Calculus and Analytic Geometry

39b $$\int \sin^n u \cos^m u \, du = \frac{\sin^{n+1} u \cos^{m-1} u}{n+m} + \frac{m-1}{n+m} \int \sin^n u \cos^{m-2} u \, du \quad \text{if } m \neq -n$$

40 $$\int u \sin u \, du = \sin u - u \cos u + C$$

41 $$\int u \cos u \, du = \cos u + u \sin u + C$$

42 $$\int u^n \sin u \, du = -u^n \cos u + n \int u^{n-1} \cos u \, du$$

43 $$\int u^n \cos u \, du = u^n \sin u - n \int u^{n-1} \sin u \, du$$

FORMS INVOLVING $\sqrt{u^2 \pm a^2}$

44 $$\int \sqrt{u^2 \pm a^2} \, du = \frac{u}{2} \sqrt{u^2 \pm a^2} \pm \frac{a^2}{2} \ln|u + \sqrt{u^2 \pm a^2}| + C$$

45 $$\int \frac{du}{\sqrt{u^2 \pm a^2}} = \ln|u + \sqrt{u^2 \pm a^2}| + C$$

46 $$\int \frac{\sqrt{u^2 + a^2}}{u} \, du = \sqrt{u^2 + a^2} - a \ln\left(\frac{a + \sqrt{u^2 + a^2}}{u}\right) + C$$

47 $$\int \frac{\sqrt{u^2 - a^2}}{u} \, du = \sqrt{u^2 - a^2} - a \sec^{-1} \frac{u}{a} + C$$

48 $$\int u^2 \sqrt{u^2 \pm a^2} \, du = \frac{u}{8} (2u^2 \pm a^2) \sqrt{u^2 \pm a^2} - \frac{a^4}{8} \ln|u + \sqrt{u^2 \pm a^2}| + C$$

49 $$\int \frac{u^2 \, du}{\sqrt{u^2 \pm a^2}} = \frac{u}{2} \sqrt{u^2 \pm a^2} \mp \frac{a^2}{2} \ln|u + \sqrt{u^2 \pm a^2}| + C$$

50 $$\int \frac{du}{u^2 \sqrt{u^2 \pm a^2}} = \mp \frac{\sqrt{u^2 \pm a^2}}{a^2 u} + C$$

51 $$\int \frac{\sqrt{u^2 \pm a^2}}{u^2} \, du = -\frac{\sqrt{u^2 \pm a^2}}{u} + \ln|u + \sqrt{u^2 \pm a^2}| + C$$

52 $$\int \frac{du}{(u^2 \pm a^2)^{3/2}} = \frac{\pm u}{a^2 \sqrt{u^2 \pm a^2}} + C$$

53 $$\int (u^2 \pm a^2)^{3/2} \, du = \frac{u}{8} (2u^2 \pm 5a^2) \sqrt{u^2 \pm a^2} + \frac{3a^4}{8} \ln|u + \sqrt{u^2 \pm a^2}| + C$$

FORMS INVOLVING $\sqrt{a^2 - u^2}$

54 $$\int \sqrt{a^2 - u^2} \, du = \frac{u}{2} \sqrt{a^2 - u^2} + \frac{a^2}{2} \sin^{-1} \frac{u}{a} + C$$

55 $$\int \frac{\sqrt{a^2 - u^2}}{u} \, du = \sqrt{a^2 - u^2} - a \ln\left|\frac{a + \sqrt{a^2 - u^2}}{u}\right| + C$$

56 $$\int \frac{u^2 \, du}{\sqrt{a^2 - u^2}} = -\frac{u}{2} \sqrt{a^2 - u^2} + \frac{a^2}{2} \sin^{-1} \frac{u}{a} + C$$

57 $$\int u^2 \sqrt{a^2 - u^2} \, du = \frac{u}{8} (2u^2 - a^2) \sqrt{a^2 - u^2} + \frac{a^4}{8} \sin^{-1} \frac{u}{a} + C$$

58 $$\int \frac{du}{u^2 \sqrt{a^2 - u^2}} = -\frac{\sqrt{a^2 - u^2}}{a^2 u} + C$$

59 $$\int \frac{\sqrt{a^2 - u^2}}{u^2} \, du = -\frac{\sqrt{a^2 - u^2}}{u} - \sin^{-1} \frac{u}{a} + C$$

60 $$\int \frac{du}{u \sqrt{a^2 - u^2}} = -\frac{1}{a} \ln\left|\frac{a + \sqrt{a^2 - u^2}}{u}\right| + C$$

61 $$\int \frac{du}{(a^2 - u^2)^{3/2}} = \frac{u}{a^2 \sqrt{a^2 - u^2}} + C$$

62 $$\int (a^2 - u^2)^{3/2} \, du = \frac{u}{8} (5a^2 - 2u^2) \sqrt{a^2 - u^2} + \frac{3a^4}{8} \sin^{-1} \frac{u}{a} + C$$

EXPONENTIAL AND LOGARITHMIC FORMS

63 $$\int u e^u \, du = (u - 1) e^u + C$$

64 $$\int u^n e^u \, du = u^n e^u - n \int u^{n-1} e^u \, du$$

65 $$\int \ln u \, du = u \ln u - u + C$$

66 $$\int u^n \ln u \, du = \frac{u^{n+1}}{n+1} \ln u - \frac{u^{n+1}}{(n+1)^2} + C$$

67 $$\int e^{au} \sin bu \, du = \frac{e^{au}}{a^2 + b^2} (a \sin bu - b \cos bu) + C$$

68 $$\int e^{au} \cos bu \, du = \frac{e^{au}}{a^2 + b^2} (a \cos bu + b \sin bu) + C$$

(Continued inside back cover)

Calculus and Analytic Geometry

THIRD EDITION

C. H. EDWARDS, Jr
The University of Georgia, Athens

DAVID E. PENNEY
The University of Georgia, Athens

PRENTICE HALL
Englewood Cliffs, New Jersey, 07632

Library of Congress Cataloging-in-Publication Data

Edwards, C. H. (Charles Henry),
Calculus and analytic geometry/C. H. Edwards, Jr., David E. Penney. — 3rd ed.
p. cm.
Includes bibliographical references.
ISBN 0-13-111204-X:
1. Calculus. 2. Geometry, Analytic. I. Penney, David E. II. Title.
QA303.E223 1990
515'. 15—dc20 89-22949
CIP

Design: Maureen Eide/Florence Silverman
Illustrator: Ron Weickart/Network Graphics
Manufacturing buyer: Paula Massenaro
Production: Nicholas Romanelli

Calculus and Analytic Geometry, Third Edition
C. H. Edwards, Jr. and David E. Penney

A Division of Simon & Schuster
Englewood Cliffs, New Jersey 07632

Printed in the United States of America
10 9 8 7 6 5 4 3 2 1

ISBN 0-13-111204-X

Prentice-Hall International (UK) Limited, *London*
Prentice-Hall of Australia Pty. Limited, *Sydney*
Prentice-Hall of Canada, Inc., *Toronto*
Prentice-Hall Hispanoamericana, S.A., *Mexico*
Prentice-Hall of India Private Limited, *New Delhi*
Prentice-Hall of Japan, Inc., *Tokyo*
Simon & Schuster Asia Pte. Ltd., *Singapore*
Editora Prentice-Hall do Brasil, Ltda., *Rio de Janeiro*

Contents

Preface

For three centuries calculus has served as the prinicpal quantitative language of science and technology, and thereby has helped to shape the world in which we live. Today it remains a vibrant and living subject. Indeed, a wider range of students than ever before find a knowledge of the basic concepts of calculus necessary for their chosen courses of study.

Our goal in preparing this revision was to make the ideas of calculus more attractive and accessible to these many students for whom calculus is a keystone of their education. This edition is somewhat leaner and (we hope) crisper than its predecessors. Its text is about seventy-five pages shorter than that of the second edition, but we believe no one will find his or her favorite topic in the second edition missing from this one. Our main editorial technique has been the judicious pruning of excess foliage rather than the removal of whole trees.

This edition (like its predecessors) was written with five related objectives in constant view: *concreteness*, *readability*, *motivation*, *applicability*, and *accuracy*.

CONCRETENESS

The power of calculus is impressive in its precise answers to realistic questions and problems. In the necessary conceptual development of the subject, we keep in sight the central question: How does one actually *compute* it? We place special emphasis on concrete examples, applications, and problems that serve both to highlight the development of the theory and to demonstrate the remarkable versatility of calculus in the investigation of important scientific questions.

READABILITY

Difficulties in learning mathematics often are complicated by language difficulties. Our writing style stems from the belief that crisp exposition, both intuitive and precise, makes mathematics more accessible—and hence more readily learned—with no loss of rigor. We hope our language is clear and attractive to students and that they can and actually will read it, thereby enabling the instructor to concentrate class time on the less routine aspects of teaching calculus.

MOTIVATION

Our exposition is centered around examples of the use of calculus to solve real problems of interest to real people. In selecting such problems for examples and exercises, we took the view that stimulating interest and motivating effective study go hand in hand. We attempt to make it clear to students how the knowledge gained with each new concept or technique will be worth the effort expended. In theoretical discussions, especially, we try to provide an intuitive picture of the goal before we set off in pursuit of it.

APPLICATIONS

Its diverse applications are what attract many students to calculus, and realistic applications provide valuable motivation and reinforcement for all students. This book is well-known for the broad range of applications that we include, but it is neither necessary nor desirable that the course cover all the applications in the book. Each section or subsection that may be omitted without loss of continuity is marked with an asterisk. This provides flexibility for each instructor to determine his or her own flavor and emphasis.

ACCURACY

Our coverage of calculus is complete (though we hope it is somewhat less than encyclopedic). Like its predecessors, this edition was subjected to a comprehensive reviewing process to help ensure accuracy. With regard to the selection and sequence of mathematical topics, our approach is traditional. However, close examination of the treatment of standard topics will uncover evidence of our interest in the current movement to revitalize the teaching of calculus. We continue to favor an intuitive approach that emphasizes both conceptual understanding and care in the formulation of definitions and statements of theorems. Some proofs that may be omitted at the discretion of the instructor are placed at the ends of sections. Others (such as the proofs of the intermediate value theorem and of the integrability of continuous functions) are deferred to the book's appendices. In this way we leave ample room for variation in seeking the proper balance between rigor and intuition.

THIRD EDITION FEATURES

In preparing this edition, we have benefited from many valuable comments and suggestions from users of the first two editions. This revision was so pervasive that the individual changes are too numerous to be detailed in a preface, but the following paragraphs summarize those that may be of widest interest.

Additional Problems The number of problems has steadily increased since the first edition, and now totals over 6000. Most of the new problems are practice exercises that have been inserted near the beginnings of problem sets to insure that students gain sufficient confidence and computational skill before moving on to the more conceptual problems that constitute the real goal of calculus. However, we have continued to add occasional new conceptual and applied problems.

New Examples and Computational Details In many sections throughout this edition, we have inserted a simpler first example or have replaced existing examples with ones that are computationally simpler. Moreover, we have inserted an additional line or two of computational detail in many of the worked-out examples to make them easier for student readers to follow. The purpose of these computational changes is to make the computations themselves less of a barrier to conceptual understanding.

Optional Computer Applications We have included a dozen and a half optional computer notes for supplementary reading (or at least perusal) by students who might be motivated by computer applications. Each of these notes appears at the end of a section (following the problems) and uses some aspect of modern computational technology to illustrate the principal ideas of the section. These notes are completely optional and are never referred to in the text proper. Most of them apply very simple BASIC programming to calculus problems, but others range in content from illustrations of the use of handheld calculators with graphics capabilities to applications of symbolic algebra systems (such as MACSYMA, Maple, or Mathematica). The purpose of these notes is to stimulate interest in calculus in the rapidly increasing population of students who already are interested in computers. Those who would like to pursue computer applications further may consult the following books:

C. H. Edwards, Jr., *Calculus and the Personal Computer*, Englewood Cliffs, N. J.: Prentice Hall, 1986.

C. H. Edwards, Jr., *A Calculus Companion: The Personal Computer*, Englewood Cliffs, N. J.: Prentice Hall, 1990.

The latter is suitable for use in a computer lab that is conducted in association with a standard calculus course, perhaps meeting weekly. It can also be used as a basis for computer assignments that students will complete outside of class, for individual study or for supplementary projects.

Introductory Chapters Chapters 1 and 2 have been streamlined for a leaner and quicker start on calculus. Chapter 1 now consists of just three sections dealing with fundamental ideas about functions and graphs. Auxiliary ideas (e.g., inverse functions) now are deferred until later when they are actually needed. Chapter 2 on limits begins with a section on tangent lines to motivate the official introduction of limits in Section 2.2. The review of elementary trigonometry now appears as Appendix A (complete with an exercise set) at the back of the book, and can be used as a Chapter 2 insert where appropriate. The material on trigonometric limits is now delayed until it is needed in Chapter 3. Proofs of the limit laws are now included for reference in Appendix B.

Differentiation Chapters We have substantially reordered the sequence of topics in Chapters 3 and 4, with the objective of building student confidence by introducing topics more nearly in order of increasing difficulty. The chain rule now appears earlier (in Section 3.3) and we cover the basic techniques for differentiating algebraic functions before discussing maxima and minima in Sections 3.5 and 3.6. Implicit differentiation and related rates have been combined in a single section (Section 3.8). The mean value theorem and its applications are deferred to Chapter 4. Section 4.6 on higher derivatives and concavity has been streamlined, and applied problems using the second derivative test have been added.

Integration Chapters New and simpler examples, together with enhanced artwork, have been inserted throughout Chapters 5 and 6. Many instructors now believe that first applications of integration ought not be confined to the standard area and volume computations; Section 6.5 is an optional new section that introduces separable differential equations. The material on centroids and the theorems of Pappus has been moved to Chapter 16 (Multiple Integrals) where it can be treated in a more natural context.

Transcendental Functions Chapter 7 now begins with a more intuitive introduction to exponential and logarithmic functions that is based on the precalculus idea of a logarithm as "the power to which the base a must be raised to get the number x." On this basis, Section 7.1 carries out a low-key review of the laws of exponents and of logarithms, and investigates informally the differentiation of exponential and logarithmic functions. This informal discussion, together with much-needed review of precalculus material, is intended to provide students with a conceptual foundation for the "official" treatments of these functions that appear in Sections 7.2 and 7.3. Many of the lengthier applications that formerly appeared in Chapters 7 and 8 have been deleted, but sufficiently many remain to amply illustrate the role of transcendental functions in the real world.

Techniques of Integration The first four sections (through integration by parts) of Chapter 9 are essentially unchanged, but the remainder of the chapter has been reorganized for the benefit of those instructors who feel that methods of formal integration now require less emphasis, in view of modern techniques for both numerical and symbolic integration. The method of partial fractions now appears in Section 9.5 (immediately following integration by parts in Section 9.4). Trigonometric substitutions and integrals involving quadratic polynomials follow in Sections 9.6 and 9.7. This rearrangement of Chapter 9 makes it more convenient to stop wherever the instructor desires.

Multivariable Topics Section 14.5 is a unified treatment of curvature and acceleration for both plane curves and space curves (which were treated separately in the second edition). The discussion of directional derivatives in Section 15.8 has been simplified considerably. Section 16.5 (Applications of Double Integrals) now contains the theorems of Pappus that formerly were discussed in Chapter 6.

Differential Equations Many calculus instructors now believe that differential equations should be seen as early and as often as possible. The very simplest differential equations (of the form $y' = f(x)$) appear in an optional subsection at the end of Section 4.8 (Antiderivatives). Section 6.5 is a new section that illustrates applications of integration to the solution of separable differential equations. However, these are optional sections, and the instructor can delay them until Chapter 18 (on differential equations) is covered. Chapter 18 has been revised substantially, and now ends with Section 18.6 on elementary numerical methods. Some of the lengthier applications in earlier editions have been deleted but are now included in Edwards and

Penney, *Elementary Differential Applications with Applications*, second edition (Englewood Cliffs, N.J.: Prentice Hall, 1989).

Computer Graphics The ability of students to visualize surfaces and graphs of functions of two variables should be enhanced by the computer graphics that appear in Chapters 14 through 16. All of these computer-generated figures either are new or were reworked for the third edition. We are indebted to John K. Edwards for the "hard copy" computer graphics and to Roy Myers for the photographs of monitor screens generated by his Calculus 3–D Function Plotter.

Historical Comments Both authors are fond of the history of mathematics, and believe that it can favorably influence our teaching of mathematics. For this reason numerous historical comments appear in the text. However, we have resisted the temptation to insert full-fledged historical notes. Instructors who would like to include more historical material in their courses are invited to consult Edwards, *The Historical Development of the Calculus* (New York: Springer-Verlag, 1979).

SUPPLEMENTARY MATERIAL

Answers to most of the odd-numbered problems appear in the back of the book. Solutions to most problems (other than those odd-numbered ones for which an answer alone is sufficient) are available in the *Instructor's Manual*. A subset of this manual, containing solutions to problems numbered 1, 5, 9, 13, . . . is available as a *Student Manual*. A collection of some 1400 additional problems suitable for use as test questions, the *Calculus Test Item File*, is available for use by instructors. Finally, an *Instructor's Annotated Edition* of the text itself is available to those who are teaching from this book. A computer diskette that accompanies the instructor's edition includes memos of possible interest to instructors as well as the BASIC programs that appear in the optional computer notes in the text itself. A variety of additional supplements are provided by the publisher.

Acknowledgments

All experienced textbook authors know the value of critical reviewing during the preparation and revision of a manuscript. In our work on various editions of this book we have benefited greatly from the advice (and frequently the consent) of the following exceptionally able reviewers:

Leon E. Arnold, Delaware County Community College
H.L. Bentley, University of Toledo
Michael L. Berry, West Virginia Wesleyan College
William Blair, Northern Illinois University
George Cain, Georgia Institute of Technology
Wil Clarke, Atlantic Union College
Peter Colwell, Iowa State University
James W. Daniel, University of Texas at Austin
Robert Devaney, Boston University
Dianne H. Haber, Westfield State College
John C. Higgins, Brigham Young University
W. Cary Huffman, Loyola University of Chicago
Lois E. Knouse, Le Tourneau College
Morris Kalka, Tulane University
Catherine Lilly, Westfield State College
Barbara Moses, Bowling Green University
Barbara L. Osofsky, Rutgers University—New Brunswick
James P. Qualey, Jr., University of Colorado
Thomas Roe, South Dakota State University
Lawrence Runyan, Shoreline Community College
John Spellman, Southwest Texas State University
Virginia Taylor, University of Lowell
Samuel A. Truitt, Jr., Middle Tennessee State University
Robert Urbanski, Middlesex County College
Robert Whiting, Villanova University

Many of the best improvements that have been made must be credited to colleagues and users of the first two editions throughout the country (and abroad). We are grateful to all those, especially students, who have written to us, and hope they will continue to do so. We also believe that the quality of the finished book itself is adequate testimony to the skill, diligence, and talent of an exceptional staff at Prentice Hall; we owe special thanks to Bob Sickles, mathematics editor; Nick Romanelli, production editor; Maureen Eide and Florence Silverman, designers; and Ron Weickart, illustrator. Finally, we again are unable to thank Alice Fitzgerald Edwards and Carol Wilson Penney adequately for their continued assistance, encouragement, support, and patience.

Athens, Georgia C.H.E., Jr. / D.E.P.

Calculus and Analytic Geometry

chapter one

Prelude to Calculus

1.1 Functions and Real Numbers

Calculus is one of the supreme accomplishments of the human intellect. This mathematical discipline stems largely from the seventeenth-century investigations of Isaac Newton (1642–1727) and Gottfried Wilhelm Leibniz (1646–1716). Yet some of its ideas date back as far as Archimedes (287–212 B.C.), and originated in cultures as diverse as those of Greece, Egypt, Babylonia, India, China, and Japan. Many of the scientific discoveries that have shaped our civilization during the past three centuries would have been impossible without the use of calculus.

The principal objective of calculus is the analysis of problems of change and motion. These problems are fundamental because we live in a world of ceaseless change, filled with bodies in motion and with phenomena of ebb and flow. Consequently, calculus remains a vibrant subject, and today this body of computational technique continues to serve as the principal quantitative language of science and technology.

Much of calculus involves the use of real numbers or variables to describe changing quantities, and the use of functions to describe relationships between different variables. In this initial section we first review briefly the notation and terminology of real numbers, then discuss functions in more detail.

REAL NUMBERS

The **real numbers** are already familiar to you; they are just those numbers ordinarily used in most measurements. The mass, speed, temperature, and charge of a body are measured by real numbers. Real numbers can be represented by **terminating** or **nonterminating** decimal expansions. Any terminating decimal can be written in nonterminating form by adding zeros:

$$\tfrac{3}{8} = 0.375 = 0.375000000\ldots.$$

Any **repeating** nonterminating decimal, such as

$$\tfrac{7}{22} = 0.31818181818\ldots,$$

represents a **rational** number, one that is the quotient of two integers. Conversely, every rational number is represented by a repeating decimal expansion like the ones displayed above. The decimal expansion of an **irrational** number (one that is not rational), such as

$$\sqrt{2} = 1.414213562\ldots \quad \text{or} \quad \pi = 3.141592653589793\ldots,$$

is both nonterminating and nonrepeating.

The geometric representation of real numbers as points on the **real line** $\boldsymbol{R}$ should also be familiar to you. Each real number is represented by precisely one point of $\boldsymbol{R}$, and each point of $\boldsymbol{R}$ represents precisely one real number. By convention, the positive numbers lie to the right of zero and the negative numbers to its left, as indicated in Fig. 1.1.

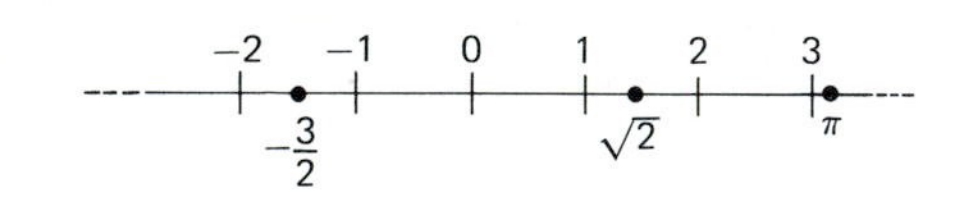

1.1 The real line $\boldsymbol{R}$

The following properties of inequalities of real numbers are fundamental and are often used.

$$\begin{aligned} &\text{If } a < b \text{ and } b < c, \text{ then } a < c. \\ &\text{If } a < b, \text{ then } a + c < b + c. \\ &\text{If } a < b \text{ and } c > 0, \text{ then } ac < bc. \\ &\text{If } a < b \text{ and } c < 0, \text{ then } ac > bc. \end{aligned} \tag{1}$$

The last two statements mean that an inequality is preserved when its members are multiplied by a *positive* number but *reversed* when they are multiplied by a *negative* number.

ABSOLUTE VALUE

The (nonnegative) distance along the real line between zero and the real number a is the **absolute value** of a, written $|a|$. Equivalently,

$$|a| = \begin{cases} a & \text{if } a \geqq 0; \\ -a & \text{if } a < 0. \end{cases} \tag{2}$$

The notation $a \geqq 0$ means that a is *either* greater than zero *or* equal to zero. Note that (2) implies that $|a| \geqq 0$ for every real number a, while $|a| = 0$ if and only if $a = 0$. For example,

$$|4| = 4, \qquad |-3| = 3, \qquad |0| = 0, \quad \text{and} \quad |\sqrt{2} - 2| = 2 - \sqrt{2},$$

the latter being true because $2 > \sqrt{2}$. Thus $\sqrt{2} - 2 < 0$, and hence

$$|\sqrt{2} - 2| = -(\sqrt{2} - 2) = 2 - \sqrt{2}.$$

The following properties of absolute values are frequently used.

$$
\begin{aligned}
|a| &= |-a| = \sqrt{a^2} \geqq 0, \\
|ab| &= |a|\,|b|, \\
-|a| &\leqq a \leqq |a|, \quad \text{and} \\
|a| &< b \quad \text{if and only if} \quad -b < a < b.
\end{aligned}
\tag{3}
$$

The **distance** between the real numbers a and b is defined to be $|a - b|$ (or $|b - a|$; there's no difference). This distance is simply the length of the line segment of the real line $\boldsymbol{R}$ with endpoints a and b, as indicated in Fig. 1.2.

The properties of inequalities and absolute values here imply the following important fact.

|b − a|
or |a − b|
a
b

1.2 The distance between a and b

> ***Triangle Inequality***
>
> For all real numbers a and b,
>
> $$|a + b| \leqq |a| + |b|. \tag{4}$$

Proof We add the inequalities $-|a| \leqq a \leqq |a|$ and $-|b| \leqq b \leqq |b|$. This gives us

$$-(|a| + |b|) \leqq a + b \leqq |a| + |b|. \tag{5}$$

But the properties in (3) imply that

$$|c| \leqq d \quad \text{if and only if} \quad -d \leqq c \leqq d. \tag{6}$$

We set $c = a + b$ and $d = |a| + |b|$, and it follows that (4) and (5) are equivalent. ■

INTERVALS

Suppose that S is a set (collection) of real numbers. It is common to describe S by means of the notation

$$S = \{x : C\}$$

where C is a condition that's true for those numbers x in S and false for those numbers x not in S. The most important sets of real numbers in calculus are **intervals.** If $a < b$, then the **open interval** (a, b) is defined to be the set

$$(a, b) = \{x : a < x < b\}$$

of real numbers, and the **closed interval** $[a, b]$ is

$$[a, b] = \{x : a \leqq x \leqq b\}.$$

Thus a closed interval contains its endpoints, while an open interval does not. We also use the **half-open intervals**

$$[a, b) = \{x : a \leqq x < b\} \quad \text{and} \quad (a, b] = \{x : a < x \leqq b\}.$$

Thus the open interval $(1, 3)$ is the set of those real numbers x such that $1 < x < 3$; the closed interval $[-1, 2]$ is the set of those real numbers x such that $-1 \leqq x \leqq 2$; the half-open interval $(-1, 2]$ is the set of those real numbers x such that $-1 < x \leqq 2$. In Fig. 1.3 we show examples of such

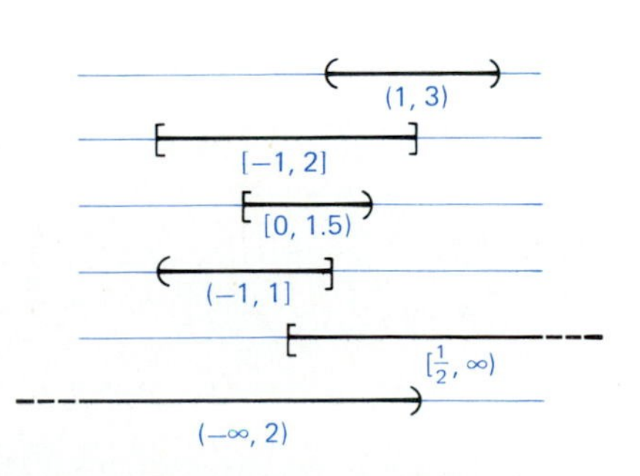

1.3 Some examples of intervals of real numbers

intervals, as well as some **unbounded intervals,** which have forms such as

$$[a, \infty) = \{x : x \geqq a\},$$
$$(-\infty, a] = \{x : x \leqq a\},$$
$$(a, \infty) = \{x : x > a\}, \quad \text{and}$$
$$(-\infty, a) = \{x : x < a\}.$$

We emphasize that the symbol ∞, denoting infinity, is merely a notational convenience and does *not* represent a real number—the real line $\boldsymbol{R}$ does *not* have "endpoints at infinity." The use of this symbol is justified by the brief and natural descriptions of the sets $[\pi, \infty)$ and $(-\infty, 2)$ for the sets of all real numbers x such that $x \geqq \pi$ and $x < 2$, respectively.

FUNCTIONS

The key to the mathematical analysis of a geometric or scientific situation is often the recognition of relationships between the variables that describe the situation. Such a relationship may be a formula that expresses one variable as a function of another. For example, the area A of a circle of radius r (see Fig. 1.4) is given by $A = \pi r^2$. The volume V and surface area S of a sphere of radius r are given by

$$V = \tfrac{4}{3}\pi r^3 \quad \text{and} \quad S = 4\pi r^2,$$

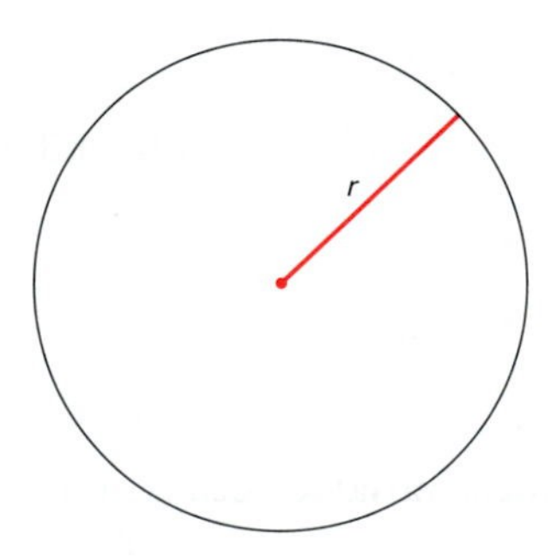

1.4 Circle, area $A = \pi r^2$, circumference $C = 2\pi r$

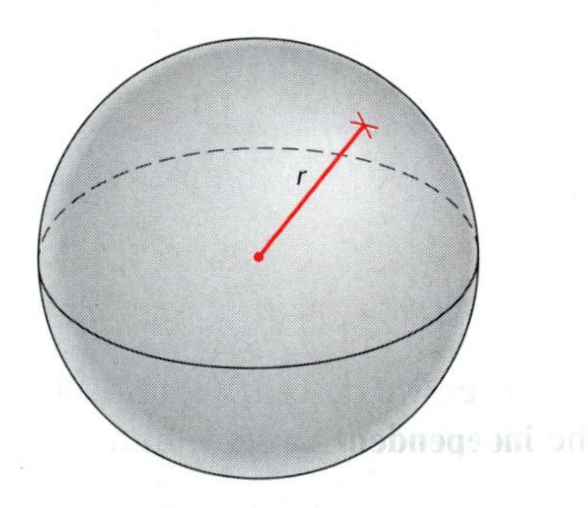

1.5 Sphere, volume $V = \frac{4}{3}\pi r^3$, surface area $S = 4\pi r^2$

respectively (see Fig. 1.5). After t seconds a body that has been dropped from rest has fallen a distance

$$s = \tfrac{1}{2}gt^2$$

feet and has speed $v = gt$ feet per second, where $g \approx 32$ ft/s^2 is gravitational acceleration. The volume V (in liters) of 3 g of carbon dioxide (CO_2) at 27°C is given in terms of its pressure p (in atmospheres) by $V = 1.68/p$. These are all examples of real-valued functions of a real variable.

Definition of Function

A real-valued **function** f defined on a set D of real numbers is a rule that assigns to each number x in D exactly one real number, denoted by $f(x)$.

The set D of all those numbers x for which $f(x)$ is defined is called the **domain** (or **domain of definition**) of the function f. The number $f(x)$, read "f of x," is called the **value** of the function at the number or point x. The set of all values $y = f(x)$ is called the **range** of f. That is, the range of f is the set

$$\{y : y = f(x) \quad \text{for some } x \text{ in } D\}.$$

In this section we will be more concerned with the domain of a function than with its range.

Often a function is described by means of a formula that specifies how to compute the number $f(x)$ in terms of the number x. The symbol $f(\ \ )$ may be regarded as an operation that is to be performed whenever a value from the domain of f is inserted between the parentheses.

EXAMPLE 1 The formula

$$f(x) = x^2 + x - 3 \tag{7}$$

describes the rule of a function f having as its domain the entire real line $\boldsymbol{R}$. Some typical values of f are $f(-2) = -1$, $f(0) = -3$, and $f(3) = 9$. Some other values of the function f are

$$\begin{aligned} f(4) &= (4)^2 + (4) - 3 = 17, \\ f(c) &= c^2 + c - 3, \\ f(2 + h) &= (2 + h)^2 + (2 + h) - 3 = (4 + 4h + h^2) + (2 + h) - 3 \\ &= 3 + 5h + h^2, \quad \text{and} \\ f(-t^2) &= (-t^2)^2 + (-t^2) - 3 = t^4 - t^2 - 3. \end{aligned}$$

When we describe the function f by writing a formula $y = f(x)$, we call x the **independent variable** and y the **dependent variable** because the value of y depends—through f—upon the choice of x. As x changes, or varies, then so does y, and the way that y varies with x is determined by the rule of the function f. For example, if f is the function of Equation (7), then $y = -1$ when $x = -2$, $y = -3$ when $x = 0$, and $y = 9$ when $x = 3$.

Sometimes it is useful to visualize the dependence of the value $y = f(x)$ on x by thinking of a function as a kind of machine that accepts as input a number x, then produces as output the number $f(x)$, perhaps displayed or printed. See Fig. 1.6.

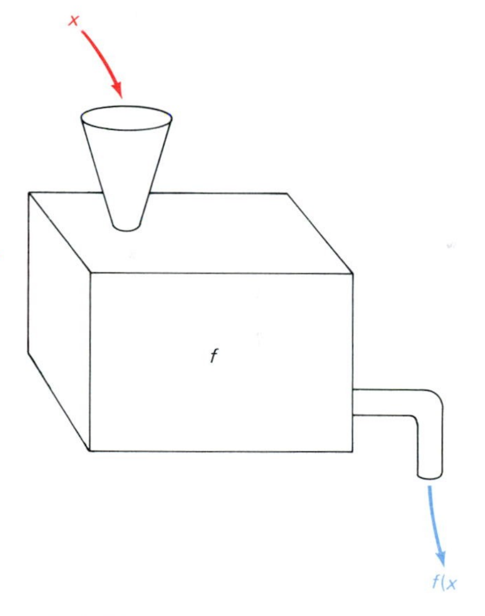

1.6 A "function machine"

One such machine is the familiar pocket calculator with a square root key. When a nonnegative number x is entered and this key is pressed, the calculator displays (an approximation to) the number $\sqrt{x}$. Note that the domain of this *square root function* $f(x) = \sqrt{x}$ is the set of all nonnegative real numbers, because no negative number has a real square root. Its range is also the set of all nonnegative real numbers, because the symbol $\sqrt{x}$ always denotes the *nonnegative* square root of x. The calculator illustrates its knowledge of the domain by displaying an adverse reaction if we ask it to calculate the square root of a negative number (unless it's a more sophisticated calculator like the HP-28S, which knows about complex numbers).

Not every function has a rule expressible as a simple one-part formula like $f(x) = \sqrt{x}$. For example, if we write

$$f(x) = \begin{cases} x^2 & \text{if } x \geqq 0, \\ -x & \text{if } x < 0, \end{cases}$$

then we have defined a perfectly good function with domain $\boldsymbol{R}$. Some of its values are $f(-3) = 3$, $f(0) = 0$, and $f(2) = 4$. The function in Example 2 is defined initially by means of a verbal description rather than by means of formulas.

EXAMPLE 2 For each real number x, let $f(x)$ denote the greatest integer that is less than or equal to x. For instance, $f(-0.5) = -1, f(0) = 0, f(3) = 3$, and $f(\pi) = 3$. If n is an integer, then $f(x) = n$ for every x in the half-open interval $[n, n + 1)$. This function f is called the **greatest integer function** and is often denoted by

$$f(x) = [\![x]\!]. \tag{8}$$

Thus $[\![-0.5]\!] = -1$, $[\![0]\!] = 0$, $[\![3]\!] = 3$, and $[\![\pi]\!] = 3$. Note that while $[\![x]\!]$ is defined for all real x, the range of the greatest integer function consists only of the set of all integers.

What is the domain of a function supposed to be when it is not specified? This is a common situation, and it occurs when we give a function f by writing only its formula $y = f(x)$. If no domain is given, we agree for convenience that the domain D is the set of all real numbers x for which the expression $f(x)$ makes sense. For example, the domain of $f(x) = 1/x$ is the set of all nonzero real numbers (because $1/x$ is defined exactly when $x \neq 0$).

EXAMPLE 3 Find the domain of the function g with formula

$$g(x) = \frac{1}{\sqrt{2x + 4}}.$$

Solution In order that the square root $\sqrt{2x + 4}$ be defined, it is necessary that $2x + 4 \geqq 0$. This holds when $2x \geqq -4$, and thus when $x \geqq -2$. In order that the reciprocal $1/\sqrt{2x + 4}$ be defined, we also require that $\sqrt{2x + 4} \neq 0$, and thus that $x \neq -2$. Hence the domain of g is the interval $D = (-2, \infty)$.

EXAMPLE 4 If the function f has the formula

$$f(x) = \frac{25 - (3 - 2x)^2}{16 + 9x^2}, \tag{9}$$

for what values of x is it true that $f(x) > 0$?

Solution The denominator in (9) is positive for all x, so $f(x) > 0$ if and only if the numerator is positive,

$$25 - (3 - 2x)^2 > 0.$$

This is so if and only if $(3 - 2x)^2 < 25$, so we need to solve the inequality

$$-5 < 3 - 2x < 5. \tag{10}$$

Using the properties of inequalities listed in (1), we proceed much as if we were solving an equation for x: We attempt to isolate the variable x. Proceeding one step at a time, we find that

$$-8 < -2x < 2 \qquad \text{(subtracting 3)},$$

$$-2 < 2x < 8 \qquad \text{(multiplying by } -1\text{)},$$

$$-1 < x < 4 \qquad \text{(dividing by 2)}.$$

Thus $f(x) > 0$ if and only if x lies in the open interval $(-1, 4)$.

FUNCTIONS AND APPLICATIONS

The investigation of an applied problem often hinges on the definition of a function that captures the essence of a geometrical or physical situation. The next two examples illustrate this process.

EXAMPLE 5 A rectangular box with a square base has volume 125. Express its total surface area A as a function of the edge length x of its base.

Solution The first step is to draw a sketch and label the relevant dimensions. Figure 1.7 shows a rectangular box with square base of edge length x and height y (its other dimension). We are given that the volume of the box is

$$V = x^2y = 125. \tag{11}$$

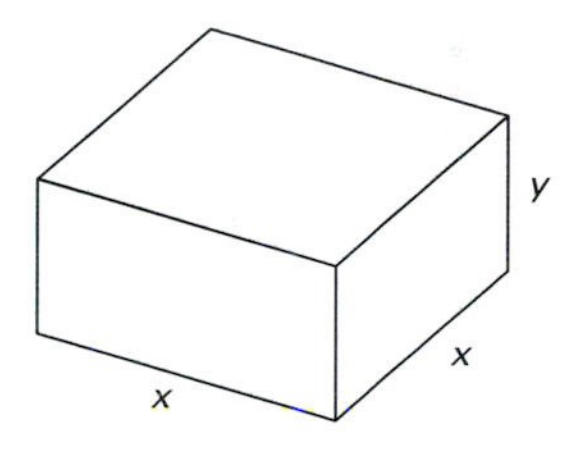

1.7 The box of Example 3

The top and bottom of the box each have area x^2 and each of its four vertical sides has area xy, so its total surface area is

$$A = 2x^2 + 4xy. \tag{12}$$

Note, however, that this is a formula for A in terms of the *two* variables x and y rather than a function of the *single* variable x. To eliminate y, and thereby obtain A in terms of x alone, we solve Eq. (11) for $y = 125/x^2$, then substitute this result in Eq. (12) to obtain

$$A = 2x^2 + (4x)\frac{125}{x^2} = 2x^2 + \frac{500}{x}.$$

Thus the surface area is given as a function of the edge length x by

$$A(x) = 2x^2 + \frac{500}{x}, \qquad 0 < x < \infty. \tag{13}$$

It is necessary to specify the domain because negative values of x make sense in the *formula* in (13) but do not belong in the domain of the *function A*. Because every $x > 0$ determines such a box, the domain does in fact include all positive numbers.

COMMENT In Example 5 our goal was to express the dependent variable A as a *function* of the independent variable x. Initially, the geometric situation provided us instead with

1. The *formula* in (12) expressing A in terms of both x and the additional variable y, and

2. The *relation* in (11) between x and y, which we used to eliminate y, thereby obtaining A as a function of x alone.

We will see that this is a common pattern in many different applied problems, such as the one that follows.

THE ANIMAL PEN PROBLEM A rectangular pen is to be built, using an existing wall as one of its four sides. The fence for the other three sides costs \$5 per foot, and \$1 per foot must be spent to paint the portion of the wall that forms the fourth side of the pen. If a total of \$180 is to be spent, what dimensions will maximize the area of the resulting pen?

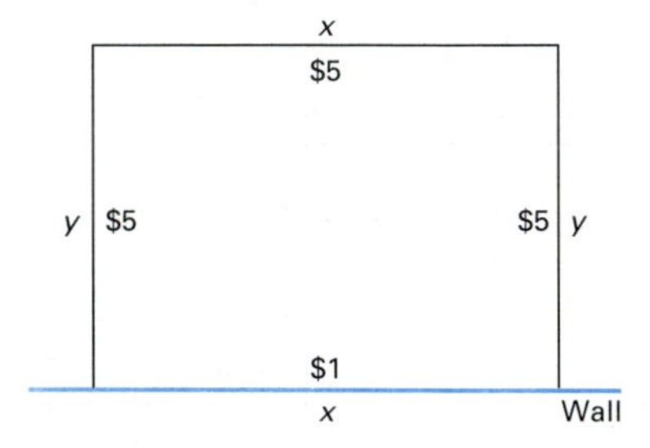

1.8 The animal pen

Figure 1.8 shows the animal pen with its dimensions x and y labeled, along with the cost per foot of each of its four sides. When confronted with a verbally stated applied problem such as the animal pen problem, anyone's first question is: How on earth do we get started on it? The function concept is the key to getting a handle on such a situation. If we can express the quantity to be maximized—the dependent variable—as a function of some independent variable, then we have something tangible to do: Find the maximum value attained by this function.

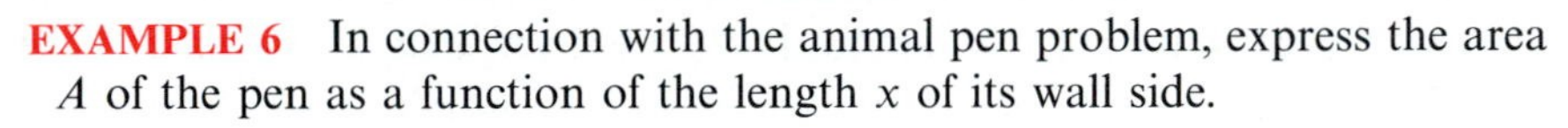

EXAMPLE 6 In connection with the animal pen problem, express the area A of the pen as a function of the length x of its wall side.

Solution The area A of the rectangular pen with length x and width y is

$$A = xy. \tag{14}$$

When we multiply the length of each side in Fig. 1.8 by its cost per foot, then add the results, we find that the total cost C of the pen is given by

$$C = x + 5y + 5x + 5y,$$

so

$$6x + 10y = 180 \tag{15}$$

because we are given $C = 180$. Choosing x to be the independent variable, we use the relation in (15) to eliminate the additional variable y from the area formula in (14). When we solve Eq. (15) for y and substitute the result

$$y = \tfrac{1}{10}(180 - 6x) = \tfrac{3}{5}(30 - x) \tag{16}$$

in (14), we obtain the desired function

$$A(x) = \tfrac{3}{5}(30x - x^2)$$

that expresses the area A as a function of the length x.

In addition to this formula for the function A, we must also specify its domain. Only if $x > 0$ will actual rectangles be produced, but we find it convenient to include the value $x = 0$ as well. This value of x corresponds to a "degenerate rectangle" with base zero and height

$$y = \tfrac{3}{5}(30) = 18$$

(a consequence of Eq. (16)). For similar reasons, we have the restriction $y \geqq 0$. Because

$$y = \tfrac{3}{5}(30 - x),$$

it follows that $x \leqq 30$. Thus the complete definition of the area function is

$$A(x) = \tfrac{3}{5}(30x - x^2), \qquad 0 \leqq x \leqq 30. \tag{17}$$

Example 6 illustrates an important part of the solution of a typical problem involving applications of mathematics. The domain of a function is a necessary part of its definition, and for each function we must specify the domain of values of the independent variable. In applications, we use the values of the independent variable that are relevant to the problem at hand.

NUMERICAL INVESTIGATION

Armed with the result of Example 6, we might attack the animal pen problem by calculating a table of values of the area function $A(x)$ in (17). Such a table is shown in Fig. 1.9. The data in this table suggest strongly that the maximum area is $A = 135$ ft^2, attained with side length $x = 15$ ft, in which case Eq. (16) yields $y = 9$ ft. This conjecture appears to be corroborated by the more refined data shown in Fig. 1.10.

x	$A(x)$
0	0
5	75
10	120
15	135 ←
20	120
25	75
30	0

1.9 Area $A(x)$ of a pen with side length x

Thus it appears that the animal pen with maximal area (costing \$180) is $x = 15$ ft long and $y = 9$ ft wide. The tables in Figs. 1.9 and 1.10 show only *integral* values of x, however, and there remains the possibility that the length x of the pen of maximal area is *not* an integer. Consequently, numerical tables alone do not settle the matter. A new mathematical idea is needed in order to *prove* that $A(15) = 135$ is the maximum value of

$$A(x) = \tfrac{3}{5}(30x - x^2), \qquad 0 \leqq x \leqq 30$$

for *all* x in its domain. We will attack this problem again in Section 1.3 after a review of rectangular coordinates in Section 1.2.

x	$A(x)$
10	120
11	125.4
12	129.6
13	132.6
14	134.4
15	135 ←
16	134.4
17	132.6
18	129.6
19	125.4
20	120

1.10 Further indication that $x = 15$ yields maximal area $A = 135$

NOTE Many scientific calculators allow the user to program a given function for repeated evaluation, and thereby to compute painlessly tables like those in Figs. 1.9 and 1.10. For example, Fig. 1.11 shows the display of a Casio fx-7000G calculator that is programmed to evaluate $A(x) = (0.6)(30x - x^2)$ when x is entered. Figure 1.12 shows the display of an HP-28S calculator that is

1.11 A calculator programmed to evaluate

$A(x) = (0.6)(30x - x^2)$

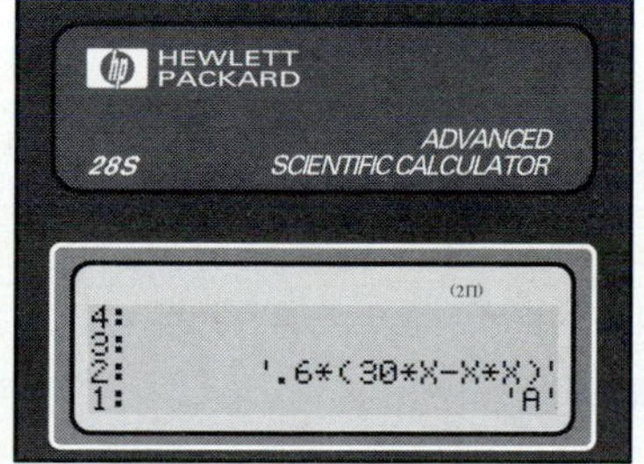

1.12 A calculator prepared to store the expression

$A = .6*(30*x - x*x)$

ready to store the expression

```
'0.6*(30*x - x*x)'
```

in the variable A, to be evaluated for different values of x.

1.1 PROBLEMS

Simplify each expression in Problems 1–10 by writing it without using absolute value symbols.

1. $|3 - 17|$

2. $|-3| - |17|$

3. $|-0.25 - \frac{1}{4}|$

4. $|5| - |-7|$

5. $|(-5)(4 - 9)|$

6. $\dfrac{|-6|}{|4| + |-2|}$

7. $|(-3)^3|$

8. $|3 - \sqrt{3}|$

9. $|\pi - \frac{22}{7}|$

10. $-|7 - 4|$

Solve each of the inequalities in Problems 11–20. Write each solution set in terms of intervals.

11. $-3 < 2x + 5 < 7$

12. $4 \leqq 3x - 5 \leqq 10$

13. $-6 \leqq 5 - 2x < 2$

14. $3 < 1 - 5x < 7$

15. $|3 - 2x| < 5$

16. $|5x + 3| \leqq 4$

17. $|1 - 3x| > 2$

18. $1 < |7x - 1| < 3$

19. $2 \leqq |4 - 5x| \leqq 4$

20. $\dfrac{1}{2x + 1} > 3$

In Problems 21–24, find each of the following values:
(a) $f(-a)$; (b) $f(a^{-1})$; (c) $f(\sqrt{a})$; (d) $f(a^2)$

21. $f(x) = \dfrac{1}{x}$

22. $f(x) = x^2 + 5$

23. $f(x) = \dfrac{1}{x^2 + 5}$

24. $f(x) = \sqrt{1 + x^2 + x^4}$

In each of Problems 25–30, find all values of a such that $g(a) = 5$.

25. $g(x) = 3x + 4$

26. $g(x) = \dfrac{1}{2x - 1}$

27. $g(x) = \sqrt{x^2 + 16}$

28. $g(x) = x^3 - 3$

29. $g(x) = \sqrt[3]{x + 25}$

30. $g(x) = 2x^2 - x + 4$

In each of Problems 31–36, compute the quantity $f(a + h) - f(a)$.

31. $f(x) = 3x - 2$

32. $f(x) = 1 - 2x$

33. $f(x) = x^2$

34. $f(x) = x^2 + 2x$

35. $f(x) = \dfrac{1}{x}$

36. $f(x) = \dfrac{2}{x + 1}$

In each of Problems 37–40, find the range of values of the given function.

37. $f(x) = \dfrac{x}{|x|}$ if $x \neq 0$, while $f(0) = 0$.

38. $f(x) = [\![3x]\!]$, where $[\![x]\!]$ is the largest integer not exceeding x.

39. $f(x) = (-1)^{[\![x]\!]}$

40. $f(x)$ is the first-class postage (in cents) for a letter mailed in the United States and weighing x ounces, $0 < x < 12$. In 1990 the postage rate for such a letter was 25¢ for the first ounce plus 20¢ for each additional ounce or fraction thereof.

In each of Problems 41–55, find the largest domain (of real numbers) on which the given formula determines a function.

41. $f(x) = 10 - x^2$

42. $f(x) = x^3 + 5$

43. $f(t) = \sqrt{t^2}$

44. $g(t) = (\sqrt{t})^2$

45. $f(x) = \sqrt{3x - 5}$

46. $g(t) = \sqrt[3]{t + 4}$

47. $f(t) = \sqrt{1 - 2t}$

48. $g(x) = \dfrac{1}{(x + 2)^2}$

49. $f(x) = \dfrac{2}{3 - x}$

50. $g(t) = \left(\dfrac{2}{3 - t}\right)^{1/2}$

51. $f(x) = \sqrt{x^2 + 9}$

52. $h(z) = \dfrac{1}{\sqrt{4 - z^2}}$

53. $f(x) = (4 - \sqrt{x})^{1/2}$

54. $f(x) = \left(\dfrac{x + 1}{x - 1}\right)^{1/2}$

55. $g(t) = \dfrac{t}{|t|}$

56. Express the area A of a square as a function of its perimeter P.

57. Express the circumference C of a circle as a function of its area A.

58. Express the volume V of a sphere as a function of its surface area S.

59. Given: 0°C is the same as 32°F and a change of 1°C is the same as a change of 1.8°F. Express the Celsius temperature C as a function of the Fahrenheit temperature F.

60. Show that if a rectangle has base x and perimeter 100, then its area A is given by the function

$$A(x) = x(50 - x), \qquad 0 \leqq x \leqq 50.$$

61. A rectangle with base of length x is inscribed in a circle of radius 2. Express the area A of the rectangle as a function of x.

62. An oil field containing 20 wells has been producing 4000 barrels of oil daily. For each new well that is drilled, the daily production of each well decreases by 5 barrels. Write the total daily production of the oil field as a function of the number x of new wells drilled.

63. Suppose that a rectangular box has volume 324 in.^{3} and a square base with edge length x inches. The material for the bottom of the box costs 2 cents per square inch and the material for its top and four sides costs 1 cent per square inch. Express the total cost of the box as a function of x.

64. A rectangle of fixed perimeter 36 is rotated about one of its sides S to generate a right circular cylinder. Express the volume V of this cylinder as a function of the length x of the side S.

65. A right circular cylinder has volume 1000 in.^{3} and the radius of its base is x inches. Express the total surface area A of the cylinder as a function of x.

66. A rectangular box has total surface area 600 in.^{2} and a square base with edge length x inches. Express the volume V of the box as a function of x.

67. An open-topped box is to be made from a square piece of cardboard with edge length 50 in. First, four small squares each having edge length x inches are cut from the four corners of the cardboard, as indicated in Fig. 1.13.

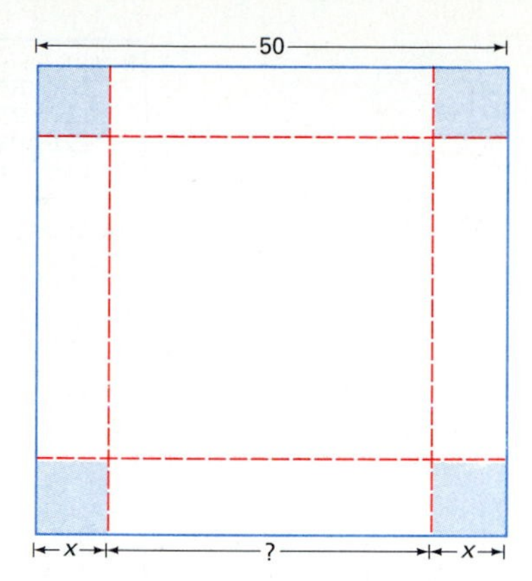

1.13 Fold the edges up to make a box

Then the four resulting flaps are turned up—fold along the dotted lines—to form the four sides of the box, which will thus have a square base and a depth of x inches. Express its volume V as a function of x.

68. Continuing Problem 60, investigate numerically the area of a rectangle with perimeter 100. What dimensions (length and width) appear to maximize the area of such a rectangle?

69. Determine numerically the number of new oil wells that should be drilled in order to maximize the total daily production of the oil field of Problem 62.

70. Investigate numerically the total surface area A of the rectangular box of Example 5. Assuming that both $x \geqq 1$ and $y \geqq 1$, what dimensions x and y appear to minimize A?

*1.1 Optional Computer Application

Repetitive computations of the sort required to produce tables of values of functions (as in Figs. 1.9 and 1.10) can be tedious and boring. This kind of labor can be lightened by the use of the pocket computers that are now widely available and relatively inexpensive (to say nothing of the microcomputers and terminals of mainframe computers that are available to students on many campuses). A surprising amount can be accomplished with a very modest knowledge of BASIC programming, such as a student might acquire from an hour's study of a typical manual for a pocket computer.

The program listed in Fig. 1.14 was used to compute the tables of values of the area function

$$A(x) = \tfrac{3}{5}(30x - x^2), \qquad 0 \leqq x \leqq 30$$

that appear in Figs. 1.9 and 1.10.

When lines 150 and 160 are executed, the computer asks you to supply the endpoints of the desired interval $[a, b]$ and the number n of subintervals

```
100 REM--Program TABULATE
110 REM--Tabulates the values of f(x) on the interval
120 REM--[a,b]  by increments of  h = (b-a)/n.  The
130 REM--function  f(x)  must be edited into line 250.
140 REM
150       INPUT "Endpoints a,b"; A,B
160       INPUT "Number of subintervals"; N
170       LET H = (B - A)/N
180       PRINT "x", "f(x)"
190       PRINT
200       LET X = A
210 REM
220 REM--Loop to calculate values:
230 REM
240       FOR I = 0 TO N
250          F = (3/5)*(30*X - X*X)
260          PRINT X, F
270          X = X + H
280       NEXT I
290 REM
300       END
```

1.14 Program TABULATE

into which you wish it subdivided for the purpose of tabulation. The input

```
Endpoints a, b? 0, 30
Number of subintervals? 6
```

produced the data shown in Fig. 1.9, while the input

```
Endpoints a, b? 10, 20
Number of subintervals? 10
```

produced the data of Fig. 1.10.

In order to use this program to tabulate values of another function $f(x)$, you need only edit the formula for $f(x)$ in line 250. The REM statements are merely comments ("REMarks") and may be omitted when you key in the program. That is, only lines 150–200 and 240–270 need be entered. The BASIC programming language varies slightly from one computer to another. For instance, with some the semicolons in lines 150 and 160 should be replaced with commas.

Exercise 1: Load and run the program listed in Fig. 1.14, making whatever corrections are necessary for its proper execution on the computer you are using. Compare your results with the data shown in Figs. 1.9 and 1.10.

Exercises 2–4: Use the program above to carry out the numerical investigations of Problems 68–70 in this section.

1.2 The Coordinate Plane and Straight Lines

Imagine the flat, featureless two-dimensional plane of Euclid's geometry. Install a copy of the real number line $\boldsymbol{R}$, with the line horizontal and the positive numbers to the right. Add another copy of $\boldsymbol{R}$ perpendicular to the first, with the two lines crossing where zero is located on each. The vertical line should have the positive numbers above and the negative numbers below the horizontal line, as in Fig. 1.15. The horizontal line is called the **x-axis** and the vertical line the **y-axis.**

With these added features, we call the plane the **coordinate plane,** because it is now possible to locate any point there by a pair of numbers called the *coordinates of the point.* If P is a point in the plane, draw perpendiculars from P to the coordinate axes, as shown in Fig. 1.16. One perpendicular meets the x-axis in the **x-coordinate** of P, labeled x_1 in Fig. 1.16. The other meets the y-axis in the **y-coordinate** y_1 of P. The pair of numbers (x_1, y_1), in that order, is called the **coordinate pair** for P, or simply the **coordinates** of P. In concise notation, we speak of "the point $P(x_1, y_1)$." The numbers x_1 and y_1 are also called the **abscissa** and **ordinate,** respectively, of the point P.

This coordinate system is called the **rectangular coordinate system,** or sometimes the **Cartesian coordinate system** (because its use in geometry was popularized, beginning in the 1630s, by the French mathematician and philosopher René Descartes (1596–1650)). The plane, thus coordinatized, is sometimes denoted by $\boldsymbol{R}^2$ because of the two copies of $\boldsymbol{R}$ used, and is sometimes called the **Cartesian plane.**

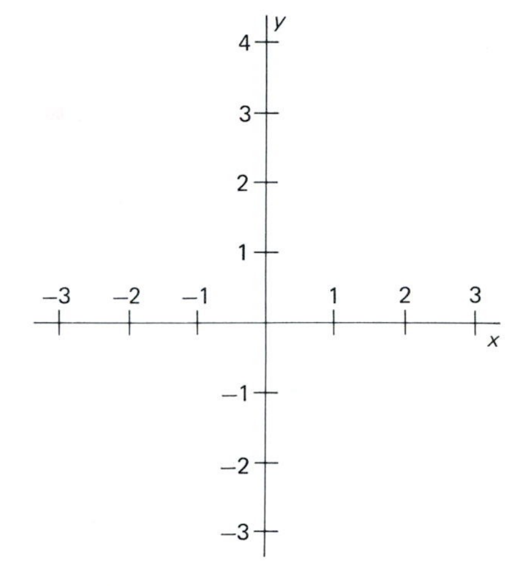

1.15 The coordinate plane

Rectangular coordinates are easy to use because $P(x_1, y_1)$ and $Q(x_2, y_2)$ denote the same point when and only when $x_1 = x_2$ and $y_1 = y_2$. Thus when you know that P and Q are different points, you may conclude that P and Q have different abscissas or different ordinates (or both).

The point of symmetry (0, 0) where the coordinate axes cross is called the **origin.** The points on the x-axis all have coordinates of the form $(x, 0)$, and, while the *real number* x is not the same as the geometric point $(x, 0)$, there are situations in which it is useful to think of the two as the same. Similar remarks apply to points $(0, y)$ on the y-axis.

The concept of distance in the plane is based on the **Pythagorean theorem:** If ABC is a right triangle with right angle C and hypotenuse c, as in Fig. 1.17, then

$$c^2 = a^2 + b^2. \tag{1}$$

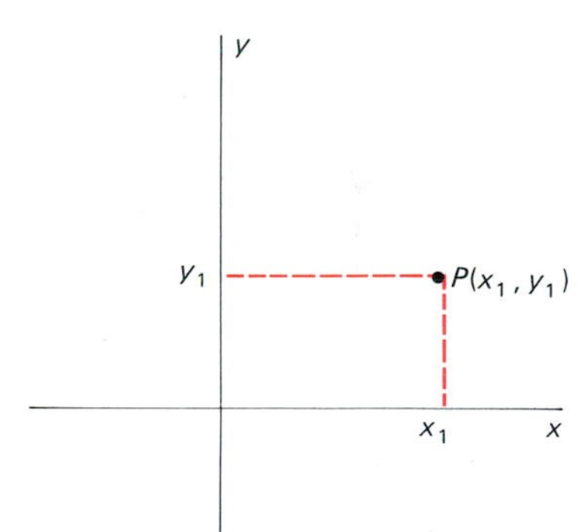

1.16 The point P has rectangular coordinates (x_1, y_1)

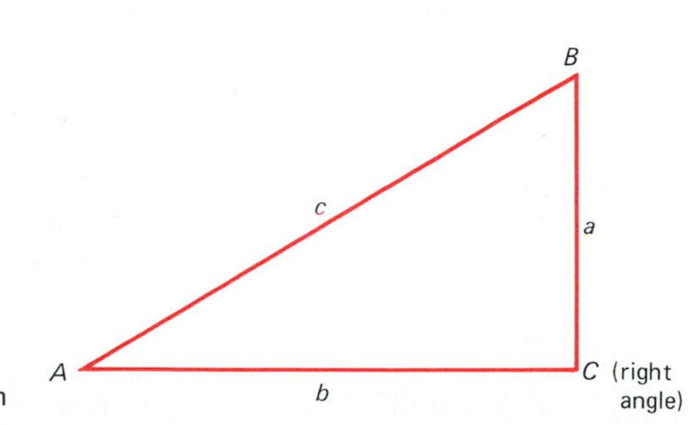

1.17 The Pythagorean theorem

The converse of the Pythagorean theorem is also true—that is, if the three sides of a given triangle satisfy the Pythagorean relation in (1), then the angle opposite side c is a right angle.

The *distance* $d(P_1, P_2)$ between the points P_1 and P_2 is, by definition, the length of the straight line segment joining P_1 and P_2. The following formula gives $d(P_1, P_2)$ in terms of the coordinates of the two points.

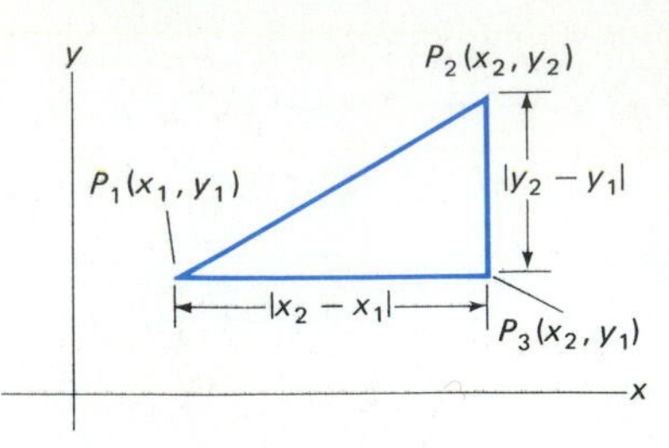

1.18 Use this triangle to deduce the distance formula.

> ***Distance Formula***
> The **distance** between $P_1(x_1, y_1)$ and $P_2(x_2, y_2)$ is
> $$d(P_1, P_2) = \sqrt{(x_2 - x_1)^2 + (y_2 - y_1)^2}. \tag{2}$$

Proof If $x_1 \neq x_2$ and $y_1 \neq y_2$, then the formula in (2) follows from the Pythagorean theorem. Use the right triangle with vertices P_1, P_2, and $P_3(x_2, y_1)$ shown in Fig. 1.18.

If $x_1 = x_2$, then P_1 and P_2 lie in a vertical line. In this case

$$d(P_1, P_2) = |y_2 - y_1| = \sqrt{(y_2 - y_1)^2}.$$

This agrees with the formula in (2) because $x_1 = x_2$. The remaining case, in which $y_1 = y_2$, is similar. ■

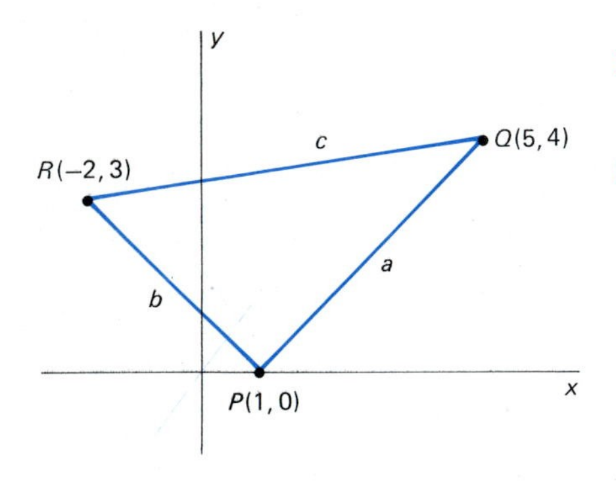

1.19 Is this a right triangle? See Example 1.

EXAMPLE 1 Show that the triangle PQR with vertices $P(1, 0)$, $Q(5, 4)$, and $R(-2, 3)$ is a right triangle. (This triangle is shown in Fig. 1.19.)

Solution The distance formula gives

$$a^2 = [d(P, Q)]^2 = (5 - 1)^2 + (4 - 0)^2 = 32,$$
$$b^2 = [d(P, R)]^2 = (-2 - 1)^2 + (3 - 0)^2 = 18, \quad \text{and}$$
$$c^2 = [d(Q, R)]^2 = (-2 - 5)^2 + (3 - 4)^2 = 50.$$

Because $a^2 + b^2 = c^2$, the *converse* of the Pythagorean theorem implies that RPQ is a right angle. The right angle is at P because P is at the vertex opposite the longest side QR.

Another application of the distance formula is an expression for the coordinates of the midpoint M of the line segment P_1P_2 with endpoints P_1 and P_2, shown in Fig. 1.20. Recall from geometry that M is the one (and only) point of the line segment P_1P_2 equally distant from P_1 and P_2. The following formula tells us that the coordinates of M are the *averages* of the corresponding coordinates of P_1 and P_2.

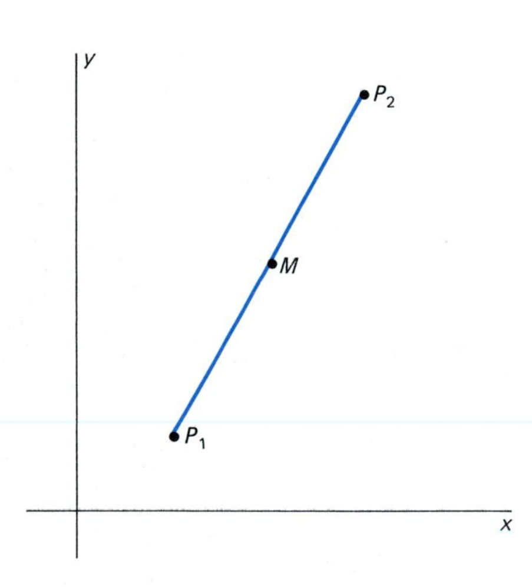

1.20 The midpoint M

> ***Midpoint Formula***
> The **midpoint** of the line segment with endpoints $P_1(x_1, y_1)$ and $P_2(x_2, y_2)$ is the point $M(\bar{x}, \bar{y})$ with coordinates
> $$\bar{x} = \tfrac{1}{2}(x_1 + x_2), \qquad \bar{y} = \tfrac{1}{2}(y_1 + y_2). \tag{3}$$

Proof If you substitute the coordinates of P_1, M, and P_2 in the distance formula, you find that $d(P_1, M) = d(P_2, M)$. All that remains is to show that M is on the line segment P_1P_2. We ask you to do this in Problem 31. ■

STRAIGHT LINES AND SLOPE

We need to define the *slope* of a straight line, a measure of its rate of rise or fall from left to right. Given a line L in the plane, with L not vertical, choose two points $P_1(x_1, y_1)$ and $P_2(x_2, y_2)$ on L. Consider the **increments**

Δx and Δy (read "delta x" and "delta y") in the x- and y-coordinates from P_1 to P_2. These are defined to be

$$\Delta x = x_2 - x_1 \quad \text{and} \quad \Delta y = y_2 - y_1. \tag{4}$$

Civil engineers would call Δy the **rise** from P_1 to P_2 and Δx the **run** from P_1 to P_2, as shown in Fig. 1.21. The **slope** m of the nonvertical line L is then the ratio of the rise to the run; that is,

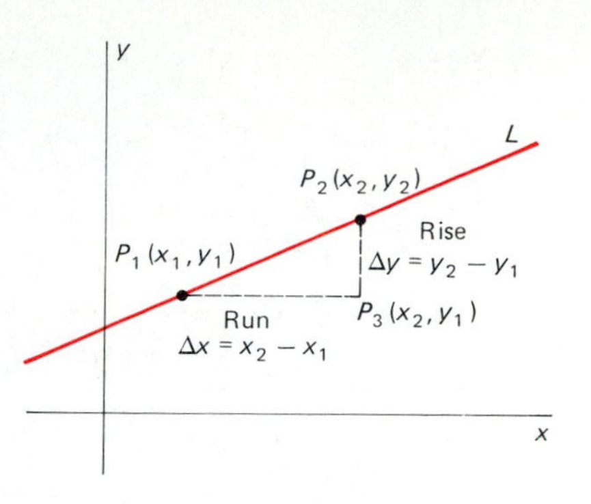

1.21 The slope of a straight line

$$m = \frac{\Delta y}{\Delta x} = \frac{y_2 - y_1}{x_2 - x_1}. \tag{5}$$

This, too, is exactly the definition in civil engineering, and in a surveying text you are likely to find the memory aid

$$\text{"slope} = \frac{\text{rise}}{\text{run}}\text{."}$$

Recall that corresponding sides of similar (that is, equal-angled) triangles have equal ratios. Now note that if $P_3(x_3, y_3)$ and $P_4(x_4, y_4)$ are two other points on L, then the similarity of the triangles in Fig. 1.22 implies that

$$\frac{y_4 - y_3}{x_4 - x_3} = \frac{y_2 - y_1}{x_2 - x_1}.$$

Hence the slope m as defined in Equation (5) does *not* depend upon the particular choice of P_1 and P_2.

If the line L is horizontal, then $\Delta y = 0$, and in this case Equation (5) gives $m = 0$. If L is vertical, then $\Delta x = 0$; the slope of L is *not defined* in this case. Thus we have the following statements:

Horizontal lines have slope zero.
Vertical lines have no slope at all.

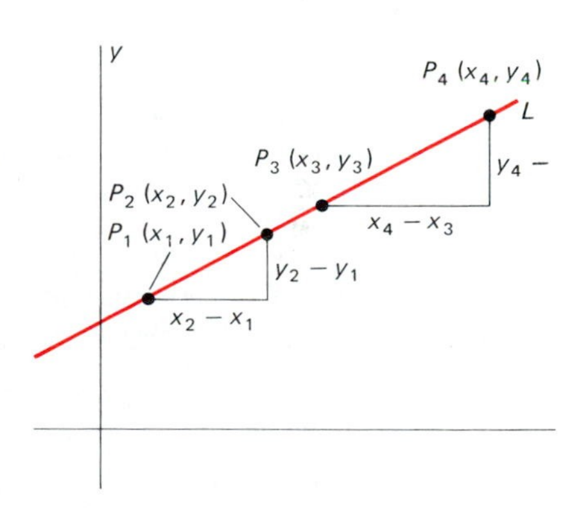

1.22 The result of the slope computation does not depend on which two points of L are used.

EXAMPLE 2 (a) The slope of the line through the points $(3, -2)$ and $(-1, 4)$ is

$$m = \frac{4 - (-2)}{(-1) - 3} = \frac{6}{-4} = -\frac{3}{2}.$$

(b) The points $(3, -2)$ and $(7, -2)$ have the same y-coordinate. Therefore, the line through them is horizontal and thus has slope $m = 0$.

(c) The points $(3, -2)$ and $(3, 4)$ have the same x-coordinate. Thus the line through them is vertical. Hence its slope is undefined.

EQUATIONS OF STRAIGHT LINES

Our immediate goal is to be able to write equations of given straight lines. That is, if L is a straight line, we wish to construct a mathematical sentence about points (x, y) in the plane, a sentence that is true when (x, y) is on L and false when (x, y) is not on L. Of course, this mathematical sentence is merely an equation involving the variables x and y together with some constants determined by the particular line L. To write this equation the concept of the slope of L is almost essential.

Suppose, then, that $P_0(x_0, y_0)$ is a fixed point on the nonvertical line L of slope m. Let $P(x, y)$ be any *other* point on L. We apply Equation (5) with P and P_0 in place of P_1 and P_2 to find that

$$m = \frac{y - y_0}{x - x_0};$$

that is,

$$y - y_0 = m(x - x_0). \tag{6}$$

Because the point (x_0, y_0) satisfies Equation (6), as does every other point of L, and because no other point of the plane can do so, Equation (6) is indeed an equation for the given line L.

The Point-Slope Equation

The point $P(x, y)$ lies on the line with slope m through the fixed point (x_0, y_0) if and only if its coordinates satisfy the equation

$$y - y_0 = m(x - x_0). \tag{6}$$

EXAMPLE 3 Write an equation for the straight line L through the points $P_1(1, -1)$ and $P_2(3, 5)$.

Solution The slope m of L may be obtained from the two given points:

$$m = \frac{5 - (-1)}{3 - 1} = 3.$$

Either P_1 or P_2 will do for the fixed point. We use $P_1(1, -1)$. Then, with the aid of (6), the point-slope equation of L is

$$y + 1 = 3(x - 1).$$

If simplification is appropriate, we write $3x - y = 4$.

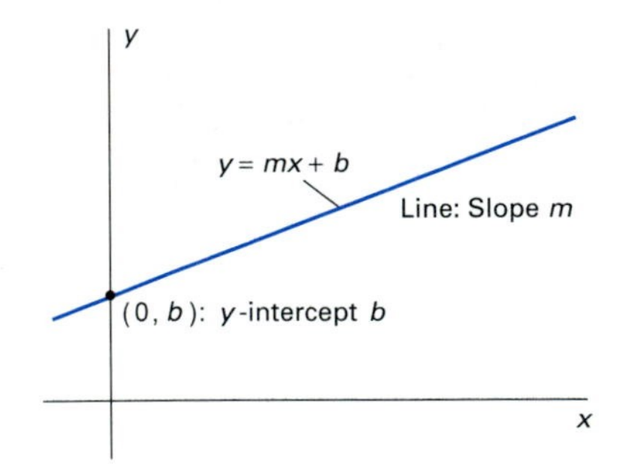

1.23 The straight line with equation $y = mx + b$ has slope m and y-intercept b.

Equation (6) can be written in the form

$$y = mx + b \tag{7}$$

where $b = y_0 - mx_0$ is a constant. Because $y = b$ when $x = 0$, the **y-intercept** of L is the point $(0, b)$ shown in Fig. 1.23. Equations (6) and (7) are different forms of the equation of a straight line.

The Slope-Intercept Equation

The point $P(x, y)$ lies on the line with slope m and y-intercept b, if and only if its coordinates satisfy the equation

$$y = mx + b. \tag{7}$$

Note that Equations (6) and (7) each can be written in the form of the general linear equation

$$Ax + By = C \tag{8}$$

where A, B, and C are constants. Conversely, if $B \neq 0$, then Equation (8) can be written in the form of Equation (7) by division of each term by B. Therefore, Equation (8) represents a straight line with its slope being the coefficient of x *after* solution of the equation for y. If $B = 0$, then (8) reduces to the equation of a vertical line: $x = K$ (where K is a constant). If $A = 0$, then (8) reduces to the equation of a horizontal line: $y = H$ (H a constant). Thus we see that Equation (8) is always an equation of a straight line unless $A = B = 0$. Conversely, every straight line in the plane—even a vertical one—has an equation of the form in (8).

PARALLEL AND PERPENDICULAR LINES

If the line L is not horizontal, it must cross the x-axis. Then its **angle of inclination** is the angle ϕ measured counterclockwise from the positive x-axis to L. It follows that $0° < \phi < 180°$ if ϕ is measured in degrees. The angle of inclination of a horizontal line is $\phi = 0°$. Figure 1.24 makes it clear that this angle ϕ and the slope m of a nonvertical line are related by the equation

$$m = \frac{\Delta y}{\Delta x} = \tan \phi. \tag{9}$$

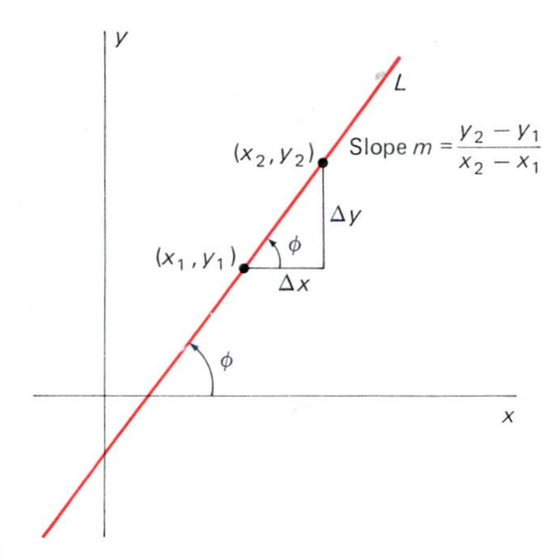

1.24 How is the angle of inclination ϕ related to the slope m?

This is true because if ϕ is an acute angle in a right triangle, then $\tan \phi$ is the ratio of the leg opposite ϕ to the leg adjacent to ϕ (a *leg* of a right triangle is either of its two shorter sides).

It is equally clear that two lines are parallel if and only if they have the same angle of inclination. So it follows from (9) that two parallel and nonvertical lines have the same slope and that two lines with the same slope must be parallel. This completes the proof of Theorem 1.

> ***Theorem 1** Slopes of Parallel Lines*
>
> Two nonvertical lines are parallel if and only if they have the same slope.

Theorem 1 can also be proved without use of the tangent function. The two lines shown in Fig. 1.25 are parallel if and only if the two right triangles are similar, in which case the two slopes are equal.

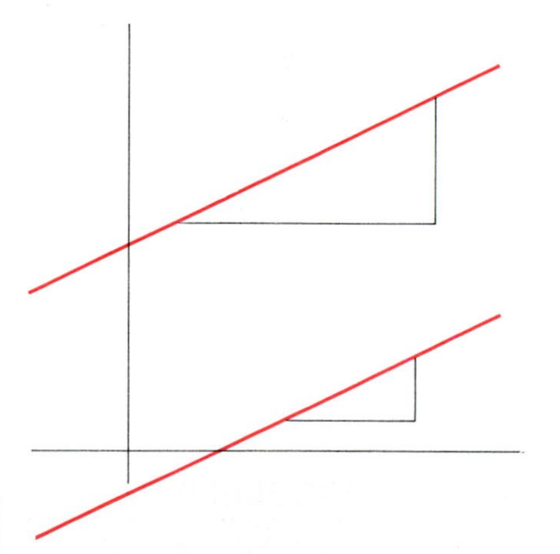

1.25 Two parallel lines

EXAMPLE 4 Write an equation of the line L that passes through the point $P(3, -2)$ and is parallel to the line L' with equation $x + 2y = 6$.

Solution When we solve the equation of L' for y, we get $y = -\frac{1}{2}x + 3$. So L' has slope $m = -\frac{1}{2}$. Because L has the same slope, its point-slope equation is then

$$y + 2 = -\tfrac{1}{2}(x - 3);$$

that is, $x + 2y = -1$.

> ***Theorem 2** Slopes of Perpendicular Lines*
>
> Two lines L_1 and L_2 with slopes m_1 and m_2, respectively, are perpendicular if and only if
>
> $$m_1 m_2 = -1. \tag{10}$$
>
> That is, the slope of each is the *negative reciprocal* of the slope of the other.

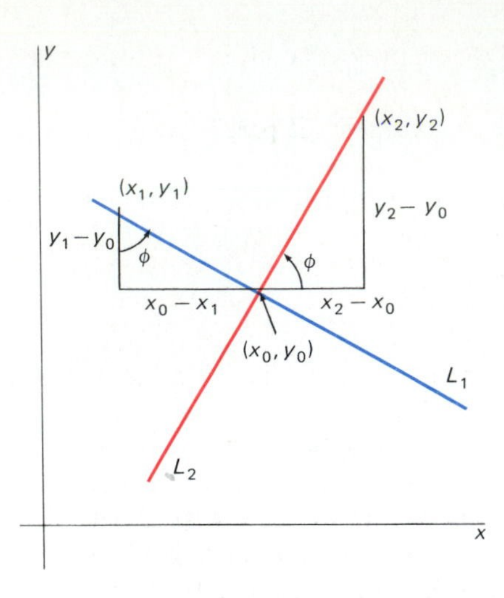

1.26 Illustration of the proof of Theorem 2

Proof If the two lines are perpendicular and the slope of each exists, then neither is horizontal or vertical. Thus the situation is like the one shown in Fig. 1.26, with the two lines intersecting at the point (x_0, y_0). It is easy to see that the two right triangles are similar, so equality of ratios of corresponding sides yields

$$m_2 = \frac{y_2 - y_0}{x_2 - x_0} = \frac{x_0 - x_1}{y_1 - y_0} = -\frac{x_1 - x_0}{y_1 - y_0} = -\frac{1}{m_1}.$$

Thus (10) holds if the two lines are perpendicular. This argument can be reversed to prove the converse—that the lines are perpendicular if $m_1 m_2 = -1$. ■

EXAMPLE 5 Write an equation of the line L through the point $P(3, -2)$ that is perpendicular to the line L' with equation $x + 2y = 6$.

Solution As we saw in Example 4, the slope of L' is $m' = -\frac{1}{2}$. By Theorem 2, the slope of L is $m = -1/m' = 2$, so L has the point-slope equation

$$y + 2 = 2(x - 3);$$

equivalently, $2x - y - 8 = 0$.

Finally, note that the *sign* of the slope m of the line L indicates whether L runs upward or downward as your eyes move from left to right. If $m > 0$, then the angle of inclination ϕ of L must be an acute angle, because $m = \tan \phi$. In this case, L "runs upward." If $m < 0$, then ϕ is obtuse, so that L "runs downward." Figure 1.27 shows the geometry behind these observations.

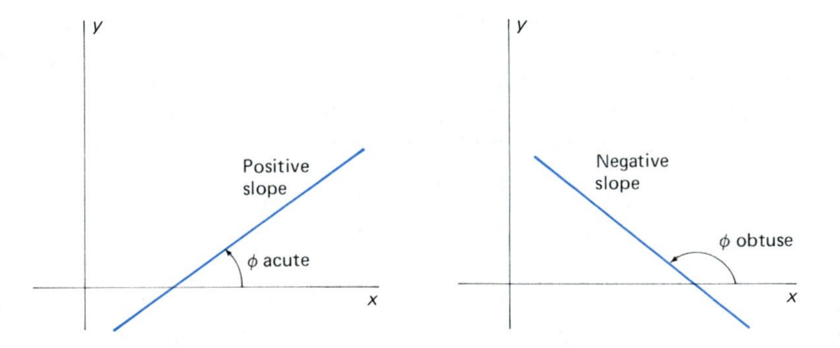

1.27 Positive and negative slope; effect on ϕ

1.2 PROBLEMS

Three points A, B, and C lie on a single straight line if and only if the slope of AB is equal to that of BC. In each of Problems 1–4, plot the three given points, then determine whether or not they lie on a single line.

1. $A(-1, -2)$, $B(2, 1)$, $C(4, 3)$
2. $A(-2, 5)$, $B(2, 3)$, $C(8, 0)$
3. $A(-1, 6)$, $B(1, 2)$, $C(4, -2)$
4. $A(-3, 2)$, $B(1, 6)$, $C(8, 14)$

In Problems 5 and 6, use the concept of slope to show that the points A, B, C, and D are the vertices of a parallelogram.

5. $A(-1, 3)$, $B(5, 0)$, $C(7, 4)$, $D(1, 7)$
6. $A(7, -1)$, $B(-2, 2)$, $C(1, 4)$, $D(10, 1)$

In Problems 7 and 8, show that the given points A, B, and C are the vertices of a right triangle.

7. $A(-2, -1)$, $B(2, 7)$, $C(4, -4)$
8. $A(6, -1)$, $B(2, 3)$, $C(-3, -2)$

In each of Problems 9–13, find the slope m and y-intercept b of the line with the given equation. Then sketch the line.

9. $2x = 3y$
10. $x + y = 1$
11. $2x - y + 3 = 0$
12. $3x + 4y = 6$
13. $2x = 3 - 5y$

In Problems 14–23, write an equation of the straight line L that is described.

14. L is vertical and has x-intercept 7.
15. L is horizontal and passes through $(3, -5)$.
16. L has x-intercept 2 and y-intercept -3.
17. L passes through $(2, -3)$ and $(5, 3)$.
18. L passes through $(-1, -4)$ and has slope $\frac{1}{2}$.
19. L passes through $(4, 2)$ and has angle of inclination $135°$.
20. L has slope 6 and y-intercept 7.
21. L passes through $(1, 5)$ and is parallel to the line with equation $2x + y = 10$.
22. L passes through $(-2, 4)$ and is perpendicular to the line with equation $x + 2y = 17$.
23. L is the perpendicular bisector of the line segment that has endpoints $(-1, 2)$ and $(3, 10)$.
24. Find the perpendicular distance from the point $(2, 1)$ to the line with equation $y = x + 1$.
25. Find the perpendicular distance between the parallel lines $y = 5x + 1$ and $y = 5x + 9$.
26. The points $A(-1, 6)$, $B(0, 0)$, and $C(3, 1)$ are three consecutive vertices of a parallelogram. Find the fourth vertex. (What happens if the word "consecutive" is omitted here?)
27. Prove that the two diagonals of the parallelogram of Problem 26 bisect each other.
28. Show that the points $A(-1, 2)$, $B(3, -1)$, $C(6, 3)$, and $D(2, 6)$ are the vertices of a *rhombus*—a parallelogram with all sides of equal length. Then prove that the diagonals of this rhombus are perpendicular to each other.
29. The points $A(2, 1)$, $B(3, 5)$, and $C(7, 3)$ are the vertices of a triangle. Prove that the line joining the midpoints of AB and BC is parallel to AC.
30. A median of a triangle is a line joining a vertex to the midpoint of the opposite side. Prove that the three medians of the triangle of Problem 29 intersect in a single point.
31. Complete the proof of the midpoint formula in Equation (3). It is necessary to show that the point M actually lies on the segment P_1P_2. One way to do this is to show that the slope of P_1M is equal to the slope of MP_2.
32. Let $P(x_0, y_0)$ be a point of the circle with center $C(0, 0)$ and radius r. Then the tangent line to the circle at P is perpendicular to the radius CP. Prove that the equation of this tangent line is $x_0x + y_0y = r^2$.
33. The Fahrenheit temperature F and the absolute temperature K satisfy a linear equation. Given that $K = 273.16$ when $F = 32$ and that $K = 373.16$ when $F = 212$, express K in terms of F. What is the value of F when $K = 0$?
34. The length L (in centimeters) of a copper rod is a linear function of its Celsius temperature C. If $L = 124.942$ when $C = 20$ and $L = 125.134$ when $C = 110$, express L in terms of C.
35. The owner of a grocery store finds that he can sell 980 gal of milk each week at \$1.69 per gallon and 1220 gal of milk each week at \$1.49 per gallon. Assume a linear relationship between price and sales. How many gallons would he then expect to sell each week at \$1.56 per gallon?

1.3 Graphs of Equations and Functions

We saw in Section 1.2 that the points (x, y) satisfying the linear equation $Ax + By = C$ form a very simple set: a straight line. By contrast, the set of points (x, y) that satisfy the equation

$$x^4 - 4x^3 + 3x^2 + 2x^2y^2 = y^2 + 4xy^2 - y^4$$

is the exotic curve shown in Fig. 1.28 (though the connection between the curve and the equation is by no means obvious). But both the straight line and the complicated curve are examples of graphs.

> ***Definition*** *Graph of an Equation*
> The **graph** of an equation in two variables x and y is the set of all points (x, y) in the plane that satisfy the equation.

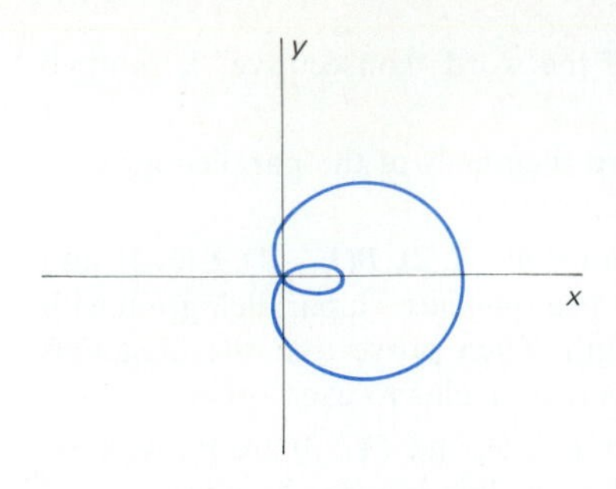

1.28 The graph of the equation $x^4 - 4x^3 + 3x^2 + 2x^2y^2 = y^2 + 4xy^2 - y^4$

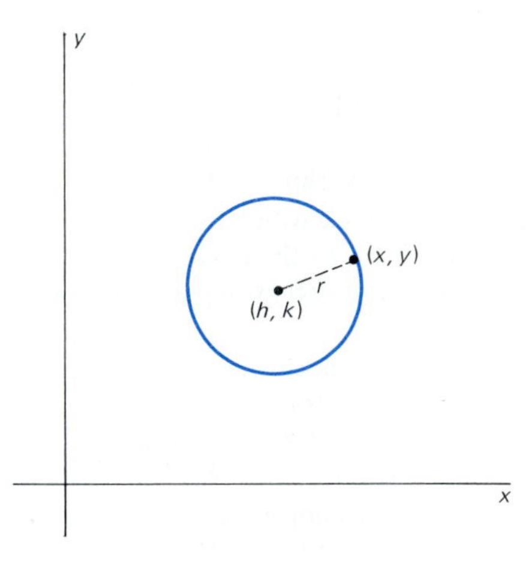

1.29 A translated circle

For instance, the distance formula tells us that the graph of the equation

$$x^2 + y^2 = r^2 \tag{1}$$

is the circle of radius r centered at the origin (0, 0). More generally, the graph of the equation

$$(x - h)^2 + (y - k)^2 = r^2 \tag{2}$$

is the circle of radius r with center (h, k). This follows immediately from the distance formula because the distance between the points (x, y) and (h, k) in Fig. 1.29 is r.

EXAMPLE 1 The equation of the circle with center (3, 4) and radius 10 is

$$(x - 3)^2 + (y - 4)^2 = 100,$$

which may also be written in the form

$$x^2 + y^2 - 6x - 8y - 75 = 0.$$

We may regard the general circle with equation in (2) as a *translate* of the origin-centered circle with equation in (1): Each point of the former is obtained by translating (moving) each point of the xy-plane h units to the right and k units upward. (A negative value of h corresponds to a translation to the left; a negative value of k means a downward translation.) Observe that the equation in (2) of the translated circle with center (h, k) is obtained from Equation (1) upon replacing x by $x - h$ and y by $y - k$. We will see that this principle applies to arbitrary curves:

> When the graph of an equation is translated h units to the right and k units upward, the equation of the translated curve is obtained from the original equation by replacement of x by $x - h$ and of y by $y - k$.

Observe that the equation of a translated circle in (2) can be written in the general form

$$x^2 + y^2 + ax + by = c. \tag{3}$$

What, then, can we do when we encounter an equation already of the form in (3)? We first recognize that it is an equation of a circle; moreover, we can recapture its center and radius by the technique of *completing the square.* To do so, note that

$$x^2 + ax = \left(x + \frac{a}{2}\right)^2 - \frac{a^2}{4},$$

which shows that $x^2 + ax$ can be made into a perfect square by adding to it the square of *half* the coefficient of x.

EXAMPLE 2 Find the center and radius of the circle having the equation

$$x^2 + y^2 - 4x + 6y = 12.$$

Solution We complete the square separately in each of the two variables x and y. This gives

$$(x^2 - 4x + 4) + (y^2 + 6y + 9) = 12 + 4 + 9;$$

$$(x - 2)^2 + (y + 3)^2 = 25.$$

Hence the circle has center $(2, -3)$ and radius 5.

GRAPHS OF FUNCTIONS

The graph of a function is a special case of the graph of an equation.

> ***Definition*** *Graph of a Function*
> The **graph** of a function f is the graph of the equation $y = f(x)$.

Thus the graph of the function f is the set of all points in the plane having the form $(x, f(x))$, where x is in the domain of f. Because the second coordinate of such a point is uniquely determined by its first coordinate, we obtain the following useful principle.

No *vertical* line can intersect the graph of a function in more than one point.

Alternatively,

Each vertical line through a point in the domain of a function meets its graph in exactly one point.

If you examine Fig. 1.28 (at the beginning of this section), you will see from these remarks that the graph shown there cannot be the graph of a *function*, although it *is* the graph of an equation.

EXAMPLE 3 Construct the graph of the absolute value function $f(x) = |x|$.

Solution Recall that

$$|x| = \begin{cases} x & \text{if } x \geqq 0; \\ -x & \text{if } x < 0. \end{cases}$$

So the graph of $y = |x|$ consists of the right half of the line $y = x$ together with the left half of the line $y = -x$, as shown in Fig. 1.30.

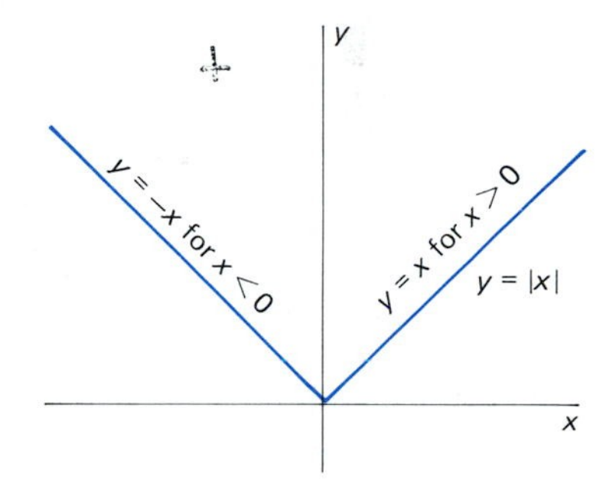

1.30 The graph of the absolute value function $y = |x|$

EXAMPLE 4 Sketch the graph of the reciprocal function

$$f(x) = \frac{1}{x}.$$

Solution Let's examine four cases.

- When x is large positive, $f(x)$ is small and positive.
- When x is small positive, $f(x)$ is large and positive.
- When x is small negative (negative and close to zero), $f(x)$ is large and negative.
- When x is large negative (negative but $|x|$ is large), $f(x)$ is small and negative (negative and close to zero).

To get started with the graph, we can plot a few points, such as $(1, 1)$, $(-1, -1)$, $(10, 0.1)$, $(0.1, 10)$, $(-10, -0.1)$, and $(-0.1, -10)$. The rest of the information displayed here suggests that the actual graph is much like the one shown in Fig. 1.31.

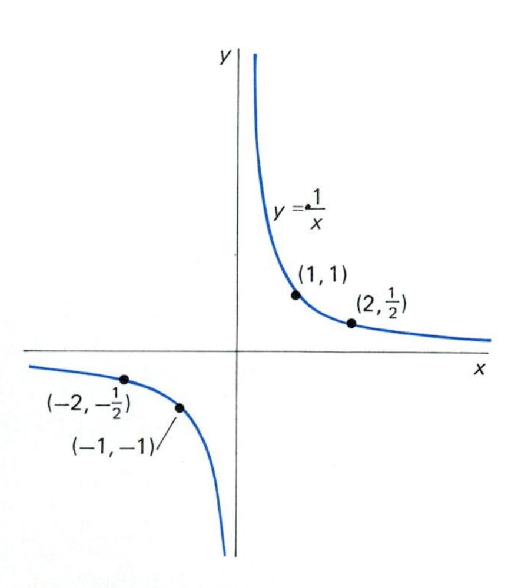

1.31 The graph of the reciprocal function $y = 1/x$

Figure 1.31 exhibits a "gap" or "discontinuity" in the graph of $y = 1/x$ at $x = 0$. Indeed, it might be (and is) called an *infinite discontinuity* because y increases without bound as x approaches zero from the right, while y decreases without bound as x approaches zero from the left. This phenomenon is signaled by the presence of denominators that are zero at certain values of x, as in the case of the functions

$$f(x) = \frac{1}{1-x} \quad \text{and} \quad f(x) = \frac{1}{x^2}$$

that we ask you to graph in the problems.

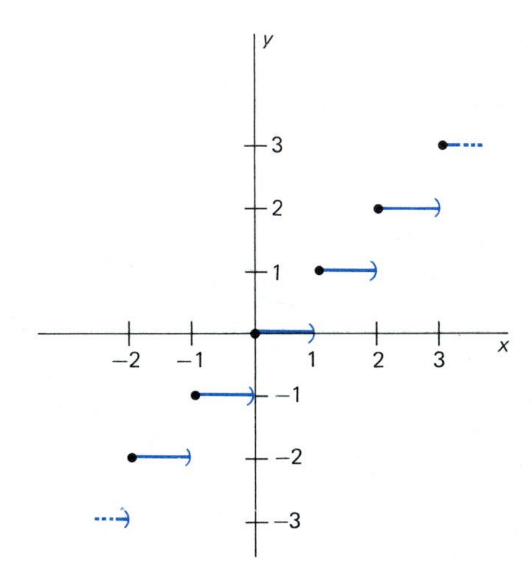

1.32 The graph of the greatest integer ("staircase") function $f(x) = [\![x]\!]$

EXAMPLE 5 Figure 1.32 shows the graph of the greatest integer function $f(x) = [\![x]\!]$ of Example 2 in Section 1.1. Note the "jumps" that occur at integral values of x.

EXAMPLE 6 Graph the function with formula

$$f(x) = x - [\![x]\!] - \tfrac{1}{2}.$$

Solution Recall that $[\![x]\!] = n$, where n is the greatest integer not exceeding x: $n \leqq x < n + 1$. Hence if n is an integer, then

$$f(n) = n - n - \tfrac{1}{2} = -\tfrac{1}{2}.$$

This implies that the point $(n, -\frac{1}{2})$ lies on the graph for each integer n. Next, if $n \leqq x < n + 1$ (where, again, n is an integer), then

$$f(x) = x - n - \tfrac{1}{2}.$$

Because $y = x - n - \frac{1}{2}$ has as its graph a straight line of slope 1, it follows that the graph of f takes the form shown in Fig. 1.33. This *sawtooth function* is another example of a discontinuous function. The values of x where the value of $f(x)$ makes a jump are called **points of discontinuity** of the function f. Thus the points of discontinuity of the sawtooth function are the integers: As x approaches the integer n from the left, the value of $f(x)$ approaches $+\frac{1}{2}$, but $f(x)$ abruptly jumps to the value $-\frac{1}{2}$ when $x = n$. A precise definition of continuity and discontinuity for functions is given in Section 2.4.

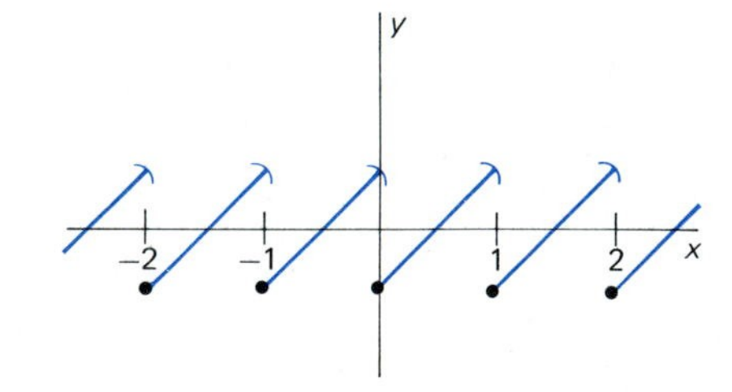

1.33 The graph of the sawtooth function

$$f(x) = x - [\![x]\!] - \tfrac{1}{2}$$

PARABOLAS

The graph of a *quadratic* function of the form

$$f(x) = ax^2 + bx + c \qquad (a \neq 0) \tag{4}$$

is a *parabola* whose shape resembles that of the particular parabola in the following example.

EXAMPLE 7 Construct the graph of the parabola $y = x^2$.

Solution We plot some points in a short table of values.

x	-3	-2	-1	0	1	2	3
$y = x^2$	9	4	1	0	1	4	9

When we draw a smooth curve through these points, we obtain the curve shown in Fig. 1.34.

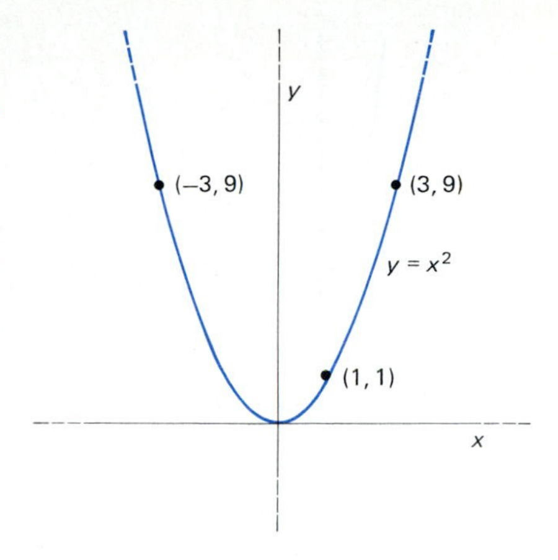

1.34 The graph of the parabola $y = x^2$

The parabola $y = -x^2$ would look similar except that it would open downward instead of upward. More generally, the graph of the equation

$$y = ax^2 \tag{5}$$

is a parabola with its *vertex* at the origin (provided that $a \neq 0$). This parabola opens upward if $a > 0$ and downward if $a < 0$. (For the time being, we may regard the vertex of a parabola as the point at which it "changes direction." The vertex of a parabola of the form $y = ax^2$ ($a \neq 0$) will always be at the origin. A precise definition of *vertex* of a parabola appears in Chapter 10.)

EXAMPLE 8 Construct the graphs of the functions $f(x) = \sqrt{x}$ and $g(x) = -\sqrt{x}$.

Solution After plotting and connecting points as in Example 7, we obtain the parabola $y^2 = x$ shown in Fig. 1.35. Note that the parabola opens to the right. The upper half is the graph of $f(x) = \sqrt{x}$, while the lower half is the graph of $g(x) = -\sqrt{x}$. Thus the union of the graphs of these *two* functions is the graph of the *single* equation $y^2 = x$. More generally, the graph of the equation

$$x = by^2 \tag{6}$$

is a parabola with its vertex at the origin (provided that $b \neq 0$). This parabola opens to the right (as in Fig. 1.35) if $b > 0$, but it opens to the left if $b < 0$.

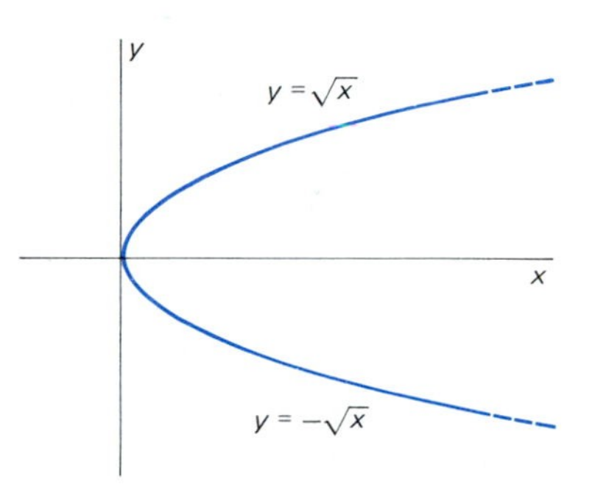

1.35 The graph of the parabola $x = y^2$

Figure 1.36 shows a parabola that has the shape of the "standard parabola" of Example 7, but its vertex is located at the point (h, k). In the indicated uv-coordinate system, the equation of this parabola is $v = u^2$, analogous to Equation (5) with $a = 1$. But the uv-coordinates and xy-coordinates are related as follows:

$$u = x - h, \qquad v = y - k.$$

Hence the xy-coordinate equation of this parabola is

$$y - k = (x - h)^2. \tag{7}$$

Thus when the parabola $y = x^2$ is translated h units to the right and k units upward, the equation in (7) of the translated parabola is obtained by replacing x by $x - h$ and y by $y - k$. This is another instance of the *translation principle* that we observed in connection with circles.

More generally, the graph of any equation of the form

$$y = ax^2 + bx + c \qquad (a \neq 0) \tag{8}$$

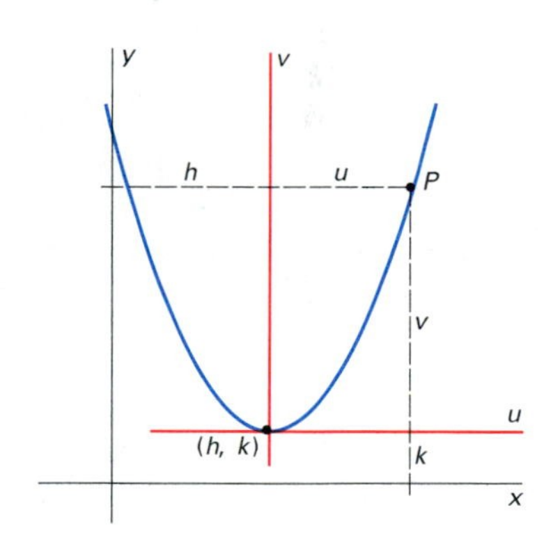

1.36 A translated parabola

can be recognized as a translated parabola by first completing the square in x to obtain an equation of the form

$$y - k = a(x - h)^2. \tag{9}$$

The graph of this equation is a parabola with vertex at (h, k). The size of $|a|$ determines the width of the parabola, while the sign of a determines whether the parabola opens upward or downward.

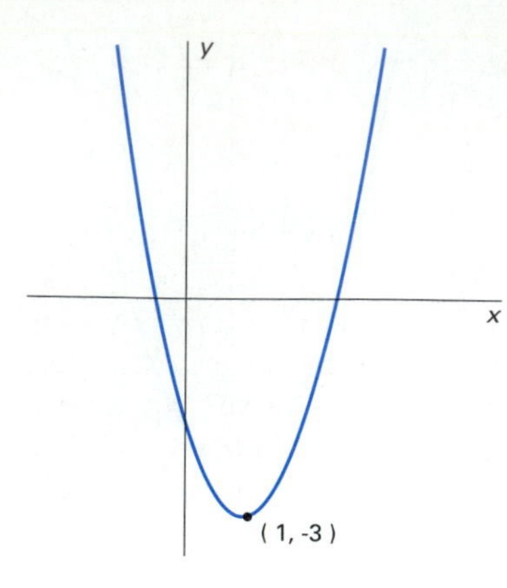

1.37 The parabola $y = 2x^2 - 4x - 1$

EXAMPLE 9 Determine the shape of the graph of the equation

$$y = 2x^2 - 4x - 1. \tag{10}$$

Solution If we complete the square in x, the equation in (10) takes the form

$$y = 2(x^2 - 2x + 1) - 3; \qquad y + 3 = 2(x - 1)^2.$$

Hence the graph of Equation (10) is the parabola shown in Fig. 1.37. It opens upward and its vertex is at $(1, -3)$.

APPLICATIONS OF QUADRATIC FUNCTIONS

In Section 1.1 we saw that a certain type of applied problem may call for us to find the maximum or minimum value attained by a certain function f. If the function f is a quadratic function as in (4), then the graph of $y = f(x)$ will be a parabola. In this case the maximum (or minimum) value of $f(x)$ corresponds to the highest (or lowest) point of the parabola—that is, to its vertex, which we can locate by completing the square as in Example 9.

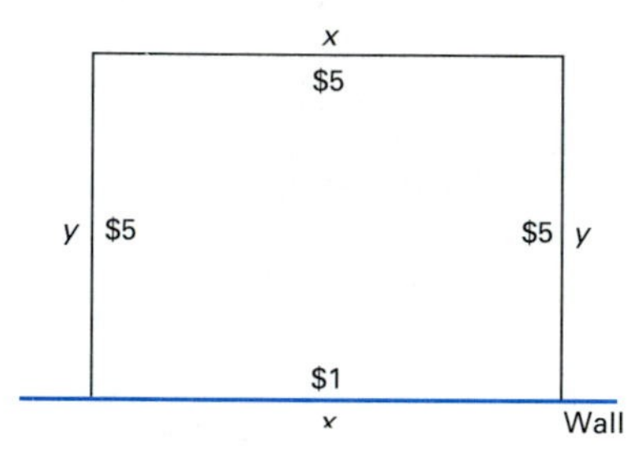

1.38 The animal pen

For instance, recall the animal pen problem of Section 1.1. In Example 6 there we saw that the area A of the pen (see Fig. 1.38) is given as a function of its base length x by

$$A(x) = \tfrac{3}{5}(30x - x^2), \qquad 0 \leqq x \leqq 30. \tag{11}$$

Completing the square, we get

$$\begin{aligned} A &= -\tfrac{3}{5}(x^2 - 30x) = -\tfrac{3}{5}(x^2 - 30x + 225 - 225) \\ &= -\tfrac{3}{5}(x^2 - 30x + 225) + 135; \end{aligned}$$

that is,

$$A - 135 = -\tfrac{3}{5}(x - 15)^2. \tag{12}$$

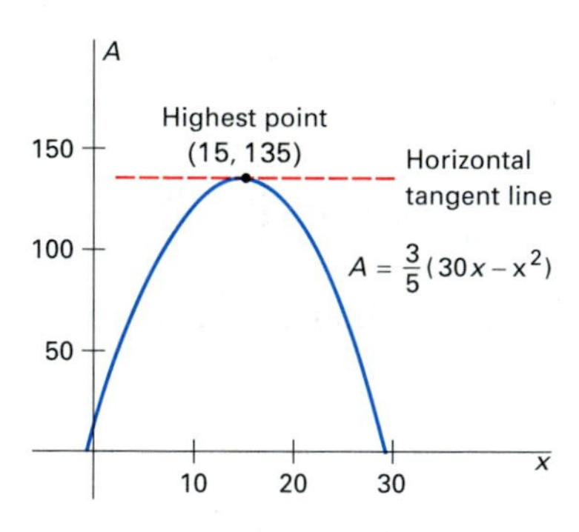

1.39 The graph of $A(x) = \frac{3}{5}(30x - x^2)$ for $0 \leqq x \leqq 30$

It follows from Equation (12) that the graph of Equation (11) is the parabola shown in Fig. 1.39, which opens downward from its vertex (15, 135). This *proves* that the maximum value of $A(x)$ on the interval $[0, 30]$ is the value $A(15) = 135$, as our numerical investigation at the end of Section 1.1 *suggested*. Alternatively, it suffices to glance at Equation (12) in the form

$$A(x) = 135 - \tfrac{3}{5}(x - 15)^2$$

to see that the maximum possible value of $135 - \frac{3}{5}u^2$ is 135, and that this occurs when $u = x - 15 = 0$ or, what is the same thing here, when $x = 15$.

NOTE In Fig. 1.39 we have used different scales along the two axes: a 10-unit interval on the x-axis has the same length as a 50-unit interval on the y-axis. In an applied problem in which there is no physical relation between the units used to measure the independent and dependent variables, there is no compelling reason to use the same scale on the two axes. As in this example, you should choose the scale for convenience in sketching the graph. Only in a geometric problem involving angles or distances will you find it necessary to use the same scale on the two axes.

The technique of completing the square is quite limited; it can be used to find maximum or minimum values only of *quadratic* functions. One of our goals in calculus is to develop a more general technique that can also be applied to a far wider variety of functions.

The basis of this more general technique lies in the following observation: Visual inspection of the graph of

$$A(x) = \tfrac{3}{5}(30x - x^2)$$

in Fig. 1.39 suggests that the tangent line to the curve at its highest point is horizontal. If we *knew* that the tangent line to a graph at its highest point must be horizontal, then our problem would reduce to showing that (15, 135) is the only point of the graph of $y = A(x)$ where the tangent line is horizontal.

But what is meant by the *tangent line* to an arbitrary curve? We pursue this question in Section 2.1. The answer will open the door to the possibility of finding maximum and minimum values of virtually arbitrary functions.

1.3 PROBLEMS

Sketch each of the translated circles in Problems 1–6. Indicate the center and radius of each.

1. $x^2 + y^2 = 4x$

2. $x^2 + y^2 + 6y = 0$

3. $x^2 + y^2 + 2x + 2y = 2$

4. $x^2 + y^2 + 10x - 20y + 100 = 0$

5. $2x^2 + 2y^2 + 2x - 2y = 1$

6. $9x^2 + 9y^2 - 6x - 12y = 11$

Sketch each of the translated parabolas in Problems 7–12. Indicate the vertex of each.

7. $y = x^2 - 6x + 9$

8. $y = 16 - x^2$

9. $y = x^2 + 2x + 4$

10. $2y = x^2 - 4x + 8$

11. $y = 5x^2 + 20x + 23$

12. $y = x - x^2$

The graph of the equation $(x - h)^2 + (y - k)^2 = C$ is a circle if $C > 0$, is the single point (h, k) if $C = 0$, and contains *no* points if $C < 0$. (Why?) Find the graphs of the equations in Problems 13–16. If the graph is a circle, give its center and radius.

13. $x^2 + y^2 - 6x + 8y = 0$

14. $x^2 + y^2 - 2x + 2y + 2 = 0$

15. $x^2 + y^2 + 2x + 6y + 20 = 0$

16. $2x^2 + 2y^2 - 2x + 6y + 5 = 0$

Sketch the graphs of the functions in Problems 17–40, taking into account the domain of definition of each function, and plotting points as necessary.

17. $f(x) = 2 - 5x, \quad -1 \leqq x \leqq 1$

18. $f(x) = 2 - 5x, \quad 0 \leqq x < 2$

19. $f(x) = 10 - x^2$

20. $f(x) = 1 + 2x^2$

21. $f(x) = x^3$

22. $f(x) = x^4$

23. $f(x) = \sqrt{4 - x^2}$

24. $f(x) = -\sqrt{9 - x^2}$

25. $f(x) = \sqrt{x^2 - 9}$

26. $f(x) = \dfrac{1}{1 - x}$

27. $f(x) = \dfrac{1}{x + 2}$

28. $f(x) = \dfrac{1}{x^2}$

29. $f(x) = \dfrac{1}{(x - 1)^2}$

30. $f(x) = \dfrac{|x|}{x}$

31. $f(x) = \dfrac{1}{2x + 3}$

32. $f(x) = \dfrac{1}{(2x + 3)^2}$

33. $f(x) = \sqrt{1 - x}$

34. $f(x) = \dfrac{1}{\sqrt{1 - x}}$

35. $f(x) = \dfrac{1}{\sqrt{2x + 3}}$

36. $f(x) = |2x - 2|$

37. $f(x) = |x| + x$

38. $f(x) = |x - 3|$

39. $f(x) = |2x + 5|$

40. $f(x) = \begin{cases} |x| & \text{if } x < 0; \\ x^2 & \text{if } x \geqq 0. \end{cases}$

Graph the functions given in Problems 41–46, indicating any points of discontinuity.

41. $f(x) = \begin{cases} 0 & \text{if } x < 0; \\ 1 & \text{if } x \geqq 0 \end{cases}$

42. $f(x) = 1$ if x is an integer; otherwise, $f(x) = 0$

43. $f(x) = [\![2x]\!]$

44. $f(x) = \dfrac{x - 1}{|x - 1|}$

45. $f(x) = [\![x]\!] - x$

46. $f(x) = [\![x]\!] + [\![-x]\!] + 1$

In each of Problems 47–50, make use of the method of completing the square to graph the appropriate function, and thereby determine the maximum or minimum value requested.

47. If a ball is thrown straight upward with initial velocity 96 ft/s, then its height t seconds later is $y = 96t - 16t^2$ (ft).

Determine the maximum height the ball attains by constructing the graph of y as a function of t.

48. Find the maximum possible area of the rectangle with perimeter 100 described in Problem 60 of Section 1.1.

49. Find the maximum possible value of the product of two positive numbers whose sum is 50.

50. In Problem 62 of Section 1.1, you were asked to express the daily production of a certain oil field as a function $P = f(x)$ of the number x of new oil wells drilled. Now construct the graph of f and use it to find the value of x that maximizes P.

CHAPTER 1 REVIEW: Definitions, Concepts, Results

Use the list below as a guide to concepts that you may need to review.

1. Rational and irrational numbers
2. The real line
3. Properties of inequalities
4. Absolute value of a real number
5. Properties of the absolute value function
6. The triangle inequality
7. Open and closed intervals
8. Solving inequalities
9. The definition of function
10. The domain of a function
11. Independent and dependent variables
12. The coordinate plane
13. The Pythagorean theorem
14. The distance formula
15. The midpoint formula
16. The slope of a straight line
17. Point-slope equation of a line
18. Slope-intercept equation of a line
19. Slope relationship of parallel lines
20. Slope relationship of perpendicular lines
21. The graph of an equation
22. The graph of a function
23. Equations and translates of circles
24. Parabolas and graphs of quadratic functions

CHAPTER 1 MISCELLANEOUS PROBLEMS

Express the solutions of the inequalities in Problems 1–10 in terms of intervals.

1. $3x + 5 < -10$
2. $2x + 17 \geqq 33$
3. $2 - 5x < 17 + 2x$
4. $0 < 3x + 4 < 20$
5. $-3 < 1 - 2x < -1$
6. $x^2 + 1 > 2x$
7. $-7 \leqq 1 - 4x < 3$
8. $3 < |2x - 5| \leqq 6$
9. $-2 \leqq \dfrac{3}{4x - 1} \leqq 4$
10. $\dfrac{2}{|3x - 4|} < 1$

In each of Problems 11–20, find the domain of definition of the function with the given formula.

11. $f(x) = \sqrt{x - 4}$
12. $f(x) = \dfrac{1}{2 - x}$
13. $f(x) = \dfrac{1}{x^2 - 9}$
14. $f(x) = \dfrac{x}{x^2 + 1}$
15. $f(x) = (1 + \sqrt{x})^3$
16. $f(x) = \dfrac{x + 1}{x^2 - 2x}$
17. $f(x) = \sqrt{2 - 3x}$
18. $f(x) = \dfrac{1}{\sqrt{9 - x^2}}$
19. $f(x) = (x - 2)(4 - x)$
20. $f(x) = \sqrt{(x - 2)(4 - x)}$

21. In accord with Boyle's law, the pressure p (lb/in.^{2}) and volume v (in.^{3}) of a certain gas satisfy the condition $pv = 800$. What is the range of possible values of the pressure, given $100 \leqq v \leqq 200$?

22. The relationship between the Fahrenheit temperature F and the Celsius temperature C is given by

$$F = 32 + \tfrac{9}{5}C.$$

If the temperature on a certain day ranged from a low of 70°F to a high of 90°F, what was the range of the temperature in degrees Celsius?

23. An electrical circuit contains a battery supplying E volts in series with a resistance of R ohms, as shown in Fig. 1.40. Then the current of I amperes that flows in the circuit satisfies Ohm's law, $E = IR$ If $E = 100$ and $25 < R < 50$, what is the range of possible values of I?

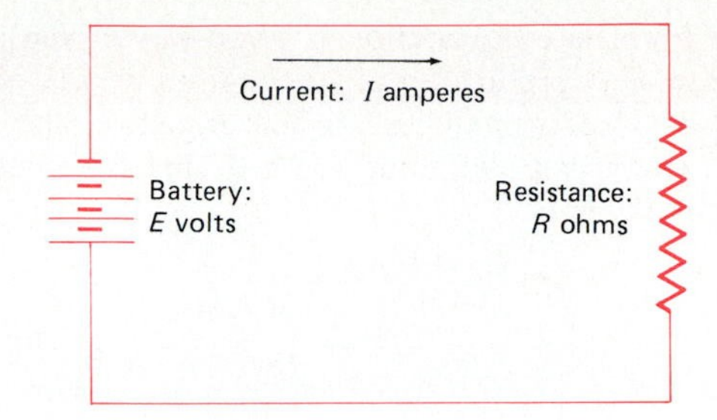

1.40 A simple electric circuit

24. The period T (seconds) of a simple pendulum of length L (ft) is given by $T = 2\pi\sqrt{L/32}$. If $3 < L < 4$, what is the range of possible values of T?

25. Express the volume V of a cube as a function of its total surface area S.

26. The height of a certain right circular cylinder is equal to its radius. Express its total surface area A (including both ends) as a function of its volume.

27. Express the area A of an equilateral triangle as a function of its perimeter P.

28. A piece of wire 100 in. long is cut into two pieces of lengths x and $100 - x$. The first piece is bent into the shape of a square and the second is bent into the shape of a circle. Express as a function of x the sum A of the areas of the square and the circle.

In each of Problems 29–34, write an equation of the straight line L described.

29. L passes through $(-3, 5)$ and $(1, 13)$.

30. L passes through $(4, -1)$ and has slope -3.

31. L has slope $\frac{1}{2}$ and y-intercept -5.

32. L passes through $(2, -3)$ and is parallel to the line with equation $3x - 2y = 4$.

33. L passes through $(-3, 7)$ and is perpendicular to the line with equation $y - 2x = 10$.

34. L is the perpendicular bisector of the segment joining $(1, -5)$ and $(3, -1)$.

Sketch the graphs of the equations and functions given in Problems 35–44.

35. $2x - 5y = 7$

36. $|x - y| = 1$

37. $x^2 + y^2 = 2x$

38. $x^2 + y^2 = 4y - 6x + 3$

39. $y = 2x^2 - 4x - 1$

40. $y = 4x - x^2$

41. $f(x) = \dfrac{1}{x + 5}$

42. $f(x) = \dfrac{1}{4 - x^2}$

43. $f(x) = |x - 3|$

44. $f(x) = |x - 3| + |x + 2|$

45. Apply the triangle inequality twice to show that

$$|a + b + c| \leqq |a| + |b| + |c|$$

for arbitrary real numbers a, b, and c.

46. Write $a = (a - b) + b$ to deduce from the triangle equality that

$$|a| - |b| \leqq |a - b|$$

for arbitrary real numbers a and b.

47. Solve the inequality $x^2 - x - 6 > 0$. (*Suggestion:* Conclude from the factorization $x^2 - x - 6 = (x - 3)(x + 2)$ that the quantities $x - 3$ and $x + 2$ are either both positive or both negative. Consider the two cases separately to conclude that the solution set is $(-\infty, -2) \cup (3, +\infty)$.)

Use the method of Problem 47 to solve the inequalities in Problems 48–50.

48. $x^2 - 3x + 2 < 0$

49. $x^2 - 2x - 8 > 0$

50. $2x \geqq 15 - x^2$

51. Prove that

$$\max\{x, y\} = \tfrac{1}{2}(x + y + |x - y|)$$

for arbitrary real numbers x and y.

52. Derive a formula like the one in Problem 51 for $\min\{x, y\}$.

53. Outline Chapter 1.

chapter two

Limits of Functions

2.1 Tangent Lines and the Derivative—A First Look

At the end of Section 1.3 we saw that certain applied problems raise the question of what is meant by the *tangent line* L at a point P of a general curve $y = f(x)$. In this section we will see that this question leads to the introduction of a new function called the *derivative* of f. In turn, the derivative involves the new concept of *limits*. Our use of limits here will be quite intuitive and informal. Indeed, the main purpose of this section is to provide background and motivation for the treatment of limits in the subsequent sections of Chapter 2. We will initiate a detailed study of derivatives in Chapter 3.

To begin our investigation of tangent lines, we first need to decide how to *define* the tangent line L at an arbitrary point P of a curve $y = f(x)$. Our intuitive idea is this: The tangent line L should be the straight line through P having the same direction as the curve does there. Because the direction of a line is determined by its slope, our plan for defining the tangent line amounts to finding an appropriate "slope-prediction formula"—one that will give the proper slope of the desired tangent line.

Our first example illustrates this approach in the case of one of the simplest of all nonstraight curves, the parabola with equation $y = x^2$.

EXAMPLE 1 Determine the slope of the tangent line to the parabola $y = x^2$ at the point $P(a, a^2)$.

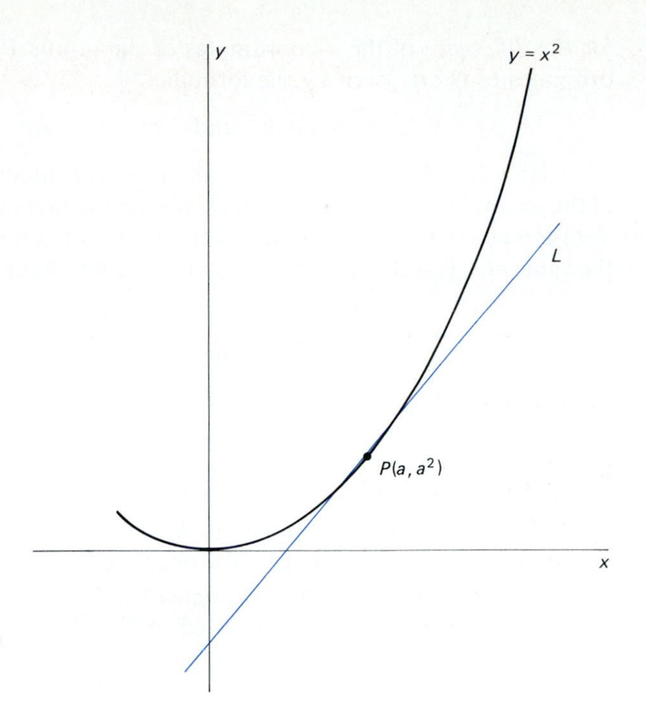

2.1 The tangent line at P should have the same direction as the curve does there.

Solution In Fig. 2.1, we show the parabola $y = x^2$ with a typical point $P(a, a^2)$ marked on it. The figure also shows the result of a visual guess about the position of the desired tangent line L. Our problem is to find the slope of L.

We cannot immediately calculate the slope of L because we know the coordinates of only *one* point $P(a, a^2)$ on the line L. Hence we begin with another line whose slope we *can* compute. Figure 2.2 shows the **secant line** K that passes through the point P and nearby point $Q(b, b^2)$ of the parabola $y = x^2$. If we write

$$h = b - a \neq 0$$

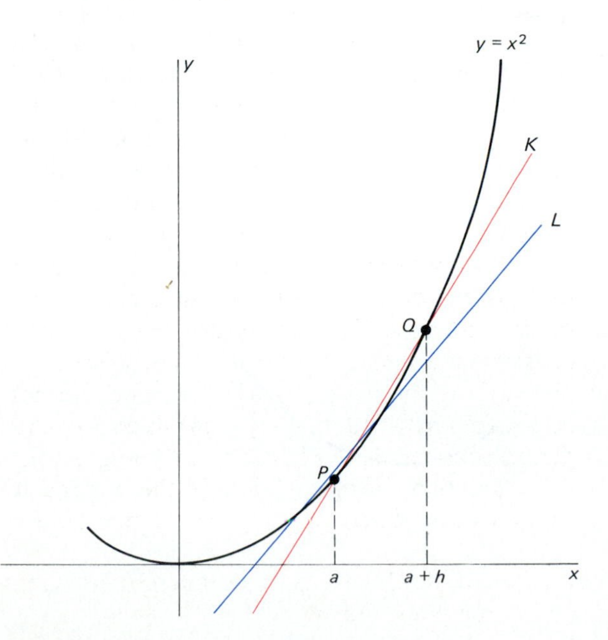

2.2 The secant line K passes through two points —P and Q—which we can use to compute its slope.

for the *difference* of the x-coordinates of the points P and Q, then the coordinates of Q are given by the formulas

$$b = a + h \quad \text{and} \quad b^2 = (a + h)^2.$$

Now $Q \neq P$, so we can use the definition of slope to compute the slope of the secant line K through P and Q. We denote this slope with the function notation $m(h)$ because the slope m actually is a function of h; if you change the value of h, you change the line K, and thereby change its slope. Therefore,

$$m(h) = \frac{b^2 - a^2}{b - a} = \frac{(a + h)^2 - a^2}{(a + h) - a} = \frac{2ah + h^2}{h}. \tag{1}$$

Because h is nonzero, we may cancel it in the last fraction. Thus we find that

$$m(h) = 2a + h. \tag{2}$$

Now imagine what happens as the point Q draws nearer and nearer to the point P. This corresponds to h approaching zero. The line K still passes through P and Q, but it pivots around the fixed point P. As h approaches zero, the secant line K comes closer to coinciding with the tangent line L. This phenomenon is suggested in Fig. 2.3, which shows the secant line K almost indistinguishable from the tangent line L.

Our idea is that the tangent line L must, by definition, lie in the limiting position of the secant line K. To see precisely what this means, we can examine what happens to the slope of K as K pivots into coincidence with L:

As h approaches zero,
Q approaches P, and so
K approaches L; meanwhile,
the slope of K approaches the slope of L.

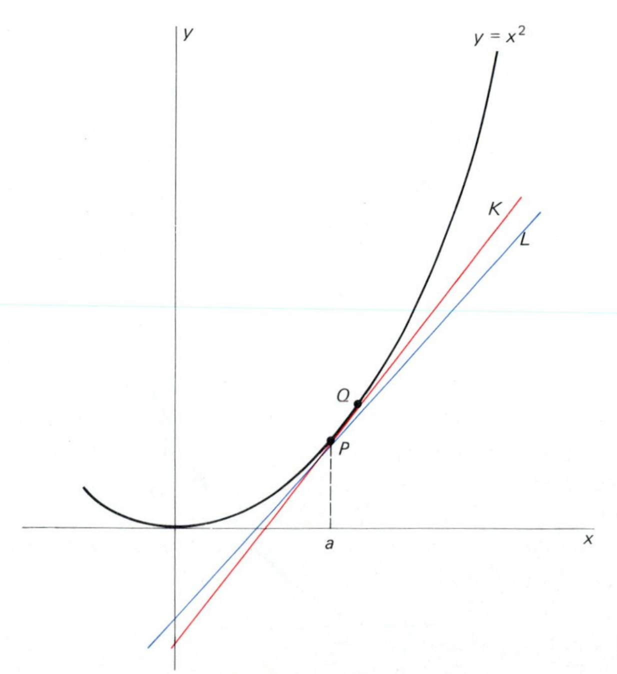

2.3 As $h \to 0$, Q approaches P, and K moves into coincidence with the tangent line L.

But we found in Equation (2) that the slope of the secant line K is

$$m(h) = 2a + h.$$

If we think of numerically smaller and smaller values of h—such as $h = 0.1$, $h = 0.01$, and $h = 0.001$—then it is clear that

$$m(h) = 2a + h \quad \text{approaches } 2a \text{ as } h \text{ approaches zero.}$$

Consequently, we *must* define the tangent line L to be that straight line through P with slope

$$m = 2a.$$

In this circumstance we say that $m = 2a$ is the **limit of $m(h)$** as h approaches zero, and we write

$$m = \lim_{h \to 0} m(h) = \lim_{h \to 0} (2a + h) = 2a. \tag{3}$$

The phrase "$h \to 0$" is read "h approaches zero," and as an alternative to the form in Equation (3) we will sometimes write

$$2a + h \to 2a \quad \text{as} \quad h \to 0.$$

The result $m = 2a$ in (3) is a "slope predictor" for tangent lines to the parabola $y = x^2$. We can use it to write the *equation* of the tangent line to the parabola $y = x^2$ at any desired point (a, a^2).

EXAMPLE 2 To find the equation of the line tangent to the parabola $y = x^2$ at the point (2, 4), we take $a = 2$ in Equation (3) and find that the slope of the tangent is $m = 4$. The point-slope equation of the tangent line is therefore

$$y - 4 = 4(x - 2);$$

its slope-intercept form is $y = 4x - 4$.

The general case $y = f(x)$ is scarcely more complicated than the special case $y = x^2$. Suppose that $y = f(x)$ is given and that we want to find the slope of the tangent line L to its graph at the point $(a, f(a))$. As shown in Fig. 2.4, let K be the secant line passing through the point $P(a, f(a))$ and a nearby point $Q(a + h, f(a + h))$ on the graph. The slope of this secant line K is

$$m_{\text{sec}} = m(h) = \frac{\Delta y}{\Delta x} = \frac{f(a + h) - f(a)}{h} \qquad (\text{for } h \neq 0). \tag{4}$$

We now force Q to approach P along the graph of f by making h approach zero. Suppose that m_{sec} approaches the number m as h gets closer and closer to zero. Then the *tangent line* L to the curve $y = f(x)$ at the point $P(a, f(a))$ is, *by definition*, the line through P whose slope is this number m.

To describe the fact that m_{sec} approaches m as h approaches zero, we call m the **limit of m_{sec}** as h approaches zero, and we write

$$m = \lim_{h \to 0} m_{\text{sec}} = \lim_{h \to 0} \frac{f(a + h) - f(a)}{h}. \tag{5}$$

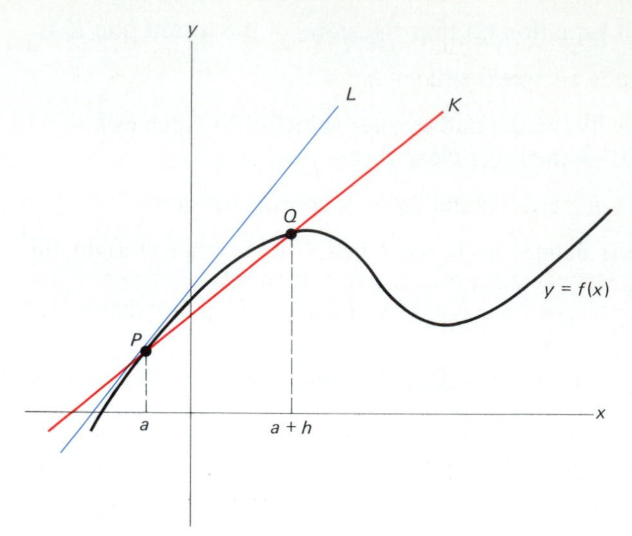

2.4 As $h \to 0$, $Q \to P$, and the slope of K approaches the slope of the tangent line L.

The slope m in Equation (5) depends both on the function f and on the number a. We indicate this by writing

$$m = f'(a) = \lim_{h \to 0} \frac{f(a + h) - f(a)}{h}. \tag{6}$$

Note that f' is a *function* with independent variable a in (6). To return to the standard symbol x for the independent variable, we simply replace a by x. (Nothing was said about the particular value of a, so it really changes nothing to write x in place of a.) So here is the definition of the new function f':

$$f'(x) = \lim_{h \to 0} \frac{f(x + h) - f(x)}{h}. \tag{7}$$

This new function f' is defined for all values of x such that the limit in (7) exists. It is called the **derivative** of the original function f, and the process of computing the formula for f' (read "f prime") is called **differentiation** of f.

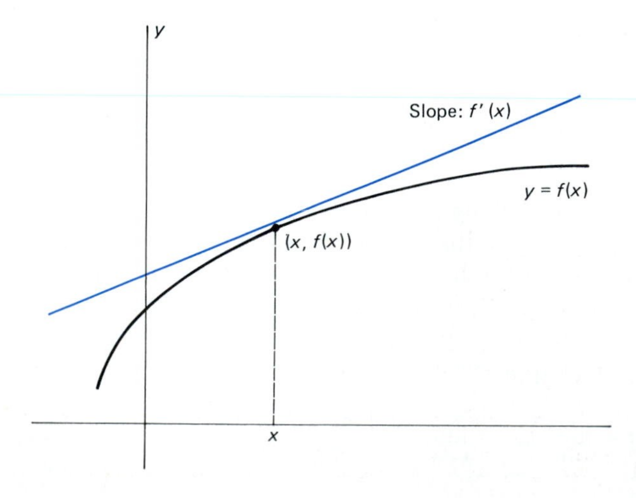

2.5 The slope of the tangent at $(x, f(x))$ is $f'(x)$.

Our discussion leading up to Equation (7) shows that the derivative of f has the following important geometric interpretation:

$f'(x)$ is the slope of the line tangent to the curve $y = f(x)$ at the point $(x, f(x))$.

This interpretation is illustrated in Fig. 2.5.

EXAMPLE 3 Upon replacing a by x in (3), we see that the slope of the tangent line to the parabola $y = x^2$ at the point (x, x^2) is $m = 2x$. In the language of derivatives:

$$\text{If} \quad f(x) = x^2, \quad \text{then} \quad f'(x) = 2x. \tag{8}$$

What's important is the *idea* a formula like the one in (8) expresses, not the particular notation used. Thus it follows immediately from Equation (8) that if $g(t) = t^2$, then $g'(t) = 2t$. In short, the derivative of the squaring function is the doubling function.

In order for us to find the derivative of a given function f, the definition

$$f'(x) = \lim_{h \to 0} \frac{f(x + h) - f(x)}{h}$$

in (7) calls for us to carry out the following four steps.

1. Recall the definition of $f'(x)$.
2. Substitute the given function f.
3. Make algebraic substitutions until Step 4 can be carried out.
4. Recognize the value of the limit as $h \to 0$.

We should remember that x may be thought of as a *constant* throughout this computation—it is h that is the variable in this four-step process.

EXAMPLE 4 The derivative of the function

$$f(x) = 3x + 2$$

surely *must* be the constant function $f'(x) = 3$, because the line $y = 3x + 2$ of slope $m = 3$ *should* be its own tangent line at every point. To check this guess by using the four-step process above, we find that

$$f'(x) = \lim_{h \to 0} \frac{f(x + h) - f(x)}{h} = \lim_{h \to 0} \frac{[3(x + h) + 2] - [3x + 2]}{h}$$

$$= \lim_{h \to 0} \frac{3h}{h} = \lim_{h \to 0} 3 = 3.$$

Therefore,

$$f'(x) = 3$$

(a constant function). In the final step we conclude that the limit is 3 because the constant 3 is "not going anywhere" as $h \to 0$.

Note that the symbol $\lim_{h \to 0}$ signifies an operation to be performed, so we must continue to write it until the final step—when the operation of taking the limit *is* performed.

With computations like those in Examples 1 and 4, it is just as easy to find—once and for all—the derivative of an arbitrary *quadratic function* $f(x) = ax^2 + bx + c$ (where a, b, and c are constants).

> ***Theorem*** *Differentiation of Quadratic Functions*
> If $f(x) = ax^2 + bx + c$, then
>
> $$f'(x) = 2ax + b. \tag{9}$$

Proof The four-step process for applying the definition of the derivative yields

$$\begin{aligned} f'(x) &= \lim_{h \to 0} \frac{f(x+h) - f(x)}{h} \\ &= \lim_{h \to 0} \frac{[a(x+h)^2 + b(x+h) + c] - [ax^2 + bx + c]}{h} \\ &= \lim_{h \to 0} \frac{2axh + bh + ah^2}{h} = \lim_{h \to 0} (2ax + b + ah). \end{aligned}$$

Thus we find that

$$f'(x) = 2ax + b$$

because the value of ah approaches zero as $h \to 0$. ■

In geometric terms, this theorem tells us that the slope-prediction formula for curves of the form $y = ax^2 + bx + c$ is

$$m = 2ax + b. \tag{10}$$

How would one find the point $P(c, c^2)$ on the parabola $y = x^2$ closest to the point (3, 0)? Intuitively, the line segment N with endpoints (3, 0) and (c, c^2) should be perpendicular or *normal* to the graph at the point (c, c^2), as in Fig. 2.6. This warrants a precise definition: The line N through the point

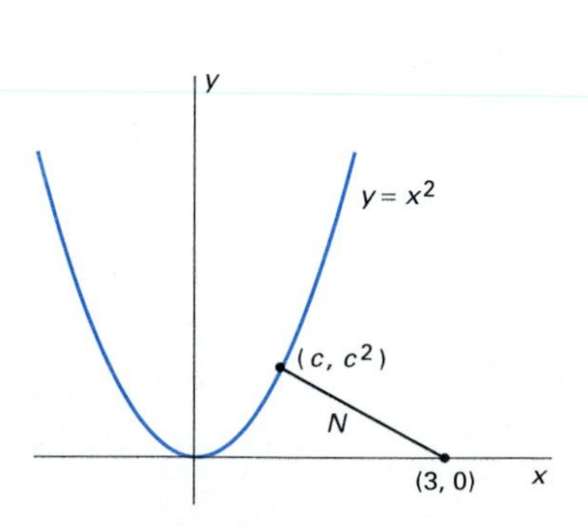

2.6 The normal line N from the point (3, 0) to the point (c, c^2) on the parabola $y = x^2$

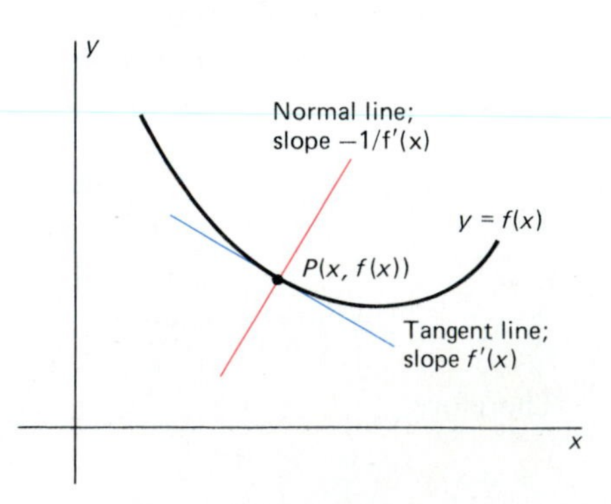

2.7 The tangent and normal lines to a curve at a point

P on the curve $y = f(x)$ is said to be **normal** to the curve at P provided that N is perpendicular to the tangent line at P, as in Fig. 2.7. By Theorem 2 in Section 1.2, its slope is $-1/f'(c)$ provided that $f'(c) \neq 0$. Thus with the aid of the formula in (10), we can write equations of normal lines at points of quadratic curves as well as equations of tangent lines.

Now that the derivative formula in (9) is available, we can simply apply it to differentiate specific quadratic functions instead of retracing the steps in the proof of the theorem. For example, if

$$f(x) = 2x^2 - 3x + 5,$$

then we can immediately write

$$f'(x) = (2)(2x) - 3 = 4x - 3.$$

EXAMPLE 5 Write equations for both the tangent line and the normal line to the parabola $y = 2x^2 - 3x + 5$ at the point $P(-1, 10)$.

Solution We use the derivative just computed. The slope of the tangent line at $(-1, 10)$ is

$$f'(-1) = (4)(-1) - 3 = -7.$$

Hence the point-slope equation of the desired tangent line is

$$y - 10 = -7(x + 1).$$

The normal line has slope $m = -1/(-7)$, so its point-slope equation is

$$y - 10 = \tfrac{1}{7}(x + 1).$$

EXAMPLE 6 Differentiate the function

$$f(x) = \frac{1}{x + 1}.$$

Solution No formula for this derivative has yet been established, so we have no recourse but to carry out the four steps that use the definition of the derivative. We get

$$\begin{aligned}
f'(x) &= \lim_{h \to 0} \frac{f(x + h) - f(x)}{h} \\
&= \lim_{h \to 0} \frac{1}{h}\left(\frac{1}{x + h + 1} - \frac{1}{x + 1}\right) \\
&= \lim_{h \to 0} \frac{1}{h}\left(\frac{(x + 1) - (x + h + 1)}{(x + h + 1)(x + 1)}\right) \\
&= \lim_{h \to 0} \frac{1}{h} \cdot \frac{-h}{(x + h + 1)(x + 1)} \\
&= \lim_{h \to 0} \frac{-1}{(x + h + 1)(x + 1)}.
\end{aligned}$$

It is apparent that $\lim_{h \to 0} (x + h + 1) = x + 1$. Hence we finally obtain the derivative,

$$f'(x) = -\frac{1}{(x + 1)^2}.$$

In conclusion, we apply the derivative to wrap up our continuing discussion of the animal pen problem of Section 1.1. In Example 6 there, we found that the area A of the pen (see Fig. 2.8) is given as a function of its base length x by

$$A(x) = \tfrac{3}{5}(30x - x^2) = -\tfrac{3}{5}x^2 + 18x, \qquad 0 \leqq x \leqq 30. \tag{11}$$

Our problem is to find the maximum value of $A(x)$ for x in the closed interval $[0, 30]$.

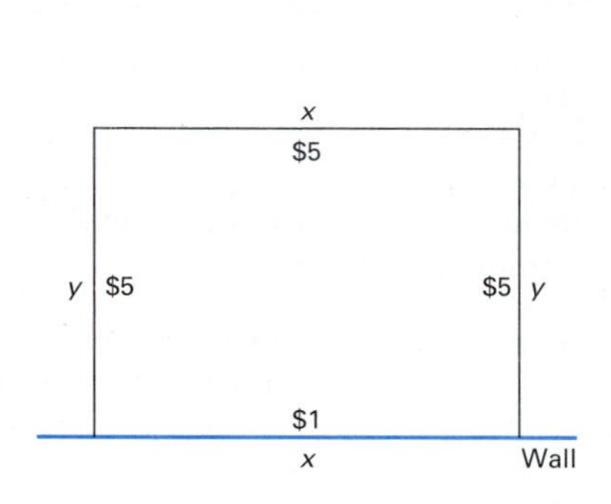

2.8 The animal pen

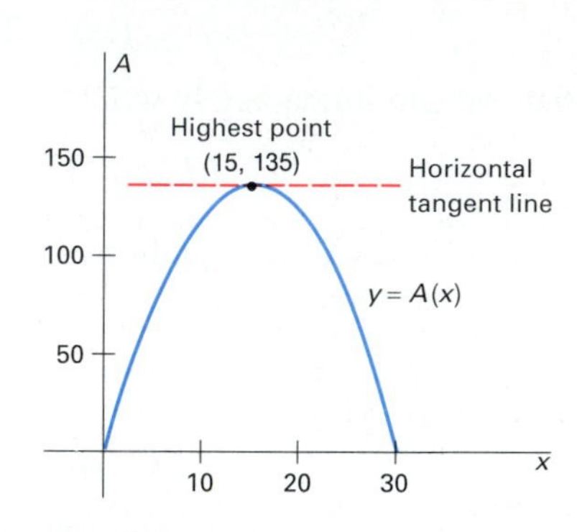

2.9 The graph of $y = A(x)$, $0 \leqq x \leqq 30$

Let us accept as intuitively obvious (we shall give a rigorous proof in Chapter 3) the fact that the maximum value of $A(x)$ occurs at a point where the tangent line to the graph of $y = A(x)$ is horizontal, as in Fig. 2.9. Then we can apply the theorem on differentiation of quadratic functions (see Equation (9)) to find this maximum point. The slope of the tangent line at an arbitrary point $(x, A(x))$ is given by

$$m = A'(x) = -\tfrac{3}{5}(2x) + (18) = -\tfrac{6}{5}x + 18.$$

We ask when $m = 0$, and find that this happens when

$$-\tfrac{6}{5}x + 18 = 0,$$

and thus when $x = 15$. In agreement with the result obtained by algebraic methods in Section 1.3, we find that the maximum possible area of the pen is

$$A(15) = \tfrac{3}{5}(30 \cdot 15 - 15^2) = 135$$

square feet.

2.1 PROBLEMS

In each of Problems 1–14, apply the theorem on differentiation of quadratic functions (see Equation (9)) to write the derivative $f'(x)$ of the given function f.

1. $f(x) = 5$

2. $f(x) = x$

3. $f(x) = x^2$

4. $f(x) = 1 - 2x^2$

5. $f(x) = 4x - 5$

6. $f(x) = 7 - 3x$

7. $f(x) = 2x^2 - 3x + 4$

8. $f(x) = 5 - 3x - x^2$

9. $f(x) = 2x(x + 3)$

10. $f(x) = 3x(5 - x)$

11. $f(x) = 2x - \left(\dfrac{x}{10}\right)^2$

12. $f(x) = 4 - (3x + 2)^2$

13. $f(x) = (2x + 1)^2 - 4x$

14. $f(x) = (2x + 3)^2 - (2x - 3)^2$

In Problems 15–20, find all points of the curve $y = f(x)$ at which the tangent line is horizontal.

15. $y = 10 - x^2$

16. $y = 10x - x^2$

17. $y = x^2 - 2x + 1$

18. $y = x^2 + x - 2$

19. $y = x - \left(\frac{x}{10}\right)^2$

20. $y = x(100 - x)$

Apply the definition of the derivative—that is, follow the four-step process of this section—to find $f'(x)$ for each of the functions in Problems 21–32. Then write an equation for the straight line tangent to the graph of the curve $y = f(x)$ at the point $(2, f(2))$.

21. $f(x) = 3x - 1$

22. $f(x) = x^2 - x - 2$

23. $f(x) = 2x^2 - 3x + 5$

24. $f(x) = 70x - x^2$

25. $f(x) = (x - 1)^2$

26. $f(x) = x^3$

27. $f(x) = \frac{1}{x}$

28. $f(x) = x^4$

29. $f(x) = \frac{1}{x^2}$

30. $f(x) = x^2 + \frac{3}{x}$

31. $f(x) = \frac{2}{x - 1}$

32. $f(x) = \frac{x}{x - 1}$

In the remaining problems, use Equation (9) to find the derivative of a quadratic function if necessary. In Problems 33–35, write equations for the tangent line and the normal line to the curve $y = f(x)$ at the point P given in the problem.

33. $y = x^2$; $P(-2, 4)$

34. $y = 5 - x - 2x^2$; $P(-1, 4)$

35. $y = 2x^2 + 3x - 5$; $P(2, 9)$

36. Prove that the line tangent to the parabola $y = x^2$ at the point (x_0, y_0) intersects the x-axis at the point $(x_0/2, 0)$.

37. If a ball is thrown straight upward with initial velocity 96 ft/s, then its height t seconds later is $y(t) = 96t - 16t^2$ feet. Determine the maximum height the ball attains by finding when $y'(t) = 0$.

38. Find the maximum possible area of the rectangle with perimeter 100 described in Problem 60 of Section 1.1.

39. Find the maximum possible value of the product of two positive numbers whose sum is 50.

40. If a projectile is fired at an angle of 45° from the horizontal, with initial position the origin in the xy-plane and with an initial velocity of $100\sqrt{2}$ feet per second, then its trajectory is the part of the parabola $y = x - (x/25)^2$ for which $y \geqq 0$.

(a) How far does the projectile travel (horizontally) before it hits the ground?

(b) What is the maximum height above the ground that the projectile attains?

41. One of the two lines that pass through the point (3, 0) and are tangent to the parabola $y = x^2$ is the x-axis itself. Find an equation for the *other* line. (*Suggestion:* Sketch the parabola and the other line. Let (a, a^2) be the point at which this other line is tangent to the parabola; first find a.)

42. Write equations for the two straight lines through the point (2, 5) that are tangent to the parabola $y = 4x - x^2$. (See the suggestion for Problem 41.)

43. Between Examples 4 and 5, we raised—but did not answer—the question of the location of the point on the graph of $y = x^2$ closest to the point (3, 0). It's now time for you to find that point. (The suggestion for Problem 41, appropriately modified, will be helpful. Note that the cubic equation you obtain has one solution apparent upon inspection.)

2.2 Limits and the Limit Laws

In Section 2.1 we defined the derivative $f'(a)$ of the function f at the number a to be

$$f'(a) = \lim_{h \to 0} \frac{f(a + h) - f(a)}{h}. \tag{1}$$

The picture that motivated this definition appears in Fig. 2.10, with $a + h$ relabeled as x (so that $h = x - a$). We see that x approaches a as h approaches zero, so (1) can be written in the form

$$f'(a) = \lim_{x \to a} \frac{f(x) - f(a)}{x - a}. \tag{2}$$

Thus the computation of $f'(a)$ amounts to the determination of the limit, as x approaches a, of the function

$$G(x) = \frac{f(x) - f(a)}{x - a}. \tag{3}$$

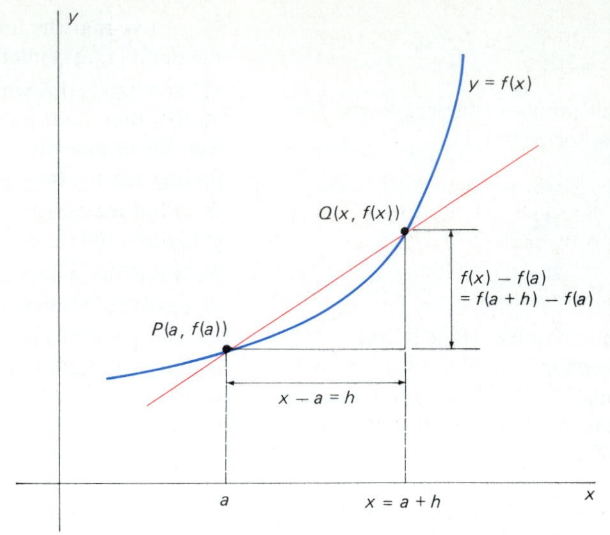

2.10 The derivative can be defined in this way:

$$f'(a) = \lim_{x \to a} \frac{f(x) - f(a)}{x - a}$$

To prepare for the differentiation of more complicated functions than we could manage in Section 2.1, we now turn our attention to the meaning of the statement

$$\lim_{x \to a} F(x) = L. \tag{4}$$

This is read "the limit of $F(x)$, as x approaches a, is L." We shall sometimes write (4) in the concise form

$$F(x) \to L \quad \text{as} \quad x \to a.$$

To discuss the limit of F at a does not require that F be defined there. Neither its value there, nor the possibility that it has no assigned value at a, is of importance. All we require is that F be defined in some **deleted neighborhood** of a—that is, in a set obtained by deleting the single point a from some open interval containing a. For instance, it would suffice for F to be defined on the open intervals $(a - 0.5, a)$ and $(a, a + 0.5)$, but not necessarily defined at the point a itself. Note that this is exactly the situation in the case of a function like the one in Equation (3), which is defined *except at a*.

The following statement gives the meaning of Equation (4) in intuitive language.

Idea of the Limit

We say that the number L is the *limit* of $F(x)$ as x approaches a provided that the number $F(x)$ can be made as close to L as one pleases merely by choosing x sufficiently near, though not equal to, the number a.

What this means, very roughly, is that $F(x)$ tends to get closer and closer to L as x gets closer and closer to a. Once we decide how close to L we want $F(x)$ to be, it is necessary that $F(x)$ be that close to L for *all* x

sufficiently close to (but not equal to) a. The actual definition of the limit makes this requirement precise.

Definition of the Limit

We say that the number L is the *limit* of $F(x)$ as x approaches a provided that, given any number $\varepsilon > 0$, there exists a number $\delta > 0$ such that

$$|F(x) - L| < \varepsilon$$

for all x such that

$$0 < |x - a| < \delta.$$

Figure 2.11 is a geometric illustration of the meaning of this definition. The points on the graph of $y = F(x)$ that satisfy the inequality $|F(x) - L| < \varepsilon$ are those that lie between the two (horizontal) lines $y = L - \varepsilon$ and $y = L + \varepsilon$. The points on this graph that satisfy the inequality $|x - a| < \delta$ are those that lie between the two (vertical) lines $x = a - \delta$ and $x = a + \delta$. Consequently the definition implies that $\lim_{x \to a} F(x) = L$ if and only if the following is true. Suppose that the two horizontal lines $y = L - \varepsilon$ and $y = L + \varepsilon$ (with $\varepsilon > 0$) are given in advance. Then it is possible to choose two vertical lines $x = a - \delta$ and $x = a + \delta$ (with $\delta > 0$) with the following property: Every point on the graph of $y = F(x)$ (with $x \neq a$) that lies between the two vertical lines must also lie between the two horizontal lines.

An inspection of Fig. 2.11 suggests the likelihood that the closer together are the two horizontal lines, the closer together the two vertical lines will have to be. This is the meaning of "making $F(x)$ closer to L by making x closer to a."

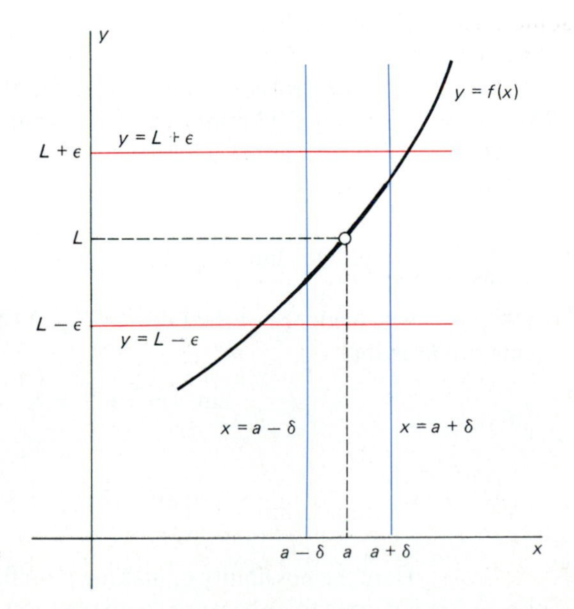

2.11 Geometric illustration of the limit definition

In some cases in which $\lim_{x \to a} F(x) = L$, we will find that $F(x)$ does not get *steadily* closer to L as x moves steadily toward a; this is perfectly permissible, as you can see either by a careful examination of the definition of limit or by study of Fig. 2.11.

In this section we explore the "idea of the limit," mainly through the investigation of specific examples. We defer further discussion of the precise definition of the limit to Appendix B.

x	x^2	(Values rounded)
2.1	4.410	
1.9	3.610	
2.01	4.040	
1.99	3.960	
2.001	4.004	
1.999	3.996	
2.0001	4.0004	
1.9999	3.9996	
2.00001	4.0000	
1.99999	4.0000	

2.12 What happens to x^2 as x approaches 2?

EXAMPLE 1 Evaluate $\lim_{x \to 2} x^2$.

Investigation Note that we call this an "investigation" rather than a solution. Our method here is useful for gathering preliminary information about what the value of a limit might be. This method may reinforce our intuitive guess as to what the value of a limit should be, but it does not constitute a proof.

We form a table of values of x^2, using values of x approaching $a = 2$. Such a table is shown in Fig. 2.12. We use regularly changing values of x because that makes the behavior exhibited in the table easier to understand. We use values of x that are "nice" in the decimal system for the same reason.

In any case, examine the table—read down the column for x, because *down* is the table's direction for "approaches"—to see what happens to the values of x^2. Despite the fact that we are using only special values of x, the table surely provides convincing evidence that

$$\lim_{x \to 2} x^2 = 4.$$

IMPORTANT *Note that we did not substitute the value $x = 2$ into the function $F(x) = x^2$ to obtain the value 4 of the limit.* While such substitution produces the correct answer in this particular case, in many important limits it produces either no answer at all or else an incorrect answer.

x	$\dfrac{x-1}{x+2}$	(Values rounded)
3.1	0.41176	
3.01	0.40120	
3.001	0.40012	
3.0001	0.40001	
3.00001	0.40000	
3.000001	0.40000	

2.13 Investigating the limit in Example 2

EXAMPLE 2 Evaluate $\lim_{x \to 3} \dfrac{x-1}{x+2}$.

Investigation Your natural guess is that the value of the limit is plainly $\frac{2}{5} = 0.4$. The data shown in Fig. 2.13 reinforce this guess. If you experiment with other sequences of values of x approaching 3, the results will support the guess that the limit is, indeed, 0.4.

x	$\sqrt{x^2+9}$	(Values rounded)
−4.1	5.080354	
−4.01	5.008004	
−4.001	5.000800	
−4.0001	5.000080	
−4.00001	5.000008	
−4.000001	5.000001	
−4.0000001	5.000000	

2.14 The behavior of $\sqrt{x^2+9}$ as $x \to -4$

EXAMPLE 3 Evaluate $\lim_{x \to -4} \sqrt{x^2 + 9}$.

Investigation The evidence of the table in Fig. 2.14 is certainly suggestive. It seems obvious that

$$\lim_{x \to -4} \sqrt{x^2 + 9} = 5.$$

EXAMPLE 4 Evaluate $\lim_{x \to 0} \dfrac{\sqrt{x + 25} - 5}{x}$.

Investigation Here the possibility of making a preliminary guess by substituting $x = 0$ is not open to us because the expression in question is meaningless when $x = 0$. But the table in Fig. 2.15 certainly indicates that

$$\lim_{x\to 0} \frac{\sqrt{x+25}-5}{x} = 0.1.$$

x	$\frac{\sqrt{x+25}-5}{x}$ (Values rounded)
10	0.09161
1	0.09902
0.1	0.09990
0.01	0.09999
0.001	0.10000
0.0001	0.10000

2.15 Numerical data for Example 4

The numerical investigations in Examples 2–4 are incomplete in that each of the associated tables shows values of the function $F(x)$ on only one side of the point $x = a$. But in order that $\lim_{x\to a} F(x) = L$, it is necessary that $F(x)$ approaches L *both* as x approaches a from the left *and* as x approaches a from the right. If $F(x)$ approaches different values as x approaches a from different sides, then $\lim_{x\to a} F(x)$ does not exist. In Section 2.3 we will discuss *one-sided limits* in more detail.

EXAMPLE 5 Investigate $\lim_{x\to 0} F(x)$ if

$$F(x) = \frac{x}{|x|} = \begin{cases} 1 & \text{if } x > 0; \\ -1 & \text{if } x < 0. \end{cases}$$

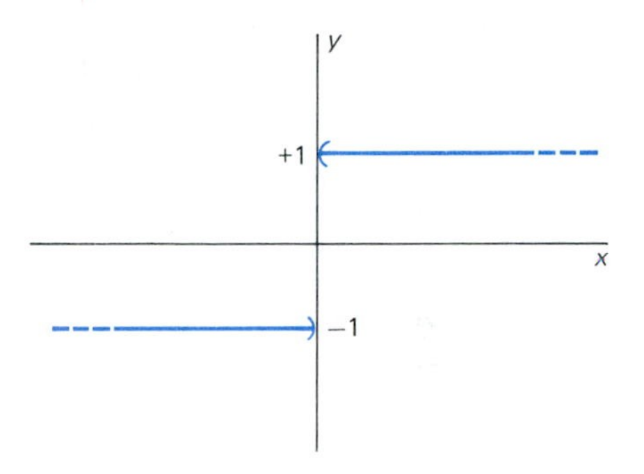

2.16 The graph of $f(x) = \dfrac{x}{|x|}$

Solution From the graph of F shown in Fig. 2.16, it is apparent that $F(x) \to 1$ as $x \to 0$ from the right, while $F(x) \to -1$ as $x \to 0$ from the left. In particular, there are positive values of x as close as you please to zero such that $F(x) = 1$ and negative values of x equally close to zero such that $F(x) = -1$. Hence $F(x)$ cannot be made as close as you please to any *single* value of L merely by choosing x sufficiently close to zero. Therefore,

$$\lim_{x\to 0} \frac{x}{|x|} \qquad \text{does not exist.}$$

In the next example the value obtained by substituting $x = a$ in $F(x)$ to find $\lim_{x\to a} F(x)$ is incorrect.

EXAMPLE 6 Evaluate $\lim_{x\to 0} F(x)$ where

$$F(x) = \begin{cases} 1 & \text{if } x \neq 0; \\ 0 & \text{if } x = 0. \end{cases}$$

The graph of F is shown in Fig. 2.17.

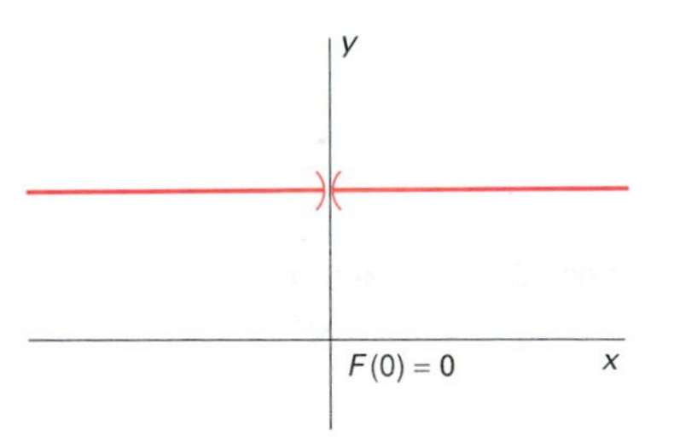

2.17 The graph of the function F of Example 6

Solution The fact that $F(x) = 1$ for *every* value of x in any deleted neighborhood of zero implies that

$$\lim_{x\to 0} F(x) = 1.$$

But the value of the limit is *not* equal to the functional value $F(0) = 0$.

Numerical investigations like those in Examples 1–4 provide us with an intuitive feeling for limits and often suggest the correct value of a limit. But most limit computations are based neither on merely suggestive numerical estimates nor on direct applications of the definition of limit. Instead, such computations are most easily and naturally performed with the aid of the *limit laws* that we give next. The proofs of these laws, which *are* based on the earlier definition of limit, are included in Appendix B.

THE LIMIT LAWS

Constant Law
If $f(x) = C$, a *constant* function, then

$$\lim_{x \to a} f(x) = \lim_{x \to a} C = C. \tag{5}$$

Addition Law
If both limits

$$\lim_{x \to a} F(x) = L \quad \text{and} \quad \lim_{x \to a} G(x) = M$$

exist, then

$$\lim_{x \to a} [F(x) \pm G(x)] = \lim_{x \to a} F(x) \pm \lim_{x \to a} G(x) = L + M. \tag{6}$$

(The limit of the sum is the sum of the limits; the limit of the difference is the difference of the limits.)

Product Law
If both limits

$$\lim_{x \to a} F(x) = L \quad \text{and} \quad \lim_{x \to a} G(x) = M$$

exist, then

$$\lim_{x \to a} [F(x)G(x)] = \left[\lim_{x \to a} F(x)\right]\left[\lim_{x \to a} G(x)\right] = LM. \tag{7}$$

(The limit of the product is the product of the limits.)

Quotient Law
If both limits

$$\lim_{x \to a} F(x) = L \quad \text{and} \quad \lim_{x \to a} G(x) = M$$

exist *and* $M \neq 0$, then

$$\lim_{x \to a} \frac{F(x)}{G(x)} = \frac{\lim_{x \to a} F(x)}{\lim_{x \to a} G(x)} = \frac{L}{M}. \tag{8}$$

(The limit of the quotient is the quotient of the limits, provided that the limit of the denominator is not zero.)

Root Law

If n is a positive integer, and $a > 0$ if n is even, then

$$\lim_{x \to a} \sqrt[n]{x} = \sqrt[n]{a}. \tag{9}$$

The case $n = 1$ of the root law is the virtual tautology

$$\lim_{x \to a} x = a. \tag{10}$$

The next two examples show how the limit laws can be used to evaluate limits of polynomials and rational functions.

EXAMPLE 7

$$\begin{aligned}\lim_{x \to 3} (x^2 + 2x + 4) &= \lim_{x \to 3} x^2 + \lim_{x \to 3} 2x + \lim_{x \to 3} 4 \\ &= \left(\lim_{x \to 3} x\right)^2 + 2\left(\lim_{x \to 3} x\right) + \lim_{x \to 3} 4 \\ &= (3)^2 + (2)(3) + 4 = 19.\end{aligned}$$

EXAMPLE 8

$$\begin{aligned}\lim_{x \to 3} \frac{2x + 5}{x^2 + 2x + 4} &= \frac{\lim_{x \to 3} (2x + 5)}{\lim_{x \to 3} (x^2 + 2x + 4)} \\ &= \frac{(2)(3) + 5}{(3)^2 + (2)(3) + 4} = \frac{11}{19}.\end{aligned}$$

Note in each of these two examples that we systematically applied the limit laws until we could simply substitute $\lim_{x \to a} x = 3$ at the final step. In this way we can find the limit as $x \to a$ of any polynomial. In the case of a quotient of polynomials the procedure is much the same, except that the limit may fail to exist at a point where the denominator is zero.

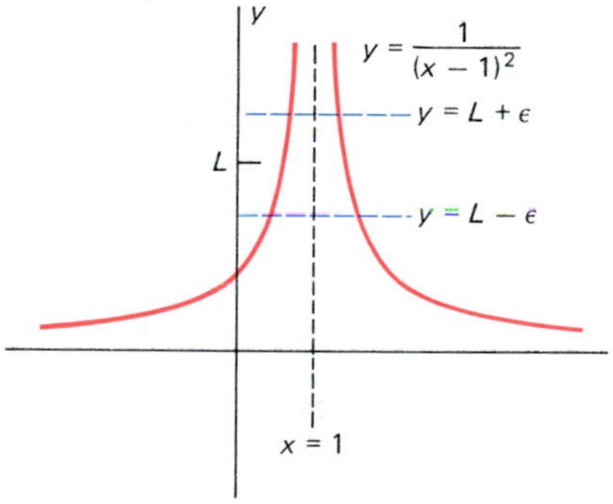

EXAMPLE 9 Investigate $\displaystyle\lim_{x \to 1} \frac{1}{(x - 1)^2}$.

Solution Because $\lim_{x \to 1} (x - 1)^2 = 0$, it follows that $1/(x - 1)^2$ can be made arbitrarily large by choosing x sufficiently close to 1. Hence $1/(x - 1)^2$ cannot approach any (finite) number as x approaches 1. Therefore, the limit in this example does not exist. You can see the geometric reason if you examine the graph of $y = 1/(x - 1)^2$ in Fig. 2.18. If x is near 1, the points on the graph *cannot* be trapped between two horizontal lines $y = L - \varepsilon$ and $y = L + \varepsilon$ bracketing any proposed limit L.

2.18 The graph of $y = 1/(x - 1)^2$

EXAMPLE 10 Investigate $\lim_{x \to 2} \frac{x^2 - 4}{x^2 + x - 6}$.

Solution We cannot apply the quotient law immediately because the denominator approaches zero as x approaches 2. If the numerator were approaching some number other than zero, then the limit would fail to exist (as in Example 9). But the numerator is approaching zero as $x \to 2$, so there is a possibility that a factor of the numerator can be canceled with a factor of the denominator before evaluating the limit. Indeed,

$$\lim_{x \to 2} \frac{x^2 - 4}{x^2 + x - 6} = \lim_{x \to 2} \frac{(x - 2)(x + 2)}{(x - 2)(x + 3)}$$

$$= \lim_{x \to 2} \frac{x + 2}{x + 3} = \frac{4}{5}.$$

We are allowed to cancel the factor $x - 2$ because it is nonzero: $x \neq 2$ when we evaluate the limit as x approaches 2.

Now let us reconsider the limit of Example 3. It is tempting to write

$$\lim_{x \to -4} \sqrt{x^2 + 9} = \sqrt{\lim_{x \to -4} (x^2 + 9)}$$

$$= \sqrt{(-4)^2 + 9} = \sqrt{25} = 5. \tag{11}$$

But is it correct simply to "move the limit inside the radical" in (11)? To analyze this question, let us write

$$f(x) = \sqrt{x} \quad \text{and} \quad g(x) = x^2 + 9.$$

Then the function that appears in (11) is a *function of a function*,

$$f(g(x)) = \sqrt{g(x)} = \sqrt{x^2 + 9}.$$

Hence our question is whether

$$\lim_{x \to a} f(g(x)) = f\left(\lim_{x \to a} g(x)\right).$$

The next limit law answers this question in the affirmative provided that the "outside function" f meets a certain condition; if so, then the limit of the combination function $f(g(x))$ as $x \to a$ may be found by substituting the limit of $g(x)$ as $x \to a$ into the function f.

Substitution Law

Suppose that

$$\lim_{x \to a} g(x) = L \quad \text{and that} \quad \lim_{x \to L} f(x) = f(L).$$

Then

$$\lim_{x \to a} f(g(x)) = f\left(\lim_{x \to a} g(x)\right) = f(L). \tag{12}$$

Thus the condition under which (12) holds is that the limit of the *outer* function f not only exists at $x = L$, but is also equal to the "expected" value of f—namely, $f(L)$. In particular, because

$$\lim_{x \to -4} (x^2 + 9) = 25 \quad \text{and} \quad \lim_{x \to 25} \sqrt{x} = \sqrt{25} = 5,$$

this condition is satisfied in (11), and hence the computations shown there are valid.

In this section we will use only the following special case of the substitution law. With $f(x) = x^{1/n}$, where n is a positive integer, Equation (12) takes the form

$$\lim_{x \to a} \sqrt[n]{g(x)} = \sqrt[n]{\lim_{x \to a} g(x)}, \tag{13}$$

under the assumption that the limit of $g(x)$ exists as $x \to a$ (and is positive if n is even). With $g(x) = x^m$, where m is a positive integer, Equation (13) in turn yields

$$\lim_{x \to a} x^{m/n} = a^{m/n}, \tag{14}$$

again with the condition that $a > 0$ if n is even. Equations (13) and (14) may be regarded as generalized root laws. The following example illustrates the use of these special cases of the substitution law.

EXAMPLE 11

$$\begin{aligned}
\lim_{x \to 4} (3x^{3/2} + 20\sqrt{x})^{1/3} &= \left(\lim_{x \to 4} (3x^{3/2} + 20\sqrt{x}) \right)^{1/3} && \text{(applying (13))} \\
&= \left(\lim_{x \to 4} 3x^{3/2} + \lim_{x \to 4} 20\sqrt{x} \right)^{1/3} && \text{(sum law)} \\
&= [(3)(4^{3/2}) + 20\sqrt{4}]^{1/3} && \text{(applying (14))} \\
&= (24 + 40)^{1/3} = \sqrt[3]{64} = 4.
\end{aligned}$$

Our discussion of limits began with the derivative. If we write the definition of the derivative in the form

$$f'(a) = \lim_{x \to a} \frac{f(x) - f(a)}{x - a},$$

the limit involved is similar to those in the preceding examples in that x is changing while other numbers (such as a) are held fixed. But when we use the definition of f' in the form

$$f'(x) = \lim_{h \to 0} \frac{f(x + h) - f(x)}{h},$$

it is important to remember that x plays the role of a *constant* and that h is the variable approaching zero.

Our next example illustrates an algebraic device often used in "preparing" functions before taking limits. This device can be applied when roots are present and resembles the simple computation

$$\frac{1}{\sqrt{3} - \sqrt{2}} = \frac{1}{\sqrt{3} - \sqrt{2}} \cdot \frac{\sqrt{3} + \sqrt{2}}{\sqrt{3} + \sqrt{2}} = \frac{\sqrt{3} + \sqrt{2}}{3 - 2} = \sqrt{3} + \sqrt{2}.$$

EXAMPLE 12 Differentiate $f(x) = \sqrt{x}$.

Solution

$$f'(x) = \lim_{h \to 0} \frac{\sqrt{x+h} - \sqrt{x}}{h}. \tag{15}$$

To prepare the fraction for evaluation of the limit, we first multiply the numerator and denominator by $\sqrt{x+h} + \sqrt{x}$:

$$f'(x) = \lim_{h \to 0} \frac{\sqrt{x+h} - \sqrt{x}}{h} \cdot \frac{\sqrt{x+h} + \sqrt{x}}{\sqrt{x+h} + \sqrt{x}}$$

$$= \lim_{h \to 0} \frac{(x+h) - x}{h(\sqrt{x+h} + \sqrt{x})} = \lim_{h \to 0} \frac{1}{\sqrt{x+h} + \sqrt{x}}.$$

Thus

$$f'(x) = \frac{1}{2\sqrt{x}}. \tag{16}$$

(In the final step we use the sum, quotient, and root laws.)

Note that if we equate the right-hand sides in (15) and (16) and take $x = 25$, we then get the limit of Example 4:

$$\lim_{h \to 0} \frac{\sqrt{25+h} - 5}{h} = \frac{1}{10}.$$

A final property of limits that we will need is the fact that taking limits preserves inequalities between functions.

Squeeze Law

Suppose that $f(x) \leqq g(x) \leqq h(x)$ in some deleted neighborhood of a and also that

$$\lim_{x \to a} f(x) = L = \lim_{x \to a} h(x).$$

Then

$$\lim_{x \to a} g(x) = L$$

as well.

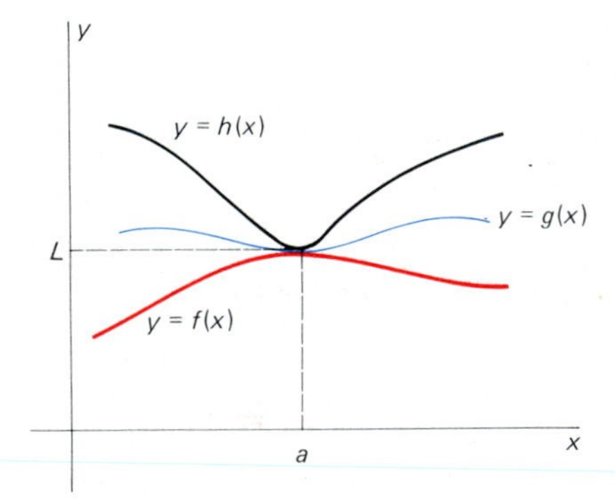

2.19 How the squeeze law works

Figure 2.19 illustrates not only how and why the squeeze law works but also how it got its name. The idea is that $g(x)$ is trapped between $f(x)$ and $h(x)$; the latter both approach L, so $g(x)$ must approach L as well.

EXAMPLE 13 Apply the squeeze law to prove that

$$\lim_{x \to 0} \frac{x^2}{1 + (1 + x^4)^{5/2}} = 0.$$

Solution Because $1 + (1 + x^4)^{5/2} \geqq 1$ for all x, we see that

$$0 \leqq \frac{x^2}{1 + (1 + x^4)^{5/2}} \leqq x^2$$

for all x. We now apply the squeeze law with $f(x) \equiv 0$ and $h(x) = x^2$. Because $f(x)$ and $h(x)$ both approach zero as $x \to 0$, it follows that the given function, trapped between them, also must approach zero as $x \to 0$.

2.2 PROBLEMS

In Problems 1–30, use the limit laws of this section to evaluate those limits that exist.

1. $\lim_{x\to 0} (3x^2 + 7x - 12)$

2. $\lim_{x\to 2} (4x^2 - x + 5)$

3. $\lim_{x\to -1} (2x - x^5)$

4. $\lim_{x\to -2} (x^2 - 2)^5$

5. $\lim_{x\to 3} \dfrac{(x-1)^7}{(2x-5)^4}$

6. $\lim_{x\to 1} \dfrac{x+1}{x^2 - x - 2}$

7. $\lim_{x\to -1} \dfrac{x+1}{x^2 - x - 2}$

8. $\lim_{t\to 2} \dfrac{t^2 + 2t - 5}{t^3 - 2t}$

9. $\lim_{t\to 3} \dfrac{t^2 - 9}{t - 3}$

10. $\lim_{y\to 3} \dfrac{1}{y-3}\left(\dfrac{1}{y} - \dfrac{1}{3}\right)$

11. $\lim_{x\to 3} (x^2 - 1)^{3/2}$

12. $\lim_{t\to -4} \sqrt{\dfrac{t+8}{25 - t^2}}$

13. $\lim_{z\to 8} \dfrac{z^{2/3}}{z - \sqrt{2z}}$

14. $\lim_{t\to 2} \sqrt[3]{3t^3 + 4t - 5}$

15. $\lim_{x\to 0} \dfrac{\sqrt{x+4} - 2}{x}$

16. $\lim_{h\to 0} \dfrac{1}{h}\left(\dfrac{1}{2+h} - \dfrac{1}{2}\right)$

17. $\lim_{h\to 0} \dfrac{1}{h}\left(\dfrac{1}{\sqrt{9+h}} - \dfrac{1}{3}\right)$

18. $\lim_{x\to 2} \dfrac{(x-2)^2}{x^4 - 16}$

19. $\lim_{x\to 3} |1 - x|$

20. $\lim_{x\to -5} |3x - 2|$

21. $\lim_{x\to 4} \dfrac{x-4}{\sqrt{x} - 2}$

22. $\lim_{x\to 9} \dfrac{3 - \sqrt{x}}{9 - x}$

23. $\lim_{x\to 1} \dfrac{x^2 + x - 2}{x^2 - 4x + 3}$

24. $\lim_{x\to -1/2} \dfrac{4x^2 - 1}{4x^2 + 8x + 3}$

25. $\lim_{x\to 4} \dfrac{x^2 - 16}{2 - \sqrt{x}}$

26. $\lim_{x\to 5} \left(\dfrac{2x^2 + 2x + 4}{6x - 3}\right)^{1/3}$

27. $\lim_{x\to 1} \sqrt{(x-2)^2}$

28. $\lim_{x\to -4} \sqrt[3]{(x+1)^6}$

29. $\lim_{x\to -2} \sqrt[3]{\dfrac{x+2}{(x-2)^2}}$

30. $\lim_{x\to 0} \dfrac{1}{x}(\sqrt{1+x} - \sqrt{1-x})$

Use a calculator to investigate numerically the limits in Problems 31–36.

31. $\lim_{x\to 0} \dfrac{(1+x)^2 - 1}{x}$

32. $\lim_{x\to 1} \dfrac{x^4 - 1}{x - 1}$

33. $\lim_{x\to 0} \dfrac{\sqrt{x+9} - 3}{x}$

34. $\lim_{x\to 4} \dfrac{x^{3/2} - 8}{x - 4}$

35. $\lim_{x\to 0} \dfrac{1}{x}\left(\dfrac{1}{x+5} - \dfrac{1}{5}\right)$

36. $\lim_{x\to 8} \dfrac{x^{2/3} - 4}{x - 8}$

In each of Problems 37–42, use the method of Example 12—where appropriate—to find the derivative $f'(x)$ of $f(x)$.

37. $f(x) = \dfrac{1}{\sqrt{x}}$

38. $f(x) = \dfrac{1}{2x + 3}$

39. $f(x) = \dfrac{x}{2x+1}$

40. $f(x) = \sqrt{3x + 1}$

41. $f(x) = \dfrac{x^2}{x+1}$

42. $f(x) = \dfrac{1}{\sqrt{x+4}}$

In each of Problems 43–46, use the squeeze law (as in Example 13) to prove that $g(x) \to 0$ as $x \to 0$.

43. $g(x) = \dfrac{|x|}{1 + x^2}$

44. $g(x) = \dfrac{x^2}{1 + |x|^{1/2}}$

45. $g(x) = \dfrac{\sqrt{x+1} - 1}{5x^2 + 2}$

46. $g(x) = \dfrac{|x+1| - |x-1|}{1 + \sqrt{1 + |x+2|^3}}$

47. Apply the product law to prove that $\lim_{x\to a} x^n = a^n$ if n is a positive integer.

48. Suppose that $p(x) = b_n x^n + \cdots + b_1 x + b_0$ is a polynomial. Prove that $\lim_{x\to a} p(x) = p(a)$.

49. Let $f(x) = [\![x]\!]$ be the greatest integer function. For what values of a does $\lim_{x\to a} f(x)$ exist?

50. Suppose that there exists a number M such that $|f(x)| \leq M$ for all x. Apply the squeeze law to show that $\lim_{x\to 0} xf(x) = 0$.

51. Let $r(x) = p(x)/q(x)$, where $p(x)$ and $q(x)$ are polynomials. Suppose that $p(a) \neq 0$ and that $q(a) = 0$. Prove that $\lim_{x\to a} r(x)$ does not exist. (*Suggestion:* Consider $\lim_{x\to a} q(x)r(x)$.)

52. Let $g(x) = 1 + [\![x]\!] + [\![-x]\!]$. Show that the limit of $g(x)$ as $x \to 3$ cannot be obtained by substitution of 3 for x in the formula for g.

53. Let $h(x) = x - (0.01)[\![100x]\!]$. Show that $h(x) \to 0$ as $x \to 0$, but that $h(x)$ does not approach zero "steadily" as $x \to 0$. First establish what the phrase "approaches steadily" ought to mean.

54. Repeat Problem 53 with

$$h(x) = \frac{1}{[\![1/x]\!]} - x.$$

The numerical investigation of limits as in Examples 1–4 of this section requires much repetitive calculation of functional values. This is an activity that computers perform efficiently. Figure 2.20 lists a program for investigating limits of the function $f(x)$ specified on the right-hand side in line 250.

```
100 REM--Program LIMIT
110 REM--Approximates the limit of the function  f(x)
120 REM--as  x --> a.  The function  f(x)  must be
130 REM--edited into line 250 before the program is run.
140 REM
150        DEFDBL  A, F, H, X
160        INPUT "Point a"; A
170        INPUT "Initial h"; H
180        INPUT "Number of values"; N
190        D$ = "##.#####      ##.#####"
200 REM
210 REM--Loop to calculate values:
220 REM
230        FOR I = 0 TO N
240            X = A + H
250            F = (SQR(X + 25) - 5)/X
260            PRINT USING D$; X,F
270            H = H/10
280        NEXT I
290 REM
300        END
```

2.20 Listing of Program LIMIT

The point a (where the limit of $f(x)$ is desired) is entered at line 160, and different sequences of values of x approaching a can be generated with different choices of h in line 170. The effect of lines 190 and 260 is to direct that the values of x and $f(x)$ be displayed rounded to five decimal places; you may change this if you wish.

Line 250 as listed in Fig. 2.20 corresponds to the function $f(x) = (\sqrt{x+25} - 5)/x$ of Example 4. Four "runs" of this program with $a = 0$ and different values of h gave the following results.

```
Point a? 0
Initial h? 1
Number of values? 5
 1.00000        0.09902
 0.10000        0.09990
 0.01000        0.09999
 0.00100        0.10000
 0.00010        0.10000
 0.00001        0.10000
```

```
Point a? 0
Initial h? 3
Number of values? 5
 3.00000        0.09717
 0.30000        0.09970
 0.03000        0.09997
 0.00300        0.10000
 0.00030        0.10000
 0.00003        0.10000
```

```
Point a? 0
Initial h? -1
Number of values? 5
-1.00000        0.10102
-0.10000        0.10010
-0.01000        0.10001
-0.00100        0.10000
-0.00010        0.10000
-0.00001        0.10000
```

```
Point a? 0
Initial h? -4
Number of values? 5
-4.00000        0.10436
-0.40000        0.10040
-0.04000        0.10004
-0.00400        0.10000
-0.00040        0.10000
-0.00004        0.10000
```

These results corroborate the fact that

$$\lim_{x \to 0} \frac{\sqrt{x+25}-5}{x} = 0.1.$$

Use Program LIMIT to investigate similarly the limits that appear in Exercises 1–8 below. An alternative to the use of a full-fledged computer is a hand-held calculator with a "function memory." For instance, with the Tandy PC-6 one can simply enter the formula

$$F = (\text{SQR}(X + 25) - 5)/X$$

in function memory to define the function $f(x) = (\sqrt{x+25} - 5)/x$. When the (CALC) key is pressed the query X? is displayed. After a value of x is entered, the corresponding value $f(x)$ of the function is calculated and displayed. Obviously, one can investigate $\lim_{x \to a} f(x)$ by entering successive values of x that approach a.

Exercises 1–6: The limits of Problems 31–36 of this section.

Exercise 7: $\lim_{x \to 0} \frac{\sin x}{x} = 1.$

(NOTE The BASIC function SIN(X) gives the sine of an angle of X *radians*.)

Exercise 8: $\lim_{x \to 0} (1 + x)^{1/x} \approx 2.718.$

(NOTE This limit defines the famous number $e \approx 2.71828$ that plays an important role in calculus (beginning in Chapter 7). The BASIC notation for $(1 + x)^{1/x}$ is usually $(1 + X) \wedge (1/X)$.)

2.3 One-Sided Limits

In Example 5 of Section 2.2 we examined the function

$$f(x) = \frac{x}{|x|} = \begin{cases} 1 & \text{if } x > 0; \\ -1 & \text{if } x < 0. \end{cases}$$

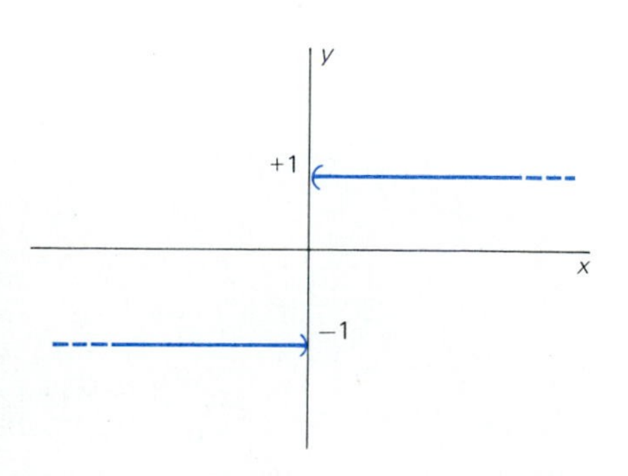

2.21 The graph of $f(x) = \frac{x}{|x|}$ again

The graph of $y = f(x)$ is shown in Fig. 2.21. We argued that the limit of $f(x) = x/|x|$ as $x \to 0$ does not exist because $f(x)$ approaches $+1$ as x approaches zero from the right, while $f(x) \to -1$ as x approaches zero from the left. A natural way of describing this situation is to say that at $x = 0$ the *right-hand limit* of $f(x)$ is $+1$ and the *left-hand limit* of $f(x)$ is -1.

In this section we define and investigate these one-sided limits. Their definitions will be stated initially in the informal language used to describe the "idea of the limit" in Section 2.2. To define the right-hand limit of $f(x)$ at $x = a$, we must assume that f is defined on an open interval immediately to the right of a; to define the left-hand limit, we must assume that f is defined on an open interval just to the left of a.

The Right-Hand Limit of a Function

Suppose that f is defined on an open interval (a, c). Then we say that the number L is the **right-hand limit** of $f(x)$ as x approaches a, and we write

$$\lim_{x \to a^+} f(x) = L, \tag{1}$$

provided that the number $f(x)$ can be made as close to L as one pleases merely by choosing the point x in (a, c) sufficiently near the number a.

We may describe the right-hand limit in (1) by saying that $f(x) \to L$ as $x \to a^+$, or as x approaches a from the right; the symbol a^+ denotes the right-hand, or positive, side of a. More precisely, (1) means that, given $\varepsilon > 0$, there exists $\delta > 0$ such that

$$|f(x) - L| < \varepsilon \tag{2}$$

for all x such that

$$a < x < a + \delta. \tag{3}$$

See Fig. 2.22 for a geometric interpretation of these one-sided limits.

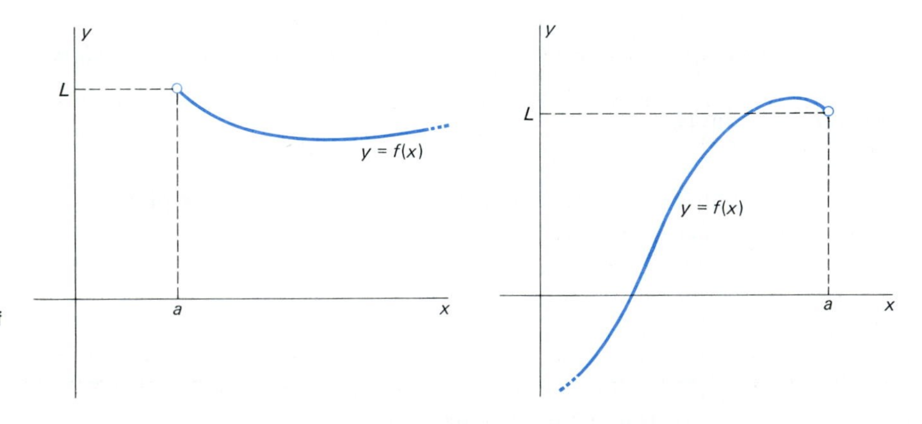

2.22 (a) The right-hand limit of $f(x)$ is L.
(b) The left-hand limit of $f(x)$ is L.

The Left-Hand Limit of a Function

Suppose that f is defined on an open interval (c, a). Then we say that the number L is the **left-hand limit** of $f(x)$ as x approaches a, and we write

$$\lim_{x \to a^-} f(x) = L, \tag{4}$$

provided that the number $f(x)$ can be made as close to L as one pleases merely by choosing the point x in (c, a) sufficiently near the number a.

Note that a consequence of each definition is that the value of $f(a)$ itself is not relevant to the existence or value of the one-sided limits, just as it is not relevant to the existence or value of the two-sided limit.

We may describe the left-hand limit in (4) by saying that $f(x) \to L$ as $x \to a^-$, or as x approaches a from the left; the symbol a^- denotes the left, or negative, side of a. We get a precise definition of the left-hand limit in (4) merely by changing the open interval in (3) to $a - \delta < x < a$.

Our preliminary discussion of the function $f(x) = x/|x|$ amounts to saying that its one-sided limits at $x = 0$ are

$$\lim_{x \to 0^+} \frac{x}{|x|} = 1 \quad \text{and} \quad \lim_{x \to 0^-} \frac{x}{|x|} = -1.$$

In Example 5 of Section 2.2, we argued further that, because these two limits are not equal, *the* two-sided limit of $x/|x|$ as $x \to 0$ does not exist. The following theorem is readily proved on the basis of precise definitions of all the limits involved.

Theorem *One-Sided and Two-Sided Limits*

Suppose that the function f is defined on a deleted neighborhood of the point a. Then the limit $\lim_{x \to a} f(x)$ exists and is equal to the number L if and only if the one-sided limits $\lim_{x \to a+} f(x)$ and $\lim_{x \to a-} f(x)$ both exist and both are equal to the number L.

This theorem will be particularly useful in proving that certain (two-sided) limits do *not* exist.

EXAMPLE 1 The graph of the greatest integer function $f(x) = [\![x]\!]$ is shown in Fig. 2.23. It should be apparent that if a is not an integer, then

$$\lim_{x \to a^+} [\![x]\!] = \lim_{x \to a^-} [\![x]\!] = \lim_{x \to a} [\![x]\!] = [\![a]\!].$$

On the other hand, if $a = n$, an integer, then

$$\lim_{x \to n^-} [\![x]\!] = n - 1 \quad \text{and} \quad \lim_{x \to n^+} [\![x]\!] = n.$$

Because these left-hand and right-hand limits are not equal, it follows from the theorem that the limit of $f(x) = [\![x]\!]$ does not exist as x approaches an integer n.

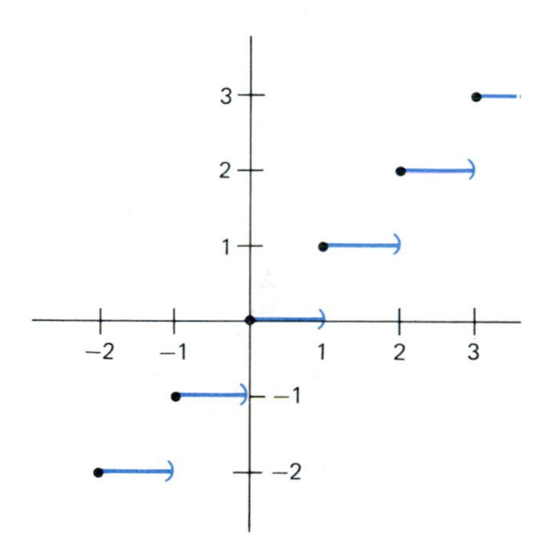

2.23 The graph of the greatest integer function $f(x) = [\![x]\!]$

EXAMPLE 2 According to the root law in Section 2.2,

$$\lim_{x \to a} \sqrt{x} = \sqrt{a} \qquad \text{if} \quad a > 0.$$

But the limit of $f(x) = \sqrt{x}$ as $x \to 0^-$ is not defined because the square root of a negative number is undefined, and hence f is undefined on a deleted neighborhood of zero. What we *can* say at $a = 0$ is that

$$\lim_{x \to 0^+} \sqrt{x} = 0,$$

although the left-hand limit $\lim_{x \to 0^-} \sqrt{x}$ does not exist.

To each of the limit laws stated in Section 2.2, there correspond two *one-sided limit laws*—a right-hand version and a left-hand version. We apply these one-sided limit laws wherever possible to evaluate one-sided limits.

EXAMPLE 3 Upon application of the appropriate one-sided limit theorems, we find that

$$\lim_{x\to 3^-}\left(\frac{x^2}{x^2+1}+\sqrt{9-x^2}\right)=\frac{\lim_{x\to 3^-} x^2}{\lim_{x\to 3^-}(x^2+1)}+\sqrt{\lim_{x\to 3^-}(9-x^2)}$$

$$=\frac{9}{9+1}+\sqrt{0}=\frac{9}{10}.$$

Note that the two-sided limit as $x \to 3$ is not defined here because $\sqrt{9-x^2}$ is not defined when $x > 3$.

Recall that the derivative $f'(x)$ of the function $f(x)$ is defined by

$$f'(x)=\lim_{h\to 0}\frac{f(x+h)-f(x)}{h} \tag{5}$$

provided that this (two-sided) limit exists, in which case we say that the function f is **differentiable** at the point x. The following example shows that it is possible for a function to be defined everywhere but fail to be differentiable at certain points of its domain.

EXAMPLE 4 Show that the function $f(x) = |x|$ is *not* differentiable at $x = 0$.

Solution When $x = 0$, we have

$$\frac{f(x+h)-f(x)}{h}=\frac{|h|}{h}=\begin{cases}-1 & \text{if } h<0;\\ +1 & \text{if } h>0.\end{cases}$$

Hence the left-hand limit of the quotient is -1, while the right-hand limit is $+1$. Therefore, the *two-sided* limit in (5) does not exist, so $f(x) = |x|$ is not differentiable at $x = 0$.

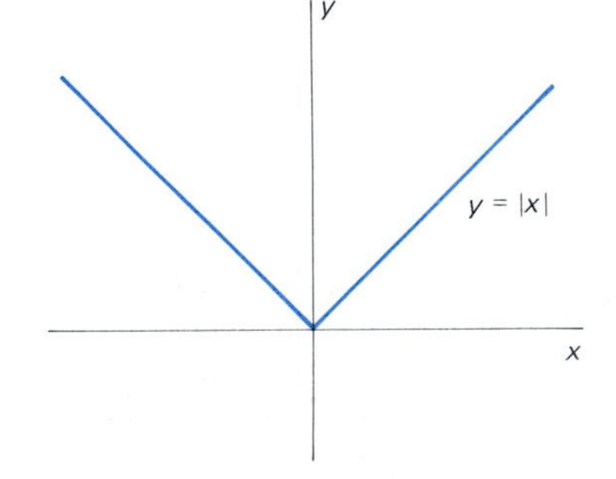

2.24 The graph of $f(x) = |x|$ has a corner point at (0, 0)

Figure 2.24 shows the graph of the function $f(x) = |x|$ of Example 4. The graph has a sharp corner at the point (0, 0), and this explains why there can be no tangent line there—no single straight line is a good approximation to the shape of the graph at (0, 0). On the other hand, the figure makes it evident that $f'(x)$ exists if $x \neq 0$. Indeed,

$$f'(x)=\begin{cases}-1 & \text{if } x<0;\\ +1 & \text{if } x>0.\end{cases}$$

In Problem 36 we ask you to derive this result directly from the definition of the derivative, without appeal to Fig. 2.24.

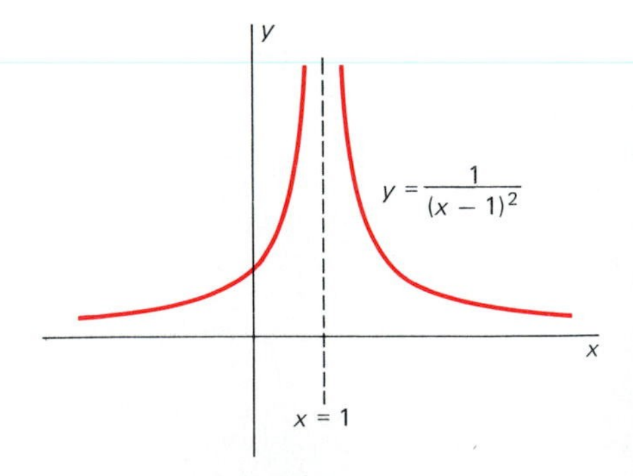

2.25 The graph of the function $f(x) = 1/(x-1)^2$

INFINITE LIMITS

In Example 9 of Section 2.2, we investigated the function $f(x) = 1/(x-1)^2$; the graph of f is shown in Fig. 2.25. The value of $f(x)$ *increases without bound* (that is, eventually exceeds any preassigned number) as x approaches 1 either from the right or from the left. This situation is sometimes described by writing

$$\lim_{x \to 1^-} \frac{1}{(x-1)^2} = +\infty = \lim_{x \to 1^+} \frac{1}{(x-1)^2}, \tag{6}$$

and we say that each of these one-sided limits equals "plus infinity."

CAUTION Merely to write the expression

$$\lim_{x \to 1^+} \frac{1}{(x-1)^2} = +\infty \tag{7}$$

does not mean that there exists an "infinite real number" denoted by $+\infty$—there does not! Neither does it mean that the limit on the left-hand side in (7) exists—it does not! To the contrary, the expression in (7) is simply a convenient way of saying *why* the right-hand limit of $1/(x-1)^2$ as $x \to 1^+$ does not exist: It is because the quantity $1/(x-1)^2$ increases without bound as $x \to 1^+$.

With similar provisos we may write

$$\lim_{x \to 1} \frac{1}{(x-1)^2} = +\infty \tag{8}$$

despite the fact that the (two-sided) limit of $1/(x-1)^2$ as $x \to 1$ does not exist. Our understanding is that the expression in (8) is only a convenient way of saying that the limit in (8) fails to exist because $1/(x-1)^2$ increases without bound as $x \to 1$ from either side.

Now consider the function $f(x) = 1/x$; its graph is shown in Fig. 2.26. Note that $1/x$ increases without bound as x approaches zero from the right but decreases without bound—eventually becomes less than any preassigned negative number—as x approaches zero from the left. We therefore write

$$\lim_{x \to 0^-} \frac{1}{x} = -\infty \quad \text{and} \quad \lim_{x \to 0^+} \frac{1}{x} = +\infty. \tag{9}$$

There is no shorthand for the two-sided limit in this case. We may say only that

$$\lim_{x \to 0} \frac{1}{x} \qquad \text{does not exist.}$$

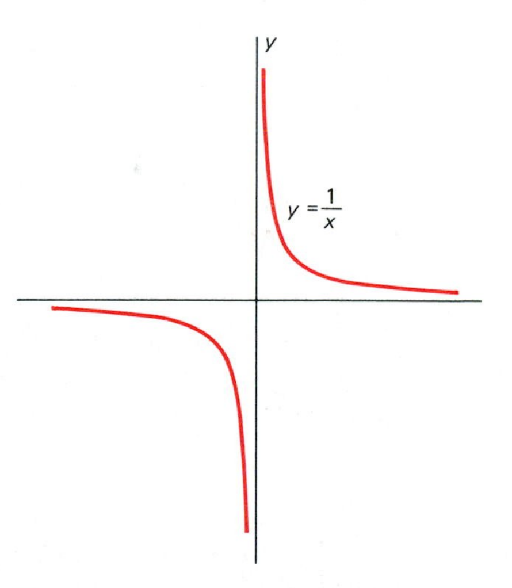

2.26 The graph of the function $f(x) = 1/x$

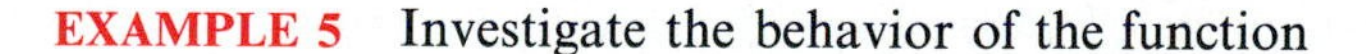

EXAMPLE 5 Investigate the behavior of the function

$$f(x) = \frac{2x+1}{x-1}$$

near the point $x = 1$, where its limit does not exist.

Solution First we look at the behavior of $f(x)$ just to the right of the number 1. If x is greater than 1 but still close to 1, then $2x + 1$ is close to 3, while $x - 1$ is a small *positive* number. In this case the quotient $(2x + 1)/(x - 1)$ will be a large positive number, and the closer x is to 1, the larger this positive quotient will be. For such x, $f(x)$ increases without bound as x approaches 1 from the

x	$\frac{2x+1}{x-1}$	x	$\frac{2x+1}{x-1}$
1	32	0.9	−28
01	302	0.99	−298
001	3002	0.999	−2998
0001	30002	0.9999	−29998

2.27 The behavior of $f(x) = \dfrac{2x+1}{x-1}$ for x near 1

right. That is,

$$\lim_{x \to 1^+} \frac{2x+1}{x-1} = +\infty, \tag{10}$$

as the data in Fig. 2.27 also suggest.

If x is less than 1 but close to 1, then $2x + 1$ is still close to 3, but now $x - 1$ is a *negative* number close to zero. In this case the quotient $(2x + 1)/(x - 1)$ will be a (numerically) large negative number, and decrease without bound as $x \to 1^-$. Hence we conclude that

$$\lim_{x \to 1^-} \frac{2x+1}{x-1} = -\infty. \tag{11}$$

The results in (10) and (11) provide a concise description of the behavior of $f(x) = (2x + 1)/(x - 1)$ near the point $x = 1$. Finally, to remain consistent with the theorem on one-sided and two-sided limits, we say in this case that

$$\lim_{x \to 1} \frac{2x+1}{x-1} \quad \text{does not exist.}$$

2.3 PROBLEMS

Use one-sided limit laws to find the limits in Problems 1–20 or to determine that they do not exist.

1. $\displaystyle\lim_{x \to 0^+} (3 - \sqrt{x})$

2. $\displaystyle\lim_{x \to 0^+} (4 + 3x^{3/2})$

3. $\displaystyle\lim_{x \to 1^-} \sqrt{x - 1}$

4. $\displaystyle\lim_{x \to 4^-} \sqrt{4 - x}$

5. $\displaystyle\lim_{x \to 2^+} \sqrt{x^2 - 4}$

6. $\displaystyle\lim_{x \to 3^+} \sqrt{9 - x^2}$

7. $\displaystyle\lim_{x \to 5^-} \sqrt{x(5 - x)}$

8. $\displaystyle\lim_{x \to 2^-} x\sqrt{4 - x^2}$

9. $\displaystyle\lim_{x \to 4^+} \sqrt{\frac{4x}{x - 4}}$

10. $\displaystyle\lim_{x \to -3^+} \sqrt{6 - x - x^2}$

11. $\displaystyle\lim_{x \to 5^-} \frac{x - 5}{|x - 5|}$

12. $\displaystyle\lim_{x \to -4^+} \frac{16 - x^2}{\sqrt{16 - x^2}}$

13. $\displaystyle\lim_{x \to 3^+} \frac{\sqrt{x^2 - 6x + 9}}{x - 3}$

14. $\displaystyle\lim_{x \to 2^+} \frac{x - 2}{x^2 - 5x + 6}$

15. $\displaystyle\lim_{x \to 2^+} \frac{2 - x}{|x - 2|}$

16. $\displaystyle\lim_{x \to 7^-} \frac{7 - x}{|x - 7|}$

17. $\displaystyle\lim_{x \to 1^+} \frac{1 - x^2}{1 - x}$

18. $\displaystyle\lim_{x \to 0^-} \frac{x}{x - |x|}$

19. $\displaystyle\lim_{x \to 5^+} \frac{\sqrt{(5 - x)^2}}{5 - x}$

20. $\displaystyle\lim_{x \to -4^-} \frac{4 + x}{\sqrt{(4 + x)^2}}$

For each of the functions in Problems 21–30, there is exactly one point a where the right-hand and left-hand limits of $f(x)$ both fail to exist. Describe (as in Example 5) the behavior of $f(x)$ for x near a.

21. $f(x) = \dfrac{1}{x - 1}$

22. $f(x) = \dfrac{2}{3 - x}$

23. $f(x) = \dfrac{x - 1}{x + 1}$

24. $f(x) = \dfrac{2x - 5}{5 - x}$

25. $f(x) = \dfrac{1 - x^2}{x + 2}$

26. $f(x) = \dfrac{1}{(x - 5)^2}$

27. $f(x) = \dfrac{|1 - x|}{(1 - x)^2}$

28. $f(x) = \dfrac{x + 1}{x^2 + 6x + 9}$

29. $f(x) = \dfrac{x - 2}{4 - x^2}$

30. $f(x) = \dfrac{x - 1}{x^2 - 3x + 2}$

In Problems 31–35, $[\![x]\!]$ denotes the greatest integer function. First find the limits $\displaystyle\lim_{x \to n^-} f(x)$ and $\displaystyle\lim_{x \to n^+} f(x)$ for each integer n (the answer will normally be in terms of n); then sketch the graph of f.

31. $f(x) = 2$ if x is not an integer; $f(x) = 2 + (-1)^x$ if x is an integer.

32. $f(x) = x$ if x is not an integer; $f(x) = 0$ otherwise.

33. $f(x) = (-1)^{[\![x]\!]}$

34. $f(x) = \left[\!\!\left[\dfrac{x}{2}\right]\!\!\right]$

35. $f(x) = 1 + [\![x]\!] - x$

36. Given $f(x) = |x|$, use the definition of the derivative to find $f'(x)$ in the two separate cases $x > 0$ and $x < 0$.

2.4 Continuous Functions

When we stated the substitution law for limits in Section 2.2, we needed for its validity the condition

$$\lim_{x \to a} f(x) = f(a) \tag{1}$$

on the outer function in the composition $f(g(x))$. (In Section 2.2 we wrote L in place of a, but of course this is not important.) A function that satisfies the condition in (1) is said to be *continuous* at the number a.

> ***Definition of Continuity at a Point***
> Suppose that the function f is defined in a neighborhood of a. We say that f is **continuous at a** provided that $\lim_{x \to a} f(x)$ exists and, moreover, that the value of this limit is $f(a)$. In other words, f is continuous at a provided that Equation (1) holds.

Briefly, continuity of f at a means that the limit of f at a is equal to its value there. Another way to put it is this: The limit of f at a is its "expected" value—the value that you would assign if you knew the values of f in a deleted neighborhood of a and you knew f to be "predictable." Alternatively, continuity of f at a means this: When x is close to a, then $f(x)$ is close to $f(a)$.

Analysis of the definition of continuity shows us this: In order to be continuous at the point a, the function must satisfy the following three conditions:

1. The function f must be defined at a (so that $f(a)$ exists).
2. The limit of $f(x)$ as x approaches a must exist.
3. The numbers in parts 1 and 2 must be equal:

$$\lim_{x \to a} f(x) = f(a).$$

If any one of these conditions is not satisfied, then f is not continuous at a.

EXAMPLE 1 If $f(x) = x^2$, then

$$\lim_{x \to a} f(x) = \lim_{x \to a} x^2 = a^2 = f(a)$$

for every real number a. Therefore, the squaring function is continuous at every point of the real line $\boldsymbol{R}$.

Suppose that the function f is defined on an open interval or a union of open intervals. Then we say simply that f is **continuous** if it is continuous at each point of its domain of definition. The definition of continuity in terms of limits makes precise the intuitive idea of a function with a graph that is a connected curve—connected in the sense that it can be traced with a "continuous" motion of the pencil without lifting the pencil from the paper.

If the function f is *not* continuous at a, we say that it is **discontinuous** there, or that a is a **discontinuity** of f. Intuitively, a discontinuity of f is a

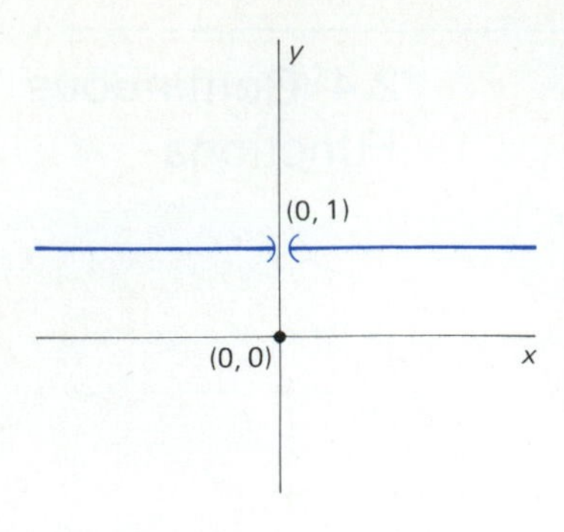

2.28 A discontinuity of f at $x = 0$

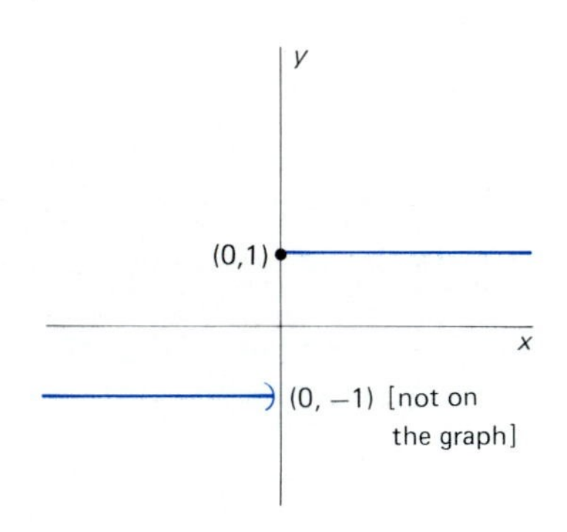

2.29 A discontinuity of g at $x = 0$

point where its graph has a "gap" or "jump" of some sort. For example, the function

$$f(x) = \begin{cases} 1 & \text{if } x \neq 0; \\ 0 & \text{if } x = 0 \end{cases}$$

(its graph appears in Fig. 2.28) jumps in value from 1 to 0 and back as x increases through the value 0. And f is not continuous at $x = 0$ because, as noted in Example 6 of Section 2.2,

$$\lim_{x \to 0} f(x) = 1 \neq 0 = f(0);$$

the limit of f at 0 is not equal to its value there.

Also, the function

$$g(x) = \begin{cases} 1 & \text{if } x \geqq 0; \\ -1 & \text{if } x < 0 \end{cases}$$

is discontinuous at $x = 0$. As its graph (in Fig. 2.29) shows, there is a jump from -1 to 1 at $x = 0$. We saw in Example 5 of Section 2.2 that the limit of $g(x)$ as $x \to 0$ does not exist.

The graph of

$$h(x) = \begin{cases} \dfrac{1}{x} & \text{if } x \neq 0; \\ 0 & \text{if } x = 0 \end{cases}$$

appears in Fig. 2.30. This function has what might be called an "infinite discontinuity" at $x = 0$. Finally, the greatest integer function $f(x) = [\![x]\!]$, which has the graph shown in Fig. 2.31, is continuous at every *noninteger* point but discontinuous at every integer point.

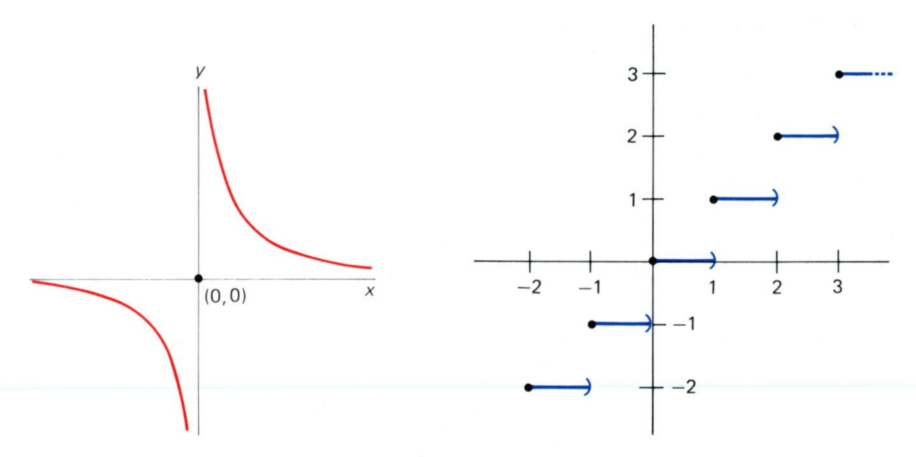

2.30 The function h has an infinite discontinuity at $x = 0$

2.31 The graph of the greatest integer function $f(x) = [\![x]\!]$

COMBINATIONS OF FUNCTIONS

We are more interested in functions that *are* continuous; most (though not all) of the functions we study in calculus are continuous on their natural domains of definition. The reason is that many varied and complex functions can be "put together" using quite simple "building blocks." Here we discuss some of the ways of combining functions to obtain new ones.

Suppose that f and g are functions and that c is a fixed real number. The **(scalar) multiple** cf, the **sum** $f + g$, the **difference** $f - g$, the **product** $f \cdot g$, and the **quotient** f/g are the new functions determined by these formulas:

$$(cf)(x) = cf(x), \tag{2}$$

$$(f + g)(x) = f(x) + g(x), \tag{3}$$

$$(f - g)(x) = f(x) - g(x), \tag{4}$$

$$(f \cdot g)(x) = f(x) \cdot g(x), \quad \text{and} \tag{5}$$

$$\left(\frac{f}{g}\right)(x) = \frac{f(x)}{g(x)}. \tag{6}$$

The combinations in (3)–(5) are defined for every number x that lies both in the domain of f and in the domain of g. In (6) we must also require that $g(x) \neq 0$.

For example, let $f(x) = x^2 + 1$ and $g(x) = x - 1$. Then

$$(3f)(x) = 3(x^2 + 1),$$

$$(f + g)(x) = (x^2 + 1) + (x - 1) = x^2 + x,$$

$$(f - g)(x) = (x^2 + 1) - (x - 1) = x^2 - x + 2,$$

$$(f \cdot g)(x) = (x^2 + 1)(x - 1) = x^3 - x^2 + x - 1, \quad \text{and}$$

$$\left(\frac{f}{g}\right)(x) = \frac{x^2 + 1}{x - 1} \qquad (x \neq 1).$$

If $f(x) = \sqrt{1 - x}$ for $x \leqq 1$ and $g(x) = \sqrt{1 + x}$ for $x \geqq -1$, then the sum or product of f and g is defined where *both* f and g are defined. In this case the domain would be the closed interval $[-1, 1]$.

It follows readily from the limit laws in Section 2.2 that *any sum or product of continuous functions is continuous.* That is, if the functions f and g are continuous at $x = a$, then so are $f + g$ and $f \cdot g$. It follows in turn that *every* ***polynomial function***

$$p(x) = b_n x^n + b_{n-1} x^{n-1} + \cdots + b_1 x + b_0$$

is continuous at each point of the real line; in short, every polynomial is continuous everywhere.

If $p(x)$ and $g(x)$ are polynomials, then the quotient law for limits and the continuity of polynomials imply that

$$\lim_{x \to a} \frac{p(x)}{q(x)} = \frac{\lim_{x \to a} p(x)}{\lim_{x \to a} q(x)} = \frac{p(a)}{q(a)}$$

provided that $q(a) \neq 0$. Thus every **rational function** $f(x) = p(x)/q(x)$ is continuous wherever it is defined—that is, wherever the denominator polynomial is nonzero.

EXAMPLE 2 Suppose that

$$f(x) = \frac{x - 2}{x^2 - 3x + 2}. \tag{7}$$

We factor the denominator: $x^2 - 3x + 2 = (x - 1)(x - 2)$. This shows that f is not defined either at $x = 1$ or at $x = 2$. Thus the rational function f defined in (7) is continuous except at these two points. In the case $x \neq 2$, we may cancel the factor $x - 2$ to obtain a new function, an "enhanced" version $\bar{f}$ of f, with formula

$$\bar{f}(x) = \frac{x-2}{(x-1)(x-2)} = \frac{1}{x-1}. \tag{8}$$

This function $\bar{f}$ agrees with f for all x other than $x = 1$ (where neither is defined) and $x = 2$. But $\bar{f}$ *is* defined at $x = 2$; its value there is 1. This tells us that although the original function f in (7) is discontinuous at both $x = 1$ and $x = 2$, it can be given the value $f(2) = 1$ in order to "remove" the discontinuity at $x = 2$. Thus $x = 1$ is the only point where f cannot be defined so as to be continuous—clearly, the one-sided limits of $1/(x - 1)$ as $x \to 1$ do not exist.

COMPOSITION OF FUNCTIONS

Some important functions cannot be constructed in any of the ways suggested above. One example is the function

$$k(x) = \sqrt{1 - x^2}, \qquad -1 \leqq x \leqq 1.$$

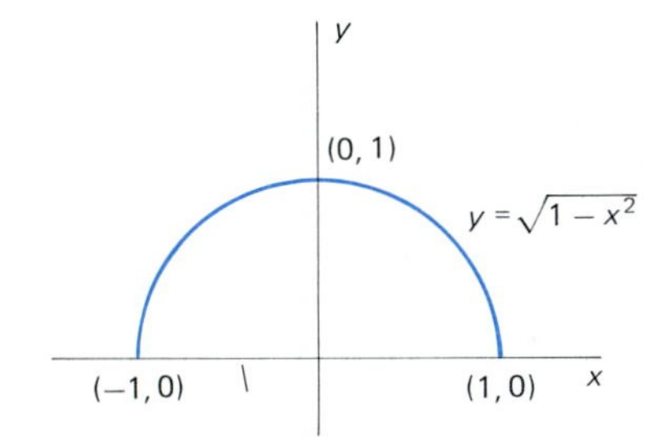

2.32 The graph of k

The graph of k is the upper half of the circle with equation $x^2 + y^2 = 1$, as shown in Fig. 2.32.

Certainly, $f(x) = \sqrt{x}$ and $g(x) = 1 - x^2$ are simple functions. But they cannot be added, subtracted, multiplied, or divided to produce $k(x)$. On the other hand, if you begin with x, let g act upon it,

$$x \xrightarrow[g]{} 1 - x^2 = g(x),$$

and next use the *output* of g as the *input* for f,

$$x \xrightarrow[g]{} 1 - x^2 \xrightarrow[f]{} \sqrt{1 - x^2} = f(g(x)),$$

then the *net* effect is that x is transformed into $\sqrt{1 - x^2}$. This is exactly how k is supposed to act on the input x.

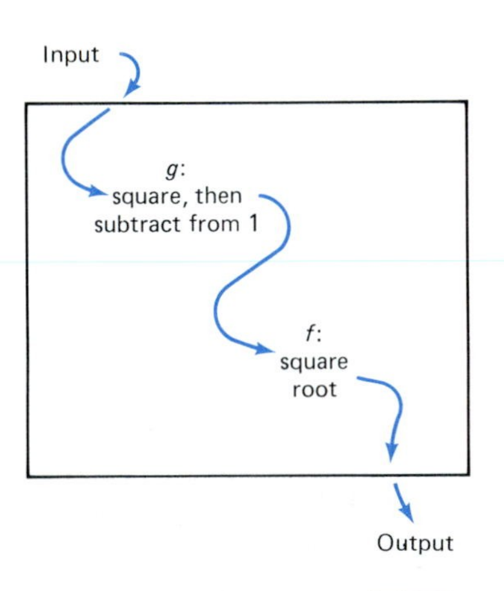

2.33 The action of $k(x) = \sqrt{1 - x^2}$

When you think of a function as a sort of calculator, one that accepts values of x and produces values of the function, then the way we have combined $f(x) = \sqrt{x}$ and $g(x) = 1 - x^2$ to produce $k(x)$ could be represented as in Fig. 2.33. Moreover, a natural way to write k in terms of f and g would be this:

$$k(x) = f(g(x)).$$

(Pronounce the right-hand side "f of g of x.") The expression $g(x)$ appears as the input or argument of f to signify that the *value* $g(x)$ is to be acted on by the function f. This is exactly what we need to produce the effect of the function k.

The order in which the two functions g and f are to be applied is this: *first g, then f*. Although the notation $k(x) = f(g(x))$ shows f appearing typographically first, you should remember that it means that f is to be applied *last*.

The way in which f and g are combined here to form k is called the *composition* of f and g and is perhaps the most important way in which two functions can be combined. The reason is that many human activities are performed in the same way. Certain operations take place in a certain order and each step acts on the result of the previous step: It's much like assembling an automobile, making a pizza, or learning calculus.

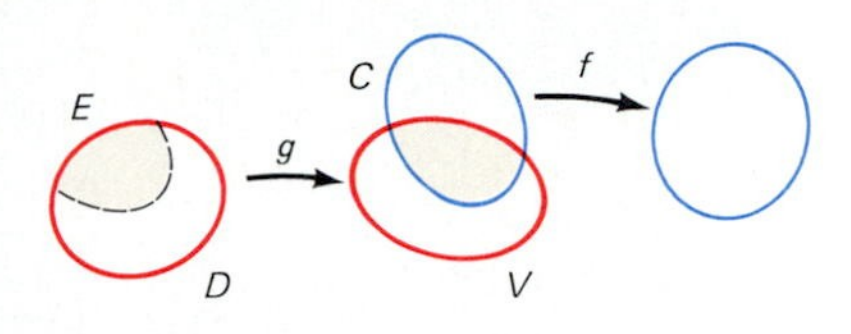

2.34 The domain E of $f(g)$

What about the domain of $f(g(x))$? Suppose that g has domain D and range V and that f has domain C that at least partly overlaps V, as we show in Fig. 2.34. Then the **composition** of f and g is denoted by $f(g)$ (read this "f of g"), and its value at x is given by

$$\{f(g)\}(x) = f(g(x)). \tag{9}$$

The domain E of $f(g)$ is the largest set of real numbers for which the rule in (9) makes sense: all numbers in D that are assigned values in C by g.

Another common notation for the composition of f and g is $f \circ g$, with a small circle between f and g.

The composition notation is very different from the product notation. The product of the two functions f and g is written $f \cdot g$, or simply fg. But the composition of f and g may involve no multiplication whatsoever and is written $f(g)$ or $f \circ g$.

A sure way to see that products differ from compositions is this: The product of f and g is the same whether it is f or g that is written first. But when we reverse the order of composition, we almost always obtain a different function. For example, with the functions $f(x) = \sqrt{x}$ and $g(x) = 1 - x^2$ of our earlier discussion, we find that

$$g(f(x)) = 1 - (\sqrt{x})^2 = 1 - x, \qquad x \geqq 0.$$

This is not at all the same as

$$f(g(x)) = \sqrt{1 - x^2}, \qquad -1 \leqq x \leqq 1.$$

EXAMPLE 3 Find $f(g)$ and $g(f)$, given $f(x) = x^3$ and $g(x) = x - 2$.

Solution First, $f(g)$ has the rule

$$f(g(x)) = f(x - 2) = (x - 2)^3.$$

The domain of g is all of $\boldsymbol{R}$. Because every possible value of g lies in the domain (also $\boldsymbol{R}$) of f, the domain of $f(g)$ is the set $\boldsymbol{R}$ of all real numbers.

Next, $g(f)$ has the rule

$$g(f(x)) = g(x^3) = x^3 - 2,$$

and its domain is also $\boldsymbol{R}$.

Note that the two compositions in Example 3 yield *different* functions. Even though $f(g)$ and $g(f)$ have the same domain, $f(g(3)) = 1 \neq 25 = g(f(3))$, so $f(g)$ and $g(f)$ yield different values at $x = 3$ (and at almost all other values of x as well!). Consequently, $f(g)$ and $g(f)$ are different. Moreover, each is different from the product fg, for

$$(fg)(x) = f(x)g(x) = x^3(x - 2),$$

and $(fg)(3) = 27$, which is neither equal to 1 nor to 25.

Sometimes we are given a function h and need to find two simpler functions f and g whose composition is h; that is, such that $h(x) = f(g(x))$. We can think of f and g as "machines," as illustrated in Fig. 2.35. The "inner" function g accepts x as input and produces the output $u = g(x)$. The "outer" function f represents the final step in manufacturing the result $h(x)$; it accepts the output $u = g(x)$ as its input and produces the ultimate output $f(u) = f(g(x)) = h(x)$. Thus the sequence of operations is

$$x \xrightarrow[g]{} g(x) = u \xrightarrow[f]{} f(u) = f(g(x)).$$

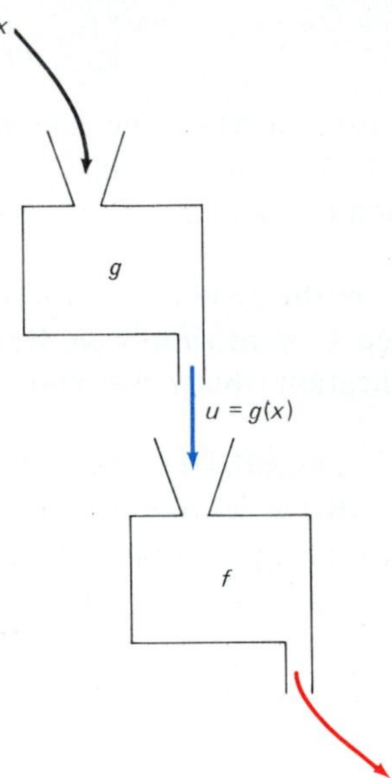

2.35 The composition of f and g

EXAMPLE 4 Given the function $h(x) = (x^2 + 4)^{3/2}$, find two functions f and g such that $h(x) = f(g(x))$.

Solution It is technically correct, but useless, simply to let $g(x) = x$ and $f(u) = (u^2 + 4)^{3/2}$. We seek a nontrivial solution here. So we observe that to calculate $(x^2 + 4)^{3/2}$, we first must calculate $x^2 + 4$. So we choose $g(x) = x^2 + 4$ as the inner function. The last step is to raise $g(x)$ to the power $\frac{3}{2}$, so we take $f(u) = u^{3/2}$ as the outer function. Thus if

$$f(x) = x^{3/2} \quad \text{and} \quad g(x) = x^2 + 4,$$

then $f(g(x)) = f(x^2 + 4) = (x^2 + 4)^{3/2} = h(x)$.

Alternatively, we could have decided first to take $f(x) = x^{3/2}$, because the last operation in calculating a value of $(x^2 + 4)^{3/2}$ is to raise a number to the power $\frac{3}{2}$. In this case we ask what $u = g(x)$ should be in order that

$$f(u) = u^{3/2} = (x^2 + 4)^{3/2} = h(x).$$

If we raise each of these terms to the power $\frac{2}{3}$, we get $u = x^2 + 4$. Therefore, we choose $g(x) = x^2 + 4$.

The following theorem implies that functions built by forming compositions of continuous functions are themselves continuous.

Theorem 1 *Continuity of Compositions*

The composition of two continuous functions is continuous. More precisely, if g is continuous at a and f is continuous at $g(a)$, then $f(g)$ is continuous at a.

Proof The continuity of g at a means that $g(x) \to g(a)$ as $x \to a$, and the continuity of f at $g(a)$ implies that $f(g(x)) \to f(g(a))$ as $g(x) \to g(a)$. Hence the substitution law for limits (Section 2.2) yields

$$\lim_{x \to a} f(g(x)) = f\left(\lim_{x \to a} g(x)\right) = f(g(a)),$$

as desired. ■

Recall from the root law in Section 2.2 that

$$\lim_{x \to a} \sqrt[n]{x} = \sqrt[n]{a}$$

under the conditions that n is an integer and that $a > 0$ if n is even. Thus the nth root function $f(x) = \sqrt[n]{x}$ is continuous everywhere if n is odd, and continuous for $x > 0$ if n is even.

We may combine this result with Theorem 1. Then we see that a root of a continuous function is continuous wherever it is defined. That is, the composition

$$g(x) = [f(x)]^{1/n}$$

is continuous at a if f is, assuming that $f(a) > 0$ if n is even (so that $\sqrt[n]{f(a)}$ will be defined).

EXAMPLE 5 Show that the function

$$f(x) = \left(\frac{x - 7}{x^2 + 2x + 2}\right)^{2/3}$$

is continuous on the whole real line.

Solution Note first that the denominator

$$x^2 + 2x + 2 = (x + 1)^2 + 1$$

is never zero. Hence the rational function $r(x) = (x - 7)/(x^2 + 2x + 2)$ is defined and continuous everywhere. It then follows from Theorem 1 and the continuity of the cube root function that

$$f(x) = ([r(x)]^2)^{1/3}$$

is continuous everywhere. Hence (for example)

$$\lim_{x \to -1} \left(\frac{x - 7}{x^2 + 2x + 2}\right)^{2/3} = f(-1) = (-8)^{2/3} = 4.$$

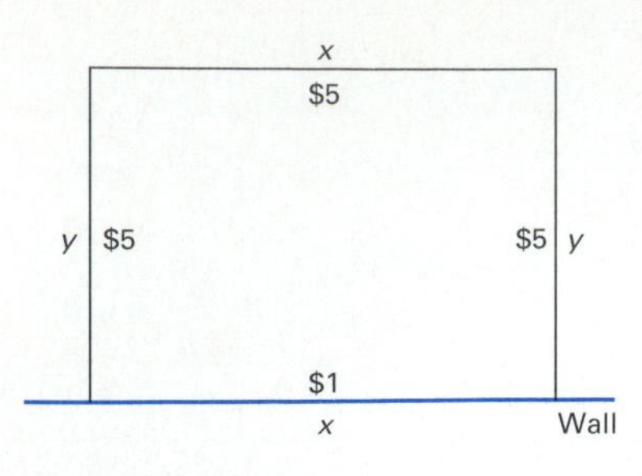

2.36 The animal pen

CONTINUOUS FUNCTIONS ON CLOSED INTERVALS

An applied problem often involves a function with domain a *closed interval.* For example, in the animal pen problem of Section 1.1, we found that the area A of the rectangular pen in Fig. 2.36 was expressed as a function of its base length x by

$$A = f(x) = \tfrac{3}{5}x(30 - x).$$

Although this formula for f is meaningful for all x, only values in the closed interval $[0, 30]$ correspond to actual rectangles, so only such values are pertinent to the animal pen problem.

The function f defined on the closed interval $[a, b]$ is said to be **continuous on** $[a, b]$ provided that it is continuous at each point of the open interval (a, b) *and* that

$$\lim_{x \to a^+} f(x) = f(a) \quad \text{and} \quad \lim_{x \to b^-} f(x) = f(b).$$

The last two conditions mean that at each endpoint, the value of the function is equal to its limit from within the interval. For instance, every polynomial is continuous on every closed interval. The square root function $f(x) = \sqrt{x}$ is continuous on the closed interval $[0, 1]$, although f is not defined for $x < 0$.

Continuous functions defined on closed intervals are very special. Example: Every such function has the *intermediate value property* of Theorem 2. (A proof of this theorem is given in Appendix D.) We suggested earlier that continuity of a function is related to the possibility of tracing its graph without lifting the pencil from the paper. Theorem 2 expresses this fact precisely.

> ***Theorem 2*** *Intermediate Value Property*
> Suppose that the function f is continuous on the closed interval $[a, b]$. Then $f(x)$ assumes every intermediate value between $f(a)$ and $f(b)$. Thus if K is any number between $f(a)$ and $f(b)$, then there exists at least one number c in (a, b) such that $f(c) = K$.

Figure 2.37 shows the graph of a typical continuous function f with domain the closed interval $[a, b]$. The number K is located on the y-axis, somewhere between $f(a)$ and $f(b)$. In the figure we have $f(a) < f(b)$, but this is not important. The horizontal line through K must cut the graph of f somewhere, and the x-coordinate of the point where graph and line meet yields the value of c. This number c is the one whose existence is guaranteed by the intermediate value property of the continuous function f.

Thus the intermediate value property implies that each horizontal line meeting the y-axis between $f(a)$ and $f(b)$ must cut the graph of the continuous function f somewhere. This is a way of saying that the graph has no gaps or jumps, suggesting that the idea of being able to trace such a graph without lifting the pencil from the paper is accurate.

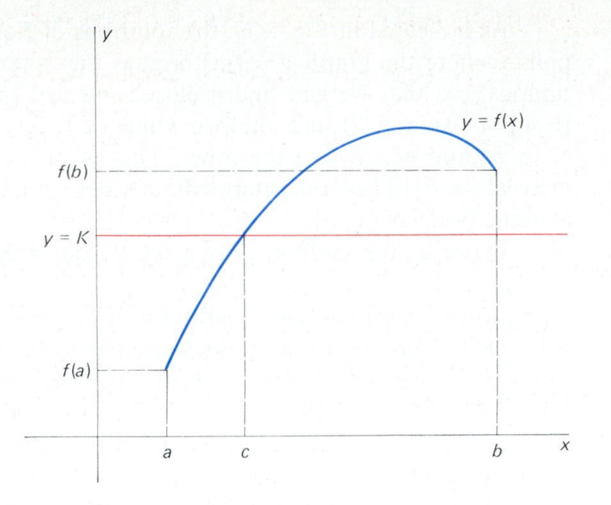

2.37 The continuous function f attains the intermediate value K at $x = c$.

EXAMPLE 6 The discontinuous function defined on $[-1, 1]$ as

$$f(x) = \begin{cases} 0 & \text{if} \quad x < 0, \\ 1 & \text{if} \quad x \geqq 0 \end{cases}$$

does *not* attain the intermediate value 0.5. See Fig. 2.38.

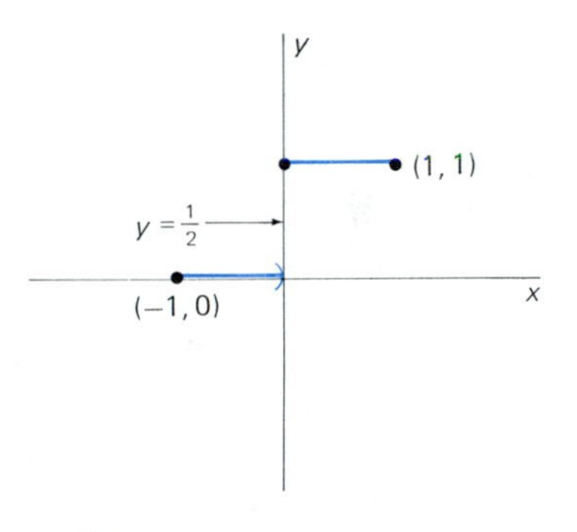

2.38 This discontinuous function does not have the intermediate value property.

EXAMPLE 7 Show that the equation

$$x^3 + x - 1 = 0 \tag{10}$$

has a solution between 0 and 1.

Solution The function $f(x) = x^3 + x - 1$ is continuous on $[0, 1]$ because it is a polynomial. Next, $f(0) = -1$ and $f(1) = +1$. So every number between -1 and $+1$ is a value of f on $(0, 1)$. In particular,

$$f(0) < 0 < f(1),$$

so the intermediate value property of f implies that it attains the value 0 at some number c between $x = 0$ and $x = 1$. That is, $f(c) = 0$ for some number c in $(0, 1)$. In other words, $c^3 + c - 1 = 0$. Therefore, the equation $x^3 + x - 1 = 0$ has a solution in $(0, 1)$.

*THE METHOD OF BISECTION

Example 7 raises a natural question. It's one thing to know that the equation $x^3 + x - 1 = 0$ has a solution in $(0, 1)$; it's another thing altogether to find it. Can the intermediate value property of continuous functions help us here?

The answer is that it does; the intermediate value property is the basis for the *method of bisection*, a natural technique for obtaining numerical approximations to solutions of equations of the form

$$f(x) = 0. \tag{11}$$

Of course, any equation in one variable can easily be put into the form in (11), so the method is quite general.

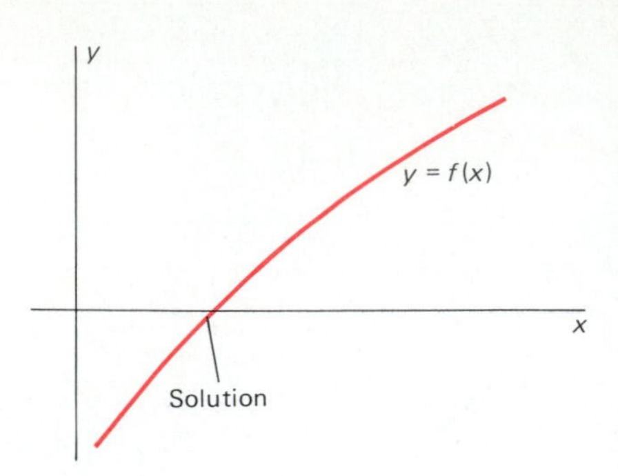

2.39 A root of the equation $f(x) = 0$

As indicated in Fig. 2.39, the solutions of Equation (11) are simply the points where the graph $y = f(x)$ crosses the x-axis. Suppose that f is continuous and that we can find a closed interval $[a, b]$ (such as the interval $[0, 1]$ of Example 7) such that the value of $f(x)$ is positive at one end point of $[a, b]$ and negative at the other. That is, $f(x)$ *changes sign* on the closed interval $[a, b]$. Then the intermediate value property ensures that $f(x) = 0$ at some point of $[a, b]$.

Let m be the midpoint of $[a, b]$. If $f(m) = 0$, we are done. Otherwise, $f(x)$ must change sign on either $[a, m]$ or $[m, b]$. If $f(x)$ changes sign on $[a, m]$, then $[a, m]$ contains a solution of $f(x) = 0$; we can bisect $[a, m]$ and continue this process for as many steps as may be needed to find the solution to the desired accuracy. We proceed in the same way if $f(x)$ changes sign on $[m, b]$. Theoretically, there is no limit to the accuracy we can attain with sufficient patience, because the length of the interval containing the desired solution halves at each stage.

The data shown in Fig. 2.40 result from carrying the method of bisection through eight steps for the equation

$$f(x) = x^3 + x - 1 = 0.$$

At the last step we see that the root lies between

$$\tfrac{87}{128} \approx 0.6797 \quad \text{and} \quad \tfrac{175}{256} \approx 0.6836.$$

Thus the root is approximately 0.68, accurate to two decimal places. By continuing this work through sufficiently many steps, you can verify that the first four decimal places of the root are, in fact, 0.6823.

x	$f(x)$ (rounded)	$f(x)$ changes sign on:	Average of end points:
0	$-1.0000 < 0$		
1	$1.0000 > 0$	[0, 1]	0.5000
1/2	$-0.3750 < 0$	[1/2, 1]	0.7500
3/4	$0.1719 > 0$	[1/2, 3/4]	0.6250
5/8	$-0.1309 < 0$	[5/8, 3/4]	0.6875
11/16	$0.0125 > 0$	[5/8, 11/16]	0.6563
21/32	$-0.0611 < 0$	[21/32, 11/16]	0.6719
43/64	$-0.0248 < 0$	[43/64, 11/16]	0.6797
87/128	$-0.0063 < 0$	[87/128, 11/16]	0.6836
175/256	$0.0030 > 0$	[87/128, 175/256]	0.6816

2.40 Result of the method of bisection for the equation $x^3 + x - 1 = 0$ on the interval [0, 1]

CONTINUITY AND DIFFERENTIABILITY

We have seen that a wide variety of functions are continuous. Although there do exist continuous functions so exotic that they are nowhere differentiable, the following theorem tells us that every differentiable function is continuous.

> ***Theorem 3*** *Differentiability Implies Continuity*
> Suppose that f is defined in a neighborhood of a. If f is differentiable at a, then f is continuous at a.

Proof Because $f'(a)$ exists, the product law for limits yields

$$\begin{aligned}\lim_{x\to a}[f(x)-f(a)] &= \lim_{x\to a}(x-a)\cdot\frac{f(x)-f(a)}{x-a}\\ &= \left(\lim_{x\to a}(x-a)\right)\left(\lim_{x\to a}\frac{f(x)-f(a)}{x-a}\right)\\ &= 0\cdot f'(a)=0.\end{aligned}$$

Thus, $\lim_{x\to a} f(x) = f(a)$, so f is continuous at a. ■

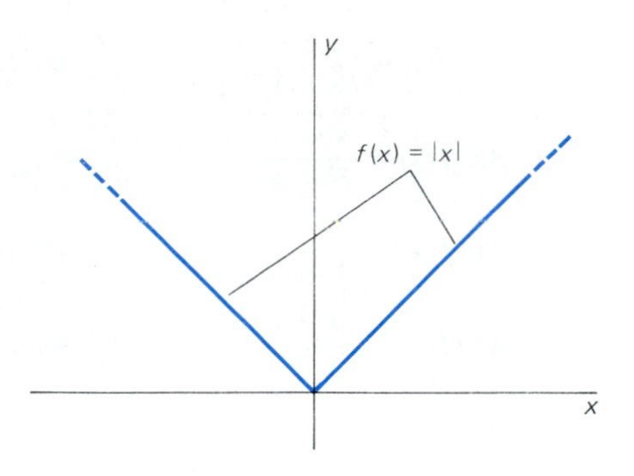

2.41 The graph of $f(x) = |x|$

Although it is complicated to exhibit a nowhere differentiable but everywhere continuous function, the absolute value function $f(x) = |x|$ provided us with a simple example of an everywhere continuous function that is not differentiable at the one point $x = 0$ (as we saw in Example 4 of Section 2.3). The "corner point" on the graph evident in Fig. 2.41 indicates that $y = |x|$ has no tangent line at (0, 0), and thus that $f(x) = |x|$ cannot be differentiable there.

2.4 PROBLEMS

In Problems 1–5, find $f + g$, $f \cdot g$, and f/g, and give the domain of each of these new functions.

1. $f(x) = x + 1, \quad g(x) = x^2 + 2x - 3$

2. $f(x) = \dfrac{1}{x-1}, \quad g(x) = \dfrac{1}{2x+1}$

3. $f(x) = \sqrt{x}, \quad g(x) = \sqrt{x-2}$

4. $f(x) = \sqrt{x+1}, \quad g(x) = \sqrt{5-x}$

5. $f(x) = \sqrt{x^2-1}, \quad g(x) = 1/\sqrt{4-x^2}$

In Problems 6–10, find $f(g(x))$ and $g(f(x))$.

6. $f(x) = 1 - x^2, \quad g(x) = 2x + 3$

7. $f(x) = -17, \quad g(x) = |x|$

8. $f(x) = \sqrt{x^2-3}, \quad g(x) = x^2 + 3$

9. $f(x) = x^2 + 1, \quad g(x) = \dfrac{1}{x^2+1}$

10. $f(x) = x^3 - 4, \quad g(x) = \sqrt[3]{x+4}$

In each of Problems 11–20, find a function of the form $f(x) = x^k$ (you must specify k) and a function g such that $f(g(x)) = h(x)$.

11. $h(x) = (2 + 3x)^2$

12. $h(x) = (4 - x)^3$

13. $h(x) = \sqrt{2x - x^3}$

14. $h(x) = (1 + x^4)^{17}$

15. $h(x) = (5 - x^2)^{3/2}$

16. $h(x) = \sqrt[3]{(4x-6)^4}$

17. $h(x) = \dfrac{1}{x+1}$

18. $h(x) = \dfrac{1}{1+x^2}$

19. $h(x) = \dfrac{1}{\sqrt{x+10}}$

20. $h(x) = \dfrac{1}{(1+x+x^2)^3}$

In each of Problems 21–36, tell where the function with the given formula is continuous. Recall that when the domain of a function is not specified, it is the set of all real numbers for which the formula of the function is meaningful.

21. $f(x) = 2x + \sqrt[3]{x}$

22. $f(x) = x^2 + \dfrac{1}{x}$

23. $f(x) = \dfrac{1}{x+3}$

24. $f(x) = \dfrac{5}{5-x}$

25. $f(x) = \dfrac{1}{x^2+1}$

26. $f(x) = \dfrac{1}{x^2-1}$

27. $f(x) = \dfrac{x-5}{|x-5|}$

28. $f(x) = \dfrac{x^2+x+1}{x^2+1}$

29. $f(x) = \dfrac{x^2+4}{x-2}$

30. $f(x) = \sqrt[4]{4+x^4}$

31. $f(x) = \sqrt[3]{\dfrac{x+1}{x-1}}$

32. $f(x) = \sqrt[3]{3-x^3}$

33. $f(x) = \dfrac{3}{x^2-x}$

34. $f(x) = \sqrt{9-x^2}$

35. $f(x) = \dfrac{x}{\sqrt{4-x^2}}$

36. $f(x) = \sqrt{\dfrac{1-x^2}{4-x^2}}$

In Problems 37–46, find the points where the given function is not defined. For each such point a, tell whether a value of $f(a)$ can be assigned in such a way that f is continuous at a.

37. $f(x) = \dfrac{x}{(x+3)^2}$

38. $f(x) = \dfrac{x}{x^2 - 1}$

39. $f(x) = \dfrac{x-2}{x^2-4}$

40. $f(x) = \dfrac{x+1}{x^2 - x - 6}$

41. $f(x) = \dfrac{1}{1 - |x|}$

42. $f(x) = \dfrac{|x-1|}{(x-1)^3}$

43. $f(x) = \dfrac{x-17}{|x-17|}$

44. $f(x) = \dfrac{x^2+5x+6}{x+2}$

45. $f(x) = \begin{cases} -x & \text{if } x < 0; \\ x^2 & \text{if } x > 0 \end{cases}$

46. $f(x) = \begin{cases} x+1 & \text{if } x < 1; \\ 3-x & \text{if } x > 1 \end{cases}$

In each of Problems 47–52, apply the intermediate value property of continuous functions to show that the given equation has a solution in the given interval.

47. $x^2 - 5 = 0$ on $[2, 3]$

48. $x^3 + x + 1 = 0$ on $[-1, 0]$

49. $x^3 - 3x^2 + 1 = 0$ on $[0, 1]$

50. $x^3 = 5$ on $[1, 2]$

51. $x^4 + 2x - 1 = 0$ on $[0, 1]$

52. $x^5 - 5x^3 + 3 = 0$ on $[-3, -2]$

In each of Problems 53–54, show that the given equation has three distinct roots by calculating the values of the left-hand side at $x = -3, -2, -1, 0, 1, 2$, and 3 and then applying the intermediate value property of continuous functions on appropriate closed intervals.

53. $x^3 - 4x + 1 = 0$

54. $x^3 - 3x^2 + 1 = 0$

55. Apply the intermediate value property of continuous functions to show that every positive number a has a square root. That is, given $a > 0$, prove that there exists a number r such that $r^2 = a$.

56. Apply the intermediate value property to prove that every real number has a cube root.

57. Determine where the function $f(x) = [\![x/3]\!]$ is continuous.

58. Determine where the function $f(x) = x - [\![x]\!]$ is continuous.

59. Suppose that $f(x) = 0$ if x is a rational number, while $f(x) = 1$ if x is irrational. Prove that f is discontinuous at every real number.

60. Suppose that $f(x) = 0$ if x is a rational number, while $f(x) = x^2$ if x is irrational. Prove that f is continuous only at the single point $x = 0$.

61. Use the method of bisection to approximate the solution of $x^4 + 2x - 1 = 0$ in $[0, 1]$ to two-place accuracy.

62. Use the method of bisection to approximate the solution of $x^5 + x = 1$ in $[0, 1]$ to three-place accuracy.

*2.4 Optional Computer Application

With a programmable calculator or one with a function memory, one can search for a solution of an equation $f(x) = 0$ by a process of "educated guessing" that amounts to an "intelligent" bisection method. For example, consider the equation

$$f(x) = x^3 + x - 1 = 0$$

of Example 7.

With a Casio *fx*-7000G the self-explanatory program

```
Lbl 1
''X''?→X
X*X*X + X - 1
Goto 1
```

repeatedly calls for a new value of x to be entered, then displays the value of $f(x) = x^3 + x - 1$. With an HP-28S, the reverse Polish program

```
<< DUP SQ 1 + * 1 - >>
```

calculates $f(x)$ when x is entered.

With the Tandy PC-6 one needs only to store the formula

```
F = X*X*X + X - 1
```

in function memory. Then when the (CALC) key is pressed and a value of X is entered, the corresponding value of F is displayed. To approximate the root of our equation in the interval [0, 1], we begin with

```
X? 0          F = -1.00000
X? 1          F - +1.00000
X? 0.5        F = -0.37500
```

as in the method of bisection. Now it is apparent that the root we seek is in the interval [0.5, 1] and is somewhat closer to 0.5 than to 1.0. We therefore try $x = 0.6$ and 0.7 next:

```
X? 0.6        F = -0.18400
X? 0.7        F = +0.04300.
```

Because 0.0 is about four-fifths of the way from -0.184 to $+0.043$, we continue with

```
X? 0.68       F = -0.00557
X? 0.69       F = +0.01851.
```

Now it is clear that the solution is much closer to 0.68 than to 0.69, and the further data

```
X? 0.681      F = -0.00318
X? 0.682      F = -0.00079
X? 0.682      F = +0.00161
```

indicate that the solution is about one-third of the way from 0.682 to 0.683. Finally, the values

```
X? 0.6823     F = -0.00007
X? 0.6824     F = +0.00017
```

tell us that the solution is $x \approx 0.6823$ accurate to four decimal places.

With a personal computer the TABULATE program listed in the Section 1.1 Optional Computer Application (page 11) can be a powerful tool for approximating solutions in this manner. We simply tabulate values of $f(x)$ on successively smaller intervals, with the interval at each stage suggested by the results obtained at the previous stage of the process. For the equation $x^3 + x - 1 = 0$ we obtained the following results in this way.

```
ENDPOINTS A,B? 0.5, 1.0
NUMBER OF SUBINTERVALS? 5
x              f(x)
0.5         -0.37500
0.6         -0.18400   ←
0.7         +0.04300
0.8         +0.31200
0.9         +0.62900
1.0         +1.00000
```

```
ENDPOINTS A,B? 0.65, 0.70
NUMBER OF SUBINTERVALS? 5
x            f(x)
0.65        -0.07538
0.66        -0.05250
0.67        -0.02924
0.68        -0.00557
0.69        +0.01851    ⟵
0.70        +0.04300

ENDPOINTS A,B? 0.680, 0.685
NUMBER OF SUBINTERVALS? 5
x            f(x)
0.680       -0.00557
0.681       -0.00318
0.682       -0.00079
0.683       +0.00161    ⟵
0.684       +0.00401
0.685       +0.00642

ENDPOINTS A,B? 0.6820, 0.6825
NUMBER OF SUBINTERVALS? 5
x            f(x)
0.6820      -0.00079
0.6821      -0.00055
0.6822      -0.00031
0.6823      -0.00007
0.6824      +0.00017    ⟵
0.6825      +0.00041
```

Note that at each stage we have chosen the *half*-subinterval on which $f(x)$ changed sign in the previous stage. In this manner we gain an additional decimal place of accuracy at each stage, and find after four tabulations that $x \approx 0.6823$ to four decimal places.

In each of Exercises 1–5, use one of the methods illustrated here to find (accurate to four decimal places) the solution of the given equation in the given interval.

Exercise 1: $x^3 - 3x^2 + 1 = 0$; $[0, 1]$

Exercise 2: $x^3 - 3x^2 + 1 = 0$; $[2, 3]$

Exercise 3: $x^2 - 2 = 0$; $[1, 2]$ (to find the positive square root of 2)

Exercise 4: $x^3 - 5 = 0$; $[1, 2]$ (to find the cube root of 5)

Exercise 5: $x^5 + x^4 = 100$; $[2, 3]$

CHAPTER 2 REVIEW: Definitions, Concepts, Results

Use the list below as a guide to concepts that you may need to review.

1. Limit of $f(x)$ as x approaches a
2. Limit laws: constant, addition, product, quotient, root, substitution, squeeze
3. Definition of the derivative of a function
4. Four-step process for finding the derivative

5. The derivative of $f(x) = ax^2 + bx + c$
6. Tangent and normal lines to the graph of a function
7. Right-hand and left-hand limits
8. Relation between one-sided and two-sided limits
9. Infinite limits
10. Composition of functions
11. Continuity of a function at a point
12. Continuity of polynomials and rational functions
13. Continuity of compositions of functions
14. Continuity of a function on a closed interval
15. Intermediate value property of continuous functions
16. Differentiability implies continuity
17. Method of bisection

CHAPTER 2 MISCELLANEOUS PROBLEMS

Apply the limit laws to evaluate the limits in Problems 1–30 or to show that the indicated limit does not exist, as appropriate.

1. $\lim_{x\to 0} (x^2 - 3x + 4)$

2. $\lim_{x\to -1} (3 - x + x^3)$

3. $\lim_{x\to 2} (4 - x^2)^{10}$

4. $\lim_{x\to 1} (x^2 + x - 1)^{17}$

5. $\lim_{x\to 2} \dfrac{1 + x^2}{1 - x^2}$

6. $\lim_{x\to 3} \dfrac{2x}{x^2 - x - 3}$

7. $\lim_{x\to 1} \dfrac{x^2 - 1}{1 - x}$

8. $\lim_{x\to -2} \dfrac{x + 2}{x^2 + x - 2}$

9. $\lim_{t\to -3} \dfrac{t^2 + 6t + 9}{9 - t^2}$

10. $\lim_{x\to 0} \dfrac{4x - x^3}{3x + x^2}$

11. $\lim_{x\to 3} (x^2 - 1)^{2/3}$

12. $\lim_{x\to 2} \sqrt{\dfrac{2x^2 + 1}{2x}}$

13. $\lim_{x\to 3} \left(\dfrac{5x + 1}{x^2 - 8}\right)^{3/4}$

14. $\lim_{x\to 1} \dfrac{x^4 - 1}{x^2 + 2x - 3}$

15. $\lim_{x\to 7} \dfrac{\sqrt{x + 2} - 3}{x - 7}$

16. $\lim_{x\to 1^+} (x - \sqrt{x^2 - 1})$

17. $\lim_{x\to -4} \dfrac{1}{x + 4}\left(\dfrac{1}{\sqrt{13 + x}} - \dfrac{1}{3}\right)$

18. $\lim_{x\to 1^+} \dfrac{1 - x}{|1 - x|}$

19. $\lim_{x\to 2^+} \dfrac{2 - x}{\sqrt{4 - 4x + x^2}}$

20. $\lim_{x\to -2^-} \dfrac{x + 2}{|x + 2|}$

21. $\lim_{x\to 4^+} \dfrac{x - 4}{|x - 4|}$

22. $\lim_{x\to 3^-} \sqrt[3]{x^2 - 9}$

23. $\lim_{x\to 2^+} \sqrt{4 - x^2}$

24. $\lim_{x\to -3} \dfrac{x}{(x + 3)^2}$

25. $\lim_{x\to 2} \dfrac{x + 2}{(x - 2)^2}$

26. $\lim_{x\to 1^-} \dfrac{x}{x - 1}$

27. $\lim_{x\to 3^+} \dfrac{x}{x - 3}$

28. $\lim_{x\to 1^-} \dfrac{x - 2}{x^2 - 3x + 2}$

29. $\lim_{x\to 1^-} \dfrac{x + 1}{(x - 1)^3}$

30. $\lim_{x\to 5^+} \dfrac{25 - x^2}{x^2 - 10x + 25}$

Apply the formula for the derivative of $f(x) = ax^2 + bx + c$ to differentiate the functions in Problems 31–36. Then write an equation for the tangent line to the curve $y = f(x)$ at the point $(1, f(1))$.

31. $f(x) = 3 + 2x^2$

32. $f(x) = x - 5x^2$

33. $f(x) = 3x^2 + 4x - 5$

34. $f(x) = 1 - 2x - 3x^2$

35. $f(x) = (x - 1)(2x + 1)$

36. $f(x) = \dfrac{x}{3} - \left(\dfrac{x}{4}\right)^2$

In each of Problems 37–43, apply the definition of the derivative directly to find $f'(x)$.

37. $f(x) = 2x^2 + 3x$

38. $f(x) = x - x^3$

39. $f(x) = \dfrac{1}{3 - x}$

40. $f(x) = \dfrac{1}{2x + 1}$

41. $f(x) = x - \dfrac{1}{x}$

42. $f(x) = \dfrac{x}{x + 1}$

43. $f(x) = \dfrac{x + 1}{x - 1}$

44. Find the derivative of

$$f(x) = 3x - x^2 + |2x + 3|$$

at the points where it is differentiable. Find the point where it is *not* differentiable. Sketch the graph of f.

45. Write equations of the two lines through $(3, 4)$ that are tangent to the parabola $y = x^2$. (*Suggestion:* Let (a, a^2) denote either point of tangency; first solve for a.)

46. Write an equation of the circle with center $(2, 3)$ that is tangent to the line $x + y + 3 = 0$.

47. Suppose that $f(x) = 1 + x^2$. Find g so that

$$f(g(x)) = 1 + x^2 - 2x^3 + x^4.$$

48. Suppose that $g(x) = 1 + \sqrt{x}$. Find f so that

$$f(g(x)) = 3 + 2\sqrt{x} + x.$$

In Problems 49–52, find g such that $f(g(x)) = h(x)$.

49. $f(x) = 2x + 3, \quad h(x) = 2x + 5$

50. $f(x) = x + 1, \quad h(x) = x^3$

51. $f(x) = x^2, \quad h(x) = x^4 + 1$

52. $f(x) = \dfrac{1}{x}, \quad h(x) = x^5$

Explain why each of the functions in Problems 53–56 is continuous wherever it is defined by the given formula. For each point a where f is not defined by the formula, tell whether a value can be assigned to $f(a)$ in such a way to make f continuous at a.

53. $f(x) = \dfrac{1 - x}{1 - x^2}$

54. $f(x) = \dfrac{1 - x}{(2 - x)^2}$

55. $f(x) = \dfrac{x^2 + x - 2}{x^2 + 2x - 3}$

56. $f(x) = \dfrac{|x^2 - 1|}{x^2 - 1}$

57. Apply the intermediate value property of continuous functions to prove that the equation $x^5 + x = 1$ has a solution.

58. Apply the intermediate value property of continuous functions to prove that the equation $x^3 - 3x^2 + 1 = 0$ has three different solutions.

59. Show that there is a number x between 0 and $\pi/2$ (radians) such that $x = \cos x$.

60. Show that there is a number x between $\pi/2$ and π (radians) such that $\tan x = -x$. (*Suggestion:* First sketch the graphs of $y = \tan x$ and $y = -x$.)

61. (*Challenge*) Let f be defined as follows:

If x is irrational, then $f(x) = 0$.

If $x = 0$ then $f(x) = 1$.

If x is a nonzero rational number, write it as a fraction p/q in lowest terms with $q > 0$. Then $f(x) = 1/q$.

(a) Show that f is discontinuous at a if a is a rational number.

(b) Show that f is continuous at a if a is an irrational number. (*Suggestion:* Note first that, given $\varepsilon > 0$, there are only finitely many positive integers q such that $1/q \geqq \varepsilon$. Then show that for each fixed positive integer q_0, the neighborhood $(a - 1, a + 1)$ of a contains only finitely many rational numbers of the form p/q_0.)

chapter three

The Derivative

3.1 The Derivative and Rates of Change

Our preliminary investigation of tangent lines and derivatives in Section 2.1 led to the study of limits in the remainder of Chapter 2. Armed with this knowledge of limits, we are now prepared to study derivatives more fully.

In Section 2.1 we introduced the derivative $f'(x)$ as the slope of the tangent line to the graph of the function f at the point $(x, f(x))$. More precisely, we were motivated by the geometry to *define* the tangent line to the graph at the point $P(a, f(a))$ to be the straight line through P with slope

$$m = \lim_{h \to 0} \frac{f(a + h) - f(a)}{h}, \tag{1}$$

as indicated in Fig. 3.1. If we replace the arbitrary number a in (1) with the independent variable x, we get a new function f', the *derivative* of the original function f.

Definition of the Derivative

The **derivative** of the function f is the function f' defined by

$$f'(x) = \lim_{h \to 0} \frac{f(x + h) - f(x)}{h} \tag{2}$$

for all x such that this limit exists.

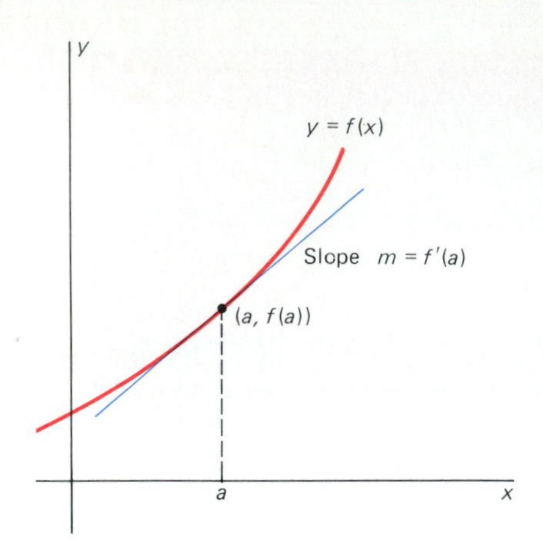

3.1 The geometric motivation for the definition of the derivative

The last phrase of the definition requires, in particular, that f must be defined in a neighborhood of x.

We emphasized in Section 2.1 that we hold x *fixed* in (2) while h approaches zero. When we are specifically interested in the value of the derivative f' at $x = a$, we sometimes rewrite (2) in the form

$$f'(a) = \lim_{h \to 0} \frac{f(a+h) - f(a)}{h} = \lim_{x \to a} \frac{f(x) - f(a)}{x - a}. \tag{3}$$

The second limit in (3) is obtained from the first by writing $x = a + h$, $h = x - a$, and noting that $x \to a$ as $h \to 0$. The statement that these equivalent limits exist is sometimes abbreviated as "$f'(a)$ exists." In this case we say that the function f is **differentiable** at $x = a$. Finally, the process of finding the derivative f' is called **differentiation** of f.

In Sections 2.1 and 2.2, we used several examples to illustrate the process of differentiating a given function f by direct evaluation of the limit in (2). This involves carrying out these four steps:

1. Write the definition in (2) of the derivative.
2. Substitute the expressions $f(x + h)$ and $f(x)$ as determined by the particular function f.
3. Simplify the result by algebraic methods to make Step 4 possible.
4. Evaluate the limit—typically, by application of the limit laws.

EXAMPLE 1 Differentiate $f(x) = \dfrac{x}{x+3}$.

Solution

$$f'(x) = \lim_{h \to 0} \frac{f(x+h) - f(x)}{h} = \lim_{h \to 0} \frac{1}{h}\left(\frac{x+h}{x+h+3} - \frac{x}{x+3}\right)$$

$$= \lim_{h \to 0} \frac{(x+h)(x+3) - x(x+h+3)}{h(x+h+3)(x+3)} = \lim_{h \to 0} \frac{3h}{h(x+h+3)(x+3)}$$

$$= \frac{3}{\left(\lim_{h \to 0} (x+h+3)\right)\left(\lim_{h \to 0} (x+3)\right)}.$$

Therefore,

$$f'(x) = \frac{3}{(x+3)^2}.$$

Even when the function f is rather simple, this process for computing f' directly from the definition of the derivative can be rather tedious. Also, Step 3 may require considerable ingenuity. Moreover, it would be very repetitious to continue to rely on the four-step process above. To avoid tedium, we want a fast, easy, short method for computing $f'(x)$.

That method is the main subject of this chapter: the development of systematic methods ("rules") for differentiating those functions that occur most frequently. Such functions include polynomials, rational functions, the trigonometric functions $\sin x$ and $\cos x$, and combinations of such functions.

Once these general differentiation rules have been established, they can be applied formally, almost mechanically, to compute derivatives. Only rarely should recourse to the definition of the derivative be necessary.

An example of a "differentiation rule" is the theorem of Section 2.1 on differentiation of quadratic functions:

$$\text{If} \quad f(x) = ax^2 + bx + c, \quad \text{then} \quad f'(x) = 2ax + b. \tag{4}$$

Once we know this rule, we need never again apply the definition of the derivative to differentiate a quadratic function. For example, if $f(x) = 3x^2 - 4x + 5$, we can apply (4) to write immediately

$$f'(x) = (2)(3x) + (-4) = 6x - 4.$$

Similarly, if $g(t) = 2t - 5t^2$, then

$$g'(t) = (2) + (2)(-5t) = 2 - 10t.$$

It makes *no* difference what the name of the function is, or whether we write x or t for the independent variable. This flexibility is valuable—in general, it is such adaptability that makes mathematics applicable to virtually every other branch of human knowledge. So you should learn every differentiation rule in a form independent of the notation used to state it.

We will develop additional differentiation rules in Sections 3.2–3.4. First, however, we introduce some new notation and a new interpretation of the derivative.

DIFFERENTIAL NOTATION

An important alternative notation for the derivative originates from the early custom of writing Δx in place of h (because $h = \Delta x$ is an increment in x) and

$$\Delta y = f(x + \Delta x) - f(x)$$

for the resulting change (or increment) in y. The slope of the secant line K of Fig. 3.2 is then

$$m_{\text{sec}} = \frac{\Delta y}{\Delta x},$$

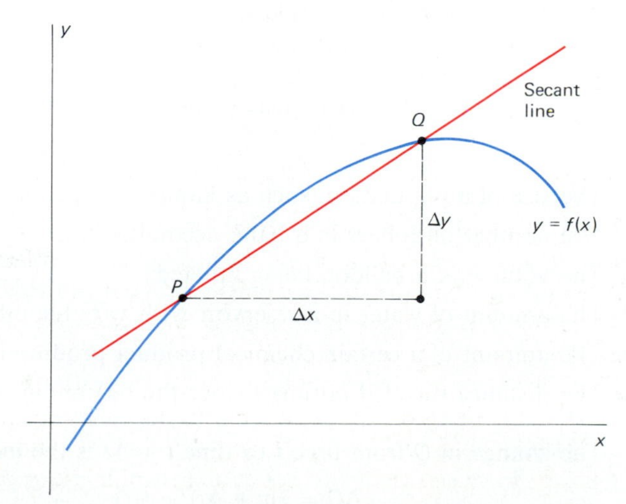

3.2 Genesis of the *dy*/*dx* notation

and the slope of the tangent line is

$$m = \frac{dy}{dx} = \lim_{\Delta x \to 0} \frac{\Delta y}{\Delta x}. \tag{5}$$

Hence, if $y = f(x)$, we often write

$$\frac{dy}{dx} = f'(x). \tag{6}$$

The two symbols $f'(x)$ and dy/dx for the derivative of the function $y = f(x)$ are used interchangeably in mathematics and its applications, so familiarity with both is necessary. It is also important to remember that dy/dx is a single symbol representing the derivative and is *not* the quotient of two separate "differential" quantities dy and dx.

EXAMPLE 2 If $y = ax^2 + bx + c$, then the derivative in (4) in differential notation takes the form

$$\frac{dy}{dx} = 2ax + b.$$

Consequently,

$$\text{if} \quad y = 3x^2 - 4x + 5, \quad \text{then} \quad \frac{dy}{dx} = 6x - 4;$$

$$\text{if} \quad z = 2t - 5t^2, \quad \text{then} \quad \frac{dz}{dt} = 2 - 10t.$$

The letter d in this notation is for the word "differential." Note that whether we write dy/dx or dz/dt, the dependent variable appears "upstairs," the independent variable "downstairs."

RATE OF CHANGE

In Section 2.1 we introduced the derivative of a function as the slope of the tangent line to its graph. Here we introduce the equally important interpretation of the derivative of a function as the function's rate of change with respect to the independent variable.

We begin with the instantaneous rate of change of a function having time t as its independent variable. Suppose that Q is a quantity that varies with time t, and write $Q = f(t)$ for its value at time t. For example, Q might be

1. The size of a population (such as kangaroos, people, or bacteria);
2. The number of dollars in a bank account;
3. The volume of a balloon being inflated;
4. The amount of water in a reservoir with variable inflow and outflow;
5. The amount of a certain chemical product produced in a reaction; or
6. The distance traveled in time t since the beginning of a journey.

The change in Q from time t to time $t + \Delta t$ is the **increment**

$$\Delta Q = f(t + \Delta t) - f(t).$$

The **average rate of change** of Q (per unit of time) is, by definition, the ratio of the change ΔQ in Q to the change Δt in t: It is thus the quotient

$$\frac{\Delta Q}{\Delta t} = \frac{f(t + \Delta t) - f(t)}{\Delta t}. \tag{7}$$

We define the **instantaneous rate of change** of Q (per unit of time) to be the limit of this average rate as $\Delta t \to 0$. That is, the instantaneous rate of change of Q is

$$\lim_{\Delta t \to 0} \frac{\Delta Q}{\Delta t} = \lim_{\Delta t \to 0} \frac{f(t + \Delta t) - f(t)}{\Delta t}. \tag{8}$$

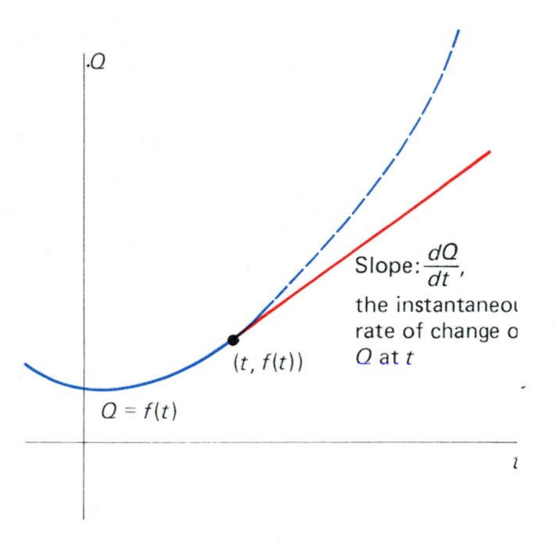

3.3 The relation between the tangent line at $(t, f(t))$ and the instantaneous rate of change of f at t

But the latter limit is simply the derivative $f'(t)$. So we see that the instantaneous rate of change of $Q = f(t)$ is the derivative

$$\frac{dQ}{dt} = f'(t). \tag{9}$$

The concept of the instantaneous rate of change agrees with our intuitive idea of what it should be. If we think of Q as changing with time, but then suddenly, at time t, continuing in the direction of its graph at time t without subsequent curvature, the graph of Q would appear as shown in Fig. 3.3. The broken line in that figure is meant to indicate how $Q = f(t)$ might normally behave. But if Q were to follow the straight line, that would be "change at a constant rate." Because the straight line is tangent to the graph of Q, we can interpret dQ/dt as the instantaneous rate of change of Q at time t. In brief:

> The instantaneous rate of change of $Q = f(t)$ at time t is equal to the slope of the tangent line to the curve $Q = f(t)$ at the point $(t, f(t))$.

There are important additional conclusions that we can draw. Because a positive slope corresponds to a rising tangent line and a negative slope to a falling tangent line, we say that

$$\begin{aligned} &Q \text{ is increasing at time } t \quad \text{if} \quad \frac{dQ}{dt} > 0; \\ &Q \text{ is decreasing at time } t \quad \text{if} \quad \frac{dQ}{dt} < 0. \end{aligned} \tag{10}$$

NOTE The meaning of the phrase "$Q = f(t)$ is increasing *over* (or *during*) *the time interval from* $t = a$ to $t = b$" should be intuitively clear. What we have in (10) is a way to make precise what is meant by "$Q = f(t)$ is increasing *at time* t"—that is, at the instant t.

EXAMPLE 3 A cylindrical tank with vertical axis is initially filled with 600 gal of water. This tank takes 60 min to empty after a drain in its bottom is opened. Suppose that the drain is opened at time $t = 0$. Suppose that the volume V of water remaining in the tank after t minutes is

$$V(t) = \tfrac{1}{6}(60 - t)^2 = 600 - 20t + \tfrac{1}{6}t^2$$

gallons. Find the instantaneous rate at which the water is flowing out of the tank at time $t = 15$ (min) *and* at time $t = 45$ (min).

Solution The rate of change of the volume of water in the tank is given by the derivative

$$\frac{dV}{dt} = -20 + \frac{1}{3}t.$$

At the instants $t = 15$ and $t = 45$ we obtain

$$V'(15) = -20 + \tfrac{1}{3}(15) = -15$$

and

$$V'(45) = -20 + \tfrac{1}{3}(45) = -5.$$

The units here are gal/min (gallons per minute). The fact that $V'(15)$ and $V'(45)$ are negative is consistent with the observation that V is a decreasing function of t. One way to indicate this in normal conversation is to say that after 15 min, the water is flowing *out* of the tank at 15 gal/min; after 45 min, the water is flowing *out* at 5 gal/min. The instantaneous rate of change of V at $t = 15$ is -15 and the instantaneous rate of change of V at $t = 45$ is -5 in gallons per minute. The units could be predicted in advance because $\Delta V/\Delta t$ is a ratio of gallons to minutes; thus its limit $V'(t) = dV/dt$ must be expressed in the same units.

Our examples of functions up to this point have been restricted to those having either formulas or verbal descriptions. In science and engineering we often work with tables of values obtained from observations or empirical measurements. Our next example shows how the instantaneous rate of change of such a tabulated function can be estimated.

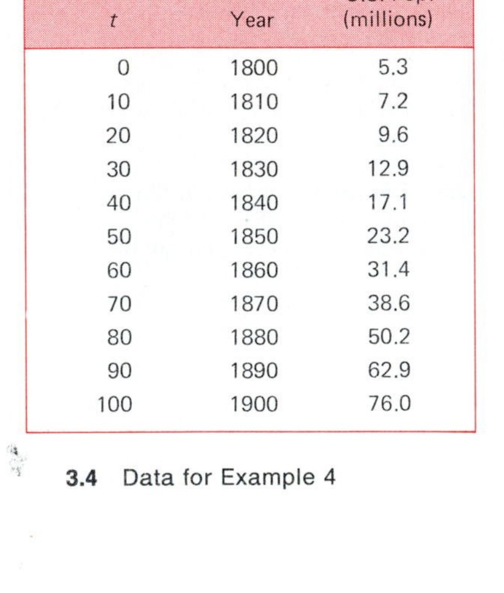

t	Year	U.S. Pop. (millions)
0	1800	5.3
10	1810	7.2
20	1820	9.6
30	1830	12.9
40	1840	17.1
50	1850	23.2
60	1860	31.4
70	1870	38.6
80	1880	50.2
90	1890	62.9
100	1900	76.0

3.4 Data for Example 4

EXAMPLE 4 The table in Fig. 3.4 gives the U.S. population P (in millions) at 10-year intervals in the nineteenth century. Estimate the instantaneous rate of population growth in 1850.

Solution We take $t = 0$ (years) in 1800, so that $t = 50$ corresponds to the year 1850. In Fig. 3.5 we have plotted the data of our example and added a freehand sketch of a smooth curve that fits these data.

However it may be obtained, a curve that fits the data ought to be a good approximation to the true graph of the unknown function $P = f(t)$.

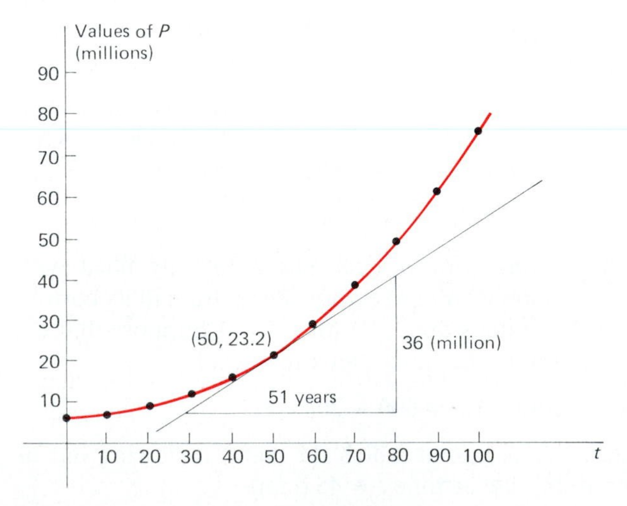

3.5 A smooth curve that fits the data well

The instantaneous rate of change dP/dt in 1850 is the slope of the tangent line at the point (50, 23.2). We draw the tangent as accurately as we can by visual inspection, then measure the base and height of the triangle of Fig. 3.5. Thus we approximate the slope of the tangent at $t = 50$ as

$$\frac{dP}{dt} \approx \frac{36}{51} \approx 0.71$$

millions of people per year (in 1850). Although there was no national census in 1851, we would expect the population of the United States then to have been approximately $23.2 + 0.7 = 23.9$ million people.

VELOCITY AND ACCELERATION

Suppose that a particle moves along a straight line, with its location s at time t given by its **position function** $s = f(t)$. Thus we make the line of motion a coordinate axis with an origin and a positive direction; $f(t)$ is merely the coordinate of the moving particle at time t, as in Fig. 3.6.

3.6 The particle in motion is at the point $s = f(t)$ at time t.

Think of the time interval from t to $t + \Delta t$. The particle moves from position $f(t)$ to position $f(t + \Delta t)$. Its displacement is then the increment

$$\Delta s = f(t + \Delta t) - f(t).$$

We calculate the average velocity of the particle during this time interval exactly as you calculate your average speed on your last long motor trip: Divide the distance by the time to obtain an average speed in miles per hour. So in this case we divide the displacement of the particle by the elapsed time to obtain the **average velocity**

$$\bar{v} = \frac{\Delta s}{\Delta t} = \frac{f(t + \Delta t) - f(t)}{\Delta t}. \tag{11}$$

(The overbar is a standard symbol in mathematics and statistics, usually connoting an average of some sort.) We define the **instantaneous velocity** v of the particle at the time t to be the limit of the average velocity $\bar{v}$ as $\Delta t \to 0$. That is,

$$v = \lim_{\Delta t \to 0} \frac{\Delta s}{\Delta t} = \lim_{\Delta t \to 0} \frac{f(t + \Delta t) - f(t)}{\Delta t}. \tag{12}$$

We recognize the limit on the right in (12)—it is the definition of the derivative of f at time t. Therefore, the velocity of the moving particle at time t is simply

$$v = \frac{ds}{dt} = f'(t). \tag{13}$$

Thus *velocity is instantaneous rate of change of position.* The velocity of a moving particle may be positive or negative, depending on whether it is moving in the positive or negative direction along the line of motion. The **speed** of the particle is defined to be the absolute value $|v|$ of the velocity.

EXAMPLE 5 If the position function of the particle is $s(t) = 5t^2 + 100$, then its velocity at time t is $v(t) = s'(t) = 10t$. For instance, at time $t = 10$ its position is $s(10) = (5)(10^2) + 100 = 600$ and its velocity at that time is $v(10) = (10)(10) = 100$. If the units on the s-axis were in meters and time t were measured in seconds, we would say this: At time $t = 10$ s, the particle would be located 600 m to the right of the origin, moving to the right at 100 m/s.

The case of vertical motion under the influence of constant gravity is of special interest. If a particle is projected straight upward from height s_0 (feet) above the ground, at time $t = 0$ (seconds), and with initial velocity v_0 (feet per second), *and* if air resistance is negligible, then its height s (in feet above the ground) at time t is given by

$$s = -\tfrac{1}{2}gt^2 + v_0t + s_0. \tag{14}$$

Here g denotes the acceleration due to the force of gravity. Near the surface of the earth, g is nearly constant, so we assume that it is exactly constant, and at the surface of the earth $g \approx 32$ ft/s^2, or $g \approx 980$ cm/s^2.

If we differentiate s with respect to time t, we obtain the velocity of the particle at time t:

$$v = \frac{ds}{dt} = -gt + v_0. \tag{15}$$

The **acceleration** of the particle is, by definition, the instantaneous rate of change (derivative) of its velocity:

$$a = \frac{dv}{dt} = -g. \tag{16}$$

Your intuition and probably your experience tell you that a body so projected upward will reach its maximum height at the instant that its velocity vanishes—when $v = 0$. (We shall see why this is true in Section 3.5.)

EXAMPLE 6 Find the maximum height attained by a ball that is thrown straight upward from the ground with initial velocity $v_0 = +96$ ft/s. Also find the velocity with which it hits the ground upon its return.

Solution We use Equation (15), with $v_0 = 96$ and $g = 32$. We find the velocity of the ball at time t to be

$$v = -32t + 96$$

(while it remains aloft; this formula is clearly not valid outside a certain closed interval). The ball attains its maximum height when $v = 0$; that is, when

$$-32t + 96 = 0.$$

This occurs when $t = 3$ (seconds). Upon substitution of this value of t in the altitude function in (14), taking $s_0 = 0$, we find that the maximum height of the ball is

$$s_{max} = s(3) = (-16)(3^2) + (96)(3) = 144 \quad \text{(ft)}.$$

The ball returns to the ground when $s(t) = 0$. The equation

$$s(t) = -16t^2 + 96t = -16t(t - 6) = 0$$

has the two solutions $t = 0$ and $t = 6$, and we now know that the formulas in this example are valid on the interval $0 \leqq t \leqq 6$. The velocity with which the ball hits the ground is

$$v(6) = -32(6) + 96 = -96 \quad \text{(ft/s)}.$$

The derivative of any function—not merely a function of time—may be interpreted as its instantaneous rate of change with respect to the independent variable. If $y = f(x)$, then the **average rate of change** of y (per unit change in x) on the interval $[x, x + \Delta x]$ is the quotient

$$\frac{\Delta y}{\Delta x} = \frac{f(x + \Delta x) - f(x)}{\Delta x}.$$

The **instantaneous rate of change of y with respect to x** is the limit as $\Delta x \to 0$ of the average rate of change. Thus the instantaneous rate of change of y with respect to x is

$$\lim_{\Delta x \to 0} \frac{\Delta y}{\Delta x} = \frac{dy}{dx} = f'(x). \tag{17}$$

The following example illustrates the fact that a dependent variable may sometimes be expressed as two different functions of two different independent variables. The derivatives of these functions are then rates of change of the dependent variable with respect to the two different independent variables.

EXAMPLE 7 A square with edge length x centimeters has area $A = x^2$, so the derivative of A with respect to x,

$$\frac{dA}{dx} = 2x, \tag{18}$$

is the rate of change of its area A with respect to x. (See the computations in Fig. 3.7.) The units for dA/dx are square centimeters *per centimeter*. Now suppose that the edge length of the square is increasing with time: $x = 5t$, with time t in seconds. Then the area of the square at time t is

$$A = (5t)^2 = 25t^2.$$

The derivative of A with respect to t is

$$\frac{dA}{dt} = (2)(25t) = 50t; \tag{19}$$

this is the rate of change of A with respect to time t, with units of square centimeters *per second*. For instance, when $t = 10$ (so that $x = 50$), the values of the two derivatives of A in (18) and (19) are

$$\left.\frac{dA}{dx}\right|_{x=50} = (2)(50) = 100 \quad (\text{cm}^2/\text{cm})$$

and

$$\left.\frac{dA}{dt}\right|_{t=10} = (50)(10) = 500 \quad (\text{cm}^2/\text{s}).$$

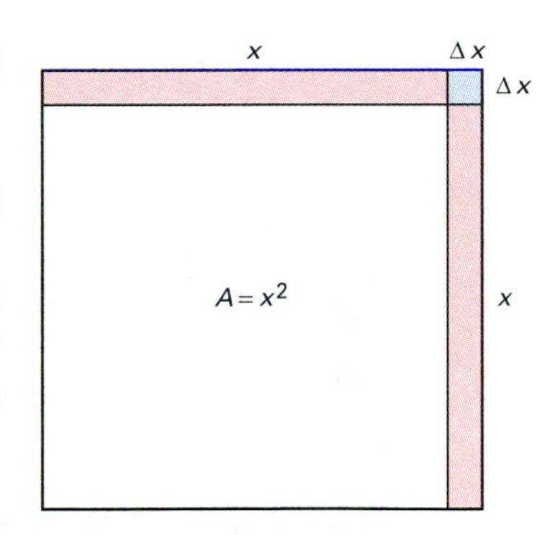

3.7 The square of Example 7:

$$A + \Delta A = (x + \Delta x)^2;$$
$$\Delta A = 2x\,\Delta x + (\Delta x)^2;$$
$$\frac{\Delta A}{\Delta x} = 2x + \Delta x;$$
$$\frac{dA}{dx} = 2x.$$

The notation dA/dt for the derivative suffers from the minor inconvenience of not providing a "place" to substitute a particular value of t, such as $t = 10$. The last lines of Example 7 illustrate one way around this difficulty.

3.1 PROBLEMS

In each of Problems 1–10, find the indicated derivative by using the differentiation rule in (4):

$$\text{If} \quad f(x) = ax^2 + bx + c, \quad \text{then} \quad f'(x) = 2ax + b.$$

1. $f(x) = 4x - 5$; find $f'(x)$.
2. $g(t) = 100 - 16t^2$; find $g'(t)$.
3. $h(z) = z(25 - z)$; find $h'(z)$.
4. $f(x) = 16 - 49x$; find $f'(x)$.
5. $y = 2x^2 + 3x - 17$; find dy/dx.
6. $x = 16t - 100t^2$; find dx/dt.
7. $z = 5u^2 + 3u$; find dz/du.
8. $v = 5y(100 - y)$; find dv/dy.
9. $x = -5y^2 + 17y + 300$; find dx/dy.
10. $u = 7t^2 + 13t$; find du/dt.

In each of Problems 11–20, apply the definition of the derivative (as in Example 1) to find $f'(x)$.

11. $f(x) = 2x - 1$
12. $f(x) = 2 - 3x$
13. $f(x) = x^2 + 5$
14. $f(x) = 3 - 2x^2$
15. $f(x) = \dfrac{1}{2x + 1}$
16. $f(x) = \dfrac{1}{3 - x}$
17. $f(x) = \sqrt{2x + 1}$
18. $f(x) = \dfrac{1}{\sqrt{x + 1}}$
19. $f(x) = \dfrac{x}{1 - 2x}$
20. $f(x) = \dfrac{x + 1}{x - 1}$

In each of Problems 21–25, the position function $s = f(t)$ of a particle moving in a straight line is given. Find its location s when its velocity v is zero.

21. $s = 100 - 16t^2$
22. $s = -16t^2 + 160t + 25$
23. $s = -16t^2 + 80t - 1$
24. $s = 100t^2 + 50$
25. $s = 100 - 20t - 5t^2$

In each of Problems 26–29, the height $s(t)$ (in feet, at time t seconds) of a ball thrown vertically upward is given. Find the maximum height that the ball attains.

26. $s = -16t^2 + 160t$
27. $s = -16t^2 + 64t$
28. $s = -16t^2 + 128t + 25$
29. $s = -16t^2 + 96t + 50$
30. The Celsius temperature C is given in terms of Fahrenheit temperature F by $C = \frac{5}{9}(F - 32)$. Find the rate of change of C with respect to F and also the rate of change of F with respect to C.
31. Find the rate of change of the area A of a circle with respect to its circumference.
32. A stone dropped into a pond at time $t = 0$ s causes a circular ripple that travels out from the point of impact at 5 ft/s. At what rate (in ft^2/s) is the area within the circle increasing when $t = 10$?
33. A car is traveling at 100 ft/s when the driver suddenly applies the brakes ($s = 0$, $t = 0$). The position function of the skidding car is $s = 100t - 5t^2$. How far and for how long does the car skid before coming to a stop?
34. A water bucket containing 10 gal of water develops a leak at time $t = 0$, and the volume V of water in the bucket t seconds later is given by

$$V = 10\left(1 - \frac{t}{100}\right)^2$$

until the bucket is empty at time $t = 100$.

(a) At what rate is water leaking from the bucket after exactly 1 min has passed?

(b) When is the instantaneous rate of change of V equal to the average rate of change of V from $t = 0$ to $t = 100$?

35. A certain population of rodents numbers

$$P = 100[1 + (0.3)t + (0.04)t^2]$$

at time t months.

(a) How long does it take for this population to double its initial ($t = 0$) size?

(b) What is the rate of growth of the population when $P = 200$?

36. The following data describe the growth of the population P (in thousands) of a certain city during the 1970s. Use the graphical method of Example 4 to estimate its rate of growth in 1975.

Year	1970	1972	1974	1976	1978	1980
P	265	293	324	358	395	437

37. The following data give the distance s in feet traveled by an accelerating car (that starts from rest at time $t = 0$) in the first t seconds. Use the graphical method of Example

4 to estimate its speed (in mi/h) when $t = 20$ and again when $t = 40$.

t	0	10	20	30	40	50	60
s	0	224	810	1655	2686	3850	5109

In Problems 38–43, use the fact (proved in Section 3.2) that the derivative of $y = ax^3 + bx^2 + cx + d$ is $dy/dx = 3ax^2 + 2bx + c$.

38. Prove that the rate of change of the volume of a cube with respect to its edge length is equal to half the surface area of the cube.

39. Show that the rate of change of the volume of a sphere with respect to its radius is equal to its surface area.

40. The height of a changing cylinder is always twice its radius. Show that the rate of change of its volume with respect to its radius is equal to its total surface area.

41. A spherical balloon with an initial radius of 5 in. begins to leak at time $t = 0$, and its radius t seconds later is $r = (60 - t)/12$ in. At what rate (in in.^{3}/s) is air leaking from the balloon when $t = 30$?

42. The volume V (in liters) of 3 g of CO_2 at 27°C is given in terms of its pressure p (in atmospheres) by the formula $V = 1.68/p$. What is the rate of change of V with respect to p when $p = 2$ (atm)? (Use the fact that the derivative of $f(x) = c/x$ is $f'(x) = -c/x^2$ if c is a constant; you can establish this by using the definition of the derivative if you wish.)

43. As a snowball with an initial radius of 12 cm melts, its radius decreases at a constant rate. It begins to melt when $t = 0$ (hours) and takes 12 h to disappear.

(a) What is its rate of change of volume when $t = 6$?

(b) What is its average rate of change of volume from $t = 3$ to $t = 9$?

44. A ball thrown vertically upward at time $t = 0$ (s) with initial velocity 96 ft/s and with initial height 112 ft has height function $y = -16t^2 + 96t + 112$.

(a) What is the maximum height attained by the ball?

(b) When and with what impact speed does it hit the ground?

45. A spaceship approaching touchdown on a distant planet has height y (meters) at time t (seconds) given by $y = 100 - 100t + 25t^2$. When and with what speed does it hit the ground?

46. A certain city has population (in thousands) given by

$$P = 100[1 + (0.04)t + (0.003)t^2],$$

with t in years and with $t = 0$ corresponding to 1970.

(a) What was the rate of change of P in 1975?

(b) What was the average rate of change of P from 1973 to 1978?

3.2 Basic Differentiation Rules

In this section we begin our development of formal rules for finding the derivative f' of the function f:

$$f'(x) = \lim_{h \to 0} \frac{f(x + h) - f(x)}{h}. \tag{1}$$

Some alternative notation for derivatives will be helpful.

When we interpreted the derivative in Section 3.1 as a rate of change, we found it useful to employ the dependent–independent variable notation

$$y = f(x), \qquad \Delta x = h, \qquad \Delta y = f(x + \Delta x) - f(x). \tag{2}$$

This led to the "differential notation"

$$\frac{dy}{dx} = \lim_{\Delta x \to 0} \frac{\Delta y}{\Delta x} = \lim_{\Delta x \to 0} \frac{f(x + \Delta x) - f(x)}{\Delta x} \tag{3}$$

for the derivative. When you use this notation, it is important to remember that the symbol dy/dx is simply another notation for the derivative $f'(x)$; it is *not* the quotient of two separate entities dy and dx.

A third notation is sometimes used for the derivative $f'(x)$; it is $Df(x)$. Here, think of D as a "machine" that "operates" on the function f to produce

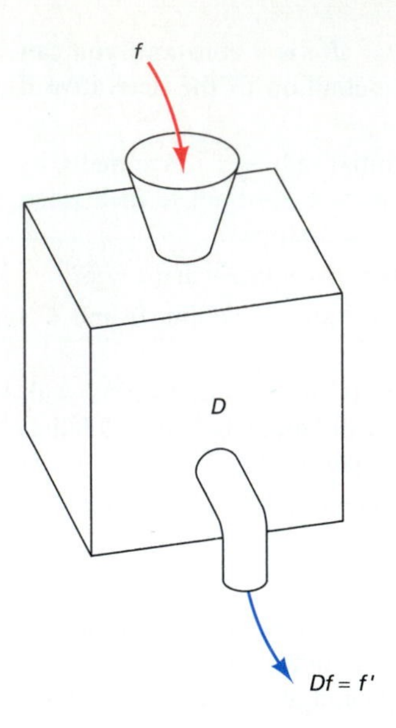

3.8 The "differentiation machine" D

the derivative function Df, as indicated in Fig. 3.8. Thus we can write the derivative of $y = f(x) = x^2$ in any of three ways:

$$f'(x) = \frac{dy}{dx} = Dx^2 = 2x.$$

These three notations for the derivative—the function notation $f'(x)$, the differential notation dy/dx, and the operator notation $Df(x)$—are used interchangeably in mathematical and scientific writing, so familiarity with each is necessary.

DERIVATIVES OF POLYNOMIALS

Our first differentiation rule says that *the derivative of a constant function is identically zero.* This is geometrically obvious, because the graph of a constant function is a horizontal straight line and therefore has slope zero at each point.

> ***Theorem 1*** *Derivative of a Constant*
> If $f(x) = c$ (a constant) for all x, then $f'(x) = 0$ for all x. That is,
>
> $$\frac{dc}{dx} = Dc = 0. \tag{4}$$

Proof Because $f(x + h) = f(x) = c$, we see that

$$f'(x) = \lim_{h \to 0} \frac{f(x+h) - f(x)}{h} = \lim_{h \to 0} \frac{c - c}{h} = \lim_{h \to 0} \frac{0}{h} = 0. \quad \blacksquare$$

As motivation for the next rule, consider the following list of derivatives, all computed in Chapter 2.

$$Dx = 1 \qquad \text{(Theorem, Section 2.1)}$$

$$Dx^2 = 2x \qquad \text{(Theorem, Section 2.1)}$$

$$Dx^3 = 3x^2 \qquad \text{(Problem 26, Section 2.1)}$$

$$Dx^4 = 4x^3 \qquad \text{(Problem 28, Section 2.1)}$$

$$Dx^{-1} = -x^{-2} \qquad \text{(Problem 27, Section 2.1)}$$

$$Dx^{-2} = -2x^{-3} \qquad \text{(Problem 29, Section 2.1)}$$

$$Dx^{1/2} = \tfrac{1}{2}x^{-1/2} \qquad \text{(Example 12, Section 2.2)}$$

$$Dx^{-1/2} = -\tfrac{1}{2}x^{-3/2} \qquad \text{(Problem 37, Section 2.2)}$$

Each of these formulas fits the simple pattern

$$Dx^n = nx^{n-1}. \tag{5}$$

As an inference from the instances listed here, (5) is only a conjecture. But many discoveries in mathematics are made by detecting such patterns, then proving that they hold universally.

Eventually, we shall see that the formula in (5), called the **power rule,** is valid for all real numbers n. At this time we give a proof only for the case in which the exponent n is a *positive integer.* We need the *binomial formula* from high-school algebra, which says that

$$(a+b)^n = a^n + na^{n-1}b + \frac{n(n-1)}{2\cdot 1}a^{n-2}b^2$$
$$+\cdots+\frac{n(n-1)\cdots(n-k+1)}{k(k-1)\cdots 3\cdot 2\cdot 1}a^{n-k}b^k$$
$$+\cdots+nab^{n-1}+b^n \tag{6}$$

if n is a positive integer. The cases $n=2$ and $n=3$ are the familiar formulas

$$(a+b)^2 = a^2+2ab+b^2$$

and

$$(a+b)^3 = a^3+3a^2b+3ab^2+b^3.$$

There are $n+1$ terms on the right-hand side in (6), but all we need to know to prove the power rule is the exact form of the first two terms and the fact that all the other terms include b^2 as a factor.

Theorem 2 *Power Rule for n a Positive Integer*
If n is a positive integer and $f(x)=x^n$, then $f'(x)=nx^{n-1}$.

Proof The binomial formula gives

$$(x+h)^n = x^n + nx^{n-1}h + Q$$

where Q is a sum of $n-1$ terms. From Equation (6) we see that each of these $n-1$ terms involves a numerical coefficient, a nonnegative power of x, and h raised to the second power or higher. This means that we can factor h^2 from Q, resulting in $Q=h^2R$, where R contains no negative powers of x or of h. Thus

$$(x+h)^n - x^n = nx^{n-1}h + h^2R.$$

Therefore,

$$f'(x) = \lim_{h\to 0}\frac{(x+h)^n-x^n}{h} = \lim_{h\to 0}\frac{nx^{n-1}h+h^2R}{h}$$
$$= \lim_{h\to 0}(nx^{n-1}+hR) = nx^{n-1}.$$

The last limit exists because R is a sum of $n-1$ terms, each of which is the product of a real number, a nonnegative power of x, and a nonnegative power of h. So we see that the power rule holds for all x as well as for every positive exponent n. ■

For example, $Dx^{17}=17x^{16}$ and $Dx^{1000}=1000x^{999}$. Nor need we always use x as the independent variable. We may write the power rule with other symbols for the independent variable:

$$Dt^m = mt^{m-1}, \qquad D\xi^r = r\xi^{r-1}, \qquad Dy^n = ny^{n-1}.$$

To use the power rule to differentiate polynomials, we need to know how to differentiate linear combinations. A **linear combination** of the functions f and g is a function of the form $af + bg$, where a and b are constants. It follows from the sum and product laws for limits that

$$\lim_{x \to c} [af(x) + bg(x)] = a\left(\lim_{x \to c} f(x)\right) + b\left(\lim_{x \to c} g(x)\right) \tag{7}$$

provided that the two limits on the right in (7) both exist. The formula in (7) is called the **linearity property** of the limit operation. It implies an analogous linearity property of differentiation.

Theorem 3 *Derivative of a Linear Combination*

If f and g are differentiable functions and a and b are fixed real numbers, then

$$D[af(x) + bg(x)] = aDf(x) + bDg(x). \tag{8}$$

With $u = f(x)$ and $v = g(x)$, this takes the form

$$\frac{d(au + bv)}{dx} = a\frac{du}{dx} + b\frac{dv}{dx}. \tag{8'}$$

Proof The linearity property of limits immediately gives

$$\begin{aligned} D[af(x) + bg(x)] &= \lim_{h \to 0} \frac{[af(x+h) + bg(x+h)] - [af(x) + bg(x)]}{h} \\ &= a\left(\lim_{h \to 0} \frac{f(x+h) - f(x)}{h}\right) + b\left(\lim_{h \to 0} \frac{g(x+h) - g(x)}{h}\right) \\ &= aDf(x) + bDg(x). \quad \blacksquare \end{aligned}$$

Now take $a = c$ and $b = 0$ in Equation (8). The result is

$$D[cf(x)] = cDf(x); \tag{9}$$

alternatively,

$$\frac{d(cu)}{dx} = c\frac{du}{dx}. \tag{9'}$$

That is, *the derivative of a constant multiple of a function is the same constant multiple of its derivative.*

Next, take $a = b = 1$ in Equation (8). We find that

$$D[f(x) + g(x)] = Df(x) + Dg(x). \tag{10}$$

In differential notation,

$$\frac{d(u + v)}{dx} = \frac{du}{dx} + \frac{dv}{dx}. \tag{10'}$$

Thus *the derivative of the sum of two functions is the sum of their derivatives.* Similarly, for differences we have

$$\frac{d(u - v)}{dx} = \frac{du}{dx} - \frac{dv}{dx}. \tag{11}$$

Repeated application of (10) to a sum of a finite number of differentiable functions gives

$$\frac{d(u_1 + u_2 + \cdots + u_n)}{dx} = \frac{du_1}{dx} + \frac{du_2}{dx} + \cdots + \frac{du_n}{dx}. \tag{12}$$

When we apply (9) and (12) and the power rule to the polynomial

$$p(x) = a_n x^n + a_{n-1} x^{n-1} + \cdots + a_2 x^2 + a_1 x + a_0,$$

we find the derivative *as fast as we can write it*; it is

$$p'(x) = n a_n x^{n-1} + (n-1) a_{n-1} x^{n-2} + \cdots + 3 a_3 x^2 + 2 a_2 x + a_1. \tag{13}$$

With this result, it becomes a routine matter to write an equation for a tangent line to the graph of a polynomial.

EXAMPLE 1 Find the tangent line to the curve $y = 2x^3 - 7x^2 + 3x + 4$ at the point $(1, 2)$.

Solution We compute the derivative as in (13) and find it to be

$$\frac{dy}{dx} = (2)(3x^2) - (7)(2x) + 3 = 6x^2 - 14x + 3.$$

We substitute $x = 1$ in the derivative and find that the slope of the tangent line at $(1, 2)$ is $m = -5$. So the point-slope equation of the tangent line is

$$y - 2 = -5(x - 1).$$

EXAMPLE 2 The volume V in cubic centimeters of a quantity of water varies with changing temperature T. For T between 0 and 30°C, the relationship is almost exactly

$$V = V_0(1 + \alpha T + \beta T^2 + \gamma T^3),$$

where V_0 is the volume at 0°C, and the three constants have the values

$$\alpha = -0.06427 \times 10^{-3}, \qquad \beta = 8.5053 \times 10^{-6}, \qquad \gamma = -6.7900 \times 10^{-8}.$$

The rate of change of volume with respect to temperature is

$$\frac{dV}{dT} = V_0(\alpha + 2\beta T + 3\gamma T^2).$$

Suppose that $V_0 = 10^5$ cm^3 and that $T = 20$°C. Substitution of these numerical data in the formulas for V and dV/dT yields $V \approx 100{,}157$ cm^3 and $dV/dT \approx 19.4$ cm^3/°C. We may conclude that, at $T = 20$°C, this amount of water should increase in volume by approximately 19.4 cm^3 for each rise of 1°C in temperature. By comparison, direct substitution into the formula for V shown here yields

$$V(20.5) - V(19.5) \approx 19.4445 \text{ cm}^3.$$

THE PRODUCT AND QUOTIENT RULES

It might be natural to conjecture that the derivative of a product $f(x)g(x)$ is the product of the derivatives. This is *false!* For example, if $f(x) = g(x) = x$, then

$$D[f(x)g(x)] = Dx^2 = 2x$$

while

$$[Df(x)] \cdot [Dg(x)] = (Dx) \cdot (Dx) = 1 \cdot 1 = 1.$$

In general, the derivative of a product is *not* merely the product of the derivatives. The following theorem tells what it *is*.

Theorem 4 *The Product Rule*

If f and g are differentiable at x, then fg is differentiable at x, and

$$D[f(x)g(x)] = f'(x)g(x) + f(x)g'(x). \tag{14}$$

With $u = f(x)$ and $v = g(x)$, this **product rule** takes the form

$$\frac{d(uv)}{dx} = v\frac{du}{dx} + u\frac{dv}{dx}. \tag{14'}$$

When it is clear what the independent variable is, we can write the product rule even more briefly:

$$(uv)' = u'v + uv'. \tag{14''}$$

Proof We use an "add and subtract" device.

$$\begin{aligned}
D[f(x)g(x)] &= \lim_{h\to 0}\frac{f(x+h)g(x+h) - f(x)g(x)}{h} \\
&= \lim_{h\to 0}\frac{f(x+h)g(x+h) - f(x)g(x+h) + f(x)g(x+h) - f(x)g(x)}{h} \\
&= \lim_{h\to 0}\frac{f(x+h)g(x+h) - f(x)g(x+h)}{h} \\
&\quad + \lim_{h\to 0}\frac{f(x)g(x+h) - f(x)g(x)}{h} \\
&= \left(\lim_{h\to 0}\frac{f(x+h) - f(x)}{h}\right)\left(\lim_{h\to 0} g(x+h)\right) \\
&\quad + f(x)\left(\lim_{h\to 0}\frac{g(x+h) - g(x)}{h}\right) \\
&= f'(x)g(x) + f(x)g'(x).
\end{aligned}$$

In this proof we used the sum and product laws for limits, the definitions of $f'(x)$ and $g'(x)$, and the fact that

$$\lim_{h\to 0} g(x+h) = g(x).$$

This last equation holds because g is differentiable and therefore continuous at x, by Theorem 3 in Section 2.4. ■

In words, the product rule says that *the derivative of the product of two functions is the derivative of the first factor times the second factor, plus the first factor times the derivative of the second factor.*

EXAMPLE 3 Find the derivative of

$$f(x) = (1 - 4x^3)(3x^2 - 5x + 2)$$

without first multiplying the two factors.

Solution

$$\begin{aligned} D[(1 - 4x^3)(3x^2 - 5x + 2)] \\ &= [D(1 - 4x^3)](3x^2 - 5x + 2) + (1 - 4x^3)[D(3x^2 - 5x + 2)] \\ &= (-12x^2)(3x^2 - 5x + 2) + (1 - 4x^3)(6x - 5) \\ &= -60x^4 + 80x^3 - 24x^2 + 6x - 5. \end{aligned}$$

We can apply the product rule repeatedly to find the derivative of a product of three or more differentiable functions $u_1, u_2, \ldots, u_n$ of x. For example,

$$\begin{aligned} D[u_1u_2u_3] &= [D(u_1u_2)] \cdot u_3 + (u_1u_2) \cdot Du_3 \\ &= [(Du_1)u_2 + u_1(Du_2)]u_3 + (u_1u_2)Du_3 \\ &= (Du_1)u_2u_3 + u_1(Du_2)u_3 + u_1u_2(Du_3). \end{aligned}$$

Note that the derivative of each factor in the original product is multiplied by the other two factors, and then the three results are added. This is, indeed, the general result:

$$\begin{aligned} D(u_1u_2 \cdots u_n) = (Du_1)u_2 \cdots u_n + u_1(Du_2)u_3 \cdots u_n \\ + \cdots + u_1u_2 \cdots u_{n-1}(Du_n), \end{aligned} \tag{15}$$

where the sum in (15) has one term corresponding to each of the n factors in $u_1u_2 \cdots u_n$. It is easy to establish this result (Problem 62) if we use the following strategy for a *proof by induction on n.*

Principle of Mathematical Induction

Suppose that $P(n)$ is a statement meaningful for each positive integer n. In order to prove that $P(n)$ is true for all such integers, it is sufficient to establish two facts:

1. The statement $P(n)$ is true when $n = 1$ (that is, $P(1)$ is true).
2. Whenever the statement $P(n)$ is true for $n = k$, a fixed but unspecified positive integer, then $P(n)$ also holds when $n = k + 1$ (that is, if $P(k)$ is true then $P(k + 1)$ is also true).

EXAMPLE 4 Use the principle of mathematical induction to establish the power rule for n a positive integer.

Solution We are to show that $Dx^n = nx^{n-1}$ for all positive integers n. This statement is clearly true when $n = 1$, because $Dx^1 = 1 \cdot x^0$.

Assume that the statement holds when $n = k$; that is, assume that $Dx^k = kx^{k-1}$ for some integer $k \geqq 1$. The product rule then gives

$$Dx^{k+1} = D(x \cdot x^k) = (Dx) \cdot x^k + x \cdot (Dx^k).$$

Our assumption that the power rule holds for exponent k now lets us continue these computations, and we find that

$$Dx^{k+1} = 1 \cdot x^k + x \cdot kx^{k-1} = x^k + kx^k = (k+1)x^k.$$

This is the correct form of the power rule in the case $n = k + 1$. Thus we satisfy both conditions of the induction principle, and we conclude that $Dx^n = nx^{n-1}$ for every integer $n \geqq 1$.

Our next theorem tells us how to find the derivative of the reciprocal of a function if we know the derivative of the function itself.

Theorem 5 *The Reciprocal Rule*

If f is differentiable at x and $f(x) \neq 0$, then

$$D\frac{1}{f(x)} = -\frac{f'(x)}{[f(x)]^2}. \tag{16}$$

With $u = f(x)$, the reciprocal rule takes the form

$$\frac{d}{dx}\left(\frac{1}{u}\right) = -\frac{1}{u^2} \cdot \frac{du}{dx}. \tag{16$'$}$$

If there can be no doubt what the independent variable is, we may write

$$\left(\frac{1}{u}\right)' = -\frac{u'}{u^2}. \tag{16$''$}$$

Proof As in the proof of Theorem 4, we use the limit laws, the definition of the derivative, and the fact that a function is continuous wherever it is differentiable (by Theorem 3 of Section 2.4). Moreover, note that $f(x + h) \neq 0$ for h near zero because $f(x) \neq 0$ and f is continuous at x (Problem 16, Appendix B). Therefore,

$$\begin{aligned} D\frac{1}{f(x)} &= \lim_{h \to 0} \frac{1}{h}\left(\frac{1}{f(x+h)} - \frac{1}{f(x)}\right) \\ &= \lim_{h \to 0} \frac{f(x) - f(x+h)}{hf(x+h)f(x)} \\ &= -\left(\lim_{h \to 0} \frac{1}{f(x+h)f(x)}\right)\left(\lim_{h \to 0} \frac{f(x+h) - f(x)}{h}\right) \\ &= -\frac{f'(x)}{[f(x)]^2}. \quad \blacksquare \end{aligned}$$

EXAMPLE 5 With $f(x) = x^2 + 1$ in (16) we get

$$D\frac{1}{x^2+1} = -\frac{D(x^2+1)}{(x^2+1)^2} = -\frac{2x}{(x^2+1)^2}.$$

We now combine the reciprocal rule with the power rule for positive integral exponents to establish the power rule for negative integral exponents.

Theorem 6 *Power Rule for n a Negative Integer*

If n is a negative integer, then $Dx^n = nx^{n-1}$.

Proof Let $m = -n$, so that m is a positive integer. Then

$$Dx^n = D\frac{1}{x^m} = -\frac{D(x^m)}{(x^m)^2} = -\frac{mx^{m-1}}{x^{2m}} = (-m)x^{(-m)-1} = nx^{n-1}. \quad \blacksquare$$

This proof also shows that the rule of Theorem 6 holds exactly when the function being differentiated is defined: when $x \neq 0$. As an illustration, $Dx^{-7} = -7x^{-8}$ for all $x \neq 0$.

Now we apply the product and reciprocal rules to get a rule for differentiation of the quotient of two functions.

Theorem 7 *The Quotient Rule*

If f and g are differentiable at x and $g(x) \neq 0$, then f/g is differentiable at x and

$$D\frac{f(x)}{g(x)} = \frac{f'(x)g(x) - f(x)g'(x)}{[g(x)]^2}. \tag{17}$$

With $u = f(x)$ and $v = g(x)$, this rule takes the form

$$\frac{d}{dx}\left(\frac{u}{v}\right) = \frac{v\dfrac{du}{dx} - u\dfrac{dv}{dx}}{v^2}. \tag{17'}$$

If it is clear what the independent variable is, the quotient rule may also be written in the form

$$\left(\frac{u}{v}\right)' = \frac{u'v - uv'}{v^2}. \tag{17''}$$

Proof We apply the product rule to the factorization

$$\frac{f(x)}{g(x)} = f(x) \cdot \frac{1}{g(x)}.$$

This gives

$$\begin{aligned} D\frac{f(x)}{g(x)} &= (Df(x)) \cdot \frac{1}{g(x)} + f(x) \cdot D\frac{1}{g(x)} \\ &= \frac{f'(x)}{g(x)} + f(x)\left(-\frac{g'(x)}{[g(x)]^2}\right) \\ &= \frac{f'(x)g(x) - f(x)g'(x)}{[g(x)]^2}. \quad \blacksquare \end{aligned}$$

Note that the numerator in (17) is *not* the derivative of the product of f and g. And the minus sign means that the *order* of terms in the numerator is important.

EXAMPLE 6 Find dz/dt given

$$z = \frac{1 - t^3}{1 + t^4}.$$

Solution With t rather than x as the independent variable, the quotient rule gives

$$\frac{dz}{dt} = \frac{[D(1 - t^3)](1 + t^4) - (1 - t^3)[D(1 + t^4)]}{(1 + t^4)^2}$$

$$= \frac{(-3t^2)(1 + t^4) - (1 - t^3)(4t^3)}{(1 + t^4)^2} = \frac{t^6 - 4t^3 - 3t^2}{(1 + t^4)^2}.$$

3.2 PROBLEMS

Apply the differentiation rules of this section to find the derivatives of the functions given in Problems 1–40.

1. $f(x) = 3x^2 - x + 5$

2. $g(t) = 1 - 3t^2 - 2t^4$

3. $f(x) = (2x + 3)(3x - 2)$

4. $g(x) = (2x^2 - 1)(x^3 + 2)$

5. $h(x) = (x + 1)^3$

6. $g(t) = (4t - 7)^2$

7. $f(y) = y(2y - 1)(2y + 1)$

8. $f(x) = 4x^4 - \dfrac{1}{x^2}$

9. $g(x) = \dfrac{1}{x + 1} - \dfrac{1}{x - 1}$

10. $f(t) = \dfrac{1}{4 - t^2}$

11. $h(x) = \dfrac{3}{x^2 + x + 1}$

12. $f(x) = \dfrac{1}{1 - \dfrac{2}{x}}$

13. $g(t) = (t^2 + 1)(t^3 + t^2 + 1)$

14. $f(x) = (2x^3 - 3)(17x^4 - 6x + 2)$

15. $g(z) = \dfrac{1}{2z} - \dfrac{1}{3z^2}$

16. $f(x) = \dfrac{2x^3 - 3x^2 + 4x - 5}{x^2}$

17. $g(y) = 2y(3y^2 - 1)(y^2 + 2y + 3)$

18. $f(x) = \dfrac{x^2 - 4}{x^2 + 4}$

19. $g(t) = \dfrac{t - 1}{t^2 + 2t + 1}$

20. $u(x) = \dfrac{1}{(x + 2)^2}$

21. $v(t) = \dfrac{1}{(t - 1)^3}$

22. $h(x) = \dfrac{2x^3 + x^2 - 3x + 17}{2x - 5}$

23. $g(x) = \dfrac{3x}{x^3 + 7x - 5}$

24. $f(t) = \dfrac{1}{\left[t + \dfrac{1}{t}\right]^2}$

25. $g(x) = \dfrac{\dfrac{1}{x} - \dfrac{2}{x^2}}{\dfrac{2}{x^3} - \dfrac{3}{x^4}}$

26. $f(x) = \dfrac{x^3 - \dfrac{1}{x^2 + 1}}{x^4 + \dfrac{1}{x^2 + 1}}$

27. $y = x^3 - 6x^5 + \frac{3}{2}x^{-4} + 12$

28. $y = \dfrac{3}{x} - \dfrac{4}{x^2} - 5$

29. $y = \dfrac{5 - 4x^2 + x^5}{x^3}$

30. $y = \dfrac{2x - 3x^2 + 2x^4}{5x^2}$

31. $y = 3x - \dfrac{1}{4x^2}$

32. $y = \dfrac{1}{x(x^2 + 2x + 2)}$

33. $y = \dfrac{x}{x - 1} + \dfrac{x + 1}{3x}$

34. $y = \dfrac{1}{1 - 4x^{-2}}$

35. $y = \dfrac{x^3 - 4x + 5}{x^2 + 9}$

36. $y = x^2\left(2x^3 - \dfrac{3}{4x^4}\right)$

37. $y = \dfrac{2x^2}{3x - \dfrac{4}{5x^4}}$

38. $y = \dfrac{4}{(x^2 - 3)^2}$

39. $y = \dfrac{x^2}{x + 1}$

40. $y = \dfrac{x + 10}{x^2}$

In each of Problems 41–50, write an equation of the tangent line to the curve $y = f(x)$ at the given point P on the curve. Express the answer in the form $ax + by = c$.

41. $y = x^3$; $P(2, 8)$

42. $y = 3x^2 - 4$; $P(1, -1)$

43. $y = \dfrac{1}{x - 1}$; $P(2, 1)$

44. $y = 2x - \dfrac{1}{x}$; $P(0.5, -1)$

45. $y = x^3 + 3x^2 - 4x - 5$; $P(1, -5)$

46. $y = \left(\dfrac{1}{x} - \dfrac{1}{x^2}\right)^{-1}$; $P(2, 4)$

47. $y = \dfrac{3}{x^2} - \dfrac{4}{x^3}$; $P(-1, 7)$

48. $y = \dfrac{3x - 2}{3x + 2}$; $P(2, 0.5)$

49. $y = \dfrac{3x^2}{x^2 + x + 1}$; $P(-1, 3)$

50. $y = \dfrac{6}{1 - x^2}$; $P(2, -2)$

51. Apply the formula of Example 2 to answer the following two questions.
(a) If 1000 cm^3 of water at 0°C is heated, does it initially expand or contract?
(b) What is the rate (in cubic centimeters per degree Celsius) at which it initially contracts or expands?

52. Susan's weight in pounds is given by the formula $W = 2 \times 10^9/R^2$, where R is her distance in miles from the center of the earth. What is the rate of change of W with respect to R when $R = 3960$ mi? If Susan climbs a mountain beginning at sea level, at what rate in ounces per (vertical) miles does her weight initially decrease?

53. The conical tank shown in Fig. 3.9 has radius 160 cm and height 800 cm. Water is running out a small hole in the bottom of the tank. When the height h of water in the tank is 600 cm, what is the rate of change of its volume V with respect to h?

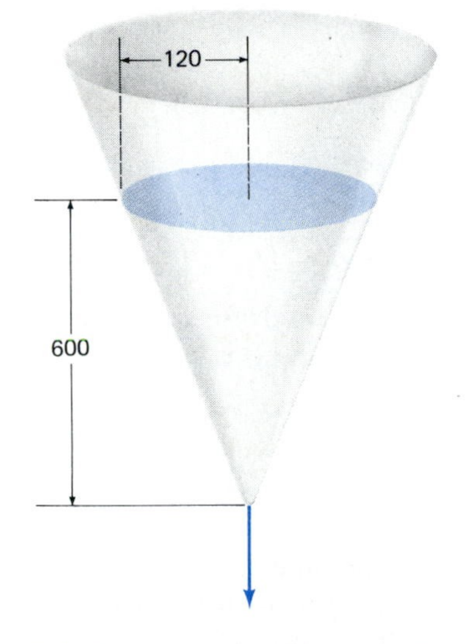

3.9 The leaky tank of Problem 53

54. Find the intercepts of the straight line that is tangent to the curve $y = x^3 + x^2 + x$ at the point (1, 3).

55. Find the line through the point (1, 5) that is tangent to the curve $y = x^3$.

56. Find *two* lines through the point (2, 8) that are tangent to the curve $y = x^3$.

57. Prove that no straight line can be tangent to the curve $y = x^2$ at two different points.

58. Find the two straight lines of slope -2 that are tangent to the curve $y = 1/x$.

59. Let n be a fixed but unspecified integer, $n \geqq 2$. Find the x-intercept of the line that is tangent to the curve $y = x^n$ at the point $P(x_0, y_0)$.

60. Prove that the curve $y = x^5 + 2x$ has no horizontal tangent line. What is the smallest slope that a tangent to this curve can have?

61. Apply Equation (15) with $n = 3$ and $u_1 = u_2 = u_3 = f(x)$ to show that

$$D([f(x)]^3) = 3[f(x)]^2 f'(x).$$

62. Use the principal of mathematical induction to prove the result in Equation (15).

63. Apply Equation (15) to show that

$$D([f(x)]^n) = n[f(x)]^{n-1} f'(x)$$

if n is a positive integer and $f'(x)$ exists.

64. Use the result of Problem 63 to compute $D(x^2 + x + 1)^{100}$.

65. Find $g'(x)$ given $g(x) = (x^3 - 17x + 35)^{17}$.

66. Find constants a, b, c, and d so that the curve $y = ax^3 + bx^2 + cx + d$ has horizontal tangent lines at the points (0, 1) and (1, 0).

3.3 The Chain Rule

In Section 3.2 we saw how to differentiate polynomials and rational functions. But one often needs to differentiate *powers* of such functions. For instance, the product rule yields

$$\begin{aligned} D[f(x)]^2 &= D[f(x) \cdot f(x)] \\ &= f'(x) \cdot f(x) + f(x) \cdot f'(x) = 2f(x)f'(x) \end{aligned}$$

if f is a differentiable function. If

$$y = [f(x)]^3 \tag{1}$$

then another application of the product rule yields

$$\frac{dy}{dx} = D(f(x) \cdot [f(x)]^2) = f'(x) \cdot [f(x)]^2 + f(x) \cdot D[f(x)]^2$$

$$= f'(x)[f(x)]^2 + f(x)[2f(x)f'(x)].$$

Hence

$$\frac{dy}{dx} = 3[f(x)]^2 f'(x). \tag{2}$$

Observe that the derivative of $[f(x)]^3$ is not simply the quantity $3[f(x)]^2$ that one might naively guess in analogy with the correct formula $Dx^3 = 3x^2$. There is an additional factor of $f'(x)$ whose origin may be explained by writing $y = [f(x)]^3$ in the form

$$y = u^3 \quad \text{with} \quad u = f(x).$$

Then

$$\frac{dy}{dx} = D[f(x)]^3,$$

$$\frac{dy}{du} = 3u^2 = 3[f(x)]^2, \quad \text{and} \tag{3}$$

$$\frac{du}{dx} = f'(x),$$

so the derivative formula in (2) takes the form

$$\frac{dy}{dx} = \frac{dy}{du} \cdot \frac{du}{dx}. \tag{4}$$

The **chain rule** tells us that this last formula holds for *any* two differentiable functions $y = g(u)$ and $u = f(x)$. The formula in (2) is simply the special case of Equation (4) with $g(u) = u^3$.

EXAMPLE 1 If

$$y = (3x + 5)^{17},$$

it would be impractical to write the binomial expansion of the 17th power of $3x + 5$: The result would be a polynomial with 18 terms, and some of the coefficients would have 14 digits. But if we write

$$y = u^{17} \quad \text{with} \quad u = 3x + 5,$$

then

$$\frac{dy}{du} = 17u^{16} \quad \text{and} \quad \frac{du}{dx} = 3.$$

Hence the chain rule in (4) yields

$$D(3x + 5)^{17} = \frac{dy}{dx} = \frac{dy}{du} \cdot \frac{du}{dx} = (17u^{16})(3)$$

$$= 17(3x + 5)^{16}(3) = 51(3x + 5)^{16}.$$

The formula in (4) is one that, once learned, is truly unlikely to be forgotten. Although dy/du and du/dx are *not* fractions—they are merely symbols representing the derivatives $g'(u)$ and $f'(x)$—it is just though they *were* fractions, with the du in the first "canceling" the du in the second. Of course, such "cancelation" no more proves the chain rule than cancelling d's proves that

$$\frac{dy}{dx} = \frac{y}{x} \quad \text{(an absurdity).}$$

It is nevertheless an excellent way to *remember* the chain rule. Such manipulations with differentials are so suggestive (even when invalid) that they played a substantial role in the early development of calculus in the seventeenth and eighteenth centuries. For thus were produced formulas that were later proved valid (as well as some formulas that were incorrect).

Although the formula in (4) is a memorable statement of the chain rule in differential notation, it suffers from the defect of not specifying where each of the derivatives is evaluated. This is better said in the notation in which derivatives are expressed as functions. Let us write

$$y = g(u), \qquad u = f(x), \qquad y = h(x) = g(f(x)). \tag{5}$$

Then

$$\frac{du}{dx} = f'(x), \qquad \frac{dy}{dx} = h'(x),$$

and

$$\frac{dy}{du} = g'(u) = g'(f(x)). \tag{6}$$

Substitution of these derivatives in (4) now recasts the chain rule in the form

$$h'(x) = g'(f(x)) \cdot f'(x). \tag{7}$$

In this form the chain rule gives the derivative of the *composition* of two functions in terms of *their* derivatives.

Theorem *The Chain Rule*

Suppose that f is differentiable at x and that g is differentiable at $f(x)$. Then the composition $h = g(f)$ is differentiable at x, and its derivative there is

$$D[g(f(x))] = g'(f(x)) \cdot f'(x). \tag{8}$$

An important note: While the derivative of $h = g(f)$ is a product of the derivatives of f and g, these two derivatives are evaluated at *different* points. For g' is evaluated at $f(x)$, while f' is evaluated at x. For a particular number $x = x_0$, (7) tells us that

$$h'(x_0) = g'(u_0)f'(x_0), \quad \text{where} \quad u_0 = f(x_0). \tag{9}$$

For example, if $h(x) = g(f(x))$ where f and g are differentiable functions, with

$$f(2) = 17, \qquad f'(2) = -3, \quad \text{and} \quad g'(17) = 5,$$

then the chain rule gives

$$h'(2) = g'(f(2)) \cdot f'(2) = g'(17) \cdot f'(2) = (5)(-3) = -15.$$

OUTLINE OF PROOF To *outline* a proof of the chain rule, suppose that we are given $y = g(u)$, $u = f(x)$, and want to compute the derivative

$$\frac{dy}{dx} = \lim_{\Delta x \to 0} \frac{\Delta y}{\Delta x} = \lim_{\Delta x \to 0} \frac{g(f(x + \Delta x)) - g(f(x))}{\Delta x}. \tag{10}$$

The differential form of the chain rule suggests the factorization

$$\frac{\Delta y}{\Delta x} = \frac{\Delta y}{\Delta u}\frac{\Delta u}{\Delta x}. \tag{11}$$

The product law of limits then gives

$$\frac{dy}{dx} = \lim_{\Delta x \to 0} \frac{\Delta y}{\Delta u}\frac{\Delta u}{\Delta x} = \left(\lim_{\Delta u \to 0} \frac{\Delta y}{\Delta u}\right)\left(\lim_{\Delta x \to 0} \frac{\Delta u}{\Delta x}\right) = \frac{dy}{du}\frac{du}{dx}. \tag{12}$$

This will suffice to prove the chain rule *provided that*

$$\Delta u = f(x + \Delta x) - f(x) \tag{13}$$

is a *nonzero* quantity that approaches zero as $\Delta x \to 0$. Certainly, $\Delta u \to 0$ as $\Delta x \to 0$ because f is differentiable and thus continuous. But it is quite possible that Δu is zero for some (even all) nonzero values of Δx. In such a case, the proposed factorization

$$\frac{\Delta y}{\Delta x} = \frac{\Delta y}{\Delta u}\frac{\Delta u}{\Delta x}$$

would involve the *invalid* step of division by zero. Thus our proof is incomplete. A complete proof of the chain rule is given at the end of this section.

If we substitute $f(x) = u$ and $f'(x) = du/dx$ in (8), we get the hybrid form

$$D_x g(u) = g'(u)\frac{du}{dx} \tag{14}$$

of the chain rule that often is the most useful for purely computational purposes. The subscript x in D_x specifies that $g(u)$ is being differentiated with respect to x, rather than with respect to u (as would be understood if we wrote $Dg(u)$ without any subscript).

Let us set $g(u) = u^n$ in (14), where n is an integer. Because $g'(u) = nu^{n-1}$, we thereby obtain the *chain rule version*

$$D_x u^n = nu^{n-1}\frac{du}{dx} \tag{15}$$

of the power rule. If $u = f(x)$ is a differentiable function, then Equation (15) implies that

$$D[f(x)]^n = n[f(x)]^{n-1}f'(x). \tag{16}$$

(If $n - 1 < 0$, we must add the proviso that $f(x) \neq 0$ in order that the right-hand side in (16) is meaningful.) We refer to this chain rule version of the power rule as the **generalized power rule.**

EXAMPLE 2 To differentiate

$$y = \frac{1}{(2x^3 - x + 7)^2},$$

we first write

$$y = (2x^3 - x + 7)^{-2}$$

in order to apply the generalized power rule in (16) with $n = -2$. This gives

$$\frac{dy}{dx} = (-2)(2x^3 - x + 7)^{-3} D(2x^3 - x + 7)$$

$$= (-2)(2x^3 - x + 7)^{-3}(6x^2 - 1) = \frac{2(1 - 6x^2)}{(2x^3 - x + 7)^3}.$$

EXAMPLE 3 Find the derivative $h'(z)$ of the function

$$h(z) = \left(\frac{z - 1}{z + 1}\right)^5.$$

Solution The key to applying the generalized power rule is observing *what* the given function is a power *of*. Here,

$$h(z) = u^5, \quad \text{where} \quad u = \frac{z - 1}{z + 1},$$

and z is the independent variable instead of x. Hence we apply first (15), then the quotient rule, to get

$$h'(z) = 5u^4 \frac{du}{dz} = 5\left(\frac{z - 1}{z + 1}\right)^4 D_z\left(\frac{z - 1}{z + 1}\right)$$

$$= 5\left(\frac{z - 1}{z + 1}\right)^4 \cdot \frac{(1)(z + 1) - (z - 1)(1)}{(z + 1)^2}$$

$$= 5\left(\frac{z - 1}{z + 1}\right)^4 \cdot \frac{2}{(z + 1)^2} = \frac{10(z - 1)^4}{(z + 1)^6}.$$

The importance of the chain rule goes far beyond the power function differentiations illustrated in Examples 1–3. In later sections we will learn how to differentiate exponential, logarithmic, and trigonometric functions. Each time we learn a new differentiation formula—for the derivative $g'(x)$ of a new function $g(x)$—the formula in (14) immediately provides us with the chain rule version

$$D_x g(u) = g'(u) D_x u$$

of that formula. The step from the power rule $Dx^n = nx^{n-1}$ to the generalized power rule $D_x u^n = nu^{n-1} D_x u$ is our first instance of this general phenomenon.

RATE OF CHANGE APPLICATIONS

Suppose that the physical or geometric quantity p depends on the quantity q, which in turn depends on time t. Then the *dependent* variable p is a function both of the *intermediate* variable q and of the *independent* variable t, and the derivatives that appear in the chain rule formula

$$\frac{dp}{dt} = \frac{dp}{dq} \cdot \frac{dq}{dt}$$

are rates of change (as in Section 3.1) of these variables with respect to one another. For instance, suppose that a spherical balloon is being inflated or deflated. Then its volume V and its radius r are changing with time t, and

$$\frac{dV}{dt} = \frac{dV}{dr} \cdot \frac{dr}{dt}.$$

Remember that a positive derivative signals an increasing quantity and that a negative derivative signals a decreasing quantity.

EXAMPLE 4 A spherical balloon is being inflated, and its radius r is increasing at the rate of 0.2 in./s when $r = 5$ in. At what rate is the volume V of the balloon increasing at that instant?

Solution Given $dr/dt = 0.2$ in./s when $r = 5$ in., we want to find dV/dt. Because the volume of the balloon is

$$V = \tfrac{4}{3}\pi r^3,$$

we see that $dV/dr = 4\pi r^2$. So the chain rule gives

$$\frac{dV}{dt} = \frac{dV}{dr} \cdot \frac{dr}{dt} = 4\pi r^2 \frac{dr}{dt} = 4\pi(5)^2(0.2) \approx 62.83$$

in.^{3}/s at the instant mentioned in the example.

Observe that in Example 4 we did not need to know r explicitly as a function of t. On the other hand, suppose we are told that after t seconds the radius (in inches) of an inflating balloon is $r = 3 + (0.2)t$ (at least until the balloon bursts). Then the volume of the balloon is

$$V = \frac{4}{3}\pi r^3 = \frac{4}{3}\pi\left(3 + \frac{t}{5}\right)^3,$$

so dV/dt is given explicitly as a function of t by

$$\frac{dV}{dt} = \frac{4}{3}\pi(3)\left(3 + \frac{t}{5}\right)^2\left(\frac{1}{5}\right) = \frac{4}{5}\pi\left(3 + \frac{t}{5}\right)^2.$$

EXAMPLE 5 Imagine a spherical raindrop falling through water vapor in the air. Suppose that the vapor adheres to its surface in such a way that the time rate of increase of the mass M of the droplet is proportional to the surface area S of the droplet. If the drop starts its fall with a radius that is effectively zero, and $r = 1$ mm after 20 s, when is the radius 3 mm?

Solution We are given

$$\frac{dM}{dt} = kS$$

where k is some constant that depends upon atmospheric conditions. Now

$$M = \tfrac{4}{3}\rho\pi r^3 \quad \text{and} \quad S = 4\pi r^2$$

where ρ is the density of water. Hence the chain rule gives

$$4\pi k r^2 = kS = \frac{dM}{dt} = \frac{dM}{dr} \cdot \frac{dr}{dt};$$

that is,

$$4\pi k r^2 = 4\pi\rho r^2 \frac{dr}{dt}.$$

This implies that

$$\frac{dr}{dt} = \frac{k}{\rho}, \qquad \text{a constant.}$$

So the radius of the droplet grows at a *constant* rate. Thus if it takes 20 s for r to grow to 1 mm, it will take 1 minute for the radius to grow to 3 mm.

PROOF OF THE CHAIN RULE

To prove the chain rule, we need to show that if f is differentiable at a and g is differentiable at $f(a)$, then

$$\lim_{h\to 0} \frac{g(f(a+h)) - g(f(a))}{h} = g'(f(a)) \cdot f'(a). \tag{17}$$

If the quantities h and

$$k(h) = f(a+h) - f(a) \tag{18}$$

are nonzero, then we can write the difference quotient on the left-hand side in (17) as

$$\frac{g(f(a+h)) - g(f(a))}{h} = \frac{g(f(a)+k(h)) - g(f(a))}{k(h)} \cdot \frac{k(h)}{h}. \tag{19}$$

To investigate the first factor on the right-hand side in (19), we define a new function ϕ as follows:

$$\phi(k) = \begin{cases} \dfrac{g(f(a)+k) - g(f(a))}{k} & \text{if } \ k \neq 0; \\ g'(f(a)) & \text{if } \ k = 0. \end{cases} \tag{20}$$

By the definition of the derivative of g, we see from (20) that ϕ is continuous at $k = 0$; that is,

$$\lim_{k\to 0} \phi(k) = g'(f(a)). \tag{21}$$

Next,

$$\lim_{h\to 0} k(h) = \lim_{h\to 0} [f(a+h) - f(a)] = 0 \tag{22}$$

because f is continuous at $x = a$, and $\phi(0) = g'(f(a))$. It therefore follows from (21) that

$$\lim_{h\to 0} \phi(k(h)) = g'(f(a)). \tag{23}$$

We are now ready to assemble all this information. Note from (19) that if $h \neq 0$, then

$$\frac{g(f(a+h)) - g(f(a))}{h} = \phi(k(h)) \cdot \frac{f(a+h) - f(a)}{h} \tag{24}$$

even if $k(h) = 0$, for in this case both sides in (24) are zero. Hence the product rule for limits yields

$$\lim_{h\to 0} \frac{g(f(a+h)) - g(f(a))}{h} = \lim_{h\to 0} \phi(k(h)) \cdot \frac{f(a+h) - f(a)}{h}$$

$$= g'(f(a)) \cdot f'(a),$$

a consequence of Equation (23) and the definition of the derivative of the function f. We have therefore established the chain rule in the form in Equation (17).

3.3 PROBLEMS

Find dy/dx in Problems 1–12.

1. $y = (3x + 4)^5$
2. $y = (2 - 5x)^3$
3. $y = \dfrac{1}{3x - 2}$
4. $y = \dfrac{1}{(2x + 1)^3}$
5. $y = (x^2 + 3x + 4)^3$
6. $y = (7 - 2x^3)^{-4}$
7. $y = (2 - x)^4(3 + x)^7$
8. $y = (x + x^2)^5(1 + x^3)^2$
9. $y = \dfrac{x + 2}{(3x - 4)^3}$
10. $y = \dfrac{(1 - x^2)^3}{(4 + 5x + 6x^2)^2}$
11. $y = [1 + (1 + x)^3]^4$
12. $y = [x + (x + x^2)^{-3}]^{-5}$

In each of Problems 13–20, express the derivative dy/dx in terms of x.

13. $y = (u + 1)^3$ and $u = \dfrac{1}{x^2}$
14. $y = \dfrac{1}{2u} - \dfrac{1}{3u^2}$ and $u = 2x + 1$
15. $y = (1 + u^2)^3$ and $u = (4x - 1)^2$
16. $y = u^5$ and $u = \dfrac{1}{3x - 2}$
17. $y = u(1 - u)^3$ and $u = \dfrac{1}{x^4}$
18. $y = \dfrac{u}{u + 1}$ and $u = \dfrac{x}{x + 1}$
19. $y = u^2(u - u^4)^3$ and $u = \dfrac{1}{x^2}$
20. $y = \dfrac{u}{(2u + 1)^4}$ and $u = x - \dfrac{2}{x}$

In each of Problems 21–26, identify a function u of x and an integer $n \neq 1$ such that $f(x) = u^n$. Then compute $f'(x)$.

21. $f(x) = (2x - x^2)^3$
22. $f(x) = \dfrac{1}{2 + 5x^3}$
23. $f(x) = \dfrac{1}{(1 - x^2)^4}$
24. $f(x) = (x^2 - 4x + 1)^3$
25. $f(x) = \left(\dfrac{x + 1}{x - 1}\right)^7$
26. $f(x) = \dfrac{(x^2 + x + 1)^7}{(x + 1)^4}$

Differentiate each of the functions given in Problems 27–36.

27. $g(y) = y + (2y - 3)^5$
28. $h(z) = z^2(z^2 + 4)^3$
29. $F(s) = \left(s - \dfrac{1}{s^2}\right)^3$
30. $G(t) = \left[t^2 + \left(1 + \dfrac{1}{t}\right)\right]^2$
31. $f(u) = (1 + u)^3(1 + u^2)^4$
32. $g(w) = (w^2 - 3w + 4)(w + 4)^5$
33. $h(v) = \left[v - \left(1 - \dfrac{1}{v}\right)^{-1}\right]^{-2}$
34. $p(t) = \left(\dfrac{1}{t} + \dfrac{1}{t^2} + \dfrac{1}{t^3}\right)^{-4}$
35. $F(z) = \dfrac{1}{(3 - 4z + 5z^5)^{10}}$
36. $G(x) = \{1 + [x + (x^2 + x^3)^4]^5\}^6$

In each of Problems 37–44, dy/dx can be found in two ways—one way using the chain rule, the other way without using it. Use both techniques to find dy/dx, then compare the answers (they should agree!).

37. $y = (x^3)^4 = x^{12}$
38. $y = x = \left(\dfrac{1}{x}\right)^{-1}$
39. $y = (x^2 - 1)^2 = x^4 - 2x^2 + 1$
40. $y = (1 - x)^3 = 1 - 3x + 3x^2 - x^3$
41. $y = (x + 1)^4 = x^4 + 4x^3 + 6x^2 + 4x + 1$
42. $y = (x + 1)^{-2} = \dfrac{1}{x^2 + 2x + 1}$
43. $y = (x^2 + 1)^{-1} = \dfrac{1}{x^2 + 1}$
44. $y = (x^2 + 1)^2 = (x^2 + 1)(x^2 + 1)$

In Section 3.7 we will establish that $D \sin x = \cos x$ (provided that x is in radian measure). Use this fact and the chain rule to find the derivatives of the functions given in Problems 45–48.

45. $f(x) = \sin(x^3)$
46. $g(t) = (\sin t)^3$
47. $g(z) = (\sin 2z)^3$
48. $k(u) = \sin(1 + \sin u)$
49. At what rate is the area of a circle increasing when its radius is 10 in. and is increasing at the rate of 2 in./s?
50. At what rate is the radius of a circle decreasing when its area is 75π in.^{2} and is decreasing at the rate of 2π in.^{2}/s?
51. At what rate is the area of a square increasing when each edge is 10 in. and is increasing at 2 in./s?

52. At what rate is the area of an equilateral triangle increasing when each edge is 10 in. and is increasing at 2 in./s?

53. A cubical block of ice is melting in such a way that each edge decreases steadily by 2 in. every hour. At what rate is its volume decreasing when its edge is 10 in.?

54. Find $f'(-1)$ given $f(y) = h(g(y))$, $g(-1) = 2$, $h'(2) = -1$, and $g'(-1) = 7$.

55. Given: $G(t) = f(h(t))$, $h(1) = 4$, $f'(4) = 3$, and $h'(1) = -6$. Find $G'(1)$.

56. Suppose that $f(0) = 0$ and that $f'(0) = 1$. Calculate the value of the derivative of $f(f(f(x)))$ at $x = 0$.

57. Air is being pumped into a spherical balloon in such a way that its radius is increasing at the rate of $dr/dt = 1$ cm/s. What is the time rate of increase, in cm^3/s, of the balloon's volume when its radius r is 10 cm?

58. Suppose that the air is being pumped into the balloon of Problem 57 at the constant rate of $200\pi\ \text{cm}^3/\text{s}$. What is the time rate of increase of the radius when $r = 5$ cm?

59. Air is escaping from a spherical balloon at the constant rate of $300\pi\ \text{cm}^3/\text{s}$. What is the radius of the balloon when its radius is decreasing at 3 cm/s?

60. A spherical hailstone is losing mass by melting uniformly over its surface as it falls. At a certain time, its radius is 2 cm and its volume is decreasing at the rate of $0.1\ \text{cm}^3/\text{s}$. How fast is its radius decreasing at that time?

61. A spherical snowball is melting in such a way that the rate of decrease of its volume is proportional to its surface area. At 10 A.M. its volume was 500 in.^3, and at 11 A.M. its volume was 250 in.^3. When did the snowball finish melting? (See Example 5.)

62. A cubical block of ice with edge 20 in. begins to melt at 8 A.M. Each edge decreases steadily thereafter, and at 4 P.M. is 8 in. What was the rate of change of the volume of the block of ice at 12 noon?

63. Suppose that u is a function of v, that v is a function of w, that w is a function of x, and that all these functions are differentiable. Explain why it follows from the chain rule that

$$\frac{du}{dx} = \frac{du}{dv} \cdot \frac{dv}{dw} \cdot \frac{dw}{dx}.$$

3.4 Derivatives of Algebraic Functions

In Section 3.3 we saw that the chain rule yields the differentiation formula

$$D[f(x)]^n = n[f(x)]^{n-1}f'(x)$$

if f is a differentiable function and the exponent n is an integer. According to Theorem 1 of this section, this **generalized power rule** holds not only when the exponent is an integer, but also when it is a rational number $r = p/q$ (where p and q are integers). Recall that rational powers are defined in terms of integral roots and powers as follows:

$$u^{p/q} = \sqrt[q]{u^p} = (\sqrt[q]{u})^p.$$

Theorem 1 *Generalized Power Rule*

If f is differentiable at x and r is a rational number, then

$$D[f(x)]^r = r[f(x)]^{r-1}f'(x) \tag{1}$$

at all points where the right-hand side in (1) is meaningful: the points where $f(x) \neq 0$ if $r - 1 < 0$ and $f(x) > 0$ if an even root is involved.

In terms of the variable $u = f(x)$, we can write the generalized power rule in the form

$$D_x u^r = \frac{d(u^r)}{dx} = ru^{r-1}\frac{du}{dx}. \tag{1'}$$

Recall that the subscript x in D_x specifies that u^r is being differentiated with respect to x, rather than with respect to u (as would be understood if we wrote Du^r). When $u = f(x) = x$, we see that the rule in (1′) takes the familiar form

$$Dx^r = rx^{r-1}, \tag{2}$$

the power rule for rational exponents. For example,

$$Dx^{1/2} = \tfrac{1}{2}x^{-1/2}, \qquad Dx^{3/2} = \tfrac{3}{2}x^{1/2}, \qquad Dx^{2/3} = \tfrac{2}{3}x^{-1/3},$$

and so forth.

To differentiate a function involving roots (or radicals) we first "prepare" it for an application of the generalized power rule: We rewrite it as a power function with fractional exponents. The first four examples illustrate this technique.

EXAMPLE 1 If $y = 5\sqrt{x^3} - \dfrac{2}{\sqrt[3]{x}}$, then

$$y = 5x^{3/2} - 2x^{-1/3},$$

so

$$\frac{dy}{dx} = (5)(\tfrac{3}{2})x^{1/2} - (2)(-\tfrac{1}{3})x^{-4/3}$$

$$= \tfrac{15}{2}x^{1/2} + \tfrac{2}{3}x^{-4/3} = \frac{15}{2}\sqrt{x} + \frac{2}{3\sqrt[3]{x^4}}$$

provided that $x > 0$.

EXAMPLE 2 With $f(x) = 3 - 5x$ and $r = 7$, the generalized power rule yields

$$D(3-5x)^7 = 7(3-5x)^6 D(3-5x) = 7(3-5x)^6(-5) = -35(3-5x)^6.$$

EXAMPLE 3 With $f(x) = 2x^2 - 3x + 5$ and $r = \frac{1}{2}$, the generalized power rule gives

$$D\sqrt{2x^2 - 3x + 5} = D(2x^2 - 3x + 5)^{1/2}$$

$$= \tfrac{1}{2}(2x^2 - 3x + 5)^{-1/2}D(2x^2 - 3x + 5)$$

$$= \frac{4x - 3}{2\sqrt{2x^2 - 3x + 5}}.$$

EXAMPLE 4 If $y = [5x + \sqrt[3]{(3x-1)^4}]^{10}$, then the formula in (1′) with $u = 5x + (3x-1)^{4/3}$ gives

$$\frac{dy}{dx} = 10u^9\frac{du}{dx}$$

$$= 10[5x + (3x-1)^{4/3}]^9 D[5x + (3x-1)^{4/3}]$$

$$= 10[5x + (3x-1)^{4/3}]^9[D(5x) + D(3x-1)^{4/3}]$$

$$= 10[5x + (3x-1)^{4/3}]^9[5 + \tfrac{4}{3}(3x-1)^{1/3}(3)]$$

$$= 10[5x + (3x-1)^{4/3}]^9[5 + 4\sqrt[3]{3x-1}].$$

Observe that in Example 4 we applied the generalized power rule twice in succession—first in the form $D_x u^{10} = 10u^9 D_x u$, a second time in computing $D_x u$.

PROOF OF THE GENERALIZED POWER RULE

In Section 3.3 we verified that the formula

$$D[f(x)]^r = r[f(x)]^{r-1} f'(x)$$

of Equation (1) holds when the exponent r is an *integer*. In extending the generalized power rule from integral to rational exponents, the most difficult step is the case in which $f(x) = x$ (so that $f'(x) \equiv 1$) and $r = 1/q$, where q is a *positive* integer:

$$Dx^{1/q} = \frac{1}{q}\, x^{(1/q)-1}. \tag{3}$$

Because $x^{1/q} = \sqrt[q]{x}$, it is natural to call the formula in (3) the **root rule** of differentiation. The chain rule version of the root rule is

$$D_x u^{1/q} = \frac{1}{q}\, u^{(1/q)-1} D_x u \tag{3'}$$

with u a differentiable function of x.

The root rule is established (independently) in Theorem 3 of this section. Assuming it for now, we can complete the proof of the generalized power rule

$$D[f(x)]^{p/q} = \frac{p}{q}\,[f(x)]^{(p/q)-1} f'(x)$$

for the rational exponent $r = p/q$ in the following way. First we write

$$[f(x)]^{p/q} = \{[f(x)]^p\}^{1/q} = u^{1/q}$$

with q a positive integer and $u = [f(x)]^p$. Then

$$D_x u = p[f(x)]^{p-1} f'(x) \tag{4}$$

by the integral case of the generalized power rule (for p is an integer). It follows that

$$\begin{aligned}
D_x[f(x)]^{p/q} &= D_x u^{1/q} \\
&= \frac{1}{q}\, u^{(1/q)-1} D_x u && \text{(by Eq. (3'))} \\
&= \frac{1}{q}\, \{[f(x)]^p\}^{(1/q)-1} p[f(x)]^{p-1} f'(x) && \text{(using (4))} \\
&= \frac{p}{q}\,[f(x)]^{p((1/q)-1)+p-1} f'(x) \\
&= \frac{p}{q}\,[f(x)]^{(p/q)-1} f'(x),
\end{aligned}$$

as desired (subject to the restrictions mentioned in the statement of Theorem 1). In Chapter 7 we will see that the generalized power rule continues to hold even when the exponent r in (1) is an *irrational* number (such as $\sqrt{2}$ or π).

DERIVATIVES OF INVERSE FUNCTIONS

We sometimes encounter pairs of functions that are natural opposites of one another, in the sense that each undoes the result of applying the other. We show a schematic of this phenomenon in Fig. 3.10. But actual examples may be more informative, so in Example 5 we give some pairs of functions that are *inverses* of each other. Recall that the domain of every function is, unless otherwise specified, the set of all real numbers for which its formula is meaningful.

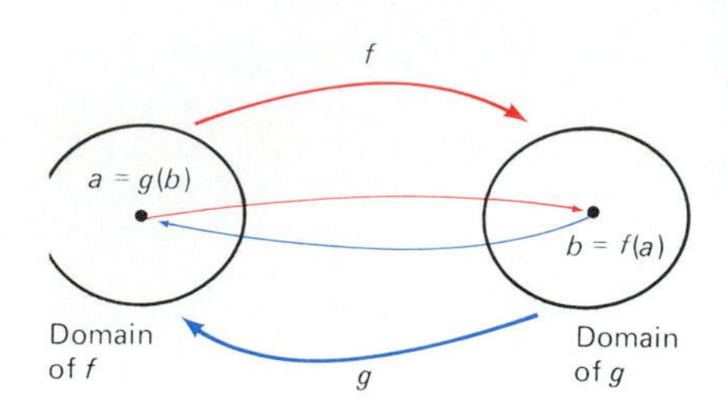

3.10 The action of inverse functions f and g

EXAMPLE 5 The following are some pairs of inverse functions.

$$f(x) = x + 1 \quad \text{and} \quad g(x) = x - 1$$

Adding 1 and subtracting 1 are inverse operations; doing either undoes the other.

$$f(x) = 2x \quad \text{and} \quad g(x) = \frac{x}{2}$$

Doubling and halving are inverse operations.

$$f(x) = \frac{1}{x} \quad \text{and} \quad g(x) = \frac{1}{x}$$

A function can be its own inverse.

$$\begin{cases} f\colon \ (0, \infty) \to \mathbf{R} \\ \text{by } f(x) = x^2 \end{cases} \quad \text{and} \quad \begin{cases} g\colon \ (0, \infty) \to \mathbf{R} \\ \text{by } g(x) = \sqrt{x} \end{cases}$$

Squaring and taking the square root are inverse operations when only positive numbers are involved.

Each pair f and g of functions given in the example has the property that

$$f(g(x)) = x \quad \text{and} \quad g(f(x)) = x. \tag{5}$$

When this pair of equations holds for all values of x in the appropriate domains, then we say that f and g are *inverses* of each other.

More precisely, we say that the two functions f and g are **inverse functions,** or are **inverses** of each other, provided that

1. The range of values of each function is the domain of definition of the other, and
2. The relations in (5) hold for all x in the domains of g and f, respectively.

Care is required in specifying the domains of f and g to insure that the condition in (1) is satisfied. For example, the rule $f(x) = x^2$ makes sense for all real x, but it is not the same as the function

$$f(x) = x^2, \qquad x > 0$$

that is the inverse of $g(x) = \sqrt{x}$, $x > 0$. On the other hand, the functions

$$f(x) = x^3 \quad \text{and} \quad g(x) = x^{1/3}$$

are inverses of each other with no restriction on their natural domains.

Our interest in inverse function pairs at this point stems from the following general principle: When we know the derivative of either of two inverse functions, then we can use the inverse function relationship between them to *discover* the derivative of the other of the two functions. The following example illustrates this principle with the inverse functions $f(x) = x^2$ and $g(x) = \sqrt{x}$, each defined for $x > 0$.

EXAMPLE 6 Suppose that we begin with the fact that

$$Dx^2 = 2x.$$

If $u = g(x) = \sqrt{x}$, then the first relation in (5) is simply the fact that

$$u^2 = x. \tag{6}$$

If we *assume* that $u = g(x)$ is differentiable, then differentiation of each side in (6)—using the chain rule and the fact that $D_u u^2 = 2u$ on the left—yields the equation

$$2u\frac{du}{dx} = 1. \tag{7}$$

We now solve for du/dx, and find that

$$\frac{du}{dx} = \frac{1}{2u} = \frac{1}{2\sqrt{x}} = \frac{1}{2}x^{-1/2}.$$

Because $u = g(x) = \sqrt{x}$, we thus have deduced, from the fact that $Dx^2 = 2x$, the fact that $Dx^{1/2} = \frac{1}{2}x^{-1/2}$.

Note that the derivation in Example 6 is *not* a complete proof that $Dx^{1/2} = \frac{1}{2}x^{-1/2}$, because it is based on the assumption that $g(x) = x^{1/2}$ is differentiable. What it *does* show is that *if* g is differentiable, *then* $g'(x) = \frac{1}{2}x^{-1/2}$.

In order to use the method of Example 6, we need to know when and whether a differentiable function f has an inverse function g that also is differentiable. The following theorem answers this question, and ordinarily is proved in an advanced calculus course.

> ***Theorem 2*** *Differentiation of an Inverse Function*
> Suppose that the function f is defined on the open interval I and that $f'(x) > 0$ for all x in I. Then f has an inverse function g, the function g is also differentiable, and
>
> $$g'(x) = \frac{1}{f'(g(x))} \tag{8}$$
>
> for all x in the domain of g.

COMMENT This theorem is also true when the condition $f'(x) > 0$ is replaced by the condition $f'(x) < 0$. Assuming the fact that g is differentiable, the derivative formula in (8) can be derived by differentiating each side in the inverse function relation

$$f(g(x)) = x.$$

When we differentiate each side, of course using the chain rule on the left-hand side, the result is

$$f'(g(x)) \cdot g'(x) = 1.$$

When we solve this equation for $g'(x)$, the result is that in Equation (8).

The formula in (8) is easy to remember in differential notation. Let us write $x = f(y)$ and $y = g(x)$. Then $dy/dx = g'(x)$ and $dx/dy = f'(y)$. So (8) becomes the seemingly inevitable formula

$$\frac{dy}{dx} = \frac{1}{\dfrac{dx}{dy}}. \tag{8'}$$

In using Equation (8′) it is important to remember that dy/dx is to be evaluated at x, while dx/dy is to be evaluated at the corresponding value of y; namely, $y = g(x)$.

We can now use the fact that $f(x) = x^q$ and $g(x) = x^{1/q}$ are mutual inverses to compute the derivative of $x^{1/q}$.

Theorem 3 *The Root Rule*

$$Dx^{1/q} = \frac{1}{q}x^{(1/q)-1}$$

if $x \neq 0$, and with the proviso that $x > 0$ if the positive integer q is even.

Proof If $y = x^{1/q}$ and $x = y^q$, then Equation (8′) gives

$$\frac{dy}{dx} = \frac{1}{\dfrac{dx}{dy}} = \frac{1}{qy^{q-1}} = \frac{1}{q(x^{1/q})^{q-1}},$$

so that

$$\frac{dy}{dx} = \frac{1}{q}x^{(1/q)-1},$$

as desired. ■

EXAMPLE 7 The root rule gives $Dx^{1/3} = \frac{1}{3}x^{-2/3}$, which increases without bound as $x \to 0$. How is this information related to the graph of $y = x^{1/3}$? The graph appears in Fig. 3.11. The derivative dy/dx does not exist at $x = 0$, so the curve does not have a tangent line (as defined in Section 2.1) at the origin. Nevertheless, in this case we may regard the vertical line $x = 0$ as the tangent line to $y = x^{1/3}$ at (0, 0). Examples such as this motivate the following definition.

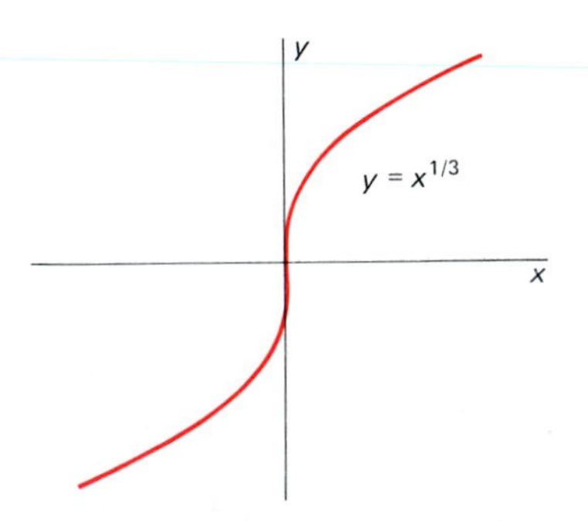

3.11 The graph of the cube root function

Definition *Vertical Tangent Line*

The curve $y = f(x)$ has a **vertical tangent line** at the point $(a, f(a))$ provided that $|f'(x)| \to +\infty$ as $x \to a$.

Note that f must be defined at $x = a$ if its graph is to have a tangent line (vertical or not!) there.

If f is defined (and differentiable) on only one side of $x = a$, we mean in this definition that $|f'(x)| \to +\infty$ as x approaches a from that side.

EXAMPLE 8 Find the points on the curve

$$y = f(x) = x\sqrt{1 - x^2}, \qquad -1 \leqq x \leqq 1$$

at which the tangent line is either horizontal or vertical.

Solution We differentiate using the product rule (primarily) and the chain rule (secondarily):

$$f'(x) = (1 - x^2)^{1/2} + \frac{x}{2}(1 - x^2)^{-1/2}(-2x)$$

$$= (1 - x^2)^{-1/2}[(1 - x^2) - x^2] = \frac{1 - 2x^2}{\sqrt{1 - x^2}}.$$

Then $f'(x) = 0$ only when the numerator $1 - 2x^2$ is zero—that is, when $x = \pm 1/\sqrt{2}$. Because $f(\pm 1/\sqrt{2}) = \pm 0.5$, the curve has a horizontal tangent line at each of the two points $(1/\sqrt{2}, 0.5)$ and $(-1/\sqrt{2}, -0.5)$.

We observe also that the denominator $\sqrt{1 - x^2}$ approaches zero as $x \to -1^+$ and as $x \to +1^-$. Because $f(\pm 1) = 0$, we see that the curve has a vertical tangent line at each of the two points $(1, 0)$ and $(-1, 0)$. The graph of f is shown in Fig. 3.12.

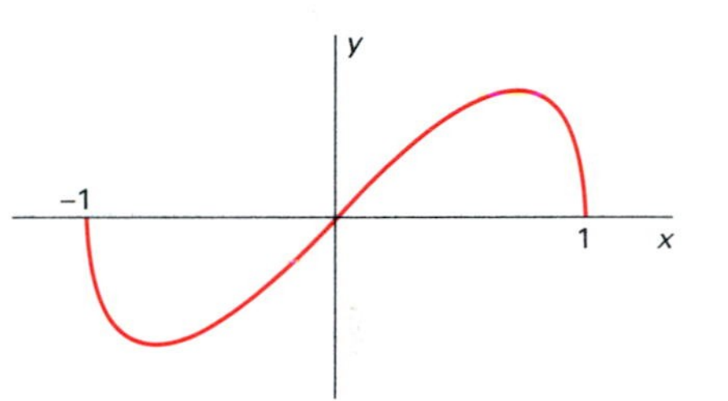

3.12 The graph of $f(x) = x\sqrt{1 - x^2}$, $-1 \leqq x \leqq 1$

3.4 PROBLEMS

Differentiate the functions given in Problems 1–34.

1. $f(x) = \sqrt{x^3 + 1}$

2. $f(x) = \dfrac{1}{(x^4 + 3)^2}$

3. $f(x) = \sqrt{2x^2 + 1}$

4. $f(x) = \dfrac{x}{\sqrt{1 + x^4}}$

5. $f(t) = \sqrt{2t^3}$

6. $g(t) = \sqrt{\dfrac{1}{3t^5}}$

7. $f(x) = (2x^2 - x + 7)^{3/2}$

8. $g(z) = (3z^2 - 4)^{97}$

9. $g(x) = \dfrac{1}{(x - 2x^3)^{4/3}}$

10. $f(t) = [t^2 + (1 + t)^4]^5$

11. $f(x) = x\sqrt{1 - x^2}$

12. $g(x) = \sqrt{\dfrac{2x + 1}{x - 1}}$

13. $f(t) = \sqrt{\dfrac{t^2 + 1}{t^2 - 1}}$

14. $h(y) = \left(\dfrac{y + 1}{y - 1}\right)^{17}$

15. $f(x) = \left(x - \dfrac{1}{x}\right)^3$

16. $g(z) = \dfrac{z^2}{\sqrt{1 + z^2}}$

17. $f(v) = \dfrac{\sqrt{v + 1}}{v}$

18. $h(x) = \left(\dfrac{x}{1 + x^2}\right)^{5/3}$

19. $f(x) = \sqrt[3]{1 - x^2}$

20. $g(x) = \sqrt{x + \sqrt{x}}$

21. $f(x) = x(3 - 4x)^{1/2}$

22. $g(t) = \dfrac{1}{t^2}[t - (1 + t^2)^{1/2}]$

23. $f(x) = (1 - x^2)(2x + 4)^{4/3}$

24. $f(x) = (1 - x)^{1/2}(2 - x)^{1/3}$

25. $g(t) = \left(1 + \dfrac{1}{t}\right)^2(3t^2 + 1)^{1/2}$

26. $f(x) = x(1 + 2x + 3x^2)^{10}$

27. $f(x) = \dfrac{2x - 1}{(3x + 4)^5}$

28. $h(z) = (z - 1)^4(z + 1)^6$

29. $f(x) = \dfrac{(2x + 1)^{1/2}}{(3x + 4)^{1/3}}$

30. $f(x) = (1 - 3x^4)^5(4 - x)^{1/3}$

31. $h(y) = \dfrac{(1 + y)^{1/2} + (1 - y)^{1/2}}{y^{5/3}}$

32. $f(x) = (1 - x^{1/3})^{1/2}$

33. $g(t) = [t + (t + t^{1/2})^{1/2}]^{1/2}$

34. $f(x) = x^3\left(1 - \dfrac{1}{x^2 + 1}\right)^{1/2}$

For each curve given in Problems 35–40, find all points on the graph where the tangent line is either horizontal or vertical.

35. $y = x^{2/3}$

36. $y = x\sqrt{4 - x^2}$

37. $y = x^{1/2} - x^{3/2}$

38. $y = \dfrac{1}{\sqrt{9 - x^2}}$

39. $y = \dfrac{x}{\sqrt{1 - x^2}}$

40. $y = \sqrt{(1 - x^2)(4 - x^2)}$

41. The period of oscillation P (seconds) of a simple pendulum of length L feet is given by $P = 2\pi\sqrt{L/g}$, where $g = 32$ ft/s^2. Find the rate of change of P with respect to L when $P = 2$.

42. Find the rate of change of the volume V of sphere with respect to its surface area S when its radius is 10.

43. Find the two points on the circle $x^2 + y^2 = 1$ at which the slope of the tangent line is -2.

44. Find the two points on the circle $x^2 + y^2 = 1$ at which the slope of the tangent line is 3.

45. Find a line through the point $P(18, 0)$ that is normal to the parabola $y = x^2$ at some point $Q(a, a^2)$. (*Suggestion:* You will obtain a cubic equation in the unknown a. Find by inspection a small integral root r. The cubic polynomial is then the product of $a - r$ and a quadratic polynomial; you can find the quadratic by division of $a - r$ into the cubic.)

46. Find three distinct lines through the point (3, 10) that are normal to the parabola $y = x^2$. See the suggestion for Problem 45.

47. Find two lines through the point (0, 2.5) that are normal to the curve $y = x^{2/3}$.

48. (*Challenge*) Suppose that f' is continuous on the set of all real numbers, that $f'(0) \geqq 0$, and that $f = f^{-1}$. Prove that $f(x) = x$ for all x, or show that this assertion is false by finding a different function with the given properties.

49. Consider the cubic equation

$$x^3 = 3x + 8.$$

If we differentiate each side with respect to x, we obtain

$$3x^2 = 3,$$

which has the two solutions $x = 1$ and $x = -1$. But neither of these is a solution of the original equation. What went wrong? After all, in several examples and theorems of this section we *appeared* to differentiate both sides of an equation. Explain carefully why differentiation of both sides of Equation (6) is valid and why the differentiation in this problem is not.

50. If $y = x^{3/5}$, then $y^5 = x^3$. Differentiate both sides of the latter equation with respect to x, then solve for dy/dx.

3.5 Maxima and Minima of Functions on Closed Intervals

In applications we often need to find the maximum or minimum value that a specified quantity can attain. The animal pen problem stated in Section 1.1 is a simple yet typical example of an applied maximum-minimum problem. We saw there that the animal pen problem is equivalent to the purely mathematical problem of finding the maximum value attained by the function

$$f(x) = \tfrac{3}{5}x(30 - x)$$

on the closed interval $[0, 30]$.

Definition *Maximum and Minimum Values*

If c is in the closed interval $[a, b]$, then $f(c)$ is called the **maximum value** of $f(x)$ on $[a, b]$ if $f(c) \geqq f(x)$ for all x in $[a, b]$. Similarly, if d is in $[a, b]$, then the value $f(d)$ is called the **minimum value** of $f(x)$ on $[a, b]$ if $f(d) \leqq f(x)$ for all x in $[a, b]$.

Thus the maximum value of $f(x)$ is simply a value of f no smaller than any other value of $f(x)$. The point of the next theorem is not merely that there is a number M such that $M \geqq f(x)$ for all x in $[a, b]$, but that if f is continuous and its domain is a closed interval, then M is actually a *value* of f.

> ***Theorem 1*** *Maximum Value Property*
> If f is continuous on the closed interval $[a, b]$, then there exists a number c in $[a, b]$ such that $f(c)$ is the maximum value of f on $[a, b]$.

By applying this result to the function $g(x) = -f(x)$, we conclude that $f(x)$ must also attain a minimum value at some point of the interval. In short, a continuous function on a closed interval attains both a maximum value and a minimum value at points of the interval. Hence we see it is the *continuity* of the function

$$f(x) = \tfrac{3}{5}x(30 - x)$$

on the *closed* interval $[0, 30]$ that guarantees that the maximum value of f exists and is attained at some point of the interval $[0, 30]$.

Suppose that the function f is defined on the interval I. The following examples show that if *either* f is not continuous *or* I is not closed, then f may fail to attain maximum and minimum values at points of I. Thus both hypotheses in Theorem 1 are necessary.

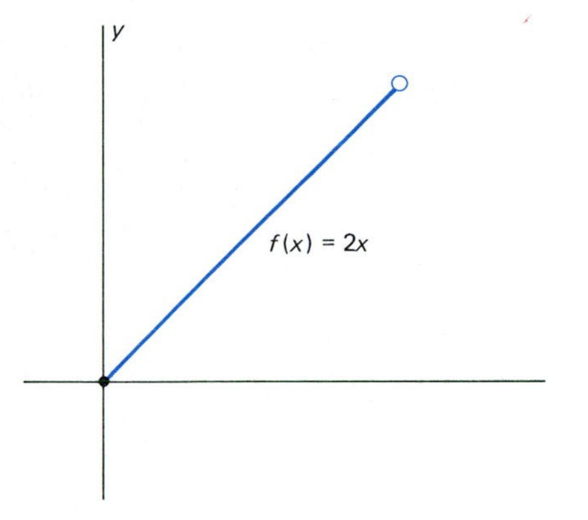

3.13 The graph of the function of Example 1

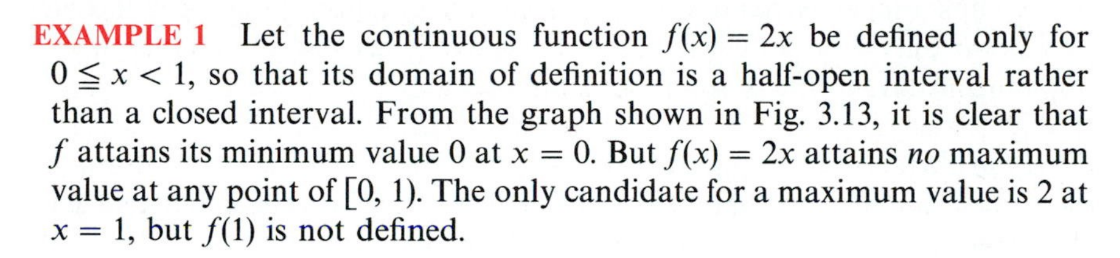

EXAMPLE 1 Let the continuous function $f(x) = 2x$ be defined only for $0 \leqq x < 1$, so that its domain of definition is a half-open interval rather than a closed interval. From the graph shown in Fig. 3.13, it is clear that f attains its minimum value 0 at $x = 0$. But $f(x) = 2x$ attains *no* maximum value at any point of $[0, 1)$. The only candidate for a maximum value is 2 at $x = 1$, but $f(1)$ is not defined.

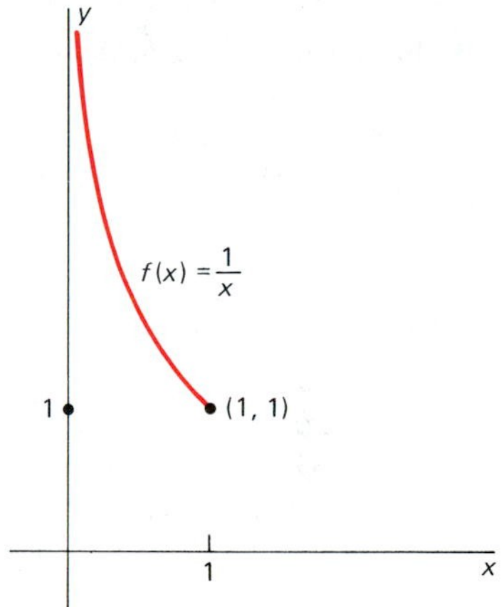

3.14 The graph of the function of Example 2

EXAMPLE 2 The function f defined on the closed interval $[0, 1]$ with formula

$$f(x) = \begin{cases} \dfrac{1}{x} & \text{if} \quad 0 < x \leqq 1, \\ 1 & \text{if} \quad x = 0 \end{cases}$$

is not continuous on $[0, 1]$ because $\lim\limits_{x \to 0^+} (1/x)$ does not exist (see Fig. 3.14). This function does attain its minimum value of 1 at $x = 0$ and also at $x = 1$. But it attains no maximum value on $[0, 1]$ because $1/x$ can be made arbitrarily large by choosing x positive and very close to zero.

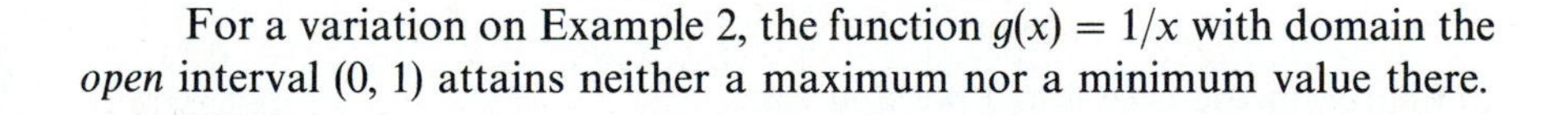

For a variation on Example 2, the function $g(x) = 1/x$ with domain the *open* interval $(0, 1)$ attains neither a maximum nor a minimum value there.

LOCAL MAXIMA AND MINIMA

Once we know that the continuous function f *does* attain maximum and minimum values on the closed interval $[a, b]$, the remaining question is this: Exactly *where* are these values located? In Section 2.1 we solved the animal pen problem on the basis of the geometrically motivated assumption that the function $f(x) = \frac{3}{5}x(30 - x)$ attains its maximum value on $[0, 30]$ at an interior point of that interval, one at which the tangent line is horizontal.

Theorems 2 and 3 of this section provide a rigorous basis for the method we used there.

We say that the value $f(c)$ is a **local maximum value** of the function f if $f(x) \leqq f(c)$ for all x sufficiently near c. More precisely, if this inequality holds for all x that are simultaneously in the domain of f and in some open interval containing c, then $f(c)$ is a local maximum of f. Similarly, we say that the value $f(c)$ is a **local minimum value** of f if $f(x) \geqq f(c)$ for all x sufficiently near c.

As the graph in Fig. 3.15 shows, a local maximum is a point such that no nearby points on the graph are higher and a local minimum is one such that no nearby points on the graph are lower. A **local extremum** of f is a value of f that is either a local maximum or a local minimum.

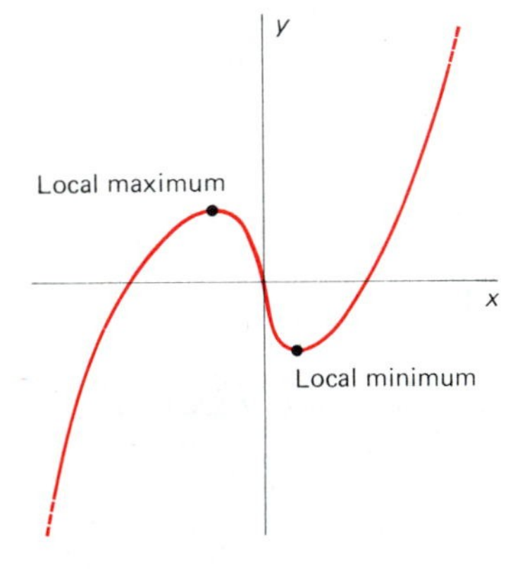

3.15 Local extrema

> ***Theorem 2*** *Local Maxima and Minima*
>
> **If f is differentiable at c and is defined on an open interval containing c, and if $f(c)$ is either a local maximum value or a local minimum value of f, then $f'(c) = 0$.**

Thus a local extremum of a *differentiable* function on an *open* interval can occur only at a point where the derivative is zero and, therefore, where the tangent line to the graph is horizontal.

Proof of Theorem 2 Suppose, for instance, that $f(c)$ is a local maximum value of f. The assumption that $f'(c)$ exists means that the right-hand and left-hand limits

$$\lim_{h \to 0^+} \frac{f(c+h) - f(c)}{h} \quad \text{and} \quad \lim_{h \to 0^-} \frac{f(c+h) - f(c)}{h}$$

both exist and are equal to $f'(c)$.

If $h > 0$, then

$$\frac{f(c+h) - f(c)}{h} \leqq 0$$

because $f(c) \geqq f(c + h)$ for all small positive values of h. Hence, by a one-sided version of the squeeze property for limits (Section 2.2), the inequality above will be preserved when we take the limit as $h \to 0$. We thus find that

$$f'(c) = \lim_{h \to 0^+} \frac{f(c+h) - f(c)}{h} \leqq \lim_{h \to 0^+} 0 = 0.$$

Similarly, in the case $h < 0$, we find that

$$\frac{f(c+h) - f(c)}{h} \geqq 0.$$

Therefore,

$$f'(c) = \lim_{h \to 0^-} \frac{f(c+h) - f(c)}{h} \geqq \lim_{h \to 0^-} 0 = 0.$$

Because both $f'(c) \leqq 0$ and $f'(c) \geqq 0$, we may conclude that $f'(c) = 0$. This establishes Theorem 2. ■

BEWARE The converse of Theorem 2 is false. That is, the fact that $f'(c) = 0$ is *not enough* to imply that $f(c)$ is a local extremum. For example, consider the function $f(x) = x^3$. Its derivative $f'(x) = 3x^2$ vanishes at $x = 0$. But a glance at its graph, shown in Fig. 3.16, shows us that $f(0) = 0$ is *not* a local extremum of x^3.

Thus the equation $f'(c) = 0$ is a *necessary* condition for $f(c)$ to be a local maximum or minimum value for a function f differentiable on an open interval. It is *not* a *sufficient* condition. For $f'(x)$ may well be zero at points other than local maxima and minima. We shall give sufficient conditions for local maxima and minima in Chapter 4.

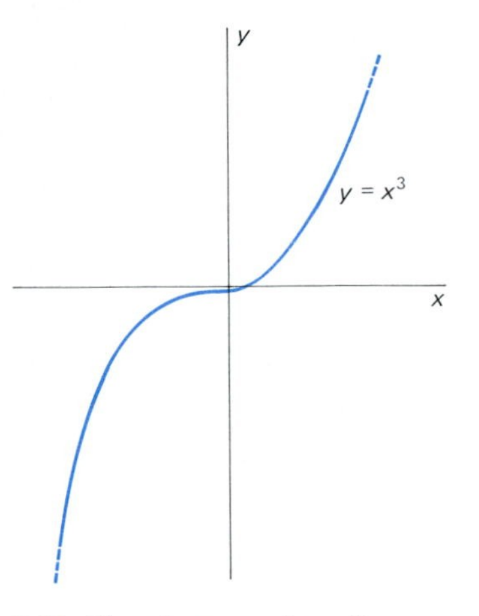

3.16 No extremum at $x = 0$ even though the derivative is zero there

ABSOLUTE MAXIMA AND MINIMA

In most sorts of optimization problems, we are less interested in the local extrema (as such) than in the global or *absolute* maximum and minimum values attained by a given continuous function. If f is a function with domain D, we call $f(c)$ the **absolute maximum value** of f on D provided that $f(c) \geqq f(x)$ for *all* x in D. Briefly, $f(c)$ is the largest value of f on D. It should be clear how the global or absolute minimum of f is to be defined. The graph in Fig. 3.17 illustrates some local and global extrema. Note that every global extremum is, of course, local as well; on the other hand, the graph shows some local extrema that are not global.

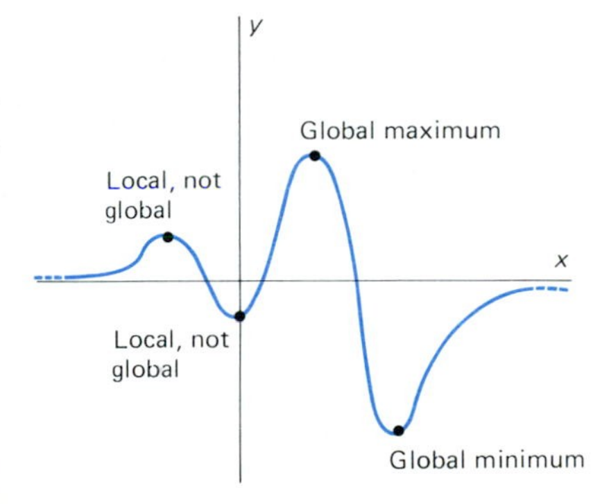

3.17 Some extrema are global; others are merely local.

Theorem 3 tells us that the absolute maximum and absolute minimum values of the continuous function f on the closed interval $[a, b]$ each occur either at one of the end points a or b or at a critical point of f. The number c in the domain of f is called a **critical point** of f if either

(1) $f'(c) = 0$, or

(2) $f'(c)$ does not exist.

> ***Theorem 3*** *Absolute Maxima and Minima*
>
> Suppose that $f(c)$ is the absolute maximum (or absolute minimum) value of the continuous function f on the closed interval $[a, b]$. Then c is either a critical point of f or one of the endpoints a and b.

Proof This follows almost immediately from Theorem 2. If c is not an endpoint of $[a, b]$, then $f(c)$ is a local extremum of f on the open interval (a, b). In this case Theorem 2 implies that $f'(c) = 0$, provided that f is differentiable at c. ■

As a consequence of Theorem 2, we can find the (absolute) maximum and minimum values of f on the closed interval $[a, b]$ as follows.

1. First locate the critical points of f in $[a, b]$.
2. Then find the value of f at each of these critical points *and* at the two endpoints a and b.

The largest of these values will automatically be the absolute maximum of f; the smallest, the absolute minimum. We call this procedure the **closed interval maximum–minimum method.**

EXAMPLE 3 Find the maximum and minimum values of

$$f(x) = 2x^3 - 3x^2 - 12x + 15$$

on the closed interval $[0, 3]$.

Solution The derivative of f is

$$f'(x) = 6x^2 - 6x - 12 = 6(x - 2)(x + 1).$$

So the critical points of f are the solutions of the equation

$$6(x - 2)(x + 1) = 0$$

and the numbers c for which $f'(c)$ does not exist. There are none of the latter, so the critical points of f occur at $x = -1$ and at $x = 2$. The first of these is not in the domain of f; we discard it, and thus the only critical point of f in $[0, 3]$ is $x = 2$. Including the two endpoints, our list of possibilities for a maximum or minimum consists of $x = 0, 2$, and 3. We evaluate the function f at each:

$$f(0) = 15, \quad \longleftarrow \quad \text{absolute maximum}$$

$$f(2) = -5, \quad \longleftarrow \quad \text{absolute minimum}$$

$$f(3) = 6.$$

Therefore, the maximum value of f on $[0, 3]$ is $f(0) = 15$ and the minimum value is $f(2) = -5$.

In Example 3 the function f was differentiable everywhere. Examples 4 and 5 illustrate the case of an extremum at a critical point where the function is not differentiable.

EXAMPLE 4 Find the maximum and minimum values of the function $f(x) = 3 - |x - 2|$ on the interval $[1, 4]$.

Solution If $x \leqq 2$, then $x - 2 \leqq 0$. If so, then

$$f(x) = 3 - (2 - x) = x + 1.$$

If $x \geqq 2$, then $x - 2 \geqq 0$, so

$$f(x) = 3 - (x - 2) = 5 - x.$$

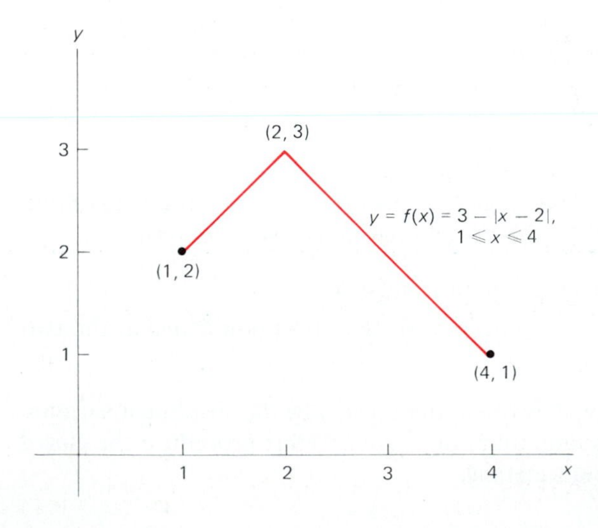

3.18 Graph of the function of Example 4

Consequently, the graph of f looks like the one shown in Fig. 3.18. The only critical point of f in $[1, 4]$ is the point $x = 2$, because $f'(x)$ takes on only the two values $+1$ and -1 (so is never zero), and $f'(2)$ does not exist. (Why?) Evaluation of f at this critical point and at the two endpoints yields

$$f(1) = 2,$$
$$f(2) = 3, \quad \longleftarrow \quad \text{absolute maximum}$$
$$f(4) = 1. \quad \longleftarrow \quad \text{absolute minimum}$$

EXAMPLE 5 Find the maximum and minimum values of

$$f(x) = 5x^{2/3} - x^{5/3}$$

on the closed interval $[-1, 4]$.

Solution Differentiation of f yields

$$f'(x) = \frac{10}{3}x^{-1/3} - \frac{5}{3}x^{2/3} = \frac{5}{3}x^{-1/3}(2 - x) = \frac{5(2 - x)}{3x^{1/3}}.$$

Hence f has two critical points in the interval: $x = 2$, where $f'(x) = 0$, and $x = 0$, where f' does not exist (the graph of f has a vertical tangent line at $(0, 0)$). When we evaluate f at these two critical points and at the two endpoints, we get

$$f(-1) = 6, \quad \longleftarrow \quad \text{absolute maximum}$$
$$f(0) = 0, \quad \longleftarrow \quad \text{absolute minimum}$$
$$f(2) = (5)(2^{2/3}) - 2^{5/3} \approx 4.76,$$
$$f(4) = (5)(4^{2/3}) - 4^{5/3} \approx 2.52.$$

Thus the maximum value $f(-1) = 6$ occurs at an endpoint. The minimum value $f(0) = 0$ occurs at a point where f is not differentiable.

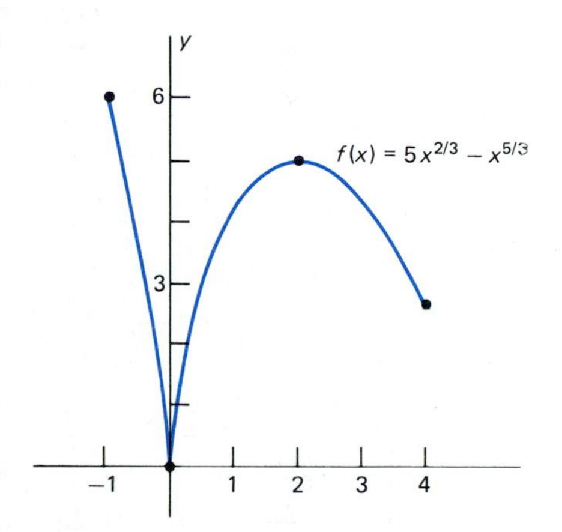

3.19 Graph of the function of Example 5

By using a "smart" calculator or pocket computer with graphics capabilities, you can verify that the graph of f looks as it is shown in Fig. 3.19. (If you don't have such a calculator, an efficient method of sketching graphs with the aid of calculus is presented in Chapter 4.) But in the usual case of a continuous function having only finitely many critical points in a given closed interval, the closed interval maximum–minimum method suffices to determine its maximum and minimum values without requiring any detailed knowledge of the graph of the function.

3.5 PROBLEMS

In each of Problems 1–10, state whether the given function attains a maximum value or a minimum value (or both) on the given interval. (*Suggestion:* Begin by sketching a graph of the function.)

1. $f(x) = 1 - x, \quad -1 \leqq x < 1$
2. $f(x) = 2x + 1, \quad -1 \leqq x < 1$
3. $f(x) = |x|, \quad -1 < x < 1$
4. $f(x) = \dfrac{1}{\sqrt{x}}, \quad 0 < x \leqq 1$
5. $f(x) = |x - 2|, \quad 1 < x \leqq 4$
6. $f(x) = 5 - x^2, \quad -1 \leqq x < 2$
7. $f(x) = x^3 + 1, \quad -1 \leqq x \leqq 1$

8. $f(x) = \dfrac{1}{x^2 + 1}, \quad -\infty < x < \infty$

9. $f(x) = \dfrac{1}{x(1 - x)}, \quad 2 \leqq x \leqq 3$

10. $f(x) = \dfrac{1}{x(1 - x)}, \quad 0 < x < 1$

In each of Problems 11–40, find the maximum and minimum values attained by the given function on the indicated closed interval.

11. $f(x) = 3x + 2; \quad [-2, 3]$
12. $g(x) = 4 - 3x; \quad [-1, 5]$
13. $h(x) = 4 - x^2; \quad [1, 3]$
14. $f(x) = x^2 + 3; \quad [0, 5]$
15. $g(x) = (x - 1)^2; \quad [-1, 4]$
16. $h(x) = x^2 + 4x + 7; \quad [-3, 0]$
17. $f(x) = x^3 - 3x; \quad [-2, 4]$
18. $g(x) = 2x^3 - 9x^2 + 12x; \quad [0, 4]$
19. $h(x) = x + \dfrac{4}{x}; \quad [1, 4]$
20. $f(x) = x^2 + \dfrac{16}{x}; \quad [1, 3]$
21. $f(x) = 3 - 2x; \quad [-1, 1]$
22. $f(x) = x^2 - 4x + 3; \quad [0, 2]$
23. $f(x) = 5 - 12x - 9x^2; \quad [-1, 1]$
24. $f(x) = 2x^2 - 4x + 7; \quad [0, 2]$
25. $f(x) = x^3 - 3x^2 - 9x + 5; \quad [-2, 4]$
26. $f(x) = x^3 + x; \quad [-1, 2]$
27. $f(x) = 3x^5 - 5x^3; \quad [-2, 2]$
28. $f(x) = |2x - 3|; \quad [1, 2]$
29. $f(x) = 5 + |7 - 3x|; \quad [1, 5]$
30. $f(x) = |x + 1| + |x - 1|; \quad [-2, 2]$
31. $f(x) = 50x^3 - 105x^2 + 72x; \quad [0, 1]$
32. $f(x) = 2x + \dfrac{1}{2x}; \quad [1, 4]$
33. $f(x) = \dfrac{x}{x + 1}; \quad [0, 3]$
34. $f(x) = \dfrac{x}{x^2 + 1}; \quad [0, 3]$
35. $f(x) = \dfrac{1 - x}{x^2 + 3}; \quad [-2, 5]$
36. $f(x) = 2 - \sqrt[3]{x}; \quad [-1, 8]$
37. $f(x) = x\sqrt{1 - x^2}; \quad [-1, 1]$
38. $f(x) = x\sqrt{4 - x^2}; \quad [0, 2]$
39. $f(x) = x(2 - x)^{1/3}; \quad [1, 3]$
40. $f(x) = x^{1/2} - x^{3/2}; \quad [0, 4]$

41. Suppose that $f(x) = Ax + B$ is a linear function. Explain why the maximum and minimum values of f on a closed interval $[a, b]$ must occur at the endpoints of the interval.

42. Suppose that f is continuous on $[a, b]$, differentiable on (a, b), and $f'(x)$ is never zero at any point of (a, b). Explain why the maximum and minimum values of f must occur at the endpoints of the interval $[a, b]$.

43. Explain why every real number is a critical point of the greatest integer function $f(x) = [\![x]\!]$.

44. Prove that every quadratic function

$$f(x) = ax^2 + bx + c \qquad (a \neq 0)$$

has exactly one critical point on the real line.

45. Explain why the cubic polynomial function

$$f(x) = ax^3 + bx^2 + cx + d \quad (a \neq 0)$$

can have either two, one, or no critical points on the real line. Produce examples illustrating each of the three cases.

46. Define $f(x)$ to be the distance from x to the nearest integer. What are the critical points of f?

3.6 Applied Maximum–Minimum Problems

This section is devoted to applied maximum–minimum problems (like the animal pen problem of Section 1.1) for which the closed-interval maximum–minimum method of Section 3.5 can be used. When we confront such a problem, there is an important first step: We must determine the quantity to be maximized or minimized. This quantity will be the dependent variable in the solution.

This dependent variable must then be expressed as a function of an independent variable, one that "controls" the values of the dependent variable. If the domain of values of the independent variable—those that are pertinent to the applied problem—is a closed interval, then we may proceed with the closed interval maximum–minimum method. This plan of attack can be summarized in the following steps.

1. *Find the quantity to be maximized or minimized.* This quantity, which you should describe with a word or short phrase and label with a letter, will be your dependent variable. Because it is a *dependent* variable, it depends on something else; that will be your independent variable. We call the independent variable x in what follows.

2. *Express the dependent variable as a function of the independent variable.* Use the information in the problem to write the dependent variable as a function of x. By all means, draw a figure or diagram and *label the variables;* this is usually the best way to find the relationship between the dependent and independent variables. Use auxiliary variables if they help, but not too many, for you must eliminate them in the long run. You *must* express the dependent variable in terms of the *single* independent variable x and various constants before you can compute any derivatives. Find the domain of this function as well as its formula. Force the domain to be a closed interval if possible—if it's an open interval, adjoin the endpoints if you can.

3. *Apply calculus to find the critical points.* Compute the derivative f' of the function f found in Step 2. Use the derivative to find the critical points of type (1) (where $f'(x) = 0$) and those of type (2) (where $f'(x)$ does not exist).

4. *Identify the extrema.* Evaluate f at each critical point in its domain *and* at the two endpoints. The values you obtain will tell you which is the absolute maximum and which is the absolute minimum. Of course, either or both of these may occur at more than one point.

5. *Answer the question posed in the problem.* That is, interpret your results. The answer to the original problem may be something other than merely the largest (or smallest) possible value of f. Give a precise answer to the specific question originally asked.

Observe how this five-step process is followed in our first example.

EXAMPLE 1 A farmer has 200 yards of fence to be used in constructing three sides of a rectangular pen; an existing long straight wall will be used for the fourth side. What dimensions will maximize the area of the pen?

Solution We want to maximize the area A of the pen shown in Fig. 3.20. To get a formula for the *dependent* variable A, we observe that the area of a rectangle is the product of its base and its height. So we let x denote the length of each of the two sides of the pen perpendicular to the wall. We also let y denote the length of the side parallel to the wall. Then we can write a *formula* for A:

$$A = xy.$$

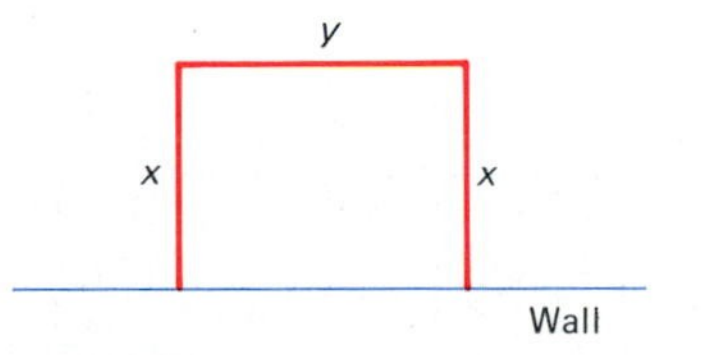

3.20 The rectangular pen of Example 1

Now we need to write A as a *function* of either x or y. Because all 200 yards of fence are to be used, this means that

$$2x + y = 200, \quad \text{so that} \quad y = 200 - 2x. \tag{1}$$

We choose to express y in terms of x merely because the algebra is simpler. In any case, we substitute this value of y in the formula $A = xy$ to obtain

$$A(x) = x(200 - 2x) = 200x - 2x^2. \tag{2}$$

This equation expresses the dependent variable A as a function of the independent variable x.

Before proceeding, we need the domain of the function A. To obtain an actual rectangular pen, we need $x > 0$ and $y > 0$; the latter, by Equation (1), implies that $x < 100$. Hence we obtain the open interval $0 < x < 100$. But to apply the closed interval maximum–minimum method, we need a closed interval. In this case we may adjoin the endpoints to $(0, 100)$ to get the *closed* interval $[0, 100]$. The values $x = 0$ and $x = 100$ correspond to "degenerate" pens of area zero. Because zero is certainly not the maximum value of A, there is no harm in thus enlarging the domain of the function A.

Now we compute the derivative of the function A in (2):

$$\frac{dA}{dx} = 200 - 4x.$$

Now

$$\frac{dA}{dx} \quad \text{exists for all } x,$$

so the only critical points of the function A occur when

$$\frac{dA}{dx} = 0;$$

that is, when

$$200 - 4x = 0.$$

So $x = 50$ is the only interior critical point. Including the endpoints, the extrema of A can occur only at $x = 0$, 50, and 100. We evaluate A at each:

$$A(0) = 0,$$

$$A(50) = 5000, \quad \longleftarrow \quad \text{absolute maximum}$$

$$A(100) = 0.$$

Thus the maximal area is $A(50) = 5000$ square yards. From (1) we find that $y = 100$ when $x = 50$. Therefore, the pen of maximal area should have the two sides perpendicular to the wall of length 50 yards each, and the side parallel to the wall of length 100 yards.

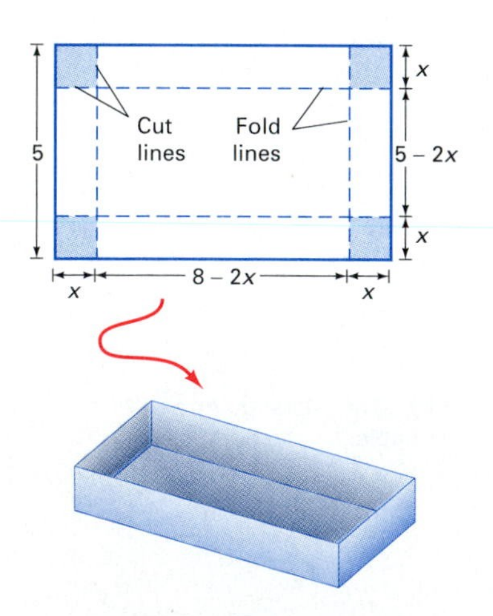

3.21 Making the box of Example 2

EXAMPLE 2 A piece of sheet metal is rectangular, 5 ft wide and 8 ft long. Congruent squares are to be cut from its four corners. The resulting piece of metal is to be folded and welded to form a box with an open top, as shown in Fig. 3.21. How should this be done to get a box of largest possible volume?

Solution The quantity to be maximized—the dependent variable—is the volume V of the box to be constructed. The shape and thus the volume of the box are determined by the length x of the edge of each corner square removed. Hence x is a natural choice for the independent variable.

To write the volume V as a function of x, note that the finished box will have height x, and its base will measure $8 - 2x$ feet by $5 - 2x$ feet. So

$$V = V(x) = x(5 - 2x)(8 - 2x) = 4x^3 - 26x^2 + 40x.$$

The procedure described in this example will produce an actual box only if $0 < x < 2.5$. But we make the domain the *closed* interval $[0, 2.5]$ in order to have the assurance that a maximum of $V(x)$ exists and to apply the closed interval maximum–minimum method. The values $x = 0$ and $x = 2.5$ correspond to "degenerate" boxes of zero volume, so adjoining these points to $(0, 2.5)$ will affect neither the location of the absolute maximum nor its value.

Now we compute the derivative of V:

$$V'(x) = 12x^2 - 52x + 40 = 4(3x - 10)(x - 1).$$

Because

$$V'(x) \text{ exists for all } x,$$

there are no critical points of type (2). So we find all critical points of V by solving the equation

$$V'(x) = 0;$$

that is,

$$4(3x - 10)(x - 1) = 0.$$

The solutions of this equation are $x = 1$ and $x = \frac{10}{3}$. We discard the latter because it does not lie in the domain $[0, 2.5]$ of V. So we examine these values of V:

$$\begin{aligned} V(0) &= 0, \\ V(1) &= 18, \quad \longleftarrow \quad \text{absolute maximum} \\ V(2.5) &= 0. \end{aligned}$$

Thus the maximum value of $V(x)$ on $[0, 2.5]$ is $V(1) = 18$. The answer to the question posed in the example is this: The squares cut from the corners should be of edge length 1 ft each. The resulting box will measure 6 ft by 3 ft by 1 ft, and its volume will be 18 ft^3.

For our next application of the closed interval maximum–minimum method, let us consider a typical problem in business management. Suppose that x units of a certain product are to be manufactured at a total cost of $C(x)$ dollars. We make the simple (and not always valid) assumption that the cost function $C(x)$ is the sum of two terms:

- A constant term a representing the fixed cost of acquiring and maintaining production facilities ("overhead"), and
- A variable term representing the additional cost of making x units at, for example, b dollars each.

Then $C(x)$ will be given by

$$C(x) = a + bx. \tag{3}$$

We also assume that the number of units that can be sold is a linear function of the selling price p, so that $x = m - np$ where m and n are positive constants. The minus sign indicates, naturally enough, that an increase in selling price will result in a decrease in sales. If we solve this last equation for p, we get the so-called price function

$$p = p(x) = A - Bx \tag{4}$$

(A and B are also constants).

The quantity to be maximized is profit, given here by the profit function $P(x)$, and which is of course equal to the sales revenue minus the production costs. Thus

$$P = P(x) = xp(x) - C(x). \tag{5}$$

EXAMPLE 3 Suppose that the cost of publishing a small book is $10,000 to set up the (annual) press run, plus $8 for each book actually printed. The publisher sold 7000 copies last year at $13 each, but this year sales dropped to 5000 copies when the price was raised to $15 per copy. Assume that as many as 10,000 copies can be printed in a single press run. How many copies should be printed, and what should be the selling price of each copy, to maximize the year's profit on this book?

Solution The dependent variable to be maximized is the profit P. As independent variable we choose the number x of copies to be printed, with $0 \leqq x \leqq 10{,}000$. The given cost information then implies that

$$C(x) = 10{,}000 + 8x.$$

Now we substitute the data $x = 7000$ when $p = 13$ in Equation (4), as well as the data $x = 5000$ when $p = 15$. We obtain the equations

$$A - 7000B = 13, \qquad A - 5000B = 15.$$

When we solve these equations simultaneously, we find that $A = 20$ and $B = 0.001$. Hence the price function is

$$p = p(x) = 20 - \frac{x}{1000},$$

and thus the profit function is

$$P(x) = x\left(20 - \frac{x}{1000}\right) - (10{,}000 + 8x).$$

We expand and collect terms to obtain

$$P(x) = 12x - \frac{x^2}{1000} - 10{,}000, \qquad 0 \leqq x \leqq 10{,}000.$$

Now

$$\frac{dP}{dx} = 12 - \frac{x}{500},$$

and

$$\frac{dP}{dx} \quad \text{exists for all } x.$$

So the only critical point of the function P occurs when

$$\frac{dP}{dx} = 0;$$

that is, when

$$12 - \frac{x}{500} = 0; \qquad x = (12)(500) = 6000.$$

We check this value of x with the endpoints to find the maximum profit:

$$P(0) = -10{,}000,$$

$$P(6000) = 26{,}000, \longleftarrow \quad \text{absolute maximum}$$

$$P(10{,}000) = 10{,}000.$$

Therefore, the maximum possible annual profit of \$26,000 results from printing 6000 copies of the book. Each copy should be sold for \$14, because

$$p = 20 - \frac{6000}{1000} = 14.$$

EXAMPLE 4 Suppose that a cylindrical can with radius r and height h is to be made. The top and bottom must be of copper, which will cost 2 cents per square inch. The curved side is to be made of aluminum, which will cost 1 cent per square inch. What dimensions will maximize the volume of the can if its total cost is to be 300π cents?

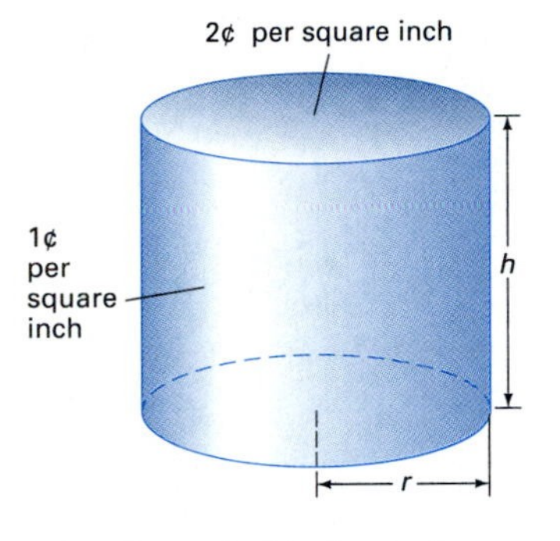

3.22 The cylindrical can of Example 4

Solution We need to maximize the volume V of the can, which we can compute if we know its radius r and its height h, shown in Fig. 3.22. With such dimensions, we find that

$$V = \pi r^2 h, \tag{6}$$

but we need to express V as a function of r alone (or, less likely, as a function of h alone).

Because each has radius r, the circular top and bottom of the can will each have area πr^2 square inches, so the area of copper used will be $2\pi r^2$ and its cost will be $4\pi r^2$ cents. The curved side of the can will have area $2\pi rh$ square inches, so the area of aluminum used will be the same and the cost of the aluminum will be $2\pi rh$ cents.

The total cost of the can is obtained by adding the cost of the copper to the cost of the aluminum, and therefore

$$4\pi r^2 + 2\pi rh = 300\pi. \tag{7}$$

We can eliminate h in (6) by solving Equation (7) for

$$h = \frac{300\pi - 4\pi r^2}{2\pi r} = \frac{1}{r}(150 - 2r^2). \tag{8}$$

Hence

$$V = V(r) = (\pi r^2)\frac{1}{r}(150 - 2r^2) = 2\pi(75r - r^3). \tag{9}$$

To determine the domain of definition of V, we note from (7) that $4\pi r^2 < 300\pi$, so $r < \sqrt{75}$ for an actual can as specified; with $r = \sqrt{75} = 5\sqrt{3}$ we get a degenerate can with height $h = 0$. With $r = 0$ we obtain *no* value of h in (8), and therefore no can, but $V(r)$ is nevertheless continuous at $r = 0$. Consequently, we can take the closed interval $[0, 5\sqrt{3}]$ as the domain of V.

Calculating the derivative, we get

$$V'(r) = 2\pi(75 - 3r^2) = 6\pi(25 - r^2).$$

Because $V(r)$ is a polynomial,

$$V'(r) \qquad \text{exists for all values of } r,$$

so we obtain all critical points by solving the equation

$$V'(r) = 0;$$

that is,

$$6\pi(25 - r^2) = 0.$$

We discard the solution -5 as it does not lie in the domain of V. Thus we obtain only the single critical point $r = 5$ in $(0, 5\sqrt{3})$. Now

$$V(0) = 0,$$

$$V(5) = 500\pi, \quad \longleftarrow \quad \text{absolute maximum}$$

$$V(5\sqrt{3}) = 0.$$

Thus the can of maximum volume has radius $r = 5$ in., and Equation (8) yields its height to be $h = 20$ in.

EXAMPLE 5 (*A Sawmill Problem*) Suppose that you need to cut a beam with maximal rectangular cross section from a circular log of radius 1 ft. (This is the geometric problem of finding the rectangle of greatest area that can be inscribed in a circle of radius 1.) What is the shape and cross-sectional area of such a beam?

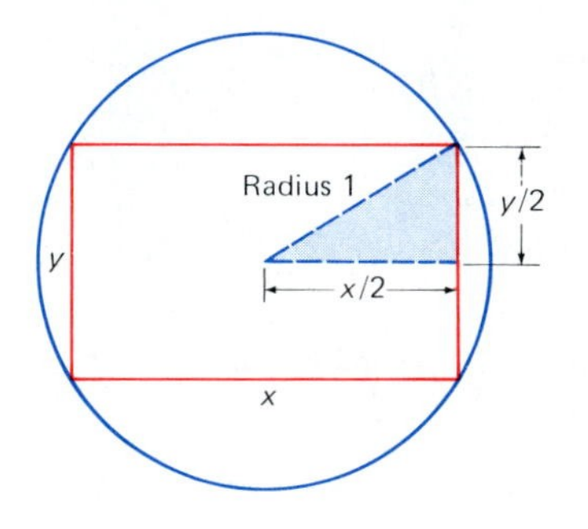

3.23 A sawmill problem—Example 5

Solution Let x and y, as indicated in Fig. 3.23, denote the base and height of the inscribed rectangle, respectively. Apply the Pythagorean theorem to the small shaded right triangle in the figure. This yields the equation

$$y = 2\left[1 - \left(\frac{x}{2}\right)^2\right]^{1/2} = (4 - x^2)^{1/2}.$$

The area of the inscribed rectangle is $A = xy$. And A now may be expressed as a function of x alone:

$$A = A(x) = x(4 - x^2)^{1/2}.$$

The practical domain of definition of A is $(0, 2)$ and there is no harm in adjoining the endpoints, so we take $[0, 2]$ as the domain. Next,

$$\frac{dA}{dx} = 1 \cdot (4 - x^2)^{1/2} + \frac{x}{2}(4 - x^2)^{-1/2}(-2x) = \frac{4 - 2x^2}{(4 - x^2)^{1/2}}.$$

Note that $A'(2)$ does not exist, but this causes no trouble, because differentiability at the endpoints is not assumed in Theorem 3 of Section 3.5. Hence we need only solve the equation

$$A'(x) = 0;$$

that is,

$$\frac{4 - 2x^2}{\sqrt{4 - x^2}} = 0.$$

Now a fraction can be zero only when its numerator is zero and its denominator is *not*, so $A'(x) = 0$ when $4 - 2x^2 = 0$. So the only critical point of

A in the open interval $(0, 2)$ is $x = \sqrt{2}$. We evaluate A here and at the two endpoints to find that

$$A(0) = 0,$$

$$A(\sqrt{2}) = 2, \quad \longleftarrow \quad \text{absolute maximum}$$

$$A(2) = 0.$$

Therefore, the beam with rectangular cross section of maximal area is square, with edge of length $\sqrt{2}$ feet, and with cross-sectional area 2 ft^2.

In Problem 43 at the conclusion of this section we ask you to maximize the cross-sectional area of the four planks that can be cut from the four pieces of log that remain after cutting the square beam (see Fig. 3.24).

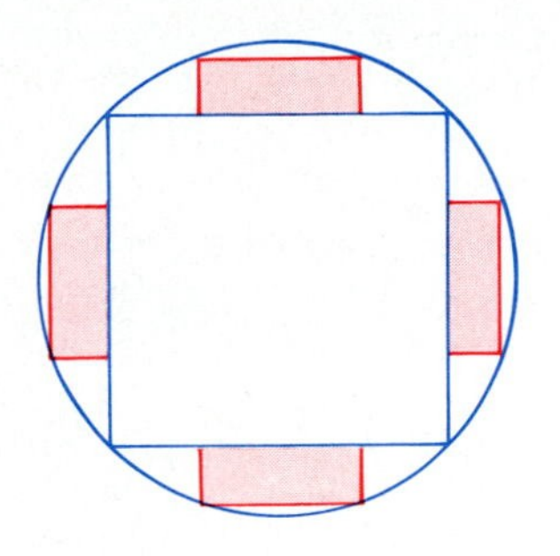

3.24 Cut four more beams after cutting one large beam.

PLAUSIBILITY

You should always check your answers for plausibility. In Example 5, the cross-sectional area of the log from which the beam is to be cut is $\pi \approx 3.14 \text{ ft}^2$. The beam of maximal cross-sectional area 2 ft^2 thus uses a little less than 64% of the log. This *is* plausible. Had the fraction been an extremely inefficient 3% or a wildly optimistic 98%, you should search for an error in arithmetic, algebra, calculus, or logic (as you should if the fraction had been −15% or 150%!). Check the results of Examples 1 through 4 for plausibility now.

DIMENSIONS

Another way to check answers is by dimensional analysis. Work the problem with unspecified constants in place of the numbers actually given. In Example 5, it would be good practice to find instead the beam of maximal rectangular cross section that can be cut from a circular log of radius R. You could always substitute the given value $R = 1$ at the conclusion of the problem. A brief solution to this problem might go as follows:

Dimension of beam: Base x, height y.

Area of beam: $A = xy$.

Draw a diameter of the log that connects the lower left-hand corner of the beam to the upper right-hand corner in Fig. 3.23. This diameter has length $2R$, so by the Pythagorean theorem,

$$x^2 + y^2 = 4R^2; \qquad y = (4R^2 - x^2)^{1/2}.$$

Area of beam:

$$A = A(x) = x(4R^2 - x^2)^{1/2}, \qquad 0 \leqq x \leqq 2R.$$

$$A'(x) = (4R^2 - x^2)^{1/2} + \frac{x}{2}(4R^2 - x^2)^{-1/2}(-2x)$$

$$= \frac{4R^2 - 2x^2}{(4R^2 - x^2)^{1/2}}.$$

$A'(x)$ does not exist when $x = 2R$, but that's an endpoint; we'll check it separately.

$A'(x) = 0$ when $x = R\sqrt{2}$ (ignore the negative root; it's not in the domain of A).

$$A(0) = 0,$$

$$A(R\sqrt{2}) = 2R^2, \quad \longleftarrow \quad \text{absolute maximum}$$

$$A(2) = 0.$$

Now you can check the results for dimensional accuracy. The value of x that maximizes A is a length (R) multiplied by a pure numerical constant ($\sqrt{2}$), so x has the dimensions of length—that's correct; had it been anything else, we should search for the error. Moreover, the maximum cross-sectional area of the beam is $2R^2$, the product of a pure number and the square of a length, therefore having the dimensions of area. This, too, is correct.

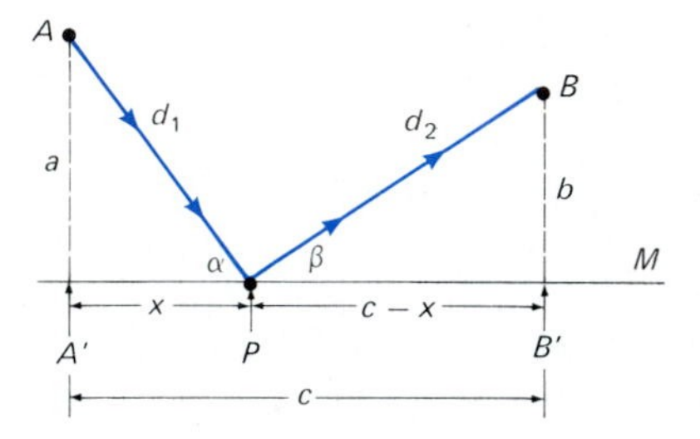

3.25 Reflection at P of a light ray by a mirror M

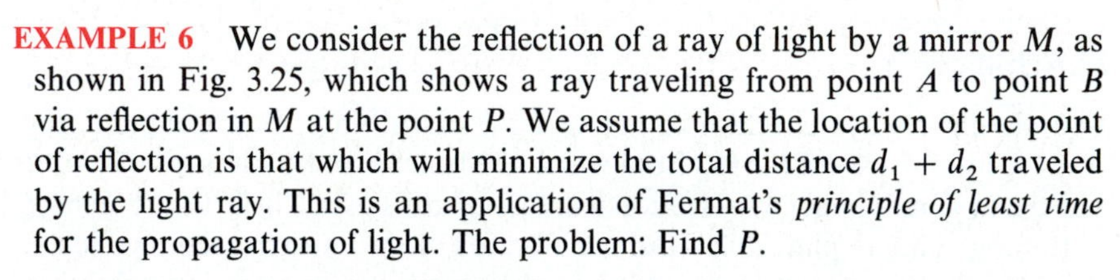

EXAMPLE 6 We consider the reflection of a ray of light by a mirror M, as shown in Fig. 3.25, which shows a ray traveling from point A to point B via reflection in M at the point P. We assume that the location of the point of reflection is that which will minimize the total distance $d_1 + d_2$ traveled by the light ray. This is an application of Fermat's *principle of least time* for the propagation of light. The problem: Find P.

Solution Drop perpendiculars from A and B to the plane of the mirror M. Denote the feet of these perpendiculars by A' and B', as in Fig. 3.25. Let a, b, c, and x denote the lengths of the segments AA', BB', $A'B'$, and $A'P$, respectively. Then $c - x$ is the length of the segment PB', and so—by the Pythagorean theorem—the distance to be minimized is

$$d_1 + d_2 = f(x) = (x^2 + a^2)^{1/2} + ([c - x]^2 + b^2)^{1/2}. \tag{10}$$

We may choose as the domain of f the interval $[0, c]$ because the minimum of f must occur somewhere within that interval. (To see why, examine the picture you get if x is *not* in that interval.)

Then

$$f'(x) = \frac{x}{(x^2 + a^2)^{1/2}} + \frac{(c - x)(-1)}{([c - x]^2 + b^2)^{1/2}}. \tag{11}$$

Because

$$f'(x) = \frac{x}{d_1} - \frac{c - x}{d_2}, \tag{12}$$

we find that any horizontal tangent to the graph of f must occur over the point x determined by the equation

$$\frac{x}{d_1} = \frac{c - x}{d_2}. \tag{13}$$

At such a point, $\cos\alpha = \cos\beta$, where α is the angle of the incident light ray and β is the angle of the reflected ray (see Fig. 3.25 again). But both α and β lie between 0 and $\pi/2$, and thus we find that $\alpha = \beta$. In other words, the angle of incidence is equal to the angle of reflection, a familiar principle from physics.

The computation in Example 6 has an alternative interpretation that is interesting, if somewhat whimsical. Figure 3.26 shows a feedlot 200 ft long

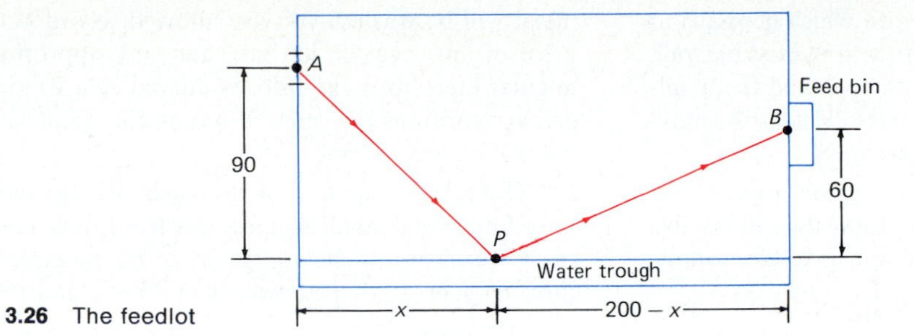

3.26 The feedlot

with a water trough along one edge, and a feed bin located on an adjacent edge. A cow enters the gate at the point A 90 ft from the water trough. She walks straight to point P at the water trough, gets a drink, and then walks straight to the feed bin at B, 60 ft from the trough. Where along the water trough should she select the point P in order to minimize the total distance she travels?

In comparing Figs. 3.25 and 3.26, you see that the cow's problem is to minimize the distance function f in (10) with the numerical values $a = 90$, $b = 60$, and $c = 200$. Substituting these values—together with

$$d_1 = (x^2 + a^2)^{1/2} \quad \text{and} \quad d_2 = [(c - x)^2 + b^2]^{1/2}$$

—in (13), we get

$$\frac{x}{(x^2 + 8100)^{1/2}} = \frac{200 - x}{[(200 - x)^2 + 3600]^{1/2}}.$$

We square both sides, clear the equation of fractions, and simplify; the result is

$$x^2[(200 - x)^2 + 3600] = (200 - x)^2(x^2 + 8100);$$

$$3600x^2 = 8100(200 - x)^2 \qquad \text{(Why?)};$$

$$60x = 90(200 - x);$$

$$150x = 18{,}000;$$

$$x = 120.$$

Thus the cow should proceed directly to the point P located 120 ft down the water trough.

These examples indicate that the closed interval maximum–minimum method is applicable to a wide range of particular problems. Indeed, applied optimization problems that initially seem as different as light rays and cows may turn out to have essentially identical mathematical models. This is only one illustration of the power of generality that calculus exploits so effectively.

3.6 PROBLEMS

1. Find two positive real numbers x and y such that their sum is 50 and their product is as large as possible.

2. Find the maximum possible area of a rectangle of perimeter 200 ft.

3. A rectangle with horizontal and vertical sides has one vertex at the origin, one on the positive x-axis, one on the positive y-axis, and its fourth vertex in the first quadrant on the line with equation $2x + y = 100$. What is the maximum possible area of such a rectangle?

4. A farmer has 600 m of fencing with which he plans to enclose a rectangular pen adjacent to a long existing wall. He will use the wall for one side of the pen and the available fencing for the remaining three sides. What is the maximum area that he can enclose in this way?

5. A rectangular box has a square base with edge at least 1 inch long. It has no top, and the total area of its five sides is 300 in.^{2}. What is the maximum possible volume of such a box?

6. If x is in the interval $[0, 1]$, then $x - x^2$ is not negative. What is the maximum value it can have on that interval?

7. The sum of two positive numbers is 48. What is the smallest possible value of the sum of their squares?

8. A rectangle of fixed perimeter 36 is rotated about one of its sides to generate a right circular cylinder. What is the maximum possible volume of the cylinder?

9. The sum of two nonnegative numbers is 10. Find the minimum possible value of the sum of their cubes.

10. The strength of a rectangular beam is proportional to the product of the width and the square of the height of its cross section. What shape beam should be cut from a cylindrical log of radius r to achieve the greatest possible strength?

11. A farmer has 600 yards of fencing with which to build a rectangular corral having two internal dividers both parallel to two of the sides of the corral. What is the maximum total area of such a corral?

12. Find the maximum possible volume of a right circular cylinder if its total surface area—including both circular ends—is 150π.

13. Find the maximum possible area of a rectangle having a diagonal of length 16.

14. A rectangle has a line of fixed length L reaching from one vertex to the midpoint of one of the far sides. What is the maximum possible area of such a rectangle?

15. The volume of 1 kg of water at a temperature T between 0°C and 30°C is approximately

$$V = 999.87 - (0.06426)T + (0.0085043)T^2 - (0.0000679)T^3$$

cubic centimeters. At what temperature does water have its maximum density?

16. What is the maximum possible area of a rectangle with a base that lies on the x-axis and with two upper vertices that lie on the graph of the equation $y = 4 - x^2$?

17. A rectangular box has a square base whose edge has length at least 1 cm, and its total surface area is 600 cm^2. What is the largest possible volume that such a box can have?

18. From 300π square inches of sheet metal, a cylindrical can with a bottom but no top is to be made. No sheet metal will be wasted; you are allowed to order a circular piece of any size for its base and any appropriate rectangular piece to make into its curved side so long as the given conditions are met. What is the greatest possible volume of such a can?

19. Three large squares of tin, each of edge length 1 m, have four equal small squares cut from their corners. All twelve resulting small squares are to be the same size. The three large cross-shaped pieces are then folded and welded to make boxes with no tops, and the twelve small squares are used to make two cubes. How should this be done to maximize the total volume of all five boxes?

20. Suppose that a rectangular box with a square base is to be made using two different materials. The material for the top and four sides of the box costs \$1/ft^2; the material for its base costs \$2/ft^2. Find the dimensions of such a box of greatest possible volume subject to the condition that \$144 is spent for the material to make it.

21. A piece of wire 80 in. long is cut into two or fewer pieces. Each piece is bent into the shape of a square. How should this be done to minimize the sum of the area of the square(s)? To maximize it?

22. A wire of length 100 cm is to be cut into two pieces. One piece is bent into a circle, the other into a square. How should the cut be made to maximize the sum of the areas of the square and the circle? How should it be done to minimize that sum?

23. A farmer has 600 m of fencing with which she plans to enclose a rectangular pasture adjacent to a long existing wall. She plans to build one fence parallel to the wall, two to form the ends of the enclosure, and a fourth (parallel to the two ends of the enclosure) to divide it equally. What is the maximum area that she can enclose in this way?

24. A rectangular outdoor pen is to be added to an animal house with a corner notch, as shown in Fig. 3.27. If 85 m of new fence is available, what should be the dimensions

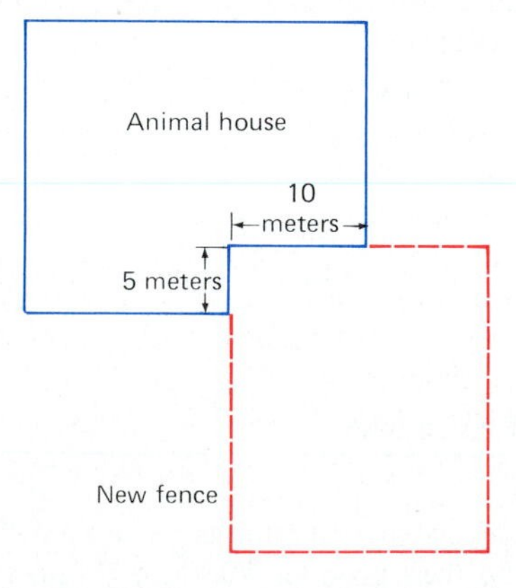

3.27 The rectangular pen of Problem 24

of the pen to maximize its area? Of course, no fence need be used along the walls of the animal house.

25. Suppose that a post office can accept a package for mailing by parcel post only if the sum of its length and its girth (the circumference of its cross section) is at most 100 in. What is the maximum volume of a rectangular box with square cross section that can be mailed?

26. Repeat Problem 25, except use a package that is cylindrical; its cross section is circular.

27. A printing company has eight presses, each of which can print 3600 copies per hour. It costs \$5.00 to set up each press for a run, and $10 + 6n$ dollars to run n presses for one hour. How many presses should be used in order to print 50,000 copies of a poster most profitably?

28. A farmer wants to hire workers to pick 900 bushels of beans. Each worker can pick 5 bushels per hour and is paid \$1.00 per bushel. The farmer must also pay a supervisor \$10 per hour while the picking is in progress, and he has miscellaneous additional expenses of \$8 for each worker hired. How many workers should he hire to minimize the total cost? What will then be the cost per bushel picked?

29. The heating and cooling costs for a certain uninsulated house are \$500 per year, but with $x \leq 10$ in. of insulation the costs would be $1000/(2 + x)$ dollars per year. It will cost \$150 for each inch (thickness) of insulation installed. How many inches of insulation should be installed in order to minimize the *total* (initial plus annual) costs over a 10-year period? What will then be the annual savings resulting from this optimal insulation?

30. A concessionaire had been selling 5000 wieners each game night at 50 cents each. When she raised the price to 70 cents each, sales dropped to 4000 per night. Assume a linear relationship between price and sales. If she has fixed costs of \$1000 per night and each wiener costs her 25 cents, what price will maximize her nightly profit?

31. A commuter train carries 600 passengers each day from a suburb to a city. It costs \$1.50 per person to ride the train. It is found that 40 fewer people will ride the train for each 5-cent increase in the fare, 40 more for each 5-cent decrease. What fare should be charged to make the largest possible revenue?

32. Find the shape of the cylinder of maximum volume that can be inscribed in a sphere of radius R. Show that the ratio of the height of the cylinder to its radius is $\sqrt{2}$, and that the ratio of the volume of the sphere to that of the maximal cylinder is $\sqrt{3}$.

33. Find the dimensions of the right circular cylinder of greatest volume that can be inscribed in a right circular cone of radius R and height H.

34. In a circle of radius 1 is inscribed a trapezoid with the longer of its parallel sides coincident with a diameter of the circle. What is the maximum possible area of such a trapezoid? (*Suggestion:* A positive quantity is maximized when its square is maximized.)

35. Show that the rectangle with maximum perimeter that can be inscribed in a circle is a square.

36. Find the dimensions of the rectangle (with vertical and horizontal sides) of maximum area that can be inscribed in the ellipse with equation $x^2/25 + y^2/9 = 1$. The ellipse is shown in Fig. 3.28.

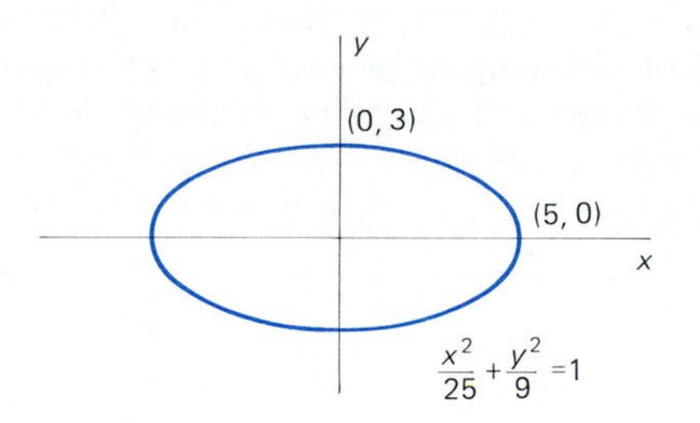

3.28 The ellipse of Problem 34

37. A right circular cone of radius r and height h has slant height $L = (r^2 + h^2)^{1/2}$. What is the maximum possible volume of a cone with slant height 10?

38. Two vertical poles are each 10 ft tall and stand 10 ft apart. Find the length of the shortest rope that can reach from the top of one pole to a point between them on the ground and then to the top of the other pole.

39. The sum of two nonnegative numbers is 16. Find the maximum and the minimum possible value of the sum of their cube roots.

40. A straight wire 60 cm long is bent into the shape of an L. What is the shortest possible distance between the two ends of the bent wire?

41. What is the shortest possible distance from a point of the parabola $y = x^2$ to the point $(0, 1)$?

42. Given: There is exactly one point on the graph of $y = (3x - 4)^{1/3}$ closest to the origin. Find it. (*Suggestion:* Solve the equation you obtain by inspection.)

43. Find the dimensions that maximize the cross-sectional area of the four planks that can be cut from the four pieces of the circular log of Example 5, the pieces that remain after cutting a square beam from the log (see Fig. 3.23).

44. Assume that the strength of a rectangular beam is proportional to the product of its width and the *cube* of its depth. Find the dimensions of the strongest rectangular beam that can be cut from a cylindrical log of radius R.

45. A small island is 2 km offshore in a large lake. A woman on the island can row her boat 10 km/h and can run at a speed of 20 km/h. Where should she land her boat in order to most quickly reach a village on the straight shoreline of the lake and 6 km down the shoreline from the point nearest the island?

46. A factory is located on one bank of a straight river that is 2000 m wide. On the opposite bank but 4500 m downstream is a power station from which the factory must draw its electricity. Assume that it costs three times as much per meter to lay an underwater cable as to lay a cable over the ground. What path should a cable connecting the power station to the factory take to minimize the cost of laying the cable?

47. A company has plants that are located (in an appropriate coordinate system) at the points $A(0, 1)$, $B(0, -1)$, and $C(3, 0)$. The company plans to construct a distribution center at the point $P(x, 0)$. What should be the value of x to minimize the sum of the distances of P from A, B, and C? (The object is to minimize transportation costs from the plants to the center.)

48. Light travels with speed c in air and with a somewhat slower speed v in water. (The constant c is approximately 3×10^{10} cm/s; the ratio $n = c/v$, known as the **index of refraction** of water, depends on the color of the light, but is approximately 1.33.) In Fig. 3.29 we show the path of a light ray traveling from point A in air to point B in water, with what appears to be a sudden change in direction as the ray moves through the air–water interface.

(a) Write the time T required for the ray to travel from A to B in terms of the variable x and the constants a, b, c, s, and v, all of which have been defined or are shown in the figure.

(b) Show that the equation $T'(x) = 0$ for minimizing T is equivalent to the condition

$$\frac{\sin \alpha}{\sin \beta} = \frac{c}{v} = n.$$

This is **Snell's law:** The ratio of the sines of the angles of incidence and refraction is equal to the index of refraction.

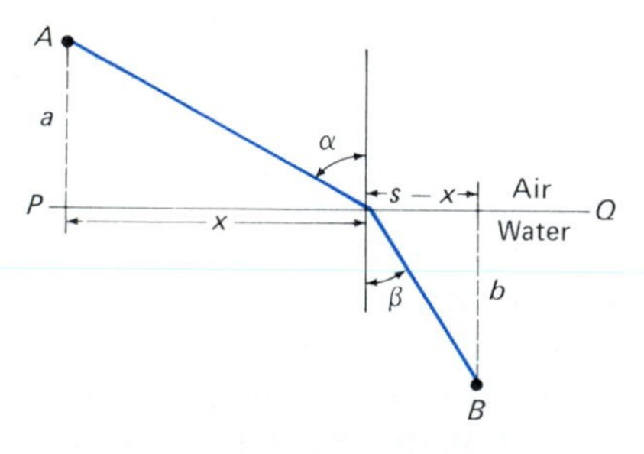

3.29 Snell's law gives the path of refracted light.

49. The mathematics of Snell's law is applicable to situations other than the refraction of light. Figure 3.30 shows a geological fault running from west to east, separating two towns at A and B. Assume that A is a miles north of the fault, that B is b miles south of the fault, and that B is L miles east of A. We want to build a road from A to B. Because of differences in terrain, the cost of construction is C_1 (in millions of dollars per mile) north of the fault and C_2 south of it. Where should the point P be placed to minimize the total cost of road construction?

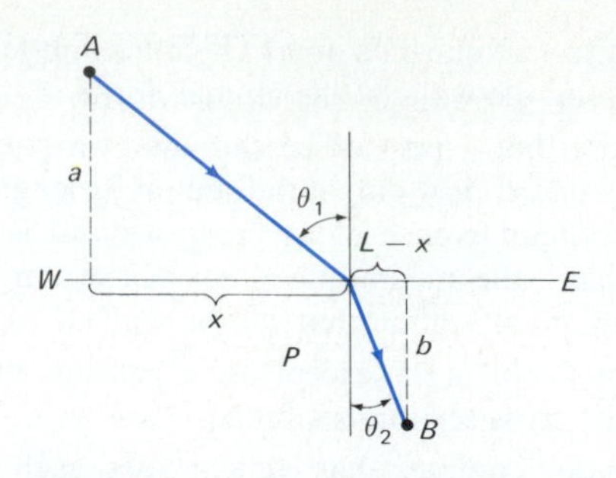

3.30 Building a road from A to B

(a) Using the notation in the figure, show that the cost is minimal when $C_1 \sin \theta_1 = C_2 \sin \theta_2$.

(b) Take the case $a = b = C_1 = 1$, $C_2 = 2$, and $L = 4$. Show that the equation in part (a) is equivalent to

$$f(x) = 3x^4 - 24x^3 + 51x^2 - 32x + 64 = 0.$$

To approximate the desired solution of this equation, calculate $f(0)$, $f(1)$, $f(2)$, $f(3)$, and $f(4)$. You should find that $f(3)$ is positive while $f(4)$ is negative. Interpolate between $f(3)$ and $f(4)$ to approximate the root of this equation.

50. The sum of the surface area of a cube and a sphere is 1000 in.^{2} What should be their dimensions in order to minimize the sum of their volumes? To maximize it?

51. A kite frame is to be made of six pieces of wood as shown in Fig. 3.31. The four outer pieces with the indicated lengths have already been cut. What should be the lengths of the inner struts in order to maximize the area of the kite?

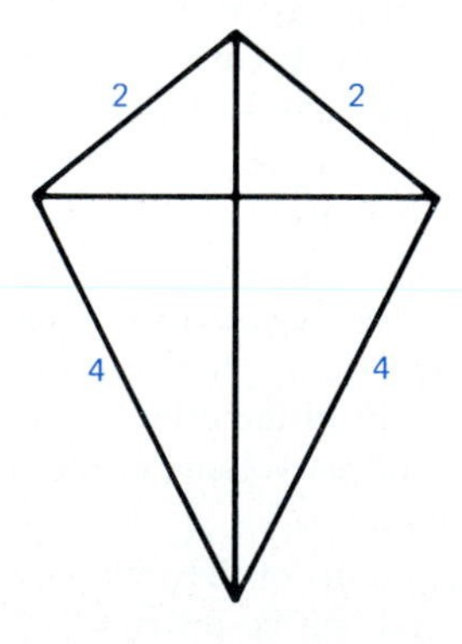

3.31 The kite frame

3.7 Derivatives of Sines and Cosines

In this section we begin our study of the calculus of trigonometric functions, concentrating on the sine and cosine functions. The definitions and the elementary properties of trigonometric functions are reviewed in Appendix A.

When we write $\sin\theta$ (or $\cos\theta$) in calculus, we mean the sine (or cosine) of an angle of θ *radians*. Recall the fundamental relation between radian measure and degree measure of angles:

$$\text{There are } \pi \text{ radians in 180 degrees.} \tag{1}$$

Figure 3.32 shows radian–degree conversions for some frequently occurring angles.

Radians	Degrees
0	0
$\pi/6$	30
$\pi/4$	45
$\pi/3$	60
$\pi/2$	90
$2\pi/3$	120
$3\pi/4$	135
$5\pi/6$	150
π	180
$3\pi/2$	270
2π	360
4π	720

3.32 Some radian–degree conversions

Figure 3.33 shows the geometric meaning of radian measure. An angle of θ radians placed at the center of a circle of radius r subtends a circular arc of length

$$s = r\theta \qquad (\theta \text{ in radians}), \tag{2}$$

and the area of the circular sector bounded by this angle is

$$A = \tfrac{1}{2}r^2\theta \qquad (\theta \text{ in radians}). \tag{3}$$

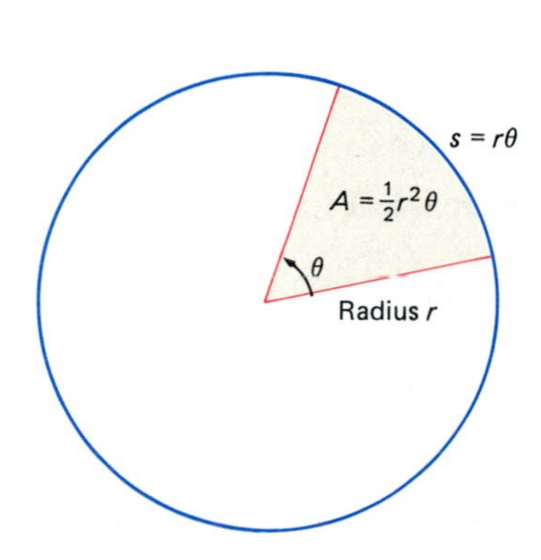

3.33 Area of a sector and arc length for a circle

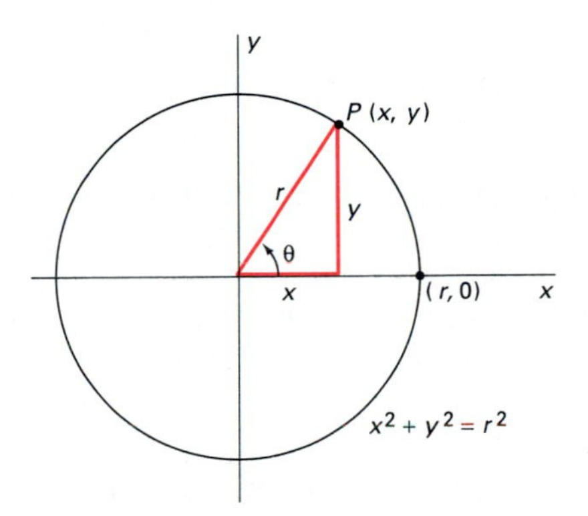

3.34 Definition of $\sin\theta$ and $\cos\theta$

Figure 3.34 shows an angle of θ radians measured counterclockwise from the positive x-axis together with the intersection $P(x, y)$ of its terminal side with the circle of radius r. Recall that the sine and cosine of θ may be defined as follows:

$$\cos\theta = \frac{x}{r}, \qquad \sin\theta = \frac{y}{r}. \tag{4}$$

The derivatives of the sine and cosine functions depend on the limits

$$\lim_{\theta\to 0}\frac{\sin\theta}{\theta} = 1, \qquad \lim_{\theta\to 0}\frac{1-\cos\theta}{\theta} = 0 \tag{5}$$

that we establish at the end of this section.

Theorem 1 *Derivatives of Sines and Cosines*
The functions $f(x) = \sin x$ and $g(x) = \cos x$ are differentiable for all x, and

$$D \sin x = \cos x, \tag{6}$$

$$D \cos x = -\sin x. \tag{7}$$

Proof To differentiate $f(x) = \sin x$, we begin with the definition of the derivative,

$$f'(x) = \lim_{h \to 0} \frac{f(x+h) - f(x)}{h} = \lim_{h \to 0} \frac{\sin(x+h) - \sin x}{h}.$$

Next we apply the addition formula for the sine and the limit laws to get

$$\begin{aligned} f'(x) &= \lim_{h \to 0} \frac{(\sin x \cos h + \sin h \cos x) - \sin x}{h} \\ &= \lim_{h \to 0} \left[(\cos x) \frac{\sin h}{h} - (\sin x) \frac{1 - \cos h}{h} \right] \\ &= (\cos x)\left(\lim_{h \to 0} \frac{\sin h}{h} \right) - (\sin x)\left(\lim_{h \to 0} \frac{1 - \cos h}{h} \right). \end{aligned}$$

The limits in (5) now yield

$$f'(x) = (\cos x)(1) - (\sin x)(0) = \cos x,$$

which proves (6). The proof of (7) is quite similar (see Problem 62). ■

The following four examples illustrate the application of Equations (6) and (7) in conjunction with the general differentiation formulas of Sections 3.2–3.4 to differentiate various combinations of trigonometric functions.

EXAMPLE 1

$$D(x^2 \sin x) = (Dx^2)(\sin x) + (x^2)(D \sin x) = 2x \sin x + x^2 \cos x.$$

EXAMPLE 2

$$\begin{aligned} D(\cos^3 t) &= D[(\cos t)^3] = 3(\cos t)^2 D(\cos t) \\ &= 3(\cos^2 t)(-\sin t) = -3 \cos^2 t \sin t. \end{aligned}$$

EXAMPLE 3 If $y = \dfrac{\cos x}{1 - \sin x}$, then

$$\begin{aligned} \frac{dy}{dx} &= \frac{(D \cos x)(1 - \sin x) - (\cos x)[D(1 - \sin x)]}{(1 - \sin x)^2} \\ &= \frac{(-\sin x)(1 - \sin x) - (\cos x)(-\cos x)}{(1 - \sin x)^2} \\ &= \frac{-\sin x + \sin^2 x + \cos^2 x}{(1 - \sin x)^2} = \frac{-\sin x + 1}{(1 - \sin x)^2}; \end{aligned}$$

$$\frac{dy}{dx} = \frac{1}{1 - \sin x}.$$

EXAMPLE 4 If $g(t) = (2 - 3 \cos t)^{3/2}$, then

$$g'(t) = \tfrac{3}{2}(2 - 3 \cos t)^{1/2} D(2 - 3 \cos t) = \tfrac{3}{2}(2 - 3 \cos t)^{1/2}(3 \sin t);$$

$$g'(t) = \tfrac{9}{2}(2 - 3 \cos t)^{1/2} \sin t.$$

It is easy to differentiate the other four trigonometric functions because each of them may be defined in terms of the sine and cosine functions. For example,

$$\tan x = \frac{\sin x}{\cos x},$$

so

$$\begin{aligned} D \tan x &= \frac{(D \sin x)(\cos x) - (\sin x)(D \cos x)}{(\cos x)^2} \\ &= \frac{(\cos x)(\cos x) - (\sin x)(-\sin x)}{\cos^2 x} \\ &= \frac{\cos^2 x + \sin^2 x}{\cos^2 x} = \frac{1}{\cos^2 x}; \end{aligned}$$

$$D \tan x = \sec^2 x \tag{8}$$

because $\sec x = 1/\cos x$. In a similar way, we can derive the formulas

$$D \cot x = -\csc^2 x, \tag{9}$$

$$D \sec x = \sec x \tan x \tag{10}$$

$$D \csc x = -\csc x \cot x. \tag{11}$$

Although we will have occasion to differentiate these other four trigonometric functions, more extensive discussion of them is deferred to Section 8.2.

By now it is a familiar fact that the chain rule in the form

$$D_x g(u) = g'(u) \frac{du}{dx} \tag{12}$$

provides a chain rule version of each new differentiation formula that we learn. For the sine and cosine functions these chain rule versions are

$$D_x \sin u = (\cos u) \frac{du}{dx} \tag{13}$$

and

$$D_x \cos u = (-\sin u) \frac{du}{dx} \tag{14}$$

where u denotes a differentiable function of x. The cases in which $u = kx$ (where k is a constant) are worth noting specifically:

$$D_x \sin kx = k \cos kx \quad \text{and} \quad D_x \cos kx = -k \sin kx. \tag{15}$$

The formulas in (15) provide an explanation of why radian measure is more natural than degree measure. Because it follows from (1) that an angle of degree measure x has radian measure $\pi x/180$, the "sine of an angle of x degrees" is a *new* and *different* function with formula

$$\sin x^\circ = \sin \frac{\pi x}{180},$$

expressed in terms of the standard (radian measure) sine function on the right-hand side. Hence the first formula in (15) yields

$$D \sin x^\circ = \frac{\pi}{180} \cos \frac{\pi x}{180},$$

so

$$D \sin x^\circ \approx (0.01745) \cos x^\circ.$$

The necessity of using the approximate value 0.01745 here, and indeed its very presence, is one reason why radians instead of degrees are used in the calculus of trigonometric functions.

EXAMPLE 5 If $y = 2 \sin 10t + 3 \cos \pi t$, then

$$\frac{dy}{dt} = 20 \cos 10t - 3\pi \sin \pi t.$$

EXAMPLE 6

$$\begin{aligned} &D(\sin^2 3x \cos^4 5x) \\ &\quad = [D(\sin 3x)^2](\cos^4 5x) + (\sin^2 3x)[D(\cos 5x)^4] \\ &\quad = 2(\sin 3x)(D \sin 3x) \cdot (\cos^4 5x) + (\sin^2 3x) \cdot 4(\cos 5x)^3(D \cos 5x) \\ &\quad = 2(\sin 3x)(3 \cos 3x)(\cos^4 5x) + (\sin^2 3x)(4 \cos^3 5x)(-5 \sin 5x) \\ &\quad = 6 \sin 3x \cos 3x \cos^4 5x - 20 \sin^2 3x \sin 5x \cos^3 5x. \end{aligned}$$

EXAMPLE 7 Differentiate $f(x) = \cos\sqrt{x}$.

Solution If $u = \sqrt{x}$, then $du/dx = 1/(2\sqrt{x})$, so (14) yields

$$\begin{aligned} D_x \cos\sqrt{x} &= D_x \cos u = (-\sin u)\frac{du}{dx} \\ &= -(\sin\sqrt{x})\frac{1}{2\sqrt{x}} = -\frac{\sin\sqrt{x}}{2\sqrt{x}}. \end{aligned}$$

Alternatively, we can carry out this computation without introducing the auxiliary variable u:

$$D \cos\sqrt{x} = (-\sin\sqrt{x})D(\sqrt{x}) = -\frac{\sin\sqrt{x}}{2\sqrt{x}}.$$

EXAMPLE 8 Differentiate

$$y = \sin^2(2x - 1)^{3/2} = [\sin(2x - 1)^{3/2}]^2.$$

Solution Here, $y = u^2$, where $u = \sin(2x - 1)^{3/2}$, so

$$\begin{aligned} \frac{dy}{dx} &= 2u\frac{du}{dx} \\ &= 2[\sin(2x - 1)^{3/2}]D_x[\sin(2x - 1)^{3/2}] \\ &= 2[\sin(2x - 1)^{3/2}][\cos(2x - 1)^{3/2}]D(2x - 1)^{3/2} \\ &= 2[\sin(2x - 1)^{3/2}][\cos(2x - 1)^{3/2}]\tfrac{3}{2}(2x - 1)^{1/2}(2) \\ &= 6(2x - 1)^{1/2}[\sin(2x - 1)^{3/2}][\cos(2x - 1)^{3/2}]. \end{aligned}$$

The following two examples illustrate the applications of trigonometric functions to rate of change and maximum–minimum problems.

EXAMPLE 9 A rocket is launched vertically and is tracked by a radar station located on the ground 5 mi from the launch pad. Suppose that the elevation angle θ of the line of sight to the rocket is increasing at 3° per second when $\theta = 60°$. What is the velocity of the rocket at this instant?

Solution First we convert the given data from degrees into radians. Because there are $\pi/180$ radians in 1 degree, the rate of increase of θ becomes

$$\frac{3\pi}{180} = \frac{\pi}{60}$$

radians per second at the instant when

$$\theta = \frac{60\pi}{180} = \frac{\pi}{3}$$

radians. From Fig. 3.35 we see that the height y (in miles) of the rocket is

$$y = 5 \tan\theta = \frac{5 \sin\theta}{\cos\theta}.$$

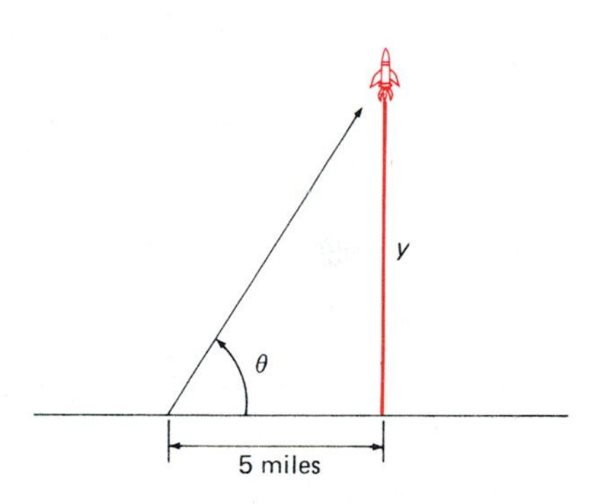

3.35 Tracking an ascending rocket

Hence its velocity is

$$\begin{aligned}\frac{dy}{dt} &= \frac{dy}{d\theta}\cdot\frac{d\theta}{dt}\\ &= 5\,\frac{(\cos\theta)(\cos\theta) - (\sin\theta)(-\sin\theta)}{\cos^2\theta}\cdot\frac{d\theta}{dt}\\ &= 5(\sec^2\theta)\,\frac{d\theta}{dt}.\end{aligned}$$

Because $\sec(\pi/3) = 2$, the velocity of the rocket is

$$\frac{dy}{dt} = (5)(2^2)\left(\frac{\pi}{60}\right) = \frac{\pi}{3}$$

miles per second, about 3770 mi/h, at the instant when $\theta = 60°$.

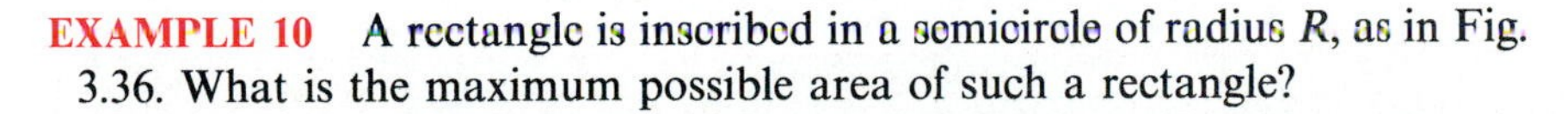

EXAMPLE 10 A rectangle is inscribed in a semicircle of radius R, as in Fig. 3.36. What is the maximum possible area of such a rectangle?

Solution If we denote the length of *half* the base of the rectangle by x and its height by y, then its area is $A = 2xy$. We see in Fig. 3.36 that the right triangle has hypotenuse R, the radius of the circle. So

$$x = R\cos\theta \quad \text{and} \quad y = R\sin\theta. \tag{16}$$

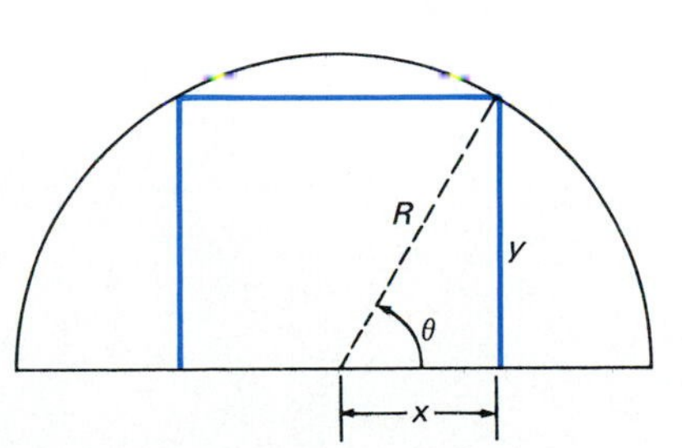

3.36 The rectangle of Example 10

Each value of θ between 0 and $\pi/2$ corresponds to a possible inscribed rectangle; the values $\theta = 0$ and $\theta = \pi/2$ yield degenerate rectangles.

We substitute the data in (16) in the formula $A = 2xy$ to obtain the area

$$A = A(\theta) = 2(R\cos\theta)(R\sin\theta) = 2R^2\cos\theta\sin\theta \tag{17}$$

as a function of θ on the closed interval $[0, \pi/2]$. To find the critical points, we differentiate:

$$\frac{dA}{d\theta} = 2R^2(-\sin\theta\sin\theta + \cos\theta\cos\theta)$$
$$= 2R^2(\cos^2\theta - \sin^2\theta). \tag{18}$$

Because $dA/d\theta$ always exists, we have critical points only if

$$\cos^2\theta - \sin^2\theta = 0; \qquad \sin^2\theta = \cos^2\theta; \qquad \tan^2\theta = 1; \qquad \tan\theta = \pm 1.$$

The only value of θ in $[0, \pi/2]$ such that $\tan\theta = \pm 1$ is $\theta = \pi/4$.

Upon evaluation of $A(\theta)$ at each of the possibilities $\theta = 0, \pi/4, \pi/2$ (the endpoints and the critical point), we find that

$$A(0) = 0,$$
$$A\left(\frac{\pi}{4}\right) = 2R^2\left(\frac{1}{\sqrt{2}}\right)\left(\frac{1}{\sqrt{2}}\right) = R^2, \longleftarrow \quad \text{absolute maximum}$$
$$A\left(\frac{\pi}{2}\right) = 0.$$

Thus the area of the largest inscribed rectangle is R^2; its dimensions are $2x = R\sqrt{2}$ and $y = R/\sqrt{2}$.

TRIGONOMETRIC LIMITS

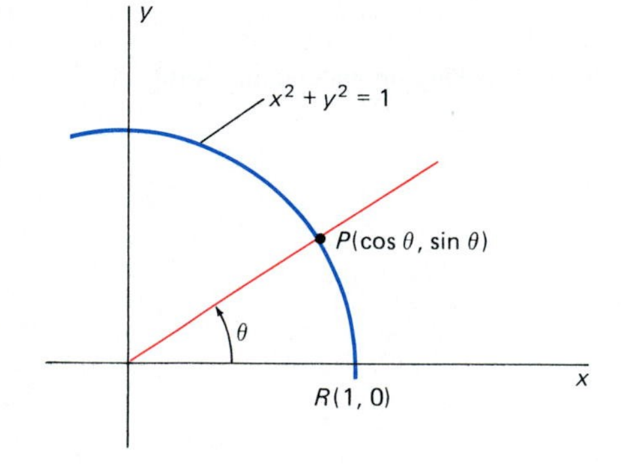

3.37 An angle θ

We want to show first that the sine and cosine functions are continuous. We begin with Fig. 3.37, which shows an angle θ with its vertex at the origin, its initial side along the positive x-axis, and its terminal side intersecting the unit circle at the point P. By the definition of the sine and cosine functions, the coordinates of P are $P(\cos\theta, \sin\theta)$. It is geometrically obvious that, as $\theta \to 0$, the point $P(\cos\theta, \sin\theta)$ approaches the point $R(1, 0)$. Hence $\cos\theta \to 1$ and $\sin\theta \to 0$ as $\theta \to 0^+$. A similar picture gives the same results as $\theta \to 0^-$, so we see that

$$\lim_{\theta\to 0} \cos\theta = 1 \quad \text{and} \quad \lim_{\theta\to 0} \sin\theta = 0. \tag{19}$$

But $\cos\theta = 1$ and $\sin\theta = 0$, so these limits imply that the functions $\cos\theta$ and $\sin\theta$ are, by definition, continuous at the point $\theta = 0$. We can use this fact to prove that they are continuous everywhere.

> ***Theorem 2*** *Continuity of Sine and Cosine*
> The functions $f(x) = \sin x$ and $g(x) = \cos x$ are continuous functions of x on the whole real line.

Proof We give the proof only for $\sin x$; the proof for $\cos x$ is similar. We want to show that

$$\lim_{x\to a} \sin x = \sin a$$

for every real number a. So let a be a fixed but arbitrary real number. If we write $x = a + h$, so that $h = x - a$, then $h \to 0$ as $x \to a$, so it will be sufficient to show that

$$\lim_{h\to 0} \sin(a+h) = \sin a.$$

But the addition formula for the sine function yields

$$\lim_{h\to 0} \sin(a+h) = \lim_{h\to 0} (\sin a \cos h + \cos a \sin h)$$

$$= (\sin a)\left(\lim_{h\to 0} \cos h\right) + (\cos a)\left(\lim_{h\to 0} \sin h\right) = \sin a$$

as desired; we used the limits in (19) in the last step. ■

The limit of $(\sin\theta)/\theta$ as $\theta \to 0$ plays a special role in the calculus of trigonometric functions. A calculator set in *radian mode* provides us with the numerical evidence shown in Fig. 3.38. This table strongly suggests that the limit of $(\sin\theta)/\theta$ is 1 as $\theta \to 0$. The next theorem replaces this suggestion with certainty.

θ	$\dfrac{\sin\theta}{\theta}$	(Values rounded)
1.0	0.84147	
0.1	0.99833	
0.01	0.99998	
0.001	1.00000	
0.0001	1.00000	

3.38 The numerical data suggest that $\lim_{\theta\to 0} \dfrac{\sin\theta}{\theta} = 1$.

Theorem 3 *The Basic Trigonometric Limit*

$$\lim_{\theta\to 0} \frac{\sin\theta}{\theta} = 1. \tag{20}$$

Proof Figure 3.39 shows the angle θ, the triangles OPQ and ORS, and the circular sector OPR that contains the triangle OPQ and is contained in the triangle ORS. Hence

$$\text{area}(\triangle OPQ) < \text{area}(\text{sector } OPR) < \text{area}(\triangle ORS).$$

In terms of θ, this means that

$$\frac{1}{2}\sin\theta\cos\theta < \frac{1}{2}\theta < \frac{1}{2}\tan\theta = \frac{\sin\theta}{2\cos\theta}.$$

Here we use the standard formula for the area of a triangle to obtain the area of OPQ and ORS. We also use the fact that the area of a circular sector

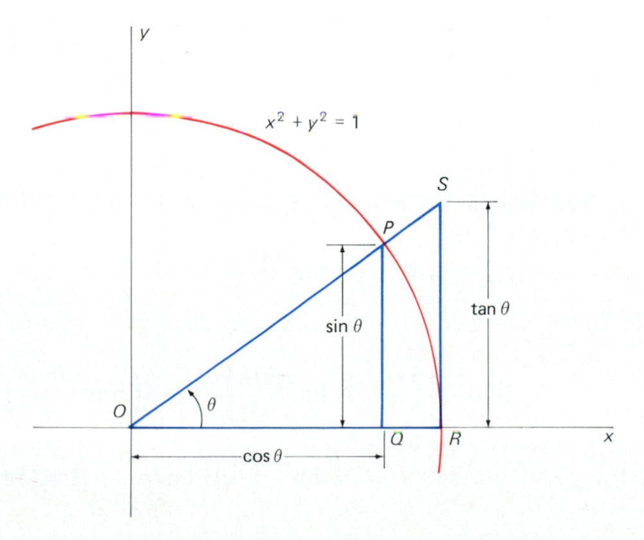

3.39 Aid to the proof of Theorem 3

in a circle of radius r is $A = \frac{1}{2}r^2\theta$ if the sector is subtended by an angle of θ radians (in this special case we have $r = 1$).

Now we let $\theta \to 0$, and consider first the case in which θ approaches zero through positive values. Hence we may suppose that $0 < \theta < \pi/2$. We divide each member of the inequality above by the positive number $\frac{1}{2}\sin\theta$ to obtain

$$\cos\theta < \frac{\theta}{\sin\theta} < \frac{1}{\cos\theta}.$$

We take reciprocals, which reverses the inequalities, and thus

$$\cos\theta < \frac{\sin\theta}{\theta} < \frac{1}{\cos\theta}.$$

Now we apply the squeeze law of limits with

$$f(\theta) = \cos\theta, \qquad g(\theta) = \frac{\sin\theta}{\theta}, \quad \text{and} \quad h(\theta) = \frac{1}{\cos\theta}.$$

Because it is clear from Equation (19) that $f(\theta)$ and $h(\theta)$ both approach 1 as $\theta \to 0^+$, so does $g(\theta) = (\sin\theta)/\theta$. This geometric argument shows that $(\sin\theta)/\theta \to 1$ for *positive* values of θ approaching zero. But the same result follows for negative values of θ because $\sin(-\theta) = -\sin\theta$. So we have proved (20). ■

As in the following two examples, many other trigonometric limits can be reduced to the one in Theorem 3.

EXAMPLE 11 Show that

$$\lim_{\theta\to 0} \frac{1-\cos\theta}{\theta} = 0. \tag{21}$$

Solution

$$\begin{aligned}
\lim_{\theta\to 0} \frac{1-\cos\theta}{\theta} &= \lim_{\theta\to 0} \frac{1-\cos\theta}{\theta}\cdot\frac{1+\cos\theta}{1+\cos\theta} \\
&= \lim_{\theta\to 0} \frac{\sin^2\theta}{\theta(1+\cos\theta)} \\
&= \left(\lim_{\theta\to 0} \frac{\sin\theta}{\theta}\right)\left(\lim_{\theta\to 0} \frac{\sin\theta}{1+\cos\theta}\right) \\
&= (1)\left(\frac{0}{1+1}\right) = 0.
\end{aligned}$$

Note that in the last step we used all the limits in Equations (19) and (20).

EXAMPLE 12 Evaluate $\displaystyle\lim_{x\to 0} \frac{\tan 3x}{x}$.

Solution

$$\begin{aligned}
\lim_{x\to 0} \frac{\tan 3x}{x} &= 3\left(\lim_{x\to 0} \frac{\tan 3x}{3x}\right) = 3\left(\lim_{u\to 0} \frac{\tan u}{u}\right) \qquad (u = 3x) \\
&= 3\left(\lim_{u\to 0} \frac{\sin u}{u}\right)\left(\lim_{u\to 0} \frac{1}{\cos u}\right) = (3)(1)\left(\frac{1}{1}\right) = 3.
\end{aligned}$$

We used the fact that $\tan u = (\sin u)/(\cos u)$ as well as some of the limits in (19) and (20).

EXAMPLE 13 Figure 3.40 shows the graph of the function f given by

$$f(x) = \begin{cases} x \sin \dfrac{1}{x} & \text{if } x \neq 0; \\ 0 & \text{if } x = 0. \end{cases}$$

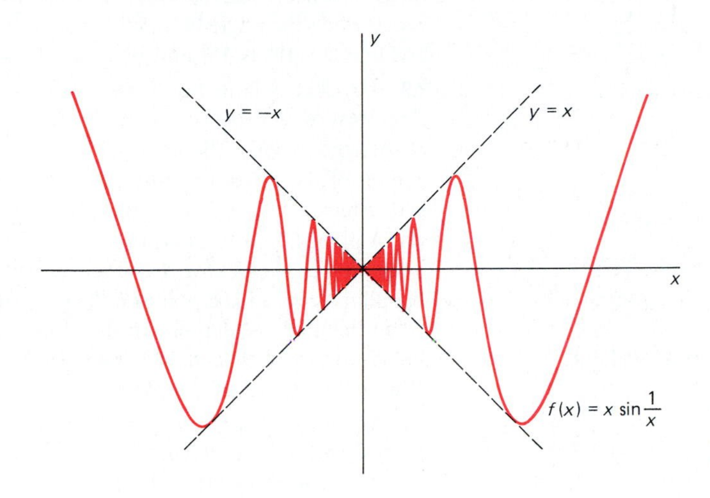

3.40 The graph of an interesting function

As $x \to 0$, $1/x$ increases without bound, so because of the periodicity of the sine function, $f(x)$ oscillates infinitely often between the values $-x$ and x. Because $|\sin(1/x)| \leqq 1$ for all x, we see that

$$-|x| \leqq x \sin \frac{1}{x} \leqq |x|$$

for all x. Moreover, $|x| \to 0$ as $x \to 0$, so it follows from the squeeze law of limits that

$$\lim_{x \to 0} x \sin \frac{1}{x} = 0.$$

Thus f is continuous at $x = 0$. In Problem 77 we ask you to show that f is *not* differentiable at $x = 0$.

3.7 PROBLEMS

Differentiate the functions given in Problems 1–20.

1. $f(x) = 3 \sin^2 x$

2. $f(x) = 2 \cos^4 x$

3. $f(x) = x \cos x$

4. $f(x) = \sqrt{x} \sin x$

5. $f(x) = \dfrac{\sin x}{x}$

6. $f(x) = \dfrac{\cos x}{\sqrt{x}}$

7. $f(x) = \sin x \cos^2 x$

8. $f(x) = \cos^3 x \sin^2 x$

9. $g(t) = (1 + \sin t)^4$

10. $g(t) = (2 - \cos^2 t)^3$

11. $g(t) = \dfrac{1}{\sin t + \cos t}$

12. $g(t) = \dfrac{\sin t}{1 + \cos t}$

13. $f(x) = 2x \sin x - 3x^2 \cos x$

14. $f(x) = x^{1/2}\cos x - x^{-1/2}\sin x$

15. $f(x) = \cos 2x \sin 3x$ **16.** $f(x) = \cos 5x \sin 7x$

17. $g(t) = t^3 \sin^2 2t$ **18.** $g(t) = \sqrt{t}\cos^3 3t$

19. $g(t) = (\cos 3t + \cos 5t)^{5/2}$

20. $g(t) = \dfrac{1}{\sqrt{\sin^2 t + \sin^2 3t}}$

Find dy/dx in Problems 21–40.

21. $y = \sin^2(\sqrt{x})$ **22.** $y = \dfrac{\cos 2x}{x}$

23. $y = x^2\cos(3x^2 - 1)$ **24.** $y = \sin^3(x^4)$

25. $y = (\sin 2x)(\cos 3x)$ **26.** $y = \dfrac{x}{\sin 3x}$

27. $y = \dfrac{\cos 3x}{\sin 5x}$ **28.** $y = \sqrt{\cos\sqrt{x}}$

29. $y = \sin^2(x^2)$ **30.** $y = \cos^3(x^3)$

31. $y = \sin 2\sqrt{x}$ **32.** $y = \cos 3\sqrt[3]{x}$

33. $y = x\sin x^2$ **34.** $y = x^2\cos\left(\dfrac{1}{x}\right)$

35. $y = \sqrt{x}\sin\sqrt{x}$ **36.** $y = (\sin x - \cos x)^2$

37. $y = \sqrt{x}(x - \cos x)^3$ **38.** $y = \sqrt{x}\sin\sqrt{x + \sqrt{x}}$

39. $y = \cos(\sin x^2)$ **40.** $y = \sin(1 + \sqrt{\sin x})$

Find the limits in Problems 41–60.

41. $\lim_{\theta\to 0}\dfrac{\theta^2}{\sin\theta}$ **42.** $\lim_{\theta\to 0}\dfrac{\sin^2\theta}{\theta^2}$

43. $\lim_{\theta\to 0}\dfrac{1 - \cos\theta}{\theta^2}$ **44.** $\lim_{\theta\to 0}\dfrac{\tan\theta}{\theta}$

45. $\lim_{x\to 0}\dfrac{2x}{(\sin x) - x}$ **46.** $\lim_{\theta\to 0}\dfrac{\sin(2\theta^2)}{\theta^2}$

47. $\lim_{x\to 0}\dfrac{\sin 5x}{x}$ **48.** $\lim_{x\to 0}\dfrac{\sin 2x}{x\cos 3x}$

49. $\lim_{x\to 0}\dfrac{\sin x}{\sqrt{x}}$ **50.** $\lim_{x\to 0}\dfrac{1 - \cos 2x}{x}$

51. $\lim_{x\to 0}\dfrac{1}{x}\sin\dfrac{x}{3}$ **52.** $\lim_{x\to 0}\dfrac{(\sin 3x)^2}{x^2\cos x}$

53. $\lim_{x\to 0}\dfrac{1 - \cos x}{\sin x}$ **54.** $\lim_{x\to 0}\dfrac{\tan 3x}{\tan 5x}$

55. $\lim_{x\to 0} x\sec x\csc x$ **56.** $\lim_{\theta\to 0}\dfrac{\sin 2\theta}{\theta}$

57. $\lim_{\theta\to 0}\dfrac{1 - \cos\theta}{\theta\sin\theta}$ **58.** $\lim_{\theta\to 0}\dfrac{\sin^2\theta}{\theta}$

59. $\lim_{x\to 0} x\cot x$ **60.** $\lim_{x\to 0}\dfrac{\tan 2x}{3x}$

61. Derive the differentiation formulas in (9)–(11).

62. Use the definition of the derivative to show directly that $g'(x) = -\sin x$ if $g(x) = \cos x$.

63. If a projectile is fired from ground level with initial velocity v_0 and inclination angle α, then its horizontal range (ignoring air resistance) is

$$R = \tfrac{1}{16}v_0^2 \sin\alpha\cos\alpha.$$

What value of α maximizes R?

64. A weather balloon is rising vertically and is being observed from a point on the ground 300 ft from the spot directly beneath the balloon. At what rate is the balloon rising when the angle between the ground and the observer's line of sight is 45° and is increasing at 1° per second?

65. A rocket is launched vertically upward from a point 2 mi west of an observer on the ground. What is the speed of the rocket when the angle of elevation (from the horizontal) of the observer's line of sight to the rocket is 50° and is increasing at 5° per second?

66. A plane flying at an altitude of 25,000 ft has an inoperative airspeed indicator. In order to determine his speed, the pilot sights a fixed point on the ground. At the moment when the angle of depression (from the horizontal) is 65°, he observes that this angle is increasing at 1.5° per second. What is the speed of the plane?

67. An observer on the ground sights an approaching plane flying at constant speed and at an altitude of 20,000 ft. From her point of view, the plane's angle of elevation is increasing at 0.5° per second when the angle is 60°. What is the plane's speed?

68. Find the largest possible area A of a rectangle inscribed in the unit circle by maximizing A as a function of the angle θ indicated in Fig. 3.41.

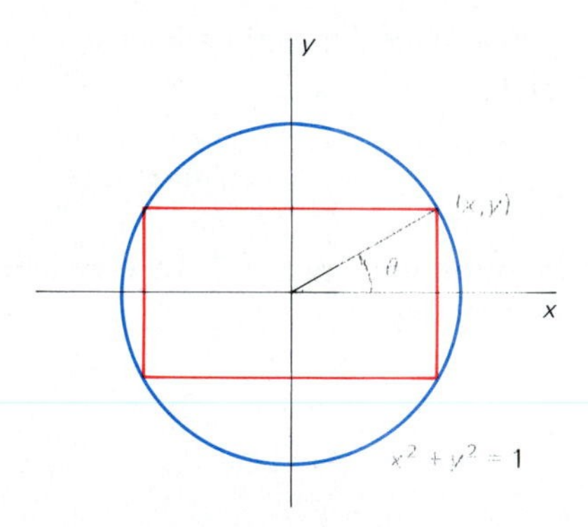

3.41 A rectangle inscribed in the unit circle

69. A water trough is to be made from a long strip of tin 6 ft wide by bending up at an angle θ a 2-ft strip on each side, as shown in Fig. 3.42. What should be the value of the angle θ to maximize the cross-sectional area, and thus the volume, of the trough?

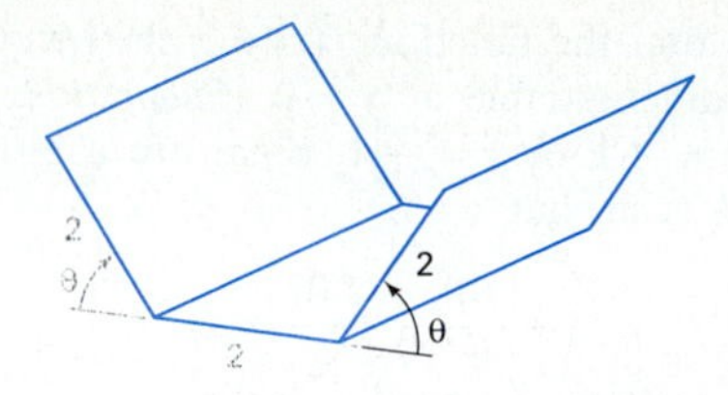

3.42 The water trough of Problem 69

70. A circular area of radius 20 m is surrounded by a walkway, and a light is placed atop a lamp post at the center. At what height should the light be placed to illuminate the walkway most strongly? The intensity of illumination I of a surface is given by $I = (k \sin \theta)/D^2$, where D is the distance from the light source to the surface, θ is the angle at which light strikes the surface, and k is a positive constant.

71. Find the minimum possible volume V of a cone in which a sphere of given radius R is inscribed. Minimize V as a function of the angle θ indicated in Fig. 3.43.

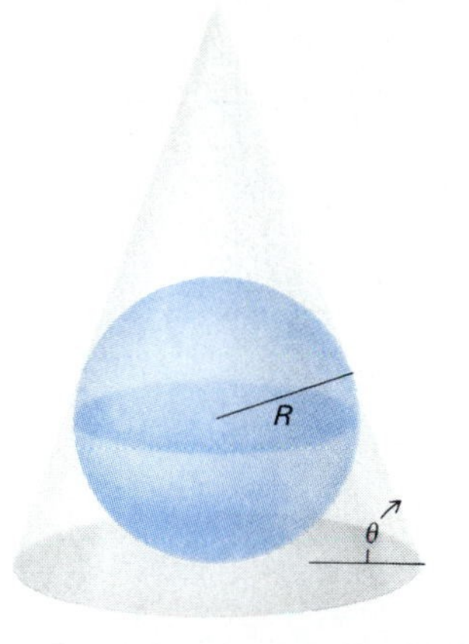

3.43 Finding the smallest cone containing a fixed sphere

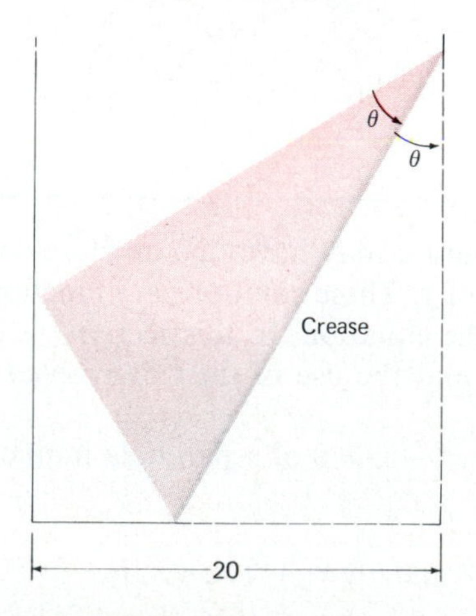

3.44 Fold a piece of paper; make the crease of minimal length.

72. A very long rectangular piece of paper is 20 cm wide. The bottom right-hand corner is folded along the crease shown as a heavy line in Fig. 3.44, so that the corner just touches the left-hand side of the page. How should this be done so that the crease is as short as possible?

73. Find the maximum possible area A of a trapezoid inscribed in a semicircle of radius 1, as shown in Fig. 3.45. Begin by expressing A as a function of θ.

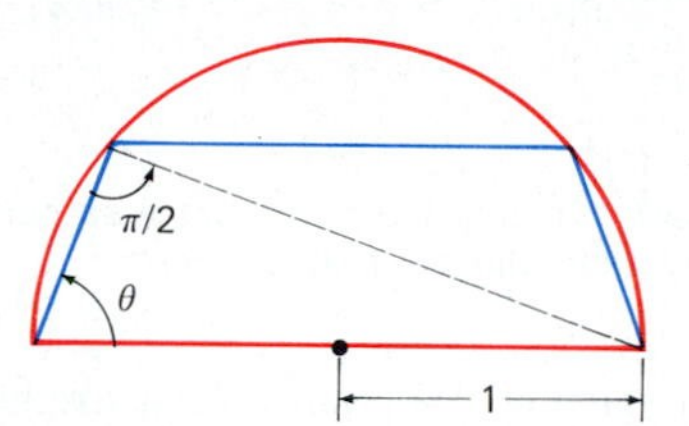

3.45 A trapezoid inscribed in a circle

74. A six-sided beam is to be cut from a circular log with diameter 30 cm, so that its cross section is as shown in Fig. 3.46; the beam is highly symmetrical, with only two different internal angles α and β. Show that the area of the cross section is maximal when it is a regular hexagon, with equal sides and angles (corresponding to $\alpha = \beta = 2\pi/3$). Note that $\alpha + 2\beta = 2\pi$. (Why?)

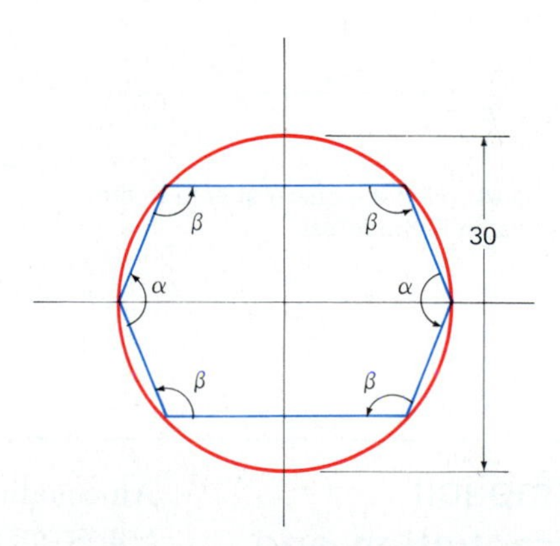

3.46 A hexagonal beam cut from a circular log

75. Consider a circular arc of given length s and with its end points on the x-axis. Show that the area A bounded by this arc and the x-axis is maximal when the circular arc is in the shape of a semicircle. (*Suggestion:* Express A in terms of the angle θ subtended by the arc at the center of the circle, as shown in Fig. 3.47. Show that A is maximal when $\theta = \pi$.)

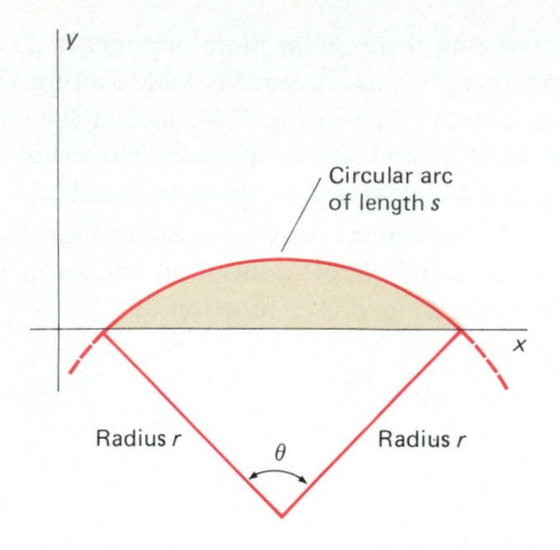

3.47 Finding the maximum area bounded by a circular arc and its chord

76. A hiker starting at a point P on a straight road wants to reach a forest cabin that is 2 km from the point Q that itself is 3 km down the road from P, as shown in Fig. 3.48. She can walk 8 km/h along the road but only 3 km/h through the forest. She wants to minimize the time required to reach the cabin. How far down the road should she walk first before setting off through the forest straight for the cabin? (*Suggestion:* Use the angle between the road and the path she takes through the forest as the independent variable.)

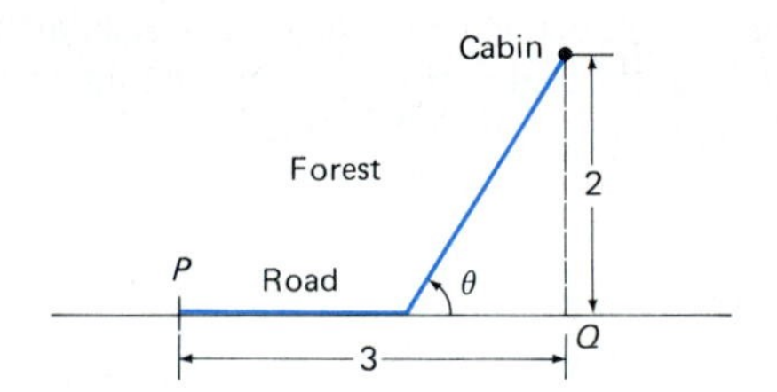

3.48 Find the quickest path to the cabin in the forest.

77. Prove that the function $f(x) = x\sin(1/x)$ of Example 13 is *not* differentiable at $x = 0$. (*Suggestion:* Show that whether $z = +1$ or $z = -1$, there are arbitrarily small values of h such that

$$\frac{f(h) - f(0)}{h} = z.)$$

78. Let $f(x) = x^2\sin(1/x)$ for $x \neq 0$; $f(0) = 0$ (the graph of f is shown in Fig. 3.49). Apply the definition of the derivative to prove that f is differentiable at $x = 0$ and that $f'(0) = 0$.

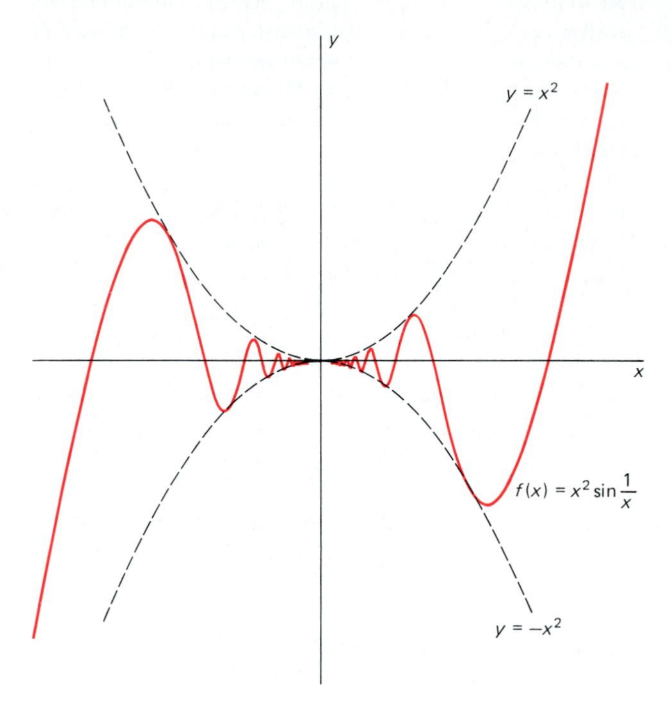

3.49 Graph of the function of Problem 78

3.8 Implicit Differentiation and Related Rates

An equation in two variables x and y may have one or more solutions for y in terms of x or for x in terms of y. These solutions are functions that are said to be **implicitly defined** by the equation. In this section we discuss the differentiation of such functions and the use of their derivatives in solving certain rate of change problems.

For example, the equation $y^2 - x = 0$ of a parabola implicitly defines two continuous functions of x:

$$y = \sqrt{x} \quad \text{and} \quad y = -\sqrt{x}.$$

Each has domain the half line $x \geqq 0$. The graphs of these two functions are the upper and lower branches of the parabola shown in Fig. 3.50. The whole parabola cannot be the graph of a function of x because no vertical line can meet the graph of a function in more than one point.

The equation of the unit circle, $x^2 + y^2 = 1$, implicitly defines four functions (among others):

$$y = +\sqrt{1 - x^2} \quad \text{for} \quad x \text{ in } [-1, 1],$$
$$y = -\sqrt{1 - x^2} \quad \text{for} \quad x \text{ in } [-1, 1],$$
$$x = +\sqrt{1 - y^2} \quad \text{for} \quad y \text{ in } [-1, 1], \quad \text{and}$$
$$x = -\sqrt{1 - y^2} \quad \text{for} \quad y \text{ in } [-1, 1].$$

The four graphs are highlighted against the four unit circles in Fig. 3.51.

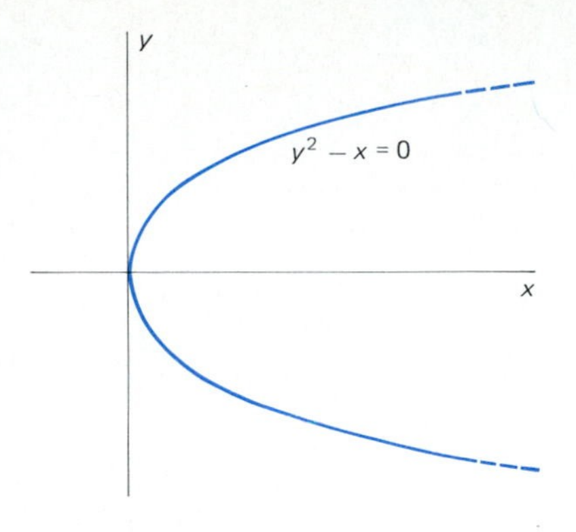

3.50 The parabola $y^2 - x = 0$

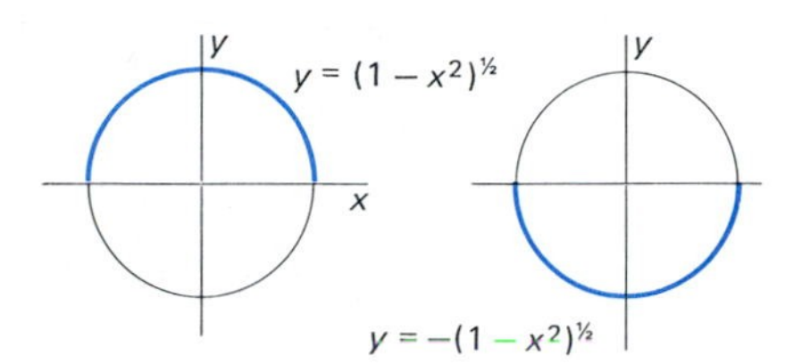

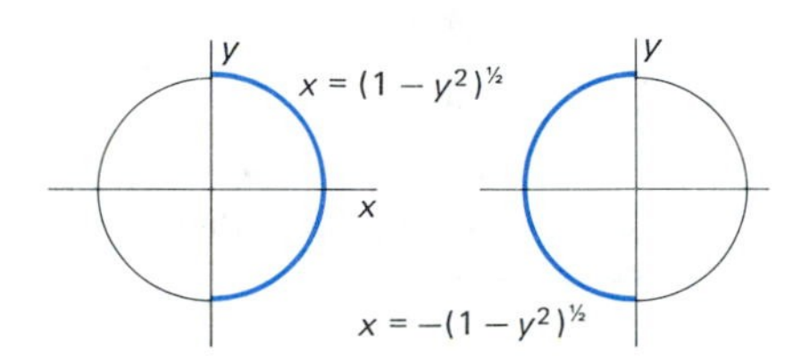

3.51 Continuous functions defined implicitly by $x^2 + y^2 = 1$

The equation $2x^2 + 2y^2 + 3 = 0$ implicitly defines *no* function, because this equation has no real solution (x, y).

In advanced calculus one studies conditions that will guarantee when an implicitly defined function is actually differentiable. Here we will proceed on the assumption that our implicitly defined functions are differentiable at most points in their domains. (The functions with the graphs shown in Fig. 3.51 are *not* differentiable at the endpoints of their domains.) This is a reasonable assumption whenever the equation we are working with is "nice," such as a polynomial equation in two variables.

When we assume differentiability, we can use the chain rule to differentiate the given equation, thinking of x as the independent variable. We can then solve the resulting equation for the derivative $dy/dx = f'(x)$ of the implicitly defined function f. This process is called **implicit differentiation.**

EXAMPLE 1 Use implicit differentiation to find the derivative of a differentiable function $y = f(x)$ implicitly defined by the equation

$$x^2 + y^2 = 1.$$

Solution The equation $x^2 + y^2 = 1$ is to be thought of as an *identity* that implicitly defines y as a function of x. Because $x^2 + y^2$ is then a function of x, it has the same derivative as the constant function 1 on the other side of the identity. Thus we may differentiate both sides of $x^2 + y^2 = 1$ with respect to x and equate the results. We obtain

$$2x + 2y\frac{dy}{dx} = 0.$$

In this step, it is essential to remember that y is a function of x, so that $D_x(y^2) = 2yD_xy$.

Then we solve for

$$\frac{dy}{dx} = -\frac{x}{y}. \tag{1}$$

It may be surprising to see a formula for dy/dx containing both x and y, but such a formula can be just as useful as one containing only x. For example, the formula in (1) tells us that the slope of the tangent line to the circle $x^2 + y^2 = 1$ at the point $(\frac{3}{5}, \frac{4}{5})$ is

$$\left.\frac{dy}{dx}\right|_{(0.6,0.8)} = -\frac{0.6}{0.8} = -0.75.$$

Note that if $y = \pm\sqrt{1 - x^2}$, then

$$\frac{dy}{dx} = \frac{-x}{\pm\sqrt{1 - x^2}} = -\frac{x}{y},$$

in agreement with Equation (1). Thus (1) simultaneously gives us the derivatives of both the functions $y = +\sqrt{1 - x^2}$ and $y = -\sqrt{1 - x^2}$ defined implicitly by the equation $x^2 + y^2 = 1$.

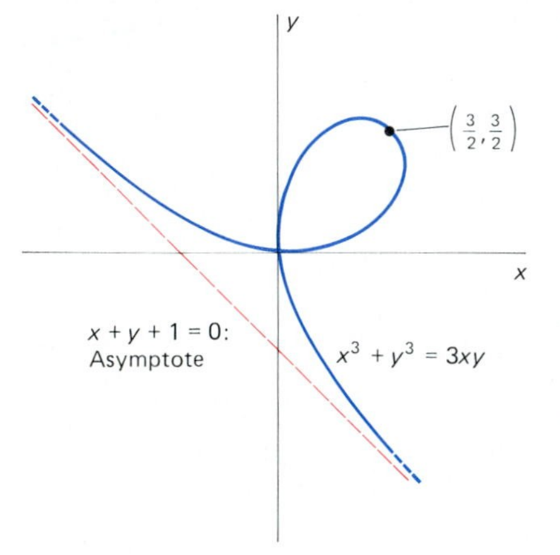

3.52 The folium of Descartes; see Example 2.

EXAMPLE 2 The *folium of Descartes* is the graph of the equation

$$x^3 + y^3 = 3xy. \tag{2}$$

This curve was originally proposed by René Descartes as a challenge to Pierre Fermat (1601–1665) to find its tangent line at an arbitrary point. The graph of the folium of Descartes appears in Fig. 3.52; it has the line $x + y + 1 = 0$ as an asymptote. We indicate in Problem 24 of Section 13.1 how this graph can be constructed. Here we want to find the slope of its tangent line.

Solution We use implicit differentiation (rather than the *ad hoc* methods with which Fermat met Descartes' challenge). We differentiate both sides in (2) and find that

$$3x^2 + 3y^2\frac{dy}{dx} = 3y + 3x\frac{dy}{dx}.$$

We solve for the derivative,

$$\frac{dy}{dx} = \frac{y - x^2}{y^2 - x}. \tag{3}$$

For instance, at the point $(\frac{3}{2}, \frac{3}{2})$ of the folium, the slope of the tangent line is

$$\left.\frac{dy}{dx}\right|_{(1.5,1.5)} = \frac{1.5 - (1.5)^2}{(1.5)^2 - 1.5} = -1,$$

and this result agrees with our intuition about the figure. At the point $(\frac{2}{3}, \frac{4}{3})$, Equation (3) gives

$$\left.\frac{dy}{dx}\right|_{(2/3,4/3)} = \frac{(\frac{4}{3}) - (\frac{2}{3})^2}{(\frac{4}{3})^2 - (\frac{2}{3})} = \frac{4}{5}.$$

Thus the equation of the tangent line at this point is

$$y - \tfrac{4}{3} = \tfrac{4}{5}(x - \tfrac{2}{3});$$

alternatively, $4x - 5y + 4 = 0$.

RELATED RATES

A **related rates problem** involves two or more quantities that vary with time, and an equation that expresses some relationship between these quantities. Typically, the values of these quantities at some instant are given, together with all their time rates of change but one. The problem is usually to find the time rate of change that is *not* given, at some instant specified in the problem. One common method for solving such a problem is to begin by implicit differentiation of the equation that relates the given quantities.

For example, suppose that $x(t)$ and $y(t)$ are the x- and y-coordinates at time t of a point moving around the circle with equation

$$x^2 + y^2 = 25. \tag{4}$$

Let us use the chain rule to differentiate both sides of this equation *with respect to time t*. This produces the equation

$$2x\frac{dx}{dt} + 2y\frac{dy}{dt} = 0. \tag{5}$$

If the values of x, y, and dx/dt are known at a certain instant t, then Equation (5) can be solved for the value of dy/dt then. Note that it is *not* necessary to know x and y as functions of t. Indeed, it is common for a related rates problem to contain insufficient information to express x and y as functions of t.

For instance, suppose that we are given $x = 3$, $y = 4$, and $dx/dt = 12$ at a certain instant. Substitution of these values in Equation (5) yields

$$(2)(3)(12) + (2)(4)\frac{dy}{dt} = 0,$$

so we find that $dy/dt = -9$ at the same instant.

EXAMPLE 3 A rocket is launched vertically and is tracked by a radar station, which is located on the ground 3 mi from the launch site. What is the vertical speed of the rocket at the instant when its distance from the radar station is 5 mi and this distance is increasing at the rate of 5000 mi/h?

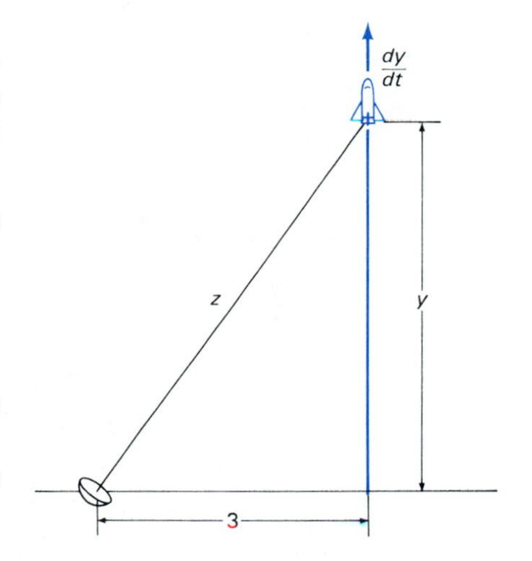

3.53 The rocket of Example 3

Solution Figure 3.53 illustrates this situation. We denote the altitude of the rocket by y and its distance from the radar station by z. We are given

$$\frac{dz}{dt} = 5000 \quad \text{when} \quad z = 5$$

(units are miles, hours, and miles per hour). We want to find dy/dt at this instant.

We apply the Pythagorean theorem to the right triangle in the figure and obtain

$$y^2 + 9 = z^2$$

as a relation between y and z. From this we see that $y = 4$ when $z = 5$. Implicit differentiation next gives

$$2y\frac{dy}{dt} = 2z\frac{dz}{dt}.$$

We substitute the data $y = 4$, $z = 5$, and $dz/dt = 5000$. Thus we find that

$$\frac{dy}{dt} = 6250 \quad \text{(mi/h)}$$

at the instant in question.

Example 3 illustrates the following steps in the solution of a typical related rates problem of the sort that involves a geometric situation:

1. Draw a diagram and label as variables the various changing quantities involved in the problem.
2. Record the values of the variables and their rates of change, as given in the problem.
3. Read from the diagram an equation that relates the important variables of the problem.
4. Differentiate this equation implicitly with respect to time t.
5. Substitute the given numerical data in the resulting equation, then solve for the unknown.

WARNING The most common error to be avoided is the premature substitution of the given data before differentiating implicitly. If, in Example 3, we had substituted $z = 5$ to begin with, our equation would have been $y^2 + 9 = 25$, and implicit differentiation would give the absurd result that $dy/dt = 0$.

In the next example, we use similar triangles (rather than the Pythagorean theorem) to discover the needed relation between the variables.

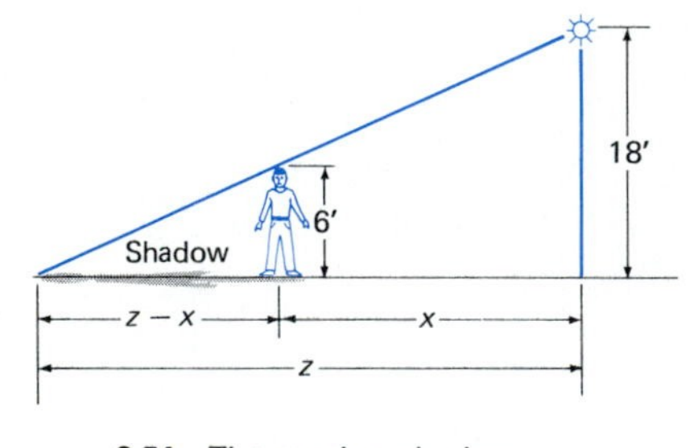

3.54 The moving shadow

EXAMPLE 4 A man 6 ft tall walks with a speed of 8 ft/s away from a street light atop an 18-ft pole. How fast is the tip of his shadow moving along the ground when he is 100 ft from the light pole?

Solution Let x be the man's distance from the pole and z the distance of the tip of his shadow from the base of the pole. Note that, although x and z are functions of t, we do *not* attempt to obtain explicit formulas for either.

It's given that $dx/dt = 8$ (ft/s); we want to find dz/dt when $x = 100$ (ft). We equate ratios of corresponding sides of the two similar triangles of Fig. 3.54 and find that

$$\frac{z}{18} = \frac{z - x}{6}.$$

Thus,

$$2z = 3x.$$

Implicit differentiation now gives

$$2\frac{dz}{dt} = 3\frac{dx}{dt}.$$

We substitute $dx/dt = 8$, and find that

$$\frac{dz}{dt} = \frac{3}{2} \cdot \frac{dx}{dt} = \frac{3}{2} \cdot (8) = 12.$$

So the tip of the man's shadow is moving at 12 ft/s.

Example 4 is somewhat unusual in that the answer is independent of the man's distance from the light pole—the given value $x = 100$ is superfluous. The next example is a related rates problem with two relationships between the variables, not quite so unusual.

EXAMPLE 5 Two radar stations at A and B, with B 6 mi east of A, are tracking a ship. At a certain instant, the ship is 5 mi from A, and this distance is increasing at the rate of 28 mi/h. At the same instant the ship is also 5 mi from B, while this distance is increasing at only 4 mi/h. Where is the ship, how fast is it moving, and in what direction is it moving?

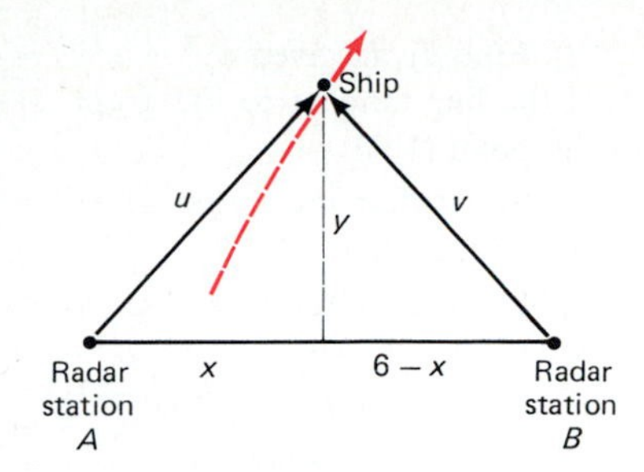

3.55 Radar stations tracking a ship

Solution With the distances indicated in Fig. 3.55, we find—again with the aid of the Pythagorean theorem—that

$$x^2 + y^2 = u^2 \quad \text{and} \quad (6 - x)^2 + y^2 = v^2.$$

We are given these data: $u = v = 5$, $du/dt = 28$, and $dv/dt = 4$ at the instant in question. Because the ship is equally distant from A and B, it is clear that $x = 3$. So $y = 4$. Thus the ship is 3 mi east and 4 mi north of A.

We differentiate implicitly the two equations above, and we obtain

$$2x\frac{dx}{dt} + 2y\frac{dy}{dt} = 2u\frac{du}{dt}$$

and

$$-2(6 - x)\frac{dx}{dt} + 2y\frac{dy}{dt} = 2v\frac{dv}{dt}.$$

When we substitute the numerical data given and deduced, we find that

$$3\frac{dx}{dt} + 4\frac{dy}{dt} = 140 \quad \text{and} \quad -3\frac{dx}{dt} + 4\frac{dy}{dt} = 20.$$

These equations are easy to solve: $dx/dt = dy/dt = 20$. Therefore, the ship is sailing northeast at a speed of

$$\sqrt{(20)^2 + (20)^2} = 20\sqrt{2}$$

miles per hour . . . if the figure is correct! For a mirror image along the line AB will reflect *another* ship, 3 mi east and 4 mi *south* of A, sailing *southeast* at a speed of $20\sqrt{2}$ mi/h.

The lesson? Figures are important, helpful, often essential—and potentially misleading. Try to avoid taking anything for granted when you draw a figure. In this example there would be no real problem, for each radar station could certainly determine whether the ship was generally to the north or to the south.

3.8 PROBLEMS

In Problems 1–10, find dy/dx by implicit differentiation.

1. $x^2 - y^2 = 1$

2. $xy = 1$

3. $16x^2 + 25y^2 = 400$

4. $x^3 + y^3 = 1$

5. $\sqrt{x} + \sqrt{y} = 1$

6. $x^2 + xy + y^2 = 9$

7. $x^{2/3} + y^{2/3} = 1$

8. $(x - 1)y^2 = x + 1$

9. $x^2(x - y) = y^2(x + y)$

10. $(x^2 + y^2)^2 = 4xy$

In each of Problems 11–20, first find dy/dx by implicit differentiation. Then write an equation of the tangent line to the graph of the equation at the given point.

11. $x^2 + y^2 = 25$; $(3, -4)$

12. $xy = -8$; $(4, -2)$

13. $x^2y = x + 2$; $(2, 1)$

14. $x^{1/4} + y^{1/4} = 4$; $(16, 16)$

15. $xy^2 + x^2y = 2$; $(1, -2)$

16. $\dfrac{1}{x + 1} + \dfrac{1}{y + 1} = 1$; $(1, 1)$

17. $12(x^2 + y^2) = 25xy$; $(3, 4)$

18. $x^2 + xy + y^2 = 7$; $(3, -2)$

19. $\dfrac{1}{x^3} + \dfrac{1}{y^3} = 2$; $(1, 1)$

20. $(x^2 + y^2)^3 = 8x^2y^2$; $(1, -1)$

21. Find dy/dx given $xy^3 - x^5y^2 = 4$. Then find the slope of the line tangent to the graph of the given equation at the point (1, 2).

22. Show that the graph of $xy^5 + x^5y = 1$ has no horizontal tangents.

23. Show that there are no points on the graph of the equation

$$x^3 + y^3 = 3xy - 1$$

at which the tangent line is horizontal.

24. Find all points on the graph of the equation

$$x^4 + y^4 + 2 = 4xy^3$$

at which the tangent line is horizontal.

25. Find all points on the graph of

$$x^2 + y^2 = 4x + 4y$$

at which the tangent line is horizontal.

26. Find the first quadrant points of the folium of Example 2 at which the tangent line is either horizontal ($dy/dx = 0$) or vertical (where $dx/dy = 1/(dy/dx) = 0$).

27. The graph of the equation

$$x^2 - xy + y^2 = 9$$

is the rotated ellipse shown in Fig. 3.56. Find the tangent lines to this curve at the two points where it intersects the x-axis, and show that these lines are parallel.

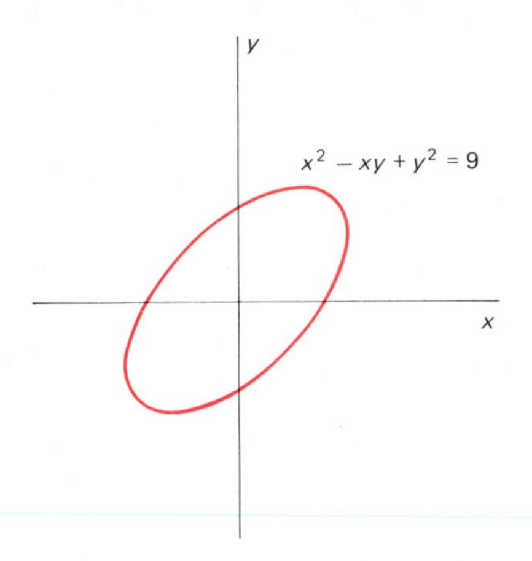

3.56 The rotated ellipse of Problem 27

28. Find the points on the curve

$$x^2 - xy + y^2 = 9$$

where the tangent line is horizontal ($dy/dx = 0$) and where it is vertical ($dx/dy = 0$).

29. The lemniscate with equation

$$(x^2 + y^2)^2 = x^2 - y^2$$

is shown in Fig. 3.57. Find by implicit differentiation the four points on the lemniscate where the tangent line is horizontal. Then find the two points where the tangent line is vertical—that is, where $dx/dy = 1/(dy/dx) = 0$.

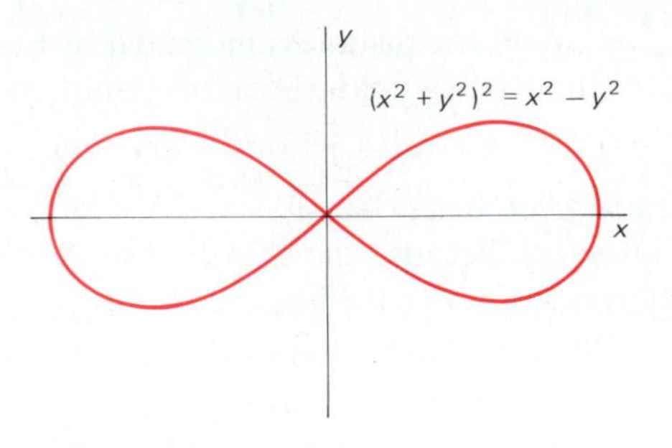

3.57 The lemniscate of Problem 29

30. Water is being collected from a block of ice with a square base. The water is produced because the ice is melting in such a way that each edge of the base of the block is decreasing at 2 in./h, while the height of the block is decreasing at 3 in./h. What is the rate of flow of water into the collecting pan when the base has edge length 20 in. and the height of the block is 15 in.? Make the simplifying assumption that the water and ice have the same density.

31. Sand being emptied from a hopper at the rate of 10 ft^3/s forms a conical pile whose height is always twice its radius. At what rate is the radius of the pile increasing when its height is 5 ft?

32. Suppose that water is being emptied from a spherical tank of radius 10 ft. If the depth of water in the tank is 5 ft and is decreasing at the rate of 3 ft/s, at what rate is the radius of the top surface of the water decreasing?

33. A circular oil slick of uniform thickness is caused by a spill of 1 m^3 of oil. The thickness of the oil slick is decreasing at the rate of 0.1 cm/h. At what rate is the radius of the slick increasing when it is 8 m?

34. Suppose that an ostrich 5 ft tall is walking at a speed of 4 ft/s directly toward a street light 10 ft high. How fast is the tip of the ostrich's shadow moving along the ground? At what rate is the ostrich's shadow decreasing in length?

35. The width of a rectangle is half its length. At what rate is its area increasing if its width is 10 cm and is increasing at 0.5 cm/s?

36. At what rate is the area of an equilateral triangle increasing if its base is 10 cm long and is increasing at 0.5 cm/s?

37. A gas balloon is being filled at the rate of 100π cubic centimeters of gas per second. At what rate is the radius of the balloon increasing when the radius is 10 cm?

38. The volume V (in cubic inches) and pressure P (in pounds per square inch) of a certain gas sample satisfy the

equation $PV = 1000$. At what rate is the volume of the sample changing if the pressure is 100 lb/in.^{2} and is increasing at the rate of 2 lb/in.^{2} per second?

39. A kite in the air at an altitude of 400 ft is being blown horizontally at the rate of 10 ft/s away from the person holding the kite string at ground level. At what rate is the string being payed out when 500 ft of string are already out?

40. A weather balloon is rising vertically and is being observed from a point on the ground 300 ft from the spot directly beneath the balloon. At what rate is the balloon rising when the angle between the ground and the observer's line of sight is 45° and is increasing at 1° per second?

41. An airplane flying horizontally at an altitude of 3 mi and at a speed of 480 mi/h passes directly above an observer on the ground. How fast is the distance from the observer to the airplane increasing 30 s later?

42. The volume V of water in a partially filled spherical tank of radius a is $V = \frac{1}{3}\pi y^2(3a - y)$, where y is the maximum depth of the water. Suppose that water is being drained from a spherical tank of radius 5 ft at the rate of 100 gal/min. Find the rate at which the depth y of water is decreasing when

(a) $y = 7$ (ft);

(b) $y = 3$ (ft).

(NOTE One gallon of water occupies a volume of approximately 0.1337 ft^3.)

43. Repeat Problem 42, except use a tank that is hemispherical, flat side on top, with radius 10 ft.

44. A swimming pool is 50 ft long and 20 ft wide. Its depth varies uniformly from 2 ft at the shallow end to 12 ft at the deep end. Suppose that the pool is being filled at the rate of 1000 gal/min. At what rate is the depth of water at the deep end increasing when it is 6 ft? (NOTE One gallon of water occupies a volume of approximately 0.1337 ft^3.)

45. A ladder 41 ft long has been leaning against a vertical wall. It begins to slip, so that its top slides down the wall while its bottom moves along the ground; the bottom moves at a constant speed of 10 ft/s. How fast is the top of the ladder moving when it is 9 ft above the ground?

46. The base of a rectangle is increasing at 4 cm/s while its height is decreasing at 3 cm/s. At what rate is its area changing when its base is 20 cm and its height is 12 cm?

47. The height of a cone is decreasing at 3 cm/s while its radius is increasing at 2 cm/s. When the radius is 4 cm and the height is 6 cm, is the volume of the cone increasing or is it decreasing? At what rate is the volume changing then?

48. A square is expanding. When each edge is 10 in., its area is increasing at 120 in.^{2}/s. At what rate is the length of each edge changing then?

49. A rocket is launched vertically and is tracked by a radar station located on the ground 4 mi from the launch site. What is the vertical speed of the rocket at the instant when its distance from the radar station is 5 mi and this distance is increasing at the rate of 3600 mi/h?

50. Two straight roads intersect at right angles. At 10 A.M. a car passes through the intersection headed due east at 30 mi/h. At 11 A.M. a truck passes through the intersection heading due north at 40 mi/h. Assume that the two vehicles maintain the given speeds and directions. At what rate are they separating at 1 P.M.?

51. A 10-ft ladder is leaning against a wall. The bottom of the ladder begins to slide away from the wall at a speed of 1 mi/h. Find the rate at which the top of the ladder is moving when it is (a) 4 ft above the ground; (b) 1 in. above the ground.

52. Two ships are sailing toward the same small island. One ship, the Pinta, is east of the island and is sailing due west at 15 mi/h. The other ship, the Niña, is north of the island and is sailing due south at 20 mi/h. At a certain time the Pinta is 30 mi from the island and the Niña is 40 mi from it. Are the two ships drawing closer together or farther apart at that time? At what rate?

53. At time $t = 0$, a single-engine military jet is flying due east at 12 mi/min. At the same altitude and 208 mi directly ahead of it, still at time $t = 0$, is a commercial jet, flying due north at 8 mi/min. When are the two planes closest to each other? What is the minimum distance between them?

54. A ship with a long anchor chain is anchored in 11 fathoms of water. The anchor chain is being wound in at the rate of 10 fathoms per minute, causing the ship to move toward the spot directly above the anchor resting on the seabed. The hawsehole—the point of contact between ship and chain—is located 1 fathom above the waterline. At what speed is the ship moving when there are exactly 13 fathoms of chain still out?

55. A water tank is in the shape of a cone with vertical axis and vertex downward. The tank's radius is 3 ft and the tank is 5 ft high. At first the tank is full of water, but at time $t = 0$ (in seconds), a small hole at the vertex is opened and the water begins to drain. When the height of water in the tank has dropped to 3 ft, the water is flowing out at 2 ft^3/s. At what rate, in feet per second, is the water level dropping then?

56. When a spherical tank with a radius of 10 ft contains water with a maximum depth of y feet, the volume of water in the tank is $V = (\pi/3)(30y^2 - y^3)$ cubic feet. If the tank is being filled at the rate of 200 gal/min, how fast is the water level rising when $y = 5$ (ft)? (NOTE One gallon of water occupies a volume of approximately 0.1337 ft^3.)

57. A water bucket is shaped like the frustum of a cone with height 2 ft, base radius 6 in., and top radius 12 in. Water is leaking from the bucket at 10 in.^{3}/min. At what rate is the water level falling when the depth of water in the bucket is 1 ft? (NOTE The volume of a conical frustum with height h and base radii a and b is

$$V = \frac{\pi h}{3}(a^2 + ab + b^2).)$$

58. Suppose that the radar stations A and B of Example 5 are now 12.6 mi apart. At a certain instant, a ship is 10.4 mi from A and its distance from A is increasing at 19.2 mi/h. At the same instant, its distance from B is 5 mi and this distance is decreasing at 0.6 mi/h. Find the location, speed, and direction of motion of the ship.

59. An airplane climbing at an angle of 45° passes directly over a ground radar station at an altitude of 1 mi. A later reading shows that the distance from the radar station to the plane is 5 mi and is increasing at 7 mi/min. What is the speed of the plane (in mi/h)?

60. The water tank of Problem 56 is completely full when a plug at its bottom is removed. According to Torricelli's law, the water drains in such a way that $dV/dt = -k\sqrt{y}$ where k is a positive empirical constant.

(a) Find dy/dt as a function of y.

(b) Find the depth of water when the water level is falling the *least* rapidly. (You will need to compute the derivative of dy/dt with respect to y.)

61. Sand is pouring from a pipe at the rate of 120π ft^3/s. The falling sand forms a conical pile on the ground; the altitude of the cone is always one-third the radius of its base. How fast is the altitude increasing when the pile is 20 ft high?

62. A man 6 ft tall walks at 5 ft/s along one edge of a road 30 ft wide. On the other edge of the road is a light atop a pile 18 ft high. How fast is the length of the man's shadow (on the horizontal ground) increasing when he is 40 ft beyond the point directly across the road from the pole?

63. (*Challenge*) As a follow-up to Problem 23, show that the graph of the equation $x^3 + y^3 = 3xy - 1$ consists of the straight line $x + y + 1 = 0$ together with the isolated point (1, 1).

3.9 Successive Approximations and Newton's Method

The quadratic formula is one tool for obtaining an exact solution of any second-degree equation $ax^2 + bx + c = 0$. Formulas are known for the exact solution of third- and fourth-degree equations, but they are rarely used because they are complicated. And it has been proved that there is *no* general formula possible for giving the roots of an arbitrary fifth (or higher)-degree polynomial equation in terms of algebraic operations. Thus the exact solution (for all its roots) of an equation such as

$$x^5 - 3x^3 + x^2 - 23x + 19 = 0$$

may be quite difficult, or even—as a practical matter—impossible.

Actually, there is a question about what it means to solve even so simple an equation as

$$x^2 - 2 = 0. \tag{1}$$

The positive exact solution is $x = \sqrt{2}$. But the number $\sqrt{2}$ is irrational and hence cannot be expressed as a terminating or repeating decimal. Thus if we mean by a *solution* an exact decimal value for x, even Equation (1) can be solved only approximately.

Over two millennia ago, ancient Babylonian mathematicians devised an effective way of generating a sequence of better and better approximations to $\sqrt{A}$, the square root of a given positive number A. Here is the *Babylonian square root method:* We begin with a first guess x_0 for the value of $\sqrt{A}$. For $\sqrt{2}$, we might guess $x_0 = 1.5$. If x_0 is too large—that is, if $x_0 > \sqrt{A}$—then

$$\frac{A}{x_0} < \frac{A}{\sqrt{A}} = \sqrt{A},$$

so A/x_0 is too small an estimate of $\sqrt{A}$. Similarly, if x_0 is too small (if $x_0 < \sqrt{A}$), then A/x_0 is too large an estimate of $\sqrt{A}$; that is, $A/x_0 > \sqrt{A}$.

Thus in each case one of the two numbers x_0 and A/x_0 is an underestimate of $\sqrt{A}$ and the other is an overestimate. The Babylonian idea was that we should get a better estimate of $\sqrt{A}$ by *averaging* x_0 and A/x_0. This yields a first approximation

$$x_1 = \frac{1}{2}\left(x_0 + \frac{A}{x_0}\right) \tag{2}$$

to $\sqrt{A}$. But why not repeat this process? We can average x_1 and A/x_1 to get a second approximation x_2, average x_2 and A/x_2 to get x_3, and so on. By repeating this process, we generate a sequence

$$x_1,\ x_2,\ x_3,\ x_4, \ldots$$

which we have every right to hope will consist of better and better approximations to $\sqrt{A}$.

Specifically, having calculated the nth approximation x_n, we calculate the next one by means of the *iterative formula*

$$x_{n+1} = \frac{1}{2}\left(x_n + \frac{A}{x_n}\right). \tag{3}$$

In other words, we plow each approximation to $\sqrt{A}$ back into the right-hand side in (3) to calculate the next approximation. This is an *iterative* process—the words *iteration* and *iterative* are derived from the Latin *iterare*, to plow again.

Suppose we find that after sufficiently many steps in this iteration, $x_{n+1} \approx x_n$ to the number of decimal places we are retaining in our computations. Then Equation (3) yields

$$x_n \approx x_{n+1} = \frac{1}{2}\left(x_n + \frac{A}{x_n}\right) = \frac{1}{2x_n}(x_n^2 + A),$$

so $2x_n^2 \approx x_n^2 + A$, and hence $x_n^2 \approx A$ to some degree of accuracy.

EXAMPLE 1 With $A = 2$ we begin with the crude first guess $x_0 = 1$ to $\sqrt{2}$. Then successive application of the formula in (3) yields

$$x_1 = \frac{1}{2}\left(1 + \frac{2}{1}\right) = \frac{3}{2} = 1.5,$$

$$x_2 = \frac{1}{2}\left(\frac{3}{2} + \frac{2}{3/2}\right) = \frac{17}{12} \approx 1.416666667,$$

$$x_3 = \frac{1}{2}\left(\frac{17}{12} + \frac{2}{17/12}\right) = \frac{577}{408} \approx 1.414215686,$$

$$x_4 = \frac{1}{2}\left(\frac{577}{408} + \frac{2}{577/408}\right) = \frac{665{,}857}{470{,}832} \approx 1.414213562,$$

rounding results to nine decimal places. It happens that x_4 gives $\sqrt{2}$ accurate to all nine places!

The Babylonian iteration defined in (3) is a method for generating a sequence of approximations to a root $r = \sqrt{A}$ of the particular equation

$x^2 - A = 0$. There are numerous iterative methods in current use for approximating a root r of an equation $f(x) = 0$, and one of the most effective is Newton's method, which we describe after Example 2. First we need a brief discussion of convergence and interpolation.

CONVERGENCE

We say that the sequence of approximations $x_1, x_2, x_3, \ldots$ *converges* to the number r provided that we can make x_n as close to r as we please merely by choosing n sufficiently large. More precisely, for any given $\varepsilon > 0$, there exists a positive integer N such that $|x_n - r| < \varepsilon$ for all $n \geqq N$. As a practical matter this means, as illustrated in Example 1, that for any positive integer k, x_n and r will agree to k or more decimal places once n becomes sufficiently large.

LINEAR INTERPOLATION

Any approximation method requires an initial approximation x_0 to the solution of the equation $f(x) = 0$ that we wish to solve. Suppose that a and b are two points such that $f(a)$ and $f(b)$ differ in sign and f is continuous on the interval $[a, b]$. Then we know from the intermediate value property of continuous functions (Section 2.4) that the equation $f(x) = 0$ has at least one solution in (a, b). If, in addition, the curve $y = f(x)$ either is steadily rising or is steadily falling on $[a, b]$, then it follows that the equation $f(x) = 0$ has *exactly one* solution x_* in (a, b).

A common way to obtain an initial approximation x_0 to x_* is by **linear interpolation.** In this method we choose x_0 as the point where the line segment joining $(a, f(a))$ and $(b, f(b))$ meets the x-axis. From the similar triangles in Fig. 3.58, where $f(a)$ is negative and $f(b)$ is positive, we obtain the proportion

$$\frac{x_0 - a}{-f(a)} = \frac{b - x_0}{f(b)}. \tag{4}$$

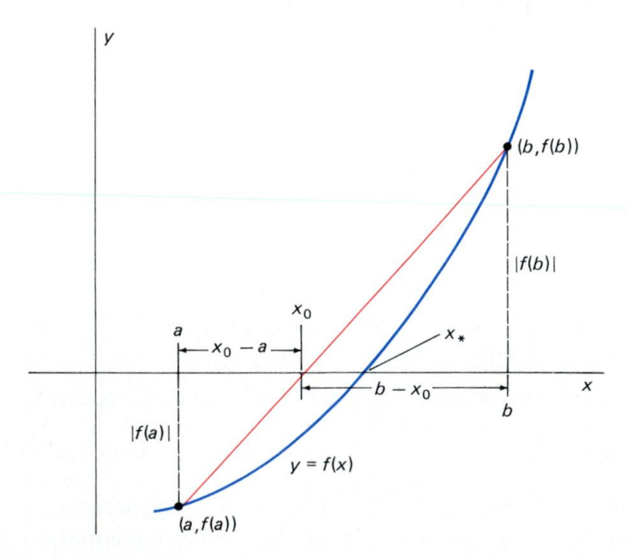

3.58 Obtaining x_0 by linear interpolation

It is then easy to solve this equation for

$$x_0 = \frac{af(b) - bf(a)}{f(b) - f(a)}. \tag{4'}$$

The case $f(a) > 0 > f(b)$ leads to the same formula for the interpolated value x_0. In practice it will probably be more convenient to set up the proportion than to remember the formula in (4′).

EXAMPLE 2 A large cork ball has radius 1 ft, and its density is $\frac{1}{4}$ that of water. Archimedes' law of buoyancy implies that when the ball floats in water, $\frac{1}{4}$ of its volume ($\frac{1}{4}$ of $4\pi/3$, which is $\pi/3$) is submerged. Find an equation determining the depth x to which the cork ball sinks, and use linear interpolation to make an initial approximation to its solution.

Solution The volume of a spherical segment of height h and radius r (the part beneath the waterline in Fig. 3.59) is given by the known formula

$$V = \frac{\pi x}{6}(3r^2 + x^2).$$

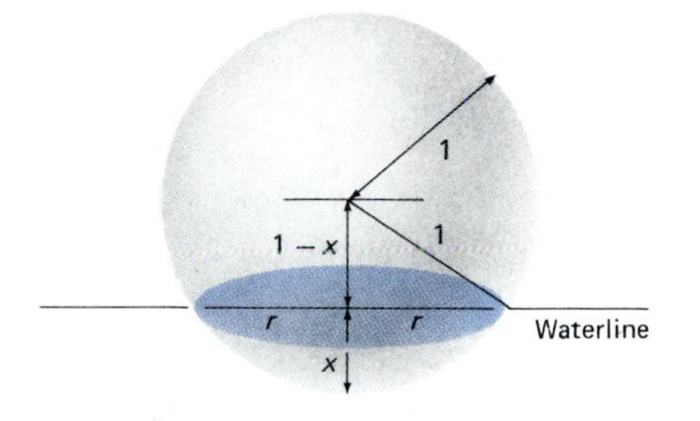

3.59 The floating cork ball

From the right triangle in Fig. 3.59, we see that

$$r^2 + (1 - x)^2 = 1^2,$$

so that $r^2 = 2x - x^2$. We combine Archimedes' law and the volume formula above to obtain

$$\frac{\pi x^2}{3}(3 - x) = \frac{\pi}{3}; \quad \text{that is,} \quad 3x^2 - x^3 = 1.$$

So we must solve the equation

$$f(x) = x^3 - 3x^2 + 1 = 0. \tag{5}$$

Equation (5) has no rational solutions; we cannot expect to solve it by entirely elementary methods. We note, however, that $f(0) = 1$ and $f(1) = -1$, so there is a root between 0 and 1. This is the solution we seek, because it is physically evident that the desired depth is between 0 and 1. With $a = 0$, $b = 1$, $f(a) = 1$, and $f(b) = -1$, the proportion in (4) is

$$\frac{x_0 - 0}{1} = \frac{1 - x_0}{1},$$

with solution $x_0 = 0.5$. This will be our initial approximation to the root of Equation (5) that lies in (0, 1).

NEWTON'S METHOD

In Fig. 3.60 we illustrate Newton's method of constructing a rapidly convergent sequence of successive approximations to a root x_* of the equation $f(x) = 0$. The tangent line at $(x_n, f(x_n))$ is used to construct a better approximation x_{n+1} to x_* as follows. Begin at the point x_n on the x-axis. Go vertically up (or down) to the point $(x_n, f(x_n))$ on the curve $y = f(x)$. Then follow the tangent line L there to the point where L meets the x-axis. That point will be x_{n+1}.

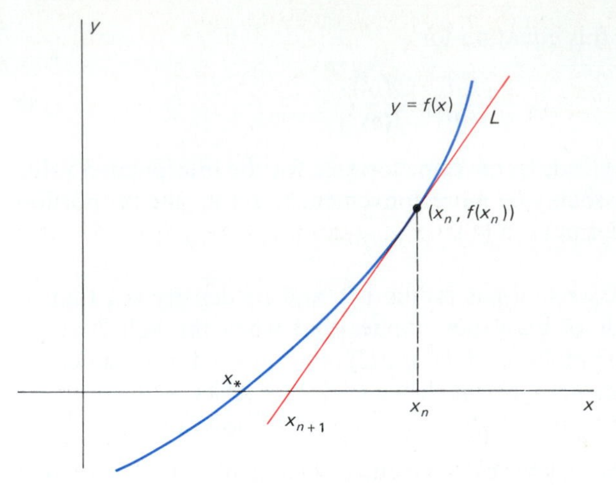

3.60 Geometry of the iteration of Newton's method

Here is a formula for x_{n+1}. We obtain it by computing the slope of the line L in two ways: from the derivative and from the two-point definition of slope. Thus

$$f'(x_n) = \frac{f(x_n) - 0}{x_n - x_{n+1}},$$

and we solve easily for x_{n+1}:

$$x_{n+1} = x_n - \frac{f(x_n)}{f'(x_n)}. \tag{6}$$

This equation is the **iterative formula** for Newton's method, so called because in about 1669, Newton introduced an algebraic procedure (rather than the geometric construction above) that is equivalent to the iterative use of Equation (6). Newton's first example was the cubic equation $x^3 - 2x - 5 = 0$, for which he found the root $x_* \approx 2.0946$ (as we ask you to do in Problem 18).

Suppose now that we want to apply Newton's method to solve the equation

$$f(x) = 0 \tag{7}$$

with an accuracy of k decimal places. Remember that an equation must be written precisely in the form in (7) in order to use the formula in (6). If we reach the point in our iteration at which x_n and x_{n+1} agree to k decimal places, it then follows that

$$x_n \approx x_{n+1} = x_n - \frac{f(x_n)}{f'(x_n)}; \qquad 0 \approx -\frac{f(x_n)}{f'(x_n)}; \qquad f(x_n) \approx 0.$$

Thus we have found an approximate root $x_n \approx x_{n+1}$ of the equation in (7). In practice, then, we retain k decimal places in our computations and persist until $x_n = x_{n+1}$ to this degree of accuracy. (We do not consider here the possibility of roundoff error, an important subject in numerical analysis.)

EXAMPLE 3 Use Newton's method to find $\sqrt{2}$ accurate to nine decimal places.

Solution More generally, consider the square root of the positive number A as the positive root of the equation

$$f(x) = x^2 - A = 0.$$

Because $f'(x) = 2x$, Equation (6) gives the iterative formula

$$x_{n+1} = x_n - \frac{x_n^2 - A}{2x_n} = \frac{1}{2}\left(x_n + \frac{A}{x_n}\right). \tag{8}$$

Thus we have derived the Babylonian iterative formula as a special case of Newton's method. The use of (8) with $A = 2$ therefore yields the values x_1, x_2, x_3, and x_4 that we calculated in Example 1, and upon another iteration we find that

$$x_5 = \frac{1}{2}\left(x_4 + \frac{2}{x_4}\right) \approx 1.414213562,$$

in agreement with x_4 to nine decimal places. The very rapid convergence here is an important characteristic of Newton's method. As a general rule (with some exceptions), each iteration doubles the number of decimal places of accuracy.

EXAMPLE 4 Use Newton's method to solve the cork ball equation

$$f(x) = x^3 - 3x^2 + 1 = 0.$$

Solution Here $f'(x) = 3x^2 - 6x$, so the iterative formula in (6) becomes

$$x_{n+1} = x_n - \frac{x_n^3 - 3x_n^2 + 1}{3x_n^2 - 6x_n}. \tag{9}$$

With $x_0 = 0.5$, this gives the values

$$x_0 = 0.5,$$

$$x_1 = 0.5 - \frac{(-0.5)^3 - (3)(0.5)^2 + 1}{(3)(0.5)^2 - (6)(0.5)} \approx 0.6667,$$

$$x_2 = 0.6528, \qquad x_3 = 0.6527, \qquad x_4 = 0.6527.$$

Thus we obtain the root $x_* = 0.6527$, retaining only four decimal places.

The equation $x^3 - 3x^2 + 1 = 0$ has two additional solutions. This is because the continuous function $f(x) = x^3 - 3x^2 + 1$ has the intermediate value property on each of the intervals $[-1, 0]$, $[0, 1]$, and $[2, 3]$. The starting values of x_0 shown in the adjacent table are estimated by linear interpolation. Beginning with $x_0 = -0.25$ and, subsequently, with $x_0 = 2.75$, the iteration in (9) produces the two sequences

x	$f(x)$	
−1	−3	$x_0 = -0.25$
0	1	
0	1	$x_0 = +0.50$
1	−1	
2	−3	$x_0 = +2.75$
3	1	

$$x_0 = -0.25, \qquad x_0 = 2.75,$$

$$x_1 = -0.7222, \qquad x_1 = 2.8939,$$

$$x_2 = -0.5626, \qquad x_2 = 2.8795,$$

$$x_3 = -0.5331, \qquad x_3 = 2.8794,$$

$$x_4 = -0.5321, \qquad x_4 = 2.8794.$$

$$x_5 = -0.5321.$$

Thus the other two roots are -0.5321 and 2.8794 (to four decimal places).

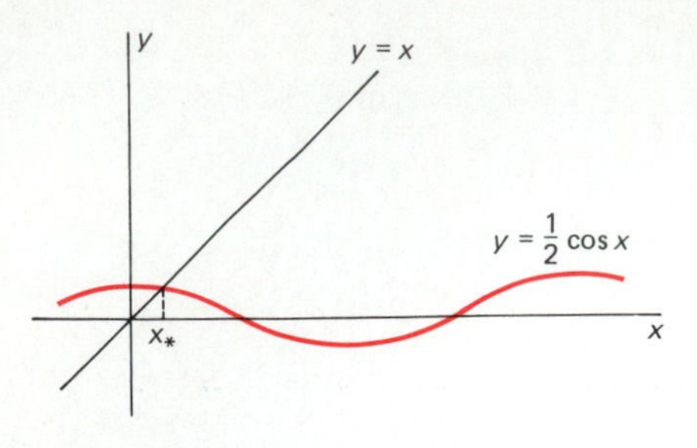

3.61 Solving the equation $x = \frac{1}{2}\cos x$

EXAMPLE 5 Figure 3.61 indicates that the equation

$$x = \tfrac{1}{2}\cos x \tag{10}$$

has a solution $x_* \approx 0.5$. To apply Newton's method to approximate x_*, we rewrite Equation (10) in the form

$$f(x) = 2x - \cos x = 0.$$

Because $f'(x) = 2 + \sin x$, the iterative formula of Newton's method is

$$x_{n+1} = x_n - \frac{2x_n - \cos x_n}{2 + \sin x_n}.$$

Beginning with $x_0 = 0.5$ and retaining five decimal places, this formula yields

$$x_1 = 0.45063, \qquad x_2 = 0.45018, \qquad x_3 = 0.45018.$$

Thus the root is 0.45018 to five decimal places.

Newton's method is one for which "the proof is in the pudding." If it works, it's obvious that it does, and everything's fine. When Newton's method fails, it may do so spectacularly. For example, suppose that we want to solve the equation

$$f(x) = x^{1/3} = 0.$$

Note that $x_* = 0$ is the only solution. The iterative formula in (6) becomes

$$x_{n+1} = x_n - \frac{(x_n)^{1/3}}{\frac{1}{3}(x_n)^{-2/3}} = x_n - 3x_n = -2x_n.$$

If we being with $x_0 = 1$, Newton's method yields $x_1 = -2$, $x_2 = +4$, $x_3 = -8$, and so on. Figure 3.62 indicates why our "approximations" are not converging.

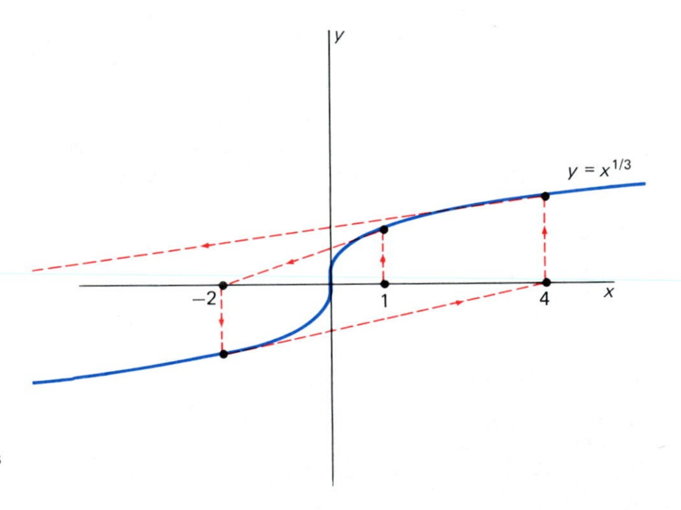

3.62 A failure of Newton's method

When Newton's method fails to provide a sequence of approximations that converge to a solution, the implication is that some other method should be used. For instance, the method of bisection discussed in Section 2.4 can always be used as a last resort. But Newton's method is much faster (far fewer iterations are required), easier to program, and therefore preferable when it succeeds.

3.9 PROBLEMS

In each of Problems 1–20, use Newton's method to find the solution of the given equation $f(x) = 0$ in the indicated interval $[a, b]$ accurate to four decimal places. Choose the initial estimate of the solution by using the interpolation formula in (4′).

1. $x^2 - 5 = 0$; $[2, 3]$ (to find the positive square root of 5)

2. $x^3 - 2 = 0$; $[1, 2]$ (to find the cube root of 2)

3. $x^5 - 100 = 0$; $[2, 3]$ (to find the fifth root of 100)

4. $x^{3/2} - 10 = 0$; $[4, 5]$ (to find $10^{2/3}$)

5. $x^2 + 3x - 1 = 0$; $[0, 1]$

6. $x^3 + 4x - 1 = 0$; $[0, 1]$

7. $x^6 + 7x^2 - 4 = 0$; $[-1, 0]$

8. $x^3 + 3x^2 + 2x = 10$; $[1, 2]$

9. $x - \cos x = 0$; $[0, 2]$

10. $x^2 - \sin x = 0$; $[0.5, 1.0]$

11. $4x - \sin x = 4$; $[1, 2]$

12. $5x + \cos x = 5$; $[0, 1]$

13. $x^5 + x^4 = 100$; $[2, 3]$

14. $x^5 + 2x^4 + 4x = 5$; $[0, 1]$

15. $x + \tan x = 0$; $[2, 3]$

16. $x + \tan x = 0$; $[11, 12]$

17. $x^3 - 10 = 0$; $[2, 3]$

18. $x^3 - 2x - 5 = 0$; $[2, 3]$ (Newton's own example)

19. $x^5 - 5x - 10 = 0$; $[1, 2]$

20. $x^5 = 32$; $[0, 5]$

21. (a) Show that Newton's method applied to the equation $x^3 - a = 0$ yields the iteration

$$x_{n+1} = \frac{1}{3}\left(2x_n + \frac{a}{x_n^2}\right)$$

for approximating the cube root of a.
(b) Use this iteration to find $\sqrt[3]{2}$ accurate to five decimal places.

22. (a) Show that Newton's method yields the iteration

$$x_{n+1} = \frac{1}{k}\left[(k-1)x_n + \frac{a}{(x_n)^{k-1}}\right]$$

for approximating the kth root of the positive number a.
(b) Use this iteration to find $\sqrt[10]{100}$ accurate to five decimal places.

23. Equation (10) has the special form $x = G(x)$ where $G(x) = \frac{1}{2}\cos x$. For an equation of this form, the iterative formula

$$x_{n+1} = G(x_n)$$

sometimes produces a sequence $x_1, x_2, x_3, \ldots$ of approximations that converge to a root. In the case of Equation (10), this *repeated substitution* formula is simply

$$x_{n+1} = \tfrac{1}{2}\cos x_n.$$

Begin with $x_0 = 0.5$ as in Example 5 and retain five decimal places in your computation of the solution of Equation (10). (CHECK You should find that $x_4 \approx 0.45018$.)

24. See Problem 23 for a discussion of the method of repeated substitution. The equation

$$x^4 = x + 1$$

has a solution between $x = 1$ and $x = 2$. Use the initial guess $x_0 = 1.5$ and the method of repeated substitution to discover that one of the solutions of this equation is $x_* \approx 1.220744$. Iterate using the formula

$$x_{n+1} = (x_n + 1)^{1/4}.$$

Then compare the result with what happens when you iterate using the formula

$$x_{n+1} = x_n^4 - 1.$$

25. See Problem 23 for a discussion of the method of repeated substitution. The cork ball equation

$$x^3 - 3x^2 + 1 = 0$$

has a solution between $x = 0$ and $x = 1$. To apply the method of repeated substitution to this equation, it may be written either in the form

$$x = 3 - \frac{1}{x^2}$$

or in the form

$$x = (3x^2 - 1)^{1/3}.$$

If you begin with $x_0 = 0.5$ in the hope of finding the nearby solution $x_* = 0.6527$ of the cork ball equation, using each of the iterative formulas above, you will observe some of the drawbacks of the method. Describe what goes wrong.

26. Show that Newton's method applied to the equation

$$\frac{1}{x} - a = 0$$

yields the iterative formula

$$x_{n+1} = 2x_n - a(x_n)^2,$$

and thus provides a method for approximating the reciprocal $1/a$ without performing any divisions. The reason for using such a method is that in most high-speed computers, the operation of division is more time-consuming than even several additions and multiplications.

27. Prove that the equation $x^5 + x = 1$ has exactly one real solution, then use Newton's method to find it with three places correct on the right of the decimal.

In each of Problems 28–30, use Newton's method to find all real roots of the given equation with two digits correct to the right of the decimal. (*Suggestion:* In order to determine the number of roots and their approximate location, graph the left-hand and right-hand sides of each equation and observe about where the graphs cross.)

28. $x^2 = \cos x$

29. $x = 2 \sin x$

30. $\cos x = -\frac{1}{5}x$ (There are exactly three solutions, as indicated in Fig. 3.63.)

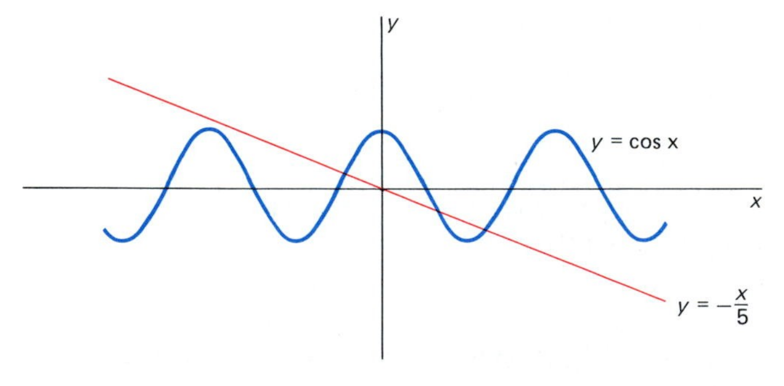

3.63 Solving the equation of Problem 30

31. Prove that the equation $x^7 - 3x^3 + 1 = 0$ has at least one solution. Then use Newton's method to find one solution.

32. Use Newton's method to approximate $\sqrt[3]{5}$ to three-place accuracy.

33. Use Newton's method to find the value of x for which $x^3 = \cos x$.

34. Use Newton's method to find the smallest positive value of x for which $x = \tan x$.

35. In Problem 49 of Section 3.6, we dealt with the problem of minimizing the cost of joining by a road two points lying on opposite sides of a geological fault. This problem led to the equation

$$f(x) = 3x^4 - 24x^3 + 51x^2 - 32x + 64 = 0.$$

Use Newton's method to find to four-place accuracy the root of this equation lying in the interval (3, 4).

36. A moon of a certain planet has an elliptical orbit with eccentricity 0.5, and its period of revolution about the planet is 100 days. If the moon is at the position $(a, 0)$ when $t = 0$, then, as illustrated in Fig. 3.64, the central angle θ after t days is given by *Kepler's equation*

$$\frac{2\pi t}{100} = \theta - (0.5) \sin \theta.$$

Use Newton's method to solve for θ when $t = 17$ (days). Take $\theta_0 = 1.5$ (radians), and calculate the first two approximations θ_1 and θ_2. Express θ_2 in degrees as well.

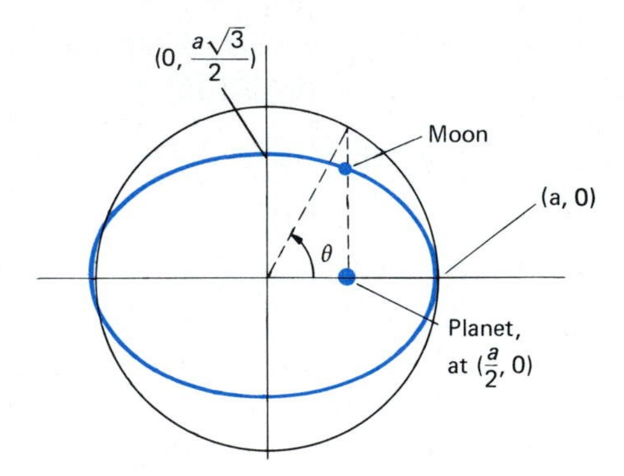

3.64 The elliptical orbit of Problem 36

37. A great problem of Archimedes was that of using a plane to cut a sphere into two segments with volumes having a given (preassigned) ratio. Archimedes showed that the volume of a segment of height h of a sphere of radius a is

$$V = \tfrac{1}{3}\pi h^2(3a - h).$$

If a plane at distance x from the center of the unit sphere cuts it into two segments, one with twice the volume of the other, show that

$$3x^3 - 9x + 2 = 0.$$

Then use Newton's method to find x accurate to four decimal places.

38. The equation

$$f(x) = x^3 - 4x + 1 = 0$$

has three distinct real roots. Locate them by calculating the value of f for $x = -3, -2, -1, 0, 1, 2,$ and 3. Then use Newton's method to approximate each of the three roots to four-place accuracy.

39. The equation $x + \tan x = 0$ is important in a variety of applications—for example, in the study of the diffusion of heat. It has a sequence $\alpha_1, \alpha_2, \alpha_3, \ldots$ of positive roots, with the nth one slightly larger than $(n - 0.5)\pi$. Use Newton's method to compute α_1 and α_2 to three-place accuracy.

*3.9 Optional Computer Application

Iterative processes such as Newton's method virtually cry out for the facility of the computer to perform the repetitive computations. Program NEWTON (listed in Fig. 3.65) implements Newton's method to solve the cork ball equation $x^3 - 3x^2 + 1 = 0$.

```
100 REM--Program NEWTON
110 REM--Applies Newton's method to solve the equation
120 REM--f(x) = 0.  The formulas FCTN for the function
130 REM--and DERIV for its derivative f'(x) must be
140 REM--edited into lines 230 and 240 before execution.
150 REM
160        INPUT "INITIAL GUESS X0"; X
170        N = 0
180        PRINT " N", "     X" : PRINT
190 REM
200 REM--Newton's iteration:
210 REM
220        PRINT N, X
230        FCTN  = X*X*X - 3*X*X + 1
240        DERIV = 3*X*X - 6*X
250        XNEW  = X - FCTN/DERIV
260        X = XNEW  :  N = N + 1
270        IF INKEY$ = "" THEN 270
280        GOTO 220
290 REM
300 REM--Press Ctrl/Break to stop.
310        END
```

3.65 Listing of Program NEWTON

Note that the function $f(x) = x^3 - 3x^2 + 1$ and its derivative $f'(x) = 3x^2 - 6x$ appear in lines 230 and 240; these can be changed to adapt the program to solving a different equation. Newton's iterative formula itself appears in line 250. When the program is run, line 160 prompts the user to enter an initial guess x_0. The effect of line 270 is to pause after each iteration. When any key is pressed, another approximation is printed. When several successive approximations agree, the user may interrupt the program in order to stop execution. Three runs of Program NEWTON with three different initial guesses gave the following results, showing the three solutions of the cork ball equation.

```
RUN
INITIAL GUESS XO?  1
 N                X

 0              1
 1               .6666666
 2               .6527778
 3               .6527036
 4               .6527036
 5               .6527036
```

```
RUN
INITIAL GUESS XO?  3
 N                  X
 0                 3
 1                 2.888889
 2                 2.879452
 3                 2.879385
 4                 2.879385
 5                 2.879385

RUN
INITIAL GUESS XO?  -1
 N                  X
 0                -1
 1                 -.6666666
 2                 -.5486111
 3                 -.5323902
 4                 -.532089
 5                 -.532089
 6                 -.532089
```

Calculators are now available that can differentiate symbolic expressions. The following HP-28S program exploits this capability.

```
<< ''Store expression F''
   HALT DROP 'X'      PURGE
   F  'X' ∂ 'DF'      STO
   ''Store initial X''
   HALT   DROP  X
   WHILE   1  0  >
   REPEAT
      DUP  DUP  'X'  STO
      F  EVAL  DF  EVAL
      /  -  HALT
   END  >>
```

Line 1 prompts the user to store the formula $f(x)$. For instance, to solve the cork ball equation, we store the algebraic expression

```
'X*X*X - 3*X*X + 1'
```

as the variable F, then press (CONT) to continue. Then line 3 calculates the derivative

```
'3*X*X - 6*X'
```

and stores it as DF. Line 4 prompts us to store the initial guess X. The WHILE-REPEAT loop carries out the Newton iteration repeatedly until we press (KILL) to halt program execution. After each iteration we simply press (CONT) to see the next approximation (together, for comparison, with the preceding two approximations).

Exercises 1–20: Use one of the programs shown here, or one of your own devising, to solve Problems 1–20 of Section 3.9.

Exercise 21: The cubic equation

$$4x^3 - 42x^2 - 19x - 28 = 0$$

has a single real solution. Find it (accurate to four decimal places). First try the initial guess $x_0 = 0$; be prepared for at least 25 iterations. Then try initial guesses $x_0 = 10$ and $x_0 = 100$.

CHAPTER 3 REVIEW: Formulas, Concepts, Definitions

Differentiation Formulas

$$D_x(cu) = c\frac{du}{dx} \qquad D_x(u^r) = ru^{r-1}\frac{du}{dx}$$

$$D_x(u+v) = \frac{du}{dx} + \frac{dv}{dx} \qquad D_x \sin u = (\cos u)\frac{du}{dx}$$

$$D_x(uv) = u\frac{dv}{dx} + v\frac{du}{dx} \qquad D_x \cos u = (-\sin u)\frac{du}{dx}$$

$$D_x\frac{u}{v} = \frac{v\dfrac{du}{dx} - u\dfrac{dv}{dx}}{v^2} \qquad D_x g(u) = g'(u)\frac{du}{dx}$$

Use the list below as a guide to concepts that you may need to review.

1. Definition of the derivative
2. Average rate of change of a function
3. Instantaneous rate of change of a function
4. Position function; velocity and acceleration
5. Differential, function, and operator notation for derivatives
6. The binomial formula
7. The power rule
8. Linearity of differentiation
9. The product rule
10. The quotient rule
11. Principle of mathematical induction
12. The chain rule
13. The generalized power rule
14. Local maxima and minima
15. $f'(c) = 0$ as a necessary condition for local extrema
16. Critical points
17. The closed interval maximum–minimum method
18. Differentiation of inverse functions
19. Vertical tangent lines
20. Steps in the solution of an applied maximum–minimum problem
21. Derivatives of the sine and cosine functions
22. Implicitly defined functions
23. Implicit differentiation
24. Solution of related rates problems
25. Linear interpolation
26. Newton's method

CHAPTER 3 MISCELLANEOUS PROBLEMS

Find dy/dx in Problems 1–35.

1. $y = x^2 + \dfrac{3}{x^2}$
2. $y^2 = x^2$
3. $y = \sqrt{x} + \dfrac{1}{\sqrt[3]{x}}$
4. $y = (x^2 + 4x)^{5/2}$
5. $y = (x-1)^7(3x+2)^9$
6. $y = \dfrac{x^4 + x^2}{x^2 + x + 1}$
7. $y = \left(3x - \dfrac{1}{2x^2}\right)^4$
8. $y = x^{10}\sin 10x$
9. $xy = 9$
10. $y = \sqrt{\dfrac{1}{5x^6}}$
11. $y = \dfrac{1}{\sqrt{(x^3 - x)^3}}$
12. $y = \sqrt[3]{2x+1}\sqrt[5]{3x-2}$
13. $y = \dfrac{1}{1+u^2}$ where $u = \dfrac{1}{1+x^2}$
14. $x^3 = \sin^2 y$
15. $y = (\sqrt{x} + \sqrt[3]{2x})^{7/3}$
16. $y = \sqrt{3x^5 - 4x^2}$
17. $y = \dfrac{u+1}{u-1}$, where $u = \sqrt{x+1}$
18. $y = \sin(2\cos 3x)$
19. $x^2y^2 = x + y$
20. $y = \sqrt{1 + \sin\sqrt{x}}$
21. $y = \sqrt{x + \sqrt{2x + \sqrt{3x}}}$
22. $y = \dfrac{x + \sin x}{x^2 + \cos x}$
23. $\sqrt[3]{x} + \sqrt[3]{y} = 4$
24. $x^3 + y^3 = xy$

25. $y = (1 + 2u)^3$ where $u = \dfrac{1}{(1 + x)^3}$

26. $y = \cos^2(\sin^2 x)$

27. $y = \sqrt{\dfrac{\sin^2 x}{1 + \cos x}}$

28. $y = (1 + \sqrt{x})^3(1 - 2\sqrt[3]{x})^4$

29. $y = \dfrac{\cos 2x}{\sqrt{\sin 3x}}$

30. $x^3 - x^2y + xy^2 - y^3 = 4$

31. $y = \sin^3 2x \cos^2 3x$

32. $y = [1 + (2 + 3x)^{-3/2}]^{2/3}$

33. $y = \sin^5\left(x + \dfrac{1}{x}\right)$

34. $\sqrt{x + y} = \sqrt[3]{x - y}$

35. $y = \cos^3(\sqrt[3]{x^4 + 1})$

In each of Problems 36–39, find the tangent line to the given curve at the indicated point.

36. $y = \dfrac{x + 1}{x - 1}$; $(0, -1)$

37. $x = \sin 2y$; $(1, \pi/4)$

38. $x^2 - 3xy + 2y^2 = 0$; $(2, 1)$

39. $y^3 = x^2 + x$; $(0, 0)$

40. If a hemispherical bowl with radius 1 ft is filled with water to a depth of x inches, the volume of liquid in the bowl is

$$V = \frac{\pi}{3}(36x^2 - x^3)$$

cubic inches. If the water flows out a hole at the bottom of the bowl at the rate of 36π in.^{3}/s, how fast is x decreasing when $x = 6$ (in.)?

41. Falling sand forms a conical sandpile with height that always remains twice its radius r while both are increasing. If sand is falling onto the pile at the rate of 25π ft^3/min, how fast is r increasing when $r = 5$ (ft)?

Find the limits in Problems 42–47.

42. $\displaystyle\lim_{x\to 0} \frac{x - \tan x}{\sin x}$

43. $\displaystyle\lim_{x\to 0} x \cot 3x$

44. $\displaystyle\lim_{x\to 0} \frac{\sin 2x}{\sin 5x}$

45. $\displaystyle\lim_{x\to 0} x^2 \csc 2x \cot 2x$

46. $\displaystyle\lim_{x\to 0} x^2 \sin\frac{1}{x^2}$

47. $\displaystyle\lim_{x\to 0^+} \sqrt{x}\sin\frac{1}{x}$

In each of Problems 48–53, identify two functions f and g such that $h(x) = f(g(x))$, then apply the chain rule to find $h'(x)$.

48. $h(x) = \sqrt[3]{x + x^4}$

49. $h(x) = \dfrac{1}{\sqrt{x^2 + 25}}$

50. $h(x) = \sqrt{\dfrac{x}{x^2 + 1}}$

51. $h(x) = \sqrt[3]{(x - 1)^5}$

52. $h(x) = \dfrac{(x + 1)^{10}}{(x - 1)^{10}}$

53. $h(x) = \cos(x^2 + 1)$

54. The period T of oscillation (seconds) of a simple pendulum of length L (ft) is given by $T = 2\pi\sqrt{L/32}$. What is the rate of change of T with respect to L when $L = 4$ (ft)?

55. What is the rate of change of the volume $V = 4\pi r^3/3$ of a sphere with respect to its surface area $S = 4\pi r^2$?

56. What is an equation for the straight line through $(1, 0)$ that is tangent to the graph of

$$h(x) = x + \frac{1}{x}$$

at a point in the first quadrant?

57. A rocket is launched vertically upward from a point 2 miles west of an observer on the ground. What is the speed of the rocket when the angle of elevation (from the horizontal) of the observer's line of sight to the rocket is 50° and is increasing at 5° per second?

58. An oil field containing 20 wells has been producing 4000 barrels of oil daily. For each new well drilled, the daily production of each well decreases by 5 barrels. How many new wells should be drilled in order to maximize the total daily production of the oil field?

59. A triangle is inscribed in a circle of radius R with one side of the triangle coincident with a diameter of the circle. What, in terms of R, is the maximum possible area of such a triangle?

60. Five rectangular pieces of sheet metal measure 210 by 336 cm each. Equal squares are to be cut from all their corners, and the resulting five cross-shaped pieces of metal are to be folded and welded to form five boxes without tops. The 20 little squares that remain are to be assembled in groups of four into five larger squares, and these five larger squares are to be assembled into a cubical box with no top. What is the maximum possible total volume of the six boxes that are constructed in this way?

61. A mass of clay of volume V is formed into two spheres. For what distribution of clay is the total surface area of the two spheres a maximum? A minimum?

62. A right triangle has legs of lengths 3 and 4 m. What is the maximum possible area of a rectangle inscribed in the triangle in the "obvious" way—with one corner at the triangle's right angle, two adjacent sides of the triangle lying on the triangle's legs, and the opposite corner on the hypotenuse?

63. What is the maximum possible volume of a right circular cone inscribed in a sphere of radius R?

64. A farmer has 400 ft of fencing with which to build a rectangular corral. He will use part or even all of an existing straight wall 100 ft long as part of the perimeter of the corral. What is the maximum area that he can enclose?

65. In one simple model of the spread of a contagious disease among members of a population of M people, the incidence of the disease, measured as the number of new cases per day, is given in terms of the number x of individuals already infected by

$$R(x) = kx(M - x) = kMx - kx^2,$$

where k is some positive constant. How many individuals in the population are infected when the incidence R is the greatest?

66. Three sides of a trapezoid have length L, a constant. What should be the length of the fourth side if the trapezoid is to have maximum area?

67. A box with no top must have a base twice as long as it is wide, and the total surface area of the box is to be 54 ft^2. What is the maximum possible volume of such a box?

68. A small right circular cone is inscribed in a larger one, as shown in Fig 3.66. The larger cone has fixed radius R and fixed altitude H. What is the largest fraction of the volume of the larger cone that the smaller one can occupy?

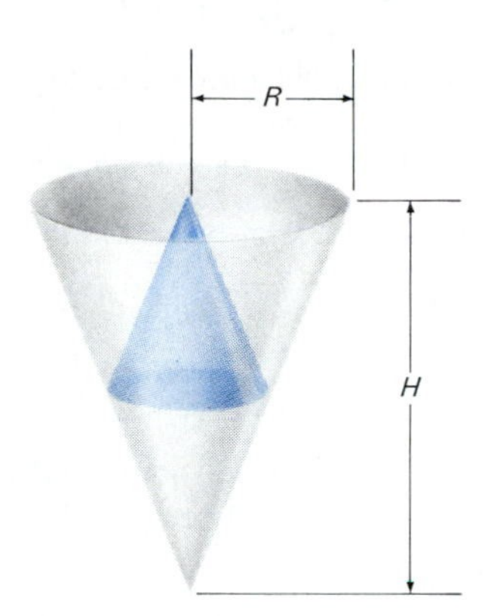

3.66 A small cone inscribed in a larger one

69. Two vertices of a trapezoid are at $(-2, 0)$ and $(2, 0)$ and the other two lie on the semicircle $x^2 + y^2 = 4$, $y \geqq 0$. What is the maximum possible area of the trapezoid? (NOTE The area of a trapezoid of height h and with bases b_1 and b_2 is $A = h(b_1 + b_2)/2$.)

70. Suppose that f is a differentiable function defined on the whole real number line **R** and that the graph of f contains a point $Q(x, y)$ closest to the point $P(x_0, y_0)$ not on the graph. Show that

$$f'(x) = -\frac{x - x_0}{y - y_0}$$

at Q. Conclude that the segment PQ is perpendicular to the tangent line to the curve at Q. (*Suggestion:* Minimize the square of the distance PQ.)

71. Use the result of Problem 70 to show that the minimum distance from the point (x_0, y_0) to a point of the straight line $Ax + By + C = 0$ is

$$\frac{|Ax_0 + By_0 + C|}{\sqrt{A^2 + B^2}}.$$

72. A race track is to be built in the shape of two parallel and equal straightaways connected by semicircles on each end. The length of the track, one lap, is to be exactly 5 km. What should be its design to maximize the rectangular area within it, as shown in Fig. 3.67?

3.67 Design the race track to maximize the shaded area.

73. Two towns are located on the straight shore of a lake. Their nearest distance to points on the shore are 1 mi and 2 mi, respectively, and these points on the shore are 6 mi apart. Where should a fishing pier be located to minimize the total amount of paving necessary to build a straight road from each town to the pier?

74. A hiker finds herself in a forest 2 km from a long straight road. She wants to walk to her cabin 10 km away in the forest, and it's also 2 km from the road. She can walk 8 km/h along the road but only 3 km/h through the forest. So she decides to walk first to the road, then along the road, and finally through the forest to the cabin (see Fig 3.68). What should be the angle θ shown in the figure in order to minimize the total time required for the hiker to reach her cabin? How much time is saved in comparison with the straight route through the forest?

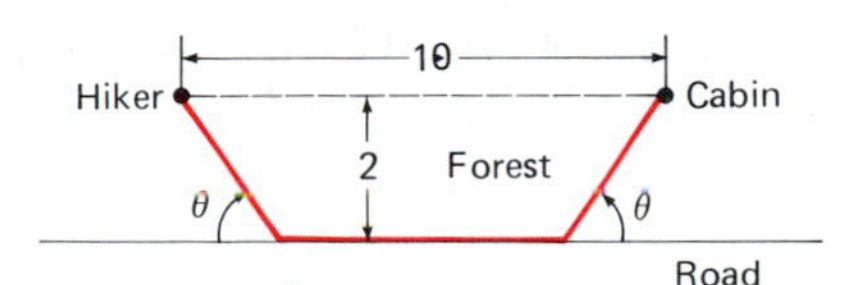

3.68 The hiker's quickest path to the cabin

75. When an arrow is shot from the origin with initial velocity v and initial angle of inclination α (from the horizontal x-axis, which represents the ground), then its trajectory is the curve

$$y = mx - \frac{16}{v^2}(1 + m^2)x^2$$

where $m = \tan \alpha$.

(a) Find the maximum height reached by the arrow (in terms of m and v).

(b) For what value of m (and hence, for what α) does the arrow travel the greatest horizontal distance?

76. A projectile is fired with initial velocity v and angle of elevation θ from the base of a plane inclined at 45° from the horizontal, as shown in Fig. 3.69. The range of the projectile, measured up this slope, is given by

$$R = \frac{v^2}{16}\sqrt{2}(\cos\theta \sin\theta - \cos^2\theta).$$

What value of θ maximizes R?

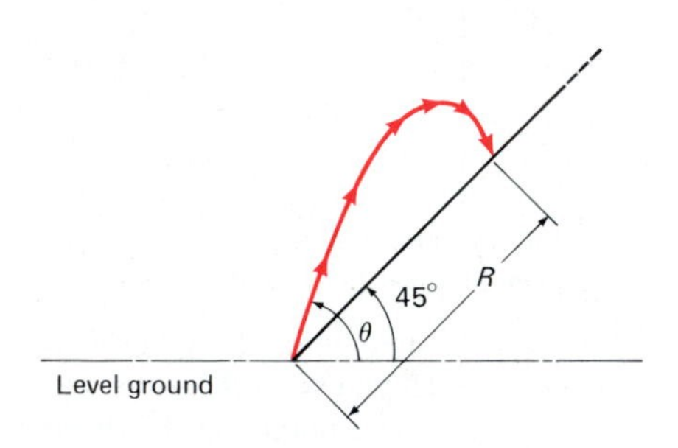

3.69 A projectile fired uphill

In each of Problems 77–88, use Newton's method to find the solution of the given equation $f(x) = 0$ in the indicated interval $[a, b]$ accurate to four decimal places.

77. $x^2 - 7 = 0$; $[2, 3]$ (to find the positive square root of 7)

78. $x^3 - 3 = 0$; $[1, 2]$ (to find the cube root of 3)

79. $x^5 - 75 = 0$; $[2, 3]$ (to find the fifth root of 75)

80. $x^{4/3} - 10 = 0$; $[5, 6]$ (to approximate $10^{3/4}$)

81. $x^3 - 3x - 1 = 0$; $[-1, 0]$

82. $x^3 - 4x - 1 = 0$; $[-1, 0]$

83. $x^6 + 7x^2 - 4 = 0$; $[0, 1]$

84. $x^3 - 3x^2 + 2x + 10 = 0$; $[-2, -1]$

85. $x + \cos x = 0$; $[-2, 0]$

86. $x^2 + \sin x = 0$; $[-1.0, -0.5]$

87. $4x - \sin x + 4 = 0$; $[-2, -1]$

88. $5x - \cos x + 5 = 0$; $[-1, 0]$

89. Find the depth to which a wooden ball with radius 2 ft sinks in water if its density is one-third that of water. A useful formula appears in Problem 37 of Section 3.9.

90. The equation $x^2 + 1 = 0$ has no real solutions. Try finding a solution by using Newton's method and report what happens. Use the initial estimate $x_0 = 2$.

91. At the beginning of Section 3.9 we mentioned the fifth-degree equation

$$x^5 - 3x^3 + x^2 - 23x + 19 = 0,$$

which has at least one and no more than five distinct real solutions. Find *all* real solutions by Newton's method.

92. The equation

$$\tan x = \frac{1}{x}$$

has a sequence $\alpha_1, \alpha_2, \alpha_3, \ldots$ of positive roots, with α_n slightly larger than $(n - 1)\pi$. Use Newton's method to approximate α_1 and α_2 to three-place accuracy.

93. Criticize the following "proof" that 3 = 2. Begin by writing

$$x^3 = x \cdot (x^2) = x^2 + x^2 + \cdots + x^2 \qquad (x \text{ summands}).$$

Differentiate to obtain

$$3x^2 = 2x + 2x + \cdots + 2x \qquad (x \text{ summands still}).$$

Thus $3x^2 = 2x^2$, and "therefore" 3 = 2.

If we substitute $z = x + h$ in the definition of the derivative, the result is

$$f'(x) = \lim_{z \to x} \frac{f(z) - f(x)}{z - x}.$$

Use this formula in Problems 94–95, together with the formula

$$a^3 - b^3 = (a - b)(a^2 + ab + b^2)$$

for factoring the difference of two cubes.

94. Show that

$$Dx^{3/2} = \lim_{z \to x} \frac{z^{3/2} - x^{3/2}}{z - x} = \frac{3}{2}x^{1/2}.$$

(*Suggestion:* Factor the numerator as a difference of cubes and the denominator as a difference of squares.)

95. Prove that

$$Dx^{2/3} = \lim_{z \to x} \frac{z^{2/3} - x^{2/3}}{z - x} = \frac{2}{3}x^{-1/3}.$$

(*Suggestion:* Factor the numerator as a difference of squares and the denominator as a difference of cubes.)

96. A rectangular block with square base is being squeezed in such a way that its height y is decreasing at the rate of 2 in./min while its volume remains constant. At what rate is the edge x of its base increasing when $x = 30$ and $y = 20$ (in.)?

97. Air is being pumped into a spherical balloon at the constant rate of 10 in.^{3}/s. At what rate is the surface area of the balloon increasing when its radius is 5 in.?

98. A ladder 10 ft long is leaning against a wall. If the bottom of the ladder slides away from the wall at the constant rate of 1 mi/h, how fast (mi/h) is the top of the ladder moving when it is 0.01 ft above the ground?

99. A water tank has the shape of an inverted cone (axis vertical and vertex downward) with a top radius of 5 ft and height 10 ft. The water is flowing out of the tank, through a hole at the vertex, at the rate of 50 ft^3/min. What is the time rate of change of the depth of the water in the tank at the instant when it is 6 ft deep?

100. Plane A is flying west toward an airport at an altitude of 2 mi; plane B is flying south toward the same airport at an altitude of 3 mi. When both planes are 2 mi (ground distance) from the airport, the speed of plane A is 500 mi/h and the distance between the two planes is decreasing at 600 mi/h. What is the speed of plane B?

101. A water tank is shaped so that the volume of water in the tank is $V = 2y^{3/2}$ cubic inches when its depth is y inches. If water flows out a hole at the bottom at the rate of $3\sqrt{y}$ cubic inches per minute, at what rate does the water level in the tank fall? Can you think of a practical application of such a water tank?

102. Water is being poured into the conical tank of Problem 99 at the rate of 50 ft^3/min and is draining out the hole at the bottom at the rate of $10\sqrt{y}$ cubic feet per minute, where y is the depth of water in the tank.

(a) At what rate is the water level rising when the water is 5 ft deep?

(b) Suppose that the tank is initially empty, water is poured in at 25 ft^3/min, and water continues to drain at $10\sqrt{y}$ cubic feet per minute. What is the maximum depth attained by the water?

103. Let L be a straight line passing through the fixed point $P(x_0, y_0)$ and also tangent to the parabola $y = x^2$ at the point $Q(a, a^2)$.

(a) Show that $a^2 - 2ax_0 + y_0 = 0$.

(b) Apply the quadratic formula to show that if $y_0 < x_0^2$ (that is, if P lies below the parabola), then there are two possible values for a and thus two lines through P that are tangent to the parabola.

(c) Similarly, show that if $y_0 > x_0^2$ (P lies above the parabola), then no line through P is also tangent to the parabola.

104. (*Challenge*) What is the shape of the tank of Problem 101? That is, assume that the tank is obtained by rotating about the y-axis the graph of $y = f(x)$, a nonnegative function such that $f(0) = 0$. Find a formula for f.

chapter four

Applications of Derivatives and Antiderivatives

4.1 Introduction

In Chapter 3 you learned to differentiate a wide variety of algebraic and trigonometric functions. You saw that derivatives have such diverse applications as maximum–minimum problems, related rates problems, and the solution of equations by Newton's method. The further applications of differentiation that we discuss in this chapter all depend ultimately upon a single fundamental question. Suppose that $y = f(x)$ is a differentiable function defined on the closed interval $[a, b]$ of length $\Delta x = b - a$. Then the *increment* Δy in the value of $f(x)$ as x changes from $x = a$ to $x = b = a + \Delta x$ is

$$\Delta y = f(b) - f(a). \tag{1}$$

The question is this: How is the increment Δy related to the derivative—the rate of change—of the function f at the points of the interval $[a, b]$?

An *approximate* answer is given in Section 4.2. If the function continued throughout the interval with the same rate of change $f'(a)$ that it had at $x = a$, then the change in its value would be $f'(a)(b - a) = f'(a)\,\Delta x$. This observation motivates the tentative approximation

$$\Delta y \approx f'(a)\,\Delta x. \tag{2}$$

A precise answer to the question above is provided by the mean value theorem of Section 4.3. This theorem implies that the exact increment is given by

$$\Delta y = f'(c)\,\Delta x \qquad (3)$$

for some point c in (a, b). The mean value theorem is the central theoretical result of differential calculus, and is also the key to many of the more advanced applications of derivatives.

In Section 4.8 we introduce the operation of **antidifferentiation**—the inverse of differentiation. If $g'(x) = f(x)$, then the function g is called an **antiderivative** of f. The importance of antiderivatives results partly from the fact that scientific laws often specify the *rates of change* of quantities. The quantities themselves are then found by antidifferentiation. For example, in Section 4.9 we begin with the given acceleration of a moving particle. From this we find its velocity and then its position function, using successive antidifferentiation.

4.2 Increments, Differentials, and Linear Approximation

Sometimes we need a quick and simple estimate of the change in $f(x)$ that results from a given change in x. Write y for $f(x)$, and suppose that the change in the independent variable is the *increment* Δx, so that x changes from its original value to the new value $x + \Delta x$. The actual change in the value of y is the **increment** Δy, computed by subtracting the old value of y from its new value:

$$\Delta y = f(x + \Delta x) - f(x). \qquad (1)$$

The increments Δx and Δy are represented geometrically in Fig. 4.1.

Now we compare the actual increment Δy with the change that *would* occur in y *if* it continued to change at the *fixed* rate $f'(x)$ as x changes to $x + \Delta x$. This hypothetical change in y is the **differential**

$$dy = f'(x)\,\Delta x. \qquad (2)$$

As Fig. 4.2 shows, dy is the change in height of a point that moves along the tangent line at the point $(x, f(x))$, rather than along the curve $y = f(x)$.

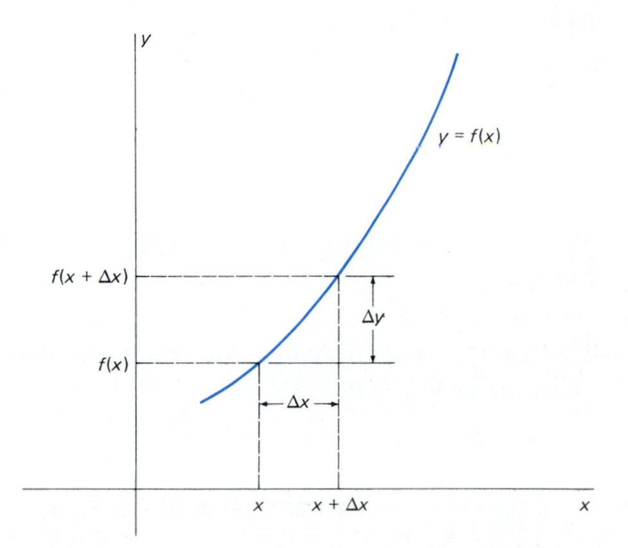

4.1 The increments Δx and Δy

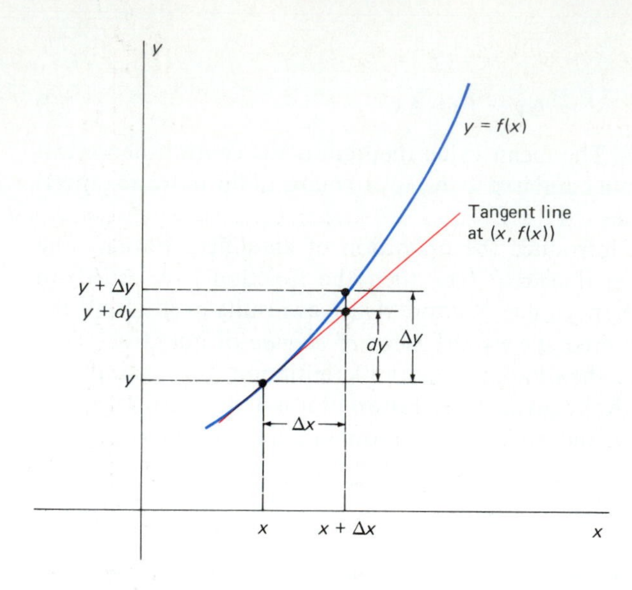

4.2 The estimate dy of the actual increment Δy

Think of x as fixed. Then (2) shows that the differential dy is a *linear* function of the increment Δx. For this reason, dy is called the **linear approximation** to the actual increment Δy. We can approximate $f(x + \Delta x)$ by substituting dy for Δy:

$$f(x + \Delta x) = y + \Delta y \approx y + dy.$$

Because $y = f(x)$ and $dy = f'(x)\,\Delta x$, this gives the **linear approximation formula**

$$f(x + \Delta x) \approx f(x) + f'(x)\,\Delta x. \tag{3}$$

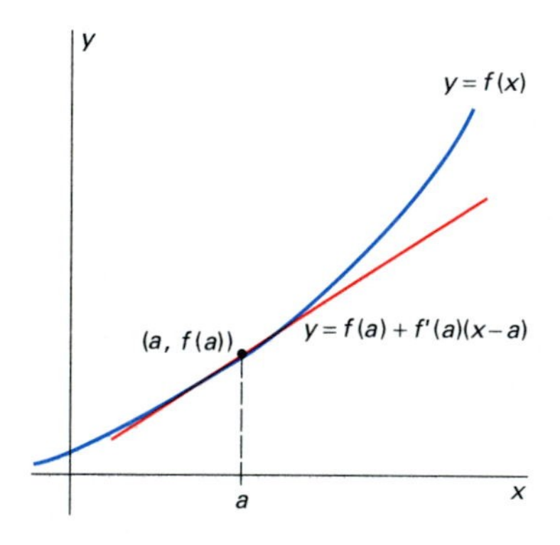

4.3 The linear approximation $y = f(a) + f'(a)(x - a)$

The idea is that this last approximation is a "good" one, at least when Δx is relatively small. If we combine (1), (2), and (3), we see that

$$\Delta y \approx f'(x)\,\Delta x = dy. \tag{4}$$

Thus the differential $dy = f'(x)\,\Delta x$ is an approximation to the actual increment $\Delta y = f(x + \Delta x) - f(x)$.

If we replace x with a in (3), we get the approximation

$$f(a + \Delta x) \approx f(a) + f'(a)\,\Delta x. \tag{5}$$

If we now write $\Delta x = x - a$, so that $x = a + \Delta x$, the result is

$$f(x) \approx f(a) + f'(a)(x - a). \tag{6}$$

This is the **linear approximation to the function** $f(x)$ **near the point** $x = a$. See Fig. 4.3.

EXAMPLE 1 Find the linear approximation to the function $f(x) = \sqrt{1 + x}$ near the point $a = 0$.

Solution Note that $f(0) = 1$ and

$$f'(x) = \frac{1}{2}(1 + x)^{-1/2} = \frac{1}{2\sqrt{1 + x}},$$

so $f'(0) = \frac{1}{2}$. Hence Equation (6) with $a = 0$ yields

$$\sqrt{1 + x} \approx 1 + \tfrac{1}{2}x. \tag{7}$$

It is important to realize that the linear approximation in (7) is likely to be accurate only if x is close to zero. For instance, the approximations

$$\sqrt{1.1} \approx 1 + \tfrac{1}{2}(0.1) = 1.05 \quad \text{and} \quad \sqrt{1.03} \approx 1 + \tfrac{1}{2}(0.03) = 1.015$$

are accurate to two and three decimal places (rounded), respectively. But

$$\sqrt{3} \approx 1 + \tfrac{1}{2}(2) = 2$$

is a very poor approximation to $\sqrt{3} \approx 1.73205$.

The approximation $\sqrt{1 + x} \approx 1 + \frac{1}{2}x$ is a special case of the approximation

$$(1 + x)^k \approx 1 + kx \tag{8}$$

(k is a constant, x is near zero), one with numerous applications. The derivation of (8) is similar to Example 1 (Problem 39).

EXAMPLE 2 Use the linear approximation formula to approximate $(122)^{2/3}$. Note that

$$(125)^{2/3} = [(125)^{1/3}]^2 = (5)^2 = 25.$$

Solution We need to approximate a particular value of $x^{2/3}$, so our strategy is to apply (6) with $f(x) = x^{2/3}$. We first note that $f'(x) = \frac{2}{3}x^{-1/3}$. We choose $a = 125$ because we know the exact values

$$f(125) = (125)^{2/3} = 25 \quad \text{and} \quad f'(125) = \tfrac{2}{3}(125)^{-1/3} = \tfrac{2}{15}$$

and because 125 is "close" to 122. Then the linear approximation in (6) to $f(x) = x^{2/3}$ near $a = 125$ takes the form

$$x^{2/3} \approx 25 + \tfrac{2}{15}(x - 125).$$

With $x = 122$ we get

$$(122)^{2/3} \approx 25 + \tfrac{2}{15}(-3) = 24.6;$$

that is, $(122)^{2/3}$ is approximately 24.6. The actual value of $(122)^{2/3}$ is about 24.59838 (the digits given are correct), so the formula in (6) gives us a relatively good approximation.

EXAMPLE 3 A hemispherical bowl of radius 10 in. is filled with water to a depth of x inches. The volume V of water in the bowl (in cubic inches) is given by the formula

$$V = \frac{\pi}{3}(30x^2 - x^3)$$

—see Fig. 4.4. (You will be able to derive this formula after you study Chapter 6). Suppose that you *measure* the depth of water in the bowl, and you find it to be 5 in. with a maximum possible measurement error of $\frac{1}{16}$ in. Estimate the maximum error in the calculated volume of water in the bowl.

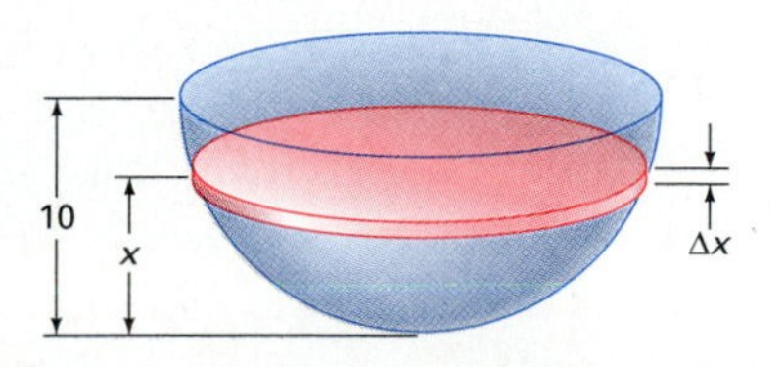

4.4 The bowl of Example 3

Solution The error ΔV in the calculated volume V caused by an error Δx in the measured depth x is approximately equal to the differential

$$dV = V'(x)\,\Delta x = \frac{\pi}{3}(60x - 3x^2)\,\Delta x = \pi(20x - x^2)\,\Delta x.$$

We take $x = 5$ and $\Delta x = \pm\frac{1}{16}$, and obtain

$$dV = \pi[(20)(5) - (5)^2](\pm\tfrac{1}{16}) \approx \pm 14.73$$

cubic inches. With $x = 5$, the formula for V gives $V(5) \approx 654.50$ in.^{3}, but we see now that this may be in error by almost 15 in.^{3} either way.

The **absolute error** in a measured or approximated value is defined to be the remainder when the approximate value is subtracted from the true value. The **relative error** is the ratio of the absolute error to the true value. Thus in Example 2, a relative error in the measured depth x of

$$\frac{\Delta x}{x} = \frac{\frac{1}{16}}{5} = 0.0125 = 1.25\%$$

leads to a relative error in the estimated volume of

$$\frac{dV}{V} = \frac{14.73}{654.50} = 0.0225 = 2.25\%.$$

The relationship between these two relative errors is of some interest. The formulas for dV and V in Example 3 give

$$\frac{dV}{V} = \frac{\pi(20x - x^2)\,\Delta x}{\frac{1}{3}\pi(30x^2 - x^3)} = \frac{3(20 - x)}{30 - x} \cdot \frac{\Delta x}{x}.$$

In particular, when $x = 5$, this gives

$$\frac{dV}{V} = (1.80)\frac{\Delta x}{x}.$$

Hence, in order to approximate the volume of water in the bowl with a relative error of at most 0.5%, we would need to measure the depth with a relative error of at most (0.5%)/1.8, thus with a relative error of less than 0.3%.

THE ERROR IN LINEAR APPROXIMATION

Now we take up the question of how closely the differential dy approximates the actual increment Δy. It is apparent in Fig. 4.2 that the smaller is Δx, the closer are the corresponding points on the curve $y = f(x)$ and its tangent line. Because the difference in the heights of two such points is the value of $\Delta y - dy$ determined by a particular choice of Δx, we conclude that $\Delta y - dy \to 0$ as $\Delta x \to 0$.

But even more is true: As $\Delta x \to 0$, the difference $\Delta y - dy$ is small *even in comparison with* Δx. For

$$\frac{\Delta y - dy}{\Delta x} = \frac{f(x + \Delta x) - f(x) - f'(x)\,\Delta x}{\Delta x} = \frac{f(x + \Delta x) - f(x)}{\Delta x} - f'(x).$$

That is,

$$\frac{\Delta y - dy}{\Delta x} = \varepsilon(\Delta x) \tag{9}$$

where, by the definition of the derivative $f'(x)$, we see that $\varepsilon = \varepsilon(\Delta x)$ is a function of Δx that approaches zero as $\Delta x \to 0$. We multiply both sides in (9) by Δx, write ε for $\varepsilon(\Delta x)$, and obtain

$$\Delta y = dy + \varepsilon \cdot \Delta x \tag{10}$$

where $\varepsilon \to 0$ as $\Delta x \to 0$. If Δx is "very small," so that ε is also "very small," we might well describe the product $\varepsilon \cdot \Delta x = \Delta y - dy$ as "very *very* small." These concepts and quantities are illustrated in Fig. 4.5.

4.5 The error $\varepsilon \cdot \Delta x$ in the linear approximation $\Delta y \approx f'(x)\,\Delta x$

EXAMPLE 4 Suppose that $y = x^3$. Verify that $\varepsilon = (\Delta y - dy)/\Delta x$ approaches zero as $\Delta x \to 0$.

Solution Simple computations give

$$\Delta y = (x + \Delta x)^3 - x^3 = 3x^2\,\Delta x + 3x(\Delta x)^2 + (\Delta x)^3$$

and $dy = 3x^2\,\Delta x$. Hence

$$\varepsilon = \frac{\Delta y - dy}{\Delta x} = \frac{3x(\Delta x)^2 + (\Delta x)^3}{\Delta x} = 3x\,\Delta x + (\Delta x)^2,$$

which obviously approaches zero as $\Delta x \to 0$.

DIFFERENTIALS

The linear approximation formula in (3) is often written with dx in place of Δx, in the form

$$f(x + dx) \approx f(x) + f'(x)\,dx. \tag{11}$$

In this case dx is an independent variable, called the **differential** of x, while x is fixed. Thus the differentials of x and y are defined to be

$$dx = \Delta x \quad \text{and} \quad dy = f'(x)\,\Delta x = f'(x)\,dx. \tag{12}$$

With this definition it follows immediately that

$$\frac{dy}{dx} = \frac{f'(x)\,dx}{dx} = f'(x),$$

in agreement with the notation we have been using. Indeed, Leibniz originated differential notation by visualizing infinitesimal increments dx and dy as in Fig. 4.6, with their ratio dy/dx being the slope of the tangent line.

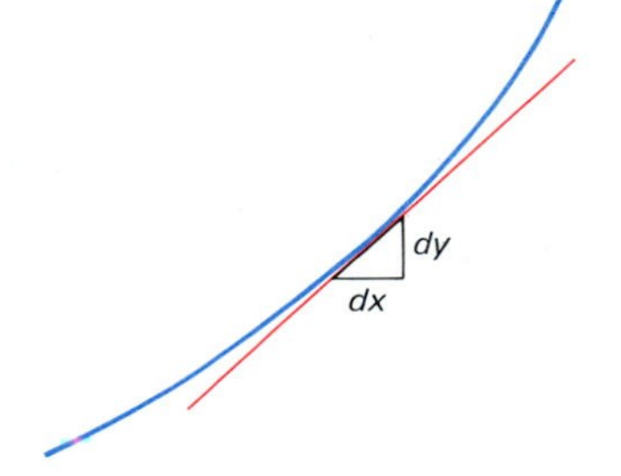

4.6 The slope of the tangent line as the ratio of the infinitesimals *dy* and *dx*

Differential notation also provides us with a convenient way to write derivative formulas. For suppose that $z = g(u)$, so that $dz = g'(u)\,du$. For particular choices of the function g, we get the formulas

$$d(u^n) = nu^{n-1}\,du, \tag{13}$$

$$d(\sin u) = (\cos u)\,du, \tag{14}$$

and so on. This allows us to write differentiation rules in "differential form," without the need to identify the independent variable. The sum, product, and quotient rules take the forms

$$d(u + v) = du + dv, \tag{15}$$

$$d(uv) = u\,dv + v\,du, \quad \text{and} \tag{16}$$

$$d\left(\frac{u}{v}\right) = \frac{v\,du - u\,dv}{v^2}. \tag{17}$$

If $u = f(x)$ and $z = g(u)$, we may substitute $du = f'(x)\,dx$ in the formula $dz = g'(u)\,du$. This gives

$$dz = g'(f(x)) \cdot f'(x)\,dx. \tag{18}$$

This is the differential form of the chain rule

$$Dg(f(x)) = g'(f(x)) \cdot f'(x).$$

Thus the chain rule appears here as though it were the result of mechanical manipulations of the differential notation. This compatibility with the chain rule is one reason for the extraordinary usefulness of differential notation in calculus.

4.2 PROBLEMS

In Problems 1–16, write dy in terms of x and dx.

1. $y = 3x^2 - \dfrac{4}{x^2}$

2. $y = 2\sqrt{x} - \dfrac{3}{\sqrt[3]{x}}$

3. $y = x - \sqrt{4 - x^3}$

4. $y = \dfrac{1}{x - \sqrt{x}}$

5. $y = 3x^2(x - 3)^{3/2}$

6. $y = \dfrac{x}{x^2 - 4}$

7. $y = x(x^2 + 25)^{1/4}$

8. $y = \dfrac{1}{(x^2 - 1)^{4/3}}$

9. $y = \cos\sqrt{x}$

10. $y = x^2 \sin x$

11. $y = \sin 2x \cos 2x$

12. $y = \cos^3 3x$

13. $y = \dfrac{\sin 2x}{3x}$

14. $y = \dfrac{\cos x}{\sqrt{x}}$

15. $y = \dfrac{1}{1 - x \sin x}$

16. $y = (1 + \cos 2x)^{3/2}$

In each of Problems 17–24, find the linear approximation (as in Example 1) to the given function f near the point $a = 0$.

17. $f(x) = \dfrac{1}{1 - x}$

18. $f(x) = \dfrac{1}{\sqrt{1 + x}}$

19. $f(x) = (1 + x)^2$

20. $f(x) = (1 - x)^3$

21. $f(x) = (1 - 2x)^{3/2}$

22. $f(x) = \dfrac{1}{(1 + 3x)^{2/3}}$

23. $f(x) = \sin x$

24. $f(x) = \cos x$

In each of Problems 25–34, find—as in Example 2—a linear approximation to the indicated number.

25. $\sqrt[3]{25}$

26. $\sqrt{102}$

27. $(15)^{1/4}$

28. $\sqrt{80}$

29. $(65)^{-2/3}$

30. $(80)^{3/4}$

31. $\cos 43°$

32. $\sin 32°$

33. $\sin 88°$

34. $\cos 62°$

In Problems 35–38, compute the differential of each side of the given equation, regarding x and y as *dependent* variables (as if both were functions of some third unspecified variable). Then solve for dy/dx.

35. $x^2 + y^2 = 1$

36. $x^{2/3} + y^{2/3} = 4$

37. $x^3 + y^3 = 3xy$

38. $x \sin y = 1$

39. Assuming the fact (to be established in Chapter 7) that $Dx^k = kx^{k-1}$ for any real constant k, derive the linear approximation formula $(1 + x)^k \approx 1 + kx$ for x near zero.

In each of Problems 40–47, use linear approximation to estimate the change in the given quantity.

40. The circumference of a circle, if its radius is increased from 10 to 10.5 in.

41. The area of a square if its edge is decreased from 10 to 9.8 in.

42. The surface area of a sphere, if its radius is increased from 5 to 5.2 in.

43. The volume of a cylinder, if both its height and its radius are decreased from 15 to 14.7 cm.

44. The volume of a conical sandpile of height 7 in. and radius 14 in. if its height is increased to 7.1 in.

45. The range

$$R = \tfrac{1}{16}v^2 \sin 2\theta$$

of a shell fired at inclination angle $\theta = 45°$, if its initial velocity v is increased from 80 to 81 ft/s.

46. The range

$$R = \tfrac{1}{16}v^2 \sin 2\theta$$

of a projectile fired with initial velocity $v = 80$ ft/s, if its initial inclination angle θ is increased from 45° to 46°.

47. The wattage $W = RI^2$ of a floodlight with resistance $R = 10$ ohms, if the current I is increased from 3 to 3.1 amperes.

48. The equatorial radius of the earth is approximately 3960 miles. Suppose that a wire is wrapped tightly around the earth at the equator. Approximately how much must this wire be lengthened if it is to be strung on poles 10 ft above the ground, all the way around the earth? Use the linear approximation formula!

49. The radius of a spherical ball is measured as 10 in., with a maximum error of $\frac{1}{16}$ in. What is the maximum resulting error in its calculated volume?

50. With what accuracy must the radius of the ball of Problem 49 be measured to ensure an error of at most 1 in.^{3} in its calculated volume?

51. The radius of a hemispherical dome is measured as 100 m with a maximum error of 1 cm. What is the maximum resulting error in its calculated surface area?

52. With what accuracy must the radius of a hemispherical dome be measured in order to have an error of at most 0.01% in its calculated surface area?

4.3 The Mean Value Theorem and Applications

The significance of the *sign* of the first derivative is simple but crucial:

f is increasing on an interval where $f'(x) > 0$;

f is decreasing on an interval where $f'(x) < 0$.

Geometrically, this means that where $f'(x) > 0$, the graph of f is rising, as you scan it from left to right. Where $f'(x) < 0$, the graph is falling. We can make the terms *increasing* and *decreasing* more precise, as follows.

Definition *Increasing and Decreasing Functions*

The function f is **increasing** on the interval I if, for each two numbers u and v in I with $u < v$,

$$f(u) < f(v).$$

That is,

$$u < v \quad \text{implies} \quad f(u) < f(v).$$

The function f is **decreasing** on I provided that

$$u < v \quad \text{implies} \quad f(u) > f(v)$$

for any two points u and v in I.

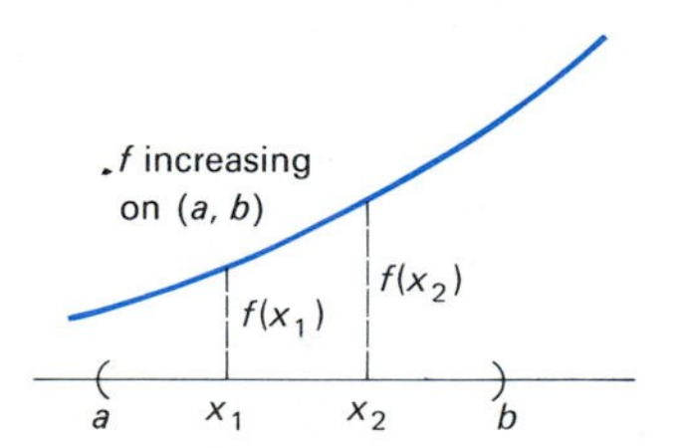

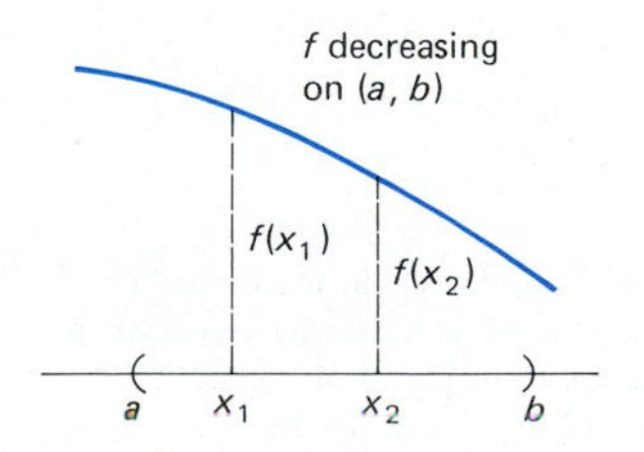

4.7 An increasing function and a decreasing function

Figure 4.7 illustrates this definition.

Note that we speak of a function as increasing or decreasing *on an interval*, not at a single point. Nevertheless, if we consider the sign of f' at a single point, we get a useful intuitive picture of the significance of the sign of the derivative. This is because the derivative $f'(x)$ is the slope of the tangent line at the point $(x, f(x))$ on the graph of f. If $f'(x) > 0$, then the tangent line has positive slope. Therefore it rises as you scan from left to right. Intuitively, it seems obvious that a rising tangent corresponds to a rising graph and thus to an increasing function. Similarly, we expect to see a falling graph where $f'(x)$ is negative. Figure 4.8 shows a pair of graphs illustrating this intuitive idea. One caution: In order to determine whether a function f is increasing or decreasing, we must examine the sign of f' on a whole interval, not merely at a single point (see Problem 45).

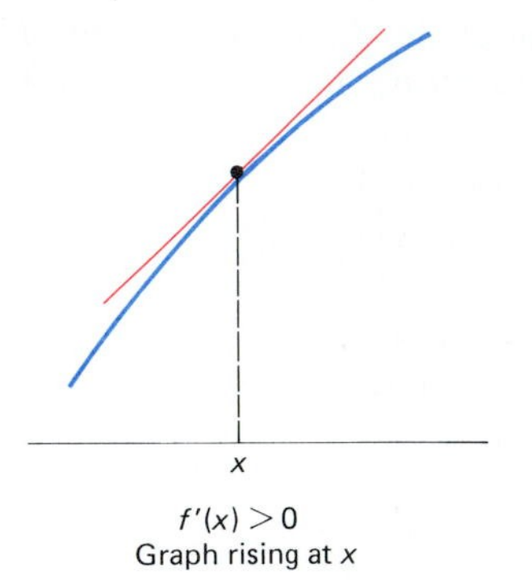

$f'(x) > 0$
Graph rising at x

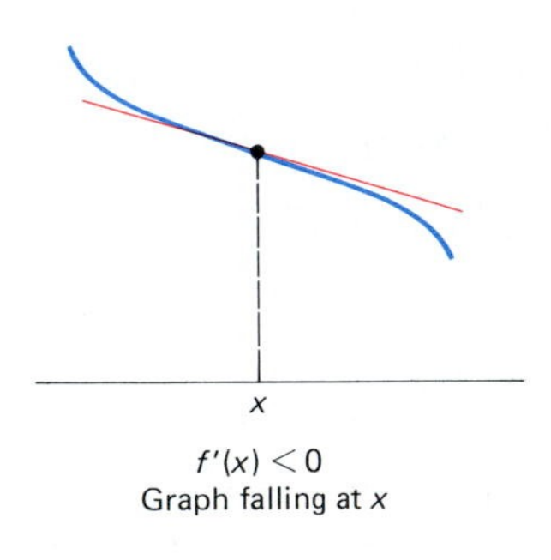

$f'(x) < 0$
Graph falling at x

4.8 A graph rising at x and a graph falling at x

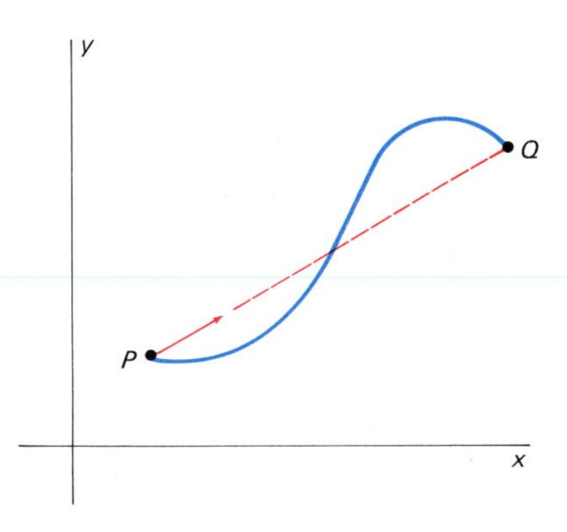

4.9 Can you sail from P to Q without ever sailing—even for an instant—in the direction PQ (the direction of the arrow)?

THE MEAN VALUE THEOREM

Although pictures of rising and falling graphs are suggestive, they provide no actual proof of the significance of the sign of the derivative. To establish the connection between the graph's rising or falling and the sign of the derivative, we need the **mean value theorem,** stated later in this section. This theorem is the principal theoretical tool of differential calculus, and we shall see that it has many important applications.

As an introduction to the mean value theorem, we pose the following question. Suppose that P and Q are two points in the plane, with Q lying generally to the east of P, as shown in Fig. 4.9. Is it possible to sail a boat from P to Q, sailing always roughly east, without *ever* (even for an instant) sailing in the exact direction from P to Q? That is, can we sail from P to Q without our instantaneous line of motion ever being parallel to the line PQ?

The mean value theorem says that the answer to this question is no; there will always be at least one instant when we are sailing parallel to the line PQ, no matter which path we choose.

Here is a mathematical paraphrase: Let the path of the sailboat be the graph of a differentiable function $y = f(x)$ with endpoints $P(a, f(a))$ and $Q(b, f(b))$. Then we say that there must be some point on this graph where the tangent line to the curve is parallel to the line PQ joining its end points. This is a *geometric interpretation* of the mean value theorem.

But the slope of the tangent line at the point $(c, f(c))$, shown in Fig. 4.10, is $f'(c)$, while the slope of the line PQ is

$$\frac{f(b) - f(a)}{b - a}.$$

We may think of this last quotient as the average (or *mean*) value of the slope of f. The mean value theorem guarantees that there is a point c in (a, b) for which the tangent line at $(c, f(c))$ is indeed parallel to the line PQ. In the language of algebra, there's a number c in (a, b) such that

$$f'(c) = \frac{f(b) - f(a)}{b - a}. \tag{1}$$

We first give a preliminary result, a lemma to expedite the proof of the mean value theorem. This lemma is called **Rolle's theorem,** after the Frenchman Michel Rolle (1652–1719), who discovered it in 1690. Rolle studied the emerging subject of calculus in his youth but later renounced it; he argued that it was based on logical fallacies, and today he is remembered only for a single theorem that bears his name. It is ironic that his theorem plays an important role in the rigorous proofs of several calculus theorems.

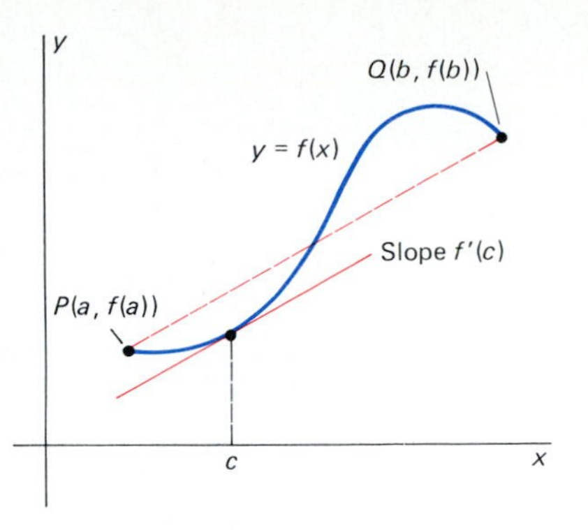

4.10 The sailboat problem in mathematical terminology

> ***Rolle's Theorem***
>
> Suppose that the function f is continuous on the closed interval $[a, b]$ and is differentiable on its interior (a, b). If $f(a) = 0 = f(b)$, then there exists some number c in (a, b) such that $f'(c) = 0$.

Thus between each pair of zeros of a differentiable function, there is *at least one* point at which the tangent line is horizontal. Some possible pictures of the situation are indicated in Fig. 4.11.

Proof Because f is continuous on $[a, b]$, it must attain both a maximum and a minimum value on $[a, b]$ (by the maximum value property of Section 3.5). If f has any positive values at all, consider its maximum value $f(c)$. Now c is not an end point of $[a, b]$ because $f(a) = 0$ and $f(b) = 0$. Therefore, c is a point of (a, b). But we know also that f is differentiable at c, so it follows from Theorem 2 of Section 3.5 that $f'(c) = 0$.

Similarly, if f has any negative values, we may consider its minimum value $f(c)$ and conclude that $f'(c) = 0$.

If f has neither positive nor negative values, then f is identically zero on $[a, b]$, and it follows that $f'(c) = 0$ for *every* c in (a, b).

Thus we see that the conclusion of Rolle's theorem is justified in every case. ■

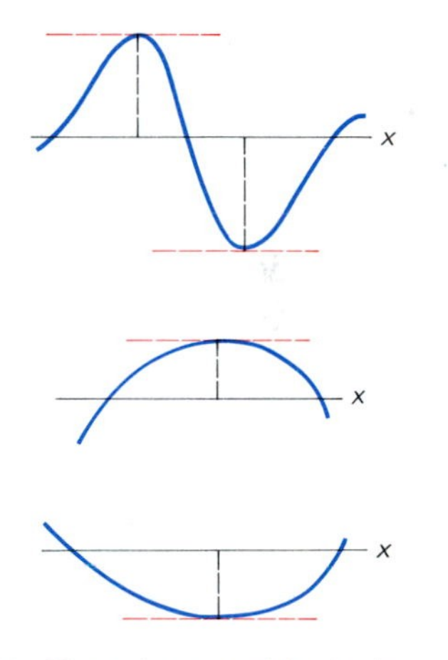

4.11 The existence of the horizontal tangent is a consequence of Rolle's theorem.

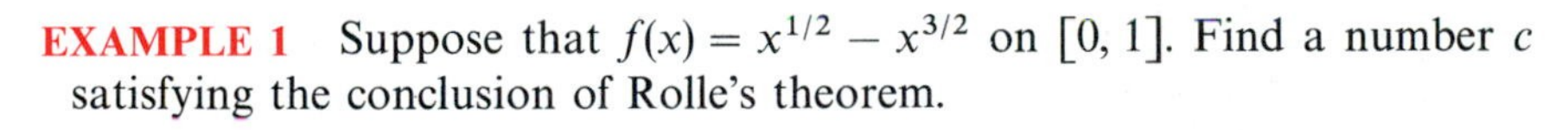

EXAMPLE 1 Suppose that $f(x) = x^{1/2} - x^{3/2}$ on $[0, 1]$. Find a number c satisfying the conclusion of Rolle's theorem.

Solution Note that f is continuous on $[0, 1]$ and differentiable on $(0, 1)$. Because of the presence of the term $x^{1/2}$, f is *not* differentiable at $x = 0$, but this is irrelevant. Also $f(0) = 0 = f(1)$, so all hypotheses of Rolle's theorem are satisfied. Finally,

$$f'(x) = \tfrac{1}{2}x^{-1/2} - \tfrac{3}{2}x^{1/2} = \tfrac{1}{2}x^{-1/2}(1 - 3x),$$

so we see that $f'(c) = 0$ for $c = \frac{1}{3}$.

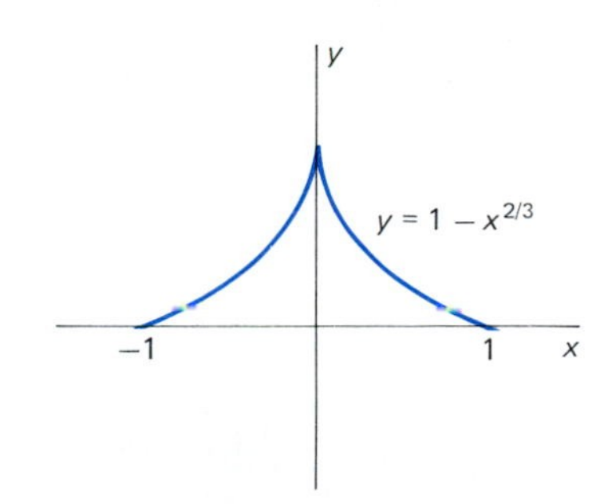

4.12 The function of Example 2; Rolle's hypotheses are not satisfied.

EXAMPLE 2 Suppose that $f(x) = 1 - x^{2/3}$ on $[-1, 1]$. Then f satisfies the hypotheses of Rolle's theorem *except* for the fact that $f'(0)$ does not exist. It is clear from the graph of f (Fig. 4.12) that there is *no* point where the tangent line is horizontal. Indeed,

$$f'(x) = -\frac{2}{3}x^{-1/3} = -\frac{2}{3x^{1/3}},$$

so $f'(x) \neq 0$ for $x \neq 0$, and we see that $|f'(x)| \to \infty$ as $x \to 0$, so the graph of f has a vertical tangent line—rather than a horizontal one—at the point $(0, 1)$.

Now we are ready to state formally and prove the mean value theorem.

> ***The Mean Value Theorem***
>
> Suppose that the function f is continuous on the closed interval $[a, b]$ and differentiable on the open interval (a, b). Then
>
> $$f(b) - f(a) = f'(c)(b - a) \qquad (2)$$
>
> for some number c in (a, b).

COMMENT Because Equation (2) is equivalent to Equation (1), the conclusion of the mean value theorem is that there must be at least one point on the curve $y = f(x)$ at which the tangent line is parallel to the line joining its end points $P(a, f(a))$ and $Q(b, f(b))$.

Motivation for the Proof We consider the auxiliary function ϕ suggested by Fig. 4.13. The value $\phi(x)$ is, by definition, the vertical height difference over x of the point $(x, f(x))$ on the curve and the corresponding point on the line PQ. It appears that a point on the curve $y = f(x)$, where the tangent line is parallel to PQ, corresponds to a maximum or minimum of ϕ. It's clear also that $\phi(a) = 0 = \phi(b)$, so Rolle's theorem can be applied to the function ϕ on $[a, b]$. So our plan for proving the mean value theorem is this: First, we obtain a formula for the function ϕ. Second, we locate the point c such that $\phi'(c) = 0$. Finally, we show that this number c is exactly the number needed to satisfy the conclusion of the theorem in Equation (2).

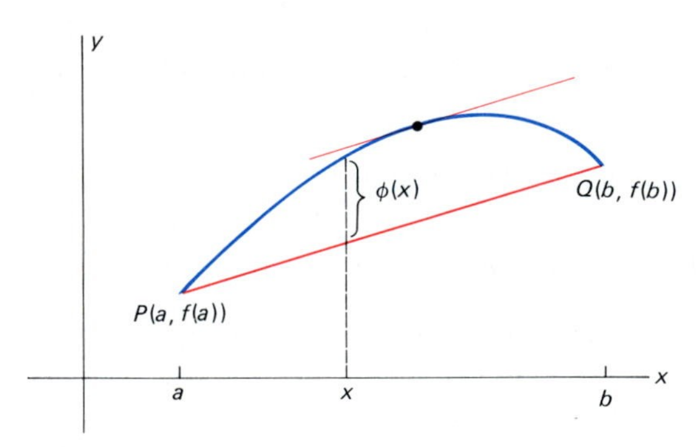

4.13 The construction of the auxiliary function ϕ

Proof of the Mean Value Theorem Because the line PQ passes through $(a, f(a))$ with slope

$$m = \frac{f(b) - f(a)}{b - a},$$

the point-slope formula for the equation of a straight line gives us the following equation for PQ:

$$y = y_{\text{line}} = f(a) + m(x - a).$$

Thus

$$\phi(x) = y_{\text{curve}} - y_{\text{line}} = f(x) - f(a) - m(x - a).$$

You may verify by direct substitution that $\phi(a) = 0 = \phi(b)$. And, as ϕ is certainly continuous on $[a, b]$ and differentiable on (a, b), we may apply Rolle's theorem to it. Thus there is a point c somewhere in the open interval (a, b) at which $\phi'(c) = 0$. But

$$\phi'(x) = f'(x) - m = f'(x) - \frac{f(b) - f(a)}{b - a}.$$

Because $\phi'(c) = 0$, we may conclude that

$$0 = f'(c) - \frac{f(b) - f(a)}{b - a}.$$

That is,

$$f(b) - f(a) = f'(c)(b - a). \quad ■$$

Observe that the proof of the mean value theorem is an application of Rolle's theorem, while Rolle's theorem is the special case of the mean value theorem in which $f(a) = 0 = f(b)$.

EXAMPLE 3 Suppose that we drive from Kristiansand, Norway to Oslo—a road distance of almost exactly 350 km—in a time of exactly 4 h, from time $t = 0$ to time $t = 4$. Let $f(t)$ denote the distance we have traveled at time t, and assume that f is a differentiable function. Then the mean value theorem implies that

$$350 = f(4) - f(0) = f'(c)(4 - 0) = 4f'(c)$$

and thus that

$$f'(c) = \tfrac{350}{4} = 87.5$$

at some instant c in $(0, 4)$. But $f'(c)$ is our *instantaneous* velocity at time $t = c$ and 87.5 km/h is our *average* velocity for the trip. Thus the mean value theorem implies that we must have an instantaneous velocity of exactly 87.5 km/h at least once during the trip.

CONSEQUENCES OF THE MEAN VALUE THEOREM

The first of these consequences is the *non*trivial converse of the trivial fact that the derivative of a constant function is identically zero. That is, we prove that there can be *no* unknown exotic function that is nonconstant but which has derivative identically zero. In Corollaries 1–3 we assume that f and g are continuous on the closed interval $[a, b]$ and differentiable in its interior.

Corollary 1 *Functions with Zero Derivative*
If $f'(x) = 0$ for all x in (a, b), then f is a constant function on $[a, b]$. That is, there exists a constant C such that $f(x) = C$ for all x in $[a, b]$.

Proof Apply the mean value theorem to the function f on the interval $[a, x]$ where x is a fixed but arbitrary point of the interval $(a, b]$. We find that

$$f(x) - f(a) = f'(c)(x - a)$$

for some number c between a and x. But f' is always zero on the interval (a, b), so $f'(c) = 0$. Thus $f(x) - f(a) = 0$, so $f(x) = f(a)$.

But this last equation holds for *all* x in $(a, b]$. Therefore, $f(x) = f(a)$ for all x in $(a, b]$; indeed, for all x in $[a, b]$. That is, $f(x)$ has the fixed value $C = f(a)$. This establishes Corollary 1. ■

The ease and simplicity of this proof suggest that the mean value theorem is a powerful tool.

Corollary 1 is usually applied in a different but equivalent form, which we state and prove next.

Corollary 2 *Functions with Equal Derivatives*

Suppose that $f'(x) = g'(x)$ for all x in the open interval (a, b). Then f and g differ by a constant on $[a, b]$. That is, there exists a constant K such that

$$f(x) = g(x) + K$$

for all x in $[a, b]$.

Proof Given the hypotheses, let $h(x) = f(x) - g(x)$. Then

$$h'(x) = f'(x) - g'(x) = 0$$

for all x in (a, b) because $f'(x) = g'(x)$ for such x. So, by Corollary 1, $h(x)$ is a constant K on $[a, b]$. That is, $f(x) - g(x) = K$ for all x in $[a, b]$, and therefore

$$f(x) = g(x) + K$$

for all x in $[a, b]$. This establishes Corollary 2. ■

The following consequence of the mean value theorem verifies the remarks about increasing and decreasing functions with which we opened this section.

Corollary 3 *Increasing and Decreasing Functions*

If $f'(x) > 0$ for all x in (a, b), then f is an increasing function on $[a, b]$. If $f'(x) < 0$ for all x in (a, b), then f is a decreasing function on $[a, b]$.

Proof Suppose, for example, that $f'(x) > 0$ for all x in (a, b). We need to show the following: If u and v are points of $[a, b]$ with $u < v$, then $f(u) < f(v)$. We apply the mean value theorem to f, but on the closed interval $[u, v]$. This gives

$$f(v) - f(u) = f'(c)(v - u)$$

for some c in (u, v). Because $v > u$ and because, by hypothesis, $f'(c) > 0$, it follows that

$$f(v) - f(u) > 0; \quad \text{that is,} \quad f(u) < f(v),$$

as we wanted to show. The proof is similar in the case that $f'(x)$ is negative on (a, b). ■

EXAMPLE 4 Where is the function $f(x) = x^2 - 4x + 5$ increasing and where is it decreasing?

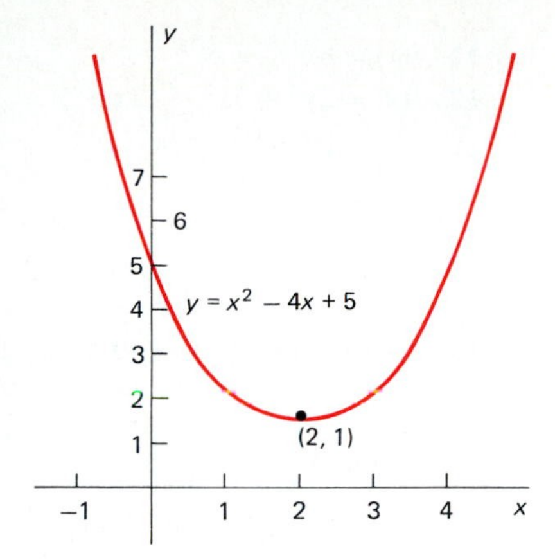

4.14 The parabola of Example 4

Solution The derivative of f is $f'(x) = 2x - 4$. Clearly $f'(x) > 0$ if $x > 2$, while $f'(x) < 0$ if $x < 2$. Hence f is decreasing on $(-\infty, 2)$ and increasing on $(2, +\infty)$. Together with the observations that $f(2) = 1$ and $f'(2) = 0$, we have enough information to sketch a qualitatively accurate graph of f. It appears in Fig. 4.14.

EXAMPLE 5 Show that the equation $x^3 + x - 1 = 0$ has exactly one (real) solution.

Solution Let $f(x) = x^3 + x - 1$. Because $f(0) = -1 < 0$, $f(1) = +1 > 0$, and f is continuous (everywhere), the intermediate value property guarantees that $f(x)$ has at *least* one zero in $[0, 1]$. But

$$f'(x) = 3x^2 + 1 > 0$$

for all x, so Corollary 3 implies that f is increasing on the whole real line. Thus $f(x)$ cannot have more than one zero. (Why?)

EXAMPLE 6 Determine the open intervals of the x-axis on which the function

$$f(x) = 3x^4 - 4x^3 - 12x^2 + 5$$

is increasing and those on which it is decreasing.

Solution The derivative of f is

$$\begin{aligned} f'(x) &= 12x^3 - 12x^2 - 24x \\ &= 12x(x^2 - x - 2) = 12x(x + 1)(x - 2). \end{aligned}$$

The critical points $x = -1, 0, 2$ separate the x-axis into the open intervals $(-\infty, -1)$, $(-1, 0)$, $(0, 2)$, and $(2, +\infty)$. In the following table we have recorded the sign of each of the three factors of $f'(x)$ on each of these four open intervals. We note the resulting sign of $f'(x)$ on each interval, then apply Corollary 3. It follows that f is decreasing on $(-\infty, -1)$ and on $(0, 2)$, while f is increasing on $(-1, 0)$ and on $(2, +\infty)$.

Interval	$x + 1$	$12x$	$x - 2$	$f'(x)$	f
$(-\infty, -1)$	Neg.	Neg.	Neg.	Neg.	Decreasing
$(-1, 0)$	Pos.	Neg.	Neg.	Pos.	Increasing
$(0, 2)$	Pos.	Pos.	Neg.	Neg.	Decreasing
$(2, +\infty)$	Pos.	Pos.	Pos.	Pos.	Increasing

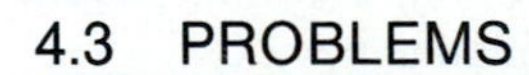

4.3 PROBLEMS

For each of the functions in Problems 1–14, determine (as in Example 6) the open intervals on the x-axis on which it is increasing as well as those on which it is decreasing.

1. $f(x) = 3x + 2$

2. $f(x) = 4 - 5x$

3. $f(x) = 4 - x^2$

4. $f(x) = 4x^2 + 8x + 13$

5. $f(x) = 6x - 2x^2$

6. $f(x) = x^3 - 12x + 17$

7. $f(x) = x^4 - 2x^2 + 1$

8. $f(x) = \dfrac{x}{x + 1}$

9. $f(x) = 3x^4 + 4x^3 - 12x^2$

10. $f(x) = x\sqrt{x^2 + 1}$

11. $f(x) = 8x^{1/3} - x^{4/3}$

12. $f(x) = 2x^3 + 3x^2 - 12x + 5$

13. $f(x) = \dfrac{(x-1)^2}{x^2 - 3}$

14. $f(x) = x^2 + \dfrac{16}{x^2}$

In Problems 15–18, show that the given function satisfies the hypotheses of Rolle's theorem on the indicated interval $[a, b]$, and find all numbers c in $[a, b]$ that satisfy the conclusion of the theorem.

15. $f(x) = x^2 - 2x$ on $[0, 2]$

16. $f(x) = 9x^2 - x^4$ on $[-3, 3]$

17. $f(x) = \dfrac{1 - x^2}{1 + x^2}$ on $[-1, 1]$

18. $f(x) = 5x^{2/3} - x^{5/3}$ on $[0, 5]$

In Problems 19–21, show that the given function f does not satisfy the conclusion of Rolle's theorem on the indicated interval. Which of the hypotheses does it fail to satisfy?

19. $f(x) = 1 - |x|$ on $[-1, 1]$

20. $f(x) = 1 - (2 - x)^{2/3}$ on $[1, 3]$

21. $f(x) = x^4 + x^2$ on $[0, 1]$

In Problems 22–26, show that the given function f satisfies the hypotheses of the mean value theorem on the indicated interval, and find all numbers c in (a, b) that satisfy the conclusion of that theorem.

22. $f(x) = x^3$ on $[-1, 1]$

23. $f(x) = 3x^2 + 6x - 5$ on $[-2, 1]$

24. $f(x) = \sqrt{x - 1}$ on $[2, 5]$

25. $f(x) = (x - 1)^{2/3}$ on $[1, 2]$

26. $f(x) = x + \dfrac{1}{x}$ on $[1, 2]$

In Problems 27–30, show that the given function satisfies neither the hypotheses nor the conclusion of the mean value theorem on the indicated interval.

27. $f(x) = |x - 2|$ on $[1, 4]$

28. $f(x) = 1 + |x - 1|$ on $[0, 3]$

29. $f(x) = [\![x]\!]$ (the greatest integer function) on $[-1, 1]$

30. $f(x) = 3x^{2/3}$ on $[-1, 1]$

In each of Problems 31–33, show that the given equation has exactly one solution in the indicated interval.

31. $x^5 + 2x - 3 = 0$ in $[0, 1]$

32. $x^{10} = 1000$ in $[1, 2]$

33. $x^4 - 3x = 20$ in $[2, 3]$

34. Show that the function $f(x) = x^{2/3}$ does not satisfy the hypotheses of the mean value theorem on $[-1, 27]$, but nevertheless there is a number c in $(-1, 27)$ such that

$$f'(c) = \frac{f(27) - f(-1)}{27 - (-1)}.$$

35. Prove that the function

$$f(x) = (1 + x)^{3/2} - \tfrac{3}{2}x - 1$$

is increasing on $(0, +\infty)$. Conclude that

$$(1 + x)^{3/2} > 1 + \tfrac{3}{2}x$$

for all $x > 0$.

36. Suppose that f' is a constant function on the interval $[a, b]$. Prove that f must be a linear function (a function having a straight line as its graph).

37. Suppose that $f'(x)$ is a polynomial of degree $n - 1$ on the interval $[a, b]$. Prove that $f(x)$ must be a polynomial of degree n on $[a, b]$.

38. Suppose that there are k different points of $[a, b]$ at which the differentiable function f vanishes (is zero). Prove that f' must vanish on at least $k - 1$ points of $[a, b]$.

39. (a) Apply the mean value theorem to $f(x) = \sqrt{x}$ on $[100, 101]$ to show that

$$\sqrt{101} = 10 + \frac{1}{2\sqrt{c}}$$

for some number c in $(100, 101)$.

(b) Show that if $100 < c < 101$, then $10 < \sqrt{c} < 10.5$, and use this fact to conclude from part (a) that $10.0475 < \sqrt{101} < 10.0500$.

40. Prove that the equation $x^7 + x^5 + x^3 + 1 = 0$ has exactly one real solution.

41. (a) Show that $D\tan^2 x = D\sec^2 x$ on the open interval $(-\pi/2, \pi/2)$.

(b) Conclude that there exists a constant C such that

$$\tan^2 x = \sec^2 x + C$$

for all x in $(-\pi/2, \pi/2)$. Then evaluate C.

42. Explain why the mean value theorem does not apply to the function $f(x) = |x|$ on the interval $[-1, 2]$.

43. Suppose that the function f is differentiable on the interval $[-1, 2]$, with $f(-1) = -1$ and $f(2) = 5$. Prove that there is a point on the graph of f at which the tangent line is parallel to the line with equation $y = 2x$.

44. Let $f(x) = x^4 - x^3 + 7x^2 + 3x - 11$. Prove that the graph of f has at least one horizontal tangent line.

45. Let the function g be defined as follows:

$$g(x) = \frac{x}{2} + x^2 \sin\frac{1}{x}$$

for $x \neq 0$; $g(0) = 0$.

(a) Apply the result of Problem 78 in Section 3.7 to show that $g'(0) = \frac{1}{2} > 0$.

(b) Perhaps with the aid of Fig. 3.49, sketch the graph of g near $x = 0$. Is g increasing on any open interval containing $x = 0$? (*Answer:* No.)

46. Suppose that f is increasing on every closed interval $[a, b]$ provided that $2 \leqq a < b$. Prove that f is increasing on the unbounded open interval $(2, +\infty)$. Note that the principle you discover was used implicitly in Example 4 of this section.

4.4 The First Derivative Test

Suppose that f is a function differentiable on the open interval I with a local extremum somewhere in I. Theorem 2 of Section 3.5 tells us this: The extremum must occur at the sort of critical point at which $f'(x) = 0$. But if $f'(c) = 0$ for c in I, it does *not* follow that there is an extremum at c. What we need is a way of finding, if $f'(c) = 0$, whether $f(c)$ is a local maximum value of f, a local minimum value, or neither.

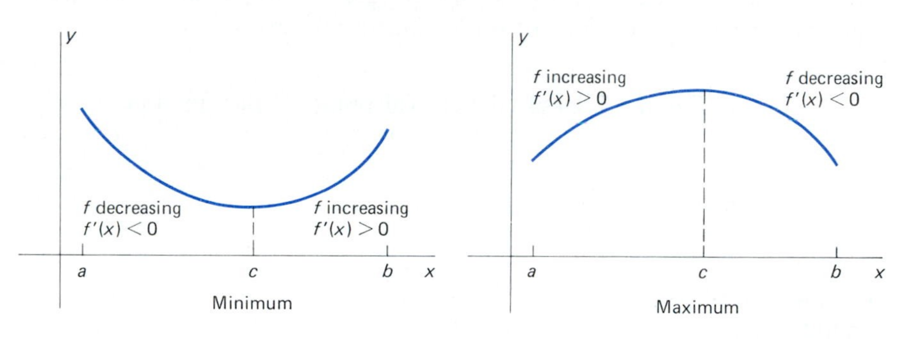

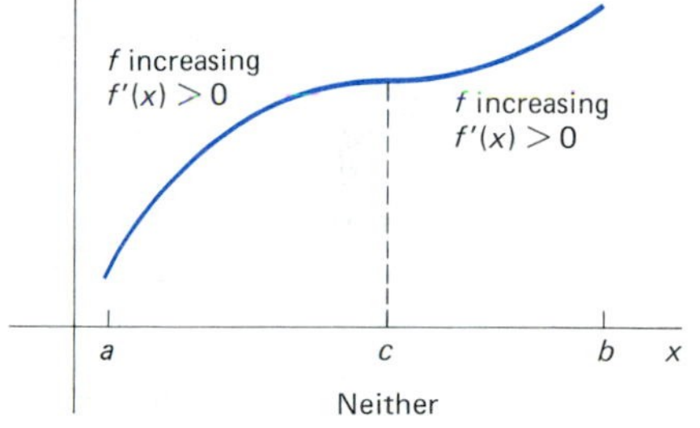

4.15 The first derivative test

Figure 4.15 suggests how such distinctions might be developed. If f is decreasing to the left of c and increasing to the right, then $f(c)$ should be a local minimum value of f. On the other hand, if f is increasing to the left of c and decreasing to the right, then $f(c)$ should be a local maximum. Moreover, with the aid of Corollary 3 to the mean value theorem, we can tell where f is increasing and where it is decreasing; such behavior is determined by the sign of $f'(x)$. In the following test for local maxima and minima, we denote by L and R the left and right open subintervals, respectively, into which the point c separates the given open interval I. See Fig. 4.16.

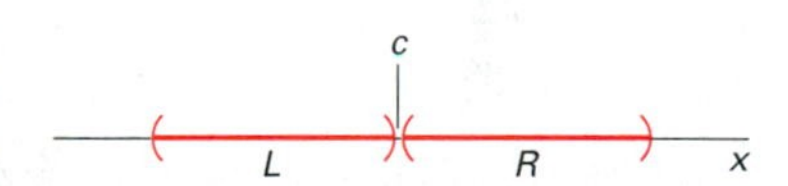

4.16 The left and right parts of the interval I with the point c deleted

Theorem *The First Derivative Test*

Suppose that the function f is continuous on the open interval I and is differentiable there except possibly at the point c.

1. If $f'(x) < 0$ on the left subinterval L and $f'(x) > 0$ on the right subinterval R, then $f(c)$ is the minimum value of f on I.
2. If $f'(x) > 0$ on L and $f'(x) < 0$ on R, then $f(c)$ is the maximum value of f on I.
3. If $f'(x) > 0$ on both L and R, or if $f'(x) < 0$ on both L and R, then $f(c)$ is neither a maximum nor a minimum value for f.

Thus $f(c)$ is a local extremum if the first derivative $f'(x)$ *changes sign* as x increases through c, and the direction of this sign change determines whether $f(c)$ is a local maximum or a local minimum. A good way to remember parts 1 and 2 of the first derivative test is simply to visualize Fig. 4.16.

Proof of the Theorem We will prove part 1; the other two parts are similar. Let z be a point of the interval I other than c. If $z < c$, then the fact that $f'(x) < 0$ on L implies that f is decreasing on $[z, c]$, and it follows that $f(z) > f(c)$. This result holds for all z in the left-hand subinterval L, so $f(z) > f(c)$ for all z in L. On the other hand, if $z > c$ then the fact that $f'(x) > 0$ on R implies that f is increasing on $[c, z]$, so it follows that $f(z) > f(c)$ for all z in R.

Therefore we have shown that $f(c) < f(z)$ for all points z of I other than c itself, and this means that $f(c)$ is the minimum value of $f(x)$ on I. The idea of this proof is illustrated in Fig. 4.17(a), which shows f decreasing to the left of c and increasing to the right. Figure 4.17(b) similarly illustrates part 2 of the Theorem, and Fig. 4.17(c) illustrates part 3. ■

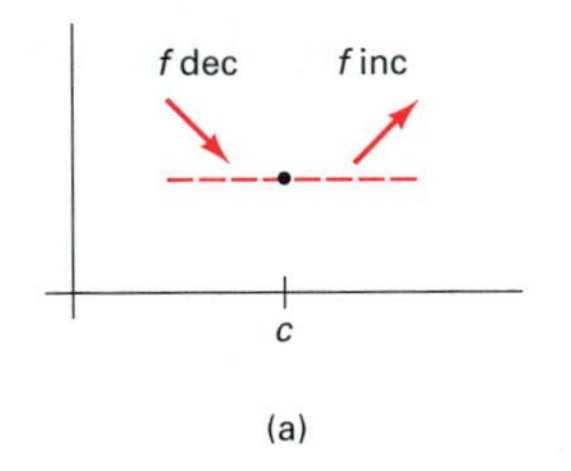

(a)

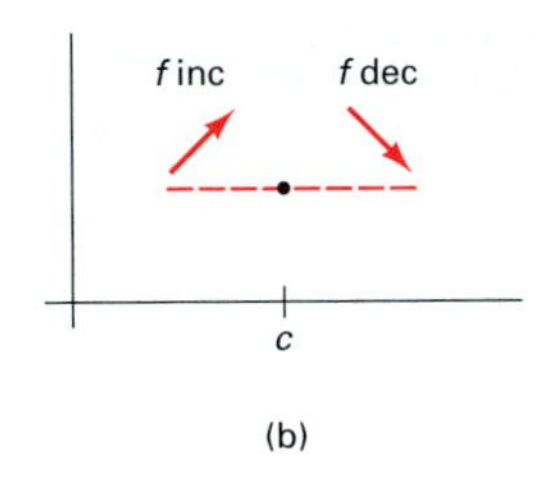

(b)

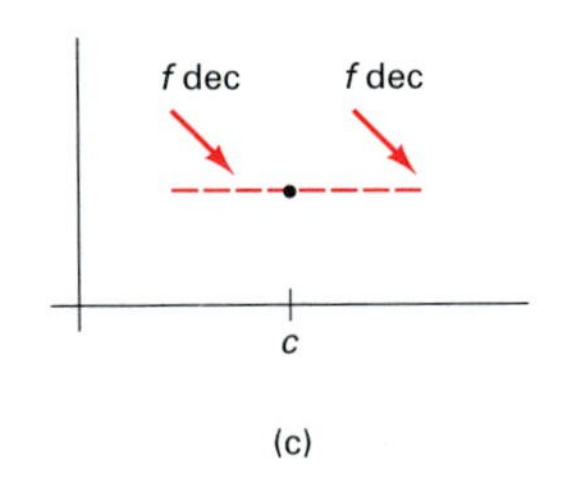

(c)

4.17 The three cases in the first derivative test

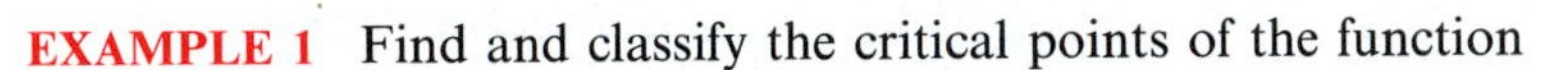

EXAMPLE 1 Find and classify the critical points of the function

$$f(x) = 2x^3 - 3x^2 - 36x + 7.$$

Solution The derivative is

$$f'(x) = 6x^2 - 6x - 36 = 6(x + 2)(x - 3),$$

so the critical points (where $f'(x) = 0$) are $x = -2$ and $x = 3$. These two points separate the x-axis into the three open intervals $(-\infty, -2)$, $(-2, 3)$, and $(3, +\infty)$. The next table shows the sign of $f'(x)$ in each of these intervals.

Interval	$x + 2$	$x - 3$	$f'(x)$	f
$(-\infty, -2)$	Neg.	Neg.	Pos.	Increasing
$(-2, 3)$	Pos.	Neg.	Neg.	Decreasing
$(3, +\infty)$	Pos.	Pos.	Pos.	Increasing

Thus $f'(x)$ is positive to the left and negative to the right of the critical point $x = -2$, so by part 2 of the theorem, $f(-2)$ is a local maximum value of f. At the critical point $x = 3$, $f(x)$ is negative to the left and positive to the right, so $f(3)$ is a local minimum value by part 1.

OPEN INTERVAL MAXIMUM–MINIMUM PROBLEMS

In Section 3.6 we discussed applied maximum–minimum problems in which the values of the dependent variable are given by a function defined on a closed interval. Sometimes, though, the function f describing the variable to be maximized is defined on an *open* interval (a, b), and we cannot "close" the interval by adjoining its end points. The typical reason is that $|f(x)| \to +\infty$ as x approaches a or b. But if f has only a single critical point in (a, b), then the first derivative test may tell us that $f(c)$ is the desired extremum, and determine whether it is a maximum or a minimum value.

EXAMPLE 2 Find the (absolute) minimum value of

$$f(x) = x + \frac{4}{x}$$

for $0 < x < +\infty$.

Solution The derivative is

$$f'(x) = 1 - \frac{4}{x^2} = \frac{x^2 - 4}{x^2}.$$

First we solve

$$f'(x) = 0.$$

The roots of the equation

$$\frac{x^2 - 4}{x^2} = 0$$

are $x = -2$ and $x = 2$. But only $x = 2$ is in the open interval $(0, +\infty)$. Because $f'(x) < 0$ to the left of $x = 2$ and $f'(x) > 0$ to its right, the first derivative test implies that $f(2) = 4$ is a local minimum value. Now we note also that $f(x) \to +\infty$ as either $x \to 0^+$ or $x \to +\infty$. Hence the graph of f must resemble the one shown in Fig. 4.18, and we see that $f(2) = 4$ is the absolute minimum value of $f(x)$ on the entire interval $(0, +\infty)$.

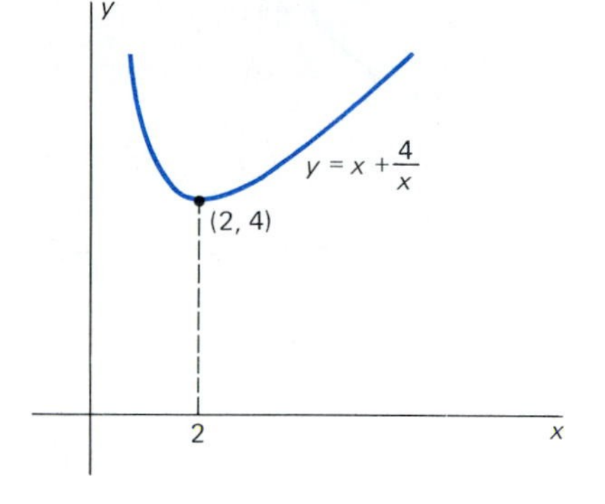

4.18 The graph of the function of Example 2

EXAMPLE 3 A cylindrical can with a volume of 125 in.^{3} (about 2 liters) is to be made by cutting its top and bottom from metal squares and forming its curved side by bending a rectangular sheet of metal to match its ends. What radius r and height h of the can will minimize the amount of material required?

Solution We assume that the corners cut from the two squares, shown in Fig. 4.19, are wasted, but that there is no other waste. As the figure shows, the area of the total amount of sheet metal required is

$$A = 8r^2 + 2\pi rh.$$

The volume of the resulting can is then

$$V = \pi r^2 h = 125,$$

so that $h = 125/\pi r^2$. Hence A is given as a function of r by

$$A(r) = 8r^2 + 2\pi r \cdot \frac{125}{\pi r^2} = 8r^2 + \frac{250}{r}.$$

Now r can have any positive value, so $A(r)$ is defined on the open interval $(0, +\infty)$. But $A \to +\infty$ as $r \to 0$ and as $r \to +\infty$. So we cannot use the closed-interval maximum–minimum method. But we *can* use the first derivative test.

The derivative of A is

$$\frac{dA}{dr} = 16r - \frac{250}{r^2} = \frac{16}{r^2}\left(r^3 - \frac{125}{8}\right).$$

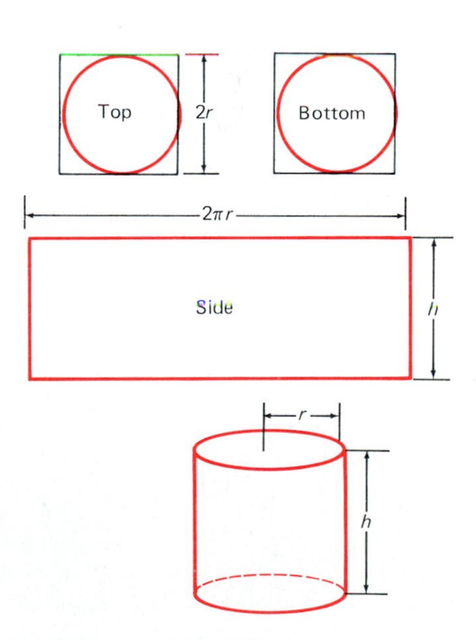

4.19 The parts to make the cylindrical can of Example 3

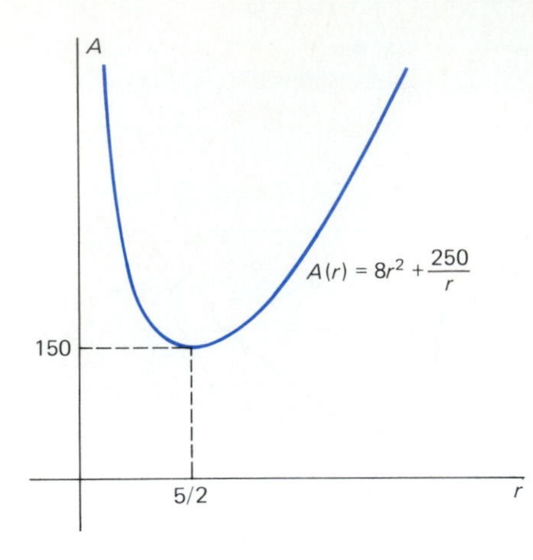

4.20 Graph of the function of Example 3

Thus the only critical point in $(0, +\infty)$ is at

$$r = \sqrt[3]{\frac{125}{8}} = \frac{5}{2} = 2.5.$$

It is clear that $dA/dr < 0$ if $0 < r < 2.5$, and that $dA/dr > 0$ if $r > 2.5$. Therefore, the first derivative test implies that the minimum value of $A(r)$ on $(0, +\infty)$ is

$$A(2.5) = (8)(2.5)^2 + \frac{250}{2.5} = 150.$$

Considering the fact that $A \to +\infty$ as $r \to 0^+$ and as $r \to +\infty$, we see that the graph of $A(r)$ on $(0, +\infty)$ looks like the one shown in Fig. 4.20. This clinches the fact that $A(2.5) = 150$ is an *absolute* minimum value. Therefore, we minimize the amount of material required by making a can with radius $r = 2.5$ in. and height

$$h = \frac{125}{\pi(2.5)^2} \approx 6.37$$

inches. The total amount of material used is 150 in.^{2}

EXAMPLE 4 Find the length of the longest rod that can be carried horizontally around the corner from a hall 2 m wide into one that is 4 m wide.

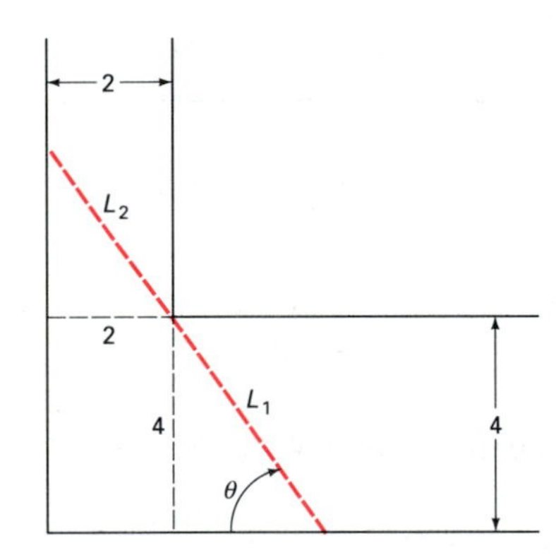

4.21 Carrying a rod around a corner

Solution The desired length will actually be the *minimum* length L of the dashed line in Fig. 4.21, which represents the rod being carried around the corner. From the two similar triangles in the figure we see that

$$L = L_1 + L_2 = 4 \csc \theta + 2 \sec \theta,$$

and thus that

$$L = L(\theta) = \frac{4}{\sin \theta} + \frac{2}{\cos \theta}.$$

The domain of L is the open interval $0 < \theta < \pi/2$. Clearly $L \to +\infty$ as either $\theta \to 0^+$ or $\theta \to (\pi/2)^-$.

We differentiate:

$$\frac{dL}{d\theta} = -\frac{4 \cos \theta}{\sin^2 \theta} + \frac{2 \sin \theta}{\cos^2 \theta} = \frac{2 \sin^3 \theta - 4 \cos^3 \theta}{\sin^2 \theta \cos^2 \theta}$$

so $dL/d\theta = 0$ when

$$2 \sin^3 \theta = 4 \cos^3 \theta, \qquad \tan \theta = \sqrt[3]{2},$$

and thus when $\theta \approx 0.90$ rad.

It's clear that $dL/d\theta < 0$ when $\theta < 0.90$, and that $dL/d\theta > 0$ when $0 > 0.90$. So the graph of L resembles the one shown in Fig. 4.22. This means that the minimum value of L, and therefore the maximum length of the rod in question, is about

$$L(0.90) = \frac{4}{\sin(0.90)} + \frac{2}{\cos(0.90)},$$

approximately 8.32 m.

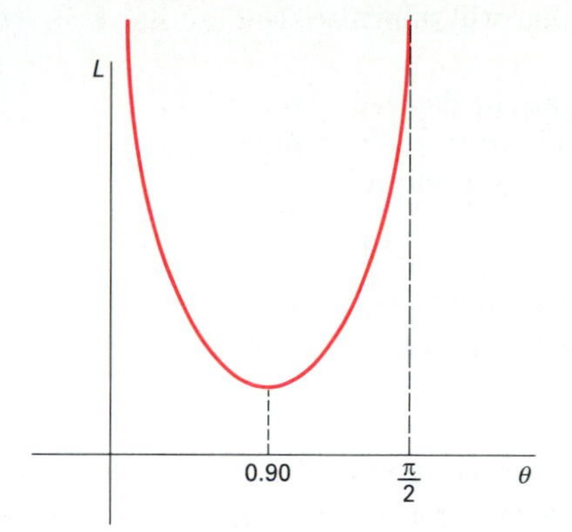

4.22 Graph of $L(\theta)$ (Example 4)

A final remark: When a function $f(x)$ has only a single critical point c in an open interval I, the first derivative test may apply to tell us either that $f(c)$ is the absolute minimum value of $f(x)$ on I (part 1 of the Theorem) or that $f(c)$ is the absolute maximum (part 2). But it is good practice to verify your conclusion by sketching the graph as we did in Examples 2–4.

4.4 PROBLEMS

Apply the first derivative test to classify each critical point of the functions in Problems 1–16.

1. $f(x) = x^2 - 4x + 5$

2. $f(x) = 6x - x^2$

3. $f(x) = x^3 - 3x^2 + 5$

4. $f(x) = x^3 - 3x + 5$

5. $f(x) = x^3 - 3x^2 + 3x + 5$

6. $f(x) = 2x^3 + 3x^2 - 36x + 17$

7. $f(x) = 10 + 60x + 9x^2 - 2x^3$

8. $f(x) = 27 - x^3$

9. $f(x) = x^4 - 2x^2$

10. $f(x) = 3x^5 - 5x^3$

11. $f(x) = x + \dfrac{1}{x}$

12. $f(x) = x + \dfrac{9}{x}$

13. $f(x) = x^2 + \dfrac{2}{x}$

14. $f(x) = x^2 + \dfrac{8}{x}$

15. $f(x) = 3 - x^{2/3}$

16. $f(x) = 4 + x^{1/3}$

In the following applied maximum–minimum problems, use the first derivative test to verify your solution.

17. Determine two real numbers with difference 20 and minimum possible product.

18. A long rectangular sheet of metal strip is to be made into a rain gutter by turning up two sides at right angles to the remaining center strip. The rectangular cross section of the gutter is to have area 18 in.^{2}. Find the minimum possible width of the strip.

19. Find the point (x, y) on the line $2x + y = 3$ that is closest to the point $(3, 2)$.

20. A closed rectangular box with volume 576 in.^{3} is to be made so its bottom is a rectangle with length twice its width. Find the dimensions of the box that will minimize its total surface area.

21. Repeat Problem 20, but use an open-topped box with volume 972 in.^{3}.

22. An open-topped cylindrical pot is to have volume 125 in.^{3}. What dimensions will minimize the total amount of material used in making this pot (neglect thickness and wastage)?

23. An open-topped cylindrical pot is to have volume 250 in.^{3}. The material for its bottom costs 4 cents/in.^{2}, while that for its curved side costs 2 cents/in.^{2} What dimensions will minimize the total cost of this pot?

24. Find the point (x, y) on the parabola $y = 4 - x^2$ that is closest to the point $(3, 4)$. (*Suggestion:* The cubic equation that you should obtain has a small integer as one of its roots.)

25. Show that the rectangle with area 100 and minimum perimeter is a square.

26. Show that the rectangular solid with square base and volume 1000 with least total surface area is a cube.

27. A box with square base and open top is to have volume 62.5 in.^{3} Neglect the thickness of the material used to make the box, and find the dimensions that will minimize the amount of material used.

28. A tin can in the shape of a right circular cylinder is to have volume 16π cm^3. What should be its radius r and height h to minimize its total surface area (including top and bottom)?

29. The metal used to make the top and bottom of a cylindrical can costs 4 cents/in.^{2}, while the metal used for the sides costs 2 cents/in.^{2} The volume of the can is to be exactly 100 in.^{3} What should be the dimensions of the can to minimize its cost?

30. The pages of a book are each to contain 30 in.^{2} of print, and each page must have 2-in. margins at top and bottom and 1-in. margins at each side. What is the minimum possible area of such a page?

31. What point or points on the curve $y = x^2$ are nearest the point (0, 2)? (*Suggestion:* The square of a distance is minimized exactly when the distance itself is minimized.)

32. What is the length of the shortest line segment lying wholly in the first quadrant with its endpoints on the coordinate axes and tangent to the graph of $y = 1/x$?

33. A rectangle has area 64 in.^{2} A straight line is to be drawn from one corner of the rectangle to the midpoint of one of the two more distant sides. What is the minimum possible length of such a line?

34. An oil can is to have volume 1000 in.^{3} and is to be shaped like a cylinder with a flat bottom but capped by a hemisphere. Neglect the thickness of the material of the can, and find the dimensions that will minimize the amount of material needed to make it.

35. Find the length L of the longest rod that can be carried horizontally around a corner from a corridor 2 m wide into one 4 m wide. Do this by *minimizing* the length of the rod, indicated in Fig. 4.23 by a dashed line, by minimizing its square as a function of x.

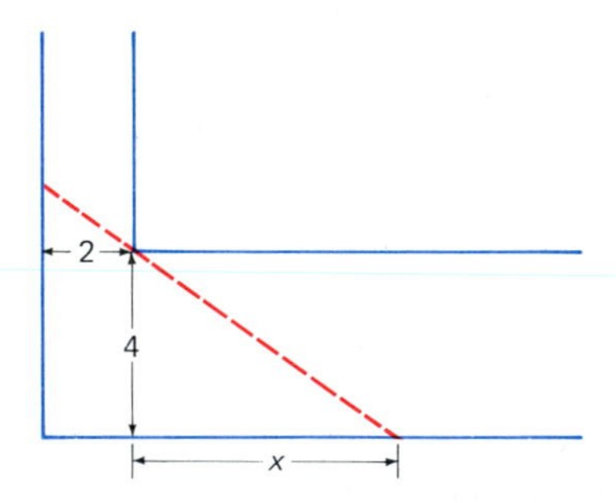

4.23 Carrying a rod around a corner (Problem 35)

36. Find the length of the shortest ladder that will reach from the ground, over a wall 8 ft high, to the side of a building 1 ft behind the wall. That is, minimize the length $L = L_1 + L_2$ shown in Fig. 4.24.

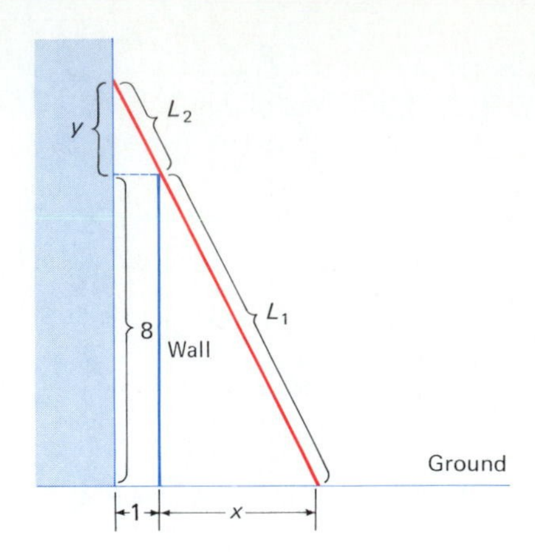

4.24 The ladder of Problem 36

37. A sphere with fixed radius a is inscribed in a pyramid with a square base so that the sphere touches the base of the pyramid and also each of its four sides. Show that the minimum possible volume of the pyramid is $8/\pi$ times the volume of the sphere. (*Suggestion:* Use the two right triangles in Fig. 4.25 to show that the volume of the pyramid is

$$V = V(y) = \frac{4a^2y^2}{3(y - 2a)}.$$

This can be done easily with the aid of the angle θ and *without* the formula for $\tan(\theta/2)$.)

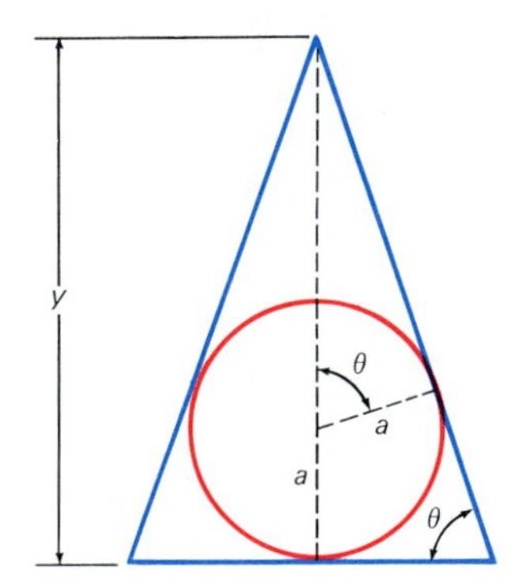

4.25 Cross section through the centers of the sphere and pyramid of Problem 37

38. Two noisy discothèques, one of them four times as noisy as the other, are located on opposite ends of a block 1000 ft long. What is the quietest point on the block between the two discos? Use the fact that the intensity of noise at a point removed from its source is proportional to the noisiness and inversely proportional to the square of the distance from the source.

39. A floored tent with fixed volume V is to be shaped like a pyramid with a square base and congruent sides, as in Fig. 4.26. What should be its height y and base edge $2x$

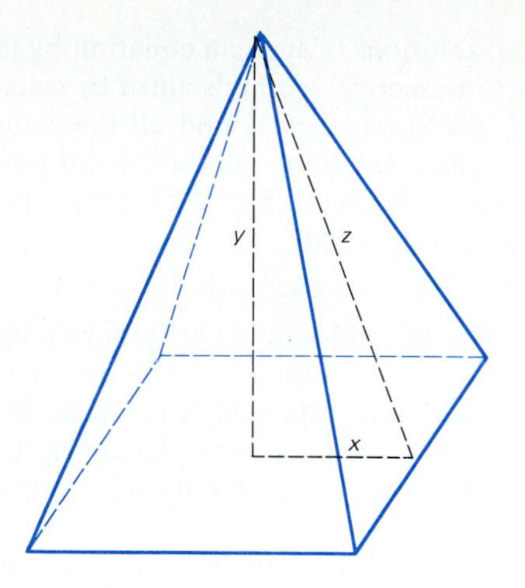

4.26 The tent of Problem 39

in order to minimize its total surface area (including its floor)?

40. Suppose that the distance from the building to the wall in Problem 36 is a and the height of the wall is b. Show that the minimal length of the ladder is

$$L_{\text{min}} = (a^{2/3} + b^{2/3})^{3/2}.$$

4.5 Graphs of Polynomials

We can construct a reasonably accurate graph of the polynomial function

$$f(x) = a_n x^n + a_{n-1} x^{n-1} + \cdots + a_2 x^2 + a_1 x + a_0 \tag{1}$$

by assembling the following information.

1. The critical points of f. These are the points on the graph where the tangent line is horizontal, because a polynomial has no critical points of type 2 (where f' does not exist).
2. The intervals on which f is increasing and those on which it is decreasing.
3. The behavior of $f(x)$ as $x \to +\infty$ and as $x \to -\infty$.

To carry out the task in item 3, we write $f(x)$ in the form

$$f(x) = x^n\left(a_n + \frac{a_{n-1}}{x} + \cdots + \frac{a_1}{x^{n-1}} + \frac{a_0}{x^n}\right).$$

Thus we conclude that the behavior of $f(x)$ as $x \to \pm\infty$ is much the same as that of its *leading term* $a_n x^n$, because the terms with powers of x in the denominator all approach zero as $x \to \pm\infty$. In particular, if $a_n > 0$, then

$$\lim_{x \to \infty} f(x) = +\infty, \tag{2}$$

meaning that $f(x)$ increases without bound as $x \to +\infty$. Also,

$$\lim_{x \to -\infty} f(x) = \begin{cases} +\infty & \text{if } n \text{ is even;} \\ -\infty & \text{if } n \text{ is odd.} \end{cases} \tag{3}$$

If $a_n < 0$, simply reverse the signs on the right-hand sides in (2) and (3).

Every polynomial, such as $f(x)$ in (1), is differentiable everywhere. So the critical points of $f(x)$ are the roots of the polynomial equation $f'(x) = 0$—that is,

$$na_n x^{n-1} + (n-1)a_{n-1}x^{n-2} + \cdots + 2a_2 x + a_1 = 0. \tag{4}$$

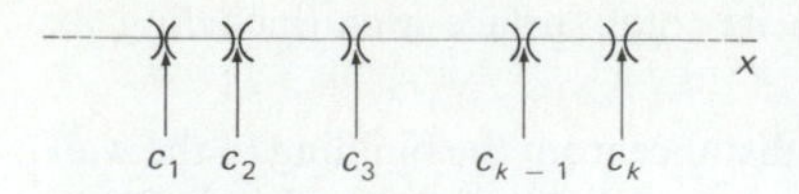

4.27 The zeros of f' divide the x-axis into intervals on each of which f' does not change sign.

Sometimes one can find all (real) solutions of such an equation by factoring, but sometimes one must resort to numerical methods aided by calculator or computer. But suppose that we have somehow found *all* the solutions $c_1, c_2, \ldots, c_k$ of Equation (4). Then these solutions are the critical points of f. If they are arranged in increasing order, as in Fig. 4.27, they separate the x-axis into the finite number of open intervals

$$(-\infty, c_1),\ (c_1, c_2),\ (c_2, c_3),\ \ldots,\ (c_{k-1}, c_k),\ (c_k, +\infty)$$

that also appear in the figure. The intermediate value property applied to $f'(x)$ tells us that f' can change sign only at the critical points of f, so that f' has only one sign on each of these open intervals. It is typical for $f'(x)$ to be negative on some intervals and positive on others. Moreover, it's easy to find the sign of f' on any such interval I: We need only substitute *any* convenient point of I into $f'(x)$.

Once we know the sign of f' on each of these intervals, we know where f is increasing and where it is decreasing. We then apply the first derivative test to find which of the critical values are local maxima, which are local minima, and which are neither—merely places where the tangent line is horizontal. With this information, the knowledge of the behavior of f as $x \to \pm\infty$, and the fact that f is continuous, we can sketch its graph. We plot the critical points $(c_i, f(c_i))$ and connect them with a smooth curve that is consistent with our other data.

It may also be helpful to plot the y-intercept $(0, f(0))$ and also any x-intercepts that are easy to find. But we recommend (until inflection points are introduced in Section 4.6) that you plot *only* these points—critical points and intercepts—and rely otherwise on the increasing–decreasing behavior of f.

EXAMPLE 1 Sketch the graph of $f(x) = x^3 - 27x$.

Solution Because the leading term is x^3, we see that

$$\lim_{x\to+\infty} f(x) = +\infty \quad\text{and}\quad \lim_{x\to-\infty} f(x) = -\infty.$$

Moreover, as

$$f'(x) = 3x^2 - 27 = 3(x+3)(x-3),$$

we see that the critical points where $f'(x) = 0$ are $x = -3$ and $x = 3$. The corresponding points on the graph of f are $(-3, 54)$ and $(3, -54)$. The critical points separate the x-axis into the three open intervals

$$(-\infty, -3), \qquad (-3, 3), \quad\text{and}\quad (3, +\infty).$$

The next table shows the behavior of f on each of these intervals.

Interval	$x+3$	$x-3$	$f'(x)$	f
$(-\infty, -3)$	Neg.	Neg.	Pos.	Increasing
$(-3, 3)$	Pos.	Neg.	Neg.	Decreasing
$(3, +\infty)$	Pos.	Pos.	Pos.	Increasing

We use this information to plot the critical points and the intercepts $(0, 0)$, $(3\sqrt{3}, 0)$, and $(-3\sqrt{3}, 0)$, and thus obtain the graph sketched in Fig. 4.28.

In the figure we use plus and minus signs to mark the sign of $f'(x)$ in each interval. This makes it clear that $(-3, 54)$ is a local maximum and that $(3, -54)$ is a local minimum.

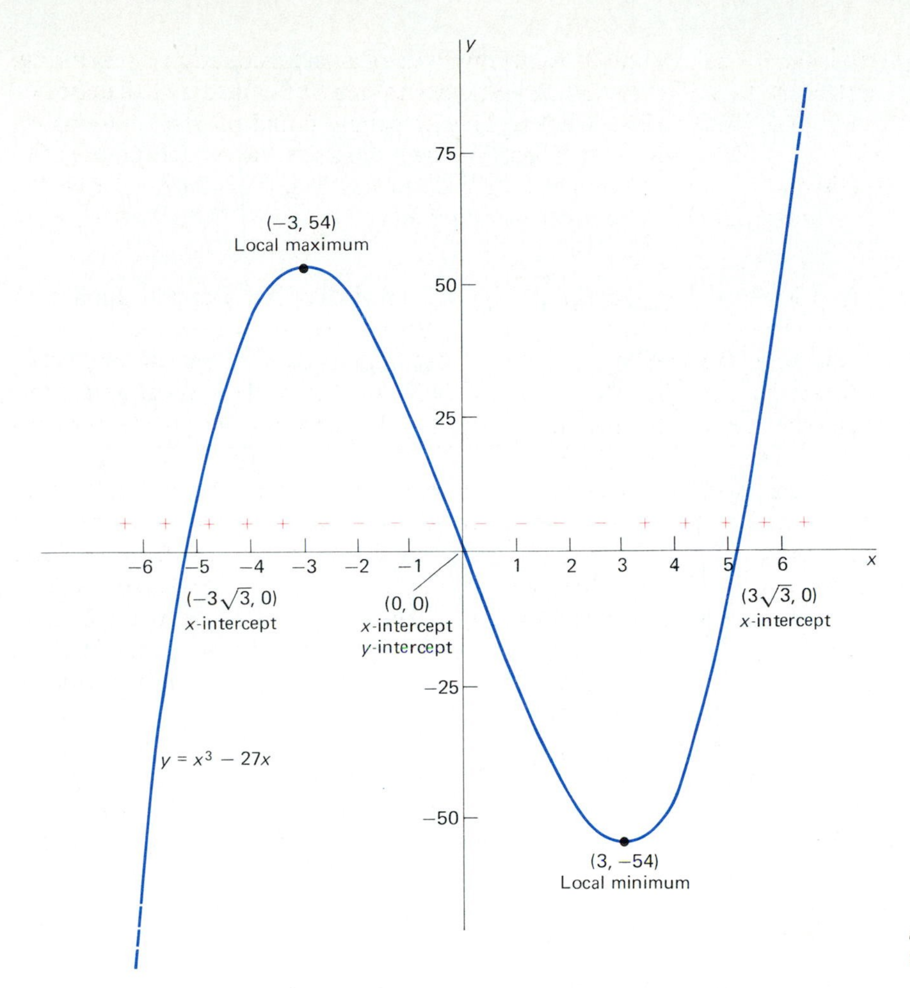

4.28 Graph of the function of Example 1

EXAMPLE 2 Sketch the graph of $f(x) = 8x^5 - 5x^4 - 20x^3$.

Solution Because

$$f'(x) = 40x^4 - 20x^3 - 60x^2 = 20x^2(x + 1)(2x - 3),$$

the critical points where $f'(x) = 0$ are -1, 0, and 1.5. The behavior of f on the four open intervals determined by these critical points is indicated in the next table.

Interval	$x + 1$	x^2	$2x - 3$	$f'(x)$	f
$(-\infty, -1)$	Neg.	Pos.	Neg.	Pos.	Increasing
$(-1, 0)$	Pos.	Pos.	Neg.	Neg.	Decreasing
$(0, 1.5)$	Pos.	Pos.	Neg.	Neg.	Decreasing
$(1.5, +\infty)$	Pos.	Pos.	Pos.	Pos.	Increasing

The points on the graph that correspond to the critical points are $(-1, 7)$, $(0, 0)$, and $(1.5, -32.06)$ (the last ordinate is an approximation).

We write $f(x)$ in the form

$$f(x) = x^3(8x^2 - 5x - 20)$$

in order to use the quadratic formula to find the intercepts. They turn out to be $(-1.30, 0)$ and $(1.92, 0)$ (approximations again), and the origin $(0, 0)$.

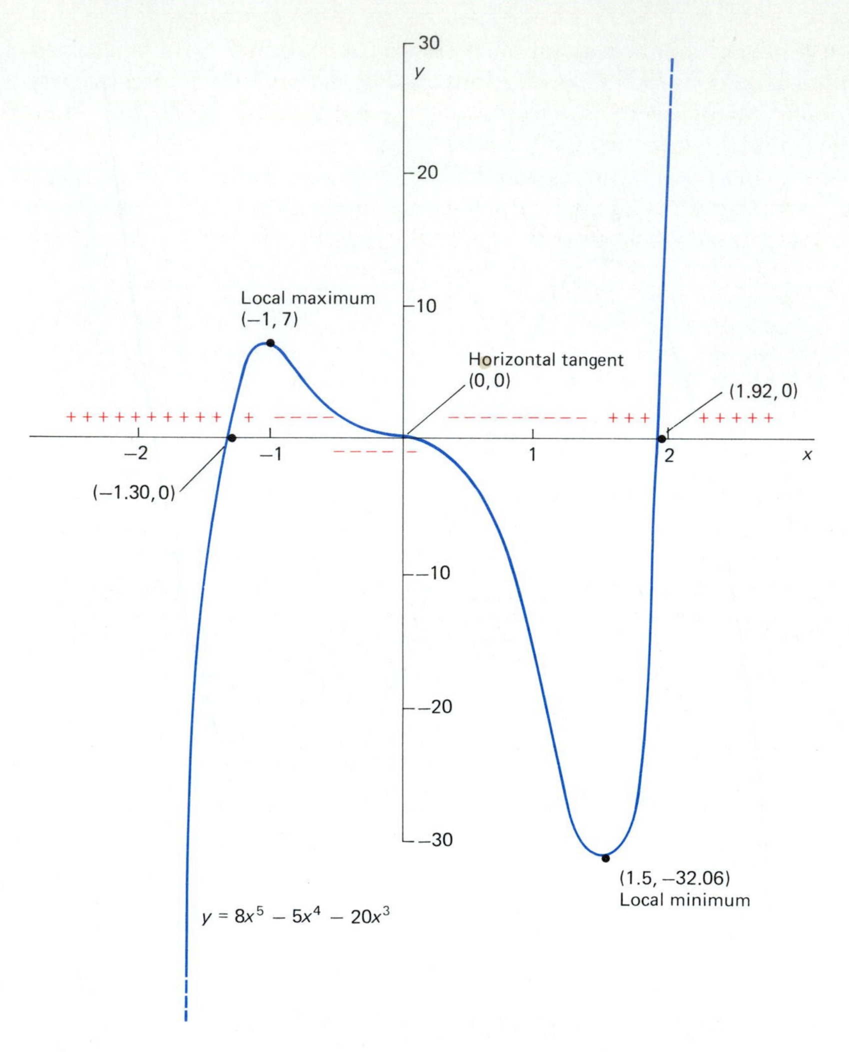

4.29 Graph of the function of Example 2

The latter is also the y-intercept. We apply the first derivative test to the increasing-decreasing behavior shown in the table. It follows that $(-1, 7)$ is a local maximum, $(1.5, -32.06)$ is a local minimum, and $(0, 0)$ is neither. The graph looks like the one shown in Fig. 4.29.

In our next example, the function is not a polynomial. Nevertheless, the methods of this section suffice for sketching its graph.

EXAMPLE 3 Sketch the graph of

$$f(x) = x^{2/3}(x^2 - 2x - 6) = x^{8/3} - 2x^{5/3} - 6x^{2/3}.$$

Solution The derivative of f is

$$f'(x) = \frac{8}{3}x^{5/3} - \frac{10}{3}x^{2/3} - \frac{12}{3}x^{-1/3} = \frac{2}{3}x^{-1/3}(4x^2 - 5x - 6)$$

$$= \frac{2(4x + 3)(x - 2)}{3x^{1/3}}.$$

The tangent line is horizontal at the two critical points $x = -0.75$ and $x = 2$, where the numerator in the last fraction is zero (and the denominator is not). In addition, because of the presence of the term $x^{1/3}$ in the denominator, $|f'(x)| \to +\infty$ as $x \to 0$. Thus $x = 0$ (a critical point because f is not differentiable there) is a point where the tangent line is vertical. These three critical points give the points $(-0.75, -3.25)$, $(0, 0)$, and $(2, -9.52)$ on the graph (we use approximations where appropriate). The following table summarizes our analysis of the sign of $f'(x)$ on each of the open intervals into which these critical points divide the x-axis.

Interval	$4x + 3$	$x^{-1/3}$	$x - 2$	$f'(x)$	f
$(-\infty, -0.75)$	Neg.	Neg.	Neg.	Neg.	Decreasing
$(-0.75, 0)$	Pos.	Neg.	Neg.	Pos.	Increasing
$(0, 2)$	Pos.	Pos.	Neg.	Neg.	Decreasing
$(2, +\infty)$	Pos.	Pos.	Pos.	Pos.	Increasing

The first derivative test now shows local minima at $(-0.75, -3.25)$ and $(2, -9.52)$ and a local maximum at $(0, 0)$. Although $f'(0)$ does not exist, f is continuous at $x = 0$, so it is continuous everywhere.

We use the quadratic formula to find the x-intercepts. In addition to the origin, they occur where $x^2 - 2x - 6 = 0$, thus they are located at $(1 - \sqrt{7}, 0)$ and at $(1 + \sqrt{7}, 0)$. We then plot the approximations $(-1.65, 0)$ and $(3.65, 0)$. Finally, we note that $f(x) \to +\infty$ as $x \to \pm\infty$. So the graph has the shape shown in Fig. 4.30.

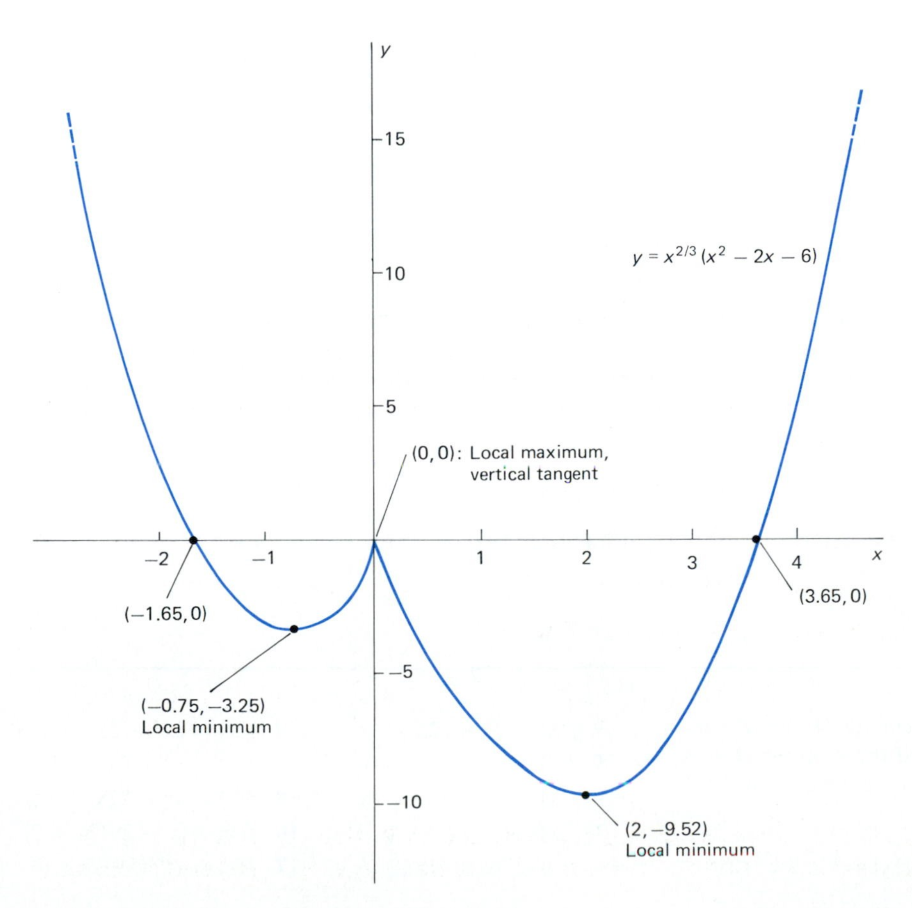

4.30 The technique is valid for nonpolynomial functions, as in Example 3.

An important application of curve-sketching techniques is to the solution of an equation of the form

$$f(x) = 0. \tag{5}$$

The real (as opposed to complex) solutions of this equation are simply the x-intercepts of the graph of $y = f(x)$. Hence by sketching this graph with reasonable accuracy we can glean information about the number of real solutions of Equation (5) and even their approximate locations.

To illustrate this approach, let us consider a cubic equation of the special form

$$f(x) = x^3 + px + q = 0, \tag{6}$$

special in that it contains no second-degree term (see Problem 39 for the technique of eliminating the second-degree term from a cubic equation). The derivative of f is

$$f'(x) = 3x^2 + p. \tag{7}$$

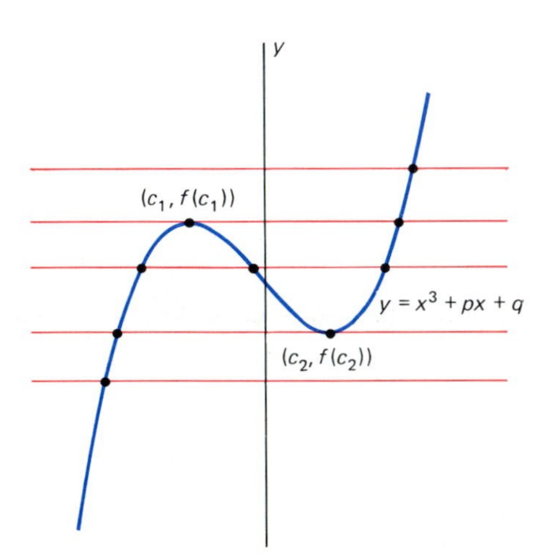

4.31 Real solutions of the cubic equation in (6)

If $p \geqq 0$, then $f'(x) > 0$ everywhere, with the single exception that $f'(0) = 0$ if $p = 0$. It follows that if $p \geqq 0$, then f is increasing on the entire real line. Because f takes on arbitrary large negative and positive values and has the intermediate value property, we may conclude that Equation (6) has precisely one real solution.

The more interesting case is the one in which $p < 0$. In this case we see from (7) that f has two distinct critical points,

$$c_1 = -\sqrt{-p/3} \quad \text{and} \quad c_2 = \sqrt{-p/3}, \tag{8}$$

with $c_1 < 0 < c_2$. In Problem 35 we ask you to show that $f(c_1) > f(c_2)$. It follows that the curve $y = f(x)$ has the shape shown in Fig. 4.31. The five horizontal lines represent all the possibilities for the location of the x-axis relative to the curve. We may conclude from this figure that Equation (6) has

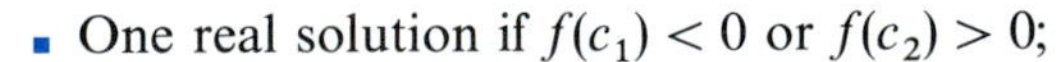

- One real solution if $f(c_1) < 0$ or $f(c_2) > 0$;
- Two real solutions if $f(c_1) = 0$ or $f(c_2) = 0$;
- Three real solutions if $f(c_1) > 0 > f(c_2)$.

The assumption that the cubic equation in (6) contained no term of the form ax^2 simplified this discussion, but a similar analysis can be carried out with an arbitrary cubic equation

$$f(x) = x^3 + ax^2 + bx + c = 0 \tag{9}$$

to determine whether it has one, two, or three real solutions.

4.5 PROBLEMS

For each of the functions in Problems 1–34, find the intervals on which it is increasing and those on which it is decreasing. Sketch the graph of the function and label the local maxima and minima.

1. $f(x) = 3x^2 - 6x + 5$

2. $f(x) = 5 - 8x - 2x^2$

3. $f(x) = x^3 - 12x$

4. $f(x) = 2x^3 + 3x^2 - 12x$

5. $f(x) = x^3 - 6x^2 + 9x$

6. $f(x) = x^3 + 6x^2 + 9x$

7. $f(x) = x^3 + 3x^2 + 9x$

8. $f(x) = x^3 - 27x$

9. $f(x) = (x - 1)^2(x + 2)^2$

10. $f(x) = (x - 2)^2(2x + 3)^2$

11. $f(x) = 3\sqrt{x} - x\sqrt{x}$

12. $f(x) = x^{2/3}(5 - x)$

13. $f(x) = 3x^5 - 5x^3$

14. $f(x) = x^4 + 4x^3$

15. $f(x) = x^4 - 8x^2 + 7$

16. $f(x) = \dfrac{1}{x}$

17. $f(x) = 2x^2 - 3x - 9$

18. $f(x) = 6 - 5x - 6x^2$

19. $f(x) = 2x^3 + 3x^2 - 12x$

20. $f(x) = x^3 + 4x$

21. $f(x) = 50x^3 - 105x^2 + 72x$

22. $f(x) = x^3 - 3x^2 + 3x - 1$

23. $f(x) = 3x^4 - 4x^3 - 12x^2 + 8$

24. $f(x) = x^4 - 2x^2 + 1$

25. $f(x) = 3x^5 - 20x^3$

26. $f(x) = 3x^5 - 25x^3 + 60x$

27. $f(x) = 2x^3 + 3x^2 + 6x$

28. $f(x) = x^4 - 4x^3$

29. $f(x) = 8x^4 - x^8$

30. $f(x) = 1 - x^{1/3}$

31. $f(x) = x^{1/3}(4 - x)$

32. $f(x) = x^{2/3}(x^2 - 16)$

33. $f(x) = x(x - 1)^{2/3}$

34. $f(x) = x^{1/3}(2 - x)^{2/3}$

35. Verify that if c_1 and c_2 are the critical points in (8) of the cubic equation in (6), then $f(c_1) > f(c_2)$.

The *discriminant* of the special cubic equation

$$x^3 + px + q = 0 \qquad (p < 0)$$

is defined to be $D = 4p^3 + 27q^2$. The following three problems, together with the discussion at the end of this section, show that this cubic equation has one real solution if $D > 0$, two real solutions if $D = 0$, and three real solutions if $D < 0$.

36. Prove that $D > 0$ if and only if either $f(c_1) < 0$ or $f(c_2) > 0$.

37. Prove that $D = 0$ if and only if either $f(c_1) = 0$ or $f(c_2) = 0$.

38. Prove that $D < 0$ if and only if $f(c_1) > 0 > f(c_2)$.

39. Prove that the substitution $x = z - k$, where $k = a/3$, transforms the general cubic equation

$$x^3 + ax^2 + bx + c = 0$$

into the special cubic equation

$$z^3 + pz + q = 0,$$

where $p = b - 2ak + 3k^2$ and $q = c - bk + ak^2 + k^3$. Consequently, our results on the number of roots of a special cubic equation can be applied to a general cubic equation.

40. Show by constructing a graph that the equation

$$x^5 - 5x^3 - 20x + 17 = 0$$

has precisely three real solutions.

41. The computer-generated graph in Fig. 4.32 shows how the curve

$$y = [x(x - 1)(2x - 1)]^2$$

looks in any "reasonable scale" with integral units of measurement on the y-axis. Use the methods of this section to show that the graph really has the appearance shown in Fig. 4.33, with critical points at

$$0, \quad \tfrac{1}{2}, \quad \tfrac{1}{6}(3 \pm \sqrt{3}), \quad \text{and} \quad 1.$$

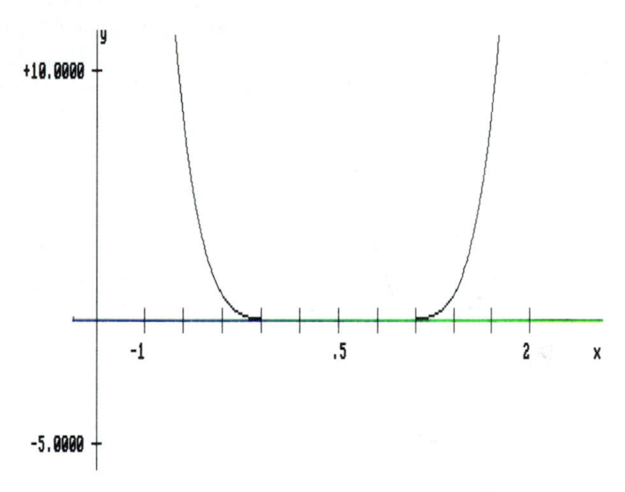

4.32 The graph $y = [x(1 - x)(1 - 2x)]^2$ on a "reasonable" scale

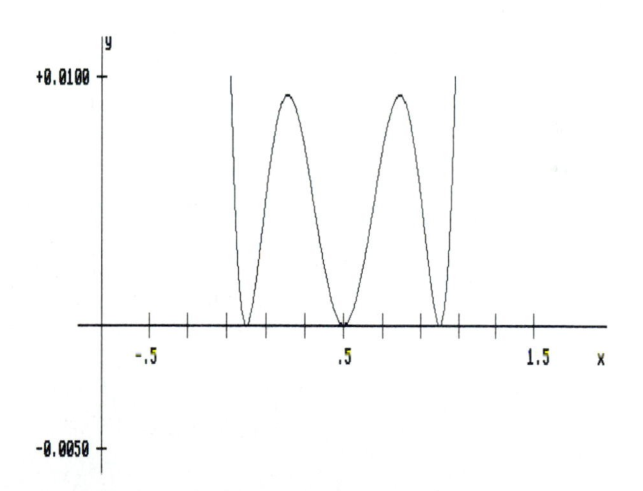

4.33 The graph $y = [x(1 - x)(1 - 2x)]^2$ on a finer scale

4.6 Higher Derivatives and Concavity

We saw in Section 4.3 that the sign of the first derivative f' tells whether the graph of the function f is rising or falling. In this section we shall see that the sign of the *second* derivative of f, the derivative of f', tells which way the curve $y = f(x)$ is *bending*, upward or downward.

HIGHER DERIVATIVES

The **second derivative** of f is denoted by f'', and its value at x is

$$f''(x) = D(f'(x)) = D(Df(x)) = D^2f(x).$$

The derivative of f'' is the **third derivative** f''' of f, with

$$f'''(x) = D(f''(x)) = D(D^2f(x)) = D^3f(x).$$

The third derivative is also denoted by $f^{(3)}$. More generally, the result of beginning with the function f and differentiating n times in succession is the ***n*th derivative** $f^{(n)}$ of f, with $f^{(n)}(x) = D^nf(x)$.

If $y = f(x)$, then the first n derivatives may be written as

$$D_xy,\ D_x^2y,\ D_x^3y,\ \ldots,\ D_x^ny$$

or

$$y',\ y'',\ y''',\ \ldots,\ y^{(n)},$$

or finally as

$$\frac{dy}{dx},\ \frac{d^2y}{dx^2},\ \frac{d^3y}{dx^3},\ \ldots,\ \frac{d^ny}{dx^n}$$

in differential notation. The history of the curious use of superscripts in the differential notation for higher derivatives involves the metamorphosis

$$\frac{d}{dx}\left(\frac{dy}{dx}\right) \to \frac{d}{dx}\frac{dy}{dx} \to \frac{(d)^2y}{(dx)^2} \to \frac{d^2y}{dx^2}.$$

EXAMPLE 1 Find the first four derivatives of

$$f(x) = 2x^3 + \frac{1}{x^2} + 16x^{7/2}.$$

Solution Write

$$f(x) = 2x^3 + x^{-2} + 16x^{7/2}. \qquad \text{Then}$$

$$f'(x) = 6x^2 - 2x^{-3} + 56x^{5/2} = 6x^2 - \frac{2}{x^3} + 56x^{5/2},$$

$$f''(x) = 12x + 6x^{-4} + 140x^{3/2} = 12x + \frac{6}{x^4} + 140x^{3/2},$$

$$f'''(x) = 12 - 24x^{-5} + 210x^{1/2} = 12 - \frac{24}{x^5} + 210\sqrt{x},$$

$$f^{(4)}(x) = 120x^{-6} + 105x^{-1/2} = \frac{120}{x^6} + \frac{105}{\sqrt{x}}.$$

The next example shows how higher derivatives of implicitly defined functions may be found.

EXAMPLE 2 Find the second derivative $y''(x)$ of the function $y = y(x)$ given

$$x^2 - xy + y^2 = 9.$$

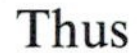

Solution Let us write y' for $y'(x)$ and y'' for $y''(x)$. Then implicit differentiation of the given equation with respect to x gives

$$2x - y - xy' + 2yy' = 0, \quad \text{so} \quad y' = \frac{y - 2x}{2y - x}.$$

We obtain y'' by differentiating implicitly again, using the quotient rule. After that, we make substitutions for y' by using the result just found.

$$y'' = D_x\left(\frac{y - 2x}{2y - x}\right) = \frac{(y' - 2)(2y - x) - (y - 2x)(2y' - 1)}{(2y - x)^2}$$

$$= \frac{3xy' - 3y}{(2x - y)^2} = \frac{3x\dfrac{y - 2x}{2y - x} - 3y}{(2y - x)^2}.$$

Thus

$$y'' = -\frac{6(x^2 - xy + y^2)}{(2y - x)^3}.$$

The original equation $x^2 - xy + y^2 = 9$ may now be used for one final simplification:

$$y'' = -\frac{54}{(2y - x)^3}.$$

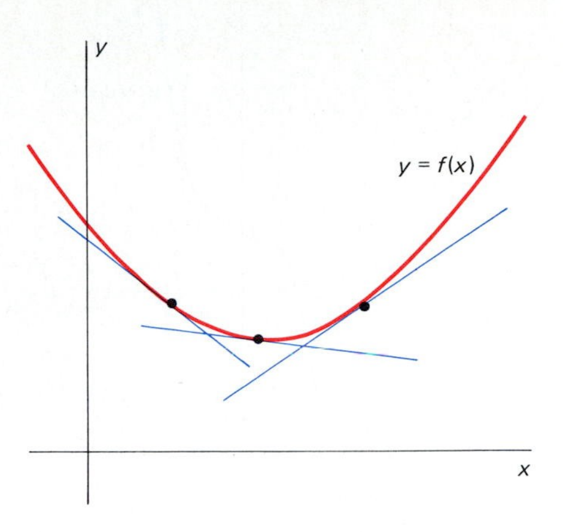

4.34 The graph is bending upward (concave upward).

THE SECOND DERIVATIVE TEST

Now we investigate the significance of the *sign* of the second derivative. Suppose first that $f''(x) > 0$ on the interval I. Then the first derivative f' is an increasing function on I because *its* derivative $f''(x)$ is positive. Thus, as we scan the graph $y = f(x)$ from left to right, we see the tangent line turning in a counterclockwise direction, as shown in Fig. 4.34. This situation may be described by saying that the curve $y = f(x)$ is **bending upward.** Note that a curve can be bending upward without rising, as illustrated in Fig. 4.35.

On the other hand, if $f''(x) < 0$ on the interval I, then the first derivative f' is decreasing on I, so the tangent line turns clockwise as x increases. In this case we say that the curve $y = f(x)$ is **bending downward.** Figures 4.36 and 4.37 show two ways this can happen.

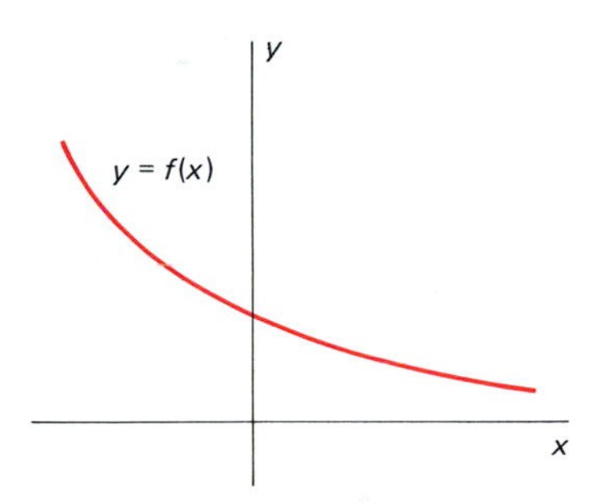

4.35 Another graph bending upward (concave upward)

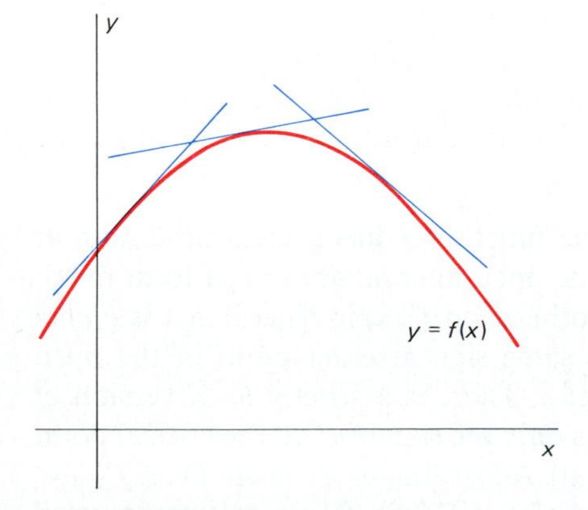

4.36 A graph bending downward (concave downward)

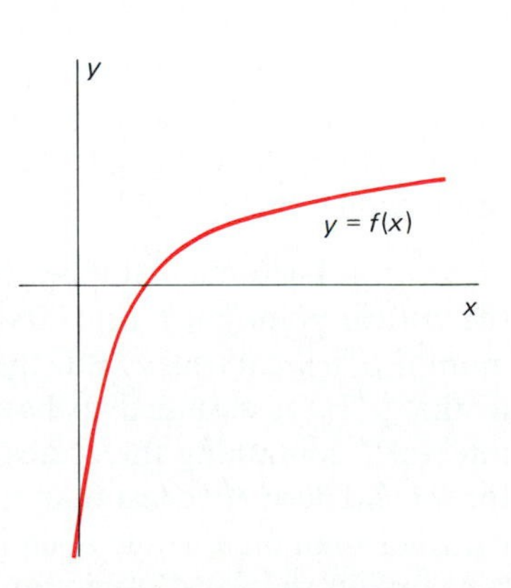

4.37 Another graph bending downward (concave downward)

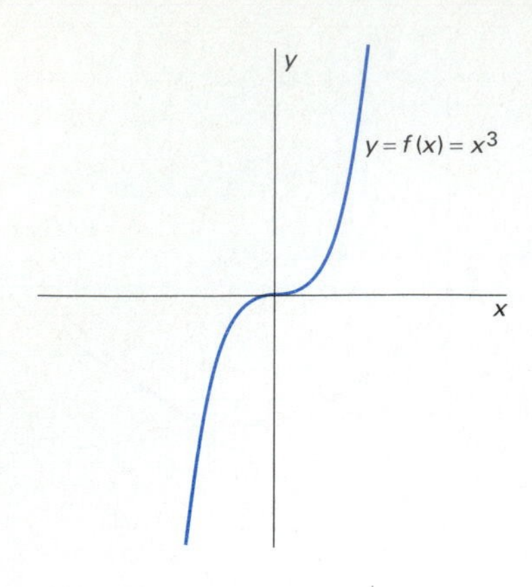

4.38 Although $f'(0) = 0$, $f(0)$ is not an extremum.

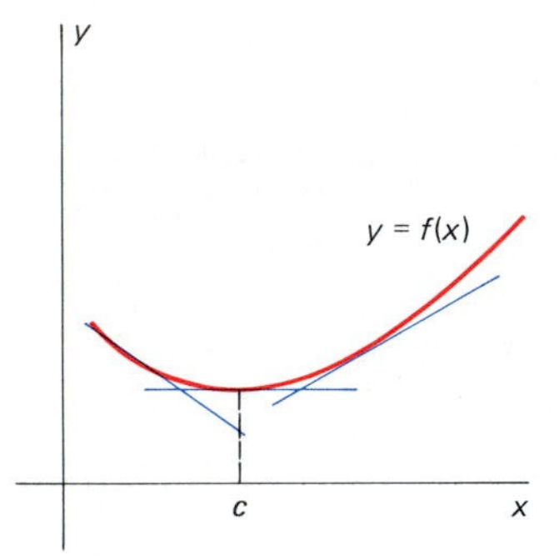

$f''(x) > 0$;
tangent turning counterclockwise;
graph concave up;
local minimum at c.

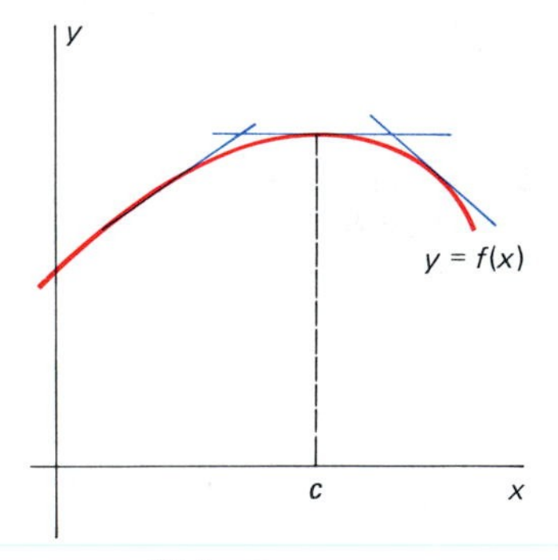

$f''(x) < 0$;
tangent turning clockwise;
graph concave down;
local maximum at c.

4.39 The second derivative test (Theorem 1)

EXAMPLE 3 In Example 1 of Section 4.5 we constructed the graph of the function $f(x) = x^3 - 27x$, for which

$$f''(x) = 6x.$$

Because $f''(x) < 0$ for $x < 0$, while $f''(x) > 0$ for $x > 0$, we conclude that the curve $y = x^3 - 27x$ is bending downward on the interval $(-\infty, 0)$ but is bending upward on $(0, +\infty)$. Observe that this conclusion agrees with the appearance of the graph in Fig. 4.28 on page 183.

We know that a local extremum of a differentiable function f can occur only at a critical point c where $f'(c) = 0$, so the tangent line at the point $(c, f(c))$ on the curve $y = f(x)$ is horizontal. But the example $f(x) = x^3$, for which $x = 0$ is a critical point but not an extremum (Fig. 4.38) shows that the *necessary condition* $f'(c) = 0$ is *not* a sufficient condition to conclude that $f(c)$ is an extreme value of the function f.

Now suppose not only that $f'(c) = 0$ but also that the curve $y = f(x)$ is bending upward on some open interval containing the critical point $x = c$. Then it is apparent from Fig. 4.39 that $f(c)$ is a local minimum value. Similarly, $f(c)$ is a local maximum value if $f'(c) = 0$ while $y = f(x)$ is bending downward on some open interval about c. But the *sign* of the second derivative $f''(x)$ tells us whether $y = f(x)$ is bending upward or downward, and therefore provides us with a *sufficient* condition for a local extremum.

Theorem 1 *Second Derivative Test*

Suppose that the function f is twice differentiable on the open interval I containing the critical point c at which $f'(c) = 0$. Then

1. If $f''(x) > 0$ on I, then $f(c)$ is the minimum value of f on I.
2. If $f''(x) < 0$ on I, then $f(c)$ is the maximum value of f on I.

Proof We will prove only part 1. If $f''(x) > 0$ on I, then it follows that the first derivative f' is an increasing function on I. Because $f'(c) = 0$, we may conclude that $f'(x) < 0$ for $x < c$ in I, and that $f'(x) > 0$ for $x > c$ in I. Consequently, the first derivative test of Section 4.4 implies that $f(c)$ is the minimum value of f on I. ■

REMARK 1 Rather than memorizing verbatim the conditions in parts 1 and 2 of Theorem 1, it is easier and more reliable to remember the second derivative test by visualizing continuously turning tangent lines and pictures like those shown in Fig. 4.39.

REMARK 2 Theorem 1 implies that the function f has a local minimum at the critical point c if $f''(x) > 0$ on some open interval about c, a local maximum if $f''(x) < 0$ near c. But the hypothesis on $f''(x)$ in Theorem 1 is *global* in that $f''(x)$ is assumed to have the same sign at *every* point of the open interval I containing the critical point c. There is a strictly *local* version of the second derivative test that involves only the sign of f'' *at* the critical point c (rather than on a whole open interval). According to Problem 68, if $f'(c) = 0$, then $f(c)$ is a local minimum value of f if $f''(c) > 0$, a local maximum if $f''(c) < 0$.

REMARK 3 Note that the second derivative test says *nothing* about what happens if $f''(c) = 0$ at the critical point c. Consider the three functions $f(x) = x^4$, $f(x) = -x^4$, and $f(x) = x^3$. For each of these, $f'(0) = 0$ and $f''(0) = 0$. But their graphs, shown in Fig. 4.40, demonstrate that *anything* can happen at such a point.

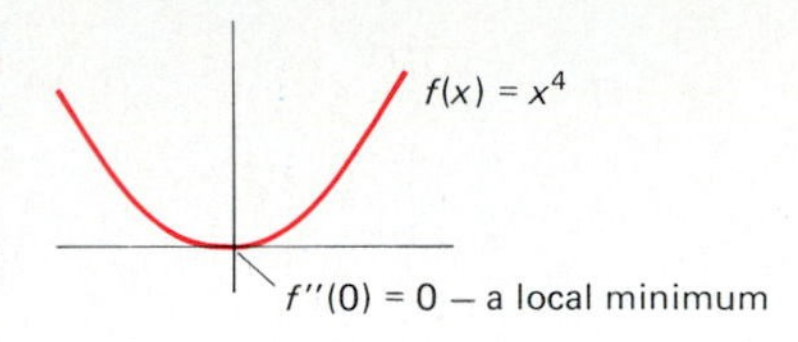

REMARK 4 Suppose that we want to maximize or minimize the function f on the open interval I, and we find that f has only one critical point in I, a number c at which $f'(c) = 0$. If $f''(x)$ has the same sign at all points of I, then Theorem 1 implies that $f(c)$ is an *absolute* extremum of f on I (a minimum of $f''(x) > 0$, a maximum if $f''(x) < 0$). This absolute interpretation of the second derivative test is useful in applied open interval maximum–minimum problems.

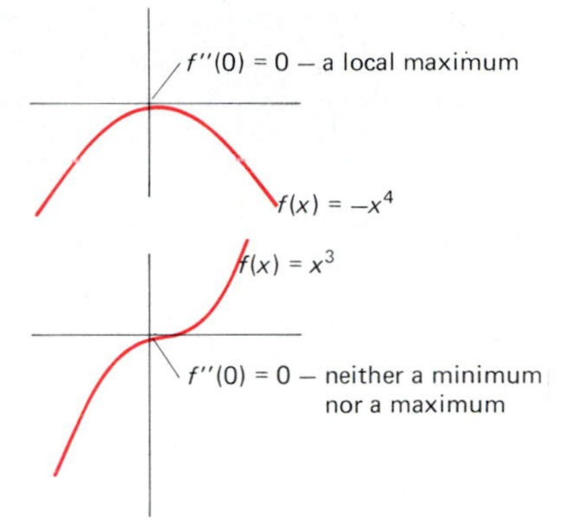

4.40 No conclusion possible if $f'(c) = 0 = f''(c)$

EXAMPLE 3 continued Consider again the function $f(x) = x^3 - 27x$, for which

$$f'(x) = 3x^2 - 27 \quad \text{and} \quad f''(x) = 6x.$$

Then f has two critical points: $x = -3$ and $x = +3$. Obviously $f''(x) < 0$ for x near -3, so $f(-3) = 54$ is a local maximum value of f. Also, $f''(x) > 0$ for x near 3, so $f(3) = -54$ is a local minimum value of f.

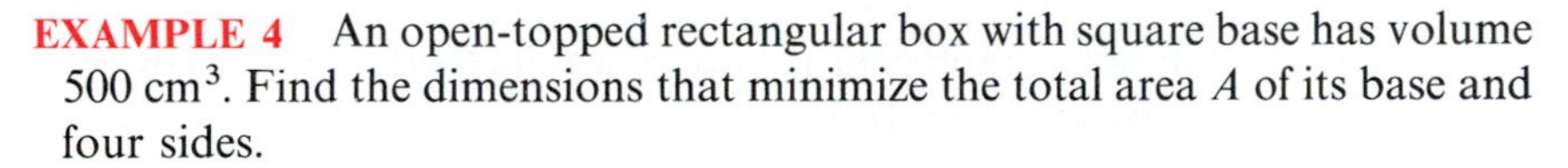

EXAMPLE 4 An open-topped rectangular box with square base has volume 500 cm^3. Find the dimensions that minimize the total area A of its base and four sides.

Solution If we denote by x the edge length of the square base and by y the height of the box (as in Fig. 4.41), then the volume of the box is

$$V = x^2y = 500 \tag{1}$$

and the total area of its base and four sides is

$$A = x^2 + 4xy. \tag{2}$$

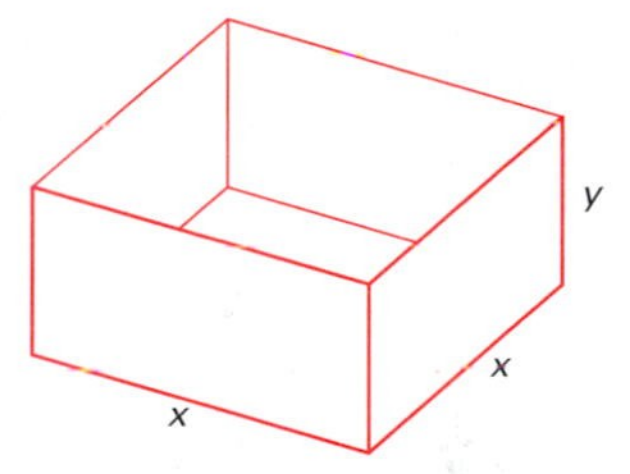

4.41 The open-topped box of Example 4

When we solve (1) for $y = 500/x^2$ and substitute in (2), we get the area function

$$A(x) = x^2 + \frac{2000}{x}$$

defined on the open interval $(0, +\infty)$. The first derivative is

$$A'(x) = 2x - \frac{2000}{x^2}. \tag{3}$$

The equation $A'(x) = 0$ yields $x^3 = 1000$, so the only critical point of A in $(0, +\infty)$ is $x = 10$. To investigate this critical point we calculate the second derivative

$$A''(x) = 2 + \frac{4000}{x^3}. \tag{4}$$

Because it is clear that $A''(x) > 0$ on $(0, +\infty)$, it follows from the second derivative test and Remark 4 that $A(10) = 300$ is the absolute minimum value of $A(x)$ on $(0, +\infty)$. Finally, because $y = 500/x^2$, $y = 5$ when $x = 10$.

Therefore this absolute minimum corresponds to a box with base 10 cm by 10 cm and of height 5 cm.

CONCAVITY AND CURVE SKETCHING

A comparison of Fig. 4.34 with Fig. 4.36 suggests that the question of whether the curve $y = f(x)$ is bending upward or downward is closely related to the question of whether it lies above or below its tangent lines. The latter question refers to the important property of *concavity*.

> ***Definition of Concavity***
>
> Suppose that the function f is differentiable at the point a and that L is the tangent line to the graph $y = f(x)$ at $(a, f(a))$. Then the function f (or its graph) is said to be
>
> 1. **Concave upward** at a if on some open interval containing a, the graph of f lies *above* L;
> 2. **Concave downward** at a if on some open interval containing a, the graph of f lies *below* L.

Figure 4.42 shows a graph concave upward at $(a, f(a))$; Fig. 4.43 shows a graph concave downward at $(a, f(a))$.

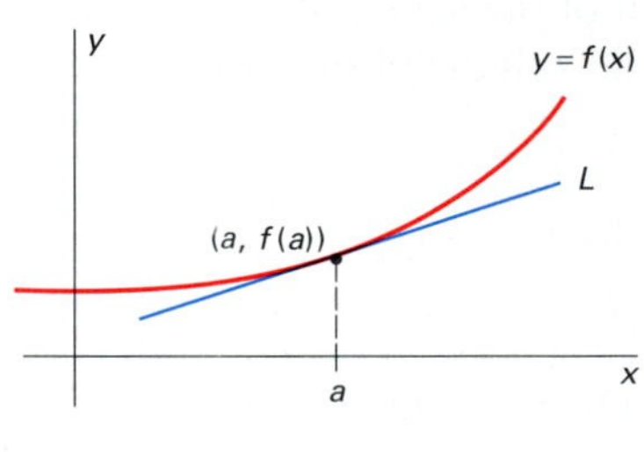

4.42 At $x = a$, f is concave upward.

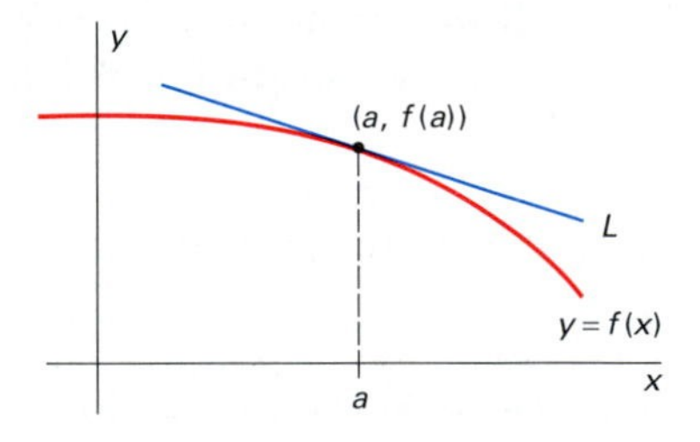

4.43 At $x = a$, f is concave downward.

The next theorem establishes the connection between concavity and the sign of the second derivative. That connection is the one suggested by our discussion of bending.

> ***Theorem 2*** *Test for Concavity*
>
> Suppose that the function f is twice differentiable on the open interval I.
>
> 1. If $f''(x) > 0$ on I, then f is concave upward at each point of I.
> 2. If $f''(x) < 0$ on I, then f is concave downward at each point of I.

A proof of Theorem 2 based on the second derivative test is given at the end of this section.

NOTE The significance of the sign of the *first* derivative must not be confused with the significance of the sign of the *second* derivative. The possibilities illustrated in Figs. 4.44–4.47 show that the signs of f' and f'' are independent of each other.

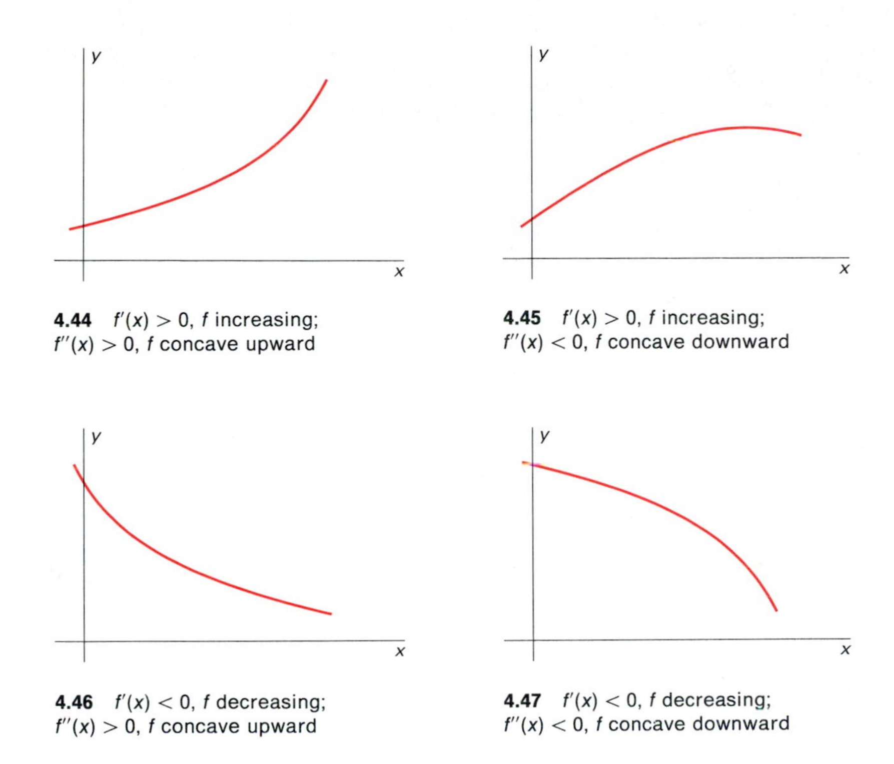

4.44 $f'(x) > 0$, f increasing; $f''(x) > 0$, f concave upward

4.45 $f'(x) > 0$, f increasing; $f''(x) < 0$, f concave downward

4.46 $f'(x) < 0$, f decreasing; $f''(x) > 0$, f concave upward

4.47 $f'(x) < 0$, f decreasing; $f''(x) < 0$, f concave downward

Observe that the test for concavity in Theorem 2 says nothing about the case in which $f''(x) = 0$. A point where the second derivative vanishes *may or may not* be a point where the function changes from concave upward on one side to concave downward on the other. A point where the concavity of a function *does* change in this manner is called an *inflection point.* More precisely, the point $x = a$ is an **inflection point** of the function provided that f is concave upward on one side of a, concave downward on the other side, and continuous at $x = a$. We also refer to $(a, f(a))$ as an inflection point on the graph of f.

Theorem 3 *Inflection Point Test*

The point a is an inflection point of the function f provided there is an open interval I containing a such that f is continuous on I and, for points x in I, either

1. $f''(x) > 0$ if $x < a$ and $f''(x) < 0$ if $x > a$, or
2. $f''(x) < 0$ if $x < a$ and $f''(x) > 0$ if $x > a$.

The fact that a point where the second derivative changes sign is an inflection point follows immediately from Theorem 2 and the definition of inflection point. NOTE At the inflection point a itself, either $f''(a) = 0$ or $f''(a)$ does not exist. Thus one finds inflection points of f by examining the critical points of f'. Some of the various possibilities are indicated in Fig. 4.48. Note how we mark the intervals of upward concavity and downward concavity with small cups opening upward and downward, respectively.

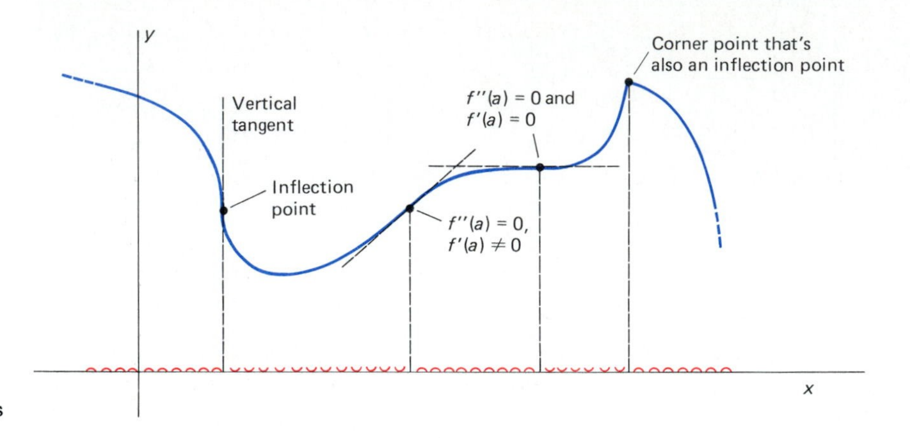

4.48 Some inflection points

EXAMPLE 5 Sketch the graph of $f(x) = 8x^5 - 5x^4 - 20x^3$, indicating local extrema, inflection points, and concave structure.

Solution We sketched this curve in Example 2 of Section 4.5; see Fig. 4.29 for the graph. In that example we found the first derivative to be

$$f'(x) = 40x^4 - 20x^3 - 60x^2 = 20x^2(x + 1)(2x - 3),$$

so the critical points are $x = -1$, 0, and 1.5. The second derivative is

$$f''(x) = 160x^3 - 60x^2 - 120x = 20x(8x^2 - 3x - 6).$$

When we compute $f''(x)$ at each critical point, we find that

$$f''(-1) = -100 < 0, \qquad f''(0) = 0, \qquad \text{and} \qquad f''(1.5) = 225 > 0.$$

Hence $f''(x) < 0$ near the critical point $x = -1$ and $f''(x) > 0$ near the critical point $x = 1.5$, so the second derivative test tells us that f has a local maximum at $x = -1$ and a local minimum at $x = 1.5$. The second derivative test is not enough to determine the behavior of f at $x = 0$.

Because f'' exists everywhere, the possible inflection points are the solutions of the equation

$$f''(x) = 0; \quad \text{that is,} \quad 20x(8x^2 - 3x - 6) = 0.$$

Clearly, one solution is $x = 0$. To find the other two, we use the quadratic formula to solve the equation

$$8x^2 - 3x - 6 = 0.$$

This gives

$$x = \tfrac{1}{16}(3 \pm \sqrt{201}),$$

so $x \approx 1.07$ and $x \approx -0.70$ are possible inflection points along with $x = 0$.

We take these two approximations as sufficiently exact and write

$$f''(x) = 20x(8x^2 - 3x - 6) = 160x(x^2 - \tfrac{3}{8}x - \tfrac{3}{4})$$
$$= 160x(x + 0.70)(x - 1.07).$$

This lets us analyze the concave structure of f in a manner like our analysis of increasing–decreasing behavior in Section 4.5. To do this we construct a table, using the open intervals into which the zeros of f'' separate the x-axis.

Interval	$x + 0.70$	$160x$	$x - 1.07$	$f''(x)$	f
$(-\infty, -0.70)$	Neg.	Neg.	Neg.	Neg.	Concave down
$(-0.70, 0)$	Pos.	Neg.	Neg.	Pos.	Concave up
$(0, 1.07)$	Pos.	Pos.	Neg.	Neg.	Concave down
$(1.07, +\infty)$	Pos.	Pos.	Pos.	Pos.	Concave up

From the table we see that the direction of concavity of f changes at each of the points $x = -0.70$, $x = 0$, and $x = 1.07$. So these three points are indeed inflection points. This information is shown in the graph sketched in Fig. 4.49.

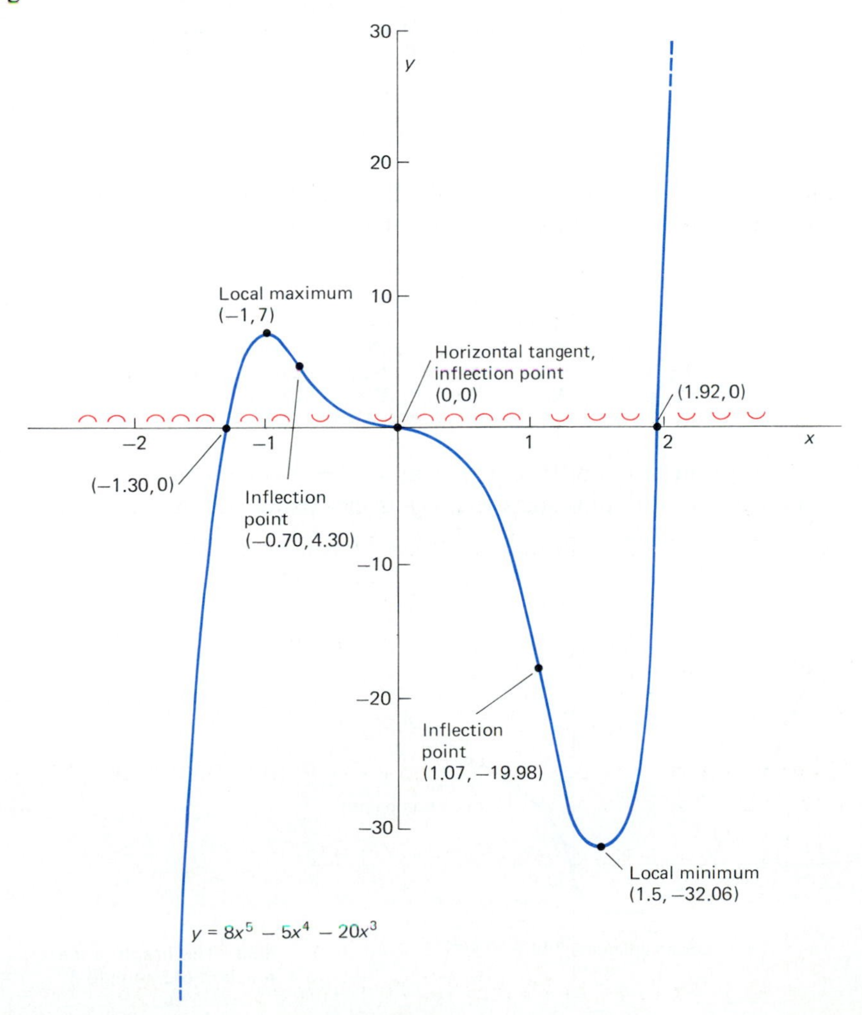

4.49 The graph of the function of Example 5

EXAMPLE 6 Sketch the graph of $f(x) = 4x^{1/3} + x^{4/3}$. Indicate local extrema, inflection points, and concave structure.

Solution First,

$$f'(x) = \frac{4}{3}x^{-2/3} + \frac{4}{3}x^{1/3} = \frac{4(x+1)}{3x^{2/3}},$$

so the critical points are $x = -1$ (where the tangent line is horizontal) and $x = 0$ (where it is vertical). Next,

$$f''(x) = -\frac{8}{9}x^{-5/3} + \frac{4}{9}x^{-2/3} = \frac{4(x-2)}{9x^{5/3}},$$

so the possible inflection points are $x = 2$ (where $f''(x) = 0$) and $x = 0$ (where $f''(x)$ does not exist).

To determine where f is increasing and decreasing, we construct the table shown next.

Interval	$x + 1$	$x^{2/3}$	$f'(x)$	f
$(-\infty, -1)$	Neg.	Pos.	Neg.	Decreasing
$(-1, 0)$	Pos.	Pos.	Pos.	Increasing
$(0, +\infty)$	Pos.	Pos.	Pos..	Increasing

Thus f is decreasing when $x < -1$ and increasing when $x > -1$.

To determine the concavity of f, we construct a table to find the sign of $f''(x)$ on each of the intervals separated by its zeros. It shows that f is concave downward on (0, 2) and concave upward for $x < 0$ and for $x > 2$.

Interval	$x^{5/3}$	$x - 2$	$f''(x)$	f
$(-\infty, 0)$	Neg.	Neg.	Pos.	Concave upward
$(0, 2)$	Pos.	Neg.	Neg.	Concave downward
$(2, +\infty)$	Pos.	Pos.	Pos.	Concave upward

We note also that $f(x) \to +\infty$ as $x \to \pm\infty$, and we mark the intervals on the x-axis with plus signs where f is increasing, minus signs where it is

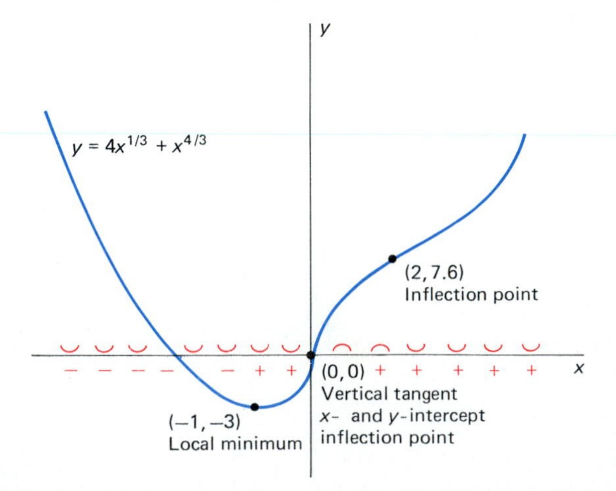

4.50 The graph of the function of Example 6

decreasing, cups opening upward where f is concave upward, and cups opening downward where f is concave downward. We plot (at least approximately) the points on the graph of f corresponding to the zeros and discontinuities of f' and f''; these are $(-1, -3)$, $(0, 0)$, and $(2, 6\sqrt[3]{2})$. Finally we use all this information to draw the smooth curve shown in Fig. 4.50.

Proof of Theorem 2 We will prove only part 1—the proof of part 2 is similar. Given a fixed point a of the open interval I where $f''(x) > 0$, we want to show that the graph $y = f(x)$ lies above the tangent line at $(a, f(a))$. The tangent line in question has equation

$$y = T(x) = f(a) + f'(a)(x - a). \tag{5}$$

Consider the auxiliary function

$$g(x) = f(x) - T(x) \tag{6}$$

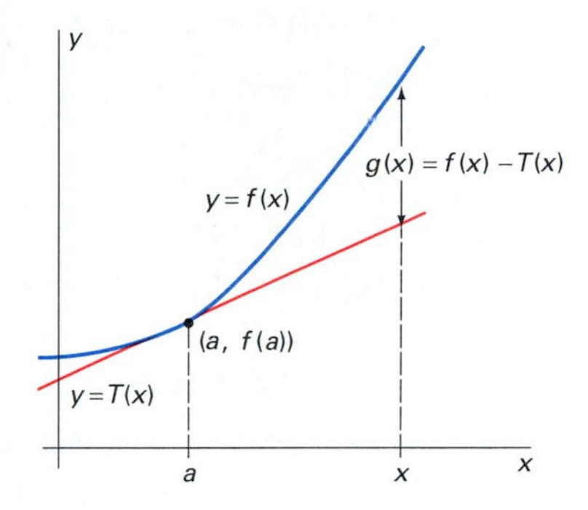

4.51 Illustrating the proof of Theorem 2

illustrated in Fig. 4.51. Note first that $g(a) = g'(a) = 0$, so $x = a$ is a critical point of g. Moreover, (5) implies that $T'(x) \equiv f'(a)$ and that $T''(x) \equiv 0$, so

$$g''(x) = f''(x) - T''(x) = f''(x) > 0$$

at each point of I. Therefore, the second derivative test implies that $g(a) = 0$ is the minimum value of $g(x) = f(x) - T(x)$ on I. It follows that the curve $y = f(x)$ lies above the tangent line $y = T(x)$. ■

4.6 PROBLEMS

Calculate the first three derivatives of the functions given in Problems 1–15.

1. $f(x) = 2x^4 - 3x^3 + 6x - 17$

2. $f(x) = 2x^5 + x^{3/2} - \dfrac{1}{2x}$

3. $f(x) = \dfrac{2}{(2x - 1)^2}$

4. $g(t) = t^2 + \sqrt{t + 1}$

5. $g(t) = (3t - 2)^{4/3}$

6. $f(x) = x\sqrt{x + 1}$

7. $h(y) = \dfrac{y}{y + 1}$

8. $f(x) = (1 + \sqrt{x})^3$

9. $g(t) = \dfrac{1}{2t^{1/2}} - \dfrac{3}{(1 - t)^{1/3}}$

10. $h(z) = \dfrac{z^2}{z^2 + 4}$

11. $f(x) = \sin 3x$

12. $f(x) = \cos^2 2x$

13. $f(x) = \sin x \cos x$

14. $f(x) = x^2 \cos x$

15. $f(x) = \dfrac{\sin x}{x}$

In each of Problems 16–22, calculate y' and y'' assuming that y is defined implicitly as a function of x by the given equation. Primes denote derivatives with respect to x.

16. $x^2 + y^2 = 4$

17. $x^2 + xy + y^2 = 3$

18. $x^{1/3} + y^{1/3} = 1$

19. $y^3 + x^2 + x = 5$

20. $\dfrac{1}{x} + \dfrac{1}{y} = 2$

21. $\sin y = xy$

22. $\sin^2 x + \cos^2 y = 1$

Apply the second derivative test to find the local maxima and local minima of the functions given in Problems 23–33, and apply the inflection point test to find all inflection points.

23. $f(x) = x^2 - 4x + 3$

24. $f(x) = 5 - 6x - x^2$

25. $f(x) = x^3 - 3x + 1$

26. $f(x) = x^3 - 3x^2$

27. $f(x) = x^3$

28. $f(x) = x^4$

29. $f(x) = x^5 + 2x$

30. $f(x) = x^4 - 8x^2$

31. $f(x) = x^2(x - 1)^2$

32. $f(x) = x^3(x + 2)^2$

33. $f(x) = (x - 1)^2(x - 2)^3$

In each of Problems 34–43, rework the indicated problem from Section 4.4, using now the second derivative test to verify that you have found the desired absolute maximum or minimum value.

34. Problem 18

35. Problem 19

36. Problem 20

37. Problem 21

38. Problem 22

39. Problem 23

40. Problem 26

41. Problem 27

42. Problem 28

43. Problem 29

Sketch the graphs of the functions in Problems 44–58, indicating all critical points and inflection points. Apply the second derivative test at each critical point. Show the correct concave structure in your sketches, and indicate the behavior of $f(x)$ as $x \to \pm\infty$.

44. $f(x) = 12x - x^3$

45. $f(x) = 2x^3 - 3x^2 - 12x + 3$

46. $f(x) = 3x^4 - 4x^3 - 5$

47. $f(x) = 6 + 8x^2 - x^4$

48. $f(x) = 3x^5 - 5x^3$

49. $f(x) = 3x^4 - 4x^3 - 12x^2 - 1$

50. $f(x) = 3x^5 - 25x^3 + 60x$

51. $f(x) = x^3(1 - x)^4$

52. $f(x) = (x - 1)^2(x + 2)^3$

53. $f(x) = 1 + x^{1/3}$

54. $f(x) = 2 - (x - 3)^{1/3}$

55. $f(x) = (x + 3)\sqrt{x}$

56. $f(x) = x^{2/3}(5 - 2x)$

57. $f(x) = (4 - x)\sqrt[3]{x}$

58. $f(x) = x^{1/3}(6 - x)^{2/3}$

59. Find all the nonzero derivatives of $f(x) = (x + 1)^5$.

60. Suppose that $f(x) = x^n$, where n is a positive integer. Show by induction that

$$f^{(n)}(x) = n! = n(n - 1) \cdots 3 \cdot 2 \cdot 1.$$

61. Conclude from the result of Problem 60 that if $f(x)$ is a polynomial of degree n, then $f^{(k)}(x) \equiv 0$ if $k > n$.

62. (a) Calculate the first four derivatives of $f(x) = \sin x$.
(b) Conclude that $D^{n+4} \sin x = D^n \sin x$ if n is a positive integer.

63. Suppose that $z = g(y)$ and that $y = f(x)$. Show that

$$\frac{d^2z}{dx^2} = \frac{d^2z}{dy^2}\left(\frac{dy}{dx}\right)^2 + \frac{dz}{dy} \cdot \frac{d^2y}{dx^2}.$$

64. Prove that the graph of a quadratic polynomial has no inflection points.

65. Prove that the graph of a cubic polynomial (one of degree 3) has exactly one inflection point.

66. Prove that the graph of a polynomial function of degree 4 has either no inflection point or exactly two inflection points.

67. Suppose that the pressure P (in atmospheres), volume V (in cubic centimeters), and temperature T (in degrees kelvin) of n moles of carbon dioxide satisfies van der Waals' equation

$$\left(P + \frac{n^2a}{V^2}\right)(V - nb) = nRT$$

where a, b, and R are empirical constants. The following experiment was carried out in order to find the values of these constants.

One mole of CO_2 was compressed at the constant temperature $T = 304$°K. The measured pressure–volume (PV) data were then plotted as in Fig. 4.52, with the PV curve showing a *horizontal* inflection point at $V = 128.1$, $P = 72.8$. Use this information to calculate a, b, and R. (*Suggestion:* Solve van der Waals' equation for P and then calculate dP/dV and d^2P/dV^2.)

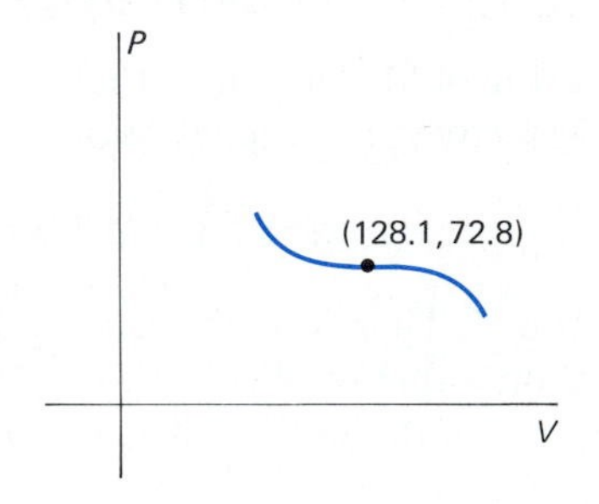

4.52 A problem involving van der Waals' equation

68. Suppose that the function f is differentiable on an open interval containing the point c at which $f'(c) = 0$, and that the second derivative

$$f''(c) = \lim_{h \to 0} \frac{f'(c + h) - f'(c)}{h} = \lim_{h \to 0} \frac{f'(c + h)}{h}$$

exists.

(a) If $f''(c) > 0$, reason that $f'(c + h)$ and h have the same sign if $h \neq 0$ is sufficiently small. Hence apply the first derivative test to show in this case that $f(c)$ is a local minimum value of f.

(b) If $f''(c) < 0$, show similarly that $f(c)$ is a local maximum value of f.

4.7 Curve Sketching and Asymptotes

Before we turn to scientific applications of differentiation, we want to extend the limit concept to include infinite limits and limits at infinity. This extension adds a powerful new weapon to our arsenal of curve-sketching techniques, the notion of an *asymptote* to a curve—a straight line that the curve approaches arbitrarily closely in a sense we soon make exact.

Recall from Section 2.3 that $f(x)$ is said to **increase without bound,** or **become infinite,** as x approaches a, and we write

$$\lim_{x \to a} f(x) = +\infty, \tag{1}$$

provided that $f(x)$ can be made arbitrarily large by choosing x sufficiently close (though not equal) to a. More precisely, (1) means that given $M > 0$, there exists $\delta > 0$ such that

$$0 < |x - a| < \delta \quad \text{implies} \quad f(x) > M.$$

For example, it is apparent that

$$\lim_{x \to 2} \frac{1}{(x-2)^2} = +\infty$$

because $(x - 2)^2$ is positive and is approaching zero as $x \to 2$.

The statement that $f(x)$ **decreases without bound,** or **becomes negatively infinite,** as $x \to a$, written

$$\lim_{x \to a} f(x) = -\infty, \tag{2}$$

has an analogous definition.

One-sided versions of (1) and (2) make sense as well. For instance, if n is an *odd* integer, then it is apparent that

$$\lim_{x \to 2^-} \frac{1}{(x-2)^n} = -\infty \quad \text{while} \quad \lim_{x \to 2^+} \frac{1}{(x-2)^n} = +\infty,$$

because $(x - 2)^n$ is negative when x is to the left of 2 and positive when x is to the right of 2.

We say that the line $x = a$ is a **vertical asymptote** for the curve $y = f(x)$ provided that *either*

$$\lim_{x \to a^-} f(x) = \pm\infty \tag{3a}$$

or

$$\lim_{x \to a^+} f(x) = \pm\infty \tag{3b}$$

or both. It is usually the case that both one-sided limits are infinite, rather than only one, and if so we write

$$\lim_{x \to a} f(x) = \pm\infty. \tag{3c}$$

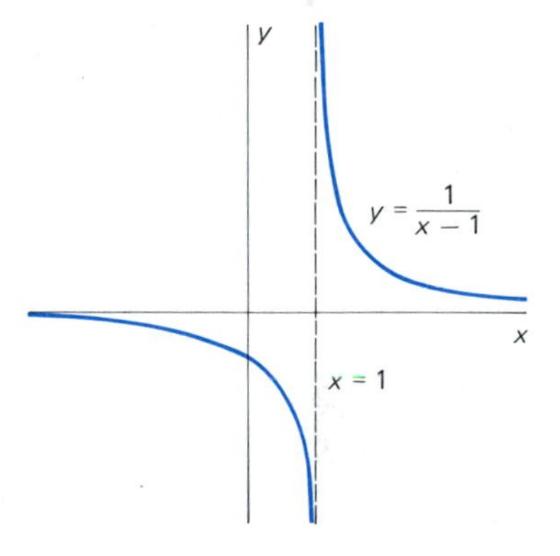

4.53 The graph of $y = 1/(x - 1)$

The geometric significance of a vertical asymptote is illustrated by the graphs of $y = 1/(x - 1)$ and $y = 1/(x - 1)^2$ in Figs. 4.53 and 4.54. In each case, as $x \to 1$ and $f(x) \to \pm\infty$, the corresponding point $(x, f(x))$ on the curve approaches the vertical asymptote $x = 1$.

Figure 4.55 shows the graph of a function having left-hand limit zero at $x = 1$. But the right-hand limit there is $+\infty$, and this is the reason that the line $x = 1$ is also a vertical asymptote for this graph. In Fig. 4.56 the right-hand limit does not even exist, but because the left-hand limit at $x = 1$ is $-\infty$, the vertical line $x = 1$ is again a vertical asymptote.

The most typical occurrence of a vertical asymptote is for a rational function $f(x) = p(x)/q(x)$ at a point $x = a$ where $q(a) = 0$ but $p(a) \neq 0$. Several examples appear shortly.

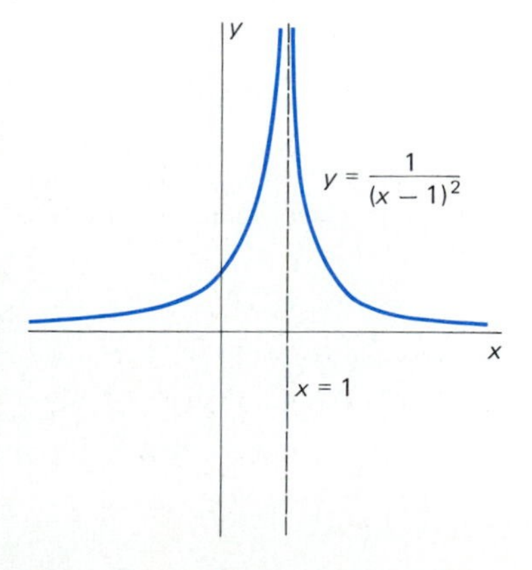

4.54 The graph of $y = 1/(x - 1)^2$

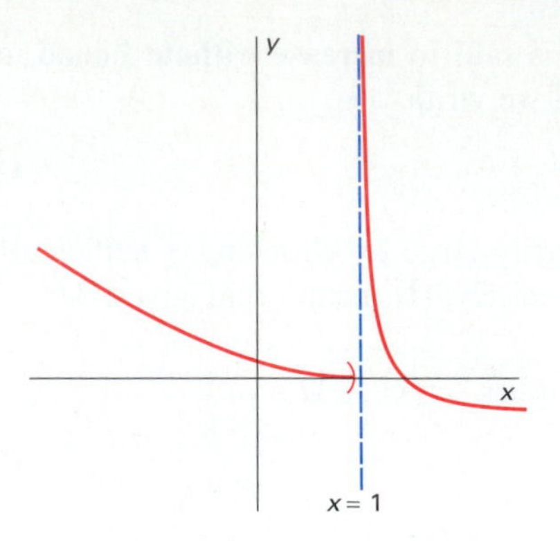

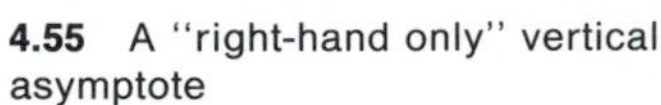

4.55 A "right-hand only" vertical asymptote

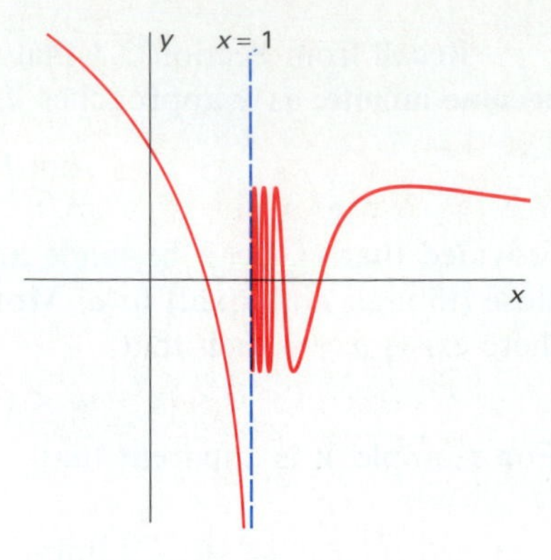

4.56 The behavior of the graph to its *left* produces the vertical asymptote.

In Section 4.5 we mentioned infinite limits at infinity in connection with the behavior of a polynomial as $x \to \pm\infty$. There is also such a thing as a finite limit at infinity. We say that $f(x)$ **approaches the number** L **as** x **increases without bound,** and write

$$\lim_{x \to +\infty} f(x) = L, \tag{4}$$

provided that $|f(x) - L|$ can be made arbitrarily small (close to zero) by choosing x sufficiently large. That is, given $\varepsilon > 0$, there exists $M > 0$ such that

$$x > M \quad \text{implies} \quad |f(x) - L| < \varepsilon. \tag{5}$$

The statement that

$$\lim_{x \to -\infty} f(x) = L$$

has a definition of similar form—merely replace the condition $x > M$ by the condition $x < -M$.

The analogues for limits at infinity of the limit laws of Section 2.2 all hold, including in particular the sum, product, and quotient laws. In addition, it is not difficult to show that if

$$\lim_{x \to +\infty} f(x) = L \quad \text{and} \quad \lim_{x \to +\infty} g(x) = \pm\infty,$$

then

$$\lim_{x \to +\infty} \frac{f(x)}{g(x)} = 0.$$

It follows from this result that

$$\lim_{x \to +\infty} \frac{1}{x^k} = 0 \tag{6}$$

for any choice of the positive rational number k.

Using (6) and the limit laws, limits at infinity of rational functions are easy to evaluate. The general method is this: First divide each term in both numerator and denominator by the highest power of x that appears in any of the terms. Then apply the limit laws.

EXAMPLE 1 Find

$$\lim_{x\to+\infty} f(x) \quad \text{if} \quad f(x) = \frac{3x^3 - x}{2x^3 + 7x^2 - 4}.$$

Solution We begin by dividing each term in numerator and denominator by x^3. Thus

$$\lim_{x\to+\infty} \frac{3x^3 - x}{2x^3 + 7x^2 - 4} = \lim_{x\to+\infty} \frac{3 - \dfrac{1}{x^2}}{2 + \dfrac{7}{x} - \dfrac{4}{x^3}} = \frac{\lim\limits_{x\to+\infty}\left(3 - \dfrac{1}{x^2}\right)}{\lim\limits_{x\to+\infty}\left(2 + \dfrac{7}{x} - \dfrac{4}{x^3}\right)}$$

$$= \frac{3 - 0}{2 + 0 - 0} = 1.5.$$

Note that the same computation, but with $x \to -\infty$, also gives the result

$$\lim_{x\to-\infty} f(x) = 1.5.$$

EXAMPLE 2 Find $\lim\limits_{x\to+\infty} (\sqrt{x+a} - \sqrt{x})$.

Solution We use the familiar "divide and multiply" technique.

$$\lim_{x\to+\infty} (\sqrt{x+a} - \sqrt{x}) = \lim_{x\to+\infty} (\sqrt{x+a} - \sqrt{x}) \cdot \frac{\sqrt{x+a} + \sqrt{x}}{\sqrt{x+a} + \sqrt{x}}$$

$$= \lim_{x\to+\infty} \frac{a}{\sqrt{x+a} + \sqrt{x}} = 0.$$

The geometric meaning of the statement

$$\lim_{x\to+\infty} f(x) = L$$

is that the point $(x, f(x))$ on the curve $y = f(x)$ approaches the horizontal line $y = L$ as $x \to +\infty$. In particular, with the numbers M and ε of the condition in (5), the part of the curve for which $x > M$ lies between the horizontal lines $y = L - \varepsilon$ and $y = L + \varepsilon$ (see Fig. 4.57). Therefore, we say that the line $y = L$

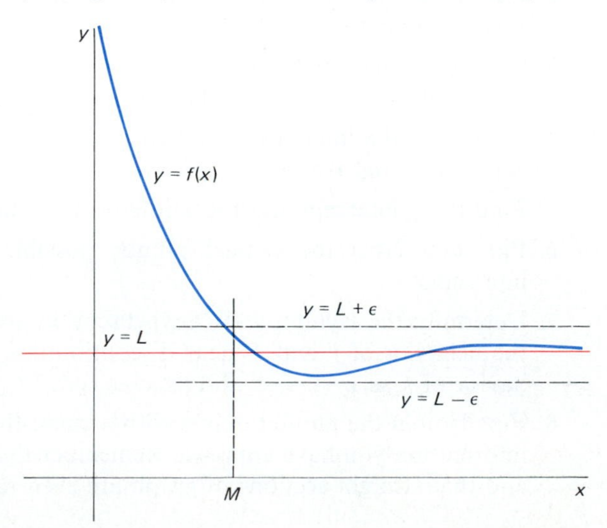

4.57 Geometry of the definition of horizontal asymptote

is a **horizontal asymptote** for the curve $y = f(x)$ if either

$$\lim_{x \to +\infty} f(x) = L \quad \text{or} \quad \lim_{x \to -\infty} f(x) = L.$$

EXAMPLE 3 Sketch the graph of $f(x) = x/(x - 2)$. Indicate any horizontal or vertical asymptotes.

Solution First we note that $x = 2$ is a vertical asymptote because $|f(x)| \to +\infty$ as $x \to 2$. Also,

$$\lim_{x \to \pm\infty} \frac{x}{x - 2} = \lim_{x \to \pm\infty} \frac{1}{1 - \dfrac{2}{x}} = 1.$$

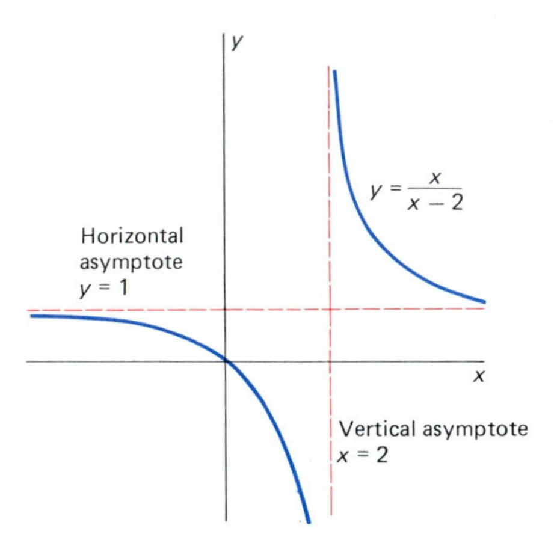

4.58 The graph for Example 3

So $y = 1$ is a horizontal asymptote. The first two derivatives of f are

$$f'(x) = -\frac{2}{(x - 2)^2} \quad \text{and} \quad f''(x) = \frac{4}{(x - 2)^2}.$$

Note that neither $f'(x)$ nor $f''(x)$ vanishes anywhere, so the function f has no critical points and no inflection points. Because $f'(x) < 0$ for $x \neq 2$, we see that $f(x)$ is decreasing on the open intervals $(-\infty, 2)$ and $(2, +\infty)$. Because $f''(x) < 0$ for $x < 2$ while $f''(x) > 0$ for $x > 2$, we also see that f is concave downward on $(-\infty, 2)$ and concave upward on $(2, +\infty)$. The graph of f appears in Fig. 4.58.

The curve-sketching techniques of Sections 4.5 and 4.6, amalgamated with those of the present section, can be summarized as a list of steps. If you follow these steps, loosely rather than rigidly, you will obtain a qualitatively accurate sketch of the graph of a given function f.

1. Solve the equation $f'(x) = 0$ and also find where $f'(x)$ does not exist. This gives the critical points of f. Note whether the tangent line is horizontal, vertical, or nonexistent.
2. Determine the intervals on which f is increasing and those on which it is decreasing.
3. Solve the equation $f''(x) = 0$ and also find where $f''(x)$ does not exist. These will be the *possible* inflection points of the graph.
4. Determine the intervals on which f is concave upward and those on which it is concave downward.
5. Find the y-intercept and the x-intercepts (if any) of the graph.
6. Plot and label the critical points, possible inflection points, and intercepts.
7. Determine the asymptotes (if any), discontinuities (if any), and *especially* the behavior of f and f' near discontinuities. Also determine the behavior of f as $x \to +\infty$ and as $x \to -\infty$.
8. Finally, join the plotted points with a curve that is consistent with the information you have amassed. Remember that corner points are rare and that straight sections of graph are even rarer.

Of course, you may follow these steps in any convenient order and omit any that present formidable computational difficulties. Many examples will require fewer than all eight steps; see Example 3. But our next example requires them all!

EXAMPLE 4 Sketch the graph of

$$f(x) = \frac{2 + x - x^2}{(x-1)^2}.$$

Solution We notice immediately that

$$\lim_{x \to 1} f(x) = +\infty,$$

because the numerator approaches 2 as $x \to 1$ while the denominator approaches zero through *positive* values. So the line $x = 1$ is a vertical asymptote. Also,

$$\lim_{x \to \pm\infty} \frac{2 + x - x^2}{(x-1)^2} = \lim_{x \to \pm\infty} \frac{\dfrac{2}{x^2} + \dfrac{1}{x} - 1}{\left(1 - \dfrac{1}{x}\right)^2} = -1,$$

so the line $y = -1$ is a horizontal asymptote.

Next we apply the quotient rule and simplify to find that

$$f'(x) = \frac{x - 5}{(x-1)^3}.$$

So the only critical point in the domain of f is $x = 5$, and we plot the point $(5, f(5)) = (5, -1.125)$ on a convenient cooordinate plane and mark the horizontal tangent at $(5, f(5))$. To determine the increasing–decreasing behavior of f, we use both the critical point $x = 5$ and the point $x = 1$ (where f is not defined) to separate the x-axis into open intervals. Here are the results.

Interval	$(x-1)^3$	$x-5$	$f'(x)$	f
$(-\infty, 1)$	Neg.	Neg.	Pos.	Increasing
$(1, 5)$	Pos.	Neg.	Neg.	Decreasing
$(5, +\infty)$	Pos.	Pos.	Pos.	Increasing

After some simplifications, we find the second derivative to be

$$f''(x) = \frac{2(7 - x)}{(x-1)^4}.$$

The only possible inflection point is at $x = 7$, corresponding to the point $(7, -\frac{10}{9})$ on the graph. We use both $x = 7$ and $x = 1$ (where f'' is not defined) to separate the x-axis into open intervals. The concave structure of the graph can be deduced with the aid of the next table.

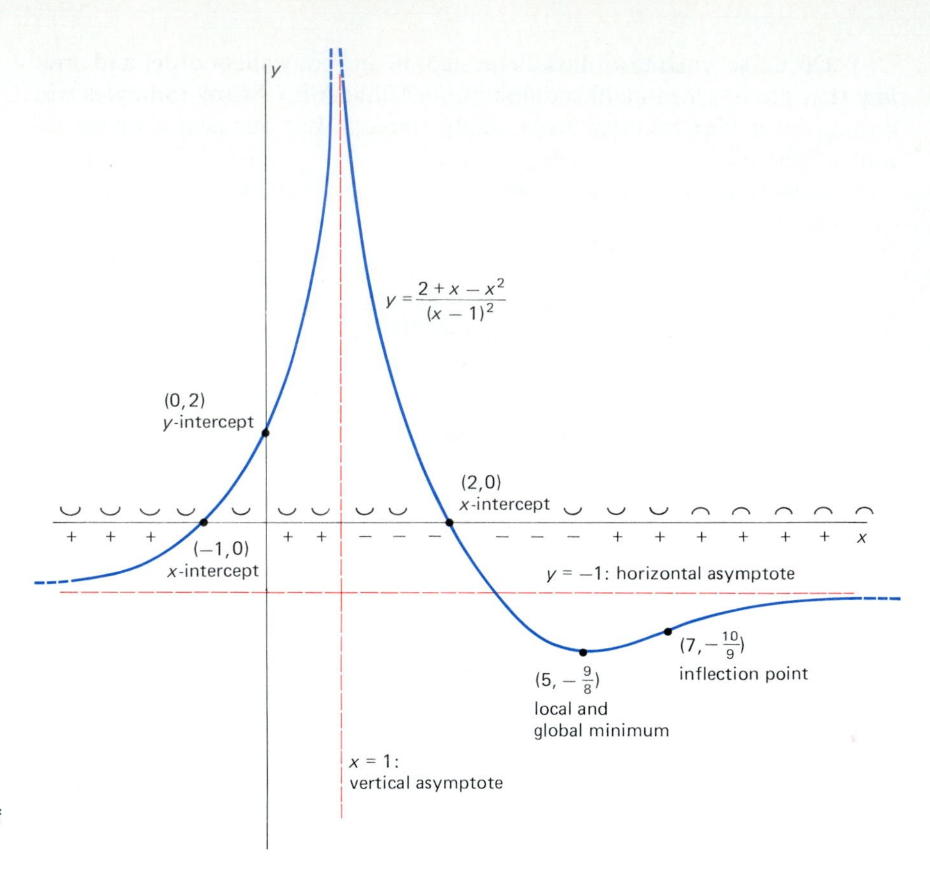

4.59 Graphing the function of Example 4

Interval	$(x-1)^4$	$7-x$	$f''(x)$	f
$(-\infty, 1)$	Pos.	Pos.	Pos.	Concave upward
$(1, 7)$	Pos.	Pos.	Pos.	Concave upward
$(7, +\infty)$	Pos.	Neg.	Neg.	Concave downward

The y-intercept of f is $(0, 2)$, and the equation $2 + x - x^2 = 0$ readily yields the x-intercepts $(-1, 0)$ and $(2, 0)$. We plot these intercepts, sketch the asymptotes, and finally sketch the graph with the aid of the two tables above—their information now appears in Fig. 4.59 along the x-axis, with plus signs above the intervals where f is increasing, minus signs where f is decreasing, cups opening upward where f is concave upward, and cups opening downward where f is concave downward.

Asymptotes may be inclined, not merely horizontal or vertical. The nonvertical line $y = mx + b$ is an **asymptote** for the curve $y = f(x)$ provided that either

$$\lim_{x \to +\infty} [f(x) - (mx + b)] = 0 \tag{7a}$$

or

$$\lim_{x \to -\infty} [f(x) - (mx + b)] = 0 \tag{7b}$$

(or both). These conditions mean that as $x \to +\infty$ or as $x \to -\infty$ (or both), the vertical distance between the point $(x, f(x))$ on the curve and the point $(x, mx + b)$ on the line approaches zero.

If $f(x) = p(x)/q(x)$ is a rational function with the degree of p greater by 1 than that of q, then by long division of $q(x)$ into $p(x)$ we find that $f(x)$ has the form

$$f(x) = mx + b + g(x),$$

where

$$\lim_{x \to \pm\infty} g(x) = 0.$$

Thus the nonvertical line $y = mx + b$ will be an asymptote for the graph of $y = f(x)$.

EXAMPLE 5 Sketch the graph of

$$f(x) = \frac{x^2 + x - 1}{x - 1}.$$

Solution The long division suggested above takes this form:

$$\begin{array}{r} x + 2 \\ x - 1 \,\overline{)\, x^2 + x - 1} \\ \underline{x^2 - x} \\ 2x - 1 \\ \underline{2x - 2} \\ 1 \end{array}$$

Thus

$$f(x) = x + 2 + \frac{1}{x - 1}.$$

So $y = x + 2$ is an asymptote for the curve. Also,

$$\lim_{x \to 1} |f(x)| = +\infty,$$

so $x = 1$ is a vertical asymptote. The first two derivatives of f are

$$f'(x) = 1 - \frac{1}{(x - 1)^2} = \frac{x(x - 2)}{(x - 1)^2}$$

and

$$f''(x) = \frac{2}{(x - 1)^3}.$$

It follows that f has critical points at $x = 0$ and at $x = 2$ but no inflection points. The sign of f' tells us that f is increasing on $(-\infty, 0)$ and on $(2, +\infty)$ and decreasing on $(0, 1)$ and on $(1, 2)$. Examination of $f''(x)$ reveals that f is concave downward on $(-\infty, 1)$ and concave upward on $(1, +\infty)$. In particular, $f(0) = 1$ is a local maximum value and $f(2) = 5$ is a local minimum value. So the graph of f looks much like the one in Fig. 4.60.

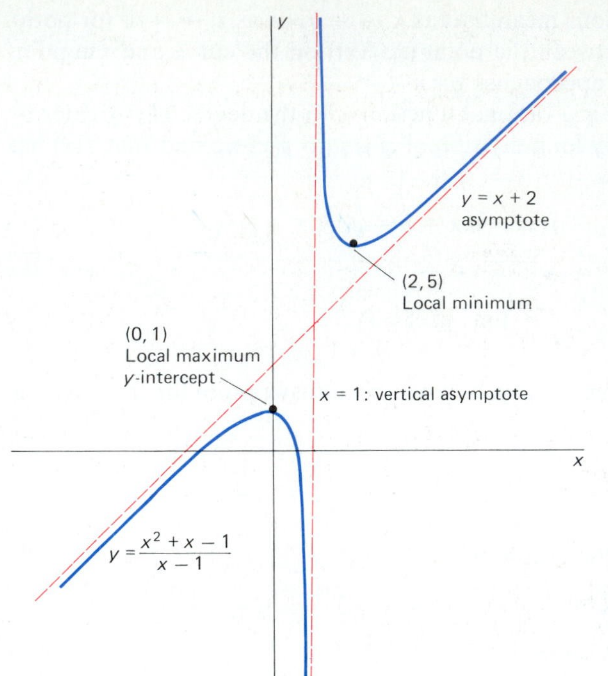

4.60 A function with asymptote the line $y = x + 2$

4.7 PROBLEMS

Investigate the limits in Problems 1–16.

1. $\lim_{x \to +\infty} \frac{x}{x+1}$

2. $\lim_{x \to -\infty} \frac{x^2+1}{x^2-1}$

3. $\lim_{x \to 1} \frac{x^2+x-2}{x-1}$

4. $\lim_{x \to 1} \frac{x^2-x-2}{x-1}$

5. $\lim_{x \to +\infty} \frac{2x^2-1}{x^2-3x}$

6. $\lim_{x \to -\infty} \frac{x^2+3x}{x^3-5}$

7. $\lim_{x \to -1} \frac{x^2+2x+1}{(x+1)^2}$

8. $\lim_{x \to +\infty} \frac{5x^3-2x+1}{7x^3+4x^2-2}$

9. $\lim_{x \to 4} \frac{x-4}{\sqrt{x}-2}$

10. $\lim_{x \to +\infty} \frac{2x+1}{x-x^{3/2}}$

11. $\lim_{x \to -\infty} \frac{8-\sqrt[3]{x}}{2+x}$

12. $\lim_{x \to +\infty} \frac{2x^2-17}{x^3-2x+27}$

13. $\lim_{x \to +\infty} \sqrt{\frac{4x^2-x}{x^2+9}}$

14. $\lim_{x \to -\infty} \frac{\sqrt[3]{x^3-8x+1}}{3x-4}$

15. $\lim_{x \to -\infty} (\sqrt{x^2+2x}-x)$

16. $\lim_{x \to -\infty} (2x-\sqrt{4x^2-5x})$

Sketch the graphs of the functions in Problems 17–42. Identify and label all extrema, inflection points, intercepts, and asymptotes. Show the concave structure clearly, as well as the behavior of the graph for $|x|$ large and for x near any discontinuities of the function.

17. $f(x) = \frac{2}{x-3}$

18. $f(x) = \frac{4}{5-x}$

19. $f(x) = \frac{3}{(x+2)^2}$

20. $f(x) = -\frac{4}{(3-x)^2}$

21. $f(x) = \frac{1}{(2x-3)^3}$

22. $f(x) = \frac{x+1}{x-1}$

23. $f(x) = \frac{x^2}{x^2+1}$

24. $f(x) = \frac{2x}{x^2+1}$

25. $f(x) = \frac{1}{x^2-9}$

26. $f(x) = \frac{x}{4-x^2}$

27. $f(x) = \frac{1}{x^2+x-6}$

28. $f(x) = \frac{2x^2+1}{x^2-2x}$

29. $f(x) = x + \frac{1}{x}$

30. $f(x) = 2x + \frac{1}{x^2}$

31. $f(x) = \frac{x^2}{x-1}$

32. $f(x) = \frac{2x^3-5x^2+4x}{x^2-2x+1}$

33. $f(x) = \frac{1}{(x-1)^2}$

34. $f(x) = \frac{1}{x^2-4}$

35. $f(x) = \frac{x}{x+1}$

36. $f(x) = \frac{1}{(x+1)^3}$

37. $f(x) = \dfrac{1}{x^2 - x - 2}$

38. $f(x) = \dfrac{1}{(x - 1)(x + 1)^2}$

39. $f(x) = \dfrac{x^2 - 4}{x}$

40. $f(x) = \dfrac{x}{x^2 - 1}$

41. $f(x) = \dfrac{x^3 - 4}{x^2}$

42. $f(x) = \dfrac{x^2 + 1}{x - 2}$

43. Suppose that

$$f(x) = x^2 + \frac{2}{x}.$$

Note that

$$\lim_{x \to \pm\infty} [f(x) - x^2] = 0,$$

so the curve $y = f(x)$ approaches the parabola $y = x^2$ as $x \to \pm\infty$. Use this observation to make an accurate sketch of the graph of f.

44. Use the method of Problem 43 to make an accurate sketch of the graph of

$$f(x) = x^3 - \frac{12}{x - 1}.$$

4.8 Antiderivatives

The language of change is the natural language for the statement of most scientific laws and principles. For example, Newton's law of cooling says that the *rate of change* of the temperature T of a body is proportional to the difference between T and the temperature of the surrounding medium. That is,

$$\frac{dT}{dt} = k(A - T) \tag{1}$$

where k is a positive constant and A, normally assumed constant, is the surrounding temperature. Similarly, the *rate of change* of a population P with constant birth and death rates is proportional to the size of the population:

$$\frac{dP}{dt} = kP \qquad (k \text{ constant}). \tag{2}$$

Torricelli's law implies that the *rate of change* of the volume V of water in a draining tank is proportional to the square root of the depth y of the water; that is,

$$\frac{dV}{dt} = -k\sqrt{y} \qquad (k \text{ constant}). \tag{3}$$

Thus mathematical models of real-world situations frequently involve equations containing *derivatives* of unknown functions. Such equations, including those in (1)–(3), are called **differential equations.**

The simplest kind of differential equation has the form

$$\frac{dy}{dx} = f(x),$$

where f is a given (known) function and the function y of x is unknown. The process of finding a function from its derivative is the opposite of differentiation and is therefore called **antidifferentiation.** If we can find a function $Y(x)$ having $y(x)$ as its derivative—

$$Y'(x) = y(x)$$

—we call Y an *antiderivative* of y.

Definition *Antiderivative*
An **antiderivative** of the function f is a function F such that

$$F'(x) = f(x)$$

wherever f is defined.

Function $f(x)$	*Antiderivative* $F(x)$
1	x
$2x$	x^2
x^3	$\frac{1}{4}x^4$
$\cos x$	$\sin x$
$\sin 2x$	$-\frac{1}{2}\cos 2x$

The adjacent table shows some examples of functions and antiderivatives.

Observe that if a function has one antiderivative, then it has many antiderivatives. (By contrast, a function can have only one derivative.) While $F(x) = x^3$ is an antiderivative of $f(x) = 3x^2$, so are the functions $G(x) = x^3 + 17$, $H(x) = x^3 + \pi$, and $K(x) = x^3 - \sqrt{2}$. Indeed, $x^3 + C$ is an antiderivative of $3x^2$ for *any* choice of the constant C.

More generally, if $F(x)$ is an antiderivative of $f(x)$ then so is $F(x) + C$ for any constant C. The converse of this statement is more subtle: If $F(x)$ is one antiderivative of $f(x)$ *on the interval I*, then *every* antiderivative of $f(x)$ on I is of the form $F(x) + C$. This follows immediately from Corollary 2 of the mean value theorem in Section 4.3, according to which two functions with the same derivative on an interval differ only by a constant. Hence the next theorem describes the set of *all* antiderivatives of f.

Theorem 1 *The Most General Antiderivative*
If $F'(x) = f(x)$ at each point of the open interval I, then every antiderivative G of f on I has the form

$$G(x) = F(x) + C \tag{4}$$

where C is a constant.

Thus if F is one antiderivative of f on I, then the *most general* antiderivative of f has the form $F(x) + C$ given in (4). We have noted the different antiderivatives x^3, $x^3 + 17$, $x^2 + \pi$, and $x^3 - \sqrt{2}$ of the function $f(x) = 3x^2$. Each of these has the form in (4) with $F(x) = x^3$ and each is said to be a **particular** antiderivative of $f(x) = 3x^2$.

When we want to emphasize the fact that antidifferentiation is the opposite (or inverse) of differentiation, we use the notation

$$D^{-1}f(x) = F(x) + C \tag{5a}$$

or

$$D_x^{-1}f(x) = F(x) + C \tag{5b}$$

for the most general antiderivative of f. Thus

$$D^{-1}f(x) = F(x) + C \quad \text{if and only if} \quad DF(x) = f(x),$$

where D denotes the operation of differentiation. For example, we may write $D^{-1}3x^2 = x^3 + C$ to state concisely the form of the most general antiderivative of $3x^2$.

Every differentiation formula yields an immediate antidifferentiation formula. We ordinarily write antidifferentiation formulas in terms of most general antiderivatives, as in the following theorem.

Theorem 2 *Antidifferentiation Formulas*

The most general antiderivative of

$$f(x) = x^r \ (r \neq -1) \quad \text{is} \quad F(x) = \frac{x^{r+1}}{r+1} + C; \tag{6}$$

$$f(x) = \cos x \quad \text{is} \quad F(x) = \sin x + C; \tag{7}$$

$$f(x) = \sin x \quad \text{is} \quad F(x) = -\cos x + C. \tag{8}$$

Each such formula may be checked by differentiation of the right-hand side. This is the sure-fire way to check any antidifferentiation: To verify that F is an antiderivative of f, compute F' and see whether it is equal to f.

The linearity of the operation of differentiation implies immediately that antidifferentiation is linear in the following sense. If F and G are antiderivatives of f and g, respectively, and a is a constant, then the most general antiderivative of

$$af(x) \quad \text{is} \quad aF(x) + C; \tag{9}$$

$$f(x) + g(x) \quad \text{is} \quad F(x) + G(x) + C. \tag{10}$$

In essence, we may antidifferentiate a sum of terms by antidifferentiating each of them individually.

EXAMPLE 1 Find the most general antiderivative of

$$f(x) = x^3 + 3\sqrt{x} - \frac{4}{x^2}.$$

Solution To prepare the function f for antidifferentiation, we write it in the form

$$f(x) = x^3 + 3x^{1/2} - 4x^{-2}.$$

Then Equations (6), (9), and (10) yield

$$F(x) = \frac{x^4}{4} + 3\left(\frac{x^{3/2}}{3/2}\right) - 4\left(\frac{x^{-1}}{-1}\right) + C = \frac{1}{4}x^4 + 2x^{3/2} + \frac{4}{x} + C$$

for the most general antiderivative F of f.

A common technique of antidifferentiation is the application of the chain rule in reverse. The chain rule in the form

$$D_x g(u) = g'(u)\frac{du}{dx}$$

yields the following result.

> ***Theorem 3*** *Chain Rule Antidifferentiation*
> If u is a differentiable function of x and $g(u)$ is a differentiable function of u, then the most general antiderivative of
>
> $$f(x) = g'(u)\frac{du}{dx} \quad \text{is} \quad F(x) = g(u) + C, \tag{11}$$
>
> where u is expressed in terms of x on the right-hand side.

When we combine the formula in (11) with the antidifferentiation formulas of Theorem 2, we find that the most general antiderivative of

$$f(x) = u^r\frac{du}{dx}\ (r \neq -1) \quad \text{is} \quad F(x) = \frac{u^{r+1}}{r+1} + C; \tag{12}$$

$$f(x) = (\cos u)\frac{du}{dx} \quad \text{is} \quad F(x) = \sin u + C; \tag{13}$$

$$f(x) = (\sin u)\frac{du}{dx} \quad \text{is} \quad F(x) = -\cos u + C. \tag{14}$$

The key to applying the formulas in Equations (12)–(14) is to spot the appropriate function $u(x)$ such that $f(x)$ is the product of a function of u and the derivative du/dx. We then need only antidifferentiate that function of u.

EXAMPLE 2 Find the most general antiderivative of

$$f(x) = (x + 1)^{10}.$$

Solution If

$$u = x + 1, \quad \text{so that} \quad \frac{du}{dx} = 1,$$

then

$$f(x) = (x + 1)^{10}(1) = u^{10}\frac{du}{dx}.$$

Hence the formula in (12) yields

$$F(x) = \frac{u^{11}}{11} + C = \frac{1}{11}(x + 1)^{11} + C.$$

for the most general antiderivative F of f. The resubstitution of $x + 1$ for u is essential because the symbol u does not appear in the original problem.

EXAMPLE 3 Find the most general antiderivative of

$$f(x) = \frac{20}{(4 - 5x)^3}.$$

Solution Rewriting f as

$$f(x) = 20(4 - 5x)^{-3}$$

suggests that we try

$$u = 4 - 5x, \quad \text{so that} \quad \frac{du}{dx} = -5.$$

In order to match the pattern in Equation (12), we need for the derivative $du/dx = -5$ to appear explicitly in the formula for f:

$$f(x) = (-4) \cdot (4 - 5x)^{-3}(-5) = (-4)u^{-3}\frac{du}{dx}.$$

It now follows that the most general antiderivative of $f(x)$ is given by

$$F(x) = (-4)\frac{u^{-2}}{-2} + C = 2u^{-2} + C = \frac{2}{(4 - 5x)^2} + C.$$

At this point in our studies, antidifferentiation involves educated guessing backed up by our experience in finding derivatives. The following example illustrates how we may make more effective guesses: Once the proper function $u(x)$ for use in Equation (11) has been identified, the remainder of the computation often can be carried out quite systematically.

EXAMPLE 4 Find the most general antiderivative of

$$f(x) = 2x\sqrt{7 + 5x^2}.$$

Solution Because

$$f(x) = 2x \cdot (7 + 5x^2)^{1/2},$$

we try

$$u = 7 + 5x^2, \quad \text{so that} \quad \frac{du}{dx} = 10x.$$

At this point we see $2x$ rather than $10x$ as a factor in $f(x)$. But as a general rule, we can always adjust for a "discrepancy" that requires only a *constant* multiplicative factor. Here we need only write

$$2x = \frac{1}{5}(10x) = \frac{1}{5} \cdot \frac{du}{dx},$$

and thus

$$f(x) = \left(\frac{1}{5} \cdot \frac{du}{dx}\right)u^{1/2} = \frac{1}{5} \cdot u^{1/2}\frac{du}{dx}.$$

It now follows that the most general antiderivative of $f(x)$ is given by

$$F(x) = \frac{1}{5} \cdot \frac{u^{3/2}}{3/2} + C = \frac{2}{15}u^{3/2} + C = \frac{2}{15}(7 + 5x^2)^{3/2} + C.$$

EXAMPLE 5 Find the most general antiderivative of

$$f(t) = \sqrt{t}(2t - 3)^2.$$

Solution The natural substitution

$$u = 2t - 3, \qquad \frac{du}{dt} = 2$$

is useless because of the presence of the factor $\sqrt{t}$. Instead, we simply expand the binomial first:

$$f(t) = \sqrt{t}(4t^2 - 12t + 9) = 4t^{5/2} - 12t^{3/2} + 9t^{1/2}.$$

It is now clear that the desired antiderivative is

$$F(t) = \tfrac{8}{7}t^{7/2} - \tfrac{24}{5}t^{5/2} + 6t^{3/2} + C.$$

If k is a constant, then the familiar differentiation formulas

$$D_x \cos kx = -k \sin kx, \qquad D_x \sin kx = k \cos kx$$

imply that the most general antiderivative of

$$f(x) = \cos kx \quad \text{is} \quad F(x) = \frac{1}{k} \sin kx + C; \tag{15}$$

$$f(x) = \sin kx \quad \text{is} \quad F(x) = -\frac{1}{k} \cos kx + C. \tag{16}$$

Note that (15) and (16) amount to the special case $u = kx$ in (13) and (14).

EXAMPLE 6 The most general antiderivative of

$$f(x) = 2 \cos 3x + 5 \sin 4x$$

is

$$\begin{aligned} F(x) &= 2(\tfrac{1}{3} \sin 3x) + 5(-\tfrac{1}{4} \cos 4x) + C \\ &= \tfrac{2}{3} \sin 3x - \tfrac{5}{4} \cos 4x + C. \end{aligned}$$

EXAMPLE 7 Find the most general antiderivative of

$$h(t) = 3 \sin^4 t \cos t.$$

Solution The more complicated factor $\sin^4 t = (\sin t)^4$ prompts us to try

$$u = \sin t, \quad \text{so that} \quad \frac{du}{dt} = \cos t.$$

Then

$$h(t) = 3u^4 \frac{du}{dt},$$

so we actually have a *power rule* antidifferentiation. The most general antiderivative is

$$H(t) = 3 \cdot \frac{u^5}{5} + C = \frac{3}{5} \sin^5 t + C.$$

VERY SIMPLE DIFFERENTIAL EQUATIONS

The technique of antidifferentiation can often be used to solve a differential equation of the special form

$$\frac{dy}{dx} = f(x) \tag{17}$$

in which the dependent variable y does not appear on the right-hand side. In this case the **general solution** of (17) is the most general antiderivative

$$y = F(x) + C \tag{18}$$

of the function $f(x)$. Any function $y = y(x)$ satisfying (17) must be of the form in (18) for some value of the constant C—because if $y' = f$, then y and F have the same derivative f, so $y(x)$ and $F(x)$ can differ only by a constant.

A differential equation of the form in (17) often appears in conjunction with an **initial condition** of the form

$$y(x_0) = y_0. \tag{19}$$

This condition specifies the value $y = y_0$ that the solution function $y(x)$ must have at $x = x_0$. Once we have found the general solution in (18), we can determine the value of the constant C by substituting the information that $y = y_0$ when $x = x_0$. With this specific value of C, (18) then gives the **particular solution** of (17) that satisfies the initial condition in (19). The combination of a differential equation (as in (17)) with an initial condition (as in (19)) is called an **initial value problem.**

EXAMPLE 8 Find the particular solution of the differential equation

$$\frac{dy}{dx} = 2x + 3 \tag{20}$$

that satisfies the initial condition $y(1) = 2$.

Solution The most general antiderivative of $f(x) = 2x + 3$ is

$$F(x) = x^2 + 3x + C.$$

Hence the *general solution* of the differential equation in (2) is given by

$$y(x) = x^2 + 3x + C. \tag{21}$$

To determine the value of C in (21), we substitute the initial data $y = 2$ when $x = 1$:

$$2 = (1)^2 + (3)(1) + C.$$

It follows that $C = -2$, so the desired *particular solution* of (20) is given by

$$y(x) = x^2 + 3x - 2. \tag{22}$$

REMARK The method used in Example 7 may be described as "antidifferentiating each side of the differential equation with respect to x." That is, we begin with (2) and write

$$D_x^{-1}\left(\frac{dy}{dx}\right) = D_x^{-1}(2x + 3),$$

$$y = x^2 + 3x + C,$$

supplying in the final step the constant C by which two functions with the same derivative may differ.

4.8 PROBLEMS

Find the most general antiderivatives of the functions in Problems 1–46.

1. $f(x) = 3x^2 + 2x + 1$

2. $g(t) = 3t^4 + 5t - 6$

3. $h(x) = 1 - 2x^2 + 3x^3$

4. $j(t) = -\dfrac{1}{t^2}$

5. $f(x) = \dfrac{3}{x^3} + 2x^{3/2} - 1$

6. $h(x) = x^{5/2} - \dfrac{5}{x^4} - \sqrt{x}$

7. $f(t) = \frac{3}{2}t^{1/2} + 7$

8. $h(x) = \dfrac{2}{x^{3/4}} - \dfrac{3}{x^{2/3}}$

9. $f(x) = \sqrt[3]{x^2} + \dfrac{4}{\sqrt[4]{x^5}}$

10. $g(x) = 2x^{3/2} - \dfrac{1}{x^{1/2}}$

11. $f(x) = 4x^3 - 4x + 6$

12. $g(t) = \frac{1}{4}t^5 - 6t^{-2}$

13. $h(x) = 7$

14. $f(x) = 4x^{2/3} - \dfrac{5}{x^{1/3}}$

15. $g(x) = (x + 1)^4$

16. $h(t) = (t + 1)^{10}$

17. $f(x) = \dfrac{1}{(x - 10)^7}$

18. $g(z) = \sqrt{z + 1}$

19. $f(x) = \sqrt{x}(1 - x)^2$

20. $f(x) = x^{1/3}(x + 1)^3$

21. $f(x) = \dfrac{2x^4 - 3x^3 + 5}{7x^2}$

22. $f(x) = \dfrac{(3x + 4)^2}{\sqrt{x}}$

23. $f(t) = (9t + 11)^5$

24. $h(z) = \dfrac{1}{(3z + 10)^7}$

25. $h(x) = \dfrac{7}{(x + 77)^2}$

26. $g(x) = \dfrac{3}{\sqrt{(x - 1)^3}}$

27. $f(t) = 6t(3t^2 - 1)^7$

28. $f(x) = x\sqrt{1 - x^2}$

29. $f(x) = 2x\sqrt{x^2 + 1}$

30. $h(t) = t(1 - 3t^2)^{10}$

31. $f(x) = \dfrac{3x}{\sqrt{x^2 + 4}}$

32. $g(x) = x\sqrt{(2x^2 + 3)^3}$

33. $f(x) = x^2(x^3 + 2)^{1/3}$

34. $h(t) = \dfrac{3t^2}{\sqrt{2t^3 + 1}}$

35. $f(x) = x^7(x^8 + 9)^{10}$

36. $f(x) = x^2(x^7 - x^5)$

37. $f(x) = (x^2 + 1)^3$

38. $f(x) = x\left(\sqrt{x} - \dfrac{1}{\sqrt{x}}\right)^2$

39. $f(x) = 5 \cos 10x - 10 \sin 5x$

40. $f(x) = 2 \cos \pi x + 3 \sin \pi x$

41. $g(t) = 3 \cos \pi t + \cos 3\pi t$

42. $h(t) = 4 \sin 2\pi t - 2 \sin 4\pi t$

43. $f(x) = \sin^5 x \cos x$

44. $f(x) = \cos^3 x \sin x$

45. $f(x) = \sin^3 2x \cos 2x$

46. $f(x) = \cos^4 3x \sin 3x$

47. Suppose that $f(x) = \sin x \cos x$.

(a) Substitute $u = \sin x$ to obtain the antiderivative

$$F_1(x) = \tfrac{1}{2} \sin^2 x + C_1.$$

(b) Substitute $u = \cos x$ to obtain the antiderivative

$$F_2(x) = -\tfrac{1}{2} \cos^2 x + C_2.$$

Reconcile the answers in parts (a) and (b). What is the relation between the constants C_1 and C_2?

48. Show that

$$F_1(x) = \frac{1}{1 - x} \quad \text{and} \quad F_2(x) = \frac{x}{1 - x}$$

are both antiderivatives of $f(x) = 1/(1 - x)^2$. What is the relationship between $F_1(x)$ and $F_2(x)$?

In each of Problems 49–62, find a function $y = y(x)$ satisfying the given differential equation and the prescribed initial condition.

49. $\dfrac{dy}{dx} = 2x + 1$; $y(0) = 3$

50. $\dfrac{dy}{dx} = (x - 2)^3$; $y(2) = 1$

51. $\dfrac{dy}{dx} = \sqrt{x}$; $y(4) = 0$

52. $\dfrac{dy}{dx} = \dfrac{1}{x^2}$; $y(1) = 5$

53. $\dfrac{dy}{dx} = \dfrac{1}{\sqrt{x + 1}}$; $y(2) = -1$

54. $\dfrac{dy}{dx} = x\sqrt{x^2 + 9}$; $y(-4) = 0$

55. $\dfrac{dy}{dx} = 3x^3 + \dfrac{2}{x^2}$; $y(1) = 1$

56. $\dfrac{dy}{dx} = x^4 - 3x + \dfrac{3}{x^3}$; $y(1) = -1$

57. $\dfrac{dy}{dx} = (x - 1)^3$; $y(0) = 2$

58. $\dfrac{dy}{dx} = \sqrt{x + 5}$; $y(4) = -3$

59. $\dfrac{dy}{dx} = \dfrac{1}{\sqrt{x - 13}}$; $y(17) = 2$

60. $\dfrac{dy}{dx} = (2x + 3)^{3/2}$; $y(3) = 100$

61. $\dfrac{dy}{dx} = x\sqrt{1 - x^2}$; $y(1) = 0$

62. $\dfrac{dy}{dx} = \dfrac{2x}{\sqrt{3x^2 + 4}}$; $y(2) = 3$

4.9 Velocity and Acceleration

Antidifferentiation is the tool that enables us, in many important cases, to analyze the motion of a particle (or "mass point") in terms of the forces acting on it. We consider first the motion of a particle moving along a straight line under the influence of a *constant* acceleration force. If we regard the line of motion as the x-axis, then—as in Section 3.1—the motion of the particle is described by its **position function**

$$x = f(t) \tag{1}$$

giving its x-coordinate at time t. Generally, the function $f(t)$ will be unknown to begin with, and our problem will be to find a formula for it, using such data as the initial position, the initial velocity, and the constant acceleration of the particle.

Recall from Section 3.1 that the *velocity* $v(t)$ of the moving particle is the derivative of its position function,

$$v(t) = f'(t); \quad \text{that is,} \quad v = \frac{dx}{dt}, \tag{2}$$

and that its *acceleration* $a(t)$ is the derivative of its velocity,

$$a(t) = v'(t) = f''(t); \quad \text{that is,} \quad a = \frac{dv}{dt} = \frac{d^2x}{dt^2}. \tag{3}$$

So if the acceleration is *constant*, we begin with the equation

$$\frac{dv}{dt} = a \qquad (a \text{ constant}). \tag{4}$$

The antiderivatives v and at of the expressions on each side of this equation can differ only by a constant, so we conclude that

$$v = at + C_1. \tag{5}$$

The constant C_1 is ordinarily evaluated by substitution of $t = 0$ in *both* sides in (5); this gives

$$v_0 = v(0) = a \cdot 0 + C_1 = C_1,$$

so C_1 turns out to be the *initial velocity* v_0. Hence the velocity $v = dx/dt$ of the particle at time t is

$$\frac{dx}{dt} = v(t) = at + v_0. \tag{6}$$

To find the position function $x(t)$, we antidifferentiate each side in Equation (6). The two antiderivatives x of dx/dt and $\frac{1}{2}at^2 + v_0 t$ of $at + v_0$ again can differ only by a constant, so it follows that

$$x(t) = \tfrac{1}{2}at^2 + v_0 t + C_2 \tag{7}$$

is the position of the particle at time t. We evaluate the constant C_2 in Equation (7) by substituting $t = 0$ in both sides of (7). This gives

$$x_0 = x(0) = \tfrac{1}{2}a \cdot 0^2 + v_0 \cdot 0 + C_2,$$

so that $C_2 = x_0$, the *initial position* of the particle. Thus the position function of the particle is

$$x(t) = \tfrac{1}{2}at^2 + v_0 t + x_0. \tag{8}$$

WARNING The formulas in (6) and (8) are valid only in the case of *constant* acceleration a. They do *not* apply to problems in which the acceleration varies.

EXAMPLE 1 The skid marks made by an automobile indicate that its brakes were fully applied for a distance of 160 ft before it came to a stop. Suppose that it is known that the car in question has a constant deceleration of 20 ft/s^2 under the conditions of the skid. How fast was the car traveling when its brakes were first applied?

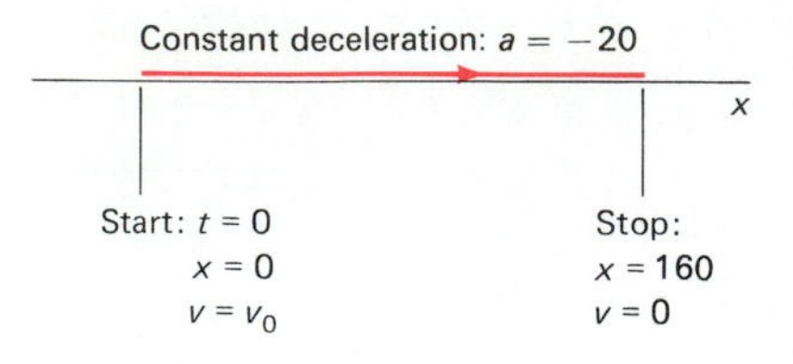

4.61 Skid marks 160 feet long

Solution The introduction of a convenient coordinate system is often crucial to the successful solution of a physical problem. Here we take the x-axis as positively oriented in the direction of motion of the car. We choose the origin so that $x_0 = 0$ when $t = 0$, as indicated in Fig. 4.61. In this coordinate system, the car's velocity v will be a decreasing function of time t (in seconds), so that its acceleration is $a = -20$ (ft/s^2) rather than $a = +20$. Hence we begin with the constant acceleration equation

$$\frac{dv}{dt} = -20.$$

Antidifferentiation gives

$$v(t) = -20t + v_0,$$

where (as in Equation (6)) the constant is the initial velocity v_0. Hence

$$\frac{dx}{dt} = -20t + v_0,$$

and another antidifferentiation yields the position function

$$x(t) = -10t^2 + v_0 t + x_0 = -10t^2 + v_0 t,$$

because we have chosen a coordinate system in which the initial position is $x_0 = 0$.

The fact that the skid marks are 160 ft long tells us that $x = 160$ when the car comes to a stop; that is, that

$$x = 160 \quad \text{when} \quad v = 0.$$

Substitution of these values in the velocity and position equations above then yields the two simultaneous equations

$$-20t + v_0 = 0, \qquad -10t^2 + v_0 t = 160.$$

We now solve these for v_0 and t to find the initial velocity v_0 *and* the duration t of the car's skid. If we multiply the first equation by $-t$ and add the result to the second equation, we find that $10t^2 = 160$, so that $t = 4$ when the car comes to a stop. It follows that

$$v_0 = 20t = (20)(4) = 80$$

feet per second, or about 55 mi/h, was the velocity of the car when the brakes were first applied.

VERTICAL MOTION WITH CONSTANT GRAVITATIONAL ACCELERATION

One common application of Equations (4), (6), and (8) involves vertical motion near the surface of the earth. A particle in such motion is subject to a downward acceleration whose magnitude is denoted by g, approximately 32 ft/s^2.

If we neglect air resistance, we may assume that this acceleration of gravity is the only outside influence on the moving particle. Moreover, if the motion involved is fairly close to the surface of the earth, we also may assume that g remains constant. (If you need more accurate values for g, you may use $g = 32.16$ ft/s^2 in the fps system, $g = 980$ cm/s^2 in the cgs system, or $g = 9.8$ m/s^2 in the mks system.)

Because we deal with vertical motion here, it is natural to choose the y-axis as the coordinate system for position. If we choose the upward direction as the positive direction, then the effect of gravity on the particle is to *decrease* its height and also to *decrease* its velocity $v = dy/dt$, so we see that the acceleration of the particle is

$$a = \frac{dv}{dt} = -g = -32$$

feet per second per second. Equations (6) and (8) then become

$$v(t) = -32t + v_0 \tag{6'}$$

and

$$y(t) = -16t^2 + v_0 t + y_0. \tag{8'}$$

Here y_0 is the initial height of the particle in feet, v_0 is its initial velocity in feet per second, and time t is measured in seconds.

EXAMPLE 2 Suppose that a bolt was fired vertically upward from a powerful crossbow at ground level, and that it struck the ground 48 s later. If air resistance may be neglected, find the initial velocity of the bolt and the maximum altitude it reached.

Solution We set up the coordinate system illustrated in Fig. 4.62, with ground level corresponding to $y = 0$, with the bolt fired at time $t = 0$ (seconds), and with the positive direction being the upward direction. Units on the y-axis are, of course, in feet.

We are given that $y = 0$ when $t = 48$. We lack any information about the initial velocity v_0. But we may use Equations (6′) and (8′) because we have set up a coordinate system in which the acceleration of gravity acts in the negative direction, and thus

$$y(t) = -16t^2 + v_0 t + y_0$$
$$= -16t^2 + v_0 t$$

and

$$v(t) = -32t + v_0.$$

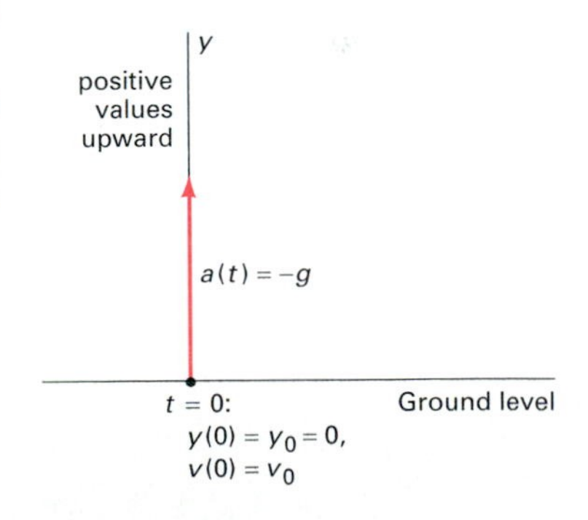

4.62 A bolt fired straight upward from a crossbow

We use the information that $y = 0$ when $t = 48$ in the first equation:

$$0 = -16 \cdot 48^2 + 48v_0, \quad \text{and thus} \quad v_0 = 16 \cdot 48 = 768$$

feet per second. To find the maximum altitude of the bolt, we maximize $y(t)$ by finding when its derivative is zero—in other words, the bolt reaches its maximum altitude when its velocity is zero:

$$-32t + v_0 = 0,$$

so at maximum altitude, $t = v_0/32 = 24$. At that time, the bolt has reached its maximum altitude of

$$y_{\max} = y(24) = -16 \cdot 24^2 + 768 \cdot 24 = 9216$$

feet—almost 2 mi.

The result seems contrary to experience. We must conclude that air resistance cannot always be neglected, particularly not in problems involving long journeys at high velocity.

*NEWTON'S INVERSE-SQUARE LAW OF GRAVITATION

Because we have already used whatever calculus is necessary in deriving the two key formulas in Equations (6) and (8), such problems as that in Example 2—though interesting and potentially important—challenge only one's skills at arithmetic. In order to discuss some more exciting problems, perhaps involving motion of objects (meteors? spacecraft?) at considerable distance from the earth, we need to know Newton's second law of motion and his inverse-square law of gravitation. According to the law of motion, the force F acting on a particle of mass m is proportional to the acceleration a that it produces,

$$F = ma. \tag{9}$$

According to the inverse-square law of gravitation, the gravitational force of attraction between two point masses m and M located at a distance r apart is given by

$$F = \frac{GMm}{r^2} \tag{10}$$

where G is a certain empirical constant. The formula also applies if either (or both) of the two masses is not merely a point mass, but a homogeneous sphere. In this case, the distance r is measured between the centers of the spheres. (We shall derive this result in Chapter 16.)

Let M denote the mass of the earth and R its radius. We obtain the gravitational acceleration $a = g$ of a particle of mass m at the surface of the earth by using Equations (9) and (10) simultaneously:

$$ma = mg = \frac{GMm}{r^2} = \frac{GMm}{R^2},$$

so that

$$g = \frac{GM}{R^2}. \tag{11}$$

We shall use Equation (11) from time to time to remove the necessity for the actual determination of G in our problems and examples.

We mentioned that Equation (10) holds for point masses, and that we shall later see that it also holds for spheres. It appears that Newton was aware of the inverse-square law of gravitation for point masses before 1670, but was unable to show that it applies to massive objects (such as spherical planets) until some years later. This may have been one reason for the delay until 1687 of the publication of his *Philosophiae Naturalis Principia Math-*

ematica (*Mathematical Principles of Natural Philosophy*), the founding document of modern exact science.

EXAMPLE 3 If a woman has enough "spring" in her legs to jump vertically to a height of 4 ft on the earth, how high could she jump on the moon? Use the fact that the mass $\bar{M}$ and the radius $\bar{R}$ of the moon are given in terms of the mass M and radius R of the earth by $\bar{M} \approx (0.0123)M$ and $\bar{R} \approx (0.2725)R$. Because her mass is (presumably) the same anywhere in the universe, it is also reasonable to suppose that she attains the same initial velocity before liftoff on the moon as on the earth.

Solution First we need to find the initial velocity v_0 required to jump 4 ft high on the earth. Then we can calculate how high she can jump on the moon with the same initial velocity.

On the earth, we use the standard coordinate system of Example 2 and the standard equations (8′)

$$y(t) = -16t^2 + v_0 t + y_0 \quad \text{and (6′)} \quad v(t) = -32t + v_0.$$

In the woman's jump on the earth, we have $v = v_0$ when $t = 0$, $y_0 = 0$ (because $y = 0$ corresponds to ground level), and $v = 0$ when $y = 4$. So we substitute $v = 0$ in (6′), and find that the time to reach her maximum height of 4 ft is $t = v_0/32$ s.

Hence Equation (8′) gives

$$4 = -16\left(\frac{v_0}{32}\right)^2 + v_0\left(\frac{v_0}{32}\right) = \frac{v_0^2}{64}.$$

Thus $v_0 = +16$ ft/s.

Now we shift our attention to the moon. In order to find the correct versions of Equations (6′) and (8′) for the moon, we need to find the gravitational acceleration $\bar{g}$ at its surface. But Equation (11) tells us that

$$\begin{aligned}\bar{g} &= \frac{G\bar{M}}{\bar{R}^2} = G\frac{(0.0123)M}{(0.2725R)^2} = (0.1656)\frac{GM}{R^2}\\ &= (0.1656)g = (0.1656)(32) = 5.3\end{aligned}$$

feet per second per second (approximately).

On the moon, then,

$$\frac{dv}{dt} = -5.3,$$

$$v(t) = \frac{dy}{dt} = (-5.3)t + v_0, \tag{6″}$$

and

$$y(t) = (-2.65)t^2 + v_0 t + y_0. \tag{8″}$$

In the moon jump, we have $y_0 = 0$; also $v_0 = 16$, from our previous work with the earth jump. In addition, $v = 0$ at maximum height y. So (6″) gives

$$0 = (-5.3)t + 16$$

then, and thus $t \approx 3.02$ (s) at her maximum height y_{max}. We substitute this value of t in (8″) and obtain

$$y_{max} \approx (-2.65)(3.02)^2 + (16)(3.02) \approx 24.15$$

feet as the height she can jump on the moon. Note that she is off the ground over 6 s during the moon jump but only 1 s during the earth jump.

4.9 PROBLEMS

In Problems 1–10, find the position function of a moving particle with the given acceleration $a(t)$, initial position $x_0 = x(0)$, and initial velocity $v_0 = v(0)$.

1. $a(t) = 50, \quad v_0 = 10, \quad x_0 = 20$
2. $a(t) = -20, \quad v_0 = -15, \quad x_0 = 5$
3. $a(t) = 3t, \quad v_0 = 5, \quad x_0 = 0$
4. $a(t) = 2t + 1, \quad v_0 = -7, \quad x_0 = 4$
5. $a(t) = 4(t + 3)^2, \quad v_0 = -1, \quad x_0 = 1$
6. $a(t) = \dfrac{3}{\sqrt{t + 4}}, \quad v_0 = -1, \quad x_0 = 1$
7. $a(t) = \dfrac{1}{(t + 1)^3}, \quad v_0 = 0, \quad x_0 = 0$
8. $a(t) = \sqrt{t}, \quad v_0 = 0, \quad x_0 = 0$
9. $a(t) = \sin t, \quad v_0 = 0, \quad x_0 = 2$
10. $a(t) = 3 \cos 2\pi t, \quad v_0 = 0, \quad x_0 = 2$

Problems 11–27 deal with vertical motion near the surface of the earth with air resistance considered negligible. Use $g = 32$ ft/s^2 for the magnitude of the gravitational acceleration.

11. A ball is thrown straight upward from the ground with initial velocity 96 ft/s. How high does it rise and how long is it aloft?

12. A ball thrown straight upward from the ground reaches a maximum height of 400 ft. What was its initial velocity?

13. A stone is dropped into a well and hits the bottom 3 s later. How deep is the well?

14. A ball is thrown straight upward alongside a tree. It just reaches the top of the tree and then falls back to the ground; it remains aloft for 4 s. How tall is the tree?

15. A ball is thrown upward with an initial velocity of 48 ft/s from the top of a building 160 ft tall, then falls to the ground at the base of the building. How long does the ball remain aloft, and with what speed does it strike the ground?

16. A ball is dropped from the top of a building 576 ft high. With what velocity should a second ball be thrown straight downward 3 s later in order that the two balls hit the ground simultaneously?

17. A ball is dropped from the top of the Empire State Building, 960 ft above 34th Street. How long does it take for the ball to reach the street, and with what velocity does it hit the street?

18. An arrow is shot straight upward from the ground with initial velocity 320 ft/s.

(a) How high is the arrow after exactly 3 s have elapsed?

(b) At what time is the arrow exactly 1200 ft above the ground?

(c) How many seconds after its release does the arrow strike the ground?

19. A ball thrown upward from the ground reaches a maximum height of 225 ft. What was its initial velocity?

20. A rock is dropped into a well in which the water surface is 256 ft below the ground. How long does it take the rock to reach the water surface? How fast is the rock moving as it penetrates the water surface?

21. A ball is dropped from the top of a building 400 ft high. How long does it take to reach the ground? With what velocity does it strike the ground?

22. A ball is thrown straight downward from the top of a tall building. The initial speed of the ball is 25 ft/s. It hits the ground with a speed of 153 ft/s. How tall is the building?

23. A ball is thrown straight upward from ground level with an initial speed of 160 ft/s. What is the maximum height that the ball attains?

24. A ball is dropped from the top of a tall building h feet high. At the same time a second ball is thrown upward from ground level from a point directly below the first ball. What should be the velocity given the second ball so that they meet at the halfway point, where each has altitude $h/2$?

25. A baseball is thrown straight downward with an initial speed of 40 ft/s from the top of the Washington Monument, 555 ft high. How long does it take to reach the ground, and with what speed does it strike the ground?

26. A rock is dropped from an initial height of h feet above the surface of the earth. Show that the speed with which the rock strikes the surface is $\sqrt{2gh}$.

27. A bomb is dropped from a balloon hovering at an altitude of 800 ft. From directly below the balloon, a projectile is fired straight upward toward the bomb exactly 2 s after the bomb is released. With what initial speed should the projectile be fired in order to hit the bomb at an altitude of exactly 400 ft?

28. A car's brakes are applied when it is moving at 60 mi/h and provide a constant deceleration of 40 ft/s^2. How far does the car travel before coming to a stop?

29. A car traveling at 60 mi/h skids 176 ft after its brakes are applied. If the deceleration provided by the braking system is constant, what is its value?

30. A spacecraft is in free fall toward the surface of the moon at a speed of 1000 mi/h. Its retrorockets, when fired, provide a deceleration of 20,000 mi/h per hour. At what height above the surface should the retrorockets be turned on to insure a "soft touchdown" ($v = 0$ at impact)? Ignore the effect of the moon's gravitational field.

31. If you can throw a ball vertically upward to a maximum height of 144 ft on the earth, how high could you throw it on the moon? (See Example 3.)

32. Arthur C. Clarke's *The Wind from the Sun* (1963) deals with the solar yacht Diana. This spacecraft was propelled by the "solar wind," its 2-mi^2 aluminized sail providing it with an acceleration of $0.001g = 0.032\ \text{ft/s}^2$. If the Diana starts from rest, calculate its distance x traveled (in miles) and its velocity v (mi/h) after 1 min, 1 h, and 1 day.

CHAPTER 4 REVIEW: Definitions, Concepts, Results

Use the list below as a guide to concepts that you may need to review.

1. Increment Δy
2. Differential dy
3. Linear approximation formula
4. Differentiation rules in differential form
5. Increasing and decreasing functions
6. Significance of the sign of the derivative
7. Rolle's theorem
8. The mean value theorem
9. Consequences of the mean value theorem
10. First derivative test
11. Open interval maximum–minimum problems
12. Graphing of polynomials
13. Calculation of higher derivatives
14. Concave-upward and concave-downward functions
15. Test for concavity
16. Second derivative test
17. Inflection points
18. Inflection point test
19. Infinite limits
20. Vertical asymptotes
21. Limits as $x \to \pm\infty$
22. Horizontal asymptotes
23. Curve-sketching strategies
24. Antidifferentiation and antiderivatives
25. The most general antiderivative of a function
26. Antidifferentiation formulas
27. Antidifferentiation using the chain rule
28. Velocity and acceleration
29. Solution of problems involving constant acceleration
30. Newton's inverse-square law of gravitation

CHAPTER 4 MISCELLANEOUS PROBLEMS

In Problems 1–6, write dy in terms of x and dx.

1. $y = (4x - x^2)^{3/2}$

2. $y = 8x^3\sqrt{x^2 + 9}$

3. $y = \dfrac{x+1}{x-1}$

4. $y = \sin(x^2)$

5. $y = x^2 \cos\sqrt{x}$

6. $y = \dfrac{x}{\sin 2x}$

In Problems 7–12, estimate the indicated number by linear approximation.

7. $\sqrt[3]{1005}$

8. $\sqrt[3]{62}$

9. $26^{3/2}$

10. $\sqrt[5]{30}$

11. $\sqrt[4]{17}$

12. $\sqrt[10]{1000}$

In Problems 13–18, estimate by linear approximation the change in the indicated quantity.

13. The volume $V = s^3$ of a cube, if its side length s is increased from 5 in. to 5.1 in.

14. The area of a circle, if its radius is decreased from 10 cm to 9.8 cm.

15. The volume of a sphere, if its radius is increased from 5 to 5.1 in.

16. The volume $V = 1000/P$ cubic inches of a gas, if the pressure P is decreased from 100 lb/in.^2 to 99 lb/in.^2.

17. The period of oscillation $T = 2\pi\sqrt{L/32}$ of a pendulum, if its length L is increased from 2 ft to 25 in. (The units in this problem are feet and seconds.)

18. The lifetime $L = (10^{30})/(E^{13})$ in hours of a light bulb with applied voltage E volts, if the voltage is increased from 110 V to 111 V. Compare with the actual decrease of the lifetime.

If the mean value theorem applies to the function f on the interval $[a, b]$, it assures the existence of a solution c in the interval (a, b) of the equation

$$f'(c) = \frac{f(b) - f(a)}{b - a}.$$

In Problems 19–24, a function f and an interval $[a, b]$ are given. Verify that the hypotheses of the mean value theorem are satisfied for f on $[a, b]$. Then use the equation shown here to find the value of the number c.

19. $f(x) = x - \frac{1}{x}$; $[1, 3]$

20. $f(x) = x^3 + x - 4$; $[-2, 3]$

21. $f(x) = x^3$; $[-1, 2]$

22. $f(x) = x^3$; $[-2, 1]$

23. $f(x) = \frac{11}{5}x^5$; $[-1, 2]$

24. $f(x) = \sqrt{x}$; $[0, 4]$

Sketch the graphs of the functions in Problems 25–29. Indicate the local maxima and minima of each function and the intervals on which it is increasing or decreasing.

25. $f(x) = x^2 - 6x + 4$

26. $f(x) = 2x^3 - 3x^2 - 36x$

27. $f(x) = 3x^5 - 5x^3 + 60x$

28. $f(x) = (3 - x)\sqrt{x}$

29. $f(x) = (1 - x)\sqrt[3]{x}$

30. Show that the equation $x^5 + x = 5$ has exactly one real solution.

Calculate the first three derivatives of the functions in Problems 31–40.

31. $f(x) = x^3 - 2x$

32. $f(x) = (x + 1)^{100}$

33. $g(t) = \frac{1}{t} - \frac{1}{2t + 1}$

34. $h(y) = \sqrt{3y - 1}$

35. $f(t) = 2t^{3/2} - 3t^{4/3}$

36. $g(x) = \frac{1}{x^2 + 9}$

37. $h(t) = \frac{t + 2}{t - 2}$

38. $f(z) = \sqrt[3]{z} + \frac{3}{\sqrt[5]{z}}$

39. $g(x) = \sqrt[3]{5 - 4x}$

40. $g(t) = \frac{8}{(3 - t)^{3/2}}$

In each of Problems 41–44, calculate y' and y'' (primes denote derivatives with respect to x) under the assumption that y is defined implicitly as a function of x by the given equation.

41. $x^{1/3} + y^{1/3} = 1$

42. $2x^2 - 3xy + 5y^2 = 25$

43. $y^5 - 4y + 1 = \sqrt{x}$

44. $\sin(xy) = xy$

Sketch the graphs of the functions in Problem 45–64, indicating all critical points, inflection points, and asymptotes; show the concave structure clearly.

45. $f(x) = x^4 - 32x$

46. $f(x) = 18x^2 - x^4$

47. $f(x) = x^6 - 2x^4$

48. $f(x) = x\sqrt{x - 3}$

49. $f(x) = x\sqrt[3]{4 - x}$

50. $f(x) = \frac{x - 1}{x + 2}$

51. $f(x) = \frac{x^2 + 1}{x^2 - 4}$

52. $f(x) = \frac{x}{x^2 - x - 2}$

53. $f(x) = \frac{2x^2}{x^2 - x - 2}$

54. $f(x) = \frac{x^3}{x^2 - 1}$

55. $f(x) = 3x^4 - 4x^3$

56. $f(x) = x^4 - 2x^2$

57. $f(x) = \frac{x^2}{x^2 - 1}$

58. $f(x) = x^3 - 12x$

59. $f(x) = -10 + 6x^2 - x^3$

60. $f(x) = \frac{x}{1 + x^2}$; note that

$$f'(x) = -\frac{(x - 1)(x + 1)}{(x^2 + 1)^2}$$

and that

$$f''(x) = \frac{2x(x^2 - 3)}{(x^2 + 1)^3}$$

61. $f(x) = x^3 - 3x$

62. $f(x) = x^4 - 12x^2$

63. $f(x) = x^3 + x^2 - 5x + 3$

64. $f(x) = \frac{1}{x} + \frac{1}{x^2}$

Find the most general antiderivatives of the functions given in Problems 65–76.

65. $f(x) = \frac{x^5 - 2x + 5}{x^3}$

66. $f(x) = \sqrt{x}(1 + \sqrt{x})^3$

67. $f(x) = (1 - 3x)^9$

68. $f(x) = \frac{7}{(2x + 3)^3}$

69. $f(x) = \sqrt[3]{9 + 4x}$

70. $f(x) = \frac{24}{\sqrt{6x + 7}}$

71. $f(x) = x^3(1 + x^4)^5$

72. $f(x) = 3x^2\sqrt{4 + x^3}$

73. $f(x) = x\sqrt[3]{1 - x^2}$

74. $f(x) = \frac{3x}{\sqrt{1 + 3x^2}}$

75. $f(x) = 7\cos 5x - 5\sin 7x$

76. $f(x) = 5\sin^3 4x \cos 4x$

In each of Problems 77–82, find a function $y = f(x)$ satisfying the given differential equation and the prescribed initial condition.

77. $\frac{dy}{dx} = 3x^2 + 2x$, $y(0) = 5$

78. $\dfrac{dy}{dx} = 3\sqrt{x}, \quad y(4) = 20$

79. $\dfrac{dy}{dx} = (2x + 1)^5, \quad y(0) = 2$

80. $\dfrac{dy}{dx} = \dfrac{2}{\sqrt{x + 5}}, \quad y(4) = 3$

81. $\dfrac{dy}{dx} = \dfrac{1}{\sqrt[3]{x}}, \quad y(1) = 1$

82. $\dfrac{dy}{dx} = 1 - \cos x, \quad y(0) = 0$

83. The function

$$f(x) = \frac{1}{x^2 + 2x + 2}$$

has a maximum value, and only one. Find it.

84. Suppose that we need to manufacture a cylindrical pot, without a top, with a volume of 1 ft^3. The cylindrical part of the pot is to be made of aluminum, the bottom of copper. Copper is five times as expensive as aluminum. What dimensions will minimize the total cost of the pot?

85. A wall 8 ft high stands 4 ft away from a tall building. What is the length of the shortest ladder that will lean over the wall and touch the building? Choose as independent variable the angle that the ladder makes with the ground.

86. A certain freight train has a constant deceleration of 1 ft/s^2 when its brakes are fully applied. Suppose that it is traveling at 60 mi/h. What is the minimum time in which it can come to a stop? How far will it travel during that time?

87. When its brakes are fully applied, a certain automobile has constant deceleration of 22 ft/s^2. If its initial velocity is 90 mi/h, how long will it take to come to a stop? How many feet will it travel during that time?

88. In Hal Clement's novel *Mission of Gravity* (serialized under the title *Heavy Planet*), much of the action takes place in the polar regions of the planet Mesklin, where the acceleration of gravity is 22,500 ft/s^2. A stone is dropped near the north pole of Mesklin from a height of 450 ft. How long does it remain aloft? With what speed does it strike the ground?

89. An automobile is traveling along the x-axis in the positive direction. At time $t = 0$ its brakes are fully applied, and the car experiences a constant deceleration of 40 ft/s^2 while skidding. The car skids for 180 ft before coming to a stop. What was its initial velocity?

90. If a car starts from rest with an acceleration of 8 ft/s^2, how far has it traveled by the time it reaches a speed of 60 mi/h?

91. Consider a planet where a ball dropped from a height of 20 ft hits the ground in 2 s. If the ball is dropped from the top of a 200-ft building on this planet, how long will it take to reach the ground? With what speed will it hit?

92. A man can throw a ball from the earth's surface straight upward to a maximum height of 144 ft.

(a) How high could he throw it on the planet of Problem 91?

(b) How high could he throw it in the polar regions of Mesklin (see Problem 88)?

93. Suppose that a car skids 44 ft if its velocity is 30 mi/h when the brakes are fully applied. Assuming the same constant deceleration, how far will it skid if its velocity is 60 mi/h when the brakes are fully applied?

94. Suppose that $f(x)$ is a cubic polynomial having exactly three distinct real zeros. Prove that the two zeros of $f'(x)$ are real and distinct.

95. Suppose that it costs $1 + (0.0003)v^{3/2}$ dollars per mile to operate a truck at v miles per hour. If there are additional costs (such as the driver's pay) of $10 per hour, what speed will minimize the total cost of a 1000-mi trip?

96. The numbers $a_1, a_2, \ldots, a_n$ are fixed. Find a simple formula for the number x such that the sum of the squares of the distances of x from the n fixed numbers is as small as possible.

97. Sketch the curve $y^2 = x(x - 1)(x - 2)$, indicating that it consists of two pieces—one bounded and the other unbounded—and has two horizontal tangent lines, three vertical tangent lines, and two inflection points. (*Suggestion:* Note that the curve is symmetric about the x-axis, and begin by determining the intervals on which the product $x(x - 1)(x - 2)$ is positive. Compute dy/dx and d^2y/dx^2 by implicit differentiation.)

98. The graph of the velocity of a model rocket fired at time $t = 0$ is shown in Fig. 4.63.

(a) At what time was the fuel exhausted?

(b) At what time did the parachute open?

(c) At what time did the rocket reach its maximum altitude?

(d) At what time did the rocket land?

(e) How high did the rocket go?

(f) How high was the pole on which the rocket landed?

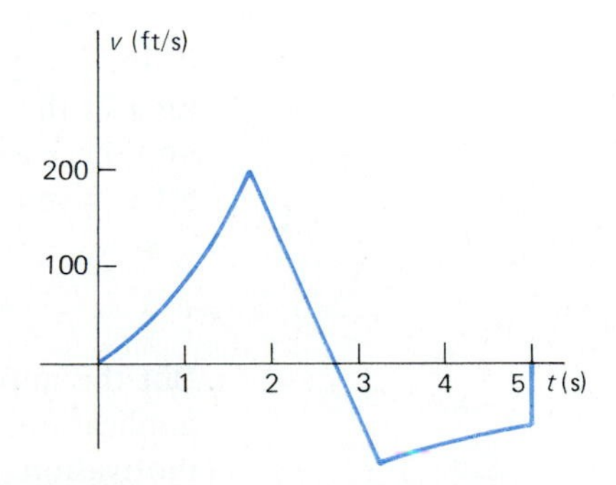

4.63 Rocket velocity graph for Problem 98

chapter five

The Integral

5.1 Introduction

The earlier chapters have dealt with **differential calculus,** which is one of two closely related parts of *the* calculus. Differential calculus is centered on the concept of the *derivative.* Recall that the original motivation for the derivative was the problem of defining tangent lines to graphs of functions and calculating the slopes of such lines. By contrast, the importance of the derivative stems from its applications to diverse problems that may seem, upon initial inspection, to have little connection with tangent lines to graphs.

Integral calculus is based on the concept of the *integral.* The definition of the integral is motivated by the problem of defining and calculating the area of the region lying between the graph of a positive-valued function f and the x-axis over a closed interval $[a, b]$. The area of the region R of Fig. 5.1 is given by the **integral** of f from a to b, denoted by the symbol

$$\int_a^b f(x)\,dx.$$

But the importance of the integral, like that of the derivative, is due to its applications in many problems that may appear unrelated to its original motivation.

The principal theorem of this chapter is the *fundamental theorem of calculus* in Section 5.5. It provides a vital connection between the operations

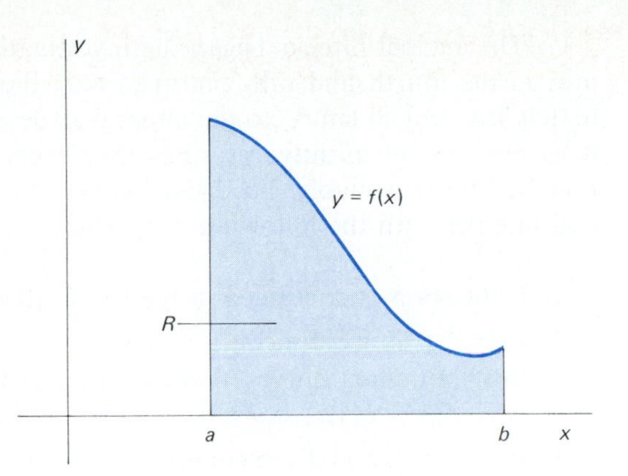

5.1 The problem of computing the area of R motivates integral calculus.

of differentiation and integration and also a method of computing values of integrals. In this chapter we concentrate on the definition and the basic computational properties of the integral, and reserve applications (other than those closely connected with area computations) to Chapter 6.

5.2 Elementary Area Computations

Perhaps everyone's first contact with the concept of area is the formula $A = bh$, which gives the area A of a rectangle as the product of its base b and its height h. We next learn that the area of a triangle is half the product of its base and height. This follows because any triangle can be split into two *right* triangles, and every right triangle is exactly half a rectangle (see Fig. 5.2).

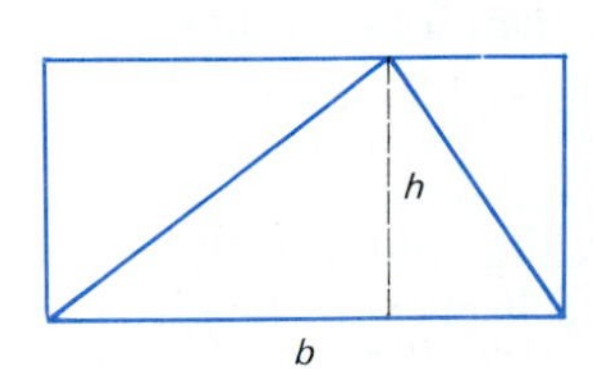

5.2 The formula for the area of a triangle follows with the aid of this figure.

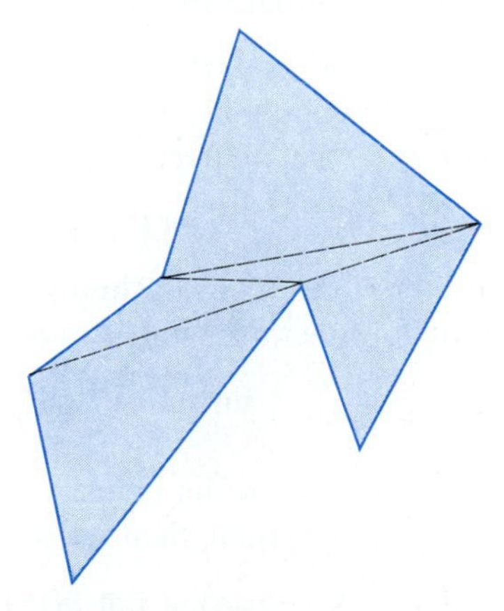

5.3 Every polygon can be represented as the union of nonoverlapping triangles.

Given the formula $A = \frac{1}{2}bh$ for the area of a triangle, we can find the area of an arbitrary polygonal figure (a bounded plane set bounded by a closed "curve" consisting of finitely many straight line segments). The reason is that any polygonal figure can be divided into nonoverlapping triangles, as shown in Fig. 5.3, and the area of the polygonal figure is then the sum of the area of these triangles. This approach to area dates back several thousand years to the ancient civilizations of Egypt and Babylonia.

The ancient Greeks began the investigation of areas of *curvilinear* figures in the fourth and fifth centuries B.C.; the approach they devised ultimately led—much later—to the integral as described in Section 5.1. We take it as obvious on intuitive grounds that every reasonably nice plane set S bounded by reasonably nice closed curves has an area $a(S)$, a nonnegative real number with the following properties:

1. If the set S is contained in the set T, then $a(S) \leqq a(T)$.
2. If S and T are nonoverlapping sets—they intersect only in points of their boundary curves, if at all—then the area of their union $S \cup T$ is $a(S \cup T) = a(S) + a(T)$.
3. If the sets S and T are congruent—have the same size and shape—then $a(S) = a(T)$.

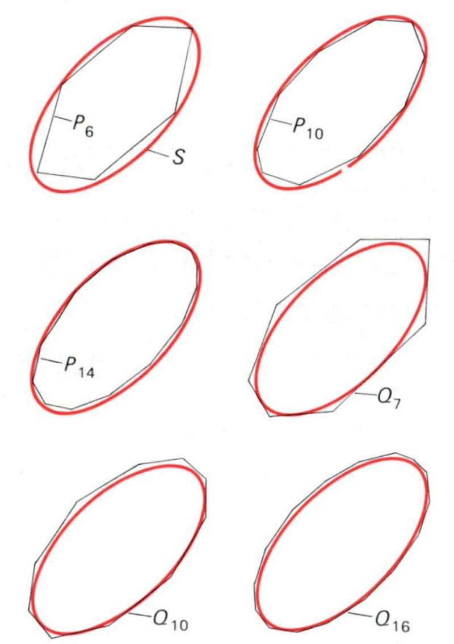

5.4 Estimating the area $a(S)$ of the set S by using areas of inscribed and circumscribed polygons

To investigate the area of a curvilinear set S, we might inscribe in it a sequence of polygons $P_1, P_2, P_3, \ldots$ with increasing areas that gradually fill or "exhaust" the set S. We can then attempt to compute the limit $\lim_{n\to\infty} a(P_n)$ of the areas of these inscribed polygons. Recall from the definition of the limit of a sequence (Section 3.9) that

$$A = \lim_{n\to\infty} a(P_n)$$

means that $a(P_n)$ can be made as close to A as we please merely by choosing n sufficiently large.

Suppose that we also circumscribe about S a sequence of polygons $Q_1, Q_2, Q_3, \ldots$ with decreasing areas that gradually "close down" on the set S, and attempt to compute the limit $\lim_{n\to\infty} a(Q_n)$ of the areas of these circumscribed polygons. Our earlier inscribed polygons and these circumscribed polygons should enable us, as Fig. 5.4 suggests, to obtain arbitrarily good estimates of the actual area $a(S)$. For by Property 1 of area above, we see that

$$a(P_n) \leqq a(S) \leqq a(Q_n)$$

for all n. By the squeeze law for limits of sequences (analogous to the squeeze law for limits of functions, stated in Section 2.2), it then follows that

$$\lim_{n\to\infty} a(P_n) \leqq a(S) \leqq \lim_{n\to\infty} a(Q_n),$$

provided that both these limits exist. If, moreover, the values of these two limits turn out to be equal, then the area of the set S itself must be given by

$$a(S) = \lim_{n\to\infty} a(P_n) = \lim_{n\to\infty} a(Q_n). \tag{1}$$

THE NUMBER π

In the third century B.C., Archimedes, the greatest mathematician of antiquity, used an approach similar to that outlined above to derive the famous estimate

$$\tfrac{223}{71} = 3\tfrac{10}{71} < \pi < 3\tfrac{1}{7} = \tfrac{22}{7}.$$

The number π may be defined as the ratio of the circumference of a circle to its diameter; an alternative definition is the ratio of the area of a circle to the square of its radius. It may also be defined as the area of the unit circle $x^2 + y^2 \leqq 1$. With this last definition, the problem of computing π is that of finding the area of the unit circle.

Let P_n and Q_n be n-sided regular polygons, with P_n inscribed in the unit circle and Q_n circumscribed around it, as shown in Fig. 5.5. Because the polygons are regular, all their sides and angles are equal, so all we need to find is the area of *one* of the triangles that we've shown making up P_n and *one* of those making up Q_n.

Let α_n be the central angle be subtended by *half* of one of the sides. The angle α_n is the same whether we work with P_n or Q_n. In degrees,

$$\alpha_n = \frac{360°}{2n} = \frac{180°}{n}.$$

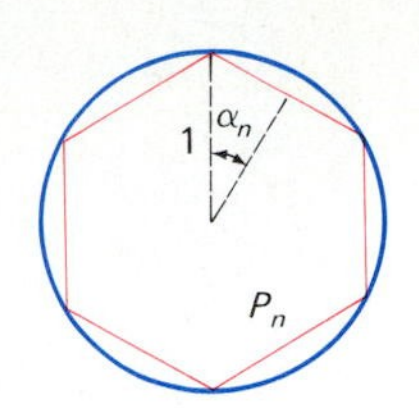

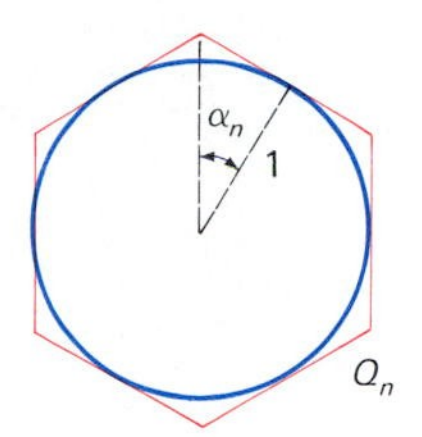

5.5 Estimating π by using inscribed and circumscribed regular polygons and the unit circle

We can read various dimensions and proportions from Fig. 5.5. For example, we see that the area of P_n is given by

$$a(P_n) = n \cdot 2 \cdot \frac{1}{2} \sin \alpha_n \cos \alpha_n = \frac{n}{2} \sin 2\alpha_n = \frac{n}{2} \sin\left(\frac{360°}{n}\right) \tag{2}$$

and that

$$a(Q_n) = n \cdot 2 \cdot \frac{1}{2} \tan \alpha_n = n \tan\left(\frac{180°}{n}\right). \tag{3}$$

We substitute selected values of n in the formulas in (2) and (3) and thereby obtain the entries of the table shown in Fig. 5.6. Because $a(P_n) \leqq \pi \leqq a(Q_n)$ for all n, we see that $\pi \approx 3.14159$ to five decimal places. We should remark that Archimedes' reasoning was *not* "circular"—he used a direct method for computing the sines and cosines in (2) and (3) that does not depend upon a prior knowledge of the value of π (see Problem 36).

n	$a(P_n)$	$a(Q_n)$
6	2.598076	3.464102
12	3.000000	3.215390
24	3.105829	3.159660
48	3.132629	3.146086
96	3.139350	3.142715
180	3.140955	3.141912
360	3.141433	3.141672
720	3.141553	3.141613
1440	3.141583	3.141598
2880	3.141590	3.141594
5760	3.141592	3.141593

5.6 Data in estimating π (rounded to six-place accuracy)

SUMMATION NOTATION

For convenient application of the method of inscribed and circumscribed polygons to compute areas of rather general figures, we need a compact notation for sums of many numbers. We use the symbol $\sum_{i=1}^{n} a_i$ to abbreviate the sum of the n numbers $a_1, a_2, a_3, \ldots, a_n$, so that

$$\sum_{i=1}^{n} a_i = a_1 + a_2 + a_3 + \cdots + a_n.$$

The Σ (Greek capital sigma) indicates the sum of the terms a_i as the **summation index** i runs through integer values from 1 to n. For example, the sum of the squares of the first ten positive integers is

$$\sum_{i=1}^{10} i^2 = 1^2 + 2^2 + 3^2 + \cdots + 10^2$$
$$= 1 + 4 + 9 + \cdots + 100 = 385.$$

The particular symbol used for the summation index is immaterial; note that

$$\sum_{i=1}^{10} i^2 = \sum_{k=1}^{10} k^2 = \sum_{r=1}^{10} r^2 = 385.$$

The simple rules of summation

$$\sum_{i=1}^{n} ca_i = c \sum_{i=1}^{n} a_i \tag{4}$$

and

$$\sum_{i=1}^{n} (a_i + b_i) = \left(\sum_{i=1}^{n} a_i\right) + \left(\sum_{i=1}^{n} b_i\right) \tag{5}$$

are easy to verify by writing out each sum in full.

Note also that if $a_i = a$ (a constant) for $i = 1, 2, \ldots, n$, then (5) yields

$$\sum_{i=1}^{n} (a + b_i) = \sum_{i=1}^{n} a + \sum_{i=1}^{n} b_i = \underbrace{(a + a + \cdots + a)}_{(n \text{ times})} + \sum_{i=1}^{n} b_i,$$

so that

$$\sum_{i=1}^{n} (a + b_i) = na + \sum_{i=1}^{n} b_i.$$

In particular,

$$\sum_{i=1}^{n} 1 = n.$$

The sum of the kth powers of the first n positive integers,

$$\sum_{i=1}^{n} i^k = 1^k + 2^k + 3^k + \cdots + n^k,$$

often occurs in area computations. The values of this sum for $k = 1, 2, 3,$ and 4 are given by the following formulas. They may be established by mathematical induction (see Problems 27–30).

$$\sum_{i=1}^{n} i = \frac{n(n+1)}{2} = \frac{1}{2}n^2 + \frac{1}{2}n. \tag{6}$$

$$\sum_{i=1}^{n} i^2 = \frac{n(n+1)(2n+1)}{6} = \frac{1}{3}n^3 + \frac{1}{2}n^2 + \frac{1}{6}n. \tag{7}$$

$$\sum_{i=1}^{n} i^3 = \frac{n^2(n+1)^2}{4} = \frac{1}{4}n^4 + \frac{1}{2}n^3 + \frac{1}{4}n^2. \tag{8}$$

$$\sum_{i=1}^{n} i^4 = \frac{1}{5}n^5 + \frac{1}{2}n^4 + \frac{1}{3}n^3 - \frac{1}{30}n. \tag{9}$$

Indeed, in the general case it is known that

$$\sum_{i=1}^{n} i^k = \frac{1}{n+1}n^{k+1} + \frac{1}{2}n^k + f(n), \tag{10}$$

where $f(n)$ is a polynomial in the variable n having degree *less* than k.

EXAMPLE 1 Use the summation formulas in (6)–(9) to evaluate the sum

$$\sum_{i=1}^{10} (i^2 + 2i) = 3 + 8 + 15 + \cdots + 120.$$

Solution

$$\sum_{i=1}^{10} (i^2 + 2i) = \sum_{i=1}^{10} i^2 + 2 \sum_{i=1}^{10} i$$

$$= \frac{(10)(11)(21)}{6} + 2\,\frac{(10)(11)}{2} = 495.$$

EXAMPLE 2 Evaluate

$$\lim_{n\to\infty} \frac{1^2 + 2^2 + 3^2 + \cdots + n^2}{n^3}.$$

Solution Using the formula in (7) we obtain

$$\lim_{n\to\infty} \frac{1^2 + 2^2 + 3^2 + \cdots + n^2}{n^3} = \lim_{n\to\infty} \frac{1}{n^3}\left(\frac{1}{3}n^3 + \frac{1}{2}n^2 + \frac{1}{6}n\right)$$

$$= \lim_{n\to\infty} \left(\frac{1}{3} + \frac{1}{2n} + \frac{1}{6n^2}\right) = \frac{1}{3}.$$

We use the fact that the terms $1/2n$ and $1/6n^2$ approach zero as $n \to \infty$.

AREAS UNDER GRAPHS

Let f be a continuous and positive-valued function defined on the closed interval $[a, b]$. Suppose that we want to calculate the **area A under the graph of f from a to b.** That is, A is the area of the region R that is bounded above by the curve $y = f(x)$, below by the x-axis, on the left by the vertical line $x = a$, and on the right by the vertical line $x = b$. This is the region shown in Fig. 5.7.

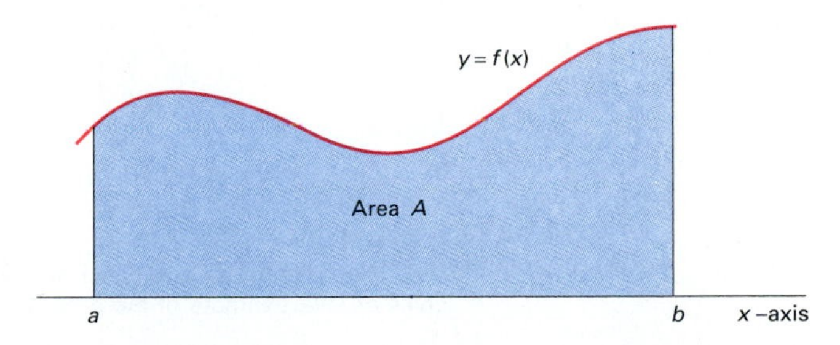

5.7 The area under the graph of $y = f(x)$ from $x = a$ to $x = b$

Our plan is to get high and low (but close) estimates of the area $A = a(R)$ by using certain special inscribed and circumscribed polygons. These polygons will be unions of rectangles with vertical sides, each having its base on the x-axis and within the interval $[a, b]$. We use such polygons because they are particularly easy to work with.

We begin with a fixed positive integer n, which will be the number of the inscribed rectangles *and* the number of circumscribed rectangles. Given n, we *partition* or subdivide the interval $[a, b]$ into n subintervals in the following way. Choose points $x_0, x_1, x_2, \ldots, x_n$ such that

$$a = x_0 < x_1 < x_2 < \cdots < x_n = b.$$

Then the subintervals will be

$$[x_0, x_1],\ [x_1, x_2],\ \ldots,\ [x_{n-1}, x_n].$$

These subintervals all have the same length; the length of each is the length of the typical ith subinterval, which is

$$\Delta x = x_i - x_{i-1} = \frac{b-a}{n},$$

$1 \leqq i \leqq n$. The point x_i will be

$$x_i = x_0 + i \cdot \Delta x = a + \frac{i}{n}(b - a).$$

Symbol	Pronounced
$x_i^\flat$	x_i-flat
$x_i^\#$	x_i-sharp

Now let $x_i^\flat$ be a point of $[x_{i-1}, x_i]$ at which f attains its minimum value, so that $f(x_i^\flat)$ is the minimum value of $f(x)$ on the ith subinterval $[x_{i-1}, x_i]$. The rectangle with base $[x_{i-1}, x_i]$ and height $f(x_i^\flat)$ then lies *within* the region R. (See Fig. 5.8.) The union of these rectangles for $i = 1, 2, \ldots, n$ is the **inscribed rectangular polygon** P_n associated with our partition of $[a, b]$ into n equal subintervals. Its area is the sum of the areas of the n rectangles with base length Δx and heights $f(x_1^\flat), f(x_2^\flat), \ldots, f(x_n^\flat)$. So

$$a(P_n) = \sum_{i=1}^{n} f(x_i^\flat)\,\Delta x. \tag{11}$$

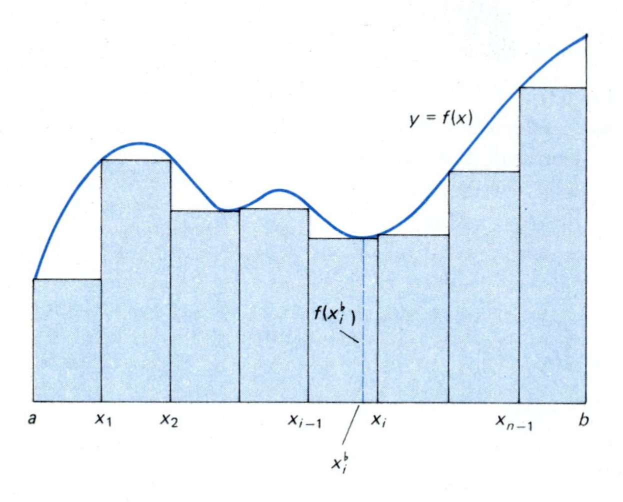

5.8 The area of the inscribed polygon P_n is an underestimate of the area $A = a(R)$.

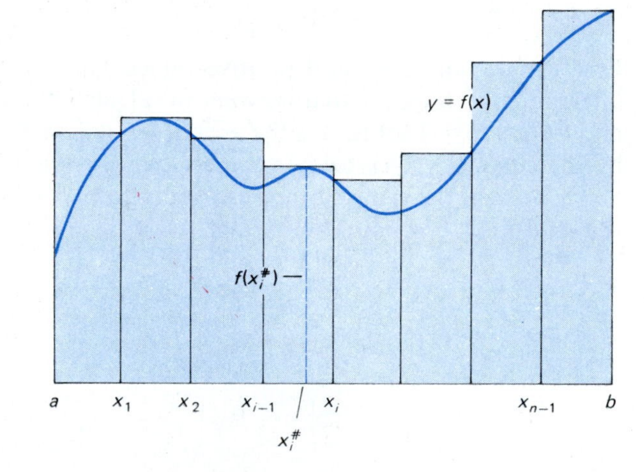

5.9 The rectangular circumscribed polygon gives an overestimate of the area $A = a(R)$.

Next, let $x_i^\#$ be a point of $[x_{i-1}, x_i]$ such that $f(x_i^\#)$ is the maximum value of f on the ith subinterval $[x_{i-1}, x_i]$. For each i $(1 \leqq i \leqq n)$, consider the rectangle with base $[x_{i-1}, x_i]$, height $f(x_i^\#)$, and area $f(x_i^\#)\,\Delta x$. (See Fig. 5.9.) The union of these n rectangles contains the region R under the graph; it is the **circumscribed rectangular polygon** Q_n associated with the partition of $[a, b]$ into n equal subintervals. Its area is

$$a(Q_n) = \sum_{i=1}^{n} f(x_i^\#)\,\Delta x. \tag{12}$$

Because the region R contains P_n but is contained in Q_n, we conclude from (11) and (12) that its area $A = a(R)$ satisfies the inequalities

$$\sum_{i=1}^{n} f(x_i^\flat)\,\Delta x \leqq A \leqq \sum_{i=1}^{n} f(x_i^\#)\,\Delta x. \tag{13}$$

We could therefore attempt to estimate the value of A by computing (for a fixed value of n) the values of the two sums in (11) and (12).

Figure 5.10 suggests that if n is very large, and hence Δx is very small, then the areas $a(P_n)$ and $a(Q_n)$ of the inscribed and circumscribed rectangular polygons will be very close to $A = a(R)$. For the difference in the heights of the inscribed and circumscribed rectangles over $[x_{i-1}, x_i]$ is $f(x_i^{\#}) - f(x_i^{\flat})$, which should be small when Δx is small; intuitively, the difference between the two sums in (13) will be small when n is very large. Indeed, the assumption that f is continuous is enough to ensure that this is so—in Section 5.3 we shall state a theorem which implies that

$$A = \lim_{n\to\infty} \sum_{i=1}^{n} f(x_i^{\flat})\,\Delta x = \lim_{n\to\infty} \sum_{i=1}^{n} f(x_i^{\#})\,\Delta x. \tag{14}$$

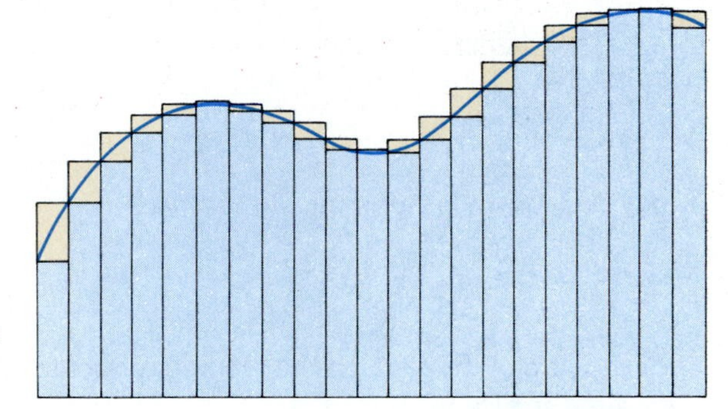

5.10 The areas of the inscribed and circumscribed polygons should be nearly equal when Δx is small.

If f is an increasing function on $[a, b]$, then for each i we have $x_i^{\#} = x_i$, the right-hand endpoint of the ith subinterval $[x_{i-1}, x_i]$. But if f is a decreasing function, then $x_i^{\flat} = x_i$ for each $i = 1, 2, 3, \ldots, n$. In each case it follows from (14) that the area A under $y = f(x)$ from $x = a$ to $x = b$ is given by

$$A = \lim_{n\to\infty} \sum_{i=1}^{n} f(x_i)\,\Delta x. \tag{15}$$

If the interval $[a, b]$ is divided into n subintervals all of the same length, then

$$\Delta x = \frac{b-a}{n} \tag{16}$$

and

$$x_i = a + i\,\Delta x \qquad (i = 0, 1, 2, \ldots, n). \tag{17}$$

EXAMPLE 3 Find the area A under the graph of $f(x) = x^3$ from $x = 0$ to $x = 2$.

Solution If we subdivide the interval $[0, 2]$ into n equal subintervals, then (16) and (17) give

$$\Delta x = \frac{2}{n} \quad \text{and} \quad x_i = 0 + i\left(\frac{2}{n}\right) = \frac{2i}{n}.$$

Therefore,

$$\sum_{i=1}^{n} f(x_i)\,\Delta x = \sum_{i=1}^{n} (x_i)^3\,\Delta x = \sum_{i=1}^{n} \left(\frac{2i}{n}\right)^3\left(\frac{2}{n}\right) = \frac{16}{n^4}\sum_{i=1}^{n} i^3.$$

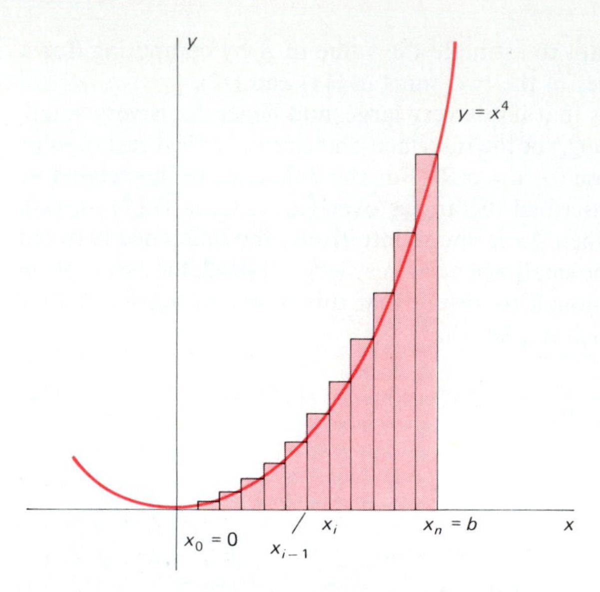

5.11 An area computation using circumscribed rectangles

Next, the formula in (8) for $\sum i^3$ yields

$$\sum_{i=1}^{n} f(x_i)\,\Delta x = \frac{16}{n^4}\left(\frac{1}{4}n^4 + \frac{1}{2}n^3 + \frac{1}{4}n^2\right) = 4 + \frac{8}{n} + \frac{4}{n^2}$$

for the sum of the areas of the rectangles shown in Fig. 5.11. When we take the limit in (15) we get

$$A = \lim_{n\to\infty}\left(4 + \frac{8}{n} + \frac{4}{n^4}\right) = 4$$

because the terms $8/n$ and $4/n^2$ approach zero as $n \to +\infty$. Note that the answer $A = 4$ is *not* an approximation; it is *exact*.

EXAMPLE 4 Find the area A under the graph of $f(x) = 100 - 3x^2$ from $x = 1$ to $x = 5$.

Solution With $a = 1$ and $b = 5$, the formulas in (16) and (17) give

$$\Delta x = \frac{4}{n} \quad\text{and}\quad x_i = 1 + i\,\Delta x = 1 + \frac{4i}{n},$$

so

$$\begin{aligned}
\sum_{i=1}^{n} f(x_i)\,\Delta x &= \sum_{i=1}^{n}\left[100 - 3\left(1 + \frac{4i}{n}\right)^2\right]\left(\frac{4}{n}\right)\\
&= \sum_{i=1}^{n}\left[97 - \frac{24i}{n} - \frac{48i^2}{n^2}\right]\left(\frac{4}{n}\right)\\
&= \frac{388}{n}\sum_{i=1}^{n} 1 - \frac{96}{n^2}\sum_{i=1}^{n} i - \frac{192}{n^3}\sum_{i=1}^{n} i^2\\
&= \frac{388}{n}(n) - \frac{96}{n^2}\left(\frac{1}{2}n^2 + \frac{1}{2}n\right) - \frac{192}{n^3}\left(\frac{1}{3}n^3 + \frac{1}{2}n^2 + \frac{1}{6}n\right)\\
&= 276 - \frac{144}{n} - \frac{32}{n^2}.
\end{aligned}$$

We applied the formulas in (6) and (7) in the next-to-last step. Consequently, the limit in (15) yields

$$A = \lim_{n\to\infty}\left(276 - \frac{144}{n} - \frac{32}{n^2}\right) = 276$$

for the desired area, because the terms $144/n$ and $32/n^2$ approach zero as $n \to \infty$.

5.2 PROBLEMS

Use the formulas in (6)–(9) to find the sums in Problems 1–10.

1. $\sum_{i=1}^{10} (4i - 3)$

2. $\sum_{j=1}^{8} (5 - 2j)$

3. $\sum_{i=1}^{10} (3i^2 + 1)$

4. $\sum_{k=1}^{6} (2k - 3k^2)$

5. $\sum_{r=1}^{8} (r - 1)(r + 2)$

6. $\sum_{i=1}^{5} (i^3 - 3i + 2)$

7. $\sum_{i=1}^{6} (i^4 - i^3)$

8. $\sum_{k=1}^{10} (2k - 1)^2$

9. $\sum_{i=1}^{1000} i^2$

10. $\sum_{i=1}^{100} i^3$

Use the method of Example 2 to evaluate the limits in Problems 11–13.

11. $\lim_{n\to\infty} \frac{1 + 2 + 3 + \cdots + n}{n^2}$

12. $\lim_{n\to\infty} \frac{1^3 + 2^3 + 3^3 + \cdots + n^3}{n^4}$

13. $\lim_{n\to\infty} \frac{1^4 + 2^4 + 3^4 + \cdots + n^4}{n^5}$

Use the formulas in Equations (6)–(9) to derive compact formulas in terms of n for the sums in Problems 14–16.

14. $\sum_{i=1}^{n} (2i - 1)$

15. $\sum_{i=1}^{n} (2i - 1)^2$

16. $\sum_{i=1}^{n} (n^2 - i^2)^2$

In each of Problems 17–26, use the method of Examples 3 and 4 to find the area under the graph of f from a to b.

17. $f(x) = 2x + 5;\quad a = 0,\quad b = 3$

18. $f(x) = 13 - 3x;\quad a = 0,\quad b = 4$

19. $f(x) = x^2;\quad a = 0,\quad b = 1$

20. $f(x) = x^4;\quad a = 0,\quad b = 3$

21. $f(x) = 3x^2 + 2;\quad a = 1,\quad b = 5$

22. $f(x) = 10 - x^2;\quad a = 1,\quad b = 3$

23. $f(x) = 3x^2 + 5x + 2;\quad a = 2,\quad b = 4$

24. $f(x) = 4x^3 + 2x;\quad a = 0,\quad b = 3$

25. $f(x) = x^2;\quad a = 0,\quad b = b$

26. $f(x) = x^3;\quad a = 0,\quad b = b$

27. Establish Equation (6) by mathematical induction.

28. Establish Equation (7) by mathematical induction.

29. Derive Equation (6) by adding the equations

$$\sum_{i=1}^{n} i = 1 + 2 + 3 + \cdots + n$$

and

$$\sum_{i=1}^{n} i = n + (n - 1) + (n - 2) + \cdots + 1.$$

30. Write the n equations obtained by substituting the values $k = 1, 2, 3, \ldots, n$ into the identity

$$(k + 1)^3 - k^3 = 3k^2 + 3k + 1.$$

Add these n equations and use their sum to deduce Equation (7) from Equation (6).

31. Derive the formula $A = \frac{1}{2}bh$ for the area of a right triangle by using circumscribed rectangular polygons to find the area under the graph of $f(x) = hx/b$ from $x = 0$ to $x = b$.

In Problems 32 and 33, let A denote the area and C the circumference of a circle of radius r and let A_n and C_n denote the area and perimeter, respectively, of a regular n-sided polygon inscribed in this circle.

32. Show that

$$A_n = nr^2 \sin\left(\frac{\pi}{n}\right)\cos\left(\frac{\pi}{n}\right) \quad\text{and that}\quad C_n = 2nr\sin\left(\frac{\pi}{n}\right).$$

33. Deduce that $A = \frac{1}{2}rC$ by taking the limit of A_n/C_n as $n \to \infty$. Then, under the assumption that $A = \pi r^2$, deduce that $C = 2\pi r$.

34. Use the method of Example 2 to deduce from the formula in (10) that

$$\lim_{n\to\infty} \frac{1^k + 2^k + 3^k + \cdots + n^k}{n^{k+1}} = \frac{1}{k + 1}$$

if k is a positive integer.

35. Use circumscribed rectangular polygons and the result of Problem 34 to show that the area under the graph of $f(x) = x^k$ (k is a positive integer) from $x = 0$ to $x = b$ is

$$A = \frac{1}{k+1} b^{k+1}.$$

36. The formulas in Equations (2) and (3) imply that the areas of the inscribed and circumscribed regular n-sided polygons are

$$a(P_n) = n \sin \alpha_n \cos \alpha_n \quad \text{and} \quad a(Q_n) = \frac{n \sin \alpha_n}{\cos \alpha_n},$$

where $\alpha_n = 180°/n$. Thus $\alpha_6 = 30°$, so $\sin \alpha_6 = \frac{1}{2}$ and $\cos \alpha_6 = \frac{1}{2}\sqrt{3}$. Now $\alpha_{12} = \frac{1}{2}\alpha_6$, so the half-angle identities yield

$$\sin \alpha_{12} = \sqrt{\tfrac{1}{2}(1 - \cos \alpha_6)} = \tfrac{1}{2}\sqrt{2 - \sqrt{3}},$$

$$\cos \alpha_{12} = \sqrt{\tfrac{1}{2}(1 + \cos \alpha_6)} = \tfrac{1}{2}\sqrt{2 + \sqrt{3}}.$$

Once $\sin \alpha_{12}$ and $\cos \alpha_{12}$ have been calculated, the half-angle identities can be applied again to calculate $\sin \alpha_{24}$ and $\cos \alpha_{24}$, and so on.

(a) Use this approach—not touching the sine or cosine keys on your calculator—to verify the entries in Fig. 5.6 for $n = 6$, 12, 24, 48, and 96. This is how far Archimedes got.

(b) Verify that your results for $n = 96$ imply Archimedes' inequality

$$3\tfrac{10}{71} < \pi < 3\tfrac{1}{7}.$$

5.3 Riemann Sums and the Integral

In the preceding section we saw that the area A under the graph from $x = a$ to $x = b$ of the *continuous, positive-valued* function f satisfies the inequalities

$$\sum_{i=1}^{n} f(x_i^\flat)\,\Delta x \leqq A \leqq \sum_{i=1}^{n} f(x_i^\#)\,\Delta x, \tag{1}$$

where $f(x_i^\flat)$ and $f(x_i^\#)$ are the minimum and maximum values of f on the ith subinterval $[x_{i-1}, x_i]$ of a partition of $[a, b]$ into n equal subintervals each of length Δx. The two approximating sums in (1) are both of the form

$$\sum_{i=1}^{n} f(x_i^*)\,\Delta x, \tag{2}$$

where x_i^* denotes an arbitrary point of the ith subinterval $[x_{i-1}, x_i]$ (see Fig. 5.12). Sums of the form in (2) appear as approximations in a wide range of applications and also form the basis for the definition of the integral. Motivated by our discussion of area in Section 5.2, we want to define the integral of f from a to b as some sort of limit as $\Delta x \to 0$ of sums such as the one in (2). Our goal is to begin with a fairly general function f and define a computable real number I (its integral), which—in the special case when f is continuous and positive-valued on $[a, b]$—will equal the area under the graph of $y = f(x)$.

We begin with a function f on $[a, b]$ *not* necessarily either continuous or positive-valued. A **partition** P of $[a, b]$ is a collection of subintervals

$$[x_0, x_1],\ [x_1, x_2],\ [x_2, x_3],\ \ldots,\ [x_{n-1}, x_n]$$

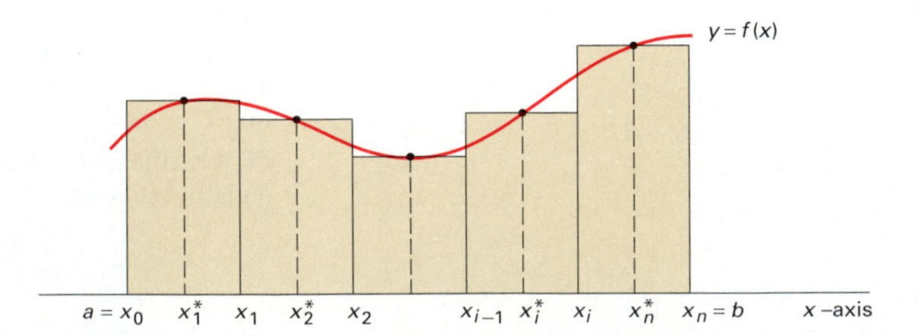

5.12 The Riemann sum in (2) as a sum of areas of rectangles

of $[a, b]$ such that

$$a = x_0 < x_1 < x_2 < x_3 < \cdots < x_{n-1} < x_n = b.$$

The **mesh** of the partition P is the largest of the lengths

$$\Delta x_i = x_i - x_{i-1}$$

of the subintervals in P and is denoted by $|P|$. To get a sum such as (2) we need a point x_i^* of the ith subinterval for each, i, $1 \leqq i \leqq n$. A collection of points

$$S = \{x_1^*, x_2^*, x_3^*, \ldots, x_n^*\}$$

with x_i^* in $[x_{i-1}, x_i]$ (for each i) is called a **selection** for the partition P.

Definition *Riemann Sum*

Let f be a function defined on the interval $[a, b]$. If P is a partition of $[a, b]$ and S is a selection for P, then the Riemann sum for f determined by P and S is

$$R = \sum_{i=1}^{n} f(x_i^*)\, \Delta x_i. \tag{3}$$

We also say that this Riemann sum is **associated with** the partition P.

The German mathematician G. F. B. Riemann (1826–1866) provided a rigorous definition of the integral. Various special types of "Riemann sums" had appeared in area and volume computations since the time of Archimedes, but it was Riemann who framed the definition above in its full generality.

The point x_i^* in (3) is an arbitrary point of the ith subinterval $[x_{i-1}, x_i]$. That is, it can be *any* point of this subinterval. But when we actually compute Riemann sums, we usually choose the points of the selection S in some systematic manner, as in the following example.

EXAMPLE 1 Let $f(x) = x^2$ on $[0, 2]$. We are going to calculate three different Riemann sums associated with the partition of $[0, 2]$ into $n = 10$ equal subintervals, each having length

$$\Delta x_i = \Delta x = \tfrac{2}{10} = \tfrac{1}{5}.$$

The endpoints of these subintervals are the integral multiples of $\frac{1}{5}$ that appear in the inequalities

$$0 < \tfrac{1}{5} < \tfrac{2}{5} < \tfrac{3}{5} < \cdots < \tfrac{8}{5} < \tfrac{9}{5} < \tfrac{10}{5} = 2.$$

The ith subinterval is $[x_{i-1}, x_i]$, where $x_i = i/5$ for $i = 0, 1, 2, \ldots, 9, 10$. We shall select x_i^* in the same way for each of these ten subintervals.

(a) If, as in Fig. 5.13(a), we choose

$$x_i^* = x_{i-1} = \frac{i-1}{5},$$

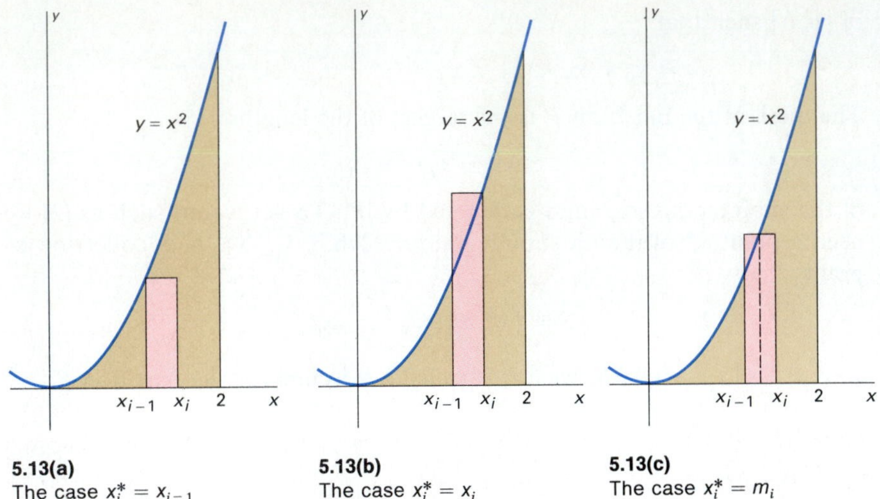

5.13(a)
The case $x_i^* = x_{i-1}$

5.13(b)
The case $x_i^* = x_i$

5.13(c)
The case $x_i^* = m_i$

the *left-hand endpoint* of $[x_{i-1}, x_i]$, then (3) is the Riemann sum

$$R_{\text{left}} = \sum_{i=1}^{n} f(x_{i-1})\,\Delta x_i = \sum_{i=1}^{n} \left(\frac{i-1}{5}\right)^2 \left(\frac{1}{5}\right)$$

$$= \frac{1}{125}(0^2 + 1^2 + 2^2 + \cdots + 9^2) = \frac{285}{125} = \frac{57}{25} = 2.28.$$

(b) If we choose $x_i^* = x_i = i/5$, the *right-hand endpoint* of $[x_{i-1}, x_i]$, then (3) is the Riemann sum

$$R_{\text{right}} = \sum_{i=1}^{n} f(x_i)\,\Delta x_i = \sum_{i=1}^{n} \left(\frac{i}{5}\right)^2 \left(\frac{1}{5}\right)$$

$$= \frac{1}{125}(1^2 + 2^2 + 3^2 + \cdots + 10^2)$$

$$= \frac{385}{125} = \frac{77}{25} = 3.08.$$

This is illustrated in Fig. 5.13(b).

(c) If, as in Fig. 5.13(c), we choose x_i^* to be the *midpoint*

$$m_i = \frac{1}{2}(x_{i-1} + x_i) = \frac{1}{2}\left(\frac{i-1}{5} + \frac{i}{5}\right) = \frac{1}{10}(2i - 1),$$

then (3) is the Riemann sum

$$R_{\text{mid}} = \sum_{i=1}^{n} f(m_i)\,\Delta x_i = \sum_{i=1}^{n} \left(\frac{2i-1}{10}\right)^2 \left(\frac{1}{5}\right)$$

$$= \frac{1}{500}(1^2 + 3^2 + 5^2 + \cdots + 19^2) = \frac{1330}{500} = 2.66.$$

The Riemann sum in (3) can be interpreted geometrically, as in Fig. 5.14. On each subinterval $[x_{i-1}, x_i]$, we construct a rectangle with width Δx_i and "height" $f(x_i^*)$. If $f(x_i^*) > 0$, then this rectangle stands above the x-axis,

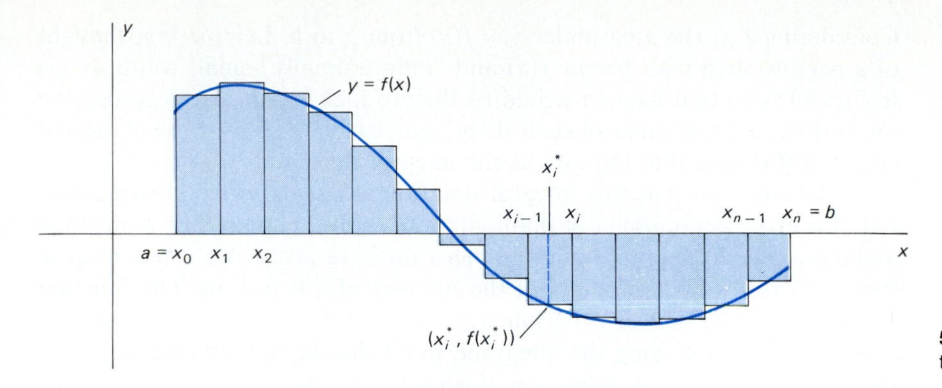

5.14 A geometric interpretation of the Riemann sum in (3)

while if $f(x_i^*) < 0$, it lies below the x-axis. The Riemann sum R is then the sum of the **signed** areas of these rectangles—that is, the sum of the areas of those rectangles that lie above the x-axis *minus* the sum of the areas of those that lie below the x-axis.

If the widths Δx_i of these rectangles are all very small—that is, if the mesh $|P|$ is very small—then it appears that the Riemann sum R will closely approximate the area from a to b under $y = f(x)$ above the x-axis, minus the area below the x-axis. This suggests that the integral of f from a to b be defined by taking the limit of the Riemann sums as the mesh $|P|$ approaches zero:

$$I = \lim_{|P| \to 0} \sum_{i=1}^{n} f(x_i^*) \, \Delta x_i. \tag{4}$$

The actual definition of the integral is obtained by saying precisely what it means for this limit to exist. Briefly, it means that if $|P|$ is sufficiently small, then *all* the Riemann sums associated with P are very close to the number I.

Definition *The (Definite) Integral*

The **(definite) integral of the function** f **from** a **to** b is the number

$$I = \lim_{|P| \to 0} \sum_{i=1}^{n} f(x_i^*) \, \Delta x_i \tag{4}$$

provided that this limit exists, in which case we say that f is **integrable** on $[a, b]$. The meaning of (4) is that, for each number $\varepsilon > 0$, there exists a number $\delta > 0$ such that

$$\left| I - \sum_{i=1}^{n} f(x_i^*) \, \Delta x_i \right| < \varepsilon$$

for every Riemann sum associated with any partition P of $[a, b]$ for which $|P| < \delta$.

The customary notation for the integral of f from a to b, due to the German mathematician and philosopher G. W. Leibniz (1646–1716), is

$$I = \int_a^b f(x) \, dx = \lim_{|P| \to 0} \sum_{i=1}^{n} f(x_i^*) \, \Delta x_i. \tag{5}$$

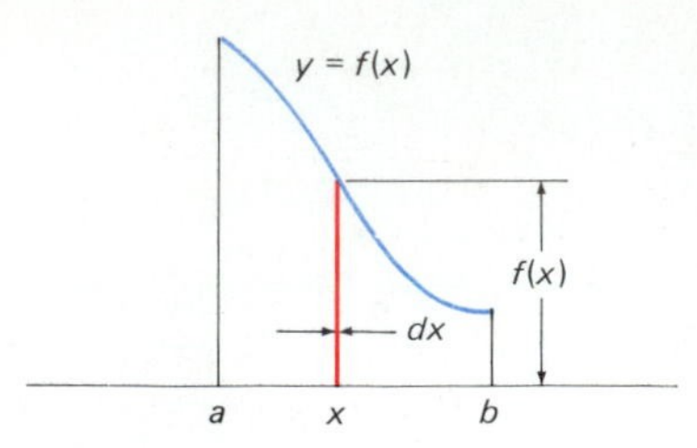

5.15 Origin of Leibniz's notation for the integral

Considering I as the area under $y = f(x)$ from a to b, Leibniz first thought of a narrow strip with height $f(x)$ and "infinitesimally" small width dx (as in Fig. 5.15), so that its area would be the product $f(x)\,dx$. He regarded the integral as a sum of areas of such strips, denoting this sum by the elongated capital S (for *sum*) that appears as the integral sign in (5).

We shall see that this integral notation is not only highly suggestive, but also exceedingly useful in manipulations with integrals. The numbers a and b are called the **lower limit** and **upper limit,** respectively, of the integral; they are merely the endpoints of the interval of integration. The function $f(x)$ that appears between the integral sign and dx is called the **integrand.** The symbol dx following the integrand in (5) should, for the time being, be thought of as simply an indication of what the independent variable is. Like the index in a summation, the independent variable x itself is merely a dummy variable, and if f is integrable on $[a, b]$ we may write

$$\int_a^b f(x)\,dx = \int_a^b f(t)\,dt = \int_a^b f(u)\,du.$$

The definition given of the definite integral applies only if $a < b$, but it is convenient to include the cases $a = b$ and $a > b$ as well. The integral is *defined* in these cases as follows:

$$\int_a^a f(x)\,dx = 0 \tag{6}$$

and

$$\int_a^b f(x)\,dx = -\int_b^a f(x)\,dx, \tag{7}$$

provided that the right-hand integral exists. Thus *interchanging the limits of integration reverses the sign of the integral.*

Not every function is integrable. Suppose that c is a point of $[a, b]$ such that $f(x) \to +\infty$ as $x \to c$. If $[x_{k-1}, x_k]$ is the subinterval of the partition P that contains c, then the Riemann sum in (3) can be made arbitrarily large by choosing x_k^* sufficiently close to c. For our purposes, however, it is sufficient to know that every continuous function is integrable. The following theorem is proved in Appendix C.

Theorem 1 *Existence of the Integral*
If the function f is continuous on $[a, b]$, then f is integrable on $[a, b]$.

Although we omit the details, it is not difficult to show that the definition of the integral can be reformulated in terms of sequences of Riemann sums, as follows.

Theorem 2 *The Integral as a Limit of a Sequence*
The function f is integrable on $[a, b]$ with integral I if and only if

$$\lim_{n\to\infty} R_n = I \tag{8}$$

for every sequence $\{R_n\}_1^\infty$ of Riemann sums associated with a sequence of partitions $\{P_n\}_1^\infty$ of $[a, b]$ such that $|P_n| \to 0$ as $n \to +\infty$.

This reformulation is advantageous because it is easier to visualize a specific sequence of Riemann sums than the vast totality of all possible Riemann sums. In the case of a continuous function f known to be integrable (by the existence theorem above), the situation can be simplified even more by using only Riemann sums associated with partitions consisting of subintervals all having the same length,

$$\Delta x_1 = \Delta x_2 = \cdots = \Delta x_n = \frac{b-a}{n} = \Delta x.$$

Such a partition of $[a, b]$ into equal subintervals is called a **regular** partition of $[a, b]$.

Any Riemann sum associated with a regular partition can be written in the form

$$\sum_{i=1}^{n} f(x_i^*)\,\Delta x, \tag{9}$$

where the absence of a subscript in Δx signifies that the sum is associated with a regular partition. In such a case, the conditions $|P| \to 0$, $\Delta x \to 0$, and $n \to +\infty$ are equivalent, so the integral of a *continuous* function can be defined quite simply:

$$\int_a^b f(x)\,dx = \lim_{n\to\infty} \sum_{i=1}^{n} f(x_i^*)\,\Delta x = \lim_{\Delta x \to 0} \sum_{i=1}^{n} f(x_i^*)\,\Delta x. \tag{10}$$

Therefore, because we are concerned for the most part only with integrals of continuous functions, in our subsequent discussions we will employ only regular partitions.

EXAMPLE 2 Use Riemann sums to compute

$$\int_0^b x^2\,dx$$

where b is a fixed positive real number.

Solution We use a regular partition of $[0, b]$ into n subintervals all of the same length, so that

$$\Delta x = \frac{b}{n} \quad \text{and} \quad x_i = \frac{ib}{n}.$$

The computations are simplest if we choose $x_i^* = x_i$ for each $i = 1, 2, 3, \ldots, n$. Then the limit in (10) gives

$$\int_0^b x^2\,dx = \lim_{n\to\infty} \sum_{i=1}^{n} x_i^2\,\Delta x = \lim_{n\to\infty} \sum_{i=1}^{n} \left(\frac{ib}{n}\right)^2 \left(\frac{b}{n}\right)$$

$$= \lim_{n\to\infty} \frac{b^3}{n^3} \sum_{i=1}^{n} i^2 = \lim_{n\to\infty} \frac{b^3}{n^3}\left(\frac{1}{3}n^3 + \frac{1}{2}n^2 + \frac{1}{6}n\right)$$

(using Equation (7) in Section 5.2)

$$= \lim_{n\to\infty} b^3\left(\frac{1}{3} + \frac{1}{2n} + \frac{1}{6n^2}\right);$$

therefore,

$$\int_0^b x^2\,dx = \frac{1}{3}b^3$$

because the terms $b^3/2n$ and $b^3/6n^2$ (with b fixed) approach zero as $n \to +\infty$.

For example, with $b = 2$ the result of Example 2 yields

$$\int_0^2 x^2\,dx = \tfrac{8}{3} \approx 2.6667.$$

Thus our third Riemann sum in Example 1, $R_{\text{mid}} = 2.66$, was quite close to the true value of the integral. Note also that (7) yields

$$\int_2^0 x^2\,dx = -\int_0^2 x^2\,dx = -\tfrac{8}{3},$$

with the sign of the integral changed when its upper and lower limits are interchanged.

5.3 PROBLEMS

In each of Problems 1–8, compute the Riemann sum

$$\sum_{i=1}^{n} f(x_i^*)\,\Delta x$$

for the indicated function and a regular partition of the given interval into n equal subintervals. Use $x_i^* = x_i$, the right-hand endpoint of the ith subinterval $[x_{i-1}, x_i]$.

1. $f(x) = x^2$ on $[0, 1]$; $n = 5$
2. $f(x) = x^3$ on $[0, 1]$; $n = 5$
3. $f(x) = \dfrac{1}{x}$ on $[1, 6]$; $n = 5$
4. $f(x) = \sqrt{x}$ on $[0, 5]$; $n = 5$
5. $f(x) = 2x + 1$ on $[1, 4]$; $n = 6$
6. $f(x) = x^2 + 2x$ on $[1, 4]$; $n = 6$
7. $f(x) = x^3 - 3x$ on $[1, 4]$; $n = 5$
8. $f(x) = 1 + 2\sqrt{x}$ on $[2, 3]$; $n = 5$

9–16. Repeat each of Problems 1–8, except with $x_i^* = x_{i-1}$, the left-hand endpoint.

17–24. Repeat each of Problems 1–8, except with $x_i^* = (x_{i-1} + x_i)/2$, the midpoint of the ith subinterval.

25. Work Problem 3 with $x_i^* = (3x_{i-1} + 2x_i)/5$.
26. Work Problem 4 with $x_i^* = (x_{i-1} + 2x_i)/3$.

Use the method of Example 2 to evaluate the integrals in Problems 27 and 28.

27. $\displaystyle\int_0^2 x\,dx$
28. $\displaystyle\int_0^4 x^3\,dx$

In Problems 29–32, evaluate the given integral by computing

$$\lim_{n\to\infty} \sum_{i=1}^{n} f(x_i^*)\,\Delta x$$

for a regular partition of the interval of integration. Take $x_i^* = x_i$ in each case.

29. $\displaystyle\int_0^3 (2x + 1)\,dx$
30. $\displaystyle\int_1^5 (4 - 3x)\,dx$
31. $\displaystyle\int_0^3 (3x^2 + 1)\,dx$
32. $\displaystyle\int_0^4 (x^3 - x)\,dx$
33. Show by the method of Problems 29–32 that
$$\int_0^b x\,dx = \tfrac{1}{2}b^2$$
if $b > 0$.
34. Show by the method of Problems 29–32 that
$$\int_0^b x^3\,dx = \tfrac{1}{4}b^4$$
if $b > 0$.
35. Let $f(x) = x$, and let $\{x_0, x_1, x_2, \ldots, x_n\}$ be an arbitrary partition of the closed interval $[a, b]$. For each i ($1 \leqq i \leqq n$), let $x_i^* = (x_{i-1} + x_i)/2$. Then show that
$$\sum_{i=1}^{n} x_i^*\,\Delta x_i = \tfrac{1}{2}b^2 - \tfrac{1}{2}a^2.$$
Explain why this computation proves that
$$\int_a^b x\,dx = \frac{b^2 - a^2}{2}.$$

36. Suppose that f is a continuous function on $[a, b]$ and that k is a constant. Use Riemann sums to prove that

$$\int_a^b kf(x)\,dx = k\int_a^b f(x)\,dx.$$

37. Suppose that $f(x) = c$, a constant. Use Riemann sums to prove that

$$\int_a^b c\,dx = c(b - a).$$

*5.3 Optional Computer Application

The BASIC program shown in Fig. 5.16 can be used to compute Riemann sums for the function $f(x) = x^2$ of Examples 1 and 2. Lines 140–150 call for you to enter the endpoints of the interval $[a, b]$ and the desired number n of subintervals. Each has length

$$\Delta x = \frac{b - a}{n},$$

denoted by H in line 160. The running Riemann sum S begins with value zero (line 170) before any terms have been added.

```
100 REM--Program RIEMANN
110 REM--Computes Riemann sums for the function
120 REM--f(x) which must be defined in line 230.
130 REM
140       INPUT "Endpoints a,b"; A,B
150       INPUT "Number of subintervals"; N
155 REM
160       H = (B - A)/N       'Delta x
170       S = 0               'S is running sum
180       X = A + H/2         'Point in 1st interval
190 REM
200 REM--Loop to add terms of Riemann sum:
210 REM
220       FOR I = 1 TO N
230          F = X*X          'Value f(x) of function
240          S = S + F*H      'Add new term to S
250          X = X + H        'Point of next interval
260       NEXT I
270 REM
280       PRINT "Riemann sum = "; S
290       END
```

5.16 Listing of Program RIEMANN

Line 180 defines the *midpoint* of the first subinterval. Each time the value of X is updated (line 250), the midpoint of the next subinterval is selected. In order to use right endpoints instead of midpoints, we would need only change $X = A + H/2$ in line 180 to $X = A + H$.

After the Ith pass through the FOR-NEXT loop in lines 220–260, S is the value of the sum of the first I terms of the Riemann sum. When all N terms have been added, the result is printed (or displayed) upon execution of line 280. The results

```
Endpoints a,b? 0,2
Number of subintervals? 10
Riemann sum =   2.66
```

```
Endpoints a,b? 0,2
Number of subintervals? 20
Riemann sum =   2.665001

Endpoints a,b? 0,2
Number of subintervals? 40
Riemann sum =   2.666249
```

corroborate the exact value

$$\int_0^2 x^2\,dx = \frac{8}{3} = 2.666666666\ldots$$

that we found in Example 2. Note the increased accuracy obtained by increasing the number n of subintervals.

In order to compute Riemann sums for a different function $f(x)$, it is necessary only to alter line 230 accordingly. For example, to approximate the value of the integral

$$\int_1^2 \frac{1}{x}\,dx \approx 0.693147,$$

we would insert $F = 1/X$ in line 230. (We will see in Chapter 7 that the exact value of this integral is the natural logarithm of the number 2.)

Exercise 1: Use the program above (and the indicated modifications) to check our numerical results in Example 1.

Exercise 2: Use the program above (with appropriate modifications) to compute the three Riemann sums as in Example 1, except use the function $f(x) = 1/x$ on the interval $[1, 2]$ with $n = 10$ equal subintervals.

5.4 Evaluation of Integrals

The evaluation of integrals using Riemann sums as in Section 5.3 is tedious and time-consuming. Fortunately, we will seldom find it necessary to evaluate an integral in this way. In 1666, Isaac Newton, while still a student at Cambridge University, discovered a much more efficient way of doing it. A few years later Gottfried Wilhelm Leibniz, working with a different approach, discovered this method independently.

Newton's key idea was that in order to evaluate the *number*

$$\int_a^b f(x)\,dx,$$

we should first introduce the *function* $A(x)$ defined as follows:

$$A(x) = \int_a^x f(t)\,dt. \tag{1}$$

Note that the independent variable x appears in the *upper limit* of the integral in (1); the dummy variable t is used in the integrand simply to avoid duplication of notation. If f is positive-valued and $x > a$, then $A(x)$ is the area under the curve $y = f(x)$ over the interval $[a, x]$, as in Fig. 5.17.

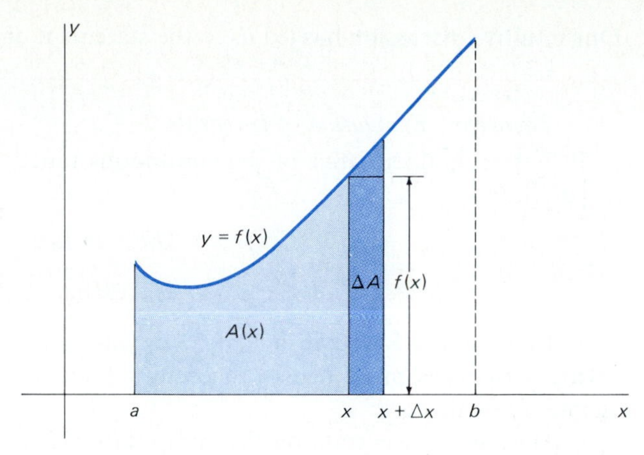

5.17 The area function $A(x)$

It is apparent from Fig. 5.17 that $A(x)$ increases as x increases. When x increases by Δx, A increases bv the area ΔA of the narrow strip in Fig. 5.17 with base $[x, x + \Delta x]$. If Δx is very small, it appears that the area of this strip is very close to the area $f(x)\,\Delta x$ of the rectangle with base $[x, x + \Delta x]$ and height $f(x)$. Thus

$$\Delta A \approx f(x)\,\Delta x; \qquad \frac{\Delta A}{\Delta x} \approx f(x). \tag{2}$$

Moreover, the figure makes it plausible that we get equality in the limit as $\Delta x \to 0$:

$$\frac{dA}{dx} = \lim_{\Delta x \to 0} \frac{\Delta A}{\Delta x} = f(x).$$

That is,

$$A'(x) = f(x), \tag{3}$$

so *the derivative of the area function* $A(x)$ *is the curve's height function* $f(x)$. In other words, (3) implies that $A(x)$ is an *antiderivative* of $f(x)$.

Now suppose that $G(x)$ is any other antiderivative of $f(x)$—perhaps one found by the methods of Section 4.8. Then

$$A(x) = G(x) + C \tag{4}$$

because two antiderivatives of the same function can differ only by a constant. Also,

$$A(a) = \int_a^a f(t)\,dt = 0 \tag{5}$$

and

$$A(b) = \int_a^b f(t)\,dt = \int_a^b f(x)\,dx, \tag{6}$$

so it finally follows that

$$\int_a^b f(x)\,dx = A(b) - A(a) = [G(b) + C] - [G(a) + C],$$

and thus

$$\int_a^b f(x)\,dx = G(b) - G(a).$$

Our intuitive discussion has led us to the statement of the following theorem.

Theorem *Evaluation of Integrals*
If G is an antiderivative of the continuous function f on the interval $[a, b]$, then

$$\int_a^b f(x)\,dx = G(b) - G(a). \tag{7}$$

In Section 5.5 we will fill in the details in the discussion above, thus giving a rigorous proof of this theorem (which is part of the fundamental theorem of calculus).

Here we concentrate on the computational applications of this theorem. The difference $G(b) - G(a)$ is customarily abbreviated as $[G(x)]_a^b$, so the theorem implies that

$$\int_a^b f(x)\,dx = \Big[G(x)\Big]_a^b \tag{8}$$

if G is any antiderivative of the continuous function f on the interval $[a, b]$. Thus if we can find the antiderivative G, we can quickly evaluate the integral *without* recourse to the paraphernalia of limits of Riemann sums.

For example, if $f(x) = x^n$ with $n \neq -1$, then an antiderivative of f is

$$G(x) = \frac{x^{n+1}}{n+1},$$

so (8) yields

$$\int_a^b x^n\,dx = \left[\frac{x^{n+1}}{n+1}\right]_a^b = \frac{b^{n+1} - a^{n+1}}{n+1}. \tag{9}$$

Contrast the immediacy of this result with the complexity of the computations of Example 2 in Section 5.3.

Similarly, if $f(x) = \cos x$, then $G(x) = D^{-1}\cos x = \sin x$, so

$$\int_a^b \cos x\,dx = \Big[\sin x\Big]_a^b = \sin b - \sin a. \tag{10}$$

Also,

$$\int_a^b \sin x\,dx = \Big[-\cos x\Big]_a^b = \cos a - \cos b. \tag{11}$$

EXAMPLE 1 Here are some immediate applications of the evaluation theorem.

$$\int_0^2 x^5\,dx = \Big[\tfrac{1}{6}x^6\Big]_0^2 = \tfrac{64}{6} - 0 = \tfrac{32}{3}.$$

$$\int_1^9 (2x - x^{-1/2} - 3)\,dx = \Big[x^2 - 2x^{1/2} - 3x\Big]_1^9 = 52.$$

$$\int_0^1 (2x+1)^3\,dx = \Big[\tfrac{1}{8}(2x+1)^4\Big]_0^1 = \tfrac{1}{8}(81 - 1) = 10.$$

$$\int_0^{\pi/2} \sin 2x\,dx = \Big[-\tfrac{1}{2}\cos 2x\Big]_0^{\pi/2}$$

$$= -\tfrac{1}{2}(\cos \pi - \cos 0) = 1.$$

We have not shown the details of finding the antiderivatives, but you can (and should) check each of these results by showing that the derivative of the function within the brackets on the right is equal to the integrand on the left. In Example 2 we show the details.

EXAMPLE 2 Evaluate $\int_1^5 \sqrt{3x+1}\,dx$.

Solution If

$$u = 3x + 1, \quad \text{then} \quad \frac{du}{dx} = 3.$$

Hence

$$\sqrt{3x+1} = \frac{1}{3}\cdot(3x+1)^{1/2}(3) = \frac{1}{3}u^{1/2}\frac{du}{dx}.$$

Thus an antiderivative of $\sqrt{3x+1}$ is of the form

$$\frac{1}{3}\cdot\frac{u^{3/2}}{\frac{3}{2}} + C = \frac{2}{9}(3x+1)^{3/2} + C.$$

With

$$f(x) = \sqrt{3x+1} \quad \text{and} \quad G(x) = \tfrac{2}{9}(3x+1)^{3/2},$$

the evaluation formula in (8) therefore yields

$$\int_1^5 \sqrt{3x+1}\,dx = \Big[\tfrac{2}{9}(3x+1)^{3/2}\Big]_1^5$$
$$= \tfrac{2}{9}(16^{3/2} - 4^{3/2}) = \tfrac{2}{9}(4^3 - 2^3) = \tfrac{112}{9}.$$

If the derivative $F'(x)$ of the function $F(x)$ is continuous, then the evaluation theorem, with $F'(x)$ in place of $f(x)$ and $F(x)$ in place of $G(x)$, yields

$$\int_a^b F'(x)\,dx = \Big[F(x)\Big]_a^b = F(b) - F(a). \tag{12}$$

Here is an immediate application.

EXAMPLE 3 Suppose that an animal population $P(t)$ initially numbers $P(0) = 100$, and its rate of growth after t months is observed to be given by the formula

$$P'(t) = 10 + t + (0.06)t^2.$$

What is the population after 10 months?

Solution By the formula in (12), we know that

$$P(10) - P(0) = \int_0^{10} P'(t)\,dt = \int_0^{10} (10 + t + (0.06)t^2)\,dt$$
$$= \Big[10t + \tfrac{1}{2}t^2 + (0.02)t^3\Big]_0^{10} = 170.$$

Thus $P(10) = 100 + 170 = 270$ individuals.

EXAMPLE 4 Evaluate

$$\lim_{n\to\infty}\sum_{i=1}^{n}\frac{2i}{n^2}$$

by recognizing this limit as the value of an integral.

Solution If we write

$$\sum_{i=1}^{n}\frac{2i}{n^2}=\sum_{i=1}^{n}\left(\frac{2i}{n}\right)\left(\frac{1}{n}\right),$$

we recognize that we have a Riemann sum for the function $f(x) = 2x$, associated with a partition of the interval $[0, 1]$ into n equal subintervals. The ith point of subdivision is $x_i = i/n$ and $\Delta x = 1/n$. Hence it follows from the definition of the integral and from the evaluation theorem that

$$\lim_{n\to\infty}\sum_{i=1}^{n}\frac{2i}{n^2}=\lim_{n\to\infty}\sum_{i=1}^{n}2x_i\,\Delta x=\lim_{n\to\infty}\sum_{i=1}^{n}f(x_i)\,\Delta x$$

$$=\int_0^1 f(x)\,dx=\int_0^1 2x\,dx.$$

Therefore,

$$\lim_{n\to\infty}\sum_{i=1}^{n}\frac{2i}{n^2}=\Big[x^2\Big]_0^1=1.$$

BASIC PROPERTIES OF INTEGRALS

Elementary proofs of the properties stated below are outlined in Problems 38–40 at the end of this section. We assume that f is integrable on $[a, b]$.

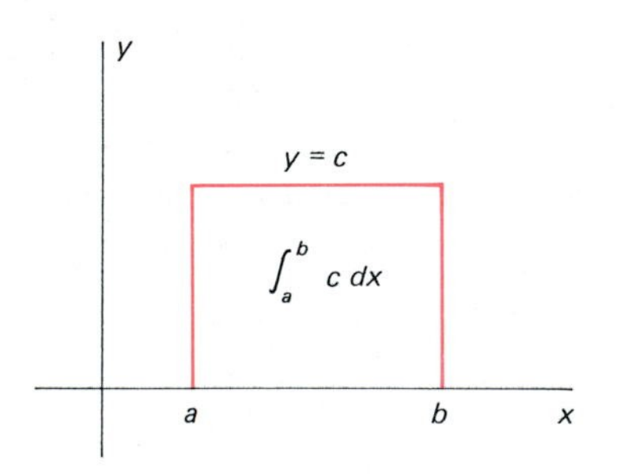

5.18 The integral of a constant is the area of a rectangle.

Integral of a Constant

$$\int_a^b c\,dx = c(b-a).$$

This is intuitively obvious because the area represented by the integral is simply a rectangle with base $b - a$ and height c (see Fig. 5.18).

Constant Multiple Property

$$\int_a^b cf(x)\,dx = c\int_a^b f(x)\,dx.$$

Thus a constant can be "moved across the integral sign." For example,

$$\int_0^{\pi/2} 2\sin x\,dx = 2\int_0^{\pi/2}\sin x\,dx = 2\Big[-\cos x\Big]_0^{\pi/2} = 2.$$

Interval Union Property

If $a < c < b$, then

$$\int_a^b f(x)\,dx = \int_a^c f(x)\,dx + \int_c^b f(x)\,dx.$$

Figure 5.19 indicates the plausibility of this property, which allows manipulations such as this:

$$\int_{-1}^{3} 2|x|\,dx = \int_{-1}^{0} -2x\,dx + \int_{0}^{3} 2x\,dx = \Big[-x^2\Big]_{-1}^{0} + \Big[x^2\Big]_{0}^{3}$$

$$= [0 - (-1)] + [9 - 0] = 10.$$

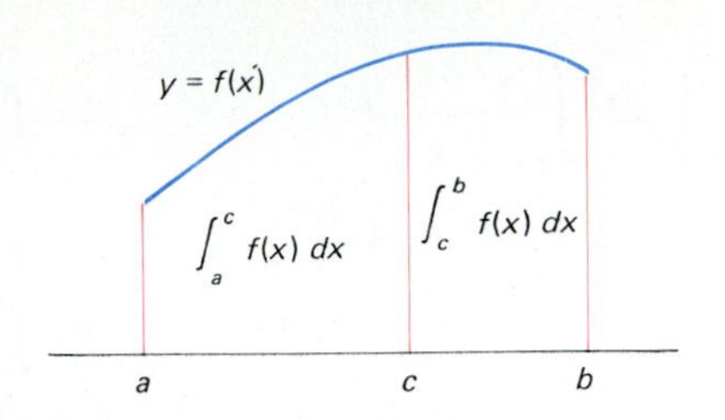

5.19 The way the interval union property works

> ***Comparison Property***
>
> If $m \leqq f(x) \leqq M$ for all x in $[a, b]$, then
>
> $$m(b-a) \leqq \int_a^b f(x)\,dx \leqq M(b-a).$$

The plausibility of this property is indicated in Fig. 5.20. Note that m and M are not necessarily the minimum and maximum values of $f(x)$ on $[a, b]$. For an example, suppose that $2 \leqq x \leqq 3$. Then

$$5 \leqq x^2 + 1 \leqq 10,$$

so

$$m = \frac{1}{10} \leqq \frac{1}{x^2+1} \leqq \frac{1}{5} = M.$$

Hence the comparison property yields

$$\frac{1}{10} \leqq \int_2^3 \frac{1}{x^2+1}\,dx \leqq \frac{1}{5}.$$

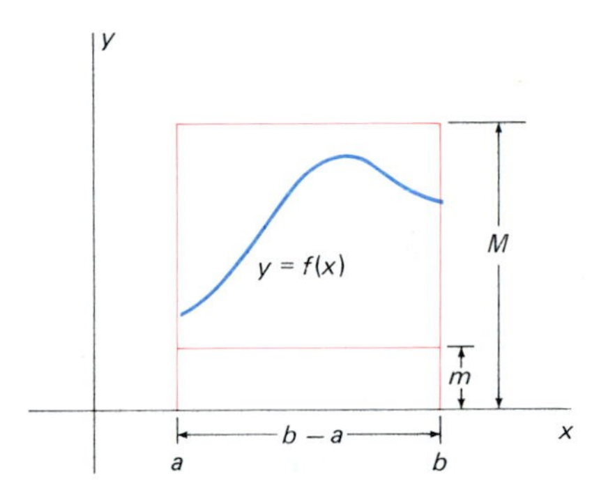

5.20 Plausibility of the comparison property

The properties of integrals stated here are frequently used in computations and will be applied in the proof of the fundamental theorem of calculus in Section 5.5.

5.4 PROBLEMS

Apply the evaluation theorem of this section to evaluate the integrals in Problems 1–30.

1. $\int_0^1 (3x^2 + 2\sqrt{x} + 3\sqrt[3]{x})\,dx$

2. $\int_1^3 \frac{6}{x^2}\,dx$

3. $\int_0^1 x^3(1+x)^2\,dx$

4. $\int_{-2}^{-1} \frac{1}{x^4}\,dx$

5. $\int_0^1 (x^4 - x^3)\,dx$

6. $\int_1^2 (x^4 - x^3)\,dx$

7. $\int_{-1}^0 (x+1)^3\,dx$

8. $\int_1^3 \frac{x^4+1}{x^2}\,dx$

9. $\int_0^4 \sqrt{x}\,dx$

10. $\int_1^4 x^{-1/2}\,dx$

11. $\int_{-1}^2 (3x^2 + 2x + 4)\,dx$

12. $\int_0^1 x^{99}\,dx$

13. $\int_{-1}^1 x^{99}\,dx$

14. $\int_0^4 (7x^{5/2} - 5x^{3/2})\,dx$

15. $\int_1^3 (x-1)^5\,dx$

16. $\int_1^2 (x^2+1)^3\,dx$

17. $\int_{-1}^0 (2x+1)^3\,dx$

18. $\int_1^3 \frac{10}{(2x+3)^2}\,dx$

19. $\int_1^8 x^{2/3}\,dx$

20. $\int_1^9 (1+\sqrt{x})^2\,dx$

21. $\int_0^1 (x^2 - 3x + 4)\,dx$

22. $\int_1^4 \sqrt{3t}\,dt$

23. $\int_1^9 \left(\sqrt{x} - \frac{2}{\sqrt{x}}\right)dx$

24. $\int_2^3 \frac{du}{u^2}$ $\left(\text{Note the abbreviation for } \frac{1}{u^2}\,du.\right)$

25. $\displaystyle\int_1^4 \frac{x^2 - 1}{\sqrt{x}}\,dx$

26. $\displaystyle\int_1^4 (t^2 - 2)\sqrt{t}\,dt$

27. $\displaystyle\int_4^7 \sqrt{3x + 4}\,dx$

28. $\displaystyle\int_0^{\pi/2} \cos 2x\,dx$

29. $\displaystyle\int_0^{\pi/4} \sin x \cos x\,dx$

30. $\displaystyle\int_0^{\pi} \sin^2 x \cos x\,dx$

In each of Problems 31–36, evaluate the given limit by first recognizing the indicated sum as a Riemann sum associated with a regular partition of [0, 1] and then evaluating the corresponding integral.

31. $\displaystyle\lim_{n\to\infty} \sum_{i=1}^{n} \left(\frac{2i}{n} - 1\right)\frac{1}{n}$

32. $\displaystyle\lim_{n\to\infty} \sum_{i=1}^{n} \frac{i^2}{n^3}$

33. $\displaystyle\lim_{n\to\infty} \frac{1 + 2 + 3 + \cdots + n}{n^2}$

34. $\displaystyle\lim_{n\to\infty} \frac{1^3 + 2^3 + 3^3 + \cdots + n^3}{n^4}$

35. $\displaystyle\lim_{n\to\infty} \frac{\sqrt{1} + \sqrt{2} + \sqrt{3} + \cdots + \sqrt{n}}{n\sqrt{n}}$

36. $\displaystyle\lim_{n\to\infty} \sum_{i=1}^{n} \frac{1}{n} \sin\frac{\pi i}{n}$

37. Evaluate the integral

$$\int_0^5 \sqrt{25 - x^2}\,dx$$

by interpreting it as the area under the graph of an appropriate function.

38. Use sequences of Riemann sums to establish the constant multiple property.

39. Use sequences of Riemann sums to establish the interval union property of the integral. Note that if R_n' and R_n'' are Riemann sums for f on the intervals $[a, c]$ and $[c, b]$, respectively, then $R_n = R_n' + R_n''$ is a Riemann sum for f on $[a, b]$.

40. Use Riemann sums to establish the comparison property for integrals. Show first that if $m \leq f(x) \leq M$ for all x in $[a, b]$, then

$$m(b - a) \leq R \leq M(b - a)$$

for every Riemann sum R for f on $[a, b]$.

41. Suppose that a tank initially contains 1000 gallons of water and that the rate of change of its volume after draining for t minutes is $V'(t) = (0.8)t - 40$ (in gallons per minute). How much water does the tank contain after it has been draining for a half hour?

42. Suppose that the population of a city in 1960 was 125 (thousand) and that its rate of growth t years later was $P'(t) = 8 + (0.5)t + (0.03)t^2$ (in thousands per year). What was its population in 1980?

43. Find a lower and an upper bound for

$$\int_1^2 \frac{1}{x}\,dx$$

by using the comparison property and a subdivision of [1, 2] into five equal subintervals.

44. Find a lower and an upper bound for

$$\int_0^1 \frac{1}{x^2 + 1}\,dx$$

by using the comparison property and a subdivision of [0, 1] into five equal subintervals.

5.5 The Fundamental Theorem of Calculus

Newton and Leibniz are generally credited with the invention of calculus in the latter part of the seventeenth century. Actually, others had earlier calculated areas essentially equivalent to integrals and tangent line slopes essentially equivalent to derivatives. The great accomplishment of Newton and Leibniz was the discovery and computational exploitation of the inverse relationship between differentiation and integration. This relationship is embodied in the **fundamental theorem of calculus.** One part of this theorem is the evaluation theorem in Section 5.4: In order to evaluate

$$\int_a^b f(x)\,dx,$$

it suffices to find an antiderivative of f on $[a, b]$. The other part of the fundamental theorem tells us that doing so is usually possible, at least in theory: Every continuous function has an antiderivative.

THE AVERAGE VALUE OF A FUNCTION

As a preliminary to the proof of the fundamental theorem of calculus, and also as a matter of independent interest, we investigate the notion of *average value*. For example, let the temperature T during a particular 24-hour day be described by $T = f(t)$, $0 \leqq t \leqq 24$. We might define the average temperature $\bar{T}$ for the day as the (ordinary arithmetical) average of the hourly temperatures:

$$\bar{T} = \tfrac{1}{24} \sum_{i=1}^{24} f(i) = \tfrac{1}{24} \sum_{i=1}^{24} f(t_i)$$

where $t_i = i$. If we subdivided the day into n equal subintervals $[t_{i-1}, t_i]$ rather than into 24 one-hour intervals, we would obtain the more general average

$$\bar{T} = \frac{1}{n} \sum_{i=1}^{n} f(t_i).$$

The larger n is, the closer would we expect $\bar{T}$ to be to the "true" average temperature for the entire day. It is therefore plausible to define the true average temperature by letting n increase without bound. This gives

$$\bar{T} = \lim_{n \to \infty} \frac{1}{n} \sum_{i=1}^{n} f(t_i).$$

The right-hand side resembles a Riemann sum, and we can make it into a Riemann sum by introducing the factor

$$\Delta t = \frac{b - a}{n}$$

where $a = 0$ and $b = 24$. Then

$$\begin{aligned}
\bar{T} &= \lim_{n \to \infty} \frac{1}{b - a} \sum_{i=1}^{n} f(t_i) \frac{b - a}{n} \\
&= \frac{1}{b - a} \lim_{n \to \infty} \sum_{i=1}^{n} f(t_i)\, \Delta t \\
&= \frac{1}{b - a} \int_a^b f(t)\, dt = \frac{1}{24} \int_0^{24} f(t)\, dt.
\end{aligned}$$

under the assumption that f is continuous, so that the Riemann sums converge to the integral as $n \to \infty$. This example motivates the following definition.

Definition *Average Value of a Function*

Suppose that the function f is integrable on $[a, b]$. Then the **average value** $\bar{y}$ of $y = f(x)$ on $[a, b]$ is

$$\bar{y} = \frac{1}{b - a} \int_a^b f(x)\, dx. \tag{1}$$

For example, the average value of $f(x) = x^2$ on $[0, 2]$ is

$$\bar{y} = \tfrac{1}{2}\int_0^2 x^2\,dx = \tfrac{1}{2}\Big[\tfrac{1}{3}x^3\Big]_0^2 = \tfrac{4}{3}.$$

EXAMPLE 1 The mean daily temperature in degrees Fahrenheit in Athens, Georgia, t months after July 15, is closely approximated by

$$T = 61 + 18\cos\frac{\pi t}{6}.$$

Find the average temperature between September 15 ($t = 2$) and December 15 ($t = 5$).

Solution Equation (1) gives

$$\bar{T} = \frac{1}{5-2}\int_2^5 \left(61 + 18\cos\frac{\pi t}{6}\right)dt$$

$$= \frac{1}{3}\left[61t + \frac{(6)(18)}{\pi}\sin\frac{\pi t}{6}\right]_2^5 \approx 57°\text{F}.$$

The following theorem tells us that every continuous function on a closed interval *attains* its average value at some point of the interval.

Average Value Theorem

If f is continuous on $[a, b]$, then

$$f(\bar{x}) = \frac{1}{b-a}\int_a^b f(x)\,dx \tag{2}$$

for some point $\bar{x}$ of $[a, b]$.

Proof Let $m = f(c)$ be the minimum value of f on $[a, b]$ and let $M = f(d)$ be its maximum value there. Then, by the comparison property of Section 5.4,

$$m = f(c) \leqq \bar{y} = \frac{1}{b-a}\int_a^b f(x)\,dx \leqq f(d) = M.$$

Because f is continuous, we can now apply the intermediate value property. The number $\bar{y}$ is between the two values m and M of f, and consequently $\bar{y}$ itself must be a value of f. Specifically, $\bar{y} = f(\bar{x})$ for some number $\bar{x}$ between a and b. This yields (2). ■

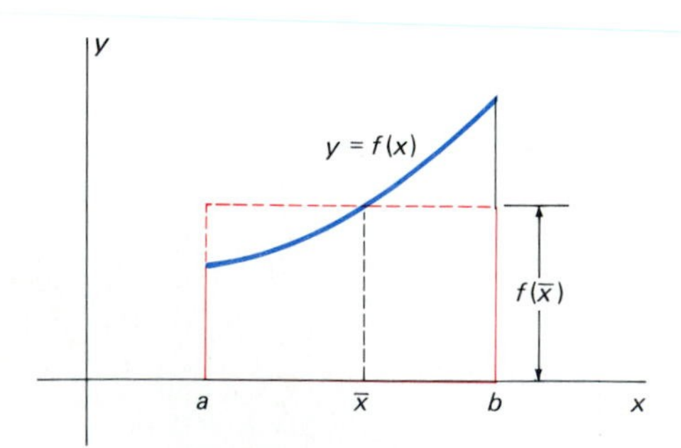

5.21 Rectangle illustrating the average value theorem

Note that (2) can be rewritten in the form

$$\int_a^b f(x)\,dx = f(\bar{x})(b-a).$$

If f is positive-valued on $[a, b]$, this equation implies that the area under $y = f(x)$ over $[a, b]$ is equal to the area of a rectangle with base $b - a$ and height $f(\bar{x})$, as indicated in Fig. 5.21.

THE FUNDAMENTAL THEOREM

We state the fundamental theorem of calculus in two parts. The first part is the fact that every function f continuous on an interval I has an antiderivative on I. In particular, an antiderivative of f can be obtained by integrating f in a certain way. Intuitively, in the case $f(x) > 0$, we let $F(x)$ denote the area under the graph of f from a fixed point a of I to x, a point of I with $x > a$. We shall prove that $F'(x) = f(x)$. We show the construction of the function F in Fig. 5.22. More precisely, we define the function F as follows:

$$F(x) = \int_a^x f(t)\,dt,$$

where we use the dummy variable t in the integrand to avoid confusion with the upper limit x. The proof that $F'(x) = f(x)$ will be independent of the supposition that $x > a$.

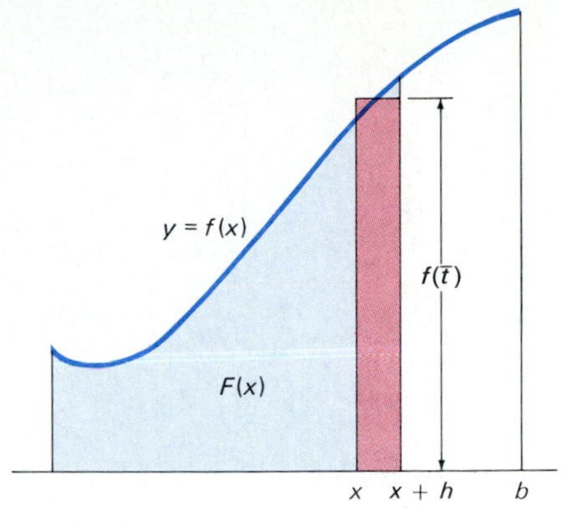

5.22 The area function F is an antiderivative of f.

The Fundamental Theorem of Calculus

Let f be a continuous function defined on $[a, b]$.

1. If the function F is defined on $[a, b]$ by
$$F(x) = \int_a^x f(t)\,dt, \tag{3}$$
then F is an antiderivative of f. That is, $F'(x) = f(x)$ for x in (a, b).
2. If G is any antiderivative of f on $[a, b]$, then
$$\int_a^b f(x)\,dx = \Big[G(x)\Big]_a^b = G(b) - G(a). \tag{4}$$

Proof of Part 1 By the definition of the derivative,

$$F'(x) = \lim_{h\to 0} \frac{F(x+h) - F(x)}{h}$$
$$= \lim_{h\to 0} \frac{1}{h}\left(\int_a^{x+h} f(t)\,dt - \int_a^x f(t)\,dt\right).$$

But

$$\int_a^{x+h} f(t)\,dt = \int_a^x f(t)\,dt + \int_x^{x+h} f(t)\,dt$$

by the interval union property of Section 5.4. Thus

$$F'(x) = \lim_{h\to 0} \frac{1}{h}\int_x^{x+h} f(t)\,dt.$$

The average value theorem of this section tells us that

$$\frac{1}{h}\int_x^{x+h} f(t)\,dt = f(\bar{t})$$

for some number $\bar{t}$ in $[x, x + h]$. Finally, we note that $\bar{t} \to x$ as $h \to 0$. Thus, because f is continuous, we see that

$$F'(x) = \lim_{h \to 0} \frac{1}{h} \int_x^{x+h} f(t)\, dt = \lim_{h \to 0} f(\bar{t}) = \lim_{\bar{t} \to x} f(\bar{t}) = f(x).$$

Hence the function F defined in (3) is, indeed, an antiderivative of f. ■

Proof of Part 2 Here we apply Part 1 to give a proof of the evaluation theorem in Section 5.4. If G is *any* antiderivative of f, then—because it and the function F of Part 1 are both antiderivatives of f on the interval $[a, b]$—we know that

$$G(x) = F(x) + C$$

on $[a, b]$ for some constant C. To evaluate C, we substitute $x = a$ and obtain

$$C = G(a) - F(a) = G(a)$$

because

$$F(a) = \int_a^a f(t)\, dt = 0.$$

Hence $G(x) = F(x) + G(a)$; in other words,

$$F(x) = G(x) - G(a)$$

for all x in $[a, b]$. With $x = b$ this gives

$$G(b) - G(a) = F(b) = \int_a^b f(x)\, dx,$$

which establishes Equation (4). ■

Sometimes the fundamental theorem of calculus is interpreted as a statement to the effect that differentiation and integration are *inverse processes*. Part 1 may be written in the form

$$\frac{d}{dx}\left(\int_a^x f(t)\, dt\right) = f(x)$$

if f is continuous on an interval containing a and x. That is, if we first integrate the function f (with *variable* upper limit of integration x) and then differentiate with respect to x, the result is the function f again. So differentiation "cancels" the effect of integration of continuous functions.

Moreover, part 2 of the fundamental theorem may be written in the form

$$\int_a^x G'(t)\, dt = G(x) - G(a)$$

if we assume that G' is continuous. If so, this equation means that if we first differentiate the function G, then integrate the result from a to x, the result can differ from the original function G by, at worst, the *constant* $G(a)$. If a is chosen so that $G(a) = 0$, this means that integration "cancels" the effect of differentiation.

For our first application of the fundamental theorem of calculus, we use it to establish the **linearity property** of the integral:

$$\int_a^b [\alpha f(x) + \beta g(x)]\, dx = \alpha \int_a^b f(x)\, dx + \beta \int_a^b g(x)\, dx \tag{5}$$

if α and β are constants and the functions f and g are continuous on $[a, b]$. For example,

$$\int_0^1 (3x^2 - 2x\sqrt{x^2+1})\, dx = 3\int_0^1 x^2\, dx - 2\int_0^1 x\sqrt{x^2+1}\, dx.$$

This linearity property enables us to "divide and conquer," to split a relatively complicated integral into simpler ones.

The linearity of integration follows—via the fundamental theorem of calculus—from the linearity of differentiation. The linearity formula in (5) is a unified statement of the following two properties of integration.

$$\int_a^b cf(x)\, dx = c\int_a^b f(x)\, dx \tag{6}$$

if c is a constant, and

$$\int_a^b [f(x) + g(x)]\, dx = \int_a^b f(x)\, dx + \int_a^b g(x)\, dx. \tag{7}$$

The formula in (6) is merely the constant multiple property of Section 5.4. To prove (7), let F be an antiderivative of f and let G be an antiderivative of g. Then

$$D_x[F(x) + G(x)] = f(x) + g(x)$$

because differentiation is linear. This means that $F + G$ is an antiderivative of $f + g$, which in turn implies that

$$\begin{aligned}\int_a^b [f(x) + g(x)]\, dx &= \Big[F(x) + G(x)\Big]_a^b \\ &= [F(b) + G(b)] - [F(a) + G(a)] \\ &= [F(b) - F(a)] + [G(b) - G(a)] \\ &= \Big[F(x)\Big]_a^b + \Big[G(x)\Big]_a^b = \int_a^b f(x)\, dx + \int_a^b g(x)\, dx.\end{aligned}$$

Note that we have invoked the fundamental theorem of calculus twice in this derivation.

One consequence of linearity is the fact that integration preserves inequalities between functions. That is, if f and g are continuous functions with $f(x) \leqq g(x)$ for all x in $[a, b]$, then

$$\int_a^b f(x)\, dx \leqq \int_a^b g(x)\, dx. \tag{8}$$

To prove this, let $h(x) = g(x) - f(x)$. Then $h(x) \geqq 0$ on $[a, b]$, so it follows from the comparison property of Section 5.4 that

$$\int_a^b h(x)\, dx \geqq 0.$$

Therefore,

$$\int_a^b g(x)\,dx - \int_a^b f(x)\,dx = \int_a^b h(x)\,dx \geqq 0,$$

which is equivalent to (8).

The verification of the following consequence of (8) is left for the problems:

$$\left|\int_a^b f(x)\,dx\right| \leqq \int_a^b |f(x)|\,dx. \tag{9}$$

Examples 1 and 2 of Section 5.4 illustrate the use of part 2 of the fundamental theorem in the evaluation of integrals. Additional examples appear in the problems and in Section 5.6. The next example illustrates the necessity of splitting an integral into a sum of integrals when its integrand has different antiderivative formulas on different intervals.

EXAMPLE 2 Evaluate $\int_{-1}^{2} |x^3 - x|\,dx$.

Solution We note that $x^3 - x \geqq 0$ on $[-1, 0]$, that $x^3 - x \leqq 0$ on $[0, 1]$, and that $x^3 - x \geqq 0$ on $[1, 2]$. So we write

$$\begin{aligned}\int_{-1}^{2} |x^3 - x|\,dx &= \int_{-1}^{0} (x^3 - x)\,dx + \int_{0}^{1} (x - x^3)\,dx + \int_{1}^{2} (x^3 - x)\,dx \\ &= \left[\tfrac{1}{4}x^4 - \tfrac{1}{2}x^2\right]_{-1}^{0} + \left[\tfrac{1}{2}x^2 - \tfrac{1}{4}x^4\right]_{0}^{1} + \left[\tfrac{1}{4}x^4 - \tfrac{1}{2}x^2\right]_{1}^{2} \\ &= \tfrac{1}{4} + \tfrac{1}{4} + \left[2 - \left(-\tfrac{1}{4}\right)\right] = \tfrac{11}{4} = 2.75.\end{aligned}$$

The first part of the fundamental theorem of calculus says that the derivative of an integral with respect to its upper limit is equal to the value of the integrand *at* the upper limit. For example, if

$$y = \int_0^x t^3 \sin t\,dt,$$

then

$$\frac{dy}{dx} = x^3 \sin x.$$

The following example is a bit more complicated in that the upper limit of the integral is a function of the independent variable.

EXAMPLE 3 Find $h'(x)$ given

$$h(x) = \int_0^{x^2} t^3 \sin t\,dt.$$

Solution Let $y = h(x)$ and $u = x^2$. Then

$$y = \int_0^u t^3 \sin t\,dt,$$

so

$$\frac{dy}{du} = u^3 \sin u$$

by the fundamental theorem of calculus. Then the chain rule yields

$$h'(x) = \frac{dy}{dx} = \frac{dy}{du} \cdot \frac{du}{dx} = (u^3 \sin u)(2x) = 2x^7 \sin x^2.$$

5.5 PROBLEMS

In Problems 1–11, find the average value of the given function on the specified interval.

1. $f(x) = x^4$; $[0, 2]$
2. $g(x) = \sqrt{x}$; $[1, 4]$
3. $h(x) = 3x^2\sqrt{x^3 + 1}$; $[0, 2]$
4. $f(x) = 8x$; $[0, 4]$
5. $g(x) = 8x$; $[-4, 4]$
6. $h(x) = x^2$; $[-4, 4]$
7. $f(x) = x^3$; $[0, 5]$
8. $g(x) = x^{-1/2}$; $[1, 4]$
9. $f(x) = \sqrt{x + 1}$; $[0, 3]$
10. $g(x) = \sin 2x$; $[0, \pi/2]$
11. $f(x) = \cos^2 x$; $[0, \pi]$

Evaluate the integrals in Problems 12–30.

12. $\int_{-1}^{2} (4 - 3x + 2x^2)\, dx$
13. $\int_{-1}^{3} dx$ (Here, dx stands for $1\, dx$.)
14. $\int_{1}^{2} (y^5 - 1)\, dy$
15. $\int_{1}^{4} \frac{dx}{\sqrt{9x^3}}$
16. $\int_{-1}^{1} (x^3 + 2)^2\, dx$
17. $\int_{1}^{3} \frac{3t - 5}{t^4}\, dt$
18. $\int_{-2}^{-1} \frac{x^2 - x + 3}{\sqrt[3]{x}}\, dx$
19. $\int_{0}^{\pi} \sin x \cos x\, dx$
20. $\int_{-1}^{2} |x|\, dx$
21. $\int_{1}^{2} \left(t - \frac{1}{2t}\right)^2 dt$
22. $\int_{-1}^{1} \frac{x^2 - 4}{x + 2}\, dx$
23. $\int_{0}^{\sqrt{\pi}} x \cos x^2\, dx$
24. $\int_{0}^{2} |x - \sqrt{x}|\, dx$
25. $\int_{-2}^{2} |x^2 - 1|\, dx$
26. $\int_{0}^{\pi/3} \sin 3x\, dx$
27. $\int_{2}^{7} \sqrt{x + 2}\, dx$
28. $\int_{5}^{10} \frac{dx}{\sqrt{x - 1}}$
29. $\int_{0}^{3} x\sqrt{x^2 + 16}\, dx$
30. $\int_{0}^{4} \frac{x\, dx}{\sqrt{9 + x^2}}$

In each of Problems 31–35, apply the fundamental theorem of calculus to find the derivative of the given function.

31. $f(x) = \int_{-1}^{x} (t^2 + 1)^{17}\, dt$
32. $g(t) = \int_{0}^{t} (x^2 + 25)^{1/2}\, dx$
33. $h(z) = \int_{2}^{z} (u - 1)^{1/3}\, du$
34. $A(x) = \int_{1}^{x} \frac{1}{t}\, dt$
35. $f(x) = \int_{x}^{10} \left(t + \frac{1}{t}\right) dt$

In Problems 36–39, $G(x)$ is the integral of the given function $f(t)$ over the specified interval of the form $[a, x]$, $x > a$. Apply the first part of the fundamental theorem of calculus to find $G'(x)$.

36. $f(t) = \frac{t}{t^2 + 1}$; $[2, x]$
37. $f(t) = \sqrt{t + 4}$; $[0, x]$
38. $f(t) = \sin^3 t$; $[0, x]$
39. $f(t) = \sqrt{t^3 + 1}$; $[1, x]$

In Problems 40–46, differentiate the function f by first writing $f(x)$ in the form $g(u)$, where u is the upper limit of the integral.

40. $f(x) = \int_{0}^{x^2} \sqrt{1 + t^3}\, dt$
41. $f(x) = \int_{2}^{3x} \sin t^2\, dt$
42. $f(x) = \int_{0}^{\sin x} \sqrt{1 - t^2}\, dt$
43. $f(x) = \int_{0}^{x^2} \sin t\, dt$
44. $f(x) = \int_{1}^{\sin x} (t^2 + 1)^3\, dt$
45. $f(x) = \int_{1}^{x^2 + 1} \frac{dt}{t}$
46. $f(x) = \int_{1}^{x^5} \sqrt{1 + t^2}\, dt$

47. The fundamental theorem of calculus *seems* to say that

$$\int_{-1}^{1} \frac{dx}{x^2} = \left[-\frac{1}{x}\right]_{-1}^{1} = -2,$$

in apparent contradiction to the fact that $1/x^2$ is always positive. What's wrong here?

48. Note that $\pm f(x) \leq |f(x)|$, and thereby deduce the inequality in (9) from the one in Equation (8).

49. Apply part 2 of the fundamental theorem of calculus to establish the interval union property (in Section 5.4) of the integral.

50. Prove that the average rate of change

$$\frac{f(b) - f(a)}{b - a}$$

of the differentiable function f on $[a, b]$ is equal to the average value of its derivative on $[a, b]$.

51. If a ball is dropped from a height of 400 ft, find its average height and its average velocity between the time it is dropped and the time it strikes the ground.

52. Find the average value on [0, 10] of the animal population $P(t) = 100 + 10t + (0.5)t^2 + (0.02)t^3$ of Example 3 in Section 5.4.

53. Suppose that a 5000-gal water tank takes 10 minutes to drain and that after t minutes, the amount of water remaining in the tank is $V(t) = 50(10 - t)^2$ gallons. What is the average amount of water in the tank during the time it drains?

54. On a certain day the temperature t hours past midnight was

$$T(t) = 80 + 10 \sin\left(\frac{\pi}{12}(t - 10)\right).$$

What was the average temperature between noon and 6 P.M.?

5.6 Integration by Substitution

We can write the computational part of the fundamental theorem of calculus in the form

$$\int_a^b f(x)\,dx = \Big[D^{-1}f(x)\Big]_a^b. \tag{1}$$

Because of this relationship between integration and antidifferentiation, it is customary to write

$$\int f(x)\,dx = D^{-1}f(x) = F(x) + C \tag{2}$$

if $F'(x) = f(x)$. The expression

$$\int f(x)\,dx,$$

with no limits on the integral sign, is called the **indefinite integral** of the function f, in contrast with the **definite integral,** which has upper and lower limits. Thus the indefinite integral of f is simply the most general antiderivative of f, and indefinite integration is simply antidifferentiation. For example,

$$\int (3x^2 - 4)\,dx = x^3 - 4x + C$$

and

$$\int (t - \cos t)\,dt = \tfrac{1}{2}t^2 - \sin t + C.$$

Note that the indefinite integral is a *function* (actually, a collection of functions), while the definite integral is a *number*. The relationship between definite and indefinite integration is obtained by rewriting Equation (1) as

$$\int_a^b f(x)\,dx = \left[\int f(x)\,dx\right]_a^b. \tag{3}$$

In the notation of indefinite integrals, the antidifferentiation formulas in Equations (6)–(10) of Section 4.8 take the forms

$$\int cf(x)\,dx = c\int f(x)\,dx \qquad (c \text{ is a constant}), \tag{4}$$

$$\int [f(x) + g(x)]\,dx = \int f(x)\,dx + \int g(x)\,dx, \tag{5}$$

$$\int x^r\,dx = \frac{1}{r+1}x^{r+1} + C \qquad (\text{if } r \neq -1), \tag{6}$$

$$\int \cos x\, dx = \sin x + C, \tag{7}$$

$$\int \sin x\, dx = -\cos x + C. \tag{8}$$

A common but very important sort of indefinite integral takes the form

$$\int f(g(x))g'(x)\, dx.$$

If we write $u = g(x)$, then the differential of u is $du = g'(x)\, dx$, so a purely mechanical substitution gives the *tentative* formula

$$\int f(g(x))g'(x)\, dx = \int f(u)\, du. \tag{9}$$

One of the beauties of differential notation is that the formula in (9) is not only plausible, but in fact true—with the understanding that u is to be replaced by $g(x)$ after the indefinite integration on the right-hand side in (9) has been performed. Indeed, Equation (9) is merely an indefinite integral version of the chain rule. For if $F'(x) = f(x)$, then

$$D_x F(g(x)) = F'(g(x))g'(x) = f(g(x))g'(x)$$

by the chain rule, so that

$$\begin{aligned}
\int f(g(x))g'(x)\, dx &= \int F'(g(x))g'(x)\, dx = \int D_x[F(g(x))]\, dx \\
&= F(g(x)) + C = F(u) + C \qquad (u = g(x)) \\
&= \int f(u)\, du.
\end{aligned}$$

The formula in (9) is the basis for the powerful technique of indefinite **integration by substitution.** It may be used whenever the integrand function is recognizable in the form $f(g(x))g'(x)$.

EXAMPLE 1 Find $\int x^2\sqrt{x^3 + 9}\, dx$.

Solution Note that x^2 is, to within a constant factor, the derivative of $x^3 + 9$. We can therefore substitute

$$u = x^3 + 9, \qquad du = 3x^2\, dx.$$

The constant factor 3 can be supplied if we compensate by multiplying the integral by $\frac{1}{3}$. This gives

$$\begin{aligned}
\int x^2\sqrt{x^3 + 9}\, dx &= \frac{1}{3}\int (x^3 + 9)^{1/2} 3x^2\, dx \\
&= \frac{1}{3}\int u^{1/2}\, du = \frac{1}{3} \cdot \frac{u^{3/2}}{\frac{3}{2}} + C \\
&= \frac{2}{9} u^{3/2} + C = \frac{2}{9}(x^3 + 9)^{3/2} + C.
\end{aligned}$$

Another valid way to carry out the same substitution is to solve $du = 3x^2\, dx$ for $x^2\, dx = \frac{1}{3}\, du$, and then write

$$\int (x^3 + 9)^{1/2}\, dx = \int u^{1/2} \cdot \tfrac{1}{3}\, du = \tfrac{1}{3}\int u^{1/2}\, du.$$

Three items worth noting appear upon examination of the solution to Example 1:

- The differential dx along with the rest of the integrand is "transformed," or replaced, in terms of u and du.
- Once the actual integration has been performed, the constant C of integration is added.
- A final resubstitution is necessary to write the answer in terms of the original variable x.

The method of integration by substitution can also be used with definite integrals. Only one additional step is required—evaluation of the final antiderivative at the original limits of integration. For example, the substitution of Example 1 gives

$$\int_0^3 x^2\sqrt{x^3+9}\,dx = \tfrac{1}{3}\int_*^{**} u^{1/2}\,du = \tfrac{1}{3}\Big[\tfrac{2}{3}u^{3/2}\Big]_*^{**}$$
$$= \tfrac{2}{9}\Big[(x^3+9)^{3/2}\Big]_0^3$$
$$= \tfrac{2}{9}(216-27) = 42.$$

The limits * and ** on u are so indicated because they weren't calculated—there was no need to know them—and because it would be only a coincidence if they were the same as the limits on x.

But sometimes it is more convenient to determine the limits of integration with respect to the new variable u. With the substitution $u = x^3 + 9$ of this example, the lower limit $x = 0$ corresponds to $u = 9$ and the upper limit $x = 3$ corresponds to $u = 36$. Hence we may alternatively write

$$\int_0^3 x^2\sqrt{x^3+9}\,dx = \tfrac{1}{3}\int_9^{36} u^{1/2}\,du = \tfrac{1}{3}\Big[\tfrac{2}{3}u^{3/2}\Big]_9^{36} = 42.$$

The following theorem tells how to transform the limits $x = a$ and $x = b$ under the substitution $u = g(x)$. The new lower limit is $u = g(a)$ and the new upper limit is $u = g(b)$, *whether or not* $g(b)$ is greater than $g(a)$.

Theorem *Definite Integration by Substitution*

Suppose that the function g has a continuous derivative on $[a, b]$ and that f is continuous on the set $g([a, b])$. Let $u = g(x)$. Then

$$\int_a^b f(g(x))g'(x)\,dx = \int_{g(a)}^{g(b)} f(u)\,du. \tag{10}$$

Proof Choose an antiderivative F of f, so that $F' = f$. Then, by the chain rule,

$$D[F(g(x))] = F'(g(x))g'(x) = f(g(x))g'(x).$$

Therefore,

$$\int_a^b f(g(x))g'(x)\,dx = \Big[F(g(x))\Big]_a^b = F(g(b)) - F(g(a))$$
$$= \Big[F(u)\Big]_{u=g(a)}^{g(b)} = \int_{g(a)}^{g(b)} f(u)\,du.$$

We used the fundamental theorem to obtain the first and last equalities in this argument. ■

EXAMPLE 2 Evaluate $\displaystyle\int_4^9 \frac{\sqrt{x}\,dx}{(30 - x^{3/2})^2}$.

Solution Note that $30 - x^{3/2}$ is nonzero on [4, 9], so the integrand is continuous there. We substitute

$$u = 30 - x^{3/2}, \quad \text{so that} \quad du = -\tfrac{3}{2}x^{1/2}\,dx.$$

We solve the last equation for the expression in the numerator of the integrand:

$$\sqrt{x}\,dx = -\tfrac{2}{3}\,du.$$

To obtain the correct limits of integration for u, note that

$$\text{if} \quad x = 4, \quad \text{then} \quad u = 22;$$

$$\text{if} \quad x = 9, \quad \text{then} \quad u = 3.$$

Hence our substitution gives

$$\int_4^9 \frac{\sqrt{x}\,dx}{(30 - x^{3/2})^2} = \int_{22}^3 \frac{1}{u^2}\left(-\frac{2}{3}\right)du = \frac{2}{3}\int_3^{22} \frac{1}{u^2}\,du = \frac{2}{3}\left[-\frac{1}{u}\right]_3^{22}$$

$$= \frac{2}{3}\left(-\frac{1}{22} + \frac{1}{3}\right) = \frac{19}{99}.$$

EXAMPLE 3 Evaluate $\displaystyle\int_0^{\pi/4} \sin^3 2t \cos 2t\,dt$.

Solution We substitute

$$u = \sin 2t, \quad \text{so that} \quad du = 2\cos 2t\,dt.$$

Then $u = 0$ when $t = 0$; $u = 1$ when $t = \pi/4$. Hence

$$\int_0^{\pi/4} \sin^3 2t \cos 2t\,dt = \tfrac{1}{2}\int_0^1 u^3\,du = \tfrac{1}{2}\left[\tfrac{1}{4}u^4\right]_0^1 = \tfrac{1}{8}.$$

5.6 PROBLEMS

Use the given substitution to compute the most general antiderivative of $f(x)$ in Problems 1–12.

1. $f(x) = x^3\sqrt{1 + x^4}$; $u = x^4$

2. $f(x) = \sin^2 x \cos x$; $u = \sin x$

3. $f(x) = \dfrac{1}{\sqrt{x}(1 + \sqrt{x})^2}$; $u = 1 + \sqrt{x}$

4. $f(x) = \dfrac{1}{\sqrt{x}(1 + \sqrt{x})^2}$; $u = \sqrt{x}$

5. $f(x) = x^2 \cos 4x^3$; $u = 4x^3$

6. $f(x) = x(x + 1)^{14}$; $u = x + 1$

7. $f(x) = x(x^2 + 1)^{14}$; $u = x^2 + 1$

8. $f(x) = x^3 \cos x^4$; $u = x^4$

9. $f(x) = x\sqrt{4 - x}$; $u = 4 - x$

10. $f(x) = \dfrac{x + 2x^3}{(x^4 + x^2)^3}$; $u = x^4 + x^2$

11. $f(x) = \dfrac{2x^3}{\sqrt{1 + x^4}}$; $u = x^4$

12. $f(x) = (2x + 1)(x^2 + x)^{-1/2}$; $u = x^2 + x$

Evaluate the integrals in Problems 13–36.

13. $\int (4x-3)^5\,dx$

14. $\int x\sqrt{x^2-1}\,dx$

15. $\int x\sqrt{2-3x^2}\,dx$

16. $\int 3t(1-2t^2)^{10}\,dt$

17. $\int \frac{x^2\,dx}{(x^3+5)^4}$

18. $\int \frac{t\,dt}{\sqrt{2t^2+1}}$

19. $\int x^2\sqrt[3]{2-4x^3}\,dx$

20. $\int \frac{(x+1)\,dx}{(x^2+2x+5)^2}$

21. $\int \sin\frac{t}{3}\,dt$

22. $\int \frac{\cos\sqrt{x}}{\sqrt{x}}\,dx$

23. $\int_{-1}^{-2} \frac{dt}{(t+3)^3}$

24. $\int_0^4 x\sqrt{x^2+9}\,dx$

25. $\int_0^4 \frac{dx}{\sqrt{2x+1}}$

26. $\int_{-1}^{1} \frac{(x+1)\,dx}{\sqrt{x^2+2x+2}}$

27. $\int_0^8 t\sqrt{t+1}\,dt$ (Try $u=t+1$.)

28. $\int_0^{\pi/2} \sin x\cos x\,dx$

29. $\int_0^{\pi/6} \sin 2x\cos^3 2x\,dx$

30. $\int_0^{\sqrt{\pi}} x\sin\frac{1}{2}x^2\,dx$

31. $\int_0^{\pi/2} (1+\sin t)^{3/2}\cos t\,dt$

32. $\int_1^4 \frac{(1+\sqrt{x})^4}{\sqrt{x}}\,dx$

33. $\int \frac{x^3-1}{(x^4-4x)^{2/3}}\,dx$

34. $\int \frac{1}{x^3}\left(1+\frac{1}{x^2}\right)^{5/3}dx$

35. $\int (2-t^2)\sqrt[4]{6t-t^3}\,dt$

36. $\int \frac{2-x^2}{(x^3-6x+1)^5}\,dx$

In each of Problems 37–40, verify the given formula by differentiation of the right-hand side.

37. $\int \frac{dx}{x^2\sqrt{x^2+a^2}} = -\frac{\sqrt{x^2+a^2}}{a^2x}+C$

38. $\int (x^2-a^2)^{3/2}\,dx = \frac{1}{4}x(x^2-a^2)^{3/2}-\frac{3}{4}a^2\int\sqrt{x^2-a^2}\,dx$

39. $\int \sin^3 x\,dx = \frac{1}{3}\cos^3 x-\cos x+C$

40. $\int \frac{x}{(1-x^2)^2}\,dx = \frac{x^2}{2(1-x^2)}+C$

41. Substitute $u=1-x^2$ to show that

$$\int \frac{x}{(1-x^2)^2}\,dx = \frac{1}{2(1-x^2)}+C.$$

Is this answer consistent with the formula of Problem 40? Explain.

42. Substitute $\sin^3 x = (\sin x)(1-\cos^2 x)$ to derive the formula of Problem 39.

43. Suppose that f is an **odd** function, meaning that $f(-x)=-f(x)$ for all x. Substitute $u=-x$ in the integral

$$\int_{-a}^{0} f(x)\,dx \quad\text{to show that}\quad \int_{-a}^{a} f(x)\,dx = 0$$

if f is continuous on $[-a,a]$.

44. If f is an **even** function, meaning that $f(-x)=f(x)$ for all x, use the method of Problem 43 to show that

$$\int_{-a}^{a} f(x)\,dx = 2\int_0^a f(x)\,dx$$

if f is continuous on $[-a,a]$.

5.7 Computing Areas by Integration

In Section 5.2 we discussed the area A under the graph of a positive-valued continuous function f on the interval $[a,b]$. This discussion motivated our definition in Section 5.3 of the integral of f from a to b, with the result that

$$A = \int_a^b f(x)\,dx \tag{1}$$

by definition.

Here we consider the more general problem of finding the area of a region bounded by the graphs of *two* functions. Suppose that the functions f and g are continuous on $[a,b]$ and that $f(x)\geqq g(x)$ for all x in $[a,b]$. We are interested in the area A of the region R shown in Fig. 5.23, in which R is bounded by the graphs of f and g and by the vertical lines $x=a$ and $x=b$.

In order to approximate A, we consider a regular partition of $[a,b]$ into n equal subintervals, each with length $\Delta x=(b-a)/n$. If ΔA_i denotes the area of the region between the graphs of f and g and lying above the

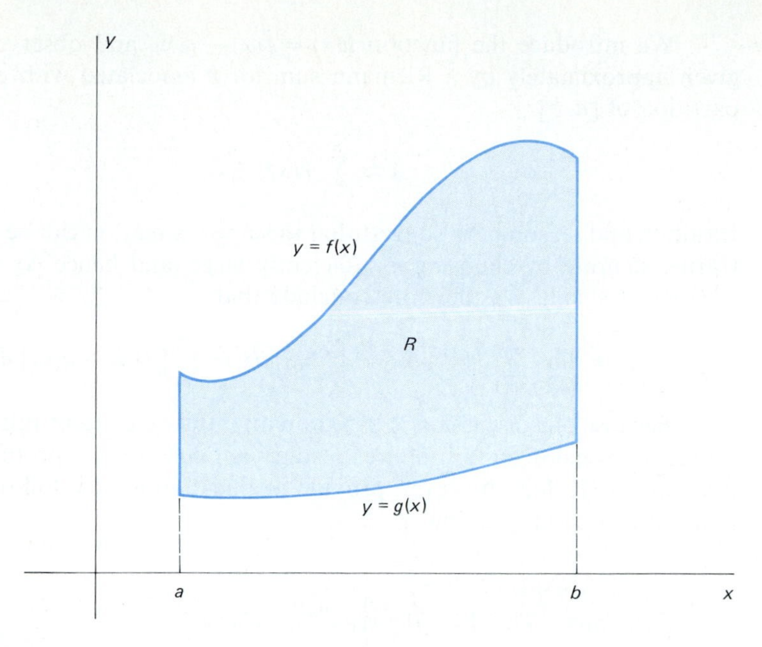

5.23 A region between two graphs

ith subinterval $[x_{i-1}, x_i]$, and x_i^* is an arbitrary number chosen in that subinterval (all this for $i = 1, 2, 3, \ldots, n$), then ΔA_i is approximately equal to the area of a rectangle with height $f(x_i^*) - g(x_i^*)$ and width Δx (see Fig. 5.24). Hence

$$\Delta A_i \approx [f(x_i^*) - g(x_i^*)]\,\Delta x,$$

so

$$A = \sum_{i=1}^{n} \Delta A_i \approx \sum_{i=1}^{n} [f(x_i^*) - g(x_i^*)]\,\Delta x.$$

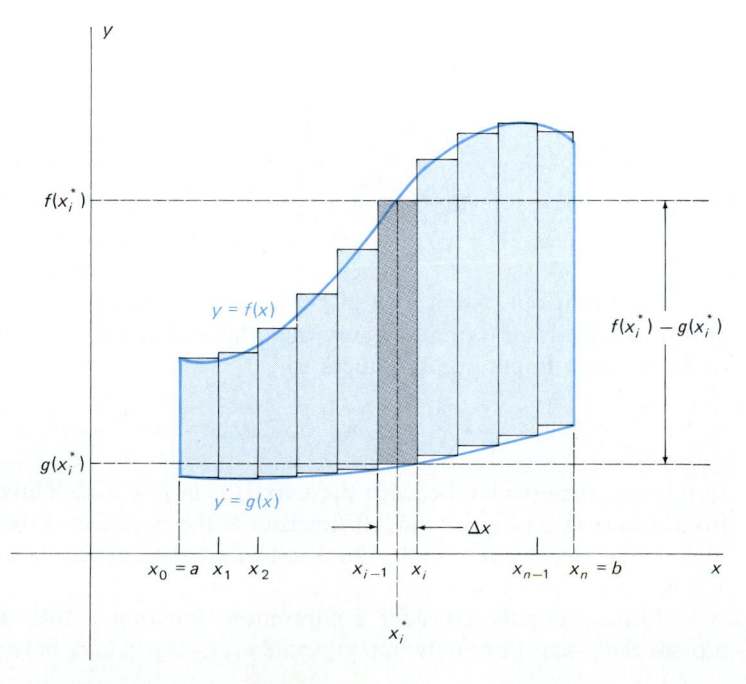

5.24 A partition of $[a, b]$ divides R into vertical strips that we approximate with rectangular strips

We introduce the function $h(x) = f(x) - g(x)$, and observe that A is given approximately by a Riemann sum for h associated with our regular partition of $[a, b]$:

$$A \approx \sum_{i=1}^{n} h(x_i^*)\, \Delta x.$$

Intuition and reason both suggest that this approximation can be made arbitrarily accurate by choosing n sufficiently large (and hence $\Delta x = (b - a)/n$ sufficiently small). We therefore conclude that

$$A = \lim_{\Delta x \to 0} \sum_{i=1}^{n} h(x_i^*)\, \Delta x = \int_a^b h(x)\, dx = \int_a^b [f(x) - g(x)]\, dx.$$

Because our discussion is based on an intuitive concept rather than on a precise logical definition of area, it does *not* constitute a proof of the formula above. It does, however, provide justification for the following *definition* of the area in question.

Definition *The Area Between Two Curves*

Let f and g be continuous with $f(x) \geqq g(x)$ for x in $[a, b]$. Then the **area** A of the region bounded by the curves $y = f(x)$ and $y = g(x)$ and the vertical lines $x = a$ and $x = b$ is

$$A = \int_a^b [f(x) - g(x)]\, dx. \tag{2}$$

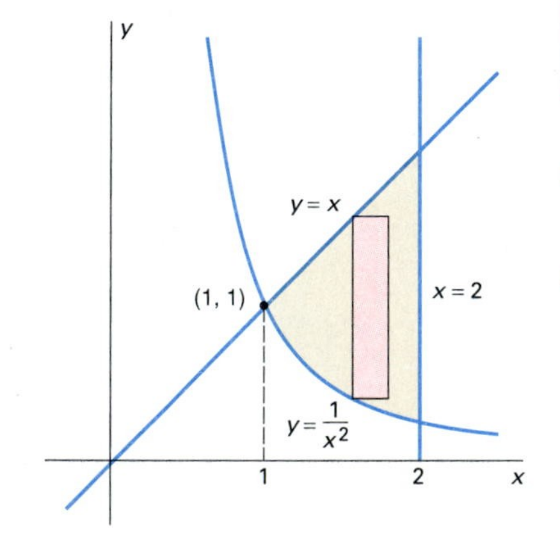

5.25 The region of Example 1

EXAMPLE 1 Find the area of the region shown in Fig. 5.25, bounded by the lines $y = x$ and $x = 2$ and the curve $y = 1/x^2$.

Solution Here the top curve is $y = f(x) = x$ and the bottom curve is $y = g(x) = 1/x^2$, while $a = 1$ and $b = 2$. Hence the formula in (2) yields

$$A = \int_1^2 \left[x - \frac{1}{x^2}\right] dx = \left[\frac{1}{2}x^2 + \frac{1}{x}\right]_1^2$$

$$= \left(2 + \frac{1}{2}\right) - \left(\frac{1}{2} + 1\right) = 1.$$

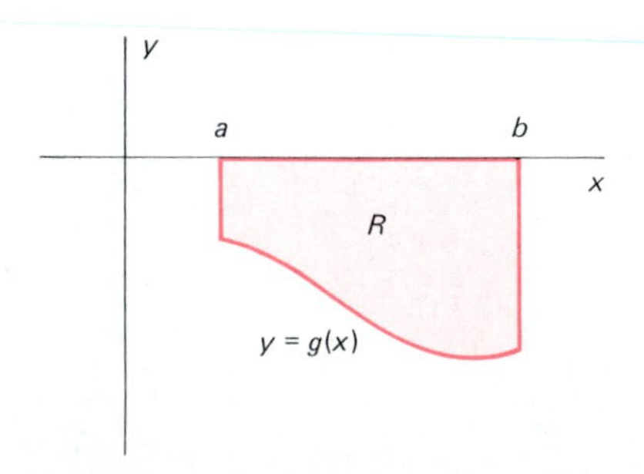

5.26 The integral gives the negative of the geometric area for a region lying below the x-axis.

Note that our earlier formula in (1) is the special case of (2) in which $g(x)$ is identically zero on $[a, b]$. On the other hand, if $f(x) \equiv 0$ and $g(x) \leqq 0$ on $[a, b]$, then Equation (2) reduces to

$$A = -\int_a^b g(x)\, dx \quad \text{or} \quad \int_a^b g(x)\, dx = -A.$$

In this case the area lies beneath the x-axis, as in Fig. 5.26. Thus the integral from a to b of a negative-valued function is the *negative* of the area of the region bounded by its graph, the x-axis, and the vertical lines $x = a$ and $x = b$.

More generally, consider a continuous function f with a graph that crosses the x-axis at finitely many points $c_1, c_2, c_3, \ldots, c_k$ between a and b,

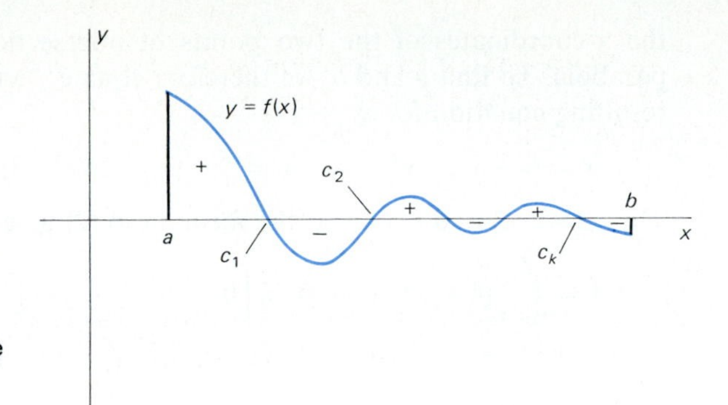

5.27 The integral computes the area above the x-axis *minus* the area below the x-axis.

as shown in Fig. 5.27. We write

$$\int_a^b f(x)\,dx = \int_a^{c_1} f(x)\,dx + \int_{c_1}^{c_2} f(x)\,dx + \cdots + \int_{c_k}^{b} f(x)\,dx.$$

Thus we see that

$$\int_a^b f(x)\,dx$$

is equal to the area under $y = f(x)$ *above* the x-axis *minus* the area over $y = f(x)$ *below* the x-axis.

The following *heuristic* (suggestive, though nonrigorous) way of setting up integral formulas like that in (2) is sometimes useful. Consider the vertical strip of area lying above the interval $[x, x + dx]$, shown shaded in Fig. 5.28, where we have written

$$y_{\text{top}} = f(x) \quad \text{and} \quad y_{\text{bot}} = g(x)$$

for the top and bottom boundary curves. We think of the length dx of the interval $[x, x + dx]$ as being so small that this strip may be regarded as a rectangle with width dx and height $y_{\text{top}} - y_{\text{bot}}$, so that its area is

$$dA = [y_{\text{top}} - y_{\text{bot}}]\,dx.$$

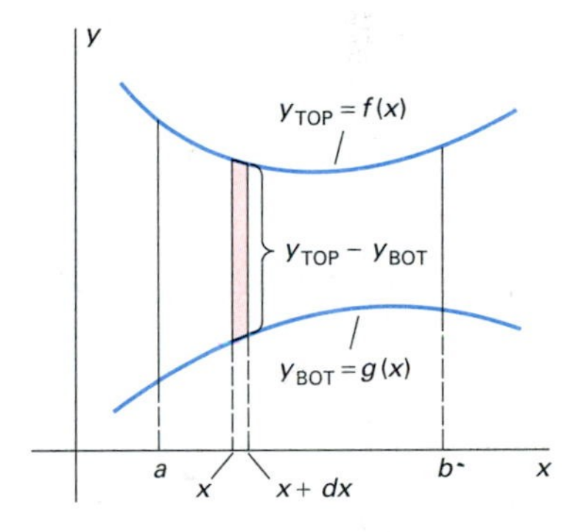

5.28 Heuristic approach to setting up area integrals

Think now of the region over $[a, b]$ between $y_{\text{top}} = f(x)$ and $y_{\text{bot}} = g(x)$ as being made up of many such vertical strips. Its area may be regarded as a sum of areas of such rectangular strips. If we write $\int$ for *sum*, this gives the formula

$$A = \int dA = \int_a^b [y_{\text{top}} - y_{\text{bot}}]\,dx.$$

This heuristic approach bypasses the subscript notation associated with Riemann sums. Nevertheless, it *is not and should not be regarded as a complete* derivation of the formula. It is best used only as a convenient memory device. For instance, in the figures for illustrative examples, we shall often show a strip of width dx as a visual aid in properly setting up the correct integral.

EXAMPLE 2 Find the area A of the region R bounded by the line $y = x$ and the parabola $y = 6 - x^2$.

Solution The region R is shown in Fig. 5.29. It is clear that we should use the formula in (2), taking $f(x) = 6 - x^2$ and $g(x) = x$. The limits a and b will be

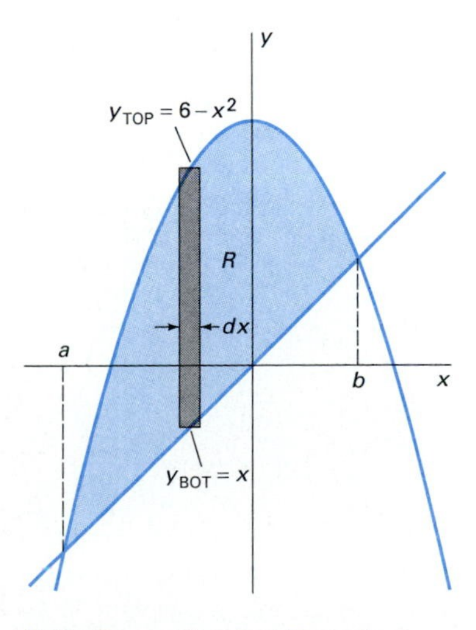

5.29 The region R of Example 2

the x-coordinates of the two points of intersection of the line and the parabola. To find a and b, we therefore equate $f(x)$ and $g(x)$ and solve the resulting equation for x:

$$x = 6 - x^2; \quad x^2 + x - 6 = 0; \quad (x-2)(x+3) = 0; \quad x = -3, 2.$$

Thus $a = -3$ and $b = 2$, so the formula in (2) gives

$$A = \int_{-3}^{2} (6 - x^2 - x)\,dx = \left[6x - \tfrac{1}{3}x^3 - \tfrac{1}{2}x^2\right]_{-3}^{2}$$

$$= [(6)(2) - \tfrac{1}{3}(2)^3 - \tfrac{1}{2}(2)^2] - [(6)(-3) - \tfrac{1}{3}(-3)^3 - \tfrac{1}{2}(-3)^2] = \tfrac{125}{6}.$$

Our next example shows that it is sometimes necessary to subdivide a region before applying the formula in Equation (2).

EXAMPLE 3 Find the area A of the region R bounded by the line $y = \frac{1}{2}x$ and the parabola $y^2 = 8 - x$.

Solution The region R is shown in Fig. 5.30. The points of intersection $(-8, -4)$ and $(4, 2)$ are found by equating $y = \frac{1}{2}x$ and $y = \pm\sqrt{8-x}$ and then solving for x. The lower boundary of R is given by $y_{\text{bot}} = -\sqrt{8-x}$ on $[-8, 8]$. But the upper boundary of R is given by

$$y_{\text{top}} = \tfrac{1}{2}x \quad \text{on } [-8, 4], \qquad y_{\text{top}} = +\sqrt{8-x} \quad \text{on } [4, 8].$$

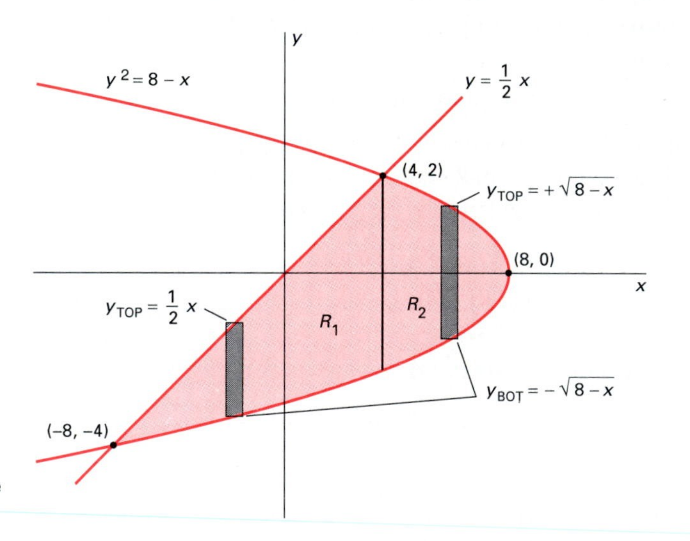

5.30 In Example 3, we split the region R into R_1 and R_2.

We must therefore subdivide R into the two regions R_1 and R_2 as indicated in Fig. 5.30. Then the formula in (2) gives

$$A = \int_{-8}^{4} (\tfrac{1}{2}x + \sqrt{8-x})\,dx + \int_{4}^{8} 2\sqrt{8-x}\,dx$$

$$= \left[\tfrac{1}{4}x^2 - \tfrac{2}{3}(8-x)^{3/2}\right]_{-8}^{4} + \left[-\tfrac{4}{3}(8-x)^{3/2}\right]_{4}^{8}$$

$$= [(\tfrac{16}{4} - \tfrac{16}{3}) - (\tfrac{64}{4} - \tfrac{128}{3})] + [0 + \tfrac{32}{3}] = 36.$$

The region of Example 3 appears simpler if it is considered to be bounded by graphs of functions of y rather than by functions of x. Figure 5.31 shows a region R bounded by the curves $x = f(y)$ and $x = g(y)$, with $f(y) \geqq g(y)$ for y in $[c, d]$, and by the horizontal lines $y = c$ and $y = d$. To approximate the area A of R, we begin with a regular partition of $[c, d]$ into n subintervals each with length Δy. We choose a point y_i^* in the ith subinterval $[y_{i-1}, y_i]$ for each $i = 1, 2, 3, \ldots, n$. The strip of R lying opposite $[y_{i-1}, y_i]$ is approximated by a rectangle with width Δy and length $f(y_i^*) - g(y_i^*)$. Hence

$$A \approx \sum_{i=1}^{n} [f(y_i^*) - g(y_i^*)]\,\Delta y.$$

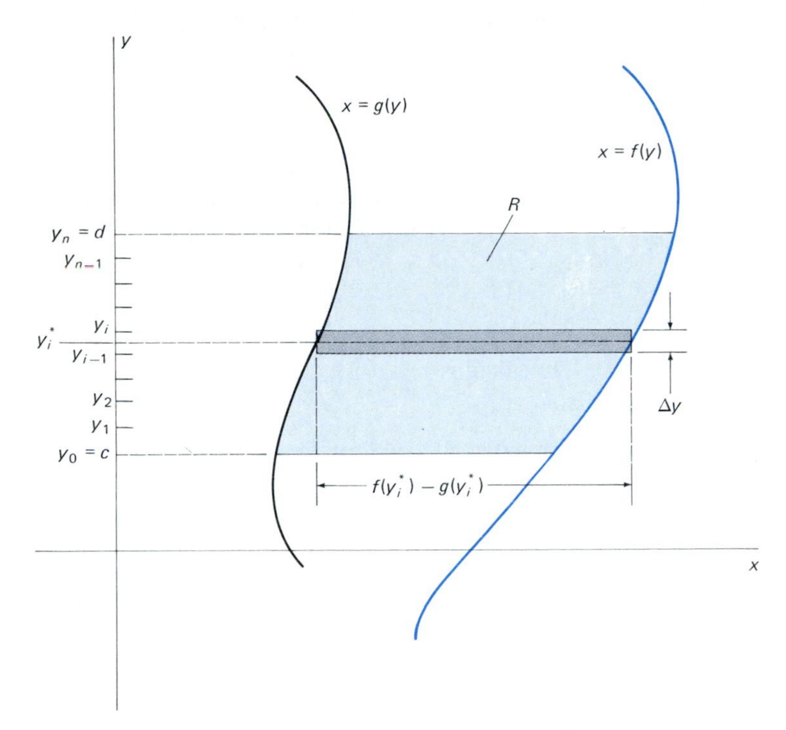

5.31 Finding area using an integral with respect to y

Recognition of this sum as a Riemann sum for the integral

$$\int_c^d [f(y) - g(y)]\,dy$$

motivates the following definition.

Definition *The Area Between Two Curves*

Let f and g be continuous with $f(y) \geqq g(y)$ for y in $[c, d]$. Then the **area** A of the region bounded by the curves $x = f(y)$ and $x = g(y)$ and the horizontal lines $y = c$ and $y = d$ is

$$A = \int_c^d [f(y) - g(y)]\,dy. \tag{3}$$

In a more advanced course, one would now prove that the formulas in (2) and (3) yield the same area A for a region that can be described both in the manner shown in Fig. 5.22 and in the manner shown in Fig. 5.31.

Let us write

$$x_{\text{right}} = f(y) \quad \text{and} \quad x_{\text{left}} = g(y)$$

for the right and left boundary curves of the region in Fig. 5.31. Then Equation (3) takes the form

$$A = \int_c^d [x_{\text{right}} - x_{\text{left}}]\, dy.$$

Comparison of Example 3 with the one following illustrates the advantage of choosing the "right" variable of integration—the one that makes the resulting computations simpler.

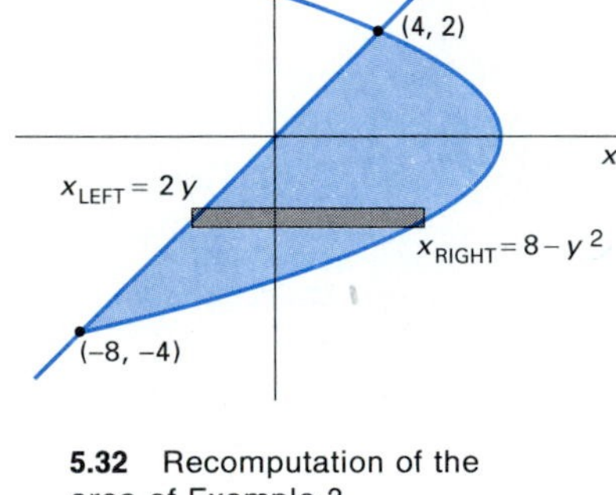

5.32 Recomputation of the area of Example 3

EXAMPLE 4 Integrate with respect to y to find the area of the region R of Example 3.

Solution We see from Fig. 5.32 that the formula in (3) applies with $x_{\text{right}} = f(y) = 8 - y^2$ and $x_{\text{left}} = g(y) = 2y$ for y in $[-4, 2]$. This gives

$$A = \int_{-4}^{2} [(8 - y^2) - 2y]\, dy = \left[8y - \tfrac{1}{3}y^3 - y^2\right]_{-4}^{2} = 36.$$

EXAMPLE 5 Use calculus to derive the formula $A = \pi r^2$ for the area of a circle of radius r.

Solution We begin with the *definition* (in Section 5.2) of the number π as the area of the *unit* circle $x^2 + y^2 \leqq 1$. Then, with the aid of Fig. 5.33, we may write

$$\pi = 4 \int_0^1 \sqrt{1 - x^2}\, dx, \tag{4}$$

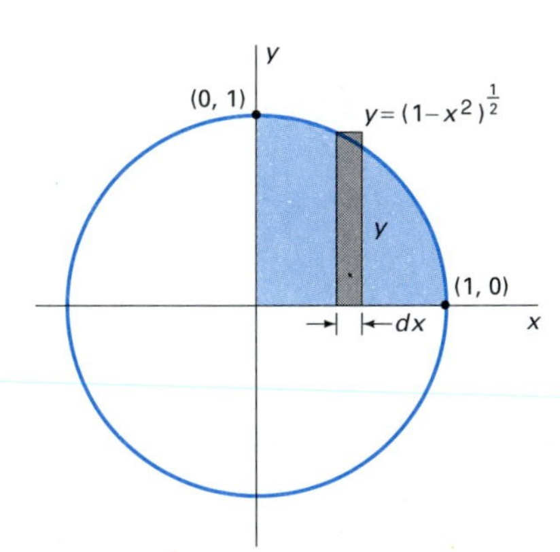

5.33 The number π is four times the shaded area.

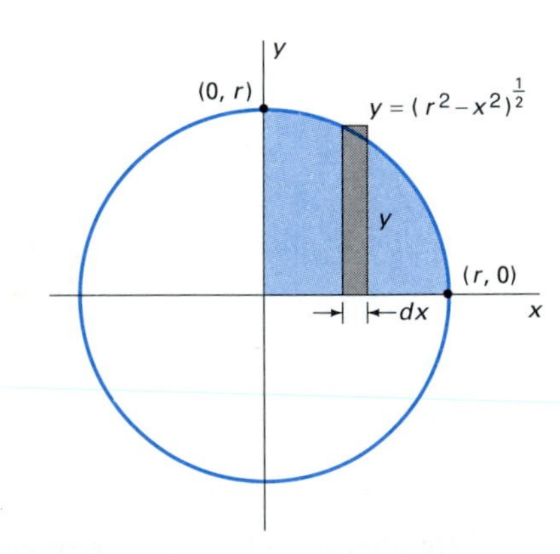

5.34 The shaded area can be written as an integral

because the integral in (4) is, by the formula in (1), the area of the first quadrant of the circle. We apply the formula in (1) to the first quadrant of the circle of radius r of Fig. 5.34, and we find the total area A of that circle to be

$$A = 4\int_0^r \sqrt{r^2 - x^2}\,dx = 4r\int_0^r \sqrt{1-(x/r)^2}\,dx$$

$$= 4r\int_0^1 \sqrt{1-u^2}\,r\,du$$

$$\left(\text{Substitution: } u = \frac{x}{r},\ dx = r\,du.\right)$$

$$= 4r^2\int_0^1 \sqrt{1-u^2}\,du = 4r^2\int_0^1 \sqrt{1-x^2}\,dx.$$

So, by Equation (4), $A = \pi r^2$.

5.7 PROBLEMS

Find the areas shown in Problems 1–10.

1.

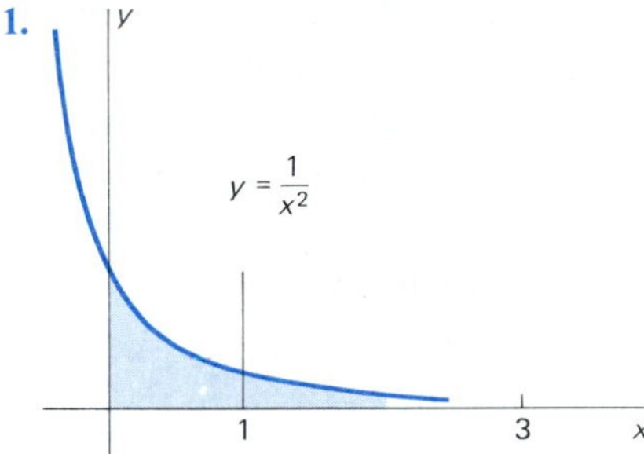

2.

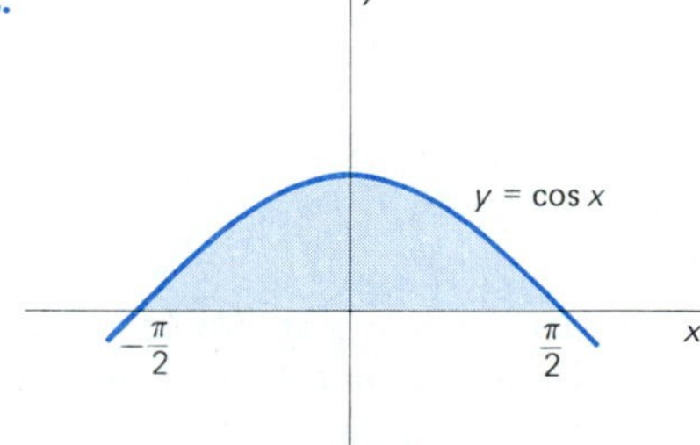

3.

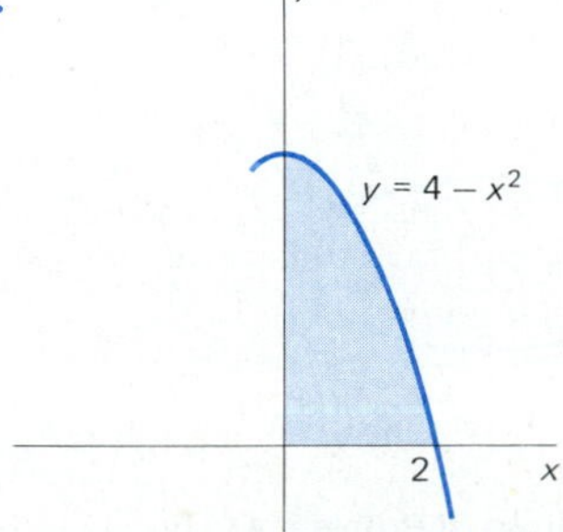

4.

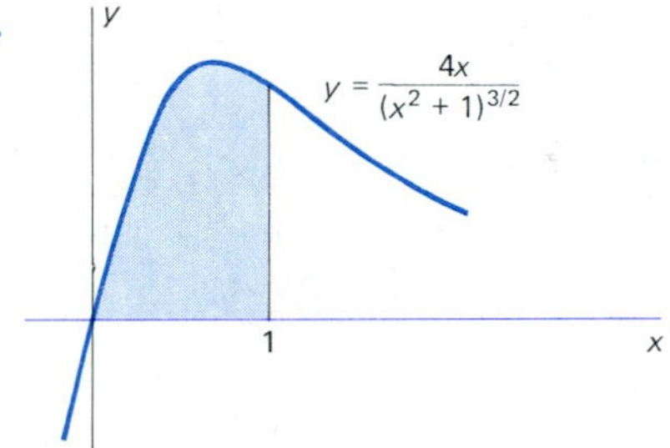

5.

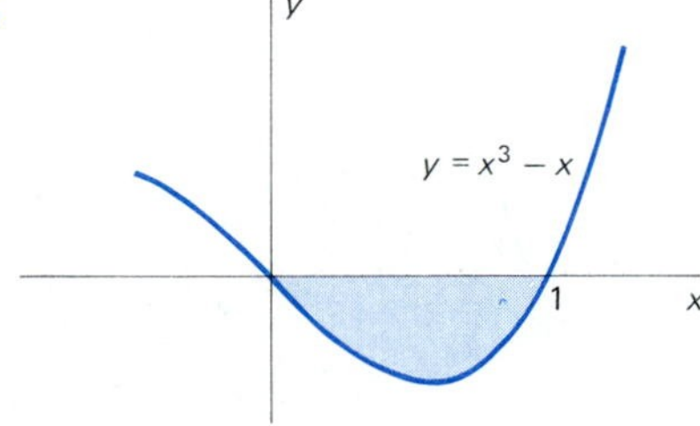

6.

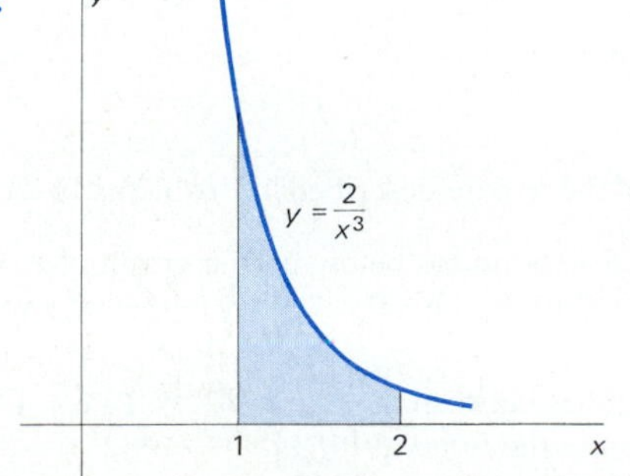

7. 

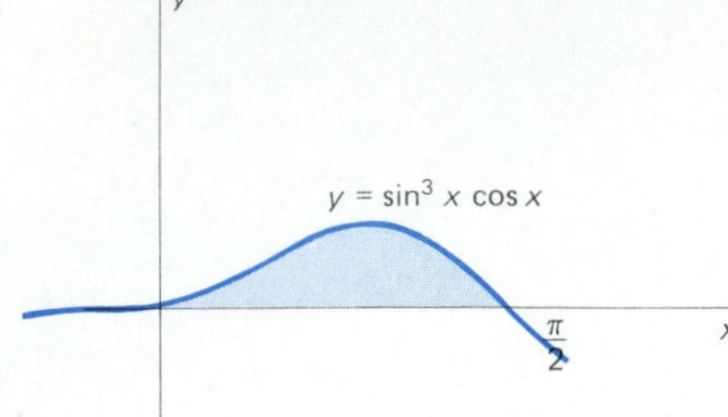

8.

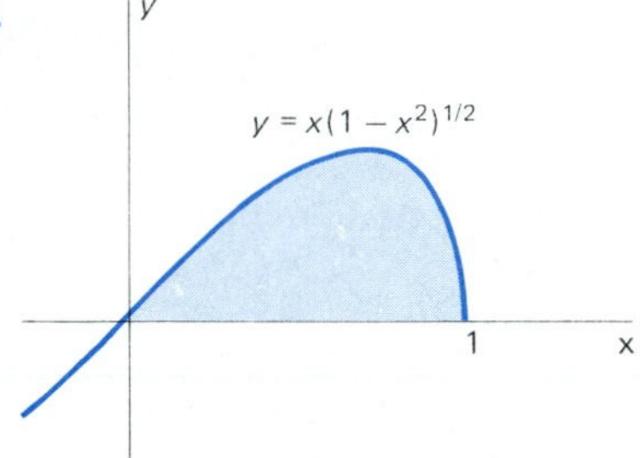

9.

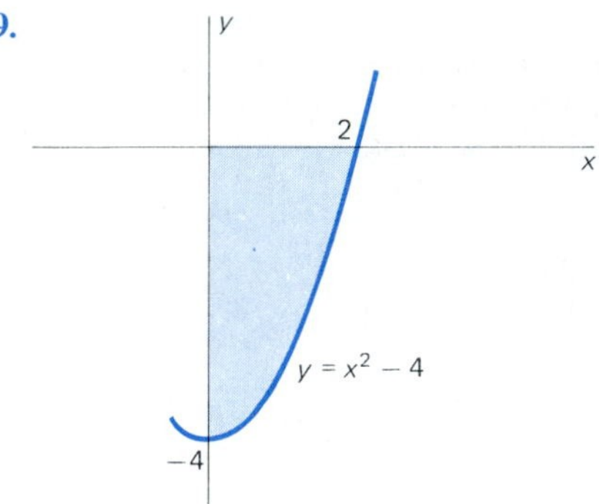

10.

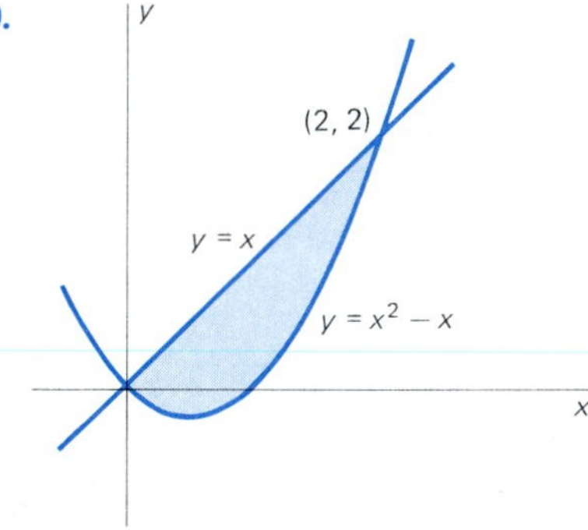

Find the areas of the regions described in Problems 11–20.

11. The region R is bounded below by the graph of $y = x^3$ and above by the graph of $y = x$ over the interval $[0, 1]$.

12. The region R lies between the graph of $y = 1/(x + 1)^2$ and the x-axis over the interval $[1, 3]$.

13. The region R is bounded above by the graph of $y = x^3$ and below by the graph of $y = x^4$ over the interval $[0, 1]$.

14. The region R is bounded above by the graph of $y = x^2$ and below by the horizontal line $y = -1$ over the interval $[-1, 2]$.

15. The region R is bounded above by the graph of $y = 1/(x + 1)^3$ and below by the x-axis over the interval $[0, 2]$.

16. The region R is bounded above by the graph of $y = 4x - x^2$ and below by the x-axis.

17. The region R is bounded on the left by the graph of $x = y^2$ and on the right by the vertical line $x = 4$.

18. The region R lies between the graphs of $y = x^4 - 4$ and $y = 3x^2$.

19. The region R lies between the graphs of $x = 8 - y^2$ and $x = y^2 - 8$.

20. The region R lies between the graphs of $y = x^{1/3}$ and $y = x^3$.

In each of Problems 21–42, sketch the region bounded by the given curves, then find its area.

21. $y = 0, \quad y = 25 - x^2$

22. $y = x^2, \quad y = 4$

23. $y = x^2, \quad y = 8 - x^2$

24. $x = 0, \quad x = 16 - y^2$

25. $x = y^2, \quad x = 25$

26. $x = y^2, \quad x = 32 - y^2$

27. $y = x^2, \quad y = 2x$

28. $y = x^2, \quad x = y^2$

29. $y = x^2, \quad y = x^3$

30. $y = 2x^2, \quad y = 5x - 3$

31. $x = 4y^2, \quad x + 12y + 5 = 0$

32. $y = x^2, \quad y = 3 + 5x - x^2$

33. $x = 3y^2, \quad x = -y^2 + 12y - 5$

34. $y = x^2, \quad y = 4(x - 1)^2$

35. $x = y^2 - 2y - 2, \quad x = -2y^2 + y + 4$

36. $y = x^4, \quad y = 32 - x^4$

37. $y = x^3, \quad y = 32\sqrt{x}$

38. $y = x^3, \quad y = 2x - x^2$

39. $y = x^2, \quad y = x^{2/3}$

40. $y^2 = x, \quad y^2 = 2(x - 3)$

41. $y = x^3, \quad y = 2x^3 + x^2 - 2x$

42. $y = x^3, \quad x + y = 0, \quad y = x + 6$

43. The *ellipse* $x^2/a^2 + y^2/b^2 = 1$ is shown in Fig. 5.35. Use the method of Example 4 to show that the area of the region it bounds is $A = \pi ab$, a pleasing generalization of the area formula for the circle.

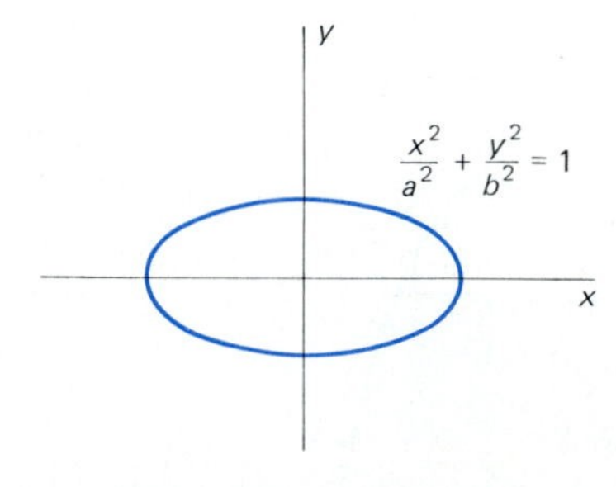

5.35 The ellipse of Problem 43

44. Let A and B be the points of intersection of the parabola $y = x^2$ and the line $y = x + 2$, and let C be the point on the parabola where the tangent line is parallel to the graph of $y = x + 2$. Show that the area of the parabolic segment cut off from the parabola by the line (as in Fig. 5.36) is four-thirds the area of triangle ABC.

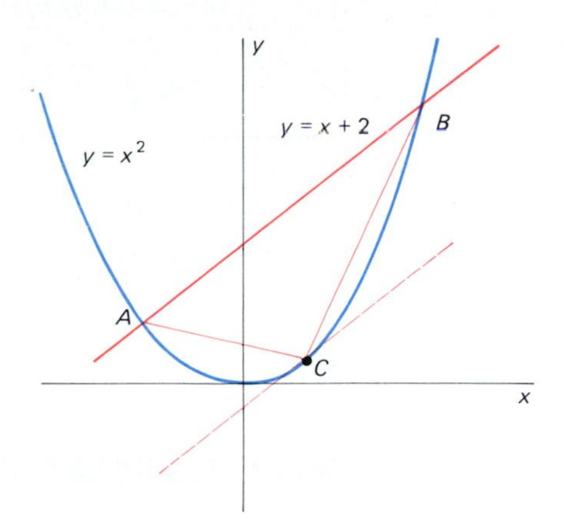

5.36 The parabolic segment of Problem 44

45. Find the area of the unbounded region shaded in Fig. 5.37, regarding it as the limit as $b \to +\infty$ of the region bounded by $y = 1/x^2$, $y = 0$, $x = 1$, and $x = b > 1$.

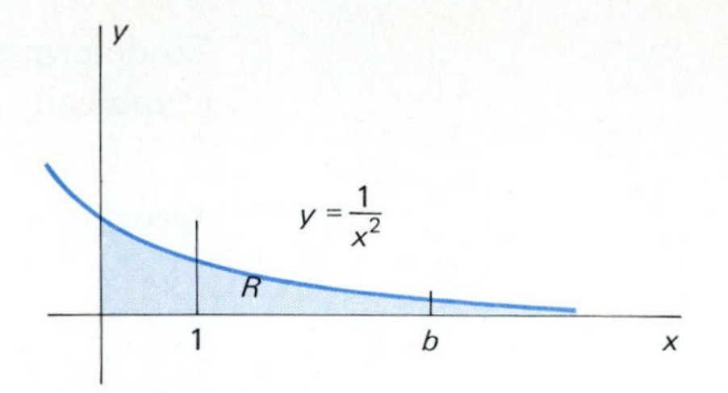

5.37 The unbounded region of Problem 45

46. Find the total area of the bounded regions that are bounded by the x-axis and the curve $y = 2x^3 - 2x^2 - 12x$.

47. Suppose that the quadratic function $f(x) = px^2 + qx + r$ is never negative on $[a, b]$. Show that the area under the graph of f from a to b is

$$A = \tfrac{1}{3}h[f(a) + 4f(m) + f(b)]$$

where $h = \frac{1}{2}(b - a)$ and $m = \frac{1}{2}(a + b)$. (*Suggestion:* By horizontal translation of this region, you may assume that $a = -h$, $m = 0$, and $b = h$.)

In Problems 48–50, the area $A(u)$ between the graph of the positive and continuous function f and the x-axis over the interval $0 \leqq x \leqq u$ is given. Find $f(x)$.

48. $A(u) = u^5$

49. $A(u) = \frac{2}{3}u^{3/2}$

50. $A(u) = [f(u)]^2$

*5.7 Optional Computer Application

The following HP-28S program can be used to compute the area A of the region R bounded by the curves $y_{\text{top}} = f(x)$ and $y_{\text{bot}} = g(x)$.

```
<< YTOP  YBOT  -
   'X'  A  B  3  →LIST
   .000001  ∫  DROP >>
```

Before this program is executed, the algebraic expressions $f(x)$ and $g(x)$ must be stored in the variables YTOP and YBOT, and the lower and upper limits of integration must be stored in the variables A and B. Then the first line of this program forms the integrand function YTOP − YBOT, and the second line assembles the three-element list {'X', A, B} specifying the variable of integration and the lower and upper limits. The last line of the program then integrates numerically with an error tolerance of $10^{-6} = 0.000001$.

For example, suppose that R is bounded above by the line $y = 4 - 2x$ and below by the parabola $y = 6x^2 - 7x$. Assuming that the program has been stored as AREA, our first step is to enter and store the functions.

```
'4-2*X  [ENTER]
'YTOP  [STO]
'6*X*X-7*X  [ENTER]
'YBOT  [STO]
```

To determine A and B, we plot both curves to find their intersection points graphically.

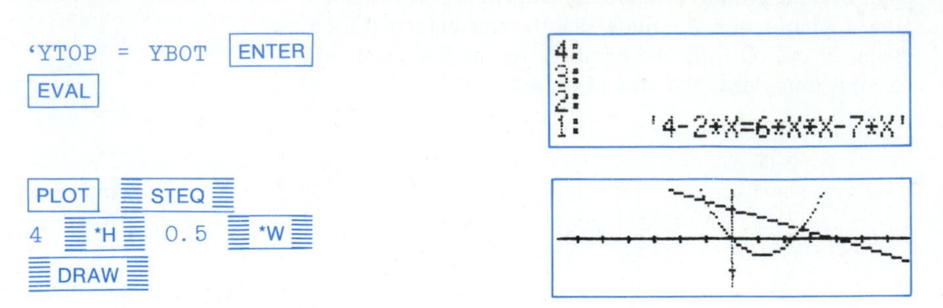

We now use the cursor and INS keys to digitize the approximate intersection points.

Finally, we use the HP-28S's equation solver to find and store the desired limits of integration.

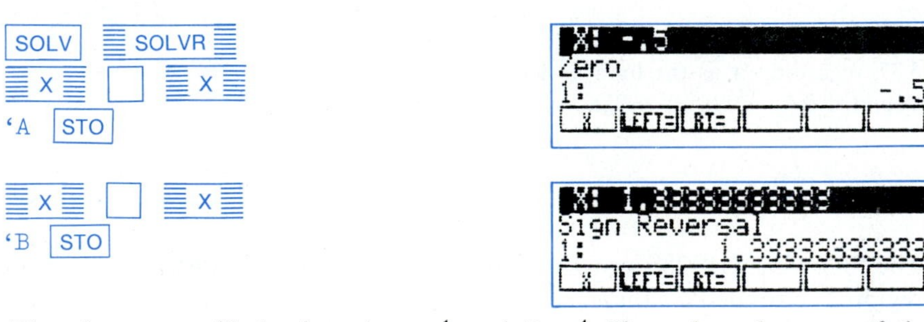

Thus it appears likely that $A = -\frac{1}{2}$ and $B = \frac{4}{3}$. If so, then the area of the region R is given by the integral

$$A = \int_{-1/2}^{4/3} [(4 - 2x) - (6x^2 - 7x)]\, dx.$$

When we press AREA to evaluate this integral, we get the result

$$A = 6.16203703703.$$

Inspection of the integral itself suggests that A should be an integral multiple of

$$\frac{1}{2^3 \cdot 3^3} = \frac{1}{216}.$$

Sure enough, multiplication by 216 yields $216A = 1331$, so our final calculator result is $A = 1331/216$. Anyone who wishes may verify this result by evaluating the integral analytically.

Exercises: Use an appropriate calculator or computer to solve some of Problems 11–42 of Section 5.7.

5.8 Numerical Integration

The fundamental theorem of calculus,

$$\int_a^b f(x)\,dx = \Big[G(x)\Big]_a^b,$$

can be used to evaluate the integral only if a convenient formula for the antiderivative G of f can be found. But there are simple functions with antiderivatives that are not elementary functions. An **elementary function** is one that can be expressed in terms of polynomial, trigonometric, exponential, and logarithmic functions by means of finite combinations of sums, differences, products, quotients, roots, and function composition.

The problem is that elementary functions can have nonelementary antiderivatives. For example, it is known that the elementary function $f(x) = (1 + x^2)^{1/3}$ has no elementary antiderivative. Consequently, we cannot use the fundamental theorem of calculus to evaluate an integral such as

$$\int_0^1 (1 + x^2)^{1/3}\,dx.$$

In this section we discuss the use of Riemann sums to numerically approximate integrals that cannot conveniently be evaluated exactly, whether or not nonelementary functions are involved. Given a continuous function f on $[a, b]$ with an integral to be approximated, consider a regular partition P of $[a, b]$ into n subintervals, each with length $\Delta x = (b - a)/n$. Then the value of any Riemann sum of the form

$$S = \sum_{i=1}^{n} f(x_i^*)\,\Delta x \tag{1}$$

may be taken as an approximation to the actual value of the integral $\int_a^b f(x)\,dx$.

With $x_i^* = x_{i-1}$ and with $x_i^* = x_i$ in (1), we get the *left-endpoint approximation* L_n and the *right-endpoint approximation* R_n to the definite integral $\int_a^b f(x)\,dx$, associated with the regular partition of $[a, b]$ into n equal subintervals. Thus

$$L_n = \sum_{i=1}^{n} f(x_{i-1})\,\Delta x \tag{2}$$

and

$$R_n = \sum_{i=1}^{n} f(x_i)\,\Delta x. \tag{3}$$

The notation for L_n and R_n may be simplified by writing y_i for $f(x_i)$.

Definition *Endpoint Approximations*

The **left-endpoint approximation** L_n and the **right-endpoint approximation** R_n to $\int_a^b f(x)\,dx$ with $\Delta x = (b - a)/n$ are

$$L_n = (\Delta x)(y_0 + y_1 + y_2 + \cdots + y_{n-1}) \tag{2}$$

and

$$R_n = (\Delta x)(y_1 + y_2 + y_3 + \cdots + y_n). \tag{3}$$

We illustrate the numerical techniques of this section by approximating the integral

$$\int_0^1 \frac{4}{1+x^2}\,dx. \tag{4}$$

In Chapter 8 we will study the inverse tangent function $y = \arctan x$ (y is the angle between $-\pi/2$ and $\pi/2$ such that $\tan y = x$), and we will show there that the derivative of $A(x) = \arctan x$ is

$$A'(x) = \frac{1}{1+x^2}.$$

This implies that

$$\int_0^1 \frac{1}{1+x^2}\,dx = \Big[\arctan x\Big]_0^1 = \arctan 1 - \arctan 0 = \frac{\pi}{4}.$$

Hence the true value of the integral in (4) is

$$\pi = 3.14159\;26\;5358\;9793\ldots.$$

This last fact will provide us with an independent check of the accuracy of our approximations.

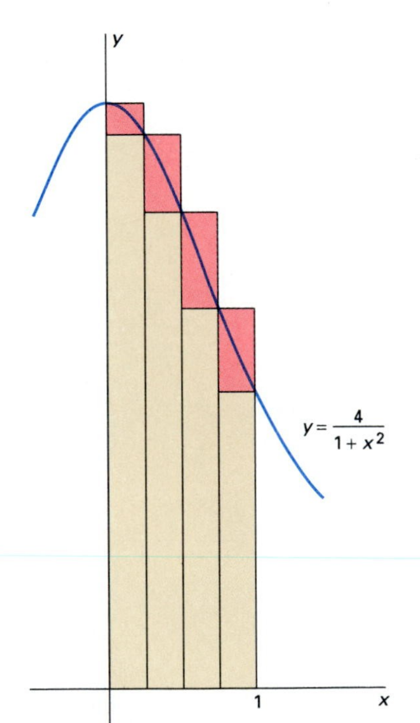

5.38 The left-hand and right-hand approximations to

$$\int_0^1 \frac{4}{1+x^2}\,dx$$

EXAMPLE 1 Calculate the left-endpoint and right-endpoint approximations to the integral in (4) with $n = 10$ and $\Delta x = 0.1$.

Solution Note first that $f(x) = 4/(1+x^2)$ is a decreasing function on $[0, 1]$. As indicated in Fig. 5.38, it follows that L_{10} is the area of a circumscribed polygon, while R_{10} is the area of an inscribed polygon. Hence L_{10} will be a high estimate of the integral and R_{10} a low estimate. We use a hand-held calculator and record only the final value of each approximation, because writing the individual terms and (subsequently) their sum would vitiate the eight- or ten-place internal accuracy of such a calculator. Because

$$x_0 = 0.0,\; x_1 = 0.1,\; x_2 = 0.2,\; \ldots,\; x_9 = 0.9, \quad \text{and} \quad x_{10} = 1.0,$$

we obtain

$$L_{10} = (0.1)\left(\frac{4}{1.00} + \frac{4}{1.01} + \frac{4}{1.04} + \cdots + \frac{4}{1.81}\right)$$
$$\approx 3.239925990 \approx 3.24$$

and

$$R_{10} = (0.1)\left(\frac{4}{1.01} + \frac{4}{1.04} + \frac{4}{1.09} + \cdots + \frac{4}{2.00}\right)$$
$$\approx 3.039925990 \approx 3.04.$$

Note that the average

$$\tfrac{1}{2}(L_{10} + R_{10}) \approx \tfrac{1}{2}(3.239925990 + 3.039925990)$$
$$= 3.139925990 \approx 3.14$$

is much closer to the true value of π than is either L_{10} or R_{10}. The technique of averaging upper and lower estimates plays an important role in numerical integration.

Now π lies between R_{10} and L_{10}, so these computations show that $\pi = 3.14 \pm 0.10$ if rounded to two decimal places (3.14 is the two-place average of R_{10} and L_{10}).

As Example 1 illustrates, the *average* of the endpoint sums L_n and R_n ordinarily is a considerably more accurate approximation to the integral than is either by itself. The average $T_n = \frac{1}{2}(L_n + R_n)$ of the left-endpoint and right-endpoint approximations is called the *trapezoidal approximation* to $\int_a^b f(x)\,dx$ associated with the partition of $[a, b]$ into n equal subintervals. Written in full, we see that

$$T_n = \tfrac{1}{2}(L_n + R_n) = \frac{\Delta x}{2} \sum_{i=1}^{n} [f(x_{i-1}) + f(x_i)];$$

that is,

$$T_n = \frac{\Delta x}{2} [f(x_0) + 2f(x_1) + 2f(x_2) + \cdots + 2f(x_{n-2}) + 2f(x_{n-1}) + f(x_n)].$$

Note the 1–2–2– · · · –2–2–1 pattern of the coefficients.

Definition *Trapezoidal Approximation*

The **trapezoidal approximation** to

$$\int_a^b f(x)\,dx \quad \text{with} \quad \Delta x = \frac{b-a}{n}$$

is

$$T_n = \frac{\Delta x}{2}(y_0 + 2y_1 + 2y_2 + \cdots + 2y_{n-2} + 2y_{n-1} + y_n). \tag{5}$$

Figure 5.39 shows where T_n gets its name. The partition points x_0, $x_1, x_2, \ldots, x_n$ are used to build trapezoids from the x-axis to the graph of the function. The trapezoid over the ith subinterval $[x_{i-1}, x_i]$ has altitude Δx and its parallel bases have lengths $f(x_{i-1})$ and $f(x_i)$. So its area is

$$\frac{\Delta x}{2} [f(x_{i-1}) + f(x_i)] = \frac{\Delta x}{2} (y_{i-1} + y_i).$$

Comparison of this with Equation (5) shows that T_n is merely the sum of the areas of the n trapezoids shown in Fig. 5.40.

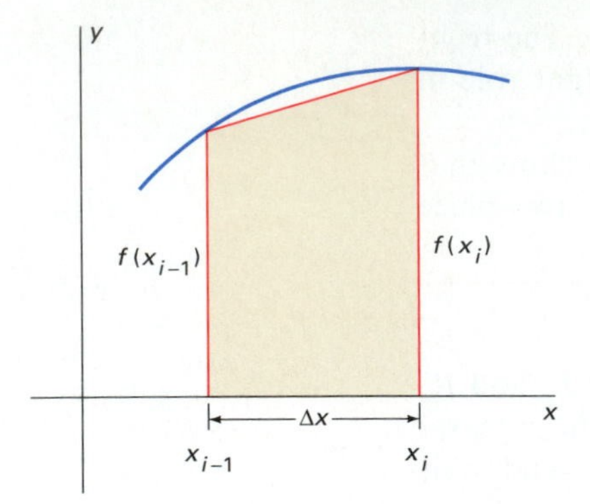

5.39 The area of the trapezoid is

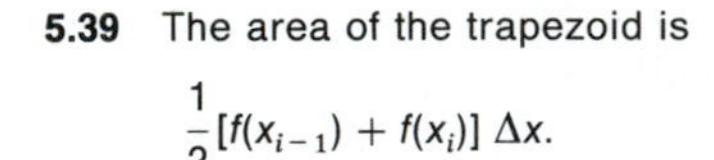

$\frac{1}{2}[f(x_{i-1}) + f(x_i)]\,\Delta x.$

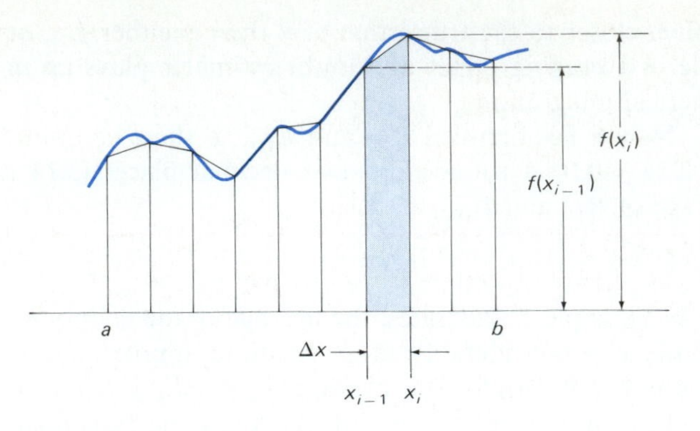

5.40 Geometry of the trapezoidal approximation

EXAMPLE 2 Calculate the trapezoidal approximation to the integral in (7) with $n = 10$ and $\Delta x = 0.1$.

Solution Because we have already calculated L_{10} and R_{10} in Example 1, we can simply write

$$T_{10} = \tfrac{1}{2}(L_{10} + R_{10}) \approx \tfrac{1}{2}(3.239952990 + 3.039925990),$$

so

$$T_{10} \approx 3.139925990.$$

Had L_{10} and R_{10} not been available, we would have written

$$T_{10} = \frac{0.1}{2}\left[1 \cdot \frac{4}{1.00} + 2 \cdot \frac{4}{1.01} + 2 \cdot \frac{4}{1.04} + \cdots + 2 \cdot \frac{4}{1.81} + 1 \cdot \frac{4}{2.00}\right]$$

$$\approx 3.139925990.$$

Another useful approximation to $\int_a^b f(x)\,dx$ is the *midpoint approximation* M_n. It is the Riemann sum obtained by choosing the point x_i^* in $[x_{i-1}, x_i]$ to be its midpoint $m_i = \frac{1}{2}(x_{i-1} + x_i)$. Thus

$$M_n = \sum_{i=1}^{n} f(m_i)\,\Delta x = (\Delta x)[f(m_1) + f(m_2) + \cdots + f(m_n)]. \tag{6}$$

Because m_1 is the midpoint of $[x_0, x_1]$, it is sometimes convenient to write $y_{1/2}$ for $f(m_1)$, $y_{3/2}$ for $f(m_2)$, and so on.

Definition *Midpoint Approximation*

The **midpoint approximation** to

$$\int_a^b f(x)\,dx$$

with $\Delta x = (b - a)/n$ is

$$M_n = (\Delta x)(y_{1/2} + y_{3/2} + y_{5/2} + \cdots + y_{n-1/2}). \tag{6}$$

EXAMPLE 3 Calculate the midpoint approximation to the integral in (4) with $n = 10$ and $\Delta x = 0.1$.

Solution We still have the integral

$$\int_0^1 \frac{4}{1 + x^2}\,dx$$

to be approximated. But now $m_1 = 0.05$, $m_2 = 0.15$, $m_3 = 0.25, \ldots,$ and $m_{10} = 0.95$. So

$$M_{10} = (0.1)\left[\frac{4}{1.0025} + \frac{4}{1.0225} + \frac{4}{1.0625} + \cdots + \frac{4}{1.9025}\right]$$

$$\approx 3.142425986 \approx 3.14.$$

Thus M_{10} gives π rounded to two decimal places.

The midpoint approximation in (6) is sometimes called the **tangent line** approximation, because the area of the rectangle with base $[x_{i-1}, x_i]$ and height $f(m_i)$ is also the area of another approximating figure. As shown in Fig. 5.41, we draw a tangent segment to the graph of f, tangent at the point $(m_i, f(m_i))$ of its graph, and use that segment for one side of a trapezoid somewhat like the method of the trapezoidal rule. The trapezoid and the rectangle mentioned above have the same area, and so the value of M_n is the sum of the areas of the trapezoids like the one in Fig. 5.41.

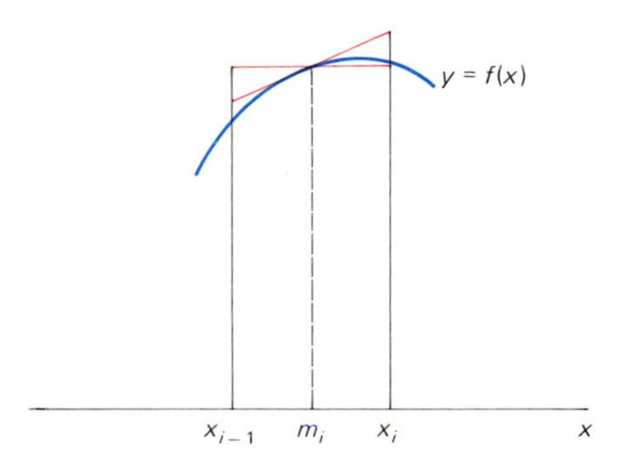

5.41 The midpoint or tangent rule

When we compare the results of Examples 2 and 3, we see that the midpoint approximation $M_{10} \approx 3.1424$ is somewhat closer to $\pi \approx 3.1416$ than is the trapezoidal approximation $T_{10} \approx 3.1399$. Indeed, the area of the trapezoid associated with the midpoint approximation is generally closer to the true value of

$$\int_{x_{i-1}}^{x_i} f(x)\,dx$$

than is the area of the trapezoid associated with the trapezoidal approximation. Figure 5.42 also shows this, in that the midpoint error E_m, shown in red in the figure, is generally smaller than the trapezoidal error E_t, shown in blue in the figure. Figure 5.42 also indicates that if $y = f(x)$ is concave downward, then M_n will be an overestimate and T_n will be an underestimate of $\int_a^b f(x)\,dx$. If the graph is concave upward, then the situation will be reversed.

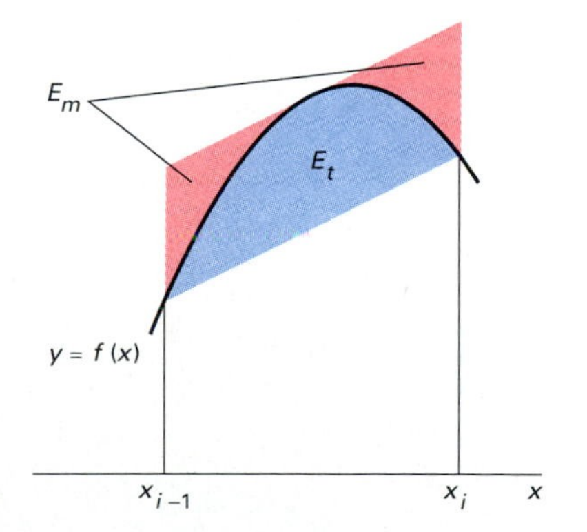

5.42 Comparison of the midpoint rule error E_m with the trapezoidal rule error E_t

Such observations motivate the consideration of a weighted average of M_n and T_n, with M_n weighted more heavily than T_n, for improving our numerical estimates of the definite integral. The particular weighted average

$$S_{2n} = \tfrac{1}{3}(2M_n + T_n) = \tfrac{2}{3}M_n + \tfrac{1}{3}T_n \tag{7}$$

is called *Simpson's approximation* to $\int_a^b f(x)\,dx$. The reason for the subscript $2n$ is that it's most natural to regard S_{2n} as being associated with a partition of $[a, b]$ into an *even* number $2n$ of equal subintervals with the endpoints

$$a = x_0 < x_1 < x_2 < \cdots < x_{2n-2} < x_{2n-1} < x_{2n} = b.$$

The midpoint and trapezoidal approximations associated with the n subintervals

$$[x_0, x_2],\ [x_2, x_4],\ [x_4, x_6],\ \ldots,\ [x_{2n-4}, x_{2n-2}],\ [x_{2n-2}, x_{2n}]$$

that all have the same length $2\,\Delta x$ can then be written in the form

$$M_n = (2\,\Delta x)(y_1 + y_3 + y_5 + \cdots + y_{2n-1})$$

and

$$T_n = \frac{2\,\Delta x}{2}(y_0 + 2y_2 + 2y_4 + \cdots + 2y_{2n-2} + y_{2n}).$$

We substitute these formulas for M_n and T_n in (7), and find—after a bit of algebra—that

$$S_{2n} = \frac{\Delta x}{3}(y_0 + 4y_1 + 2y_2 + 4y_3 + 2y_4 + \cdots + 2y_{2n-2} + 4y_{2n-1} + y_{2n}). \tag{8}$$

Definition *Simpson's Approximation*

Simpson's approximation to $\int_a^b f(x)\,dx$ with $\Delta x = (b - a)/2n$, associated with a partition of $[a, b]$ into an even number $2n$ of equal subintervals, is the sum S_{2n} given in Equation (8).

Note the 1–4–2–4–2–$\cdots$–4–2–4–1 pattern of coefficients in Simpson's approximation.

EXAMPLE 4 Calculate Simpson's approximation S_{10} to the integral

$$\int_0^1 \frac{4}{1 + x^2}\,dx$$

of Equation (4).

Solution Working with 10 subintervals of $[0, 1]$ and with $\Delta x = 0.1$, we obtain

$$S_{10} = \frac{0.1}{3}\left(1 \cdot \frac{4}{1.00} + 4 \cdot \frac{4}{1.01} + 2 \cdot \frac{4}{1.04} + 4 \cdot \frac{4}{1.09} + \cdots + 2 \cdot \frac{4}{1.64} + 4 \cdot \frac{4}{1.81} + 1 \cdot \frac{4}{2.00}\right);$$

then a calculator yields

$$S_{10} \approx 3.141592616.$$

Simpson's approximation correctly gives the first *seven* decimal places of π!

EXAMPLE 5 Calculate Simpson's approximation S_{20} to the integral of Example 4.

Solution Because we have already calculated T_{10} and M_{10} in Examples 2 and 3, we can use Equation (7) immediately:

$$S_{20} = \tfrac{2}{3}M_{10} + \tfrac{1}{3}T_{10} \approx \tfrac{2}{3}(3.142425986) + \tfrac{1}{3}(3.139925990)$$

so that

$$S_{20} \approx 3.141592654,$$

the correct value of π rounded to *nine* decimal places!

Although we have defined Simpson's approximation S_{2n} as a weighted average of the midpoint and trapezoidal approximations, it has an important interpretation in terms of **parabolic approximations** to the curve $y = f(x)$. Beginning with the partition of $[a, b]$ into $2n$ equal subintervals, we define the parabolic function

$$p_i(x) = A_i + B_i x + C_i x^2$$

on $[x_{2i-2}, x_{2i}]$. We choose the coefficients A_i, B_i, and C_i so that $p_i(x)$ agrees with $f(x)$ at the three points x_{2i-2}, x_{2i-1}, and x_{2i} (see Fig. 5.43). This can be done by solving the three equations

$$\begin{aligned} A_i + B_i x_{2i-2} + C_i(x_{2i-2})^2 &= f(x_{2i-2}), \\ A_i + B_i x_{2i-1} + C_i(x_{2i-1})^2 &= f(x_{2i-1}), \\ A_i + B_i x_{2i} + C_i(x_{2i})^2 &= f(x_{2i}) \end{aligned}$$

in the three unknowns A_i, B_i, and C_i. A routine (though tedious) algebraic computation—see Problem 47 in Section 5.7—then shows that

$$\int_{x_{2i-2}}^{x_{2i}} p_i(x)\,dx = \frac{\Delta x}{3}(y_{2i-2} + 4y_{2i-1} + y_{2i}).$$

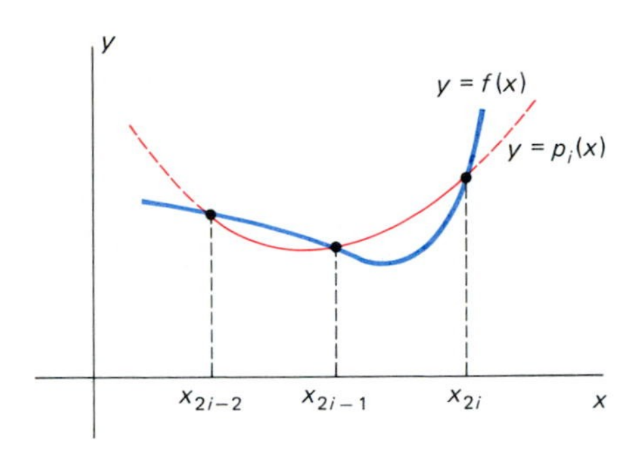

5.43 The parabolic approximation $y = p_i(x)$ to $y = f(x)$ on $[x_{2i-2}, x_{2i}]$

We now approximate $\int_a^b f(x)\,dx$ by replacing $f(x)$ by $p_i(x)$ on the interval $[x_{2i-2}, x_{2i}]$ for $i = 1, 2, 3, \ldots, n$. This gives

$$\begin{aligned} \int_a^b f(x)\,dx &= \sum_{i=1}^{n} \int_{x_{2i-2}}^{x_{2i}} f(x)\,dx \approx \sum_{i=1}^{n} \int_{x_{2i-2}}^{x_{2i}} p_i(x)\,dx \\ &= \sum_{i=1}^{n} \frac{\Delta x}{3}(y_{2i-2} + 4y_{2i-1} + y_{2i}) \\ &= \frac{\Delta x}{3}(y_0 + 4y_1 + 2y_2 + 4y_3 + \cdots \\ &\qquad + 4y_{2n-3} + 2y_{2n-2} + 4y_{2n-1} + y_{2n}). \end{aligned}$$

Thus the parabolic approximation described here results in Simpson's approximation S_{2n} to $\int_a^b f(x)\,dx$.

The numerical methods of this section are especially useful for approximating integrals of functions that are available only in graphical or in tabular form. This is often the case with functions derived from empirical data or from experimental measurements.

EXAMPLE 6 Suppose that the graph in Fig. 5.44 shows the velocity recorded by instruments on board a submarine traveling under the polar ice cap directly toward the North Pole. Use the trapezoidal approximation and Simpson's approximation to estimate the distance $s = \int_a^b v(t)\,dt$ traveled by the submarine during the 10-h period from $t = 0$ to $t = 10$.

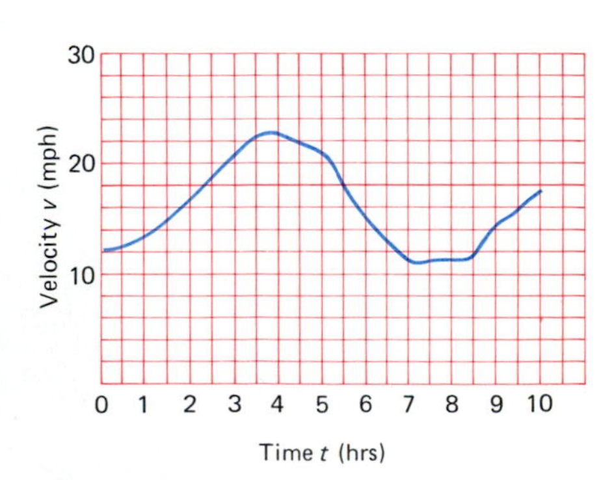

5.44 Velocity graph for the submarine of Example 6

Solution We read the following data from the graph.

t	0	1	2	3	4	5	6	7	8	9	10	h
v	13	14	17	21	23	21	15	11	11	14	17	mi/h

Using the trapezoidal approximation with $n = 10$ and $\Delta x = 1$, we obtain

$$s = \int_0^{10} v(t)\,dt \approx \tfrac{1}{2}[13 + (2)(14 + 17 + 21 + 23 + 21 + 15 + 11 + 11 + 14) + 17] = 162$$

miles. Using Simpson's approximation with $2n = 10$ and $\Delta x = 1$, we obtain

$$s = \int_0^{10} v(t)\,dt \approx \tfrac{1}{3}[13 + 4 \cdot 14 + 2 \cdot 17 + 4 \cdot 21 + 2 \cdot 23 + 4 \cdot 21 + 2 \cdot 15 + 4 \cdot 11 + 2 \cdot 11 + 4 \cdot 14 + 17] = 162$$

miles as an estimate of the distance traveled by the submarine during this 10-h period.

ERROR ESTIMATES

The trapezoidal and Simpson approximations are widely used for numerical integration, and there are *error estimates* that can be used to predict in advance the maximum possible error in a particular approximation. The trapezoidal error ET_n and the Simpson's error ES_n are defined by means of the equations

$$\int_a^b f(x)\,dx = T_n + ET_n \tag{9}$$

and

$$\int_a^b f(x)\,dx = S_n + ES_n \qquad (n \text{ even}). \tag{10}$$

Thus $|ET_n|$ is the numerical difference between the actual value of the integral and the trapezoidal approximation with n subintervals, while $|ES_n|$ is the numerical difference between the actual integral and Simpson's approximation.

The following two theorems are proved in numerical analysis textbooks.

Theorem 1 *Trapezoidal Error Estimate*
Suppose that the second derivative f'' is continuous on $[a, b]$ and that $|f''(x)| \leqq M_2$ for all x in $[a, b]$. Then

$$|ET_n| \leqq \frac{M_2(b-a)^3}{12n^2}. \tag{11}$$

Theorem 2 *Simpson's Error Estimate*
Suppose that the fourth derivative $f^{(4)}$ is continuous on $[a, b]$ and that $|f^{(4)}(x)| \leqq M_4$ for all x in $[a, b]$. If n is even, then

$$|ES_n| \leqq \frac{M_4(b-a)^5}{180n^4}. \tag{12}$$

EXAMPLE 7 In Chapter 7 we will see that the natural logarithm of the number 2 is the value of the integral

$$\ln 2 = \int_1^2 \frac{dx}{x}.$$

Estimate the errors in the trapezoidal and Simpson approximations to this integral using $n = 10$ subintervals. (Its actual value is approximately 0.693147.)

Solution With $f(x) = 1/x$ we calculate

$$f'(x) = -\frac{1}{x^2}, \qquad f'''(x) = -\frac{6}{x^4},$$

$$f''(x) = \frac{2}{x^3}, \qquad f^{(4)}(x) = \frac{24}{x^5}.$$

The maximum numerical values on $[1, 2]$ occur at $x = 1$, so we may take $M_2 = 2$ and $M_4 = 24$ in (11) and (12). From (11) we see that

$$|ET_{10}| \leqq \frac{2 \cdot 1^3}{12 \cdot 10^2} \approx 0.0016667, \tag{13}$$

so we would expect the trapezoidal approximation T_{10} to be accurate to at least two decimal places. From (12) we see that

$$|ES_{10}| \leqq \frac{24 \cdot 1^5}{180 \cdot 10^4} \approx 0.000013, \tag{14}$$

so we would expect Simpson's approximation to be accurate to at least four decimal places.

It turns out that $T_{10} \approx 0.693771$ and that $S_{10} \approx 0.691350$, so the actual errors in these approximations (in comparison with the actual value 0.693147 of the integral) are $ET_{10} \approx 0.000624$ and $ES_{10} \approx 0.000003$. Note that the actual errors are somewhat smaller than their estimates calculated in (13) and (14). It is fairly typical of numerical integration that the trapezoidal and Simpson's approximations computed in practice are somewhat more accurate than the "worst case" estimates provided in Theorems 1 and 2.

5.8 PROBLEMS

In Problems 1–6, first calculate the endpoint approximations L_n and R_n to the given integral. Use the indicated number of subintervals and round answers to two decimal places. Then calculate the average T_n of L_n and R_n and compare T_n with the exact value of the integral.

1. $\int_0^4 x\,dx, \quad n = 4$

2. $\int_1^2 x^2\,dx, \quad n = 5$

3. $\int_0^1 \sqrt{x}\,dx, \quad n = 5$

4. $\int_1^3 \frac{1}{x^2}\,dx, \quad n = 4$

5. $\int_0^{\pi/2} \cos x\,dx, \quad n = 3$

6. $\int_0^{\pi} \sin x\,dx, \quad n = 4$

7–12. Calculate the midpoint approximations to the integrals in Problems 1–6, respectively, using the indicated number of subintervals. In each case compare M_n with the exact value of the integral.

In Problems 13–20, calculate both the trapezoidal approximation T_n and Simpson's approximation S_n to the given interval. Use the indicated number of subintervals and round answers to four decimal places. In Problems 13–16, also compare these approximations with the exact value of the integral.

13. $\int_1^3 x^2\,dx, \quad n = 4$ **14.** $\int_1^4 x^3\,dx, \quad n = 4$

15. $\int_2^4 \frac{1}{x^3}\,dx, \quad n = 4$ **16.** $\int_0^1 \sqrt{1 + x}\,dx, \quad n = 4$

17. $\int_0^2 \sqrt{1 + x^3}\,dx, \quad n = 6$ **18.** $\int_0^3 \frac{1}{1 + x^4}\,dx, \quad n = 6$

19. $\int_1^5 \sqrt[3]{1 + x^2}\,dx, \quad n = 8$ **20.** $\int_0^1 \frac{\sin x}{x}\,dx, \quad n = 10$

(NOTE Make the integrand in Problem 20 continuous by assuming that its value at $x = 0$ is its limit there,

$$\lim_{x \to 0} \frac{\sin x}{x} = 1.)$$

In Problems 21 and 22, calculate (a) the trapezoidal approximation and (b) Simpson's approximation to

$$\int_a^b f(x)\,dx,$$

where f is the given tabulated function.

21.

x	$a = 1.00$	1.25	1.50	1.75	2.00	2.25	$2.50 = b$
$f(x)$	3.43	2.17	0.38	1.87	2.65	2.31	1.97

22.

x	$a = 0$	1	2	3	4	5	6	7	8	9	$10 = b$
$f(x)$	23	8	-4	12	35	47	53	50	39	29	5

23. The graph in Fig. 5.45 shows the measured rate of water flow (in gal/min) into a tank during a 10-min period. Using ten subintervals in each case, estimate the total amount of water flowing into the tank during this period using (a) the trapezoidal approximation and (b) Simpson's approximation.

24. The graph in Fig. 5.46 shows the daily mean temperatures recorded during a winter month at a certain location. Using ten subintervals in each case, estimate the average

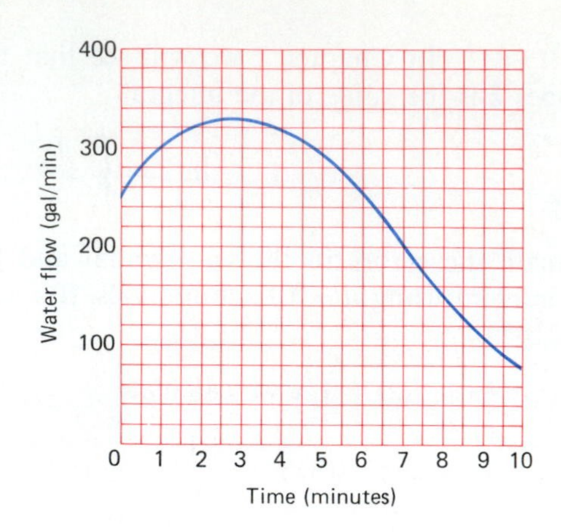

5.45 Water flow graph for Problem 23

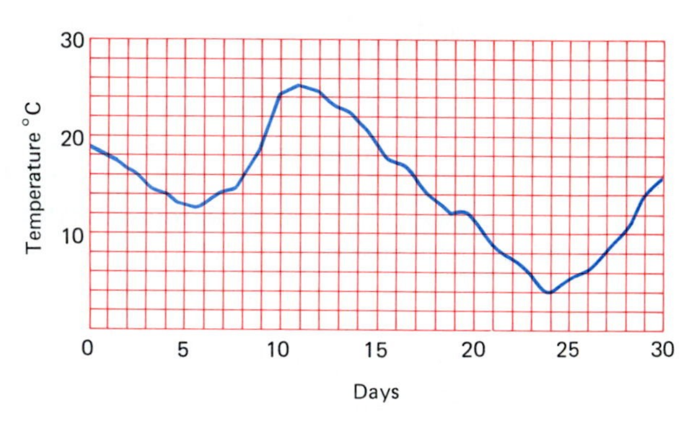

5.46 Temperature graph for Problem 24

temperature during the month using (a) the trapezoidal approximation and (b) Simpson's approximation.

25. Figure 5.47 shows a tract of land with measurements in feet. A surveyor has measured its width w at 50-ft intervals (the values of x shown below), with the following results.

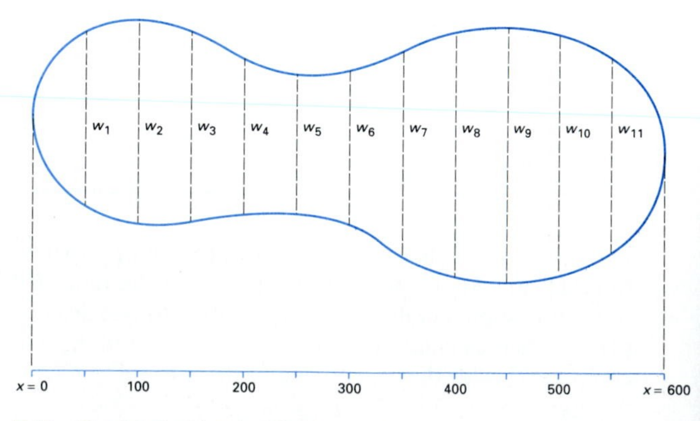

5.47 The tract of land of Problem 25

x	0	50	100	150	200	250	300
w	0	165	192	146	63	42	84

x	350	400	450	500	550	600
w	155	224	270	267	215	0

Use (a) the trapezoidal approximation and (b) Simpson's approximation to estimate the acreage of this tract. NOTE An acre is 4840 yd^2.

26. The base for natural logarithms is a certain number e. In Chapter 7 we will see that

$$\int_1^e \frac{1}{x}\,dx = 1.$$

Approximate the integrals

$$\int_1^{2.7} \frac{1}{x}\,dx \quad \text{and} \quad \int_1^{2.8} \frac{1}{x}\,dx$$

with sufficient accuracy to show that $2.7 < e < 2.8$.

Problems 27 and 28 deal with the integral

$$\ln 2 = \int_1^2 \frac{1}{x}\,dx$$

of Example 7.

27. Use the trapezoidal error estimate to determine how large n must be in order to guarantee that T_n differs from ln 2 by at most 0.0005.

28. Use the Simpson's error estimate to determine how large n must be in order to guarantee that S_n differs from ln 2 by at most 0.000005.

29. Deduce from the error estimate for Simpson's approximation the following: If $p(x)$ is a polynomial of degree at most 3, then Simpson's approximation with $n = 2$ subintervals gives the *exact* value of the integral $\int_a^b p(x)\,dx$.

30. Use the result of Problem 29 to calculate (without explicit integration) the area of the region shown in Fig. 5.48. (*Answer:* 1331/216.)

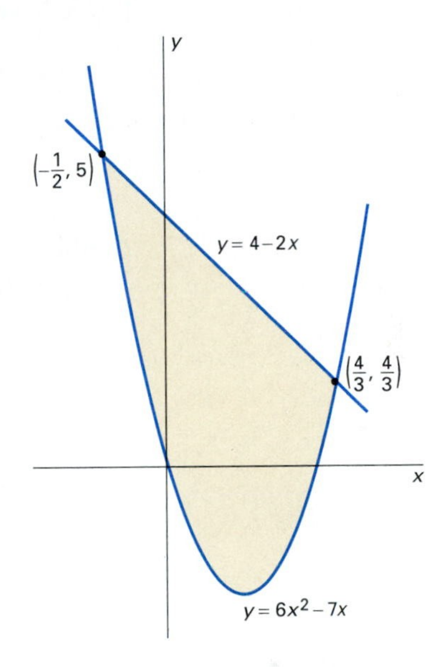

5.48 The region of Problem 30

*5.8 Optional Computer Application

Numerical integration is an area in which computer methods are obviously natural and extremely helpful. Figure 5.49 shows Program TRAPZOID, written to compute trapezoidal approximations to the integral

$$\ln 2 = \int_1^2 \frac{1}{x}\,dx \approx 0.693147$$

of Example 7.

The function $f(x) = 1/x$ is defined in line 140, and the endpoints $a = 1$ and $b = 2$ are prescribed in line 150. Line 160 calls for the desired number n of subintervals to be entered. The calculation of the sum

$$S = f(a) + 2f(x_1) + 2f(x_2) + \cdots + 2f(x_{n-1}) + f(b)$$

is begun in line 180 with the two endpoint terms. The interior point terms are added in the FOR-NEXT loop of lines 210–240, and finally S is multiplied by $\Delta x/2$ in line 260.

```
100 REM--Program TRAPZOID
110 REM--Computes trapezoidal approximation to the
120 REM--integral of  f(x)  from  x = a  to  x = b.
130 REM
140      DEF FNF(X) = 1/X          'The function  f(x)
150      A = 1  :  B = 2           'Limits of integration
155 REM
160      INPUT "Number of subintervals"; N
170      H = (B - A)/N             'Delta x
180      S = FNF(A) + FNF(B)       'Start with the endpts
190      X = A + H                 'First interior point
200 REM
210      FOR I = 1 TO N-1
220          S = S + 2*FNF(X)      'Add  2  times  f(x)
230          X = X + H             'Next interior point
240      NEXT I
250 REM
260      S = (H/2)*S               'Multiply by  h/2
270      PRINT "Sum = "; S         'Display result
280      END
```

5.49 Program TRAPZOID

It is common practice to compute several approximations successively, doubling the value of n for each new approximation. In this way Program TRAPZOID produces the following results.

```
RUN
Number of subintervals? 10
Sum =  .6937714

RUN
Number of subintervals? 20
Sum =  .6933036

RUN
Number of subintervals? 40
Sum =  .6931863
```

Thus we find that $\ln 2 \approx 0.693$, the correct value rounded to three decimal places.

Only slight alterations are needed to convert Program TRAPZOID into Program SIMPSON, which is listed in Fig. 5.50. Note how the

$$1\text{–}4\text{–}2\text{–}4\text{–}\cdots\text{–}2\text{–}4\text{–}1$$

pattern of coefficients is generated in lines 215–220.

Using Program SIMPSON we get these results.

```
RUN
EVEN number of subintervals? 10
Sum =  .6931503

RUN
EVEN number of subintervals? 20
Sum =  .6931476

RUN
EVEN number of subintervals? 40
Sum =  .6931475
```

The results are quite consistent with the correct six-place value of ln 2.

```
100 REM--Program SIMPSON
110 REM--Computes Simpson's approximation to the
120 REM--integral of  f(x)  from  x = a  to  x = b.
130 REM
140      DEF FNF(X) = 1/X          'The function  f(x)
150      A = 1  :  B = 2           'Limits of integration
155 REM
160      INPUT "EVEN number of subintervals"; N
170      H = (B - A)/N             'Delta x
180      S = FNF(A) + FNF(B)       'Start with the endpts
190      X = A + H                 'First interior point
195      E = 1                     'Alternating sign
200 REM
210      FOR I = 1 TO N-1
215      K = 3 + E : E = -E        '4-2-4---2-4 pattern
220         S = S + K*FNF(X)       'Add next term
230         X = X + H              'Next interior point
240      NEXT I
250 REM
260      S = (H/3)*S               'Multiply by  h/3
270      PRINT "Sum = "; S         'Display result
280      END
```

5.50 Program SIMPSON

Exercises 1–4: Apply Programs TRAPZOID and SIMPSON to the integrals of Problems 17–20 of this section. Keep increasing the number of subintervals until you are confident of the result accurate to five decimal places.

Exercise 5: Add the line

```
135   DEFDBL  A, B, F, H, S, X:  REM  Double precision
```

to Program SIMPSON, then use the integral

$$\pi = \int_0^1 \frac{4}{1 + x^2}\,dx$$

to approximate the number π accurate to as many digits as possible with your computer.

CHAPTER 5 REVIEW: Definitions, Concepts, Results

Use the list below as a guide to concepts that you may need to review.

1. Properties of area
2. Summation notation
3. The area under the graph of f from a to b
4. Inscribed and circumscribed rectangular polygons
5. A partition of $[a, b]$
6. The mesh of a partition
7. A Riemann sum associated with a partition
8. The (definite) integral of f from a to b
9. Existence of the integral of a continuous function
10. The integral as the limit of a sequence of Riemann sums
11. Regular partitions and
$$\lim_{\Delta x \to 0} \sum_{i=1}^{n} f(x_i^*)\,\Delta x$$
12. Upper and lower Riemann sums
13. The constant multiple, interval union, and comparison properties of integrals
14. Evaluation of definite integrals using antiderivatives
15. The average value of $f(x)$ on the interval $[a, b]$
16. The average value theorem
17. The fundamental theorem of calculus
18. Linearity of integration
19. Indefinite integrals
20. The method of integration by substitution

21. Transforming the limits in integration by substitution

22. The area between $y = f(x)$ and $y = g(x)$ by integration with respect to x

23. The area between $x = f(y)$ and $x = g(y)$ by integration with respect to y

24. The right-endpoint and left-endpoint approximations

25. The trapezoidal approximation

26. The midpoint approximation

27. Simpson's approximation

28. Error estimates for the trapezoidal and Simpson's approximations

CHAPTER 5 MISCELLANEOUS PROBLEMS

Find the sums in Problems 1–4.

1. $\sum_{i=1}^{100} 17$

2. $\sum_{k=1}^{100} \left(\frac{1}{k} - \frac{1}{k+1}\right)$

3. $\sum_{n=1}^{10} (3n-2)^2$

4. $\sum_{n=1}^{16} \sin \frac{n\pi}{2}$

In each of Problems 5–7, find the limit of the given Riemann sum associated with a regular partition of the indicated interval $[a, b]$. First express it as an integral from a to b, then evaluate that integral.

5. $\lim_{n\to\infty} \sum_{i=1}^{n} \frac{\Delta x}{\sqrt{x_i^*}}$; $[1, 2]$

6. $\lim_{n\to\infty} \sum_{i=1}^{n} [(x_i^*)^2 - 3x_i^*]\,\Delta x$; $[0, 3]$

7. $\lim_{n\to\infty} \sum_{i=1}^{n} 2\pi x_i^* \sqrt{1 + (x_i^*)^2}\,\Delta x$; $[0, 1]$

8. Evaluate

$$\lim_{n\to\infty} \frac{1}{n^{11}} (1^{10} + 2^{10} + 3^{10} + \cdots + n^{10})$$

by expressing this limit as an integral over $[0, 1]$.

9. Use Riemann sums to prove that if $f(x) \equiv c$ (a constant), then

$$\int_a^b f(x)\,dx = c(b-a).$$

10. Use Riemann sums to prove that if f is continuous on $[a, b]$ and $f(x) \geqq 0$ for all x in $[a, b]$, then

$$\int_a^b f(x)\,dx \geqq 0.$$

11. Use the comparison property of integrals (Section 5.4) to prove that

$$\int_a^b f(x)\,dx > 0$$

if f is a continuous function with $f(x) > 0$ on $[a, b]$.

Evaluate the integrals in Problems 12–25.

12. $\int_0^1 (1-x^2)^3\,dx$

13. $\int \left(\sqrt{2x} - \frac{1}{\sqrt{3x^3}}\right) dx$

14. $\int \frac{(1 - \sqrt[3]{x})^2}{\sqrt{x}}\,dx$

15. $\int \frac{4 - x^3}{2x^2}\,dx$

16. $\int_0^1 \frac{dt}{(3-2t)^2}$

17. $\int \sqrt{x} \cos x\sqrt{x}\,dx$

18. $\int_0^2 x^2\sqrt{9 - x^3}\,dx$

19. $\int \frac{1}{t^2} \sin \frac{1}{t}\,dt$

20. $\int_1^2 \frac{2t+1}{\sqrt{t^2+t}}\,dt$

21. $\int \frac{\sqrt[3]{u}}{(1 + u^{4/3})^3}\,du$

22. $\int_0^{\pi/4} \frac{\sin t}{\sqrt{\cos t}}\,dt$

23. $\int_1^4 \frac{(1+\sqrt{t})^2}{\sqrt{t}}\,dt$

24. $\int \frac{\sqrt[3]{1 - (1/u)}}{u^2}\,du$

25. $\int \frac{\sqrt{4x^2 - 1}}{x^4}\,dx$

Find the areas of the regions bounded by the curves given in Problems 26–32.

26. $y = x^3$, $x = -1$, $y = 1$

27. $y = x^4$ and $y = x^5$

28. $y^2 = x$ and $3y^2 = x + 6$

29. $y = x^4$ and $y = 2 - x^2$

30. $y = x^4$ and $y = 2x^2 - 1$

31. $y = (x-2)^2$ and $y = 10 - 5x$

32. $y = x^{2/3}$ and $y = 2 - x^2$

33. Evaluate the integral

$$\int_0^2 \sqrt{2x - x^2}\,dx$$

by interpreting it as the area of a region.

34. Evaluate the integral

$$\int_1^5 \sqrt{6x - 5 - x^2}\,dx$$

by interpreting it as the area of a region.

35. Find a function f such that

$$x^2 = 1 + \int_1^x \sqrt{1 + [f(t)]^2}\,dt$$

for all $x > 1$. (*Suggestion:* Differentiate both sides of the equation with the aid of the fundamental theorem of calculus.)

36. Show that $G'(x) = \phi(h(x))h'(x)$ if

$$G(x) = \int_a^{h(x)} \phi(t)\,dt.$$

37. Use right-endpoint and left-endpoint approximations to estimate

$$\int_0^1 \sqrt{1 + x^2}\,dx$$

with error not exceeding 0.05.

38. Calculate the trapezoidal and Simpson's approximations to

$$\int_0^{\pi} \sqrt{1 - \cos x}\,dx$$

with six subintervals. For comparison, use an appropriate half-angle identity to calculate the exact value of this integral.

39. Calculate the midpoint and trapezoidal approximations to

$$\int_1^2 \frac{dx}{x + x^2}$$

with $n = 5$ subintervals. Explain why the exact value of the integral lies between these two approximations.

In Problems 40–42, let $\{x_0, x_1, x_2, \ldots, x_n\}$ be a partition of $[a, b]$ where $0 < a < b$.

40. For $i = 1, 2, 3, \ldots, n$, let x_i^* be given by

$$(x_i^*)^2 = \tfrac{1}{3}[(x_{i-1})^2 + x_{i-1}x_i + (x_i)^2].$$

Show first that $x_{i-1} < x_i^* < x_i$. Then use the algebraic identity

$$(c - d)(c^2 + cd + d^2) = c^3 - d^3$$

to show that

$$\sum_{i=1}^{n} (x_i^*)^2\,\Delta x_i = \tfrac{1}{3}(b^3 - a^3).$$

Explain why this computation proves that

$$\int_a^b x^2\,dx = \tfrac{1}{3}(b^3 - a^3).$$

41. Let $x_i^* = \sqrt{x_{i-1}x_i}$ for $i = 1, 2, 3, \ldots, n$. Show that

$$\sum_{i=1}^{n} \frac{\Delta x_i}{(x_i^*)^2} = \frac{1}{a} - \frac{1}{b}.$$

Then explain why this computation proves that

$$\int_a^b \frac{dx}{x^2} = \frac{1}{a} - \frac{1}{b}.$$

42. Define x_i^* by means of the equation

$$(x_i^*)^{1/2}(x_i - x_{i-1}) = \tfrac{2}{3}[(x_i)^{3/2} - (x_{i-1})^{3/2}].$$

Show that $x_{i-1} < x_i^* < x_i$, then use this selection for the given partition to prove that

$$\int_a^b \sqrt{x}\,dx = \tfrac{2}{3}(b^{3/2} - a^{3/2}).$$

43. In Section 5.8 we derived Equation (8) for Simpson's approximation by giving the midpoint approximation "weight" $\frac{2}{3}$ and the trapezoidal approximation "weight" $\frac{1}{3}$. Why not some *other* weights? Read the discussion of parabolic approximations following Example 5 of Section 5.8, then work Problem 44 of Section 5.7. This should provide most of the answer.

chapter six

Applications of the Integral

6.1 Setting Up Integral Formulas

In Section 5.3 we defined the integral of the function f from a to b as a limit of Riemann sums. Specifically, one begins with a partition P of the interval $[a, b]$. A selection of numbers x_i^* in each subinterval of P produces a Riemann sum associated with the partition P; this is a number of the form

$$\sum_{i=1}^{n} f(x_i^*)\,\Delta x_i.$$

Finally, the integral of f on $[a, b]$ is defined to be the limit of such sums as the mesh $|P|$ of P approaches zero. That is,

$$\int_a^b f(x)\,dx = \lim_{|P|\to 0} \sum_{i=1}^{n} f(x_i^*)\,\Delta x_i. \tag{1}$$

The wide applicability of the definite integral arises from the fact that many geometric and physical quantities can be approximated arbitrarily closely by Riemann sums. Such approximations lead to integral formulas for the computation of such quantities.

For example, our discussion in Sections 5.2 and 5.7 of the area from a to b under the graph of the positive-valued continuous function f can be summarized as follows. Begin with a regular partition of $[a, b]$ into n subintervals, each with length $\Delta x = (b - a)/n$. For each i ($i = 1, 2, 3, \ldots, n$), select a point x_i^* in the ith subinterval $[x_{i-1}, x_i]$. Let ΔA_i denote the area below the graph of f over this subinterval. Then this area is given approximately by

$$\Delta A_i \approx f(x_i^*)\,\Delta x.$$

Figure 6.1 shows why this approximation should be good and why it should improve as the mesh of the partition approaches zero.

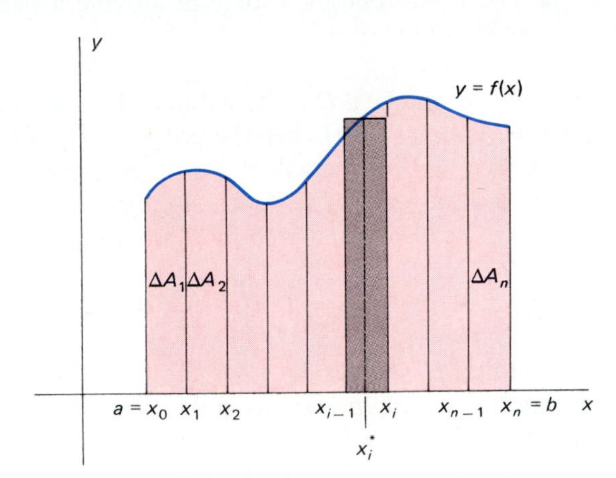

6.1 Approximating an area by means of a Riemann sum

It follows that the total area A under the graph of f is given approximately by

$$A = \sum_{i=1}^{n} \Delta A_i \approx \sum_{i=1}^{n} f(x_i^*)\,\Delta x. \tag{2}$$

The point is that the approximating sum on the right is a Riemann sum for f on $[a, b]$. Moreover:

1. It is intuitively evident that the Riemann sum in (2) approaches the actual area A as $n \to +\infty$.
2. By the definition of the integral, this Riemann sum approaches $\int_a^b f(x)\,dx$ as $n \to +\infty$.

These observations justify the *definition* of the area A by means of the formula

$$A = \int_a^b f(x)\,dx. \tag{3}$$

This justification of the area formula in (3) illustrates the following general method of setting up integral formulas. Suppose that we want to compute a certain quantity Q, where Q is associated with the interval $[a, b]$ in such a way that subintervals of $[a, b]$ correspond to specific portions of Q. Also assume that if $\Delta Q_1, \Delta Q_2, \ldots, \Delta Q_n$ are the portions of Q corresponding to the subintervals of a partition of $[a, b]$ into n subintervals, then

$$Q = \sum_{i=1}^{n} \Delta Q_i. \tag{4}$$

Such assumptions hold if, for example, Q is:

- An *area* lying over the interval $[a, b]$;
- The *distance* traveled by a moving particle during the time interval $a \leqq t \leqq b$;
- The *volume* of water flowing into a tank during the time interval $a \leqq t \leqq b$; or
- The *work* done by a force in moving a particle from the point $x = a$ to the point $x = b$.

For example, if Q is the volume of water that flows into a tank during the time interval $[a, b]$, then the portion ΔQ_i of Q corresponding to the subinterval $[t_{i-1}, t_i]$ is the volume of water that flows into the tank during this subinterval of time. In any such situation it will generally turn out that the quantity Q can be computed by evaluating a certain integral,

$$Q = \int_a^b f(x)\,dx \quad \text{or} \quad Q = \int_a^b f(t)\,dt, \tag{5}$$

depending on whether x or t is the independent variable. The key step in setting up such an integral formula is finding the particular function f to be integrated.

The usual way of finding f is to approximate the portion ΔQ_i of the quantity Q that corresponds to the subinterval $[x_{i-1}, x_i]$. We are still thinking of a regular partition of $[a, b]$ into n equal subintervals of length Δx, or into subintervals $[t_{i-1}, t_i]$ of length Δt in case t is the independent variable. Suppose we find that

$$\Delta Q_i \approx f(x_i^*)\,\Delta x \tag{6}$$

for some point x_i^* in $[x_{i-1}, x_i]$. What follows now is the approximation

$$Q = \sum_{i=1}^{n} \Delta Q_i \approx \sum_{i=1}^{n} f(x_i^*)\,\Delta x. \tag{7}$$

The right-hand sum in (7) is a Riemann sum that approaches the integral

$$\int_a^b f(x)\,dx \quad \text{as} \quad n \to +\infty.$$

If it is also evident—for geometric or physical reasons, for example—that this Riemann sum must approach the quantity Q as $n \to +\infty$, then (7) justifies our setting up the integral formula

$$Q = \int_a^b f(x)\,dx. \tag{5}$$

The crucial step in this process is the determination of the particular function f in the approximation in (6):

$$\Delta Q_i \approx f(x_i^*)\,\Delta x.$$

If Q is the area under a curve, then (as we have seen) $f(x)$ should be the height of the curve. If Q is the distance traveled by a moving particle, then (as we soon shall see) $f(t)$ should be its velocity function. If Q is the volume of water that flows into a tank during a specified time interval, then $f(t)$

should be the rate of water flow (in appropriate units, such as liters per second). This chapter is devoted largely to ways of setting up integral formulas like (5) in a variety of specific situations. The applications that we consider include computations of distances, volumes, work, lengths of curves, and areas of surfaces of revolution.

Observe that the integral in (5) results from the summation in (7) when we make the following replacements:

$$\sum_{i=1}^{n} \quad \text{becomes} \quad \int_a^b,$$

$$x_i^* \quad \text{becomes} \quad x,$$

$$\Delta x \quad \text{becomes} \quad dx.$$

EXAMPLE 1 Suppose that water is pumped into an initially empty tank. It is known that the rate of flow of water into the tank at time t (in seconds) is $50 - t$ liters per second. How much water flows into the tank during the first 30 s?

Solution We want to compute the amount Q of water that flows into the tank during the time interval $[0, 30]$. Think of a regular partition of $[0, 30]$ into n equal subintervals each of length $\Delta t = 30/n$.

Next choose a point t_i^* in the ith subinterval $[t_{i-1}, t_i]$. The rate of water flow between time t_{i-1} and time t_i is approximately $50 - t_i^*$ liters per second, so the amount ΔQ_i of water that flows into the tank during this subinterval of time is given approximately by

$$\Delta Q_i \approx (50 - t_i^*)\,\Delta t$$

(in liters). Therefore, the total amount Q we seek is given approximately by

$$Q = \sum_{i=1}^{n} \Delta Q_i \approx \sum_{i=1}^{n} (50 - t_i^*)\,\Delta t$$

(still in liters). We recognize the Riemann sum on the right, and—most important—we see that the function involved in this sum is $f(t) = 50 - t$. Hence we may conclude that

$$Q = \lim_{n\to\infty} \sum_{i=1}^{n} (50 - t_i^*)\,\Delta t = \int_0^{30} (50 - t)\,dt = \left[50t - \tfrac{1}{2}t^2\right]_0^{30}$$

$$= 1050 \quad \text{(liters)}.$$

EXAMPLE 2 Calculate Q if

$$Q = \lim_{n\to\infty} \sum_{i=1}^{n} m_i\sqrt{25 - (m_i)^2}\,\Delta x,$$

where (for each $i = 1, 2, \dots, n$) m_i denotes the midpoint of the ith subinterval $[x_{i-1}, x_i]$ of a partition of the interval $[3, 4]$ into n equal subintervals each of length $\Delta x = 1/n$.

Solution We recognize the given sum as a Riemann sum (with $x_i^* = m_i$) for the function $f(x) = x\sqrt{25 - x^2}$. Hence

$$Q = \int_3^4 x\sqrt{25 - x^2}\,dx = \left[-\tfrac{1}{3}(25 - x^2)^{3/2}\right]_3^4 = \tfrac{1}{3}(16^{3/2} - 9^{3/2}) = \tfrac{37}{3}.$$

DISTANCE AND VELOCITY

Consider a particle traveling along a (directed) straight line with velocity $v = f(t)$ at time t. We want to compute the **net distance** s it travels between time $t = a$ and time $t = b$; that is, the distance between its initial position at time $t = a$ and its final position at time $t = b$.

We begin with a regular partition of $[a, b]$ into n equal subintervals, each of length $\Delta t = (b - a)/n$. If t_i^* is an arbitrary point of the ith subinterval $[t_{i-1}, t_i]$, then during this subinterval the velocity of the particle is approximately $f(t_i^*)$. Hence the distance Δx_i traveled by the particle between time $t = t_{i-1}$ and time $t = t_i$ is given approximately by

$$\Delta s_i \approx f(t_i^*)\,\Delta t.$$

Consequently, the net distance s is given approximately by

$$s = \sum_{i=1}^{n} \Delta s_i \approx \sum_{i=1}^{n} f(t_i^*)\,\Delta t. \tag{8}$$

We recognize the right-hand approximation as a Riemann sum for the integral $\int_a^b f(t)\,dt$, and we reason that this approximation should approach the actual net distance as $n \to +\infty$. Thus we conclude that

$$s = \int_a^b f(t)\,dt = \int_a^b v\,dt. \tag{9}$$

Note that the *net distance is the definite integral of the velocity.*

If the velocity function $v = f(t)$ has both positive and negative values for t in $[a, b]$, then forward and backward distances cancel when we compute the net distance s given by the integral formula in (9). The **total distance** traveled, irrespective of direction, may be found by evaluating the integral of the *absolute value* of the velocity. Thus the total distance is given by

$$\int_a^b |f(t)|\,dt = \int_a^b |v|\,dt. \tag{10}$$

This integral may be computed by integrating separately over the subintervals where v is positive and those where v is negative, then adding the absolute values of the results. Note that this is exactly the same procedure used to find the area between the graph of a function and the x-axis when the function takes on both positive and negative values.

EXAMPLE 3 Suppose that the velocity of a moving particle is $v = 30 - 2t$ ft/s. Find both the net distance and the total distance it travels between the times $t = 0$ and $t = 20$ s.

Solution For net distance, we use the formula in Equation (9) and find that

$$s = \int_0^{20} (30 - 2t)\,dt = \Big[30t - t^2\Big]_0^{20} = 200$$

(ft). To find the total distance traveled, we note that v is positive if $t < 15$, negative if $t > 15$. Because

$$\int_0^{15} (30 - 2t)\,dt = \Big[30t - t^2\Big]_0^{15} = 225 \quad \text{(ft)}$$

and

$$\int_{15}^{20} (2t - 30)\, dt = \Big[t^2 - 30t\Big]_{15}^{20} = 25 \quad \text{(ft)},$$

we see that the particle travels 225 ft forward, then 25 ft backward, for a total distance traveled of 250 ft.

6.1 PROBLEMS

In each of Problems 1–10, compute both the net distance and the total distance traveled between time $t = a$ and $t = b$ by a particle moving along a line with the given velocity function $v = f(t)$.

1. $v = -32; \quad a = 0, \quad b = 10$

2. $v = 2t + 10; \quad a = 1, \quad b = 5$

3. $v = 4t - 25; \quad a = 0, \quad b = 10$

4. $v = |2t - 5|; \quad a = 0, \quad b = 5$

5. $v = 4t^3; \quad a = -2, \quad b = 3$

6. $v = t - \dfrac{1}{t^2}; \quad a = 0.1, \quad b = 1$

7. $v = \sin 2t; \quad a = 0, \quad b = \pi/2$

8. $v = \cos 2t; \quad a = 0, \quad b = \pi/2$

9. $v = \cos \pi t; \quad a = -1, \quad b = 1$

10. $v = \sin t + \cos t; \quad a = 0, \quad b = \pi$

In each of Problems 11–19, x_i^* denotes an arbitrary point, and m_i the midpoint, of the ith subinterval $[x_{i-1}, x_i]$ of a regular partition of the indicated interval $[a, b]$ into n equal subintervals of length Δx. Evaluate the given limit by computing the value of the appropriate related integral.

11. $\lim_{n\to\infty} \sum_{i=1}^{n} 2x_i^* \, \Delta x; \quad a = 0, \quad b = 1$

12. $\lim_{n\to\infty} \sum_{i=1}^{n} \dfrac{\Delta x}{(x_i^*)^2}; \quad a = 1, \quad b = 2$

13. $\lim_{n\to\infty} \sum_{i=1}^{n} (\sin \pi x_i^*)\, \Delta x; \quad a = 0, \quad b = 1$

14. $\lim_{n\to\infty} \sum_{i=1}^{n} [3(x_i^*)^2 - 1]\, \Delta x; \quad a = -1, \quad b = 3$

15. $\lim_{n\to\infty} \sum_{i=1}^{n} x_i^* \sqrt{(x_i^*)^2 + 9}\, \Delta x; \quad a = 0, \quad b = 4$

16. $\lim_{n\to\infty} \sum_{i=1}^{n} (x_i)^2\, \Delta x; \quad a = 2, \quad b = 4$

17. $\lim_{n\to\infty} \sum_{i=1}^{n} (2m_i - 1)\, \Delta x; \quad a = -1, \quad b = 3$

18. $\lim_{n\to\infty} \sum_{i=1}^{n} \sqrt{2m_i + 1}\, \Delta x; \quad a = 0, \quad b = 4$

19. $\lim_{n\to\infty} \sum_{i=1}^{n} \dfrac{m_i}{\sqrt{(m_i)^2 + 16}}\, \Delta x; \quad a = -3, \quad b = 0$

In Problems 20–23 the notation is the same as in Problems 11–19. Express the given limit as an integral involving the function f.

20. $\lim_{n\to\infty} \sum_{i=1}^{n} 2\pi x_i^* f(x_i^*)\, \Delta x; \quad a = 1, \quad b = 4$

21. $\lim_{n\to\infty} \sum_{i=1}^{n} [f(x_i^*)]^2\, \Delta x; \quad a = -1, \quad b = 1$

22. $\lim_{n\to\infty} \sum_{i=1}^{n} \sqrt{1 + [f(x_i^*)]^2}\, \Delta x; \quad a = 0, \quad b = 10$

23. $\lim_{n\to\infty} \sum_{i=1}^{n} 2\pi m_i \sqrt{1 + [f(m_i)]^2}\, \Delta x; \quad a = -2, \quad b = 3$

24. If a particle is thrown straight upward from the ground with an initial velocity of 160 ft/s, then its velocity after t seconds is $v = -32t + 160$ ft/s, and it attains its maximum height when $t = 5$ seconds (and $v = 0$). Use the formula in (9) to compute this maximum height. Check your answer by the methods of Section 4.9.

25. Suppose that the rate of flow of water into an initially empty tank is $100 - 3t$ gal/min at time t (in minutes). How much water flows into the tank during the interval from $t = 10$ to $t = 20$ min?

26. Suppose that the birth rate in a certain city t years after 1960 was $13 + t$ thousands of births per year. Set up and evaluate an appropriate integral to compute the total number of births that occurred between 1960 and 1980.

27. Assume that the city of Problem 26 had a death rate of $5 + t/2$ thousands per year t years after 1960. If the population of the city was 125,000 in 1960, what was its population in 1980? Consider both births and deaths in this problem.

28. The average daily rainfall in a certain locale is $r(t)$ inches per day at time t (days), $0 \leqq t \leqq 365$. Begin with a regular partition of the interval $[0, 365]$ and derive the formula

$$R = \int_0^{365} r(t)\, dt$$

for the average total annual rainfall R.

29. Suppose, in connection with Problem 28, that

$$r(t) = a - b \cos \frac{2\pi t}{365}$$

where a and b are constants to be determined. If the average daily rainfall on January 1 ($t = 0$) is 0.1 in. and the average daily rainfall on July 1 ($t = 182.5$) is 0.5 in., what is the average total annual rainfall in this locale?

30. Suppose that the rate of flow of water into a tank is $r(t)$ gallons per minute at time t. Use the method of Example 1 to derive the formula

$$Q = \int_a^b r(t)\,dt$$

for the amount of water that flows into the tank between times $t = a$ and $t = b$.

31. Evaluate

$$\lim_{n\to\infty} \frac{\sqrt[3]{1} + \sqrt[3]{2} + \sqrt[3]{3} + \cdots + \sqrt[3]{n}}{n^{4/3}}$$

by first finding a function f such that the limit is equal to

$$\int_0^1 f(x)\,dx.$$

32. A spherical ball has radius 1 ft and, at distance r from its center, its density is $100(1 + r)$ pounds per cubic foot. Use Riemann sums to find a function $f(r)$ such that the weight of the ball is

$$W = \int_0^1 f(r)\,dr$$

(in pounds). Then compute W by evaluating this integral. (*Suggestion:* The surface area of a sphere of radius r is given by the formula $A = 4\pi r^2$. Given a partition $0 = r_0 < r_1 < \cdots < r_n = 1$ of $[0, 1]$, estimate the weight ΔW_i of the spherical shell $r_{i-1} \leqq r \leqq r_i$ of the ball.)

6.2 Volumes by the Method of Cross Sections

In this section we show how to use integrals to calculate the volumes of certain solids or regions in space. We begin with the intuitive idea that volume has properties analogous to the properties of area listed in Section 5.2. That is, we assume that every reasonably nice bounded solid region R has a volume $v(R)$, a nonnegative real number with the following properties:

1. If the solid region R is contained in the solid region S, then $v(R) \leqq v(S)$.
2. If R and S are nonoverlapping solid regions, then the volume of their union is $v(R \cup S) = v(R) + v(S)$.
3. If the solid regions R and S are congruent (have the same size and shape), then $v(R) = v(S)$.

The **method of cross sections** is a way of computing the volume of a solid when that solid is described in terms of its cross sections in planes perpendicular to a fixed *reference line L*. This reference line will ordinarily be either the x-axis or the y-axis. In a specific problem we attempt to choose as reference line that axis which makes the resulting computations simpler.

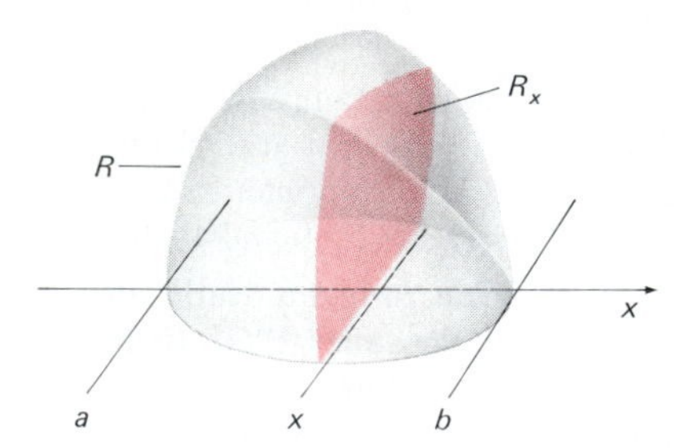

6.2 R_x is the cross section of R in the plane perpendicular to the x-axis at x.

We first consider the case in which L is the x-axis. Suppose that the solid R with volume $V = v(R)$ lies opposite the interval $[a, b]$ on the x-axis. That is, a plane perpendicular to the x-axis intersects the solid if and only if this plane meets the x-axis in a point of $[a, b]$, as indicated in Fig. 6.2. Let R_x denote the intersection of R with the perpendicular plane that meets the x-axis at the point x of $[a, b]$. We call R_x the (plane) **cross section** of the solid R at x.

This situation is especially simple if all the cross sections of R are congruent to one another and, in addition, are parallel translates of each other. In this case the solid R is called a **cylinder** with **bases** R_a and R_b and height $h = b - a$. If R_a and R_b are circles, then R is the familiar **circular cylinder.**

In order to calculate volumes by integration, we need a fourth property of volume adjoined to the three listed earlier:

4. The volume of any cylinder, circular or not, is the product of its height and the area of its base.

This is a generalization of the familiar formula $V = Ah = \pi r^2 h$ for the volume of a circular cylinder with radius r and height h.

Now for each x in $[a, b]$, let $A(x)$ denote the area of the cross section R_x of the solid R:

$$A(x) = a(R_x). \tag{1}$$

We shall assume that the shape of R is sufficiently simple that this **cross-sectional area function** A is continuous, and therefore integrable.

To set up an integral formula for $V = v(R)$, we begin with a regular partition of $[a, b]$ into n equal subintervals each with the usual length $\Delta x = (b - a)/n$. Let R_i denote the slab or slice of the solid R that lies opposite the ith subinterval $[x_{i-1}, x_i]$, as shown in Fig. 6.3. We denote the volume of this ith slice of R by $\Delta V_i = v(R_i)$, so that

$$V = \sum_{i=1}^{n} \Delta V_i.$$

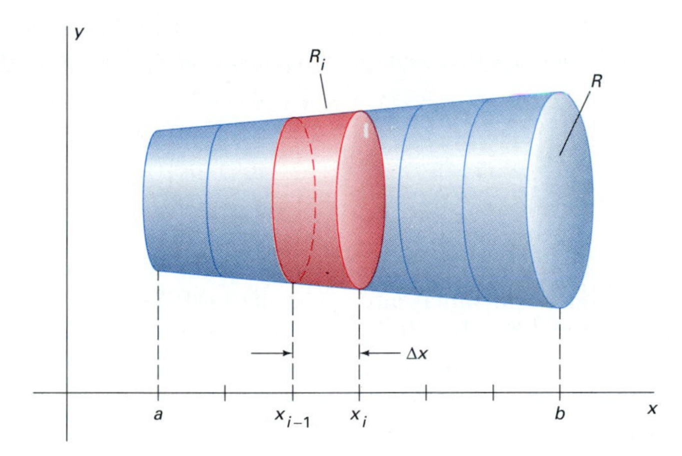

6.3 Planes through the partition points $x_0, x_1, \ldots, x_n$ partition the solid R into slabs $R_1, R_2, \ldots, R_n$.

To approximate ΔV_i, we select an arbitrary point x_i^* in $[x_{i-1}, x_i]$, and consider the *cylinder* C_i with height Δx and whose base is the cross section $R_{x_i^*}$ of R *at* x_i^*. Figure 6.4 suggests that if Δx is small, then $v(C_i)$ is a good

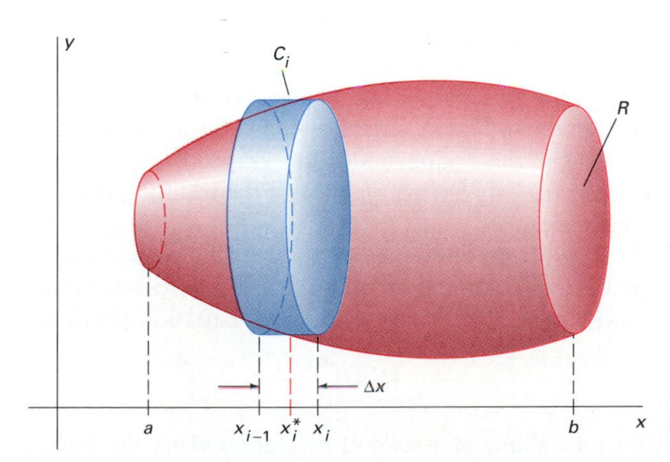

6.4 The slab R_i is approximated by the cylinder C_i of volume $a(R_{x_i^*})\,\Delta x$.

approximation to $\Delta V_i = v(R_i)$:

$$\Delta V_i \approx v(C_i) = a(R_{x_i^*})\,\Delta x = A(x_i^*)\,\Delta x,$$

using the fourth property of volume listed earlier.

Then we add the volumes of these approximating cylinders for $i = 1, 2, 3, \ldots, n$. We find that

$$V = \sum_{i=1}^{n} \Delta V_i \approx \sum_{i=1}^{n} A(x_i^*)\,\Delta x.$$

We recognize the approximating sum on the right as a Riemann sum that approaches $\int_a^b A(x)\,dx$ as $n \to +\infty$. But this sum should also approach the actual volume V as $n \to +\infty$. This justifies the following *definition* of the volume of a solid R in terms of its cross-sectional area function $A(x)$.

Definition *Volume by Cross Sections*

If the solid R lies opposite the interval $[a, b]$ on the x-axis and has continuous cross-sectional area function $A(x)$, then its volume $V = v(R)$ is

$$V = \int_a^b A(x)\,dx. \tag{2}$$

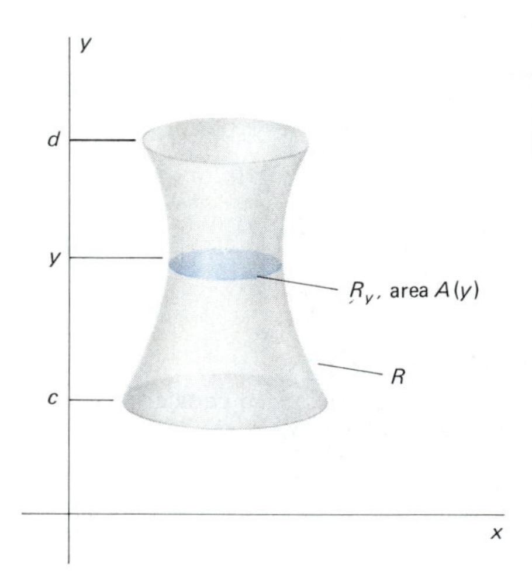

6.5 $A(y)$ is the area of the cross section R_y in the plane perpendicular to the y-axis at the point y.

The formula in (2) is also known as **Cavalieri's principle,** after the Italian mathematician Bonaventura Cavalieri (1598–1647), who systematically exploited the fact that the volume of a solid is determined by the areas of its cross sections perpendicular to a given reference line.

In the case of a solid R that lies opposite the interval $[c, d]$ on the y-axis, we denote by $A(y)$ the area of its cross section R_y in the plane perpendicular to the y-axis at the point y of $[c, d]$ (see Fig. 6.5). A similar discussion, beginning with a regular partition of $[c, d]$, leads to the volume formula

$$V = \int_c^d A(y)\,dy. \tag{3}$$

SOLIDS OF REVOLUTION

An important special case of the formula in (2) gives the volume of a **solid of revolution.** For example, let the solid R be obtained by revolving around the x-axis the region under the graph of $y = f(x)$ over the interval $[a, b]$, where $f(x) \geqq 0$. Such a region and the resulting solid of revolution are shown in Fig. 6.6.

Because the solid R is obtained by revolution, each cross section of R at x is a circular disk of radius $f(x)$. The cross-sectional area function is then $A(x) = \pi[f(x)]^2$, so the formula in Equation (2) yields

$$V = \int_a^b \pi y^2\,dx = \int_a^b \pi[f(x)]^2\,dx \tag{4}$$

for the **volume of a solid of revolution about the x-axis.**

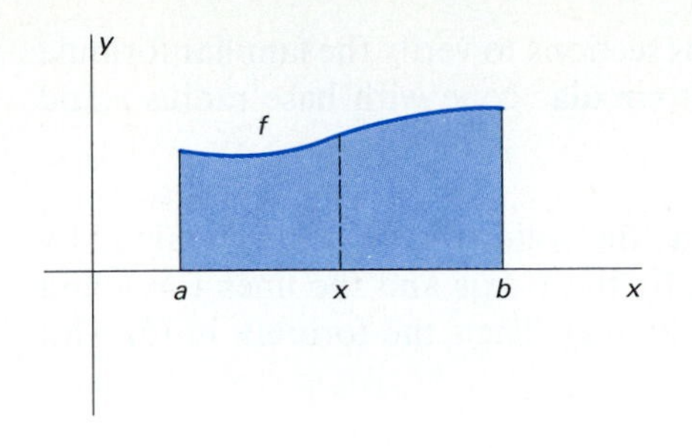

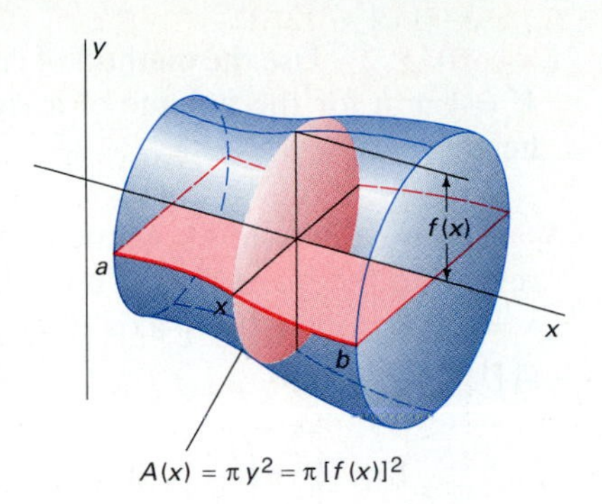

6.6 Volume of revolution around the x-axis

NOTE In the expression $\pi y^2\,dx$, the differential dx tells us what the independent variable is. You *must* express y (and any other dependent variables) in terms of x in order to perform the indicated integration.

By a similar argument, if the region bounded by the curve $x = g(y)$, the y-axis, and the horizontal lines $y = c$ and $y = d$ is rotated about the y-axis, then the volume of the resulting solid of revolution (Fig. 6.7) is

$$V = \int_c^d \pi x^2\,dy = \int_c^d \pi[g(y)]^2\,dy. \tag{5}$$

Here it is essential to express x (and any other dependent variables) in terms of the independent variable y before attempting to antidifferentiate in (5).

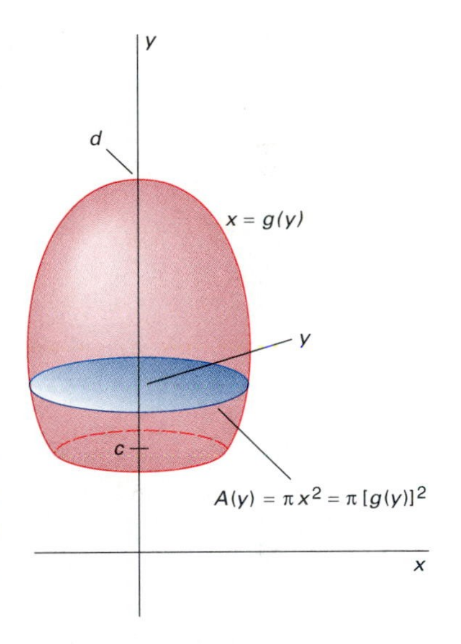

6.7 Volume of revolution around the y-axis

EXAMPLE 1 Use the method of cross sections to verify the familiar formula $V = \frac{4}{3}\pi R^3$ for the volume of a sphere of radius R.

Solution We think of the sphere as the solid of revolution obtained by revolving the semicircular plane region of Fig. 6.8 around the x-axis. This is the region bounded above by the semicircle $y = \sqrt{R^2 - x^2}$ $(-R \leqq x \leqq R)$ and below by the interval $[-R, R]$ on the x-axis. To use the formula in (4), we take $f(x) = \sqrt{R^2 - x^2}$, $a = -R$, and $b = R$. This gives

$$V = \int_{-R}^{R} \pi(\sqrt{R^2 - x^2})^2\,dx = \pi \int_{-R}^{R} (R^2 - x^2)\,dx$$

$$= \pi\Big[R^2 x - \tfrac{1}{3}x^3\Big]_{-R}^{R} = \tfrac{4}{3}\pi R^3.$$

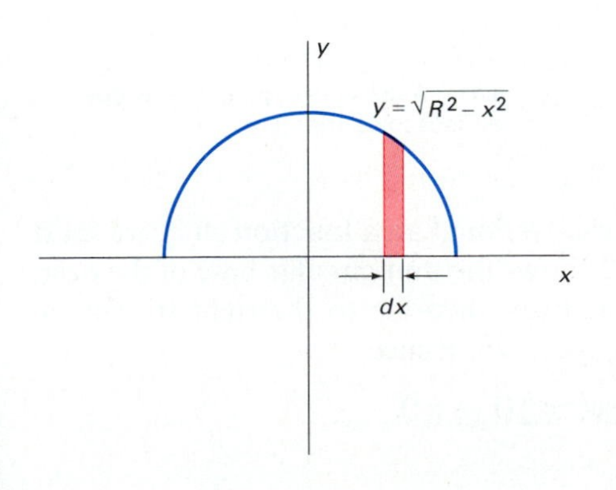

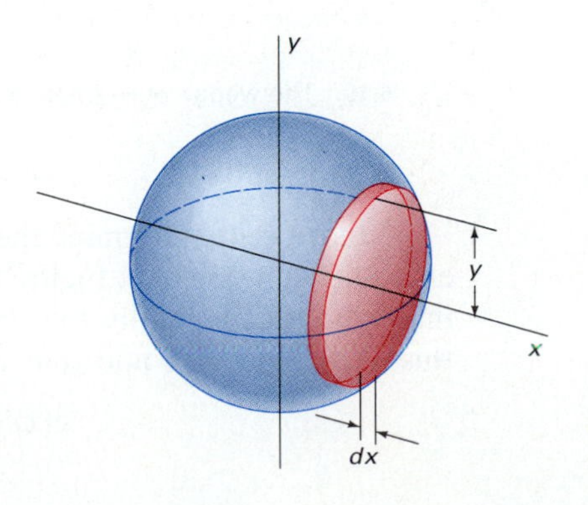

6.8 A sphere generated by rotation of a semicircular region

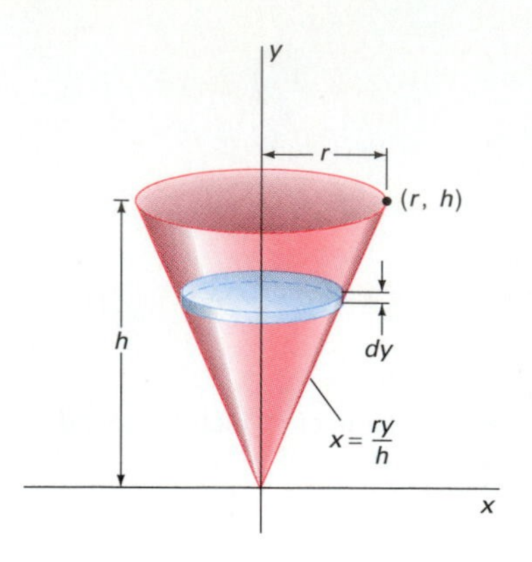

6.9 Generating a cone by rotation (see Example 2)

EXAMPLE 2 Use the method of cross sections to verify the familiar formula $V = \frac{1}{3}\pi r^2 h$ for the volume of a right circular cone with base radius r and height h.

Solution We may think of the cone as the solid of revolution obtained by revolving the plane region bounded by the y-axis and the lines $y = h$ and $x = ry/h$ around the y-axis (as in Fig. 6.9). Then the formula in (5) with $g(y) = ry/h$ gives

$$V = \int_0^h \pi\left(\frac{ry}{h}\right)^2 dy = \frac{\pi r^2}{h^2}\int_0^h y^2\,dy = \frac{\pi r^2}{h^2}\left[\frac{1}{3}y^3\right]_0^h = \frac{1}{3}\pi r^2 h.$$

EXAMPLE 3 Find the volume of the "wedge" that is cut from a circular cylinder with unit radius and height by a plane that passes through a diameter of the bottom base of the cylinder and through a point on the circumference of its top.

Solution The cylinder and wedge are shown in Fig. 6.10. To form such a wedge, fill a cylindrical glass with cider, then drink slowly until half the bottom of the glass is exposed; the remaining cider forms the wedge.

We choose as reference line and x-axis the line through the "edge of the wedge"—the original diameter of the base of the cylinder. It's easy to verify with similar triangles that each cross section of the wedge perpendicular to the diameter is an isosceles right triangle. One of these triangles is shown in Fig. 6.11. We denote by y the equal base and height of this triangle.

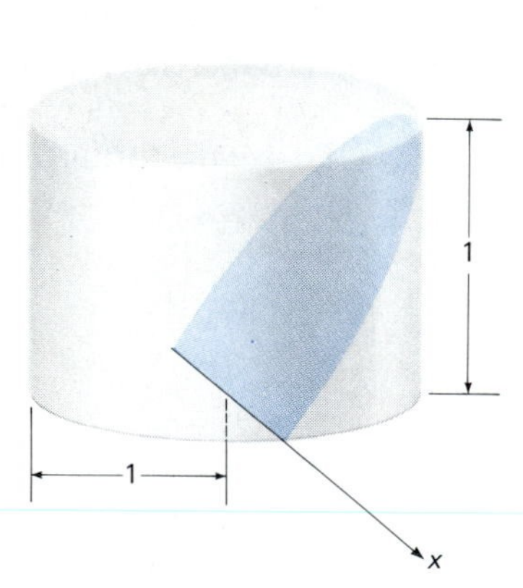

6.10 The wedge of Example 3

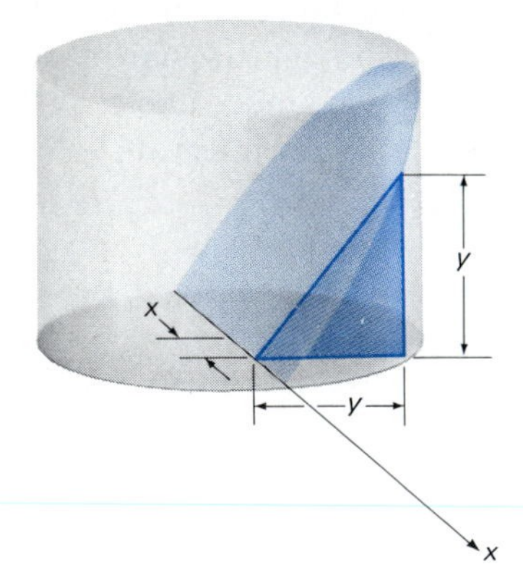

6.11 A cross section of the wedge—an isosceles triangle

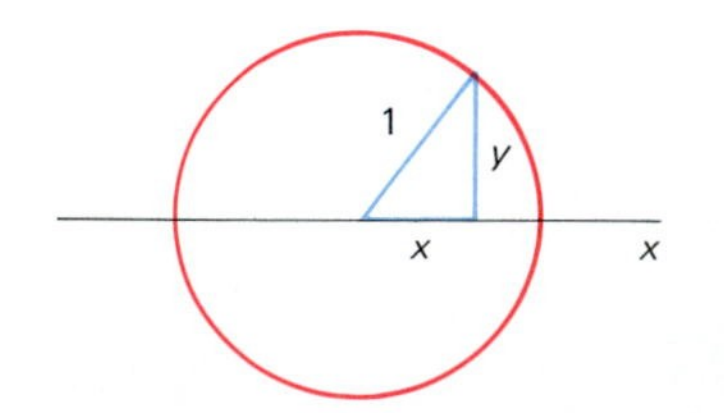

6.12 The base of the cylinder

In order to determine the cross-sectional area function $A(x)$, we must express y in terms of x. Figure 6.12 shows the unit circular base of the original cylinder. We apply the Pythagorean theorem to the right triangle in this figure and thus find that $y = \sqrt{1 - x^2}$. Hence

$$A(x) = \tfrac{1}{2}y^2 = \tfrac{1}{2}(1 - x^2),$$

so the formula in (2) gives

$$V = \int_{-1}^{1} A(x)\,dx = 2\int_{0}^{1} A(x)\,dx \qquad \text{(by symmetry)}$$

$$= 2\int_{0}^{1} \tfrac{1}{2}(1 - x^2)\,dx = \left[x - \tfrac{1}{3}x^3\right]_0^1 = \tfrac{2}{3}$$

for the volume of the wedge.

A useful habit is the checking of answers for plausibility whenever convenient. For example, we may compare a given solid with one whose volume is already known. Because the volume of the original cylinder in Example 3 is π, we have found that the wedge occupies the fraction

$$\frac{V_{\text{wedge}}}{V_{\text{cyl}}} = \frac{\frac{2}{3}}{\pi} \approx 21\%$$

of the volume of the cylinder. A glance at Fig. 6.10 indicates that this is plausible. An error in our computation would likely have given an unbelievable answer, thereby revealing the existence of error.

The wedge of Example 3 has an ancient history. Its volume was first calculated in the third century B.C. by Archimedes, to whom also is originally due the formula $V = \frac{4}{3}\pi r^3$ for the volume of a sphere. His work on the wedge is found in a manuscript that was rediscovered in 1906 after having been lost for centuries. Archimedes used a method of exhaustion for volume similar to that discussed for areas in Section 5.2.

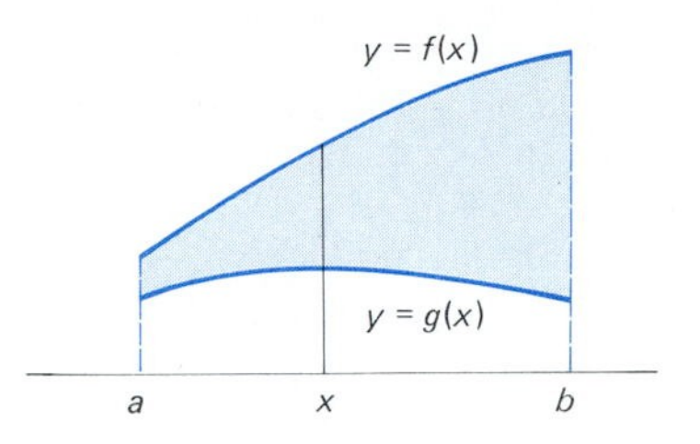

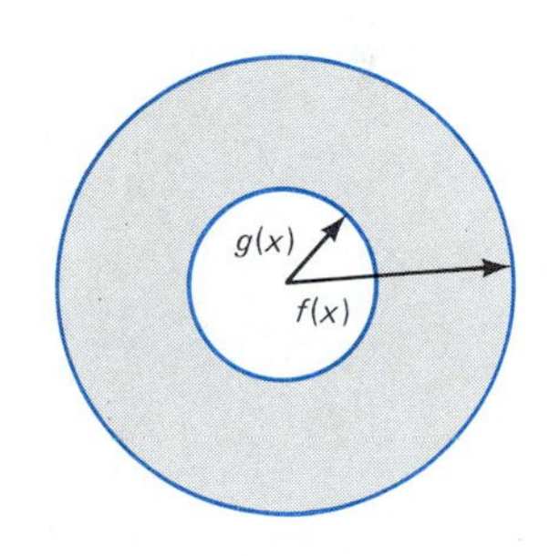

6.13 The region between two positive graphs is rotated about the x-axis. Cross sections are annular rings.

Sometimes we need to calculate the volume of a solid generated by revolution of a plane region lying between two given curves. Suppose that $f(x) > g(x) > 0$ for x in the interval $[a, b]$, and that the solid R is generated by revolving the region between $y = f(x)$ and $y = g(x)$ about the x-axis. Then the cross section at x is an **annular ring** bounded by two circles, as shown in Fig. 6.13. The ring has inner radius $r_{\text{in}} = g(x)$ and outer radius $r_{\text{out}} = f(x)$, so the formula for the cross-sectional area of R at x is

$$A(x) = \pi(r_{\text{out}})^2 - \pi(r_{\text{in}})^2 = \pi[(y_{\text{top}})^2 - (y_{\text{bot}})^2] = \pi\{[f(x)]^2 - [g(x)]^2\},$$

where, as usual, we write $y_{\text{top}} = f(x)$ and $y_{\text{bot}} = g(x)$ for the top and bottom boundary curves of the plane region. Therefore, the formula in (2) yields

$$V = \int_a^b \pi[(y_{\text{top}})^2 - (y_{\text{bot}})^2]\,dx = \int_a^b \pi\{[f(x)]^2 - [g(x)]^2\}\,dx \qquad (6)$$

for the volume V of the solid.

Similarly, if $f(y) > g(y) > 0$ for $c \leqq y \leqq d$, then the volume of the solid obtained by revolving the region between $x_{\text{right}} = f(y)$ and $x_{\text{left}} = g(y)$ about the y-axis is

$$V = \int_c^d \pi[(x_{\text{right}})^2 - (x_{\text{left}})^2]\,dy = \int_c^d \pi\{[f(y)]^2 - [g(y)]^2\}\,dy. \qquad (7)$$

EXAMPLE 4 Consider the plane region shown in Fig. 6.14, bounded by the curves $y^2 = x$ and $y = x^3$, which intersect at the points (0, 0) and (1, 1). If this

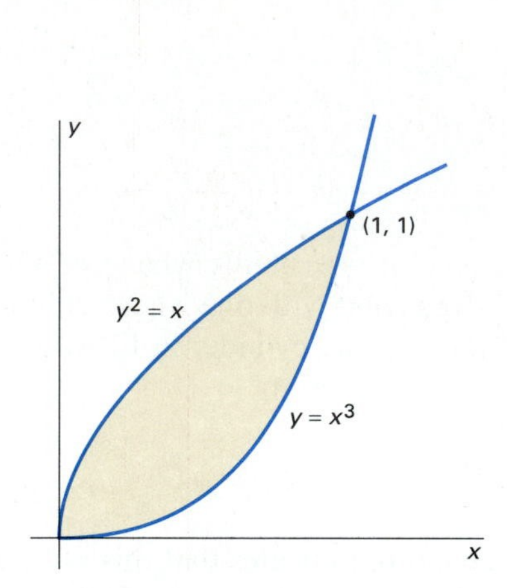

6.14 The plane region of Example 4

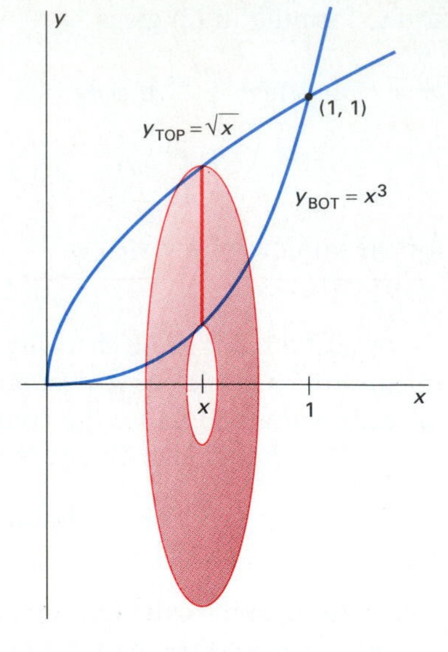

6.15 Revolution about the x-axis

region is revolved about the x-axis (Fig. 6.15), then the formula in (6) with

$$y_{\text{top}} = \sqrt{x}, \qquad y_{\text{bot}} = x^3$$

gives

$$V = \int_0^1 \pi[(\sqrt{x})^2 - (x^3)^2]\,dx = \int_0^1 \pi(x - x^6)\,dx = \pi\Big[\tfrac{1}{2}x^2 - \tfrac{1}{7}x^7\Big]_0^1 = \tfrac{5}{14}\pi$$

for the volume of revolution.

If the same region is revolved around the y-axis (Fig. 6.16), then each cross section perpendicular to the y-axis is an annular ring with outer radius $x_{\text{right}} = y^{1/3}$ and inner radius $x_{\text{left}} = y^2$. Hence the formula in (7) gives volume of revolution

$$V = \int_0^1 \pi[(y^{1/3})^2 - (y^2)^2]\,dy = \int_0^1 \pi(y^{2/3} - y^4)\,dy = \pi\Big[\tfrac{3}{5}y^{5/3} - \tfrac{1}{5}y^5\Big]_0^1 = \tfrac{2}{5}\pi.$$

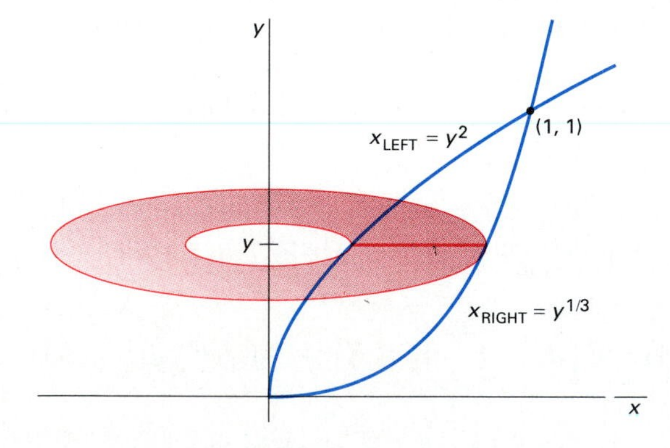

6.16 Revolution about the y-axis

EXAMPLE 5 Suppose instead that the plane region of Example 4 (Fig. 6.14) is revolved about the vertical line $x = -1$, as indicated in Fig. 6.17. Then

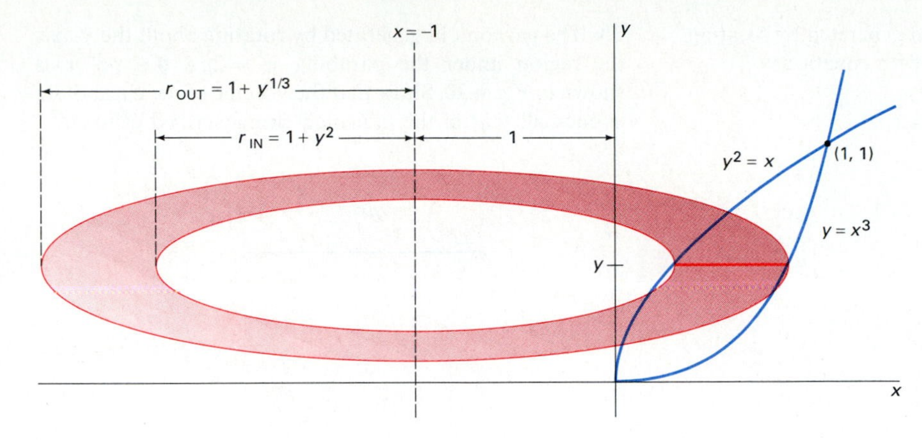

6.17 The annular ring of Example 5

each cross section of the resulting solid is an annular ring with outer radius

$$r_{\text{out}} = 1 + x_{\text{right}} = 1 + y^{1/3}$$

and inner radius

$$r_{\text{in}} = 1 + x_{\text{left}} = 1 + y^2.$$

The area of such a cross section is

$$A(y) = \pi(1 + y^{1/3})^2 - \pi(1 + y^2)^2 = \pi(2y^{1/3} + y^{2/3} - 2y^2 - y^4),$$

so the volume of the resulting solid of revolution is

$$V = \int_0^1 \pi(2y^{1/3} + y^{2/3} - 2y^2 - y^4)\, dy$$

$$= \pi\left[\tfrac{3}{2}y^{4/3} + \tfrac{3}{5}y^{5/3} - \tfrac{2}{3}y^3 - \tfrac{1}{5}y^5\right]_0^1 = \tfrac{37}{30}\pi.$$

6.2 PROBLEMS

In each of Problems 1–24, find the volume of the solid that is generated by rotating about the indicated axis the plane region bounded by the given curves.

1. $y = x^2$, $y = 0$, $x = 1$; about the x-axis

2. $y = \sqrt{x}$, $y = 0$, $x = 4$; about the x-axis

3. $y = x^2$, $y = 4$, $x = 0$ (first quadrant only); about the y-axis

4. $y = 1/x$, $y = 0$, $x = 0.1$, $x = 1$; about the x-axis

5. $y = \sin x$ on $[0, \pi]$, $y = 0$; about the x-axis

6. $y = 9 - x^2$, $y = 0$; about the x-axis

7. $y = x^2$, $x = y^2$; about the x-axis

8. $y = x^2$, $y = 4x$; about the line $x = 5$

9. $y = x^2$, $y = 8 - x^2$; about the x-axis

10. $x = y^2$, $x = y + 6$; about the y-axis

11. $y = 1 - x^2$, $y = 0$; about the x-axis

12. $y = x - x^3$, $y = 0$ $(0 \leqq x \leqq 1)$; about the x-axis

13. $y = 1 - x^2$, $y = 0$; about the y-axis

14. $y = 6 - x^2$, $y = 2$; about the x-axis

15. $y = 6 - x^2$, $y = 2$; about the y-axis

16. $y = 1 - x^2$, $y = 0$; about the vertical line $x = 2$

17. $y = x - x^3$, $y = 0$ $(0 \leqq x \leqq 1)$; about the horizontal line $y = -1$

18. $y = 4$, $x = 0$, $y = x^2$; about the x-axis

19. $y = 4$, $x = 0$, $y = x^2$; about the y-axis

20. $x = 16 - y^2$, $x = 0$, $y = 0$; about the x-axis

21. $y = x^2$, $x = y^2$; about the line $y = -2$

22. $y = x^2$, $y = 8 - x^2$; about the line $y = -1$

23. $y = x^2$, $x = y^2$; about the line $x = 3$

24. $y = x^2$, $y = 8 - x^2$; about the line $x = 4$

25. Find the volume of the solid generated by rotating the region bounded by the parabolas $y^2 = x$ and $y^2 = 2(x - 3)$ about the x-axis.

26. Find the volume of the ellipsoid generated by rotating the region bounded by the ellipse with equation

$$\left(\frac{x}{a}\right)^2 + \left(\frac{y}{b}\right)^2 = 1$$

about the x-axis (see Fig. 6.18).

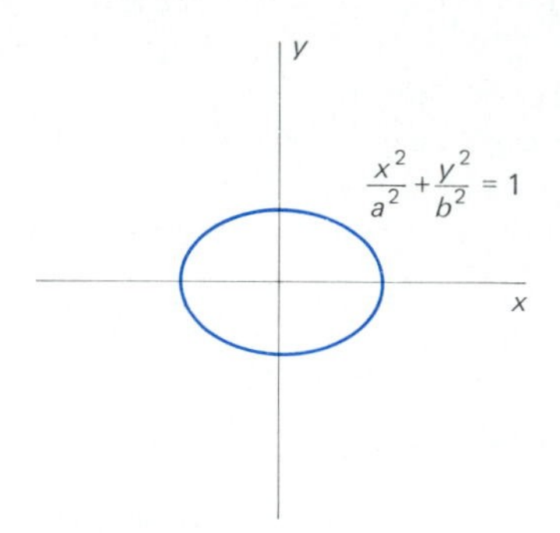

6.18 The ellipse of Problems 26 and 27

27. Repeat Problem 26, except rotate the elliptical region about the y-axis.

28. Find the volume of the unbounded solid generated by rotating the unbounded region of Fig. 6.19 about the x-axis. This is the region between the graph of $y = 1/x^2$ and the x-axis for $x \geqq 1$. (METHOD Compute the volume from $x = 1$ to $x = b$ where $b > 1$. Then find the limit of this volume as $b \to +\infty$.)

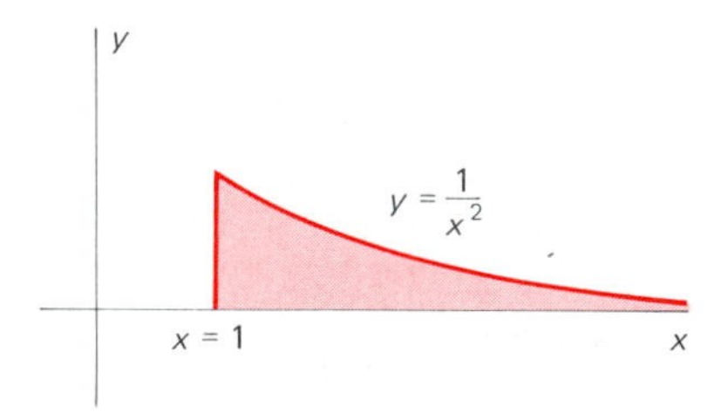

6.19 The unbounded plane region of Problem 28

29. The base of a certain solid is a circle with diameter AB of length $2a$. Find the volume of the solid if each cross section perpendicular to AB is a square.

30. The base of a certain solid is a circle with diameter AB of length $2a$. Find the volume of the solid if each cross section perpendicular to AB is a semicircle.

31. The base of a certain solid is a circle with diameter AB of length $2a$. Find the volume of the solid if each cross section perpendicular to AB is an equilateral triangle.

32. The base of a certain solid is the region in the xy-plane bounded by the parabolas $y = x^2$ and $x = y^2$. Find the volume of this solid if every cross section perpendicular to the x-axis is a square with base in the xy-plane.

33. The paraboloid generated by rotating about the x-axis the region under the parabola $y^2 = 2px$, $0 \leqq x \leqq h$, is shown in Fig. 6.20. Show that the volume of the paraboloid is one-half that of the indicated circumscribed cylinder.

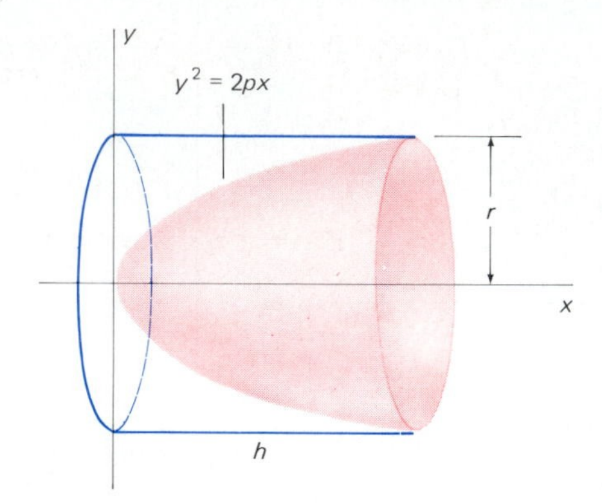

6.20 The paraboloid and cylinder of Problem 33

34. A pyramid has height h and square base with area A. Show that its volume is $V = \frac{1}{3}Ah$. Note that each cross section parallel to the base is a square.

35. Repeat Problem 34, except make the base a triangle with area A.

36. Find the volume that remains after a hole of radius 3 is bored through the center of a solid sphere of radius 5.

37. Two horizontal circular cylinders each have radius a and their axes intersect at right angles. Find the volume of their solid of intersection. (*Suggestion:* Draw the usual xy-coordinate axes and add a z-axis "coming out of the paper" toward you. Imagine the first cylinder with its axis coinciding with the x-axis and the second with its axis coinciding with the z-axis. But draw only the quarter of the first cylinder that lies in front of the xy-plane and above the xz-plane. Draw only the quarter of the second that lies to the right of the yz-plane and above the xz-plane. Their intersection is one-eighth the total volume and is now relatively easy to sketch.)

38. Figure 6.21 shows a "spherical segment" of height h cut off from a sphere of radius r by a horizontal plane.

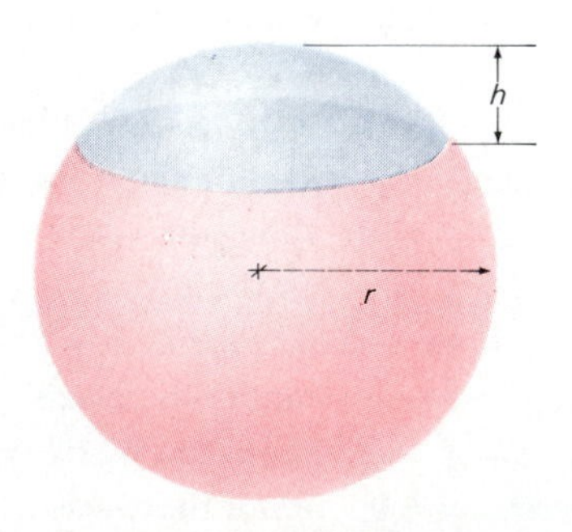

6.21 A spherical segment

Show that its volume is

$$V = \tfrac{1}{3}\pi h^2(3r - h).$$

39. A doughnut-shaped solid called a *torus* is generated by revolving about the y-axis the circular disk

$$(x - b)^2 + y^2 \leqq a^2$$

centered at the point $(b, 0)$ where $0 < a < b$. Show that the volume of this torus is $V = 2\pi^2 a^2 b$. (*Suggestion:* Note that each cross section perpendicular to the y-axis is an annular ring, and recall that

$$\int_0^a \sqrt{a^2 - y^2}\, dy = \tfrac{1}{4}\pi a^2$$

because the integral represents the area of a quarter-circle of radius a.)

40. The summit of a hill is 100 ft higher than the surrounding level terrain, and each horizontal cross section of the hill is circular. The following table gives the radius r (in feet) for selected values of the height h (in feet) above the surrounding terrain. Use Simpson's approximation to estimate the volume of the hill.

h	0	25	50	75	100
r	60	55	50	35	0

41. *Newton's Wine Barrel.* Consider a barrel with the shape of the solid generated by revolving around the x-axis the region under the parabola

$$y = R - kx^2, \qquad -\tfrac{1}{2}h \leqq x \leqq \tfrac{1}{2}h.$$

(a) Show that the radius of each end of the barrel is $r = R - \delta$, where $4\delta = kh^2$.

(b) Then show that the volume of the barrel is

$$V = \tfrac{1}{3}\pi h(2R^2 + r^2 - \tfrac{2}{5}\delta^2).$$

42. *The Clepsydra, or Water Clock.* Consider a water tank whose lateral surface is generated by rotating the curve $y = kx^4$ about the y-axis (k is a positive constant).

(a) Compute $V(y)$, the volume of water in the tank as a function of its depth y.

(b) Suppose that water drains from the tank through a small hole at its bottom. Use the chain rule and Torricelli's law (Equation (3) in Section 4.8) to show that the water level in this tank falls at a *constant* rate.

43. A contractor wants to bid on the job of leveling a 60-ft hill. She knows from previous experience that it will cost \$3.30 per cubic yard of material in the hill. The table below, based on surveying data, shows areas of horizontal cross sections of the hill at 10-ft height intervals. Use (a) the trapezoidal rule, and (b) Simpson's approximation, to estimate how much this job should cost. Round each answer to the nearest hundred dollars.

Height x (ft)	0	10	20	30	40	50	60
Area (ft^2)	1513	882	381	265	151	50	0

44. Water evaporates from an open bowl at a rate proportional to the area of the surface of the water. Show that whatever the shape of the bowl, the water level will drop at a constant rate.

45. A frustum of a right circular cone has height h and volume V. Its base is a circle with radius R and its top is a circle with radius r. Apply the method of cross sections to show that

$$V = \tfrac{1}{3}\pi h(r^2 + rR + R^2).$$

46. The base of a solid is the region in the first quadrant bounded by the graphs of $y = x$ and $y = x^2$. Each cross section perpendicular to the line $y = x$ is a square. Find the volume of the solid.

*6.2 Optional Computer Application

Program WASHER, shown in Fig. 6.22, uses midpoint Riemann sums to approximate a volume of revolution around the x-axis. The top boundary curve $y_{\text{top}} = f(x)$ and the bottom boundary curve $y_{\text{bot}} = g(x)$ of the plane region being revolved are defined in lines 130 and 140. When an integer n is entered in response to line 170, the interval $[a, b]$ is subdivided into n subintervals each of length $\Delta x = (b - a)/n$, and the FOR-NEXT loop of lines 220–270 computes the sum VOL of the volumes of the corresponding slices or "washers." In line 250 the volume

$$\Delta V = \pi[(y_{\text{top}})^2 - (y_{\text{bot}})^2]\,\Delta x$$

is added to the previous value of VOL.

```
100 REM--Program WASHER
110 REM--Volume of revolution around x-axis
120 REM
130       DEF FNF(X) = X^.5             'Ytop
140       DEF FNG(X) = X^3              'Ybot
150       PI = 3.141593
160       A = 0   :   B = 1             'Endpoints
165 REM
170       INPUT "No of subintervals"; N
180       DX = (B - A)/N                'Delta x
190       X = A + DX/2                  'First midpoint
200       VOL = 0                       'Running sum
210 REM
220       FOR I = 1 TO N
230           YT = FNF(X)    :   YB = FNG(X)
240           AREA = PI*(YT*YT - YB*YB)
250           VOL  = VOL + AREA*DX
260           X = X + DX
270       NEXT I
280 REM
290       PRINT "Sum  = "; VOL
300       END
```

6.22 Listing of Program WASHER

Exercise 1: Use this program with $n = 10, 100, \ldots$ to check numerically the first result in Example 4.

Exercise 2: Alter program WASHER so that it approximates a volume of revolution around the y-axis. Then check numerically the second result in Example 4.

6.3 Volume by the Method of Cylindrical Shells

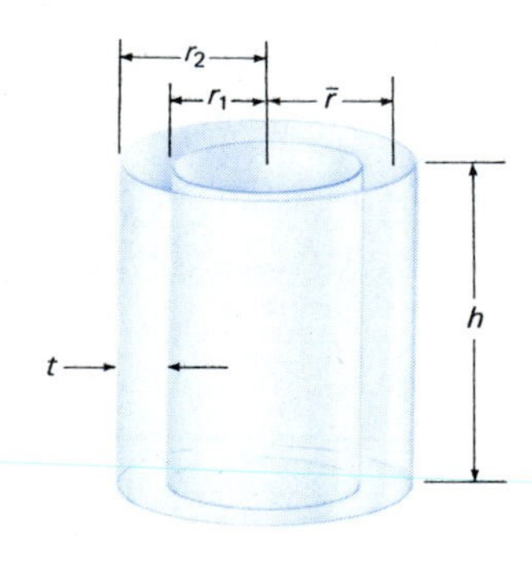

6.23 A cylindrical shell

The method of cross sections of Section 6.2 is a technique of approximating a solid with a stack of thin slabs or slices. In the case of a solid of revolution, these slices are circular disks or annular rings. The **method of cylindrical shells** is a second way of computing volumes of solids of revolution. It is a technique of approximating a solid of revolution with a collection of thin cylindrical shells, and it frequently leads to simpler computations.

A **cylindrical shell** is a region bounded by two concentric circular cylinders of the same height h. If, as in Fig. 6.23, the inner cylinder has radius r_1 and the outer one has radius r_2, then we can write $\bar{r} = (r_1 + r_2)/2$ for the **average radius** of the cylindrical shell and $t = r_2 - r_1$ for its **thickness.** We then get the volume of the cylindrical shell by subtracting the volume of the inner cylinder from that of the outer one. So the shell has volume

$$V = \pi r_2^2 h - \pi r_1^2 h = 2\pi \frac{r_1 + r_2}{2}(r_2 - r_1)h = 2\pi \bar{r} t h. \tag{1}$$

In words, the volume of the shell is the product of 2π, its average radius, its thickness, and its height. Another way to remember this formula is by noting that the volume of the shell is closely approximated by multiplying its surface area by its thickness.

Now suppose that we want to find the volume V of revolution generated by revolving around the y-axis the region under $y = f(x)$ from $x = a$ to $x = b$. We assume, as indicated in Fig. 6.24, that $0 \leqq a < b$ and that $f(x)$ is continuous and nonnegative on $[a, b]$. The solid will resemble the one shown in Fig. 6.24.

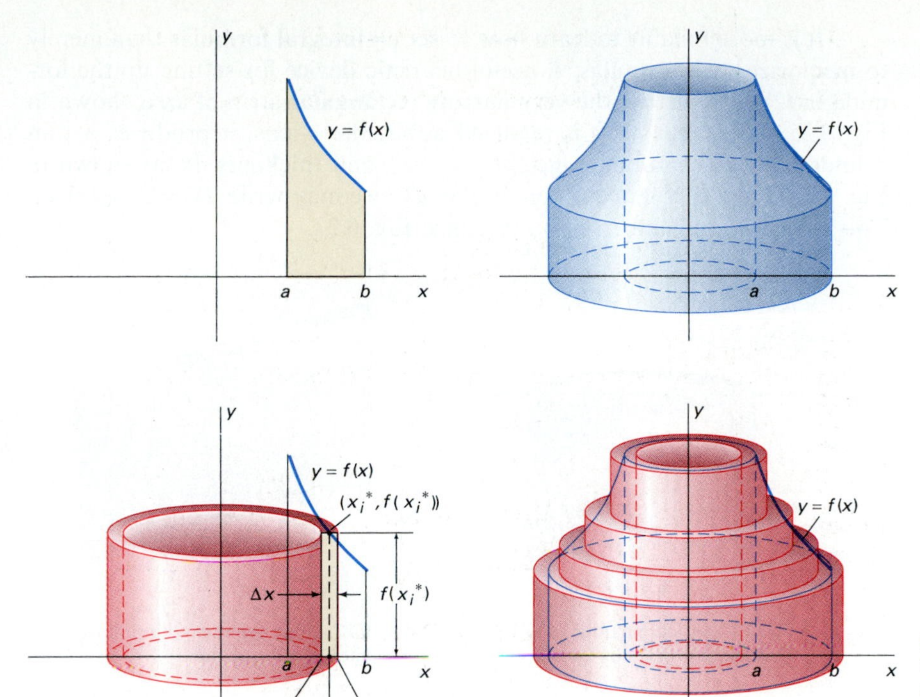

6.24 A solid of revolution—note the hole through its center—and a way to approximate it with nested cylindrical shells

To find V, we begin with a regular partition of $[a, b]$ into n equal subintervals each of length $\Delta x = (b - a)/n$. Let x_i^* denote the midpoint of the ith subinterval $[x_{i-1}, x_i]$. Consider the rectangle with base $[x_{i-1}, x_i]$ and height $f(x_i^*)$. When this rectangle is revolved around the y-axis, it sweeps out a cylindrical shell like the one in Fig. 6.24, with average radius x_i^*, thickness Δx, and height $f(x_i^*)$. This cylindrical shell approximates the solid with volume ΔV_i that is obtained by revolving the region under $y = f(x)$ and over $[x_{i-1}, x_i]$, and thus the formula in (1) gives

$$\Delta V_i \approx 2\pi x_i^* f(x_i^*)\, \Delta x.$$

We add the volumes of the n cylindrical shells determined by the partition. This sum should approximate V because—as Fig. 6.24 suggests—the union of the shells physically approximates the solid of revolution. Thus we obtain the approximation

$$V = \sum_{i=1}^{n} \Delta V_i \approx \sum_{i=1}^{n} 2\pi x_i^* f(x_i^*)\, \Delta x.$$

This approximation to the volume V is a Riemann sum that approaches

$$\int_a^b 2\pi x f(x)\, dx \quad \text{as} \quad \Delta x \to 0,$$

so it appears that the volume of the solid of revolution is given by

$$V = \int_a^b 2\pi x f(x)\, dx. \tag{2}$$

A complete discussion would require a proof that this formula gives the same volume as that *defined* by the method of cross sections in Section 6.2 (see Appendix E).

It is more reliable to learn how to set up integral formulas than merely to memorize such formulas. A useful heuristic device for setting up the formula in (2) is to picture the very narrow rectangular strip of area shown in Fig. 6.25. When this strip is revolved about the y-axis, it produces a thin cylindrical shell of radius x, height $y = f(x)$, and thickness dx, as shown in Fig. 6.26. So, if its volume is denoted by dV, we may write $dV = 2\pi x f(x)\,dx$. This is easy to remember if you visualize Fig. 6.27.

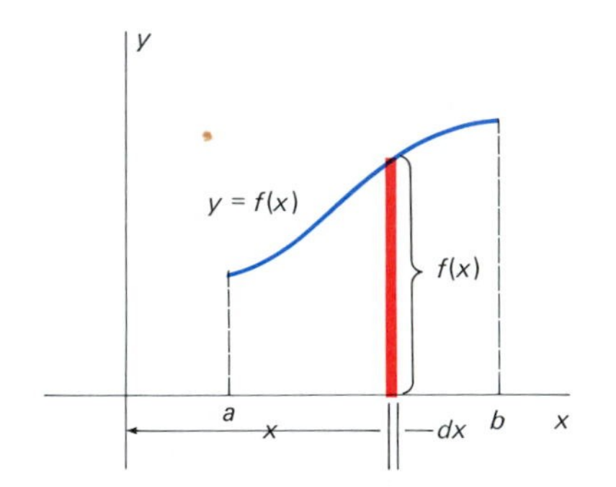

6.25 Heuristic device for setting up the formula in Equation (2)

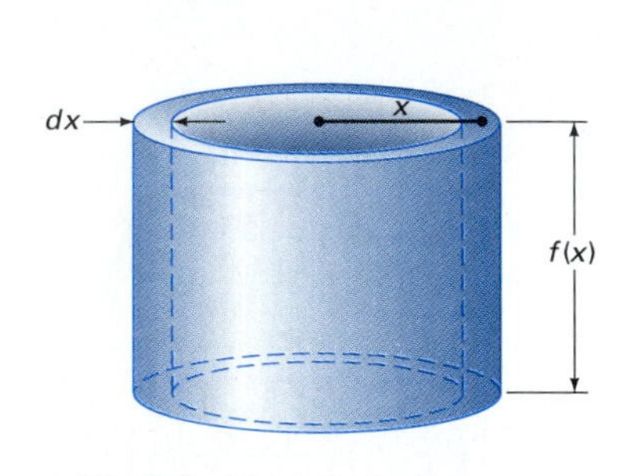

6.26 Cylindrical shell of infinitesimal thickness

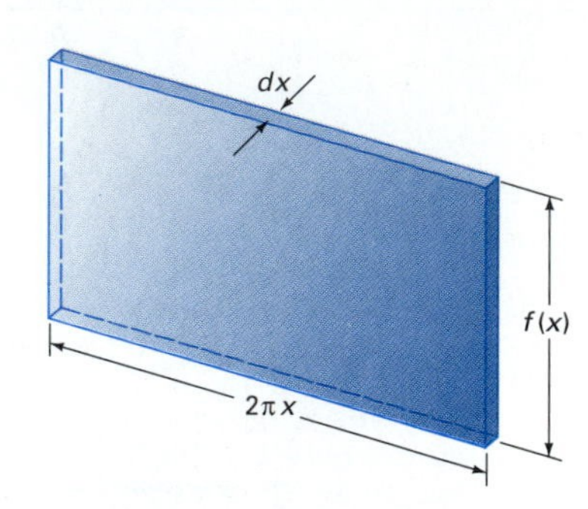

6.27 Infinitesimal cylindrical shell, flattened out

We think of V as a sum of very many such volumes, nested concentrically about the axis of revolution and forming the solid itself. This makes it natural to write

$$V = \int dV = \int_a^b 2\pi x y\,dx = \int_a^b 2\pi x f(x)\,dx.$$

Do not forget to express y (and any other dependent variables) in terms of the independent variable x (identified by the differential dx) before integrating.

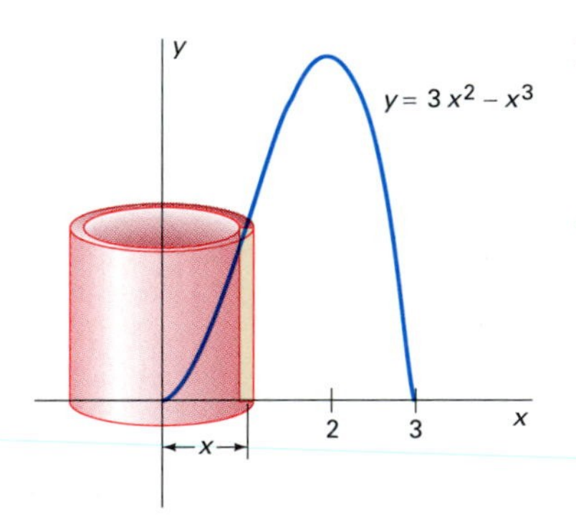

6.28 The region of Example 1: Rotate it about the y-axis.

EXAMPLE 1 Find the volume of the solid generated by revolving around the y-axis the region under $y = 3x^2 - x^3$ from $x = 0$ to $x = 3$. This region is shown in Fig. 6.28.

Solution Here it would be impractical to use the method of cross sections, because a cross section perpendicular to the y-axis is an annular ring and finding its inner and outer radii would require solution of the equation $y = 3x^2 - x^3$ for x in terms of y. We would prefer to avoid this troublesome task, and the formula in (2) provides us with an alternative: We take $f(x) = 3x^2 - x^3$, $a = 0$, and $b = 3$. It immediately follows that

$$V = \int_0^3 2\pi x(3x^2 - x^3)\,dx = 2\pi \int_0^3 (3x^3 - x^4)\,dx$$

$$= 2\pi\left[\tfrac{3}{4}x^4 - \tfrac{1}{5}x^5\right]_0^3 = \tfrac{243}{10}\pi.$$

EXAMPLE 2 Find the volume of the solid that remains after boring a circular hole of radius a through the center of a solid sphere of radius $r > a$ (see Fig. 6.29).

Solution We think of the sphere of radius r as being generated by revolving the right half of the circular disk $x^2 + y^2 = r^2$ around the y-axis, and we

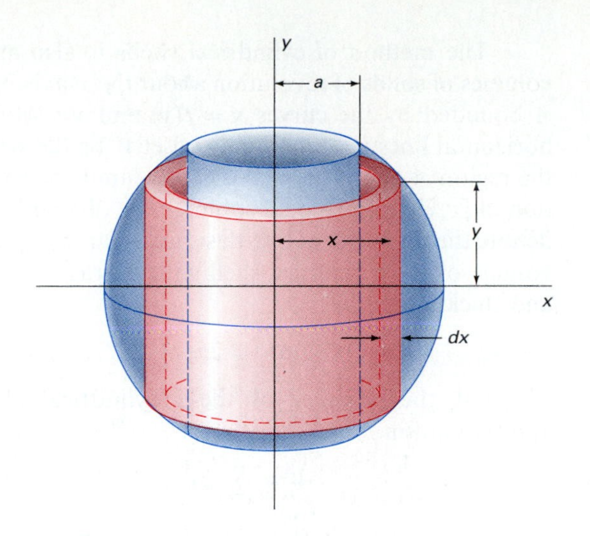

6.29 The sphere-with-hole of Example 2

think of the hole as vertical and with its centerline lying on the y-axis. Then the upper *half* of the solid in question is generated by revolving the region shaded in Fig. 6.30 around the y-axis. This is the region under the graph of $y = (r^2 - x^2)^{1/2}$ (and over the x-axis) from $x = a$ to $x = r$. The volume of the entire sphere-with-hole is then double that of the upper half, and the formula in (2) gives

$$V = 2\int_a^r 2\pi x(r^2 - x^2)^{1/2}\,dx = 4\pi\Big[-\tfrac{1}{3}(r^2 - x^2)^{3/2}\Big]_a^r,$$

so that

$$V = \tfrac{4}{3}\pi(r^2 - a^2)^{3/2}.$$

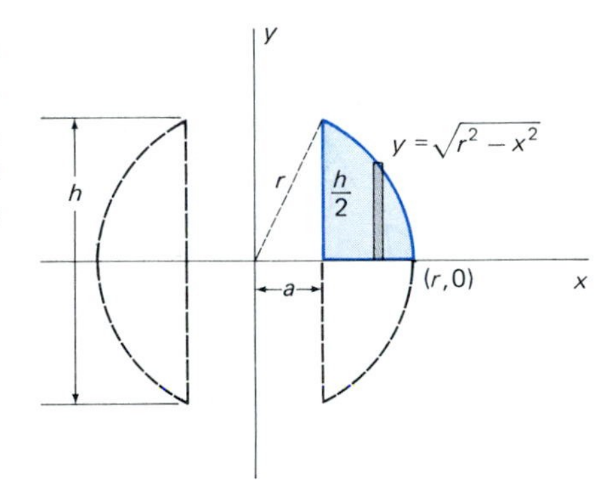

6.30 Middle cross section of the sphere-with-hole

A way to check an answer such as this is to test it in some extreme cases. If $a = 0$, which corresponds to boring no hole at all, our result reduces to the volume $V = \frac{4}{3}\pi r^3$ of the entire sphere. If $a = r$, which corresponds to using a drill bit as large as the sphere, then $V = 0$; this, too, is correct.

Now let A denote the region between the curves $y = f(x)$ and $y = g(x)$ over the interval $[a, b]$, where $0 \leqq a < b$ and $g(x) \leqq f(x)$ for x in $[a, b]$. Such a region is shown in Fig. 6.31. When A is rotated about the y-axis, it generates a solid of revolution. Suppose that we want to find the volume V of this solid. A development similar to that of Equation (2) leads to the approximation

$$V \approx \sum_{i=1}^{n} 2\pi x_i^*[f(x_i^*) - g(x_i^*)]\,\Delta x,$$

from which we may conclude that

$$V = \int_a^b 2\pi x[f(x) - g(x)]\,dx. \tag{3}$$

Thus

$$V = \int_a^b 2\pi x[y_{\text{top}} - y_{\text{bot}}]\,dx \tag{3}$$

with the usual notation $y_{\text{top}} = f(x)$, $y_{\text{bot}} = g(x)$.

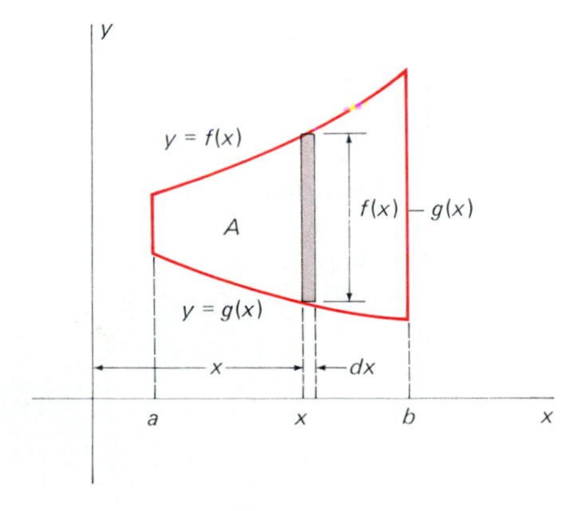

6.31 The region A between the graphs of f and g over $[a, b]$ is to be rotated about the y-axis.

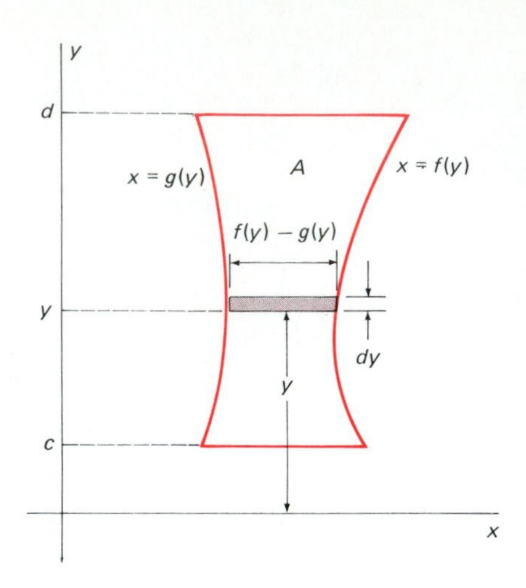

6.32 The region A is to be rotated about the x-axis.

The method of cylindrical shells is also an effective way to compute volumes of solids of revolution about the x-axis. Figure 6.32 shows the region A bounded by the curves $x = f(y)$ and $x = g(y)$ for $c \leqq y \leqq d$ and by the horizontal lines $y = c$ and $y = d$. Let V be the volume obtained by revolving the region A about the x-axis. To compute V, we begin with a regular partition of $[c, d]$ into n equal subintervals of length $\Delta y = (d - c)/n$ each. Let y_i^* denote the midpoint of the ith subinterval $[y_{i-1}, y_i]$ of the partition. Then the volume of the cylindrical shell with average radius y_i^*, height $f(y_i^*) - g(y_i^*)$, and thickness Δy is

$$\Delta V_i = 2\pi y_i^*[f(y_i^*) - g(y_i^*)]\,\Delta y.$$

We add the volumes of these cylindrical shells and thus obtain the approximation

$$V \approx \sum_{i=1}^{n} 2\pi y_i^*[f(y_i^*) - g(y_i^*)]\,\Delta y.$$

We recognize the right-hand side as a Riemann sum for an integral with respect to y from c to d, and so conclude that the volume of the solid of revolution is given by

$$V = \int_c^d 2\pi y[f(y) - g(y)]\,dy. \tag{4}$$

Thus

$$V = \int_c^d 2\pi y[x_{\text{right}} - x_{\text{left}}]\,dy \tag{4'}$$

where $x_{\text{right}} = f(y)$ and $x_{\text{left}} = g(y)$. NOTE In the formulas in (3′) and (4′), the integrand must be expressed in terms of the variable of integration specified by the differential.

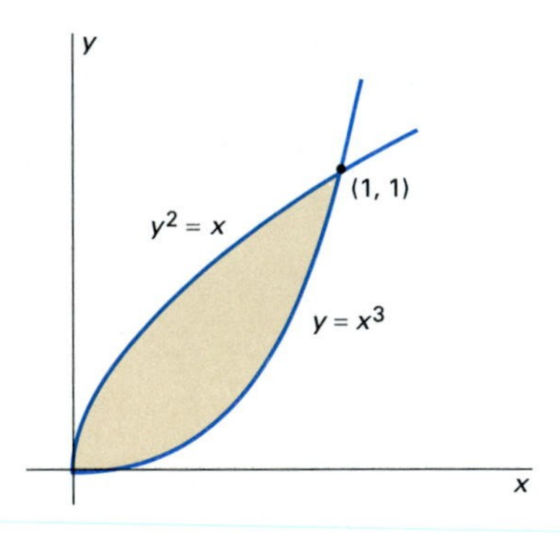

6.33 The region of Example 3

EXAMPLE 3 Consider the region in the first quadrant bounded by the curves $y^2 = x$ and $y = x^3$, shown in Fig. 6.33. Use the method of cylindrical shells to compute the volume of the solids obtained by revolving this region first about the y-axis, second about the x-axis.

Solution It is best to use cylindrical shells as in Figs. 6.34 and 6.35, rather than memorized formulas, to set up the appropriate integrals. Thus the volume of revolution around the y-axis (Fig. 6.34) is given by

$$\begin{aligned} V &= \int_0^1 2\pi x(y_{\text{top}} - y_{\text{bot}})\,dx = \int_0^1 2\pi x(\sqrt{x} - x^3)\,dx \\ &= \int_0^1 2\pi(x^{3/2} - x^4)\,dx = 2\pi\left[\tfrac{2}{5}x^{5/2} - \tfrac{1}{5}x^5\right]_0^1 = \tfrac{2}{5}\pi. \end{aligned}$$

The volume of revolution around the x-axis (see Fig. 6.35) is given by

$$\begin{aligned} V &= \int_0^1 2\pi y(x_{\text{right}} - x_{\text{left}})\,dy = \int_0^1 2\pi y(y^{1/3} - y^2)\,dy \\ &= \int_0^1 2\pi(y^{4/3} - y^3)\,dy \\ &= 2\pi\left[\tfrac{3}{7}y^{7/3} - \tfrac{1}{4}y^4\right]_0^1 = \tfrac{5}{14}\pi. \end{aligned}$$

These answers are the same, of course, as those we obtained using the method of cross sections in Example 4 of Section 6.2.

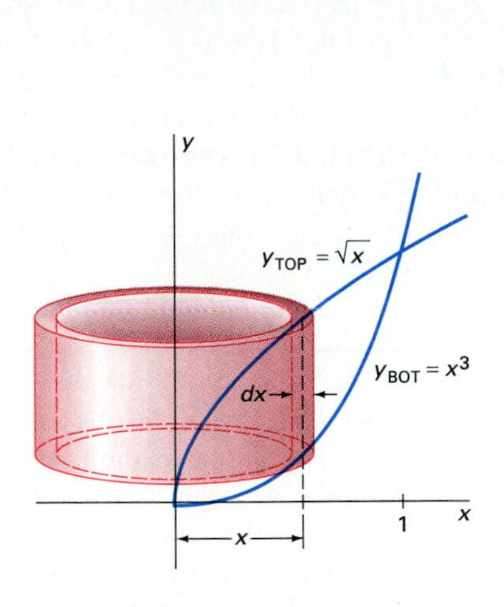

6.34 Revolution about the y-axis

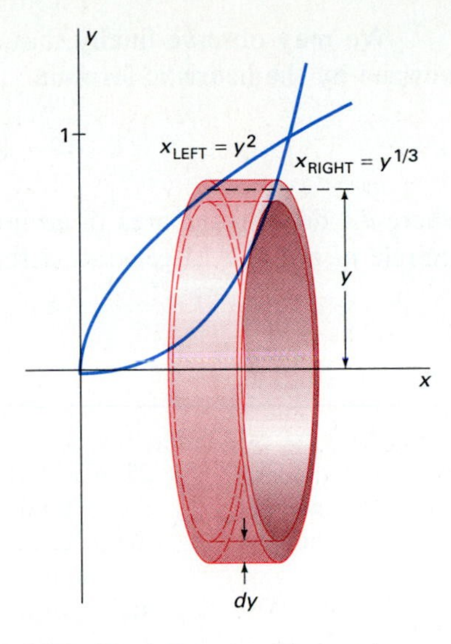

6.35 Revolution about the x-axis

EXAMPLE 4 Suppose instead that the region of Example 3 is rotated about the vertical line $x = -1$ (Fig. 6.36). Then the area element

$$dA = (y_{\text{top}} - y_{\text{bot}})\, dx = (\sqrt{x} - x^3)\, dx$$

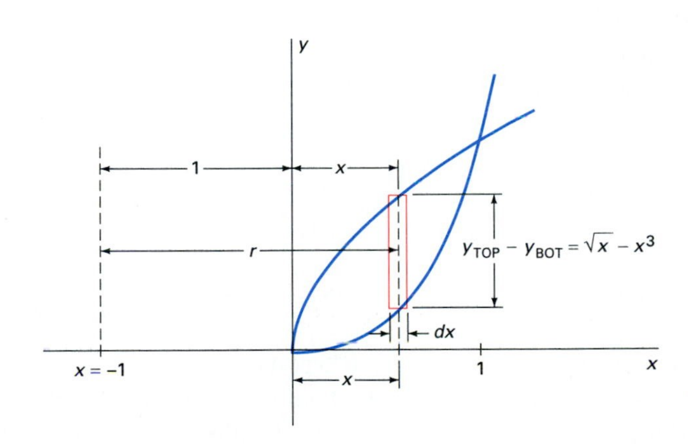

6.36 Revolution about the line $x = -1$

is revolved through a circle of radius $r = 1 + x$. Hence the volume of the resulting cylindrical shell is

$$\begin{aligned} dV = &= 2\pi r\, dA = 2\pi(1 + x)(x^{1/2} - x^3)\, dx \\ &= 2\pi(x^{1/2} + x^{3/2} - x^3 - x^4)\, dx, \end{aligned}$$

so the volume of the resulting solid of revolution is

$$\begin{aligned} V &= \int_0^1 2\pi(x^{1/2} + x^{3/2} - x^3 - x^4)\, dx \\ &= 2\pi\left[\tfrac{2}{3}x^{3/2} + \tfrac{2}{5}x^{5/2} - \tfrac{1}{4}x^4 - \tfrac{1}{5}x^5\right]_0^1 = \tfrac{37}{30}\pi, \end{aligned}$$

as we found using the method of cross sections in Example 5 of Section 6.2.

We may observe finally that the method of cylindrical shells is summarized by the heuristic formula

$$V = \int_{*}^{**} 2\pi r \, dA,$$

where dA denotes the area of an infinitesimal strip that is revolved through a circle of radius r to generate a thin cylindrical shell.

6.3 PROBLEMS

In each of Problems 1–28, use the method of cylindrical shells to find the volume of the solid generated by revolving about the indicated axis the region bounded by the given curves.

1. $y = x^2$, $y = 0$, $x = 2$; about the y-axis
2. $x = y^2$, $x = 4$; about the y-axis
3. $y = 25 - x^2$, $y = 0$; about the y-axis
4. $y = 2x^2$, $y = 8$; about the y-axis
5. $y = x^2$, $y = 8 - x^2$; about the y-axis
6. $x = 9 - y^2$, $x = 0$; about the x-axis
7. $x = y$, $x + 2y = 3$, $y = 0$; about the x-axis
8. $y = x^2$, $y = 2x$; about the line $y = 5$
9. $y = 2x^2$, $y^2 = 4x$; about the x-axis
10. $y = 3x - x^2$, $y = 0$; about the y-axis
11. $y = 4x - x^3$, $y = 0$; about the y-axis
12. $x = y^3 - y^4$, $x = 0$; about the line $y = -2$
13. $y = x - x^3$, $y = 0$ $(0 \leqq x \leqq 1)$; about the y-axis
14. $x = 16 - y^2$, $x = 0$, $y = 0$ $(0 \leqq y \leqq 4)$; about the x-axis
15. $y = x - x^3$, $y = 0$ $(0 \leqq x \leqq 1)$; about the vertical line $x = 2$
16. $y = x^3$, $y = 0$, $x = 2$; about the y-axis
17. $y = x^3$, $y = 0$, $x = 2$; about the vertical line $x = 3$
18. $y = x^3$, $y = 0$, $x = 2$; about the x-axis
19. $y = x^2$, $y = 0$, $x = -1$, $x = 1$; about the vertical line $x = 2$
20. $y = x^2$, $y = x$ $(0 \leqq x \leqq 1)$; about the y-axis
21. $y = x^2$, $y = x$ $(0 \leqq x \leqq 1)$; about the x-axis
22. $y = x^2$, $y = x$ $(0 \leqq x \leqq 1)$; about the horizontal line $y = 2$
23. $y = x^2$, $y = x$ $(0 \leqq x \leqq 1)$; about the vertical line $x = -1$
24. $x = y^2$, $x = 2 - y^2$; about the x-axis
25. $x = y^2$, $x = 2 - y^2$; about the horizontal line $y = 1$
26. $y = 4x - x^2$, $y = 0$; about the y-axis
27. $y = 4x - x^2$, $y = 0$; about the vertical line $x = -1$
28. $y = x^2$, $x = y^2$; about the horizontal line $y = -1$

29. Verify the formula for the volume of a cone by using the method of cylindrical shells. Apply the method to the figure generated by revolving the triangular region with vertices $(0, 0)$, $(r, 0)$, and $(0, h)$ about the y-axis.

30. Use the method of cylindrical shells to compute the volume of the paraboloid of Problem 33 in Section 6.2.

31. Use the method of cylindrical shells to find the volume of the ellipsoid obtained by revolving the elliptical region bounded by the graph of the equation

$$\left(\frac{x}{a}\right)^2 + \left(\frac{y}{b}\right)^2 = 1$$

about the y-axis.

32. Use the method of cylindrical shells to derive the formula given in Problem 38 of Section 6.2 for the volume of a spherical segment.

33. Use the method of cylindrical shells to compute the volume of the torus of Problem 39 in Section 6.2. (*Suggestion:* Substitute u for $x - b$ in the integral given by the formula in Equation (2).)

34. (a) Find the volume of the solid generated by revolving the region bounded by the curves $y = x^2$ and $y = x + 2$ about the line $x = -2$.
(b) Repeat (a), but revolve the region about the line $x = 3$.

35. Find the volume of the solid generated by revolving the circular disk $x^2 + y^2 \leqq a^2$ about the vertical line $x = -a$.

36. (a) Verify by differentiation that

$$\int x \sin x \, dx = \sin x - x \cos x + C.$$

(b) Find the volume of the solid obtained by rotating about the y-axis the area under $y = \sin x$ from $x = 0$ to $x = \pi$.

37. In Example 2 of this section, we found that the volume remaining after a hole of radius a is bored through the center of a sphere of radius $r > a$ is

$$V = \tfrac{4}{3}\pi(r^2 - a^2)^{3/2}.$$

(a) Express the volume V of this formula *without* use of

the hole radius a; use instead the hole height h. (*Suggestion:* Use the right triangle in Fig. 6.30.)

(b) What is remarkable about the answer to part (a)?

38. The plane region R is bounded above and on the right by $y = 25 - x^2$, on the left by the y-axis, and below by the x-axis. A paraboloid is generated by revolving R around the y-axis. Then a vertical hole of radius 3 and centered along the y-axis is bored through the paraboloid. Find the volume of the solid that remains using (a) the method of cross sections and (b) the method of cylindrical shells.

*6.3 Optional Computer Application

Progrram SHELL, shown in Fig. 6.37, uses midpoint Riemann sums to approximate a volume of revolution around the y-axis. Note that lines 100–200 of this program are identical to lines 100–200 of Program WASHER (Optional Computer Application 6.2). The FOR-NEXT loop in lines 220–270 sums the volumes of n cylindrical shells corresponding to a subdivision of $[a, b]$ into n equal subintervals. In line 250 the volume

$$\Delta V = 2\pi x(y_{\text{top}} - y_{\text{bot}})\,\Delta x$$

is added to the previous value of VOL.

```
100 REM--Program SHELL
110 REM--Volume of revolution around y-axis
120 REM
130       DEF FNF(X) = X^.5         'Ytop
140       DEF FNG(X) = X^3          'Ybot
150       PI = 3.141593
160       A = 0  :  B = 1           'Endpoints
165 REM
170       INPUT "No of subintervals"; N
180       DX = (B - A)/N            'Delta x
190       X = A + DX/2              'First midpoint
200       VOL = 0                   'Running sum
210 REM
220       FOR I = 1 TO N
230           YT = FNF(X)   :  YB = FNG(X)
240           AREA = 2*PI*X*(YT - YB)
250           VOL  = VOL + AREA*DX
260           X = X + DX
270       NEXT I
280 REM
290       PRINT "Sum  = "; VOL
300       END
```

6.37 Listing of Program SHELL

Exercise 1: Use this program with $n = 10, 100, \ldots$ to check numerically the first result in Example 3.

Exercise 2: Alter Program SHELL so that it approximates a volume of revolution around the x--axis, then check numerically the second result in Example 3.

6.4 Arc Length and Surface Area of Revolution

A **smooth arc** is the graph of a smooth function defined on a closed interval; a **smooth function** f on $[a, b]$ is one with its derivative f' continuous on $[a, b]$. The continuity of f' rules out the possibility of corner points on the graph of f, points where the direction of the tangent line changes abruptly.

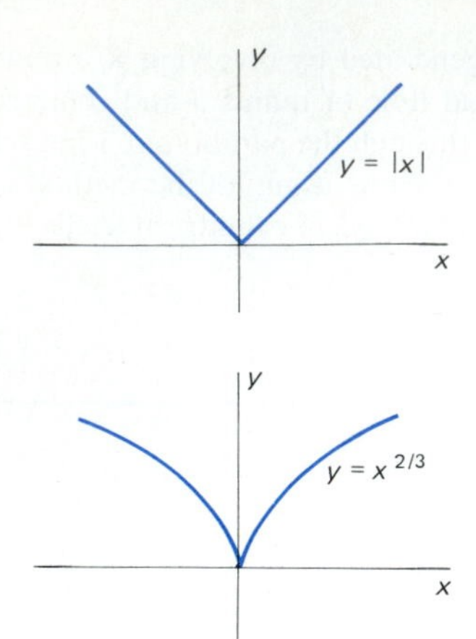

6.38 Graphs having corner points

The graphs of $f(x) = |x|$ and $g(x) = x^{2/3}$ are shown in Fig. 6.38; neither is smooth because each has a corner point at the origin.

To investigate the length of a smooth arc, we begin with the length of a straight line segment, which is simply the distance between its end points. Then, given a smooth arc C, we pose the following question: If C were a thin wire and we straightened it without stretching it, how long would the resulting straight wire be? The answer is what we would call the *length* of C.

To approximate the length s of the smooth arc C, we can inscribe in C a polygonal arc—one made up of straight line segments—and then calculate the length of this polygonal arc. We proceed in the following way, under the assumption that C is the graph of a smooth function f defined on the closed interval $[a, b]$. Consider a regular partition of $[a, b]$ into n subintervals each having length Δx. Let P_i denote the point $(x_i, f(x_i))$ on the arc C corresponding to the ith subdivision point x_i. Our polygonal arc "inscribed in" C is then the union of the line segments $P_0P_1, P_1P_2, P_2P_3, \ldots, P_{n-1}P_n$. So an approximation to the length s of C is

$$s \approx \sum_{i=1}^{n} |P_{i-1}P_i|, \tag{1}$$

the sum of the lengths of these line segments (see Fig. 6.39).

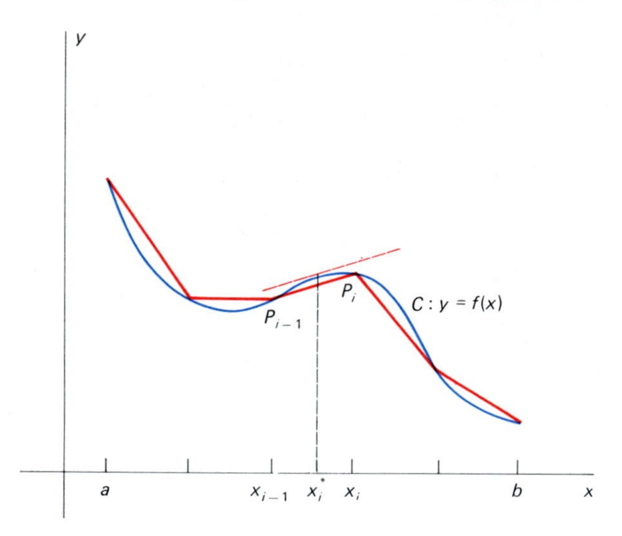

6.39 A polygonal arc inscribed in the smooth curve C

The length of $P_{i-1}P_i$ is

$$|P_{i-1}P_i| = [(x_i - x_{i-1})^2 + (f(x_i) - f(x_{i-1}))^2]^{1/2}.$$

We apply the mean value theorem to the function f on the interval $[x_{i-1}, x_i]$, and thereby conclude the existence of a point x_1^* in this interval such that

$$f(x_i) - f(x_{i-1}) = f'(x_i^*)(x_i - x_{i-1}).$$

Hence

$$\begin{aligned} |P_{i-1}P_i| &= \left[1 + \left(\frac{f(x_i) - f(x_{i-1})}{x_i - x_{i-1}}\right)^2\right]^{1/2}(x_i - x_{i-1}) \\ &= \{1 + [f'(x_i^*)]]^2\}^{1/2}\,\Delta x, \end{aligned}$$

where $\Delta x = x_i - x_{i-1}$.

We next substitute this expression for $|P_{i-1}P_i|$ into (1), which gives the approximation

$$s \approx \sum_{i=1}^{n} \sqrt{1 + [f'(x_i^*)]^2}\, \Delta x.$$

This sum is a Riemann sum for the function $\sqrt{1 + [f'(x)]^2}$ on $[a, b]$, and therefore—because f' is continuous—approaches the integral

$$\int_a^b \sqrt{1 + [f'(x)]^2}\, dx$$

as $\Delta x \to 0$. But our approximation ought also to approach the actual length s as $\Delta x \to 0$. On this basis we *define* the **length** s of the smooth arc C to be

$$s = \int_a^b \sqrt{1 + [f'(x)]^2}\, dx = \int_a^b \sqrt{1 + \left(\frac{dy}{dx}\right)^2}\, dx. \tag{2}$$

In the case of a smooth arc given as a graph $x = g(y)$ for y in $[c, d]$, a similar discussion beginning with a regular partition of $[c, d]$ leads to the formula

$$s = \int_c^d \sqrt{1 + [g'(y)]^2}\, dy = \int_c^d \sqrt{1 + \left(\frac{dx}{dy}\right)^2}\, dy \tag{3}$$

for its length. The length of a more general curve, such as a circle, can be computed by subdividing it into finitely many smooth arcs, then applying to each of these arcs whichever of the formulas in (2) and (3) is required.

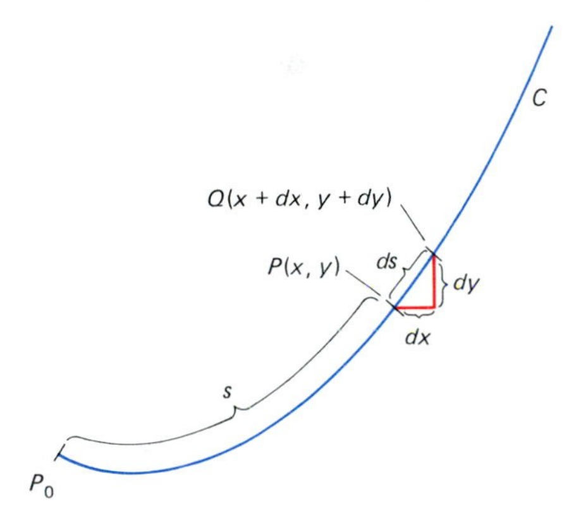

6.40 Heuristic development of the arc length formula

There is a convenient symbolic device that we can employ to remember both the formulas in (2) and (3) simultaneously. We think of two nearby points $P(x, y)$ and $Q(x + dx, y + dy)$ on the smooth arc C, and denote by ds the length of the arc joining P and Q. Imagine that P and Q are so close that ds is, for all practical purposes, equal to the length of the straight line segment PQ. Then the Pythagorean theorem, applied to the small right triangle in Fig. 6.40, gives

$$ds = \sqrt{(dx)^2 + (dy)^2} \tag{4}$$

$$= \sqrt{1 + \left(\frac{dy}{dx}\right)^2}\, dx \tag{4'}$$

$$= \sqrt{1 + \left(\frac{dx}{dy}\right)^2}\, dy. \tag{4''}$$

Thinking of the entire length s of C as the sum of small pieces like ds, we write

$$s = \int ds. \tag{5}$$

Then formal (symbolic) substitution of the expressions in (4′) and (4″) for ds in Equation (5) yields the formulas in (2) and (3); only the limits of integration remain to be determined.

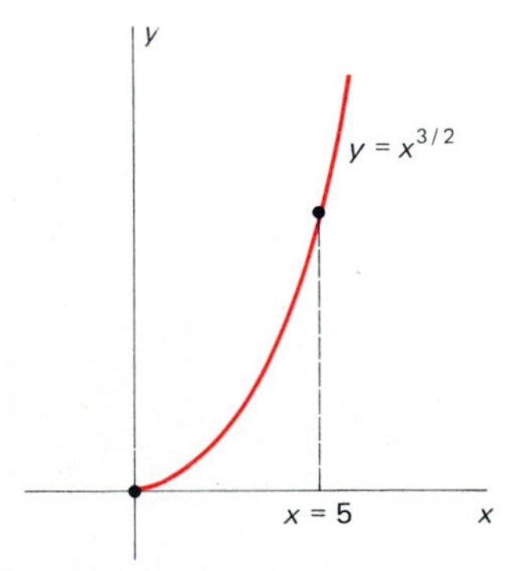

6.41 The semicubical parabola of Example 1

EXAMPLE 1 Find the length of the "semicubical parabola" $y = x^{3/2}$ on $[0, 5]$, shown in Fig. 6.41.

Solution We first compute the integrand in (2):

$$\left[1+\left(\frac{dy}{dx}\right)^2\right]^{1/2} = [1+(\tfrac{3}{2}x^{1/2})^2]^{1/2}$$

$$= (1+\tfrac{9}{4}x)^{1/2} = \tfrac{1}{2}(4+9x)^{1/2}.$$

Hence the length of the arc $y = x^{3/2}$ over the interval $[0, 5]$ is

$$s = \int_0^5 \tfrac{1}{2}(4+9x)^{1/2}\,dx = \left[\tfrac{1}{27}(4+9x)^{3/2}\right]_0^5 = \tfrac{335}{27} \approx 12.41.$$

As a plausibility check, the endpoints of the arc are (0, 0) and $(5, 5\sqrt{5})$, so the straight line segment connecting these points has length $5\sqrt{6} \approx 12.25$, which is as it should be, somewhat less than the calculated length of the arc.

EXAMPLE 2 Find the length of the curve

$$x = \frac{1}{6}y^3 + \frac{1}{2y}, \qquad 1 \leqq y \leqq 2.$$

Solution Here y is the natural independent variable, so we use the arc length formula in (3). First we calculate

$$1+\left(\frac{dx}{dy}\right)^2 = 1+\left(\frac{1}{2}y^2 - \frac{1}{2y^2}\right)^2$$

$$= 1 + \frac{1}{4}y^4 - \frac{1}{2} + \frac{1}{4y^4} = \frac{1}{4}y^4 + \frac{1}{2} + \frac{1}{4y^4}$$

$$= \left(\frac{1}{2}y^2 + \frac{1}{2y^2}\right)^2.$$

Thus we can "get out from under the radical" in (3):

$$s = \int_c^d \sqrt{1+\left(\frac{dx}{dy}\right)^2}\,dy = \int_1^2 \left(\frac{1}{2}y^2 + \frac{1}{2y^2}\right)dy$$

$$= \left[\frac{1}{6}y^3 - \frac{1}{2y}\right]_1^2 = \frac{17}{12}.$$

EXAMPLE 3 A manufacturer needs to make corrugated metal sheets 36 in. wide with cross sections in the shape of the curve

$$y = \tfrac{1}{2}\sin \pi x, \qquad 0 \leqq x \leqq 36,$$

shown in Fig. 6.42. What is the width of the flat sheets the manufacturer should use to produce these corrugated sheets?

6.42 The corrugated sheet in the shape of $y = \frac{1}{2}\sin x$

Solution If

$$f(x) = \tfrac{1}{2}\sin \pi x, \quad \text{then} \quad f'(x) = \frac{\pi}{2}\cos \pi x.$$

Hence the formula in (2) yields the arc length of the graph of f over $[0, 36]$ to be

$$s = \int_0^{36} \sqrt{1 + \left(\frac{\pi}{2}\right)^2 \cos^2 \pi x}\, dx = 36 \int_0^1 \sqrt{1 + \left(\frac{\pi}{2}\right)^2 \cos^2 \pi x}\, dx.$$

It turns out that these integrals cannot be evaluated in terms of elementary functions. Because of this, we cannot apply the fundamental theorem of calculus. So we estimate their values with the aid of Simpson's approximation (Section 5.8). Both with $n = 6$ and with $n = 12$ subintervals we find that

$$\int_0^1 \sqrt{1 + \left(\frac{\pi}{2}\right)^2 \cos^2 \pi x}\, dx \approx 1.46$$

(in.). Therefore the manufacturer should use flat sheets each having width approximately $(36)(1.46) \approx 52.6$ in.

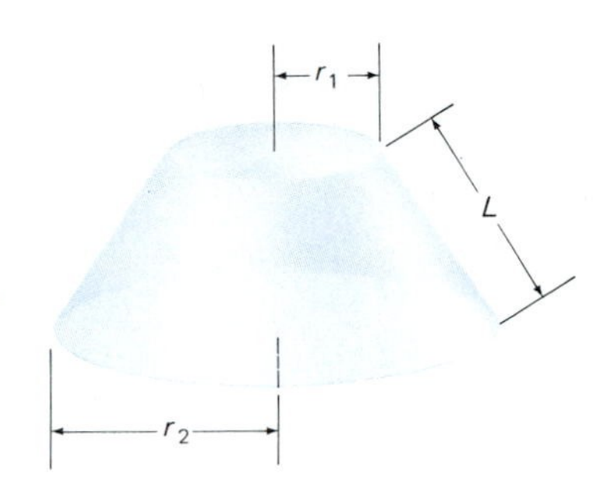

6.43 A frustum of a cone. The slant height is L.

AREAS OF SURFACES OF REVOLUTION

A **surface of revolution** is one obtained by revolving an arc or curve about an axis lying in its plane. The surface of a cylinder or of a sphere and the curved surface of a cone are important examples of surfaces of revolution.

Our basic approach to the area of such a surface is this: First we inscribe a polygonal arc in the curve to be revolved. We then regard the area of the surface generated by revolving the polygonal arc as an approximation to the surface generated by revolving the original curve. Because a surface generated by revolving a polygonal arc about an axis consists of frusta (sections) of cones, we can calculate its area in a reasonably simple way.

This approach to surface area originated with Archimedes. For example, it's the method he used to establish the formula $A = 4\pi r^2$ for the surface area of a sphere of radius r.

We will need the formula

$$A = 2\pi \bar{r} L \tag{6}$$

for the curved surface area of a frustum of a cone with average radius $\bar{r} = \frac{1}{2}(r_1 + r_2)$ and *slant height* L, as shown in Fig. 6.43. The formula in (6) follows from the formula

$$A = \pi r L \tag{7}$$

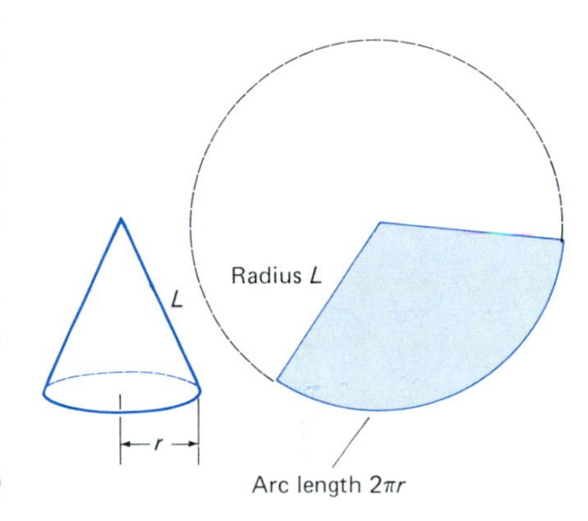

6.44 Surface area of a cone: Cut along L, then unroll the cone onto the circular sector.

for the area of a conical surface with base radius r and slant height L, as shown in Fig. 6.44. It is easy to derive the formula in (7) by "unrolling" the conical surface onto a sector of a circle of radius L (also in Fig. 6.44) because the area of this sector is

$$A = \frac{2\pi r}{2\pi L} \cdot \pi L^2 = \pi r L.$$

To derive the formula in (6) from the one in (7), we think of the frustum as the lower section of a cone with slant height $L_2 = L + L_1$, as indicated in Fig. 6.45. Then subtraction of the area of the upper conical section from that of the entire cone gives

$$A = \pi r_2 L_2 - \pi r_1 L_1 = \pi r_2(L + L_1) - \pi r_1 L_1 = \pi(r_2 - r_1)L_1 + \pi r_2 L$$

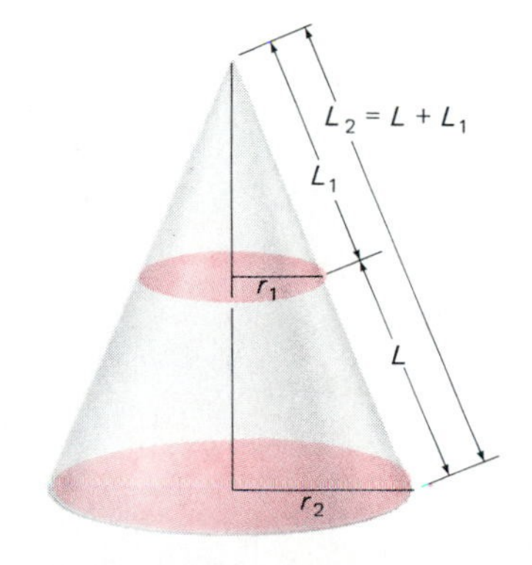

6.45 Derivation of the formula in (6)

for the area of the frustum. But the similar right triangles in Fig. 6.45 yield the proportion

$$\frac{r_1}{L_1} = \frac{r_2}{L_2} = \frac{r_2}{L + L_1},$$

from which we find that $(r_2 - r_1)L_1 = r_1 L$. Hence the area of the frustum is

$$A = \pi r_1 L + \pi r_2 L = 2\pi \bar{r} L$$

where $\bar{r} = \frac{1}{2}(r_1 + r_2)$. So we have verified the formula in (6).

Now suppose that the surface S has area A and is generated by revolving around the x-axis the smooth arc $y = f(x)$, $a \leqq x \leqq b$; suppose also that $f(x)$ is never negative on $[a, b]$. To approximate A we begin with a regular partition of $[a, b]$ into n equal subintervals each of length Δx. As in our discussion of arc length leading to Equation (2), let P_i denote the point $(x_i, f(x_i))$ on the arc. Then, as before, the line segment $P_{i-1}P_i$ has length

$$L_i = |P_{i-1}P_i| = \{1 + [f'(x_i^*)]^2\}^{1/2}\, \Delta x$$

for some point x_i^* in the ith subinterval $[x_{i-1}, x_i]$.

The conical frustum obtained by revolving the segment $P_{i-1}P_i$ around the x-axis has slant height L_i and, as shown in Fig. 6.46, average radius

$$\bar{r}_i = \tfrac{1}{2}[f(x_{i-1}) + f(x_i)].$$

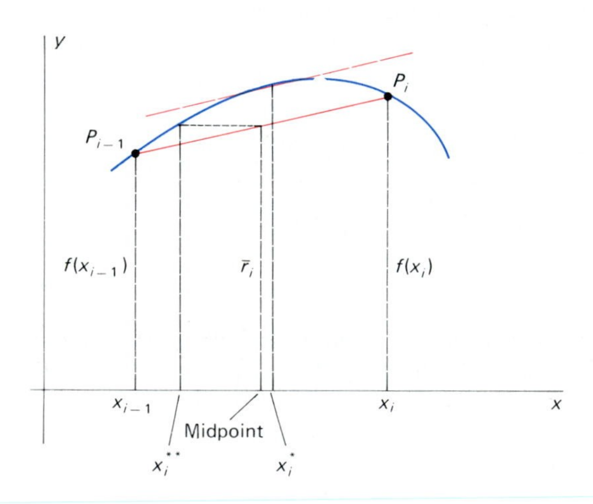

6.46 Approximation of a surface area of revolution by the surface of a frustum of a cone

Because $\bar{r}_i$ lies between the values $f(x_{i-1})$ and $f(x_i)$, the intermediate value property of continuous functions (Section 2.4) yields a point x_i^{**} in $[x_{i-1}, x_i]$ such that $\bar{r}_i = f(x_i^{**})$. By the formula in (6), the area of this conical frustum is therefore

$$2\pi \bar{r}_i L_i = 2\pi f(x_i^{**})\{1 + [f'(x_i^*)]^2\}^{1/2}\, \Delta x.$$

We add the areas of these conical frusta for $i = 1, 2, 3, \ldots, n$. This gives the approximation

$$A \approx \sum_{i=1}^{n} 2\pi f(x_i^{**})\{1 + [f'(x_i^*)]^2\}^{1/2}\, \Delta x.$$

If x_i^* and x_i^{**} were the *same* point of the ith subinterval $[x_{i-1}, x_i]$, then this approximation would be a Riemann sum for the integral

$$\int_a^b 2\pi f(x)\sqrt{1 + [f'(x)]^2}\, dx.$$

Even though the numbers x_i^* and x_i^{**} are generally unequal, it still follows (from a result stated in Appendix E) that our approximation approaches the integral above as $\Delta x \to 0$.

We therefore *define* the **area** A of the surface generated by revolving the smooth arc $y = f(x)$, $a \leqq x \leqq b$, around the x-axis by the integral formula

$$A = \int_a^b 2\pi f(x)\sqrt{1 + [f'(x)]^2}\, dx. \tag{8}$$

If we write y for $f(x)$ and ds for $\sqrt{1 + (dy/dx)^2}\, dx$, as in the formula in (4′), then the formula in (8) can be abbreviated to

$$A = \int_a^b 2\pi y\, ds \qquad (x\text{-axis}). \tag{9}$$

This abbreviated formula is conveniently remembered by thinking of $dA = 2\pi y\, ds$ as the area of the narrow frustum obtained by revolving the tiny arc ds around the x-axis in a circle of radius y, as in Fig. 6.47.

If our smooth arc being revolved around the x-axis is given by $x = g(y)$, $c \leqq y \leqq d$, then an approximation based on a regular partition of $[c, d]$ leads to the area formula

$$A = \int_c^d 2\pi y\sqrt{1 + [g'(y)]^2}\, dy. \tag{10}$$

Note that Equation (10) can be obtained by making the formal substitution $ds = \sqrt{1 + (dx/dy)^2}\, dy$ of (4″) in the abbreviated formula in (9) for surface area of revolution, then replacing a and b there by the correct limits of integration.

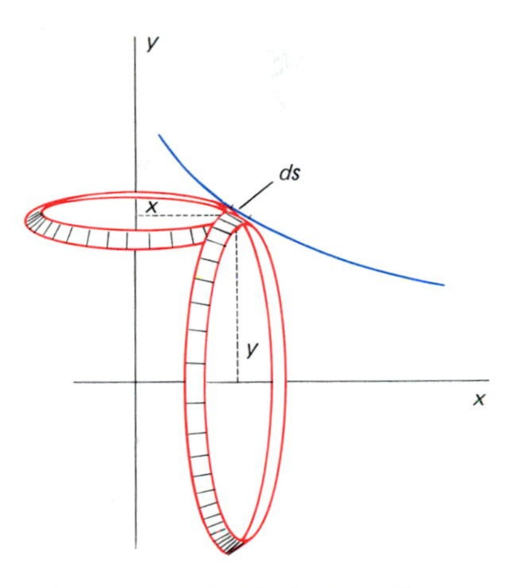

6.47 The arc ds may be rotated about either the x-axis or the y-axis.

Now let us consider the surface generated by revolving a smooth arc about the y-axis rather than about the x-axis. In Fig. 6.47 we see that the average radius of the narrow frustum obtained by revolving the tiny arc ds is now x instead of y. This suggests the abbreviated formula

$$A = \int_a^b 2\pi x\, ds \qquad (y\text{-axis}) \tag{11}$$

for a surface area of revolution about the y-axis. If the smooth arc is given by $y = f(x)$, $a \leqq x \leqq b$, then the symbolic substitution $ds = \sqrt{1 + (dy/dx)^2}\, dx$ gives

$$A = \int_a^b 2\pi x\sqrt{1 + [f'(x)]^2}\, dx. \tag{12}$$

But if the smooth arc is presented in the form $x = g(y)$, $c \leqq y \leqq d$, then the symbolic substitution $ds = \sqrt{1 + (dx/dy)^2}\, dy$ in (11) gives

$$A = \int_c^d 2\pi g(y)\sqrt{1 + [g'(y)]^2}\, dy. \tag{13}$$

The formulas in (12) and (13) may be verified by using approximations similar to the one leading to Equation (8).

Thus we have *four* formulas for areas of surfaces of revolution, summarized in the table in Fig. 6.48. Which of these formulas is appropriate for computing the area of a given surface depends on two factors:

1. Whether the smooth arc that generates the surface is presented in the form $y = f(x)$ or in the form $x = g(y)$, and
2. Whether this arc is to be revolved about the x-axis or about the y-axis.

Description of curve C	Axis of revolution: x-axis	Axis of revolution: y-axis
$y = f(x)$, $a \leqslant x \leqslant b$	$\int_a^b 2\pi f(x)\sqrt{1 + [f'(x)]^2}\,dx$ (8)	$\int_a^b 2\pi x\sqrt{1 + [f'(x)]^2}\,dx$ (10)
$x = g(y)$, $c \leqslant y \leqslant d$	$\int_c^d 2\pi y\sqrt{1 + [g'(y)]^2}\,dy$ (12)	$\int_c^d 2\pi g(y)\sqrt{1 + [g'(y)]^2}\,dy$ (13)

6.48 Area formulas for surfaces of revolution

But memorizing the four formulas in the table is unnecessary. We suggest that you instead remember the abbreviated formulas in (9) and (11) in conjunction with Fig. 6.47 and make either the substitution

$$y = f(x), \qquad ds = \sqrt{1 + \left(\frac{dy}{dx}\right)^2}\,dx$$

or the substitution

$$x = g(y), \qquad ds = \sqrt{1 + \left(\frac{dx}{dy}\right)^2}\,dy,$$

depending on whether the smooth arc is presented as a function of x or as a function of y. It may also be helpful to note that each of these four surface area formulas is of the form

$$A = \int_{*}^{**} 2\pi r\,ds \tag{14}$$

where r denotes the radius of the circle around which the arc length element ds is revolved.

In earlier sections we cautioned you to identify the independent variable by examination of the differential, and to express every dependent variable in terms of the independent variable before antidifferentiating. That is, either express everything, including ds, in terms of x (and dx) or everything in terms of y (and dy).

6.49 The paraboloid of Example 4

EXAMPLE 4 Find the area of the paraboloid shown in Fig. 6.49, obtained by revolving the parabolic arc $y = x^2$, $0 \leqq x \leqq \sqrt{2}$, around the y-axis.

Solution Following the suggestion that precedes the example, we get

$$A = \int_{*}^{**} 2\pi x\, ds = \int_a^b 2\pi x \sqrt{1 + \left(\frac{dy}{dx}\right)^2}\, dx = \int_0^{\sqrt{2}} 2\pi x\sqrt{1 + (2x)^2}\, dx$$

$$= \int_0^{\sqrt{2}} \frac{\pi}{4}(1 + 4x^2)^{1/2}(8x)\, dx = \left[\tfrac{1}{6}\pi(1 + 4x^2)^{3/2}\right]_0^{\sqrt{2}} = \tfrac{13}{3}\pi.$$

Note that the decision of which abbreviated formula—(9) or (11)—should be used is determined by the axis of revolution. By contrast, the decision whether the variable of integration should be x or y is determined by the way in which the smooth arc is given: as a function of x or as a function of y. In some problems either x or y may be used as the variable of integration, but the integral is usually much simpler to evaluate if you make the correct choice. Experience is very helpful here.

6.4 PROBLEMS

In Problems 1–10, set up and simplify the integral that gives the length of the given smooth arc; do not evaluate the integral.

1. $y = x^2, \quad 0 \leqq x \leqq 1$
2. $y = x^{5/2}, \quad 1 \leqq x \leqq 3$
3. $y = 2x^3 - 3x^2, \quad 0 \leqq x \leqq 2$
4. $y = x^{4/3}, \quad -1 \leqq x \leqq 1$
5. $y = 1 - x^2, \quad 0 \leqq x \leqq 100$
6. $x = 4y - y^2, \quad 0 \leqq y \leqq 1$
7. $x = y^4, \quad -1 \leqq y \leqq 2$
8. $x^2 = y, \quad 1 \leqq y \leqq 4$
9. $xy = 1, \quad 1 \leqq x \leqq 2$
10. $x^2 + y^2 = 4, \quad 0 \leqq x \leqq 2$

In Problems 11–20, set up and simplify the integral that gives the surface area of revolution generated by rotation of the given smooth arc about the given axis; do not evaluate the integral.

11. $y = x^2, \quad 0 \leqq x \leqq 4;$ the x-axis
12. $y = x^2, \quad 0 \leqq x \leqq 4;$ the y-axis
13. $y = x - x^2, \quad 0 \leqq x \leqq 1;$ the x-axis
14. $y = x^2, \quad 0 \leqq x \leqq 1;$ the line $y = 4$
15. $y = x^2, \quad 0 \leqq x \leqq 1;$ the line $x = 2$
16. $y = x - x^3, \quad 0 \leqq x \leqq 1;$ the x-axis
17. $y = \sqrt{x}, \quad 1 \leqq x \leqq 4;$ the x-axis
18. $y = \sqrt{x}, \quad 1 \leqq x \leqq 4;$ the y-axis
19. $y = x^{3/2}, \quad 1 \leqq x \leqq 4;$ the line $x = -1$
20. $y = x^{5/2}, \quad 1 \leqq x \leqq 4;$ the line $y = -2$

Find the lengths of the smooth arcs in Problems 21–28.

21. $y = \frac{2}{3}(x^2 + 1)^{3/2}$ from $x = 0$ to $x = 2$
22. $x = \frac{2}{3}(y - 1)^{3/2}$ from $y = 1$ to $y = 5$
23. $y = \dfrac{1}{6}x^3 + \dfrac{1}{2x}$ from $x = 1$ to $x = 3$
24. $x = \dfrac{1}{8}y^4 + \dfrac{1}{4y^2}$ from $y = 1$ to $y = 2$
25. $8x^2y - 2x^6 = 1$ from $(1, 3/8)$ to $(2, 129/32)$
26. $12xy - 4y^4 = 3$ from $(7/12, 1)$ to $(67/24, 2)$
27. $y^3 = 8x^2$ from $(1, 2)$ to $(8, 8)$
28. $(y - 3)^2 = 4(x + 2)^3$ from $(-1, 5)$ to $(2, 19)$

In each of Problems 29–35, find the area of the surface of revolution generated by revolving the given curve about the indicated axis.

29. $y = \sqrt{x}, \quad 0 \leqq x \leqq 1;$ the x-axis
30. $y = x^3, \quad 1 \leqq x \leqq 2;$ the x-axis
31. $y = \dfrac{1}{5}x^5 + \dfrac{1}{12x^3}, \quad 1 \leqq x \leqq 2;$ the y-axis
32. $x = \dfrac{1}{8}y^4 + \dfrac{1}{4y^2}, \quad 1 \leqq y \leqq 2;$ the x-axis
33. $y^3 = 3x, \quad 0 \leqq x \leqq 9;$ the y-axis
34. $y = \frac{2}{3}x^{3/2}, \quad 1 \leqq x \leqq 2;$ the y-axis
(*Suggestion:* Make the substitution $u = 1 + x$.)
35. $y = (2x - x^2)^{1/2}, \quad 0 \leqq x \leqq 2;$ the x-axis
36. Prove that the length of one arch of the sine curve $y = \sin x$ is equal to half the circumference of the ellipse

$2x^2 + y^2 = 2$. (*Suggestion:* Substitute $x = \cos\theta$ into the arc length integral for the ellipse.)

37. Use Simpson's approximation with $n = 6$ subintervals to estimate the length of the sine arch of Problem 36.

38. Use Simpson's approximation with $n = 10$ subintervals to estimate the length of the parabola $y = x^2$ from $x = 0$ to $x = 1$.

39. Verify the formula in (6) for the area of a conical frustum. Think of the frustum as being generated by revolving the line segment from $(r_1, 0)$ to (r_2, h) around the y-axis.

40. By considering a sphere of radius r as a surface of revolution, derive the formula $A = 4\pi r^2$ for its surface area.

41. Find the total length of the *astroid* shown in Fig. 6.50. The equation of its graph is $x^{2/3} + y^{2/3} = 1$.

42. Find the area of the surface generated by revolving the astroid of Problem 41 about the y-axis.

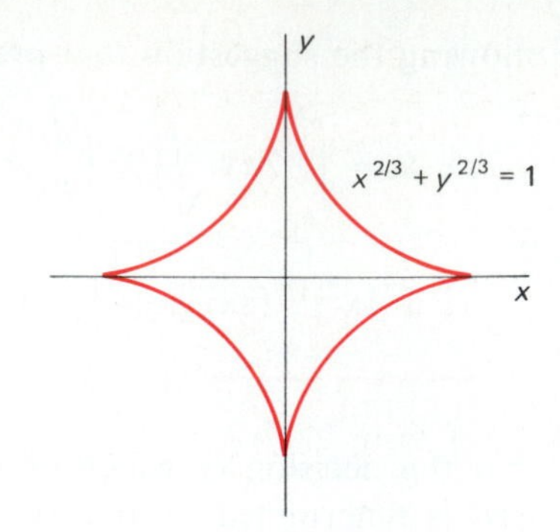

6.50 The astroid of Problems 41 and 42

6.5 Separable Differential Equations[†]

We remarked in Section 4.8 that mathematical models of changing real-world phenomena frequently involve differential equations—that is, equations containing *derivatives* of unknown functions. A *first-order* **differential equation** is one that can be written in the form

$$\frac{dy}{dx} = F(x, y) \tag{1}$$

where x denotes the independent variable and y the unknown function. A **solution** of (1) is a function $y = y(x)$ such that $y'(x) = F(x, y(x))$ for all x in some appropriate interval I.

EXAMPLE 1 If $y = x^{3/2}$ then

$$\frac{dy}{dx} = \frac{3}{2}x^{1/2} = \frac{3}{2}\cdot\frac{x^{3/2}}{x} = \frac{3y}{2x}$$

if $x > 0$. Hence the function $y(x) = x^{3/2}$ is a solution (for $x > 0$) of the differential equation

$$\frac{dy}{dx} = \frac{3y}{2x}.$$

The differential equation in (1) is said to be **separable** provided that its right-hand side is the product of a function of x and a function of y. If so, the equation takes the special form

$$\frac{dy}{dx} = g(x)\phi(y). \tag{2}$$

[†] The study of differential equations can be deferred if the instructor desires.

In this case the variables x and y can be "separated"—isolated on opposite sides of an equation—by writing informally the (literally) differential equation

$$f(y)\,dy = g(x)\,dx, \tag{3}$$

which we understand to be concise notation for the equation

$$f(y)\frac{dy}{dx} = g(x) \tag{4}$$

that we get upon multiplying each side in (2) by $f(y) = 1/\phi(y)$.

METHOD OF SOLUTION

One of the most important applications of integration is to the solution of differential equations. We show now that the solution of the differential equation in (4) reduces to the evaluation of two integrals. Integration of each side of (4) with respect to x yields

$$\int f(y(x))\frac{dy}{dx}\,dx = \int g(x)\,dx + C \tag{5}$$

because two antiderivatives of the same function can differ only by a constant. If we make the substitution

$$y = y(x), \qquad dy = \frac{dy}{dx}\,dx$$

in the left-hand side in (5), we get

$$\int f(y)\,dy = \int g(x)\,dx + C, \tag{6}$$

just as though we integrated each side of $f(y)\,dy = g(x)\,dx$ with respect to its own variable. If the indefinite integrals

$$F(y) = \int f(y)\,dy \quad \text{and} \quad G(x) = \int g(x)\,dx \tag{7}$$

can be evaluated, then (6) takes the form

$$F(y) = G(x) + C. \tag{8}$$

Finally, we can hope to solve (8) algebraically for an *explicit* solution $y = y(x)$ of the original differential equation. If not, it may be equally satisfactory to solve (8) for $x = x(y)$. If this, too, is impossible, we will generally be satisfied with the solution in the form in (8), which we call an *implicit solution* of the original differential equation. The reason for the name is that (8) presents the solution in implicitly defined form.

Observe that $y(x)$ will involve the arbitrary constant C. A solution of a differential equation involving an arbitrary constant is called a **general solution** of the differential equation. A general solution actually describes an infinite collection of different solutions of the differential equation, because different values for the constant C yield different solutions.

EXAMPLE 2 Find a general solution of the differential equation

$$\frac{dy}{dx} = \sqrt{xy} \qquad (x,\ y > 0). \tag{9}$$

Solution When we separate the variables and integrate as in (6), we get

$$\int y^{-1/2}\,dy = \int x^{1/2}\,dx + C; \qquad 2y^{1/2} = \tfrac{2}{3}x^{3/2} + C. \tag{10}$$

We can now solve for y to obtain the general solution

$$y(x) = (\tfrac{1}{3}x^{3/2} + \tfrac{1}{2}C)^2. \tag{11}$$

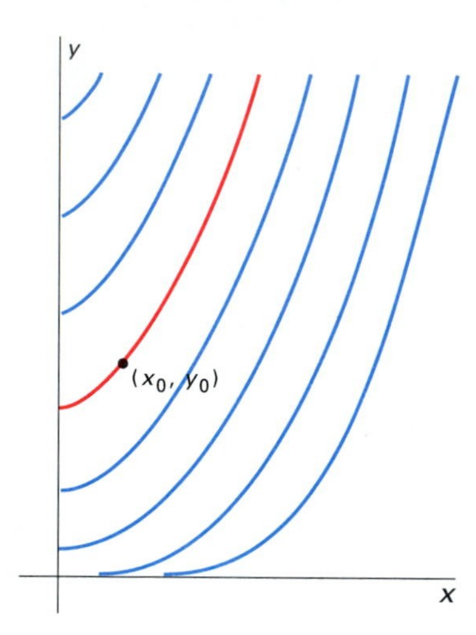

6.51 Solution curves of $y' = \sqrt{xy}$.

For each fixed value of the constant C, the graph of the solution $y(x)$ in (11) is a *solution curve* of the differential equation $y' = \sqrt{xy}$ (primes will denote derivatives with respect to x unless otherwise noted). These solution curves fill the first quadrant of the xy-plane, as indicated in Fig. 6.51. Through any point (x_0, y_0) in the first quadrant there passes exactly one of the solution curves—the one corresponding to the constant

$$C = 2(y_0)^{1/2} - \tfrac{2}{3}(x_0)^{3/2}$$

obtained by substituting $x = x_0$ and $y = y_0$ in Equation (10).

A specific solution of a differential equation—such as one obtained by specifying the value of the constant C in a general solution—is called a **particular solution** of the differential equation. Often we seek a particular solution satisfying an **initial condition** of the form

$$y(x_0) = y_0. \tag{12}$$

In this case we want to solve the **initial value problem**

$$\frac{dy}{dx} = F(x, y), \qquad y(x_0) = y_0, \tag{13}$$

consisting of a differential equation *and* an initial condition.

It is proved in differential equations texts that if $F(x, y)$ is a sufficiently well-behaved function of x and y, then the initial value problem in (13) has exactly one solution $y = y(x)$ defined for x near x_0. If the differential equation is separable, then we can attempt to find this solution as follows.

- First separate the variables and integrate to find a general solution.
- Then impose the initial condition $y(x_0) = y_0$ to evaluate the constant C and thereby determine the desired particular solution.

EXAMPLE 3 Solve the initial value problems

(a) $\dfrac{dy}{dx} = \sqrt{xy}, \quad y(0) = 1;$

(b) $\dfrac{dy}{dx} = \sqrt{xy}, \quad y(1) = 4.$

Solution In Example 2 we found the implicit solution

$$2y^{1/2} = \tfrac{2}{3}x^{3/2} + C \tag{10}$$

that yields the general solution

$$y(x) = (\tfrac{1}{3}x^{3/2} + \tfrac{1}{2}C)^2$$

of the differential equation $dy/dx = \sqrt{xy}$.

To solve the initial value problem in part (a), we substitute $x = 0$ and $y = 1$ in (10). This yields the information that $C = 2$. Hence the desired particular solution is given by

$$y(x) = (\tfrac{1}{3}x^{3/2} + 1)^2 = \tfrac{1}{9}(x^{3/2} + 3)^2.$$

To solve the initial value problem of part (b), we substitute $x = 1$ and $y = 4$ in (10):

$$2 \cdot 4^{1/2} = \tfrac{2}{3} \cdot 1^{1/2} + C.$$

It follows that $C = 4 - \frac{2}{3} = \frac{10}{3}$, so the desired particular solution is

$$y(x) = (\tfrac{1}{3}x^{3/2} + \tfrac{5}{3})^2 = \tfrac{1}{9}(x^{3/2} + 5)^2.$$

POPULATION GROWTH

A differential equation of the form

$$\frac{dP}{dt} = kP^{\alpha} \tag{14}$$

is frequently used to model the growth of an animal population that measures $P(t)$ individuals at time t (we regard the continuous function $P(t)$ as a sufficiently accurate approximation to the actual discrete population). The proportionality constant k ordinarily must be determined by experiment, while the value of the exponent α depends on the assumptions made about the way in which the population changes. For example, the case of *constant* birth and death rates corresponds to the value $\alpha = 1$ and yields the *natural* population growth equation $dP/dt = kP$. In the following example we take $\alpha = 0.5$ (mainly to make the calculations simpler).

EXAMPLE 4 Suppose that a lake is stocked initially with $P(0) = 100$ fish and that the fish population $P(t)$ satisfies thereafter the differential equation

$$\frac{dP}{dt} = k\sqrt{P} \qquad (k \text{ constant}). \tag{15}$$

If after $t = 6$ months there are 169 fish in the lake, how many will there be after 1 year (that is, when $t = 12$)?

Solution We will use the fact that $P(6) = 169$ to find the value of k. But first we must solve the differential equation. We separate the variables in (15) and integrate:

$$\int P^{-1/2}\,dP = \int k\,dt + C; \qquad 2\sqrt{P} = kt + C. \tag{16}$$

Now $P = 100$ when $t = 0$. Simultaneous substitution of these values in (16) gives $C = 2\sqrt{100} = 20$, so

$$2\sqrt{P} = kt + 20.$$

Next, substitution of $t = 6$ and $P = 169$ yields the value $k = 1$, so $2\sqrt{P} = t + 20$. Hence the fish population is given by

$$P(t) = \tfrac{1}{4}(t + 20)^2$$

after t months. Finally, the number of fish in the lake after 1 year is

$$P(12) = \tfrac{1}{4} \cdot 32^2 = 256.$$

TORRICELLI'S LAW

Suppose that a water tank has a hole with area a at its bottom and that water is leaking from the hole. Denote by $y(t)$ the depth (in feet) of water in the tank at time t (in seconds) and by $V(t)$ the volume of water (in ft^3) in the tank then. It is plausible—as well as true under ideal conditions—that the velocity of the stream of water exiting through the hole is

$$v = \sqrt{2gy} \qquad (g = 32 \text{ ft/s}^2), \tag{17}$$

which is the velocity a drop of water would acquire in falling freely from the surface of the water to the hole.

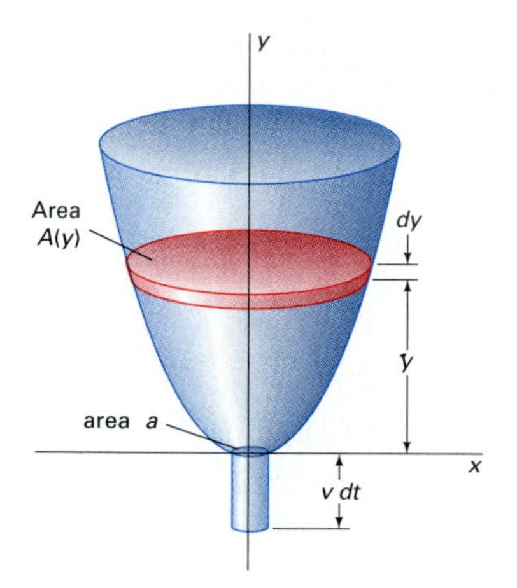

6.52 Derivation of Torricelli's law

As indicated in Fig. 6.52, the amount of water that leaves through the bottom hole during a tiny time interval dt amounts to a cylinder with base area a and height $v\,dt$. Hence the resulting change dV in the volume of water in the tank is given by

$$dV = -av\,dt = -a\sqrt{2gy}\,dt. \tag{18}$$

On the other hand, if $A(y)$ denotes the horizontal cross-sectional area of the tank at height y, then

$$dV = A(y)\,dy \tag{19}$$

as usual. Comparing (18) and (19), we conclude that $y(t)$ satisfies the differential equation

$$A(y)\frac{dy}{dt} = -a\sqrt{2gy}. \tag{20}$$

In some applications this is a very convenient form of Torricelli's law (Section 4.8). In other situations you may find that you prefer to work with the differential equation in (18) in the form

$$\frac{dV}{dt} = -a\sqrt{2gy}$$

or, if the area of the bottom hole is unknown, the form

$$\frac{dV}{dt} = -c\sqrt{y}$$

where $c = a\sqrt{2g}$ is a positive constant.

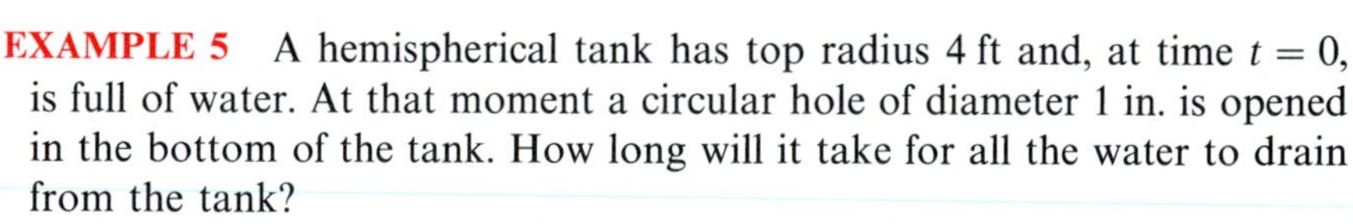

EXAMPLE 5 A hemispherical tank has top radius 4 ft and, at time $t = 0$, is full of water. At that moment a circular hole of diameter 1 in. is opened in the bottom of the tank. How long will it take for all the water to drain from the tank?

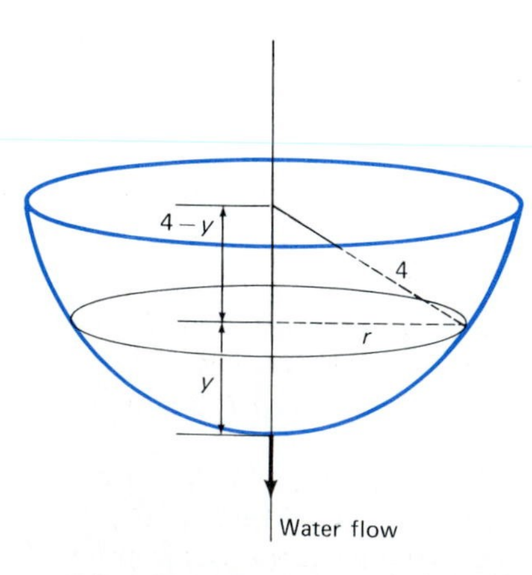

6.53 Draining a hemispherical tank

Solution From the right triangle in Fig. 6.53, we see that

$$A(y) = \pi r^2 = \pi[16 - (4 - y)^2] = \pi(8y - y^2).$$

With $g = 32$ ft/s², Equation (20) takes the form

$$\pi(8y - y^2)\frac{dy}{dt} = -\pi\left(\tfrac{1}{24}\right)^2\sqrt{64y};$$

$$\int (8y^{1/2} - y^{3/2})\,dy = -\int \tfrac{1}{72}\,dt + C;$$

$$\tfrac{16}{3}y^{3/2} - \tfrac{2}{5}y^{5/2} = -\tfrac{1}{72}t + C.$$

Now $y(0) = 4$, so

$$C = \tfrac{16}{3} \cdot 4^{3/2} - \tfrac{2}{5} \cdot 4^{5/2} = \tfrac{448}{15}.$$

The tank is empty when $y = 0$, thus when

$$t = 72 \cdot \tfrac{448}{15} \approx 2150$$

seconds; that is, about 35 min 50 s. So it takes slightly less than 36 minutes for the tank to drain.

6.5 PROBLEMS

Find general solutions (implicit if necessary) of the differential equations in Problems 1–10.

1. $\dfrac{dy}{dx} = 2x\sqrt{y}$

2. $\dfrac{dy}{dx} = 2xy^2$

3. $\dfrac{dy}{dx} = x^2y^3$

4. $\dfrac{dy}{dx} = (xy)^{3/2}$

5. $\dfrac{dy}{dx} = 2x\sqrt{y-1}$

6. $\dfrac{dy}{dx} = 4x^3(y-4)^2$

7. $\dfrac{dy}{dx} = \dfrac{1+\sqrt{x}}{1+\sqrt{y}}$

8. $\dfrac{dy}{dx} = \dfrac{x+x^3}{y+y^3}$

9. $\dfrac{dy}{dx} = \dfrac{x^2+1}{x^2(3y^2+1)}$

10. $\dfrac{dy}{dx} = \dfrac{(x^3-1)y^3}{x^2(2y^3-3)}$

Solve the initial value problems in Problems 11–20.

11. $\dfrac{dy}{dx} = y^2, \quad y(0) = 1$

12. $\dfrac{dy}{dx} = \sqrt{y}, \quad y(0) = 4$

13. $\dfrac{dy}{dx} = \dfrac{1}{4y^3}, \quad y(0) = 1$

14. $\dfrac{dy}{dx} = \dfrac{1}{x^2y}, \quad y(1) = 2$

15. $\dfrac{dy}{dx} = \sqrt{xy^3}, \quad y(0) = 4$

16. $\dfrac{dy}{dx} = \dfrac{x}{y}, \quad y(3) = 5$

17. $\dfrac{dy}{dx} = -\dfrac{x}{y}, \quad y(12) = -5$

18. $y^2\dfrac{dy}{dx} = x^2 + 2x + 1, \quad y(1) = 2$

19. $\dfrac{dy}{dx} = 3x^2y^2 - y^2, \quad y(0) = 1$

20. $\dfrac{dy}{dx} = 2xy^3(2x^2+1), \quad y(1) = 1$

21. Suppose that the fish population $P(t)$ in a lake is attacked by disease at time $t = 0$, with the result that

$$\frac{dP}{dt} = -k\sqrt{P} \qquad (k > 0)$$

thereafter. If there were initially 900 fish in the lake and 441 were left after 6 weeks, how long did it take all the fish in the lake to die?

22. Prove that the solution of the initial value problem

$$\frac{dP}{dt} = k\sqrt{P}, \qquad P(0) = P_0,$$

is given by

$$P(t) = (\tfrac{1}{2}kt + \sqrt{P_0})^2.$$

23. Suppose that the population of a certain city satisfies the differential equation of Problem 22.

(a) If $P = 100{,}000$ in 1970 and $P = 121{,}000$ in 1980, what will the population be in the year 2000?

(b) When will the population be 200,000?

24. Consider a prolific breed of rabbits whose population $P(t)$ satisfies the initial value problem

$$\frac{dP}{dt} = kP^2, \qquad P(0) = P_0,$$

where k is a positive constant. Prove that

$$P(t) = \frac{P_0}{1 - kP_0t}.$$

25. In Problem 24, suppose that $P_0 = 2$ and that there are 4 rabbits after 3 months. What happens in the next 3 months?

26. Suppose that a motorboat is traveling at $v = 40$ ft/s when its motor is cut off at time $t = 0$. Thereafter, its deceleration due to water resistance is given by $dv/dt = -kv^2$, where k is a positive constant.

(a) Solve this differential equation to show that the speed of the boat after t seconds is $v = 40/(1 + 40kt)$ ft/s.

(b) If the boat's speed after 10 s is 20 ft/s, how long does it take to slow to 5 ft/s?

27. At time $t = 0$, the bottom plug (at the vertex) of a full conical water tank 16 ft high is removed. After 1 h the water in the tank is 9 ft deep. When will the tank be empty?

28. A tank is shaped like a vertical cylinder. It initially contains water to a depth of 9 ft, and a bottom plug is pulled at time $t = 0$(h). After 1 h the depth has dropped to 4 ft. How long does it take for all the water to drain?

29. Suppose that the tank of Problem 28 has a radius of 3 ft and that its bottom hole is circular with radius 1 in. How long will it take the water (initially 9 ft deep) to drain completely?

30. A cylindrical tank with length 5 ft and radius 3 ft is situated with its axis horizontal. If a circular bottom hole with a radius of 1 in. is opened, and the tank is initially half full of benzene, how long will it take for the liquid to drain completely?

31. A spherical tank with radius 4 ft is full of gasoline when a circular bottom hole with radius 1 in. is opened. How long will be required for all the gasoline to drain from the tank?

32. *The Clepsydra, or Water Clock* A 12-h water clock is to be designed with the dimensions shown in Fig. 6.54, shaped like the surface obtained by revolving the curve $y = f(x)$ around the y-axis. What should be the equation of this curve, *and* what should be the radius of the circular bottom hole in order that the water level will fall at the *constant* rate of 4 in./h?

33. Suppose that a cylindrical tank initially containing V_0 gallons of water drains through a bottom hole in T minutes. Use Torricelli's law to show that the volume of water in the tank after $t \leqq T$ minutes will be

$$V(t) = V_0\left(1 - \frac{t}{T}\right)^2.$$

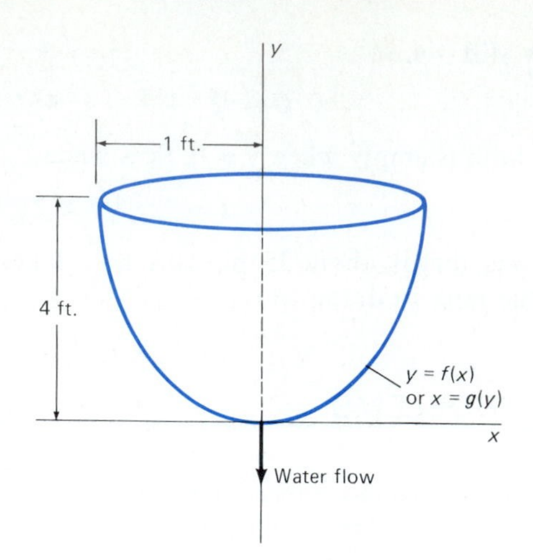

6.54 The clepsydra

6.6 Force and Work

The concept of *work* is introduced to measure the cumulative effect of a force in moving a body from one position to another. In the simplest case, a particle is moved along a straight line by the action of a *constant* force. In this case, the work done by the force is defined to be the product of the force and the distance through which it acts. Thus if the constant force has magnitude F and the particle is moved through the distance d, the work done is given by

$$W = F \cdot d. \tag{1}$$

For example, if a constant horizontal force of 50 newtons (N) is applied to push a heavy box a distance of 10 m along a rough floor (see Fig. 6.55), then the work done by the force is

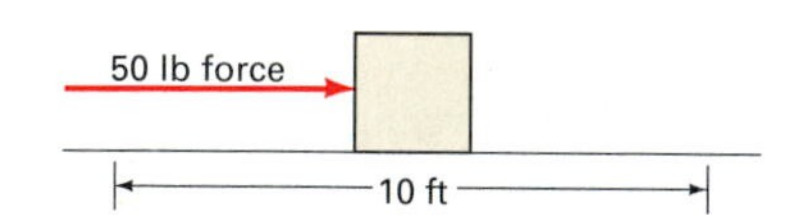

6.55 A 50-lb force does 500 ft-lb of work in pushing a box 10 ft.

$$W = (50)(10) = 500$$

newton-meters. Note the units; because of its very definition, work units are always products of force units and distance units. For another example, in order to lift a weight of 75 lb a vertical distance of 5 ft, a constant force of 75 lb must be applied, and the work done by this force is

$$W = (75)(5) = 375$$

ft-lb.

In this section we use the integral to generalize the definition of work to the case in which a particle is moved along a straight line by a *variable* force. Given a **force function** $F(x)$ defined at each point x of the straight line segment $[a, b]$, we want to define the work W done by this variable force in pushing the particle from the point $x = a$ to the point $x = b$ (see Fig. 6.56).

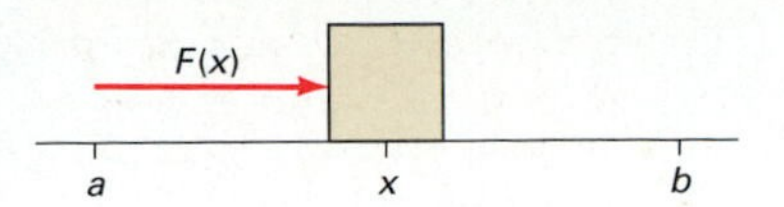

6.56 A variable force pushing a particle from a to b

We begin with the usual subdivision of the interval $[a, b]$ into n subintervals all having the same length $\Delta x = (b - a)/n$. For each $i = 1, 2, 3, \ldots, n$, let x_i^* be an arbitrary point of the ith subinterval $[x_{i-1}, x_i]$. The key idea is to approximate the actual work ΔW_i done by the **variable** force $F(x)$ in moving the particle from x_{i-1} to x_i by the work $F(x_i^*)\,\Delta x$ (force × distance) in moving a particle the distance Δx from x_{i-1} to x_i (see Fig. 6.57). Thus

$$\Delta W_i \approx F(x_i^*)\,\Delta x. \tag{2}$$

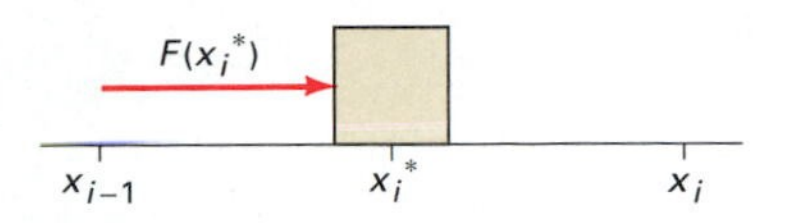

6.57 The *constant* force $F(x_i^*)$ acting through the ith subinterval

We approximate the total work W by summing from $i = 1$ to $i = n$:

$$W = \sum_{i=1}^{n} \Delta W_i \approx \sum_{i=1}^{n} F(x_i^*)\,\Delta x. \tag{3}$$

But the final sum in (3) is a Riemann sum for $F(x)$ on the interval $[a, b]$, and as $n \to +\infty$ (and $\Delta x \to 0$), such sums approach the *integral* of $F(x)$ from $x = a$ to $x = b$. We therefore are motivated to *define* the **work** W done by the force $F(x)$ in moving the particle from $x = a$ to $x = b$ to be

$$W = \int_a^b F(x)\,dx. \tag{4}$$

The following heuristic way of setting up the formula in (4) is useful in setting up integrals for work problems in general. Imagine that dx is so small a number that the value of F does not change appreciably on the tiny interval from x to $x + dx$. Then the work done by the force in moving a particle from x to $x + dx$ should be very close to

$$dW = F(x)\,dx.$$

The natural additive property of work then implies that we could obtain the total work W by adding these tiny elements of work, so that

$$W = \int_{*}^{**} dW = \int_a^b F(x)\,dx.$$

ELASTIC SPRINGS

Consider a spring with left end held fixed and right end free to move along the x-axis. We assume that the right end is at the origin $x = 0$ when the spring has its **natural length;** that is, when the spring is in its rest position, neither compressed nor stretched by outside forces.

According to **Hooke's law** for elastic springs, the force $F(x)$ that must be exerted on the spring to hold its right end at the point x is proportional to the displacement x of the right end from its rest position. That is,

$$F(x) = kx \tag{5}$$

for some positive constant k. The constant k is a characteristic of the particular spring under study and is called the **spring constant.**

Figure 6.58 shows the arrangement of such a spring along the x-axis. The spring is being held with its right end at position x on the x-axis by a force $F(x)$. The figure shows the situation for $x > 0$, so that the spring is

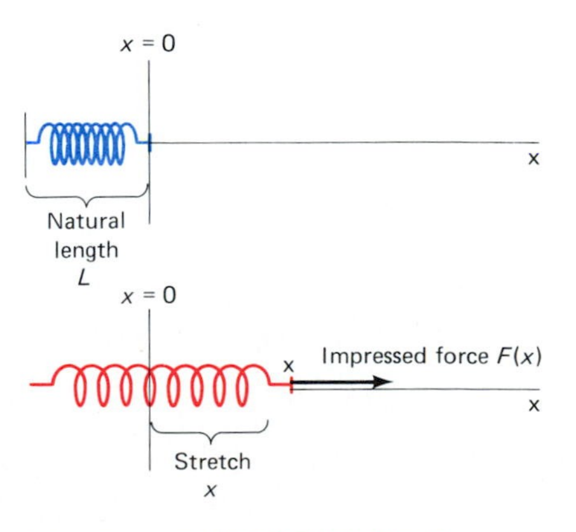

6.58 The stretch x is proportional to the impressed force F.

stretched. The force that the spring exerts on its right-hand end is directed to the left, so—as the figure shows—the external force $F(x)$ must act to the right. The right is the positive direction here, so $F(x)$ must be a positive number. Because x and $F(x)$ have the same sign, k must be positive too. You can check that k is also positive in the case $x < 0$.

EXAMPLE 1 Suppose that a spring has a natural length of 1 ft and that a force of 10 lb is required to hold it compressed to a length of 6 in. How much work is done in stretching the spring from its natural length to a total length of 2 ft?

Solution To move the free end from $x = 0$ (the natural length position) to $x = 1$ (stretched by 1 ft), we must exert a variable force $F(x)$ determined by Hooke's law. We are given that $F = -10$ (lb) when $x = -0.5$ (ft), so Equation (5), $F = kx$, implies that the spring constant for this spring is $k = 20$ (lb/ft). Thus $F(x) = 20x$, and so—using the formula in (4)—we find that the work done in stretching this spring in the manner given is

$$W = \int_0^1 20x\,dx = \Big[10x^2\Big]_0^1 = 10 \quad \text{(ft-lb).}$$

*WORK AGAINST GRAVITY

According to Newton's law of gravitation, the force that must be exerted on a body to hold it at a distance r from the center of the earth is inversely proportional to r^2. That is, if $F(r)$ denotes the holding force, then

$$F(r) = \frac{k}{r^2} \tag{6}$$

for some positive constant k. The value of this force at the surface of the earth, where $r = R \approx 4000$ mi, is called the **weight** of the body.

Given the weight $F(R)$ of a particular body, we can find the corresponding value of k by using Equation (6):

$$k = R^2F(R).$$

The work that must be done to lift the body vertically from the surface (where $r = R$) to a distance R_1 from the center of the earth is then

$$W = \int_R^{R_1} \frac{k}{r^2}\,dr. \tag{7}$$

If distance is measured in miles and force in pounds, then this integral gives the work in mile-pounds. This is a very unconventional unit of work; we shall multiply by 5280 (ft/mi) to convert any such result into foot-pounds.

EXAMPLE 2 (*Satellite Launch*) How much work must be done in order to lift a 1000-lb satellite vertically from the surface of the earth to an orbit 1000 mi above the surface? Take $R = 4000$ (mi) as the radius of the earth.

Solution Because $F = 1000$ (lb) when $r = R = 4000$ (mi), we find from Equation (6) that

$$k = (4000)^2(1000) = 16 \times 10^9 \quad \text{(mi}^2\text{-lb).}$$

Then by the formula in (7), the work done is

$$W = \int_{4000}^{5000} \frac{k}{r^2}\,dr = \left[-\frac{k}{r}\right]_{4000}^{5000}$$

$$= (16 \times 10^9)\left(\frac{1}{4000} - \frac{1}{5000}\right) = 8 \times 10^5 \quad \text{(mi-lb)}.$$

We multiply by 5280 and write the answer as

$$4.224 \times 10^9 = 4{,}224{,}000{,}000 \quad \text{(ft-lb.)}$$

We can also express the answer in terms of the power that the launch rocket must provide. **Power** is the rate at which work is done. For instance, 1 **horsepower** (hp) is defined to be 33,000 ft-lb/min. If it is known that the ascent to orbit takes 15 min and that only 2% of the power generated by the rocket is effective in lifting the satellite (the rest is used to lift the rocket and its fuel), we can convert the answer in Example 2 to horsepower. The *average* power that the rocket engine must produce during the 15-min ascent is

$$P = \frac{(50)(4.224 \times 10^9)}{(15)(33{,}000)} \approx 426{,}667 \quad \text{(hp)}.$$

The term 50 in the numerator comes from the fact that because of the 2% "efficiency" of the rocket, the total power must be multiplied by $1/(0.02) = 50$.

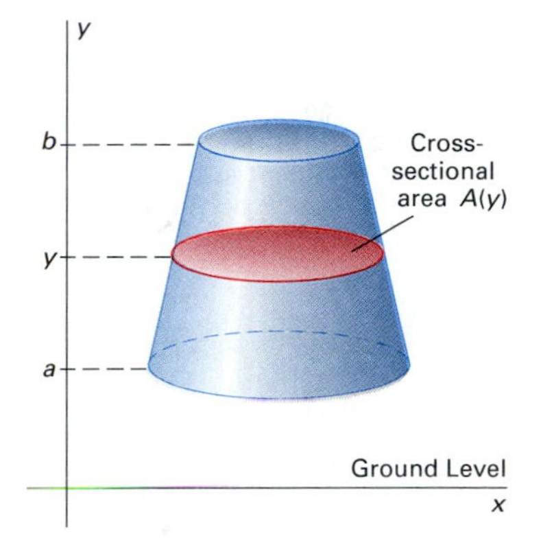

6.59 An above-ground tank

WORK TO FILL A TANK

Examples 1 and 2 are applications of the formula in (4) for calculating the work done by a variable force in moving a particle a certain distance. Another common type of force-work problem involves the summation of work done by constant forces that act through different distances. For example, consider the problem of pumping a fluid from ground level up into an above-ground tank, like the one shown in Fig. 6.59.

A convenient way to think of how the tank is filled is this: Think of thin horizontal layers of fluid, each lifted from the ground to its final position in the tank. No matter how the fluid actually behaves as the tank is filled, this simple way of thinking about the process gives a way to compute the work done in the filling process. When we think of filling the tank in this way, different layers of fluid must be lifted different distances to reach their eventual positions in the tank.

Suppose that the bottom of the tank is at height $y = a$ and that its top is at height $y = b$. Let $A(y)$ be the cross-sectional area of the tank at height y. Consider a regular partition of $[a, b]$ into n equal subintervals of length Δy. Then the volume of the horizontal slice of the tank that corresponds to the ith subinterval $[y_{i-1}, y_i]$ is

$$\Delta V_i = \int_{y_{i-1}}^{y_i} A(y)\,dy = A(y_i^*)\,\Delta y$$

for some number y_i^* in $[y_{i-1}, y_i]$; this is a consequence of the average value theorem for integrals (Section 5.5). If ρ is the density of the fluid (for instance, in pounds per cubic foot), then the force required to lift this slice from the ground to its final position in the tank is

$$F_i = \rho \, \Delta V_i = \rho A(y_i^*) \, \Delta y.$$

What about the distance through which this force must act? The fluid in question is lifted from ground level to the level of the subinterval $[y_{i-1}, y_i]$, so every particle of the fluid is lifted at least the distance y_{i-1} and at most the distance y_i. Hence the work ΔW_i needed to lift this ith slice of fluid satisfies the inequalities

$$F_i y_{i-1} \leqq \Delta W_i \leqq F_i y_i;$$

that is,

$$\rho y_{i-1} A(y_i^*) \, \Delta y \leqq \Delta W_i \leqq \rho y_i A(y_i^*) \, \Delta y.$$

Now we add these inequalities for $i = 1, 2, 3, \ldots, n$ and find that the total work $W = \Delta W_i$ satisfies the inequalities

$$\sum_{i=1}^{n} \rho y_{i-1} A(y_i^*) \, \Delta y \leqq W \leqq \sum_{i=1}^{n} \rho y_i A(y_i^*) \, \Delta y.$$

If the three points y_{i-1}, y_i, and y_i^* of $[y_{i-1}, y_i]$ were the same, then both the last two sums would be Riemann sums for the function $f(y) = \rho y A(y)$ on $[a, b]$. Altthough the three points are not the same, it still follows—from a result stated in Appendix E—that both sums approach

$$\int_a^b \rho y A(y) \, dy \quad \text{as} \quad \Delta y \to 0.$$

The squeeze law of limits therefore gives the formula

$$W = \int_a^b \rho y A(y) \, dy. \tag{8}$$

This is the work W done in pumping fluid of density ρ from the ground into a tank with horizontal cross-sectional area $A(y)$, located between heights $y = a$ and $y = b$ above the ground.

A quick heuristic way to set up the formula in (8), and many variants of it, is to think of a thin horizontal slice of fluid with volume $dV = A(y) \, dy$ and weight $\rho \, dV = \rho A(y) \, dy$. The work required to lift this slice a distance y is

$$dW = y \cdot \rho \, dV = \rho y A(y) \, dy,$$

so the total work required to fill the tank is

$$W = \int_{*}^{**} dW = \int_a^b \rho y A(y) \, dy,$$

because the horizontal slices lie between $y = a$ and $y = b$.

EXAMPLE 3 Suppose that it took 20 years to construct the great pyramid of Khufu at Gizeh. This pyramid is 500 ft high and has a square base with edge length 750 ft. Suppose also that the pyramid is made of rock with density $\rho = 120$ lb/ft^3. Finally, suppose that each laborer did 160 ft-lb per hour of effective work in lifting rocks from ground level to their final position in the pyramid and worked 12 h daily for 330 days each year. How many laborers would have been required to construct the pyramid?

Solution We assume a constant labor force throughout the 20-year period of construction. We think of the pyramid as being made up of thin horizontal slabs of rock, each slab lifted (just like a slice of water) from ground level to its ultimate height. Hence we can use the formula in (8) to compute the work W required.

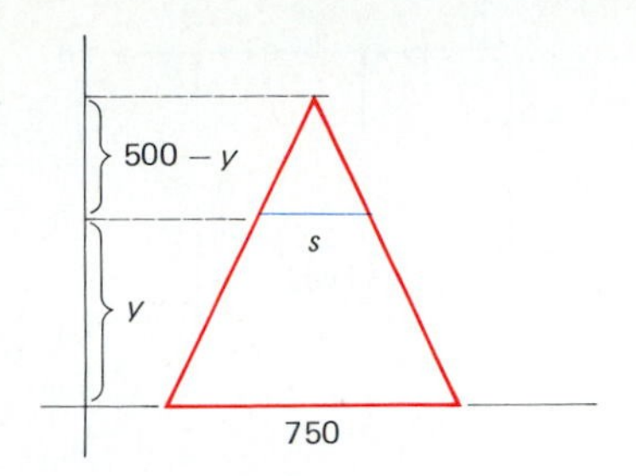

6.60 Vertical cross section of Khufu's pyramid

Figure 6.60 shows a vertical cross section of the pyramid. The horizontal cross section at height y is a square with edge s. From the simillar triangles in Fig. 6.60 we see that

$$\frac{s}{750} = \frac{500 - y}{500}, \quad \text{so that} \quad s = \frac{3}{2}(500 - y).$$

Hence the cross-sectional area at height y is

$$A(y) = \tfrac{9}{4}(500 - y)^2.$$

The formula in (8) therefore gives

$$\begin{aligned} W &= \int_0^{500} 120 \cdot y \cdot \tfrac{9}{4}(500 - y)^2 \, dy \\ &= 270 \int_0^{500} (250{,}000y - 1000y^2 + y^3) \, dy \\ &= 270\Big[125{,}000y^2 - \tfrac{1000}{3}y^3 + \tfrac{1}{4}y^4\Big]_0^{500}, \end{aligned}$$

so that $W \approx 1.406 \times 10^{12}$ ft-lb.

Because each laborer does

$$(160)(12)(330)(20) \approx 1.267 \times 10^7$$

ft-lb of work, the construction of the pyramid would—under our assumptions—have required

$$\frac{1.406 \times 10^{12}}{1.267 \times 10^7},$$

or about 111,000, laborers.

Suppose now that the tank shown in Fig. 6.61 is already filled with a liquid of density ρ pounds per cubic foot, and we want to pump all this liquid from the tank up to the level $y = h$ above the top of the tank. We imagine a thin horizontal slice of liquid at height y. If its thickness is dy, then its volume is $dV = A(y)\,dy$, so its weight is $\rho\,dV = \rho A(y)\,dy$. This slice must be lifted the distance $h - y$, so the work done to lift the slice is

$$dW = (h - y)\rho\,dV = \rho(h - y)A(y)\,dy.$$

Hence the total amount of work done on all the liquid originally in the tank is

$$W = \int_a^b \rho(h - y)A(y)\,dy. \tag{9}$$

In Problem 14 we ask you to use Riemann sums to set up this integral.

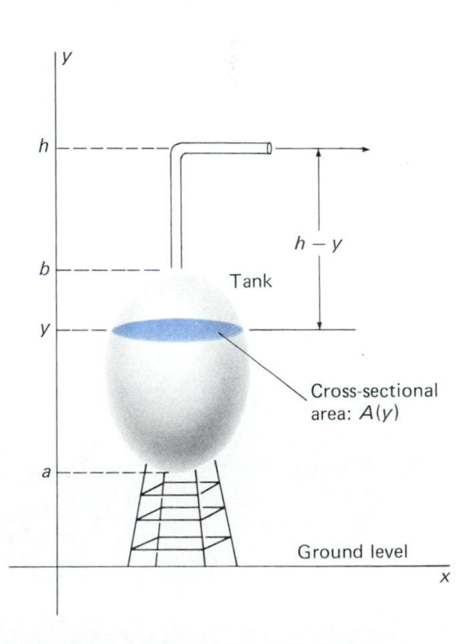

6.61 Pumping liquid from a tank to a higher level

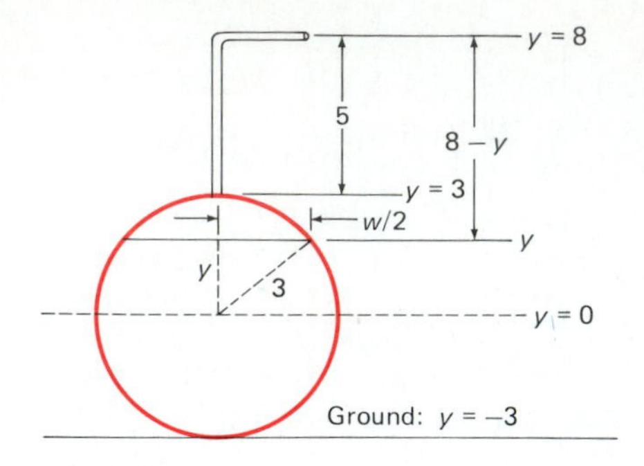

6.62 End view of the cylindrical tank of Example 4

EXAMPLE 4 A cylindrical tank of radius 3 ft and length 10 ft is lying on its side on horizontal ground. If this tank initially is full of gasoline weighing 40 lb/ft^3, how much work is done in pumping this gasoline to a point 5 ft above the top of the tank?

Solution Figure 6.62 shows an end view of the tank. In order to exploit circular symmetry, we choose $y = 0$ at the *center* of the circular vertical section, so the tank lies between $y = -3$ and $y = 3$. A horizontal cross section of the tank that meets the y-axis at y is a rectangle of length 10 ft and width w. From the right triangle in Fig. 6.62, we see that

$$\tfrac{1}{2}w = (9 - y^2)^{1/2},$$

so the area of this cross section is

$$A(y) = 10w = 20(9 - y^2)^{1/2}.$$

This cross section must be lifted from its initial position y to the final position $5 + 3 = 8$, so it is to be lifted the distance $8 - y$. Thus the formula in (9) with $\rho = 40$, $a = -3$, and $b = 3$ yields

$$\begin{aligned} W &= \int_{-3}^{3} (40)(8 - y)(20)(9 - y^2)^{1/2}\,dy \\ &= 6400 \int_{-3}^{3} (9 - y^2)^{1/2}\,dy - 800 \int_{-3}^{3} y(9 - y^2)^{1/2}\,dy. \end{aligned}$$

We attack the two integrals separately. First,

$$\int_{-3}^{3} y(9 - y^2)^{1/2}\,dy = \Big[-\tfrac{1}{3}(9 - y^2)^{3/2}\Big]_{-3}^{3} = 0.$$

Second,

$$\int_{-3}^{3} (9 - y^2)^{1/2}\,dy = \tfrac{1}{2}\pi(3)^2 = \tfrac{9}{2}\pi$$

because the integral is simply the area of a semicircle of radius 3. Hence

$$W = 6400 \cdot \tfrac{9}{2}\pi = 28{,}800\pi;$$

that is, approximately 90,478 ft-lb.

As in Example 4, you may use as needed in the problems the integral

$$\int_{0}^{a} (a^2 - x^2)^{1/2}\,dx = \tfrac{1}{4}\pi a^2 \tag{10}$$

that corresponds to the area of a quarter-circle of radius a.

FORCE EXERTED BY A LIQUID

The **pressure** p at depth h in a liquid is the force per unit area exerted by the liquid at that depth, and is given by

$$p = \rho h \tag{11}$$

where ρ is the (weight) density of the liquid. For example, at a depth of 10 ft in water, for which $\rho = 62.4$ lb/ft^3, the pressure is $(62.4)(10) = 624$ lb/ft^2.

Hence if a thin flat plate with area 5 ft^2 is suspended in a horizontal position at a depth of 10 ft in water, then the water exerts a downward force of $(624)(5) = 3120$ lb on the top face of the plate and an equal upward force on its bottom face.

It is an important fact that at a given depth in a liquid the pressure is the same in all directions. But if a flat plate is submerged in a vertical position in the liquid, then the pressure on the face of the plate is not constant, because by (11) it increases with increasing depth. Consequently, the total force exerted on the vertical plate must be computed by integration.

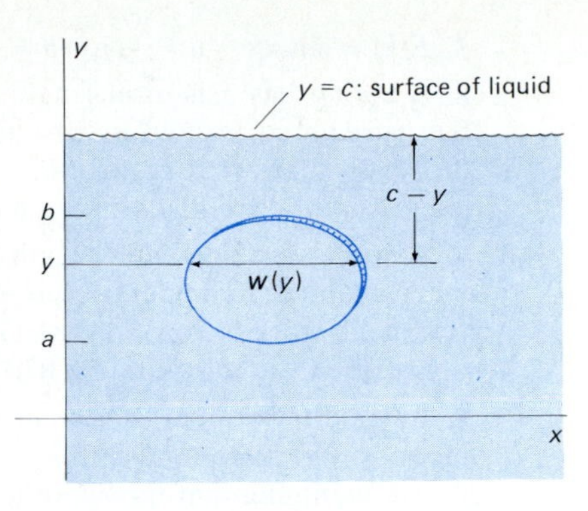

6.63 A thin plate suspended vertically in a liquid

Consider a vertical flat plate submerged in a liquid of density ρ as indicated in Fig. 6.63. The surface of the liquid is the line $y = c$, and the plate lies opposite the interval $a \leqq y \leqq b$. The width of the plate at depth $c - y$ is some function of y, which we denote by $w(y)$.

To compute the total force F exerted by the liquid on either face of this plate, we begin with a regular partition of $[a, b]$ into n equal subintervals of length Δy, and denote by y_i^* the midpoint of the subinterval $[y_{i-1}, y_i]$. The horizontal strip of the plate opposite this ith subinterval is approximated by a rectangle with width $w(y_i^*)$ and height Δy, and its average depth in the liquid is $c - y_i^*$. Hence the force ΔF_i exerted by the liquid on this horizontal strip of the plate is given approximately by

$$\Delta F_i \approx \rho(c - y_i^*)w(y_i^*)\,\Delta y, \tag{12}$$

and the total force on the entire plate is given approximately by

$$F = \sum_{i=1}^{n} \Delta F_i \approx \sum_{i=1}^{n} \rho(c - y_i^*)w(y_i^*)\,\Delta y.$$

We obtain the exact value of F by taking the limit of this Riemann sum as $\Delta y \to 0$:

$$F = \int_a^b \rho(c - y)w(y)\,dy. \tag{13}$$

EXAMPLE 5 A cylindrical tank 8 ft in diameter is lying on its side and is half full of oil with density $\rho = 75$ lb/ft^3. Find the force exerted by the oil on one end of the tank.

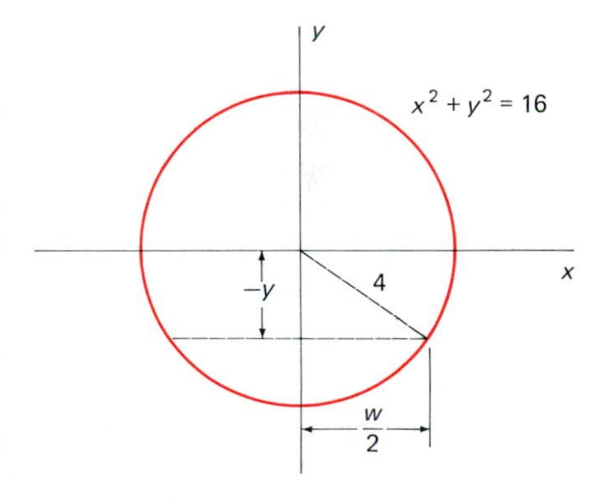

6.64 View of one end of the oil tank

Solution We locate the y-axis as indicated in Fig. 6.64, so that the surface of the oil is at the level $y = 0$. The oil corresponds to the interval $-4 \leqq y \leqq 0$, and from the right triangle in the figure we see that the width of the oil at depth $-y$ is $w(y) = 2(16 - y^2)^{1/2}$. Hence the formula in (13) gives

$$F = \int_{-4}^{0} 75(-y)[2(16 - y^2)^{1/2}]\,dy = 75\Big[\tfrac{2}{3}(16 - y^2)^{3/2}\Big]_{-4}^{0} = 3200 \quad \text{(lb)}.$$

6.6 PROBLEMS

In each of Problems 1–5, find the work done by the given force $F(x)$ in moving a particle along the x-axis from $x = a$ to $x = b$.

1. $F(x) = 10; \quad a = -2, \quad b = 1$
2. $F(x) = 3x - 1; \quad a = 1, \quad b = 5$
3. $F(x) = \dfrac{10}{x^2}; \quad a = 1, \quad b = 10$
4. $F(x) = -3\sqrt{x}; \quad a = 0, \quad b = 4$

5. $F(x) = \sin \pi x; \quad a = -1, \quad b = 1$

6. A spring has a natural length of 1 ft, and a force of 10 lb is required to hold it stretched to a total length of 2 ft. How much work is done in compressing this spring from its natural length to a length of 6 in.?

7. A spring has a natural length of 2 ft, and a force of 15 lb is required to hold it compressed to a length of 18 in. How much work is done in stretching this spring from its natural length to a length of 3 ft?

8. Apply the formula in Equation (4) to compute the amount of work done in lifting a 100-lb weight a height of 10 ft, assuming that this work is done against the constant force of gravity.

9. Compute the amount of work (in foot-pounds) done in lifting a 1000-lb weight from an orbit 1000 mi above the earth's surface to one 2000 mi above the earth's surface. Use the value of k given in Example 2.

10. A cylindrical tank is resting on the ground with its axis vertical; it has radius 5 ft and height 10 ft. Use the formula in (8) to compute the amount of work done in filling this tank with water pumped in from ground level. (Use $\rho = 62.4$ lb/ft^3 for the density of water.)

11. A conical tank is resting on its base—which is at ground level—and with its axis vertical. The tank has radius 5 ft and height 10 ft. Compute the work done in filling this tank with water ($\rho = 62.4$ lb/ft^3) pumped in from ground level.

12. Repeat Problem 11, except that now the tank is upended: Its vertex is at ground level and its base is 10 ft above the ground (but its axis is still vertical).

13. A tank with its lowest point 10 ft above ground has the shape of a cup obtained by rotating the parabola $x^2 = 5y$, $-5 \leqq x \leqq 5$, around the y-axis. The units on the coordinate axes are also in feet. How much work is done in filling this tank with oil weighing 50 lb/ft^3 if the oil is pumped in from ground level?

14. Suppose that the tank of Fig. 6.61 is filled with fluid of density ρ and that all this fluid is to be pumped from the tank to the level $y = h$ above the top of the tank. Use Riemann sums, as in the derivation of the formula in Equation (8), to obtain the formula

$$W = \int_a^b \rho(h - y)A(y)\, dy$$

for the work required to do this.

15. Use the formula of Problem 14 to find the amount of work done in pumping the water in the tank of Problem 10 to a height of 5 ft above the top of the tank.

16. Gasoline at a service station is stored in a cylindrical tank buried on its side, with the highest part of the tank 5 ft below the surface. The tank is 6 ft in diameter and 10 ft long. The density of gasoline is 45 lb/ft^3. Assume that the filler cap of each automobile gas tank is 2 ft above the ground. How much work is done in emptying all the gasoline from this tank, which is initially full, into automobiles?

17. Consider a spherical water tank of radius 10 ft, with center 50 ft above the ground. How much work is required to fill this tank by pumping water up from ground level? (*Suggestion:* It may simplify your computations to take $y = 0$ at the center of the tank and think of the distance each horizontal slice of water must be lifted.)

18. A hemispherical tank of radius 10 ft is located with its flat side down atop a tower 60 ft high. How much work is required to fill it with oil weighing 50 lb/ft^3 if the oil is to be pumped into the tank from ground level?

19. Water is being drawn from a well 100 ft deep, using a bucket that scoops up 100 lb of water. The bucket is being pulled up at the rate of 2 ft/s, but it has a hole in its bottom through which water leaks out at the rate of 0.5 lb/s. How much work is done in pulling the bucket to the top? Neglect the weight of the bucket, the weight of the rope, and the work done in overcoming friction. (*Suggestion:* Take $y = 0$ at the level of water in the well, so that $y = 100$ at ground level. Let $\{y_0, y_1, y_2, \ldots, y_n\}$ be a subdivision of $[0, 100]$ into n equal subintervals. Estimate the amount of work ΔW_i required to raise the bucket from y_{i-1} to y_i. Then set up the sum $W = \Sigma\, \Delta W_i$ and proceed to the appropriate integral by letting $n \to +\infty$.

20. A rope 100 ft long and weighing 0.25 lb per linear foot hangs from the edge of a very tall building. How much work is required to pull this rope to the top of the building?

21. Suppose that we plug the hole in the leaking bucket of Problem 19. How much work is done in lifting the mended bucket, full of water, to the surface, using the rope of Problem 20? Ignore friction and the weight of the bucket, but allow for the weight of the rope.

22. Consider a volume V of gas in a cylinder fitted with a piston at one end, where the pressure p of the gas is a function $p(V)$ of its volume. Such a cylinder is shown in Fig. 6.65. Let A be the area of the face of the piston. Then the force exerted on the piston by gas in the cylinder is $F = pA$. Assume that the gas expands from volume V_1 to volume V_2. Show that the work done by the force F is then given by

$$W = \int_{V_1}^{V_2} p(V)\, dV.$$

(*Suggestion:* If x is the length of the cylinder (from its fixed end to the face of the piston), then $F = Ap(Ax)$. Apply the

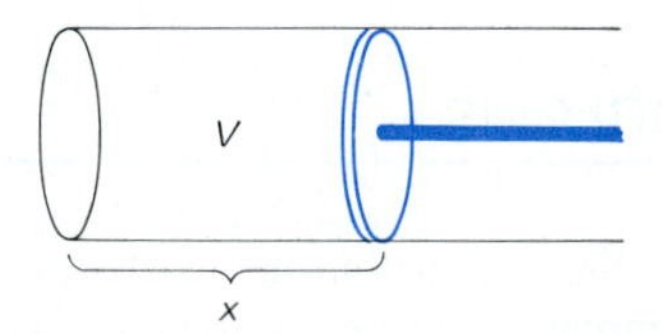

6.65 A cylinder fitted with a piston

formula in Equation (4) of this section, and make the substitution $V = Ax$ in the resulting integral.)

23. The pressure p and volume V of the steam in a small steam engine satisfy the condition $pV^{1.4} = c$ (a constant). In one cycle, the steam expands from a volume $V_1 = 50$ in.^{3} to $V_2 = 500$ in.^{3} with an initial pressure of 200 lb/in.^{2} Use the formula of Problem 22 to compute the work, in foot-pounds, done by this engine in each such cycle.

24. A tank is in the shape of a hemisphere of radius 60 ft, resting on its flat base with the curved surface on top. It is filled with alcohol weighing 40 lb/ft^3. How much work is done in pumping all the alcohol to the level of the top of the tank?

25. A tank is in the shape of the surface generated by rotating the graph of $y = x^4$, $0 \leqq x \leqq 1$, about the y-axis. The tank is initially full of oil weighing 60 lb/ft^3. The units on the coordinate axes are in feet. Find how much work is done in pumping all the oil to the level of the top of the tank.

26. A cylindrical tank of radius 3 ft and length 20 ft is lying on its side on horizontal ground. Gasoline weighing 40 lb/ft^3 is available at ground level and is to be pumped into the tank. Find the work required to fill the tank.

27. A spherical storage tank has radius 12 ft. The base of the tank is at ground level. Find the amount of work done in filling the tank with oil weighing 50 lb/ft^3 if all the oil is initially at ground level.

28. A 20-lb monkey is attached to a 50-ft chain weighing 0.5 lb per (linear) foot. The other end of the chain is attached to the 40-ft-high ceiling of the monkey's cage. Find the amount of work done by the monkey in climbing up her chain to the ceiling.

29. A kite is flying at a height of 500 ft above the ground. Suppose that the kite string weighs $\frac{1}{16}$ oz per (linear) foot, and (for simplicity) that its string is stretched in a straight line that makes an angle of 45° with the ground. How much work was done by the wind in lifting the string from ground level up to its flying position?

30. A spherical tank has radius R and its center is at distance $H > R$ above the ground. A liquid of weight density ρ is available at ground level. Show that the work required to pump the initially empty tank full of this liquid is the same as lifting the full tank the distance H.

31. A water trough 10 ft long has a square cross section that is 2 ft wide. If the trough is full of water ($\rho =$ 62.4 lb/ft^3), find the force exerted by the water on one end of the trough.

32. Repeat Problem 31 in the case in which the cross section of the trough is an equilateral triangle with edge 3 ft.

33. Repeat Problem 31 in the case in which the cross section of the trapezoid is 3 ft high, 2 ft wide at the bottom, and 4 ft wide at the top.

34. Find the force on one end of the cylindrical tank of Example 5 if it is filled with oil weighing 50 lb/ft^3. Remember that

$$\int_0^a (a^2 - y^2)^{1/2}\,dy = \tfrac{1}{4}\pi a^2$$

because the integral represents the area of a quarter-circle of radius a.

In each of Problems 35–38, a gate in the vertical face of a dam is described. Find the total force of water on this gate if its top is 10 ft beneath the surface of the water.

35. A square of edge 5 ft with its top parallel to the water surface.

36. A circle with radius 3 ft.

37. An isosceles triangle 5 ft high and 8 ft wide at the top.

38. A semicircle with radius 4 ft and with its diameter also its top edge and parallel to the water surface.

39. Suppose that the dam of Fig. 6.66 is $L = 200$ ft long and $T = 30$ ft thick at its base. Find the force of water on the dam if the water is 100 ft deep and the *slanted* end of the dam faces the water.

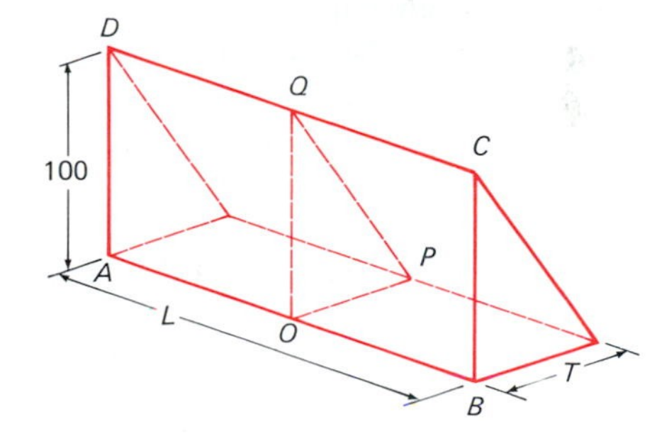

6.66 View of a model of a dam

CHAPTER 6 REVIEW: Definitions, Concepts, Results

Use the list below as a guide to concepts that you may need to review.

1. The general method of setting up an integral formula for a quantity by approximating it and then recognizing the approximation as a Riemann sum corresponding to the desired integral: If the interval $[a, b]$ is subdivided into n subintervals all of the same length $\Delta x = (b - a)/n$ and x_i^* denotes a point of the ith subinterval, then

$$\lim_{n\to\infty} \sum_{i=1}^{n} f(x_i^*)\,\Delta x = \int_a^b f(x)\,dx.$$

2. Net distance traveled as the integral of velocity:

$$s = \int_a^b v(t)\,dt.$$

3. The method of cross sections for computing volumes:

$$V = \int_a^b A(x)\,dx$$

where $A(x)$ denotes the area of a slice with infinitesimal thickness dx.

4. The volume of a solid of revolution by the method of cross sections, with the cross sections being either disks or annular rings.

5. The volume of a solid of revolution by the method of cylindrical shells:

$$V = \int_*^{**} 2\pi r\,dA$$

where r denotes the radius of the circle through which the area element dA is revolved.

6. The arc length of a smooth arc described either in the form $y = f(x)$, $a \leqq x \leqq b$, or in the form $x = g(y)$, $c \leqq y \leqq d$:

$$s = \int_*^{**} ds \qquad (ds = \sqrt{(dx)^2 + (dy)^2})$$

where

$$ds = \sqrt{1 + [f'(x)]^2}\,dx \qquad \text{for } y = f(x);$$

$$ds = \sqrt{1 + [g'(y)]^2}\,dy \qquad \text{for } x = g(y).$$

7. The area of the surface of revolution generated by revolving a smooth arc, given in either of the forms $y = f(x)$ or $x = g(y)$, about either the x-axis or the y-axis:

$$A = \int_*^{**} 2\pi r\,ds$$

where r denotes the radius of the circle through which the arc length element ds is revolved.

8. The work done by a force function in moving a particle along a straight line segment:

$$W = \int_a^b F(x)\,dx$$

if the force $F(x)$ acts from $x = a$ to $x = b$.

9. Hooke's law and the work done in stretching or compressing an elastic spring.

10. Work done against the force of gravity.

11. Work done in filling a tank, or in pumping the liquid in a tank to another level:

$$W = \int_a^b \rho h(y) A(y)\,dy$$

where $h(y)$ denotes the vertical distance the horizontal fluid slice of volume $dV = A(y)\,dy$, at height y, must be lifted.

12. The force exerted by liquid on the face of a submerged vertical plate:

$$F = \int_*^{**} \rho h\,dA$$

where h denotes the depth of the horizontal area element dA beneath the surface of the fluid of density ρ.

13. Solution of the separable differential equation

$$\frac{dy}{dx} = g(x)\phi(y)$$

by integration:

$$\int f(y)\,dy = \int g(x)\,dx + C \qquad \left(f(y) = \frac{1}{\phi(y)}\right).$$

CHAPTER 6 MISCELLANEOUS PROBLEMS

In Problems 1–3, find both the net distance and the total distance traveled between times $t = a$ and $t = b$ by a particle moving along a line with the given velocity function $v = f(t)$.

1. $v = t^2 - t - 2;\quad a = 0,\quad b = 3$

2. $v = |t^2 - 4|;\quad a = 1,\quad b = 4$

3. $v = \pi \sin \dfrac{\pi}{2}(2t - 1);\quad a = 0,\quad b = 1.5$

In Problems 4–8, a solid extends along the x-axis from $x = a$ to $x = b$, and its cross-sectional area at x is $A(x)$. Find its volume.

4. $A(x) = x^3;\quad a = 0,\quad b = 1$

5. $A(x) = x^{1/2};\quad a = 1,\quad b = 4$

6. $A(x) = x^3;\quad a = 1,\quad b = 2$

7. $A(x) = \pi(x^2 - x^4);\quad a = 0,\quad b = 1$

8. $A(x) = x^{100};\quad a = -1,\quad b = 1$

9. Suppose that rainfall begins at time $t = 0$ and that the rate after t hours is $(t + 6)/12$ in./h. How many inches of rain fall during the first 12 h?

10. The base of a certain solid is the region in the first quadrant bounded by the curves $y = x^3$ and $y = 2x - x^2$. Find its volume, if each cross section perpendicular to the x-axis is a square with one edge in the base of the solid.

11. Find the volume of the solid generated by revolving the first-quadrant region of Problem 10 about the x-axis.

12. Find the volume of the solid generated by revolving the region bounded by $y = 2x^4$ and $y = x^2 + 1$ around (a) the x-axis; (b) the y-axis.

13. A wire is made of copper (density 8.5 g/cm^3) and is shaped like a helix that spirals around the x-axis from $x = 0$ to $x = 20$ cm. Each cross section of this wire perpendicular to the x-axis is a circular disk of radius 0.25 cm. What is the total mass of the wire?

14. Derive the formula $V = \frac{1}{3}\pi h(r_1^2 + r_1 r_2 + r_2^2)$ for the volume of a frustum of a cone with height h and base radii r_1 and r_2.

15. Suppose that the point P lies on a line perpendicular to the xy-plane at the origin O, with $|OP| = h$. Consider the "elliptical cone" consisting of all points on line segments from P to points on and within the ellipse

$$\left(\frac{x}{a}\right)^2 + \left(\frac{y}{b}\right)^2 = 1.$$

Show that the volume of this elliptical cone is $V = \frac{1}{3}\pi abh$.

16. Figure 6.67 shows the region R bounded by the ellipse $(x/a)^2 + (y/b)^2 = 1$ and the line $x = a - h$, where $0 < h < a$. Revolution of R around the x-axis generates a "segment of an ellipsoid" with radius r, height h, and volume V. Show that

$$r^2 = \frac{b^2(2ah - h^2)}{a^2} \quad \text{and that} \quad V = \frac{1}{3}\pi r^2 h \frac{3a - h}{2a - h}.$$

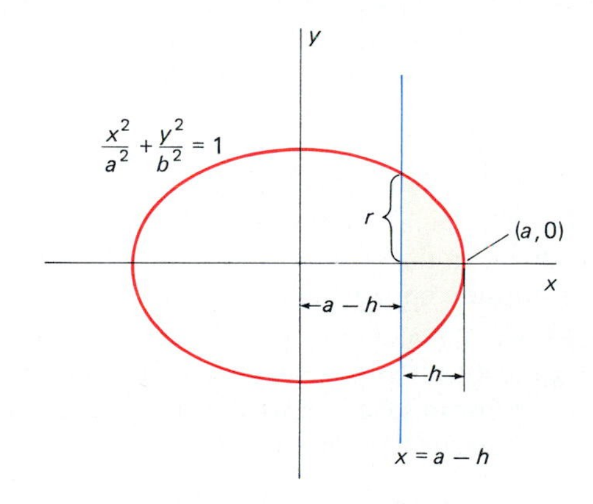

6.67 A segment of an ellipsoid

17. Figure 6.68 shows the region R bounded by the hyperbola $(x/a)^2 - (y/b)^2 = 1$ and the line $x = a + h$, where $h > 0$. Revolution of R about the x-axis generates a "segment of a hyperboloid" with radius r, height h, and volume V. Show that

$$r^2 = \frac{b^2(2ah + h^2)}{a^2} \quad \text{and that} \quad V = \frac{1}{3}\pi r^2 h \frac{3a + h}{2a + h}.$$

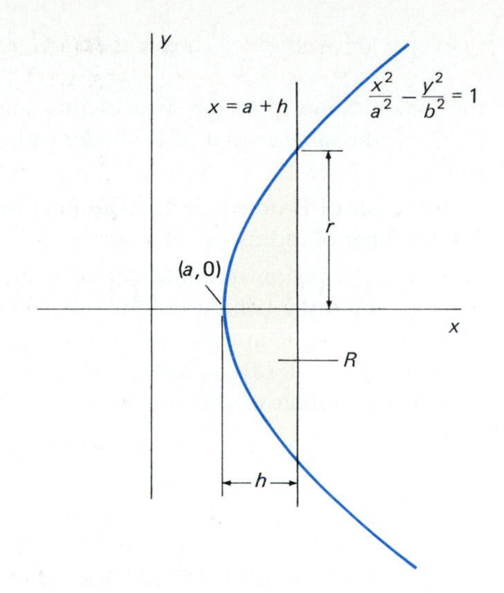

6.68 The region R of Problem 17

In Problems 18–20, the function $f(x)$ is nonnegative and continuous for $x \leqq 1$. When the region lying under $y = f(x)$ from $x = 1$ to $x = t$ is revolved around the indicated axis, the volume of tthe resulting solid is $V(t)$. Find the function $f(x)$.

18. $V(t) = \pi\left(1 - \frac{1}{t}\right)$; the x-axis

19. $V(t) = \frac{1}{6}\pi[(1 + 3t)^2 - 16]$; the x-axis

20. $V(t) = \frac{2}{9}\pi[(1 + 3t^2)^{3/2} - 8]$; the y-axis

21. Use the integral formula of Problem 36 in Section 6.3 to find the volume of the solid generated by revolving the first-quadrant region bounded by $y = x$ and $y = \sin(\pi x/2)$ around the y-axis.

22. Use the method of cylindrical shells to find the volume of the solid generated by revolving the region bounded by $y = x^2$ and $y = x + 2$ around the line $x = -2$.

23. Find the length of the curve $y = \frac{1}{3}x^{3/2} - x^{1/2}$ from $x = 1$ to $x = 4$.

24. Find the area of the surface generated by revolving the curve of Problem 23 around (a) the x-axis; (b) the y-axis.

25. Find the length of the curve $y = \frac{3}{8}(y^{4/3} - 2y^{2/3})$ from $y = 1$ to $y = 8$.

26. Find the area of the surface generated by revolving the curve of Problem 25 around (a) the x-axis; (b) y-axis.

27. Find the area of the surface generated by revolving the curve of Problem 23 around the line $x = 1$.

28. If $-r < a < b < r$, then a "spherical zone" of "height" $h = b - a$ is generated by revolving the circular arc

$$y = (r^2 - x)^{1/2}, \qquad a \leqq x \leqq b$$

around the x-axis. Show that the area of this spherical zone is $A = 2\pi rh$, the same as that of a cylinder with radius r and height h.

29. Apply the result of Problem 28 to show that the surface area of a sphere of radius r is $A = 4\pi r^2$.

30. Let R denote the region bounded by the curves $y = 2x^3$ and $y^2 = 4x$. Find the volumes of the solids obtained by revolving the region R around (a) the x-axis; (b) the y-axis; (c) the line $y = -1$; (d) the line $x = 2$. In each case use both the method of cross sections and the method of cylindrical shells.

In Problems 31–42, find the general solution of the given differential equation. If an initial condition is given, find the corresponding particular solution.

31. $\dfrac{dy}{dx} = 2x + \cos x; \quad y(0) = 0$

32. $\dfrac{dy}{dx} = 3\sqrt{x} + \dfrac{1}{\sqrt{x}}; \quad y(1) = 10$

33. $\dfrac{dy}{dx} = (y + 1)^2$

34. $\dfrac{dy}{dx} = \sqrt{y + 1}$

35. $\dfrac{dy}{dx} = 3x^2y^2; \quad y(0) = 1$

36. $\dfrac{dy}{dx} = \sqrt[3]{xy}; \quad y(1) = 1$

37. $x^2y^2\dfrac{dy}{dx} = 1$

38. $\sqrt{xy}\,\dfrac{dy}{dx} = 1$

39. $\dfrac{dy}{dx} = y^2 \cos x; \quad y(0) = 1$

40. $\dfrac{dy}{dx} = \sqrt{y} \sin x; \quad y(0) = 4$

41. $\dfrac{dy}{dx} = \dfrac{y^2(1 - \sqrt{x})}{x^2(1 - \sqrt{y})}$

42. $\dfrac{dy}{dx} = \dfrac{\sqrt{y}(x + 1)^3}{\sqrt{x}(y + 1)^3}$

43. Find the natural length L of a spring if five times as much work is required to stretch it from a length of 2 ft to a length of 5 ft, as is required to stretch it from a length of 2 ft to a length of 3 ft.

44. A steel beam weighing 1000 lb hangs from a 50-ft cable; the cable weighs 5 lb per linear foot. How much work is done in winding in 25 ft of the cable with a windlass?

45. A spherical tank of radius R (in feet) is initially full of oil weighing ρ pounds per cubic foot. Find the total work done in pumping all the oil from the sphere to a height of $2R$ above the top of the tank.

46. How much work is done by a colony of ants in building a conical anthill with height and diameter both 1 ft, using sand initially at ground level and with a density of 150 lb/ft^3?

47. Below the surface of the earth the gravitational attraction is directly proportional to the distance from the earth's center. Suppose that a straight cylindrical hole with radius 1 ft is dug from the surface of the earth (radius 3960 mi) to its center. Assume that the earth has a uniform density of 350 lb/ft^3. How much work, in foot-pounds, is done in lifting a 1-lb weight from the bottom of this hole to its top?

48. How much work is done in digging the hole of Problem 47; that is, in lifting the material it initially contained to the earth's surface?

49. Suppose that a dam is shaped like a trapezoid with height 100 ft, 300 ft long at the top and 200 ft long at the bottom. When the water level behind the dam is level with its top, what is the total force that the water exerts on the dam?

50. Suppose that a dam has the same top and bottom lengths as in Problem 49 and the same vertical height of 100 ft, but that its face toward the water is slanted at an angle of 30° from the vertical. Now what is the total force of water pressure on the dam?

51. For $c > 0$, the graphs of $y = c^2x^2$ and $y = c$ bound a plane region. Revolve this region about the horizontal line $y = -1/c$ to form a solid. For what value of c is the volume of this solid maximal? Minimal?

chapter seven

Exponential and Logarithmic Functions

7.1 Introduction

Up to this point our study of calculus has been concentrated largely on power functions and on the algebraic functions that are obtained from power functions by forming finite combinations and compositions. This chapter is devoted to the calculus of exponential and logarithmic functions. This introductory section gives a brief and somewhat informal overview of these nonalgebraic functions, partly from the perspective of precalculus mathematics. In addition to reviewing the laws of exponents and the laws of logarithms, our purpose here is to provide some intuitive background for the systematic treatment of exponential and logarithmic functions that begins in Section 7.2.

An *exponential function* is one of the form $f(x) = a^x$. Note that x is the variable; the number a is a constant. Thus an exponential function is "a constant raised to a variable power," whereas the power function $p(x) = x^k$ is "a variable raised to a constant power."

In precalculus courses, the *logarithmic function* with base a is introduced as the inverse of the exponential function $f(x) = a^x$. That is, $\log_a x$ is "the power to which a must be raised to get x," so that

$$y = \log_a x \quad \text{means that} \quad a^y = x.$$

But there are some fundamental questions that underlie these seemingly straightforward definitions of exponential and logarithmic functions.

In elementary algebra a *rational* power of the positive real number a is defined in terms of integral roots and powers. There we learn that if p and q are integers (with $q > 0$), then

$$a^{p/q} = \sqrt[q]{a^p} = (\sqrt[q]{a})^p.$$

The following **laws of exponents** are then established for all *rational* exponents r and s:

$$a^{r+s} = a^r a^s, \qquad (a^r)^s = a^{rs},$$
$$a^{-r} = \frac{1}{a^r}, \qquad (ab)^r = a^r b^r. \tag{1}$$

Moreover, recall that

$$a^0 = 1$$

for every positive real number a. A typical calculation using the laws of exponents is

$$\frac{(2^2)^3(3)^{-4}}{(2)^{-2}} = \frac{2^6 \cdot 2^2}{3^4} = \frac{2^8}{3^4} = \frac{256}{81}.$$

In applications we often need to work with irrational exponents as well as with rational ones. For example, consider a bacteria population $P(t)$ that increases by the same factor in any two time intervals of the same length and, in particular, doubles every hour. Suppose that the initial population is $P(0) = 1$ million. In 3 h, P will increase by a factor of $2 \cdot 2 \cdot 2 = 2^3$, so that $P(3) = 2^3$ (million). If k is the factor by which P increases in $\frac{1}{3}$ h, then in 1 h P will increase by a factor of $k \cdot k \cdot k = k^3 = 2$, so

$$k = 2^{1/3} \quad \text{and} \quad P(\tfrac{1}{3}) = 2^{1/3} \quad \text{(million)}.$$

More generally, if p and q are positive integers, then $P(1/q) = 2^{1/q}$; in p/q hours, P will increase p times by a factor of $2^{1/q}$ each time, so

$$P\left(\frac{p}{q}\right) = (2^{1/q})^p = 2^{p/q}.$$

Thus $P(t) = 2^t$ if t is a rational number. But because time is not restricted to rational values alone, we surely ought to conclude that $P(t) = 2^t$ for *all* $t > 0$.

But what is meant by an expression such as $2^{\sqrt{2}}$ or 2^π involving an irrational exponent? To find the value of 2^π we might work with (rational) finite decimal approximations to the irrational number $\pi = 3.1415926\ldots$. A calculator gives the (rounded) values

$$\begin{aligned}
2^{3.1} &\approx 8.5742\\
2^{3.14} &\approx 8.8152\\
2^{3.141} &\approx 8.8214\\
2^{3.1415} &\approx 8.8244\\
2^{3.14159} &\approx 8.8250\\
2^{3.141592} &\approx 8.8250\\
2^{3.1415926} &\approx 8.8250.
\end{aligned}$$

Thus it appears that $2^\pi \approx 8.8250$, rounded to four decimal places. (A ten-place calculator actually approximates 2^π by computing instead $2^{3.141592654} \approx 8.824977830$ with a rational exponent.)

But the real question is this: What is the precise *definition* of the number a^x if the exponent x is irrational? One answer is to define the number a^x using limits:

$$a^x = \lim_{n \to \infty} a^{r_n}, \tag{2}$$

where $r_1, r_2, r_3, \ldots$ is a sequence of *rational* numbers (like the finite decimal approximations to π) having x as its limit:

$$\lim_{n \to \infty} r_n = x.$$

In order to carry out this approach we would need to show all of the following:

1. The limit in (2) always exists, and its value is independent of the particular sequence of rational numbers used to approximate x. (This implies that $f(x) = a^x$ is a properly defined *function.*)
2. The laws of exponents in (1) hold for irrational exponents as well as for rational exponents.
3. The function $f(x) = a^x$ is continuous, indeed differentiable.

All of this can be done, but the complete program requires much in the way of detailed analysis of limit processes.

Fortunately, there is an alternative approach that employs concepts of calculus to circumvent this painstaking work with limits. In this section we will assume the laws of exponents and of logarithms, and proceed to investigate informally the derivatives of the exponential and logarithmic functions. What we learn will tell us how (in subsequent sections) to formulate precise definitions of these functions. These "official" definitions can finally be used to establish rather simply the laws of exponents and of logarithms, as well as other useful properties of the exponential and logarithmic functions.

Let us investigate the function $f(x) = a^x$ in the case $a > 1$. We first note that $a^r > 1$ if r is a positive rational number. If $r < s$, another rational number, then the laws of exponents give

$$a^r < a^r a^{s-r} = a^s,$$

so $f(r) < f(s)$ whenever $r < s$. That is, $f(x) = a^x$ is an *increasing* function of the rational exponent r. If we plot points on the graph of $y = a^x$ for rational values x using a typical fixed value of a such as $a = 2$, we obtain a graph like the one in Fig. 7.1. This graph is shown with a dotted curve to suggest that it is densely filled with "holes" corresponding to irrational values of x. In Section 7.4 we shall show that the "holes" in this graph can be filled to obtain the graph of an increasing and continuous function f that is defined for all real x and such that $f(r) = a^r$ for every rational number r. We therefore write $f(x) = a^x$ for all x and call f the **exponential function with base a.**

In elementary algebra the common logarithm function is introduced as the *inverse function* (Section 3.4) of the exponential function 10^x with base

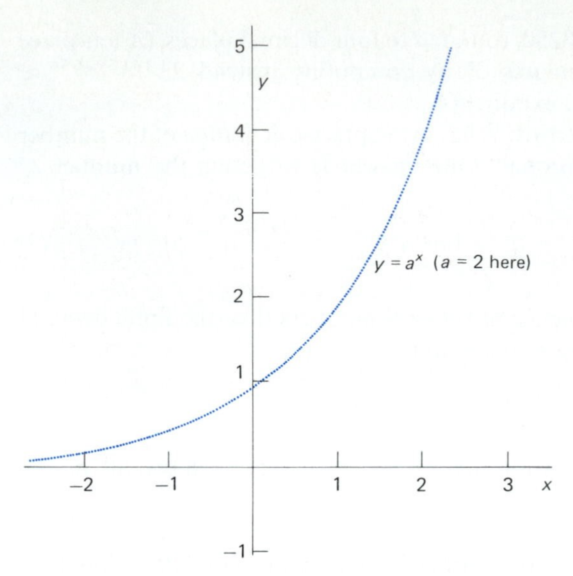

7.1 The graph of $y = a^x$ has "holes" if only rational values of x are used.

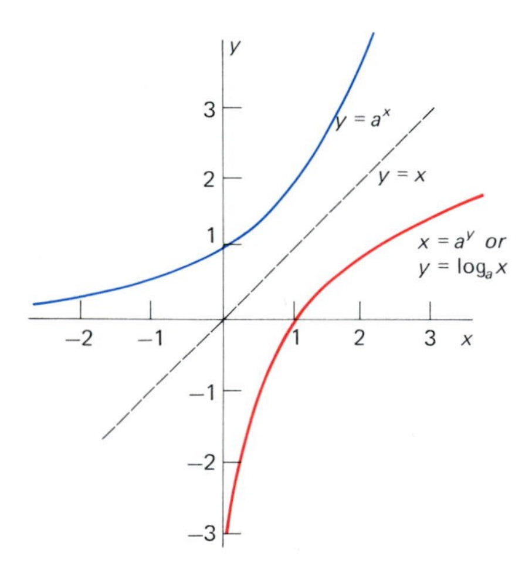

7.2 The graph of $x = a^y$ is the graph of the inverse function $\log_a x$ of the exponential function a^x. The case $a > 1$ is shown here.

10. That is, $y = \log_{10} x$ is the power to which 10 must be raised to obtain x, so that

$$y = \log_{10} x \quad \text{if and only if} \quad 10^y = x.$$

Thus $\log_{10} 1000 = 3$ because $1000 = 10^3$, and $\log_{10}(0.1) = -1$ because $0.1 = 10^{-1}$. Similarly, the **base *a* logarithm function $\log_a x$** is the inverse function of the exponential function $f(x) = a^x$ with base $a > 1$. That is,

$$y = \log_a x \quad \text{if and only if} \quad a^y = x. \tag{3}$$

Thus $\log_2 16 = 4$ because $2^4 = 16$, and $\log_3 9 = 2$ because $3^2 = 9$.

Because interchanging x and y in $a^y = x$ yields $y = a^x$, it follows from (3) that the graph of $y = \log_a x$ is the reflection in the line $y = x$ of the graph of $y = a^x$ and therefore has the shape shown in Fig. 7.2. Because $a^0 = 1$, it also follows that

$$\log_a 1 = 0,$$

so the intercepts in the figure are independent of the choice of $a > 1$. Note also that $\log_a x$ is defined *only* for $x > 0$.

The inverse function relationship between $\log_a x$ and a^x can be used to deduce, from the laws of exponents in (1), the following **laws of logarithms:**

$$\begin{aligned} \log_a xy &= \log_a x + \log_a y, \\ \log_a \frac{1}{x} &= -\log_a x, \\ \log_a \frac{x}{y} &= \log_a x - \log_a y, \\ \log_a x^y &= y \log_a x. \end{aligned} \tag{4}$$

We will verify these laws of logarithms in Section 7.2. A typical calculation using the laws of logarithms is

$$\log_{10}\left(\frac{3000}{17}\right) = \log_{10} 3000 - \log_{10} 17$$

$$= \log_{10}(3)(10^3) - \log_{10} 17$$

$$= \log_{10} 3 + 3\log_{10} 10 - \log_{10} 17$$

$$\approx 0.47712 + (3)(1) - 1.23045;$$

$$\log_{10}\left(\frac{3000}{17}\right) \approx 2.24667.$$

We used a calculator to find the values $\log_{10} 3 \approx 0.47712$ and $\log_{10} 17 \approx 1.23045$.

THE DERIVATIVES

To compute the derivative of the exponential function $f(x) = a^x$, we begin with the definition of the derivative and then use the first law of exponents in (1) to simplify. This gives

$$f'(x) = D_x a^x = \lim_{h \to 0} \frac{f(x+h) - f(x)}{h} = \lim_{h \to 0} \frac{a^{x+h} - a^x}{h}$$

$$= \lim_{h \to 0} \frac{a^x a^h - a^x}{h} \qquad \text{(by the law of exponents)}$$

$$= a^x \left(\lim_{h \to 0} \frac{a^h - 1}{h} \right) \qquad \text{(because } a^x \text{ is "constant" with respect to } h\text{).}$$

Under the assumption that $f(x) = a^x$ is indeed differentiable, it follows that the limit

$$m(a) = \lim_{h \to 0} \frac{a^h - 1}{h} \tag{5}$$

exists. Though its value $m(a)$ depends on a, it is a constant as far as x is concerned. Thus we find that the derivative of a^x is a *constant multiple* of a^x itself,

$$D_x a^x = m(a) \cdot a^x. \tag{6}$$

Because $a^0 = 1$, we see from (6) that the constant $m(a)$ is the slope of the tangent line to the curve $y = a^x$ at the point where $x = 0$.

The numerical data shown in Fig. 7.3 suggest that $m(2) \approx 0.693$ and $m(3) \approx 1.099$ (to three places). The tangent lines with these slopes are shown in Fig. 7.4. Thus it appears that

$$D_x 2^x \approx (0.693)2^x \quad \text{and} \quad D_x 3^x \approx (1.099)3^x. \tag{7}$$

We would like somehow to avoid awkward numerical factors like those in Equation (7). It seems plausible that the value $m(a)$ defined in (5) is a continuous function of a. If so, then, because $m(2) < 1$ and $m(3) > 1$, the intermediate value property implies that $m(e) = 1$ (exactly) for some number e

h	$\frac{2^h - 1}{h}$	$\frac{3^h - 1}{h}$
0.1	0.718	1.161
0.01	0.696	1.105
0.001	0.693	1.099
0.0001	0.693	1.099

7.3 Investigating the values of $m(2)$ and $m(3)$

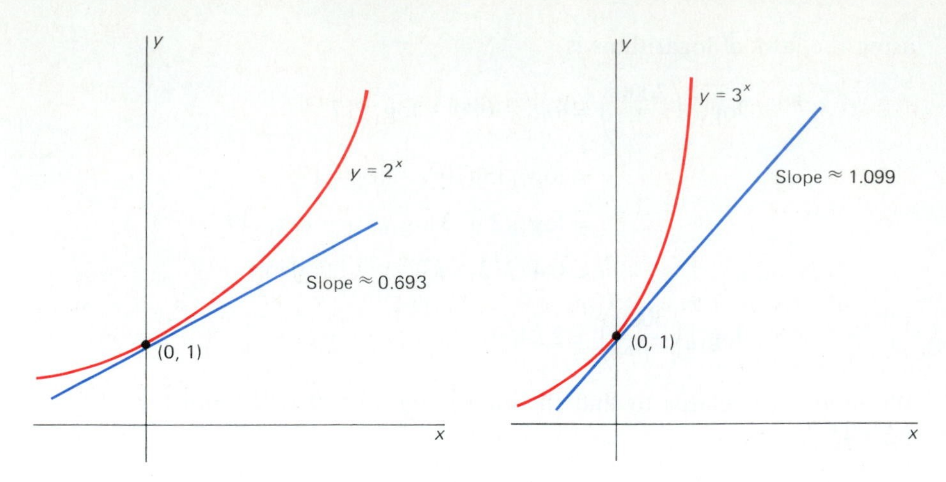

7.4 The graphs $y = 2^x$ and $y = 3^x$

between 2 and 3. If we use this particular number e as base, it then follows from Equation (6) that the derivative of the resulting exponential function $f(x) = e^x$ is given by

$$D_x e^x = e^x. \tag{8}$$

Thus the function e^x is its own derivative! For this reason we call $f(x) = e^x$ the **natural exponential function.** Its graph is shown in Fig. 7.5.

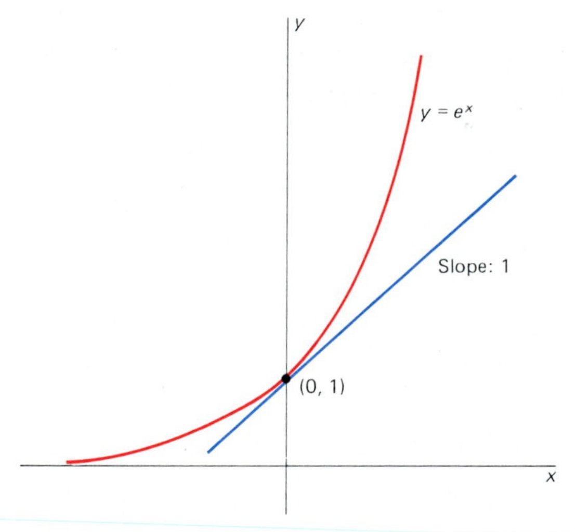

7.5 The graph $y = e^x$

In Section 7.3 we will see that the number e is given by the limit

$$e = \lim_{n \to \infty} \left(1 + \frac{1}{n}\right)^n. \tag{9}$$

n	$\left(1+\frac{1}{n}\right)^n$
10	2.594
100	2.705
1,000	2.717
10,000	2.718
100,000	2.718

7.6 Numerical estimate of the number e

Let us investigate this limit numerically. With a calculator we obtain the values in the table of Fig. 7.6. The evidence suggests (although it does not prove) that $e \approx 2.718$ to three decimal places. This number e is one of the most important special numbers in mathematics, and its value to 20 places is

$$e \approx 2.71828\,1828\,459045\,23536.$$

EXAMPLE 1 Find dy/dx if $y = x^2e^x$.

Solution The formula in (8) and the product rule yield

$$\frac{dy}{dx} = (Dx^2)e^x + x^2(De^x) = (2x)e^x + x^2(e^x) = x(2 + x)e^x.$$

The logarithm function $g(x) = \log_e x$ with base e is called the **natural logarithm function.** It is commonly denoted by the special logarithm symbol ln, so that

$$\ln x = \log_e x. \tag{10}$$

The natural logarithm function is the inverse of the natural exponential function, so

$$e^{\ln x} = x \quad \text{for all} \quad x > 0 \tag{11a}$$

and

$$\ln(e^x) = x \quad \text{for all} \quad x. \tag{11b}$$

In order to find the derivative of the function

$$u = \ln x,$$

we begin with Equation (11a) in the form

$$e^u = x.$$

Because this last equation is actually an identity (for $x > 0$), the derivative of the left-hand side with respect to x is also identically equal (for $x > 0$) to the derivative of the right-hand side with respect to x:

$$D_x e^u = D_x x.$$

With the aid of the chain rule we differentiate, and find that

$$D_u(e^u)\frac{du}{dx} = 1.$$

But $D_u(e^u) = e^u$ by (8), so

$$e^u \frac{du}{dx} = 1,$$

and finally,

$$\frac{du}{dx} = \frac{1}{e^u} = \frac{1}{e^{\ln x}} = \frac{1}{x}.$$

Hence the derivative du/dx of the natural logarithm function $u = \ln x$ is given by

$$D_x \ln x = \frac{1}{x}. \tag{12}$$

Thus $\ln x$ is the hitherto missing function with derivative $1/x$.

Just as with exponentials, the derivative of a logarithm function with base other than e involves an inconvenient numerical factor. For instance, we will see in Section 7.4 that

$$D_x \log_{10} x \approx \frac{0.4343}{x}. \tag{13}$$

The contrast between (12) and (13) illustrates one way in which base e logarithms are "natural."

EXAMPLE 2 Find the derivative of

$$y = \frac{\ln x}{x}.$$

Solution The formula in (12) and the quotient rule yield

$$\frac{dy}{dx} = \frac{(D \ln x)(x) - (\ln x)(Dx)}{x^2} = \frac{(1/x)(x) - (\ln x)(1)}{x^2} = \frac{1 - \ln x}{x^2}.$$

EXAMPLE 3 To find the derivative of $\ln x^2$ we can apply a law of logarithms to simplify before differentiation. Thus

$$D_x(\ln x^2) = D_x(2 \ln x) = 2(D_x \ln x) = \frac{2}{x}.$$

In the preliminary discussion of this section, we have introduced in an informal manner

- The number $e \approx 2.71828$,
- The natural exponential function e^x, and
- The natural logarithm function $\ln x$.

Our investigation here of the derivatives of e^x and $\ln x$—given by the formulas in (8) and (12)—should be regarded as provisional, pending a more complete discussion of these new functions in Sections 7.2 and 7.3. In any case, this brief preview of how the theory of exponential and logarithmic functions fits together should help you better understand the systematic development in the following two sections.

In subsequent sections of this chapter we shall see that the functions e^x and $\ln x$ play a vital role in the quantitative analysis of a wide range of natural phenomena—including population growth, radioactive decay, spread of epidemics, growth of investments, diffusion of pollutants, and motion with the effect of resistance taken into account.

7.1 PROBLEMS

Use the laws of exponents to simplify the expressions in Problems 1–10. Then write each answer as an integer.

1. $(2^3)(2^4)$
2. $(3^2)(3^3)$
3. $(2^2)^3$
4. $2^{(2^3)}$
5. $(3^5)(3^{-5})$
6. $(10^{10})(10^{-10})$
7. $(2^{12})^{1/3}$
8. $(3^6)^{1/2}$
9. $(4^5)(2^{-6})$
10. $(6^5)(3^{-5})$

Note that $\log_{10} 100 = \log_{10} 10^2 = 2 \log_{10} 10 = 2$. Use a similar technique to evaluate (without use of a calculator) each of the expressions in Problems 11–16.

11. $\log_2 16$
12. $\log_3 27$
13. $\log_5 125$
14. $\log_7 49$

15. $\log_{10} 1000$

16. $\log_{12} 144$

Use the laws of logarithms to express each of the natural logarithms in Problems 17–26 in terms of the three numbers ln 2, ln 3, and ln 5.

17. $\ln 8$

18. $\ln 9$

19. $\ln 6$

20. $\ln 15$

21. $\ln 72$

22. $\ln 200$

23. $\ln \frac{8}{27}$

24. $\ln \frac{12}{25}$

25. $\ln \frac{27}{40}$

26. $\ln \frac{1}{90}$

27. Which is larger, $2^{(3^4)}$ or $(2^3)^4$?

28. Evaluate $\log_{0.5} 16$.

29. By inspection, find two values of x such that $x^2 = 2^x$.

30. Show that the number $\log_2 3$ is irrational. (*Suggestion:* Assume to the contrary that $\log_2 3 = p/q$, where p and q are positive integers, and then express the consequence of this assumption in exponential form. Can an integral power of 2 equal an integral power of 3?)

In Problems 31–40, solve for x *without* using a calculator.

31. $2^x = 64$

32. $10^{-x} = 0.001$

33. $10^{-x} = 100$

34. $(3^x)^2 = 81$

35. $x^x = x^2$ (Find *all* solutions!)

36. $\log_x 16 = 2$

37. $\log_3 x = 4$

38. $e^{5x} = 7$

39. $3e^x = 3$

40. $2e^{-7x} = 5$

Find dy/dx in each of Problems 41–54.

41. $y = xe^x$

42. $y = x^3 e^x$

43. $y = \sqrt{x}\, e^x$

44. $y = \dfrac{1}{x} e^x$

45. $y = \dfrac{e^x}{x^2}$

46. $y = \dfrac{e^x}{\sqrt{x}}$

47 $y = x \ln x$

48. $y = x^2 \ln x$

49. $y = \sqrt{x} \ln x$

50. $y = \dfrac{\ln x}{\sqrt{x}}$

51. $y = \dfrac{x}{e^x}$

52. $y = e^x \ln x$

53. $y = \ln x^3$ [This means $\ln(x^3)$.]

54. $y = \ln \sqrt[3]{x}$

55. Use the chain rule to deduce from $De^x = e^x$ that

$$D_x(e^{kx}) = ke^{kx}$$

if k is a constant.

Apply the formula in Problem 55 to find the derivatives of the functions in Problems 56–58.

56. $f(x) = e^{3x}$

57. $f(x) = e^{x/10}$

58. $f(x) = e^{-10x}$

59. Substitute $2 = e^{\ln 2}$ in $P(t) = 2^t$, then apply the chain rule to show that $P'(t) = 2^t \ln 2$.

Use the method of Problem 59 to find the derivatives of the functions in Problems 60–63.

60. $P(t) = 10^t$

61. $P(t) = 3^t$

62. $P(t) = 2^{3t}$

63. $P(t) = 2^{-t}$

64. If a population $P(t)$ of bacteria (in millions) at time t (in hours) doubles every hour and initially (at time $t = 0$) the population is 1 (million), then its population at time $t \geqq 0$ is given by $P(t) = 2^t$. Use the result of Problem 59 to find the instantaneous rate of growth (in millions of bacteria per hour) of this population (a) at time $t = 0$; (b) after 4 h; that is, at time $t = 4$. (NOTE The natural logarithm of 2 is $\ln 2 \approx 0.69315$.)

7.2 The Natural Logarithm

We now begin our systematic development of the properties of exponential and logarithmic functions. Although we gave an informal overview of this material in Section 7.1, we now begin anew; in this section, we shall include the important details.

It is simplest to make the definition of the natural logarithm our initial step. Guided by the results in Section 7.1, we want to define $\ln x$ for $x > 0$ in such a way that

$$\ln 1 = 0 \quad \text{and} \quad D \ln x = \frac{1}{x}. \tag{1}$$

To do so, we recall part 1 of the fundamental theorem of calculus (Section 5.5), according to which

$$D_x \int_a^x f(t)\,dt = f(x)$$

if f is a continuous function. In order that $\ln x$ satisfy the equations in (1), we take $a = 1$ and $f(t) = 1/t$.

> ***Definition*** *The Natural Logarithm*
>
> The **natural logarithm** $\ln x$ of the positive number x is defined to be
>
> $$\ln x = \int_1^x \frac{1}{t}\,dt. \tag{2}$$

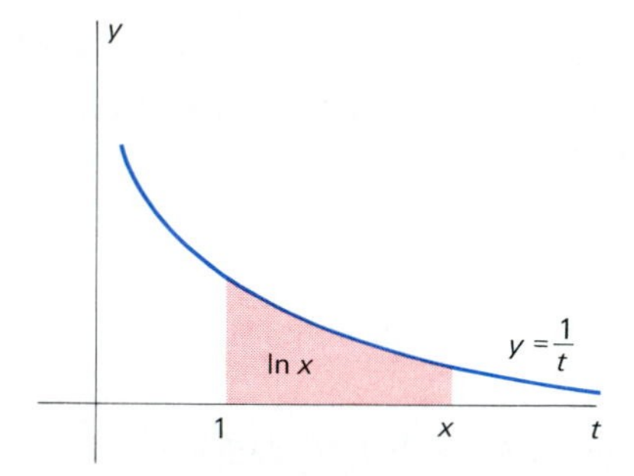

7.7 The natural logarithm function defined by means of an integral

Note that $\ln x$ is *not* defined for $x \leqq 0$. Geometrically, $\ln x$ is the area under the graph of $y = 1/t$ from $t = 1$ to $t = x$ if $x > 1$ (this is the situation shown in Fig. 7.7), the negative of this area if $0 < x < 1$, and 0 if $x = 1$. The fact that $D_x \ln x = 1/x$ follows immediately from the fundamental theorem of calculus. By Theorem 3 in Section 2.4, the fact that the function $\ln x$ is differentiable implies that it is continuous for $x > 0$.

Because $D_x \ln x = 1/x > 0$ for $x > 0$, we see that $\ln x$ must be an increasing function. Because its second derivative

$$D_x^2 \ln x = D_x \frac{1}{x} = -\frac{1}{x^2}$$

is negative for $x > 0$, it follows from Theorem 2 in Section 4.6 that the graph of $y = \ln x$ is everywhere concave downward. Later in this section we shall use the laws of logarithms to show that

$$\lim_{x \to 0^+} \ln x = -\infty \tag{3}$$

and

$$\lim_{x \to \infty} \ln x = +\infty. \tag{4}$$

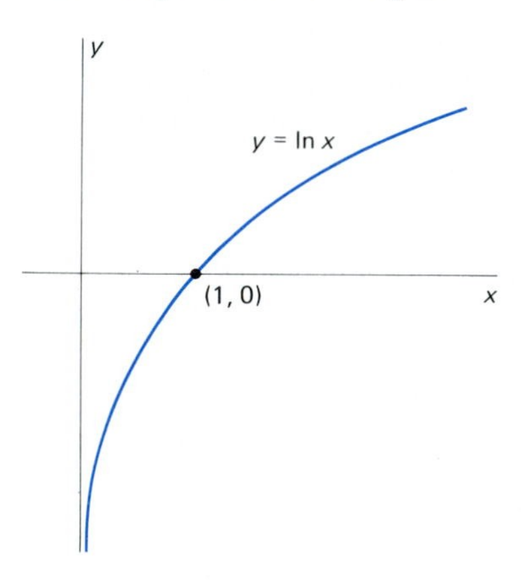

7.8 The graph of the natural logarithm function

When we assemble all these facts, we see that the graph of the equation $y = \ln x$ has the shape shown in Fig. 7.8.

When we combine the formula $D \ln x = 1/x$ with the chain rule, we obtain the differentiation formula

$$D_x \ln u = \frac{D_x u}{u} = \frac{1}{u} \cdot \frac{du}{dx}. \tag{5}$$

Here, $u = u(x)$ denotes a positive-valued differentiable function of x.

EXAMPLE 1 If $y = \ln(x^2 + 1)$, then the formula in (5) with $u = x^2 + 1$ gives

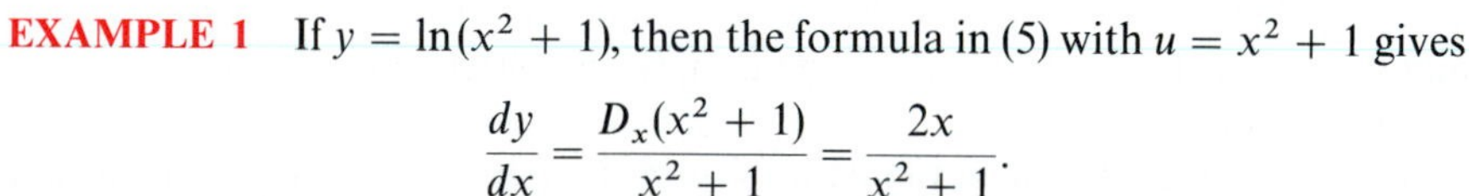

$$\frac{dy}{dx} = \frac{D_x(x^2 + 1)}{x^2 + 1} = \frac{2x}{x^2 + 1}.$$

If $y = \ln\sqrt{x^2 + 1}$, then with $u = \sqrt{x^2 + 1}$ we obtain

$$\frac{dy}{dx} = \frac{D_x\sqrt{x^2 + 1}}{\sqrt{x^2 + 1}} = \frac{1}{\sqrt{x^2 + 1}} \cdot \frac{x}{\sqrt{x^2 + 1}} = \frac{x}{x^2 + 1}.$$

Note that these two derivatives are consistent with the fact that

$$\ln\sqrt{x^2 + 1} = \tfrac{1}{2}\ln(x^2 + 1)$$

(by a law of logarithms that we have not yet rigorously proved).

EXAMPLE 2 To differentiate $f(x) = (\ln x)^2$, we use the chain rule with $u = \ln x$. This gives

$$D(\ln x)^2 = D_x(u^2) = 2u\frac{du}{dx} = 2(\ln x)\, D \ln x = \frac{2}{x}\ln x.$$

The function $f(x) = \ln|x|$ is defined for all $x \neq 0$. If $x > 0$, then $|x| = x$ and, in this case,

$$f'(x) = D \ln x = \frac{1}{x}.$$

On the other hand, if $x < 0$, then $|x| = -x$, so

$$f'(x) = D \ln(-x) = \frac{-1}{-x} = \frac{1}{x}.$$

In the second computation, we used (5) with $u = -x$. Thus we have shown that

$$D \ln|x| = \frac{1}{x} \qquad (x \neq 0) \tag{6}$$

whether x is positive or negative.

When we combine (6) with the chain rule, we obtain the formula

$$D_x \ln|u| = \frac{D_x u}{u} = \frac{1}{u}\cdot\frac{du}{dx}, \tag{7}$$

valid wherever the function $u = u(x)$ is both differentiable and nonzero.

The formula in (7) is equivalent to the integral formula

$$\int \frac{g'(x)}{g(x)}\, dx = \ln|g(x)| + C,$$

or simply

$$\int \frac{du}{u} = \ln|u| + C. \tag{8}$$

EXAMPLE 3 If $x > 0$, then the formula in (8) with

$$u = 4x + 3, \qquad du = 4\, dx$$

yields

$$\int \frac{dx}{4x+3} = \frac{1}{4}\int \frac{4}{4x+3}\, dx = \frac{1}{4}\int \frac{du}{u} = \frac{1}{4}\ln|u| + C$$

$$= \frac{1}{4}\ln|4x+3| + C = \frac{1}{4}\ln(4x+3) + C.$$

EXAMPLE 4 With $u = \cos x$, the formula in (7) yields

$$D_x \ln|\cos x| = \frac{D \cos x}{\cos x} = \frac{-\sin x}{\cos x} = -\tan x.$$

EXAMPLE 5 With $u = x^2 - 1$, the formula in (8) gives

$$\int \frac{2x}{x^2 - 1}\,dx = \ln|x^2 - 1| + C = \begin{cases} \ln(x^2 - 1) + C & \text{if } |x| > 1; \\ \ln(1 - x^2) + C & \text{if } |x| < 1. \end{cases}$$

Note the dependence on whether $|x| > 1$ or $|x| < 1$. For example,

$$\int_2^4 \frac{2x}{x^2 - 1}\,dx = \Big[\ln(x^2 - 1)\Big]_2^4 = \ln 15 - \ln 3 = \ln 5,$$

while

$$\int_{1/3}^{3/4} \frac{2x}{x^2 - 1}\,dx = \Big[\ln(1 - x^2)\Big]_{1/3}^{3/4} = \ln\frac{7}{16} - \ln\frac{8}{9} = \ln\frac{63}{128}.$$

We now use our ability to differentiate logarithms to establish the laws of logarithms.

Theorem *Laws of Logarithms*

If x and y are positive numbers and r is a rational number, then

$$\ln xy = \ln x + \ln y; \tag{9}$$

$$\ln\left(\frac{1}{x}\right) = -\ln x; \tag{10}$$

$$\ln\left(\frac{x}{y}\right) = \ln x - \ln y; \tag{11}$$

$$\ln(x^r) = r\ln x. \tag{12}$$

Proof of (9) We temporarily fix y; thus we may regard x as the independent variable and y as a constant. Then

$$D_x \ln xy = \frac{D_x(xy)}{xy} = \frac{y}{xy} = \frac{1}{x} = D_x \ln x.$$

Thus $\ln xy$ and $\ln x$ have the same derivative with respect to x. We antidifferentiate each and conclude that

$$\ln xy = \ln x + C$$

for some constant C. To evaluate C, we substitute $x = 1$ in both sides of the last equation. The fact that $\ln 1 = 0$ then implies that $C = \ln y$, and this establishes Equation (9).

Proof of (10) We differentiate $\ln(1/x)$:

$$D\left(\ln\frac{1}{x}\right) = \frac{-1/x^2}{1/x} = -\frac{1}{x} = D(-\ln x).$$

Thus $\ln(1/x)$ and $-\ln x$ have the same derivative. Hence antidifferentiation gives

$$\ln\left(\frac{1}{x}\right) = -\ln x + C$$

where C is a constant. We substitute $x = 1$ in this last equation. Because $\ln 1 = 0$, it follows that $C = 0$, and this proves (10).

Proof of (11) Because $x/y = x \cdot (1/y)$, Equation (11) follows immediately from Equations (9) and (10).

Proof of (12) We know that $Dx^r = rx^{r-1}$ if r is rational. So

$$D_x(\ln x^r) = \frac{rx^{r-1}}{x^r} = \frac{r}{x} = D_x(r \ln x).$$

Antidifferentiation then gives

$$\ln(x^r) = r \ln x + C$$

for some constant C. As before, substitution of $x = 1$ then gives $C = 0$, which proves (12). In Section 7.4 we will show that (12) holds whether or not r is rational. ■

Note that the proofs of (9), (10), and (12) are all quite similar—we differentiate the left-hand side, apply the fact that two functions with the same derivative (on an interval) differ by a constant C (on that interval), and evaluate C using the fact that $\ln 1 = 0$.

The laws of logarithms can often be used to simplify an expression prior to differentiating it, as in the following two examples.

EXAMPLE 6 If $f(x) = \ln(3x + 1)^{17}$ then

$$f'(x) = D_x \ln(3x + 1)^{17} = D_x[17 \ln(3x + 1)] = 17 \frac{3}{3x + 1} = \frac{51}{3x + 1}.$$

EXAMPLE 7 Find dy/dx given

$$y = \ln \frac{\sqrt{x^2 + 1}}{\sqrt[3]{x^3 + 1}}.$$

Solution Immediate differentiation would require using the quotient rule and the chain rule (*several* times), all the while working with a complicated fraction. Our work is made much easier if we first use the laws of logarithms to simplify the formula for y:

$$\begin{aligned} y &= \ln[(x^2 + 1)^{1/2}] - \ln[(x^3 + 1)^{1/3}] \\ &= \tfrac{1}{2} \ln(x^2 + 1) - \tfrac{1}{3} \ln(x^3 + 1). \end{aligned}$$

Finding dy/dx is now no trouble:

$$\frac{dy}{dx} = \frac{1}{2}\left(\frac{2x}{x^2 + 1}\right) - \frac{1}{3}\left(\frac{3x^2}{x^3 + 1}\right) = \frac{x}{x^2 + 1} - \frac{x^2}{x^3 + 1}.$$

Now we establish the limits of $\ln x$ as $x \to 0^+$ and as $x \to +\infty$, as stated in (3) and (4). Because $\ln 2$ is the area under $y = 1/x$ from $x = 1$ to $x = 2$, we can inscribe and circumscribe a pair of rectangles, as shown in Fig. 7.9, and conclude that

$$\tfrac{1}{2} < \ln 2 < 1.$$

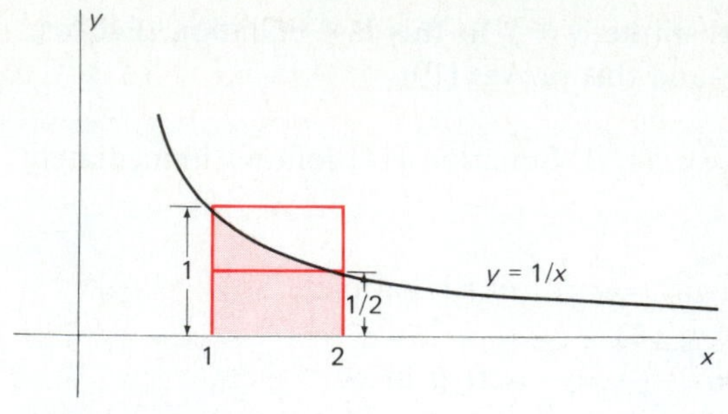

7.9 Using rectangles to estimate ln 2

The law of logarithms in Equation (12) then gives

$$\ln(2^n) = n \ln 2 > \frac{n}{2}.$$

Next suppose that $M > 0$. Let n be an integer greater than $2M$. If $x > 2^n$, then—because $\ln x$ is an increasing function—it follows that

$$\ln x > \ln(2^n) > \frac{n}{2} > \frac{2M}{2} = M.$$

Thus we can make $\ln x$ as large as we please by choosing x sufficiently large. This proves (4):

$$\lim_{x \to \infty} \ln x = +\infty.$$

To prove (3), we use the law of logarithms in Equation (10) to write

$$\lim_{x \to 0^+} \ln x = -\lim_{x \to 0^+} \ln\left(\frac{1}{x}\right) = -\lim_{y \to \infty} \ln y = -\infty,$$

by taking $y = 1/x$ and applying the result in (4).

The graph of the natural logarithm function (Fig. 7.8) suggests that though $\ln x \to +\infty$ as $x \to +\infty$, it does so rather slowly. Indeed, the function $\ln x$ increases more slowly than any positive integral power of x. By this we mean that

$$\lim_{x \to \infty} \frac{\ln x}{x^n} = 0 \tag{13}$$

if n is a fixed positive integer. To prove this, note first that

$$\ln x = \int_1^x \frac{dt}{t} \leqq \int_1^x \frac{dt}{t^{1/2}} = 2(x^{1/2} - 1),$$

because $1/t \leqq 1/t^{1/2}$ if $t \geqq 1$. Hence if $x > 1$, then

$$0 < \frac{\ln x}{x^n} \leqq \frac{\ln x}{x} \leqq \frac{2}{x^{1/2}} - \frac{2}{x}.$$

The last expression on the right approaches zero as $x \to +\infty$. Equation (13) follows by the squeeze law for limits. Moreover, this argument shows that the positive integer n in (13) may be replaced by any rational number $k \geqq 1$.

Because $\ln x$ is an increasing function, the intermediate value property implies that the curve $y = \ln x$ crosses the horizontal line $y = 1$ precisely once. The point of intersection has as its abscissa the important number $e \approx 2.71828$ mentioned in Section 7.1 (see Fig. 7.10).

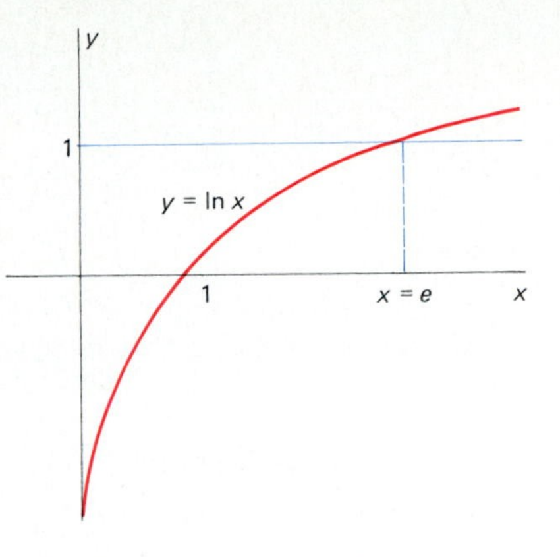

7.10 The fact that $\ln e = 1$ is expressed graphically here.

Definition of e

The number e is the unique real number such that

$$\ln e = 1. \tag{14}$$

The letter e has been used to denote the number with natural logarithm 1 ever since this number was introduced by the Swiss mathematician Leonhard Euler (1707–1783); he used e for "exponential."

EXAMPLE 8 Sketch the graph of

$$f(x) = \frac{\ln x}{x}, \qquad x > 0.$$

Solution First we compute the derivative

$$f'(x) = \frac{(1/x)(x) - (\ln x)(1)}{x^2} = \frac{1 - \ln x}{x^2}.$$

Thus the only critical point is where $\ln x = 1$; that is, where $x = e$. Because $f'(x) > 0$ if $x < e$ (the same as $\ln x < 1$) and $f'(x) < 0$ if $x > e$ (the same as $\ln x > 1$), we see that f is increasing if $x < e$ and decreasing if $x > e$. Hence f has a local maximum at $x = e$.

The second derivative of f is

$$f''(x) = \frac{(-1/x)(x^2) - (1 - \ln x)(2x)}{x^4} = \frac{2 \ln x - 3}{x^3},$$

so the only inflection point of the graph is where $\ln x = \frac{3}{2}$; that is, at $x = e^{3/2} \approx 4.48$.

Because

$$\lim_{x \to 0^+} \frac{\ln x}{x} = -\infty \quad \text{and} \quad \lim_{x \to \infty} \frac{\ln x}{x} = 0$$

—consequences of (3) and (13), respectively—we conclude that the graph of f looks like the one shown in Fig. 7.11.

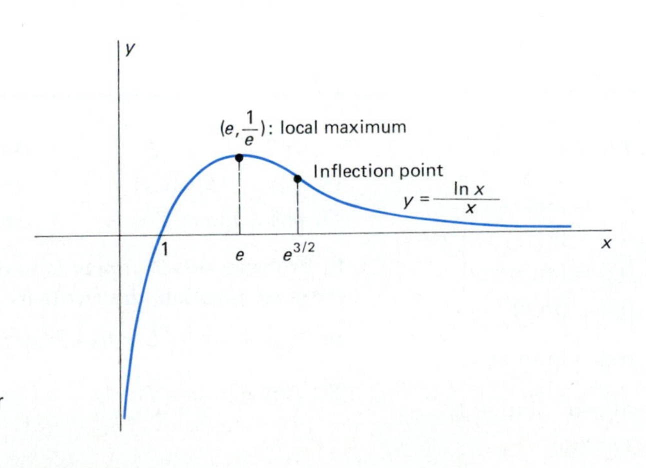

7.11 The graph for Example 8

LOGARITHMS AND EXPERIMENTAL DATA

Certain empirical data can be explained by assuming that the observed dependent variable y is a **power** function of the independent variable x. In other words, y is described by a mathematical model of the form

$$y = kx^m$$

where k and m are constants. If so, the laws of logarithms imply that

$$\ln y = \ln k + m \ln x.$$

The experimenter then plots values of $\ln y$ against values of $\ln x$. If the power function model is valid, the resulting data points will lie on a straight line with slope m and y-intercept $\ln k$, as shown in Fig. 7.12. The usefulness of this technique is that it is easy to see whether or not the data lie on a straight line, and—if they do—it is easy to measure the slope and y-intercept of the line and thereby find the values of k and m.

7.12 Plotting the logarithms of data may reveal a hidden relationship.

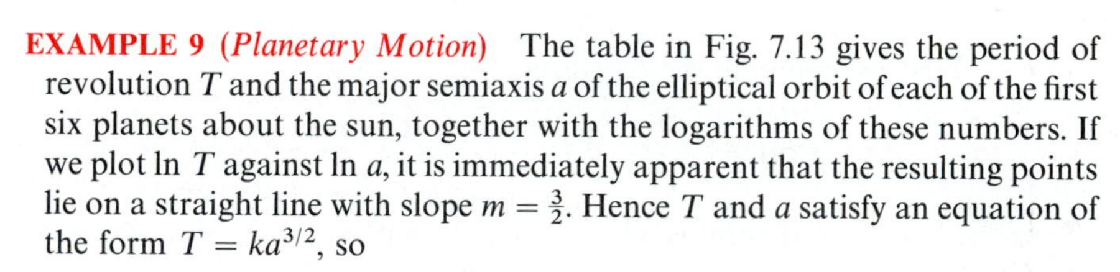

EXAMPLE 9 (*Planetary Motion*) The table in Fig. 7.13 gives the period of revolution T and the major semiaxis a of the elliptical orbit of each of the first six planets about the sun, together with the logarithms of these numbers. If we plot $\ln T$ against $\ln a$, it is immediately apparent that the resulting points lie on a straight line with slope $m = \frac{3}{2}$. Hence T and a satisfy an equation of the form $T = ka^{3/2}$, so

$$T^2 = Ca^3.$$

Planet	T (in days)	a (in 10^6 km)	$\ln T$	$\ln a$
Mercury	87.97	58	4.48	4.06
Venus	224.70	108	5.41	4.68
Earth	365.26	149	5.90	5.00
Mars	686.98	228	6.53	5.43
Jupiter	4332.59	778	8.37	6.66
Saturn	10759.20	1426	9.28	7.26

7.13 Data for Example 9

This means that the square of the period T is proportional to the cube of the major semiaxis a. This is Kepler's third law of planetary motion, which he discovered empirically in 1619.

7.2 PROBLEMS

Differentiate the functions given in Problems 1–18.

1. $f(x) = \ln(3x - 1)$

2. $f(x) = \ln(4 - x^2)$

3. $f(x) = \ln\sqrt{1 + 2x}$

4. $f(x) = \ln[(1 + x)^3]$

5. $f(x) = \ln\sqrt[3]{x^3 - x}$

6. $f(x) = \ln(\sin^2 x)$

7. $f(x) = \cos(\ln x)$

8. $f(x) = (\ln x)^3$

9. $f(x) = \dfrac{1}{\ln x}$

10. $f(x) = \ln(\ln x)$

11. $f(x) = \ln(x\sqrt{x^2 + 1})$

12. $g(t) = t^{3/2}\ln(t + 1)$

13. $f(x) = \ln(\cos x)$

14. $f(x) = \ln(2 \sin x)$

15. $f(t) = t^2 \ln(\cos t)$

16. $f(x) = \sin(\ln 2x)$

17. $g(t) = t(\ln t)^2$

18. $g(t) = \sqrt{t}\,[\cos(\ln t)]^2$

In Problems 19–28, apply laws of logarithms to simplify the given function, then write its derivative.

19. $f(x) = \ln[(2x + 1)^3(x^2 - 4)^4]$

20. $f(x) = \ln\sqrt{\dfrac{1 - x}{1 + x}}$

21. $f(x) = \ln \sqrt{\dfrac{4 - x^2}{9 + x^2}}$

22. $f(x) = \ln \dfrac{\sqrt{4x - 7}}{(3x - 2)^3}$

23. $f(x) = \ln \dfrac{x + 1}{x - 1}$

24. $f(x) = x^2 \ln \dfrac{1}{2x + 1}$

25. $g(t) = \ln \dfrac{t^2}{t^2 + 1}$

26. $f(x) = \ln \dfrac{\sqrt{x + 1}}{(x - 1)^3}$

27. $f(x) = \ln \dfrac{\sin x}{x}$

28. $f(x) = \ln \dfrac{\sin x}{\cos x}$

In Problems 29–32, find dy/dx by implicit differentiation.

29. $y = x \ln y$

30. $y = (\ln x)(\ln y)$

31. $xy = \ln(\sin y)$

32. $xy + x^2(\ln y)^2 = 4$

Evaluate the indefinite integrals in Problems 33–50.

33. $\displaystyle\int \frac{dx}{2x - 1}$

34. $\displaystyle\int \frac{dx}{3x + 5}$

35. $\displaystyle\int \frac{x}{1 + 3x^2}\,dx$

36. $\displaystyle\int \frac{x^2}{4 - x^3}\,dx$

37. $\displaystyle\int \frac{x + 1}{2x^2 + 4x + 1}\,dx$

38. $\displaystyle\int \frac{\cos x}{1 + \sin x}\,dx$

39. $\displaystyle\int \frac{1}{x}(\ln x)^2\,dx$

40. $\displaystyle\int \frac{1}{x \ln x}\,dx$

41. $\displaystyle\int \frac{1}{x + 1}\,dx$

42. $\displaystyle\int \frac{x}{1 - x^2}\,dx$

43. $\displaystyle\int \frac{2x + 1}{x^2 + x + 1}\,dx$

44. $\displaystyle\int \frac{x + 1}{x^2 + 2x + 3}\,dx$

45. $\displaystyle\int \frac{\ln x}{x}\,dx$

46. $\displaystyle\int \frac{\ln(x^3)}{x}\,dx$

47. $\displaystyle\int \frac{\sin 2x}{1 - \cos 2x}\,dx$

48. $\displaystyle\int \frac{dx}{x(\ln x)^2}$

49. $\displaystyle\int \frac{x^2 - 2x}{x^3 - 3x^2 + 1}\,dx$

50. $\displaystyle\int \frac{dx}{\sqrt{x}(1 + \sqrt{x})}$ (*Suggestion:* Let $u = 1 + \sqrt{x}$.)

Apply (13) to evaluate the limits in Problems 51–56.

51. $\displaystyle\lim_{x \to \infty} \frac{\ln(x^{1/2})}{x}$

52. $\displaystyle\lim_{x \to \infty} \frac{\ln(x^3)}{x^2}$

53. $\displaystyle\lim_{x \to \infty} \frac{\ln x}{x^{1/2}}$ (*Suggestion:* Substitute $x = u^2$.)

54. $\displaystyle\lim_{x \to 0^+} x \ln x$ (*Suggestion:* Substitute $x = 1/u$).

55. $\displaystyle\lim_{x \to 0^+} x^{1/2} \ln x$

56. $\displaystyle\lim_{x \to \infty} \frac{(\ln x)^2}{x}$

57. Prove: If $x \geqq 1$, then

$$\ln(x + \sqrt{x^2 - 1}) = -\ln(x - \sqrt{x^2 - 1}).$$

58. Find a formula for $f^{(n)}(x)$ given $f(x) = \ln x$.

59. The heart rate R (in beats per minute) and weight W (in pounds) of various mammals were measured, with the results shown in Fig. 7.14. Use the method of Example 9 to find a relation between the two of the form $R = kW^m$.

W	R
25	131
67	103
127	88
175	81
240	75
975	53

7.14 Data for Problem 59

60. During the adiabatic expansion of a certain diatomic gas, its volume V (in liters) and pressure P (in atmospheres) were measured, with the results shown in Fig. 7.15. Use the method of Example 9 to find a relation between V and P for the form $P = kV^m$.

V	P
1.46	28.3
2.50	13.3
3.51	8.3
5.73	4.2
7.26	3.0

7.15 Data for Problem 60

61. Substitute $y = x^p$ and then apply (13) to show that

$$\lim_{x \to \infty} \frac{\ln x}{x^p} = 0$$

if $0 < p < 1$.

62. Deduce from Problem 61 that

$$\lim_{x \to \infty} \frac{(\ln x)^k}{x} = 0 \qquad \text{if} \quad k > 0.$$

63. Substitute $y = 1/x$ and then apply (13) to show that

$$\lim_{x \to 0^+} x^k \ln x = 0 \qquad \text{if} \quad k > 0.$$

Use the limits in the preceding problems to sketch the graphs, for $x > 0$, of the functions given in Problems 64–67.

64. $y = x \ln x$

65. $y = x^2 \ln x$

66. $y = \sqrt{x} \ln x$

67. $y = \dfrac{\ln x}{\sqrt{x}}$

68. Problem 26 of Section 5.8 calls for showing, by numerical integration, that

$$\int_1^{2.7} \frac{dx}{x} < 1 < \int_1^{2.8} \frac{dx}{x}.$$

Explain carefully why this result proves that $2.7 < e < 2.8$.

69. If n moles of an ideal gas expand at *constant* temperature T, then its pressure and volume satisfy the equation $pV = nRT$ (n and R are constants). With the aid of Problem 22 in Section 6.6, show that the work W done by the gas in expanding from volume V_1 to volume V_2 is

$$W = nRT \ln \frac{V_2}{V_1}.$$

70. "Gabriel's trumpet" is obtained by revolving the curve $y = 1/x$, $x \geqq 1$, around the x-axis. Let A_b denote its surface area from $x = 1$ to $x = b$. Show that $A_b \geqq 2\pi \ln b$, so—as a consequence—$A_b \to +\infty$ as $b \to +\infty$. Thus the surface area of Gabriel's trumpet is infinite. Is its volume finite or infinite?

71. According to the prime number theorem, which was conjectured by the great German mathematician Carl Friedrich Gauss in 1792 (when he was 15 years old) but not proved until over a century later (independently, in 1896, by Jacques Hadamard and C. J. de la Vallée Poussin), the number of primes between the large positive numbers a and b ($a < b$) is given to a close approximation by the integral

$$\int_a^b \frac{1}{\ln x}\, dx.$$

The midpoint and trapezoidal approximations with $n = 1$ subinterval provide an underestimate and an overestimate of the value of this integral. (Why?) Calculate them with $a = 90{,}000$ and $b = 100{,}000$. The actual number of primes in this range is 879.

*7.2 Optional Computer Application

The fact that the number e satisfies (by definition) the relation

$$\int_1^e \frac{dx}{x} = 1$$

suggests the possibility of using numerical integration to investigate the value of e. For this purpose we can employ Program SIMPSON in the Optional Computer Application that follows Section 5.8. It is convenient first to replace the single line

```
150    A = 1 : B = 2     ‘Limits of integration’
```

there with the two lines

```
150    A = 1     ‘Lower limit of integration’
155    INPUT “Upper limit b”; B
```

so that we can approximate

$$\int_1^b \frac{dx}{x}$$

with different values of b.

Our strategy is to "bracket" e between closer and closer approximations. First, the following results of the modified Program SIMPSON with $n = 50$ subintervals indicate that $2.7 < e < 2.8$:

```
RUN
Upper limit b? 2
EVEN number of subintervals? 50
SUM = .6931473
```

```
RUN
Upper limit b? 3
EVEN number of subintervals? 50
Sum = 1.098613

RUN
Upper limit b? 2.7
EVEN number of subintervals? 50
SUM = .9932514

RUN
Upper limit b? 2.8
EVEN number of subintervals? 50
Sum = 1.029619
```

Next, the following results with $n = 100$ subintervals refine our estimate to $2.718 < e < 2.719$:

```
RUN
Upper limit b? 2.71
EVEN number of subintervals? 100
SUM = .9969487

RUN
Upper limit b? 2.72
EVEN number of subintervals? 100
Sum = 1.000632

RUN
Upper limit b? 2.718
EVEN number of subintervals? 100
SUM = .999897

RUN
Upper limit b? 2.719
EVEN number of subintervals? 100
Sum = 1.000265
```

The final results shown next indicate that $e \approx 2.71828$ (rounded to five decimal places):

```
RUN
Upper limit b? 2.7182
EVEN number of subintervals? 200
SUM = .9999682

RUN
Upper limit b? 2.7183
EVEN number of subintervals? 200
Sum = 1.000006

RUN
Upper limit b? 2.71828
EVEN number of subintervals? 200
SUM = .9999986
```

```
RUN
Upper limit b? 2.71929
EVEN number of subintervals? 200
Sum = 1.000004
```

7.3 The Exponential Function

We saw in Section 7.2 that the natural logarithm function $\ln x$ is continuous and increasing for $x > 0$, and that it attains arbitrarily large positive and negative values (because of the limits in (3) and (4) of Section 7.2). It follows that $\ln x$ has an inverse function that is defined for all x. To see this, let y be any (fixed) real number whatsoever. If a and b are positive numbers such that $\ln a < y < \ln b$, then the intermediate value property gives a number $x > 0$, with x between a and b, such that $\ln x = y$. Because ln is an increasing function, there is only *one* such number x with $\ln x = y$. This inverse function of ln is called the *natural exponential function* and is denoted by exp.

Definition *The Natural Exponential Function*

The (natural) **exponential function** exp is defined for all x as follows:

$$\exp x = y \quad \text{if and only if} \quad \ln y = x. \tag{1}$$

Thus $\exp x$ is simply that (positive) number whose natural logarithm is x. It is an immediate consequence of (1) that

$$\ln(\exp x) = x \quad \text{for all} \quad x, \tag{2}$$

and that

$$\exp(\ln y) = y \quad \text{for all} \quad y > 0. \tag{3}$$

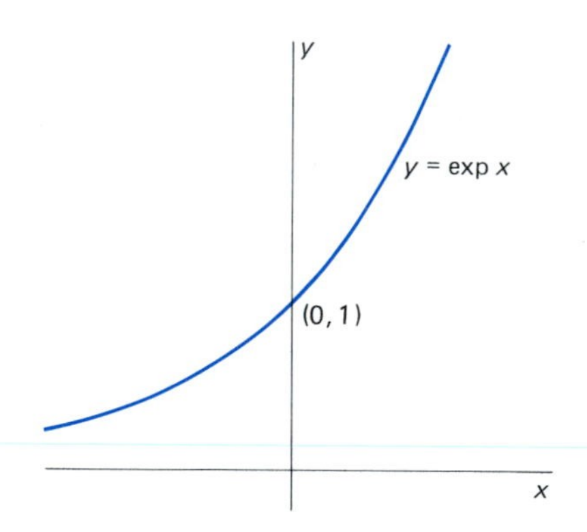

7.16 The graph of the exponential function, exp

Just like the graphs of $y = a^x$ and $y = \log_a x$ discussed informally in Section 7.1, the fact that $\exp x$ and $\ln x$ are inverse functions implies that the graphs $y = \exp x$ and $y = \ln x$ are reflections of each other in the line $y = x$ (like those in Fig. 7.2). Therefore, the graph of the exponential function looks like the one shown in Fig. 7.16. In particular, $\exp x$ is positive-valued for all x, and

$$\exp 0 = 1, \tag{4}$$

$$\lim_{x \to \infty} \exp x = +\infty, \quad \text{and} \tag{5}$$

$$\lim_{x \to -\infty} \exp x = 0. \tag{6}$$

These facts follow from the equation $\ln 1 = 0$ and from the limits in (3) and (4) in Section 7.2.

Recall that the number $e \approx 2.71828$ was defined in Section 7.2 as the number that has natural logarithm 1. If r is any rational number, it follows that

$$\ln(e^r) = r \ln e = r.$$

But Equation (1) implies that $\ln(e^r) = r$ if and only if

$$\exp r = e^r.$$

Thus exp x is equal to e^x (e raised to the power x) if x is a rational number. We therefore *define* e^x for x irrational as well as rational by

$$e^x = \exp x. \tag{7}$$

This is our first instance of irrational powers.

Equation (7) is the reason for calling exp the natural exponential function. With this notation, Equations (1) through (3) become

$$e^x = y \quad \text{if and only if} \quad \ln y = x, \tag{8}$$

$$\ln(e^x) = x \quad \text{for all} \quad x, \quad \text{and} \tag{9}$$

$$e^{\ln x} = x \quad \text{for all} \quad x > 0. \tag{10}$$

To justify Equation (7), we should show that powers of e satisfy the laws of exponents. We can do this immediately.

Theorem *Laws of Exponents*

If x and y are real numbers and r is rational, then:

$$e^x e^y = e^{x+y}, \tag{11}$$

$$e^{-x} = \frac{1}{e^x}, \quad \text{and} \tag{12}$$

$$(e^x)^r = e^{rx}. \tag{13}$$

Proof The laws of logarithms and Equation (9) give

$$\ln(e^x e^y) = \ln(e^x) + \ln(e^y) = x + y = \ln(e^{x+y}).$$

Thus (11) follows from the fact that ln is an increasing function, and therefore is one-to-one. Similarly,

$$\ln([e^x]^r) = r \ln(e^x) = rx = \ln(e^{rx}).$$

So (13) follows in the same way. The proof of (12) is almost identical. In Section 7.4, we will see that the restriction that r be rational in (13) is actually unnecessary. That is,

$$(e^x)^y = e^{xy}$$

for *all* real numbers x and y. ■

Because e^x is the inverse of the differentiable and increasing function $\ln x$, it follows from Theorem 2 in Section 3.4 that e^x is differentiable and therefore also continuous. We may thus differentiate both sides of the equation (actually, the *identity*)

$$\ln(e^x) = x$$

with respect to x. Let $u = e^x$. Then this equation becomes

$$\ln u = x,$$

and the derivatives also must be equal:

$$\frac{1}{u} \cdot \frac{du}{dx} = 1 \qquad (\text{because } u > 0).$$

So

$$\frac{du}{dx} = u = e^x;$$

that is,

$$D_x e^x = e^x, \tag{14}$$

as we indicated in Section 7.1.

If u denotes a differentiable function of x, then (14) in combination with the chain rule gives

$$D_x e^u = e^u \frac{du}{dx}. \tag{15}$$

The corresponding integration formula is

$$\int e^u \, du = e^u + C. \tag{16}$$

The special case of (15) with $u = kx$ (k constant) is worth noting:

$$D_x e^{kx} = ke^{kx}.$$

For instance, $D_x e^{5x} = 5e^{5x}$.

EXAMPLE 1 Find dy/dx given $y = e^{\sqrt{x}}$.

Solution With $u = \sqrt{x}$, the formula in (15) gives

$$\frac{dy}{dx} = e^{\sqrt{x}} D_x(\sqrt{x}) = e^{\sqrt{x}}\left(\frac{1}{2} x^{-1/2}\right) = \frac{e^{\sqrt{x}}}{2\sqrt{x}}.$$

EXAMPLE 2 If $y = x^2 e^{-2x^3}$, then the formula in (15) and the product rule yield

$$\begin{aligned}\frac{dy}{dx} = (Dx^2)e^{-2x^3} + x^2 D(e^{-2x^3}) &= 2xe^{-2x^3} + x^2 e^{-2x^3} D(-2x^3)\\ &= 2xe^{-2x^3} + x^2 e^{-2x^3}(-6x^2).\end{aligned}$$

Therefore,

$$\frac{dy}{dx} = (2x - 6x^4)e^{-2x^3}.$$

EXAMPLE 3 Find $\int xe^{-3x^2} \, dx$.

Solution We substitute

$$u = -3x^2, \quad \text{so that} \quad du = -6x \, dx.$$

Then we have $x \, dx = -\frac{1}{6} \, du$, and hence we obtain

$$\int xe^{-3x^2} \, dx = -\tfrac{1}{6} \int e^u \, du = -\tfrac{1}{6} e^u + C = -\tfrac{1}{6} e^{-3x^2} + C.$$

ORDER OF MAGNITUDE

The exponential function is remarkable for its high rate of increase with increasing x. In fact, e^x increases more rapidly as $x \to +\infty$ than *any* fixed power of x. In the language of limits,

$$\lim_{x\to\infty} \frac{x^k}{e^x} = 0 \quad \text{for any fixed} \quad k > 0. \tag{17}$$

Alternatively,

$$\lim_{x\to\infty} \frac{e^x}{x^k} = +\infty \quad \text{for any fixed} \quad k > 0. \tag{17'}$$

Because we have not yet defined x^k for k irrational, we prove (17) for the case k rational; once we know that (for $x > 1$) the power function x^k is an increasing function of k for all k, then the general case will follow.

We begin by taking logarithms. We find that

$$\ln\left(\frac{e^x}{x^k}\right) = x - k \ln x = \left(\frac{x}{\ln x} - k\right) \ln x.$$

Because we know (from the formula in (13) of Section 7.2) that $x/(\ln x) \to +\infty$ as $x \to +\infty$, this makes it clear that

$$\lim_{x\to\infty} \ln\left(\frac{e^x}{x^k}\right) = +\infty.$$

Hence $e^x/x^k \to +\infty$ as $x \to +\infty$, so we have proved both (17) and (17′). The table in Fig. 7.17 illustrates the limit in (17). Although both $x^5 \to +\infty$ and $e^x \to +\infty$ as $x \to +\infty$, we see that e^x (the hare) increases so much more rapidly than x^5 (the tortoise) that $x^5/e^x \to 0$.

x	x^5	e^x	x^5/e^x
10	1.00×10^5	2.20×10^4	4.54×10^0
20	3.20×10^6	4.85×10^8	6.60×10^{-3}
30	2.43×10^7	1.07×10^{13}	2.27×10^{-6}
40	1.02×10^8	2.35×10^{17}	4.35×10^{-10}
50	3.13×10^8	5.18×10^{21}	6.03×10^{-14}
	$\downarrow$	$\downarrow$	$\downarrow$
	∞	∞	0

7.17 Orders of magnitude of x^5 and e^x

The formulas in (17) and (17′) may be used to evaluate certain limits using the laws of limits. For example,

$$\lim_{x\to\infty} \frac{2e^x - 3x}{3e^x + x^3} = \lim_{x\to\infty} \frac{2 - 3xe^{-x}}{3 + x^3e^{-x}} = \frac{2-0}{3+0} = \frac{2}{3}.$$

EXAMPLE 4 Sketch the graph of $f(x) = xe^{-x}$.

Solution From (17) we see that $f(x) \to 0$ as $x \to +\infty$, while $f(x) \to -\infty$ as $x \to -\infty$. Because

$$f'(x) = e^{-x} - xe^{-x} = e^{-x}(1 - x),$$

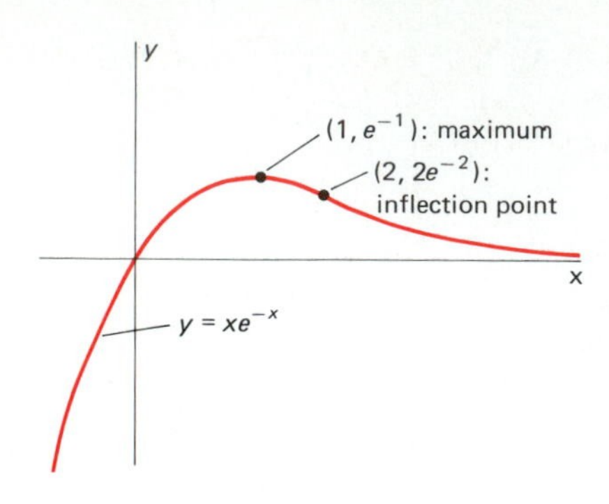

7.18 Graph of the function of Example 4

the only critical point of f is $x = 1$, at which $y = e^{-1} \approx 0.37$. Moreover,

$$f''(x) = -e^{-x}(1 - x) + e^{-x}(-1) = e^{-x}(x - 2),$$

so the only inflection point occurs at $x = 2$, where $y = 2e^{-2} \approx 0.27$. Hence the graph of f resembles the one in Fig. 7.18.

THE NUMBER e AS A LIMIT

We now establish the following limit expression for the exponential function:

$$e^x = \lim_{n \to \infty} \left(1 + \frac{x}{n}\right)^n. \tag{18}$$

We begin by differentiating $\ln t$ using the definition of the derivative in combination with the fact that we already know that the derivative is $1/t$. Thus

$$\begin{aligned} \frac{1}{t} = D \ln t &= \lim_{h \to 0} \frac{\ln(t + h) - \ln t}{h} \\ &= \lim_{h \to 0} \frac{1}{h} \ln\left(\frac{t + h}{t}\right) \\ &= \lim_{h \to 0} \ln\left[\left(1 + \frac{h}{t}\right)^{1/h}\right] \qquad \text{(by laws of logarithms)} \\ &= \ln\left[\lim_{h \to 0} \left(1 + \frac{h}{t}\right)^{1/h}\right] \qquad \text{(by continuity of the logarithm function).} \end{aligned}$$

The substitution $n = 1/h$ allows us to write

$$\frac{1}{t} = \ln\left[\lim_{n \to \infty} \left(1 + \frac{1}{nt}\right)^n\right].$$

Then the substitution $x = 1/t$ gives

$$x = \ln\left[\lim_{n \to \infty} \left(1 + \frac{x}{n}\right)^n\right].$$

Now Equation (18) follows, because $x = \ln y$ implies $e^x = y$. With $x = 1$, we obtain the following important expression of e as a limit:

$$e = \lim_{n \to \infty} \left(1 + \frac{1}{n}\right)^n. \tag{19}$$

7.3 PROBLEMS

Differentiate the functions in Problems 1–30.

1. $f(x) = e^{2x}$

2. $f(x) = e^{3x-1}$

3. $f(x) = e^{x^2} = \exp(x^2)$

4. $f(x) = e^{4-x^3}$

5. $f(x) = e^{1/x^2}$

6. $f(x) = x^2 e^{x^3}$

7. $g(t) = te^{\sqrt{t}}$

8. $g(t) = (e^{2t} + e^{3t})^7$

9. $g(t) = (t^2 - 1)e^{-t}$

10. $g(t) = \sqrt{e^t - e^{-t}}$

11. $g(t) = e^{\cos t}$

12. $f(x) = xe^{\sin x}$

13. $f(x) = \cos(1 - e^{-x})$

14. $f(x) = \sin^2(e^{-x})$

15. $f(x) = \ln(x + e^{-x})$

16. $f(x) = e^x \cos 2x$

17. $f(x) = e^{-2x} \sin 3x$

18. $g(t) = \ln(te^{t^2})$

19. $g(t) = 3(e^t - \ln t)^5$

20. $g(t) = \sin(e^t) \cos(e^{-t})$

21. $f(x) = \dfrac{2 + 3x}{e^{4x}}$

22. $g(t) = \dfrac{1 + e^t}{1 - e^t}$

23. $g(t) = \dfrac{1 - e^{-t}}{t}$

24. $f(x) = e^{-1/x}$

25. $f(x) = \dfrac{1 - x}{e^x}$

26. $f(x) = e^{\sqrt{x}} + e^{-\sqrt{x}}$

27. $f(x) = e^{(e^x)}$

28. $f(x) = \sqrt{e^{2x} + e^{-2x}}$

29. $f(x) = \sin(2e^x)$

30. $f(x) = \cos(e^x + e^{-x})$

In Problems 31–35, find dy/dx by implicit differentiation.

31. $xe^y = y$

32. $\sin(e^{xy}) = x$

33. $e^x + e^y = e^{xy}$

34. $x = ye^y$

35. $e^{x-y} = xy$

Find the antiderivatives indicated in Problems 36–53.

36. $\int e^{3x}\,dx$

37. $\int e^{1-2x}\,dx$

38. $\int xe^{x^2}\,dx$

39. $\int x^2 e^{3x^3-1}\,dx$

40. $\int \sqrt{x}\, e^{2x\sqrt{x}}\,dx$

41. $\int \dfrac{e^{2x}}{1 + e^{2x}}\,dx$

42. $\int (\cos x)e^{\sin x}\,dx$

43. $\int (\sin 2x)e^{1-\cos 2x}\,dx$

44. $\int (e^x + e^{-x})^2\,dx$

45. $\int \dfrac{x + e^{2x}}{x^2 + e^{2x}}\,dx$

46. $\int e^{2x+3}\,dx$

47. $\int te^{-t^2/2}\,dt$

48. $\int x^2 e^{1-x^3}\,dx$

49. $\int \dfrac{e^{\sqrt{x}}}{\sqrt{x}}\,dx$

50. $\int \dfrac{e^{1/t}}{t^2}\,dt$

51. $\int \dfrac{e^x}{1 + e^x}\,dx$

52. $\int \exp(x + e^x)\,dx$

53. $\int \sqrt{x}\,\exp(-\sqrt{x^3})\,dx$

Apply the formula in (18) to evaluate (in terms of the exponential function) the limits in Problems 54–58.

54. $\lim_{n\to\infty} \left(1 - \dfrac{1}{n}\right)^n$

55. $\lim_{n\to\infty} \left(1 + \dfrac{2}{n}\right)^n$

56. $\lim_{n\to\infty} \left(1 + \dfrac{2}{3n}\right)^n$

57. $\lim_{h\to 0} (1 + h)^{1/h}$

58. $\lim_{h\to 0} (1 + 2h)^{1/h}$ (*Suggestion:* Substitute $k = 2h$.)

Evaluate the limits in Problems 59–62 by applying the fact that

$$\lim_{x\to\infty} x^k e^{-x} = 0.$$

59. $\lim_{x\to\infty} \dfrac{e^x}{x}$

60. $\lim_{x\to\infty} \dfrac{e^x}{x^{1/2}}$

61. $\lim_{x\to\infty} \dfrac{e^{(x^{1/2})}}{x}$

62. $\lim_{x\to\infty} x^2 e^{-x}$

In Problems 63–65, sketch the graph of the given equation. Show and label all extrema, inflection points, and asymptotes; show the concave structure clearly.

63. $y = x^2 e^{-x}$

64. $y = x^3 e^{-x}$

65. $y = \exp(-x^2)$

66. Find the area under the graph of $y = e^x$ from $x = 0$ to $x = 1$.

67. Find the volume generated by revolving the region of Problem 66 about the x-axis.

68. Let R be the plane figure bounded below by the x-axis, above by the graph of $y = \exp(-x^2)$, and on the sides by the vertical lines at $x = 0$ and $x = 1$. Find the volume generated by rotating R about the y-axis.

69. Find the length of the curve $y = (e^x + e^{-x})/2$ from $x = 0$ to $x = 1$.

70. Find the area of the surface generated by revolving the curve of Problem 69 about the x-axis.

71. Prove that the equation $e^{-x} = x - 1$ has a single solution. Then use Newton's method to find it to three-place accuracy.

72. If a plant releases an amount A of pollutant into a canal at time $t = 0$, then the resulting concentration of pollutant at time t in the water at a town on the canal a distance x_0 downstream of the plant is

$$C(t) = \frac{A}{\sqrt{k\pi t}} \exp\left(-\frac{x_0^2}{4kt}\right)$$

where k is a certain constant. Show that the maximum concentration at the town is

$$C_{\text{max}} = \frac{A}{x_0}\sqrt{\frac{2}{\pi e}}.$$

73. Sketch the graph of $f(x) = x^n e^{-x}$ for $x \geqq 0$ (n is a fixed but arbitrary positive integer). In particular, show that the maximum value of f is $f(n) = n^n e^{-n}$.

74. Approximate the number e as follows. First apply Simpson's approximation with $n = 2$ subintervals to the integral

$$\int_0^1 e^x\,dx = e - 1$$

to obtain the approximation $5e - 4\sqrt{e} - 7 \approx 0$. Then solve for e.

75. Suppose that $f(x) = x^n e^{-x}$, where n is a fixed but arbitrary positive integer. Conclude from Problem 73 that the numbers $f(n - 1)$ and $f(n + 1)$ are each less than $f(n) = n^n e^{-n}$. Deduce from this that

$$\left(1 + \frac{1}{n}\right)^n < e < \left(1 - \frac{1}{n}\right)^{-n}$$

Substitute $n = 1024$ to show that $2.716 < e < 2.720$. Note that $1024 = 2^{10}$, so that a^{1024} can be computed easily with almost any calculator by entering a and then squaring a ten times in succession.

76. Suppose that the quadratic equation $am^2 + bm + c = 0$ has the two real roots m_1 and m_2, and suppose that C_1 and C_2 are arbitrary constants. Show that the function

$$y = y(x) = C_1 \exp(m_1 x) + C_2 \exp(m_2 x)$$

satisfies the differential equation $ay'' + by' + cy = 0$.

77. Use the result of Problem 76 to find a solution $y = y(x)$ of the differential equation $y'' + y' - 2y = 0$ such that $y(0) = 5$ and $y'(0) = 2$.

*7.3 Optional Computer Application

Here we use the limit

$$e = \lim_{n \to \infty} \left(1 + \frac{1}{n}\right)^n \tag{19}$$

to investigate numerically the number e. Assuming that this limit exists, we can "accelerate" the convergence by calculating the quantity $(1 + 1/n)^n$ only for each power $n = 2^k$ of 2 instead of for every positive integer n. That is, we consider the sequence $n = 1, 2, 4, 8, \ldots, 2^k, \ldots$ instead of the sequence $n = 1, 2, 3, 4, \ldots$ of all positive integers:

$$e = \lim_{k \to \infty} \left(1 + \frac{1}{2^k}\right)^{2^k}. \tag{20}$$

This approach has the advantage that

$$\left(1 + \frac{1}{2^k}\right)^{2^k}$$

can be calculated simply by beginning with $1 + 1/2^k$ and *squaring* k times in succession, because

$$(x^2)^2 = x^4, \qquad (x^4)^2 = x^8, \qquad (x^8)^2 = x^{16},$$

and so forth.

Program EBYSQARE (listed in Fig. 7.19) approximates the limit in (20) in this way, and produces the results shown in the table on the left. We see once again that $e \approx 2.71828$.

n	$\left(1 + \frac{1}{2^n}\right)^{2^n}$
2	2.44141
4	2.63793
6	2.69734
8	2.71299
10	2.71696
12	2.71795
14	2.71820
16	2.71826
18	2.71828
20	2.71828

```
100 REM--Program EBYSQARE
110 REM--Approximates  e  as  [1 + 1/2^n]^(2^n)
120 REM
130       DEFDBL E, X
140       X = 1/2                    'With  n = 1
145 REM--Loop:
150       FOR N = 1 TO 20
160           E = 1 + X              'E is  1 + 1/2^n
170           FOR J = 1 TO N         'This subloop
180               E = E*E            ' squares  E
190           NEXT J                 ' n  times.
200           IF INT(N/2) = N/2 THEN
                 PRINT USING "##      #.#####"; N,E
210           X = X/2                'Next  1/2^n
220       NEXT N
225 REM
230       END
```

7.19 Program EBYSQARE

7.4 General Exponential and Logarithmic Functions

The natural exponential function e^x and the natural logarithm function $\ln x$ are often called the exponential and logarithm with *base e*. We now define general exponential and logarithm functions, having the forms a^x and $\log_a x$, with base a positive number $a \neq 1$. But it is now convenient to reverse the order of treatment in Sections 7.2 and 7.3, so we first consider the general exponential function.

If r is a rational number, then one of the laws of exponents (Equation (13) in Section 7.3) gives

$$a^r = (e^{\ln a})^r = e^{r \ln a}.$$

We therefore *define* arbitrary powers (rational *and* irrational) of the positive number a in this way:

$$a^x = e^{x \ln a} \tag{1}$$

for all x. Then $f(x) = a^x$ is called the **exponential function with base *a*.** Note that $a^x > 0$ for all x and that $a^0 = e^0 = 1$ for all $a > 0$.

The *laws of exponents* for general exponentials follow almost immediately from the definition in (1) and the laws of exponents for the natural exponential function.

$$a^x a^y = a^{x+y}, \tag{2}$$

$$a^{-x} = \frac{1}{a^x}, \quad \text{and} \tag{3}$$

$$(a^x)^y = a^{xy} \tag{4}$$

for all x and y. To prove the first formula, we write

$$\begin{aligned} a^x a^y &= e^{x \ln a}\, e^{y \ln a} \\ &= e^{(x \ln a) + (y \ln a)} = e^{(x+y)\ln a} = a^{x+y}. \end{aligned}$$

To derive (4), note first from (1) that $\ln a^x = x \ln a$. Then

$$(a^x)^y = e^{y \ln(a^x)} = e^{xy \ln a} = a^{xy}.$$

Observe that this follows for all real numbers x and y, so the restriction that r be rational in the formula $(e^x)^r = e^{rx}$ (see Equation (13) in Section 7.3) has now been removed.

If $a > 1$, so that $\ln a > 0$, then Equations (5) and (6) in Section 7.3 immediately give us the results

$$\lim_{a \to \infty} a^x = +\infty \quad \text{and} \quad \lim_{a \to -\infty} a^x = 0. \tag{5}$$

The values of these two limits are interchanged if $0 < a < 1$, for then $\ln a < 0$.

Because

$$D_x a^x = D_x e^{x \ln a} = a^x \ln a \tag{6}$$

is positive for all x if $a > 1$, we see that—in this case—a^x is an increasing function of x. If $a > 1$, then the graph of $y = a^x$ resembles that of $f(x) = e^x$. For example, examine the graph of $y = 2^x$ shown in Fig. 7.20. But if $0 < a < 1$, then $\ln a < 0$, and it then follows from (6) that a^x is a decreasing function. In this case, the graph of $y = a^x$ will look like the one shown in Fig. 7.21.

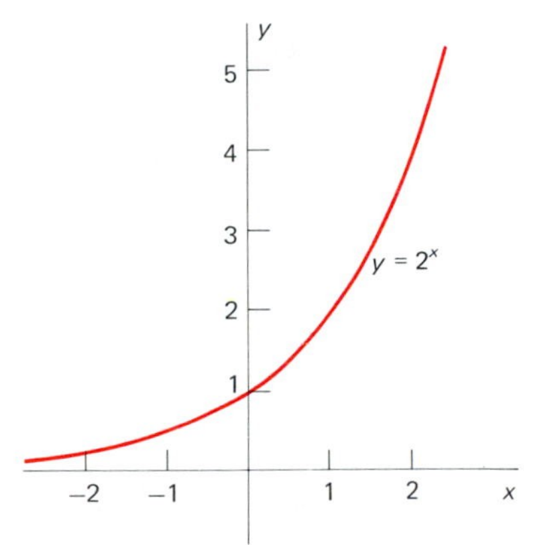

7.20 The graph of $y = 2^x$

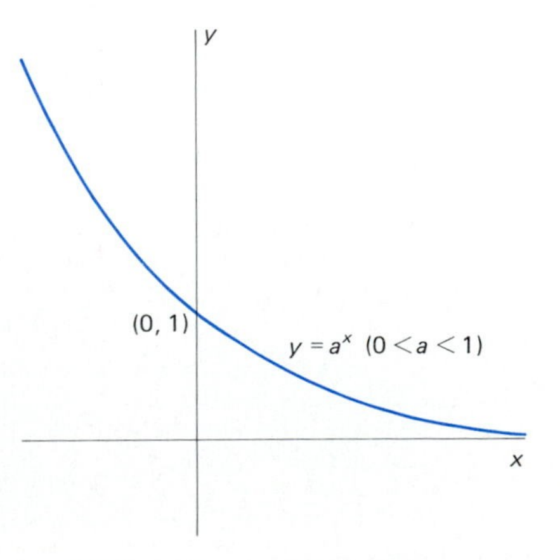

7.21 The graph of $y = a^x$ is decreasing and concave upward if $0 < a < 1$.

If $u = u(x)$ is a differentiable function of x, then (6) combined with the chain rule gives

$$D_x a^u = (a^u \ln a)\frac{du}{dx}. \tag{7}$$

The corresponding integral formula is

$$\int a^u\,du = \frac{a^u}{\ln a} + C. \tag{8}$$

But rather than using these general formulas, it usually is simpler to rely solely on the definition in (1), as in the following examples.

EXAMPLE 1 To differentiate $f(x) = 3^{x^2}$, we may first write

$$3^{x^2} = (e^{\ln 3})^{x^2} = e^{x^2 \ln 3}.$$

Then

$$D_x 3^{x^2} = D_x e^{x^2 \ln 3} = e^{x^2 \ln 3} D_x(x^2 \ln 3) = 3^{x^2}(\ln 3)(2x).$$

EXAMPLE 2 Find $\displaystyle\int \frac{10^{\sqrt{x}}}{\sqrt{x}}\,dx.$

Solution We first write

$$10^{\sqrt{x}} = (e^{\ln 10})^{\sqrt{x}} = e^{\sqrt{x}\ln 10}.$$

Then

$$\begin{aligned}\int \frac{10^{\sqrt{x}}}{\sqrt{x}}\,dx &= \int \frac{e^{\sqrt{x}\ln 10}}{\sqrt{x}}\,dx \\ &= \int \frac{2e^u}{\ln 10}\,du \qquad \left(u = \sqrt{x}\ln 10, \quad du = \frac{\ln 10}{2\sqrt{x}}\,dx\right) \\ &= \frac{2e^u}{\ln 10} + C = \frac{2}{\ln 10}\,10^{\sqrt{x}} + C.\end{aligned}$$

Whether or not the exponent r is rational, the **general power function** $f(x) = x^r$ is now defined for $x > 0$ by

$$x^r = e^{r \ln x}.$$

We may now prove the power rule of differentiation for an *arbitrary* (constant) exponent, as follows:

$$D_x x^r = D_x(e^{r \ln x}) = e^{r \ln x} D_x(r \ln x) = x^r \cdot \frac{r}{x} = rx^{r-1}.$$

For example, we now know that

$$Dx^{\pi} = \pi x^{\pi - 1} \approx (3.14159)x^{2.14159}.$$

If $a > 1$, then the general exponential function a^x is continuous and increasing for all x, and attains all positive values (by an argument similar to that in the first paragraph of Section 7.3, using (6) and the limits in (5) above). It therefore has an inverse function that is defined for all $x > 0$. This inverse function of a^x is called the **logarithm function with base *a*** and is denoted by $\log_a x$. Thus

$$y = \log_a x \quad \text{if and only if} \quad x = a^y. \tag{9}$$

Note that the logarithm function with base e is the natural logarithm function $\log_e x = \ln x$.

The following *laws of logarithms* are easy to derive from the laws of exponents in Equations (2) through (4).

$$\log_a xy = \log_a x + \log_a y, \tag{10}$$

$$\log_a\left(\frac{1}{x}\right) = -\log_a x, \tag{11}$$

$$\log_a x^y = y \log_a x. \tag{12}$$

These formulas hold for any positive base $a \neq 1$ and for all positive values of x and y; in (12), y may be negative or zero as well.

Logarithms with one base are related to logarithms with another base, and the relationship is most easily expressed by the formula

$$(\log_a b)(\log_b c) = \log_a c. \tag{13}$$

This formula holds for all values of a, b, and c for which it makes sense—the bases a and b are positive numbers other than 1 and c is positive. The proof of this formula is outlined in Problem 53. Note also how easy it is to remember the formula in (13)—it is as if some arcane cancellation law held.

If we take $c = a$ in (13), this gives

$$(\log_a b)(\log_b a) = 1, \tag{14}$$

which in turn, with $b = e$, gives

$$\ln a = \frac{1}{\log_a e}. \tag{15}$$

If in (13) we replace a with e, b with a, and c with x, we obtain

$$(\log_e a)(\log_a x) = \log_e x,$$

so

$$\log_a x = \frac{\log_e x}{\log_e a} = \frac{\ln x}{\ln a}. \tag{16}$$

On most calculators, the "log" button denotes common (base 10) logarithms: $\log x = \log_{10} x$. In BASIC, however, only the natural logarithm LOG(X) appears explicitly. To get $\log_{10} x$ we write LOG(X)/LOG(10).

Differentiation in (16) yields

$$D_x \log_a x = \frac{1}{x \ln a} = \frac{\log_a e}{x}. \tag{17}$$

For example,

$$D \log_{10} x = \frac{\log_{10} e}{x} \approx \frac{0.4343}{x}.$$

If we now reason as in Equation (6) of Section 7.2, the chain rule yields the general formula

$$D_x \log_a |u| = \frac{1}{u \ln a} \cdot \frac{du}{dx} = \frac{\log_a e}{u} \cdot \frac{du}{dx} \qquad (u \neq 0) \tag{18}$$

if u is a differentiable function of x. For instance,

$$D_x \log_2 \sqrt{x^2+1} = \frac{1}{2} D_x \log_2 (x^2+1)$$

$$= \frac{1}{2} \cdot \frac{\log_2 e}{x^2+1}(2x) \approx \frac{(1.4427)x}{x^2+1}.$$

We used here the fact that $\log_2 e = 1/(\ln 2)$ by Equation (15).

LOGARITHMIC DIFFERENTIATION

The derivatives of certain functions are most conveniently found by first differentiating their logarithms. This process—called **logarithmic differentiation**—involves the following steps for finding $f'(x)$.

1. Given: $\quad y = f(x).$
2. Take *natural* logarithms, then simplify using laws of logarithms: $\quad \ln y = \ln f(x).$
3. Now differentiate with respect to x: $\quad \dfrac{1}{y} \cdot \dfrac{dy}{dx} = D_x[\ln f(x)].$
4. Finally multiply both sides by $y = f(x)$: $\quad \dfrac{dy}{dx} = f(x)\, D_x[\ln f(x)].$

REMARK If $f(x)$ is not positive-valued everywhere, Steps 1 and 2 should be replaced by $y = |f(x)|$ and $\ln y = \ln|f(x)|$, respectively. The differentiation in Step 3 then leads to the result $dy/dx = f(x)\, D_x[\ln|f(x)|]$ in Step 4. In practice, we need not be overly concerned in advance with the sign of $f(x)$, because the appearance of what seems to be the logarithm of a negative quantity will signal the fact that absolute values should be taken.

EXAMPLE 3 Find dy/dx given

$$y = \frac{\sqrt{(x^2+1)^3}}{\sqrt[3]{(x^3+1)^4}}.$$

Solution The laws of logarithms give

$$\ln y = \ln \frac{(x^2+1)^{3/2}}{(x^3+1)^{4/3}} = \frac{3}{2}\ln(x^2+1) - \frac{4}{3}\ln(x^3+1).$$

So differentiation with respect to x gives

$$\frac{1}{y} \cdot \frac{dy}{dx} = \frac{3}{2} \cdot \frac{2x}{x^2+1} - \frac{4}{3} \cdot \frac{3x^2}{x^3+1} = \frac{3x}{x^2+1} - \frac{4x^2}{x^3+1}.$$

Finally, to solve for dy/dx, we multiply both sides by

$$y = (x^2+1)^{3/2}(x^3+1)^{-4/3},$$

and we obtain

$$\frac{dy}{dx} = \left(\frac{3x}{x^2+1} - \frac{4x^2}{x^3+1}\right)\frac{(x^2+1)^{3/2}}{(x^3+1)^{4/3}}.$$

EXAMPLE 4 Find dy/dx given $y = x^{x+1}$.

Solution If $y = x^{x+1}$, then

$$\ln y = \ln(x^{x+1}) = (x+1)\ln x,$$

$$\frac{1}{y}\cdot\frac{dy}{dx} = (1)\ln x + (x+1)\left(\frac{1}{x}\right) = 1 + \frac{1}{x} + \ln x.$$

And now multiplication of both sides by $y = x^{x+1}$ gives

$$\frac{dy}{dx} = \left(1 + \frac{1}{x} + \ln x\right)x^{x+1}.$$

7.4 PROBLEMS

In Problems 1–24, find the derivative of the given function $f(x)$.

1. 10^x
2. $2^{1/x^2}$
3. $\dfrac{3^x}{4^x}$
4. $\log_{10}\cos x$
5. $7^{\cos x}$
6. $2^x 3^{(x^2)}$
7. $2^{x\sqrt{x}}$
8. $\log_{100}(10^x)$
9. $2^{\ln x}$
10. $7^{(8^x)}$
11. 17^x
12. $2^{\sqrt{x}}$
13. $10^{1/x}$
14. $3^{\sqrt[3]{1-x^2}}$
15. $2^{(2^x)}$
16. $\log_2 x$
17. $\log_3\sqrt{x^2+4}$
18. $\log_{10}(e^x)$
19. $\log_3(2^x)$
20. $\log_{10}(\log_{10}x)$
21. $\log_2(\log_3 x)$
22. $\pi^x + x^\pi + \pi^\pi$
23. $\exp(\log_{10}x)$
24. $\pi^{(x^3)}$

Evaluate the integrals given in Problems 25–32.

25. $\int 3^{2x}\,dx$
26. $\int x(10^{-x^2})\,dx$
27. $\int \dfrac{2^{\sqrt{x}}}{\sqrt{x}}\,dx$
28. $\int \dfrac{10^{1/x}}{x^2}\,dx$
29. $\int x^2 7^{x^3+1}\,dx$
30. $\int \dfrac{dx}{x\log_{10}x}$
31. $\int \dfrac{\log_2 x}{x}\,dx$
32. $\int (2^x)3^{(2^x)}\,dx$

In Problems 33–52, find dy/dx by logarithmic differentiation.

33. $y = \sqrt{(x^2-4)\sqrt{2x+1}}$
34. $y = \dfrac{(3-x^2)^{1/3}}{(x^4+1)^{1/4}}$
35. $y = 2^x$
36. $y = x^x$
37. $y = x^{\ln x}$
38. $y = (1+x)^{1/x}$
39. $y = \left[\dfrac{(x+1)(x+2)}{(x^2+1)(x^2+2)}\right]^{1/3}$
40. $y = \sqrt{x+1}\,\sqrt[3]{x+2}\,\sqrt[4]{x+3}$
41. $y = (\ln x)^{\sqrt{x}}$
42. $y = (3+2^x)^x$
43. $y = \dfrac{(1+x^2)^{3/2}}{(1+x^3)^{4/3}}$
44. $y = (x+1)^x$
45. $y = (x^2+1)^{x^2}$
46. $y = \left(1+\dfrac{1}{x}\right)^x$
47. $y = (\sqrt{x})^{\sqrt{x}}$
48. $y = x^{\sin x}$
49. $y = e^x$
50. $y = (\ln x)^{\ln x}$
51. $y = x^{(e^x)}$
52. $y = (\cos x)^x$
53. Prove the formula in Equation (13). *Suggestion:* Let $x = \log_a b$, $y = \log_b c$, and $z = \log_a c$. Then show that $a^z = a^{xy}$, and conclude that $z = xy$.
54. Suppose that u and v are differentiable functions of x. Show by logarithmic differentiation that

$$D_x(u^v) = v(u^{v-1})\frac{du}{dx} + (u^v\ln u)\frac{dv}{dx}.$$

Interpret the two terms on the right in relation to each of the two special cases: (i) u is a constant; (ii) v is a constant.

55. Suppose that $a > 0$. Show that

$$\lim_{x\to\infty} a^{1/x} = 1$$

by examining $\ln(a^{1/x})$. It follows that

$$\lim_{n\to\infty} a^{1/n} = 1.$$

Test this conclusion by entering some positive number in your calculator and then pressing the square root key repeatedly. Make a table of the results of two such experiments.

56. Show that

$$\lim_{n \to \infty} n^{1/n} = 1$$

by showing that

$$\lim_{x \to \infty} x^{1/x} = 1.$$

Use the method of Problem 55.

57. Show that

$$\lim_{x \to \infty} \frac{x^x}{e^x} = +\infty$$

by examining $\ln(x^x/e^x)$. Thus x^x increases even faster than the exponential function e^x as $x \to +\infty$.

58. (a) Show that the equation $2^x = x^{10}$ has the same positive solutions as the equation $(\ln x)/x = (\ln 2)/10$.

(b) Conclude from the graph of $y = (\ln x)/x$ (Example 8 in Section 7.2) that the equation $2^x = x^{10}$ has exactly two positive solutions.

(c) Show that one of these two solutions is between 1 and 2, while the other is between 50 and 60. Then use Newton's method to approximate each solution with at least two-place accuracy.

59. Consider the function

$$f(x) = \frac{1}{1 + 2^{1/x}} \qquad \text{for} \quad x \neq 0.$$

Show that the left-hand and right-hand limits of $f(x)$ at $x = 0$ both exist but are unequal.

60. Find dy/dx if $y = \log_x 2$.

61. Suppose that $y = uvw/pqr$ where u, v, w, p, q, and r are nonzero differentiable functions. Show by logarithmic differentiation that

$$y' = (y)\left(\frac{u'}{u} + \frac{v'}{v} + \frac{w'}{w} - \frac{p'}{p} - \frac{q'}{q} - \frac{r'}{r}\right)$$

(primes denote derivatives with respect to x). Is the generalization—for arbitrary finite numbers of factors in numerator and denominator—obvious?

7.5 Natural Growth and Decay

Consider a population that numbers $P(t)$ persons—or animals, bacteria, or any sort of entity—at time t. We assume that this population has a constant birth rate β and a constant death rate δ. Roughly speaking, this means that during any one-year period, βP births and δP deaths occur.

But because P changes during the course of that year, some allowance must be made for changes in the number of births and the number of deaths. To be more precise, we think of a very brief time interval from t to $t + \Delta t$. For very small values of Δt, the value of $P = P(t)$ will change by such a small amount during the time interval $[t, t + \Delta t]$ that we can regard P as "almost constant." We require that the number of births and deaths during this time interval be given with sufficient accuracy by these approximations:

$$\begin{aligned} &\text{The number of births is approximately } \beta P(t)\,\Delta t; \\ &\text{The number of deaths is approximately } \delta P(t)\,\Delta t. \end{aligned} \qquad (1)$$

What we mean when we say that the birth rate is β and the death rate is δ is this: The ratio to Δt of the errors in these approximations both approach zero as $\Delta t \to 0$.

We would like to use the information in (1) to deduce, if possible, the form of the function $P(t)$ that describes the population in question. Our strategy begins with finding the **time rate of change** of P. Hence we consider the increment

$$\Delta P = P(t + \Delta t) - P(t)$$

of P during the time interval $[t, t + \Delta t]$. Because ΔP is simply the number of births minus the number of deaths, we find from (1) that

$$\Delta P = P(t + \Delta t) - P(t) \approx \beta P(t)\,\Delta t - \delta P(t)\,\Delta t.$$

Therefore,

$$\frac{\Delta P}{\Delta t} = \frac{P(t + \Delta t) - P(t)}{\Delta t} \approx (\beta - \delta)P(t).$$

The quotient on the left-hand side approaches the derivative $P'(t)$ as $\Delta t \to 0$ and, by the assumption following (1), it also approaches the right-hand side $(\beta - \delta)P(t)$. Hence, when we take the limit as $\Delta t \to 0$, we get the differential equation

$$P'(t) = (\beta - \delta)P(t);$$

that is,

$$\frac{dP}{dt} = kP, \qquad \text{where} \quad k = \beta - \delta. \tag{2}$$

This differential equation may be regarded as a *mathematical model* of the changing population.

The differential equation

$$\frac{dx}{dt} = kx \qquad (k \text{ a constant}) \tag{3}$$

serves as the mathematical model for an extraordinarily wide range of natural phenomena. It is easy to solve; we first write it in the form

$$\frac{1}{x} \cdot \frac{dx}{dt} = k$$

—that is,

$$D_t(\ln x) = D_t(kt).$$

We then antidifferentiate and obtain

$$\ln x = kt + C,$$

under the assumption that $x > 0$. Note that the same result is obtained formally if we first "separate the variables" in (3) and then integrate:

$$\frac{dx}{x} = k\,dt; \qquad \int \frac{dx}{x} = \int k\,dt; \qquad \ln x = kt + C.$$

In either case, we next apply the exponential function to both sides of this equation to solve for x. This gives

$$x = e^{\ln x} = e^{kt + C} = Ae^{kt}.$$

Here, $A = e^C$ is a constant that remains to be determined. But we see that A will simply be the value of x when $t = 0$, and thus $A = x(0) = x_0$. Therefore, the solution of the differential equation in (3) with the *initial value* $x(0) = x_0$ is

$$x(t) = x_0 e^{kt}. \tag{4}$$

As a result, Equation (3) is often called the **exponential growth equation** or the **natural growth equation.** We see from (4) that with $x_0 > 0$, the solution $x(t)$ is an increasing function if $k > 0$ and a decreasing function if $k < 0$.

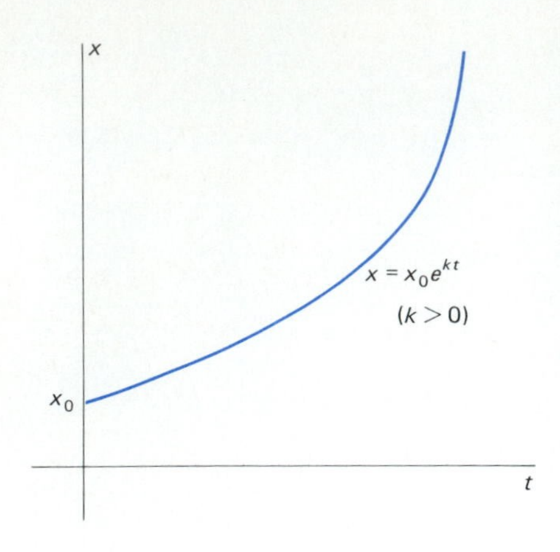

7.22 Solution of the exponential growth equation for $k > 0$

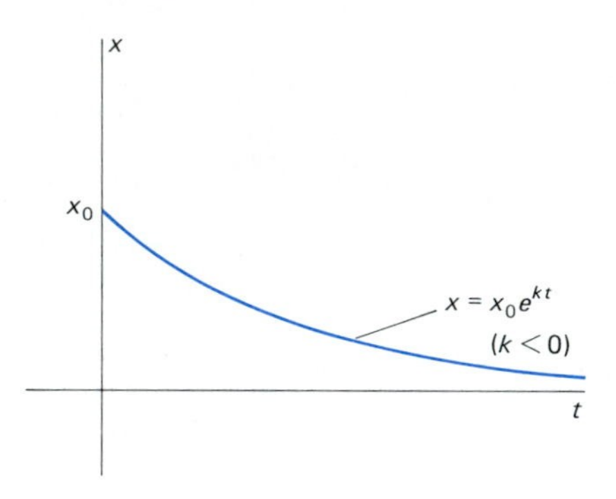

7.23 Solution of the exponential growth equation—now actually a *decay* equation—for the case $k < 0$

These cases are shown in Figs. 7.22 and 7.23, respectively. The remainder of this section concerns examples of natural phenomena for which this differential equation serves as a mathematical model.

POPULATION GROWTH

When we compare Equations (2), (3), and (4), we see that a population $P(t)$ with constant birth rate β and constant death rate δ is given by

$$P(t) = P_0e^{kt} \tag{5}$$

where $P_0 = P(0)$ and $k = \beta - \delta$. If t is measured in years, then k is called the **annual growth rate,** whether it be positive, negative, or zero. Its value is often given as a percentage (which is its actual decimal value multiplied by 100). If k is not too large, this value is fairly close to the actual percentage increase (or decrease) of the population each year.

EXAMPLE 1 According to an Associated Press release of March, 1987, the world population had then reached 5 billion persons, and was increasing then at the rate of 380 thousand persons each day. Let us assume constant birth and death rates. We would like to find the answers to these questions:

1. What is the annual growth rate k?
2. How long will it take for the world's population to double?
3. What will the world's population be in the year 2000?
4. When will the world's population be 50 billion (some demographers believe that this is the maximum for which the planet can provide food)?

Solution We measure the world's population $P(t)$ in billions and measure time t in years. We take $t = 0$ to correspond to 1987, so that $P_0 = 5$. The fact that P is increasing by 380 thousand, or 0.00038 billion, persons per day at time $t = 0$ means that

$$P'(0) = (0.00038)(365.25) \approx 0.1388$$

billions per year. From Equation (2) we now obtain

$$k = \left[\frac{1}{P} \cdot \frac{dP}{dt}\right]_{t=0} = \frac{P'(0)}{P(0)} = \frac{0.1388}{5} \approx 0.0278.$$

Thus the population was growing at the rate of about 2.78% per year in 1987.

We use this value of k to conclude that the world population at time t should be

$$P(t) = 5e^{(0.0278)t}.$$

For example, $t = 13$ yields

$$P(13) = 5e^{(0.0278)(13)} \approx 7.17 \text{ (billions)}$$

for the population in the year 2000.

To find when the population will be 10 billion, we solve the equation

$$10 = 5e^{(0.0278)t}$$

(by first taking the natural logarithm of each side) for

$$t = \frac{\ln 2}{0.0278} \approx 25 \quad \text{(years)},$$

which corresponds to the year 2012. Finally, under the assumptions here, the world population will reach 50 billion when

$$50 = 5e^{(0.0278)t}; \qquad t = \frac{\ln 10}{0.0278} \approx 83;$$

that is, in the year 2070.

RADIOACTIVE DECAY

Consider a sample of material that contains $N(t)$ atoms of a certain radioactive isotope at time t. It has been observed that a constant fraction of these radioactive atoms will spontaneously decay (into atoms of another element or another isotope of the same element) during each unit of time. Consequently the sample behaves exactly like a population with a constant death rate but with no births occurring. To write a model for $N(t)$, we use Equation (2) with N in place of P, with $k > 0$ in place of δ, and with $\beta = 0$. We thus get the differential equation

$$\frac{dN}{dt} = -kN. \tag{6}$$

From the solution in (4) of Equation (3), with k replaced by $-k$, we conclude that

$$N(t) = N_0 e^{-kt} \tag{7}$$

where $N_0 = N(0)$, the number of radioactive atoms present in the sample at time $t = 0$.

The value of the *decay constant* k depends on the particular isotope with which we are dealing. If k is large, then the isotope decays rapidly, while if k is near zero, the isotope decays quite slowly and thus may be a relatively persistent factor in its environment. The decay constant k is often specified in terms of another empirical parameter, the *half-life* of the isotope, because this parameter is more convenient. The **half-life** τ of a radioactive isotope is the time required for *half* of it to decay. To find the relationship between k and τ, we set

$$t = \tau \quad \text{and} \quad N = \tfrac{1}{2}N_0$$

in Equation (7), so that

$$\tfrac{1}{2}N_0 = N_0 e^{-k\tau}. \tag{8}$$

When we solve for τ, we find that

$$\tau = \frac{\ln 2}{k}. \tag{9}$$

Note that the concept of half-life is meaningful—the value of τ depends *only* on k and thus depends only on the particular isotope involved. It does *not* depend on the amount of that isotope present.

EXAMPLE 2 The half-life of radioactive carbon ^{14}C is about 5700 years. A specimen of charcoal found at Stonehenge turns out to contain 63% as much ^{14}C as a sample of present-day charcoal. What is the age of the sample?

Solution The key to the method of *radiocarbon dating* is that living organic matter maintains a constant level of ^{14}C by "breathing" air (or by consuming organic matter that does so). Because air contains ^{14}C along with the much more common stable isotope ^{12}C of carbon, mostly in the gas CO_2, the same percentage permeates all life, because organic processes seem to make no distinction between the two isotopes. But when a living organism dies, it ceases its metabolism of carbon, and the process of radioactive decay begins to deplete its ^{14}C content. The fraction of ^{14}C in the air remains roughly constant because new ^{14}C is being generated constantly by the action of cosmic rays on nitrogen atoms in the upper atmosphere, and this generation has long been in a steady-state equilibrium with the loss of ^{14}C through radioactive decay.

For our Stonehenge example, we take $t = 0$ as the time of death of the tree from which the charcoal was made. From Equation (8) we know that

$$\tfrac{1}{2}N_0 = N_0 e^{-5700k}$$

so

$$k = \frac{\ln 2}{\tau} = \frac{\ln 2}{5700} \approx 0.0001216.$$

We are given that $N = (0.63)N_0$ now, so we solve the equation

$$(0.63)N_0 = N_0 e^{-kt}$$

with this value of k and thus find that

$$t = -\frac{\ln(0.63)}{0.0001216} \approx 3800$$

years. Thus the sample is about 3800 years old, and if it has any connection with the builders of Stonehenge, our computations suggest that this observatory, monument, or temple—whichever it may be—dates from almost 1800 B.C.

EXAMPLE 3 According to one cosmological theory, there were equal amounts of the two uranium isotopes ^{235}U and ^{238}U at the creation of the universe in the "big bang." At present there are 137.7 ^{238}U atoms for each atom of ^{235}U. Using the half-lives

4.51 billion years for ^{238}U,
0.71 billion years for ^{235}U,

calculate the age of the universe.

Solution Let $N_8(t)$ and $N_5(t)$ be the numbers of ^{238}U and ^{235}U atoms, respectively, at time t (in billions of years after the creation of the universe). Then

$$N_8(t) = N_0 e^{-kt} \quad \text{and} \quad N_5(t) = N_0 e^{-ct}$$

where N_0 is the original number of atoms of each isotope. Also

$$k = \frac{\ln 2}{4.51} \quad \text{and} \quad c = \frac{\ln 2}{0.71},$$

a consequence of Equation (9). We divide the equation for N_8 by the equation for N_5, and find that when t has the value corresponding to "now," then

$$137.7 = \frac{N_8}{N_5} = e^{(c-k)t}.$$

Finally, we solve this equation for

$$t = \frac{\ln 137.7}{[(1/0.71) - (1/4.51)] \ln 2} \approx 5.99.$$

Thus our estimate of the age of the universe is about 6 billion years, which is at least of the same order of magnitude as recent estimates of about 15 billion years.

CONTINUOUSLY COMPOUNDED INTEREST

Consider a savings account that is opened with an initial deposit of A_0 dollars and which thenceforth earns interest at the annual rate r. If there are $A(t)$ dollars in the account at time t and the interest is compounded at time $t + \Delta t$, this means that $rA(t)\,\Delta t$ dollars in interest are added to the account then. So

$$A(t + \Delta t) = A(t) + rA(t)\,\Delta t,$$

and thus

$$\frac{\Delta A}{\Delta t} = \frac{A(t + \Delta t) - A(t)}{\Delta t} = rA(t).$$

Continuous compounding of interest is the situation that results by taking the limit as $\Delta t \to 0$, so that

$$\frac{dA}{dt} = rA. \tag{10}$$

This is an exponential growth equation with solution

$$A(t) = A_0 e^{rt}. \tag{11}$$

For example, if $A_0 = \$1000$ is invested at 6% annual interest rate compounded continuously, then $r = 0.06$, and Equation (11) gives

$$A(1) = 1000e^{(0.06)(1)} = \$1061.84$$

for the amount present after one year. Thus the "effective" annual interest rate is 6.184%. Most people are aware that the more often interest is compounded, the more rapidly their savings grow, but bank advertisements sometimes tend to overemphasize this advantage. For example, 6% compounded *monthly* multiplies your investment by

$$1 + \frac{0.06}{12} = 1.005$$

at the end of each month, so an initial investment of \$1000 would grow in one year to

$$(1000)(1.005)^{12} = \$1061.68,$$

only 16 cents less than would be yielded by continuous compounding.

*DRUG ELIMINATION

In many cases the amount $A(t)$ of a certain drug in the bloodstream, measured by the excess over the natural level of the drug, will decline at a rate proportional to the excess amount. That is,

$$\frac{dA}{dt} = -\lambda A, \qquad \text{so that} \quad A(t) = A_0 e^{-\lambda t}. \tag{12}$$

The parameter λ is called the *elimination constant* of the drug, and $T = 1/\lambda$ is called the *elimination time.*

EXAMPLE 4 The elimination time for alcohol varies from one person to another. If a person's "sobering time" $T = 1/\lambda$ is 2.5 h, how long will it take for the excess bloodstream alcohol concentration to be reduced from 0.10% to 0.02%?

Solution We assume that the normal concentration of alcohol in the blood is zero, so that any amount is an excess amount. In this problem, we have $\lambda = 1/2.5 = 0.4$, so Equation (12) yields

$$0.02 = (0.10)e^{-(0.4)\lambda}.$$

Thus

$$t = -\frac{\ln(0.2)}{0.4} \approx 4.02 \quad \text{(hours)}.$$

*SALES DECLINE

Marketing studies for certain products show that if advertising for a particular product is halted and other market conditions remain unchanged—we mean such things as number and promotion of competing products, their prices, and so on—then the sales of the unadvertised product will decline at a rate that is proportional at any time t to the current sales S. That is,

$$\frac{dS}{dt} = -\lambda S, \qquad \text{so that} \quad S(t) = S_0 e^{-\lambda t}. \tag{13}$$

As usual, S_0 denotes the initial value of the sales, which we take to be sales in the last advertising month. If we take months as the natural units for time t, then $S(t)$ denotes the number of sales t months after advertising is halted, and λ might be called the *sales decay constant.*

*LINGUISTICS

Consider a basic list of N_0 words in use in a given language at time $t = 0$. Let $N(t)$ denote the number of these words that are still in use at time t—those that have neither disappeared from the language nor been replaced by noncognates. According to one theory in linguistics, the rate of decrease of N is proportional to N. That is,

$$\frac{dN}{dt} = -\lambda N, \qquad \text{so that} \quad N(t) = N_0 e^{-\lambda t}. \tag{14}$$

If t is measured in millennia (standard in linguistics), then $k = e^{-\lambda}$ is the fraction of the words in the original list that survive for 1000 years.

7.5 PROBLEMS

1. Suppose that \$1000 is deposited in a savings account that pays 8% annual interest compounded continuously. At what rate (in dollars per year) is this account earning interest after 5 years? After 20 years?

2. (Population Growth) A certain city had a population of 25,000 in 1960 and a population of 30,000 in 1970. Assume that its population will continue to grow exponentially at a constant rate. What population can its city planners expect in the year 2000?

3. (Population Growth) In a certain culture of bacteria, the number of bacteria increased sixfold in 10 h. How long did it take for their number to double?

4. (Radiocarbon Dating) Carbon extracted from an ancient skull contained only one-sixth as much radioactive ^{14}C as carbon extracted from present-day bone. How old is the skull?

5. (Radiocarbon Dating) Carbon taken from a purported relic of the time of Christ contained 4.6×10^{10} atoms of ^{14}C per gram. Carbon extracted from a present-day specimen of the same substance contained 5.0×10^{10} atoms of ^{14}C per gram. Compute the approximate age of the relic. What is your opinion as to its authenticity?

6. An amount A is invested for t years at an annual interest rate r that is compounded n times over these years at equal intervals.

(a) Explain why the amount that has accrued after t years is

$$A_{r,n} = A\left(1 + \frac{rt}{n}\right)^n.$$

(b) Conclude from the limit in Equation (18) of Section 7.3 that

$$\lim_{n \to \infty} A_{r,n} = Ae^{rt},$$

in agreement with Equation (11) in this section.

7. If an investment of A_0 dollars returns A_1 dollars after one year, the **effective annual interest rate** r is defined by means of the equation

$$A_1 = (1 + r)A_0.$$

Banks often advertise that they increase the effective interest rates on their customers' savings accounts by increasing the frequency of compounding. Calculate the effective annual interest rate if a 9% annual interest rate is compounded (a) quarterly; (b) monthly; (c) weekly; (d) daily; (e) continuously.

8. (Continuously Compounded Interest) Upon the birth of their first child, a couple deposited \$5000 in a savings account that draws 6% interest compounded continuously. The interest payments are allowed to accumulate. How much will the account contain when the child is ready to go to college at age 18?

9. (Continuously Compounded Interest) Suppose that you discover in your attic an overdue library book on which your great-great-grandfather owed a fine of 30 cents exactly 100 years ago. If an overdue fine grows exponentially at a 5% annual interest rate compounded continuously, how much would you have to pay if you returned the book today?

10. (Drug Elimination) Suppose that sodium pentobarbitol is used to anesthetize a dog; the dog is anesthetized when its bloodstream contains at least 45 mg of sodium pentobarbitol per kilogram of the dog's body weight. Suppose also that sodium pentobarbitol is eliminated exponentially from the dog's bloodstream, with a half-life of 5 h. What single dose should be administered in order to anesthetize a 50-kg dog for 1 h?

11. (Sales Decline) Advertising of a certain product has been discontinued; the company plans to resume advertising when sales have declined to 75% of their initial rate. (This phenomenon actually occurred when the Sony Corporation first introduced home videotape recorders in the United States in 1976.) If after 1 week without advertising, sales have declined to 95% of their initial rate, when should the company expect to resume advertising?

12. (Linguistics) The English language evolves in such a way that 77% of all words disappear (or are replaced by noncognates) every 1000 years. Of a basic list of words used by Chaucer in A.D. 1400, what percentage should we expect to find still in use today?

13. (Half-life) The half-life of radioactive cobalt is 5.27 years. Suppose that a nuclear accident has left the level of cobalt radiation in a certain region at 100 times the level acceptable for human habitation. How long will it be before the region is again habitable? (Ignore the likely presence of other radioactive substances.)

14. Suppose that a mineral body formed in an ancient cataclysm—perhaps the formation of the earth itself—originally contained the uranium isotope ^{238}U (which has a half-life of 4.51×10^9 years) but no lead, the end product of the radioactive decay of ^{238}U. If today the ratio of ^{238}U atoms to lead atoms in the mineral body is 0.9, when did the cataclysm occur?

15. A certain moon rock was found to contain equal numbers of potassium and argon atoms. Assume that all the argon is the result of radioactive decay of potassium (its half-life is about 1.28×10^9 years) and that 1 of every 9 potassium atom disintegrations yields an argon atom. What is the age of the rock, measured from the time it contained only potassium?

16. If a body is cooling in a medium with constant temperature A, then—according to Newton's law of cooling—the rate of change of the temperature T of the body is proportional to $T - A$. A pitcher of buttermilk initially at 25°C is to be cooled by setting it out on the front porch, where the temperature is 0°C. Suppose that the temperature of the buttermilk has dropped to 15°C after 20 min. When will it be at 5°C?

17. When sugar is dissolved in water, the amount A that remains undissolved after t minutes satisfies the differential equation $dA/dt = -kA$ $(k > 0)$. If 25% of the sugar dissolves in 1 min, how long does it take for half the sugar to dissolve?

18. The intensity of light I at a depth x meters below the surface of a lake satisfies the differential equation $dI/dx = -(1.4)I$.

(a) At what depth is the intensity half the intensity I_0 at the surface (where $x = 0$)?

(b) What is the intensity at a depth of 10 m (as a fraction of I_0)?

(c) At what depth will the intensity be 1% of its value at the surface?

19. The barometric pressure p (in inches of mercury) at an altitude x miles above sea level satisfies the differential equation $dp/dx = -(0.2)p$; $p(0) = 29.92$.

(a) Calculate the barometric pressure at 10,000 ft and again at 30,000 ft.

(b) Without prior conditioning, few people can survive when the pressure drops to less than 15 in. of mercury. How high is that?

*7.6 Linear First-Order Differential Equations and Applications

A **first-order differential equation** is one in which only the first derivative (not higher derivatives) of the dependent variable appears. It is called **linear** if it can be written in the form

$$\frac{dx}{dt} = ax + b, \tag{1}$$

where a and b denote functions of the independent variable t. In this section we discuss applications in the special case in which the coefficients a and b are *constants*.

Equation (1) is separable, so we can separate the variables as in Section 6.5 and immediately integrate. Assuming that $ax + b > 0$, we get

$$\int \frac{a\,dx}{ax + b} = \int a\,dt; \qquad \ln(ax + b) = at + C.$$

Then exponentiation gives

$$ax + b = Ke^{at},$$

where $K = e^C$. When we substitute $t = 0$ and denote the resulting value of x by x_0, we find that $K = ax_0 + b$. So

$$ax + b = (ax_0 + b)e^{at}.$$

Finally, we solve this equation for the solution $x = x(t)$ of Equation (1):

$$x(t) = \left(x_0 + \frac{b}{a}\right)e^{at} - \frac{b}{a}. \tag{2}$$

In this development, we have assumed that $ax + b > 0$, but the formula in (2) also gives the correct solution in the case $ax + b < 0$ (see Problem 17). In Problem 21 we outline a method by which Equation (1) may be solved when the coefficients a and b are functions of t rather than constants, but the solution in (2) for the constant-coefficient case will be sufficient for the following applications.

POPULATION GROWTH WITH IMMIGRATION

Consider a population $P(t)$ with constant birth and death rates (β and δ, respectively), as in Section 7.5, but also with a constant immigration rate of I persons per year entering the country. To account for the immigration, our derivation of Equation (2) in Section 7.5 must be amended as follows:

$$\begin{aligned} P(t + \Delta t) - P(t) &= \{\text{births}\} - \{\text{deaths}\} + \{\text{immigrants}\} \\ &\approx \beta P(t)\,\Delta t - \delta P(t)\,\Delta t + I\,\Delta t, \end{aligned}$$

so

$$\frac{P(t + \Delta t) - P(t)}{\Delta t} \approx (\beta - \delta)P(t) + I.$$

We take limits as $\Delta t \to 0$ and thus obtain the linear first order differential equation

$$\frac{dP}{dt} = kP + I \tag{3}$$

with constant coefficients $k = \beta - \delta$ and I. According to the formula in (2), the solution of (3) is

$$P(t) = P_0 e^{kt} + \frac{I}{k}(e^{kt} - 1). \tag{4}$$

The first term on the right-hand side is the effect of natural population growth and the second term is the effect of immigration.

EXAMPLE 1 Consider the U.S. population with $P_0 = 222$ million in 1980 ($t = 0$). Suppose that we ask about the effect of allowing immigration at the rate of half a million people per year for the next 20 years, assuming a natural growth rate of 1% annually, so that $k = 0.01$. Then

$$P_0 e^{kt} = 222e^{(0.01)(20)} \approx 271.2 \quad \text{(million)}$$

and

$$\frac{I}{k}(e^{kt} - 1) = \frac{0.5}{0.01}(e^{(0.01)(20)} - 1) \approx 11.1 \quad \text{(million)}.$$

Thus the effect of the immigration would be to increase the U.S. population in the year 2000 from 271.2 million to 282.3 million.

SAVINGS ACCOUNT WITH CONTINUOUS DEPOSITS

Consider the savings account of Section 7.5, containing A_0 dollars initially and earning interest at the annual rate r compounded continuously. In addition, we now suppose that deposits are added to this account at the rate of Q dollars per year. To simplify the mathematical model, we assume that these deposits are made continuously rather than (for instance) monthly. We may then regard the amount $A(t)$ in the account at time t as a "population" of dollars, having a natural annual growth rate r and with "immigration" (deposits) at the rate of Q dollars annually. Then by merely changing the notation in Equations (3) and (4), we get the differential equation

$$\frac{dA}{dt} = rA + Q \tag{5}$$

with solution

$$A(t) = A_0 e^{rt} + \frac{Q}{r}(e^{rt} - 1). \tag{6}$$

EXAMPLE 2 Suppose that you wish to arrange, at the time of her birth, for your daughter to have \$20,000 available for college expenses at age 18. You plan to do so by making frequent small—essentially continuous—deposits in a savings account, at the rate of Q dollars per year. This account will accumulate 6% annual interest compounded continuously. What should Q be so that you achieve your goal?

Solution With $A_0 = 0$ and $r = 0.06$, we are asking for the value of Q so that Equation (6) yields the result

$$A(18) = 20{,}000.$$

That is, we are to find Q so that

$$20{,}000 = \frac{Q}{0.06}(e^{(0.06)(18)} - 1).$$

When we solve this equation, we find that $Q \approx 617.07$. Thus you should deposit \$617.07 per year, or about \$51.42 per month, in order to have \$20,000 in the account after 18 years. You may wish to verify that your total deposits will be \$11,107.23 and that the total interest accumulated will be \$8892.77.

COOLING AND HEATING

According to Newton's law of cooling (or heating!), the time rate of change of the temperature T of a body is proportional to the difference between T and the temperature A of its surroundings. We will always assume that A is constant. We may translate this law into the language of differential equations by writing

$$\frac{dT}{dt} = -k(T - A). \tag{7}$$

Here k is a positive constant; the minus sign is required to make $T'(t)$ negative when T exceeds A (and it correctly makes $T'(t)$ positive when A exceeds T).

From Equation (2) we obtain the solution of Equation (7):

$$T(t) = A + (T_0 - A)e^{-kt}. \tag{8}$$

EXAMPLE 3 A 5-lb roast, initially at 50°F, is put into a 375°F oven when $t = 0$; it's found that the temperature $T(t)$ of the roast is 125°F when $t = 75$ (min). When will the roast be medium rare, a temperature of 150°F?

Solution Although we could simply substitute $A = 375$ and $T_0 = 50$ in Equation (8), let us instead solve explicitly the differential equation

$$\frac{dT}{dt} = -k(T - 375) = k(375 - T)$$

that we get from Equation (7). We separate the variables and integrate to obtain

$$\int \frac{dT}{375 - T} = \int k\,dt; \qquad -\ln(375 - T) = kt + C.$$

When $t = 0$, $T = T_0 = 50$. Substitution of these data now yields the value $C = -\ln 325$, so

$$-\ln(375 - T) = kt - \ln 325, \qquad 375 - T = 325e^{-kt},$$

and hence

$$T = T(t) = 375 - 325e^{-kt}.$$

We also know that $T = 125$ when $t = 75$. It follows that

$$k = \tfrac{1}{75}\ln\tfrac{325}{250} \approx 0.0035.$$

So all we need do is solve the equation

$$150 = 375 - 325e^{-kt}.$$

We find that t is approximately 105, so the roast should remain in the oven for about another 30 min.

DIFFUSION OF INFORMATION

Let $N(t)$ denote the number of people (in a fixed population P) who by time t have heard a certain piece of information that is being spread by the mass media. Under certain common conditions, the time rate of increase of N will be proportional to the number of people who have not yet heard the piece of information. Thus

$$\frac{dN}{dt} = k(P - N). \tag{9}$$

If $N(0) = 0$, the solution of Equation (9) is

$$N(t) = P(1 - e^{-kt}). \tag{10}$$

If P and some later value $N(t_1)$ are known, we can then solve for k and thereby determine $N(t)$ for all t. Problem 15 illustrates this situation.

ELIMINATION OF POLLUTANTS

In the following example, we envision a lake that has been polluted, perhaps by factories operating on its shores. Suppose that the pollution is halted, perhaps by a legal order, perhaps by improved technology. We ask how long it will take for natural processes to reduce the pollutant concentration in the lake to an acceptable level.

EXAMPLE 4 Consider a lake with a volume of 8 billion ft^3 and an initial pollutant concentration of 0.25%. Suppose that an inflowing river brings in 500 million ft^3 of water daily with a (low) pollutant concentration of 0.05%, and that an outflowing river also removes 500 million ft^3 of the lake water daily. We make the simplifying assumption that the water in the lake, including that removed by the second river, is perfectly mixed at all times. If so, how long will it take to reduce the pollutant concentration in the lake to 0.10%?

Solution Let $x(t)$ denote the amount of pollutants in the lake after t days, measured in millions of cubic feet. Then $x_0 = (0.0025)(8000) = 20$. We want to know when $x = (0.001)(8000) = 8$.

We construct a mathematical model for this situation by estimating the increment in x during a short time interval Δt. Thus

$$\begin{aligned} x(t + \Delta t) - x(t) &= \{\text{pollutant in}\} - \{\text{pollutant out}\} \\ &\approx (0.0005)(500)\,\Delta t - \frac{x(t)}{8000}\,500\,\Delta t \\ &= \frac{1}{4}\Delta t - \frac{x}{16}\Delta t. \end{aligned}$$

It follows that

$$\frac{dx}{dt} = \frac{1}{4} - \frac{x}{16}.$$

With $x_0 = 20$, Equation (2) gives the solution

$$x(t) = 4 + 16e^{-t/16}.$$

We can find when $x(t) = 8$ by solving the equation

$$8 = 4 + 16e^{-t/16},$$

and this gives

$$t = 16 \ln 4 \approx 22.2 \quad \text{(days)}.$$

7.6 PROBLEMS

In each of Problems 1–10, use the method of derivation of Equation (2), rather than the equation itself, to find the solution of the given differential equation satisfying the indicated initial condition.

1. $\dfrac{dy}{dx} = y + 1;\quad y(0) = 1$

2. $\dfrac{dy}{dx} = 2 - y;\quad y(0) = 3$

3. $\dfrac{dy}{dx} = 2y - 3;\quad y(0) = 2$

4. $\dfrac{dy}{dx} = \dfrac{1}{4} - \dfrac{y}{16};\quad y(0) = 20$

5. $\dfrac{dx}{dt} = 2(x - 1);\quad x(0) = 0$

6. $\dfrac{dx}{dt} = 2 - 3x;\quad x(0) = 4$

7. $\dfrac{dx}{dt} = 5(x + 2);\quad x(0) = 25$

8. $\dfrac{dx}{dt} = -3 - 4x;\quad x(0) = -5$

9. $\dfrac{dv}{dt} = 10(10 - v);\quad v(0) = 0$

10. $\dfrac{dv}{dt} = -5(10 - v);\quad v(0) = -10$

11. A certain city had a population of 1.5 million in 1980. Assume that it grows continuously at a 4% annual rate and also absorbs 50,000 newcomers per year. What will be its population in the year 2000?

12. A cake is removed from an oven at 210°F and left to cool at room temperature, which is 70°F. After 30 min the temperature of the cake is 140°F. When will it be 100°F?

13. Payments are made on a mortgage (original loan) of P_0 dollars continuously at the constant rate of c dollars per month. Let $P(t)$ denote the balance (amount still owed) after t months, and let r denote the monthly interest rate paid by the borrower. (For example, $r = 0.06/12 = 0.005$ if the annual interest rate is 6%.) Derive the differential equation

$$\frac{dP}{dt} = rP - c, \qquad P(0) = P_0.$$

14. An auto loan of \$3600 is to be paid off continuously over a period of 36 months. Apply the result of Problem 13 to determine the monthly payment required if the annual interest rate is (a) 12%; (b) 18%.

15. A certain piece of dubious information about phenylethylamine in the drinking water began to spread one day in a city with a population of 100,000. Within a week, 10,000 people had heard this rumor. Assuming that the rate of increase of the number who have heard the rumor is proportional to the number who have not yet heard it, how long will it be until half the population of the city has heard the rumor?

16. A tank contains 1000 liters of a solution consisting of 50 lb of salt dissolved in water. Pure water is pumped into the tank t the rate of 5 liter/s and the mixture—kept uniform by stirring—is pumped out at the same rate. After how many seconds will only 10 lb of salt remain in the tank?

17. Derive the solution in (2) of Equation (1) under the assumption that $ax + b < 0$.

18. Suppose that a body moves through a resisting medium with resistance proportional to its velocity v, so that $dv/dt = -kv$.

(a) Show that its velocity $v(t)$ and position $x(t)$ at time t are given by

$$v = v_0 e^{-kt} \quad \text{and} \quad x = x_0 + \frac{v_0}{k}(1 - e^{-kt}).$$

(b) Conclude that the body travels only a *finite* distance v_0/k.

19. Suppose that a motorboat is moving at 40 ft/s when its motor suddenly quits and that 10 s later the boat has slowed to 20 ft/s. Assume, as in Problem 18, that the resistance it encounters while coasting is proportional to its velocity. How far will the motorboat coast in all?

20. The acceleration of a certain sports car is proportional to the difference between 250 km/h and the velocity of the sports car. If this machine can accelerate from rest to 100 km/h in 10 s, how long will it take for it to accelerate from rest to 200 km/h?

21. Consider the linear first order differential equation

$$x'(t) + p(t)x(t) = q(t)$$

with variable coefficients. Let $P(t)$ be an antiderivative of $p(t)$. Multiply both sides of the given equation by $e^{P(t)}$, and note that the left-hand side of the resulting equation is $D_t[e^{P(t)}x(t)]$. Conclude by antidifferentiation that

$$x(t) = e^{-P(t)}\left[\int e^{P(t)}q(t)\,dt + C\right].$$

22. Use the method of Problem 21 to derive the solution

$$x(t) = x_0 e^{-at} + b\,\frac{e^{ct} - e^{-at}}{a + c}$$

of the differential equation $dx/dt + ax = be^{ct}$ (under the assumption that $a + c \neq 0$).

23. A 30-year-old woman accepts an engineering position with a starting salary of \$30,000 per year. Her salary S increases exponentially, with

$$S(t) = 30e^{(0.05)t}$$

thousand dollars after t years. Meanwhile, 12% of her salary is deposited continuously in a retirement account, which accumulates interest at an annual rate of 6%, compounded continuously.

(a) Estimate ΔA in terms of Δt to derive this equation for the amount $A(t)$ in her retirement account at time t:

$$\frac{dA}{dt} - (0.06)A = (3.6)e^{(0.05)t}.$$

(b) Apply the result of Problem 22 to compute $A(40)$, the amount available for her retirement at age 70.

CHAPTER 7 REVIEW: Definitions, Concepts, Results

Use this list as a guide to concepts that you may need to review.

1. The laws of exponents
2. The laws of logarithms
3. The definition of the natural logarithm function
4. The graph of $y = \ln x$
5. The definition of the number e
6. The definition of the natural exponential function
7. The inverse function relationship between $\ln x$ and e^x
8. The graphs of $y = e^x$ and $y = e^{-x}$

9. Differentiation of $\ln u$ and e^u where u is a differentiable function of x
10. The order of magnitude of $(\ln x)/x^k$ and x^k/e^x as $x \to \infty$
11. The number e as a limit
12. Definition of general exponential and logarithm functions
13. Differentiation of a^u and $\log_a u$
14. Logarithmic differentiation
15. Solution of the differential equation $dx/dt = kx$
16. Natural population growth
17. Radioactive decay and radiocarbon dating
18. Solution of a linear first order differential equation with constant coefficients
19. Solution of separable first order differential equations
20. Evaluation of the constant of integration in an initial value problem

CHAPTER 7 MISCELLANEOUS PROBLEMS

Differentiate the given function $f(x)$ in Problems 1–24.

1. $\ln 2\sqrt{x}$
2. $e^{-2\sqrt{x}}$
3. $\ln(x - e^x)$
4. $10^{\sqrt{x}}$
5. $\ln(2^x)$
6. $\log_{10}(\sin x)$
7. $x^3 e^{-1/x^2}$
8. $x(\ln x)^2$
9. $(\ln x)[\ln(\ln x)]$
10. $\exp(10^x)$
11. $2^{\ln x}$
12. $\ln\left(\dfrac{e^x + e^{-x}}{e^x - e^{-x}}\right)$
13. $e^{(x+1)/(x-1)}$
14. $\ln(\sqrt{1+x}\,\sqrt[3]{2+x^2})$
15. $\ln[(x-1)/(3-4x^2)]^{3/2}$
16. $\sin(\ln x)$
17. $\exp([1 + \sin^2 x]^{1/2})$
18. $\dfrac{x}{(\ln x)^2}$
19. $\ln(3^x \sin x)$
20. $(\ln x)^x$
21. $x^{1/x}$
22. $x^{\sin x}$
23. $(\ln x)^{\ln x}$
24. $(\sin x)^{\cos x}$

Evaluate the indefinite integrals in Problems 25–36.

25. $\displaystyle\int \frac{dx}{1-2x}$
26. $\displaystyle\int \frac{\sqrt{x}}{1+x^{3/2}}\,dx$
27. $\displaystyle\int \frac{3-x}{1+6x-x^2}\,dx$
28. $\displaystyle\int \frac{e^x - e^{-x}}{e^x + e^{-x}}\,dx$
29. $\displaystyle\int \frac{\sin x}{2+\cos x}\,dx$
30. $\displaystyle\int \frac{e^{-1/x^2}}{x^3}\,dx$
31. $\displaystyle\int \frac{10^{\sqrt{x}}}{\sqrt{x}}\,dx$
32. $\displaystyle\int \frac{1}{x(\ln x)^2}\,dx$
33. $\displaystyle\int e^x\sqrt{1+e^x}\,dx$
34. $\displaystyle\int \frac{1}{x}\sqrt{1+\ln x}\,dx$
35. $\displaystyle\int 2^x 3^x\,dx$
36. $\displaystyle\int \frac{dx}{x^{1/3}(1+x^{2/3})}$

Solve the initial value problems in Problems 37–44.

37. $\dfrac{dx}{dt} = 2t;\quad x(0) = 17$
38. $\dfrac{dx}{dt} = 2x;\quad x(0) = 17$
39. $\dfrac{dx}{dt} = e^t;\quad x(0) = 2$
40. $\dfrac{dx}{dt} = e^x;\quad x(0) = 2$
41. $\dfrac{dx}{dt} = 3x - 2;\quad x(0) = 3$
42. $\dfrac{dx}{dt} = x^2 t^2;\quad x(0) = -1$
43. $\dfrac{dx}{dt} = x\cos t;\quad x(0) = \sqrt{2}$
44. $\dfrac{dx}{dt} = \sqrt{x};\quad x(1) = 0$

Sketch the graphs of the equations given in Problems 45–49.

45. $y = e^{-x}\sqrt{x}$
46. $y = x - \ln x$
47. $y = \sqrt{x} - \ln x$
48. $y = x(\ln x)^2$
49. $y = e^{-1/x}$

50. Find the length of the curve $y = \frac{1}{2}x^2 - \frac{1}{4}\ln x$ from $x = 1$ to $x = e$.

51. A grain warehouse holds B bushels of grain, which is deteriorating in such a way that only $B \cdot 2^{-t/12}$ bushels will be salable after t months. Meanwhile, the market price is increasing linearly; after t months it will be $2 + t/12$ dollars per bushel. After how many months should the grain be sold in order to maximize the revenue obtained?

52. Suppose that you have borrowed \$1000 at 10% annual interest compounded continuously to plant timber on a tract of land. Your agreement is to repay the loan, plus interest, when the timber is cut and sold. If the cut timber can be sold after t years for $800\exp(\frac{1}{2}\sqrt{t})$ dollars, when should you cut and sell in order to maximize your profit?

53. Suppose that blood samples from 1000 students are to be tested for a certain disease known to occur in 1% of the population. Each test costs \$5, so it would cost \$5000 to test the samples individually. Suppose, however, that "lots" made up of x samples each are formed by pooling individual samples, and that these lots are tested first (for \$5 each). Only in case a lot tests positive—the probability

of this is $1 - (0.99)^x$—will the x samples used to make up this lot be tested individually.

(a) Show that the total expected number of tests is

$$f(x) = \frac{1000}{x}[(1)(0.99)^x + (x+1)(1-(0.99)^x)]$$

$$= 1000 + \frac{1000}{x} - (1000)(0.99)^x \quad \text{if } x \geqq 2.$$

(b) Show that the value of x that minimizes $f(x)$ is a root of the equation

$$x = \frac{(0.99)^{-x/2}}{[\ln(100/99)]^{1/2}}.$$

Because the last denominator is approximately 0.1, it may be convenient to solve instead the simpler equation $x = (10)(0.99)^{-x/2}$. Whichever you choose, the method of repeated substitution (beginning with $x_0 = 10$) is very effective.

(c) From the results in parts (a) and (b), compute the cost of using this batch method to test the original 1000 samples.

54. Deduce from Problem 63 in Section 7.2 that

$$\lim_{x \to 0^+} x^x = 1.$$

55. Show that

$$\lim_{x \to 0} \frac{\ln(1+x)}{x} = 1$$

by considering the value of $D \ln x$ for $x = 1$. Thus show that $\ln(1+x) \approx x$ if x is very close to zero.

56. Prove that

$$\lim_{h \to 0} \frac{1}{h}(a^h - 1) = \ln a$$

by considering the definition of the derivative of a^x at $x = 0$. Substitute $h = 1/n$ to obtain

$$\ln a = \lim_{n \to \infty} n(a^{1/n} - 1).$$

Approximate $\ln 2$ by taking $n = 1024 = 2^{10}$ and using only the square root key (10 times) on your calculator.

57. Suppose that the fish population in a lake is attacked by disease at time $t = 0$, with the result that

$$\frac{dP}{dt} = -3\sqrt{P}$$

thereafter, with time t in weeks. Initially, there are $P_0 = 900$ fish in the lake. How long will it take for all of the fish to die?

58. A race car sliding along a level surface is decelerated by frictional forces proportional to its speed. Suppose that it decelerates at 2 m/s^2 and travels a total distance of 1800 m. What was its initial velocity? See Problem 18 in Section 7.6.

59. A home mortgage of \$120,000 is to be paid off continuously over a period of 25 years. Apply the result of Problem 13 in Section 7.6 to determine the monthly payment required if the annual interest rate is (a) 8%; (b) 12%.

60. A powerboat weighs 32,000 lb, and its motor provides a thrust of 5000 lb. Assume that the water resistance is 100 lb for each foot per second of the boat's speed. Then the velocity $v(t)$ of the boat (in ft/s^2) of the boat at time t (in seconds) satisfies the differential equation

$$1000\frac{dv}{dt} = 5000 - 100v.$$

Find the maximum velocity that the boat can attain if it starts from rest.

61. The temperature within a certain freezer is -16°C, and the room temperature is a constant 20°C. At 11 P.M. the power goes off, and at 6 A.M. the next morning the temperature in the freezer has risen to -10°C. At what time will the temperature in the freezer reach the critical value of 0°C if the power remains off?

62. Suppose that the action of fluorocarbons depletes the ozone in the upper atmosphere by 0.25% annually, so that the amount A of ozone in the atmosphere satisfies the differential equation

$$\frac{dA}{dt} = -\frac{1}{400}A \quad (t \text{ in years}).$$

(a) What percentage of the original amount A_0 of atmospheric ozone will remain 25 years from now?

(b) How long will it take for the amount of atmospheric ozone to be reduced to half its initial amount?

63. A car starts from rest and travels along a straight road. Its engine provides a constant acceleration of a feet per second per second. Air resistance and road friction cause a deceleration of ρ feet per second per second for every foot per second of the car's velocity.

(a) Show that after t seconds the car's velocity is

$$v = \frac{a}{\rho}(1 - e^{-\rho t}).$$

(b) If $a = 17.6$ ft/s^2 and $\rho = 0.1$, find v when $t = 10$, and also find the limiting velocity as $t \to \infty$. Give each answer in *miles per hour*.

chapter eight

Trigonometric and Hyperbolic Functions

8.1 Introduction

The function f is called an **algebraic function** provided that $y = f(x)$ satisfies an equation of the form

$$a_n(x)y^n + a_{n-1}(x)y^{n-1} + \cdots + a_1(x)y + a_0(x) = 0$$

where the coefficients $a_0(x), a_1(x), \ldots, a_n(x)$ are polynomials in x. For example, because the equation $y^2 - p(x) = 0$ has the form shown above, we see that the square root of the polynomial $p(x)$ is an algebraic function. The equation $q(x)y - p(x) = 0$ also has the necessary form, so we see that a rational function (a quotient of polynomials) is also an algebraic function.

A function that is *not* algebraic is said to be **transcendental.** The natural logarithm function $\ln x$ and the exponential function e^x are examples of transcendental functions. In this chapter we study the remaining transcendental functions of elementary character—the trigonometric and the hyperbolic functions. These functions have extensive scientific applications and also provide the basis for certain important methods of integration (these methods are discussed in Chapter 9).

8.2 Derivatives and Integrals of Trigonometric Functions

In Section 3.7 we saw that the derivatives of the sine and cosine functions are given by

$$D_x \sin u = (\cos u)\frac{du}{dx}, \qquad D_x \cos u = (-\sin u)\frac{du}{dx}, \tag{1}$$

where u denotes a differentiable function of x. Recall that when differentiating and integrating trigonometric functions, we use radian measure of angles exclusively.

The other four trigonometric functions—the tangent, cotangent, secant, and cosecant—may be defined in terms of the sine and cosine:

$$\tan x = \frac{\sin x}{\cos x}, \qquad \cot x = \frac{\cos x}{\sin x},$$
$$\sec x = \frac{1}{\cos x}, \qquad \csc x = \frac{1}{\sin x}. \tag{2}$$

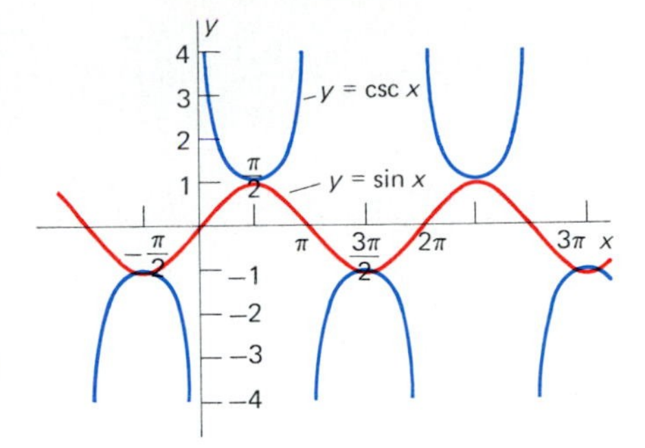

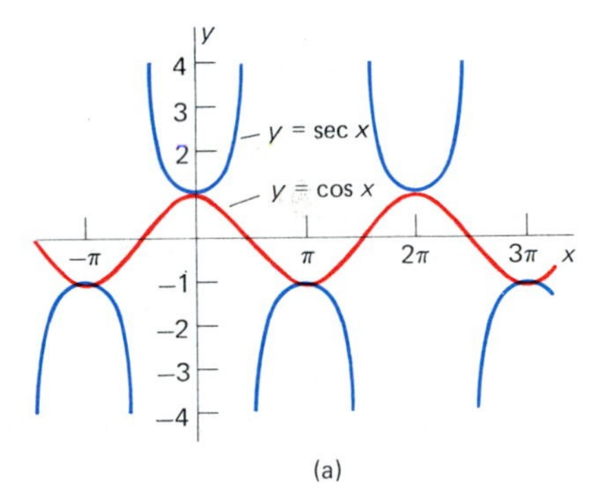

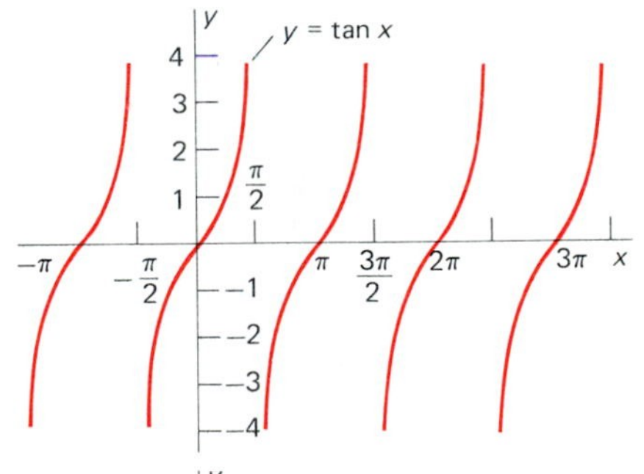

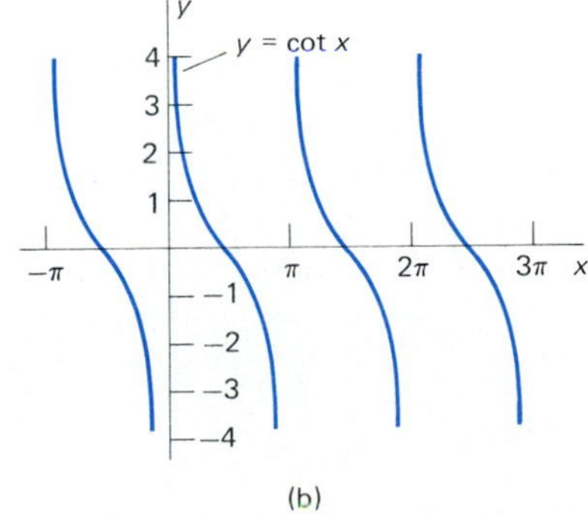

8.1 Graphs of the six trigonometric functions

These formulas are valid except where the denominators are zero. Thus $\tan x$ and $\sec x$ are undefined when x is an odd multiple of $\pi/2$, while $\cot x$ and $\csc x$ are undefined when x is an integral multiple of π. The graphs of the six trigonometric functions appear in Fig. 8.1. There we show the sine and its reciprocal, the cosecant, in the same coordinate plane; we also pair the cosine with the secant and the tangent with the cotangent.

The graphs in Fig. 8.1 suggest that all six trigonometric functions are periodic, and this is indeed true. The sine and cosine functions each have period 2π, and it follows from (2) that the other four must repeat their values in cycles of length 2π as well. For example, $\tan(x + 2\pi) = \tan x$. But the **period** of the tangent and cotangent functions is the length of the shortest interval of repetition, so for these two functions the period is π, because $\tan(x + \pi) = \tan x$ for all (meaningful) values of x. This relation, too, is suggested in Fig. 8.1. The secant and cosecant functions have period 2π.

If we begin with the derivatives in (1) and the definitions in (2), we can easily compute the derivatives of the latter four trigonometric functions. All we need is the quotient rule. The results (as listed in Section 3.7) are

$$D \tan x = \sec^2 x, \qquad D \cot x = -\csc^2 x,$$
$$D \sec x = \sec x \tan x, \qquad D \csc x = -\csc x \cot x.$$

When we combine these results with the chain rule, we get the four differentiation formulas

$$D_x \tan u = (\sec^2 u)\frac{du}{dx}, \tag{3}$$

$$D_x \cot u = (-\csc^2 u)\frac{du}{dx}, \tag{4}$$

$$D_x \sec u = (\sec u \tan u)\frac{du}{dx}, \tag{5}$$

$$D_x \csc u = (-\csc u \cot u)\frac{du}{dx}. \tag{6}$$

The patterns in (1) and in (3) through (6) make them easy to memorize. The formulas in (4) and (6) are the "cofunction analogues" of those in (3) and (5), respectively. Note also that the derivative formulas for the three cofunctions are those involving minus signs.

EXAMPLE 1 The following are some derivatives involving trigonometric functions.

(a) $D\tan(2x^3) = [\sec^2 2x^3]D_x(2x^3) = 6x^2\sec^2 2x^3.$

(b) $$\begin{aligned} D\cot^3(2x) = D(\cot 2x)^3 &= [3(\cot 2x)^2]D_x(\cot 2x) \\ &= [3\cot^2(2x)][-\csc^2(2x)]D_x(2x) \\ &= -6\cot^2 2x \csc^2 2x. \end{aligned}$$

(c) $D\sec\sqrt{x} = [\sec\sqrt{x}\tan\sqrt{x}]D_x(\sqrt{x}) = \dfrac{\sec\sqrt{x}\tan\sqrt{x}}{2\sqrt{x}}.$

(d) $$\begin{aligned} D\csc^{1/2}x &= \tfrac{1}{2}(\csc x)^{-1/2}D_x(\csc x) \\ &= \frac{-\csc x\cot x}{2(\csc x)^{1/2}} = -\frac{1}{2}\csc^{1/2}x\cot x. \end{aligned}$$

By now we know that to every derivative formula there corresponds an integral formula. The integral versions of (1) and (3) through (6) are these:

$$\int \cos u\,du = \sin u + C, \tag{7}$$

$$\int \sin u\,du = -\cos u + C, \tag{8}$$

$$\int \sec^2 u\,du = \tan u + C, \tag{9}$$

$$\int \csc^2 u\,du = -\cot u + C, \tag{10}$$

$$\int \sec u\tan u\,du = \sec u + C, \tag{11}$$

$$\int \csc u\cot u\,du = -\csc u + C. \tag{12}$$

EXAMPLE 2 Compute the antiderivative: $\int \sec^2 3x\,dx.$

Solution With $u = 3x$ and $du = 3\,dx$, the formula in (9) gives

$$\int \sec^2 3x\,dx = \tfrac{1}{3}\int \sec^2 u\,du = \tfrac{1}{3}\tan u + C = \tfrac{1}{3}\tan 3x + C.$$

EXAMPLE 3 Find the antiderivative:

$$\int \frac{1}{\sqrt{x}}\csc\sqrt{x}\cot\sqrt{x}\,dx.$$

Solution With

$$u = \sqrt{x} \quad \text{we have} \quad du = \frac{1}{2\sqrt{x}}\,dx.$$

Hence the formula in (12) gives

$$\int \frac{\csc\sqrt{x}\cot\sqrt{x}}{\sqrt{x}}\,dx = 2\int \csc u \cot u\,du$$
$$= -2\csc u + C = -2\csc\sqrt{x} + C.$$

To integrate $\tan x$, the substitution

$$u = \cos x, \qquad du = -\sin x\,dx$$

gives

$$\int \tan x\,dx = \int \frac{\sin x}{\cos x}\,dx = -\int \frac{1}{u}\,du = -\ln|u| + C,$$

and thus

$$\int \tan x\,dx = -\ln|\cos x| + C = \ln|\sec x| + C. \tag{13}$$

In the last step we used the fact that $|\sec x| = 1/|\cos x|$.

Similarly,

$$\int \cot x\,dx = \ln|\sin x| + C = -\ln|\csc x| + C. \tag{14}$$

The first person to integrate $\sec x$ may well have spent much time doing so. First we must "prepare" the function for integration:

$$\sec x = \frac{1}{\cos x} = \frac{\cos x}{\cos^2 x} = \frac{\cos x}{1 - \sin^2 x}.$$

Now

$$\frac{1}{1+x} + \frac{1}{1-x} = \frac{2}{1-x^2}.$$

Similarly, working backwards, we have

$$\frac{2\cos x}{1-\sin^2 x} = \frac{\cos x}{1+\sin x} + \frac{\cos x}{1-\sin x}.$$

Therefore

$$\int \sec x\,dx = \frac{1}{2}\int \left(\frac{\cos x}{1+\sin x} + \frac{\cos x}{1-\sin x}\right)dx$$
$$= \frac{1}{2}(\ln|1+\sin x| - \ln|1-\sin x|) + C.$$

It is customary to perform some algebraic simplifications of this last result:

$$\int \sec x\,dx = \frac{1}{2}\ln\left|\frac{1+\sin x}{1-\sin x}\right| + C = \frac{1}{2}\ln\left|\frac{(1+\sin x)^2}{1-\sin^2 x}\right| + C$$
$$= \ln\left|\frac{(1+\sin x)^2}{\cos^2 x}\right|^{1/2} + C = \ln\left|\frac{1+\sin x}{\cos x}\right| + C$$
$$= \ln|\sec x + \tan x| + C.$$

Of course, after verifying by differentiation that

$$\int \sec x \, dx = \ln|\sec x + \tan x| + C, \tag{15}$$

one can always "derive" this result by an unmotivated trick:

$$\int \sec x \, dx = \int (\sec x) \frac{\tan x + \sec x}{\sec x + \tan x} \, dx = \int \frac{\sec x \tan x + \sec^2 x}{\sec x + \tan x} \, dx$$
$$= \ln|\sec x + \tan x| + C.$$

A similar technique yields

$$\int \csc x \, dx = -\ln|\csc x + \cot x| + C. \tag{16}$$

EXAMPLE 4 The substitution

$$u = \tfrac{1}{2}x, \qquad du = \tfrac{1}{2}dx$$

gives

$$\int_0^{\pi/2} \sec\frac{x}{2} \, dx = 2\int_0^{\pi/4} \sec u \, du = 2\Big[\ln|\sec u + \tan u|\Big]_0^{\pi/4}$$
$$= 2\ln(1 + \sqrt{2}) \approx 1.76275.$$

To integrate $\sin^2 x$ and $\cos^2 x$ we use the half-angle identities (in Equations (10) and (11) in Appendix A)

$$\sin^2 x = \tfrac{1}{2}(1 - \cos 2x), \qquad \cos^2 x = \tfrac{1}{2}(1 + \cos 2x). \tag{17}$$

EXAMPLE 5 Find $\int \sin^2 3x \, dx$.

Solution The first identity in (17)—with $3x$ in place of x—yields

$$\int \sin^2 3x \, dx = \int \tfrac{1}{2}(1 - \cos 6x) \, dx = \tfrac{1}{2}(x - \tfrac{1}{6}\sin 6x) + C$$
$$= \tfrac{1}{12}(6x - \sin 6x) + C.$$

To integrate $\tan^2 x$ and $\cot^2 x$ we use the identities

$$1 + \tan^2 x = \sec^2 x, \qquad 1 + \cot^2 x = \csc^2 x. \tag{18}$$

The first of these follows from the fundamental identity $\sin^2 x + \cos^2 x = 1$ upon division of both sides by $\cos^2 x$. To obtain the second formula in (18), divide both sides of the fundamental identity by $\sin^2 x$.

EXAMPLE 6 Compute the antiderivative: $\int \cot^2 3x \, dx$.

Solution By using the second identity in (18) we obtain

$$\int \cot^2 3x \, dx = \int (\csc^2 3x - 1) \, dx$$
$$= \int (\csc^2 u - 1)(\tfrac{1}{3}\, du) \qquad (\text{with } u = 3x)$$
$$= \tfrac{1}{3}(-\cot u - u) + C = -\tfrac{1}{3}\cot 3x - x + C.$$

*PERIODIC PHENOMENA AND SIMPLE HARMONIC MOTION

As we saw in Chapter 7, exponential functions are important in the study of natural growth and decay processes—population growth, radioactive decay, and the like. The trigonometric functions are equally important in the analysis of periodic phenomena in nature, such as the ebb and flow of the seasons and the tides, and the oscillatory motions that occur in vibrating mechanical systems. These applications are so remarkably diverse partly because of the truth of the following theorem.

Theorem *The Equation* $x'' = -k^2x$

The function $x = x(t)$ satisfies the second order differential equation

$$\frac{d^2x}{dt^2} = -k^2x \qquad (k > 0, k \text{ constant}) \tag{19}$$

if and only if there exist constants A and B such that

$$x(t) = A\cos kt + B\sin kt. \tag{20}$$

Proof Suppose first that $x = x(t)$ has the form given in Equation (20). Then

$$\frac{dx}{dt} = -kA\sin kt + kB\cos kt \tag{21}$$

and

$$\frac{d^2x}{dt^2} = -k^2A\cos kt - k^2B\sin kt = -k^2(A\cos kt + B\sin kt) = -k^2x.$$

Thus the function in (20) satisfies the given differential equation in (19). This proves the *if* part of the theorem. Problem 88 is an outline of the *only if* part—the fact that if a function $x = x(t)$ satisfies the differential equation in (19), then it *must* be of the form in (20). ■

If a particle moves along the x-axis with acceleration given by Equation (19), then the theorem implies that its position function $x = x(t)$ is of the form in (20). Motion of this type is called **simple harmonic motion.** The constants A and B may be determined in terms of the particle's initial position $x_0 = x(0)$ and initial velocity $v_0 = x'(0)$, as follows. Substitution of $t = 0$ in (20) gives $A = x_0$, and substitution of $t = 0$ in (21) gives $B = v_0/k$. Hence the position function of the particle is

$$x = x(t) = x_0\cos kt + \frac{v_0}{k}\sin kt \tag{22}$$

The expression on the right-hand side in Equation (20) can be rewritten in a way that makes it easier to visualize the graph of $x = x(t)$. To do this, we choose positive constants C and α as indicated in Fig. 8.2—that is, so that

$$C\cos k\alpha = A \quad \text{and} \quad C\sin k\alpha = B. \tag{23}$$

Here is one way to find C and α. First, the sum of the squares of the left-hand sides in (23) is C^2, so

$$C = \sqrt{A^2 + B^2}. \tag{24}$$

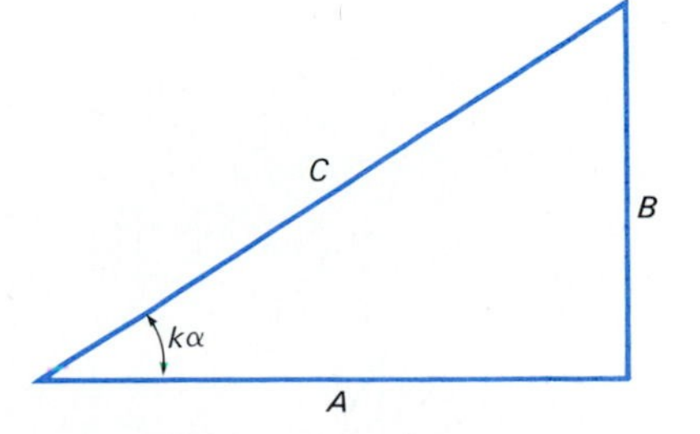

8.2 The constants in Equation (23)

Then we form quotients using the two equations in (23) and find that

$$\tan k\alpha = \frac{B}{A}. \tag{25}$$

The signs of A and B determine which quadrant $k\alpha$ lies in—the first quadrant if both are positive, the third if both are negative, the second if $A < 0 < B$, and the fourth if $B < 0 < A$. Then substitution of the formulas in (24) and (25) in Equation (20) yields

$$\begin{aligned} x(t) &= (C \cos k\alpha) \cos kt + (C \sin k\alpha) \sin kt \\ &= C(\cos kt \cos k\alpha + \sin kt \sin k\alpha). \end{aligned}$$

Thus

$$x(t) = C \cos k(t - \alpha). \tag{26}$$

With $k\beta = k\alpha - \pi/2$, the addition formula for cosines gives the alternative form

$$x(t) = C \sin k(t - \beta). \tag{27}$$

The graph of Equation (27) exhibits the same oscillatory behavior as the curve $x = \sin t$. But the effect of the coefficient C is to stretch the graph in the vertical x-direction, so that x has maximum value $+C$ and minimum value $-C$.

Because of the constant k, the function $x(t) = C \sin k(t - \beta)$ has period $2\pi/k$ rather than 2π, for

$$\begin{aligned} x\left(t + \frac{2\pi}{k}\right) &= C \sin k\left(t + \frac{2\pi}{k} - \beta\right) = C \sin(kt - k\beta + 2\pi) \\ &= C \sin k(t - \beta) = x(t). \end{aligned}$$

Finally, the effect of the constant β is to translate the graph β units to the right of the curve $x = C \sin kt$. When we assemble all this information, we see that the graph of $x(t)$ in (27) resembles the one shown in Fig. 8.3.

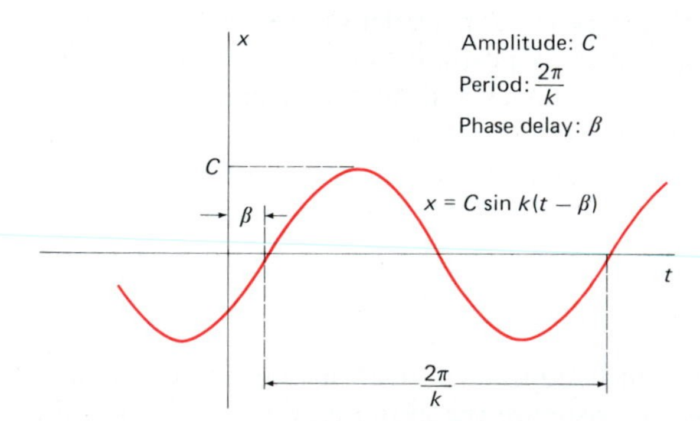

8.3 Amplitude, period, and phase delay

Simple harmonic motion with the position function in (27) may be described as having

$$\text{amplitude} \quad C, \qquad \text{period} \quad \frac{2\pi}{k}, \qquad \text{phase delay} \quad \beta. \tag{28}$$

EXAMPLE 7 Sketch the graph of $x = 3\sin(2t - \pi/2)$.

Solution We write $x = 3\sin 2(t - \pi/4)$, which shows that the function x has amplitude 3, period $2\pi/k = \pi$, and phase delay $\beta = \pi/4$, so the graph looks like the one in Fig. 8.4.

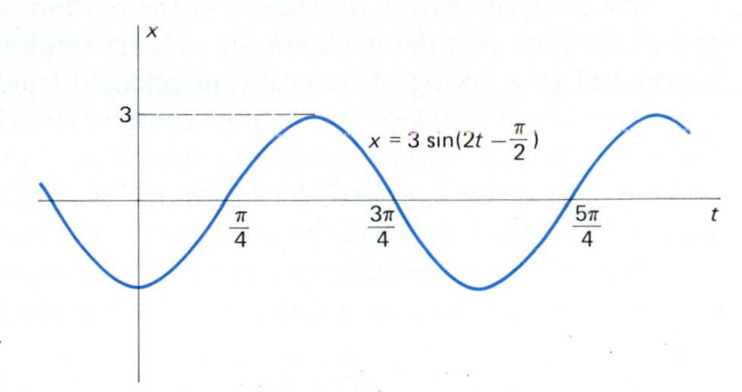

8.4 The graph for Example 7

EXAMPLE 8 Sketch the graph of $x = \cos 2t + \sin 2t$.

Solution Here we have the form in (20) with $A = B = 1$ and $k = 2$. But instead of applying the general relations in (24) and (25), it generally is simpler in a numerical example to proceed as follows. We first factor out $C = (A^2 + B^2)^{1/2}$, and write

$$x(t) = \cos 2t + \sin 2t = \sqrt{2}\left[\frac{1}{\sqrt{2}}\sin 2t + \frac{1}{\sqrt{2}}\cos 2t\right]$$

$$= \sqrt{2}\left(\sin 2t \cos\frac{\pi}{4} + \cos 2t \sin\frac{\pi}{4}\right).$$

By the addition formula for the sine, it now follows that

$$x(t) = \sqrt{2}\sin 2(t + \tfrac{1}{8}\pi).$$

Thus the curve in question has amplitude $C = \sqrt{2}$, period $2\pi/k = \pi$, and phase delay $\beta = -\pi/8$. Hence its graph looks like the one in Fig. 8.5.

Some typical examples of simple harmonic motion are the motion of a mass attached to a vibrating spring, the motion of an air molecule in a sound

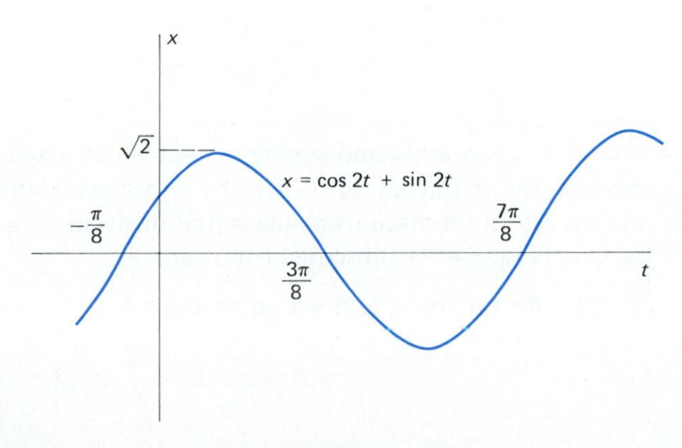

8.5 The function of Example 8

wave, the motion of a point on a vibrating musical string (hence the term *harmonic*), and the motion of a piston in an engine.

*MASS AND SPRING

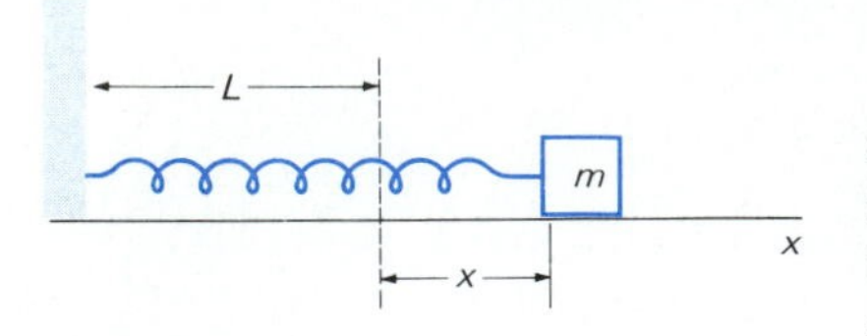

8.6 A simple mass-and-spring system

We conclude with a discussion of the motion of a mass attached to a spring. Suppose that the mass m rests on a frictionless horizontal plane and is attached to a spring of natural (unstretched) length L. We assume that the spring obeys Hooke's law: If the spring is stretched or compressed an amount x, then it exerts a restorative force $F = -cx$, where c is a positive constant (the *spring constant*). As shown in Fig. 8.6, the spring has one end attached to a fixed wall and the other end to the mass m.

We set up a coordinate system in which the mass moves along the x-axis and with the mass at position $x = 0$ when the spring is at its natural length.

Now suppose that the mass is set in motion with position function $x(t)$, so that $x(t)$ measures the amount by which the spring is stretched (the case $x > 0$) or compressed (tthe case $x < 0$). Combining Newton's second law $F = ma = mx''(t)$ and Hooke's law $F = -cx$, we find that

$$mx''(t) = -cx(t),$$

and thus

$$x''(t) = -k^2x(t)$$

where $k = \sqrt{c/m}$. Thus the position function of the mass satisfies the differential equation in (19), and so by the theorem above, its motion is simple harmonic:

$$x(t) = A\cos kt + B\sin kt.$$

The constants A and B are determined by the initial position x_0 and the initial velocity v_0 of the mass (as in Equation (22)). From (28) we see that the period of motion of the mass is

$$\frac{2\pi}{k} = 2\pi\sqrt{m/c}, \tag{29}$$

which depends only upon the mass m and the spring constant c and is independent of x_0 and v_0. The **frequency** f of this simple harmonic motion—the number of cycles completed per second—is the reciprocal of the period:

$$f = \frac{k}{2\pi} = \frac{1}{2\pi}\sqrt{c/m}. \tag{30}$$

EXAMPLE 9 A mass-and-spring system is constructed as in the above discussion; the frequency turns out, by experiment, to be $f = 1/\pi$. The mass is then set into motion from the initial position $x_0 = 1$ with initial velocity $v_0 = 2$. Describe its subsequent motion.

Solution By Equation (22) we know that

$$x(t) = \cos kt + \frac{2}{k}\sin kt.$$

From Equation (30) we see that $f = 1/\pi$ implies that $k = 2$, so

$$x(t) = \cos 2t + \sin 2t = \sqrt{2} \sin 2(t + \tfrac{1}{8}\pi).$$

In the last step we used the result of Example 8. Thus we see that the amplitude of the motion is $\sqrt{2}$ and that the mass first reaches the position $x = \sqrt{2}$ when $t = \pi/8$ and first reaches the position $x = -\sqrt{2}$ when $t = 5\pi/8$.

8.2 PROBLEMS

Find the derivatives of the functions given in Problems 1–26.

1. $f(x) = \sin(2x + 3)$

2. $f(x) = \cos(x^2 - 1)$

3. $f(x) = \tan\left(\dfrac{2x}{3}\right)$

4. $f(x) = \cot\sqrt{x}$

5. $f(x) = \sec\left(\dfrac{1}{x}\right)$

6. $f(x) = \csc e^x$

7. $f(x) = \tan^3 4x$

8. $f(x) = \sec^2 \sqrt[3]{x}$

9. $f(x) = \ln(\sin x)$

10. $f(x) = e^{\cot 2x}$

11. $f(x) = \tan^2(\ln x)$

12. $f(x) = \sec(\csc x)$

13. $f(x) = \ln|\csc x + \cot x|$

14. $f(x) = \tan^3 x \sec^3 x$

15. $f(x) = \sin^3(\csc x)$

16. $f(x) = \dfrac{\cos^2 x}{1 - \sin x}$

17. $f(x) = \dfrac{\tan^2 x}{1 + \sec x}$

18. $f(x) = x^3 \tan^{3/2}\left(\dfrac{1}{x^2}\right)$

19. $f(x) = e^{\sin x} \sin x$

20. $f(x) = (\cot x) \ln|\cos x|$

21. $g(x) = (\sin x)^{1/3} - \sin(x^{1/3})$

22. $h(x) = \ln(\ln \sec x)$

23. $f(x) = \dfrac{1 + \tan x}{1 + \sec x}$

24. $g(x) = \sec(7 \ln x)$

25. $h(t) = \cot^3(1 + t^4)$

26. $f(x) = \cos(e^{-x \ln x})$

In each of Problems 27–30, find dy/dx by implicit differentiation. Then find the tangent line to the graph of the equation at the indicated point P.

27. $\tan xy = 1$; $P(\tfrac{1}{2}\sqrt{\pi}, \tfrac{1}{2}\sqrt{\pi})$

28. $\tan x + \cot y = 2$; $P(\pi/4, \pi/4)$

29. $\sin^2 x + \cos^2 y = 1$; $P(\pi/4, \pi/4)$

30. $6(x \sin y + y \sin x) = \pi$; $P(\pi/6, \pi/6)$

Evaluate the integrals in Problems 31–60.

31. $\displaystyle\int \sec^2 \frac{x}{2}\, dx$

32. $\displaystyle\int \csc^2 x \cot x\, dx$

33. $\displaystyle\int \cos^2 3x\, dx$

34. $\displaystyle\int \frac{\tan^2 \sqrt{x}}{\sqrt{x}}\, dx$

35. $\displaystyle\int \frac{dx}{\csc x}$

36. $\displaystyle\int \frac{dx}{\sin 2x \tan 2x}$

37. $\displaystyle\int \frac{\sec x \tan x}{1 + \sec x}\, dx$

38. $\displaystyle\int \tan 3x \sec^2 3x\, dx$

39. $\displaystyle\int x \sec(x^2) \tan(x^2)\, dx$

40. $\displaystyle\int \frac{\tan^2 x}{\sec x}\, dx$

41. $\displaystyle\int \frac{e^{\sin 2x}}{\sec 2x}\, dx$

42. $\displaystyle\int \frac{e^{\tan x}}{\cos^2 x}\, dx$

43. $\displaystyle\int \frac{\sin x}{\cos^2 x}\, dx$

44. $\displaystyle\int \frac{1}{x} \sec(\ln x)\, dx$

45. $\displaystyle\int \frac{e^x \sec^2 e^x}{\tan e^x}\, dx$

46. $\displaystyle\int \cos^2 2x\, dx$

47. $\displaystyle\int \sin^5 x \cos x\, dx$

48. $\displaystyle\int e^{\tan x} \sec^2 x\, dx$

49. $\displaystyle\int \frac{\sin 2x}{\cos^5 2x}\, dx$

50. $\displaystyle\int \sin x \cos x\, dx$

51. $\displaystyle\int \sin^{10} x \cos x\, dx$

52. $\displaystyle\int \tan^7 x \sec^2 x\, dx$

53. $\displaystyle\int \frac{\sin x}{(1 + \cos x)^5}\, dx$

54. $\displaystyle\int \tan^6 x \sec^2 x\, dx$

55. $\displaystyle\int e^{2x} \cos e^{2x}\, dx$

56. $\displaystyle\int (\cos x) \sec(\sin x)\, dx$

57. $\displaystyle\int 2^{\sec x} \sec x \tan x\, dx$

58. $\displaystyle\int \sec^2 5x\, dx$

59. $\displaystyle\int \frac{1}{x} \cos(\ln x)\, dx$

60. $\displaystyle\int \frac{\sin x - \cos x}{\cos x + \sin x}\, dx$

61. Find the critical points and inflection points of the function $f(x) = \tan x$, and verify that its graph looks as indicated in Fig. 8.1.

62. Find the critical points and inflection points of the function $f(x) = \sec x$, and verify that its graph looks as indicated in Fig. 8.1.

63. Sketch the graph of $y = \sin x + \cos x$.

64. Sketch the graph of the function $f(x) = e^{-x} \cos x$ for $x \geqq 0$. This curve oscillates between the two curves $y = e^{-x}$ and $y = -e^{-x}$. Where are its local maxima and minima?

65. Find the area of the region between the curves $y = \tan^2 x$ and $y = \sec^2 x$ from $x = 0$ to $x = \pi/4$.

66. Find the volume generated by rotating the region under $y = \sec x$ from $x = 0$ to $x = \pi/4$ around the x-axis.

67. Find the volume generated by revolving the region under $y = \tan(\pi x^2/4)$ from $x = 0$ to $x = 1$ around the y-axis.

68. Find the length of the curve $y = \ln(\cos x)$ from $x = 0$ to $x = \pi/4$.

69. Show by evaluating the integral in two different ways that

$$\int \cot x \csc^2 x \, dx = -\tfrac{1}{2}\cot^2 x + C_1 = -\tfrac{1}{2}\csc^2 x + C_2.$$

Are these two answers consistent with one another?

70. Show by evaluating the integral in two different ways that

$$\int \sin x \cos x \, dx = \tfrac{1}{2}\sin^2 x + C_1 = -\tfrac{1}{2}\cos^2 x + C_2.$$

Are these two answers consistent with one another?

71. Sketch the graphs of $y = 1/x$ and $y = \tan x$ on the same coordinate plane. Then use Newton's method to find the least two positive solutions of the equation

$$\frac{1}{x} = \tan x.$$

72. Figure 8.7 shows a belt of total length L stretched tightly around two pulleys with radii R and r. The distance between the centers of the pulleys is x.

(a) Show that $x = (R - r)\sec\alpha$ and that

$$L = 2\pi R + 2(R - r)[(\tan\alpha) - \alpha]. \tag{31}$$

(b) Find x if $R = 3$ ft, $r = 1$ ft, and $L = 25$ ft. *Suggestion:* First use Newton's method to solve Equation (31) for α.

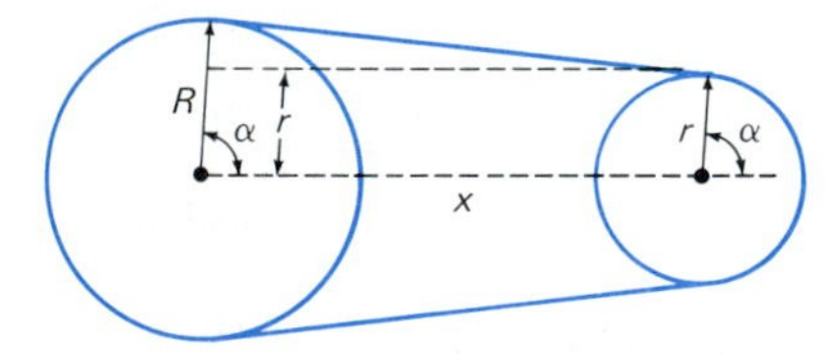

8.7 The belt and pulleys of Problem 72

Problems 73–79 deal with trigonometric graphs as in Examples 7 and 8. Show at least one cycle of the graph of the given function.

73. $x = 2\sin\pi t$

74. $x = 2\cos 3\pi t$

75. $x = \sin\pi(t - 1)$

76. $x = 2\cos(2t - \pi)$

77. $x = 3\sin\dfrac{\pi}{2}(4t - 1)$

78. $x = \sin t + \sqrt{3}\cos t$

79. $x = 3\sin\pi t + 4\cos\pi t$

80. In a certain city the average temperature $f(t)$ on the tth day of the year (with $t = 1$ on January 1) is given by a formula of the form

$$f(t) = A + B\sin k(t - \alpha).$$

(a) What must k be for the period of $f(t)$ to be 365 days?

(b) The minimum daily average temperature of 50°F occurs on January 20 and the maximum of 80°F occurs half a year later. What are the values of A, B, and α?

(c) On what days of the year is the average temperature 60°F? Repeat for 70°F.

Problems 81–87 deal with a mass on a spring, undergoing simple harmonic motion with position function $x(t)$.

81. What is the amplitude of the motion if its frequency is 2 Hz (cycles per second), $x(0) = 1$, and $x'(0) = 0$?

82. What is the amplitude of the motion if its frequency is $2/\pi$ Hz, $x(0) = 0$, and $x'(0) = -4$?

83. What is the amplitude of the motion if its period is π seconds and $x(0) = x'(0) = 1$?

84. What is the period of the motion if its amplitude is 4, $x(0) = 0$, and $x'(0) = 2$?

85. What is the maximum velocity of the mass if the amplitude of the motion is 5 and its period is 2 s?

86. What is the amplitude of the motion if its frequency is $1/\pi$ Hz and the maximum velocity of the mass is 10?

87. Find the period and amplitude of the motion if $x(0) = 3$, $x'(0) = 4$, and $x''(0) = -3$.

88. This is an outline of the proof of the *only if* part of the theorem in this section.

(a) Suppose that $f(t)$ is twice differentiable and that $f''(t) = -k^2 f(t)$. Let the new function $g(t)$ be defined as follows:

$$g(t) = f(t) - f(0)\cos kt - \frac{1}{k}f'(0)\sin kt.$$

Prove that $g''(t) = -k^2 g(t)$ and that $g(0) = g'(0) = 0$.

(b) Let $H(t) = [g'(t)]^2 + k^2[g(t)]^2$. Prove that $H'(t)$ is identically zero. Conclude that $H(t)$ is identically zero. Why does this complete the proof?

8.3 Trigonometric Functions

If the function f is one-to-one on its domain of definition, then it has an inverse function f^{-1}. The domain of f^{-1} is the range of values of f, and

$$f^{-1}(x) = y \quad \text{if and only if} \quad f(y) = x. \tag{1}$$

For example, the exponential function e^x is increasing, and therefore one-to-one, on the whole real line, and it attains all positive values. Hence it has an inverse function, which we already know to be $\ln x$, and this inverse function is defined for all $x > 0$.

Here we want to define the inverses of the trigonometric functions. We must, however, confront the fact that the trigonometric functions fail to be one-to-one because each has period π or 2π. For example, $\sin y = 0.5$ if y is either $\pi/6$ plus any multiple of 2π or $5\pi/6$ plus any multiple of 2π. Consequently, we *cannot* define $y = \sin^{-1} x$, the inverse of the sine function, by saying simply that y is that number such that $\sin y = x$. There are *many* such values of y, and we must specify just which particular one of these is to be used.

We do this by suitably restricting the domain of the sine function. Because the function $\sin x$ is increasing on $[-\pi/2, \pi/2]$ and its range of values is $[-1, 1]$, for each x in $[-1, 1]$ there is a *single* number y in $[-\pi/2, \pi/2]$ such that $\sin y = x$. This observation leads to the following definition of the **inverse sine** (or **arcsine**) **function,** denoted by $\sin^{-1} x$ or $\arcsin x$.

Definition

The **inverse sine function** is defined as follows:

$$y = \sin^{-1} x \quad \text{if and only if} \quad \sin y = x \tag{2}$$

where $-1 \leqq x \leqq 1$ and $-\pi/2 \leqq y \leqq \pi/2$.

Thus, if x is between -1 and $+1$ (inclusive), then $\sin^{-1} x$ is that number y between $-\pi/2$ and $\pi/2$ such that $\sin y = x$. Even more briefly, $\arcsin x$ is the angle (in radians) nearest zero whose sine is x. For instance,

$$\sin^{-1} 1 = \frac{\pi}{2}, \qquad \sin^{-1} 0 = 0, \qquad \sin^{-1}(-1) = -\frac{\pi}{2},$$

and $\sin^{-1} 2$ does not exist.

Note that the symbol -1 in the notation $\sin^{-1} x$ is *not an exponent*—it does *not* mean $(\sin x)^{-1}$.

Because interchanging x and y in $\sin y = x$ yields $y = \sin x$, it follows from (2) that the graph of $y = \sin^{-1} x$ is the reflection in the line $y = x$ of the graph of $y = \sin x$, $-\pi/2 \leqq x \leqq \pi/2$, and therefore looks like the graph in Fig. 8.8.

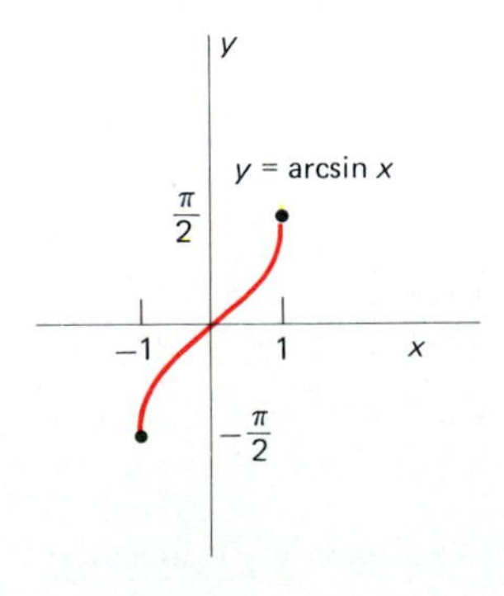

8.8 The graph of $y = \arcsin x$

It follows from Equation (2) that

$$\sin(\sin^{-1} x) = x \qquad \text{if} \quad -1 \leqq x \leqq 1, \tag{3a}$$

$$\sin^{-1}(\sin x) = x \qquad \text{if} \quad -\frac{\pi}{2} \leqq x \leqq \frac{\pi}{2}. \tag{3b}$$

Because the derivative of $\sin x$ is positive for $-\pi/2 < x < \pi/2$, it follows from Theorem 2 in Section 3.4 that $\sin^{-1} x$ is differentiable on $(-1, 1)$. We may therefore differentiate both sides of the identity in (3a), though we begin by writing it in the form

$$\sin y = x$$

where $y = \sin^{-1} x$. This gives

$$(\cos y)\frac{dy}{dx} = 1.$$

So

$$\frac{dy}{dx} = \frac{1}{\cos y} = \frac{1}{\sqrt{1 - \sin^2 y}} = \frac{1}{\sqrt{1 - x^2}}.$$

We are correct in taking the positive square root in this computation because $\cos y > 0$ for $-\pi/2 < y < \pi/2$. Thus

$$D \sin^{-1} x = \frac{1}{\sqrt{1 - x^2}} \tag{4}$$

provided that $-1 < x < 1$. When we combine this result with the chain rule, we get

$$D_x \sin^{-1} u = \frac{1}{\sqrt{1 - u^2}} \cdot \frac{du}{dx} \tag{5}$$

if u is a differentiable function with values in the interval $(-1, 1)$.

The definition of the inverse cosine is similar, except that we begin by restricting the cosine function to the interval $[0, \pi]$, where it is a decreasing function. Thus the **inverse cosine** (or **arccosine**) **function** is defined by means of the rule

$$y = \cos^{-1} x \quad \text{if and only if} \quad \cos y = x \tag{6}$$

where $-1 \leqq x \leqq 1$ and $0 \leqq y \leqq \pi$. Thus $\cos^{-1} x$ is the angle in $[0, \pi]$ whose cosine is x. For instance,

$$\cos^{-1} 1 = 0, \qquad \cos^{-1} 0 = \frac{\pi}{2}, \qquad \cos^{-1}(-1) = \pi.$$

We may compute the derivative of $\cos^{-1} x$, also written arccos x, by differentiation of both sides of the identity

$$\cos(\cos^{-1} x) = x \qquad (-1 < x < 1).$$

The computations are similar to those for $D \sin^{-1} x$, and lead to the result

$$D \cos^{-1} x = -\frac{1}{\sqrt{1 - x^2}}.$$

And if u denotes a differentiable function with values in $(-1, 1)$, the chain rule then gives

$$D_x \cos^{-1} u = -\frac{1}{\sqrt{1 - u^2}} \cdot \frac{du}{dx}. \tag{7}$$

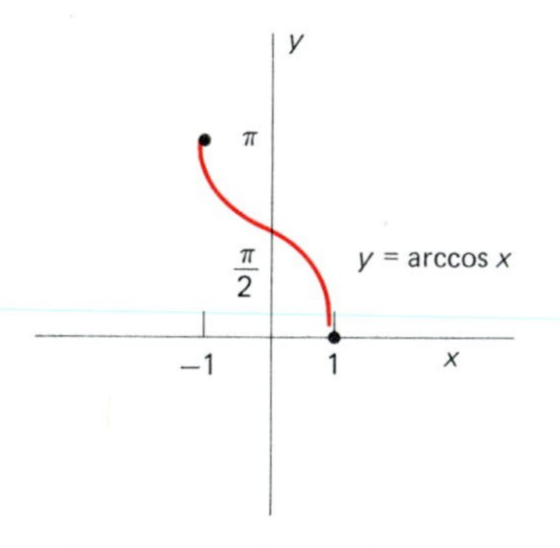

8.9 The graph of $y = \arccos x$

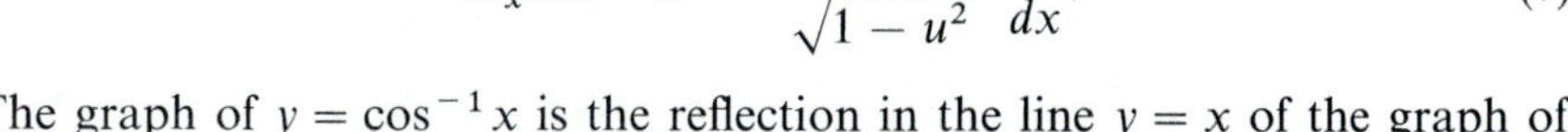

The graph of $y = \cos^{-1} x$ is the reflection in the line $y = x$ of the graph of $y = \cos x$, $0 \leqq x \leqq \pi$, and is shown in Fig. 8.9.

EXAMPLE 1 Suppose that $y = \sin^{-1} x^2$. Then (5) with $u = x^2$ gives

$$\frac{dy}{dx} = \frac{2x}{\sqrt{1 - x^4}}.$$

The tangent function is increasing on the *open* interval $(-\pi/2, \pi/2)$—it is not defined at the end points $-\pi/2$ and $\pi/2$—and its range of values is the whole real line. We may consequently define the **inverse tangent** (or **arctangent) function,** denoted by $\tan^{-1} x$ or by $\arctan x$, as the inverse of $y = \tan x$, $-\pi/2 < x < \pi/2$.

Definition

The **inverse tangent function** is defined as follows:

$$y = \tan^{-1} x \quad \text{if and only if} \quad \tan y = x, \tag{8}$$

where $-\pi/2 < y < \pi/2$.

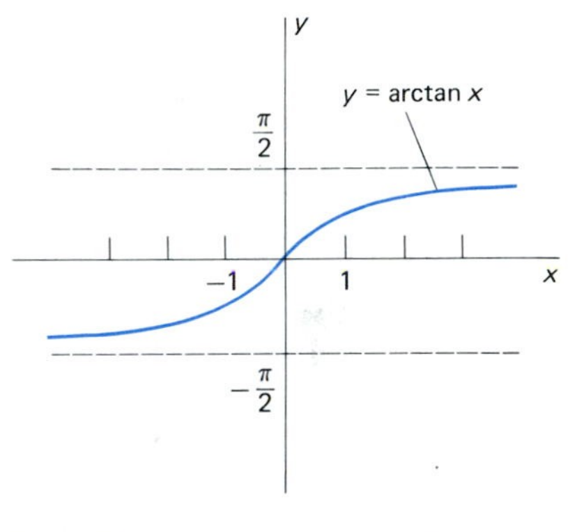

8.10 The inverse tangent function has domain all real x.

Because the tangent function attains all real values, $\tan^{-1} x$ is defined for all real numbers x; $\tan^{-1} x$ is that number y between $-\pi/2$ and $\pi/2$ such that $\tan y = x$. Alternatively, $\arctan x$ is the angle nearest zero whose tangent is x. The graph of $y = \tan^{-1} x$ is the reflection in the line $y = x$ of the graph of $y = \tan x$, $-\pi/2 < x < \pi/2$, and is shown in Fig. 8.10.

It follows from (8) that

$$\tan(\tan^{-1} x) = x \qquad \text{for all} \quad x, \tag{9a}$$

$$\tan^{-1}(\tan x) = x \qquad \text{if} \quad -\pi/2 < x < \pi/2. \tag{9b}$$

Because the derivative of $\tan x$ is positive for all x in the interval $(-\pi/2, \pi/2)$, it follows from Theorem 2 in Section 3.4 that $\tan^{-1} x$ is differentiable for all x. We may therefore differentiate both sides of the identity in (9a). First, we write that identity in the form

$$\tan y = x$$

where $y = \tan^{-1} x$. Then

$$(\sec^2 y)\frac{dy}{dx} = 1.$$

So

$$\frac{dy}{dx} = \frac{1}{\sec^2 y} = \frac{1}{1 + \tan^2 y} = \frac{1}{1 + x^2}.$$

Thus

$$D \tan^{-1} x = \frac{1}{1 + x^2}, \tag{10}$$

and if u is any differentiable function of x, then

$$D_x \tan^{-1} u = \frac{1}{1 + u^2} \cdot \frac{du}{dx}. \tag{11}$$

The definition of the inverse cotangent function is similar, except that we begin by restricting the cotangent function to the interval $(0, \pi)$, where it is a decreasing function attaining all real values. Thus the **inverse cotangent** (or **arccotangent) function** is defined as

$$y = \cot^{-1} x \quad \text{if and only if} \quad \cot y = x \tag{12}$$

where x is any real number and $0 < y < \pi$. Then differentiation of the identity $\cot(\cot^{-1} x) = x$ leads, as in the derivation of Equation (10), to

$$D \cot^{-1} x = -\frac{1}{1 + x^2}.$$

If u is a differentiable function of x, then the chain rule gives

$$D_x \cot^{-1} u = -\frac{1}{1 + u^2} \cdot \frac{du}{dx}. \tag{13}$$

EXAMPLE 2 A mountain climber on one edge of a deep canyon 800 ft wide sees a large rock fall from the opposite edge at time $t = 0$. As he watches the rock plummeting downward, he notices that his eyes first move slowly, then faster, then more slowly again. Let α denote the angle of depression of his line of sight below the horizontal. At what angle α would the rock *seem* to be moving the most rapidly? That is, when would $d\alpha/dt$ be maximal?

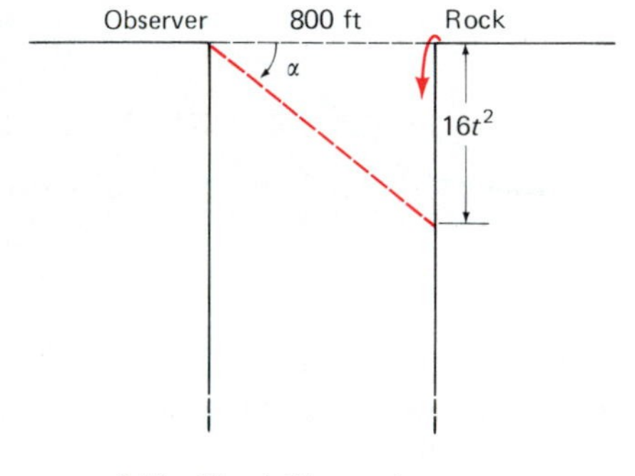

8.11 The falling rock

Solution From our study of constant acceleration in Section 4.9, we know that the rock will fall $16t^2$ feet in the first t seconds. We refer to Fig. 8.11 and see that the value of α at time t will be

$$\alpha = \alpha(t) = \tan^{-1}\left(\frac{16t^2}{800}\right) = \tan^{-1}\left(\frac{t^2}{50}\right).$$

Hence

$$\frac{d\alpha}{dt} = \frac{1}{1 + (t^2/50)^2} \cdot \frac{2t}{50} = \frac{100t}{t^4 + 2500}.$$

To find when $d\alpha/dt$ is maximal, we find when *its* derivative is zero.

$$\frac{d}{dt}\left(\frac{d\alpha}{dt}\right) = \frac{100(t^4 + 2500) - 100t(4t^3)}{(t^4 + 2500)^2} = \frac{100(2500 - 3t^4)}{(t^4 + 2500)^2}.$$

So $d^2\alpha/dt^2$ is zero when $3t^4 = 2500$; that is, when

$$t = (2500/3)^{1/4} \approx 5.37 \text{ s}.$$

This is the value of t when $d\alpha/dt$ is maximal, and at this time we have $t^2 = 50/\sqrt{3}$. So the angle at this time is

$$\alpha = \arctan\left(\frac{1}{50} \cdot \frac{50}{\sqrt{3}}\right) = \arctan\left(\frac{1}{\sqrt{3}}\right) = \frac{\pi}{6}.$$

So the *apparent* speed of the falling rock is greatest when the climber's line of sight is 30° below the horizontal. The speed of the rock then is $32t$ with $t \approx 5.37$ and thus is about 172 ft/s.

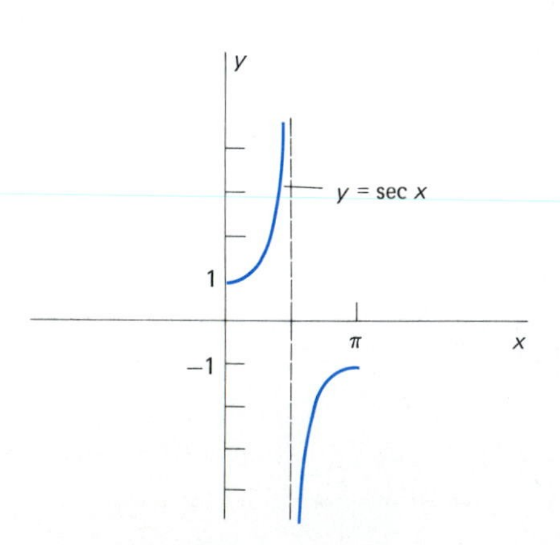

8.12 Restriction of the secant function to the union of the two intervals $[0, \pi/2)$ and $(\pi/2, \pi]$

Figure 8.12 shows that the secant function is increasing on each of the intervals $[0, \pi/2)$ and $(\pi/2, \pi]$. On the union of these two intervals the secant function attains all real values y such that $|y| \geqq 1$. We may therefore define the **inverse secant** (or **arcsecant**) **function,** denoted by $\sec^{-1} x$ or by $\operatorname{arcsec} x$, by restricting the secant function to the union of the intervals $[0, \pi/2)$ and $(\pi/2, \pi]$.

Definition

The **inverse secant function** is defined as follows:

$$y = \sec^{-1} x \quad \text{if and only if} \quad \sec y = x \tag{14}$$

where $|x| \geqq 1$ and $0 \leqq y \leqq \pi$.

The graph of $y = \sec^{-1} x$ is the reflection in the line $y = x$ of the graph of $y = \sec x$, suitably restricted to the intervals $0 \leqq x < \pi/2$ and $\pi/2 < x \leqq \pi$. The graph appears in Fig. 8.13. It follows from the definition of the inverse secant that

$$\sec(\sec^{-1} x) = x \qquad \text{if} \quad |x| \geqq 1, \tag{15a}$$

$$\sec^{-1}(\sec x) = x \qquad \text{for} \quad x \text{ in } [0, \pi/2) \cup (\pi/2, \pi]. \tag{15b}$$

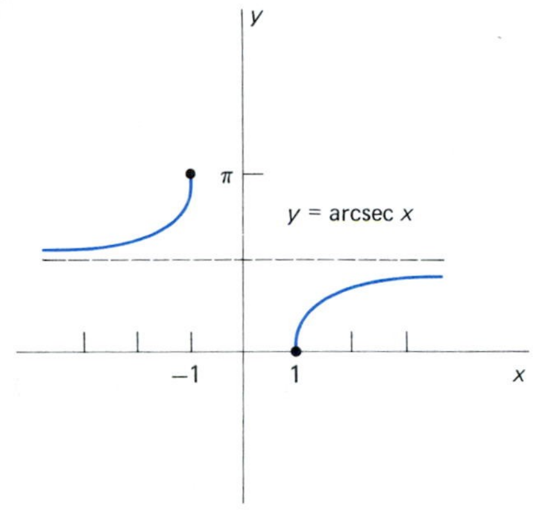

8.13 The graph of $y = \sec^{-1} x = \text{arcsec } x$

Following the now familiar pattern, we find $D \sec^{-1} x$ by differentiating both sides of (15a) in the form

$$\sec y = x$$

where $y = \sec^{-1} x$. This yields

$$(\sec y \tan y)\frac{dy}{dx} = 1,$$

so

$$\frac{dy}{dx} = \frac{1}{\sec y \tan y} = \frac{1}{\pm x\sqrt{x^2 - 1}},$$

because $\tan y = \pm\sqrt{\sec^2 y - 1} = \pm\sqrt{x^2 - 1}$.

To obtain the correct choice of sign here, note what happens in the two cases $x > 1$ and $x < -1$. In the first case, $0 < y < \pi/2$ and $\tan y > 0$, so we choose the plus sign. If $x < -1$ then $\pi/2 < y < \pi$ and $\tan y < 0$, so we take the minus sign. Thus

$$D \sec^{-1} x = \frac{1}{|x|\sqrt{x^2 - 1}} \qquad (|x| > 1). \tag{16}$$

If u is a differentiable function of x with values that exceed 1 in magnitude, then by the chain rule we have

$$D_x \sec^{-1} u = \frac{1}{|u|\sqrt{u^2 - 1}} \cdot \frac{du}{dx}. \tag{17}$$

EXAMPLE 3 The function $\sec^{-1} e^x$ is defined if $x > 0$, for then $e^x > 1$. Then by (17),

$$D_x \sec^{-1} e^x = \frac{e^x}{|e^x|\sqrt{e^{2x} - 1}} = \frac{1}{\sqrt{e^{2x} - 1}}$$

because $|e^x| = e^x$ for all x.

The **inverse cosecant** (or **arccosecant**) **function** is the inverse of the function $y = \csc x$ where x is restricted to the union of the intervals $[-\pi/2, 0)$ and $(0, \pi/2]$. Thus

$$y = \csc^{-1} x \quad \text{if and only if} \quad \csc y = x \tag{18}$$

where $|x| \geqq 1$ and $-\pi/2 < y < \pi/2$. Its derivative formula, which has a derivation similar to that of the inverse secant function, is

$$D_x \csc^{-1} u = -\frac{1}{|u|\sqrt{u^2 - 1}} \cdot \frac{du}{dx}. \tag{19}$$

It is noteworthy that the derivatives of the six inverse trigonometric functions are all simple *algebraic* functions. As a consequence, inverse trigonometric functions frequently appear when we integrate algebraic functions. Observe also that the derivatives of $\cos^{-1} x$, $\cot^{-1} x$, and $\csc^{-1} x$ differ only in sign from the derivatives of their respective cofunctions. For this reason only the arcsine, arctangent, and arcsecant functions are necessary for integration, and only these three are in common use. That is, one need commit to memory the integral formulas only for the latter three functions. They follow immediately from (5), (11), and (17), and may be written in the forms that follow.

$$\int \frac{du}{\sqrt{1 - u^2}} = \sin^{-1} u + C, \tag{20}$$

$$\int \frac{du}{1 + u^2} = \tan^{-1} u + C, \tag{21}$$

$$\int \frac{du}{u\sqrt{u^2 - 1}} = \sec^{-1} |u| + C. \tag{22}$$

It is easy to verify that the absolute value on the right side in (22) follows from the one in (17). Remember also that because $\sec^{-1}|u|$ is undefined unless $|u| \geqq 1$, the definite integral

$$\int_a^b \frac{du}{u\sqrt{u^2 - 1}}$$

is meaningful only when the limits a and b are both at least 1 or both at most -1.

EXAMPLE 4 It follows immediately from Equation (21) that

$$\int_0^1 \frac{dx}{1 + x^2} = \Big[\tan^{-1} x\Big]_0^1 = \tan^{-1} 1 - \tan^{-1} 0 = \frac{\pi}{4}.$$

EXAMPLE 5 The substitution

$$u = 3x, \qquad du = 3\,dx$$

gives

$$\int \frac{1}{1 + 9x^2}\,dx = \frac{1}{3}\int \frac{3}{1 + (3x)^2}\,dx$$

$$= \frac{1}{3}\int \frac{du}{1 + u^2} = \frac{1}{3}\tan^{-1} u + C = \frac{1}{3}\tan^{-1} 3x + C.$$

EXAMPLE 6 The substitution

$$u = \tfrac{1}{2}x, \qquad du = \tfrac{1}{2}\,dx$$

gives

$$\int \frac{1}{\sqrt{4-x^2}}\,dx = \int \frac{1}{2\sqrt{1-(x/2)^2}}\,dx$$

$$= \int \frac{1}{\sqrt{1-u^2}}\,du = \arcsin u + C = \arcsin\left(\frac{x}{2}\right) + C.$$

EXAMPLE 7 The substitution

$$u = x\sqrt{2}, \qquad du = \sqrt{2}\,dx$$

gives

$$\int_1^{\sqrt{2}} \frac{1}{x\sqrt{2x^2-1}}\,dx = \int_{\sqrt{2}}^{2} \frac{1}{u\sqrt{u^2-1}}\,du$$

$$= \Big[\sec^{-1}|u|\Big]_{\sqrt{2}}^{2} = \sec^{-1}2 - \sec^{-1}\sqrt{2}$$

$$= \frac{\pi}{3} - \frac{\pi}{4} = \frac{\pi}{12}.$$

8.3 PROBLEMS

Find the values indicated in each of Problems 1–4.

1. (a) $\sin^{-1}(\frac{1}{2})$
(b) $\sin^{-1}(-\frac{1}{2})$
(c) $\sin^{-1}(\frac{1}{2}\sqrt{2})$
(d) $\sin^{-1}(-\frac{1}{2}\sqrt{3})$

2. (a) $\cos^{-1}(\frac{1}{2})$
(b) $\cos^{-1}(-\frac{1}{2})$
(c) $\cos^{-1}(\frac{1}{2}\sqrt{2})$
(d) $\cos^{-1}(-\frac{1}{2}\sqrt{3})$

3. (a) $\tan^{-1}0$
(b) $\tan^{-1}1$
(c) $\tan^{-1}(-1)$
(d) $\tan^{-1}\sqrt{3}$

4. (a) $\sec^{-1}1$
(b) $\sec^{-1}(-1)$
(c) $\sec^{-1}2$
(d) $\sec^{-1}(-\sqrt{2})$

Differentiate the functions in Problems 5–26.

5. $f(x) = \sin^{-1}(x^{100})$

6. $f(x) = \arctan(e^x)$

7. $f(x) = \sec^{-1}(\ln x)$

8. $f(x) = \ln(\tan^{-1}x)$

9. $f(x) = \arcsin(\tan x)$

10. $f(x) = x \arctan x$

11. $f(x) = \sin^{-1}e^x$

12. $f(x) = \arctan\sqrt{x}$

13. $f(x) = \cos^{-1}x + \sec^{-1}\left(\frac{1}{x}\right)$

14. $f(x) = \cot^{-1}\left(\frac{1}{x^2}\right)$

15. $f(x) = \csc^{-1}x^2$

16. $f(x) = \arccos\left(\frac{1}{\sqrt{x}}\right)$

17. $f(x) = \frac{1}{\arctan x}$

18. $f(x) = (\arcsin x)^2$

19. $f(x) = \tan^{-1}(\ln x)$

20. $f(x) = \operatorname{arcsec}\sqrt{x^2+1}$

21. $f(x) = \tan^{-1}e^x + \cot^{-1}e^{-x}$

22. $f(x) = \exp(\arcsin x)$

23. $f(x) = \sin(\arctan x)$

24. $f(x) = \sec(\sec^{-1}e^x)$

25. $f(x) = \frac{\arctan x}{(1+x^2)^2}$

26. $f(x) = (\sin^{-1}2x^2)^{-2}$

In each of Problems 27–30, find dy/dx by implicit differentiation. Then find the tangent line to the graph of the equation at the indicated point P.

27. $\tan^{-1}x + \tan^{-1}y = \frac{\pi}{2}$; $P(1, 1)$

28. $\sin^{-1}x + \sin^{-1}y = \frac{\pi}{2}$; $P(\frac{1}{2}, \frac{1}{2}\sqrt{3})$

29. $(\sin^{-1}x)(\sin^{-1}y) = \frac{\pi^2}{16}$; $P(\frac{1}{2}\sqrt{2}, \frac{1}{2}\sqrt{2})$

30. $(\sin^{-1}x)^2 + (\sin^{-1}y)^2 = \frac{5\pi^2}{36}$; $P(\frac{1}{2}, \frac{1}{2}\sqrt{3})$

Evaluate or antidifferentiate, as appropriate, in Problems 31–55.

31. $\displaystyle\int_0^1 \frac{dx}{1+x^2}$

32. $\displaystyle\int_0^{1/2} \frac{dx}{\sqrt{1-x^2}}$

33. $\displaystyle\int_{\sqrt{2}}^{2} \frac{dx}{x\sqrt{x^2-1}}$

34. $\displaystyle\int_{-2}^{-2/\sqrt{3}} \frac{dx}{x\sqrt{x^2-1}}$

35. $\displaystyle\int_0^3 \frac{dx}{9+x^2}$

36. $\displaystyle\int_0^{\sqrt{12}} \frac{dx}{\sqrt{16-x^2}}$

37. $\displaystyle\int \frac{dx}{\sqrt{1-4x^2}}$

38. $\displaystyle\int \frac{dx}{9x^2+4}$

39. $\displaystyle\int \frac{dx}{x\sqrt{x^2-25}}$

40. $\displaystyle\int \frac{dx}{x\sqrt{4x^2-9}}$

41. $\displaystyle\int \frac{e^x}{1+e^{2x}}\,dx$

42. $\displaystyle\int \frac{x^2}{x^6+25}\,dx$

43. $\displaystyle\int \frac{dx}{x\sqrt{x^6-25}}$

44. $\displaystyle\int \frac{\sqrt{x}}{1+x^3}\,dx$

45. $\displaystyle\int \frac{dx}{\sqrt{x(1-x)}}$

46. $\displaystyle\int \frac{\sec x \tan x}{1+\sec^2 x}\,dx$

47. $\displaystyle\int \frac{x^{49}}{1+x^{100}}\,dx$

48. $\displaystyle\int \frac{x^4}{(1-x^{10})^{1/2}}\,dx$

49. $\displaystyle\int \frac{1}{x[1+(\ln x)^2]}\,dx$

50. $\displaystyle\int \frac{\arctan x}{1+x^2}\,dx$

51. $\displaystyle\int_0^1 \frac{1}{1+(2x-1)^2}\,dx$

52. $\displaystyle\int_0^1 \frac{x^3}{1+x^4}\,dx$

53. $\displaystyle\int_1^e \frac{dx}{x[1-(\ln x)^2]^{1/2}}$

54. $\displaystyle\int_1^2 \frac{dx}{x(x^2-1)^{1/2}}$

55. $\displaystyle\int_1^3 \frac{dx}{2x^{1/2}(1+x)}$ (*Suggestion:* Let $u = x^{1/2}$.)

56. Conclude from the formula $D\cos^{-1}x = -D\sin^{-1}x$ that $\sin^{-1}x + \cos^{-1}x = \pi/2$ if $0 \leqq x \leqq 1$.

57. Show that $D\sec^{-1}x = D\cos^{-1}(1/x)$ if $x \geqq 1$, and conclude that $\sec^{-1}x = \cos^{-1}(1/x)$ if $x \geqq 1$. This fact can be used to find arcsecants on a calculator that has the key for the arccosine function, usually written "inv cos" or "$\cos^{-1}$."

58. (a) Deduce from the addition formula for tangents (Problem 10 in Appendix A) that

$$\arctan x + \arctan y = \arctan\frac{x+y}{1-xy}$$

provided that $xy < 1$.

(b) Apply part (a) to show that each of the following numbers is equal to $\pi/4$:

(i) $\arctan(\frac{1}{2}) + \arctan(\frac{1}{3})$

(ii) $2\arctan(\frac{1}{3}) + \arctan(\frac{1}{7})$

(iii) $\arctan(\frac{120}{119}) - \arctan(\frac{1}{239})$

(iv) $4\arctan(\frac{1}{5}) - \arctan(\frac{1}{239})$

59. A billboard *parallel* to a highway is to be 12 ft high, and its bottom will be 4 ft above the eye level of a passing motorist. How far from the highway should the billboard be placed in order to maximize the angle it subtends at the motorist's eyes?

60. Use inverse trigonometric functions to prove that the vertical angle subtended by a rectangular painting on a wall is greatest when the painting is hung with its center at the level of the observer's eyes.

61. Show that the circumference of a circle of radius a is $2\pi a$ by finding the length of the circular arc

$$y = \sqrt{a^2 - x^2}$$

from $x = 0$ to $x = a/\sqrt{2}$ and then multiplying by 8.

62. Find the volume generated by revolving the area under $y = 1/(1+x^4)$ from $x = 0$ to $x = 1$ around the x-axis.

63. The unbounded region R is bounded on the left by the y-axis, below by the x-axis, and above by the graph of $y = 1/(1+x^2)$. Show that the area of R is finite by evaluating

$$\lim_{a\to\infty}\int_0^a \frac{dx}{1+x^2}.$$

64. A building 250 ft high is equipped with an external elevator. The elevator starts at the top at time $t = 0$ and descends at the constant rate of 25 ft/s. You are watching the elevator from a window that is 100 ft above the ground, in a building 50 ft from the elevator. At what height does the elevator appear to you to be moving the fastest?

65. Suppose that the function f is defined for all x with $|x| > 1$, and has the property that

$$f'(x) = \frac{1}{x\sqrt{x^2-1}}$$

for all x.

(a) Explain why there exist two constants A and B such that

$$f(x) = \operatorname{arcsec} x + A \quad \text{if} \quad x > 1;$$

$$f(x) = -\operatorname{arcsec} x + B \quad \text{if} \quad x < -1.$$

(b) Determine the values of A and B so that $f(2) = 1 = f(-2)$. Then sketch the graph of $y = f(x)$.

66. The arctangent is the only inverse trigonometric function included in many versions of BASIC and FORTRAN, so it is necessary in programming to express $\sin^{-1}x$ and $\sec^{-1}x$ in terms of the arctangent. Show each of the following.

(a) If $|x| < 1$, then

$$\sin^{-1}x = \arctan\left(\frac{x}{\sqrt{1-x^2}}\right).$$

(b) If $x > 1$, then

$$\sec^{-1}x = \arctan(\sqrt{x^2-1}).$$

(c) If $x < -1$, then

$$\sec^{-1}x = \pi - \arctan(\sqrt{x^2-1}).$$

8.4 Hyperbolic Functions

The **hyperbolic cosine** and the **hyperbolic sine** of the real number x are denoted by $\cosh x$ and $\sinh x$ and are defined to be

$$\cosh x = \frac{e^x + e^{-x}}{2}, \qquad \sinh x = \frac{e^x - e^{-x}}{2}. \tag{1}$$

These particular combinations of familiar exponentials are useful in certain applications of calculus and are also helpful in evaluating certain integrals. The other four hyperbolic functions—the hyperbolic tangent, cotangent, secant, and cosecant—are defined in terms of $\cosh x$ and $\sinh x$ by analogy with trigonometry:

$$\tanh x = \frac{\sinh x}{\cosh x} = \frac{e^x - e^{-x}}{e^x + e^{-x}},$$

$$\coth x = \frac{\cosh x}{\sinh x} = \frac{e^x + e^{-x}}{e^x - e^{-x}} \qquad (x \neq 0); \tag{2}$$

$$\operatorname{sech} x = \frac{1}{\cosh x} = \frac{2}{e^x + e^{-x}},$$

$$\operatorname{csch} x = \frac{1}{\sinh x} = \frac{2}{e^x - e^{-x}} \qquad (x \neq 0). \tag{3}$$

The trigonometric terminology and notation for these hyperbolic functions stem from the fact that they satisfy a list of identities that, apart from an occasional difference of sign, much resemble the familiar trigonometric identities:

$$\cosh^2 x - \sinh^2 x = 1; \tag{4}$$

$$1 - \tanh^2 x = \operatorname{sech}^2 x; \tag{5}$$

$$\coth^2 x - 1 = \operatorname{csch}^2 x; \tag{6}$$

$$\sinh(x + y) = \sinh x \cosh y + \cosh x \sinh y; \tag{7}$$

$$\cosh(x + y) = \cosh x \cosh y + \sinh x \sinh y; \tag{8}$$

$$\sinh 2x = 2 \sinh x \cosh x; \tag{9}$$

$$\cosh 2x = \cosh^2 x + \sinh^2 x; \tag{10}$$

$$\cosh^2 x = \tfrac{1}{2}(\cosh 2x + 1); \tag{11}$$

$$\sinh^2 x = \tfrac{1}{2}(\cosh 2x - 1). \tag{12}$$

The identities in (4), (7), and (8) follow directly from the definitions of $\cosh x$ and $\sinh x$. For example,

$$\begin{aligned} \cosh^2 x - \sinh^2 x &= \tfrac{1}{4}(e^x + e^{-x})^2 - \tfrac{1}{4}(e^x - e^{-x})^2 \\ &= \tfrac{1}{4}(e^{2x} + 2 + e^{-2x}) - \tfrac{1}{4}(e^{2x} - 2 + e^{-2x}) = 1 \end{aligned}$$

The other identities above may be derived from (4), (7), and (8) in ways that parallel the derivations of the corresponding trigonometric identities.

The trigonometric functions are sometimes called *circular* functions because the point $(\cos\theta, \sin\theta)$ lies on the circle $x^2 + y^2 = 1$ for all θ. Similarly, the identity in (4) tells us that the point $(\cosh\theta, \sinh\theta)$ lies on the hyperbola

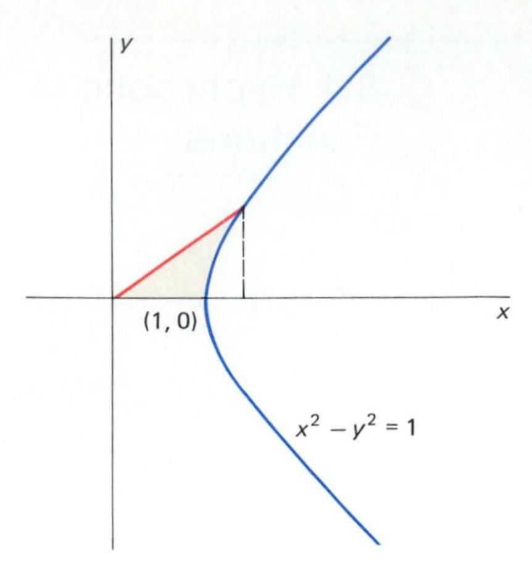

8.14 Relation of the hyperbolic cosine and hyperbolic sine to the hyperbola $x^2 - y^2 = 1$

$x^2 - y^2 = 1$, and this is the reason for the name *hyperbolic* function (see Fig. 8.14).

The graphs of $y = \cosh x$ and $y = \sinh x$ are easily constructed. Add (for cosh) or subtract (for sinh) the ordinates of the graphs of $y = \frac{1}{2}e^x$ and $y = \frac{1}{2}e^{-x}$. The graphs of the other four hyperbolic functions can then be constructed by dividing ordinates. The graphs of all six are shown in Fig. 8.15.

These graphs show a striking difference between the hyperbolic functions and the ordinary trigonometric functions: None of the hyperbolic functions is periodic. They do, however, have even–odd properties like the circular functions. The two functions cosh and sech are even, because

$$\cosh(-x) = \cosh x \quad \text{and} \quad \operatorname{sech}(-x) = \operatorname{sech} x$$

for all x. The other four hyperbolic functions are odd:

$$\sinh(-x) = -\sinh x, \qquad \tanh(-x) = -\tanh x,$$

and so on.

The formulas for the derivatives of the hyperbolic functions parallel those for the trigonometric functions, with occasional sign differences. For example,

$$D \cosh x = D(\tfrac{1}{2}e^x + \tfrac{1}{2}e^{-x}) = \tfrac{1}{2}e^x - \tfrac{1}{2}e^{-x} = \sinh x.$$

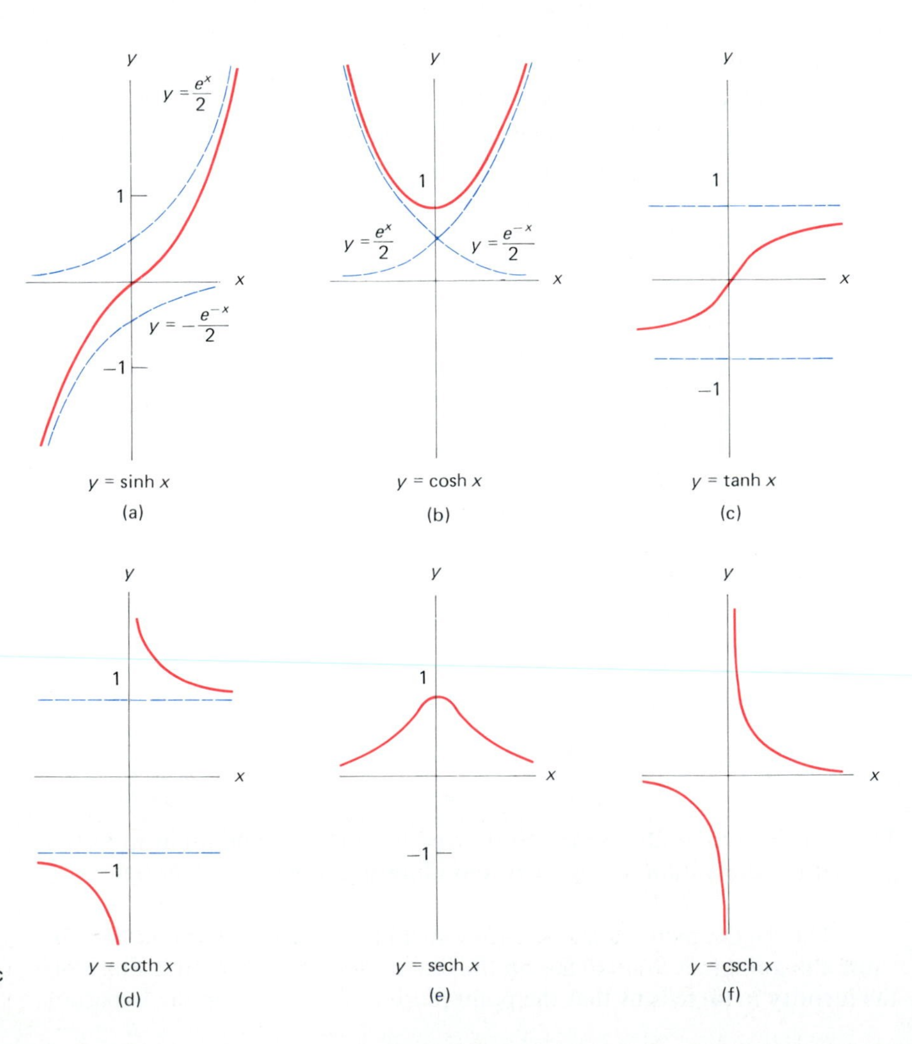

8.15 Graphs of the six hyperbolic functions

The chain rule then gives

$$D_x \cosh u = (\sinh u)\frac{du}{dx} \tag{13}$$

if u is a differentiable function of x. The other five differentiation formulas are

$$D_x \sinh u = (\cosh u)\frac{du}{dx}, \tag{14}$$

$$D_x \tanh u = (\operatorname{sech}^2 u)\frac{du}{dx}, \tag{15}$$

$$D_x \coth u = (-\operatorname{csch}^2 u)\frac{du}{dx}, \tag{16}$$

$$D_x \operatorname{sech} u = (-\operatorname{sech} u \tanh u)\frac{du}{dx}, \tag{17}$$

$$D_x \operatorname{csch} u = (-\operatorname{csch} u \coth u)\frac{du}{dx}. \tag{18}$$

The formula in (14) is derived exactly as (13) is. Then the formulas in (15) through (18) follow from (13) and (14) with the aid of the quotient rule and the identities in (5) and (6).

As indicated in the next example, the differentiation of hyperbolic functions using the formulas in (13) through (18) is very similar to the differentiation of trigonometric functions.

EXAMPLE 1

(a) $D \cosh 2x = 2 \sinh 2x$.

(b) $D \sinh^2 x = 2 \sinh x \cosh x$.

(c) $D(x \tanh x) = \tanh x + x \operatorname{sech}^2 x$.

(d) $D \operatorname{sech}(x^2) = -2x \operatorname{sech} x^2 \tanh x^2$.

The antiderivative versions of the differentiation formulas in (13) through (18) are the following integral formulas.

$$\int \sinh u \, du = \cosh u + C, \tag{19}$$

$$\int \cosh u \, du = \sinh u + C, \tag{20}$$

$$\int \operatorname{sech}^2 u \, du = \tanh u + C, \tag{21}$$

$$\int \operatorname{csch}^2 u \, du = -\coth u + C, \tag{22}$$

$$\int \operatorname{sech} u \tanh u \, du = -\operatorname{sech} u + C, \tag{23}$$

$$\int \operatorname{csch} u \coth u \, du = -\operatorname{csch} u + C. \tag{24}$$

The integrals in the following example illustrate the fact that simple hyperbolic integrals may be treated in much the same way as simple trigonometric integrals (Section 8.2).

EXAMPLE 2

(a) $\int \sinh^2 x\,dx = \int \frac{1}{2}(\cosh 2x - 1)\,dx = \frac{1}{4}\sinh 2x - \frac{1}{2}x + C.$

(b) $\int \cosh^2 x \sinh x\,dx = \frac{1}{3}\cosh^3 x + C.$

(c) $\displaystyle\int \frac{\operatorname{sech}^2\sqrt{x}}{\sqrt{x}}\,dx = 2\int (\operatorname{sech}^2\sqrt{x})\frac{dx}{2\sqrt{x}} = 2\tanh\sqrt{x} + C.$

(d)
$$\begin{aligned}\int_0^1 \tanh^2 x\,dx &= \int_0^1 (1 - \operatorname{sech}^2 x)\,dx\\ &= \Big[x - \tanh x\Big]_0^1 = 1 - \tanh 1\\ &= 1 - \frac{e - e^{-1}}{e + e^{-1}} = \frac{2}{e^2 + 1} \approx 0.238406.\end{aligned}$$

(e) $\displaystyle\int \coth x\,dx = \int \frac{\cosh x}{\sinh x}\,dx = \ln|\sinh x| + C.$

8.4 PROBLEMS

Find the derivatives of the functions in Problems 1–14.

1. $f(x) = \cosh(3x - 2)$
2. $f(x) = \sinh\sqrt{x}$
3. $f(x) = x^2\tanh\left(\frac{1}{x}\right)$
4. $f(x) = \operatorname{sech} e^{2x}$
5. $f(x) = \coth^3 4x$
6. $f(x) = \ln \sinh 3x$
7. $f(x) = e^{\operatorname{csch} x}$
8. $f(x) = \cosh \ln x$
9. $f(x) = \sin(\sinh x)$
10. $f(x) = \tan^{-1}(\tanh x)$
11. $f(x) = \sinh x^4$
12. $f(x) = \sinh^4 x$
13. $f(x) = \dfrac{1}{x + \tanh x}$
14. $f(x) = \cosh^2 x - \sinh^2 x$

Evaluate the integrals in Problems 15–28.

15. $\int x \sinh x^2\,dx$
16. $\int \cosh^2 3u\,du$
17. $\int \tanh^2 3x\,dx$
18. $\displaystyle\int \frac{\operatorname{sech}\sqrt{x}\tanh\sqrt{x}}{\sqrt{x}}\,dx$
19. $\int \sinh^2 2x \cosh 2x\,dx$
20. $\int \tanh 3x\,dx$
21. $\displaystyle\int \frac{\sinh x}{\cosh^3 x}\,dx$
22. $\int \sinh^4 x\,dx$
23. $\int \coth x \operatorname{csch}^2 x\,dx$
24. $\int \operatorname{sech} x\,dx$
25. $\displaystyle\int \frac{\sinh x}{1 + \cosh x}\,dx$
26. $\displaystyle\int \frac{\sinh \ln x}{x}\,dx$
27. $\displaystyle\int \frac{1}{(e^x + e^{-x})^2}\,dx$
28. $\displaystyle\int \frac{e^x + e^{-x}}{e^x - e^{-x}}\,dx$

29. Apply the definition in (1) to prove the identity in (7).

30. Derive the identities in (5) and (6) from the identity in (4).

31. Deduce the identities in (10) and (11) from the identity in (8).

32. Suppose that A and B are constants. Show that the function

$$x = A\cosh kt + B\sinh kt$$

is a solution of the differential equation

$$x''(t) = k^2 x(t).$$

33. Find the length of the curve $y = \cosh x$ over the interval $[0, a]$.

34. Find the volume of the solid obtained by revolving around the x-axis the area under $y = \sinh x$ from $x = 0$ to $x = \pi$.

35. Show that the area $A(\theta)$ of the shaded sector in Fig. 8.14 is $\theta/2$. This corresponds to the fact that the area of the sector of the unit circle between the positive x-axis and the radius to the point $(\cos\theta, \sin\theta)$ is $\theta/2$. *Suggestion:* Note first that

$$A(\theta) = \tfrac{1}{2}\cosh\theta\sinh\theta - \int_1^{\cosh\theta}\sqrt{x^2 - 1}\,dx.$$

Then use the fundamental theorem of calculus to show that $A'(\theta) = \frac{1}{2}$ for all θ.

*8.5 Inverse Hyperbolic Functions

A glance at Fig. 8.15 in Section 8.4 shows that the hyperbolic sine and tangent functions are increasing on the whole real line. Thus each is one-to-one on the set of all real numbers. The hyperbolic cotangent and cosecant functions are also one-to-one where they are defined; each of the latter has domain the set of all real numbers other than zero. Thus each of these four hyperbolic functions has an inverse function whose domain is in each case the range of values of the original function:

$$y = \sinh^{-1} x \quad \text{if} \quad x = \sinh y, \quad \text{all } x; \tag{1}$$

$$y = \tanh^{-1} x \quad \text{if} \quad x = \tanh y, \quad -1 < x < 1; \tag{2}$$

$$y = \coth^{-1} x \quad \text{if} \quad x = \coth y, \quad |x| > 1; \tag{3}$$

$$y = \operatorname{csch}^{-1} x \quad \text{if} \quad x = \operatorname{csch} y, \quad x \neq 0. \tag{4}$$

The hyperbolic cosine and secant functions are one-to-one on the half-line $x \geqq 0$, where cosh is increasing and sech is decreasing. Hence we define

$$y = \cosh^{-1} x \quad \text{if} \quad x = \cosh y \tag{5}$$

where $x \geqq 1$ and $y \geqq 0$, and

$$y = \operatorname{sech}^{-1} x \quad \text{if} \quad x = \operatorname{sech} y \tag{6}$$

where $0 < x \leqq 1$ and $y \geqq 0$. The graphs of the six inverse hyperbolic functions may be obtained by reflecting through the line $y = x$ the graphs of the hyperbolic functions (restricted to the half line $x \geqq 0$ in the cases of cosh and sech). We show these graphs in Fig. 8.16.

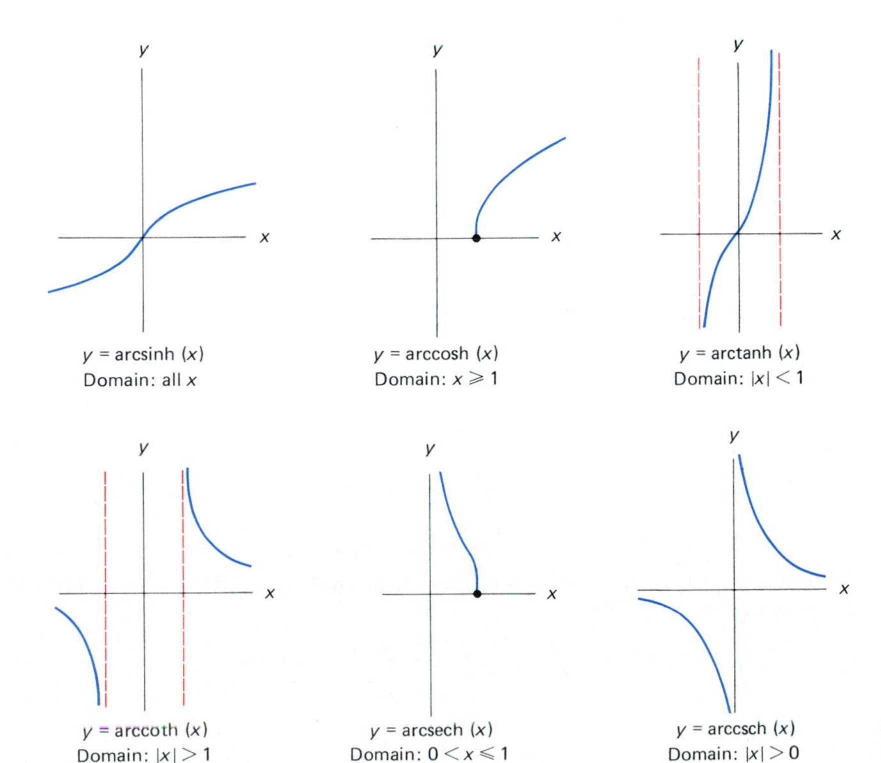

8.16 The inverse hyperbolic functions

Here are the derivatives of the six inverse hyperbolic functions:

$$D \sinh^{-1} x = \frac{1}{\sqrt{x^2 + 1}}, \tag{7}$$

$$D \cosh^{-1} x = \frac{1}{\sqrt{x^2 - 1}}, \tag{8}$$

$$D \tanh^{-1} x = \frac{1}{1 - x^2}, \tag{9}$$

$$D \coth^{-1} x = \frac{1}{1 - x^2}, \tag{10}$$

$$D \operatorname{sech}^{-1} x = -\frac{1}{x\sqrt{1 - x^2}}, \tag{11}$$

$$D \operatorname{csch}^{-1} x = -\frac{1}{|x|\sqrt{1 - x^2}}. \tag{12}$$

One can derive these formulas by the standard method of finding the derivative of an inverse function given the derivative of the function itself. The only requirement is that the inverse function is known in advance to be differentiable. For example, to differentiate $\tanh^{-1} x$, we begin with the relation

$$\tanh(\tanh^{-1} x) = x$$

which follows from the definition in (2), and substitute $u = \tanh^{-1} x$. Then, because the equation above is actually an identity,

$$D_x \tanh u = Dx = 1,$$

so that

$$(\operatorname{sech}^2 u)\frac{du}{dx} = 1.$$

Thus

$$D \tanh^{-1} x = \frac{du}{dx} = \frac{1}{\operatorname{sech}^2 u} = \frac{1}{1 - \tanh^2 u}$$

$$= \frac{1}{1 - \tanh^2(\tanh^{-1} x)} = \frac{1}{1 - x^2}.$$

This establishes the formula in (9). One can use similar methods to verify the formulas for the derivatives of the other five inverse hyperbolic functions.

The hyperbolic functions are defined in terms of the natural exponential function e^x, so it's no surprise to find that their inverses may be expressed in terms of $\ln x$. In fact,

$$\sinh^{-1} x = \ln(x + \sqrt{x^2 + 1}) \qquad \text{for all} \quad x; \tag{13}$$

$$\cosh^{-1} x = \ln(x + \sqrt{x^2 - 1}) \qquad \text{for all} \quad x \geqq 1; \tag{14}$$

$$\tanh^{-1} x = \frac{1}{2}\ln\left(\frac{1 + x}{1 - x}\right) \qquad \text{for} \quad |x| < 1; \tag{15}$$

$$\coth^{-1} x = \frac{1}{2} \ln\left(\frac{x+1}{x-1}\right) \quad \text{for} \quad |x| > 1; \tag{16}$$

$$\operatorname{sech}^{-1} x = \ln\left(\frac{1 + \sqrt{1 - x^2}}{x}\right) \quad \text{if} \quad 0 < x \leqq 1; \tag{17}$$

$$\operatorname{csch}^{-1} x = \ln\left(\frac{1}{x} + \frac{\sqrt{1 + x^2}}{|x|}\right) \quad \text{if} \quad x \neq 0. \tag{18}$$

Each of these identities may be established by showing that each side has the same derivative, and also that the two sides agree for at least one value of x in every interval in their respective domains. For example,

$$D \ln(x + \sqrt{x^2 + 1}) = \frac{1 + x(x^2 + 1)^{-1/2}}{x + (x^2 + 1)^{1/2}} = \frac{1}{\sqrt{x^2 + 1}} = D \sinh^{-1} x.$$

So

$$\sinh^{-1} x = \ln(x + \sqrt{x^2 + 1}) + C.$$

But $\sinh^{-1}(0) = 0 = \ln(0 + \sqrt{0 + 1})$. This implies that $C = 0$ and thus establishes the formula in (13). It is not quite so easy to show that $C = 0$ in the proofs of (16) and (18); see Problems 25 and 26.

The formulas in (13) through (18) may be used to calculate inverse hyperbolic function values. This is convenient if you own a calculator with a repertoire that does not include the inverse hyperbolic functions, or if you are programming in a language such as BASIC, most forms of which do not include these functions.

EXAMPLE 1 Suppose that a body is dropped at time $t = 0$ from a height y_0 above the ground. Assume that it is acted on both by gravity and by air resistance proportional to v^2 (where v is its velocity). According to Problem 30, its height t seconds after release will be

$$y = y_0 - \frac{1}{\rho} \ln(\cosh t\sqrt{\rho g}) \tag{19}$$

where ρ is the proportionality constant for the resistance of the air. Assume that $y_0 = 2000$ ft, that $g = 32$ ft/s^2, and that $\rho = 0.001$. How long will it take for this body to fall to the ground?

Solution If we set $y = 0$ in Equation (19) and then solve for t, we get

$$t = \frac{1}{\sqrt{\rho g}} \cosh^{-1}(\exp(\rho y_0)).$$

Then the formula in (14) gives

$$t = \frac{1}{\sqrt{\rho g}} \ln[\exp(\rho y_0) + \sqrt{\exp(2\rho y_0) - 1}].$$

We substitute the data given in the problem and find that

$$t = \frac{1}{\sqrt{0.032}} \ln(e^2 + \sqrt{e^4 - 1}) \approx 15.03$$

seconds will be the time required for the body to fall to the ground.

The differentiation formulas in (7) through (12) may, in the usual way, be written as the following integral formulas.

$$\int \frac{du}{\sqrt{u^2+1}} = \sinh^{-1} u + C. \tag{20}$$

$$\int \frac{du}{\sqrt{u^2-1}} = \cosh^{-1} u + C. \tag{21}$$

$$\int \frac{du}{1-u^2} = \begin{cases} \tanh^{-1} u + C & \text{if } |u| < 1; \quad (22a) \\ \coth^{-1} u + C & \text{if } |u| > 1; \quad (22b) \end{cases}$$

$$= \frac{1}{2} \ln \left| \frac{1+u}{1-u} \right| + C. \tag{22c}$$

$$\int \frac{du}{u\sqrt{1-u^2}} = -\operatorname{sech}^{-1} u + C. \tag{23}$$

$$\int \frac{du}{u\sqrt{1+u^2}} = -\operatorname{csch}^{-1} |u| + C. \tag{24}$$

The distinction between the two cases $|u| < 1$ and $|u| > 1$ in (22) results from the fact that the inverse hyperbolic tangent is defined for $|x| < 1$, while the inverse hyperbolic cotangent is defined for $|x| > 1$.

EXAMPLE 2

$$\int \frac{dx}{\sqrt{4x^2+1}} = \frac{1}{2}\int \frac{du}{\sqrt{u^2+1}} \qquad (u = 2x, \quad dx = \tfrac{1}{2}\,du)$$

$$= \frac{1}{2} \sinh^{-1} 2x + C.$$

EXAMPLE 3

$$\int_0^{1/2} \frac{dx}{1-x^2} = \Big[\tanh^{-1} x\Big]_0^{1/2}$$

$$= \frac{1}{2}\left[\ln\left|\frac{1+x}{1-x}\right|\right]_0^{1/2} = \frac{1}{2}\ln 3.$$

EXAMPLE 4

$$\int_2^5 \frac{dx}{1-x^2} = \Big[\coth^{-1} x\Big]_2^5 = \frac{1}{2}\left[\ln\left|\frac{1+x}{1-x}\right|\right]_2^5$$

$$= \frac{1}{2}\left[\ln\left(\frac{6}{4}\right) - \ln 3\right] = -\frac{1}{2}\ln 2.$$

8.5 PROBLEMS

Find the derivatives of the functions in Problems 1–10.

1. $f(x) = \sinh^{-1} 2x$
2. $f(x) = \cosh^{-1}(x^2+1)$
3. $f(x) = \tanh^{-1}\sqrt{x}$
4. $f(x) = \coth^{-1}\sqrt{x^2+1}$
5. $f(x) = \operatorname{sech}^{-1}\left(\dfrac{1}{x}\right)$
6. $f(x) = \operatorname{csch}^{-1} e^x$
7. $f(x) = (\sinh^{-1} x)^{3/2}$
8. $f(x) = \sinh^{-1}(\ln x)$

9. $f(x) = \ln(\tanh^{-1} x)$

10. $f(x) = \dfrac{1}{\tanh^{-1} 3x}$

Evaluate the integrals in Problems 11–20.

11 $\displaystyle\int \frac{dx}{\sqrt{x^2 + 9}}$

12. $\displaystyle\int \frac{dy}{\sqrt{4y^2 - 9}}$

13. $\displaystyle\int_{1/2}^{1} \frac{dx}{4 - x^2}$

14. $\displaystyle\int_{5}^{10} \frac{dx}{4 - x^2}$

15. $\displaystyle\int \frac{dx}{x\sqrt{4 - 9x^2}}$

16. $\displaystyle\int \frac{dx}{x\sqrt{x^2 + 25}}$

17. $\displaystyle\int \frac{e^x}{\sqrt{e^{2x} + 1}}\,dx$

18. $\displaystyle\int \frac{x}{\sqrt{x^4 - 1}}\,dx$

19. $\displaystyle\int \frac{1}{\sqrt{1 - e^{2x}}}\,dx$

20. $\displaystyle\int \frac{\cos x}{\sqrt{1 + \sin^2 x}}\,dx$

21. Establish the formula for $D \sinh^{-1} x$ in Equation (7).

22. Establish the formula for $D \operatorname{sech}^{-1} x$ in Equation (11).

23. Prove the formula in Equation (15) by differentiation of both sides.

24. Establish the formula in Equation (13) by solving the equation

$$x = \sinh y = \tfrac{1}{2}(e^y - e^{-y})$$

for y in terms of x.

25. Establish the formula in Equation (16) by solving the equation

$$x = \coth y = \frac{e^y + e^{-y}}{e^y - e^{-y}}$$

for y in terms of x.

26. (a) Differentiate both sides of the formula in Equation (16) to show that they differ by a constant C.
(b) Then prove that $C = 0$ by using the definition of $\coth x$ to show that $\coth^{-1} 2 = \frac{1}{2} \ln 3$.

27. (a) Differentiate both sides of the formula in Equation (18) to show that the two sides differ by a constant C.
(b) Then prove that $C = 0$ by using the definition of $\operatorname{csch} x$ to show that $\operatorname{csch}^{-1} 1 = \ln(1 + \sqrt{2})$.

28. Suppose that the body of Example 1 is dropped from the top of a 500-ft building. After how long and with what speed does it strike the ground?

29. Assume that $\rho = 0.075$ (in the formula of Example 1) for a paratrooper falling with open parachute. If she jumps from an altitude of 10,000 ft and opens her parachute immediately, after how long and with what speed will she land?

30. (a) Beginning with the initial value problem

$$\frac{dv}{dt} = -g + \rho v^2, \qquad v(0) = 0,$$

first separate variables, then use Equation (9) to derive the solution

$$v = -\sqrt{g/\rho}\,\tanh t\sqrt{\rho g}.$$

(b) Integrate again to obtain Equation (19) in this section.

CHAPTER 8 REVIEW: Definitions and Formulas

Use the list below as a guide to concepts that you may need to review.

1. The derivatives of the six trigonometric functions, and the corresponding integral formulas
2. The definitions of the six inverse trigonometric functions
3. The derivatives of the inverse trigonometric functions
4. The integral formulas corresponding to the derivatives of the inverse sine, tangent, and secant functions
5. The definitions and derivatives of the hyperbolic functions
6. The definitions of the inverse hyperbolic functions

CHAPTER 8 MISCELLANEOUS PROBLEMS

Differentiate the functions in Problems 1–20.

1. $f(x) = \sin\sqrt{x}$

2. $f(x) = x \cos\left(\dfrac{1}{x}\right)$

3. $f(x) = \tan(\ln x)$

4. $f(x) = \sec^2(3x + 1)$

5. $f(x) = (\csc 2x + \cot 2x)^5$

6. $f(x) = \sqrt{\tan x^2}$

7. $f(x) = \ln(\sec^2 3x)$

8. $f(x) = \cot e^{3x}$

9. $f(x) = \exp(\arctan x)$

10. $f(x) = \ln(\sin^{-1} x)$

11. $f(x) = \arcsin\sqrt{x}$

12. $f(x) = x \sec^{-1} x^2$

13. $f(x) = \tan^{-1}(1 + x^2)$

14. $f(x) = \sin^{-1}\sqrt{1 - x^2}$

15. $f(x) = e^x \sinh e^x$

16. $f(x) = \ln \cosh x$

17. $f(x) = \tanh^2 3x + \operatorname{sech}^2 3x$

18. $f(x) = \sinh^{-1}\sqrt{x^2 - 1}$

19. $f(x) = \cosh^{-1}\sqrt{x^2 + 1}$

20. $f(x) = \tanh^{-1}(1 - x^2)$

Evaluate the integrals in Problems 21–44.

21. $\int x \sec^2 x^2\, dx$

22. $\int e^x \sec e^x \tan e^x\, dx$

23. $\int \csc^2 2x \cot 2x\, dx$

24. $\int \dfrac{\cot^2\sqrt{x}}{\sqrt{x}}\, dx$

25. $\int \dfrac{\tan^2(1/x)}{x^2}\, dx$

26. $\int \tan^3 2x \sec^2 2x\, dx$

27. $\int x \tan(x^2 + 1)\, dx$

28. $\int \dfrac{e^x}{\sqrt{1 - e^x}}\, dx$

29. $\int \dfrac{e^x}{\sqrt{1 - e^{2x}}}\, dx$

30. $\int \dfrac{x}{1 + x^4}\, dx$

31. $\int \dfrac{1}{\sqrt{9 - 4x^2}}\, dx$

32. $\int \dfrac{1}{9 + 4x^2}\, dx$

33. $\int \dfrac{x^2}{1 + x^6}\, dx$

34. $\int \dfrac{\cos x}{1 + \sin^2 x}\, dx$

35. $\int \dfrac{1}{x\sqrt{4x^2 - 1}}\, dx$

36. $\int \dfrac{1}{x\sqrt{x^4 - 1}}\, dx$

37. $\int \dfrac{1}{\sqrt{e^{2x} - 1}}\, dx$

38. $\int x^2 \cosh x^3\, dx$

39. $\int \dfrac{\sinh\sqrt{x}}{\sqrt{x}}\, dx$

40. $\int \operatorname{sech}^2(3x - 2)\, dx$

41. $\int \dfrac{\arctan x}{1 + x^2}\, dx$

42. $\int \dfrac{1}{\sqrt{4x^2 - 1}}\, dx$

43. $\int \dfrac{1}{\sqrt{4x^2 + 9}}\, dx$

44. $\int \dfrac{x}{\sqrt{x^4 + 1}}\, dx$

45. Find the volume generated by revolving the region under $y = (1 - x^4)^{1/2}$ from $x = 0$ to $x = 1/\sqrt{2}$ around the y-axis.

46. Find the volume generated by revolving the region under $y = (x^4 + 1)^{-1/2}$ from $x = 0$ to $x = 1$ around the y-axis.

47. Sketch the graph of the curve $y = 5 \cos 2x + 12 \sin 2x$. Show at least one complete cycle.

48. A mass on the end of a spring oscillates vertically with a period of 2π seconds. At time $t = 0$ the mass is 4 ft from the equilibrium point ($y = 0$) and is approaching it at 3 ft/s. What is the amplitude of the motion of this mass?

49. Use the formulas in Equations (14) through (17) of Section 8.5 to show each of the following.

(a) $\coth^{-1} x = \tanh^{-1}\left(\dfrac{1}{x}\right)$;

(b) $\operatorname{sech}^{-1} x = \cosh^{-1}\left(\dfrac{1}{x}\right)$.

50. Show that $x''(t) = k^2 x(t)$ if

$$x(t) = A \cosh kt + B \sinh kt,$$

where A and B are constants. Determine A and B if:

(a) $x(0) = 1$, $x'(0) = 0$; (b) $x(0) = 0$, $x'(0) = 1$.

51. Use Newton's method to find the least positive solution of the equation $\cos x \cosh x = 1$. Begin by sketching the graphs $y = \cos x$ and $y = \operatorname{sech} x$.

52. (a) Verify by differentiation that

$$\int \sec x\, dx = \sinh^{-1}(\tan x) + C.$$

(b) Show similarly that

$$\int \operatorname{sech} x\, dx = \tan^{-1}(\sinh x) + C.$$

53. Assume that the earth is a solid sphere of uniform density, with mass M and radius $R = 3960$ mi. For a particle of mass m *within* the earth at distance r from the center of the earth, the gravitational force attracting m toward the center is $F_r = -GM_r m/r^2$, where M_r is the mass of the part of the earth within a sphere of radius r.

(a) Show that $F_r = -(GMm/R^3)r$.

(b) Now suppose that a hole is drilled straight through the center of the earth, connecting two antipodal points on its surface. Let a particle of mass m be dropped at time $t = 0$ into this hole with initial speed zero, and let $r(t)$ be its distance from the center of the earth at time t. Conclude from Newton's second law and part (a) that $r''(t) = -k^2 r(t)$, where $k^2 = GM/R^3 = g/R$ and $g = 32.2$ ft/s^2.

(c) Deduce from part (b) that the particle undergoes simple harmonic motion back and forth between the ends of the hole, with period approximately 84 min.

(d) With what speed (in mi/h) does the particle pass through the earth's center?

chapter nine

Techniques of Integration

9.1 Introduction

In the past three chapters we have seen that many geometric and physical quantities can be expressed as definite integrals. The fundamental theorem of calculus reduces the problem of calculating the definite integral $\int_a^b f(x)\,dx$ to that of finding an antiderivative $G(x)$ of $f(x)$. Once this is accomplished, then

$$\int_a^b f(x)\,dx = \Big[G(x)\Big]_a^b = G(b) - G(a).$$

But as yet we have relied largely on trial-and-error methods for finding the required antiderivative $G(x)$. Sometimes a knowledge of elementary derivative formulas, perhaps in combination with a simple substitution, will allow us to integrate a given function. This approach can, however, be inefficient and time-consuming, especially in view of the following surprising fact: There exist simple-looking integrals, such as

$$\int e^{-x^2}\,dx, \qquad \int \frac{\sin x}{x}\,dx, \quad \text{and} \quad \int \sqrt{1 + x^4}\,dx,$$

that cannot be evaluated in terms of finite combinations of the familiar algebraic and elementary transcendental functions. For example, the antiderivative

$$H(x) = \int_0^x e^{-t^2}\,dt$$

of e^{-x^2} has no finite expression in terms of elementary functions. Any attempt to find such an expression will, therefore, inevitably be unsuccessful.

The presence of such examples indicates that we cannot hope to reduce integration to a routine process like differentiation. In fact, finding antiderivatives is an art, depending upon experience and practice. Nevertheless, there are a number of techniques whose systematic use can substantially reduce our dependence upon chance and intuition alone. This chapter deals with some of these systematic techniques of integration.

9.2 Integral Tables and Simple Substitutions

Integration would be a fairly simple matter if we had a list of integral formulas, an *integral table*, in which we could locate any integral that we ever needed to evaluate. But the diversity of integrals that we encounter in practice is too great for such an all-inclusive integral table to be practical. It is more sensible to print, or memorize, a short table of integrals of the sort seen frequently and to learn techniques by which the range of applicability of this short table can be extended. We may begin with the list of integrals in Fig. 9.1, which are familiar from earlier chapters; each is equivalent to one of the basic derivative formulas.

A table of approximately 110 integral formulas appears inside the covers of this book. Still more extensive integral tables are readily available. For example, the volume of *Standard Mathematical Tables* edited by William H. Beyer and published annually by the Chemical Rubber Company in Cleveland, contains a list of over 700 integral formulas. But even such a lengthy table can be expected to include only a small fraction of the integrals we may need to evaluate. Thus it is necessary to learn techniques for deriving new formulas and for transforming a given integral either into one that's already familiar or into one appearing in an accessible table.

$$\int u^n\,du = \frac{u^{n+1}}{n+1} + C \quad [n \neq -1] \tag{1}$$

$$\int \frac{du}{u} = \ln|u| + C \tag{2}$$

$$\int e^u\,du = e^u + C \tag{3}$$

$$\int \cos u\,du = \sin u + C \tag{4}$$

$$\int \sin u\,du = -\cos u + C \tag{5}$$

$$\int \sec^2 u\,du = \tan u + C \tag{6}$$

$$\int \csc^2 u\,du = -\cot u + C \tag{7}$$

$$\int \sec u \tan u\,du = \sec u + C \tag{8}$$

$$\int \csc u \cot u\,du = -\csc u + C \tag{9}$$

$$\int \frac{du}{\sqrt{1-u^2}} = \sin^{-1} u + C \tag{10}$$

$$\int \frac{du}{1+u^2} = \tan^{-1} u + C \tag{11}$$

$$\int \frac{du}{u\sqrt{u^2-1}} = \sec^{-1}|u| + C \tag{12}$$

9.1 A short table of integrals

The principal such method is the *method of substitution*, which we first considered in Section 5.6. Recall that if

$$\int f(u)\,du = F(u) + C,$$

then

$$\int f(g(x))g'(x)\,dx = F(g(x)) + C.$$

Thus the substitution

$$u = g(x), \qquad du = g'(x)\,dx$$

transforms the integral $\int f(g(x))g'(x)\,dx$ into the simpler integral $\int f(u)\,du$. The key to making this simplification lies in spotting the composition $f(g(x))$ in the given integrand. In order to convert this integrand into a function of u alone, it is necessary that the remaining factor be a constant multiple of the derivative $g'(x)$ of the inside function $g(x)$. In this case we replace $f(g(x))$ by the simpler $f(u)$ and also $g'(x)\,dx$ by du. The preceding three chapters contain numerous illustrations of this method of substitution, and the problems for the present section provide an opportunity to review it.

EXAMPLE 1 Find $\displaystyle\int \frac{1}{x}(1 + \ln x)^5\,dx.$

Solution We need to spot *both* the inner function $g(x)$ *and* its derivative $g'(x)$. If we choose $g(x) = 1 + \ln x$, then $g'(x) = 1/x$. Hence the given integral is of the form discussed above with $f(u) = u^5$, $u = 1 + \ln x$, and $du = dx/x$. Therefore,

$$\int \frac{1}{x}(1 + \ln x)^5\,dx = \int u^5\,du = \frac{1}{6}u^6 + C = \frac{1}{6}(1 + \ln x)^6 + C.$$

EXAMPLE 2 Find $\displaystyle\int \frac{x}{1 + x^4}\,dx$.

Solution Here it is not so clear what the inside function is. But comparison with the integral formula in (11) (in Fig. 9.1) suggests that we try the substitution

$$u = x^2, \qquad du = 2x\,dx.$$

We take advantage of the factor $x\,dx = \frac{1}{2}\,du$ that is available in the integrand and make these computations:

$$\int \frac{x\,dx}{1 + x^4} = \frac{1}{2}\int \frac{du}{1 + u^2} = \frac{1}{2}\tan^{-1}u + C = \frac{1}{2}\tan^{-1}x^2 + C.$$

Note that the substitution $u = x^2$ would have been of little use had the integrand been either $1/(1 + x^4)$ or $x^2/(1 + x^4)$.

Example 2 illustrates the device of using a substitution to convert a given integral into a familiar one. Often an integral that does not appear in any integral table can be transformed into one that does by using the techniques of this chapter. In the following example we employ an appropriate substitution to "reconcile" the given integral with the standard integral formula

$$\int \frac{u^2}{\sqrt{a^2 - u^2}}\,du = \frac{a^2}{2}\sin^{-1}\left(\frac{u}{a}\right) - \frac{u}{2}\sqrt{a^2 - u^2} + C, \tag{13}$$

which is Formula 56 (inside the front cover).

EXAMPLE 3 Find $\displaystyle\int \frac{x^2}{\sqrt{25 - 16x^2}}\,dx$.

Solution In order that $25 - 16x^2$ be equal to $a^2 - u^2$ in (1), we take $a = 5$ and $u = 4x$. Then $du = 4\,dx$ and so $dx = \frac{1}{4}\,du$. This gives

$$\begin{aligned}\int \frac{x^2}{\sqrt{25 - 16x^2}}\,dx &= \int \frac{(u/4)^2(1/4)}{\sqrt{25 - u^2}}\,du \\ &= \frac{1}{64}\int \frac{u^2}{\sqrt{25 - u^2}}\,du \\ &= \frac{1}{64}\left[\frac{25}{2}\sin^{-1}\left(\frac{u}{5}\right) - \frac{u}{2}\sqrt{25 - u^2}\,\right] + C \\ &= \frac{25}{128}\sin^{-1}\left(\frac{4x}{5}\right) - \frac{x}{32}\sqrt{25 - 16x^2} + C.\end{aligned}$$

In Section 9.6 we will see how integral formulas like the one in (13) can be derived.

9.2 PROBLEMS

Evaluate the integrals in Problems 1–30.

1. $\int (2-3x)^4\,dx$

2. $\int \frac{1}{(1+2x)^2}\,dx$

3. $\int x^2\sqrt{2x^3-4}\,dx$

4. $\int \frac{5t}{5+2t^2}\,dt$

5. $\int \frac{3x}{\sqrt[3]{2x^2+3}}\,dx$

6. $\int x\sec^2 x^2\,dx$

7. $\int \frac{\cot\sqrt{y}\,\csc\sqrt{y}}{\sqrt{y}}\,dy$

8. $\int \sin \pi(2x+1)\,dx$

9. $\int (1+\sin\theta)^5\cos\theta\,d\theta$

10. $\int \frac{\sin 2x}{4+\cos 2x}\,dx$

11. $\int e^{-\cot x}\csc^2 x\,dx$

12. $\int \frac{e^{\sqrt{x+4}}}{\sqrt{x+4}}\,dx$

13. $\int \frac{(\ln t)^{10}}{t}\,dt$

14. $\int \frac{t}{\sqrt{1-9t^2}}\,dt$

15. $\int \frac{1}{\sqrt{1-9t^2}}\,dt$

16. $\int \frac{e^{2x}}{1+e^{2x}}\,dx$

17. $\int \frac{e^{2x}}{1+e^{4x}}\,dx$

18. $\int \frac{e^{\arctan x}}{1+x^2}\,dx$

19. $\int \frac{3x}{\sqrt{1-x^4}}\,dx$

20. $\int \sin^3 2x\cos 2x\,dx$

21. $\int \tan^4 3x\sec^2 3x\,dx$

22. $\int \frac{1}{1+4t^2}\,dt$

23. $\int \frac{\cos\theta}{1+\sin^2\theta}\,d\theta$

24. $\int \frac{\sec^2\theta}{1+\tan\theta}\,d\theta$

25. $\int \frac{(1+\sqrt{x})^4}{\sqrt{x}}\,dx$

26. $\int t^{-1/3}\sqrt{t^{2/3}-1}\,dt$

27. $\int \frac{1}{(1+t^2)\arctan t}\,dt$

28. $\int \frac{\sec 2x\tan 2x}{(1+\sec 2x)^{3/2}}\,dx$

29. $\int \frac{1}{\sqrt{e^{2x}-1}}\,dx$

30. $\int \frac{x}{\sqrt{\exp(2x^2)-1}}\,dx$

In Problems 31–35, make the indicated substitution to evaluate the given integral.

31. $\int x^2\sqrt{x-2}\,dx;\quad u=x-2$

32. $\int \frac{x^2}{\sqrt{x+3}}\,dx;\quad u=x+3$

33. $\int \frac{x}{\sqrt{2x+3}}\,dx;\quad u=2x+3$

34. $\int x\sqrt[3]{x-1}\,dx;\quad u=x-1$

35. $\int \frac{x}{\sqrt[3]{x+1}}\,dx;\quad u=x+1$

In each of Problems 36–50, evaluate the given integral by first transforming it into one of those listed in the integral table inside the covers of this book.

36. $\int \frac{1}{100+9x^2}\,dx$

37. $\int \frac{1}{100-9x^2}\,dx$

38. $\int \sqrt{9-4x^2}\,dx$

39. $\int \sqrt{4+9x^2}\,dx$

40. $\int \frac{1}{\sqrt{16x^2+9}}\,dx$

41. $\int \frac{x^2}{\sqrt{16x^2+9}}\,dx$

42. $\int \frac{x^2}{\sqrt{25+16x^2}}\,dx$

43. $\int x^2\sqrt{25-16x^2}\,dx$

44. $\int x\sqrt{4-x^4}\,dx$

45. $\int e^x\sqrt{9+e^{2x}}\,dx$

46. $\int \frac{\cos x}{(\sin^2 x)\sqrt{1+\sin^2 x}}\,dx$

47. $\int \frac{\sqrt{x^4-1}}{x}\,dx$

48. $\int \frac{e^{3x}}{\sqrt{25+16e^{2x}}}\,dx$

49. $\int \frac{(\ln x)^2}{x}\sqrt{1+(\ln x)^2}\,dx$

50. $\int x^8\sqrt{4x^6-1}\,dx$

51. The substitution $u=x^2$, $x=\sqrt{u}$, and $dx=du/(2\sqrt{u})$ appears to lead to this result:

$$\int_{-1}^{1} x^2\,dx = \tfrac{1}{2}\int_{1}^{1}\sqrt{u}\,du = 0.$$

Do you believe this result? If not, why not?

52. Use the fact that $x^2 + 4x + 5 = (x + 2)^2 + 1$ to evaluate

$$\int \frac{1}{x^2 + 4x + 5}\,dx.$$

53. Use the fact that $1 - (x - 1)^2 = 2x - x^2$ to evaluate

$$\int \frac{1}{\sqrt{2x - x^2}}\,dx.$$

9.3 Trigonometric Integrals

The substitution $u = \sin x$, $du = \cos x\,dx$ gives

$$\int \sin^3 x \cos x\,dx = \int u^3\,du = \tfrac{1}{4}u^4 + C = \tfrac{1}{4}\sin^4 x + C.$$

This type of substitution can be used to evaluate an integral of the form

$$\int \sin^m x \cos^n x\,dx \tag{1}$$

in the first of the following two cases.

- *Case 1:* At least one of the two numbers m and n is an *odd positive integer*. If so, the other may be any real number.
- *Case 2:* Both m and n are *nonnegative even integers.*

Suppose, for example, that $m = 2k + 1$ is an odd positive integer. Then we split off one $\sin x$ factor and use the identity $\sin^2 x = 1 - \cos^2 x$ to express the remaining factor $\sin^{m-1} x$ in terms of $\cos x$, as follows:

$$\begin{aligned}\int \sin^m x \cos^n x\,dx &= \int \sin^{m-1} x \cos^n x \sin x\,dx\\ &= \int (\sin^2 x)^k \cos^n x \sin x\,dx\\ &= \int (1 - \cos^2 x)^k \cos^n x \sin x\,dx.\end{aligned}$$

Now the substitution $u = \cos x$, $du = -\sin x\,dx$ yields

$$\int \sin^m x \cos^n x\,dx = -\int (1 - u^2)^k u^n\,du.$$

The exponent $k = (m - 1)/2$ is a nonnegative integer because m is an odd positive integer. Thus the factor $(1 - u^2)^k$ of the integrand is a polynomial in the variable u, and so its product with u^n is easy to integrate.

In essence, this method consists of peeling off one copy of $\sin x$ (if m is odd) and then converting the remaining sines into cosines. If it is n that is odd, then we can split off one copy of $\cos x$ and convert the remaining cosines into sines.

EXAMPLE 1

(a) $$\begin{aligned}\int \sin^3 x \cos^2 x\,dx &= \int (1 - \cos^2 x)\cos^2 x \sin x\,dx\\ &= \int (u^4 - u^2)\,du \qquad (u = \cos x)\\ &= \tfrac{1}{5}u^5 - \tfrac{1}{3}u^3 + C = \tfrac{1}{5}\cos^5 x - \tfrac{1}{3}\cos^3 x + C.\end{aligned}$$

(b) $\int \cos^5 x\,dx = \int (1 - \sin^2 x)^2 \cos x\,dx$

$$= \int (1 - u^2)^2\,du \qquad (u = \sin x)$$

$$= \int (1 - 2u^2 + u^4)\,du = u - \tfrac{2}{3}u^3 + \tfrac{1}{5}u^5 + C$$

$$= \sin x - \tfrac{2}{3}\sin^3 x + \tfrac{1}{5}\sin^5 x + C.$$

In Case 2 of the sine–cosine integral in (1), with both m and n nonnegative even integers, we use the half-angle formulas

$$\sin^2\theta = \tfrac{1}{2}(1 - \cos 2\theta) \tag{2a}$$

and

$$\cos^2\theta = \tfrac{1}{2}(1 + \cos 2\theta) \tag{2b}$$

to halve the even powers of $\sin x$ and $\cos x$. Repetition of this process with the resulting powers of $\cos 2x$—if necessary—leads to integrals involving odd powers, and we have seen how to handle these in Case 1.

EXAMPLE 2 Use of the formulas in (2a) and (2b) gives

$$\int \sin^2 x \cos^2 x\,dx = \int \tfrac{1}{2}(1 - \cos 2x)\tfrac{1}{2}(1 + \cos 2x)\,dx$$

$$= \tfrac{1}{4}\int (1 - \cos^2 2x)\,dx = \tfrac{1}{4}\int \left[1 - \tfrac{1}{2}(1 + \cos 4x)\right]dx$$

$$= \tfrac{1}{8}\int (1 - \cos 4x)\,dx = \tfrac{1}{8}x - \tfrac{1}{32}\sin 4x + C.$$

In the third step we have used Equation (2b) with $\theta = 2x$.

EXAMPLE 3 Here we apply Equation (2b), first with $\theta = 3x$, then with $\theta = 6x$.

$$\int \cos^4 3x\,dx = \int \tfrac{1}{4}(1 + \cos 6x)^2\,dx$$

$$= \tfrac{1}{4}\int (1 + 2\cos 6x + \cos^2 6x)\,dx$$

$$= \tfrac{1}{4}\int \left(\tfrac{3}{2} + 2\cos 6x + \tfrac{1}{2}\cos 12x\right)dx$$

$$= \tfrac{3}{8}x + \tfrac{1}{12}\sin 6x + \tfrac{1}{96}\sin 12x + C.$$

An integral of the form

$$\int \tan^m x \sec^n x\,dx \tag{3}$$

can be routinely evaluated in either of the following two cases.

- *Case 1:* m is an *odd positive integer.*
- *Case 2:* n is an *even positive integer.*

In Case 1, we split off the factor $\sec x \tan x$ to form, along with dx, the differential $\sec x \tan x\,dx$ of $\sec x$. We then use the identity $\tan^2 x = \sec^2 x - 1$ to convert the remaining even power of $\tan x$ into powers of $\sec x$. This prepares the integrand for the substitution $u = \sec x$.

EXAMPLE 4

$$\begin{aligned}\int \tan^3 x \sec^3 x\,dx &= \int (\sec^2 x - 1)\sec^2 x \sec x \tan x\,dx \\ &= \int (u^4 - u^2)\,du \qquad (u = \sec x) \\ &= \tfrac{1}{5}u^5 - \tfrac{1}{3}u^3 + C = \tfrac{1}{5}\sec^5 x - \tfrac{1}{3}\sec^3 x + C.\end{aligned}$$

To evaluate the integral in (3) in Case 2, we split off $\sec^2 x$ to form, along with dx, the differential of $\tan x$. Use of the identity $\sec^2 x = 1 + \tan^2 x$ to convert the remaining even power of $\sec x$ to powers of $\tan x$ then prepares the integrand for the substitution $u = \tan x$.

EXAMPLE 5

$$\begin{aligned}\int \sec^6 2x\,dx &= \int (1 + \tan^2 2x)^2 \sec^2 2x\,dx \\ &= \tfrac{1}{2}\int (1 + \tan^2 2x)^2 (2\sec^2 2x)\,dx \\ &= \tfrac{1}{2}\int (1 + u^2)^2\,du \qquad (u = \tan 2x) \\ &= \tfrac{1}{2}\int (1 + 2u^2 + u^4)\,du \\ &= \tfrac{1}{2}u + \tfrac{1}{3}u^3 + \tfrac{1}{10}u^5 + C \\ &= \tfrac{1}{2}\tan 2x + \tfrac{1}{3}\tan^3 2x + \tfrac{1}{10}\tan^5 2x + C.\end{aligned}$$

Similar methods are effective with integrals of the form

$$\int \csc^m x \cot^n x\,dx.$$

The method of Case 1 succeeds with the integral $\int \tan^n x\,dx$ only when n is an odd positive integer, but there is another approach that works equally well whether n be even *or* odd. We split off the factor $\tan^2 x$ and replace it by $\sec^2 x - 1$. This gives

$$\begin{aligned}\int \tan^n x\,dx &= \int (\tan^{n-2} x)(\sec^2 x - 1)\,dx \\ &= \int \tan^{n-2} x \sec^2 x\,dx - \int \tan^{n-2} x\,dx.\end{aligned}$$

We integrate what we can and find that

$$\int \tan^n x\,dx = \frac{\tan^{n-1} x}{n-1} - \int \tan^{n-2} x\,dx. \tag{4}$$

Equation (4) is our first example of a **reduction formula.** Its use reduces the original exponent from n to $n - 2$. Repeated application of (4) leads eventually either to

$$\int \tan^2 x\,dx = \int (\sec^2 x - 1)\,dx = \tan x - x + C$$

or to

$$\begin{aligned}\int \tan x\,dx &= \int \frac{\sin x}{\cos x}\,dx \\ &= -\ln|\cos x| + C = \ln|\sec x| + C.\end{aligned}$$

EXAMPLE 6 Two applications of the formula in (4) give

$$\begin{aligned}\int \tan^6 x\,dx &= \tfrac{1}{5}\tan^5 x - \int \tan^4 x\,dx\\ &= \tfrac{1}{5}\tan^5 x - \left(\tfrac{1}{3}\tan^3 x - \int \tan^2 x\,dx\right)\\ &= \tfrac{1}{5}\tan^5 x - \tfrac{1}{3}\tan^3 x + \tan x - x + C.\end{aligned}$$

Finally, in the case of a trigonometric integral involving tangents, cosecants, and so on, a last resort is to express the integrand wholly in terms of sines and cosines. Simplification may then yield an integrable expression.

9.3 Problems

Evaluate the integrals in Problems 1–35.

1. $\int \sin^3 x\,dx$

2. $\int \sin^4 x\,dx$

3. $\int \sin^2\theta\cos^3\theta\,d\theta$

4. $\int \sin^3 t\cos^3 t\,dt$

5. $\int \cos^5 x\,dx$

6. $\int \frac{\sin t}{\cos^3 t}\,dt$

7. $\int \frac{\sin^3 x}{\sqrt{\cos x}}\,dx$

8. $\int \sin^3 3\phi\cos^4 3\phi\,d\phi$

9. $\int \sin^5 2z\cos^2 2z\,dz$

10. $\int \sin^{3/2} x\cos^3 x\,dx$

11. $\int \frac{\sin^3 4x}{\cos^2 4x}\,dx$

12. $\int \cos^6 4\theta\,d\theta$

13. $\int \sec^4 t\,dt$

14. $\int \tan^3 x\,dx$

15. $\int \cot^3 2x\,dx$

16. $\int \tan\theta\sec^4\theta\,d\theta$

17. $\int \tan^5 2x\sec^2 2x\,dx$

18. $\int \cot^3 x\csc^2 x\,dx$

19. $\int \csc^6 2t\,dt$

20. $\int \frac{\sec^4 t}{\tan^2 t}\,dt$

21. $\int \frac{\tan^3\theta}{\sec^4\theta}\,d\theta$

22. $\int \frac{\cot^3 x}{\csc^2 x}\,dx$

23. $\int \frac{\tan^3 t}{\sqrt{\sec t}}\,dt$

24. $\int \frac{1}{\cos^4 2x}\,dx$

25. $\int \frac{\cot\theta}{\csc^3\theta}\,d\theta$

26. $\int \sin^2 3\alpha\cos^2 3\alpha\,d\alpha$

27. $\int \cos^3 5t\,dt$

28. $\int \tan^4 x\,dx$

29. $\int \cot^4 3t\,dt$

30. $\int \tan^2 2t\sec^4 2t\,dt$

31. $\int \sin^5 2t\cos^{3/2} 2t\,dt$

32. $\int \cot^2\xi\csc^{3/2}\xi\,d\xi$

33. $\int \frac{\tan x + \sin x}{\sec x}\,dx$

34. $\int \frac{\cot x + \csc x}{\sin x}\,dx$

35. $\int \frac{\cot x + \csc^2 x}{1 - \cos^2 x}\,dx$

36. Derive a reduction formula, one analogous to that in Equation (4), for $\int \cot^n x\,dx$.

37. Find $\int \tan x\sec^4 x\,dx$ in two different ways. Then show that your two results are equivalent.

38. Find $\int \cot^3 x\,dx$ in two different ways. Then show that your two results are equivalent.

Problems 39–42 are applications of the trigonometric identities

$$\sin\alpha\sin\beta = \tfrac{1}{2}[\cos(\alpha - \beta) - \cos(\alpha + \beta)],$$

$$\sin\alpha\cos\beta = \tfrac{1}{2}[\sin(\alpha - \beta) + \sin(\alpha + \beta)],$$

$$\cos\alpha\cos\beta = \tfrac{1}{2}[\cos(\alpha - \beta) + \cos(\alpha + \beta)].$$

39. Find $\int \sin 3x\cos 5x\,dx$.

40. Find $\int \sin 2x\sin 4x\,dx$.

41. Find $\int \cos x\cos 4x\,dx$.

42. Suppose that m and n are positive integers with $m \neq n$. Show that

(a) $\int_0^{2\pi} \sin mx\sin nx\,dx = 0;$

(b) $\int_0^{2\pi} \cos mx\sin nx\,dx = 0;$

(c) $\int_0^{2\pi} \cos mx\cos nx\,dx = 0.$

43. Substitute $\sec x\csc x = (\sec^2 x)/(\tan x)$ to derive the formula

$$\int \sec x\csc x\,dx = \ln|\tan x| + C.$$

44. Show that

$$\csc x = \frac{1}{2\sin(x/2)\cos(x/2)}$$

and then apply the result of Problem 43 to derive the formula

$$\int \csc x\,dx = \ln|\tan(x/2)| + C.$$

45. Substitute $x = (\pi/2) - u$ in the integral formula of Problem 44 to show that

$$\int \sec x\,dx = \ln\left|\cot\left(\frac{\pi}{4} - \frac{x}{2}\right)\right| + C.$$

46. Use appropriate trigonometric identities to deduce from the result of Problem 45 that

$$\int \sec x\,dx = \ln|\sec x + \tan x| + C.$$

9.4 Integration by Parts

One reason for transforming one given integral into another is to produce an integral that is easier to evaluate. There are two general ways to accomplish this. We have seen the first, integration by substitution. The second is *integration by parts.*

The formula for integration by parts is a simple consequence of the product rule for derivatives

$$D_x(uv) = v\frac{du}{dx} + u\frac{dv}{dx}.$$

If we write this formula in the form

$$u(x)v'(x) = D_x[u(x)v(x)] - v(x)u'(x), \tag{1}$$

then antidifferentiation gives

$$\int u(x)v'(x)\,dx = u(x)v(x) - \int v(x)u'(x)\,dx. \tag{2}$$

This is the formula for **integration by parts.** With $du = u'(x)\,dx$ and $dv = v'(x)\,dx$, the formula in (2) becomes

$$\int u\,dv = uv - \int v\,du. \tag{3}$$

To apply the integration by parts formula to a given integral, we must first factor its integrand into two "parts," u and dv, the latter including the differential dx. We try to choose these parts in accord with two principles:

1. The antiderivative $v = \int dv$ is easy to find.
2. The new integral $\int v\,du$ is easier to compute than the original integral $\int u\,dv$.

An effective strategy is to choose for dv the most complicated factor that can readily be integrated. The other part u then is differentiated to find du.

We begin with two examples in which we have little flexibility in choosing the parts u and dv.

EXAMPLE 1 Find $\int \ln x\,dx$.

Solution Here there is little alternative to the natural choice $u = \ln x$ and $dv = dx$. It is helpful to systematize the procedure of integration by parts by writing u, dv, du, and v in an array like this:

$$\text{Let} \qquad u = \ln x \quad \text{and} \quad dv = dx.$$

$$\text{Then} \quad du = \frac{1}{x}\,dx \quad \text{and} \quad v = x.$$

The first line specifies the choice of u and dv; the second line is computed from the first. Then the formula in (3) gives

$$\int \ln x\,dx = x \ln x - \int dx = x \ln x - x + C.$$

COMMENT 1 The constant of integration appears only at the last step. We know that once we have found one antiderivative, any other may be obtained by adding a constant C to the one we have found.

COMMENT 2 In computing $v = \int dv$, we ordinarily take the constant of integration to be zero. Had we written $v = x + C_1$ in Example 1, the answer would have been

$$\int \ln x\,dx = (x + C_1) \ln x - \int \left(1 + \frac{C_1}{x}\right) dx$$

$$= x \ln x + C_1 \ln x - (x + C_1 \ln x) + C = x \ln x - x + C$$

as before, so introducing the extra constant C_1 makes no difference.

EXAMPLE 2 Find $\int \arcsin x\,dx$.

Solution Again, there is only one choice for u and dv.

$$\text{Let} \qquad u = \arcsin x, \qquad dv = dx.$$

$$\text{Then} \quad du = \frac{dx}{\sqrt{1 - x^2}}, \qquad v = x.$$

Then the formula in (3) gives

$$\int \arcsin x\,dx = x \arcsin x - \int \frac{x}{\sqrt{1 - x^2}}\,dx$$

$$= x \arcsin x + \sqrt{1 - x^2} + C.$$

EXAMPLE 3 Find $\int xe^{-x}\,dx$.

Solution Here we have some flexibility. Suppose that we try

$$u = e^{-x}, \qquad dv = x\,dx$$

so that

$$du = -e^{-x}\,dx, \qquad v = \tfrac{1}{2}x^2.$$

Then integration by parts gives

$$\int xe^{-x}\,dx = \tfrac{1}{2}x^2 e^{-x} + \tfrac{1}{2}\int x^2 e^{-x}\,dx.$$

The new integral on the right looks more troublesome than the one we started with! Let us begin anew.

$$\text{Let} \quad u = x, \qquad dv = e^{-x}\,dx.$$

$$\text{Then} \quad du = dx, \qquad v = -e^{-x}.$$

Now integration by parts gives

$$\int xe^{-x}\,dx = -xe^{-x} + \int e^{-x}\,dx = -xe^{-x} - e^{-x} + C.$$

Integration by parts can also be applied to definite integrals. We integrate Equation (1) from $x = a$ to $x = b$ and apply the fundamental theorem of calculus. This gives

$$\int_a^b u(x)v'(x)\,dx = \int_a^b D_x[u(x)v(x)]\,dx - \int_a^b v(x)u'(x)\,dx$$

$$= \Big[u(x)v(x)\Big]_a^b - \int_a^b v(x)u'(x)\,dx.$$

In the notation of (3) this is

$$\int_{x=a}^{x=b} u\,dv = \Big[uv\Big]_{x=a}^{x=b} - \int_{x=a}^{x=b} v\,du, \tag{4}$$

though we must not forget that u and v are functions of x. For example, with $u = x$ and $dv = e^{-x}\,dx$, as in Example 3, we obtain

$$\int_0^1 xe^{-x}\,dx = \Big[-xe^{-x}\Big]_0^1 + \int_0^1 e^{-x}\,dx = -e^{-1} + \Big[-e^{-x}\Big]_0^1 = 1 - \frac{2}{e}.$$

EXAMPLE 4 Find $\int x^2e^{-x}\,dx$.

Solution If we choose $u = x^2$, then $du = 2x\,dx$, so we will reduce the exponent of x by this choice.

$$\text{Let} \quad u = x^2, \qquad dv = e^{-x}\,dx.$$

$$\text{Then} \quad du = 2x\,dx, \qquad v = -e^{-x}.$$

Then integration by parts gives

$$\int x^2e^{-x}\,dx = -x^2e^{-x} + 2\int xe^{-x}\,dx.$$

We substitute the result $\int xe^{-x}\,dx = -xe^{-x} - e^{-x}$ of Example 3, and this yields

$$\int x^2e^{-x}\,dx = -x^2e^{-x} - 2xe^{-x} - 2e^{-x} + C$$

$$= -(x^2 + 2x + 2)e^{-x} + C.$$

In effect, we have annihilated the original factor x^2 by integrating by parts twice in succession (because Example 3 was itself an integration by parts).

EXAMPLE 5 Find $\int e^{2x}\sin 3x\,dx$.

Solution This is another example in which repeated integration by parts succeeds, but with a different twist.

$$\text{Let} \quad u = \sin 3x, \qquad dv = e^{2x}\,dx.$$
$$\text{Then} \quad du = 3\cos 3x\,dx, \qquad v = \tfrac{1}{2}e^{2x}.$$

Therefore,

$$\int e^{2x}\sin 3x\,dx = \tfrac{1}{2}e^{2x}\sin 3x - \tfrac{3}{2}\int e^{2x}\cos 3x\,dx.$$

At first it might appear that little progress has been made, for the integral on the right is surely as difficult as the one on the left. We ignore this objection and try again, applying integration by parts to the new integral:

$$\text{Let} \quad u = \cos 3x, \qquad dv = e^{2x}\,dx.$$
$$\text{Then} \quad du = -3\sin 3x\,dx, \qquad v = \tfrac{1}{2}e^{2x}.$$

Now we find that

$$\int e^{2x}\cos 3x\,dx = \tfrac{1}{2}e^{2x}\cos 3x + \tfrac{3}{2}\int e^{2x}\sin 3x\,dx.$$

When we substitute this result in the previous equation, we discover that

$$\int e^{2x}\sin 3x\,dx = \tfrac{1}{2}e^{2x}\sin 3x - \tfrac{3}{4}e^{2x}\cos 3x - \tfrac{9}{4}\int e^{2x}\sin 3x\,dx.$$

So we're back where we started. Or are we? In fact we are *not*, because we actually can *solve* the last equation for the desired integral! We move the right-hand integral here to the left-hand side of the equation. This gives

$$\tfrac{13}{4}\int e^{2x}\sin 3x\,dx = \tfrac{1}{4}e^{2x}(2\sin 3x - 3\cos 3x) + C_1,$$

so

$$\int e^{2x}\sin 3x\,dx = \tfrac{1}{13}e^{2x}(2\sin 3x - 3\cos 3x) + C.$$

EXAMPLE 6 Find a reduction formula for $\int \sec^n x\,dx$.

Solution The idea is that n is a (large) positive integer, and that we want to express the given integral in terms of a lower power of $\sec x$. The easiest power of $\sec x$ to integrate is $\sec^2 x$, so we proceed as follows.

$$\text{Let} \quad u = \sec^{n-2} x, \qquad dv = \sec^2 x\,dx.$$
$$\text{Then} \quad du = (n-2)\sec^{n-2} x \tan x\,dx, \qquad v = \tan x.$$

This gives

$$\begin{aligned}\int \sec^n x\,dx &= \sec^{n-2} x \tan x - (n-2)\int \sec^{n-2} x \tan^2 x\,dx \\ &= \sec^{n-2} x \tan x - (n-2)\int (\sec^{n-2} x)(\sec^2 x - 1)\,dx.\end{aligned}$$

Hence

$$\int \sec^n x\,dx = \sec^{n-2} x \tan x - (n-2)\int \sec^n x\,dx + (n-2)\int \sec^{n-2} x\,dx.$$

We solve this equation for the desired integral and find that

$$\int \sec^n x\,dx = \frac{\sec^{n-2} x \tan x}{n-1} + \frac{n-2}{n-1}\int \sec^{n-2} x\,dx. \tag{5}$$

This is the desired reduction formula. For example, if we take $n = 3$ in this formula, we find that

$$\begin{aligned}\int \sec^3 x\,dx &= \tfrac{1}{2}\sec x \tan x + \tfrac{1}{2}\int \sec x\,dx \\ &= \tfrac{1}{2}\sec x \tan x + \tfrac{1}{2}\ln|\sec x + \tan x| + C.\end{aligned} \tag{6}$$

In the last step we used Equation (15) of Section 8.2,

$$\int \sec x\,dx = \ln|\sec x + \tan x| + C.$$

The reason for using the reduction formula in (5) is that repeated application must yield one of the two elementary integrals $\int \sec x\,dx$ and $\int \sec^2 x\,dx$. For instance, with $n = 4$ we get

$$\begin{aligned}\int \sec^4 x\,dx &= \tfrac{1}{3}\sec^2 x \tan x + \tfrac{2}{3}\int \sec^2 x\,dx \\ &= \tfrac{1}{3}\sec^2 x \tan x + \tfrac{2}{3}\tan x + C,\end{aligned} \tag{7}$$

and with $n = 5$ we get

$$\begin{aligned}\int \sec^5 x\,dx &= \tfrac{1}{4}\sec^3 x \tan x + \tfrac{3}{4}\int \sec^3 x\,dx \\ &= \tfrac{1}{4}\sec^3 x \tan x + \tfrac{3}{8}\sec x \tan x + \tfrac{3}{8}\ln|\sec x + \tan x| + C,\end{aligned} \tag{8}$$

using in the last step the formula in Equation (6).

9.4 PROBLEMS

Use integration by parts to compute the integrals in Problems 1–35.

1. $\int xe^{2x}\,dx$

2. $\int x^2e^{2x}\,dx$

3. $\int t \sin t\,dt$

4. $\int t^2 \sin t\,dt$

5. $\int x \cos 3x\,dx$

6. $\int x \ln x\,dx$

7. $\int x^3 \ln x\,dx$

8. $\int e^{3z} \cos 3z\,dz$

9. $\int \arctan x\,dx$

10. $\int \frac{\ln x}{x^2}\,dx$

11. $\int y^{1/2} \ln y\,dy$

12. $\int x \sec^2 x\,dx$

13. $\int (\ln t)^2\,dt$

14. $\int t(\ln t)^2\,dt$

15. $\int x\sqrt{x+3}\,dx$

16. $\int x^3\sqrt{1-x^2}\,dx$

17. $\int x^5\sqrt{x^3+1}\,dx$

18. $\int \sin^2\theta\,d\theta$

19. $\int \csc^3\theta\,d\theta$

20. $\int \sin(\ln t)\,dt$

21. $\int x^2 \arctan x\,dx$

22. $\int \ln(1+x^2)\,dx$

23. $\int \sec^{-1}\sqrt{x}\,dx$

24. $\int x \tan^{-1}\sqrt{x}\,dx$

25. $\int \tan^{-1}\sqrt{x}\,dx$

26. $\int x^2 \sin 4x\,dx$

27. $\int x \csc^2 x\,dx$

28. $\int x \arctan x\,dx$

29. $\int x^3 \cos x^2\,dx$

30. $\int e^{-3x} \sin 4x\,dx$

31. $\int \frac{\ln x}{x^{3/2}}\,dx$

32. $\int \frac{x^7}{(1+x^4)^{3/2}}\,dx$

33. $\int x \cosh x\,dx$

34. $\int e^x \cosh x\,dx$

35. $\int x^2 \sinh x \, dx$

36. Use the method of cylindrical shells to find the volume generated by rotating the area under $y = \cos x$, for $0 \leqq x \leqq \pi/2$, around the y-axis.

37. Find the volume of the solid obtained by revolving the area bounded by the graph of $y = \ln x$, the x-axis, and the vertical line $y = e$ about the x-axis.

38. Use integration by parts to evaluate

$$\int 2x \arctan x \, dx,$$

with $dv = 2x$, but let $v = x^2 + 1$ rather than $v = x^2$. Is there any reason why one cannot choose v in this way?

39. Use integration by parts to evaluate

$$\int xe^x \cos x \, dx.$$

40. Use integration by parts to evaluate

$$\int \sin 3x \cos x \, dx.$$

Derive the reduction formulas given in Problems 41–46.

41. $\int x^n e^x \, dx = x^n e^x - n \int x^{n-1} e^x \, dx$

42. $\displaystyle\int x^n e^{-x^2} \, dx = -\frac{1}{2} x^{n-1} e^{-x^2} + \frac{n-1}{2} \int x^{n-2} e^{-x^2} \, dx$

43. $\int (\ln x)^n \, dx = x(\ln x)^n - n \int (\ln x)^{n-1} \, dx$

44. $\int x^n \cos x \, dx = x^n \sin x - n \int x^{n-1} \sin x \, dx$

45. $\displaystyle\int \sin^n x \, dx = -\frac{\sin^{n-1} x \cos x}{n} + \frac{n-1}{n} \int \sin^{n-2} x \, dx$

46. $\displaystyle\int \cos^n x \, dx = \frac{\cos^{n-1} x \sin x}{n} + \frac{n-1}{n} \int \cos^{n-2} x \, dx$

Use appropriate reduction formulas from the preceding list to evaluate the integrals in Problems 47–49.

47. $\int_0^1 x^3 e^x \, dx$ **48.** $\int_0^1 x^5 \exp(-x^2) \, dx$

49. $\int_1^e (\ln x)^3 \, dx$

50. Apply the reduction formula in Problem 45 to show that for each positive integer n,

$$\int_0^{\pi/2} \sin^{2n} x \, dx = \frac{\pi}{2} \cdot \frac{1}{2} \cdot \frac{3}{4} \cdot \frac{5}{6} \cdots \frac{2n-1}{2n}$$

and

$$\int_0^{\pi/2} \sin^{2n+1} x \, dx = \frac{2}{3} \cdot \frac{4}{5} \cdot \frac{6}{7} \cdot \frac{8}{9} \cdots \frac{2n}{2n+1}.$$

51. Derive the formula

$$\int \ln(x+10) \, dx = (x+10)\ln(x+10) - x + C$$

in three different ways:

(a) By substituting $u = x + 10$ and applying the result of Example 1.

(b) By integrating by parts with $u = \ln(x+10)$ and $dv = dx$, noting that

$$\frac{x}{x+10} = 1 - \frac{10}{x+10}.$$

(c) By integrating by parts with $u = \ln(x+10)$ and $dv = dx$, but with $v = x + 10$.

52. Derive the formula

$$\int x^3 \tan^{-1} x \, dx = \tfrac{1}{4}(x^4 - 1)\tan^{-1} x - \tfrac{1}{12}x^3 + \tfrac{1}{4}x + C$$

by integrating by parts with $u = \tan^{-1} x$ and $v = \frac{1}{4}(x^4 - 1)$.

53. Let

$$J_n = \int_0^1 x^n e^x \, dx$$

for each integer $n \geqq 0$.

(a) Show that

$$J_0 = 1 - \frac{1}{e} \quad \text{and that} \quad J_n = nJ_{n-1} - \frac{1}{e}$$

for $n \geqq 1$.

(b) Deduce by mathematical induction that

$$J_n = n! - \frac{n!}{e} \sum_{k=0}^{n} \frac{1}{k!}$$

for each integer $n \geqq 0$.

(c) Explain why $J_n \to 0$ as $n \to +\infty$.

(d) Conclude that

$$e = \lim_{n \to \infty} \sum_{k=0}^{n} \frac{1}{k!}.$$

54. Let m and n be positive integers. Derive the reduction formula

$$\int x^m (\ln x)^n \, dx = \frac{x^{m+1}}{m+1} (\ln x)^n - \frac{n}{m+1} \int x^m (\ln x)^{n-1} \, dx.$$

55. According to an advertisement in the March 1984 issue of the *American Mathematical Monthly* for the computer algebra program MACSYMA®,

> An engineer working for a major aerospace company needed to evaluate the following integral dealing with turbulence and boundary layers:
>
> $$\int (k \ln x - 2x^3 + 3x^2 + b)^4 \, dx.$$

He had worked on this problem for more than three weeks with pencil and paper, always arriving at a different result. He was never sure which of the many results he had come upon was correct.

In less than 10 seconds after entering the problem in the computer, MACSYMA gave him the correct answer, not in numerical terms, but in symbolic terms that gave him real insight into the physical nature of the problem.

(Reprinted with the permission of the Computer-Aided Mathematics Group, Symbolics, Inc., Eleven Cambridge Center, Cambridge, MA 02142.) Explain how you could use the reduction formula of Problem 54 to find the engineer's integral (but don't actually do it). Can you see any reason why it should take three weeks?

9.5 Rational Functions and Partial Fractions

In this section we show how every rational function can be integrated in terms of elementary functions. Recall that a rational function $R(x)$ is one that can be expressed as a quotient of two polynomials. That is,

$$R(x) = \frac{P(x)}{Q(x)} \tag{1}$$

where $P(x)$ and $Q(x)$ are polynomials. The **method of partial fractions** involves decomposing $R(x)$ into a sum of terms:

$$R(x) = \frac{P(x)}{Q(x)} = p(x) + F_1(x) + F_2(x) + \ldots + F_k(x), \tag{2}$$

where $p(x)$ is a polynomial and each expression $F_i(x)$ is a fraction that can be integrated by the methods of earlier sections.

For example, one can verify (by finding a common denominator on the right) that

$$\frac{x^3 - 1}{x^3 + x} = 1 - \frac{1}{x} + \frac{x - 1}{x^2 + 1}. \tag{3}$$

It follows that

$$\int \frac{x^3 - 1}{x^3 + x}\,dx = \int \left(1 - \frac{1}{x} + \frac{x}{x^2 + 1} - \frac{1}{x^2 + 1}\right) dx$$

$$= x - \ln|x| + \frac{1}{2}\ln(x^2 + 1) - \tan^{-1}x + C.$$

Of course, the key to this simple integration lies in finding the decomposition given in Equation (3). That such a decomposition exists and the technique of finding it is what the method of partial fractions is about.

According to a theorem proved in advanced algebra, every rational function can be written in the form in (2) with each of the $F_i(x)$ being either a fraction of the form

$$\frac{A}{(ax + b)^n} \tag{4}$$

or one of the form

$$\frac{Bx + C}{(ax^2 + bx + c)^n}, \tag{5}$$

where the quadratic polynomial $ax^2 + bx + c$ is **irreducible,** meaning that it is not a product of linear factors with real coefficients. This is the same as saying that the equation $ax^2 + bx + c = 0$ has no real roots, and the quadratic formula tells us that this is the case exactly when its discriminant is negative: $b^2 - 4ac < 0$.

Fractions of the form in (4) and (5) are called **partial fractions,** and the sum in (2) is called the **partial fraction decomposition** of $R(x)$. Thus (3) gives the partial fraction decomposition of $(x^3 - 1)/(x^3 + x)$. A partial fraction of the form in (4) may be integrated immediately, and we will see in Section 9.7 how to integrate one of the form in (5).

The first step in finding the partial fraction decomposition of $R(x)$ is finding the polynomial $p(x)$ in (2). It turns out that $p(x) \equiv 0$ provided that the degree of the numerator $P(x)$ is *less than* that of the denominator $Q(x)$; in this case the rational fraction $R(x) = P(x)/Q(x)$ is said to be **proper.** If $R(x)$ is not proper, then $p(x)$ may be found by ordinary division of $Q(x)$ into $P(x)$, as in the following example.

EXAMPLE 1 Find $\displaystyle\int \frac{x^3 + x^2 + x - 1}{x^2 + 2x + 2}\,dx.$

Solution Long division of denominator into numerator may be carried out as follows.

$$\begin{array}{rl} & x \quad - 1 \quad \longleftarrow p(x) \text{ (quotient)} \\ x^2 + 2x + 2 \,\big) & x^3 + \;x^2 + \;x - 1 \\ & x^3 + 2x^2 + 2x \\ \hline & \quad -x^2 - \;x - 1 \\ & \quad -x^2 - 2x - 2 \\ \hline & \qquad\qquad x + 1 \quad \longleftarrow r(x) \text{ (remainder)} \end{array}$$

As in arithmetic,

$$\text{Fraction} = \text{quotient} + \frac{\text{remainder}}{\text{divisor}}.$$

Thus

$$\frac{x^3 + x^2 + x - 1}{x^2 + 2x + 2} = (x - 1) + \frac{x + 1}{x^2 + 2x + 2}.$$

And hence

$$\begin{aligned} \int \frac{x^3 + x^2 + x - 1}{x^2 + 2x + 2}\,dx &= \int \left(x - 1 + \frac{x + 1}{x^2 + 2x + 2}\right) dx \\ &= \tfrac{1}{2}x^2 - x + \tfrac{1}{2}\ln(x^2 + 2x + 2) + C. \end{aligned}$$

By using long division as in Example 1, any rational function $R(x)$ can be written as a sum of a polynomial $p(x)$ and a *proper* rational fraction,

$$R(x) = p(x) + \frac{r(x)}{Q(x)}.$$

To see how to integrate an arbitrary rational function, we therefore need only see how to find the partial fraction decomposition of a proper rational function.

To obtain such a decomposition, the first step is to factor the denominator $Q(x)$ into a product of linear factors (those of the form $ax + b$) and irreducible quadratic factors (those of the form $ax^2 + bx + c$ with $b^2 - 4ac < 0$). This is always possible in principle but may be difficult in practice. But once this factorization of $Q(x)$ has been found, the partial fraction decomposition may be obtained by routine algebraic methods (described below). Each linear or irreducible quadratic factor of $Q(x)$ leads to one or more partial fractions of the forms in (4) and (5).

LINEAR FACTORS

Let $R(x) = P(x)/Q(x)$ be a *proper* rational fraction, and suppose that the linear factor $ax + b$ occurs n times in the factorization of $Q(x)$. That is, $(ax + b)^n$ is the highest power of $ax + b$ that divides "evenly" into $Q(x)$. In this case we call n the **multiplicity** of the factor $ax + b$.

Rule 1 *Linear Factor Partial Fractions*

The part of the partial fraction decomposition of $R(x)$ corresponding to the linear factor $ax + b$ of multiplicity n is a sum of n partial fractions, specifically

$$\frac{A_1}{ax+b} + \frac{A_2}{(ax+b)^2} + \cdots + \frac{A_n}{(ax+b)^n}, \tag{6}$$

where $A_1, A_2, \ldots, A_n$ are constants.

If *all* the factors of $Q(x)$ are linear, then the partial fraction decomposition of $R(x)$ is a sum of expressions like the one in (6). The situation is especially simple if each of these linear factors is *nonrepeated*—that is, if each has multiplicity $n = 1$. In this case, the expression in (6) reduces to its first term, and the partial fraction decomposition of $R(x)$ is a sum of such terms. The solutions in the following examples illustrate how the constant numerators can be determined.

EXAMPLE 2 Find $\displaystyle\int \frac{5}{(2x+1)(x-2)}\,dx$.

Solution The linear factors in the denominator are distinct, so we seek a partial fraction decomposition of the form

$$\frac{5}{(2x+1)(x-2)} = \frac{A}{2x+1} + \frac{B}{x-2}.$$

To find the constants A and B, we multiply both sides of this identity by the left-hand (common) denominator $(2x + 1)(x - 2)$. The result is

$$5 = A(x-2) + B(2x+1) = (A+2B)x + (-2A+B).$$

We next equate coefficients of x and of 1 on the two sides of this equation. This yields the simultaneous equations

$$\begin{aligned} A + 2B &= 0, \\ -2A + B &= 5 \end{aligned}$$

that we readily solve for $A = -2$, $B = 1$. Hence

$$\begin{aligned} \int \frac{5}{(2x+1)(x-2)}\,dx &= \int \left(\frac{-2}{2x+1} + \frac{1}{x-2} \right) dx \\ &= -\ln|2x+1| + \ln|x-2| + C = \ln\left|\frac{x-2}{2x+1}\right| + C. \end{aligned}$$

EXAMPLE 3 Find $\displaystyle\int \frac{4x^2 - 3x - 4}{x^3 + x^2 - 2x}\,dx.$

Solution The rational function to be integrated is proper, so we immediately proceed to factor its denominator.

$$x^3 + x^2 - 2x = x(x^2 + x - 2) = x(x-1)(x+2).$$

We are dealing with three nonrepeated linear factors, so the partial fraction decomposition is of the form

$$\frac{4x^2 - 3x - 4}{x^3 + x^2 - 2x} = \frac{A}{x} + \frac{B}{x-1} + \frac{C}{x+2}.$$

To find the constants A, B, and C, we multiply both sides of this equation by the common denominator $x(x-1)(x+2)$ and find thereby that

$$4x^2 - 3x - 4 = A(x-1)(x+2) + Bx(x+2) + Cx(x-1). \tag{7}$$

Then we collect coefficients of like powers of x on the right:

$$4x^2 - 3x - 4 = (A + B + C)x^2 + (A + 2B - C)x + (-2A).$$

Now two polynomials are equal only if the coefficients of corresponding powers of x are the same, and so we may conclude that

$$\begin{aligned} A + B + C &= 4, \\ A + 2B - C &= -3, \\ -2A &= -4. \end{aligned}$$

We solve these simultaneous equations and thus find that $A = 2$, $B = -1$, and $C = 3$.

There is an alternative way of finding A, B, and C that is especially convenient in the case of nonrepeated linear factors. Substitute the values $x = 0$, $x = 1$, and $x = -2$ (the zeros of the linear factors of the denominator) in Equation (7). Substitution of $x = 0$ in (7) immediately gives $-4 = -2A$, so that $A = 2$. Substitution of $x = 1$ in (7) gives $-3 = 3B$, so $B = -1$. Substitution of $x = -2$ gives $18 = 6C$, so $C = 3$.

With these values of $A = 2$, $B = -1$, and $C = 3$, however obtained, we find that

$$\int \frac{4x^2 - 3x - 4}{x^3 + x^2 - 2x}\,dx = \int \left(\frac{2}{x} - \frac{1}{x - 1} + \frac{3}{x + 2}\right) dx$$

$$= 2 \ln|x| - \ln|x - 1| + 3 \ln|x + 2| + C.$$

Laws of logarithms allow us to write this antiderivative in the more compact form

$$\int \frac{4x^2 - 3x - 4}{x^3 + x^2 - 2x}\,dx = \ln\left|\frac{x^2(x + 2)^3}{x - 1}\right| + C.$$

EXAMPLE 4 Find $\displaystyle\int \frac{x^3 - 4x - 1}{x(x - 1)^3}\,dx.$

Solution Here we have a linear factor of multiplicity $n = 3$. According to Rule 1, the partial fraction decomposition of the integrand has the form

$$\frac{x^3 - 4x - 1}{x(x - 1)^3} = \frac{A}{x} + \frac{B}{x - 1} + \frac{C}{(x - 1)^2} + \frac{D}{(x - 1)^3}.$$

To find the constants A, B, C, and D, we multiply both sides of this equation by $x(x - 1)^3$. We find that

$$x^3 - 4x - 1 = A(x - 1)^3 + Bx(x - 1)^2 + Cx(x - 1) + Dx.$$

We expand, then collect coefficients of like powers of x on the right-hand side. This yields

$$x^3 - 4x - 1 = (A + B)x^3 + (-3A - 2B + C)x^2$$
$$+ (3A + B - C + D)x - A.$$

Then we equate coefficients of like powers of x on each side of this equation. We get the four simultaneous equations

$$\begin{aligned} A + B \qquad\qquad &= 1, \\ -3A - 2B + C \qquad &= 0, \\ 3A + B - C + D &= -4, \\ -A \qquad\qquad\qquad &= -1. \end{aligned}$$

The last equation gives $A = 1$, then the first equation gives $B = 0$. Next, the second equation gives $C = 3$. Substitution of these three values in the third equation finally gives $D = -4$. Hence

$$\int \frac{x^3 - 4x - 1}{x(x - 1)^3}\,dx = \int \left(\frac{1}{x} + \frac{3}{(x - 1)^2} - \frac{4}{(x - 1)^3}\right) dx$$

$$= \ln|x| - \frac{3}{x - 1} + \frac{2}{(x - 1)^2} + C.$$

QUADRATIC FACTORS

Suppose that $R(x) = P(x)/Q(x)$ is a proper rational fraction, and suppose that the irreducible quadratic factor $ax^2 + bx + c$ occurs n times in the factorization of $Q(x)$. That is, $(ax^2 + bx + c)^n$ is the highest power of $ax^2 + bx + c$ that divides evenly into $Q(x)$.

Rule 2 *Quadratic Factor Partial Fractions*

The part of the partial fraction decomposition of $R(x)$ corresponding to the irreducible quadratic factor $ax^2 + bx + c$ of multiplicity n is a sum of n partial fractions. It has the form

$$\frac{B_1x + C_1}{ax^2 + bx + c} + \frac{B_2x + C_2}{(ax^2 + bx + c)^2} + \cdots + \frac{B_nx + C_n}{(ax^2 + bx + c)^n}, \tag{8}$$

where $B_1, B_2, \ldots, B_n, C_1, C_2, \ldots, C_n$ are constants.

If $Q(x)$ has both linear and irreducible quadratic factors, then the partial fraction decomposition of $R(x)$ is the sum of the expressions of the form in (6) corresponding to the linear factors, plus the sum of the expressions of the form in (8) corresponding to the quadratic factors. In the case of an irreducible quadratic factor of multiplicity $n = 1$, the expression in (8) reduces to just its first term.

The most important case is that of a nonrepeated quadratic factor of the "sum of squares" form $x^2 + k^2$ (where k is a positive constant). The corresponding partial fraction $(Bx + C)/(x^2 + k^2)$ is readily integrated using the familiar integrals

$$\int \frac{x}{x^2 + k^2}\,dx = \frac{1}{2}\ln(x^2 + k^2) + C,$$

$$\int \frac{1}{x^2 + k^2}\,dx = \frac{1}{k}\arctan\frac{x}{k} + C.$$

The integration of more general partial fractions involving irreducible quadratic factors will be discussed in Section 9.7.

EXAMPLE 5 Find $\displaystyle\int \frac{5x^3 - 3x^2 + 2x - 1}{x^4 + x^2}\,dx.$

Solution The denominator $x^4 + x^2 = x^2(x^2 + 1)$ has both a quadratic factor and a repeated linear factor. The partial fraction decomposition takes the form

$$\frac{5x^3 - 3x^2 + 2x - 1}{x^4 + x^2} = \frac{A}{x} + \frac{B}{x^2} + \frac{Cx + D}{x^2 + 1}.$$

We multiply both sides by $x^4 + x^2$ and obtain

$$\begin{aligned} 5x^3 - 3x^2 + 2x - 1 &= Ax(x^2 + 1) + B(x^2 + 1) + (Cx + D)x^2 \\ &= (A + C)x^3 + (B + D)x^2 + Ax + B. \end{aligned}$$

As before, we equate coefficients of like powers of x. This yields the four simultaneous equations

$$\begin{aligned} A \quad + C \quad &= 5, \\ B \quad + D &= -3, \\ A \quad &= 2, \\ B \quad &= -1. \end{aligned}$$

These equations are easily solved for $A = 2$, $B = -1$, $C = 3$, and $D = -2$. Thus

$$\int \frac{5x^3 - 3x^2 + 2x - 1}{x^4 + x^2}\,dx = \int \left(\frac{2}{x} - \frac{1}{x^2} + \frac{3x - 2}{x^2 + 1}\right) dx$$

$$= 2\ln|x| + \frac{1}{x} + \frac{3}{2}\int \frac{2x\,dx}{x^2 + 1} - 2\int \frac{dx}{x^2 + 1}$$

$$= 2\ln|x| + \frac{1}{x} + \frac{3}{2}\ln(x^2 + 1) - 2\tan^{-1}x + C.$$

*APPLICATIONS TO DIFFERENTIAL EQUATIONS

The following example illustrates the use of partial fractions to solve certain types of separable differential equations.

EXAMPLE 6 Suppose that at time $t = 0$, half of a population of 100 thousand persons has heard a certain rumor, and that the number $P(t)$ of those who know it is then increasing at the rate of 1 thousand persons per day. If $P(t)$ satisfies the differential equation

$$\frac{dP}{dt} = kP(100 - P) \tag{9}$$

(with P in thousands of persons and t in days), determine how many people will know the rumor after $t = 30$ days.

Solution The differential equation in (9) is the model for a simple but widely used assumption: The rate at which the rumor spreads is proportional to the number of contacts between those who know the rumor and those who have not yet heard it. That is, dP/dt is jointly proportional to P and to $100 - P$; thus dP/dt is proportional to the product of P and $100 - P$.

To find the constant of proportionality, we substitute the given values $P(0) = 50$ and $P'(0) = 1$ in (9). This yields the equation

$$1 = k(50)(100 - 50).$$

It follows that $k = 0.0004$, so the differential equation in (9) takes the form

$$\frac{dP}{dt} = (0.0004)P(100 - P). \tag{10}$$

Separation of the variables leads to

$$\int \frac{dP}{P(100 - P)} = \int 0.0004\,dt.$$

Then the partial fraction decomposition

$$\frac{100}{P(100 - P)} = \frac{1}{P} + \frac{1}{100 - P}$$

yields

$$\int \left(\frac{1}{P} + \frac{1}{100 - P}\right) dP = \int 0.04\,dt;$$

$$\ln P - \ln(100 - P) = (0.04)t + C. \tag{11}$$

Substitution of the initial data $P = 50$ when $t = 0$ now gives $C = 0$, so (11) takes the form

$$\ln \frac{P}{100 - P} = (0.04)t, \quad \text{so} \quad \frac{P}{100 - P} = e^{(0.04)t}.$$

We readily solve this last equation for the solution

$$P(t) = \frac{100e^{(0.04)t}}{1 + e^{(0.04)t}}. \tag{12}$$

Hence the number of people who know the rumor after 30 days is $P(30) \approx 76.85$ thousand persons.

The method of Example 6 can be used to solve any differential equation of the form

$$\frac{dx}{dt} = k(x - a)(x - b) \tag{13}$$

(where a, b, and k are constants). As Problems 47–51 indicate, this differential equation serves as a mathematical model for a wide variety of natural phenomena.

9.5 PROBLEMS

Find the integrals in Problems 1–35.

1. $\int \frac{x^2}{x + 1}\,dx$

2. $\int \frac{x^3}{2x - 1}\,dx$

3. $\int \frac{1}{x^3 - 3x}\,dx$

4. $\int \frac{x}{x^2 + 4x}\,dx$

5. $\int \frac{1}{x^2 + x - 6}\,dx$

6. $\int \frac{x^3}{x^2 + x - 6}\,dx$

7. $\int \frac{1}{x^3 + 4x}\,dx$

8. $\int \frac{1}{(x + 1)(x^2 + 1)}\,dx$

9. $\int \frac{x^4}{x^2 + 4}\,dx$

10. $\int \frac{1}{(x^2 + 1)(x^2 + 4)}\,dx$

11. $\int \frac{x - 1}{x + 1}\,dx$

12. $\int \frac{2x^3 - 1}{x^2 + 1}\,dx$

13. $\int \frac{x^2 + 2x}{(x + 1)^2}\,dx$

14. $\int \frac{2x - 4}{x^2 - x}\,dx$

15. $\int \frac{1}{x^2 - 4}\,dx$

16. $\int \frac{x^4}{x^2 + 4x + 4}\,dx$

17. $\int \frac{x + 10}{2x^2 + 5x - 3}\,dx$

18. $\int \frac{x + 1}{x^3 - x^2}\,dx$

19. $\int \frac{x^2 + 1}{x^3 + 2x^2 + x}\,dx$

20. $\int \frac{x^2 + x}{x^3 - x^2 - 2x}\,dx$

21. $\int \frac{4x^3 - 7x}{x^4 - 5x^2 + 4}\,dx$

22. $\int \frac{2x^2 + 3}{x^4 - 2x^2 + 1}\,dx$

23. $\int \frac{x^2}{(x + 2)^3}\,dx$

24. $\int \frac{x^2 + x}{(x^2 - 4)(x + 4)}\,dx$

25. $\int \frac{1}{x^3 + x}\,dx$

26. $\int \frac{6x^3 - 18x}{(x^2 - 1)(x^2 - 4)}\,dx$

27. $\int \frac{x + 4}{x^3 + 4x}\,dx$

28. $\int \frac{4x^4 + x + 1}{x^5 + x^4}\,dx$

29. $\int \frac{x}{(x + 1)(x^2 + 1)}\,dx$

30. $\int \frac{x^2 + 2}{(x^2 + 1)^2}\,dx$

31. $\int \frac{x^2 - 10}{2x^4 + 9x^2 + 4}\,dx$

32. $\int \frac{x^2}{x^4 - 1}\,dx$

33. $\int \frac{x^3 + x^2 + 2x + 3}{x^4 + 5x^2 + 6}\,dx$

34. $\int \frac{x^2 + 4}{(x^2 + 1)^2(x^2 + 2)}\,dx$

35. $\int \frac{x^4 + 3x^2 - 4x + 5}{(x - 1)^2(x^2 + 1)}\,dx$

In Problems 36–39, make a preliminary substitution before using the method of partial fractions.

36. $\int \frac{\cos\theta}{\sin^2\theta - \sin\theta - 6}\,d\theta$

37. $\int \frac{e^{4t}}{(e^{2t} - 1)^3}\,dt$

38. $\int \frac{\sec^2 t}{\tan^3 t + \tan^2 t}\,dt$

39. $\int \frac{1 + \ln t}{t(3 + 2\ln t)^2}\,dt$

40. The plane region R is bounded by the curve

$$y^2 = x^2 \frac{1-x}{1+x}, \qquad 0 \leqq x \leqq 1.$$

Find the volume generated by revolving R around the x-axis.

Solve the initial value problems in 41–46.

41. $\dfrac{dx}{dt} = x - x^2$; $x(0) = 2$

42. $\dfrac{dx}{dt} = 10x - x^2$; $x(0) = 1$

43. $\dfrac{dx}{dt} = 1 - x^2$; $x(0) = 3$

44. $\dfrac{dx}{dt} = 9 - 4x^2$; $x(0) = 0$

45. $\dfrac{dx}{dt} = x^2 + 5x + 6$; $x(0) = 5$

46. $\dfrac{dx}{dt} = 2x^2 + x - 15$; $x(0) = 10$

47. Suppose that the population $P(t)$ (in millions) of a certain country satisfies the differential equation

$$\frac{dP}{dt} = kP(200 - P) \qquad (k \text{ constant}).$$

In 1940 its population was 100 million and was then growing at the rate of 1 million per year. Determine this country's predicted population in the year 2000.

48. Suppose that a community contains 15,000 people who are susceptible to a certain contagious disease. At time $t = 0$ the number $N(t)$ of people who have the disease is 5000 and is increasing then at the rate of 500 per day. Assume that $N'(t)$ is proportional to the product of the numbers of those who have the disease and those who do not. How long will it take for another 5000 people to contract the disease?

49. As a certain salt dissolves in a solvent, the number $x(t)$ of grams of the salt in solution after t seconds satisfies the differential equation

$$\frac{dx}{dt} = (0.8)x - (0.004)x^2.$$

(a) If $x = 50$ when $t = 0$, how long will it take for an additional 50 g of the salt to dissolve?

(b) What is the maximum amount of the salt that will ever dissolve in the solvent?

50. A population $P(t)$ (t in months) of small rodents satisfies the differential equation

$$\frac{dP}{dt} = (0.001)P^2 - kP \qquad (k \text{ constant}).$$

If $P(0) = 100$ and $P'(0) = 8$, how long will it take for this population to double to 200 rodents?

51. Consider an animal population $P(t)$ (t in years) that satisfies the differential equation

$$\frac{dP}{dt} = kP^2 - (0.01)P \qquad (k \text{ constant}).$$

Suppose also that $P(0) = 200$ and that $P'(0) = 2$.

(a) When is $P = 1000$?

(b) When does doomsday occur?

*9.5 Optional Computer Application

A number of symbolic computation programs—such as Derive®, MACSYMA®, Maple®, Mathematica®, and REDUCE®—are now available to perform algebraic manipulations and to compute derivatives and antiderivatives symbolically. For example, the Maple command

```
int( 100 /( (x + 1)^3 * (x^2 + 100) ) , x );
```

produces the immediate response

```
   50       1          200       1        9700
- ----- ----------- - ------- ------- - --------- ln(x + 1)
   101            2    10201   x + 1    1030301
         (x + 1)

     4850         2             2990
  + --------- ln(x  + 100) - --------- arctan(1/10 x)
    1030301                   1030301
```

Can you explain why the integral

$$\int \frac{100}{(x+1)^3(x^2+100)}\,dx$$

should be of the form

$$\frac{A}{(x+1)^2} + \frac{B}{x+1} + C\ln(x+1) + D\ln(x^2+100) + \tan^{-1}\left(\frac{x}{10}\right)?$$

Maple can also solve differential equations. The command

```
dsolve( { diff(P(t),t) = 0.0004*P(t)*(100 - P(t)) , P(0) = 50 } , P(t) );
```

to solve Equation (10) in this section (with $P(0) = 50$) produces the result

```
                                              1
P(t) = - 100. ------------------------------------------
              - 1. - 1.000000000 exp(- .04000000000 t)
```

Does this agree with the solution we found in Equation (12)?

9.6 Trigonometric Substitution

The method of *trigonometric substitution* is often effective in dealing with integrals when the integrands involve certain algebraic expressions such as $(a^2 - u^2)^{1/2}$, $(u^2 - a^2)^{3/2}$, and $1/(a^2 + u^2)^2$. There are three basic trigonometric substitutions:

If the integral involves	*then substitute*	*and use the identity*
$a^2 - u^2$	$u = a\sin\theta$	$1 - \sin^2\theta = \cos^2\theta$
$a^2 + u^2$	$u = a\tan\theta$	$1 + \tan^2\theta = \sec^2\theta$
$u^2 - a^2$	$u = a\sec\theta$	$\sec^2\theta - 1 = \tan^2\theta$

By the substitution $u = a\sin\theta$ we mean, more precisely, the *inverse* trigonometric substitution

$$\theta = \sin^{-1}\frac{u}{a}, \qquad -\frac{\pi}{2} \leqq \theta \leqq \frac{\pi}{2},$$

where $|u| \leqq a$. Suppose, for example, that an integral involves the expression $(a^2 - u^2)^{1/2}$. Then this substitution yields

$$(a^2 - u^2)^{1/2} = (a^2 - a^2\sin^2\theta)^{1/2} = (a^2\cos^2\theta)^{1/2} = a\cos\theta.$$

We choose the nonnegative square root in the last step because $\cos\theta \geqq 0$ for $-\pi/2 \leqq \theta \leqq \pi/2$. Thus the troublesome factor $(a^2 - u^2)^{1/2}$ becomes $a\cos\theta$, and meanwhile, $du = a\cos\theta\,d\theta$. If the trigonometric integral that results from this substitution can be evaluated using the methods of Section 9.3, the result will normally involve $\theta = \sin^{-1}(u/a)$ and trigonometric functions of θ. The final step will be to express the answer in terms of the original variable. For this purpose the values of the various trigonometric functions can be read from the right triangle shown in Fig. 9.2, which contains an angle θ such that $\sin \theta = u/a$ (if u is negative, then θ is negative).

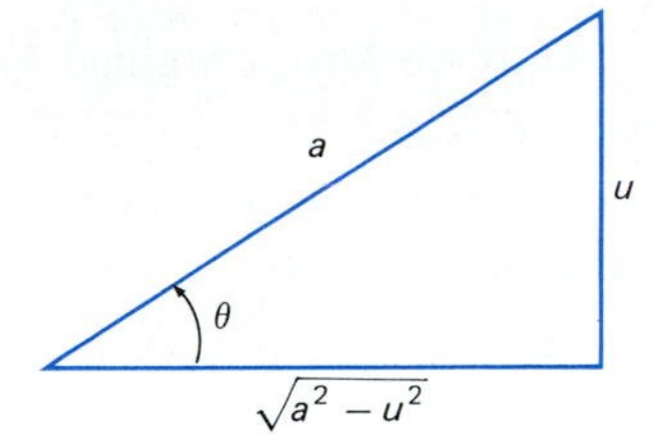

9.2 The reference triangle for the substitution $u = a \sin \theta$

EXAMPLE 1 Evaluate $\displaystyle\int \frac{x^3}{\sqrt{1 - x^2}}\, dx$ where $|x| < 1$.

Solution Here $a = 1$ and $u = x$, so we substitute

$$x = \sin \theta, \qquad dx = \cos \theta \, d\theta.$$

This gives

$$\int \frac{x^3}{\sqrt{1 - x^2}}\, dx = \int \frac{\sin^3 \theta \cos \theta}{\sqrt{1 - \sin^2 \theta}}\, d\theta$$

$$= \int \sin^3 \theta \, d\theta = \int (\sin \theta)(1 - \cos^2 \theta)\, d\theta$$

$$= \tfrac{1}{3} \cos^3 \theta - \cos \theta + C.$$

Because $\cos \theta = (1 - \sin^2 \theta)^{1/2} = \sqrt{1 - x^2}$, our final answer is

$$\int \frac{x^3}{\sqrt{1 - x^2}}\, dx = \frac{1}{3}(1 - x^2)^{3/2} - \sqrt{1 - x^2} + C.$$

The following example illustrates the use of trigonometric substitution to find integrals like those appearing in Formulas 44–62 of the table inside the front cover.

EXAMPLE 2 Find $\int \sqrt{a^2 - u^2}\, du, \quad |u| < a.$

Solution The substitution $u = a \sin \theta$, $du = a \cos \theta \, d\theta$ gives

$$\int \sqrt{a^2 - u^2}\, du = \int \sqrt{a^2 - a^2 \sin^2 \theta}\, (a \cos \theta)\, d\theta$$

$$= \int a^2 \cos^2 \theta \, d\theta = \tfrac{1}{2} a^2 \int (1 + \cos 2\theta)\, d\theta$$

$$= \tfrac{1}{2} a^2 (\theta + \tfrac{1}{2} \sin 2\theta) + C = \tfrac{1}{2} a^2 (\theta + \sin \theta \cos \theta) + C.$$

(We used the identity $\sin 2\theta = 2 \sin \theta \cos \theta$ in the last step.) Now from Fig. 9.2 we see that

$$\sin \theta = \frac{u}{a}, \qquad \cos \theta = \frac{\sqrt{a^2 - u^2}}{a}.$$

Hence

$$\int \sqrt{a^2 - u^2}\, du = \frac{1}{2}a^2\left(\sin^{-1}\frac{u}{a} + \frac{u}{a}\cdot\frac{\sqrt{a^2 - u^2}}{a}\right) + C.$$

$$= \frac{u}{2}\sqrt{a^2 - u^2} + \frac{a^2}{2}\sin^{-1}\frac{u}{a} + C.$$

Thus we have obtained Formula 54 in the table.

By the substitution $u = a \tan \theta$ in an integral involving $a^2 + u^2$ is meant the substitution

$$\theta = \tan^{-1}\frac{u}{a}, \qquad -\frac{\pi}{2} < \theta < \frac{\pi}{2}.$$

Note that in this case

$$\sqrt{a^2 + u^2} = \sqrt{a^2 + a^2\tan^2\theta} = \sqrt{a^2\sec^2\theta} = a \sec \theta$$

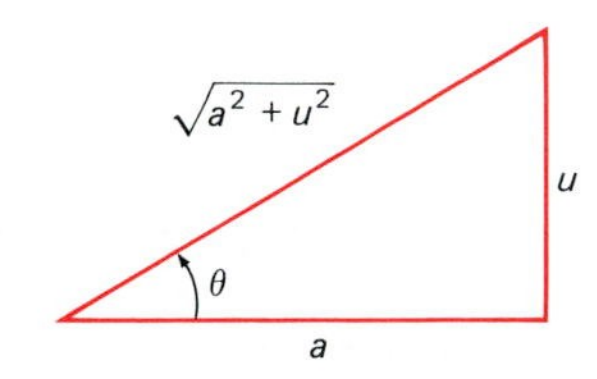

9.3 The substitution $u = a \tan \theta$

under the assumption that $a > 0$. We take the positive square root in the last step here because $\sec \theta > 0$ for $-\pi/2 < \theta < \pi/2$. The values of the various trigonometric functions of θ under this substitution can be read from the right triangle of Fig. 9.3, which shows a (positive or negative) acute angle θ such that $\tan \theta = u/a$.

EXAMPLE 3 Find $\displaystyle\int \frac{1}{(4x^2 + 9)^2}\, dx.$

Solution The factor $4x^2 + 9$ corresponds to $u^2 + a^2$ with $u = 2x$ and $a = 3$. Hence the substitution $u = a \tan \theta$ amounts to

$$2x = 3 \tan \theta, \qquad x = \tfrac{3}{2}\tan \theta, \quad \text{and} \quad dx = \tfrac{3}{2}\sec^2\theta\, d\theta.$$

This gives

$$\int \frac{1}{(4x^2 + 9)^2}\, dx = \int \frac{\frac{3}{2}\sec^2\theta}{(9\tan^2\theta + 9)^2}\, d\theta$$

$$= \frac{3}{2}\int \frac{\sec^2\theta}{(9\sec^2\theta)^2}\, d\theta = \frac{1}{54}\int \frac{1}{\sec^2\theta}\, d\theta$$

$$= \frac{1}{54}\int \cos^2\theta\, d\theta = \frac{1}{108}(\theta + \sin\theta\cos\theta) + C.$$

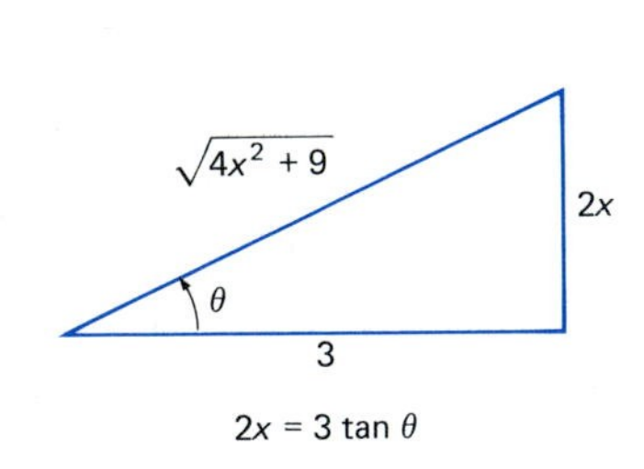

9.4 The reference triangle for Example 3

The actual integration in the last step is the same as in Example 2. Now $\theta = \tan^{-1}(2x/3)$, and the triangle of Fig. 9.4 gives

$$\sin\theta = \frac{2x}{\sqrt{4x^2 + 9}}, \qquad \cos\theta = \frac{3}{\sqrt{4x^2 + 9}}.$$

Hence

$$\int \frac{1}{(4x^2 + 9)^2}\, dx = \frac{1}{108}\left[\tan^{-1}\left(\frac{2x}{3}\right) + \frac{2x}{\sqrt{4x^2 + 9}}\cdot\frac{3}{\sqrt{4x^2 + 9}}\right] + C$$

$$= \frac{1}{108}\tan^{-1}\left(\frac{2x}{3}\right) + \frac{x}{18(4x^2 + 9)} + C.$$

By the substitution $u = a \sec \theta$ in an integral involving $u^2 - a^2$ is meant the substitution

$$\theta = \sec^{-1}\frac{u}{a}, \qquad 0 \leqq \theta \leqq \pi,$$

where $|u| \geqq a$ (because of the domain and range of the inverse secant function). Then

$$\sqrt{u^2 - a^2} = \sqrt{a^2 \sec^2\theta - a^2} = \sqrt{a^2 \tan^2\theta} = \pm a \tan \theta.$$

Here we must take the plus sign if $u > a$, so that $0 < \theta < \pi/2$ and $\tan \theta > 0$. If $u < -a$, so that $\pi/2 < \theta < \pi$ and $\tan \theta < 0$, we take the minus sign above. In either case the values of the various trigonometric functions of θ can be read from the right triangle in Fig. 9.5.

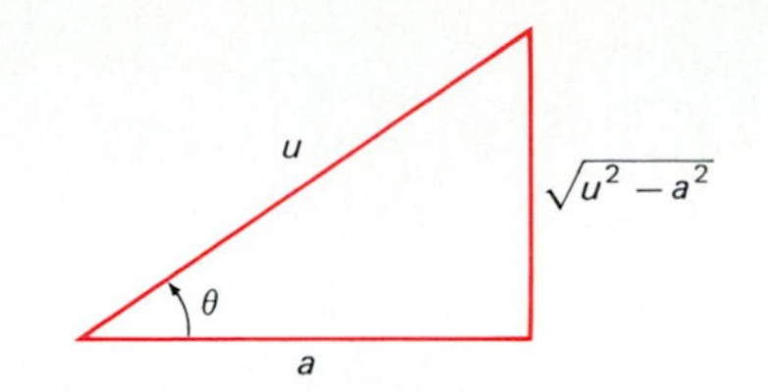

9.5 The reference triangle for the substitution $u = a \sec \theta$

EXAMPLE 4 Find $\displaystyle\int \frac{\sqrt{x^2 - 25}}{x}\,dx, \quad x > 5.$

Solution We substitute $x = 5 \sec \theta$, $dx = 5 \sec \theta \tan \theta \, d\theta$. Then

$$\sqrt{x^2 - 25} = \sqrt{25(\sec^2\theta - 1)} = 5 \tan \theta,$$

because $x > 5$ implies $0 < \theta < \pi/2$, so that $\tan \theta > 0$. Hence this substitution gives

$$\begin{aligned}
\int \frac{\sqrt{x^2 - 25}}{x}\,dx &= \int \frac{5 \tan \theta}{5 \sec \theta}(5 \sec \theta \tan \theta)\,d\theta) \\
&= 5 \int \tan^2\theta \, d\theta = 5 \int (\sec^2\theta - 1)\,d\theta \\
&= 5 \tan \theta - 5\theta + C = \sqrt{x^2 - 25} - 5 \sec^{-1}\left(\frac{x}{5}\right) + C.
\end{aligned}$$

Hyperbolic substitutions may be used in a similar way, and to the same effect, as trigonometric substitutions. The three basic hyperbolic substitutions—which are not ordinarily memorized—are listed here for reference.

If the integral involves	*then substitute*	*and use the identity*
$a^2 - u^2$	$u = a \tanh \theta$	$1 - \tanh^2\theta = \operatorname{sech}^2\theta$
$a^2 + u^2$	$u = a \sinh \theta$	$1 + \sinh^2\theta = \cosh^2\theta$
$u^2 - a^2$	$u = a \cosh \theta$	$\cosh^2\theta - 1 = \sinh^2\theta$

EXAMPLE 5 Find $\displaystyle\int \frac{1}{\sqrt{x^2 - 1}}\,dx, \quad x > 1.$

Solution For purposes of comparison, we evaluate this integral both by trigonometric substitution and by hyperbolic substitution. The trigonometric substitution

$$x = \sec \theta, \qquad dx = \sec \theta \tan \theta \, d\theta, \qquad \tan \theta = \sqrt{x^2 - 1}$$

gives

$$\int \frac{1}{\sqrt{x^2-1}}\,dx = \int \frac{\sec\theta\tan\theta}{\tan\theta}\,d\theta = \int \sec\theta\,d\theta$$

$$= \ln|\sec\theta + \tan\theta| + C \quad \text{(Equation (15), Section 8.2)}$$

$$= \ln\left|x + \sqrt{x^2-1}\right| + C.$$

With the hyperbolic substitution $x = \cosh\theta$, $dx = \sinh\theta\,d\theta$, we have

$$\sqrt{x^2-1} = \sqrt{\cosh^2\theta - 1} = \sinh\theta.$$

We take the positive square root here because $x > 1$ implies that $\theta = \cosh^{-1}x > 0$ and thus that $\sinh\theta > 0$. Hence

$$\int \frac{1}{\sqrt{x^2-1}}\,dx = \int \frac{\sinh\theta}{\sinh\theta}\,d\theta = \int 1\,d\theta = \theta + C = \cosh^{-1}x + C.$$

The two results appear to differ, but Equation (14) in Section 8.5 shows that they are equivalent.

9.6 PROBLEMS

Use trigonometric substitutions to evaluate the integrals in Problems 1–36.

1. $\displaystyle\int \frac{1}{\sqrt{16-x^2}}\,dx$

2. $\displaystyle\int \frac{1}{\sqrt{4-9x^2}}\,dx$

3. $\displaystyle\int \frac{1}{x^2\sqrt{4-x^2}}\,dx$

4. $\displaystyle\int \frac{1}{x^2\sqrt{x^2-25}}\,dx$

5. $\displaystyle\int \frac{x^2}{\sqrt{16-x^2}}\,dx$

6. $\displaystyle\int \frac{x^2}{\sqrt{9-4x^2}}\,dx$

7. $\displaystyle\int \frac{1}{(9-16x^2)^{3/2}}\,dx$

8. $\displaystyle\int \frac{1}{(25+16x^2)^{3/2}}\,dx$

9. $\displaystyle\int \frac{\sqrt{x^2-1}}{x^2}\,dx$

10. $\displaystyle\int x^3\sqrt{4-x^2}\,dx$

11. $\displaystyle\int x^3\sqrt{9+4x^2}\,dx$

12. $\displaystyle\int \frac{x^3}{\sqrt{x^2+25}}\,dx$

13. $\displaystyle\int \frac{(1-4x^2)^{1/2}}{x}\,dx$

14. $\displaystyle\int \frac{1}{\sqrt{1+x^2}}\,dx$

15. $\displaystyle\int \frac{1}{\sqrt{9+4x^2}}\,dx$

16. $\displaystyle\int \sqrt{1+4x^2}\,dx$

17. $\displaystyle\int \frac{x^2}{\sqrt{25-x^2}}\,dx$

18. $\displaystyle\int \frac{x^3}{\sqrt{25-x^2}}\,dx$

19. $\displaystyle\int \frac{x^2}{\sqrt{1+x^2}}\,dx$

20. $\displaystyle\int \frac{x^3}{\sqrt{1+x^2}}\,dx$

21. $\displaystyle\int \frac{x^2}{\sqrt{4+9x^2}}\,dx$

22. $\displaystyle\int (1-x^2)^{3/2}\,dx$

23. $\displaystyle\int \frac{1}{(1+x^2)^{3/2}}\,dx$

24. $\displaystyle\int \frac{1}{(4-x^2)^2}\,dx$

25. $\displaystyle\int \frac{1}{(4-x^2)^3}\,dx$

26. $\displaystyle\int \frac{1}{(4x^2+9)^3}\,dx$

27. $\displaystyle\int \sqrt{9+16x^2}\,dx$

28. $\displaystyle\int (9+16x^2)^{3/2}\,dx$

29. $\displaystyle\int \frac{1}{x}\sqrt{x^2-25}\,dx$

30. $\displaystyle\int \frac{1}{x}\sqrt{9x^2-16}\,dx$

31. $\displaystyle\int x^2\sqrt{x^2-1}\,dx$

32. $\displaystyle\int \frac{x^2}{\sqrt{4x^2-9}}\,dx$

33. $\displaystyle\int \frac{1}{(4x^2-1)^{3/2}}\,dx$

34. $\displaystyle\int \frac{1}{x^2\sqrt{4x^2-9}}\,dx$

35. $\displaystyle\int \frac{\sqrt{x^2-5}}{x^2}\,dx$

36. $\displaystyle\int (4x^2-5)^{3/2}\,dx$

Use hyperbolic substitutions to evaluate the following integrals.

37. $\displaystyle\int \frac{1}{\sqrt{25+x^2}}\,dx$

38. $\displaystyle\int \sqrt{1+x^2}\,dx$

39. $\displaystyle\int \frac{1}{x^2}\sqrt{x^2-4}\,dx$

40. $\displaystyle\int \frac{1}{\sqrt{1+9x^2}}\,dx$

41. $\int x^2\sqrt{1 + x^2}\,dx$

42. Compute the arc length of the parabola $y = x^2$ over the interval $0 \leqq x \leqq 1$.

43. Compute the area of the surface obtained by revolving the parabolic arc of Problem 42 around the x-axis.

44. Show that the length of one arch of the sine curve $y = \sin x$ is equal to half the circumference of the ellipse $x^2 + \frac{1}{2}y^2 = 1$. (*Suggestion:* Substitute $x = \cos\theta$ in the arc length integral for the ellipse.)

45. Compute the arc length of the curve $y = \ln x$ for $1 \leqq x \leqq 2$.

46. Compute the area of the surface obtained by revolving the curve of Problem 45 around the y-axis.

47. A torus is obtained by revolving the circle

$$(x - b)^2 + y^2 = a^2 \qquad (0 < a \leqq b)$$

around the y-axis. Show that its surface area is $4\pi^2 ab$.

48. Find the area under the curve $y = \sqrt{9 + x^2}$ from $x = 0$ to $x = 4$.

49. Find the area of the surface obtained by revolving the curve $y = \sin x$, $0 \leqq x \leqq \pi$, around the x-axis.

50. An ellipsoid of revolution is obtained by revolving the ellipse $x^2/a^2 + y^2/b^2 = 1$ around the x-axis. Suppose that $a > b$, and show then that the ellipsoid has surface area

$$A = 2\pi ab\left[\frac{b}{a} + \frac{a}{c}\sin^{-1}\left(\frac{c}{a}\right)\right]$$

where $c^2 = a^2 - b^2$. Assume that $a \approx b$, so that $c \approx 0$ and $\sin^{-1}(c/a) \approx c/a$. Conclude that $A \approx 4\pi a^2$.

51. Suppose that $b > a$ for the ellipsoid of revolution of Problem 50. Show that its surface area is then

$$A = 2\pi ab\left[\frac{b}{a} + \frac{a}{c}\ln\left(\frac{b + c}{a}\right)\right]$$

where $c^2 = b^2 - a^2$. Use the fact that $\ln(1 + x) \approx x$ if $x \approx 0$, and thereby conclude that $A \approx 4\pi a^2$ if $a \approx b$.

52. A road is to be built from the point (2, 1) to the point (5, 3), following the path of the parabola

$$y = -1 + 2\sqrt{x - 1}.$$

Calculate the length of this road (the units are in miles). (*Suggestion:* Substitute $x = \sec^2\theta$ in the arc length integral.)

53. Suppose that the cost of the road in the previous problem is $\sqrt{x}$ million dollars per mile. Calculate the total cost of the road.

54. A kite is flying at a height of 500 ft at a horizontal distance of 100 ft from the string-holder on the ground. The kite string weighs 1/16 oz/ft and is hanging in the shape of the parabola $y = x^2/20$ joining the string-holder at (0, 0) to the kite at (100, 500). Calculate the work (in ft-lb) done in lifting the kite string from the ground to its present position.

9.7 Integrals Involving Quadratic Polynomials

An integral involving a square root or negative power of a quadratic polynomial $ax^2 + bx + c$ can often be simplified by the process of *completing the square.* For example,

$$x^2 + 2x + 2 = (x + 1)^2 + 1,$$

and hence the substitution $u = x + 1$, $du = dx$ yields

$$\int \frac{1}{x^2 + 2x + 2}\,dx = \int \frac{1}{u^2 + 1}\,du = \tan^{-1}u + C = \tan^{-1}(x + 1) + C.$$

In general, the object is to convert $ax^2 + bx + c$ into either a sum or a difference of two squares—either $u^2 \pm a^2$ or $a^2 - u^2$—so that the method of trigonometric substitution may then be used. To see how this works in practice, suppose first that $a = 1$, so that the quadratic in question is of the form $x^2 + bx + c$. The sum $x^2 + bx$ of the first two terms can be completed to a perfect square by adding $b^2/4$, the square of half the coefficient of x, and in turn subtracting $b^2/4$ from the constant term c. This gives

$$x^2 + bx + c = \left(x^2 + bx + \frac{b^2}{4}\right) + \left(c - \frac{b^2}{4}\right)$$
$$= \left(x + \frac{b}{2}\right)^2 + \left(c - \frac{b^2}{4}\right).$$

With $u = x + \frac{1}{2}b$, our result above is of the form $u^2 + A^2$ or $u^2 - A^2$ (depending on the sign of $c - \frac{1}{4}b^2$). If the coefficient a of x^2 is not 1, we factor it out to begin with and proceed as before:

$$ax^2 + bx + c = a\left(x^2 + \frac{b}{a}x + \frac{c}{a}\right).$$

EXAMPLE 1 Find $\displaystyle\int \frac{1}{9x^2 + 6x + 5}\,dx.$

Solution Our first step is completion of the square.

$$\begin{aligned} 9x^2 + 6x + 5 = 9(x^2 + \tfrac{2}{3}x) + 5 &= 9(x^2 + \tfrac{2}{3}x + \tfrac{1}{9}) - 1 + 5 \\ &= 9(x + \tfrac{1}{3})^2 + 4 = (3x + 1)^2 + 2^2. \end{aligned}$$

Hence

$$\begin{aligned} \int \frac{1}{9x^2 + 6x + 5}\,dx &= \int \frac{1}{(3x + 1)^2 + 4}\,dx \\ &= \frac{1}{3}\int \frac{1}{u^2 + 4}\,du \qquad (u = 3x + 1) \\ &= \frac{1}{6}\int \frac{1/2}{(u/2)^2 + 1}\,du = \frac{1}{6}\int \frac{1}{v^2 + 1}\,dv \qquad \left(v = \frac{1}{2}u\right) \\ &= \frac{1}{6}\tan^{-1} v + C = \frac{1}{6}\tan^{-1}\left(\frac{u}{2}\right) + C \\ &= \frac{1}{6}\tan^{-1}\left(\frac{3x + 1}{2}\right) + C. \end{aligned}$$

EXAMPLE 2 Find $\displaystyle\int \frac{1}{\sqrt{9 + 16x - 4x^2}}\,dx.$

Solution First we complete the square:

$$\begin{aligned} 9 + 16x - 4x^2 = 9 - 4(x^2 - 4x) &= 9 - 4(x^2 - 4x + 4) + 16 \\ &= 25 - 4(x - 2)^2. \end{aligned}$$

Hence

$$\begin{aligned} \int \frac{1}{\sqrt{9 + 16x - 4x^2}}\,dx &= \int \frac{1}{\sqrt{25 - 4(x - 2)^2}}\,dx \\ &= \frac{1}{5}\int \frac{1}{\sqrt{1 - \frac{4}{25}(x - 2)^2}}\,dx \\ &= \frac{1}{2}\int \frac{1}{\sqrt{1 - u^2}}\,du \qquad \left(u = \frac{2(x - 2)}{5}\right) \\ &= \frac{1}{2}\sin^{-1} u + C = \frac{1}{2}\sin^{-1}\frac{2(x - 2)}{5} + C. \end{aligned}$$

An alternative approach is to make the trigonometric substitution

$$2(x-2) = 5\sin\theta, \qquad 2\,dx = 5\cos\theta\,d\theta$$

immediately after completing the square. This yields

$$\int \frac{1}{\sqrt{9+16x-4x^2}}\,dx = \int \frac{1}{\sqrt{25-4(x-2)^2}}\,dx = \int \frac{\frac{5}{2}\cos\theta}{\sqrt{25-25\sin^2\theta}}\,d\theta$$

$$= \frac{1}{2}\int 1\,d\theta = \frac{1}{2}\theta + C$$

$$= \frac{1}{2}\arcsin\frac{2(x-2)}{5} + C.$$

An integral involving a quadratic expression can sometimes be split into two simpler integrals. The next two examples illustrate this technique.

EXAMPLE 3 Find $\displaystyle\int \frac{2x+3}{9x^2+6x+5}\,dx.$

Solution Because $D(9x^2+6x+5) = 18x+6$, this would be a simpler integral if the numerator $2x+3$ were a constant multiple of $18x+6$. Our strategy is to write

$$2x+3 = A(18x+6) + B,$$

so we can split the given integral into a sum of two integrals, one of which has numerator $18x+6$ in its integrand. By matching coefficients, we find that $A = \frac{1}{9}$ and $B = \frac{7}{3}$. Hence

$$\int \frac{2x+3}{9x^2+6x+5}\,dx = \frac{1}{9}\int \frac{18x+6}{9x^2+6x+5}\,dx + \frac{7}{3}\int \frac{1}{9x^2+6x+5}\,dx.$$

The first integral on the right is a logarithm, and the second is given in Example 1. Thus

$$\int \frac{2x+3}{9x^2+6x+5}\,dx = \frac{1}{9}\ln(9x^2+6x+5) + \frac{7}{18}\tan^{-1}\left(\frac{3x+1}{2}\right) + C.$$

Alternatively, we could first complete the square in the denominator. The substitution $u = 3x+1$, $x = \frac{1}{3}(u-1)$, $dx = \frac{1}{3}\,du$ then gives

$$\int \frac{2x+3}{(3x+1)^2+4}\,dx = \int \frac{\frac{2}{3}(u-1)+3}{u^2+4}\cdot\frac{1}{3}\,du$$

$$= \frac{1}{9}\int \frac{2u}{u^2+4}\,du + \frac{7}{9}\int \frac{1}{u^2+4}\,du$$

$$= \frac{1}{9}\ln(u^2+4) + \frac{7}{18}\tan^{-1}\left(\frac{u}{2}\right) + C$$

$$= \frac{1}{9}\ln(9x^2+6x+5) + \frac{7}{18}\tan^{-1}\left(\frac{3x+1}{2}\right) + C.$$

EXAMPLE 4 Find $\displaystyle\int \frac{4-x}{(2x-x^2)^2}\,dx$ given $|x-1|<1$.

Solution Because $D_x(2x-x^2) = 2-2x$, we first write

$$\begin{aligned}\int \frac{4-x}{(2x-x^2)^2}\,dx &= \frac{1}{2}\int \frac{2-2x}{(2x-x^2)^2}\,dx + 3\int \frac{1}{(2x-x^2)^2}\,dx \\ &= -\frac{1}{2(2x-x^2)} + 3\int \frac{1}{(2x-x^2)^2}\,dx.\end{aligned}$$

To find the remaining integral, we complete the square:

$$2x - x^2 = -(x^2-2x) = -(x^2-2x+1-1) = 1-(x-1)^2.$$

This suggests the substitution $x - 1 = \sin\theta$, so that

$$dx = \cos\theta\,d\theta \quad \text{and} \quad 1-(x-1)^2 = \cos^2\theta.$$

This substitution gives

$$\begin{aligned}\int \frac{1}{(2x-x^2)^2}\,dx &= \int \frac{1}{[1-(x-1)^2]^2}\,dx \\ &= \int \frac{\cos\theta}{(\cos^2\theta)^2}\,d\theta = \int \sec^3\theta\,d\theta \\ &= \frac{1}{2}\sec\theta\tan\theta + \frac{1}{2}\int \sec\theta\,d\theta \\ &\qquad \text{(by Equation (6) in Section 9.4)} \\ &= \frac{1}{2}\sec\theta\tan\theta + \frac{1}{2}\ln|\sec\theta + \tan\theta| + C \\ &= \frac{1}{2}\left(\frac{x-1}{2x-x^2}\right) + \frac{1}{2}\ln\left|\frac{x}{\sqrt{2x-x^2}}\right| + C.\end{aligned}$$

9.6 The reference triangle for Example 4

Here we have read the values of $\sec\theta$ and $\tan\theta$ from the right triangle in Fig. 9.6.

When we combine all our results above, we finally obtain the answer:

$$\begin{aligned}\int \frac{4-x}{(2x-x^2)^2}\,dx &= -\frac{1}{2(2x-x^2)} + \frac{3(x-1)}{2(2x-x^2)} + \frac{3}{2}\ln\left|\frac{x}{\sqrt{2x-x^2}}\right| + C \\ &= \frac{3x-4}{2(2x-x^2)} + \frac{3}{4}\ln\left|\frac{x^2}{2x-x^2}\right| + C.\end{aligned}$$

The method of Example 4 can be used to evaluate a general integral having the form

$$\int \frac{Ax+B}{(ax^2+bx+c)^n}\,dx, \tag{1}$$

where n is a positive integer. By splitting such an integral into two simpler ones and by completing the square in the quadratic expression in the denominator, the problem of evaluating the integral in (1) can be reduced to that of computing

$$\int \frac{1}{(a^2 \pm u^2)^n}\,du. \tag{2}$$

If the sign in (2) is the plus sign, then the substitution $u = a\tan\theta$ transforms the integral into the form

$$\int \cos^m\theta\,d\theta$$

(see Problem 35). This integral can be handled by the methods of Section 9.3 or by using the reduction formula

$$\int \cos^k\theta\,d\theta = \frac{1}{k}\cos^{k-1}\theta\sin\theta + \frac{k-1}{k}\int \cos^{k-2}\theta\,d\theta$$

of Problem 46 in Section 9.4.

If the sign in (2) is the minus sign, then the substitution $u = a\sin\theta$ transforms the integral into the form

$$\int \sec^m\theta\,d\theta$$

(see Problem 36). This integral may be evaluated with the aid of the reduction formula

$$\int \sec^k\theta\,d\theta = \frac{1}{k-1}\sec^{k-2}\theta\tan\theta + \frac{k-2}{k-1}\int \sec^{k-2}\theta\,d\theta$$

(Equation (5) in Section 9.4).

9.7 PROBLEMS

Find the integrals in Problems 1–34.

1. $\displaystyle\int \frac{1}{x^2+4x+5}\,dx$

2. $\displaystyle\int \frac{2x+5}{x^2+4x+5}\,dx$

3. $\displaystyle\int \frac{5-3x}{x^2+4x+5}\,dx$

4. $\displaystyle\int \frac{x+1}{(x^2+4x+5)^2}\,dx$

5. $\displaystyle\int \frac{1}{\sqrt{3-2x-x^2}}\,dx$

6. $\displaystyle\int \frac{x+3}{\sqrt{3-2x-x^2}}\,dx$

7. $\displaystyle\int x\sqrt{3-2x-x^2}\,dx$

8. $\displaystyle\int \frac{1}{4x^2+4x-3}\,dx$

9. $\displaystyle\int \frac{3x+2}{4x^2+4x-3}\,dx$

10. $\displaystyle\int \sqrt{4x^2+4x-3}\,dx$

11. $\displaystyle\int \frac{1}{x^2+4x+13}\,dx$

12. $\displaystyle\int \frac{1}{\sqrt{2x-x^2}}\,dx$

13. $\displaystyle\int \frac{1}{3+2x-x^2}\,dx$

14. $\displaystyle\int x\sqrt{8+2x-x^2}\,dx$

15. $\displaystyle\int \frac{2x-5}{x^2+2x+2}\,dx$

16. $\displaystyle\int \frac{2x-1}{4x^2+4x-15}\,dx$

17. $\displaystyle\int \frac{x}{\sqrt{5+12x-9x^2}}\,dx$

18. $\displaystyle\int (3x-2)\sqrt{9x^2+12x+8}\,dx$

19. $\displaystyle\int (7-2x)\sqrt{9+16x-4x^2}\,dx$

20. $\displaystyle\int \frac{2x+3}{\sqrt{x^2+2x+5}}\,dx$

21. $\displaystyle\int \frac{x+4}{(6x-x^2)^{3/2}}\,dx$

22. $\displaystyle\int \frac{x-1}{(x^2+1)^2}\,dx$

23. $\displaystyle\int \frac{2x+3}{(4x^2+12x+13)^2}\,dx$

24. $\displaystyle\int \frac{x^3}{(1-x^2)^4}\,dx$

25. $\displaystyle\int \frac{3x-1}{x^2+x+1}\,dx$

26. $\displaystyle\int \frac{3x-1}{(x^2+x+1)^2}\,dx$

27. $\displaystyle\int \frac{1}{(x^2-4)^2}\,dx$

28. $\displaystyle\int (x-x^2)^{3/2}\,dx$

29. $\displaystyle\int \frac{x^2+1}{x^3+x^2+x}\,dx$

30. $\displaystyle\int \frac{x^2+2}{(x^2+1)^2}\,dx$

31. $\displaystyle\int \frac{2x^2+3}{x^4-2x^2+1}\,dx$

32. $\int \frac{x^2+4}{(x^2+1)^2(x^2+2)}\,dx$

33. $\int \frac{3x+1}{(x^2+2x+5)^2}\,dx$

34. $\int \frac{x^3-2x}{(x^2+2x+2)}\,dx$

35. Show that the substitution $u = a\tan\theta$ gives

$$\int \frac{1}{(a^2+u^2)^n}\,du = \frac{1}{a^{2n-1}}\int \cos^{2n-2}\theta\,d\theta.$$

36. Show that the substitution $u = a\sin\theta$ gives

$$\int \frac{1}{(a^2-u^2)^n}\,du = \frac{1}{a^{2n-1}}\int \sec^{2n-2}\theta\,d\theta.$$

37. A road is to be built joining the points (0, 0) and (3, 2), following the path of the circle with equation

$$(4x+4)^2 + (4y-19)^2 = 377.$$

Find the length of this road, with distance measured in miles.

38. Suppose that the road of Problem 37 costs $10/(1+x)$ million dollars per mile.

(a) Calculate its total cost.

(b) With the same cost per mile, calculate the total cost of a straight line road from (0, 0) to (3, 2). You should find that it is *more* expensive than the *longer* circular road!

In Problems 39–41, factor the denominator by first noting by inspection a root r of the denominator and then employing long division by $x - r$. Finally, use the method of partial fractions to aid in finding the indicated antiderivative.

39. $\int \frac{3x+2}{x^3+x^2-2}\,dx$

40. $\int \frac{1}{x^3+8}\,dx$

41. $\int \frac{x^4+2x^2}{x^3-1}\,dx$

42. (a) Find constants a and b such that

$$x^4 + 1 = (x^2+ax+1)(x^2+bx+1).$$

(b) Prove that

$$\int_0^1 \frac{x^2+1}{x^4+1}\,dx = \frac{\pi}{2\sqrt{2}}.$$

Hint: If u and v are positive numbers with $uv = 1$, then

$$\arctan u + \arctan v = \frac{\pi}{2}.$$

43. Factor $x^4 + x^2 + 1$ as in Problem 42. Then evaluate

$$\int \frac{x^3+2x}{x^4+x^2+1}\,dx.$$

*9.8 Rationalizing Substitutions

In the following example we make a substitution that eliminates a radical. This is an example of a *rationalizing* substitution.

EXAMPLE 1 Find $\int x^2\sqrt{x+1}\,dx$.

Solution Let $u = x + 1$. Then $x = u - 1$ and $dx = du$. This substitution then yields

$$\begin{aligned}\int x^2\sqrt{x+1}\,dx &= \int (u-1)^2\sqrt{u}\,du = \int (u^2-2u+1)\sqrt{u}\,du\\ &= \int (u^{5/2} - 2u^{3/2} + u^{1/2})\,du\\ &= \tfrac{2}{7}u^{7/2} - \tfrac{4}{5}u^{5/2} + \tfrac{2}{3}u^{3/2} + C\\ &= \tfrac{2}{7}(x+1)^{7/2} - \tfrac{4}{5}(x+1)^{5/2} + \tfrac{2}{3}(x+1)^{3/2} + C.\end{aligned}$$

The method of Example 1 succeeds with any integral of the form

$$\int p(x)\sqrt{ax+b}\,dx \tag{1}$$

where $p(x)$ is a polynomial. In fact, either the substitution $u = ax + b$ or the substitution $u = (ax+b)^{1/2}$ may be used. More generally, the substitution $u^n = f(x)$ may succeed when the integrand involves $[f(x)]^{1/n}$. It *always* succeeds in the case of an integral of the form

$$\int p(x) \sqrt[n]{\frac{ax+b}{cx+d}}\, dx, \tag{2}$$

where $p(x)$ is a polynomial. In particular, the substitution

$$u^n = \frac{ax+b}{cx+d}$$

converts (2) into the integral of a rational function of u (see Problem 21), which can then be integrated by the method of partial fractions.

EXAMPLE 2 Find $\displaystyle\int \frac{\sqrt{y}}{\sqrt{c-y}}\, dy.$

COMMENT This integral arises in the solution of the problem of determining the shape of a wire joining two fixed points, down which a bead slides in the least possible time from the upper point O to the lower point P. This is the famous *brachistochrone problem* (from the Greek *brachistos*, shortest, and *chronos*, time).

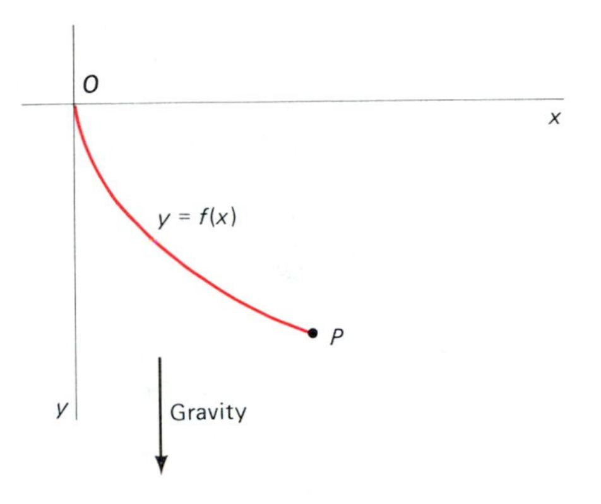

9.7 The brachistochrone problem: Find the shape of the graph of $y = f(x)$ to minimize the time it takes a bead to slide down the graph from O to P.

We take the y-axis pointing downward from O, as shown in Fig. 9.7. An analysis of the physics of the situation, together with an advanced technique from the calculus of variations, shows that the optimal shape $y = f(x)$ of the wire—the shape that gives the bead the minimum time of transit—satisfies the differential equation

$$\left[1 + \left(\frac{dy}{dx}\right)^2\right] y(x) = c \qquad (c \text{ constant}),$$

so that

$$\frac{dy}{dx} = \frac{\sqrt{c-y}}{\sqrt{y}}.$$

Thus

$$x = x(y) = \int \frac{\sqrt{y}}{\sqrt{c-y}}\, dy.$$

So we can solve the brachistochrone problem if we can perform the antidifferentiation indicated in Example 2, and thus obtain x as a function of y; this will give us the formula for the inverse of the desired function f.

Solution We use the substitution suggested above for an integral of the form in (2). If $u^2 = y/(c - y)$, then

$$y = \frac{cu^2}{1+u^2} \quad \text{and} \quad dy = \frac{2cu}{(1+u^2)^2}\, du.$$

Hence

$$x = \int \frac{2cu^2}{(1+u^2)^2}\, du = \int \frac{2c \tan^2\theta \sec^2\theta}{(1+\tan^2\theta)^2}\, d\theta \qquad (u = \tan\theta)$$

$$= 2c \int \frac{\tan^2\theta}{\sec^2\theta}\, d\theta = 2c \int \sin^2\theta\, d\theta.$$

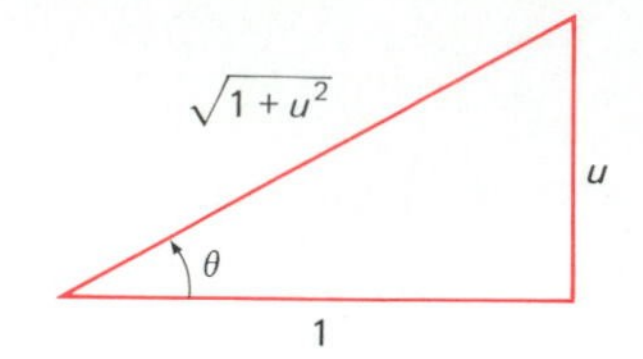

9.8 The reference triangle for the substitution $u = \tan\theta$

And so

$$x = c(\theta - \sin\theta\cos\theta). \tag{3}$$

The constant of integration is zero because $y = u = \theta = 0$ when $x = 0$.
Now, with the aid of Fig. 9.8, we retrace our steps.

$$x = x(y) = \int \frac{\sqrt{y}}{\sqrt{c - y}}\,dy = c(\theta - \sin\theta\cos\theta) = c\tan^{-1}u - \frac{cu}{u^2 + 1},$$

and therefore

$$x = c\tan^{-1}\left(\frac{\sqrt{y}}{\sqrt{c - y}}\right) - \sqrt{cy - y^2}.$$

This *is* a solution, but its form tells us little about the actual shape of the curve of quickest descent. In Section 13.1 we will see what this curve looks like. For reference then, we record the fact that

$$y = \frac{cu^2}{1 + u^2} = \frac{c\tan^2\theta}{1 + \tan^2\theta} = c\sin^2\theta. \tag{4}$$

Equations (3) and (4) express x and y in terms of the substitution variable θ. This is a *parametric* description of the curve of quickest descent.

In the case of an integral involving roots of the variable x, we may try the substitution $x = u^n$, with n chosen to be the smallest integer that will rationalize all the roots.

EXAMPLE 3 Find $\displaystyle\int \frac{1}{x^{1/2} + x^{1/3}}\,dx.$

Solution The least common multiple of 2 and 3 (from the exponents $\frac{1}{2}$ and $\frac{1}{3}$) is 6, so we try the substitution $x = u^6$, $dx = 6u^5\,du$. This gives

$$\begin{aligned}
\int \frac{1}{x^{1/2} + x^{1/3}}\,dx &= \int \frac{6u^5}{u^3 + u^2}\,du = 6\int \frac{u^3}{u + 1}\,du \\
&= 6\int \left(u^2 - u + 1 - \frac{1}{u + 1}\right)du \\
&\qquad\qquad \text{(by long division)} \\
&= 2u^3 - 3u^2 + 6u - 6\ln|u + 1| + C \\
&= 2x^{1/2} - 3x^{1/3} + 6x^{1/6} - 6\ln|1 + x^{1/6}| + C.
\end{aligned}$$

RATIONAL FUNCTIONS OF sin θ AND cos θ

In Section 9.3 we saw how to evaluate certain integrals of the form

$$\int R(\sin\theta, \cos\theta)\,d\theta, \tag{5}$$

where $R(\sin\theta, \cos\theta)$ is a rational function of $\sin\theta$ and $\cos\theta$. That is, $R(\sin\theta, \cos\theta)$ is a quotient of polynomials in the two "variables" $\sin\theta$ and

$\cos\theta$. The special substitution

$$u = \tan\frac{\theta}{2} \tag{6}$$

can be used to find *any* integral of the form in (5).

In order to carry out the substitution indicated in (6), we must express $\sin\theta$, $\cos\theta$, and $d\theta$ in terms of u and du. Note first that

$$\theta = 2\tan^{-1}u, \qquad \text{so that} \qquad d\theta = \frac{2du}{1+u^2}. \tag{7}$$

From the triangle in Fig. 9.9 we see that

$$\sin\frac{\theta}{2} = \frac{u}{\sqrt{1+u^2}}, \qquad \cos\frac{\theta}{2} = \frac{1}{\sqrt{1+u^2}}.$$

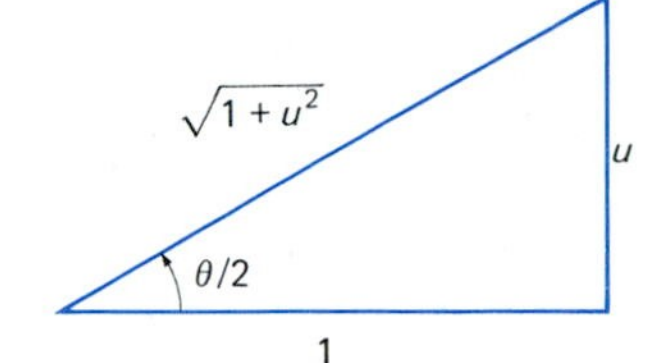

9.9 The spectral rationalizing substitution $u = \tan(\theta/2)$

Hence

$$\sin\theta = 2\sin\frac{\theta}{2}\cos\frac{\theta}{2} = \frac{2u}{1+u^2}, \tag{8}$$

$$\cos\theta = \cos^2\frac{\theta}{2} - \sin^2\frac{\theta}{2} = \frac{1-u^2}{1+u^2}. \tag{9}$$

It should be clear that these substitutions will convert the integral in (5) into an integral of a rational function of u. The latter can then be evaluated by the methods of Section 9.5.

In Section 8.2 we used elaborate algebraic manipulations to show that

$$\int \sec\theta\, d\theta = \ln|\sec\theta + \tan\theta| + C.$$

We can now use the substitution $u = \tan(\theta/2)$ to integrate $\sec\theta$ more systematically.

EXAMPLE 4 Find $\int \sec\theta\, d\theta$.

Solution Equations (7) through (9) give

$$\begin{aligned}\int \sec\theta\, d\theta &= \int \frac{1}{\cos\theta}\, d\theta = \int \frac{1+u^2}{1-u^2}\cdot\frac{2}{1+u^2}\, du \\ &= \int \frac{2}{1-u^2}\, du = \int \left(\frac{1}{1-u} + \frac{1}{1+u}\right) du \\ &= \ln\left|\frac{1+u}{1-u}\right| + C = \ln\left|\frac{1+\tan(\theta/2)}{1-\tan(\theta/2)}\right| + C.\end{aligned}$$

Elementary trigonometric identities may now be used to show that this result is the same as our earlier one (see Problem 28).

EXAMPLE 5 Suppose that $a > b > 0$. Find

$$\int \frac{1}{a + b\cos\theta}\, d\theta.$$

Solution We make the substitutions of Equations (7) through (9) and thus find that

$$\begin{aligned}
\int \frac{1}{a + b\cos\theta}\,d\theta &= \int \frac{1}{a + b\dfrac{1-u^2}{1+u^2}} \cdot \frac{2}{1+u^2}\,du \\
&= \int \frac{2}{(a+b) + (a-b)u^2}\,du \\
&= \frac{2}{a-b}\int \frac{1}{c^2 + u^2}\,du \quad \left(\text{where } c = \left(\frac{a+b}{a-b}\right)^{1/2}\right) \\
&= \frac{2/c}{a-b}\tan^{-1}\left(\frac{u}{c}\right) + C.
\end{aligned}$$

Thus

$$\int \frac{1}{a + b\cos\theta}\,d\theta = \frac{2}{\sqrt{a^2 - b^2}}\tan^{-1}\left(\sqrt{\frac{a-b}{a+b}}\tan\frac{\theta}{2}\right) + C. \tag{10}$$

9.8 PROBLEMS

Find the integrals in Problems 1–20.

1. $\int x^3\sqrt{3x-2}\,dx$
2. $\int x^3\sqrt[3]{x^2+1}\,dx$
3. $\int \frac{1}{1+\sqrt{x}}\,dx$
4. $\int \frac{1}{x^{1/2} - x^{1/4}}\,dx$
5. $\int \frac{x^3}{(x^2-1)^{4/3}}\,dx$
6. $\int x^2(x-1)^{3/2}\,dx$
7. $\int \frac{1-\sqrt{x}}{1+\sqrt[4]{x}}\,dx$
8. $\int \frac{\sqrt[3]{x}}{1+\sqrt{x}}\,dx$
9. $\int \frac{x^5}{\sqrt{x^3+1}}\,dx$
10. $\int x^7\sqrt[3]{x^4+1}\,dx$
11. $\int \frac{1}{1+x^{2/3}}\,dx$
12. $\int \sqrt{\frac{1+x}{1-x}}\,dx$
13. $\int \frac{1}{1+\sqrt{x+4}}\,dx$
14. $\int \frac{\sqrt[3]{x+1}}{x}\,dx$
15. $\int \frac{1}{1+\sin\theta}\,d\theta$
16. $\int \frac{1}{(1-\cos\theta)^2}\,d\theta$
17. $\int \frac{1}{\sin\theta + \cos\theta}\,d\theta$
18. $\int \frac{1}{\sin\phi + \cos\phi + 2}\,d\phi$
19. $\int \frac{\sin\theta}{2+\cos\theta}\,d\theta$
20. $\int \frac{\sin\theta - \cos\theta}{\sin\theta + \cos\theta}\,d\theta$

21. Prove that if $p(x)$ is a polynomial, then the substitution $u^n = (ax+b)/(cx+d)$ transforms the integral

$$\int p(x)\left(\frac{ax+b}{cx+d}\right)^{1/n}dx$$

into the integral of a rational function of u.

22. Rework Example 2 using the trigonometric substitution $y = c\sin^2\theta$ instead of the algebraic substitution used in the text.

23. Find the area bounded by the loop of the curve

$$y^2 = x^2(1-x), \qquad 0 \leqq x \leqq 1.$$

24. Find the area bounded by the loop of the curve

$$y^2 = \frac{1-x}{1+x}x^2, \qquad 0 \leqq x \leqq 1.$$

25. Find the length of the arc $y = 2\sqrt{x}$, $0 \leqq x \leqq 1$. (*Suggestion:* Substitute $x = \tan^2\theta$ in the arc length integral.)

26. Use the result of Example 5 to find

$$\int \frac{A + B\cos\theta}{a + b\cos\theta}\,d\theta \qquad (a > b > 0).$$

Do this by dividing the denominator into the numerator—think of the denominator as a polynomial in the variable $\cos\theta$.

27. Show that

$$\int \frac{1}{a + b\cos\theta}\,d\theta = \frac{1}{\sqrt{b^2 - a^2}} \times \ln\left|\frac{\sqrt{b-a}\tan(\theta/2) + \sqrt{b+a}}{\sqrt{b-a}\tan(\theta/2) - \sqrt{b+a}}\right| + C$$

if $0 < a < b$.

28. Use the trigonometric identity

$$\tan\frac{\theta}{2} = \left(\frac{1 - \cos\theta}{1 + \cos\theta}\right)^{1/2}$$

to derive our earlier formula

$$\int \sec\theta\,d\theta = \ln|\sec\theta + \tan\theta| + C$$

from the solution to Example 4.

29. (a) Use the method of Example 4 to show that

$$\int \csc\theta\,d\theta = \ln\left|\tan\frac{\theta}{2}\right| + C.$$

(b) Use trigonometric identities to derive the formula

$$\int \csc\theta\,d\theta = \ln|\csc\theta - \cot\theta| + C$$

from part (a).

30. Substitute $x = u^2$ to find

$$\int \sqrt{1 + \sqrt{x}}\,dx.$$

31. Substitute $u^2 = 1 + e^{2x}$ to find

$$\int \sqrt{1 + e^{2x}}\,dx.$$

32. Find the area A of the surface obtained by revolving the curve $y = \frac{2}{3}x^{3/2}$, $3 \leqq x \leqq 8$, around the x-axis. (*Suggestion:* Substitute $x = u^2$ in the surface area integral.) (*Answer:* $A \approx 732.39$.)

CHAPTER 9 SUMMARY

When you confront the problem of evaluating a particular integral, you must first decide which of the several methods of this chapter to try. There are only two *general* methods of integration—integration by substitution (Section 9.2) and integration by parts (Section 9.4). These are the analogues for integration of the chain rule and product rule, respectively, for differentiation.

Look first at the given integral to see if you can spot a substitution that transforms it into an elementary or familiar integral or one likely to be found in an integral table. If the integrand is an unfamiliar product of two functions, one of which is easily differentiated and the other easily integrated, then an attempt to integrate by parts is indicated.

Beyond these two quite general methods, the chapter deals with a number of *special* methods. In the case of a conspicuously trigonometric integral, the simple "split-off" methods of Section 9.3 may succeed. Recall that reduction formulas (like Equation (5) and Problems 45 and 46 in Section 9.4) are available for integrating an integral power of a single trigonometric function. As a last resort, change to sines and cosines and remember that any rational function of $\sin\theta$ and $\cos\theta$ can be integrated using the substitution $u = \tan(\theta/2)$ of Section 9.8. The result of this substitution is a rational function of u.

Any integral of a rational function (quotient of two polynomials) can be evaluated by the method of partial fractions (Section 9.5). If the degree of the numerator is not less than that of the denominator—that is, if the rational function is not proper—first use long division to express it as the sum of a polynomial (easily integrated) and a proper rational fraction. Then decompose the latter into partial fractions. The partial fractions corresponding to linear factors are easily integrated, and those corresponding to irreducible quadratic factors can be integrated by completing the square and making (if necessary) a trigonometric substitution. As we explained in Section 9.7, the trigonometric integrals that result can always be evaluated.

In the case of an integral involving

$$\sqrt{ax^2 + bx + c},$$

you should first complete the square (Section 9.7) and then rationalize the integral by making an appropriate trigonometric substitution (Section 9.6). This will leave you with a trigonometric integral. Finally, an integral involving $[f(x)]^{1/n}$ is sometimes rationalized by the special substitution $u^n = f(x)$ of Section 9.8.

CHAPTER 9 MISCELLANEOUS PROBLEMS

Evaluate the integrals in Problems 1–100.

1. $\displaystyle\int \frac{1}{\sqrt{x}(1+x)}\,dx$

2. $\displaystyle\int \frac{\sec^2 t}{1+\tan t}\,dt$

3. $\displaystyle\int \sin x \sec x\,dx$

4. $\displaystyle\int \frac{\csc x \cot x}{1+\csc^2 x}\,dx$

5. $\displaystyle\int \frac{\tan\theta}{\cos^2\theta}\,d\theta$

6. $\displaystyle\int \csc^4 x\,dx$

7. $\displaystyle\int x\tan^2 x\,dx$

8. $\displaystyle\int x^2\cos^2 x\,dx$

9. $\displaystyle\int x^5\sqrt{2-x^3}\,dx$

10. $\displaystyle\int \frac{1}{\sqrt{x^2+4}}\,dx$

11. $\displaystyle\int \frac{x^2}{\sqrt{25+x^2}}\,dx$

12. $\displaystyle\int (\cos x)\sqrt{4-\sin^2 x}\,dx$

13. $\displaystyle\int \frac{1}{x^2-x+1}\,dx$

14. $\displaystyle\int \sqrt{x^2+x+1}\,dx$

15. $\displaystyle\int \frac{5x+31}{3x^2-4x+11}\,dx$

16. $\displaystyle\int \frac{x^4+1}{x^2+2}\,dx$

17. $\displaystyle\int \frac{1}{5+4\cos\theta}\,d\theta$

18. $\displaystyle\int \frac{\sqrt{x}}{1+x}\,dx$

19. $\displaystyle\int \frac{\cos x}{\sqrt{4-\sin^2 x}}\,dx$

20. $\displaystyle\int \frac{\cos 2x}{\cos x}\,dx$

21. $\displaystyle\int \frac{\tan x}{\ln(\cos x)}\,dx$

22. $\displaystyle\int \frac{x^7}{\sqrt{1-x^4}}\,dx$

23. $\displaystyle\int \ln(1+x)\,dx$

24. $\displaystyle\int x\sec^{-1}x\,dx$

25. $\displaystyle\int \sqrt{x^2+9}\,dx$

26. $\displaystyle\int \frac{x^2}{\sqrt{4-x^2}}\,dx$

27. $\displaystyle\int \sqrt{2x-x^2}\,dx$

28. $\displaystyle\int \frac{4x-2}{x^3-x}\,dx$

29. $\displaystyle\int \frac{x^4}{x^2-2}\,dx$

30. $\displaystyle\int \frac{\sec x\tan x}{\sec x+\sec^2 x}\,dx$

31. $\displaystyle\int \frac{x}{(x^2+2x+2)^2}\,dx$

32. $\displaystyle\int \frac{x^{1/3}}{x^{1/2}+x^{1/4}}\,dx$

33. $\displaystyle\int \frac{1}{1+\cos 2\theta}\,d\theta$

34. $\displaystyle\int \frac{\sec x}{\tan x}\,dx$

35. $\displaystyle\int \sec^3 x\tan^3 x\,dx$

36. $\displaystyle\int x^2\tan^{-1}x\,dx$

37. $\displaystyle\int x(\ln x)^3\,dx$

38. $\displaystyle\int \frac{1}{x\sqrt{1+x^2}}\,dx$

39. $\displaystyle\int e^x\sqrt{1+e^{2x}}\,dx$

40. $\displaystyle\int \frac{x}{\sqrt{4x-x^2}}\,dx$

41. $\displaystyle\int \frac{1}{x^3\sqrt{x^2-9}}\,dx$

42. $\displaystyle\int \frac{x}{(7x+1)^{17}}\,dx$

43. $\displaystyle\int \frac{4x^2+x+1}{4x^3+x}$

44. $\displaystyle\int \frac{4x^3-x+1}{x^3+1}\,dx$

45. $\displaystyle\int \tan^2 x\sec x\,dx$

46. $\displaystyle\int \frac{x^2+2x+2}{(x+1)^3}\,dx$

47. $\displaystyle\int \frac{x^4+2x+2}{x^5+x^4}\,dx$

48. $\displaystyle\int \frac{8x^2-4x+7}{(x^2+1)(4x+1)}\,dx$

49. $\displaystyle\int \frac{3x^5-x^4+2x^3-12x^2-2x+1}{(x^3-1)^2}\,dx$

50. $\displaystyle\int \frac{x}{x^4+4x^2+8}\,dx$

51. $\displaystyle\int \frac{1}{4+5\cos\theta}\,d\theta$

52. $\displaystyle\int \frac{(1 + x^{2/3})^{3/2}}{x^{1/3}}\,dx$

53. $\displaystyle\int \frac{(\arcsin x)^2}{\sqrt{1 - x^2}}\,dx$

54. $\displaystyle\int \frac{1}{x^{3/2}(1 + x^{1/3})}\,dx$

55. $\displaystyle\int \tan^3 z\,dz$

56. $\displaystyle\int \sin^2\omega \cos^4\omega\,d\omega$

57. $\displaystyle\int \frac{xe^{x^2}}{1 + e^{2x^2}}\,dx$

58. $\displaystyle\int \frac{\cos^3 x}{\sqrt{\sin x}}\,dx$

59. $\displaystyle\int x^3 \exp(-x^2)\,dx$

60. $\displaystyle\int \sin\sqrt{x}\,dx$

61. $\displaystyle\int \frac{\arcsin x}{x^2}\,dx$

62. $\displaystyle\int \sqrt{x^2 - 9}\,dx$

63. $\displaystyle\int x^2\sqrt{1 - x^2}\,dx$

64. $\displaystyle\int x\sqrt{2x - x^2}\,dx$

65. $\displaystyle\int \frac{x - 2}{4x^2 + 4x + 1}\,dx$

66. $\displaystyle\int \frac{2x^2 - 5x - 1}{x^3 - 2x^2 - x + 2}\,dx$

67. $\displaystyle\int \frac{e^{2x}}{e^{2x} - 1}\,dx$

68. $\displaystyle\int \frac{\cos x}{\sin^2 x - 3\sin x + 2}\,dx$

69. $\displaystyle\int \frac{2x^3 + 3x^2 + 4}{(x + 1)^4}\,dx$

70. $\displaystyle\int \frac{\sec^2 x}{\tan^2 x + 2\tan x + 2}\,dx$

71. $\displaystyle\int \frac{x^3 + x^2 + 2x + 1}{x^4 + 2x^2 + 1}\,dx$

72. $\displaystyle\int \frac{3 + \cos\theta}{2 - \cos\theta}\,d\theta$

73. $\displaystyle\int x^5\sqrt{x^3 - 1}\,dx$

74. $\displaystyle\int \frac{1}{2 + 2\cos\theta + \sin\theta}\,d\theta$

75. $\displaystyle\int \frac{\sqrt{1 + \sin x}}{\sec x}\,dx$

76. $\displaystyle\int \frac{1}{x^{2/3}(1 + x^{2/3})}\,dx$

77. $\displaystyle\int \frac{\sin x}{\sin 2x}\,dx$

78. $\displaystyle\int \sqrt{1 + \cos t}\,dt$

79. $\displaystyle\int \sqrt{1 + \sin t}\,dt$

80. $\displaystyle\int \frac{\sec^2 t}{1 - \tan^2 t}\,dt$

81. $\displaystyle\int \ln(x^2 + x + 1)\,dx$

82. $\displaystyle\int e^x \sin^{-1}(e^x)\,dx$

83. $\displaystyle\int \frac{\arctan x}{x^2}\,dx$

84. $\displaystyle\int \frac{x^2}{\sqrt{x^2 - 25}}\,dx$

85. $\displaystyle\int \frac{x^3}{(x^2 + 1)^2}\,dx$

86. $\displaystyle\int \frac{1}{x\sqrt{6x - x^2}}\,dx$

87. $\displaystyle\int \frac{3x + 2}{(x^2 + 4)^{3/2}}\,dx$

88. $\displaystyle\int x^{3/2}\ln x\,dx$

89. $\displaystyle\int \frac{(1 + \sin^2 x)^{1/2}}{\sec x \csc x}\,dx$

90. $\displaystyle\int \frac{\exp(\sqrt{\sin x})}{(\sec x)\sqrt{\sin x}}\,dx$

91. $\displaystyle\int xe^x \sin x\,dx$

92. $\displaystyle\int x^2 \exp(x^{3/2})\,dx$

93. $\displaystyle\int \frac{\arctan x}{(x - 1)^3}\,dx$

94. $\displaystyle\int \ln(1 + \sqrt{x})\,dx$

95. $\displaystyle\int \frac{2x + 3}{\sqrt{3 + 6x - 9x^2}}\,dx$

96. $\displaystyle\int \frac{1}{2 + 2\sin\theta + \cos\theta}\,d\theta$

97. $\displaystyle\int \frac{\sin^3\theta}{\cos\theta - 1}\,d\theta$

98. $\displaystyle\int x^{3/2}\tan^{-1}(x^{1/2})\,dx$

99. $\displaystyle\int \sec^{-1}\sqrt{x}\,dx$

100. $\displaystyle\int x\left(\frac{1 - x^2}{1 + x^2}\right)^{1/2} dx$

101. Find the area of the surface generated by revolving the curve $y = \cosh x$, $0 \leqq x \leqq 1$, around the x-axis.

102. Find the length of the curve $y = e^{-x}$, $0 \leqq x \leqq 1$.

103. (a) Find the area A_b of the surface generated by revolving the curve $y = e^{-x}$, $0 \leqq x \leqq b$, around the x-axis.
(b) Find $\lim_{b\to\infty} A_b$.

104. (a) Find the area A_b of the surface generated by revolving the curve $y = 1/x$, $1 \leqq x \leqq b$, around the x-axis.
(b) Find $\lim_{b\to\infty} A_b$.

105. Find the area of the surface generated by revolving the curve

$$y = \sqrt{x^2 - 1}, \qquad 1 \leqq x \leqq 2,$$

around the x-axis.

106. (a) Derive the reduction formula

$$\int x^m(\ln x)^n\,dx = \frac{1}{m + 1}x^{m+1}(\ln x)^n - \frac{n}{m + 1}\int x^m(\ln x)^{n-1}\,dx.$$

(b) Evaluate $\displaystyle\int_1^e x^3(\ln x)^3\,dx$.

107. Derive the reduction formula

$$\int \sin^m x \cos^n x\,dx = -\frac{1}{m + n}\sin^{m-1}x\cos^{n+1}x + \frac{m - 1}{m + n}\int \sin^{m-2}x\cos^n x\,dx.$$

108. Use the reduction formula of Problem 107 here and Problem 46 in Section 9.4 to evaluate

$$\int_0^{\pi/2} \sin^6 x \cos^5 x \, dx.$$

109. Find the area bounded by the curve $y^2 = x^5(2 - x)$, $0 \leqq x \leqq 2$. (*Suggestion:* Substitute $x = 2\sin^2\theta$, and then use the result of Problem 50 in Section 9.4.)

110. Show that

$$\int_0^1 \frac{t^4(1-t)^4}{1+t^2}\, dt = \frac{22}{7} - \pi.$$

111. Evaluate

$$\int_0^1 t^4(1-t)^4 \, dt,$$

then apply the result of Problem 110 to conclude that

$$\tfrac{22}{7} - \tfrac{1}{630} < \pi < \tfrac{22}{7} - \tfrac{1}{1260}.$$

Thus $3.1412 < \pi < 3.1421$.

112. Find the length of the curve $y = \frac{4}{5}x^{5/4}$, $0 \leqq x \leqq 1$.

113. Find the length of the curve $y = \frac{4}{3}x^{3/4}$, $1 \leqq x \leqq 4$.

114. An initially empty water tank is shaped like a cone with vertical axis, vertex at the bottom, 9 ft deep, and top radius 4.5 ft. Beginning at time $t = 0$, water is poured into this tank at 50 ft^3/min. Meanwhile, water leaks out a hole at the bottom at the rate of $10\sqrt{y}$ cubic feet per minute where y is the depth of the water in the tank. (This is consistent with Torricelli's law.) How long does it take to fill the tank?

115. (a) Evaluate $\displaystyle\int \frac{1}{1 + e^x + e^{-x}}\, dx.$

(b) Explain why your substitution in (a) suffices to integrate any rational function of e^x.

116. (a) The equation $x^3 + x + 1 = 0$ has one real root r. Use Newton's method to find it, accurate to at least two places.

(b) Use long division to find the irreducible quadratic factor of $x^3 + x + 1$.

(c) Use the factorization of part (b) to evaluate

$$\int_0^1 \frac{1}{x^3 + x + 1}\, dx.$$

117. Evaluate $\displaystyle\int \frac{1}{1 + e^x}\, dx.$

118. The integral

$$\int \frac{1 + 2x^2}{x^5(1 + x^2)^3}\, dx = \int \frac{x + 2x^3}{(x^4 + x^2)^3}\, dx$$

would require solving eleven equations in eleven unknowns if the method of partial fractions were used to evaluate it. Use the substitution $u = x^4 + x^2$ to evaluate it much more simply.

119. Evaluate

$$\int \sqrt{\tan\theta}\, d\theta.$$

Suggestion: First substitute $u = \tan\theta$. Then substitute $u = x^2$. Finally use the method of partial fractions; see Problem 42 in Section 9.7.)

chapter ten

Polar Coordinates and Conic Sections

10.1 Analytic Geometry and the Conic Sections

Plane analytic geometry, the main topic of this chapter, is the use of algebra and calculus to study the properties of curves in the plane. The ancient Greeks used deductive reasoning and the methods of axiomatic Euclidean geometry to study lines, circles, and the **conic sections** (parabolas, ellipses, and hyperbolas). The properties of conic sections have played an important role in diverse scientific applications since the 17th century, when Kepler discovered—and Newton explained—the fact that the orbits of planets and other bodies in the solar system are conic sections.

The French mathematicians Descartes and Fermat, working almost independently of one another, initiated analytic geometry in 1637. The central idea of analytic geometry is the correspondence between an equation $F(x, y) = 0$ and the set or **locus** (typically, a curve) of all those points (x, y) in the plane with coordinates that satisfy this equation.

The idea is this: Given a geometric locus or curve, its properties can be derived algebraically or analytically from its defining equation $F(x, y) = 0$. For example, suppose that the equation of a given curve turns out to be the linear equation

$$Ax + By = C, \tag{1}$$

where A, B, and C are constants with $B \neq 0$. This equation may be written in the form

$$y = mx + b \tag{2}$$

where $m = -A/B$ and $b = C/B$. But (2) is the slope-intercept equation (Section 1.2) of the straight line with slope m and y-intercept b. Hence the given curve is this straight line. In the following example we use this approach to show that a certain geometrically described locus is a particular straight line.

EXAMPLE 1 Prove that the set of all points equally distant from the points (1, 1) and (5, 3) is the perpendicular bisector of the line segment joining these two points.

Solution The typical point $P(x, y)$ is equally distant from (1, 1) and (5, 3) if and only if

$$(x-1)^2 + (y-1)^2 = (x-5)^2 + (y-3)^2;$$

$$(x^2 - 2x + 1) + (y^2 - 2y + 1) = (x^2 - 10x + 25) + (y^2 - 6y + 9);$$

$$2x + y = 8. \tag{3}$$

Thus the given locus is the straight line in (3) with slope -2. The straight line through (1, 1) and (5, 3) has equation

$$y - 1 = \tfrac{1}{2}(x - 1) \tag{4}$$

and slope $\frac{1}{2}$. Because the product of the slopes of these two lines is -1, it follows (from Theorem 2 in Section 1.2) that they are perpendicular lines. If we solve Equations (3) and (4) simultaneously, we find that the intersection of these lines is, indeed, the midpoint (3, 2) of the given line segment. Thus the locus described is the perpendicular bisector of this line segment.

The circle with center (h, k) and radius r is described geometrically as the set of all points $P(x, y)$ whose distance from (h, k) is r. The distance formula then gives

$$(x - h)^2 + (y - k)^2 = r^2 \tag{5}$$

as the equation of this circle. In particular, if $h = k = 0$, then (5) takes the simple form

$$x^2 + y^2 = r^2. \tag{6}$$

We can see directly from this equation, without further reference to the definition, that a circle centered at the origin has the following symmetry properties:

- *Symmetry about the x-axis:* The equation of the curve is unaltered when y is replaced by $-y$.
- *Symmetry about the y-axis:* The equation of the curve is unaltered when x is replaced by $-x$.
- *Symmetry with respect to the origin:* The equation of the curve is unaltered when x is replaced by $-x$ and y is replaced by $-y$.
- *Symmetry about the* 45°*-line* $y = x$*:* The equation is unaltered when x and y are interchanged.

The relationship between Equations (5) and (6) is an illustration of the *translation principle* stated informally in Section 1.3. Imagine a translation (or "slide") of the plane that moves the point (x, y) to the new position

$(x + h, y + k)$. Under such a translation, a curve C would be moved to a new curve. The equation of the new curve is easily obtained from the old equation—we simply replace x by $x - h$ and y by $y - k$. Conversely, we can recognize a translated circle from its equation: Any equation of the form

$$x^2 + y^2 + Ax + By + C = 0 \tag{7}$$

can be rewritten in the form

$$(x - h)^2 + (y - k)^2 = p$$

by completing the square as in Example 2 of Section 1.3. Thus the graph of Equation (7) is either a circle (if $p > 0$), a single point (if $p = 0$), or empty (if $p < 0$). We use this approach in the following example to discover that the locus described is a particular circle.

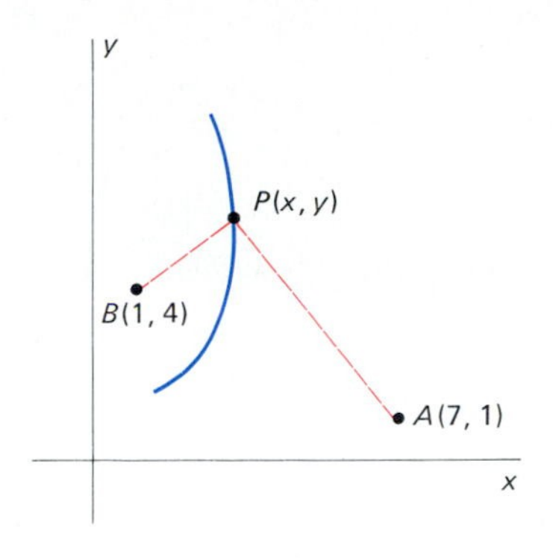

10.1 The locus of Example 2

EXAMPLE 2 Determine the locus of a point $P(x, y)$ if its distance $|AP|$ from $A(7, 1)$ is twice its distance $|BP|$ from $B(1, 4)$.

Solution The points A, B, and P appear in Fig. 10.1, along with a curve through P representing the given locus. Because

$$|AP|^2 = 4|BP|^2 \qquad (\text{from } |AP| = 2|BP|),$$

we get the equation

$$(x - 7)^2 + (y - 1)^2 = 4[(x - 1)^2 + (y - 4)^2].$$

Hence

$$3x^2 + 3y^2 + 6x - 30y + 18 = 0;$$

$$x^2 + y^2 + 2x - 10y = -6;$$

$$(x + 1)^2 + (y - 5)^2 = 20.$$

Thus the locus is a circle with center $(-1, 5)$ and radius $r = \sqrt{20} = 2\sqrt{5}$.

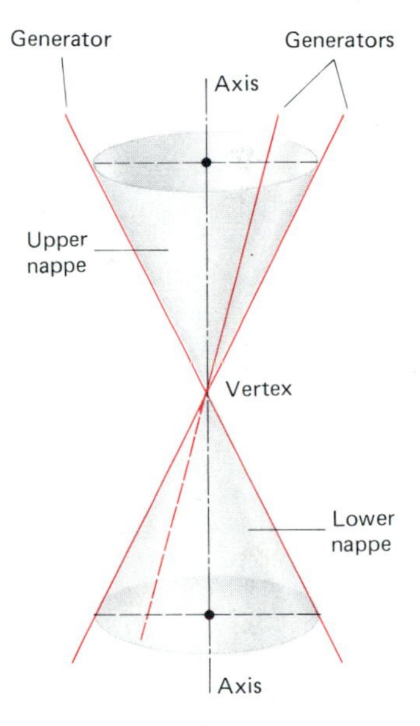

10.2 A cone of two nappes

CONIC SECTIONS

The phrase *conic sections* stems from the fact that these are the curves in which a plane intersects a cone. The cone used is a right circular cone with two *nappes* extending infinitely far in both directions, as in Fig. 10.2. There are three types of conic sections, as illustrated in Fig. 10.3. If the cutting

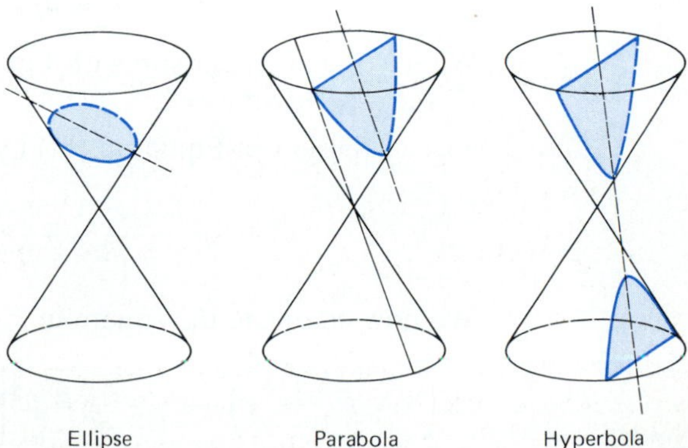

10.3 The conic sections

plane is parallel to some generator of the cone, then the curve of intersection is a *parabola.* Otherwise, it is either a single closed curve—an *ellipse*–or a *hyperbola* with two **branches.**

In Section 14.6 we shall use the methods of three-dimensional analytic geometry to show that if an appropriate xy-coordinate system is set up in the intersecting plane, then the equations of the three conic sections take the following forms:

$$\text{Parabola:} \qquad y^2 = kx; \tag{8}$$

$$\text{Ellipse:} \qquad \frac{x^2}{a^2} + \frac{y^2}{b^2} = 1; \tag{9}$$

$$\text{Hyperbola:} \qquad \frac{x^2}{a^2} - \frac{y^2}{b^2} = 1. \tag{10}$$

In Sections 10.4 through 10.6 we will discuss these conic sections on the basis of definitions that are "two-dimensional"—they will not require the three-dimensional setting of a cone and an intersecting plane. The following example illustrates one such approach to the conic sections.

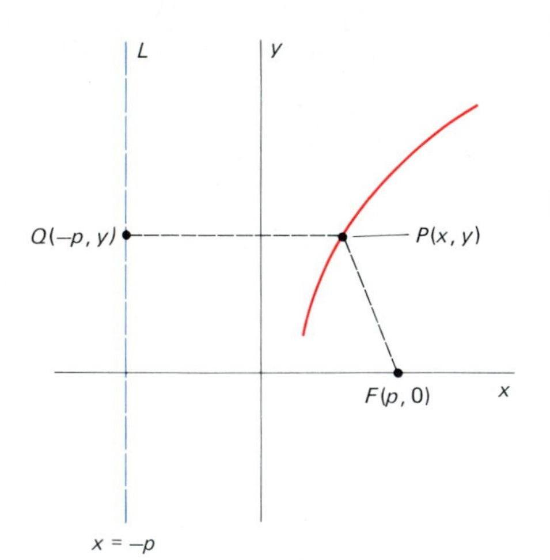

10.4 A figure for Example 3

EXAMPLE 3 Let e be a given positive number (*not* to be confused with the natural logarithm base). Determine the locus of a point $P(x, y)$ if its distance from the fixed point $F(p, 0)$ is e times its distance from the vertical line L with equation $x = -p$. (See Fig. 10.4.)

Solution Let PQ be the perpendicular from P to the line L. Then the condition

$$|PF| = e|PQ|$$

takes the analytic form

$$\sqrt{(x-p)^2 + y^2} = e|x - (-p)|.$$

That is,

$$(x^2 - 2px + p^2) + y^2 = e^2(x^2 + 2px + p^2),$$

so

$$x^2(1 - e^2) - 2p(1 + e^2)x + y^2 = -p^2(1 - e^2). \tag{11}$$

- *Case 1:* $e = 1$. Then (11) reduces to

$$y^2 = 4px. \tag{12}$$

We see upon comparison with Equation (8) that the locus of P is a *parabola* if $e = 1$.

- *Case 2:* $e < 1$. Division of Equation (11) by $1 - e^2$ yields

$$x^2 - 2p\frac{1 + e^2}{1 - e^2}x + \frac{y^2}{1 - e^2} = -p^2.$$

We now complete the square in x. The result is

$$\left(x - p\frac{1 + e^2}{1 - e^2}\right)^2 + \frac{y^2}{1 - e^2} = p^2\left[\left(\frac{1 + e^2}{1 - e^2}\right)^2 - 1\right] = a^2.$$

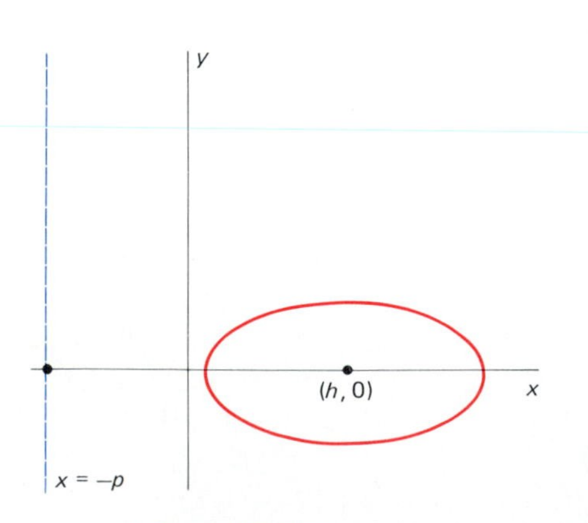

10.5 An ellipse: $e < 1$.

This equation has the form

$$\frac{(x-h)^2}{a^2} + \frac{y^2}{b^2} = 1, \tag{13}$$

where

$$h = +p\,\frac{1+e^2}{1-e^2} \quad \text{and} \quad b^2 = a^2(1-e^2). \tag{14}$$

When we compare Equations (9) and (13), we see that if $e < 1$ then the locus of P is an *ellipse* with (0, 0) translated to $(h, 0)$, as illustrated in Fig. 10.5.

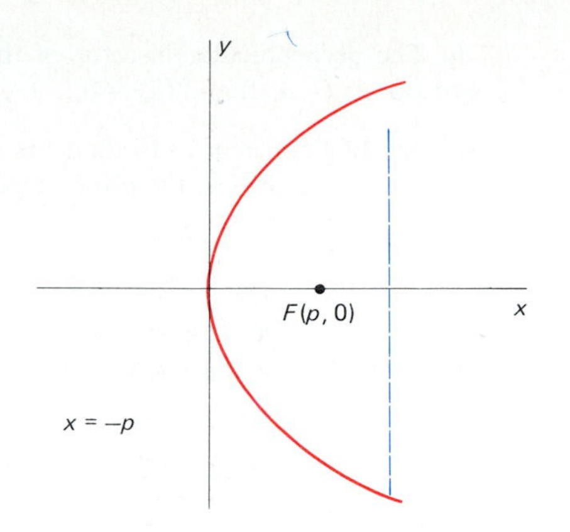

10.6 A parabola: $e = 1$.

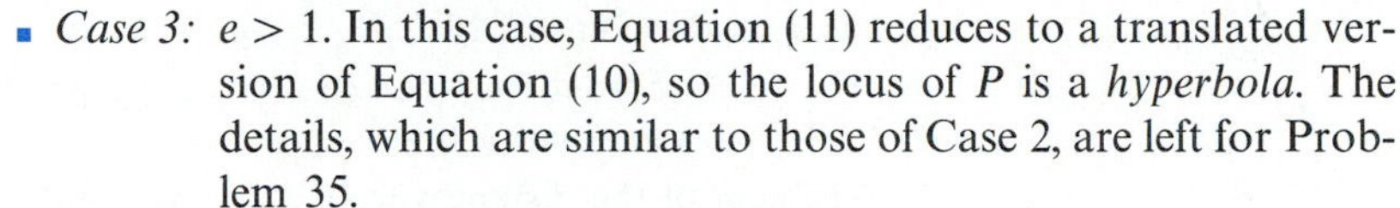

- *Case 3:* $e > 1$. In this case, Equation (11) reduces to a translated version of Equation (10), so the locus of P is a *hyperbola*. The details, which are similar to those of Case 2, are left for Problem 35.

Thus the locus in Example 3 is a *parabola* if $e = 1$, an *ellipse* if $e < 1$, and a *hyperbola* if $e > 1$. The number e is called the *eccentricity* of the conic section. The point $F(p, 0)$ is commonly called its **focus** in the parabolic case. Figure 10.6 shows the parabola of Case 1, and Fig. 10.7 illustrates the hyperbola of Case 3.

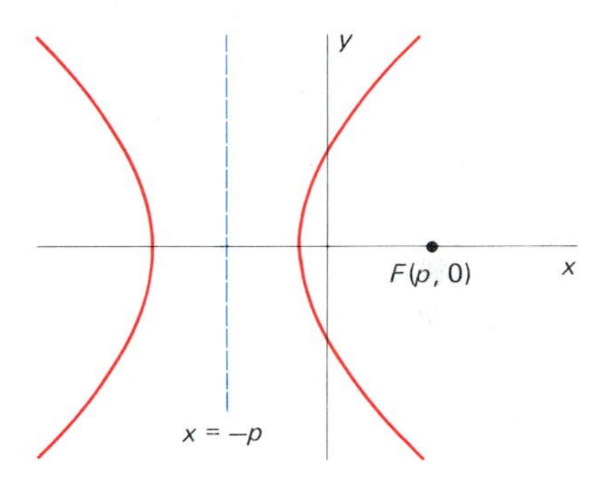

10.7 A hyperbola: $e > 1$.

If we begin with Equations (8) through (10), we can derive the general characteristics of the three conic sections shown in Figs. 10.5 through 10.7. For example, in the case of the parabola with the equation in (8), we see that the curve passes through the origin, that $x \geqq 0$ at each of its points, that $y \to \pm\infty$ as $x \to \infty$, and that the graph is symmetric about the x-axis (because the curve is unchanged when y is replaced by $-y$).

In the case of the ellipse with the equation in (9), we see that the graph must be symmetric about both axes. At each point (x, y) of the graph, we must have $|x| \leqq a$ and $|y| \leqq b$. The graph intersects the axes at the four points $(\pm a, 0)$ and $(0, \pm b)$.

Finally, the hyperbola with the equation in (10), or its alternative form

$$y = \pm\frac{b}{a}\sqrt{x^2 - a^2},$$

is symmetric about both axes. It meets the x-axis at the two points $(\pm a, 0)$, has one branch consisting of points with $x \geqq a$, and has another branch where $x \leqq -a$. And $|y| \to \infty$ as $|x| \to \infty$.

10.1 PROBLEMS

In each of Problems 1–6, write an equation of the specified straight line.

1. The line through the point $(1, -2)$ that is parallel to the line with equation $x + 2y = 5$.

2. The line through the point $(-3, 2)$ that is perpendicular to the line with equation $3x - 4y = 7$.

3. The line that is tangent to the circle $x^2 + y^2 = 25$ at the point $(3, -4)$.

4. The line that is tangent to the curve $y^2 = x + 3$ at the point $(6, -3)$.

5. The line that is perpendicular to the curve $x^2 + 2y^2 = 6$ at the point $(2, -1)$.

6. The perpendicular bisector of the line segment with end points $(-3, 2)$ and $(5, -4)$.

In each of Problems 7–16, find the center and radius of the circle described in the given equation.

7. $x^2 + 2x + y^2 = 4$
8. $x^2 + y^2 - 4y = 5$
9. $x^2 + y^2 - 4x + 6y = 3$
10. $x^2 + y^2 + 8x - 6y = 0$
11. $4x^2 + 4y^2 - 4x = 3$
12. $4x^2 + 4y^2 + 12y = 7$
13. $2x^2 + 2y^2 - 2x + 6y = 13$
14. $9x^2 + 9y^2 - 12x = 5$
15. $9x^2 + 9y^2 + 6x - 24y = 19$
16. $36x^2 + 36y^2 - 48x - 108y = 47$

In each of Problems 17–20, show that the graph of the given equation consists either of a single point or of no points.

17. $x^2 + y^2 - 6x - 4y + 13 = 0$
18. $2x^2 + 2y^2 + 6x + 2y + 5 = 0$
19. $x^2 + y^2 - 6x - 10y + 84 = 0$
20. $9x^2 + 9y^2 - 6x - 6y + 11 = 0$

In each of Problems 21–24, write the equation of the specified circle.

21. The circle with center $(-1, -2)$ that passes through the point $(2, 3)$.
22. The circle with center $(2, -2)$ that is tangent to the line $y = x + 4$.
23. The circle with center $(6, 6)$ that is tangent to the line $y = 2x - 4$.
24. The circle that passes through the points $(4, 6)$, $(-2, -2)$, and $(5, -1)$.

In each of Problems 25–30, derive the equation of the set of all points $P(x, y)$ satisfying the given condition. Then sketch the graph of the equation.

25. The point $P(x, y)$ is equally distant from the two points $(3, 2)$ and $(7, 4)$.
26. The distance from P to the point $(-2, 1)$ is one-half its distance from $(4, -2)$.
27. The point P is three times as far from the point $(-3, 2)$ as it is from the point $(5, 10)$.
28. The distance from P to the line $x = -3$ is equal to its distance from the point $(3, 0)$.
29. The sum of the distances from P to the points $(4, 0)$ and $(-4, 0)$ is 10.
30. The sum of the distances from P to the points $(0, 3)$ and $(0, -3)$ is 10.
31. Find all lines through the point $(2, 1)$ that are tangent to the parabola $y = x^2$.
32. Find all lines through the point $(-1, 2)$ that are normal to the parabola $y = x^2$.
33. Find all lines that are normal to the curve $xy = 4$ and are simultaneously parallel to the line $y = 4x$.
34. Find all lines that are tangent to the curve $y = x^3$ and are also parallel to the line $3x - y = 5$.
35. Suppose that $e > 1$. Show that Equation (11) of this section can be written in the form

$$\frac{(x-h)^2}{a^2} - \frac{y^2}{a^2} = 1.$$

thus showing that its graph is a hyperbola. Find a, b, and h in terms of p and e.

10.2 Polar Coordinates

A familiar way to locate a point in the plane is by specifying its rectangular coordinates (x, y)—that is, by giving its abscissa x and ordinate y relative to given perpendicular axes. In some problems it is more convenient to locate a point by means of its *polar coordinates*. The polar coordinates of a point give its position relative to a fixed reference point O (the **pole**) and a given ray beginning at O (the **polar axis**).

For convenience, we begin with a given xy-coordinate system, then take the origin as the pole and the nonnegative x-axis as the polar axis. Given the pole O and the polar axis, the point P with **polar coordinates** r and θ, written as the ordered pair (r, θ), is located as follows. First find the terminal side of the angle θ, given in radians, where θ is measured counterclockwise (if $\theta > 0$) from the positive x-axis (the polar axis) as its initial side. If $r \geqq 0$, then P is on this terminal side at the distance r from the origin. If $r < 0$, then P lies on the ray opposite the terminal side at the distance

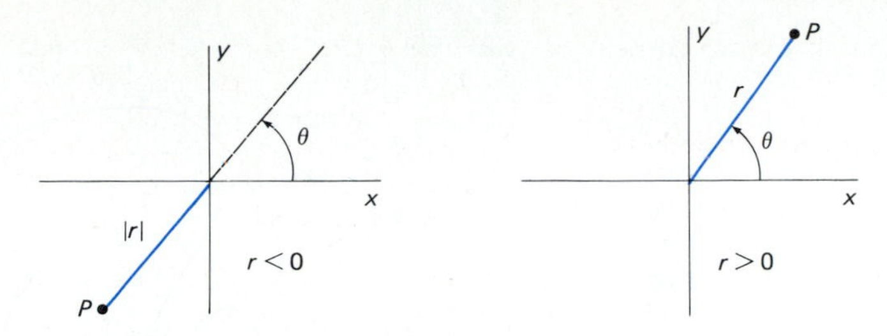

10.8 The difference between the two cases $r > 0$ and $r < 0$.

$|r| = -r > 0$ from the pole. The **radial coordinate** r can be described as the *directed* distance of P from the pole along the terminal side of the angle θ. Thus, if r is positive, the point P lies in the same quadrant as θ, while if r is negative, then P lies in the opposite quadrant. If $r = 0$, the angle θ does not matter; the polar coordinates $(0, \theta)$ represent the origin whatever the **angular coordinate** θ might be. Of course, the origin or pole is the only point for which $r = 0$. The cases $r > 0$ and $r < 0$ are illustrated separately in Fig. 10.8.

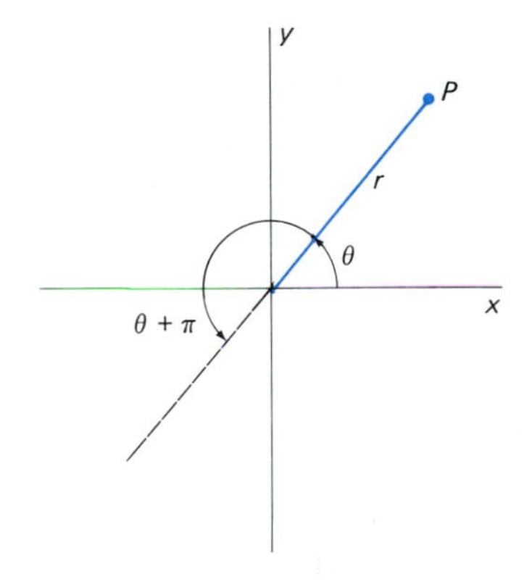

10.9 The polar coordinates (r, θ) and $(-r, \theta + \pi)$ represent the same point P.

Polar coordinates differ from rectangular coordinates in that any point has more than one representation in polar coordinates. For example, the polar coordinates (r, θ) and $(-r, \theta + \pi)$ represent the same point P, as shown in Fig. 10.9. More generally, this same point P has the polar coordinates $(r, \theta + n\pi)$ for any even integer n and also the coordinates $(-r, \theta + n\pi)$ for any odd integer n. Thus the polar coordinate pairs

$$\left(2, \frac{\pi}{3}\right), \quad \left(-2, \frac{4\pi}{3}\right), \quad \left(2, \frac{7\pi}{3}\right), \quad \text{and} \quad \left(-2, -\frac{2\pi}{3}\right)$$

all represent the point P shown in Fig. 10.10. The rectangular coordinates of P are $(1, \sqrt{3})$.

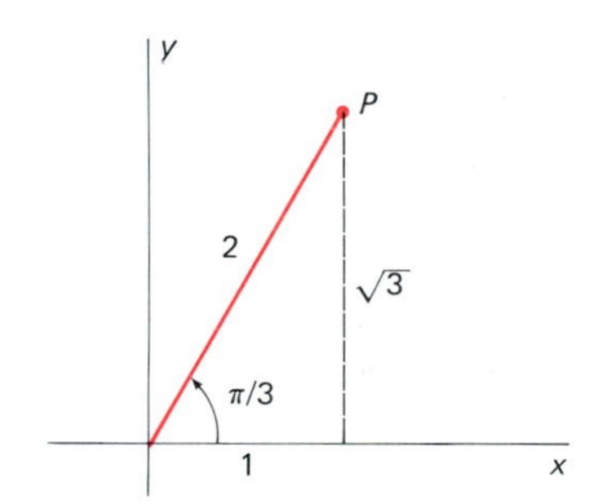

10.10 The point P can be described in many ways in polar coordinates.

Some curves have simpler equations in polar coordinates than in rectangular coordinates; this is an important reason for the usefulness of polar coordinates. The **graph** of an equation in the polar coordinate variables r and θ is the set of all those points P such that P has some pair of polar coordinates (r, θ) satisfying the given equation. The graph of an equation $r = f(\theta)$ can be constructed by computing a table of values of r against θ, then plotting the corresponding points (r, θ) on polar coordinate graph paper.

EXAMPLE 1 Construct the graph of the equation $r = 2 \sin \theta$.

Solution Figure 10.11 shows a table of values of r as a function of θ. The corresponding points are plotted in Fig. 10.12, using the rays at multiples of $\pi/6$ and the circles (centered at the pole) of radii 1 and 2 to locate these points. A visual inspection of the smooth curve connecting these points suggests that it is a circle of radius 1. Let us assume for the moment that this is so. Note then that the point $P(r, \theta)$ moves *once around this circle counterclockwise* as θ increases from 0 to π, then moves around the circle *a second time* as θ increases from π to 2π. This is because the negative values of r for θ between π and 2π give—in this example—the same geometric points as the positive values of r found when θ is between 0 and π. (Why?)

θ	r
0	0.00
$\pi/6$	1.00
$\pi/3$	1.73
$\pi/2$	2.00
$2\pi/3$	1.73
$5\pi/6$	1.00
π	0.00
$7\pi/6$	−1.00
$4\pi/3$	−1.73
$3\pi/2$	−2.00
$5\pi/3$	−1.73
$11\pi/6$	−1.00
2π	0.00
	(data rounded)

10.11 Values of $r = 2 \sin \theta$

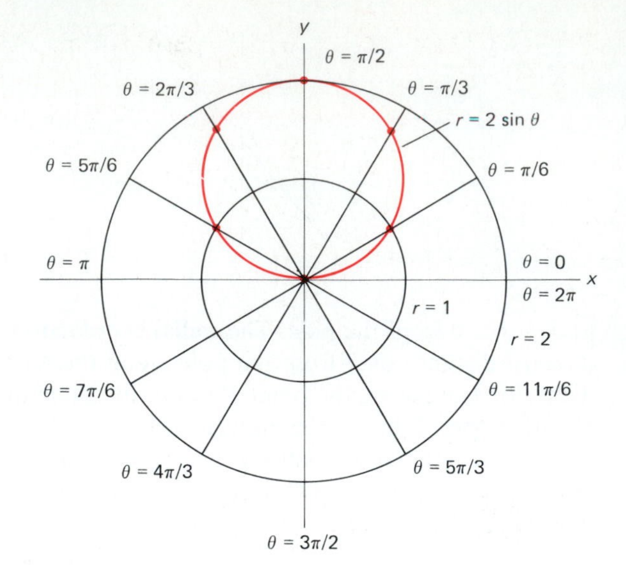

10.12 The graph of the polar equation $r = 2 \sin \theta$

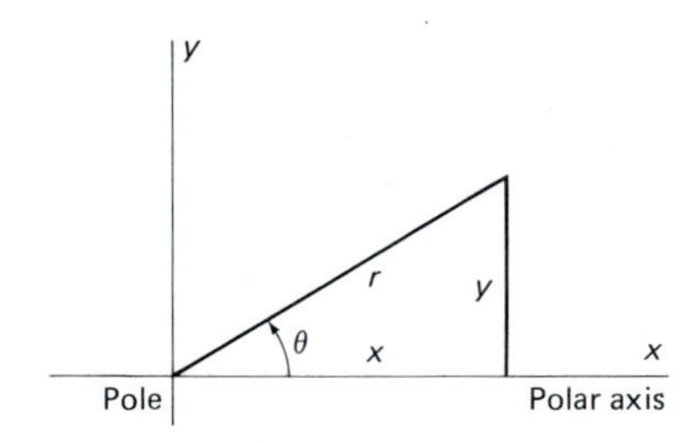

10.13 Read Equations (1) and (2)—conversions between polar and rectangular coordinates—from this figure.

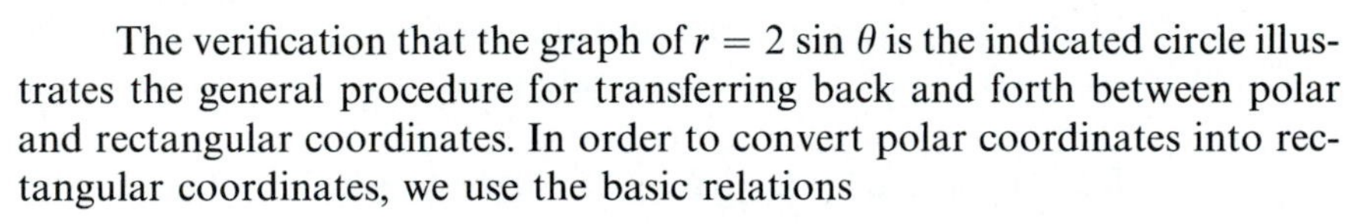

The verification that the graph of $r = 2 \sin \theta$ is the indicated circle illustrates the general procedure for transferring back and forth between polar and rectangular coordinates. In order to convert polar coordinates into rectangular coordinates, we use the basic relations

$$x = r \cos \theta, \qquad y = r \sin \theta \tag{1}$$

that we read from the right triangle of Fig. 10.13. In the opposite direction we have

$$r^2 = x^2 + y^2, \qquad \tan \theta = \frac{y}{x} \quad \text{if } x \neq 0. \tag{2}$$

Some care is required in the correct choice of θ here. If $x > 0$, then (x, y) lies in either the first or fourth quadrant, so that $-\pi/2 < \theta < \pi/2$, which is the range of the inverse tangent function. But if $x < 0$, then (x, y) lies in the second or third quadrant. In this case an appropriate choice for the angle is $\theta = \pi + \tan^{-1}(y/x)$. In any event, the signs of x and y in (1) with $r > 0$ indicate the quadrant in which θ lies.

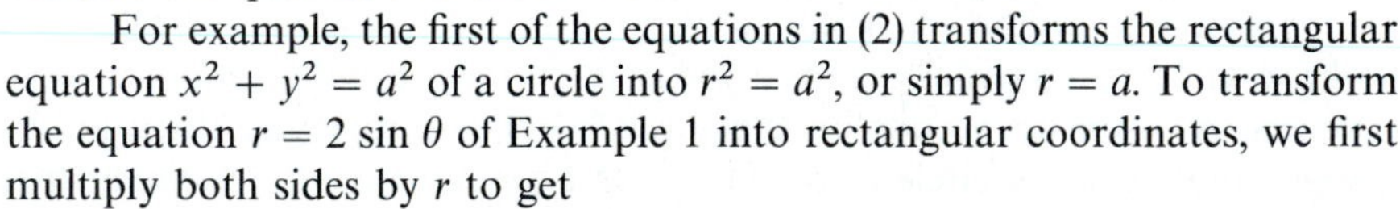

For example, the first of the equations in (2) transforms the rectangular equation $x^2 + y^2 = a^2$ of a circle into $r^2 = a^2$, or simply $r = a$. To transform the equation $r = 2 \sin \theta$ of Example 1 into rectangular coordinates, we first multiply both sides by r to get

$$r^2 = 2r \sin \theta.$$

Equations (1) and (2) now give

$$x^2 + y^2 = 2y.$$

Finally, after we complete the square, we have

$$x^2 + (y - 1)^2 = 1,$$

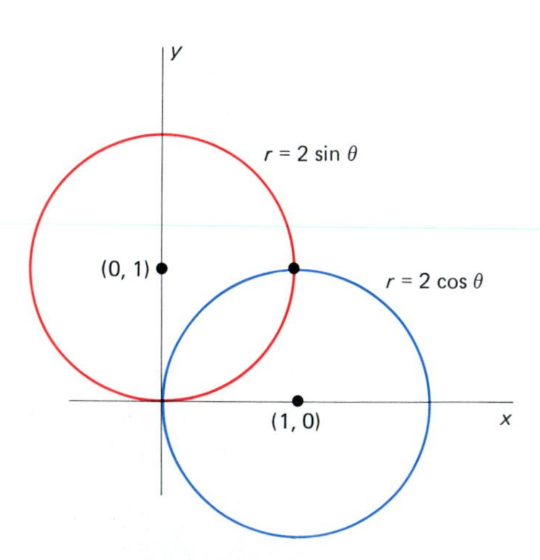

10.14 The graphs of the circles with equations in (3)

the rectangular coordinates equation of a circle with center (0, 1) and radius 1. More generally, the graphs of the equations

$$r = 2a \sin \theta \quad \text{and} \quad r = 2a \cos \theta \tag{3}$$

are circles of radius a centered, respectively, at the points $(0, a)$ and $(a, 0)$, as illustrated (with $a = 1$) in Fig. 10.14.

Substitution of the equations in (1) transforms the rectangular equation $ax + by = c$ of a straight line into

$$ar \cos \theta + br \sin \theta = c.$$

Let us take $a = 1$ and $b = 0$. Then we see that the polar coordinates equation of the vertical line $x = c$ is $r = c \sec \theta$, as we can deduce directly from Fig. 10.15.

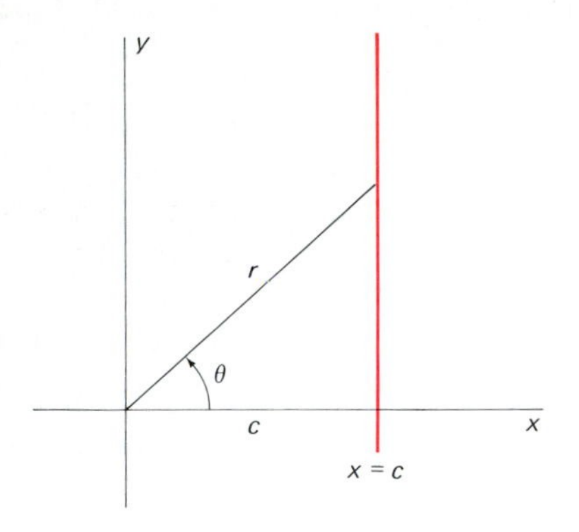

10.15 Finding the polar equation of the vertical line $x = c$

EXAMPLE 2 Sketch the graph of the polar equation

$$r = 2 + 2 \sin \theta.$$

Solution If we scan the second column of the table in Fig. 10.11, mentally adding 2 to each entry for r, we see that

- r increases from 2 to 4 as θ increases from 0 to $\pi/2$;
- r decreases from 4 to 2 as θ increases from $\pi/2$ to π;
- r decreases from 2 to 0 as θ increases from π to $3\pi/2$;
- r increases from 0 to 2 as θ increases from $3\pi/2$ to 2π.

This information tells us that the graph resembles the curve shown in Fig. 10.16. This heart-shaped graph is called a **cardioid.** The graphs of the equations

$$r = a(1 \pm \sin \theta) \quad \text{and} \quad r = a(1 \pm \cos \theta)$$

are all cardioids, differing only in size (determined by a), axis of symmetry (horizontal or vertical), and the direction in which the cusp at the pole points.

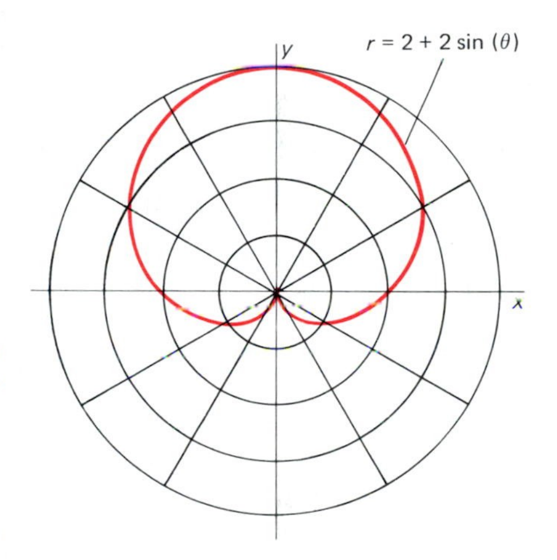

10.16 A cardioid

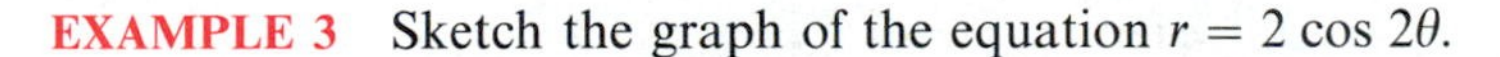

EXAMPLE 3 Sketch the graph of the equation $r = 2 \cos 2\theta$.

Solution Rather than constructing a table of values of r as a function of θ and then plotting points, let us reason qualitatively. Note first that $r = 0$ if θ is an odd multiple of $\pi/4$. Also, $|r| = 2$ if θ is an even multiple of $\pi/4$. We begin by plotting all these points.

Then think of what happens as θ increases from one odd multiple of $\pi/4$ to the next. For instance, if θ goes from $\pi/4$ to $3\pi/4$, then 2θ increases from $\pi/2$ to $3\pi/2$. So $r = 2 \cos 2\theta$ decreases from 0 to -2, then increases back to 0. The graph for such values of θ forms a loop, starting at the origin and lying between the rays $\theta = 5\pi/4$ and $\theta = 7\pi/4$. Something similar happens between any two consecutive odd multiples of $\pi/4$, so the entire graph consists of four loops—see Fig. 10.17. The arrows in the figure indicate the direction in which the point $P(r, \theta)$ moves along the curve as θ increases.

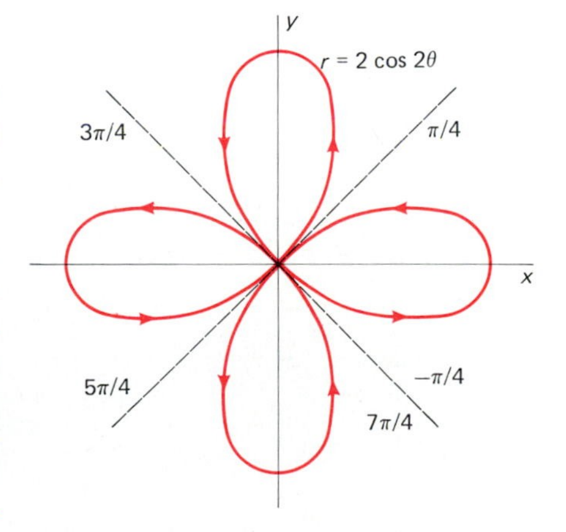

10.17 A four-leaved rose

The curve of Example 3 is called a *four-leaved rose*. The equations $r = a \cos n\theta$ and $r = a \sin n\theta$ represent "roses" with $2n$ "leaves" or loops if n is even and $n \geqq 2$, and with n loops if n is odd and $n \geqq 3$.

The four-leaved rose exhibits several types of symmetry. The following are some *sufficient* conditions for symmetry in polar coordinates.

- *For symmetry about the x-axis:* The equation is unaltered when θ is replaced by $-\theta$.
- *For symmetry about the y-axis:* The equation is unaltered when θ is replaced by $\pi - \theta$.
- *For symmetry with respect to the origin:* The equation is unchanged when r is replaced by $-r$.

Because $\cos 2\theta = \cos(-2\theta) = \cos 2(\pi - \theta)$, the equation $r = 2 \cos 2\theta$ of the four-leaved rose satisfies the first two symmetry conditions, and therefore its graph is symmetric about both the x-axis and the y-axis. Thus it is also symmetric about the origin. Nevertheless, this equation does *not* satisfy the third condition, the one for symmetry about the origin. This illustrates that while the symmetry conditions given here are *sufficient* for the symmetries described, they are not *necessary* conditions.

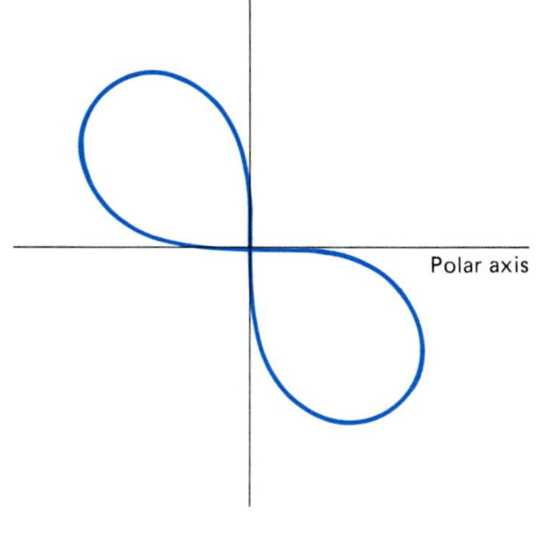

10.18 The lemiscate $r^2 = -4 \sin 2\theta$

EXAMPLE 4 Figure 10.18 shows the lemniscate with equation

$$r^2 = -4 \sin 2\theta.$$

To see why it has loops only in the second and fourth quadrants, we examine a table of values of $-4 \sin 2\theta$.

θ	2θ	$-4 \sin 2\theta$
$0 < \theta < \frac{\pi}{2}$	$0 < 2\theta < \pi$	Negative
$\frac{\pi}{2} < \theta < \pi$	$\pi < 2\theta < 2\pi$	Positive
$\pi < \theta < \frac{3\pi}{2}$	$2\pi < 2\theta < 3\pi$	Negative
$\frac{3\pi}{2} < \theta < 2\pi$	$3\pi < 2\theta < 4\pi$	Positive

When θ lies in the first or the third quadrant, the quantity $-4 \sin 2\theta$ is negative, so that equation $r^2 = -4 \sin 2\theta$ cannot be satisfied for any real value of r.

Example 3 illustrates a peculiarity of graphs of polar equations, one that is caused by the fact that a single point has multiple representations in polar coordinates. For the point with polar coordinates $(2, \pi/2)$ clearly lies on the four-leaved rose, but these coordinates do *not* satisfy the equation $r = 2 \cos 2\theta$. This means that a point may have one pair of polar coordinates that satisfy a given equation and others that do not. Hence we must be careful to understand this: The graph of a polar equation consists of all those points having *at least one* polar coordinate representation satisfying the given equation.

Another result of the multiplicity of polar coordinates is that the simultaneous solution of two polar equations does not always give all the points of intersection of their graphs. For instance, consider the circles $r = 2\sin\theta$ and $r = 2\cos\theta$ shown in Fig. 10.14. The origin is clearly a point of intersection of these two circles. Its polar representation $(0, \pi)$ satisfies the equation $r = 2\sin\theta$, and its representation $(0, \pi/2)$ satisfies the other equation $r = 2\cos\theta$. But the origin has no *single* polar representation that satisfies both equations simultaneously! If we think of θ as increasing uniformly with time, then the corresponding moving points on the two circles pass through the origin at different times. Hence the origin cannot be discovered as a point of intersection of the two circles by solving their equations simultaneously.

As a consequence of the phenomenon illustrated by this example, the only way we can be certain of finding *all* points of intersection of two curves in polar coordinates is by graphing both curves.

EXAMPLE 5 Find all points of intersection of the graphs of the equation $r = 1 + \sin\theta$ and $r^2 = 4\sin\theta$.

Solution The graph of $r = 1 + \sin\theta$ is a scaled-down version of the cardioid of Example 2. In Problem 42 we ask you to show that the graph of $r^2 = 4\sin\theta$ is the figure-eight curve shown with the cardioid in Fig. 10.19. The figure shows four points of intersection, marked A, B, C, and O. Can we find all four points with algebra?

Given the two equations, we begin by eliminating r. Because

$$(1 + \sin\theta)^2 = r^2 = 4\sin\theta,$$

it follows that

$$\sin^2\theta - 2\sin\theta + 1 = 0, \qquad (\sin\theta - 1)^2 = 0,$$

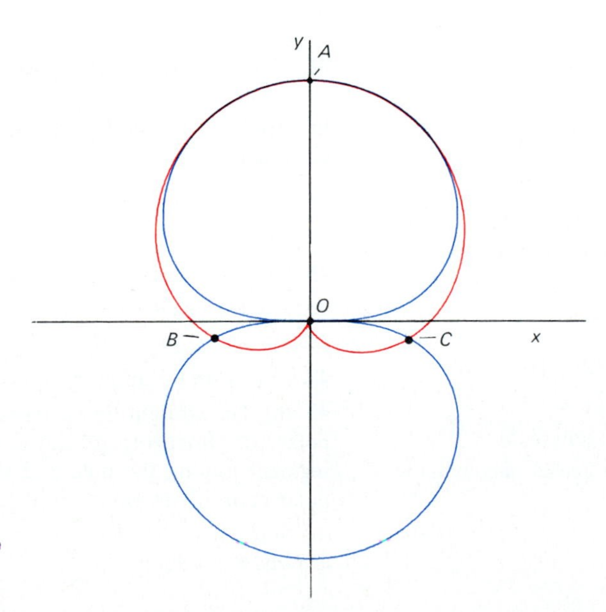

10.19 The cardioid $r = 1 + \sin\theta$ and the figure eight $r^2 = 4\sin\theta$ meet in four points.

and thus that $\sin \theta = 1$. So θ must be an angle of the form $\pi/2 \pm 2n\pi$, where n is an integer. All points on the cardioid and all points on the figure-eight curve are produced by letting θ increase from 0 to 2π, so $\theta = \pi/2$ will produce all the solutions that we can possibly obtain by the method. The only such point is $A(2, \pi/2)$, and the other three points of intersection are detected only when the two equations are graphed.

10.2 PROBLEMS

1. Plot the points with the given polar coordinates and then find the rectangular coordinates of each.

(a) $(1, \pi/4)$ (b) $(-2, 2\pi/3)$
(c) $(1, -\pi/3)$ (d) $(3, 3\pi/2)$
(e) $(2, 9\pi/4)$ (f) $(-2, -7\pi/6)$
(g) $(2, 5\pi/6)$

2. Find two polar coordinate representations, one with $r > 0$ and the other with $r < 0$, for the points with these rectangular coordinates.

(a) $(-1, -1)$ (b) $(\sqrt{3}, -1)$
(c) $(2, 2)$ (d) $(-1, \sqrt{3})$
(e) $(\sqrt{2}, -\sqrt{2})$ (f) $(-3, \sqrt{3})$

In each of Problems 3–10, express the given rectangular equations in polar form.

3. $x = 4$ **4.** $y = 6$
5. $x = 3y$ **6.** $x^2 + y^2 = 25$
7. $xy = 1$ **8.** $x^2 - y^2 = 1$
9. $y = x^2$ **10.** $x + y = 4$

In each of Problems 11–18, express the given polar equation in rectangular coordinates.

11. $r = 3$ **12.** $\theta = 3\pi/4$
13. $r = -5 \cos \theta$ **14.** $r = \sin 2\theta$
15. $r = 1 - \cos 2\theta$ **16.** $r = 2 + \sin \theta$
17. $r = 3 \sec \theta$ **18.** $r^2 = \cos 2\theta$

For each of the curves described in Problems 19–28, write equations in both rectangular and polar coordinates.

19. The vertical line through (2, 0)
20. The horizontal line through (1, 3)
21. The line with slope -1 through $(2, -1)$
22. The line with slope $+1$ through (4, 2)
23. The line through the points (1, 3) and (3, 5)
24. The circle with center (3, 0) that passes through the origin
25. The circle with center $(0, -4)$ that passes through the origin
26. The circle with center (3, 4) and radius 5
27. The circle with center (1, 1) that passes through the origin
28. The circle with center $(5, -2)$ that passes through the point (1, 1)

Sketch the graphs of the polar equations in Problems 29–42. Indicate any symmetry about either axis or the origin.

29. $r = 2 \cos \theta$
30. $r = 2 \sin \theta + 2 \cos \theta$ (circle)
31. $r = 1 + \cos \theta$ (cardioid)
32. $r = 1 - \sin \theta$ (cardioid)
33. $r = 2 + 4 \cos \theta$ (limaçon)
34. $r = 4 + 2 \cos \theta$ (limaçon)
35. $r^2 = 4 \sin 2\theta$ (lemniscate)
36. $r^2 = 4 \cos 2\theta$ (lemniscate)
37. $r = 2 \sin 2\theta$ (four-leaved rose)
38. $r = 3 \sin 3\theta$ (three-leaved rose)
39. $r = 3 \cos 3\theta$ (three-leaved rose)
40. $r = 3\theta$ (spiral of Archimedes)
41. $r = 2 \sin 5\theta$ (five-leaved rose)
42. $r^2 = 4 \sin \theta$ (figure eight)

In Problems 43–48, find all points of intersection of the given curves.

43. $r = 2, \quad r = \cos \theta$
44. $r = \sin \theta, \quad r^2 = 3 \cos^2 \theta$
45. $r = \sin \theta$ and $r = \cos 2\theta$
46. $r = 1 + \cos \theta$ and $r = 1 - \sin \theta$
47. $r = 1 - \cos \theta$ and $r^2 = 4 \cos \theta$
48. $r^2 = 4 \sin \theta$ and $r^2 = 4 \cos \theta$

49. (a) The straight line L passes through the point with polar coordinates (p, α) and is perpendicular to the line segment joining the pole and the point (p, α). Write the polar coordinates equation of L.
(b) Show that the rectangular coordinates equation of L is $x \cos \alpha + y \sin \alpha = p$.

50. Show that the graph of the rectangular coordinates equation

$$x^2 + y^2 = (x + x^2 + y^2)^2$$

is the cardioid with polar equation $r = 1 - \cos\theta$.

51. Did you wonder if the cusp shown in the cardioid of Example 2 is really there? After all, it's conceivable that the cardioid has a *horizontal* tangent at the origin. Here is one way to be sure about the actual shape of a polar graph. Begin with the equation

$$r = 2 + 2\sin\theta$$

of the cardioid of Example 2. Use the equations in (1) to write

$$x = r\cos\theta = 2\cos\theta + 2\sin\theta\cos\theta$$

and

$$y = r\sin\theta = 2\sin\theta + 2\sin^2\theta.$$

Compute $dx/d\theta$, $dy/d\theta$, and

$$\frac{dx}{dy} = \frac{dx/d\theta}{dy/d\theta}$$

as functions of θ. Compute the limit of dx/dy as θ approaches $3\pi/2$ by using l'Hôpital's rule (Section 11.1). You will obtain the limit zero, and thus the cardioid has a vertical tangent at the origin; Fig. 10.16 is correctly drawn.

*10.2 Optional Computer Application

The BASIC program shown in Fig. 10.20 was written to calculate the data needed to plot the three-leaved rose $r = 4\sin 3\theta$. The function $f(\theta) = 4\sin 3\theta$ is defined (with T in place of θ) in line 120, and the x- and y-coordinates of the points on the curve are calculated in line 210.

```
100 REM--Program POLAR
110 REM
120       DEF FNF(T) = 4*SIN(3*T)
130       INPUT "Initial theta"; A
140       INPUT "Final   theta"; B
150       INPUT "No. of intervals"; N
160       T = A  :  H = (B - A)/N
170       PRINT "  X        Y"
180 REM
190       FOR I = 0 TO N
200           R = FNF(T)
210           X = R*COS(T) : Y = R*SIN(T)
220           PRINT USING "##.##  ##.##";X,Y
230           T = T + H
240       NEXT I
245 REM
250       END
```

10.20 Listing of Program POLAR

Exercise 1: Run Program POLAR with $A = 0$, $B = 3.141593$, and $N = 18$. Then plot the points it produces.

Exercise 2: Alter the program to generate the data to plot the four-leaved rose $r = 3\cos 2\theta$.

Exercise 3: Alter the program to generate the data for plotting the figure-eight curve $r^2 = 4\sin\theta$. You will need to take into account the fact that $4\sin\theta$ is negative for certain values of θ.

10.3 Area in Polar Coordinates

The graph of the polar coordinates equation $r = f(\theta)$ may bound an area, as does the cardioid $r = 2(1 + \sin\theta)$ of Fig. 10.16 in Section 10.2. To calculate this area, we may find it more convenient to work directly in polar coordinates, rather than change to rectangular coordinates.

To see how to set up an area integral in polar coordinates, we consider the region R of Fig. 10.21. This region is bounded by the two radial lines $\theta = \alpha$ and $\theta = \beta$ and by the curve $r = f(\theta)$, $\alpha \leqq \theta \leqq \beta$. To approximate the area A of R, we begin with a regular partition

$$\alpha = \theta_0 < \theta_1 < \theta_2 < \cdots < \theta_n = \beta$$

of the interval $[\alpha, \beta]$ into n equal subintervals, each having length $\Delta\theta = (\beta - \alpha)/n$, and select a point $\theta_i{}^*$ in the ith subinterval $[\theta_{i-1}, \theta_i]$ for $i = 1, 2, \ldots, n$.

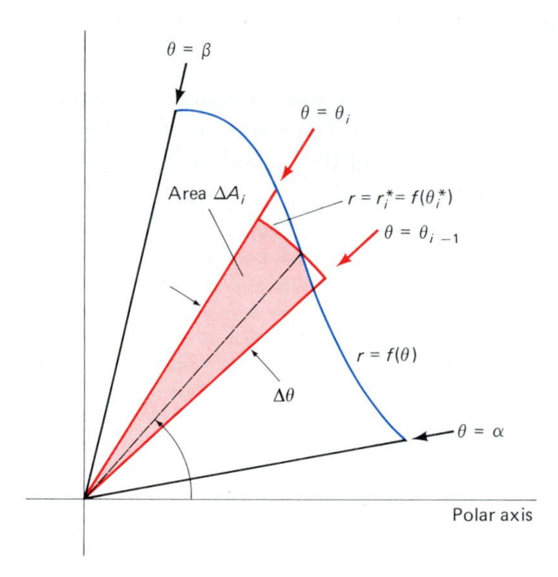

10.21 We obtain the area formula from Riemann sums.

Let ΔA_i denote the area of the sector bounded by the lines $\theta = \theta_{i-1}$ and $\theta = \theta_i$ and the curve $r = f(\theta)$. We see from Fig. 10.21 that for small values of $\Delta\theta$, ΔA_i is approximately equal to the area of the *circular* sector with radius $r_i^* = f(\theta_i^*)$ and bounded by the same angles. That is,

$$\Delta A_i \approx \tfrac{1}{2}(r_i^*)^2\,\Delta\theta = \tfrac{1}{2}[f(\theta_i^*)]^2\,\Delta\theta.$$

We add the areas of these sectors for $i = 1, 2, \ldots, n$ and thereby find that

$$A = \sum_{i=1}^{n} \Delta A_i \approx \sum_{i=1}^{n} \tfrac{1}{2}[f(\theta_i^*)]^2\,\Delta\theta.$$

The right-hand sum is a Riemann sum for the integral

$$\tfrac{1}{2}\int_{\alpha}^{\beta} [f(\theta)]^2\,d\theta.$$

Hence, if f is continuous, the value of this integral is the limit, as $\Delta\theta \to 0$, of the sum shown above. We therefore conclude that *the area A of the* *region R bounded by the lines $\theta = \alpha$ and $\theta = \beta$ and the curve $r = f(\theta)$ is*

$$A = \frac{1}{2}\int_{\alpha}^{\beta} [f(\theta)]^2 \, d\theta. \tag{1}$$

The infinitesimal sector shown in Fig. 10.22, with radius r, central angle $d\theta$, and area $dA = \frac{1}{2}r^2 \, d\theta$, serves as a useful visual device for remembering the formula in (1) in the abbreviated form

$$A = \int_{\alpha}^{\beta} \tfrac{1}{2}r^2 \, d\theta. \tag{2}$$

EXAMPLE 1 Find the area of the region bounded by the limaçon with equation $r = 3 + 2\cos\theta$, $0 \leqq \theta \leqq 2\pi$, shown in Fig. 10.23.

Solution We could apply the formula in (2) with $\alpha = 0$ and $\beta = 2\pi$. Here, instead, we will make use of symmetry. We will calculate the area of the upper half of the region, then double the result. Note that the infinitesimal sector shown in Fig. 10.23 sweeps out the upper half of the limaçon as θ increases from 0 to π (see Fig. 10.24). Hence

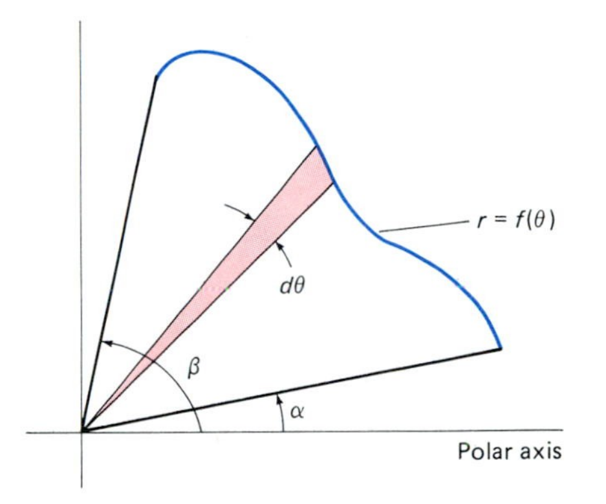

10.22 Heuristic derivation of the area formula in polar coordinates

$$A = 2\int_0^{\pi} \tfrac{1}{2}r^2 \, d\theta = \int_0^{\pi} (3 + 2\cos\theta)^2 \, d\theta$$
$$= \int_0^{\pi} (9 + 12\cos\theta + 4\cos^2\theta) \, d\theta.$$

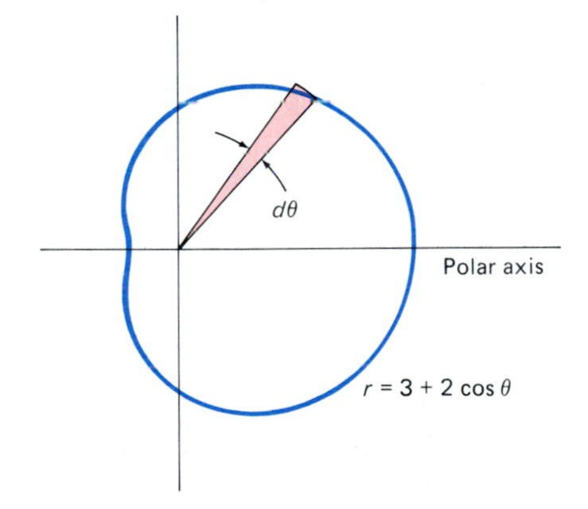

10.23 The limaçon of Example 1

10.24 Infinitesimal sectors from $\theta = 0$ to $\theta = \pi$

Because

$$4\cos^2\theta = 4\,\frac{1 + \cos 2\theta}{2} = 2 + 2\cos 2\theta,$$

we now get

$$A = \int_0^{\pi} (11 + 12\cos\theta + 2\cos 2\theta) \, d\theta$$
$$= \Big[11\theta + 12\sin\theta + \sin 2\theta\Big]_0^{\pi} = 11\pi.$$

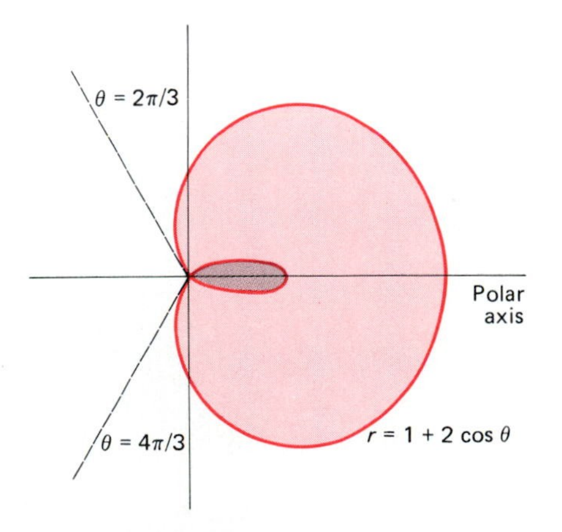

10.25 The limaçon of Example 2

EXAMPLE 2 Find the area bounded by each loop of the limaçon with equation $r = 1 + 2\cos\theta$. This limaçon is shown in Fig. 10.25.

Solution The equation $1 + 2\cos\theta = 0$ has two solutions for θ between 0 and 2π: $\theta = 2\pi/3$ and $\theta = 4\pi/3$. The upper half of the outer loop of the limaçon corresponds to values of θ between 0 and $2\pi/3$, where r is positive. Because the curve is symmetric about the x-axis, we can find the total area A_1 bounded by the outer loop by integrating from 0 to $2\pi/3$ and then doubling. Thus

$$A_1 = 2\int_0^{2\pi/3} \tfrac{1}{2}(1 + 2\cos\theta)^2\,d\theta = \int_0^{2\pi/3} (1 + 4\cos\theta + 4\cos^2\theta)\,d\theta$$

$$= \int_0^{2\pi/3} (3 + 4\cos\theta + 2\cos 2\theta)\,d\theta$$

$$= \Big[3\theta + 4\sin\theta + \sin 2\theta\Big]_0^{2\pi/3} = 2\pi + \tfrac{3}{2}\sqrt{3}.$$

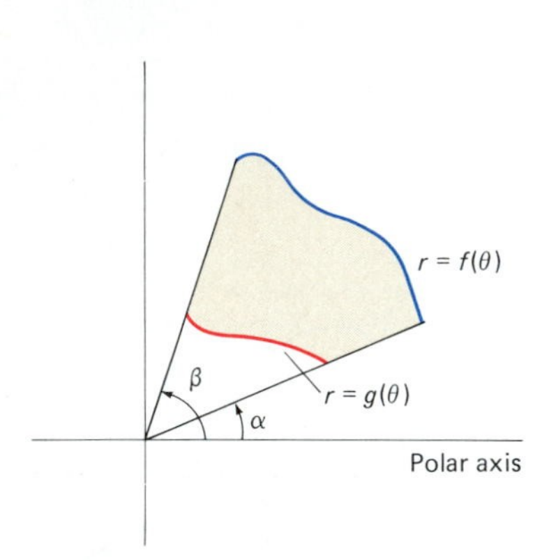

10.26 The area between the graphs of f and g

The inner loop of the limaçon corresponds to values of θ between $2\pi/3$ and $4\pi/3$, where r is negative. Hence the area bounded by the inner loop is

$$A_2 = \tfrac{1}{2}\int_{2\pi/3}^{4\pi/3} (1 + 2\cos\theta)^2\,d\theta$$

$$= \tfrac{1}{2}\Big[3\theta + 4\sin\theta + \sin 2\theta\Big]_{2\pi/3}^{4\pi/3} = \pi - \tfrac{3}{2}\sqrt{3}.$$

The area of the region lying *between* the two loops of the limaçon is then

$$A = A_1 - A_2 = (2\pi + \tfrac{3}{2}\sqrt{3}) - (\pi - \tfrac{3}{2}\sqrt{3}) = \pi + 3\sqrt{3}.$$

Now consider two curves $r = f(\theta)$ and $r = g(\theta)$, with $f(\theta) \geqq g(\theta)$ for $\alpha \leqq \theta \leqq \beta$. Then the area of the region bounded by these curves and the rays $\theta = \alpha$ and $\theta = \beta$ (shown in Fig. 10.26) may be found by subtracting the area bounded by the inner curve from that bounded by the outer curve. That is, the area A between the two curves is given by

$$A = \tfrac{1}{2}\int_\alpha^\beta [f(\theta)]^2\,d\theta - \tfrac{1}{2}\int_\alpha^\beta [g(\theta)]^2\,d\theta$$
$$= \tfrac{1}{2}\int_\alpha^\beta \{[f(\theta)]^2 - [g(\theta)]^2\}\,d\theta. \tag{3}$$

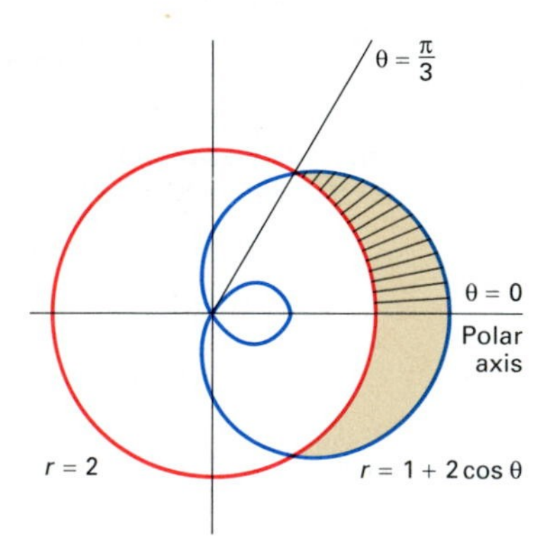

10.27 The region of Example 3

With r_{outer} for the outer curve and r_{inner} for the inner curve, we get the abbreviated formula

$$A = \tfrac{1}{2}\int_\alpha^\beta [(r_{\text{outer}})^2 - (r_{\text{inner}})^2]\,d\theta \tag{4}$$

for the area of the region shown in Fig. 10.26.

EXAMPLE 3 Find the area A of the region that lies within the limaçon $r = 1 + 2\cos\theta$ and outside the circle $r = 2$.

Solution The circle and limaçon are shown in Fig. 10.27, with the area A between them shaded. The points of intersection of the circle and limaçon are given by

$$1 + 2\cos\theta = 2, \qquad \text{so} \quad \cos\theta = \tfrac{1}{2},$$

and the figure shows that we should choose the solutions $\theta = \pm\pi/3$. These two values of θ are the needed limits of integration, and when we use the formula in (3) we find that

$$A = \frac{1}{2}\int_{-\pi/3}^{\pi/3} [(1 + 2\cos\theta)^2 - 2^2]\,d\theta$$

$$= \int_0^{\pi/3} (4\cos\theta + 4\cos^2\theta - 3)\,d\theta$$

(because of symmetry about the polar axis)

$$= \int_0^{\pi/3} (4\cos\theta + 2\cos 2\theta - 1)\,d\theta$$

$$= \Big[4\sin\theta + \sin 2\theta - \theta\Big]_0^{\pi/3} = \frac{5}{2}\sqrt{3} - \frac{\pi}{3}.$$

10.3 PROBLEMS

In each of Problems 1–10, find the area bounded by the given curve.

1. $r = 2\cos\theta$
2. $r = 4\sin\theta$
3. $r = 1 + \cos\theta$
4. $r = 2 - 2\sin\theta$
5. $r = 2 - \cos\theta$
6. $r = 3 + 2\sin\theta$
7. $r = -4\cos\theta$
8. $r = 5(1 + \sin\theta)$
9. $r = 3 - \cos\theta$
10. $r = 2 + \sin\theta + \cos\theta$

In each of Problems 11–18, find the area bounded by one loop of the given curve.

11. $r = 2\cos 2\theta$
12. $r = 3\sin 3\theta$
13. $r = 2\cos 4\theta$
14. $r = \sin 5\theta$
15. $r^2 = 4\sin 2\theta$
16. $r^2 = 4\cos 2\theta$
17. $r^2 = 4\sin\theta$
18. $r = 6\cos 6\theta$

In each of Problems 19–30, find the area of the region described.

19. Inside $r = 2\sin\theta$ and outside $r = 1$.

20. Inside both $r = 4\cos\theta$ and $r = 2$.

21. Inside both $r = \cos\theta$ and $r = \sqrt{3}\sin\theta$.

22. Inside $r = 2 + \cos\theta$ and outside $r = 2$.

23. Inside $r = 3 + 2\sin\theta$ and outside $r = 4$.

24. Inside $r^2 = 2\cos 2\theta$ and outside $r = 1$.

25. Inside both $r^2 = \cos 2\theta$ and $r^2 = \sin 2\theta$.

26. Inside the large loop and outside the small loop of $r = 1 - 2\sin\theta$.

27. Inside $r = 2(1 + \cos\theta)$ and outside $r = 1$.

28. Inside the figure-eight curve $r^2 = 4\cos\theta$ and outside $r = 1 - \cos\theta$.

29. Inside both $r = 2\cos\theta$ and $r = 2\sin\theta$.

30. Inside $r = 2 + 2\sin\theta$ and outside $r = 3$.

31. Find the area of the circle $r = \sin\theta + \cos\theta$ by integration in polar coordinates. Check the answer by writing the equation of the circle in rectangular coordinates, finding its radius, and then using the familiar area formula.

32. Find the area of the region that lies interior to all three circles $r = 1$, $r = 2\cos\theta$, and $r = 2\sin\theta$.

33. The *spiral of Archimedes*, shown in Fig. 10.28, has the simple equation $r = a\theta$ (a is a constant). Let A_n denote the area bounded by the nth turn of the spiral, where $2\pi(n-1) \leqq \theta \leqq 2\pi n$, and the portion of the polar axis joining its end points. For each $n \geqq 2$, let $R_n = A_n - A_{n-1}$ denote the area between the $(n-1)$st and the nth turns. Then derive the following results of Archimedes:

(a) $A_1 = \frac{1}{3}\pi(2\pi a)^2$,
(b) $A_2 = \frac{7}{12}\pi(4\pi a)^2$,
(c) $R_2 = 6A_1$,
(d) $R_{n+1} = nR_2$ for $n \geqq 2$.

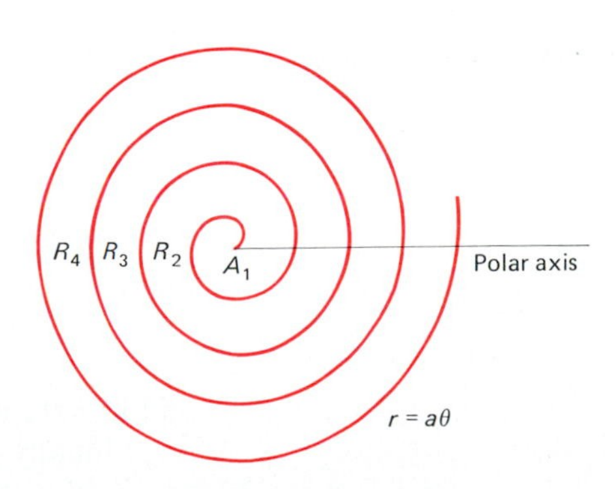

10.28 The spiral of Archimedes

34. Two circles each have radius a, and each passes through the center of the other. Find the area of the region lying within both circles.

10.4 The Parabola

The case $e = 1$ of Example 3 in Section 10.1 is motivation for this formal definition.

> ***Definition*** *The Parabola*
> A **parabola** is the set of all points P in the plane that are equally distant from a fixed point F (called the **focus** of the parabola) and a fixed line L (called its **directrix**) not containing F.

If the focus of the parabola is $F(p, 0)$ and its directrix is the vertical line $x = -p$, $p > 0$, then it follows from Equation (12) of Section 10.1 that the equation of this parabola is

$$y^2 = 4px. \tag{1}$$

When we replace x by $-x$, both in the equation and in the discussion that precedes it, we get the equation of the parabola with focus at $(-p, 0)$ and with directrix the vertical line $x = +p$. The new parabola has equation

$$y^2 = -4px. \tag{2}$$

The old and new parabolas appear in Fig. 10.29.

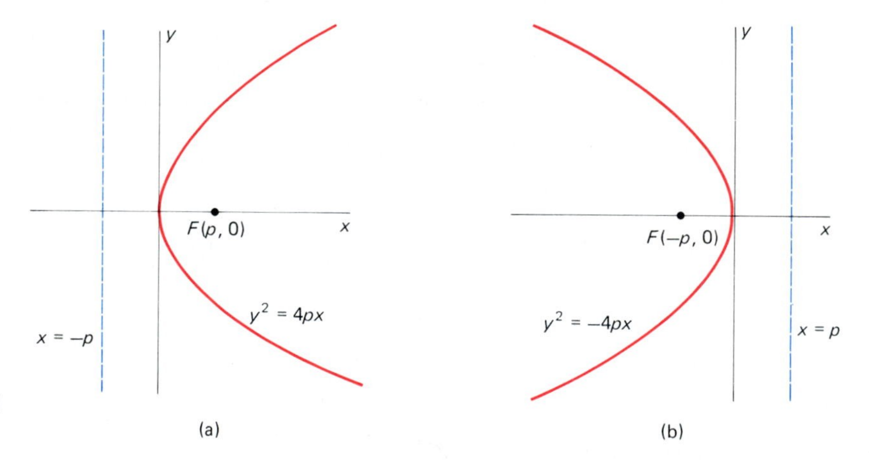

10.29 Two parabolas with vertical directrices

We could also interchange x and y in Equation (1). This would give the equation of a parabola with focus at $(0, p)$ and directrix the horizontal line $y = -p$. This parabola opens upward. Its equation is

$$x^2 = 4py, \tag{3}$$

and it is shown in Fig. 10.30(a).

Finally, we replace y with $-y$ in Equation (3). This gives the equation

$$x^2 = -4py \tag{4}$$

of a parabola opening downward, with focus $(0, -p)$, and with directrix $y = p$. It appears in Fig. 10.30(b).

Each of the parabolas discussed so far is symmetric about one of the coordinate axes. The line about which a parabola is symmetric is called its

axis. The point of a parabola midway between its focus and its directrix is called its **vertex**. Each parabola we discussed in connection with Equations (1) through (4) has its vertex at the origin (0, 0).

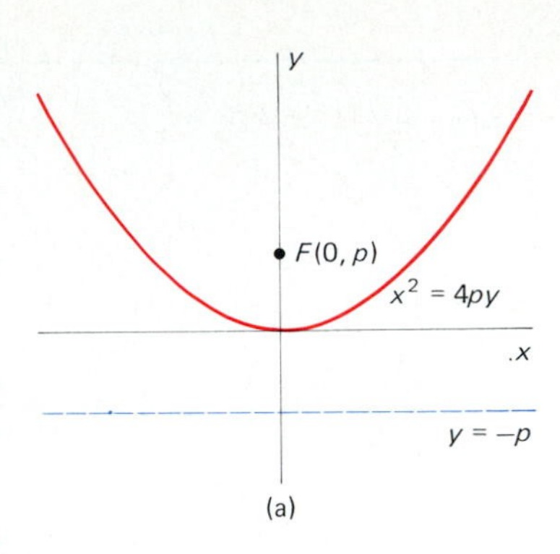

EXAMPLE 1 Determine the focus, directrix, axis, and vertex of the parabola $x^2 = 12y$.

Solution We write the given equation as $x^2 = (4)(3)(y)$. In this form it matches Equation (3) with $p = 3$. Hence the given parabola has its focus at (0, 3), and its directrix is the horizontal line $y = -3$. The y-axis is its axis of symmetry, and the parabola opens upward from its vertex at the origin.

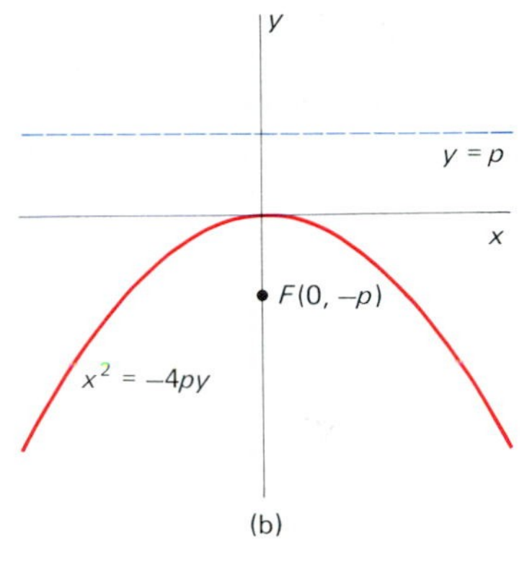

10.30 Two parabolas with horizontal directrices

Suppose that we begin with the parabola of Equation (1), and translate it in such a way that its vertex moves to the point (h, k). Then the translated parabola has equation

$$(y - k)^2 = 4p(x - h). \tag{1a}$$

The new parabola has focus $F(p + h, k)$ and its directrix is the vertical line $x = -p + h$, as indicated in Fig. 10.31. Its axis is the horizontal line $y = k$.

We can obtain the translates of the other three parabolas in (2) through (4) in the same way. If the vertex is moved from the origin to the point (h, k), then the three equations take these forms:

$$(y - k)^2 = -4p(x - h), \tag{2a}$$

$$(x - h)^2 = 4p(y - k), \tag{3a}$$

$$(x - h)^2 = -4p(y - k). \tag{4a}$$

Now note that Equations (1a) and (2a) both take the general form

$$y^2 + Ax + By + C = 0 \qquad (A \neq 0), \tag{5}$$

while Equations (3a) and (4a) both take the general form

$$x^2 + Ax + By + C = 0 \qquad (B \neq 0). \tag{6}$$

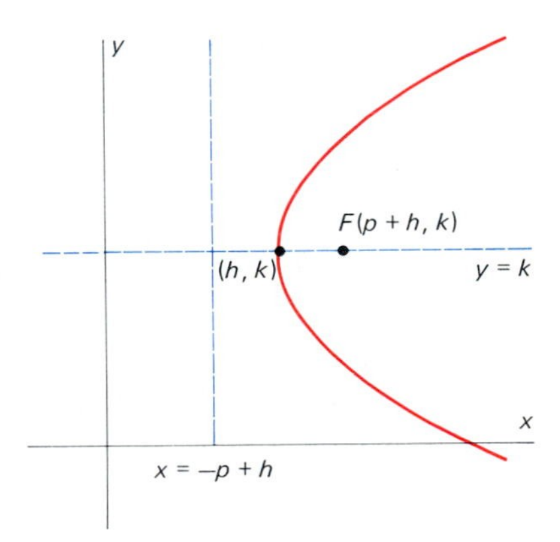

10.31 A translation of the parabola $y^2 = 4px$

What is significant about these last two equations is what they have in common: Each is linear in one of the coordinate variables and quadratic in the other. In fact, *any* such equation can be reduced to one of the standard forms in (1a) through (4a) by completing the square in the coordinate variable that appears quadratically. This means that the graph of any equation of the form of either (5) or (6) is a parabola. The features of the parabola can be read from the standard form of its equation.

EXAMPLE 2 Determine the graph of the equation

$$4y^2 - 8x - 12y + 1 = 0.$$

Solution This equation is linear in x and quadratic in y. We divide through by the coefficient of y^2 and then collect all terms involving y on one side of the equation:

$$y^2 - 3y = 2x - \tfrac{1}{4}.$$

Then we complete the square in the variable y and thus find that

$$y^2 - 3y + \tfrac{9}{4} = 2x - \tfrac{1}{4} + \tfrac{9}{4} = 2x + 2.$$

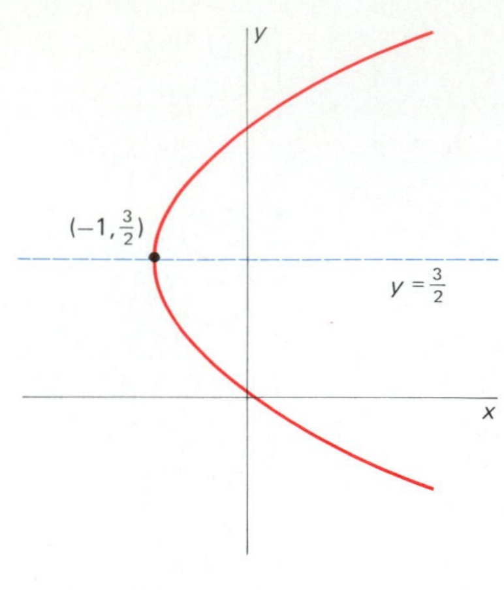

10.32 The parabola of Example 2

The final step is to write the terms on the right involving x in the form $4p(x - h)$:

$$(y - \tfrac{3}{2})^2 = (4)(\tfrac{1}{2})(x + 1).$$

This equation has the form of Equation (1a), with $p = \frac{1}{2}$, $h = -1$, and $k = \frac{3}{2}$. Thus the graph is a parabola that opens to the right from the vertex at $(-1, \frac{3}{2})$. It focus is at $(-\frac{1}{2}, \frac{3}{2})$, its directrix is the vertical line $x = -\frac{3}{2}$, and its axis is the horizontal line $y = \frac{3}{2}$, as indicated in Fig. 10.32.

An important property of the parabola, known as its **reflection property,** has many practical applications. This property involves the fact that the tangent line to the parabola $y^2 = 4px$ at any point $P(x_0, y_0)$ makes equal angles with the horizontal line through P and the line PF from P to the focus F of the parabola. That is, $\alpha = \beta$ in Fig. 10.33. To verify this, note first that the derivative of $y = 2\sqrt{px}$ is $dy/dx = \sqrt{p/x}$, so

$$\tan \beta = \sqrt{\frac{p}{x_0}}.$$

The slope of the line PF is

$$\tan \gamma = \frac{y_0 - 0}{x_0 - p} = \frac{2\sqrt{px_0}}{x_0 - p}.$$

Now $\alpha = \gamma - \beta$ because γ and $\alpha + \beta$ are supplementary to the same angle (see Fig. 10.33). It follows that

$$\begin{aligned}\tan \alpha = \tan(\gamma - \beta) &= \frac{\tan \gamma - \tan \beta}{1 + \tan \gamma \tan \beta} \\ &= \frac{[2\sqrt{px_0}/(x_0 - p)] - (\sqrt{p}/\sqrt{x_0})}{1 + [2\sqrt{px_0}/(x_0 - p)](\sqrt{p}/\sqrt{x_0})} \cdot \frac{(x_0 - p)\sqrt{x_0}}{(x_0 - p)\sqrt{x_0}} \\ &= \frac{2\sqrt{px_0} - \sqrt{p}(x_0 - p)}{(x_0 - p)\sqrt{x_0} + 2\sqrt{px_0}\sqrt{p}} = \frac{\sqrt{p}(x_0 + p)}{\sqrt{x_0}(x_0 + p)} = \sqrt{\frac{p}{x_0}} = \tan \beta.\end{aligned}$$

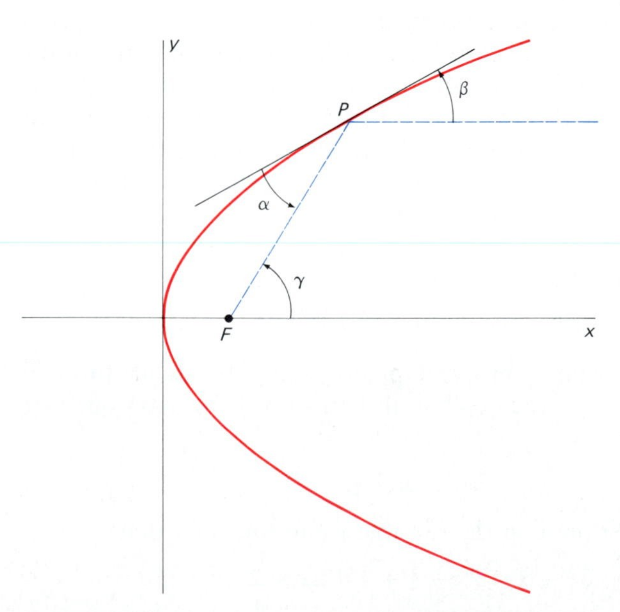

10.33 The reflection property of the parabola: $\alpha = \beta$.

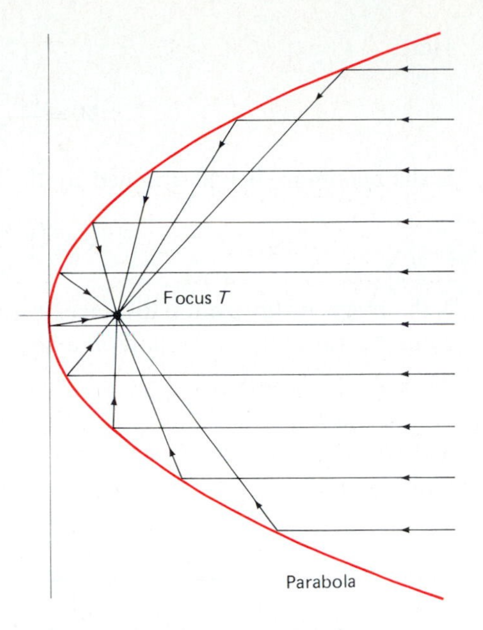

10.34 Incident rays parallel to the axis reflect through the focus.

Because α and β are acute angles, the fact that $\tan \alpha = \tan \beta$ implies that $\alpha = \beta$.

The reflection property is exploited in the design of parabolic mirrors. Such a mirror has the shape of the surface obtained by revolving a parabola around its axis of symmetry. Because of the law of reflection for light rays, a beam of incoming rays parallel to the axis will then be focused at the point T, as shown in Fig. 10.34. The reflection property is also exploited "in reverse"—rays emanating from the focus will be reflected in a beam parallel to the axis, thus keeping the light beam intense. Parabolic mirrors are used in visual and radio telescopes, radar antennas, searchlights, automobile headlights, microphone systems, satellite ground stations, and solar heating devices.

Galileo discovered in the early seventeenth century that the trajectory of a projectile fired from a gun is a parabola (under the assumption that air resistance is ignored). Suppose that the projectile is fired at time $t = 0$ from the origin with initial velocity v_0 at an angle α of inclination from the horizontal x-axis. Then the initial velocity splits into the components

$$v_{0x} = v_0 \cos \alpha \quad \text{and} \quad v_{0y} = v_0 \sin \alpha$$

as indicated in Fig. 10.35. The fact that the projectile continues to move horizontally with *constant* speed v_{0x}, together with Equation (8) in Section 4.9, implies that its x- and y-coordinates after t seconds are

$$x = (v_0 \cos \alpha)t, \tag{7}$$

$$y = -\frac{1}{2}gt^2 + (v_0 \sin \alpha)t. \tag{8}$$

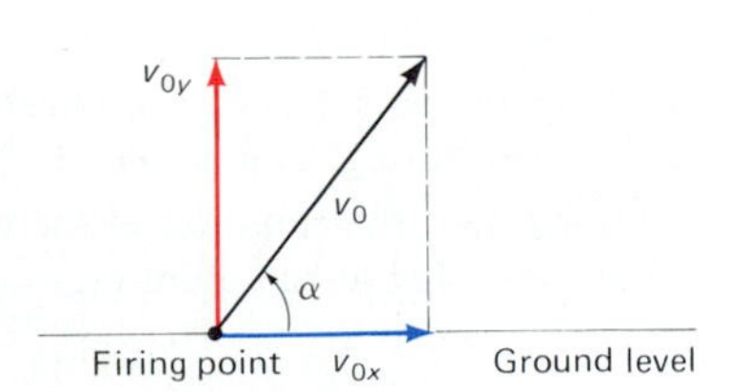

10.35 Resolution of the initial velocity v_0 into horizontal and vertical components

By substituting $t = x/(v_0 \cos \alpha)$ from (7) in (8) and then completing the square, we can derive (as in Problem 24) an equation of the form

$$y - M = -4p(x - \tfrac{1}{2}R)^2. \tag{9}$$

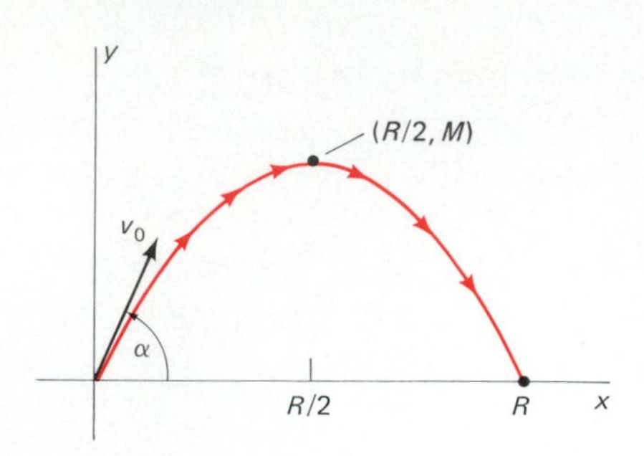

10.36 The trajectory of the projectile, showing its maximum altitude and its range

Here,

$$M = \frac{v_0^2 \sin^2 \alpha}{2g} \tag{10}$$

is the maximum height attained by the projectile, and

$$R = \frac{v_0^2 \sin 2\alpha}{g} \tag{11}$$

is the **range,** or horizontal distance, it travels before returning to the ground. Thus its trajectory is the parabola shown in Fig. 10.36.

10.4 PROBLEMS

In each of Problems 1–5, find the equation and sketch the graph of the parabola with vertex V and focus F.

1. $V(0, 0)$, $F(3, 0)$
2. $V(0, 0)$, $F(0, -2)$
3. $V(2, 3)$, $F(2, 1)$
4. $V(-1, -1)$, $F(-3, -1)$
5. $V(2, 3)$, $F(0, 3)$

In each of Problems 6–10, find the equation and sketch the graph of the parabola with the given focus and directrix.

6. $F(1, 2)$, $x = -1$
7. $F(0, -3)$, $y = 0$
8. $F(1, -1)$, $x = 3$
9. $F(0, 0)$, $y = -2$
10. $F(-2, 1)$, $x = -4$

In each of Problems 11–18, sketch the parabola with the given equation. Show and label its vertex, focus, axis, and directrix.

11. $y^2 = 12x$
12. $x^2 = -8y$
13. $y^2 = -6x$
14. $x^2 = 7y$
15. $x^2 - 4x - 4y = 0$
16. $y^2 - 2x + 6y + 15 = 0$
17. $4x^2 + 4x + 4y + 13 = 0$
18. $4y^2 - 12y + 9x = 0$

19. Prove that the point of the parabola $y^2 = 4px$ closest to its focus is its vertex.

20. Find the equation of the parabola with vertical axis that passes through the points (2, 3), (4, 3), and (6, −5).

21. Show that the equation of the tangent line to the parabola $y^2 = 4px$ at the point (x_0, y_0) is

$$2px - y_0 y + 2px_0 = 0.$$

Conclude that this tangent line intersects the x-axis at the point $(-x_0, 0)$. This fact provides a quick method for constructing a tangent line to a parabola at a given point.

22. A comet has a parabolic orbit with the sun at its focus. When the comet is $100\sqrt{2}$ million miles from the sun, the line from the sun to the comet makes an angle of 45° with the axis of the parabola, as shown in Fig. 10.37. What will be the minimum distance between the comet and the sun? *Suggestion:* Write the equation of the parabola with the origin at the focus and then use the result of Problem 19.

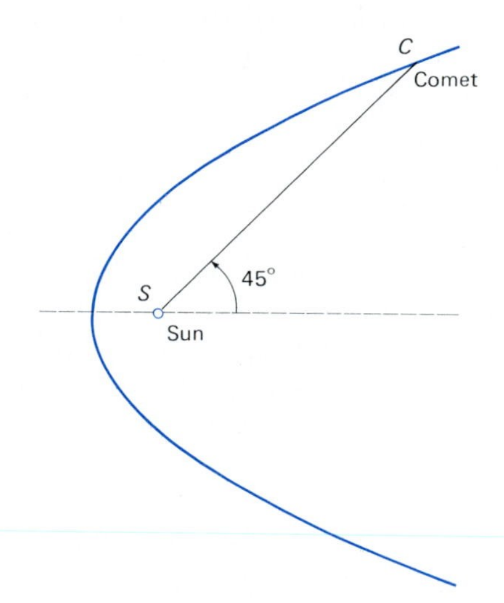

10.37 The comet of Problem 22 in parabolic orbit about the sun

23. Suppose that the angle of Problem 22 increases from 45° to 90° in 3 days. How much longer will be required for the comet to reach its point of closest approach to the sun? Assume that the line segment from the sun to the comet sweeps out area at a constant rate (Kepler's second law).

24. Use Equations (7) and (8) to derive Equation (9) with the values of M and R given by Equations (10) and (11).

25. Deduce from Equation (11) that, given a fixed initial velocity v_0, the maximum range of the projectile is $R_{max} = v_0^2/g$ and is attained when $\alpha = 45°$.

In Problems 26–28, assume that a projectile is fired from the origin with initial velocity $v_0 = 160$ ft/s and with initial angle of inclination α. Use $g = 32$ ft/s^2.

26. If $\alpha = 45°$, find the range of the projectile and the maximum height it attains.

27. For what value(s) of α is the range $R = 400$ ft?

28. Find the range R of the projectile and the length of time it remains above the ground if (a) $\alpha = 30°$; (b) $\alpha = 60°$.

29. Prove that the plane curve with equation

$$\sqrt{x} + \sqrt{y} = \sqrt{a}$$

is a parabola.

10.5 The Ellipse

An ellipse is a conic section with eccentricity e less than 1, as in Example 3 of Section 10.1.

> ***Definition*** *The Ellipse*
> Suppose that $0 < e < 1$, and let F be a fixed point and L a fixed line not containing F. The **ellipse** with **eccentricity** e, **focus** F, and **directrix** L is the set of all points P such that the distance $|PF|$ is e times the (perpendicular) distance from P to the line L.

The equation of the ellipse is especially simple if F is the point $(c, 0)$ on the x-axis and L is the vertical line $x = c/e^2$. The case $c > 0$ is shown in Fig. 10.38. If Q is the point $(c/e^2, y)$, then PQ is the perpendicular from $P(x, y)$ to L. The condition $|PF| = e|PQ|$ gives

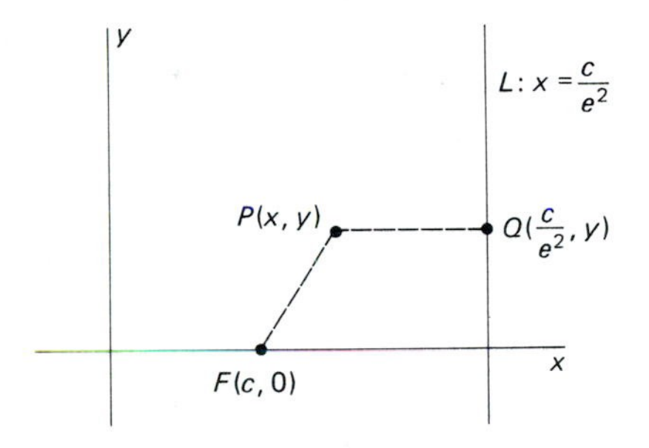

10.38 Ellipse: focus F. directrix L, eccentricity e

$$(x - c)^2 + y^2 = e^2\left(x - \frac{c}{e^2}\right)^2,$$

$$x^2 - 2cx + c^2 + y^2 = e^2x^2 - 2cx + \frac{c^2}{e^2},$$

$$x^2(1 - e^2) + y^2 = c^2\left(\frac{1}{e^2} - 1\right) = \frac{c^2}{e^2}(1 - e^2).$$

Thus

$$x^2(1 - e^2) + y^2 = a^2(1 - e^2)$$

where

$$a = \frac{c}{e}. \tag{1}$$

Division of both sides of the next-to-last equation by $a^2(1 - e^2)$ gives

$$\frac{x^2}{a^2} + \frac{y^2}{a^2(1 - e^2)} = 1.$$

Finally, with the aid of the fact that $e < 1$, we may let

$$b^2 = a^2(1 - e^2) = a^2 - c^2. \tag{2}$$

Then the equation of the ellipse with focus $(c, 0)$ and directrix $x = c/e^2 = a/e$ takes the pleasant form

$$\frac{x^2}{a^2} + \frac{y^2}{b^2} = 1. \tag{3}$$

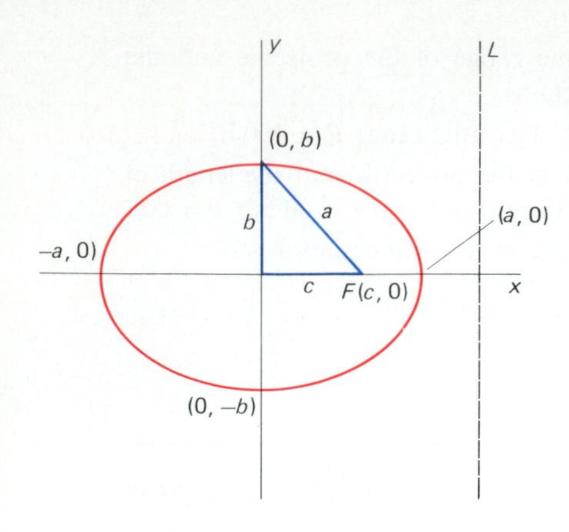

10.39 The parts of an ellipse

We see from Equation (3) that this ellipse is symmetric about both coordinate axes. Its x-intercepts are $(\pm a, 0)$ and its y-intercepts are $(0, \pm b)$. The points $(\pm a, 0)$ are called the **vertices** of the ellipse, and the line segment joining them is called its **major axis.** The line segment joining $(0, b)$ with $(0, -b)$ is called the **minor axis** (note from (2) that $b < a$). The alternative form

$$a^2 = b^2 + c^2 \tag{4}$$

of (2) is the Pythagorean relation for the right triangle of Fig. 10.39. Indeed, visualization of this triangle is an excellent way to remember Equation (4). The numbers a and b are the lengths of the major and minor **semiaxes,** respectively.

Because $a = c/e$, the directrix of the ellipse in (3) is $x = a/e$. If we had begun instead with focus $(-c, 0)$ and directrix $x = -a/e$, we would still have obtained Equation (3), because only the squares of a and c are involved in its derivation. Thus the ellipse in (3) has *two* foci, $(c, 0)$ and $(-c, 0)$, and *two* directrices, $x = a/e$ and $x = -a/e$. These are shown in Fig. 10.40.

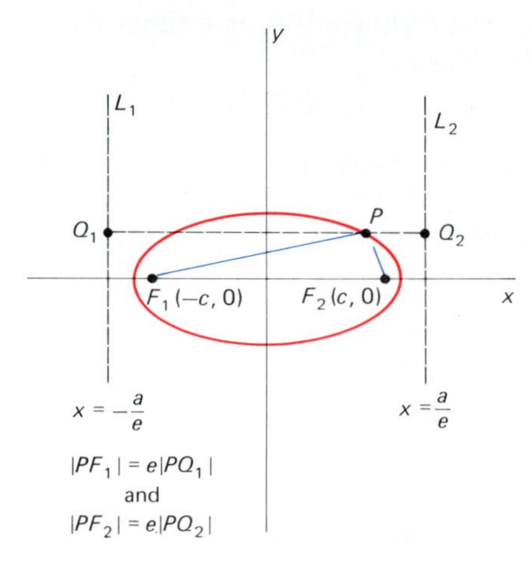

10.40 The ellipse as a conic section: Two foci, two directrices

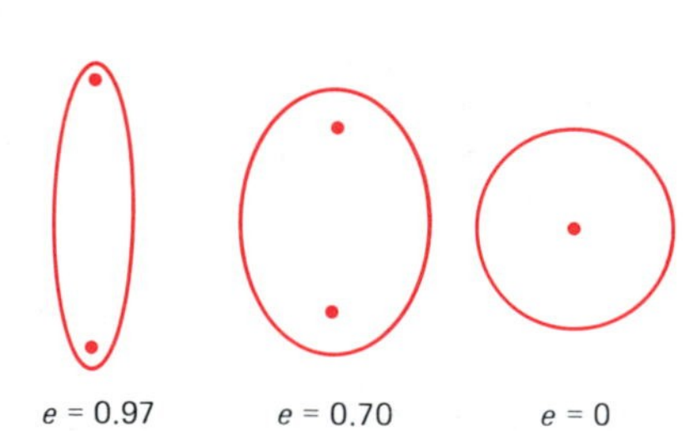

10.41 The relation between the eccentricity of an ellipse and its shape

The larger the eccentricity $e < 1$ is, the more elongated is the ellipse. If $e = 0$ then Equation (2) gives $b = a$, so Equation (3) reduces to the equation of a circle of radius a. Thus a circle may be regarded as an ellipse with eccentricity zero. Compare the three cases shown in Fig. 10.41.

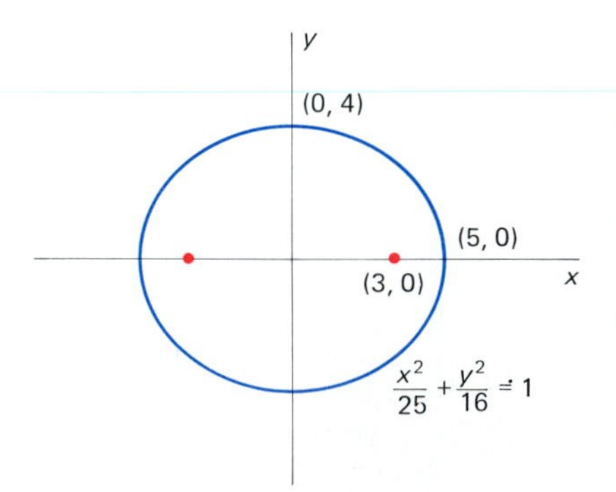

10.42 The ellipse of Example 1

EXAMPLE 1 Find the equation of the ellipse with foci $(\pm 3, 0)$ and vertices $(\pm 5, 0)$.

Solution We are given $c = 3$ and $a = 5$, so Equation (2) gives $b = 4$. Thus Equation (3) gives

$$\frac{x^2}{25} + \frac{y^2}{16} = 1$$

for the desired equation. This ellipse is shown in Fig. 10.42.

If an ellipse has its two foci on the y-axis, such as $F_1(0, c)$ and $F_2(0, -c)$, then its equation is

$$\frac{x^2}{b^2} + \frac{y^2}{a^2} = 1, \tag{5}$$

and it is still true that

$$a^2 = b^2 + c^2. \tag{4}$$

But now the major axis of length $2a$ is vertical, and the minor axis of length $2b$ is horizontal. The derivation of Equation (5) is similar to that of Equation (3); see Problem 23. Figure 10.43 shows the case of an ellipse with vertical major axis.

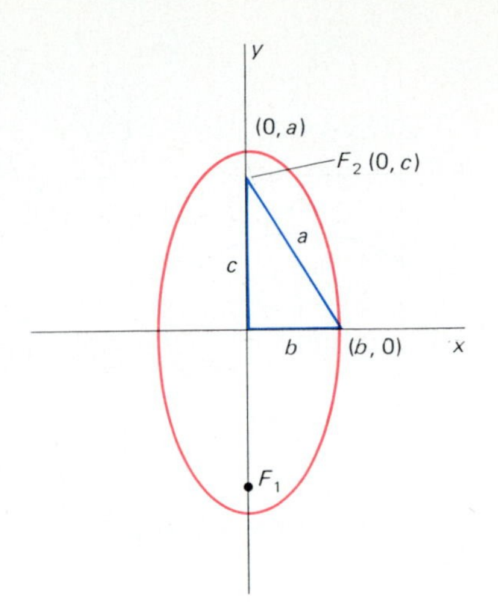

10.43 An ellipse with vertical major axis

In practice, there is little chance of confusing Equations (3) and (5). The equation or given data will make it clear whether the major axis of the ellipse is horizontal or vertical. Just use the equation to read the ellipse's intercepts. The two that are farthest from the origin are the end points of the major axis, and the other two are the end points of the minor axis. The two foci lie on the major axis, each at distance c from the center of the ellipse—which will be the origin if the equation of the ellipse has the form of either of Equations (3) or (5).

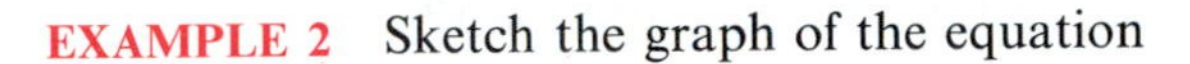

EXAMPLE 2 Sketch the graph of the equation

$$\frac{x^2}{16} + \frac{y^2}{25} = 1$$

Solution The x-intercepts are $(\pm 4, 0)$; the y-intercepts are $(0, \pm 5)$. So the major axis is vertical. We take $a = 5$ and $b = 4$ in Equation (4) and find that $c = 3$. The foci are thus at $(0, \pm 3)$. Hence this ellipse has the appearance of the one shown in Fig. 10.44.

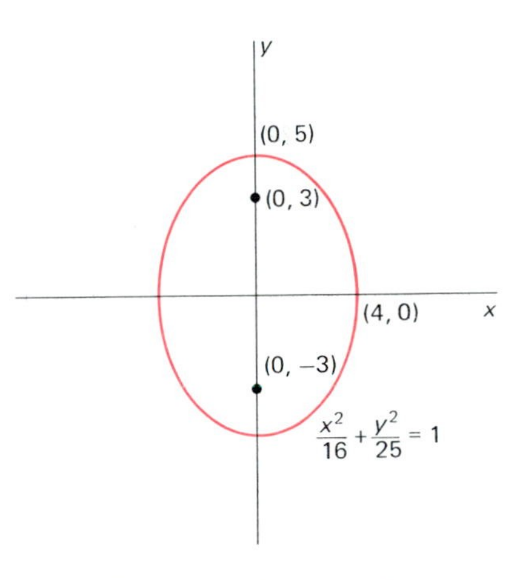

10.44 The ellipse of Example 2

Any equation of the form

$$Ax^2 + Cy^2 + Dx + Ey + F = 0 \tag{6}$$

in which the coefficients A and C of the squared terms are *both nonzero* and have the *same sign*, may be reduced to the form

$$A(x - h)^2 + C(y - k)^2 = G$$

by completing the square in x and y. We may assume that A and B are both positive. Then if $G < 0$, there are no points satisfying (6), and the graph is the empty set. If $G = 0$, there is exactly one point on the locus—the single point (h, k). And if $G > 0$, we can divide both sides of the last equation by G and get an equation that resembles one of these two:

$$\frac{(x - h)^2}{a^2} + \frac{(y - k)^2}{b^2} = 1, \tag{7a}$$

$$\frac{(x - h)^2}{b^2} + \frac{(y - k)^2}{a^2} = 1. \tag{7b}$$

Which equation to choose? The one consistent with the condition $a \geqq b > 0$. Finally note that either of the equations in (7) is the equation of a translated ellipse. Thus, apart from the exceptional cases already noted, the graph of Equation (6) is an ellipse if $AC > 0$.

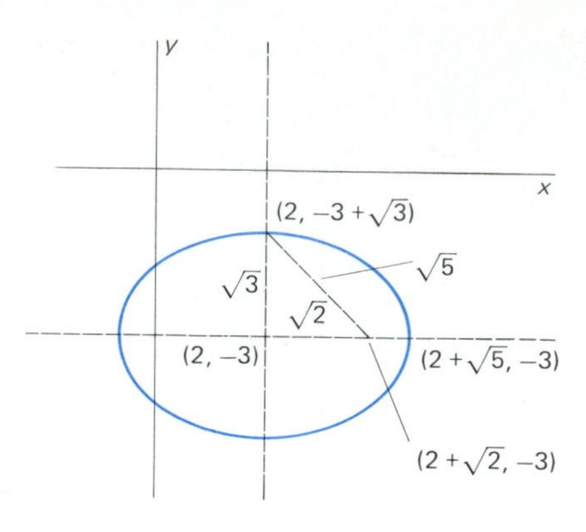

10.45 The ellipse of Example 3

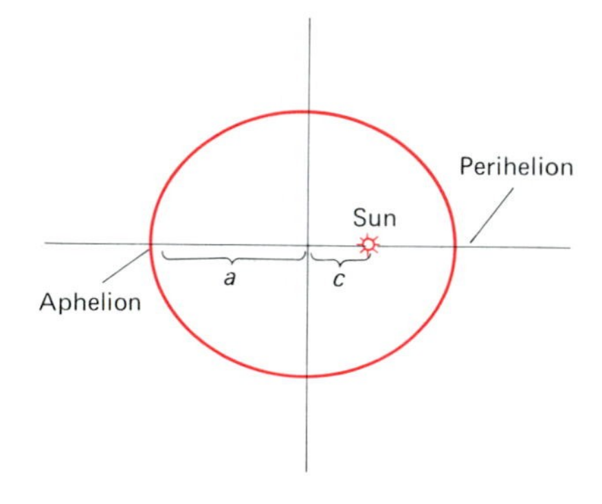

10.46 The orbit of the earth with its eccentricity exaggerated

EXAMPLE 3 Determine the graph of the equation

$$3x^2 + 5y^2 - 12x + 30y + 42 = 0.$$

Solution We collect terms containing x, terms containing y, and complete the square in each variable. This gives

$$3(x^2 - 4x) + 5(y^2 + 6y) = -42,$$

$$3(x^2 - 4x + 4) + 5(y^2 + 6y + 9) = 15,$$

$$\frac{(x-2)^2}{5} + \frac{(y+3)^2}{3} = 1.$$

Thus the given equation is that of a translated ellipse. The center is at $(2, -3)$, the ellipse has horizontal major semiaxis of length $a = \sqrt{5}$, and the minor semiaxis has length $b = \sqrt{3}$. The distance from the center to each focus is $c = \sqrt{2}$, and the eccentricity is $e = c/a = \sqrt{2/5}$. This ellipse appears in Fig. 10.45.

EXAMPLE 4 The orbit of the earth is an ellipse with the sun at one focus. The planet's maximum distance from the sun is 94.56 million miles and its minimum distance is 91.45 million miles. What are the major and minor semiaxes of the earth's orbit, and what is its eccentricity?

Solution As Fig. 10.46 shows, we have

$$a + c = 94.56 \quad \text{and} \quad a - c = 91.45,$$

with units in millions of miles. From these equations we conclude that $a = 93.00$, that $c = 1.56$, and then that

$$b = \sqrt{(93.00)^2 - (1.56)^2} \approx 92.99$$

million miles. Finally,

$$e = \frac{1.56}{93.00} \approx 0.017,$$

a number relatively close to zero. This means that the earth's orbit is nearly circular. Indeed, the major and minor semiaxes are so nearly equal that, on any usual scale, the earth's orbit would appear to be a perfect circle. But the difference between uniform circular motion and the earth's actual motion has some important aspects, including the facts that the sun is 1.56 million miles off center, and—as we shall see in Chapter 13—the orbital speed of the earth is not constant.

EXAMPLE 5 One of the most famous of all comets is Halley's comet, named for Edmund Halley (1656–1742, a disciple of Newton). By studying the records of the paths of earlier comets, Halley deduced that the comet of 1682 was the same one that had been sighted in 1607, in 1531, in 1456, and in 1066 (an omen at the Battle of Hastings). In 1682 he predicted that the comet would return in 1759, in 1835, and in 1910; he was correct each time. The period of Halley's comet is about 76 years—it can vary a couple of years in either direction because of perturbations of its orbit by Jupiter. The orbit of this comet is an ellipse with the sun at one focus. In terms of astronomical units (1 A.U. is the earth's mean distance from the sun), the major and minor semiaxes of this elliptical orbit are 18.09 A.U. and 4.56 A.U., respectively. What are the maximum and minimum distances from the sun of Halley's comet?

Solution We are given that $a = 18.09$ (all units will be in A.U.) and that $b = 4.56$, so

$$c = \sqrt{(18.09)^2 - (4.56)^2} \approx 17.51.$$

Hence its maximum distance from the sun is $a + c \approx 35.60$ A.U. and its minimum distance is $a - c \approx 0.58$ A.U. The eccentricity of its orbit is

$$e = \frac{17.51}{18.09} \approx 0.97,$$

a very eccentric orbit.

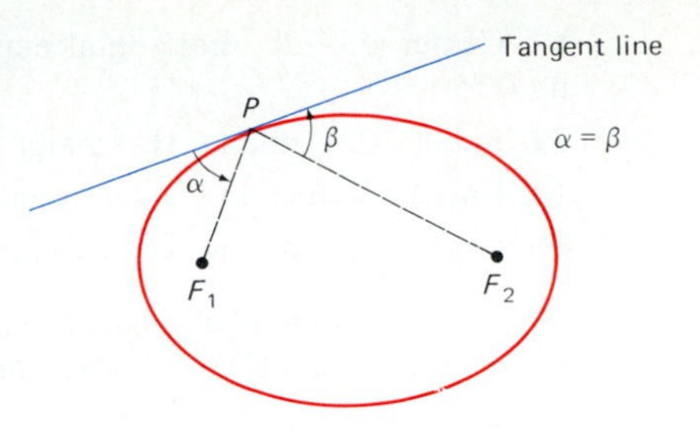

10.47 The reflection property: $\alpha = \beta$.

The *reflection property* of the ellipse is the fact that at a point P of an ellipse, the tangent line makes equal angles with the two lines PF_1 and PF_2 from P to the two foci of the ellipse (see Fig. 10.47). This property is the basis of the "whispering gallery" phenomenon, which can be observed (for example) in the whispering gallery of the U.S. Senate. Suppose that the ceiling of a large room is shaped like half an ellipsoid of the sort obtained by revolving an ellipse about its major axis. Sound waves, like light waves, are reflected at equal angles of incidence and reflection. Thus if two diplomats are standing in quiet conversation near one focus of the ellipsoidal surface, a reporter standing near the other focus would be able to eavesdrop on their conversation even though standing 50 ft away, and even if the diplomats' conversation were inaudible to others in the same room.

Billiard tables are sometimes manufactured in the shape of an ellipse. The one we saw had the foci plainly marked for the convenience of enthusiasts of this unusual game.

A more serious application of the reflection property of ellipses is the nonintrusive kidney stone treatment called *shockwave lithotripsy*. An ellipsoidal reflector with a transducer at one focus is positioned (outside the patient's body) so that the offending kidney stone is located at the other focus. The stone then is pulverized by reflected shock waves emanating from the transducer. For further details, see the COMAP *Newsletter* **20** (November 1986).

An alternative definition of the ellipse with foci F_1 and F_2 and major axis of length $2a$ is this: It is the locus of a point P such that the sum of the distances $|PF_1|$ and $|PF_2|$ is the constant $2a$ (Problem 26). This fact gives a convenient way of drawing the ellipse, making use of two tacks placed at F_1 and F_2, a string of length $2a$, and a pencil (see Fig. 10.48).

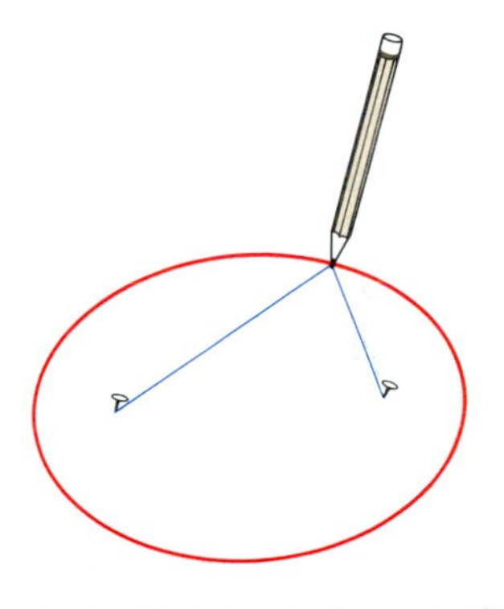

10.48 One way to draw an ellipse

10.5 PROBLEMS

In each of Problems 1–15, find an equation of the ellipse specified.

1. Vertices $(\pm 4, 0)$ and $(0, \pm 5)$
2. Foci $(\pm 5, 0)$ and major semiaxis 13
3. Foci $(0, \pm 8)$ and major semiaxis 17
4. Center $(0, 0)$, vertical major axis 12, minor axis 8
5. Foci $(\pm 3, 0)$, eccentricity 0.75
6. Foci $(0, \pm 4)$, eccentricity 2/3
7. Center $(0, 0)$, horizontal major axis 20, eccentricity 0.5
8. Center $(0, 0)$, horizontal minor axis 10, eccentricity 0.5
9. Foci $(\pm 2, 0)$, directrices $x = \pm 8$
10. Foci $(0, \pm 4)$, directrices $y = \pm 9$
11. Center $(2, 3)$, horizontal axis 8, vertical axis 4

12. Center $(1, -2)$, horizontal major axis 8, eccentricity 0.75

13. Foci $(-2, 1)$ and $(4, 1)$, major axis 10

14. Foci $(-3, 0)$ and $(-3, 4)$, minor axis 6

15. Foci $(-2, 2)$ and $(4, 2)$, eccentricity 1/3

Sketch the graphs of the equations in Problems 16–20. Indicate centers, foci, and lengths of axes.

16. $4x^2 + y^2 = 16$

17. $4x^2 + 9y^2 = 144$

18. $4x^2 + 9y^2 = 24x$

19. $9x^2 + 4y^2 - 32y + 28 = 0$

20. $2x^2 + 3y^2 + 12x - 24y + 60 = 0$

21. The orbit of the comet Kahoutek is an ellipse of extreme eccentricity $e = 0.999925$; the sun is at one focus of this ellipse. The minimum distance between Kahoutek and the sun is 0.13 A.U. What is the maximum distance between Kahoutek and the sun?

22. The orbit of the planet Mercury is an ellipse with eccentricity $e = 0.206$. Its maximum and minimum distances from the sun are 0.467 and 0.307 A.U., respectively. What are the major and minor semiaxes of the orbit of Mercury? Do you feel that "nearly circular" is an accurate description of the orbit of Mercury?

23. Derive Equation (5) for an ellipse if its foci lie on the y-axis.

24. Show that the tangent line to the ellipse

$$\frac{x^2}{a^2} + \frac{y^2}{b^2} = 1$$

at the point $P(x_0, y_0)$ of that ellipse has equation

$$\frac{xx_0}{a^2} + \frac{yy_0}{b^2} = 1.$$

25. Use the result of Problem 24 to establish the reflection property of the ellipse. (*Suggestion:* Let m be the slope of the normal line to the ellipse at $P(x_0, y_0)$ and let m_1 and m_2 be the slopes of the lines PF_1 and PF_2, respectively. Show that

$$\frac{m - m_1}{1 + m_1 m} = \frac{m_2 - m}{1 + m_2 m},$$

and use the identity for $\tan(A - B)$.)

26. Given $F_1(-c, 0)$ and $F_2(c, 0)$ and $a > c > 0$, prove that the ellipse $x^2/a^2 + y^2/b^2 = 1$ (with $b^2 = a^2 - c^2$) is the locus of a point P such that $|PF_1| + |PF_2| = 2a$.

27. Find the equation of the ellipse with horizontal and vertical axes that passes through the points $(-1, 0)$, $(3, 0)$, $(0, 2)$, and $(0, -2)$.

28. Derive an equation for the ellipse with foci $(3, -3)$, $(-3, 3)$, and major axis of length 10. Note that the foci of this ellipse lie on neither a vertical line nor a horizontal line.

10.6 The Hyperbola

A hyperbola is defined in the same way as is an ellipse, except that the eccentricity e of a hyperbola is greater than 1.

> ***Definition*** *The Hyperbola*
>
> Suppose that $e > 1$, and let F be a fixed point and L a fixed line not containing F. Then the **hyperbola** with **eccentricity** e, **focus** F, and **directrix** L is the set of all points P such that the distance $|PF|$ is e times the (perpendicular) distance from P to the line L.

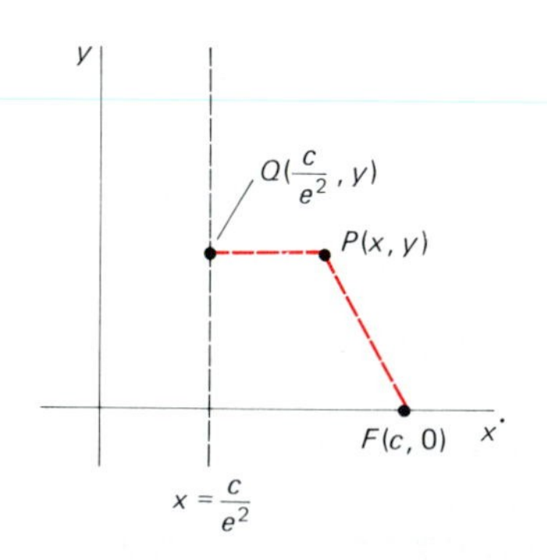

10.49 The definition of the hyperbola

As with the ellipse, the equation of a hyperbola is simplest if F is the point $(c, 0)$ on the x-axis and L is the vertical line $x = c/e^2$. The case $c > 0$ is shown in Fig. 10.49. If Q is the point $(c/e^2, y)$, then PQ is the perpendicular from $P(x, y)$ to L. The condition $|PF| = e|PQ|$ gives

$$(x - c)^2 + y^2 = e^2\left(x - \frac{c}{e^2}\right)^2,$$

$$x^2 - 2cx + c^2 + y^2 = e^2x^2 - 2cx + \frac{c^2}{e^2},$$

$$(e^2 - 1)x^2 - y^2 = c^2\left(1 - \frac{1}{e^2}\right) = \frac{c^2}{e^2}(e^2 - 1).$$

Thus

$$(e^2 - 1)x^2 - y^2 = a^2(e^2 - 1)$$

where

$$a = \frac{c}{e}. \tag{1}$$

Division of both sides of the next-to-last equation by $a^2(e^2 - 1)$ gives

$$\frac{x^2}{a^2} - \frac{y^2}{a^2(e^2 - 1)} = 1.$$

To simplify this equation, we let

$$b^2 = a^2(e^2 - 1) = c^2 - a^2. \tag{2}$$

This is permissible because $e > 1$. So the equation of the hyperbola with focus $(c, 0)$ and directrix $x = c/e^2 = a/e$ takes the form

$$\frac{x^2}{a^2} - \frac{y^2}{b^2} = 1. \tag{3}$$

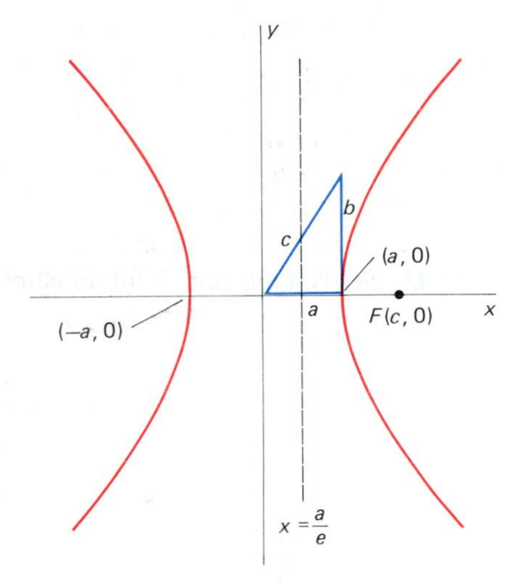

10.50 A hyperbola has two *branches.*

The minus sign on the left-hand side is the only difference between the equation of a hyperbola and that of an ellipse. Of course, Equation (2) differs from the relation

$$b^2 = a^2(1 - e^2) = a^2 - c^2$$

for the case of the ellipse.

The hyperbola of Equation (3) is clearly symmetric about both coordinate axes, and has x-intercepts $(\pm a, 0)$. But it has no y-intercept. If we rewrite Equation (3) in the form

$$y = \pm\frac{b}{a}\sqrt{x^2 - a^2}, \tag{4}$$

then we see that there are points on the graph only if $|x| \geqq a$. Hence the hyperbola has two **branches,** as shown in Fig. 10.50. We also see from Equation (4) that $|y| \to \infty$ as $|x| \to \infty$.

The x-intercepts $V_1(-a, 0)$ and $V_2(a, 0)$ are the **vertices** of the hyperbola, and the line segment joining them is its **transverse axis.** The line segment joining $W_1(0, -b)$ and $W_2(0, b)$ is its **conjugate axis.** The alternative form

$$c^2 = a^2 + b^2 \tag{5}$$

of Equation (2) is the Pythagorean relation for the right triangle shown in Fig. 10.50.

The lines $y = \pm bx/a$ that pass through the **center** (0, 0) and the opposite vertices of the rectangle in Fig. 10.51 are **asymptotes** of the two branches of the hyperbola in both directions. That is, if

$$y_1 = \frac{bx}{a} \quad \text{and} \quad y_2 = \frac{b}{a}\sqrt{x^2 - a^2},$$

then

$$\lim_{x \to \infty} (y_1 - y_2) = 0 = \lim_{x \to -\infty} (y_1 - (-y_2)). \tag{6}$$

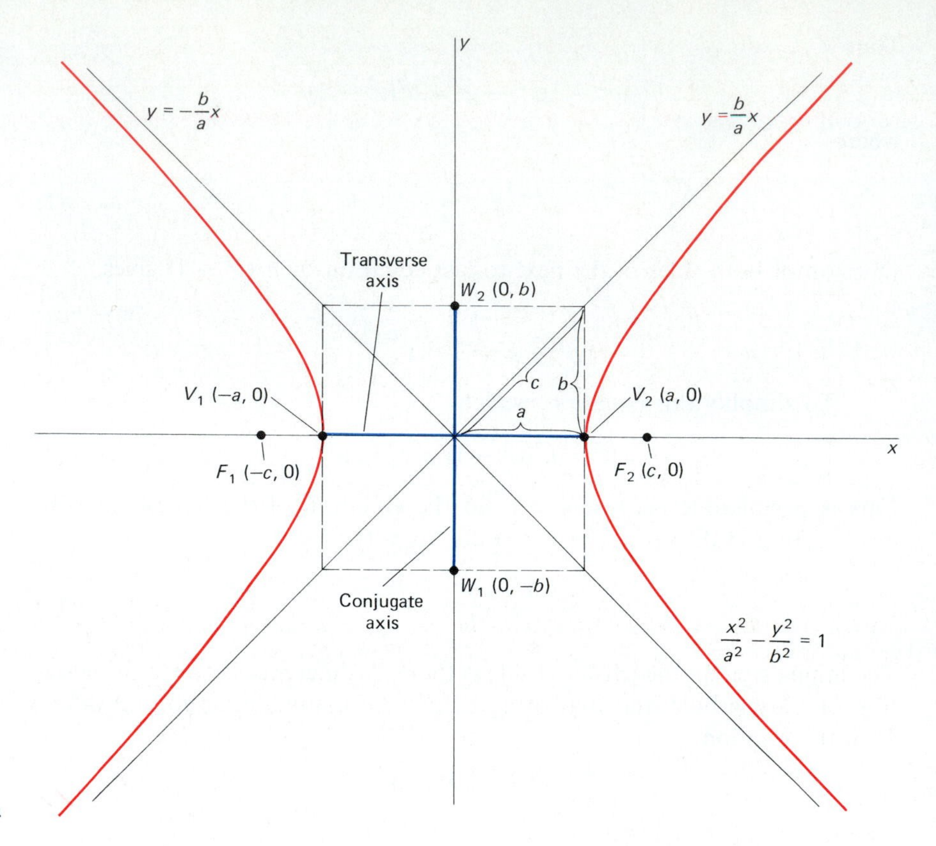

10.51 The parts of a hyperbola

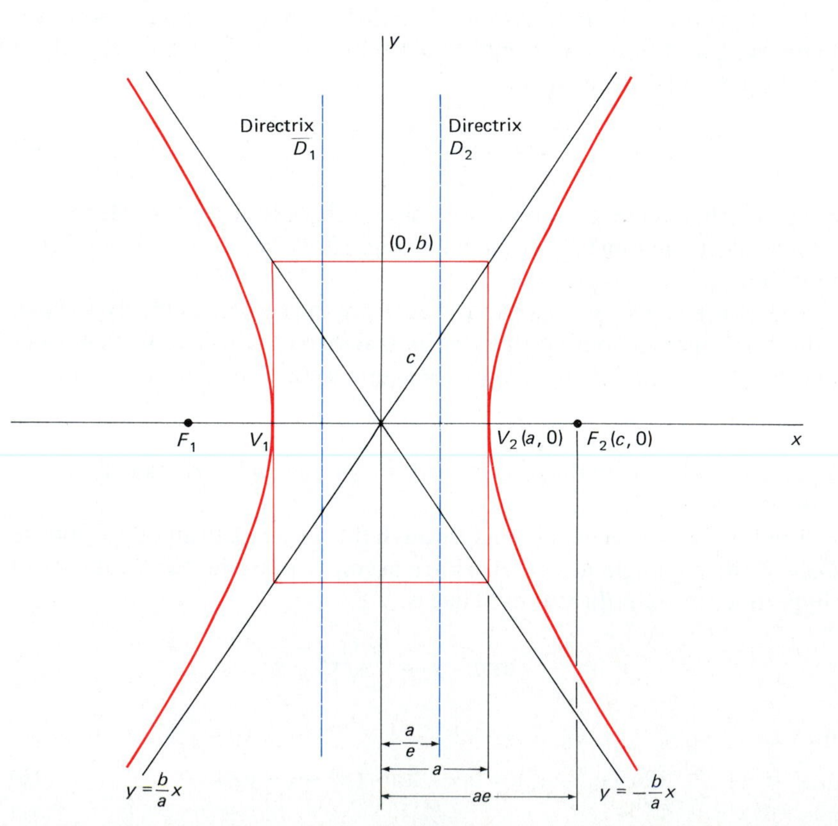

10.52 The relations between the parts of a hyperbola

To verify the first limit, note that

$$\lim_{x\to+\infty} \frac{b}{a}(x - \sqrt{x^2 - a^2}) = \lim_{x\to+\infty} \frac{b}{a} \cdot \frac{(x - \sqrt{x^2 - a^2})(x + \sqrt{x^2 - a^2})}{x + \sqrt{x^2 - a^2}}$$

$$= \lim_{x\to+\infty} \frac{b}{a} \cdot \frac{a^2}{x + \sqrt{x^2 - a^2}} = 0.$$

Just as in the case of the ellipse, the hyperbola with focus $(c, 0)$ and directrix $x = a/e$ also has focus $(-c, 0)$ and directrix $x = -a/e$, as shown both in Figs. 10.51 and 10.52. Because $c = ae$ by (1), the foci $(\pm ae, 0)$ and the directrices $x = \pm a/e$ take the same forms in terms of a and e for both the hyperbola $(e > 1)$ and the ellipse $(e < 1)$.

If we interchange x and y in Equation (3), we obtain

$$\frac{y^2}{a^2} - \frac{x^2}{b^2} = 1. \tag{7}$$

This hyperbola has foci at $(0, \pm c)$. They, and its transverse axis, lie on the y-axis. Its asymptotes are $y = \pm ax/b$, and its graph generally resembles the one in Fig. 10.53.

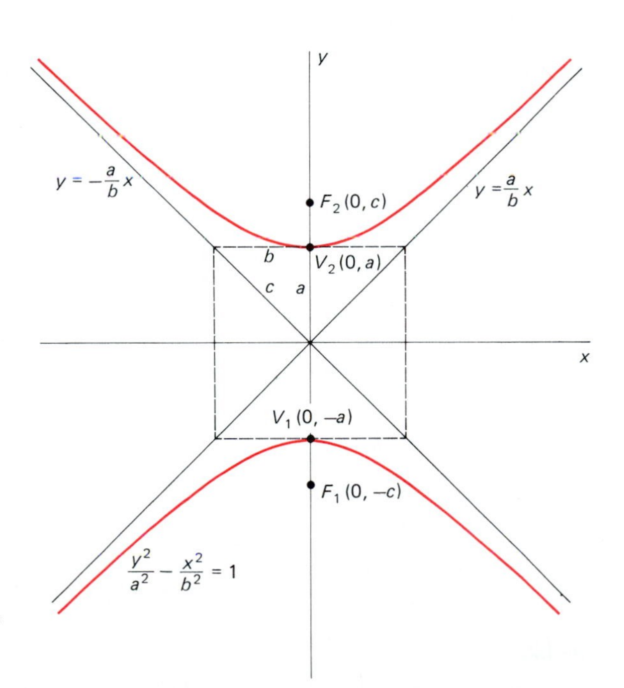

10.53 The hyperbola of Equation (7) has horizontal directrices.

When we studied the ellipse, we saw that its orientation—whether the major axis is horizontal or vertical—is determined by the relative sizes of a and b. In the case of the hyperbola, the situation is quite different, for the relative sizes of a and b make no such difference. The direction in which the hyperbola opens—horizontal as in Fig. 10.52 or vertical as in Fig. 10.53—is determined by the signs of the terms involving x^2 and y^2.

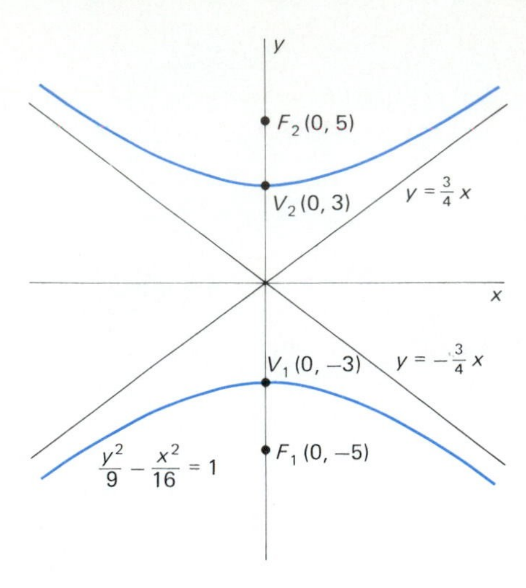

10.54 The hyperbola of Example 1

EXAMPLE 1 Sketch the graph of the hyperbola with equation

$$\frac{y^2}{9} - \frac{x^2}{16} = 1$$

Solution This is an equation of the form in (7), so the hyperbola opens vertically. Because $a = 3$ and $b = 4$, we find that $c = 5$ by using Equation (5): $c^2 = a^2 + b^2$. Thus the vertices are $(0, \pm 3)$, the foci are the two points $(0, \pm 5)$, and the asymptotes are the two lines $y = \pm 3x/4$. This hyperbola appears in Fig. 10.54.

EXAMPLE 2 Find the equation of the hyperbola with foci $(\pm 10, 0)$ and asymptotes $y = \pm 4x/3$.

Solution Because $c = 10$, we have

$$a^2 + b^2 = 100 \quad \text{and} \quad \frac{b}{a} = \frac{4}{3}.$$

Thus $b = 8$ and $a = 6$, and the equation of the hyperbola is

$$\frac{x^2}{36} - \frac{y^2}{64} = 1.$$

As we noted in Section 10.5, any equation of the form

$$Ax^2 + Cy^2 + Dx + Ey + F = 0 \tag{8}$$

with A and C nonzero can be reduced to the form

$$A(x - h)^2 + C(y - k)^2 = G$$

by completing the square in x and y. Now suppose that the coefficients A and C of the quadratic terms have *opposite signs;* for example, suppose that $A = p^2$ and $C = -q^2$. The equation above then becomes

$$p^2(x - h)^2 - q^2(y - k)^2 = G. \tag{9}$$

If $G = 0$, then factorization of the difference of squares on the left yields the equations

$$p(x - h) + q(y - k) = 0, \qquad p(x - h) - q(y - k) = 0$$

of two straight lines through (h, k) with slopes $\pm p/q$. If $G \neq 0$, then division of Equation (9) by G gives an equation that looks either like

$$\frac{(x - h)^2}{a^2} - \frac{(y - k)^2}{b^2} = 1 \qquad (\text{if } G > 0)$$

or like

$$\frac{(y - k)^2}{a^2} - \frac{(x - h)^2}{b^2} = 1 \qquad (\text{if } G < 0).$$

Thus if $AC < 0$ in Equation (8), the graph is either a pair of intersecting straight lines or a hyperbola.

EXAMPLE 3 Determine the graph of the equation

$$9x^2 - 4y^2 - 36x + 8y = 4.$$

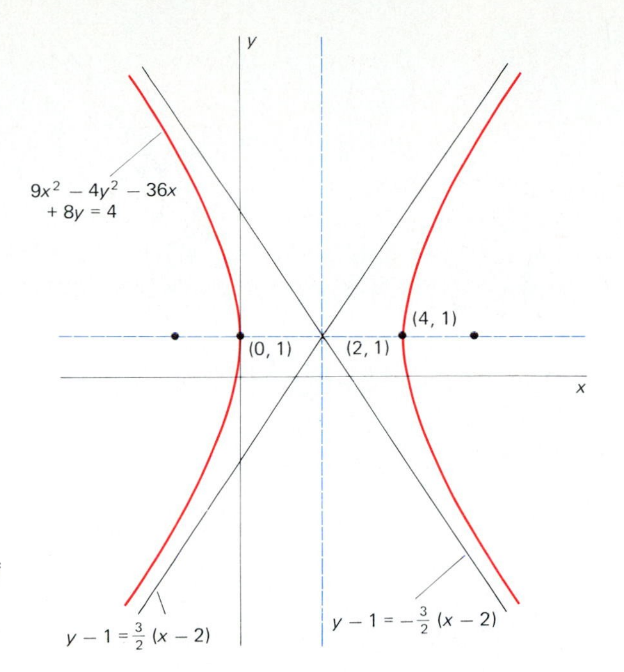

10.55 The hyperbola of Example 3, a translate of the hyperbola $x^2/4 - y^2/9 = 1$

Solution We collect the terms containing x, those containing y, and complete the square in each variable. We find that

$$9(x - 2)^2 - 4(y - 1)^2 = 36,$$

so

$$\frac{(x - 2)^2}{4} - \frac{(y - 1)^2}{9} = 1.$$

Hence the graph is a hyperbola with horizontal transverse axis and center (2, 1). Because $a = 2$ and $b = 3$, we find that $c = \sqrt{13}$. The vertices of the hyperbola are (0, 1) and (4, 1) and its foci are the two points $(2 \pm \sqrt{13}, 1)$. Its asymptotes are the two lines

$$y - 1 = \pm\tfrac{3}{2}(x - 2),$$

translates of the asymptotes $y = \pm 3x/2$ of the hyperbola $x^2/4 - y^2/9 = 1$. Figure 10.55 shows the graph of the translated hyperbola.

The *reflection property* of the hyperbola takes the same form as for the ellipse. If P is a point on a hyperbola, then the two lines PF_1 and PF_2 from P to the two foci make equal angles with the tangent line at P. In Fig. 10.56 this means that $\alpha = \beta$.

For an application of this reflection property, consider a mirror shaped like one branch of a hyperbola and made reflective on its outer (convex) surface. Then an incoming light ray aimed toward one focus will be reflected toward the other focus, as shown in Fig. 10.57. Figure 10.58 indicates the design of a reflecting telescope that makes use of the reflection properties of the parabola and the hyperbola. The parallel incoming light rays first are reflected by the parabola toward its focus at F. Then they are intercepted by an auxiliary hyperbolic mirror with foci at E and F and reflected into the eyepiece located at E.

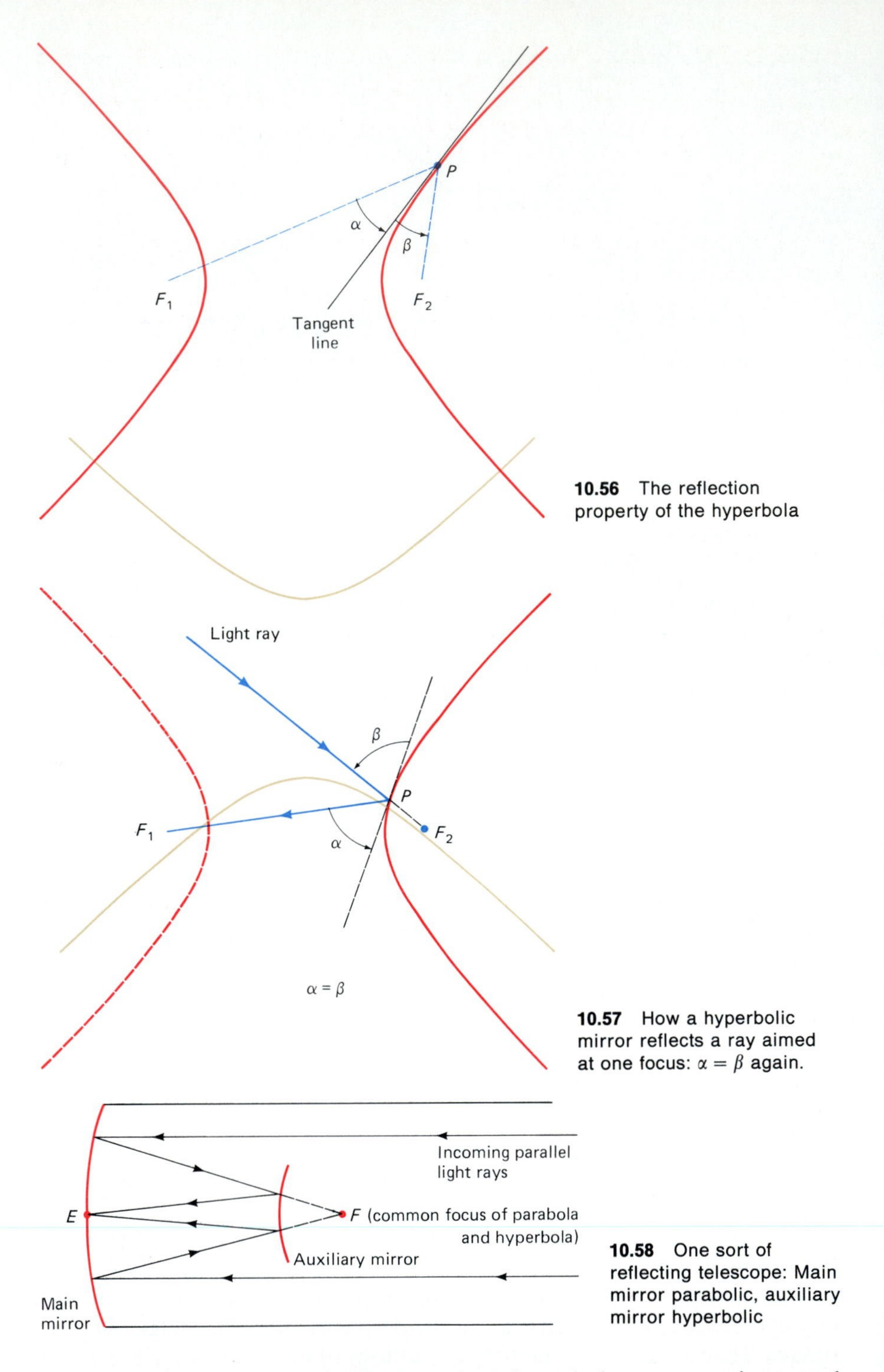

10.56 The reflection property of the hyperbola

10.57 How a hyperbolic mirror reflects a ray aimed at one focus: $\alpha = \beta$ again.

10.58 One sort of reflecting telescope: Main mirror parabolic, auxiliary mirror hyperbolic

The following example illustrates how hyperbolas are sometimes used to determine the position of ships at sea.

EXAMPLE 4 A ship lies at sea due east of point A on a long north–south coastline. Simultaneous signals are transmitted by a radio station at A and by one at B on the coast 200 miles south of A. The ship receives the signal from A 500 μs (microseconds) before it receives the one from B. Assume that the speed of radio signals is 980 ft/μs. How far out at sea is the ship?

Solution The situation is diagrammed in Fig. 10.59.

The difference between the distances of the ship S from A and B is

$$|SB| - |SA| = (500)(980)$$

ft; that is, 92.8 mi. Thus (by Problem 24) the ship lies on a hyperbola with foci A and B, and with $c = 100$ we have

$$a = \tfrac{1}{2}(92.8) = 46.4 \quad \text{and} \quad b = \sqrt{(100)^2 - (46.4)^2} \approx 88.6.$$

In the coordinate system of Fig. 10.59, the hyperbola has equation

$$\frac{y^2}{(46.4)^2} - \frac{x^2}{(88.6)^2} = 1.$$

We substitute $y = 100$ because the ship is due east of A. Thus we find that the ship's distance from the coastline is $x \approx 169.1$ miles.

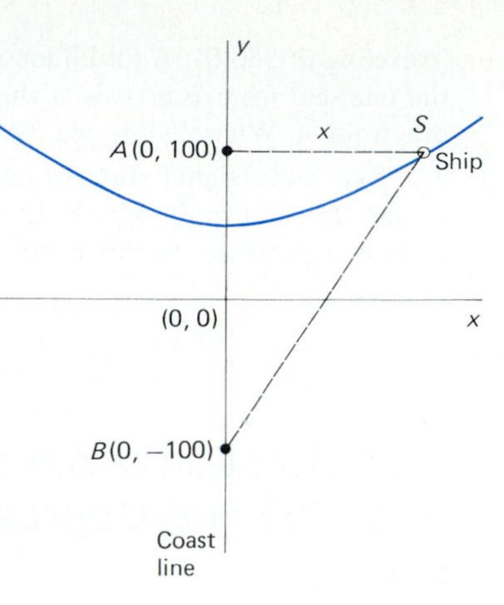

10.59 A navigation problem (Example 4)

10.6 PROBLEMS

In each of Problems 1–14, find an equation of the hyperbola described there.

1. Foci $(\pm 4, 0)$, vertices $(\pm 1, 0)$

2. Foci $(0, \pm 3)$, vertices $(0, \pm 2)$

3. Foci $(\pm 5, 0)$, asymptotes $y = \pm 3x/4$

4. Vertices $(\pm 3, 0)$, asymptotes $y = \pm 3x/4$

5. Vertices $(0, \pm 5)$, asymptotes $y = \pm x$

6. Vertices $(\pm 3, 0)$, eccentricity $e = 5/3$

7. Foci $(0, \pm 6)$, eccentricity $e = 2$

8. Vertices $(\pm 4, 0)$ and passing through $(8, 3)$

9. Foci $(\pm 4, 0)$, directrices $x = \pm 1$

10. Foci $(0, \pm 9)$, directrices $y = \pm 4$

11. Center $(2, 2)$, horizontal transverse axis of length 6, eccentricity $e = 2$

12. Center $(-1, 3)$, vertices $(-4, 3)$ and $(2, 3)$, foci $(-6, 3)$ and $(4, 3)$

13. Center $(1, -2)$, vertices $(1, 1)$ and $(1, -5)$, asymptotes $3x - 2y = 7$ and $3x + 2y = -1$

14. Focus $(8, -1)$, asymptotes $3x - 4y = 13$ and $3x + 4y = 5$

Sketch the graph of the equation given in each of Problems 15–20; indicate centers, foci, and asymptotes.

15. $x^2 - y^2 - 2x + 4y = 4$

16. $x^2 - 2y^2 + 4x = 0$

17. $y^2 - 3x^2 - 6y = 0$

18. $x^2 - y^2 - 2x + 6y = 9$

19. $9x^2 - 4y^2 + 18x + 8y = 31$

20. $4y^2 - 9x^2 - 18x - 8y = 41$

21. Show that the graph of the equation

$$\frac{x^2}{15 - c} - \frac{y^2}{c - 6} = 1 \qquad \text{is:}$$

(a) a hyperbola with foci $(\pm 3, 0)$ if $6 < c < 15$;
(b) an ellipse if $c < 6$.
Identify the graph in the case $c > 15$.

22. Establish that the tangent line to the hyperbola

$$\frac{x^2}{a^2} - \frac{y^2}{b^2} = 1$$

at the point $P(x_0, y_0)$ has equation

$$\frac{xx_0}{a^2} - \frac{yy_0}{b^2} = 1.$$

23. Use the result of Problem 22 to establish the reflection property of the hyperbola. (See the suggestion for Problem 25 in Section 10.5.)

24. Suppose that $0 < a < c$, and let $b = (c^2 - a^2)^{1/2}$. Show that the hyperbola $x^2/a^2 - y^2/b^2 = 1$ is the locus of a point P such that the *difference* between the distances $|PF_1|$ and $|PF_2|$ is equal to $2a$, where F_1 and F_2 are the two foci of the hyperbola.

25. Derive an equation for the hyperbola with foci at $(\pm 5, \pm 5)$ and vertices at $(\pm 3/\sqrt{2}, \pm 3/\sqrt{2})$. Use the difference definition of a hyperbola implied by Problem 24.

26. Two radio signal stations at A and B lie on an east–west line with A 100 mi west of B. A plane is flying west on a line 50 mi north of the line AB. Radio signals are sent (traveling at 980 ft/μs) simultaneously from A and B, and the one sent from B arrives at the plane 400 μs before the one from A. Where is the plane?

27. Two radio signal stations are located as in Problem 26 and transmit radio signals that travel at the same speed as in that problem. In this problem, however, it is known only that the plane is generally somewhere north of the line AB, that the signal sent from B arrives 400 μs before the one sent from A, and that the signal sent from A and reflected by the plane takes a total of 600 μs to reach B. Where is the plane?

10.7 Rotation of Axes and Second-Degree Curves

In the preceding three sections we have studied the second degree equation

$$Ax^2 + Cy^2 + Dx + Ey + F = 0, \tag{1}$$

which contains no xy-term. We found that its graph is always a conic section, apart from exceptional cases of the following types:

$$2x^2 + 3y^2 = -1 \quad \text{(no locus)},$$
$$2x^2 + 3y^2 = 0 \quad \text{(a single point)},$$
$$(2x - 1)^2 = 0 \quad \text{(a line)},$$
$$(2x - 1)^2 = 1 \quad \text{(two parallel lines)},$$
$$x^2 - y^2 = 0 \quad \text{(two intersecting lines)}.$$

We may therefore say that the graph of Equation (1) is a conic section, possibly **degenerate.** If either A or C is zero (but not both), then the graph is a parabola. It is an ellipse if $AC > 0$ (by the discussion of Equation (6) in Section 10.5), a hyperbola if $AC < 0$ (by the discussion of Equation (8) in Section 10.6).

Let us assume that $AC \neq 0$. Then we can determine the particular conic section represented by Equation (1) by completing squares. That is, we write (1) in the form

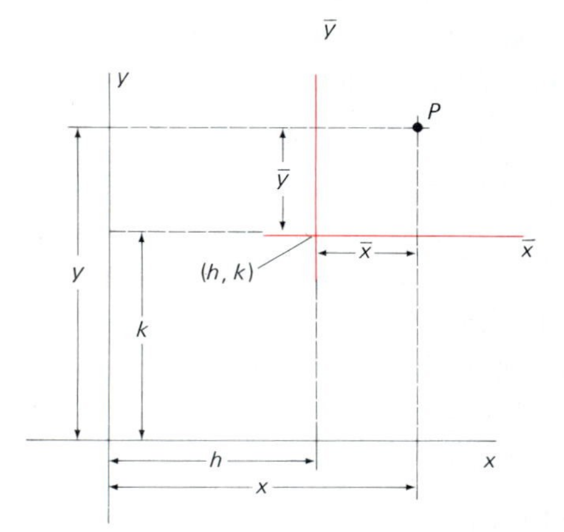

10.60 A translation of coordinates

$$A(x - h)^2 + C(y - k)^2 = G. \tag{2}$$

This equation can be simplified further by a **translation of coordinates** to a new $\bar{x}\bar{y}$-coordinate system centered at the point (h, k) in the old xy-system. The geometry of this change in coordinates is shown in Fig. 10.60. The relation between the old and the new coordinates is

$$\left.\begin{aligned} \bar{x} &= x - h, \\ \bar{y} &= y - k \end{aligned}\right\} \quad \text{or} \quad \left\{\begin{aligned} x &= \bar{x} + h, \\ y &= \bar{y} + k. \end{aligned}\right. \tag{3}$$

In the new $\bar{x}\bar{y}$-coordinate system, Equation (2) takes the simpler form

$$A\bar{x}^2 + C\bar{y}^2 = G, \tag{2$'$}$$

from which it is clear whether we have an ellipse, a hyperbola, or a degenerate case.

We now turn to the general second degree equation

$$Ax^2 + Bxy + Cy^2 + Dx + Ey + F = 0. \tag{4}$$

Note the presence of the "cross product," or xy-, term. In order to recognize its graph, we need to change to a new $x'y'$-coordinate system obtained by a **rotation of axes.**

We get the $x'y'$-axes from the xy-axes by a rotation through an angle α in the counterclockwise direction. In the notation of Fig. 10.61, we have

$$x = OQ = OP\cos(\phi + a) \quad \text{and} \quad y = PQ = OP\sin(\phi + \alpha). \tag{5}$$

Similarly,

$$x' = OR = OP\cos\phi \quad \text{and} \quad y' = PR = OP\sin\phi. \tag{6}$$

Recall the addition formulas

$$\cos(\phi + \alpha) = \cos\phi\cos\alpha - \sin\phi\sin\alpha,$$

$$\sin(\phi + \alpha) = \sin\phi\cos\alpha + \cos\phi\sin\alpha.$$

With the aid of these identities and the substitution of the equations in (6) in the equations in (5), we obtain this result.

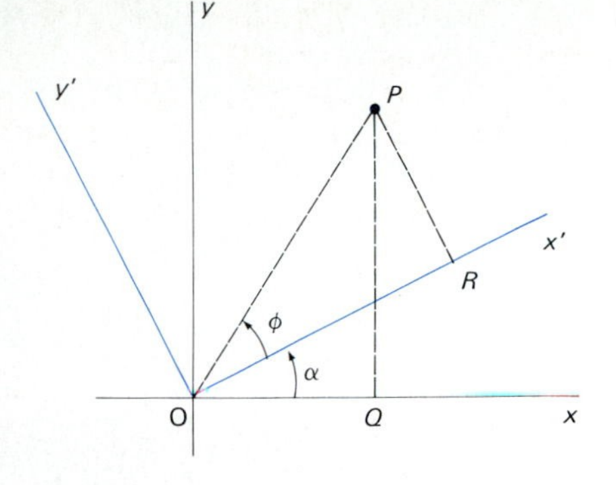

10.61 A rotation of coordinates through the angle α

Equations for Rotation of Axes

$$x = x'\cos\alpha - y'\sin\alpha,$$
$$y = x'\sin\alpha + y'\cos\alpha. \tag{7}$$

These equations express the old xy-coordinates of the point P in terms of its new $x'y'$-coordinates and the rotation angle α. The following example illustrates how the equations in (7) may be used to transform the equation of a curve from xy-coordinates into the rotated $x'y'$-coordinates.

EXAMPLE 1 The xy-axes are rotated through an angle of $\alpha = 45°$. Find the equation of the curve $2xy = 1$ in the new $x'y'$-coordinate system.

Solution Because $\cos 45° = \sin 45° = 1/\sqrt{2}$, the equations in (7) yield

$$x = \frac{x' - y'}{\sqrt{2}} \quad \text{and} \quad y = \frac{x' + y'}{\sqrt{2}}.$$

The original equation $2xy = 1$ thus becomes

$$(x')^2 - (y')^2 = 1.$$

So, in the $x'y'$-coordinate system, we have a hyperbola with $a = b = 1$, $c = \sqrt{2}$, and foci $(\pm\sqrt{2}, 0)$. In the original xy-coordinates, its foci are $(1, 1)$ and $(-1, -1)$ and its asymptotes are the x- and y-axes. This hyperbola is shown in Fig. 10.62. (A hyperbola of this form, one which has equation $xy = k$, is called a **rectangular** hyperbola because its asymptotes are perpendicular.)

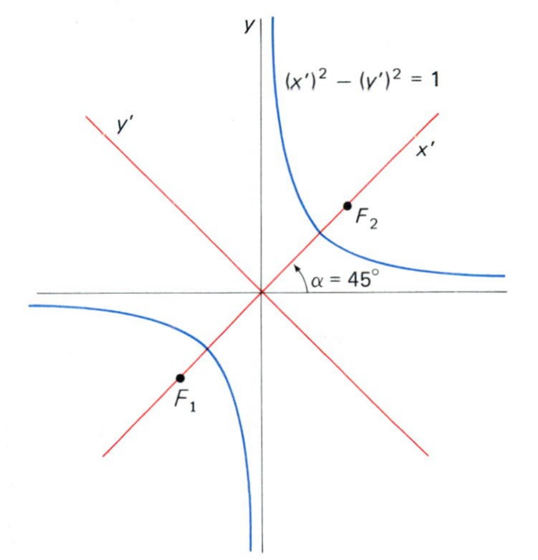

10.62 The graph of $2xy = 1$ is a hyperbola.

Example 1 strongly suggests that the cross-product term Bxy of Equation (4) may disappear upon rotation of the coordinate axes. One can, indeed, always choose an appropriate angle of rotation so that, in the new coordinate system, there is no cross-product term.

To determine the appropriate rotation angle, we substitute the equations in (7) for x and y in the general second degree equation in (4). We obtain the following new second degree equation:

$$A'(x')^2 + B'x'y' + C'(y')^2 + D'x' + E'y' + F' = 0. \tag{8}$$

The new coefficients are given in terms of the old ones and the angle α by the following equations.

$$
\begin{aligned}
A' &= A\cos^2\alpha + B\cos\alpha\sin\alpha + C\sin^2\alpha,\\
B' &= B(\cos^2\alpha - \sin^2\alpha) + 2(C - A)\sin\alpha\cos\alpha,\\
C' &= A\sin^2\alpha - B\sin\alpha\cos\alpha + C\cos^2\alpha,\\
D' &= D\cos\alpha + E\sin\alpha,\\
E' &= -D\sin\alpha + E\cos\alpha, \quad \text{and}\\
F' &= F.
\end{aligned}
\tag{9}
$$

Now suppose that an equation of the form in (4) is given, with $B \neq 0$. We simply choose α so that $B' = 0$ in the list of new coefficients in (9). Then Equation (8) will have no cross-product term, and we can identify and sketch the curve with little trouble in the $x'y'$-coordinate system. But is it really easy to choose such an angle α?

It is. We recall that

$$\cos 2\alpha = \cos^2\alpha - \sin^2\alpha \quad \text{and} \quad \sin 2\alpha = 2\sin\alpha\cos\alpha.$$

So the equation for B' in (9) may be written

$$B' = B\cos 2\alpha + (C - A)\sin 2\alpha.$$

We can cause B' to be zero by choosing as α that (unique) acute angle such that

$$\cot 2\alpha = \frac{A - C}{B}. \tag{10}$$

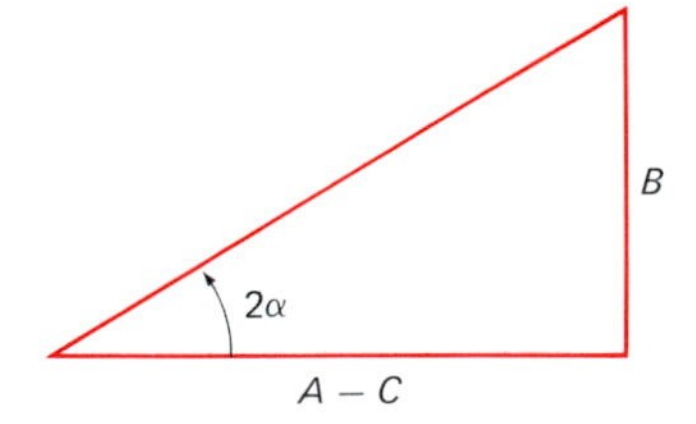

10.63 Finding $\sin\alpha$ and $\cos\alpha$ given $\cot 2\alpha = (A - C)/B$

If we plan to use the equations in (9) to calculate the coefficients in the transformed Equation (8), we will need the values of $\sin\alpha$ and $\cos\alpha$ that follow from Equation (10). It is sometimes convenient to calculate these values directly from $\cot 2\alpha$, as follows. From the right triangle in Fig. 10.63, we can read the numerical value of $\cos 2\alpha$. Because the cosine and cotangent are both positive in the first quadrant and both negative in the second quadrant, we give $\cos 2\alpha$ the same sign as $\cot 2\alpha$. Then we use the half-angle formulas to get $\sin\alpha$ and $\cos\alpha$:

$$\sin\alpha = \left(\frac{1 - \cos 2\alpha}{2}\right)^{1/2}, \qquad \cos\alpha = \left(\frac{1 + \cos 2\alpha}{2}\right)^{1/2} \tag{11}$$

Once we have the values of $\sin\alpha$ and $\cos\alpha$, we can compute the coefficients in the resulting Equation (8) by means of the equations in (9). Alternatively, it's frequently simpler to get Equation (8) directly by substituting the equations in (7), with the numerical values of $\sin\alpha$ and $\cos\alpha$ obtained as before, in Equation (4). If we are using a calculator or a computer that has an inverse tangent function provided, then we can calculate α more briefly by using the formula

$$\alpha = \frac{\pi}{4} - \frac{1}{2}\tan^{-1}\left(\frac{A - C}{B}\right), \tag{12}$$

which follows from Equation (10) and the observation that $\cot^{-1}x = \dfrac{\pi}{2} - \tan^{-1}x$.

EXAMPLE 2 Determine the graph of the equation

$$73x^2 - 72xy + 52y^2 - 30x - 40y - 75 = 0.$$

Solution We begin with Equation (10) and find that $\cot 2\alpha = -\frac{7}{24}$, so that $\cos 2\alpha = -\frac{7}{25}$. Thus

$$\sin\alpha = \left(\frac{1-(-\frac{7}{25})}{2}\right)^{1/2} = \frac{4}{5}, \qquad \cos\alpha = \left(\frac{1+(-\frac{7}{25})}{2}\right)^{1/2} = \frac{3}{5}.$$

Then, with $A = 73$, $B = -72$, $C = 52$, $D = -30$, $E = -40$, and $F = -75$, the equations in (9) yield

$$A' = 25, \qquad D' = -50,$$
$$B' = 0 \quad \text{(this was our goal)}, \qquad E' = 0,$$
$$C' = 100, \qquad F' = -75.$$

Consequently the equation in the new $x'y'$-coordinate system, obtained by rotation through an angle of $\alpha = \arcsin(\frac{4}{5})$ (approximately 53.13°), is

$$25(x')^2 + 100(y')^2 - 50x' = 75.$$

Alternatively, we could have obtained this equation by substituting

$$x = \tfrac{3}{5}x' - \tfrac{4}{5}y', \qquad y = \tfrac{4}{5}x' + \tfrac{3}{5}y'$$

in the original equation.

By completing the square in x' we finally obtain

$$25(x'-1)^2 + 100(y')^2 = 100,$$

which we put into the standard form

$$\frac{(x'-1)^2}{4} + \frac{(y')^2}{1} = 1.$$

Thus the original curve is an ellipse with major semiaxis 2, minor semiaxis 1, and center (1, 0) in the $x'y'$-coordinate system, as shown in Fig. 10.64.

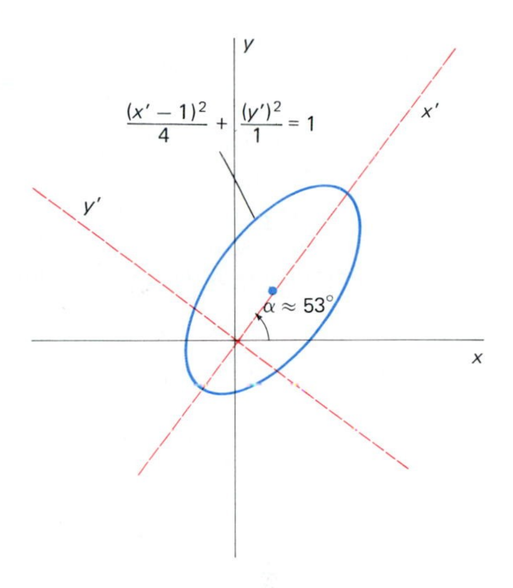

10.64 The ellipse of Example 2

CLASSIFICATION OF CONICS

Example 2 illustrates the general procedure for finding the graph of a second degree equation. First, if there is a cross-product term, rotate axes to eliminate it. Then translate axes (if necessary) to reduce the equation to the standard form of a parabola, ellipse, or hyperbola (or a degenerate case of one of these).

There *is* a test by which the nature of the curve may be discovered without actually carrying out these transformations. This test stems from the fact (see Problem 14) that, whatever the angle α of rotation, the equations in (9) imply that

$$(B')^2 - 4A'C' = B^2 - 4AC. \tag{13}$$

Thus the **discriminant** $B^2 - 4AC$ is an *invariant* under any rotation of axes. If α is chosen so that $B' = 0$, then the left-hand side in Equation (13) is simply $-4A'C'$. Because A' and C' are the coefficients of the squared terms, our earlier discussion of Equation (1) now applies. It follows that the graph will be

1. A *parabola* if $B^2 - 4AC = 0$,
2. An *ellipse* if $B^2 - 4AC < 0$,
3. A *hyperbola* if $B^2 - 4AC > 0$.

Of course, degenerate cases may occur.

Here are some examples.

1. $x^2 + 2xy + y^2 = 1$ is a (degenerate) parabola.
2. $x^2 + xy + y^2 = 1$ is an ellipse.
3. $x^2 + 3xy + y^2 = 1$ is a hyperbola.

10.7 PROBLEMS

In each of Problems 1–6, the graph of the given equation is a translated conic section, possibly degenerate. Give its equation in standard position in the appropriate $x'y'$-coordinate system, and give the origin (in terms of xy-coordinates) of the new coordinate system.

1. $2x^2 + y^2 - 8x - 6y + 13 = 0$
2. $x^2 + 3y^2 + 6x + 12y + 18 = 0$
3. $9x^2 - 16y^2 - 18x - 32y - 151 = 0$
4. $y^2 - 4x - 4y - 8 = 0$
5. $2x^2 + 3y^2 - 8x - 18y + 35 = 0$
6. $4x^2 - y^2 - 8x - 4y = 0$

In each of Problems 7–20, the graph of the given equation is a rotated conic section (possibly degenerate). Identify it; give the counterclockwise angle α of rotation that yields a rotated $x'y'$-system in which the graph is in standard position, and give the transformed equation in $x'y'$-coordinates.

7. $3x^2 + 2xy + 3y^2 = 1$
8. $x^2 + 6xy + y^2 = 1$
9. $4x^2 + 4xy + y^2 = 20$
10. $9x^2 + 4xy + 6y^2 = 40$
11. $4x^2 + 6xy - 4y^2 = 5$
12. $19x^2 + 6xy + 11y^2 = 40$
13. $22x^2 + 12xy + 17y^2 = 26$
14. $9x^2 + 12xy + 4y^2 = 13$
15. $52x^2 + 72xy + 73y^2 = 100$
16. $9x^2 + 24xy + 16y^2 = 0$
17. $33x^2 + 8xy + 18y^2 = 68$
18. $40x^2 + 36xy + 25y^2 = 52$
19. $119x^2 + 240xy - 119y^2 = 0$
20. $313x^2 + 120xy + 194y^2 = 0$

In each of Problems 21–26, identify the graph of the given equation by carrying out first a rotation and then a translation. Give the angle of rotation, the translated center of coordinates in the rotated coordinate system, and the final transformed equation.

21. $34x^2 - 24xy + 41y^2 - 40x - 30y - 25 = 0$
22. $41x^2 - 24xy + 34y^2 + 20x - 140y + 125 = 0$
23. $23x^2 - 72xy + 2y^2 + 140x + 20y - 75 = 0$
24. $9x^2 + 24xy + 16y^2 - 170x - 60y + 245 = 0$
25. $161x^2 + 480xy - 161y^2 - 510x - 272y = 0$
26. $144x^2 - 120xy + 25y^2 - 65x - 156y - 169 = 0$
27. Solve the equations in (7) to show that the rotated coordinates (x', y') are given in terms of the original coordinates by
$$x' = x \cos \alpha + y \sin \alpha, \qquad y' = -x \sin \alpha + y \cos \alpha.$$
28. Use the equations in (9) to verify Equation (13).
29. Prove that the sum $A + C$ of the coefficients of x^2 and y^2 in Equation (4) is invariant under rotation. That is, show that $A' + C' = A + C$ for any rotation through an angle α.
30. Use the equations in (9) to prove that any rotation of axes transforms the equation $x^2 + y^2 = r^2$ into the equation $(x')^2 + (y')^2 = r^2$.
31. Consider the equation
$$x^2 + Bxy - y^2 + Dx + Ey + F = 0.$$
Prove that there is a rotation of axes such that $A' = 0 = C'$ in the resulting equation. *Suggestion:* Find the angle α for which $A' = 0$. Then apply the result of Problem 29. What can you conclude about the graph of the given equation?

32. Suppose that $B^2 - 4AC < 0$, so that the equation

$$Ax^2 + Bxy + Cy^2 = 1$$

represents an ellipse. Prove that its area is

$$\pi ab = \frac{2\pi}{\sqrt{4AC - B^2}}$$

where a and b are the lengths of its semiaxes.

33. Show that the equation $27x^2 + 37xy + 17y^2 = 1$ represents an ellipse, and then find the points of the ellipse that are closest to and farthest from the origin.

34. Show that the equation $x^2 + 14xy + 49y^2 = 100$ represents a parabola (possibly degenerate), and then find the point of this parabola that is closest to the origin.

*10.7 Optional Computer Application

The program listed in Fig. 10.65 was written (in IBM-PC BASIC) to calculate the new coefficients that result when the axes are rotated to eliminate the xy-term in the equation

$$Ax^2 + Bxy + Cy^2 + Dx + Ey + F = 0.$$

```
100 REM--Program ROTATE
105 REM
110 REM--Rotates the general second-degree curve
115 REM
120 REM    Ax^2 + Bxy + Cy^2 + Dx + Ey + F = 0
125 REM
130 REM--so as to eliminate the cross-product term.
140 REM
150        INPUT "A, B, C"; A, B, C
160        INPUT "D, E";    D, E
170        INPUT "F";       F
180        PI = 3.141593
190 REM
200        IF B = 0 THEN  ALPHA = 0
                    ELSE  ALPHA = PI/4 - (ATN((A-C)/B))/2
210        COSA = COS(ALPHA)  :  SINA = SIN(ALPHA)
220 REM
230        A1 = A*COSA^2 + B*COSA*SINA + C*SINA^2
240        B1 = B*(COSA^2 - SINA^2) + 2*(C - A)*SINA*COSA
250        C1 = A*SINA^2 - B*SINA*COSA + C*COSA^2
260        D1 =  D*COSA + E*SINA
270        E1 = -D*SINA + E*COSA
280 REM
290        PRINT : PRINT "ALPHA = "; 180*ALPHA/PI ; "DEG"
295        PRINT
300        PRINT "NEW COEFFICIENTS:"
305        PRINT
310        PRINT " A = "; A1
320        PRINT " B = "; B1
330        PRINT " C = "; C1
340        PRINT " D = "; D1
350        PRINT " E = "; E1
360        PRINT " F = "; F
365 REM
370        END
```

10.65 Listing of Program ROTATE

Lines 150–170 call for the values of the coefficients A, B, C, D, E, and F to be entered. The angle α = ALPHA is then calculated in line 200 using the formula in Equation (12). Then the new coefficients are calculated in lines 230–270 using the rotation formulas in (9). A run of this program with the data of Example 2 produces the following output.

```
RUN
A, B, C? 73, -72, 52
D, E? -30, -40
F? -75

ALPHA = 53.13011 DEGREES

NEW COEFFICIENTS:
A = 25
B = 1.716614E-05
C = 100
D = -50
E = 5.722046E-06
F = -75
```

The "slightly nonzero" values for B and E result from roundoff error.

Exercises: Use this program—rewritten if necessary for a computer that accepts only single-character variables—to solve Problems 7–26 of Section 10.7.

CHAPTER 10 REVIEW: Properties of Conic Sections

The parabola with focus $(p, 0)$ and directrix $x = -p$ has eccentricity $e = 1$ and equation $y^2 = 4px$. The table on the right compares the properties of an ellipse and a hyperbola, each with foci $(\pm c, 0)$ and major axis of length $2a$.

	Ellipse	*Hyperbola*
Eccentricity	$e = \frac{c}{a} < 1$	$e = \frac{c}{a} > 1$
a, b, c relation	$a^2 = b^2 + c^2$	$c^2 = a^2 + b^2$
Equation	$\frac{x^2}{a^2} + \frac{y^2}{b^2} = 1$	$\frac{x^2}{a^2} - \frac{y^2}{b^2} = 1$
Vertices	$(\pm a, 0)$	$(\pm a, 0)$
y-intercepts	$(0, \pm b)$	None
Directrices	$x = \pm\frac{a}{e}$	$x = \pm\frac{a}{e}$
Asymptotes	None	$y = \pm\frac{bx}{a}$

Use the list below as a guide to additional concepts that you may need to review.

1. The relationship between rectangular and polar coordinates
2. The graph of an equation in polar coordinates
3. The area formula in polar coordinates
4. Translation of coordinates
5. Equations and procedure for rotation of axes
6. Use of the discriminant to classify the graph of a second degree equation

CHAPTER 10 MISCELLANEOUS PROBLEMS

Sketch the graphs of the equations in Problems 1–30. If the graph in 1–18 is a conic section, label its center, foci, and vertices.

1. $x^2 + y^2 - 2x + 2y = 2$
2. $x^2 + y^2 = x + y$
3. $x^2 + y^2 - 6x + 2y + 9 = 0$
4. $y^2 = 4(x + y)$
5. $x^2 = 8x - 2y - 20$
6. $x^2 + 2y^2 - 2x + 8y + 8 = 0$
7. $9x^2 + 4y^2 = 36x$
8. $x^2 - y^2 = 2x - 2y - 1$
9. $y^2 - 2x^2 = 4x + 2y + 3$
10. $9y^2 - 4x^2 = 8x + 18y + 31$
11. $x^2 + 2y^2 = 4x + 4y - 12$
12. $x^2 + 2xy + y^2 + 1 = 0$
13. $x^2 + 2xy - y^2 = 7$
14. $xy + 8 = 0$
15. $3x^2 - 22xy + 3y^2 = 4$

16. $x^2 - 6xy + y^2 = 4$

17. $9x^2 - 24xy + 16y^2 = 20x + 15y$

18. $7x^2 + 48xy - 7y^2 + 25 = 0$

19. $r = -2\cos\theta$

20. $\cos\theta + \sin\theta = 0$

21. $r = \dfrac{1}{\sin\theta - \cos\theta}$

22. $r\sin^2\theta = \cos\theta$

23. $r = 3\csc\theta$

24. $r = 2(\cos\theta - 1)$

25. $r^2 = 4\cos\theta$

26. $r\theta = 1$

27. $r = 3 - 2\sin\theta$

28. $r = \dfrac{1}{1 + \cos\theta}$

29. $r = \dfrac{4}{2 + \cos\theta}$

30. $r = \dfrac{4}{1 - 2\cos\theta}$

In each of Problems 31–38, find the area of the region described.

31. Inside both $r = 2\sin\theta$ and $r = 2\cos\theta$

32. Inside $r^2 = 4\cos\theta$

33. Inside $r = 3 - 2\sin\theta$ and outside $r = 4$

34. Inside $r^2 = 2\sin 2\theta$ and outside $r = 2\sin\theta$

35. Inside $r = 2\sin 2\theta$ and outside $r = \sqrt{2}$

36. Inside $r = 3\cos\theta$ and outside $r = 1 + \cos\theta$

37. Inside $r = 1 + \cos\theta$ and outside $r = \cos\theta$

38. Between the loops of $r = 1 - 2\sin\theta$

39. Find a polar coordinates equation of the circle that passes through the origin and is centered at the point with polar coordinates (p, α).

40. Find a simple equation of the parabola with focus the origin and directrix the line $y = x + 4$. Recall from Chapter 3, Miscellaneous Problem 71, that the distance from the point (x_0, y_0) to the line $Ax + By + C = 0$ is

$$\frac{|Ax_0 + By_0 + C|}{\sqrt{A^2 + B^2}}.$$

41. A *diameter* of an ellipse is a chord through its center. Find the maximum and minimum lengths of diameters of the ellipse with equation

$$\frac{x^2}{a^2} + \frac{y^2}{b^2} = 1$$

42. Use calculus to prove that the ellipse of Problem 41 is normal to the coordinate axis at each of its four vertices.

43. The parabolic arch of a bridge has base width b and height h at its center. Write its equation, choosing the origin on the ground at the left end of the arch.

44. Use methods of calculus to find the points on the ellipse

$$\frac{x^2}{a^2} + \frac{y^2}{b^2} = 1$$

that are nearest to and farthest from: (a) the center $(0, 0)$; (b) the focus $(c, 0)$.

45. Consider a line segment QR that contains a point P such that $|QP| = a$ and $|PR| = b$. Suppose that Q is constrained to move on the y-axis, while R must remain on the x-axis. Prove that the locus of P is an ellipse.

46. Suppose that $a > 0$ and that F_1 and F_2 are two fixed points in the plane with $|F_1F_2| > 2a$. Imagine a point P that moves in such a way that $|PF_2| = 2a + |PF_1|$. Prove that the locus of P is one branch of a hyperbola with foci F_1 and F_2. Then—as a consequence—explain how to construct points on a hyperbola by drawing appropriate circles centered at its foci.

47. Let Q_1 and Q_2 be two points on the parabola $y^2 = 4px$. Let P be the point of the parabola at which the tangent line is parallel to Q_1Q_2. Prove that the horizontal line through P bisects the segment Q_1Q_2.

48. Determine the locus of a point P such that the product of its distances from the two fixed points $F_1(-a, 0)$ and $F_2(a, 0)$ is a^2.

49. Find the eccentricity of the conic section

$$3x^2 - y^2 + 12x + 9 = 0.$$

50. Find the area bounded by the loop of the *strophoid* $r = \sec\theta - 2\cos\theta$ shown in Fig. 10.66.

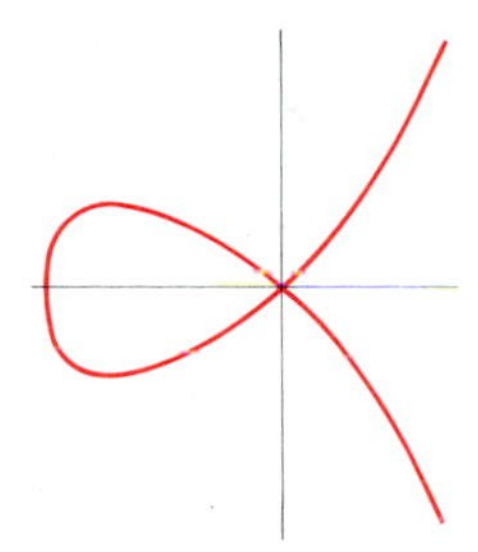

10.66 Strophoid

51. Find the area bounded by the loop of the *folium of Descartes* $x^3 + y^3 = 3xy$ shown in Fig. 10.67. *Suggestion:*

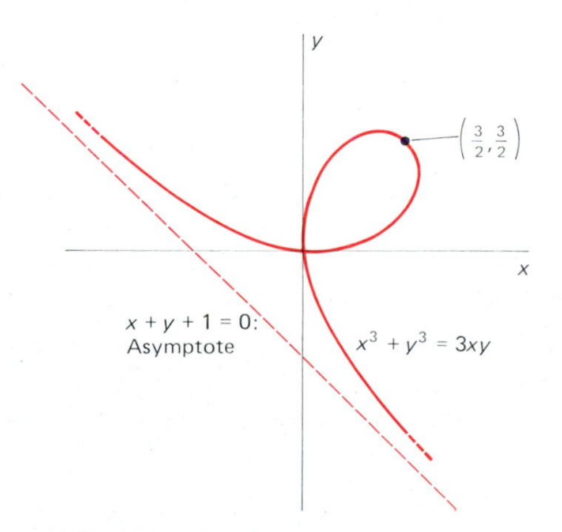

10.67 The folium of Descartes
$x^3 + y^3 = 3xy$

Change to polar coordinates, then substitute $u = \tan\theta$ to evaluate the area integral.

52. Use the method of Problem 51 to find the area bounded by the first quadrant loop (similar to the folium of Problem 51) of the curve $x^5 + y^5 = 5x^2y^2$.

53. Show that the graph of the equation

$$2929x^2 - 3456xy + 1921y^2 - 9000x - 12{,}000y - 15{,}625 = 0$$

is the ellipse shown in Fig. 10.68. This ellipse has its vertices at (15, 20) and $(-0.6, -0.8)$ and its major axis is inclined at the angle $\tan^{-1}(\frac{4}{3})$ from the horizontal.

54. Suppose that $0 < c < \pi$. Prove that the graph of the equation

$$\sin^{-1} x + \sin^{-1} y = c$$

is an ellipse rotated (from standard position) through an angle of 45°. (*Suggestion:* If $u = \sin^{-1} x$ and $v = \sin^{-1} y$, then $u + v = c$. Expand the left-hand side of the equation $\cos(u + v) = \cos c = k$.)

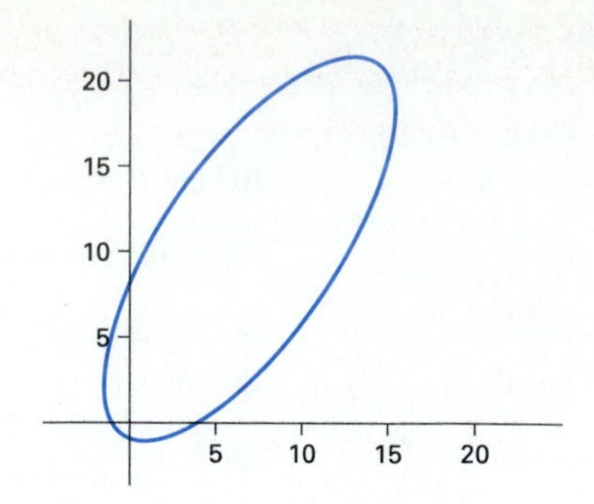

10.68 The ellipse of Problem 53

chapter eleven

Indeterminate Forms, Taylor's Formula, and Improper Integrals

11.1 Indeterminate Forms and l'Hôpital's Rule

An *indeterminate form* is a certain type of expression with a limit that is not evident by inspection. If

$$\lim_{x \to a} f(x) = 0 = \lim_{x \to a} g(x),$$

then we say that the quotient $f(x)/g(x)$ has the **indeterminate form** 0/0 at $x = a$. For example, to differentiate the trigonometric functions (Section 3.7), we needed to know that

$$\lim_{x \to 0} \frac{\sin x}{x} = 1. \tag{1}$$

Here, $f(x) = \sin x$ and $g(x) = x$. Because $\sin x$ and x both approach zero as $x \to 0$, the quotient $(\sin x)/x$ has the indeterminate form 0/0 at $x = 0$. Consequently, we had to use a special geometric argument to find the limit in (1)—see Theorem 3 of Section 3.7. Indeed, something of this sort happens whenever we compute a derivative, because the quotient

$$\frac{f(x) - f(a)}{x - a},$$

whose limit as $x \to a$ is the derivative $f'(a)$, has the indeterminate form 0/0 at $x = a$.

Sometimes the limit of an indeterminate form can be found by a special algebraic manipulation or construction, as in our earlier computations of derivatives. Often, however, it is more convenient to apply a rule that appeared in the first calculus textbook ever published, in 1696, by the Marquis de l'Hôpital. L'Hôpital was a French nobleman who had hired the Swiss mathematician John Bernoulli as his calculus tutor, and "l'Hôpital's rule" is actually due to Bernoulli.

Theorem 1 *L'Hôpital's Rule*

Suppose that the functions f and g are differentiable in a deleted neighborhood of the point a, and that $g'(x)$ is nonzero in that neighborhood. Suppose also that

$$\lim_{x\to a} f(x) = 0 = \lim_{x\to a} g(x).$$

Then

$$\lim_{x\to a} \frac{f(x)}{g(x)} = \lim_{x\to a} \frac{f'(x)}{g'(x)} \tag{2}$$

provided that the limit on the right either exists (as a finite number) or is $\pm\infty$.

In essence, the rule says that if $f(x)/g(x)$ has the indeterminate form 0/0 at $x = a$, then—subject to a few mild restrictions—this quotient has the same limit at $x = a$ as does the quotient $f'(x)/g'(x)$ of *derivatives*. The proof appears at the end of this section.

EXAMPLE 1 Find $\displaystyle\lim_{x\to 0} \frac{e^x - 1}{\sin 2x}$.

Solution The fraction whose limit we seek has the indeterminate form 0/0 at $x = 0$. The numerator and denominator are clearly differentiable in some deleted neighborhood of $x = 0$, and the derivative of the denominator is certainly nonzero if the neighborhood is small enough (if $0 < |x| < \pi/4$). So l'Hôpital's rule applies, and

$$\lim_{x\to 0} \frac{e^x - 1}{\sin 2x} = \lim_{x\to 0} \frac{e^x}{2\cos 2x} = \frac{1}{2}.$$

If it turns out that the quotient $f'(x)/g'(x)$ is again indeterminate, then l'Hôpital's rule may be applied a second (or third, or . . .) time, as in the next example. When the rule is applied repeatedly, however, the conditions for its applicability must be checked at each stage.

EXAMPLE 2 Find $\displaystyle\lim_{x\to 1} \frac{1 - x + \ln x}{1 + \cos \pi x}$.

Solution

$$\lim_{x\to 1}\frac{1-x+\ln x}{1+\cos \pi x}=\lim_{x\to 1}\frac{-1+1/x}{-\pi\sin \pi x}\qquad\text{(still of the form 0/0)}$$

$$=\lim_{x\to 1}\frac{x-1}{\pi x\sin \pi x}\qquad\text{(algebraic simplification)}$$

$$=\lim_{x\to 1}\frac{1}{\pi\sin \pi x+\pi^2 x\cos \pi x}\qquad\text{(l'Hôpital's rule again)}$$

$$=-\frac{1}{\pi^2}\qquad\text{(by inspection).}$$

Because the final limit exists, so do the previous ones; the existence of the final limit in (2) implies the existence of the first.

When you need to apply l'Hôpital's rule repeatedly in this way, you need only keep differentiating the numerator and denominator separately until at least one of them has a nonzero finite limit. At that point you can recognize the limit of the quotient by inspection, as in the final step in Example 2.

EXAMPLE 3 Find $\lim_{x\to 0}\dfrac{\sin x}{x+x^2}$.

Solution If we simply applied l'Hôpital's rule twice in succession, the result would be the *incorrect* computation

$$\lim_{x\to 0}\frac{\sin x}{x+x^2}=\lim_{x\to 0}\frac{\cos x}{1+2x}$$

$$=\lim_{x\to 0}\frac{-\sin x}{2}=0.\qquad\textbf{(Wrong!)}$$

The reason this answer is wrong is that $(\cos x)/(1+2x)$ is *not* an indeterminate form. Thus l'Hôpital's rule must not be applied to it. The correct computation is

$$\lim_{x\to 0}\frac{\sin x}{x+x^2}=\lim_{x\to 0}\frac{\cos x}{1+2x}=\frac{\lim_{x\to 0}\cos x}{\lim_{x\to 0}(1+2x)}=\frac{1}{1}=1.$$

The point of Example 3 is to issue a warning: Verify the hypotheses of l'Hôpital's rule *before* applying it. It is an oversimplification to say that l'Hôpital's rule "works when you need it and doesn't work when you don't," but there is still much truth in this statement.

INDETERMINATE FORMS INVOLVING ∞

L'Hôpital's rule has several variations. In addition to the fact that the limit in Equation (2) is allowed to be infinite, the real number a in l'Hôpital's rule may be replaced by either $+\infty$ or $-\infty$. For example,

$$\lim_{x\to\infty}\frac{f(x)}{g(x)}=\lim_{x\to\infty}\frac{f'(x)}{g'(x)}\qquad(3)$$

provided the other hypotheses are satisfied. In particular, to use Equation (3), we must first verify that

$$\lim_{x \to \infty} f(x) = 0 = \lim_{x \to \infty} g(x)$$

and that the right-hand limit in Equation (3) exists. The proof of this version of l'Hôpital's rule is outlined in Problem 50.

L'Hôpital's rule may also be used when $f(x)/g(x)$ has the **indeterminate form** ∞/∞. This means that

$$\lim_{x \to a} f(x) \quad \text{is either} \quad +\infty \quad \text{or} \quad -\infty$$

and

$$\lim_{x \to a} g(x) \quad \text{is either} \quad +\infty \quad \text{or} \quad -\infty.$$

The proof of this extension of the rule is more difficult, and will be omitted. For a proof, see, for example, A. E. Taylor and W. R. Mann, *Advanced Calculus*, third edition (New York: John Wiley, 1983), page 107.

EXAMPLE 4 Find $\displaystyle\lim_{x \to \infty} \frac{e^x}{x^2 + x}$.

Solution The quotients $e^x/(x^2 + x)$ and $e^x/(2x + 1)$ each have the indeterminate form ∞/∞, so two applications of l'Hôpital's rule yield

$$\lim_{x \to \infty} \frac{e^x}{x^2 + x} = \lim_{x \to \infty} \frac{e^x}{2x + 1} = \lim_{x \to \infty} \frac{e^x}{2} = +\infty.$$

Remember that l'Hôpital's rule "allows" the final result to be an infinite limit.

EXAMPLE 5

$$\lim_{x \to \infty} \frac{\ln x}{x^{1/2}} = \lim_{x \to \infty} \frac{1/x}{\frac{1}{2}x^{-1/2}} = \lim_{x \to \infty} \frac{2}{x^{1/2}} = 0.$$

CAUCHY'S MEAN VALUE THEOREM

In order to prove l'Hôpital's rule, we need a generalization of the mean value theorem. This generalization is due to the French mathematician Augustin-Louis Cauchy (1789–1857), who used it in the early nineteenth century to give rigorous proofs of several calculus results not previously established firmly.

Theorem 2 *Cauchy's Mean Value Theorem*

Suppose that the functions f and g are continuous on the closed interval $[a, b]$ and differentiable on (a, b). Then there exists a number c in (a, b) such that

$$[f(b) - f(a)]g'(c) = [g(b) - g(a)]f'(c). \tag{4}$$

REMARK 1 To see that this theorem is, indeed, a generalization of the (ordinary) mean value theorem, we take $g(x) = x$. Then $g'(x) = 1$, and the conclusion in (4) reduces to the fact that

$$f(b) - f(a) = (b - a)f'(c)$$

for some number c in (a, b).

REMARK 2 Equation (4) has a geometric interpretation like that of the ordinary mean value theorem. Let us think of the equations $x = g(t)$, $y = f(t)$ as describing the motion of a point $P(x, y)$ moving along a curve C in the xy-plane (see Fig. 11.1) as t increases from a to b. That is, $P(x, y) = P(g(t), f(t))$ is the location of the point P at time t. Under the assumption that $g(b) \neq g(a)$, the slope of the line L connecting the end points of the curve C is

$$m = \frac{f(b) - f(a)}{g(b) - g(a)}. \tag{5}$$

On the other hand, if $g'(c) \neq 0$, then the chain rule gives

$$\frac{dy}{dx} = \frac{dy/dt}{dx/dt} = \frac{f'(c)}{g'(c)} \tag{6}$$

for the slope of the tangent line to the curve C at the point $(g(c), f(c))$. But if $g(b) \neq g(a)$ and $g'(c) \neq 0$, then Equation (4) may be written as

$$\frac{f(b) - f(a)}{g(b) - g(a)} = \frac{f'(c)}{g'(c)}, \tag{7}$$

so the two slopes in (5) and (6) are equal. Thus Cauchy's mean value theorem implies that (under our assumptions) there is a point on the curve C where the tangent line is *parallel* to the line joining the endpoints of C. Of course, this is exactly what the (ordinary) mean value theorem says for an explicitly

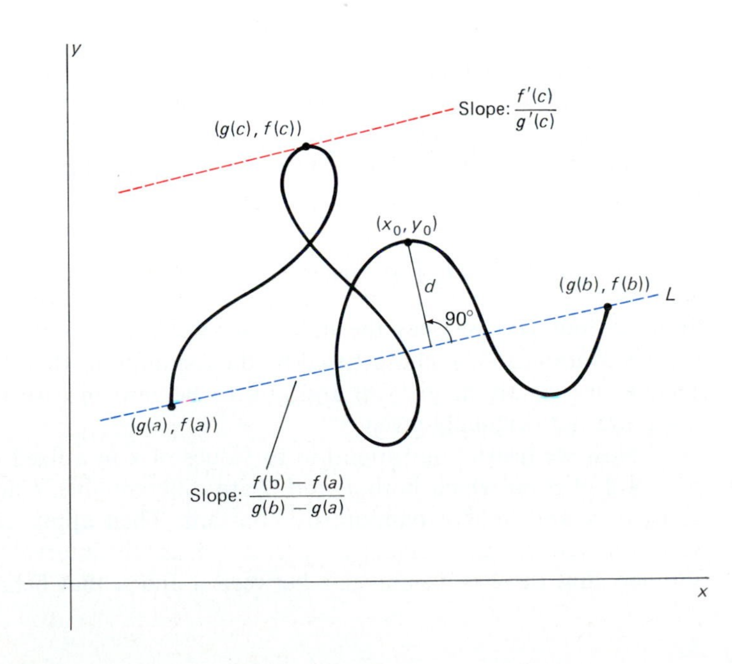

11.1 The idea of Cauchy's mean value theorem

defined curve $y = f(x)$. This geometric interpretation motivates the following proof of Cauchy's mean value theorem.

Proof The line L through the endpoints in Fig. 11.1 has point-slope equation

$$y - f(a) = \frac{f(b) - f(a)}{g(b) - g(a)}[x - g(a)],$$

which can be rewritten in the form $Ax + By + C = 0$ with

$$A = f(b) - f(a), \qquad B = -[g(b) - g(a)], \quad \text{and}$$
$$C = f(a)[g(b) - g(a)] - g(a)[f(b) - f(a)]. \tag{8}$$

According to Miscellaneous Problem 71 at the end of Chapter 3, the (perpendicular) distance from the point (x_0, y_0) to the line L is

$$d = \frac{|Ax_0 + By_0 + C|}{\sqrt{A^2 + B^2}}.$$

Figure 11.1 suggests that the point $(g(c), f(c))$ will maximize this distance d for points on the curve C.

We therefore are motivated to define the auxiliary function

$$\phi(t) = Ag(t) + Bf(t) + C, \tag{9}$$

with the constants A, B, and C as defined in (8). Thus $\phi(t)$ is essentially a constant multiple of the distance from $(g(t), f(t))$ to the line L in Fig. 11.1.

Now $\phi(a) = \phi(b) = 0$ (why?), so Rolle's theorem implies the existence of a number c in (a, b) such that

$$\phi'(c) = Ag'(c) + Bf'(c) = 0. \tag{10}$$

We substitute the values of A and B from (8) in (10) and obtain the equation

$$[f(b) - f(a)]g'(c) - [g(b) - g(a)]f'(c) = 0.$$

This is the same as Equation (4) in the conclusion of Cauchy's mean value theorem, and the proof is complete. Note: Whereas the assumptions that $g(b) \neq g(a)$ and $g'(c) \neq 0$ were needed for our geometric interpretation of the theorem, they were not used in its proof—only in the motivation for the method of proof. ■

PROOF OF L'HÔPITAL'S RULE

Suppose that $f(x)/g(x)$ has the indeterminate form 0/0 at $x = a$. We may invoke continuity of f and g to allow the assumption that $f(a) = 0 = g(a)$. That is, we simply define $f(a)$ and $g(a)$ to be zero in case their values at $x = a$ are not originally given.

Now we restrict our attention to values of x in a fixed deleted neighborhood of a on which both f and g are differentiable. Choose one such value of x and hold it temporarily constant. Then apply Cauchy's mean value theorem on the interval $[a, x]$. (If $x < a$, use the interval $[x, a]$ instead.) We find that there is a number z between a and x that behaves as c does in Equation (4). Hence, by virtue of Equation (4), we obtain the equation

$$\frac{f(x)}{g(x)} = \frac{f(x) - f(a)}{g(x) - g(a)} = \frac{f'(z)}{g'(z)}.$$

Now z depends upon x, but z is trapped between x and a, so that z must approach a as x does (see Fig. 11.2). We conclude that

$$\lim_{x \to a} \frac{f(x)}{g(x)} = \lim_{z \to a} \frac{f'(z)}{g'(z)} = \lim_{x \to a} \frac{f'(x)}{g'(x)},$$

under the assumption that the right-hand limit exists. ■

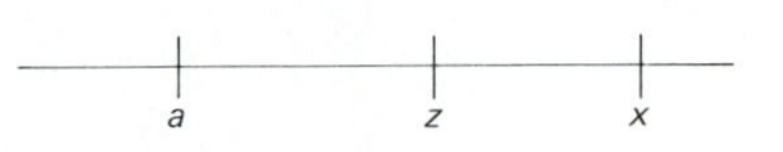

11.2 One case in the proof of l'Hôpital's rule

11.1 PROBLEMS

Find the limits in Problems 1–48.

1. $\lim_{x \to 1} \frac{x - 1}{x^2 - 1}$

2. $\lim_{x \to \infty} \frac{3x - 4}{2x - 5}$

3. $\lim_{x \to \infty} \frac{2x^2 - 1}{5x^2 + 3x}$

4. $\lim_{x \to 0} \frac{e^{3x} - 1}{x}$

5. $\lim_{x \to 0} \frac{\sin x^2}{x}$

6. $\lim_{x \to 0^+} \frac{1 - \cos\sqrt{x}}{x}$

7. $\lim_{x \to 1} \frac{x - 1}{\sin x}$

8. $\lim_{x \to 0} \frac{1 - \cos x}{x^3}$

9. $\lim_{x \to 0} \frac{e^x - x - 1}{x^2}$

10. $\lim_{z \to \pi/2} \frac{1 + \cos 2z}{1 - \sin 2z}$

11. $\lim_{u \to 0} \frac{u \arctan u}{1 - \cos u}$

12. $\lim_{x \to 0} \frac{x - \arctan x}{x^3}$

13. $\lim_{x \to \infty} \frac{\ln x}{x^{1/10}}$

14. $\lim_{r \to \infty} \frac{e^r}{(r + 1)^4}$

15. $\lim_{x \to 10} \frac{\ln(x - 9)}{x - 10}$

16. $\lim_{t \to \infty} \frac{t^2 + 1}{t \ln t}$

17. $\lim_{x \to 0} \frac{e^x + e^{-x} - 2}{x \sin x}$

18. $\lim_{x \to (\pi/2)^-} \frac{\tan x}{\ln \cos x}$

19. $\lim_{x \to 0} \frac{\sin 3x}{\tan 5x}$

20. $\lim_{x \to 0} \frac{e^x - e^{-x}}{x}$

21. $\lim_{x \to 1} \frac{x^3 - 1}{x^2 - 1}$

22. $\lim_{x \to 2} \frac{x^3 - 8}{x^4 - 16}$

23. $\lim_{x \to \infty} \frac{x + \sin x}{3x + \cos x}$

24. $\lim_{x \to \infty} \frac{1}{x}\sqrt{x^2 + 4}$

25. $\lim_{x \to 0} \frac{2^x - 1}{3^x - 1}$

26. $\lim_{x \to \infty} \frac{2^x}{3^x}$

27. $\lim_{x \to \infty} \frac{\sqrt{x^2 - 1}}{\sqrt{4x^2 - x}}$

28. $\lim_{x \to \infty} \frac{\sqrt{x^3 + x}}{\sqrt{2x^3 - 4}}$

29. $\lim_{x \to 0} \frac{\ln(1 + x)}{x}$

30. $\lim_{x \to \infty} \frac{\ln(\ln x)}{x \ln x}$

31. $\lim_{x \to 0} \frac{2e^x - x^2 - 2x - 2}{x^3}$

32. $\lim_{x \to 0} \frac{\sin x - \tan x}{x^3}$

33. $\lim_{x \to 0} \frac{1 - \cosh x}{x^2}$

34. $\lim_{x \to 0} \frac{\sinh 3x}{x}$

35. $\lim_{x \to \pi/2} \frac{2x - \pi}{\tan 2x}$

36. $\lim_{x \to \pi/2} \frac{\sec x}{\tan x}$

37. $\lim_{x \to 2} \frac{x - 2 \cos \pi x}{x^2 - 4}$

38. $\lim_{x \to 1/2} \frac{2x - \sin \pi x}{4x^2 - 1}$

39. $\lim_{x \to 0} \frac{\arctan 2x}{\arctan 3x}$

40. $\lim_{x \to \infty} \frac{\arctan 2x}{\arctan 3x}$

41. $\lim_{x \to 0} \frac{\exp(x^3) - 1}{x - \sin x}$

42. $\lim_{x \to 0} \frac{\sqrt{1 + 3x} - 1}{x}$

43. $\lim_{x \to 0} \frac{\sqrt[3]{1 + 4x} - 1}{x}$

44. $\lim_{x \to 0} \frac{\sqrt{3 + 2x} - \sqrt{3 + x}}{x}$

45. $\lim_{x \to 0} \frac{\sqrt[3]{1 + x} - \sqrt[3]{1 - x}}{x}$

46. $\lim_{x \to \pi/4} \frac{1 - \tan x}{4x - \pi}$

47. $\lim_{x \to 0} \frac{\ln(1 + x^2)}{e^x - \cos x}$

48. $\lim_{x \to 2} \frac{x^5 - 5x^2 - 12}{x^{10} - 500x - 24}$

49. Suppose that f is a twice-differentiable function. Use l'Hôpital's rule to prove these two results:

(a) $\lim_{h \to 0} \frac{f(x + h) - f(x - h)}{2h} = f'(x);$

(b) $\lim_{h \to 0} \frac{f(x + h) - 2f(x) + f(x - h)}{h^2} = f''(x).$

50. Establish the 0/0 version of l'Hôpital's rule in the case $a = \infty$. (*Suggestion:* Let $F(t) = f(1/t)$ and $G(t) = g(1/t)$. Then show that

$$\lim_{x \to \infty} \frac{f(x)}{g(x)} = \lim_{t \to 0^+} \frac{F(t)}{G(t)}$$

$$= \lim_{t \to 0^+} \frac{F'(t)}{G'(t)} = \lim_{x \to \infty} \frac{f'(x)}{g'(x)},$$

using l'Hôpital's rule in the case $a = 0$.)

11.2 Additional Indeterminate Forms

If $\lim_{x \to a} f(x) = 0$ and $\lim_{x \to a} g(x) = \infty$, we say that the product $f(x)g(x)$ has the **indeterminate form** $0 \cdot \infty$ at $x = a$. To find the limit of $f(x)g(x)$ at $x = a$, we can change the problem to one of the forms 0/0 or ∞/∞ in this way:

$$f(x)g(x) = \frac{f(x)}{1/g(x)} = \frac{g(x)}{1/f(x)}.$$

Then l'Hôpital's rule can be applied if its other hypotheses are satisfied.

EXAMPLE 1 Find $\lim_{x \to \infty} x \ln\left(\frac{x-1}{x+1}\right)$.

Solution This has the indeterminate form $\infty \cdot 0$, so we write

$$\lim_{x \to \infty} x \ln\left(\frac{x-1}{x+1}\right) = \lim_{x \to \infty} \frac{\ln[(x-1)/(x+1)]}{1/x}.$$

The latter limit has the form 0/0, so we can apply l'Hôpital's rule. First we note that

$$D \ln[(x-1)/(x+1)] = \frac{2}{x^2 - 1}.$$

Thus

$$\lim_{x \to \infty} x \ln\left(\frac{x-1}{x+1}\right) = \lim_{x \to \infty} \frac{2/(x^2-1)}{-1/x^2} = \lim_{x \to \infty} \frac{-2}{1 - (1/x^2)} = -2.$$

If $\lim_{x \to a} f(x) = \infty = \lim_{x \to a} g(x)$, then we say that $f(x) - g(x)$ has the **indeterminate form** $\infty - \infty$. To then evaluate

$$\lim_{x \to a} [f(x) - g(x)],$$

we try by algebraic manipulations to convert $f(x) - g(x)$ into a form of type 0/0 or ∞/∞, to which l'Hôpital's rule can then be applied. If $f(x)$ or $g(x)$ is a fraction, we sometimes can do this by finding a common denominator. In most cases, however, subtler methods are required. Example 2 illustrates the technique of finding a common denominator, while Example 3 illustrates a factoring technique that is sometimes effective.

EXAMPLE 2

$$\lim_{x \to 0} \left(\frac{1}{x} - \frac{1}{\sin x}\right) = \lim_{x \to 0} \frac{(\sin x) - x}{x \sin x} \qquad \text{(form 0/0)}$$

$$= \lim_{x \to 0} \frac{(\cos x) - 1}{\sin x + x \cos x} \qquad (\textit{still } 0/0)$$

$$= \lim_{x \to 0} \frac{-\sin x}{2 \cos x - x \sin x} = 0.$$

EXAMPLE 3

$$\begin{aligned}\lim_{x\to\infty}(\sqrt{x^2+3x}-x) &= \lim_{x\to\infty} x\left(\sqrt{1+\frac{3}{x}}-1\right) \qquad \text{(form } \infty\cdot 0\text{)}\\ &= \lim_{x\to\infty}\frac{\sqrt{1+3/x}-1}{1/x} \qquad \text{(form 0/0 now)}\\ &= \lim_{x\to\infty}\frac{\frac{1}{2}(1+3/x)^{-1/2}(-3/x^2)}{-1/x^2}\\ &= \lim_{x\to\infty}\frac{\frac{3}{2}}{\sqrt{1+3/x}}=\frac{3}{2}.\end{aligned}$$

THE INDETERMINATE FORMS 0^0, ∞^0, AND 1^∞

Suppose that we need to find the limit of a quantity

$$y=[f(x)]^{g(x)}$$

where the limits of f and g as $x\to a$ are such that one of the indeterminate forms 0^0, ∞^0, or 1^∞ is produced. We first compute the natural logarithm

$$\ln y=\ln([f(x)]^{g(x)})=g(x)\ln f(x).$$

For each of the three indeterminate cases mentioned here, $g(x)\ln f(x)$ has the form $0\cdot\infty$, so we can use our previous methods to find $L=\lim_{x\to a}\ln y$. Then

$$\begin{aligned}\lim_{x\to a}[f(x)]^{g(x)} &= \lim_{x\to a} y=\lim_{x\to a}\exp(\ln y)\\ &= \exp\left(\lim_{x\to a}\ln y\right)=e^L,\end{aligned}$$

because the exponential function is continuous. Thus we have the following four steps for finding the limit of $[f(x)]^{g(x)}$ as $x\to a$:

1. Let $y=[f(x)]^{g(x)}$.
2. Simplify $\ln y=g(x)\ln f(x)$.
3. Evaluate $L=\lim_{x\to a}\ln y$.
4. Conclude that $\lim_{x\to a}[f(x)]^{g(x)}=e^L$.

EXAMPLE 4 Find $\lim_{x\to 0}(\cos x)^{1/x^2}$.

Solution Here we have the indeterminate form 1^∞. If we let $y=(\cos x)^{1/x^2}$, then

$$\ln y=\ln[(\cos x)^{1/x^2}]=\frac{\ln\cos x}{x^2}.$$

As $x \to 0$, $\cos x \to 1$, and so $\ln \cos x \to 0$; consequently we are dealing with the indeterminate form 0/0. Hence two applications of l'Hôpital's rule yield

$$\lim_{x\to 0} \ln y = \lim_{x\to 0} \frac{\ln \cos x}{x^2} = \lim_{x\to 0} \frac{-\tan x}{2x} \qquad (0/0 \text{ form})$$

$$= \lim_{x\to 0} \frac{-\sec^2 x}{2} = -\frac{1}{2}.$$

Consequently,

$$\lim_{x\to 0} (\cos x)^{1/x^2} = e^{-1/2} = \frac{1}{\sqrt{e}}.$$

EXAMPLE 5 Find $\lim_{x\to 0^+} x^{\tan x}$.

Solution This is an indeterminate of the form 0^0. If $y = x^{\tan x}$, then

$$\ln y = (\tan x)(\ln x) = \frac{\ln x}{\cot x}.$$

Now we have the indeterminate form ∞/∞, and l'Hôpital's rule yields

$$\lim_{x\to 0^+} \ln y = \lim_{x\to 0^+} \frac{\ln x}{\cot x} = \lim_{x\to 0^+} \frac{1/x}{-\csc^2 x} = -\lim_{x\to 0^+} \frac{\sin^2 x}{x}$$

$$= -\lim_{x\to 0^+} \left(\frac{\sin x}{x}\right)(\sin x) = (-1)(0) = 0.$$

Therefore, $\lim_{x\to 0^+} x^{\tan x} = e^0 = 1.$

Although $a^0 = 1$ for any *nonzero* constant a, the form 0^0 *is* indeterminate—the limit is not necessarily 1 (see Problem 37). On the other hand, the form 0^∞ is not indeterminate if it is defined; if so, its limit is zero. For example,

$$\lim_{x\to 0^+} x^{1/x} = 0.$$

11.2 PROBLEMS

Find the limits in Problems 1–34.

1. $\lim_{x\to 0} x \cot x$

2. $\lim_{x\to 0} \left(\frac{1}{x} - \cot x\right)$

3. $\lim_{x\to 0} \frac{1}{x} \ln\left(\frac{7x+8}{4x+8}\right)$

4. $\lim_{x\to 0^+} (\sin x)(\ln \sin x)$

5. $\lim_{x\to 0} x^2 \csc^2 2x$

6. $\lim_{x\to \infty} e^{-x} \ln x$

7. $\lim_{x\to \infty} x(e^{1/x} - 1)$

8. $\lim_{x\to 2} \left(\frac{1}{x-2} - \frac{1}{\ln(x-1)}\right)$

9. $\lim_{x\to 0^+} x \ln x$

10. $\lim_{x\to \pi/2} (\tan x)(\cos 3x)$

11. $\lim_{x\to \pi} (x - \pi) \csc x$

12. $\lim_{x\to \infty} e^{-x^2}(x - \sin x)$

13. $\lim_{x\to 0^+} \left(\frac{1}{\sqrt{x}} - \frac{1}{\sin x}\right)$

14. $\lim_{x\to 0} \left(\frac{1}{x} - \frac{1}{e^x - 1}\right)$

15. $\lim_{x\to 1^+} \left(\frac{x}{x^2 + x - 2} - \frac{1}{x-1}\right)$

16. $\lim_{x\to \infty} (\sqrt{x+1} - \sqrt{x})$

17. $\lim_{x\to 0} \left(\frac{1}{x} - \frac{1}{\ln(1+x)}\right)$

18. $\lim_{x\to \infty} (\sqrt{x^2 + x} - \sqrt{x^2 - x})$

19. $\lim_{x\to\infty} (\sqrt[3]{x^3 + 2x + 5} - x)$

20. $\lim_{x\to 0^+} x^x$

21. $\lim_{x\to 0^+} x^{\sin x}$

22. $\lim_{x\to\infty} \left(1 + \frac{1}{x}\right)^x$

23. $\lim_{x\to\infty} (\ln x)^{1/x}$

24. $\lim_{x\to\infty} \left(1 - \frac{1}{x^2}\right)^x$

25. $\lim_{x\to 0} \left(\frac{\sin x}{x}\right)^{1/x^2}$

26. $\lim_{x\to 0^+} (1 + 2x)^{1/3x}$

27. $\lim_{x\to\infty} \left(\cos \frac{1}{x^2}\right)^{x^4}$

28. $\lim_{x\to 0^+} (\sin x)^{\sec x}$

29. $\lim_{x\to 0^+} (x + \sin x)^x$

30. $\lim_{x\to\pi/2} (\tan x - \sec x)$

31. $\lim_{x\to 1} x^{1/(1-x)}$

32. $\lim_{x\to 1^+} (x - 1)^{\ln x}$

33. $\lim_{x\to 2^+} \left(\frac{1}{\sqrt{x^2 - 4}} - \frac{1}{x - 2}\right)$

34. $\lim_{x\to\infty} (\sqrt[5]{x^5 - 3x^4 + 17} - x)$

35. Use l'Hôpital's rule to establish these two limits:

(a) $\lim_{h\to 0} (1 + hx)^{1/h} = e^x$;

(b) $\lim_{n\to\infty} \left(1 + \frac{x}{n}\right)^n = e^x$.

36. Sketch the graph of $y = x^{1/x}$, $x > 0$.

37. Let $f(x) = \exp(-1/x^2)$ and $g(x) = \cos x - 1$, so that $[f(x)]^{g(x)}$ is indeterminate of the form 0^0 as $x \to 0$. Show that $[f(x)]^{g(x)} \to \sqrt{e}$ as $x \to 0$.

38. Let n be a fixed positive integer and let $p(x)$ be the polynomial

$$p(x) = x^n + a_1x^{n-1} + a_2x^{n-2} + \cdots + a_{n-1}x + a_n;$$

the numbers $a_1, a_2, \ldots, a_n$ are fixed real numbers. Prove that

$$\lim_{n\to\infty} ([p(x)]^{1/n} - x) = \frac{a_1}{n}.$$

39. According to Problem 50 in Section 9.6, the surface area of the ellipsoid obtained by revolving the ellipse

$$\frac{x^2}{a^2} + \frac{y^2}{b^2} = 1 \qquad (a > b)$$

around the x-axis is

$$A = 2\pi ab\left[\frac{b}{a} + \frac{a}{c}\sin^{-1}\left(\frac{c}{a}\right)\right]$$

where $c^2 = a^2 - b^2$. Use l'Hôpital's rule to show that $\lim_{b\to a} A = 4\pi a^2$, the area of a sphere of radius a.

40. Consider a long thin rod with heat diffusivity k and coinciding with the x-axis. Suppose that at time $t = 0$ the temperature at x is $A/2\varepsilon$ if $-\varepsilon \leqq x \leqq \varepsilon$ and is zero if $|x| > \varepsilon$. Then it turns out that the temperature $T(x, t)$ of the rod at the point x at time $t > 0$ is given by

$$T(x, t) = \frac{A}{\varepsilon\sqrt{4\pi kt}} \int_0^\varepsilon \exp\left(-\frac{(x - u)^2}{4kt}\right) du.$$

Use l'Hôpital's rule to show that

$$\lim_{\varepsilon\to 0} T(x, t) = \frac{A}{\sqrt{4\pi kt}} \exp\left(-\frac{x^2}{4kt}\right).$$

This is the temperature resulting from an initial "hot spot" at the origin.

41. Explain why

$$\lim_{x\to 0^+} (\ln x)^{1/x} \neq 0.$$

42. Let α be a fixed real number.

(a) Evaluate (in terms of α) the 0^0 indeterminate form

$$\lim_{x\to 0} \left[\exp\left(-\frac{1}{x^2}\right)\right]^{\alpha x^2}.$$

(Note that l'Hôpital's rule is not needed.) Thus the indeterminate form 0^0 may have as its limit any positive real number. Explain why.

(b) Can its limit be zero, negative, or infinite? Explain.

11.3 Taylor's Formula and Polynomial Approximations

The definitions of the various elementary transcendental functions leave it unclear how to compute their values precisely, except at a few isolated points. For example,

$$\ln x = \int_1^x \frac{1}{t}\,dt \qquad (x > 0)$$

by definition, so obviously $\ln 1 = 0$, but no other value of $\ln x$ is obvious. The natural exponential function is the inverse of $\ln x$, so it is clear that $e^0 = 1$, but it is not at all clear how to compute e^x for $x \neq 0$. Indeed, even such an innocent-looking expression as $\sqrt{x}$ is not computable (precisely, and in a finite number of steps) unless x happens to be the square of a rational number.

On the other hand, *any* value of a polynomial

$$P(x) = c_0 + c_1x + c_2x^2 + \cdots + c_nx^n$$

with known coefficients $c_0, c_1, c_2, \ldots, c_n$ is easy to calculate—only addition and multiplication are required. The goal of this section is to use the fact that polynomial values are so readily computable to help us calculate approximate values of functions like $\ln x$ and e^x.

Suppose that we want to calculate (or at least closely approximate) a specific value $f(x_0)$ of a given function f. It would suffice to find a polynomial $P(x)$ with a graph that is very close to that of f on some interval containing x_0. For then we could use the value $P(x_0)$ as an approximation to the actual value of $F(x_0)$. Once we know how to find such an approximating polynomial $P(x)$, we can then ask how accurately $P(x_0)$ approximates the desired value $f(x_0)$.

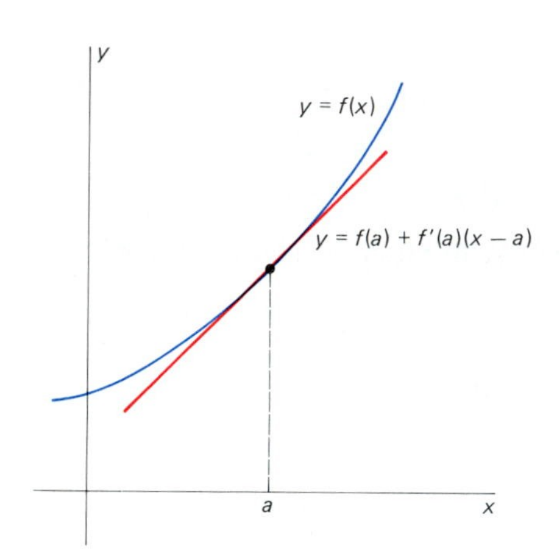

11.3 The tangent line at $(a, f(a))$ is the best linear approximation to f near a.

The simplest example of polynomial approximation is the linear approximation

$$f(x) \approx f(a) + f'(a)(x - a)$$

obtained by writing $\Delta x = x - a$ in Equation (3) of Section 4.2. The graph of the first degree polynomial

$$P_1(x) = f(a) + f'(a)(x - a) \tag{1}$$

is the tangent line to the curve $y = f(x)$ at the point $(a, f(a))$; see Fig. 11.3. Note that this first degree polynomial agrees with f and with its first derivative at $x = a$. That is,

$$P_1(a) = f(a) \quad \text{and} \quad P_1'(a) = f'(a).$$

EXAMPLE 1 Suppose that $f(x) = \ln x$ and that $a = 1$. Then $f(1) = 0$ and $f'(1) = 1$, so $P_1(x) = x - 1$. Hence we expect that $\ln x \approx x - 1$ for x near 1. With $x = 1.1$, we find that

$$P_1(1.1) = 0.100000 \qquad \text{while} \quad \ln(1.1) = 0.095310 \quad \text{(rounded).}$$

The error in this approximation is about 5%.

To better approximate $\ln x$ near $x = 1$, let us look for a second degree polynomial

$$P_2(x) = c_0 + c_1x + c_2x^2$$

that has not only the same value and the same first derivative as does f at $x = 1$, but also has the same second derivative there: $P_2''(1) = f''(1) = -1$. To satisfy these conditions, we must have

$$P_2(1) = c_2 + c_1 + c_0 = 0,$$

$$P_2'(1) = 2c_2 + c_1 = 1,$$

$$P_2''(1) = 2c_2 = -1.$$

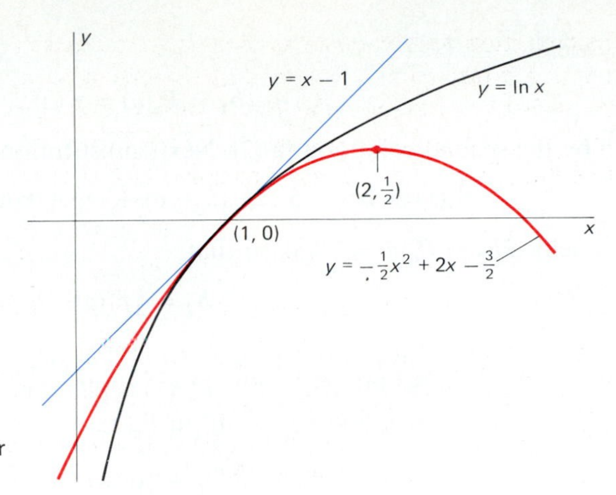

11.4 The linear and parabolic approximations to $y = \ln x$ near the point (1, 0)

When we solve these equations, we find that $c_0 = -\frac{3}{2}$, $c_1 = 2$, and $c_2 = -\frac{1}{2}$, so

$$P_2(x) = -\tfrac{1}{2}x^2 + 2x - \tfrac{3}{2}.$$

With $x = 1.1$ we find that $P_2(1.1) = 0.095000$, which is accurate to three decimal places because $\ln 1.1 \approx 0.095310$. The graph of $y = -\frac{1}{2}x^2 + 2x - \frac{3}{2}$ is a parabola through (1, 0) with the same value, slope, *and curvature* there as $y = \ln x$; see Fig. 11.4.

The tangent line and the parabola used in the computations of Example 1 illustrate one general approach to polynomial approximation. In order to approximate the function $f(x)$ near $x = a$, we look for an nth-degree polynomial

$$P_n(x) = c_0 + c_1x + c_2x^2 + \cdots + c_nx^n$$

such that its value at a and the values of its first n derivatives at a agree with the corresponding values of f. That is, we require that

$$\begin{aligned} P_n(a) &= f(a), \\ P_n'(a) &= f'(a), \\ P_n''(a) &= f''(a), \\ &\vdots \\ P_n^{(n)}(a) &= f^{(n)}(a). \end{aligned} \tag{2}$$

These $n + 1$ conditions can be used to determine the values of the $n + 1$ coefficients $c_0, c_1, \ldots, c_n$.

The algebra involved is much simpler, however, if we begin with $P_n(x)$ expressed as an nth degree polynomial in powers of $x - a$ rather than in powers of x:

$$P_n(x) = b_0 + b_1(x - a) + b_2(x - a)^2 + \cdots + b_n(x - a)^n. \tag{3}$$

Then substitution of $x = a$ in (3) yields $b_0 = P_n(a) = f(a)$ by the first condition in (2). Substitution of $x = a$ in

$$P_n'(x) = b_1 + 2b_2(x - a) + 3b_3(x - a)^2 + \cdots + nb_n(x - a)^{n-1}$$

yields

$$b_1 = P_n'(a) = f'(a)$$

by the second condition in (2). Next, substitution of $x = a$ in

$$P_n''(x) = 2b_2 + 3 \cdot 2b_3(x - a) + \cdots + n(n-1)b_n(x - a)^{n-2}$$

yields $2b_2 = P_n''(a) = f''(a)$, so that

$$b_2 = \tfrac{1}{2}f''(a).$$

We continue in this process to find $b_3, b_4, \ldots, b_n$. In general, the constant term in the kth derivative $P_n^{(k)}(x)$ is $k!b_k$, because it is the kth derivative of the kth degree term $b_k(x - a)^k$ in $P_n(x)$:

$$P_n^{(k)}(x) = k!b_k + \{\text{powers of } x - a\}.$$

So when we substitute $x = a$ in $P_n^{(k)}(x)$, we find that

$$k!b_k = P_n^{(k)}(a) = f^{(k)}(a),$$

and thus that

$$b_k = \frac{f^{(k)}(a)}{k!} \tag{4}$$

for $k = 1, 2, 3, \ldots, n$.

Indeed, the formula in (4) also holds for $k = 0$ if we use the common convention that $0! = 1$ and agree that the zeroth derivative $g^{(0)}$ of the function g is just g itself. With such conventions, our computations establish the following theorem.

Theorem 1 *The nth-Degree Taylor Polynomial*
Suppose that the first n derivatives of the function $f(x)$ exist at $x = a$. Let $P_n(x)$ be the nth-degree polynomial

$$P_n(x) = \sum_{k=0}^{n} \frac{f^{(k)}(a)}{k!}(x - a)^k = f(a) + f'(a)(x - a) + \frac{f''(a)}{2!}(x - a)^2 + \cdots + \frac{f^{(n)}(a)}{n!}(x - a)^n. \tag{5}$$

Then the values of $P_n(x)$ and its first n derivatives agree, at $x = a$, with the values of f and its first n derivatives there. That is, the equations in (2) all hold.

The polynomial given in Equation (5) is called the **nth degree Taylor polynomial of the function f at the point** $x = a$. Note that $P_n(x)$ is a polynomial in powers of $x - a$ rather than in powers of x. In order to use $P_n(x)$ effectively for the approximation of $f(x)$ near a, we shall need to be able to compute the values of the derivatives $f(a)$, $f'(a)$, $f''(a)$, and so on, all the way to $f^{(n)}(a)$.

The line $y = P_1(x)$ is simply the tangent line to the curve $y = f(x)$ at the point $(a, f(a))$. Thus $y = f(x)$ and $y = P_1(x)$ have the same slope at this point. Now recall from Section 4.6 that the second derivative $f''(a)$ measures the way the curve $y = f(x)$ is bending as it passes through $(a, f(a))$. Let us therefore call $f''(a)$ the "concavity" of $y = f(x)$ at $(a, f(a))$. Then, because $P_2''(a) = f''(a)$, we see that $y = P_2(x)$ has the same value, the same slope, *and* the same concavity at $(a, f(a))$ as does $y = f(x)$. Moreover, $P_3(x)$ and $f(x)$ will also have the same rate of change of concavity at $(a, f(a))$. Such observations suggest that the larger n is, the more closely the nth-degree Taylor polynomial will approximate $f(x)$ for x near a.

EXAMPLE 2 Find the nth degree Taylor polynomial of $f(x) = \ln x$ at $a = 1$.

Solution The first few derivatives of $f(x) = \ln x$ are

$$f'(x) = \frac{1}{x}, \qquad f''(x) = -\frac{1}{x^2}, \qquad f^{(3)}(x) = \frac{2}{x^3},$$

$$f^{(4)}(x) = -\frac{3!}{x^4}, \qquad f^{(5)}(x) = \frac{4!}{x^5}.$$

The pattern is clear:

$$f^{(k)}(x) = (-1)^{k-1}\frac{(k-1)!}{x^k} \qquad \text{for} \quad k \geqq 1.$$

Hence $f^{(k)}(1) = (-1)^{k-1}(k-1)!$, so Equation (5) gives

$$P_n(x) = (x-1) - \frac{1}{2}(x-1)^2 + \frac{1}{3}(x-1)^3$$
$$- \frac{1}{4}(x-1)^4 + \cdots + \frac{(-1)^{n-1}}{n}(x-1)^n.$$

With $n = 2$ we obtain the quadratic polynomial

$$P_2(x) = (x-1) - \tfrac{1}{2}(x-1)^2 = -\tfrac{1}{2}x^2 + 2x - \tfrac{3}{2};$$

we saw its graph in Fig. 11.4. With the third-degree Taylor polynomial

$$P_3(x) = (x-1) - \tfrac{1}{2}(x-1)^2 + \tfrac{1}{3}(x-1)^3$$

we can go a step further in approximating $\ln(1.1)$: The value $P_3(1.1) = 0.095333\ldots$ is correct to four decimal places.

Note that in the common case $a = 0$, the nth-degree Taylor polynomial in (5) reduces to

$$P_n(x) = f(0) + f'(0)x + \frac{f''(0)}{2!}x^2 + \cdots + \frac{f^{(n)}(0)}{n!}x^n. \tag{5'}$$

EXAMPLE 3 Find the nth-degree Taylor polynomial for $f(x) = e^x$ at $a = 0$.

Solution This is the easiest of all Taylor polynomials to compute because $f^{(k)}(x) = e^x$ for all $k \geqq 0$. Hence $f^{(k)}(0) = 1$ for all $k \geqq 0$, so (5′) yields

$$P_n(x) = 1 + x + \frac{x^2}{2!} + \frac{x^3}{3!} + \cdots + \frac{x^n}{n!}.$$

$x = 0.1$

n	$P_n(x)$	e^x	$e^x - P_n(x)$
0	1.000000	1.105171	0.105171
1	1.100000	1.105171	0.005171
2	1.105000	1.105171	0.000171
3	1.105167	1.105171	0.000004
4	1.105171	1.105171	0.000000085
5	1.105171	1.105171	0.000000001

$x = 0.5$

n	$P_n(x)$	e^x	$e^x - P_n(x)$
0	1.000000	1.648721	0.648721
1	1.500000	1.648721	0.148721
2	1.625000	1.648721	0.023721
3	1.645833	1.648721	0.002888
4	1.648438	1.648721	0.000284
5	1.648698	1.648721	0.000023

[all data rounded]

11.5 Approximating $y = e^x$ with Taylor polynomials at $a = 0$

The first few Taylor polynomials of the natural exponential function at $a = 0$ are, therefore,

$$P_0(x) = 1,$$

$$P_1(x) = 1 + x,$$

$$P_2(x) = 1 + x + \tfrac{1}{2}x^2,$$

$$P_3(x) = 1 + x + \tfrac{1}{2}x^2 + \tfrac{1}{6}x^3,$$

$$P_4(x) = 1 + x + \tfrac{1}{2}x^2 + \tfrac{1}{6}x^3 + \tfrac{1}{24}x^4,$$

$$P_5(x) = 1 + x + \tfrac{1}{2}x^2 + \tfrac{1}{6}x^3 + \tfrac{1}{24}x^4 + \tfrac{1}{120}x^5.$$

The table in Fig. 11.5 shows how these polynomials approximate $f(x) = e^x$ for $x = 0.1$ and for $x = 0.5$. Note that—at least for these two values of x—the closer x is to $a = 0$, the more rapidly $P_n(x)$ appears to approach $f(x)$ as n increases.

The closeness with which $P_n(x)$ approximates $f(x)$ is measured by the difference

$$R_n(x) = f(x) - P_n(x),$$

for which

$$f(x) = P_n(x) + R_n(x). \tag{6}$$

This difference $R_n(x)$ is called the ***n*th-degree remainder for $f(x)$ at $x = a$.** It is the *error* made if the value $f(x)$ is replaced by the approximation $P_n(x)$.

The theorem that lets us estimate the error or remainder $R_n(x)$ is called **Taylor's formula,** after Brook Taylor, a follower of Newton, who introduced "Taylor polynomials" in an article published in 1715. The particular expression for $R_n(x)$ that we give next is called the *Lagrange form* for the remainder because it first appeared in 1797 in a book written by the French mathematician Joseph Louis Lagrange (1736–1813).

Theorem 2 *Taylor's Formula*

Suppose that the $(n + 1)$st derivative of the function f exists on an interval containing the points a and b. Then

$$f(b) = f(a) + f'(a)(b - a) + \frac{f''(a)}{2!}(b - a)^2 + \frac{f^{(3)}(a)}{3!}(b - a)^3 + \cdots + \frac{f^{(n)}(a)}{n!}(b - a)^n + \frac{f^{(n+1)}(z)}{(n + 1)!}(b - a)^{n+1} \quad (7)$$

for some number z between a and b.

The proof of Taylor's formula is given at the end of this section. If we replace b by x in (7), we get the nth-degree **Taylor's formula with remainder** at $x = a$,

$$f(x) = f(a) + f'(a)(x - a) + \frac{f''(a)}{2!}(x - a)^2 + \frac{f^{(3)}(a)}{3!}(x - a)^3 + \cdots + \frac{f^{(n)}(a)}{n!}(x - a)^n + \frac{f^{(n+1)}(z)}{(n + 1)!}(x - a)^{n+1} \quad (8)$$

where z is some number between a and x. Thus the nth-degree remainder term is

$$R_n(x) = \frac{f^{(n+1)}(z)}{(n + 1)!}(x - a)^{n+1}, \quad (9)$$

which is easy to remember—it's the same as the *last* term of $P_{n+1}(x)$, except that $f^{(n+1)}(a)$ is replaced by $f^{(n+1)}(z)$.

Ordinarily, the exact value of z is unknown. One effective way to skirt this difficulty is to estimate $f^{(n+1)}(z)$; we normally seek an overestimate, a number M such that

$$|f^{(n+1)}(z)| \leqq M$$

for all z between a and x. If we can find such a number M, then

$$|f(x) - P_n(x)| = |R_n(x)| \leqq \frac{M|x - a|^{n+1}}{(n + 1)!}. \quad (10)$$

The fact that $(n + 1)!$ grows very rapidly as n increases is often helpful in showing that $R_n(x)$ is small.

For a particular x, we may be able to show that $\lim_{n \to \infty} R_n(x) = 0$. It will then follow that

$$f(x) = \lim_{n \to \infty} P_n(x).$$

In such a case we can then approximate $f(x)$ with *any* desired degree of accuracy simply by choosing n sufficiently large.

EXAMPLE 4 Estimate the accuracy of the approximation

$$\ln(1.1) \approx 0.095333\ldots$$

obtained in Example 2.

Solution Recall that if $f(x) = \ln x$, then

$$f^{(k)}(x) = (-1)^{k-1}\frac{(k-1)!}{x^k},$$

so that

$$f^{(k)}(1) = (-1)^{k-1}(k-1)!.$$

Hence the third-degree Taylor formula *with remainder* at $a = 1$ is

$$\ln x = (x-1) - \frac{1}{2}(x-1)^2 + \frac{1}{3}(x-1)^3 - \frac{3!}{4!z^4}(x-1)^4$$

for some z between $a = 1$ and x. With $x = 1.1$ this gives

$$\ln(1.1) = 0.095333\ldots - \frac{0.0001}{4z^4}$$

with z between $a = 1$ and $x = 1.1$. The largest possible numerical value of the remainder term is obtained with $z = 1$,

$$\frac{0.0001}{4(1)^4} = 0.000025.$$

It follows that

$$0.095308 < \ln(1.1) < 0.095333,$$

so we can say that $\ln(1.1) = 0.0953$ to four-place accuracy.

EXAMPLE 5 Use a Taylor polynomial for $f(x) = e^x$ at $a = 0$ (as in Example 3) to approximate the number e to five-place accuracy.

Solution Because $f^{(k)}(x) = e^x$ for all k, the nth-degree Taylor polynomial with remainder is

$$e^x = 1 + x + \frac{x^2}{2!} + \frac{x^3}{3!} + \cdots + \frac{x^n}{n!} + \frac{e^z}{(n+1)!}x^{n+1} \tag{11}$$

for some z between $a = 0$ and x. With $x = 1$ we find that

$$e = 2 + \frac{1}{2!} + \frac{1}{3!} + \cdots + \frac{1}{n!} + \frac{e^z}{(n+1)!}$$

with z between 0 and 1. Thus $e^0 < e^z < e^1$. But we already know (Problem 68 in Section 7.2) that $e < 3$, so it follows that $1 < e^z < 3$. We can therefore achieve at least five-place accuracy by choosing n so large that $3/(n+1)! < 0.000005$; that is, so that

$$(n+1)! > \frac{3}{0.000005} = 600{,}000.$$

With either a table of factorials or a calculator, we find that the first factorial greater than 600,000 is $10! = 3{,}628{,}800$. We therefore take $n = 9$ and find that

$$e = 2 + \frac{1}{2!} + \frac{1}{3!} + \cdots + \frac{1}{9!} + R_9 = 2.7182815 + R_9$$

where

$$0 < R_9 = \frac{e^z}{10!} < \frac{3}{10!} = 0.000000827.$$

Thus $2.7182815 < e < 2.7182824$, so the familiar $e = 2.71828$ is indeed the correct value rounded to five decimal places.

In Problems 29 and 30 we ask you to derive (with $a = 0$) the Taylor polynomial approximations

$$\cos x \approx 1 - \frac{x^2}{2!} + \frac{x^4}{4!} - \cdots + (-1)^n \frac{x^{2n}}{(2n)!} \tag{12}$$

and

$$\sin x \approx x - \frac{x^3}{3!} + \frac{x^5}{5!} - \cdots + (-1)^n \frac{x^{2n+1}}{(2n+1)!}. \tag{13}$$

In each case it can be shown that (with x fixed) the remainder term approaches zero as $n \to +\infty$. Figures 11.6 and 11.7 illustrate the increasingly better approximations that we get with larger values of n.

Because of the factor $(x - a)^{n+1}$ in the remainder term $R_n(x)$, we see that (with n fixed) the closer x is to a, the better $P_n(x)$ approximates $f(x)$. To approximate $f(x)$ with the least labor, we naturally choose a as the nearest point to x at which we already know the value of f and the required derivatives. For example, to compute the sine of 5° (which is $\pi/36$ radians), we

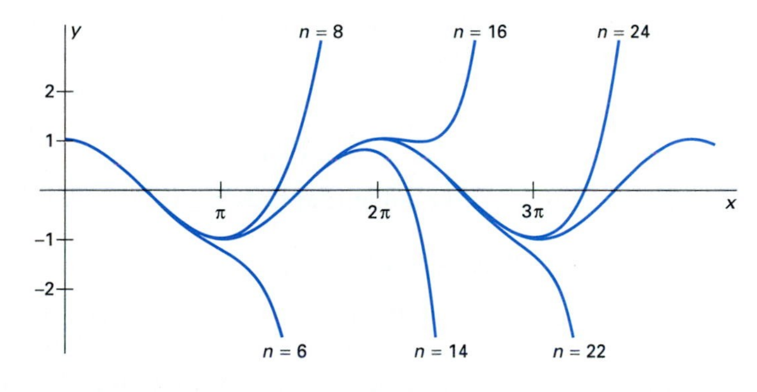

11.6 Approximations to cos x by some of its Taylor polynomials; n is the degree of the polynomial.

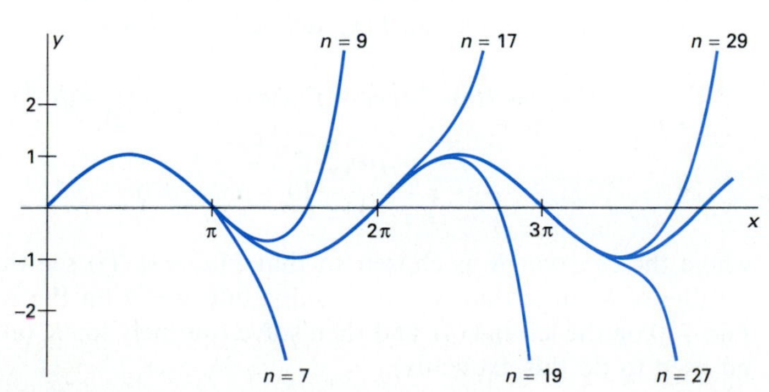

11.7 Approximations to sin x by some of its Taylor polynomials; n is the degree of the polynomial.

choose $a = 0$. But to compute the sine of $50°$ ($5\pi/18$ radians), it's much better to use $a = \pi/4$.

EXAMPLE 6 Show that the values of $\sin x$ for angles between $40°$ and $50°$ can be computed with three-place accuracy by using the approximation

$$\sin x \approx \frac{1}{\sqrt{2}}\left[1 + \left(x - \frac{\pi}{4}\right) - \frac{1}{2}\left(x - \frac{\pi}{4}\right)^2\right].$$

Solution Let $f(x) = \sin x$. Because $f'(x) = \cos x$, $f''(x) = -\sin x$, and $f^{(3)}(x) = -\cos x$, the second-degree Taylor polynomial with remainder for $\sin x$ at $a = \pi/4$ is

$$\sin x = \sin\frac{\pi}{4} + \left(\cos\frac{\pi}{4}\right)\left(x - \frac{\pi}{4}\right) - \frac{1}{2!}\left(\sin\frac{\pi}{4}\right)\left(x - \frac{\pi}{4}\right)^2 + R_2(x)$$

$$= \frac{1}{\sqrt{2}}\left[1 + \left(x - \frac{\pi}{4}\right) - \frac{1}{2}\left(x - \frac{\pi}{4}\right)^2\right] + R_2(x),$$

where

$$R_2(x) = -\frac{\cos z}{3!}\left(x - \frac{\pi}{4}\right)^3$$

for some number z between $\pi/4$ and x. Now $|\cos z| \leqq 1$ for any z, and

$$\left|x - \frac{\pi}{4}\right| \leqq \frac{\pi}{36} < 0.1$$

for the angles in question. Hence

$$|R_2(x)| < \frac{(0.1)^3}{6} \approx 0.000167 < 0.0002.$$

Thus the given polynomial will indeed give three-place accuracy, exactly as desired. For example,

$$\sin(50°) = \sin\left(\frac{\pi}{4} + \frac{\pi}{36}\right) \approx \frac{1}{\sqrt{2}}\left[1 + \frac{\pi}{36} - \frac{1}{2}\left(\frac{\pi}{36}\right)^2\right] = 0.766$$

to three places. The true value is approximately 0.766044443.

PROOF OF TAYLOR'S FORMULA

Several different proofs of Taylor's formula are known, but none of them seems very well motivated—each requires some "trick" to begin the proof. The trick we employ here (suggested by C. R. MacCluer) is to begin by introducing an auxiliary function $F(x)$, defined as follows:

$$F(x) = f(b) - f(x) - f'(x)(b - x) - \frac{f''(x)}{2!}(b - x)^2$$

$$- \cdots - \frac{f^{(n)}(x)}{n!}(b - x)^n - K(b - x)^{n+1}, \tag{14}$$

where the *constant* K is chosen so that $F(a) = 0$. To see that there *is* such a value of K, note that we could substitute $x = a$ on the right and $F(x) = F(a) = 0$ on the left in (14), and then solve routinely for K (although we have no need to do this explicitly).

Observe now that (14) makes it quite obvious that $F(b) = 0$, as well. Therefore, Rolle's Theorem (Section 4.3) implies that

$$F'(z) = 0 \tag{15}$$

for some point z of the open interval (a, b) (under the assumption that $a < b$). To see what (15) means, we differentiate (14) and find that

$$\begin{aligned} F'(x) = &-f'(x) + [f'(x) - f^{(2)}(x)(b-x)] \\ &+ \left[f^{(2)}(x)(b-x) - \frac{1}{2!}f^{(3)}(x)(b-x)^2\right] \\ &+ \left[\frac{1}{2!}f^{(3)}(x)(b-x)^2 - \frac{1}{3!}f^{(4)}(x)(b-x)^3\right] \\ &+ \cdots + \left[\frac{1}{(n-1)!}f^{(n)}(x)(b-x)^{n-1} - \frac{1}{n!}f^{(n+1)}(x)(b-x)^n\right] \\ &+ (n+1)K(b-x)^n. \end{aligned}$$

Upon careful inspection of this result, we see that all terms except the final two cancel in pairs. Thus the sum "telescopes" to give

$$F'(x) = (n+1)K(b-x)^n - \frac{f^{(n+1)}(x)}{n!}(b-x)^n. \tag{16}$$

Hence (15) means that

$$(n+1)K(b-z)^n - \frac{f^{(n+1)}(z)}{n!}(b-z)^n = 0.$$

Consequently we can cancel $(b-z)^n$ and solve for

$$K = \frac{f^{(n+1)}(z)}{(n+1)!}. \tag{17}$$

Finally we return to (14) and substitute $x = a$, $F(x) = 0$, and the value of K given in (17). The result is the equation

$$\begin{aligned} 0 = f(b) - f(a) - f'(a)(b-a) - \frac{f''(a)}{2!}(b-a)^2 \\ - \cdots - \frac{f^{(n)}(a)}{n!}(b-a)^n - \frac{f^{(n+1)}(z)}{(n+1)!}(b-a)^{n+1}, \end{aligned}$$

which is equivalent to the desired Taylor's formula in (7). ∎

11.3 PROBLEMS

The case $a = 0$ of Taylor's formula with remainder is often called *Maclaurin's formula*, after the Scottish mathematician Colin Maclaurin, who used it as a basic tool in a calculus book he published in 1742. In Problems 1–10, find Maclaurin's formula with remainder for the given function and given value of n.

1. $f(x) = e^{-x}$, $n = 5$

2. $f(x) = \sin x$, $n = 4$

3. $f(x) = \cos x$, $n = 4$

4. $f(x) = \dfrac{1}{1-x}$, $n = 4$

5. $f(x) = \sqrt{1+x}$, $n = 3$

6. $f(x) = \ln(1+x)$, $n = 4$

7. $f(x) = \tan x$, $n = 3$

8. $f(x) = \arctan x$, $n = 2$

9. $f(x) = \sin^{-1} x, \quad n = 2$

10. $f(x) = x^3 - 3x^2 + 5x - 7, \quad n = 4$

In Problems 11–16, find the Taylor polynomial with remainder using the given values of a and n.

11. $f(x) = e^x; \quad a = 1, \quad n = 4$

12. $f(x) = \cos x; \quad a = \pi/4, \quad n = 3$

13. $f(x) = \sin x; \quad a = \pi/6, \quad n = 3$

14. $f(x) = \sqrt{x}; \quad a = 100, \quad n = 3$

15. $f(x) = 1/(x-4)^2; \quad a = 5, \quad n = 5$

16. $f(x) = \tan x; \quad a = \pi/4, \quad n = 4$

In Problems 17–20, determine the number of decimal places of accuracy the given approximation formula yields for $|x| \leqq 0.1$.

17. $e^x \approx 1 + x + \frac{1}{2}x^2 + \frac{1}{6}x^3 + \frac{1}{24}x^4$

18. $\sin x \approx x - \frac{1}{6}x^3 + \frac{1}{120}x^5$

19. $\ln(1 + x) \approx x - \frac{1}{2}x^2 + \frac{1}{3}x^3 - \frac{1}{4}x^4$

20. $\sqrt{1 + x} \approx 1 + \frac{1}{2}x - \frac{1}{8}x^2$

21. Show that the approximation in Problem 17 gives e^x to within 0.001 if $|x| \leqq 0.5$. Then compute $\sqrt[3]{e}$ accurate to two decimal places.

22. For what values of x is the approximation

$$\sin x \approx x - \tfrac{1}{6}x^3$$

accurate to five decimal places?

23. (a) Show that the values of the cosine function for angles between 40° and 50° can be computed with five-place accuracy using the approximation

$$\cos x \approx \frac{1}{\sqrt{2}}\left[1 - \left(x - \frac{\pi}{4}\right) - \frac{1}{2}\left(x - \frac{\pi}{4}\right)^2 + \frac{1}{6}\left(x - \frac{\pi}{4}\right)^3\right].$$

(b) Show that this approximation yields eight-place accuracy for angles between 44° and 46°.

In Problems 24–28, calculate the indicated number with the required accuracy using Taylor's formula for an appropriate function at the given value of a.

24. $\sin 10°; \quad a = 0,$ four decimal places

25. $\cos 35°; \quad a = \pi/6,$ four decimal places

26. $e^{1/4}; \quad a = 0,$ four decimal places

27. $\sin 62°; \quad a = \pi/3,$ six decimal places

28. $\sqrt{105}; \quad a = 100,$ three decimal places

29. Show that the Taylor polynomial of degree $2n$ for $f(x) = \cos x$ at $a = 0$ is

$$\sum_{k=0}^{n} (-1)^k \frac{x^{2k}}{(2k)!} = 1 - \frac{x^2}{2!} + \frac{x^4}{4!} + \cdots + (-1)^n \frac{x^{2n}}{(2n)!}.$$

30. Show that the Taylor polynomial of degree $2n + 1$ for $f(x) = \sin x$ at $a = 0$ is

$$\sum_{k=0}^{n} (-1)^k \frac{x^{2k+1}}{(2k+1)!} = x - \frac{x^3}{3!} + \frac{x^5}{5!} - \cdots + (-1)^n \frac{x^{2n+1}}{(2n+1)!}.$$

11.4 Improper Integrals

To show the existence of the definite integral, we have relied until now on the existence theorem stated in Section 5.3. This is the theorem that guarantees the existence of the definite integral $\int_a^b f(x)\,dx$ provided that the function f is *continuous* on the closed (and bounded) interval $[a, b]$. Certain applications in calculus, however, lead naturally to the formulation of integrals in which either

1. The interval of integration is not bounded; it has one of the forms

$$[a, +\infty), \quad (-\infty, a], \quad \text{or} \quad (-\infty, +\infty); \quad \text{or}$$

2. The integrand has an infinite discontinuity at some point c:

$$\lim_{x \to c} f(x) = \pm\infty.$$

An example of Case 1 is the integral

$$\int_1^{\infty} \frac{1}{x^2}\,dx.$$

11.8 The shaded area cannot be measured using our earlier techniques.

A geometric interpretation of this integral is the area of the unbounded region (shaded in Fig. 11.8) that lies between the curve $y = 1/x^2$ and the x-axis and to the right of the vertical line $x = 1$. An example of Case 2 is the integral

$$\int_0^1 \frac{1}{\sqrt{x}}\,dx.$$

This integral may be interpreted as the area of the unbounded region (shaded in Fig. 11.9) that lies under the curve $y = 1/\sqrt{x}$ from $x = 0$ to $x = 1$.

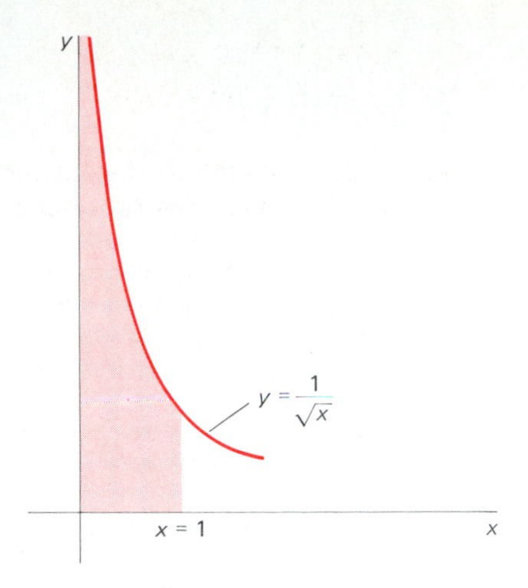

11.9 Another area that must be measured with an improper integral.

Such integrals are called **improper** integrals. The natural interpretation of an improper integral is the area of an unbounded region. The surprise is that such an area can nevertheless be finite, and this section is meant to show how to find such areas—that is, how to evaluate improper integrals.

To see why improper integrals require special care, let us consider the integral

$$\int_{-1}^1 \frac{1}{x^2}\,dx,$$

which is improper because its integrand $f(x) = 1/x^2$ is unbounded as $x \to 0$, and thus is *not* continuous at $x = 0$. If we blindly applied the fundamental theorem of calculus, we would get

$$\int_{-1}^1 \frac{1}{x^2}\,dx = \left[-\frac{1}{x}\right]_{-1}^1 = (-1) - (+1) = -2. \qquad \textbf{(Wrong!)}$$

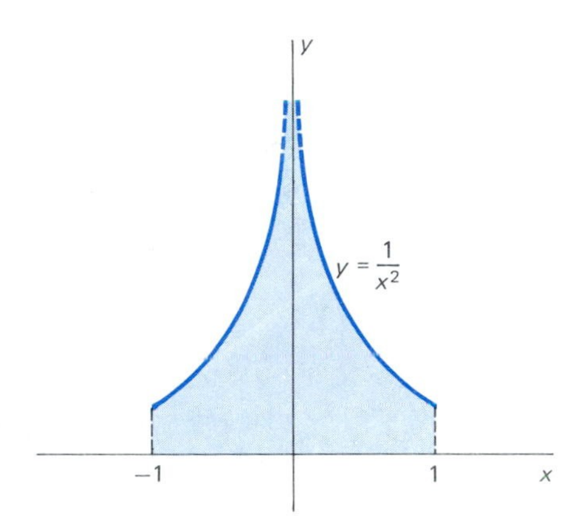

11.10 The area under $y = 1/x^2$, $-1 \leqq x \leqq 1$

The negative answer is obviously incorrect, because the area shown in Fig. 11.10 lies above the x-axis and hence cannot be negative. This simple example emphasizes that we cannot ignore the hypotheses—*continuous* function and *bounded* interval—of the fundamental theorem of calculus.

INFINITE LIMITS OF INTEGRATION

Suppose that the function f is continuous and nonnegative on the unbounded interval $[a, +\infty)$. Then, for any fixed $t > a$, the area $A(t)$ of the region under $y = f(x)$ from $x = a$ to $x = t$ (shaded in Fig. 11.11) is given by the (ordinary) definite integral

$$A(t) = \int_a^t f(x)\,dx.$$

Suppose now that we let $t \to +\infty$, and find that the limit of $A(t)$ exists. Then we may regard this limit as the area of the unbounded region lying under $y = f(x)$ and over $[a, +\infty)$. For f continuous on $[a, +\infty)$, we therefore *define*

$$\int_a^\infty f(x)\,dx = \lim_{t\to\infty} \int_a^t f(x)\,dx \tag{1}$$

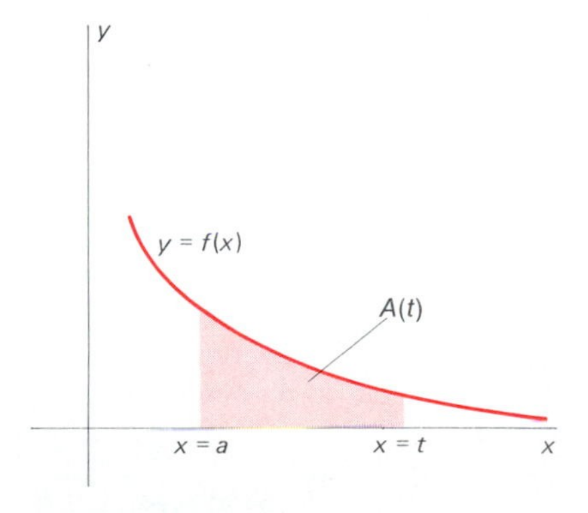

11.11 The shaded area $A(t)$ exists provided f is continuous.

provided that this limit exists (as a finite number). In this case we say that the improper integral on the left **converges;** otherwise, we say that it **diverges.** If $f(x)$ is nonnegative on $[a, +\infty)$, then the limit in Equation (1) either exists or is infinite, and in the latter case we write

$$\int_a^{\infty} f(x)\,dx = +\infty$$

and say that the improper integral **diverges to infinity.**

If the function f has both positive and negative values on $[a, +\infty)$, then the improper integral can diverge *by oscillation*; that is, without diverging to infinity. This occurs with $\int_0^{\infty} \sin x\,dx$, because it is easy to verify that $\int_0^t \sin x\,dx$ is zero if t is an even multiple of π but is 2 if t is an odd multiple of π. Thus $\int_0^t \sin x\,dx$ oscillates between 0 and 2 as $t \to \infty$, and so the limit in (1) does not exist.

We handle an infinite lower limit of integration similarly: We define

$$\int_{-\infty}^{b} f(x)\,dx = \lim_{t \to -\infty} \int_t^b f(x)\,dx \tag{2}$$

provided that the limit exists. If the function f is continuous on the whole real line, we define

$$\int_{-\infty}^{\infty} f(x)\,dx = \int_{-\infty}^{c} f(x)\,dx + \int_c^{\infty} f(x)\,dx \tag{3}$$

for any convenient choice of c, provided that both improper integrals on the right converge. Note that $\int_{-\infty}^{\infty} f(x)\,dx$ is *not* necessarily equal to

$$\lim_{t \to \infty} \int_{-t}^{t} f(x)\,dx$$

(see Problem 28).

It makes no difference what value of c is used in (3) because, if $c < d$, then

$$\begin{aligned}\int_{-\infty}^{c} f(x)\,dx + \int_c^{\infty} f(x)\,dx &= \int_{-\infty}^{c} f(x)\,dx + \int_c^{d} f(x)\,dx + \int_d^{\infty} f(x)\,dx \\ &= \int_{-\infty}^{d} f(x)\,dx + \int_d^{\infty} f(x)\,dx,\end{aligned}$$

under the assumption that the limits involved all exist.

EXAMPLE 1 Investigate the improper integrals

$$\text{(a)} \int_1^{\infty} \frac{1}{x^2}\,dx \quad \text{and} \quad \text{(b)} \int_{-\infty}^{0} \frac{1}{\sqrt{1-x}}\,dx.$$

Solution

$$\begin{aligned}\text{(a)} \int_1^{\infty} \frac{1}{x^2}\,dx &= \lim_{t \to \infty} \int_1^t \frac{1}{x^2}\,dx \\ &= \lim_{t \to \infty} \left[-\frac{1}{x}\right]_1^t = \lim_{t \to \infty} \left(-\frac{1}{t} + 1\right) = 1.\end{aligned}$$

Thus this improper integral converges to 1, and this is the area of the region shown in Fig. 11.8.

(b) $$\int_{-\infty}^{0} \frac{1}{\sqrt{1-x}}\,dx = \lim_{t\to-\infty} \int_{t}^{0} \frac{1}{\sqrt{1-x}}\,dx = \lim_{t\to-\infty} \Big[-2\sqrt{1-x}\Big]_{t}^{0}$$
$$= \lim_{t\to-\infty} (2\sqrt{1-t}-2) = +\infty.$$

Thus the second improper integral of the example diverges to $+\infty$.

EXAMPLE 2 Investigate the improper integral

$$\int_{-\infty}^{\infty} \frac{1}{1+x^2}\,dx.$$

Solution The choice $c = 0$ in Equation (3) gives

$$\int_{-\infty}^{\infty} \frac{1}{1+x^2}\,dx = \int_{-\infty}^{0} \frac{1}{1+x^2}\,dx + \int_{0}^{\infty} \frac{1}{1+x^2}\,dx$$
$$= \lim_{s\to-\infty} \int_{s}^{0} \frac{1}{1+x^2}\,dx + \lim_{t\to\infty} \int_{0}^{t} \frac{1}{1+x^2}\,dx$$
$$= \lim_{s\to-\infty} \Big[\tan^{-1}x\Big]_{s}^{0} + \lim_{t\to\infty} \Big[\tan^{-1}x\Big]_{0}^{t}$$
$$= \lim_{s\to-\infty} (-\tan^{-1}s) + \lim_{t\to\infty} (\tan^{-1}t) = \frac{\pi}{2} + \frac{\pi}{2} = \pi.$$

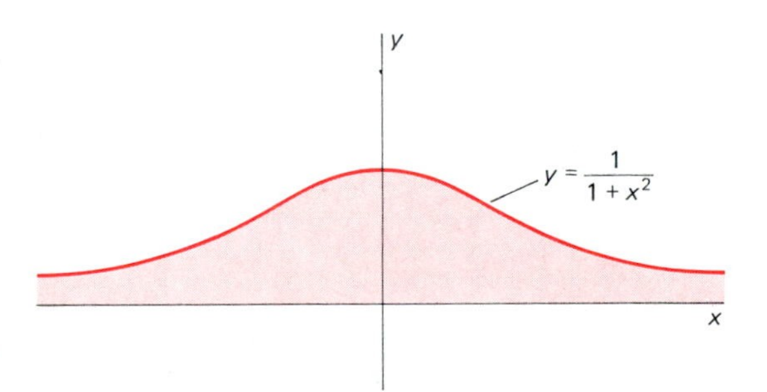

11.12 The area measured by the integral of Example 2.

The shaded region in Fig. 11.12 is a geometric interpretation of the integral of Example 2.

INFINITE INTEGRANDS

Suppose that the function f is continuous and nonnegative on $[a, b)$ but that $f(x) \to +\infty$ as $x \to b^-$. The graph of such a function appears in Fig. 11.13. The area $A(t)$ of the region lying under $y = f(x)$ from $x = a$ to $x = t < b$ is the value of the (ordinary) definite integral

$$A(t) = \int_{a}^{t} f(x)\,dx.$$

If the limit of $A(t)$ as $t \to b^-$ exists, then this limit may be regarded as the area of the (unbounded) region under $y = f(x)$ from $x = a$ to $x = b$. For f

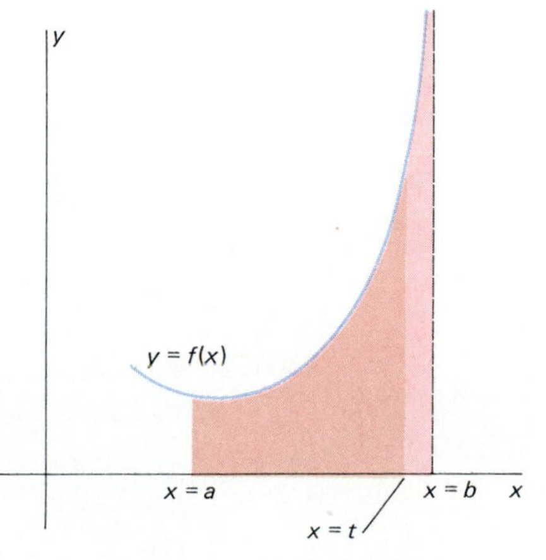

11.13 An improper integral of the second type: $f(x) \to \infty$ as $x \to b^-$.

continuous on $[a, b)$, we therefore *define*

$$\int_a^b f(x)\,dx = \lim_{t \to b^-} \int_a^t f(x)\,dx, \tag{4}$$

provided that this limit exists (as a finite number), in which case we say that the improper integral on the left **converges;** otherwise we say that it **diverges.** If

$$\int_a^b f(x)\,dx = \lim_{t \to b^-} \int_a^t f(x)\,dx = \infty,$$

then we say that the improper integral **diverges to infinity.**

If f is continuous on $(a, b]$ but the limit of $f(x)$ as $x \to a^+$ is infinite, then we *define*

$$\int_a^b f(x)\,dx = \lim_{t \to a^+} \int_t^b f(x)\,dx, \tag{5}$$

provided that the limit exists. If f is continuous at every point of $[a, b]$ except for the point c in (a, b) and one or both one-sided limits of f at c are infinite, then we *define*

$$\int_a^b f(x)\,dx = \int_a^c f(x)\,dx + \int_c^b f(x)\,dx \tag{6}$$

provided that both improper integrals on the right converge.

EXAMPLE 3 Investigate the improper integrals

$$\text{(a)} \int_0^1 \frac{1}{\sqrt{x}}\,dx \qquad \text{and} \qquad \text{(b)} \int_1^2 \frac{1}{(x-2)^2}\,dx.$$

Solution (a) The integrand $1/\sqrt{x}$ becomes infinite as $x \to 0^+$, so

$$\int_0^1 \frac{1}{\sqrt{x}}\,dx = \lim_{t \to 0^+} \int_t^1 \frac{1}{\sqrt{x}}\,dx = \lim_{t \to 0^+} \Big[2\sqrt{x}\Big]_t^1$$

$$= \lim_{t \to 0^+} 2(1 - \sqrt{t}) = 2.$$

Thus the area of the unbounded region shown in Fig. 11.9 is 2.

(b) Here the integrand becomes infinite as x approaches the right-hand endpoint, so

$$\int_1^2 \frac{1}{(x-2)^2}\,dx = \lim_{t \to 2^-} \int_1^t \frac{1}{(x-2)^2}\,dx$$

$$= \lim_{t \to 2^-} \left[-\frac{1}{x-2}\right]_1^t = \lim_{t \to 2^-} \left(-1 - \frac{1}{t-2}\right) = +\infty.$$

Hence this improper integral diverges to infinity. It follows that the improper integral

$$\int_1^3 \frac{1}{(x-2)^2}\,dx = \int_1^2 \frac{1}{(x-2)^2}\,dx + \int_2^3 \frac{1}{(x-2)^2}\,dx$$

also diverges, because not both of the right-hand improper integrals converge. (It turns out that the second one also diverges to $+\infty$.)

EXAMPLE 4 Investigate the improper integral

$$\int_0^2 \frac{1}{(2x-1)^{2/3}}\,dx.$$

Solution This improper integral corresponds to the region shaded in Fig. 11.14. The integrand has an infinite discontinuity at the point $c = \frac{1}{2}$ within the interval of integration, so we write

$$\int_0^2 \frac{1}{(2x-1)^{2/3}}\,dx = \int_0^{1/2} \frac{1}{(2x-1)^{2/3}}\,dx + \int_{1/2}^2 \frac{1}{(2x-1)^{2/3}}\,dx$$

and investigate separately the two improper integrals on the right. We find that

$$\begin{aligned}\int_0^{1/2} \frac{1}{(2x-1)^{2/3}}\,dx &= \lim_{t\to(1/2)^-} \int_0^t \frac{1}{(2x-1)^{2/3}}\,dx \\ &= \lim_{t\to(1/2)^-} \left[\frac{3}{2}(2x-1)^{1/3}\right]_0^t \\ &= \lim_{t\to(1/2)^-} \frac{3}{2}\left[(2t-1)^{1/3} - (-1)^{1/3}\right] = \frac{3}{2},\end{aligned}$$

and

$$\begin{aligned}\int_{1/2}^2 \frac{1}{(2x-1)^{2/3}}\,dx &= \lim_{t\to(1/2)^+} \int_t^2 \frac{1}{(2x-1)^{2/3}}\,dx \\ &= \lim_{t\to(1/2)^+} \left[\frac{3}{2}(2x-1)^{1/3}\right]_t^2 \\ &= \lim_{t\to(1/2)^+} \frac{3}{2}\left[(3)^{1/3} - (2t-1)^{1/3}\right] = \frac{3}{2}\sqrt[3]{3}.\end{aligned}$$

Therefore,

$$\int_0^2 \frac{1}{(2x-1)^{1/3}}\,dx = \frac{3}{2}(1 + \sqrt[3]{3}).$$

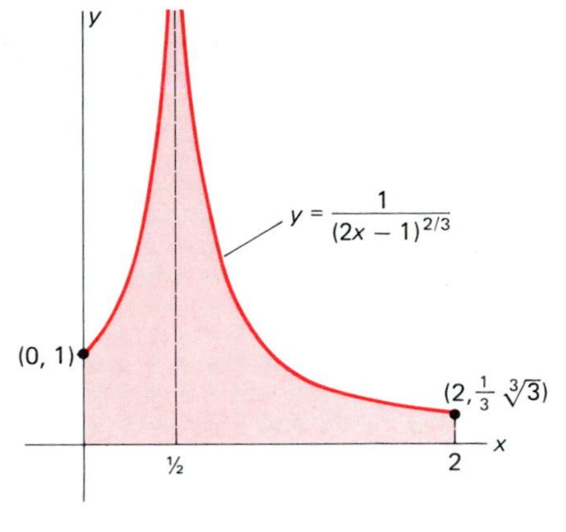

11.14 The region of Example 4

Special functions in advanced mathematics frequently are defined by means of improper integrals. An important example is the **gamma function** $\Gamma(t)$ that the prolific Swiss mathematician Leonhard Euler (1707–1783) introduced to "interpolate" the factorial function $n!$. The gamma function is defined for all real numbers $t > 0$ as follows:

$$\Gamma(t) = \int_0^\infty x^{t-1}e^{-x}\,dx. \tag{7}$$

This definition gives a continuous function of t such that

$$\Gamma(n+1) = n! \tag{8}$$

if n is a positive integer (see Problems 29 and 30).

EXAMPLE 5 Find the volume V of the unbounded solid obtained by revolving around the y-axis the region under the curve $y = \exp(-x^2)$, $x \geqq 0$.

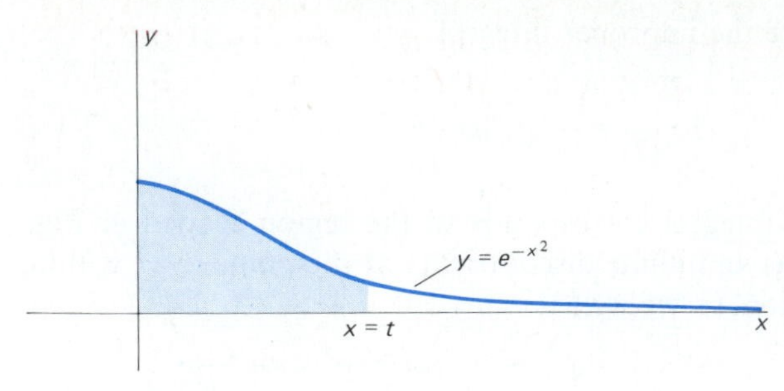

11.15 *All* the region under the graph is rotated about the y-axis; this produces an unbounded solid.

Solution The region in question is shown in Fig. 11.15. Let V_t denote the volume generated by revolving the shaded part of this region between $x = 0$ and $x = t > 0$. Then the method of cylindrical shells gives

$$V_t = \int_0^t 2\pi x \exp(-x^2)\,dx.$$

We get the whole volume V by letting $t \to +\infty$:

$$V = \lim_{t\to\infty} V_t = \int_0^\infty 2\pi x \exp(-x^2)\,dx = \lim_{t\to\infty} \Big[-\pi \exp(-x^2)\Big]_0^t = \pi.$$

*ESCAPE VELOCITY

We saw in Section 6.6 how to compute the work W_r required to lift a body of mass m from the surface of a planet (of mass M and radius R) to a distance $r > R$ from the center of the planet. According to Equation (7) in that section, the answer is

$$W_r = \int_R^r \frac{GMm}{x^2}\,dx.$$

So the work required to move the mass m "infinitely far" from the planet is

$$W = \lim_{r\to\infty} W_r = \int_R^\infty \frac{GMm}{x^2}\,dx = \lim_{r\to\infty}\left[-\frac{GMm}{x}\right]_R^r = \frac{GMm}{R}.$$

Suppose that the mass is projected straight upward from the planet's surface with initial velocity v, as in Jules Verne's novel *From the Earth to the Moon* (1865), in which a spacecraft was fired from an immense cannon. Then the initial kinetic energy $\frac{1}{2}mv^2$ is available to supply this work—by conversion into potential energy. From the equation

$$\frac{1}{2}mv^2 = \frac{GMm}{R},$$

we find that

$$v = \sqrt{\frac{2GM}{R}}.$$

Substitution of appropriate numerical values for the constants G, M, and R yields the value

$$v \approx 11{,}175 \text{ m/s} \approx 25{,}000 \text{ mi/h}$$

for the *escape velocity* from the earth.

*PRESENT VALUE OF A PERPETUITY

Consider a "perpetual annuity," under which you and your heirs (and theirs, ad infinitum) will be paid A dollars annually. The question we pose is this: What is the fair market value of such an annuity? What should you pay to purchase it?

At continuous annual interest rate r, a dollar deposited in a savings account would grow to e^{rt} dollars in t years. Hence e^{-rt} dollars deposited now would yield $1 after t years. Consequently, the **present value** of the amount you (and your heirs) will receive between time $t = 0$ (the present) and time $t = T > 0$ is defined as the integral

$$P_T = \int_0^T Ae^{-rt}\,dt.$$

Hence the total present value of the perpetual annuity is

$$P = \lim_{T\to\infty} P_T = \int_0^\infty Ae^{-rt}\,dt = \lim_{T\to\infty}\left[-\frac{A}{r}e^{-rt}\right]_0^T = \frac{A}{r}.$$

Thus $A = rP$. For instance, at an annual interest rate of 8% ($r = 0.08$), you should be able to purchase for $P = (\$10{,}000)/(0.08) = \$125{,}000$ a perpetuity that pays you (and your heirs) an annual sum of $10,000.

11.4 PROBLEMS

Determine whether or not the improper integrals in Problems 1–24 converge, and evaluate those that do converge.

1. $\displaystyle\int_4^\infty \frac{1}{x^{3/2}}\,dx$

2. $\displaystyle\int_1^\infty \frac{1}{x^{2/3}}\,dx$

3. $\displaystyle\int_0^4 \frac{1}{x^{3/2}}\,dx$

4. $\displaystyle\int_0^8 \frac{1}{x^{2/3}}\,dx$

5. $\displaystyle\int_1^\infty \frac{1}{x+1}\,dx$

6. $\displaystyle\int_3^\infty \frac{1}{\sqrt{x+1}}\,dx$

7. $\displaystyle\int_5^\infty \frac{1}{(x-1)^{3/2}}\,dx$

8. $\displaystyle\int_0^4 \frac{1}{\sqrt{4-x}}\,dx$

9. $\displaystyle\int_0^9 \frac{1}{(9-x)^{3/2}}\,dx$

10. $\displaystyle\int_0^3 \frac{1}{(x-3)^2}\,dx$

11. $\displaystyle\int_{-\infty}^{-2} \frac{1}{(x+1)^3}\,dx$

12. $\displaystyle\int_{-\infty}^{0} \frac{1}{\sqrt{4-x}}\,dx$

13. $\displaystyle\int_{-1}^{8} \frac{1}{x^{1/3}}\,dx$

14. $\displaystyle\int_{-4}^{4} \frac{1}{(x+4)^{2/3}}\,dx$

15. $\displaystyle\int_2^\infty \frac{1}{(x-1)^{1/3}}\,dx$

16. $\displaystyle\int_{-\infty}^{\infty} \frac{x}{(x^2+4)^{3/2}}\,dx$

17. $\displaystyle\int_{-\infty}^{\infty} \frac{x}{x^2+4}\,dx$

18. $\displaystyle\int_0^\infty e^{-(x+1)}\,dx$

19. $\displaystyle\int_0^1 \frac{\exp(\sqrt{x})}{\sqrt{x}}\,dx$

20. $\displaystyle\int_0^2 \frac{x}{x^2-1}\,dx$

21. $\displaystyle\int_1^\infty \frac{1}{x\ln x}\,dx$

22. $\displaystyle\int_0^\infty \sin^2 x\,dx$

23. $\int_0^\infty xe^{-2x}\,dx$

24. $\int_0^\infty e^{-x}\sin x\,dx$

Problems 25–27 deal with *Gabriel's horn*, the surface obtained by revolving the curve $y = 1/x$, $x \geqq 1$, around the x-axis (see Fig. 11.16).

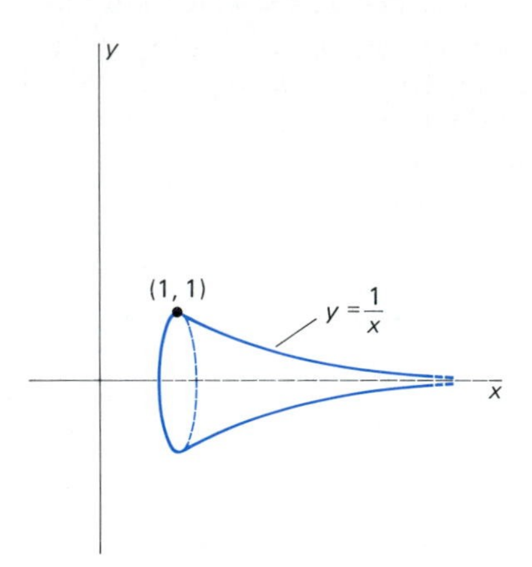

11.16 Gabriel's horn

25. Show that the area under the curve $y = 1/x$, $x \geqq 1$, is infinite.

26. Show that the volume of revolution enclosed by Gabriel's horn is finite, and compute it.

27. Show that the surface area of Gabriel's horn is infinite. *Suggestion:* Let A_t denote the surface area from 1 to $t > 1$. Prove that $A_t > 2\pi \ln t$. In any case, the implication is that we could fill Gabriel's horn with a finite amount of paint (Problem 26), but no finite amount suffices to paint its surface.

28. Show that

$$\int_{-\infty}^{\infty} \frac{1+x}{1+x^2}\,dx$$

diverges, but that

$$\lim_{t\to\infty} \int_{-t}^{t} \frac{1+x}{1+x^2}\,dx = \pi.$$

29. Let n be a fixed positive integer. Begin with the integral in Equation (7), the one where the gamma function is defined, and integrate by parts to prove that $\Gamma(n+1) = n\Gamma(n)$.

30. (a) Prove that $\Gamma(1) = 1$.

(b) Use the results of part (a) and Problem 29 to prove by mathematical induction that $\Gamma(n+1) = n!$ if n is a positive integer.

31. Use the substitution $x = e^{-u}$ and the fact that $\Gamma(n+1) = n!$ to prove that if m and n are fixed but arbitrary positive integers, then

$$\int_0^1 x^m(\ln x)^n\,dx = \frac{n!(-1)^n}{(n+1)^{n+1}}.$$

32. Consider a perpetual annuity under which you and your heirs will be paid at the rate of $10 + t$ thousand dollars per year t years hence. Thus you will receive \$20 thousand 10 years hence, your heir will receive \$110 thousand 100 years hence, and so on. Show that the present value at an annual interest rate of 10% of this perpetuity is

$$P = \int_0^\infty (10+t)e^{-t/10}\,dt,$$

and then evaluate this improper integral.

33. A "semi-infinite" uniform rod occupies the nonnegative x-axis ($x \geqq 0$) and has linear density ρ; that is, a segment of length dx has mass $\rho\,dx$. Show that the force of gravitational attraction that the rod exerts on a point mass m at $(-a, 0)$ is

$$F = \int_0^\infty \frac{Gm\rho}{(a+x)^2}\,dx = \frac{Gm\rho}{a}.$$

34. A rod of linear density ρ occupies the entire y-axis. A point mass m is located at $(a, 0)$ on the x-axis, as indicated in Fig. 11.17. Show that the total (horizontal) gravitational attraction that the rod exerts on m is

$$F = \int_{-\infty}^{\infty} \frac{Gm\rho\cos\theta}{r^2}\,dy = \frac{2Gm\rho}{a},$$

where $r^2 = a^2 + y^2$ and $\cos\theta = \dfrac{a}{r}$.

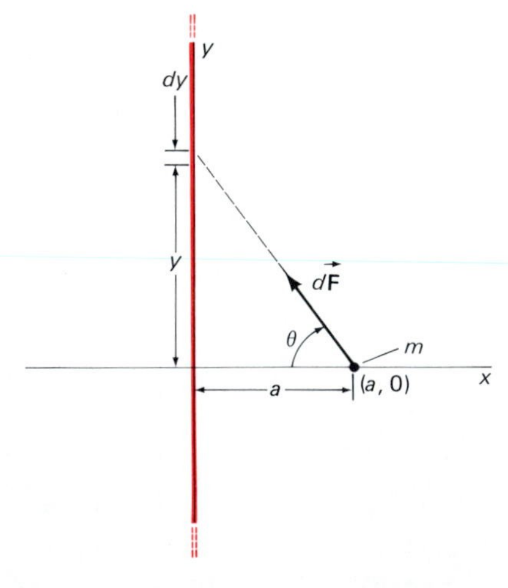

11.17 Gravitational attraction exerted on a point pass by an infinite rod

CHAPTER 11 REVIEW: Definitions, Concepts, Results

Use the list below as a guide to concepts that you may need to review.

1. L'Hôpital's rule and the indeterminate forms 0/0, ∞/∞, $0 \cdot \infty$, $\infty - \infty$; 0^0, ∞^0, and 1^∞
2. Cauchy's mean value theorem
3. The nth-degree Taylor polynomial of the function f at the point $x = a$
4. Taylor's formula with remainder
5. Use of Taylor's formula to approximate values of functions
6. Definition and evaluation of improper integrals
7. Evaluation of an improper integral of a function having an infinite discontinuity within the interval of integration

CHAPTER 11 MISCELLANEOUS PROBLEMS

Find the limits in Problems 1–15.

1. $\lim_{x\to 2} \dfrac{x-2}{x^2-4}$

2. $\lim_{x\to 0} \dfrac{\sin 2x}{x}$

3. $\lim_{x\to \pi} \dfrac{1+\cos x}{(x-\pi)^2}$

4. $\lim_{x\to 0} \dfrac{x-\sin x}{x^3}$

5. $\lim_{t\to 0} \dfrac{t \arctan t - \sin^2 t}{t^6}$

6. $\lim_{x\to \infty} \dfrac{\ln(\ln x)}{\ln x}$

7. $\lim_{x\to 0} (\cot x) \ln(1+x)$

8. $\lim_{x\to 0^+} (e^{1/x}-1) \tan x$

9. $\lim_{x\to 0} \left(\dfrac{1}{x^2} - \dfrac{1}{1-\cos x}\right)$

10. $\lim_{x\to \infty} \left(\dfrac{x^2}{x+2} - \dfrac{x^3}{x^2+3}\right)$

11. $\lim_{x\to \infty} (\sqrt{x^2-x+1} - \sqrt{x})$

12. $\lim_{x\to \infty} x^{1/x}$

13. $\lim_{x\to \infty} (e^{2x}-2x)^{1/x}$

14. $\lim_{x\to \infty} [1-\exp(-x^2)]^{1/x^2}$

15. $\lim_{x\to \infty} x\left[\left(1+\dfrac{1}{x}\right)^x - e\right]$

(*Suggestion:* Let $u = 1/x$ and take the limit as $u \to 0^+$.)

Determine whether or not the improper integrals in Problems 16–22 converge, and evaluate those that do converge.

16. $\displaystyle\int_{-4}^{0} \frac{1}{\sqrt{x+4}}\,dx$

17. $\displaystyle\int_{2}^{6} \frac{1}{(x-2)^{3/2}}\,dx$

18. $\displaystyle\int_{-\infty}^{\infty} \frac{x}{\sqrt{x^2+9}}\,dx$

19. $\displaystyle\int_{e}^{\infty} \frac{1}{x(\ln x)^2}\,dx$

20. $\displaystyle\int_{-\infty}^{\infty} \sin x\,dx$

21. $\displaystyle\int_{0}^{\infty} x^2 e^{-x}\,dx$

22. $\displaystyle\int_{0}^{2} \frac{1}{(x-1)^{2/3}}\,dx$

In Problems 23–31, find the Taylor formula with remainder for $f(x)$ with the given values of a and n.

23. $f(x) = e^{x+1}$; $a = 0$, $n = 5$
24. $f(x) = xe^x$; $a = 0$, $n = 3$
25. $f(x) = \exp(-x^2)$; $a = 0$, $n = 3$
26. $f(x) = \sqrt[3]{x-8}$; $a = 0$, $n = 3$
27. $f(x) = \sqrt[3]{x}$; $a = 125$, $n = 3$
28. $f(x) = \tan^{-1} x$; $a = 1$, $n = 3$
29. $f(x) = x^{-1/2}$; $a = 1$, $n = 4$
30. $f(x) = (x+2)^{3/2}$; $a = 2$, $n = 3$
31. $f(x) = x^4$; $a = 1$, $n = 5$
32. Use Taylor's formula to calculate $\sqrt[3]{120}$ accurate to three decimal places.
33. Use Taylor's formula to calculate $(1000)^{0.1}$ accurate to five decimal places.

In Problems 34–35, determine the number of decimal places of accuracy the given approximation formula yields for $|x| \leqq 1$.

34. $\cos x \approx 1 - \frac{1}{2}x^2 + \frac{1}{24}x^4$

35. $\tan^{-1} x \approx x - \frac{1}{3}x^3$

36. Show that the approximation

$$\sin x \approx x - \frac{x^3}{3!} + \frac{x^5}{5!} - \frac{x^7}{7!} + \frac{x^9}{9!}$$

can be used to calculate $\sin x$ accurate to four decimal places for all values of x that correspond to angles between 0° and 90°.

37. Show that the approximation of Problem 36 is accurate to 15 decimal places for values of x that correspond to angles between 0° and 10°.

38. According to Problem 51 in Section 9.6, the surface area of the ellipsoid obtained by revolving the ellipse with equation $x^2/a^2 + y^2/b^2 = 1$ $(a < b)$ around the x-axis is

$$A = 2\pi ab\left[\frac{b}{a} + \frac{a}{c}\ln\left(\frac{b+c}{a}\right)\right]$$

where $c^2 = b^2 - a^2$. Use l'Hôpital's rule to show that

$$\lim_{b \to a} A = 4\pi a^2.$$

39. Given:

$$\int_0^\infty e^{-x^2}\,dx = \frac{1}{2}\sqrt{\pi}.$$

Deduce that $\Gamma(\frac{1}{2}) = \sqrt{\pi}$.

40. Recall that $\Gamma(x + 1) = x\Gamma(x)$ if $x > 0$. Suppose that n is a positive integer. Use Problem 39 to establish that

$$\Gamma\left(n + \frac{1}{2}\right) = \frac{1 \cdot 3 \cdot 5 \cdots (2n - 1)}{2^n}\sqrt{\pi}.$$

41. (a) Suppose that $k > 1$. Show by integration by parts that

$$\int_0^\infty x^k \exp(-x^2)\,dx = \frac{k-1}{2}\int_0^\infty x^{k-2}\exp(-x^2)\,dx.$$

(b) Suppose that n is a positive integer. Prove that

$$\int_0^\infty x^{n-1}\exp(-x^2)\,dx = \frac{1}{2}\Gamma\left(\frac{n}{2}\right).$$

42. Prove as follows that the number e is irrational. First suppose to the contrary that $e = p/q$, where p and q are positive integers. Write

$$\frac{p}{q} = e = 1 + 1 + \frac{1}{2!} + \frac{1}{3!} + \cdots + \frac{1}{q!} + R_q$$

where $0 < R_q < 3/(q + 1)!$. (Why?) Then show that multiplication of both sides of this equation by $q!$ would lead to the contradiction that one side of the result is an integer but the other side is not.

chapter twelve

Infinite Series

12.1 Introduction

In the fifth century B.C., the Greek philosopher Zeno proposed the following paradox: In order for a runner to travel a given distance, the runner must first travel halfway, then half the remaining distance, then half the distance that yet remains, and so on ad infinitum. But, Zeno argued, it is clearly impossible for a runner to accomplish these infinitely many steps in a finite period of time, so motion from one point to another must be impossible.

Zeno's paradox suggests the infinite subdivision of $[0, 1]$ indicated in Fig. 12.1. There is one subinterval of length $1/2^n$ for each integer $n = 1, 2, 3, \ldots$. If the length of the interval is the sum of the lengths of the subintervals into which it is divided, then it would appear that

$$1 = \frac{1}{2} + \frac{1}{4} + \frac{1}{8} + \frac{1}{16} + \cdots + \frac{1}{2^n} + \cdots,$$

with infinitely many terms somehow adding up to 1. On the other hand, the formal infinite sum

$$1 + 2 + 3 + \cdots + n + \cdots$$

of all the positive integers seems meaningless—it does not appear to add up to *any* (finite) value.

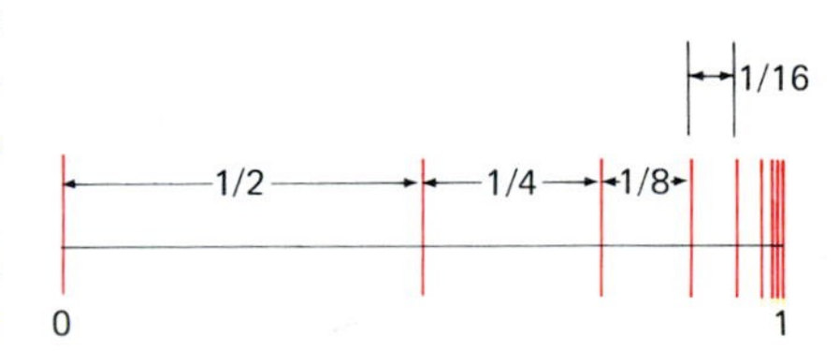

12.1 Subdivision of an interval to illustrate Zeno's paradox

The question is this: What, if anything, is meant by the sum of an *infinite* collection of numbers? This chapter explores conditions under which an *infinite* sum

$$a_1 + a_2 + a_3 + \cdots + a_n + \cdots,$$

known as an *infinite series*, is meaningful. We shall discuss methods for computing the sum of an infinite series, and applications of the algebra and calculus of infinite series. Infinite series are important in science and mathematics because many functions either arise most naturally in the form of infinite series, or have infinite series representations (such as the Taylor series of Section 12.7) that are useful for numerical computations.

12.2 Infinite Sequences

An **(infinite) sequence** of real numbers is a function whose domain of definition is the set of all positive integers. Thus if s is a sequence, then to each positive integer n there corresponds a real number $s(n)$. Ordinarily, a sequence is most conveniently described by listing its values in order, beginning with $s(1)$:

$$s(1),\ s(2),\ s(3),\ \ldots,\ s(n),\ \ldots.$$

With subscript notation rather than function notation, we may write

$$s_1,\ s_2,\ s_3,\ \ldots,\ s_n,\ \ldots \tag{1}$$

for this list of values. The values in this list are the **terms** of the sequence; s_1 is the first term, s_2 the second term, s_n the ***n*th term.**

We use the notation $\{s_n\}_{n=1}^{\infty}$, or simply $\{s_n\}$, as an abbreviation for the **ordered** list in (1), and we may refer to the sequence by saying simply "the sequence $\{s_n\}$." When a particular sequence is so described, the nth term s_n is generally (though not always) given by a formula in terms of its subscript n. In this case, listing the first few terms of the sequence often helps us to see it more concretely.

EXAMPLE 1 The following table lists explicitly the first four terms of each of several sequences.

$\{s_n\}_1^{\infty}$	$s_1,\ s_2,\ s_3,\ \ldots$
$\left\{\dfrac{1}{n}\right\}_1^{\infty}$	$1, \dfrac{1}{2}, \dfrac{1}{3}, \dfrac{1}{4}, \ldots$
$\left\{\dfrac{1}{10^n}\right\}_1^{\infty}$	$0.1, 0.01, 0.001, 0.0001, \ldots$
$\left\{\dfrac{1}{n!}\right\}_1^{\infty}$	$1, \dfrac{1}{2}, \dfrac{1}{6}, \dfrac{1}{24}, \ldots$
$\left\{\sin\dfrac{n\pi}{2}\right\}_1^{\infty}$	$1, 0, -1, 0, \ldots$
$\{1 + (-1)^n\}_1^{\infty}$	$0, 2, 0, 2, \ldots$

EXAMPLE 2 The *Fibonacci sequence* $\{F_n\}$ is defined as follows:

$$F_1 = 1, \quad F_2 = 1, \quad \text{and} \quad F_{n+1} = F_n + F_{n-1} \quad \text{for} \quad n \geqq 2.$$

The first ten terms of the Fibonacci sequence are

$$1,\ 1,\ 2,\ 3,\ 5,\ 8,\ 13,\ 21,\ 34,\ 55.$$

This is a *recursively defined* sequence—after the initial values are given, each term is defined in terms of its predecessors.

The limit of a sequence is defined in much the same way as the limit of an ordinary function (Section 2.2).

Definition *Limit of a Sequence*

We say that the sequence $\{s_n\}$ **converges** to the real number L, or has **limit** L, and we write

$$\lim_{n \to \infty} s_n = L, \tag{2}$$

provided that s_n can be made as close to L as we please by choosing n sufficiently large. That is, given any number $\varepsilon > 0$, there exists an integer N such that

$$|s_n - L| < \varepsilon \qquad \text{for all} \quad n \geqq N. \tag{3}$$

EXAMPLE 3 Prove that $\displaystyle\lim_{n \to \infty} \frac{1}{n} = 0.$

Proof We need to show this: To each positive number ε, there corresponds an integer N such that, for all $n \geqq N$,

$$\left|\frac{1}{n} - 0\right| = \frac{1}{n} < \varepsilon.$$

It suffices to choose any fixed integer $N > 1/\varepsilon$. For example, merely let $N = 1 + [\![1/\varepsilon]\!]$. Then $n \geqq N$ implies that

$$\frac{1}{n} \leqq \frac{1}{N} < \varepsilon,$$

as desired.

The limit laws stated in Section 2.2 for limits of functions have natural analogues for limits of sequences. Their proofs are based on techniques similar to those used in Appendix B.

Theorem 1 *Limit Laws for Sequences*

If the limits

$$\lim_{n\to\infty} a_n = A \quad \text{and} \quad \lim_{n\to\infty} b_n = B$$

exist (so that A and B are real numbers), then:

1. $\lim_{n\to\infty} ca_n = cA$ (c any real number);
2. $\lim_{n\to\infty} (a_n + b_n) = A + B$;
3. $\lim_{n\to\infty} a_n b_n = AB$;
4. $\lim_{n\to\infty} \dfrac{a_n}{b_n} = \dfrac{A}{B}$.

In this last case we must assume also that $B \neq 0$ and that $b_n \neq 0$ for all sufficiently large values of n.

Theorem 2 *Substitution Law for Sequences*

If $\lim_{n\to\infty} a_n = A$ and the function f is continuous at $x = A$, then

$$\lim_{n\to\infty} f(a_n) = f(A).$$

Theorem 3 *Squeeze Law for Sequences*

If $a_n \leqq b_n \leqq c_n$ for all n and

$$\lim_{n\to\infty} a_n = L = \lim_{n\to\infty} c_n,$$

then $\lim_{n\to\infty} b_n = L$ as well.

These theorems can be used to compute limits of many sequences formally, without recourse to the definition. For example, if k is a positive integer and c is a constant, then Example 3 and the product law (Theorem 1, part 3) give

$$\lim_{n\to\infty} \frac{c}{n^k} = c \cdot 0 \cdot 0 \cdots 0 = 0.$$

EXAMPLE 4 Show that $\displaystyle\lim_{n\to\infty} \frac{(-1)^n \cos n}{n^2} = 0$.

Solution This result follows from the squeeze law and the fact that $1/n^2 \to 0$ as $n \to \infty$, because

$$-\frac{1}{n^2} \leqq \frac{(-1)^n \cos n}{n^2} \leqq \frac{1}{n^2}.$$

EXAMPLE 5 Show that if $a > 0$, then $\lim_{n\to\infty} \sqrt[n]{a} = 1$.

Solution We apply the substitution law with $f(x) = a^x$ and $A = 0$. Because $1/n \to 0$ as $n \to \infty$ and f is continuous at $x = 0$, this gives

$$\lim_{n\to\infty} a^{1/n} = a^0 = 1.$$

EXAMPLE 6 The limit laws and the continuity of $f(x) = \sqrt{x}$ at $x = 4$ yield

$$\lim_{n\to\infty} \sqrt{\frac{4n-1}{n+1}} = \sqrt{\lim_{n\to\infty} \frac{4-(1/n)}{1+(1/n)}} = \sqrt{4} = 2.$$

EXAMPLE 7 Show that if $|r| < 1$, then $\lim_{n\to\infty} r^n = 0$.

Solution Because $|r^n| = |(-r)^n|$, we may assume that $0 < r < 1$. Then $1/r = 1 + a$ with $a > 0$, so the binomial formula yields

$$\frac{1}{r^n} = (1+a)^n = 1 + na + \{\text{positive terms}\} > 1 + na,$$

so

$$0 < r^n < \frac{1}{1+na}.$$

Now $1/(1+na) \to 0$ as $n \to \infty$. Therefore, the squeeze law implies that $r^n \to 0$ as $n \to \infty$.

Let f be a function defined for every real number $x \geqq 1$, and $\{a_n\}$ a sequence such that $f(n) = a_n$ for every positive integer n. Then it follows from the definitions of limits of functions and sequences that

$$\textit{if} \quad \lim_{x\to\infty} f(x) = L, \quad \textit{then} \quad \lim_{n\to\infty} a_n = L. \tag{4}$$

Note that the converse of the statement in (4) is generally false. For example,

$$\lim_{n\to\infty} \sin \pi n = 0 \quad \text{but} \quad \lim_{x\to\infty} \sin \pi x \text{ does not exist.}$$

Because of (4) we can use **l'Hôpital's rule for sequences:** If $a_n = f(n)$, $b_n = g(n)$, and $f(x)/g(x)$ has the indeterminate form ∞/∞ as $x \to \infty$, then

$$\lim_{n\to\infty} \frac{a_n}{b_n} = \lim_{x\to\infty} \frac{f(x)}{g(x)} = \lim_{x\to\infty} \frac{f'(x)}{g'(x)}, \tag{5}$$

provided that f and g satisfy the other hypotheses of l'Hôpital's rule, including the assumption that the right-hand limit exists.

EXAMPLE 8 Show that $\lim_{n\to\infty} \frac{\ln n}{n} = 0$.

Solution The function $(\ln x)/x$ is defined for all $x \geqq 1$ and agrees with the given sequence when $x = n$, a positive integer. Because $(\ln x)/x$ has the indeterminate form ∞/∞ as $x \to \infty$, l'Hôpital's rule gives

$$\lim_{n\to\infty} \frac{\ln n}{n} = \lim_{x\to\infty} \frac{\ln x}{x} = \lim_{x\to\infty} \frac{1/x}{1} = 0.$$

EXAMPLE 9 Show that $\lim_{n \to \infty} \sqrt[n]{n} = 1$.

Solution First we note that

$$\ln \sqrt[n]{n} = \ln n^{1/n} = \frac{\ln n}{n} \to 0 \quad \text{as} \quad n \to \infty$$

by Example 8. By the substitution law with $f(x) = e^x$, this gives

$$\lim_{n \to \infty} n^{1/n} = \lim_{n \to \infty} \exp(\ln n^{1/n}) = e^0 = 1.$$

EXAMPLE 10 Find $\lim_{n \to \infty} \dfrac{3n^3}{e^{2n}}$.

Solution We apply l'Hôpital's rule repeatedly, although we must be careful at each intermediate step to verify that we still have an indeterminate form. Thus we find that

$$\lim_{n \to \infty} \frac{3n^3}{e^{2n}} = \lim_{x \to \infty} \frac{3x^3}{e^{2x}} = \lim_{x \to \infty} \frac{9x^2}{2e^{2x}}$$

$$= \lim_{x \to \infty} \frac{18x}{4e^{2x}} = \lim_{x \to \infty} \frac{18}{8e^{2x}} = 0.$$

*BOUNDED MONOTONE SEQUENCES

The set of all *rational* numbers has by itself all the most familiar elementary algebraic properties of the entire real number system. In order to guarantee the existence of irrational numbers, we must assume in addition a "completeness property" of the real numbers. Otherwise, the real line might have "holes" where the irrational numbers ought to be. One way of stating this completeness property involves the convergence of an important type of sequence.

The sequence $\{a_n\}_1^\infty$ is said to be **monotone increasing** if

$$a_1 \leqq a_2 \leqq a_3 \leqq \cdots \leqq a_n \leqq \cdots$$

and **monotone decreasing** if

$$a_1 \geqq a_2 \geqq a_3 \geqq \cdots \geqq a_n \geqq \cdots.$$

It is **monotone** if it is *either* monotone increasing *or* monotone decreasing. The sequence $\{a_n\}$ is **bounded** if there is a number M such that $|a_n| \leqq M$ for all n. The following assertion may be taken as an axiom for the real number system.

Bounded Monotone Sequence Property

Every bounded monotone infinite sequence converges (that is, has a finite limit).

Suppose, for example, that the monotone increasing sequence $\{a_n\}_1^\infty$ is bounded above by a number M, meaning that $a_n \leqq M$ for all n. Because it is also bounded below (by a_1, for instance), the bounded monotone sequence property implies that

$$\lim_{n\to\infty} a_n = A \qquad \text{for some real number} \quad A \leqq M.$$

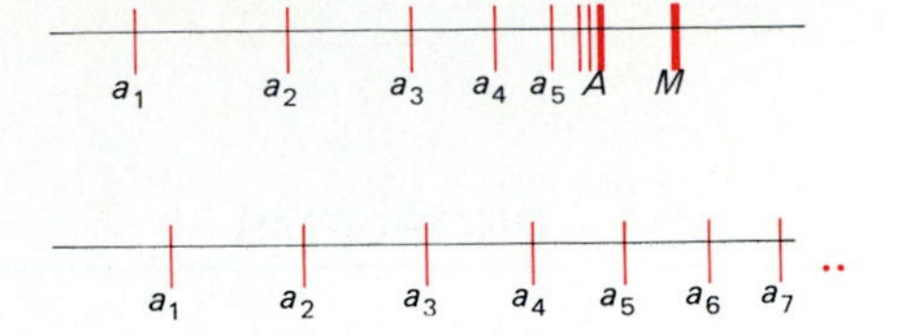

12.2 A bounded increasing sequence and an increasing sequence that is not bounded above.

If the monotone increasing sequence $\{a_n\}$ is *not* bounded above (by any number), then it follows that

$$\lim_{n\to\infty} a_n = +\infty$$

(Problem 38). These two alternatives are illustrated in Fig. 12.2.

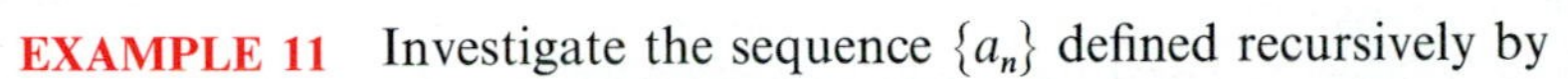

EXAMPLE 11 Investigate the sequence $\{a_n\}$ defined recursively by

$$a_1 = \sqrt{2}, \qquad a_{n+1} = \sqrt{2 + a_n} \qquad \text{for} \quad n \geqq 1. \tag{6}$$

Solution The first four terms of $\{a_n\}$ are

$$\sqrt{2},\ \sqrt{2+\sqrt{2}},\ \sqrt{2+\sqrt{2+\sqrt{2}}},\ \sqrt{2+\sqrt{2+\sqrt{2+\sqrt{2}}}}$$

If the sequence $\{a_n\}$ has a limit A, then A would seem to be the natural interpretation of the value of the infinite expression

$$\sqrt{2+\sqrt{2+\sqrt{2+\sqrt{2+\cdots}}}}.$$

We will apply the bounded monotone sequence property to show that this limit does indeed exist. Note first that $a_1 < 2$. If we assume inductively that $a_n < 2$ for some $n \geqq 1$, then it follows that

$$(a_{n+1})^2 = 2 + a_n < 4, \qquad \text{so} \quad a_{n+1} < 2.$$

Thus the sequence $\{a_n\}$ is bounded above by the number 2. We observe next that

$$(a_{n+1})^2 - (a_n)^2 = (2 + a_n) - (a_n)^2 = (2 - a_n)(1 + a_n) > 0$$

because $2 - a_n$ and $1 + a_n$ are both positive. Because a_{n+1} and a_n are positive, it follows that $a_{n+1} > a_n$ for all $n \geqq 1$. Thus $\{a_n\}$ is a monotone increasing sequence.

Therefore, the bounded monotone sequence property implies that the sequence $\{a_n\}$ has a limit A. It does not tell us what the number A is. But now that we know that the limit A of the sequence $\{a_n\}$ exists, we can write

$$A = \lim_{n\to\infty} a_{n+1} = \lim_{n\to\infty} \sqrt{2 + a_n} = \sqrt{2 + A},$$

and thus

$$A^2 = 2 + A.$$

The roots of this equation are -1 and 2. It is clear that $A > 0$, so we finally conclude that

$$\lim_{n\to\infty} a_n = 2.$$

To indicate what the bounded monotone sequence property has to do with the "completeness" property of the real numbers, we outline in Problem 42 a proof, using this property, of the existence of the number $\sqrt{2}$. In Problems 43 and 44, we outline a proof of the equivalence of the bounded monotone sequence property and another common statement of the completeness of the real numbers—the *least upper bound property.*

12.2 PROBLEMS

In Problems 1–35, determine whether or not the sequence $\{a_n\}$ converges, and find its limit if it does converge.

1. $a_n = \dfrac{2n}{5n-3}$
2. $a_n = \dfrac{1-n^2}{2+3n^2}$
3. $a_n = \dfrac{n^2-n+7}{2n^3+n^2}$
4. $a_n = \dfrac{n^3}{10n^2+1}$
5. $a_n = 1 + \left(\dfrac{9}{10}\right)^n$
6. $a_n = 2 - \left(-\dfrac{1}{2}\right)^n$
7. $a_n = 1 + (-1)^n$
8. $a_n = \dfrac{1+(-1)^n}{\sqrt{n}}$
9. $a_n = \dfrac{1+(-1)^n\sqrt{n}}{(\frac{3}{2})^n}$
10. $a_n = \dfrac{\sin n}{3^n}$
11. $a_n = \dfrac{\sin^2 n}{\sqrt{n}}$
12. $a_n = \sqrt{\dfrac{2+\cos n}{n}}$
13. $a_n = n \sin \pi n$
14. $a_n = n \cos \pi n$
15. $a_n = \pi^{-(\sin n)/n}$
16. $a_n = 2^{\cos n\pi}$
17. $a_n = \dfrac{\ln n}{\sqrt{n}}$
18. $a_n = \dfrac{\ln 2n}{\ln 3n}$
19. $a_n = \dfrac{(\ln n)^2}{n}$
20. $a_n = n \sin\left(\dfrac{1}{n}\right)$
21. $a_n = \dfrac{\tan^{-1} n}{n}$
22. $a_n = \dfrac{n^3}{e^{n/10}}$
23. $a_n = \dfrac{2^n+1}{e^n}$
24. $a_n = \dfrac{\sinh n}{\cosh n}$
25. $a_n = \left(1+\dfrac{1}{n}\right)^n$
26. $a_n = (2n+5)^{1/n}$
27. $a_n = \left(\dfrac{n-1}{n+1}\right)^n$
28. $a_n = (0.001)^{-1/n}$
29. $a_n = \sqrt[n]{2^{n+1}}$
30. $a_n = \left(1-\dfrac{2}{n^2}\right)^n$
31. $a_n = \left(\dfrac{2}{n}\right)^{3/n}$
32. $a_n = (-1)^n(n^2+1)^{1/n}$
33. $a_n = \left(\dfrac{2-n^2}{3+n^2}\right)^n$
34. $a_n = \dfrac{(\frac{2}{3})^n}{1-n^{1/n}}$
35. $a_n = \dfrac{(\frac{2}{3})^n}{(\frac{1}{2})^n + (\frac{9}{10})^n}$

36. Suppose that $\lim_{n\to\infty} a_n = A$. Prove that $\lim_{n\to\infty} |a_n| = |A|$.

37. Prove that $\{(-1)^n a_n\}$ diverges if $\lim_{n\to\infty} a_n = A \neq 0$.

38. Suppose that $\{a_n\}$ is a monotone increasing sequence that is not bounded. Prove that $\lim_{n\to\infty} a_n = +\infty$.

39. Suppose that $A > 0$. Given x_1 arbitrary, define the sequence $\{x_n\}$ recursively as follows:

$$x_{n+1} = \frac{1}{2}\left(x_n + \frac{A}{x_n}\right).$$

Prove that if $L = \lim_{n\to\infty} x_n$ exists, then $L = \sqrt{A}$.

40. Let $\{F_n\}$ be the Fibonacci sequence of Example 2. Assume that

$$\tau = \lim_{n\to\infty} \frac{F_{n+1}}{F_n}$$

exists; prove that $\tau = \frac{1}{2}(1+\sqrt{5})$.

41. Let the sequence $\{a_n\}$ be defined recursively by

$$a_1 = 2, \qquad a_{n+1} = \tfrac{1}{2}(a_n + 4) \quad \text{for} \quad n \geqq 1.$$

(a) Prove by induction on n that $a_n < 4$ for each n and that $\{a_n\}$ is a monotone increasing sequence.
(b) Find the limit of this sequence.

42. For each positive integer n, let a_n be the largest integral multiple of $1/10^n$ such that $a_n^2 \leqq 2$.
(a) Prove that $\{a_n\}$ is a bounded and monotone increasing sequence, so that $A = \lim_{n\to\infty} a_n$ exists.
(b) Prove that if $A^2 > 2$, then $a_n^2 > 2$ for n sufficiently large.
(c) Prove that if $A^2 < 2$, then $a_n^2 < B$ for some number $B < 2$ and all sufficiently large n.
(d) Conclude that $A^2 = 2$.

Problems 43 and 44 deal with the **least upper bound** property of the real numbers: If the nonempty set S of real numbers has an upper bound, then S has a least upper bound. The number M is an **upper bound** for S if $x \leqq M$ for all x in S. The upper bound L of S is a **least upper bound** for S if no number smaller than L is an upper bound for S.

43. Prove that the least upper bound property implies the bounded monotone sequence property. (*Suggestion:* If $\{a_n\}$ is a bounded and monotone increasing sequence and A is the least upper bound of $\{a_n\}$, prove that $A = \lim_{n\to\infty} a_n$.)

44. Prove that the bounded monotone sequence property implies the least upper bound property. (*Suggestion:* For each positive integer n, let a_n be the least multiple of $1/10^n$ that is an upper bound of the set S. Prove that $\{a_n\}$ is a bounded and monotone decreasing sequence, and then that $A = \lim_{n\to\infty} a_n$ is a least upper bound for S.)

12.3 Convergence of Infinite Series

An **infinite series** is an expression of the form

$$\sum_{n=1}^{\infty} a_n = a_1 + a_2 + a_3 + \cdots + a_n + \cdots \tag{1}$$

where $\{a_n\}$ is an infinite sequence of real numbers. The number a_n is called the ***n*th term** of the series. The symbol $\sum_{n=1}^{\infty} a_n$ is simply an abbreviation—compact notation—for the right-hand side in (1). An example of an infinite series is the series

$$\sum_{n=1}^{\infty} \frac{1}{2^n} = \frac{1}{2} + \frac{1}{4} + \frac{1}{8} + \cdots + \frac{1}{2^n} + \cdots$$

that was mentioned in Section 12.1. The nth term of this particular infinite series is $a_n = 1/2^n$.

To say what is meant by such an infinite sum, we introduce the partial sums of the infinite series in (1). The ***n*th partial sum** S_n of the series is the sum of its first n terms:

$$S_n = a_1 + a_2 + a_3 + \cdots + a_n. \tag{2}$$

Thus each infinite series has associated with it an infinite **sequence of partial sums**

$$S_1, S_2, S_3, \ldots, S_n, \ldots.$$

We define the sum of the infinite series to be the limit of its sequence of partial sums, provided that this limit exists.

Definition *Sum of an Infinite Series*

We say that the infinite series

$$\sum_{n=1}^{\infty} a_n \qquad \textbf{converges} \text{ (or is } \textbf{convergent)}$$

with **sum** S provided that the limit of its sequence of partial sums,

$$S = \lim_{n \to \infty} S_n, \tag{3}$$

exists (and is finite). Otherwise, we say that the series **diverges** (or is **divergent**). If a series diverges, then it has no sum.

Thus an infinite sum is a limit of finite sums,

$$S = \sum_{n=1}^{\infty} a_n = \lim_{N \to \infty} \sum_{n=1}^{N} a_n,$$

provided that this limit exists.

EXAMPLE 1 Show that the series

$$\sum_{n=1}^{\infty} \left(\tfrac{1}{2}\right)^n = \tfrac{1}{2} + \tfrac{1}{4} + \tfrac{1}{8} + \tfrac{1}{16} + \cdots$$

converges, and find its sum.

Solution The first four partial sums are

$$S_1 = \tfrac{1}{2}, \qquad S_2 = \tfrac{3}{4}, \qquad S_3 = \tfrac{7}{8}, \qquad S_4 = \tfrac{15}{16}.$$

It seems likely that $S_n = (2^n - 1)/2^n$, and indeed this follows easily by induction on n, because

$$\begin{aligned} S_{n+1} = S_n + \frac{1}{2^{n+1}} &= \frac{2^n - 1}{2^n} + \frac{1}{2^{n+1}} \\ &= \frac{2^{n+1} - 2 + 1}{2^{n+1}} = \frac{2^{n+1} - 1}{2^{n+1}}. \end{aligned}$$

Hence the sum of the given series is

$$S = \lim_{n \to \infty} S_n = \lim_{n \to \infty} \frac{2^n - 1}{2^n} = \lim_{n \to \infty} \left(1 - \frac{1}{2^n}\right) = 1.$$

EXAMPLE 2 Show that the series

$$\sum_{n=1}^{\infty} (-1)^{n+1} = 1 - 1 + 1 - 1 + \cdots$$

diverges.

Solution The sequence of partial sums of the given series is

$$1, \quad 0, \ 1, \ 0, \ 1, \ \ldots,$$

which has no limit. Therefore, the series diverges.

EXAMPLE 3 Show that the infinite series

$$\sum_{n=1}^{\infty} \frac{1}{4n^2 - 1} = \tfrac{1}{3} + \tfrac{1}{15} + \tfrac{1}{35} + \tfrac{1}{63} \cdots$$

converges, and find its sum.

Solution We need a formula for the nth partial sum S_n so that we can evaluate its limit as $n \to \infty$. To find such a formula, we begin with the observation that

$$a_n = \frac{1}{4n^2 - 1} = \frac{1}{2}\left(\frac{1}{2n - 1} - \frac{1}{2n + 1}\right).$$

It follows that

$$\begin{aligned} S_n = \frac{1}{2}\bigg[\left(1 - \frac{1}{3}\right) &+ \left(\frac{1}{3} - \frac{1}{5}\right) \\ &+ \left(\frac{1}{5} - \frac{1}{7}\right) + \cdots + \left(\frac{1}{2n - 1} - \frac{1}{2n + 1}\right)\bigg] \\ = \frac{1}{2}\left(1 - \frac{1}{2n + 1}\right) &= \frac{n}{2n + 1}. \end{aligned}$$

And hence

$$\sum_{n=1}^{\infty} \frac{1}{4n^2 - 1} = \lim_{n \to \infty} \frac{n}{2n + 1} = \frac{1}{2}.$$

The sum for S_n in Example 3 is called a *telescoping* sum and provides us with a way to find the sums of certain series. The series in Examples 1 and 2 are examples of a more common sort of series, the *geometric series.*

Definition *Geometric Series*

The series $\sum_{n=0}^{\infty} a_n$ is said to be a **geometric series** if each term after the first is a fixed multiple of the term immediately before it. That is, there is a number r, called the **ratio** of the series, such that

$$a_{n+1} = ra_n \qquad \text{for all} \quad n \geqq 0.$$

Thus every geometric series takes the form

$$a_0 + ra_0 + r^2a_0 + r^3a_0 + \cdots = \sum_{n=0}^{\infty} r^n a_0. \tag{4}$$

Note that it is convenient to begin the summation at $n = 0$, and thus we regard the sum

$$S_n = a_0(1 + r + r^2 + \cdots + r^n)$$

of the first $n + 1$ terms as the nth partial sum of the series.

EXAMPLE 4 The infinite series

$$\sum_{n=0}^{\infty} \frac{2}{3^n} = 2 + \frac{2}{3} + \frac{2}{9} + \cdots + \frac{2}{3^n} + \cdots$$

is a geometric series with first term $a_0 = 2$ and ratio $r = \frac{1}{3}$.

Theorem 1 *Sum of a Geometric Series*

If $|r| < 1$, then the geometric series in (4) converges, and its sum is

$$S = \sum_{n=0}^{\infty} r^n a_0 = \frac{a_0}{1 - r}. \tag{5}$$

If $|r| \geqq 1$ and $a_0 \neq 0$, then the geometric series diverges.

Proof If $r = 1$, then $S_n = (n + 1)a_0$, so the series certainly diverges. If $r = -1$, then it diverges by an argument like the one in Example 2. So we suppose that $|r| \neq 1$. Then the elementary identity

$$1 + r + r^2 + \cdots + r^n = \frac{1 - r^{n+1}}{1 - r}$$

follows after multiplication of each side by $1 - r$. Hence

$$S_n = a_0(1 + r + r^2 + \cdots + r^n) = a_0\left(\frac{1}{1 - r} - \frac{r^{n+1}}{1 - r}\right).$$

If $|r| < 1$, then $r^{n+1} \to 0$ as $n \to \infty$, by Example 7 in Section 12.2. So in this case

$$S = \lim_{n \to \infty} a_0\left(\frac{1}{1 - r} - \frac{r^{n+1}}{1 - r}\right) = \frac{a_0}{1 - r}.$$

If $|r| > 1$, then $\lim_{n\to\infty} r^{n+1}$ does not exist, so $\lim_{n\to\infty} S_n$ does not exist. This establishes the theorem. ■

EXAMPLE 5 With $a_0 = 1$ and $r = -\frac{1}{2}$, we find that

$$1 - \tfrac{1}{2} + \tfrac{1}{4} - \tfrac{1}{8} + \cdots = \sum_{n=0}^{\infty} \left(-\tfrac{1}{2}\right)^n = \frac{1}{1 - \left(-\frac{1}{2}\right)} = \tfrac{2}{3}.$$

The next theorem implies that the operations of addition and of multiplication by a constant can be carried out termwise in the case of *convergent* series. Because the sum of an infinite series is the limit of its sequence of partial sums, this theorem follows immediately from the limit laws for sequences (Theorem 1 in Section 12.2).

> ***Theorem 2*** *Termwise Addition and Multiplication*
> If the series $A = \sum a_n$ and $B = \sum b_n$ converge to the indicated sums and c is a constant, then the series $\sum (a_n + b_n)$ and $\sum ca_n$ also converge, with sums
>
> **1.** $\sum (a_n + b_n) = A + B$;
> **2.** $\sum ca_n = cA$.

The geometric series formula in (5) may be used to find the rational number represented by a given infinite repeating decimal.

EXAMPLE 6

$$0.55555\ldots = \tfrac{5}{10} + \tfrac{5}{100} + \tfrac{5}{1000} + \cdots = \sum_{n=0}^{\infty} \tfrac{5}{10}\left(\tfrac{1}{10}\right)^n$$

$$= \frac{\frac{5}{10}}{1 - \frac{1}{10}} = \tfrac{5}{10} \cdot \tfrac{10}{9} = \tfrac{5}{9}.$$

In a more complicated example, we may wish to use the "termwise algebra" of Theorem 2:

$$0.728\,28\,28\ldots = \frac{7}{10} + \frac{28}{10^3} + \frac{28}{10^5} + \frac{28}{10^7} + \cdots$$

$$= \frac{7}{10} + \frac{28}{10^3}\left(1 + \frac{1}{10^2} + \frac{1}{10^4} + \cdots\right)$$

$$= \frac{7}{10} + \frac{28}{1000}\sum_{n=0}^{\infty}\left(\frac{1}{10^2}\right)^n$$

$$= \frac{7}{10} + \frac{28}{1000}\left(\frac{1}{1 - \frac{1}{100}}\right)$$

$$= \frac{7}{10} + \frac{28}{1000}\left(\frac{100}{99}\right)$$

$$= \frac{7}{10} + \frac{28}{990} = \frac{721}{990}$$

It should be clear that this technique can be used to show that every repeating infinite decimal represents a rational number; consequently, the decimal expansions of irrational numbers such as π, e, and $\sqrt{2}$ must be nonrepeating as well as infinite. Conversely, if p and q are integers with $q \neq 0$, then long division of q into p yields a repeating decimal expansion for the rational number p/q because such a division can yield at each stage only q possible different remainders.

EXAMPLE 7 Suppose that Paul and Mary toss a fair six-sided die in turn until one of them wins by getting the first six. If Paul tosses first, calculate the probability that it is he who wins.

Solution Because the die is fair, the probability that Paul gets a six on the first round is $\frac{1}{6}$. The probability that he gets the game's first six on the second round is $(\frac{5}{6})^2(\frac{1}{6})$—the product of the probability $(\frac{5}{6})^2$ that neither Paul nor Mary rolls a six in the first round and the probability $\frac{1}{6}$ that Paul rolls a six in the second round. Paul's probability p of getting the first six in the game is the *sum* of his probabilities of getting it in the first round, in the second round, in the third round, and so on. Hence

$$\begin{aligned} p &= \tfrac{1}{6} + (\tfrac{5}{6})^2(\tfrac{1}{6}) + (\tfrac{5}{6})^2(\tfrac{5}{6})^2(\tfrac{1}{6}) + \cdots \\ &= \tfrac{1}{6}[1 + (\tfrac{5}{6})^2 + (\tfrac{5}{6})^4 + \cdots] \\ &= \tfrac{1}{6} \cdot \frac{1}{1 - (\frac{5}{6})^2} = \tfrac{1}{6} \cdot \tfrac{36}{11} = \tfrac{6}{11}. \end{aligned}$$

Observe that because he tosses first, Paul has more than the fair probability $\frac{1}{2}$ of getting the first six and thus winning the game.

The following theorem is often useful in showing that a given series does *not* converge.

Theorem 3 *The nth Term Test for Divergence*

If either

$$\lim_{n \to \infty} a_n \neq 0$$

or this limit does not exist, then the infinite series $\sum a_n$ diverges.

Proof We want to show under the stated hypothesis that the series $\sum a_n$ diverges. It suffices to show that *if* the series $\sum a_n$ does converge, then $\lim_{n \to \infty} a_n = 0$. So suppose that $\sum a_n$ converges with sum $S = \lim_{n \to \infty} S_n$, where

$$S_n = a_1 + a_2 + a_3 + \cdots + a_n.$$

We note that $a_n = S_n - S_{n-1}$, so that

$$\lim_{n \to \infty} a_n = \lim_{n \to \infty} (S_n - S_{n-1}) = \lim_{n \to \infty} S_n - \lim_{n \to \infty} S_{n-1} = S - S = 0.$$

Consequently, if $\lim_{n \to \infty} a_n \neq 0$, then the series $\sum a_n$ diverges. ■

EXAMPLE 8 The series

$$\sum_{n=1}^{\infty} (-1)^{n-1}n^2 = 1 - 4 + 9 - 16 + 25 - \cdots$$

diverges because $\lim_{n\to\infty} a_n$ does not exist, while the series

$$\sum_{n=1}^{\infty} \frac{n}{3n+1} = \frac{1}{4} + \frac{2}{7} + \frac{3}{10} + \frac{4}{13} + \cdots$$

diverges because

$$\lim_{n\to\infty} \frac{n}{3n+1} = \frac{1}{3} \neq 0.$$

WARNING The converse of Theorem 3 is *false*! The condition

$$\lim_{n\to\infty} a_n = 0$$

is necessary *but not sufficient* for convergence of the series

$$\sum_{n=1}^{\infty} a_n.$$

That is, a series may satisfy the condition $a_n \to 0$ as $n \to \infty$ and yet diverge. An important example of a divergent series with terms that approach zero is the **harmonic series**

$$\sum_{n=1}^{\infty} \frac{1}{n} = 1 + \frac{1}{2} + \frac{1}{3} + \frac{1}{4} + \frac{1}{5} + \cdots. \tag{6}$$

Theorem 4

The harmonic series diverges.

Proof Because each term of the harmonic series is positive, its sequence of partial sums $\{S_n\}$ is monotone increasing. We shall prove that

$$\lim_{n\to\infty} S_n = +\infty,$$

and thus that the harmonic series diverges, by showing that there are arbitrarily large partial sums. Consider the closed interval $[0, k]$ on the x-axis, where k is a positive integer. On its subinterval $[0, 1]$, imagine a rectangle of height 1 with that subinterval as its base. On the subinterval $[1, 2]$, imagine a rectangle of height $\frac{1}{2}$ having that subinterval as its base. Build a rectangle of height $\frac{1}{3}$ over the subinterval $[2, 3]$, a rectangle of height $\frac{1}{4}$ over the subinterval $[3, 4]$, and so on. The last rectangle will have the subinterval $[k-1, k]$ as its base and will have height $1/k$. These rectangles are shown in Fig. 12.3.

Next, note that the total area of all these k rectangles is the kth partial sum of the harmonic series:

$$S_k = 1 + \frac{1}{2} + \frac{1}{3} + \cdots + \frac{1}{k}.$$

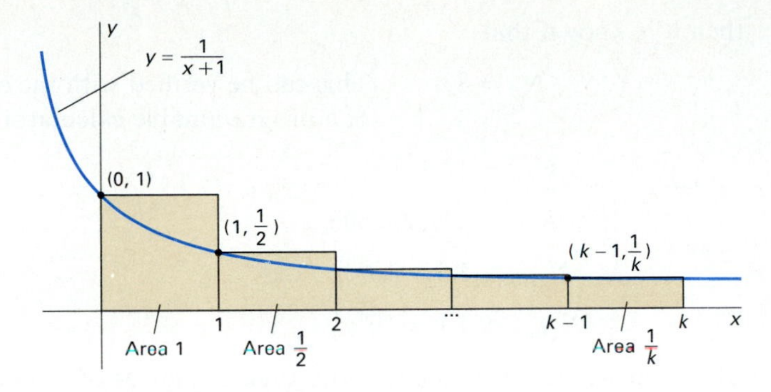

12.3 Idea of the proof of Theorem 4

Finally, the graph of the function

$$f(x) = \frac{1}{x+1}$$

—also shown in Fig. 12.3—passes through the upper left-hand corner point of each rectangle. Because f is decreasing for all $x \geqq 0$, the graph of f is never above the top of any rectangle on the corresponding subinterval. So the sum of the areas of the k rectangles is greater than the area under the graph of f over $[0, k]$. That is,

$$S_k \geqq \int_0^k \frac{1}{x+1}\,dx = \Big[\ln(x+1)\Big]_0^k = \ln(k+1).$$

But

$$\lim_{k\to\infty} \ln(k+1) = +\infty,$$

so $\ln(k+1)$ takes on arbitrarily large positive values with increasing k. Consequently, because $S_k \geqq \ln(k+1)$ for all $k \geqq 1$, S_k also takes on arbitrarily large positive values. Therefore $\lim_{k\to\infty} S_k = +\infty$, and hence the harmonic series diverges. ■

If the sequence of partial sums of the series $\sum a_n$ diverges to infinity, then we say that the series **diverges to infinity,** and we write

$$\sum_{n=1}^{\infty} a_n = \infty.$$

The series $\sum (-1)^{n+1}$ of Example 2 is one that diverges but does not diverge to infinity. In the nineteenth century it was common to say that such a series was divergent by oscillation; today we say merely that it diverges.

Our proof of Theorem 4 shows that

$$\sum_{n=1}^{\infty} \frac{1}{n} = \infty.$$

But the partial sums of the harmonic series diverge to infinity very slowly. If N_A denotes the smallest integer such that

$$\sum_{n=1}^{N_A} \frac{1}{n} \geqq A,$$

then it is known that

$$N_5 = 83, \qquad \text{(This can be verified with the aid of a programmable calculator.)}$$

$$N_{10} = 12{,}367,$$

$$N_{20} = 272{,}400{,}600,$$

$$N_{100} \approx 1.5 \times 10^{43}, \quad \text{and}$$

$$N_{1000} \approx 1.1 \times 10^{434}.$$

Thus you would need to add more than a quarter of a billion terms of the harmonic series to get a partial sum over 20. At this point the next few terms would each be approximately $0.000\,000\,004 = 4 \times 10^{-9}$. The number of terms you'd have to add to reach 1000 is far larger than the estimated number of elementary particles in the entire universe (10^{80}). If you enjoy such large numbers, see the article "Partial sums of infinite series, and how they grow," by R. P. Boas, Jr., in *American Mathematical Monthly* **84** (1977), 237–248.

The following theorem says that if two infinite series have the same terms from some point on, then either both series converge or both series diverge. The proof is left for Problem 43.

> ***Theorem 5*** *Series That Are Eventually the Same*
> If there exists a positive integer k such that $a_n = b_n$ for all $n > k$, then the series $\sum a_n$ and $\sum b_n$ either both converge or both diverge.

It follows that a *finite* number of terms can be changed, deleted from, or adjoined to an infinite series without altering its convergence or divergence (although the *sum* of a convergent series will generally be changed by such alterations). In particular, taking $b_n = 0$ for $n \leqq k$ and $b_n = a_n$ for $n > k$, we see that the series

$$\sum_{n=1}^{\infty} a_n \quad \text{and the series} \quad \sum_{n=k+1}^{\infty} a_n$$

obtained by deleting its first k terms either both converge or both diverge.

12.3 PROBLEMS

In Problems 1–29, determine whether each given infinite series converges or diverges. If it converges, find its sum.

1. $1 + \frac{1}{3} + \frac{1}{9} + \cdots + (\frac{1}{3})^n + \cdots$
2. $1 + e^{-1} + e^{-2} + \cdots + e^{-n} + \cdots$
3. $1 + 3 + 5 + 7 + \cdots + (2n - 1) + \cdots$
4. $\frac{1}{2} + \frac{1}{2^{1/2}} + \frac{1}{2^{1/3}} + \cdots + \frac{1}{2^{1/n}} + \cdots$
5. $1 - 2 + 4 - 8 + \cdots + (-2)^n + \cdots$
6. $1 - \frac{1}{4} + \frac{1}{16} - \cdots + (-\frac{1}{4})^n + \cdots$
7. $4 + \frac{4}{3} + \frac{4}{9} + \frac{4}{27} + \cdots + \frac{4}{3^n} + \cdots$
8. $\frac{1}{3} + \frac{2}{9} + \frac{4}{27} + \cdots + \frac{2^{n-1}}{3^n} + \cdots$
9. $1 + (1.01) + (1.01)^2 + (1.01)^3 + \cdots + (1.01)^n + \cdots$
10. $1 + \frac{1}{2^{1/2}} + \frac{1}{3^{1/3}} + \cdots + \frac{1}{n^{1/n}} + \cdots$

11. $\sum_{n=0}^{\infty} \frac{(-1)^n n}{n+1}$

12. $\sum_{n=1}^{\infty} \left(\frac{e}{10}\right)^n$

13. $\sum_{n=0}^{\infty} (-1)^n \left(\frac{3}{e}\right)^n$

14. $\sum_{n=0}^{\infty} \frac{3^n - 2^n}{4^n}$

15. $\sum_{n=1}^{\infty} (\sqrt{2})^{1-n}$

16. $\sum_{n=1}^{\infty} \left(\frac{2}{n} - \frac{1}{2^n}\right)$

17. $\sum_{n=1}^{\infty} \frac{n}{10n + 17}$

18. $\sum_{n=1}^{\infty} \frac{\sqrt{n}}{\ln(n+1)}$

19. $\sum_{n=1}^{\infty} (5^{-n} - 7^{-n})$

20. $\sum_{n=0}^{\infty} \frac{1}{1 + (\frac{9}{10})^n}$

21. $\sum_{n=1}^{\infty} \left(\frac{e}{\pi}\right)^n$

22. $\sum_{n=1}^{\infty} \left(\frac{\pi}{e}\right)^n$

23. $\sum_{n=0}^{\infty} \left(\frac{100}{99}\right)^n$

24. $\sum_{n=0}^{\infty} \left(\frac{99}{100}\right)^n$

25. $\sum_{n=0}^{\infty} \frac{1 + 2^n + 3^n}{5^n}$

26. $\sum_{n=0}^{\infty} \frac{1 + 2^n + 5^n}{3^n}$

27. $\sum_{n=0}^{\infty} \frac{7 \cdot 5^n + 3 \cdot 11^n}{13^n}$

28. $\sum_{n=1}^{\infty} 2^{1/n}$

29. $\sum_{n=1}^{\infty} [(\frac{7}{11})^n - (\frac{3}{5})^n]$

30. Use the method of Example 6 to verify that

(a) $0.666\,666\,666\ldots = \frac{2}{3}$;

(b) $0.111\,111\,111\ldots = \frac{1}{9}$;

(c) $0.249\,999\,999\ldots = \frac{1}{4}$;

(d) $0.999\,999\,999\ldots = 1$.

In Problems 31–35, find the rational number represented by the given repeating decimal.

31. 0.474747...

32. 0.252525...

33. 0.123123123...

34. 0.3377 3377 3377...

35. 3.14159 14159 14159...

In each of Problems 36–40, use the method of Example 3 to find a formula for the nth partial sum S_n, and then compute the sum of the infinite series. (They all *do* converge.)

36. $\sum_{n=1}^{\infty} \frac{1}{n(n+1)}$

37. $\sum_{n=1}^{\infty} \ln\left(\frac{n+1}{n}\right)$

38. $\sum_{n=0}^{\infty} \frac{4}{16n^2 - 8n - 3}$

39. $\sum_{n=2}^{\infty} \frac{2}{n^2 - 1}$

40. $\frac{1}{1 \cdot 3} - \frac{1}{2 \cdot 4} + \frac{1}{3 \cdot 5} - \frac{1}{4 \cdot 6} + \cdots$

41. Prove: If $\sum a_n$ diverges and c is a nonzero constant, then $\sum ca_n$ diverges.

42. Suppose that $\sum a_n$ converges and that $\sum b_n$ diverges. Prove that $\sum (a_n + b_n)$ diverges.

43. Let S_n and T_n denote the nth partial sums of $\sum a_n$ and $\sum b_n$, respectively. Suppose that $a_n = b_n$ for all $n > k$. Show that $S_n - T_n = S_k - T_k$ for $n > k$. Hence prove Theorem 5.

44. Suppose that $0 < x \leqq 1$. Integrate both sides of the identity

$$\frac{1}{1+t} = 1 - t + t^2 - t^3 + \cdots + (-1)^n t^n + \frac{(-1)^{n+1} t^{n+1}}{1+t}$$

from $t = 0$ to $t = x$ to show that

$$\ln(1 + x) = x - \frac{x^2}{2} + \frac{x^3}{3} - \cdots + (-1)^n \frac{x^{n+1}}{n+1} + R_n$$

where $\lim_{n \to \infty} R_n = 0$. Hence conclude that

$$\ln(1 + x) = \sum_{n=1}^{\infty} (-1)^{n+1} \frac{x^n}{n}.$$

if $0 < x \leqq 1$.

45. Criticize the following "proof" that $2 = 1$. Substitution of $x = 1$ in the result of Problem 44 gives the fact that

$$\ln 2 = 1 - \tfrac{1}{2} + \tfrac{1}{3} - \tfrac{1}{4} + \cdots.$$

If

$$S = 1 + \tfrac{1}{2} + \tfrac{1}{3} + \tfrac{1}{4} + \cdots,$$

then

$$\ln 2 = S - 2(\tfrac{1}{2} + \tfrac{1}{4} + \tfrac{1}{6} + \tfrac{1}{8} + \cdots)$$
$$= S - S = 0.$$

Hence $2 = e^{\ln 2} = e^0 = 1$.

46. A ball has *bounce coefficient* $r < 1$ if, when it is dropped from height h, it bounces back to a height of rh. Suppose that such a ball is dropped from the initial height a and subsequently bounces infinitely many times. Use a geometric series to show that the total up-and-down distance it travels in all its bouncing is

$$D = a\frac{1+r}{1-r}.$$

Note that D is *finite*.

47. A ball with bounce coefficient $r = 0.64$ (see Problem 46) is dropped from an initial height of $a = 4$ ft. Use a geometric series to compute the total time required for it to complete its infinitely many bounces. The time required for a ball to drop h feet (from rest) is $\sqrt{2h/g}$ seconds where $g = 32$ ft/s^2.

48. Suppose that the government spends \$1 billion and that each recipient spends 90% of the dollars that he or she receives. In turn, the secondary recipients spend 90% of the dollars they receive, and so on. How much total spending results from the original injection of \$1 billion into the economy?

49. A tank initially contains a mass M_0 of air. Each stroke of a vacuum pump removes 5% of the air in the container. Compute

(a) the mass M_n of air remaining in the tank after n strokes of the pump;

(b) $\lim_{n \to \infty} M_n$.

50. If Paul and Mary toss a fair coin in turn until one of them wins by getting the first head, calculate for each the probability that he or she wins the game.

51. If Peter, Paul, and Mary toss a fair coin in turn until one of them wins by getting the first head, calculate for each the probability that he or she wins the game. Check your answer by verifying that the sum of the three probabilities is 1.

52. If Peter, Paul, and Mary roll a fair die in turn until one of them wins by getting the first six, calculate for each the probability that he or she wins the game. Check your answer by verifying that the sum of the three probabilities is 1.

53. A pane of glass of a certain material reflects half the incident light, absorbs one-fourth, and transmits one-fourth. A window is made of two panes of this glass separated by a small space, as shown in Fig. 12.4. What fraction of the incident light I is transmitted by the double window?

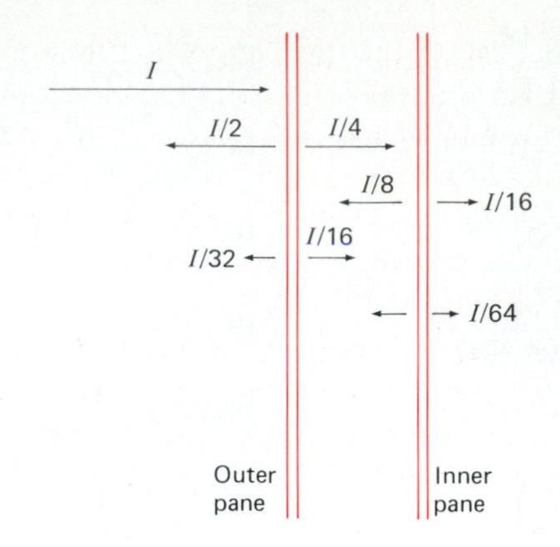

12.4 Diagram of the double-pane window of Problem 53

54. Criticize this evaluation of the sum of an infinite series:

Let $x = 1 - 2 + 4 - 8 + 16 - 32 + 64 - \cdots$.
Then $2x = 2 - 4 + 8 - 16 + 32 - 64 + \cdots$.
Add the equations to obtain $3x = 1$. Thus $x = \frac{1}{3}$, and "therefore"

$$1 - 2 + 4 - 8 + 16 - 32 + 64 - \cdots = \tfrac{1}{3}.$$

*12.3 Optional Computer Application

Figure 12.5 is a listing of Program GEOMETRC. This program computes the successive partial sums of the geometric series

$$a + ar + ar^2 + ar^3 + \cdots + ar^n + ar^{n+1} + \cdots$$

after the initial term a and the ratio r are entered at lines 140 and 150. In line 220 the old term is multiplied by r to form the new term, then in line 230 the new term is added to the old sum to form the new sum.

```
100 REM--Program GEOMETRC
110 REM--Computes the sum of a geometric series
120 REM--with input initial term A and ratio R.
130 REM
140      INPUT "Initial term"; A
150      INPUT "Common ratio"; R
160      T = A                      'Initial term
170      S = T                      'Initial sum
180      EPS  = .000001             'Tolerance
190 REM
200 REM--Loop:       'Until terms suff. small
210      PRINT S
220      T = R*T                    'New term
230      S = S + T                  'New  sum
240      IF ABS(T) > EPS THEN GOTO 210
250 REM
260      PRINT USING "Sum = ##.####"; S
270      END
```

12.5 Listing of Program GEOMETRC

Line 240 directs that terms continue to be added until they no longer exceed EPS = 0.000 001. This type of "stopping rule" is common in series summation programs, but it can well be satisfied without the limit (if any) actually having been reached. Therefore, when such a program displays the result "SUM = . . . ," we should be cautious in believing the computer merely because it prints what *we* told it to print!

Exercise: Run Program GEOMETRC with initial term $a = 1$ and with ratios $r = 0.2$, 0.5, 0.75, 0.9, and 0.095. How does the apparent rate of convergence—as measured by the number of partial sums displayed before stopping—depend upon the value of r? What happens if $r > 1$?

12.4 The Integral Test

Given an infinite series $\sum a_n$, it is the exception rather than the rule when a simple formula for its nth partial sum S_n can be found and used directly to determine whether the series converges or diverges. There are, however, several *convergence tests* that involve the *terms* of an infinite series rather than its partial sums. Such a test, when successful, will tell us whether or not the series converges. Once we know that a series $\sum a_n$ does converge, it is then a separate matter actually to find its sum S. It may be necessary to approximate S by adding sufficiently many terms; in this case we shall need to know how many terms are required for the desired accuracy.

In this section and the one that follows it, we concentrate our attention on **positive term series**—that is, those with terms that are all positive. If $a_n > 0$ for all n, then

$$S_1 < S_2 < S_3 < \cdots < S_n < \cdots,$$

so the sequence $\{S_n\}$ of partial sums of the series is monotone increasing. Hence there are just two possibilities. If the sequence $\{S_n\}$ is *bounded*—there exists a number M such that $S_n \leqq M$ for all n—then the bounded monotone sequence property (Section 12.2) implies that $S = \lim_{n\to\infty} S_n$ exists, so the series $\sum a_n$ *converges.* Otherwise, it diverges to infinity (by Problem 38 in Section 12.2).

A similar alternative holds for improper integrals. Suppose that the function f is continuous and positive-valued for $x \geqq 1$. Then it follows (from Problem 35) that the improper integral

$$\int_1^\infty f(x)\,dx = \lim_{b\to\infty} \int_1^b f(x)\,dx \tag{1}$$

either converges (the limit is a real number) or diverges to infinity (the limit is $+\infty$). This analogy between positive-term series and improper integrals of positive functions is the key to the **integral test.** We compare the behavior of the series $\sum a_n$ with that of the improper integral in (1), where f is an appropriately selected function. (Among other things, we require that $f(n) = a_n$ for all n.)

Theorem 1 *The Integral Test*
Suppose that $\sum a_n$ is a positive-term series and that f is a positive-valued, decreasing, continuous function for $x \geqq 1$. If $f(n) = a_n$ for all integers $n \geqq 1$, then the series and the improper integral

$$\sum_{n=1}^{\infty} a_n \quad \text{and} \quad \int_1^{\infty} f(x)\,dx$$

either both converge or both diverge.

Proof Because f is a decreasing function, the rectangular polygon with area

$$S_n = a_1 + a_2 + \cdots + a_n$$

shown in Fig. 12.6 contains the region under $y = f(x)$ from $x = 1$ to $x = n + 1$. Hence

$$\int_1^{n+1} f(x)\,dx \leqq S_n. \tag{2}$$

Similarly, the rectangular polygon with area

$$S_n - a_1 = a_2 + a_3 + \cdots + a_n$$

shown in Fig. 12.7 is contained in the region under $y = f(x)$ from $x = 1$ to

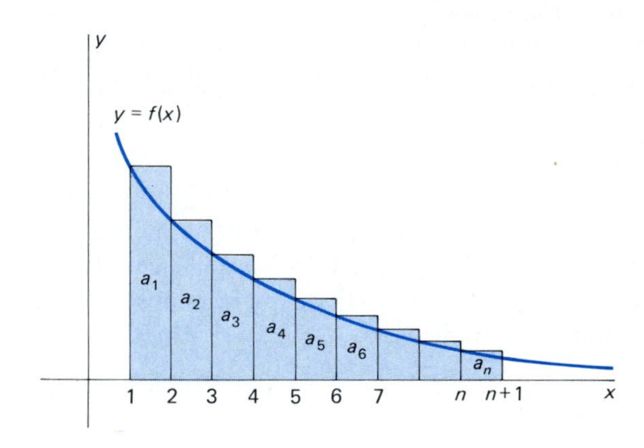

12.6 Underestimating the partial sums with an integral

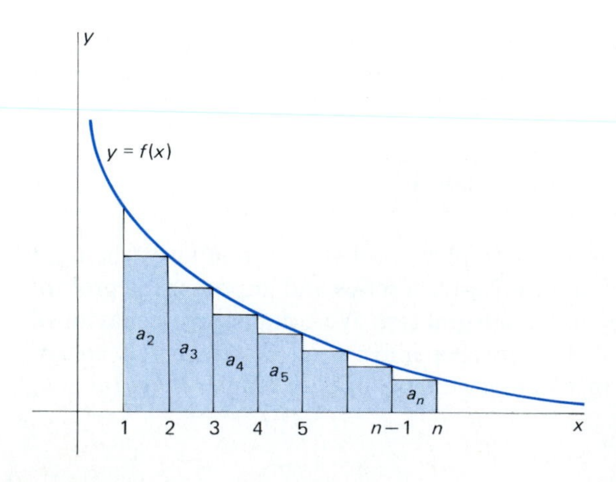

12.7 Overestimating the partial sums with an integral

$x = n$. Hence

$$S_n - a_1 \leqq \int_1^n f(x)\,dx. \tag{3}$$

Suppose first that the improper integral $\int_1^\infty f(x)\,dx$ diverges (necessarily to $+\infty$). Then

$$\lim_{n\to\infty} \int_1^{n+1} f(x)\,dx = +\infty,$$

so it follows from (2) that $\lim_{n\to\infty} S_n = +\infty$ as well, and hence the infinite series $\sum a_n$ likewise diverges.

On the other hand, suppose that the improper integral $\int_1^\infty f(x)\,dx$ converges, with (finite) value I. Then (3) implies that

$$S_n \leqq a_1 + \int_1^n f(x)\,dx \leqq a_1 + I,$$

so the monotone increasing sequence $\{S_n\}_1^\infty$ is bounded. Thus the infinite series

$$\sum a_n = \lim_{n\to\infty} S_n$$

converges also. Hence we have shown that the infinite series and the improper integral either both converge or both diverge. ■

EXAMPLE 1 We used a version of the integral test to prove in Section 12.3 that the harmonic series

$$\sum_{n=1}^{\infty} \frac{1}{n} = 1 + \tfrac{1}{2} + \tfrac{1}{3} + \tfrac{1}{4} + \cdots$$

diverges. Note that using the test as stated in Theorem 1 is a little simpler: We note that $f(x) = 1/x$ is positive, continuous, decreasing for $x \geqq 1$, and that $f(n) = 1/n$ for n a positive integer. Now

$$\int_1^\infty \frac{1}{x}\,dx = \lim_{b\to\infty} \int_1^b \frac{1}{x}\,dx = \lim_{b\to\infty} \Big[\ln x\Big]_1^b$$

$$= \lim_{b\to\infty} (\ln b - \ln 1) = +\infty.$$

Thus the improper integral diverges and, therefore, so does the harmonic series.

The harmonic series is the case $p = 1$ of the ***p*-series**

$$\sum_{n=1}^{\infty} \frac{1}{n^p} = 1 + \frac{1}{2^p} + \frac{1}{3^p} + \cdots + \frac{1}{n^p} + \cdots. \tag{4}$$

Whether the p-series converges or diverges depends upon the value of p.

EXAMPLE 2 Show that the p-series converges if $p > 1$ but diverges if $0 < p \leqq 1$.

Solution The case $p = 1$ has already been settled in Example 1. If $p > 0$ but $p \neq 1$, then the function $f(x) = 1/x^p$ satisfies the conditions of the integral test, and

$$\int_1^\infty \frac{1}{x^p}\,dx = \lim_{b\to\infty}\int_1^b \frac{1}{x^p}\,dx = \lim_{b\to\infty}\left[-\frac{1}{(p-1)x^{p-1}}\right]_1^b$$

$$= \lim_{b\to\infty}\frac{1}{p-1}\left(1 - \frac{1}{b^{p-1}}\right).$$

If $p > 1$, then

$$\int_1^\infty \frac{1}{x^p}\,dx = \frac{1}{p-1} < \infty,$$

so the integral and the series both converage. But if $0 < p < 1$, then

$$\int_1^\infty \frac{1}{x^p}\,dx = \lim_{b\to\infty}\frac{1}{1-p}(b^{1-p} - 1) = \infty,$$

and in this case the integral and the series both diverge.

As specific examples, the series

$$\sum_{n=1}^\infty \frac{1}{n^2} = 1 + \frac{1}{2^2} + \frac{1}{3^2} + \cdots + \frac{1}{n^2} + \cdots$$

converges ($p = 2$), while the series

$$\sum_{n=1}^\infty \frac{1}{\sqrt{n}} = 1 + \frac{1}{\sqrt{2}} + \frac{1}{\sqrt{3}} + \cdots + \frac{1}{\sqrt{n}} + \cdots$$

diverges ($p = \frac{1}{2}$).

Now suppose that the positive-term series $\sum a_n$ converges by the integral test, and we wish to approximate its sum by adding sufficiently many of its initial terms. The difference between the sum S and the nth partial sum S_n is the **remainder**

$$R_n = S - S_n = a_{n+1} + a_{n+2} + a_{n+3} + \cdots. \tag{5}$$

This remainder is the error made when the actual sum S is estimated by using in its place the partial sum S_n.

Theorem 2 *Integral Test Remainder Estimate*

Suppose that the infinite series and improper integral

$$\sum_{n=1}^\infty a_n \quad \text{and} \quad \int_1^\infty f(x)\,dx$$

satisfy the hypotheses of the integral test, and suppose in addition that both converge. Then

$$\int_{n+1}^\infty f(x)\,dx \leqq R_n \leqq \int_n^\infty f(x)\,dx \tag{6}$$

where R_n is the remainder given in (5).

Proof From Fig. 12.8 we see that

$$\int_k^{k+1} f(x)\,dx \leqq a_k \leqq \int_{k-1}^{k} f(x)\,dx$$

for $k = n + 1, n + 2, \ldots$. We add these inequalities for all such values of k, and the result is the inequality in (6), because

$$\sum_{k=n+1}^{\infty} \int_k^{k+1} f(x)\,dx = \int_{n+1}^{\infty} f(x)\,dx$$

and

$$\sum_{k=n+1}^{\infty} \int_{k-1}^{k} f(x)\,dx = \int_n^{\infty} f(x)\,dx. \quad \blacksquare$$

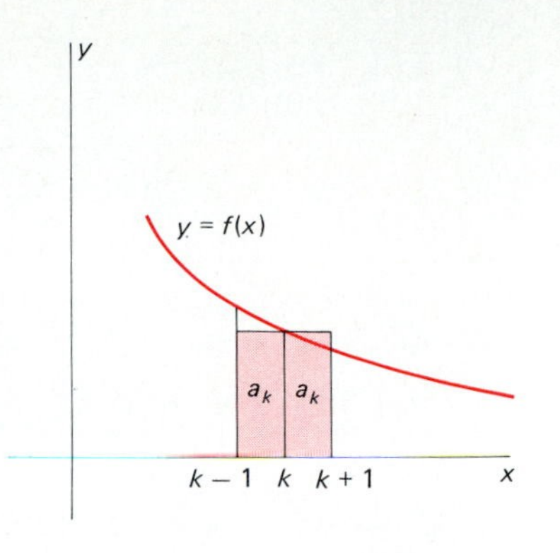

12.8 Establishing the integral test remainder estimate

If we substitute $R_n = S - S_n$, then it follows from (6) that the sum S of the series satisfies the inequality

$$S_n + \int_{n+1}^{\infty} f(x)\,dx \leqq S \leqq S_n + \int_n^{\infty} f(x)\,dx. \tag{7}$$

If the nth partial sum S_n is known and the difference

$$\int_n^{n+1} f(x)\,dx$$

between the two integrals is small, then (7) provides an accurate estimate of the actual sum S.

EXAMPLE 3 In Section 12.8 we will see that the exact sum of the p-series with $p = 2$ is $\pi^2/6$, thus giving the beautiful formula

$$\frac{\pi^2}{6} = 1 + \frac{1}{2^2} + \frac{1}{3^2} + \frac{1}{4^2} + \cdots. \tag{8}$$

Approximate the number π by applying the integral test remainder estimate with $n = 50$ to the series in Equation (8).

Solution The sum of the first 50 terms in (8) is, to the accuracy shown,

$$\sum_{n=1}^{50} \frac{1}{n^2} = 1.6251327.$$

You can add these 50 terms one by one on a pocket calculator in a few minutes, but this is precisely the sort of thing a computer or programmable calculator is really good for.

Because

$$\int_a^{\infty} \frac{1}{x^2}\,dx = \lim_{b\to\infty} \left[-\frac{1}{x}\right]_a^b = \frac{1}{a},$$

the integral test remainder estimate with $f(x) = 1/x^2$ gives $\frac{1}{51} \leqq R_{50} \leqq \frac{1}{50}$, so (7) yields

$$\frac{1}{51} + 1.6251327 \leqq \frac{\pi^2}{6} \leqq \frac{1}{50} + 1.6251327.$$

We multiply by 6, extract the square root, and round to four-place accuracy. The result is that

$$3.1414 < \pi < 3.1418.$$

With $n = 200$, a programmable calculator gives

$$\sum_{n=1}^{200} \frac{1}{n^2} = 1.6399465,$$

so

$$\frac{1}{201} + 1.6399465 \leqq \frac{\pi^2}{6} \leqq \frac{1}{200} + 1.6399465.$$

This leads to the inequality $3.14158 < \pi < 3.14161$, and it follows that $\pi = 3.1416$ rounded to four decimal places.

EXAMPLE 4 Show that the series

$$\sum_{n=2}^{\infty} \frac{1}{n(\ln n)^2}$$

converges, and find how many terms you would need to add to find its sum accurate to within 0.01.

Solution We begin the sum at $n = 2$ because $\ln 1 = 0$. Let $f(x) = x(\ln x)^2$. Then

$$\int_2^{\infty} \frac{1}{x(\ln x)^2}\,dx = \lim_{b\to\infty} \left[-\frac{1}{\ln x}\right]_2^b = \frac{1}{\ln 2} < \infty,$$

so the series does converge by the integral test. We want to choose n sufficiently large that $R_n < 0.01$. The right-hand inequality in (6) gives

$$R_n \leqq \int_n^{\infty} \frac{1}{x(\ln x)^2}\,dx = \frac{1}{\ln n}.$$

Hence we need

$$\frac{1}{\ln n} \leqq 0.01, \qquad \ln n \geqq 100, \qquad n \geqq e^{100} \approx 2.7 \times 10^{43}.$$

This is a far larger number of terms than any conceivable computer could add within the expected lifetime of the universe. But accuracy to within 0.05 would require only that $n \geqq e^{20} \approx 4.85 \times 10^8$, fewer than half a billion terms—well within the range of a modern computer.

12.4 PROBLEMS

In Problems 1–28, use the integral test to test the given series for convergence.

1. $\sum_{n=1}^{\infty} \frac{n}{n^2+1}$

2. $\sum_{n=1}^{\infty} \frac{n}{e^{n^2}}$

3. $\sum_{n=1}^{\infty} \frac{1}{\sqrt{n+1}}$

4. $\sum_{n=1}^{\infty} \frac{1}{(n+1)^{4/3}}$

5. $\sum_{n=1}^{\infty} \frac{1}{n^2+1}$

6. $\sum_{n=1}^{\infty} \frac{1}{n(n+1)}$

7. $\sum_{n=2}^{\infty} \frac{1}{n \ln n}$

8. $\sum_{n=1}^{\infty} \frac{\ln n}{n}$

9. $\sum_{n=1}^{\infty} \frac{1}{2^n}$

10. $\sum_{n=1}^{\infty} \frac{n}{e^n}$

11. $\sum_{n=1}^{\infty} \frac{n^2}{e^n}$

12. $\sum_{n=1}^{\infty} \frac{1}{17n - 13}$

13. $\sum_{n=1}^{\infty} \frac{\ln n}{n^2}$

14. $\sum_{n=1}^{\infty} \frac{n+1}{n^2}$

15. $\sum_{n=1}^{\infty} \frac{n}{n^4 + 1}$

16. $\sum_{n=1}^{\infty} \frac{1}{n^3 + n}$

17. $\sum_{n=1}^{\infty} \frac{2n + 5}{n^2 + 5n + 7}$

18. $\sum_{n=1}^{\infty} \ln\left(\frac{n+1}{n}\right)$

19. $\sum_{n=1}^{\infty} \ln\left(1 + \frac{1}{n^2}\right)$

20. $\sum_{n=1}^{\infty} \frac{2^{1/n}}{n^2}$

21. $\sum_{n=1}^{\infty} \frac{n}{4n^2 + 5}$

22. $\sum_{n=1}^{\infty} \frac{n}{(4n^2 + 5)^{3/2}}$

23. $\sum_{n=2}^{\infty} \frac{1}{n\sqrt{\ln n}}$

24. $\sum_{n=2}^{\infty} \frac{1}{n(\ln n)^2}$

25. $\sum_{n=1}^{\infty} \frac{1}{4n^2 + 9}$

26. $\sum_{n=1}^{\infty} \frac{n+1}{n + 100}$

27. $\sum_{n=1}^{\infty} \frac{n}{n^4 + 2n^2 + 1}$

28. $\sum_{n=1}^{\infty} \frac{1}{(n+1)^3}$

In each of Problems 29–31, tell why the integral test does *not* apply to the given infinite series.

29. $\sum_{n=1}^{\infty} \frac{(-1)^n}{n}$

30. $\sum_{n=1}^{\infty} e^{-n} \sin n$

31. $\sum_{n=1}^{\infty} \frac{2 + \sin n}{n^2}$

In each of Problems 32–34, find the least positive integer n such that the remainder R_n in Theorem 2 is less than E.

32. $\sum_{n=1}^{\infty} \frac{1}{n^2}$; $E = 0.0001$

33. $\sum_{n=1}^{\infty} \frac{1}{n^2}$; $E = 0.00005$

34. $\sum_{n=1}^{\infty} \frac{1}{n^6}$; $E = 2 \times 10^{-11}$

35. Suppose that the function f is continuous and positive-valued for $x \geqq 1$. Let

$$b_n = \int_1^n f(x)\, dx$$

for $n = 1, 2, 3, \ldots$.

(a) Suppose that the monotone increasing sequence $\{b_n\}$ is bounded, so that $B = \lim_{n \to \infty} b_n$ exists. Prove that

$$\int_1^{\infty} f(x)\, dx = B.$$

(b) Prove that if the sequence $\{b_n\}$ is not bounded, then

$$\int_1^{\infty} f(x)\, dx = +\infty.$$

36. Show that the series

$$\sum_{n=2}^{\infty} \frac{1}{n(\ln n)^p}$$

converges if $p > 1$ and diverges if $p \leqq 1$.

37. Use the integral test remainder estimate to find

$$\sum_{n=1}^{\infty} \frac{1}{n^5}$$

accurate to three decimal places.

38. Using the integral test remainder estimate, how many terms are needed to approximate

$$\sum_{n=1}^{\infty} \frac{1}{n^{3/2}}$$

with two-place accuracy?

39. Deduce from the inequalities in (2) and (3) with the function $f(x) = 1/x$ that

$$\ln n \leqq 1 + \frac{1}{2} + \frac{1}{3} + \cdots + \frac{1}{n} \leqq 1 + \ln n$$

for $n = 1, 2, 3, \ldots$. If a computer adds one million terms of the harmonic series per second, how long will it take for the partial sum to reach 50?

40. (a) Let

$$c_n = 1 + \frac{1}{2} + \frac{1}{3} + \cdots + \frac{1}{n} - \ln n$$

for $n = 1, 2, 3, \ldots$. Deduce from Problem 39 that $0 \leqq c_n \leqq 1$ for all n.

(b) Note that

$$\int_n^{n+1} \frac{1}{x}\, dx \geqq \frac{1}{n+1}.$$

Conclude that the sequence $\{c_n\}$ is monotone decreasing. Therefore the sequence $\{c_n\}$ converges. The number

$$\gamma = \lim_{n \to \infty} \left(1 + \frac{1}{2} + \frac{1}{3} + \cdots + \frac{1}{n} - \ln n\right) \approx 0.57722$$

is known as **Euler's constant.**

41. It is known that

$$\sum_{n=1}^{\infty} \frac{1}{n^4} = \frac{\pi^4}{90}.$$

Use the integral test remainder estimate and the first ten terms of this series to show that $\pi = 3.1416$ rounded to four decimal places.

The program listed in Fig. 12.9 computes the first 100 partial sums of a p-series $\sum 1/n^p$ after the value of p is entered at line 130. Line 190 calculates the new term T, which then is added to the old sum to form the new sum S (line 200). The effect of line 210 is to direct that every 10th partial sum be printed.

```
100 REM--Program P-SERIES
110 REM--Sums a p-series when p is input.
120 REM
130      INPUT "Value of p"; P
140      N = 1                      'Initial N
150      S = 0                      'Initial sum
160 REM
170 REM--Loop:
180      FOR I = 1 TO 100
190          T = 1/(N^P)            'New term
200          S = S + T              'New sum
210          IF INT(I/10) = I/10
                THEN PRINT N, S
220          N = N + 1              'New N
230      NEXT I
240 REM
250      END
```

12.9 Listing of Program P-SERIES

Exercise 1: Use Program P-SERIES to verify the computations in Example 3.

Exercise 2: Apply this program to the series in Problem 41 to calculate π accurate to six decimal places.

12.5 Comparison Tests for Positive-Term Series

With the integral test we attempt to determine whether or not an infinite series converges by comparing it with an improper integral. The methods of this section involve comparing the terms of the *positive-term* series $\sum a_n$ with those of another positive-term series $\sum b_n$ whose convergence or divergence is already known. We have already developed two families of *reference series* for the role of the known series $\sum b_n$; these are the geometric series of Section 12.3 and the p-series of Section 12.4. They are well adapted for our new purposes because their convergence or divergence is quite easy to determine: The geometric series $\sum r^n$ converges if $|r| < 1$ and diverges if $|r| \geqq 1$, and the p-series $\sum 1/n^p$ converges for $p > 1$ and diverges if $0 < p \leqq 1$.

Let $\sum a_n$ and $\sum b_n$ be *positive-term* series. Then we say that the series $\sum b_n$ **dominates** the series $\sum a_n$ provided that $a_n \leqq b_n$ for all n. The following theorem says that the positive-term series $\sum a_n$ converges if it is dominated by a convergent series and diverges if it dominates a divergent series.

Theorem 1 *Comparison Test*

Suppose that $\sum a_n$ and $\sum b_n$ are positive-term series. Then

1. $\sum a_n$ converges if $\sum b_n$ converges and $a_n \leqq b_n$ for all n;

2. $\sum a_n$ diverges if $\sum b_n$ diverges and $a_n \geqq b_n$ for all n.

Proof Denote the nth partial sums of the series $\sum a_n$ and $\sum b_n$ by S_n and T_n, respectively. Then $\{S_n\}$ and $\{T_n\}$ are monotone increasing sequences. To prove part (1), suppose that $\sum b_n$ converges, so that $T = \lim_{n\to\infty} T_n$ exists (T is a real number). Then the fact that $a_n \leqq b_n$ for all n implies that $S_n \leqq T_n \leqq T$ for all n. Thus the sequence $\{S_n\}$ of partial sums of $\sum a_n$ is bounded and monotone increasing and therefore converges. Thus $\sum a_n$ converges.

Part (2) is merely a restatement of part (1). If the series $\sum a_n$ converged, then the fact that $\sum a_n$ dominates $\sum b_n$ would imply—by part (1), with a_n and b_n interchanged—that $\sum b_n$ converged. But $\sum b_n$ diverges, so it follows that $\sum a_n$ must also diverge. ■

We know by Theorem 5 in Section 12.3 that the convergence or divergence of an infinite series is not affected by the addition or deletion of finitely many terms. Consequently, the conditions $a_n \leqq b_n$ and $a_n \geqq b_n$ in the two parts of the comparison test really need hold only for all $n \geqq k$ where k is some fixed positive integer. Thus we can say that the positive-term series $\sum a_n$ converges if it is "eventually dominated" by the convergent positive-term series $\sum b_n$.

EXAMPLE 1 Because

$$\frac{1}{n(n+1)(n+2)} < \frac{1}{n^3}$$

for all $n \geqq 1$, the series

$$\sum_{n=1}^{\infty} \frac{1}{n(n+1)(n+2)} = \frac{1}{1\cdot 2\cdot 3} + \frac{1}{2\cdot 3\cdot 4} + \frac{1}{3\cdot 4\cdot 5} + \cdots$$

is dominated by the series $\sum 1/n^3$, which is a convergent p-series with $p = 3$. Both are positive-term series, and hence the series $\sum 1/n(n+1)(n+2)$ converges by part (1) of the comparison test.

EXAMPLE 2 Because

$$\frac{1}{\sqrt{2n-1}} > \frac{1}{\sqrt{2n}}$$

for all $n \geqq 1$, the positive-term series

$$\sum_{n=1}^{\infty} \frac{1}{\sqrt{2n-1}} = 1 + \frac{1}{\sqrt{3}} + \frac{1}{\sqrt{5}} + \frac{1}{\sqrt{7}} + \cdots$$

dominates the series

$$\sum_{n=1}^{\infty} \frac{1}{\sqrt{2n}} = \frac{1}{\sqrt{2}} \sum_{n=1}^{\infty} \frac{1}{n^{1/2}}.$$

But $\sum 1/n^{1/2}$ is a divergent p-series with $p = \frac{1}{2}$, and a constant nonzero multiple of a divergent series diverges, so part (2) of the comparison test shows that the series $\sum 1/\sqrt{2n-1}$ also diverges.

EXAMPLE 3 Test the series

$$\sum_{n=0}^{\infty} \frac{1}{n!} = 1 + \frac{1}{1!} + \frac{1}{2!} + \frac{1}{3!} + \cdots$$

for convergence.

Solution We note first that if $n \geqq 1$, then

$$\begin{aligned} n! &= n(n-1)(n-2)\cdots 3 \cdot 2 \cdot 1 \\ &\geqq 2 \cdot 2 \cdot 2 \cdots 2 \cdot 2 \cdot 1; \quad \text{(same number of factors)} \end{aligned}$$

that is, $n! \geqq 2^{n-1}$ for $n \geqq 1$. Thus

$$\frac{1}{n!} \leqq \frac{1}{2^{n-1}} \qquad \text{for} \quad n \geqq 1,$$

so the series

$$\sum_{n=0}^{\infty} \frac{1}{n!} \quad \text{is dominated by the series} \quad 1 + \sum_{n=1}^{\infty} \frac{1}{2^{n-1}} = 1 + \sum_{n=0}^{\infty} \frac{1}{2^n},$$

which is a convergent geometric series (after the first term). Both are positive-term series, and therefore by the comparison test the series

$$\sum_{n=0}^{\infty} \frac{1}{n!}$$

converges. In Section 12.7 we will see that the sum of this series is the number e, so that

$$e = 1 + \frac{1}{1!} + \frac{1}{2!} + \frac{1}{3!} + \cdots + \frac{1}{n!} + \cdots.$$

Indeed, this series provides perhaps the simplest way of showing that

$$e \approx 2.71828\ 18284\ 59045\ 23536.$$

Suppose that $\sum a_n$ is a positive-term series such that $a_n \to 0$ as $n \to \infty$. Then, in connection with the nth term divergence test of Section 12.3, the series $\sum a_n$ has at least a *chance* of converging. How do we choose an appropriate positive-term series $\sum b_n$ to compare it with? A good idea is to pick b_n as a *simple* function of n, simpler than a_n, but such that a_n and b_n approach zero at the same rate as $n \to \infty$. If the formula for a_n is a fraction, we can try discarding all but the terms of largest magnitude in its numerator and denominator. For example, if

$$a_n = \frac{3n^2 + n}{n^4 + \sqrt{n}},$$

then we reason that n is small in comparison with $3n^2$ and $\sqrt{n}$ is small in comparison with n^4 when n is quite large. This suggests that we choose $b_n = 3n^2/n^4 = 3/n^2$. The series $\sum 3/n^2$ converges ($p = 2$), but when we attempt to compare $\sum a_n$ and $\sum b_n$, we find that $a_n \geqq b_n$ (rather than $a_n \leqq b_n$). Consequently, the comparison test does not apply immediately—the fact that $\sum a_n$ dominates a convergent series does *not* imply that $\sum a_n$ itself converges. The following theorem provides a convenient way of handling such a situation.

Theorem 2 *Limit Comparison Test*
Suppose that $\sum a_n$ and $\sum b_n$ are positive-term series. If the limit

$$L = \lim_{n \to \infty} \frac{a_n}{b_n}$$

exists and $0 < L < \infty$, then either both series converge or both series diverge.

Proof Choose two fixed positive numbers P and Q such that $P < L < Q$. Then $P < a_n/b_n < Q$ for n sufficiently large, and so

$$Pb_n < a_n < Qb_n$$

for all sufficiently large values of n. If $\sum b_n$ converges, then $\sum a_n$ is eventually dominated by the convergent series $Q \sum b_n$, so part (1) of the comparison test implies that $\sum a_n$ also converges. If $\sum b_n$ diverges, then $\sum a_n$ eventually dominates the divergent series $P \sum b_n$, so part (2) of the comparison test implies that $\sum a_n$ also diverges. Thus the convergence of either series implies the convergence of the other. ■

EXAMPLE 4 With

$$a_n = \frac{3n^2 + n}{n^4 + \sqrt{n}} \quad \text{and} \quad b_n = \frac{1}{n^2},$$

(motivated by the discussion preceding Theorem 2), we find that

$$\lim_{n \to \infty} \frac{a_n}{b_n} = \lim_{n \to \infty} \frac{3n^4 + n^3}{n^4 + \sqrt{n}} = \lim_{n \to \infty} \frac{3 + 1/n}{1 + n^{-7/2}} = 3.$$

Because $\sum 1/n^2$ is a convergent p-series ($p = 2$), the limit comparison test tells us that the series

$$\sum_{n=1}^{\infty} \frac{3n^2 + n}{n^4 + \sqrt{n}}$$

also converges.

EXAMPLE 5 Test for convergence: $\displaystyle\sum_{n=1}^{\infty} \frac{1}{2n + \ln n}$.

Solution Because $\lim_{n \to \infty} (\ln n)/n = 0$ by l'Hôpital's rule, we note that $\ln n$ is very small in comparison with $2n$ when n is large. We therefore take $a_n = 1/(2n + \ln n)$ and—ignoring the constant coefficient 2—we take $b_n = 1/n$. Then we find that

$$\lim_{n \to \infty} \frac{a_n}{b_n} = \lim_{n \to \infty} \frac{n}{2n + \ln n} = \lim_{n \to \infty} \frac{1}{2 + \dfrac{\ln n}{n}} = \frac{1}{2}.$$

Because the harmonic series $\sum 1/n$ diverges, it follows that the given series $\sum a_n$ also diverges.

It is important to realize that if $L = \lim(a_n/b_n)$ is either 0 or ∞, then the limit comparison test does not apply. (See Problem 42 for a discussion of what conclusions may sometimes be drawn in these cases.) Note, for example, that if $a_n = 1/n^2$ and $b_n = 1/n$, then $\lim(a_n/b_n) = 0$. But in this case $\sum a_n$ converges while $\sum b_n$ diverges.

We close our discussion of positive-term series with the observation that the sum of a convergent *positive*-term series is not altered by grouping or rearranging its terms. For example, let $\sum a_n$ be a convergent positive-term series and consider

$$\sum_{n=1}^{\infty} b_n = (a_1 + a_2 + a_3) + (a_4) + (a_5 + a_6) + \cdots.$$

That is, the new series has $b_1 = a_1 + a_2 + a_3$, $b_2 = a_4$, $b_3 = a_5 + a_6$, and so on. Then every partial sum T_n of $\sum b_n$ is equal to some partial sum $S_{n'}$ of $\sum a_n$. Because $\{S_n\}$ is a monotone increasing sequence with limit $S = \sum a_n$, it follows easily that $\{T_n\}$ is a monotone increasing sequence with the same limit. Thus $\sum b_n = S$ as well. The argument is more subtle if terms of $\sum a_n$ are "moved out of place," as in

$$\sum_{n=1}^{\infty} b_n = a_2 + a_1 + a_4 + a_3 + a_6 + a_5 + \cdots,$$

but the same conclusion holds: Any rearrangement of a convergent *positive*-term series also converges, and to the same sum.

Similarly, it is easy to prove that any grouping or rearrangement of a divergent *positive*-term series also diverges. But these observations all fail in the case of an infinite series having both positive and negative terms. For example, the series $\sum (-1)^n$ diverges, but it has the convergent grouping

$$(-1 + 1) + (-1 + 1) + (-1 + 1) + \cdots = 0 + 0 + 0 + \cdots = 0.$$

It follows from Problem 44 in Section 12.3 that

$$\ln 2 = 1 - \tfrac{1}{2} + \tfrac{1}{3} - \tfrac{1}{4} + \tfrac{1}{5} - \cdots,$$

but the rearrangement

$$1 + \tfrac{1}{3} - \tfrac{1}{2} + \tfrac{1}{5} + \tfrac{1}{7} - \tfrac{1}{4} + \tfrac{1}{9} + \tfrac{1}{11} - \tfrac{1}{6} + \cdots$$

converges to $\frac{3}{2}\ln 2$. The series for ln 2 above even has rearrangements that converge to zero, and others that diverge to $+\infty$.

12.5 PROBLEMS

Use comparison tests to determine whether the infinite series in Problems 1–35 converge or diverge.

1. $\displaystyle\sum_{n=1}^{\infty} \frac{1}{n^2 + n + 1}$

2. $\displaystyle\sum_{n=1}^{\infty} \frac{n^3 + 1}{n^4 + 2}$

3. $\displaystyle\sum_{n=1}^{\infty} \frac{1}{n + \sqrt{n}}$

4. $\displaystyle\sum_{n=1}^{\infty} \frac{1}{n + n^{3/2}}$

5. $\displaystyle\sum_{n=1}^{\infty} \frac{1}{1 + 3^n}$

6. $\displaystyle\sum_{n=1}^{\infty} \frac{10n^2}{n^4 + 1}$

7. $\displaystyle\sum_{n=1}^{\infty} \frac{10n^2}{n^3 - 1}$

8. $\displaystyle\sum_{n=1}^{\infty} \frac{n^2 - n}{n^4 + 2}$

9. $\displaystyle\sum_{n=1}^{\infty} \frac{1}{\sqrt{37n^3 + 3}}$

10. $\displaystyle\sum_{n=1}^{\infty} \frac{1}{\sqrt{n^2 + 1}}$

11. $\displaystyle\sum_{n=1}^{\infty} \frac{\sqrt{n}}{n^2 + n}$

12. $\displaystyle\sum_{n=1}^{\infty} \frac{1}{3 + 5^n}$

13. $\displaystyle\sum_{n=2}^{\infty} \frac{1}{\ln n}$

14. $\displaystyle\sum_{n=2}^{\infty} \frac{1}{n - \ln n}$

15. $\displaystyle\sum_{n=1}^{\infty} \frac{\sin^2 n}{n^2 + 1}$

16. $\displaystyle\sum_{n=1}^{\infty} \frac{\cos^2 n}{3^n}$

17. $\displaystyle\sum_{n=1}^{\infty} \frac{n + 2^n}{n + 3^n}$

18. $\displaystyle\sum_{n=0}^{\infty} \frac{1}{2^n + 3^n}$

19. $\displaystyle\sum_{n=2}^{\infty} \frac{1}{n^2 \ln n}$

20. $\displaystyle\sum_{n=1}^{\infty} \frac{1}{n^{1+\sqrt{n}}}$

21. $\displaystyle\sum_{n=1}^{\infty} \frac{\ln n}{n^2}$

22. $\displaystyle\sum_{n=1}^{\infty} \frac{\arctan n}{n}$

23. $\displaystyle\sum_{n=1}^{\infty} \frac{\sin^2(1/n)}{n^2}$

24. $\displaystyle\sum_{n=1}^{\infty} \frac{e^{1/n}}{n}$

25. $\displaystyle\sum_{n=2}^{\infty} \frac{\ln n}{e^n}$

26. $\displaystyle\sum_{n=1}^{\infty} \frac{n^2 + 2}{n^3 + 3n}$

27. $\displaystyle\sum_{n=1}^{\infty} \frac{n^{3/2}}{n^2 + 4}$

28. $\displaystyle\sum_{n=1}^{\infty} \frac{1}{n \cdot 2^n}$

29. $\displaystyle\sum_{n=1}^{\infty} \frac{3}{4 + \sqrt{n}}$

30. $\displaystyle\sum_{n=1}^{\infty} \frac{n^2 + 1}{e^n(n + 1)^2}$

31. $\displaystyle\sum_{n=1}^{\infty} \frac{2n^2 - 1}{n^2 \cdot 3^n}$

32. $\displaystyle\sum_{n=1}^{\infty} \frac{1}{\sqrt[3]{2n^4 + 1}}$

33. $\displaystyle\sum_{n=1}^{\infty} \frac{2 + \sin n}{n^2}$

34. $\displaystyle\sum_{n=1}^{\infty} \frac{\ln n}{n^3}$

35. $\displaystyle\sum_{n=1}^{\infty} \frac{(n + 1)^n}{n^{n+1}}$ (*Suggestion:* $\displaystyle\lim_{n\to\infty}\left(1 + \frac{1}{n}\right)^n = e$.)

36. (a) Prove that $\ln n < n^{1/8}$ for all sufficiently large values of n.
(b) Conclude that $\sum 1/(\ln n)^8$ diverges.

37. Prove that if $\sum a_n$ is a convergent positive-term series, then $\sum (a_n/n)$ converges.

38. Suppose that $\sum a_n$ is a convergent positive-term series and that $\{c_n\}$ is a sequence of positive numbers with limit zero. Prove that $\sum a_n c_n$ converges.

39. Use the result of Problem 38 to prove that if $\sum a_n$ and $\sum b_n$ are convergent positive-term series, then $\sum a_n b_n$ converges.

40. Prove that the series

$$\sum_{n=1}^{\infty} \frac{1}{1 + 2 + 3 + \cdots + n}$$

converges.

41. Use the result of Problem 40 in Section 12.4 to prove that the series

$$\sum_{n=1}^{\infty} \left(1 + \frac{1}{2} + \frac{1}{3} + \cdots + \frac{1}{n}\right)^{-1}$$

diverges.

42. Adapt the proof of the limit-comparison test to prove the following two results.
(a) Suppose that $\sum a_n$ and $\sum b_n$ are positive-term series and that $\sum b_n$ converges. If

$$L = \lim_{n\to\infty} \frac{a_n}{b_n} = 0,$$

then $\sum a_n$ converges.
(b) Suppose that $\sum a_n$ and $\sum b_n$ are positive-term series and that $\sum b_n$ diverges. If

$$L = \lim_{n\to\infty} \frac{a_n}{b_n} = +\infty,$$

then $\sum a_n$ diverges.

12.6 Alternating Series and Absolute Convergence

In Sections 12.4 and 12.5, we concentrated on positive-term series. Now we discuss infinite series that have both positive and negative terms. An important example is a series with terms that are alternately positive and negative. An **alternating series** is an infinite series of the form

$$\sum_{n=1}^{\infty} (-1)^{n+1}a_n = a_1 - a_2 + a_3 - a_4 + \cdots \tag{1}$$

or of the form $\sum_{n=1}^{\infty} (-1)^n a_n$, where $a_n > 0$ for all n. For example, the series

$$\sum_{n=1}^{\infty} \frac{(-1)^{n+1}}{n} = 1 - \tfrac{1}{2} + \tfrac{1}{3} - \tfrac{1}{4} + \cdots$$

is an alternating series. The following theorem shows that this series converges because the sequence of absolute values of its terms is monotone decreasing with limit zero.

> ***Theorem 1*** *Alternating Series Test*
> If $a_n > a_{n+1} > 0$ for all n and $\lim_{n\to\infty} a_n = 0$, then the alternating series in (1) converges.

Proof We consider first the "even" partial sums $S_2, S_4, S_6, \ldots, S_{2n}$. We may write

$$S_{2n} = (a_1 - a_2) + (a_3 - a_4) + \cdots + (a_{2n-1} - a_{2n}).$$

Because $a_k - a_{k+1} \geqq 0$ for all k, we see that the sequence $\{S_{2n}\}$ is monotone increasing. Also, because

$$S_{2n} = a_1 - (a_2 - a_3) - \cdots - (a_{2n-2} - a_{2n-1}) - a_{2n},$$

we see that $S_{2n} \leqq a_1$ for all n, so the monotone increasing sequence $\{S_{2n}\}$ is bounded above. Hence the limit

$$S = \lim_{n\to\infty} S_{2n}$$

exists by the bounded monotone sequence property of Section 12.2. It remains only to see that the "odd" partial sums $S_1, S_3, S_5, \ldots$ also converge to S. But $S_{2n+1} = S_{2n} + a_{2n+1}$ and $\lim_{n\to\infty} a_{2n+1} = 0$, so

$$\lim_{n\to\infty} S_{2n+1} = \lim_{n\to\infty} S_{2n} + \lim_{n\to\infty} a_{2n+1} = S.$$

Thus $\lim_{n\to\infty} S_n = S$, and therefore the series converges. ■

EXAMPLE 1 The series

$$\sum_{n=1}^{\infty} \frac{(-1)^{n+1}}{2n-1} = 1 - \tfrac{1}{3} + \tfrac{1}{5} - \tfrac{1}{7} + \cdots$$

satisfies the conditions of Theorem 1 and therefore converges. The alternating series test does not tell us its sum, but we will see in Section 12.7 that the sum of this series is $\pi/4$.

EXAMPLE 2 The series

$$\sum_{n=1}^{\infty} (-1)^{n+1} \frac{n}{2n-1} = 1 - \frac{2}{3} + \frac{3}{5} - \frac{4}{7} + \frac{5}{9} - \cdots$$

is an alternating series, and

$$a_n = \frac{n}{2n-1} > \frac{n+1}{2n+1} = a_{n+1}$$

for all n because

$$n(2n+1) > (2n-1)(n+1) = 2n^2 + n - 1 \quad \text{if} \quad n \geqq 1.$$

But

$$\lim_{n \to \infty} a_n = 0.5 \neq 0,$$

so the alternating series test does not apply. Indeed, this series diverges by the nth term divergence test.

If a series converges by the alternating series test, the following theorem shows how to approximate its sum with any desired degree of accuracy—*if* you have some method (a computer?) for adding enough of its terms.

> ***Theorem 2*** *Alternating Series Error Estimate*
> Suppose that the series $\sum (-1)^{n+1} a_n$ satisfies the conditions of the alternating series test and therefore converges. Let S denote its sum. Denote by $R_n = S - S_n$ the error made in replacing S by the nth partial sum S_n. Then R_n has the same sign as the next term $(-1)^{n+2} a_{n+1}$ of the series, and
>
> $$0 < |R_n| < a_{n+1}. \qquad (2)$$

In particular, *the sum S of a convergent alternating series lies between any two consecutive partial sums.* This follows from the proof of Theorem 1, where we saw that $\{S_{2n}\}$ is an increasing sequence while $\{S_{2n+1}\}$ is a decreasing sequence, each converging to S. The resulting inequalities

$$S_{2n-1} > S > S_{2n} = S_{2n-1} - a_{2n}$$

and

$$S_{2n} < S < S_{2n+1} = S_{2n} + a_{2n+1}$$

imply the inequality in (2).

EXAMPLE 3 In Section 12.7 we will see that

$$e^x = \sum_{n=0}^{\infty} \frac{x^n}{n!}$$

for all x, and thus that

$$\frac{1}{e} = e^{-1} = 1 - 1 + \frac{1}{2!} - \frac{1}{3!} + \frac{1}{4!} - \cdots.$$

Use this alternating series to compute e^{-1} accurate to four decimal places.

Solution We want $|R_n| < 1/(n+1)! \geqq 0.00005$. The least value of n for which this inequality holds is $n = 7$. Then

$$e^{-1} = 1 - 1 + \frac{1}{2!} - \frac{1}{3!} + \frac{1}{4!} - \frac{1}{5!} + \frac{1}{6!} - \frac{1}{7!} + R_7$$

$$= 0.367857 + R_7.$$

(We are carrying six places because we want four-place accuracy in the final answer.) Now the inequality in (2) gives

$$0 < R_7 < \frac{1}{8!} < 0.000025.$$

Thus

$$0.367857 < e^{-1} < 0.367882.$$

When we take reciprocals, we find that

$$2.718263 < e < 2.718448.$$

Thus $e^{-1} = 0.3679$ rounded to four places and also $e = 2.718$ rounded to three places.

ABSOLUTE CONVERGENCE

The series

$$\sum_{n=1}^{\infty} \frac{(-1)^{n+1}}{n} = 1 - \tfrac{1}{2} + \tfrac{1}{3} - \tfrac{1}{4} + \cdots$$

converges, but if we simply add the absolute values of its terms, we get the *divergent* series

$$1 + \tfrac{1}{2} + \tfrac{1}{3} + \tfrac{1}{4} + \cdots.$$

On the other hand, the convergent series

$$\sum_{n=1}^{\infty} \frac{(-1)^n}{2^n} = 1 - \tfrac{1}{2} + \tfrac{1}{4} - \tfrac{1}{8} + \cdots$$

has the property that the associated positive-term series

$$1 + \tfrac{1}{2} + \tfrac{1}{4} + \tfrac{1}{8} + \tfrac{1}{16} + \cdots$$

also converges. Our next theorem tells us this: If a series of *positive* terms converges, then we may insert minus signs in front of any of the terms—every other one, for instance—and the resulting series will also converge.

Theorem 3 *Absolute Convergence Implies Convergence*

If the series $\sum |a_n|$ converges, then so does the series $\sum a_n$.

Proof Suppose that the series $\sum |a_n|$ converges. Note that

$$0 \leqq a_n + |a_n| \leqq 2|a_n|$$

for all n. Let $b_n = a_n + |a_n|$. It then follows from the comparison test that the positive-term series $\sum b_n$ converges, for it is dominated by the convergent series $\sum 2|a_n|$. Because it is easy to verify that the termwise difference of two convergent series also converges, we now see that the series

$$\sum a_n = \sum (b_n - |a_n|) = \sum b_n - \sum |a_n|$$

converges ■

Thus we have another convergence test, one not limited to positive-term series: Given the series $\sum a_n$, test the series $\sum |a_n|$ for convergence. If the latter series converges, then so does the former. This phenomenon motivates us to make the following definition.

Definition *Absolute Convergence*
The series $\sum a_n$ is said to **converge absolutely** (and is called **absolutely convergent**) provided that the series

$$\sum |a_n| = |a_1| + |a_2| + |a_3| + \cdots + |a_n| + \cdots$$

converges.

Thus Theorem 3 may be rephrased as follows: *If a series converges absolutely, then it converges.* The two examples preceding Theorem 3 show that a convergent series may either converge absolutely or fail to do so:

$$1 - \tfrac{1}{2} + \tfrac{1}{4} - \tfrac{1}{8} + \tfrac{1}{16} - \cdots$$

is an absolutely convergent series because

$$1 + \tfrac{1}{2} + \tfrac{1}{4} + \tfrac{1}{8} + \tfrac{1}{16} + \cdots$$

converges, while

$$1 - \tfrac{1}{2} + \tfrac{1}{3} - \tfrac{1}{4} + \tfrac{1}{5} - \cdots$$

is an example of a series that, though convergent, is *not* absolutely convergent. A series that converges but does not converge absolutely is said to be **conditionally convergent.**

Consequently, the terms *absolutely convergent*, *conditionally convergent*, and *divergent* are simultaneously all-inclusive and mutually exclusive.

Note that there is some advantage in the application of Theorem 3, because to apply it we test the *positive*-term series $\sum |a_n|$ for convergence—and we have a variety of tests, such as comparison tests or the integral test, designed for use on positive-term series.

Note also that absolute convergence of the series $\sum a_n$ means convergence of *another* series $\sum |a_n|$, and the two sums will generally differ. For example, with $a_n = (-\frac{1}{3})^n$ the formula for the sum of a geometric series gives

$$\sum_{n=0}^{\infty} a_n = \sum_{n=0}^{\infty} \left(-\frac{1}{3}\right)^n = \frac{1}{1 - (-\frac{1}{3})} = \frac{3}{4},$$

while

$$\sum_{n=0}^{\infty} |a_n| = \sum_{n=0}^{\infty} \left(\frac{1}{3}\right)^n = \frac{1}{1 - (\frac{1}{3})} = \frac{3}{2}.$$

EXAMPLE 4 Discuss the convergence of the series

$$\sum_{n=1}^{\infty} \frac{\cos n}{n^2} = \cos 1 + \frac{\cos 2}{4} + \frac{\cos 3}{9} + \cdots.$$

Solution Let $a_n = (\cos n)/n^2$. Then

$$|a_n| = \frac{|\cos n|}{n^2} \leqq \frac{1}{n^2}$$

for all $n \geqq 1$. Hence the positive-term series $\sum |a_n|$ converges by the comparison test, because it is dominated by the convergent p-series $\sum (1/n^2)$. Thus the given series is absolutely convergent, and therefore it converges by Theorem 3.

One reason for the importance of absolute convergence is the fact (proved in advanced calculus) that the terms of an absolutely convergent series may be regrouped or rearranged without changing the sum of the series. As we suggested at the end of Section 12.5, this is *not* true of conditionally convergent series.

THE RATIO AND ROOT TESTS

Each of our next two convergence tests involves a way of measuring the rate of growth or decrease of the sequence $\{a_n\}$ of terms of a series in order to determine whether $\sum a_n$ diverges or converges absolutely.

Theorem 4 *The Ratio Test*
Suppose that the limit

$$\rho = \lim_{n \to \infty} \left| \frac{a_{n+1}}{a_n} \right| \tag{3}$$

either exists or is infinite. Then the infinite series $\sum a_n$ of nonzero terms

1. Converges absolutely if $\rho < 1$;
2. Diverges if $\rho > 1$.

If $\rho = 1$, the ratio test is inconclusive.

Proof If $\rho < 1$, choose a (fixed) number r with $\rho < r < 1$. Then (3) implies that there exists an integer N such that $|a_{n+1}| \leqq r|a_n|$ for all $n \geqq N$. It follows that

$$|a_{N+1}| \leqq r|a_N|,$$
$$|a_{N+2}| \leqq r|a_{N+1}| \leqq r^2|a_N|,$$
$$|a_{N+3}| \leqq r|a_{N+2}| \leqq r^3|a_N|,$$

and—in general—that

$$|a_{N+k}| \leqq r^k|a_N| \qquad \text{for} \quad k \geqq 0.$$

Hence the series

$$|a_N| + |a_{N+1}| + |a_{N+2}| + \cdots$$

is dominated by the geometric series

$$|a_N|(1 + r + r^2 + r^3 + \cdots),$$

and the latter converges because $r < 1$. Thus the series $\sum |a_n|$ converges, so the series $\sum a_n$ converges absolutely.

If $\rho > 1$, then (3) implies that there exists a positive integer N such that $|a_{n+1}| > |a_n|$ for all $n \geqq N$. It follows that $|a_n| > |a_N| > 0$ for all $n > N$. Thus the sequence $\{a_n\}$ cannot approach zero as $n \to \infty$, and consequently—by the nth term divergence test—the series $\sum a_n$ diverges. ■

To see that $\sum a_n$ may either converge or diverge if $\rho = 1$, consider the divergent series $\sum (1/n)$ and the convergent series $\sum (1/n^2)$. You should verify that for each of these series, the value of the ratio ρ is 1.

EXAMPLE 5 Consider the series

$$\sum_{n=1}^{\infty} \frac{(-1)^n 2^n}{n!} = -2 + \frac{4}{2!} - \frac{8}{3!} + \frac{16}{4!} - \cdots.$$

Then

$$\rho = \lim_{n\to\infty} \left| \frac{a_{n+1}}{a_n} \right| = \lim_{n\to\infty} \left| \frac{(-1)^{n+1}2^{n+1}/(n+1)!}{(-1)^n 2^n/n!} \right|$$

$$= \lim_{n\to\infty} \frac{2}{n+1} = 0.$$

Because $\rho < 1$, the series converges absolutely.

EXAMPLE 6 Test for convergence:

$$\sum_{n=1}^{\infty} \frac{n}{2^n}.$$

Solution We have

$$\rho = \lim_{n\to\infty} \left| \frac{a_{n+1}}{a_n} \right| = \lim_{n\to\infty} \frac{(n+1)/2^{n+1}}{n/2^n}$$

$$= \lim_{n\to\infty} \frac{n+1}{2n} = \frac{1}{2}.$$

Because $\rho < 1$, this series converges (absolutely).

EXAMPLE 7 Test for convergence:

$$\sum_{n=1}^{\infty} \frac{3^n}{n^2}.$$

Solution Here we have

$$\rho = \lim_{n\to\infty} \left| \frac{a_{n+1}}{a_n} \right| = \lim_{n\to\infty} \frac{3^{n+1}/(n+1)^2}{3^n/n^2} = \lim_{n\to\infty} \frac{3n^2}{(n+1)^2} = 3.$$

In this case $\rho > 1$, so the given series diverges.

Theorem 5 *The Root Test*
Suppose that the limit

$$\rho = \lim_{n\to\infty} \sqrt[n]{|a_n|} \tag{4}$$

either exists or is infinite. Then the infinite series $\sum a_n$

1. Converges absolutely if $\rho < 1$;
2. Diverges if $\rho > 1$.

If $\rho = 1$, the root test is inconclusive.

Proof If $\rho < 1$, choose a (fixed) number r such that $\rho < r < 1$. Then $|a_n|^{1/n} < r$, and hence $|a_n| < r^n$, for n sufficiently large. Thus the series $\sum |a_n|$ is eventually dominated by the convergent geometric series $\sum r^n$. Therefore $\sum |a_n|$ converges, and so the series $\sum a_n$ converges absolutely.

If $\rho > 1$, then $|a_n|^{1/n} > 1$, and hence $|a_n| > 1$, for n sufficiently large. Therefore the nth term divergence test implies that the series $\sum a_n$ diverges. ■

The ratio test is generally simpler to apply than the root test and therefore is ordinarily the one to try first. But there are certain series for which the root test succeeds while the ratio test fails.

EXAMPLE 8 Consider the series

$$\sum_{n=0}^{\infty} \frac{1}{2^{n+(-1)^n}} = \tfrac{1}{2} + \tfrac{1}{1} + \tfrac{1}{8} + \tfrac{1}{4} + \tfrac{1}{32} + \tfrac{1}{16} + \cdots.$$

Then $a_{n+1}/a_n = 2$ if n is even while $a_{n+1}/a_n = \frac{1}{8}$ if n is odd. So the limit required for the ratio test does not exist. But

$$\lim_{n\to\infty} |a_n|^{1/n} = \lim_{n\to\infty} \left|\frac{1}{2^{n+(-1)^n}}\right|^{1/n} = \lim_{n\to\infty} \frac{1}{2}\left|\frac{1}{2^{(-1)^n/n}}\right| = \frac{1}{2},$$

so the given series converges by the root test. (Its convergence also follows from the fact that it is a rearrangement of the positive-term convergent geometric series $\sum 1/2^n$.)

12.6 PROBLEMS

Determine whether or not the alternating series in Problems 1–10 converge.

1. $\sum_{n=1}^{\infty} \frac{(-1)^{n+1}}{n^2}$
2. $\sum_{n=1}^{\infty} \frac{(-1)^{n+1} n}{3n+2}$
3. $\sum_{n=1}^{\infty} (-1)^n \frac{n}{n^2+1}$
4. $\sum_{n=2}^{\infty} (-1)^n \frac{n}{\ln n}$
5. $\sum_{n=2}^{\infty} (-1)^n \frac{1}{\ln n}$
6. $\sum_{n=1}^{\infty} \frac{(-1)^n}{\sqrt{2n+1}}$
7. $\sum_{n=0}^{\infty} \frac{(-1)^n}{n!}$
8. $\sum_{n=1}^{\infty} (-1)^n \frac{(1.01)^n}{n^4+1}$
9. $\sum_{n=1}^{\infty} \frac{(-1)^n}{n^{1/n}}$
10. $\sum_{n=1}^{\infty} (-1)^n \frac{n!}{(2n)!}$

Determine whether the series in Problems 11–32 converge absolutely, converge conditionally, or diverge.

11. $\sum_{n=1}^{\infty} \frac{(-1)^{n+1}}{2^n}$

12. $\sum_{n=1}^{\infty} \frac{1}{n^2+1}$

13. $\sum_{n=1}^{\infty} \frac{(-1)^n \ln n}{n}$

14. $\sum_{n=1}^{\infty} \frac{1}{n^n}$

15. $\sum_{n=1}^{\infty} \left(\frac{10}{n}\right)^n$

16. $\sum_{n=1}^{\infty} \frac{3^n}{n!n}$

17. $\sum_{n=0}^{\infty} \frac{(-10)^n}{n!}$

18. $\sum_{n=1}^{\infty} (-1)^n \frac{n!}{n^n}$

19. $\sum_{n=1}^{\infty} (-1)^n \left(\frac{n}{n+1}\right)^n$

20. $\sum_{n=1}^{\infty} \frac{n!n^2}{(2n)!}$

21. $\sum_{n=1}^{\infty} \left(\frac{\ln n}{n}\right)^n$

22. $\sum_{n=0}^{\infty} (-1)^n \frac{2^{3n}}{7^n}$

23. $\sum_{n=0}^{\infty} (-1)^n(\sqrt{n+1} - \sqrt{n})$

24. $\sum_{n=1}^{\infty} n(\frac{3}{4})^n$

25. $\sum_{n=2}^{\infty} \left[\ln\left(\frac{1}{n}\right)\right]^n$

26. $\sum_{n=0}^{\infty} \frac{(n!)^2}{(2n)!}$

27. $\sum_{n=1}^{\infty} (-1)^n \frac{3^n}{n(2^n+1)}$

28. $\sum_{n=1}^{\infty} (-1)^n \frac{\arctan n}{n}$

29. $\sum_{n=1}^{\infty} (-1)^n \frac{n!}{1 \cdot 3 \cdot 5 \cdots (2n-1)}$

30. $\sum_{n=1}^{\infty} (-1)^n \frac{1 \cdot 3 \cdot 5 \cdots (2n-1)}{1 \cdot 4 \cdot 7 \cdots (3n-2)}$

31. $\sum_{n=0}^{\infty} \frac{(n+2)!}{3^n(n!)^2}$

32. $\sum_{n=1}^{\infty} (-1)^n \frac{n^n}{3^{n^2}}$

In each of Problems 33–36, find the least positive integer n such that $|R_n| = |S - S_n| < 0.0005$, so that the nth partial sum S_n of the given alternating series approximates its sum S accurate to three decimal places.

33. $\sum_{n=1}^{\infty} \frac{(-1)^{n+1}}{n}$

34. $\sum_{n=1}^{\infty} \frac{(-1)^{n+1}}{n^2}$

35. $\sum_{n=0}^{\infty} \frac{(-1)^n}{3^n}$

36. $\sum_{n=1}^{\infty} \frac{(-1)^n}{n^n}$

In each of Problems 37–40, approximate the sum of the given series accurate to three decimal places.

37. $e^{-1/2} = \sum_{n=0}^{\infty} \frac{(-1)^n}{n!2^n}$

38. $\cos 1 = \sum_{n=0}^{\infty} \frac{(-1)^n}{(2n)!}$

39. $\ln(1.1) = \sum_{n=1}^{\infty} (-1)^{n+1} \frac{(0.1)^n}{n}$

40. $\sum_{n=1}^{\infty} \frac{(-1)^{n+1}}{n^5}$

41. Approximate the sum of the series

$$\frac{\pi^2}{12} = \sum_{n=1}^{\infty} \frac{(-1)^{n+1}}{n^2}$$

with error less than 0.01. Use the corresponding partial sum and error estimate to verify that $3.13 < \pi < 3.15$.

42. Prove that $\sum |a_n|$ diverges if the series $\sum a_n$ diverges.

43. Give an example of a pair of convergent series $\sum a_n$ and $\sum b_n$ such that $\sum a_n b_n$ diverges.

44. (a) Suppose that r is a (fixed) number with $|r| < 1$. Use the ratio test to prove that the series

$$\sum_{n=0}^{\infty} nr^n$$

converges. Let S denote its sum.

(b) Show that

$$(1-r)S = \sum_{n=1}^{\infty} r^n$$

Show how to conclude that

$$\sum_{n=0}^{\infty} nr^n = \frac{r}{(1-r)^2}.$$

12.7 Taylor Series

Up to this point we have concentrated on infinite series with *constant* terms. Much of the practical importance of infinite series, however, derives from the fact that many functions have useful representations as infinite series with *variable* terms. A number of such series representations can be derived from Taylor's formula (Section 11.3).

Suppose, then, that f is a function with continuous derivatives of *all* orders in a neighborhood of the point a and that n is an arbitrary positive integer. Taylor's formula then implies that

$$f(x) = P_n(x) + R_n(x), \tag{1}$$

where

$$P_n(x) = f(a) + f'(a)(x-a) + f''(a)\frac{(x-a)^2}{2!} + \cdots + f^{(n)}(a)\frac{(x-a)^n}{n!} \tag{2}$$

and

$$R_n(x) = \frac{f^{(n+1)}(z)}{(n+1)!}(x-a)^{n+1} \tag{3}$$

where z is some number between a and x.

Now suppose that for some particular *fixed* value of x, we can show that

$$\lim_{n\to\infty} R_n(x) = 0. \tag{4}$$

Then it follows from Equation (1) that

$$f(x) = \lim_{n\to\infty} P_n(x) = \lim_{n\to\infty}\left(\sum_{k=0}^{n} \frac{f^{(k)}(a)}{k!}(x-a)^k\right), \tag{5}$$

and thus

$$f(x) = \sum_{k=0}^{\infty} \frac{f^{(k)}(a)}{k!}(x-a)^k$$

$$= f(a) + f'(a)(x-a) + \frac{f''(a)}{2!}(x-a)^2 + \cdots + \frac{f^{(n)}(a)}{n!}(x-a)^n + \cdots. \tag{6}$$

The infinite series in Equation (6) is called the **Taylor series** of the function f at $x = a$. Note that the nth-degree Taylor polynomial $P_n(x)$ is the sum of the first $n+1$ terms of the Taylor series and thus is the $(n+1)$st term in the associated sequence of partial sums of the series in (6).

Now suppose that Equation (4) holds—that, for a particular value of x, $R_n(x) \to 0$ as $n \to \infty$. Then, by Equation (5), we can compute the value of $f(x)$ with any desired accuracy by adding enough terms of the Taylor series of f at a. For example, from the problems and examples of Section 11.3, we know the following Taylor formulas for the exponential and trigonometric functions:

$$e^x = 1 + x + \frac{x^2}{2!} + \frac{x^3}{3!} + \cdots + \frac{x^n}{n!} + \frac{e^z}{(n+1)!}x^{n+1}.$$

$$\cos x = 1 - \frac{x^2}{2!} + \frac{x^4}{4!} - \cdots + (-1)^n\frac{x^{2n}}{(2n)!} + (-1)^{n+1}\frac{\cos z}{(2n+2)!}x^{2n+2}.$$

$$\sin x = x - \frac{x^3}{3!} + \frac{x^5}{5!} - \cdots + (-1)^n\frac{x^{2n+1}}{(2n+1)!} + (-1)^{n+1}\frac{\cos z}{(2n+3)!}x^{2n+3}.$$

In each case z is a number between 0 and x.

Because z is between 0 and x, it follows that $0 < e^z \leqq e^{|x|}$ in Taylor's formula for e^x. In the formulas for the sine and cosine functions, $0 \leqq |\cos z| \leqq 1$. Therefore, the fact that

$$\lim_{n\to\infty} \frac{x^n}{n!} = 0$$

for all x (see Problem 27) implies that $\lim_{n\to\infty} R_n(x) = 0$ in all three cases above. This gives the following Taylor series:

$$e^x = \sum_{n=0}^{\infty} \frac{x^n}{n!} = 1 + x + \frac{x^2}{2!} + \frac{x^3}{3!} + \frac{x^4}{4!} + \cdots, \tag{7}$$

$$\cos x = \sum_{n=0}^{\infty} (-1)^n \frac{x^{2n}}{(2n)!} = 1 - \frac{x^2}{2!} + \frac{x^4}{4!} - \frac{x^6}{6!} + \cdots, \tag{8}$$

$$\sin x = \sum_{n=0}^{\infty} (-1)^n \frac{x^{2n+1}}{(2n+1)!} = x - \frac{x^3}{3!} + \frac{x^5}{5!} - \frac{x^7}{7!} + \cdots. \tag{9}$$

These series are all examples of the case $a = 0$ of Taylor series. The series

$$f(x) = f(0) + f'(0)x + \frac{f''(0)}{2!}x^2 + \frac{f^{(3)}(0)}{3!}x^3 + \cdots \tag{10}$$

is sometimes called the **Maclaurin series** of the function f.

The formulas in (7) through (9) are *identities* that hold for all values of x. Consequently, new series can be derived by substitution as in the following two examples.

EXAMPLE 1 The substitution $x = -t^2$ in (7) yields

$$e^{-t^2} = 1 - t^2 + \frac{t^4}{2!} - \frac{t^6}{3!} + \cdots + (-1)^n \frac{t^{2n}}{n!} + \cdots.$$

EXAMPLE 2 The substitution $x = 2t$ in (9) gives

$$\sin 2t = 2t - \tfrac{4}{3}t^3 + \tfrac{4}{15}t^5 - \tfrac{8}{315}t^7 + \cdots.$$

*THE NUMBER π

In Section 5.2 we described how Archimedes used polygons inscribed in and circumscribed about the unit circle to show that $3\frac{10}{71} < \pi < 3\frac{1}{7}$. With the aid of large electronic computers, π has been computed to many millions of decimal places. We describe now some of the methods that have been used for such computations. (For a chronicle of mankind's perennial fascination with the number π, see Howard Eves, *An Introduction to the History of Mathematics*, fourth edition (Boston: Allyn and Bacon, 1976), pages 96–102.)

We begin with the elementary algebraic identity

$$\frac{1}{1-x} = 1 - x + x^2 - x^3 + \cdots + (-1)^{k-1}x^{k-1} + \frac{(-1)^k x^k}{1+x}, \tag{11}$$

which can be verified by multiplying both sides by $1 + x$. We substitute t^2 for x and $n + 1$ for k, and thus find that

$$\frac{1}{1+t^2} = 1 - t^2 + t^4 - t^6 + \cdots + (-1)^n t^{2n} + \frac{(-1)^{n+1}t^{2n+2}}{1+t^2}.$$

Because $D_t \tan^{-1} t = 1/(1 + t^2)$, integration of both sides of this last equation from $t = 0$ to $t = x$ gives

$$\tan^{-1} x = x - \frac{x^3}{3} + \frac{x^5}{5} - \frac{x^7}{7} + \cdots + (-1)^n \frac{x^{2n+1}}{2n+1} + R_{2n+1} \tag{12}$$

where

$$|R_{2n+1}| = \left| \int_0^x \frac{t^{2n+2}}{1+t^2}\,dt \right| \leq \left| \int_0^x t^{2n+2}\,dt \right| = \frac{|x|^{2n+3}}{2n+3}. \tag{13}$$

This estimate of the error makes it clear that

$$\lim_{n \to \infty} R_{2n+1} = 0$$

if $|x| \leqq 1$. Hence we obtain the Taylor series for the inverse tangent function:

$$\tan^{-1} x = \sum_{n=0}^{\infty} (-1)^n \frac{x^{2n+1}}{2n+1} = x - \frac{x^3}{3} + \frac{x^5}{5} - \frac{x^7}{7} + \cdots, \tag{14}$$

valid for $-1 \leqq x \leqq 1$.

If we substitute $x = 1$ in Equation (14), we obtain *Leibniz's series*

$$\frac{\pi}{4} = 1 - \frac{1}{3} + \frac{1}{5} - \frac{1}{7} + \cdots$$

Though this is a beautiful series, it is not an effective way to compute π. But the error estimate in (13) shows that the formula in (12) is effective for the calculation of $\tan^{-1} x$ if $|x|$ is small. For example, if $x = \frac{1}{5}$, the fact that

$$\frac{1}{9 \cdot 5^9} \approx 0.000000057$$

implies that the approximation

$$\begin{aligned} \alpha &= \tan^{-1}(\tfrac{1}{5}) \\ &\approx \tfrac{1}{5} - \tfrac{1}{3}(\tfrac{1}{5})^3 + \tfrac{1}{5}(\tfrac{1}{5})^5 - \tfrac{1}{7}(\tfrac{1}{5})^7 \end{aligned}$$

is accurate to six decimal places.

Let us begin with $\alpha = \tan^{-1}(\frac{1}{5})$. The addition formula for the tangent function can be used to show (Problem 24) that

$$\tan\left(\frac{\pi}{4} - 4\alpha\right) = -\frac{1}{239}.$$

Hence

$$\frac{\pi}{4} = 4 \tan^{-1}\left(\frac{1}{5}\right) - \tan^{-1}\left(\frac{1}{239}\right). \tag{15}$$

In 1706, John Machin used the formula in (15) to calculate the first 100 decimal places of π. In Problem 26 we ask you to use it to show that $\pi = 3.14159$ to five decimal places. In 1844 the lightning calculator Zacharias Dase of Germany computed the first 200 decimal places of π using the related formula

$$\frac{\pi}{4} = \tan^{-1}\left(\frac{1}{2}\right) + \tan^{-1}\left(\frac{1}{5}\right) + \tan^{-1}\left(\frac{1}{8}\right), \tag{16}$$

and you might enjoy verifying this formula (see Problem 25). A recent computation of 1 million decimal places of π used the formulas

$$\begin{aligned} \frac{\pi}{4} &= 12 \tan^{-1}\left(\frac{1}{18}\right) + 8 \tan^{-1}\left(\frac{1}{57}\right) - 5 \tan^{-1}\left(\frac{1}{239}\right) \\ &= 6 \tan^{-1}\left(\frac{1}{8}\right) + 2 \tan^{-1}\left(\frac{1}{57}\right) + \tan^{-1}\left(\frac{1}{239}\right). \end{aligned}$$

For derivations of these and similar formulas, with further discussion of the computations of the number π, see the article "An algorithm for the calculation of π" by George Miel in *American Mathematical Monthly*, **86** (1979), 694–697. Although no practical application is likely to require more than ten or twelve decimal places of π, these computations provide dramatic evidence of the power of Taylor's formula. Moreover, the number π continues to serve as a challenge both to human ingenuity and to the accuracy and efficiency of modern electronic computers. For an account of how investigations of the Indian mathematical genius Srinivasa Ramanujan (1887–1920) have led recently to the computation of over 100 million decimal places of π, see the article "Ramanujan and pi" in the February 1988 issue of *Scientific American*.

12.7 PROBLEMS

In each of Problems 1–8, find the Maclaurin series of the given function f by substitution in one of the known series in Equations (7) through (9).

1. $f(x) = e^{-x}$

2. $f(x) = e^{2x}$

3. $f(x) = e^{-3x}$

4. $f(x) = \exp(x^3)$

5. $f(x) = \cos 2x$

6. $f(x) = \sin \dfrac{x}{2}$

7. $f(x) = \sin x^2$

8. $f(x) = \cos\sqrt{x}$

In each of Problems 9–22, find the Taylor series (Equation (6)) of the given function at the indicated point a.

9. $f(x) = \ln(1 + x)$, $a = 0$

10. $f(x) = \dfrac{1}{1-x}$, $a = 0$

11. $f(x) = e^{-x}$, $a = 0$

12. $f(x) = \cosh x$, $a = 0$

13. $f(x) = \ln x$, $a = 1$

14. $f(x) = \sin x$, $a = \dfrac{\pi}{2}$

15. $f(x) = \cos x$, $a = \dfrac{\pi}{4}$

16. $f(x) = e^{2x}$, $a = 0$

17. $f(x) = \sinh x$, $a = 0$

18. $f(x) = \dfrac{1}{(1-x)^2}$, $a = 0$

19. $f(x) = \dfrac{1}{x}$, $a = 1$

20. $f(x) = \cos x$, $a = \dfrac{\pi}{2}$

21. $f(x) = \sin x$, $a = \dfrac{\pi}{4}$

22. $f(x) = \sqrt{1 + x}$, $a = 0$

23. Use the Taylor series for the cosine and sine functions to verify that $\cos(-x) = \cos x$ and $\sin(-x) = -\sin x$ for all x.

24. Beginning with $\alpha = \tan^{-1}(\frac{1}{5})$, use the addition formula

$$\tan(A + B) = \frac{\tan A + \tan B}{1 - \tan A \tan B}$$

to show in turn that:

(a) $\tan 2\alpha = \frac{5}{12}$;

(b) $\tan 4\alpha = \frac{120}{119}$;

(c) $\tan(\pi/4 - 4\alpha) = -\frac{1}{239}$.

25. Apply the addition formula for the tangent function to verify the formula in Equation (16).

26. Use the formulas in (12), (13), and (15) to show that $\pi = 3.14159$ to five decimal places. *Suggestion:* Compute $\arctan(\frac{1}{5})$ and $\arctan(\frac{1}{239})$ each with error less than 0.5×10^{-7}. You can do this by applying Equation (11) with $n = 4$ and with $n = 1$. Carry out your computations to seven decimal places and keep track of errors.

27. Prove that $\lim_{n\to\infty} x^n/n! = 0$ if x is a fixed real number. (*Suggestion:* Choose an integer k with $k > |2x|$, and let $L = |x|^k/k!$. Then show that

$$\frac{|x|^n}{n!} < \frac{L}{2^{n-k}}$$

if $n > k$.)

28. (a) Suppose that t is a positive real number but is otherwise arbitrary. Integrate both sides in (10) from $x = 0$ to $x = t$ to show that

$$\ln(1 + t) = t - \frac{t^2}{2} + \frac{t^3}{3} - \cdots + (-1)^{k-1}\frac{t^k}{k} + R_k$$

where

$$|R_k| \leqq \frac{t^{k+1}}{k+1}.$$

(b) Conclude that

$$\ln(1 + t) = t - \frac{t^2}{2} + \frac{t^3}{3} - \cdots = \sum_{k=1}^{\infty} (-1)^{k-1}\frac{t^k}{k}$$

if $0 \leqq t \leqq 1$.

29. Use the result of part (a) of Problem 28 to calculate $\ln(\frac{4}{3})$ accurate to three decimal places.

30. Differentiate both sides of the identity

$$\frac{1}{1-x} = 1 + x + x^2 + x^3 + \cdots + x^n + \frac{x^{n+1}}{1-x}$$

to obtain the result

$$\frac{1}{(1-x)^2} = 1 + 2x + 3x^2 + 4x^3 + \cdots + nx^{n-1} + R_n$$

where, assuming that $|x| < 1$, $R_n \to 0$ as $n \to \infty$. Thus if $-1 < x < 1$, then

$$\frac{1}{(1-x)^2} = \sum_{n=0}^{\infty} (n+1)x^n.$$

31. Substitute t^2 for x in the first identity of Problem 30, then integrate from $t = 0$ to $t = x$ to obtain the result

$$\tanh^{-1} x = x + \frac{x^3}{3} + \frac{x^5}{5} + \cdots + \frac{x^{2n+1}}{2n+1} + R_n$$

where $R_n \to 0$ as $n \to +\infty$ (under the assumption that $|x| < 1$). Thus if $-1 < x < 1$, then

$$\tanh^{-1} x = \sum_{n=0}^{\infty} \frac{x^{2n+1}}{2n+1}.$$

32. Use the formula $\ln 2 = \ln(\frac{5}{4}) + 2\ln(\frac{6}{5}) + \ln(\frac{10}{9})$ to calculate $\ln 2$ accurate to three decimal places. See part (a) of Problem 28 for the Taylor's formula for $\ln(1 + x)$.

*12.7 Optional Computer Application

The program listed in Fig. 12.10 was written to sum the Taylor series for e^x when a value of x is entered. For instance, with $x = 1$ the resulting display

```
Value of x? 1
   N =  15
 Sum = 2.718281828
```

indicates that 15 terms of the exponential series have produced the value of e accurate to 9 decimal places.

The key step in Program POWERSER is the conversion of the old term $T_{old} = x^n/n!$ of the exponential series

$$e^x = 1 + x + \frac{x^2}{2!} + \frac{x^3}{3!} + \cdots + \frac{x^n}{n!} + \frac{x^{n+1}}{(n+1)!} + \cdots$$

```
100 REM--Program POWERSER
110 REM--Computes sum of approp. power series
120 REM
130      DEFDBL X,R,S,T
140      INPUT "Value of x"; X
150      N = 0   :   T = 1           'Initial term
160      S = T                       'Initial sum
170      EPS = 1E-11                 'Tolerance
180 REM
190 REM--Loop:                       'Until T < EPS
200      R = X/(N + 1)               'R = Tnew/Told
210      T = R*T                     'Next term
220      S = S + T                   'Next sum
230      N = N + 1                   'Next power
240      IF ABS(T) => EPS THEN GOTO 200
250 REM
260      PRINT USING "  N = ###"; N
270      PRINT USING "Sum = #.#########"; S
280      END
```

12.10 Listing of Program POWERSER

into the new term $T_{\text{new}} = x^{n+1}/(n+1)!$. The factorization

$$T_{\text{new}} = \frac{x^{n+1}}{(n+1)!} = \frac{x}{n+1} \cdot \frac{x^n}{n!} = \frac{x}{n+1} \cdot T_{\text{old}}$$

provides the factor $R = x/(n+1)$ defined in line 200 and used in line 210.

Exercise 1: Explain why with just the two alterations

```
200    R = -X*X/((N+1)*(N+2))
230    N = N + 2
```

the program in Fig. 12.10 produces values of $\cos x$ instead of e^x. Verify that it gives $\cos(\pi/4)$ accurate to nine decimal places if the value $\pi/4 \approx 0.7853\,9816\,3397$ is entered.

Exercise 2: Explain why with just the alterations

```
150    N = 1  :  T = X
200    R = -X*X/((N+1)*(N+2))
230    N = N + 2
```

the program in Fig. 12.10 produces values of $\sin x$ instead of e^x. Verify that it gives $\sin(\pi/6)$ accurate to nine decimal places if the value $\pi/6 \approx 0.5235\,9877\,5598$ is entered.

12.8 Power Series

The most important infinite series representations of functions are those whose terms are constant multiples of (successive) integral powers of the independent variable x—that is, series that resemble "infinite polynomials." For example, from Theorem 1 of Section 12.3 (with x in place of r) we know that the geometric series

$$\frac{1}{1-x} = 1 + x + x^2 + x^2 + \cdots + x^n + \cdots \tag{1}$$

represents (that is, *converges to*) the function $f(x) = 1/(1-x)$ when $|x| < 1$. In Section 12.7, we derived the Taylor series

$$e^x = \sum_{n=0}^{\infty} \frac{x^n}{n!} = 1 + x + \frac{x^2}{2!} + \frac{x^3}{3!} + \frac{x^4}{4!} + \cdots, \tag{2}$$

$$\cos x = \sum_{n=0}^{\infty} (-1)^n \frac{x^{2n}}{(2n)!} = 1 - \frac{x^2}{2!} + \frac{x^4}{4!} - \frac{x^6}{6!} + \cdots, \tag{3}$$

$$\sin x = \sum_{n=0}^{\infty} (-1)^n \frac{x^{2n+1}}{(2n+1)!} = x - \frac{x^3}{3!} + \frac{x^5}{5!} - \frac{x^7}{7!} + \cdots. \tag{4}$$

There we used Taylor's formula with remainder to show that these series converge (for all x) to the functions e^x, $\cos x$, and $\sin x$, respectively. Thus we did not then require the convergence tests of this chapter.

The infinite series above all have the form

$$\sum_{n=0}^{\infty} a_n x^n = a_0 + a_1 x + a_2 x^2 + \cdots + a_n x^n + \cdots \tag{5}$$

with constant *coefficients* $a_0, a_1, a_2, \ldots$. An infinite series of this form is called a **power series** in (powers of) x. In order that the initial terms of the two sides of Equation (5) agree, we adopt here the convention that $x^0 = 1$ even if $x = 0$.

The power series in (5) obviously converges when $x = 0$. In general, it will converge for some nonzero values of x and diverge for others. Because of the way in which powers of x are involved, the ratio test is particularly effective in deciding for which values of x a power series converges. Assume that the limit

$$\rho = \lim_{n \to \infty} \left| \frac{a_{n+1}}{a_n} \right| \tag{6}$$

exists. This is the limit that we need if we want to apply the ratio test to the series $\sum a_n$ of constants. To apply the ratio test to the power series in (5), we write $u_n = a_n x^n$ and compute the limit

$$\lim_{n \to \infty} \left| \frac{u_{n+1}}{u_n} \right| = \lim_{n \to \infty} \left| \frac{a_{n+1} x^{n+1}}{a_n x^n} \right| = \rho |x|. \tag{7}$$

If $\rho = 0$, we see that $\sum a_n x^n$ converges absolutely for all x. If $\rho = +\infty$, we see that $\sum a_n x^n$ diverges for all $x \neq 0$. If ρ is a positive real number, we see from (7) that $\sum a_n x^n$ converges absolutely for all x such that $\rho|x| < 1$; that is, when

$$|x| < R = \frac{1}{\rho} = \lim_{n \to \infty} \left| \frac{a_n}{a_{n+1}} \right|. \tag{8}$$

In the case the ratio test also implies that $\sum a_n x^n$ diverges if $|x| > R$, but the test is inconclusive when $x = \pm R$. We have therefore proved the following theorem, under the additional hypothesis that the limit in (6) exists. In Problems 43 and 44 we outline a proof that does not require this additional hypothesis.

Theorem 1 *Convergence of Power Series*

If $\sum a_n x^n$ is a power series, then either

1. The series converges absolutely for all x, or
2. The series converges only when $x = 0$, or
3. There exists a number $R > 0$ such that $\Sigma\, a_n x^n$ converges absolutely if $|x| < R$ and diverges if $|x| > R$.

The number R of Case 3 is called the **radius of convergence** of the power series $\sum a_n x^n$. We shall write $R = \infty$ in Case 1 and $R = 0$ in Case 2. The set of all real numbers x for which the series converges is called its **interval of convergence.** If $0 < R < \infty$, then the interval of convergence is one of the intervals

$$(-R, R), \quad (-R, R], \quad [-R, R), \quad \text{or} \quad [-R, R].$$

When we substitute either of the end points $x = \pm R$ in the series $\sum a_n x^n$, we obtain an infinite series with constant terms whose convergence must be determined separately. Because these will be numerical series, the earlier tests of this chapter are appropriate.

EXAMPLE 1 Find the interval of convergence of the series

$$\sum_{n=1}^{\infty} \frac{x^n}{n \cdot 3^n}.$$

Solution With $u_n = x_n/(n \cdot 3^n)$ we find that

$$\lim_{n\to\infty} \left|\frac{u_{n+1}}{u_n}\right| = \lim_{n\to\infty} \left|\frac{x^{n+1}/([n+1] \cdot 3^{n+1})}{x^n/(n \cdot 3^n)}\right| = \lim_{n\to\infty} \frac{n|x|}{3(n+1)} = \frac{|x|}{3}.$$

Now $|x|/3 < 1$ provided that $|x| < 3$, so the ratio test implies that the given series converges absolutely if $|x| < 3$ and diverges if $|x| > 3$. When $x = 3$ we have the divergent harmonic series $\sum (1/n)$, and when $x = -3$ we have the convergent alternating series $\sum (-1)^n/n$. Thus the interval of convergence of the given power series is $[-3, 3)$.

EXAMPLE 2 Find the interval of convergence of

$$\sum_{n=0}^{\infty} \frac{2^n x^n}{n!}.$$

Solution With $u_n = 2^n x^n/n!$, we find that

$$\lim_{n\to\infty} \left|\frac{u_{n+1}}{u_n}\right| = \lim_{n\to\infty} \left|\frac{2^{n+1}x^{n+1}/(n+1)!}{2^n x^n/n!}\right| = \lim_{n\to\infty} \frac{2|x|}{n+1} = 0$$

for all x. Hence the ratio test implies that the power series converges for all x, and its interval of convergence is $(-\infty, \infty)$, the entire real line.

EXAMPLE 3 Find the interval of convergence of the series

$$\sum_{n=0}^{\infty} n^n x^n.$$

Solution With $u_n = n^n x^n$ we find that

$$\lim_{n\to\infty} \left|\frac{u_{n+1}}{u_n}\right| = \lim_{n\to\infty} \left|\frac{(n+1)^{n+1}x^{n+1}}{n^n x^n}\right| = \lim_{n\to\infty} (n+1)\left(1 + \frac{1}{n}\right)^n |x| = +\infty$$

for all $x \neq 0$, because

$$\lim_{n\to\infty} \left(1 + \frac{1}{n}\right)^n = e.$$

Thus the given series diverges for all $x \neq 0$, and its interval of convergence consists of the single point $x = 0$.

EXAMPLE 4 Use the ratio test to verify that the Taylor series for $\cos x$ in Equation (3) converges for all x.

Solution With $u_n = (-1)^n x^{2n}/(2n)!$ we find that

$$\lim_{n\to\infty}\left|\frac{u_{n+1}}{u_n}\right| = \lim_{n\to\infty}\left|\frac{(-1)^{n+1}x^{2n+2}/(2n+2)!}{(-1)^n x^{2n}/(2n)!}\right|$$

$$= \lim_{n\to\infty}\frac{x^2}{(2n+1)(2n+2)} = 0$$

for all x, so the series converges for all x.

IMPORTANT In Example 4, the ratio test tells us only that the series for $\cos x$ converges to *some* number, *not* necessarily the particular number $\cos x$. The argument of Section 12.7, using the Taylor formula with remainder, is required to establish that the sum of the series is actually $\cos x$.

An infinite series of the form

$$\sum_{n=0}^{\infty} a_n(x-c)^n = a_0 + a_1(x-c) + a_2(x-c)^2 + \cdots \tag{9}$$

(where c is a constant) is called a **power series in** (powers of) $x - c$. By the same reasoning that leads to Theorem 1, with x^n replaced by $(x-c)^n$ throughout, we conclude that either

1. The series in (9) converges absolutely for all x; or
2. The series converges only when $x - c = 0$—that is, when $x = c$; or
3. There exists a number $R > 0$ such that the series in (9) converges absolutely if $|x - c| < R$ and diverges if $|x - c| > R$.

As in the case of a power series with $c = 0$, the number R is called its **radius of convergence,** and the **interval of convergence** of the series $\sum a_n(x-c)^n$ is the set of all numbers x for which it converges. As before, when $0 < R < \infty$ the convergence of the series at the end points $x = c - R$ and $x = c + R$ of its interval of convergence must be checked separately.

EXAMPLE 5 Determine the interval of convergence of the series

$$\sum_{n=1}^{\infty} (-1)^n \frac{(x-2)^n}{n \cdot 4^n}.$$

Solution We let $u_n = (-1)^n(x-2)^n/(n \cdot 4^n)$. Then

$$\lim_{n\to\infty}\left|\frac{u_{n+1}}{u_n}\right| = \lim_{n\to\infty}\left|\frac{(-1)^{n+1}(x-2)^{n+1}/[(n+1)\cdot 4^{n+1}]}{(-1)^n(x-2)^n/(n\cdot 4^n)}\right|$$

$$= \lim_{n\to\infty}\frac{|x-2|}{4}\cdot\frac{n}{n+1} = \frac{|x-2|}{4}.$$

Hence the given series converges when $|x-2|/4 < 1$; that is, when $|x-2| < 4$, so the radius of convergence is $R = 4$. Because $c = 2$, the series converges when $-2 < x < 6$ and diverges if either $x < -2$ or $x > 6$. When $x = -2$ the series reduces to the divergent harmonic series, and when $x = 6$ it reduces to the convergent alternating series $\sum (-1)^n/n$. Thus the interval of convergence of the given power series is $(-2, 6]$.

POWER SERIES REPRESENTATION OF FUNCTIONS

Power series provide us with an important tool for computing (or approximating) values of functions. Suppose that the series $\sum a_n x^n$ converges to the value $f(x)$; that is,

$$f(x) = a_0 + a_1 x + a_2 x^2 + \cdots + a_n x^n + \cdots$$

for each x in the interval of convergence of the power series. Then we call $\sum a_n x^n$ a **power series representation** of $f(x)$. For example, the geometric series $\sum x^n$ in (1) is a power series representation of the function $f(x) = 1/(1 - x)$ on the interval $(-1, 1)$.

In Section 12.7 we saw how Taylor's formula with remainder can often be used to find a power series representation of a given function. Recall that the nth-degree Taylor's formula for $f(x)$ at $x = a$ is

$$f(x) = f(a) + f'(a)(x - a) + \frac{f''(a)}{2!}(x - a)^2 + \cdots + \frac{f^{(n)}(a)}{n!}(x - a)^n + R_n(x). \tag{10}$$

The remainder $R_n(x)$ is given by

$$R_n(x) = \frac{f^{(n+1)}(z)}{(n + 1)!}(x - a)^{n+1}$$

where z is some number between a and x. If we let $n \to \infty$ in (10), we obtain the following theorem.

Theorem 2 *Taylor Series Representation*

Suppose that the function f has derivatives of all orders on some interval containing a and also that

$$\lim_{n\to\infty} R_n(x) = 0 \tag{11}$$

for each x in that interval. Then

$$f(x) = \sum_{n=0}^{\infty} \frac{f^{(n)}(a)}{n!}(x - a)^n \tag{12}$$

for each x in the interval.

The power series in (12) is the **Taylor series** of the function f **at** $x = a$ (or *in powers of* $x - a$, or *with center* a). If $a = 0$, we obtain the power series

$$f(x) = \sum_{n=1}^{\infty} \frac{f^{(n)}(0)}{n!} x^n = f(0) + f'(0)x + \frac{f''(0)}{2!}x^2 + \cdots, \tag{13}$$

commonly called the **Maclaurin series** of f. Thus the power series in Equations (2) through (4) at the beginning of this section are the Maclaurin series of the functions e^x, $\cos x$, and $\sin x$.

Upon replacing x by $-x$ in the Maclaurin series for e^x, we obtain

$$e^{-x} = 1 - x + \frac{x^2}{2!} - \frac{x^3}{3!} + \cdots + (-1)^n \frac{x^n}{n!} + \cdots.$$

Let us add the series for e^x and e^{-x} and divide by 2. This gives

$$\cosh x = \frac{e^x + e^{-x}}{2} = \frac{1}{2}\left(1 + x + \frac{x^2}{2!} + \frac{x^3}{3!} + \frac{x^4}{4!} + \cdots\right)$$
$$+ \frac{1}{2}\left(1 - x + \frac{x^2}{2!} - \frac{x^3}{3!} + \frac{x^4}{4!} - \cdots\right),$$

so that

$$\cosh x = 1 + \frac{x^2}{2!} + \frac{x^4}{4!} + \frac{x^6}{6!} + \cdots.$$

Similarly,

$$\sinh x = x + \frac{x^3}{3!} + \frac{x^5}{5!} + \frac{x^7}{7!} + \cdots.$$

Note the strong resemblance with the series for $\sin x$ and $\cos x$.

Upon replacement of x by $-x^2$ in the series for e^x, we obtain

$$e^{-x^2} = \sum_{n=0}^{\infty} (-1)^n \frac{x^{2n}}{n!} = 1 - x^2 + \frac{x^4}{2!} - \frac{x^6}{3!} + \cdots.$$

Because this power series converges to $\exp(-x^2)$ for all x, it must be the Maclaurin series for $\exp(-x^2)$ (see Problem 40). Think how tedious it would be to compute the derivatives of $\exp(-x^2)$ needed to write its Maclaurin series directly from (13).

THE BINOMIAL SERIES

The following example gives one of the most famous and useful of all series, the *binomial series*, which was discovered by Newton in the 1660s. It is the infinite series generalization of the (finite) binomial formula of elementary algebra.

EXAMPLE 6 Suppose that α is a nonzero real number. Show that the Maclaurin series of $f(x) = (1 + x)^\alpha$ is

$$(1 + x)^\alpha = 1 + \sum_{n=1}^{\infty} \frac{\alpha(\alpha - 1)(\alpha - 2)\cdots(\alpha - n + 1)}{n!} x^n$$
$$= 1 + \alpha x + \frac{\alpha(\alpha - 1)}{2!} x^2 + \frac{\alpha(\alpha - 1)(\alpha - 2)}{3!} x^3 + \cdots. \qquad (14)$$

Also determine the interval of convergence of this **binomial series.**

Solution To derive the series itself, we simply list the derivatives of $f(x) = (1 + x)^\alpha$:

$$f(x) = (1 + x)^\alpha,$$
$$f'(x) = \alpha(1 + x)^{\alpha - 1},$$
$$f''(x) = \alpha(\alpha - 1)(1 + x)^{\alpha - 2}$$
$$f^{(3)}(x) = \alpha(\alpha - 1)(\alpha - 2)(1 + x)^{\alpha - 3},$$
$$\vdots$$
$$f^{(n)}(x) = \alpha(\alpha - 1)(\alpha - 2)\cdots(\alpha - n + 1)(1 + x)^{\alpha - n}.$$

Thus

$$f^{(n)}(0) = \alpha(\alpha - 1)(\alpha - 2) \cdots (\alpha - n + 1).$$

Substitution of this value of $f^{(n)}(0)$ in the Maclaurin series formula in (13) gives the binomial series in (14).

To determine the interval of convergence of the binomial series, we let

$$u_n = \frac{\alpha(\alpha - 1)(\alpha - 2) \cdots (\alpha - n + 1)}{n!} x^n.$$

We find that

$$\begin{aligned} \lim_{n\to\infty} \left| \frac{u_{n+1}}{u_n} \right| &= \lim_{n\to\infty} \left| \frac{\alpha(\alpha - 1)(\alpha - 2) \cdots (\alpha - n)x^{n+1}/(n + 1)!}{\alpha(\alpha - 1)(\alpha - 2) \cdots (\alpha - n + 1)x^n/n!} \right| \\ &= \lim_{n\to\infty} \left| \frac{(\alpha - n)x}{n + 1} \right| = |x|. \end{aligned}$$

Hence the ratio test shows that the binomial series converges absolutely if $|x| < 1$ and diverges if $|x| > 1$. Its convergence at the end points $x = \pm 1$ depends upon the value of α; we shall not pursue this problem. Problem 41 outlines a proof that the sum of the binomial series actually is $(1 + x)^\alpha$ if $|x| < 1$.

If $\alpha = k$, a positive integer, then the coefficient of x^n in (14) vanishes for $n > k$, and the binomial series reduces to the binomial formula

$$(1 + x)^k = \sum_{n=0}^{k} \frac{k!}{n!(k - n)!} x^n.$$

Otherwise (14) is an infinite series. For example, with $\alpha = \frac{1}{2}$, we obtain

$$\begin{aligned} \sqrt{1 + x} &= 1 + \frac{\frac{1}{2}}{1!} x + \frac{(\frac{1}{2})(-\frac{1}{2})}{2!} x^2 \\ &\quad + \frac{(\frac{1}{2})(-\frac{1}{2})(-\frac{3}{2})}{3!} x^3 + \frac{(\frac{1}{2})(-\frac{1}{2})(-\frac{3}{2})(-\frac{5}{2})}{4!} x^4 + \cdots \\ &= 1 + \tfrac{1}{2}x - \tfrac{1}{8}x^2 + \tfrac{1}{16}x^3 - \tfrac{5}{128}x^4 + \cdots. \end{aligned} \tag{15}$$

If we replace x by $-x$ and take $\alpha = -\frac{1}{2}$, we get the series

$$\begin{aligned} \frac{1}{\sqrt{1 - x}} &= 1 + \frac{(-\frac{1}{2})}{1!}(-x) + \frac{(-\frac{1}{2})(-\frac{3}{2})}{2!}(-x)^2 \\ &\quad + \cdots + \frac{1 \cdot 3 \cdot 5 \cdots (2n - 1)}{n! \cdot 2^n} x^n + \cdots, \end{aligned}$$

which in summation notation takes the form

$$\frac{1}{\sqrt{1 - x}} = 1 + \sum_{n=1}^{\infty} \frac{1 \cdot 3 \cdot 5 \cdots (2n - 1)}{2 \cdot 4 \cdot 6 \cdots (2n)} x^n. \tag{16}$$

Sometimes it is inconvenient to compute the repeated derivatives of a function in order to find its Taylor series. Another common method of deriving new power series is by the differentiation and integration of known power series. Suppose that a power series representation of the function $f(x)$ is known. Then the following theorem (we leave its proof to advanced calculus) implies that the function $f(x)$ may be differentiated by separately differentiating the individual terms in its power series. That is, the power series obtained by termwise differentiation converges to the derivative $f'(x)$. Similarly, a function can be integrated by termwise integration of its power series.

Theorem 3 *Termwise Differentiation and Integration*

Suppose that the function f has a power series representation

$$f(x) = \sum_{n=0}^{\infty} a_n x^n = a_0 + a_1 x + a_2 x^2 + a_3 x^3 + \cdots$$

with nonzero radius of convergence R. Then f is differentiable on $(-R, R)$ and

$$f'(x) = \sum_{n=1}^{\infty} n a_n x^{n-1} = a_1 + 2a_2 x + 3a_3 x^2 + 4a_4 x^3 + \cdots. \quad (17)$$

Also,

$$\int_0^x f(t)\,dt = \sum_{n=0}^{\infty} \frac{a_n x^{n+1}}{n+1} = a_0 x + \tfrac{1}{2}a_1 x^2 + \tfrac{1}{3}a_2 x^3 + \cdots \quad (18)$$

for each x in $(-R, R)$. Moreover, the power series in (17) and (18) have the same radius of convergence R.

REMARK Although the proof of Theorem 3 is omitted, we observe that the radius of convergence of the series in (17) is

$$\begin{aligned} R &= \lim_{n\to\infty} \left| \frac{n a_n}{(n+1)a_{n+1}} \right| \\ &= \left(\lim_{n\to\infty} \frac{n}{n+1} \right) \left(\lim_{n\to\infty} \left| \frac{a_n}{a_{n+1}} \right| \right) \\ &= \lim_{n\to\infty} \left| \frac{a_n}{a_{n+1}} \right|. \end{aligned}$$

Thus (by the formula in (8)) the power series for $f(x)$ and the power series for $f'(x)$ have the same radius of convergence (under the assumption that the limit above exists).

EXAMPLE 7 Termwise differentiation of the geometric series for $f(x) = 1/(1-x)$ yields

$$\begin{aligned} \frac{1}{(1-x)^2} &= D_x\left(\frac{1}{1-x}\right) = D_x(1 + x + x^2 + x^3 + \cdots) \\ &= 1 + 2x + 3x^2 + 4x^3 + \cdots. \end{aligned}$$

Thus

$$\frac{1}{(1-x)^2} = \sum_{n=1}^{\infty} n x^{n-1} = \sum_{n=0}^{\infty} (n+1)x^n.$$

The series converges to $1/(1-x)^2$ if $-1 < x < 1$.

EXAMPLE 8 Replacement of x by $-t$ in the geometric series of Example 7 gives

$$\frac{1}{1+t} = 1 - t + t^2 - t^3 + \cdots + (-1)^n t^n + \cdots.$$

Because $D_t \ln(1+t) = 1/(1+t)$, termwise integration from $t = 0$ to $t = x$ now gives

$$\ln(1+x) = \int_0^x \frac{1}{1+t}\,dt$$

$$= \int_0^x (1 - t + t^2 - \cdots + (-1)^n t^n + \cdots)\,dt;$$

$$\ln(1+x) = x - \frac{1}{2}x^2 + \frac{1}{3}x^3 - \frac{1}{4}x^4 + \cdots + \frac{(-1)^{n-1}}{n}x^n + \cdots \tag{19}$$

if $|x| < 1$.

EXAMPLE 9 Find a power series representation for the arctangent function.

Solution Because $D_t \tan^{-1} t = 1/(1+t^2)$, termwise integration of the series

$$\frac{1}{1+t^2} = 1 - t^2 + t^4 - t^6 + t^8 - \cdots$$

gives

$$\tan^{-1} x = \int_0^x \frac{1}{1+t^2}\,dt = \int_0^x (1 - t^2 + t^4 - t^6 + t^8 - \cdots)\,dt.$$

Therefore,

$$\tan^{-1} x = x - \tfrac{1}{3}x^3 + \tfrac{1}{5}x^5 - \tfrac{1}{7}x^7 + \tfrac{1}{9}x^9 - \cdots \tag{20}$$

if $-1 < x < 1$.

EXAMPLE 10 Find a power series representation for the arcsine function.

Solution First we substitute t^2 for x in Equation (16). This yields

$$\frac{1}{\sqrt{1-t^2}} = 1 + \sum_{n=1}^{\infty} \frac{1 \cdot 3 \cdot 5 \cdots (2n-1)}{2 \cdot 4 \cdot 6 \cdots (2n)} t^{2n}$$

if $|t| < 1$. Because $D_t \sin^{-1} t = 1/(1-t^2)^{1/2}$, termwise integration of this series from $t = 0$ to $t = x$ gives

$$\sin^{-1} x = \int_0^x \frac{1}{\sqrt{1-t^2}}\,dt = x + \sum_{n=1}^{\infty} \frac{1 \cdot 3 \cdot 5 \cdots (2n-1)}{2 \cdot 4 \cdot 6 \cdots (2n)} \cdot \frac{x^{2n+1}}{2n+1} \tag{21}$$

if $|x| < 1$. Problem 42 shows how to use this series for $\sin^{-1} x$ to derive the series

$$\frac{\pi^2}{6} = 1 + \frac{1}{2^2} + \frac{1}{3^2} + \frac{1}{4^2} + \cdots + \frac{1}{n^2} + \cdots$$

that was used in Example 3 of Section 12.4 to approximate the number π.

Theorem 3 has this important consequence: If the two power series $\sum a_n x^n$ and $\sum b_n x^n$ both converge and, for all x with $|x| < R$ $(R > 0)$, $\sum a_n x^n = \sum b_n x^n$, then $a_n = b_n$ for all n. In particular, the Taylor series of a function is its unique power series representation (if any). See Problem 40.

12.8 PROBLEMS

Find the interval of convergence of each of the power series in Problems 1–20.

1. $\sum_{n=1}^{\infty} \frac{1}{n} x^n$

2. $\sum_{n=0}^{\infty} \frac{(-1)^n}{n^2+1} x^n$

3. $\sum_{n=1}^{\infty} (-1)^n n^2 x^n$

4. $\sum_{n=1}^{\infty} n! x^n$

5. $\sum_{n=1}^{\infty} \frac{(-1)^n x^{2n}}{2n-1}$

6. $\sum_{n=1}^{\infty} \frac{n x^n}{5^n}$

7. $\sum_{n=0}^{\infty} (5x-3)^n$

8. $\sum_{n=1}^{\infty} \frac{(2x-1)^n}{n^4+16}$

9. $\sum_{n=1}^{\infty} \frac{2^n (x-3)^n}{n^2}$

10. $\sum_{n=1}^{\infty} \frac{n!}{n^n} x^n$ (Do not test the endpoints; it diverges at each.)

11. $\sum_{n=1}^{\infty} \frac{(2n)!}{n!} x^n$

12. $\sum_{n=1}^{\infty} \frac{1 \cdot 3 \cdot 5 \cdots (2n+1)}{n!} x^n$ (Do not test the endpoints; it diverges at each.)

13. $\sum_{n=1}^{\infty} \frac{n^3 (x+1)^n}{3^n}$

14. $\sum_{n=1}^{\infty} (-1)^n \frac{(x-2)^n}{n^2}$

15. $\sum_{n=1}^{\infty} \frac{(3-x)^n}{n^3}$

16. $\sum_{n=1}^{\infty} (-1)^n \frac{10^n}{n!} (x-10)^n$

17. $\sum_{n=1}^{\infty} \frac{n!}{2^n} (x-5)^n$

18. $\sum_{n=1}^{\infty} \frac{(-1)^n}{n \cdot 10^n} (x-2)^n$

19. $\sum_{n=0}^{\infty} x^{(2^n)}$

20. $\sum_{n=0}^{\infty} \left(\frac{x^2+1}{5}\right)^n$

In each of Problems 21–30, use power series established in this section to find a power series representation of the given function. Then determine the radius of convergence of the resulting series.

21. $f(x) = x^2 e^{-3x}$

22. $f(x) = \frac{1}{10+x}$

23. $f(x) = \sin x^2$

24. $f(x) = \cos^2 x = \frac{1}{2}(1 + \cos 2x)$

25. $f(x) = \sqrt[3]{1-x}$

26. $f(x) = (1+x^2)^{3/2}$

27. $f(x) = (1+x)^{-3}$

28. $f(x) = \frac{1}{\sqrt{9+x^3}}$

29. $f(x) = \frac{\ln(1+x)}{x}$

30. $f(x) = \frac{x - \arctan x}{x^3}$

In each of Problems 31–36, find a power series representation for the given function $f(x)$ using termwise integration.

31. $f(x) = \int_0^x \sin t^3 \, dt$

32. $f(x) = \int_0^x \frac{\sin t}{t} \, dt$

33. $f(x) = \int_0^x \exp(-t^3) \, dt$

34. $f(x) = \int_0^x \frac{\arctan t}{t} \, dt$

35. $f(x) = \int_0^x \frac{1 - \exp(-t^2)}{t^2} \, dt$

36. $\tanh^{-1} x = \int_0^x \frac{1}{1-t^2} \, dt$

37. Deduce from the arctangent series (Example 9) that

$$\pi = \frac{6}{\sqrt{3}} \sum_{n=0}^{\infty} \frac{(-1)^n}{2n+1} \left(\frac{1}{3}\right)^n.$$

Then use this alternating series to show that $\pi = 3.14$ accurate to two decimal places.

38. Substitute the Maclaurin series for $\sin x$, and then assume the validity of termwise integration of the resulting series, to derive the formula

$$\int_0^\infty e^{-t} \sin xt \, dt = \frac{x}{1+x^2} \qquad (|x| < 1).$$

Use the fact that

$$\int_0^\infty t^n e^{-t} \, dt = \Gamma(n+1) = n!$$

(Section 11.4).

39. (a) Deduce from the Maclaurin series for e^t that

$$\frac{1}{x^x} = \sum_{n=0}^{\infty} \frac{(-1)^n}{n!} (x \ln x)^n.$$

(b) Assuming the validity of termwise integration of the series in part (a), use the integral formula of Problem 31 in Section 11.4 to conclude that

$$\int_0^1 \frac{1}{x^x} \, dx = \sum_{n=1}^{\infty} \frac{1}{n^n}.$$

40. Suppose that $f(x)$ is represented by the power series

$$\sum_{n=0}^{\infty} a_n x^n$$

for all x in some open interval centered at $x = 0$. Show by repeated termwise differentiation of the series, substituting $x = 0$ each time, that $a_n = f^{(n)}(0)/n!$ for all n. Thus the only power series in x that represents a function at and near $x = 0$ is its Maclaurin series.

41. (a) Consider the binomial series

$$f(x) = \sum_{n=0}^{\infty} \frac{\alpha(\alpha-1)(\alpha-2)\cdots(\alpha-n+1)}{n!} x^n,$$

which converges (to *something*) if $|x| < 1$. Compute the derivative $f'(x)$ by termwise differentiation, and show that it satisfies the differential equation

$$(1+x)f'(x) = \alpha f(x).$$

(b) Solve the differential equation in part (a) to obtain $f(x) = C(1+x)^\alpha$ for some constant C. Finally, show that $C = 1$. Thus the binomial series converges to $(1+x)^\alpha$ if $|x| < 1$.

42. (a) Show by direct integration that

$$\int_0^1 \frac{\arcsin x}{\sqrt{1-x^2}}\,dx = \frac{\pi^2}{8}.$$

(b) Use the result of Problem 50 in Section 9.4 to show that

$$\int_0^1 \frac{x^{2n+1}}{\sqrt{1-x^2}}\,dx = \frac{2\cdot 4\cdot 6\cdots(2n)}{1\cdot 3\cdot 5\cdots(2n+1)}.$$

(c) Substitute the series of Example 10 for $\arcsin x$ in the integral of part (a), then use the integral of part (b) to integrate termwise. Conclude that

$$\int_0^1 \frac{\arcsin x}{\sqrt{1-x^2}}\,dx = 1 + \frac{1}{3^2} + \frac{1}{5^2} + \cdots.$$

(d) Note that

$$\sum_{n=1}^{\infty} \frac{1}{n^2} = \sum_{n=1}^{\infty} \frac{1}{(2n-1)^2} + \sum_{n=1}^{n} \frac{1}{(2n)^2}.$$

Use this information and parts (a) and (c) to show that

$$\sum_{n=1}^{\infty} \frac{1}{n^2} = \frac{\pi^2}{6}.$$

43. Prove that if the power series $\sum a_n x^n$ converges for some $x = x_0 \neq 0$, then it converges absolutely for all x such that $|x| < |x_0|$. *Suggestion:* Conclude from the fact that $\lim_{n\to\infty} a_n x_0^n = 0$ that $|a_n x^n| \leqq |x/x_0|^n$ for n sufficiently large. Thus the series $\sum |a_n x^n|$ is eventually dominated by the geometric series $\sum |x/x_0|^n$, which converges if $|x| < |x_0|$.

44. Suppose that the power series $\sum a_n x^n$ converges for some but not all nonzero values of x. Let S be the set of real numbers for which the series converges absolutely.

(a) Conclude from Problem 43 that the set S is bounded above.

(b) Let R be the least upper bound of the set S (see Problem 44 in Section 12.2). Then show that $\sum a_n x^n$ converges absolutely if $|x| < R$ and diverges if $|x| > R$. Explain why this proves Theorem 1 without the additional hypothesis that $\lim_{n\to\infty} |a_{n+1}/a_n|$ exists.

12.9 Power Series Computations

In Section 12.7 we discussed the use of Taylor polynomials to approximate numerical values of functions and integrals. Power series are often used for these same purposes. In the case of an *alternating* power series, it may be simpler to work with the alternating series error estimate (Theorem 2 in Section 12.6) than with a Taylor formula remainder term.

EXAMPLE 1 Use the binomial series

$$\sqrt{1+x} = 1 + \tfrac{1}{2}x - \tfrac{1}{8}x^2 + \tfrac{1}{16}x^3 - \tfrac{5}{128}x^4 + \cdots$$

to approximate $\sqrt{105}$ and estimate the accuracy.

Solution The series above is, after the first term, an alternating series. Hence

$$\begin{aligned}\sqrt{105} &= \sqrt{100+5} = 10\sqrt{1+0.05}\\ &= 10[1 + \tfrac{1}{2}(0.05) - \tfrac{1}{8}(0.05)^2 + \tfrac{1}{16}(0.05)^3 + E]\\ &\approx 10(1.02469531 + E) = 10.2469531 + 10E\end{aligned}$$

where E is negative and

$$|E| < \tfrac{5}{128}(0.05)^4 < 0.0000003.$$

Consequently,

$$10.246952 < \sqrt{105} < 10.246954,$$

so $\sqrt{105} = 10.24695$ to five decimal places.

Suppose that we had been asked in advance to approximate $\sqrt{105}$ accurate to five decimal places. A convenient way to do this is to continue writing terms of the series until it is clear that they have become too small in magnitude to affect the fifth decimal place. A good rule of thumb is to use in the computations two more decimal places than required. Thus we use seven decimal places in this case and get

$$\begin{aligned}\sqrt{105} &= 10(1 + 0.05)^{1/2}\\ &\approx 10(1 + 0.025 - 0.0003125 + 0.0000078 - 0.0000002 + \cdots)\\ &\approx 10.2469510 \approx 10.24695.\end{aligned}$$

EXAMPLE 2 Approximate

$$\int_0^1 \frac{1 - \cos x}{x^2}\,dx$$

accurate to five decimal places.

Solution We replace $\cos x$ by its Maclaurin series and get

$$\begin{aligned}\int_0^1 \frac{1 - \cos x}{x^2}\,dx &= \int_0^1 \frac{1}{x^2}\left(\frac{x^2}{2!} - \frac{x^4}{4!} + \frac{x^6}{6!} - \cdots\right)dx\\ &= \int_0^1 \left(\frac{1}{2!} - \frac{x^2}{4!} + \frac{x^4}{6!} - \frac{x^6}{8!} + \cdots\right)dx\\ &= \frac{1}{2!} - \frac{1}{4!3} + \frac{1}{6!5} - \frac{1}{8!7} + \cdots.\end{aligned}$$

Because this last series is a convergent alternating series, it follows that

$$\int_0^1 \frac{1 - \cos x}{x^2}\,dx = \frac{1}{2!} - \frac{1}{4!3} + \frac{1}{6!5} + E \approx 0.4863889 + E,$$

where E is negative and

$$|E| < \frac{1}{8!7} < 0.0000036.$$

Therefore,

$$\int_0^1 \frac{1 - \cos x}{x^2}\,dx \approx 0.48639$$

rounded to five decimal places.

EXAMPLE 3 The binomial series with $\alpha = \frac{1}{3}$ gives

$$(1 + x^2)^{1/3} = 1 + \tfrac{1}{3}x^2 - \tfrac{1}{9}x^4 + \tfrac{5}{81}x^6 - \tfrac{10}{243}x^8 + \cdots,$$

which alternates after its first term. Use the first five terms of this series to estimate the value of

$$\int_0^{1/2} \sqrt[3]{1 + x^2}\, dx.$$

Solution Termwise integration of the series above yields

$$\begin{aligned}
\int_0^{1/2} \sqrt[3]{1 + x^2}\, dx &= \int_0^{1/2} \left(1 + \tfrac{1}{3}x^2 - \tfrac{1}{9}x^4 + \tfrac{5}{81}x^6 - \tfrac{10}{243}x^8 + \cdots\right) dx \\
&= \left[x + \tfrac{1}{9}x^3 - \tfrac{1}{45}x^5 + \tfrac{5}{567}x^7 - \tfrac{10}{2187}x^9 + \cdots\right]_0^{1/2} \\
&= \tfrac{1}{2} + \tfrac{1}{9}(\tfrac{1}{2})^3 - \tfrac{1}{45}(\tfrac{1}{2})^5 + \tfrac{5}{567}(\tfrac{1}{2})^7 - \tfrac{10}{2187}(\tfrac{1}{2})^9 + \cdots.
\end{aligned}$$

Because this last series is a convergent alternating series, it follows that

$$\int_0^{1/2} \sqrt[3]{1 + x^2}\, dx = \tfrac{1}{2} + \tfrac{1}{9}(\tfrac{1}{2})^3 - \tfrac{1}{45}(\tfrac{1}{2})^5 + \tfrac{5}{567}(\tfrac{1}{2})^7 + E$$

where E is negative and

$$|E| < \tfrac{10}{2187}(\tfrac{1}{2})^9 < 0.000009.$$

Therefore,

$$0.513254 < \int_0^{1/2} \sqrt[3]{1 + x^2}\, dx < 0.513264,$$

so

$$\int_0^{1/2} \sqrt[3]{1 + x^2}\, dx = 0.51326 \pm 0.00001.$$

THE ALGEBRA OF POWER SERIES

The following theorem, which we state without proof, implies that power series may be added and multiplied much like polynomials. The guiding principle is that of collecting coefficients of like powers of x.

> ***Theorem 1*** *Adding and Multiplying Power Series*
> Let $\sum a_n x^n$ and $\sum b_n x^n$ be power series with nonzero radii of convergence. Then
>
> $$\sum_{n=0}^{\infty} a_n x^n + \sum_{n=0}^{\infty} b_n x^n = \sum_{n=0}^{\infty} (a_n + b_n) x^n \tag{1}$$
>
> and
>
> $$\begin{aligned}\left(\sum_{n=0}^{\infty} a_n x^n\right)\left(\sum_{n=0}^{\infty} b_n x^n\right) &= \sum_{n=0}^{\infty} c_n x^n \\ &= a_0 b_0 + (a_0 b_1 + a_1 b_0)x + (a_0 b_2 + a_1 b_1 + a_2 b_0)x^2 + \cdots,\end{aligned} \tag{2}$$
>
> where
>
> $$c_n = a_0 b_n + a_1 b_{n-1} + \cdots + a_{n-1} b_1 + a_n b_0. \tag{3}$$
>
> The series in (1) and (2) converge for any x that lies interior to the intervals of convergence of both $\sum a_n x^n$ and $\sum b_n x^n$.

Thus if $\sum a_n x^n$ and $\sum b_n x^n$ are power series representations of the functions $f(x)$ and $g(x)$, respectively, then the product power series $\sum c_n x^n$ found by "ordinary multiplication" and collection of terms is a power series representation of the function $f(x)g(x)$. This fact can also be used to divide one power series by another, *provided* it is known in advance that the quotient has a power series representation.

EXAMPLE 4 Assume that the tangent function has a power series representation $\tan x = \sum a_n x^n$. Use the Maclaurin series for $\sin x$ and $\cos x$ to find a_0, a_1, a_2, and a_3.

Solution We multiply series to obtain

$$\begin{aligned}\sin x &= \tan x \cos x \\ &= (a_0 + a_1 x + a_2 x^2 + a_3 x^3 + \cdots)\left(1 - \frac{x^2}{2} + \frac{x^4}{24} - \cdots\right) \\ &= a_0 + a_1 x + \left(a_2 - \frac{1}{2}a_0\right)x^2 + \left(a_3 - \frac{1}{2}a_1\right)x^3 + \cdots.\end{aligned}$$

But because

$$\sin x = x - \tfrac{1}{6}x^3 + \tfrac{1}{120}x^5 - \cdots,$$

comparison of coefficients gives the equations

$$\begin{aligned} a_0 \qquad\qquad\qquad &= 0, \\ a_1 \qquad\qquad &= 1, \\ -\tfrac{1}{2}a_0 \qquad + a_2 \qquad &= 0, \\ -\tfrac{1}{2}a_1 \qquad + a_3 &= -\tfrac{1}{6}. \end{aligned}$$

Then we find that $a_0 = 0$, $a_1 = 1$, $a_2 = 0$, and $a_3 = \frac{1}{3}$. So

$$\tan x = x + \tfrac{1}{3}x^3 + \cdots.$$

Things are not always as they first appear. The continuation of the tangent series is

$$\tan x = x + \tfrac{1}{3}x^3 + \tfrac{2}{15}x^5 + \tfrac{17}{315}x^7 + \cdots.$$

For the general form of the nth coefficient, see page 204 of K. Knopp's *Theory and Application of Infinite Series* (New York: Hafner Press, 1971). You may check that the first few terms above agree with the result of the ordinary long division indicated next.

$$\begin{array}{r} x + \tfrac{1}{3}x^3 + \tfrac{2}{15}x^5 + \cdots \\ (1 - \tfrac{1}{2}x^2 + \tfrac{1}{24}x^4 - \cdots)\overline{\big) \, x - \tfrac{1}{3}x^3 + \tfrac{1}{120}x^5 + \cdots} \end{array}$$

POWER SERIES AND INDETERMINATE FORMS

According to Theorem 1 in Section 12.8, a power series is differentiable and therefore continuous within its interval of convergence. It follows that

$$\lim_{x \to c} \sum_{n=0}^{\infty} a_n (x - c)^n = a_0. \tag{4}$$

The following two examples illustrate the use of this simple observation to find the limit of the indeterminate form $f(x)/g(x)$. The technique is first to substitute power series representations for $f(x)$ and $g(x)$.

EXAMPLE 5 Find $\displaystyle\lim_{x\to 0}\frac{\sin x - \arctan x}{x^2\ln(1+x)}$.

Solution The power series of Equations (4), (19), and (20) in Section 12.8 give

$$\begin{aligned}\sin x - \arctan x &= (x - \tfrac{1}{6}x^3 + \tfrac{1}{120}x^5 - \cdots) - (x - \tfrac{1}{3}x^3 + \tfrac{1}{5}x^5 - \cdots)\\ &= \tfrac{1}{6}x^3 - \tfrac{23}{120}x^5 + \cdots\end{aligned}$$

and

$$\begin{aligned}x^2\ln(1+x) &= x^2(x - \tfrac{1}{2}x^2 + \tfrac{1}{3}x^3 - \cdots)\\ &= x^3 - \tfrac{1}{2}x^4 + \tfrac{1}{3}x^5 - \cdots.\end{aligned}$$

Hence

$$\begin{aligned}\lim_{x\to 0}\frac{\sin x - \arctan x}{x^2\ln(1+x)} &= \lim_{x\to 0}\frac{\tfrac{1}{6}x^3 - \tfrac{23}{120}x^5 + \cdots}{x^3 - \tfrac{1}{2}x^4 + \cdots}\\ &= \lim_{x\to 0}\frac{\tfrac{1}{6} - \tfrac{23}{120}x^2 + \cdots}{1 - \tfrac{1}{2}x + \cdots} = \frac{1}{6}.\end{aligned}$$

EXAMPLE 6 Find $\displaystyle\lim_{x\to 1}\frac{\ln x}{x-1}$.

Solution We first replace x by $x - 1$ in the power series for $\ln(1+x)$ used in Example 5. This gives us

$$\ln x = (x-1) - \tfrac{1}{2}(x-1)^2 + \tfrac{1}{3}(x-1)^3 - \cdots.$$

Hence

$$\begin{aligned}\lim_{x\to 1}\frac{\ln x}{x-1} &= \lim_{x\to 1}\frac{(x-1) - \tfrac{1}{2}(x-1)^2 + \tfrac{1}{3}(x-1)^3 - \cdots}{x-1}\\ &= \lim_{x\to 1}\left[1 - \tfrac{1}{2}(x-1) + \tfrac{1}{3}(x-1)^2 - \cdots\right] = 1.\end{aligned}$$

The method of Examples 5 and 6 provides a useful alternative to l'Hôpital's rule, especially when repeated differentiation of numerator and denominator is inconvenient or too time-consuming.

12.9 PROBLEMS

In each of Problems 1–10, use an infinite series to approximate the indicated number accurate to three decimal places.

1. $\sqrt[3]{65}$

2. $(630)^{1/4}$

3. $\sin(0.5)$

4. $e^{-0.2}$

5. $\tan^{-1}(0.5)$

6. $\ln(1.1)$

7. $\sin\left(\dfrac{\pi}{10}\right)$

8. $\cos\left(\dfrac{\pi}{20}\right)$

9. $\sin 10^\circ$

10. $\cos 5^\circ$

In each of Problems 11–20, use an infinite series to approximate the value of the given integral accurate to three decimal places.

11. $\int_0^1 \frac{\sin x}{x}\,dx$

12. $\int_0^1 \frac{\sin x}{\sqrt{x}}\,dx$

13. $\int_0^{0.5} \frac{\arctan x}{x}\,dx$

14. $\int_0^1 \sin x^2\,dx$

15. $\int_0^{0.1} \frac{\ln(1+x)}{x}\,dx$

16. $\int_0^{0.5} \frac{1}{\sqrt{1+x^4}}\,dx$

17. $\int_0^{0.5} \frac{1-e^{-x}}{x}$

18. $\int_0^{0.5} \sqrt{1+x^3}\,dx$

19. $\int_0^1 \exp(-x^2)\,dx$

20. $\int_0^{0.5} \frac{1}{1+x^5}\,dx$

In each of Problems 21–26, use power series rather than l'Hôpital's rule to evaluate the given limit.

21. $\lim_{x\to 0} \frac{1+x-e^x}{x^2}$

22. $\lim_{x\to 0} \frac{x-\sin x}{x^3 \cos x}$

23. $\lim_{x\to 0} \frac{1-\cos x}{x(e^x-1)}$

24. $\lim_{x\to 0} \frac{e^x-e^{-x}-2x}{x-\arctan x}$

25. $\lim_{x\to 0} \left(\frac{1}{x}-\frac{1}{\sin x}\right)$

26. $\lim_{x\to 1} \frac{\ln(x^2)}{x-1}$

27. Derive the geometric series by long division of $1-x$ into 1.

28. Derive the series for $\tan x$ listed in Example 4 by long division of the Taylor series of $\cos x$ into the Taylor series of $\sin x$.

29. Derive the geometric series representation of $1/(1-x)$ by finding $a_0, a_1, a_2, \ldots$ such that

$$(1-x)(a_0+a_1x+a_2x^2+a_3x^3+\cdots)=1.$$

30. Derive the first five coefficients in the binomial series for $\sqrt{1+x}$ by finding a_0, a_1, a_2, a_3, and a_4 such that

$$(a_0+a_1x+a_2x^2+a_3x^3+a_4x^4+\cdots)^2=1+x.$$

31. Use the method of Example 4 to find the coefficients a_0, a_1, a_2, a_3, and a_4 in the series

$$\sec x=\frac{1}{\cos x}=\sum_{n=0}^{\infty} a_nx^n.$$

32. Multiply the geometric series for $1/(1-x)$ and the series for $\ln(1-x)$ to show that if $|x|<1$, then

$$-\frac{1}{1-x}\ln(1-x)=x+(1+\tfrac{1}{2})x^2+(1+\tfrac{1}{2}+\tfrac{1}{3})x^3$$
$$+(1+\tfrac{1}{2}+\tfrac{1}{3}+\tfrac{1}{4})x^4+\cdots.$$

33. Take the logarithmic series

$$\ln(1+x)=x-\tfrac{1}{2}x^2+\tfrac{1}{3}x^3-\tfrac{1}{4}x^4+\cdots$$

as known. Find the first four coefficients in the series for e^x by finding a_0, a_1, a_2, and a_3 such that

$$1+x=e^{\ln(1+x)}$$
$$=\sum_{n=0}^{\infty} a_n(x-\tfrac{1}{2}x^2+\tfrac{1}{3}x^3-\tfrac{1}{4}x^4+\cdots)^n.$$

This is exactly how the power series for e^x was first discovered (by Newton)!

34. Use the method of Example 4 to show that

$$\frac{x}{\sin x}=1+\frac{1}{6}x^2+\frac{7}{360}x^4+\cdots.$$

35. Show that long division of power series gives

$$\frac{2+x}{1+x+x^2}=2-x-x^2+2x^3-x^4-x^5$$
$$+2x^6-x^7-x^8+\cdots.$$

Also show that the radius of convergence of this series is $R=1$.

CHAPTER 12 REVIEW: Definitions, Concepts, Results

Use the list below as a guide to concepts that you may need to review.

1. Definition of limit of a sequence
2. The limit laws for sequences
3. The bounded monotone sequence property
4. Definition of the sum of an infinite series
5. Formula for the sum of a geometric series
6. The *n*th term test for divergence
7. Divergence of the harmonic series
8. The integral test
9. Convergence of *p*-series
10. The comparison and limit comparison tests
11. The alternating series test
12. Absolute convergence: Definition *and* the fact that it implies convergence

13. The ratio test

14. The root test

15. Power series; radius and interval of convergence

16. Taylor series of the elementary transcendental functions

17. The binomial series

18. Termwise differentiation and integration of power series

19. Use of power series to approximate values of functions and integrals

20. Product of two power series

21. Use of power series to evaluate indeterminate forms

CHAPTER 12 MISCELLANEOUS PROBLEMS

In Problems 1–15, determine whether or not the sequence $\{a_n\}$ converges, and find its limit if it does converge.

1. $a_n = \dfrac{n^2+1}{n^2+4}$

2. $a_n = \dfrac{8n-7}{7n-8}$

3. $a_n = 10 - (0.99)^n$

4. $a_n = n \sin \pi n$

5. $a_n = \dfrac{1+(-1)^n\sqrt{n}}{n+1}$

6. $a_n = \sqrt{\dfrac{1+(-0.5)^n}{n+1}}$

7. $a_n = \dfrac{\sin 2n}{n}$

8. $a_n = 2^{-(\ln n)/n}$

9. $a_n = (-1)^{\sin(n\pi/2)}$

10. $a_n = \dfrac{(\ln n)^3}{n^2}$

11. $a_n = \dfrac{1}{n}\sin\dfrac{1}{n}$

12. $a_n = \dfrac{n-e^n}{n+e^n}$

13. $a_n = \dfrac{\sinh n}{n}$

14. $a_n = \left(1+\dfrac{2}{n}\right)^{2n}$

15. $a_n = (2n^2+1)^{1/n}$

Determine whether each infinite series in Problems 16–30 converges or diverges.

16. $\displaystyle\sum_{n=1}^{\infty} \frac{(n^2)!}{n^n}$

17. $\displaystyle\sum_{n=1}^{\infty} (-1)^n \frac{\ln n}{n^2}$

18. $\displaystyle\sum_{n=0}^{\infty} \frac{3^n}{2^n+4^n}$

19. $\displaystyle\sum_{n=0}^{\infty} \frac{n!}{e^{n^2}}$

20. $\displaystyle\sum_{n=1}^{\infty} \frac{\sin(1/n)}{n^{3/2}}$

21. $\displaystyle\sum_{n=0}^{\infty} \frac{(-2)^n}{3^n+1}$

22. $\displaystyle\sum_{n=1}^{\infty} 2^{-(2/n^2)}$

23. $\displaystyle\sum_{n=2}^{\infty} \frac{(-1)^n n}{(\ln n)^3}$

24. $\displaystyle\sum_{n=1}^{\infty} \frac{(-1)^n}{10^{1/n}}$

25. $\displaystyle\sum_{n=1}^{\infty} \frac{n^{1/2}+n^{1/3}}{n^2+n^3}$

26. $\displaystyle\sum_{n=1}^{\infty} (-1)^n \frac{1}{n^{[1+(1/n)]}}$

27. $\displaystyle\sum_{n=1}^{\infty} (-1)^n \frac{\arctan n}{\sqrt{n}}$

28. $\displaystyle\sum_{n=1}^{\infty} n \sin\frac{1}{n}$

29. $\displaystyle\sum_{n=3}^{\infty} \frac{1}{n(\ln n)(\ln \ln n)}$

30. $\displaystyle\sum_{n=3}^{\infty} \frac{1}{n(\ln n)(\ln \ln n)^2}$

Find the interval of convergence of each of the power series in Problems 31–40.

31. $\displaystyle\sum_{n=0}^{\infty} \frac{2^n x^n}{n!}$

32. $\displaystyle\sum_{n=0}^{\infty} \frac{(3x)^n}{2^{n+1}}$

33. $\displaystyle\sum_{n=1}^{\infty} \frac{(x-1)^n}{n(3^n)}$

34. $\displaystyle\sum_{n=0}^{\infty} \frac{(2x-3)^n}{4^n}$

35. $\displaystyle\sum_{n=1}^{\infty} \frac{(-1)^n x^n}{4n^2-1}$

36. $\displaystyle\sum_{n=0}^{\infty} \frac{(2x-1)^n}{n^2+1}$

37. $\displaystyle\sum_{n=0}^{\infty} \frac{n!x^{2n}}{10^n}$

38. $\displaystyle\sum_{n=2}^{\infty} \frac{x^n}{\ln n}$

39. $\displaystyle\sum_{n=0}^{\infty} \frac{1+(-1)^n}{2(n!)} x^n$

40. $\displaystyle\sum_{n=1}^{\infty} \left(1+\frac{1}{n}\right)^n (x-1)^n$

Find the set of values of x for which each series in Problems 41–43 converges.

41. $\displaystyle\sum_{n=1}^{\infty} (x-n)^n$

42. $\displaystyle\sum_{n=1}^{\infty} (\ln x)^n$

43. $\displaystyle\sum_{n=0}^{\infty} \frac{e^{nx}}{n!}$

44. Find the rational number that has repeated decimal expansion 2.7 1828 1828 1828 . . . .

45. Give an example of two convergent numerical series $\sum a_n$ and $\sum b_n$ such that the series $\sum a_n b_n$ diverges.

46. Prove that if $\sum a_n$ is a convergent positive-term series then $\sum a_n^2$ converges.

47. Let the sequence $\{a_n\}$ be defined recursively by

$$a_1 = 1, \qquad a_{n+1} = 1 + \frac{1}{1+a_n} \quad \text{if} \quad n \geqq 1.$$

The limit of the sequence $\{a_n\}$ is the value of the *continued fraction*

$$1 + \cfrac{1}{2 + \cfrac{1}{2 + \cfrac{1}{2 + \cdots}}}.$$

Assuming that $A = \lim_{n\to\infty} a_n$ exists, prove that $A = \sqrt{2}$.

48. Let $\{F_n\}_1^\infty$ be the Fibonacci sequence of Example 2 in Section 12.2.

(a) Prove that $0 < F_n \leqq 2^n$ for all $n \geqq 1$, and hence conclude that the power series

$$F(x) = \sum_{n=1}^{\infty} F_n x^n$$

converges if $|x| < \frac{1}{2}$.

(b) Show that $(1 - x - x^2)F(x) = x$, so that $F(x) = x/(1 - x - x^2)$.

49. We say that the *infinite product* indicated by

$$\prod_{n=1}^{\infty} (1 + a_n) = (1 + a_1)(1 + a_2)(1 + a_3) \cdots$$

converges provided that the infinite series

$$S = \sum_{n=1}^{\infty} \ln(1 + a_n)$$

converges, in which case the value of the infinite product is e^S. Use the integral test to prove that

$$\prod_{n=1}^{\infty} \left(1 + \frac{1}{n}\right)$$

diverges.

50. (See Problem 49.) Prove that the infinite product

$$\prod_{n=1}^{\infty} \left(1 + \frac{1}{n^2}\right)$$

converges, and use the integral test remainder estimate to approximate its value. The actual value of this infinite product is known to be $(\sinh \pi)/\pi = 3.67608$.

In each of Problems 51–55, use infinite series to approximate the indicated number accurate to three decimal places.

51. $(1.5)^{1/5}$

52. $\ln(1.2)$

53. $\int_0^1 \exp(-x^2)\,dx$

54. $\int_0^{0.5} \sqrt[3]{1 + x^4}\,dx$

55. $\displaystyle\int_0^{0.5} \frac{1 - e^{-x}}{x}\,dx$

56. Substitute the Maclaurin series for $\sin x$ into that for e^x to obtain

$$e^{\sin x} = 1 + x + \tfrac{1}{2}x^2 - \tfrac{1}{8}x^4 + \cdots.$$

57. Substitute the Maclaurin series for the cosine and then integrate termwise to derive the formula

$$\int_0^{\infty} e^{-t^2} \cos 2xt\,dt = \frac{\sqrt{\pi}}{2} e^{-x^2}.$$

Use the reduction formula

$$\int_0^{\infty} t^{2n} e^{-t^2}\,dt = \frac{2n - 1}{2} \int_0^{\infty} t^{2n-2} e^{-t^2}\,dt$$

derived in Miscellaneous Problem 41 of Chapter 11. It should be noted that the validity of this improper termwise integration is subject to verification.

58. Prove that

$$\tanh^{-1} x = \int_0^x \frac{1}{1 - t^2}\,dt = \sum_{n=0}^{\infty} \frac{x^{2n+1}}{2n + 1}$$

if $|x| < 1$.

59. Prove that

$$\sinh^{-1} x = \int_0^x \frac{1}{(1 + t^2)^{1/2}}\,dt$$

$$= \sum_{n=0}^{\infty} (-1)^n \frac{1 \cdot 3 \cdot 5 \cdots (2n - 1)}{2 \cdot 4 \cdot 6 \cdots (2n)} \cdot \frac{x^{2n+1}}{2n + 1}$$

if $|x| < 1$.

60. Suppose that $\tan y = \sum a_n y^n$. Determine a_0, a_1, a_2, and a_3 by substituting the inverse tangent series (Equation (14) in Section 12.7) into the equation

$$x = \tan(\tan^{-1} x) = \sum a_n (\tan^{-1} x)^n.$$

61. According to *Stirling's series*, the value of $n!$ for n large is given to a close approximation by

$$n! = \sqrt{2\pi n}\left(\frac{n}{e}\right)^n e^{\mu(n)}$$

where

$$\mu(n) = \frac{1}{12n} - \frac{1}{360n^3} + \frac{1}{1260n^5}.$$

Substitute $\mu(n)$ into Maclaurin's series for e^x to show that

$$e^{\mu(n)} = 1 + \frac{1}{12n} + \frac{1}{288n^2} - \frac{139}{51{,}840n^3} + \cdots$$

Can you show that the next term in the last series is $-571/(2{,}488{,}320n^4)$?

62. Define

$$T(n) = \int_0^{\pi/4} \tan^n x\,dx$$

for $n \geqq 0$.

(a) Show by "reduction" of the integral that

$$T(n + 2) = \frac{1}{n + 1} - T(n)$$

for $n \geqq 0$.

(b) Conclude that $T(n) \to 0$ as $n \to +\infty$.

(c) Show that $T_0 = \pi/4$ and that $T_1 = \frac{1}{2}\ln 2$.

(d) Prove by induction on n that

$$T(2n) = (-1)^{n+1}\left(1 - \frac{1}{3} + \frac{1}{5} - \cdots \pm \frac{1}{2n-1} - \frac{\pi}{4}\right).$$

(e) Conclude from (b) and (d) that

$$1 - \frac{1}{3} + \frac{1}{5} - \cdots = \frac{\pi}{4}.$$

(f) Prove by induction on n that

$$T(2n+1) = \frac{1}{2}(-1)^n\left(1 - \frac{1}{2} + \frac{1}{3} - \cdots \pm \frac{1}{n} - \ln 2\right).$$

(g) Conclude from (b) and (f) that

$$1 - \tfrac{1}{2} + \tfrac{1}{3} - \tfrac{1}{4} + \cdots = \ln 2.$$

chapter thirteen

Parametric Curves and Vectors in the Plane

13.1 Parametric Curves

Until now we have encountered *curves* mainly as graphs of equations. An equation of the form $y = f(x)$ or one of the form $x = g(y)$ determines a curve by giving one of the coordinate variables explicitly as a function of the other. An equation of the form $F(x, y) = 0$ may also determine a curve, but here each variable is given implicitly as a function of the other.

Another important sort of curve is the trajectory of a point moving in the plane. The motion of the point can be described by giving its position $(x(t), y(t))$ at time t. Such a description involves expressing both the rectangular coordinate variables x and y as functions of a third variable, or **parameter,** t, rather than as functions of one another. In this context a *parameter* is an independent variable (not a constant, as is sometimes meant in popular usage). This approach motivates the following definition.

> ***Definition*** *Parametric Curve*
> A **parametric curve** C in the plane is a pair of functions
> $$x = f(t), \qquad y = g(t), \tag{1}$$
> that give x and y as continuous functions of the real number t (the parameter) in some interval I.

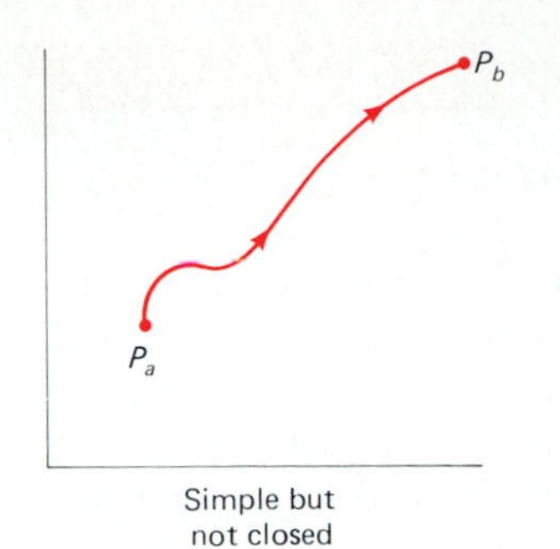

Simple but not closed

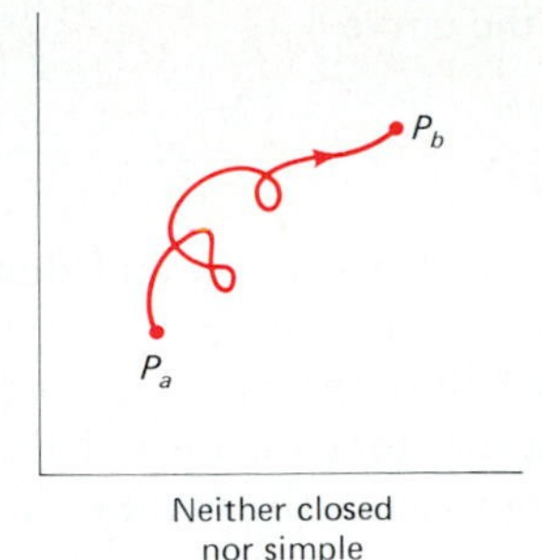

Neither closed nor simple

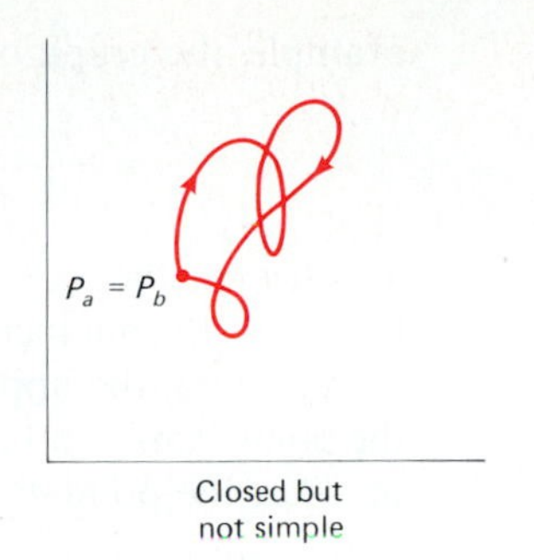

Closed but not simple

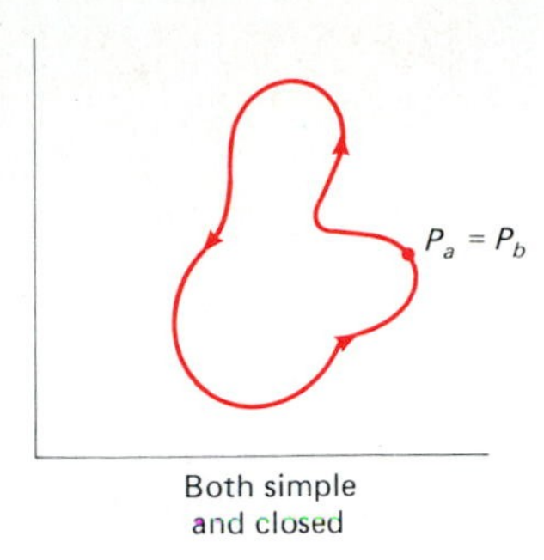

Both simple and closed

13.1 Parametric curves may be simple or not, closed or not.

Each value of the parameter t gives a point $(f(t), g(t))$, and the set of all such points is the **graph** of the curve C. Often the distinction between the curve—the pair of **coordinate functions** f and g—and the graph is not made. Therefore we shall sometimes refer interchangeably to the curve and to its graph. The two equations in (1) are called the **parametric equations** of the curve.

In most cases the interval I will be a closed interval of the form $[a, b]$, and in such cases the two points $P_a(f(a), g(a))$ and $P_b(f(b), g(b))$ are called the **end points** of the curve. If these two points coincide, then we say that the curve C is **closed.** If no distinct pair of values of t, except possibly for the values $t = a$ and $t = b$, give rise to the same point on the graph of the curve, then the curve C is non-self-intersecting, and we say that the curve is **simple.** These concepts are illustrated in Fig. 13.1.

The graph of a parametric curve may be sketched by plotting enough points to indicate its likely shape. Sometimes one can eliminate the parameter t and thus obtain an equation in x and y. This equation may give us more information about the shape of the curve.

EXAMPLE 1 Determine the graph of the curve

$$x = \cos t, \qquad y = \sin t, \qquad 0 \leqq t \leqq 2\pi.$$

Solution Figure 13.2 shows a table of values of x and y that correspond to multiples of $\pi/4$ for the parameter t. These values give the eight points highlighted in Fig. 13.2, all of which lie on the unit circle. This suggests that the graph is, in fact, the unit circle. To verify this, we note that the fundamental identity of trigonometry gives

$$x^2 + y^2 = \cos^2 t + \sin^2 t = 1,$$

so every point of the graph lies on the unit circle with equation $x^2 + y^2 = 1$. Conversely, the point of the circle with angular (polar) coordinate t is the point $(\cos t, \sin t)$ of the graph. Thus the graph is precisely the unit circle.

t	x	y
0	1	0
$\pi/4$	$1/\sqrt{2}$	$1/\sqrt{2}$
$\pi/2$	0	1
$3\pi/4$	$-1/\sqrt{2}$	$1/\sqrt{2}$
π	-1	0
$5\pi/4$	$-1/\sqrt{2}$	$-1/\sqrt{2}$
$3\pi/2$	0	-1
$7\pi/4$	$1/\sqrt{2}$	$-1/\sqrt{2}$
2π	1	0

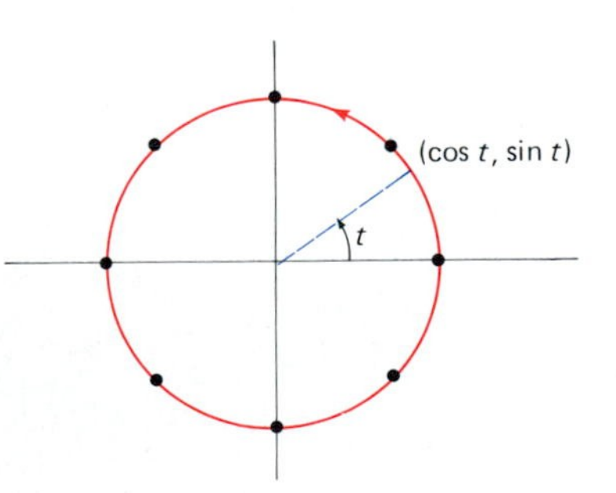

13.2 Table of values for Example 1 and the graph

What is lost in the process in Example 1 is the information about how the graph is produced as t goes from 0 to 2π. But this is easy to determine by inspection. As t increases from 0 to 2π, the point $(\cos t, \sin t)$ begins at $(1, 0)$ and travels counterclockwise around the circle, ending at $(1, 0)$ when $t = 2\pi$.

A given figure in the plane may be the graph of different curves. Speaking more loosely, a given curve may have different **parametrizations.** For example, the graph of the curve

$$x = \frac{1 - t^2}{1 + t^2}, \qquad y = \frac{2t}{1 + t^2}, \qquad -\infty < t < +\infty$$

also lies on the unit circle, because we find that $x^2 + y^2 = 1$ here, too. If t begins at 0 and increases, then the point $P(x(t), y(t))$ begins at (1, 0) and travels along the upper half of the circle. If t begins at 0 and decreases, then the point $P(x(t), t(t))$ travels along the lower half. As t approaches either $+\infty$ or $-\infty$, the point P approaches the point $(-1, 0)$. Thus the graph consists of the unit circle with the single point $(-1, 0)$ deleted. A slight modification of the curve of Example 1,

$$x = \cos t, \qquad y = \sin t, \qquad -\pi < t < \pi,$$

is a different parametrization of this graph.

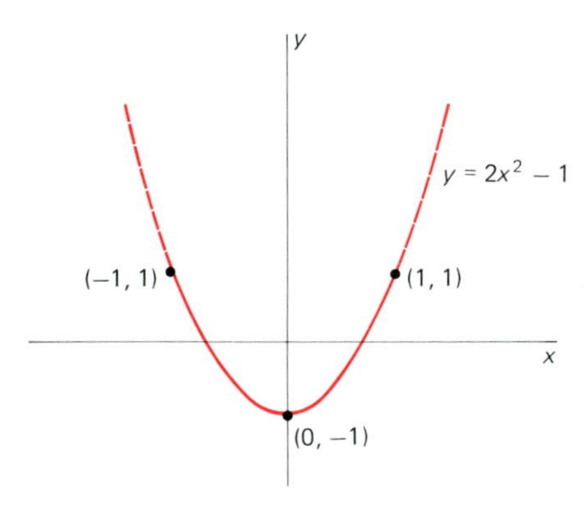

13.3 The curve of Example 2 is part of a parabola.

EXAMPLE 2 Eliminate the parameter to determine the graph of the parametric curve

$$x = t - 1, \qquad y = 2t^2 - 4t + 1, \qquad 0 \leqq t \leqq 2.$$

Solution We substitute $t = x + 1$ (from the equation for x) into the equation for y. This yields

$$y = 2(x + 1)^2 - 4(x + 1) + 1 = 2x^2 - 1$$

for $-1 \leqq x \leqq 1$. Thus the graph of the given curve is a portion of the parabola $y = 2x^2 - 1$, as shown in Fig. 13.3. As t increases from 0 to 2, the point $(t - 1, 2t^2 - 4t + 1)$ travels along the parabola from $(-1, 1)$ to $(1, 1)$.

The same parabolic arc can be reparametrized with

$$x = \sin t, \qquad y = 2\sin^2 t - 1.$$

Now, as t increases, the point $(\sin t, 2\sin^2 t - 1)$ travels back and forth along the parabola between the two points $(-1, 1)$ and $(1, 1)$, rather like the bob of a pendulum.

Examples 1 and 2 involve parametric curves in which we can eliminate the parameter and thus obtain an explicit equation $y = f(x)$. Conversely, any explicitly presented curve $y = f(x)$ can be viewed as a parametric curve by writing

$$x = t, \qquad y = f(t),$$

with t running through the values in the original domain of f. For example, the straight line

$$y - y_0 = m(x - x_0)$$

with slope m and passing through the point (x_0, y_0) may be parametrized by

$$x = t, \qquad y = y_0 + m(t - x_0).$$

An even simpler parametrization is obtained with $x - x_0 = t$. This gives

$$x = x_0 + t, \qquad y = y_0 + mt.$$

The use of parametric equations $x = x(t)$ and $y = y(t)$ is most advantageous when elimination of the parameter is either impossible or would lead to an equation $y = f(x)$ considerably more complicated than the original parametric equations. This often happens when the curve is a geometric locus or the path of a point moving under specified conditions.

EXAMPLE 3 The curve traced by a point P on the edge of a rolling circle is called a **cycloid.** The circle is to roll along a straight line without slipping or stopping. You will see a cycloid if you watch a patch of bright paint on the tire of a bicycle crossing your path from left to right. Find parametric equations for the cycloid if the line along which the circle rolls is the x-axis, the circle is above the x-axis but always tangent to it, and the point P begins at the origin.

Solution Evidently, the cycloid consists of a series of arches. As parameter t we take the angle (in radians) through which the circle has turned since it began with P at the origin. This is the angle TCP in Fig. 13.4.

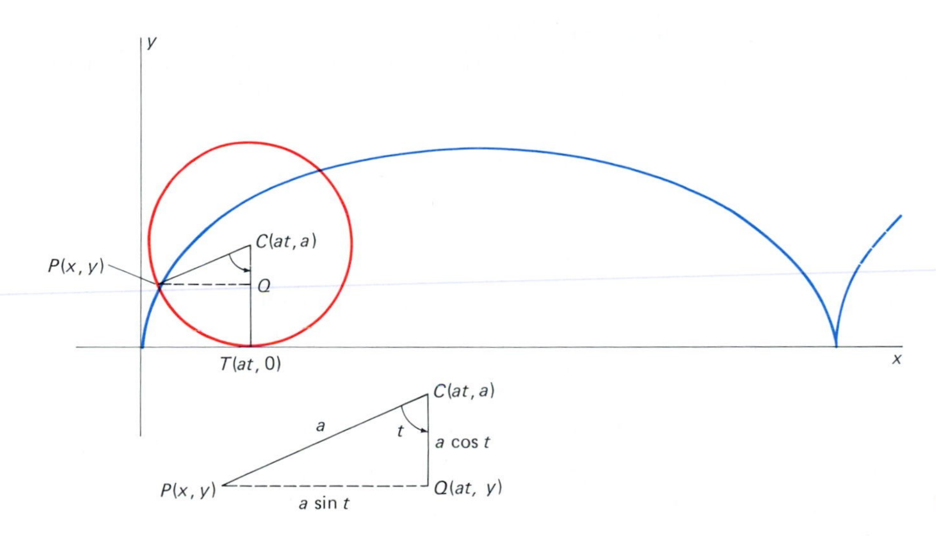

13.4 The cycloid

The distance the circle has rolled is $|OT|$, so this is also the length of the circumference subtended by the angle TCP. Thus $|OT| = at$ if a is the radius of the circle, so the center C of the rolling circle has coordinates (at, a) at time t. The right triangle CPQ of Fig. 13.4 provides us with the relations

$$at - x = a \sin t \quad \text{and} \quad a - y = a \cos t.$$

Therefore, the parametric equations of the cycloid, the path of the moving point P, are

$$x = a(t - \sin t), \qquad y = a(1 - \cos t). \tag{2}$$

In Example 2 of Section 9.8 (Equations (3) and (4) of that section), we obtained the parametrization

$$x = c(\theta - \sin\theta\cos\theta), \qquad y = c\sin^2\theta \tag{3}$$

of the brachistochrone—the curve of quickest descent from one point to a lower point. But the substitutions $t = 2\theta$, $a = c/2$, together with the trigonometric identities $\sin 2\theta = 2 \sin \theta \cos \theta$ and $\sin^2 \theta = \frac{1}{2}(1 - \cos 2\theta)$, transform the equations in (3) into those in (2). In June 1696, John Bernoulli proposed the brachistochrone problem as a public challenge, with a six-month deadline (later extended to Easter 1697 at Leibniz's request). Newton received Bernoulli's challenge on January 29, 1697 (new style), probably after a hard day at the office, and the next day communicated his solution—the curve of minimal descent time is an arc of an inverted cycloid—to the Royal Society of London.

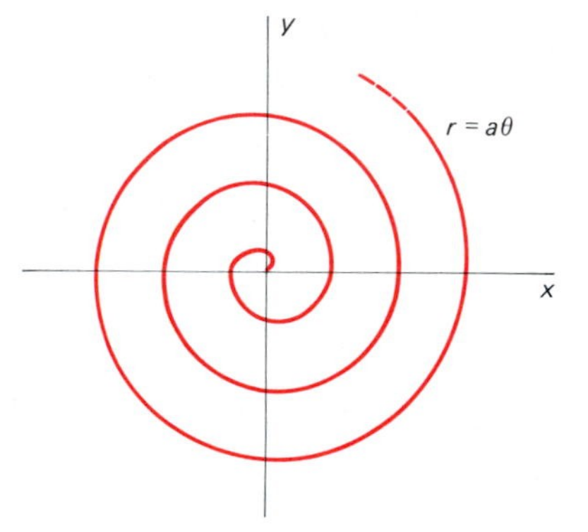

13.5 The spiral of Archimedes

A curve given in polar coordinates by the equation $r = f(\theta)$ can be regarded as a parametric curve with parameter θ. To see this, we recall that the equations $x = r \cos \theta$ and $y = r \sin \theta$ allow us to change from polar to rectangular coordinates. We replace r with $f(\theta)$, and this gives the parametric equations

$$x = f(\theta) \cos \theta, \qquad y = f(\theta) \sin \theta, \tag{4}$$

expressing x and y in terms of the parameter θ.

For example, the spiral of Archimedes shown in Fig. 13.5 has the polar coordinates equation $r = a\theta$. The equations in (4) give the spiral the parametrization

$$x = a\theta \cos \theta, \qquad y = a\theta \sin \theta.$$

TANGENT LINES TO PARAMETRIC CURVES

The parametric curve $x = f(t)$, $y = g(t)$ is called **smooth** if the derivatives $f'(t)$ and $g'(t)$ are continuous and never simultaneously zero. In some neighborhood of each point of its graph, a smooth parametric curve can be described in one or possibly both of the forms $y = F(x)$ and $x = G(y)$. To see why this is so, suppose (for example) that $f'(t) > 0$ on the interval I. Then $f(t)$ is an increasing function on t, and therefore has a inverse function $t = \phi(x)$ there. Substitution of $t = \phi(x)$ into the equation $y = g(t)$ then gives

$$y = g(\phi(x)) = F(x).$$

We can use the chain rule to compute the slope dy/dx of the tangent line to a smooth parametric curve. Differentiation of $y = F(x)$ with respect to t yields

$$\frac{dy}{dt} = \frac{dy}{dx} \cdot \frac{dx}{dt},$$

so

$$\frac{dy}{dx} = \frac{dy/dt}{dx/dt} = \frac{g'(t)}{f'(t)} \tag{5}$$

at any point where $f'(t) \neq 0$. The tangent line is vertical at a point where $f'(t) = 0$ but $g'(t) \neq 0$.

Equation (5) gives $y' = dy/dx$ as a function of t. Another differentiation with respect to t, again using the chain rule, results in the formulas

$$\frac{dy'}{dt} = \frac{dy'}{dx} \cdot \frac{dx}{dt},$$

so

$$\frac{d^2y}{dx^2} = \frac{dy'}{dx} = \frac{dy'/dt}{dx/dt}. \tag{6}$$

EXAMPLE 4 Calculate dy/dx and d^2y/dx^2 for the cycloid with the parametric equations in (2).

Solution We begin with

$$x = a(t - \sin t), \qquad y = a(1 - \cos t). \tag{2}$$

Then Equation (5) gives

$$\frac{dy}{dx} = \frac{dy/dt}{dx/dt} = \frac{a \sin t}{a(1 - \cos t)} = \frac{\sin t}{1 - \cos t}. \tag{7}$$

This derivative is zero when t is an odd multiple of π, so the tangent line is horizontal at the midpoint of each arch of the cycloid. The end points of the arches correspond to even multiples of π, where both the numerator and the denominator in (7) vanish. These are isolated points (called *cusps*) at which the cycloid fails to be a smooth curve.

Next, Equation (6) yields

$$\frac{d^2y}{dx^2} = \frac{(\cos t)(1 - \cos t) - (\sin t)(\sin t)}{(1 - \cos t)^2 \cdot a(1 - \cos t)} = -\frac{1}{a(1 - \cos t)^2}.$$

Because $d^2y/dx^2 < 0$ for all t (except for the isolated even multiples of π), this shows that each arch of the cycloid is concave downward, as shown in Fig. 13.4.

The slope dy/dx can also be computed in terms of polar coordinates. Given a polar coordinates curve $r = f(\theta)$, we use the parametrization

$$x = f(\theta) \cos \theta, \qquad y = f(\theta) \sin \theta$$

shown in (4). Then Equation (5), with θ in place of t, gives

$$\frac{dy}{dx} = \frac{dy/d\theta}{dx/d\theta} = \frac{f'(\theta) \sin \theta + f(\theta) \cos \theta}{f'(\theta) \cos \theta - f(\theta) \sin \theta} \tag{8}$$

or, alternatively, denoting $f'(\theta) = dr/d\theta$ by r',

$$\frac{dy}{dx} = \frac{r' \sin \theta + r \cos \theta}{r' \cos \theta - r \sin \theta}. \tag{9}$$

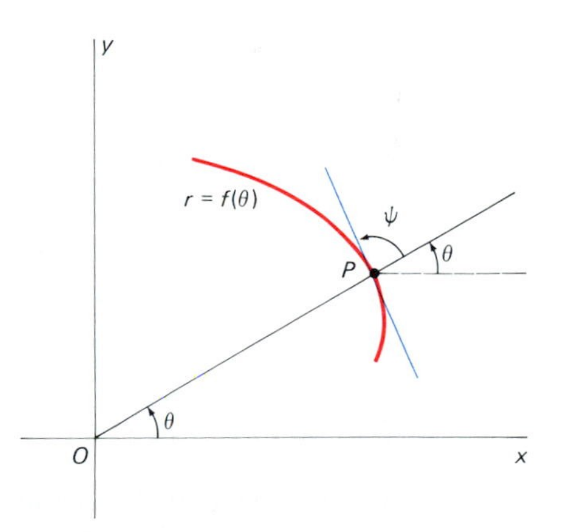

13.6 The interpretation of the angle ψ (see the formula in (10))

The formula in (9) has the following useful consequence. Let ψ denote the angle between the tangent line at P and the radius OP (extended) from the origin, as in Fig. 13.6. Then

$$\cot \psi = \frac{1}{r} \frac{dr}{d\theta} \qquad (0 \leqq \psi \leqq \pi). \tag{10}$$

In Problem 26 we indicate how Equation (10) can be derived from Equation (9).

EXAMPLE 5 Consider the *logarithmic spiral* with polar equation $r = e^{\theta}$. Show that $\psi = \pi/4$ at every point of the spiral, and write the equation of its tangent line at the point $(e^{\pi/2}, \pi/2)$.

Solution Because $dr/d\theta = e^\theta$, Equation (10) tells us that $\cot\psi = e^\theta/e^\theta = 1$. Thus $\psi = \pi/4$. When $\theta = \pi/2$, Equation (9) gives

$$\frac{dy}{dx} = \frac{e^{\pi/2}\sin(\pi/2) + e^{\pi/2}\cos(\pi/2)}{e^{\pi/2}\cos(\pi/2) - e^{\pi/2}\sin(\pi/2)} = -1.$$

But when $\theta = \pi/2$, we have $x = 0$ and $y = e^{\pi/2}$. Thus it follows that the equation of the desired tangent line is

$$y - e^{\pi/2} = -x; \qquad \text{that is,} \quad x + y = e^{\pi/2}.$$

The line and the spiral appear in Fig. 13.7.

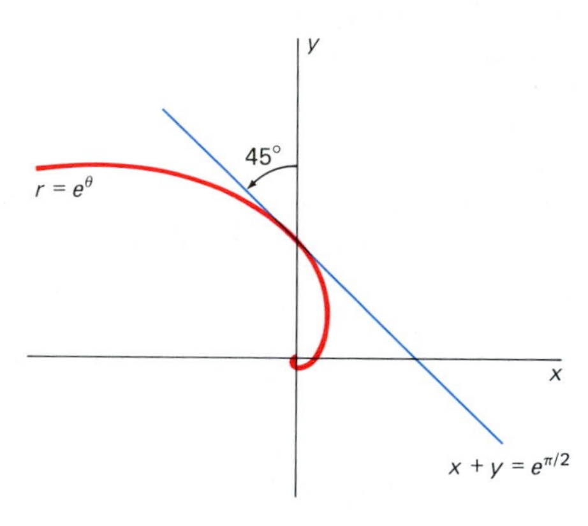

13.7 The angle ψ is always 45° for the logarithmic spiral.

13.1 PROBLEMS

In Problems 1–12, eliminate the parameter and then sketch the curve.

1. $x = t + 1, \quad y = 2t - 1$

2. $x = t^2 + 1, \quad y = 2t^2 - 1$

3. $x = t^2, \quad y = t^3$

4. $x = \sqrt{t}, \quad y = 3t - 2$

5. $x = t + 1, \quad y = 2t^2 - t - 1$

6. $x = t^2 + 3t, \quad y = t - 2$

7. $x = e^t, \quad y = 4e^{2t}$

8. $x = 2e^t, \quad y = 2e^{-t}$

9. $x = 5\cos t, \quad y = 3\sin t$

10. $x = 2\cosh t, \quad y = 3\sinh t$

11. $x = \sec t, \quad y = \tan t$

12. $x = \cos 2t, \quad y = \sin t$

Do *two* things in each of Problems 13–17:

(a) First write the equation of the line tangent to the given parametric curve at the point corresponding to the given value of t.

(b) Then calculate d^2y/dx^2 in order to determine whether the curve is concave upward or concave downward at this point.

13. $x = 2t^2 + 1, \quad y = 3t^3 + 2; \quad t = 1$

14. $x = \cos^3 t, \quad y = \sin^3 t; \quad t = \pi/4$

15. $x = t\sin t, \quad y = t\cos t; \quad t = \pi/2$

16. $x = e^t, \quad y = e^{-t}; \quad t = 0$

17. $x = \dfrac{3t}{1 + t^3}, \quad y = \dfrac{3t^2}{1 + t^3}; \quad t = 1$

In each of Problems 18–21, find the angle ψ between the radius OP and the tangent line at the point P that corresponds to the given value of θ.

18. $r = 1/\theta, \quad \theta = 1$

19. $r = \exp(\theta\sqrt{3}), \quad \theta = \pi/2$

20. $r = 1 - \cos\theta, \quad \theta = \pi/3$

21. $r = \sin 3\theta, \quad \theta = \pi/6$

22. Eliminate t to determine the graph of the parametric curve $x = t^2$, $y = t^3 - 3t$, $-2 \leqq t \leqq 2$. Note that $t = \sqrt{x}$ if $t > 0$, while $t = -\sqrt{x}$ if $t < 0$. Your sketch should contain a loop and should be symmetric about the x-axis.

23. The curve C is determined by the parametric equations $x = e^{-t}$, $y = e^{2t}$. Calculate dy/dx and d^2y/dx^2 directly from these parametric equations. Conclude that C is concave upward at every point. Then sketch C.

24. The graph of the folium of Descartes with rectangular equation $x^3 + y^3 = 3xy$ appears in Fig. 13.8. Parametrize its loop as follows: Let P be the point of intersection of

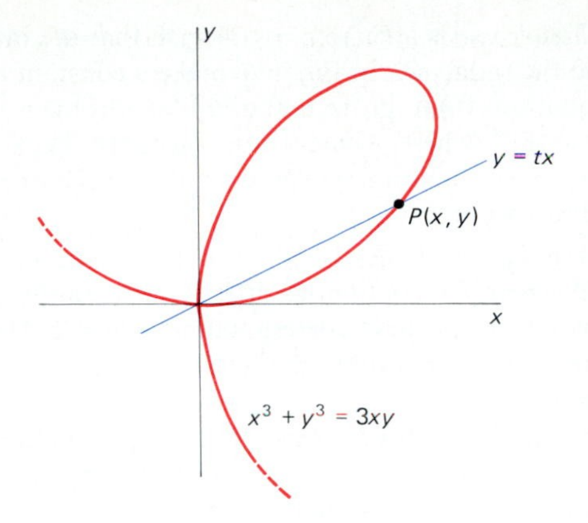

13.8 The loop of the folium of Descartes (Problem 24)

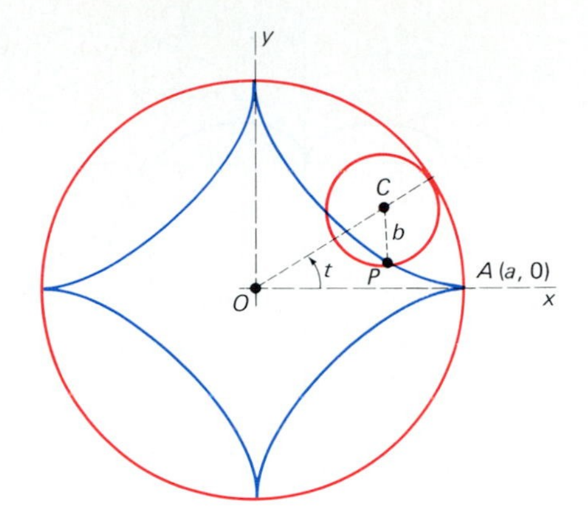

13.9 The hypocycloid of Problem 28

the line $y = tx$ with the loop; then solve for the coordinates x and y of P in terms of t.

25. Parametrize the parabola $y^2 = 4px$ by expressing x and y as functions of the slope m of the tangent line at the point $P(x, y)$ of the parabola.

26. Let P be a point of the curve with polar equation $r = f(\theta)$ and let ψ be the angle between the extended radius OP and the tangent line at P. Let α be the angle of inclination of this tangent line, measured counterclockwise from the horizontal. Then $\psi = \alpha - \theta$. Verify the formula in (10) by substituting $\tan \alpha = dy/dx$ from the formula in (9) and $\tan \theta = y/x = (\sin \theta)/(\cos \theta)$ in the identity

$$\cot \psi = \frac{1}{\tan(\alpha - \theta)} = \frac{1 + \tan \alpha \tan \theta}{\tan \alpha - \tan \theta}.$$

27. Let P_0 be the highest point of the circle of Fig. 13.4—the circle that generates the cycloid of Example 3. Show that the line through P_0 and the point P of the cycloid (the point P is also shown in Fig. 13.4) is tangent to the cycloid at P. This fact gives a geometric construction of the tangent line to the cycloid.

28. A circle of radius b rolls without slipping inside a circle of radius $a > b$. The path of a point P fixed on the circumference of the rolling circle is called a *hypocycloid*. Let P begin its journey at $A(a, 0)$, shown in Fig. 13.9, and let t be the angle AOC, where O is the center of the large circle and C the center of the rolling circle. Show that the coordinates of P are given by the parametric equations

$$x = (a - b) \cos t + b \cos\left(\frac{a - b}{b} t\right),$$

$$y = (a - b) \sin t - b \sin\left(\frac{a - b}{b} t\right).$$

29. If $b = a/4$ in Problem 28, show that the parametric equations of the hypocycloid reduce to

$$x = a \cos^3 t, \qquad y = a \sin^3 t.$$

30. (a) Prove that the hypocycloid of Problem 29 is the graph of the equation $x^{2/3} + y^{2/3} = 1$.

(b) Find all points of this hypocycloid where its tangent line is either horizontal or vertical, and the intervals on which it is concave upward and those on which it is concave downward.

(c) Sketch this hypocycloid.

31. Consider a point P on the spiral of Archimedes, the curve shown in Fig. 13.10 with polar equation $r = a\theta$. Archimedes viewed the path of P as compounded of two motions, one with speed a directly away from the origin O and another a circular motion with unit angular speed around O. This suggests Archimedes' result that the line PQ in the figure is tangent to the spiral at P. Prove that this is indeed true.

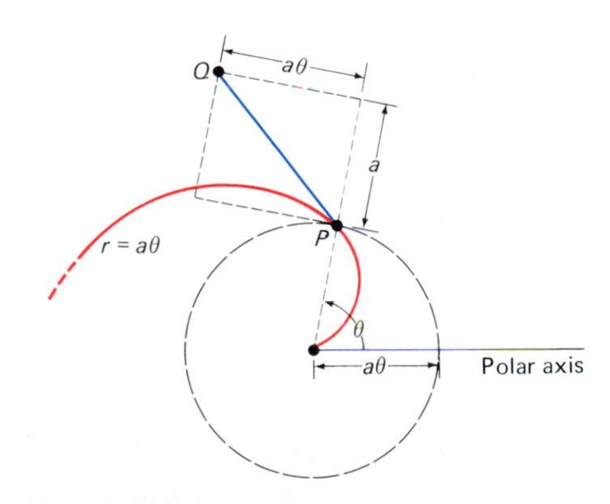

13.10 The segment PQ is tangent to the spiral (a result of Archimedes).

32. (a) Deduce from Equation (7) that if t is not a multiple of 2π, then the slope of the tangent line at the corresponding point of the cycloid is $\cot(t/2)$.

(b) Conclude that at the cusp point of the cycloid where t is a multiple of 2π, the cycloid has a vertical tangent line.

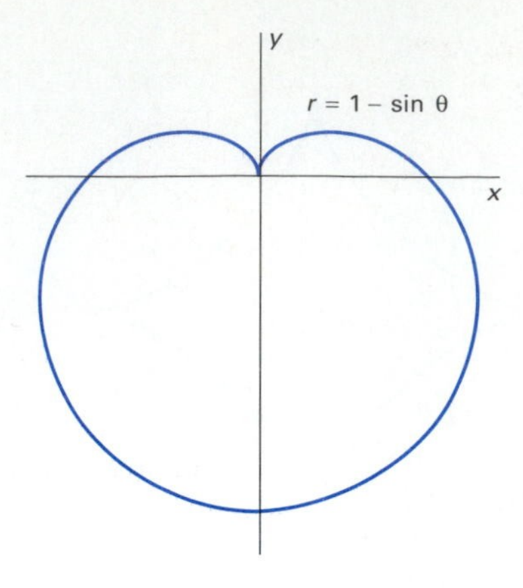

13.11 The cardioid of Problem 34

33. A *loxodrome* is a curve $r = f(\theta)$ such that the tangent line and the radius OP in Fig. 13.6 make a constant angle. Use Equation (10) to prove that every loxodrome is of the form $r = Ae^{k\theta}$, where A and k are constants. Thus every loxodrome is a logarithmic spiral similar to the one considered in Example 5.

34. Let a curve be described in polar coordinates by $r = f(\theta)$, where f is continuous. If $f(\alpha) = 0$, then the origin is the point of the curve corresponding to $\theta = \alpha$. Deduce from the parametrization $x = f(\theta)\cos\theta$, $y = f(\theta)\sin\theta$ that the tangent line to the curve at this point makes the angle α with the positive x-axis. For example, the *cardioid* $r = f(\theta) = 1 - \sin\theta$ shown in Fig. 13.11 is tangent to the y-axis at the origin. And, indeed, $f(\pi/2) = 0$: The y-axis is the line $\theta = \alpha = \pi/2$.

13.2 Integral Computations with Parametric Curves

In Chapter 6 we discussed the computation of a variety of geometric quantities associated with the graph $y = f(x)$ of a nonnegative function on the interval $[a, b]$. These included:

Area under the curve:
$$A = \int_a^b y\,dx. \tag{1}$$

Volume of revolution around the x-axis:
$$V_x = \int_a^b \pi y^2\,dx. \tag{2a}$$

Volume of revolution around the y-axis:
$$V_y = \int_a^b 2\pi xy\,dx. \tag{2b}$$

Arc length of the curve:
$$s = \int_0^s ds = \int_a^b [1 + (y')^2]^{1/2}\,dx. \tag{3}$$

Area of surface of revolution about x-axis:
$$A_x = \int_a^b 2\pi y\,ds. \tag{4a}$$

Area of surface of revolution about y-axis:
$$A_y = \int_a^b 2\pi x\,ds. \tag{4b}$$

Of course, we substitute $y = f(x)$ in each of these integrals before integrating from $x = a$ to $x = b$.

We now want to compute these same quantities for a smooth parametric curve

$$x = f(t), \qquad y = g(t), \qquad \alpha \leqq t \leqq \beta.$$

To ensure that this parametric curve looks like the graph of a function defined on some interval $[a, b]$ of the x-axis, we assume that $f(t)$ *is either an increasing or a decreasing function of t on* $[\alpha, \beta]$, *and that* $g(t) \geqq 0$. If $a = f(\alpha)$

and $b = f(\beta)$, then the curve will be traced from left to right as t increases, while if $a = f(\beta)$ and $b = f(\alpha)$, it is traced from right to left. (See Fig. 13.12.)

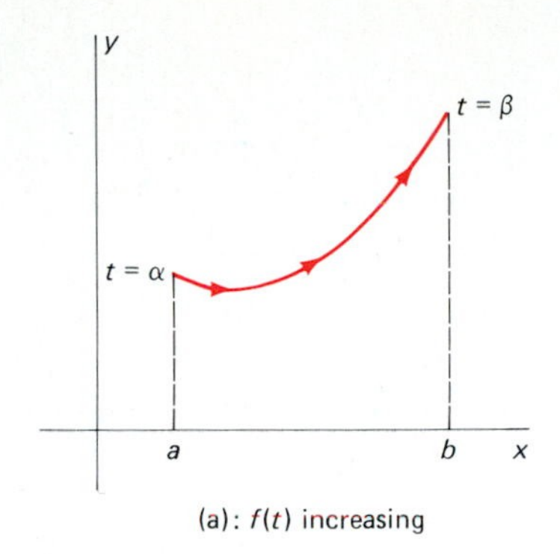

(a): $f(t)$ increasing

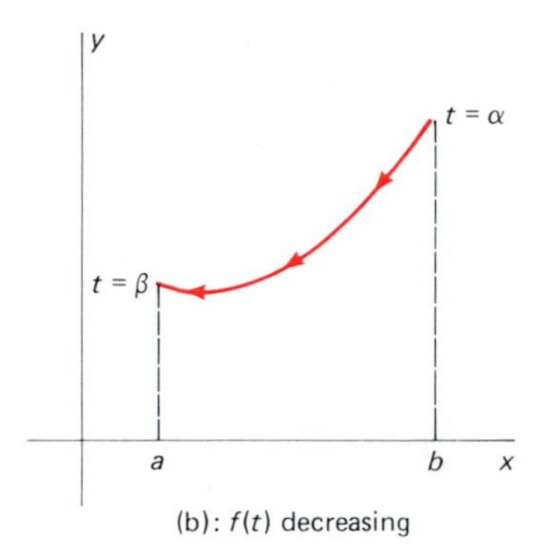

(b): $f(t)$ decreasing

13.12 Tracing a parametrized curve.

Any one of the quantities A, V_x, V_y, s, A_x, and A_y may then be computed by making the substitutions

$$x = f(t), \qquad y = g(t),$$
$$dx = f'(t)\,dt, \qquad dy = g'(t)\,dt, \quad \text{and} \tag{5}$$
$$ds = \sqrt{[f'(t)]^2 + [g'(t)]^2}\,dt$$

in the appropriate one of the integral formulas in (1) through (4). In the case of the formulas in (1) and (2), which involve dx, we then integrate from $t = \alpha$ to $t = \beta$ in the case shown in part (a) of Fig. 13.12, and from $t = \beta$ to $t = \alpha$ in the case shown in part (b). Thus the proper choice of limits on t corresponds to traversing the curve *from left to right*. For example,

$$A = \int_\alpha^\beta g(t)f'(t)\,dt \qquad \text{if} \quad f(\alpha) < f(\beta),$$

while

$$A = \int_\beta^\alpha g(t)f'(t)\,dt \qquad \text{if} \quad f(\beta) < f(\alpha).$$

The validity of this method of evaluating the integrals in (1) and (2) follows from the theorem on integration by substitution in Section 5.6.

In the case of the formulas in (3) and (4), which involve ds, we integrate from $t = \alpha$ to $t = \beta$ in either case. To see why this is so, recall from Equation (5) in Section 13.1 that $dy/dx = g'(t)/f'(t)$ if $f'(t) \neq 0$ on $[\alpha, \beta]$. Hence

$$s = \int_a^b \sqrt{1 + \left(\frac{dy}{dx}\right)^2}\,dx = \int_{f^{-1}(a)}^{f^{-1}(b)} \sqrt{1 + \left[\frac{g'(t)}{f'(t)}\right]^2}\,f'(t)\,dt.$$

Because $f'(t) > 0$ if $f(\alpha) = a$ and $f(\beta) = b$, while $f'(t) < 0$ if $f(\alpha) = b$ and $f(\beta) = a$, it follows that

$$s = \int_\alpha^\beta \sqrt{1 + \left[\frac{g'(t)}{f'(t)}\right]^2}\,|f'(t)|\,dt.$$

so

$$s = \int_\alpha^\beta \sqrt{[f'(t)]^2 + [g'(t)]^2}\,dt = \int_\alpha^\beta \sqrt{\left(\frac{dx}{dt}\right)^2 + \left(\frac{dy}{dt}\right)^2}\,dt. \tag{6}$$

This formula, derived under the assumption that $f'(t) \neq 0$ on $[\alpha, \beta]$, may be taken as the *definition* of arc length for an arbitrary smooth parametric curve. Similarly, the area of a surface of revolution is defined for smooth parametric curves as the result of first making the substitutions of (5) in the formula in (4a) or (4b), and then integrating from $t = \alpha$ to $t = \beta$.

The infinitesimal triangle of Fig. 13.13 serves as a convenient device for remembering the substitution

$$ds = \sqrt{[f'(t)]^2 + [g'(t)]^2}\,dt.$$

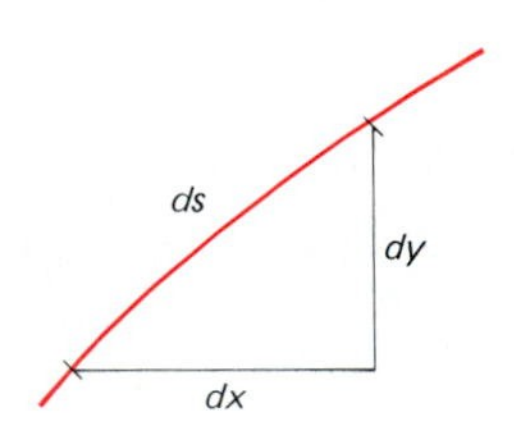

13.13 Nearly a right triangle for dx and dy close to zero

When the Pythagorean theorem is applied to this right triangle, it leads to the symbolic manipulation

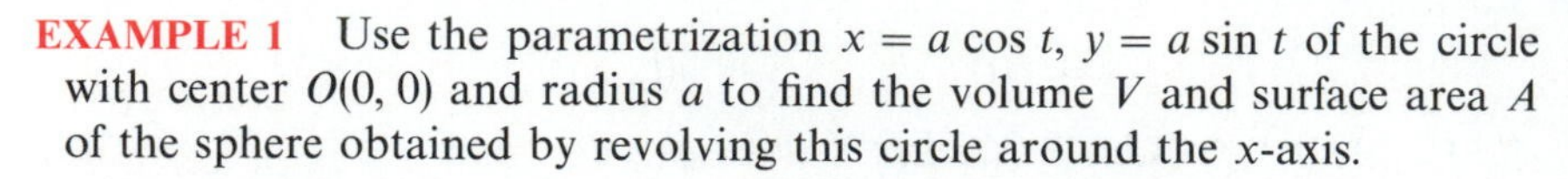

$$ds = \sqrt{(dx)^2 + (dy)^2} = \sqrt{\left(\frac{dx}{dt}\right)^2 + \left(\frac{dy}{dt}\right)^2}\, dt = \sqrt{[f'(t)]^2 + [g'(t)]^2}\, dt. \quad (7)$$

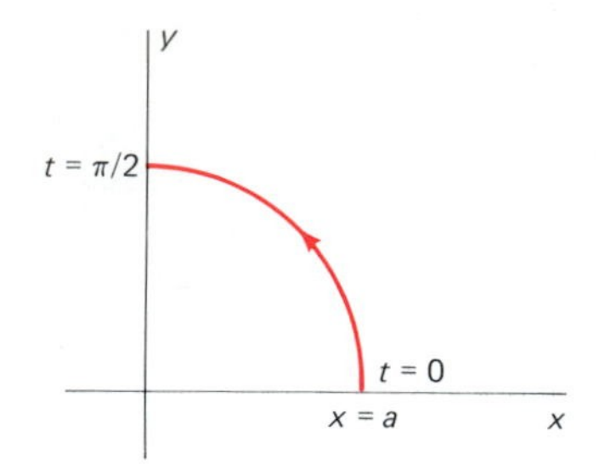

13.14 The quarter circle of Example 1

EXAMPLE 1 Use the parametrization $x = a \cos t$, $y = a \sin t$ of the circle with center $O(0, 0)$ and radius a to find the volume V and surface area A of the sphere obtained by revolving this circle around the x-axis.

Solution Half of the sphere is obtained by revolving the first quadrant of the circle, which is shown in Fig. 13.14. The left-to-right direction along the curve is from $t = \pi/2$ to $t = 0$, so the formula in (2a) gives

$$V = 2\int_{\pi/2}^{0} \pi(a \sin t)^2(-a \sin t\, dt) = 2\pi a^3 \int_0^{\pi/2} (1 - \cos^2 t) \sin t\, dt$$

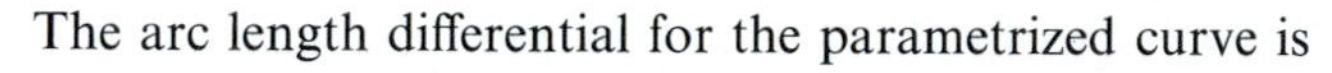

$$= 2\pi a^3\Big[-\cos t + \tfrac{1}{3}\cos^3 t\Big]_0^{\pi/2} = \tfrac{4}{3}\pi a^3.$$

The arc length differential for the parametrized curve is

$$ds = \sqrt{(-a \sin t)^2 + (a \cos t)^2}\, dt = a\, dt.$$

Hence the formula in (4a) gives

$$A = 2\int_0^{\pi/2} 2\pi(a \sin t)(a\, dt) = 4\pi a^2 \int_0^{\pi/2} \sin t\, dt = 4\pi a^2\Big[-\cos t\Big]_0^{\pi/2} = 4\pi a^2.$$

Of course, the results of Example 1 are familiar. By contrast, the following example requires the methods of this section.

EXAMPLE 2 Find the area under and the arc length of the cycloidal arch

$$x = a(t - \sin t), \qquad y = a(1 - \cos t), \qquad 0 \leqq t \leqq 2\pi.$$

Solution We use the identity $1 - \cos t = 2 \sin^2(t/2)$ and the result of Problem 50 in Section 9.4,

$$\int_0^{\pi} \sin^{2n} u\, du = \pi \cdot \frac{1}{2} \cdot \frac{3}{4} \cdot \frac{5}{6} \cdots \frac{2n-1}{2n}.$$

Because $dx = a(1 - \cos t)\, dt$ and the left-to-right direction along the curve is from $t = 0$ to $t = 2\pi$, the formula for area in (1) gives

$$A = \int_0^{2\pi} a(1 - \cos t) \cdot a(1 - \cos t)\, dt = a^2 \int_0^{2\pi} (1 - \cos t)^2\, dt$$

$$= 4a^2 \int_0^{2\pi} \sin^4 \frac{t}{2}\, dt = 8a^2 \int_0^{\pi} \sin^4 u\, du \qquad \left(u = \frac{t}{2}\right)$$

$$= 8a^2 \cdot \pi \cdot \frac{1}{2} \cdot \frac{3}{4} = 3\pi a^2$$

for the area under one arch of the cycloid. The arc length differential is

$$ds = \sqrt{a^2(1 - \cos t)^2 + (a \sin t)^2}\, dt = a\sqrt{2(1 - \cos t)}\, dt$$

$$= 2a \sin\left(\frac{t}{2}\right) dt,$$

so the formula in (3) gives

$$s = \int_0^{2\pi} 2a \sin\frac{t}{2}\, dt = \left[-4a \cos\frac{t}{2}\right]_0^{2\pi} = 8a$$

for the length of one arch of the cycloid.

PARAMETRIC POLAR COORDINATES

Suppose that a parametric curve is determined by giving its polar coordinates

$$r = r(t), \qquad \theta = \theta(t), \qquad \alpha \leqq t \leqq \beta$$

as functions of the parameter t. Then this curve is described in rectangular coordinates by means of the parametric equations

$$x(t) = r(t)\cos\theta(t), \qquad y(t) = r(t)\sin\theta(t), \qquad \alpha \leqq t \leqq \beta,$$

giving x and y as functions of t. The latter parametric equations may then be used in the integral formulas in Equations (1) through (4).

To compute ds, we first calculate the derivatives

$$\frac{dx}{dt} = (\cos\theta)\frac{dr}{dt} - (r\sin\theta)\frac{d\theta}{dt},$$

$$\frac{dy}{dt} = (\sin\theta)\frac{dr}{dt} + (r\cos\theta)\frac{d\theta}{dt}.$$

Upon substituting these expressions for dx/dt and dy/dt in (7) and making algebraic simplifications, we find that the arc length differential in parametric polar coordinates is

$$ds = \sqrt{\left(\frac{dr}{dt}\right)^2 + \left(r\frac{d\theta}{dt}\right)^2}\, dt. \tag{8}$$

In the case of a curve with explicit polar coordinate equation $r = f(\theta)$, we may use θ itself as the parameter. Then (8) takes the simpler form

$$ds = \sqrt{\left(\frac{dr}{d\theta}\right)^2 + r^2}\, d\theta. \tag{9}$$

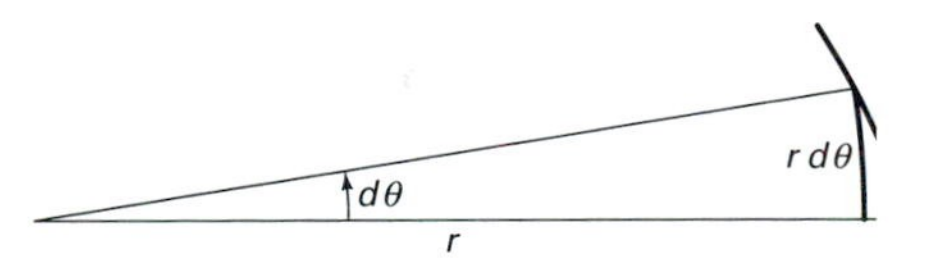

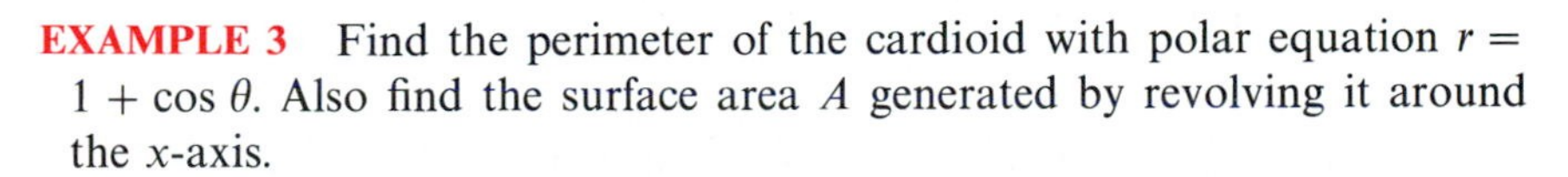

EXAMPLE 3 Find the perimeter of the cardioid with polar equation $r = 1 + \cos\theta$. Also find the surface area A generated by revolving it around the x-axis.

Solution This cardioid is shown in Fig. 13.15. The formula in (9) gives

$$ds = \sqrt{(-\sin\theta)^2 + (1 + \cos\theta)^2}\, d\theta = \sqrt{2(1 + \cos\theta)}\, d\theta$$

$$= \sqrt{4\cos^2\left(\frac{\theta}{2}\right)}\, d\theta.$$

Hence $ds = 2\cos(\theta/2)\, d\theta$ on the upper half of the cardioid, where $0 \leqq \theta \leqq \pi$ and $\cos(\theta/2) \geqq 0$. Therefore,

$$s = 2\int_0^{\pi} 2\cos\frac{\theta}{2}\, d\theta = 8\left[\sin\frac{\theta}{2}\right]_0^{\pi} = 8.$$

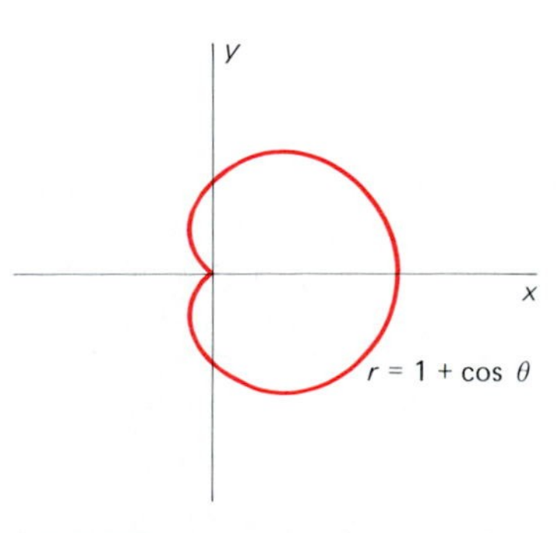

13.15 The cardioid of Example 3

The surface area of revolution is

$$A = \int_{*}^{**} 2\pi y\, ds = \int_{\theta=0}^{\pi} 2\pi(r \sin \theta)\, ds$$

$$= \int_{0}^{\pi} 2\pi(1 + \cos \theta)(\sin \theta) \cdot 2 \cos \frac{\theta}{2}\, d\theta$$

$$= 16\pi \int_{0}^{\pi} \cos^4 \frac{\theta}{2} \sin \frac{\theta}{2}\, d\theta = 16\pi \left[-\frac{2}{5} \cos^5 \frac{\theta}{2} \right]_0^{\pi} = \frac{32\pi}{5}.$$

13.2 PROBLEMS

In each of Problems 1–6, find the area of the region that lies between the given parametric curve and the x-axis.

1. $x = t^3, \quad y = 2t^2 + 1; \quad -1 \leqq t \leqq 1$
2. $x = e^{3t}, \quad y = e^{-t}; \quad 0 \leqq t \leqq \ln 2$
3. $x = \cos t, \quad y = \sin^2 t; \quad 0 \leqq t \leqq \pi$
4. $x = 2 - 3t, \quad y = e^{2t}; \quad 0 \leqq t \leqq 1$
5. $x = \cos t, \quad y = e^t; \quad 0 \leqq t \leqq \pi$
6. $x = 1 - e^t, \quad y = 2t + 1; \quad 0 \leqq t \leqq 1$

In Problems 7–10, find the volume obtained by revolving around the x-axis the region described in the given problem.

7. Problem 1
8. Problem 2
9. Problem 3
10. Problem 5

In Problems 11–16, find the arc length of the given curve.

11. $x = 2t, \quad y = \frac{2}{3}t^{3/2}; \quad 5 \leqq t \leqq 12$
12. $x = \frac{1}{2}t^2, \quad y = \frac{1}{3}t^3; \quad 0 \leqq t \leqq 1$
13. $x = \sin t - \cos t, \quad y = \sin t + \cos t; \quad \frac{\pi}{4} \leqq t \leqq \frac{\pi}{2}$
14. $x = e^t \sin t, \quad y = e^t \cos t; \quad 0 \leqq t \leqq \pi$
15. $r = e^{\theta/2}; \quad 0 \leqq \theta \leqq 4\pi$
16. $r = \theta; \quad 2\pi \leqq \theta \leqq 4\pi$

In Problems 17–22, find the area of the surface of revolution generated by revolving the given curve around the indicated axis.

17. $x = 1 - t, \quad y = 2t^{1/2}, \quad 1 \leqq t \leqq 4;$ the x-axis
18. $x = 2t^2 + \frac{1}{t}, \quad y = 8t^{1/2}, \quad 1 \leqq t \leqq 2;$ the x-axis
19. $x = t^3, \quad y = 2t + 3, \quad -1 \leqq t \leqq 1;$ the y-axis
20. $x = 2t + 1, \quad y = t^2 + t, \quad 0 \leqq t \leqq 3;$ the y-axis
21. $r = 4 \sin \theta;$ the x-axis
22. $r = e^{\theta}, \quad 0 \leqq \theta \leqq \frac{\pi}{2};$ the y-axis
23. Find the volume generated by revolving the region under the cycloidal arch of Example 2 around the x-axis.
24. Find the area of the surface generated by revolving the cycloidal arch of Example 2 around the x-axis.
25. Use the parametrization $x = a \cos t$, $y = b \sin t$ to find
(a) the area bounded by the ellipse $x^2/a^2 + y^2/b^2 = 1$;
(b) the volume of the ellipsoid generated by revolving this ellipse around the x-axis.
26. Find the area bounded by the loop of the parametric curve $x = t^2$, $y = t^3 - 3t$ of Problem 22 in Section 13.1.
27. Use the parametrization $x = t \cos t$, $y = t \sin t$ of the Archimedean spiral to find the arc length of the first full turn of this spiral (corresponding to $0 \leqq t \leqq 2\pi$).
28. The circle $(x - b)^2 + y^2 = a^2$ with radius $a < b$ and center $(b, 0)$ can be parametrized by

$$x = b + a \cos t, \qquad y = a \sin t, \qquad 0 \leqq t \leqq 2\pi.$$

Find the surface area of the torus obtained by revolving this circle around the y-axis.

29. The *astroid* (or four-cusped hypocycloid) with equation $x^{2/3} + y^{2/3} = 1$ arose in connection with Problems 28–30 of Section 13.1 and is shown in Fig. 13.9 at the end of that section. It has the parametrization

$$x = a \cos^3 t, \qquad y = a \sin^3 t, \qquad 0 \leqq t \leqq 2\pi.$$

Find the area of the region bounded by the astroid.

30. Find the total length of the astroid of Problem 29.
31. Find the area of the surface obtained by revolving the astroid of Problem 29 around the x-axis.
32. Find the area of the surface generated by revolving the lemniscate $r^2 = 2a^2 \cos 2\theta$ around the y-axis. (*Suggestion:* Use the formula in (9) and note that $r\, dr = -2a^2 \sin 2\theta\, d\theta$.)
33. Figure 13.16 shows the graph of the parametric curve

$$x = t^2\sqrt{3}, \qquad y = 3t - \tfrac{1}{3}t^3;$$

the shaded region is bounded by the part of the curve for which $-3 \leqq t \leqq 3$. Find its area.

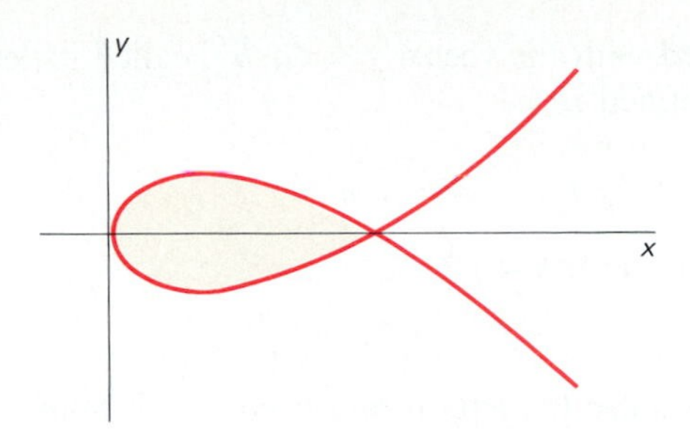

13.16 The parametric curve of Problems 33–36

34. Find the arc length of the loop in Problem 33.

35. Find the volume of the solid obtained by revolving the shaded region in Fig. 13.16 around the x-axis.

36. Find the surface area of revolution generated by revolving the loop of Fig. 13.16 around the x-axis.

37. (a) With reference to Problem 24 and Fig. 13.8 in Section 13.1, show that the arc length of the first quadrant loop of the folium of Descartes is

$$s = 6\int_0^1 \frac{(1 + 4t^2 - 4t^3 - 4t^5 + 4t^6 + t^8)^{1/2}}{(1 + t^3)^2}\,dt.$$

(b) If you have a programmable calculator or computer, apply Simpson's approximation to obtain $s \approx 4.9175$.

13.3 Vectors in the Plane

A physical quantity such as length, temperature, or mass can be specified in terms of a single real number, its magnitude. Such a quantity is called a **scalar.** Other physical entities, such as force and velocity, possess both magnitude and direction; these entities are called **vector quantities,** or simply **vectors.**

For example, to specify the velocity of a moving point in the plane, we must give both the rate at which it moves (its speed) and the direction of that motion. The combination is the **velocity vector** of the moving point. It is convenient to represent this velocity vector by an arrow, located at the current position of the moving point on its trajectory, as shown in Fig. 13.17.

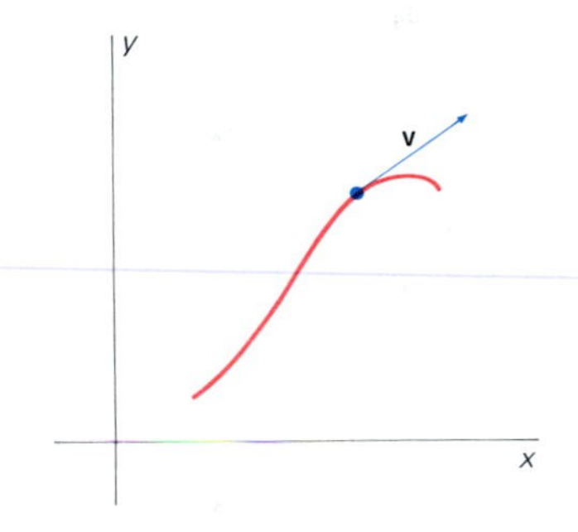

13.17 A velocity vector may be represented by an arrow.

Although the arrow or directed line segment carries the desired information—both magnitude (its length) and direction—it is a pictorial object rather than a mathematical object. We shall see that the following formal definition of a vector captures the essence of magnitude in combination with direction.

> ***Definition of Vector***
>
> A **vector** $\mathbf{v}$ in the Cartesian plane is an ordered pair of real numbers, of the form $\langle a, b\rangle$. We write $\mathbf{v} = \langle a, b\rangle$ and call a and b the **components** of the vector $\mathbf{v}$.

The directed line segment $\overrightarrow{OP}$ from the origin O to the point $P(a, b)$ is one geometric representation of the vector $\mathbf{v}$. For this reason, the vector $\mathbf{v} = \langle a, b\rangle$ is called the **position vector** of the point $P(a, b)$. In fact, the relationship between $\mathbf{v} = \langle a, b\rangle$ and $P(a, b)$ is so close that, in certain contexts, it is convenient to confuse the two deliberately—to regard $\mathbf{v}$ and P as the same mathematical object. In other contexts the distinction between the two is important—in particular, *any* directed line segment with the same direction and magnitude (length) as $\overrightarrow{OP}$ is a representation of $\mathbf{v}$, as Fig. 13.18 suggests. What is important about the vector $\mathbf{v}$ is not usually *where* it is, but how long it is and which way it points.

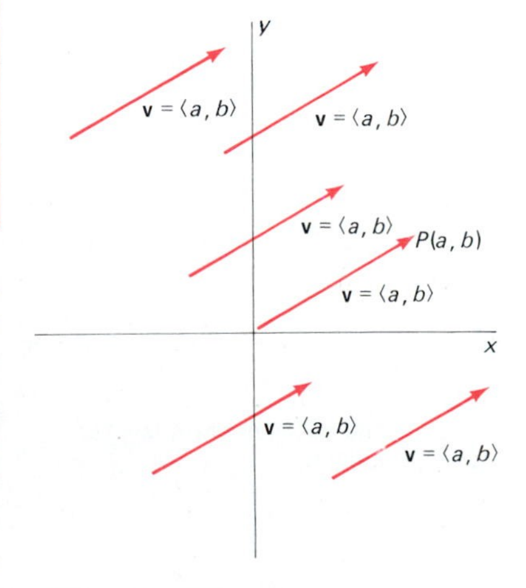

13.18 All these arrows represent the same vector $\mathbf{v} = \langle a, b\rangle$.

The magnitude associated with the vector $\mathbf{v} = \langle a, b \rangle$, called its **length**, is denoted by $v = |\mathbf{v}|$ and is defined to be

$$v = |\mathbf{v}| = |\langle a, b \rangle| = \sqrt{a^2 + b^2}. \tag{1}$$

For example, the length of the vector $\mathbf{v} = \langle 1, -2 \rangle$ is

$$|\mathbf{v}| = |\langle 1, -2 \rangle| = \sqrt{(1)^2 + (-2)^2} = \sqrt{5}.$$

The notation $v = |\mathbf{v}|$ is used because the length of a vector is in many ways analogous to the absolute value of a real number.

The only vector with length zero is the **zero vector** with both components zero, denoted by $\mathbf{0} = \langle 0, 0 \rangle$. The zero vector is also unique in having no specific direction.

The operations of addition and multiplication of real numbers have analogues for vectors. We shall define each of these operations of "vector algebra" in terms of components of vectors and then give a geometric interpretation in terms of arrows.

Definition *Addition of Vectors*

The **sum** $\mathbf{u} + \mathbf{v}$ of the two vectors $\mathbf{u} = \langle u_1, u_2 \rangle$ and $\mathbf{v} = \langle v_1, v_2 \rangle$ is the vector

$$\mathbf{u} + \mathbf{v} = \langle u_1 + v_1, u_2 + v_2 \rangle. \tag{2}$$

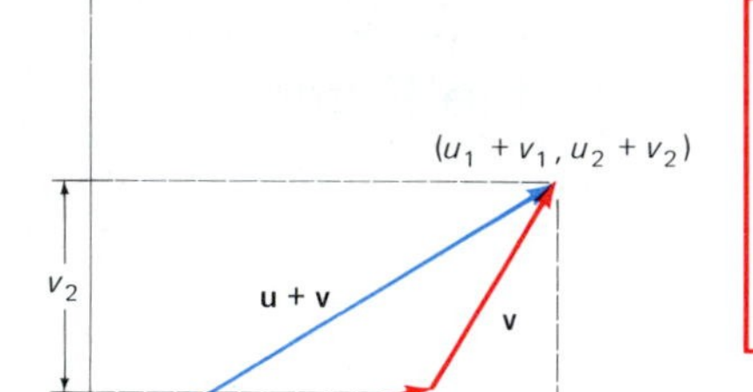

13.19 The triangle law is a geometric interpretation of vector addition.

For example, the sum of the vectors $\mathbf{u} = \langle 4, 3 \rangle$ and $\mathbf{v} = \langle -5, 2 \rangle$ is

$$\langle 4, 3 \rangle + \langle -5, 2 \rangle = \langle 4 + (-5), 3 + 2 \rangle = \langle -1, 5 \rangle.$$

Thus we add vectors by adding corresponding components—that is, by *componentwise addition.* The geometric interpretation of vector addition is the **triangle law of addition,** illustrated in Fig. 13.19, where the labeled lengths indicate why this interpretation is valid. An equivalent interpretation is the **parallelogram law of addition** illustrated in Fig. 13.20.

It is natural to write $2\mathbf{u} = \mathbf{u} + \mathbf{u}$. But if $\mathbf{u} = \langle u_1, u_2 \rangle$, then

$$2\mathbf{u} = \mathbf{u} + \mathbf{u} = \langle u_1, u_2 \rangle + \langle u_1, u_2 \rangle = \langle 2u_1, 2u_2 \rangle.$$

This suggests that multiplication of a vector by a scalar (real number) also be defined in a componentwise manner.

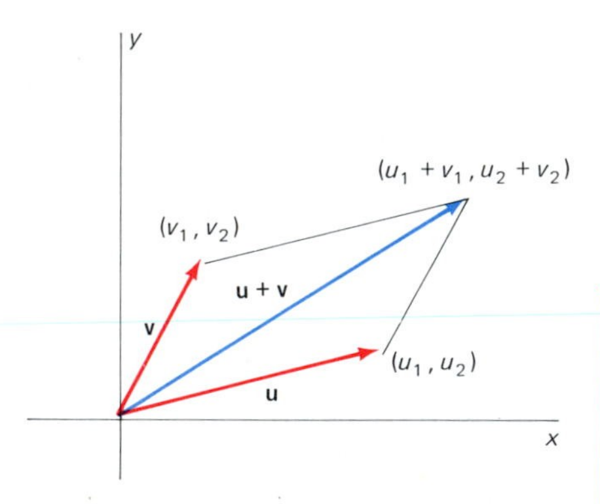

13.20 The parallelogram law for vector addition

Definition *Multiplication of a Vector by a Scalar*

If $\mathbf{u} = \langle u_1, u_2 \rangle$ and c is a real number, then the **scalar multiple** $c\mathbf{u}$ is the vector

$$c\mathbf{u} = \langle cu_1, cu_2 \rangle. \tag{3}$$

Note that

$$|c\mathbf{u}| = \sqrt{(cu_1)^2 + (cu_2)^2} = |c|\sqrt{(u_1)^2 + (u_2)^2} = |c| \cdot |\mathbf{u}|.$$

Thus the length of $c\mathbf{u}$ is $|c|$ times that of $\mathbf{u}$.

The negative of the vector **u** is the vector

$$-\mathbf{u} = (-1)\mathbf{u} = \langle -u_1, -u_2 \rangle,$$

with the same length as **u** but the opposite direction. We say that the two nonzero vectors **u** and **v** have

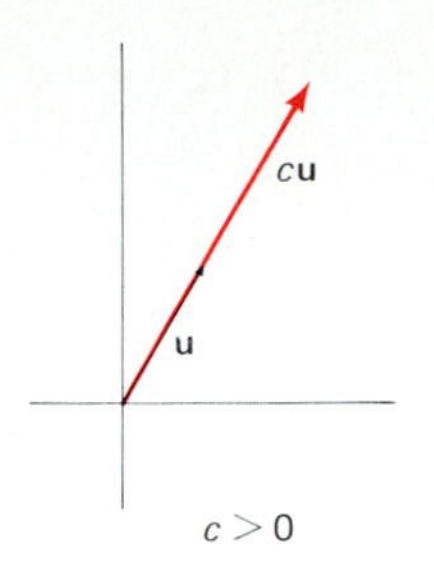

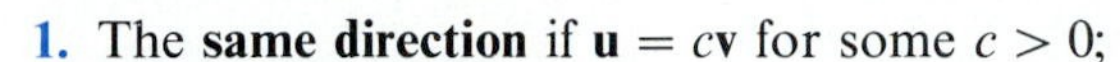

1. The **same direction** if $\mathbf{u} = c\mathbf{v}$ for some $c > 0$;
2. **Opposite directions** if $\mathbf{u} = c\mathbf{v}$ for some $c < 0$.

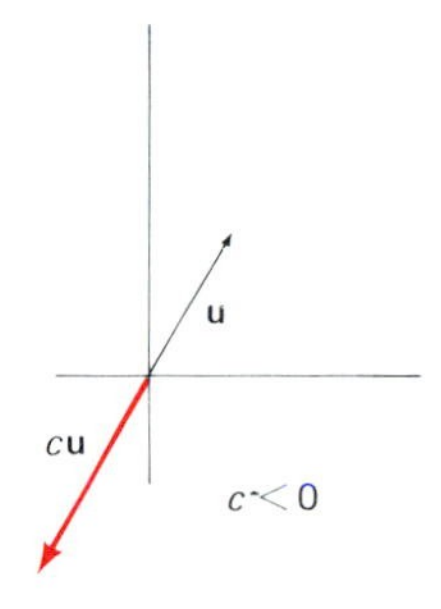

13.21 The vector $c\mathbf{u}$ may have the same direction as **u** or the opposite direction.

The geometric interpretation of scalar multiplication is that $c\mathbf{u}$ is the vector with length $|c| \cdot |\mathbf{u}|$, with the same direction as **u** if $c > 0$ but the opposite direction if $c < 0$. See Fig. 13.21.

The **difference** $\mathbf{u} - \mathbf{v}$ of the vectors $\mathbf{u} = \langle u_1, u_2 \rangle$ and $\mathbf{v} = \langle v_1, v_2 \rangle$ is defined to be

$$\mathbf{u} - \mathbf{v} = \mathbf{u} + (-\mathbf{v}) = \langle u_1 - v_1, u_2 - v_2 \rangle. \tag{4}$$

If we think of $\langle u_1, u_2 \rangle$ and $\langle v_1, v_2 \rangle$ as position vectors of the points P and Q, respectively, then $\mathbf{u} - \mathbf{v}$ may be represented by the arrow $\overrightarrow{QP}$ from Q to P. We may therefore write

$$\mathbf{u} - \mathbf{v} = \overrightarrow{OP} - \overrightarrow{OQ} = \overrightarrow{QP},$$

as illustrated in Fig. 13.22.

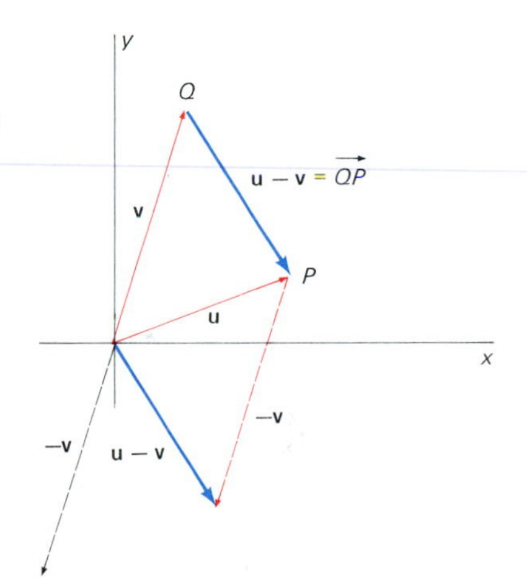

13.22 Geometric interpretation of the difference $\mathbf{u} - \mathbf{v}$

EXAMPLE 1 Suppose that $\mathbf{u} = \langle 4, -3 \rangle$ and $\mathbf{v} = \langle -2, 3 \rangle$. Find $|\mathbf{u}|$ and the vectors $\mathbf{u} + \mathbf{v}$, $\mathbf{u} - \mathbf{v}$, $3\mathbf{u} - 2\mathbf{v}$, and $2\mathbf{u} + 4\mathbf{v}$.

Solution

$$|\mathbf{u}| = \sqrt{4^2 + (-3)^2} = \sqrt{25} = 5.$$

$$\mathbf{u} + \mathbf{v} = \langle 4 + (-2), -3 + 3 \rangle = \langle 2, 0 \rangle.$$

$$\mathbf{u} - \mathbf{v} = \langle 4 - (-2), -3 - 3 \rangle = \langle 6, -6 \rangle.$$

$$3\mathbf{u} = \langle 3 \cdot (4), 3 \cdot (-3) \rangle = \langle 12, -9 \rangle.$$

$$-2\mathbf{v} = \langle -2 \cdot (-2), -2 \cdot (3) \rangle = \langle 4, -6 \rangle.$$

$$2\mathbf{u} + 4\mathbf{v} = \langle 2 \cdot (4) + 4 \cdot (-2), 2 \cdot (-3) + 4 \cdot (3) \rangle = \langle 0, 6 \rangle.$$

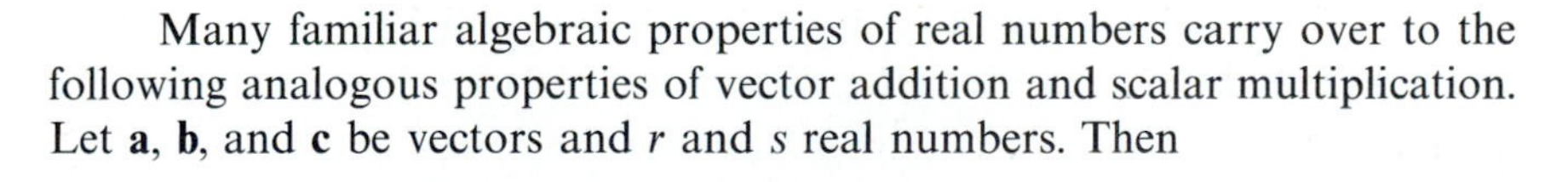

Many familiar algebraic properties of real numbers carry over to the following analogous properties of vector addition and scalar multiplication. Let **a**, **b**, and **c** be vectors and r and s real numbers. Then

$$
\begin{aligned}
&1. \quad \mathbf{a} + \mathbf{b} = \mathbf{b} + \mathbf{a}, \\
&2. \quad \mathbf{a} + (\mathbf{b} + \mathbf{c}) = (\mathbf{a} + \mathbf{b}) + \mathbf{c}, \\
&3. \quad r(\mathbf{a} + \mathbf{b}) = r\mathbf{a} + r\mathbf{b}, \qquad (5) \\
&4. \quad (r + s)\mathbf{a} = r\mathbf{a} + s\mathbf{a}, \\
&5. \quad (rs)\mathbf{a} = r(s\mathbf{a}) = s(r\mathbf{a}).
\end{aligned}
$$

These identities are verified by working with components. For example, if $\mathbf{a} = \langle a_1, a_2 \rangle$ and $\mathbf{b} = \langle b_1, b_2 \rangle$, then

$$r(\mathbf{a} + \mathbf{b}) = r\langle a_1 + b_1, a_2 + b_2 \rangle = \langle r(a_1 + b_1), r(a_2 + b_2) \rangle$$
$$= \langle ra_1 + rb_1, ra_2 + rb_2 \rangle = \langle ra_1, ra_2 \rangle + \langle rb_1, rb_2 \rangle = r\mathbf{a} + r\mathbf{b}.$$

The proofs of the other four identities in (5) are left as exercises.

THE UNIT VECTORS i AND j

A **unit** vector is one with length 1. If $\mathbf{a} = \langle a_1, a_2 \rangle \neq \mathbf{0}$ then

$$\mathbf{u} = \frac{\mathbf{a}}{|\mathbf{a}|} \tag{6}$$

is the unit vector having the same direction as $\mathbf{a}$, because

$$|\mathbf{u}| = \sqrt{\left(\frac{a_1}{|\mathbf{a}|}\right)^2 + \left(\frac{a_2}{|\mathbf{a}|}\right)^2} = \frac{1}{|\mathbf{a}|}\sqrt{a_1^2 + a_2^2} = 1.$$

For example, if $\mathbf{a} = \langle 3, -4 \rangle$, then $|\mathbf{a}| = 5$. Thus $\langle \frac{3}{5}, -\frac{4}{5} \rangle$ is a unit vector that has the same direction as $\mathbf{a}$.

Two particular unit vectors play a special role. They are the vectors

$$\mathbf{i} = \langle 1, 0 \rangle \quad \text{and} \quad \mathbf{j} = \langle 0, 1 \rangle.$$

The first points in the positive x-direction, the second in the positive y-direction. Together they provide a useful alternative notation for vectors. For if $\mathbf{a} = \langle a_1, a_2 \rangle$, then

$$\mathbf{a} = \langle a_1, 0 \rangle + \langle 0, a_2 \rangle = a_1\langle 1, 0 \rangle + a_2\langle 0, 1 \rangle = a_1\mathbf{i} + a_2\mathbf{j}. \tag{7}$$

Thus every vector is a **linear combination** of $\mathbf{i}$ and $\mathbf{j}$. The usefulness of this notation is based on the fact that such linear combinations of $\mathbf{i}$ and $\mathbf{j}$ may be manipulated as if they were ordinary sums. For example, if

$$\mathbf{a} = a_1\mathbf{i} + a_2\mathbf{j} \quad \text{and} \quad \mathbf{b} = b_1\mathbf{i} + b_2\mathbf{j},$$

then

$$\mathbf{a} + \mathbf{b} = (a_1\mathbf{i} + a_2\mathbf{j}) + (b_1\mathbf{i} + b_2\mathbf{j}) = (a_1 + b_1)\mathbf{i} + (a_2 + b_2)\mathbf{j}.$$

Also,

$$c\mathbf{a} = c(a_1\mathbf{i} + a_2\mathbf{j}) = (ca_1)\mathbf{i} + (ca_2)\mathbf{j}.$$

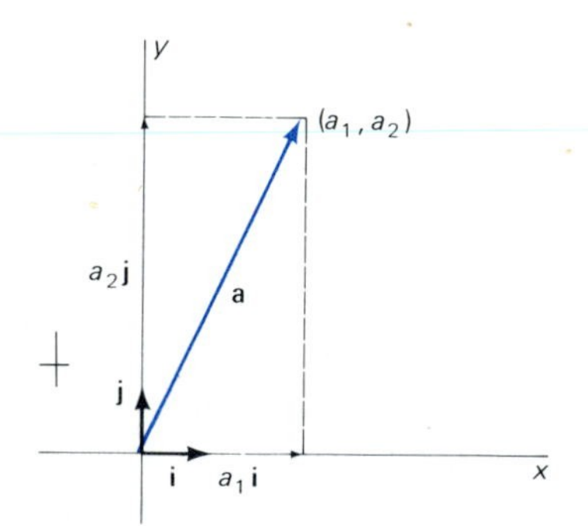

13.23 Resolution of $\mathbf{a} = \langle a_1, a_2 \rangle$ into its horizontal and vertical components

EXAMPLE 2 Suppose that $\mathbf{a} = 2\mathbf{i} - 3\mathbf{j}$ and $\mathbf{b} = 3\mathbf{i} + 4\mathbf{j}$. Express $5\mathbf{a} - 3\mathbf{b}$ in terms of $\mathbf{i}$ and $\mathbf{j}$.

Solution

$$5\mathbf{a} - 3\mathbf{b} = 5(2\mathbf{i} - 3\mathbf{j}) - 3(3\mathbf{i} + 4\mathbf{j})$$
$$= (10 - 9)\mathbf{i} + (-15 - 12)\mathbf{j} = \mathbf{i} - 27\mathbf{j}.$$

The formula in (7) expresses the vector $\mathbf{a} = \langle a_1, a_2 \rangle$ as the sum of a horizontal vector $a_1\mathbf{i}$ and a vertical vector $a_2\mathbf{j}$, as Fig. 13.23 shows. The decomposition or "resolution" of a vector into its horizontal and vertical components is an important technique in the study of vector quantities. For example, a force $\mathbf{F}$ may be decomposed into its horizontal and vertical components $F_1\mathbf{i}$ and $F_2\mathbf{j}$, respectively. The physical effect of the single force $\mathbf{F}$ is the same as the combined effect of the separate forces $F_1\mathbf{i}$ and $F_2\mathbf{j}$. (This is an instance of the empirically verifiable parallelogram law of addition of forces.) Because of this decomposition, many two-dimensional problems can be reduced to one-dimensional problems, the latter solved, and the two results combined (again by vector methods) to give the solution of the original problem.

PERPENDICULAR VECTORS AND THE DOT PRODUCT

Two nonzero vectors $\mathbf{a}$ and $\mathbf{b}$ are called **perpendicular** if, when they are represented as position vectors $\mathbf{a} = \overrightarrow{OP}$ and $\mathbf{b} = \overrightarrow{OQ}$, the line segments $\overrightarrow{OP}$ and $\overrightarrow{OQ}$ are perpendicular. We need a simple way to determine whether or not two given vectors are perpendicular.

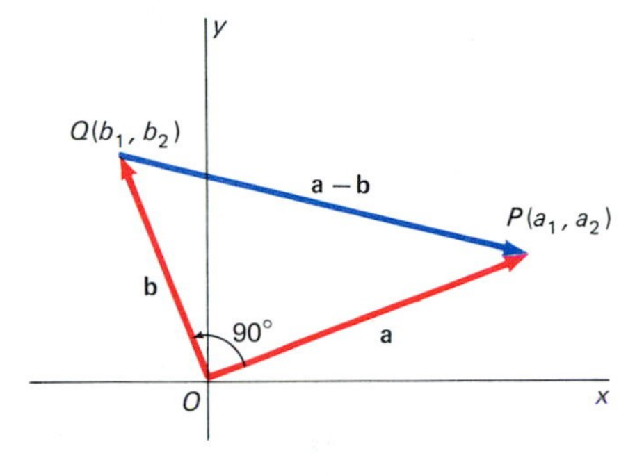

13.24 Two perpendicular vectors

Figure 13.24 shows our two nonzero vectors

$$\mathbf{a} = \langle a_1, a_2 \rangle \quad \text{and} \quad \mathbf{b} = \langle b_1, b_2 \rangle. \tag{8}$$

Their difference

$$\mathbf{a} - \mathbf{b} = \langle a_1 - b_1, a_2 - b_2 \rangle \tag{9}$$

is represented by the third side $\overrightarrow{QP}$ of the triangle OPQ. By the Pythagorean theorem and its converse, this triangle has a right angle at O if and only if

$$|\mathbf{a}|^2 + |\mathbf{b}|^2 = |\mathbf{a} - \mathbf{b}|^2;$$

$$\begin{aligned}(a_1^2 + a_2^2) + (b_1^2 + b_2^2) &= (a_1 - b_1)^2 + (a_2 - b_2)^2 \\ &= (a_1^2 - 2a_1b_1 + b_1^2) + (a_2^2 - 2a_2b_2 + b_2^2);\end{aligned}$$

the latter is true if and only if

$$a_1b_1 + a_2b_2 = 0. \tag{10}$$

This last condition is the condition for perpendicularity of the vectors $\mathbf{a}$ and $\mathbf{b}$, and we have proved the following theorem.

Theorem *Test for Perpendicular Vectors*

The two nonzero vectors $\mathbf{a} = \langle a_1, a_2 \rangle$ and $\mathbf{b} = \langle b_1, b_2 \rangle$ are perpendicular if and only if

$$a_1b_1 + a_2b_2 = 0. \tag{10}$$

Note that it is consistent with (10) to regard the zero vector $\mathbf{0} = \langle 0, 0 \rangle$ as perpendicular to every vector.

The particular combination of the components of $\mathbf{a}$ and $\mathbf{b}$ in (10) has important applications and is called the *dot product* of $\mathbf{a}$ and $\mathbf{b}$.

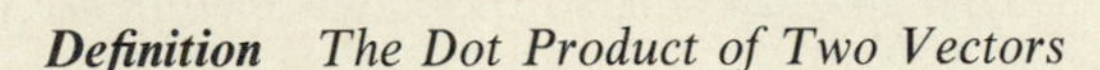

Definition *The Dot Product of Two Vectors*

The **dot product** $\mathbf{a} \cdot \mathbf{b}$ of the two vectors $\mathbf{a} = \langle a_1, a_2 \rangle$ and $\mathbf{b} = \langle b_1, b_2 \rangle$ is the real number

$$\mathbf{a} \cdot \mathbf{b} = a_1b_1 + a_2b_2. \tag{11}$$

Note that the dot product of two vectors is not a vector, but a *scalar*. For this reason it is sometimes called the *scalar product* of vectors. For example,

$$\langle 2, -3 \rangle \cdot \langle -4, 2 \rangle = (2)(-4) + (-3)(2) = -14.$$

The following algebraic properties of the dot product are easy to verify by componentwise calculations.

$$\begin{aligned} &1. \quad \mathbf{a} \cdot \mathbf{a} = |\mathbf{a}|^2, \\ &2. \quad \mathbf{a} \cdot \mathbf{b} = \mathbf{b} \cdot \mathbf{a}, \\ &3. \quad \mathbf{a} \cdot (\mathbf{b} + \mathbf{c}) = \mathbf{a} \cdot \mathbf{b} + \mathbf{a} \cdot \mathbf{c}, \\ &4. \quad (r\mathbf{a}) \cdot \mathbf{b} = r(\mathbf{a} \cdot \mathbf{b}) = \mathbf{a} \cdot (r\mathbf{b}). \end{aligned} \tag{12}$$

For example,

$$\mathbf{a} \cdot \mathbf{a} = (a_1)^2 + (a_2)^2 = |\mathbf{a}|^2$$

and

$$\begin{aligned} \mathbf{a} \cdot (\mathbf{b} + \mathbf{c}) &= \langle a_1, a_2 \rangle \cdot \langle b_1 + c_1, b_2 + c_2 \rangle = a_1(b_1 + c_1) + a_2(b_2 + c_2) \\ &= (a_1 b_1 + a_2 b_2) + (a_1 c_1 + a_2 c_2) = \mathbf{a} \cdot \mathbf{b} + \mathbf{a} \cdot \mathbf{c}. \end{aligned}$$

Because $\mathbf{i} \cdot \mathbf{i} = \mathbf{j} \cdot \mathbf{j} = 1$ and $\mathbf{i} \cdot \mathbf{j} = \mathbf{j} \cdot \mathbf{i} = 0$, application of the properties in (12) gives

$$\begin{aligned} \mathbf{a} \cdot \mathbf{b} &= (a_1\mathbf{i} + a_2\mathbf{j}) \cdot (b_1\mathbf{i} + b_2\mathbf{j}) \\ &= (a_1 b_1)\mathbf{i} \cdot \mathbf{i} + (a_1 b_2 + a_2 b_1)\mathbf{i} \cdot \mathbf{j} + (a_2 b_2)\mathbf{j} \cdot \mathbf{j} \\ &= a_1 b_1 + a_2 b_2, \end{aligned}$$

in accord with the definition of $\mathbf{a} \cdot \mathbf{b}$. Thus dot products of vectors expressed in terms of $\mathbf{i}$ and $\mathbf{j}$ can be computed by "ordinary" algebra.

The test for perpendicularity is best remembered in terms of the dot product.

> ***Corollary*** *Test for Perpendicular Vectors*
> The two nonzero vectors $\mathbf{a}$ and $\mathbf{b}$ are perpendicular if and only if $\mathbf{a} \cdot \mathbf{b} = 0$.

EXAMPLE 3 Determine whether or not the following pairs of vectors are perpendicular.

(a) $\langle 3, 4 \rangle$ and $\langle 8, -6 \rangle$;

(b) $\langle 2, 3 \rangle$ and $\langle 4, -3 \rangle$.

Solution (a) $\langle 3, 4 \rangle \cdot \langle 8, -6 \rangle = 24 - 24 = 0$, so these two vectors *are* perpendicular.

(b) $\langle 2, 3 \rangle \cdot \langle 4, -3 \rangle = 8 - 9 = -1 \neq 0$, so $\langle 2, 3 \rangle$ and $(4, -3\rangle$ are *not* perpendicular.

In Section 14.1 we shall see that $\mathbf{a} \cdot \mathbf{b} = |\mathbf{a}|\,|\mathbf{b}| \cos \theta$, where θ is the angle between the vectors **a** and **b**. The case $\theta = \pi/2$ gives the corollary above.

13.3 PROBLEMS

In each of Problems 1–8, find $|\mathbf{a}|$, $|-2\mathbf{b}|$, $|\mathbf{a} - \mathbf{b}|$, $\mathbf{a} + \mathbf{b}$, and $3\mathbf{a} - 2\mathbf{b}$. Also determine whether or not **a** and **b** are perpendicular.

1. $\mathbf{a} = \langle 1, -2 \rangle$, $\mathbf{b} = \langle -3, 2 \rangle$
2. $\mathbf{a} = \langle 3, 4 \rangle$, $\mathbf{b} = \langle -4, 3 \rangle$
3. $\mathbf{a} = \langle -2, -2 \rangle$, $\mathbf{b} = \langle -3, -4 \rangle$
4. $\mathbf{a} = -2\langle 4, 7 \rangle$, $\mathbf{b} = -3\langle -4, -2 \rangle$
5. $\mathbf{a} = \mathbf{i} + 3\mathbf{j}$, $\mathbf{b} = 2\mathbf{i} - 5\mathbf{j}$
6. $\mathbf{a} = 2\mathbf{i} - 5\mathbf{j}$, $\mathbf{b} = \mathbf{i} - 6\mathbf{j}$
7. $\mathbf{a} = 4\mathbf{i}$, $\mathbf{b} = -7\mathbf{j}$
8. $\mathbf{a} = -\mathbf{i} - \mathbf{j}$, $\mathbf{b} = 2\mathbf{i} + 2\mathbf{j}$

In each of Problems 9–12, find a unit vector **u** with the same direction as the given vector **a**. Express **u** in terms of **i** and **j**. Also find a unit vector **v** with the direction opposite that of **a**.

9. $\mathbf{a} = \langle -3, -4 \rangle$
10. $\mathbf{a} = \langle 5, -12 \rangle$
11. $\mathbf{a} = 8\mathbf{i} + 15\mathbf{j}$
12. $\mathbf{a} = 7\mathbf{i} - 24\mathbf{j}$

In each of Problems 13–16, find the vector **a**, expressed in terms of **i** and **j**, that is represented by the arrow $\overrightarrow{PQ}$ in the plane.

13. $P = (3, 2)$, $Q = (3, -2)$
14. $P = (-3, 5)$, $Q = (-3, 6)$
15. $P = (-4, 7)$, $Q = (4, -7)$
16. $P = (1, -1)$, $Q = (-4, -1)$

In each of Problems 17–20, determine whether or not the given vectors **a** and **b** are perpendicular.

17. $\mathbf{a} = \langle 6, 0 \rangle$, $\mathbf{b} = \langle 0, -7 \rangle$
18. $\mathbf{a} = 3\mathbf{j}$, $\mathbf{b} = 3\mathbf{i} - \mathbf{j}$
19. $\mathbf{a} = 2\mathbf{i} - \mathbf{j}$, $\mathbf{b} = 4\mathbf{j} + 8\mathbf{i}$
20. $\mathbf{a} = 8\mathbf{i} + 10\mathbf{j}$, $\mathbf{b} = 15\mathbf{i} - 12\mathbf{j}$

21. Find a vector that has the same direction as $5\mathbf{i} - 7\mathbf{j}$ and
(a) three times its length;
(b) one-third its length.

22. Find a vector that has the opposite direction from $-3\mathbf{i} + 5\mathbf{j}$ and
(a) four times its length;
(b) one-fourth its length.

23. Find a vector of length 5 with
(a) the same direction as $7\mathbf{i} - 3\mathbf{j}$;
(b) the opposite direction from $8\mathbf{i} + 5\mathbf{j}$.

24. For what numbers c are the vectors $\langle c, 2 \rangle$ and $\langle c, -8 \rangle$ perpendicular?

25. For what numbers c are the vectors $2c\mathbf{i} - 4\mathbf{j}$ and $3\mathbf{i} + c\mathbf{j}$ perpendicular?

26. Given the three points $A(2, 3)$, $B(-5, 7)$, and $C(1, -5)$, verify by direct computation of the vectors and their sum that $\overrightarrow{AB} + \overrightarrow{BC} + \overrightarrow{CA} = \mathbf{0}$.

In each of Problems 27–32, give a componentwise proof of the indicated property of vector algebra. Take $\mathbf{a} = \langle a_1, a_2 \rangle$ and $\mathbf{b} = \langle b_1, b_2 \rangle$ throughout.

27. $\mathbf{a} + (\mathbf{b} + \mathbf{c}) = (\mathbf{a} + \mathbf{b}) + \mathbf{c}$
28. $(r + s)\mathbf{a} = r\mathbf{a} + s\mathbf{a}$
29. $(rs)\mathbf{a} = r(s\mathbf{a})$
30. $\mathbf{a} \cdot \mathbf{b} = \mathbf{b} \cdot \mathbf{a}$
31. $(r\mathbf{a}) \cdot \mathbf{b} = r(\mathbf{a} \cdot \mathbf{b})$
32. If $\mathbf{a} + \mathbf{b} = \mathbf{a}$, then $\mathbf{b} = \mathbf{0}$.

In Problems 33–35, assume the following fact: If an airplane flies with velocity vector $\mathbf{v}_a$ relative to the air and the velocity of the wind is **w**, then the velocity vector of the plane relative to the ground is

$$\mathbf{v}_g = \mathbf{v}_a + \mathbf{w}$$

(see Fig. 13.25). The **apparent velocity** vector is $\mathbf{v}_a$, while $\mathbf{v}_g$ is called the **true velocity** vector.

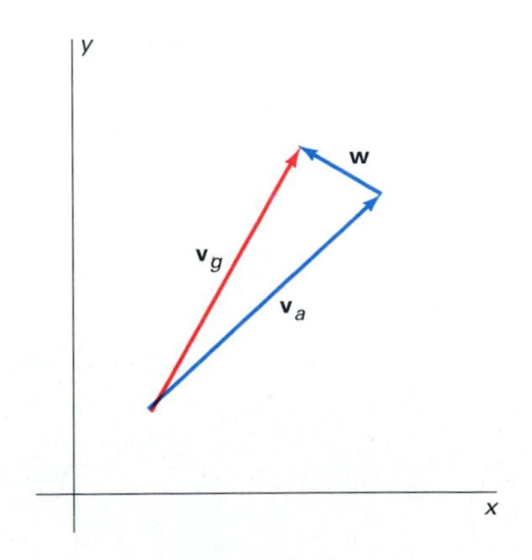

13.25 Apparent velocity: $\mathbf{v}_a$
Wind: **w**
True velocity: $\mathbf{v}_g = \mathbf{v}_a + \mathbf{w}$

33. Suppose that the wind is blowing from the northeast at 50 mi/h, and the pilot wishes to fly due east at 500 mi/h. What should the plane's apparent velocity vector be?

34. Repeat Problem 33 with the phrase *due east* replaced by *due west*.

35. Repeat Problem 33 in the case that the pilot wishes to fly northwest at 500 mi/h.

36. In the triangle ABC, let M_1 and M_2 be the midpoints of AB and AC, respectively. Show that $\overrightarrow{M_1M_2} = \frac{1}{2}\overrightarrow{BC}$. Conclude that the line segment joining the midpoints of two sides of a triangle is parallel to the third side.

37. Prove that the diagonals of a parallelogram $ABCD$ bisect each other. (*Suggestion:* If M_1 and M_2 denote the midpoints of AC and BD, respectively, show that $\overrightarrow{OM_1} = \overrightarrow{OM_2}$.

38. Use vectors to prove that the midpoints of the four sides of an arbitrary quadrilateral are the vertices of a parallelogram.

39. Prove that the vector $\mathbf{n} = a\mathbf{i} + b\mathbf{j}$ is perpendicular to the line with equation $ax + by + c = 0$. (*Suggestion:* If $P_1(x_1, y_1)$ and $P_2(x_2, y_2)$ are two points on the line, show that $\mathbf{n} \cdot \overrightarrow{P_1P_2} = 0$.)

40. Figure 13.26 shows the vector $\mathbf{a}_\perp$ that is obtained by rotating the vector $\mathbf{a} = a_1\mathbf{i} + a_2\mathbf{j}$ through a counterclockwise angle of 90°. Show that

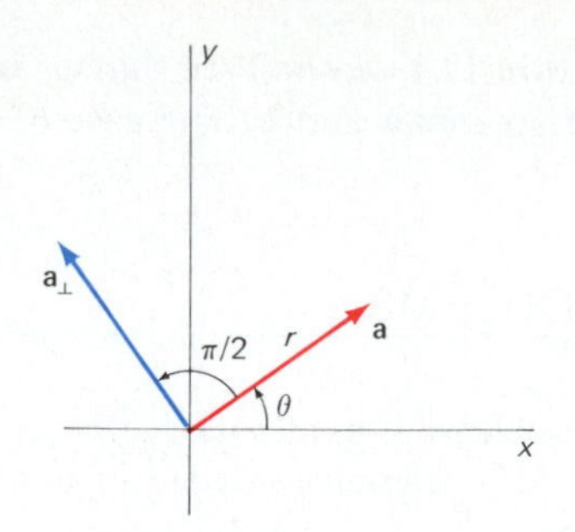

13.26 Rotate **a** counterclockwise 90° to obtain $\mathbf{a}_\perp$.

$$\mathbf{a}_\perp = -a_2\mathbf{i} + a_1\mathbf{j}.$$

Suggestion: Begin by writing

$$\mathbf{a} = (r\cos\theta)\mathbf{i} + (r\sin\theta)\mathbf{j}.$$

41. Is the dot product associative? That is, if **a**, **b**, and **c** are vectors, does the equation $\mathbf{a} \cdot (\mathbf{b} \cdot \mathbf{c}) = (\mathbf{a} \cdot \mathbf{b}) \cdot \mathbf{c}$ always hold? Does this question make any sense at all?

13.4 Motion and Vector-Valued Functions

We now use vectors to study the motion of a point in the plane. If the coordinates of the moving point at time t are given by the parametric equations $x = f(t)$, $y = g(t)$, then the vector

$$\mathbf{r}(t) = f(t)\mathbf{i} + g(t)\mathbf{j}, \quad \text{or} \quad \mathbf{r} = x\mathbf{i} + y\mathbf{j}, \tag{1}$$

is called the **position vector** of the point. Equation (1) determines a **vector-valued function** $\mathbf{r} = \mathbf{r}(t)$ that associates with the number t the position vector $\mathbf{r}(t)$ of the moving point. In the bracket notation for vectors, a vector-valued function is an ordered pair of real-valued functions: $\mathbf{r}(t) = \langle f(t), g(t) \rangle$.

Much of the calculus of (ordinary) real-valued functions applies to vector-valued functions. To begin with, the **limit** of a vector-valued function $\mathbf{r} = \langle f, g \rangle$ is defined as follows:

$$\lim_{x \to a} \mathbf{r}(t) = \left\langle \lim_{x \to a} f(t), \lim_{x \to a} g(t) \right\rangle = \mathbf{i}\left(\lim_{x \to a} f(t)\right) + \mathbf{j}\left(\lim_{x \to a} g(t)\right), \tag{2}$$

provided that the limits in the latter two expressions exist. Thus we take limits of vector-valued functions by taking limits of their component functions.

We say that $\mathbf{r} = \mathbf{r}(t)$ is **continuous** at the number a provided that

$$\lim_{t \to a} \mathbf{r}(t) = \mathbf{r}(a).$$

This amounts to saying that **r** is continuous at a if and only if its component functions f and g are continuous at a.

The derivative $\mathbf{r}'(t)$ of the vector-valued function $\mathbf{r}(t)$ is defined in almost exactly the same way as the derivative of a real-valued function. Specifically,

$$\mathbf{r}'(t) = \lim_{h \to 0} \frac{\mathbf{r}(t+h) - \mathbf{r}(t)}{h}, \tag{3}$$

provided that this limit exists. A geometric note: Figure 13.27 suggests that $\mathbf{r}'(t)$ will be tangent to the curve swept out by $\mathbf{r}$ if $\mathbf{r}'(t)$ is attached to the curve at the point of evaluation. Note also that because $|\mathbf{r}(t+h) - \mathbf{r}(t)|$ is the distance from the point with position vector $\mathbf{r}(t)$ to the point with position vector $\mathbf{r}(t+h)$, the quotient

$$\frac{|\mathbf{r}(t+h) - \mathbf{r}(t)|}{h}$$

is equal to the average speed of a particle that travels from $\mathbf{r}(t)$ to $\mathbf{r}(t+h)$ in time h. Consequently, the limit in (3) yields both the direction of motion and the instantaneous speed of a particle moving with position vector $\mathbf{r}(t)$ along a curve.

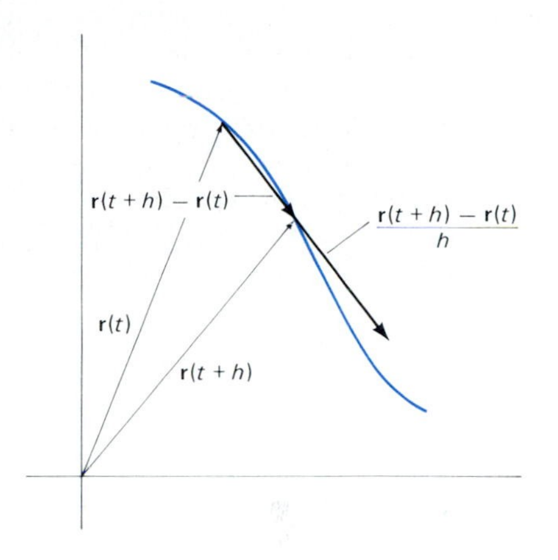

13.27 Geometry of the derivative of a vector-valued function.

Our next result implies the simple *but important* fact that $\mathbf{r}'(t)$ can be calculated by componentwise differentiation. We shall also denote derivatives by

$$\mathbf{r}'(t) = D_t\mathbf{r}(t) = \frac{d\mathbf{r}}{dt}.$$

Theorem 1 *Componentwise Differentiation*

Suppose that

$$\mathbf{r}(t) = \langle f(t), g(t) \rangle = f(t)\mathbf{i} + g(t)\mathbf{j},$$

where both f and g are differentiable functions. Then

$$\mathbf{r}'(t) = \langle f'(t), g'(t) \rangle = f'(t)\mathbf{i} + g'(t)\mathbf{j}. \tag{4}$$

That is, if $\mathbf{r} = x\mathbf{i} + y\mathbf{j}$, then

$$\frac{d\mathbf{r}}{dt} = \frac{dx}{dt}\mathbf{i} + \frac{dy}{dt}\mathbf{j}.$$

Proof We simply take the limit in Equation (3) by taking limits of components. We find that

$$\begin{aligned}
\mathbf{r}'(t) &= \lim_{h \to 0} \frac{\mathbf{r}(t+h) - \mathbf{r}(t)}{h} \\
&= \lim_{h \to 0} \frac{f(t+h)\mathbf{i} + g(t+h)\mathbf{j} - f(t)\mathbf{i} - g(t)\mathbf{j}}{h} \\
&= \left(\lim_{h \to 0} \frac{f(t+h) - f(t)}{h}\right)\mathbf{i} + \left(\lim_{h \to 0} \frac{g(t+h) - g(t)}{h}\right)\mathbf{j} \\
&= f'(t)\mathbf{i} + g'(t)\mathbf{j}. \quad \blacksquare
\end{aligned}$$

In Equation (5) of Section 13.1, we saw that the tangent line to the parametric curve $x = f(t)$, $y = g(t)$ has slope $g'(t)/f'(t)$ and therefore is parallel to the line through (0, 0) and $(f'(t), g'(t))$. Hence, by Theorem 1, the derivative vector $\mathbf{r}'(t)$ may be visualized as a tangent vector to the curve at the point $(f(t), g(t))$. (See Fig. 13.28.)

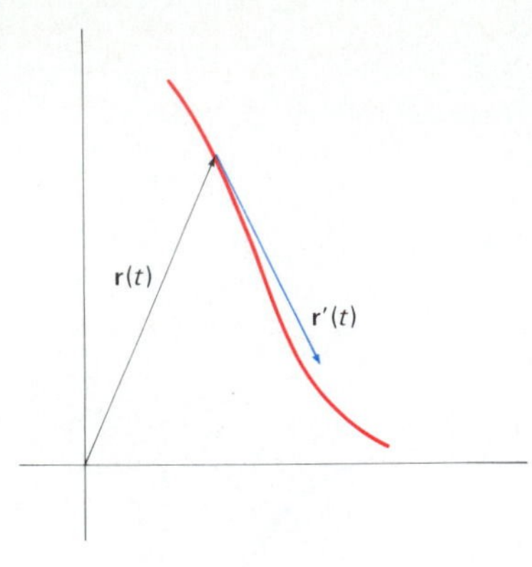

13.28 The derivative vector is tangent to the curve at the point of evaluation.

Our next theorem tells us that the formulas for computing derivatives of sums and products of vector-valued functions are formally similar to those for real-valued functions.

Theorem 2 *Differentiation Formulas*

Let $\mathbf{u}(t)$ and $\mathbf{v}(t)$ be differentiable vector-valued functions. Let $h(t)$ be a differentiable real-valued function and let c be a (constant) scalar. Then

1. $D_t[\mathbf{u}(t) + \mathbf{v}(t)] = \mathbf{u}'(t) + \mathbf{v}'(t)$,
2. $D_t[c\mathbf{u}(t)] = c\mathbf{u}'(t)$,
3. $D_t[h(t)\mathbf{u}(t)] = h'(t)\mathbf{u}(t) + h(t)\mathbf{u}'(t)$,
4. $D_t[\mathbf{u}(t)\cdot\mathbf{v}(t)] = \mathbf{u}'(t)\cdot\mathbf{v}(t) + \mathbf{u}(t)\cdot\mathbf{v}'(t)$.

Proof We shall prove part (4) and leave the other parts as exercises. If

$$\mathbf{u}(t) = \langle f_1(t), f_2(t)\rangle \quad \text{and} \quad \mathbf{v}(t) = \langle g_1(t), g_2(t)\rangle,$$

then

$$\mathbf{u}(t)\cdot\mathbf{v}(t) = f_1(t)g_1(t) + f_2(t)g_2(t).$$

Hence the product rule for ordinary real-valued functions gives

$$\begin{aligned} D_t[\mathbf{u}(t)\cdot\mathbf{v}(t)] &= D_t[f_1(t)g_1(t) + f_2(t)g_2(t)] \\ &= [f_1'(t)g_1(t) + f_2'(t)g_2(t)] + [f_1(t)g_1'(t) + f_2(t)g_2'(t)] \\ &= \mathbf{u}'(t)\cdot\mathbf{v}(t) + \mathbf{u}(t)\cdot\mathbf{v}'(t). \quad \blacksquare \end{aligned}$$

Now we can discuss the motion of a point with position vector $\mathbf{r}(t) = f(t)\mathbf{i} + g(t)\mathbf{j}$. Its **velocity vector** $\mathbf{v}(t)$ and **acceleration vector** $\mathbf{a}(t)$ are defined as follows:

$$\left.\begin{aligned} \mathbf{v}(t) &= \mathbf{r}'(t) = f'(t)\mathbf{i} + g'(t)\mathbf{j}, \\ \mathbf{v} &= \frac{d\mathbf{r}}{dt} = \frac{dx}{dt}\mathbf{i} + \frac{dy}{dt}\mathbf{j}; \end{aligned}\right\} \tag{5}$$

$$\left.\begin{aligned} \mathbf{a}(t) &= \mathbf{v}'(t) = f''(t)\mathbf{i} + g''(t)\mathbf{j}, \\ \mathbf{a} &= \frac{d^2\mathbf{r}}{dt^2} = \frac{d^2x}{dt^2}\mathbf{i} + \frac{d^2y}{dt^2}\mathbf{j}. \end{aligned}\right\} \tag{6}$$

The moving point also has a **speed** $v(t)$ and **scalar acceleration** $a(t)$. These are merely the lengths of the corresponding velocity and acceleration vectors. Thus

$$v(t) = |\mathbf{v}(t)| = \sqrt{\left(\frac{dx}{dt}\right)^2 + \left(\frac{dy}{dt}\right)^2} \tag{7}$$

and

$$a(t) = |\mathbf{a}(t)| = \sqrt{\left(\frac{d^2x}{dt^2}\right)^2 + \left(\frac{d^2y}{dt^2}\right)^2}. \tag{8}$$

Do *not* assume that the scalar acceleration a and the derivative dv/dt of the speed are equal. Although this holds part of the time for motion in a straight line, it is almost *never* true for two-dimensional motion.

EXAMPLE 1 A moving particle has position vector

$$\mathbf{r}(t) = t\mathbf{i} + t^2\mathbf{j}.$$

Find the velocity and acceleration vectors and the speed and the scalar acceleration when $t = 2$.

Solution Because $\mathbf{r}(2) = 2\mathbf{i} + 4\mathbf{j}$, the particle is at the point $(2, 4)$ on the parabola $y = x^2$ when $t = 2$. Its velocity vector is

$$\mathbf{v}(t) = \mathbf{i} + 2t\mathbf{j}.$$

So $\mathbf{v}(2) = \mathbf{i} + 4\mathbf{j}$ and $v(t) = |\mathbf{v}(2)| = \sqrt{17}$. Note that the *velocity* $\mathbf{v}(2)$ is a *vector*, while the *speed* $v(2)$ is a *scalar*.

The acceleration vector is $\mathbf{a} = 2\mathbf{j}$, so the acceleration of the particle is the (constant) vector $2\mathbf{j}$. In particular, when $t = 2$, its acceleration is $2\mathbf{j}$ and its scalar acceleration is $a = 2$. Figure 13.29 shows the trajectory of the particle with the vectors $\mathbf{v}(2)$ and $\mathbf{a}(2)$ attached at the point corresponding to $t = 2$.

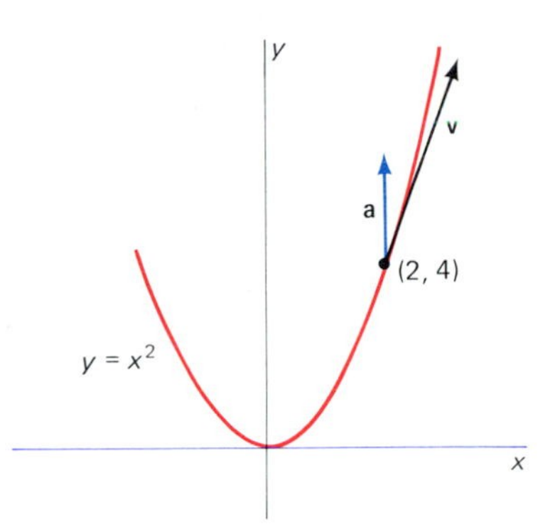

13.29 The velocity and acceleration vectors at $t = 2$ (see Example 1).

Integrals of vector-valued functions are defined by analogy with the definition of an integral of a real-valued function:

$$\int_a^b \mathbf{r}(t)\,dt = \lim_{\Delta t \to 0} \sum_{i=1}^{n} \mathbf{r}(t_i^*)\,\Delta t, \tag{9}$$

where t_i^* is a point of the ith subinterval of a subdivision of $[a, b]$ into n equal subintervals each of length $\Delta t = (b - a)/n$.

If $\mathbf{r}(t) = f(t)\mathbf{i} + g(t)\mathbf{j}$ is continuous, then—by taking limits componentwise—we get

$$\begin{aligned}\int_a^b \mathbf{r}(t)\,dt &= \lim_{\Delta t \to 0} \sum_{i=1}^{n} \mathbf{r}(t_i^*)\,\Delta t \\ &= \mathbf{i}\left(\lim_{\Delta t \to 0} \sum_{i=1}^{n} f(t_i^*)\,\Delta t\right) + \mathbf{j}\left(\lim_{\Delta t \to 0} \sum_{i=1}^{n} g(t_i^*)\,\Delta t\right).\end{aligned}$$

This gives the result that

$$\int_a^b \mathbf{r}(t)\,dt = \mathbf{i}\left(\int_a^b f(t)\,dt\right) + \mathbf{j}\left(\int_a^b g(t)\,dt\right). \tag{10}$$

Thus *a vector-valued function may be integrated componentwise.*

Now suppose that $\mathbf{R}(t)$ is an *antiderivative* of $\mathbf{r}(t)$, meaning that $\mathbf{R}'(t) = \mathbf{r}(t)$. That is, if $\mathbf{R}(t) = F(t)\mathbf{i} + G(t)\mathbf{j}$, then

$$\mathbf{R}'(t) = F'(t)\mathbf{i} + G'(t)\mathbf{j} = f(t)\mathbf{i} + g(t)\mathbf{j} = \mathbf{r}(t).$$

Then componentwise integration yields

$$\begin{aligned}\int_a^b \mathbf{r}(t)\,dt &= \mathbf{i}\left(\int_a^b f(t)\,dt\right) + \mathbf{j}\left(\int_a^b g(t)\,dt\right) = \mathbf{i}\Big[F(t)\Big]_a^b + \mathbf{j}\Big[G(t)\Big]_a^b \\ &= [F(b)\mathbf{i} + G(b)\mathbf{j}] - [F(a)\mathbf{i} + G(a)\mathbf{j}].\end{aligned}$$

Thus the *fundamental theorem of calculus* for vector-valued functions takes the form

$$\int_a^b \mathbf{r}(t)\,dt = \Big[\mathbf{R}(t)\Big]_a^b = \mathbf{R}(b) - \mathbf{R}(a) \tag{11}$$

where $\mathbf{R}'(t) = \mathbf{r}(t)$.

Vector integration is the basis for one method of navigation. If a submarine is cruising beneath the icecap at the North Pole, and thus can use neither visual nor radio methods to determine its position, here is an alternative: Build a sensitive gyroscope–accelerometer combination and install it in the submarine. The device continuously measures the sub's acceleration vector, beginning at the time $t = 0$ when its position $\mathbf{r}(0)$ and velocity $\mathbf{v}(0)$ are known. Because $\mathbf{v}'(t) = \mathbf{a}(t)$, Equation (10) gives

$$\int_0^t \mathbf{a}(t)\,dt = \Big[\mathbf{v}(t)\Big]_0^t = \mathbf{v}(t) - \mathbf{v}(0)$$

so that

$$\mathbf{v}(t) = \mathbf{v}(0) + \int_0^t \mathbf{a}(t)\,dt.$$

Thus the velocity at every time t is known. Similarly, because $\mathbf{r}'(t) = \mathbf{v}(t)$, a second integration gives

$$\mathbf{r}(t) = \mathbf{r}(0) + \int_0^t \mathbf{v}(t)\,dt$$

for the position of the ship at every time t. On-board computers can be programmed to carry out these integrations (perhaps using Simpson's approximation), and thus continuously provide captain and crew with the submarine's position and velocity.

Indefinite integrals of vector-valued functions may also be computed. If $\mathbf{R}'(t) = \mathbf{r}(t)$, then every antiderivative of $\mathbf{r}(t)$ is of the form $\mathbf{R}(t) + \mathbf{C}$ for some constant vector $\mathbf{C}$. We therefore write

$$\int \mathbf{r}(t)\,dt = \mathbf{R}(t) + \mathbf{C} \qquad \text{if} \quad \mathbf{R}'(t) = \mathbf{r}(t), \tag{12}$$

on the basis of a componentwise computation similar to the one leading to Equation (11).

EXAMPLE 2 Suppose that a moving point has initial position $\mathbf{r}(0) = 2\mathbf{i}$, initial velocity $\mathbf{v}(0) = \mathbf{i} - \mathbf{j}$, and acceleration $\mathbf{a}(t) = 2\mathbf{i} + 6t\mathbf{j}$. Find its position and velocity at time t.

Solution Because $\mathbf{a}(t) = \mathbf{v}'(t)$, Equation (12) gives

$$\mathbf{v}(t) = \int \mathbf{a}(t)\,dt + \mathbf{C} = \int (2\mathbf{i} + 6t\mathbf{j})\,dt + \mathbf{C} = 2t\mathbf{i} + 3t^2\mathbf{j} + \mathbf{C}.$$

To evaluate $\mathbf{C}$, we use the fact that $\mathbf{v}(0)$ is known. We substitute $t = 0$ in both sides of the last equation and find that $\mathbf{C} = \mathbf{v}(0) = \mathbf{i} - \mathbf{j}$. So

$$\mathbf{v}(t) = (2t\mathbf{i} + 3t^2\mathbf{j}) + (\mathbf{i} - \mathbf{j}) = (2t + 1)\mathbf{i} + (3t^2 - 1)\mathbf{j}.$$

Next,

$$\mathbf{r}(t) = \int \mathbf{v}(t)\,dt + \mathbf{C}^* = (t^2 + t)\mathbf{i} + (t^3 - t)\mathbf{j} + \mathbf{C}^*.$$

Again we substitute $t = 0$, and find that $\mathbf{C}^* = \mathbf{r}(0) = 2\mathbf{i}$. Hence

$$\mathbf{r}(t) = (t^2 + t + 2)\mathbf{i} + (t^3 - t)\mathbf{j}$$

is the positive vector of the moving point.

*MOTION OF PROJECTILES

Suppose that a projectile is launched from the point (x_0, y_0) with y_0 denoting its initial height above the surface of the earth. Let α be the angle of inclination from the horizontal of its initial velocity vector $\mathbf{v}_0$, as in Fig. 13.30. Then its initial position vector is

$$\mathbf{r}_0 = x_0\mathbf{i} + y_0\mathbf{j}, \tag{13a}$$

and from Fig. 13.30 we see that

$$\mathbf{v}_0 = (v_0 \cos \alpha)\mathbf{i} + (v_0 \sin \alpha)\mathbf{j} \tag{13b}$$

where $v_0 = |\mathbf{v}_0|$ is the initial speed of the projectile.

We suppose that the motion takes place sufficiently close to the surface that we may assume the earth is flat and gravity perfectly uniform. Then, if we also ignore air resistance, the acceleration of the projectile is

$$\mathbf{a} = \frac{d\mathbf{v}}{dt} = -g\mathbf{j}$$

where $g \approx 32 \text{ ft/s}^2$. Antidifferentiation gives

$$\mathbf{v} = -gt\mathbf{j} + \mathbf{C}_1.$$

Put $t = 0$ in both sides of this last equation. This shows that $\mathbf{C}_1 = \mathbf{v}_0$, and thus that

$$\mathbf{v} = \frac{d\mathbf{r}}{dt} = -gt\mathbf{j} + \mathbf{v}_0.$$

Another antidifferentiation gives

$$\mathbf{r} = -\tfrac{1}{2}gt^2\mathbf{j} + \mathbf{v}_0 t + \mathbf{C}_2.$$

Now substitution of $t = 0$ gives $\mathbf{C}_2 = \mathbf{r}_0$, so the position vector of the projectile at time t is

$$\mathbf{r}(t) = -\tfrac{1}{2}gt^2\mathbf{j} + \mathbf{v}_0 t + \mathbf{r}_0. \tag{14}$$

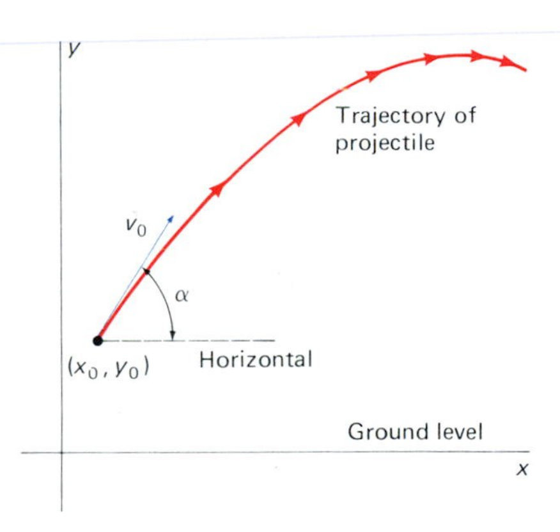

13.30 Trajectory of a projectile launched at the angle α

Equations (13a) and (13b) now give

$$\mathbf{r}(t) = [(v_0 \cos \alpha)t + x_0]\mathbf{i} + [-\tfrac{1}{2}gt^2 + (v_0 \sin \alpha)t + y_0]\mathbf{j},$$

so the parametric equations of the trajectory of the projectile are

$$x = (v_0 \cos \alpha)t + x_0, \tag{15}$$

$$y = -\tfrac{1}{2}gt^2 + (v_0 \sin \alpha)t + y_0. \tag{16}$$

EXAMPLE 3 An airplane is flying horizontally at an altitude of 1600 ft, in order to pass directly over snowbound cattle on the ground. Its speed is a constant 150 mi/h (220 ft/s). At what angle of sight ϕ (between the horizontal and the direct line to the target) should a bale of hay be released in order to hit the target?

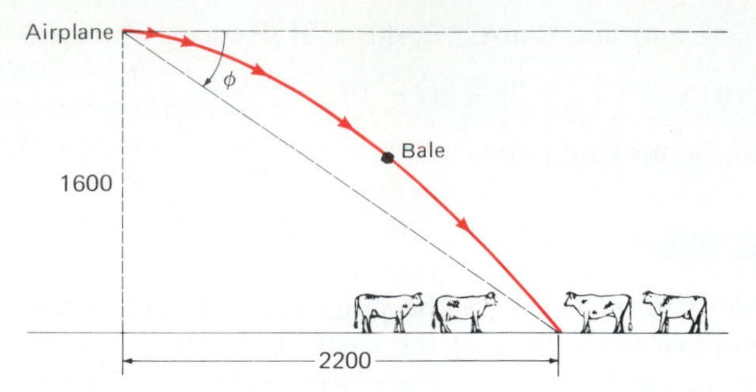

13.31 Trajectory of the hay bale (see Example 3).

Solution See Fig. 13.31. We take $x_0 = 0$ where the bale of hay is released at time $t = 0$. Then $y_0 = 1600$ (ft), $v_0 = 220$ (ft/s), and $\alpha = 0$. Then Equations (15) and (16) give

$$x = 220t, \qquad y = -16t^2 + 1600.$$

From the second of these equations we find that $t = 10$ (s) when the bale of hay hits the ground ($y = 0$). It has then traveled a horizontal distance of

$$x = (220)(10) = 2200$$

ft. Hence the required angle of sight is

$$\phi = \tan^{-1}\left(\frac{1600}{2200}\right) \approx 36^\circ.$$

13.4 PROBLEMS

In Problems 1–8, find the values of $\mathbf{r}'(t)$ and $\mathbf{r}''(t)$ for the indicated value of t.

1. $\mathbf{r}(t) = 3\mathbf{i} - 2\mathbf{j}; \quad t = 1$

2. $\mathbf{r}(t) = t^2\mathbf{i} - t^3\mathbf{j}; \quad t = 2$

3. $\mathbf{r}(t) = e^{2t}\mathbf{i} + e^{-t}\mathbf{j}; \quad t = 0$

4. $\mathbf{r}(t) = \mathbf{i}\cos t + \mathbf{j}\sin t; \quad t = \pi/4$

5. $\mathbf{r}(t) = 3\mathbf{i}\cos 2\pi t + 3\mathbf{j}\sin 2\pi t; \quad t = 3/4$

6. $\mathbf{r}(t) = 5\mathbf{i}\cos t + 4\mathbf{j}\sin t; \quad t = \pi$

7. $\mathbf{r}(t) = \mathbf{i}\sec t + \mathbf{j}\tan t; \quad t = 0$

8. $\mathbf{r}(t) = (2t + 3)\mathbf{i} + (6t - 5)\mathbf{j}; \quad t = 10$

Calculate the integrals in Problems 9–12.

9. $\displaystyle\int_0^{\pi/4} (\mathbf{i}\sin t + 2\mathbf{j}\cos t)\,dt$

10. $\displaystyle\int_1^e \left(\frac{1}{t}\mathbf{i} - \mathbf{j}\right) dt$

11. $\displaystyle\int_0^2 t^2(1 + t^3)^{3/2}\mathbf{i}\,dt$

12. $\displaystyle\int_0^1 (\mathbf{i}e^t - \mathbf{j}t\exp(-t^2))\,dt$

In Problems 13–15, apply Theorem 2 to compute the derivative $D_t[\mathbf{u}(t)\cdot\mathbf{v}(t)]$.

13. $\mathbf{u}(t) = 3t\mathbf{i} - \mathbf{j}, \quad \mathbf{v}(t) = 2\mathbf{i} - 5t\mathbf{j}$

14. $\mathbf{u}(t) = t\mathbf{i} + t^2\mathbf{j}, \quad \mathbf{v}(t) = t^2\mathbf{i} - t\mathbf{j}$

15. $\mathbf{u}(t) = \langle \cos t, \sin t\rangle, \quad \mathbf{v}(t) = \langle \sin t, -\cos t\rangle$

In Problems 16–20, find $\mathbf{v}(t)$ and $\mathbf{r}(t)$ corresponding to the given values of the acceleration $\mathbf{a}(t)$, initial velocity $\mathbf{v}_0$, and initial position $\mathbf{r}_0$.

16. $\mathbf{a} = \mathbf{0}, \quad \mathbf{r}_0 = 2\mathbf{i} + 3\mathbf{j}, \quad \mathbf{v}_0 = -2\mathbf{j}$

17. $\mathbf{a} = 2\mathbf{j}, \quad \mathbf{r}_0 = 0, \quad \mathbf{v}_0 = \mathbf{i}$

18. $\mathbf{a} = \mathbf{i} - \mathbf{j}, \quad \mathbf{r}_0 = \mathbf{j}, \quad \mathbf{v}_0 = \mathbf{i} + \mathbf{j}$

19. $\mathbf{a} = t\mathbf{i} + t^2\mathbf{j}, \quad \mathbf{r}_0 = \mathbf{i}, \quad \mathbf{v}_0 = \mathbf{0}$

20. $\mathbf{a} = -\mathbf{i}\sin t + \mathbf{j}\cos t, \quad \mathbf{r}_0 = \mathbf{i}, \quad \mathbf{v}_0 = \mathbf{j}$

Problems 21–26 deal with a projectile fired from the origin (so that $x_0 = y_0 = 0$) with initial speed v_0 and initial angle of inclination α. The *range* R of the projectile is the horizontal distance it travels before returning to the ground.

21. If $\alpha = 45^\circ$, what value of v_0 gives a range of 1 mi?

22. If $\alpha = 60°$ and $R = 1$ mi, what is the maximum height attained by the projectile?

23. Deduce from Equations (15) and (16) the fact that

$$R = \tfrac{1}{16}v_0^2 \sin \alpha \cos \alpha.$$

24. Given the initial speed v_0, find the angle α that maximizes the range R. Use the result of Problem 23.

25. Suppose that $v_0 = 160$ ft/s. Find the maximum height y_m and the range R of the projectile if
(a) $\alpha = 30°$;
(b) $\alpha = 45°$;
(c) $\alpha = 60°$.

26. The projectile of Problem 25 is to be fired at a target 600 ft away, and there is a hill 300 ft high midway between the gun site and this target. At what initial angle of inclination should the projectile be fired?

A projectile is to be fired horizontally from the top of a 400-ft cliff at a target 1 mi from the base of the cliff. What should be the initial velocity of the projectile?

28. A bomb is dropped (initial speed zero) from a helicopter hovering at a height of 800 ft. A projectile is fired from a gun located on the ground 800 ft west of the point directly beneath the helicopter. The intent is for the projectile to intercept the bomb at a height of exactly 400 ft. If the projectile is fired at the same instant that the bomb is dropped, what should be its initial velocity and angle of inclination?

Suppose, more realistically, that the projectile of Problem 28 is fired 1 s after the bomb is dropped. What should be its initial velocity and angle of inclination?

30. An artillery gun with a muzzle velocity of 1000 ft/s is located atop a seaside cliff 500 ft high. At what initial inclination angle (or angles) should it fire a projectile in order to hit a ship at sea 20,000 ft from the base of the cliff?

Suppose that the vector-valued functions $\mathbf{u}(t)$ and $\mathbf{v}(t)$ both have limits as $t \to a$. Prove that
(a) $\lim_{t\to a} (\mathbf{u}(t) + \mathbf{v}(t)) = \lim_{t\to a} \mathbf{u}(t) + \lim_{t\to a} \mathbf{v}(t)$;
(b) $\lim_{t\to a} (\mathbf{u}(t) \cdot \mathbf{v}(t)) = \left(\lim_{t\to a} \mathbf{u}(t)\right) \cdot \left(\lim_{t\to a} \mathbf{v}(t)\right).$

32. Prove part (1) of Theorem 2.

33. Prove part (2) of Theorem 2.

34. Suppose that both the vector-valued function $\mathbf{r}(t)$ and the real-valued function $h(t)$ are differentiable. Deduce the chain rule for vector-valued functions,

$$D_t[\mathbf{r}(h(t))] = h'(t)\mathbf{r}'(h(t)),$$

in componentwise fashion from the ordinary chain rule.

35. A point moves with constant speed, so that its velocity vector $\mathbf{v}$ satisfies the condition

$$|\mathbf{v}|^2 = \mathbf{v} \cdot \mathbf{v} = C \qquad \text{(a constant)}.$$

Prove that the velocity and acceleration vectors of the point are always perpendicular to each other.

36. A point moves on a circle with center at the origin. Use the dot product to show that the position and velocity vectors of the moving point are always perpendicular.

37. A point moves on the hyperbola $x^2 - y^2 = 1$, with position vector

$$\mathbf{r}(t) = \mathbf{i} \cosh \omega t + \mathbf{j} \sinh \omega t$$

(the number ω is a constant). Prove that $\mathbf{a}(t) = c\mathbf{r}(t)$ where c is a positive constant. What sort of external force would produce this sort of motion?

38. Suppose that a point moves on the ellipse $x^2/a^2 + y^2/b^2 = 1$ with position vector $\mathbf{r}(t) = \mathbf{i}a \cos \omega t + \mathbf{j}b \sin \omega t$, ω a constant. Prove that $\mathbf{a}(t) = c\mathbf{r}(t)$, where c is a negative constant. To what sort of external force $\mathbf{F}(t)$ does this motion correspond?

39. A point moves in the plane with constant acceleration vector $\mathbf{a} = a\mathbf{j}$. Prove that its path is a parabola or a straight line.

40. Suppose that a particle is subject to no force, so its acceleration vector $\mathbf{a}(t)$ is identically zero. Prove that the particle travels with constant speed along a straight line (Newton's first law of motion).

41. (Uniform circular motion) Consider a particle that moves counterclockwise around the circle with center (0, 0) and radius r at a constant angular speed of ω radians per second (see Fig. 13.32). If its initial position is $(r, 0)$, then its position vector is

$$\mathbf{r}(t) = \mathbf{i}r \cos \omega t + \mathbf{j}r \sin \omega t.$$

(a) Show that the velocity vector of the particle is tangent to the circle and that the speed of the particle is

$$v(t) = |\mathbf{v}(t)| = r\omega.$$

(b) Show that the acceleration vector $\mathbf{a}$ of the particle is directed opposite to $\mathbf{r}$ and that

$$a(t) = |\mathbf{a}(t)| = r\omega^2.$$

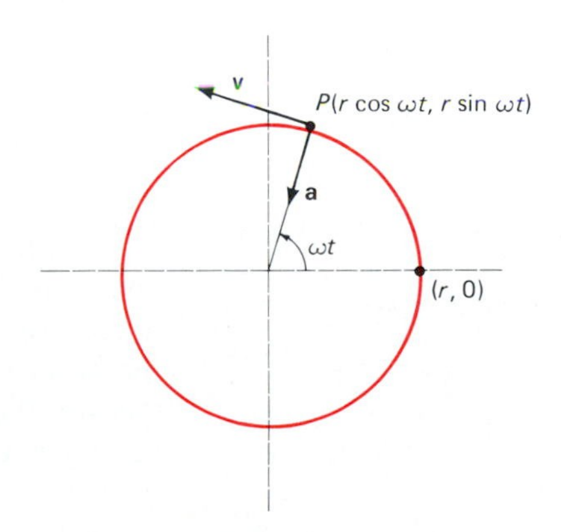

13.32 Uniform circular motion

*13.5 Orbits of Planets and Satellites

Ancient Greek astronomers and mathematicians developed an elaborate mathematical model to account for the complicated motions of the sun, moon, and six planets (then known) as viewed from the Earth. A combination of uniform circular motions was used to describe the motion of each body about the Earth—if the Earth is placed at the origin, then each body *does* orbit the Earth.

In this system, it was typical for a planet P to travel uniformly around a small circle (the *epicycle*) with center C, which in turn traveled uniformly around a circle centered at the Earth, labeled E in Fig. 13.33. The radii of the circles and the angular speeds of P and C around them were chosen to match the observed motion of the planet as closely as possible. For greater accuracy, one could use secondary circles. In fact, several circles were required for each body in the solar system. The theory of epicycles reached its definitive form in Ptolemy's *Almagest* of the second century A.D.

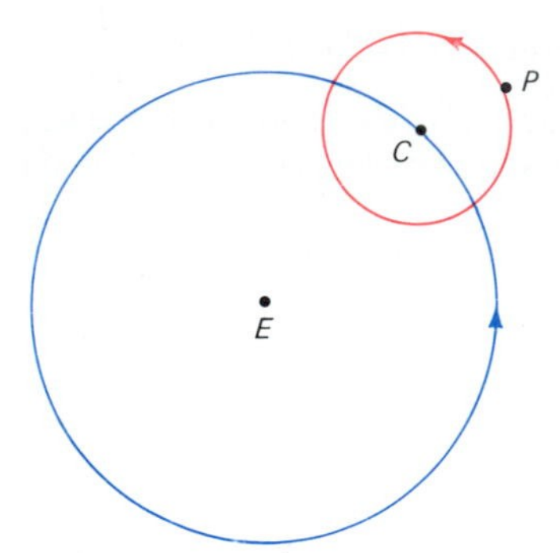

13.33 The small circle is the epicycle.

In 1543, Copernicus altered Ptolemy's approach by placing the center of each primary circle at the sun rather than at the Earth. This change was of much greater philosophical than mathematical importance. For, contrary to popular belief, this *heliocentric system* was *not* simpler than Ptolemy's geocentric system. Indeed, Copernicus's system actually required more circles.

It was Johann Kepler (1571–1630) who finally got rid of all these circles. On the basis of a detailed analysis of a lifetime of planetary observations by the Danish astronomer Tycho Brahe, Kepler stated the following three propositions, now known as **Kepler's laws of planetary motion:**

1. The orbit of each planet is an ellipse with the sun at one focus.
2. The radius vector from the sun to a planet sweeps out area at a constant rate.
3. The *square* of the period of revolution of a planet is proportional to the *cube* of the major semiaxis of its elliptical orbit.

In his *Principia Mathematica* (1687) Newton showed that Kepler's laws follow from the basic principles of mechanics ($F = ma$, and so on) and the inverse-square law of gravitational attraction. His success in using mathematics to explain natural phenomena ("I now demonstrate the frame of the System of the World") inspired confidence that the universe could be understood, and perhaps even mastered. This new confidence permanently altered humanity's perception of itself and of its place in the scheme of things.

In this section we show how Kepler's laws can be derived as indicated above. To begin with, set up a coordinate system in which the sun is located at the origin in the plane of motion of the planet. Let $r = r(t)$ and $\theta = \theta(t)$ be the polar coordinates at time t of the planet as it moves in its orbit about the sun. We want first to split the planet's position, velocity, and acceleration vectors $\mathbf{r}$, $\mathbf{v}$, and $\mathbf{a}$ into *radial* and *transverse* components. To do so, we introduce at each point (r, θ) of the plane (the origin excepted) the *unit* vectors

$$\mathbf{u}_r = \mathbf{i}\cos\theta + \mathbf{j}\sin\theta, \qquad \mathbf{u}_\theta = -\mathbf{i}\sin\theta + \mathbf{j}\cos\theta. \tag{1}$$

If we substitute $\theta = \theta(t)$, then $\mathbf{u}_r$ and $\mathbf{u}_\theta$ become functions of t. The **radial** unit vector $\mathbf{u}_r$ always points directly away from the origin; the **transverse** unit vector $\mathbf{u}_\theta$ is obtained from $\mathbf{u}_r$ by a 90° counterclockwise rotation, as shown in Fig. 13.34.

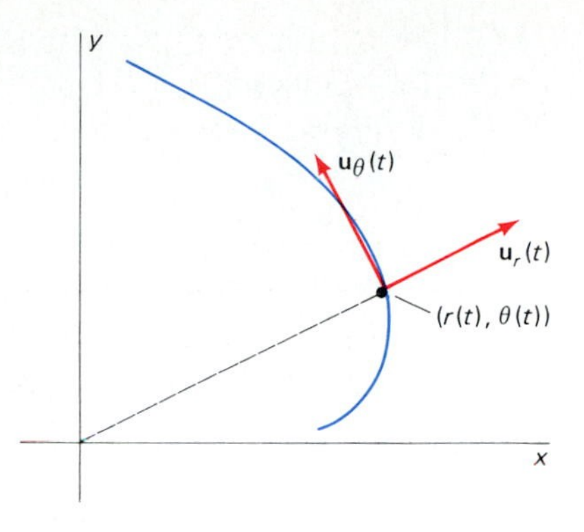

13.34 The radial and transverse unit vectors $\mathbf{u}_r$ and $\mathbf{u}_\theta$

In Problem 6 we ask you to verify, by componentwise differentiation of the equations in (1), that

$$\frac{d\mathbf{u}_r}{dt} = \mathbf{u}_\theta \frac{d\theta}{dt} \quad \text{and} \quad \frac{d\mathbf{u}_\theta}{dt} = -\mathbf{u}_r \frac{d\theta}{dt}. \tag{2}$$

The position vector $\mathbf{r}$ points directly away from the origin and has length $|\mathbf{r}| = r$, so

$$\mathbf{r} = r\mathbf{u}_r. \tag{3}$$

Differentiation of both sides in Equation (3) with respect to t gives the result

$$\mathbf{v} = \frac{d\mathbf{r}}{dt} = \mathbf{u}_r \frac{dr}{dt} + r\frac{d\mathbf{u}_r}{dt}.$$

We use the first equation in (2) and find that the planet's velocity vector is

$$\mathbf{v} = \mathbf{u}_r \frac{dr}{dt} + r\frac{d\theta}{dt}\mathbf{u}_\theta. \tag{4}$$

Thus we have expressed the velocity $\mathbf{v}$ in terms of the radial vector $\mathbf{u}_r$ and the transverse vector $\mathbf{u}_\theta$.

We differentiate this last equation, and find that

$$\mathbf{a} = \frac{d\mathbf{v}}{dt} = \left(\mathbf{u}_r \frac{d^2r}{dt^2} + \frac{dr}{dt}\frac{d\mathbf{u}_r}{dt}\right)$$
$$+ \left(\frac{dr}{dt}\frac{d\theta}{dt}\mathbf{u}_\theta + r\frac{d^2\theta}{dt^2}\mathbf{u}_\theta + r\frac{d\theta}{dt}\frac{d\mathbf{u}_\theta}{dt}\right).$$

Then, by using the equations in (2) and collecting the coefficients of $\mathbf{u}_r$ and $\mathbf{u}_\theta$ (Problem 7), we obtain the decomposition

$$\mathbf{a} = \left[\frac{d^2r}{dt^2} - r\left(\frac{d\theta}{dt}\right)^2\right]\mathbf{u}_r + \left[\frac{1}{r}\frac{d}{dt}\left(r^2\frac{d\theta}{dt}\right)\right]\mathbf{u}_\theta \tag{5}$$

of the acceleration vector into its radial and transverse components.

Let M denote the mass of the sun and m the mass of the orbiting planet. The inverse-square law of gravitation in vector form is

$$\mathbf{F} = m\mathbf{a} = -\frac{GMm}{r^2}\mathbf{u}_r,$$

so the acceleration of the planet *also* is given by

$$\mathbf{a} = -\frac{\mu}{r^2}\mathbf{u}_r \tag{6}$$

where $\mu = GM$. We equate the transverse components in (5) and (6), and thus obtain

$$\frac{1}{r}\cdot\frac{d}{dt}\left(r^2\frac{d\theta}{dt}\right) = 0.$$

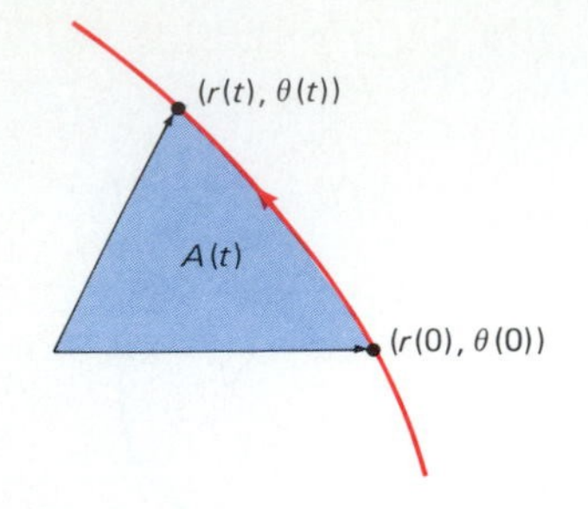

13.35 Area swept out by the radius vector

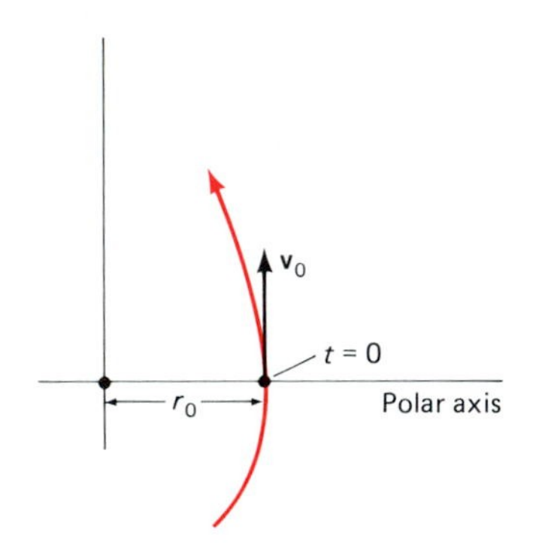

13.36 Setup of the coordinate system for the derivation of Kepler's first law

We drop the factor $1/r$, then antidifferentiate both sides. We find that

$$r^2 \frac{d\theta}{dt} = h \qquad (h \text{ a constant}). \tag{7}$$

We know from Section 10.3 that if $A(t)$ denotes the area swept out by the planet's radius vector from time 0 to time t (see Fig. 13.35), then

$$A(t) = \int_{*}^{**} \frac{1}{2} r^2 \, d\theta = \int_0^t \frac{1}{2} r^2 \frac{d\theta}{dt} \, dt.$$

Now we apply the fundamental theorem of calculus, which yields

$$\frac{dA}{dt} = \frac{1}{2} r^2 \frac{d\theta}{dt}. \tag{8}$$

When we compare Equations (7) and (8), we see that

$$\frac{dA}{dt} = \frac{h}{2}. \tag{9}$$

Because $h/2$ is a constant, we have derived Kepler's second law: The radius vector from sun to planet sweeps out area at a constant rate.

Now we derive Kepler's first law. Choose coordinate axes so that at time $t = 0$ the planet is on the polar axis and is at its closest point of approach to the sun, with initial position vector $\mathbf{r}_0$ and initial velocity vector $\mathbf{v}_0$, as in Fig. 13.36. By Equation (4),

$$\mathbf{v}_0 = r_0 \theta'(0) \mathbf{u}_\theta = v_0 \mathbf{j} \tag{10}$$

because, when $t = 0$, $\mathbf{u}_\theta = \mathbf{j}$ and $dr/dt = 0$ (as r is minimal then).

From Equation (6) we have

$$\frac{d\mathbf{v}}{dt} = -\frac{\mu}{r^2} \mathbf{u}_r = -\frac{\mu}{r^2} \left(-\frac{1}{d\theta/dt} \cdot \frac{d\mathbf{u}_\theta}{dt} \right)$$

$$= \frac{\mu}{r^2 (d\theta/dt)} \frac{d\mathbf{u}_\theta}{dt} = \frac{\mu}{h} \cdot \frac{d\mathbf{u}_\theta}{dt},$$

by using the second equation in (2) in the first step and then Equation (7).

Next we antidifferentiate and find that

$$\mathbf{v} = \frac{\mu}{h} \mathbf{u}_\theta + \mathbf{C}.$$

To find $\mathbf{C}$, we substitute $t = 0$. This yields $\mathbf{u}_\theta = \mathbf{j}$ and $\mathbf{v} = v_0 \mathbf{j}$. We get

$$\mathbf{C} = v_0 \mathbf{j} - \frac{\mu}{h} \mathbf{j} = \left(v_0 - \frac{\mu}{h} \right) \mathbf{j}.$$

Consequently,

$$\mathbf{v} = \frac{\mu}{h} \mathbf{u}_\theta + \left(v_0 - \frac{\mu}{h} \right) \mathbf{j}. \tag{11}$$

Now we take the dot product of each side in (11) with the unit vector $\mathbf{u}_\theta$. Remember that

$$\mathbf{u}_\theta \cdot \mathbf{v} = r \frac{d\theta}{dt} \quad \text{and} \quad \mathbf{u}_\theta \cdot \mathbf{j} = \cos \theta,$$

consequences of Equations (4) and (1), respectively. The result is

$$r\frac{d\theta}{dt} = \frac{\mu}{h} + \left(v_0 - \frac{\mu}{h}\right)\cos\theta.$$

But $r(d\theta/dt) = h/r$ by Equation (7). It follows that

$$\frac{h}{r} = \frac{\mu}{h} + \left(v_0 - \frac{\mu}{h}\right)\cos\theta.$$

We solve for r, and find that

$$r = \frac{h^2/\mu}{1 + [(v_0 h/\mu) - 1]\cos\theta}.$$

Next, from Equations (7) and (10) we see that

$$h = r_0 v_0, \tag{12}$$

so finally we may write the equation of the planet's orbit:

$$r = \frac{(r_0 v_0)^2/\mu}{1 + (r_0 v_0^2/\mu - 1)\cos\theta} = \frac{pe}{1 + e\cos\theta}. \tag{13}$$

In Problem 17 we ask you to show that (13) is the polar coordinates equation of a conic section with focus at the origin and with eccentricity

$$e = \frac{r_0 v_0^2}{GM} - 1. \tag{14}$$

Because the nature of a conic section is determined by its eccentricity, we see that the planet's orbit is:

- A circle if $r_0 v_0^2 = GM$,
- An ellipse if $GM < r_0 v_0^2 < 2GM$,
- A parabola if $r_0 v_0^2 = 2GM$,
- A hyperbola if $r_0 v_0^2 > 2GM$.

(15)

This comprehensive description of the situation is a generalization of Kepler's first law. Of course, the elliptical case is the one that holds for planets in the solar system.

Let us consider further the case in which the orbit is an ellipse with major semiaxis a and minor semiaxis b. (The case in which the ellipse is actually a circle, with $a = b$, will be a special case of this discussion.) In Problem 16 we ask you to show that the constant

$$pe = \frac{h^2}{GM}$$

used in (13) satisfies the equations

$$pe = a(1 - e^2) = a\left(1 - \frac{a^2 - b^2}{a^2}\right) = \frac{b^2}{a}. \tag{16}$$

We equate these two expressions for pe and find that

$$h^2 = GM\frac{b^2}{a}.$$

Now let T denote the period of revolution of the planet in its elliptical orbit. Then from Equation (9) we see that the area of the ellipse is $A = \frac{1}{2}hT = \pi ab$, and thus that

$$T^2 = \frac{4\pi^2 a^2 b^2}{h^2} = \frac{4\pi^2 a^2 b^2}{GMb^2/a},$$

so that

$$T^2 = \gamma a^3 \tag{17}$$

where $\gamma = 4\pi^2/GM$ is a constant. This is Kepler's third law.

EXAMPLE 1 The period of revolution of Mercury in its elliptical orbit about the sun is 87.97 days, while that of Earth is 365.26 days. Compute the major semiaxis (in astronomical units) of the orbit of Mercury.

Solution The major semiaxis of Earth's orbit is, by definition, 1 A.U. So Equation (17) gives the value of the constant $\gamma = (365.26)^2$ (day^2/A.U.^{3}). Hence the major semiaxis of the orbit of Mercury is

$$a = \left(\frac{T^2}{\gamma}\right)^{1/3} = \left(\frac{(87.97)^2}{(365.26)^2}\right)^{1/3} \approx 0.387 \quad \text{(A.U.)}.$$

As yet we have considered only planets in orbits about the sun. But Kepler's laws and the equations of this section apply to bodies in orbit about any common central mass, so long as they move solely under the influence of *its* gravitational attraction. Examples include satellites (artificial or natural) orbiting the Earth or the moons of Jupiter. All we need to know is the value of the constant $\mu = GM$ for the central body. For Earth, we can compute μ by beginning with the values

$$g = \frac{GM}{R^2} \approx 32.16$$

ft/s^2 for surface gravitational acceleration and the radius 3960 (mi) of the Earth—which we assume to be spherical. Then

$$\mu = GM = R^2 g = (3960)^2\left(\frac{32.16}{5280}\right) \approx 95{,}500 \quad (\text{mi}^3/\text{s}^2).$$

EXAMPLE 2 A communications relay satellite is to be placed in a circular orbit about the earth and to have a period of revolution of 24 h. This is a *synchronous* orbit, in which the satellite appears to be stationary in the sky. Assume that the earth's natural moon has a period of 27.32 days in a circular orbit of radius 238,850 mi. What should be the radius of the satellite's orbit?

Solution Equation (17), when applied to the moon, yields

$$(27.32)^2 = k(238{,}850)^3.$$

For the stationary satellite that has period $T = 1$ (day), it yields $1^2 = kr^3$, where r is the radius of the synchronous orbit. To eliminate k, we divide the second of these equations by the first, and find that

$$r^3 = \frac{(238{,}850)^3}{(27.32)^2}.$$

Thus r is approximately 26,330 mi. The radius of the earth is about 3960 mi, so the satellite will be 22,370 mi above the surface.

13.5 PROBLEMS

For each of the parametric curves in Problems 1–5, express its velocity and acceleration vectors in terms of $\mathbf{u}_r$ and $\mathbf{u}_\theta$; that is, in the form $A(t)\mathbf{u}_r + B(t)\mathbf{u}_\theta$.

1. $r = a, \quad \theta = t$
2. $r = 2\cos t, \quad \theta = t$
3. $r = t, \quad \theta = t$
4. $r = e^t, \quad \theta = 2t$
5. $r = 3\sin 4t, \quad \theta = 2t$

6. Derive both equations in (2) by differentiation of the equations in (1).

7. Derive Equation (5) by differentiating Equation (4).

8. Consider a body in an elliptical orbit with major and minor semiaxes a and b and period of revolution T.

(a) Deduce from Equation (4) that $v = r(d\theta/dt)$ when the body is nearest to and farthest from its focus.

(b) Then apply Kepler's second law to conclude that $v = 2\pi ab/rT$ at the body's nearest and farthest points.

In each of Problems 9–12, apply the equation of part (b) of Problem 8 to compute the speed (in miles per second) of the given body at the nearest and farthest points of its orbit. Convert 1 A.U., the major semiaxis of Earth's orbit, into 92,956,000 mi.

9. The planet Mercury: $a = 0.387$ A.U., $e = 0.206$, $T = 87.97$ days

10. The planet Earth: $e = 0.0167$, $T = 365.26$ days

11. The moon: $a = 238{,}900$ miles, $e = 0.055$, $T = 27.32$ days

12. An earth satellite: $a = 10{,}000$ mi, $e = 0.5$

13. Assuming the earth to be a sphere with radius 3960 mi, find the altitude above the earth's surface of a satellite in a circular orbit that has a period of revolution of 1 hour.

14. Given the fact that Jupiter's period of (almost) circular revolution about the sun is 11.86 years, calculate the distance of Jupiter from the sun.

15. Suppose that an earth satellite in elliptical orbit varies in altitude from 100 to 1000 mi above the earth's surface (presumed spherical). Find this satellite's period of revolution.

16. Substitute $\theta = 0$ and $\theta = \pi$ in Equation (13) to deduce that $pe = a(1 - e^2)$.

17. Figure 13.37 shows a conic section with eccentricity e and focus at the origin. Derive Equation (13) from the defining relation $|OP| = e|PQ|$ of the conic section.

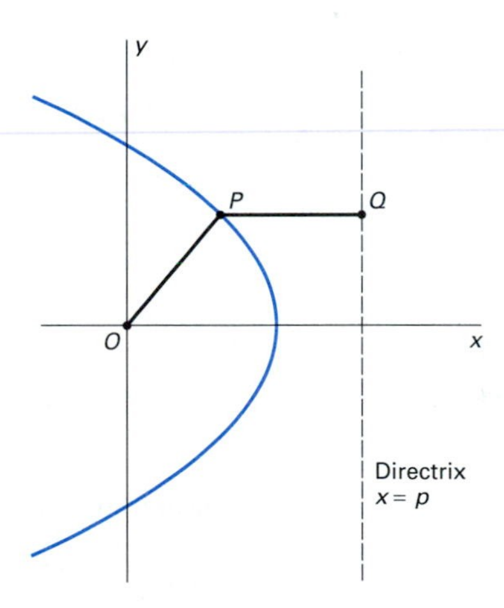

13.37 A conic section with eccentricity e: $|OP| = e|PQ|$.

CHAPTER 13 REVIEW: Definitions and Concepts

Use the list below as a guide to concepts that you may need to review.

1. Definition of a parametric curve; of a smooth parametric curve
2. The slope of the tangent line to a smooth parametric curve (both in rectangular and in polar coordinates)
3. Integral computations with parametric curves (Equations (1) through (4) in Section 13.2)
4. Arc length of a parametric curve
5. Vectors: their definition, length, addition, multiplication by scalars, and dot product
6. Test for perpendicular vectors

7. Vector-valued functions, velocity and acceleration vectors

8. Componentwise differentiation and integration of vector-valued functions

9. The equations of motion of a projectile

10. Kepler's three laws of planetary motion

11. The radial and transverse unit polar vectors

12. Polar decomposition of velocity and acceleration vectors

13. Outline of the derivation of Kepler's laws from Newton's law of gravitation

CHAPTER 13 MISCELLANEOUS PROBLEMS

In Problems 1–5, eliminate the parameter and sketch the curve.

1. $x = 2t^3 - 1, \quad y = 2t^3 + 1$
2. $x = \cosh t, \quad y = \sinh t$
3. $x = 2 + \cos t, \quad y = 1 - \sin t$
4. $x = \cos^4 t, \quad y = \sin^4 t$
5. $x = 1 + t^2, \quad y = t^3$

In Problems 6–10, write an equation of the tangent line to the given curve at the indicated point.

6. $x = t^2, \quad y = t^3; \quad t = 1$
7. $x = 3 \sin t, \quad y = 4 \cos t; \quad t = \pi/4$
8. $x = e^t, \quad y = e^{-t}; \quad t = 0$
9. $r = \theta; \quad \theta = \pi/2$
10. $r = 1 + \sin \theta; \quad \theta = \pi/3$

In each of Problems 11–14, find the area of the region between the given curve and the x-axis.

11. $x = 2t + 1, \quad y = t^2 + 3; \quad -1 \leqq t \leqq 2$
12. $x = e^t, \quad y = e^{-t}; \quad 0 \leqq t \leqq 10$
13. $x = 3 \sin t, \quad y = 4 \cos t; \quad 0 \leqq t \leqq \pi/2$
14. $x = \cosh t, \quad y = \sinh t; \quad 0 \leqq t \leqq 1$

In each of Problems 15–19, find the arc length of the given curve.

15. $x = t^2, \quad y = t^3; \quad 0 \leqq t \leqq 1$
16. $x = \ln(\cos t), \quad y = t; \quad 0 \leqq t \leqq \pi/4$
17. $x = 2t, \quad y = t^3 + 1/(3t); \quad 1 \leqq t \leqq 2$
18. $r = \sin \theta; \quad 0 \leqq \theta \leqq \pi$
19. $r = \sin^3(\theta/3); \quad 0 \leqq \theta \leqq \pi$

In each of Problems 20–24, find the area of the surface generated by revolving the given curve around the x-axis.

20. $x = t^2 + 1, \quad y = 3t; \quad 0 \leqq t \leqq 2$
21. $x = 4t^{1/2}, \quad y = \frac{1}{3}t^3 + \frac{1}{2}t^{-2}; \quad 1 \leqq t \leqq 4$
22. $r = 4 \cos \theta$
23. $r = \exp(\theta/2), \quad 0 \leqq \theta \leqq \pi$
24. $x = e^t \cos t, \quad y = e^t \sin t; \quad 0 \leqq \theta \leqq \pi/2$

25. Consider the rolling circle of radius a that was used to generate the cycloid in Example 3 of Section 13.1. Suppose that this circle is the rim of a disk, and let Q be a point of this disk at distance $b < a$ from its center. Find parametric equations for the curve traced by Q as the circle rolls along the x-axis; assume that Q begins at the point $(0, a - b)$. Sketch this curve, which is called a *trochoid*.

26. If the smaller circle of Problem 28 in Section 13.1 rolls around the *outside* of the larger circle, the path of the point P is called an *epicycloid*. Show that it has parametric equations

$$x = (a + b) \cos t - b \cos\left(\frac{a + b}{b} t\right),$$

$$y = (a + b) \sin t - b \sin\left(\frac{a + b}{b} t\right).$$

27. Suppose that $b = a$ in Problem 26. Show that the epicycloid is then the cardioid $r = 2a(1 - \cos \theta)$ translated a units to the right.

28. Find the area of the surface generated by revolving the lemniscate $r^2 = 2a^2 \cos 2\theta$ around the x-axis.

29. Find the volume generated by revolving around the y-axis the area under the cycloid

$$x = a(t - \sin t), \qquad y = a(1 - \cos t), \qquad 0 \leqq t \leqq 2\pi.$$

30. Show that the length of one arch of the hypocycloid of Problem 28 in Section 13.1 is $s = 8b(a - b)/a$.

31. Let ABC be an isosceles triangle with $|AB| = |AC|$. Let M be the midpoint of BC. Use the dot product to show that AM and BC are perpendicular.

32. Use the dot product to show that the diagonals of a rhombus (a parallelogram with all four sides equal) are perpendicular to each other.

33. The acceleration of a certain particle is

$$\mathbf{a} = \mathbf{i} \sin t - \mathbf{j} \cos t.$$

Assume that the particle begins at time $t = 0$ at the point $(0, 1)$, and has initial velocity $\mathbf{v}_0 = -\mathbf{i}$. Show that its path is a circle.

34. A particle moves in an attracting central force field, with force proportional to the distance from the origin. This implies that the particle's acceleration is given by

$\mathbf{a} = -\omega^2\mathbf{r}$, where $\mathbf{r}$ is the position vector of the particle. Assume that the particle's initial position is $\mathbf{r}_0 = p\mathbf{i}$ and that its initial velocity is $\mathbf{v}_0 = q\omega\mathbf{j}$. Show that the trajectory of the particle is the ellipse with equation $x^2/p^2 + y^2/q^2 = 1$. (*Suggestion:* Apply the theorem of Section 8.2.)

35. At time $t = 0$, a ground target is 160 ft from a gun and is moving directly away from it with a constant speed of 80 ft/s. If the muzzle velocity of the gun is 320 ft/s, at what angle of elevation α should it be fired in order to strike the moving target?

36. Suppose that a gun with muzzle velocity v_0 is located at the foot of a hill with a 30° slope. At what angle of elevation (from the horizontal) should the gun be fired in order to maximize its range, as measured up the hill?

37. Assume that a body has an elliptical orbit

$$r = \frac{pe}{1 + e\cos\theta}$$

and satisfies Kepler's second law in the form

$$r^2\frac{d\theta}{dt} = h \qquad \text{(a constant).}$$

Then deduce from Equation (5) in Section 13.5 that its acceleration vector $\mathbf{a}$ is equal to $(k/r^2)\mathbf{u}_r$ for some constant k. This shows that Newton's inverse-square law of gravitation follows from Kepler's first and second laws.

chapter fourteen

Vectors, Curves, and Surfaces in Space

14.1 Rectangular Coordinates and Three-Dimensional Vectors

In the first thirteen chapters we have discussed many aspects of the calculus of functions of a *single* variable. The geometry of such functions is two-dimensional because the graph of a function of a single variable is a curve in the plane. Most of the remainder of this book deals with the calculus of functions of *several* (two or more) independent variables. The geometry of functions of two variables is three-dimensional because the graphs of such functions are generally surfaces in space.

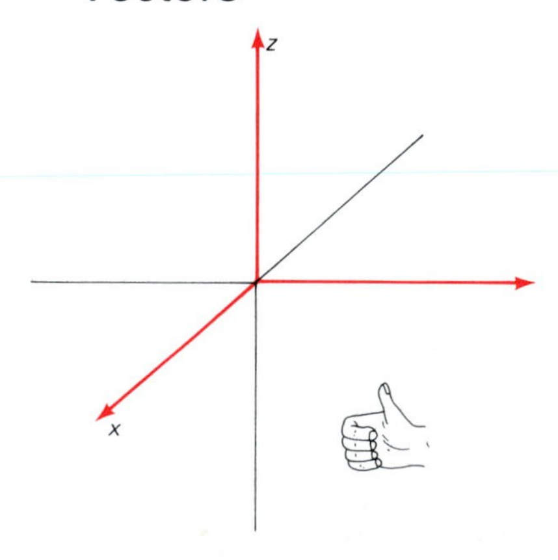

14.1 The right-handed rectangular coordinate system

Rectangular coordinates in the plane may be generalized in a natural way to rectangular coordinates in space. A point in space is determined by giving its location relative to three mutually perpendicular **coordinate axes** passing through the origin O. We shall always draw the x-, y-, and z-axes as shown in Fig. 14.1, with arrows indicating the positive direction along each axis. With this configuration of axes, our rectangular coordinate system is said to be **right-handed:** If the curled fingers of the right hand point in the direction of a 90° rotation from the positive x-axis to the positive y-axis, then the thumb points in the direction of the positive z-axis. If the x- and y-axes were interchanged, then we would have a left-handed coordinate system. These two coordinate systems are different in that it is impossible to bring one into coincidence with the other by means of rotations and translations. Similarly, the L- and D-alanine molecules shown in Fig. 14.2

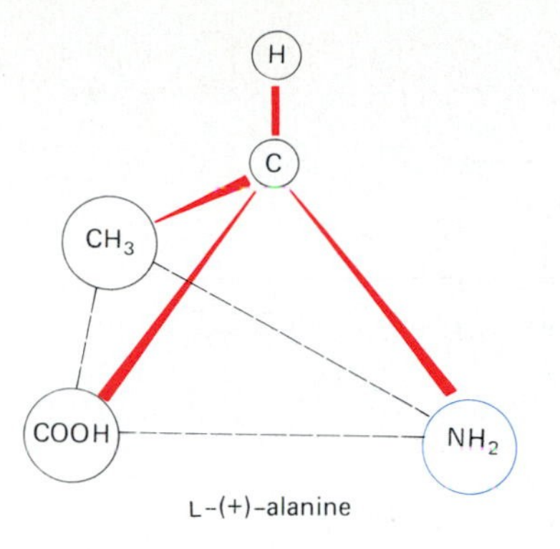

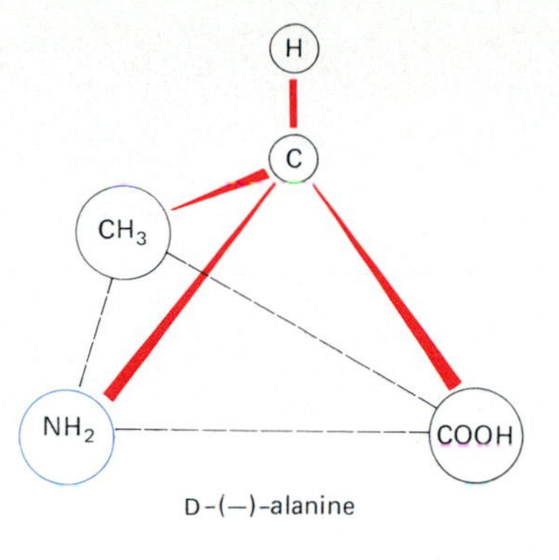

14.2 The stereoisomers of the amino acid alanine are physically and biologically different.

are different; you can digest the left-handed ("levo") version but not the right-handed ("dextro") version. In this book we shall use right-handed coordinate systems exclusively and always draw the x-, y-, and z-axes with the orientation shown in Fig. 14.1.

The three coordinate axes taken in pairs determine the three **coordinate planes:**

- The (horizontal) xy-plane, where $z = 0$;
- The (vertical) yz-plane, where $x = 0$; and
- The (vertical) xz-plane, where $y = 0$.

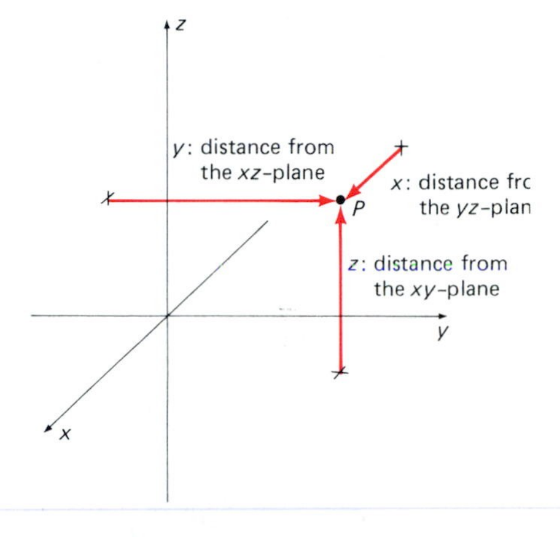

14.3 Locating the point P with rectangular coordinates

The point P in space is said to have **rectangular coordinates** (x, y, z) if (see Fig. 14.3):

- x is its signed distance from the yz-plane;
- y is its signed distance from the xz-plane; and
- z is its signed distance from the xy-plane.

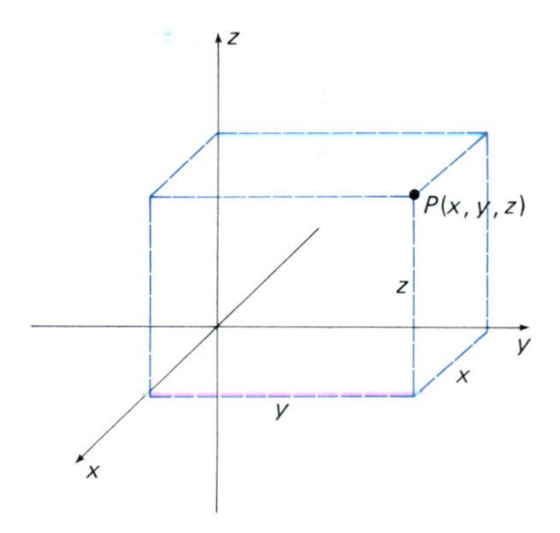

14.4 "Completing the box" to show P with the illusion of the third dimension

In this case we may describe the location of the point P by simply calling it "the point $P(x, y, z)$." There is a natural one-to-one correspondence between ordered triples (x, y, z) of real numbers and points P in space; this correspondence is called a **rectangular coordinate system** in space. In Fig. 14.4 the point P is located in the **first octant**—the eighth of space in which all three rectangular coordinates are positive.

If we apply the Pythagorean theorem to the right triangles P_1QR and P_1RP_2 in Fig. 14.5, we get

$$|P_1P_2|^2 = |RP_2|^2 + |P_1R|^2 = |RP_2|^2 + |QR|^2 + |P_1Q|^2$$
$$= (x_1 - x_2)^2 + (y_1 - y_2)^2 + (z_1 - z_2)^2.$$

Thus the **distance formula** for the **distance** $|P_1P_2|$ between the points P_1 and P_2 is

$$|P_1P_2| = \sqrt{(x_1 - x_2)^2 + (y_1 - y_2)^2 + (z_1 - z_2)^2}. \qquad (1)$$

For example, the distance between the points $P_1(1, 3, -2)$ and $P_2(4, -3, 1)$ is

$$|P_1P_2| = \sqrt{(4 - 1)^2 + (-3 - 3)^2 + (1 + 2)^2} = \sqrt{54} \approx 7.34847.$$

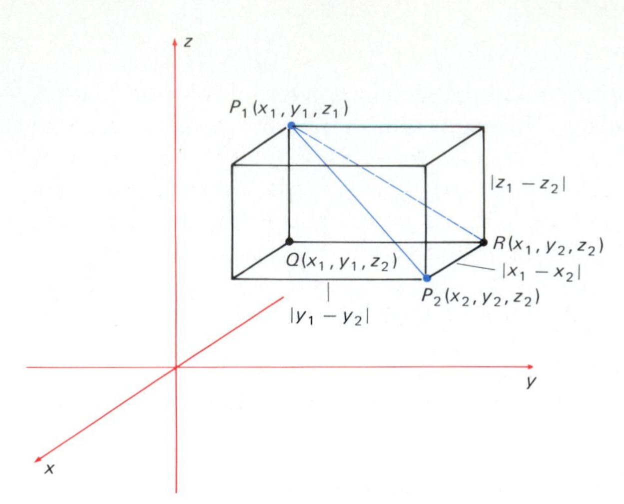

14.5 The distance between P_1 and P_2 is the length of the long diagonal of the box.

In Problem 45 we ask you to apply the distance formula in (1) to show that the **midpoint** M of the line segment joining $P_1(x_1, y_1, z_1)$ and $P_2(x_2, y_2, z_2)$ is

$$M\left(\frac{x_1 + x_2}{2}, \frac{y_1 + y_2}{2}, \frac{z_1 + z_2}{2}\right). \tag{2}$$

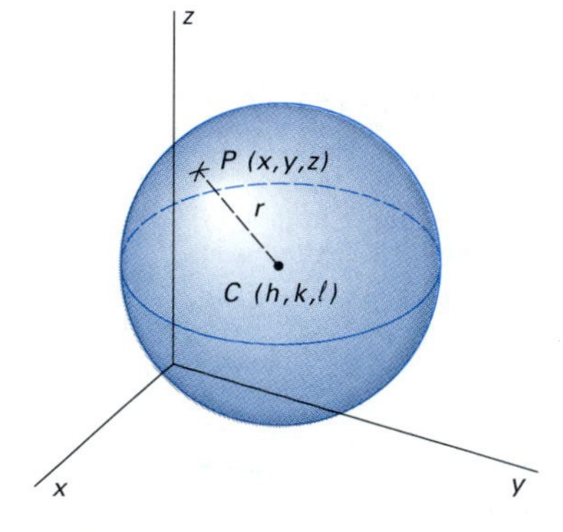

14.6 The sphere with center (h, k, ℓ) and radius r

The **graph** of an equation in three variables x, y, and z is the set of all points in space with rectangular coordinates that satisfy the equation. In general, the graph of an equation in three variables will be a *two-dimensional surface* in $\boldsymbol{R}^3$ (three-dimensional space with rectangular coordinates). For example, let $C(h, k, l)$ be a fixed point. Then the graph of the equation

$$(x - h)^2 + (y - k)^2 + (z - l)^2 = r^2 \tag{3}$$

is the set of all points $P(x, y, z)$ at distance $r > 0$ from the fixed point C. This means that Equation (3) is the equation of the **sphere with radius r and center** $C(h, k, l)$ shown in Fig. 14.6. Moreover, given an equation of the form

$$x^2 + y^2 + z^2 + Ax + By + Cz + D = 0,$$

we can attempt—by completing the square in each variable—to write it in the form in (3), and thereby show that its graph is a sphere.

EXAMPLE 1 Determine the graph of the equation

$$x^2 + y^2 + z^2 + 4x + 2y - 6z - 2 = 0.$$

Solution We complete the square in each variable. The equation then takes the form

$$(x^2 + 4x + 4) + (y^2 + 2y + 1) + (z^2 - 6z + 9) = 16.$$

That is,

$$(x + 2)^2 + (y + 1)^2 + (z - 3)^2 = 4^2.$$

Thus the given equation has as its graph a sphere with radius 4 and center $(-2, -1, 3)$.

VECTORS IN SPACE

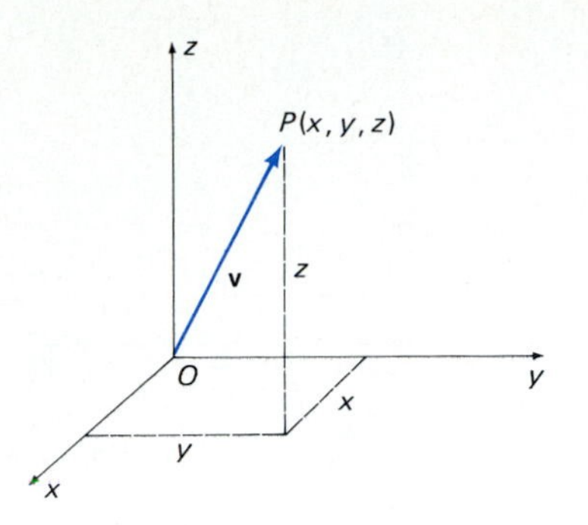

14.7 The segment OP is a realization of the vector $\mathbf{v} = OP$.

The discussion of vectors in the plane in Section 13.3 may be repeated almost verbatim for vectors in space. The major difference is that a vector in space has three components rather than two. The vector determined by the point $P(x, y, z)$ is its **position vector** $\mathbf{v} = \overrightarrow{OP} = \langle x, y, z \rangle$, which is represented pictorially (Fig. 14.7) by the arrow from the origin O to P (or by any parallel translate of this arrow). The distance formula in (1) gives

$$|\mathbf{v}| = \sqrt{x^2 + y^2 + z^2} \tag{4}$$

for the **length** of the vector $\mathbf{v} = \langle x, y, z \rangle$.

The vector $\overrightarrow{AB}$ represented by the arrow (Fig. 14.8) from $A(a_1, a_2, a_3)$ to $B(b_1, b_2, b_3)$ is defined to be

$$\overrightarrow{AB} = \langle b_1 - a_1, b_2 - a_2, b_3 - a_3 \rangle,$$

and its length is simply the distance between the two points A and B.

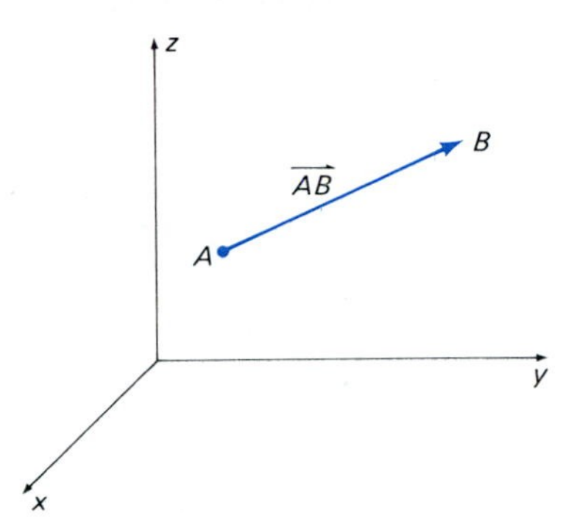

14.8 The segment AB is an instance of the vector $\overrightarrow{AB}$.

We define addition and scalar multiplication of vectors exactly as in Section 13.3, taking into account that our vectors now have three components instead of two: The **sum** of the vectors $\mathbf{a} = \langle a_1, a_2, a_3 \rangle$ and $\mathbf{b} = \langle b_1, b_2, b_3 \rangle$ is the vector

$$\mathbf{a} + \mathbf{b} = \langle a_1 + b_1, a_2 + b_2, a_3 + b_3 \rangle. \tag{5}$$

Because $\mathbf{a}$ and $\mathbf{b}$ lie in a plane (though perhaps not the xy-plane) if their initial points coincide, addition of vectors obeys the same **parallelogram law** as in the two-dimensional case (see Fig. 14.9).

If c is a real number, then the **scalar multiple** $c\mathbf{a}$ is the vector

$$c\mathbf{a} = \langle ca_1, ca_2, ca_3 \rangle. \tag{6}$$

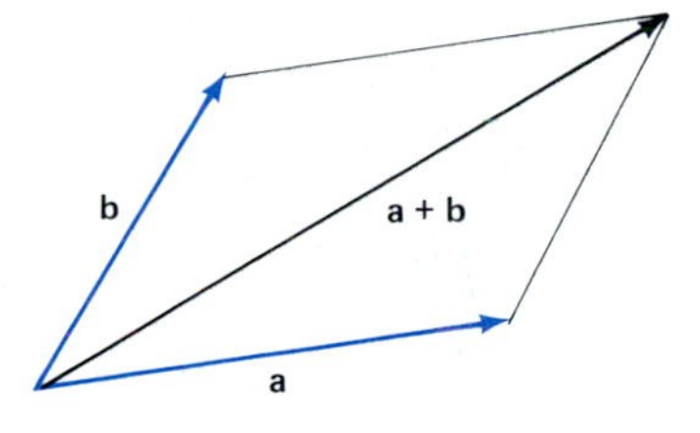

14.9 The parallelogram law for addition of vectors

The length of $c\mathbf{a}$ is $|c|$ times the length of $\mathbf{a}$, and $c\mathbf{a}$ has the same direction as $\mathbf{a}$ if $c > 0$ but the opposite direction if $c < 0$. The following algebraic properties of vector addition and scalar multiplication are easy to establish—they follow from computations with components, just as in Section 13.3.

$$\begin{aligned}
\mathbf{a} + \mathbf{b} &= \mathbf{b} + \mathbf{a}, \\
\mathbf{a} + (\mathbf{b} + \mathbf{c}) &= (\mathbf{a} + \mathbf{b}) + \mathbf{c}, \\
r(\mathbf{a} + \mathbf{b}) &= r\mathbf{a} + r\mathbf{b}, \\
(r + s)\mathbf{a} &= r\mathbf{a} + s\mathbf{a}, \\
(rs)\mathbf{a} &= r(s\mathbf{a}) = s(r\mathbf{a}).
\end{aligned} \tag{7}$$

EXAMPLE 2 If $\mathbf{a} = \langle 3, 4, 12 \rangle$ and $\mathbf{b} = \langle -4, 3, 0 \rangle$, then

$$\begin{aligned}
\mathbf{a} + \mathbf{b} &= \langle 3 - 4, 4 + 3, 12 + 0 \rangle = \langle -1, 7, 12 \rangle, \\
|\mathbf{a}| &= \sqrt{3^2 + 4^2 + 12^2} = \sqrt{169} = 13, \\
2\mathbf{a} &= \langle 2 \cdot 3, 2 \cdot 4, 2 \cdot 12 \rangle = \langle 6, 8, 24 \rangle, \\
2\mathbf{a} - 3\mathbf{b} &= \langle 6 + 12, 8 - 9, 24 - 0 \rangle = \langle 18, -1, 24 \rangle.
\end{aligned}$$

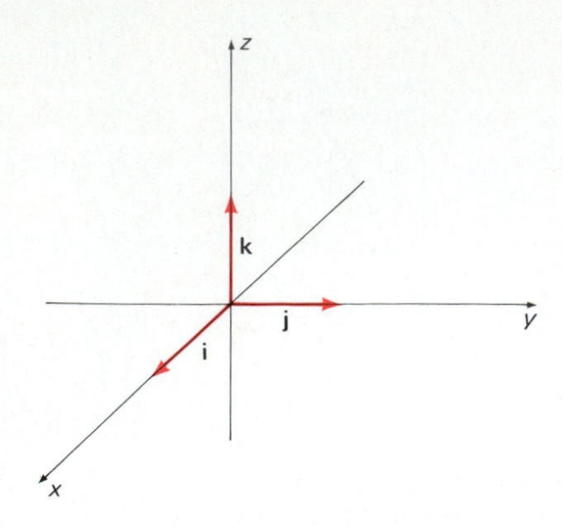

14.10 The basic unit vectors **i**, **j**, and **k**

A **unit vector** is one with length 1. Any space vector can be expressed in terms of the three **basic unit vectors**

$$\mathbf{i} = \langle 1, 0, 0 \rangle, \qquad \mathbf{j} = \langle 0, 1, 0 \rangle, \qquad \mathbf{k} = \langle 0, 0, 1 \rangle.$$

When located with their initial points at the origin, these basic unit vectors form a right-handed triple of vectors pointing in the positive directions along the three coordinate axes (as shown in Fig. 14.10).

Any space vector $\mathbf{a} = \langle a_1, a_2, a_3 \rangle$ can be written as

$$\mathbf{a} = a_1\mathbf{i} + a_2\mathbf{j} + a_3\mathbf{k}$$

in terms of the basic unit vectors. As in the two-dimensional case, the usefulness of this representation is that algebraic operations involving vectors may be carried out simply by collecting coefficients of **i**, **j**, and **k**. For example,

$$\begin{aligned}\mathbf{a} + \mathbf{b} &= (a_1\mathbf{i} + a_2\mathbf{j} + a_3\mathbf{k}) + (b_1\mathbf{i} + b_2\mathbf{j} + b_3\mathbf{k}) \\ &= (a_1 + b_1)\mathbf{i} + (a_2 + b_2)\mathbf{j} + (a_3 + b_3)\mathbf{k}.\end{aligned}$$

The **dot product** of the two vectors

$$\mathbf{a} = a_1\mathbf{i} + a_2\mathbf{j} + a_3\mathbf{k} \quad \text{and} \quad \mathbf{b} = b_1\mathbf{i} + b_2\mathbf{j} + b_3\mathbf{k}$$

is defined almost exactly as before: Multiply corresponding components, then add the results. Thus

$$\mathbf{a} \cdot \mathbf{b} = a_1b_1 + a_2b_2 + a_3b_3. \tag{8}$$

If $a_3 = 0 = b_3$, then we may think of **a** and **b** as vectors in the xy-plane. Then the definition in (8) reduces to the one given in Section 13.3 for the dot product of two vectors in the plane. The three-dimensional dot product has the same list of properties as the two-dimensional dot product, and all those shown in (9) can be established routinely by working with components.

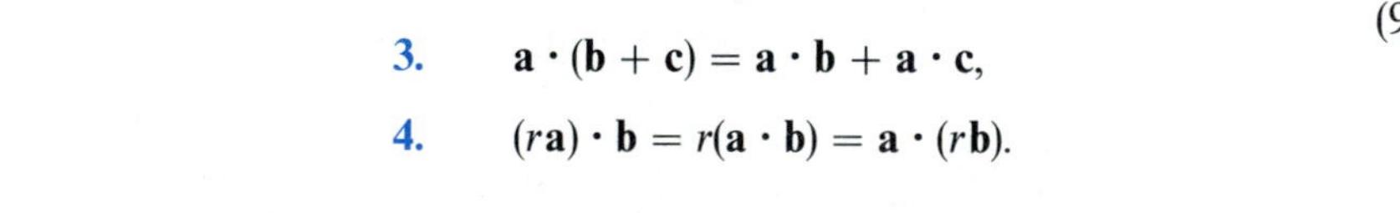

$$\begin{aligned}&\textbf{1.} \quad \mathbf{a} \cdot \mathbf{a} = |\mathbf{a}|^2, \\ &\textbf{2.} \quad \mathbf{a} \cdot \mathbf{b} = \mathbf{b} \cdot \mathbf{a}, \\ &\textbf{3.} \quad \mathbf{a} \cdot (\mathbf{b} + \mathbf{c}) = \mathbf{a} \cdot \mathbf{b} + \mathbf{a} \cdot \mathbf{c}, \\ &\textbf{4.} \quad (r\mathbf{a}) \cdot \mathbf{b} = r(\mathbf{a} \cdot \mathbf{b}) = \mathbf{a} \cdot (r\mathbf{b}).\end{aligned} \tag{9}$$

EXAMPLE 3 If $\mathbf{a} = \langle 3, 4, 12 \rangle$ and $\mathbf{b} = \langle -4, 3, 0 \rangle$, then

$$\mathbf{a} \cdot \mathbf{b} = (3)(-4) + (4)(3) + (12)(0) = -12 + 12 + 0 = 0.$$

If $\mathbf{c} = \langle 4, 5, -3 \rangle$, then

$$\mathbf{a} \cdot \mathbf{c} = (3)(4) + (4)(5) + (12)(-3) = 12 + 20 - 36 = -4.$$

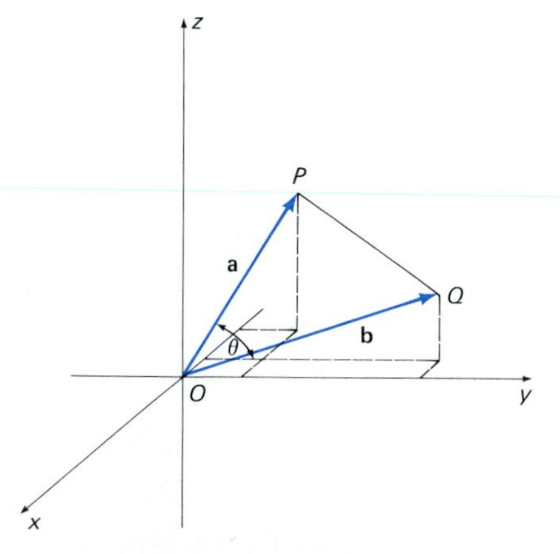

14.11 The angle θ between the vectors **a** and **b**

It is important to remember that the dot product of two *vectors* is a *scalar*—that is, a real number. For this reason the dot product is often called the *scalar product*.

The significance of the dot product resides in its geometric interpretation. Let the vectors **a** and **b** be represented by the position vectors $\overrightarrow{OP}$ and $\overrightarrow{OQ}$, respectively. Then the angle θ between **a** and **b** is the angle at O in the triangle OPQ of Fig. 14.11. We say that **a** and **b** are **parallel** if $\theta = 0$ or if

$\theta = \pi$, and that $\mathbf{a}$ and $\mathbf{b}$ are **perpendicular** if $\theta = \pi/2$. For convenience, we regard the zero vector $\mathbf{0} = \langle 0, 0, 0\rangle$ as both parallel to *and* perpendicular to *every* vector.

Theorem *Interpretation of the Dot Product*
If θ is the angle between the vectors $\mathbf{a}$ and $\mathbf{b}$, then

$$\mathbf{a} \cdot \mathbf{b} = |\mathbf{a}|\,|\mathbf{b}| \cos\theta \tag{10}$$

Proof If either $\mathbf{a} = \mathbf{0}$ or $\mathbf{b} = \mathbf{0}$, then Equation (10) follows immediately. If the vectors $\mathbf{a}$ and $\mathbf{b}$ are parallel, then $\mathbf{b} = t\mathbf{a}$ with either $t > 0$ and $\theta = 0$ or $t < 0$ and $\theta = \pi$. In either case, both sides in (10) reduce to $t|\mathbf{a}|^2$, and again Equation (10) follows.

So we turn to the general case in which the vectors $\mathbf{a} = \overrightarrow{OP}$ and $\mathbf{b} = \overrightarrow{OQ}$ are nonzero and nonparallel. Then

$$\begin{aligned}|\overrightarrow{QP}|^2 = |\mathbf{a} - \mathbf{b}|^2 = (\mathbf{a} - \mathbf{b}) \cdot (\mathbf{a} - \mathbf{b}) &= \mathbf{a} \cdot \mathbf{a} - \mathbf{a} \cdot \mathbf{b} - \mathbf{b} \cdot \mathbf{a} + \mathbf{b} \cdot \mathbf{b} \\ &= |\mathbf{a}|^2 + |\mathbf{b}|^2 - 2\mathbf{a} \cdot \mathbf{b}.\end{aligned}$$

on the other hand, $c = |\overrightarrow{QP}|$ is the side of the triangle OPQ (Fig. 14.11) opposite the angle θ included by the sides $a = |\mathbf{a}|$ and $b = |\mathbf{b}|$. Hence the law of cosines yields

$$|\overrightarrow{QP}|^2 = c^2 = a^2 + b^2 - 2ab\cos\theta = |\mathbf{a}|^2 + |\mathbf{b}|^2 - 2|\mathbf{a}|\,|\mathbf{b}|\cos\theta.$$

Finally, comparison of these two expressions for $|\overrightarrow{QP}|^2$ gives Equation (10). ■

This theorem tells us that the angle between the nonzero vectors $\mathbf{a}$ and $\mathbf{b}$ is given by

$$\cos\theta = \frac{\mathbf{a} \cdot \mathbf{b}}{|\mathbf{a}|\,|\mathbf{b}|}. \tag{11}$$

Note that this immediately implies the perpendicularity test of Section 13.3: *The two vectors* $\mathbf{a}$ *and* $\mathbf{b}$ *are perpendicular* ($\theta = \pi/2$) *if and only if* $\mathbf{a} \cdot \mathbf{b} = 0$. For instance, the vectors $\mathbf{a}$ and $\mathbf{b}$ of Example 3 are perpendicular because we found there that $\mathbf{a} \cdot \mathbf{b} = 0$.

EXAMPLE 4 Find the angles shown in the triangle of Fig. 14.12, having vertices at $A(2, -1, 0)$, $B(5, -4, 3)$, and $C(1, -3, 2)$.

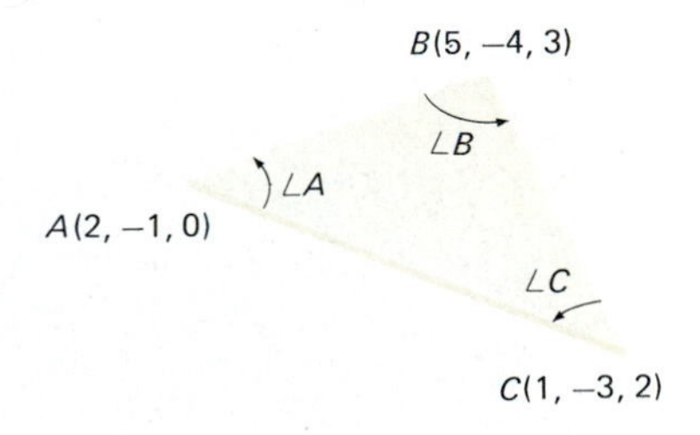

14.12 The triangle of Example 4

Solution We apply Equation (10) with $\theta = \angle A$, $\mathbf{a} = \overrightarrow{AB} = \langle 3, -3, 3\rangle$, and $\mathbf{b} = \overrightarrow{AC} = \langle -1, -2, 2\rangle$. This yields

$$\begin{aligned}\angle A &= \cos^{-1}\left(\frac{\langle 3, -3, 3\rangle \cdot \langle -1, -2, 2\rangle}{\sqrt{27}\sqrt{9}}\right) \\ &= \cos^{-1}\left(\frac{9}{\sqrt{27}\sqrt{9}}\right) \approx 54.74^\circ.\end{aligned}$$

Similarly,

$$\angle B = \cos^{-1}\left(\frac{\overrightarrow{BA} \cdot \overrightarrow{BC}}{|\overrightarrow{BA}||\overrightarrow{BC}|}\right)$$

$$= \cos^{-1}\left(\frac{\langle -3, 3, -3\rangle \cdot \langle -4, 1, -1\rangle}{\sqrt{27}\sqrt{18}}\right)$$

$$= \cos^{-1}\left(\frac{18}{\sqrt{27}\sqrt{18}}\right) \approx 35.26^\circ.$$

Then $\angle C = 180^\circ - \angle A - \angle B = 90^\circ$. As a check, note that

$$\overrightarrow{CA} \cdot \overrightarrow{CB} = \langle 1, 2, -2,\rangle \cdot \langle 4, -1, 1\rangle = 0.$$

So the angle at C is, indeed, a right angle.

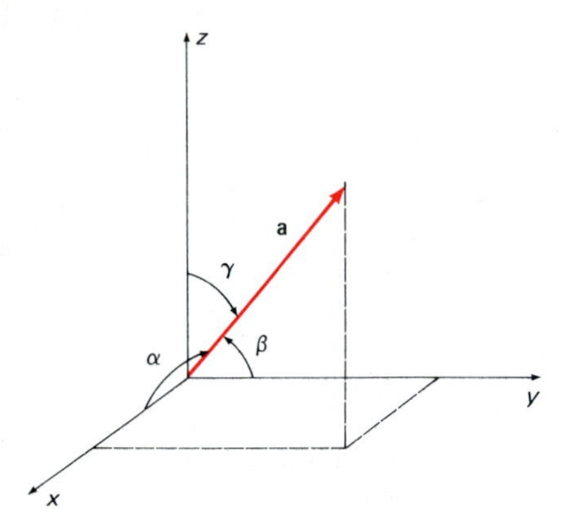

14.13 The direction angles of the vector **a**

The **direction angles** of the nonzero vector $\mathbf{a} = \langle a_1, a_2, a_3\rangle$ are the angles α, β, and γ that it makes with the vectors **i**, **j**, and **k** respectively, as shown in Fig. 14.13. The cosines of these angles, $\cos\alpha$, $\cos\beta$, and $\cos\gamma$, are called the **direction cosines** of the vector **a**. When we replace **b** in Equation (11) by **i**, **j**, and **k** in turn, we find that

$$\cos\alpha = \frac{\mathbf{a}\cdot\mathbf{i}}{|\mathbf{a}||\mathbf{i}|} = \frac{a_1}{|\mathbf{a}|},$$

$$\cos\beta = \frac{\mathbf{a}\cdot\mathbf{j}}{|\mathbf{a}||\mathbf{j}|} = \frac{a_2}{|\mathbf{a}|}, \quad \text{and} \tag{12}$$

$$\cos\gamma = \frac{\mathbf{a}\cdot\mathbf{k}}{|\mathbf{a}||\mathbf{k}|} = \frac{a_3}{|\mathbf{a}|}.$$

That is, the direction cosines of **a** are the components of the **unit vector** $\mathbf{a}/|\mathbf{a}|$ with the same direction as **a**. Consequently

$$\cos^2\alpha + \cos^2\beta + \cos^2\gamma = 1. \tag{13}$$

EXAMPLE 5 Find the direction angles of the vector $\mathbf{a} = 2\mathbf{i} + 3\mathbf{j} - \mathbf{k}$.

Solution Because $\mathbf{a} = \sqrt{14}$, the equations in (12) give

$$\alpha = \cos^{-1}\left(\frac{2}{\sqrt{14}}\right) \approx 57.69^\circ, \qquad \beta = \cos^{-1}\left(\frac{3}{\sqrt{14}}\right) \approx 36.70^\circ,$$

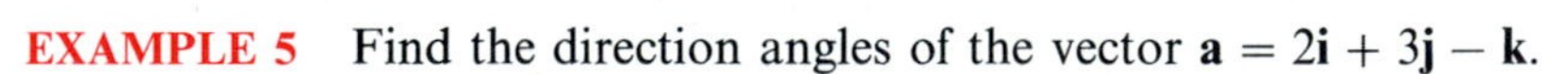

$$\text{and} \quad \gamma = \cos^{-1}\left(\frac{-1}{\sqrt{14}}\right) \approx 105.50^\circ.$$

Sometimes we need to find the component of one vector **a** in the direction of another (nonzero) vector **b**. Think of the two vectors located with the same initial point, as in Fig. 14.14. Then the (scalar) **component of a along b**, denoted by $\text{comp}_\mathbf{b}\,\mathbf{a}$, is numerically the length of the perpendicular projection of **a** onto the straight line determined by **b**. It is positive if the angle θ between **a** and **b** is acute (so that **a** and **b** point in the same general direction) and negative if $\theta > \pi/2$. Thus $\text{comp}_\mathbf{b}\,\mathbf{a} = |\mathbf{a}|\cos\theta$ in either case.

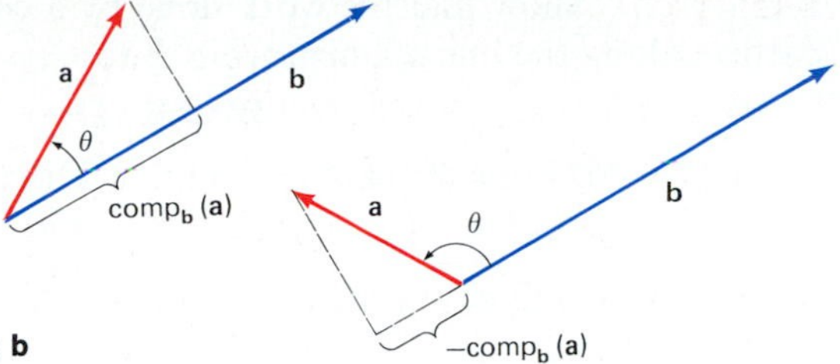

14.14 The component of **a** along **b**

Equation (10) then gives

$$\text{comp}_{\mathbf{b}}\mathbf{a} = \frac{|\mathbf{a}|\,|\mathbf{b}|\cos\theta}{|\mathbf{b}|} = \frac{\mathbf{a}\cdot\mathbf{b}}{|\mathbf{b}|}. \tag{14}$$

There is no need to memorize this formula, for—in practice—we can always read $\text{comp}_{\mathbf{b}}\mathbf{a} = |\mathbf{a}|\cos\theta$ from the figure and then apply (10) to eliminate $\cos\theta$. Note that $\text{comp}_{\mathbf{b}}\mathbf{a}$ is a scalar, not a vector.

EXAMPLE 6 Given $\mathbf{a} = \langle 4, -5, 3\rangle$ and $\mathbf{b} = \langle 2, 1, -2\rangle$, write $\mathbf{a}$ as the sum of a vector $\mathbf{a}_{||}$ parallel to $\mathbf{b}$ and a vector $\mathbf{a}_{\perp}$ perpendicular to $\mathbf{b}$.

Solution Our method of solution is motivated by the diagram in Fig. 14.15. We take

$$\mathbf{a}_{||} = (\text{comp}_{\mathbf{b}}\mathbf{a})\frac{\mathbf{b}}{|\mathbf{b}|} = \frac{\mathbf{a}\cdot\mathbf{b}}{|\mathbf{b}|^2}\,\mathbf{b} = \frac{8-5-6}{9}\,\mathbf{b}$$

$$= -\frac{1}{3}\langle 2, 1, -2\rangle = \left\langle -\frac{2}{3}, -\frac{1}{3}, \frac{2}{3}\right\rangle,$$

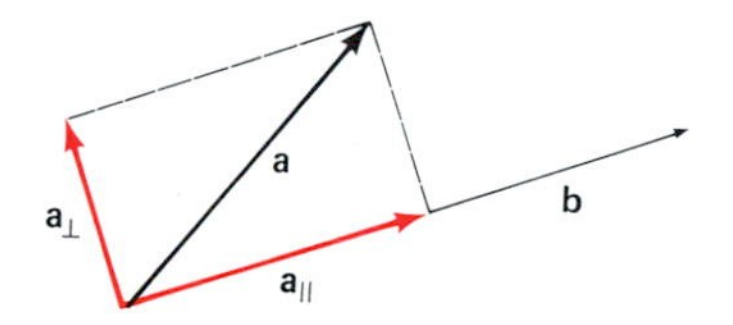

14.15 Construction of $\mathbf{a}_{||}$ and $\mathbf{a}_{\perp}$ (see Example 6)

and

$$\mathbf{a}_{\perp} = \mathbf{a} - \mathbf{a}_{||} = \langle 4, -5, 3\rangle - \left\langle -\frac{2}{3}, -\frac{1}{3}, \frac{2}{3}\right\rangle = \left\langle \frac{14}{3}, -\frac{14}{3}, \frac{7}{3}\right\rangle.$$

The diagram makes our choice of $\mathbf{a}_{||}$ plausible, and we have deliberately chosen $\mathbf{a}_{\perp}$ so that $\mathbf{a} = \mathbf{a}_{||} + \mathbf{a}_{\perp}$. To verify that the vector $\mathbf{a}_{||}$ is, indeed, parallel to $\mathbf{b}$, we simply note that it is a scalar multiple of $\mathbf{b}$. To verify that $\mathbf{a}_{\perp}$ is perpendicular to $\mathbf{b}$, we compute the dot product:

$$\mathbf{a}_{\perp}\cdot\mathbf{b} = \tfrac{28}{3} - \tfrac{14}{3} - \tfrac{14}{3} = 0.$$

Thus $\mathbf{a}_{||}$ and $\mathbf{a}_{\perp}$ have the required properties.

One important application of vector components is to the definition and computation of work. Recall that the work W done by a constant force F exerted along the line of motion in moving a particle a distance d is given by $W = Fd$. But what if the force is a constant vector $\mathbf{F}$ pointing in some direction other than the line of motion, as when a child pulls a sled against the resistance of friction (see Fig. 14.16)? Suppose that $\mathbf{F}$ moves a particle along the line segment from P to Q, and let $\mathbf{u} = \overrightarrow{PQ}$. Then the **work** W done by $\mathbf{F}$ in moving the particle along the line from P to Q is *by definition the product of the component of* $\mathbf{F}$ along $\mathbf{u} = \overrightarrow{PQ}$ and the distance moved:

$$\mathbf{W} = (\text{comp}_{\mathbf{u}}\mathbf{F})|\mathbf{u}|. \tag{15}$$

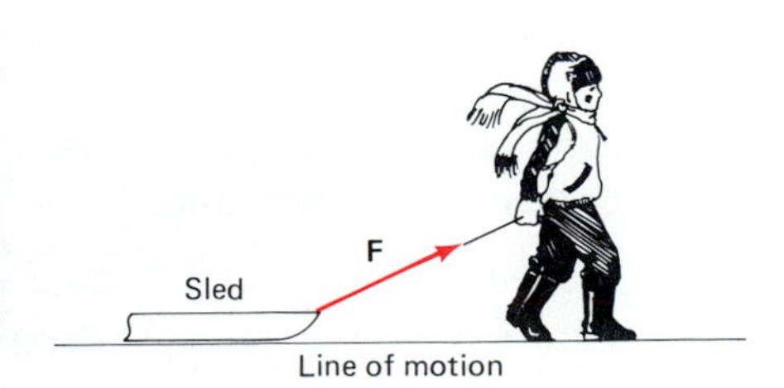

14.16 The vector force **F** is constant but acts at an angle to the line of motion (see Problem 36).

EXAMPLE 7 Show that the work done by a constant force $\mathbf{F}$ in moving a particle along the line segment from P to Q is

$$W = \mathbf{F} \cdot \mathbf{D} \tag{16}$$

where $\mathbf{D} = \overrightarrow{PQ}$ is the displacement vector. This formula is the natural vector generalization of the scalar formula $W = Fd$.

Solution We combine the formulas in (14) and (15). This gives

$$W = (\text{comp}_{\mathbf{D}}\,\mathbf{F})|\mathbf{D}| = \frac{\mathbf{F} \cdot \mathbf{D}}{|\mathbf{D}|}|\mathbf{D}| = \mathbf{F} \cdot \mathbf{D}.$$

14.1 PROBLEMS

In Problems 1–5 find: (a) $2\mathbf{a} + \mathbf{b}$, (b) $3\mathbf{a} - 4\mathbf{b}$, (c) $\mathbf{a} \cdot \mathbf{b}$, (d) $|\mathbf{a} - \mathbf{b}|$, and (e) $\mathbf{a}/|\mathbf{a}|$.

1. $\mathbf{a} = \langle 2, 5, 4 \rangle$, $\mathbf{b} = \langle 1, -2, -3 \rangle$
2. $\mathbf{a} = \langle -1, 0, 2 \rangle$, $\mathbf{b} = \langle 3, 4, -5 \rangle$
3. $\mathbf{a} = \mathbf{i} + \mathbf{j} + \mathbf{k}$, $\mathbf{b} = \mathbf{j} - \mathbf{k}$
4. $\mathbf{a} = 2\mathbf{i} - 3\mathbf{j} + 5\mathbf{k}$, $\mathbf{b} = 5\mathbf{i} + 3\mathbf{j} - 7\mathbf{k}$
5. $\mathbf{a} = 2\mathbf{i} - \mathbf{j}$, $\mathbf{b} = \mathbf{j} - 3\mathbf{k}$

6–10. Find the angle between the vectors $\mathbf{a}$ and $\mathbf{b}$ in Problems 1–5.

11–15. Find $\text{comp}_{\mathbf{a}}\,\mathbf{b}$ and $\text{comp}_{\mathbf{b}}\,\mathbf{a}$ for the vectors $\mathbf{a}$ and $\mathbf{b}$ given in Problems 1–5.

In Problems 16–20, write the equation of the indicated sphere.

16. Center (3, 1, 2), radius 5
17. Center $(-2, 1, -5)$, radius $\sqrt{7}$
18. One diameter the segment joining $(3, 5, -3)$ and (7, 3, 1)
19. Center $(4, 5, -2)$, passing through the point (1, 0, 0)
20. Center $(3, -4, 3)$, tangent to the xz-plane

In Problems 21–23, find the center and radius of the sphere having the given equation.

21. $x^2 + y^2 + z^2 + 4x - 6y = 0$
22. $x^2 + y^2 + z^2 - 8x - 9y + 10z + 40 = 0$
23. $3x^2 + 3y^2 + 3z^2 - 18z - 48 = 0$

In Problems 24–30, describe the graph of the given equation in geometric terms.

24. $x = 0$
25. $z = 10$
26. $xy = 0$
27. $xyz = 0$
28. $x^2 + y^2 + z^2 + 7 = 0$
29. $x^2 + y^2 + z^2 - 2x + 1 = 0$
30. $x^2 + y^2 + z^2 - 6x + 8y + 25 = 0$

In Problems 31–33, find the direction angles of the vector $\overrightarrow{PQ}$.

31. $P(1, -1, 0)$, $Q(3, 4, 5)$
32. $P(2, -3, 5)$, $Q(1, 0, -1)$
33. $P(-1, -2, -3)$, $Q(5, 6, 7)$

In Problems 34 and 35 find the work W done by the force $\mathbf{F}$ in moving a particle in a straight line from P to Q.

34. $\mathbf{F} = \mathbf{i} - \mathbf{k}$; $P(0, 0, 0)$, $Q(3, 1, 0)$
35. $\mathbf{F} = 2\mathbf{i} - 3\mathbf{j} + 5\mathbf{k}$; $P(5, 3, -4)$, $Q(-1, -2, 5)$
36. Suppose that the force vector in Fig. 14.16 is inclined at an angle of 30° to the ground. If the child exerts a constant force of 20 lb, how much work (in ft-lb) is done in pulling the sled a distance of 100 ft along the ground?
37. Suppose that the horizontal and vertical components of the vectors shown in Fig. 14.17 balance. How much work is done by the constant force $\mathbf{F}$ (parallel to the inclined plane) in pulling the weight mg up the inclined plane a vertical height h?

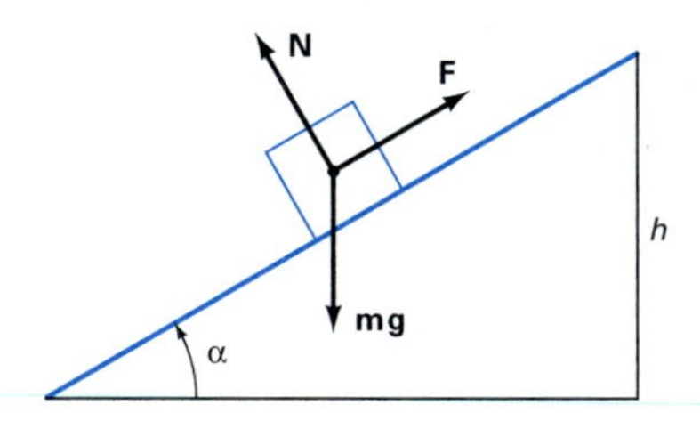

14.17 The inclined plane of Problem 37

38. Prove the **Cauchy–Schwarz inequality:**

$$|\mathbf{a} \cdot \mathbf{b}| \leqq |\mathbf{a}|\,|\mathbf{b}|$$

for all pairs of vectors $\mathbf{a}$ and $\mathbf{b}$.

39. Given two arbitrary vectors $\mathbf{a}$ and $\mathbf{b}$, prove that they satisfy the **triangle inequality**

$$|\mathbf{a} + \mathbf{b}| \leqq |\mathbf{a}| + |\mathbf{b}|.$$

(*Suggestion:* Square both sides.)

40. Prove that if **a** and **b** are arbitrary vectors, then

$$|\mathbf{a} - \mathbf{b}| \geqq |\mathbf{a}| - |\mathbf{b}|.$$

(*Suggestion:* Write $\mathbf{a} = (\mathbf{a} - \mathbf{b}) + \mathbf{b}$, then apply the triangle inequality (Problem 39).)

41. Find the area of the triangle with vertices $A(1, 1, 1)$, $B(3, -2, 3)$, and $C(3, 4, 6)$.

42. Find the three angles of the triangle of Problem 41.

43. Find the angle between any longest diagonal of a cube and any edge it meets.

44. Prove that the three points $P(0, -2, 4)$, $Q(1, -3, 5)$, and $R(4, -6, 8)$ lie on a single straight line.

45. Prove that the point M given in Formula (2) is indeed the midpoint of the segment P_1P_2. (NOTE You must prove *both* that M is equally distant from P_1 and P_2 *and* that M lies on the segment P_1P_2.)

46. Given vectors **a** and **b**, let $a = |\mathbf{a}|$ and $b = |\mathbf{b}|$. Prove that the vector $\mathbf{c} = (b\mathbf{a} + a\mathbf{b})/(a + b)$ bisects the angle between **a** and **b**.

47. Let **a**, **b**, and **c** be three vectors in the xy-plane with **a** and **b** nonzero and not parallel. Show that there exist scalars α and β such that $\mathbf{c} = \alpha\mathbf{a} + \beta\mathbf{b}$. Begin by expressing **a**, **b**, and **c** in terms of **i**, **j**, and **k**.

48. Let $ax + by + c = 0$ be the equation of the line L in the xy-plane with normal vector $\mathbf{n} = \langle a, b \rangle$. Let $P_0(x_0, y_0)$ be a point on this line and $P_1(x_1, y_1)$ a point not on L. Prove that the perpendicular distance from P_1 to L is

$$d = \frac{|\mathbf{n} \cdot \overrightarrow{P_0P_1}|}{|\mathbf{n}|} = \frac{|ax_1 + by_1 + c|}{\sqrt{a^2 + b^2}}.$$

49. Given the two points $A(3, -2, 4)$ and $B(5, 7, -1)$, write an equation in x, y, and z that says this: The point $P(x, y, z)$ is equally distant from the points A and B. Then simplify this equation and give a geometric description of the set of all such points $P(x, y, z)$.

50. Given the fixed point $A(1, 3, 5)$, the point $P(x, y, z)$, and the vector $\mathbf{n} = \mathbf{i} - \mathbf{j} + 2\mathbf{k}$, use the dot product to help you write an equation in x, y, and z that says this: **n** and $\overrightarrow{AP}$ are perpendicular. Then simplify this equation and give a geometric description of the set of all such points $P(x, y, z)$.

51. Prove that the points $(0, 0, 0)$, $(1, 1, 0)$, $(1, 0, 1)$, and $(0, 1, 1)$ are the vertices of a regular tetrahedron by showing that each of the six edges has length $\sqrt{2}$. Then use the dot product to find the angle between any two edges of the tetrahedron.

52. The methane molecule CH_4 is arranged with the four hydrogen atoms at the vertices of a regular tetrahedron and with the carbon atom at its center. Suppose that the axes and scale are chosen so that the tetrahedron is the one of Problem 51, which has center at $(\frac{1}{2}, \frac{1}{2}, \frac{1}{2})$. Find the *bond angle* between the lines from the carbon atom to two of the hydrogen atoms.

14.2 The Vector Product of Two Vectors

We often need to find a vector that is perpendicular to each of two space vectors **a** and **b**. A routine way of doing this is provided by the **vector product**, or **cross product**, $\mathbf{a} \times \mathbf{b}$ of the vectors **a** and **b**. This vector product is quite unlike the dot product $\mathbf{a} \cdot \mathbf{b}$ in that $\mathbf{a} \cdot \mathbf{b}$ is a scalar, while $\mathbf{a} \times \mathbf{b}$ is a vector.

The vector product of the vectors $\mathbf{a} = \langle a_1, a_2, a_3 \rangle$ and $\mathbf{b} = \langle b_1, b_2, b_3 \rangle$ can be defined by the formula

$$\mathbf{a} \times \mathbf{b} = \langle a_2b_3 - a_3b_2, a_3b_1 - a_1b_3, a_1b_2 - a_2b_1 \rangle. \tag{1}$$

Though this formula seems unmotivated, it has a redeeming feature: The product $\mathbf{a} \times \mathbf{b}$ is perpendicular both to **a** and to **b**, as suggested in Fig. 14.18.

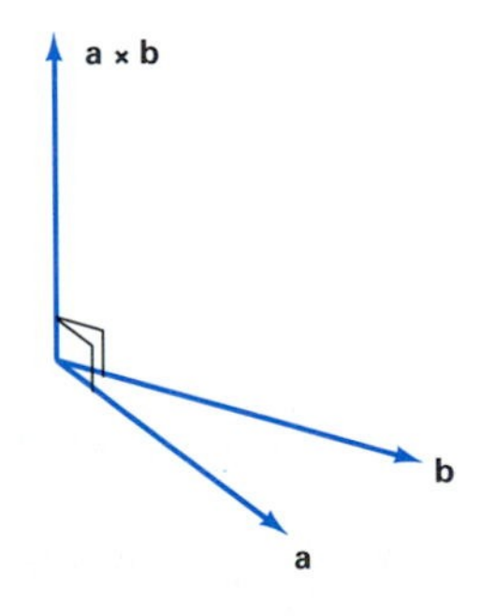

14.18 The cross product **a × b** is perpendicular both to **a** and to **b**.

> ***Theorem 1*** *Perpendicularity of the Vector Product*
> The vector product **a × b** is perpendicular both to **a** and to **b**.

Proof We show that **a × b** is perpendicular to **a** by showing that the dot product of **a** and **a × b** is zero. With the components as in Equation (1), we find that

$$\begin{aligned}\mathbf{a} \cdot (\mathbf{a} \times \mathbf{b}) &= a_1(a_2b_3 - a_3b_2) + a_2(a_3b_1 - a_1b_3) + a_3(a_1b_2 - a_2b_1)\\ &= a_1a_2b_3 - a_1a_3b_2 + a_2a_3b_1 - a_2a_1b_3 + a_3a_1b_2 - a_3a_2b_1\\ &= 0.\end{aligned}$$

A similar computation shows that $\mathbf{b} \cdot (\mathbf{a} \times \mathbf{b}) = 0$ as well, so that $\mathbf{a} \times \mathbf{b}$ is also perpendicular to the vector $\mathbf{b}$. ■

The formula in (1) need not be memorized, because there is an alternative version involving determinants that is easy to remember. Recall that a determinant of order two is defined as follows:

$$\begin{vmatrix} a_1 & a_2 \\ b_1 & b_2 \end{vmatrix} = a_1 b_2 - a_2 b_1. \tag{2}$$

For example,

$$\begin{vmatrix} 2 & -1 \\ 3 & 4 \end{vmatrix} = (2)(4) - (-1)(3) = 11.$$

A determinant of order three can be defined in terms of determinants of order two:

$$\begin{vmatrix} a_1 & a_2 & a_3 \\ b_1 & b_2 & b_3 \\ c_1 & c_2 & c_3 \end{vmatrix} = +a_1 \begin{vmatrix} b_2 & b_3 \\ c_2 & c_3 \end{vmatrix} - a_2 \begin{vmatrix} b_1 & b_3 \\ c_1 & c_3 \end{vmatrix} + a_3 \begin{vmatrix} b_1 & b_2 \\ c_1 & c_2 \end{vmatrix}. \tag{3}$$

Note that each element a_i of the first row is multiplied by the 2-by-2 "subdeterminant" obtained by deleting the row *and* column that contain a_i. Note also in (3) that signs are attached to the a_i in accord with the checkerboard pattern

$$\begin{vmatrix} + & - & + \\ - & + & - \\ + & - & + \end{vmatrix}.$$

The formula in (3) is an expansion of the 3-by-3 determinant along its first row. It also can be expanded along any other row or column. For example, its expansion along its second column is

$$\begin{vmatrix} a_1 & a_2 & a_3 \\ b_1 & b_2 & b_3 \\ c_1 & c_2 & c_3 \end{vmatrix} = -a_2 \begin{vmatrix} b_1 & b_3 \\ c_1 & c_3 \end{vmatrix} + b_2 \begin{vmatrix} a_1 & a_3 \\ c_1 & c_3 \end{vmatrix} - c_2 \begin{vmatrix} a_1 & a_3 \\ b_1 & b_3 \end{vmatrix}.$$

In linear algebra it is proved that all such expansions yield the same value for the determinant.

Although a determinant of order three can be expanded along any row or column, we shall use only expansions along the first row, as in (3). For example,

$$\begin{aligned} \begin{vmatrix} 1 & 3 & -2 \\ 2 & -1 & 4 \\ -3 & 7 & 5 \end{vmatrix} &= (1) \begin{vmatrix} -1 & 4 \\ 7 & 5 \end{vmatrix} - (3) \begin{vmatrix} 2 & 4 \\ -3 & 5 \end{vmatrix} + (-2) \begin{vmatrix} 2 & -1 \\ -3 & 7 \end{vmatrix} \\ &= (1)(-5 - 28) + (-3)(10 + 12) + (-2)(14 - 3) \\ &= -33 - 66 - 22 = -121. \end{aligned}$$

The formula in (1) for the vector product of the vectors $\mathbf{a} = a_1\mathbf{i} + a_2\mathbf{j} + a_3\mathbf{k}$ and $\mathbf{b} = b_1\mathbf{i} + b_2\mathbf{j} + b_3\mathbf{k}$ is equivalent to

$$\mathbf{a} \times \mathbf{b} = \begin{vmatrix} a_2 & a_3 \\ b_2 & b_3 \end{vmatrix} \mathbf{i} - \begin{vmatrix} a_1 & a_3 \\ b_1 & b_3 \end{vmatrix} \mathbf{j} + \begin{vmatrix} a_1 & a_2 \\ b_1 & b_2 \end{vmatrix} \mathbf{k}. \tag{4}$$

This is easy to verify by expanding the 2-by-2 determinants on the right-hand side, and noting that the three components of the right-hand side of the formula in (1) result. Motivated by Equation (4), we write

$$\mathbf{a} \times \mathbf{b} = \begin{vmatrix} \mathbf{i} & \mathbf{j} & \mathbf{k} \\ a_1 & a_2 & a_3 \\ b_1 & b_2 & b_3 \end{vmatrix}. \tag{5}$$

The "symbolic determinant" in this equation is to be evaluated by expanding along its first row, just as in Equation (3) and just as though it were an ordinary determinant with real number entries. The result of this expansion is the right-hand side of the formula in (4). Note that the components of the *first* vector **a** in **a** × **b** constitute the *second* row of the 3-by-3 determinant, while the components of the *second* vector **b** constitute the *third* row of the determinant. The order in which the vectors **a** and **b** is written is important, because we shall soon see that **a** × **b** is generally *not* equal to **b** × **a**: The vector product is *not commutative*.

The formula in (5) for the vector product is the form most convenient for computational purposes.

EXAMPLE 1 If

$$\mathbf{a} = 3\mathbf{i} - \mathbf{j} + 2\mathbf{k} \quad \text{and} \quad \mathbf{b} = 2\mathbf{i} + 2\mathbf{j} - \mathbf{k},$$

then

$$\mathbf{a} \times \mathbf{b} = \begin{vmatrix} \mathbf{i} & \mathbf{j} & \mathbf{k} \\ 3 & -1 & 2 \\ 2 & 2 & -1 \end{vmatrix} = \begin{vmatrix} -1 & 2 \\ 2 & -1 \end{vmatrix}\mathbf{i} - \begin{vmatrix} 3 & 2 \\ 2 & -1 \end{vmatrix}\mathbf{j} + \begin{vmatrix} 3 & -1 \\ 2 & 2 \end{vmatrix}\mathbf{k}$$

$$= (1 - 4)\mathbf{i} - (-3 - 4)\mathbf{j} + (6 - (-2))\mathbf{k};$$

thus

$$\mathbf{a} \times \mathbf{b} = -3\mathbf{i} + 7\mathbf{j} + 8\mathbf{k}.$$

You might now pause to verify (using the dot product) that the vector $-3\mathbf{i} + 7\mathbf{j} + 8\mathbf{k}$ is perpendicular both to **a** and to **b**.

If the vectors **a** and **b** are located with the same initial point, then Theorem 1 implies that **a** × **b** is normal to the plane determined by **a** and **b**, as indicated in Fig. 14.19. There are still two possible directions for **a** × **b**, but it turns out that if $\mathbf{a} \times \mathbf{b} \neq \mathbf{0}$, then the triple **a**, **b**, **a** × **b** is a *right-handed* triple in exactly the same sense as the triple **i**, **j**, **k**. Thus if the thumb of your right hand points in the direction of **a** × **b**, then your fingers curl in the direction of a rotation (less than 180°) from **a** to **b**.

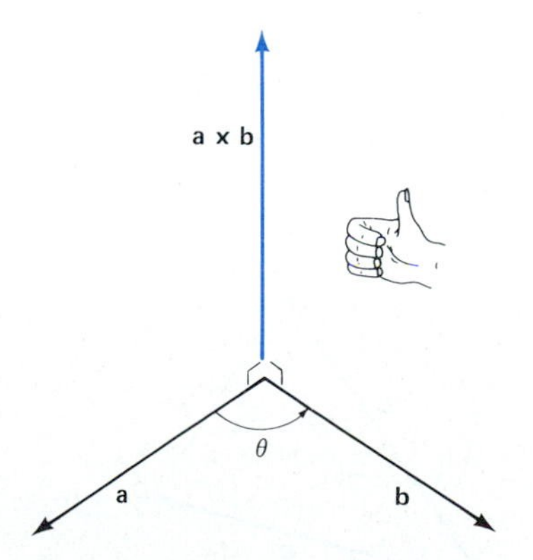

14.19 The vectors **a**, **b**, and **a** × **b** form—in that order—a right-handed triple.

Once the direction of **a** × **b** has been established, the vector product can be described in completely geometric terms by telling what the length $|\mathbf{a} \times \mathbf{b}|$ of the vector **a** × **b** is. This is given by the formula

$$|\mathbf{a} \times \mathbf{b}|^2 = |\mathbf{a}|^2 |\mathbf{b}|^2 - (\mathbf{a} \cdot \mathbf{b})^2. \tag{6}$$

This vector identity can be verified routinely (though tediously) by writing $\mathbf{a} = \langle a_1, a_2, a_3 \rangle$ and $\mathbf{b} = \langle b_1, b_2, b_3 \rangle$, computing both sides of Equation (6), and then noting that the results are equal (Problem 28).

The formula in (6) tells us what $|\mathbf{a} \times \mathbf{b}|$ is, but the following theorem reveals the *geometric significance* of the length of the vector product.

Theorem 2 *Length of the Vector Product*

Let θ be the angle between the nonzero vectors **a** and **b** (measured so that $0 \leqq \theta \leqq \pi$). Then

$$|\mathbf{a} \times \mathbf{b}| = |\mathbf{a}|\,|\mathbf{b}| \sin \theta. \tag{7}$$

Proof We begin with Equation (6) and use the fact that $\mathbf{a} \cdot \mathbf{b} = |\mathbf{a}|\,|\mathbf{b}| \cos \theta$. Thus

$$\begin{aligned}|\mathbf{a} \times \mathbf{b}|^2 &= |\mathbf{a}|^2 |\mathbf{b}|^2 - (\mathbf{a} \cdot \mathbf{b})^2 = |\mathbf{a}|^2 |\mathbf{b}|^2 - (|\mathbf{a}|\,|\mathbf{b}| \cos \theta)^2 \\ &= |\mathbf{a}|^2 |\mathbf{b}|^2 (1 - \cos^2 \theta) = |\mathbf{a}|^2 |\mathbf{b}|^2 \sin^2 \theta.\end{aligned}$$

Equation (7) now follows after we take the positive square root of both sides. (This is the correct root on the right-hand side because $\sin \theta \geqq 0$ for $0 \leqq \theta \leqq \pi$.) ■

Corollary *Parallel Vectors*

Two nonzero vectors **a** and **b** are parallel ($\theta = 0$ or $\theta = \pi$) if and only if $\mathbf{a} \times \mathbf{b} = \mathbf{0}$.

In particular, the cross product of any vector with itself is the zero vector. Also, Equation (1) shows immediately that the cross product of any vector with the zero vector is the zero vector again. Thus

$$\mathbf{a} \times \mathbf{a} = \mathbf{a} \times \mathbf{0} = \mathbf{0} \times \mathbf{a} = \mathbf{0} \tag{8}$$

for every vector **a**.

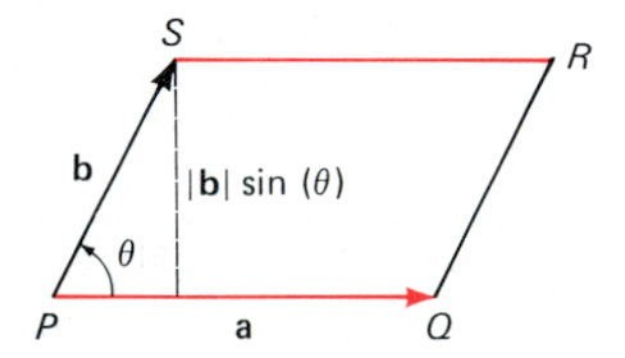

14.20 The area of the parallelogram *PQRS* is $|\mathbf{a} \times \mathbf{b}|$.

Equation (7) has an important geometric interpretation. Suppose that **a** and **b** are represented by adjacent sides of a parallelogram *PQRS*, with $\mathbf{a} = \overrightarrow{PQ}$ and $\mathbf{b} = \overrightarrow{PS}$, as indicated in Fig. 14.20. The parallelogram then has base of length $|\mathbf{a}|$ and height $|\mathbf{b}| \sin \theta$, so its area is

$$A = |\mathbf{a}|\,|\mathbf{b}| \sin \theta = |\mathbf{a} \times \mathbf{b}|. \tag{9}$$

Thus *the length of the vector product* $\mathbf{a} \times \mathbf{b}$ *is the same, numerically, as the area of the parallelogram determined by* **a** *and* **b**. It follows that the area of the triangle *PQS* in Fig. 14.21—because its area is half that of the parallelogram—is

$$\tfrac{1}{2}A = \tfrac{1}{2}|\mathbf{a} \times \mathbf{b}| = \tfrac{1}{2}|\overrightarrow{PQ} \times \overrightarrow{PS}|. \tag{10}$$

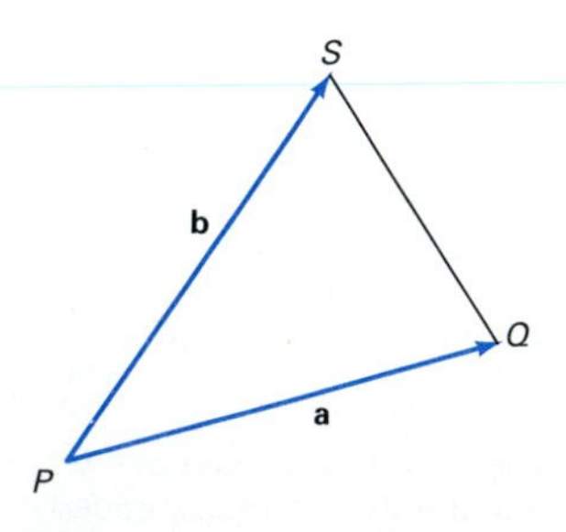

14.21 The area of $\triangle PQS$ is $A = \frac{1}{2}|\mathbf{a} \times \mathbf{b}|$.

The formula in (10) gives a quick way to compute the area of a triangle—even one in space—without the need of finding any of its angles.

EXAMPLE 2 Find the area of the triangle with vertices $A(3, 0, -1)$, $B(4, 2, 5)$, and $C(7, -2, 4)$.

Solution $\overrightarrow{AB} = \langle 1, 2, 6 \rangle$ and $\overrightarrow{AC} = \langle 4, -2, 5 \rangle$, so

$$\overrightarrow{AB} \times \overrightarrow{AC} = \begin{vmatrix} \mathbf{i} & \mathbf{j} & \mathbf{k} \\ 1 & 2 & 6 \\ 4 & -2 & 5 \end{vmatrix} = 22\mathbf{i} + 19\mathbf{j} - 10\mathbf{k}.$$

So by Equation (10), the area of triangle ABC is

$$\tfrac{1}{2}\sqrt{(22)^2 + (19)^2 + (-10)^2} = \tfrac{1}{2}\sqrt{945} \approx 15.37.$$

Now let $\mathbf{u}$, $\mathbf{v}$, $\mathbf{w}$ be a right-handed triple of mutually perpendicular *unit* vectors. The angle between any two of these is $\theta = \pi/2$, so $|\mathbf{u}| = |\mathbf{v}| = |\mathbf{w}| = \sin\theta = 1$. Thus it follows from (7) that $\mathbf{u} \times \mathbf{v} = \mathbf{w}$. When we apply this observation to the basic unit vectors $\mathbf{i}$, $\mathbf{j}$, and $\mathbf{k}$ (see Fig. 14.22), we see that

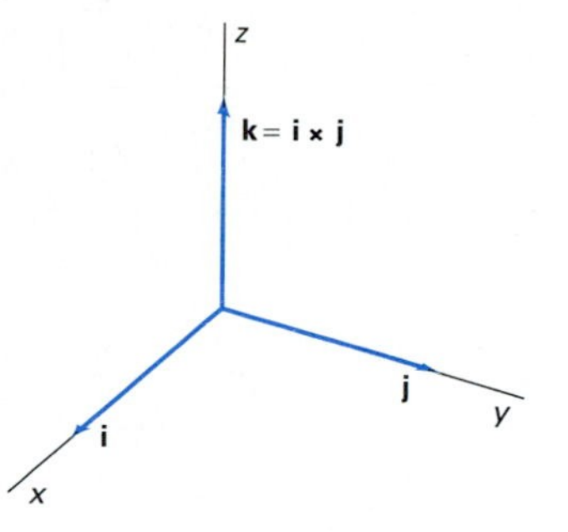

14.22 The basic unit vectors in space

$$\mathbf{i} \times \mathbf{j} = \mathbf{k}, \qquad \mathbf{j} \times \mathbf{k} = \mathbf{i}, \quad \text{and} \quad \mathbf{k} \times \mathbf{i} = \mathbf{j}. \tag{11a}$$

But

$$\mathbf{j} \times \mathbf{i} = -\mathbf{k}, \qquad \mathbf{k} \times \mathbf{j} = -\mathbf{i}, \quad \text{and} \quad \mathbf{i} \times \mathbf{k} = -\mathbf{j}. \tag{11b}$$

These observations, together with the fact that

$$\mathbf{i} \times \mathbf{i} = \mathbf{j} \times \mathbf{j} = \mathbf{k} \times \mathbf{k} = \mathbf{0}, \tag{11c}$$

also follow directly from the original definition of the cross product (in the form of Equation (5)). The products in (11a) are easily remembered in terms of the sequence

$$\mathbf{i}, \quad \mathbf{j}, \quad \mathbf{k}, \quad \mathbf{i}, \quad \mathbf{j}, \quad \mathbf{k}, \quad \ldots.$$

The product of any two consecutive unit vectors—in the order in which they appear here—equals the next one in the sequence.

Note that $\mathbf{i} \times \mathbf{j} \neq \mathbf{j} \times \mathbf{i}$. *The vector product is not commutative.* Instead, it is *anticommutative:* For any two vectors $\mathbf{a}$ and $\mathbf{b}$, $\mathbf{a} \times \mathbf{b} = -(\mathbf{b} \times \mathbf{a})$. This is the first part of the following theorem.

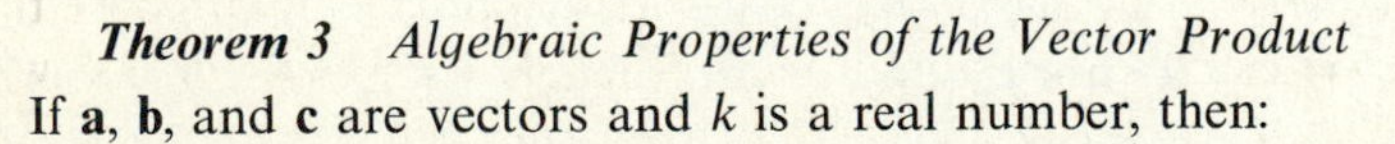

Theorem 3 *Algebraic Properties of the Vector Product*

If $\mathbf{a}$, $\mathbf{b}$, and $\mathbf{c}$ are vectors and k is a real number, then:

1. $\mathbf{a} \times \mathbf{b} = -(\mathbf{b} \times \mathbf{a});$ (12)
2. $(k\mathbf{a}) \times \mathbf{b} = \mathbf{a} \times (k\mathbf{b}) = k(\mathbf{a} \times \mathbf{b});$ (13)
3. $\mathbf{a} \times (\mathbf{b} + \mathbf{c}) = (\mathbf{a} \times \mathbf{b}) + (\mathbf{a} \times \mathbf{c});$ (14)
4. $\mathbf{a} \cdot (\mathbf{b} \times \mathbf{c}) = (\mathbf{a} \times \mathbf{b}) \cdot \mathbf{c};$ (15)
5. $\mathbf{a} \times (\mathbf{b} \times \mathbf{c}) = (\mathbf{a} \cdot \mathbf{c})\mathbf{b} - (\mathbf{a} \cdot \mathbf{b})\mathbf{c}.$ (16)

The proofs of the first four of these properties are straightforward applications of the definition of the vector product in terms of components. See Problem 24 for an outline of the proof of Equation (16).

Cross products of vectors expressed in terms of the basic unit vectors $\mathbf{i}$, $\mathbf{j}$, and $\mathbf{k}$ can be found by means of computations that closely resemble those of ordinary algebra. We simply apply the algebraic properties summarized in Theorem 3 together with the relations in (11) giving the various products of the basic unit vectors. Care must be taken to preserve the order of factors, because vector multiplication is not commutative—although, of course, one should not hesitate to use Equation (12). For example,

$$\begin{aligned}(\mathbf{i} - 2\mathbf{j} + 3\mathbf{k}) &\times (3\mathbf{i} + 2\mathbf{j} - 4\mathbf{k}) \\ &= 3(\mathbf{i} \times \mathbf{i}) + 2(\mathbf{i} \times \mathbf{j}) - 4(\mathbf{i} \times \mathbf{k}) - 6(\mathbf{j} \times \mathbf{i}) - 4(\mathbf{j} \times \mathbf{j}) + 8(\mathbf{j} \times \mathbf{k}) \\ &\quad + 9(\mathbf{k} \times \mathbf{i}) + 6(\mathbf{k} \times \mathbf{j}) - 12(\mathbf{k} \times \mathbf{k}) \\ &= 3(\mathbf{0}) + 2\mathbf{k} - 4(-\mathbf{j}) - 6(-\mathbf{k}) - 4(\mathbf{0}) + 8\mathbf{i} + 9\mathbf{j} + 6(-\mathbf{i}) - 12(\mathbf{0}) \\ &= 2\mathbf{i} + 13\mathbf{j} + 8\mathbf{k}.\end{aligned}$$

SCALAR TRIPLE PRODUCTS

Let us examine the product $\mathbf{a} \cdot (\mathbf{b} \times \mathbf{c})$ appearing in Equation (15). Note first that the expression would not make sense were the parentheses around $\mathbf{a} \cdot \mathbf{b}$, because $\mathbf{a} \cdot \mathbf{b}$ is a scalar, and thus we could not form the cross product of $\mathbf{a} \cdot \mathbf{b}$ with the vector $\mathbf{c}$. This means that we may omit the parentheses; the expression $\mathbf{a} \cdot \mathbf{b} \times \mathbf{c}$ is not ambiguous. The dot product of the vectors $\mathbf{a}$ and $\mathbf{b} \times \mathbf{c}$ is a real number, called the **scalar triple product** of the vectors $\mathbf{a}$, $\mathbf{b}$, and $\mathbf{c}$. Equation (15) implies that the operations $\cdot$ (dot) and $\times$ (cross) can be interchanged without affecting the value of the expression:

$$\mathbf{a} \cdot \mathbf{b} \times \mathbf{c} = \mathbf{a} \times \mathbf{b} \cdot \mathbf{c}$$

for all vectors $\mathbf{a}$, $\mathbf{b}$, and $\mathbf{c}$.

To compute the scalar triple product in terms of components, write $\mathbf{a} = \langle a_1, a_2, a_3 \rangle$, $\mathbf{b} = \langle b_1, b_2, b_3 \rangle$, and $\mathbf{c} = \langle c_1, c_2, c_3 \rangle$. Then

$$\mathbf{b} \times \mathbf{c} = (b_2 c_3 - b_3 c_2)\mathbf{i} - (b_1 c_3 - b_3 c_1)\mathbf{j} + (b_1 c_2 - b_2 c_1)\mathbf{k},$$

so

$$\mathbf{a} \cdot (\mathbf{b} \times \mathbf{c}) = a_1(b_2 c_3 - b_3 c_2) - a_2(b_1 c_3 - b_3 c_1) + a_3(b_1 c_2 - b_2 c_1).$$

But the expression on the right is the value of the 3-by-3 determinant

$$\mathbf{a} \cdot \mathbf{b} \times \mathbf{c} = \begin{vmatrix} a_1 & a_2 & a_3 \\ b_1 & b_2 & b_3 \\ c_1 & c_2 & c_3 \end{vmatrix}. \tag{17}$$

This is the quickest way to compute the scalar triple product.

EXAMPLE 3 If

$$\mathbf{a} = 2\mathbf{i} - 3\mathbf{k}, \qquad \mathbf{b} = \mathbf{i} + \mathbf{j} + \mathbf{k}, \quad \text{and} \quad \mathbf{c} = 4\mathbf{j} - \mathbf{k},$$

then

$$\begin{aligned}\mathbf{a} \cdot \mathbf{b} \times \mathbf{c} &= \begin{vmatrix} 2 & 0 & -3 \\ 1 & 1 & 1 \\ 0 & 4 & -1 \end{vmatrix} \\ &= +(2)\begin{vmatrix} 1 & 1 \\ 4 & -1 \end{vmatrix} - (0)\begin{vmatrix} 1 & 1 \\ 0 & -1 \end{vmatrix} + (-3)\begin{vmatrix} 1 & 1 \\ 0 & 4 \end{vmatrix} \\ &= +(2)(-5) + (-3)(4) = -22.\end{aligned}$$

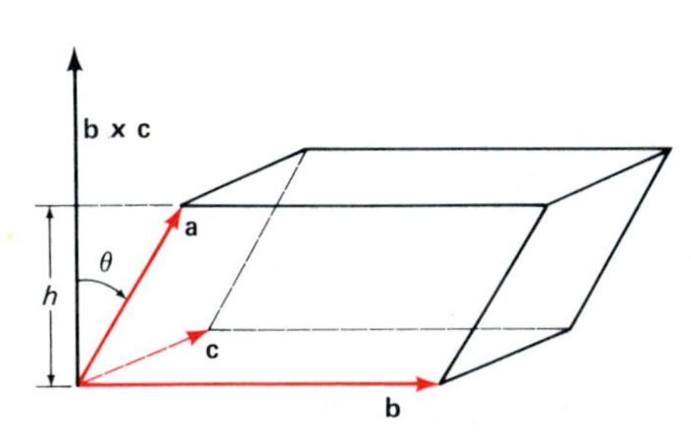

14.23 The volume of the parallelepiped is $|\mathbf{a} \cdot \mathbf{b} \times \mathbf{c}|$.

The importance of the scalar triple product for applications depends upon the following geometric interpretation. Let $\mathbf{a}$, $\mathbf{b}$, and $\mathbf{c}$ be three vectors with the same initial point. Figure 14.23 shows the parallelepiped determined by these vectors—that is, with the vectors as adjacent edges. If the vectors

a, **b**, and **c** are coplanar (lie in a single plane), then the parallelepiped is *degenerate:* Its volume is zero. Theorem 4 holds whether or not the three vectors are coplanar, but it is of most interest when they are not.

Theorem 4 *Scalar Triple Products and Volume*

The volume V of the parallelepiped determined by the vectors **a**, **b**, and **c** is the absolute value of the scalar triple product $\mathbf{a} \cdot \mathbf{b} \times \mathbf{c}$; that is,

$$V = |\mathbf{a} \cdot \mathbf{b} \times \mathbf{c}|. \tag{18}$$

Proof If the three vectors are coplanar, then **a** and $\mathbf{b} \times \mathbf{c}$ are perpendicular, so $V = |\mathbf{a} \cdot \mathbf{b} \times \mathbf{c}| = 0$. Assume that they are not coplanar. By Equation (9) the area of the base (determined by **b** and **c**) of the parallelepiped is $A = |\mathbf{b} \times \mathbf{c}|$.

Now let α be the *acute* angle between **a** and the line through $\mathbf{b} \times \mathbf{c}$ that is perpendicular to the base. Then the height of the parallelepiped is $h = |\mathbf{a}| \cos \alpha$. If θ is the angle between the vectors **a** and $\mathbf{b} \times \mathbf{c}$, then either $\theta = \alpha$ or $\theta = \pi - \alpha$. Hence $\cos \alpha = |\cos \theta|$, so

$$V = Ah = |\mathbf{b} \times \mathbf{c}|\,|\mathbf{a}| \cos \alpha = |\mathbf{a}|\,|\mathbf{b} \times \mathbf{c}|\,|\cos \theta| = |\mathbf{a} \cdot \mathbf{b} \times \mathbf{c}|.$$

Thus we have verified Equation (18). ■

EXAMPLE 4 Figure 14.24 shows both pyramid $OPQR$ and the parallelepiped determined by the vectors $\mathbf{a} = \overrightarrow{OP} = \langle 3, 2, -1 \rangle$, $\mathbf{b} = \overrightarrow{OQ} = \langle -2, 5, 1 \rangle$, and $\mathbf{c} = \overrightarrow{OR} = \langle 2, 1, 5 \rangle$. The volume of the pyramid is $V = \frac{1}{3}Ah$, where h is its height and the area of its base OPQ is *half* the area of the corresponding base of the parallelepiped. It therefore follows from (17) and (18) that V is $\frac{1}{6}$ the volume of the parallelepiped:

$$V = \tfrac{1}{6}|\mathbf{a} \cdot \mathbf{b} \times \mathbf{c}| = \tfrac{1}{6}\begin{vmatrix} 3 & 2 & -1 \\ -2 & 5 & 1 \\ 2 & 1 & 5 \end{vmatrix} = \tfrac{108}{6} = 18.$$

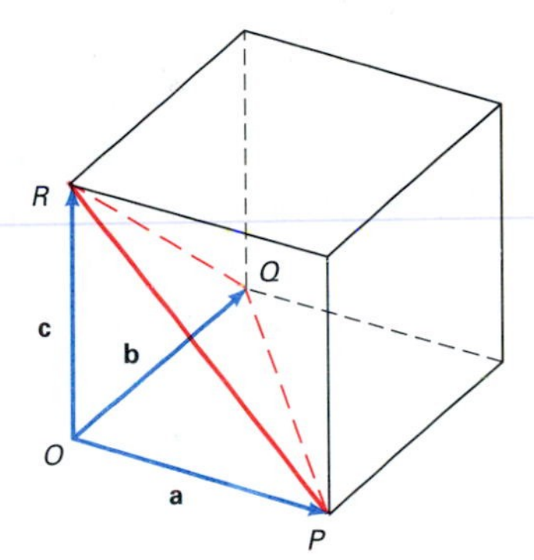

14.24 The pyramid of Example 4

EXAMPLE 5 Use the scalar triple product to show that the points $A(1, -1, 2)$, $B(2, 0, 1)$, $C(3, 2, 0)$, and $D(5, 4, -2)$ are coplanar.

Solution It is enough to show that the vectors $\overrightarrow{AB} = \langle 1, 1, -1 \rangle$, $\overrightarrow{AC} = \langle 2, 3, -2 \rangle$, and $\overrightarrow{AD} = \langle 4, 5, -4 \rangle$ are coplanar. But their scalar triple product is

$$\begin{vmatrix} 1 & 1 & -1 \\ 2 & 3 & -2 \\ 4 & 5 & -4 \end{vmatrix} = (1)(-2) - (1)(0) + (-1)(-2) = 0,$$

so Theorem 4 guarantees us that the parallelepiped determined by these three vectors has volume zero. Hence the four given points are coplanar.

The vector product occurs quite naturally in many scientific applications. For example, suppose that a body in space is free to rotate about the fixed point O. If a force **F** acts at the point P of the body, its effect is to cause rotation of the body. This effect is measured by the **torque vector** $\boldsymbol{\tau}$ defined by the relation

$$\boldsymbol{\tau} = \mathbf{r} \times \mathbf{F}$$

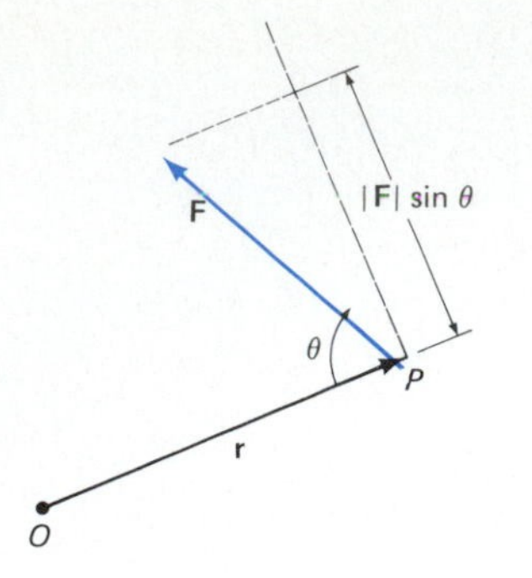

14.25 The torque vector τ is normal to both **r** and **F**.

where $\mathbf{r} = \overrightarrow{OP}$. The straight line through O determined by τ is the axis of rotation, and the length

$$|\tau| = |\mathbf{r}|\,|\mathbf{F}| \sin\theta$$

(see Fig. 14.25) is the **moment** of the force **F** about this axis.

Another example is the force exerted on a moving charged particle by a magnetic field. This force is important in the cyclotron and in the television picture tube; controlling the paths of the ions is accomplished through the interplay of electric and magnetic fields. In such circumstances, the force **F** on the particle due to a magnetic field depends upon three things: the charge q of the particle, its velocity vector **v**, and the magnetic field vector **B** at the instantaneous location of the particle. And it turns out that

$$\mathbf{F} = (q\mathbf{v}) \times \mathbf{B}.$$

14.2 PROBLEMS

Find $\mathbf{a} \times \mathbf{b}$ in Problems 1–4.

1. $\mathbf{a} = \langle 5, -1, -2\rangle, \quad \mathbf{b} = \langle -3, 2, 4\rangle$
2. $\mathbf{a} = \langle 3, -2, 0\rangle, \quad \mathbf{b} = \langle 0, 3, -2\rangle$
3. $\mathbf{a} = \mathbf{i} - \mathbf{j} + 3\mathbf{k}, \quad \mathbf{b} = -2\mathbf{i} + 3\mathbf{j} + \mathbf{k}$
4. $\mathbf{a} = 4\mathbf{i} + 2\mathbf{j} - 2\mathbf{k}, \quad \mathbf{b} = 2\mathbf{i} - 5\mathbf{j} + 5\mathbf{k}$
5. Apply Equation (5) to verify the equations in (11a).
6. Apply Equation (5) to verify the equations in (11b).
7. Prove that the vector product is not associative by calculating and comparing $\mathbf{a} \times (\mathbf{b} \times \mathbf{c})$ and $(\mathbf{a} \times \mathbf{b}) \times \mathbf{c}$ with $\mathbf{a} = \mathbf{i}$, $\mathbf{b} = \mathbf{i} + \mathbf{j}$, and $\mathbf{c} = \mathbf{i} + \mathbf{j} + \mathbf{k}$.
8. Find nonzero vectors **a**, **b**, and **c** such that $\mathbf{a} \times \mathbf{b} = \mathbf{a} \times \mathbf{c}$ but $\mathbf{b} \neq \mathbf{c}$.
9. Suppose that the three vectors **a**, **b**, and **c** are mutually perpendicular. Prove that $\mathbf{a} \times (\mathbf{b} \times \mathbf{c}) = \mathbf{0}$.
10. Find the area of the triangle with vertices $P(1, 1, 0)$, $Q(1, 0, 1)$, and $R(0, 1, 1)$.
11. Find the area of the triangle with vertices $P(1, 3, -2)$, $Q(2, 4, 5)$, and $R(-3, -2, 2)$.
12. Find the volume of the parallelepiped with adjacent edges $\overrightarrow{OP}$, $\overrightarrow{OQ}$, and $\overrightarrow{OR}$, where P, Q, and R are the points given in Problem 10.
13. (a) Find the volume of the parallelepiped with adjacent edges $\overrightarrow{OP}$, $\overrightarrow{OQ}$, and $\overrightarrow{OR}$ where P, Q, and R are the points given in Problem 11.

(b) Find the volume of the pyramid with vertices O, P, Q, and R.

14. Find a unit vector **n** perpendicular to the plane through the three points P, Q, and R of Problem 11. Then find the distance from the origin to this plane by computing $\mathbf{n} \cdot \overrightarrow{OP}$.
15. Figure 14.26 shows a polygonal plot of land, with angles and lengths measured by a surveyor. First find coordinates of each vertex, then use the cross product (as in Equation (10)) to calculate the area of the plot.

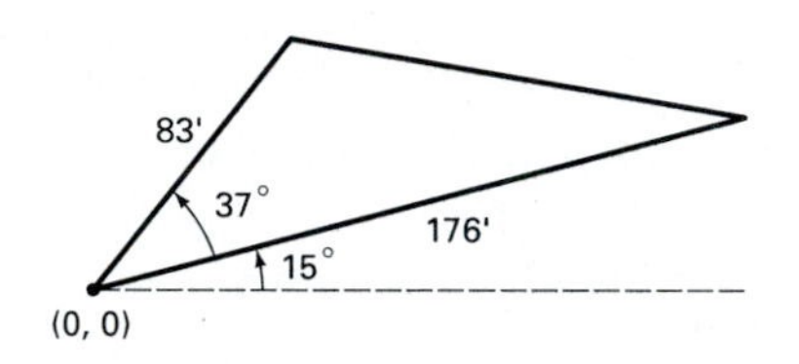

14.26

16. Repeat Problem 15 with the plot shown in Fig. 14.27.

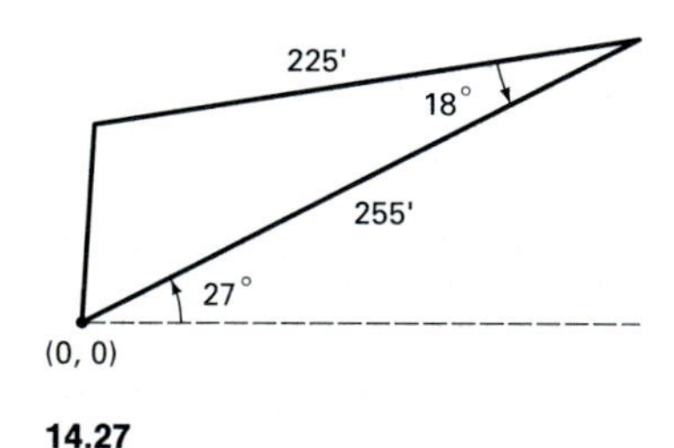

14.27

17. Repeat Problem 15 with the plot shown in Fig. 14.28. But first divide the plot into triangles.

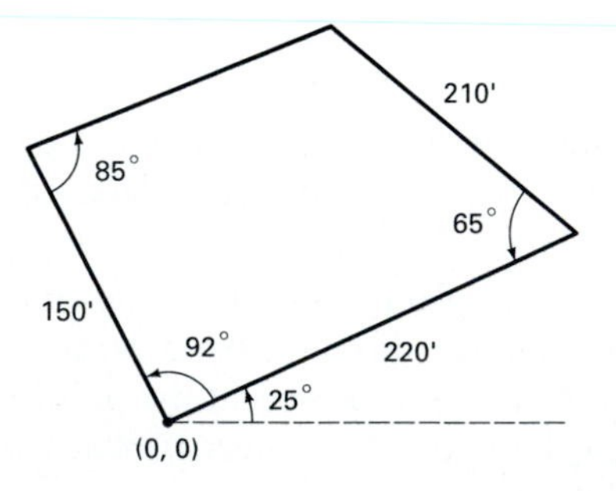

14.28

18. Repeat Problem 17 with the plot shown in Fig. 14.29.

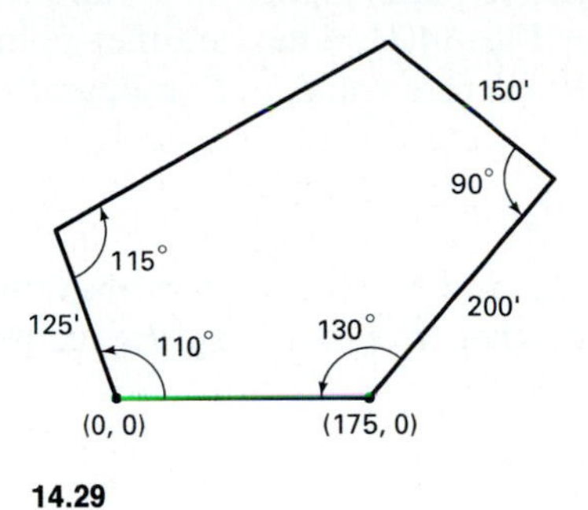

14.29

19. Apply Equation (5) to verify Equation (12), the anticommutativity of the vector product.

20. Apply Equation (17) to verify the identity for scalar triple products found in Equation (15).

21. Suppose that P and Q are points on a line L in space. Let A be a point not on L. Calculate in two ways the area of the triangle APQ to show that the perpendicular distance from A to the line L is $d = |\overrightarrow{AP} \times \overrightarrow{AQ}|/|\overrightarrow{PQ}|$. Then use this formula to compute the distance from the point $A(1, 0, 1)$ to the line through the two points $P(2, 3, 1)$ and $Q(-3, 1, 4)$. See Fig. 14.30.

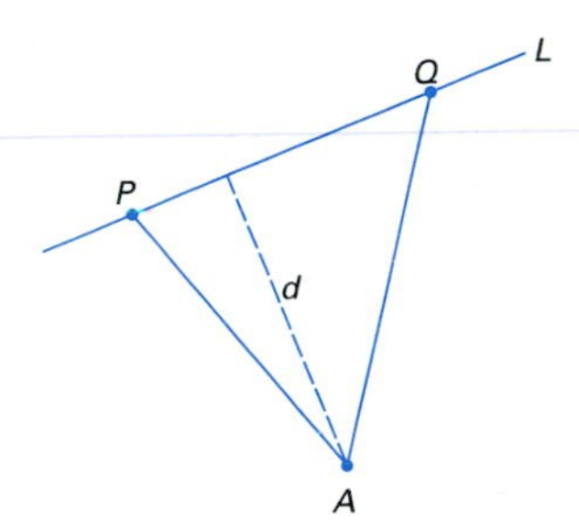

14.30 Figure for Problem 21

22. Suppose that A is a point not on the plane determined by the three points P, Q, and R. Calculate in two ways the volume of the pyramid $APQR$ to show that the perpendicular distance from A to this plane is

$$d = \frac{|\overrightarrow{AP} \cdot \overrightarrow{AQ} \times \overrightarrow{AR}|}{|\overrightarrow{PQ} \times \overrightarrow{PR}|}.$$

Use this formula to compute the distance from the point $A(1, 0, 1)$ to the plane through the points $P(2, 3, 1)$, $Q(3, -1, 4)$, and $R(0, 0, 2)$.

23. Suppose that P_1 and Q_1 are two points on the line L_1, and that P_2 and Q_2 are two points on the line L_2. If the lines L_1 and L_2 are not parallel, then the perpendicular distance d between them is the projection of $\overrightarrow{P_1P_2}$ onto a vector $\mathbf{n}$ that is perpendicular both to $\overrightarrow{P_1Q_1}$ and $\overrightarrow{P_2Q_2}$. Prove that

$$d = \frac{|\overrightarrow{P_1P_2} \cdot \overrightarrow{P_1Q_1} \times \overrightarrow{P_2Q_2}|}{|\overrightarrow{P_1Q_1} \times \overrightarrow{P_2Q_2}|}.$$

24. Use the following method to establish that the **vector triple product** $(\mathbf{a} \times \mathbf{b}) \times \mathbf{c}$ is equal to $(\mathbf{a} \cdot \mathbf{c})\mathbf{b} - (\mathbf{b} \cdot \mathbf{c})\mathbf{a}$.

(a) Let $\mathbf{I}$ be a unit vector in the direction of $\mathbf{a}$ and let $\mathbf{J}$ be a unit vector perpendicular to $\mathbf{I}$ and parallel to the plane of $\mathbf{a}$ and $\mathbf{b}$. Let $\mathbf{K} = \mathbf{I} \times \mathbf{J}$. Explain why there are scalars a_1, b_1, b_2, c_1, c_2, and c_3 such that

$$\mathbf{a} = a_1\mathbf{I}, \qquad \mathbf{b} = b_1\mathbf{I} + b_2\mathbf{J}, \qquad \mathbf{c} = c_1\mathbf{I} + c_2\mathbf{J} + c_3\mathbf{K}.$$

(b) Now show that

$$(\mathbf{a} \times \mathbf{b}) \times \mathbf{c} = -a_1b_2c_2\mathbf{I} + a_1b_2c_1\mathbf{J}.$$

(c) Finally, substitute for $\mathbf{I}$ and $\mathbf{J}$ in terms of $\mathbf{a}$ and $\mathbf{b}$.

25. By permutation of the vectors $\mathbf{a}$, $\mathbf{b}$, and $\mathbf{c}$, deduce from Problem 24 that

$$\mathbf{a} \times (\mathbf{b} \times \mathbf{c}) = (\mathbf{a} \cdot \mathbf{c})\mathbf{b} - (\mathbf{a} \cdot \mathbf{b})\mathbf{c}$$

(Equation (16)).

26. Deduce from the orthogonality properties of the vector product that the vector $(\mathbf{a} \times \mathbf{b}) \times (\mathbf{c} \times \mathbf{d})$ can be written both in the form $r_1\mathbf{a} + r_2\mathbf{b}$ and in the form $s_1\mathbf{c} + s_2\mathbf{d}$.

27. Consider the triangle in the xy-plane that has vertices $(x_1, y_1, 0)$, $(x_2, y_2, 0)$, and $(x_3, y_3, 0)$. Use the cross product to prove that the area of this triangle is *half* the *absolute value* of the determinant

$$\begin{vmatrix} 1 & 1 & 1 \\ x_1 & x_2 & x_3 \\ y_1 & y_2 & y_3 \end{vmatrix}.$$

28. Given the vectors $\mathbf{a} = \langle a_1, a_2, a_3 \rangle$ and $\mathbf{b} = \langle b_1, b_2, b_3 \rangle$, verify Equation (6),

$$|\mathbf{a} \times \mathbf{b}|^2 = |\mathbf{a}|^2|\mathbf{b}|^2 - (\mathbf{a} \cdot \mathbf{b})^2,$$

by computing each side in terms of the components of $\mathbf{a}$ and $\mathbf{b}$.

14.3 Lines and Planes in Space

A straight line in space is determined by any two points P_0 and P_1 on it. Alternatively, a line in space can be specified by giving a point P_0 on it *and* a vector, such as $\overrightarrow{P_0P_1}$, that determines the direction of the line.

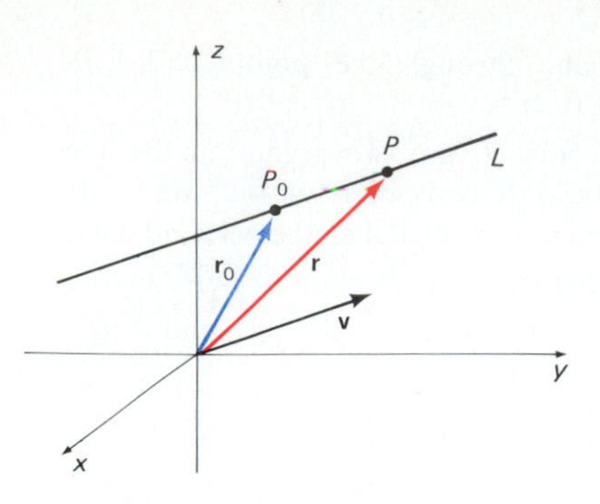

14.31 Finding the equation of the line L through the point P_0 parallel to the vector **v**.

To investigate equations describing lines in space, let us begin with a straight line L that passes through the point $P_0(x_0, y_0, z_0)$ and is parallel to the vector $\mathbf{v} = a\mathbf{i} + b\mathbf{j} + c\mathbf{k}$, as in Fig. 14.31. Then another point $P(x, y, z)$ lies on the line L if and only if the vectors $\mathbf{v}$ and $\overrightarrow{P_0P}$ are parallel, in which case

$$\overrightarrow{P_0P} = t\mathbf{v} \tag{1}$$

for some real number t. If $\mathbf{r}_0 = \overrightarrow{OP_0}$ and $\mathbf{r} = \overrightarrow{OP}$ are the position vectors of the points P_0 and P, respectively, then $\overrightarrow{P_0P} = \mathbf{r} - \mathbf{r}_0$. Hence (1) gives the *vector equation*

$$\mathbf{r} = \mathbf{r}_0 + t\mathbf{v} \tag{2}$$

describing the line L. As indicated in Fig. 14.31, $\mathbf{r}$ is the position vector of an *arbitrary* point P on the line L, and Equation (2) gives $\mathbf{r}$ in terms of the parameter t, the position vector $\mathbf{r}_0$ of a *fixed* point P_0 on L, and the fixed vector $\mathbf{v}$ that determines the direction of L.

The left- and right-hand sides are equal in Equation (2), and each side is a vector. So corresponding components are also equal. When we write the resulting equations, we get a scalar description of the line L. Because $\mathbf{r}_0 = \langle x_0, y_0, z_0 \rangle$ and $\mathbf{r} = \langle x, y, z \rangle$, Equation (2) thereby yields the three scalar equations

$$x = x_0 + at, \qquad y = y_0 + bt, \qquad z = z_0 + ct. \tag{3}$$

These are **parametric equations** of the line L through the point (x_0, y_0, z_0) parallel to the vector $\mathbf{v} = \langle a, b, c \rangle$.

EXAMPLE 1 Write parametric equations of the line L that passes through the points $P_1(1, 2, 2)$ and $P_2(3, -1, 3)$.

Solution The line L is parallel to the vector

$$\mathbf{v} = \overrightarrow{P_1P_2} = 2\mathbf{i} - 3\mathbf{j} + \mathbf{k},$$

so we take $a = 2$, $b = -3$, and $c = 1$. With P_1 as the fixed point, the equations in (3) give

$$x = 1 + 2t, \qquad y = 2 - 3t, \qquad z = 2 + t$$

as parametric equations of L. In contrast, with P_2 as the fixed point and with

$$-2\mathbf{v} = -4\mathbf{i} + 6\mathbf{j} - 2\mathbf{k}$$

as the direction vector, the equations in (3) yield the parametric equations

$$x = 3 - 4t, \qquad y = -1 + 6t, \qquad z = 3 - 2t.$$

Thus the parametric equations of a line are not unique.

Given two straight lines L_1 and L_2 with parametric equations

$$x = x_1 + a_1t, \qquad y = y_1 + b_1t, \qquad z = z_1 + c_1t \tag{4}$$

and

$$x = x_2 + a_2s, \qquad y = y_2 + b_2s, \qquad z = z_2 + c_2s, \tag{5}$$

respectively, we can see at a glance whether or not L_1 and L_2 are parallel. Because L_1 is parallel to $\mathbf{v}_1 = \langle a_1, b_1, c_1 \rangle$ and L_2 is parallel to $\mathbf{v}_2 = \langle a_2, b_2, c_2 \rangle$, it follows that the lines L_1 and L_2 are parallel if and only if the vectors $\mathbf{v}_1$ and $\mathbf{v}_2$ are scalar multiples of each other. If the two lines are not parallel, we can attempt to find a point of intersection by solving the equations

$$x_1 + a_1 t = x_2 + a_2 s \quad \text{and} \quad y_1 + b_1 t = y_2 + b_2 s$$

simultaneously for s and t. If these values satisfy the equation $z_1 + c_1 t = z_2 + c_2 s$, then we have found a point of intersection (whose coordinates are obtained by substitution of the resulting value of t in (4) or that of s in (5)). Otherwise, the two lines L_1 and L_2 do not intersect. Two nonparallel and nonintersecting lines in space are called **skew** lines.

EXAMPLE 2 The line L_1 with parametric equations

$$x = 1 + 2t, \qquad y = 2 - 3t, \qquad z = 2 + t$$

passes through the point $P_1(1, 2, 2)$ and is parallel to the vector $\mathbf{v}_1 = \langle 2, -3, 1 \rangle$. The line L_2 with parametric equations

$$x = 3 + 4t, \qquad y = 1 - 6t, \qquad z = 5 + 2t$$

passes through the point $P_2(3, 1, 5)$ and is parallel to the vector $\mathbf{v}_2 = \langle 4, -6, 2 \rangle$. Because $\mathbf{v}_2 = 2\mathbf{v}_1$, we see that L_1 and L_2 are parallel.

But are L_1 and L_2 actually different lines, or are we perhaps dealing with two different parametrizations of the same line? To answer this question, we note that $\overrightarrow{P_1P_2} = \langle 2, -1, 3 \rangle$ is not parallel to $\mathbf{v}_1 = \langle 2, -3, 1 \rangle$. Thus the point P_2 does not line on the line L_1, and hence the lines L_1 and L_2 are indeed distinct.

If the coefficients a, b, and c in (3) are all nonzero, then we can eliminate the parameter by equating the three expressions obtained by solving each equation for t. This gives

$$\frac{x - x_0}{a} = \frac{y - y_0}{b} = \frac{z - z_0}{c} \tag{6}$$

These are called the **symmetric equations** of the line L. If one or more of a or b or c is zero, this means that L lies in a plane parallel to one of the coordinate planes, and in this case the line does not have symmetric equations. For example, if $c = 0$, then L lies in the horizontal plane $z = z_0$. Of course, it is still possible to write equations for L not involving the parameter t; if $c = 0$ while a and b are nonzero, we could describe the line L as the simultaneous solution of the equations

$$\frac{x - x_0}{a} = \frac{y - y_0}{b}, \qquad z = z_0.$$

EXAMPLE 3 Find both parametric and symmetric equations of the line L through the points $P_0(3, 1, -2)$ and $P_1(4, -1, 1)$. Also find the points in which L intersects the three coordinate planes.

Solution The line L is parallel to the vector $\mathbf{v} = \overrightarrow{P_0P_1} = \langle 1, -2, 3 \rangle$, so we take $a = 1$, $b = -2$, and $c = 3$. The equations in (3) then give the parametric equations

$$x = 3 + t, \qquad y = 1 - 2t, \qquad z = -2 + 3t$$

of L, while the equations in (6) give the symmetric equations

$$\frac{x-3}{1} = \frac{y-1}{-2} = \frac{z+2}{3}.$$

To find the point at which L intersects the xy-plane, we set $z = 0$ in the symmetric equations. This gives

$$\frac{x-3}{1} = \frac{y-1}{-2} = \frac{2}{3},$$

and so $x = \frac{11}{3}$ and $y = -\frac{1}{3}$. Thus L meets the xy-plane at the point $(\frac{11}{3}, -\frac{1}{3}, 0)$. Similarly, $x = 0$ gives $(0, 7, -11)$ for the point where L meets the yz-plane, and $y = 0$ gives $(\frac{7}{2}, 0, -\frac{1}{2})$ for its intersection with the xz-plane.

PLANES IN SPACE

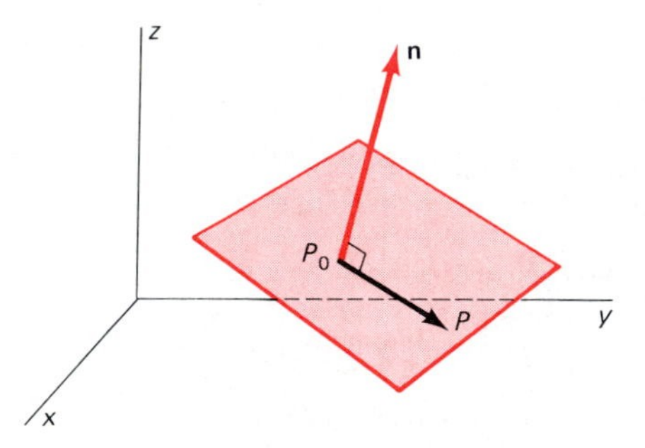

14.32 Because **n** is normal to $\mathscr{P}$, it follows that **n** is normal to $\overrightarrow{P_0P}$ for all points P in $\mathscr{P}$.

A plane $\mathscr{P}$ in space is determined by a point $P_0(x_0, y_0, z_0)$ through which $\mathscr{P}$ passes and a line through P_0 that is normal to $\mathscr{P}$. Alternatively, we may be given P_0 on $\mathscr{P}$ and a normal vector $\mathbf{n} = \langle a, b, c \rangle$ to the plane $\mathscr{P}$. Then the point $P(x, y, z)$ lies on the plane $\mathscr{P}$ if and only if the vectors $\mathbf{n}$ and $\overrightarrow{P_0P}$ are perpendicular (see Fig. 14.32), in which case $\mathbf{n} \cdot \overrightarrow{P_0P} = 0$. We write $\overrightarrow{P_0P} = \mathbf{r} - \mathbf{r}_0$, where $\mathbf{r}$ and $\mathbf{r}_0$ are the position vectors $\mathbf{r} = \overrightarrow{OP}$ and $\mathbf{r}_0 = \overrightarrow{OP_0}$ of the points P and P_0, respectively. Thus we obtain a **vector equation**

$$\mathbf{n} \cdot (\mathbf{r} - \mathbf{r}_0) = 0 \tag{7}$$

of the plane $\mathscr{P}$.

If we substitute $\mathbf{n} = \langle a, b, c \rangle$, $\mathbf{r} = \langle x, y, z \rangle$, and $\mathbf{r}_0 = \langle x_0, y_0, z_0 \rangle$ in Equation (7), we thereby obtain a **scalar equation**

$$a(x - x_0) + b(y - y_0) + c(z - z_0) = 0 \tag{8}$$

of the plane through $P_0(x_0, y_0, z_0)$ **with normal vector** $\mathbf{n} = \langle a, b, c \rangle$.

EXAMPLE 4 An equation of the plane through $P_0(-1, 5, 2)$ with normal vector $\mathbf{n} = \langle 1, -3, 2 \rangle$ is

$$(1)(x + 1) + (-3)(y - 5) + (2)(z - 2) = 0;$$

that is,

$$x - 3y + 2z = -12.$$

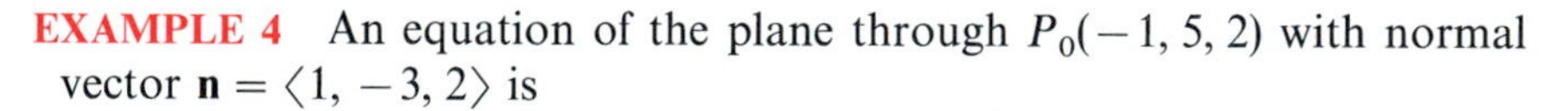

Note that the coefficients of x, y, and z in the last equation are the components of the normal vector. This is always the case, for Equation (8) can be written in the form

$$ax + by + cz = d \tag{9}$$

where $d = ax_0 + by_0 + cz_0$. Conversely, every *linear equation* in x, y, and z of the form in (9) represents a plane in space provided that the coefficients a, b, and c are not all zero. For if $c \neq 0$ (for instance), we can pick x_0 and y_0 arbitrarily and solve the equation $ax_0 + by_0 + cz_0 = d$ for z_0. With these values, Equation (9) takes the form

$$ax + by + cz = ax_0 + by_0 + cz_0$$

—that is,

$$a(x - x_0) + b(y - y_0) + c(z - z_0) = 0,$$

so this equation represents the plane through (x_0, y_0, z_0) with normal vector $\langle a, b, c \rangle$.

EXAMPLE 5 Find an equation for the plane through the three points $P(2, 4, -3)$, $Q(3, 7, -1)$, and $R(4, 3, 0)$.

Solution We want to use Equation (8), so we first need a vector **n** that is normal to the plane in question. One easy way to obtain such a normal vector is through the cross product. Let

$$\mathbf{n} = \overrightarrow{PQ} \times \overrightarrow{PR} = \begin{vmatrix} \mathbf{i} & \mathbf{j} & \mathbf{k} \\ 1 & 3 & 2 \\ 2 & -1 & 3 \end{vmatrix} = 11\mathbf{i} + \mathbf{j} - 7\mathbf{k}.$$

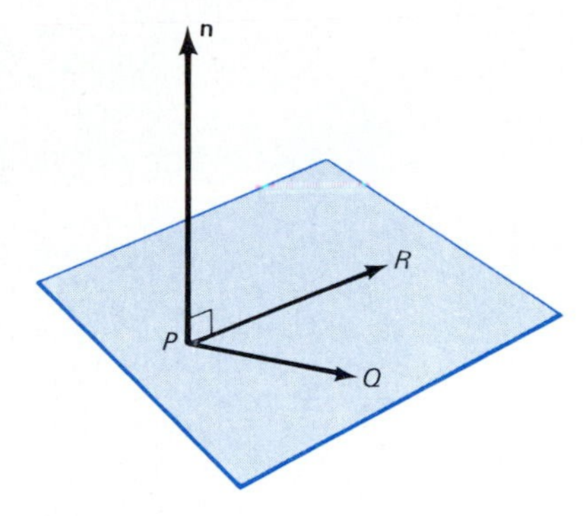

14.33 The normal vector **n** as a cross product

Because $\overrightarrow{PQ}$ and $\overrightarrow{PR}$ are in the plane, their cross product **n** is normal to the plane, as indicated in Fig. 14.33. Hence the plane has equation

$$11(x - 2) + (y - 4) - 7(z + 3) = 0.$$

After simplifications, we write it as

$$11x + y - 7z = 47.$$

Two planes with normal vectors **n** and **m** are called **parallel** provided that **n** and **m** are parallel. Otherwise, the two planes meet in a straight line. We define the angle between the two planes to be the angle between their normal vectors **n** and **m**, as in Fig. 14.34.

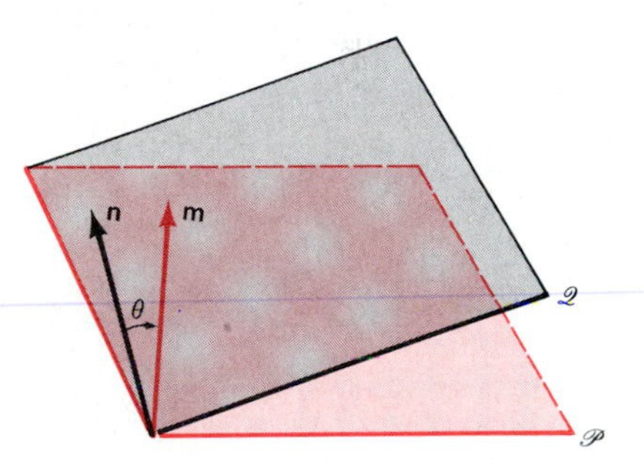

14.34 Vectors **m** and **n** normal to the planes $\mathscr{P}$ and $\mathscr{Q}$, respectively

EXAMPLE 6 Find the angle θ between the planes with equations

$$2x + 3y - z = -3 \quad \text{and} \quad 4x + 5y + z = 1.$$

Then write symmetric equations of their line of intersection L.

Solution The vectors $\mathbf{n} = \langle 2, 3, -1 \rangle$ and $\mathbf{m} = \langle 4, 5, 1 \rangle$ are normal to the two planes, so

$$\cos\theta = \frac{\mathbf{n} \cdot \mathbf{m}}{|\mathbf{n}||\mathbf{m}|} = \frac{22}{\sqrt{14}\sqrt{42}}.$$

Hence $\theta = \cos^{-1}(\frac{11}{21}\sqrt{3}) \approx 24.87^\circ$.

To determine the line L of intersection of the two planes, we need first to find a point P_0 lying on L. We can do this by substituting an arbitrarily chosen value of x into the equations of the given planes and then solving the resulting equations for y and z. With $x = 1$ we get the equations

$$2 + 3y - z = -3,$$

$$4 + 5y + z = 1.$$

The common solution is $y = -1$, $z = 2$. Thus the point $P_0(1, -1, 2)$ lies on the line L.

Next we need a vector **v** parallel to L. The normal vectors **n** and **m** to the two planes are both perpendicular to L, so their cross product will be parallel to L. Alternatively, we can find a second point P_1 on L by substituting a second value of x in the equations of the given planes and solving for y and z, as before. With $x = 5$ we obtain the equations

$$10 + 3y - z = -3,$$
$$20 + 5y + z = 1,$$

with common solution $y = -4$, $z = 1$. Thus we obtain the vector

$$\mathbf{v} = \overrightarrow{P_0P_1} = \langle 4, -3, -1 \rangle.$$

From (6) we now find the symmetric equations

$$\frac{x-1}{4} = \frac{y+1}{-3} = \frac{z-2}{-1}$$

of the line of intersection of the two given planes.

In conclusion, we note that the symmetric equations of a line L exhibit the line as an intersection of planes. For we can rewrite the equations in (6) in the form

$$\begin{aligned} b(x - x_0) - a(y - y_0) &= 0, \\ c(x - x_0) - a(z - z_0) &= 0, \\ c(y - y_0) - b(z - z_0) &= 0. \end{aligned} \tag{10}$$

These are the equations of three planes that intersect in the line L. The first has normal vector $\langle b, -a, 0 \rangle$, a vector parallel to the xy-plane. So the first plane is perpendicular to the xy-plane. Similarly, the second plane is perpendicular to the xz-plane and the third is perpendicular to the yz-plane.

The equations in (10) are symmetric equations of the line through $P_0(x_0, y_0, z_0)$ parallel to $\mathbf{v} = \langle a, b, c \rangle$. They enjoy the advantage over the equations in (6) of being meaningful whether or not the components a, b, and c of $\mathbf{v}$ are all nonzero. They have a special form, though, if one of the components is zero. If (for example) $a = 0$, then the first two equations in (10) take the form $x = x_0$. The line is then the intersection of the two planes $x = x_0$ and $c(y - y_0) = b(z - z_0)$.

14.3 PROBLEMS

In Problems 1–3, write parametric equations of the straight line through the point P that is parallel to the vector $\mathbf{v}$.

1. $P(0, 0, 0)$, $\mathbf{v} = \mathbf{i} + 2\mathbf{j} + 3\mathbf{k}$

2. $P(3, -4, 5)$, $\mathbf{v} = -2\mathbf{i} + 7\mathbf{j} + 3\mathbf{k}$

3. $P(4, 13, -3)$, $\mathbf{v} = 2\mathbf{i} - 3\mathbf{k}$

In Problems 4–6, write parametric equations of the straight line that passes through the points P_1 and P_2.

4. $P_1(0, 0, 0)$, $P_2(-6, 3, 5)$

5. $P_1(3, 5, 7)$, $P_2(6, -8, 10)$

6. $P_1(3, 5, 7)$, $P_2(6, 5, 4)$

In Problems 7–10, write an equation of the plane with normal vector $\mathbf{n}$ that passes through the point P.

7. $P(0, 0, 0)$, $\mathbf{n} = \langle 1, 2, 3 \rangle$

8. $P(3, -4, 5)$, $\mathbf{n} = \langle -2, 7, 3 \rangle$

9. $P(5, 12, 13)$, $\mathbf{n} = \mathbf{i} - \mathbf{k}$

10. $P(5, 12, 13)$, $\mathbf{n} = \mathbf{j}$

In Problems 11–16, write both parametric and symmetric equations for the indicated straight line.

11. Through $P(2, 3, -4)$ and parallel to $\mathbf{v} = \langle 1, -1, -2 \rangle$

12. Through $P(2, 5, -7)$ and $Q(4, 3, 8)$

13. Through $P(1, 1, 1)$ and perpendicular to the xy-plane

14. Through the origin and perpendicular to the plane with equation $x + y + z = 1$

15. Through $P(2, -3, 4)$ and perpendicular to the plane with equation $2x - y + 3z = 4$

16. Through $P(2, -1, 5)$ and parallel to the line with parametric equations $x = 3t$, $y = 2 + t$, $z = 2 - t$

In Problems 17–24, write an equation of the indicated plane.

17. Through $P(5, 7, -6)$ parallel to the xz-plane

18. Through $P(1, 0, -1)$ with normal vector $\mathbf{n} = \langle 2, 2, -1 \rangle$

19. Through $P(10, 4, -3)$ with normal vector $\mathbf{n} = \langle 7, 11, 0 \rangle$

20. Through $P(1, -3, 2)$ with normal vector $\mathbf{n} = \overrightarrow{OP}$

21. Through the origin and parallel to the plane with equation $3x + 4y = z + 10$

22. Through $P(5, 1, 4)$ and parallel to the plane with equation $x + y - 2z = 0$

23. Through the origin and the points $P(1, 1, 1)$ and $Q(1, -1, 3)$

24. Through the points $A(1, 0, -1)$, $B(3, 3, 2)$, and $C(4, 5, -1)$

In Problems 25–27, find the angle between the given planes.

25. $x = 10$ and $x + y + z = 0$

26. $2x - y + z = 5$ and $x + y - z = 1$

27. $x - y - 2z = 1$ and $x - y - 2z = 5$

28. Find parametric equations of the line of intersection of the planes $2x + y + z = 4$ and $3x - y + z = 3$.

29. Write symmetric equations for the line through $P(3, 3, 1)$ that is parallel to the line of Problem 28.

30. Find an equation of the plane through $P(3, 3, 1)$ that is perpendicular to the planes $x + y = 2z$ and $2x + z = 10$.

31. Find an equation of the plane through $(1, 1, 1)$ that intersects the xy-plane in the same line as does the plane $3x + 2y - z = 6$.

32. Find an equation for the plane through the point $P(1, 3, -2)$ and the line of intersection of the planes $x - y + z = 1$ and $x + y - z = 1$.

33. Find an equation of the plane through the points $P(1, 0, -1)$ and $Q(2, 1, 0)$ that is also parallel to the line of intersection of the planes $x + y + z = 5$ and $3x - y = 4$.

34. Prove that the lines $x - 1 = \frac{1}{2}(y + 1) = z - 2$ and $x - 2 = \frac{1}{3}(y - 2) = \frac{1}{2}(z - 4)$ intersect. Find an equation of the (only) plane containing them both.

35. Prove that the line of intersection of the planes $x + 2y - z = 2$ and $3x + 2y + 2z = 7$ is parallel to the line $x = 1 + 6t$, $y = 3 - 5t$, $z = 2 - 4t$. Find an equation of the plane determined by these two lines.

36. Show that the perpendicular distance D from the point $P_0(x_0, y_0, z_0)$ to the plane $ax + by + cz = d$ is

$$D = \frac{|ax_0 + by_0 + cz_0 - d|}{\sqrt{a^2 + b^2 + c^2}}.$$

(*Suggestion:* The line through P_0 perpendicular to the given plane has parametric equations $x = x_0 + at$, $y = y_0 + bt$, $z = z_0 + ct$. Let $P_1(x_1, y_1, z_1)$ be the point of this line, corresponding to $t = t_1$, at which it intersects the given plane. Solve for t_1, and then compute $D = |\overrightarrow{P_0P_1}|$.)

In Problems 37 and 38, use the formula of Problem 36 to find the distance between the given point and the given plane.

37. The origin and the plane $x + y + z = 10$

38. The point $P(5, 12, -13)$ and the plane with equation $3x + 4y + 5z = 12$

39. Prove that any two skew lines lie in parallel planes.

40. Use the formula of Problem 36 to show that the perpendicular distance D between the two parallel planes $ax + by + cz + d_1 = 0$ and $ax + by + cz + d_2 = 0$ is

$$D = \frac{|d_1 - d_2|}{\sqrt{a^2 + b^2 + c^2}}.$$

41. The line L_1 is described by the equations

$$x - 1 = 2y + 2; \qquad z = 4.$$

The line L_2 passes through the points $P(2, 1, -3)$ and $Q(0, 8, 4)$.

(a) Show that L_1 and L_2 are skew lines.

(b) Use the results of Problems 39 and 40 to find the perpendicular distance between L_1 and L_2.

14.4 Curves and Motion in Space

In Section 13.4 we used vector-valued functions to discuss curves and motion in the plane. Much of that discussion applies, with only minor changes, to curves and motion in three-dimensional space. The principal difference is that our vectors now have three components rather than two.

Think of a point that moves along a curve in space. Its position at time t can be described by *parametric equations*

$$x = f(t), \qquad y = g(t), \qquad z = h(t)$$

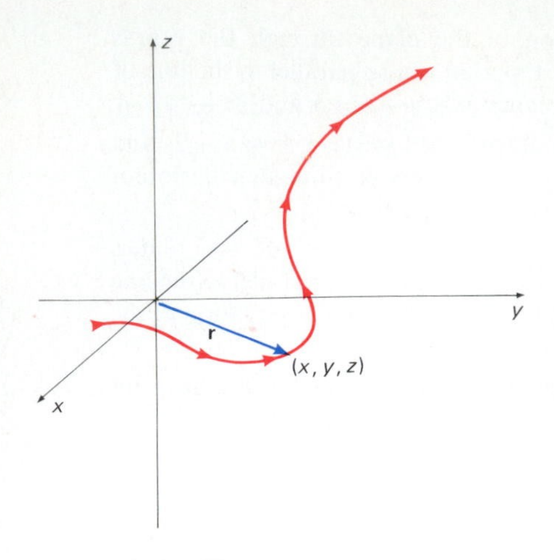

14.35 The position vector $\mathbf{r} = \langle x, y, z \rangle$ of a moving particle in space

that give its coordinates at time t. Alternatively, the location of the point can be given by its **position vector**

$$\mathbf{r} = f(t)\mathbf{i} + g(t)\mathbf{j} + h(t)\mathbf{k} = x\mathbf{i} + y\mathbf{j} + z\mathbf{k} \tag{1}$$

shown in Fig. 14.35.

A three-dimensional vector-valued function like $\mathbf{r}(t)$ can be differentiated and integrated in a componentwise manner, just like a two-dimensional vector-valued function (see Theorem 1 in Section 13.4). Thus the **velocity vector** $\mathbf{v} = \mathbf{v}(t)$ of the moving point at time t is given by

$$\begin{aligned} \mathbf{v}(t) &= \mathbf{r}'(t) = f'(t)\mathbf{i} + g'(t)\mathbf{j} + h'(t)\mathbf{k}, \\ \mathbf{v} &= \frac{d\mathbf{r}}{dt} = \frac{dx}{dt}\mathbf{i} + \frac{dy}{dt}\mathbf{j} + \frac{dz}{dt}\mathbf{k}. \end{aligned} \tag{2}$$

Its **acceleration vector** $\mathbf{a} = \mathbf{a}(t)$ is given by

$$\begin{aligned} \mathbf{a}(t) &= \mathbf{v}'(t) = f''(t)\mathbf{i} + g''(t)\mathbf{j} + h''(t)\mathbf{k}, \\ \mathbf{a} &= \frac{d\mathbf{v}}{dt} = \frac{d^2x}{dt^2}\mathbf{i} + \frac{d^2y}{dt^2}\mathbf{j} + \frac{d^2z}{dt^2}\mathbf{k}. \end{aligned} \tag{3}$$

The **speed** $v(t)$ and **scalar acceleration** $a(t)$ of the moving point are the lengths of its velocity and acceleration vectors, respectively:

$$v(t) = |\mathbf{v}(t)| = \sqrt{\left(\frac{dx}{dt}\right)^2 + \left(\frac{dy}{dt}\right)^2 + \left(\frac{dz}{dt}\right)^2} \tag{4}$$

and

$$a(t) = |\mathbf{a}(t)| = \sqrt{\left(\frac{d^2x}{dt^2}\right)^2 + \left(\frac{d^2y}{dt^2}\right)^2 + \left(\frac{d^2z}{dt^2}\right)^2}. \tag{5}$$

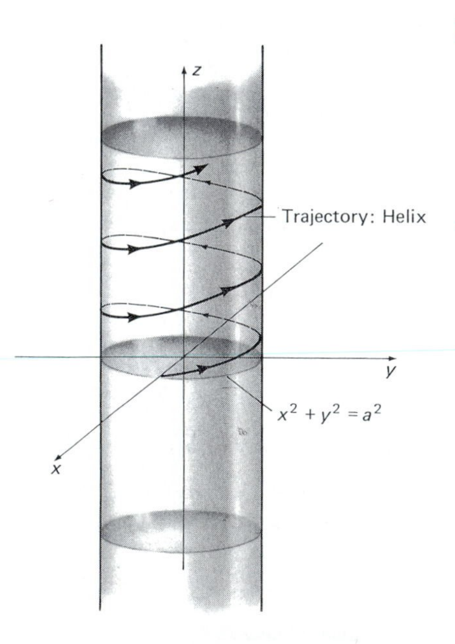

14.36 The particle of Example 1 moves in a helical path.

EXAMPLE 1 The parametric equations of a moving point are

$$x = a\cos\omega t, \qquad y = a\sin\omega t, \qquad z = bt,$$

where a, b, and ω are positive. Describe the path of the point in geometric terms. Compute its velocity, speed, and acceleration at time t.

Solution Because

$$x^2 + y^2 = a^2\cos^2\omega t + a^2\sin^2\omega t = a^2,$$

the path of the moving point lies on the **cylinder** that stands above and below the circle $x^2 + y^2 = a^2$ in the xy-plane, extending infinitely far in both directions, as indicated in Fig. 14.36. The given parametric equations for x and y tell us that the projection of the point into the xy-plane moves counterclockwise around the circle $x^2 + y^2 = a^2$ with angular speed ω. Meanwhile, because $z = bt$, the point itself is rising with vertical speed b. Its path on the cylinder is a spiral called a **helix** (also shown in Fig. 14.36).

The derivative of the position vector

$$\mathbf{r}(t) = (a\cos\omega t)\mathbf{i} + (a\sin\omega t)\mathbf{j} + (bt)\mathbf{k}$$

of the moving point is its velocity vector

$$\mathbf{v}(t) = (-a\omega\sin\omega t)\mathbf{i} + (a\omega\cos\omega t)\mathbf{j} + b\mathbf{k}. \tag{6}$$

Another differentiation gives its acceleration vector

$$\begin{aligned}\mathbf{a}(t) &= (-a\omega^2 \cos \omega t)\mathbf{i} + (-a\omega^2 \sin \omega t)\mathbf{j} \\ &= -a\omega^2(\mathbf{i} \cos \omega t + \mathbf{j} \sin \omega t).\end{aligned} \tag{7}$$

The speed of the moving point is a constant, for

$$v(t) = |\mathbf{v}(t)| = \sqrt{a^2\omega^2 + b^2}.$$

Note that the acceleration vector is a horizontal vector of length $a\omega^2$. Moreover, if we think of $\mathbf{a}(t)$ as attached to the moving point at the time t of evaluation—so that the initial point of $\mathbf{a}(t)$ is the terminal point of $\mathbf{r}(t)$—then $\mathbf{a}(t)$ points directly toward the point $(0, 0, bt)$ on the z-axis.

REMARK The helix of Example 1 is a typical trajectory of a charged particle in a constant magnetic field. Such a particle must satisfy both Newton's law $\mathbf{F} = m\mathbf{a}$ and the magnetic force law $\mathbf{F} = q\mathbf{v} \times \mathbf{B}$ mentioned in Section 14.2. Hence its velocity and acceleration vectors must satisfy the equation

$$q\mathbf{v} \times \mathbf{B} = m\mathbf{a}. \tag{8}$$

If the constant magnetic field is vertical, $\mathbf{B} = B\mathbf{k}$, then with the velocity vector of Equation (6) we find that

$$\begin{aligned}q\mathbf{v} \times \mathbf{B} &= q\begin{vmatrix} \mathbf{i} & \mathbf{j} & \mathbf{k} \\ -a\omega \sin \omega t & a\omega \cos \omega t & b \\ 0 & 0 & B \end{vmatrix} \\ &= qa\omega B(\mathbf{i} \cos \omega t + \mathbf{j} \sin \omega t).\end{aligned}$$

The acceleration vector of Equation (7) gives

$$m\mathbf{a} = -ma\omega^2(\mathbf{i} \cos \omega t + \mathbf{j} \sin \omega t).$$

When we compare these results, we see that the helix of Example 1 satisfies Equation (8) provided that

$$qa\omega B = -ma\omega^2; \qquad \text{that is,} \quad \omega = -\frac{qB}{m}.$$

For example, this equation would determine the angular speed ω for the helical trajectory of electrons ($q < 0$) in a cathode-ray tube placed in a constant magnetic field parallel to the axis of the tube (see Fig. 14.37).

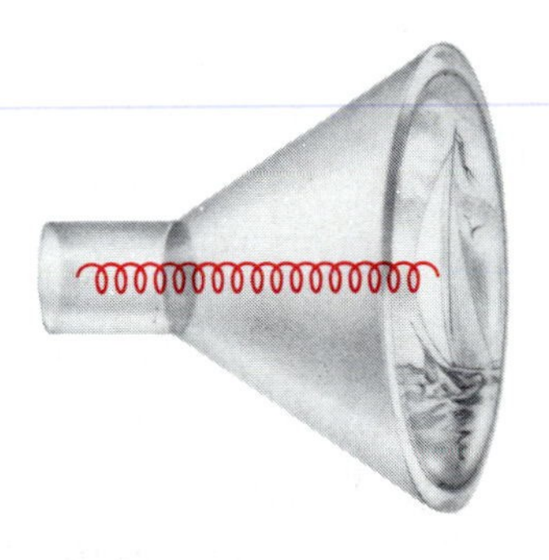

14.37 A spiraling electron in a cathode-ray tube

Differentiation of three-dimensional vector-valued functions satisfies the same formal properties that we listed for two-dimensional vector-valued functions in Theorem 2 of Section 13.4:

$$\begin{aligned}&\textbf{1.}\quad D_t[\mathbf{u}(t) + \mathbf{v}(t)] = \mathbf{u}'(t) + \mathbf{v}'(t), \\ &\textbf{2.}\quad D_t[c\mathbf{u}(t)] = c\mathbf{u}'(t), \\ &\textbf{3.}\quad D_t[h(t)\mathbf{u}(t)] = h'(t)\mathbf{u}(t) + h(t)\mathbf{u}'(t), \\ &\textbf{4.}\quad D_t[\mathbf{u}(t) \cdot \mathbf{v}(t)] = \mathbf{u}'(t) \cdot \mathbf{v}(t) + \mathbf{u}(t) \cdot \mathbf{v}'(t).\end{aligned} \tag{9}$$

Here, $\mathbf{u}(t)$ and $\mathbf{v}(t)$ are vector-valued functions with differentiable component functions and $h(t)$ is a differentiable real-valued function. In addition, the expected product rule for the derivative of a cross product also holds:

$$\textbf{5.}\quad D_t[\mathbf{u}(t) \times \mathbf{v}(t)] = \mathbf{u}'(t) \times \mathbf{v}(t) + \mathbf{u}(t) \times \mathbf{v}'(t). \tag{10}$$

Note that the order of the factors in Equation (10) *must* be preserved because the cross product is not commutative. Each of the properties in (9) and (10) can be verified routinely by componentwise differentiation, as in Section 13.4.

EXAMPLE 2 If $\mathbf{u}(t) = 3\mathbf{j} + 4t\mathbf{k}$ and $\mathbf{v}(t) = 5t\mathbf{i} - 4\mathbf{k}$, then

$$\mathbf{u}'(t) = 4\mathbf{k} \quad \text{and} \quad \mathbf{v}'(t) = 5\mathbf{i}.$$

Hence Equation (10) yields

$$\begin{aligned} D_t[\mathbf{u}(t) \times \mathbf{v}(t)] &= 4\mathbf{k} \times (5t\mathbf{i} - 4\mathbf{k}) + (3\mathbf{j} + 4t\mathbf{k}) \times 5\mathbf{i} \\ &= 20t(\mathbf{k} \times \mathbf{i}) - 16(\mathbf{k} \times \mathbf{k}) + 15(\mathbf{j} \times \mathbf{i}) + 20t(\mathbf{k} \times \mathbf{i}) \\ &= 20t\mathbf{j} - 16(\mathbf{0}) + 15(-\mathbf{k}) + 20t\mathbf{j} = 40t\mathbf{j} - 15\mathbf{k}, \end{aligned}$$

without our having to calculate the cross product $\mathbf{u}(t) \times \mathbf{v}(t)$.

EXAMPLE 3 A ball is thrown northward into the air from the origin in xyz-space (where the xy-plane represents the ground) with initial velocity vector

$$\mathbf{v}_0 = \mathbf{v}(0) = 80\mathbf{j} + 80\mathbf{k}. \tag{11}$$

In addition to gravitational acceleration, the spin of the ball causes an eastward acceleration of 2 ft/s^2, so the acceleration vector produced by the combination of gravity and the spin is

$$\mathbf{a}(t) = 2\mathbf{i} - 32\mathbf{k}. \tag{12}$$

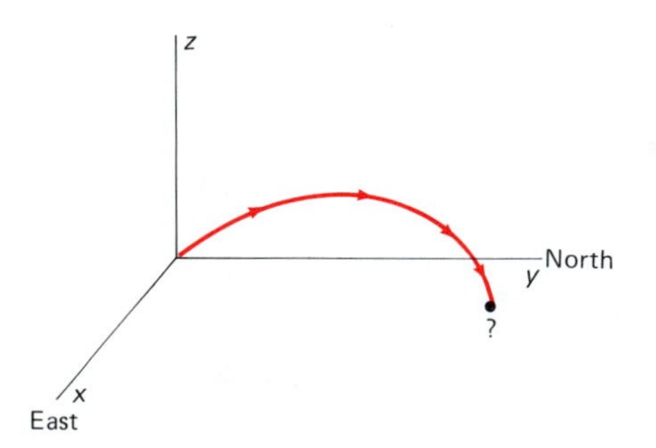

14.38 The trajectory of the ball of Example 3

First find the velocity vector $\mathbf{v}(t)$ of the ball, as well as its position vector $\mathbf{r}(t)$. Then determine where and with what speed it hits the ground. See Fig. 14.38.

Solution When we antidifferentiate $\mathbf{a}(t)$ we get

$$\begin{aligned} \mathbf{v}(t) &= \int \mathbf{a}(t)\,dt \\ &= \int (2\mathbf{i} - 32\mathbf{k})\,dt = 2t\mathbf{i} - 32t\mathbf{k} + \mathbf{c}_1. \end{aligned}$$

We substitute $t = 0$ to find that $\mathbf{c}_1 = \mathbf{v}_0 = 80\mathbf{j} + 80\mathbf{k}$, so

$$\mathbf{v}(t) = 2t\mathbf{i} + 80\mathbf{j} + (80 - 32t)\mathbf{k}.$$

Another antidifferentiation yields

$$\begin{aligned} \mathbf{r}(t) &= \int \mathbf{v}(t)\,dt = \int [2t\mathbf{i} + 80\mathbf{j} + (80 - 32t)\mathbf{k}]\,dt \\ &= t^2\mathbf{i} + 80t\mathbf{j} + (80t - 16t^2)\mathbf{k} + \mathbf{c}_2, \end{aligned}$$

and substitution of $t = 0$ gives $\mathbf{c}_2 = \mathbf{r}(0) = \mathbf{0}$. Hence the position vector of the ball is

$$\mathbf{r}(t) = t^2\mathbf{i} + 80t\mathbf{j} + (80t - 16t^2)\mathbf{k}.$$

The ball hits the ground when $z = 80t - 16t^2 = 0$, thus when $t = 5$. Its position vector then is

$$\mathbf{r}(5) = 5^2\mathbf{i} + (80)(5)\mathbf{j} = 25\mathbf{i} + 400\mathbf{j},$$

so the ball has traveled 25 ft eastward and 400 ft northward. Its velocity vector is

$$\begin{aligned} \mathbf{v}(5) &= (2)(5)\mathbf{i} + 80\mathbf{j} + [80 - (32)(5)]\mathbf{k} \\ &= 10\mathbf{i} + 80\mathbf{j} - 80\mathbf{k}, \end{aligned}$$

so its speed when it hits the ground is

$$v(5) = |\mathbf{v}(5)| = \sqrt{(10)^2 + (80)^2 + (-80)^2},$$

approximately 113.58 ft/s. Because the ball started with initial speed $v_0 = [(80)^2 + (80)^2]^{1/2} \approx 113.14$ ft/s, its eastward acceleration has slightly increased its terminal speed.

*DOES A BASEBALL PITCH REALLY CURVE?

Have you ever wondered whether a baseball pitch really curves, or whether it's some sort of optical illusion? To answer this question, let us suppose that the pitcher throws the ball toward home plate (60 ft away, as in Fig. 14.39), and gives it a spin of S revolutions per second, counterclockwise

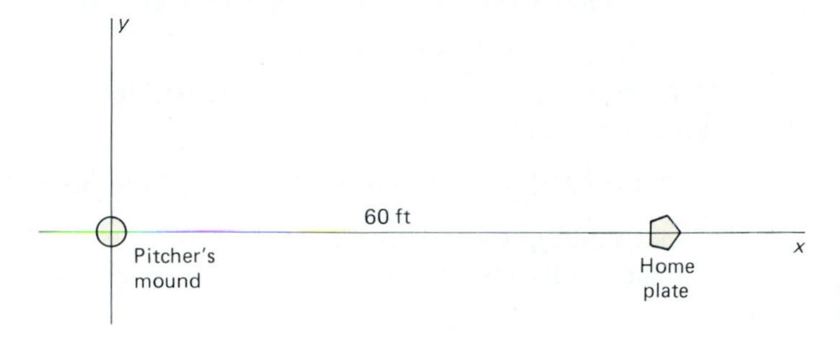

14.39 The x-axis points toward home plate.

(as viewed from above) about a vertical axis through the center of the ball. This spin is described by the *spin vector* **S** shown in Fig. 14.40; **S** points along the axis of revolution in the right-handed direction and has length S.

In aerodynamics it is shown that this spin causes a difference in air pressure on the sides of the ball toward and away from the spin, and this results in a *spin acceleration*

$$\mathbf{a}_s = c\mathbf{S} \times \mathbf{v} \tag{13}$$

of the ball (where c is an empirical constant). The total acceleration of the ball is then

$$\mathbf{a} = c\mathbf{S} \times \mathbf{v} - g\mathbf{k} \tag{14}$$

where $g \approx 32$ ft/s² is the usual gravitational acceleration.

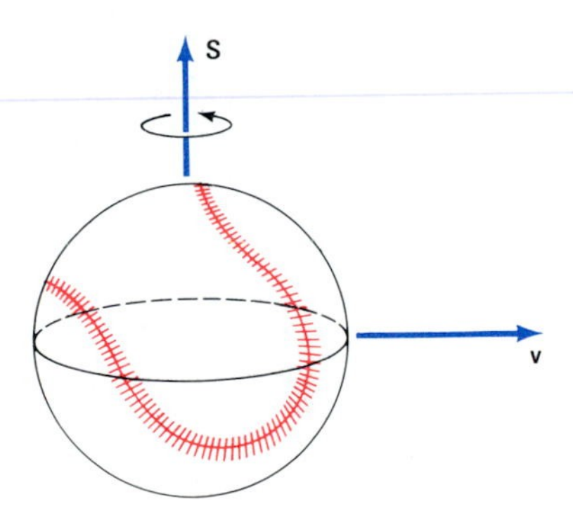

14.40 The spin and velocity vectors

With the spin vector $\mathbf{S} = S\mathbf{k}$ pointing upward as in Fig. 14.40, we find that

$$\mathbf{S} \times \mathbf{v} = \begin{vmatrix} \mathbf{i} & \mathbf{j} & \mathbf{k} \\ 0 & 0 & S \\ v_x & v_y & v_z \end{vmatrix} = -Sv_y\mathbf{i} + Sv_x\mathbf{j}.$$

For a ball pitched along the x-axis, v_x is much larger than v_y, and the approximation $\mathbf{S} \times \mathbf{v} = Sv_x\mathbf{i}$ is sufficiently accurate for our purposes. The acceleration vector of the ball may then be taken to be

$$\mathbf{a} = cSv_x\mathbf{j} - g\mathbf{k}. \tag{15}$$

Now suppose that the pitcher throws the ball from the initial position $x_0 = y_0 = 0$, $z_0 = 5$ (ft), with initial velocity vector

$$\mathbf{v}_0 = 120\mathbf{i} - 3\mathbf{j} + 4\mathbf{k} \tag{16}$$

(with components in feet per second, so that $v_0 \approx 120$ ft/s, about 82 mi/h), and with a spin of $S = 40$ rev/s. A reasonable value of c is

$$c = 0.005 \text{ ft/s}^2 \text{ per ft/s of velocity and rev/s of spin,}$$

although the precise value depends on whether the pitcher has (accidentally, of course) scuffed the ball or contaminated it with some foreign substance. Then the formula in (15) gives

$$\mathbf{a} = (0.005)(40)(120)\mathbf{j} - 32\mathbf{k} = 24\mathbf{j} - 32\mathbf{k},$$

and two successive integrations yield

$$\mathbf{v} = 120\mathbf{i} + (-3 + 24t)\mathbf{j} + (4 - 32t)\mathbf{k}$$

and

$$\mathbf{r} = 120t\mathbf{i} + (-3t + 12t^2)\mathbf{j} + (5 + 4t - 16t^2)\mathbf{k}$$

for the velocity and position vectors of the ball.

Thus the position of the ball during the half-second it takes to reach home plate is given by

$$x = 120t, \qquad y = -3t + 12t^2, \qquad z = 5 + 4t - 16t^2.$$

To see what its trajectory looks like, we calculate its position at intervals of 0.125 s; the results appear in the table with distance converted to feet and inches.

t	x	y	z
0	0	0	5 ft
0.125	15 ft	−2.25 in.	5 ft 3 in.
0.250	30 ft	0	5 ft
0.375	45 ft	6.75 in.	4 ft 3 in.
0.500	60 ft	1 ft 6 in.	3 ft

Suppose that the batter gets a "fix" on the ball and prepares to swing by observing it during the first quarter second. After 0.25 s it still appears to be straight "on target" for home plate at a height of 5 ft. During the last quarter second, however, it suddenly tails off 2 ft downward and 18 in. away from a right-handed batter.

According to an article in the July 1984 issue of *Science Digest*, a pitcher can easily give a baseball a spin of up to 30 rev/s. Although the values $c = 0.005$ and $S = 40$ may slightly exceed those attainable by a pitcher with an "unloaded" ball, our computation makes it clear that the effect is real—*not* an optical illusion.

14.4 PROBLEMS

In Problems 1–8, the position vector $\mathbf{r}(t)$ of a particle moving in space is given. Find its velocity and acceleration vectors and its speed at time t.

1. $\mathbf{r}(t) = 3\mathbf{i} - 4\mathbf{j} + 5t\mathbf{k}$
2. $\mathbf{r}(t) = 3\mathbf{i} - 4t\mathbf{j} + 5t^2\mathbf{k}$
3. $\mathbf{r}(t) = t\mathbf{i} + t^2\mathbf{j} + t^3\mathbf{k}$
4. $\mathbf{r}(t) = t^2(3\mathbf{i} + 4\mathbf{j} - 12\mathbf{k})$
5. $\mathbf{r}(t) = t\mathbf{i} + 3e^t\mathbf{j} + 4e^t\mathbf{k}$
6. $\mathbf{r}(t) = e^t\mathbf{i} + e^{2t}\mathbf{j} + e^{3t}\mathbf{k}$
7. $\mathbf{r}(t) = (3\cos t)\mathbf{i} + (3\sin t)\mathbf{j} - 4t\mathbf{k}$
8. $\mathbf{r}(t) = 12t\mathbf{i} + (5\sin 2t)\mathbf{j} - (5\cos 2t)\mathbf{k}$

In Problems 9–16, the acceleration vector $\mathbf{a}(t)$, the initial position $\mathbf{r}_0 = \mathbf{r}(0)$, and the initial velocity $\mathbf{v}_0 = \mathbf{v}(0)$ of a particle moving in xyz-space are given. Find its position vector $\mathbf{r}(t)$ at time t.

9. $\mathbf{a}(t) = 2\mathbf{i} - 4\mathbf{k};\quad \mathbf{r}_0 = \mathbf{0},\quad \mathbf{v}_0 = 10\mathbf{j}$

10. $\mathbf{a}(t) = \mathbf{i} - \mathbf{j} + 3\mathbf{k};\quad \mathbf{r}_0 = 5\mathbf{i},\quad \mathbf{v}_0 = 7\mathbf{j}$

11. $\mathbf{a}(t) = 2\mathbf{j} - 6t\mathbf{k};\quad \mathbf{r}_0 = 2\mathbf{i},\quad \mathbf{v}_0 = 5\mathbf{k}$

12. $\mathbf{a}(t) = 6t\mathbf{i} - 5\mathbf{j} + 12t^2\mathbf{k};\quad \mathbf{r}_0 = 3\mathbf{i} + 4\mathbf{j},\quad \mathbf{v}_0 = 4\mathbf{j} - 5\mathbf{k}$

13. $\mathbf{a}(t) = t\mathbf{i} + t^2\mathbf{j} + t^3\mathbf{k};\quad \mathbf{r}_0 = 10\mathbf{i},\quad \mathbf{v}_0 = 10\mathbf{j}$

14. $\mathbf{a}(t) = t\mathbf{i} + e^{-t}\mathbf{j};\quad \mathbf{r}_0 = 3\mathbf{i} + 4\mathbf{j},\quad \mathbf{v}_0 = 5\mathbf{k}$

15. $\mathbf{a}(t) = \mathbf{i}\cos t + \mathbf{j}\sin t;\quad \mathbf{r}_0 = \mathbf{j},\quad \mathbf{v}_0 = -\mathbf{i} + 5\mathbf{k}$

16. $\mathbf{a}(t) = -9(\mathbf{i}\sin 3t + \mathbf{j}\cos 3t) + 4\mathbf{k};\quad \mathbf{r}_0 = 3\mathbf{i} + 4\mathbf{j},$
$\mathbf{v}_0 = 2\mathbf{i} - 7\mathbf{k}$

17. The parametric equations of a moving point are

$$x = 3\cos 2t, \qquad y = 3\sin 2t, \qquad z = 8t.$$

Find its velocity, speed, and acceleration at time $t = 7\pi/8$.

18. Use the equations in (9) and (10) to calculate

$$D_t[\mathbf{u}(t)\cdot\mathbf{v}(t)] \quad\text{and}\quad D_t[\mathbf{u}(t)\times\mathbf{v}(t)]$$

if $\mathbf{u}(t) = \langle t, t^2, t^3\rangle$ and $\mathbf{v}(t) = \langle e^t, \cos t, \sin t\rangle$.

19. Verify the result obtained in Example 2 by first calculating $\mathbf{u}(t)\times\mathbf{v}(t)$ and then differentiating.

20. Verify that

$$D_t[\mathbf{u}(t)\times\mathbf{v}(t)] = \mathbf{u}'(t)\times\mathbf{v}(t) + \mathbf{u}(t)\times\mathbf{v}'(t).$$

21. A point moves on a sphere centered at the origin. Show that its velocity vector is always tangent to the sphere.

22. A particle moves along a space curve with constant speed. Show that its velocity and acceleration vectors are always perpendicular.

23. Find the maximum height reached by the ball in Example 3 and also its speed then.

24. The **angular momentum** $\mathbf{L}(t)$ and **torque** $\boldsymbol{\tau}(t)$ of a moving particle of mass m with position vector $\mathbf{r}(t)$ are defined to be

$$\mathbf{L}(t) = m\mathbf{r}(t)\times\mathbf{v}(t), \qquad \boldsymbol{\tau}(t) = m\mathbf{r}(t)\times\mathbf{a}(t).$$

Prove that $\mathbf{L}'(t) = \boldsymbol{\tau}(t)$. It follows that $\mathbf{L}(t)$ must be constant if $\boldsymbol{\tau} \equiv \mathbf{0}$; this is the law of the conservation of angular momentum.

25. Suppose that a particle is moving under the influence of a *central* force field $\mathbf{F} = k\mathbf{r}$, where k is a scalar function of x, y, and z. Conclude that the trajectory of the particle lies in a *fixed* plane through the origin.

26. A baseball is thrown straight upward from the ground with an initial velocity of 160 ft/s. It experiences a downward gravitational acceleration of 32 ft/s^2. Because of spin, it also experiences a (horizontal) northward acceleration of 0.1 ft/s^2; otherwise, the air has no effect on its motion. How far north of the throwing point will the ball land?

27. A baseball is hit from ground level straight down a foul line with an initial velocity of 96 ft/sec and an initial inclination angle of 15°. Because of spin it experiences a horizontal acceleration of 2 ft/s^2 perpendicular to the foul line; otherwise, the air has no effect on its motion. When the ball hits the ground, how far is it from the foul line?

28. A gun fires a shell with a muzzle velocity of 500 ft/s. While the shell is in the air, it experiences a downward (vertical) gravitational acceleration of 32 ft/s^2 and also an eastward (horizontal) Coriolis acceleration of 1/6 ft/s^2; ignore air resistance. The target is 5000 ft due north of the gun, and both the gun and the target are on level ground. Halfway between them is a hill 2000 ft high. Tell precisely how to aim the gun—both compass heading and inclination from the horizontal—so that the shell will clear the hill and hit the target.

14.5 Curvature and Acceleration

The speed of a moving point is closely related to the arc length of its trajectory. The arc length formula for parametric space curves is a natural generalization of the formula for parametric plane curves (Equation (6) in Section 13.2). The **arc length** s along the smooth curve with position vector

$$\mathbf{r}(t) = f(t)\mathbf{i} + g(t)\mathbf{j} + h(t)\mathbf{k} = x\mathbf{i} + y\mathbf{j} + z\mathbf{k} \tag{1}$$

from the point $\mathbf{r}(a)$ to the point $\mathbf{r}(b)$ is, by definition,

$$\begin{aligned} s &= \int_a^b \sqrt{[x'(t)]^2 + [y'(t)]^2 + [z'(t)]^2}\,dt \\ &= \int_a^b \sqrt{\left(\frac{dx}{dt}\right)^2 + \left(\frac{dy}{dt}\right)^2 + \left(\frac{dz}{dt}\right)^2}\,dt. \end{aligned} \tag{2}$$

From Equation (4) in Section 14.4 we see that the integrand is the speed of the moving point with position vector $\mathbf{r}(t)$, so

$$s = \int_a^b v(t)\, dt. \tag{3}$$

EXAMPLE 1 Find the arc length of one turn (from $t = 0$ to $t = 2\pi/\omega$) of the helix with parametric equations

$$x = a \cos \omega t, \qquad y = a \sin \omega t, \qquad z = bt.$$

Solution In Example 1 of Section 14.4, we found that

$$v(t) = \sqrt{a^2\omega^2 + b^2}.$$

Hence the formula in (3) gives

$$s = \int_0^{2\pi/\omega} \sqrt{a^2\omega^2 + b^2}\, dt = \frac{2\pi}{\omega}\sqrt{a^2\omega^2 + b^2}.$$

For instance, if $a = b = \omega = 1$, then $s = 2\pi\sqrt{2}$, which is $\sqrt{2}$ times the circumference of the circle in the xy-plane over which the helix lies.

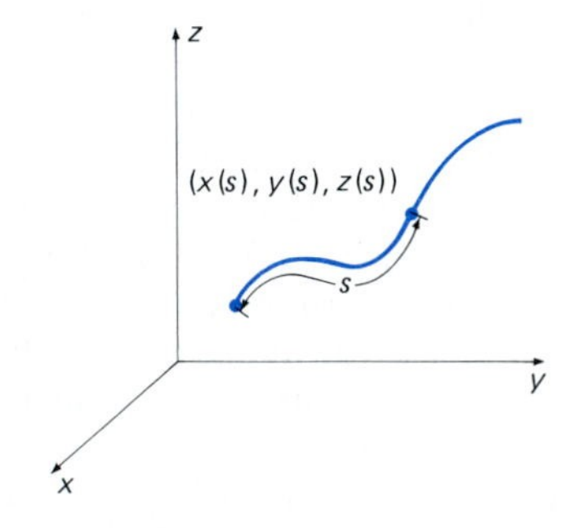

14.41 A curve parametrized by arc length s

Let $s(t)$ denote the arc length along a smooth curve from its initial point $\mathbf{r}(a)$ to the variable point $\mathbf{r}(t)$ (see Fig. 14.41). Then from the formula in (3) we obtain the **arc length function** $s(t)$ of the curve:

$$s(t) = \int_a^t v(\tau)\, d\tau. \tag{4}$$

The fundamental theorem of calculus then gives

$$\frac{ds}{dt} = v. \tag{5}$$

Thus *the speed of the moving point is the time rate of change of its arc length function.* If $v(t) > 0$ for all t, then it follows that $s(t)$ is an increasing function of t and therefore has an inverse function $t(s)$. When we replace t by $t(s)$ in the curve's original parametric equations, we obtain the **arc length parametrization**

$$x = x(s), \qquad y = y(s), \qquad z = z(s).$$

This gives the position of the moving point as a function of arc length measured along the curve from its initial point (see Fig. 14.41).

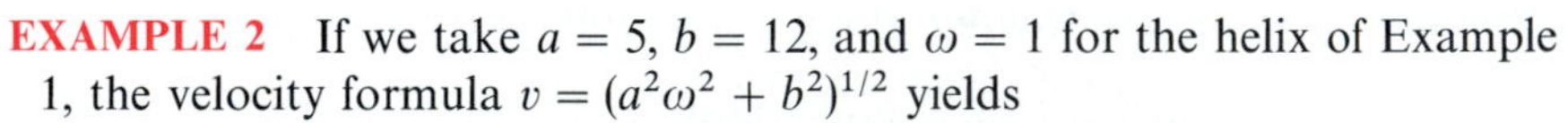

EXAMPLE 2 If we take $a = 5$, $b = 12$, and $\omega = 1$ for the helix of Example 1, the velocity formula $v = (a^2\omega^2 + b^2)^{1/2}$ yields

$$v = \sqrt{5^2 \cdot 1^2 + 12^2} = \sqrt{169} = 13.$$

Hence Equation (5) gives $ds/dt = 13$, so

$$s = 13t,$$

taking $s = 0$ when $t = 0$, and thereby measuring arc length from the natural starting point $(5, 0, 0)$. When we substitute $t = s/13$ and the numerical values of a, b, and ω in the original parametric equations of the helix, we get the arc length parametrization

$$x = 5\cos\frac{s}{13}, \qquad y = 5\sin\frac{s}{13}, \qquad z = \frac{12}{13}s$$

of the helix.

CURVATURE FOR PLANE CURVES

The word *curvature* has an intuitive meaning that we need to make precise. Most people would agree that a straight line does not curve at all, while a circle of small radius is more curved than a circle of large radius (see Fig. 14.42). This judgment may be based on a feeling that curvature is "rate of change of direction." The direction of a curve is determined by its velocity vector, so you would expect the idea of curvature to have something to do with the rate at which the velocity vector is turning.

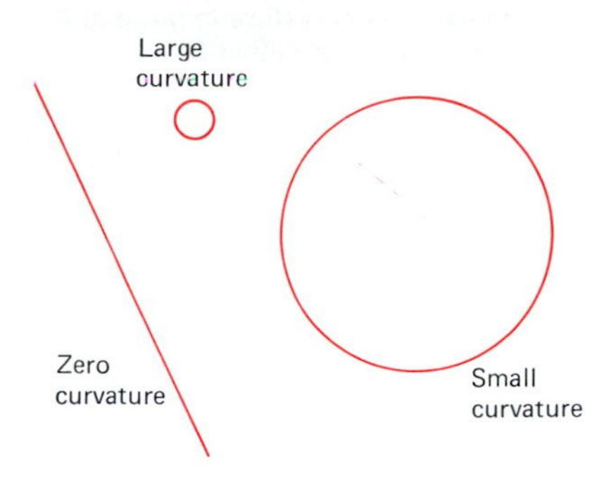

14.42 The intuitive idea of curvature

Let

$$\mathbf{r}(t) = x(t)\mathbf{i} + y(t)\mathbf{j}, \qquad a \leqq t \leqq b, \tag{6}$$

be the position vector of a smooth *plane* curve with nonzero velocity vector $\mathbf{v}(t) = \mathbf{r}'(t)$. The curve's **unit tangent vector** at the point $\mathbf{r}(t)$ is the unit vector

$$\mathbf{T}(t) = \frac{\mathbf{v}(t)}{|\mathbf{v}(t)|} = \frac{\mathbf{v}(t)}{v(t)} \tag{7}$$

where $v(t) = |\mathbf{v}(t)|$ is the speed. Now denote by ϕ the angle of inclination of $\mathbf{T}$, measured counterclockwise from the positive x-axis, as in Fig. 14.43. Then

$$\mathbf{T} = \mathbf{i}\cos\phi + \mathbf{j}\sin\phi. \tag{8}$$

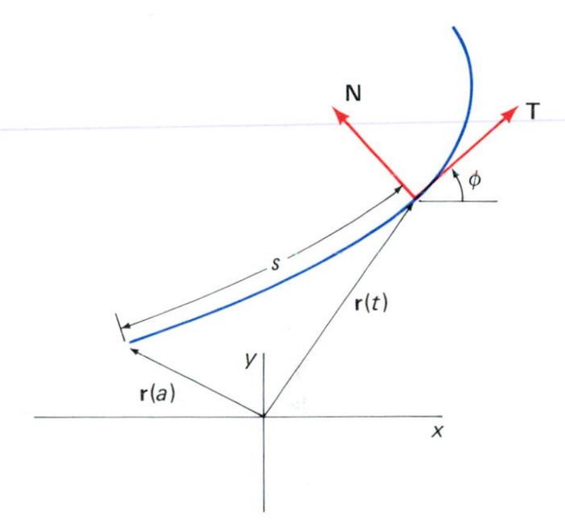

14.43 The unit tangent vector **T**

Now the unit tangent vector $\mathbf{T}$ in (8) can be expressed as a function of the arc length parameter s indicated in Fig. 14.43. Then the rate at which $\mathbf{T}$ is turning is measured by the derivative

$$\frac{d\mathbf{T}}{ds} = \frac{d\mathbf{T}}{d\phi}\cdot\frac{d\phi}{ds} = (-\mathbf{i}\sin\phi + \mathbf{j}\cos\phi)\frac{d\phi}{ds}. \tag{9}$$

Note that

$$\left|\frac{d\mathbf{T}}{ds}\right| = \left|\frac{d\phi}{ds}\right| \tag{10}$$

because the vector on the right-hand side in (9) is a unit vector.

The **curvature** at a point of a plane curve, denoted by the lowercase Greek letter kappa, is therefore defined to be

$$\kappa = \left|\frac{d\phi}{ds}\right|, \tag{11}$$

the absolute value of the rate of change of the angle ϕ with respect to arc length s. We define the curvature κ in terms of $d\phi/ds$ rather than $d\phi/dt$ because the latter depends not only upon the shape of the curve but also upon the speed of the moving point $\mathbf{r}(t)$. For a straight line the angle ϕ is a constant, so the curvature given by Equation (11) is zero. If you imagine a point moving with constant speed along a curve, the curvature is greatest at points where ϕ changes the most rapidly, such as the points P and R on

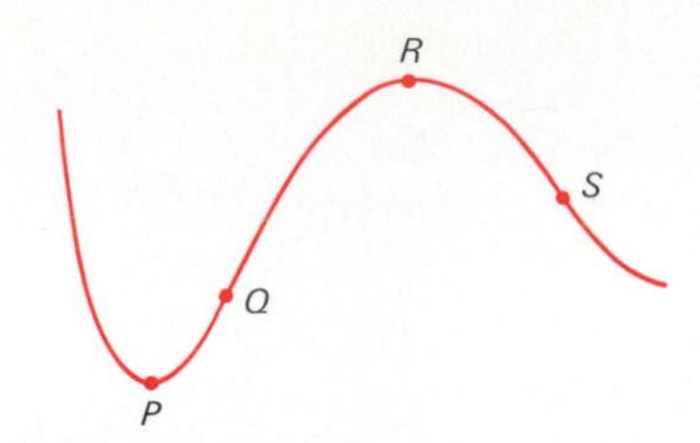

14.44 The curvature is large at P and R, small at Q and S.

the curve of Fig. 14.44. The curvature is least at points such as Q and S, where ϕ is changing the least rapidly.

We need to derive a formula that is effective in computing the curvature of a smooth parametric curve $x = x(t)$, $y = y(t)$. First we note that

$$\phi = \tan^{-1}\left(\frac{dy}{dx}\right) = \tan^{-1}\left(\frac{y'(t)}{x'(t)}\right)$$

provided $x'(t) \neq 0$. Hence

$$\begin{aligned}\frac{d\phi}{dt} &= \frac{y''x' - y'x''}{(x')^2} \div \left(1 + \left(\frac{y'}{x'}\right)^2\right)\\ &= \frac{x'y'' - x''y'}{(x')^2 + (y')^2},\end{aligned}$$

where primes denote derivatives with respect to t. Because $v = ds/dt > 0$, Equation (11) gives

$$\kappa = \left|\frac{d\phi}{ds}\right| = \left|\frac{d\phi}{dt}\cdot\frac{dt}{ds}\right| = \frac{1}{v}\left|\frac{d\phi}{dt}\right|;$$

thus

$$\kappa = \frac{|x'y'' - x''y'|}{[(x')^2 + (y')^2]^{3/2}} = \frac{|x'y'' - x''y'|}{v^3}. \tag{12}$$

At a point where $x'(t) = 0$, we know that $y'(t) \neq 0$ because the curve is smooth. Thus we will obtain the same result if we begin with the equation $\phi = \cot^{-1}(x'/y')$.

An explicitly described curve $y = f(x)$ may be regarded as a parametric curve $x = x$, $y = f(x)$. Then $x' = 1$ and $x'' = 0$, so Equation (12)—with x in place of t as the parameter—becomes

$$\kappa = \frac{|y''|}{[1 + (y')^2]^{3/2}} = \frac{|d^2y/dx^2|}{[1 + (dy/dx)^2]^{3/2}}. \tag{13}$$

EXAMPLE 3 Show that the curvature at each point of a circle of radius a is $\kappa = 1/a$.

Solution With the familiar parametrization $x = a\cos t$, $y = a\sin t$ of such a circle centered at the origin, we have

$$\begin{aligned} x' &= -a\sin t, & y' &= a\cos t,\\ x'' &= -a\cos t, & y'' &= -a\sin t.\end{aligned}$$

Hence (12) gives

$$\begin{aligned}\kappa &= \frac{|(-a\sin t)(-a\sin t) - (-a\cos t)(a\cos t)|}{[(-a\sin t)^2 + (a\cos t)^2]^{3/2}}\\ &= \frac{a^2}{a^3} = \frac{1}{a}.\end{aligned}$$

Alternatively, we could have used the formula in (13); our point of departure would be the equation $x^2 + y^2 = a^2$ of the same circle, and we would compute y' and y'' by implicit differentiation (see Problem 27).

It follows immediately from Equations (8) and (9) that

$$\mathbf{T} \cdot \frac{d\mathbf{T}}{ds} = 0,$$

so the unit tangent vector $\mathbf{T}$ and its derivative vector $d\mathbf{T}/ds$ are perpendicular. The *unit* vector $\mathbf{N}$ that points in the direction of $d\mathbf{T}/ds$ is called the **principal unit normal vector** to the curve. Because $\kappa = |d\phi/ds| = |d\mathbf{T}/ds|$ by Equation (10), it follows that

$$\frac{d\mathbf{T}}{ds} = \kappa\mathbf{N}. \tag{14}$$

Intuitively, $\mathbf{N}$ *is the unit normal vector to the curve that points in the direction in which the curve is bending.*

Suppose that P is a point on a parametrized curve where $\kappa \neq 0$. Consider the circle that is tangent to the curve at P and which has the same curvature there. The center of the circle is to lie on the concave side of the curve, that is, on the side toward which the normal vector $\mathbf{N}$ points. This circle is called the **osculating circle** (or **circle of curvature**) of the curve at the given point because it touches the curve so closely there (*osculum* is the Latin word for *kiss*). Let ρ be the radius of the osculating circle and let γ be the position vector of its center; thus $\gamma = \overrightarrow{OC}$ where C is the center of the osculating circle. Then ρ is called the **radius of curvature** of the curve at the point P, and γ is called the (vector) **center of curvature** of the curve at P. (See Fig. 14.45.)

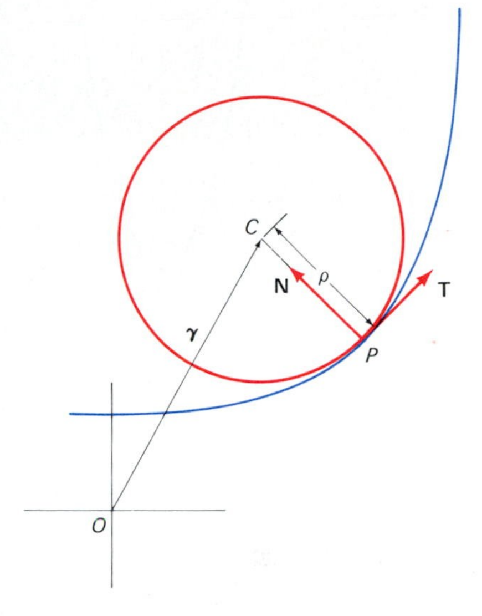

14.45 Osculating circle, radius of curvature, center of curvature

Example 3 implies that the radius of curvature is

$$\rho = \frac{1}{\kappa}, \tag{15}$$

and the fact that $|\mathbf{N}| = 1$ implies that the position vector of the center of curvature is

$$\gamma = \mathbf{r} + \rho\mathbf{N} \qquad (\mathbf{r} = \overrightarrow{OP}). \tag{16}$$

EXAMPLE 4 Determine the vectors $\mathbf{T}$ and $\mathbf{N}$, the curvature κ, and the center of curvature of the parabola $y = x^2$ at the point (1, 1).

Solution If the parabola is parametrized by $x = t$, $y = t^2$, then its position vector is $\mathbf{r}(t) = t\mathbf{i} + t^2\mathbf{j}$, so $\mathbf{v}(t) = \mathbf{i} + 2t\mathbf{j}$. The speed is $v(t) = (1 + 4t^2)^{1/2}$, so Equation (7) yields

$$\mathbf{T}(t) = \frac{\mathbf{i} + 2t\mathbf{j}}{\sqrt{1 + 4t^2}}.$$

By substituting $t = 1$, we find that the unit tangent vector at (1, 1) is

$$\mathbf{T} = \frac{1}{\sqrt{5}}\mathbf{i} + \frac{2}{\sqrt{5}}\mathbf{j}.$$

Because the parabola is concave upward at (1, 1), the principal unit normal vector is the upward-pointing unit vector

$$\mathbf{N} = -\frac{2}{\sqrt{5}}\mathbf{i} + \frac{1}{\sqrt{5}}\mathbf{j}$$

that is perpendicular to **T**. If $y = x^2$, then $dy/dx = 2x$ and $d^2y/dx^2 = 2$, so Equation (13) yields

$$\kappa = \frac{|2|}{(1 + 4x^2)^{3/2}}.$$

So at the point (1, 1) we find the curvature and radius of curvature to be

$$\kappa = \frac{2}{5\sqrt{5}} \quad \text{and} \quad \rho = \frac{5}{2}\sqrt{5},$$

respectively.

Next, Equation (16) gives the center of curvature as

$$\gamma = \langle 1, 1 \rangle + \frac{5}{2}\sqrt{5}\left\langle -\frac{2}{\sqrt{5}}, \frac{1}{\sqrt{5}} \right\rangle = \left\langle -4, \frac{7}{2} \right\rangle.$$

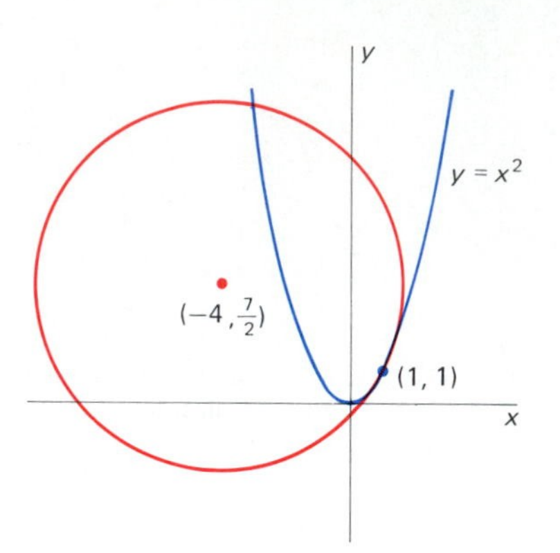

14.46 The osculating circle of Example 4

The equation of the osculating circle to the parabola is, therefore,

$$(x + 4)^2 + (y - \tfrac{7}{2})^2 = \rho^2 = \tfrac{125}{4}.$$

The parabola and this osculating circle are shown in Fig. 14.46.

CURVATURE FOR SPACE CURVES

Consider now a moving particle in space with twice-differentiable position vector $\mathbf{r}(t)$. Suppose also that the velocity vector $\mathbf{v}(t)$ is always nonzero. The **unit tangent vector** at time t is defined, as before, to be

$$\mathbf{T}(t) = \frac{\mathbf{v}(t)}{|\mathbf{v}(t)|} = \frac{\mathbf{v}}{v}, \tag{17}$$

so

$$\mathbf{v} = v\mathbf{T}. \tag{18}$$

We defined the curvature of a plane curve to be $\kappa = |d\phi/dt|$, where ϕ is the angle of inclination of **T** from the positive x-axis. For a space curve, there is no single angle that determines the direction of **T**, so we adopt the following approach (which leads to the same value for curvature when applied to a space curve that happens to lie in the xy-plane). Differentiation of the identity $\mathbf{T} \cdot \mathbf{T} = 1$ with respect to arc length gives

$$\mathbf{T} \cdot \frac{d\mathbf{T}}{ds} = 0.$$

It follows that the vectors **T** and $d\mathbf{T}/ds$ are always perpendicular.

Then we define the **curvature** κ of the curve at the point $\mathbf{r}(t)$ to be

$$\kappa = \left|\frac{d\mathbf{T}}{ds}\right| = \left|\frac{d\mathbf{T}}{dt}\frac{dt}{ds}\right| = \frac{1}{v}\left|\frac{d\mathbf{T}}{dt}\right|. \tag{19}$$

At a point where $\kappa \neq 0$, we define the **principal unit normal vector N** to be

$$\mathbf{N} = \frac{d\mathbf{T}/ds}{|d\mathbf{T}/ds|} = \frac{1}{\kappa}\frac{d\mathbf{T}}{ds}, \tag{20}$$

so

$$\frac{d\mathbf{T}}{ds} = \kappa\mathbf{N}. \tag{21}$$

Equation (21) shows that **N** has the same direction as $d\mathbf{T}/ds$, and Equation (20) shows that **N** is a unit vector. Because Equation (21) is the same as Equation (14), we see that the present definitions of κ and **N** agree with those given earlier in the two-dimensional case (see Fig. 14.47).

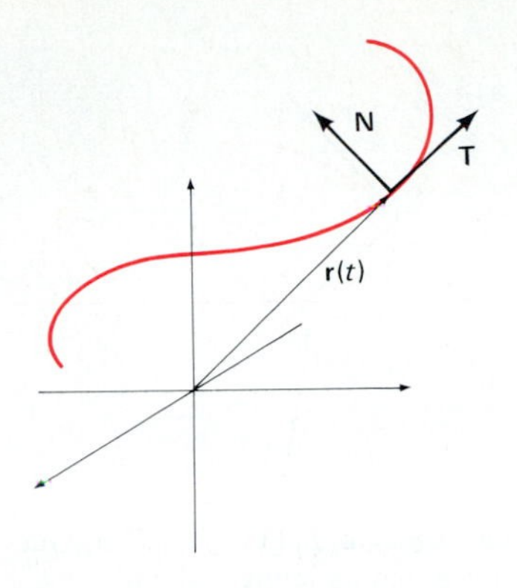

14.47 The principal unit normal vector **N** points in the direction in which the curve is turning.

EXAMPLE 5 Compute the curvature κ of the helix of Example 1, the one with parametric equations

$$x = a \cos \omega t, \qquad y = a \sin \omega t, \qquad z = bt.$$

Solution In Example 1 of Section 14.4, we computed the velocity vector

$$\mathbf{v} = \mathbf{i}(-a\omega \sin \omega t) + \mathbf{j}(a\omega \cos \omega t) + b\mathbf{k}$$

and speed

$$v = |\mathbf{v}| = \sqrt{a^2\omega^2 + b^2}.$$

Hence Equation (17) gives the unit tangent vector

$$\mathbf{T} = \frac{\mathbf{v}}{v} = \frac{\mathbf{i}(-a\omega \sin \omega t) + \mathbf{j}(a\omega \cos \omega t) + b\mathbf{k}}{\sqrt{a^2\omega^2 + b^2}}.$$

Then

$$\frac{d\mathbf{T}}{dt} = \frac{\mathbf{i}(-a\omega^2 \cos \omega t) + \mathbf{j}(-a\omega^2 \sin \omega t)}{\sqrt{a^2\omega^2 + b^2}},$$

so Equation (19) gives

$$\kappa = \frac{1}{v}\left|\frac{d\mathbf{T}}{dt}\right| = \frac{a\omega^2}{a^2\omega^2 + b^2}$$

for the curvature of the helix of Example 1. Note that the helix has constant curvature. Note also, if $b = 0$ (so that the helix reduces to a circle of radius a in the xy-plane), our result reduces to $\kappa = 1/a$, in agreement with our computation of the curvature of a circle in Example 3.

NORMAL AND TANGENTIAL COMPONENTS OF ACCELERATION

We may apply Equation (21) to analyze the meaning of the acceleration vector of a moving particle with velocity vector **v** and speed v. Then Equation (17) gives $\mathbf{v} = v\mathbf{T}$, so the acceleration vector of the particle is

$$\mathbf{a} = \frac{d\mathbf{v}}{dt} = \frac{dv}{dt}\mathbf{T} + v\frac{d\mathbf{T}}{dt} = \frac{dv}{dt}\mathbf{T} + v\frac{d\mathbf{T}}{ds}\frac{ds}{dt}.$$

But $ds/dt = v$, so Equation (21) gives us the formula

$$\mathbf{a} = \frac{dv}{dt}\mathbf{T} + \kappa v^2 \mathbf{N}. \tag{22}$$

Because **T** and **N** are unit vectors tangent and normal to the curve, respectively, Equation (22) provides a *decomposition of the acceleration vector* into its components tangent and normal to the trajectory. The **tangential component**

$$a_T = \frac{dv}{dt} \tag{23}$$

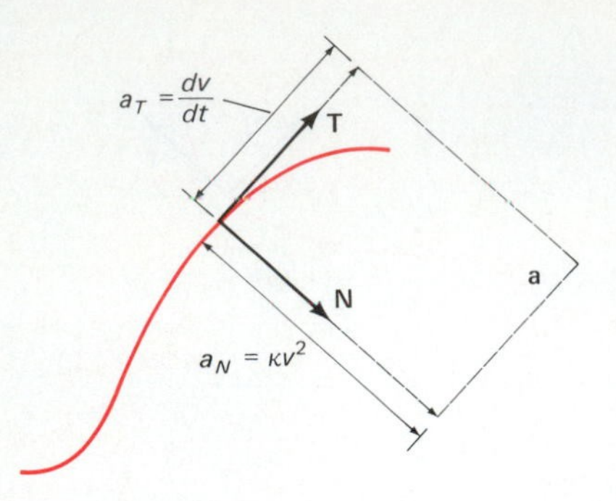

14.48 Resolution of the acceleration vector **a** into its tangential and normal components

is the rate of change of speed of the particle, while the **normal component**

$$a_N = \kappa v^2 = \frac{v^2}{\rho} \tag{24}$$

measures the rate of change of its direction of motion. The decomposition

$$\mathbf{a} = a_T\mathbf{T} + a_N\mathbf{N} \tag{25}$$

is illustrated in Fig. 14.48.

As an application of Equation (22), think of a train moving along a straight track with constant speed v, so that $a_T = 0 = a_N$ (the latter because $\kappa = 0$ for a straight line). Suppose that at time $t = 0$, the train enters a circular curve of radius ρ. At that instant, it will be *suddenly* subjected to a normal acceleration of magnitude v^2/ρ, proportional to the *square* of the speed of the train. A passenger in the train will experience a sudden jerk to the side. If v is large, the stresses may be great enough to damage the track or derail the train. It is for exactly this reason that railroads are not built with curves shaped like arcs of circles, but with *approach curves* in which the curvature, and hence the normal acceleration, builds up smoothly.

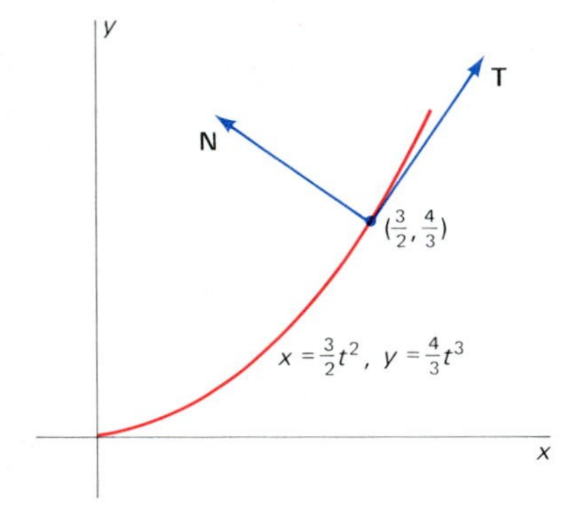

14.49 The moving particle of Example 6

EXAMPLE 6 A particle moves in the xy-plane with parametric equations

$$x = \tfrac{3}{2}t^2, \qquad y = \tfrac{4}{3}t^3.$$

Find the tangential and normal components of its acceleration vector when $t = 1$.

Solution The trajectory and the vectors **N** and **T** appear in Fig. 14.49; **N** and **T** are shown attached at the point of evaluation, at which $t = 1$. The particle has position vector

$$\mathbf{r}(t) = \tfrac{3}{2}t^2\mathbf{i} + \tfrac{4}{3}t^3\mathbf{j}$$

and thus velocity

$$\mathbf{v}(t) = 3t\mathbf{i} + 4t^2\mathbf{j}.$$

Hence its speed is

$$v(t) = \sqrt{9t^2 + 16t^4},$$

from which we calculate

$$a_T = \frac{dv}{dt} = \frac{9t + 32t^3}{\sqrt{9t^2 + 16t^4}}.$$

Thus $v = 5$ and $a_T = \frac{41}{5}$ when $t = 1$.

To use Equation (12) to compute the curvature at $t = 1$, we compute $dx/dt = 3t$, $dy/dt = 4t^2$, $d^2x/dt^2 = 3$, and $d^2y/dt^2 = 8t$. Thus at $t = 1$ we have

$$\kappa = \frac{|x'y'' - x''y'|}{v^3} = \frac{|3 \cdot 8 - 3 \cdot 4|}{5^3} = \frac{12}{125}.$$

Hence

$$a_N = \kappa v^2 = \tfrac{12}{125}(5^2) = \tfrac{12}{5}$$

when $t = 1$. As a check (Problem 28), you might compute **T** and **N** when $t = 1$ and verify that

$$\tfrac{41}{5}\mathbf{T} + \tfrac{12}{5}\mathbf{N} = \mathbf{a} = 3\mathbf{i} + 8\mathbf{j}.$$

It remains for us to see how to compute a_T, a_N, and **N** effectively in the case of a space curve. We would prefer to have formulas that explicitly involve only the vectors **r**, **v**, and **a**.

If we compute the scalar product of $\mathbf{v} = v\mathbf{T}$ with the acceleration **a** as given in Equation (22) and use the facts that $\mathbf{T} \cdot \mathbf{T} = 1$ and $\mathbf{T} \cdot \mathbf{N} = 0$, we get

$$\mathbf{v} \cdot \mathbf{a} = v\mathbf{T} \cdot \left(\frac{dv}{dt}\mathbf{T}\right) + (v\mathbf{T}) \cdot (\kappa v^2 \mathbf{N}) = v\frac{dv}{dt}.$$

It follows that

$$a_T = \frac{dv}{dt} = \frac{\mathbf{v} \cdot \mathbf{a}}{v} = \frac{\mathbf{r}'(t) \cdot \mathbf{r}''(t)}{|\mathbf{r}'(t)|}. \tag{26}$$

Similarly, when we compute the vector product of $\mathbf{v} = v\mathbf{T}$ with each side in Equation (22), we find that

$$\mathbf{v} \times \mathbf{a} = v\mathbf{T} \times \frac{dv}{dt}\mathbf{T} + v\mathbf{T} \times \kappa v^2 \mathbf{N} = \kappa v^3 \mathbf{T} \times \mathbf{N}.$$

Because κ and v are nonnegative and because $\mathbf{T} \times \mathbf{N}$ is a unit vector, we may conclude that

$$\kappa = \frac{|\mathbf{v} \times \mathbf{a}|}{v^3} = \frac{|\mathbf{r}'(t) \times \mathbf{r}''(t)|}{|\mathbf{r}'(t)|^3}. \tag{27}$$

It now follows from (24) that

$$a_N = \frac{|\mathbf{r}'(t) \times \mathbf{r}''(t)|}{|\mathbf{r}'(t)|}. \tag{28}$$

The curvature of a space curve is not often as easy to compute directly from the definition as we found in the case of the helix of Example 5. It is generally more convenient to use the formula in (27). Once **a**, **T**, a_T, and a_N have been computed, we can rewrite Equation (25) as

$$\mathbf{N} = \frac{\mathbf{a} - a_T \mathbf{T}}{a_N} \tag{29}$$

to find the principal unit normal vector.

EXAMPLE 7 Compute **T**, **N**, κ, a_T, and a_N at the point $(1, \frac{1}{2}, \frac{1}{3})$ of the twisted cubic with parametric equations

$$x = t, \qquad y = \tfrac{1}{2}t^2, \qquad z = \tfrac{1}{3}t^3.$$

Solution Differentiation of the position vector

$$\mathbf{r}(t) = \langle t, \tfrac{1}{2}t^2, \tfrac{1}{3}t^3 \rangle$$

gives

$$\mathbf{r}'(t) = \langle 1, t, t^2 \rangle$$

and

$$\mathbf{r}''(t) = \langle 0, 1, 2t \rangle.$$

When we substitute $t = 1$, we obtain

$$\mathbf{v}(1) = \langle 1, 1, 1 \rangle \qquad \text{(velocity)},$$

$$v(1) = |\mathbf{v}(1)| = \sqrt{3} \qquad \text{(speed), and}$$

$$\mathbf{a}(1) = \langle 0, 1, 2 \rangle \qquad \text{(acceleration)}$$

at the point $(1, \frac{1}{2}, \frac{1}{3})$. Then Equation (26) gives the tangential component of acceleration:

$$a_T = \frac{\mathbf{v} \cdot \mathbf{a}}{v} = \frac{3}{\sqrt{3}} = \sqrt{3}.$$

Because

$$\mathbf{v} \times \mathbf{a} = \begin{vmatrix} \mathbf{i} & \mathbf{j} & \mathbf{k} \\ 1 & 1 & 1 \\ 0 & 1 & 2 \end{vmatrix} = \langle 1, -2, 1 \rangle,$$

Equation (27) gives the curvature:

$$\kappa = \frac{|\mathbf{v} \times \mathbf{a}|}{v^3} = \frac{\sqrt{6}}{(\sqrt{3})^3} = \tfrac{1}{3}\sqrt{2}.$$

The normal component of acceleration is $a_N = \kappa v^2 = \sqrt{2}$. The unit tangent vector is

$$\mathbf{T} = \frac{\mathbf{v}}{v} = \frac{1}{\sqrt{3}} \langle 1, 1, 1 \rangle = \frac{\mathbf{i} + \mathbf{j} + \mathbf{k}}{\sqrt{3}}.$$

Finally, Equation (29) gives

$$\mathbf{N} = \frac{\mathbf{a} - a_T \mathbf{T}}{a_N} = \frac{1}{\sqrt{2}} (\langle 0, 1, 2 \rangle - \langle 1, 1, 1 \rangle)$$

$$= \frac{1}{\sqrt{2}} \langle -1, 0, 1 \rangle = \frac{-\mathbf{i} + \mathbf{k}}{\sqrt{2}}.$$

14.5 PROBLEMS

Find the arc length of each of the curves in Problems 1–6.

1. $x = 3 \sin 2t,\quad y = 3 \cos 2t,\quad z = 8t$; from $t = 0$ to $t = \pi$

2. $x = t,\quad y = t^2/\sqrt{2},\quad z = t^3/3$; from $t = 0$ to $t = 1$

3. $x = 6e^t \cos t,\quad y = 6e^t \sin t,\quad z = 17e^t$; from $t = 0$ to $t = 1$

4. $x = t^2/2,\quad y = \ln t,\quad z = t\sqrt{2}$; from $t = 1$ to $t = 2$

5. $x = 3t \sin t,\quad y = 3t \cos t,\quad z = 2t^2$; from $t = 0$ to $t = 4/5$

6. $x = 2e^t,\quad y = e^{-t},\quad z = 2t$; from $t = 0$ to $t = 1$

In Problems 7–12, find the curvature of the given plane curve at the indicated point.

7. $y = x^3$ at $(0, 0)$

8. $y = x^3$ at $(-1, -1)$

9. $y = \cos x$ at $(0, 1)$

10. $x = t - 1$, $y = t^2 + 3t + 2$, where $t = 2$

11. $x = 5 \cos t$, $y = 4 \sin t$, where $t = \pi/4$

12. $x = 5 \cosh t$, $y = 3 \sinh t$, where $t = 0$

In Problems 13–16, find the point or points of the given plane curve at which the curvature is a maximum.

13. $y = e^x$

14. $y = \ln x$

15. $x = 5 \cos t$, $y = 3 \sin t$

16. $xy = 1$

For each of the plane curves in Problems 17–21, find the unit tangent and normal vectors at the indicated point.

17. $y = x^3$ at $(-1, -1)$

18. $x = t^3$, $y = t^2$ at $(-1, 1)$

19. $x = 3 \sin 2t$, $y = 4 \cos 2t$, where $t = \pi/6$

20. $x = t - \sin t$, $y = 1 - \cos t$, where $t = \pi/2$

21. $x = \cos^3 t$, $y = \sin^3 t$, where $t = 3\pi/4$

The position vector of a particle moving in the plane is given in each of Problems 22–26. Find the tangential and normal components of the acceleration vector.

22. $\mathbf{r}(t) = 3\mathbf{i} \sin \pi t + 3\mathbf{j} \cos \pi t$

23. $\mathbf{r}(t) = (2t + 1)\mathbf{i} + (3t^2 - 1)\mathbf{j}$

24. $\mathbf{r}(t) = \mathbf{i} \cosh 3t + \mathbf{j} \sinh 3t$

25. $\mathbf{r}(t) = \mathbf{i}t \cos t + \mathbf{j}t \sin t$

26. $\mathbf{r}(t) = \langle e^t \sin t, e^t \cos t \rangle$

27. Use the formula in (13) to compute the curvature of the circle with equation $x^2 + y^2 = a^2$.

28. Verify the equation

$$\tfrac{41}{5}\mathbf{T} + \tfrac{12}{5}\mathbf{N} = 3\mathbf{i} + 8\mathbf{j}$$

given at the end of Example 6.

In each of Problems 29–31, find the equation of the osculating circle for the given plane curve at the indicated point.

29. $y = 1 - x^2$ at $(0, 1)$

30. $y = e^x$ at $(0, 1)$

31. $xy = 1$ at $(1, 1)$

Find the curvature κ of each of the space curves with position vector given in Problems 32–36.

32. $\mathbf{r}(t) = t\mathbf{i} + (2t - 1)\mathbf{j} + (3t + 5)\mathbf{k}$

33. $\mathbf{r}(t) = t\mathbf{i} + \mathbf{j} \sin t + \mathbf{k} \cos t$

34. $\mathbf{r}(t) = \langle t, t^2, t^3 \rangle$

35. $\mathbf{r}(t) = \langle e^t \cos t, e^t \sin t, e^t \rangle$

36. $\mathbf{r}(t) = \mathbf{i}t \sin t + \mathbf{j}t \cos t + \mathbf{k}t$

37–41. Find the tangential and normal components of acceleration a_T and a_N for the curves of Problems 32–36, respectively.

In each of Problems 42–45, find the unit vectors **T** and **N** for the given curve at the indicated point.

42. The curve of Problem 34 at $(1, 1, 1)$

43. The curve of Problem 33 at $(0, 0, 1)$

44. The curve of Problem 3 at $(6, 0, 17)$

45. The curve of Problem 35 at $(1, 0, 1)$

46. Find **T**, **N**, a_T, and a_N as functions of t for the helix of Example 1.

47. Find the arc length parametrization of the line

$$x = 2 + 4t, \qquad y = 1 - 12t, \qquad z = 3 + 3t$$

in terms of the arc length s measured from the initial point $(2, 1, 3)$.

48. Find the arc length parametrization of the circle

$$x = 2 \cos t, \qquad y = 2 \sin t, \qquad z = 0$$

in terms of the arc length s measured from the initial point $(2, 0, 0)$.

49. Find the arc length parametrization of the helix

$$x = 3 \cos t, \qquad y = 3 \sin t, \qquad z = 4t$$

in terms of the arc length s measured from the initial point $(3, 0, 0)$.

50. Substitute $x = t$, $y = f(t)$, and $z = 0$ in the formula in Equation (27) to verify that the curvature of the plane curve $y = f(x)$ is

$$\kappa = \frac{|f''(x)|}{[1 + (f'(x))^2]^{3/2}}.$$

51. A particle moves under the influence of a force that is always perpendicular to its direction of motion. Show that the speed of the particle must be constant.

52. Deduce from Equation (20) that

$$\kappa = \frac{\sqrt{a^2 - a_T^2}}{v^2} = \frac{[(x'')^2 + (y'')^2 - (v')^2]^{1/2}}{(x')^2 + (y')^2},$$

where primes denote differentiation with respect to t.

53. Apply the formula of Problem 52 to calculate the curvature of the curve

$$x = \cos t + t \sin t, \qquad y = \sin t - t \cos t.$$

54. Find the curvature and center of curvature of the folium of Descartes $x^3 + y^3 = 3xy$ at the point $(\frac{3}{2}, \frac{3}{2})$. Begin by calculating dy/dx and d^2y/dx^2 by implicit differentiation.

55. Determine the constants A, B, C, D, E, and F so that the curve

$$y = Ax^5 + Bx^4 + Cx^3 + Dx^2 + Ex + F$$

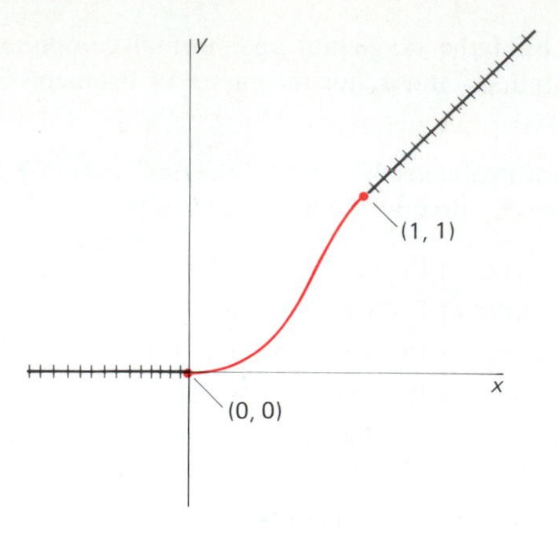

14.50 Connecting two railroad tracks (Problem 55)

does, simultaneously, all of the following:

- Joins the two points (0, 0) and (1, 1);
- Has slope 0 at (0, 0) and slope 1 at (1, 1);
- Has curvature 0 at both (0, 0) and (1, 1).

The curve in question is shown in color in Fig. 14.50. Why would this be a good curve to join the railway tracks, which are shown in black in the figure?

56. Show that the point on the curve $y = e^x$ where the curvature κ is maximal is the point where $y = 1/\sqrt{5}$.

14.6 Cylinders and Quadric Surfaces

Just as the graph of an equation $f(x, y) = 0$ is generally a curve in the xy-plane, the graph of an equation in three variables is generally a surface in space. A function F of three variables associates with each ordered triple of real numbers (x, y, z) a real number $F(x, y, z)$. The graph of the equation

$$F(x, y, z) = 0 \tag{1}$$

is the set of all points whose coordinates (x, y, z) satisfy this equation. We will refer to the graph of such an equation as a **surface.** It should be noted, however, that the graph of Equation (1) does not always agree with the intuitive notion of a surface. For example, the graph of the equation

$$(x^2 + y^2)(y^2 + z^2) = 0$$

consists of the x- and z-axes because $x^2 + y^2 = 0$ implies that $x = y = 0$ (the z-axis) and $y^2 + z^2 = 0$ implies that $y = z = 0$ (the x-axis). We leave for advanced calculus the precise definition of *surface* as well as the study of conditions sufficient to imply that the graph of Equation (1) actually is a surface.

The simplest example of a surface is a plane with linear equation $Ax + By + Cz + D = 0$. This section is devoted to examples of other simple surfaces that frequently appear in multivariable calculus.

In order to sketch a surface S, it is often helpful to examine its intersections with various planes. The **trace** of the surface S in the plane $\mathcal{P}$ is the intersection of $\mathcal{P}$ and S. For example, if S is a sphere, then the methods of elementary geometry may be used to verify that the trace of S in the plane $\mathcal{P}$ is a circle (see Fig. 14.51), provided that $\mathcal{P}$ intersects the sphere but is not merely tangent to it (Problem 49). When we want to visualize a specified surface in space, it often suffices to examine its traces in the coordinate planes and possibly a few planes parallel to them.

EXAMPLE 1 Consider the plane with equation

$$3x + 2y + 2z = 6.$$

We find its trace in the xy-plane by setting $z = 0$. The equation then reduces to the equation $3x + 2y = 6$ of a line in the xy-plane. Similarly, when we set $y = 0$ we get the line $3x + 2z = 6$ as the trace of the given plane in

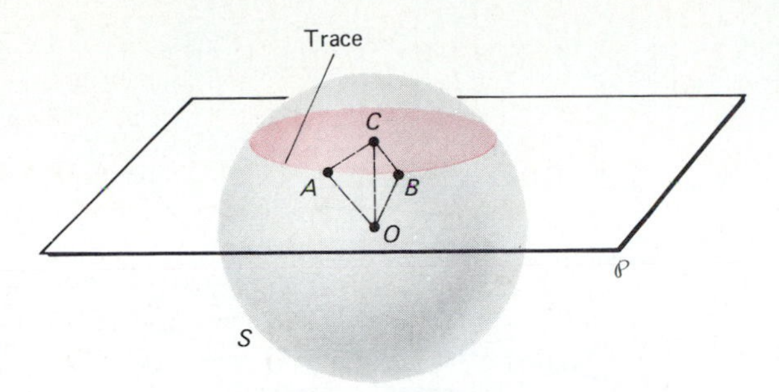

14.51 The intersection of the sphere S and the plane $\mathscr{P}$ is a circle.

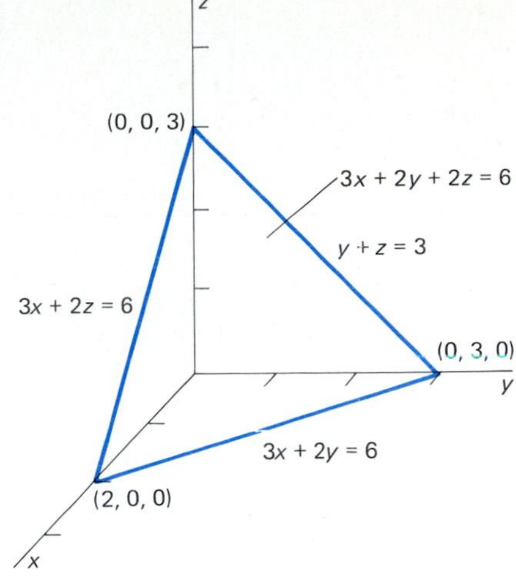

14.52 Traces of the plane $3x + 2y + 2z = 6$ in the coordinate planes

the xz-plane. To find its trace in the yz-plane we set $x = 0$, and this yields the line $y + z = 3$. Figure 14.52 shows the portions of these three trace lines that lie in the first octant. Together they give us a good picture of how the plane $3x + 2y + 2z = 6$ is situated in space.

Let C be a curve in a plane and let L be a line not parallel to the plane. Then the set of all points on all lines parallel to L that intersect C is called a **cylinder.** This is a generalization of the familiar right circular cylinder, for which the curve C is a circle and the line L is perpendicular to the plane of the circle. Figure 14.53 shows this cylinder when C is the circle $x^2 + y^2 = a^2$ in the xy-plane. The trace of this cylinder in any horizontal plane $z = c$ is a circle of radius a and with center the point $(0, 0, c)$ on the z-axis. Thus the point (x, y, c) lies on the cylinder if and only if $x^2 + y^2 = a^2$. Hence this cylinder is the graph of the equation $x^2 + y^2 = a^2$, considered as an equation in *three* variables.

The fact that the variable z does not appear explicitly in the equation $x^2 + y^2 = a^2$ means that given any point $(x_0, y_0, 0)$ on the *circle* $x^2 + y^2 = a^2$ in the xy-plane, the point (x_0, y_0, z) lies on the cylinder for any and all values of z. The set of all such points is the vertical line through the point $(x_0, y_0, 0)$. Thus the *cylinder* $x^2 + y^2 = a^2$ in space is the union of all vertical lines through points of the *circle* $x^2 + y^2 = a^2$ in the plane, as illustrated in Fig. 14.54.

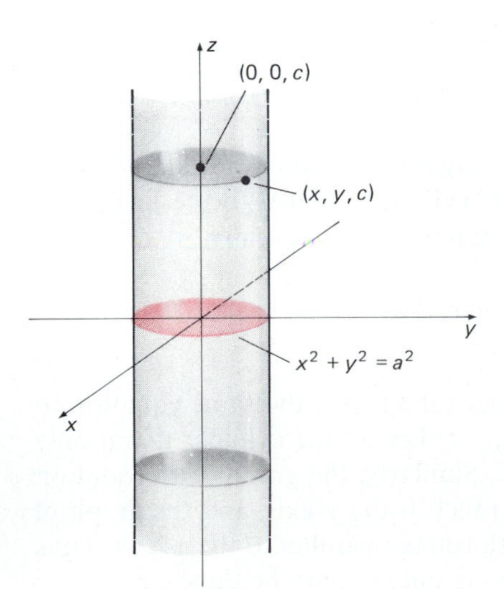

14.53 A right circular cylinder

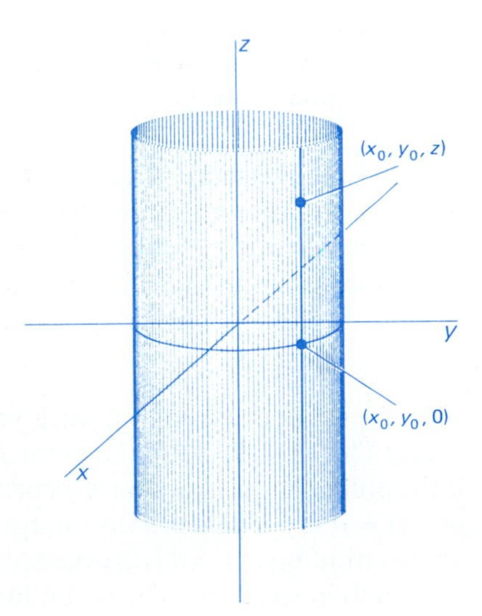

14.54 The cylinder $x^2 + y^2 = a^2$; its rulings are parallel to the z-axis.

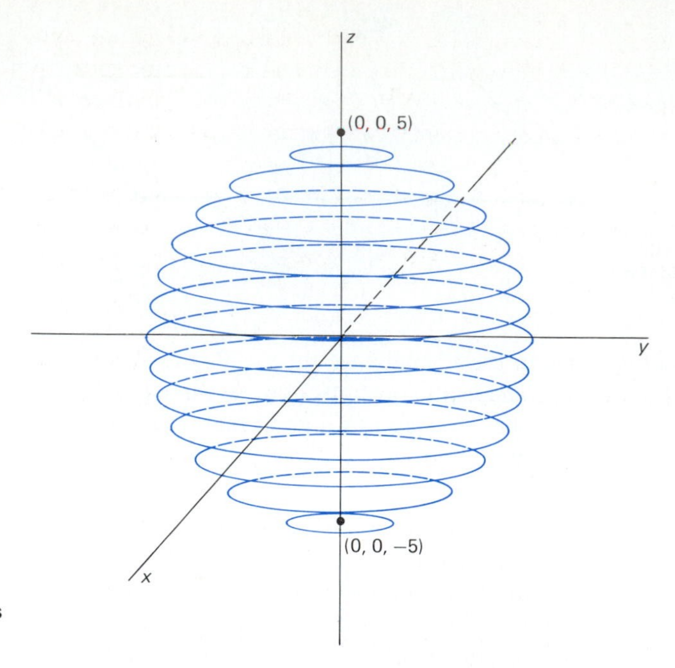

14.55 A sphere as a union of circles (and two points).

EXAMPLE 2 Consider the sphere of radius 5 with equation

$$x^2 + y^2 + z^2 = 25.$$

If $0 < |z_0| < 5$, then the cylinder

$$x^2 + y^2 = 25 - (z_0)^2$$

intersects the sphere in two circles with radius

$$r_0 = \sqrt{25 - (z_0)^2}$$

each. These circles are the traces of the sphere in the horizontal planes $z = +z_0$ and $z = -z_0$. Figure 14.55 shows the sphere presented as the union of such circles (plus the two isolated points (0, 0, 5) and (0, 0, −5)). For instance, $z = 3$ gives the circle with equation $x^2 + y^2 = 16$ in the horizontal plane $z = 3$.

Given a general cylinder generated by a plane curve C and a line L as in the definition above, the lines on the cylinder parallel to the line L are called **rulings** of the cylinder. Thus the rulings of the cylinder $x^2 + y^2 = a^2$ are vertical lines (parallel to the z-axis).

If the curve C in the xy-plane has equation

$$f(x, y) = 0, \tag{2}$$

then the cylinder through C with vertical rulings has the same equation in space. This is so because the point $P(x, y, z)$ lies on the cylinder if and only if the point $Q(x, y, 0)$ lies on the curve C. Similarly, the graph of an equation $g(x, z) = 0$ is a cylinder with rulings parallel to the y-axis, and the graph of an equation $h(y, z) = 0$ is a cylinder with rulings parallel to the x-axis. Thus the graph in space of an equation involving only two of the three coordinate variables is always a cylinder; its rulings are parallel to the axis corresponding to the *missing* variable.

EXAMPLE 3 The graph of the equation $4y^2 + 9z^2 = 36$ is the **elliptic cylinder** shown in Fig. 14.56. Its rulings are parallel to the x-axis, and its trace in every plane perpendicular to the x-axis is an ellipse with semiaxes of lengths 3 and 2 (just like the pictured ellipse $y^2/9 + z^2/4 = 1$ in the yz-plane).

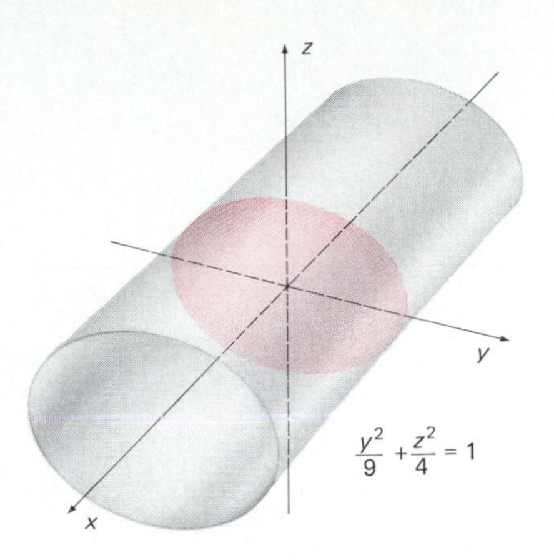

14.56 An elliptic cylinder

EXAMPLE 4 The graph of the equation $z = 4 - x^2$ is the **parabolic cylinder** shown in Fig. 14.57. Its rulings are parallel to the y-axis, and its trace in every plane perpendicular to the y-axis is a parabola that is a parallel translate of the parabola $z = 4 - x^2$ in the xz-plane.

Another way to use a plane curve C to generate a surface is to revolve the curve (in space) about a line L in its plane. This gives a **surface of revolution** with **axis** L. For example, Fig. 14.58 shows the surface generated by revolving the curve $f(x, y) = 0$ in the first quadrant of the xy-plane around

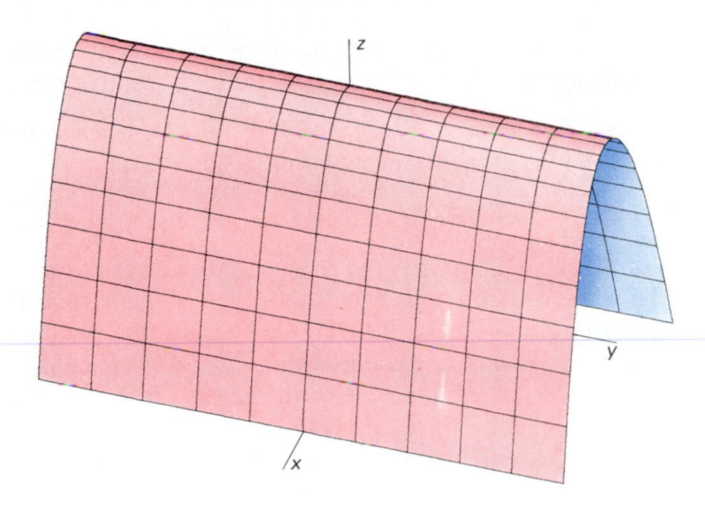

14.57 The parabolic cylinder $z = 4 - x^2$

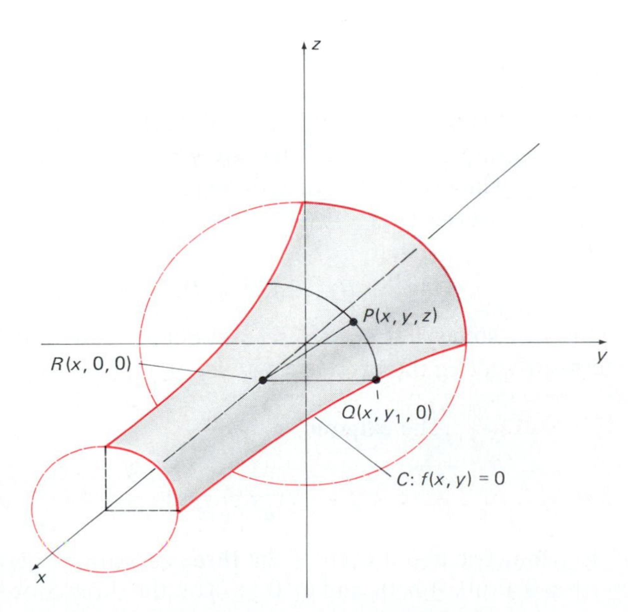

14.58 The surface generated by rotating C around the x-axis. (For clarity, only a quarter of the surface is shown.)

the x-axis. The point $P(x, y, z)$ lies on the surface of revolution if and only if the point $Q(x, y_1, 0)$ lies on the curve, where

$$y_1 = |RQ| = |RP| = \sqrt{y^2 + z^2}.$$

Thus it is necessary that $F(x, y_1) = 0$, so the equation of the indicated surface of revolution around the x-axis is

$$f(x, \sqrt{y^2 + z^2}) = 0. \tag{3}$$

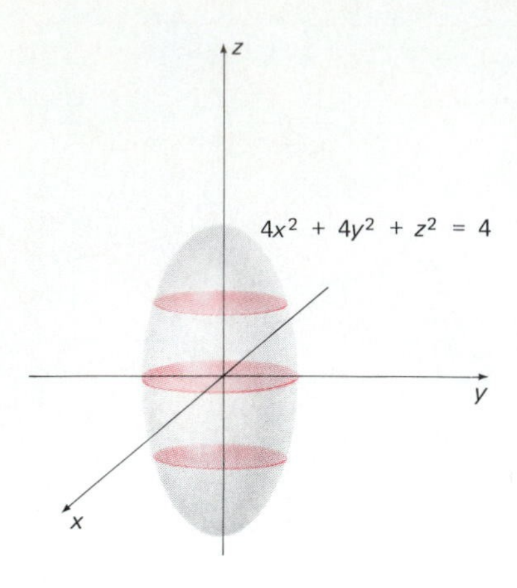

14.59 The ellipsoid of revolution of Example 5

The equations of surfaces of revolution around other coordinate axes are obtained similarly. If the first quadrant curve $f(x, y) = 0$ is revolved around the y-axis, we replace x by $(x^2 + z^2)^{1/2}$ to get the equation $f((x^2 + z^2)^{1/2}, y) = 0$ of the resulting surface of revolution. If the curve $g(y, z) = 0$ in the first quadrant of the yz-plane is revolved around the z-axis, we replace y by $(x^2 + y^2)^{1/2}$. Thus the equation of the resulting surface of revolution about the z-axis is $g((x^2 + y^2)^{1/2}, z) = 0$. These assertions are easily verified with the aid of diagrams similar to Fig. 14.58.

EXAMPLE 5 Find an equation of the **ellipsoid of revolution** obtained by revolving the ellipse $4y^2 + z^2 = 4$ around the z-axis (see Fig. 14.59).

Solution We replace y by $(x^2 + y^2)^{1/2}$ in the given equation. This yields $4x^2 + 4y^2 + z^2 = 4$ as an equation of the ellipsoid.

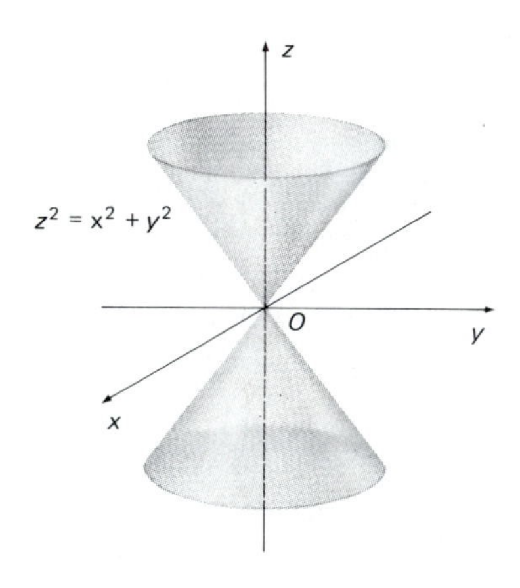

14.60 The cone of Example 6

EXAMPLE 6 Determine the graph of the equation $z^2 = x^2 + y^2$.

Solution First we rewrite the given equation in the form $z = \pm(x^2 + y^2)^{1/2}$. Thus the surface is symmetric about the xy-plane, and the upper half has equation $z = +(x^2 + y^2)^{1/2}$. This last equation is obtained from the simple equation $z = y$ by replacing y by $(x^2 + y^2)^{1/2}$. Thus the upper half of the surface is obtained by revolving the line $z = y$ (for $y \geqq 0$) around the z-axis. Thus the graph is the **cone** shown in Fig. 14.60. Its upper half has equation $z = +(x^2 + y^2)^{1/2}$, and its lower half has equation $z = -(x^2 + y^2)^{1/2}$. The whole cone $z^2 = x^2 + y^2$ is obtained by revolving the entire line $z = y$ around the z-axis.

QUADRIC SURFACES

Cones, spheres, circular and parabolic cylinders, and ellipsoids of revolution are all examples of surfaces that are graphs of second-degree equations in x, y, and z. The graph of a second-degree equation in three variables is called a **quadric surface.** We discuss here some important special cases of the equation

$$Ax^2 + By^2 + Cz^2 + Dx + Ey + Fz + G = 0. \tag{4}$$

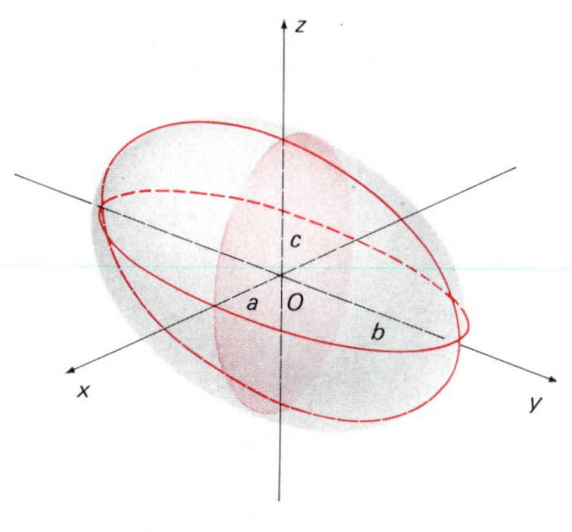

14.61 The ellipsoid of Example 7

This is a somewhat special second-degree equation in that it contains no terms involving the products xy, xz, or yz.

EXAMPLE 7 The **ellipsoid**

$$\frac{x^2}{a^2} + \frac{y^2}{b^2} + \frac{z^2}{c^2} = 1 \tag{5}$$

is symmetric about each of the three coordinate planes and has intercepts $(\pm a, 0, 0)$, $(0, \pm b, 0)$, and $(0, 0, \pm c)$ on the three coordinate axes. If $P(x, y, z)$

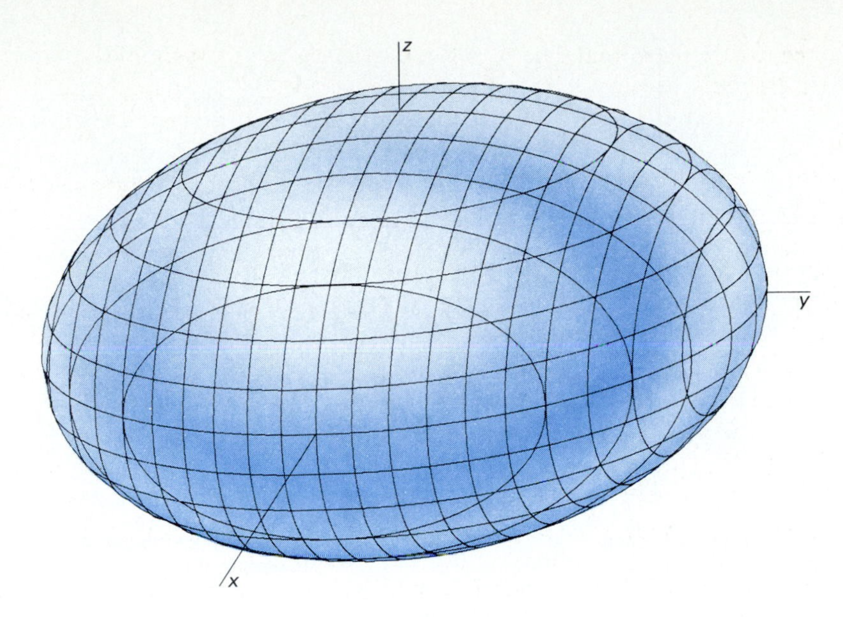

14.62 The traces of the ellipsoid $x^2/a^2 + y^2/b^2 + z^2/c^2 = 1$

is a point of this ellipsoid, then $|x| \leqq a$, $|y| \leqq b$, and $|z| \leqq c$. Each trace in a plane parallel to one of the three coordinate planes is either a single point or an ellipse. For example, if $-c < z_0 < c$, then Equation (5) can be reduced to

$$\frac{x^2}{a^2} + \frac{y^2}{b^2} = 1 - \frac{z_0^2}{c^2} > 0,$$

which is the equation of an ellipse with semiaxes $(a/c)(c^2 - z_0^2)^{1/2}$ and $(b/c)(c^2 - z_0^2)^{1/2}$. Figure 14.61 shows this ellipsoid with its semiaxes a, b, and c labeled. Figure 14.62 shows its trace ellipses in planes parallel to the three coordinate planes.

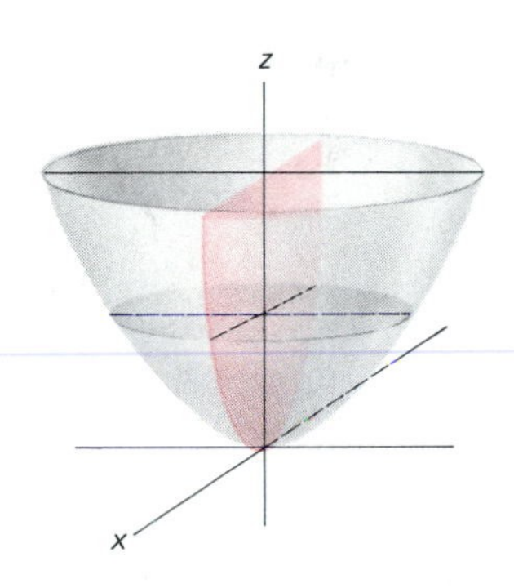

14.63 An elliptic paraboloid

EXAMPLE 8 The **elliptic paraboloid**

$$\frac{x^2}{a^2} + \frac{y^2}{b^2} = \frac{z}{c} \tag{6}$$

is shown in Fig. 14.63. Its trace in the horizontal plane $z = z_0 > 0$ is the ellipse $x^2/a^2 + y^2/b^2 = z_0/c$ with semiaxes $a(z_0/c)^{1/2}$ and $b(z_0/c)^{1/2}$. Its trace in any vertical plane is a parabola. For instance, its trace in the plane $y = y_0$ has equation $x^2/a^2 + y_0^2/b^2 = z/c$, which can be rewritten in the form $z - z_1 = k(x - x_1)^2$ by taking $z_1 = cy_0^2/b^2$ and $x_1 = 0$. The paraboloid opens upward if $c > 0$ and downward if $c < 0$. If $a = b$ the paraboloid is circular. Figure 14.64 shows the traces of a circular paraboloid in planes parallel to the xz- and yz-planes.

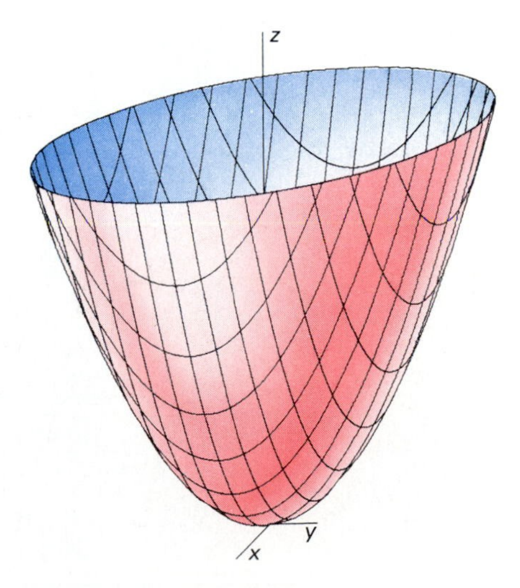

14.64 Trace parabolas of a circular paraboloid

EXAMPLE 9 The **elliptic cone**

$$\frac{x^2}{a^2} + \frac{y^2}{b^2} = \frac{z^2}{c^2} \tag{7}$$

is shown in Fig. 14.65. Its trace in the horizontal plane $z = z_0 \neq 0$ is an ellipse with semiaxes $a|z_0|/c$ and $b|z_0|/c$.

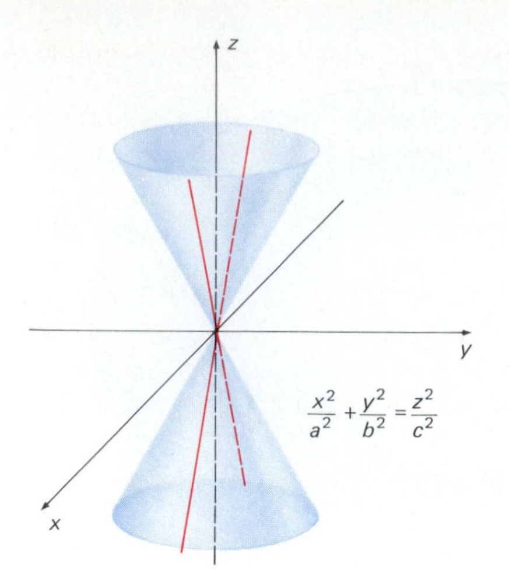

14.65 An elliptic cone (see Example 9)

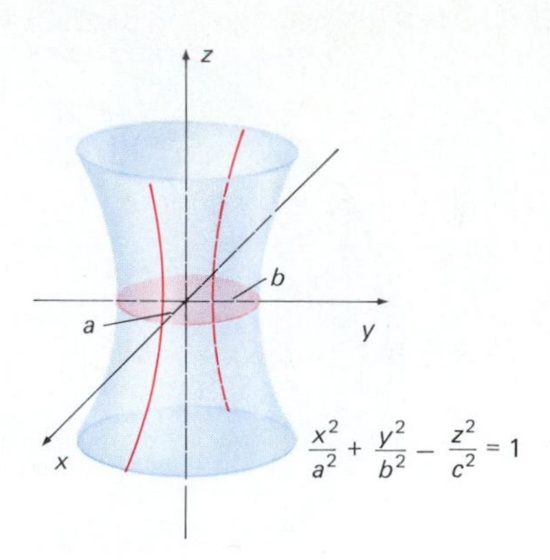

14.66 A hyperboloid of one sheet (Example 10)

EXAMPLE 10 The **hyperboloid of one sheet** with equation

$$\frac{x^2}{a^2}+\frac{y^2}{b^2}-\frac{z^2}{c^2}=1 \tag{8}$$

is shown in Fig. 14.66. Its trace in the horizontal plane $z = z_0$ is the ellipse $x^2/a^2 + y^2/b^2 = 1 + z_0^2/c^2 > 0$. Its trace in a vertical plane is a hyperbola except when the vertical plane intersects the xy-plane in a line tangent to the ellipse $x^2/a^2 + y^2/b^2 = 1$. In this special case, the trace is a degenerate hyperbola consisting of two intersecting lines. Figure 14.67 shows the traces

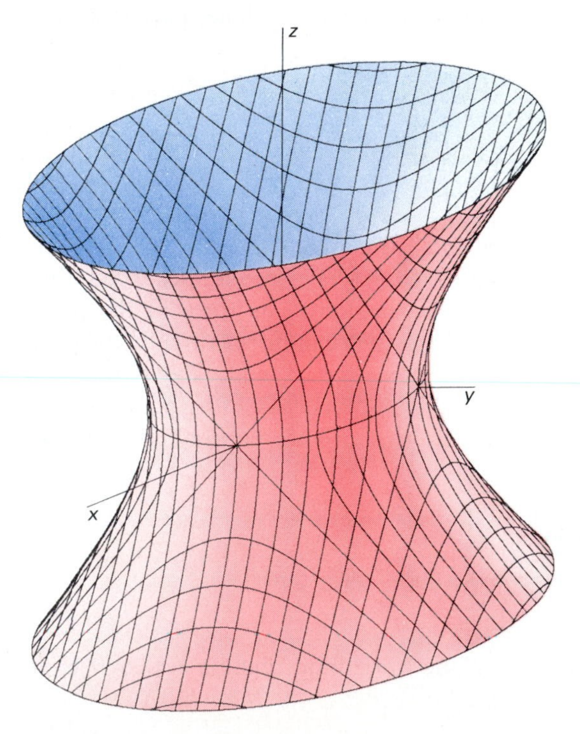

14.67 A circular hyperboloid of one sheet. Its traces in horizontal planes are circles; its traces in vertical planes are hyperbolas.

(in planes parallel to the coordinate planes) of a circular ($a = b$) hyperboloid of one sheet.

The graphs of the equations

$$\frac{y^2}{b^2} + \frac{z^2}{c^2} - \frac{x^2}{a^2} = 1 \quad \text{and} \quad \frac{x^2}{a^2} + \frac{z^2}{c^2} - \frac{y^2}{b^2} = 1$$

are also hyperboloids of one sheet, opening along the x- and y-axes, respectively.

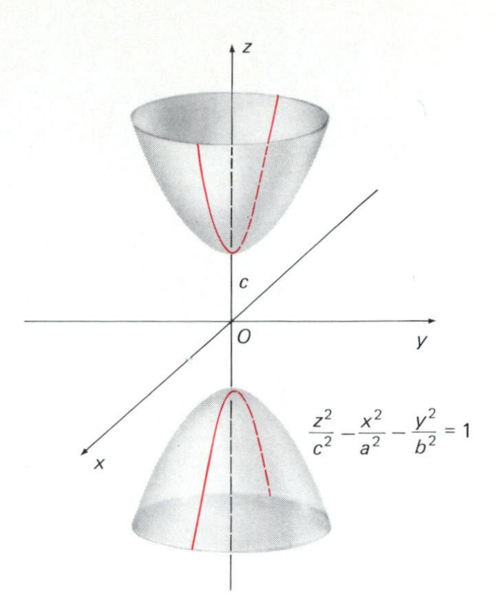

14.68 A hyperboloid of two sheets (see Example 11)

EXAMPLE 11 The **hyperboloid of two sheets** with equation

$$\frac{z^2}{c^2} - \frac{x^2}{a^2} - \frac{y^2}{b^2} = 1 \tag{9}$$

consists of two separated pieces or *sheets* (Fig. 14.68). The two sheets open along the positive and negative z-axis and intersect it at the points $(0, 0, \pm c)$. The trace of the hyperboloid in a horizontal plane $z = z_0$ with $|z_0| > c$ is the ellipse

$$\frac{x^2}{a^2} + \frac{y^2}{b^2} = \frac{z_0^2}{c^2} - 1 > 0.$$

Its trace in any vertical plane is a nondegenerate hyperbola. Figure 14.69 shows traces of a circular hyperboloid of two sheets.

The graphs of the equations

$$\frac{x^2}{a^2} - \frac{y^2}{b^2} - \frac{z^2}{c^2} = 1 \quad \text{and} \quad \frac{y^2}{b^2} - \frac{x^2}{a^2} - \frac{z^2}{c^2} = 1$$

are also hyperboloids of two sheets, opening along the x- and y-axes, respectively. Note that when the equation of a hyperboloid is written in standard form with $+1$ on the right-hand side (as in Equations (8) and (9)), the number of sheets is equal to the number of negative terms on the left-hand side.

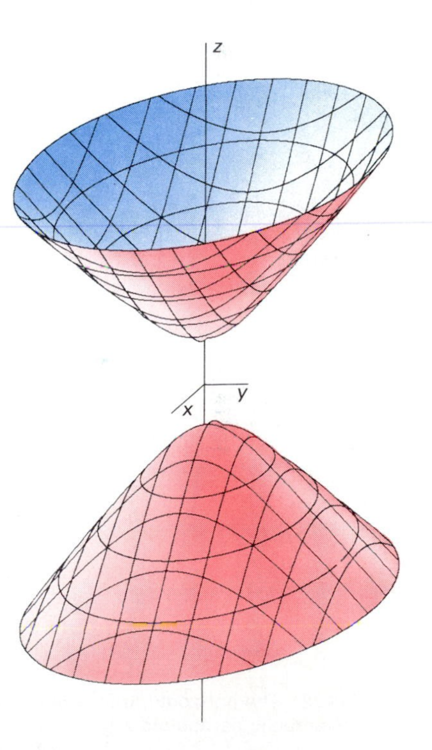

14.69 A circular hyperboloid of two sheets. Its (nondegenerate) traces in horizontal planes are circles; its traces in vertical planes are hyperbolas.

EXAMPLE 12 The **hyperbolic paraboloid**

$$\frac{y^2}{b^2} - \frac{x^2}{a^2} = \frac{z}{c} \qquad (c > 0) \tag{10}$$

is saddle-shaped, as indicated in Fig. 14.70. Its trace in the horizontal plane $z = z_0$ is a hyperbola (or two intersecting lines if $z_0 = 0$). Its trace in a vertical plane parallel to the xz-plane is a parabola that opens downward, while its trace in a vertical plane parallel to the yz-plane is a parabola that opens upward. In particular, the trace of the hyperbolic paraboloid in the xz-plane is a parabola opening downward from the origin, while its trace in the yz-plane is a parabola opening upward from the origin. Thus the origin looks like a local maximum from one direction but like a local minimum from another direction. Such a point on a surface is called a **saddle point.**

Figure 14.71 shows the parabolic traces in vertical planes of the hyperbolic paraboloid $z = y^2 - x^2$. Figure 14.72 shows its hyperbolic traces in horizontal planes.

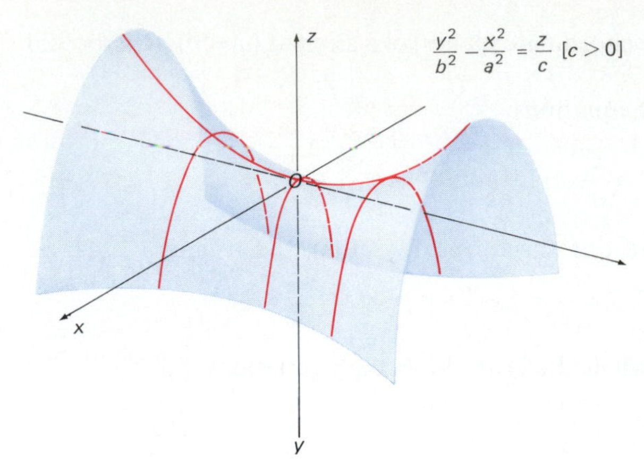

14.70 The hyperbolic paraboloid is a saddle-shaped surface.

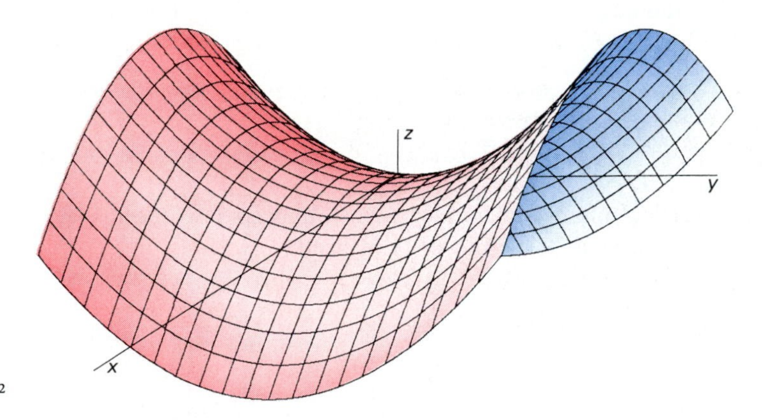

14.71 The vertical traces of the hyperbolic paraboloid $z = y^2 - x^2$

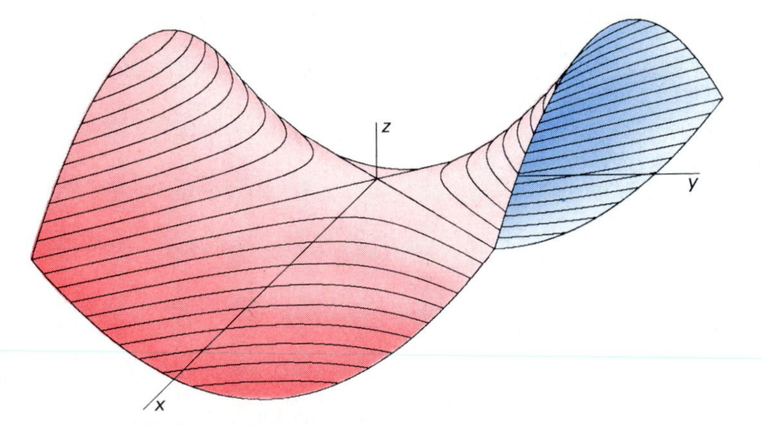

14.72 The horizontal traces of the hyperbolic paraboloid $z = y^2 - x^2$

*CONIC SECTIONS AS SECTIONS OF A CONE

The parabola, ellipse, and hyperbola that we studied in Chapter 10 were originally introduced by the ancient Greek mathematicians as plane sections (traces) of a right circular cone. Here we show that the intersection of a plane and a cone is, indeed, one of the three conic sections as defined in Chapter 10.

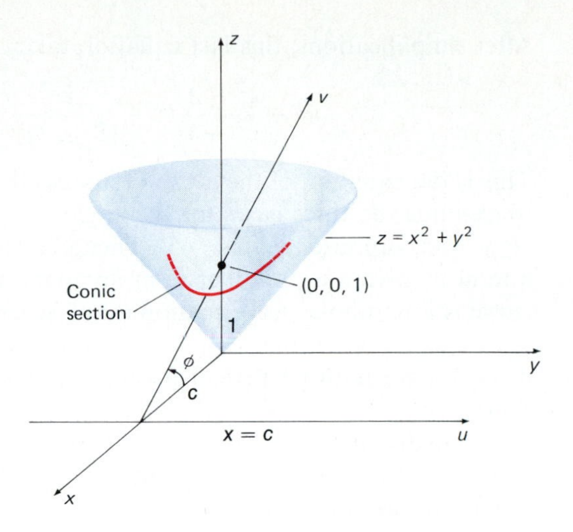

14.73 Finding an equation for a conic section

Figure 14.73 shows the cone with equation $z = (x^2 + y^2)^{1/2}$ and its intersection with a plane $\mathscr{P}$ that passes through the point (0, 0, 1) and the line $x = c > 0$ in the xy-plane. The equation of $\mathscr{P}$ is

$$z = 1 - \frac{x}{c}. \tag{11}$$

The angle between $\mathscr{P}$ and the xy-plane is $\phi = \tan^{-1}(1/c)$. We want to show that the conic section obtained by intersecting the cone and the plane is:

A parabola if $\phi = 45°$ $\quad (c = 1)$,

An ellipse if $\phi < 45°$ $\quad (c > 1)$,

A hyperbola if $\phi > 45°$ $\quad (c < 1)$.

We begin by introducing uv-coordinates in the plane $\mathscr{P}$ as follows. The u-coordinate of the point (x, y, z) of P is $u = y$. The v-coordinate of the same point is its perpendicular distance from the line $x = c$. This explains the u- and v-axes indicated in Fig. 14.73. Figure 14.74 shows the cross section in the plane $y = 0$ exhibiting the relation between v, x, and z. We see that

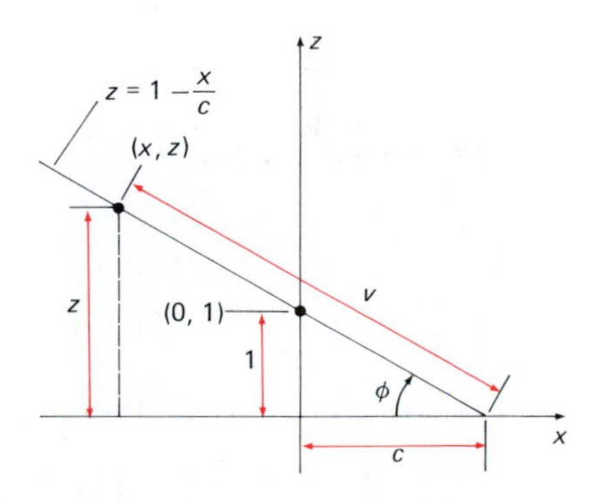

14.74 Computing coordinates in the uv-plane

$$z = v \sin \phi = \frac{v}{(1 + c^2)^{1/2}}. \tag{12}$$

Equations (11) and (12) give

$$x = c(1 - z) = c\left[1 - \frac{v}{(1 + c^2)^{1/2}}\right]. \tag{13}$$

We had $z^2 = x^2 + y^2$ for the equation of the cone. We make the following substitutions in this equation: Replace y by u, and replace z and x by the expressions on the right-hand sides of Equations (12) and (13), respectively. This yields

$$\frac{v^2}{1 + c^2} = c^2\left[1 - \frac{v}{(1 + c^2)^{1/2}}\right]^2 + u^2.$$

After simplifications, this last equation takes the form

$$u^2 + \frac{c^2 - 1}{c^2 + 1} v^2 - \frac{2c^2}{(1 + c^2)^{1/2}} v + c^2 = 0. \tag{14}$$

This is the equation of the curve of intersection in the uv-plane. We proceed to examine the three cases for the angle ϕ.

First suppose that $\phi = 45°$. Then $c = 1$, so that Equation (14) contains a term involving u^2, another term involving v, and a constant term. So the curve is a parabola; see Equation (6) of Section 10.4.

Next suppose that $\phi < 45°$. Then $c > 1$, and the coefficients of u^2 and v^2 in (14) are both positive. Thus the curve is an ellipse; see Equation (6) of Section 10.5.

Finally, if $\phi > 45°$, then $c < 1$, and the coefficients of u^2 and v^2 in Equation (14) have different signs. So the curve is a hyperbola; see Equation (8) in Section 10.6.

14.6 PROBLEMS

Describe and sketch the graphs of the equations given in Problems 1–30.

1. $3x + 2y + 10z = 20$
2. $3x + 2y = 30$
3. $x^2 + y^2 = 9$
4. $y^2 = x^2 - 9$
5. $xy = 4$
6. $z = 4x^2 + 4y^2$
7. $z^2 = 4x^2 + y^2$
8. $4x^2 + 9y^2 = 36$
9. $z = 4 - x^2 - y^2$
10. $y^2 + z^2 = 1$
11. $2z = x^2 + y^2$
12. $x = 1 + y^2 + z^2$
13. $z^2 = 4(x^2 + y^2)$
14. $y^2 = 4x$
15. $x^2 = 4z + 8$
16. $x = 9 - z^2$
17. $4x^2 + y^2 = 4$
18. $x^2 + z^2 = 4$
19. $x^2 = 4y^2 + 9z^2$
20. $x^2 - 4y^2 = z$
21. $x^2 + y^2 + 4z = 0$
22. $x = \sin y$
23. $x = 2y^2 - z^2$
24. $x^2 + 4y^2 + 2z^2 = 4$
25. $x^2 + y^2 - 9z^2 = 9$
26. $x^2 - y^2 - 9z^2 = 9$
27. $y = 4x^2 + 9z^2$
28. $y^2 + 4x^2 - 9z^2 = 36$
29. $y^2 - 9x^2 - 4z^2 = 36$
30. $x^2 + 9y^2 + 4z^2 = 36$

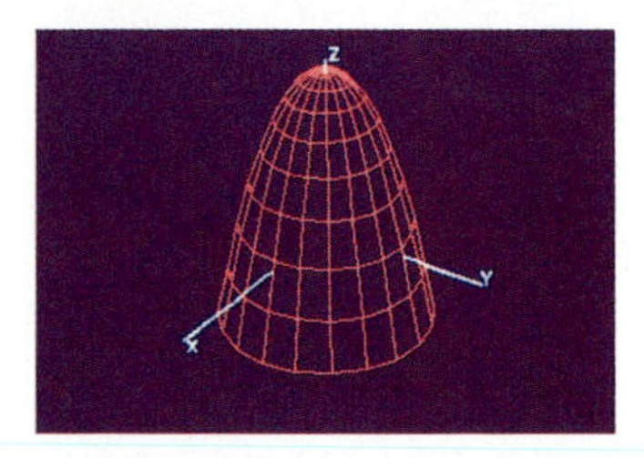

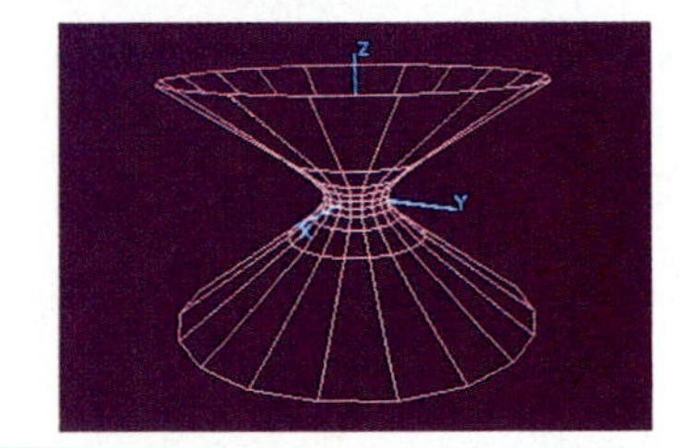

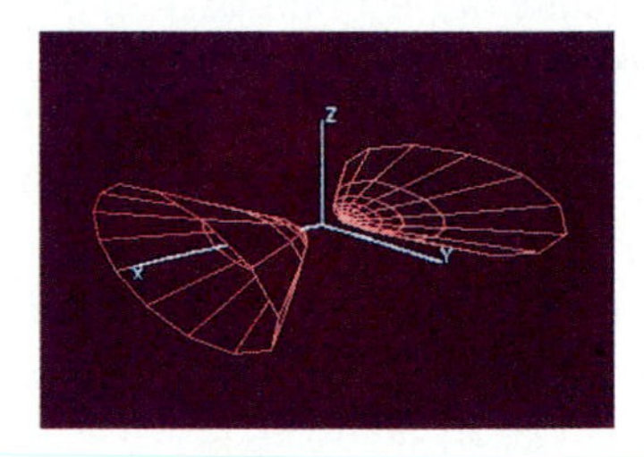

Figures for some of Problems 1–30

Each of Problems 31–40 gives the equation of a curve in one of the coordinate planes. Write an equation for the surface generated by revolving this curve about the indicated axis. Then sketch the surface.

31. $x = 2z^2$; the x-axis
32. $4x^2 + 9y^2 = 36$; the y-axis
33. $y^2 - z^2 = 1$; the z-axis
34. $z = 4 - x^2$; the z-axis
35. $y^2 = 4x$; the x-axis
36. $yz = 1$; the z-axis
37. $z = \exp(-x^2)$; the z-axis
38. $(y - z)^2 + z^2 = 1$; the z-axis
39. The line $z = 2x$; the z-axis
40. The line $z = 2x$; the x-axis

In each of Problems 41–47, describe the traces of the given surface in planes of the indicated type.

41. $x^2 + 4y^2 = 4$; in horizontal planes (those parallel to the xy-plane)

42. $x^2 + 4y^2 + 4z^2 = 4$; in horizontal planes

43. $x^2 + 4y^2 + 4z^2 = 4$; in planes parallel to the yz-plane

44. $z = 4x^2 + 9y^2$; in horizontal planes

45. $z = 4x^2 + 9y^2$; in planes parallel to the xz-plane

46. $z = xy$; in horizontal planes

47. $z = xy$; in vertical planes through the z-axis

48. Identify the surface $z = xy$ by making a suitable rotation of axes in the xy-plane (as in Section 10.7).

49. Prove that the triangles OAC and OBC in Fig. 14.51 are congruent, and thereby conclude that the trace of a sphere in an intersecting plane is a circle.

50. Prove that the projection into the yz-plane of the curve of intersection of the surfaces $x = 1 - y^2$ and $x = y^2 + z^2$ is an ellipse.

51. Show that the projection into the xy-plane of the intersection of the plane $z = 2y$ and the paraboloid $z = x^2 + y^2$ is a circle.

52. Prove that the projection into the xz-plane of the intersection of the paraboloids $y = 2x^2 + 3z^2$ and $y = 5 - 3x^2 - 2z^2$ is a circle.

53. Prove that the projection into the xy-plane of the intersection of the plane $x + y + z = 1$ and the ellipsoid $x^2 + 4y^2 + 4z^2 = 4$ is an ellipse.

54. Show that the curve of intersection of the plane $z = ky$ and the cylinder $x^2 + y^2 = 1$ is an ellipse. (*Suggestion:* Introduce uv-coordinates into the plane $z = ky$ as follows: Let the u-axis be the original x-axis and let the v-axis be the line $z = ky$, $x = 0$.)

14.7 Cylindrical and Spherical Coordinates†

Rectangular coordinates provide only one of several useful ways of describing points, curves, and surfaces in space. In this section we discuss two additional coordinate systems in three-dimensional space. Each is a generalization of polar coordinates in the plane.

Recall that the relationship between the rectangular coordinates (x, y) and the polar coordinates (r, θ) of a point in the plane is given by

$$x = r\cos\theta, \qquad y = r\sin\theta \tag{1}$$

and

$$r^2 = x^2 + y^2, \qquad \tan\theta = \frac{y}{x} \quad \text{if} \quad x \neq 0. \tag{2}$$

These relationships may be read directly from the triangle of Fig. 14.75.

The **cylindrical coordinates** (r, θ, z) of a point P in space are a natural hybrid of its polar and rectangular coordinates. We use the polar coordinates (r, θ) of the point in the plane with rectangular coordinates (x, y) and use the same z-coordinate as in rectangular coordinates. (The cylindrical coordinates of a point P in space are illustrated in Fig. 14.76.) This means that the relation between the rectangular coordinates (x, y, z) of the point P and its cylindrical coordinates (r, θ, z) is obtained by simply adjoining the identity $z = z$ to the equations in (1) and (2):

$$x = r\cos\theta, \qquad y = r\sin\theta, \qquad z = z \tag{3}$$

and

$$r^2 = x^2 + y^2, \qquad \tan\theta = \frac{y}{x}, \qquad z = z. \tag{4}$$

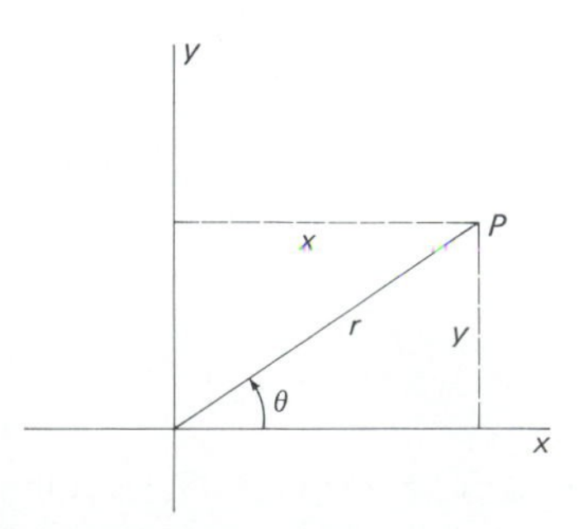

14.75 The relation between rectangular and polar coordinates in the xy-plane

†The material in this section will not be required until Section 16.7 and thus may be deferred until just before that section is covered.

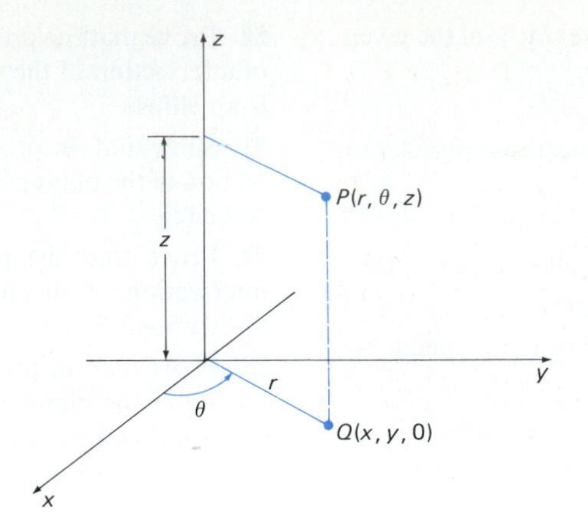

14.76 Finding the cylindrical coordinates of the point P

These equations may be used to convert from rectangular to cylindrical coordinates, and vice versa. The following table lists the rectangular and corresponding cylindrical coordinates for a few points in space.

(x, y, z)	(r, θ, z)
$(1, 0, 0)$	$(1, 0, 0)$
$(-1, 0, 0)$	$(1, \pi, 0)$
$(0, 2, 3)$	$(2, \pi/2, 3)$
$(1, 1, 2)$	$(\sqrt{2}, \pi/4, 2)$
$(1, -1, 2)$	$(\sqrt{2}, 7\pi/4, 2)$
$(-1, 1, 2)$	$(\sqrt{2}, 3\pi/4, 2)$
$(-1, -1, -2)$	$(\sqrt{2}, 5\pi/4, -2)$
$(0, -3, -3)$	$(3, 3\pi/2, -3)$

The term *cylindrical coordinates* arises from the fact that the graph in space of the equation $r = c$ (c a constant) is a cylinder of radius c symmetric about the x-axis. This suggests that cylindrical coordinates be used when working problems involving circular symmetry about the z-axis, for cylindrical coordinates exhibit a special simplicity in such cases. For instance, the sphere $x^2 + y^2 + z^2 = a^2$ and the cone $z^2 = x^2 + y^2$ have cylindrical coordinates equations $r^2 + z^2 = a^2$ and $z^2 = r^2$, respectively. It follows from our discussion of surfaces of revolution in Section 14.6 that if the curve $f(y, z) = 0$ in the yz-plane is revolved around the z-axis, then the cylindrical coordinates equation of the surface generated is $f(r, z) = 0$.

EXAMPLE 1 If the parabola $z = y^2$ is revolved around the z-axis, then we get the cylindrical coordinates equation of the paraboloid thus generated quite simply: We replace y with r. Thus the paraboloid has cylindrical equation

$$z = r^2.$$

Similarly, if the ellipse $y^2/9 + z^2/4 = 1$ is revolved around the z-axis, the cylindrical coordinates equation of the resulting ellipsoid is

$$\frac{r^2}{9} + \frac{z^2}{4} = 1.$$

EXAMPLE 2 Sketch the region that is bounded by the graphs of the cylindrical coordinates equations $z = r^2$ and $z = 8 - r^2$.

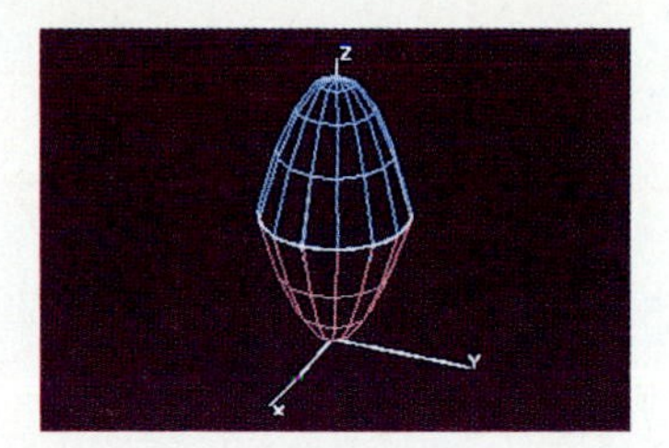

The solid of Example 2

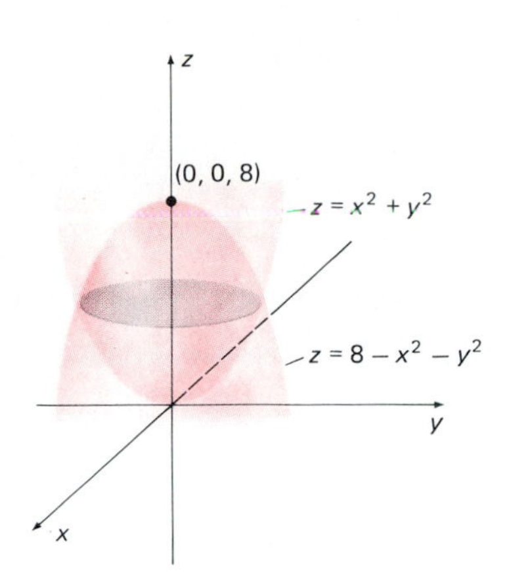

14.77 The two paraboloids of Example 2

Solution We first substitute $r^2 = x^2 + y^2$ from (4) in the given equations. Thus the two surfaces of the example have rectangular coordinate equations

$$z = x^2 + y^2 \quad \text{and} \quad z = 8 - x^2 - y^2.$$

Their graphs are the two paraboloids shown in Fig. 14.77. The region in question is bounded above by the paraboloid $z = 8 - x^2 - y^2$ and below by the paraboloid $z = x^2 + y^2$.

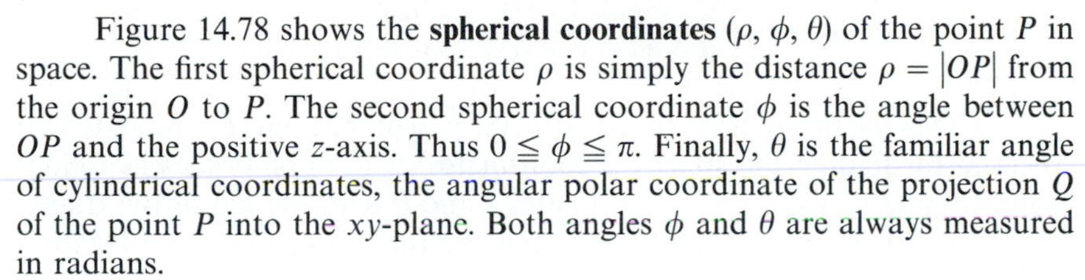

Figure 14.78 shows the **spherical coordinates** (ρ, ϕ, θ) of the point P in space. The first spherical coordinate ρ is simply the distance $\rho = |OP|$ from the origin O to P. The second spherical coordinate ϕ is the angle between OP and the positive z-axis. Thus $0 \leqq \phi \leqq \pi$. Finally, θ is the familiar angle of cylindrical coordinates, the angular polar coordinate of the projection Q of the point P into the xy-plane. Both angles ϕ and θ are always measured in radians.

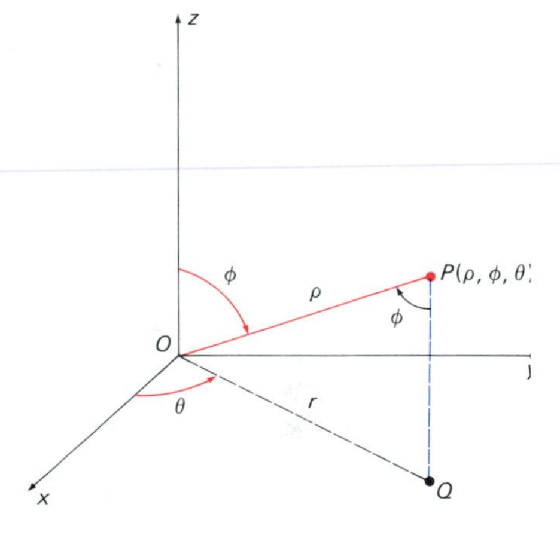

14.78 Finding the spherical coordinates of the point P

The term *spherical coordinates* is used because the graph of the equation $\rho = c$ (c a constant) is a sphere of radius c centered at the origin. Note also that $\phi = c$ (constant) describes (one nappe of) a cone if $0 < c < \pi/2$ or $\pi/2 < \phi < \pi$. The spherical coordinates equation of the xy-plane is $\phi = \pi/2$ (see Fig. 14.79).

From the right triangle OPQ of Fig. 14.78, we see that

$$r = \rho \sin \phi \quad \text{and} \quad z = \rho \cos \phi. \tag{5}$$

Indeed, these equations are most easily remembered by visualizing this triangle. Substitution of the equations in (5) into those in (3) yields

$$x = \rho \sin \phi \cos \theta, \qquad y = \rho \sin \phi \sin \theta, \qquad z = \rho \cos \phi. \tag{6}$$

These three equations give the relationship between rectangular and spherical coordinates. Also useful is the formula

$$\rho^2 = x^2 + y^2 + z^2, \tag{7}$$

a consequence of the distance formula.

It is important to note the order in which the spherical coordinates (ρ, ϕ, θ) of a point P are written—first the distance ρ of P from the origin, then the angle ϕ down from the positive z-axis, and last the counterclockwise angle θ around from the positive x-axis. The next table lists the rectangular coordinates and corresponding spherical coordinates of a few points in space.

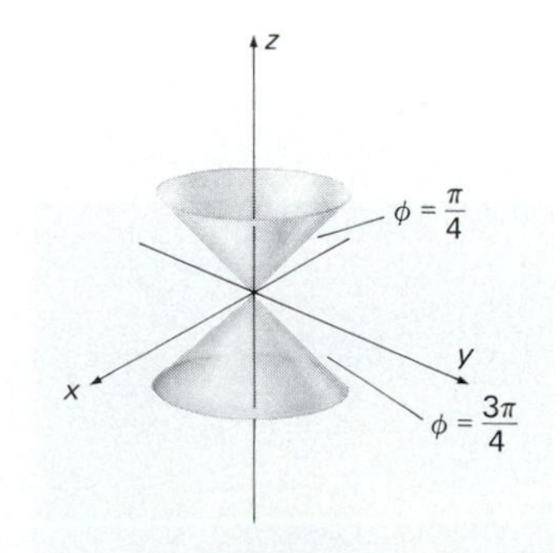

14.79 The two nappes of a 45° cone; $\phi = \pi/2$ is the xy-plane.

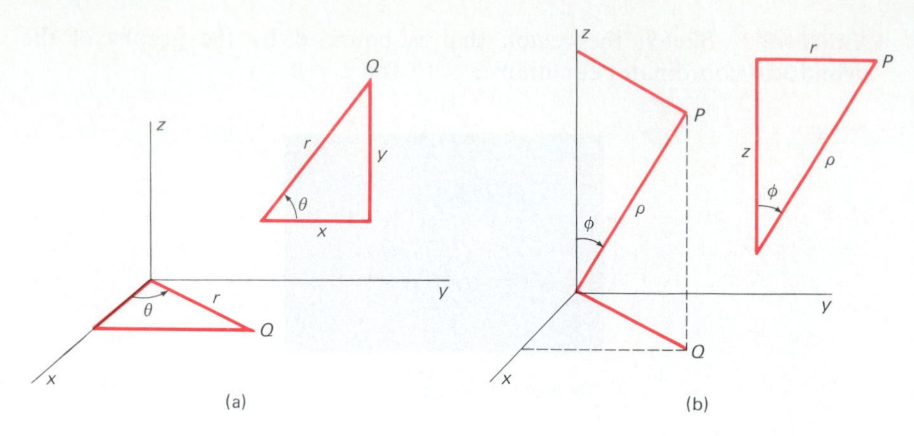

14.80 Triangles used in finding spherical coordinates

(x, y, z)	(ρ, ϕ, θ)
$(1, 0, 0)$	$(1, \pi/2, 0)$
$(0, 1, 0)$	$(1, \pi/2, \pi/2)$
$(0, 0, 1)$	$(1, 0, 0)$
$(0, 0, -1)$	$(1, \pi, 0)$
$(1, 1, \sqrt{2})$	$(2, \pi/4, \pi/4)$
$(-1, -1, \sqrt{2})$	$(2, \pi/4, 5\pi/4)$
$(1, -1, -\sqrt{2})$	$(2, 3\pi/4, 7\pi/4)$
$(1, 1, \sqrt{6})$	$(2\sqrt{2}, \pi/6, \pi/4)$

Given the rectangular coordinates (x, y, z) of the point P, one systematic method for finding the spherical coordinates (ρ, ϕ, θ) of P goes as follows. First we find its cylindrical coordinates r and θ with the aid of the triangle in Fig. 14.80(a). Then we find ρ and ϕ using the triangle in Fig. 14.80(b).

EXAMPLE 3 Find a spherical coordinates equation for the paraboloid with rectangular equation $z = x^2 + y^2$.

Solution We substitute $z = \rho \cos \phi$ from (5) and $x^2 + y^2 = r^2 = \rho^2 \sin^2 \phi$ from (6). This gives $\rho \cos \phi = \rho^2 \sin^2 \phi$. Cancellation of ρ gives $\cos \phi = \rho \sin^2 \phi$; that is, $\rho = \csc \phi \cot \phi$ as the spherical coordinates equation of the paraboloid. We get the whole paraboloid by using ϕ in the range $0 < \phi \leqq \pi/2$. Note that $\phi = \pi/2$ gives the point $\rho = 0$ that might otherwise have been lost by cancelling ρ.

EXAMPLE 4 Determine the graph of the spherical coordinates equation $\rho = 2 \cos \phi$.

Solution Multiplication by ρ gives

$$\rho^2 = 2\rho \cos \phi,$$

then substitution of $\rho^2 = x^2 + y^2 + z^2$ and $z = \rho \cos \theta$ yields

$$x^2 + y^2 + z^2 = 2z$$

as the rectangular coordinates equation of the graph. Completion of the square in z now gives

$$x^2 + y^2 + (z - 1)^2 = 1,$$

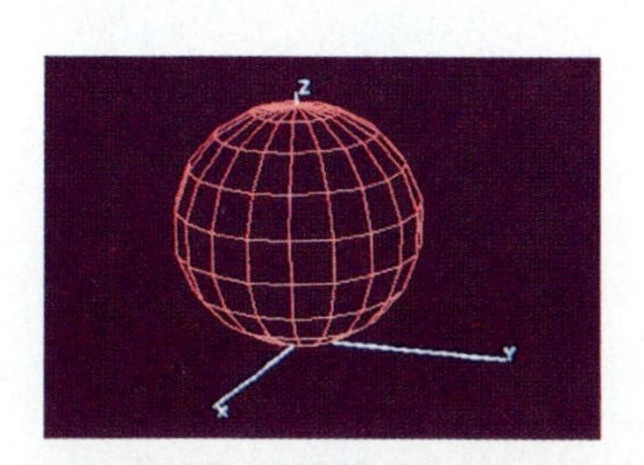

The sphere of Example 4

so the graph is a sphere with center (0, 0, 1) and radius 1. It is tangent to the xy-plane at the origin (see Fig. 14.81).

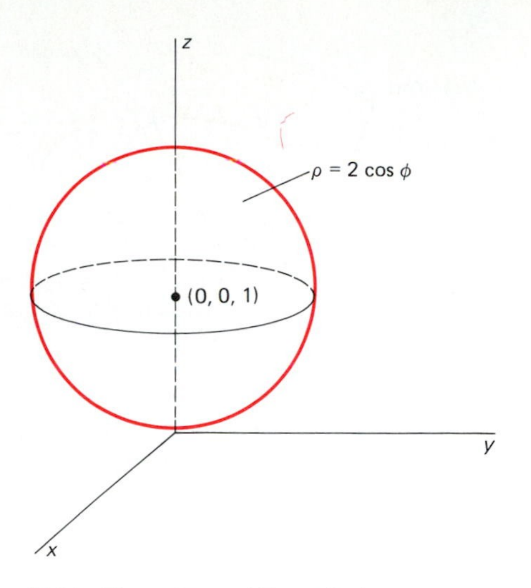

14.81 The sphere of Example 4

EXAMPLE 5 Determine the graph of the spherical coordinates equation $\rho = \sin\phi \sin\theta$.

Solution We first multiply each side by ρ and get $\rho^2 = \rho \sin\phi \sin\theta$. We then use Equations (6) and (7) and find that $x^2 + y^2 + z^2 = y$. This is a rectangular coordinates equation of a sphere with center $(0, \frac{1}{2}, 0)$ and radius $\frac{1}{2}$.

*LATITUDE AND LONGITUDE

A **great circle** of a spherical surface is a circle formed by the intersection of the surface with a plane through the center of the sphere it bounds. Thus a great circle of a spherical surface is a circle in the surface having the same radius as the sphere. Therefore, a great circle is a circle of maximum possible circumference lying on the sphere. It is easy to see that any two points on a spherical surface lie on a great circle (uniquely determined unless the two points are *antipodal*–lie on the ends of a diameter of the sphere). In the calculus of variations it is shown that the shortest distance between two such points—measured along the curved surface—is the shorter of the two arcs of the great circle containing them. The surprise is that the *shortest* distance is found by using the *largest* circle.

The spherical coordinates ϕ and θ are closely related to the latitude and longitude of points on the surface of the earth. Assume that the earth is a sphere with radius $\rho = 3960$ mi. We begin with the **prime meridian** (a **meridian** is a north–south great semicircle) through Greenwich, just outside London. This is the point marked G in Fig. 14.82.

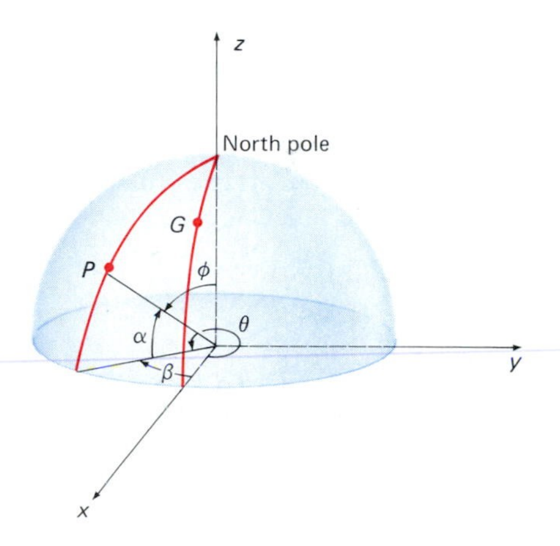

14.82 The relation between latitude, longitude, and spherical coordinates

We take the z-axis through the North Pole and the x-axis through the point where the prime meridian intersects the equator. The **latitude** α and (west) **longitude** β of a point P in the Northern Hemisphere are given by the equations

$$\alpha = 90^\circ - \phi^\circ \quad \text{and} \quad \beta = 360^\circ - \theta^\circ, \tag{8}$$

where ϕ° and θ° are the angular spherical coordinates, measured in *degrees*, of P. (That is, ϕ° and θ° denote the degree equivalents of the angles ϕ and θ, respectively, which are always measured in radians unless otherwise specified.) Thus the latitude α is measured northward from the equator and the longitude β is measured westward from the prime meridian.

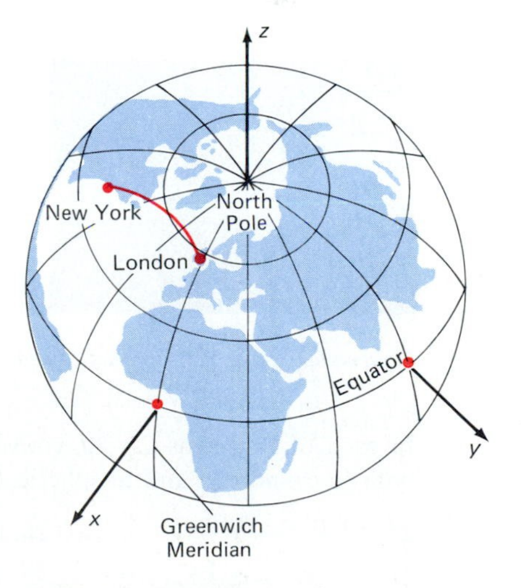

14.84 Finding the great circle distance d from New York to London

EXAMPLE 6 Find the great circle distance between New York (latitude 40.75° north, longitude 74° west) and London (latitude 51.5° north, longitude 0°). See Fig. 14.83.

Solution From the equations in (8) we find that $\phi^\circ = 49.25^\circ$, $\theta^\circ = 286^\circ$ for New York, while $\phi^\circ = 38.5^\circ$, $\phi^\circ = 360^\circ$ (or 0°) for London. Hence the spherical coordinates of New York are $\phi = (49.25/180)\pi$, $\theta = (286/180)\pi$, while those of London are $\phi = (38.5/180)\pi$, $\theta = 0$. With these values of ϕ and θ and with $\pi = 3960$ (mi), the equations in (6) give the rectangular coordinates

$$\text{New York:} \qquad P_1(826.90, -2883.74, 2584.93)$$

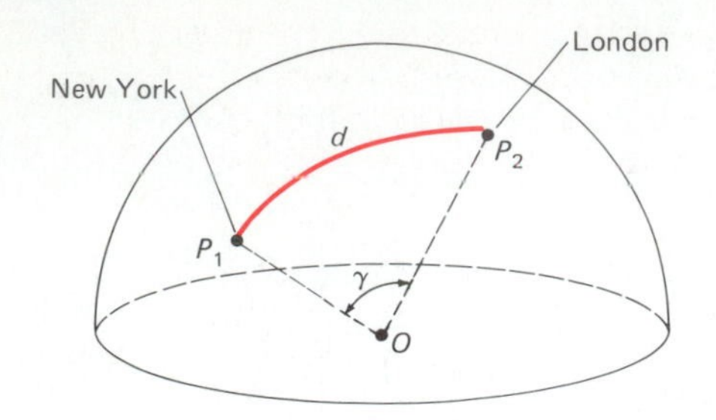

14.83 The great circle arc between New York and London

and

$$\text{London:} \qquad P_2(2465.16, 0.0, 3099.13).$$

The angle γ between the radius vectors $\overrightarrow{OP_1}$ and $\overrightarrow{OP_2}$ in Fig. 14.84 is given by

$$\cos \gamma = \frac{\overrightarrow{OP_1} \cdot \overrightarrow{OP_2}}{|\overrightarrow{OP_1}||\overrightarrow{OP_2}|}$$

$$= \frac{(826.90)(2465.16) - (2883.74)(0.0) + (2584.93)(3099.13)}{(3960)^2}$$

$$\approx 0.641.$$

Thus γ is approximately 0.875 rad. Hence the great circle distance between New York and London is close to

$$d = (3960)(0.875) \approx 3465 \quad \text{(mi)}.$$

14.7 PROBLEMS

In Problems 1–10 find both the cylindrical coordinates and the spherical coordinates of the point P having the given rectangular coordinates.

1. $P(0, 0, 5)$
2. $P(0, 0, -3)$
3. $P(1, 1, 0)$
4. $P(2, -2, 0)$
5. $P(1, 1, 1)$
6. $P(-1, 1, -1)$
7. $P(2, 1, -2)$
8. $P(-2, -1, -2)$
9. $P(3, 4, 12)$
10. $P(-2, 4, -12)$

In each of Problems 11–24, describe the graph of the given equation.

11. $r = 5$
12. $\theta = 3\pi/4$
13. $\theta = \pi/4$
14. $\rho = 5$
15. $\phi = \pi/6$
16. $\phi = 5\pi/6$
17. $\phi = \pi/2$
18. $\phi = \pi$
19. $r = 2 \sin \theta$
20. $\rho = 2 \sin \phi$
21. $\cos \theta + \sin \theta = 0$
22. $z = 10 - 3r^2$
23. $\rho \cos \phi = 1$
24. $\rho = \cot \phi$

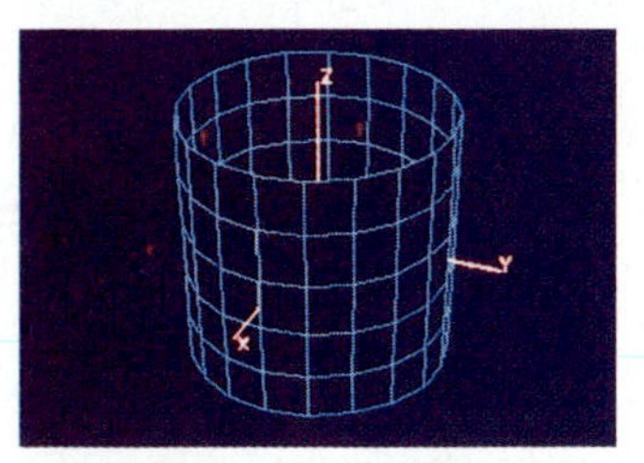

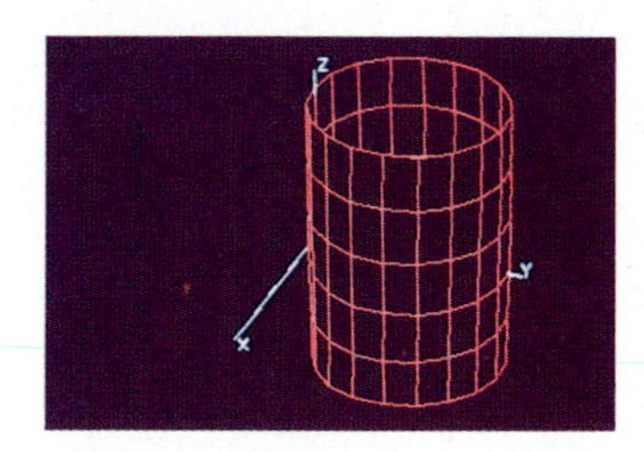

Figures for some of Problems 11–24

In each of Problems 25–30, convert the given equation both to cylindrical and to spherical coordinates.

25. $x^2 + y^2 + z^2 = 25$
26. $x^2 + y^2 = 2x$
27. $x + y + z = 1$
28. $x + y = 4$
29. $x^2 + y^2 + z^2 = x + y + z$
30. $z = x^2 - y^2$

31. The parabola $z = x^2$, $y = 0$ is rotated about the z-axis. Write a cylindrical coordinates equation for the surface thereby generated.

32. The hyperbola $y^2 - z^2 = 1$, $x = 0$ is rotated about the z-axis. Write a cylindrical coordinates equation for the surface thereby generated.

33. A sphere of radius 2 is centered at the origin. A hole of radius 1 is drilled through the sphere, with the axis of the hole lying on the z-axis. Describe the solid region that remains in
(a) cylindrical coordinates; (b) spherical coordinates.

34. Find the great circle distance from Atlanta (latitude 33.75° north, longitude 84.40° west) to San Francisco (latitude 37.78° north, longitude 122.42° west).

35. Find the great circle distance from Fairbanks (latitude 64.80° north, longitude 147.85° west) to Leningrad (latitude 59.91° north, longitude 30.43° *east* of Greenwich—alternatively, longitude 329.57° west).

36. Because Fairbanks and Leningrad are at almost the same latitude, a plane could fly from one to the other roughly along the 62nd parallel of latitude. Accurately estimate the length of such a trip.

37. In flying the great circle route from Fairbanks to Leningrad, how close to the North Pole would a plane come?

38. A right circular cone of radius R and height H is located with its vertex at the origin and its axis coincident with part of the nonnegative z-axis. Describe the solid cone in cylindrical coordinates.

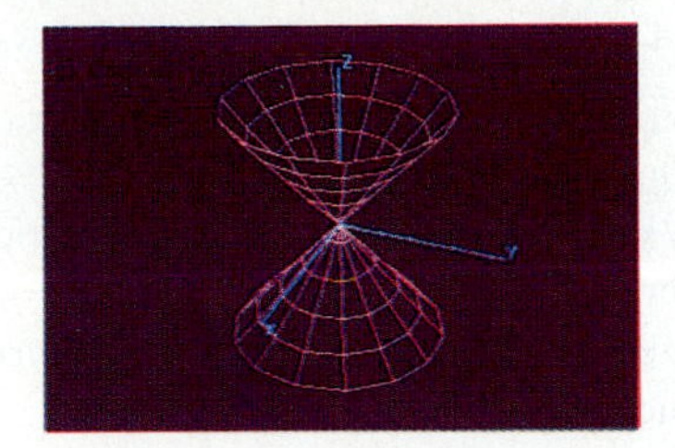

The cone of Problems 38–39

39. Describe the cone of Problem 38 in spherical coordinates.

CHAPTER 14 REVIEW: Definitions, Concepts, Results

Use the list below as a guide to concepts that you may need to review.

1. Properties of addition of vectors in space and of multiplication of vectors by scalars

2. The dot (scalar) product of vectors—definition and geometric interpretation

3. Use of the dot product to test perpendicularity of vectors and, generally, finding the angle between two vectors

4. The cross (vector) product of two vectors—definition and geometric interpretation

5. The scalar triple product of three vectors—definition and geometric interpretation

6. The parametric and symmetric equations of the straight line through a given point and parallel to a given vector

7. Equation of the plane through a given point normal to a given vector

8. The velocity and acceleration vectors of a parametric space curve

9. Arc length of a parametric space curve

10. The curvature, unit tangent vector, and principal unit normal vector of a parametric curve in the plane or in space

11. Tangential and normal components of the acceleration vector of a parametric curve

12. Equations of cylinders and of surfaces of revolution

13. The standard examples of quadric surfaces

14. Definition of the cylindrical coordinate and spherical coordinate systems, and the equations relating cylindrical and spherical coordinates to rectangular coordinates

CHAPTER 14 MISCELLANEOUS PROBLEMS

1. Suppose that M is the midpoint of the segment PQ in space and that A is another point. Show that

$$\overrightarrow{AM} = \tfrac{1}{2}(\overrightarrow{AP} + \overrightarrow{AQ}).$$

2. Let $\mathbf{a}$ and $\mathbf{b}$ be nonzero vectors. Define

$$\mathbf{a}_{||} = (\text{comp}_{\mathbf{b}}\,\mathbf{a})\frac{\mathbf{b}}{|\mathbf{b}|} \quad \text{and} \quad \mathbf{a}_{\perp} = \mathbf{a} - \mathbf{a}_{||}.$$

Prove that $\mathbf{a}_{\perp}$ is perpendicular to $\mathbf{b}$.

3. Let P and Q be different points in space. Show that the point R lies on the line through P and Q *if and only if* there exist numbers a and b such that $a + b = 1$ and $\overrightarrow{OR} = a\overrightarrow{OP} + b\overrightarrow{OQ}$. Conclude that

$$\mathbf{r}(t) = t\overrightarrow{OP} + (1 - t)\overrightarrow{OQ}$$

is a parametric equation of this line.

4. Conclude from the result of Problem 3 that the points P, Q, and R are collinear if and only if there exist numbers

a, b, and c, not all zero, such that $a + b + c = 0$ and $a\overrightarrow{OP} + b\overrightarrow{OQ} + c\overrightarrow{OR} = \mathbf{0}$.

5. Let $P(x_0, y_0)$, $Q(x_1, y_1)$, and $R(x_2, y_2)$ be points in the xy-plane. Use the cross product to show that the area of the triangle PQR is

$$A = \tfrac{1}{2}|(x_1 - x_0)(y_2 - y_0) - (x_2 - x_0)(y_1 - y_0)|.$$

6. Write both symmetric and parametric equations of the line through $P(1, -1, 0)$ parallel to $\mathbf{v} = \langle 2, -1, 3 \rangle$.

7. Write both symmetric and parametric equations of the line through $P_1(1, -1, 2)$ and $P_2(3, 2, -1)$.

8. Write an equation of the plane through $P(3, -5, 1)$ with normal vector $\mathbf{n} = \mathbf{i} + \mathbf{j}$.

9. Show that the lines with symmetric equations

$$x - 1 = 2(y + 1) = 3(z - 2)$$

and

$$x - 3 = 2(y - 1) = 3(z + 1)$$

are parallel. Then write an equation of the plane through these two lines.

10. Let the lines L_1 and L_2 have symmetric equations

$$\frac{x - x_i}{a_i} = \frac{y - y_i}{b_i} = \frac{z - z_i}{c_i}$$

for $i = 1, 2$. Show that L_1 and L_2 are skew lines if and only if

$$\begin{vmatrix} x_1 - x_2 & y_1 - y_2 & z_1 - z_2 \\ a_1 & b_1 & c_1 \\ a_2 & b_2 & c_2 \end{vmatrix} \neq 0.$$

11. Given the four points $A(2, 3, 2)$, $B(4, 1, 0)$, $C(-1, 2, 0)$, and $D(5, 4, -2)$, find an equation of the plane through A and B parallel to the line through C and D.

12. Given the points A, B, C, and D of Problem 11, find points P on the line AB and Q on the line CD such that the line PQ is perpendicular to both AB and CD. What is the perpendicular distance d between the lines AB and CD?

13. Let $P_0(x_0, y_0, z_0)$ be a point of the plane with equation

$$ax + by + cz + d = 0.$$

By projecting $\overrightarrow{OP_0}$ onto the normal vector $\mathbf{n} = \langle a, b, c \rangle$ show that the distance D from the origin to this plane is

$$D = \frac{|d|}{\sqrt{a^2 + b^2 + c^2}}.$$

14. Show that the distance D from the point $P_1(x_1, y_1, z_1)$ to the plane $ax + by + cz + d = 0$ is equal to the distance from the origin to the plane with equation

$$a(x + x_1) + b(y + y_1) + c(z + z_1) + d = 0.$$

Hence conclude from the result of Problem 13 that

$$D = \frac{|ax_1 + by_1 + cz_1 + d|}{\sqrt{a^2 + b^2 + c^2}}.$$

15. Find the perpendicular distance between the parallel planes $2x - y + 2z = 4$ and $2x - y + 2z = 13$.

16. Write an equation of the plane through the point $(1, 1, 1)$ that is normal to the twisted cubic $x = t$, $y = t^2$, $z = t^3$ at this point.

17. A particle moves in space with parametric equations $x = t$, $y = t^2$, $z = \frac{4}{3}t^{3/2}$. Find the curvature of its trajectory and the tangential and normal components of its acceleration when $t = 1$.

18. The **osculating plane** to a space curve at a point P of that curve is the plane through P that is parallel to the curve's unit tangent and principal unit normal vectors at P. Write an equation of the osculating plane to the curve of Problem 17 at the point $(1, 1, \frac{4}{3})$.

19. Show that the equation of the plane through the point $P_0(x_0, y_0, z_0)$ and parallel to the vectors $\mathbf{v}_1 = \langle a_1, b_1, c_1 \rangle$ and $\mathbf{v}_2 = \langle a_2, b_2, c_2 \rangle$ can be written in the form

$$\begin{vmatrix} x - x_0 & y - y_0 & z - z_0 \\ a_1 & b_1 & c_1 \\ a_2 & b_2 & c_2 \end{vmatrix} = 0.$$

20. Deduce from Problem 19 that the equation of the osculating plane (Problem 18) to the parametric curve $\mathbf{r}(t)$ at the point $\mathbf{r}(t_0)$ can be written in the form

$$(\mathbf{R} - \mathbf{r}(t_0)) \cdot \mathbf{r}'(t_0) \times \mathbf{r}''(t_0) = 0$$

where $\mathbf{R} = \langle x, y, z \rangle$. Note first that the vectors $\mathbf{T}$ and $\mathbf{N}$ are coplanar with $\mathbf{r}'(t)$ and $\mathbf{r}''(t)$.

21. Use the result of Problem 20 to write an equation of the osculating plane to the twisted cubic $x = t$, $y = t^2$, $z = t^3$ at the point $(1, 1, 1)$.

22. Let a parametric curve in space be described by equations $r = r(t)$, $\theta = \theta(t)$, $z = z(t)$ giving the cylindrical coordinates of a moving point on the curve for $a \leqq t \leqq b$. Use the equations relating rectangular and cylindrical coordinates to show that the arc length of the curve is

$$s = \int_a^b \left[\left(\frac{dr}{dt}\right)^2 + \left(r\frac{d\theta}{dt}\right)^2 + \left(\frac{dz}{dt}\right)^2 \right]^{1/2} dt.$$

23. A point moves on the *unit* sphere $\rho = 1$ with its spherical angular coordinates at time t given by $\phi = \phi(t)$, $\theta = \theta(t)$, $a \leqq t \leqq b$. Use the equations relating rectangular and spherical coordinates to show that the arc length of its path is

$$s = \int_a^b \left[\left(\frac{d\phi}{dt}\right)^2 + (\sin^2\phi)\left(\frac{d\theta}{dt}\right)^2 \right]^{1/2} dt.$$

24. The cross product $\mathbf{B} = \mathbf{T} \times \mathbf{N}$ of the unit tangent vector and the principal unit normal vector is the *unit binormal vector* $\mathbf{B}$ of a curve.

(a) Differentiate $\mathbf{B} \cdot \mathbf{T} = 0$ to show that $\mathbf{T}$ is perpendicular to $d\mathbf{B}/ds$.

(b) Differentiate $\mathbf{B} \cdot \mathbf{B} = 1$ to show that $\mathbf{B}$ is perpendicular to $d\mathbf{B}/ds$.

(c) Conclude from (*a*) and (*b*) that $d\mathbf{B}/ds = -\tau\mathbf{N}$ for some number τ. This number τ, called the **torsion** of the curve, measures the amount it twists at each point in space.

25. Show that the torsion of the helix of Example 1 in Section 14.4 is constant by showing that its value is

$$\tau = \frac{b\omega}{a^2\omega^2 + b^2}.$$

26. Deduce from the definition of torsion (Problem 24) that $\tau = 0$ for any curve such that $\mathbf{r}(t)$ lies in a fixed plane.

27. Write an equation in spherical coordinates of the sphere with radius 1 and center $x = 0 = y$, $z = 1$.

28. Let C be the circle in the yz-plane with radius 1 and center $y = 1$, $z = 0$. Write equations in both rectangular and cylindrical coordinates of the surface obtained by revolving C around the z-axis.

29. Let C be the curve in the yz-plane with equation $(y^2 + z^2)^2 = 2(z^2 - y^2)$. Write an equation in spherical coordinates of the surface obtained by revolving this curve around the z-axis. Then sketch this surface. (*Suggestion:* Remember that $r^2 = 2\cos 2\theta$ is the polar equation of a figure-eight curve.)

30. Let A be the area of a parallelogram in space determined by the vectors $\mathbf{a} = \overrightarrow{PQ}$ and $\mathbf{b} = \overrightarrow{RS}$. Let A' be the area of the perpendicular projection of $PQRS$ into a plane that makes an acute angle γ with the plane of $PQRS$. Assuming the fact that $A' = A\cos\gamma$ in such a situation, prove that the areas of the perpendicular projections of the parallelogram $PQRS$ into the three coordinate planes are

$$|\mathbf{i}\cdot\mathbf{a}\times\mathbf{b}|, \quad |\mathbf{j}\cdot\mathbf{a}\times\mathbf{b}|, \quad \text{and} \quad |\mathbf{k}\cdot\mathbf{a}\times\mathbf{b}|.$$

Conclude that the square of the area of a parallelogram in space is equal to the sum of the squares of the areas of its perpendicular projections into the three coordinate planes.

31. Take $\mathbf{a} = \langle a_1, a_2, a_3\rangle$ and $\mathbf{b} = \langle b_1, b_2, b_3\rangle$ in Problem 30. Show that

$$A^2 = \begin{vmatrix} a_2 & a_3 \\ b_2 & b_3 \end{vmatrix}^2 + \begin{vmatrix} a_3 & a_1 \\ b_3 & b_1 \end{vmatrix}^2 + \begin{vmatrix} a_1 & a_2 \\ b_1 & b_2 \end{vmatrix}^2.$$

32. Let C be a curve in a plane $\mathscr{P}$ that is not parallel to the z-axis. Suppose that the projection of C into the xy-plane is an ellipse. Introduce uv-coordinates in the plane $\mathscr{P}$ to prove that the curve C is itself an ellipse.

33. Conclude from Problem 32 that the intersection of a nonvertical plane and an elliptic cylinder with vertical sides is an ellipse.

34. Use the result of Problem 32 to prove that the intersection of the plane $z = Ax + By$ and the paraboloid $z = a^2x^2 + b^2y^2$ is either empty, a single point, or an ellipse.

35. Use the result of Problem 32 to prove that the intersection of the plane $z = Ax + By$ and the ellipsoid $x^2/a^2 + y^2/b^2 + z^2/c^2 = 1$ is either empty, a single point, or an ellipse.

36. Suppose that $y = f(x)$ is the graph of a function f for which f'' is continuous, and suppose also that the graph has an inflection point at $(a, f(a))$. Prove that the curvature of the graph at $x = a$ is zero.

37. Find the points on the curve $y = \sin x$ where the curvature is maximal and those where it is minimal.

38. The right branch of the hyperbola $x^2 - y^2 = 1$ may be parametrized by $x = \cosh t$, $y = \sinh t$. Find the point where its curvature is minimal.

39. Find the vectors $\mathbf{N}$ and $\mathbf{T}$ at the point of the curve $x = t\cos t$, $y = t\sin t$ at the point where $t = \pi/2$.

40. Find the points on the ellipse $x^2/a^2 + y^2/b^2 = 1$ (with $a > b$) where the curvature is maximal and those where it is minimal.

41. Suppose that the plane curve $f = f(\theta)$ is given in polar coordinates. Write r' for $f'(\theta)$ and r'' for $f''(\theta)$. Show that its curvature is given by

$$k = \frac{|r^2 + 2(r')^2 - rr''|}{[r^2 + (r')^2]^{3/2}}.$$

42. Use the formula of Problem 41 to calculate the curvature $k(\theta)$ at the point (r, θ) of the Archimedean spiral $r = \theta$. Then show that $k(\theta) \to 0$ as $\theta \to +\infty$.

43. A railway curve is to join two straight tracks, one extending due west from $(-1, -1)$, the other extending due east from $(1, 1)$. Determine A, B, and C so that the curve $y = Ax + Bx^3 + Cx^5$ joins $(-1, -1)$ and $(1, 1)$ and so that the slope and curvature of this connecting curve are zero at both its end points.

chapter fifteen

Partial Differentiation

15.1 Introduction

In this and the following two chapters, we turn our attention to the calculus of functions of more than one variable. Many real-world functions depend upon two or more variables. For example:

- In physical chemistry the ideal gas law $PV = nRT$ (where n and R are constants) is used to express any one of the variables P, V, and T as a function of the other two.
- The altitude above sea level at a particular location on the earth's surface depends upon its latitude and longitude.
- A manufacturer's profit depends upon sales, overhead costs, the cost of each raw material used, and perhaps upon additional variables.
- The amount of usable energy a solar panel can gather depends upon its efficiency, its angle of inclination to the sun's rays, the angle of elevation of the sun above the horizon, and perhaps upon other factors.

A typical application may call for us to find an extreme value of a function of several variables. For example, suppose that we want to minimize the cost of making a rectangular box with a volume of 48 ft^3, given that its front and back cost \$1/ft^2, its top and bottom cost \$2/ft^2, and its two ends cost

\$3/ft². Figure 15.1 shows such a box with length x, width y, and height z. Under the conditions given, its total cost will be

$$C = 2xz + 4xy + 6yz \quad \text{(dollars).}$$

But x, y, and z are not independent variables, because the box has fixed volume

$$V = xyz = 48.$$

We eliminate z from the first formula by using the second; because $z = 48/xy$, we find that the cost we want to minimize is given by

$$C = 4xy + \frac{288}{x} + \frac{96}{y}.$$

Because neither of the variables x and y can be expressed in terms of the other, the single-variable maximum-minimum techniques of Chapter 3 cannot be applied here. We need new optimization techniques applicable to functions of two or more independent variables. In Section 15.5 we shall return to this problem.

The problem of optimization is merely a single example. In this chapter, we shall see that the main ingredients of single-variable differential calculus—limits, derivatives and rates of change, chain rule computations, and maximum–minimum techniques—can all be generalized to functions of two or more variables.

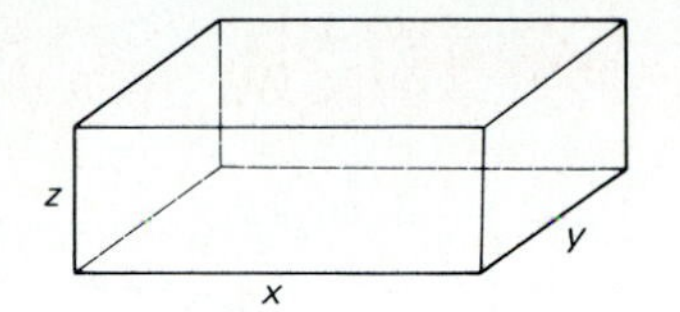

15.1 Minimizing the total cost of a box

15.2 Functions of Several Variables

Recall that a real-valued *function* is a rule or correspondence f that associates a unique real number with each element of a set D. The set D is called the *domain* of definition of the function f. The domain D has always been a subset of the real line for the functions of a single variable that we have studied up to this point. If D is a subset of the plane, then f is a function of *two* variables—for, given a point P in D, we naturally associate with P its rectangular coordinates (x, y).

> ***Definition*** *Functions of Two or Three Variables*
>
> A **function of two variables**, defined on the **domain** D in the plane, is a rule f that associates with each point (x, y) in D a real number $f(x, y)$. A **function of three variables**, defined on the **domain** D in space, is a rule f that associates with each point (x, y, z) in D a real number $f(x, y, z)$.

A function f of two (or three) variables is often defined by giving a formula that specifies $f(x, y)$ in terms of x and y (or $f(x, y, z)$ in terms of x, y, and z). In case the domain D of f is not explicitly specified, it is to be understood that D consists of all points for which the given formula is meaningful. For example, the domain of the function f with formula $f(x, y) = \sqrt{25 - x^2 - y^2}$ is the set of all (x, y) such that $25 - x^2 - y^2 \geqq 0$; that is, the circular disk $x^2 + y^2 \leqq 25$ of radius 5 centered at the origin. Similarly, the function g defined as

$$g(x, y, z) = \frac{x + y + z}{\sqrt{x^2 + y^2 + z^2}}$$

is defined at all points of space where $x^2 + y^2 + z^2 > 0$. Thus its domain consists of all points of three-dimensional space $\mathbf{R}^3$ other than the origin (0, 0, 0).

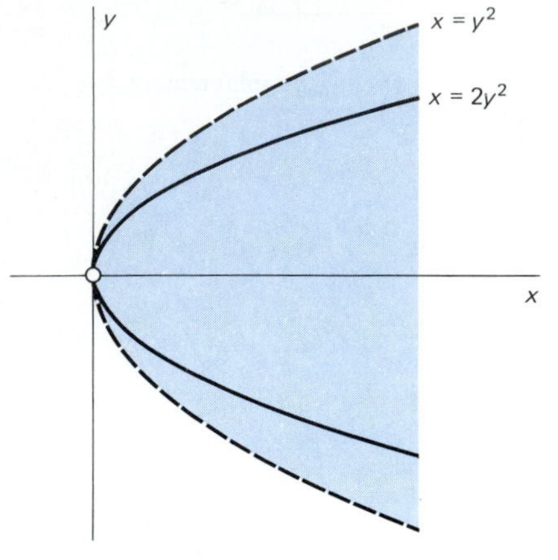

15.2 The domain of $f(x, y) = y/\sqrt{x - y^2}$

EXAMPLE 1 Find the domain of definition of the function with formula

$$f(x, y) = \frac{y}{\sqrt{x - y^2}}. \tag{1}$$

Also find the points (x, y) at which $f(x, y) = \pm 1$.

Solution In order that $f(x, y)$ be defined, it is necessary that the radicand $x - y^2$ be positive—that is, that $y^2 < x$. Hence the domain of f is the set of points lying strictly to the right of the parabola $x = y^2$. The domain is shown shaded in Fig. 15.2. The parabola itself is shown dotted to indicate that it is not included in the domain; any point for which $x = y^2$ would entail division by zero in (1).

The function $f(x, y)$ has the value ± 1 wherever

$$\frac{y}{\sqrt{x - y^2}} = \pm 1;$$

that is, $y^2 = x - y^2$, so that $x = 2y^2$. Thus $f(x, y) = \pm 1$ at each point of the parabola $x = 2y^2$ (other than its vertex (0, 0), which is not included in the domain of f).

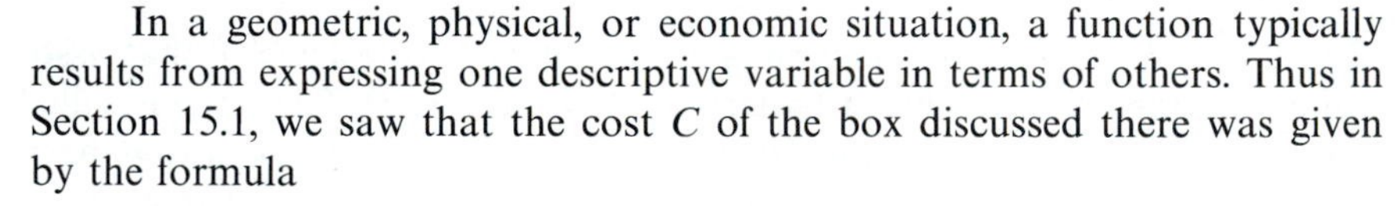

In a geometric, physical, or economic situation, a function typically results from expressing one descriptive variable in terms of others. Thus in Section 15.1, we saw that the cost C of the box discussed there was given by the formula

$$C = 4xy + \frac{288}{x} + \frac{96}{y}$$

in terms of its length x and width y. The value C of this function is a variable that depends upon the values of x and y. Hence we call C a **dependent variable**, while x and y are **independent variables**. If the temperature T at the point (x, y, z) in space is given by some formula $T = h(x, y, z)$, then the dependent variable T is a function of the three independent variables x, y, and z.

A function of four or more variables can be defined by giving a formula involving the appropriate number of independent variables. For example, if an amount A of heat is released at the origin at time $t = 0$ in a medium with diffusivity k, then—under appropriate conditions—it turns out that the temperature T at the point (x, y, z) at time $t > 0$ is given by

$$T(x, y, z, t) = \frac{A}{(4\pi kt)^{3/2}} \exp\left(-\frac{x^2 + y^2 + z^2}{4kt}\right).$$

This formula gives the temperature T as a function of the four independent variables x, y, z, and t.

We shall see that the main differences between single-variable and multivariable calculus show up with examples involving only two independent variables. Hence most of our results will be stated in terms of functions of two variables. Many of these results readily generalize by analogy to the case of three or more independent variables.

GRAPHS AND LEVEL CURVES

A function f of two variables x and y has the property that we can visualize how it "works" in terms of its graph. The **graph** of f is the graph of the equation $z = f(x, y)$. Thus the graph of f is the set of all points in space with coordinates (x, y, z) that satisfy the equation $z = f(x, y)$ (see Fig. 15.3).

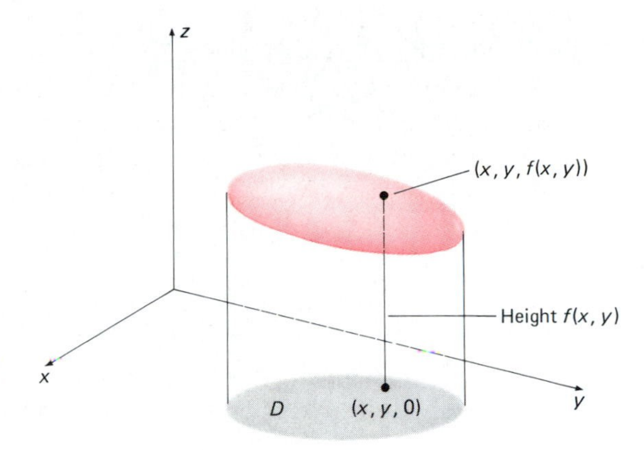

15.3 The graph of a function of two variables is frequently a surface "over" the domain of the function.

You saw several examples of such graphs in Chapter 14. For example, the graph of the function $f(x, y) = x^2 + y^2$ is the paraboloid $z = x^2 + y^2$ shown in Fig. 15.4. The graph of the function

$$g(x, y) = c\sqrt{1 - \left(\frac{x}{a}\right)^2 - \left(\frac{y}{b}\right)^2}$$

is the *upper half* of the ellipsoid $x^2/a^2 + y^2/b^2 + z^2/c^2 = 1$ and is shown in Fig. 15.5. Generally speaking, the graph of a function of two variables is a surface that lies above (or below, or both) its domain D in the xy-plane.

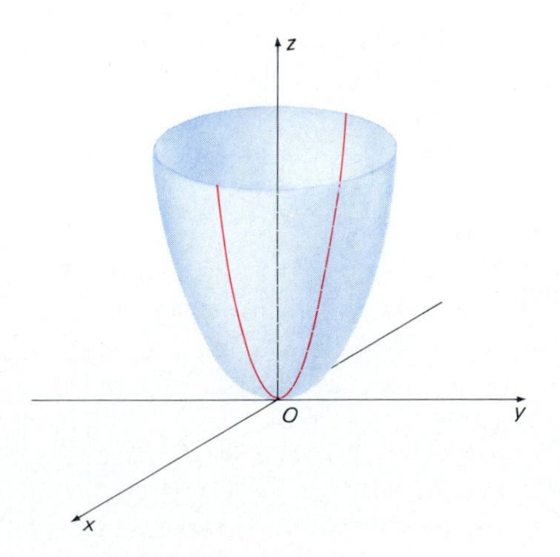

15.4 The paraboloid is (part of) the graph of the function $f(x, y) = x^2 + y^2$.

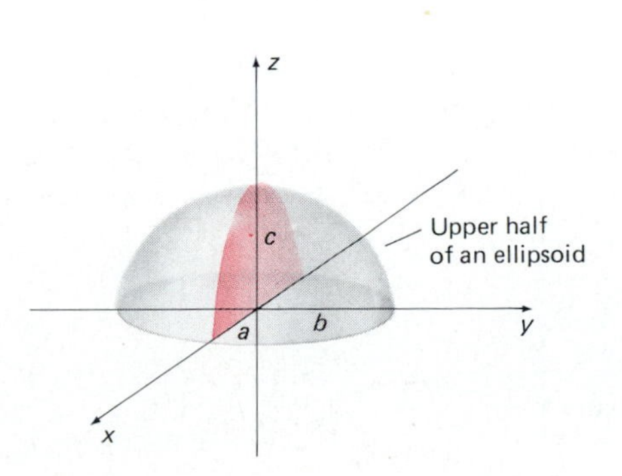

15.5 The upper half of an ellipsoid is the graph of a function of two variables.

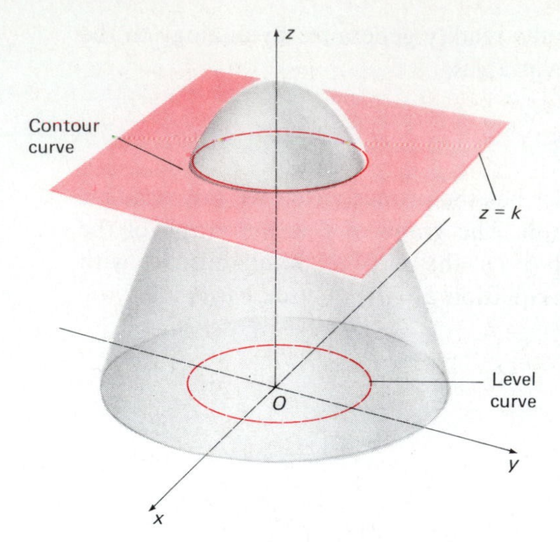

15.6 A contour curve and the corresponding level curve

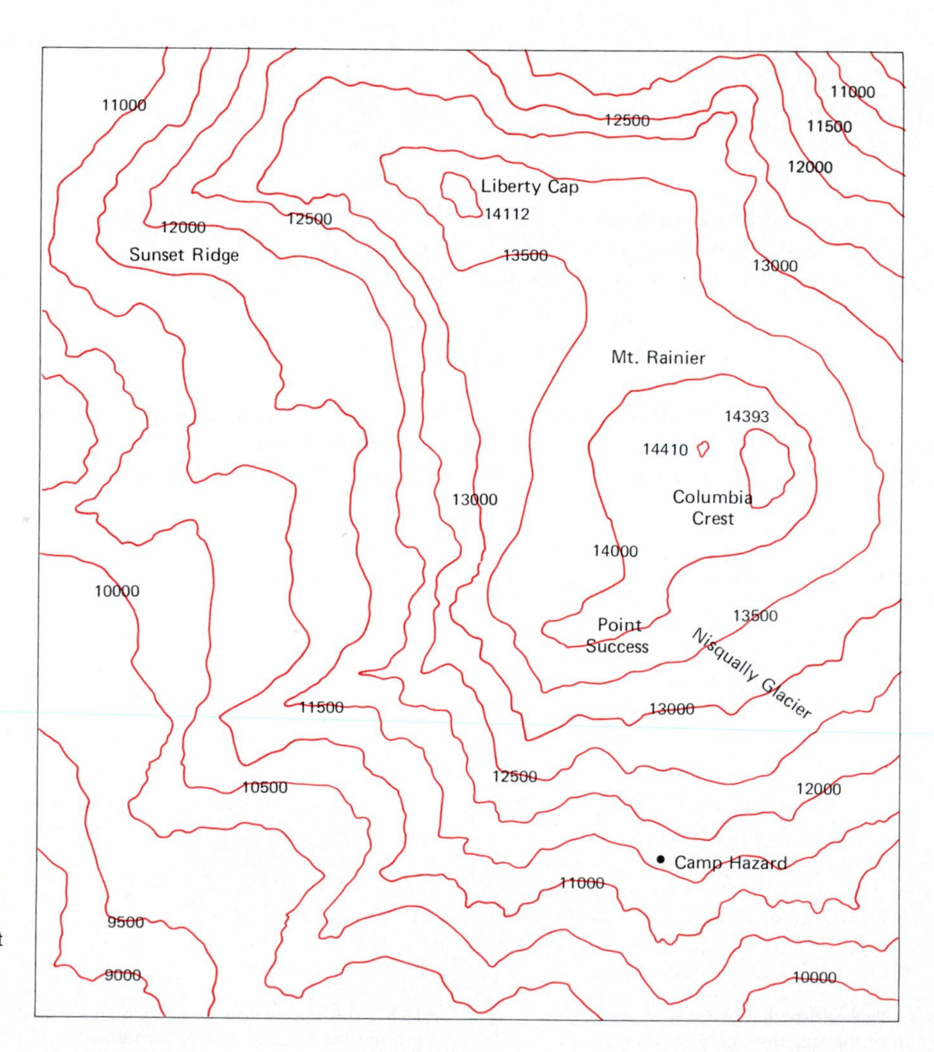

15.7 Area around Mt. Rainier, Washington, showing level curves at 500-ft intervals (Adapted from U.S. Geological Survey Map N4645-W12145/7.5 (1971))

The intersection of the horizontal plane $z = k$ with the surface $z = f(x, y)$ is called the **contour curve** of height k on the surface, as in Fig. 15.6. The vertical projection of this contour curve into the xy-plane is the **level curve** $f(x, y) = k$ of the function f. The level curves of f are simply the sets on which the value of f is constant. On a topographic map, like the one in Fig. 15.7, the level curves are the curves of constant height above sea level.

Level curves afford us a two-dimensional way of representing a three-dimensional surface $z = f(x, y)$, just as the two-dimensional map in Fig. 15.7 represents the three-dimensional mountain. We do this by drawing typical level curves of $z = f(x, y)$ in the xy-plane, labeling each with the corresponding (constant) value of z. Figure 15.8 illustrates this process for a simple hill.

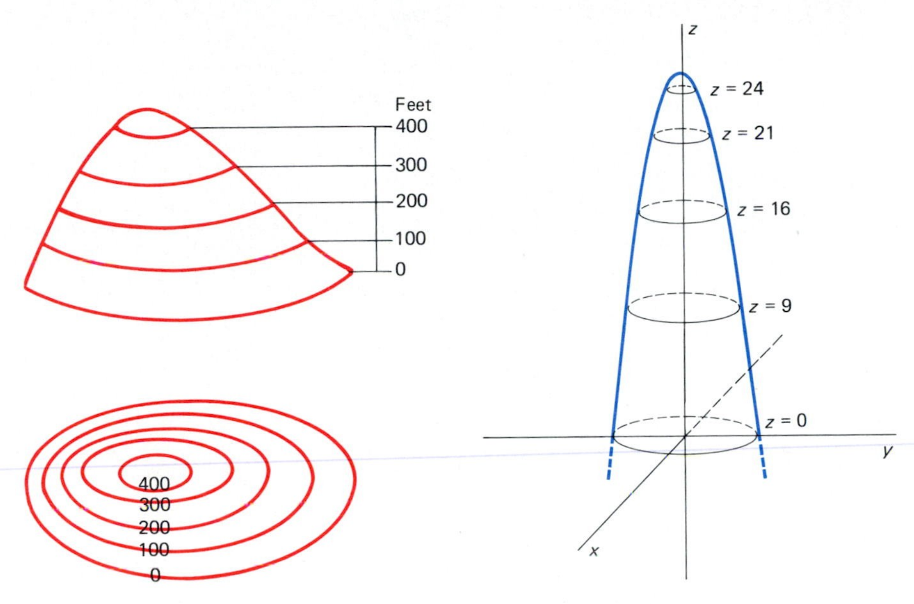

15.8 Contour curves and level curves for a hill

15.9 Contour curves on $z = 25 - x^2 - y^2$

EXAMPLE 2 Figure 15.9 shows some typical contour curves on the paraboloid $z = 25 - x^2 - y^2$. Figure 15.10 shows the corresponding level curves.

EXAMPLE 3 Figure 15.11 shows contour curves on the hyperbolic paraboloid $z = y^2 - x^2$. Figure 15.12 shows the corresponding level curves of the function $f(x, y) = y^2 - x^2$. If $z = k > 0$, then $y^2 - x^2 = k$ is a hyperbola opening along the y-axis, while if $k < 0$, it opens along the x-axis. The level curve for which $k = 0$ consists of the two straight lines $y = x$ and $y = -x$.

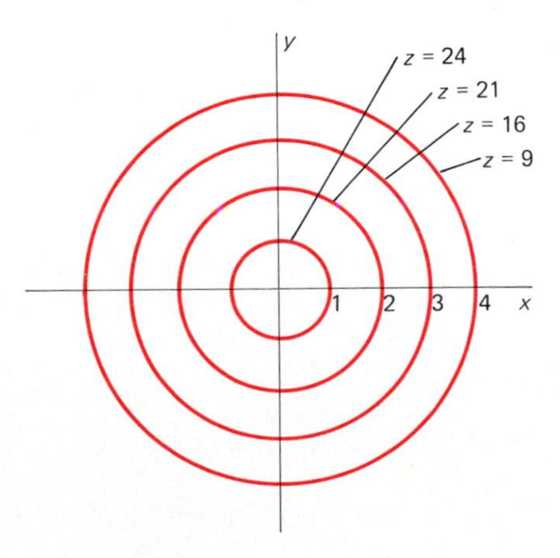

15.10 Level curves of $f(x, y) = 25 - x^2 - y^2$

The graph of a function $f(x, y, z)$ of three variables cannot be drawn in three dimensions, but we can readily visualize its **level surfaces** of the form $f(x, y, z) = k$. For example, the level surfaces of the function $f(x, y, z) = x^2 + y^2 + z^2$ are spheres centered at the origin. Thus the level surfaces of f are the sets in space on which the value $f(x, y, z)$ is constant.

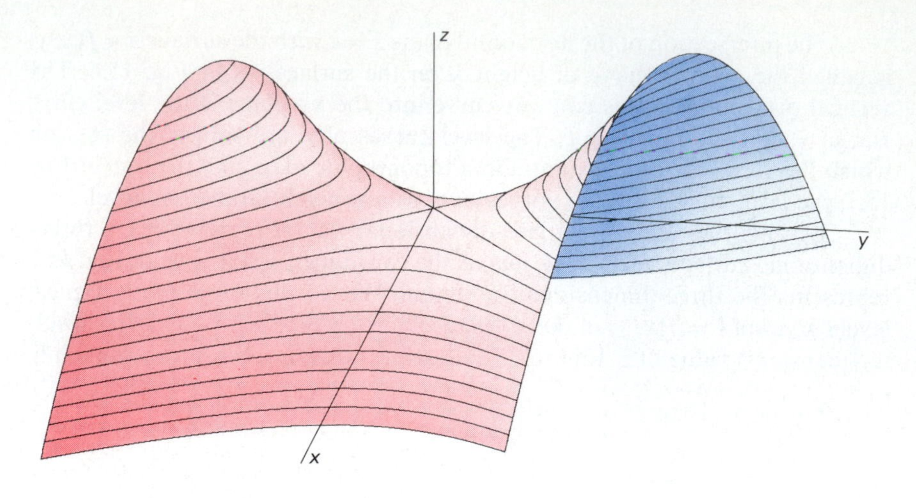

15.11 Contour curves on $z = y^2 - x^2$

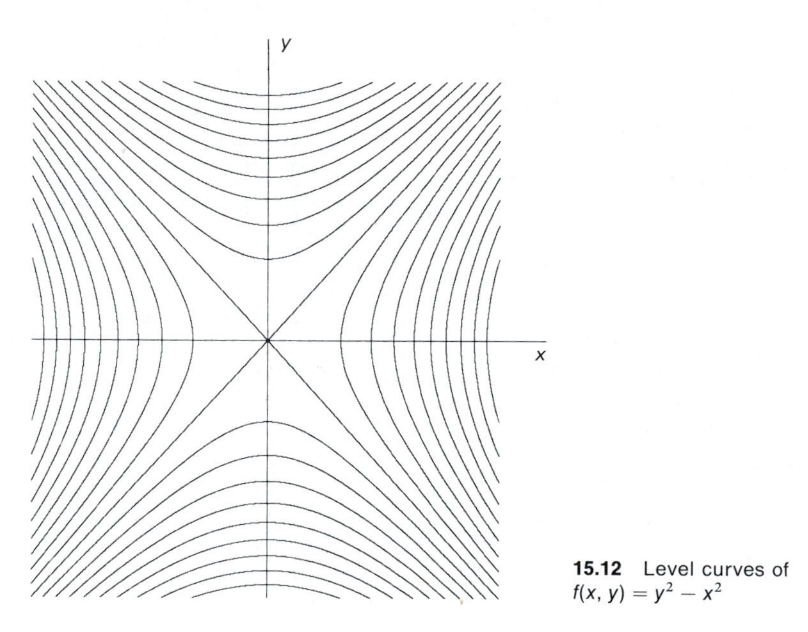

15.12 Level curves of $f(x, y) = y^2 - x^2$

If f is a temperature function, then its level curves or surfaces are called **isotherms**. A weather map often includes level curves of the ground level atmospheric pressure; these are called **isobars**. Even though you may be able to construct the graph of a function of two variables, that graph might be so complicated that information about the function (or the situation it describes) is obscure. Frequently the level curves themselves give more information—as in weather maps. For example, Fig. 15.13 shows level curves for the annual numbers of days of *high* air pollution forecast at different localities in the United States. The scale of this figure does not show local variations caused by individual cities. But a glance indicates that western Colorado, southern Georgia, and central Illinois all expect the same number (10, in this case) of high pollution days each year.

EXAMPLE 4 Figure 15.14 shows some level surfaces of the function

$$f(x, y, z) = x^2 + y^2 - z^2. \tag{2}$$

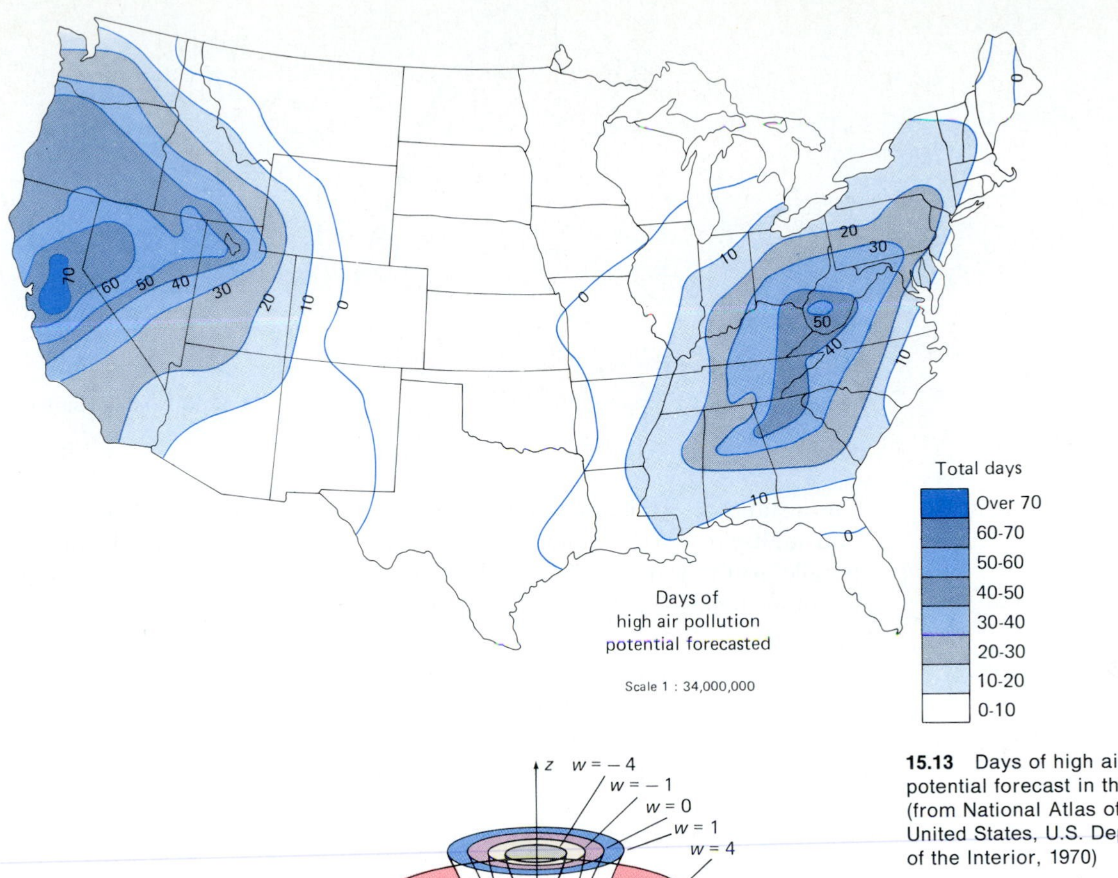

15.13 Days of high air pollution potential forecast in the United States (from National Atlas of the United States, U.S. Department of the Interior, 1970)

15.14 Some level curves of the function $w = f(x, y, z) = x^2 + y^2 - z^2$

If $k > 0$, then the graph of $x^2 + y^2 - z^2 = k$ is a hyperboloid of one sheet, while if $k < 0$, it is a hyperboloid of two sheets. The cone $x^2 + y^2 - z^2 = 0$ lies between these two types of hyperboloids.

EXAMPLE 5 The surface

$$z = \sin\sqrt{x^2 + y^2} \tag{3}$$

is symmetrical with respect to the z-axis, because Equation (3) reduces to the equation $z = \sin r$ (see Fig. 15.15) in terms of the radial coordinate $r = (x^2 + y^2)^{1/2}$ that measures perpendicular distance from the z-axis. The *surface* $z = \sin r$ is generated by revolving the curve $z = \sin x$ around the z-axis. Hence its level curves are circles centered at the origin in the xy-plane.

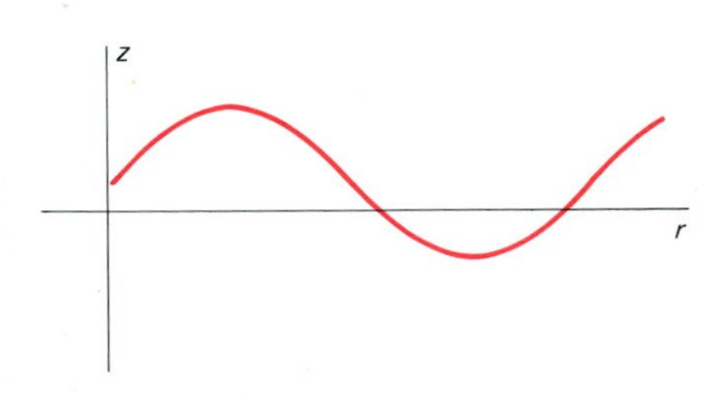

15.15 The curve $z = \sin r$

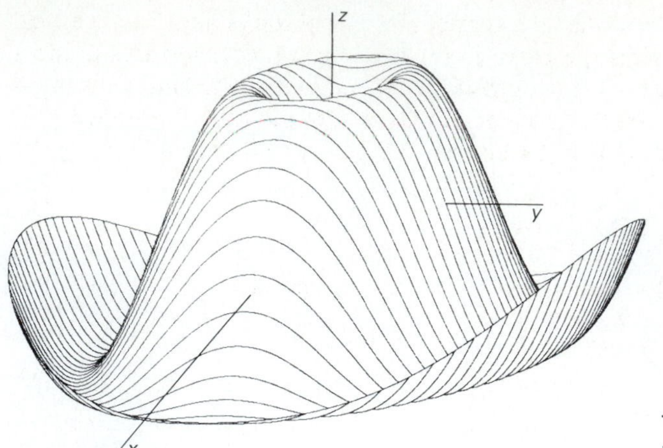

15.16 The hat surface $z = \sin\sqrt{x^2 + y^2}$

For instance, $z = 0$ if r is an integral multiple of π, while $z = \pm 1$ if r is any odd multiple of $\pi/2$. Figure 15.16 shows traces of this surface in planes parallel to the yz-plane. The "hat effect" was achieved by plotting (x, y, z) for those points (x, y) lying within an appropriate ellipse in the xy-plane.

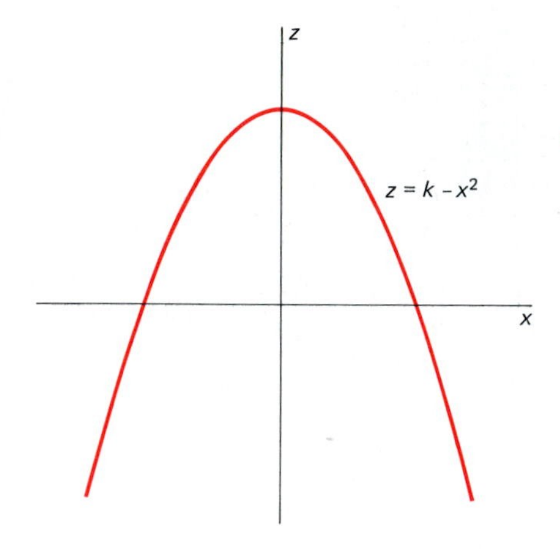

15.17 Intersection of $z = f(x, y)$ and the plane $y = y_0$

Given an arbitrary function $f(x, y)$, it can be quite a challenge to construct a picture of the surface $z = f(x, y)$. The following example illustrates some special techniques that may be useful. Additional surface-sketching techniques will appear in the remainder of this chapter.

EXAMPLE 6 Investigate the graph of the function

$$f(x, y) = \tfrac{3}{4}y^2 + \tfrac{1}{24}y^3 - \tfrac{1}{32}y^4 - x^2. \tag{4}$$

Solution The key feature in (4) is that the right-hand side is the *sum* of a function of x and a function of y. If we set $x = 0$, we get the curve

$$z = \tfrac{3}{4}y^2 + \tfrac{1}{24}y^3 - \tfrac{1}{32}y^4 \tag{5}$$

in which the surface $z = f(x, y)$ intersects the yz-plane. On the other hand, if we set $y = y_0$ in (4), we get

$$z = (\tfrac{3}{4}y_0^2 + \tfrac{1}{24}y_0^3 - \tfrac{1}{32}y_0^4) - x^2;$$

that is,

$$z = k - x^2, \tag{6}$$

which is the equation of a parabola in the xz-plane. Hence the trace of $z = f(x, y)$ in each plane $y = y_0$ is a parabola of the form in (6)—see Fig. 15.17.

We can use the techniques of Section 4.5 to sketch the curve in (5). Calculating the derivative of z with respect to y, we get

$$\frac{dz}{dy} = \frac{3}{2}y + \frac{1}{8}y^2 - \frac{1}{8}y^3 = -\frac{1}{8}y(y^2 - y - 12) = -\frac{1}{8}y(y + 3)(y - 4).$$

Hence the critical points are $y = -3$, $y = 0$, and $y = 4$. The corresponding values of z are $f(0, -3) \approx 3.09$, $f(0, 0) = 0$, and $f(0, 4) \approx 6.67$. Because $z \to -\infty$ as $y \to \pm\infty$, it follows readily that the graph of (5) looks like the one shown in Fig. 15.18.

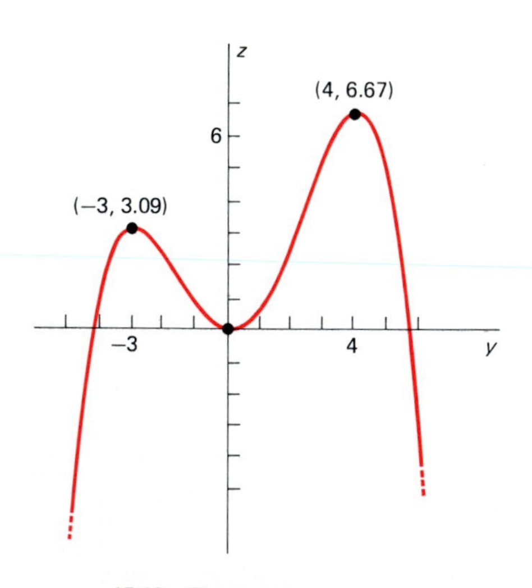

15.18 The curve $z = \frac{3}{4}y^2 + \frac{1}{24}y^3 - \frac{1}{32}y^4$

Now we can see what the surface $z = f(x, y)$ looks like. Each vertical plane $y = y_0$ intersects the curve in (5) at a single point, and this point is the vertex of a parabola that opens downward like the one in (6); this parabola is the intersection of the plane and the surface. Thus the surface $z = f(x, y)$ is generated by translating the vertex of such a parabola along the curve

$$z = \tfrac{3}{4}y^2 + \tfrac{1}{24}y^3 - \tfrac{1}{32}y^4,$$

as indicated in Fig. 15.19.

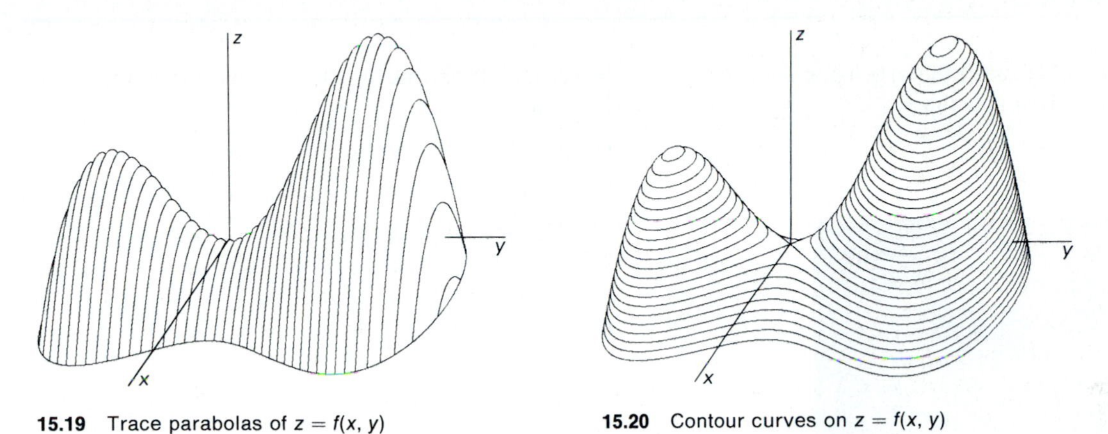

15.19 Trace parabolas of $z = f(x, y)$

15.20 Contour curves on $z = f(x, y)$

Figure 15.20 shows some typical contour curves on this surface. Note that it resembles two peaks separated by a mountain pass. To check this figure, we programmed a microcomputer to plot typical level curves of the function $f(x, y)$. The result is shown in Fig. 15.21. The nested level curves

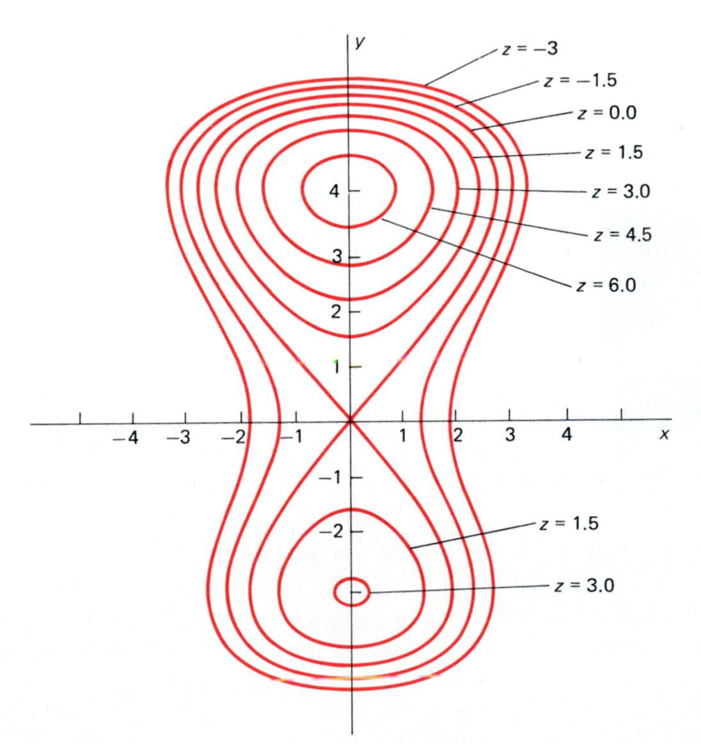

15.21 Level curves of the function $f(x, y) = \frac{3}{4}y^2 + \frac{1}{24}y^3 - \frac{1}{32}y^4 - x^2$

around the points $(0, -3)$ and $(0, 4)$ are indicative of the local maxima of $z = f(x, y)$. The level figure-eight curve through $(0, 0)$ is indicative of the *saddle point* we see in Figs. 15.19 and 15.20. Local extrema and saddle points of functions of two variables are discussed in Sections 15.5 and 15.10.

15.2 PROBLEMS

In each of Problems 1–10, state the largest possible domain of definition of the given function f.

1. $f(x, y) = \exp(-x^2 - y^2)$

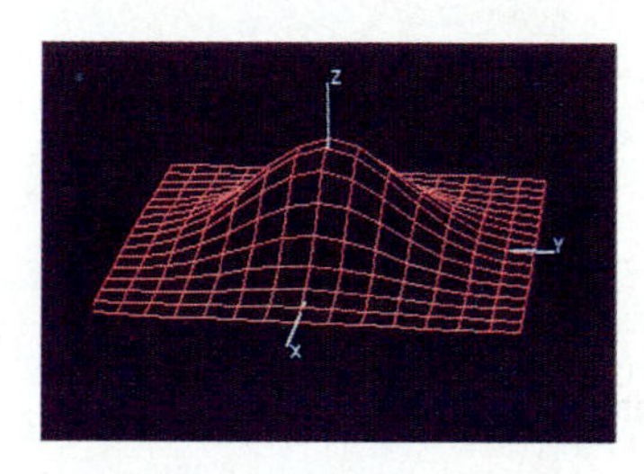

Graph of the function in Problem 1

2. $f(x, y) = \ln(x^2 - y^2 - 1)$

3. $f(x, y) = \dfrac{x + y}{x - y}$

4. $f(x, y) = \sqrt{4 - x^2 - y^2}$

5. $f(x, y) = \dfrac{1 + \sin xy}{xy}$

6. $f(x, y) = \dfrac{1 + \sin xy}{x^2 + y^2}$

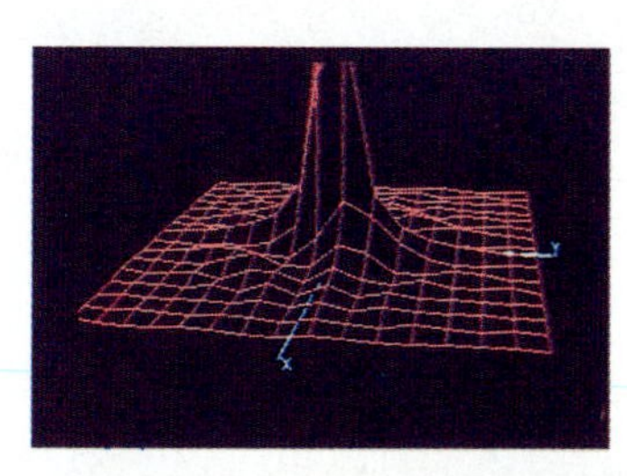

Graph of the function in Problem 6

7. $f(x, y) = \dfrac{xy}{x^2 - y^2}$

8. $f(x, y, z) = \dfrac{1}{\sqrt{z - x^2 - y^2}}$

9. $f(x, y, z) = \exp\left(\dfrac{1}{x^2 + y^2 + z^2}\right)$

10. $f(x, y, z) = \ln(xyz)$

In each of Problems 11–20, describe the graph of the function f.

11. $f(x, y) = 10$

12. $f(x, y) = x$

13. $f(x, y) = x + y$

14. $f(x, y) = \sqrt{x^2 + y^2}$

15. $f(x, y) = x^2 + y^2$

16. $f(x, y) = 4 - x^2 - y^2$

17. $f(x, y) = \sqrt{4 - x^2 - y^2}$

18. $f(x, y) = 16 - y^2$

19. $f(x, y) = 10 - \sqrt{x^2 + y^2}$

20. $f(x, y) = -\sqrt{36 - 4x^2 - 9y^2}$

In each of Problems 21–30, sketch some typical level curves of the function f.

21. $f(x, y) = x - y$

22. $f(x, y) = x^2 - y^2$

23. $f(x, y) = x^2 + 4y^2$

24. $f(x, y) = y - x^2$

25. $f(x, y) = y - x^3$

26. $f(x, y) = y - \cos x$

27. $f(x, y) = x^2 + y^2 - 4x$

28. $f(x, y) = x^2 + y^2 - 6x + 4y + 7$

29. $f(x, y) = \exp(-x^2 - y^2)$

30. $f(x, y) = \dfrac{1}{1 + x^2 + y^2}$

In each of Problems 31–36, describe the level surfaces of the function f.

31. $f(x, y, z) = x^2 + y^2 - z$

32. $f(x, y, z) = z + \sqrt{x^2 + y^2}$

33. $f(x, y, z) = x^2 + y^2 + z^2 - 4x - 2y - 6z$

34. $f(x, y, z) = z^2 - x^2 - y^2$

35. $f(x, y, z) = x^2 + 4y^2 - 4x - 8y + 17$

36. $f(x, y, z) = x^2 + z^2 + 25$

In each of Problems 37–40, the function $f(x, y)$ is the sum of a function of x and a function of y. Use the method of Example 6 to construct a sketch of the surface $z = f(x, y)$.

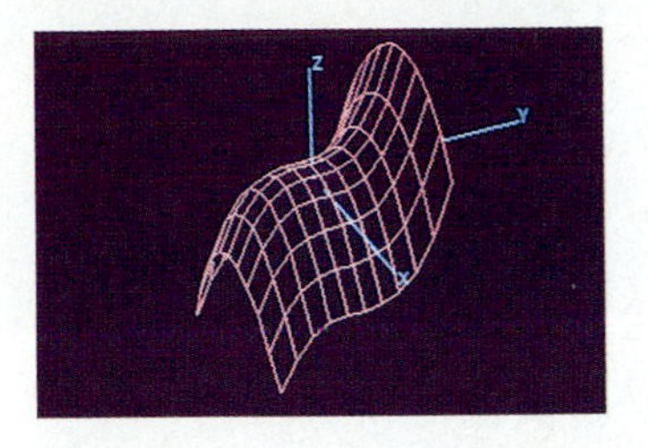

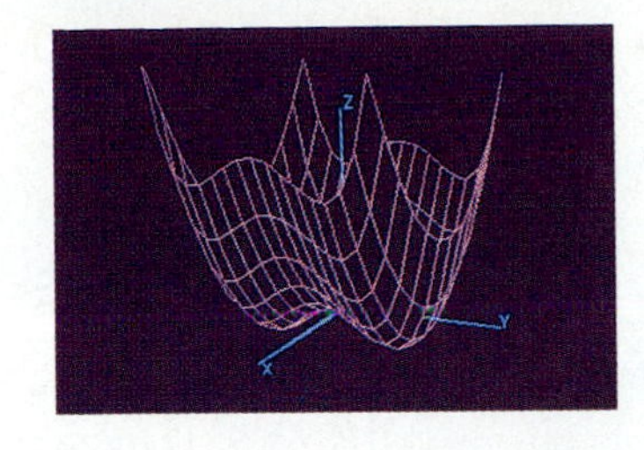

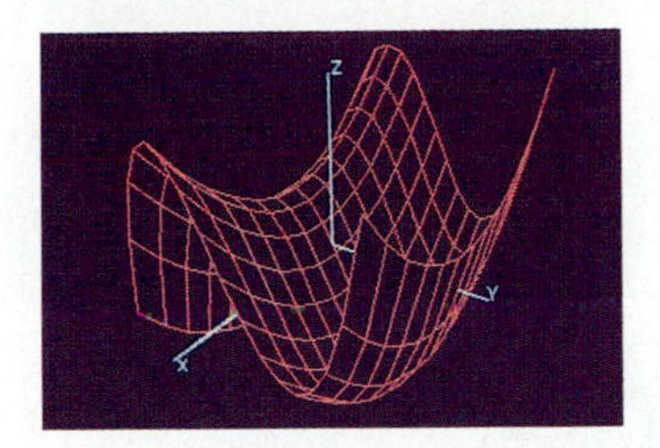

Figures for some of Problems 37–40

37. $f(x, y) = y^3 - x^2$

38. $f(x, y) = y^4 + x^2$

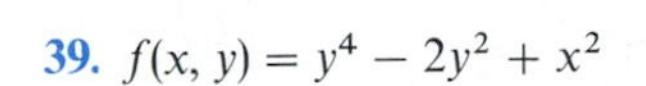

39. $f(x, y) = y^4 - 2y^2 + x^2$

40. $f(x, y) = 2y^3 - 3y^2 - 12y + x^2$

41. Figures 15.22 through 15.27 show the graphs of six functions $z = f(x, y)$. Figures 15.28 through 15.33 show level curves of the same six functions, but not in the same order. The level curves in each figure correspond to contours at equally spaced heights on the surface $z = f(x, y)$. Match each surface with its level curves.

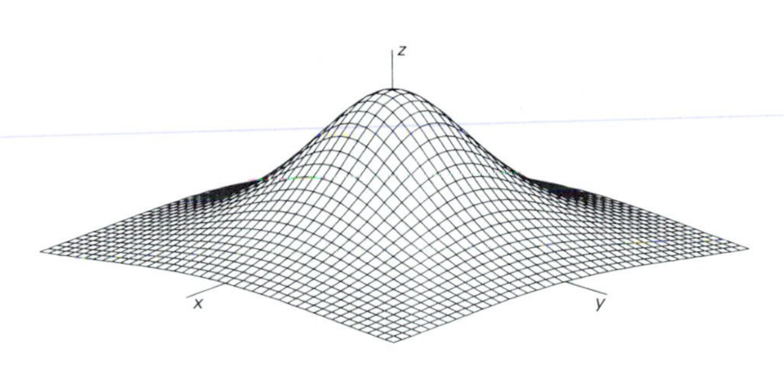

15.22 $z = \dfrac{1}{1 + x^2 + y^2}$, $|x| \leqq 2$, $|y| \leqq 2$

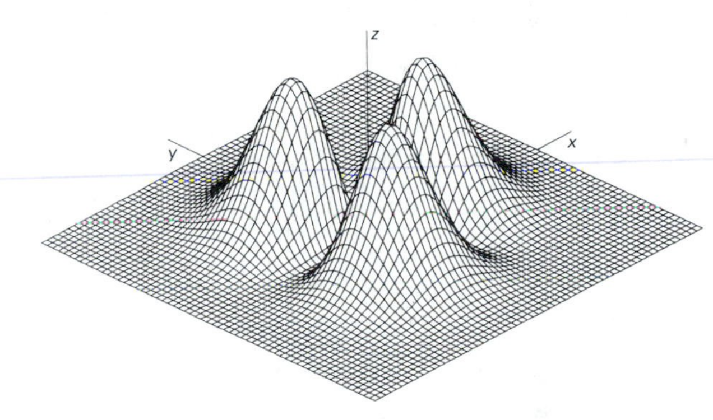

15.24 $z = r^2 \exp(-r^2) \cos^2(3\theta/2)$, $|x| \leqq 3$, $|y| \leqq 3$

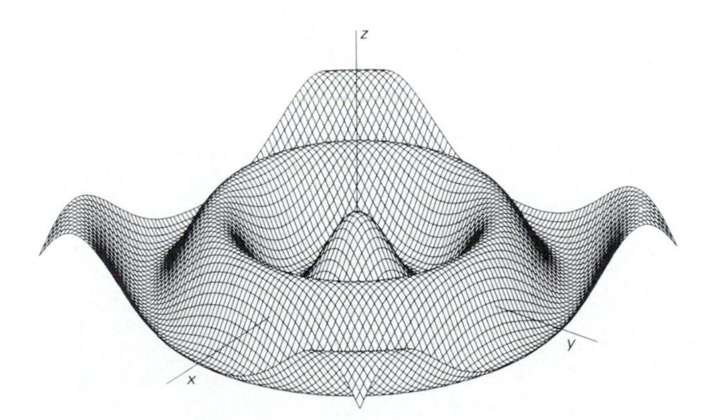

15.23 $z = \cos\sqrt{x^2 + y^2}$, $|x| \leqq 10$, $|y| \leqq 10$

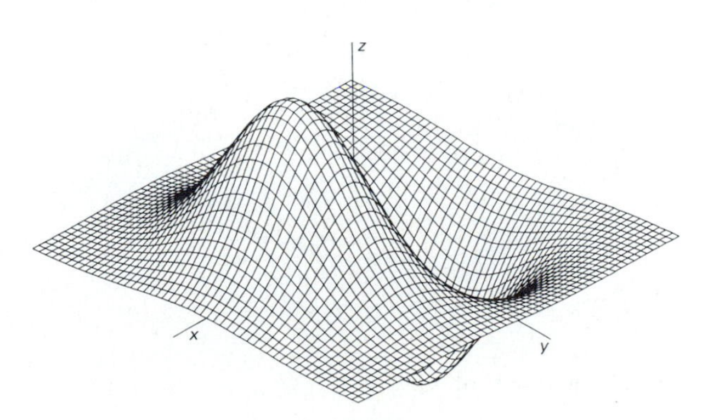

15.25 $z = x \exp(-x^2 - y^2)$, $|x| \leqq 2$, $|y| \leqq 2$

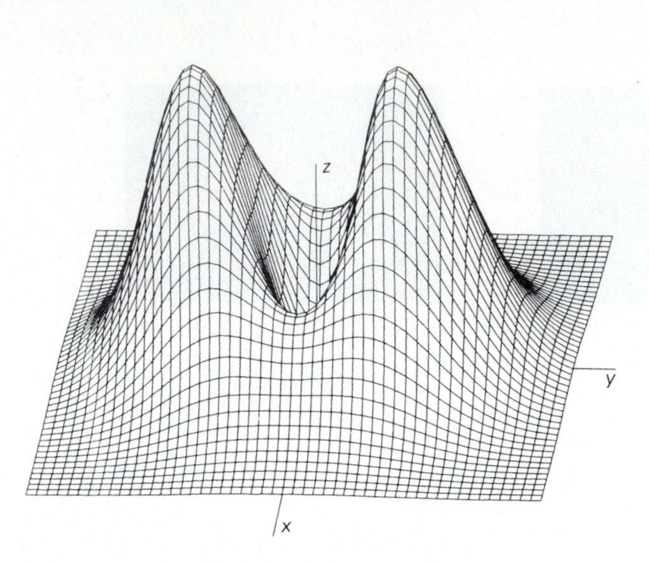

15.26 $z = 3(x^2 + 3y^2)\exp(-x^2 - y^2)$, $|x| \leqq 2.5$, $|y| \leqq 2.5$

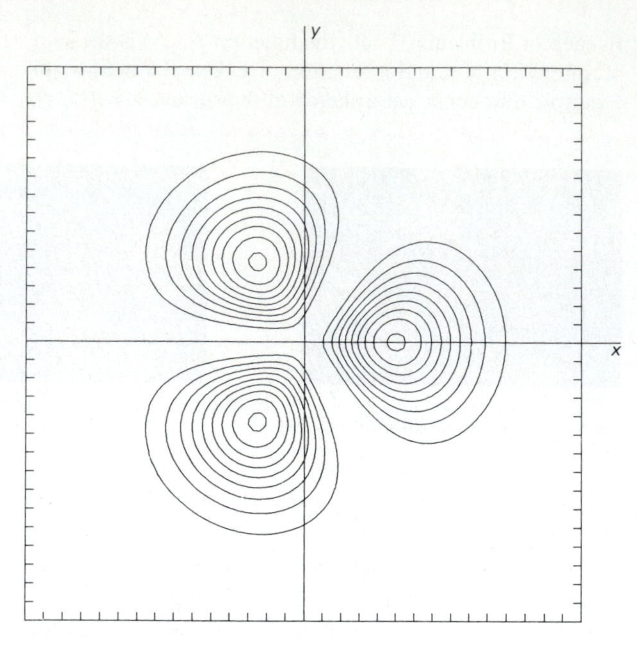

15.28

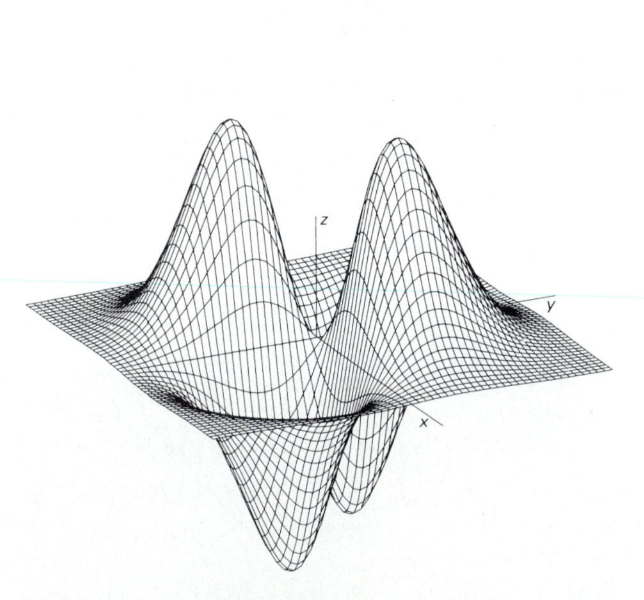

15.27 $z = xy\exp(-\frac{1}{2}(x^2 + y^2))$, $|x| \leqq 3.5$, $|y| \leqq 3.5$

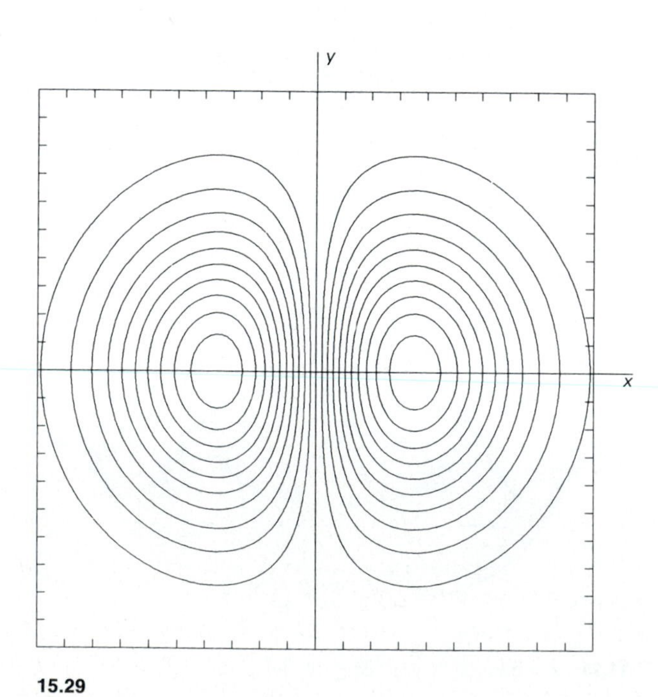

15.29

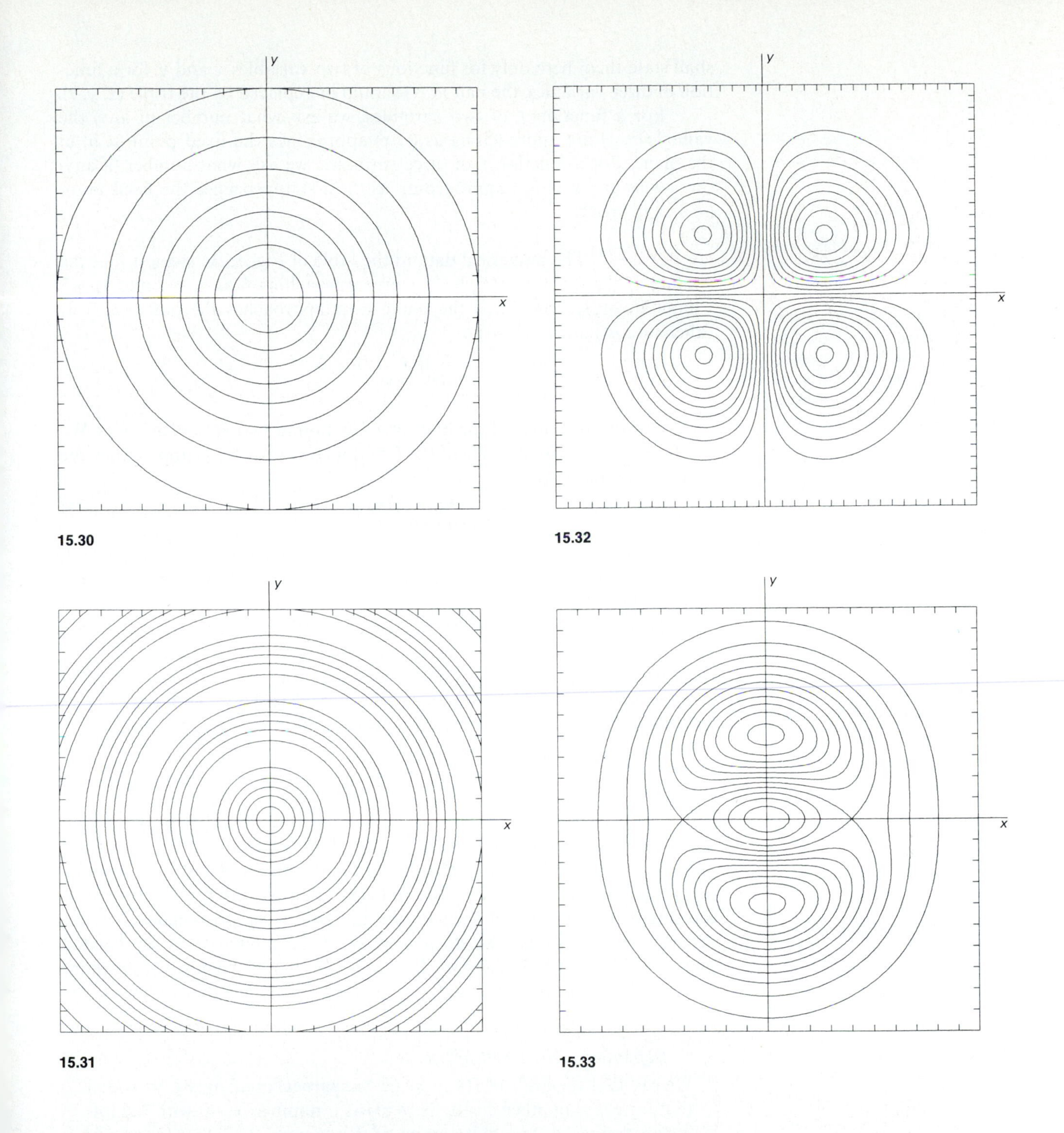

15.30

15.32

15.31

15.33

15.3 Limits and Continuity

We need limits of functions of several variables for the same reasons that we needed limits of functions of a single variable—so that we can discuss slopes and rates of change. Both the definition and the basic properties of limits of functions of several variables are essentially the same as those we stated in Section 2.2 for functions of a single variable. For simplicity, we shall state them here only for functions of two variables x and y; for a function of three variables, the pair (x, y) should be replaced by the triple (x, y, z).

For a function f of two variables, we ask what number (if any) the values $f(x, y)$ are approaching as (x, y) approaches the fixed point (a, b) in the plane. For a function f of three variables, we ask what number (if any) the values $f(x, y, z)$ are approaching as (x, y, z) approaches the fixed point (a, b, c) in space.

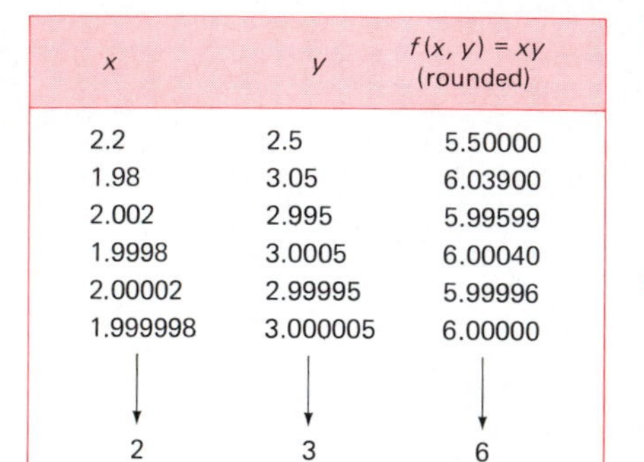

x	y	$f(x, y) = xy$ (rounded)
2.2	2.5	5.50000
1.98	3.05	6.03900
2.002	2.995	5.99599
1.9998	3.0005	6.00040
2.00002	2.99995	5.99996
1.999998	3.000005	6.00000
↓	↓	↓
2	3	6

15.34 The numerical data of Example 1

EXAMPLE 1 The numerical data in the table of Fig. 15.34 suggest that the values of the function $f(x, y) = xy$ are approaching 6 as $x \to 2$ and $y \to 3$ simultaneously; that is, as the point (x, y) approaches the point $(2, 3)$. It therefore is natural to write

$$\lim_{(x,y)\to(2,3)} xy = 6.$$

Our intuitive idea of the limit of a function of two variables is this. We say the number L is the *limit* of the function $f(x, y)$ as (x, y) approaches the point (a, b) and write

$$\lim_{(x,y)\to(a,b)} f(x, y) = L, \tag{1}$$

provided that the number $f(x, y)$ can be made as close as we please to L if the point (x, y) is chosen sufficiently close to—though not equal to—the point (a, b).

To make this idea precise, we must specify how close to L—within $\varepsilon > 0$, say—we want $f(x, y)$ to be and then how close to (a, b) the point (x, y) must be to accomplish this. We think of the point (x, y) as being close to (a, b) provided that it lies within a small square (Fig. 15.35) with center (a, b) and edge length 2δ, where δ is a small positive number. The point (x, y) lies within this square if and only if both

$$|x - a| < \delta \quad \text{and} \quad |y - b| < \delta. \tag{2}$$

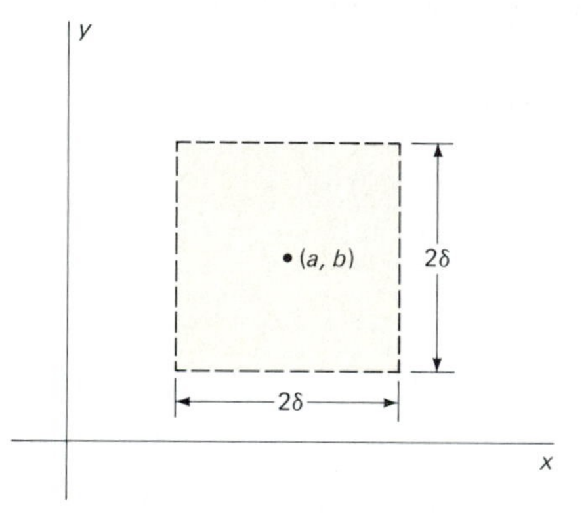

15.35 The square $|x - a| < \delta$, $|y - b| < \delta$

This observation serves as motivation for the formal definition, with two additional conditions. First, we define the limit of $f(x, y)$ as $(x, y) \to (a, b)$ *only* under the condition that the domain of definition of f contains points $(x, y) \neq (a, b)$ lying arbitrarily close to (a, b)—that is, within *every* square of the sort in Fig. 15.35 and thus within any and every preassigned positive distance of (a, b). Hence we do not speak of the limit of f at an isolated point of its domain D. Finally, we do *not* require that f be defined at the point (a, b) itself. Hence we deliberately exclude the possibility that $(x, y) = (a, b)$.

Definition *The Limit of* $f(x, y)$

We say that the **limit of** $f(x, y)$ **as** (x, y) **approaches** (a, b) is L provided that for every number $\varepsilon > 0$, there exists a number $\delta > 0$ with the following property: If (x, y) is a point of the domain of f other than (a, b) such that both

$$|x - a| < \delta \quad \text{and} \quad |y - b| < \delta, \tag{2}$$

then it follows that

$$|f(x, y) - L| < \varepsilon. \tag{3}$$

We ordinarily shall rely upon continuity rather than the formal definition of the limit to evaluate limits of functions of several variables. We say that f is **continuous at the point** (a, b) provided that $f(a, b)$ exists and $f(x, y)$ approaches $f(a, b)$ as (x, y) approaches (a, b). That is,

$$\lim_{(x,y)\to(a,b)} f(x, y) = f(a, b).$$

Thus f is continuous at (a, b) if it is defined there and its limit there is equal to its value there, exactly as in the case of a function of a single variable. The function f is **continuous on the set** D if it is continuous at each point of D.

EXAMPLE 2 Let D be the circular disk consisting of the points (x, y) such that $x^2 + y^2 \leqq 1$, and let $f(x, y) = 1$ at each point of D. Then the limit of $f(x, y)$ at each point of D is obviously 1, so f is continuous on D. But let the new function $g(x, y)$ be defined on the entire plane $\mathbf{R}^2$ as follows:

$$g(x, y) = \begin{cases} f(x, y) & \text{if } (x, y) \text{ is in } D; \\ 0 & \text{otherwise.} \end{cases}$$

(See Fig. 15.36.) Then g is *not* continuous on $\mathbf{R}^2$. For instance, the limit of $g(x, y)$ as $(x, y) \to (1, 0)$ does not exist because there exist both points within D arbitrarily close to $(1, 0)$ at which g has value 1, and points outside of D arbitrarily close to $(1, 0)$ at which g has value 0. Thus $g(x, y)$ cannot approach any single value as $(x, y) \to (1, 0)$. Because g has no limit at $(1, 0)$, it cannot be continuous there.

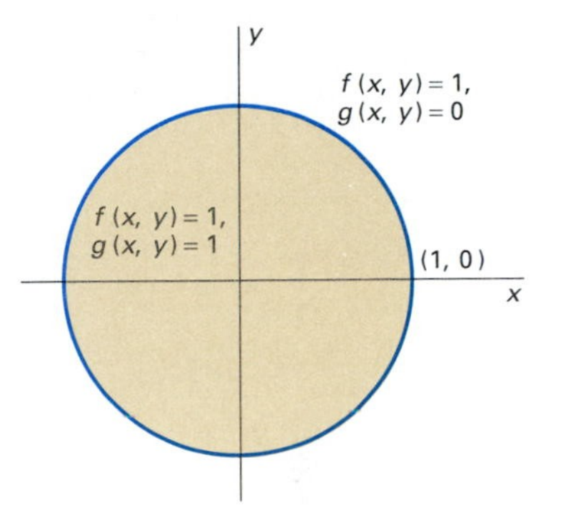

15.36 The circular disk of Example 2

The limit laws of Section 2.2 have natural analogues for functions of several variables. If

$$\lim_{(x,y)\to(a,b)} f(x, y) = L \quad \text{and} \quad \lim_{(x,y)\to(a,b)} g(x, y) = M, \tag{4}$$

then the sum, product, and quotient laws for limits are these:

$$\lim_{(x,y)\to(a,b)} [f(x, y) + g(x, y)] = L + M, \tag{5}$$

$$\lim_{(x,y)\to(a,b)} [f(x, y)g(x, y)] = LM, \quad \text{and} \tag{6}$$

$$\lim_{(x,y)\to(a,b)} \frac{f(x, y)}{g(x, y)} = \frac{L}{M} \quad \text{if} \quad M \neq 0. \tag{7}$$

EXAMPLE 3 Show that $\lim_{(x,y)\to(a,b)} xy = ab.$

Solution We take $f(x, y) = x$ and $g(x, y) = y$. Then it follows from the definition of limit that

$$\lim_{(x,y)\to(a,b)} f(x, y) = a \quad \text{and} \quad \lim_{(x,y)\to(a,b)} g(x, y) = b.$$

Hence the product law gives

$$\begin{aligned} \lim_{(x,y)\to(a,b)} xy &= \lim_{(x,y)\to(a,b)} f(x, y)g(x, y) \\ &= \left[\lim_{(x,y)\to(a,b)} f(x, y)\right]\left[\lim_{(x,y)\to(a,b)} g(x, y)\right] = ab. \end{aligned}$$

More generally, suppose that $P(x, y)$ is a polynomial in the two variables x and y, so that it can be written in the form

$$P(x, y) = \sum c_{ij}x^i y^j.$$

Then the sum and product laws imply that

$$\lim_{(x,y)\to(a,b)} P(x, y) = P(a, b).$$

An immediate but important consequence is that every polynomial in two (or more) variables is a continuous function.

Just as in the single-variable case, any composition of continuous multivariable functions is also a continuous function. For instance, suppose that the functions f and g are each continuous at (a, b) and that h is continuous at the point $(f(a, b), g(a, b))$. Then the composite function

$$H(x, y) = h(f(x, y), g(x, y))$$

is also continuous at (a, b). As a consequence, any finite combination involving sums, products, quotients, and compositions of the familiar elementary functions is continuous, except possibly at points where a denominator is zero or where the formula for the function is otherwise meaningless. This general rule suffices for the evaluation of most limits that we shall encounter.

EXAMPLE 4 By application of the limit laws we get

$$\lim_{(x,y)\to(1,2)} \left[e^{xy}\sin\frac{\pi y}{4} + xy\ln\sqrt{y-x}\right]$$

$$= \lim_{(x,y)\to(1,2)} e^{xy}\sin\frac{\pi y}{4} + \lim_{(x,y)\to(1,2)} xy\ln\sqrt{y-x}$$

$$= \left(\lim_{(x,y)\to(1,2)} e^{xy}\right)\left(\lim_{(x,y)\to(1,2)} \sin\frac{\pi y}{4}\right) + \left(\lim_{(x,y)\to(1,2)} xy\right)\left(\lim_{(x,y)\to(1,2)} \ln\sqrt{y-x}\right)$$

$$= (e^2)(1) + (2)(\ln 1) = e^2.$$

The next two examples illustrate techniques that sometimes are successful in handling cases with denominators that approach zero.

EXAMPLE 5 Show that $\displaystyle\lim_{(x,y)\to(0,0)} \frac{xy}{\sqrt{x^2+y^2}} = 0.$

Solution Let (r, θ) be the polar coordinates of the point (x, y). Then $x = r\cos\theta$ and $y = r\sin\theta$, so

$$\frac{xy}{\sqrt{x^2+y^2}} = \frac{(r\cos\theta)(r\sin\theta)}{\sqrt{r^2\cos^2\theta + r^2\sin^2\theta}}$$

$$= r\cos\theta\sin\theta \qquad \text{for} \quad r > 0.$$

Because $r = \sqrt{x^2 + y^2}$, it is clear that $r \to 0$ as x and y both approach zero. It therefore follows that

$$\lim_{(x,y)\to(0,0)} \frac{xy}{\sqrt{x^2 + y^2}} = \lim_{r\to 0} r \cos\theta \sin\theta = 0,$$

because $|\cos\theta \sin\theta| \leqq 1$ for all values of θ.

EXAMPLE 6 Show that

$$\lim_{(x,y)\to(0,0)} \frac{xy}{x^2 + y^2}$$

does not exist.

Solution Our plan is to show that $f(x, y) = xy/(x^2 + y^2)$ approaches different values as (x, y) approaches $(0, 0)$ from different directions. Suppose that (x, y) approaches $(0, 0)$ along the straight line of slope m through the origin. On this line we have $y = mx$. So on this line,

$$f(x, y) = \frac{x(mx)}{x^2 + m^2x^2} = \frac{m}{1 + m^2}$$

if $x \neq 0$. If we take $m = 1$, we see that $f(x, y) = \frac{1}{2}$ at every point of the line $y = x$ other than $(0, 0)$. If we take $m = -1$, we see that $f(x, y) = -\frac{1}{2}$ at every point of the line $y = -x$ other than $(0,0)$. Thus $f(x, y)$ approaches two different values as (x, y) approaches $(0, 0)$ along these two lines, shown in Fig. 15.37. Hence $f(x, y)$ cannot approach any *single* value as (x, y) approaches $(0, 0)$, and this implies that the limit in question cannot exist.

Figure 15.38 shows a computer-generated graph of the function $f(x, y) = xy/(x^2 + y^2)$. It consists of linear rays along each of which the polar angular coordinate θ is constant. For each number z between $-\frac{1}{2}$ and $\frac{1}{2}$ (inclusive), there are rays along which $f(x, y)$ has the constant value z. Hence we can make $f(x, y)$ approach any number we please in $[-\frac{1}{2}, \frac{1}{2}]$ by letting (x, y) approach $(0, 0)$ from the appropriate direction.

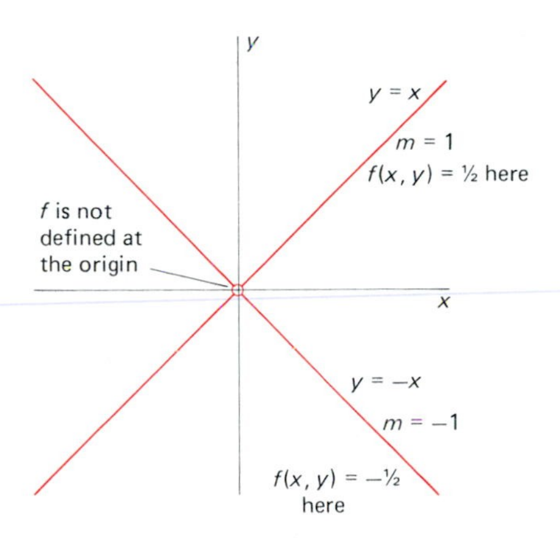

15.37 The function f of Example 6 takes on both values $+\frac{1}{2}$ and $-\frac{1}{2}$ at points arbitrarily close to the origin.

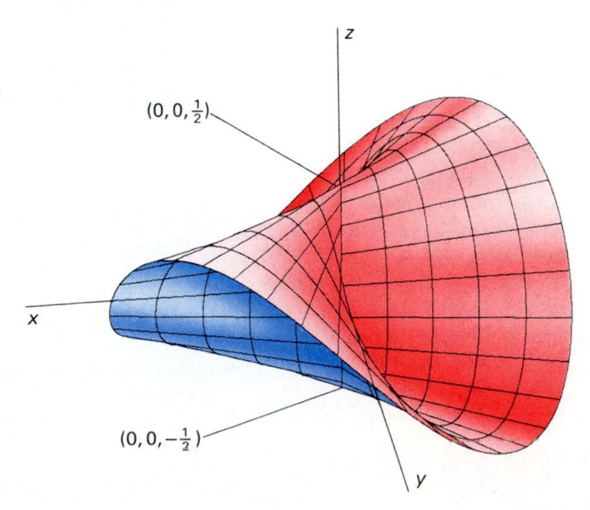

15.38 The graph of $f(x, y) = xy/(x^2 + y^2)$

In order for

$$\lim_{(x,y)\to(a,b)} f(x, y)$$

to exist, $f(x, y)$ must approach L for *any and every* mode of approach of (x, y) to (a, b). In Problem 27 we give an example of a function f such that $f(x, y) \to 0$ as $(x, y) \to (0, 0)$ along any straight line through the origin, but $f(x, y) \to 1$ as (x, y) approaches the origin along the parabola $y = x^2$. Fortunately, many important applications, including those we discuss in the remainder of this chapter, involve only functions that exhibit no such exotic behavior.

15.3 PROBLEMS

Use the limit laws and consequences of continuity to evaluate the limits in Problems 1–15.

1. $\lim_{(x,y)\to(0,0)} (7 - x^2 + 5xy)$

2. $\lim_{(x,y)\to(1,-2)} (3x^2 - 4xy + 5y^2)$

3. $\lim_{(x,y)\to(1,-1)} e^{-xy}$

4. $\lim_{(x,y)\to(0,0)} \dfrac{x + y}{1 + xy}$

5. $\lim_{(x,y)\to(0,0)} \dfrac{5 - x^2}{3 + x + y}$

6. $\lim_{(x,y)\to(2,3)} \dfrac{9 - x^2}{1 + xy}$

7. $\lim_{(x,y)\to(0,0)} \ln\sqrt{1 - x^2 - y^2}$

8. $\lim_{(x,y)\to(2,-1)} \ln\dfrac{1 + x + 2y}{3y^2 - x}$

9. $\lim_{(x,y)\to(0,0)} \dfrac{e^{xy}\sin xy}{xy}$

10. $\lim_{(x,y)\to(0,0)} \exp\left(-\dfrac{1}{x^2 + y^2}\right)$

11. $\lim_{(x,y,z)\to(1,1,1)} \dfrac{x^2 + y^2 + z^2}{1 - x - y - z}$

12. $\lim_{(x,y,z)\to(1,1,1)} (x + y + z) \ln xyz$

13. $\lim_{(x,y,z)\to(1,1,0)} \dfrac{xy - z}{\cos xyz}$

14. $\lim_{(x,y,z)\to(2,-1,3)} \dfrac{x + y + z}{x^2 + y^2 + z^2}$

15. $\lim_{(x,y,z)\to(2,8,1)} \sqrt{xy}\tan\dfrac{3\pi z}{4}$

In each of Problems 16—20, evaluate the limits

$$\lim_{h\to 0} \frac{f(x + h, y) - f(x, y)}{h}$$

and

$$\lim_{k\to 0} \frac{f(x, y + k) - f(x, y)}{k}.$$

16. $f(x, y) = x + y$

17. $f(x, y) = xy$

18. $f(x, y) = x^2 + y^2$

19. $f(x, y) = xy^2 - 2$

20. $f(x, y) = x^2y^3 - 10$

In Problems 21–23, use the method of Example 5 to verify the given limit.

21. $\lim_{(x,y)\to(0,0)} \dfrac{x^2 - y^2}{\sqrt{x^2 + y^2}} = 0$

22. $\lim_{(x,y)\to(0,0)} \dfrac{x^3 - y^3}{x^2 + y^2} = 0$

23. $\lim_{(x,y)\to(0,0)} \dfrac{x^4 + y^4}{(x^2 + y^2)^{3/2}} = 0$

24. Use the method of Example 6 to show that

$$\lim_{(x,y)\to(0,0)} \frac{x^2 - y^2}{x^2 + y^2}$$

does not exist.

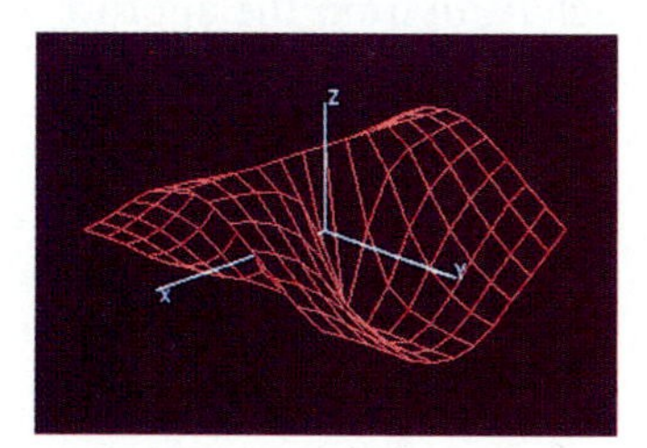

Graph of the function in Problem 24

25. Substitute spherical coordinates $x = \rho \sin\phi \cos\theta$, $y = \rho \sin\phi \sin\theta$, $z = \rho\cos\phi$ to show that

$$\lim_{(x,y,z)\to(0,0,0)} \frac{xyz}{x^2 + y^2 + z^2} = 0.$$

26. Determine whether or not

$$\lim_{(x,y,z)\to(0,0,0)} \frac{xy + xz + yz}{x^2 + y^2 + z^2}$$

exists.

27. Let $f(x, y) = 2x^2y/(x^4 + y^2)$. Show that

(a) $f(x, y) \to 0$ as $(x, y) \to (0, 0)$ along *any* (every) straight line through the origin, but that

(b) $f(x, y) \to 1$ as $(x, y) \to (0, 0)$ along the parabola $y = x^2$.

Hence conclude that the limit of $f(x, y)$ as $(x, y) \to (0, 0)$ does not exist.

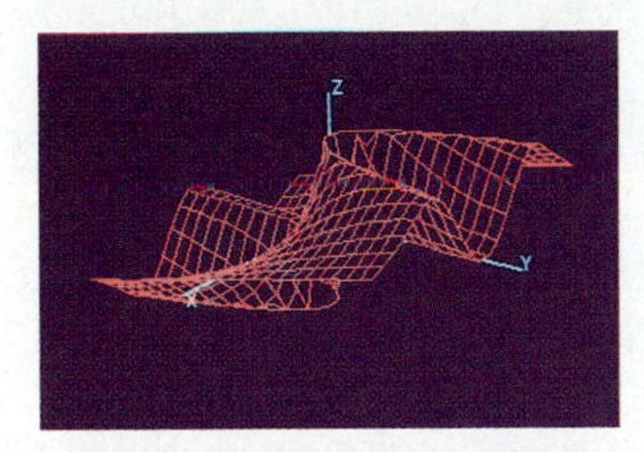

Two views of the graph of the function of Problem 27

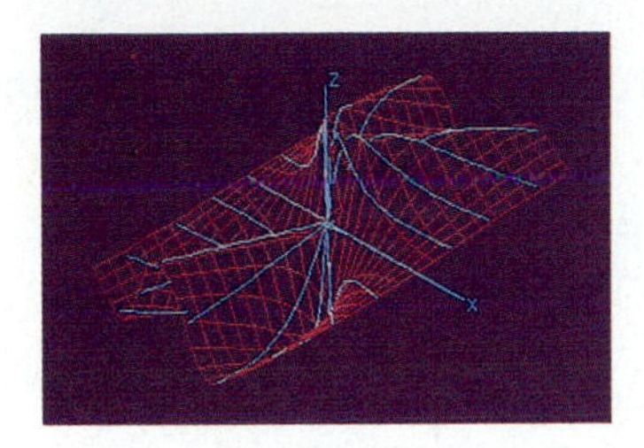

28. Suppose that $f(x, y) = (x - y)/(x^3 - y)$ except at points of the curve $y = x^3$; we *define* $f(x, y)$ to be 1 at those points. Show that f is not continuous at the point $(1, 1)$. Evaluate the limits of $f(x, y)$ as $(x, y) \to (1, 1)$ along the vertical line $x = 1$ and along the horizontal line $y = 1$. (*Suggestion:* Recall that $a^3 - b^3 = (a - b)(a^2 + ab + b^2)$.)

29. Locate and identify the extrema (local or global, maximum or minimum) of the function $f(x, y) = x^2 - x + y^2 + 2y + 1$. (NOTE The point of this problem is that you do *not* need calculus to work it.)

30. Sketch enough level curves of the function $h(x, y) = y - x^2$ to show that the function h has *no* extreme values—no local or global maxima or minima.

15.4 Partial Derivatives

Suppose that $y = f(x)$ is a function of *one* real variable. Its first derivative

$$\frac{dy}{dx} = D_x f(x) = \lim_{h \to 0} \frac{f(x + h) - f(x)}{h} \tag{1}$$

can be interpreted as the instantaneous rate of change of y with respect to x. For a function $z = f(x, y)$ of two variables, we need a similar understanding of the rate at which z changes as x and y vary (either singly or simultaneously). To reach this more complicated concept, we adopt a "divide and conquer" strategy.

First we hold y fixed and let x vary. The rate of change of z with respect to x is then denoted by $\partial z/\partial x$ and has the value

$$\frac{\partial z}{\partial x} = \lim_{h \to 0} \frac{f(x + h, y) - f(x, y)}{h}. \tag{2}$$

The value of the limit—if it exists—is called the **partial derivative of f with respect to x**. In like manner, we may hold x fixed and let y vary. The rate of change of z with respect to y is then the **partial derivative of f with respect to y**, defined to be

$$\frac{\partial z}{\partial y} = \lim_{k \to 0} \frac{f(x, y + k) - f(x, y)}{k} \tag{3}$$

for all (x, y) for which this limit exists. Note the symbol ∂ that is used instead of d to denote the partial derivatives of a function of two variables. A function of three or more independent variables has a partial derivative (defined similarly) with respect to each of its independent variables. Some other common notations for partial derivatives are

$$\frac{\partial z}{\partial x} = \frac{\partial f}{\partial x} = f_x(x, y) = D_x f(x, y) = D_1 f(x, y), \tag{4}$$

$$\frac{\partial z}{\partial y} = \frac{\partial f}{\partial y} = f_y(x, y) = D_y f(x, y) = D_2 f(x, y). \tag{5}$$

Note that if the symbol y in Equation (2) is deleted, the result is the limit in Equation (1). This means that $\partial z/\partial x$ can be calculated as an "ordinary" derivative with respect to x, simply by regarding y as a constant during the process of differentiation. Similarly, we can compute $\partial z/\partial y$ as an ordinary derivative, thinking of y as the *only* variable and treating x as a constant during the computation.

EXAMPLE 1 Compute the partial derivatives $\partial f/\partial x$ and $\partial f/\partial y$ of the function $f(x, y) = x^2 + 2xy^2 - y^3$.

Solution To compute the partial of f with respect to x, we regard y as a constant. Then we differentiate normally and find that

$$\frac{\partial f}{\partial x} = 2x + 2y^2.$$

When we regard x as a constant and differentiate with respect to y, we find that

$$\frac{\partial f}{\partial y} = 4xy - 3y^2.$$

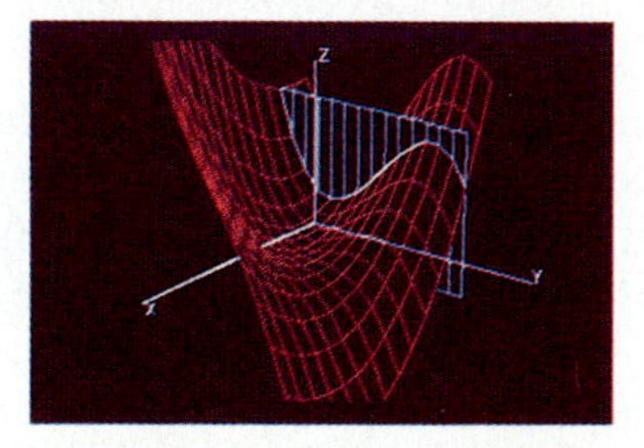

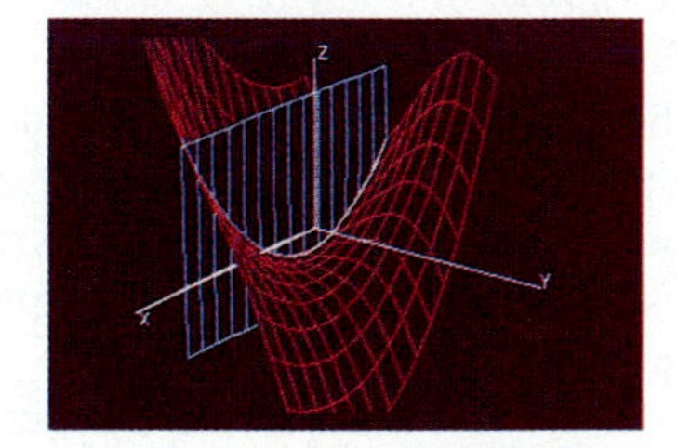

Vertical cross-sections of a graph similar to the one of Example 1

To get an intuitive feel for the meaning of partial derivatives, we can think of $f(x, y)$ as the temperature at the point (x, y) of the plane. Then $f_x(x, y)$ is the instantaneous rate of change of temperature at (x, y) per unit increase in x (with y held constant), while $f_y(x, y)$ is the rate of change of temperature per unit increase in y (with x held constant). For example, with the temperature function $f(x, y) = x^2 + 2xy^2 - y^3$ of Example 1, the rate of change of temperature at the point $(1, -1)$ is $+4°$ per unit distance in the positive x-direction and $-7°$ per unit distance in the y-direction.

EXAMPLE 2 Find $\partial z/\partial x$ and $\partial z/\partial y$ if $z = (x^2 + y^2)e^{-xy}$.

Solution Because $\partial z/\partial x$ is calculated as if it were an ordinary derivative with respect to x while y is held constant, we use the product rule. This gives

$$\frac{\partial z}{\partial x} = (2x)(e^{-xy}) + (x^2 + y^2)(-ye^{-xy}) = (2x - x^2y - y^3)e^{-xy}.$$

Because x and y appear symmetrically in the expression $z = (x^2 + y^2)e^{-xy}$, we get $\partial z/\partial y$ when we interchange x and y in the expression for $\partial z/\partial x$:

$$\frac{\partial z}{\partial y} = (2y - xy^2 - x^3)e^{-xy}.$$

You should check this result by actually differentiating z with respect to y.

EXAMPLE 3 The volume V (in cubic centimeters) of 1 mole of an ideal gas is given by

$$V = \frac{(82.06)T}{P},$$

where P is the pressure (in atmospheres) and T is the absolute temperature (in kelvin (K), where K = °C + 273). Find the rates of change of the volume of 1 mole of an ideal gas with respect to pressure and with respect to temperature with $T = 300$ K and $P = 5$ atm.

Solution The partial derivatives of V with respect to its two variables are

$$\frac{\partial V}{\partial P} = -\frac{(82.06)T}{P^2} \quad \text{and} \quad \frac{\partial V}{\partial T} = \frac{82.06}{P}.$$

With $T = 300$ K and $P = 5$ atm, we have the two values $\partial V/\partial P = -984.72\ \text{cm}^3/\text{atm}$ and $\partial V/\partial T = 16.41\ \text{cm}^3/\text{K}$. These partial derivatives allow us to estimate the effect of a change in temperature or in pressure on the volume V of gas in question, as follows. We are given $T = 300$ K and $P = 5$ atm, so the volume of gas we are dealing with is

$$V = \frac{(82.06)(300)}{5} = 4923.60\ \text{cm}^3.$$

We would expect an increase in pressure of 1 atm (with temperature held constant) to decrease the volume of gas by approximately 1 liter (1000 cm^3). An increase in temperature of 1 K (or 1 °C) would, with pressure held constant, increase the volume by about 16 cm^3.

GEOMETRIC INTERPRETATION OF PARTIAL DERIVATIVES

The partial derivatives f_x and f_y are the slopes of tangent lines to certain curves on the surface $z = f(x, y)$. Consider the point $P(a, b, f(a, b))$ on this surface. It lies directly above the point $Q(a, b, 0)$ in the xy-plane, as shown in Fig. 15.39. The vertical plane $y = b$ parallel to the xz-plane intersects the surface in the curve $z = f(x, b)$ through the point P. Along this curve, x varies

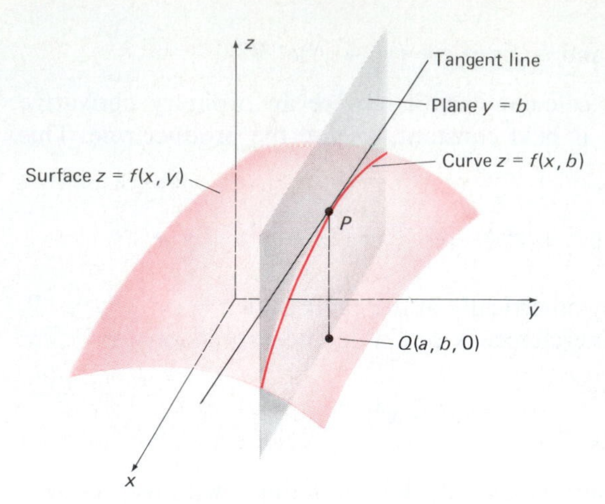

15.39 Geometric interpretation of $f_x(a, b)$ as the slope of a tangent to a *curve*

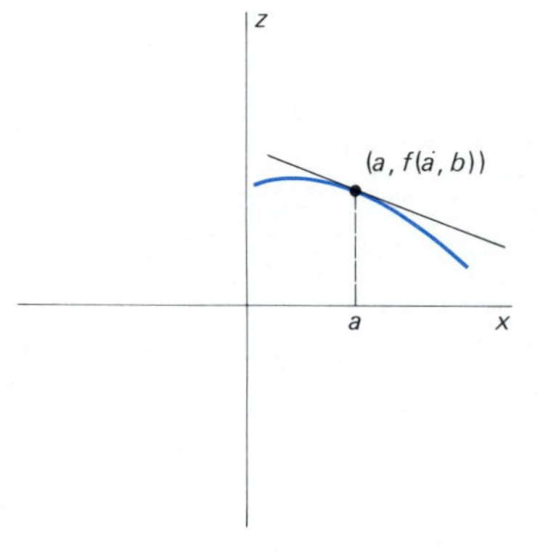

15.40 The projection of the curve and tangent line of Fig. 15.39 into the xz-plane

but y is constant; y is always equal to b because the curve lies in the vertical plane with equation $y = b$.

Let us project this curve and its tangent line at P into the xz-plane. We get exactly the same curve and tangent line, for this is a normal projection from the plane $y = b$ into a parallel plane. But because we are now working in the xz-plane, we can actually "forget" the presence of y. In effect, as indicated in Fig. 15.40, we are dealing with z as a function of the *single* variable x, and the projected curve is the graph of this function. Hence the slope of the tangent line to the original curve at the point $P(a, b, f(a, b))$ is equal to the slope of the tangent line in Fig. 15.40. But by familiar single-variable calculus, this slope is

$$\lim_{h \to 0} \frac{f(a + h, b) - f(a, b)}{h} = f_x(a, b).$$

Thus $f_x(a, b)$ is the slope of the tangent line at P to the curve formed by intersecting the plane $y = b$ parallel to the xz-plane with the surface $z = f(x, y)$.

We proceed in much the same way to get a geometric interpretation of the partial derivative of f with respect to y. As Fig. 15.41 suggests, $f_y(a, b)$ *is the slope of a line tangent to the curve of intersection of the surface $z = f(x, y)$ with the plane $x = a$ parallel to the yz-plane.*

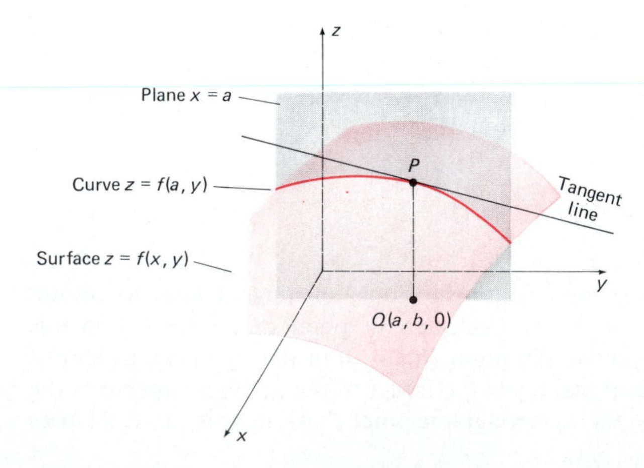

15.41 The partial derivative f_y is also the slope of a line tangent to a curve in the surface $z = f(x, y)$.

The two tangent lines we have just found determine a unique plane through the point $P(a, b, f(a, b))$. In Section 15.7 we will see that if the partial derivatives f_x and f_y are continuous functions of x and y, then this plane contains the tangent line at P to *every* smooth curve on the surface $z = f(x, y)$ passing through P. This plane is therefore (by definition) the tangent plane to the surface at P.

Definition *Tangent Plane to* $z = f(x, y)$

Suppose that the function $f(x, y)$ has continuous partial derivatives on a rectangle in the xy-plane containing (a, b) in its interior. Then the **tangent plane** to the surface $z = f(x, y)$ at the point $P(a, b, f(a, b))$ is the plane through P that contains the tangent lines to the two curves

$$z = f(x, b), \qquad y = b \tag{6}$$

and

$$z = f(a, y), \qquad x = a. \tag{7}$$

In order to write the equation of this tangent plane, all we need is a vector $\mathbf{n}$ normal to the plane. One way to get such a vector is to find the cross product of the tangent vectors to the curves in (6) and (7). Figures 15.42 and 15.43 show these two curves. As we saw above, the curve in (7) has slope $f_y(a, b)$ at P, so we can take

$$\mathbf{u} = \mathbf{j} + \mathbf{k} f_y(a, b) \tag{8}$$

as its tangent vector at P. The curve in (6) has slope $f_x(a, b)$ at P, so we can take

$$\mathbf{v} = \mathbf{i} + \mathbf{k} f_x(a, b) \tag{9}$$

as its tangent vector. Using these two tangent vectors, we obtain the normal vector

$$\mathbf{n} = \mathbf{u} \times \mathbf{v} = \begin{vmatrix} \mathbf{i} & \mathbf{j} & \mathbf{k} \\ 0 & 1 & f_y(a, b) \\ 1 & 0 & f_x(a, b) \end{vmatrix},$$

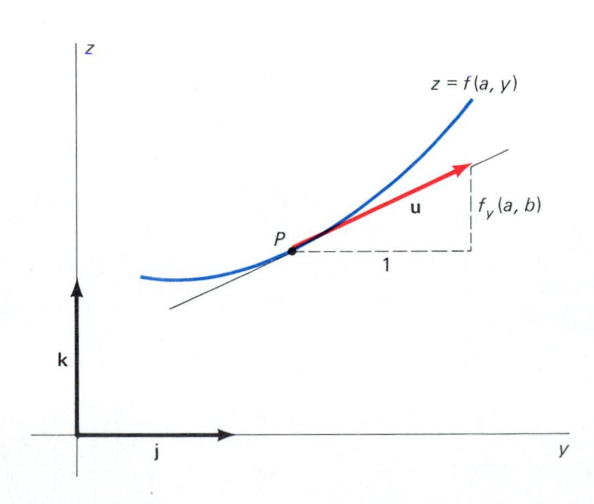

15.42 The curve $z = f(a, y)$ in the plane $x = a$

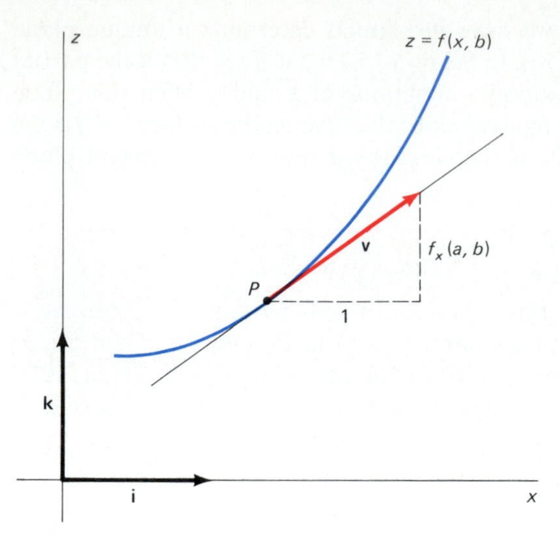

15.43 The curve $z = f(x, b)$ in the plane $y = b$

so that

$$\mathbf{n} = \mathbf{i}f_x(a, b) + \mathbf{j}f_y(a, b) - \mathbf{k}. \tag{10}$$

Note that

$$\mathbf{n} = \left\langle \frac{\partial z}{\partial x}, \frac{\partial z}{\partial y}, -1 \right\rangle \tag{11}$$

is a downward-pointing vector—its negative $-\mathbf{n}$ is the normal vector shown in Fig. 15.44.

Finally, we use the normal vector $\mathbf{n}$ of Equation (10) to find the equation of the tangent plane to the surface $z = f(x, y)$ at the point $(a, b, f(a, b))$. This

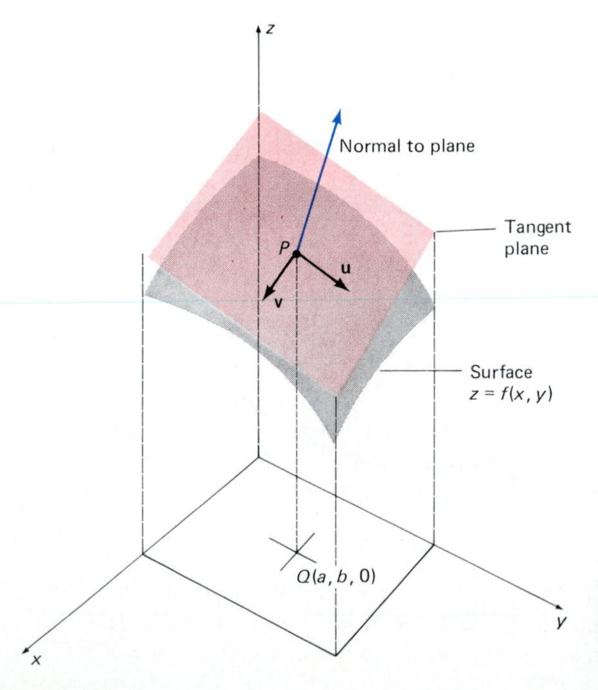

15.44 A normal vector to the tangent plane can be found by forming the cross product of two tangent vectors.

equation is

$$f_x(a, b)(x - a) + f_y(a, b)(y - b) - [z - f(a, b)] = 0. \tag{12}$$

An equivalent form of this equation is

$$z - c = \frac{\partial z}{\partial x}(x - a) + \frac{\partial z}{\partial y}(y - b) \tag{13}$$

where $c = f(a, b)$ and we remember that the partial derivatives $\partial z/\partial x$ and $\partial z/\partial y$ are evaluated at the point (a, b).

EXAMPLE 4 Write an equation of the tangent plane to the paraboloid $z = x^2 + y^2$ at the point $(2, -1, 5)$.

Solution We begin by computing $\partial z/\partial x = 2x$ and $\partial z/\partial y = 2y$. The values of these partial derivatives when $x = 2$ and $y = -1$ are 4 and -2, respectively. So Equation (12) gives

$$4(x - 2) - 2(y + 1) - (z - 5) = 0$$

(when simplified, $4x - 2y - z = 5$) as an equation of the indicated tangent plane.

HIGHER PARTIAL DERIVATIVES

The **first-order** partial derivatives f_x and f_y are themselves functions of x and y, so they may be differentiated with respect to x and y. The partial derivatives of $f_x(x, y)$ and $f_y(x, y)$ are called the **second-order partial derivatives** of f. There are four of them because there are four possibilities in the order of differentiation:

$$(f_x)_x = f_{xx} = \frac{\partial f_x}{\partial x} = \frac{\partial}{\partial x}\left(\frac{\partial f}{\partial x}\right) = \frac{\partial^2 f}{\partial x^2},$$

$$(f_x)_y = f_{xy} = \frac{\partial f_x}{\partial y} = \frac{\partial}{\partial y}\left(\frac{\partial f}{\partial x}\right) = \frac{\partial^2 f}{\partial y \partial x},$$

$$(f_y)_x = f_{yx} = \frac{\partial f_y}{\partial x} = \frac{\partial}{\partial x}\left(\frac{\partial f}{\partial y}\right) = \frac{\partial^2 f}{\partial x \partial y},$$

$$(f_y)_y = f_{yy} = \frac{\partial f_y}{\partial y} = \frac{\partial}{\partial y}\left(\frac{\partial f}{\partial y}\right) = \frac{\partial^2 f}{\partial y^2},$$

If we write $z = f(x, y)$, we can replace each occurrence of the symbol f above by z.

Note that f_{xy} is the second-order partial derivative of f with respect to x first, then y; f_{yx} is the result of differentiating with respect to x and y in the reverse order. While f_{xy} and f_{yx} are not necessarily equal, it is proved in advanced calculus that these two "mixed" second-order partial derivatives are equal if they are continuous. More precisely, if f_{xy} and f_{yx} are continuous on a circular disk centered at the point (a, b), then

$$f_{xy}(a, b) = f_{yx}(a, b). \tag{14}$$

(If f_{xy} and f_{yx} are both continuous merely at (a, b), they may well be unequal there.) Because most functions of interest to us will have second-order partial derivatives that are continuous everywhere they are defined, we will ordinarily need deal only with three distinct second-order partial derivatives rather than with four. Similarly, if $f(x, y, z)$ is a function of three variables with continuous second-order partial derivatives, then

$$\frac{\partial^2 f}{\partial x\,\partial y} = \frac{\partial^2 f}{\partial y\,\partial x}, \qquad \frac{\partial^2 f}{\partial x\,\partial z} = \frac{\partial^2 f}{\partial z\,\partial x}, \quad \text{and} \quad \frac{\partial^2 f}{\partial y\,\partial z} = \frac{\partial^2 f}{\partial z\,\partial y}.$$

Third-order and higher-order partial derivatives are defined similarly, and the order in which the differentiations are performed is unimportant as long as all derivatives involved are continuous. For example, the distinct third-order partial derivatives of a function $z = f(x, y)$ are

$$f_{xxx} = \frac{\partial}{\partial x}\left(\frac{\partial^2 f}{\partial x^2}\right) = \frac{\partial^3 f}{\partial x^3},$$

$$f_{xxy} = \frac{\partial}{\partial y}\left(\frac{\partial^2 f}{\partial x^2}\right) = \frac{\partial^3 f}{\partial y\,\partial x^2},$$

$$f_{xyy} = \frac{\partial}{\partial y}\left(\frac{\partial^2 f}{\partial y\,\partial x}\right) = \frac{\partial^3 f}{\partial y^2\,\partial x},$$

$$f_{yyy} = \frac{\partial}{\partial y}\left(\frac{\partial^2 f}{\partial y^2}\right) = \frac{\partial^3 f}{\partial y^3}.$$

EXAMPLE 5 Show that the partial derivatives of third and higher order of the function $f(x, y) = x^2 + 2xy^2 - y^3$ are constant.

Solution We find that

$$f_x(x, y) = 2x + 2y^2 \quad \text{and} \quad f_y(x, y) = 4xy - 3y^2.$$

So

$$f_{xx}(x, y) = 2, \qquad f_{xy}(x, y) = f_{yx}(x, y) = 4y, \quad \text{and} \quad f_{yy}(x, y) = 4x - 6y.$$

Finally,

$$f_{xxx}(x, y) = 0, \qquad f_{xxy}(x, y) = 0,$$

$$f_{xyy}(x, y) = 4, \quad \text{and} \quad f_{yyy}(x, y) = -6.$$

The function f is a polynomial, so all its partial derivatives are polynomials and are therefore continuous. Hence there is no need to compute any other third-order partial derivative—each is equal to one of the four above. Because the third-order partial derivatives are all constant, all higher-order partial derivatives of f are zero.

15.4 PROBLEMS

Compute the first-order partial derivatives of the functions in Problems 1–20.

1. $f(x, y) = x^4 - x^3y + x^2y^2 - xy^3 + y^4$

2. $f(x, y) = x \sin y$

3. $f(x, y) = e^x(\cos y - \sin y)$

4. $f(x, y) = x^2e^{xy}$

5. $f(x, y) = \dfrac{x + y}{x - y}$

6. $f(x, y) = \dfrac{xy}{x^2 + y^2}$

7. $f(x, y) = \ln(x^2 + y^2)$

8. $f(x, y) = (x - y)^{14}$

9. $f(x, y) = x^y$

10. $f(x, y) = \tan^{-1} xy$

11. $f(x, y, z) = x^2y^3z^4$

12. $f(x, y, z) = x^2 + y^3 + z^4$

13. $f(x, y, z) = e^{xyz}$

14. $f(x, y, z) = x^4 - 16yz$

15. $f(x, y, z) = x^2e^y \ln z$

16. $f(u, v) = (2u^2 + 3v^2) \exp(-u^2 - v^2)$

17. $f(r, s) = \dfrac{r^2 - s^2}{r^2 + s^2}$

18. $f(u, v) = e^{uv}(\cos uv + \sin uv)$

19. $f(u, v, w) = ue^v + ve^w + we^u$

20. $f(r, s, t) = (1 - r^2 - s^2 - t^2)e^{-rst}$

Verify that $z_{xy} = z_{yx}$ in Problems 21–30.

21. $z = x^2 - 4xy + 3y^2$

22. $z = 2x^3 + 5x^2y - 6y^2 + xy^4$

23. $z = x^2 \exp(-y^2)$

24. $z = xye^{-xy}$

25. $z = \ln(x + y)$

26. $z = (x^3 + y^3)^{10}$

27. $z = e^{-3x} \cos y$

28. $z = (x + y) \sec xy$

29. $z = x^2 \cosh(1/y^2)$

30. $z = \sin xy + \tan^{-1} xy$

In each of Problems 31–40, find an equation of the tangent plane to the given surface $z = f(x, y)$ at the indicated point P.

31. $z = x^2 + y^2, \quad P = (3, 4, 25)$

32. $z = \sqrt{25 - x^2 - y^2}, \quad P = (4, -3, 0)$

33. $z = \sin \dfrac{\pi xy}{2}, \quad P = (3, 5, -1)$

34. $z = \dfrac{4}{\pi} \tan^{-1} xy, \quad P = (1, 1, 1)$

35. $z = x^3 - y^3, \quad P = (3, 2, 19)$

36. $z = 3x + 4y, \quad P = (1, 1, 7)$

37. $z = xy, \quad P = (1, -1, -1)$

38. $z = \exp(-x^2 - y^2), \quad P = (0, 0, 1)$

39. $z = x^2 - 4y^2, \quad P = (5, 2, 9)$

40. $z = \sqrt{x^2 + y^2}, \quad P = (3, -4, 5)$

41. Verify that the mixed second-order partial derivatives f_{xy} and f_{yx} are indeed equal if $f(x, y) = x^m y^n$, where m and n are positive integers.

42. Suppose that $z = e^{x+y}$. Show that the result of differentiation of z first m times with respect to x, then n times with respect to y, is e^{x+y}.

43. Let $f(x, y, z) = e^{xyz}$. Calculate the distinct second-order partial derivatives of f and also the third-order partial derivative f_{xyz}.

44. Suppose that $g(x, y) = \sin xy$. Verify that $g_{xy} = g_{yx}$ and that $g_{xxy} = g_{xyx} = g_{yxx}$.

45. In physics it is shown that the temperature $u(x, t)$ at the point x and the time t of a long insulated rod lying along the x-axis satisfies the *one-dimensional heat equation*

$$\frac{\partial u}{\partial t} = k \frac{\partial^2 u}{\partial x^2} \qquad (k \text{ is a constant}).$$

Show that the function

$$u = u(x, t) = \exp(-n^2kt) \sin nx$$

satisfies the one-dimensional heat equation for any choice of the constant n.

46. The *two-dimensional heat equation* for an insulated plane is

$$\frac{\partial u}{\partial t} = k\left(\frac{\partial^2 u}{\partial x^2} + \frac{\partial^2 u}{\partial y^2}\right).$$

Show that the function

$$u = u(x, y, t) = \exp(-[m^2 + n^2]kt) \sin mx \cos ny$$

satisfies this equation for any choice of the constants m and n.

47. A string is stretched along the x-axis, fixed at each end, and then set in vibration. In physics it is shown that the displacement $y = y(x, t)$ of the point of the string at location x at time t satisfies the *one-dimensional wave equation*

$$\frac{\partial^2 y}{\partial t^2} = a^2 \frac{\partial^2 y}{\partial x^2}$$

where the constant a depends upon the density and tension of the string. Show that the following functions each satisfy the one-dimensional wave equation.

(a) $y = \sin(x + at)$

(b) $y = \cosh(3[x - at])$

(c) $y = \sin kx \cos kat \quad$ (k is a constant)

48. A steady-state temperature function $y = u(x, y)$ for a thin flat plate satisfies *Laplace's equation*

$$\frac{\partial^2 u}{\partial x^2} + \frac{\partial^2 u}{\partial y^2} = 0.$$

Determine which of the following functions satisfies Laplace's equation.

(a) $u = \ln(\sqrt{x^2 + y^2})$

(b) $u = \sqrt{x^2 + y^2}$

(c) $u = \arctan(y/x)$

(d) $u = e^{-x} \sin y$

49. The **ideal gas law** $PV = nRT$ (n is the number of moles of the gas, R is a constant) determines each of the three variables P, V, and T (pressure, volume, and absolute temperature, respectively) as functions of the other two. Show that

$$\frac{\partial P}{\partial V} \cdot \frac{\partial V}{\partial T} \cdot \frac{\partial T}{\partial P} = -1.$$

50. It is geometrically clear that every tangent plane to the cone $z^2 = x^2 + y^2$ passes through the origin. Show this by methods of calculus.

51. There is only one point at which the tangent plane to the surface $z = x^2 + 2xy + 2y - 6x + 8y$ is horizontal. Find it.

52. Show that the plane tangent to the paraboloid with equation $z = x^2 + y^2$ at the point (a, b, c) intersects the xy-plane in the line with equation $2ax + 2by = a^2 + b^2$. Then apply the formula for the distance from a point to a line to conclude that this line is tangent to the circle with equation $4x^2 + 4y^2 = a^2 + b^2$.

53. According to the van der Waals equation, 1 mole of a gas satisfies the equation

$$\left(P + \frac{a}{V^2}\right)(V - b) = (82.06)T$$

where P, V, and T are as in Example 2. For carbon dioxide, $a = 3.59 \times 10^6$ and $b = 42.7$, and V is 25,600 cm^3 when P is 1 atm and $T = 313$ K.

(a) Compute $\partial V/\partial P$ by differentiating van der Waals' equation with T held constant. Then estimate the change in volume that would result from an increase of 0.1 atm of pressure with T at 313 K.

(b) Compute $\partial V/\partial T$ by differentiating van der Waals' equation with P held constant. Then estimate the change in volume that would result from an increase of 1 K in temperature with P held at 1 atm.

54. A *minimal surface* is one that has the least surface area of all surfaces with the same boundary. Figure 15.45 shows Scherk's minimal surface. It has the equation

$$z = \ln(\cos x) - \ln(\cos y).$$

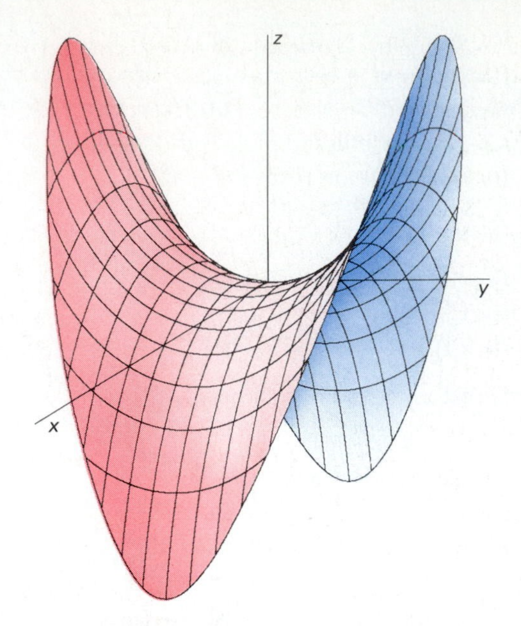

15.45 Scherk's minimal surface

It is known that a minimal surface $z = f(x, y)$ satisfies the partial differential equation

$$(1 + z_y^2)z_{xx} - zz_x z_y z_{xy} + (1 + z_x^2)z_{yy} = 0.$$

Verify this in the case of Scherk's minimal surface.

15.5 Maxima and Minima of Functions of Several Variables

The single-variable maximum–minimum techniques of Section 3.5 generalize in a natural manner to functions of several variables. We consider first a function f of two variables. Suppose that we are interested in the extreme values attained by f on a plane region R consisting of the points on and within a simple closed curve C, as in Fig. 15.46. We say that the function f attains its **absolute** (or **global**) **maximum value** M on R at the point (a, b) of R provided that

$$f(x, y) \leqq M = f(a, b)$$

for all points (x, y) of R. Similarly, f attains its **absolute** (or **global**) **minimum value** m on R at the point (c, d) provided that $f(x, y) \geqq m = f(c, d)$ for all points (x, y) of R. The following theorem, proved in advanced calculus courses, guarantees the existence of absolute maximum and minimum values in many situations of practical interest.

> ***Theorem 1*** *Existence of Extreme Values*
>
> Suppose that the function f is continuous on the region R which consists of the points on and within a simple closed curve C in the plane. Then f attains an absolute maximum value at some point (a, b) of R and attains an absolute minimum value at some point (c, d) of R.

Theorem 1 is the two-dimensional analogue of the theorem we used so much in Chapters 3 and 4: A continuous function (of one variable) defined on the closed (and bounded) interval $[a, b]$ attains its absolute maximum value at some point of $[a, b]$ (as well as its absolute minimum value at some point of $[a, b]$).

But, as in single-variable calculus, the next natural question is this: How do we actually *find* these extrema? We are interested mainly in the case in which the function f attains its absolute maximum (or minimum) value at an interior point of R. The point (a, b) of R is called an **interior point** of R provided that some circular disk centered at (a, b) lies wholly within R. The interior points of a region R of the sort described in Theorem 1 are precisely those that do *not* lie on the boundary curve C. Although this fact seems intuitively plausible, it is by no means self-evident; we omit its proof.

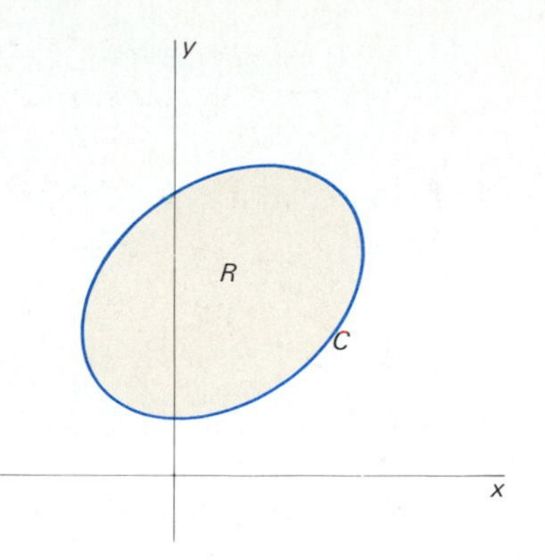

15.46 A bounded plane region R with boundary the simple closed curve C

An absolute extreme value attained by the function f at an *interior* point of R is necessarily a local extreme value. We say that $f(a, b)$ is a **local maximum value** of f if there is a circular disk D centered at (a, b) such that f is defined on D and $f(x, y) \leqq f(a, b)$ for all points (x, y) of D. If the inequality is reversed, then $f(a, b)$ is a **local minimum value** of the function f.

EXAMPLE 1 Figure 15.47 shows the graph of a certain function $f(x, y)$ defined on a region R in the xy-plane bounded by the simple closed curve C. We can think of the surface $z = f(x, y)$ as formed by starting with a stretched elastic membrane with fixed boundary C and then poking two fingers upward and two downward. The two fingers in each direction have different lengths. Looking at the four extreme values of $f(x, y)$ that result, we see

1. A local maximum that is not an absolute maximum,

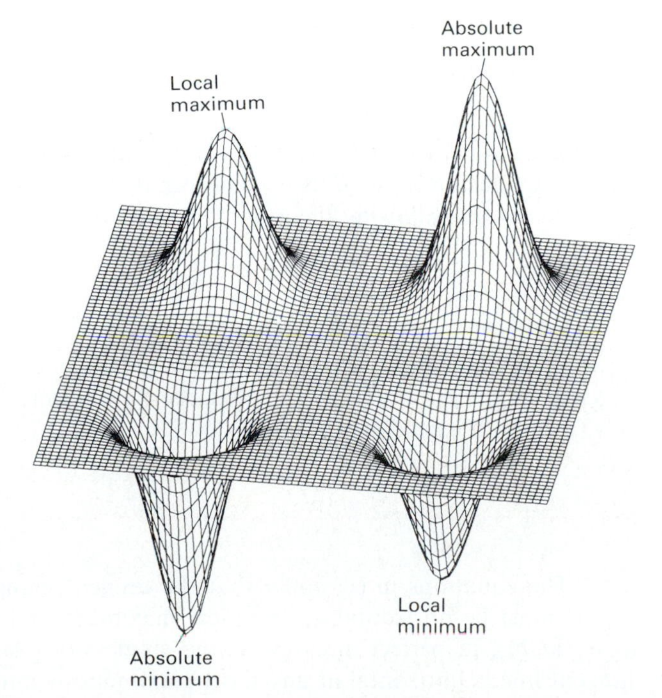

15.47 Local extrema contrasted with global extrema

2. A local maximum that is also an absolute maximum,
3. A local minimum that is not an absolute minimum,
4. A local minimum that is also an absolute minimum.

In Theorem 2 of Section 3.5, we saw that if the function f has a local maximum or local minimum value at a point $x = c$ where f is differentiable, then $f'(c) = 0$. We shall now show that an analogous event occurs in the two-dimensional situation: If $f(a, b)$ is either a local maximum value or a local minimum value of the function $f(x, y)$, then $f_x(a, b)$ and $f_y(a, b)$ are both zero, provided that these two partial derivatives both exist at the point (a, b).

Suppose, for example, that $f(a, b)$ is a local maximum value of $f(x, y)$ and that the partial derivatives $f_x(a, b)$ and $f_y(a, b)$ both exist. We look at cross sections of the graph of $z = f(x, y)$ in the same way as when we defined these partial derivatives in Section 15.4. Let

$$G(x) = f(x, b) \quad \text{and} \quad H(y) = f(a, y).$$

Because f is defined on a circular disk centered at (a, b), it follows that $G(x)$ is defined on some open interval containing the point $x = a$ and that $H(y)$ is defined on some open interval containing the point $y = b$. But we are assuming that f has a local maximum at (a, b), so it follows that $G(x)$ has a local maximum at $x = a$ and that $H(y)$ has a local maximum at $y = b$. The single-variable maximum–minimum result cited above therefore implies that $G'(a) = 0$ and that $H'(b) = 0$. But

$$G'(a) = \lim_{h \to 0} \frac{G(a + h) - G(a)}{h} = \lim_{h \to 0} \frac{f(a + h, b) - f(a, b)}{h} = f_x(a, b)$$

and

$$H'(b) = \lim_{k \to 0} \frac{H(b + k) - H(b)}{k} = \lim_{k \to 0} \frac{f(a, b + k) - f(a, b)}{k} = f_y(a, b).$$

Hence we conclude that $f_x(a, b) = 0 = f_y(a, b)$. A similar argument gives the same conclusion if $f(a, b)$ is a local minimum value of f. This discussion establishes the following theorem.

Theorem 2 *Necessary Conditions for Local Extrema*

Suppose that $f(x, y)$ attains a local maximum value or a local minimum value at the point $f(a, b)$ and that the partial derivatives $f_x(a, b)$ and $f_y(a, b)$ both exist. Then

$$f_x(a, b) = 0 = f_y(a, b). \tag{1}$$

The equations in (1) imply that the tangent plane to the surface $z = f(x, y)$ must be horizontal at any local maximum or local minimum point $(a, b, f(a, b))$, in perfect analogy to the single variable case (in which the tangent line is horizontal at any local maximum or minimum point).

EXAMPLE 2 Consider the three familiar surfaces

$$z = f(x, y) = x^2 + y^2,$$
$$z = g(x, y) = -x^2 - y^2, \quad \text{and}$$
$$z = h(x, y) = y^2 - x^2$$

shown in Fig. 15.48. In each case $\partial z/\partial x = \pm 2x$ and $\partial z/\partial y = \pm 2y$. Thus both partial derivatives vanish at the origin (0, 0) (and only there). It is clear from the figure that $f(x, y) = x^2 + y^2$ has a local minimum at (0, 0). In fact, because a square cannot be negative, it is plain that $z = x^2 + y^2$ has the global minimum value 0 at (0, 0). Similarly, $g(x, y)$ has a local (indeed, global) maximum at (0, 0), while $h(x, y)$ has neither a local minimum nor a local maximum there—the origin is a *saddle point* of h. This example shows that a point (a, b) where

$$\frac{\partial z}{\partial x} = 0 = \frac{\partial z}{\partial y}$$

may correspond to either a local maximum, a local minimum, or neither. Thus the necessary condition in (1) is *not* a sufficient condition for a local extremum.

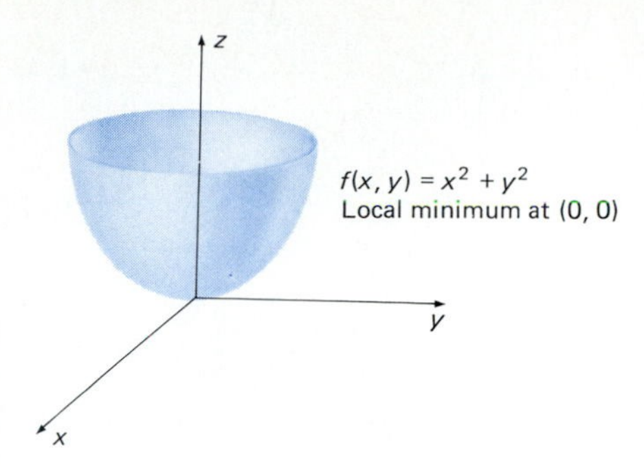

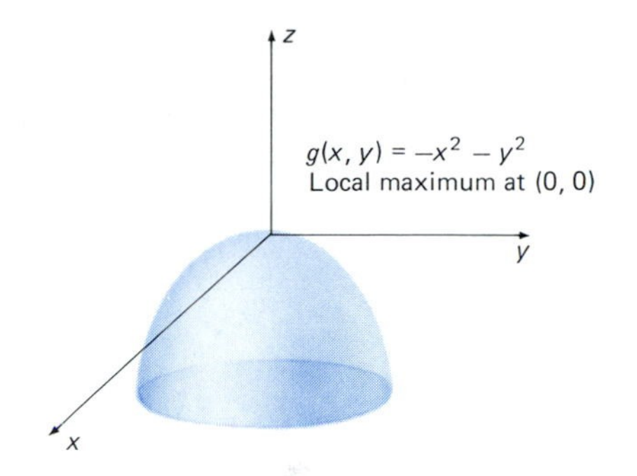

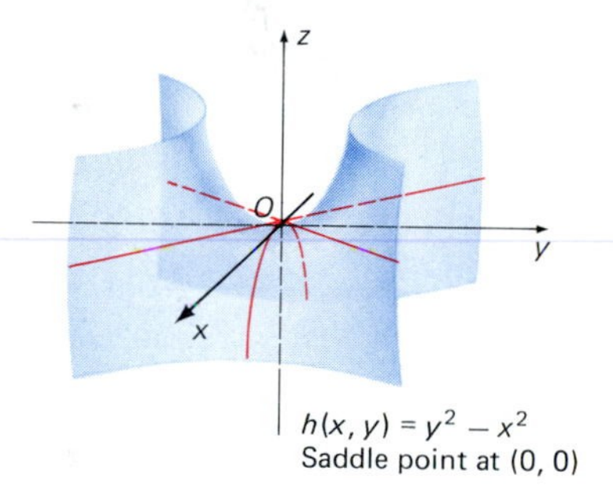

15.48 When both partial derivatives are zero, there may be a maximum, a minimum, or neither.

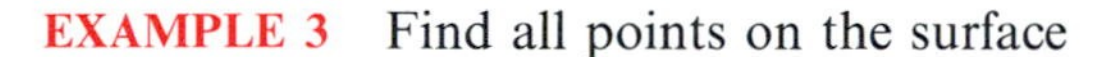

EXAMPLE 3 Find all points on the surface

$$z = \tfrac{3}{4}y^2 + \tfrac{1}{24}y^3 - \tfrac{1}{32}y^4 - x^2$$

at which the tangent plane is horizontal.

Solution We first calculate the partial derivatives $\partial z/\partial x$ and $\partial z/\partial y$:

$$\frac{\partial z}{\partial x} = -2x,$$

$$\frac{\partial z}{\partial y} = \frac{3}{2}y + \frac{1}{8}y^2 - \frac{1}{8}y^3 = -\frac{1}{8}y(y^2 - y - 12) = -\frac{1}{8}y(y + 3)(y - 4).$$

We next equate both $\partial z/\partial x$ and $\partial z/\partial y$ to zero. This yields

$$-2x = 0 \quad \text{and} \quad -\tfrac{1}{8}y(y + 3)(y - 4) = 0.$$

Simultaneous solution of these equations yields exactly three points where both partial derivatives vanish: (0, −3), (0, 0), and (0, 4). The three corresponding points on the surface where the tangent plane is horizontal are $(0, -3, \frac{99}{32})$, (0, 0, 0), and $(0, 4, \frac{20}{3})$. These three points are indicated on the graph in Fig. 15.49 of this surface. (Recall that we constructed this surface in Example 6 of Section 15.2.)

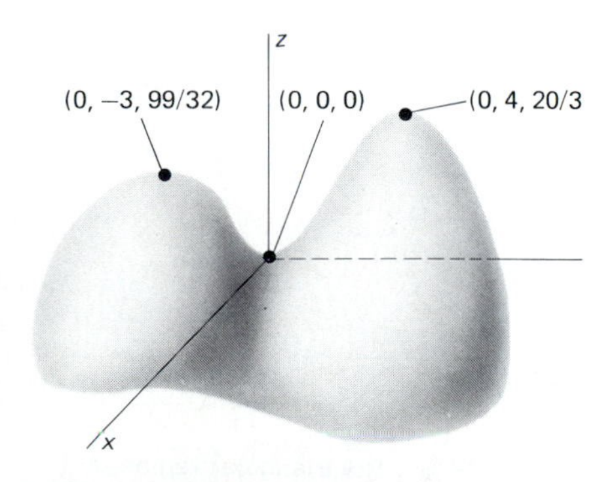

15.49 The surface of Example 3

Theorem 2 is a very useful tool for finding the absolute maximum and minimum values attained by a continuous function f on a region R of the type described in Theorem 1. If $f(a, b)$ is the absolute maximum value, for example, then (a, b) is either an interior point of R or a point of the boundary curve C. If (a, b) is an interior point and the partial derivatives $f_x(a, b)$ and $f_y(a, b)$ both exist, then Theorem 2 implies that both these partial derivatives must be zero. Thus we have the following result.

Theorem 3 *Types of Absolute Extrema*

Suppose that f is continuous on the plane region R consisting of the points on and within a simple closed curve C. If $f(a, b)$ is either the absolute maximum or the absolute minimum value of $f(x, y)$ on R, then (a, b) is either:

1. An interior point of R where
$$f_x(a, b) = 0 = f_y(a, b),$$
2. An interior point of R where not both first partial derivatives of f exist, or
3. A point of the boundary curve C.

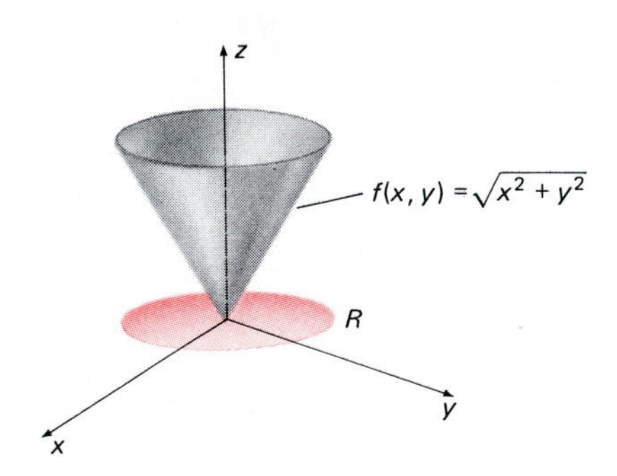

15.50 The graph of the function of Example 4

A point (a, b) where either condition (1) or (2) holds is called a **critical point** of the function f. Thus Theorem 3 says that *any extreme value of the continuous function f on the plane region R must occur at an interior critical point or at a boundary point.* Note the analogy with Theorem 3 in Section 3.5, which implies that an extreme value of a single-variable function $f(x)$ on a closed interval I must occur either at a critical point within I or at an end point (boundary point) of I.

As a consequence of Theorem 3, we can find the absolute maximum and minimum values of $f(x, y)$ on R as follows:

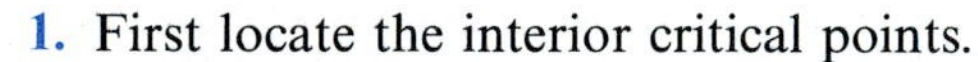

1. First locate the interior critical points.
2. Next find the possible extreme values of f on the boundary curve C.
3. Finally compare the values of f at the points found in Steps 1 and 2.

The technique to be used in Step 2 will depend upon the nature of the boundary curve C, as illustrated in Example 5.

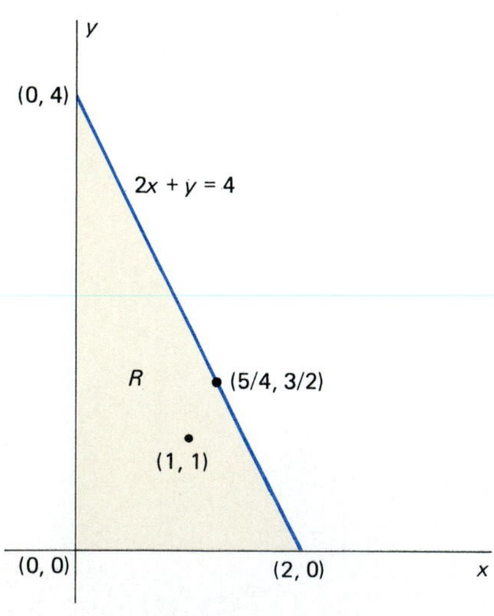

15.51 The triangular region of Example 5

EXAMPLE 4 Let $f(x, y) = \sqrt{x^2 + y^2}$ on the region R consisting of the points on and within the circle $x^2 + y^2 = 1$ in the xy-plane. The graph of f is shown in Fig. 15.50. We see that the minimum value 0 of f occurs at the origin (0, 0), where the partial derivatives f_x and f_y both fail to exist (why?), while the maximum value 1 of f on R occurs at *each and every* point of the boundary circle.

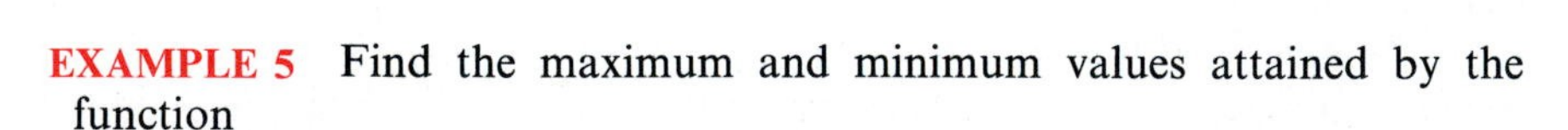

EXAMPLE 5 Find the maximum and minimum values attained by the function
$$f(x, y) = xy - x - y + 3$$
at points of the triangular region R in the xy-plane with vertices at (0, 0), (2, 0), and (0, 4).

Solution The region R is shown in Fig. 15.51. Its boundary "curve" C consists of the segment $0 \leqq x \leqq 2$ on the x-axis, the segment $0 \leqq y \leqq 4$ on the y-axis, and the part of the line $2x + y = 4$ that lies in the first quadrant.

Any interior extremum must occur at a point where

$$\frac{\partial f}{\partial x} = y - 1 \quad \text{and} \quad \frac{\partial f}{\partial y} = x - 1$$

are both zero. Hence the only interior critical point is (1, 1).

Along the edge $y = 0$, the function $f(x, y)$ takes the form

$$\alpha(x) = f(x, 0) = 3 - x, \qquad 0 \leqq x \leqq 2.$$

Because $\alpha(x)$ is a decreasing function, its extrema for $0 \leqq x \leqq 2$ occur at the endpoints $x = 0$ and $x = 2$. This gives the two possibilities (0, 0) and (2, 0) for extrema of $f(x, y)$.

Along the edge $x = 0$, $f(x, y)$ takes the form

$$\beta(y) = f(0, y) = 3 - y, \qquad 0 \leqq y \leqq 4.$$

The end points of this interval yield the possibilities (0, 0) and (0, 4) for extrema of $f(x, y)$.

On the edge $y = 4 - 2x$, we may substitute $4 - 2x$ for y in the formula for $f(x, y)$ and thus express f as a function of a single variable:

$$\begin{aligned}\gamma(x) &= x(4 - 2x) - x - (4 - 2x) + 3 \\ &= -2x^2 + 5x - 1, \qquad 0 \leqq x \leqq 2.\end{aligned}$$

To find the extreme values of $\gamma(x)$, we first calculate

$$\gamma'(x) = -4x + 5;$$

$\gamma'(x) = 0$ where $x = \frac{5}{4}$. Thus each extreme value of $\gamma(x)$ on $[0, 2]$ must occur either at the interior point $x = \frac{5}{4}$ of the interval $0 \leqq x \leqq 2$ or at one of the endpoints $x = 0$ and $x = 2$. This gives the possibilities (0, 4), $(\frac{5}{4}, \frac{3}{2})$, and (2, 0) for extrema of $f(x, y)$.

We conclude by evaluating f at each of the possible extrema points we have found. The result:

$$\begin{aligned}
&f(0, 0) = 3, &&\longleftarrow \quad \text{maximum} \\
&f(\tfrac{5}{4}, \tfrac{3}{2}) = 2.125, && \\
&f(1, 1) = 2, && \\
&f(2, 0) = 1, && \\
&f(0, 4) = -1. &&\longleftarrow \quad \text{minimum}
\end{aligned}$$

Thus the maximum value of $f(x, y)$ on the region R is $f(0, 0) = 3$ and the minimum value is $f(0, 4) = -1$.

Note the terminology used throughout this section. In Example 5, the maximum *value* of f is 3, the maximum value *occurs at* the point (0, 0) in the domain of f, and the *highest point* on the graph of f is (0, 0, 3).

In applied problems we frequently know in advance that the absolute maximum (or minimum) value of $f(x, y)$ occurs at an *interior* point of R where the partial derivatives of f both exist. In this important case, Theorem 3 tells us that we can locate every possible point at which the maximum might occur by solving (simultaneously) the two equations

$$f_x(x, y) = 0 \quad \text{and} \quad f_y(x, y) = 0. \tag{2}$$

If we are so fortunate as to find that these equations have only one simultaneous solution (x, y) interior to R, then *it* must be the location of the desired maximum. If we find that the equations in (2) have several solutions interior to R, then we simply evaluate the function f at each of them to determine which yields the largest value of $f(x,y)$ and is therefore the desired maximum point.

EXAMPLE 6 Find the highest point on the surface

$$z = 1 + \tfrac{4}{3}x^3 + 4y^3 - x^4 - y^4. \tag{3}$$

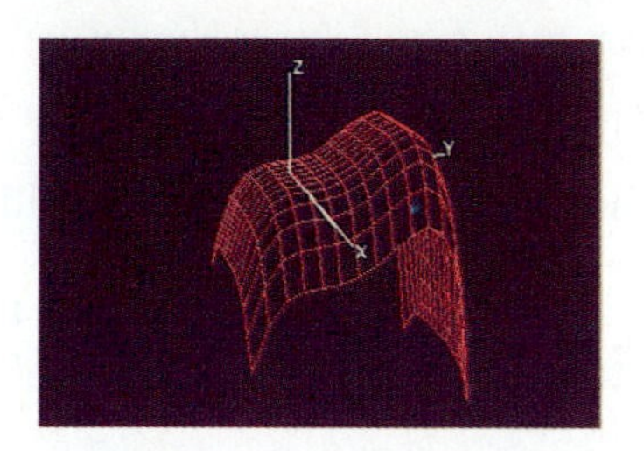

The surface of Example 6

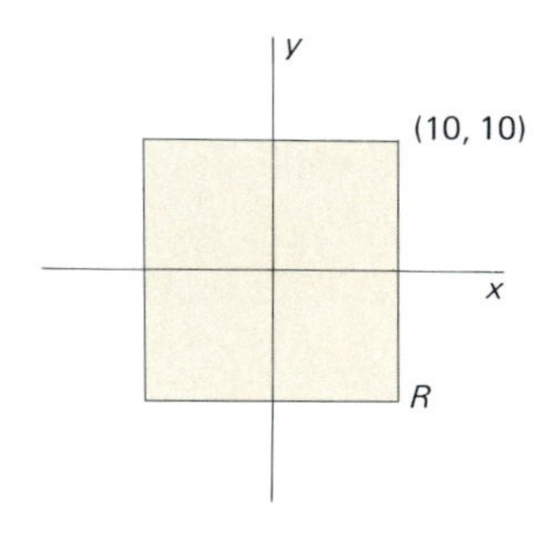

15.52 The function $z = z(x, y)$ of Example 6 is negative outside the square.

Solution Note that $z = 1$ at $(0, 0)$, so the maximum value of z (if any) is certainly positive. If we can find a region R such that z is negative at every point outside R and on the boundary of R, it will then follow that the maximum value of $z = z(x, y)$ must occur at some interior point of R. The fact that the terms of highest degree in z are of even degree and have negative coefficients suggests that such a region R exists. That is, we suspect that $z(x, y) < 0$ if either $|x|$ or $|y|$ is sufficiently large. So we let R be the square shown in Fig. 15.52, consisting of all points (x, y) such that $|x| \leqq 10$ and $|y| \leqq 10$. We claim that if the point (x, y) is *not* interior to the set R, then $z(x, y) < 0$.

The idea is that if $|x|$ or $|y|$ is large, then the corresponding negative fourth-degree term will predominate in (3). To make this precise, we rewrite (3) in the form

$$\frac{z}{x^4 + y^4} = \frac{1 + \frac{4}{3}x^3 + 4y^3}{x^4 + y^4} - 1. \tag{4}$$

Then if either $|x| \geqq 10$ or $|y| \geqq 10$, it follows that

$$\frac{1}{x^4 + y^4} \leqq \frac{1}{10^4}, \qquad \frac{\frac{4}{3}x^3}{x^4 + y^4} \leqq \frac{4}{10} \quad \text{and} \quad \frac{4y^3}{x^4 + y^4} \leqq \frac{4}{10}. \tag{5}$$

The first inequality in (5) is clear. For the other two there are several cases to consider. For instance, if $x \geqq 10$ and $0 \leqq y \leqq 10$ then

$$\frac{4y^3}{x^4 + y^4} \leqq \frac{4y^3}{x^4} \leqq \frac{4 \cdot 10^3}{10^4} = \frac{4}{10}.$$

Similar arguments establish the same inequalities in the remaining cases.

Hence it follows from (4) that

$$\frac{z}{x^4 + y^4} \leqq \frac{1}{10^4} + \frac{4}{10} + \frac{4}{10} - 1 < 0$$

if either $|x| \geqq 10$ or $|y| \geqq 10$. Therefore, z is the product of a negative number and $x^4 + y^4$. This implies that $z < 0$ if either $|x| \geqq 10$ or $|y| \geqq 10$. We therefore conclude that the (positive) maximum value of z must occur at some *interior* point of the region R shown in Fig. 15.52.

Because the partial derivatives of z with respect to x and y exist everywhere, Theorem 3 implies that we need only solve the equations $\partial z/\partial x = 0$ and $\partial z/\partial y = 0$ in (2)—that is,

$$\frac{\partial z}{\partial x} = 4x^2 - 4x^3 = 4x^2(1 - x) = 0,$$

$$\frac{\partial z}{\partial y} = 12y^2 - 4y^3 = 4y^2(3 - y) = 0.$$

If these two equations are satisfied, then

Either $x = 0$ or $x = 1$	and	Either $y = 0$ or $y = 3$.

It follows that either

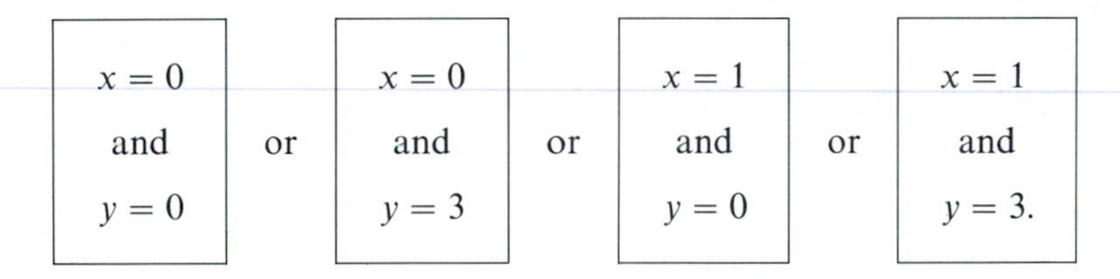

Consequently, we need only inspect the values

$$z(0, 0) = 1,$$

$$z(1, 0) = 1.333\,333\,333\ldots,$$

$$z(0, 3) = 28,$$

$$z(1, 3) = 28.333\,333\,333\ldots. \quad \longleftarrow \quad \text{maximum}$$

Thus the highest point on the surface is the point $(1, 3, \frac{85}{3})$.

APPLIED MAXIMUM–MINIMUM PROBLEMS

The analysis of a multivariable applied maximum–minimum problem involves the same general steps that we listed at the beginning of Section 3.6. Here, however, we will express the dependent variable—the quantity to be maximized or minimized—as a function $f(x, y)$ of *two* independent variables. Once we have identified the appropriate region in the xy-plane as the domain of f, the methods of this section are applicable. We often find that a preliminary step is required: If the meaningful domain of definition of f is an unbounded region, then we first restrict f to a *bounded* plane region R on which we know the desired extreme value occurs. This procedure is similar to the one we used with open interval maximum–minimum problems in Section 4.4.

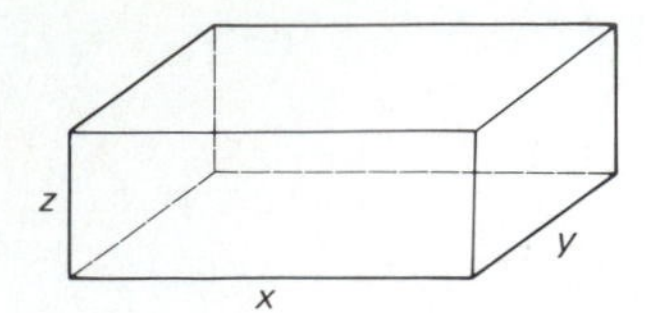

15.53 Minimizing the total cost of a box

EXAMPLE 7 Find the minimum cost of the rectangular box with volume 48 ft^3 discussed in Section 15.1. This is the box that has front and back that cost \$1/ft^2, top and bottom that cost \$2/ft^2, and two ends that cost \$3/ft^2. Such a box is shown in Fig. 15.53.

Solution In Section 15.1 we found that the cost C (in dollars) of this box is given by

$$C = 4xy + \frac{288}{x} + \frac{96}{y}$$

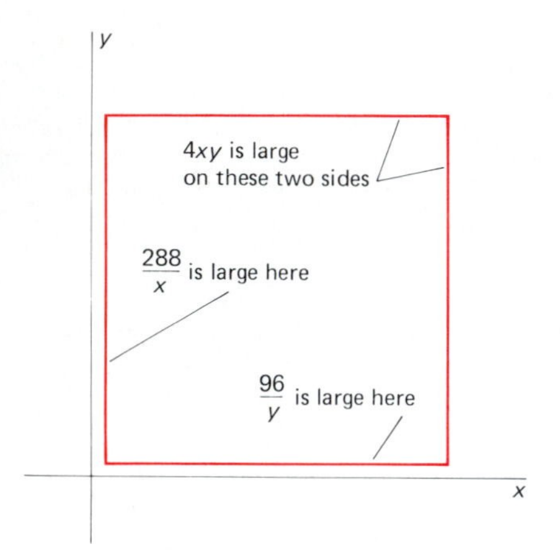

15.54 The cost function C of Example 7 takes on large positive values on the boundary of the square.

in terms of its length x and width y. Let R be a square such as the one shown in Fig. 15.54. Two sides of R are so close to the coordinate axes that $288/x > 1000$ on the side near the y-axis and $96/y > 1000$ on the side near the x-axis. Also, the square is so large that $4xy > 1000$ on each of the other two sides. This means that $C(x, y) > 1000$ at every point (x, y) of the first quadrant that lies on the boundary of the square R or outside of R. Because $C(x, y)$ attains reasonably small values within R (for instance, $C(1, 1) = 388$), it is clear that the absolute minimum of C must occur at an interior point of R. Thus while the natural domain of the cost function $C(x, y)$ is the first quadrant, we have succeeded in restricting its domain to a region R of the sort to which Theorem 3 applies.

We therefore proceed to solve the equations

$$\frac{\partial C}{\partial x} = 4y - \frac{288}{x^2} = 0,$$

$$\frac{\partial C}{\partial y} = 4x - \frac{96}{y^2} = 0.$$

We multiply the first equation by x and the second by y. This gives

$$\frac{288}{x} = 4xy = \frac{96}{y},$$

so $x = 288y/96 = 3y$. We substitute $x = 3y$ in the equation $\partial C/\partial y = 0$ and find that

$$12y - \frac{96}{y^2} = 0, \qquad \text{so that} \quad 12y^3 = 96.$$

Hence $y = \sqrt[3]{8} = 2$, so $x = 6$. Therefore, the minimum cost of this box is $C(6, 2) = 144$ (dollars). Because the volume of the box is $V = xyz = 48$, its height is $z = 48/(6 \cdot 2) = 4$ when $x = 6$ and $y = 2$. Thus the optimal box is 6 ft by 2 ft by 4 ft.

We have seen that if $f_x(a, b) = 0 = f_y(a, b)$, then $f(a, b)$ may be either a maximum value, a minimum value, or neither. In Section 15.10 we will discuss sufficient conditions that $f(a, b)$ be either a local maximum or a local minimum. These conditions involve the second-order partial derivatives of f at (a, b).

The methods of this section generalize readily to functions of three or more variables. For example, if the function $f(x, y, z)$ has a local extremum at the point (a, b, c) where its three first-order partial derivatives exist, then all three must vanish there. That is,

$$f_x(a, b, c) = f_y(a, b, c) = f_z(a, b, c) = 0. \tag{6}$$

The following example illustrates a "line-through-the-point" method that can sometimes be used to show that a point (a, b, c) where the conditions in (6) hold is neither a local maximum nor a local minimum point. (The method is also applicable to functions of two, or more than three, variables.)

EXAMPLE 8 Determine whether the function

$$f(x, y, z) = xy + yz - xz$$

has any local extrema.

Solution The necessary conditions in (5) give the equations

$$f_x(x, y, z) = y - z = 0,$$

$$f_y(x, y, z) = x + z = 0,$$

$$f_z(x, y, z) = y - x = 0.$$

We easily find that the simultaneous solution of these equations is $x = y = z = 0$. On the line $x = y = z$ through $(0, 0, 0)$, the function $f(x, y, z)$ reduces to x^2, which is minimal at $x = 0$. But on the line $x = -y = z$ it reduces to $-3x^2$, which is maximal when $x = 0$. Hence f can have neither a local maximum nor a local minimum at $(0, 0, 0)$, and therefore has no extrema (local *or* global) at all.

15.5 PROBLEMS

In each of Problems 1–12, find every point on the given surface $z = f(x, y)$ at which the tangent plane is horizontal.

1. $z = x - 3y + 5$

2. $z = 4 - x^2 - y^2$

3. $z = xy + 5$

4. $z = x^2 + y^2 + 2x$

5. $z = x^2 + y^2 - 6x + 2y + 5$

6. $z = 10 + 8x - 6y - x^2 - y^2$

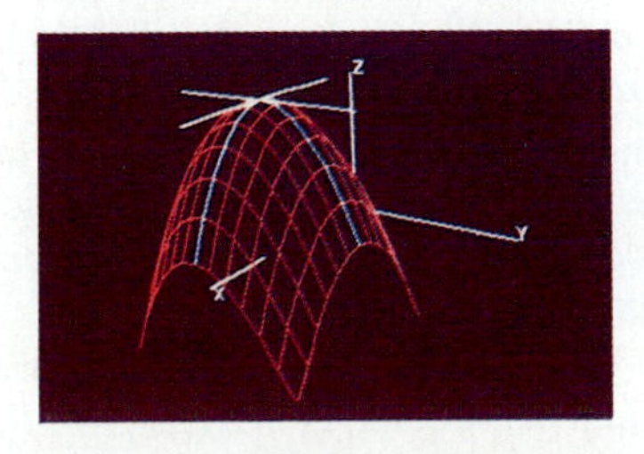

The surface of Problem 6

7. $z = x^2 + 4x + y^3$

8. $z = x^4 + y^3 - 3y$

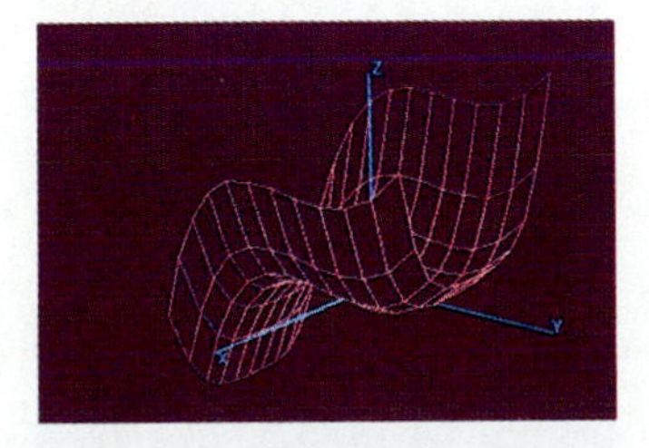

The surface of Problem 8

9. $z = 3x^2 + 12x + 4y^3 - 6y^2 + 5$

10. $z = \dfrac{1}{1 - 2x + 2y + x^2 + y^2}$

11. $z = (2x^2 + 3y^2)\exp(-x^2 - y^2)$

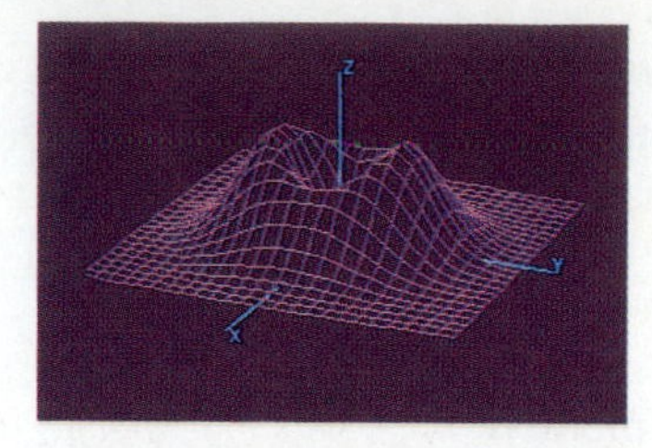

The surface of Problem 11

12. $z = 2xy\exp(-\frac{1}{8}(4x^2 + y^2))$

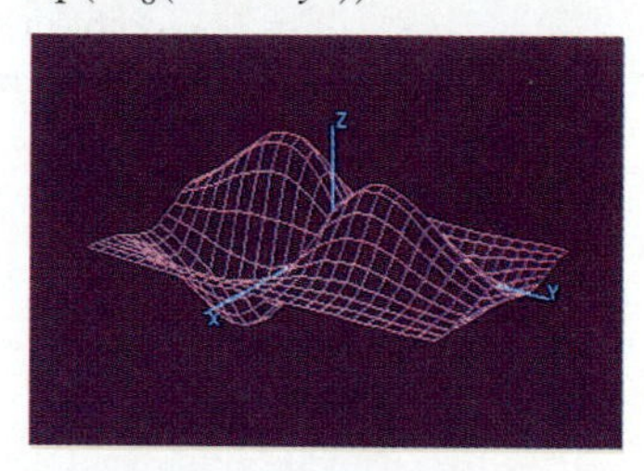

The surface of Problem 12

In each of Problems 13–20, the given function $f(x, y)$ has either an absolute minimum value or an absolute maximum value (for all x and y) but not both. Find that value.

13. $f(x, y) = x^2 - 2x + y^2 - 2y + 3$

14. $f(x, y) = 6x - 8y - x^2 - y^2$

15. $f(x, y) = 2x - x^2 + 2y^2 - y^4$

16. $f(x, y) = 3x^4 + 4x^3 + 6y^4 - 16y^3 + 12y^2 + 7$

17. $f(x, y) = 2x^2 + 8xy + y^4$

18. $f(x, y) = \dfrac{1}{10 - 2x - 4y + x^2 + y^4}$

19. $f(x, y) = \exp(2x - 4y - x^2 - y^2)$

20. $f(x, y) = (1 + x^2)\exp(-x^2 - y^2)$

In each of Problems 21–26, find the maximum and minimum values attained by the given function $f(x, y)$ on the given region R.

21. $f(x, y) = x + 2y$; R is the square with vertices at $(\pm 1, \pm 1)$.

22. $f(x, y) = x^2 + y^2 - x$; R is the square of Problem 21.

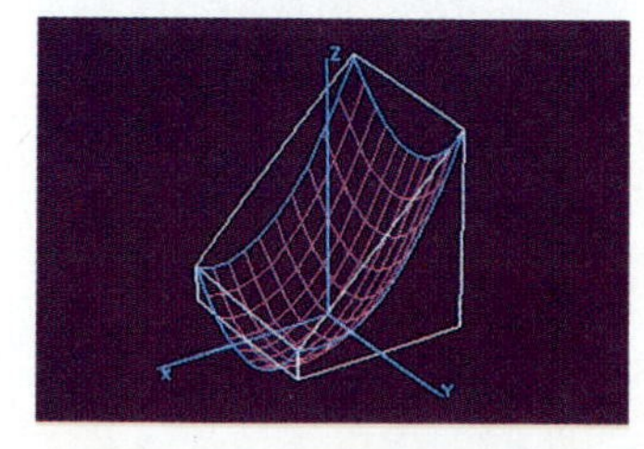

Graph for Problem 22

23. $f(x, y) = x^2 + y^2 - 2x$; R is the triangular region with vertices at (0, 0), (2, 0), and (0, 2).

24. $f(x, y) = x^2 + y^2 - x - y$; R is the region of Problem 23.

25. $f(x, y) = 2xy$; R is the circular disk $x^2 + y^2 \leq 1$.

26. $f(x, y) = xy^2$; R is the circular disk $x^2 + y^2 \leq 3$.

27. Find the dimensions x, y, and z of a rectangular box with fixed volume $V = 1000$ and minimum total surface area A.

28. Find the points on the surface $xyz = 1$ that are closest to the origin.

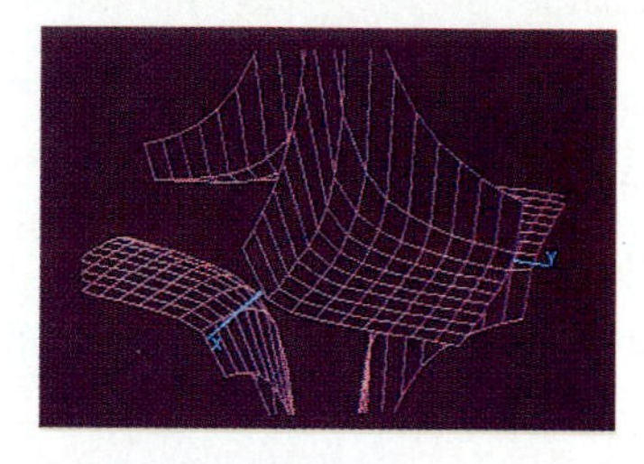

The surface xyz = 1

29. Find the dimensions of the rectangular box with maximum volume that has total surface area 600 cm^2.

30. A rectangular box without a lid is to have fixed volume 4000 cm^3. What should be its dimensions to minimize its total surface area?

31. A rectangular box is placed in the first octant, with one of its corners at the origin and three of its sides lying in the three coordinate planes. The vertex opposite the origin lies on the plane with equation $x + 2y + 3z = 6$. What is the maximum possible volume of such a box? What are the dimensions of that box?

32. The sum of three positive numbers is 120. What is the maximum possible value of their product?

33. A building in the shape of a rectangular box is to have a volume of 8000 ft^3. Annual heating and cooling costs will amount to \$2 for each square foot of top, front, and back, and \$4 for each square foot of the two end walls. What dimensions of the building will minimize these annual costs?

34. A rectangular box *without top* is to be made of materials costing \$3/ft^2 for its bottom and \$2/ft^2 for the four sides. The box is to have volume 48 ft^3. What dimensions will minimize its cost?

35. A rectangular shipping crate is to have volume 12 ft^3. Its bottom costs *twice* as much (per square foot) as its top and four sides. What dimensions will minimize the total cost of the crate?

36. Use the maximum–minimum methods of this section to find the point of the plane $2x - 3y + z = 1$ that is closest to the point $(3, -2, 1)$. (*Suggestion:* A positive quantity is minimized when its square is minimized.)

37. Find the maximum volume of a rectangular box that can be sent from a post office if the sum of its *length* and *girth* cannot exceed 108 in.

38. Repeat Problem 37 for the case of a cylindrical box—one shaped like a hatbox or a fat mailing tube.

39. A rectangular box with its base in the xy-plane is inscribed under the graph of the paraboloid $z = 1 - x^2 - y^2$, $z \geqq 0$. Find the maximum possible volume of the box. (*Suggestion:* You may assume that the sides of the box are parallel to the vertical coordinate planes, and it follows that the box is symmetrically placed around these planes.)

40. What is the maximum possible volume of a rectangular box inscribed in a *hemisphere* of radius R? You may assume that one face of the box lies in the planar base of the hemisphere.

41. A wire 120 in. long is cut into three *or fewer* pieces, and each piece is bent into the shape of a square. How should this be done in order to minimize the total area of these squares? To maximize it?

42. A lump of putty of (fixed) volume V is to be divided into three or fewer pieces and the pieces made into cubes. How should this be done to maximize the total surface area of the cubes? To minimize it?

43. Consider the function $f(x, y) = (y - x^2)(y - 3x^2)$.

(a) Show that $f_x(0, 0) = 0 = f_y(0, 0)$.

(b) Show that for every straight line $y = mx$, the function $f(x, mx)$ has a local minimum at $x = 0$.

(c) Examine the values of f at points of the parabola $y = 2x^2$ to show that f does *not* have a local minimum at (0, 0). This shows that the line-through-the-point method cannot be used to show that a point *is* a local extremum.

15.6 Increments and Differentials

Given a single-variable function $f(x)$, we defined (in Section 4.2) the *increment*

$$\Delta f = \Delta y = f(a + \Delta x) - f(a)$$

and the *differential*

$$df = dy = f'(a)\,\Delta x.$$

Each of these quantities is associated with the change in x from a to $a + \Delta x$. For a *fixed* value a, the differential df is a *linear* function of Δx. We saw that df is a good approximation to the actual change Δf in the sense that

$$\Delta f = df + \varepsilon\,\Delta x = f'(a)\,\Delta x + \varepsilon\,\Delta x,$$

where ε is a function of Δx that approaches zero as $\Delta x \to 0$. Hence

$$\Delta f - df = \varepsilon\,\Delta x$$

is "very *very* small" when Δx is "very small." The relation between Δf and df is shown in Fig. 15.55, which shows that df is the change in the height of the *tangent line* to $y = f(x)$ at $x = a$.

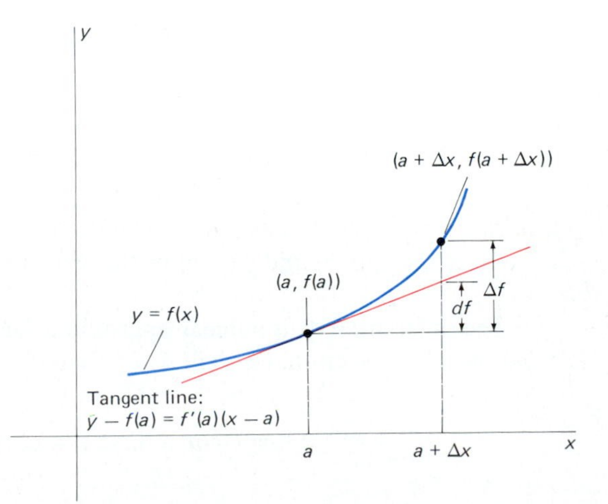

15.55 Increment and differential for a function of one variable

We define the differential of a function $f(x, y)$ of two variables in much the same way, by using the tangent *plane* to the surface $z = f(x, y)$ in a way analogous to the way we use the tangent *line* for a single-variable function. Our goal is a simple approximation to the change in the value of f between $P(a, b)$ and $Q(a + \Delta x, b + \Delta y)$; that is, the **increment**

$$\Delta f = f(a + \Delta x, b + \Delta y) - f(a, b). \tag{1}$$

We consider the tangent plane at $(a, b, f(a, b))$, shown in Fig. 15.56, with equation—by the formula in Equation (13) of Section 15.4—that may be written as

$$z - f(a, b) = f_x(a, b)(x - a) + f_y(a, b)(y - b). \tag{2}$$

The **differential** df of f at $P(a, b)$ we define to be the change in the height of

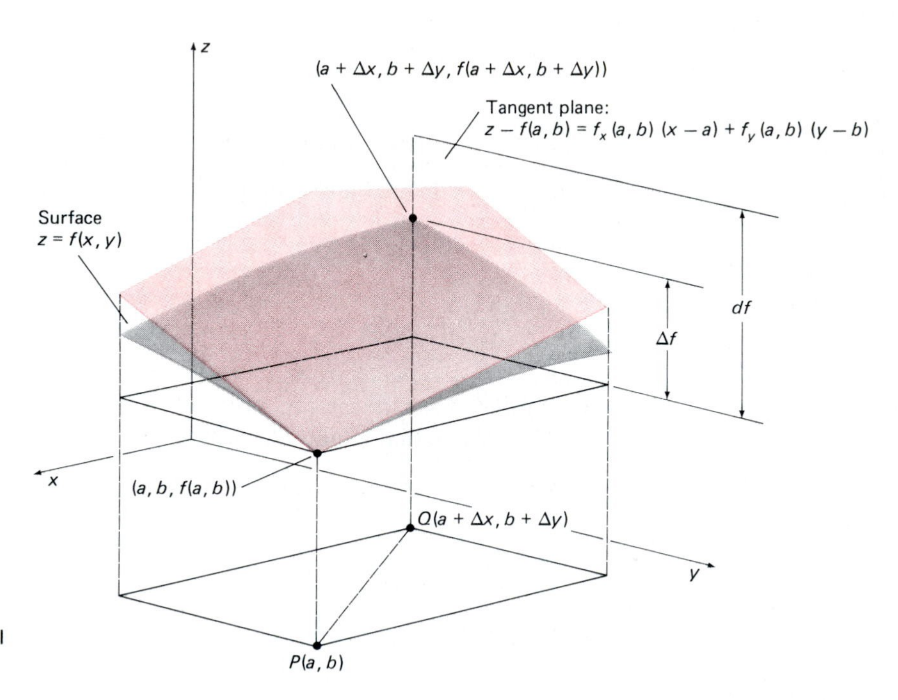

15.56 Increment and differential for $z = f(x, y)$

the tangent plane from $P(a, b)$ to $Q(a + \Delta x, b + \Delta y)$. This is simply the value of $z - f(a, b)$ that we calculate from Equation (2), taking $x = a + \Delta x$ and $y = b + \Delta y$. So

$$df = f_x(a, b)\,\Delta x + f_y(a, b)\,\Delta y. \tag{3}$$

Note that, for a and b fixed, df is a *linear* function of Δx and Δy and that the coefficients $f_x(a, b)$ and $f_y(a, b)$ in this linear function depend upon a and b.

The differential df is a **linear approximation** to the actual increment Δf. At the end of this section, we shall show that df is a very good approximation to Δf when Δx and Δy are both small, in the sense that

$$\Delta f - df = \varepsilon_1\,\Delta x + \varepsilon_2\,\Delta y, \tag{4}$$

where ε_1 and ε_2 are functions of Δx and Δy that approach zero as $\Delta x \to 0$ and $\Delta y \to 0$. Hence we write

$$\Delta f - df \approx 0, \quad \text{or} \quad \Delta f \approx df$$

when Δx and Δy are small. The approximation

$$f(a + \Delta x, b + \Delta y) = f(a, b) + \Delta f \approx f(a, b) + df;$$

$$f(a + \Delta x, b + \Delta y) \approx f(a, b) + f_x(a, b)\,\Delta x + f_y(a, b)\,\Delta y \tag{5}$$

may then be used to estimate the value of $f(a + \Delta x, b + \Delta y)$ when Δx and Δy are small and the values $f(a, b)$, $f_x(a, b)$, and $f_y(a, b)$ are all known.

EXAMPLE 1 Use Equation (5) to estimate

$$\sqrt{(2)(2.02)^3 + (2.97)^2}.$$

Note that $\sqrt{(2)(2)^3 + (3)^2} = \sqrt{25} = 5$.

Solution We take $f(x, y) = (2x^3 + y^2)^{1/2}$. Then

$$\frac{\partial f}{\partial x} = \frac{3x^2}{(2x^3 + y^2)^{1/2}} \quad \text{and} \quad \frac{\partial f}{\partial y} = \frac{y}{(2x^3 + y^2)^{1/2}}.$$

Now let $a = 2$, $b = 3$, $\Delta x = 0.02$, and $\Delta y = -0.03$. Then $f(2, 3) = 5$,

$$f_x(2, 3) = \tfrac{12}{5}, \quad \text{and} \quad f_y(2, 3) = \tfrac{3}{5}.$$

Hence Equation (5) gives

$$\begin{aligned}\sqrt{(2)(2.02)^3 + (2.97)^2} &= f(2.02, 2.97)\\ &\approx f(2, 3) + f_x(2, 3)\cdot(0.02) + f_y(2, 3)\cdot(-0.03)\\ &= 5 + \tfrac{12}{5}(0.02) + \tfrac{3}{5}(-0.03) = 5.03.\end{aligned}$$

The actual value, accurate to four decimal places, is 5.0305.

If $z = f(x, y)$, we often write dz in place of df. So the differential of the dependent variable z at the point (a, b) is $dz = f_x(a, b)\,\Delta x + f_y(a, b)\,\Delta y$. At the arbitrary point (x, y), the differential of z takes the form

$$dz = f_x(x, y)\,\Delta x + f_y(x, y)\,\Delta y.$$

More simply, we can write

$$dz = \frac{\partial z}{\partial x}\,\Delta x + \frac{\partial z}{\partial y}\,\Delta y. \tag{6}$$

It is customary to write dx for Δx and dy for Δy in this formula. When this is done, Equation (6) takes the form

$$dz = \frac{\partial z}{\partial x}\,dx + \frac{\partial z}{\partial y}\,dy. \tag{7}$$

When one uses this notation, it is important to realize that there is *no* connotation of dx and dy being "infinitesimal" or even small; dz is still simply a linear function of the ordinary real variables dx and dy, one that gives a linear approximation to the change in z when x and y are changed by the amounts dx and dy, respectively.

EXAMPLE 2 Find the differential dz given $z = x^2 + 3xy - 2y^2$. Then estimate the change in z when x increases from 3 to 3.2 and y, meanwhile, decreases from 5 to 4.9.

Solution Equation (7) gives

$$dz = \frac{\partial z}{\partial x}\, dx + \frac{\partial z}{\partial y}\, dy = (2x + 3y)\, dx + (3x - 4y)\, dy.$$

When we substitute $x = 3$, $y = 5$, $dx = 0.2$, and $dy = -0.1$, we find that

$$dz = (6 + 15)(0.2) + (9 - 20)(-0.1) = 5.3.$$

Remember that this is an approximation to the actual change

$$\Delta z = [(3.2)^2 + (3)(3.2)(4.9) - (2)(4.9)^2] - [9 + 45 - 50] = 5.26.$$

EXAMPLE 3 In Example 3 of Section 15.4, we considered a mole of an ideal gas—its volume V in cubic centimeters given in terms of its pressure P in atmospheres and temperature T in degrees kelvin by the formula $V = (82.06)T/P$. Approximate the change in its volume when P is increased from 5 atm to 5.2 atm and T is increased from 300 K to 310 K.

Solution The differential of $V = V(P, T)$ is

$$dV = \frac{\partial V}{\partial P}\, dP + \frac{\partial V}{\partial T}\, dT = -\frac{(82.06)T}{P^2}\, dP + \frac{82.06}{P}\, dT.$$

With $P = 5$, $T = 300$, $dP = 0.2$, and $dT = 10$, we compute

$$dV = -\frac{(82.06)(300)}{5^2}(0.2) + \frac{82.06}{5}(10) = -32.8 \quad (\text{cm}^3).$$

This indicates that the gas will decrease in volume by about 33 cm^3. The actual change is

$$\Delta V = \frac{(82.06)(310)}{5.2} - \frac{(82.06)(300)}{5} = 4892.0 - 4923.6 = -31.6 \quad (\text{cm}^3).$$

EXAMPLE 4 The point (1, 2) lies on the curve with equation

$$f(x, y) = 2x^3 + y^3 - 5xy = 0.$$

Approximate the nearby point on the curve with $x = 1.1$.

Solution First we calculate the differential

$$df = \frac{\partial f}{\partial x}\, dx + \frac{\partial f}{\partial y}\, dy = (6x^2 - 5y)\, dx + (3y^2 - 5x)\, dy = 0.$$

When we substitute $x = 1$, $y = 2$, and $dx = 0.1$, we find that $dy \approx 0.06$. This yields (1.1, 2.06) for the approximate coordinates of the nearby point. As a check on the accuracy of this approximation, we can substitute $x = 1.1$ in the original equation to get

$$(2)(1.1)^3 + y^3 - (5)(1.1)y = 0$$

and then solve for y using Newton's method. This technique gives $y \approx 2.05$.

Increments and differentials of functions of more than two variables are defined similarly. A function $w = f(x, y, z)$ has *increment*

$$\Delta w = \Delta f = f(x + \Delta x, y + \Delta y, z + \Delta z) - f(x, y, z)$$

and *differential*

$$dw = df = \frac{\partial f}{\partial x} \Delta x + \frac{\partial f}{\partial y} \Delta y + \frac{\partial f}{\partial z} \Delta z;$$

that is,

$$dw = w_x\, dx + w_y\, dy + w_z\, dz,$$

if (as in Equation (7)) we write dx for Δx, dy for Δy, and dz for Δz.

EXAMPLE 5 Suppose that three positive numbers, each less than 100, are rounded to the first decimal place. Estimate the largest error that this rounding might cause in computing their product xyz.

Solution The differential of $w = xyz$ is

$$dw = yz\, dx + xz\, dy + xy\, dz.$$

We substitute the largest possible values for x, y, and z: We take each to be 100. Because of the way the numbers are rounded, we take ± 0.05 for dx, dy, and dz. We then get

$$dw = (100)^2(\pm 0.05) + (100)^2(\pm 0.05) + (100)^2(\pm 0.05),$$

which shows that dw may be anywhere between -1500 and 1500. It is rather a surprise to find that an error of only 0.05 in each factor may cause an error of as much as 1500 in the product.

The differential $df = f_x\, dx + f_y\, dy$ is defined provided that the partial derivatives f_x and f_y both exist. The following theorem gives sufficient conditions that df be a good approximation to the increment Δf when Δx and Δy are small.

> ***Theorem*** *Linear Approximation*
>
> Suppose that $f(x, y)$ has continuous first order partial derivatives in a rectangular region with horizontal and vertical sides and containing the points $P(a, b)$ and $Q(a + \Delta x, b + \Delta y)$. Let
>
> $$\Delta f = f(a + \Delta x, b + \Delta y) - f(a, b)$$
>
> be the increment in Equation (1). Then
>
> $$\Delta f = f_x(a, b)\, \Delta x + f_y(a, b)\, \Delta y + \varepsilon_1\, \Delta x + \varepsilon_2\, \Delta y \quad (8)$$
>
> where ε_1 and ε_2 are functions of Δx and Δy that approach zero as $\Delta x \to 0$ and $\Delta y \to 0$.

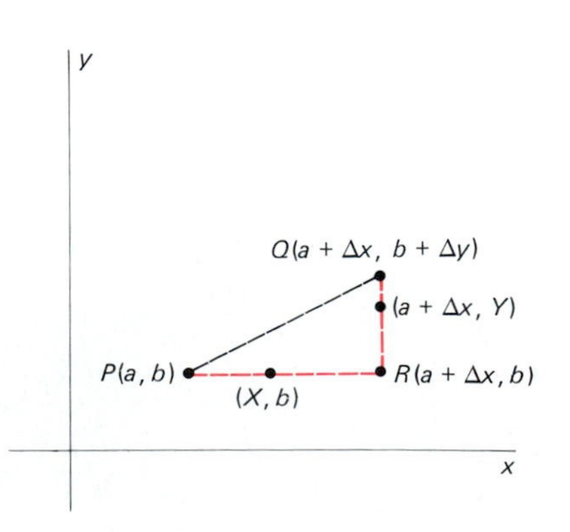

15.57 Illustration of the proof of the linear approximation theorem

Proof If R is the point $(a + \Delta x, b)$ indicated in Fig. 15.57, then

$$\begin{aligned}\Delta f &= f(Q) - f(P) = [f(R) - f(P)] + [f(Q) - f(R)] \\ &= [f(a + \Delta x, b) - f(a, b)] + [f(a + \Delta x, b + \Delta y) - f(a + \Delta x, b)]. \quad (9)\end{aligned}$$

We consider separately the two terms on the right in Equation (9). For the first term, we define the single-variable function

$$g(x) = f(x, b) \qquad \text{for} \quad x \text{ in } [a, a + \Delta x].$$

Then the mean value theorem gives

$$f(a + \Delta x, b) - f(a, b) = g(a + \Delta x) - g(x) = g'(X)\,\Delta x = f_x(X, b)\,\Delta x$$

for some number X in the open interval $(a, a + \Delta x)$.

For the second term on the right in Equation (9), we define the single-variable function

$$h(y) = f(a + \Delta x, y) \qquad \text{for} \qquad y \text{ in } [b, b + \Delta y].$$

The mean value theorem now yields

$$\begin{aligned} f(a + \Delta x, b + \Delta y) - f(a + \Delta x, b) &= h(b + \Delta y) - h(b) \\ &= h'(Y)\,\Delta y = f_y(a + \Delta x, Y)\,\Delta y \end{aligned}$$

for some number Y in the open interval $(b, b + \Delta y)$.

When we substitute these two results in Equation (9), we find that

$$\begin{aligned} \Delta f &= f_x(X, b)\Delta x + f_y(a + \Delta x, Y)\,\Delta y \\ &= [f_x(a, b) + f_x(X, b) - f_x(a, b)]\,\Delta x \\ &\quad + [f_y(a, b) + f_y(a + \Delta x, Y) - f_y(a, b)]\,\Delta y. \end{aligned}$$

So

$$\Delta f = f_x(a, b)\,\Delta x + f_y(a, b)\,\Delta y + \varepsilon_1\,\Delta x + \varepsilon_2\,\Delta y,$$

where

$$\varepsilon_1 = f_x(X, b) - f_x(a, b), \qquad \varepsilon_2 = f_y(a + \Delta x, Y) - f_y(a, b).$$

Finally, because the points (X, b) and $(a + \Delta x, Y)$ both approach (a, b) as $\Delta x \to 0$ and $\Delta y \to 0$, it follows from the continuity of f_x and f_y that ε_1 and ε_2 both approach zero as Δx and Δy approach zero. This completes the proof. ■

The function of Problem 39 illustrates the fact that a function $f(x, y)$ of two variables can have partial derivatives at a point without even being continuous there. Thus the mere existence of partial derivatives means much less for a function of two (or more) variables than does differentiability for a single-variable function. But it follows from the linear approximation theorem above that a function with partial derivatives that are *continuous* at every point within a circle is itself continuous within that circle (see Problem 40).

A function f of two variables is said to be **differentiable** at the point (a, b) if $f_x(a, b)$ and $f_y(a, b)$ both exist *and also* there exist functions ε_1 and ε_2 of Δx and Δy that approach zero as Δx and Δy do and such that Equation (8) holds. We shall have little need for this concept of differentiability, because it will always suffice for us to assume that our functions have continuous partial derivatives.

The basic theorem above can be generalized in a natural way to functions of three or more variables. For example, if $w = f(x, y, z)$, then the analogue of Equation (8) is

$$\Delta f = f_x(a, b, c)\,\Delta x + f_y(a, b, c)\,\Delta y + f_z(a, b, c)\,\Delta z$$
$$+\ \varepsilon_1\,\Delta x + \varepsilon_2\,\Delta y + \varepsilon_3\,\Delta z,$$

where ε_1, ε_2, and ε_3 all approach zero as Δx, Δy, and Δz approach zero. The proof for the three-variable case is much like the one given here for two variables.

15.6 PROBLEMS

Find the differential dw in Problems 1–16.

1. $w = 3x^2 + 4xy - 2y^3$

2. $w = \exp(-x^2 - y^2)$

3. $w = \sqrt{1 + x^2 + y^2}$

4. $w = xye^{x+y}$

5. $w = \arctan\left(\dfrac{y}{x}\right)$

6. $w = xz^2 - yx^2 + zy^2$

7. $w = \ln(x^2 + y^2 + z^2)$

8. $w = \sin xyz$

9. $w = x \tan yz$

10. $w = xye^{uv}$

11. $w = e^{-xyz}$

12. $w = \ln(1 + rs)$

13. $w = u^2 \exp(-v^2)$

14. $w = \dfrac{s + t}{s - t}$

15. $w = \sqrt{z^2 + y^2 + z^2}$

16. $w = pqr \exp(-p^2 - q^2 - r^2)$

In each of Problems 17–23, use differentials to approximate $\Delta f = f(Q) - f(P)$.

17. $f(x, y) = \sqrt{x^2 + y^2}$; $P(3, 4)$, $Q(2.97, 4.04)$

18. $f(x, y) = \sqrt{x^2 - y^2}$; $P(13, 5)$, $Q(13.2, 4.9)$

19. $f(x, y) = \dfrac{1}{1 + x + y}$; $P(3, 6)$, $Q(3.02, 6.05)$

20. $f(x, y, z) = \sqrt{xyz}$; $P(1, 3, 3)$, $Q(0.9, 2.9, 3.1)$

21. $f(x, y, z) = \sqrt{x^2 + y^2 + z^2}$; $P(3, 4, 12)$, $Q(3.03, 3.96, 12.05)$

22. $f(x, y, z) = \dfrac{xyz}{x + y + z}$; $P(2, 3, 5)$, $Q(1.98, 3.03, 4.97)$

23. $f(x, y, z) = e^{-xyz}$; $P(1, 0, -2)$, $Q(1.02, 0.03, -2.02)$

In each of Problems 24–29, use differentials to approximate the indicated number.

24. $\sqrt{26}\,\sqrt[3]{28}\,\sqrt[4]{17}$

25. $(\sqrt{15} + \sqrt{99})^2$

26. $\dfrac{\sqrt[3]{25}}{\sqrt[5]{30}}$

27. $e^{0.4} = \exp(1.1^2 - 0.9^2)$

28. The y-coordinate of the point P near $(1, 2)$ on the curve $2x^3 + 2y^3 = 9xy$, if the x-coordinate of P is 1.1.

29. The x-coordinate of the point P near $(2, 4)$ on the curve $4x^4 + 4y^4 = 17x^2y^2$, if the y-coordinate of P is 3.9.

30. The base and height of a rectangle are measured as 10 cm and 15 cm, respectively, with a possible error of as much as 0.1 cm in each. Use differentials to estimate the maximum resulting error in computing the area of the rectangle.

31. The base radius r and height h of a right circular cone are measured as 5 in. and 10 in., respectively. There is a possible error of as much as $\frac{1}{16}$ in. in each measurement. Use differentials to estimate the maximum resulting error that might occur in computing the volume of the cone.

32. The dimensions of a closed rectangular box are found by measurement to be 10 cm by 15 cm by 20 cm, but there is a possible error of 0.1 cm in each. Use differentials to estimate the maximum resulting error in computing the total surface area of the box.

33. A surveyor wants to find the area in acres (1 acre is 43,560 ft^2) of a certain field. She measures two adjacent sides, finding them to be $a = 500$ ft and $b = 700$ ft, with a possible error of as much as 1 ft in each. She finds the angle between these two sides to be $\theta = 30°$, with a possible error of as much as $0.25°$. The field is triangular, so its area is given by $A = \frac{1}{2}ab \sin \theta$. Use differentials to estimate the maximum resulting error, in acres, in computing the area of the field by this formula.

34. Use differentials to estimate the change in the volume of the gas of Example 3 if its pressure is decreased from 5 atm to 4.9 atm and its temperature is decreased from 300 K to 280 K.

35. The period of a pendulum of length L is given (approximately) by the formula $T = 2\pi\sqrt{L/g}$. Estimate the change in the period of a pendulum if its length is increased from 2 ft to 2 ft 1 in. and it is simultaneously moved from a location where g is exactly 32 ft/s^2 to one where $g = 32.2$ ft/s^2.

36. Given the pendulum of Problem 35, show that the relative error in the determination of T is half the difference of the relative errors in measuring L and g; that is, that

$$\frac{dT}{T} = \frac{1}{2}\left(\frac{dL}{L} - \frac{dg}{g}\right).$$

37. The range of a projectile fired (in vacuum) with initial velocity v_0 and inclination angle α from the horizontal is $R = \frac{1}{32}v_0^2 \sin 2\alpha$. Use differentials to approximate the change in range if v_0 is increased from 400 ft/s to 410 ft/s and α is increased from 30° to 31°.

38. A horizontal beam is supported at both ends and supports a uniform load. The deflection, or sag, at its midpoint is given by $S = k/wh^3$, where w and h are the width and height, respectively, of the beam, and k is a constant that depends on the length and composition of the beam and the amount of the load. Show that

$$dS = -S\left(\frac{1}{w}\,dw + \frac{3}{h}\,dh\right).$$

If S is 1 in. when w is 2 in. and h is 4 in., approximate the sag when $w = 2.1$ in. and $h = 4.1$ in. Compare your approximation with the actual value you compute from the formula $S = k/wh^3$.

39. Let the function f be defined on the whole xy-plane by $f(x, y) = 1$ if $x = y \neq 0$, while $f(x, y) = 0$ otherwise. Show that

(a) f is not continuous at (0, 0); but

(b) both first partial derivatives f_x and f_y exist at (0, 0).

40. Deduce from Equation (8) that under the hypotheses of the linear approximation theorem, $\Delta f \to 0$ as $\Delta x \to 0$ and $\Delta y \to 0$. What does this imply about the continuity of f at the point (a, b)?

15.7 The Chain Rule

The single-variable chain rule expresses the derivative of a composite function $f(g(t))$ in terms of the derivatives of f and g:

$$D_t f(g(t)) = f'(g(t))g'(t). \tag{1}$$

With $w = f(x)$ and $x = g(t)$, the chain rule says that

$$\frac{dw}{dt} = \frac{dw}{dx}\frac{dx}{dt}. \tag{2}$$

The simplest multivariable chain rule situation involves a function $w = f(x, y)$ where x and y are each functions of the same single variable t: $x = g(t)$ and $y = h(t)$. The composite function $f(g(t), h(t))$ is then a single-variable function of t, and the following theorem expresses its derivative in terms of the partial derivatives of f and the ordinary derivatives of g and h. We assume that the stated hypotheses hold on suitable domains such that the composite function is defined.

Theorem 1 *The Chain Rule*

Suppose that $w = f(x, y)$ has continuous first order partial derivatives, and that $x = g(t)$ and $y = h(t)$ are differentiable functions. Then w is a differentiable function of t, and

$$\frac{dw}{dt} = \frac{\partial w}{\partial x}\cdot\frac{dx}{dt} + \frac{\partial w}{\partial y}\cdot\frac{dy}{dt}. \tag{3}$$

The variable notation of the formula in (3) ordinarily will be more useful than function notation. It must be remembered, however, that the partial derivatives in (3) are to be evaluated at the point $(g(t), h(t))$, so what (3) actually says is that

$$D_t[f(g(t), h(t))] = f_x(g(t), h(t))g'(t) + f_y(g(t), h(t))h'(t). \tag{4}$$

EXAMPLE 1 Suppose that

$$w = e^{xy}, \qquad x = t^2, \quad \text{and} \quad y = t^3.$$

Then

$$\frac{\partial w}{\partial x} = ye^{xy}, \qquad \frac{\partial w}{\partial y} = xe^{xy},$$

$$\frac{dx}{dt} = 2t, \quad \text{and} \quad \frac{dy}{dt} = 3t^2.$$

So the formula in (3) yields

$$\frac{dw}{dt} = \frac{\partial w}{\partial x} \cdot \frac{dx}{dt} + \frac{\partial w}{\partial y} \cdot \frac{dy}{dt} = (ye^{xy})(2t) + (xe^{xy})(3t^2)$$

$$= (t^3 e^{t^5})(2t) + (t^2 e^{t^5})(3t^2) = 5t^4 e^{t^5}.$$

Had our purpose not been to illustrate the chain rule, we could have obtained the same result more simply by writing

$$w = e^{xy} = e^{(t^2)(t^3)} = e^{t^5}$$

and then differentiating w as a single-variable function of t.

Proof of the Chain Rule We choose a point t_0 at which we wish to compute dw/dt and write

$$a = g(t_0), \qquad b = h(t_0).$$

Let

$$\Delta x = g(t_0 + \Delta t) - g(t_0), \qquad \Delta y = h(t_0 + \Delta t) - h(t_0).$$

Then

$$g(t_0 + \Delta t) = a + \Delta x \quad \text{and} \quad h(t_0 + \Delta t) = b + \Delta y.$$

If

$$\begin{aligned} \Delta w &= f(g(t_0 + \Delta t), h(t_0 + \Delta t)) - f(g(t_0), h(t_0)) \\ &= f(a + \Delta x, b + \Delta y) - f(a, b), \end{aligned}$$

then what we need to compute is

$$\frac{dw}{dt} = \lim_{\Delta t \to 0} \frac{\Delta w}{\Delta t}.$$

The linear approximation theorem of Section 15.6 gives

$$\Delta w = f_x(a, b)\, \Delta x + f_y(a, b)\, \Delta y + \varepsilon_1\, \Delta x + \varepsilon_2\, \Delta y$$

where ε_1 and ε_2 approach zero as $\Delta x \to 0$ and $\Delta y \to 0$. We note that Δx and Δy approach zero as $\Delta t \to 0$ because the derivatives

$$\frac{dx}{dt} = \lim_{\Delta t \to 0} \frac{\Delta x}{\Delta t} \quad \text{and} \quad \frac{dy}{dt} = \lim_{\Delta t \to 0} \frac{\Delta y}{\Delta t}$$

both exist. Therefore,

$$\begin{aligned}
\frac{dw}{dt} &= \lim_{\Delta t \to 0} \frac{\Delta w}{\Delta t} \\
&= \lim_{\Delta t \to 0} \left[f_x(a, b) \frac{\Delta x}{\Delta t} + f_y(a, b) \frac{\Delta y}{\Delta t} + \varepsilon_1 \frac{\Delta x}{\Delta t} + \varepsilon_2 \frac{\Delta y}{\Delta t} \right] \\
&= f_x(a, b) \frac{dx}{dt} + f_y(a, b) \frac{dy}{dt} + 0 \frac{dx}{dt} + 0 \frac{dy}{dt};
\end{aligned}$$

hence

$$\frac{dw}{dt} = \frac{\partial w}{\partial x} \cdot \frac{dx}{dt} + \frac{\partial w}{\partial y} \cdot \frac{dy}{dt}.$$

Thus we have established the formula in (3), writing $\partial w/\partial x$ and $\partial w/\partial y$ for the partial derivatives $f_x(a, b)$ and $f_y(a, b)$ in the final step. ■

In the situation of Theorem 1, we may refer to w as the **dependent** variable, x and y as **intermediate** variables, and t as the **independent** variable. Then note that the right-hand side in Equation (3) has two terms, one for each intermediate variable, each like the right-hand side of the single-variable chain rule of Equation (2). If there are more than two intermediate variables, then there is still one term on the right-hand side for each intermediate variable. For example, if $w = f(x, y, z)$ with x, y, and z each a function of t, then the chain rule takes the form

$$\frac{dw}{dt} = \frac{\partial w}{\partial x} \cdot \frac{dx}{dt} + \frac{\partial w}{\partial y} \cdot \frac{dy}{dt} + \frac{\partial w}{\partial z} \cdot \frac{dz}{dt}. \tag{5}$$

The proof of (5) is essentially the same as the proof of (3); it uses the linear approximation theorem for three variables rather than for two variables.

EXAMPLE 2 Find dw/dt if $w = x^2 + ze^y + \sin xz$ and $x = t$, $y = t^2$, $z = t^3$.

Solution Equation (5) gives

$$\begin{aligned}
\frac{dw}{dt} &= \frac{\partial w}{\partial x} \cdot \frac{dx}{dt} + \frac{\partial w}{\partial y} \cdot \frac{dy}{dt} + \frac{\partial w}{\partial z} \cdot \frac{dz}{dt} \\
&= (2x + z \cos xz)(1) + (ze^y)(2t) + (e^y + x \cos xz)(3t^2) \\
&= 2t + (3t^2 + 2t^4)e^{t^2} + 4t^3 \cos t^4.
\end{aligned}$$

In this example we could check the result given by the chain rule by first writing w as a function of t and then computing the ordinary single-variable derivative of w with respect to t.

There may be several independent variables as well as several intermediate variables. For example, if $w = f(x, y, z)$, where $x = g(u, v)$, $y = h(u, v)$, and $z = k(u, v)$, so that

$$w = f(x, y, z) = f(g(u, v), h(u, v), k(u, v)),$$

then we have the three intermediate variables x, y, and z and the two independent variables u and v. In this case we would need to compute the *partial* derivatives $\partial w/\partial u$ and $\partial w/\partial v$ of the composite function. The general chain rule in Theorem 2 says that each partial derivative of the dependent variable w is given by a chain rule formula like (3) or (5). The only difference is that the derivatives with respect to the independent variable will be partial derivatives. For instance,

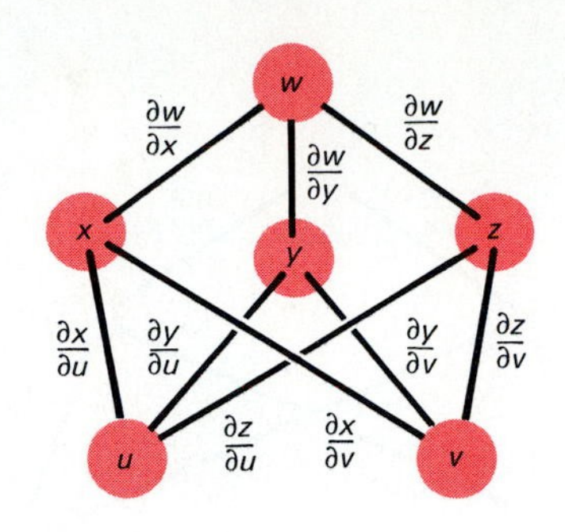

15.58 Diagram for $w = w(x, y, z)$ where $x = x(u, v)$, $y = y(u, v)$, $z = z(u, v)$

$$\frac{\partial w}{\partial u} = \frac{\partial w}{\partial x} \cdot \frac{\partial x}{\partial u} + \frac{\partial w}{\partial y} \cdot \frac{\partial y}{\partial u} + \frac{\partial w}{\partial z} \cdot \frac{\partial z}{\partial u}.$$

The "molecular model" in Fig. 15.58 illustrates this formula. The "atom" at the top represents the dependent variable w. The atoms at the next level represent the intermediate variables x, y, and z. The atoms at the bottom represent the independent variables u and v. Each "bond" in the model represents a partial derivative involving the two variables represented by the atoms joined by that bond. Finally, note that the formula displayed before this paragraph expresses $\partial w/\partial u$ as the sum of the products of the partial derivatives taken along all paths from w to u. Similarly, the sum of the products of the partial derivatives along all paths from w to v yields the correct formula

$$\frac{\partial w}{\partial v} = \frac{\partial w}{\partial x} \cdot \frac{\partial x}{\partial v} + \frac{\partial w}{\partial y} \cdot \frac{\partial y}{\partial v} + \frac{\partial w}{\partial z} \cdot \frac{\partial z}{\partial v}.$$

The following theorem describes the most general such situation.

Theorem 2 *The General Chain Rule*

Suppose that w is a function of the variables $x_1, x_2, x_3, \ldots, x_m$, and that each of these is a function of the variables $t_1, t_2, \ldots, t_n$. If all these functions have continuous first-order partial derivatives, then

$$\frac{\partial w}{\partial t_i} = \frac{\partial w}{\partial x_1} \cdot \frac{\partial x_1}{\partial t_i} + \frac{\partial w}{\partial x_2} \cdot \frac{\partial x_2}{\partial t_i} + \cdots + \frac{\partial w}{\partial x_m} \cdot \frac{\partial x_m}{\partial t_i} \tag{6}$$

for each i, $1 \leqq i \leqq n$.

Thus there is a formula in (6) for *each* of the independent variables $t_1, t_2, \ldots, t_n$, and the right-hand side of each such formula contains one typical chain rule term for each of the intermediate variables $x_1, x_2, \ldots, x_m$.

EXAMPLE 3 Suppose that

$$z = f(u, v), \qquad u = 2x + y, \qquad v = 3x - 2y.$$

Given the values $\partial z/\partial u = 3$ and $\partial z/\partial v = -2$ at the point $(u, v) = (3, 1)$, find the values $\partial z/\partial x$ and $\partial z/\partial y$ at the corresponding point $(x, y) = (1, 1)$.

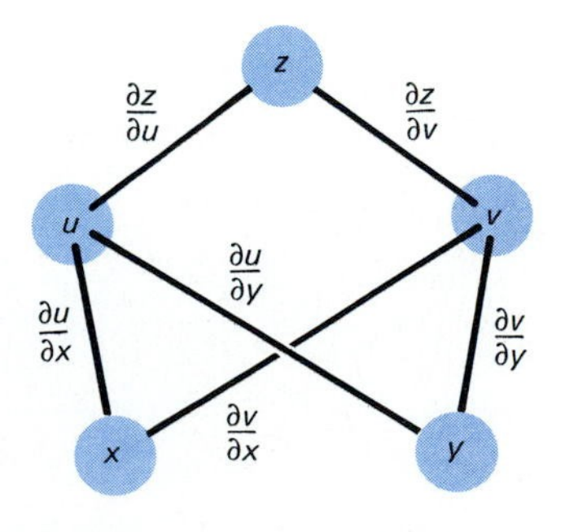

15.59 Diagram for $z = z(u, v)$ where $u = u(x, y)$ and $v = v(x, y)$

Solution The relationships between the variables are shown in Fig. 15.59. The chain rule gives

$$\frac{\partial z}{\partial x} = \frac{\partial z}{\partial u} \cdot \frac{\partial u}{\partial x} + \frac{\partial z}{\partial v} \cdot \frac{\partial v}{\partial x} = (3)(2) + (-2)(3) = 0$$

and

$$\frac{\partial z}{\partial y} = \frac{\partial z}{\partial u} \cdot \frac{\partial u}{\partial y} + \frac{\partial z}{\partial v} \cdot \frac{\partial v}{\partial y} = (3)(1) + (-2)(-2) = 7$$

at the indicated point $(x, y) = (1, 1)$.

15.60 Diagram for $w = w(x, y)$ where $x = x(r, \theta)$ and $y = y(r, \theta)$

EXAMPLE 4 Let $w = f(x, y)$, where x and y are given in polar coordinates by the equations $x = r \cos \theta$ and $y = r \sin \theta$. Calculate

$$\frac{\partial w}{\partial r}, \quad \frac{\partial w}{\partial \theta}, \quad \text{and} \quad \frac{\partial^2 w}{\partial r^2}$$

in terms of r and θ and the partial derivatives of w with respect to x and y. (See Fig. 15.60.)

Solution Here x and y are intermediate variables, while the independent variables are r and θ. First note that

$$\frac{\partial x}{\partial r} = \cos \theta, \qquad \frac{\partial y}{\partial r} = \sin \theta,$$

$$\frac{\partial x}{\partial \theta} = -r \sin \theta, \quad \text{and} \quad \frac{\partial y}{\partial \theta} = r \cos \theta.$$

Then

$$\frac{\partial w}{\partial r} = \frac{\partial w}{\partial x} \cdot \frac{\partial x}{\partial r} + \frac{\partial w}{\partial y} \cdot \frac{\partial y}{\partial r} = \frac{\partial w}{\partial x} \cos \theta + \frac{\partial w}{\partial y} \sin \theta \tag{7a}$$

and

$$\frac{\partial w}{\partial \theta} = \frac{\partial w}{\partial x} \cdot \frac{\partial x}{\partial \theta} + \frac{\partial w}{\partial y} \cdot \frac{\partial y}{\partial \theta} = -r \frac{\partial w}{\partial x} \sin \theta + r \frac{\partial w}{\partial y} \cos \theta. \tag{7b}$$

Next,

$$\frac{\partial^2 w}{\partial r^2} = \frac{\partial}{\partial r}\left(\frac{\partial w}{\partial r}\right) = \frac{\partial}{\partial r}\left(\frac{\partial w}{\partial x} \cos \theta + \frac{\partial w}{\partial y} \sin \theta\right)$$

$$= \frac{\partial w_x}{\partial r} \cos \theta + \frac{\partial w_y}{\partial r} \sin \theta,$$

where $w_x = \partial w/\partial x$ and $w_y = \partial w/\partial y$. Applying the formula in (7a) to calculate $\partial w_x/\partial r$ and $\partial w_y/\partial y$, we get

$$\frac{\partial^2 w}{\partial r^2} = \left(\frac{\partial w_x}{\partial x} \cdot \frac{\partial x}{\partial r} + \frac{\partial w_x}{\partial y} \cdot \frac{\partial y}{\partial r}\right) \cos \theta + \left(\frac{\partial w_y}{\partial x} \cdot \frac{\partial x}{\partial r} + \frac{\partial w_y}{\partial y} \cdot \frac{\partial y}{\partial r}\right) \sin \theta$$

$$= \left(\frac{\partial^2 w}{\partial x^2} \cos \theta + \frac{\partial^2 w}{\partial y\, \partial x} \sin \theta\right) \cos \theta + \left(\frac{\partial^2 w}{\partial x\, \partial y} \cos \theta + \frac{\partial^2 w}{\partial y^2} \sin \theta\right) \sin \theta.$$

Finally, because $w_{yx} = w_{xy}$, we get

$$\frac{\partial^2 w}{\partial r^2} = \frac{\partial^2 w}{\partial x^2} \cos^2 \theta + 2 \frac{\partial^2 w}{\partial x\, \partial y} \cos \theta \sin \theta + \frac{\partial^2 w}{\partial y^2} \sin^2 \theta. \tag{8}$$

EXAMPLE 5 Suppose that $w = f(u, v, x, y)$, where u and v are functions of x and y. Here x and y play dual roles as both intermediate and independent variables. The chain rule yields

$$\frac{\partial w}{\partial x} = \frac{\partial f}{\partial u}\cdot\frac{\partial u}{\partial x} + \frac{\partial f}{\partial v}\cdot\frac{\partial v}{\partial x} + \frac{\partial f}{\partial x}\cdot\frac{\partial x}{\partial x} + \frac{\partial f}{\partial y}\cdot\frac{\partial y}{\partial x}$$

$$= \frac{\partial f}{\partial u}\cdot\frac{\partial u}{\partial x} + \frac{\partial f}{\partial v}\cdot\frac{\partial v}{\partial x} + \frac{\partial f}{\partial x}$$

because $\partial x/\partial x = 1$ and $\partial y/\partial x = 0$. Similarly,

$$\frac{\partial w}{\partial y} = \frac{\partial f}{\partial u}\cdot\frac{\partial u}{\partial y} + \frac{\partial f}{\partial v}\cdot\frac{\partial v}{\partial y} + \frac{\partial f}{\partial y}.$$

Note that these results are consistent with the paths from w to x and from w to y in the molecular model shown in Fig. 15.61.

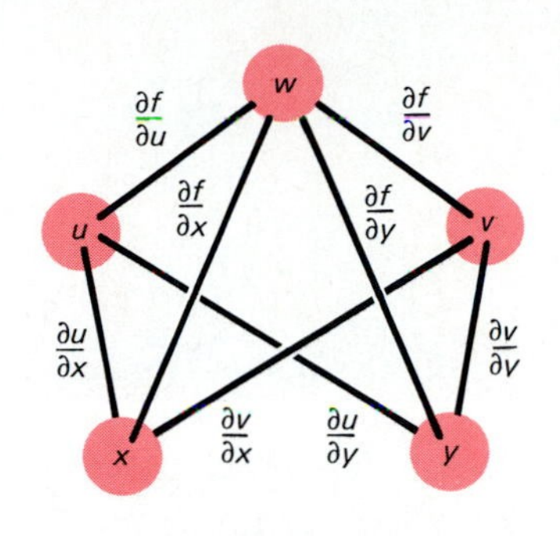

15.61 Diagram for $w = f(u, v, x, y)$ where $u = u(x, y)$ and $v = v(x, y)$

EXAMPLE 6 Consider a parametric curve $x = x(t)$, $y = y(t)$, $z = z(t)$ lying on the surface $z = f(x, y)$ in space. Recall that if

$$\mathbf{T} = \left\langle \frac{dx}{dt}, \frac{dy}{dt}, \frac{dz}{dt} \right\rangle \quad \text{and} \quad \mathbf{N} = \left\langle \frac{\partial z}{\partial x}, \frac{\partial z}{\partial y}, -1 \right\rangle,$$

then $\mathbf{T}$ is tangent to the curve and $\mathbf{N}$ is normal to the surface. Show that $\mathbf{T}$ and $\mathbf{N}$ are everywhere perpendicular.

Solution The chain rule in Equation (3) tells us that

$$\frac{dz}{dt} = \frac{\partial z}{\partial x}\cdot\frac{dx}{dt} + \frac{\partial z}{\partial y}\cdot\frac{dy}{dt}.$$

But this equation is equivalent to the vector equation

$$\left\langle \frac{\partial z}{\partial x}, \frac{\partial z}{\partial y}, -1 \right\rangle \cdot \left\langle \frac{dx}{dt}, \frac{dy}{dt}, \frac{dz}{dt} \right\rangle = 0.$$

Thus $\mathbf{N}\cdot\mathbf{T} = 0$, so $\mathbf{N}$ and $\mathbf{T}$ are indeed perpendicular.

Theorem 3 *Implicit Partial Differentiation*

Suppose that the function $F(x, y, z)$ has continuous first partial derivatives, and that the equation $F(x, y, z) = 0$ implicitly defines a function $z = f(x, y)$ that has continuous first order partial derivatives. Then

$$\frac{\partial z}{\partial x} = -\frac{F_x}{F_z} \quad \text{and} \quad \frac{\partial z}{\partial y} = -\frac{F_y}{F_z} \tag{9}$$

wherever $F_z = \partial F/\partial z \neq 0$.

Proof Because $w = F(x, y, f(x, y))$ is identically zero, differentiation with respect to x yields

$$0 = \frac{\partial w}{\partial x} = \frac{\partial F}{\partial x}\cdot\frac{\partial x}{\partial x} + \frac{\partial F}{\partial y}\cdot\frac{\partial y}{\partial x} + \frac{\partial F}{\partial z}\cdot\frac{\partial z}{\partial x}$$

$$= (F_x)(1) + (F_y)(0) + (F_z)\left(\frac{\partial z}{\partial x}\right),$$

so

$$F_x + F_z \frac{\partial z}{\partial x} = 0.$$

This gives the first formula in (9). The second is obtained similarly by differentiating w with respect to y. ■

REMARK In a specific example it is usually simpler to differentiate the equation $F(x, y, f(x, y)) = 0$ implicitly than to apply the formulas in (9).

EXAMPLE 7 Find the tangent plane at the point (1, 3, 2) to the surface with equation

$$z^3 + xz - y^2 = 1.$$

Solution Partial differentiation of the given equation with respect to x and with respect to y yields the equations

$$3z^2 \frac{\partial z}{\partial x} + z + x \frac{\partial z}{\partial x} = 0 \quad \text{and}$$

$$3z^2 \frac{\partial z}{\partial y} + x \frac{\partial z}{\partial y} - 2y = 0.$$

When we substitute $x = 1$, $y = 3$, and $z = 2$, we find that $\partial z/\partial x = -\frac{2}{13}$ and $\partial z/\partial y = \frac{6}{13}$. Hence an equation of the tangent plane in question is

$$z - 2 = -\tfrac{2}{13}(x - 1) + \tfrac{6}{13}(y - 3);$$

that is,

$$2x - 6y + 13z = 10.$$

15.7 PROBLEMS

In Problems 1–4, find dw/dt both by using the chain rule *and* by expressing w explicitly as a function of t before differentiating.

1. $w = \exp(-x^2 - y^2)$, $x = t$, $y = t^{1/2}$

2. $w = \dfrac{1}{u^2 + v^2}$, $u = \cos 2t$, $v = \sin 2t$

3. $w = \sin xyz$, $x = t$, $y = t^2$, $z = t^3$

4. $w = \ln(u + v + z)$, $u = \cos^2 t$, $v = \sin^2 t$, $z = t^2$

In Problems 5–8, find $\partial w/\partial s$ and $\partial w/\partial t$.

5. $w = \ln(x^2 + y^2 + z^2)$, $x = s - t$, $y = s + t$, $z = 2\sqrt{st}$

6. $w = pq \sin r$, $p = 2s + t$, $q = s - t$, $r = st$

7. $w = (u^2 + v^2 + z^2)^{1/2}$, $u = 3e^t \sin s$, $v = 3e^t \cos s$, $z = 4e^t$

8. $w = yz + zx + xy$, $x = s^2 - t^2$, $y = s^2 + t^2$, $z = s^2t^2$

In Problems 9–10, find $\partial r/\partial x$, $\partial r/\partial y$, and $\partial r/\partial z$.

9. $r = e^{u+v+w}$, $u = yz$, $v = xz$, $w = xy$

10. $r = uvw - u^2 - v^2 - w^2$, $u = y + z$, $v = x + z$, $w = x + y$

In Problems 11–15, find $\partial z/\partial x$ and $\partial z/\partial y$ as functions of x, y, and z, assuming that $z = f(x, y)$ satisfies the given equation.

11. $x^{2/3} + y^{2/3} + z^{2/3} = 1$

12. $x^3 + y^3 + z^3 = xyz$

13. $xe^{xy} + ye^{zx} + ze^{xy} = 3$

14. $x^5 + xy^2 + yz = 5$

15. $\dfrac{x^2}{a^2} + \dfrac{y^2}{b^2} + \dfrac{z^2}{c^2} = 1$

In Problems 16–19, use the method of Example 5 to find $\partial w/\partial x$ and $\partial w/\partial y$ as functions of x and y.

16. $w = u^2 + v^2 + x^2 + y^2, \quad u = x - y, \quad v = x + y$

17. $w = \sqrt{uvxy}, \quad u = \sqrt{x - y}, \quad v = \sqrt{x + y}$

18. $w = xy \ln(u + v), \quad u = (x^2 + y^2)^{1/3}, \quad v = (x^3 + y^3)^{1/2}$

19. $w = uv - xy, \quad u = \dfrac{x}{x^2 + y^2}, \quad v = \dfrac{y}{x^2 + y^2}$

In Problems 20–23, write an equation for the tangent plane at the point P to the surface with the given equation.

20. $x^2 + y^2 + z^2 = 9; \quad P(1, 2, 2)$

21. $x^2 + 2y^2 + 2z^2 = 14; \quad P(2, 1, -2)$

22. $x^3 + y^3 + z^3 = 5xyz; \quad P(2, 1, 1)$

23. $z^3 + (x + y)z^2 + x^2 + y^2 = 13; \quad P(2, 2, 1)$

24. Suppose that $y = g(x, z)$ satisfies the equation $F(x, y, z) = 0$ and that $F_y \neq 0$. Show that

$$\frac{\partial y}{\partial x} = -\frac{\partial F/\partial x}{\partial F/\partial y}.$$

25. Suppose that $x = h(y, z)$ satisfies the equation $F(x, y, z) = 0$ and that $F_x \neq 0$. Show that

$$\frac{\partial x}{\partial y} = -\frac{\partial F/\partial y}{\partial F/\partial x}.$$

26. Suppose that $w = f(x, y)$, $x = r\cos\theta$, and $y = r\sin\theta$. Show that

$$\left(\frac{\partial w}{\partial x}\right)^2 + \left(\frac{\partial w}{\partial y}\right)^2 = \left(\frac{\partial w}{\partial r}\right)^2 + \frac{1}{r^2}\left(\frac{\partial w}{\partial \theta}\right)^2.$$

27. Suppose that $w = f(u)$ and that $u = x + y$. Show that $\partial w/\partial x = \partial w/\partial y$.

28. Suppose that $w = f(u)$ and that $u = x - y$. Show that $\partial w/\partial x = -\partial w/\partial y$ and that

$$\frac{\partial^2 w}{\partial x^2} = \frac{\partial^2 w}{\partial y^2} = -\frac{\partial^2 w}{\partial x\,\partial y}.$$

29. Suppose that $w = f(x, y)$ where $x = u + v$ and $y = u - v$. Show that

$$\frac{\partial^2 w}{\partial x^2} - \frac{\partial^2 w}{\partial y^2} = \frac{\partial^2 w}{\partial u\,\partial v}.$$

30. Assume that $w = f(x, y)$ where $x = 2u + v$ and $y = u - v$. Show that

$$5\frac{\partial^2 w}{\partial x^2} + 2\frac{\partial^2 w}{\partial x\,\partial y} + 2\frac{\partial^2 w}{\partial y^2} = \frac{\partial^2 w}{\partial u^2} + \frac{\partial^2 w}{\partial v^2}.$$

31. Suppose that $w = f(x, y)$, $x = r\cos\theta$, and $y = r\sin\theta$. Show that

$$\frac{\partial^2 w}{\partial x^2} + \frac{\partial^2 w}{\partial y^2} = \frac{\partial^2 w}{\partial r^2} + \frac{1}{r}\frac{\partial w}{\partial r} + \frac{1}{r^2}\frac{\partial^2 w}{\partial \theta^2}.$$

(*Suggestion:* First find $\partial^2 w/\partial\theta^2$ by the method of Example 4. Then combine the result with Equations (7) and (8) of this sect on.)

32. Suppose that

$$w = \frac{1}{r} f\left(t - \frac{r}{a}\right)$$

and that $r = \sqrt{x^2 + y^2 + z^2}$. Show that

$$\frac{\partial^2 w}{\partial x^2} + \frac{\partial^2 w}{\partial y^2} + \frac{\partial^2 w}{\partial z^2} = \frac{1}{a^2}\frac{\partial^2 w}{\partial t^2}.$$

33. Suppose that $w = f(r)$ and that $r = \sqrt{x^2 + y^2 + z^2}$. Show that

$$\frac{\partial^2 w}{\partial x^2} + \frac{\partial^2 w}{\partial y^2} + \frac{\partial^2 w}{\partial z^2} = \frac{d^2 w}{dr^2} + \frac{2}{r}\frac{dw}{dr}.$$

34. Suppose that $w = f(u) + g(v)$, that $u = x - at$, and that $v = x + at$. Show that

$$\frac{\partial^2 w}{\partial t^2} = a^2\frac{\partial^2 w}{\partial x^2}.$$

35. Assume that $w = f(u, v)$ where $u = x + y$ and $v = x - y$. Show that

$$\frac{\partial w}{\partial x}\frac{\partial w}{\partial y} = \left(\frac{\partial w}{\partial u}\right)^2 - \left(\frac{\partial w}{\partial v}\right)^2.$$

36. Given: $w = f(x, y)$, $x = e^u\cos v$, and $y = e^u\sin v$. Show that

$$\left(\frac{\partial w}{\partial x}\right)^2 + \left(\frac{\partial w}{\partial y}\right)^2 = e^{-2u}\left[\left(\frac{\partial w}{\partial u}\right)^2 + \left(\frac{\partial w}{\partial v}\right)^2\right].$$

37. Assume that $w = f(x, y)$ and that there is a constant α such that $x = u\cos\alpha - v\sin\alpha$ and $y = u\sin\alpha + v\cos\alpha$. Show that

$$\left(\frac{\partial w}{\partial u}\right)^2 + \left(\frac{\partial w}{\partial v}\right)^2 = \left(\frac{\partial w}{\partial x}\right)^2 + \left(\frac{\partial w}{\partial y}\right)^2.$$

38. Suppose that $w = f(u)$, where

$$u = \frac{x^2 - y^2}{x^2 + y^2}.$$

Show that $xw_x + yw_y = 0$.

Suppose that the equation $F(x, y, z) = 0$ defines implicitly the three functions $z = f(x, y)$, $y = g(x, z)$, and $x = h(y, z)$. To keep track of the various partial derivatives, the notation

$$\left(\frac{\partial z}{\partial x}\right)_y = \frac{\partial f}{\partial x}, \qquad \left(\frac{\partial z}{\partial y}\right)_x = \frac{\partial f}{\partial y}, \tag{10a}$$

$$\left(\frac{\partial y}{\partial x}\right)_z = \frac{\partial g}{\partial x}, \qquad \left(\frac{\partial y}{\partial z}\right)_x = \frac{\partial g}{\partial z}, \tag{10b}$$

$$\left(\frac{\partial x}{\partial y}\right)_z = \frac{\partial h}{\partial y}, \qquad \left(\frac{\partial x}{\partial z}\right)_y = \frac{\partial h}{\partial z} \tag{10c}$$

is used. In short, the general symbol $(\partial w/\partial u)_v$ denotes the derivative of w with respect to u, with w regarded as a function of the independent variables u and v.

39. Using the notation in (10), show that

$$\left(\frac{\partial x}{\partial y}\right)_z \left(\frac{\partial y}{\partial z}\right)_x \left(\frac{\partial z}{\partial x}\right)_y = -1.$$

(*Suggestion:* Find the three partial derivatives on the right-hand side in terms of F_x, F_y, and F_z.)

40. Verify the result in Problem 39 for the equation

$$F(x, y, z) = x^2 + y^2 + z^2 - 1 = 0.$$

41. Verify the result in Problem 39 (with P, V, and T in place of x, y, and z) for the equation

$$F(P, V, T) = PV - nRT = 0$$

(n and R constant) that expresses the ideal gas law.

15.8 Directional Derivatives and the Gradient Vector

The change in the value of the function $w = f(x, y, z)$ from the point $P(x, y, z)$ to the nearby point $Q(x + \Delta x, y + \Delta y, z + \Delta z)$ is given by the increment

$$\Delta w = f(Q) - f(P). \tag{1}$$

The linear approximation of Section 15.6 yields

$$\Delta w \approx \frac{\partial f}{\partial x}\Delta x + \frac{\partial f}{\partial y}\Delta y + \frac{\partial f}{\partial z}\Delta z. \tag{2}$$

This approximation can be expressed concisely in terms of the **gradient vector** ∇f of the function f, which is defined to be

$$\nabla f(x, y, z) = \mathbf{i} f_x(x, y, z) + \mathbf{j} f_y(x, y, z) + \mathbf{k} f_z(x, y, z). \tag{3}$$

We also write

$$\nabla f = \left\langle \frac{\partial f}{\partial x}, \frac{\partial f}{\partial y}, \frac{\partial f}{\partial z} \right\rangle = \frac{\partial f}{\partial x}\mathbf{i} + \frac{\partial f}{\partial y}\mathbf{j} + \frac{\partial f}{\partial z}\mathbf{k}.$$

Then (2) implies that the increment $\Delta w = f(Q) - f(P)$ is given approximately by

$$\Delta w \approx \nabla f(P) \cdot \mathbf{v} \tag{4}$$

where $\mathbf{v} = \overrightarrow{PQ} = \langle \Delta x, \Delta y, \Delta z \rangle$ is the *displacement vector* from P to Q.

EXAMPLE 1 If $f(x, y, z) = x^2 + yz - 2xy - z^2$, then the definition of the gradient vector in (3) yields

$$\nabla f(x, y, z) = \frac{\partial f}{\partial x}\mathbf{i} + \frac{\partial f}{\partial y}\mathbf{j} + \frac{\partial f}{\partial z}\mathbf{k}$$

$$= (2x - 2y)\mathbf{i} + (z - 2x)\mathbf{j} + (y - 2z)\mathbf{k}.$$

For instance, the value of ∇f at the point $P(2, 1, 3)$ is

$$\nabla f(P) = \nabla f(2, 1, 3) = 2\mathbf{i} - \mathbf{j} - 5\mathbf{k}.$$

To apply Equation (4), we first calculate

$$f(P) = f(2, 1, 3) = 2^2 + 1 \cdot 3 - 2 \cdot 2 \cdot 1 - 3^2 = -6.$$

If Q is the nearby point $(1.9, 1.2, 3.1)$, then $\overrightarrow{PQ} = \mathbf{v} = (-0.1, 0.2, 0.1)$, so the approximation in (4) gives

$$f(Q) - f(P) \approx \nabla f(P) \cdot \mathbf{v} = \langle 2, -1, -5 \rangle \cdot \langle -0.1, 0.2, 0.1 \rangle = -0.9.$$

Hence $f(P) \approx -6 + (-0.9) = -6.9$. In this case we can readily calculate for comparison the exact value $f(Q) = -6.84$.

DIRECTIONAL DERIVATIVES

We know that the partial derivatives $f_x(x, y, z)$, $f_y(x, y, z)$, and $f_z(x, y, z)$ give the rates of change of $w = f(x, y, z)$ at the point $P(x, y, z)$ in the x-, y-, and z-directions, respectively. We now can use the gradient vector ∇f to calculate the rate of change of w at P in an *arbitrary* direction. Recall that a "direction" is prescribed by a *unit* vector $\mathbf{u}$.

Let Q be a point on the ray in the direction of $\mathbf{u}$ from the point P (see Fig. 15.62). The **average rate of change of w with respect to distance between P and Q** is

$$\frac{f(Q) - f(P)}{|\overrightarrow{PQ}|} = \frac{\Delta w}{\Delta s},$$

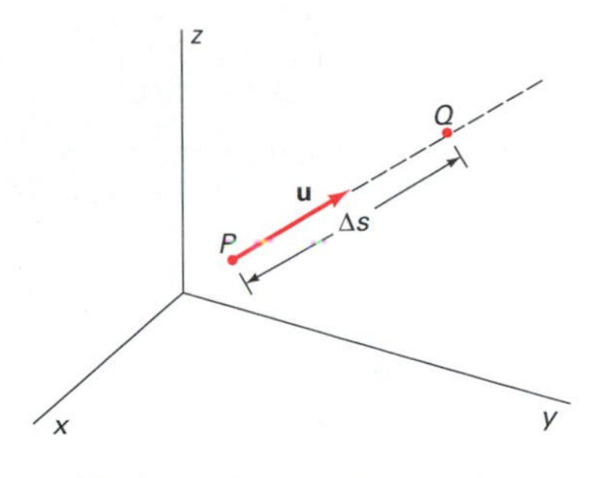

15.62 First step in computing the rate of change of $f(x, y, z)$ in the direction of the unit vector $\mathbf{u}$

where $\Delta s = |\overrightarrow{PQ}| = |\mathbf{v}|$ is the distance from P to Q. Then the approximation in (4) yields

$$\frac{\Delta w}{\Delta s} \approx \frac{\nabla f(P) \cdot \mathbf{v}}{|\mathbf{v}|} = \nabla f(P) \cdot \mathbf{u} \tag{5}$$

where $\mathbf{u} = \mathbf{v}/|\mathbf{v}|$ is the *unit* vector in the direction from P to Q. When we take the limit of the average rate of change $\Delta w/\Delta s$ as $\Delta s \to 0$, we get the *instantaneous* rate of change

$$\frac{dw}{ds} = \lim_{\Delta s \to 0} \frac{\Delta w}{\Delta s} = \nabla f(P) \cdot \mathbf{u}. \tag{6}$$

This computation motivates the *definition*

$$D_{\mathbf{u}} f(P) = \nabla f(P) \cdot \mathbf{u} \tag{7}$$

of the **directional derivative of f at $P(x, y, z)$ in the direction u.** In physics and engineering texts one may see the notation

$$\left.\frac{df}{ds}\right|_P = D_{\mathbf{u}} f(P),$$

or simply dw/ds as in (6), for the rate of change of the function $w = f(x, y, z)$ *with respect to distance* s in the direction of the unit vector $\mathbf{u}$.

REMARK Remember that the vector $\mathbf{u}$ in (7) is a *unit* vector: $|\mathbf{u}| = 1$. If $\mathbf{u} = \langle a, b, c \rangle$ then (7) implies simply that

$$D_{\mathbf{u}} f = a\frac{\partial f}{\partial x} + b\frac{\partial f}{\partial y} + c\frac{\partial f}{\partial z}. \tag{8}$$

EXAMPLE 2 Suppose that the temperature at the point (x, y, z), with distance measured in kilometers, is given by

$$w = f(x, y, z) = 10 + xy + xz + yz$$

(in degrees Celsius). Find the rate of change (in degrees per kilometer) of temperature at the point $P(1, 2, 3)$ in the direction of the vector

$$\mathbf{v} = \mathbf{i} + 2\mathbf{j} - 2\mathbf{k}.$$

Solution Because $\mathbf{v}$ is not a unit vector, we must replace it by a unit vector having the same direction before using the formulas of this section. So we take

$$\mathbf{u} = \frac{\mathbf{v}}{|\mathbf{v}|} = \left\langle \frac{1}{3}, \frac{2}{3}, -\frac{2}{3} \right\rangle.$$

The gradient vector of f is

$$\nabla f = (y + z)\mathbf{i} + (x + z)\mathbf{j} + (x + y)\mathbf{k},$$

so $\nabla f(1, 2, 3) = 5\mathbf{i} + 4\mathbf{j} + 3\mathbf{k}$. Hence (7) gives

$$D_{\mathbf{u}} f(P) = \langle 5, 4, 3 \rangle \cdot \langle \tfrac{1}{3}, \tfrac{2}{3}, -\tfrac{2}{3} \rangle = \tfrac{7}{3}$$

(degrees per kilometer) for the desired rate of change of temperature with respect to distance.

The directional derivative $D_{\mathbf{u}} f$ is closely related to a version of the multivariable chain rule. Suppose that the first-order partial derivatives of f are continuous and that

$$\mathbf{r}(t) = x(t)\mathbf{i} + y(t)\mathbf{j} + z(t)\mathbf{k}$$

is a differentiable vector-valued function. Then

$$f(\mathbf{r}(t)) = f(x(t), y(t), z(t))$$

is a differentiable function of t, and its (ordinary) derivative with respect to t is

$$D_t f(\mathbf{r}(t)) = D_t[f(x(t), y(t), z(t))]$$

$$= \frac{\partial f}{\partial x} \cdot \frac{dx}{dt} + \frac{\partial f}{\partial y} \cdot \frac{dy}{dt} + \frac{\partial f}{\partial z} \cdot \frac{dz}{dt}.$$

Hence

$$D_t f(\mathbf{r}(t)) = \nabla f(\mathbf{r}(t)) \cdot \mathbf{r}'(t), \tag{9}$$

where $\mathbf{r}'(t) = \langle x'(t), y'(t), z'(t) \rangle$ is the velocity vector of the parametric curve $\mathbf{r}(t)$. Equation (9) is the **vector chain rule.** Note that the operation on the right-hand side in (9) is the *dot* product, because the gradient of f and the derivative of $\mathbf{r}$ are both *vector*-valued functions.

If the velocity vector $\mathbf{v}(t) = \mathbf{r}'(t) \neq \mathbf{0}$, then $\mathbf{v} = v\mathbf{u}$ where $v = |\mathbf{v}|$ is the speed and $\mathbf{u} = \mathbf{v}/v$ is the unit tangent vector to the curve. Then (9) implies that

$$D_t f(\mathbf{r}(t)) = vD_{\mathbf{u}} f(\mathbf{r}(t)). \tag{10}$$

With $w = f(\mathbf{r}(t))$, $D_{\mathbf{u}} f = dw/ds$, and $v = ds/dt$, Equation (10) takes the pleasant chain-rule form

$$\frac{dw}{dt} = \frac{dw}{ds} \cdot \frac{ds}{dt}. \tag{11}$$

EXAMPLE 3 If the function

$$w = f(x, y, z) = 10 + xy + xz + yz$$

of Example 2 gives the temperature, what time rate of change (degrees per minute) will a hawk observe as it flies through $P(1, 2, 3)$ at a speed of 2 km/min, heading directly toward the point $Q(3, 4, 4)$?

Solution In Example 2 we calculated $\nabla f(P) = \langle 5, 4, 3\rangle$, and the unit vector in the direction from P to Q is

$$\mathbf{u} = \frac{\overrightarrow{PQ}}{|\overrightarrow{PQ}|} = \left\langle \frac{2}{3}, \frac{2}{3}, \frac{1}{3} \right\rangle.$$

Then

$$D_{\mathbf{u}} f(P) = \nabla f(P) \cdot \mathbf{u} = \langle 5, 4, 3\rangle \cdot \langle \tfrac{2}{3}, \tfrac{2}{3}, \tfrac{1}{3}\rangle = 7$$

(degrees per kilometer). Hence the formula in (11) yields

$$\frac{dw}{dt} = \frac{dw}{ds} \cdot \frac{ds}{dt} = \left(7 \frac{\text{deg}}{\text{km}}\right)\left(2 \frac{\text{km}}{\text{min}}\right) = 14 \frac{\text{deg}}{\text{min}}$$

as the hawk's time rate of change of temperature.

INTERPRETATION OF THE GRADIENT VECTOR

As yet we have discussed directional derivatives only for functions of three variables. For a function of two variables the formulas are analogous:

$$\nabla f(x, y) = \left\langle \frac{\partial f}{\partial x}, \frac{\partial f}{\partial y} \right\rangle = \frac{\partial f}{\partial x}\mathbf{i} + \frac{\partial f}{\partial y}\mathbf{j} \tag{12}$$

and

$$D_{\mathbf{u}} f(x, y) = \nabla f(x, y) \cdot \mathbf{u} = a\frac{\partial f}{\partial x} + b\frac{\partial f}{\partial y} \tag{13}$$

where $\mathbf{u} = \langle a, b\rangle$ is a unit vector. If α is the angle of inclination of $\mathbf{u}$ (measured counterclockwise from the positive x-axis, as in Fig. 15.63), then $a = \cos\alpha$ and $b = \sin\alpha$, so Equation (13) takes the form

$$D_{\mathbf{u}} f(x, y) = \frac{\partial f}{\partial x}\cos\alpha + \frac{\partial f}{\partial y}\sin\alpha. \tag{14}$$

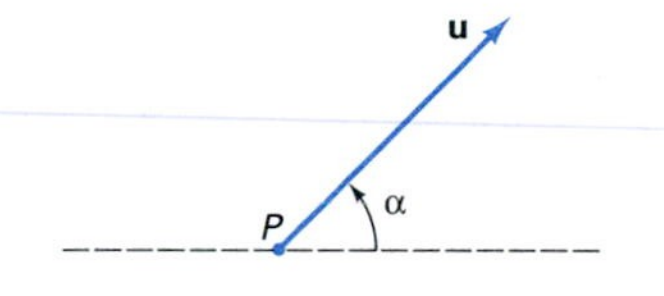

15.63 The unit vector **u** in Eq. (14)

The gradient vector ∇f has an important interpretation that involves the *maximal* directional derivative of f. If ϕ is the angle between ∇f at the point P and the unit vector $\mathbf{u}$ (Fig. 15.64), then the formula in Equation (7) gives

$$D_{\mathbf{u}} f(P) = \nabla f(P) \cdot \mathbf{u} = |\nabla f(P)| \cos\phi$$

because $|\mathbf{u}| = 1$. The maximum value of $\cos\phi$ is 1, and this occurs when $\phi = 0$. This is so when $\mathbf{u}$ is the particular unit vector $\nabla f(P)/|\nabla f(P)|$ that points in the direction of the gradient vector. In this case the formula above yields

$$D_{\mathbf{u}} f(P) = |\nabla f(P)|,$$

so that the value of the directional derivative is the length of the gradient vector. We have therefore proved the following theorem.

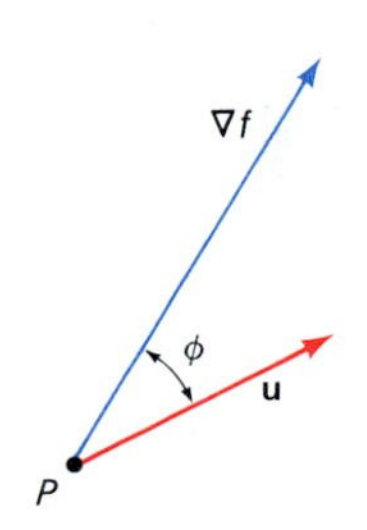

15.64 The angle ϕ between ∇f and the unit vector **u**

Theorem 1 *Significance of the Gradient Vector*
The maximum value of the directional derivative $D_{\mathbf{u}} f(P)$ is obtained when $\mathbf{u}$ is the unit vector in the direction of the gradient vector $\nabla f(P)$; that is, when $\mathbf{u} = \nabla f(P)/|\nabla f(P)|$. The value of this maximum directional derivative is $|\nabla f(P)|$, the length of the gradient vector.

Thus *the gradient vector* ∇f *points in the direction in which the function* f *increases most rapidly, and its length is the rate of increase of* f *(with respect to distance) in that direction.*

For instance, if the function f gives the temperature, then the gradient vector $\nabla f(P)$ points in the direction that a bumblebee at P should initially fly in order to get warmer the fastest.

EXAMPLE 4 Suppose that the temperature w (in degrees Celsius) at the point (x, y) is given by

$$w = f(x, y) = 10 + (0.003)x^2 - (0.004)y^2.$$

In what direction $\mathbf{u}$ should a bumblebee at the point (40, 30) initially fly in order to get warmer fastest? Find the directional derivative $D_{\mathbf{u}} f(40, 30)$ in this optimal direction $\mathbf{u}$.

Solution The gradient vector is

$$\nabla f = \frac{\partial f}{\partial x}\mathbf{i} + \frac{\partial f}{\partial y}\mathbf{j} = (0.006x)\mathbf{i} - (0.008y)\mathbf{j},$$

so

$$\nabla f(40, 30) = (0.24)\mathbf{i} - (0.24)\mathbf{j} = (0.24\sqrt{2})\mathbf{u}.$$

The unit vector

$$\mathbf{u} = \frac{\nabla f(40, 30)}{|\nabla f(40, 30)|} = \frac{\mathbf{i} - \mathbf{j}}{\sqrt{2}}$$

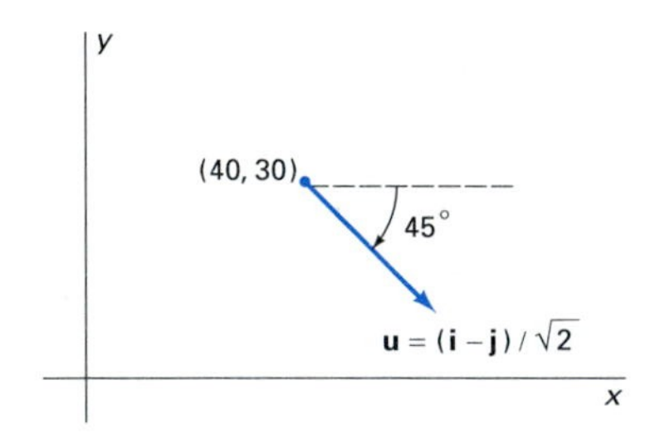

15.65 The unit vector $\mathbf{u} = \nabla f / |\nabla f|$ of Example 4

points southeast (Fig. 15.65); this is the direction in which the bumblebee should initially fly. And, according to Theorem 1, the derivative of f in this optimal direction is

$$D_{\mathbf{u}} f(40, 30) = |\nabla f(40, 30)| = (0.24)\sqrt{2} \approx 0.34$$

degrees per unit of distance.

THE GRADIENT VECTOR AS A NORMAL VECTOR

Consider the graph of the equation

$$F(x, y, z) = 0, \tag{15}$$

where F is a function with continuous first-order partial derivatives. According to the **implicit function theorem** of advanced calculus, near every point where $\nabla F \neq \mathbf{0}$—that is, at least one of the partial derivatives of F is nonzero—the graph of Equation (15) agrees with the graph of an equation of one of the forms

$$z = f(x, y), \qquad y = g(x, z), \qquad x = h(y, z).$$

Because of this, we are justified in general in referring to the graph of Equation (15) as a "surface." The gradient vector ∇F is normal to this surface, in the sense of the following theorem.

Theorem 2 *Gradient Vector as Normal Vector*

Suppose that $F(x, y, z)$ has continuous first-order partial derivatives, and let $P_0(x_0, y_0, z_0)$ be a point of the graph of the equation $F(x, y, z) = 0$ at which $\nabla F(P_0) \neq \mathbf{0}$. If $\mathbf{r}(t)$ is a differentiable curve on this surface with $\mathbf{r}(t_0) = \langle x_0, y_0, z_0 \rangle$, then

$$\nabla F(P_0) \cdot \mathbf{r}'(t_0) = 0. \tag{16}$$

Thus $\nabla F(P_0)$ is perpendicular to the tangent vector $\mathbf{r}'(t_0)$, as indicated in Fig. 15.66.

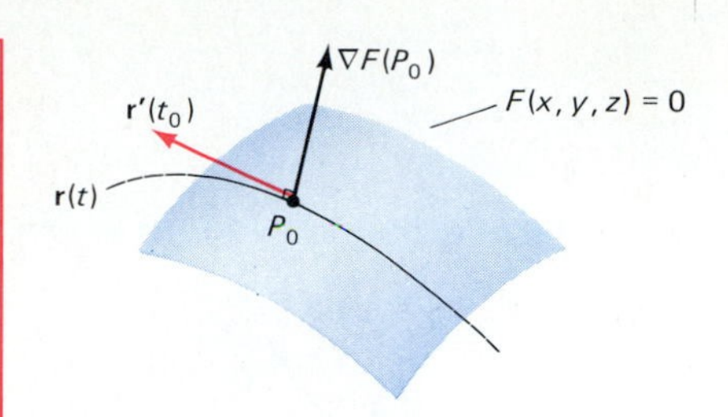

15.66 The gradient vector ∇F is normal to every curve in the surface $F(x, y, z) = 0$.

Proof The statement that $\mathbf{r}(t)$ lies on the surface $F(x, y, z) = 0$ means that $F(\mathbf{r}(t)) = 0$ for all t. Hence

$$0 = D_t F(\mathbf{r}(t_0)) = \nabla F(\mathbf{r}(t_0)) \cdot \mathbf{r}'(t_0) = \nabla F(P_0) \cdot \mathbf{r}'(t_0)$$

by the chain rule in the form of Equation (9). Therefore, the vectors $\nabla F(P_0)$ and $\mathbf{r}'(t_0)$ are perpendicular. ■

Because $\nabla F(P_0) \neq \mathbf{0}$ is perpendicular to every curve on the surface $F(x, y, z) = 0$ through the point P_0, it is a *normal vector* to the surface at P_0,

$$\mathbf{n} = \frac{\partial F}{\partial x}\mathbf{i} + \frac{\partial F}{\partial y}\mathbf{j} + \frac{\partial F}{\partial z}\mathbf{k}. \tag{17}$$

Note that if the equation $z = f(x, y)$ is rewritten in the form $F(x, y, z) = f(x, y) - z = 0$, then

$$\left\langle \frac{\partial F}{\partial x}, \frac{\partial F}{\partial y}, \frac{\partial F}{\partial z} \right\rangle = \left\langle \frac{\partial f}{\partial x}, \frac{\partial f}{\partial y}, -1 \right\rangle.$$

Thus Equation (17) agrees with the definition of normal vector that we gave in Section 15.4 (Equation (1) there).

The **tangent plane** to the surface $F(x, y, z) = 0$ at the point $P_0(x_0, y_0, z_0)$ is the plane through P_0 that is perpendicular to the normal vector $\mathbf{n}$ in Equation (13). Its equation is

$$F_x(x_0, y_0, z_0)(x - x_0) + F_y(x_0, y_0, z_0)(y - y_0) + F_z(x_0, y_0, z_0)(z - z_0) = 0. \tag{18}$$

EXAMPLE 5 Write an equation of the tangent plane to the ellipsoid $2x^2 + 4y^2 + z^2 - 45 = 0$ at the point $(2, -3, -1)$.

Solution Here we have $\nabla f(x, y, z) = \langle 4x, 8y, 2z \rangle$, so

$$\nabla F(2, -3, -1) = 8\mathbf{i} - 24\mathbf{j} - 2\mathbf{k}.$$

Equation (18) then takes the form

$$8(x - 2) - 24(y + 3) - 2(z + 1) = 0; \qquad 4x - 12y - z = 45.$$

The intersection of the two surfaces $F(x, y, z) = 0$ and $G(x, y, z) = 0$ will generally be some sort of curve. By the implicit function theorem, this curve can be represented in parametric fashion near every point where the

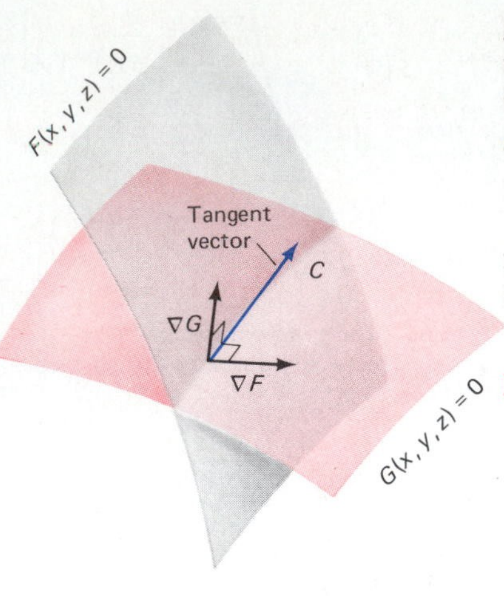

15.67 $\nabla F \times \nabla G$ is tangent to the curve C of intersection.

gradient vectors ∇F and ∇G are *not* parallel. This curve C is perpendicular to each of the normal vectors ∇F and ∇G. That is, if P is a point of C, then the tangent vector to C at P is perpendicular to each of the vectors $\nabla F(P)$ and $\nabla G(P)$, as indicated in Fig. 15.67. It follows that the vector

$$\mathbf{T} = \nabla F \times \nabla G \tag{19}$$

is tangent to the curve of intersection of the surfaces $F(x, y, z) = 0$ and $G(x, y, z) = 0$.

EXAMPLE 6 The point $P(1, -1, 2)$ lies on both the paraboloid

$$F(x, y, z) = x^2 + y^2 - z = 0$$

and the ellipsoid

$$G(x, y, z) = 2x^2 + 3y^2 + z^2 - 9 = 0.$$

Write an equation of the plane through P that is normal to their curve of intersection.

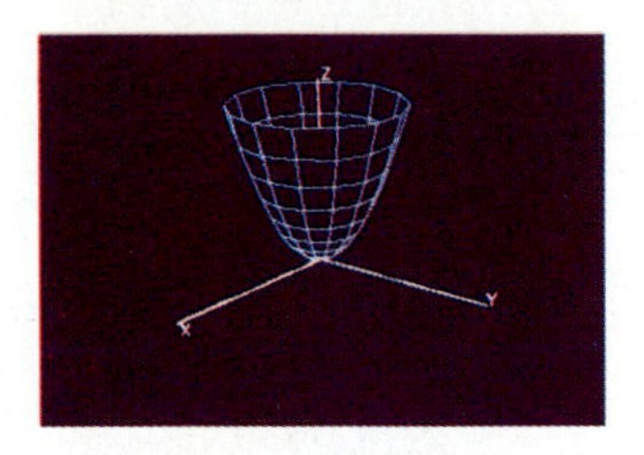

The paraboloid of Example 6

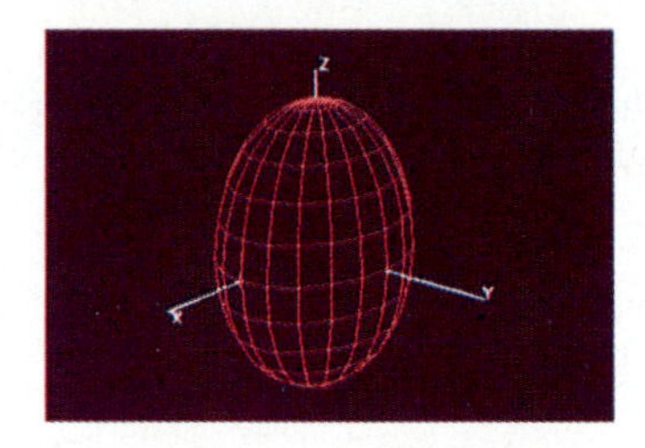

The ellipsoid of Example 6

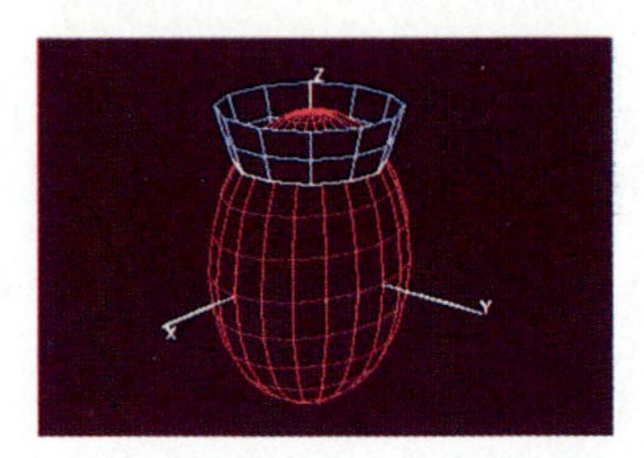

Their curve of intersection

Solution First we compute

$$\nabla F = \langle 2x, 2y, -1 \rangle \quad \text{and} \quad \nabla G = \langle 4x, 6y, 2z \rangle.$$

At $P(1, -1, 2)$ these two vectors are

$$\nabla F(1, -1, 2) = \langle 2, -2, -1 \rangle \quad \text{and} \quad \nabla G(1, -1, 2) = \langle 4, -6, 4 \rangle.$$

Hence the tangent vector to the curve of intersection of the paraboloid and the ellipsoid is

$$\mathbf{T} = \nabla F \times \nabla G = \begin{vmatrix} \mathbf{i} & \mathbf{j} & \mathbf{k} \\ 2 & -2 & -1 \\ 4 & -6 & 4 \end{vmatrix} = \langle -14, -12, -4 \rangle.$$

A slightly simpler vector parallel to $\mathbf{T}$ is $\mathbf{n} = \langle 7, 6, 2 \rangle$, and each is normal to the desired plane through $(1, -1, 2)$, so an equation of the plane is

$$7(x - 1) + 6(y + 1) + 2(z - 2) = 0;$$

that is, $7x + 6y + 2z = 5$.

A result analogous to Theorem 2 holds in two dimensions. The graph of the equation $F(x, y) = 0$ looks like a *curve* near each point at which $\nabla F \neq \mathbf{0}$, and ∇F is normal to the curve in such cases.

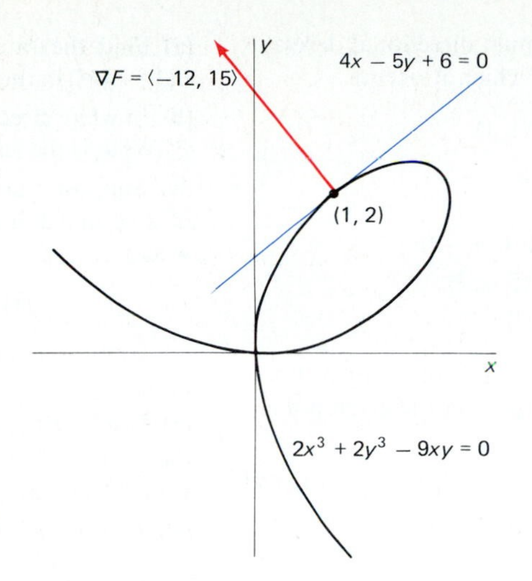

15.68 The folium and its tangent (see Example 7)

EXAMPLE 7 Write an equation of the tangent line to the folium of Descartes

$$F(x, y) = 2x^3 + 2y^3 - 9xy = 0$$

at the point (1, 2). See Fig. 15.68.

Solution The gradient of F is

$$\nabla F(x, y) = (6x^2 - 9y)\mathbf{i} + (6y^2 - 9x)\mathbf{j}.$$

So a normal vector at (1, 2) is $\nabla F(1, 2) = -12\mathbf{i} + 15\mathbf{j}$. Hence the tangent line has equation

$$-12(x - 1) + 15(y - 2) = 0;$$

that is, $4x - 5y + 6 = 0$.

15.8 PROBLEMS

In Problems 1–10, find the gradient vector ∇f at the indicated point P.

1. $f(x, y) = 3x - 7y;\quad P(17, 39)$
2. $f(x, y) = 3x^2 - 5y^2;\quad P(2, -3)$
3. $f(x, y) = \exp(-x^2 - y^2);\quad P(0, 0)$
4. $f(x, y) = \sin \frac{1}{4}\pi xy;\quad P(3, -1)$
5. $f(x, y, z) = y^2 - z^2;\quad P(17, 3, 2)$
6. $f(x, y, z) = \sqrt{x^2 + y^2 + z^2};\quad P(12, 3, 4)$
7. $f(x, y, z) = e^x \sin y + e^y \sin z + e^z \sin x;\quad P(0, 0, 0)$
8. $f(x, y, z) = x^2 - 3yz + z^3;\quad P(2, 1, 0)$
9. $f(x, y, z) = 2\sqrt{xyz};\quad P(3, -4, -3)$
10. $f(x, y, z) = (2x - 3y + 5z)^5;\quad P(-5, 1, 3)$

In Problems 11–20, find the directional derivative of f at P in the direction of $\mathbf{v}$; that is, find

$$D_{\mathbf{u}} f(P) \qquad \text{where} \quad \mathbf{u} = \frac{\mathbf{v}}{|\mathbf{v}|}.$$

11. $f(x, y) = x^2 + 2xy + 3y^2;\quad P(2, 1);\quad \mathbf{v} = \langle 1, 1\rangle$
12. $f(x, y) = e^x \sin y;\quad P(0, \pi/4);\quad \mathbf{v} = \langle 1, -1\rangle$
13. $f(x, y) = x^3 - x^2y + xy^2 + y^3;\quad P(1, -1);$ $\mathbf{v} = 2\mathbf{i} + 3\mathbf{j}$
14. $f(x, y) = \tan^{-1}\left(\dfrac{y}{x}\right);\quad P(-3, 3);\quad \mathbf{v} = 3\mathbf{i} + 4\mathbf{j}$
15. $f(x, y) = \sin x \cos y;\quad P\left(\dfrac{\pi}{3}, \dfrac{-2\pi}{3}\right);\quad \mathbf{v} = \langle 4, -3\rangle$
16. $f(x, y, z) = xy + yz + zx;\quad P(1, -1, 2);$ $\mathbf{v} = \langle 1, 1, 1\rangle$
17. $f(x, y, z) = \sqrt{xyz};\quad P(2, -1, -2);$ $\mathbf{v} = \mathbf{i} + 2\mathbf{j} - 2\mathbf{k}$
18. $f(x, y, z) = \ln(1 + x^2 + y^2 - z^2);\quad P(1, -1, 1);$ $\mathbf{v} = 2\mathbf{i} - 2\mathbf{j} + 3\mathbf{k}$
19. $f(x, y, z) = e^{xyz};\quad P(4, 0, -3);\quad \mathbf{v} = \mathbf{j} - \mathbf{k}$
20. $f(x, y, z) = (10 - x^2 - y^2 - z^2)^{1/2};\quad P(1, 1, -2);$ $\mathbf{v} = \langle 3, 4, -12\rangle$

In Problems 21–25, find the maximum directional derivative of f at P and the direction in which it occurs.

21. $f(x, y) = 2x^2 + 3xy + 4y^2$; $P(1, 1)$

22. $f(x, y) = \arctan\left(\dfrac{y}{x}\right)$; $P(1, -2)$

23. $f(x, y, z) = 3x^2 + y^2 + 4z^2$; $P(1, 5, -2)$

24. $f(x, y, z) = \exp(x - y - z)$; $P(5, 2, 3)$

25. $f(x, y, z) = (xy^2z^3)^{1/2}$; $P(2, 2, 2)$

In Problems 26–30, write an equation of the tangent line (or plane) to the given curve (or surface) at the given point P.

26. $2x^2 + 3y^2 = 35$; $P(2, 3)$

27. $x^4 + xy + y^2 = 19$; $P(2, -3)$

28. $3x^2 + 4y^2 + 5z^2 = 73$; $P(2, 2, 3)$

29. $x^{1/3} + y^{1/3} + z^{1/3} = 1$; $P(1, -1, 1)$

30. $xyz + x^2 - 2y^2 + z^3 = 14$; $P(5, -2, 3)$

31. Show that the gradient operator has the following formal properties that exhibit its close analogy with the single-variable derivative operator D.

(a) If a and b are constants, then

$$\nabla(au + bv) = a\,\nabla u + b\,\nabla v.$$

(b) $\nabla(uv) = u\,\nabla v + v\,\nabla u$.

(c) $\nabla\left(\dfrac{u}{v}\right) = \dfrac{v\,\nabla u - u\,\nabla v}{v^2}$.

(d) If n is a positive integer, then

$$\nabla u^n = nu^{n-1}\,\nabla u.$$

32. Suppose that f is a function of three independent variables x, y, and z. Show that $D_{\mathbf{i}} f = f_x$, $D_{\mathbf{j}} f = f_y$, and $D_{\mathbf{k}} f = f_z$.

33. Show that the equation of the line tangent to the conic section $Ax^2 + Bxy + Cy^2 = D$ at the point (x_0, y_0) is

$$(Ax_0)x + \tfrac{1}{2}B(y_0x + x_0y) + (Cy_0)y = D.$$

34. Show that the equation of the tangent plane to the quadric surface $Ax^2 + By^2 + Cz^2 = D$ at the point (x_0, y_0, z_0) is

$$(Ax_0)x + (By_0)y + (Cz_0)z = D.$$

35. Suppose that the temperature W (in degrees Celsius) at the point (x, y, z) in space is given by $W = 50 + xyz$.

(a) Find the rate of change of temperature with respect to distance at the point $P(3, 4, 1)$ in the direction of the vector $\mathbf{v} = \langle 1, 2, 2 \rangle$. (The units in space are in feet.)

(b) Find the maximal directional derivative $D_{\mathbf{u}}W$ at the point $P(3, 4, 1)$ and the direction $\mathbf{u}$ in which that maximum occurs.

36. Suppose that the temperature at the point (x, y, z) in space (in degrees Celsius) is given by the formula $W = 100 - x^2 - y^2 - z^2$. The units in space are in feet.

(a) Find the rate of change of temperature at the point $P(3, -4, 5)$ in the direction of the vector $\mathbf{v} = 3\mathbf{i} - 4\mathbf{j} + 12\mathbf{k}$.

(b) In what direction does W increase the most rapidly at P? What is the value of the maximal directional derivative?

37. Suppose that the altitude z (in miles above sea level) of a certain hill is described by the equation $z = f(x, y)$, where

$$f(x, y) = (0.1)(x^2 - xy + 2y^2).$$

(a) Write an equation (in the form $z = ax + by + c$) of the tangent plane at the point $P(2, 1, 0.4)$.

(b) Use $\nabla f(P)$ to approximate the altitude of the hill above the point $(2.2, 0.9)$ in the xy-plane. Compare your result with the actual altitude at this point.

38. Find an equation for the plane tangent to the paraboloid $z = 2x^2 + 3y^2$ and, simultaneously, parallel to the plane $4x - 3y - z = 10$.

39. The cone with equation $z^2 = x^2 + y^2$ and the plane with equation $2x + 3y + 4z + 2 = 0$ intersect in an ellipse. Write an equation of the plane normal to this ellipse at the point $P(3, 4, -5)$.

40. It is geometrically apparent that the highest and lowest points of the ellipse of Problem 39 at those points where its tangent line is horizontal. Find those points.

41. Show that the sphere $x^2 + y^2 + z^2 = r^2$ and the elliptical cone $z^2 = a^2x^2 + b^2y^2$ are orthogonal (that is, have perpendicular tangent planes) at every point of their intersection.

In Problems 42–45, the function $z = f(x, y)$ describes the shape of a hill: $f(P)$ is the altitude of the hill above the point $P(x, y)$ in the xy-plane. If you start at the point $(P, f(P))$ on this hill, then $D_{\mathbf{u}} f(P)$ is your rate of climb (rise per unit of horizontal distance) as you proceed in the *horizontal* direction $\mathbf{u} = a\mathbf{i} + b\mathbf{j}$. And the angle at which you climb while walking in this direction is $\gamma = \tan^{-1}(D_{\mathbf{u}} f(P))$, as shown in Fig. 15.69.

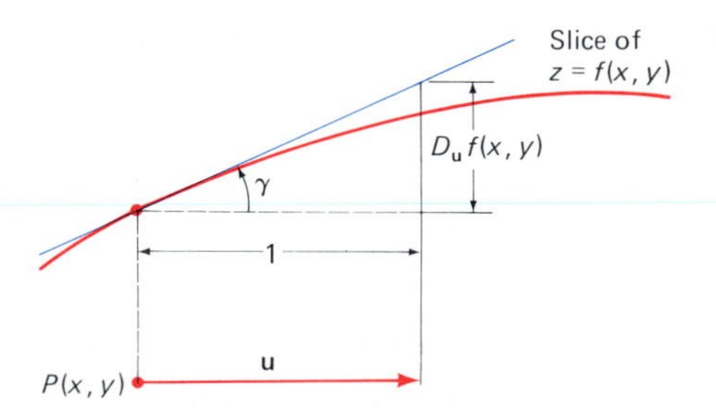

15.69 The cross section of the part of the graph above **u**

42. You are standing at the point $(-100, -100, 430)$ on a hill that has the shape of the graph of

$$z = 500 - (0.003)x^2 - (0.004)y^2.$$

What will be your rate of climb if you head northwest? At what angle from the horizontal will you be climbing?

43. Assume that the temperature at the point (x, y, z) in space is given by the function $W = 50 + xyz$, as in Problem 35. You are standing at the point $P(3, 4, 1)$ in a mountain valley shaped like the surface $z = x^2 - 2y$. Suppose that you start climbing with unit speed and a northeast compass heading. What initial rate of change of temperature do you observe?

44. (a) Suppose that a road on the hill of Problem 37 passes through the point $P(2, 1, 0.4)$. A driver passes through this point headed due east (the positive x-direction). Find her initial angle of inclination (in degrees).
(b) Repeat part (a) in the case that the driver is headed due north (the positive y-direction).
(c) Repeat part (a) in the case that the driver is headed northeast.

45. You are standing at the point (30, 20, 5) on a hill with the shape of the surface

$$z = 100 \exp(-\tfrac{1}{701}[x^2 + 3y^2]).$$

(a) In what direction (that is, with what compass heading) should you go in order to climb most steeply? At what angle from the horizontal will you initially be climbing?
(b) If instead of climbing as in part (a), you head directly west (the negative x-direction), then at what angle will you be climbing initially?

15.9 Lagrange Multipliers and Constrained Maximum–Minimum Problems

In Section 15.5 we discussed the problem of finding the maximum and minimum values attained by a function $f(x, y)$ at points of the plane region R, in the simple case in which R consists of the points on and within the simple closed curve C. We saw that any local maximum or minimum in the *interior* of R occurs at a point where $f_x = 0 = f_y$. In this section we discuss the very different matter of finding the maximum and minimum values attained by f at points of the *boundary* curve C.

If the curve C is the graph of the equation $g(x, y) = 0$, then our task is to maximize or minimize the function $f(x, y)$ subject to the **constraint** or **side condition**

$$g(x, y) = 0. \tag{1}$$

We could in principle try to solve this constraint equation for $y = \phi(x)$, and then maximize or minimize the single-variable function $f(x, \phi(x))$ by the standard method of finding where its derivative is zero. But what if it is impractical or impossible to solve Equation (1) explicitly for y in terms of x? An alternative approach that does not require that we first solve this equation is the **method of Lagrange multipliers.** It is named for its discoverer, the French mathematician Joseph Louis Lagrange (1736–1813). The method is based on the following theorem.

Theorem 1 *Lagrange Multipliers (one constraint)*

Let $f(x, y)$ and $g(x, y)$ be functions with continuous first-order partial derivatives. If the maximum (or minimum) value of f subject to the condition

$$g(x, y) = 0 \tag{1}$$

occurs at a point P where $\nabla g(P) \neq \mathbf{0}$, then

$$\nabla f(P) = \lambda \nabla g(P) \tag{2}$$

for some constant λ.

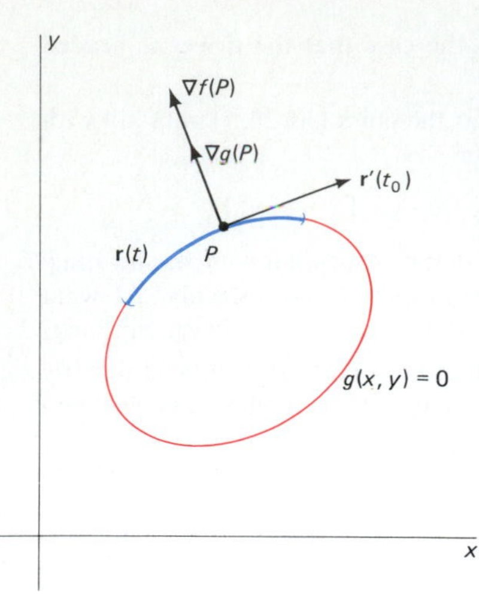

15.70 The conclusion of Theorem 1 illustrated

Proof By the implicit function theorem mentioned in Section 15.8, the fact that $\nabla g(P) \neq \mathbf{0}$ allows us to represent the curve $g(x, y) = 0$ near P by a parametric curve $\mathbf{r}(t)$, and in such fashion that $\mathbf{r}$ has a nonzero tangent vector near P. Thus $\mathbf{r}'(t) \neq \mathbf{0}$ (see Fig. 15.70). Let t_0 be the value of t such that $\mathbf{r}(t_0) = \overrightarrow{OP}$. If $f(x, y)$ attains its maximum value at P, then the composite function $f(\mathbf{r}(t))$ attains its maximum value at $t = t_0$, so

$$D_t f(\mathbf{r}(t))\Big|_{t=t_0} = \nabla f(\mathbf{r}(t_0)) \cdot \mathbf{r}'(t_0) = \nabla f(P) \cdot \mathbf{r}'(t_0) = 0. \tag{3}$$

Here we have used the vector chain rule—Equation (9) of Section 15.8.

Because $\mathbf{r}(t)$ lies on the curve $g(x, y) = 0$, the composite function $g(\mathbf{r}(t))$ is a constant function. Therefore,

$$D_t g(\mathbf{r}(t))\Big|_{t=t_0} = \nabla g(\mathbf{r}(t_0)) \cdot \mathbf{r}'(t_0) = \nabla g(P) \cdot \mathbf{r}'(t_0) = 0. \tag{4}$$

Equations (3) and (4) together imply that the vectors $\nabla f(P)$ and $\nabla g(P)$ are both perpendicular to the tangent vector $\mathbf{r}'(t_0)$. Hence $\nabla f(P)$ must be a scalar multiple of $\nabla g(P)$, and this is exactly the meaning of Equation (2). This concludes the proof of the theorem. ■

The Lagrange multiplier method based on Theorem 1 goes like this. Suppose that we want to maximize (or minimize) $z = f(x, y)$ subject to the constraint or side condition $g(x, y) = 0$. Equation (1) and the two scalar components of Equation (2) yield the three equations

$$g(x, y) = 0, \tag{1}$$

$$f_x(x, y) = \lambda g_x(x, y), \quad \text{and} \tag{2a}$$

$$f_y(x, y) = \lambda g_y(x, y). \tag{2b}$$

Thus we have three equations which we can attempt to solve for the three unknowns x, y, and λ. The points (x, y) that we find (assuming that our efforts are successful) are the only possible locations for the extrema of f subject to the constraint $g(x, y) = 0$. The associated values of λ, called **Lagrange multipliers,** may come out in the wash but are usually not of much interest to us. Finally, we calculate the value $f(x, y)$ at each of the solution points (x, y) in order to spot its maximum and minimum values.

We must bear in mind the additional possibility that the maximum or minimum (or both) values of f might occur at a point where $g_x = 0 = g_y$. The Lagrange multiplier method may fail to locate these exceptional points, but they can usually be recognized as points where the graph $g(x, y) = 0$ fails to be a smooth curve.

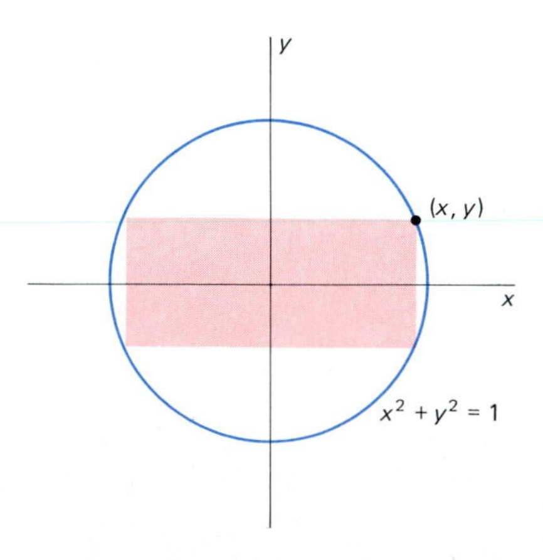

15.71 Cutting a rectangular beam from a circular log

EXAMPLE 1 Let us return to the sawmill problem of Example 5 in Section 3.6: to maximize the cross-sectional area of a rectangular beam cut from a circular log with radius 1 ft. We want to show that the optimal beam has a square cross section.

Solution With the coordinate system indicated in Fig. 15.71, we want to maximize the area $A = f(x, y) = 4xy$ subject to the constraint

$$g(x, y) = x^2 + y^2 - 1 = 0.$$

Equations (2a) and (2b) take the forms

$$4y = 2\lambda x \quad \text{and} \quad 4x = 2\lambda y.$$

It is clear that neither $x = 0$ nor $y = 0$ gives the maximum area. Hence we may divide the first equation by $4x$ and the second by $4y$ to obtain

$$\frac{\lambda}{2} = \frac{y}{x} = \frac{x}{y}.$$

We forget λ, and see that $x^2 = y^2$. We seek a solution point in the first quadrant, so we conclude that $x = 1/\sqrt{2} = y$ gives the maximum. This corresponds to a square beam of edge $\sqrt{2}$ feet, which has cross-sectional area of 2 ft^2—about 64% of the total cross-sectional area π of the log.

Note that $f(x, y) = 4xy$ attains its maximum value of 2 at both $(1/\sqrt{2}, 1/\sqrt{2})$ and $(-1/\sqrt{2}, -1/\sqrt{2})$ and its minimum value of -2 at both $(-1/\sqrt{2}, 1/\sqrt{2})$ and $(1/\sqrt{2}, -1/\sqrt{2})$. The Lagrange multiplier method actually locates all four of these points for us.

EXAMPLE 2 After the square beam of Example 1 has been cut from the circular log of radius 1 ft, let us cut four planks from the remaining pieces, each of dimensions u by $2v$, as shown in Fig. 15.72. How should this be done in order to maximize the combined cross-sectional area of all four planks, thereby using what might otherwise be scrap lumber as efficiently as possible?

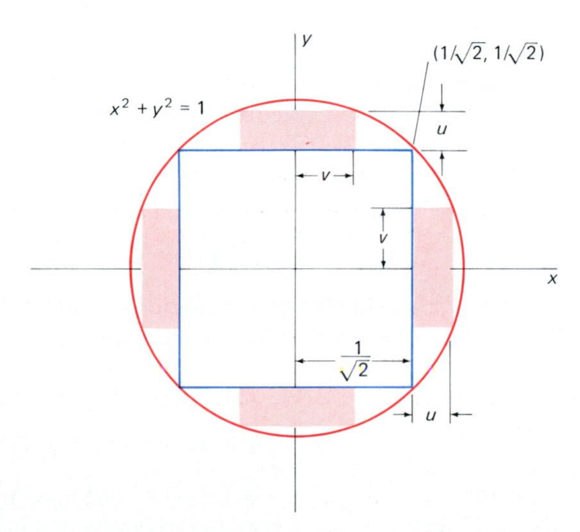

15.72 Cutting four more planks from the log

Solution Because the point $(u + 1/\sqrt{2}, v)$ lies on the unit circle, we want to maximize the function $f(u, v) = 8uv$ subject to the condition

$$g(u, v) = \left(u + \frac{1}{\sqrt{2}}\right)^2 + v^2 - 1 = 0.$$

The conditions $f_u = \lambda g_u$ and $f_v = \lambda g_v$ give us the equations

$$8v = 2\lambda\left(u + \frac{1}{\sqrt{2}}\right) \quad \text{and} \quad 8u = 2\lambda v.$$

We solve each of these equations for $\lambda/4$ and find that

$$\frac{\lambda}{4} = \frac{u}{v} = \frac{v}{u + 1/\sqrt{2}}.$$

Thus $v^2 = u(u + 1/\sqrt{2})$.

Finally we apply the one equation we've not yet used—the constraint equation $g(u, v) = 0$. We substitute our last result into that equation and get the quadratic equation

$$\left(u + \frac{1}{\sqrt{2}}\right)^2 + u\left(u + \frac{1}{\sqrt{2}}\right) - 1 = 2u^2 + \frac{3}{\sqrt{2}}u - \frac{1}{2} = 0.$$

The only positive root of this equation is $u = 0.199$ (to three-place accuracy), and this in turn means that $v = 0.424$. Thus our planks should be 0.199 ft thick and 0.848 ft wide. Their combined cross-sectional area will then be $f(0.199, 0.424) \approx 0.673 \text{ ft}^2$, approximately 21% of the original log.

LAGRANGE MULTIPLIERS IN THREE DIMENSIONS

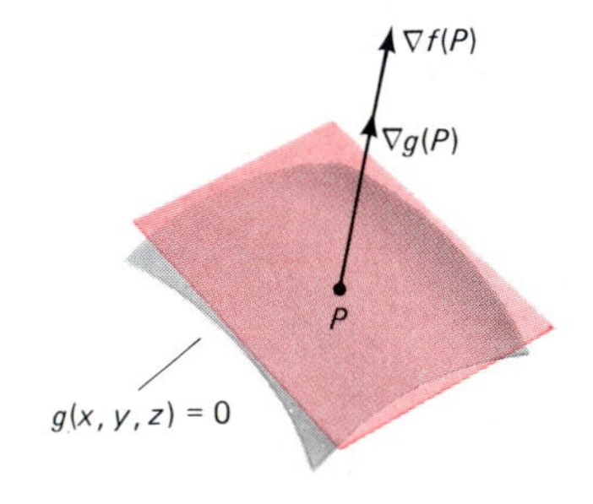

15.73 The natural generalization of Theorem 1 also holds for functions of three variables.

Now suppose that $f(x, y, z)$ and $g(x, y, z)$ have continuous first-order partial derivatives, and that we want to find the points of the *surface*

$$g(x, y, z) = 0 \tag{5}$$

at which the function $f(x, y, z)$ attains its maximum and minimum values. With functions of three rather than two variables, Theorem 1 holds precisely as we stated it earlier. We leave the details to Problem 31, but an argument similar to the proof of Theorem 1 shows that at a maximum-minimum point P of $f(x, y, z)$ on the surface in (5), the gradient vectors $\nabla f(P)$ and $\nabla g(P)$ are both normal vectors to the surface, as indicated in Fig. 15.73. It follows that

$$\nabla f(P) = \lambda \nabla g(P) \tag{6}$$

for some scalar λ. This vector equation corresponds to three scalar equations, and so we can attempt to solve simultaneously the four equations

$$g(x, y, z) = 0, \tag{5}$$

$$f_x(x, y, z) = \lambda g_x(x, y, z), \tag{6a}$$

$$f_y(x, y, z) = \lambda g_y(x, y, z), \tag{6b}$$

$$f_z(x, y, z) = \lambda g_z(x, y, z) \tag{6c}$$

for the four unknowns x, y, z, and λ. If successful, we then evaluate $f(x, y, z)$ at each of the solution points (x, y, z) to see at which it attains its maximum and minimum values. In analogy to the two-dimensional case, we also check points at which the surface $g(x, y, z) = 0$ fails to be smooth. Thus the Lagrange multiplier method with one side condition is essentially the same in dimension three as in dimension two.

EXAMPLE 3 Find the maximum volume of a rectangular box inscribed in the ellipsoid $x^2/a^2 + y^2/b^2 + z^2/c^2 = 1$ with its faces parallel to the coordinate planes.

Solution Let (x, y, z) be the vertex of the box lying in the first octant (where x, y, and z are all positive). We want to maximize the volume $V(x, y, z) = 8xyz$ subject to the constraint

$$g(x, y, z) = \frac{x^2}{a^2} + \frac{y^2}{b^2} + \frac{z^2}{c^2} - 1 = 0.$$

Equations (6a), (6b), and (6c) give

$$8yz = \frac{2\lambda x}{a^2}, \qquad 8xz = \frac{2\lambda y}{b^2}, \qquad 8xy = \frac{2\lambda z}{c^2}.$$

Part of the art of mathematics lies in pausing for a moment to find an elegant way to solve a problem, rather than rushing in headlong with brute force methods. Here, if we multiply the first equation by x, the second by y, and the third by z, we find that

$$2\lambda \frac{x^2}{a^2} = 2\lambda \frac{y^2}{b^2} = 2\lambda \frac{z^2}{c^2} = 8xyz.$$

Now $\lambda \neq 0$ because (at maximum volume) x, y, and z are nonzero. We conclude that

$$\frac{x^2}{a^2} = \frac{y^2}{b^2} = \frac{z^2}{c^2}.$$

The sum of the last three expressions is 1 because that is precisely the constraint condition in this problem. Thus each of these three expressions is equal to $\frac{1}{3}$. All three of x, y, and z are positive, and therefore

$$x = \frac{a}{\sqrt{3}}, \qquad y = \frac{b}{\sqrt{3}}, \quad \text{and} \quad z = \frac{c}{\sqrt{3}}.$$

Therefore, the box of maximum volume has volume

$$V = V_{\max} = \frac{8}{3\sqrt{3}}abc.$$

Note that the answer is dimensionally correct (the product of the three lengths a, b, and c yields a volume) and that it is also plausible—the maximum box occupies about 37% of the volume of the circumscribed ellipsoid.

PROBLEMS HAVING TWO CONSTRAINTS

Suppose that we want to find the maximum and minimum values of the function $f(x, y, z)$ at points of the curve of intersection of the two surfaces

$$g(x, y, z) = 0 \quad \text{and} \quad h(x, y, z) = 0. \tag{7}$$

This is a maximum–minimum problem with *two* constraints. The Lagrange multiplier method for such situations is based upon the following theorem.

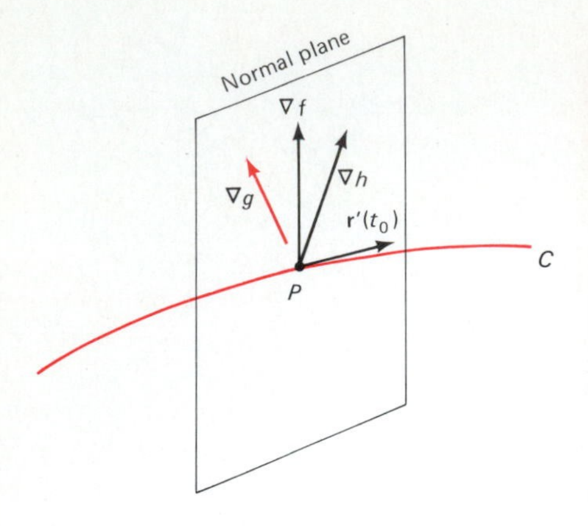

15.74 The relation between the gradient vectors in the proof of Theorem 2

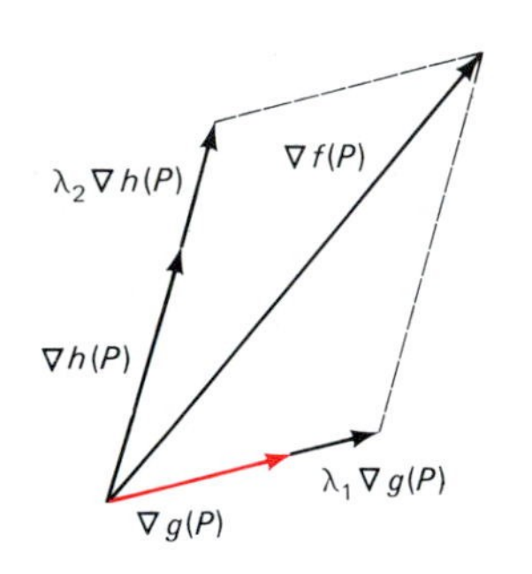

15.75 Geometry of the equation $\nabla f(P) = \lambda_1 \nabla g(P) + \lambda_2 \nabla h(P)$

Theorem 2 *Lagrange Multipliers (two constraints)*

Let $f(x, y, z)$, $g(x, y, z)$, and $h(x, y, z)$ be functions with continuous first-order partial derivatives. If the maximum (or minimum) value of f subject to the two conditions

$$g(x, y, z) = 0, \qquad h(x, y, z) = 0 \tag{7}$$

occurs at a point P where the vectors $\nabla g(P)$ and $\nabla h(P)$ are nonzero and nonparallel, then

$$\nabla f(P) = \lambda_1 \nabla g(P) + \lambda_2 \nabla h(P). \tag{8}$$

for some two constants λ_1 and λ_2.

Outline of Proof By an appropriate version of the implicit function theorem, the curve C of intersection of the two surfaces (Fig. 15.74) may be represented near P by a parametric curve $\mathbf{r}(t)$ with nonzero tangent vector $\mathbf{r}'(t)$. Let t_0 be the value of t such that $\mathbf{r}(t_0) = \overrightarrow{OP}$. We compute the derivatives at t_0 of the composite functions $f(\mathbf{r}(t))$, $g(\mathbf{r}(t))$, and $h(\mathbf{r}(t))$. We find—exactly as in the proof of Theorem 1—that

$$\nabla f(P) \cdot \mathbf{r}'(t_0) = 0, \qquad \nabla g(P) \cdot \mathbf{r}'(t_0) = 0, \quad \text{and} \quad \nabla h(P) \cdot \mathbf{r}'(t_0) = 0.$$

These three equations say that all three gradient vectors are perpendicular to the curve C at P and thus all lie in a single plane, the normal plane to the curve C at the point P.

Now $\nabla g(P)$ and $\nabla h(P)$ are nonzero and nonparallel, so $\nabla f(P)$ is the sum of its projections onto $\nabla g(P)$ and $\nabla h(P)$ (see Problem 47 of Section 14.1). As illustrated in Fig. 15.75, this fact implies Equation (8). ■

In examples we prefer to avoid subscripts by writing λ and μ for the Lagrange multipliers λ_1 and λ_2 in the statement of Theorem 2. The equations in (7) and the three scalar components of the vector equation in (8) then give rise to the five simultaneous equations

$$g(x, y, z) = 0, \tag{7a}$$

$$h(x, y, z) = 0, \tag{7b}$$

$$f_x(x, y, z) = \lambda g_x(x, y, z) + \mu h_x(x, y, z), \tag{8a}$$

$$f_y(x, y, z) = \lambda g_y(x, y, z) + \mu h_y(x, y, z), \tag{8b}$$

$$f_z(x, y, z) = \lambda g_z(x, y, z) + \mu h_z(x, y, z) \tag{8c}$$

in the five unknowns x, y, z, λ, and μ.

EXAMPLE 4 The plane $x + y + z = 12$ intersects the paraboloid $z = x^2 + y^2$ in an ellipse, as shown in Fig. 15.76. Find the highest and lowest points on this ellipse.

Solution The height of the point (x, y, z) is z, so we want to find the maximum and minimum values of

$$f(x, y, z) = z \tag{9}$$

subject to the two conditions

$$g(x, y, z) = x + y + z - 12 = 0 \tag{10}$$

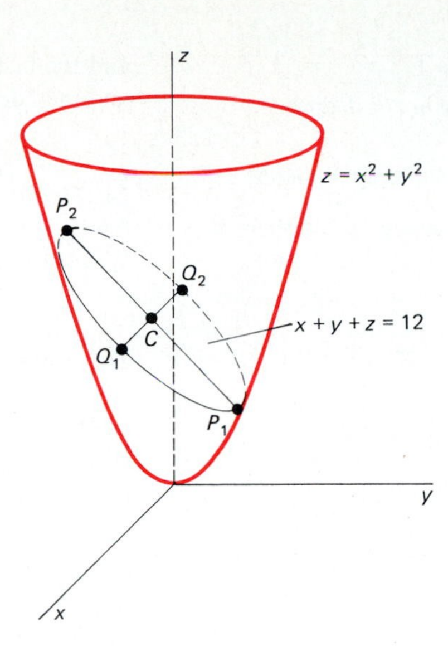

15.76 The ellipse of Example 4

and

$$h(x, y, z) = x^2 + y^2 - z = 0. \tag{11}$$

The conditions in (8a) through (8c) yield

$$0 = \lambda + 2\mu x, \tag{12a}$$

$$0 = \lambda + 2\mu y, \quad \text{and} \tag{12b}$$

$$1 = \lambda - \mu. \tag{12c}$$

If μ were zero, then (12a) would imply that $\lambda = 0$, which contradicts (12c). Hence $\mu \neq 0$, and therefore the equation

$$2\mu x = -\lambda = 2\mu y$$

implies that $x = y$. Substitution of $x = y$ in Equation (11) gives $z = 2x^2$, and then Equation (10) yields

$$2x^2 + 2x - 12 = 0;$$

$$x^2 + x - 6 = 0; \tag{13}$$

$$(x + 3)(x - 2) = 0.$$

Thus we obtain the two solutions $x = -3$ and $x = 2$. Because $y = x$ and $z = 2x^2$, the corresponding points of the ellipse are $P_1(2, 2, 8)$ and $P_2(-3, -3, 18)$. Obviously, P_1 is the lowest point and P_2 is the highest.

15.9 PROBLEMS

In each of Problems 1–10, find the maximum and minimum values—if any—of the given function f subject to the given constraint or constraints.

1. $f(x, y) = x^2 - y^2$; $x^2 + y^2 = 4$

2. $f(x, y) = x^2 + y^2$; $2x + 3y = 6$

3. $f(x, y) = xy$; $4x^2 + 9y^2 = 36$

4. $f(x, y) = 4x^2 + 9y^2$; $x^2 + y^2 = 1$

5. $f(x, y, z) = x^2 + y^2 + z^2$; $3x + 2y + z = 6$

6. $f(x, y, z) = 3x + 2y + z$; $x^2 + y^2 + z^2 = 1$

7. $f(x, y, z) = x + y + z$; $x^2 + 4y^2 + 9z^2 = 36$

8. $f(x, y, z) = xyz$; $x^2 + y^2 + z^2 = 1$

9. $f(x, y, z) = x^2 + y^2 + z^2$; $x + y + z = 1$ and $x + 2y + 3z = 6$

10. $f(x, y, z) = z$; $x^2 + y^2 = 1$ and $2x + 2y + z = 5$

In each of Problems 11–20, use Lagrange multipliers to solve the indicated problem.

11. Section 15.5, Problem 27

12. Section 15.5, Problem 28

13. Section 15.5, Problem 29

14. Section 15.5, Problem 30

15. Section 15.5, Problem 31

16. Section 15.5, Problem 32

17. Section 15.5, Problem 34

18. Section 15.5, Problem 35

19. Section 15.5, Problem 40

20. Section 15.5, Problem 41

21. Find the point or points of the surface $z = xy + 5$ closest to the origin. (*Suggestion:* Minimize the *square* of the distance.)

22. A triangle with sides x, y, and z has fixed perimeter $2s = x + y + z$. Its area A is given by Heron's formula:

$$A^2 = s(s - x)(s - y)(s - z).$$

Use the method of Lagrange multipliers to show that, among all triangles with the given perimeter, the one of largest area is equilateral.

23. Use the method of Lagrange multipliers to show that, of all triangles inscribed in the unit circle, the one of greatest area is equilateral. (*Suggestion:* Use Fig. 15.77

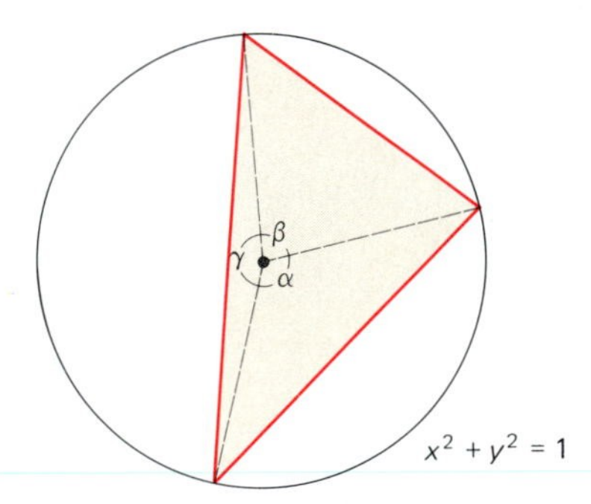

15.77 A triangle inscribed in a circle

and the fact that the area of a triangle with sides a and b and included angle θ is given by the formula

$$A = \tfrac{1}{2}ab \sin \theta.)$$

24. Find the points on the rotated ellipse $x^2 + xy + y^2 = 3$ that are closest to and farthest from the origin. (*Suggestion:* Write the Lagrange multiplier equations in the form

$$ax + by = 0 \quad \text{and} \quad cx + dy = 0.$$

These equations have a nontrivial solution *only if* $ad - bc = 0$. Use this fact to solve first for λ.)

25. Use the method of Problem 24 to find the points of the rotated hyperbola $x^2 + 12xy + 6y^2 = 130$ that are closest to the origin.

26. Find the points of the ellipse $4x^2 + 9y^2 = 36$ that are closest to the point $(1, 1)$ as well as the point or points farthest from it.

27. Find the highest and lowest points on the ellipse of intersection of the cylinder $x^2 + y^2 = 1$ and the plane $2x + y - z = 4$.

28. Apply the method of Example 4 to find the highest and lowest points on the ellipse of intersection of the cone $z^2 = x^2 + y^2$ and the plane $x + 2y + 3z = 3$.

29. Find the points on the ellipse of Problem 28 that are nearest the origin and those that are farthest from it.

30. The ice tray shown in Fig. 15.78 is to be made of material that costs 1 cent/in.^{2}. Minimize the cost function $f(x, y, z) = xy + 3xz + 7yz$ subject to the constraints that each of the 12 compartments is to be square and the total volume (ignoring partitions) is to be 12 in.^{3}.

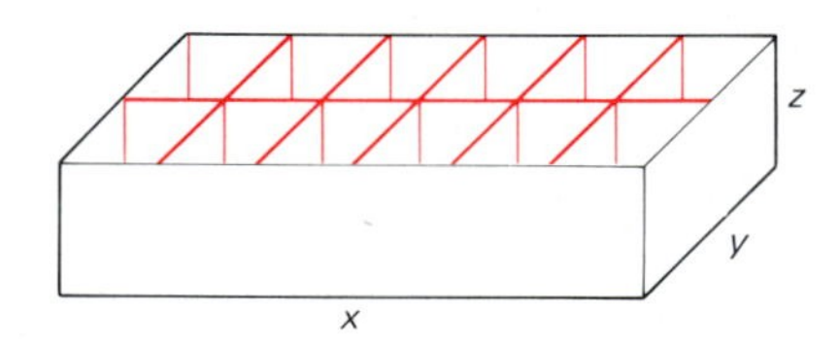

15.78 The ice tray of Problem 30

31. Prove Theorem 1 for functions of three variables by showing that each of the vectors $\nabla f(P)$ and $\nabla g(P)$ is perpendicular at P to every curve on the surface $g(x, y, z) = 0$.

32. Find the lengths of the semiaxes of the ellipse of Example 4.

15.10 The Second Derivative Test for Functions of Two Variables

We saw in Section 15.5 that in order for the differentiable function $f(x, y)$ to have either a local minimum or a local maximum at the point $P(a, b)$, it is a *necessary* condition that P be a *critical point* of f; that is, that

$$f_x(a, b) = 0 = f_y(a, b).$$

In this section we give *sufficient* conditions that f have a local extremum at a critical point. The criterion stated below involves the second-order partial derivatives of f at (a, b), and plays the role of the single-variable second derivative test (Section 4.6) for functions of two variables. To simplify the statement of this result, we shall use the following abbreviations:

$$A = f_{xx}(a, b), \qquad B = f_{xy}(a, b), \qquad C = f_{yy}(a, b), \tag{1}$$

and

$$\Delta = AC - B^2 = f_{xx}(a, b)f_{yy}(a, b) - [f_{xy}(a, b)]^2. \tag{2}$$

We will outline a proof of the following theorem at the end of this section.

> ***Theorem*** *Sufficient Conditions for Local Extrema*
>
> Let (a, b) be a critical point of the function $f(x, y)$, and suppose that f has continuous first-order and second-order partial derivatives in some circular disk centered at (a, b). Then:
>
> 1. If $\Delta > 0$ and $A > 0$, then f has a local minimum at (a, b).
> 2. If $\Delta > 0$ and $A < 0$, then f has a local maximum at (a, b).
> 3. If $\Delta < 0$, then f has neither a local minimum nor a local maximum at (a, b). Instead, it has a saddle point there.

Thus f has *either* a local maximum *or* a local minimum at the critical point (a, b) provided that the **discriminant** $\Delta = AC - B^2$ is *positive*. In this case, $A = f_{xx}(a, b)$ plays the role of the second derivative of a single-variable function: There is a local minimum at (a, b) if $A > 0$ and a local maximum if $A < 0$.

If $\Delta < 0$, then f has *neither* a local maximum *nor* a local minimum at (a, b). In this case we call (a, b) a **saddle point** for f, thinking of the appearance of the hyperbolic paraboloid $f(x, y) = x^2 - y^2$ (Fig. 15.79), a typical example of this case.

Note that the theorem is silent on the question of what happens when $\Delta = 0$. In this case, the two-variable second derivative test fails—it gives no conclusion. Moreover, at such a point (a, b), *anything* can happen, ranging

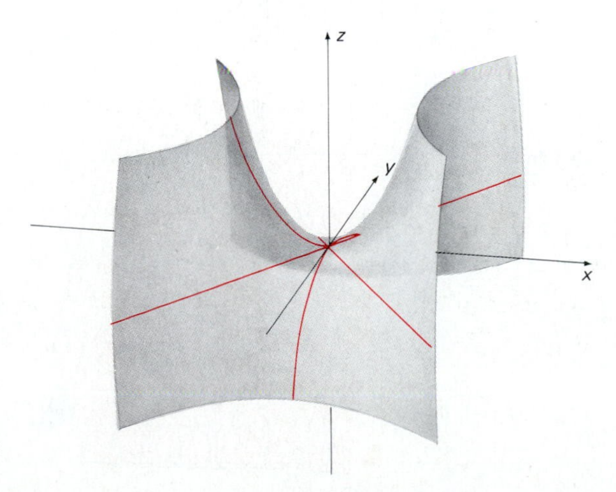

15.79 The origin is a saddle point of the surface with equation $z = x^2 - y^2$

from the local (indeed global) minimum of $f(x, y) = x^4 + y^4$ at (0, 0) to the "monkey saddle" of Example 2.

In the case of a function $f(x, y)$ with several critical points, the quantities A, B, C, and Δ must be computed separately at each critical point in order to apply the test.

The graph of Example 1

EXAMPLE 1 Locate and classify the critical points of

$$f(x, y) = 3x - x^3 - 3xy^2.$$

Solution This function is a polynomial, so all its partial derivatives exist and are continuous everywhere. When we equate its first partial derivatives to zero (in order to locate the critical points of f), we get

$$f_x(x, y) = 3 - 3x^2 - 3y^2 = 0, \qquad f_y(x, y) = -6xy = 0.$$

The second of these equations implies that one of the variables x and y must be zero, and then the first implies that the other must be -1 or 1. Thus there are four critical points: (1, 0), $(-1, 0)$, (0, 1), and $(0, -1)$.

The second-order partial derivatives of f are

$$A = f_{xx} = -6x, \qquad B = f_{xy} = -6y, \qquad C = f_{yy} = -6x.$$

Hence $\Delta = 36(x^2 - y^2)$ at each of the critical points. The table in Fig. 15.80 summarizes the situation at each of the four critical points. The graph of f is shown in Fig. 15.81.

Critical point	A	B	C	Δ	Type of extremum
(1, 0)	−6	0	−6	36	Local maximum
(−1, 0)	6	0	6	36	Local minimum
(0, 1)	0	−6	0	−36	Not an extremum
(0, −1)	0	6	0	−36	Not an extremum

15.80 Critical point analysis for the function of Example 1

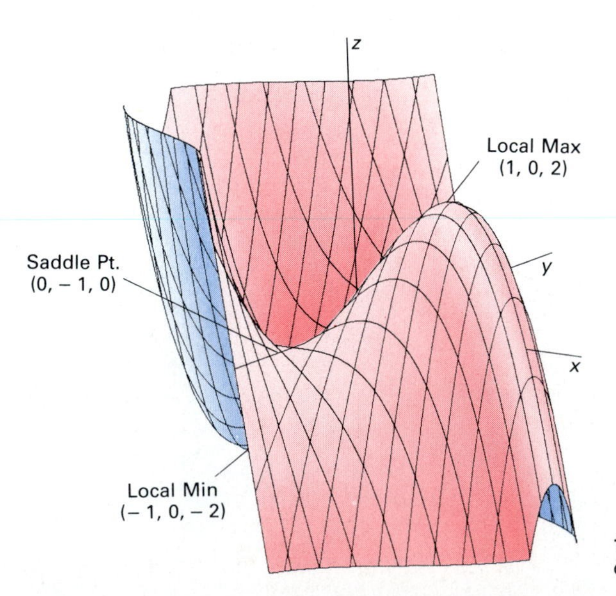

15.81 Graph of the function of Example 1

EXAMPLE 2 Find and classify the critical points of the function

$$f(x, y) = 6xy^2 - 2x^3 - 3y^4.$$

Solution Equating the first-order partial derivatives to zero, we get the equations

$$f_x(x, y) = 6y^2 - 6x^2 = 0 \quad \text{and} \quad f_y(x, y) = 12xy - 12y^3 = 0.$$

It follows that

$$x^2 = y^2 \quad \text{and} \quad xy - y^3 = 0.$$

The first of these equations gives $x = \pm y$. If $x = y$, the second equation implies that $y = 0$ or $y = 1$. If $x = -y$, the second equation implies that $y = 0$ or $y = -1$. Hence there are three critical points: (0, 0), (1, 1), and (1, −1).

The second-order partial derivatives of f are

$$A = f_{xx} = -12x, \qquad B = f_{xy} = 12y, \qquad C = f_{yy} = 12x - 36y^2.$$

Critical point	A	B	C	Δ	Type of extremum
(0, 0)	0	0	0	0	Test fails
(1, 1)	−12	+12	−24	144	Local maximum
(1, −1)	−12	−12	−24	144	Local maximum

15.82 Critical point analysis for the function of Example 2

These give the data shown in the table of Fig. 15.82. Note that the critical point test fails at (0, 0), so we must find another way to test this point. We observe that $f(x, 0) = -2x^3$ and that $f(0, y) = -3y^4$. Hence, as we move away from the origin in the

Positive x-direction: f decreases;
Negative x-direction: f increases;
Positive y-direction: f decreases;
Negative y-direction: f decreases.

Consequently, f has neither a local maximum nor a local minimum at the origin. Note (Fig. 15.83) that if a monkey sits with its rump at the origin facing the negative x-direction, then the directions in which $f(x, y)$ decreases provide places for both its tail and its two legs to hang. That's why this particular surface is called a *monkey saddle*.

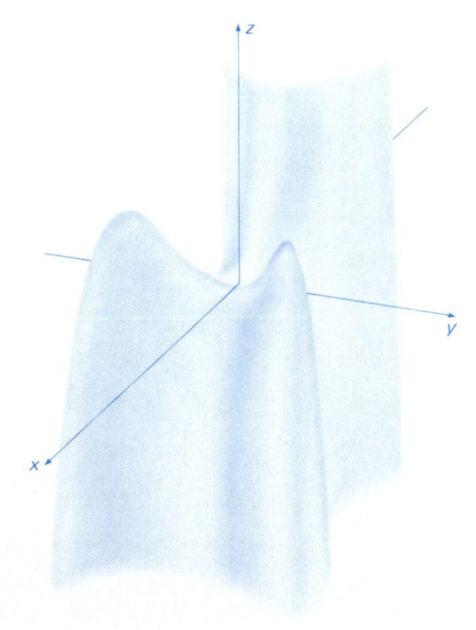

15.83 The graph of the function of Example 2 is sometimes called a monkey saddle.

EXAMPLE 3 Find and classify the critical points of the function

$$f(x, y) = \tfrac{1}{3}x^4 + \tfrac{1}{2}y^4 - 4xy^2 + 2x^2 + 2y^2 + 3.$$

Solution When we equate the first-order partial derivatives of f to zero, we obtain the equations

$$f_x(x, y) = \tfrac{4}{3}x^3 - 4y^2 + 4x = 0, \tag{3}$$

$$f_y(x, y) = 2y^3 - 8xy + 4y = 0, \tag{4}$$

which are not so easy to solve as those that appeared in Examples 1 and 2.

If we write Equation (4) in the form

$$2y(y^2 - 4x + 2) = 0,$$

we see that either $y = 0$ or

$$y^2 = 4x - 2. \tag{5}$$

If $y = 0$, then Equation (3) reduces to the equation

$$\tfrac{4}{3}x^3 + 4x = \tfrac{4}{3}x(x^2 + 3) = 0,$$

whose only solution is $x = 0$. Thus one critical point of f is $(0, 0)$.

If $y \neq 0$, we substitute $y^2 = 4x - 2$ in (3) to obtain

$$\tfrac{4}{3}x^3 - 4(4x - 2) + 4x = 0;$$

that is,

$$\tfrac{4}{3}x^3 - 12x + 8 = 0.$$

Thus we need to solve the cubic equation

$$p(x) = x^3 - 9x + 6 = 0. \tag{6}$$

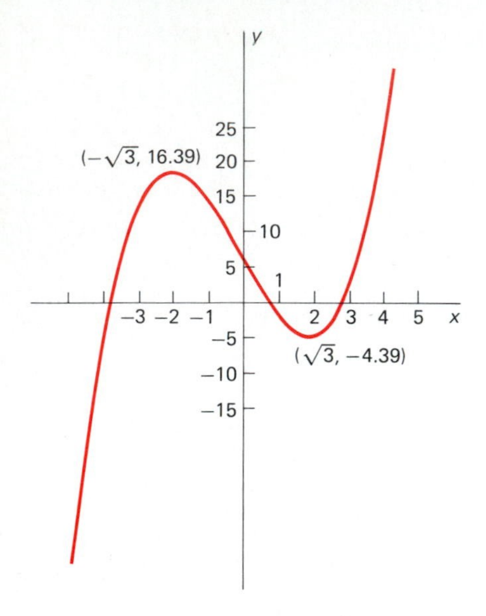

15.84 Graph of $p(x) = x^3 - 9x + 6$

As a preliminary step, we apply the methods of Section 4.5 to construct the graph of p. This graph, which appears in Fig. 15.84, indicates the existence of three real roots—very roughly, $x \approx -3$, $x \approx 1$, and $x \approx 3$.

The iteration of Newton's method (Section 3.9) for Equation (6) is

$$x_{n+1} = x_n - \frac{p(x_n)}{p'(x_n)} = x_n - \frac{(x_n)^3 - 9x_n + 6}{3(x_n)^2 - 9}.$$

Using the estimates of the root obtained graphically as initial values in the iteration, we obtain the three much improved approximations $x = -3.2899$, $x = 0.7057$, and $x = 2.5842$. We now return to Equation (5). The first of these roots yields *no* (real) value of y, the second yields $y = \pm 0.9071$, and the third yields $y = \pm 2.8874$. Thus in addition to the critical point $(0, 0)$, the function f has the four critical points

$$(0.7057, \pm 0.9071) \quad \text{and} \tag{7}$$

$$(2.5842, \pm 2.8874). \tag{8}$$

When we finally calculate the values of

$$A = f_{xx} = 4x^2 + 4, \qquad C = F_{yy} = 6y^2 - 8x + 4,$$
$$B = f_{xy} = f_{yx} = -8y, \qquad \Delta = AC - B^2$$

at each of these critical points, we find that $f(x, y)$ has a local minimum at $(0, 0)$ and at the points in (8), while those in (7) are saddle points. The level curve diagram in Fig. 15.85 shows how these five critical points fit together. Finally, we observe that the behavior of $f(x, y)$ is approximately that of $\frac{1}{3}x^4 + \frac{1}{4}y^4$ when $|x|$ or $|y|$ is large, so f must have a global minimum value but no global maximum. Consequently, because $f(0, 0) = 3$, while the value of f at the points in (8) is approximately -3.5293, the local minima at the points in (8) are actually global minima.

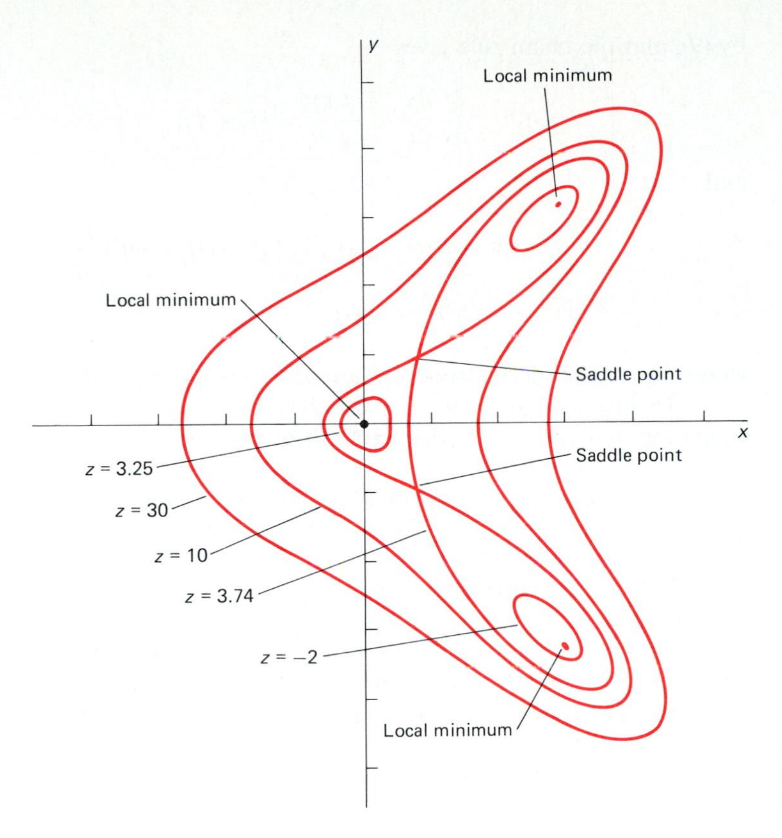

15.85 Level curves $z = -2$, 3.25, 3.74, 10, and 30 for the function $f(x, y)$ of Example 3

DISCUSSION OF THE THEOREM

A complete proof of the theorem providing sufficient conditions for local extrema of functions of two variables is best left for advanced calculus. Here, however, we provide an outline of the main ideas. Given a function $f(x, y)$ with critical point (a, b), we fix $h = \Delta x$ and $k = \Delta y$ and introduce the single-variable function

$$G(t) = f(a + ht, b + kt). \tag{9}$$

Its second-degree Taylor formula at $t = 0$ is

$$G(t) = G(0) + G'(0)t + \tfrac{1}{2}G''(0)t^2 + R, \tag{10}$$

where the remainder term is of the form

$$R = \frac{1}{3!}\,G^{(3)}(\tau)t^3 \qquad (\tau \text{ between 0 and } t).$$

With $t = 1$ we obtain

$$G(1) = G(0) + G'(0) + \tfrac{1}{2}G''(0) + R. \tag{11}$$

But

$$G(0) = f(a, b) \quad \text{and} \quad G(1) = f(a + h, b + k) \tag{12}$$

by (9), and the chain rule gives

$$G'(0) = \frac{\partial f}{\partial x}\frac{dx}{dt} + \frac{\partial f}{\partial y}\frac{dy}{dt} = hf_x + kf_y \tag{13}$$

and

$$\begin{aligned} G''(0) &= \frac{\partial}{\partial x}(hf_x + kf_y)\frac{dx}{dt} + \frac{\partial}{\partial y}(hf_x + kf_y)\frac{dy}{dt} \\ &= h^2 f_{xx} + 2hkf_{xy} + k^2 f_{yy}, \end{aligned} \tag{14}$$

where the partial derivatives of f are to be evaluated at the point (a, b).

Because $f_x(a, b) = f_y(a, b) = 0$, substitution of (12), (13), and (14) in (11) yields the two-variable Taylor expansion

$$f(a + h, b + k) = f(a, b) + \tfrac{1}{2}(Ah^2 + 2Bhk + Ck^2) + R, \tag{15}$$

where

$$A = f_{xx}(a, b), \qquad B = f_{xy}(a, b), \qquad C = f_{yy}(a, b). \tag{16}$$

It turns out that if h and k are sufficiently small, then the remainder term R is "negligible," so the behavior of $f(x, y)$ near the critical point (a, b) is determined by the behavior near $(0, 0)$ of the *quadratic form*

$$q(h, k) = Ah^2 + 2Bhk + Ck^2. \tag{17}$$

We write $\Delta = AC - B^2$ as before, and find (upon completing the square) that

$$q(h, k) = \frac{1}{A}\left[(Ah + Bk)^2 + \Delta k^2\right]. \tag{18}$$

This form makes clear the following properties of $q(h, k)$:

1. If $\Delta > 0$ and $A > 0$, then $q(h, k) > 0$ for all h and k not both zero.
2. If $\Delta > 0$ and $A < 0$, then $q(h, k) < 0$ for all h and k not both zero.
3. If $\Delta < 0$, then there are some values of h and k (not both zero) arbitrarily close to zero such that $q(h, k) > 0$, and other such values of h and k with $q(h, k) < 0$.

The assertions in (1) and (2) are obvious upon consideration of signs. The reason for (3) in the case $A > 0$, $B \neq 0$ is this: If $\Delta < 0$, then $q(h, 0) = Ah^2 > 0$ while $q(h, -Ah/B) = \Delta k^2/A < 0$.

Assuming that h and k (not both zero) are sufficiently small that R is negligible in (15), it follows from these properties of $q(h, k)$ that:

1. If $\Delta > 0$ and $A > 0$, then $f(a + h, b + k) > f(a, b)$.
2. If $\Delta > 0$ and $A < 0$, then $f(a + h, b + k) < f(a, b)$.
3. If $\Delta < 0$, then $f(a + h, b + k) > f(a, b)$ for some values of h and k arbitrarily close to zero, while $f(a + h, b + k) < f(a, b)$ for other such values of h and k.

Finally, we observe that these possibilities correspond directly to the conclusions of the theorem.

15.10 PROBLEMS

Find and classify the critical points of each of the functions in Problems 1–22.

1. $f(x, y) = 2x^2 + y^2 + 4x - 4y + 5$

2. $f(x, y) = 10 + 12x - 12y - 3x^2 - 2y^2$

3. $f(x, y) = 2x^2 - 3y^2 + 2x - 3y + 7$

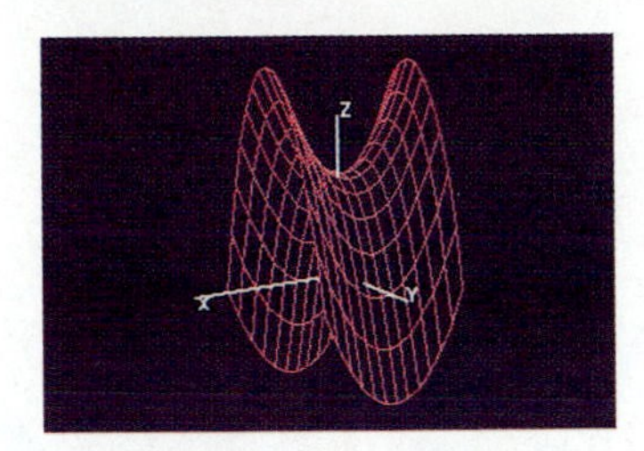

Graph for Problem 3

4. $f(x, y) = xy + 3x - 2y + 4$

5. $f(x, y) = 2x^2 + 2xy + y^2 + 4x - 2y + 1$

6. $f(x, y) = x^2 + 4xy + 2y^2 + 4x - 8y + 3$

7. $f(x, y) = x^3 + y^3 + 3xy + 3$

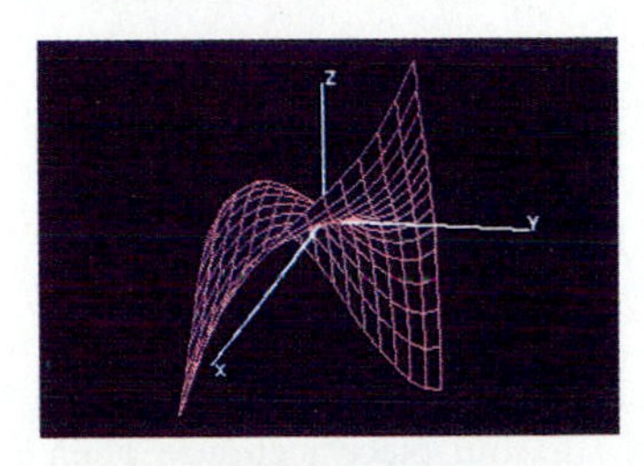

Graph for Problem 7

8. $f(x, y) = x^2 - 2xy + y^3 - y$

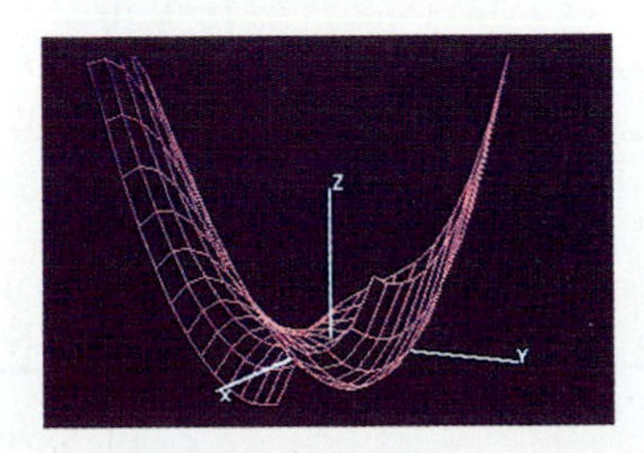

Graph of Problem 8

9. $f(x, y) = 6x - x^3 - y^3$

10. $f(x, y) = 3xy - x^3 - y^3$

11. $f(x, y) = x^4 + y^4 - 4xy$

12. $f(x, y) = x^3 + 6xy + 3y^2$

13. $f(x, y) = x^3 + 6xy + 3y^2 - 9x$

14. $f(x, y) = x^3 + 6xy + 3y^2 + 6x$

15. $f(x, y) = 3x^2 + 6xy + 2y^3 + 12x - 24y$

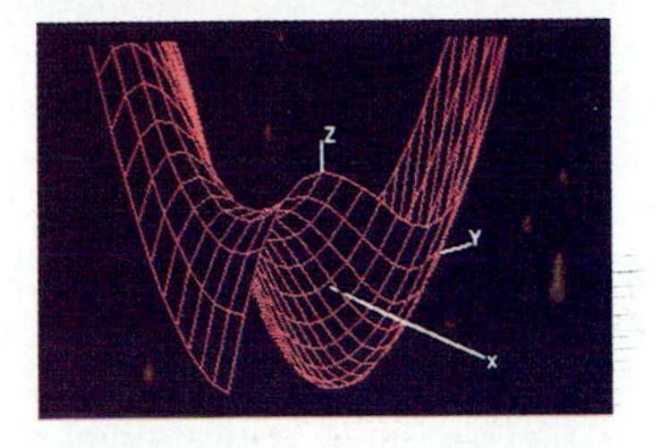

Graph for Problem 15

16. $f(x, y) = 3x^2 + 12xy + 2y^3 - 6x + 6y$

17. $f(x, y) = 4xy - 2x^4 - y^2$

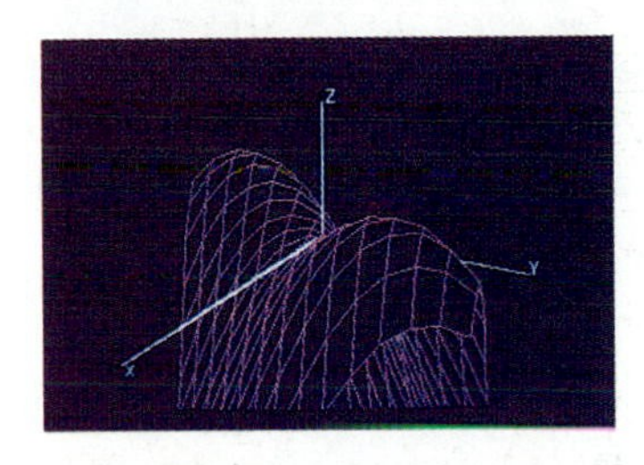

Graph for Problem 17

18. $f(x, y) = 8xy - 2x^2 - y^4$

19. $f(x, y) = 2x^3 - 3x^2 + y^2 - 12x + 10$

20. $f(x, y) = 2x^3 + y^3 - 3x^2 - 12x - 3y$

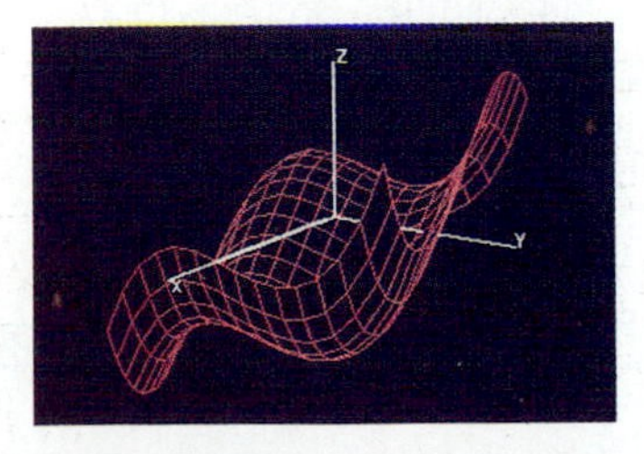

Graph for Problem 20

21. $f(x, y) = (x^2 + y^2)e^{-x}$

22. $f(x, y) = (x^2 + y^2)\exp(x^2 - y^2)$

In each of Problems 23–25, first show that $\Delta = f_{xx}f_{yy} - (f_{xy})^2$ is zero at the origin. Then classify this critical point by imagining what the surface $z = f(x, y)$ looks like.

23. $f(x, y) = x^4 + y^4$

24. $f(x, y) = x^3 + y^3$

25. $f(x, y) = \exp(-x^4 - y^4)$

26. Let $f(x, t)$ denote the *square* of the distance between a typical point of the line $x = t$, $y = t + 1$, $z = 2t$ and a typical point of the line $x = 2s$, $y = s - 1$, $z = s + 1$. Show that the single critical point of f is a local minimum. Hence find the closest points on these two skew lines.

27. Let $f(x, y)$ denote the square of the distance from $(0, 0, 2)$ to a typical point of the surface $z = xy$. Find and classify the critical points of f.

28. Show that the surface $z = (x^2 + 2y^2)\exp(1 - x^2 - y^2)$ looks like two mountain peaks joined by two ridges with a pit between them.

29. A wire 120 in. long is cut into three pieces of lengths x, y, and $120 - x - y$, and each piece is bent into the shape of a square. Let $f(x, y)$ denote the sum of the areas of these squares. Show that the single critical point of f is a local minimum. But surely it is possible to *maximize* the sum of the areas. Explain.

30. Show that the function

$$f(x, y) = xy\exp(\tfrac{1}{8}(x^2 + 4y^2))$$

has a saddle point, two local minima, and two global minima.

31. Find and classify the critical points of the function

$$f(x, y) = \sin(\tfrac{1}{2}\pi x)\sin(\tfrac{1}{2}\pi y).$$

32. Let $f(x, y) = x^3 - 3xy^2$.

(a) Show that its only critical point is $(0, 0)$ and that $\Delta = 0$ there.

(b) Show that the surface $z = x^3 - 3xy^2$ qualifies as a monkey saddle by examining the behavior of $x^3 - 3xy^2$ on straight lines through the origin. See Fig. 15.86.

33. Repeat Problem 32 with $f(x, y) = 4xy(x^2 - y^2)$. Show that near the critical point $(0, 0)$ the surface $z = f(x, y)$ qualifies as a "dog saddle." See Fig. 15.87.

34. Let $f(x, y) = xy(x^2 - y^2)/(x^2 + y^2)$. Classify the behavior of f near the critical point $(0, 0)$.

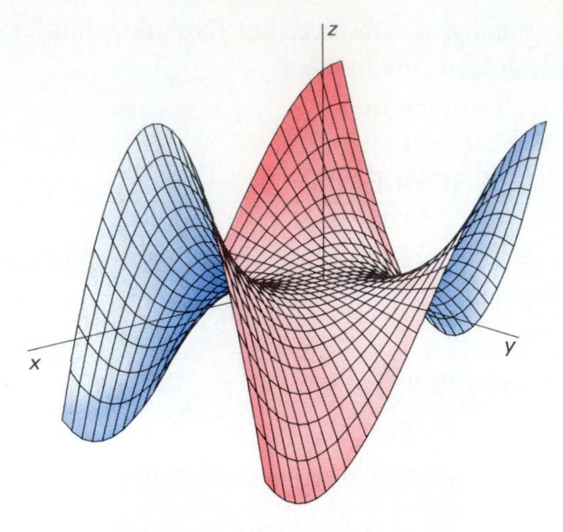

15.86 The monkey saddle of Problem 32

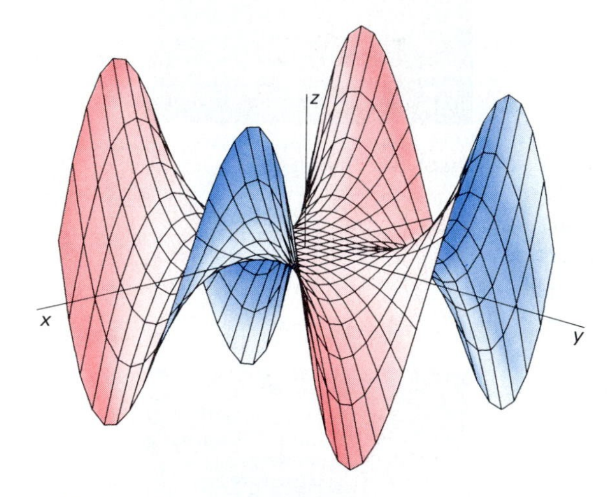

15.87 The dog saddle of Problem 33

In Problems 35–39, use Newton's method to find the critical points of f to four-place accuracy. Then classify them.

35. $f(x, y) = 2x^4 - 12x^2 + y^2 + 8x$

36. $f(x, y) = x^4 + 4x^2 - y^2 - 16x$

37. $f(x, y) = x^4 + 12xy + 6y^2 + 4x + 10$

38. $f(x, y) = x^4 + 8xy - 4y^2 - 16x + 10$

39. $f(x, y) = x^4 + 2y^4 - 12xy^2 - 20y^2$

CHAPTER 15 REVIEW: Definitions, Concepts, Results

Use the list below as a guide to concepts that you may need to review.

1. Graphs and level curves of functions of two variables

2. Limits and continuity for functions of two and three variables

3. Partial derivatives—definition and computation

4. Geometric interpretation of partial derivatives and the tangent plane to the surface $z = f(x, y)$

5. Absolute and local maxima and minima

6. Necessary conditions for a local extremum

7. Increments and differentials for functions of two and three variables

8. The linear approximation theorem

9. The chain rule for functions of several variables

10. Directional derivatives—definition and computation

11. The gradient vector and the vector chain rule

12. Significance of the length and direction of the gradient vector

13. The gradient vector as a normal vector; tangent plane to a surface $F(x, y, z) = 0$

14. Constrained maximum–minimum problems and the Lagrange multiplier method

15. Sufficient conditions for a local extremum of a function of two variables

CHAPTER 15 MISCELLANEOUS PROBLEMS

1. Use the method of Example 5 in Section 15.3 to show that

$$\lim_{(x,y)\to(0,0)} \frac{x^2y^2}{x^2+y^2} = 0.$$

2. Use spherical coordinates to show that

$$\lim_{(x,y,z)\to(0,0,0)} \frac{x^3+y^3-z^3}{x^2+y^2+z^2} = 0.$$

3. Suppose that $g(x, y) = xy/(x^2 + y^2)$ unless $(x, y) = (0, 0)$; define $g(0, 0)$ to be zero. Show that g is not continuous at $(0, 0)$.

4. Compute $g_x(0, 0)$ and $g_y(0, 0)$ for the function g of Problem 3.

5. Find a function $f(x, y)$ such that

$$f_x(x, y) = 2xy^3 + e^x \sin y$$

and

$$f_y(x, y) = 3x^2y^2 + e^x \cos y + 1.$$

6. Prove that there is *no* function f with continuous second-order partial derivatives such that $f_x(x, y) = 6xy^2$ and $f_y(x, y) = 8x^2y$.

7. Find the points on the paraboloid $z = x^2 + y^2$ at which the normal line passes through the point $(0, 0, 1)$.

8. Write an equation of the tangent plane to the surface $\sin xy + \sin yz + \sin xz = 1$ at the point $(1, \pi/2, 0)$.

9. Prove that every normal line to the cone with equation $z = (x^2 + y^2)^{1/2}$ intersects the z-axis.

10. Show that the function

$$u(x, t) = (4\pi kt)^{-1/2} \exp\left(-\frac{x^2}{4kt}\right)$$

satisfies the one-dimensional heat equation of Problem 45 in Section 15.4.

11. Show that the function

$$u(x, y, t) = (4\pi kt)^{-1} \exp\left(-\frac{x^2+y^2}{4kt}\right)$$

satisfies the two-dimensional heat equation of Problem 46 in Section 15.4.

12. Let $f(x, y) = xy(x^2 - y^2)/(x^2 + y^2)$ unless $(x, y) = (0, 0)$, in which case define $f(0, 0)$ to be zero. Show that the second-order partial derivatives f_{xx}, f_{xy}, f_{yx}, and f_{yy} all exist at $(0, 0)$ but that $f_{xy}(0, 0) \neq f_{yx}(0, 0)$.

13. Define the partial derivatives $\mathbf{r}_x$ and $\mathbf{r}_y$ of the vector-valued function $\mathbf{r}(x, y) = \mathbf{i}x + \mathbf{j}y + \mathbf{k}f(x, y)$ by componentwise partial differentiation. Then show that the vector $\mathbf{r}_x \times \mathbf{r}_y$ is normal to the surface $z = f(x, y)$.

14. An open-topped rectangular box is to have total surface area 300 in.^{2}. Find the dimensions that maximize its volume.

15. A rectangular shipping crate is to have volume 60 ft^3. Its sides cost \$1/ft^2, its top costs \$2/ft^2, and its bottom costs \$3/ft^2. What dimensions will minimize the cost of the box?

16. A pyramid is bounded by the three coordinate planes and the tangent plane to the surface $xyz = 1$ at a point in the first octant. Find the volume of this pyramid (it is independent of the point of tangency).

17. The total resistance R of two resistances R_1 and R_2 connected in parallel is given by the formula

$$\frac{1}{R} = \frac{1}{R_1} + \frac{1}{R_2}.$$

Suppose that R_1 and R_2 are measured to be 300 and 600 ohms, respectively, with a maximum error of 1% in each measurement. Use differentials to estimate the maximum error (in ohms) in the calculated value of R.

18. Consider the gas of Problem 53 in Section 15.4, a gas satisfying the van der Waals equation. Use differentials to approximate the change in its volume if P is increased from 1 atm to 1.1 atm and T is decreased from 313 K to 303 K.

19. The semiaxes a, b, and c of an ellipsoid with volume $V = \frac{4}{3}\pi abc$ are each measured with a maximum percentage error of 1%. Use differentials to estimate the maximum percentage error in the calculated value of V.

20. Two spheres have radii a and b and the distance between their centers is $c < a + b$, so that the spheres meet in a common circle. Let P be a point on this circle and let $\mathscr{P}_1$ and $\mathscr{P}_2$ be the tangent planes at P to the two spheres.

Find the angle between $\mathscr{P}_1$ and $\mathscr{P}_2$ in terms of a, b, and c. Recall that the angle between two planes is, by definition, the angle between their normal vectors.

21. Find every point on the surface of the ellipsoid $x^2 + 4y^2 + 9z^2 = 16$ at which the normal line at the point passes through the center $(0, 0, 0)$ of the ellipsoid.

22. Suppose that

$$F(x) = \int_{g(x)}^{h(x)} f(t)\,dt.$$

Show that

$$F'(x) = f(h(x))h'(x) - f(g(x))g'(x).$$

(*Suggestion:* Write $w = \int_u^v f(t)\,dt$ with $u = g(x)$ and $v = h(x)$.)

23. Suppose that $\mathbf{a}$, $\mathbf{b}$, and $\mathbf{c}$ are mutually perpendicular unit vectors, and that f is a function of the three independent variables x, y, and z. Show that

$$\nabla f = \mathbf{a}(D_{\mathbf{a}} f) + \mathbf{b}(D_{\mathbf{b}} f) + \mathbf{c}(D_{\mathbf{c}} f).$$

24. Let $\mathbf{R} = \langle \cos\theta, \sin\theta, 0\rangle$ and $\mathbf{\Theta} = \langle -\sin\theta, \cos\theta, 0\rangle$ be the polar coordinate unit vectors. Given $f(x, y, z) = w(r, \theta, z)$, show that

$$D_{\mathbf{R}} f = \frac{\partial w}{\partial r} \quad \text{and} \quad D_{\mathbf{\Theta}} f = \frac{1}{r}\frac{\partial w}{\partial \theta}.$$

Then conclude from Problem 23 that the gradient vector is given in cylindrical coordinates by

$$\nabla f = \frac{\partial w}{\partial r}\mathbf{R} + \frac{1}{r}\frac{\partial w}{\partial \theta}\mathbf{\Theta} + \frac{\partial w}{\partial z}\mathbf{k}.$$

25. Suppose that you are standing at the point with coordinates $(-100, -100, 430)$ on the hill of Problem 42 in Section 15.8.

(a) In what (horizontal) direction should you move in order to climb most steeply?

(b) What will be your resulting rate of climb (rise/run)?

26. Suppose that the blood concentration in the water at the point (x, y) is given by

$$f(x, y) = A\exp(-k[x^2 + 2y^2])$$

where A and k are positive constants. A shark always swims in the direction of ∇f. (Why?) Show that its path is a parabola $y = cx^2$. (*Suggestion:* Show that the condition that $\langle dx/dt, dy/dt\rangle$ be a multiple of ∇f implies that $x'/x = y'/2y$. Then antidifferentiate this equation.)

27. Consider a tangent plane to the surface with equation $x^{2/3} + y^{2/3} + z^{2/3} = 1$. Show that the sum of the squares of the x-, y-, and z-intercepts of this plane is 1.

28. Find the points on the ellipse $x^2/a^2 + y^2/b^2 = 1$ (with $a \neq b$) where the normal line passes through the origin.

29. (a) Show that the origin $(0, 0)$ is a critical point of the function f of Problem 12.

(b) Show that f does not have a local extremum at $(0, 0)$.

30. Find the point of the surface $z = xy + 1$ that is closest to the origin.

31. Use the method of Problem 24 in Section 15.9 to find the semiaxes of the rotated ellipse

$$73x^2 + 72xy + 52y^2 = 100.$$

32. Use the Lagrange multiplier method to show that the longest chord of the sphere $x^2 + y^2 + z^2 = 1$ has length 2. (*Suggestion:* There is no loss of generality in assuming that $(1, 0, 0)$ is one endpoint of the chord.)

33. Use the method of Lagrange multipliers, the law of cosines, and Fig. 15.77 to find the triangle of minimum perimeter inscribed in the unit circle.

34. When a current I enters two resistances R_1 and R_2 connected in parallel, it splits into two currents I_1 and I_2 (with $I = I_1 + I_2$) in such a way as to minimize the total power $R_1I_1^2 + R_2I_2^2$. Express I_1 and I_2 in terms of R_1, R_2, and I.

35. Use the method of Lagrange multipliers to find the points of the ellipse $x^2 + 2y^2 = 1$ that are closest to and farthest from the line $x + y = 2$. (*Suggestion:* Let $f(x, y, u, v)$ denote the square of the distance between the point (x, y) of the ellipse and the point (u, v) of the line.)

36. (a) Show that the maximum value of

$$f(x, y, z) = x + y + z$$

at points of the sphere $x^2 + y^2 + z^2 = a^2$ is $a\sqrt{3}$.

(b) Conclude from the result in part (a) that

$$(x + y + z)^2 \leqq 3(x^2 + y^2 + z^2)$$

for any three numbers x, y, and z.

37. Generalize the method of Problem 36 to show that

$$\left(\sum_{i=1}^{n} x_i\right)^2 \leqq n \sum_{i=1}^{n} x_i^2$$

for any n real numbers $x_1, x_2, \ldots, x_n$.

38. Find the minimum and maximum values of $f(x, y) = xy - x - y$ at points on and within the triangle with vertices $(0, 0)$, $(0, 1)$, and $(3, 0)$.

39. Find the maximum and minimum values of $f(x, y, z) = x^2 - yz$ at points of the sphere $x^2 + y^2 + z^2 = 1$.

40. Find the maximum and minimum values of $f(x, y) = x^2y^2$ at points of the ellipse $x^2 + 4y^2 = 24$.

Locate and classify the critical points (local maxima, local minima, saddle points, and other points at which the tangent plane is horizontal) of the functions in Problems 41–50.

41. $f(x, y) = x^3y - 3xy + y^2$

42. $f(x, y) = x^2 + xy + y^2 - 6x + 2$

43. $f(x, y) = x^3 - 6xy + y^3$

44. $f(x, y) = x^2y + xy^2 + x + y$

45. $f(x, y) = x^3y^2(1 - x - y)$

46. $f(x, y) = x^4 - 2x^2 + y^2 + 4y + 3$

47. $f(x, y) = e^{xy} - 2xy$

48. $f(x, y) = x^3 - y^3 + x^2 + y^2$

49. $f(x, y) = (x - y)(xy - 1)$

50. $f(x, y) = (2x^2 + y^2)\exp(-x^2 - y^2)$

51. Given data points (x_i, y_i) for $i = 1, 2, \ldots, n$, the **least squares straight line** $y = mx + b$ is the line that best fits these data in the following sense. Let $d_i = y_i - (mx_i + b)$ be the *deviation* of the predicted value $mx_i + b$ from the true value y_i. Let

$$f(m, b) = d_1^2 + d_2^2 + \cdots + d_n^2$$
$$= \sum_{i=1}^{n} [y_i - (mx_i + b)]^2$$

be the sum of the squares of the deviations. The least squares line is the one that minimizes this sum (see Fig. 15.88). Show how to choose m and b by minimizing f. (NOTE The only variables in this computation are m and b.)

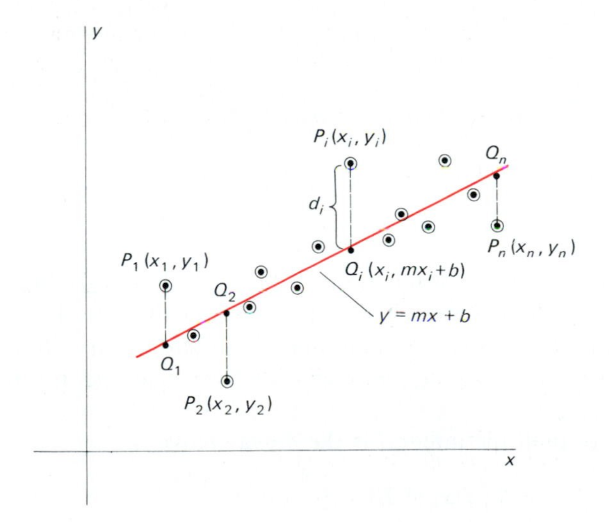

15.88 Fitting the best straight line to the data points (x_i, y_i), $1 \leqq i \leqq n$

52. *Newton's method* for solving the simultaneous equations

$$f(x, y) = 0,$$
$$g(x, y) = 0$$

is the following technique. The graph of $f(x, y) = 0$ is, generally, a curve in the xy-plane, as is the graph of $g(x, y) = 0$. The point where these curves meet is the simultaneous solution we seek. If (x_1, y_1) is a first "guess" for the simultaneous solution, then $(x_1, y_1, f(x_1, y_1))$ is a point on the surface $z = f(x, y)$ and $(x_1, y_1, g(x_1, y_1))$ is a point on the surface $z = g(x, y)$. The tangent planes at these two points generally meet the xy-plane in a pair of lines, whose intersection should be close to the desired solution (x_0, y_0) of the simultaneous equations. Let (x_2, y_2) denote the intersection of these lines, and repeat the process with (x_1, y_1) replaced by (x_2, y_2). This leads to the pair of iterative formulas

$$x_{k+1} = x_k - \frac{f(x_k, y_k)g_y(x_k, y_k) - g(x_k, y_k)f_y(x_k, y_k)}{f_x(x_k, y_k)g_y(x_k, y_k) - g_x(x_k, y_k)f_y(x_k, y_k)},$$

$$y_{k+1} = y_k - \frac{g(x_k, y_k)f_x(x_k, y_k) - f(x_k, y_k)g_x(x_k, y_k)}{f_x(x_k, y_k)g_y(x_k, y_k) - g_x(x_k, y_k)f_y(x_k, y_k)}$$

for $k = 1, 2, 3, \ldots$. Iterate these formulas to closely approximate the simultaneous solution of

$$x^2 - xy + y^2 = 3,$$
$$x^3 = y^3 - 1$$

that is close to $(x_1, y_1) = (2, 2)$.

53. Derive the iterative formulas of Problem 52.

chapter sixteen

Multiple Integrals

16.1 Double Integrals

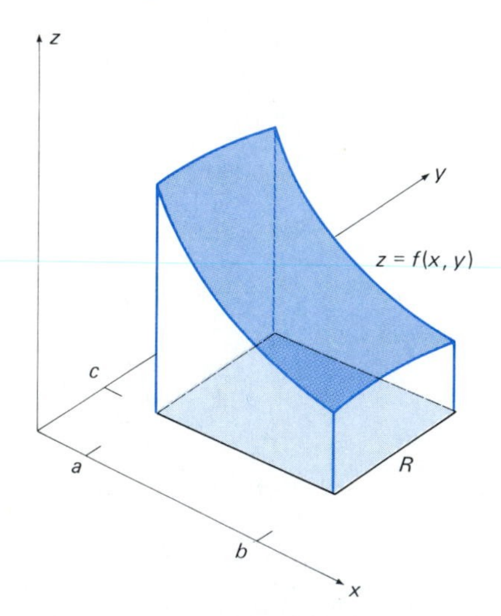

16.1 We will use a double integral to compute the volume V.

This chapter is devoted to integrals of functions of two and three variables. Such integrals are called **multiple integrals.** The applications of multiple integrals include computations of area, volume, mass, and surface area in a wider variety of situations than can be handled with the single integral of Chapters 5 and 6.

The simplest sort of multiple integral is the *double integral*

$$\iint_R f(x, y)\, dA$$

of a continuous function $f(x, y)$ over the *rectangle*

$$R = [a, b] \times [c, d] = \{(x, y) \mid a \leqq x \leqq b, c \leqq y \leqq d\}$$

in the xy-plane. Just as the definition of the single integral is motivated by the problem of computing areas, the definition of the double integral is motivated by the problem of computing the volume V of the solid of Fig. 16.1: a solid bounded above by the graph $z = f(x, y)$ of the nonnegative function f over the rectangle R in the xy-plane.

To define the *value*

$$V = \iint_R f(x, y)\, dA$$

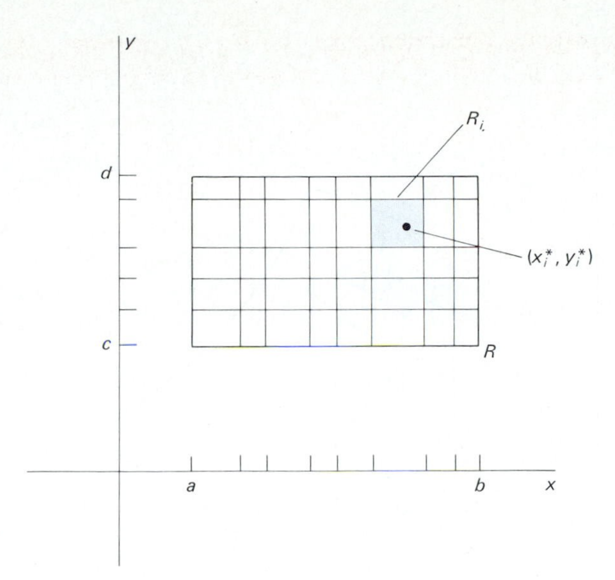

16.2 A partition $\mathscr{P}$ of the rectangle R

of such a double integral, we begin with an *approximation* to V. To obtain this approximation, the first step is to construct a **partition** $\mathscr{P}$ of R into subrectangles $R_1, R_2, \ldots, R_k$ determined by the partitions

$$a = x_0 < x_1 < x_2 < \cdots < x_m = b$$

of $[a, b]$ and

$$c = y_0 < y_1 < y_2 < \cdots < y_n = d$$

of $[c, d]$. Such a partition of R into $k = mn$ subrectangles is shown in Fig. 16.2. The order in which these rectangles are labeled makes no difference.

Next we choose an arbitrary point (x_i^*, y_i^*) of the ith subrectangle R_i for each i $(1 \leqq i \leqq k)$. The collection of points $S = \{(x_i^*, y_i^*) | 1 \leqq i \leqq k\}$ is called a **selection** for the partition $\mathscr{P} = \{R_i | 1 \leqq i \leqq k\}$. As a measure of the size of the subrectangles of the partition $\mathscr{P}$, we define its **mesh** $|\mathscr{P}|$ to be the maximum of the lengths of the diagonals of the rectangles $\{R_i\}$.

Now consider a rectangular column with base the subrectangle R_i, rising up from the xy-plane with height the value $f(x_i^*, y_i^*)$ of f at the selected point (x_i^*, y_i^*) of R_i. One such column is shown in Fig. 16.3. If ΔA_i denotes the area of R_i, then the volume of the ith column will be $f(x_i^*, y_i^*)\,\Delta A_i$. The sum of the volumes of all such columns is the **Riemann sum**

$$\sum_{i=1}^{k} f(x_i^*, y_i^*)\,\Delta A_i, \tag{1}$$

an approximation to the volume V of the solid region that lies over the rectangle R and under the graph $z = f(x, y)$.

We would expect to get the exact volume V by taking the limit of the Riemann sum in (1) as the mesh $|\mathscr{P}|$ of the partition $\mathscr{P}$ approaches zero. We therefore define the **(double) integral** of the function f over the rectangle R to be

$$\iint_R f(x, y)\,dA = \lim_{|\mathscr{P}| \to 0} \sum_{i=1}^{k} f(x_i^*, y_i^*)\,\Delta A_i, \tag{2}$$

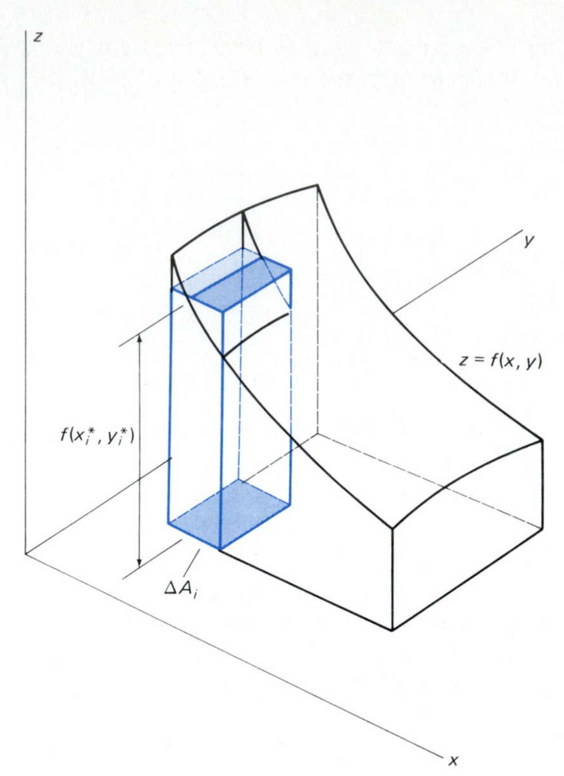

16.3 Approximating the volume under the surface by summing volumes of towers with rectangular bases

provided that this limit exists (we will make this concept more precise in Section 16.2). In advanced calculus it is proved that the limit in (2) *does* exist if f is continuous on R. In order to motivate the introduction of the Riemann sum in (1), we assumed that f was nonnegative on R, but Equation (2) serves to define the double integral whether or not f is nonnegative.

The direct evaluation of the limit in Equation (2) is generally even less practical than the direct evaluation of the limit we used in Section 5.3 to define the single-variable integral. In practice, we shall calculate double integrals over rectangles by means of the **iterated integrals** that appear in the following theorem.

Theorem *Double Integrals as Iterated Single Integrals*
Suppose that $f(x, y)$ is continuous on the rectangle $R = [a, b] \times [c, d]$. Then

$$\iint_R f(x, y)\, dA = \int_a^b \left(\int_c^d f(x, y)\, dy \right) dx = \int_c^d \left(\int_a^b f(x, y)\, dx \right) dy. \quad (3)$$

This theorem tells us how to compute a double integral by means of two successive (or *iterated*) single-variable integrations, each of which can be carried out using the fundamental theorem of calculus (if the function f is sufficiently nice).

The meaning of the parentheses in the iterated integral

$$\int_a^b \int_c^d f(x, y)\, dy\, dx = \int_a^b \left(\int_c^d f(x, y)\, dy \right) dx \quad (4)$$

is this: First we hold x constant and integrate with respect to y, from $y = c$ to $y = d$. The result of this first integration is the **partial integral of f with respect to y**, denoted by

$$\int_c^d f(x, y)\, dy,$$

and it is a function of x alone. The final step is to integrate this latter function with respect to x, from $x = a$ to $x = b$.

Similarly, the iterated integral

$$\int_c^d \int_a^b f(x, y)\, dx\, dy = \int_c^d \left(\int_a^b f(x, y)\, dx \right) dy \tag{5}$$

is calculated by first integrating from a to b with respect to x (while holding y fixed), then integrating the result from c to d with respect to y. Note that the order of integration (either first with respect to x and then with respect to y, or the reverse) is determined by the order in which the differentials dx and dy appear in the iterated integrals in (4) and (5). We always work "from the inside out." Theorem 1 guarantees that the value obtained is independent of the order of integration provided that f is continuous.

EXAMPLE 1 Compute the iterated integrals in (4) and (5) for the function $f(x, y) = 4x^3 + 6xy^2$ on the rectangle $R = [1, 3] \times [-2, 1]$.

Solution

$$\begin{aligned}
\int_1^3 \left(\int_{-2}^1 (4x^3 + 6xy^2)\, dy \right) dx &= \int_1^3 \Big[4x^3 y + 2xy^3 \Big]_{y=-2}^1 dx \\
&= \int_1^3 [(4x^3 + 2x) - (-8x^3 - 16x)]\, dx \\
&= \int_1^3 (12x^3 + 18x)\, dx \\
&= \Big[3x^4 + 9x^2 \Big]_1^3 = 312.
\end{aligned}$$

On the other hand,

$$\begin{aligned}
\int_{-2}^1 \left(\int_1^3 (4x^3 + 6xy^2)\, dx \right) dy &= \int_{-2}^1 \Big[x^4 + 3x^2y^2 \Big]_{x=1}^3 dy \\
&= \int_{-2}^1 [(81 + 27y^2) - (1 + 3y^2)]\, dy \\
&= \int_{-2}^1 (80 + 24y^2)\, dy \\
&= \Big[80y + 8y^3 \Big]_{-2}^1 = 312.
\end{aligned}$$

When we note that iterated double integrals are always evaluated from the inside out, it becomes clear that the parentheses appearing on the right-hand sides in Equations (4) and (5) are unnecessary. They are therefore generally omitted, as in the two examples that follow. When $dy\, dx$ appears in the integrand we integrate first with respect to y, while the appearance of $dx\, dy$ tells us to integrate first with respect to x.

EXAMPLE 2

$$\int_0^{\pi}\int_0^{\pi/2} \cos x \cos y \, dy \, dx = \int_0^{\pi} \Big[\cos x \sin y\Big]_{y=0}^{\pi/2} dx$$

$$= \int_0^{\pi} \cos x \, dx = \Big[\sin x\Big]_0^{\pi} = 0.$$

EXAMPLE 3

$$\int_0^1 \int_0^{\pi/2} (e^y + \sin x) \, dx \, dy = \int_0^1 \Big[xe^y - \cos x\Big]_{x=0}^{\pi/2} dy$$

$$= \int_0^1 (\tfrac{1}{2}\pi e^y + 1) \, dy$$

$$= \Big[\tfrac{1}{2}\pi e^y + y\Big]_0^1 = \frac{\pi(e-1)}{2} + 1.$$

An outline of the proof of the theorem above exhibits an instructive relationship between iterated integrals and the method of cross sections (for computing volumes) discussed in Section 6.2. First we subdivide $[a, b]$ into n equal subintervals each with length $\Delta x = (b - a)/n$, and we also subdivide $[c, d]$ into n equal subintervals each with length $\Delta y = (d - c)/n$. This gives n^2 subrectangles, each of which has area $\Delta A = \Delta x \, \Delta y$. Pick a point x_i^* in $[x_{i-1}, x_i]$ for each i, $1 \leqq i \leqq n$. Then the average value theorem for single integrals (Section 5.5) gives a point y_{ij}^* in $[y_{j-1}, y_j]$ such that

$$\int_{y_{j-1}}^{y_j} f(x_i^*, y) \, dy = f(x_i^*, y_{ij}^*) \, \Delta y.$$

This gives our selected point (x_i^*, y_{ij}^*) in the subrectangle $[x_{i-1}, x_i] \times [y_{j-1}, y_j]$. Then

$$\iint_R f(x, y) \, dA \approx \sum_{i,j=1}^{n} f(x_i^*, y_{ij}^*) \, \Delta A = \sum_{i=1}^{n} \sum_{j=1}^{n} f(x_i^*, y_{ij}^*) \, \Delta y \, \Delta x$$

$$= \sum_{i=1}^{n} \left(\sum_{j=1}^{n} \int_{y_{j-1}}^{y_j} f(x_i^*, y) \, dy \right) \Delta x$$

$$= \sum_{i=1}^{n} \left(\int_c^d f(x_i^*, y) \, dy \right) \Delta x = \sum_{i=1}^{n} A(x_i^*) \, \Delta x$$

where

$$A(x) = \int_c^d f(x, y) \, dy.$$

This last sum is a Riemann sum for the integral $\int_a^b A(x) \, dx$, so the result of our computation is that

$$\iint_R f(x, y) \, dA \approx \sum_{i=1}^{n} A(x_i^*) \, \Delta x$$

$$\approx \int_a^b A(x) \, dx = \int_a^b \left(\int_c^d f(x, y) \, dy \right) dx.$$

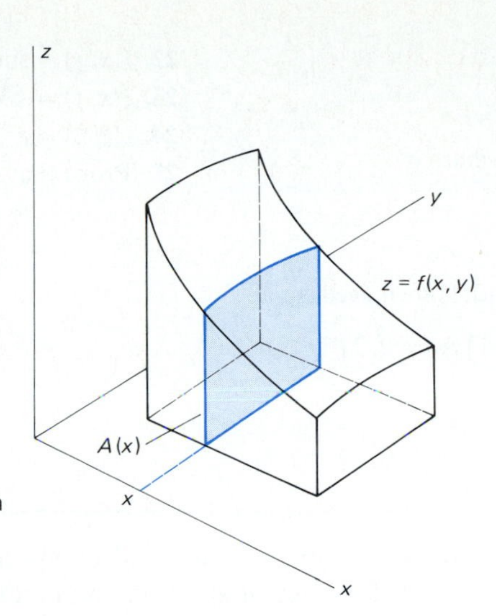

16.4 The area of the cross section "at" x is

$$A(x) = \int_c^d f(x, y)\, dy.$$

This outline can be converted into a complete proof of the theorem by showing that the approximations above become equalities when we take limits as $n \to \infty$.

In case the function f is nonnegative on R, the function $A(x)$ introduced here gives the area of the vertical cross section perpendicular to the x-axis, as shown in Fig. 16.4. Thus the iterated integral in (4) expresses the volume V as the integral from $x = a$ to $x = b$ of the cross-sectional area function $A(x)$. Similarly, the iterated integral in (5) expresses V as the integral from $y = c$ to $y = d$ of the function

$$A(y) = \int_a^b f(x, y)\, dx,$$

which gives the area of a vertical cross section in a plane perpendicular to the y-axis. (Although it is suggestive to use the notation $A(y)$ here, note that $A(y)$ and $A(x)$ are by no means the same function!)

16.1 PROBLEMS

Evaluate the iterated integrals in Problems 1–20.

1. $\int_0^2 \int_0^4 (3x + 4y)\, dx\, dy$

2. $\int_0^3 \int_0^2 x^2 y\, dx\, dy$

3. $\int_{-1}^2 \int_1^3 (2x - 7y)\, dy\, dx$

4. $\int_{-2}^1 \int_2^4 x^2 y^3\, dy\, dx$

5. $\int_0^3 \int_0^3 (xy + 7x + y)\, dx\, dy$

6. $\int_0^2 \int_2^4 (x^2 y^2 - 17)\, dx\, dy$

7. $\int_{-1}^2 \int_{-1}^2 (2xy^2 - 3x^2 y)\, dy\, dx$

8. $\int_1^3 \int_{-3}^{-1} (x^3 y - xy^3)\, dy\, dx$

9. $\int_0^{\pi/2} \int_0^{\pi/2} (\sin x \cos y)\, dx\, dy$

10. $\int_0^{\pi/2} \int_0^{\pi/2} (\cos x \sin y)\, dy\, dx$

11. $\int_0^1 \int_0^1 xe^y\, dy\, dx$

12. $\int_0^1 \int_{-2}^2 x^2 e^y\, dx\, dy$

13. $\int_0^1 \int_0^\pi e^x \sin y\, dy\, dx$

14. $\int_0^1 \int_0^1 e^{x+y}\, dx\, dy$

15. $\int_0^\pi \int_0^\pi (xy + \sin x)\, dx\, dy$

16. $\int_0^{\pi/2} \int_0^{\pi/2} (y - 1) \cos x\, dx\, dy$

17. $\int_0^{\pi/2} \int_1^e \frac{\sin y}{x}\, dx\, dy$

18. $\int_1^e \int_1^e \frac{1}{xy}\, dy\, dx$

19. $\int_0^1 \int_0^1 \left(\frac{1}{x+1} + \frac{1}{y+1}\right) dx\, dy$

20. $\displaystyle\int_1^2 \int_1^3 \left(\frac{x}{y} + \frac{y}{x}\right) dy\, dx$

In Problems 21–24, verify that the values of

$$\iint_R f(x, y)\, dA$$

given by the iterated integrals in (4) and (5) are indeed equal.

21. $f(x, y) = 2xy - 3y^2$; $R = [-1, 1] \times [-2, 2]$

22. $f(x, y) = \sin x \cos y$; $R = [0, \pi] \times [-\pi/2, \pi/2]$

23. $f(x, y) = \sqrt{x + y}$; $R = [0, 1] \times [1, 2]$

24. $f(x, y) = e^{x+y}$; $R = [0, \ln 2] \times [0, \ln 3]$

25. Prove that

$$\lim_{n\to\infty} \int_0^1 \int_0^1 x^n y^n\, dx\, dy = 0.$$

16.2 Double Integrals over More General Regions

Now we want to define and compute double integrals over regions more general than rectangles. Let the function f be defined on the plane region R, and suppose that R is **bounded**—that is, suppose that R lies within some rectangle S. To define the (double) integral of f over the region R, we begin with a partition $\mathcal{Q}$ of the rectangle S into subrectangles. Some of the rectangles of $\mathcal{Q}$ will lie wholly within R, some will lie outside of R, and some will lie partly within and partly outside R. We consider the collection $\mathcal{P} = \{R_1, R_2, \ldots, R_k\}$ of all those subrectangles of $\mathcal{Q}$ that lie *completely within* the region R. This collection $\mathcal{P}$ is called the **inner partition** of the region R determined by the partition $\mathcal{Q}$ of the rectangle S (see Fig. 16.5). By the **mesh** $|\mathcal{P}|$ of the inner partition $\mathcal{P}$ is meant the mesh of the partition $\mathcal{Q}$ that determines $\mathcal{P}$. Note that $|\mathcal{P}|$ depends not only upon $\mathcal{P}$ but upon $\mathcal{Q}$ as well.

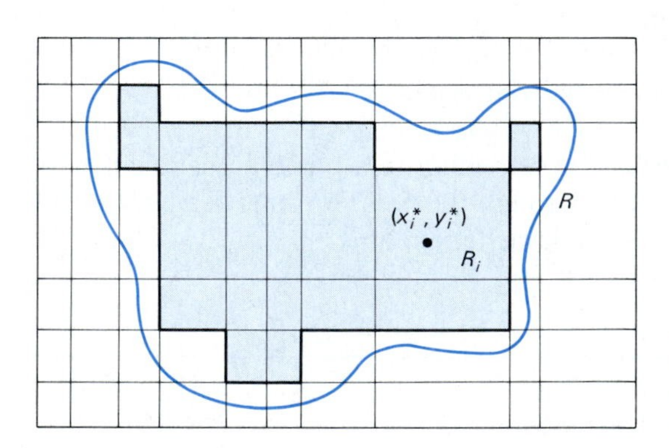

16.5 The rectangular partition of S produces an associated inner partition (shown shaded) of the region R.

Using the inner partition $\mathcal{P}$ of the region R, we can proceed in much the same way as before. By choosing an arbitrary point (x_i^*, y_i^*) in the ith subrectangle R_i of $\mathcal{P}$ for $i = 1, 2, 3, \ldots, k$, we obtain a **selection** for the inner partition $\mathcal{P}$. Let us denote by ΔA_i the area of R_i. Then this selection gives the **Riemann sum**

$$\sum_{i=1}^{k} f(x_i^*, y_i^*)\, \Delta A_i$$

associated with the inner partition $\mathcal{P}$. In case f is nonnegative on R, this Riemann sum approximates the volume of the three-dimensional region that lies under the surface $z = f(x, y)$ and above the region R in the xy-plane. We therefore define the double integral of f over the region R by taking the limit of this Riemann sum as the mesh $|\mathcal{P}|$ approaches zero. Thus

$$\iint_R f(x, y)\,dA = \lim_{|\mathcal{P}|\to 0} \sum_{i=1}^{k} f(x_i^*, y_i^*)\,\Delta A_i \tag{1}$$

provided that this limit exists in the sense of the following definition.

Definition *The Double Integral*
The **(double) integral** of the bounded function f over the plane region R is the number

$$I = \iint_R f(x, y)\,dA$$

provided that, for every $\varepsilon > 0$, there exists a number $\delta > 0$ such that

$$\left| \sum_{i=1}^{k} f(x_i^*, y_i^*)\,\Delta A_i - I \right| < \varepsilon$$

for every inner partition $\mathcal{P} = \{R_1, R_2, \ldots, R_k\}$ of R having mesh $|\mathcal{P}| < \delta$ and every selection of points (x_i^*, y_i^*) in R_i $(i = 1, 2, \ldots, k)$.

Thus the meaning of the limit in (1) is that the Riemann sum can be made arbitrarily close to $I = \iint_R f(x, y)\,dA$ by choosing the mesh of the inner partition $\mathcal{P}$ sufficiently small.

NOTE If R is a rectangle and we choose $S = R$ (so that an inner partition of R is simply a partition of R), then the definition above reduces to our earlier definition of a double integral over a rectangle. In advanced calculus it is proved that the double integral of the function f over the bounded plane region R exists provided that f is continuous on R and the *boundary* of R is reasonably well-behaved. In particular, it suffices for the boundary of R to consist of finitely many piecewise smooth simple closed curves (that is, each boundary curve consists of finitely many smooth arcs).

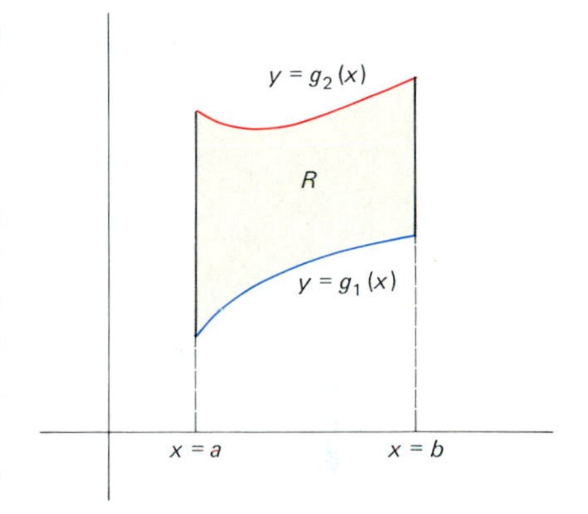

16.6 A vertically simple region R

For certain common types of regions, we can evaluate double integrals by using iterated integrals, in much the same way as when the region is a rectangle. The plane region R is called **vertically simple** if it is described by means of the inequalities

$$a \leqq x \leqq b, \qquad g_1(x) \leqq y \leqq g_2(x), \tag{2}$$

where g_1 and g_2 are continuous on $[a, b]$. Such a region appears in Fig. 16.6. The region R is called **horizontally simple** if it is described by the inequalities

$$c \leqq y \leqq d, \qquad h_1(y) \leqq x \leqq h_2(y), \tag{3}$$

where h_1 and h_2 are continuous on $[c, d]$. The region of Fig. 16.7 is horizontally simple.

The next theorem tells us how to compute by iterated integration a double integral over a region R that is either vertically simple or horizontally simple.

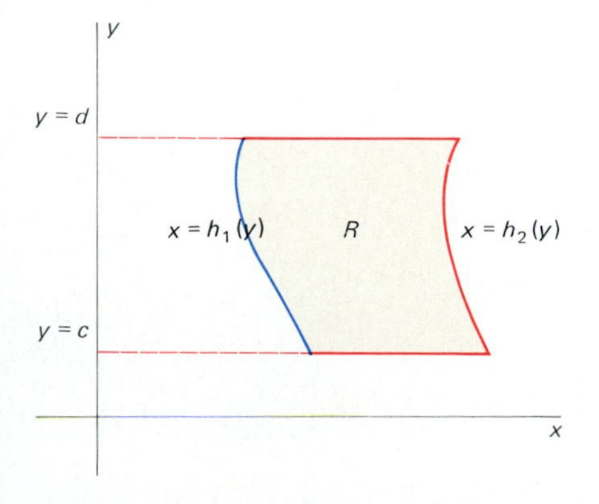

16.7 A horizontally simple region R

> ***Theorem*** *Evaluation of Double Integrals*
> Suppose that $f(x, y)$ is continuous on the region R. If R is the vertically simple region given in (2), then
> $$\iint_R f(x, y)\,dA = \int_a^b \int_{g_1(x)}^{g_2(x)} f(x, y)\,dy\,dx. \tag{4}$$
> If R is the horizontally simple region given in (3), then
> $$\iint_R f(x, y)\,dA = \int_c^d \int_{h_1(y)}^{h_2(y)} f(x, y)\,dx\,dy. \tag{5}$$

This theorem includes the theorem of Section 16.1 as a special case (when R is a rectangle), and it can be proved by a generalization of the argument we outlined there.

EXAMPLE 1 Compute in two different ways the integral

$$\iint_R xy^2\,dA,$$

where R is the first-quadrant region bounded by the two curves $y = x^2$ and $y = x^3$.

Solution *Always sketch the region R of integration before attempting to evaluate a double integral.* The region R bounded by $y = x^2$ and $y = x^3$ is shown in Fig. 16.8. This region is vertically simple with $a = 0$, $b = 1$, $g_1(x) = x^3$, and $g_2(x) = x^2$. Therefore, (4) yields

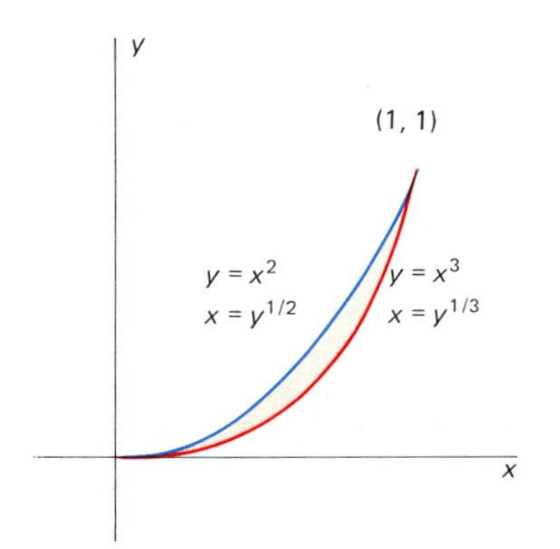

16.8 The region R of Example 1

$$\iint_R xy^2\,dA = \int_0^1 \int_{x^3}^{x^2} xy^2\,dy\,dx$$

$$= \int_0^1 \left[\tfrac{1}{3}xy^3\right]_{x^3}^{x^2} dx = \int_0^1 \left(\tfrac{1}{3}x^7 - \tfrac{1}{3}x^{10}\right) dx$$

$$= \tfrac{1}{24} - \tfrac{1}{33} = \tfrac{1}{88}.$$

The region R is also horizontally simple with $c = 0$, $d = 1$, $h_1(y) = y^{1/2}$, and $h_2(y) = y^{1/3}$, so we can reverse the order of integration, and integrate first with respect to x. Then Equation (5) gives

$$\iint_R xy^2\,dA = \int_0^1 \int_{y^{1/2}}^{y^{1/3}} xy^2\,dx\,dy = \int_0^1 \left[\tfrac{1}{2}x^2y^2\right]_{y^{1/2}}^{y^{1/3}} dy$$

$$= \int_0^1 \left(\tfrac{1}{2}y^{8/3} - \tfrac{1}{2}y^3\right) dy = \tfrac{1}{88}.$$

EXAMPLE 2 Evaluate

$$\iint_R (6x + 2y^2)\,dA,$$

where R is the region bounded by the parabola $x = y^2$ and the straight line $x + y = 2$.

Solution The region R appears in Fig. 16.9. It is both horizontally and vertically simple. If we wished to integrate first with respect to y and then with respect to x, we would need to evaluate two integrals:

$$\iint_R f(x, y)\, dA = \int_0^1 \int_{-\sqrt{x}}^{\sqrt{x}} (6x + 2y^2)\, dy\, dx + \int_1^4 \int_{-\sqrt{x}}^{2-x} (6x + 2y^2)\, dy\, dx.$$

The reason is that the "top" function $g_2(x)$ changes its formula at the point (1, 1) (see the sketch of the region R).

To avoid this extra work, we prefer to integrate in the opposite order:

$$\iint_R (6x - 2y^2)\, dA = \int_{-2}^{1} \int_{y^2}^{2-y} (6x + 2y^2)\, dx\, dy$$

$$= \int_{-2}^{1} \left[3x^2 + 2xy^2\right]_{y^2}^{2-y} dy$$

$$= \int_{-2}^{1} [3(2 - y)^2 + 2(2 - y)y^2 - 3(y^2)^2 - 2y^4]\, dy$$

$$= \int_{-2}^{1} (12 - 12y + 7y^2 - 2y^3 - 5y^4)\, dy$$

$$= \left[12y - 6y^2 + \tfrac{7}{3}y^3 - \tfrac{1}{2}y^4 - y^5\right]_{-2}^{1} = \tfrac{99}{2}.$$

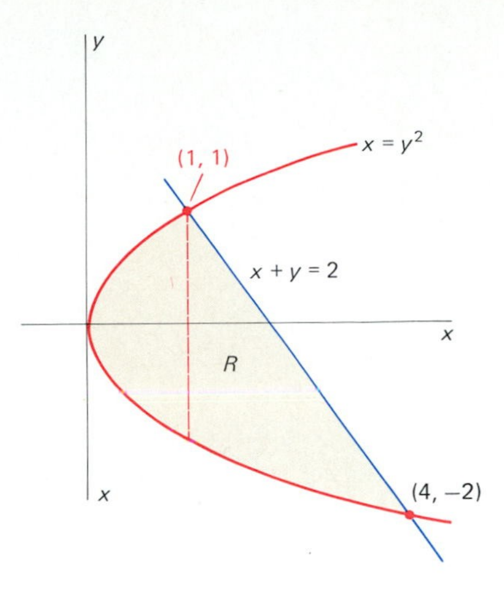

16.9 The region of Example 2

Example 2 indicates that even when the region R is both vertically and horizontally simple, it may be simpler to integrate in one order rather than the other because of the shape of R. We naturally prefer the easier route. The choice of the preferable order of integration may also be influenced by the nature of the function $f(x, y)$. For it may be difficult—or even impossible—to compute a given iterated integral, but easy to do *after reversing the order of integration.* The following example shows that the key to reversing the order of integration is this: Find (and sketch) the region R over which the integration is performed.

EXAMPLE 3 Evaluate

$$\int_0^2 \int_{y/2}^1 y\, e^{x^3}\, dx\, dy.$$

Solution We cannot integrate first with respect to x, as indicated, because it happens that $\exp(x^3)$ has no elementary antiderivative. So we try to evaluate the integral by first reversing the order of integration. To do so, we sketch the region of integration specified by the limits in the given iterated integral.

The region R is determined by the inequalities

$$\tfrac{1}{2}y \leqq x \leqq 1 \quad \text{and} \quad 0 \leqq y \leqq 2.$$

Thus all points (x, y) of R lie between the horizontal lines $y = 0$ and $y = 2$, and also between the two graphs $x = y/2$ and $x = 1$. We draw the four straight lines $y = 0$, $y = 2$, $x = y/2$, and $x = 1$ and find that the region of integration is the triangle that appears in Fig. 16.10.

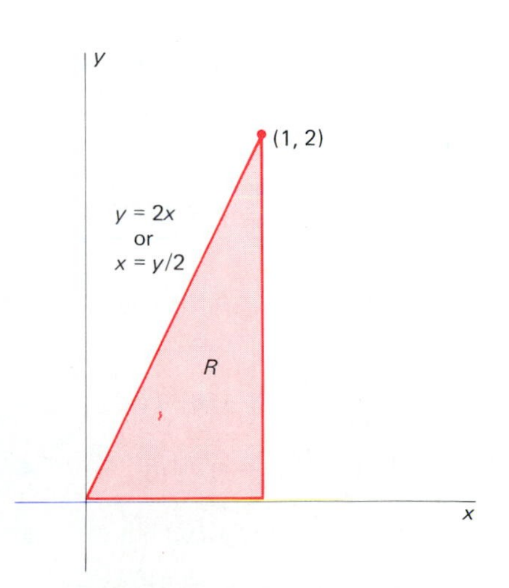

16.10 The region R of Example 3

Integrating first with respect to y, from $g_1(x) = 0$ to $g_2(x) = 2x$, we obtain

$$\begin{aligned}\int_0^2 \int_{y/2}^1 y\, e^{x^3}\, dx\, dy &= \int_0^1 \int_0^{2x} y\, e^{x^3}\, dy\, dx \\ &= \int_0^1 \left[\tfrac{1}{2}y^2\right]_0^{2x} e^{x^3}\, dx \\ &= \int_0^1 2x^2\, e^{x^3}\, dx \\ &= \left[\tfrac{2}{3}\, e^{x^3}\right]_0^1 = \tfrac{2}{3}(e - 1).\end{aligned}$$

We conclude this section by listing some useful formal properties of double integrals. Let c be a constant and f and g continuous functions on a region R on which $f(x, y)$ attains a minimum value m and a maximum value M. Let $a(R)$ denote the area of the region R. If the indicated integrals all exist, then:

$$\iint_R cf(x, y)\, dA = c \iint_R f(x, y)\, dA, \tag{6}$$

$$\iint_R [f(x, y) + g(x, y)]\, dA = \iint_R f(x, y)\, dA + \iint_R g(x, y)\, dA, \tag{7}$$

$$m \cdot a(R) \leqq \iint_R f(x, y)\, dA \leqq M \cdot a(R), \tag{8}$$

$$\iint_R f(x, y)\, dA = \iint_{R_1} f(x, y)\, dA + \iint_{R_2} f(x, y)\, dA. \tag{9}$$

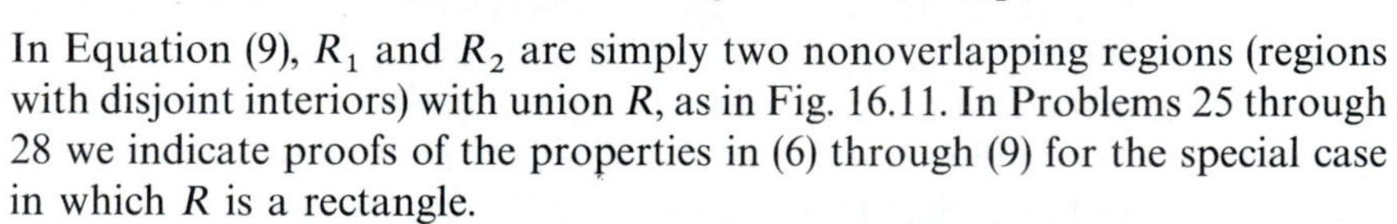

In Equation (9), R_1 and R_2 are simply two nonoverlapping regions (regions with disjoint interiors) with union R, as in Fig. 16.11. In Problems 25 through 28 we indicate proofs of the properties in (6) through (9) for the special case in which R is a rectangle.

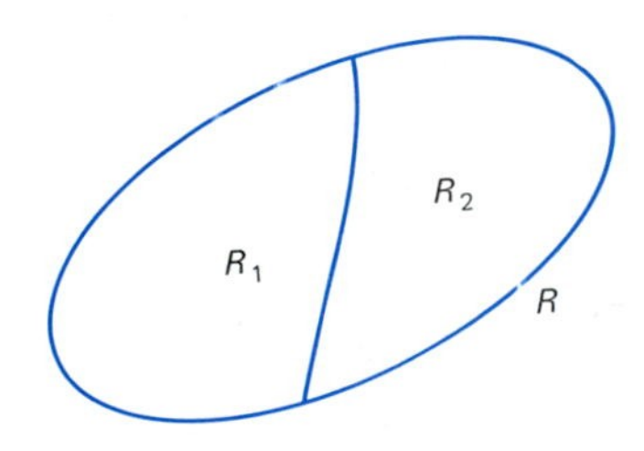

16.11 The regions of the formula in Equation (9)

The property in Equation (9) enables us to evaluate double integrals over a region R that is neither vertically nor horizontally simple. All that is necessary is to subdivide R into finitely many simple regions $R_1, R_2, \ldots, R_n$. Then we integrate over each (converting each double integral into an iterated integral as in the examples of this section) and add the results.

16.2 PROBLEMS

Evaluate the iterated integrals in Problems 1–14.

1. $\int_0^1 \int_0^x (1 + x)\, dy\, dx$

2. $\int_0^2 \int_0^{2x} (1 + y)\, dy\, dx$

3. $\int_0^1 \int_y^1 (x + y)\, dx\, dy$

4. $\int_0^2 \int_{y/2}^1 (x + y)\, dx\, dy$

5. $\int_0^1 \int_0^{x^2} xy\, dy\, dx$

6. $\int_0^1 \int_y^{\sqrt{y}} (x + y)\, dx\, dy$

7. $\int_0^1 \int_x^{\sqrt{x}} (2x - y)\, dy\, dx$

8. $\int_0^2 \int_{-\sqrt{2y}}^{\sqrt{2y}} (3x + 2y)\, dx\, dy$

9. $\int_0^1 \int_{x^4}^x (y - x)\, dy\, dx$

10. $\int_{-1}^2 \int_{-y}^{y+2} (x + 2y^2)\, dx\, dy$

11. $\int_0^1 \int_0^{x^3} e^{y/x}\, dy\, dx$

12. $\int_0^\pi \int_0^{\sin x} y\, dy\, dx$

13. $\int_0^3 \int_0^y \sqrt{y^2 + 16}\, dx\, dy$

14. $\int_1^{e^2} \int_0^{1/y} e^{xy}\, dx\, dy$

In Problems 15–24, first sketch the region of integration, then reverse the order of integration, and finally evaluate the resulting integral.

15. $\int_{-2}^{2}\int_{x^2}^{4} x^2 y \, dy \, dx$

16. $\int_{0}^{1}\int_{x^4}^{x} (x-1) \, dy \, dx$

17. $\int_{-1}^{3}\int_{x^2}^{2x+3} x \, dy \, dx$

18. $\int_{-2}^{2}\int_{y^2-4}^{4-y^2} y \, dx \, dy$

19. $\int_{0}^{2}\int_{2x}^{4x-x^2} 1 \, dy \, dx$

20. $\int_{0}^{1}\int_{y}^{1} \exp(-x^2) \, dx \, dy$

21. $\int_{0}^{\pi}\int_{x}^{\pi} \frac{\sin y}{y} \, dy \, dx$

22. $\int_{0}^{\sqrt{\pi}}\int_{y}^{\sqrt{\pi}} \sin x^2 \, dx \, dy$

23. $\int_{0}^{1}\int_{y}^{1} \frac{1}{1+x^4} \, dx \, dy$

24. $\int_{0}^{1}\int_{\tan^{-1} y}^{\pi/4} \sec x \, dx \, dy$

25. Use Riemann sums to prove (6) for the case in which R is a rectangle with its sides parallel to the coordinate axes.

26. Use iterated integrals and familiar properties of single integrals to prove (7) for the case in which R is a rectangle with its sides parallel to the coordinate axes.

27. Use Riemann sums to prove (8) for the case in which R is a rectangle with its sides parallel to the coordinate axes.

28. Use iterated integrals and familiar properties of single integrals to prove (9) if R_1 and R_2 are rectangles with their sides parallel to the coordinate axes and the right-hand edge of R_1 is the left-hand edge of R_2.

29. Use Riemann sums to prove that

$$\iint_R f(x, y) \, dA \leqq \iint_R g(x, y) \, dA$$

if $f(x, y) \leqq g(x, y)$ at each point of the region R, a rectangle with its sides parallel to the coordinate axes.

30. Suppose that the continuous function f is integrable on the plane region R and that f attains a minimum value m and a maximum value M on R. Assume also that R is *connected* in the following sense: For any two points (x_0, y_0) and (x_1, y_1) of R, there is a continuous parametric curve $\mathbf{r}(t)$ in R with $\mathbf{r}(0) = \langle x_0, y_0 \rangle$ and $\mathbf{r}(1) = \langle x_1, y_1 \rangle$. Then deduce the *average value property* of double integrals from (8):

$$\iint_R f(x, y) \, dA = f(\hat{x}, \hat{y}) \cdot \text{area}(R)$$

for some point $(\hat{x}, \hat{y})$ of R. (*Suggestion:* If $m = f(x_0, y_0)$ and $M = f(x_1, y_1)$, then you may apply the intermediate value property of the continuous function $f(\mathbf{r}(t))$.)

16.3 Area and Volume by Double Integration

In Section 16.2, our definition of $\iint_R f(x, y) \, dA$ was *motivated* by the problem of computing the volume of the solid

$$T = \{(x, y, z) \mid (x, y) \in R \text{ and } 0 \leqq z \leqq f(x, y)\}$$

that lies under the surface $z = f(x, y)$ and above the region R in the xy-plane. Such a solid appears in Fig. 16.12. Despite this geometric motivation, the actual definition of the double integral as a limit of Riemann sums does not depend upon the concept of volume. We may, therefore, turn matters around and use the double integral to *define* volume.

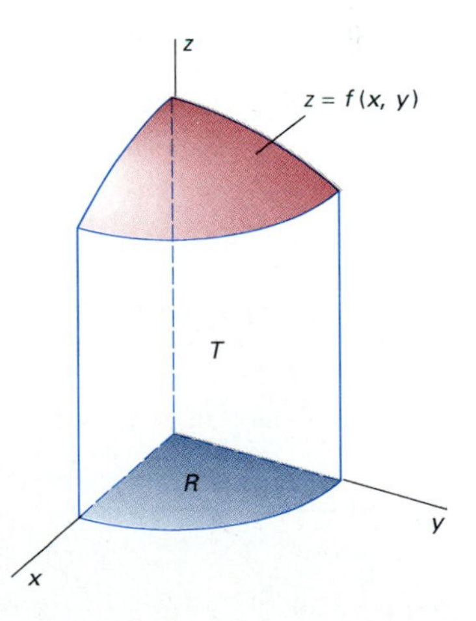

16.12 A solid region with vertical sides and base R in the xy-plane

> ***Definition*** *Volume under $z = f(x, y)$*
>
> Suppose that the function f is continuous and nonnegative on the bounded plane region R. Then the **volume** V of the solid that lies under the surface $z = f(x, y)$ and above the region R is defined to be
>
> $$V = \iint_R f(x, y) \, dA, \tag{1}$$
>
> provided that this integral exists.

It is of interest to note the connection between this definition and the cross-sections approach to volume that we discussed in Section 6.2. If, for example, the region R is vertically simple, then the volume integral in (1)

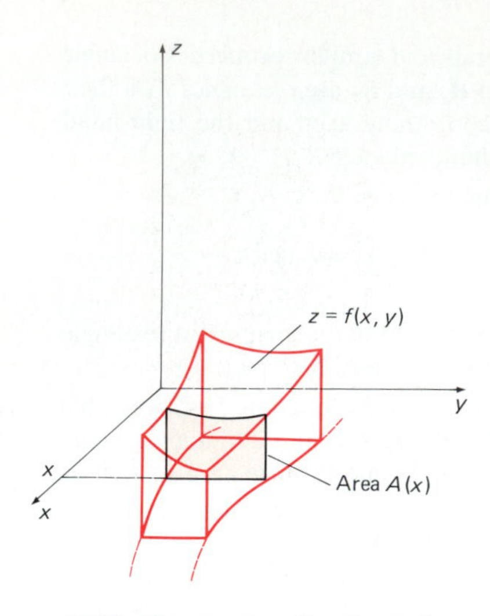

16.13 The cross-sectional area is
$$A(x) = \int_{g_1(x)}^{g_2(x)} f(x, y)\,dy.$$

takes the form

$$V = \iint_R z\,dA = \int_a^b \int_{g_1(x)}^{g_2(x)} f(x, y)\,dy\,dx$$

in terms of iterated integrals. The inner integral

$$A(x) = \int_{g_1(x)}^{g_2(x)} f(x, y)\,dy$$

is simply the area of the cross section of the solid region T in a plane perpendicular to the x-axis (see Fig. 16.13). Thus

$$V = \int_a^b A(x)\,dx,$$

and so in this case the formula in (1) reduces to "volume is the integral of cross-sectional area."

EXAMPLE 1 The rectangle R in the xy-plane consists of those points (x, y) for which $0 \leqq x \leqq 2$ and $0 \leqq y \leqq 1$. Find the volume V of the solid that lies under the surface $z = 1 + xy$ and above R.

Solution Here $f(x, y) = 1 + xy$, so the formula in (1) yields

$$V = \iint_R z\,dA = \int_0^2 \int_0^1 (1 + xy)\,dy\,dx = \int_0^2 \left[y + \tfrac{1}{2}xy^2\right]_{y=0}^1 dx$$

$$= \int_0^2 (1 + \tfrac{1}{2}x)\,dx = \left[x + \tfrac{1}{4}x^2\right]_0^2 = 3.$$

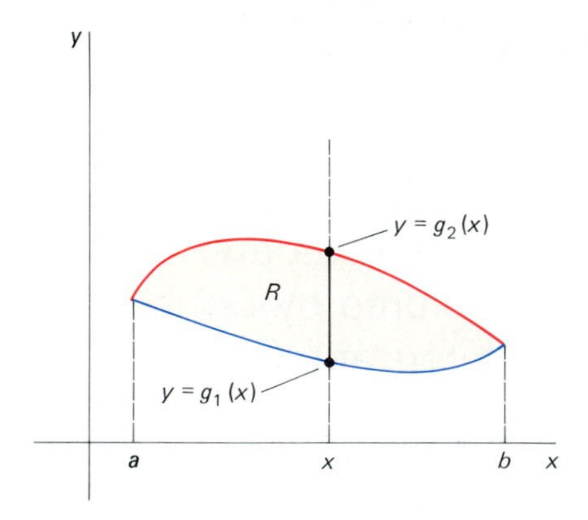

16.14 A vertically simple region

A three-dimensional region T is typically described in terms of the surfaces that bound it. The first step in applying the formula in Equation (1) to compute its volume V is to determine the region R in the xy-plane over which T lies. The second step is to determine the appropriate order of iterated integration. This may be done in the following way:

If each vertical line in the xy-plane meets R in a *single* line segment, then R is vertically simple, and you may integrate first with respect to y. The limits on y will be the y-coordinates $g_1(x)$ and $g_2(x)$ of the endpoints of this line segment (as indicated in Fig. 16.14). The limits on x will be the endpoints a and b of the interval on the x-axis onto which R projects. The theorem of Section 16.2 then gives

$$V = \iint_R f(x, y)\,dA = \int_a^b \int_{g_1(x)}^{g_2(x)} f(x, y)\,dy\,dx. \tag{2}$$

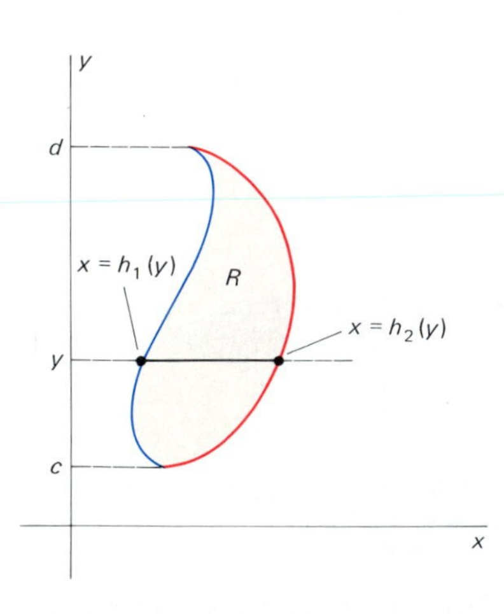

16.15 A horizontally simple region

Alternatively:

If each horizontal line in the xy-plane meets R in a *single* line segment, then R is horizontally simple, and you may integrate with respect to x first. In this case,

$$V = \iint_R f(x, y)\,dA = \int_c^d \int_{h_1(y)}^{h_2(y)} f(x, y)\,dx\,dy. \tag{3}$$

As indicated in Fig. 16.15, $h_1(y)$ and $h_2(y)$ are the x-coordinates of the endpoints of this horizontal line segment, and c and d are the endpoints of the corresponding interval on the y-axis.

If the region R is both vertically simple and horizontally simple, then you have the pleasant option of choosing the order of integration that will lead to the simpler subsequent computations. If R is neither vertically simple nor horizontally simple, then you must first subdivide R into simple regions before proceeding with iterated integration.

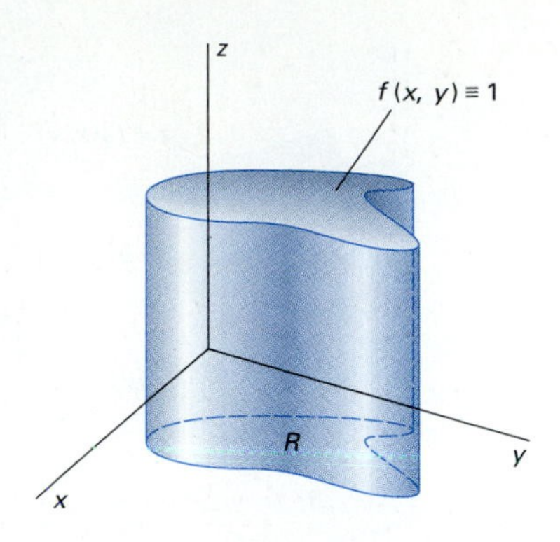

16.16 The mesa

The special case $f(x, y) \equiv 1$ in the formula of Equation (1) gives the area

$$A = a(R) = \iint_R 1\,dA = \iint_R dA \tag{4}$$

of the plane region R. In this case the solid region T resembles a desert mesa (see Fig. 16.16)—a solid cylinder with base R of area A and height 1. And the volume of any such cylinder—not necessarily circular—is the product of its height and the area of its base. In this case, the iterated integrals in (2) and (3) reduce to

$$A = \int_a^b \int_{y_{\text{bot}}}^{y_{\text{top}}} 1\,dy\,dx \quad \text{and} \quad A = \int_c^d \int_{x_{\text{left}}}^{x_{\text{right}}} 1\,dx\,dy,$$

respectively.

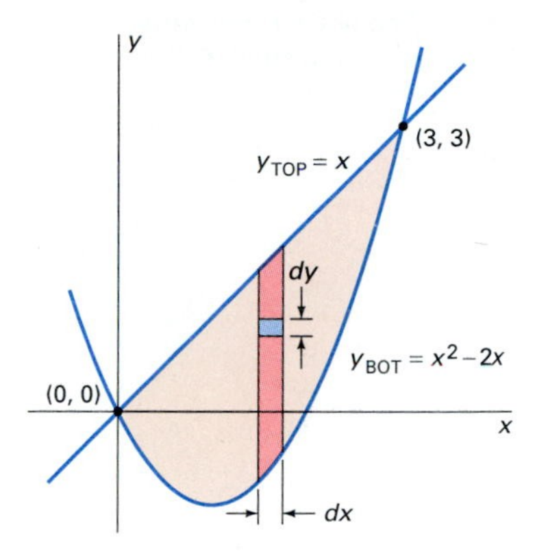

16.17 The region R of Example 2

EXAMPLE 2 Compute by double integration the area A of the region R in the xy-plane that is bounded by the line $y = x$ and the parabola $y = x^2 - 2x$.

Solution As indicated in Fig. 16.17, the line $y_{\text{top}} = x$ and the parabola $y_{\text{bot}} = x^2 - 2x$ intersect in the points (0, 0) and (3, 3). (These coordinates are easy to find by solving the equation $y_{\text{top}} = y_{\text{bot}}$.) Therefore,

$$A = \int_a^b \int_{y_{\text{bot}}}^{y_{\text{top}}} 1\,dy\,dx = \int_0^3 \int_{x^2-2x}^{x} 1\,dy\,dx = \int_0^3 \Big[y\Big]_{x^2-2x}^{x}\,dx$$
$$= \int_0^3 (3x - x^2)\,dx = \Big[\tfrac{3}{2}x^2 - \tfrac{1}{3}x^3\Big]_0^3 = \tfrac{9}{2}.$$

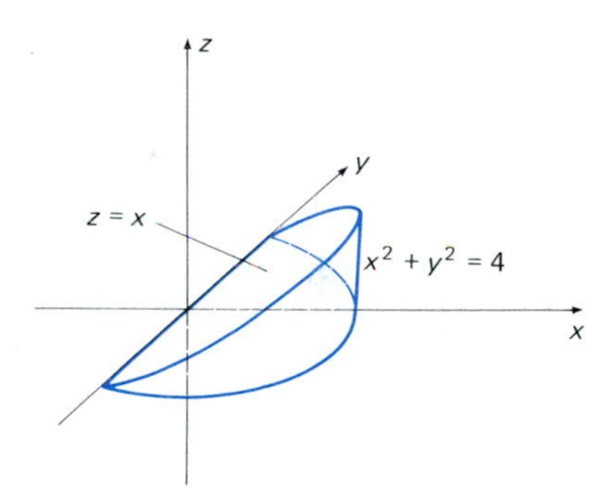

16.18 The wedge of Example 3

EXAMPLE 3 Find the volume of the wedge-shaped solid T that lies above the xy-plane, under the plane $z = x$, and within the cylinder $x^2 + y^2 = 4$ (as shown in Fig. 16.18).

Solution The base region R is a semicircle of radius 2, but by symmetry we may integrate over the first-quadrant quarter-circle alone and then double the result. A sketch of the quarter-circle (see Fig. 16.19) helps establish the limits of integration. We could integrate in either order, but integration with respect to x first gives a slightly simpler computation of the volume V:

$$V = \iint_R z\,dA = 2\int_0^2 \int_0^{\sqrt{4-y^2}} x\,dx\,dy$$
$$= 2\int_0^2 \Big[\tfrac{1}{2}x^2\Big]_0^{\sqrt{4-y^2}}\,dy = \int_0^2 (4 - y^2)\,dy$$
$$= \Big[4y - \tfrac{1}{3}y^3\Big]_0^2 = \tfrac{16}{3}.$$

You should integrate in the other order and compare the results.

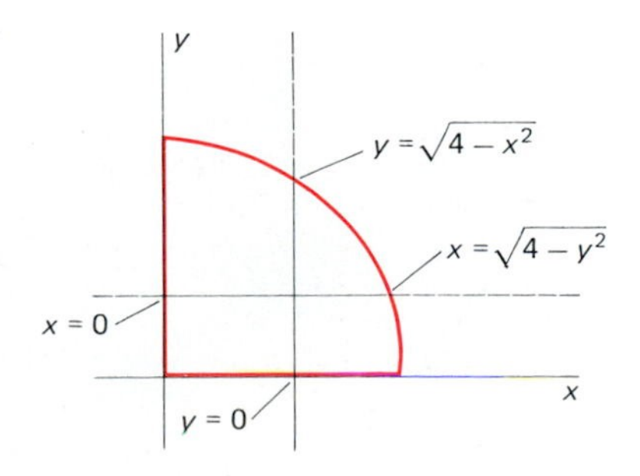

16.19 *Half* of the base R of the wedge

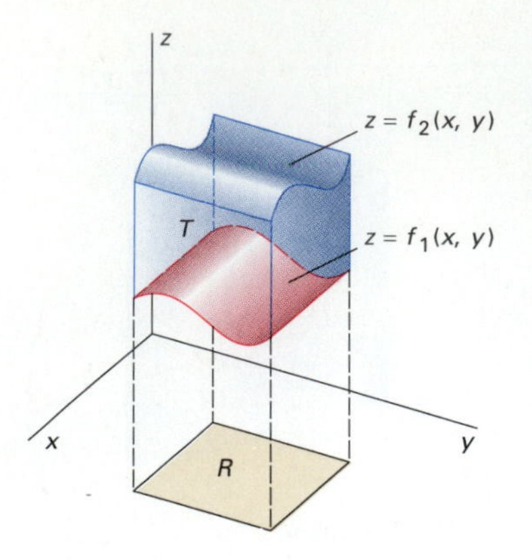

16.20 The solid T has vertical sides and is bounded above and below by surfaces.

Suppose now that the solid region T lies above the plane region R, as before, but *between* the surfaces $z = f_1(x, y)$ and $z = f_2(x, y)$, where $f_1(x, y) \leqq f_2(x, y)$ for all (x, y) in R (see Fig. 16.20). Then we get the volume V of T by subtracting the volume under $z = f_1(x, y)$ from the volume under $z = f_2(x, y)$, so

$$V = \iint_R [f_2(x, y) - f_1(x, y)]\, dA. \tag{5}$$

More briefly,

$$V = \iint_R (z_{\text{top}} - z_{\text{bot}})\, dA$$

where $z_{\text{top}} = f_2(x, y)$ describes the top surface and $z_{\text{bot}} = f_1(x, y)$ the bottom surface of T. This is a natural generalization of the formula for the area of the plane region between the curves $y = f_1(x)$ and $y = f_2(x)$ over the interval $[a, b]$.

EXAMPLE 4 Find the volume V of the solid T bounded by the planes $z = 6$ and $z = 2y$ and by the parabolic cylinders $y = x^2$ and $y = 2 - x^2$. (See Fig. 16.21.)

Solution Because the parabolic cylinders given are perpendicular to the xy-plane, the solid T has vertical sides. Thus we may think of T as lying between the planes $z_{\text{top}} = 6$ and $z_{\text{bot}} = 2y$ and above the xy-plane region R that is bounded by the parabolas $y = x^2$ and $y = 2 - x^2$. As indicated in Fig. 16.22, these parabolas intersect at the points $(-1, 1)$ and $(1, 1)$.

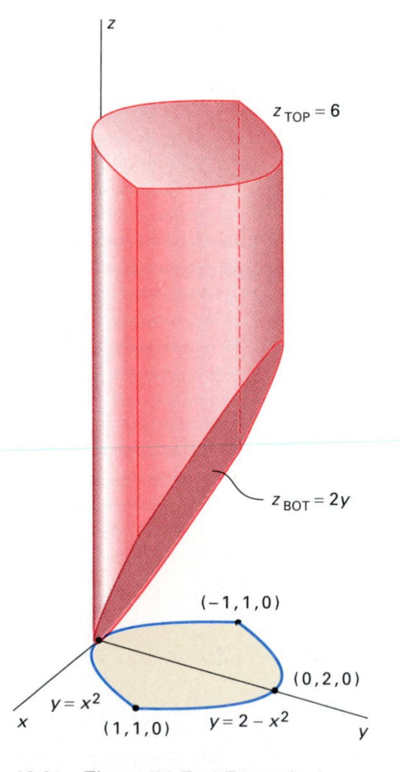

16.21 The solid T of Example 4

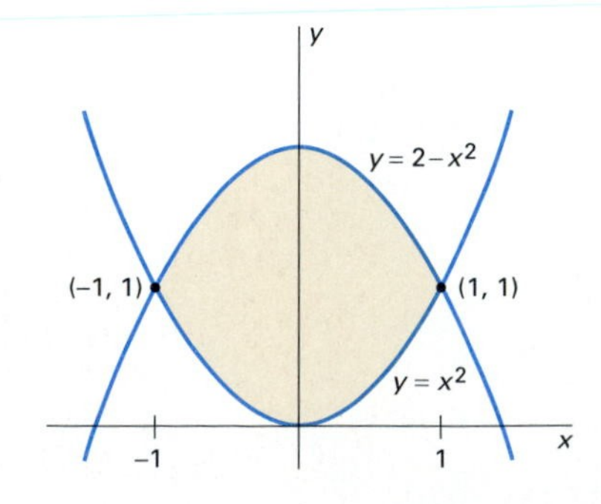

16.22 The region R of Example 4

Integrating first with respect to y (for otherwise we would need two integrals), we get

$$V = \iint_R (z_{\text{top}} - z_{\text{bot}})\,dA = \int_{-1}^{1} \int_{x^2}^{2-x^2} (6 - 2y)\,dy\,dx$$

$$= 2\int_0^1 \Big[6y - y^2\Big]_{y=x^2}^{2-x^2} dx \qquad \text{(symmetry)}$$

$$= 2\int_0^1 ([6(2 - x^2) - (2 - x^2)^2] - [6x^2 - x^4])\,dx$$

$$= 2\int_0^1 (8 - 8x^2)\,dx = 16\Big[x - \tfrac{1}{3}x^3\Big]_0^1 = \tfrac{32}{3}.$$

16.3 PROBLEMS

In each of Problems 1–10, use double integration to find the area of the region in the xy-plane bounded by the given curves.

1. $y = x, \quad y^2 = x$
2. $y = x, \quad y = x^4$
3. $y = x^2, \quad y = 2x + 3$
4. $y = 2x + 3, \quad y = 6x - x^2$
5. $y = x^2, \quad x + y = 2, \quad y = 0$
6. $y = (x - 1)^2, \quad y = (x + 1)^2, \quad y = 0$
7. $y = x^2 + 1, \quad y = 2x^2 - 3$
8. $y = x^2 + 1, \quad y = 9 - x^2$
9. $y = x, \quad y = 2x, \quad xy = 2$
10. $y = x^2, \quad y = \dfrac{2}{1 + x^2}$

In each of Problems 11–26, find the volume of the solid that lies under the surface $z = f(x, y)$ and above the region in the xy-plane bounded by the given curves.

11. $z = 1 + x + y; \quad x = 0, \quad x = 1, \quad y = 0, \quad y = 1$
12. $z = 2x + 3y; \quad x = 0, \quad x = 3, \quad y = 0, \quad y = 2$
13. $z = y + e^x; \quad x = 0, \quad x = 1, \quad y = 0, \quad y = 2$
14. $z = 3 + \cos x + \cos y; \quad x = 0, \quad x = \pi, \quad y = 0, \quad y = \pi$

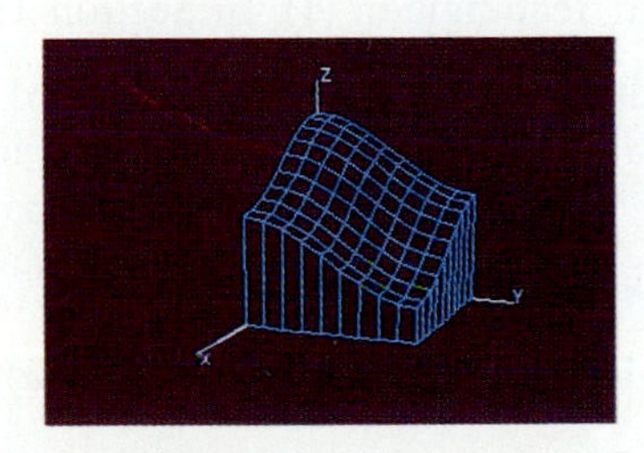

The solid of Problem 14

15. $z = x + y; \quad x = 0, \quad y = 0, \quad x + y = 1$
16. $z = 3x + 2y; \quad x = 0, \quad y = 0, \quad x + 2y = 4$
17. $z = 1 + x + y; \quad x = 1, \quad y = 0, \quad y = x^2$
18. $z = 2x + y; \quad x = 0, \quad y = 1, \quad x = \sqrt{y}$
19. $z = x^2; \quad y = x^2, \quad y = 1$
20. $z = y^2; \quad x = y^2, \quad x = 4$
21. $z = x^2 + y^2; \quad x = 0, \quad y = 0, \quad x = 1, \quad y = 2$
22. $z = 1 + x^2 + y^2; \quad y = x, \quad y = 2 - x^2$
23. $z = 9 - x - y; \quad y = 0, \quad x = 3, \quad y = 2x/3$
24. $z = 10 + y - x^2; \quad y = x^2, \quad x = y^2$
25. $z = 4x^2 + y^2; \quad x = 0, \quad y = 0, \quad 2x + y = 2$
26. $z = 2x + 3y; \quad y = x^2, \quad y = x^3$
27. Use double integration to find the volume of the tetrahedron in the first octant bounded by the coordinate planes and the plane with equation $x/a + y/b + z/c = 1$. The numbers a, b, and c are positive constants.
28. Suppose that $h > a > 0$. Show that the volume of the solid bounded by the cylinder $x^2 + y^2 = a^2$, the plane $z = 0$, and the plane $z = x + h$ is $\pi a^2 h$.
29. Find the volume of the first octant part of the solid bounded by the cylinders $x^2 + y^2 = 1$ and $y^2 + z^2 = 1$. (*Suggestion:* One order of integration is considerably easier than the other.)
30. Find the areas of the two regions bounded by the parabola $y = x^2$ and the curve $y(2x - 7) = -9$, a translated rectangular hyperbola. (*Suggestion:* $x = -1$ is one root of the cubic equation you will need to solve.)

In the following problems you may consult Chapter 9 or the integral table inside the covers of this book to find antiderivatives of such expressions as $(a^2 - u^2)^{3/2}$.

31. Find the volume of a sphere of radius a by double integration.

32. Use double integration to find the formula $V = V(a, b, c)$ for the volume of an ellipsoid with semiaxes a, b, and c.

33. Find the volume of the solid bounded by the xy-plane and the paraboloid $z = 25 - x^2 - y^2$ by evaluating a double integral.

34. Find the volume of the solid bounded by the paraboloids $z = x^2 + 2y^2$ and $z = 12 - 2x^2 - y^2$.

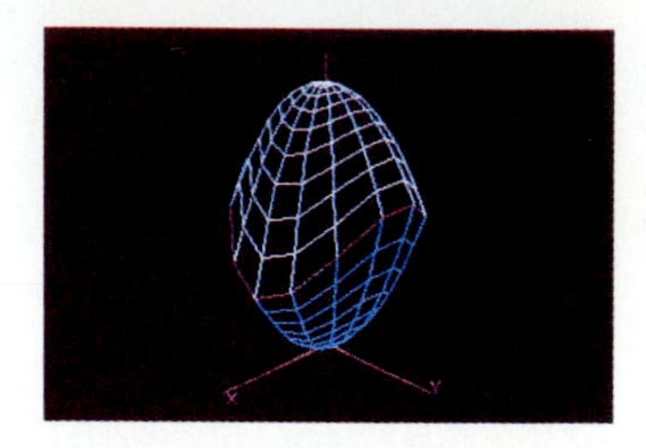

The solid of Problem 34

35. Find the volume removed when a vertical square hole of edge length R is cut directly through the center of a long horizontal cylinder of radius R.

36. Find the volume of the solid bounded by the two surfaces $z = x^2 + 3y^2$ and $z = 4 - y^2$.

37. Find the volume V of the solid T bounded by the parabolic cylinders $z = x^2$, $z = 2x^2$, $y = x^2$, and $y = 8 - x^2$.

38. Suppose that a square hole with side of length 2 is cut symmetrically through the center of a sphere of radius 2. Show that the volume removed is given by $V = \int_0^1 F(x)\,dx$ where

$$F(x) = 4\sqrt{3 - x^2} + 4(4 - x^2)\arcsin\frac{1}{\sqrt{4 - x^2}}.$$

Use Simpson's rule (or the integration key or subroutine on a calculator) to approximate the integral. Is your numerical result consistent with the exact value

$$V = \tfrac{4}{3}(19\pi + 2\sqrt{2} - 54\arctan\sqrt{2})?$$

16.4 Double Integrals in Polar Coordinates

A double integral may be easier to evaluate after it has been transformed from rectangular xy-coordinates into polar $r\theta$-coordinates. In particular, this is likely to be the case when the region R of integration is a *polar rectangle.* A **polar rectangle** is a region described in polar coordinates by the inequalities

$$a \leqq r \leqq b, \qquad \alpha \leqq \theta \leqq \beta. \tag{1}$$

This polar rectangle is shown in Fig. 16.23. If $a = 0$, it is a sector of a circular disk of radius b. If $0 < a < b$, $\alpha = 0$, and $\beta = 2\pi$, it is an annular ring of inner radius a and outer radius b. Because the area of a circular sector with radius r and central angle θ is $\frac{1}{2}r^2\theta$, the area of the polar rectangle in (1) is

$$A = \tfrac{1}{2}b^2(\beta - \alpha) - \tfrac{1}{2}a^2(\beta - \alpha) = \tfrac{1}{2}(a + b)(b - a)(\beta - \alpha) = \bar{r}\,\Delta r\,\Delta\theta, \tag{2}$$

where $\Delta r = b - a$, $\Delta\theta = \beta - \alpha$, and $\bar{r} = \frac{1}{2}(a + b)$ is the *average radius* of the polar rectangle.

16.23 A "polar rectangle"

Now suppose that we want to compute the double integral $\iint_R f(x, y)\,dA$, where R is the polar rectangle in (1). In Section 16.1 we defined the double integral as a limit of Riemann sums associated with partitions consisting of ordinary rectangles. It can also be defined in terms of *polar partitions*, made up of polar rectangles. We begin with a partition

$$a = r_0 < r_1 < r_2 < \cdots < r_m = b$$

of $[a, b]$ into m equal subintervals each of length $\Delta r = (b - a)/m$, and a partition

$$\alpha = \theta_0 < \theta_1 < \theta_2 < \cdots < \theta_n = \beta$$

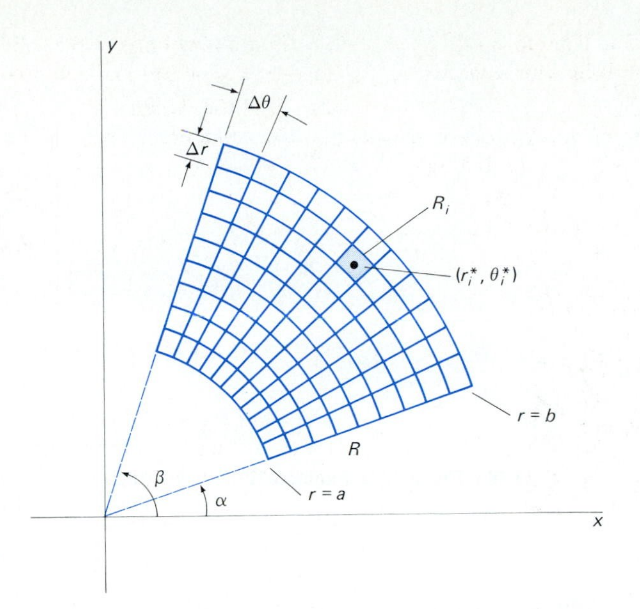

16.24 A polar partition of the polar rectangle R

of $[\alpha, \beta]$ into n equal subintervals each of length $\Delta\theta = (\beta - \alpha)/n$. This gives the **polar partition** $\mathscr{P}$ of R into the $k = mn$ polar rectangles $R_1, R_2, \ldots, R_k$ indicated in Fig. 16.24. The mesh $|\mathscr{P}|$ of this polar partition is the maximum of the lengths of the diagonals of its polar subrectangles.

Let the center point of R_i have polar coordinates (r_i^*, θ_i^*), where r_i^* is the average radius of R_i. Then the rectangular coordinates of this point are $x_i^* = r_i^* \cos \theta_i^*$ and $y_i^* = r_i^* \sin \theta_i^*$. Therefore the Riemann sum for the function $f(x, y)$ associated with the polar partition $\mathscr{P}$ is

$$\sum_{i=1}^{k} f(x_i^*, y_i^*)\,\Delta A_i,$$

where $\Delta A_i = r_i^*\,\Delta r\,\Delta\theta$ is the area of the polar rectangle R_i (in part a consequence of the formula in (2)). When we express this Riemann sum in polar coordinates, we obtain

$$\sum_{i=1}^{k} f(x_i^*, y_i^*)\,\Delta A_i = \sum_{i=1}^{k} f(r_i^* \cos \theta_i^*, r_i^* \sin \theta_i^*) r_i^*\,\Delta r\,\Delta\theta$$

$$= \sum_{i=1}^{k} g(r_i^*, \theta_i^*)\,\Delta r\,\Delta\theta,$$

where $g(r, \theta) = f(r \cos \theta, r \sin \theta) r$. This last sum is simply a Riemann sum for the double integral

$$\int_{\alpha}^{\beta}\int_{a}^{b} g(r, \theta)\,dr\,d\theta = \int_{\alpha}^{\beta}\int_{a}^{b} f(r \cos \theta, r \sin \theta)\,r\,dr\,d\theta,$$

so it finally follows that

$$\iint_R f(x, y)\,dA = \lim_{|\mathscr{P}| \to 0} \sum_{i=1}^{k} f(x_i^*, y_i^*)\,\Delta A_i$$

$$= \lim_{\Delta r, \Delta\theta \to 0} \sum_{i=1}^{k} g(r_i^*, \theta_i^*)\,\Delta r\,\Delta\theta$$

$$= \int_{\alpha}^{\beta}\int_{a}^{b} g(r, \theta)\,dr\,d\theta.$$

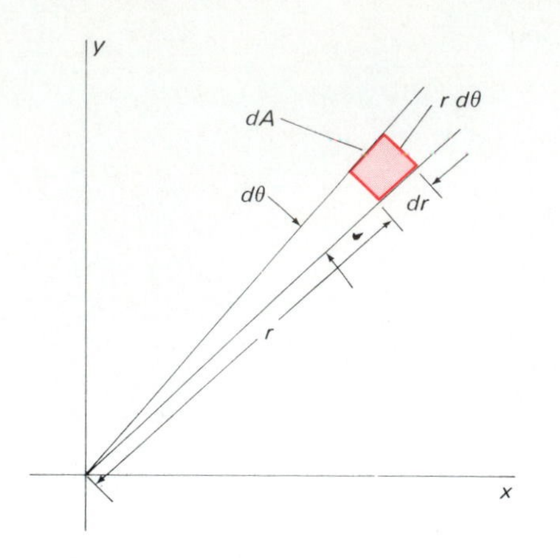

16.25 The dimensions of the small polar rectangle suggest that $dA = r\,dr\,d\theta$.

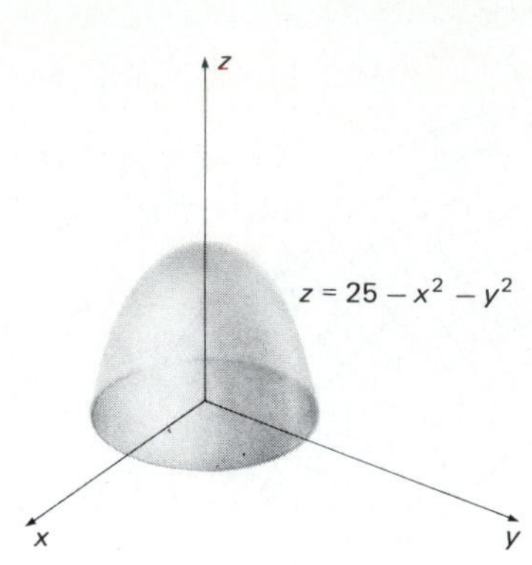

16.26 The solid of Example 1

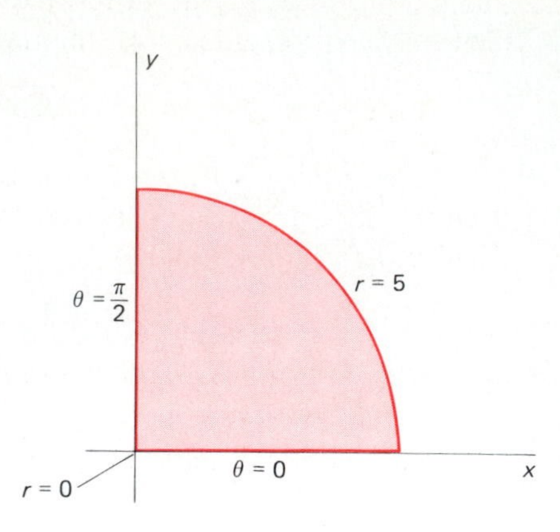

16.27 One-fourth of the domain of the integral of Example 1

That is,

$$\iint_R f(x, y)\,dA = \int_\alpha^\beta \int_a^b f(r\cos\theta, r\sin\theta)\,r\,dr\,d\theta. \tag{3}$$

Thus we formally transform a double integral over a polar rectangle of the form in (1) into polar coordinates by substituting

$$x = r\cos\theta, \qquad y = r\sin\theta, \qquad dA = r\,dr\,d\theta,$$

and inserting the appropriate limits on r and θ. In particular, *note the "extra" r on the right-hand side in Equation (3).* It may be remembered by visualizing the "infinitesimal polar rectangle" of Fig. 16.25, with "area" $dA = r\,dr\,d\theta$ (formally).

EXAMPLE 1 Find the volume V of the solid shown in Fig. 16.26. This is the figure bounded below by the xy-plane and above by the paraboloid $z = 25 - x^2 - y^2$.

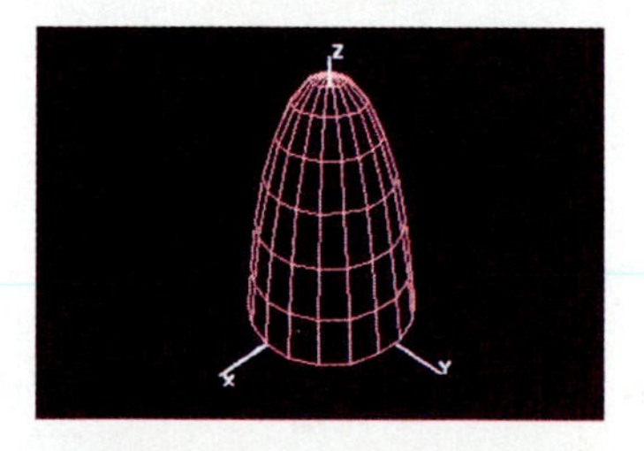

The paraboloid of Example 1

Solution The paraboloid intersects the xy-plane in the circle $x^2 + y^2 = 25$. We can compute the volume of the solid by integrating over the quarter of that circle in the first quadrant (see Fig. 16.27) and then multiplying the result by 4. Thus

$$V = 4\int_0^5 \int_0^{\sqrt{25-x^2}} (25 - x^2 - y^2)\,dy\,dx$$

There is no difficulty in performing the integration with respect to y, but then we are confronted with the integrals

$$\int \sqrt{25 - x^2}\, dx, \qquad \int x^2 \sqrt{25 - x^2}\, dx, \quad \text{and} \quad \int (25 - x^2)^{3/2}\, dx.$$

Let us instead transform the original integral into polar coordinates. Because $25 - x^2 - y^2 = 25 - r^2$ and because the first quadrant of the circular disk is described by

$$0 \leqq r \leqq 5, \qquad 0 \leqq \theta \leqq \pi/2,$$

the formula in (3) yields the volume

$$V = 4\int_0^{\pi/2}\int_0^5 (25 - r^2)\, r\, dr\, d\theta = 4\int_0^{\pi/2}\left[\frac{25}{2}r^2 - \frac{1}{4}r^4\right]_0^5 d\theta$$

$$= 4\left(\frac{625}{4}\right)\left(\frac{\pi}{2}\right) = \frac{625\pi}{2}.$$

If R is a more general region, then the double integral $\iint_R f(x, y)\, dA$ can be transformed into polar coordinates by expressing it as a limit of Riemann sums associated with "polar inner partitions" of the sort indicated in Fig. 16.28. Instead of giving the detailed derivation—a generalization of the above derivation of the formula in (3)—we shall simply give the results in two special cases of practical importance. These correspond to the two types of plane regions that play the same role in polar coordinates that horizontally simple and vertically simple regions play in rectangular coordinates.

Figure 16.29 shows a *radially simple* region R consisting of those points that have polar coordinates satisfying the inequalities

$$\alpha \leqq \theta \leqq \beta, \qquad g_1(\theta) \leqq r \leqq g_2(\theta).$$

In this case, the formula

$$\iint_R f(x, y)\, dA = \int_\alpha^\beta \int_{g_1(\theta)}^{g_2(\theta)} f(r\cos\theta, r\sin\theta)\, r\, dr\, d\theta \tag{4}$$

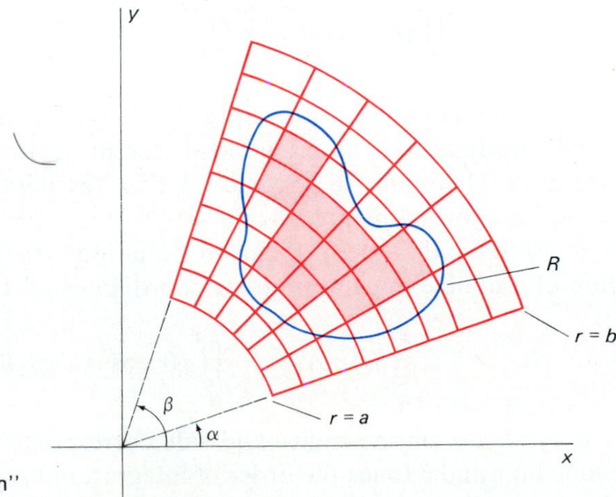

16.28 A "polar inner partition" of the region R

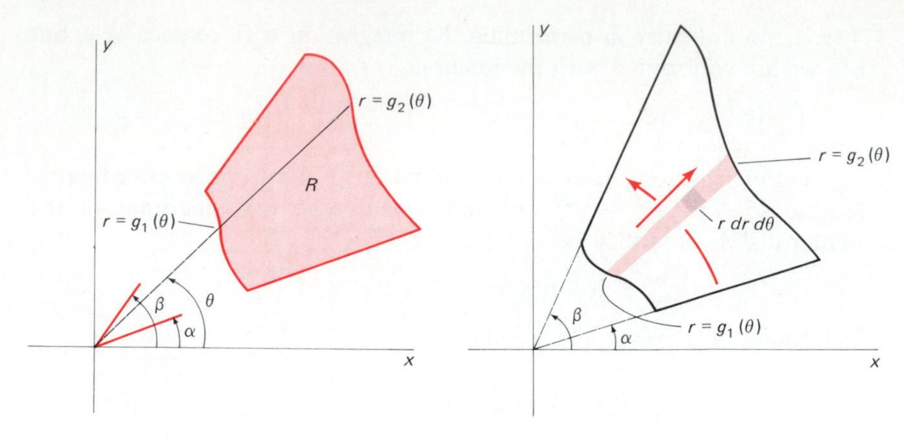

16.29 A radially simple region R

16.30 Integrating first with respect to r, then with respect to θ

gives the evaluation in polar coordinates of a double integral over R (under the usual assumption that the indicated integrals exist). Note that we integrate first with respect to r, with the limits $g_1(\theta)$ and $g_2(\theta)$ being the r-coordinates of the end points of a typical radial segment in R, as indicated in Fig. 16.29.

Figure 16.30 indicates how the iterated integral on the right-hand side in (4) can be set up in a formal way. A typical area element $dA = r\,dr\,d\theta$ is first swept radially from $r = g_1(\theta)$ to $r = g_2(\theta)$. The resulting strip is then rotated from $\theta = \alpha$ to $\theta = \beta$ to sweep out the region R. Equation (4) yields the volume formula

$$V = \int_{\alpha}^{\beta} \int_{r_{\text{inner}}}^{r_{\text{outer}}} zr\,dr\,d\theta \tag{5}$$

for the volume V of the solid lying over the region R of Fig. 16.29 and under the surface $z = f(x, y) = f(r\cos\theta, r\sin\theta)$.

Figure 16.31 shows an *angularly simple* region R described in polar coordinates by the inequalities

$$a \leqq r \leqq b, \qquad h_1(r) \leqq \theta \leqq h_2(r).$$

In this case

$$\iint_R f(x, y)\,dA = \int_a^b \int_{h_1(r)}^{h_2(r)} f(r\cos\theta, r\sin\theta)\,r\,d\theta\,dr. \tag{6}$$

Here we integrate first with respect to θ, with the limits $h_1(r)$ and $h_2(r)$ being the θ-coordinates of the end points of a typical circular arc in R, as indicated in Fig. 16.31. Figure 16.32 indicates how the iterated integral in (6) can be set up in a formal manner.

Observe that the formulas in Equations (3), (4), and (6) for the evaluation of a double integral in polar coordinates all take the form

$$\iint_R f(x, y)\,dA = \iint_S f(r\cos\theta, r\sin\theta)\,r\,dr\,d\theta. \tag{7}$$

The symbol S on the right-hand side corresponds to choosing appropriate limits on r and θ (once the order of integration has been determined) so that the region R is swept out in the manner of Figs. 16.30 and 16.32.

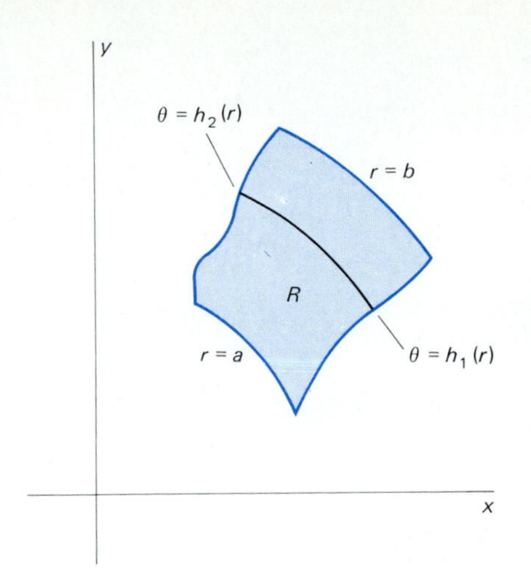

16.31 An angularly simple region

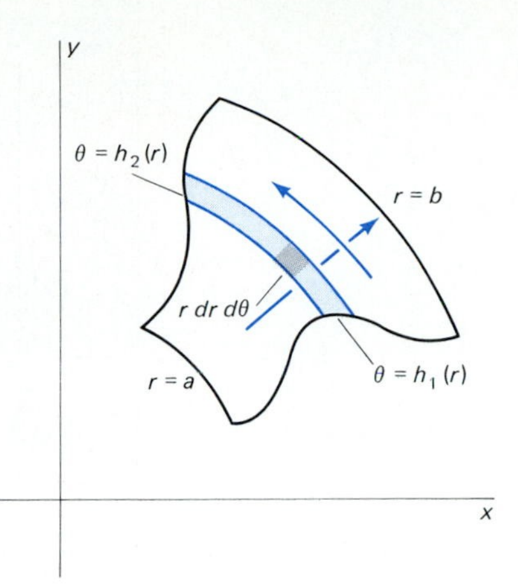

16.32 Integrating first with respect to θ, last with respect to r

With $f(x, y) \equiv 1$, the formula in (7) reduces to the formula

$$A = a(R) = \iint_S r\,dr\,d\theta \tag{8}$$

for computing the area of R by double integration in polar coordinates. Note again that the symbol S refers not to a new region in the xy-plane, but to a new description—in terms of polar coordinates—of the original region R.

EXAMPLE 2 Figure 16.33 shows the region R bounded on the inside by the circle $r = 1$ and on the outside by the limaçon $r = 2 + \cos\theta$. By following a typical radial line outward from the origin, we see that $r_{\text{inner}} = 1$ and $r_{\text{outer}} = 2 + \cos\theta$. Hence the area of R is

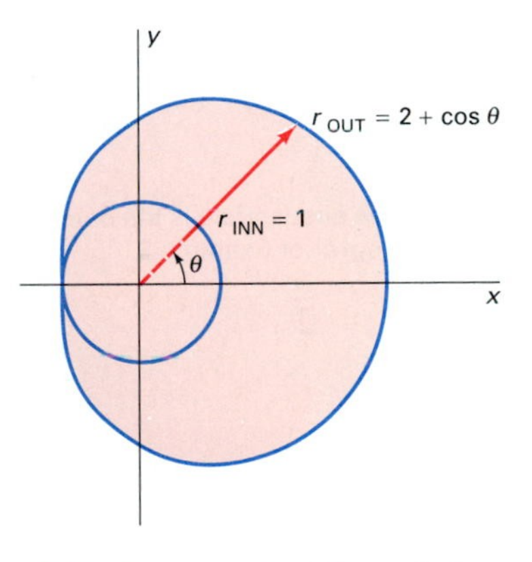

16.33 The region R of Example 2

$$\begin{aligned}
A &= \int_{\alpha}^{\beta}\int_{r_{\text{inner}}}^{r_{\text{outer}}} r\,dr\,d\theta \\
&= 2\int_0^{\pi}\int_1^{2+\cos\theta} r\,dr\,d\theta \qquad \text{(symmetry)} \\
&= 2\int_0^{\pi} \tfrac{1}{2}[(2+\cos\theta)^2 - (1)^2]\,d\theta \\
&= \int_0^{\pi} (3 + 4\cos\theta + \cos^2\theta)\,d\theta \\
&= \int_0^{\pi} (3 + 4\cos\theta + \tfrac{1}{2} + \tfrac{1}{2}\cos 2\theta)\,d\theta \\
&= \int_0^{\pi} (3 + \tfrac{1}{2})\,d\theta = \tfrac{7}{2}\pi.
\end{aligned}$$

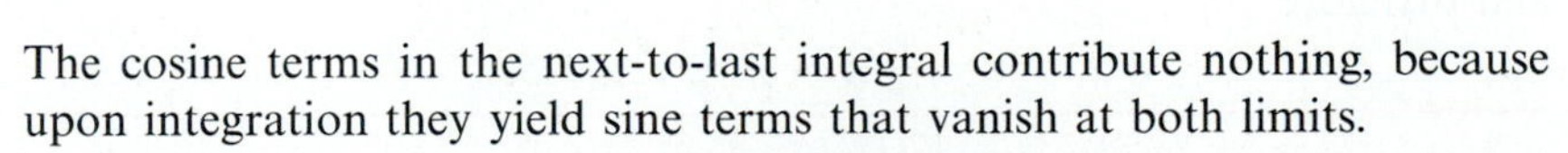

The cosine terms in the next-to-last integral contribute nothing, because upon integration they yield sine terms that vanish at both limits.

EXAMPLE 3 Find the volume of the solid region that is interior to both the sphere $x^2 + y^2 + z^2 = 4$ of radius 2 and the cylinder $(x - 1)^2 + y^2 = 1$. This is the volume of material removed when an off-center hole of radius 1 is bored just tangent to a diameter all the way through a sphere of radius 2. The upper hemisphere is shown in Fig. 16.34.

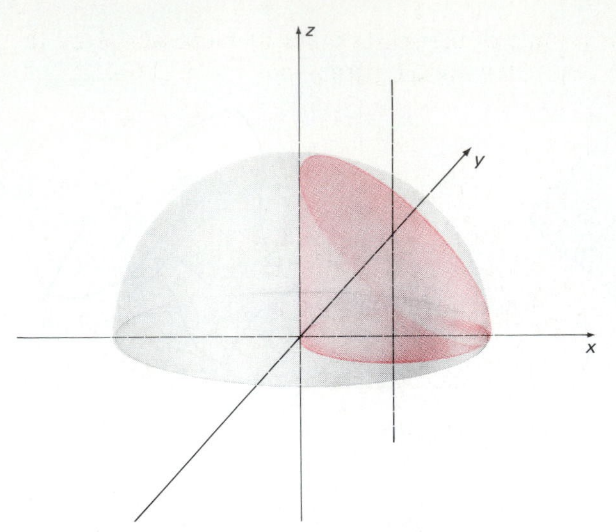

16.34 The *upper half* of the sphere-with-hole of Example 3

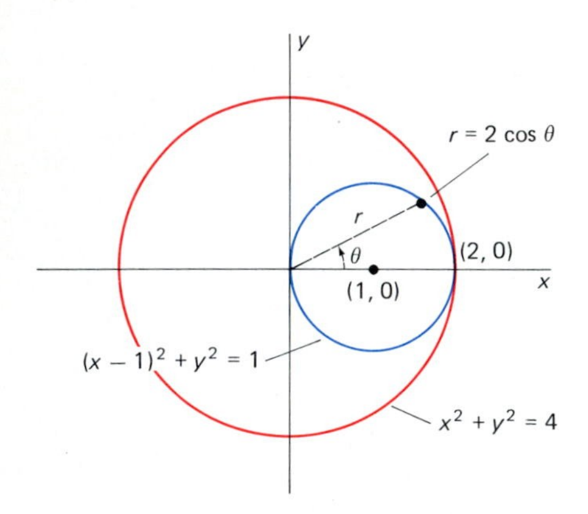

16.35 The small circle is the domain of the integral of Example 3

Solution We need to integrate the function $f(x, y) = (4 - x^2 - y^2)^{1/2}$ over the disk R bounded by the circle with center $(1, 0)$ and radius 1, shown in Fig. 16.35. The desired volume V is twice that of the part above the xy-plane, so

$$V = 2\iint_R \sqrt{4 - x^2 - y^2}\, dA.$$

But this integral would be troublesome to evaluate in rectangular coordinates, so we change to polar coordinates.

The unit circle in Fig. 16.35 is familiar from Example 2; its polar coordinates equation is $r = 2\cos\theta$. Therefore, the region R is described by the inequalities

$$0 \leqq r \leqq 2\cos\theta, \qquad -\pi/2 \leqq \theta \leqq \pi/2.$$

We shall integrate only over the upper half of R, taking advantage of the symmetry of the sphere-with-hole. This involves doubling for a second time the integral we write. So—using the formula in (4)—we find that

$$\begin{aligned} V &= 4\int_0^{\pi/2}\int_0^{2\cos\theta} (4 - r^2)^{1/2}\, r\, dr\, d\theta \\ &= 4\int_0^{\pi/2} \left[-\tfrac{1}{3}(4 - r^2)^{3/2}\right]_0^{2\cos\theta} d\theta = \tfrac{32}{3}\int_0^{\pi/2} (1 - \sin^3\theta)\, d\theta. \end{aligned}$$

Now we see from Formula (113) inside the back cover that

$$\int_0^{\pi/2} \sin^3\theta\, d\theta = \tfrac{2}{3},$$

and therefore

$$V = \tfrac{16}{3}\pi - \tfrac{64}{9} \approx 9.64405.$$

In the next example we use a polar coordinates version of the familiar volume formula

$$V = \iint_R (z_{\text{top}} - z_{\text{bot}})\, dA.$$

EXAMPLE 4 Find the volume of the solid that is bounded above by the paraboloid $z = 8 - r^2$ and below by the paraboloid $z = r^2$. These paraboloids are shown in Fig. 16.36.

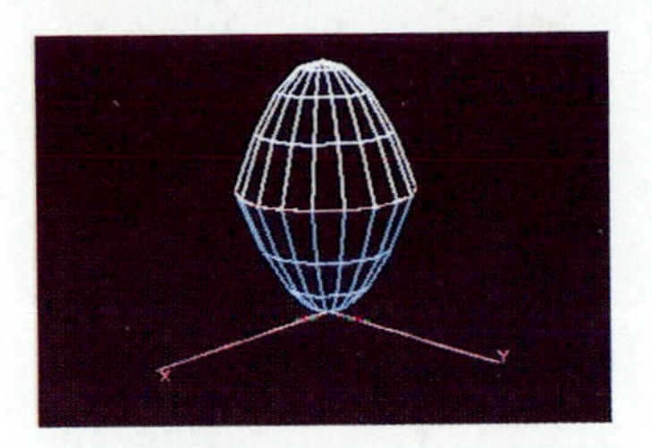

The solid of Example 4

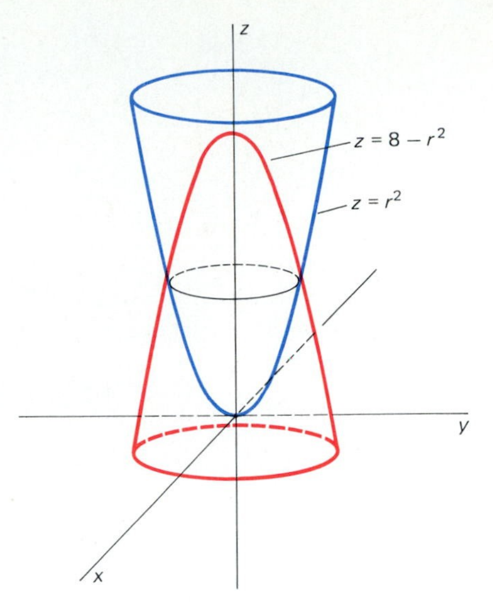

16.36 The solid of Example 4

Solution The curve of intersection of the two paraboloids is found by simultaneous solution of the equations of the two surfaces. We eliminate z to obtain

$$r^2 = 8 - r^2; \qquad \text{that is,} \quad r^2 = 4.$$

Hence the solid lies above the circular disk $r \leqq 2$, and so its volume is

$$V = \iint_{r \leqq 2} (z_{\text{top}} - z_{\text{bot}})\, dA = \int_0^{2\pi} \int_0^2 [(8 - r^2) - r^2]\, r\, dr\, d\theta$$

$$= \int_0^{2\pi} \int_0^2 (8r - 2r^3)\, dr\, d\theta = 2\pi \Big[4r^2 - \tfrac{1}{2}r^4\Big]_0^2 = 16\pi.$$

EXAMPLE 5 Here we apply a standard polar coordinates technique to show that

$$I = \int_0^{\infty} e^{-x^2}\, dx = \frac{\sqrt{\pi}}{2}. \tag{9}$$

This important improper integral converges because

$$\int_1^b e^{-x^2}\, dx \leqq \int_1^b e^{-x}\, dx \leqq \int_1^{\infty} e^{-x}\, dx = \frac{1}{e}.$$

(The first inequality is valid because $e^{-x^2} \leqq e^{-x}$ for $x \geqq 1$.) It follows that $\int_1^b e^{-x^2}\, dx$ is a bounded and increasing function of b.

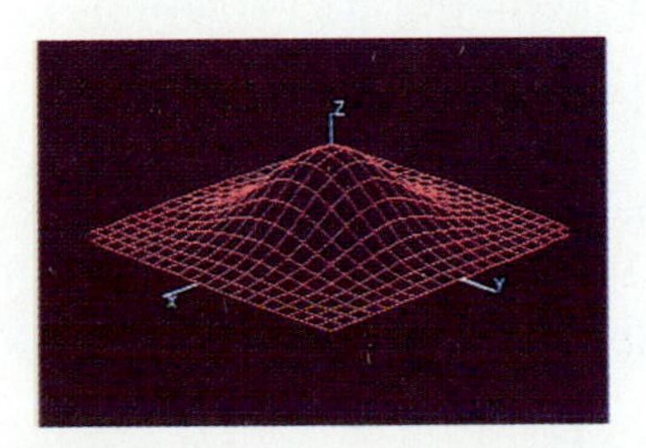

The surface $z = \exp(-x^2 - y^2)$

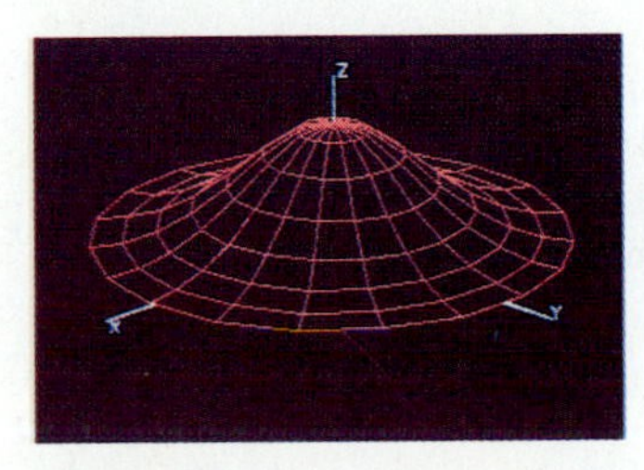

The surface $z = \exp(-r^2)$

Solution Let V_b denote the volume of the region that lies under the surface $z = e^{-x^2-y^2}$ and above the square with vertices $(\pm b, \pm b)$ in the xy-plane. Then

$$V_b = \int_{-b}^{b}\int_{-b}^{b} e^{-x^2-y^2}\,dx\,dy = \int_{-b}^{b} e^{-y^2}\left(\int_{-b}^{b} e^{-x^2}\,dx\right)dy$$

$$= \left(\int_{-b}^{b} e^{-x^2}\,dx\right)\left(\int_{-b}^{b} e^{-y^2}\,dy\right) = \left(\int_{-b}^{b} e^{-x^2}\,dx\right)^2$$

$$= 4\left(\int_{0}^{b} e^{-x^2}\,dx\right)^2.$$

It follows that the volume under $z = e^{-x^2-y^2}$ above the entire xy-plane is

$$V = \lim_{b\to\infty} V_b = \lim_{b\to\infty} 4\left(\int_0^b e^{-x^2}\,dx\right)^2 = 4\left(\int_0^\infty e^{-x^2}\,dx\right)^2 = 4I^2.$$

Now we compute V by another method—the use of polar coordinates. We take the limit, as $b \to \infty$, of the volume under $z = e^{-x^2-y^2} = e^{-r^2}$ above the circular disk with center $(0, 0)$ and radius b. This disk is described by $0 \leqq r \leqq b$, $0 \leqq \theta \leqq 2\pi$, so we obtain

$$V = \lim_{b\to\infty}\int_0^{2\pi}\int_0^b re^{-r^2}\,dr\,d\theta = \lim_{b\to\infty}\int_0^{2\pi}\left[-\tfrac{1}{2}e^{-r^2}\right]_0^b d\theta$$

$$= \lim_{b\to\infty}\int_0^{2\pi}\tfrac{1}{2}\left[1 - e^{-b^2}\right]d\theta = \lim_{b\to\infty}\pi\left[1 - e^{-b^2}\right] = \pi.$$

We equate these two values of V, and it follows that $4I^2 = \pi$. Therefore, $I = \frac{1}{2}\sqrt{\pi}$, as desired.

16.4 PROBLEMS

In Problems 1–7, find the indicated area by double integration in polar coordinates.

1. The area bounded by the circle $r = 1$
2. The area bounded by the circle $r = 3\sin\theta$
3. The area bounded by the cardioid $r = 1 + \cos\theta$
4. The area bounded by one loop of $r = 2\cos 2\theta$
5. The area within both the circles $r = 1$ and $r = 2\sin\theta$
6. The area within $r = 2 + \cos\theta$ and outside the circle $r = 2$
7. The area within the smaller loop of $r = 1 - 2\sin\theta$

In each of Problems 8–12, find by double integration in polar coordinates the volume of the solid that lies under the given surface and over the plane region R bounded by the given curve.

8. $z = x^2 + y^2$; $r = 3$
9. $z = \sqrt{x^2 + y^2}$; $r = 2$
10. $z = x^2 + y^2$; $r = 2\cos\theta$
11. $z = 10 + 2x + 3y$; $r = \sin\theta$
12. $z = a^2 - x^2 - y^2$; $r = a$

In Problems 13–18, evaluate the given integral by first changing to polar coordinates.

13. $\displaystyle\int_0^1\int_0^{\sqrt{1-y^2}} \frac{1}{1 + x^2 + y^2}\,dx\,dy$
14. $\displaystyle\int_0^1\int_0^{\sqrt{1-x^2}} \frac{1}{\sqrt{4 - x^2 - y^2}}\,dy\,dx$
15. $\displaystyle\int_0^2\int_0^{\sqrt{4-x^2}} (x^2 + y^2)^{3/2}\,dy\,dx$
16. $\displaystyle\int_0^1\int_x^1 x^2\,dy\,dx$
17. $\displaystyle\int_0^1\int_0^{\sqrt{1-y^2}} \sin(x^2 + y^2)\,dx\,dy$
18. $\displaystyle\int_1^2\int_0^{\sqrt{2x-x^2}} \frac{1}{\sqrt{x^2 + y^2}}\,dy\,dx$

In each of Problems 19–22, find the volume of the solid that is bounded above and below by the given surfaces $z = f_1(x, y)$ and $z = f_2(x, y)$ and lies over the plane region R bounded by the given curve $r = g(\theta)$.

19. $z = 1, \quad z = 3 + x + y; \quad r = 1$

20. $z = 2 + x, \quad z = 4 + 2x; \quad r = 2$

21. $z = 0, \quad z = 3 + x + y; \quad r = 2 \sin \theta$

22. $z = 0, \quad z = 1 + x; \quad r = 1 + \cos \theta$

Solve Problems 23–32 by double integration in polar coordinates.

23. Problem 31, Section 16.3

24. Problem 34, Section 16.3

25. Problem 28, Section 16.3

26. Find the volume of the wedge-shaped solid described in Example 3 of Section 16.3.

27. Find the volume bounded by the paraboloids $z = x^2 + y^2$ and $z = 4 - 3x^2 - 3y^2$.

28. Find the volume bounded by the paraboloids $z = x^2 + y^2$ and $z = 2x^2 + 2y^2 - 1$.

29. Find the volume of the "ice cream cone" bounded by the sphere $x^2 + y^2 + z^2 = a^2$ and the cone $z = (x^2 + y^2)^{1/2}$.

30. Find the volume bounded by the paraboloid $z = r^2$, the cylinder $r = 2a \sin \theta$, and the plane $z = 0$.

31. Find the volume that lies under the paraboloid $z = r^2$ and above one loop of the lemniscate with equation $r^2 = 2 \sin \theta$.

32. Find the volume that lies inside both the cylinder $x^2 + y^2 = 4$ and the ellipsoid $2x^2 + 2y^2 + z^2 = 18$.

33. If $0 < h < a$, then the plane $z = a - h$ cuts off a spherical segment of height h and radius b from the sphere $x^2 + y^2 + z^2 = a^2$.

(a) Show that $b^2 = 2ah - h^2$.

(b) Show that the volume of the spherical segement is

$$V = \tfrac{1}{6}\pi h(3b^2 + h^2).$$

34. Show by the method of Example 5 that

$$\int_0^\infty \int_0^\infty \frac{dx\, dy}{(1 + x^2 + y^2)^2} = \frac{\pi}{4}.$$

35. Find the volume of the solid torus obtained by revolving the disk $r \leqq a$ about the line $x = b > a$. (*Suggestion:* If the area element $dA = r\, dr\, d\theta$ is revolved around the line, the volume generated is $dV = 2\pi(b - x)\, dA$. Express everything in polar coordinates.)

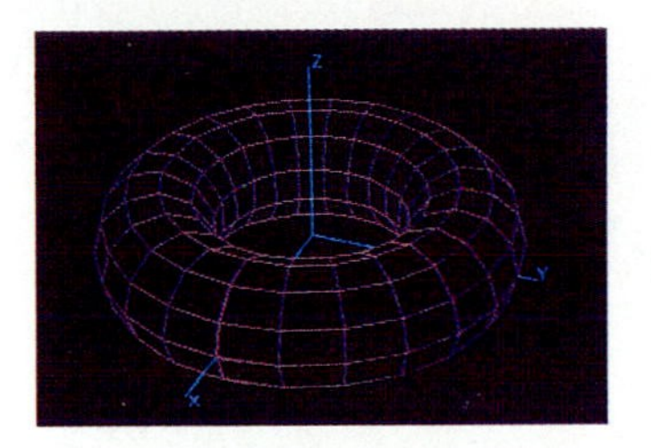

The torus of Problem 35

16.5 Applications of Double Integrals

The double integral can be used to find the mass m and the centroid $(\bar{x}, \bar{y})$ of a thin plate or plane *lamina* that occupies a bounded region R in the xy-plane. We suppose that the density of the lamina (in units of mass per unit *area*) at the point (x, y) is given by the continuous function $\rho(x, y)$.

Let $\mathscr{P} = \{R_1, R_2, R_3, \ldots, R_n\}$ be an inner partition of R and choose a point (x_i^*, y_i^*) in each subrectangle R_i (see Fig. 16.37). Then the mass of the piece of the lamina occupying R_i will be approximately $\rho(x_i^*, y_i^*)\, \Delta A_i$, where ΔA_i denotes the area $a(R_i)$ of R_i. Hence the mass of the entire lamina is given approximately by

$$m \approx \sum_{i=1}^{n} \rho(x_i^*, y_i^*)\, \Delta A_i.$$

As the mesh $|\mathscr{P}|$ of the inner partition $\mathscr{P}$ approaches zero, this Riemann sum approaches the corresponding double integral over R. We therefore *define* the **mass** m of the lamina by means of the formula

$$m = \iint_R \rho(x, y)\, dA. \tag{1}$$

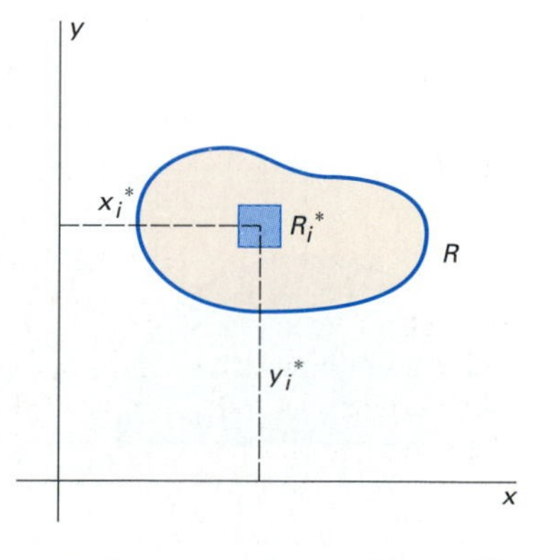

16.37 The area element $\Delta A_i = a(R_i)$

In brief,

$$m = \iint_R \rho \, dA = \iint_R dm$$

in terms of the density ρ and the element of mass

$$dm = \rho \, dA.$$

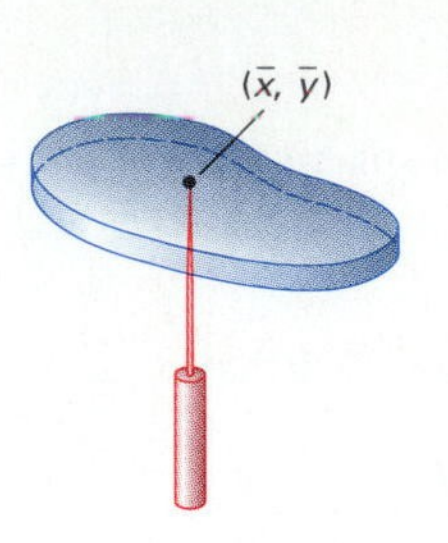

16.38 A lamina balanced on its centroid

The coordinates $(\bar{x}, \bar{y})$ of the **centroid** or center of mass of the lamina are defined to be

$$\bar{x} = \frac{1}{m} \iint_R x\rho(x, y) \, dA, \tag{2}$$

$$\bar{y} = \frac{1}{m} \iint_R y\rho(x, y) \, dA. \tag{3}$$

These formulas may be remembered in the form

$$\bar{x} = \frac{1}{m} \iint_R x \, dm, \qquad \bar{y} = \frac{1}{m} \iint_R y \, dm.$$

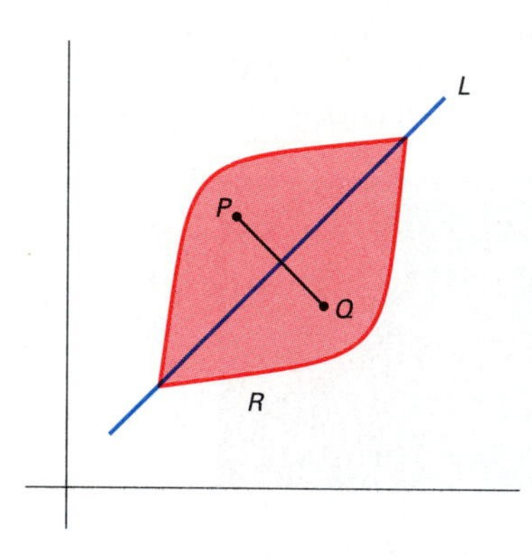

16.39 A line of symmetry

Thus $\bar{x}$ and $\bar{y}$ are the *average values* of x and y *with respect to mass* in the region R. The centroid $(\bar{x}, \bar{y})$ is the point of the lamina where it would balance horizontally if placed on the point of an icepick (see Fig. 16.38).

If the density function ρ has the *constant* value $k > 0$, then the coordinates $\bar{x}$ and $\bar{y}$ will be independent of the specific value of k (why?). In such a case we will generally take $\rho = 1$ in our computations. Moreover, in this case m will have the same numerical value as the area A of R, and $(\bar{x}, \bar{y})$ is then called the **centroid of the plane region** R.

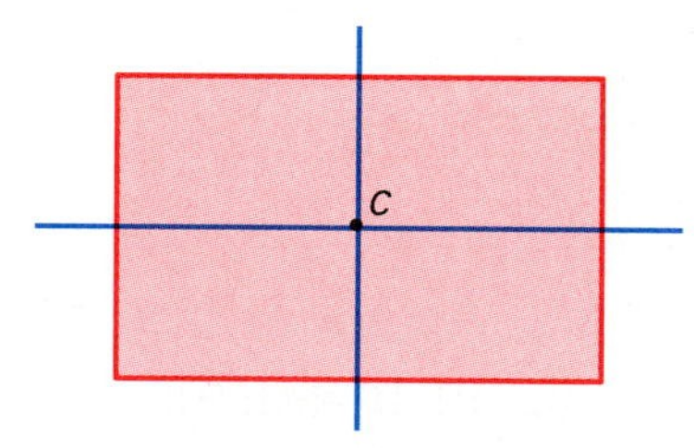

16.40 The centroid of a rectangle

Generally, we must calculate all three integrals in (1) through (3) in order to find the centroid of a lamina. But sometimes we can take advantage of the following **symmetry principle:** If the plane region R is symmetric with respect to the line L—that is, if R is carried onto itself when the plane is rotated through an angle of 180° about the line L—then the centroid of R (considered as a lamina of constant density) lies on L (see Fig. 16.39). For example, the centroid of a rectangle (Fig. 16.40) is the point where the perpendicular bisectors of its sides meet—these bisectors are also lines of symmetry.

In the case of a nonconstant density function ρ, we require (for symmetry) that ρ also be symmetric about the geometric line L of symmetry. That is, $\rho(P) = \rho(Q)$ if, as in Fig. 16.39, the points P and Q are symmetrically located with respect to L. Then the centroid of the lamina R will lie on the line L of symmetry.

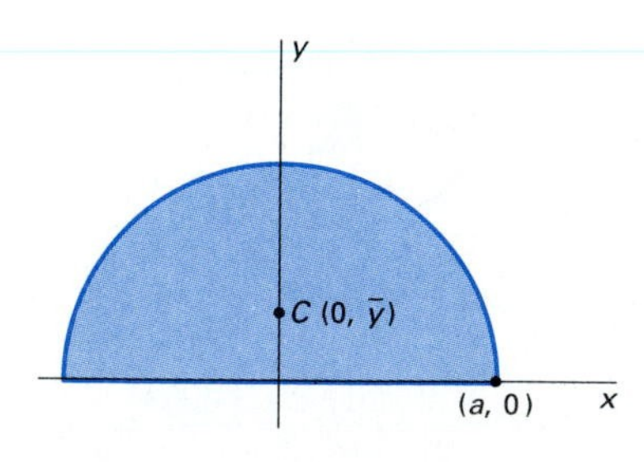

16.41 The centroid of a semicircular disk

EXAMPLE 1 Consider the semicircular disk of radius a shown in Fig. 16.41. With constant density $\rho \equiv 1$ its mass is $m = \frac{1}{2}\pi a^2$ (numerically equal to its area), and by symmetry its centroid $C(0, \bar{y})$ lies on the y-axis. Hence we need only compute

$$\bar{y} = \frac{1}{m} \iint_R y\, dm$$

$$= \frac{2}{\pi a^2} \int_0^{\pi} \int_0^{a} (r \sin \theta)\, r\, dr\, d\theta \qquad \text{(polar coordinates)}$$

$$= \frac{2}{\pi a^2} \Big[-\cos \theta\Big]_0^{\pi} \Big[\tfrac{1}{3} r^3\Big]_0^{a} = \frac{2}{\pi a^2} \cdot 2 \cdot \frac{a^3}{3} = \frac{4a}{3\pi}.$$

Thus the centroid of the semicircle is located at the point $(0, 4a/3\pi)$. Note that the computed value for $\bar{y}$ has the dimensions of length (because a is a length), as it should. Any answer having other dimensions would be suspect.

EXAMPLE 2 A lamina occupies the region bounded by the line $y = x + 2$ and the parabola $y = x^2$, as in Fig. 16.42. The density of the lamina at the point $P(x, y)$ is proportional to the square of the distance of P from the y-axis—thus $\rho(x, y) = kx^2$ (where k is a positive constant). Find the mass and centroid of the lamina.

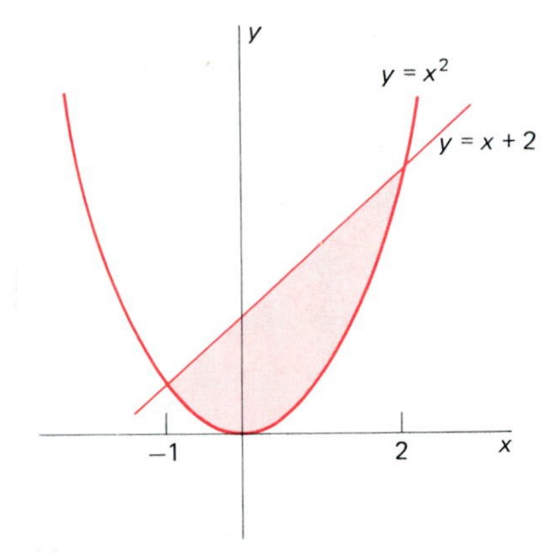

16.42 The lamina of Example 2

Solution The line and the parabola intersect in the two points $(-1, 1)$ and $(2, 4)$, so the formula in (1) gives mass

$$m = \int_{-1}^{2} \int_{x^2}^{x+2} kx^2\, dy\, dx = k \int_{-1}^{2} \Big[x^2 y\Big]_{x^2}^{x+2} dx$$

$$= k \int_{-1}^{2} (x^3 + 2x^2 - x^4)\, dx = \frac{63k}{20}.$$

Then the formulas in (2) and (3) give

$$\bar{x} = \frac{20}{63k} \int_{-1}^{2} \int_{x^2}^{x+2} kx^3\, dy\, dx = \frac{20}{63} \int_{-1}^{2} \Big[x^3 y\Big]_{x^2}^{x+2} dx$$

$$= \tfrac{20}{63} \int_{-1}^{2} (x^4 + 2x^3 - x^5)\, dx = \frac{20}{63} \cdot \frac{18}{5} = \frac{8}{7};$$

$$\bar{y} = \frac{20}{63k} \int_{-1}^{2} \int_{x^2}^{x+2} kx^2 y\, dy\, dx = \tfrac{20}{63} \int_{-1}^{2} \Big[\tfrac{1}{2} x^2 y^2\Big]_{x^2}^{x+2} dx$$

$$= \tfrac{10}{63} \int_{-1}^{2} (x^4 + 4x^3 + 4x^2 - x^6)\, dx = \tfrac{10}{63} \cdot \tfrac{531}{35} = \tfrac{118}{49}.$$

Thus the lamina of this example has mass $63k/20$ and its centroid is located at the point $(8/7, 118/49)$.

EXAMPLE 3 A lamina is shaped like the first-quadrant quarter-circle of radius a shown in Fig. 16.43. Its density is proportional to distance from the origin— that is, its density at (x, y) is $\rho(x, y) = k(x^2 + y^2)^{1/2} = kr$ (where k is a positive constant). Find its mass and centroid.

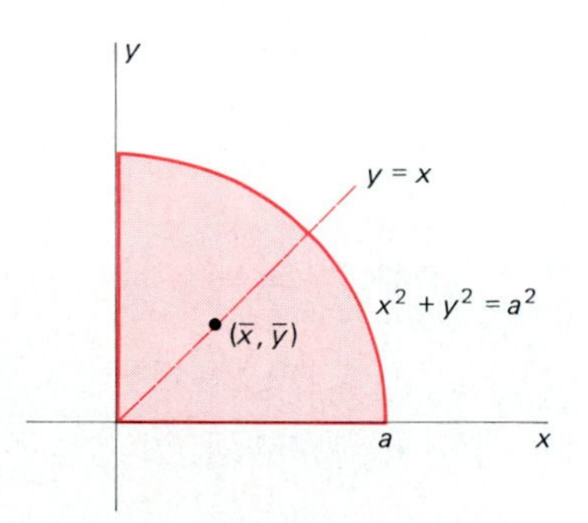

16.43 Finding mass and centroid—see Example 3.

Solution First we change to polar coordinates, for the shape of the boundary of the lamina suggests that this will make the computations simpler.

Then the formula in (1) yields the mass to be

$$m = \iint_R \rho \, dA = \int_0^{\pi/2} \int_0^a kr^2 \, dr \, d\theta$$

$$= k \int_0^{\pi/2} \left[\tfrac{1}{3} r^3 \right]_0^a d\theta = k \int_0^{\pi/2} \tfrac{1}{3} a^3 \, d\theta = \frac{k\pi a^3}{6}.$$

By symmetry of the lamina and its density function, the centroid lies on the line $y = x$. So the formula in (3) gives

$$\bar{x} = \bar{y} = \frac{1}{m} \iint_R y\rho \, dA = \frac{6}{k\pi a^3} \int_0^{\pi/2} \int_0^a kr^3 \sin\theta \, dr \, d\theta$$

$$= \frac{6}{\pi a^3} \int_0^{\pi/2} \left[\tfrac{1}{4} r^4 \sin\theta \right]_0^a d\theta = \frac{6}{\pi a^3} \cdot \frac{a^4}{4} \int_0^{\pi/2} \sin\theta \, d\theta = \frac{3a}{2\pi}.$$

Thus the given lamina has mass $\frac{1}{6} k\pi a^3$ and its centroid is located at the point $(3a/2\pi, 3a/2\pi)$.

THE THEOREMS OF PAPPUS

An important theorem relating centroids and volumes of revolution is named for the Greek mathematician who stated it during the third century A.D.

> ***First Theorem of Pappus*** *Volume of Revolution*
>
> Suppose that a plane region R is revolved about an axis in its plane, generating a solid of revolution with volume V. Assume that the axis does not intersect the interior of R. Then V is the product of the area A of R and the distance traveled by the centroid of R.

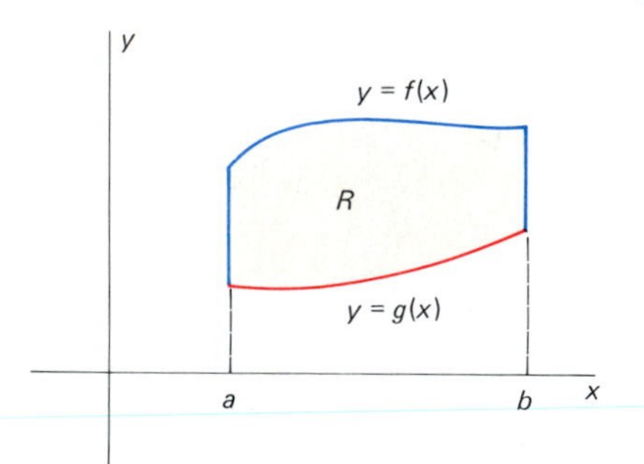

16.44 A region R between the graphs of two functions

Proof for the special case of a region like the one shown in Fig. 16.44. This is the region between the two graphs $y = f(x)$ and $y = g(x)$ for $a \leqq x \leqq b$, with the axis of revolution being the y-axis. Then, in a revolution about the y-axis, the distance traveled by the centroid of R is $d = 2\pi\bar{x}$. By the method of cylindrical shells [see Equation (3) in Section 6.3] the volume of the solid generated is

$$V = \int_a^b 2\pi x[f(x) - g(x)] \, dx = \int_a^b \int_{g(x)}^{f(x)} 2\pi x \, dy \, dx = 2\pi \iint_R x \, dA = 2\pi\bar{x} \cdot A$$

(by the formula in (2), with $\rho \equiv 1$). Thus $V = d \cdot A$, as desired. ■

EXAMPLE 4 Find the volume V of the sphere of radius a generated by revolving around the x-axis the semicircle D of Example 1.

Solution The area of D is $A = \frac{1}{2}\pi a^2$, and we found in Example 1 that $\bar{y} = 4a/3\pi$. Hence Pappus's theorem gives

$$V = 2\pi\bar{y}A = 2\pi \cdot \frac{4a}{3\pi} \cdot \frac{\pi a^2}{2} = \frac{4}{3}\pi a^3.$$

EXAMPLE 5 Consider the circular disk of Fig. 16.45, with radius a and center at the point $(b, 0)$ with $0 < a < b$. Find the volume V of the solid torus generated by revolving this disk around the y-axis.

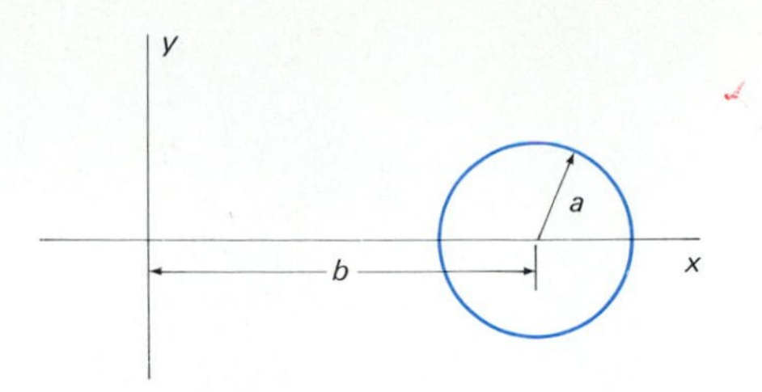

16.45 Rotate the circular disk about the y-axis to produce a torus (Example 5).

Solution The centroid of the circle is at its center $(b, 0)$, so $\bar{x} = b$. Hence the centroid is revolved through the distance $d = 2\pi b$. Consequently,

$$V = d \cdot A = (2\pi b)(\pi a^2) = 2\pi^2 a^2 b.$$

Note that this result is dimensionally correct.

Centroids of plane *curves* are defined in analogy with the method for plane regions, so we shall present this topic in less detail. It will suffice for us to treat only the case of constant density $\rho = 1$ (like a wire with unit mass per unit length). Then the centroid $(\bar{x}, \bar{y})$ of the plane curve C is defined by the formulas

$$\bar{x} = \frac{1}{s}\int_C x\,ds, \qquad \bar{y} = \frac{1}{s}\int_C y\,ds \tag{4}$$

where s is the arc length of C.

The meaning of the integrals in (4) is that of the notation of Section 6.4. That is, ds is a symbol to be replaced (before evaluation of the integral) by either

$$ds = \sqrt{1 + \left(\frac{dy}{dx}\right)^2}\,dx \quad \text{or} \quad ds = \sqrt{1 + \left(\frac{dx}{dy}\right)^2}\,dy,$$

depending on whether C is a smooth arc of the form $y = f(x)$ or one of the form $x = g(y)$. Alternatively, we may have

$$ds = \sqrt{(dx)^2 + (dy)^2} = \sqrt{[x'(t)]^2 + [y'(t)]^2}\,dt$$

if C is presented in parametric form, as in Section 13.2.

EXAMPLE 6 Let J denote the upper half of the *circle* (not the disk) of radius a and center $(0, 0)$, represented parametrically by

$$x = a\cos t, \qquad y = a\sin t, \qquad 0 \leqq t \leqq \pi.$$

The arc J is shown in Fig. 16.46. Find its centroid.

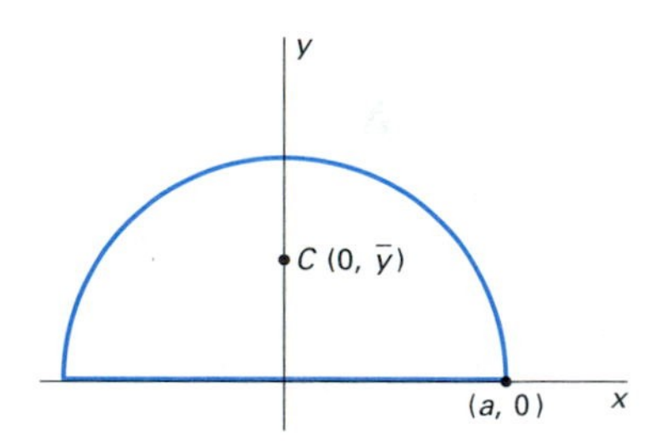

16.46 The semicircular arc of Example 6

Solution Note first that $\bar{x} = 0$ by symmetry. The arc length of J is $s = \pi a$; the arc length element is

$$ds = \sqrt{(-a\sin t\,dt)^2 + (a\cos t\,dt)^2} = a\,dt.$$

Hence the second formula in (4) yields

$$\bar{y} = \frac{1}{\pi a}\int_0^{\pi} (a\sin t)(a\,dt) = \frac{a}{\pi}\Big[-\cos t\Big]_0^{\pi} = \frac{2a}{\pi}.$$

Thus the centroid of the semicircle is located at the point $(0, 2a/\pi)$ on the y-axis. Note that the answer is both plausible and dimensionally correct.

The first theorem of Pappus has an analogue for surface area of revolution.

Second Theorem of Pappus *Surface Area of Revolution*

Let the plane curve C be revolved about an axis in its plane that does not intersect the curve. Then the area A of the surface of revolution generated is equal to the product of the length of C and the distance traveled by the centroid of C.

Proof for the special case in which C is a smooth arc described by $y = f(x)$, $a \leqq x \leqq b$, and the axis of revolution is the y-axis. The distance traveled by the centroid of C is $d = 2\pi\bar{x}$. By the formula in Equation (12) of Section 6.4, the area of the surface of revolution is

$$A = \int_a^b 2\pi x\sqrt{1 + [f'(x)]^2}\, dx = 2\pi s \cdot \frac{1}{s}\int_C x\, ds = 2\pi s\bar{x}$$

by the formula in (4). Therefore, $A = d \cdot s$, as desired. ■

EXAMPLE 7 Find the surface area A of the sphere of radius a generated by revolving around the x-axis the semicircular arc of Example 6.

Solution Because we found that $\bar{y} = 2a/\pi$ and we know that $s = \pi a$, the second theorem of Pappus gives

$$A = 2\pi\bar{y}s = 2\pi\left(\frac{2a}{\pi}\right)(\pi a) = 4\pi a^2.$$

EXAMPLE 8 Find the surface area A of the torus of Example 5.

Solution Now we think of revolving the circle (*not* the disk) of radius a centered at the point $(b, 0)$. Of course, the centroid of the circle is located at its center $(b, 0)$; this follows from the symmetry principle or can be verified by using computations like those in Example 6. Hence the distance traveled by the centroid is $d = 2\pi b$. Because the circumference of the circle is $s = 2\pi a$, the second theorem of Pappus gives

$$A = (2\pi b)(2\pi a) = 4\pi^2 ab.$$

MOMENTS OF INERTIA

Let R be a plane lamina and L a straight line that may or may not lie in the xy-plane. Then the **moment of inertia** I of R about the axis L is, by definition,

$$I = \iint_R w^2\, dm \tag{5}$$

where $w = w(x, y)$ denotes the (perpendicular) distance to L from a typical point (x, y) of R.

The most important case is that in which the axis is the z-axis, so that $w = r = (x^2 + y^2)^{1/2}$ (see Fig. 16.47). In this case we call $I_0 = I$ the **polar moment of inertia** of the lamina R. Thus the polar moment of inertia of R is defined to be

$$I_0 = \iint_R r^2\rho(x, y)\, dA = \iint_R (x^2 + y^2)\, dm. \tag{6}$$

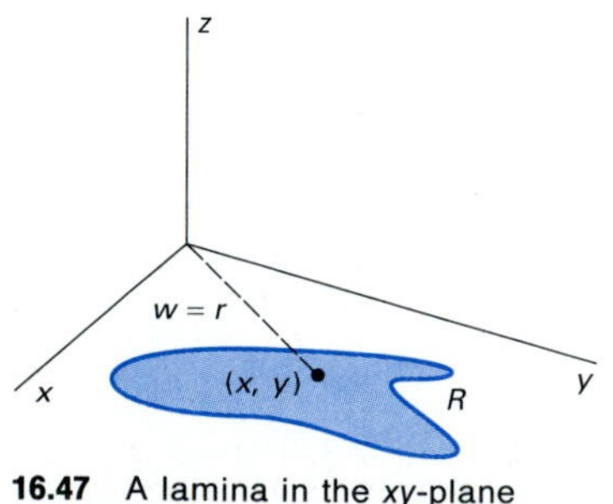

16.47 A lamina in the xy-plane in space

It follows that

$$I_0 = I_x + I_y$$

where

$$I_x = \iint_R y^2 \, dm = \iint_R y^2 \rho \, dA \tag{7}$$

and

$$I_y = \iint_R x^2 \, dm = \iint_R x^2 \rho \, dA. \tag{8}$$

Here I_x is the lamina's moment of inertia about the x-axis, while I_y is its moment of inertia about the y-axis.

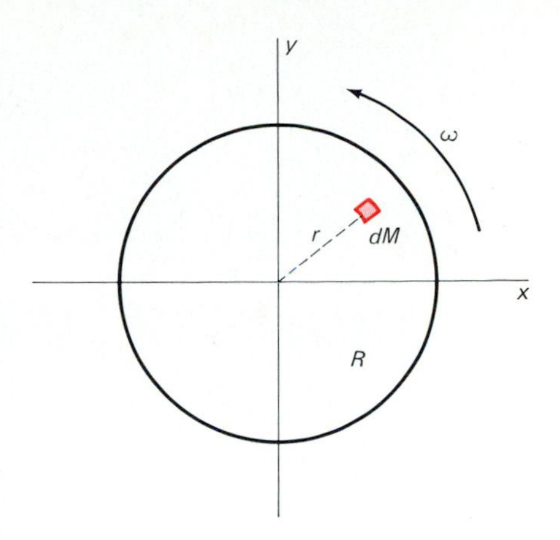

16.48 The rotating disk

An important application of moments of inertia involves *kinetic energy of rotation.* Consider a circular disk that is revolving about its center (the origin) with angular speed ω radians per second. A mass element dm at distance r from the origin is moving with (linear) velocity $v = r\omega$ (see Fig. 16.48), so the kinetic energy of this mass element is

$$\tfrac{1}{2}(dm)v^2 = \tfrac{1}{2}\omega^2 r^2 \, dm.$$

Summing by integration over the whole disk, we find that its kinetic energy due to rotation with angular speed ω is

$$\text{K.E.}_{\text{rot}} = \iint_R \tfrac{1}{2}\omega^2 r^2 \, dm = \tfrac{1}{2}\omega^2 \iint_R r^2 \, dm;$$

that is,

$$\text{K.E.}_{\text{rot}} = \tfrac{1}{2} I_0 \omega^2. \tag{9}$$

Because linear kinetic energy has the formula $\text{K.E.} = \frac{1}{2}mv^2$, the formula in (9) suggests that moment of inertia is the rotational analogue of mass.

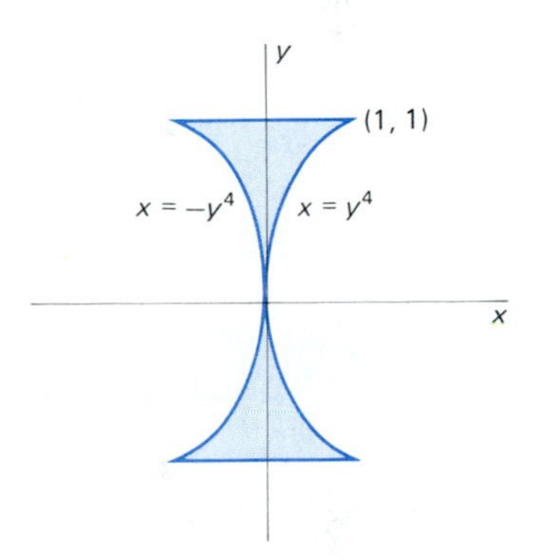

16.49 The lamina of Example 9

EXAMPLE 9 Compute I_x for a lamina of constant density $\rho = 1$ occupying the region bounded by the curves $x = \pm y^4$, $-1 \leqq y \leqq 1$. (See Fig. 16.49.)

Solution The formula in (7) gives

$$I_x = \int_{-1}^{1} \int_{-y^4}^{y^4} y^2 \, dx \, dy = \int_{-1}^{1} \left[xy^2 \right]_{-y^4}^{y^4} dy = \int_{-1}^{1} 2y^6 \, dy = \tfrac{4}{7}.$$

The region of Example 9 resembles the cross section of an I beam. It is known that the stiffness, or resistance to bending, of a horizontal beam is proportional to the moment of inertia of its cross section with respect to a horizontal axis through the centroid of the cross section. Let us compare our I beam with a rectangular beam of equal height 2 and equal area

$$A = \int_{-1}^{1} \int_{-y^4}^{y^4} 1 \, dx \, dy = \tfrac{4}{5}.$$

The cross section of such a rectangular beam is shown in Fig. 16.50. Its width will be $\frac{2}{5}$, and the moment of inertia of its cross section will be

$$I_x = \int_{-1}^{1} \int_{-1/5}^{1/5} y^2 \, dx \, dy = \tfrac{4}{15}.$$

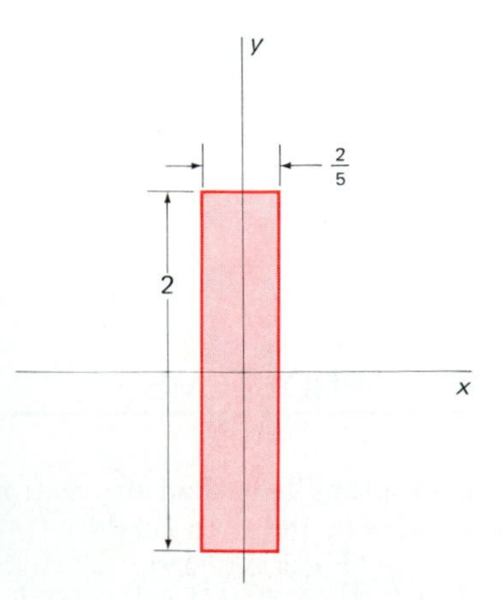

16.50 A rectangular beam for comparison with the I beam of Example 9

Because the ratio of $\frac{4}{7}$ to $\frac{4}{15}$ is $\frac{15}{7}$, we see that the I beam is more than twice as strong as a rectangular beam of the same cross-sectional area. This is why I beams are commonly used in construction.

EXAMPLE 10 Find the polar moment of inertia of a circular lamina of radius a and constant density ρ centered at the origin.

Solution The formula in (6) gives

$$I_0 = \iint_{x^2+y^2 \leqq a^2} r^2 \rho \, dA = \int_0^{2\pi} \int_0^a \rho r^3 \, dr \, d\theta = \frac{\rho \pi a^4}{2} = \frac{1}{2} m a^2,$$

where $m = \rho \pi a^2$ is the mass of the circular lamina.

Finally, the **radius of gyration** $\hat{r}$ of a lamina of mass m about an axis is defined to be

$$\hat{r} = \sqrt{\frac{I}{m}} \tag{10}$$

where I is the moment of inertia of the lamina about that axis. For example, the radii of gyration $\hat{x}$ and $\hat{y}$ about the y-axis and x-axis, respectively, are given by

$$\hat{x} = \left(\frac{I_y}{m}\right)^{1/2} \quad \text{and} \quad \hat{y} = \left(\frac{I_x}{m}\right)^{1/2}. \tag{11}$$

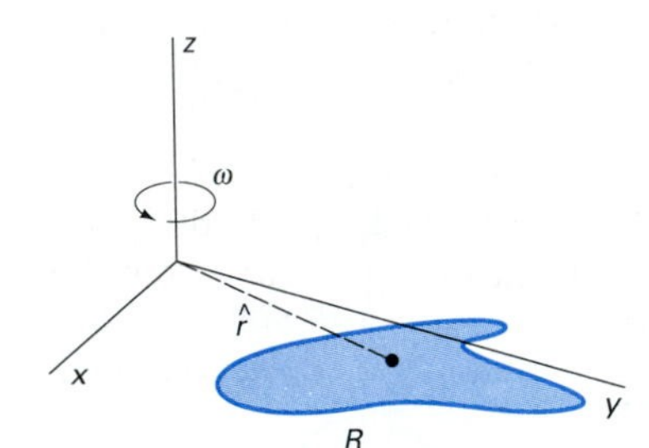

16.51 A plane lamina rotating about the z-axis

Now suppose that this lamina lies in the right half-plane $x > 0$ and is symmetric about the x-axis. If it represents the face of a racquet whose handle (considered of negligible weight) extends along the x-axis from the origin to the face, then the point $(\hat{x}, 0)$ is a plausible candidate for the tennis racquet's "sweet spot." (See Problem 56.)

The definition in (10) is motivated by consideration of a plane lamina R rotating with angular speed ω about the z-axis (see Fig. 16.51). Then (10) yields

$$I_0 = m\hat{r}^2,$$

so it follows from (9) that the kinetic energy of the lamina is

$$\text{K.E.} = \tfrac{1}{2} m (\hat{r}\omega)^2.$$

Thus the kinetic energy of the rotating lamina equals that of a single particle of mass m revolving at the distance $\hat{r}$ from the axis of revolution.

16.5 PROBLEMS

In Problems 1–10, find the centroid of the plane region bounded by the given curves.

1. $x = 0, \quad x = 4, \quad y = 0, \quad y = 6$

2. $x = 1, \quad x = 3, \quad y = 2, \quad y = 4$

3. $x = -1, \quad x = 3, \quad y = -2, \quad y = 4$

4. $x = 0, \quad y = 0, \quad x + y = 3$

5. $x = 0, \quad y = 0, \quad x + 2y = 4$

6. $y = 0, \quad y = x, \quad x + y = 2$

7. $y = 0, \quad y = x^2, \quad x = 2$

8. $y = x^2, \quad y = 9$

9. $y = 0, \quad y = x^2 - 4$

10. $x = -2, \quad x = 2, \quad y = 0, \quad y = x^2 + 1$

In each of Problems 11–30, find the mass and centroid of a plane lamina with the indicated shape and density.

11. The triangular region bounded by $x = 0$, $y = 0$, and $x + y = 1$, with $\rho(x, y) = xy$.

12. The triangular region of Problem 11, with $\rho(x, y) = x^2$.

13. The region bounded by $y = 0$ and $y = 4 - x^2$, with $\rho(x, y) = y$.

14. The region bounded by $x = 0$ and $x = 9 - y^2$, with $\rho(x, y) = x^2$.

15. The region bounded by the parabolas $y = x^2$ and $x = y^2$, with $\rho(x, y) = xy$.

16. The region of Problem 15, with $\rho(x, y) = x^2 + y^2$.

17. The region bounded by the parabolas $y = x^2$ and $y = 2 - x^2$, with $\rho(x, y) = y$.

18. The region bounded by $x = 0$, $x = e$, $y = 0$, and $y = \ln x$ for $1 \leqq x \leqq e$, with $\rho(x, y) = 1$.

19. The region bounded by $y = 0$ and $y = \sin x$ for $0 \leqq x \leqq \pi$, with $\rho(x, y) = 1$.

20. The region bounded by $y = 0$, $x = -1$, $x = 1$, and $y = \exp(-x^2)$, with $\rho(x, y) = |xy|$.

21. The square with vertices $(0, 0)$, $(0, a)$, (a, a), and $(a, 0)$, with $\rho(x, y) = x + y$.

22. The triangular region bounded by the coordinate axes and the line $x + y = a$; $\rho(x, y) = x^2 + y^2$.

23. The region bounded by $y = x^2$ and $y = 4$; $\rho(x, y) = y$.

24. The region bounded by $y = x^2$ and $y = 2x + 3$; $\rho(x, y) = x^2$.

25. The region of Problem 19; $\rho(x, y) = x$.

26. The semicircular region $x^2 + y^2 \leqq a^2$ for $y \geqq 0$; $\rho(x, y) = y$.

27. The region of Problem 26; $\rho(x, y) = r$ (the radial polar coordinate).

28. The region bounded by the cardioid $r = 1 + \cos\theta$; $\rho = r$.

29. The region within the circle $r = 2\sin\theta$ and outside the circle $r = 1$; $\rho(x, y) = y$.

30. The region within the limaçon $r = 1 + 2\cos\theta$ and outside the circle $r = 2$; $\rho(x, y) = r$.

In each of Problems 31–35, find the polar moment of inertia I_0 of the indicated lamina.

31. The region bounded by the circle $r = a$; $\rho(x, y) = r^n$, where n is a fixed positive integer.

32. The lamina of Problem 26.

33. The disk bounded by $r = 2\cos\theta$; $\rho(x, y) = k$ (a positive constant).

34. The lamina of Problem 29.

35. The region bounded by the right-hand loop of the lemniscate $r^2 = \cos 2\theta$; $\rho(x, y) = r^2$.

In each of Problems 36–40, find the radii of gyration $\hat{x}$ and $\hat{y}$ of the indicated lamina about the coordinate axes.

36. The lamina of Problem 21.

37. The lamina of Problem 23.

38. The lamina of Problem 24.

39. The lamina of Problem 27.

40. The lamina of Problem 33.

41. Find the centroid of the first quadrant of the circular disk $x^2 + y^2 \leqq r^2$ by direct computation, as in Example 1.

42. Apply the first theorem of Pappus to find the centroid of the first quadrant of the circular disk $x^2 + y^2 \leqq r^2$. Use the facts that $\bar{x} = \bar{y}$ by symmetry and that revolution of this quarter-disk about either coordinate axis gives a solid hemisphere with volume $V = \frac{2}{3}\pi r^3$.

43. Find the centroid of the arc consisting of the first-quadrant portion of the circle $x^2 + y^2 = r^2$ by direct computation, as in Example 6.

44. Apply the second theorem of Pappus to find the centroid of the quarter-circular arc of Problem 43. Note that $\bar{x} = \bar{y}$ by symmetry, and that revolution of this arc about either coordinate axis gives a hemisphere with surface area $A = 2\pi r^2$.

45. Show by direct computation that the centroid of the triangle with vertices $(0, 0)$, $(r, 0)$, and $(0, h)$ is the point $(r/3, h/3)$. Verify that this point lies on the line from the vertex $(0, 0)$ to the midpoint of the opposite side of the triangle and two-thirds of the way from the vertex to the midpoint.

46. Apply the first theorem of Pappus and the result of Problem 45 to verify the formula $V = \frac{1}{3}\pi r^2 h$ for the volume of the cone obtained by revolving the triangle about the y-axis.

47. Apply the second theorem of Pappus to show that the lateral surface area of the cone of Problem 46 is $A = \pi r L$, where $L = (r^2 + h^2)^{1/2}$ is the slant height of the cone.

48. (a) Find the centroid of the trapezoid shown in Fig. 16.52.

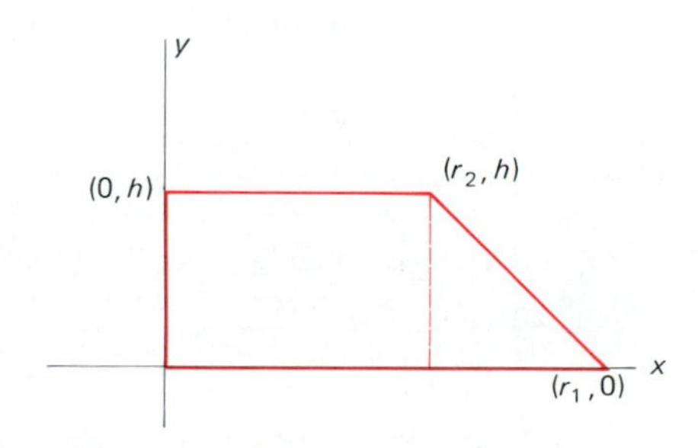

16.52 The trapezoid of Problem 48

(b) Apply the first theorem of Pappus and the result of part (a) to show that the volume of the conical frustum generated by revolving the trapezoid around the y-axis is

$$V = \frac{\pi h}{3}(r_1^2 + r_1 r_2 + r_2^2).$$

49. Apply the second theorem of Pappus to show that the lateral surface area of the conical frustum in Problem 48 is $A = \pi(r_1 + r_2)L$, where

$$L = \sqrt{(r_1 - r_2)^2 + h^2}$$

is its slant height.

50. (a) Apply the second theorem of Pappus to verify that the curved surface area of a right circular cylinder with height h and base radius r is $A = 2\pi rh$.

(b) Explain how this also follows from the result of Problem 49.

51. (a) Find the centroid of the plane region shown in Fig. 16.53, which consists of a semicircular region of radius a sitting atop a rectangular region having width $2a$ and height b.

(b) Then apply the first theorem of Pappus to find the volume generated by rotating this region about the x-axis.

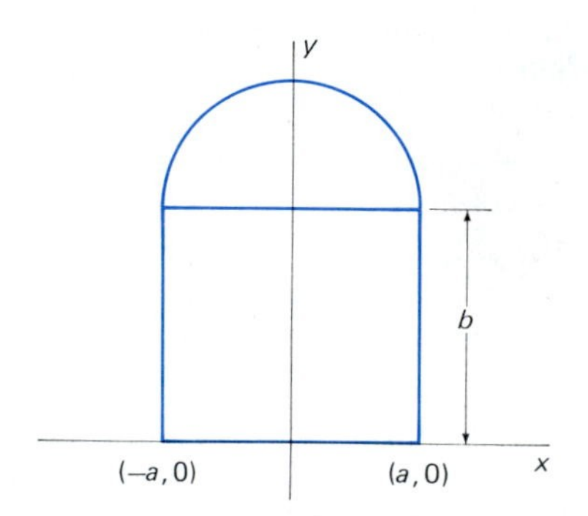

16.53 The plane region of Problem 51

52. (a) Consider the plane region of Fig. 16.54, bounded by $x^2 = 2py$, $x = 0$, and $y = h = r^2/2p$ ($p > 0$). Show that

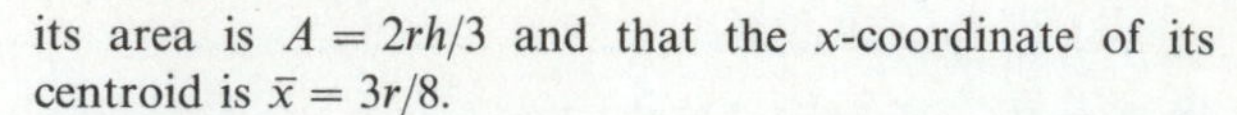

its area is $A = 2rh/3$ and that the x-coordinate of its centroid is $\bar{x} = 3r/8$.

(b) Use Pappus's theorem and the result of part (a) to show that the volume of a paraboloid of revolution with radius r and height h is $V = \frac{1}{2}\pi r^2 h$.

53. Find the centroid of the unbounded region under the graph of $y = e^{-x}$, $x > 0$.

54. A uniform plane region has centroid at (0, 0) and total mass m. Show that its moment of inertia about an axis perpendicular to the xy-plane at (x_0, y_0) is

$$I = I_0 + m(x_0^2 + y_0^2).$$

55. Suppose that a plane lamina consists of two nonoverlapping laminae. Show that its polar moment of inertia is the sum of theirs. Use this fact, together with the results of Problems 53 and 54, to find the polar moment of inertia of the T-shaped lamina of constant density $\rho = k > 0$ shown in Fig. 16.55.

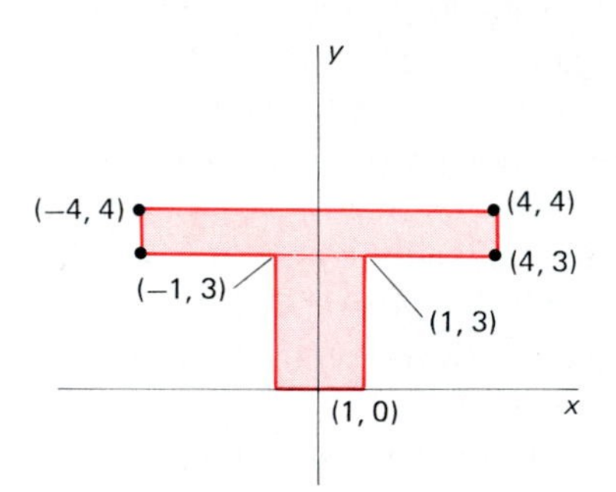

16.55 One lamina made of two simple ones—see Problem 55.

56. A racquet consists of a uniform lamina that occupies the region within the right-hand loop of $r^2 = \cos 2\theta$ on the end of a handle (assumed of negligible mass) corresponding to the interval $-1 \leqq x \leqq 0$, as in Fig. 16.56. Find the radius of gyration of the racquet about the line $x = -1$. Where is its sweet spot?

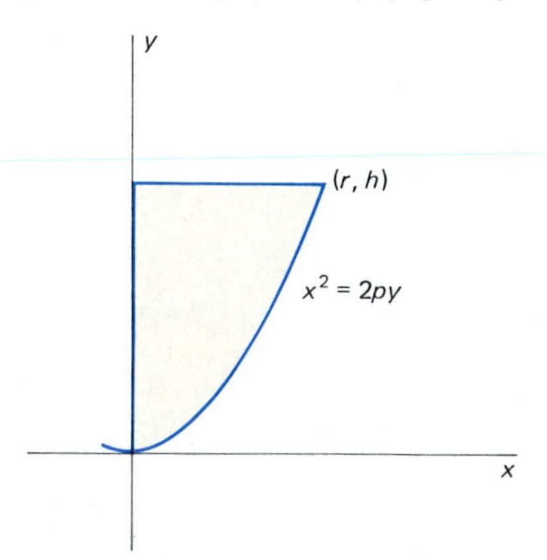

16.54 The region of Problem 52

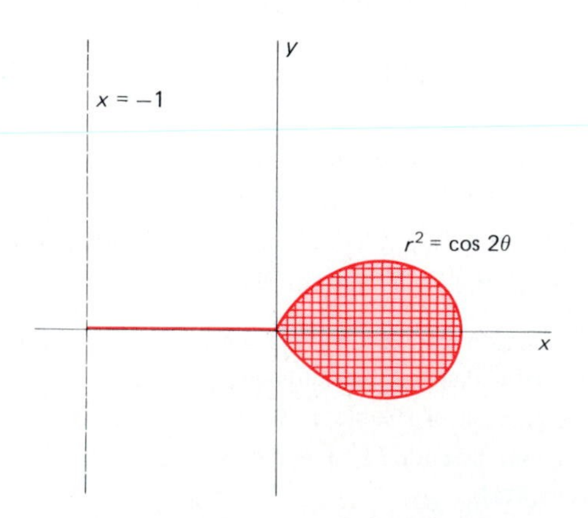

16.56 The racquet of Problem 56

16.6 Triple Integrals

The definition of the triple integral is the three-dimensional version of the definition of the double integral in Section 16.2. Let $f(x, y, z)$ be continuous on the bounded space region T, and suppose that T lies inside the rectangular block R determined by the inequalities $a \leqq x \leqq b$, $c \leqq y \leqq d$, and $p \leqq z \leqq q$. We subdivide $[a, b]$ into subintervals all of the same length Δx, $[c, d]$ into subintervals all of the same length Δy, and $[p, q]$ into subintervals all of the same length Δz. This generates a partition of R into smaller rectangular blocks (as in Fig. 16.57), each with volume $\Delta V = \Delta x \, \Delta y \, \Delta z$. Let $\mathcal{P} = \{T_1, T_2, \ldots, T_n\}$ be the collection of these smaller blocks that lie wholly within T. Then $\mathcal{P}$ is called an **inner partition** of the region T. The **mesh** $|\mathcal{P}|$ of $\mathcal{P}$ is the length of a longest diagonal of any of the blocks T_i. If (x_i^*, y_i^*, z_i^*) is an arbitrarily selected point of T_i (for each $i = 1, 2, 3, \ldots, n$), then the **Riemann sum**

$$\sum_{i=1}^{n} f(x_i^*, y_i^*, z_i^*) \, \Delta V$$

is an approximation to the triple integral of f over the region T.

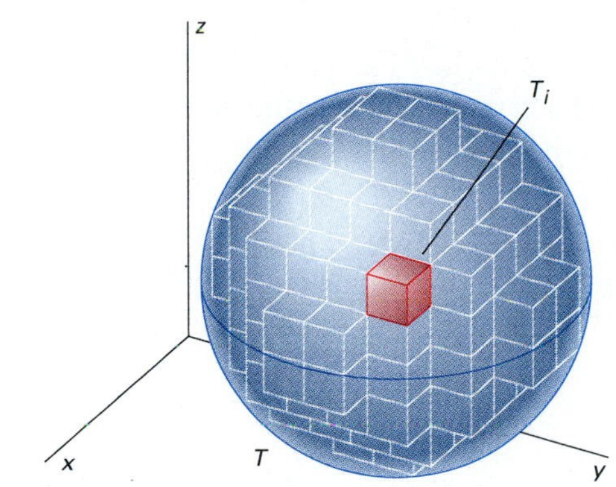

16.57 One small block in an inner partition of the bounded space region T

For example, if T is a solid body with density function f, then such a Riemann sum approximates its total mass. We define the **triple integral of f over T** by means of the equation

$$\iiint_T f(x, y, z) \, dV = \lim_{|\mathcal{P}| \to 0} \sum_{i=1}^{n} f(x_i^*, y_i^*, z_i^*) \, \Delta V. \tag{1}$$

It is proved in advanced calculus that this limit of Riemann sums exists as the mesh $|\mathcal{P}|$ approaches zero, provided that f is continuous on T and that the boundary of the region T is reasonably well-behaved. For instance, it suffices for the boundary of T to consist of finitely many pieces of smooth surfaces.

Just like double integrals, triple integrals are ordinarily computed by means of iterated integrals. If the region of integration is a rectangular block, as in Example 1, then we can integrate in any order we wish.

EXAMPLE 1 If $f(x, y, z) = xy + yz$ and T consists of those points (x, y, z) in space satisfying the inequalities $-1 \leqq x \leqq 1$, $2 \leqq y \leqq 3$, and $0 \leqq z \leqq 1$, then

$$\iiint_T f(x, y, z)\,dV = \int_{-1}^{1}\int_{2}^{3}\int_{0}^{1} (xy + yz)\,dz\,dy\,dx$$

$$= \int_{-1}^{1}\int_{2}^{3} \left[xyz + \tfrac{1}{2}yz^2\right]_{z=0}^{1} dy\,dx$$

$$= \int_{-1}^{1}\int_{2}^{3} (xy + \tfrac{1}{2}y)\,dy\,dx$$

$$= \int_{-1}^{1} \left[\tfrac{1}{2}xy^2 + \tfrac{1}{4}y^2\right]_{y=2}^{3} dx$$

$$= \int_{-1}^{1} (\tfrac{5}{2}x + \tfrac{5}{4})\,dx = \left[\tfrac{5}{4}x^2 + \tfrac{5}{4}x\right]_{-1}^{1} = \tfrac{5}{2}.$$

The applications of double integrals that we saw in earlier sections generalize immediately to triple integrals. If T is a solid body with density function $\rho(x, y, z)$, then its **mass** m is given by

$$m = \iiint_T \rho\,dV. \tag{2}$$

The case $\rho \equiv 1$ gives the **volume**

$$V = \iiint_T dV \tag{3}$$

of T. The coordinates of its **centroid** are

$$\bar{x} = \frac{1}{m}\iiint_T x\rho\,dV, \tag{4a}$$

$$\bar{y} = \frac{1}{m}\iiint_T y\rho\,dV, \quad \text{and} \tag{4b}$$

$$\bar{z} = \frac{1}{m}\iiint_T z\rho\,dV. \tag{4c}$$

The **moments of inertia** of T about the three coordinate axes are

$$I_x = \iiint_T (y^2 + z^2)\rho\,dV, \tag{5a}$$

$$I_y = \iiint_T (x^2 + z^2)\rho\,dV, \quad \text{and} \tag{5b}$$

$$I_z = \iiint_T (x^2 + y^2)\rho\,dV. \tag{5c}$$

As indicated previously, triple integrals are almost always evaluated by iterated single integration. Suppose that the nice region T is ***z*-simple,** in that each line parallel to the z-axis intersects T (if at all) in a single line segment. In effect, this means that T can be described by the inequalities

$$g_1(x, y) \leqq z \leqq g_2(x, y), \qquad (x, y) \text{ in } R,$$

where R is the vertical projection of T into the xy-plane. Then

$$\iiint_T f(x, y, z)\, dV = \iint_R \left(\int_{g_1(x,y)}^{g_2(x,y)} f(x, y, z)\, dz \right) dA. \tag{6}$$

In the formula in (6), we take $dA = dx\,dy$ or $dA = dy\,dx$, depending upon the preferable order of integration over the set R. The limits $g_1(x, y)$ and $g_2(x, y)$ for z will be the z-coordinates of the endpoints of the line segment in which the vertical line at (x, y) meets T. (See Fig. 16.58.)

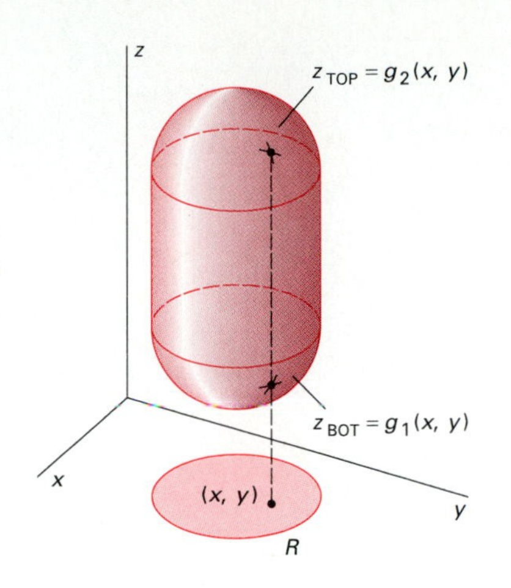

16.58 Obtaining the limits of integration for z

If the region R has the description

$$\alpha(y) \leqq x \leqq \beta(y), \qquad c \leqq y \leqq d,$$

then (integrating last with respect to y),

$$\iiint_T f(x, y, z)\, dV = \int_c^d \int_{\alpha(y)}^{\beta(y)} \int_{g_1(x,y)}^{g_2(x,y)} f(x, y, z)\, dz\, dx\, dy.$$

Thus the triple integral reduces in this case to three iterated single integrals. These can (in principle) be evaluated using the fundamental theorem of calculus.

If the solid T is bounded by the *two* surfaces $z = g_1(x, y)$ and $z = g_2(x, y)$, as in Fig. 16.59, we can find the region R of Equation (6) as follows. The equation $g_1(x, y) = g_2(x, y)$ determines a vertical cylinder that passes through the curve of intersection of the two surfaces. So the cylinder intersects the xy-plane in the boundary curve of R. For example, the solid of Example 3 below is bounded by the paraboloid $z = x^2 + y^2$ and the plane $z = y + 2$. And the graph of the equation $x^2 + y^2 = y + 2$ is a circle that bounds the region R above which the solid T lies.

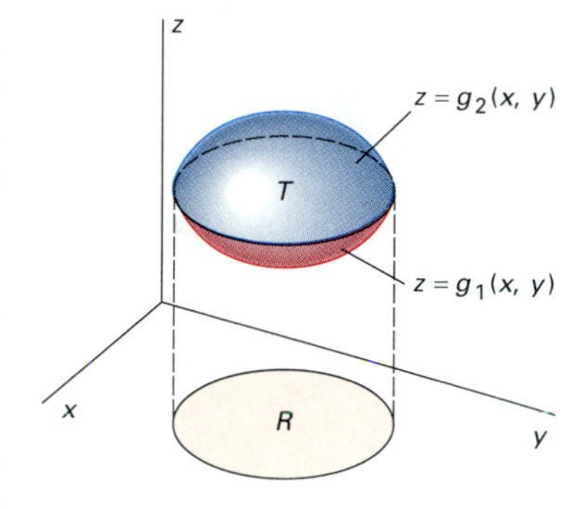

16.59 To find the boundary of R, solve the equation

$$g_1(x, y) = g_2(x, y).$$

We proceed similarly if T is either **x-simple** or **y-simple.** Such situations, as well as a z-simple solid, appear in Fig. 16.60. For example, suppose that T is y-simple, so that it has a description of the form

$$h_1(x, z) \leqq y \leqq h_2(x, z), \qquad (x, z) \text{ in } R,$$

where R is the projection of T into the xz-plane. Then

$$\iiint_T f(x, y, z)\, dV = \iint_R \left(\int_{h_1(x,z)}^{h_2(x,z)} f(x, y, z)\, dy \right) dA, \tag{7}$$

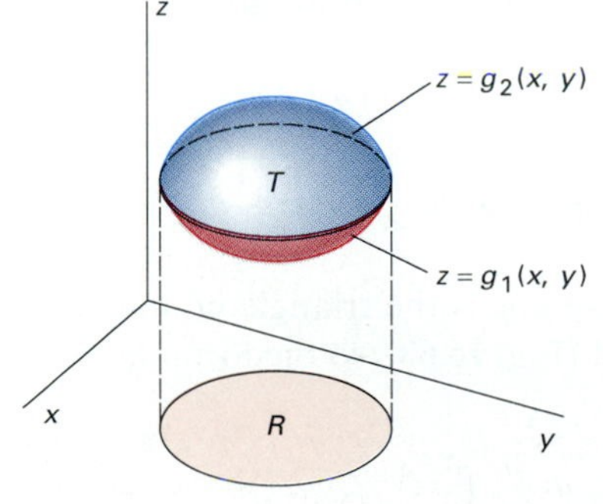

T is *z*-simple

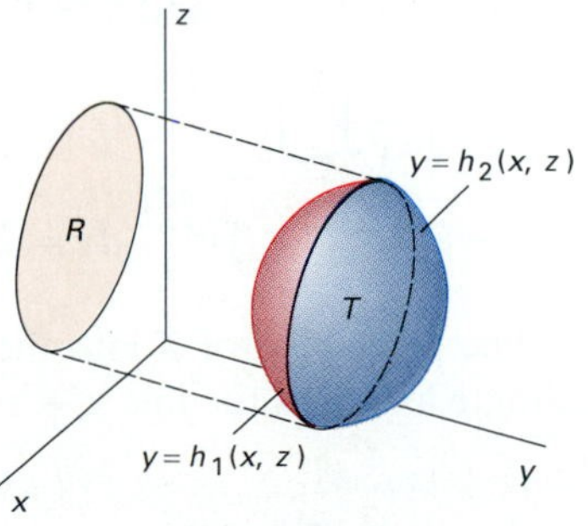

T is *y*-simple

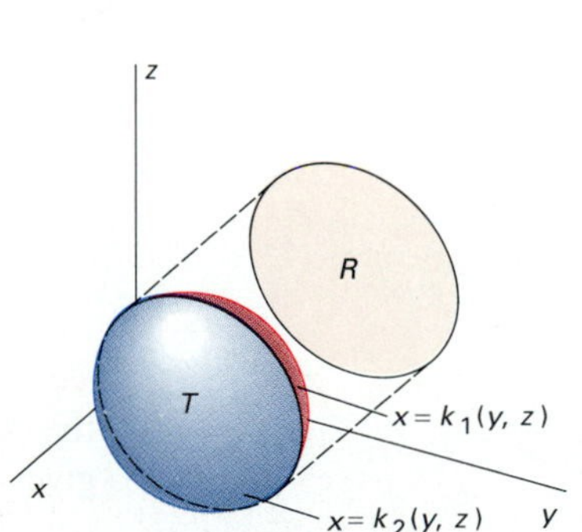

T is *x*-simple

16.60 Solids that are x-simple, y-simple, and z-simple

where $dA = dx\,dz$ or $dA = dz\,dx$, and the limits $h_1(x, z)$ and $h_2(x, z)$ are the y-coordinates of the endpoints of the line segment in which a typical line parallel to the y-axis intersects T. If T is x-simple, we have

$$\iiint_T f(x, y, z)\,dV = \iint_R \left(\int_{k_1(y,z)}^{k_2(y,z)} f(x, y, z)\,dx \right) dA \tag{8}$$

where $dA = dy\,dz$ or $dA = dz\,dy$ and R is the projection of T into the yz-plane.

EXAMPLE 2 Compute by triple integration the volume of the region T that is bounded by the parabolic cylinder $x = y^2$ and the planes $z = 0$ and $x + z = 1$. Also find the centroid of T given it has constant density $\rho = 1$.

COMMENT The three segments in Fig. 16.61 parallel to the coordinate axes indicate that the region T is simultaneously x-simple, y-simple, and z-simple. We may, therefore, integrate in any order we choose, so there are at least three solutions. Here are three computations of the volume V of T.

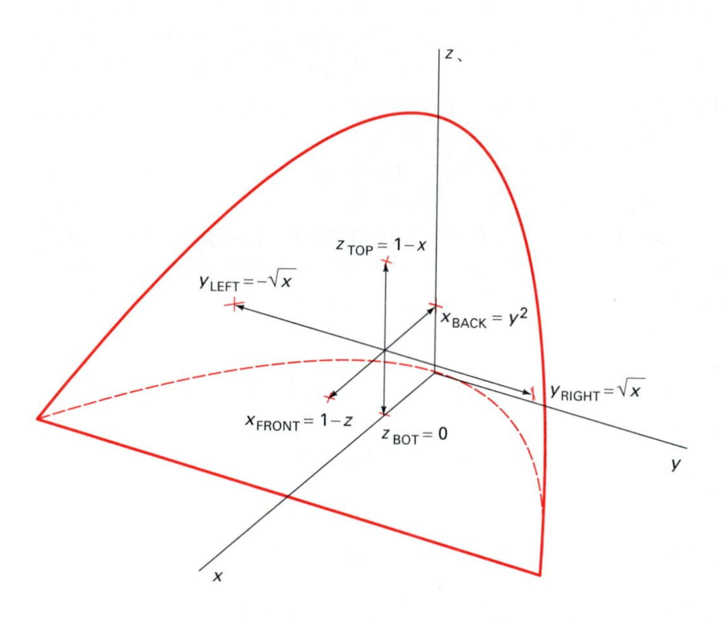

16.61 The region T of Example 2 is x-simple, y-simple, and z-simple.

Solution 1 The projection of T into the xy-plane is the region shown in Fig. 16.62, bounded by $x = y^2$ and $x = 1$. So the formula in (6) gives

$$V = \int_{-1}^{1} \int_{y^2}^{1} \int_{0}^{1-x} dz\,dx\,dy = 2 \int_0^1 \int_{y^2}^{1} (1 - x)\,dx\,dy$$

$$= 2 \int_0^1 \left[x - \tfrac{1}{2}x^2 \right]_{y^2}^{1} dy = 2 \int_0^1 \left(\tfrac{1}{2} - y^2 + \tfrac{1}{2}y^4\right) dy = \tfrac{8}{15}.$$

Solution 2 The projection of T into the xz-plane is the triangle bounded by the coordinate axes and the line $x + z = 1$ (Fig. 16.63), so the formula in (7) gives

$$V = \int_0^1 \int_0^{1-x} \int_{-\sqrt{x}}^{\sqrt{x}} dy\,dz\,dx = 2 \int_0^1 \int_0^{1-x} \sqrt{x}\,dz\,dx$$

$$= 2 \int_0^1 (x^{1/2} - x^{3/2})\,dx = \tfrac{8}{15}.$$

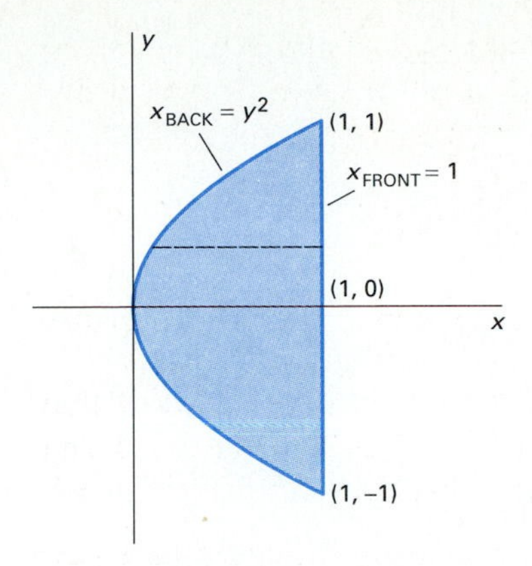

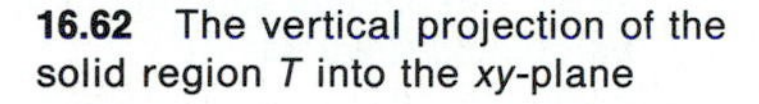

16.62 The vertical projection of the solid region T into the xy-plane

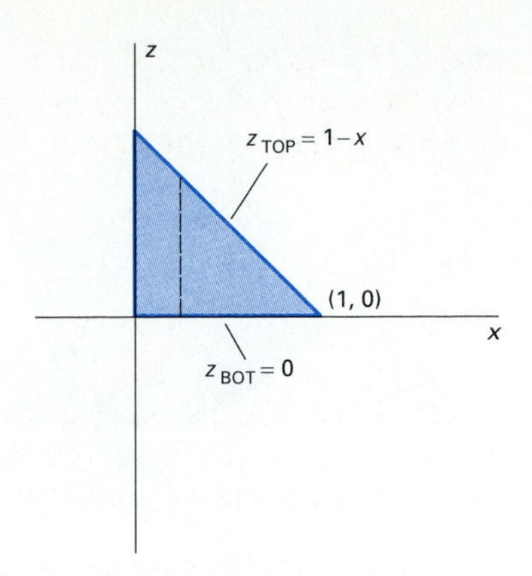

16.63 The vertical projection of the solid region T into the xz-plane

Solution 3 The projection of T into the yz-plane is bounded by the y-axis and the parabola $z = 1 - y^2$ (as in Fig. 16.64), so the formula in (8) yields

$$V = \int_{-1}^{1} \int_{0}^{1-y^2} \int_{y^2}^{1-z} dx\,dz\,dy,$$

and evaluation of this integral again gives $V = \frac{8}{15}$.

Now for the centroid of T. Because the region T is symmetric about the xz-plane, its centroid lies in this plane, and so $\bar{y} = 0$. We compute $\bar{x}$ and $\bar{z}$ by integrating first with respect to y:

$$\bar{x} = \frac{1}{V} \iiint_T x\,dV = \frac{15}{8} \int_0^1 \int_0^{1-x} \int_{-\sqrt{x}}^{\sqrt{x}} x\,dy\,dz\,dx$$

$$= \frac{15}{4} \int_0^1 \int_0^{1-x} x^{3/2}\,dz\,dx = \frac{15}{4} \int_0^1 (x^{3/2} - x^{5/2})\,dx = \frac{3}{7},$$

and similarly,

$$\bar{z} = \frac{1}{V} \iiint_T z\,dV = \frac{15}{8} \int_0^1 \int_0^{1-x} \int_{-\sqrt{x}}^{\sqrt{x}} z\,dy\,dz\,dx = \frac{2}{7}.$$

Thus the centroid of T is located at the point $(\frac{3}{7}, 0, \frac{2}{7})$.

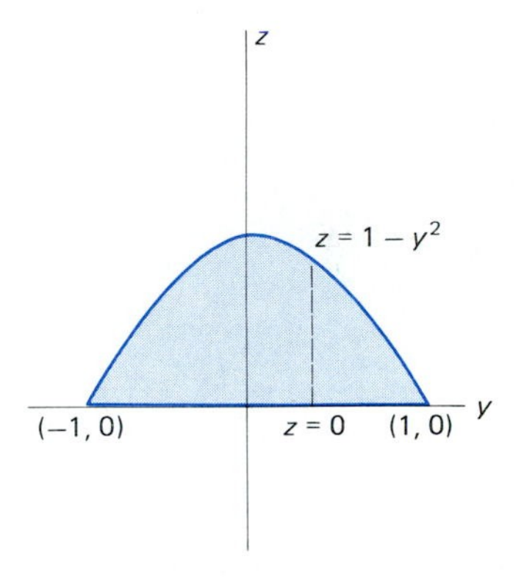

16.64 The vertical projection of the solid region T into the yz-plane

EXAMPLE 3 Find the volume of the *oblique segment of a paraboloid* bounded by the paraboloid $z = x^2 + y^2$ and the plane $z = y + 2$ (see Fig. 16.65).

Solution The given region T is z-simple, but its projection into the xy-plane is bounded by the graph of the equation $x^2 + y^2 = y + 2$, which is a translated circle. Consequently, it would be inconvenient to integrate first with respect to z.

The region T is also x-simple, so we may integrate first with respect to x. The projection of T into the yz-plane is bounded by the line $z = y + 2$ and the parabola $z = y^2$, which intersect at the points $(-1, 1)$ and $(2, 4)$,

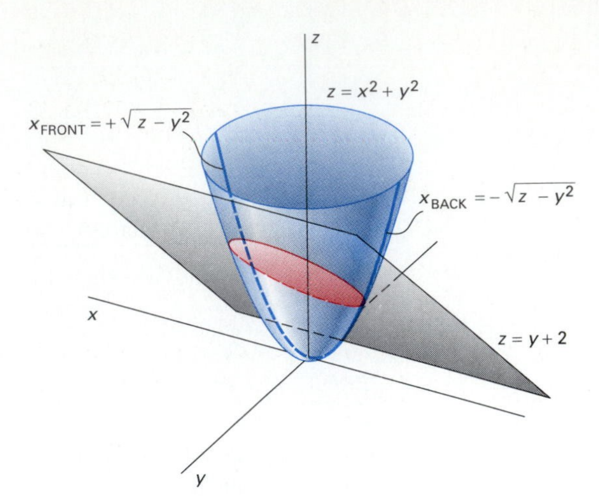

16.65 An oblique segment of a paraboloid (Example 3)

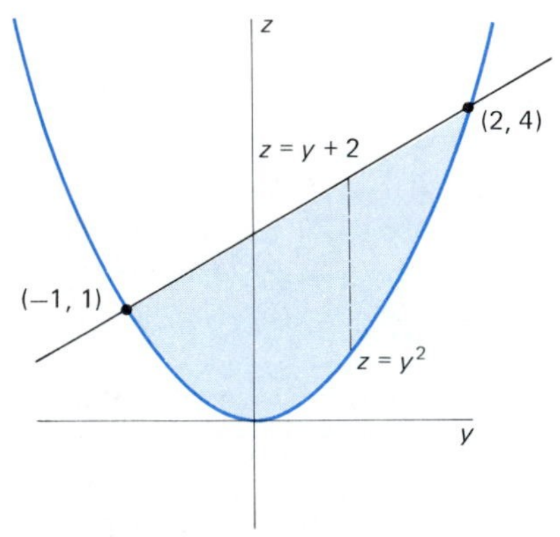

16.66 Projection of the segment of the paraboloid into the yz-plane

as shown in Fig. 16.66. The endpoints of a line segment in T parallel to the x-axis have x-coordinates $x = \pm\sqrt{z - y^2}$. Because T is symmetric about the yz-plane, we can integrate from $x = 0$ to $x = \sqrt{z - y^2}$ and double the result. Hence T has volume

$$V = 2\int_{-1}^{2}\int_{y^2}^{y+2}\int_{0}^{\sqrt{z-y^2}} dx\,dz\,dy = 2\int_{-1}^{2}\int_{y^2}^{y+2}\sqrt{z - y^2}\,dz\,dy$$

$$= 2\int_{-1}^{2}\left[\tfrac{2}{3}(z - y^2)^{3/2}\right]_{z=y^2}^{y+2} dy = \tfrac{4}{3}\int_{-1}^{2}(2 + y - y^2)^{3/2}\,dy$$

$$= \tfrac{4}{3}\int_{-3/2}^{3/2}(\tfrac{9}{4} - u^2)^{3/2}\,du \qquad \text{(completing the square; } u = y - \tfrac{1}{2})$$

$$= \tfrac{27}{4}\int_{-\pi/2}^{\pi/2}\cos^4\theta\,d\theta \qquad (u = \tfrac{3}{2}\sin\theta)$$

$$= \frac{27}{4}\cdot 2\cdot\frac{1}{2}\cdot\frac{3}{4}\cdot\frac{\pi}{2} = \frac{81\pi}{32}.$$

In the final evaluation, we have used Formula (113) (inside the back cover).

16.6 PROBLEMS

In each of Problems 1–10, compute the value of the triple integral $\iiint_T f(x, y, z)\,dV$.

1. $f(x, y, z) = x + y + z$; T is the rectangular box $0 \leqq x \leqq 2$, $0 \leqq y \leqq 3$, $0 \leqq z \leqq 1$.

2. $f(x, y, z) = xy \sin z$; T is the cube $0 \leqq x \leqq \pi$, $0 \leqq y \leqq \pi$, $0 \leqq z \leqq \pi$.

3. $f(x, y, z) = xyz$; T is the rectangular block $-1 \leqq x \leqq 3$, $0 \leqq y \leqq 2$, $-2 \leqq z \leqq 6$.

4. $f(x, y, z) = x + y + z$; T is the rectangular block of Problem 3.

5. $f(x, y, z) = x^2$; T is the tetrahedron bounded by the coordinate planes and the first octant part of the plane with equation $x + y + z = 1$.

6. $f(x, y, z) = 2x + 3y$; T is a first-octant tetrahedron as in Problem 5, except that the plane has equation $2x + 3y + z = 6$.

7. $f(x, y, z) = xyz$; T lies under the surface $z = 1 - x^2$ and above the rectangle $-1 \leqq x \leqq 1$, $0 \leqq y \leqq 2$ in the xy-plane.

8. $f(x, y, z) = 2y + z$; T lies under the surface with equation $z = 4 - y^2$ and above the rectangle $-1 \leqq x \leqq 1$, $-2 \leqq y \leqq 2$ in the xy-plane.

9. $f(x, y, z) = x + y$; T is the region between the surfaces $z = 2 - x^2$ and $z = x^2$ for $0 \leqq y \leqq 3$.

10. $f(x, y, z) = z$; T is the region between the surfaces $z = y^2$ and $z = 8 - y^2$ for $-1 \leqq x \leqq 1$.

In Problems 11–20, sketch the solid bounded by the graphs of the given equations, then find its volume by triple integration.

11. $2x + 3y + z = 6, \quad x = 0, \quad y = 0, \quad z = 0$

12. $z = y, \quad y = x^2, \quad y = 4, \quad z = 0$

13. $y + z = 4, \quad y = 4 - x^2, \quad y = 0, \quad z = 0$

14. $z = x^2 + y^2, \quad z = 0, \quad x = 0, \quad y = 0, \quad x + y = 1$

15. $z = 10 - x^2 - y^2, \quad y = x^2, \quad x = y^2, \quad z = 0$

16. $x = z^2, \quad x = 8 - z^2, \quad y = -1, \quad y = -3$

17. $z = x^2, \quad y + z = 4, \quad y = 0, \quad z = 0$

18. $z = 1 - y^2, \quad z = y^2 - 1, \quad x + z = 1, \quad x = 0$

19. $y = z^2, \quad z = y^2, \quad x + y + z = 2, \quad x = 0$

20. $y = 4 - x^2 - z^2, \quad x = 0, \quad y = 0, \quad z = 0, \quad x + z = 2$

In each of Problems 21–32, assume that the indicated solid has constant density $\rho = 1$.

21. Find the centroid of the solid of Problem 12.

22. Find the centroid of the hemisphere $x^2 + y^2 + z^2 \leqq R^2$, $z \geqq 0$.

23. Find the centroid of the solid of Problem 17.

24. Find the centroid of the solid bounded by $z = 1 - x^2$, $z = 0$, $y = -1$, and $y = 1$.

25. Find the centroid of the solid bounded by $z = \cos x$, $x = -\pi/2$, $x = \pi/2$, $y = 0$, $z = 0$, and $y + z = 1$.

26. Find the moment of inertia about the z-axis of the solid of Problem 12.

27. Find the moment of inertia about the y-axis of the solid of Problem 24.

28. Find the moment of inertia about the z-axis of the solid cylinder $x^2 + y^2 \leqq R^2$, $0 \leqq z \leqq H$.

29. Find the moment of inertia about the z-axis of the solid bounded by $x = 0$, $y = 0$, $z = 0$, and $x + y + z = 1$.

30. Find the moment of inertia about the z-axis of the cube with vertices $(\pm 0.5, 3, \pm 0.5)$ and $(\pm 0.5, 4, \pm 0.5)$.

31. Consider the solid paraboloid bounded by $z = x^2 + y^2$ and the plane $z = h > 0$. Show that its centroid lies on its axis of symmetry, two-thirds of the way from its "vertex" $(0, 0, 0)$ to its base.

32. Show that the centroid of a right circular cone lies on the axis of the cone and three-fourths of the way from the vertex to the base.

In the following problems the indicated solid has uniform density $\rho = 1$ unless otherwise indicated.

33. Find the moment of inertia about one of its edges of a cube with edge length a.

34. The first octant cube with edge length a, faces parallel to the coordinate planes, and opposite vertices $(0, 0, 0)$ and (a, a, a) has density at $P(x, y, z)$ proportional to the square of the distance from P to the origin. Find the coordinates of its centroid.

35. Find the moment of inertia about the z-axis of the cube of Problem 34.

36. The cube bounded by the coordinate planes and the planes $x = 1$, $y = 1$, and $z = 1$ has density $\rho = kz$ at the point $P(x, y, z)$ (k is a positive constant). Find its centroid.

37. Find the moment of inertia about the z-axis of the cube of Problem 36.

38. Find the moment of inertia about a diameter of a solid sphere of radius a.

39. Find the centroid of the first octant region that is interior to the two cylinders $x^2 + z^2 = 1$ and $y^2 + z^2 = 1$.

40. Find the moment of inertia about the z-axis of the solid of Problem 39.

41. Find the volume bounded by the elliptic paraboloids $z = 2x^2 + y^2$ and $z = 12 - x^2 - 2y^2$. Note that this solid projects onto a circular disk in the xy-plane.

42. Find the volume bounded by the elliptic paraboloid $y = x^2 + 4z^2$ and the plane $y = 2x + 3$.

43. Find the volume of the elliptical cone bounded by $z = (x^2 + 4y^2)^{1/2}$ and the plane $z = 1$. (*Suggestion:* Integrate first with respect to x.)

44. Find the volume of the region bounded by the paraboloid $x = y^2 + 2z^2$ and the parabolic cylinder $x = 2 - y^2$.

16.7 Integration in Cylindrical and Spherical Coordinates

Suppose that $f(x, y)$ is a continuous function defined on the z-simple region, T, which—because it is z-simple—can be described by

$$g_1(x, y) \leqq z \leqq g_2(x, y) \qquad \text{for} \quad (x, y) \text{ in } R.$$

We saw in Section 16.6 that

$$\iiint_T f(x, y, z)\, dV = \iint_R \left(\int_{g_1(x,y)}^{g_2(x,y)} f(x, y, z)\, dz \right) dA. \tag{1}$$

If the region R is described more naturally in polar coordinates than in rectangular coordinates, then it is likely that the integration over the plane region R will be simpler if it is carried out in polar coordinates.

We first express the inner partial integral in (1) in terms of r and θ by writing

$$\int_{g_1(x,y)}^{g_2(x,y)} f(x, y, z)\, dz = \int_{G_1(r,\theta)}^{G_2(r,\theta)} F(r, \theta, z)\, dz, \tag{2}$$

where

$$F(r, \theta, z) = f(r \cos \theta, r \sin \theta, z) \tag{3a}$$

and

$$G_i(r, \theta) = g_i(r \cos \theta, r \sin \theta) \tag{3b}$$

for $i = 1, 2$. Substitution of (2) in (1) with $dA = r\, dr\, d\theta$ gives

$$\iiint_T f(x, y, z)\, dV = \iint_S \left(\int_{G_1(r,\theta)}^{G_2(r,\theta)} F(r, \theta, z)\, dz \right) r\, dr\, d\theta \tag{4}$$

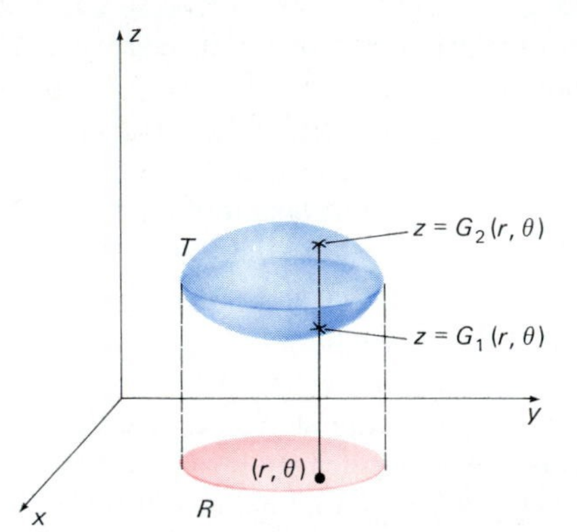

16.67 The limits on z in a triple integral in cylindrical coordinates are determined by the lower and upper surfaces.

where F, G_1, and G_2 are the functions given in (3) and S represents the appropriate limits on r and θ needed to describe the plane region R in polar coordinates (as discussed in Section 16.4). The limits on z are simply the z-coordinates (in terms of r and θ) of a typical line segment joining the lower and upper boundary surfaces of T, as indicated in Fig. 16.67.

Thus the general formula for **triple integration in cylindrical coordinates** is

$$\iiint_T f(x, y, z)\, dV = \iiint f(r \cos \theta, r \sin \theta, z)\, r\, dz\, dr\, d\theta, \tag{5}$$

with limits on z, r, and θ appropriate to describe the space region T in cylindrical coordinates. Before integrating, the variables x and y are replaced by $r \cos \theta$ and $r \sin \theta$, respectively, while z is left unchanged. The cylindrical coordinates volume element

$$dV = r\, dz\, dr\, d\theta$$

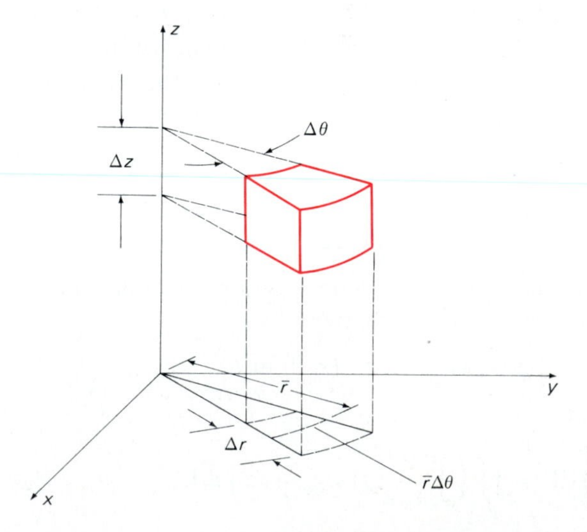

16.68 The volume of the cylindrical block is $\Delta V = \bar{r}\, \Delta z\, \Delta r\, \Delta \theta$.

may be regarded formally as the product of dz and the polar coordinates area element $dA = r\,dr\,d\theta$. It is a consequence of the formula $\Delta V = \bar{r}\,\Delta z\,\Delta r\,\Delta\theta$ for the volume of the *cylindrical block* shown in Fig. 16.68.

Integration in cylindrical coordinates is particularly useful for computations associated with solids of revolution. The solid should be placed so that the axis of revolution is the z-axis.

EXAMPLE 1 Find the centroid of the first-octant part of the solid ball bounded by the sphere $r^2 + z^2 = a^2$.

Solution The volume of the first octant of the solid ball is $V = \frac{1}{8}(\frac{4}{3}\pi a^3) = \frac{1}{6}\pi a^3$. Because $\bar{x} = \bar{y} = \bar{z}$ by symmetry, we need only calculate

$$\begin{aligned}\bar{z} &= \frac{1}{V}\iiint z\,dV = \frac{6}{\pi a^3}\int_0^{\pi/2}\int_0^a\int_0^{\sqrt{a^2-r^2}} zr\,dz\,dr\,d\theta \\ &= \frac{6}{\pi a^3}\int_0^{\pi/2}\int_0^a \frac{r}{2}(a^2 - r^2)\,dr\,d\theta \\ &= \frac{3}{\pi a^3}\int_0^{\pi/2}\left[\tfrac{1}{2}a^2r^2 - \tfrac{1}{4}r^4\right]_0^a d\theta \\ &= \frac{3}{\pi a^3}\cdot\frac{\pi}{2}\cdot\frac{a^4}{4} = \frac{3}{8}a.\end{aligned}$$

Thus the centroid is located at the point $(3a/8, 3a/8, 3a/8)$. Note that the answer is both plausible and dimensionally correct.

EXAMPLE 2 Find the volume and centroid of the solid T bounded by the paraboloid $z = b(x^2 + y^2)$ $(b > 0)$ and the plane $z = h$ $(h > 0)$.

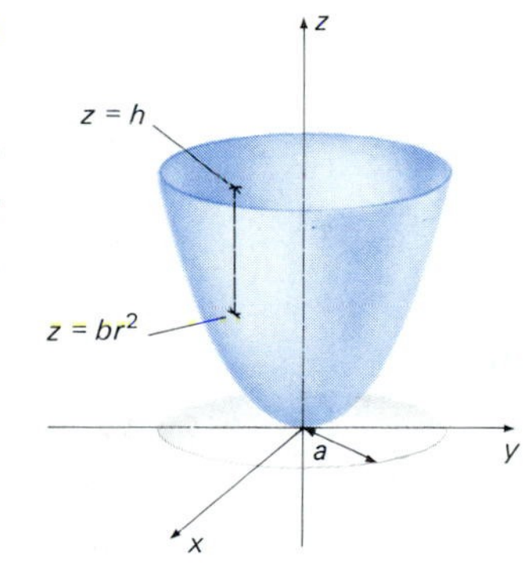

16.69 The paraboloid of Example 2

Solution Figure 16.69 makes it clear that we get the radius of the circular top by equating $z = b(x^2 + y^2) = br^2$ and $z = h$. This gives $a = \sqrt{h/b}$ for the radius of the circle over which the solid lies. Hence the formula in (4), with $f(x, y, z) \equiv 1$, gives the volume:

$$\begin{aligned}V &= \iiint_T dV = \int_0^{2\pi}\int_0^a\int_{br^2}^h r\,dz\,dr\,d\theta = \int_0^{2\pi}\int_0^a (hr - br^3)\,dr\,d\theta \\ &= 2\pi(\tfrac{1}{2}ha^2 - \tfrac{1}{4}ba^4) = \frac{\pi h^2}{2b} = \tfrac{1}{2}\pi a^2 h\end{aligned}$$

(because $a^2 = h/b$).

By symmetry, the centroid of T lies on the z-axis, so all that remains is the computation of $\bar{z}$:

$$\begin{aligned}\bar{z} &= \frac{1}{V}\iiint_T z\,dV = \frac{2}{\pi a^2 h}\int_0^{2\pi}\int_0^a\int_{br^2}^h rz\,dz\,dr\,d\theta \\ &= \frac{2}{\pi a^2 h}\int_0^{2\pi}\int_0^a (\tfrac{1}{2}h^2 r - \tfrac{1}{2}b^2 r^5)\,dr\,d\theta \\ &= \frac{4}{a^2 h}(\tfrac{1}{4}h^2a^2 - \tfrac{1}{12}b^2a^6) = \tfrac{2}{3}h,\end{aligned}$$

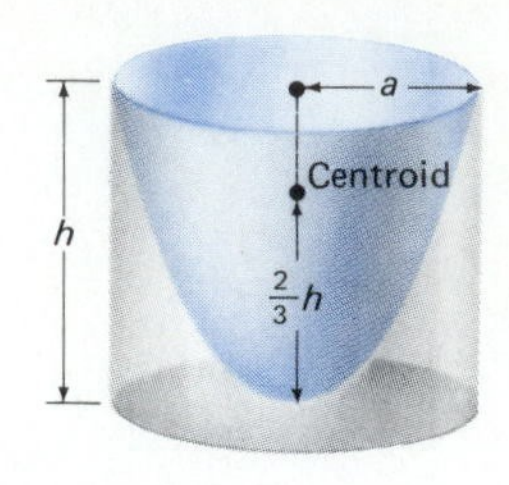

16.70 Volume and centroid of a right circular paraboloid in terms of the circumscribed cylinder

again using $a^2 = h/b$. Therefore, the centroid of T is located at the point $(0, 0, 2h/3)$. Note that this answer is both plausible and dimensionally correct.

The results of Example 2 can be summarized as follows: The volume of a right circular paraboloid is *half* that of the circumscribed cylinder (see Fig. 16.70), and its centroid lies on its axis of symmetry *two-thirds* of the way from the "vertex" at (0, 0, 0) to the top.

SPHERICAL COORDINATES

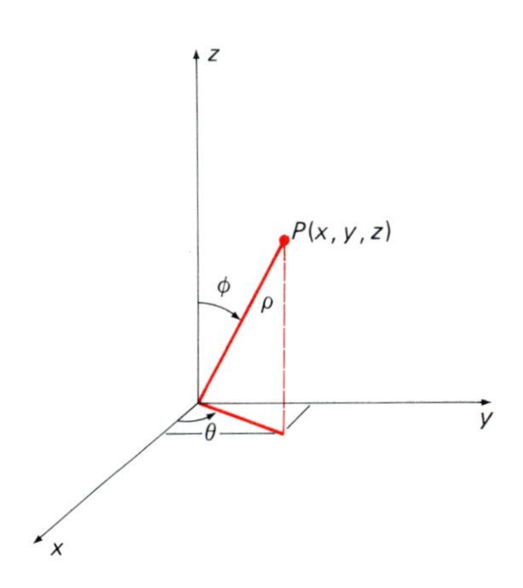

16.71 The spherical coordinates (ρ, ϕ, θ) of the point P

When the boundary surfaces of the region T of integration are spheres, cones, or other surfaces with simple descriptions in spherical coordinates, it is generally advantageous to transform the integral into spherical coordinates. Recall from Section 14.7 that the relationship between spherical coordinates (ρ, ϕ, θ) (shown in Fig. 16.71) and rectangular coordinates (x, y, z) is given by

$$x = \rho \sin \phi \cos \theta, \qquad y = \rho \sin \phi \sin \theta, \qquad z = \rho \cos \phi. \tag{6}$$

Suppose, for example, that T is the **spherical block** determined by the inequalities

$$\begin{aligned} \rho_1 &\leqq \rho \leqq \rho_2 = \rho_1 + \Delta\rho, \\ \phi_1 &\leqq \phi \leqq \phi_2 = \phi_1 + \Delta\phi, \\ \theta_1 &\leqq \theta \leqq \theta_2 = \theta_1 + \Delta\theta. \end{aligned} \tag{7}$$

As indicated by the dimensions labeled in Fig. 16.72, this spherical block is (if $\Delta\rho$, $\Delta\phi$, and $\Delta\theta$ are small) *approximately* a rectangular block with dimensions $\Delta\rho$, $\rho_1\,\Delta\phi$, and $\rho_1 \sin \phi_2\,\Delta\theta$. Thus its volume is approximately

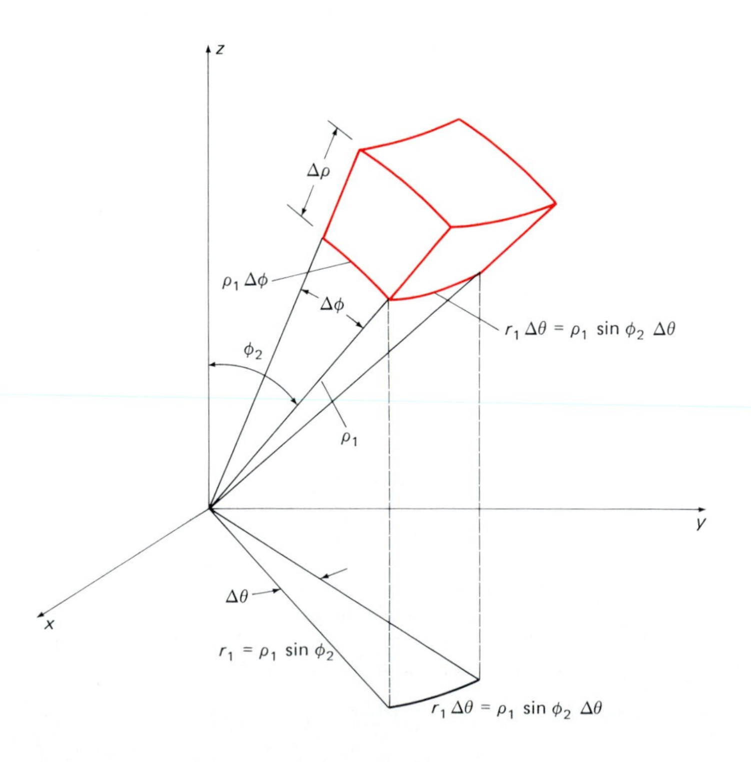

16.72 The approximate volume of the spherical block is $\rho_1^2 \sin \phi_2\,\Delta\rho\,\Delta\phi\,\Delta\theta$.

$\rho_1^2 \sin \phi_2 \, \Delta\rho \, \Delta\phi \, \Delta\theta$. It can be shown (see Problem 19 in Section 16.8) that the *exact* volume of the spherical block described in (7) is

$$\Delta V = \hat{\rho}^2 \sin \hat{\phi} \, \Delta\rho \, \Delta\phi \, \Delta\theta \tag{8}$$

for certain numbers $\hat{\rho}$ and $\hat{\phi}$ such that $\rho_1 < \hat{\rho} < \rho_2$ and $\phi_1 < \hat{\phi} < \phi_2$.

Now suppose that we partition each of the intervals $[\rho_1, \rho_2]$, $[\phi_1, \phi_2]$, and $[\theta_1, \theta_2]$ into n equal subintervals with lengths

$$\Delta\rho = \frac{\rho_2 - \rho_1}{n}, \qquad \Delta\phi = \frac{\phi_2 - \phi_1}{n}, \quad \text{and} \quad \Delta\theta = \frac{\theta_2 - \theta_1}{n},$$

respectively. This produces a **spherical partition** $\mathscr{P}$ of the spherical block T into $k = n^3$ smaller spherical blocks $T_1, T_2, T_3, \ldots, T_k$. By the formula in (8), there exists a point $(\hat{\rho}_i, \hat{\phi}_i, \hat{\theta}_i)$ of the spherical block T_i such that its volume is $\Delta V_i = \hat{\rho}_i^2 \sin \hat{\phi}_i \, \Delta\rho \, \Delta\phi \, \Delta\theta$. The mesh $|\mathscr{P}|$ of $\mathscr{P}$ is the length of the longest diagonal of any of the small spherical blocks $T_1, T_2, T_3, \ldots, T_k$.

If (x_i^*, y_i^*, z_i^*) are the rectangular coordinates of the point with spherical coordinates $(\hat{\rho}_i, \hat{\phi}_i, \hat{\theta}_i)$, then the definition of the triple integral as a limit of Riemann sums as the mesh $|\mathscr{P}|$ approaches zero gives

$$\begin{aligned} \iiint_T f(x, y, z) \, dV &= \lim_{|\mathscr{P}| \to 0} \sum_{i=1}^{k} f(x_i^*, y_i^*, z_i^*) \, \Delta V_i \\ &= \lim_{|\mathscr{P}| \to 0} \sum_{i=1}^{k} F(\hat{\rho}_i, \hat{\phi}_i, \hat{\theta}_i) \hat{\rho}_i^2 \sin \hat{\phi}_i \, \Delta\rho \, \Delta\phi \, \Delta\theta, \end{aligned} \tag{9}$$

where

$$F(\rho, \phi, \theta) = f(\rho \sin \phi \cos \theta, \rho \sin \phi \sin \theta, \rho \cos \phi) \tag{10}$$

is the result of substituting (6) in $F(x, y, z)$. But the right-hand sum in (9) is simply a Riemann sum for the triple integral

$$\int_{\theta_1}^{\theta_2} \int_{\phi_1}^{\phi_2} \int_{\rho_1}^{\rho_2} F(\rho, \phi, \theta) \rho^2 \sin \phi \, d\rho \, d\phi \, d\theta.$$

It therefore follows that

$$\iiint_T f(x, y, z) \, dV = \int_{\theta_1}^{\theta_2} \int_{\phi_1}^{\phi_2} \int_{\rho_1}^{\rho_2} F(\rho, \phi, \theta) \rho^2 \sin \phi \, d\rho \, d\phi \, d\theta. \tag{11}$$

Thus we transform the integral $\iiint_T f(x, y, z) \, dV$ into spherical coordinates by replacing the rectangular coordinate variables x, y, and z by their expressions in (6) in terms of the spherical coordinate variables ρ, ϕ, and θ. In addition, we write

$$dV = \rho^2 \sin \phi \, d\rho \, d\phi \, d\theta$$

for the volume element in spherical coordinates.

More generally, we can transform the triple integral $\iiint_T f(x, y, z) \, dV$ into spherical coordinates whenever the region T is **centrally simple;** that is, whenever it has a spherical coordinates description of the form

$$G_1(\phi, \theta) \leqq \rho \leqq G_2(\phi, \theta), \qquad \phi_1 \leqq \phi \leqq \phi_2, \quad \theta_1 \leqq \theta \leqq \theta_2. \tag{12}$$

If so, then

$$\iiint_T f(x, y, z)\, dV = \int_{\theta_1}^{\theta_2} \int_{\phi_1}^{\phi_2} \int_{G_1(\phi,\theta)}^{G_2(\phi,\theta)} F(\rho, \phi, \theta)\rho^2 \sin\phi\, d\rho\, d\phi\, d\theta. \quad (13)$$

The limits on ρ are simply the ρ-coordinates (in terms of ϕ and θ) of the endpoints of a typical radial segment joining the "inner" and "outer" parts of the boundary of T (see Fig. 16.73). Thus the general formula for **triple integration in spherical coordinates** is

$$\iiint_T f(x, y, z)\, dV$$

$$= \iiint f(\rho \sin\phi \cos\theta, \rho \sin\phi \sin\theta, \rho \cos\phi)\rho^2 \sin\phi\, d\rho\, d\phi\, d\theta, \quad (14)$$

with limits on ρ, ϕ, and θ appropriate to describe the region T in spherical coordinates.

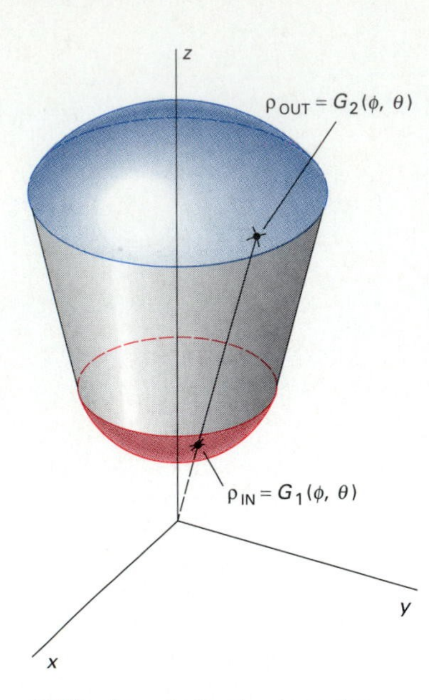

16.73 A centrally simple region

EXAMPLE 3 A solid ball T with constant density δ is bounded by the sphere $\rho = a$. Use spherical coordinates to compute its volume V and its moment of inertia I_z about the z-axis.

Solution The points of the ball are described by the inequalities

$$0 \leqq \rho \leqq a, \qquad 0 \leqq \phi \leqq \pi, \qquad 0 \leqq \theta \leqq 2\pi.$$

We take $f = F \equiv 1$ in the formula in (11) and thereby obtain

$$V = \iiint_T dV = \int_0^{2\pi} \int_0^{\pi} \int_0^{a} \rho^2 \sin\phi\, d\rho\, d\phi\, d\theta.$$

$$= \tfrac{1}{3}a^3 \int_0^{2\pi} \int_0^{\pi} \sin\phi\, d\phi\, d\theta = \tfrac{1}{3}a^3 \int_0^{2\pi} \Big[-\cos\phi\Big]_0^{\pi} d\theta$$

$$= \tfrac{2}{3}a^3 \int_0^{2\pi} d\theta = \tfrac{4}{3}\pi a^3.$$

The distance from the typical point (ρ, ϕ, θ) of the sphere to the z-axis is $r = \rho \sin\phi$, so the moment of inertia of the sphere about that axis is

$$I_z = \iiint_T r^2 \delta\, dV = \int_0^{2\pi} \int_0^{\pi} \int_0^{a} \delta \rho^4 \sin^3\phi\, d\rho\, d\phi\, d\theta$$

$$= \tfrac{1}{5}\delta a^5 \int_0^{2\pi} \int_0^{\pi} \sin^3\phi\, d\phi\, d\theta = \tfrac{2}{5}\pi\delta a^5 \int_0^{\pi} \sin^3\phi\, d\phi$$

$$= \tfrac{2}{5}\pi\delta a^5 \cdot 2 \cdot \tfrac{2}{3} = \tfrac{2}{5}ma^2$$

where $m = \frac{4}{3}\pi a^3 \delta$ is the mass of the ball. The answer is dimensionally correct because it is the product of mass and the square of a distance, and it is plausible because it implies that, for purposes of rotational inertia, the sphere acts as if its mass were concentrated 63% of the way from the axis to the equator.

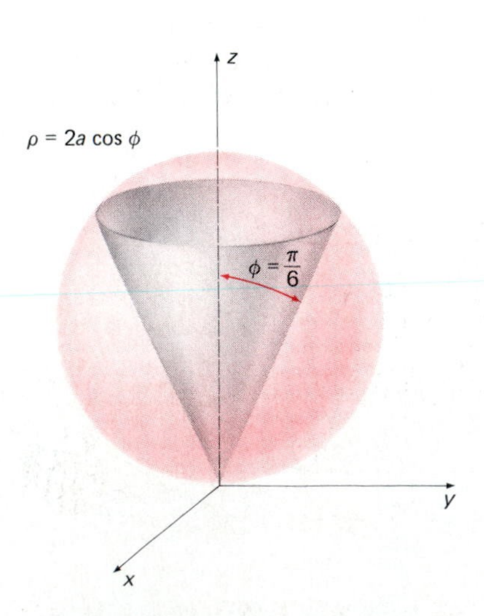

16.74 The ice cream cone of Example 4 is the part of the cone that lies within the sphere.

EXAMPLE 4 Find the volume and centroid of the "ice cream cone" C bounded by the cone $\phi = \pi/6$ and the sphere $\rho = 2a \cos\phi$ of radius a and tangent to the xy-plane at the origin. The sphere and the part of the cone within it are shown in Fig. 16.74.

Solution The ice cream cone is described by the inequalities

$$0 \leqq \theta \leqq 2\pi, \qquad 0 \leqq \phi \leqq \pi/6, \qquad 0 \leqq \rho \leqq 2a \cos \phi.$$

Using the formula in (13) to compute its volume, we get

$$\begin{aligned} V &= \int_0^{2\pi} \int_0^{\pi/6} \int_0^{2a \cos \phi} \rho^2 \sin \phi \, d\rho \, d\phi \, d\theta \\ &= \tfrac{8}{3}a^3 \int_0^{2\pi} \int_0^{\pi/6} \cos^3 \phi \sin \phi \, d\phi \, d\theta \\ &= \tfrac{16}{3}\pi a^3 \Big[-\tfrac{1}{4} \cos^4 \phi \Big]_0^{\pi/6} = \tfrac{7}{12}\pi a^3. \end{aligned}$$

Now for the centroid. It is clear by symmetry that $\bar{x} = 0 = \bar{y}$. Because $z = \rho \cos \phi$, the z-coordinate of its centroid is

$$\bar{z} = \frac{1}{V} \iiint_C z \, dV = \frac{12}{7\pi a^3} \int_0^{2\pi} \int_0^{\pi/6} \int_0^{2a \cos \phi} \rho^3 \cos \phi \sin \phi \, d\rho \, d\phi \, d\theta$$

$$= \frac{48a}{7\pi} \int_0^{2\pi} \int_0^{\pi/6} \cos^5 \phi \sin \phi \, d\phi \, d\theta = \frac{96a}{7} \left[-\frac{1}{6} \cos^6 \phi \right]_0^{\pi/6} = \frac{37a}{28}.$$

Hence the centroid of the ice cream cone is located at the point $(0, 0, 37a/28)$.

16.7 PROBLEMS

Solve Problems 1–20 by triple integration in cylindrical coordinates.

1. Find the volume of the solid bounded above by the plane $z = 4$ and below by the paraboloid $z = r^2$.

2. Find the centroid of the solid of Problem 1.

3. Derive the formula $V = 4\pi a^3/3$ for the volume of a sphere of radius a.

4. Find the moment of inertia about the z-axis of the solid sphere of Problem 3 under the assumption that it has unit density.

5. Find the volume of the region that lies inside both the sphere $x^2 + y^2 + z^2 = 4$ and the cylinder $x^2 + y^2 = 1$.

6. Find the centroid of the portion $z \geqq 0$ of the region of Problem 5.

7. Find the mass of the cylinder $0 \leqq r \leqq a$, $0 \leqq z \leqq h$ if its density at (x, y, z) is z.

8. Find the centroid of the cylinder of Problem 7.

9. Find the moment of inertia about the z-axis of the cylinder of Problem 7.

10. Find the volume of the region that lies inside both the sphere $x^2 + y^2 + z^2 = 4$ and the cylinder $x^2 + y^2 - 2x = 0$.

11. Find the volume and centroid of the region bounded by the plane $z = 0$ and the paraboloid $z = 9 - x^2 - y^2$.

12. Find the volume and centroid of the region bounded by the paraboloids $z = x^2 + y^2$ and $z = 12 - 2x^2 - 2y^2$.

13. Find the volume of the region bounded by the paraboloids $z = 2x + y^2$ and $z = 12 - x^2 - 2y^2$.

14. Find the volume of the region bounded below by the paraboloid $z = x^2 + y^2$ and above by the plane $z = 2x$.

15. Find the volume of the region bounded above by the spherical surface $x^2 + y^2 + z^2 = 2$ and below by the paraboloid $z = x^2 + y^2$.

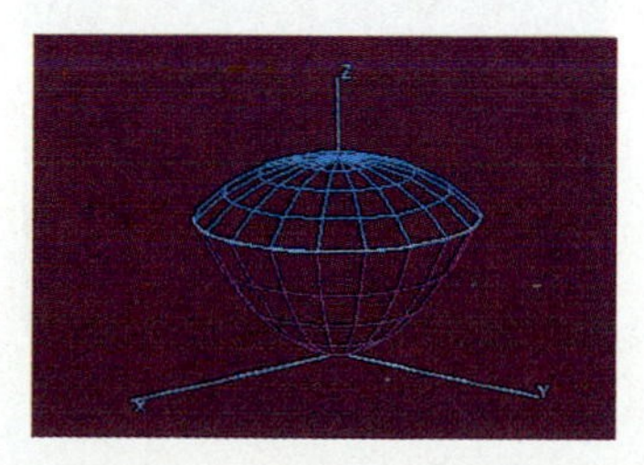

The solid of Problem 15

16. A homogeneous solid cylinder has mass m and radius a. Show that its moment of inertia about its axis of symmetry is $\frac{1}{2}ma^2$.

17. Find the moment of inertia I of a homogeneous solid cylinder about a diameter of its base. Express I in terms of the radius a, the height h, and the (constant) density δ of the cylinder.

18. Find the centroid of a homogeneous solid hemisphere of radius a.

19. Find the volume of the region bounded by the plane $z = 1$ and the cone $z = r$.

20. Show that the centroid of a homogeneous solid circular cone lies on its axis three-quarters of the way from the vertex to its base.

Solve Problems 21–30 by triple integration in spherical coordinates.

21. Find the centroid of a homogeneous solid hemisphere of radius a.

22. Find the mass and centroid of a solid hemisphere of radius a if its density δ is proportional to distance z from its base—so that $\delta = kz$ (where k is a positive constant).

23. Solve Problem 19 by triple integration in spherical coordinates.

24. Solve Problem 20 by triple integration in spherical coordinates.

25. Find the volume and centroid of the solid that lies within the sphere $\rho = a$ and above the cone $r = z$.

26. Find the moment of inertia I_z of the solid of Problem 25 under the assumption that it has constant density δ.

27. Find the moment of inertia about a tangent line of a solid homogeneous sphere of radius a and total mass m.

28. A spherical shell of mass m is bounded by the spheres $\rho = a$ and $\rho = 2a$, and its density function is $\delta = \rho^2$. Find its moment of inertia about a diameter.

29. Describe the surface $\rho = 2a \sin \phi$, and compute the volume of the region it bounds.

30. Describe the surface $\rho = 1 + \cos \phi$, and compute the volume of the region it bounds.

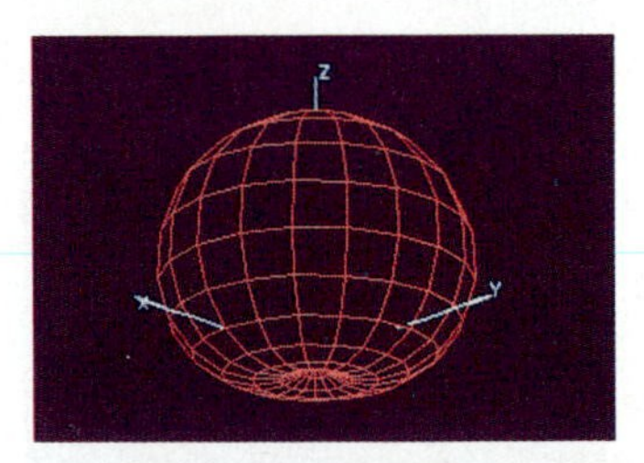

The surface of Problem 30

31. Find the moment of inertia about the x-axis of the region that lies inside both the cylinder $r = a$ and the sphere $\rho = 2a$.

32. Find the moment of inertia about the z-axis of the ice cream cone of Example 4.

33. Find the mass and centroid of the ice cream cone of Example 4 if its density is given by $\delta = z$.

34. Consider a homogeneous spherical ball of radius a centered at the origin, with density δ and mass $M = \frac{4}{3}\pi a^3 \delta$. Show that the gravitational force $\mathbf{F}$ exerted by this ball upon a point mass located at the point $(0, 0, c)$, with $c > a$ (Fig. 16.75), is the same as though all the mass of the ball were concentrated at its center $(0, 0, 0)$. That is, show that $|\mathbf{F}| = GMm/c^2$. (*Suggestion:* By symmetry you may assume that the force is vertical, so that $\mathbf{F} = F_z\mathbf{k}$. Set up the integral

$$F_z = -\int_0^{2\pi} \int_0^{a} \int_0^{\pi} \frac{GM\delta \cos \alpha}{w^2} \rho^2 \sin \phi \, d\phi \, d\rho \, d\theta$$

Change the first variable of integration from ϕ to w by using the law of cosines:

$$w^2 = \rho^2 + c^2 - 2\rho c \cos \phi.$$

Then $2w\,dw = 2\rho c \sin \phi \, d\phi$ and $w \cos \alpha + \rho \cos \phi = c$. (Why?))

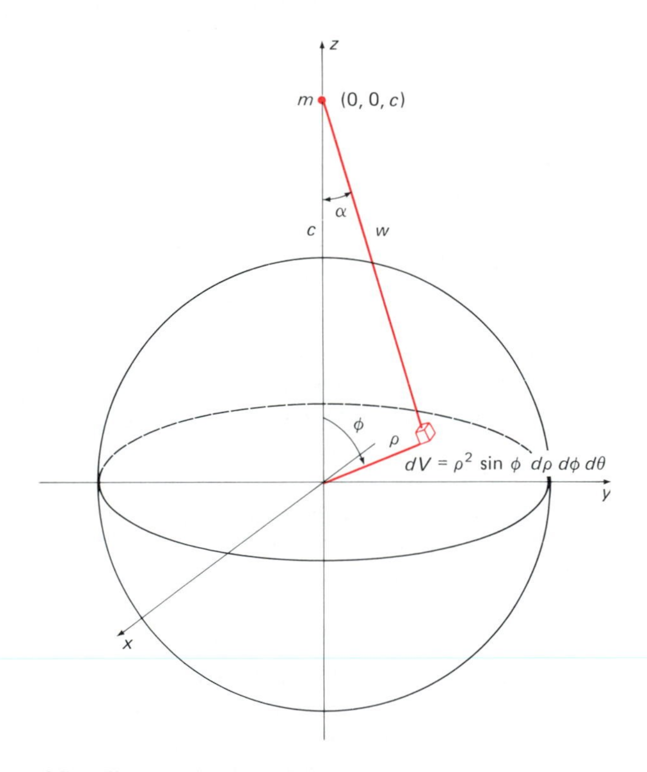

16.75 The system of Problem 34

35. Consider now the spherical shell $a \leqq r \leqq b$ with uniform density δ. Show that this shell exerts *no* force on a point mass m located at the point $(0, 0, c)$ *inside* it—that is, with $|c| < a$. The computation will be the same as in Problem 34 except for the limits of integration on ρ and w.

16.8 Surface Area

Until now our concept of a surface has been the graph $z = f(x, y)$ of a function of two variables. Occasionally, we have seen such a surface defined implicitly by an equation of the form $F(x, y, z) = 0$. Now we want to introduce the more precise concept of a parametric surface—the two-dimensional analogue of a parametric curve.

A **parametric surface** S is the image of a function or transformation $\mathbf{r}$ defined on a region R in the uv-plane and with values in xyz-space. The **image** under $\mathbf{r}$ of each point (u, v) in R is the point in xyz-space with position vector

$$\mathbf{r}(u, v) = \langle x(u, v), y(u, v), z(u, v) \rangle. \tag{1}$$

We shall assume throughout this section that the component functions of $\mathbf{r}$ have continuous partial derivatives with respect to u and v, and also that the vectors

$$\mathbf{r}_u = \frac{\partial \mathbf{r}}{\partial u} = \langle x_u, y_u, z_u \rangle = \frac{\partial x}{\partial u}\mathbf{i} + \frac{\partial y}{\partial u}\mathbf{j} + \frac{\partial z}{\partial u}\mathbf{k} \tag{2}$$

and

$$\mathbf{r}_v = \frac{\partial \mathbf{r}}{\partial v} = \langle x_v, y_v, z_v \rangle = \frac{\partial x}{\partial v}\mathbf{i} + \frac{\partial y}{\partial v}\mathbf{j} + \frac{\partial z}{\partial v}\mathbf{k} \tag{3}$$

are nonzero and nonparallel at each interior point of R. (Compare this with the definition of *smooth* parametric curve $\mathbf{r}(t)$ in Section 13.1.)

We call the variables u and v the *parameters* for the surface S (analogous to the single parameter t for a parametric curve). The graph $z = f(x, y)$ of a function may be regarded as a parametric surface with parameters x and y. In this case the transformation $\mathbf{r}$ from the xy-plane to xyz-space has the component functions

$$x = x, \qquad y = y, \qquad z = f(x, y). \tag{4}$$

Similarly, a surface given in cylindrical coordinates as the graph of $z = g(r, \theta)$ may be regarded as a parametric surface with parameters r and θ. The transformation $\mathbf{r}$ from the $r\theta$-plane to xyz-space is then given by

$$x = r \cos \theta, \qquad y = r \sin \theta, \qquad z = g(r, \theta). \tag{5}$$

A surface given in spherical coordinates by $\rho = h(\phi, \theta)$ may be regarded as a parametric surface with parameters ϕ and θ, and the corresponding transformation from the $\phi\theta$-plane to xyz-space is then given by

$$x = h(\phi, \theta) \sin \phi \cos \theta, \qquad y = h(\phi, \theta) \sin \phi \sin \theta, \qquad z = h(\phi, \theta) \cos \phi. \tag{6}$$

The concept of a parametric surface lets us treat all these special cases, and many others, with the same techniques.

Now we want to define the *surface area* of the general parametric surface given in Equation (1). We begin with an inner partition of the region R—the domain of $\mathbf{r}$ in the uv-plane—into rectangles $R_1, R_2, \ldots, R_n$, each with dimensions Δu and Δv. Let (u_i, v_i) be the lower left-hand corner of R_i (as in Fig. 16.76). The image S_i of R_i under $\mathbf{r}$ will not generally be a rectangle in xyz-space; it will look more like a *curvilinear figure* on the image surface S, with $\mathbf{r}(u_i, v_i)$ as one "vertex." (See Fig. 16.77). Let ΔS_i denote the area of this curvilinear figure S_i.

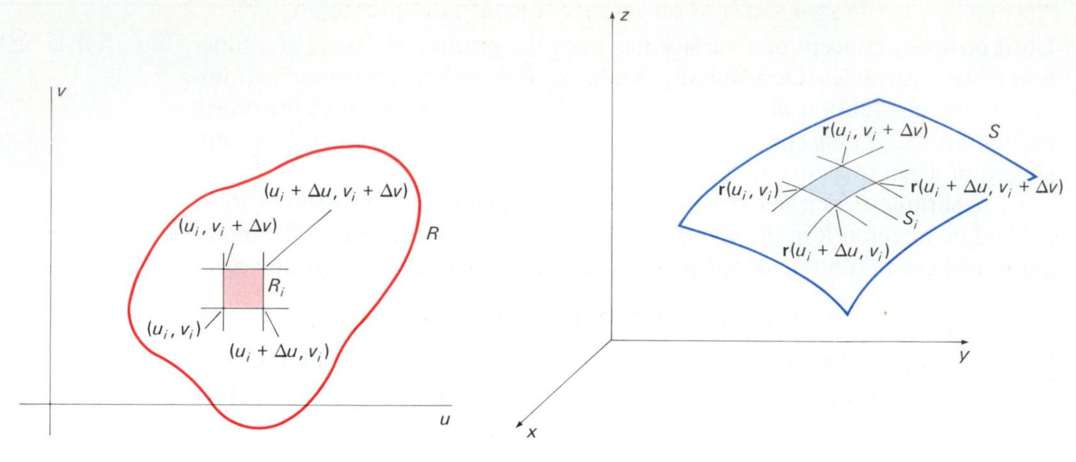

16.76 The rectangle R_i in the uv-plane

16.77 The image of R_i is a curvilinear figure.

The parametric curves $\mathbf{r}(u, v_i)$ and $\mathbf{r}(u_i, v)$—with parameters u and v, respectively—lie on the surface S and meet at the point $\mathbf{r}(u_i, v_i)$. At this point of intersection, these two curves have the tangent vectors $\mathbf{r}_u(u_i, v_i)$ and $\mathbf{r}_v(u_i, v_i)$ shown in Fig. 16.78. Hence their cross product

$$\mathbf{N}(u_i, v_i) = \mathbf{r}_u(u_i, v_i) \times \mathbf{r}_v(u_i, v_i) \tag{7}$$

is a normal vector to S at the point $\mathbf{r}(u_i, v_i)$.

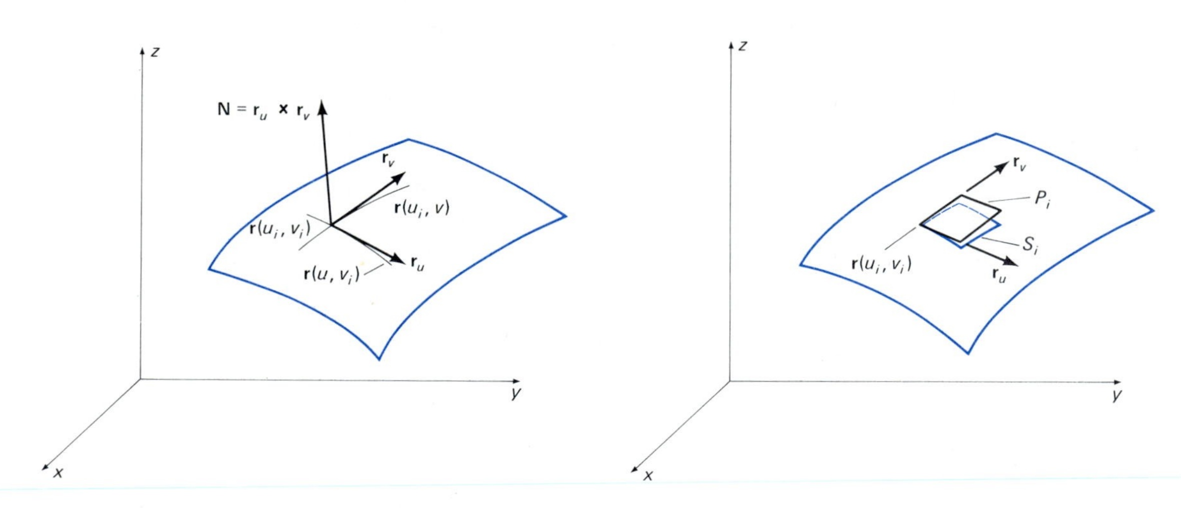

16.78 The normal **N** to the surface at $\mathbf{r}(u_i, v_i)$

16.79 The area of the actual parallelogram P_i is an approximation to the area of the curvilinear figure S_i.

Now suppose that Δu and Δv are both small. Then the area ΔS_i of the curvilinear figure S_i will be approximately equal to the area ΔP_i of the parallelogram with adjacent sides $\mathbf{r}_u(u_i, v_i)\,\Delta u$ and $\mathbf{r}_v(u_i, v_i)\,\Delta v$ (see Fig. 16.79). But the area of this parallelogram is

$$\Delta P_i = |\mathbf{r}_u(u_i, v_i)\,\Delta u \times \mathbf{r}_v(u_i, v_i)\,\Delta v| = |\mathbf{N}(u_i, v_i)|\,\Delta u\,\Delta v.$$

This means that the area $a(S)$ of the surface S is given approximately by

$$a(S) = \sum_{i=1}^{n} \Delta S_i \approx \sum_{i=1}^{n} \Delta P_i,$$

so that

$$a(S) \approx \sum_{i=1}^{n} |\mathbf{N}(u_i, v_i)| \, \Delta u \, \Delta v.$$

But this last sum is a Riemann sum for the double integral $\iint_R |\mathbf{N}(u, v)| \, du \, dv$. We are therefore motivated to *define* the **surface area** of the parametric surface S by

$$A = a(S) = \iint_R |\mathbf{N}(u, v)| \, du \, dv = \iint_R \left| \frac{\partial \mathbf{r}}{\partial u} \times \frac{\partial \mathbf{r}}{\partial v} \right| du \, dv. \tag{8}$$

In the case of the surface $z = f(x, y)$, for (x, y) in the region R in the xy-plane, the component functions of $\mathbf{r}$ are given by the equations in (4) with parameters x and y (in place of u and v). Then

$$\mathbf{N} = \frac{\partial \mathbf{r}}{\partial x} \times \frac{\partial \mathbf{r}}{\partial y} = \begin{vmatrix} \mathbf{i} & \mathbf{j} & \mathbf{k} \\ 1 & 0 & \dfrac{\partial f}{\partial x} \\ 0 & 1 & \dfrac{\partial f}{\partial y} \end{vmatrix}$$

$$= -\frac{\partial f}{\partial x}\mathbf{i} - \frac{\partial f}{\partial y}\mathbf{j} + \mathbf{k},$$

so the formula in (8) takes the special form

$$A = a(S) = \iint_R \sqrt{1 + \left(\frac{\partial f}{\partial x}\right)^2 + \left(\frac{\partial f}{\partial y}\right)^2} \, dx \, dy$$

$$= \iint_R \sqrt{1 + z_x^2 + z_y^2} \, dx \, dy. \tag{9}$$

EXAMPLE 1 Find the area of the ellipse cut from the plane $z = 2x + 2y + 1$ by the cylinder $x^2 + y^2 = 1$.

Solution Here, R is the unit circle in the xy-plane with area $\iint_R 1 \, dx \, dy = \pi$, so the formula in (9) gives area

$$A = \iint_R \sqrt{1 + z_x^2 + z_y^2} \, dx \, dy = \iint_R \sqrt{1 + 2^2 + 2^2} \, dx \, dy = \iint_R 3 \, dx \, dy = 3\pi.$$

Now consider a cylindrical coordinates surface $z = g(r, \theta)$ parametrized by the equations in (5), for (r, θ) in a region R of the $r\theta$-plane. Then the normal vector is

$$\mathbf{N} = \frac{\partial \mathbf{r}}{\partial r} \times \frac{\partial \mathbf{r}}{\partial \theta} = \begin{vmatrix} \mathbf{i} & \mathbf{j} & \mathbf{k} \\ \cos\theta & \sin\theta & \dfrac{\partial z}{\partial r} \\ -r\sin\theta & r\cos\theta & \dfrac{\partial z}{\partial \theta} \end{vmatrix}$$

$$= \mathbf{i}\left(\frac{\partial z}{\partial \theta}\sin\theta - r\frac{\partial z}{\partial r}\cos\theta\right) - \mathbf{j}\left(\frac{\partial z}{\partial \theta}\cos\theta + r\frac{\partial z}{\partial r}\sin\theta\right) + r\mathbf{k}.$$

After some simplifications, we find that

$$|\mathbf{N}| = \sqrt{r^2 + r^2\left(\frac{\partial z}{\partial r}\right)^2 + \left(\frac{\partial z}{\partial \theta}\right)^2}.$$

Thus the formula in (8) yields the formula

$$A = \iint_R \sqrt{r^2 + (rz_r)^2 + (z_\theta)^2}\, dr\, d\theta \tag{10}$$

for surface area in cylindrical coordinates.

EXAMPLE 2 Find the surface area cut from the paraboloid $z = r^2$ by the cylinder $r = 1$.

Solution The formula in (10) gives area

$$A = \int_0^{2\pi}\int_0^1 \sqrt{r^2 + r^2(2r)^2}\, dr\, d\theta = 2\pi\int_0^1 r\sqrt{1 + 4r^2}\, dr$$

$$= 2\pi\left[\tfrac{2}{3}\cdot\tfrac{1}{8}(1 + 4r^2)^{3/2}\right]_0^1 = \frac{\pi}{6}(5\sqrt{5} - 1) \approx 5.3304.$$

In Example 2, you would get the same result if you first wrote $z = x^2 + y^2$, used Equation (9), which gives

$$A = \iint_R \sqrt{1 + 4x^2 + 4y^2}\, dx\, dy,$$

and then changed to polar coordinates. In the next example it would be less convenient to begin with rectangular coordinates.

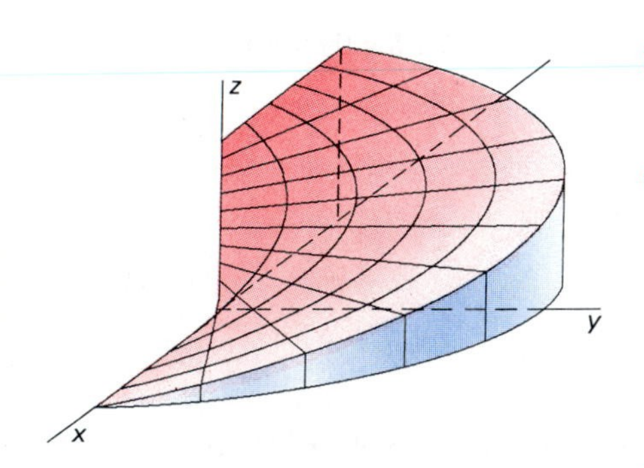

16.80 The spiral ramp of Example 3

EXAMPLE 3 Find the area of the *spiral ramp* $z = \theta$, $0 \leqq r \leqq 1$, $0 \leqq \theta \leqq \pi$. This is the upper surface of the solid shown in Fig. 16.80.

Solution The formula in Equation (10) gives

$$A = \int_0^{\pi}\int_0^1 \sqrt{r^2 + 1}\, dr\, d\theta = \frac{\pi}{2}[\sqrt{2} + \ln(1 + \sqrt{2})] \approx 3.6059;$$

we avoided a trigonometric substitution by using the table of integrals inside the front cover.

EXAMPLE 4 Find the surface area of the torus generated by revolving the circle $(x - b)^2 + z^2 = a^2$ $(a < b)$ in the xz-plane around the z-axis.

Solution With the ordinary polar coordinate θ and the angle ψ of Fig. 16.81, the torus is described for $0 \leqq \theta \leqq 2\pi$ and $0 \leqq \psi \leqq 2\pi$ by the parametric equations

$$x = r\cos\theta = (b + a\cos\psi)\cos\theta,$$
$$y = r\sin\theta = (b + a\cos\psi)\sin\theta,$$
$$z = a\sin\psi.$$

When we compute $\mathbf{N} = \mathbf{r}_\theta \times \mathbf{r}_\psi$ and simplify, we find that

$$|\mathbf{N}| = a(b + a\cos\psi).$$

Hence the general surface area formula (Equation (8)) gives area

$$A = \int_0^{2\pi}\int_0^{2\pi} a(b + a\cos\psi)\,d\theta\,d\psi = 2\pi a\Big[b\psi + a\sin\psi\Big]_0^{2\pi} = 4\pi^2 ab.$$

We obtained the same result in Section 16.5 with the aid of Pappus's theorem.

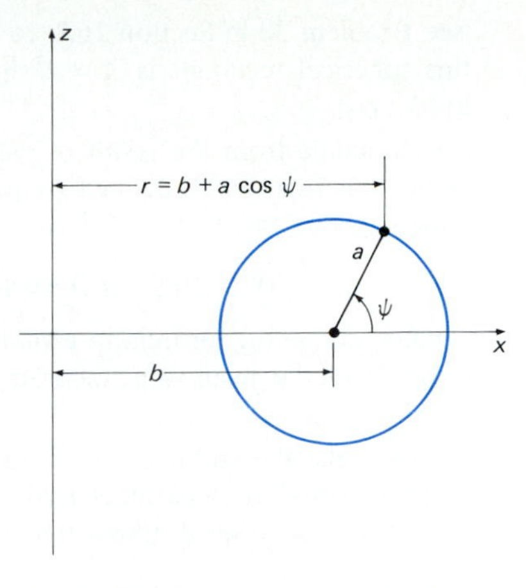

16.81 The circle that generates the torus of Example 4

16.8 PROBLEMS

1. Find the area of the portion of the plane $z = x + 3y$ that lies inside the elliptical cylinder with equation $x^2/4 + y^2/9 = 1$.

2. Find the area of the region in the plane $y = 1 + 2x + 2y$ that lies directly above the region in the xy-plane bounded by the parabolas $y = x^2$ and $x = y^2$.

3. Find the area of the part of the paraboloid $z = 9 - x^2 - y^2$ that lies above the plane $z = 5$.

4. Find the area of the part of the surface $2z = x^2$ that lies directly above the triangle in the xy-plane with vertices at $(0, 0)$, $(1, 0)$, and $(1, 1)$.

5. Find the area of the surface that is the graph of $z = x + y^2$ for $0 \leqq x \leqq 1$, $0 \leqq y \leqq 2$.

6. Find the area of that part of the surface of Problem 5 that lies above the triangle in the xy-plane with vertices at $(0, 0)$, $(0, 1)$, and $(1, 1)$.

7. Find by integration the area of the part of the plane $2x + 3y + z = 6$ that lies in the first octant.

8. Find the area of the ellipse that is cut from the plane of Problem 7 by the cylinder $x^2 + y^2 = 2$.

9. Find the area that is cut from the saddle-shaped surface $z = xy$ by the cylinder $x^2 + y^2 = 1$.

10. Find the area that is cut from the surface $z = x^2 - y^2$ by the cylinder $x^2 + y^2 = 4$.

11. Find the surface area of the part of the paraboloid $z = 16 - x^2 - y^2$ that lies above the xy-plane.

12. Show by integration that the surface area of the conical surface $z = br$ between the planes $z = 0$ and $z = h = ab$ is given by $A = \pi aL$, where L is the slant height $(a^2 + h^2)^{1/2}$ and a is the radius of the base of the cone.

13. Let the part of the cylinder $x^2 + y^2 = a^2$ between the planes $z = 0$ and $z = h$ be parametrized by $x = a\cos\theta$, $y = a\sin\theta$, $z = z$. Apply the formula in Equation (8) to show that the area of this zone is $A = 2\pi ah$.

14. Consider the meridianal zone of height $h = c - b$ lying on the sphere $r^2 + z^2 = a^2$ between the planes $z = b$ and $z = c$, where $0 \leqq b < c \leqq a$. Apply the formula in Equation (10) to show that the area of this zone is $A = 2\pi ah$.

15. Find the area of the part of the cylinder $x^2 + z^2 = a^2$ that lies within the cylinder $r^2 = x^2 + y^2 = a^2$.

16. Find the area of the part of the sphere $r^2 + z^2 = a^2$ that lies within the cylinder $r = a\sin\theta$.

17. (a) Apply the formula in Equation (8) to show that the surface area of the surface $y = f(x, z)$, for (x, z) in the region R of the xz-plane, is given by

$$A = \iint_R \sqrt{1 + (\partial f/\partial x)^2 + (\partial f/\partial z)^2}\,dx\,dz.$$

(b) State and derive a similar formula for the area of the surface $x = f(y, z)$, (y, z) in R.

18. Suppose that R is a region in the $\phi\theta$-plane. Consider the part of the sphere $\rho = a$ that corresponds to (ϕ, θ) in R, parametrized by the equations in (6) with $h(\phi, \theta) = a$. Apply (8) to show that the surface area of this part of the sphere is

$$A = \iint_R a^2 \sin\phi\,d\phi\,d\theta.$$

19. (a) Consider the "spherical rectangle" defined by $\rho = a$, $\phi_1 \leqq \phi \leqq \phi_2 = \phi_1 + \Delta\phi$, $\theta_1 \leqq \theta \leqq \theta_2 = \theta_1 + \Delta\theta$. Apply the formula of Problem 18 and the average value property (see Problem 30 in Section 16.2) to show that the area of this spherical rectangle is $A = a^2 \sin \hat{\phi}\, \Delta\phi\, \Delta\theta$ for some $\hat{\phi}$ in (ϕ_1, ϕ_2).

(b) Conclude from the result of part (a) that the volume of the spherical block defined by $\rho_1 \leqq \rho \leqq \rho_2 = \rho_1 + \Delta\rho$ and $\phi_1 \leqq \phi \leqq \phi_2$, $\theta_1 \leqq \theta \leqq \theta_2$ is

$$\Delta V = \tfrac{1}{3}(\rho_2^3 - \rho_1^3) \sin \hat{\phi}\, \Delta\phi\, \Delta\theta.$$

Finally, derive the formula in Equation (8) of Section 16.7 by applying the mean value theorem to the function $f(\rho) = \rho^3$ on the interval $[\rho_1, \rho_2]$.

20. Describe the surface $\rho = 2a \sin \phi$. Why is it called a pinched torus? It is parametrized by the equation in (6) with $h(\phi, \theta) = 2a \sin \phi$. Show that its surface area is $A = 4\pi^2 a^2$.

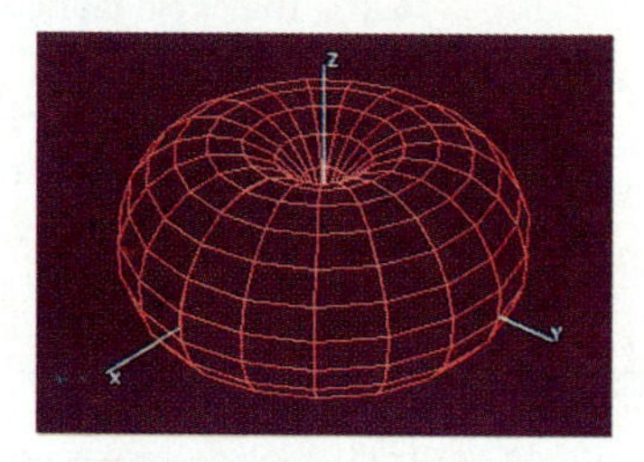

The pinched torus of Problem 20

21. The surface of revolution obtained when one revolves the curve $x = f(z)$, $a \leqq z \leqq b$, around the z-axis is parametrized in terms of θ $(0 \leqq \theta \leqq 2\pi)$ and z $(a \leqq z \leqq b)$ by $x = f(z) \cos \theta$, $y = f(z) \sin \theta$, $z = z$. From (8) derive the surface area formula

$$A = \int_0^{2\pi} \int_a^b f(z)\sqrt{1 + [f'(z)]^2}\, dz\, d\theta.$$

Note that this agrees with surface area of revolution as defined in Section 6.4.

22. Apply the formula of Problem 18 in both parts of this problem.

(a) Verify the formula $A = 4\pi a^2$ for the surface area of a sphere of radius a.

(b) Find the area of that part of a sphere of radius a and center $(0, 0, 0)$ that lies within the cone $\phi = \pi/6$.

23. Apply the result of Problem 21 to verify the formula $A = 2\pi rh$ for the lateral surface area of a right circular cylinder of radius r and height h.

24. Apply the formula in Equation (9) to verify the formula $A = 2\pi rh$ for the lateral surface area of the cylinder $x^2 + z^2 = r^2$, $0 \leqq y \leqq h$ of radius r and height h.

*16.9 Change of Variables in Multiple Integrals

We have seen in preceding sections that it is easier to evaluate certain multiple integrals by transforming them into polar or spherical coordinates. The technique of changing coordinate systems in order to evaluate a multiple integral is the multivariable analogue of substitution in a single integral. Recall from Section 5.6 that if $x = g(u)$, then

$$\int_a^b f(x)\, dx = \int_c^d f(g(u))g'(u)\, du \tag{1}$$

where $a = g(c)$ and $b = g(d)$. The method of substitution involves a "change of variables" that is tailored to the evaluation of a given integral.

Suppose that we want to evaluate the double integral $\iint_R F(x, y)\, dx\, dy$. A change of variables for this integral is determined by a **transformation** T from the uv-plane to the xy-plane; that is, a function T that associates with the point (u, v) a point $(x, y) = T(u, v)$ given by equations of the form

$$x = f(u, v), \qquad y = g(u, v). \tag{2}$$

The point (x, y) is called the **image** of the point (u, v) under the transformation T. If no two different points in the uv-plane have the same image point in the xy-plane, then the transformation T is said to be **one-to-one.** In this case it may be possible to solve the equations in (2) for u and v in terms of x and y and thus obtain the equations

$$u = h(x, y), \qquad v = k(x, y) \tag{3}$$

of the **inverse transformation** T^{-1} from the xy-plane to the uv-plane.

It is often convenient to visualize the transformation T geometrically in terms of its u-curves and v-curves. The **u-curves** of T are the images of horizontal lines in the uv-plane and the **v-curves** of T are the images of vertical lines in the uv-plane. Note that the image under T of a rectangle bounded by horizontal and vertical lines in the uv-plane is a *curvilinear figure* bounded by u-curves and v-curves in the xy-plane. This phenomenon is shown in Fig. 16.82. If we know the equations in (3) of the inverse transformation, then we can find the u-curves and the v-curves quite simply by writing the equations

$$k(x, y) = C_1 \quad \text{and} \quad h(x, y) = C_2,$$

respectively, where C_1 and C_2 are constants.

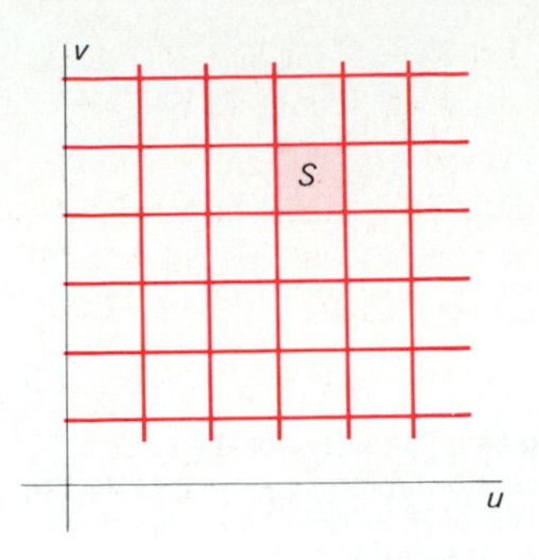

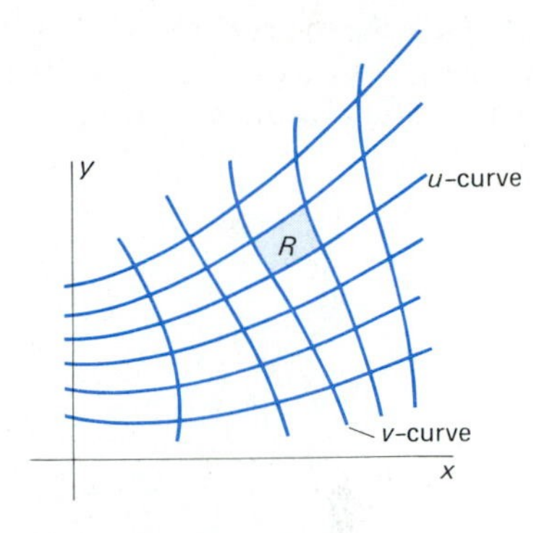

16.82 The transformation T turns the rectangle S into the curvilinear figure R.

EXAMPLE 1 Determine the u-curves and the v-curves of the transformation T whose inverse T^{-1} is specified by the equations $u = xy$, $v = x^2 - y^2$.

Solution The v-curves are the rectangular hyperbolas

$$xy = u = C_1 \qquad \text{(constant)},$$

while the u-curves are the hyperbolas

$$x^2 - y^2 = v = C_2 \qquad \text{(constant)}.$$

These two familiar families of hyperbolas are shown in Fig. 16.83.

Now we describe the change of variables in a double integral that corresponds to the transformation T specified by the equations in (2). Let the region R in the xy-plane be the image under T of the region S in the uv-plane. Let $F(x, y)$ be continuous on R. Let $\{S_1, S_2, \ldots, S_n\}$ be an inner partition of S into rectangles each having dimensions Δu by Δv. Each rectangle S_i is transformed by T into a curvilinear figure R_i in the xy-plane, as shown in Fig. 16.84. The images $\{R_1, R_2, \ldots, R_n\}$ under T of the S_i then constitute an inner partition of the region R (though into curvilinear figures rather than rectangles).

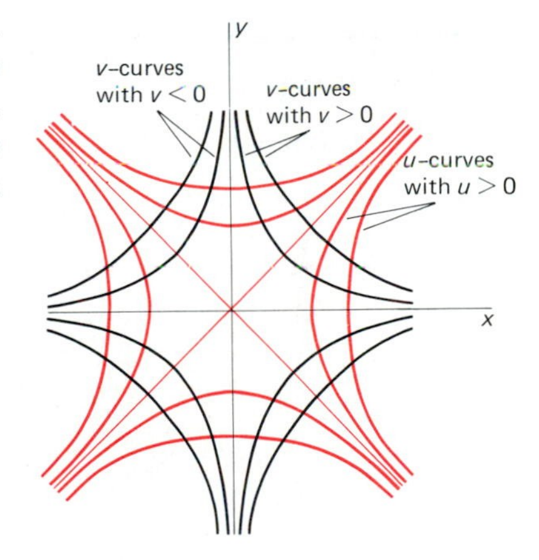

16.83 The u-curves and v-curves of Example 1

Let (u_i^*, v_i^*) be the lower left-hand corner point of S_i, and write

$$(x_i^*, y_i^*) = (f(u_i^*, v_i^*), g(u_i^*, v_i^*))$$

for its image under T. The u-curve through (x_i^*, y_i^*) has velocity vector

$$\mathbf{t}_u = \mathbf{i}f_u(u_i^*, v_i^*) + \mathbf{j}g_u(u_i^*, v_i^*) = \frac{\partial x}{\partial u}\mathbf{i} + \frac{\partial y}{\partial u}\mathbf{j},$$

while the v-curve through (x_i^*, y_i^*) has velocity vector

$$\mathbf{t}_v = \mathbf{i}f_v(u_i^*, v_i^*) + \mathbf{j}g_v(u_i^*, v_i^*) = \frac{\partial x}{\partial v}\mathbf{i} + \frac{\partial y}{\partial v}\mathbf{j}.$$

Thus we can approximate the curvilinear figure R_i by a parallelogram P_i with edges that are "copies" of the vectors $\mathbf{t}_u\,\Delta u$ and $\mathbf{t}_v\,\Delta v$. These edges and the approximating parallelogram also appear in Fig. 16.84.

Now the area ΔA_i of R_i is also approximated by the area of the parallelogram P_i, and we can compute the latter. Indeed,

$$\begin{aligned}\Delta A_i \approx a(P_i) &= |(\mathbf{t}_u\,\Delta u) \times (\mathbf{t}_v\,\Delta v)| \\ &= |\mathbf{t}_u \times \mathbf{t}_v|\,\Delta u\,\Delta v.\end{aligned}$$

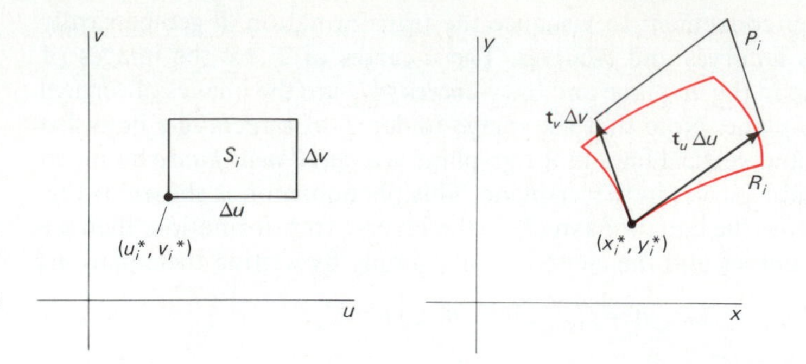

16.84 The effect of the transformation T; we estimate the area of $R_i = T(S_i)$ by computing the area of P_i.

But

$$\mathbf{t}_u \times \mathbf{t}_v = \begin{vmatrix} \mathbf{i} & \mathbf{j} & \mathbf{k} \\ \dfrac{\partial x}{\partial u} & \dfrac{\partial y}{\partial u} & 0 \\ \dfrac{\partial x}{\partial v} & \dfrac{\partial y}{\partial v} & 0 \end{vmatrix} = \begin{vmatrix} \dfrac{\partial x}{\partial u} & \dfrac{\partial x}{\partial v} \\ \dfrac{\partial y}{\partial u} & \dfrac{\partial y}{\partial v} \end{vmatrix} \mathbf{k}.$$

The two-by-two determinant on the right is called the **Jacobian** of the transformation T, after the German mathematician Carl Jacobi (1804–1851), who first investigated general changes of variables in multiple integrals. The Jacobian of the transformation T is a function of u and v, and we denote it by $J_T = J_T(u, v)$. Thus

$$J_T(u, v) = \begin{vmatrix} f_u(u, v) & f_v(u, v) \\ g_u(u, v) & g_v(u, v) \end{vmatrix}. \tag{4}$$

A common and particularly suggestive notation for the Jacobian is

$$J_T = \frac{\partial(x, y)}{\partial(u, v)}.$$

The computation above shows that the area ΔA_i of R_i is given approximately by

$$\Delta A_i \approx |J_T(u_i^*, v_i^*)| \, \Delta u \, \Delta v.$$

Therefore, when we set up Riemann sums for approximating double integrals, we find that

$$\iint_R F(x, y)\, dx\, dy \approx \sum_{i=1}^{n} F(x_i^*, y_i^*)\, \Delta A_i$$

$$\approx \sum_{i=1}^{n} F(f(u_i^*, v_i^*), g(u_i^*, v_i^*))|J_T(u_i^*, v_i^*)| \, \Delta u \, \Delta v$$

$$\approx \iint_R F(f(u, v), g(u, v))|J_T(u, v)| \, du\, dv.$$

This discussion is, in fact, an outline of a proof of the following general **change of variables** theorem. We assume that T transforms the bounded region S in the uv-plane into the bounded region R in the xy-plane, and that T is one-to-one from the interior of S to the interior of R. Suppose that the function $F(x, y)$ and the first-order partial derivatives of the component functions of T are continuous functions. Finally, in order to assure the existence of the indicated double integrals, we assume that the boundary of each of the regions S and R consists of finitely many piecewise smooth simple closed curves.

Theorem *Change of Variables*

If the transformation T with component functions $x = f(u, v)$, $y = g(u, v)$ satisfies the conditions above, then

$$\iint_R F(x, y)\, dx\, dy = \iint_S F(f(u, v), g(u, v))|J_T(u, v)|\, du\, dv. \tag{5}$$

If we write $G(u, v) = F(f(u, v), g(u, v))$, then the change of variables formula in (5) becomes

$$\iint_R F(x, y)\, dx\, dy = \iint_S G(u, v)\left|\frac{\partial(x, y)}{\partial(u, v)}\right| du\, dv. \tag{5a}$$

Thus we formally transform $\iint_R F(x, y)\, dA$ by replacing the variables x and y by $f(u, v)$ and $g(u, v)$, respectively, and writing

$$dA = \left|\frac{\partial(x, y)}{\partial(u, v)}\right| du\, dv$$

for the area element in terms of u and v. Note the analogy between (5a) and the single-variable formula in (1). In fact, if $g'(x) \neq 0$ on $[c, d]$, and we denote by α the smaller and by β the larger of the two limits c and d in (1), then it takes the form

$$\int_a^b f(x)\, dx = \int_\alpha^\beta f(g(u))|g'(u)|\, du. \tag{1a}$$

Thus the Jacobian in (5a) plays the role of the derivative $g'(u)$ in (1a).

For example, suppose that the transformation T from the $r\theta$-plane to the xy-plane is determined by the polar coordinates equations

$$x = f(r, \theta) = r\cos\theta, \qquad y = g(r, \theta) = r\sin\theta.$$

The Jacobian of T is

$$\frac{\partial(x, y)}{\partial(r, \theta)} = \begin{vmatrix} \cos\theta & -r\sin\theta \\ \sin\theta & r\cos\theta \end{vmatrix} = r > 0,$$

so Equation (5) or (5a) reduces to the familiar formula

$$\iint_R F(x, y)\, dx\, dy = \iint_S F(r\cos\theta, r\sin\theta)\, r\, dr\, d\theta.$$

Given a particular double integral $\iint_R f(x, y)\,dx\,dy$, how do we find a *productive* change of variables? One standard approach is to choose a transformation T such that the boundary of R consists of u-curves and v-curves. In case it is more convenient to express u and v in terms of x and y, we can first compute $\partial(u, v)/\partial(x, y)$ explicitly and then find the needed Jacobian $\partial(x, y)/\partial(u, v)$ from the formula

$$\frac{\partial(x, y)}{\partial(u, v)} \cdot \frac{\partial(u, v)}{\partial(x, y)} = 1. \tag{6}$$

The formula in (6) is a consequence of the chain rule (see Problem 18).

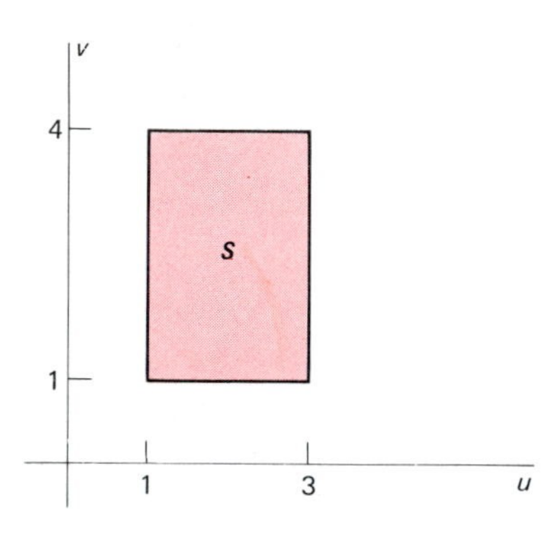

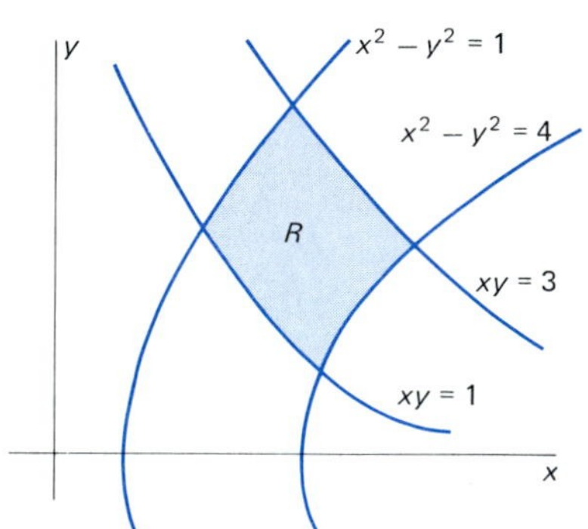

16.85 The transformation T and the new region S constructed in Example 2

EXAMPLE 2 Suppose that R is the plane region bounded by the hyperbolas

$$xy = 1, \quad xy = 3 \qquad \text{and} \qquad x^2 - y^2 = 1, \quad x^2 - y^2 = 4.$$

Find the polar moment of inertia

$$I_0 = \iint_R (x^2 + y^2)\,dx\,dy$$

of this region.

Solution The hyperbolas bounding R are v-curves and u-curves if $u = xy$ and $v = x^2 - y^2$, as in Example 1. We can most easily write $x^2 + y^2$ in terms of u and v by first noting that

$$4u^2 + v^2 = 4x^2y^2 + (x^2 - y^2)^2 = (x^2 + y^2)^2,$$

so that $x^2 + y^2 = (4u^2 + v^2)^{1/2}$. Now

$$\frac{\partial(u, v)}{\partial(x, y)} = \begin{vmatrix} y & x \\ 2x & -2y \end{vmatrix} = -2(x^2 + y^2).$$

Hence Equation (6) gives

$$\frac{\partial(x, y)}{\partial(u, v)} = -\frac{1}{2(x^2 + y^2)} = -\frac{1}{2(4u^2 + v^2)^{1/2}}.$$

We are now ready to apply the theorem, with the regions S and R as shown in Fig. 16.85. With $F(x, y) = x^2 + y^2$, the formula in (5a) gives

$$I_0 = \iint_R (x^2 + y^2)\,dx\,dy = \int_1^4 \int_1^3 (4u^2 + v^2)^{1/2} \frac{1}{2(4u^2 + v^2)^{1/2}}\,du\,dv$$

$$= \int_1^4 \int_1^3 \frac{1}{2}\,du\,dv = 3.$$

Our next example has an important application as its motivation. Consider an engine with an operating cycle that consists of alternate expansion and compression of the gas in a piston. During a cycle the point (P, V) giving the pressure and volume of this gas traces a closed curve in the PV-plane. The work done by the engine—ignoring friction and related losses—is then equal (in appropriate units) to the *area enclosed by this curve*, called the *indicator diagram* of the engine. In an ideal Carnot engine, the indicator diagram consists of two *isothermals* $xy = a$, $xy = b$ and two *adiabatics* $xy^\gamma = c$, $xy^\gamma = d$, where γ is the heat capacity ratio of the working gas in the piston. A typical value is $\gamma = 1.4$.

EXAMPLE 3 Find the area of the region R bounded by the curves $xy = 1$, $xy = 3$ and $xy^{1.4} = 1$, $xy^{1.4} = 2$ (see Fig. 16.86).

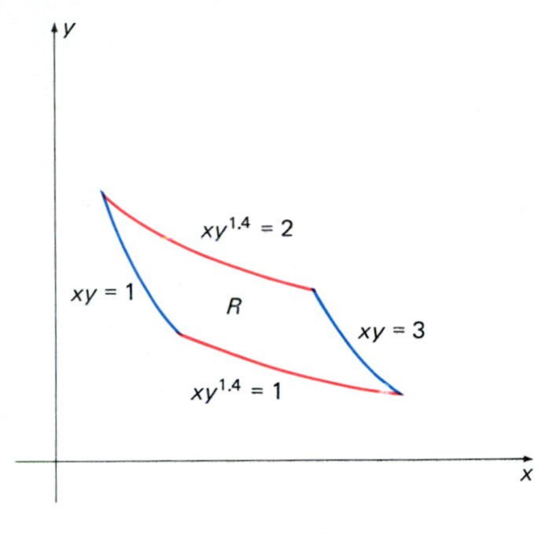

16.86 Finding the area of the region R (Example 3)

Solution To force the given curves to be u-curves and v-curves, we define our change of variables transformation by $u = xy$ and $v = xy^{1.4}$. Then

$$\frac{\partial(u, v)}{\partial(x, y)} = \begin{vmatrix} y & x \\ y^{1.4} & (1.4)xy^{0.4} \end{vmatrix} = (0.4)xy^{1.4} = (0.4)v.$$

So

$$\frac{\partial(x, y)}{\partial(u, v)} = \frac{1}{\partial(u, v)/\partial(x, y)} = \frac{2.5}{v}.$$

Consequently, the change of variables theorem gives the formula

$$A = \iint_R 1\,dx\,dy = \int_1^2 \int_1^3 \frac{2.5}{v}\,du\,dv = 5 \ln 2.$$

The change of variables formula for triple integrals is similar to the formula in (5). Let S and R be regions that correspond under the one-to-one transformation T from uvw-space to xyz-space, where the coordinate functions that comprise T are

$$x = f(u, v, w), \qquad y = g(u, v, w), \qquad z = h(u, v, w). \tag{7}$$

The Jacobian of T is

$$J_T(u, v, w) = \frac{\partial(x, y, z)}{\partial(u, v, w)} = \begin{vmatrix} \dfrac{\partial x}{\partial u} & \dfrac{\partial x}{\partial v} & \dfrac{\partial x}{\partial w} \\[2ex] \dfrac{\partial y}{\partial u} & \dfrac{\partial y}{\partial v} & \dfrac{\partial y}{\partial w} \\[2ex] \dfrac{\partial z}{\partial u} & \dfrac{\partial z}{\partial v} & \dfrac{\partial z}{\partial w} \end{vmatrix}. \tag{8}$$

Then the change of variables formula for triple integrals is

$$\iiint_R F(x, y, z)\,dx\,dy\,dz = \iiint_S G(u, v, w) \left| \frac{\partial(x, y, z)}{\partial(u, v, w)} \right| du\,dv\,dw, \tag{9}$$

where $G(u, v, w) = F(f(u, v, w), g(u, v, w), h(u, v, w))$ is the function obtained from $F(x, y, z)$ by expressing the variables x, y, and z in terms of u, v, w.

For example, if T is the spherical coordinates transformation given by

$$x = \rho \sin \phi \cos \theta, \qquad y = \rho \sin \phi \sin \theta, \qquad z = \rho \cos \phi,$$

then the Jacobian of T is

$$\frac{\partial(x, y, z)}{\partial(\rho, \phi, \theta)} = \begin{vmatrix} \sin \phi \cos \theta & \rho \cos \phi \cos \theta & -\rho \sin \phi \sin \theta \\ \sin \phi \sin \theta & \rho \cos \phi \sin \theta & \rho \sin \phi \cos \theta \\ \cos \phi & -\rho \sin \phi & 0 \end{vmatrix} = \rho^2 \sin \phi.$$

Thus (9) reduces to the familiar formula

$$\iiint_R F(x, y, z)\, dx\, dy\, dz = \iiint_S G(\rho, \phi, \theta)\rho^2 \sin\phi\, d\rho\, d\phi\, d\theta.$$

The sign is correct because $\rho^2 \sin\phi \geqq 0$ for ϕ in $[0, \pi]$.

EXAMPLE 4 Find the volume of the solid torus R obtained by revolving around the z-axis the circular disk

$$(x - b)^2 + z^2 \leqq a^2, \qquad a < b, \tag{10}$$

in the xz-plane.

Solution This is the torus of Example 4 in Section 16.8. Let us write u for the ordinary polar coordinate angle θ, v for the angle ψ of Fig. 16.81, and w for distance from the center of the circular disk described by the inequality in (10). We then define the transformation T by means of the equations

$$x = (b + w\cos v)\cos u, \qquad y = (b + w\cos v)\sin u, \qquad z = w\sin v.$$

Then the solid torus R is the image under T of the region in uvw-space described by the inequalities $0 \leqq u \leqq 2\pi$, $0 \leqq v \leqq 2\pi$, $0 \leqq w \leqq a$. By a routine computation, we find that the Jacobian of T is

$$\frac{\partial(x, y, z)}{\partial(u, v, w)} = w(b + w\cos v).$$

Hence Equation (9) with $F(x, y, z) \equiv 1$ gives

$$V = \iiint_R dx\, dy\, dz = \int_0^{2\pi}\int_0^{2\pi}\int_0^a (bw + w^2\cos v)\, dw\, du\, dv$$

$$= 2\pi \int_0^{2\pi} (\tfrac{1}{2}a^2 b + \tfrac{1}{3}a^3\cos v)\, dv = 2\pi^2 a^2 b,$$

which agrees with the value $V = (2\pi b)(\pi a^2)$ given by Pappus's theorem (Section 16.5).

16.9 PROBLEMS

In each of Problems 1–6, solve for x and y in terms of u and v and then compute the Jacobian $\partial(x, y)/\partial(u, v)$.

1. $u = x + y, \quad v = x - y$

2. $u = x - 2y, \quad v = 3x + y$

3. $u = xy, \quad v = y/x$

4. $u = 2(x^2 + y^2), \quad v = 2(x^2 - y^2)$

5. $u = x + 2y^2, \quad v = x - 2y^2$

6. $u = 2x/(x^2 + y^2), \quad v = -2y/(x^2 + y^2)$

7. Let R be the parallelogram bounded by the lines $x + y = 1$, $x + y = 2$ and $2x - 3y = 2$, $2x - 3y = 5$. Substitute $u = x + y$, $v = 2x - 3y$ to find the area

$$A = \iint_R dx\, dy$$

of R.

8. Substitute $u = xy$, $v = y/x$ to find the area of the first quadrant region bounded by the lines $y = x$, $y = 2x$ and the hyperbolas $xy = 1$, $xy = 2$.

9. Substitute $u = xy$, $v = xy^3$ to find the area of the region in the first quadrant bounded by the curves $xy = 2$, $xy = 4$ and $xy^3 = 3$, $xy^3 = 6$.

10. Find the area of the region in the first quadrant bounded by the curves $y = x^2$, $y = 2x^2$ and $x = y^2$, $x = 4y^2$. (*Suggestion:* Let $y = ux^2$ and $x = vy^2$.)

11. Use the method of Problem 10 to find the area of the region in the first quadrant bounded by the curves $y = x^3$, $y = 2x^3$ and $x = y^3$, $x = 4y^3$.

12. Let R be the region in the first quadrant bounded by the circles $x^2 + y^2 = 2x$, $x^2 + y^2 = 6x$ and the circles $x^2 + y^2 = 2y$, $x^2 + y^2 = 8y$. Use the transformation

$u = 2x/(x^2 + y^2)$, $v = 2y/(x^2 + y^2)$ to evaluate the integral

$$\iint_R (x^2 + y^2)^{-2}\, dx\, dy.$$

13. Use elliptical coordinates $x = 3r\cos\theta$, $y = 2r\sin\theta$ to find the volume of the region bounded by the xy-plane, the paraboloid $z = x^2 + y^2$, and the elliptical cylinder $x^2/9 + y^2/4 = 1$.

14. Let R be the solid ellipsoid with outer boundary surface $x^2/a^2 + y^2/b^2 + z^2/c^2 = 1$. Use the transformation $x = au$, $y = bv$, $z = cw$ to show that the volume of this ellipsoid is

$$V = \iiint_R 1\, dx\, dy\, dz = \tfrac{4}{3}\pi abc.$$

15. Find the volume of the region in the first octant that is bounded by the hyperbolic cylinders $xy = 1$, $xy = 4$; $xz = 1$, $xz = 9$; $yz = 4$, $yz = 9$. (*Suggestion:* Let $u = xy$, $v = xz$, $w = yz$, and note that $uvw = x^2y^2z^2$.)

16. Use the transformation

$$x = \frac{r}{t}\cos\theta, \qquad y = \frac{r}{t}\sin\theta, \qquad z = r^2.$$

to find the volume of the region R that lies between the paraboloids $z = x^2 + y^2$, $z = 4(x^2 + y^2)$ and also between the planes $z = 1$, $z = 4$.

17. Let R be the rotated elliptical region bounded by the graph of $x^2 + xy + y^2 = 3$. Let $x = u + v$ and $y = u - v$. Show that

$$\iint_R \exp(-x^2 - xy - y^2)\, dx\, dy = 2\iint_S \exp(-3u^2 - v^2)\, du\, dv.$$

Then substitute $u = r\cos\theta$, $v = \sqrt{3}(r\sin\theta)$ to evaluate the latter integral.

18. Derive the relation in (6) between the Jacobians of a transformation and its inverse from the chain rule and the following property of determinants:

$$\begin{vmatrix} a_1 & b_1 \\ c_1 & d_1 \end{vmatrix} \cdot \begin{vmatrix} a_2 & b_2 \\ c_2 & d_2 \end{vmatrix} = \begin{vmatrix} a_1a_2 + b_1c_2 & a_1b_2 + b_1d_2 \\ a_2c_1 + c_2d_1 & b_2c_1 + d_1d_2 \end{vmatrix}.$$

19. Change to spherical coordinates to show that, for $k > 0$,

$$\int_{-\infty}^{+\infty}\int_{-\infty}^{+\infty}\int_{-\infty}^{+\infty} (x^2 + y^2 + z^2)^{1/2} e^{-k(x^2+y^2+z^2)} \times dx\, dy\, dz = \frac{2\pi}{k^2}.$$

20. Let R be the solid ellipsoid with constant density δ and boundary surface $x^2/a^2 + y^2/b^2 + z^2/c^2 = 1$. Use ellipsoidal coordinates $x = a\rho\sin\phi\cos\theta$, $y = b\rho\sin\phi\sin\theta$, $z = c\rho\cos\phi$ to show that its mass is $M = \tfrac{4}{3}\pi\delta abc$.

21. Show that the moment of inertia of the ellipsoid of Problem 20 about the z-axis is $I_z = \tfrac{1}{5}M(a^2 + b^2)$.

CHAPTER 16 REVIEW: Definitions, Concepts, Results

Use the list below as a guide to concepts that you may need to review.

1. Definition of the double integral as a limit of Riemann sums
2. Evaluation of double integrals by iterated single integrals
3. Use of the double integral to find the volume between two surfaces above a given plane region
4. Transformation of the double integral $\iint_R f(x, y)\, dA$ into polar coordinates
5. Application of double integrals to find mass, centroids, and moments of inertia of plane laminae
6. The two theorems of Pappus
7. Definition of the triple integral as a limit of Riemann sums
8. Evaluation of triple integrals by iterated single integrals
9. Application of triple integrals to find volume, mass, centroids, and moments of inertia
10. Transformation of the triple integral $\iiint_T f(x, y, z)\, dV$ into cylindrical and spherical coordinates
11. The surface area of a parametric surface
12. The area of a surface $z = f(x, y)$ for (x, y) in the plane region R
13. The Jacobian of a transformation of coordinates
14. The transformation of a double or triple integral corresponding to a given change of variables

In each of Problems 1–5, evaluate the given integral by first reversing the order of integration.

1. $\displaystyle\int_0^1 \int_{y^{1/3}}^1 \frac{1}{(1+x^2)^{1/2}}\,dx\,dy$

2. $\displaystyle\int_0^1 \int_y^1 \frac{\sin x}{x}\,dx\,dy$

3. $\displaystyle\int_0^1 \int_x^1 \exp(-y^2)\,dy\,dx$

4. $\displaystyle\int_0^8 \int_{x^{2/3}}^4 x\cos y^4\,dy\,dx$

5. $\displaystyle\int_0^4 \int_{\sqrt{y}}^2 \frac{y}{x^3}\exp(x^2)\,dx\,dy$

6. The double integral $\displaystyle\int_0^\infty \int_x^\infty y^{-1}e^{-y}\,dy\,dx$ is an improper integral over the unbounded region in the first quadrant bounded by the lines $y = x$ and $x = 0$. Assuming the validity of reversing the order of integration, evaluate this integral by integrating first with respect to x.

7. Find the volume of the solid T that lies under the paraboloid $z = x^2 + y^2$ and over the triangle R in the xy-plane having vertices at $(0, 0, 0)$, $(1, 1, 0)$, and $(2, 0, 0)$.

8. Find by integration in cylindrical coordinates the volume bounded by the paraboloids $z = 2x^2 + 2y^2$ and $z = 48 - x^2 - y^2$.

9. Use integration in spherical coordinates to find the volume and centroid of the solid region that is inside the sphere $\rho = 3$, under the cone $\phi = \pi/3$, and above the xy-plane $\phi = \pi/2$.

10. Find the volume of the solid bounded by the elliptic paraboloids $z = x^2 + 3y^2$ and $z = 8 - x^2 - 5y^2$.

11. Find the volume bounded by the paraboloid $y = x^2 + 3z^2$ and the parabolic cylinder $y = 4 - z^2$.

12. Find the volume of the region bounded by the parabolic cylinders $z = x^2$, $z = 2 - x^2$ and the planes $y = 0$, $y + z = 4$.

13. Find the volume of the region bounded by the elliptical cylinder $y^2 + 4z^2 = 4$ and the planes $x = 0$, $x = y + 2$.

14. Show that the volume of the solid bounded by the elliptic cylinder $x^2/a^2 + y^2/b^2 = 1$ and the planes $z = 0$, $z = h + x$ (where $h > a > 0$) is $V = \pi abh$.

15. Let R be the first-quadrant region bounded by the line $y = x$ and the curve $x^4 + x^2y^2 = y^2$. Use polar coordinates to evaluate

$$\iint_R (1 + x^2 + y^2)^{-2}\,dA.$$

In each of Problems 16–20, find the mass and centroid of a plane lamina having the indicated shape and density ρ.

16. The region bounded by $y = x^2$ and $x = y^2$; $\rho = x^2 + y^2$.

17. The region bounded by $x = 2y^2$ and $y^2 = x - 4$; $\rho = y^2$.

18. The region between $y = \ln x$ and the x-axis over the interval $1 \leqq x \leqq 2$; $\rho = 1/x$.

19. The circle bounded by $r = 2\cos\theta$; $\rho = k$ (a constant)

20. The region of Problem 19; $\rho = r$

21. Use the first theorem of Pappus to find the y-coordinate of the centroid of the upper half of the ellipse $(x/a)^2 + (y/b)^2 = 1$. Employ the facts that the area of this semiellipse is $A = \pi ab/2$, while the volume of the ellipsoid it generates when rotated about the x-axis is $V = \frac{4}{3}\pi ab^2$.

22. (a) Use the first theorem of Pappus to find the centroid of the first-quadrant part of the annular ring with boundary circles $x^2 + y^2 = a^2$ and $x^2 + y^2 = b^2$, with $0 < a < b$.
(b) Show that the limiting position of this centroid as $b \to a$ is the centroid of a quarter-circular arc, as found in Problem 44 of Section 16.5

23. Find the centroid of the region in the xy-plane bounded by the x-axis and the parabola $y = 4 - x^2$.

24. Find the volume of the solid that lies under the parabolic cylinder $z = x^2$ and over the triangle in the xy-plane bounded by the x-axis, the y-axis, and the line $x + y = 1$.

25. Use cylindrical coordinates to find the volume of the ice cream cone bounded above by the sphere $x^2 + y^2 + z^2 = 5$ and below by the cone $z = 2(x^2 + y^2)^{1/2}$.

26. Find the volume and centroid of the ice cream cone bounded above by the sphere $\rho = a$ and below by the cone $\phi = \pi/3$.

27. Find the moment of inertia about its natural axis of a homogeneous solid circular cone with mass M and base radius a.

28. Find the mass of the first octant of the ball $\rho \leqq a$ if its density function is $\delta = xyz$.

29. Find the moment of inertia about the x-axis of the homogeneous solid ellipsoid with boundary surface $(x/a)^2 + (y/b)^2 + (z/c)^2 = 1$.

30. Find the volume of the region in the first octant bounded by the sphere $\rho = a$, the cylinder $r = a$, the plane $z = a$, the xz-plane, and the yz-plane.

31. Find the moment of inertia about the z-axis of the homogeneous region that lies inside both the sphere $\rho = 2$ and the cylinder $r = 2\cos\theta$.

In each of Problems 32–34, a volume is generated by revolving a plane region R around an axis. To find this volume, set up a *double* integral over R by revolving an area element dA around the indicated axis to generate a volume element dV.

32. Find the volume of the solid obtained by revolving around the y-axis the region within the circle $r = 2a\cos\theta$.

33. Find the volume of the solid obtained by revolving the region enclosed by the cardioid $r = 1 + \cos\theta$ around the x-axis.

34. Find the volume of the solid torus obtained by revolving the disk $0 \leqq r \leqq a$ about the line $x = -b$, $|b| \geqq a$.

35. This problem deals with the oblique segment of a paraboloid discussed in Example 3 of Section 16.6 and shown in Fig. 16.65 there.

(a) Show first that its centroid is at the point $C(0, \frac{1}{2}, \frac{7}{4})$.

(b) Show that the center of the elliptical upper "base" of the solid paraboloid is at the point $Q(0, \frac{1}{2}, \frac{5}{2})$.

(c) Verify that the point $V(0, \frac{1}{2}, \frac{1}{4})$ is the point where the tangent plane to the paraboloid is parallel to the upper base. The point V is called the *vertex* of the oblique segment, and the line segment VQ is its *principal axis.*

(d) Show that C lies on the principal axis *two-thirds* of the way from the vertex to the upper base. Archimedes showed that this is true for any segment cut off from a paraboloid by a plane. This was a key step in his determination of the equilibrium position of a floating right circular paraboloid, in terms of its density and dimensions. The possible positions are shown in Fig. 16.87. The principles he introduced for the solution of this problem are still important in naval architecture.

Problems 36–42 deal with average distance. The **average distance** $\bar{d}$ of the point (x_0, y_0) from the points of the plane region R with area A is defined to be

$$\bar{d} = \frac{1}{A}\iint [(x - x_0)^2 + (y - y_0)^2]^{1/2}\, dA.$$

The average distance of a point (x_0, y_0, z_0) from the points of a space region is defined analogously.

36. Show that the average distance of the points of a disk of radius a from its center is $2a/3$.

37. Show that the average distance of the point of a disk of radius a from a fixed point on its boundary is $32a/9\pi$.

38. A circle of radius 1 is interior to and tangent to a circle of radius 2. Find the average distance of the point of tangency from the points that lie between the two circles.

39. Show that the average distance of the points of a spherical ball of radius a from its center is $3a/4$.

40. Show that the average distance of the points of a spherical ball of radius a from a fixed point on its surface is $6a/5$.

41. A sphere of radius 1 is interior to and tangent to a sphere of radius 2. Find the average distance of the point of tangency from the set of all points between the two spheres.

42. A right circular cone has radius R and height H. Find the average distance of the points of the cone from its vertex.

43. Find the surface area of the part of the paraboloid $z = 10 - r^2$ that lies between the two planes $z = 1$ and $z = 6$.

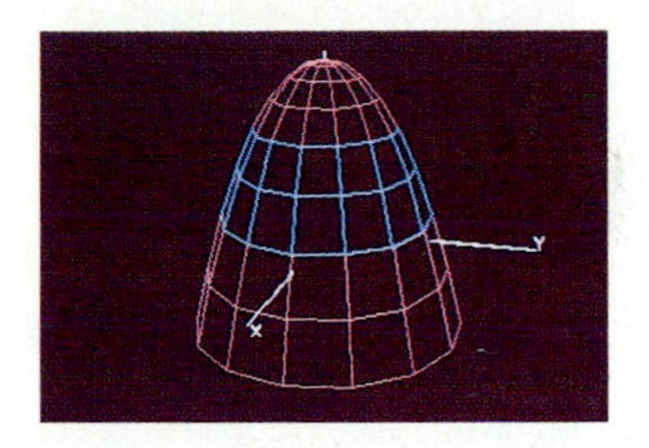

The area of Problem 43

44. Find the surface area of the part of the surface $z = y^2 - x^2$ that is inside the cylinder $x^2 + y^2 = 4$.

45. Deduce from the formula of Problem 18 in Section 16.8 that the surface area of the zone on the sphere $\rho = a$ between the planes $z = z_1$ and $z = z_2$ (where $-a \leqq z_1 < z_2 \leqq a$) is $A = 2\pi ah$ where $h = z_2 - z_1$.

46. Find the surface area of the part of the sphere $\rho = 2$ that is inside the cylinder $x^2 + y^2 = 2x$.

47. A square hole with side length 2 is cut through a cone of height and base radius 2; the center line of the hole is the axis of symmetry of the cone. Find the area of the surface removed from the cone.

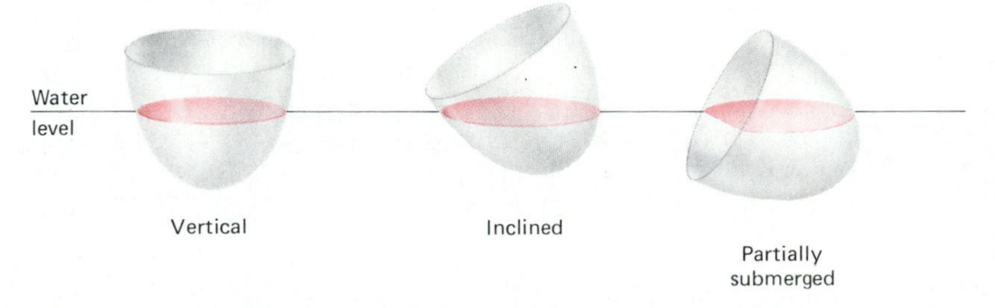

16.87 How a uniform solid paraboloid might float

48. Numerically approximate the surface area of the part of the parabolic cylinder $2z = x^2$ that lies inside the cylinder $x^2 + y^2 = 1$.

49. A "fence" of variable height $h(t)$ stands above the plane curve $(x(t), y(t))$. Thus the fence has the parametrization $x = x(t)$, $y = y(t)$, $z = z$ for $a \leqq t \leqq b$, $0 \leqq z \leqq h(t)$. Apply the formula in Equation (8) of Section 16.8 to show that the area of the fence is

$$A = \int_a^b \int_0^{h(t)} \left[\left(\frac{dx}{dt}\right)^2 + \left(\frac{dy}{dt}\right)^2\right]^{1/2} dz\, dt.$$

50. Apply the formula of Problem 49 to compute the area of the part of the cylinder $r = a \sin\theta$ that lies inside the sphere $r^2 + z^2 = a^2$.

51. Find the polar moment of inertia of the first-quadrant region of constant density δ that is bounded by the hyperbolas $xy = 1$, $xy = 3$ and $x^2 - y^2 = 1$, $x^2 - y^2 = 4$.

52. Substitute $u = x - y$ and $v = x + y$ to evaluate

$$\iint_R \exp\left(\frac{x-y}{x+y}\right) dx\, dy,$$

where R is bounded by the coordinate axes and the line $x + y = 1$.

53. Use ellipsoidal coordinates $x = a\rho \sin\phi \cos\theta$, $y = b\rho \sin\phi \sin\theta$, $z = c\rho \cos\phi$ to find the mass of the solid ellipsoid $(x/a)^2 + (y/b)^2 + (z/c)^2 \leqq 1$ if its density at the point (x, y, z) is given by $\delta = 1 - (x/a)^2 - (y/b)^2 - (z/c)^2$.

54. Let R be the first-quadrant region bounded by the lemniscates $r^2 = 3\cos 2\theta$, $r^2 = 4\cos 2\theta$ and $r^2 = 3\sin 2\theta$, $r^2 = 4\sin 2\theta$ (see Fig. 16.88). Show that its area is $A = (10 - 7\sqrt{2})/4$. (*Suggestion:* Define the transformation T from the uv-plane to the $r\theta$-plane by $r^2 = u^{1/2}\cos 2\theta$, $r^2 = v^{1/2}\sin 2\theta$.) Show first that

$$r^4 = \frac{uv}{u+v}, \qquad \theta = \frac{1}{2}\arctan\frac{u^{1/2}}{v^{1/2}}.$$

Then show that $\partial(r, \theta)/\partial(u, v) = -1/[16r(u+v)^{3/2}]$.

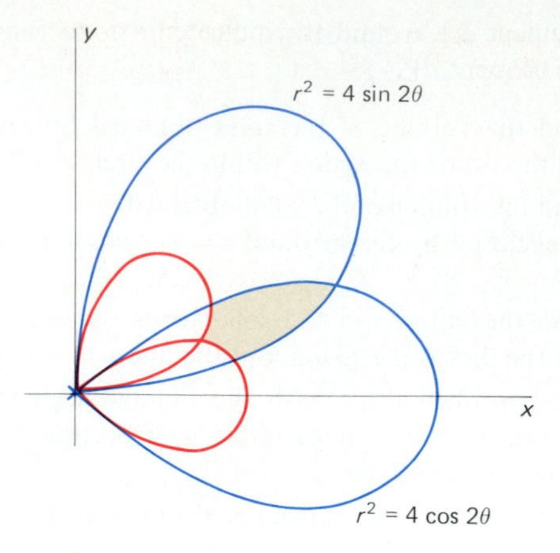

16.88 The region R of Problem 54

55. A 2 by 2 square hole is cut symmetrically through a sphere of radius $\sqrt{2}$.

(a) Show that the total surface area of the two pieces cut from the sphere is

$$A = \int_0^1 8\sqrt{2}\arcsin\left(\frac{1}{(2 - x^2)^{1/2}}\right) dx.$$

Then use Simpson's rule to approximate this integral.

(b) (Difficult!) Show that the exact value of the integral in part (a) is $A = 8\pi(\sqrt{2} - 1)$. (*Suggestion:* First integrate by parts, then substitute $x = \sin\theta$.)

56. Show that the volume enclosed by the surface

$$x^{2/3} + y^{2/3} + z^{2/3} = a^{2/3}$$

is $V = 4\pi a^3/35$. (*Suggestion:* Substitute $y = b\sin^3\theta$.)

57. Show that the volume enclosed by the surface

$$x^{1/3} + y^{1/3} + z^{1/3} = a^{1/3}$$

is $V = a^3/210$. (*Suggestion:* Substitute $y = b\sin^6\theta$.)

chapter seventeen

Vector Analysis

17.1 Vector Fields

This chapter is devoted to topics in the calculus of vector fields of importance in science and engineering. A **vector field** defined on a region T in space is a vector-valued function $\mathbf{F}$ that associates with each point (x, y, z) of T a vector

$$\mathbf{F}(x, y, z) = \mathbf{i}P(x, y, z) + \mathbf{j}Q(x, y, z) + \mathbf{k}R(x, y, z). \tag{1}$$

We may more briefly describe the vector field $\mathbf{F}$ in terms of its *component functions* P, Q, and R by writing $\mathbf{F} = \langle P, Q, R \rangle$. Note that P, Q, and R are scalar (real-valued) functions.

A **vector field** in the plane is similar except that neither z-components nor z-coordinates are involved. Thus a vector field on the plane region R is a vector-valued function $\mathbf{F}$ that associates with each point (x, y) of R a vector

$$\mathbf{F}(x, y) = \mathbf{i}P(x, y) + \mathbf{j}Q(x, y). \tag{2}$$

It is useful to be able to visualize a given vector field $\mathbf{F}$. One common way is to sketch a collection of typical vectors $\mathbf{F}(x, y)$, each represented by an arrow of length $|\mathbf{F}(x, y)|$ and placed with (x, y) as its initial point. This procedure is illustrated in the following example.

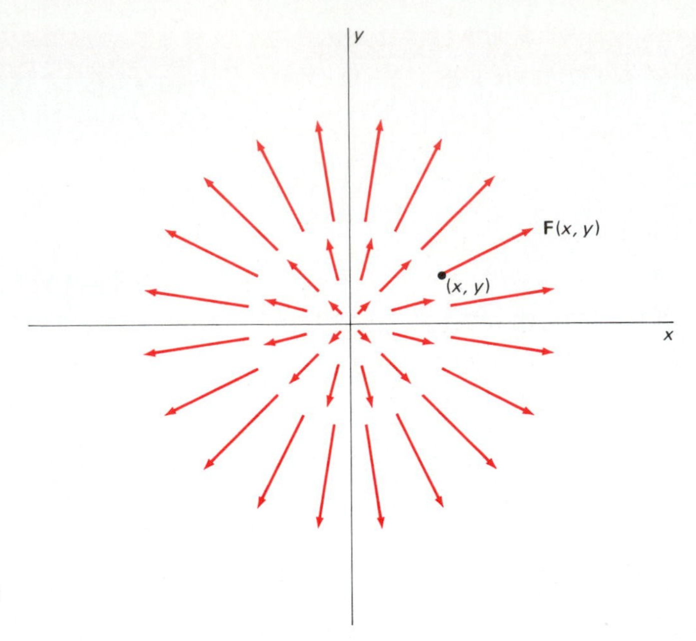

17.1 The vector field $\mathbf{F}(x, y) = x\mathbf{i} + y\mathbf{j}$

EXAMPLE 1 Describe the vector field $\mathbf{F}(x, y) = x\mathbf{i} + y\mathbf{j}$.

Solution For each point (x, y) in the plane, $\mathbf{F}(x, y)$ is simply its position vector; it points directly away from the origin and has length

$$|\mathbf{F}(x, y)| = |x\mathbf{i} + y\mathbf{j}| = \sqrt{x^2 + y^2} = r$$

equal to the distance from the origin to (x, y). Figure 17.1 shows some typical vectors representing this vector field.

Among the most important vector fields in applications are velocity vector fields. Imagine the steady flow of a fluid, such as the solar wind or the water in a river. By a *steady flow* is meant one such that the velocity vector $\mathbf{v}(x, y, z)$ of the fluid flowing through each point (x, y, z) is independent of time (although not necessarily of x, y, and z), so the pattern of the flow remains constant. Then $\mathbf{v}(x, y, z)$ is the **velocity vector field** of the fluid flow.

EXAMPLE 2 Suppose the horizontal xy-plane to be covered with a thin sheet of water that is revolving—rather like a whirlpool—about the origin with constant angular speed ω radians per second in the counterclockwise direction. Describe the associated velocity vector field.

Solution In this case we obviously have a two-dimensional vector field $\mathbf{v}(x, y)$. At each point (x, y) the water is moving tangential to the circle of radius $r = \sqrt{x^2 + y^2}$ with speed $v = r\omega$. We note that the vector field

$$\mathbf{v}(x, y) = \omega(-y\mathbf{i} + x\mathbf{j}) \tag{3}$$

has length $r\omega$, points in a generally counterclockwise direction, and that

$$\mathbf{v} \cdot \mathbf{r} = \omega(-y\mathbf{i} + x\mathbf{j}) \cdot (x\mathbf{i} + y\mathbf{j}) = 0,$$

so $\mathbf{v}$ is tangent to the circle mentioned above. The velocity field determined by Equation (3) is illustrated in Fig. 17.2.

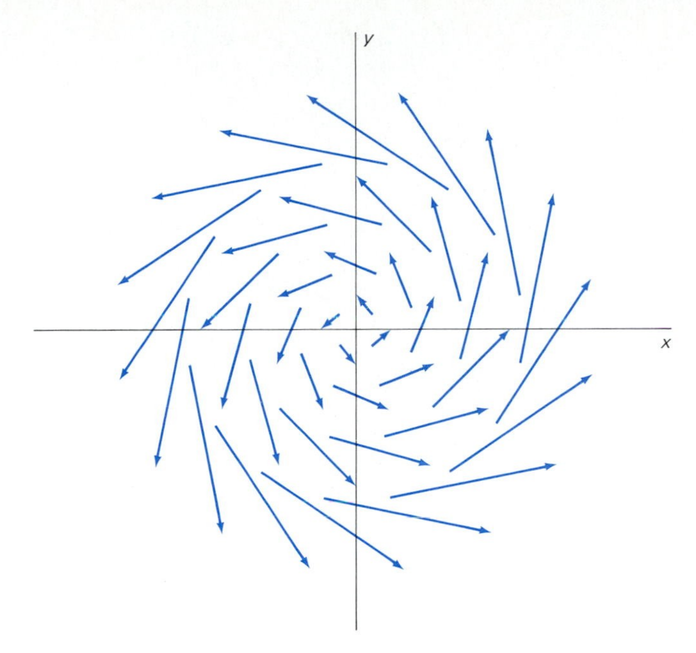

17.2 The velocity vector field $\mathbf{v}(x, y) = \omega(-y\mathbf{i} + x\mathbf{j})$, drawn with $\omega = 1$

Equally important in physical applications are *force fields*. Suppose that some circumstance (perhaps gravitational or electrical in character) causes a force $\mathbf{F}(x, y, z)$ to act on a particle when it is placed at the point (x, y, z). Then we have a force field $\mathbf{F}$. The following example deals with what is perhaps the most common force field perceived by human beings.

EXAMPLE 3 Suppose that a mass M is fixed at the origin. When a particle of unit mass is placed at the point (x, y, z) other than the origin, it is subjected to a force $\mathbf{F}(x, y, z)$ of gravitational attraction directed toward the mass M at the origin. By the inverse-square law of gravitation, the magnitude of $\mathbf{F}$ is $F = \mathrm{G}M/r^2$ where $r = \sqrt{x^2 + y^2 + z^2}$ is the length of the position vector $\mathbf{r} = x\mathbf{i} + y\mathbf{j} + z\mathbf{k}$. It follows immediately that

$$\mathbf{F}(x, y, z) = -\frac{k\mathbf{r}}{r^3} \tag{4}$$

where $k = GM$, because this vector has both the correct magnitude and the correct direction (toward the origin, for $\mathbf{F}$ is a multiple of $-\mathbf{r}$). A force field of the form in (4) is called an *inverse-square* force field. Note that $\mathbf{F}(x, y, z)$ is not defined at the origin, and that $|\mathbf{F}| \to \infty$ as $r \to 0$. Figure 17.3 illustrates an inverse-square force field.

THE GRADIENT VECTOR FIELD

In Section 15.8 we introduced the gradient vector of the real-valued function $f(x, y, z)$. It is the vector ∇f defined as follows:

$$\nabla f = \mathbf{i}\frac{\partial f}{\partial x} + \mathbf{j}\frac{\partial f}{\partial y} + \mathbf{k}\frac{\partial f}{\partial z}. \tag{5}$$

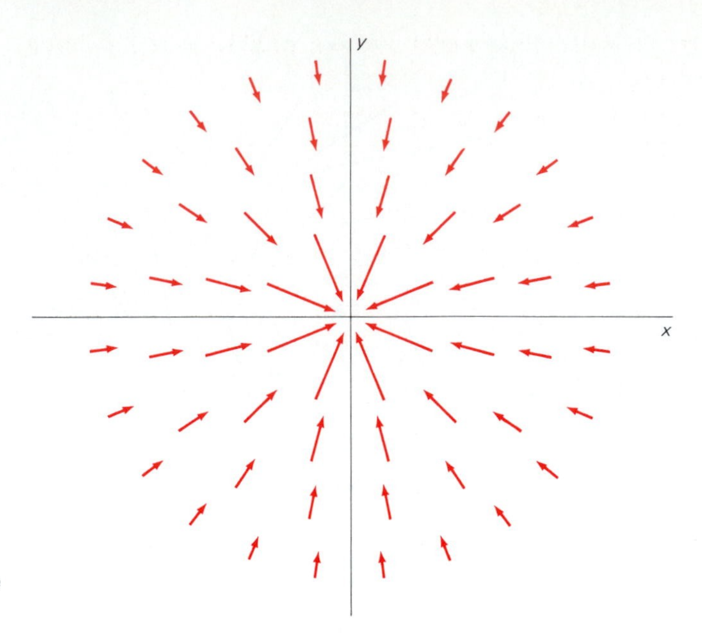

17.3 An inverse-square-law force field

The partial derivatives on the right-hand side in (5) are evaluated at the point (x, y, z). Thus $\nabla f(x, y, z)$ is a vector field—it is the **gradient vector field** of the function f. According to Theorem 1 in Section 15.8, the vector $\nabla f(x, y, z)$ points in the direction in which the maximal directional derivative of f at (x, y, z) is obtained. For example, if $f(x, y, z)$ is the temperature at the point (x, y, z) in space, then one should go in the direction $\nabla f(x, y, z)$ in order to get warmer the most quickly.

The notation in Equation (5) suggests the formal expression

$$\nabla = \mathbf{i}\frac{\partial}{\partial x} + \mathbf{j}\frac{\partial}{\partial y} + \mathbf{k}\frac{\partial}{\partial z}. \tag{6}$$

It is fruitful to think of ∇ as a *vector differential operator*. That is, ∇ is the operation which, when applied to the scalar function f, yields its gradient vector field ∇f. This operation behaves in several familiar and important ways like the operation of (single-variable) differentiation. For a familiar example of this, recall that in Chapter 15 we found the critical points of a function of several variables to be those points at which $\nabla f(x, y, z) = 0$ and those at which $\nabla f(x, y, z)$ does not exist. As a computationally useful instance, suppose that f and g are functions and a and b are constants. It then follows readily from (5) and from the linearity of partial differentiation that

$$\nabla(af + bg) = a\nabla f + b\nabla g. \tag{7}$$

Thus the gradient operator is *linear*. It also satisfies the product rule, as demonstrated in the next example.

EXAMPLE 4 Given the differentiable functions $f(x, y, z)$ and $g(x, y, z)$, show that

$$\nabla(fg) = f\,\nabla g + g\,\nabla f. \tag{8}$$

Solution We apply the definition in (5) and the product rule for partial differentiation. Thus

$$\begin{aligned}\nabla(fg) &= \mathbf{i}\,\frac{\partial(fg)}{\partial x} + \mathbf{j}\,\frac{\partial(fg)}{\partial y} + \mathbf{k}\,\frac{\partial(fg)}{\partial z}\\ &= \mathbf{i}(fg_x + gf_x) + \mathbf{j}(fg_y + gf_y) + \mathbf{k}(fg_z + gf_z)\\ &= (f)(\mathbf{i}g_x + \mathbf{j}g_y + \mathbf{k}g_z) + (g)(\mathbf{i}f_x + \mathbf{j}f_y + \mathbf{k}f_z) = f\,\nabla g + g\,\nabla f,\end{aligned}$$

as desired.

THE DIVERGENCE OF A VECTOR FIELD

Suppose that we are given the vector-valued function

$$\mathbf{F}(x, y, z) = \mathbf{i}P(x, y, z) + \mathbf{j}Q(x, y, z) + \mathbf{k}R(x, y, z)$$

with differentiable component functions P, Q, and R. Then the **divergence** of $\mathbf{F}$ is the scalar function div $\mathbf{F}$ defined as follows:

$$\operatorname{div}\mathbf{F} = \nabla\cdot\mathbf{F} = \frac{\partial P}{\partial x} + \frac{\partial Q}{\partial y} + \frac{\partial R}{\partial z}. \tag{9}$$

Of course *div* is an abbreviation for "divergence," and the alternative notation $\nabla\cdot\mathbf{F}$ is consistent with the formal expression for ∇ in Equation (6). That is,

$$\nabla\cdot\mathbf{F} = \left\langle \frac{\partial}{\partial x}, \frac{\partial}{\partial y}, \frac{\partial}{\partial z}\right\rangle\cdot\langle P, Q, R\rangle = \frac{\partial P}{\partial x} + \frac{\partial Q}{\partial y} + \frac{\partial R}{\partial z}.$$

In Section 17.7 we will see that if $\mathbf{v}$ is the velocity vector field of a steady fluid flow, then the value of div $\mathbf{v}$ at a point (x, y, z) is essentially the net rate per unit volume at which fluid mass is flowing away (or "diverging") from the point (x, y, z).

EXAMPLE 5 If the vector field $\mathbf{F}$ is given by

$$\mathbf{F}(x, y, z) = (xe^y)\mathbf{i} + (z\sin y)\mathbf{j} + (xy\ln z)\mathbf{k},$$

then $P(x, y, z) = xe^y$, $Q(x, y, z) = z\sin y$, and $R(x, y, z) = xy\ln z$. Hence (9) yields

$$\begin{aligned}\operatorname{div}\mathbf{F} &= \frac{\partial}{\partial x}(xe^y) + \frac{\partial}{\partial y}(z\sin y) + \frac{\partial}{\partial z}(xy\ln z)\\ &= e^y + z\cos y + \frac{xy}{z}.\end{aligned}$$

For instance, the value of div $\mathbf{F}$ at the point $(-3, 0, 2)$ is

$$\nabla\cdot\mathbf{F}(-3, 0, 2) = e^0 + 2\cos 0 + 0 = 3.$$

The analogues for divergence of Equations (7) and (8) are the formulas

$$\nabla\cdot(a\mathbf{F} + b\mathbf{G}) = a\,\nabla\cdot\mathbf{F} + b\,\nabla\cdot\mathbf{G} \tag{10}$$

and

$$\nabla\cdot(f\mathbf{G}) = (f)(\nabla\cdot\mathbf{G}) + (\nabla f)\cdot\mathbf{G}. \tag{11}$$

We ask you to verify these formulas in the problems. Note that the formula in (11)—in which f is a scalar function and $\mathbf{G}$ is a vector field—is consistent in that f and $\nabla \cdot \mathbf{G}$ are scalar functions, while ∇f and $\mathbf{G}$ are vector fields, so the sum on the right-hand side makes sense (and is a scalar function).

THE CURL OF A VECTOR FIELD

The **curl** of the vector field $\mathbf{F} = P\mathbf{i} + Q\mathbf{j} + R\mathbf{k}$ is the vector field curl $\mathbf{F}$ with the following definition:

$$\operatorname{curl} \mathbf{F} = \nabla \times \mathbf{F} = \begin{vmatrix} \mathbf{i} & \mathbf{j} & \mathbf{k} \\ \dfrac{\partial}{\partial x} & \dfrac{\partial}{\partial y} & \dfrac{\partial}{\partial z} \\ P & Q & R \end{vmatrix}. \tag{12}$$

Evaluation of the formal determinant expression in (12) yields

$$\operatorname{curl} \mathbf{F} = \mathbf{i}\left(\frac{\partial R}{\partial y} - \frac{\partial Q}{\partial z}\right) + \mathbf{j}\left(\frac{\partial P}{\partial z} - \frac{\partial R}{\partial x}\right) + \mathbf{k}\left(\frac{\partial Q}{\partial x} - \frac{\partial P}{\partial y}\right). \tag{13}$$

Although you may wish to memorize this formula, we recommend—because you will generally find it simpler—that in practice you set up and evaluate directly the formal determinant in (12). Our next example shows how easy this is.

EXAMPLE 6 For the vector field $\mathbf{F}$ of Example 5, the formula in (12) yields

$$\begin{aligned} \operatorname{curl} \mathbf{F} &= \begin{vmatrix} \mathbf{i} & \mathbf{j} & \mathbf{k} \\ \dfrac{\partial}{\partial x} & \dfrac{\partial}{\partial y} & \dfrac{\partial}{\partial z} \\ xe^y & z \sin y & xy \ln z \end{vmatrix} \\ &= \mathbf{i}(x \ln z - \sin y) + \mathbf{j}(-y \ln z) + \mathbf{k}(-xe^y). \end{aligned}$$

In particular, the value of curl $\mathbf{F}$ at the point $(3, \pi/2, e)$ is

$$\nabla \times \mathbf{F}(3, \pi/2, e) = 2\mathbf{i} - \frac{\pi}{2}\mathbf{j} - 3e^{\pi/2}\mathbf{k}.$$

In Section 17.7 we will see that if $\mathbf{v}$ is the velocity vector of a fluid flow, then the value of the vector curl $\mathbf{v}$ (where it is nonzero) at the point (x, y, z) determines the axis through (x, y, z) about which the fluid is rotating (or whirling or "curling").

The analogues for curl of Equations (10) and (11) are the formulas

$$\nabla \times (a\mathbf{F} + b\mathbf{G}) = a(\nabla \times \mathbf{F}) + b(\nabla \times \mathbf{G}) \tag{14}$$

and

$$\nabla \times (f\mathbf{G}) = (f)(\nabla \times \mathbf{G}) + (\nabla f) \times \mathbf{G} \tag{15}$$

that we ask you to verify in the problems.

EXAMPLE 7 If the function $f(x, y, z)$ has continuous second-order partial derivatives, show that

$$\operatorname{curl}(\operatorname{grad} f) = \mathbf{0}. \tag{16}$$

Solution Direct computation yields

$$\nabla \times \nabla f = \begin{vmatrix} \mathbf{i} & \mathbf{j} & \mathbf{k} \\ \dfrac{\partial}{\partial x} & \dfrac{\partial}{\partial y} & \dfrac{\partial}{\partial z} \\ \dfrac{\partial f}{\partial x} & \dfrac{\partial f}{\partial y} & \dfrac{\partial f}{\partial z} \end{vmatrix}$$

$$= \mathbf{i}\left(\frac{\partial^2 f}{\partial y\,\partial z} - \frac{\partial^2 f}{\partial z\,\partial y}\right) + \mathbf{j}\left(\frac{\partial^2 f}{\partial z\,\partial x} - \frac{\partial^2 f}{\partial x\,\partial z}\right) + \mathbf{k}\left(\frac{\partial^2 f}{\partial x\,\partial y} - \frac{\partial^2 f}{\partial y\,\partial x}\right).$$

Therefore

$$\nabla \times \nabla f = \mathbf{0}$$

because of the equality of continuous mixed second-order partial derivatives.

In Section 17.2 we define line integrals, which are used (for example) to compute the work done by a force field in moving a particle along a curved path. In Section 17.5 we discuss surface integrals, which are used (for example) to compute the rate at which a fluid with a known velocity vector field is moving across a given surface. The three basic integral theorems of vector calculus—Green's theorem (Section 17.4), the divergence theorem (Section 17.6), and Stokes' theorem (Section 17.7)—play much the same role for line and surface integrals that the fundamental theorem of calculus plays for ordinary single-variable integrals.

17.1 PROBLEMS

In each of Problems 1–10, illustrate the given vector field $\mathbf{F}$ by sketching several typical vectors in the field.

1. $\mathbf{F}(x, y) = \mathbf{i} + \mathbf{j}$
2. $\mathbf{F}(x, y) = 3\mathbf{i} - 2\mathbf{j}$
3. $\mathbf{F}(x, y) = x\mathbf{i} - y\mathbf{j}$
4. $\mathbf{F}(x, y) = 2\mathbf{i} + x\mathbf{j}$
5. $\mathbf{F}(x, y) = (x^2 + y^2)^{1/2}(x\mathbf{i} + y\mathbf{j})$
6. $\mathbf{F}(x, y) = (x^2 + y^2)^{-1/2}(x\mathbf{i} + y\mathbf{j})$
7. $\mathbf{F}(x, y, z) = \mathbf{j} + \mathbf{k}$
8. $\mathbf{F}(x, y, z) = \mathbf{i} + \mathbf{j} - \mathbf{k}$
9. $\mathbf{F}(x, y, z) = -x\mathbf{i} - y\mathbf{j}$
10. $\mathbf{F}(x, y, z) = x\mathbf{i} + y\mathbf{j} + z\mathbf{k}$

In each of Problems 11–20, calculate the divergence and curl of the given vector field $\mathbf{F}$.

11. $\mathbf{F}(x, y, z) = x\mathbf{i} + y\mathbf{j} + z\mathbf{k}$
12. $\mathbf{F}(x, y, z) = 3x\mathbf{i} - 2y\mathbf{j} - 4z\mathbf{k}$
13. $\mathbf{F}(x, y, z) = yz\mathbf{i} + xz\mathbf{j} + xy\mathbf{k}$
14. $\mathbf{F}(x, y, z) = x^2\mathbf{i} + y^2\mathbf{j} + z^2\mathbf{k}$
15. $\mathbf{F}(x, y, z) = xy^2\mathbf{i} + yz^2\mathbf{j} + zx^2\mathbf{k}$
16. $\mathbf{F}(x, y, z) = (2x - y)\mathbf{i} + (3y - 2z)\mathbf{j} + (7z - 3x)\mathbf{k}$
17. $\mathbf{F}(x, y, z) = (y^2 + z^2)\mathbf{i} + (x^2 + z^2)\mathbf{j} + (x^2 + y^2)\mathbf{k}$
18. $\mathbf{F}(x, y, z) = (e^{xz} \sin y)\mathbf{j} + (e^{xy} \cos z)\mathbf{k}$
19. $\mathbf{F}(x, y, z) = (x + \sin yz)\mathbf{i} + (y + \sin xz)\mathbf{j} + (z + \sin xy)\mathbf{k}$
20. $\mathbf{F}(x, y, z) = (x^2 e^{-z})\mathbf{i} + (y^3 \ln x)\mathbf{j} + (z \cosh y)\mathbf{k}$

Apply the definitions of gradient, divergence, and curl to establish the identities in Problems 21–27, where a and b denote constants, f and g denote differentiable scalar functions, and $\mathbf{F}$ and $\mathbf{G}$ denote differentiable vector fields.

21. $\nabla(af + bg) = a\,\nabla f + b\,\nabla g$
22. $\nabla \cdot (a\mathbf{F} + b\mathbf{G}) = a\,\nabla \cdot \mathbf{F} + b\,\nabla \cdot \mathbf{G}$
23. $\nabla \times (a\mathbf{F} + b\mathbf{G}) = a(\nabla \times \mathbf{F}) + b(\nabla \times \mathbf{G})$

24. $\nabla \cdot (f\mathbf{G}) = (f)(\nabla \cdot \mathbf{G}) + (\nabla f) \cdot \mathbf{G}$

25. $\nabla \times (f\mathbf{G}) = (f)(\nabla \times \mathbf{G}) + (\nabla f) \times \mathbf{G}$

26. $\nabla(f/g) = (g\,\nabla f - f\,\nabla g)/g^2$

27. $\nabla \cdot (\mathbf{F} \times \mathbf{G}) = \mathbf{G} \cdot (\nabla \times \mathbf{F}) - \mathbf{F} \cdot (\nabla \times \mathbf{G})$

Establish the identities in Problems 28–30 under the assumption that the scalar functions f and g and the vector field $\mathbf{F}$ are twice differentiable.

28. $\operatorname{div}(\operatorname{curl} \mathbf{F}) = 0$

29. $\operatorname{div}(\nabla fg) = f \operatorname{div}(\nabla g) + g \operatorname{div}(\nabla f) + 2(\nabla f) \cdot (\nabla g)$

30. $\operatorname{div}(\nabla f \times \nabla g) = 0$

Verify the identities in Problems 31–40, in which $\mathbf{a}$ is a constant vector, $\mathbf{r} = x\mathbf{i} + y\mathbf{j} + z\mathbf{k}$, and $r = |\mathbf{r}|$. Problems 33 and 34 imply that the divergence and curl of an inverse-square vector field both vanish identically.

31. $\nabla \cdot \mathbf{r} = 3$ and $\nabla \times \mathbf{r} = \mathbf{0}$

32. $\nabla \cdot (\mathbf{a} \times \mathbf{r}) = 0$ and $\nabla \times (\mathbf{a} \times \mathbf{r}) = 2\mathbf{a}$

33. $\nabla \cdot \dfrac{\mathbf{r}}{r^3} = 0$

34. $\nabla \times \dfrac{\mathbf{r}}{r^3} = \mathbf{0}$

35. $\nabla r = \dfrac{\mathbf{r}}{r}$

36. $\nabla\left(\dfrac{1}{r}\right) = -\dfrac{\mathbf{r}}{r^3}$

37. $\nabla \cdot (r\mathbf{r}) = 4r$

38. $\nabla \cdot (\nabla r) = 0$

39. $\nabla(\ln r) = \dfrac{\mathbf{r}}{r^2}$

40. $\nabla(r^{10}) = 10r^8\mathbf{r}$

17.2 Line Integrals

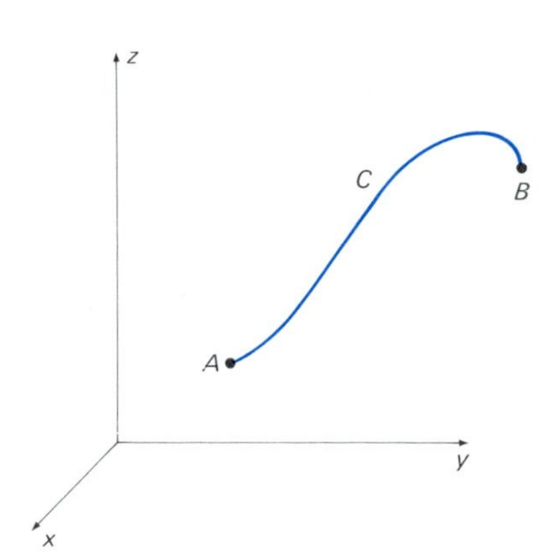

17.4 A wire of variable density in the shape of the smooth curve C

To motivate the definition of the line integral, we imagine a thin wire shaped like the smooth curve C with endpoints A and B, as in Fig. 17.4. Suppose that the wire has variable density given at the point (x, y, z) by the known continuous function $f(x, y, z)$, in units such as grams per (linear) centimeter. Let

$$x = x(t), \qquad y = y(t), \qquad z = z(t), \qquad t \text{ in } [a, b] \tag{1}$$

be a smooth parametrization of the curve C, with $t = a$ corresponding to the initial point A of the curve and $t = b$ to its terminal point B.

In order to *approximate* the total mass m of the curved wire, we begin with a partition

$$a = t_0 < t_1 < t_2 < \cdots < t_{n-1} < t_n = b$$

of $[a, b]$ into n subintervals, all having the same length $\Delta t = (b - a)/n$. These subdivision points of $[a, b]$ produce, via our parametrization, a physical subdivision of the wire into short curve segments, as shown in Fig. 17.5. We let P_i denote the point $(x(t_i), y(t_i), z(t_i))$ for $i = 0, 1, 2, \ldots, n$. Then the points $P_0, P_1, \ldots, P_n$ are the subdivision points of C.

From our study of arc length in Sections 13.2 and 14.4, we know that the arc length Δs_i of the segment of C from P_{i-1} to P_i is

$$\begin{aligned}\Delta s_i &= \int_{t_{i-1}}^{t_i} \sqrt{[x'(t)]^2 + [y'(t)]^2 + [z'(t)]^2}\,dt \\ &= \sqrt{[x'(t_i^*)]^2 + [y'(t_i^*)]^2 + [z'(t_i^*)]^2}\,\Delta t \end{aligned} \tag{2}$$

for some number t_i^* in the interval $[t_{i-1}, t_i]$. This is a consequence of the average value theorem for integrals in Section 5.5.

If we multiply the density at the point (x_i^*, y_i^*, z_i^*) by the length Δs_i of the segment of C containing it, we obtain an estimate of the mass of that segment of C. So, after we sum over all the segments, we have an estimate of the total mass of the wire:

$$m \approx \sum_{i=1}^{n} f(x(t_i^*), y(t_i^*), z(t_i^*))\,\Delta s_i.$$

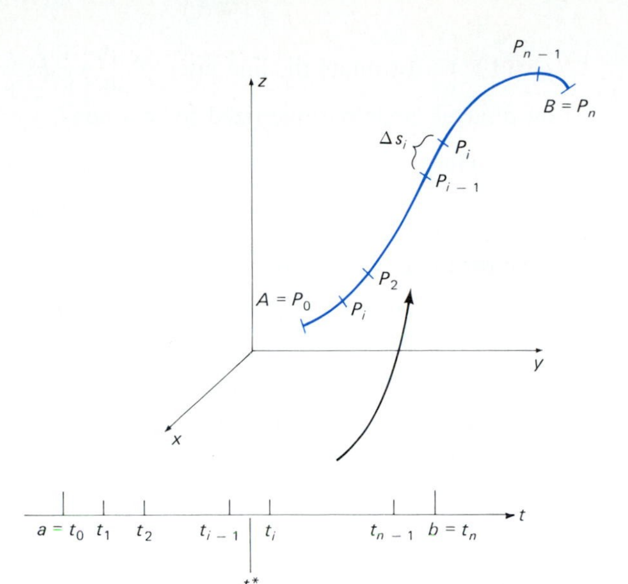

17.5 The partition of the interval $[a, b]$ determines a related partition of the curve C into short arcs.

The limit of this sum as $\Delta t \to 0$ should be the actual mass m. This is our motivation for the definition of the line integral of the function f along the curve C, denoted by

$$\int_C f(x, y, z)\, ds.$$

Definition *Line Integral with Respect to Arc Length*

Suppose that the function $f(x, y, z)$ is continuous at each point of the smooth parametric curve C from A to B, as given in (1). Then the **line integral of f along C from A to B with respect to arc length** is defined to be

$$\int_C f(x, y, z)\, ds = \lim_{\Delta t \to 0} \sum_{i=1}^{n} f(x(t_i^*), y(t_i^*), z(t_i^*))\, \Delta s_i. \tag{3}$$

When we substitute (2) into (3), we recognize the result as the limit of a Riemann sum. Therefore,

$$\int_C f(x, y, z)\, ds = \int_a^b f(x(t), y(t), z(t))\sqrt{[x'(t)]^2 + [y'(t)]^2 + [z'(t)]^2}\, dt. \tag{4}$$

Thus we may evaluate the line integral $\int_C f(x, y, z)\, ds$ by expressing everything in terms of the parameter t, including the symbolic arc length element

$$ds = \sqrt{[x'(t)]^2 + [y'(t)]^2 + [z'(t)]^2}\, dt.$$

The result—the right-hand side in (4)—is an **ordinary integral with respect to the single real variable** t.

A curve C that lies in the xy-plane may be regarded as a space curve for which $z = 0$. In this case we suppress the variable z in (4) and write

$$\int_C f(x, y)\, ds = \int_a^b f(x(t), y(t))\sqrt{[x'(t)]^2 + [y'(t)]^2}\, dt. \tag{5}$$

EXAMPLE 1 Evaluate the line integral $\int_C xy\,ds$, where C is the first-quadrant quarter circle parametrized by $x = \cos t$, $y = \sin t$, $0 \leqq t \leqq \pi/2$.

Solution Here

$$ds = \sqrt{(-\sin t)^2 + (\cos t)^2}\,dt = dt,$$

so the formula in (5) yields

$$\int_C xy\,ds = \int_0^{\pi/2} \cos t \sin t\,dt = \left[\tfrac{1}{2}\sin^2 t\right]_0^{\pi/2} = \tfrac{1}{2}.$$

Let us now return to the physical wire, and denote its density function by the more usual $\rho(x, y, z)$. The mass of a small piece of length Δs is $\Delta m = \rho\,\Delta s$, so we write

$$dm = \rho(x, y, z)\,ds$$

for its (symbolic) element of mass. Then the **mass** m of the wire and its **centroid** $(\bar{x}, \bar{y}, \bar{z})$ are defined as follows:

$$m = \int_C dm = \int_C \rho\,ds, \qquad \bar{y} = \frac{1}{m}\int_C y\,dm,$$
$$\bar{x} = \frac{1}{m}\int_C x\,dm, \qquad \bar{z} = \frac{1}{m}\int_C z\,dm. \tag{6}$$

Note the analogy with Equations (2) and (4) in Section 16.6. The **moment of inertia** of the wire about a given axis is

$$I = \int_C w^2\,dm, \tag{7}$$

where $w = w(x, y, z)$ denotes the perpendicular distance from the point (x, y, z) to the axis in question.

EXAMPLE 2 Find the centroid of a wire with density $\rho = kz$ if it has the shape of the helix C with parametrization

$$x = 3\cos t, \qquad y = 3\sin t, \qquad z = 4t, \qquad 0 \leqq t \leqq \pi.$$

Solution The mass element of the wire is

$$dm = \rho\,ds = kz\,ds = 4kt\sqrt{(-3\sin t)^2 + (3\cos t)^2 + 4^2}\,dt = 20kt\,dt.$$

Hence the formulas in (6) yield

$$m = \int_C \rho\,ds = \int_0^{\pi} 20kt\,dt = 10k\pi^2;$$

$$\bar{x} = \frac{1}{m}\int_C \rho x\,ds = \frac{1}{10k\pi^2}\int_0^{\pi} 60kt\cos t\,dt$$

$$= \frac{6}{\pi^2}\Big[\cos t + t\sin t\Big]_0^{\pi} = -\frac{12}{\pi^2} \approx -1.22;$$

$$\bar{y} = \frac{1}{m}\int_C \rho y\,ds = \frac{1}{10k\pi^2}\int_0^{\pi} 60kt\sin t\,dt$$

$$= \frac{6}{\pi^2}\Big[\sin t - t\cos t\Big]_0^{\pi} = -\frac{6}{\pi} \approx 1.91;$$

$$\bar{z} = \frac{1}{m}\int_C \rho z\,ds = \frac{1}{10k\pi^2}\int_0^{\pi} 80kt^2\,dt = \frac{8}{\pi^2}\left[\frac{1}{3}t^3\right]_0^{\pi} = \frac{8\pi}{3} \approx 8.38.$$

So the centroid of the wire is located at the point with approximate coordinates $(-1.22, 1.91, 8.38)$.

LINE INTEGRALS WITH RESPECT TO COORDINATE VARIABLES

A different type of line integral is obtained by replacing Δs_i in (3) by

$$\Delta x_i = x(t_i) - x(t_{i-1}) = x'(t_i^*)\Delta t.$$

The **line integral of** f **along** C **with respect to** x is defined to be

$$\int_C f(x, y, z)\,dx = \lim_{\Delta t \to 0} \sum_{i=1}^{n} f(x(t_i^*), y(t_i^*), z(t_i^*))\,\Delta x_i;$$

Thus

$$\int_C f(x, y, z)\,dx = \int_a^b f(x(t), y(t), z(t))x'(t)\,dt. \tag{8a}$$

Similarly, the line integrals of f along C **with respect to** y and **with respect to** z are given by

$$\int_C f(x, y, z)\,dy = \int_a^b f(x(t), y(t), z(t))y'(t)\,dt \tag{8b}$$

and

$$\int_C f(x, y, z)\,dz = \int_a^b f(x(t), y(t), z(t))z'(t)\,dt. \tag{8c}$$

The three integrals in (8) often occur together. If P, Q, and R are continuous functions of the variables x, y, and z, then we write (indeed, *define*)

$$\int_C P\,dx + Q\,dy + R\,dz = \int_C P\,dx + \int_C Q\,dy + \int_C R\,dz. \tag{9}$$

The line integrals in (8) and (9) are evaluated by expressing x, y, z, dx, dy, and dz in terms of t as determined by a suitable parametrization of the curve C. The result is an ordinary single-variable integral. For instance, if C is a parametric plane curve parametrized over the interval $[a, b]$, then

$$\int_C P\,dx + Q\,dy = \int_a^b [P(x(t), y(t))x'(t) + Q(x(t), y(t))y'(t)]\,dt.$$

EXAMPLE 3 Evaluate the line integral

$$\int_C y\,dx + z\,dy + x\,dz$$

where C is the parametric curve $x = t$, $y = t^2$, $z = t^3$, $0 \leqq t \leqq 1$.

Solution Because $dx = dt$, $dy = 2t\,dt$, and $dz = 3t^2\,dt$, substitution in terms of t yields

$$\int_C y\,dx + z\,dy + x\,dz = \int_0^1 t^2\,dt + t^3(2t\,dt) + t(3t^2\,dt)$$
$$= \int_0^1 (t^2 + 3t^3 + 2t^4)\,dt$$
$$= \left[\tfrac{1}{3}t^3 + \tfrac{3}{4}t^4 + \tfrac{2}{5}t^5\right]_0^1 = \tfrac{89}{60}.$$

There is an important difference between the line integral in (4) with respect to arc length s and the line integrals in (9) with respect to x, y, and z. Suppose that the *orientation* of the curve C (the direction in which it is traced as t increases) is reversed. Then, because of the terms $x'(t)$, $y'(t)$, and $z'(t)$ in (8), the *sign* of the line integral in (9) is changed. But this reversal of orientation does *not* change the value of the line integral in (4). We may express this by writing

$$\int_{-C} f\,ds = \int_C f\,ds, \tag{10}$$

in contrast with the formula

$$\int_{-C} P\,dx + Q\,dy + R\,dz = -\int_C P\,dx + Q\,dy + R\,dz. \tag{11}$$

Here the symbol $-C$ denotes the curve C with its orientation reversed (from B to A, rather than from A to B). It is proved in advanced calculus that for either type of line integral, two one-to-one parametrizations of the smooth curve C that *agree in orientation* will give the same value.

If the curve C consists of finitely many smooth curves joined at corner points, we say that C is **piecewise smooth.** The value of a line integral along C is then defined to be the sum of its values along the smooth segments of C.

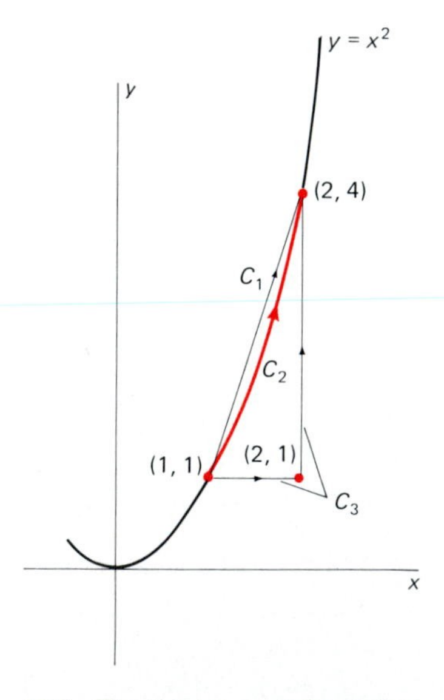

17.6 The three arcs of Example 4

EXAMPLE 4 Evaluate the line integral

$$\int_C y\,dx + 2x\,dy$$

for C each of the three curves C_1, C_2, and C_3 from $A(1, 1)$ to $B(2, 4)$ shown in Fig. 17.6.

Solution The straight line segment C_1 from A to B can be parametrized by $x = 1 + t$, $y = 1 + 3t$, $0 \leqq t \leqq 1$. So

$$\int_{C_1} y\,dx + 2x\,dy = \int_0^1 (1 + 3t)(dt) + 2(1 + t)(3\,dt) = \int_0^1 (7 + 9t)\,dt = \tfrac{23}{2}.$$

Next, the arc C_2 of the parabola $y = x^2$ from A to B has the parametrization $x = x$, $y = x^2$, $1 \leqq x \leqq 2$, so

$$\int_{C_2} y\,dx + 2x\,dy = \int_1^2 (x^2)(dx) + 2(x)(2x\,dx) = \int_1^2 5x^2\,dx = \tfrac{35}{3}.$$

Finally, along the straight line segment from (1, 1) to (2, 1) we have $y = 1$ and $dy = 0$. Along the vertical segment from (2, 1) to (2, 4) we have

$x = 2$ and $dx = 0$. Therefore,

$$\int_{C_3} y\,dx + 2x\,dy = \int_1^2 [(1)(dx) + (2x)(0)] + \int_1^4 [(y)(0) + (4)(dy)]$$
$$= \int_1^2 1\,dx + \int_1^4 4\,dy = 13.$$

Example 4 also shows that we may obtain different values for the line integral from A to B if we evaluate it along different curves from A to B. In Section 17.3 we give a sufficient condition for the line integral

$$\int_C P\,dx + Q\,dy + R\,dz$$

to have the same value for *all* smooth curves C from A to B, and thus for the integral to be *independent of path*.

LINE INTEGRALS AND WORK

Suppose now that $\mathbf{F} = P\mathbf{i} + Q\mathbf{j} + R\mathbf{k}$ is a force field that is defined on a region containing the curve C from A to B. Suppose also that C has a parametrization

$$x = x(t), \qquad y = y(t), \qquad z = z(t), \qquad t \text{ in } [a, b]$$

with a *nonzero* velocity vector

$$\mathbf{v} = \mathbf{i}\frac{dx}{dt} + \mathbf{j}\frac{dy}{dt} + \mathbf{k}\frac{dz}{dt}.$$

The speed associated with this velocity vector is

$$v = |\mathbf{v}| = \left[\left(\frac{dx}{dt}\right)^2 + \left(\frac{dy}{dt}\right)^2 + \left(\frac{dz}{dt}\right)^2\right]^{1/2}.$$

Recall that the *unit tangent vector* to the curve C is

$$\mathbf{T} = \frac{\mathbf{v}}{v} = \frac{1}{v}\left(\frac{dx}{dt}\mathbf{i} + \frac{dy}{dt}\mathbf{j} + \frac{dz}{dt}\mathbf{k}\right).$$

We want to approximate the work W done by the force field $\mathbf{F}$ in moving a particle along the curve C from A to B. Subdivide C as indicated in Fig. 17.7. Think of $\mathbf{F}$ moving the particle from P_{i-1} to P_i, two consecutive division points of C. The work ΔW_i done is approximately the product of the distance Δs_i from P_{i-1} to P_i (measured along C) and the tangential component $\mathbf{F} \cdot \mathbf{T}$ of the force $\mathbf{F}$ at a typical point $(x(t_i^*), y(t_i^*), z(t_i^*))$ between P_{i-1} and P_i. Thus

$$\Delta W_i \approx \mathbf{F}(x(t_i^*), y(t_i^*), z(t_i^*)) \cdot \mathbf{T}(t_i^*)\,\Delta s_i,$$

so the total work W is given approximately by

$$W \approx \sum_{i=1}^{n} \mathbf{F}(x(t_i^*), y(t_i^*), z(t_i^*)) \cdot \mathbf{T}(t_i^*)\,\Delta s_i.$$

This approximation suggests that we *define* the **work** W as

$$W = \int_C \mathbf{F} \cdot \mathbf{T}\,ds. \tag{12}$$

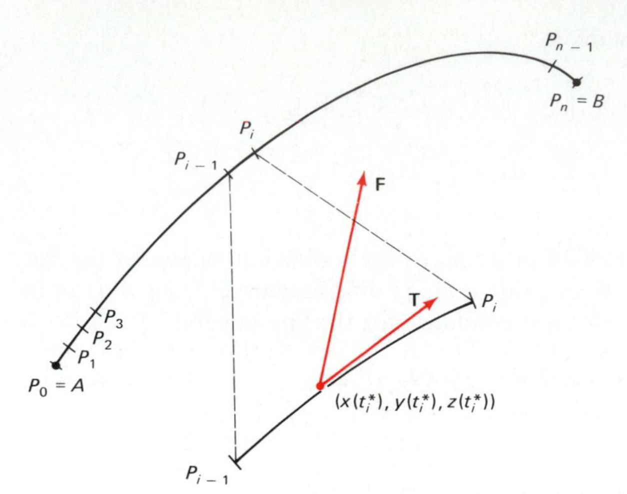

17.7 The component of **F** along C from P_{i-1} to P_i is $\mathbf{F} \cdot \mathbf{T}$.

Thus *work is the integral with respect to arc length of the tangential component of the force.* Intuitively, we may regard $dW = \mathbf{F} \cdot \mathbf{T}\, ds$ as the infinitesimal element of work done by the tangential component $\mathbf{F} \cdot \mathbf{T}$ of the force in moving the particle along the arc length element ds. The line integral in (12) is then the "sum" of all these infinitesimal elements of work.

It is customary to write formally

$$\mathbf{r} = x\mathbf{i} + y\mathbf{j} + z\mathbf{k}, \qquad d\mathbf{r} = \mathbf{i}\, dx + \mathbf{j}\, dy + \mathbf{k}\, dz,$$

and

$$\mathbf{T}\, ds = \left(\frac{dx}{ds}\mathbf{i} + \frac{dy}{ds}\mathbf{j} + \frac{dz}{ds}\mathbf{k}\right) ds = d\mathbf{r}.$$

With this notation, (12) takes the form

$$W = \int_C \mathbf{F} \cdot d\mathbf{r} \tag{13}$$

that is common in physics and engineering texts.

To evaluate the line integral in (12) or (13), we express its integrand in terms of the parameter t, as usual. Thus

$$\begin{aligned} W &= \int_C \mathbf{F} \cdot \mathbf{T}\, ds \\ &= \int_a^b (P\mathbf{i} + Q\mathbf{j} + R\mathbf{k}) \cdot \frac{1}{v}\left(\frac{dx}{dt}\mathbf{i} + \frac{dy}{dt}\mathbf{j} + \frac{dz}{dt}\mathbf{k}\right) v\, dt \\ &= \int_a^b \left(P\frac{dx}{dt} + Q\frac{dy}{dt} + R\frac{dz}{dt}\right) dt. \end{aligned}$$

Therefore,

$$W = \int_C P\, dx + Q\, dy + R\, dz. \tag{14}$$

This computation yields an important relation between the two types of line integrals we have defined in this section.

Theorem *Equivalent Line Integrals*

Suppose that the vector field $\mathbf{F} = P\mathbf{i} + Q\mathbf{j} + R\mathbf{k}$ has continuous component functions and that $\mathbf{T}$ is the unit tangent vector to the smooth curve C. Then

$$\int_C \mathbf{F} \cdot \mathbf{T}\, ds = \int_C P\, dx + Q\, dy + R\, dz. \tag{15}$$

Note that if the orientation of the curve C is reversed, then the sign of the right-hand integral is changed according to Equation (11), while the sign of the left-hand integral is changed because $\mathbf{T}$ is replaced by $-\mathbf{T}$.

EXAMPLE 5 The work done by the force field $\mathbf{F} = y\mathbf{i} + z\mathbf{j} + x\mathbf{k}$ in moving a particle from (0, 0, 0) to (1, 1, 1) along the twisted cubic $x = t$, $y = t^2$, $z = t^3$ is given by the line integral

$$W = \int_C \mathbf{F} \cdot \mathbf{T}\, ds = \int_C y\, dx + z\, dy + x\, dz,$$

and we computed the value of this integral in Example 3. Hence $W = 89/60$.

EXAMPLE 6 Find the work done by the inverse-square law force field

$$\mathbf{F}(x, y, z) = \frac{k\mathbf{r}}{r^3} = \frac{k(x\mathbf{i} + y\mathbf{j} + z\mathbf{k})}{(x^2 + y^2 + z^2)^{3/2}}$$

in moving a particle along the straight line segment C from (0, 4, 0) to (0, 4, 3).

Solution Along C we have $x = 0$, $y = 4$, and $z = z$. We choose z as the parameter. Because $dx = 0 = dy$, the formula in (14) gives

$$W = \int_C \frac{k(x\, dx + y\, dy + z\, dz)}{(x^2 + y^2 + z^2)^{3/2}} = \int_0^3 \frac{kz}{(16 + z^2)^{3/2}}\, dz = \left[\frac{-k}{(16 + z^2)^{1/2}}\right]_0^3 = \frac{k}{20}.$$

17.2 PROBLEMS

In Problems 1–5, evaluate the line integrals

$$\int_C f(x, y)\, ds, \quad \int_C f(x, y)\, dx, \quad \text{and} \quad \int_C f(x, y)\, dy$$

along the indicated parametric curve.

1. $f(x, y) = x^2 + y^2$, $x = 4t - 1$, $y = 3t + 1$, $-1 \leqq t \leqq 1$
2. $f(x, y) = x$; $x = t$, $y = t^2$, $0 \leqq t \leqq 1$
3. $f(x, y) = x + y$; $x = e^t + 1$, $y = e^t - 1$, $0 \leqq t \leqq \ln 2$
4. $f(x, y) = 2x - y$; $x = \sin t$, $y = \cos t$, $0 \leqq t \leqq \pi/2$
5. $f(x, y) = xy$; $x = 3t$, $y = t^4$, $0 \leqq t \leqq 1$
6. Evaluate

$$\int_C xy\, dx + (x + y)\, dy$$

where C is the part of the graph of $y = x^2$ from $(-1, -1)$ to (2, 4).

7. Evaluate

$$\int_C y^2\, dx + x\, dy$$

where C is the part of the graph of $x = y^3$ from $(-1, -1)$ to $(1, 1)$.

8. Evaluate

$$\int_C y\sqrt{x}\,dx + x\sqrt{x}\,dy$$

where C is the part of the graph of $y^2 = x^3$ from $(1, 1)$ to $(4, 8)$.

9. Evaluate the line integral

$$\int_C x^2 y\,dx + xy^3\,dy$$

where C consists of the line segments from $(-1, 1)$ to $(2, 1)$ and from $(2, 1)$ to $(2, 5)$.

10. Evaluate

$$\int_C (x + 2y)\,dx + (2x - y)\,dy$$

where C consists of the line segments from $(3, 2)$ to $(3, -1)$ and from $(3, -1)$ to $(-2, -1)$.

In Problems 11–15, evaluate the line integral

$$\int_C \mathbf{F} \cdot \mathbf{T}\,ds$$

along the indicated curve C.

11. $\mathbf{F} = z\mathbf{i} + x\mathbf{j} - y\mathbf{k}$; $x = t$, $y = t^2$, $z = t^3$, $0 \leqq t \leqq 1$

12. $\mathbf{F} = yz\mathbf{i} + xz\mathbf{j} + xy\mathbf{k}$; C is the straight line segment from $(2, -1, 3)$ to $(4, 2, -1)$

13. $\mathbf{F} = y\mathbf{i} - x\mathbf{j} + z\mathbf{k}$; $x = \sin t$, $y = \cos t$, $z = 2t$, $0 \leqq t \leqq \pi$

14. $\mathbf{F} = (2x + 3y)\mathbf{i} + (3x + 2y)\mathbf{j} + 3z^2\mathbf{k}$; C is the path from $(0, 0, 0)$ to $(4, 2, 3)$ that consists of three line segments parallel to the x-axis, the y-axis, and the z-axis, in that order.

15. $\mathbf{F} = yz^2\mathbf{i} + xz^2\mathbf{j} + 2xyz\mathbf{k}$; C is the path from $(-1, 2, -2)$ to $(1, 5, 2)$ consisting of three line segments parallel to the z-axis, the x-axis, and the y-axis, in that order.

16. Find $\int_C xyz\,ds$ if C is the line segment from $(1, -1, 2)$ to $(3, 2, 5)$.

17. Find $\int_C (2x + 9xy)\,ds$ where C is the curve $x = t$, $y = t^2$, $z = t^3$, $0 \leqq t \leqq 1$.

18. Evaluate $\int_C xy\,ds$ where C is the elliptical helix $x = 4\cos t$, $y = 9\sin t$, $z = 7t$, $0 \leqq t \leqq 5\pi/2$.

19. Find the centroid of a uniform thin wire shaped like the semicircle $x^2 + y^2 = a^2$, $y \geqq 0$.

20. Find the moments of inertia about the x- and y-axes of the wire of Problem 19.

21. Find the mass and centroid of a wire with constant density $\rho = k$ and shaped like the helix $x = 3\cos t$, $y = 3\sin t$, $z = 4t$, $0 \leqq t \leqq 2\pi$.

22. Find the moment of inertia about the z-axis of the wire of Example 1 of this section.

23. A wire shaped like the first-quadrant portion of the circle $x^2 + y^2 = a^2$ has density $\rho = kxy$ at the point (x, y). Find its mass, centroid, and moment of inertia about each coordinate axis.

24. Find the work done by the inverse-square force field of Example 6 in moving a particle from $(1, 0, 0)$ to $(0, 3, 4)$. Integrate first along the line segment from $(1, 0, 0)$ to $(5, 0, 0)$ and then along a path on the sphere with equation $x^2 + y^2 + z^2 = 25$. Note that the second integral is automatically zero (why?).

25. Imagine an infinitely long and uniformly charged wire that coincides with the z-axis. The electric force that it exerts on a unit charge at the point $(x, y) \neq (0, 0)$ in the xy-plane is $\mathbf{F} = k(x\mathbf{i} + y\mathbf{j})/(x^2 + y^2)$. Find the work done by $\mathbf{F}$ in moving a unit charge along the straight-line segment from:

(a) $(1, 0)$ to $(1, 1)$;

(b) $(1, 1)$ to $(0, 1)$.

26. Show that if $\mathbf{F}$ is a *constant* force field, then it does zero work on a particle that moves once uniformly counterclockwise around the unit circle in the xy-plane.

27. Show that if $\mathbf{F} = k\mathbf{r} = k(x\mathbf{i} + y\mathbf{j})$, then $\mathbf{F}$ does zero work on a particle that moves once uniformly counterclockwise around the unit circle in the xy-plane.

28. Find the work done by the force field $\mathbf{F} = -y\mathbf{i} + x\mathbf{j}$ in moving a particle counterclockwise once around the unit circle in the xy-plane.

29. Let C be a curve on the unit sphere $x^2 + y^2 + z^2 = 1$. Explain why the inverse-square force field of Example 6 does zero work in moving a particle along C.

In each of Problems 30–32, the given curve C joins the points P and Q in the xy-plane. The point P represents the top of a ten-story building and Q is a point on the ground 100 ft from the base of the building. A 150-lb person slides down a frictionless slide shaped like the curve from P to Q under the influence of gravitational force $\mathbf{F} = -150\mathbf{j}$. In each problem show that $\mathbf{F}$ does the same amount of work on the person, $W = 15{,}000$ ft-lb, as though he or she dropped straight down to the ground.

30. C is the straight line segment $y = x$ from $P(100, 100)$ to $Q(0, 0)$.

31. C is the circular arc $x = 100\sin t$, $y = 100\cos t$ from $P(0, 100)$ to $Q(100, 0)$.

32. C is the parabolic arc $y = (0.01)x^2$ from $P(100, 100)$ to $Q(0, 0)$.

17.3 Independence of Path

Let $\mathbf{F} = P\mathbf{i} + Q\mathbf{j} + R\mathbf{k}$ be a vector field with continuous component functions. By the theorem in Section 17.2, we know that

$$\int_C \mathbf{F} \cdot \mathbf{T}\, ds = \int_C P\, dx + Q\, dy + R\, dz \tag{1}$$

for any piecewise smooth curve C. Thus the two sides of Equation (1) are two different ways of writing the same line integral. In this section we discuss the question of whether this line integral has the *same value* for any *two* curves with the same endpoints (the same initial point and the same terminal point).

Definition *Independence of Path*

The line integral in (1) is said to be **independent of the path in the region** D provided that given any two points A and B of D, the integral has the same value along every piecewise smooth curve or **path** in D from A to B. In this case, we may write

$$\int_C \mathbf{F} \cdot \mathbf{T}\, ds = \int_A^B \mathbf{F} \cdot \mathbf{T}\, ds \tag{2}$$

because the value of the integral depends only on the points A and B and not on the particular choice of the path C joining them.

For a tangible interpretation of independence of path, let us think of walking along the curve C from point A to point B in the plane where a wind with velocity vector field $\mathbf{w}(x, y)$ is blowing. Suppose that when we are at (x, y) the wind exerts a force $\mathbf{F} = k\mathbf{w}(x, y)$ on us, k being a constant that depends on our size and shape (and perhaps other factors as well). Then, by the formula in Equation (12) of Section 17.2, the amount of work the wind does on us as we walk along C is given by

$$W = \int_C \mathbf{F} \cdot \mathbf{T}\, ds = k \int_C \mathbf{w} \cdot \mathbf{T}\, ds. \tag{3}$$

This is the wind's contribution to our trip from A to B. In this context, the question of independence of path is the question whether the wind's work W depends on *which* path from point A to point B we choose.

EXAMPLE 1 Suppose that there is a steady wind blowing toward the northeast with $\mathbf{w} = 10\mathbf{i} + 10\mathbf{j}$ in fps units, so its speed is $|\mathbf{w}| = 10\sqrt{2} \approx 14$ ft/s—about 10 mi/h. Assume that $k = 0.5$, so the wind exerts 0.5 lb of force for each foot per second of its velocity. Then $\mathbf{F} = 5\mathbf{i} + 5\mathbf{j}$, so the formula in (3) yields

$$W = \int_C \langle 5, 5\rangle \cdot \mathbf{T}\, ds = \int_C 5\, dx + 5\, dy \tag{4}$$

for the work done on us by the wind as we walk along C.

For instance, if C is the straight path $x = 10t$, $y = 10t$ $(0 \leqq t \leqq 1)$ from (0, 0) to (10, 10), then (4) gives

$$W = \int_0^1 (5)(10\,dt) + (5)(10\,dt) = 100 \int_0^1 1\,dt = 100$$

ft-lb of work.

On the other hand, if C is the parabolic path $y = (0.1)x^2$, $0 \leqq x \leqq 10$ from the same initial point (0, 0) to the same terminal point (10, 10), then (4) yields

$$W = \int_0^{10} (5)(dx) + (5)[(0.2)x\,dx] = \int_0^{10} (5 + x)\,dx$$

$$= \left[5x + \tfrac{1}{2}x^2\right]_0^{10} = 100$$

ft-lb of work, the same as before. Indeed, it follows from Theorem 1 of this section that the line integral in (4) is independent of path, so the wind does 100 ft-lb of work along any path from (0, 0) to (10, 10).

EXAMPLE 2 Now suppose that $\mathbf{w} = -2y\mathbf{i} + 2x\mathbf{j}$. This wind is blowing counterclockwise around the origin, as in a hurricane with its eye at the origin. With $k = 0.5$ as before, $\mathbf{F} = -y\mathbf{i} + x\mathbf{j}$, so the work integral is

$$W = \int_C \mathbf{F} \cdot \mathbf{T}\,ds = \int_C -y\,dx + x\,dy. \tag{5}$$

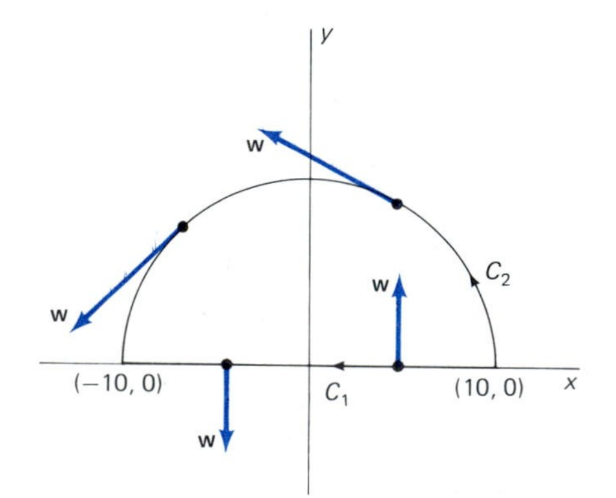

17.8 Around and through the eye of the hurricane

If we walk from (10, 0) to (−10, 0) along the straight path C_1 (through the eye of the hurricane!), then—as illustrated in Fig. 17.8—the wind is always perpendicular to our unit tangent vector $\mathbf{T}$. Hence $\mathbf{F} \cdot \mathbf{T} = 0$, and therefore

$$W = \int_{C_1} \mathbf{F} \cdot \mathbf{T}\,ds = \int_{C_1} -y\,dx + x\,dy = 0.$$

On the other hand, if we walk along the semicircular path C_2 shown in Fig. 17.8, then $\mathbf{w}$ remains tangent to our path, so $\mathbf{F} \cdot \mathbf{T} = |\mathbf{F}| = 10$ at each point. In this case,

$$W = \int_{C_2} -y\,dx + x\,dy = \int_{C_2} \mathbf{F} \cdot \mathbf{T}\,ds = (10)(10\pi) = 100\pi.$$

The fact that we get different values along different paths from (10, 0) to (−10, 0) shows that the line integral in (5) is *not* independent of path.

The following theorem tells when a given line integral is independent of path and when it is not.

> ***Theorem 1*** *Independence of Path*
>
> The line integral $\int_C \mathbf{F} \cdot \mathbf{T}\,ds$ is independent of path in the plane region D if and only if $\mathbf{F} = \nabla f$ for some function f defined on D.

Proof Suppose first that $\mathbf{F} = \nabla f = \langle \partial f/\partial x, \partial f/\partial y, \partial f/\partial z \rangle$ and that C is a path from A to B in D parametrized as usual with parameter t in $[a, b]$. Then, by Equation (1),

$$\begin{aligned}
\int_C \mathbf{F}\cdot\mathbf{T}\,ds &= \int_C \frac{\partial f}{\partial x}\,dx + \frac{\partial f}{\partial y}\,dy + \frac{\partial f}{\partial z}\,dz \\
&= \int_a^b \left(\frac{\partial f}{\partial x}\cdot\frac{dx}{dt} + \frac{\partial f}{\partial y}\cdot\frac{dy}{dt} + \frac{\partial f}{\partial z}\cdot\frac{dz}{dt}\right) dt \\
&= \int_a^b D_t(f(x(t), y(t), z(t)))\,dt \\
&= f(x(b), y(b), z(b)) - f(x(a), y(a), z(a)).
\end{aligned}$$

And therefore

$$\int_C \mathbf{F}\cdot\mathbf{T}\,ds = f(B) - f(A). \tag{6}$$

The last step follows from the fundamental theorem of calculus. Equation (6) shows that the value of the line integral depends only upon the points A and B and is therefore independent of the choice of the particular path C. This proves the *if* part of Theorem 1.

To prove the *only if* part, we suppose that the line integral is independent of path in D. Choose a *fixed* point $A_0 = A_0(x_0, y_0, z_0)$ in D, and let $B = B(x, y, z)$ be an arbitrary point in D. Given any path C from A_0 to B in D, we *define* the function f by means of the equation

$$f(x, y, z) = \int_C \mathbf{F}\cdot\mathbf{T}\,ds = \int_{(x_0,y_0,z_0)}^{(x,y,z)} \mathbf{F}\cdot\mathbf{T}\,ds. \tag{7}$$

Because of the hypothesis of independence of path, the resulting value of $f(x, y, z)$ depends only upon (x, y, z) and not upon the particular path C used. We shall omit the verification that $\nabla f = \mathbf{F}$ (see Problem 19 in Section 17.7). ■

As an application of Theorem 1, we see that the vector field $\mathbf{F} = -y\mathbf{i} + x\mathbf{j}$ of Example 2 is not the gradient of any scalar function f because $\int \mathbf{F}\cdot\mathbf{T}\,ds$ is not independent of path. More precisely, $\int \mathbf{F}\cdot\mathbf{T}\,ds$ is not independent of path in any region that either includes or encloses the origin.

Definition *Conservative Fields and Potential Functions*
The vector field $\mathbf{F}$ defined on a region D is called **conservative** provided that there exists a scalar function f defined on D such that

$$\mathbf{F} = \nabla f \tag{8}$$

at each point of D. In this case f is called a **potential function** for the vector field $\mathbf{F}$.†

Note that the formula in (6) has the form of a "fundamental theorem of calculus for line integrals," with the potential function f playing the role

† In some physical applications the scalar function f is called a *potential function* for the vector field $\mathbf{F}$ provided that $\mathbf{F} = -\nabla f$.

of an antiderivative. If the line integral $\int_C \mathbf{F} \cdot \mathbf{T}\, ds$ is known to be independent of path, then Theorem 1 guarantees that the vector field $\mathbf{F}$ is conservative and that the formula in (7) yields a potential function for $\mathbf{F}$.

EXAMPLE 3 Find a potential function for the conservative vector field

$$\mathbf{F}(x, y) = (6xy - y^3)\mathbf{i} + (4y + 3x^2 - 3xy^2)\mathbf{j}. \tag{9}$$

Solution Because we are given the information that $\mathbf{F}$ is a conservative field, the line integral $\int_C \mathbf{F} \cdot \mathbf{T}\, ds$ is independent of path by Theorem 1. Therefore, we may apply the formula in Equation (7) to find a scalar potential function for $\mathbf{F}$. Let C be the straight line path from $A(0, 0)$ to $B(x_1, y_1)$ parametrized by $x = x_1 t$, $y = y_1 t$, $0 \leqq t \leqq 1$. Then Equation (7) yields

$$\begin{aligned}
f(x_1, y_1) &= \int_A^B \mathbf{F} \cdot \mathbf{T}\, ds \\
&= \int_A^B (6xy - y^3)\, dx + (4y + 3x^2 - 3xy^2)\, dy \\
&= \int_0^1 (6x_1 y_1 t^2 - y_1^3 t^3)(x_1\, dt) + (4y_1 t + 3x_1^2 t^2 - 3x_1 y_1^2 t^3)(y_1\, dt) \\
&= \int_0^1 (4y_1^2 t + 9x_1^2 y_1 t^2 - 4x_1 y_1^3 t^3)\, dt \\
&= \Big[2y_1^2 t^2 + 3x_1^2 y_1 t^3 - x_1 y_1^3 t^4\Big]_0^1 \\
&= 2y_1^2 + 3x_1^2 y_1 - x_1 y_1^3.
\end{aligned}$$

At this point we delete the subscripts because (x_1, y_1) is an arbitrary point of the plane. Thus we obtain the potential function

$$f(x, y) = 2y^2 + 3x^2 y - xy^3$$

for the vector field $\mathbf{F}$ in (9). As a check, we can differentiate f to obtain

$$\frac{\partial f}{\partial x} = 6xy - y^3, \qquad \frac{\partial f}{\partial y} = 4y + 3x^2 - 3xy^2.$$

But how did we know in advance that the vector field $\mathbf{F}$ was conservative? The answer is provided by the next theorem, which is proved in Section 1.7 of C. H. Edwards, Jr. and D. E. Penney, *Elementary Differential Equations with Applications* (Englewood Cliffs, N.J.: Prentice-Hall, 1985, 1989).

Theorem 2 *Conservative Fields and Potential Functions*

Suppose that the functions $P(x, y)$ and $Q(x, y)$ are continuous and have continuous first-order partial derivatives in the open rectangle $R = \{(x, y) | a < x < b, c < y < d\}$. Then the vector field $\mathbf{F} = P\mathbf{i} + Q\mathbf{j}$ is conservative in R—and hence has a potential function $f(x, y)$ defined on R—if and only if

$$\frac{\partial P}{\partial y} = \frac{\partial Q}{\partial x} \tag{10}$$

at each point of R.

Observe that the vector field $\mathbf{F}$ in (9), where $P(x, y) = 6xy - y^3$ and $Q(x, y) = 4y + 3x^2 - 3xy^2$, satisfies the criterion in (10) because

$$\frac{\partial P}{\partial y} = 6x - 3y^2 = \frac{\partial Q}{\partial x}.$$

When this sufficient condition for the existence of a potential function is satisfied, the method illustrated in the next example is usually an easier way to find a potential function than the evaluation of the line integral in (7)—the method used in Example 3.

EXAMPLE 4 Given

$$P(x, y) = 6xy - y^3 \quad \text{and} \quad Q(x, y) = 4y + 3x^2 - 3xy^2,$$

note that P and Q satisfy the condition $\partial P/\partial y = \partial Q/\partial x$. Find a potential function $f(x, y)$ such that

$$\frac{\partial f}{\partial x} = 6xy - y^3 \quad \text{and} \quad \frac{\partial f}{\partial y} = 4y + 3x^2 - 3xy^2. \tag{11}$$

Solution Upon integrating the first of these two equations with respect to x, we get

$$f(x, y) = 3x^2y - xy^3 + \xi(y), \tag{12}$$

where $\xi(y)$ is an "arbitrary function" of y alone—it acts as a "constant of integration" with respect to x because its derivative with respect to x is zero. We next determine $\xi(y)$ by imposing the second condition in (11):

$$\frac{\partial f}{\partial y} = 3x^2 - 3xy^2 + \xi'(y) = 4y + 3x^2 - 3xy^2.$$

It follows that $\xi'(y) = 4y$, so $\xi(y) = 2y^2 + C$. When we set $C = 0$ and substitute the result in (12), we get the same potential function

$$f(x, y) = 3x^2y - xy^3 + 2y^2$$

that we found by entirely different methods in Example 3.

CONSERVATIVE FORCE FIELDS

Given a conservative force field $\mathbf{F}$, it is customary in physics to introduce a minus sign and write $\mathbf{F} = -\nabla V$. Then $V(x, y, z)$ is called the **potential energy** at the point (x, y, z). With $f = -V$ in Equation (6), we have

$$W = \int_A^B \mathbf{F} \cdot \mathbf{T}\, ds = V(A) - V(B), \tag{13}$$

and this means that the work W done by $\mathbf{F}$ in moving a particle from A to B is equal to the *decrease* in potential energy.

For example, a brief computation shows that

$$\nabla\left(-\frac{k}{\sqrt{x^2 + y^2 + z^2}}\right) = \frac{k(x\mathbf{i} + y\mathbf{j} + z\mathbf{k})}{(x^2 + y^2 + z^2)^{3/2}} = \mathbf{F}$$

for the inverse-square force field of Example 6 in Section 17.2. With $V = k/(x^2 + y^2 + z^2)^{1/2}$, Equation (16) then gives

$$\int_{(0,4,0)}^{(0,4,3)} \mathbf{F} \cdot \mathbf{T}\, ds = \frac{k}{(0^2 + 4^2 + 0^2)^{1/2}} - \frac{k}{(0^2 + 4^2 + 3^2)^{1/2}} = \frac{k}{20},$$

as we also found by direct integration.

Here is the reason why the expression *conservative field* is used. Suppose that a particle of mass m moves from A to B under the influence of the conservative force $\mathbf{F}$, with position vector $\mathbf{r}(t)$, $a \leqq t \leqq b$. Then Newton's law $\mathbf{F}(\mathbf{r}(t)) = m\mathbf{r}''(t) = m\mathbf{v}'(t)$ gives

$$\int_A^B \mathbf{F} \cdot \mathbf{T}\, ds = \int_a^b m\mathbf{v}'(t) \cdot \frac{\mathbf{v}(t)}{v}\, v\, dt = \int_a^b mD_t\left[\tfrac{1}{2}\mathbf{v}(t) \cdot \mathbf{v}(t)\right] dt$$

$$= \left[\tfrac{1}{2}m[v(t)]^2\right]_a^b.$$

Thus with the abbreviations v_A for $v(a)$ and v_B for $v(b)$, we see that

$$\int_A^B \mathbf{F} \cdot \mathbf{T}\, ds = \tfrac{1}{2}m(v_B)^2 - \tfrac{1}{2}m(v_A)^2. \tag{14}$$

By equating the right-hand sides in Equations (13) and (14), we get the formula

$$\tfrac{1}{2}m(v_A)^2 + V(A) = \tfrac{1}{2}m(v_B)^2 + V(B). \tag{15}$$

This is the law of **conservation of mechanical energy** for a particle moving under the influence of a *conservative* force field—the sum of its kinetic energy and potential energy remains constant.

17.3 PROBLEMS

Apply the method of Example 4 to find a potential function for each of the vector fields in Problems 1–12.

1. $\mathbf{F}(x, y) = (2x + 3y)\mathbf{i} + (3x + 2y)\mathbf{j}$
2. $\mathbf{F}(x, y) = (4x - y)\mathbf{i} + (6y - x)\mathbf{j}$
3. $\mathbf{F}(x, y) = (3x^2 + 2y^2)\mathbf{i} + (4xy + 6y^2)\mathbf{j}$
4. $\mathbf{F}(x, y) = (2xy^2 + 3x^2)\mathbf{i} + (2x^2y + 4y^3)\mathbf{j}$
5. $\mathbf{F}(x, y) = \left(x^3 + \dfrac{y}{x}\right)\mathbf{i} + (y^2 \ln x)\mathbf{j}$
6. $\mathbf{F}(x, y) = (1 + ye^{xy})\mathbf{i} + (2y + xe^{xy})\mathbf{j}$
7. $\mathbf{F}(x, y) = (\cos x + \ln y)\mathbf{i} + \left(\dfrac{x}{y} + e^y\right)\mathbf{j}$
8. $\mathbf{F}(x, y) = (x + \arctan y)\mathbf{i} + \dfrac{x + y}{1 + y^2}\mathbf{j}$
9. $\mathbf{F}(x, y) = (3x^2y^3 + y^4)\mathbf{i} + (3x^3y^2 + y^4 + 4xy^3)\mathbf{j}$
10. $\mathbf{F}(x, y) = (e^x \sin y + \tan y)\mathbf{i} + (e^x \cos y + x \sec^2 y)\mathbf{j}$
11. $\mathbf{F}(x, y) = \left(\dfrac{2x}{y} - \dfrac{3y^2}{x^4}\right)\mathbf{i} + \left(\dfrac{2y}{x^3} - \dfrac{x^2}{y^2} + \dfrac{1}{\sqrt{y}}\right)\mathbf{j}$
12. $\mathbf{F}(x, y) = \dfrac{2x^{5/2} - 3y^{5/3}}{2x^{5/2}y^{2/3}}\mathbf{i} + \dfrac{3y^{5/3} - 2x^{5/2}}{3x^{3/2}y^{5/3}}\mathbf{j}$

In each of Problems 13–16, apply the method of Example 3 to find a potential function for the indicated vector field.

13. The vector field of Problem 3
14. The vector field of Problem 4
15. The vector field of Problem 9
16. The vector field of Problem 6

In each of Problems 17–22, show that the given line integral is independent of path in the entire xy-plane, and then calculate the value of the line integral.

17. $\int_{(0,0)}^{(1,2)} (y^2 + 2xy)\,dx + (x^2 + 2xy)\,dy$

18. $\int_{(0,0)}^{(1,1)} (2x - 3y)\,dx + (2y - 3x)\,dy$

19. $\int_{(0,0)}^{(1,-1)} 2xe^y\,dx + x^2e^y\,dy$

20. $\int_{(0,0)}^{(2,\pi)} \cos y\,dx - x\sin y\,dy$

21. $\int_{(\pi/2,\pi/2)}^{(\pi,\pi)} (\sin y + y\cos x)\,dx + (\sin x + x\cos y)\,dy$

22. $\int_{(0,0)}^{(1,-1)} (e^y + ye^x)\,dx + (e^x + xe^y)\,dy$

Find a potential function for each of the conservative vector fields in Problems 23–25.

23. $\mathbf{F}(x, y, z) = yz\mathbf{i} + xz\mathbf{j} + xy\mathbf{k}$

24. $\mathbf{F}(x, y, z) = (2x - y - z)\mathbf{i} + (2y - x)\mathbf{j} + (2z - x)\mathbf{k}$

25. $\mathbf{F}(x, y, z) = (y\cos z - yze^x)\mathbf{i} + (x\cos z - ze^x)\mathbf{j} - (xy\sin z + ye^x)\mathbf{k}$

26. Let $\mathbf{F} = (-y\mathbf{i} + x\mathbf{j})/(x^2 + y^2)$ for x and y not both zero. Calculate the values of $\int_C \mathbf{F}\cdot\mathbf{T}\,ds$ along both the upper and lower halves of the circle $x^2 + y^2 = 1$ from $(1, 0)$ to $(-1, 0)$. Is there a function $f = f(x, y)$ defined for x and y not both zero such that $\nabla f = \mathbf{F}$? Why?

27. Show that if the force field $\mathbf{F} = P\mathbf{i} + Q\mathbf{j}$ is conservative, then $\partial P/\partial y = \partial Q/\partial x$. Show that the force field of Problem 26 satisfies the condition $\partial P/\partial y = \partial Q/\partial x$, but nevertheless is *not* conservative.

28. Suppose that the force field $\mathbf{F} = P\mathbf{i} + Q\mathbf{j} + R\mathbf{k}$ is conservative. Show that $\partial P/\partial y = \partial Q/\partial x$, $\partial P/\partial z = \partial R/\partial x$, and $\partial Q/\partial z = \partial R/\partial y$.

29. Apply the result of Problem 28 together with Theorem 1 to show that

$$\int_C 2xy\,dx + x^2\,dy + y^2\,dz$$

is not independent of path.

30. Let

$$\mathbf{F} = \mathbf{F}(x, y, z) = yz\mathbf{i} + (xz + y)\mathbf{j} + (xy + 1)\mathbf{k}.$$

Define the function f by

$$f(x, y, z) = \int_C \mathbf{F}\cdot\mathbf{T}\,ds$$

where C is the line segment from $(0, 0, 0)$ to (x, y, z). Determine f by evaluating this line integral, and then show that $\nabla f = \mathbf{F}$.

17.4 Green's Theorem

Green's theorem relates a line integral around a simple closed plane curve C to an ordinary double integral over the plane region R bounded by C. Suppose that the curve C is piecewise smooth—it consists of finitely many parametric arcs with continuous nonzero velocity vectors. Then C has a unit tangent vector $\mathbf{T}$ except possibly at finitely many *corner points.* The **positive** or **counterclockwise** direction along C is the one determined by a parametrization $\mathbf{r}(t)$ of C such that the region R remains on the *left* as the point $\mathbf{r}(t)$ traces the boundary curve C. That is, the vector obtained from the unit tangent vector $\mathbf{T}$ by a counterclockwise rotation through 90° always points *into* the region R, as shown in Fig. 17.9. The symbol

$$\oint_C P\,dx + Q\,dy$$

denotes a line integral along or around C in this positive direction. A reversed arrow on the circle through the integral sign indicates a line integral around C in the opposite direction, which we naturally call either the **negative** or the **clockwise** direction.

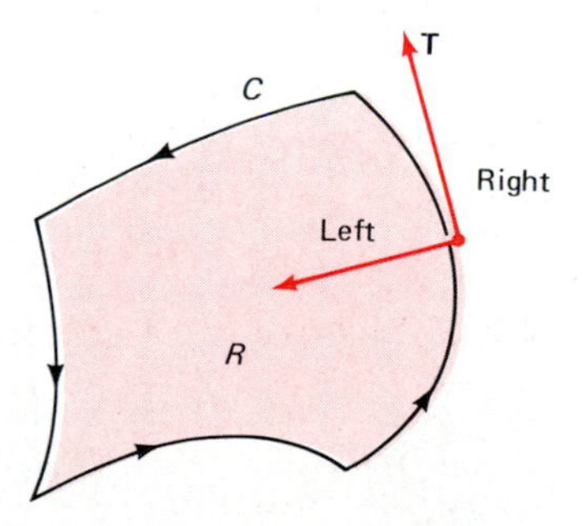

17.9 Positive orientation of the curve C: The region R within C is to the *left* of the unit tangent vector $\mathbf{T}$.

The following result first appeared (in an equivalent form) in a booklet on the applications of mathematics to electricity and magnetism, published privately in 1828 by the self-taught English mathematical physicist George Green (1793–1841).

Green's Theorem

Let C be a piecewise smooth simple closed curve that bounds the region R in the plane. Suppose that the functions $P(x, y)$ and $Q(x, y)$ are continuous and have continuous first-order partial derivatives in R. Then

$$\oint_C P\,dx + Q\,dy = \iint_R \left(\frac{\partial Q}{\partial x} - \frac{\partial P}{\partial y} \right) dA. \tag{1}$$

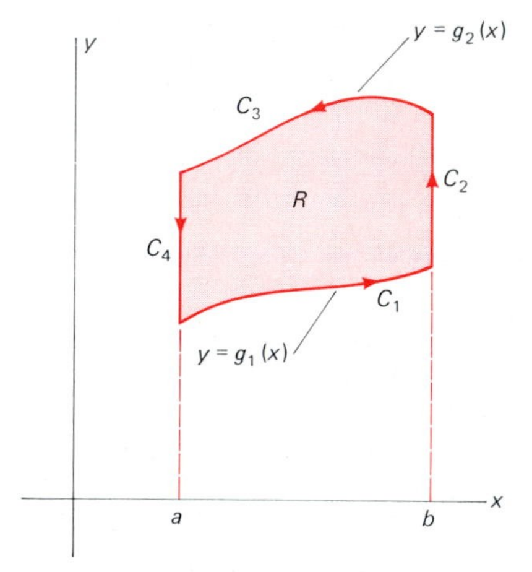

17.10 The boundary curve C is the union of the four arcs C_1, C_2, C_3, C_4.

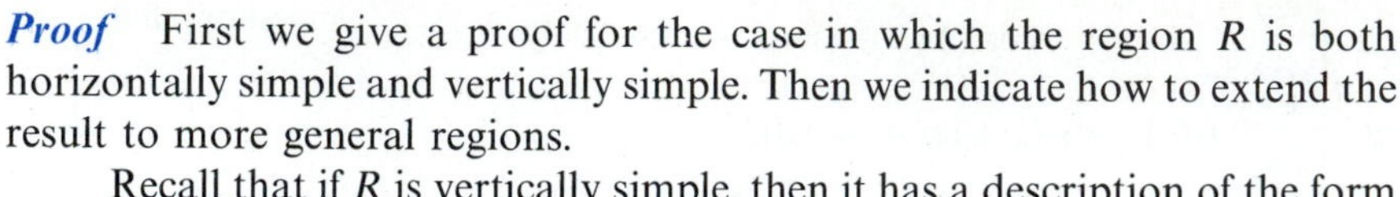

Proof First we give a proof for the case in which the region R is both horizontally simple and vertically simple. Then we indicate how to extend the result to more general regions.

Recall that if R is vertically simple, then it has a description of the form $g_1(x) \leqq y \leqq g_2(x)$, $a \leqq x \leqq b$. The boundary curve C is then the union of the four arcs C_1, C_2, C_3, and C_4 of Fig. 17.10, positively oriented as indicated there. Hence

$$\int_C P\,dx = \int_{C_1} P\,dx + \int_{C_2} P\,dx + \int_{C_3} P\,dx + \int_{C_4} P\,dx.$$

The integrals along C_2 and C_4 are both zero, because on those two curves $x(t)$ is constant, so that $dx = x'(t)\,dt = 0$. It remains to compute the integrals along C_1 and C_3.

The point $(x, g_1(x))$ traces C_1 as x increases from a to b, while the point $(x, g_2(x))$ traces C_3 as x *decreases* from b to a. Hence

$$\begin{aligned}\oint_C P\,dx &= \int_a^b P(x, g_1(x))\,dx + \int_b^a P(x, g_2(x))\,dx \\ &= -\int_a^b [P(x, g_2(x)) - P(x, g_1(x))]\,dx \\ &= -\int_a^b \int_{g_1(x)}^{g_2(x)} \frac{\partial P}{\partial y}\,dy\,dx\end{aligned}$$

by the fundamental theorem of calculus. Thus

$$\oint_C P\,dx = -\iint_R \frac{\partial P}{\partial y}\,dA. \tag{2}$$

In Problem 28 we ask you to show in a similar way that

$$\oint_C Q\,dy = +\iint_R \frac{\partial Q}{\partial x}\,dA \tag{3}$$

if the region R is horizontally simple. We then obtain Equation (1), the conclusion of Green's theorem, simply by adding Equations (2) and (3). ■

The complete proof of Green's theorem for more general regions is beyond the scope of an elementary text. But the typical region R that appears in practice can be subdivided into smaller regions $R_1, R_2, \ldots, R_k$ that are both vertically and horizontally simple. Green's theorem for the region R then follows from the fact that it applies to each of the regions $R_1, R_2, \dots, R_k$ (see Problem 29).

For example, the horseshoe-shaped region R of Fig. 17.11 can be subdivided into the regions R_1 and R_2, each of which is both horizontally simple and vertically simple. We also subdivide the boundary C of R and write $C_1 \cup D_1$ for the boundary of R_1 and $C_2 \cup D_2$ for the boundary of R_2 (see Fig. 17.11). Applying Green's theorem separately to the regions R_1 and R_2, we get

$$\oint_{C_1 \cup D_1} P\,dx + Q\,dy = \iint_{R_1} \left(\frac{\partial Q}{\partial x} - \frac{\partial P}{\partial y}\right) dA$$

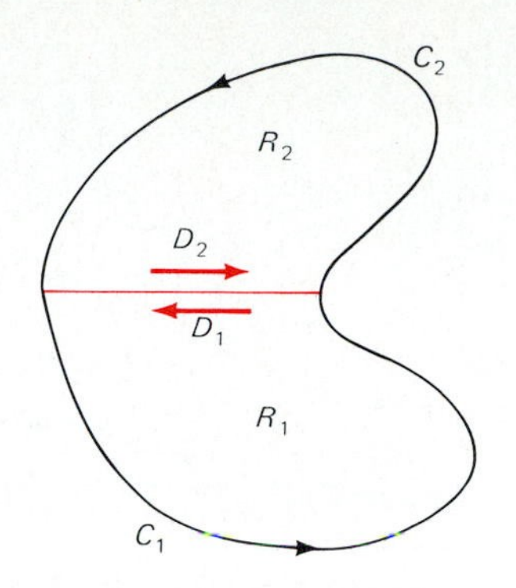

17.11 Decomposing the region R into two horizontally and vertically simple regions by means of a crosscut.

and

$$\oint_{C_2 \cup D_2} P\,dx + Q\,dy = \iint_{R_2} \left(\frac{\partial Q}{\partial x} - \frac{\partial P}{\partial y}\right) dA.$$

When we add these two equations, the result is Equation (1), Green's theorem for the region R, because the two line integrals along D_1 and D_2 cancel. This occurs because D_1 and D_2 represent the same curve with opposite orientations, so that

$$\int_{D_2} P\,dx + Q\,dy = -\int_{D_1} P\,dx + Q\,dy$$

by Equation (11) in Section 17.2. It therefore follows that

$$\int_{C_1 \cup D_1 \cup C_2 \cup D_2} P\,dx + Q\,dy = \int_{C_1 \cup C_2} P\,dx + Q\,dy = \oint_C P\,dx + Q\,dy.$$

Similarly, Green's theorem for the region shown in Fig. 17.12 could be established by subdividing it into the four simple regions as indicated.

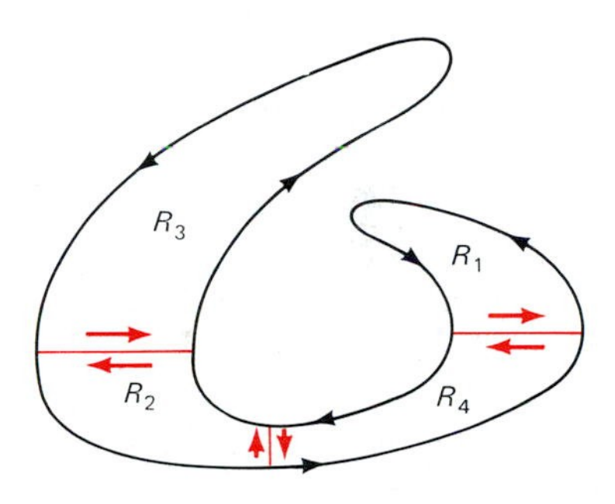

17.12 Many important regions can be decomposed into simple regions by using one or more crosscuts.

EXAMPLE 1 Use Green's theorem to evaluate the line integral

$$\oint_C (2y + \sqrt{9 + x^3})\,dx + (5x + e^{\arctan y})\,dy,$$

where C is the circle $x^2 + y^2 = 4$.

Solution With $P(x, y) = 2y + (9 + x^3)^{1/2}$ and $Q(x, y) = 5x + \exp(\arctan y)$, we see that

$$\frac{\partial Q}{\partial x} - \frac{\partial P}{\partial y} = 5 - 2 = 3.$$

Because C bounds R, a circular disk with area 4π, Green's theorem therefore implies that the given line integral is equal to

$$\iint_R 3\,dA = 3 \cdot 4\pi = 12\pi.$$

EXAMPLE 2 Evaluate the line integral

$$\oint_C 3xy\,dx + 2x^2\,dy$$

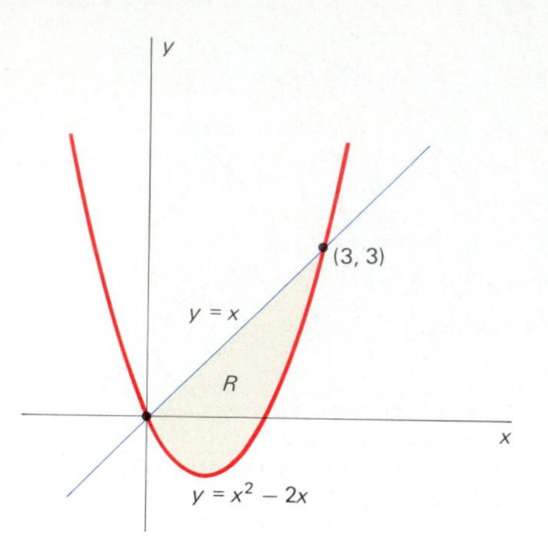

17.13 The region of Example 2

where C is the boundary of the region R shown in Fig. 17.13; it is bounded above by the line $y = x$ and below by the parabola $y = x^2 - 2x$.

Solution To evaluate the line integral directly, we would need to parametrize separately the line and the parabola. Instead, we apply Green's theorem with $P = 3xy$ and $Q = 2x^2$, so that

$$\frac{\partial Q}{\partial x} - \frac{\partial P}{\partial y} = 4x - 3x = x.$$

Then

$$\oint_C 3xy\,dx + 2x^2\,dy = \iint_R x\,dA$$

$$= \int_0^3 \int_{x^2-2x}^{x} x\,dy\,dx = \int_0^3 \Big[xy\Big]_{x^2-2x}^{x}$$

$$= \int_0^3 (3x^2 - x^3)\,dx = \Big[x^3 - \tfrac{1}{4}x^4\Big]_0^3 = \tfrac{27}{4}.$$

In Examples 1 and 2 we found the double integral easier to evaluate directly than the line integral. Sometimes the situation is reversed. The following consequence of Green's theorem illustrates the technique of evaluating a double integral $\iint_R f(x, y)\,dA$ by converting it into a line integral $\oint_C P\,dx + Q\,dy$. To do this we must be able to find functions $P(x, y)$ and $Q(x, y)$ such that $\partial Q/\partial x - \partial P/\partial y = f(x, y)$. As in the proof of the following result, this is sometimes easy.

> *Corollary to Green's Theorem*
> The area A of the region R bounded by the piecewise smooth simple closed curve C is given by
>
> $$A = \tfrac{1}{2}\oint_C -y\,dx + x\,dy = -\oint_C y\,dx = \oint_C x\,dy. \quad (4)$$

Proof With $P(x, y) = -y$ and $Q(x, y) \equiv 0$, Green's theorem gives

$$-\oint_C y\,dx = \iint_R 1\,dA = A.$$

Similarly, with $P(x, y) \equiv 0$ and $Q(x, y) = x$, we obtain $A = \oint_C x\,dy$. With $P(x, y) = -y/2$ and $Q(x, y) = x/2$, Green's theorem gives

$$\tfrac{1}{2}\oint_C -y\,dx + x\,dy = \iint_R \left(\frac{1}{2} + \frac{1}{2}\right) dA = A. \quad \blacksquare$$

EXAMPLE 3 Apply the corollary to Green's theorem to find the area A bounded by the ellipse $x^2/a^2 + y^2/b^2 = 1$.

Solution With the parametrization $x = a \cos t,\ \ y = b \sin t,\ \ 0 \leqq t \leqq 2\pi$, Equation (4) gives

$$A = \oint_{\text{ellipse}} x\, dy = \int_0^{2\pi} (a \cos t)(b \cos t\, dt)$$

$$= \tfrac{1}{2}ab \int_0^{2\pi} (1 + \cos 2t)\, dt = \pi ab.$$

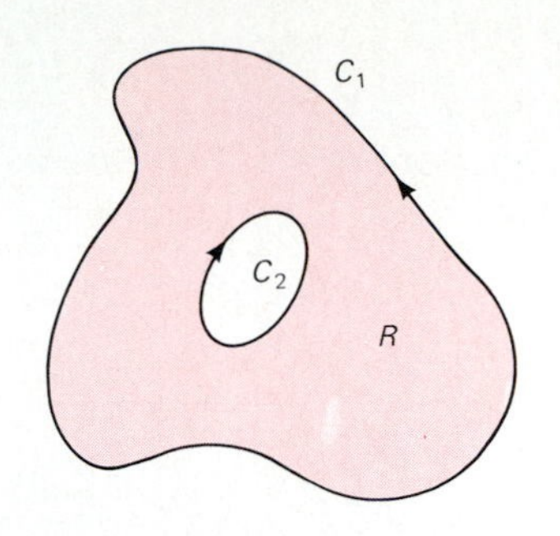

17.14 An annular region—the boundary consists of two simple closed curves, one within the other.

By using the technique of subdividing a region into simpler ones, Green's theorem may be extended to regions having boundaries that consist of two or more simple closed curves. For example, consider the annular region R of Fig. 17.14, with boundary C consisting of the two simple closed curves C_1 and C_2. The positive direction along C—the direction for which the region R always lies on the left—is counterclockwise on the outer curve C_1 but clockwise on the inner curve C_2.

We subdivide R into two regions R_1 and R_2 by using two crosscuts, as shown in Fig. 17.15. Applying Green's theorem to each of these subregions, we get

$$\iint_R (Q_x - P_y)\, dA = \iint_{R_1} (Q_x - P_y)\, dA + \iint_{R_2} (Q_x - P_y)\, dA$$

$$= \oint_{C_1} (P\, dx + Q\, dy) + \oint_{C_2} (P\, dx + Q\, dy)$$

$$= \oint_C P\, dx + Q\, dy.$$

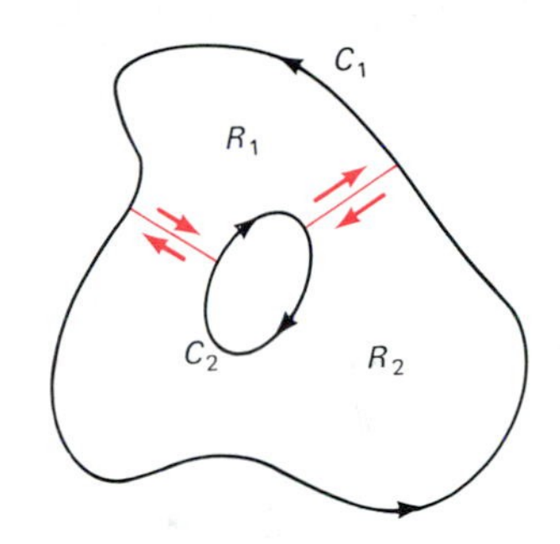

17.15 Two crosscuts convert the annular region into the union of two ordinary regions.

Thus we obtain Green's theorem for the given region R. What makes this proof work is that the opposite line integrals along the two crosscuts cancel each other.

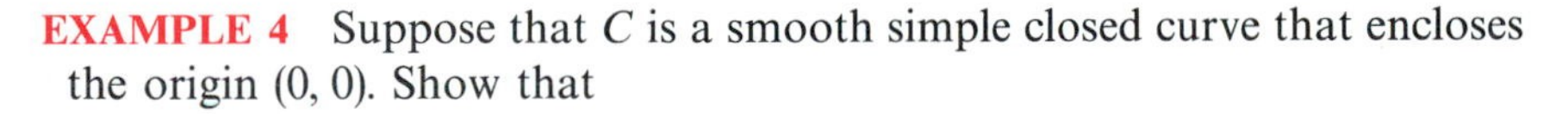

EXAMPLE 4 Suppose that C is a smooth simple closed curve that encloses the origin (0, 0). Show that

$$\oint_C \frac{-y\, dx + x\, dy}{x^2 + y^2} = 2\pi,$$

but that this integral is zero if C does *not* enclose the origin.

Solution With $P(x, y) = -y/(x^2 + y^2)$ and $Q(x, y) = x/(x^2 + y^2)$, a brief computation gives $\partial Q/\partial x - \partial P/\partial y = 0$ when x and y are not both zero. If the region R bounded by C does not contain the origin, then P and Q and their derivatives are continuous on R. Hence Green's theorem implies that the integral in question is zero.

If C does enclose the origin, then we enclose the origin in a small circle C_a interior to C (as in Fig. 17.16) and parametrize this circle by $x = a \cos t$, $y = a \sin t$. Then Green's theorem, applied to the region R between C and C_a, gives

$$\oint_C \frac{-y\, dx + x\, dy}{x^2 + y^2} + \oint_{C_a} \frac{-y\, dx + x\, dy}{x^2 + y^2} = \iint_R 0\, dA = 0.$$

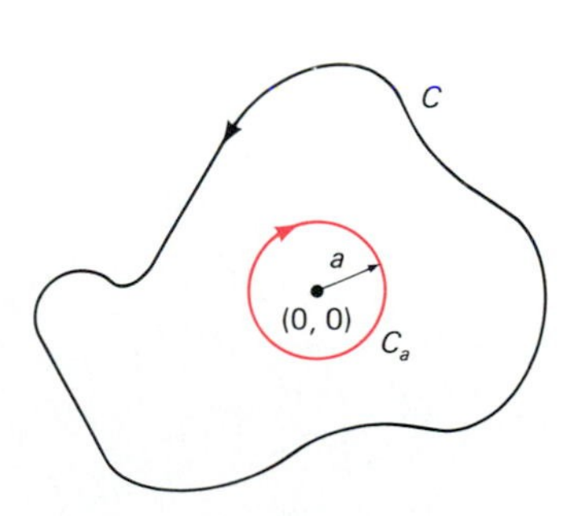

17.16 Use the small circle C_a if C encloses the origin.

Therefore,

$$\oint_C \frac{-y\,dx + x\,dy}{x^2 + y^2} = \oint_{C_a} \frac{-y\,dx + x\,dy}{x^2 + y^2}$$

$$= \int_0^{2\pi} \frac{(-a \sin t)(-a \sin t\,dt) + (a \cos t)(a \cos t\,dt)}{(a \cos t)^2 + (a \sin t)^2}$$

$$= \int_0^{2\pi} 1\,dt = 2\pi.$$

The result of Example 4 can be interpreted in terms of the polar coordinate angle $\theta = \arctan(y/x)$. Because

$$d\theta = \frac{-y\,dx + x\,dy}{x^2 + y^2},$$

the line integral of Example 4 measures the net change in θ as we go around the curve C once in a counterclockwise direction. This net change is 2π if C encloses the origin and is zero otherwise.

THE DIVERGENCE AND FLUX OF A VECTOR FIELD

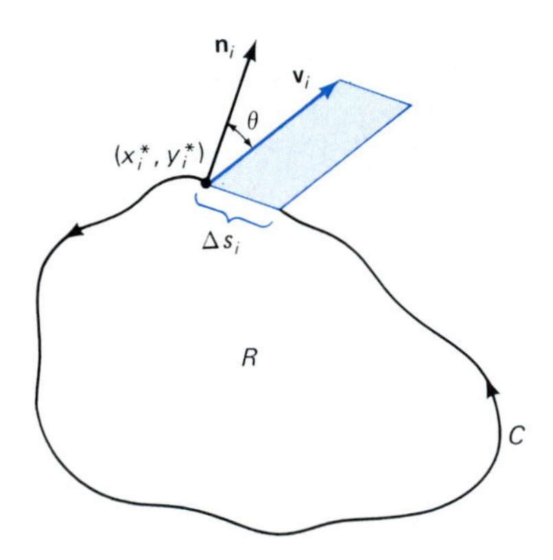

17.17 The area of the parallelogram approximates the fluid flow across Δs_i in unit time.

Now let us consider the steady flow of a thin layer of fluid in the plane (perhaps like a sheet of water spreading across the floor). Let $\mathbf{v}(x, y)$ be its velocity vector field and $\rho(x, y)$ the density of the fluid at the point (x, y). The term *steady flow* means that $\mathbf{v}$ and ρ depend only upon x and y and *not* upon time t. We want to compute the rate at which the fluid flows out of the region R bounded by a simple closed curve C (Fig. 17.17). We seek the net rate of outflow—the actual outflow minus the inflow.

Let Δs_i be a short segment of the curve C, and let (x_i^*, y_i^*) be an end point of Δs_i. Then the area of the portion of the fluid that flows out of R across Δs_i per unit time is approximately the area of the parallelogram of Fig. 17.17. This is the parallelogram spanned by the segment Δs_i and the vector $\mathbf{v}_i = \mathbf{v}(x_i^*, y_i^*)$. Suppose that $\mathbf{n}_i$ is the unit normal to C at the point (x_i^*, y_i^*), the normal pointing *out* of R. Then the area of this parallelogram is

$$(|\mathbf{v}_i| \cos \theta)\,\Delta s_i = \mathbf{v}_i \cdot \mathbf{n}_i\,\Delta s_i,$$

where θ is the angle between $\mathbf{n}_i$ and $\mathbf{v}_i$.

We multiply by the density $\rho_i = \rho(x_i^*, y_i^*)$, then add these terms over those values of i that correspond to a subdivision of the entire curve C. This gives the (net) total mass of fluid leaving R per unit of time; it is approximately

$$\sum_{i=1}^{n} \rho_i \mathbf{v}_i \cdot \mathbf{n}_i\,\Delta s_i = \sum_{i=1}^{n} \mathbf{F}_i \cdot \mathbf{n}_i\,\Delta s_i$$

where $\mathbf{F} = \rho\mathbf{v}$. The line integral around C that this sum approximates is called the *flux of the vector field* $\mathbf{F}$ *across the curve* C. Thus the flux ϕ of $\mathbf{F}$ across C is given by

$$\phi = \oint_C \mathbf{F} \cdot \mathbf{n}\,ds \tag{5}$$

where $\mathbf{n}$ is the *outer* unit normal to C.

In the present case of fluid flow with velocity vector $\mathbf{v}$, the flux ϕ of $\mathbf{F} = \rho\mathbf{v}$ is the rate at which the fluid is flowing out of R across the boundary curve C, in units of mass per unit of time. But the same terminology is used for an arbitrary vector field $\mathbf{F} = M\mathbf{i} + N\mathbf{j}$. For example, we may speak of the flux of an electric or gravitational field across a curve C.

From Fig. 17.18 we see that the outer unit normal $\mathbf{n}$ is equal to $\mathbf{T} \times \mathbf{k}$. The unit tangent $\mathbf{T}$ to the curve C is

$$\mathbf{T} = \frac{1}{v}\left(\mathbf{i}\frac{dx}{dt} + \mathbf{j}\frac{dy}{dt}\right) = \mathbf{i}\frac{dx}{ds} + \mathbf{j}\frac{dy}{ds}$$

because $v = ds/dt$. Hence

$$\mathbf{n} = \mathbf{T} \times \mathbf{k} = \left(\mathbf{i}\frac{dx}{ds} + \mathbf{j}\frac{dy}{ds}\right) \times \mathbf{k}.$$

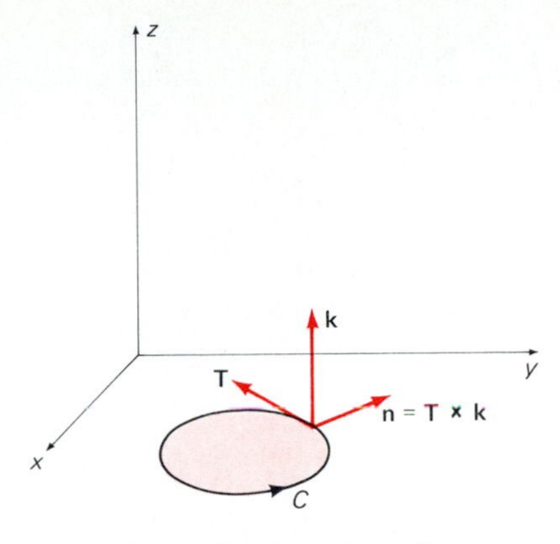

17.18 Computing the outer unit normal **n** from the unit tangent vector **T**

But $\mathbf{i} \times \mathbf{k} = -\mathbf{j}$ and $\mathbf{j} \times \mathbf{k} = \mathbf{i}$. Thus we find that

$$\mathbf{n} = \mathbf{i}\frac{dy}{ds} - \mathbf{j}\frac{dx}{ds}. \tag{6}$$

Substitution of the expression in (6) in the flux integral of (5) gives

$$\oint_C \mathbf{F} \cdot \mathbf{n}\, ds = \oint_C (M\mathbf{i} + N\mathbf{j}) \cdot \left(\mathbf{i}\frac{dy}{ds} - \mathbf{j}\frac{dx}{ds}\right) ds = \oint_C -N\, dx + M\, dy.$$

Applying Green's theorem to the last line integral with $P = -N$ and $Q = M$, we get

$$\oint_C \mathbf{F} \cdot \mathbf{n}\, ds = \iint_R \left(\frac{\partial M}{\partial x} + \frac{\partial N}{\partial y}\right) dA \tag{7}$$

for the flux of $\mathbf{F} = M\mathbf{i} + N\mathbf{j}$ across C.

The scalar function $\partial M/\partial x + \partial N/\partial y$ that appears in Equation (7) is the **divergence** of the two-dimensional vector field $\mathbf{F} = M\mathbf{i} + N\mathbf{j}$ as defined in Section 17.1 and denoted by

$$\text{div}\, \mathbf{F} = \nabla \cdot \mathbf{F} = \frac{\partial M}{\partial x} + \frac{\partial N}{\partial y}. \tag{8}$$

When we substitute (8) in (7), we obtain a **vector form of Green's theorem:**

$$\oint_C \mathbf{F} \cdot \mathbf{n}\, ds = \iint_R \nabla \cdot \mathbf{F}\, dA, \tag{9}$$

with the understanding that $\mathbf{n}$ is the *outer* unit normal to C. Thus the flux of a vector field across a simple closed curve C is equal to the double integral of its divergence over the region R bounded by C.

If the region R is bounded by a circle C_r of radius r centered at the point (x_0, y_0), then

$$\oint_{C_r} \mathbf{F} \cdot \mathbf{N}\, ds = \iint_R \nabla \cdot \mathbf{F}\, dA = (\pi r^2)\, \nabla \cdot \mathbf{F}(\bar{x}, \bar{y})$$

for some point $(\bar{x}, \bar{y})$ in R; this is a consequence of the average value property of double integrals (see Problem 30 in Section 16.2). We divide by πr^2 and then let r approach zero. Thus we find that

$$\nabla \cdot \mathbf{F}(x_0, y_0) = \lim_{r \to 0} \frac{1}{\pi r^2} \oint_{C_r} \mathbf{F} \cdot \mathbf{n}\, ds \tag{10}$$

because $(\bar{x}, \bar{y}) \to (x_0, y_0)$ as $r \to 0$.

In the case of our original fluid flow, with $\mathbf{F} = \rho \mathbf{v}$, Equation (10) implies that the value of $\nabla \cdot \mathbf{F}$ at (x_0, y_0) is a measure of the rate at which the fluid is "diverging away" from the point (x_0, y_0).

EXAMPLE 5 The vector field $\mathbf{F} = -y\mathbf{i} + x\mathbf{j}$ is the velocity field of a steady-state counterclockwise rotation about the origin. Show that the flux of $\mathbf{F}$ across any simple closed curve C is zero.

Solution This follows immediately from Equation (9) because

$$\nabla \cdot \mathbf{F} = \frac{\partial}{\partial x}(-y) + \frac{\partial}{\partial y}(x) = 0.$$

17.4 PROBLEMS

In each of Problems 1–12, apply Green's theorem to evaluate the integral $\oint P\,dx + Q\,dy$ around the specified closed curve C.

1. $P = x + y^2$, $Q = y + x^2$; C is the square with vertices $(\pm 1, \pm 1)$

2. $P = x^2 + y^2$, $Q = -2xy$; C is the boundary of the triangle bounded by the lines $x = 0$, $y = 0$, and $x + y = 1$

3. $P = y + e^x$, $Q = 2x^2 + \cos y$; C is the boundary of the triangle with vertices $(0, 0)$, $(1, 1)$, and $(2, 0)$

4. $P = x^2 - y^2$, $Q = xy$; C is the boundary of the region bounded by the line $y = x$ and the parabola $y = x^2$

5. $P = -y^2 + \exp(e^x)$, $Q = \arctan y$; C is the boundary of the region between the parabolas $y = x^2$ and $x = y^2$

6. $P = y^2$, $Q = 2x - 3y$; C is the circle $x^2 + y^2 = 9$

7. $P = x - y$, $Q = y$; C is the boundary of the region between the x-axis and the graph of $y = \sin x$ for $0 \leqq x \leqq \pi$

8. $P = e^x \sin y$, $Q = e^x \cos y$; C is the right-hand loop of the graph of the polar equation $r^2 = 4\cos\theta$

9. $P = y^2$, $Q = xy$; C is the ellipse with equation $x^2/9 + y^2/4 = 1$

10. $P = y/(1 + x^2)$, $Q = \arctan x$; C is the oval with equation $x^4 + y^4 = 1$

11. $P = xy$, $Q = x^2$; C is the first-quadrant loop of $r = \sin 2\theta$

12. $P = x^2$, $Q = -y^2$; C is the cardioid $r = 1 + \cos\theta$

In each of Problems 13–16, use the corollary to Green's theorem to find the area of the indicated region.

13. The circle bounded by $x = a\cos t$, $y = a\sin t$, $0 \leqq t \leqq 2\pi$

14. The region under one arch of the cycloid with parametric equations $x = a(t - \sin t)$, $y = a(1 - \cos t)$

15. The region bounded by the astroid $x = \cos^3 t$, $y = \sin^3 t$, $0 \leqq t \leqq 2\pi$

16. The region bounded by $y = x^2$ and $y = x^3$

17. Suppose that f is a twice-differentiable scalar function of x and y. Show that

$$\nabla^2 f = \operatorname{div}(\nabla f) = \frac{\partial^2 f}{\partial x^2} + \frac{\partial^2 f}{\partial y^2}.$$

18. Show that $f(x, y) = \ln(x^2 + y^2)$ satisfies **Laplace's equation** $\nabla^2 f = 0$ (except at the point $(0, 0)$).

19. Suppose that f and g are twice-differentiable functions. Show that $\nabla^2(fg) = f\nabla^2 g + g\nabla^2 f + 2\,\nabla f \cdot \nabla g$.

20. Suppose that the function $f(x, y)$ is twice continuously differentiable in the region R bounded by the piecewise smooth closed curve C. Prove that

$$\oint_C \frac{\partial f}{\partial x}\,dy - \frac{\partial f}{\partial y}\,dx = \iint_R \nabla^2 f\,dx\,dy.$$

21. Let R be the plane region with area A enclosed by the piecewise smooth simple closed curve C. Use Green's theorem to show that the coordinates of the centroid of R are

$$\bar{x} = \frac{1}{2A}\oint_C x^2\,dy, \qquad \bar{y} = -\frac{1}{2A}\oint_C y^2\,dx.$$

22. Use the result of Problem 21 to find the centroid of:
(a) a semicircular region of radius a;
(b) a quarter-circular region of radius a.

23. Suppose that a lamina shaped like the region of Problem 21 has constant density ρ. Show that its moments of inertia about the coordinate axes are

$$I_x = -\frac{\rho}{3}\oint_C y^3\,dx, \qquad I_y = \frac{\rho}{3}\oint_C x^3\,dy.$$

24. Use the result of Problem 23 to show that the polar moment of inertia $I_0 = I_x + I_y$ of a circular lamina of radius a, centered at the origin, and of constant density ρ, is $\frac{1}{2}Ma^2$, where M is the mass of the lamina.

25. The loop of the folium of Descartes (with equation $x^3 + y^3 = 3xy$) appears in Fig. 17.19. Apply the corollary to Green's theorem to find the area of this loop. (*Suggestion:* Set $y = tx$ to discover the parametrization $x = 3t/(1 + t^3)$, $y = 3t^2/(1 + t^3)$. To obtain the area of the loop, use values of t lying in the interval $[0, 1]$. This gives the half of the loop lying below the line $y = x$.)

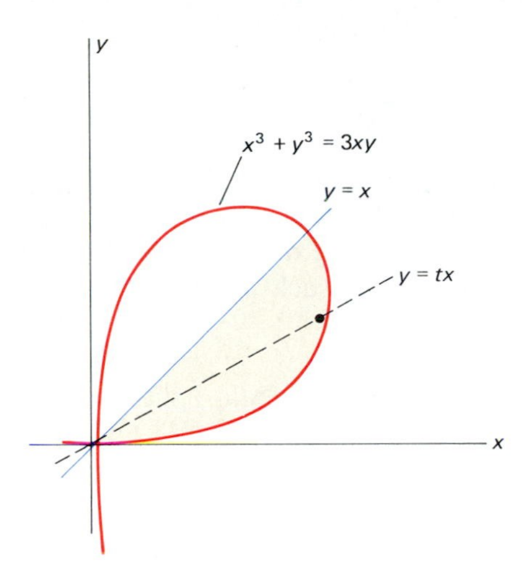

17.19 The loop of Problem 25

26. Find the area bounded by one loop of the curve $x = \sin 2t$, $y = \sin t$.

27. Let f and g be functions with continuous second-order partial derivatives in the region R bounded by the piecewise smooth simple closed curve C. Apply Green's theorem in vector form to show that

$$\oint_C f\,\nabla g \cdot \mathbf{n}\,ds = \iint_R [(f)(\nabla \cdot \nabla g) + \nabla f \cdot \nabla g]\,dA.$$

It was this formula rather than Green's theorem itself that appeared in Green's book of 1828.

28. Complete the proof of the simple case of Green's theorem by showing directly that

$$\oint_C Q\,dy = \iint_R Q_x\,dA$$

if the region R is horizontally simple.

29. Suppose that the bounded plane region R is subdivided into the nonoverlapping subregions $R_1, R_2, \ldots, R_k$. If Green's theorem, Equation (1), holds for each of these subregions, explain why it follows that Green's theorem holds for R. State carefully any assumptions that you need to make.

In Problems 30–32, find the flux of the given vector field $\mathbf{F}$ outward across the ellipse $x = 4\cos t$, $y = 3\sin t$, $0 \leqq t \leqq 2\pi$.

30. $\mathbf{F} = x\mathbf{i} + 2y\mathbf{j}$

31. $\mathbf{F} = \dfrac{x\mathbf{i} + y\mathbf{j}}{x^2 + y^2}$

32. $\mathbf{F} = -y\mathbf{i} + x\mathbf{j}$

17.5 Surface Integrals

A surface integral is to surfaces in space what a line (or "curve") integral is to curves in the plane. Consider a curved thin metal sheet shaped like the surface S. Suppose that this sheet has variable density, given at the point (x, y, z) by the known continuous function $f(x, y, z)$, in units such as grams per square centimeter of surface. We want to define the surface integral

$$\iint_S f(x, y, z)\,dS$$

in such a way that—upon evaluation—it gives the total mass of the thin metal sheet. In case $f(x, y, z) \equiv 1$, the numerical value of the integral should also equal the surface area of S.

As in Section 16.8, we assume that S is a parametric surface described by the function or transformation

$$\mathbf{r}(u, v) = \langle x(u, v), y(u, v), z(u, v)\rangle$$

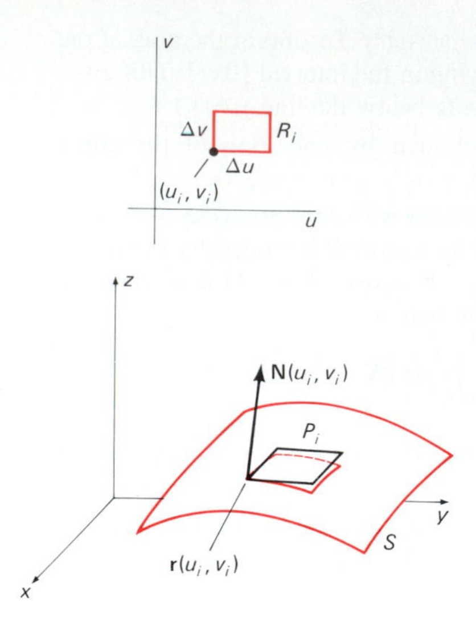

17.20 Approximating surface area with parallelograms

for (u, v) in a region D in the uv-plane. We suppose throughout that the component functions of $\mathbf{r}$ have continuous partial derivatives, and also that the vectors $\mathbf{r}_u = \partial\mathbf{r}/\partial u$ and $\mathbf{r}_v = \partial\mathbf{r}/\partial v$ are nonzero and nonparallel at each interior point of D.

Recall now how we computed the surface area A of S in Section 16.8. We began with an inner partition of D consisting of n rectangles R_1, $R_2, \ldots, R_n$, each Δu by Δv in size. The images under $\mathbf{r}$ of these rectangles are curvilinear figures filling up most of the surface S, and these pieces of S are themselves approximated by parallelograms P_i of the sort shown in Fig. 17.20. This gave us the approximation

$$A \approx \sum_{i=1}^{n} \Delta P_i = \sum_{i=1}^{n} |\mathbf{N}(u_i, v_i)|\, \Delta u\, \Delta v. \tag{1}$$

Here, $\Delta P_i = |\mathbf{N}(u_i, v_i)|\, \Delta u\, \Delta v$ is the area of the parallelogram P_i that is tangent to the surface S at the point $\mathbf{r}(u_i, v_i)$. The vector

$$\mathbf{N} = \frac{\partial \mathbf{r}}{\partial u} \times \frac{\partial \mathbf{r}}{\partial v} \tag{2}$$

is normal to S at $\mathbf{r}(u, v)$. If the surface S now has a density function $f(x, y, z)$, we can approximate the total mass m of the surface by first multiplying each parallelogram area ΔP_i in Equation (1) by the density $f(\mathbf{r}(u_i, v_i))$ at $\mathbf{r}(u_i, v_i)$, and then summing these estimates over all such parallelograms. Thus we obtain the approximation

$$m \approx \sum_{i=1}^{n} f(\mathbf{r}(u_i, v_i))\, \Delta P_i = \sum_{i=1}^{n} f(\mathbf{r}(u_i, v_i))|\mathbf{N}(u_i, v_i)|\, \Delta u\, \Delta v.$$

This approximation is a Riemann sum for the **surface integral of the function f over the surface S**, defined by

$$\begin{aligned} \iint_S f(x, y, z)\, dS &= \iint_D f(\mathbf{r}(u, v))|\mathbf{N}(u, v)|\, du\, dv \\ &= \iint_D f(\mathbf{r}(u, v))\left|\frac{\partial \mathbf{r}}{\partial u} \times \frac{\partial \mathbf{r}}{\partial v}\right| du\, dv. \end{aligned} \tag{3}$$

To evaluate the surface integral $\iint_S f(x, y, z)\, dS$, we simply use the parametrization $\mathbf{r}$ to express the variables x, y, and z in terms of u and v and formally replace the **surface area element** dS by

$$dS = |\mathbf{N}(u, v)|\, du\, dv = \left|\frac{\partial \mathbf{r}}{\partial u} \times \frac{\partial \mathbf{r}}{\partial v}\right| du\, dv. \tag{4}$$

This converts the surface integral into an *ordinary double integral* over the region D in the uv-plane.

In the important special case of a surface S described by $z = h(x, y)$, (x, y) in D, we may use x and y as the parameters (rather than u and v). The surface area element takes the form

$$dS = \sqrt{1 + (\partial h/\partial x)^2 + (\partial h/\partial y)^2}\, dx\, dy \tag{5}$$

corresponding to Equation (9) in Section 16.8. The surface integral of f over S is then given by

$$\iint_S f(x, y, z)\, dS = \iint_D f(x, y, h(x, y)) \sqrt{1 + \left(\frac{\partial h}{\partial x}\right)^2 + \left(\frac{\partial h}{\partial y}\right)^2}\, dx\, dy. \qquad (6)$$

Centroids and moments of inertia for surfaces are computed in much the same way as for curves (Section 17.2, using surface integrals in place of line integrals). For example, suppose that the surface S has density $\rho(x, y, z)$ at the point (x, y, z) and total mass m. Then the z-component $\bar{z}$ of its centroid and its moment of inertia I_z about the z-axis are given by

$$\bar{z} = \frac{1}{m} \iint_S z\rho(x, y, z)\, dS \quad \text{and} \quad I_z = \iint_S (x^2 + y^2)\rho(x, y, z)\, dS.$$

EXAMPLE 1 Find the centroid of the unit-density hemispherical surface $z = \sqrt{a^2 - x^2 - y^2}$, $x^2 + y^2 \leqq a^2$.

Solution By symmetry, $\bar{x} = 0 = \bar{y}$. A simple computation gives $\partial z/\partial x = -x/z$ and $\partial z/\partial y = -y/z$, so Equation (5) yields

$$dS = \sqrt{1 + \left(\frac{\partial z}{\partial x}\right)^2 + \left(\frac{\partial z}{\partial y}\right)^2}\, dx\, dy = \sqrt{1 + \left(\frac{x}{z}\right)^2 + \left(\frac{y}{z}\right)^2}\, dx\, dy$$

$$= \frac{1}{z}\sqrt{x^2 + y^2 + z^2}\, dx\, dy = \frac{a}{z}\, dx\, dy.$$

Hence

$$\bar{z} = \frac{1}{2\pi a^2} \iint_D z \cdot \frac{a}{z}\, dx\, dy = \frac{1}{2\pi a} \iint_D 1\, dx\, dy = \frac{a}{2}.$$

Note in the final step that D is a circle of radius a in the xy-plane; this simplifies the computation.

EXAMPLE 2 Find the moment of inertia about the z-axis of the spherical surface $x^2 + y^2 + z^2 = a^2$, assuming that it has constant density $\rho = k$.

Solution The sphere of radius a is most easily parametrized in spherical coordinates:

$$x = a \sin\phi \cos\theta, \qquad y = a \sin\phi \sin\theta, \qquad z = a \cos\phi$$

for $0 \leqq \phi \leqq \pi$ and $0 \leqq \theta \leqq 2\pi$. By Problem 18 in Section 16.8, the surface element is then $dS = a^2 \sin\phi\, d\phi\, d\theta$. We first note that

$$x^2 + y^2 = a^2 \sin^2\phi \cos^2\theta + a^2 \sin^2\phi \sin^2\theta = a^2 \sin^2\phi,$$

and it then follows that

$$I_z = \iint_S (x^2 + y^2)\rho\, dS = \int_0^{2\pi} \int_0^{\pi} k(a^2 \sin^2\phi)a^2 \sin\phi\, d\phi\, d\theta$$

$$= 2\pi k a^4 \int_0^{\pi} \sin^3\phi\, d\phi = \tfrac{8}{3}\pi k a^4 = \tfrac{2}{3} m a^2.$$

In the last step we used the fact that the mass m of the spherical surface with density k is $4\pi ka^2$. Note that the answer is dimensionally correct and numerically plausible.

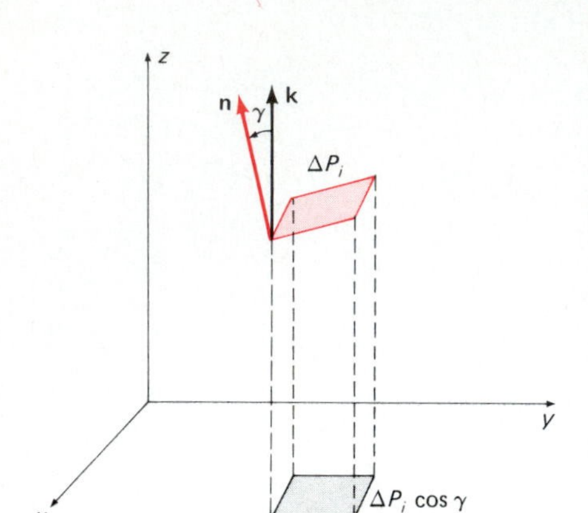

17.21 Finding the area of the projected parallelogram

The surface integral $\iint_S f(x, y, z)\, dS$ is analogous to the line integral $\int_C f(x, y)\, ds$. There is a second type of surface integral which is analogous to the line integral of the form $\int_C P\, dx + Q\, dy$. To define the surface integral

$$\iint_S f(x, y, z)\, dx\, dy,$$

with $dx\, dy$ in place of dS, we replace the parallelogram area ΔP_i in Equation (1) by the area of its projection into the xy-plane (see Fig. 17.21). To see how this works out, consider the *unit* normal to S,

$$\mathbf{n} = \frac{\mathbf{N}}{|\mathbf{N}|} = \mathbf{i}\cos\alpha + \mathbf{j}\cos\beta + \mathbf{k}\cos\gamma. \tag{7}$$

Because

$$\mathbf{N} = \begin{vmatrix} \mathbf{i} & \mathbf{j} & \mathbf{k} \\ \dfrac{\partial x}{\partial u} & \dfrac{\partial y}{\partial u} & \dfrac{\partial z}{\partial u} \\ \dfrac{\partial x}{\partial v} & \dfrac{\partial y}{\partial v} & \dfrac{\partial z}{\partial v} \end{vmatrix}$$

or—in the Jacobian notation of Section 16.9—

$$\mathbf{N} = \mathbf{i}\frac{\partial(y, z)}{\partial(u, v)} + \mathbf{j}\frac{\partial(z, x)}{\partial(u, v)} + \mathbf{k}\frac{\partial(x, y)}{\partial(u, v)},$$

we see that the components of the unit normal $\mathbf{n}$ are

$$\cos\alpha = \frac{1}{|\mathbf{N}|}\frac{\partial(y, z)}{\partial(u, v)}, \qquad \cos\beta = \frac{1}{|\mathbf{N}|}\frac{\partial(z, x)}{\partial(u, v)}, \qquad \cos\gamma = \frac{1}{|\mathbf{N}|}\frac{\partial(x, y)}{\partial(u, v)}. \tag{8}$$

From Fig. 17.21 we see that the (signed) projection of the area ΔP_i into the xy-plane is $\Delta P_i \cos\gamma$. The corresponding Riemann sum motivates the *definition*

$$\iint_S f(x, y, z)\, dx\, dy = \iint_S f(x, y, z)\cos\gamma\, dS = \iint_S f(\mathbf{r}(u, v))\frac{\partial(x, y)}{\partial(u, v)}\, du\, dv, \tag{9}$$

where the last integral is obtained by substituting for $\cos\gamma$ from (8) and for dS from (4). Similarly, we *define*

$$\iint_S f(x, y, z)\, dy\, dz = \iint_S f(x, y, z)\cos\alpha\, dS = \iint_S f(\mathbf{r}(u, v))\frac{\partial(y, z)}{\partial(u, v)}\, du\, dv \tag{10}$$

and

$$\iint_S f(x, y, z)\, dz\, dx = \iint_S f(x, y, z)\cos\beta\, dS = \iint_S f(\mathbf{r}(u, v))\frac{\partial(z, x)}{\partial(u, v)}\, du\, dv. \tag{11}$$

Note that the symbols z and x in (11) appear in the reverse of alphabetical order. It is important to write them in the correct order because $\partial(x, z)/\partial(u, v) = -\partial(z, x)/\partial(u, v)$. This implies that

$$\iint_S f(x, y, z)\, dx\, dz = -\iint_S f(x, y, z)\, dz\, dx.$$

In an ordinary *double integral*, the order in which the differentials are written simply indicates the order of integration. But in a *surface integral*, it instead indicates the order of appearance of the corresponding variables in the Jacobians in (9) through (11).

The general surface integral of the second type is the sum

$$\iint_S P\, dy\, dz + Q\, dz\, dx + R\, dx\, dy = \iint_S (P \cos \alpha + Q \cos \beta + R \cos \gamma)\, dS \quad (12)$$

$$= \iint_S \left(P \frac{\partial(y, z)}{\partial(u, v)} + Q \frac{\partial(z, x)}{\partial(u, v)} + R \frac{\partial(x, y)}{\partial(u, v)} \right) du\, dv. \quad (13)$$

Here, P, Q, and R are continuous functions of x, y, and z.

Suppose that $\mathbf{F} = P\mathbf{i} + Q\mathbf{j} + R\mathbf{k}$. Then the integrand in Equation (12) is simply $\mathbf{F} \cdot \mathbf{n}$, so we obtain the basic relation

$$\iint_S \mathbf{F} \cdot \mathbf{n}\, dS = \iint_S P\, dy\, dz + Q\, dz\, dx + R\, dx\, dy \quad (14)$$

between these two types of surface integrals. This formula is analogous to the earlier formula

$$\int_C \mathbf{F} \cdot \mathbf{T}\, ds = \int_C P\, dx + Q\, dy + R\, dz$$

for line integrals. On the other hand, the formula in (13) gives the evaluation procedure for the surface integral in (12)—substitute for x, y, z, and their derivatives in terms of u and v and then integrate over the appropriate region D in the uv-plane.

It turns out that only the *sign* of the right-hand surface integral in Equation (14) depends upon the parametrization of S. The unit normal on the left-hand side is the one provided by the parametrization of S via the equations in (8). In the case of a surface given by $z = h(x, y)$, with x and y used as the parameters u and v, this will be the *upper* normal, as you will see in the computation in the next example.

EXAMPLE 3 Suppose that S is the surface $z = h(x, y)$, (x, y) in D. Then show that

$$\iint_S P\, dy\, dz + Q\, dz\, dx + R\, dx\, dy = \iint_D \left(-P \frac{\partial z}{\partial x} - Q \frac{\partial z}{\partial y} + R \right) dx\, dy, \quad (15)$$

where P, Q, and R in the second integral are evaluated at $(x, y, h(x, y))$.

Solution This is simply a matter of computing the three Jacobians in (13) with the parameters x and y. We note first that $\partial x/\partial x = 1 = \partial y/\partial y$ and that

$\partial x/\partial y = 0 = \partial y/\partial x$. Hence

$$\frac{\partial(y, z)}{\partial(x, y)} = \begin{vmatrix} \dfrac{\partial y}{\partial x} & \dfrac{\partial y}{\partial y} \\ \dfrac{\partial z}{\partial x} & \dfrac{\partial z}{\partial y} \end{vmatrix} = -\frac{\partial z}{\partial x},$$

$$\frac{\partial(z, x)}{\partial(x, y)} = \begin{vmatrix} \dfrac{\partial z}{\partial x} & \dfrac{\partial z}{\partial y} \\ \dfrac{\partial x}{\partial x} & \dfrac{\partial x}{\partial y} \end{vmatrix} = -\frac{\partial z}{\partial y},$$

and

$$\frac{\partial(x, y)}{\partial(x, y)} = \begin{vmatrix} \dfrac{\partial x}{\partial x} & \dfrac{\partial x}{\partial y} \\ \dfrac{\partial y}{\partial x} & \dfrac{\partial y}{\partial y} \end{vmatrix} = 1.$$

Equation (15) is an immediate consequence.

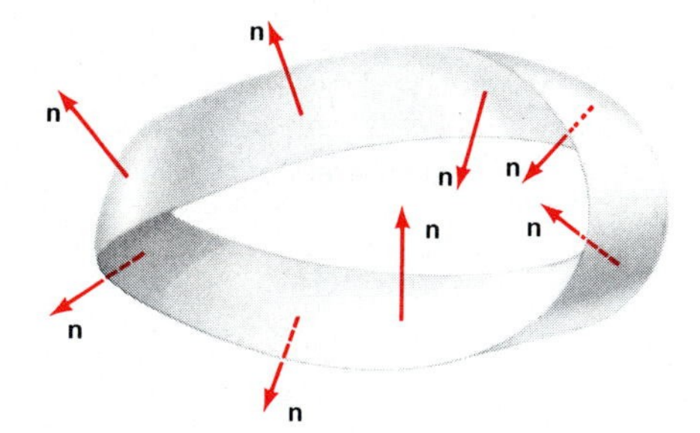

17.22 The Möbius strip is an example of a one-sided surface.

One of the most important applications of surface integrals involves the computation of the flux of a vector field. In order to define the flux of the vector field **F** across the surface S, we assume that S has a unit normal vector **n** that varies *continuously* from point to point of S. This condition excludes from our consideration one-sided (*nonorientable*) surfaces, such as the Möbius strip of Fig. 17.22. If S is a two-sided (*orientable*) surface, then there are two possible choices for **n**. For example, if S is a closed surface (such as a sphere or torus) that separates space, then we may choose for **n** either the outer normal or the inner normal. These two options are shown in Fig. 17.23. The unit normal defined by the formula in (7) may be either the outer normal or the inner normal; which of the two it is depends on how S has been parametrized.

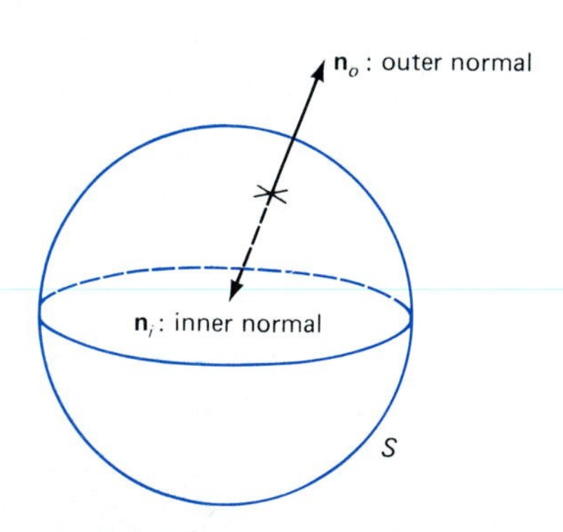

17.23 Inner and outer normals to a two-sided closed surface.

Now we turn our attention to the flux of a vector field. Suppose that we are given the vector field **F**, the orientable surface S, and a continuous unit normal vector **n** on S. We define the **flux ϕ of F across S in the direction of n** in analogy with Equation (5) of Section 17.4:

$$\phi = \iint_S \mathbf{F} \cdot \mathbf{n}\, dS. \tag{16}$$

For example, if $\mathbf{F} = \rho\mathbf{v}$ where **v** is the velocity vector field corresponding to the steady flow in space of a fluid of density ρ and **n** is the *outer* normal for a closed surface S bounding a space region T, then the flux determined by Equation (16) is the net rate of flow of the fluid *out of* T across its boundary surface S, in such units as grams per second.

A similar application is to the flow of heat, which is mathematically quite similar to the flow of a fluid. Suppose that a body has temperature $u = u(x, y, z)$ at the point (x, y, z). Experiments indicate that the flow of heat in the body is described by the heat flow vector

$$\mathbf{q} = -K\,\nabla u. \tag{17}$$

The number K—normally, but not always, a constant—is the *heat conductivity* of the body. The vector $\mathbf{q}$ points in the direction of heat flow, and its length is the rate of flow of heat across a unit area normal to $\mathbf{q}$. This flow rate is measured in such units as calories per second per square centimeter. If S is a closed surface within the body bounding a region T and $\mathbf{n}$ denotes the outer unit normal to S, then

$$\iint_S \mathbf{q} \cdot \mathbf{n}\, dS = -\iint_S K\, \nabla u \cdot \mathbf{n}\, dS \tag{18}$$

is the net rate of flow of heat (in calories per second, for example) out of the region T across its boundary surface S.

EXAMPLE 4 Calculate the flux $\iint_S \mathbf{F} \cdot \mathbf{n}\, dS$, where $\mathbf{F} = v_0\mathbf{k}$ and S is the hemispherical surface of radius a shown in Fig. 17.24, with outward unit normal vector $\mathbf{n}$.

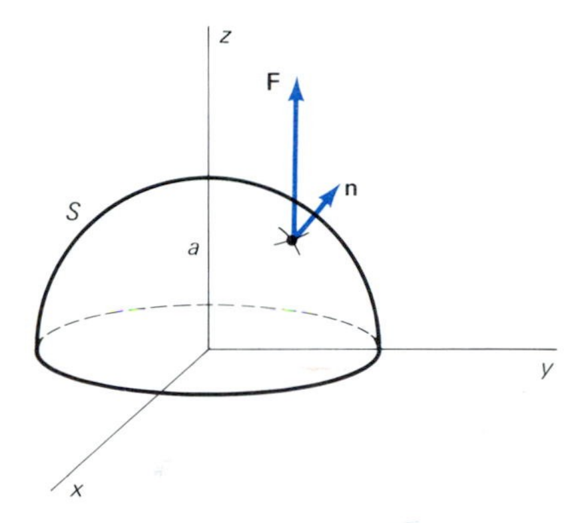

17.24 The hemisphere S of Example 4

Solution If we think of $\mathbf{F} = v_0\mathbf{k}$ as the velocity vector field of a fluid flowing upward with constant speed v_0, then the flux in question may be interpreted as the rate of flow (in cubic centimeters per second, for example) of the fluid across S. To calculate this flux, we note that

$$\mathbf{n} = \frac{x\mathbf{i} + y\mathbf{j} + z\mathbf{k}}{\sqrt{x^2 + y^2 + z^2}} = \frac{1}{a}(x\mathbf{i} + y\mathbf{j} + z\mathbf{k}).$$

Hence

$$\mathbf{F} \cdot \mathbf{n} = v_0\mathbf{k} \cdot \frac{1}{a}(x\mathbf{i} + y\mathbf{j} + z\mathbf{k}) = \frac{v_0}{a}z,$$

so

$$\iint_S \mathbf{F} \cdot \mathbf{n}\, dS = \iint_S \frac{v_0}{a} z\, dS.$$

If we introduce spherical coordinates $z = a\cos\phi$, $dS = a^2 \sin\phi\, d\phi\, d\theta$ on the sphere, we get

$$\iint_S \mathbf{F} \cdot \mathbf{n}\, dS = \frac{v_0}{a}\int_0^{2\pi}\int_0^{\pi/2} (a\cos\phi)(a^2 \sin\phi)\, d\phi\, d\theta$$

$$= 2\pi a^2 v_0 \int_0^{\pi/2} \cos\phi \sin\phi\, d\phi = 2\pi a^2 v_0 \left[\tfrac{1}{2}\sin^2\phi\right]_0^{\pi/2};$$

thus

$$\iint_S \mathbf{F} \cdot \mathbf{n}\, dS = \pi a^2 v_0.$$

Note that this result is equal to the flux of $\mathbf{F} = v_0\mathbf{k}$ across the disk $x^2 + y^2 \leqq a^2$ of area πa^2. If we think of the hemispherical region T bounded by the hemisphere S and the circular disk that forms its base, it should be no surprise that the rate of inflow of an incompressible fluid across the circular disk is equal to its rate of outflow across the hemisphere S.

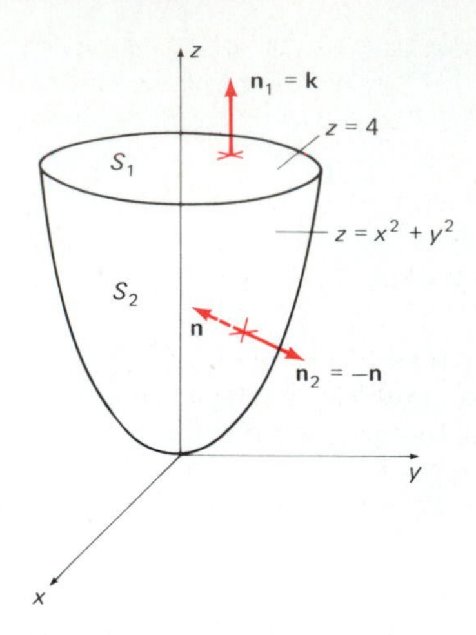

17.25 The surface of Example 5

EXAMPLE 5 Find the flux of the vector field $\mathbf{F} = x\mathbf{i} + y\mathbf{j} + 3\mathbf{k}$ out of the region T bounded by the paraboloid $z = x^2 + y^2$ and the plane $z = 4$ (see Fig. 17.25).

Solution Let S_1 denote the circular top, which has outer unit normal $\mathbf{n}_1 = \mathbf{k}$. Let S_2 be the parabolic part of this surface, with outer unit normal $\mathbf{n}_2$. The flux across S_1 is

$$\iint_{S_1} \mathbf{F} \cdot \mathbf{n}_1 \, dS = \iint_{S_1} 3 \, dS = 12\pi$$

because S_1 is a circle of radius 2.

Next, the computation of Example 3 gives

$$\mathbf{N} = \left\langle -\frac{\partial z}{\partial x}, -\frac{\partial z}{\partial y}, 1 \right\rangle = \langle -2x, -2y, 1 \rangle$$

for a normal vector to the paraboloid $z = x^2 + y^2$. Then $\mathbf{n} = \mathbf{N}/|\mathbf{N}|$ is an upper, and thus an *inner*, unit normal to the surface S_2. The unit *outer* normal is therefore $\mathbf{n}_2 = -\mathbf{n}$, opposite to the direction of $\mathbf{N} = \langle -2x, -2y, 1 \rangle$. With parameters (x, y) in the circular disk $x^2 + y^2 \leqq 4$ in the xy-plane, the surface area element is $dS = |\mathbf{N}| \, dx \, dy$. Therefore, the outward flux across S_2 is

$$\iint_{S_2} \mathbf{F} \cdot \mathbf{n}_2 \, dS = -\iint_{S_2} \mathbf{F} \cdot \mathbf{n} \, dS = -\iint_{D} \mathbf{F} \cdot \frac{\mathbf{N}}{|\mathbf{N}|} |\mathbf{N}| \, dx \, dy$$

$$= -\iint_{D} [(x)(-2x) + (y)(-2y) + (3)(1)] \, dx \, dy.$$

We change to polar coordinates and find that

$$\iint_{S_2} \mathbf{F} \cdot \mathbf{n}_2 \, dS = \int_0^{2\pi} \int_0^2 (2r^2 - 3)r \, dr \, d\theta = 2\pi \left[\tfrac{1}{2}r^4 - \tfrac{3}{2}r^2 \right]_0^2 = 4\pi.$$

Hence the total flux of $\mathbf{F}$ out of T is 16π, which is approximately 50.27.

17.5 PROBLEMS

In each of Problems 1–5, evaluate the surface integral

$$\iint_S f(x, y, z) \, dS.$$

1. $f(x, y, z) = xyz$; S is the first-octant part of the plane $x + y + z = 1$.

2. $f(x, y, z) = x + y$; S is the part of the plane $z = 2x + 3y$ that lies within the cylinder $x^2 + y^2 = 9$.

3. $f(x, y, z) = z$; S is the part of the paraboloid $z = r^2$ that lies within the cylinder $r = 2$.

4. $f(x, y, z) = z(x^2 + y^2)$; S is the hemispherical surface $\rho = 1$, $z \geqq 0$. (*Suggestion:* Use spherical coordinates.)

5. $f(x, y, z) = z^2$; S is the cylindrical surface parametrized by $x = \cos\theta$, $y = \sin\theta$, $z = z$, $0 \leqq \theta \leqq 2\pi$, $0 \leqq z \leqq 2$.

In Problems 6–10, use the formulas in (13) and (14) to evaluate the surface integral $\iint_S \mathbf{F} \cdot \mathbf{n} \, dS$, where $\mathbf{n}$ is the upward-pointing unit normal to the given surface S.

6. $\mathbf{F} = x\mathbf{i} + y\mathbf{j} + z\mathbf{k}$; S is the first-octant part of the plane $2x + 2y + z = 3$.

7. $\mathbf{F} = 2y\mathbf{i} + 3z\mathbf{k}$; S is the part of the plane $z = 3x + 2$ that lies within the cylinder $x^2 + y^2 = 4$.

8. $\mathbf{F} = z\mathbf{k}$; S is the upper half of the spherical surface $\rho = 2$. (*Suggestion:* Use spherical coordinates.)

9. $\mathbf{F} = y\mathbf{i} - x\mathbf{j}$; S is the part of the cone $z = r$ that lies within the cylinder $r = 3$.

10. $\mathbf{F} = 2x\mathbf{i} + 2y\mathbf{j} + 3\mathbf{k}$; S is the part of the paraboloid $z = 4 - x^2 - y^2$ that lies above the xy-plane.

11. The first-octant part of the spherical surface $\rho = a$ has unit density. Find its centroid.

12. The conical surface $z = r$, $r \leqq a$, has constant density $\delta = k$. Find its centroid and its moment of inertia about the z-axis.

13. The paraboloid $z = r^2$, $r \leqq a$, has constant density δ. Find its centroid and its moment of inertia about the z-axis.

14. Find the centroid of the part of the spherical surface $\rho = a$ that lies within the cone $r = z$.

15. Find the centroid of the part of the spherical surface $x^2 + y^2 + z^2 = 4$ that lies both within the cylinder $x^2 + y^2 = 2x$ and above the xy-plane.

16. Suppose that the toroidal surface of Example 4 in Section 16.8 has uniform density and total mass M. Show that its moment of inertia about the z-axis is $\frac{1}{2}M(3a^2 + 2b^2)$.

17. Let S denote the boundary of the region bounded by the coordinate planes and the plane $2x + 3y + z = 6$. Find the flux of $\mathbf{F} = x\mathbf{i} - y\mathbf{j}$ across S in the direction of its outer normal.

18. Let S denote the boundary of the solid bounded by the paraboloid $z = 4 - x^2 - y^2$ and the xy-plane. Find the flux of the vector field $\mathbf{F} = 2x\mathbf{i} + 2y\mathbf{j} + 3\mathbf{k}$ across S in the direction of the outer normal.

19. Let S denote the boundary of the solid bounded by the paraboloids $z = x^2 + y^2$ and $z = 18 - x^2 - y^2$. Find the flux of $\mathbf{F} = z^2\mathbf{k}$ across S in the direction of the outer normal.

20. Let S denote the surface $z = h(x, y)$ for (x, y) in the region D in the xy-plane, and let γ be the angle between $\mathbf{k}$ and the upper normal vector $\mathbf{N}$ to S. Prove that

$$\iint_S f(x, y, z)\, dS = \iint_D f(x, y, h(x, y)) \sec\gamma \, dx\, dy.$$

21. Consider a homogeneous thin spherical shell S of radius a centered at the origin, with density δ and total mass $M = 4\pi a^2\delta$. A particle of mass m is located at the point $(0, 0, c)$ with $c > a$. Use the method and notation of Problem 34 in Section 16.7 to show that the gravitational force of attraction between the particle and the spherical shell is

$$F = \iint_S \frac{Gm\delta}{w^2}\, dS = \frac{GMm}{c^2}.$$

17.6 The Divergence Theorem

The *divergence theorem* is for surface integrals what Green's theorem is for line integrals. It lets us convert a surface integral over a closed surface into a triple integral over the enclosed region, or vice versa. The divergence theorem is also known as *Gauss's theorem*, and in some eastern European countries it is called *Ostrogradski's theorem*. The German "prince of mathematics" Carl Friedrich Gauss (1777–1855) used it to study inverse-square force fields, while the Russian Michel Ostrogradski (1801–1861) used it to study heat flow. The work of each was done in the 1830s.

The surface S is called **piecewise smooth** if it consists of finitely many smooth parametric surfaces. It is called **closed** if it is the boundary of a bounded space region. For example, the boundary of a cube is a closed piecewise smooth surface, as are the boundary of a pyramid and the boundary of a solid cylinder.

The Divergence Theorem

Suppose that S is a closed piecewise smooth surface bounding the space region T. Let $\mathbf{F} = P\mathbf{i} + Q\mathbf{j} + R\mathbf{k}$ be a vector field with component functions that have continuous first-order partial derivatives on T. Let $\mathbf{n}$ be the *outer* unit normal to S. Then

$$\iint_S \mathbf{F}\cdot\mathbf{n}\, dS = \iiint_T \nabla\cdot\mathbf{F}\, dV. \qquad (1)$$

Equation (1) is a three-dimensional analogue of the vector form of Green's theorem that we saw in Equation (9) of Section 17.4:

$$\oint_C \mathbf{F} \cdot \mathbf{n}\, ds = \iint_R \nabla \cdot \mathbf{F}\, dA,$$

where $\mathbf{F}$ is a vector field in the plane, C is a piecewise smooth curve that bounds the plane region R, and $\mathbf{n}$ is the outer unit normal to C. Note that the left-hand side in Equation (1) is the flux of $\mathbf{F}$ across S in the direction of the outer unit normal $\mathbf{n}$. Recall that the *divergence* $\nabla \cdot \mathbf{F}$ of the vector field $\mathbf{F}$ is given in the three-dimensional case by

$$\operatorname{div} \mathbf{F} = \nabla \cdot \mathbf{F} = \frac{\partial P}{\partial x} + \frac{\partial Q}{\partial y} + \frac{\partial R}{\partial z}. \tag{2}$$

If $\mathbf{n}$ is given in terms of its direction cosines, as $\mathbf{N} = \langle \cos \alpha, \cos \beta, \cos \gamma \rangle$, then we can write the divergence theorem in scalar form:

$$\iint_S (P \cos \alpha + Q \cos \beta + R \cos \gamma)\, dS = \iiint_T \left(\frac{\partial P}{\partial x} + \frac{\partial Q}{\partial y} + \frac{\partial R}{\partial z} \right) dV. \tag{3}$$

It is best to parametrize S so that the normal vector given by the parametrization is the outer normal. For then Equation (3) can be written entirely in Cartesian form:

$$\iint_S P\, dy\, dz + Q\, dz\, dx + R\, dx\, dy = \iiint_T \left(\frac{\partial P}{\partial x} + \frac{\partial Q}{\partial y} + \frac{\partial R}{\partial z} \right) dV. \tag{4}$$

Proof of the Divergence Theorem We shall prove the theorem only in the case in which the region T is simultaneously x-simple, y-simple, and z-simple. This guarantees that every straight line parallel to a coordinate axis intersects T, if at all, in a single point or a line segment. It will suffice to derive separately the equations

$$\iint_S P\, dy\, dz = \iiint_T \frac{\partial P}{\partial x}\, dV,$$

$$\iint_S Q\, dz\, dx = \iiint_T \frac{\partial Q}{\partial y}\, dV, \quad \text{and} \tag{5}$$

$$\iint_S R\, dx\, dy = \iiint_T \frac{\partial R}{\partial z}\, dV.$$

For the sum of the equations in (5) yields Equation (4).

Because T is z-simple, it has the description

$$h_1(x, y) \leqq z \leqq h_2(x, y)$$

for (x, y) in D. As in Fig. 17.26, we denote the lower surface $z = h_1(x, y)$ of T by S_1, the upper surface $z = h_2(x, y)$ by S_2, and the lateral surface between S_1 and S_2 by S_3. In the case of some simple surfaces—such as a spherical

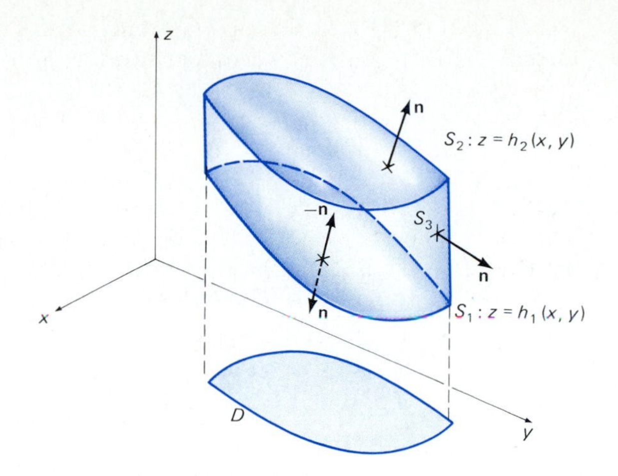

17.26 A z-simple space region bounded by the surfaces S_1, S_2, and S_3

surface—there may be no S_3 to consider. But if there is, we see that

$$\iint_{S_3} R\,dx\,dy = \iint_{S_3} R\cos\gamma\,dS = 0 \tag{6}$$

because $\gamma = 90°$ at each point of the vertical cylinder S_3.

On the upper surface S_2, the unit upper normal corresponding to the parametrization $z = h_2(x, y)$ is the given outer unit normal **n**, so the formula in Equation (15) of Section 17.5 yields

$$\iint_{S_2} R\,dx\,dy = \iint_{D} R(x, y, h_2(x, y))\,dx\,dy. \tag{7}$$

But on the lower surface S_1, the unit upper normal corresponding to the parametrization $z = h_1(x, y)$ is the inner normal $-$**n**, so we must reverse the sign. Thus

$$\iint_{S_1} R\,dx\,dy = -\iint_{D} R(x, y, h_1(x, y))\,dx\,dy. \tag{8}$$

We add Equations (6), (7), and (8). The result is that

$$\iint_{S} R\,dx\,dy = \iint_{D} [R(x, y, h_2(x, y)) - R(x, y, h_1(x, y))]\,dx\,dy$$

$$= \iint_{D} \left(\int_{h_1(x,y)}^{h_2(x,y)} \frac{\partial R}{\partial z}\,dz \right) dx\,dy.$$

Therefore,

$$\iint_{S} R\,dx\,dy = \iiint_{T} \frac{\partial R}{\partial z}\,dV.$$

This is the third of the three equations in (5), and the other two may be derived in much the same way. ■

The divergence theorem is established for more general regions by the device of subdivision of T into simpler regions, ones for which the proof

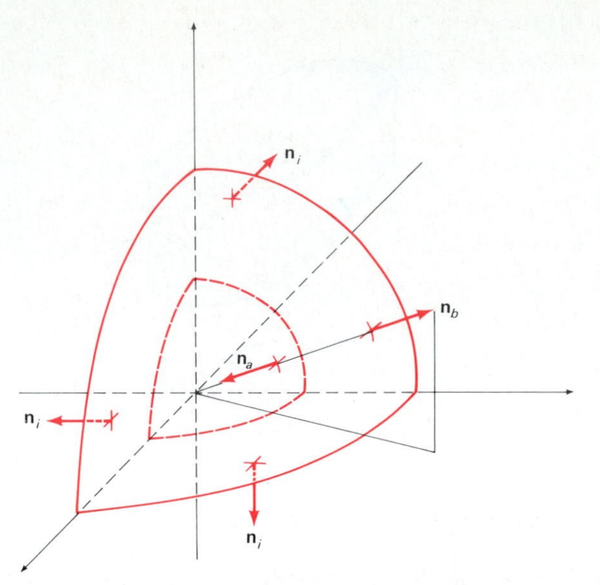

17.27 One octant of the shell between S_a and S_b

above holds. For example, suppose that T is the shell between the concentric spheres S_a and S_b of radii a and b, with $0 < a < b$. The coordinate planes separate T into eight regions $T_1, T_2, \ldots, T_8$, each shaped like the one of Fig. 17.27. Let Σ_i denote the boundary of T_i and let $\mathbf{n}_i$ be the outer unit normal to Σ_i. We apply the divergence theorem to each of these eight regions, and obtain

$$\iiint_T \nabla \cdot \mathbf{F}\, dV = \sum_{i=1}^{8} \iiint_{T_i} \nabla \cdot \mathbf{F}\, dV$$

$$= \sum_{i=1}^{8} \iint_{\Sigma_i} \mathbf{F} \cdot \mathbf{n}_i\, dS \qquad \text{(divergence theorem)}$$

$$= \iint_{S_a} \mathbf{F} \cdot \mathbf{n}_a\, dS + \iint_{S_b} \mathbf{F} \cdot \mathbf{n}_b\, dS.$$

Here we write $\mathbf{n}_a$ for the inner normal on S_a and $\mathbf{n}_b$ for the outer normal on S_b. The last equality holds because the surface integrals over the internal boundary surfaces (the ones in the coordinate planes) cancel in pairs—the normals are oppositely oriented there. As the boundary S of T is the union of the spherical surfaces S_a and S_b, it now follows that

$$\iiint_T \nabla \cdot \mathbf{F}\, dV = \iint_S \mathbf{F} \cdot \mathbf{n}\, dS.$$

This is the divergence theorem for the spherical shell T.

EXAMPLE 1 Let S be the surface (with outer unit normal $\mathbf{n}$) of the region T bounded by the planes $z = 0$, $y = 0$, $y = 2$, and the paraboloid $z = 1 - x^2$ (see Fig.17.28). Apply the divergence theorem to compute $\iint_S \mathbf{F} \cdot \mathbf{n}\, dS$ given

$$\mathbf{F} = (x + \cos y)\mathbf{i} + (y + \sin z)\mathbf{j} + (z + e^x)\mathbf{k}.$$

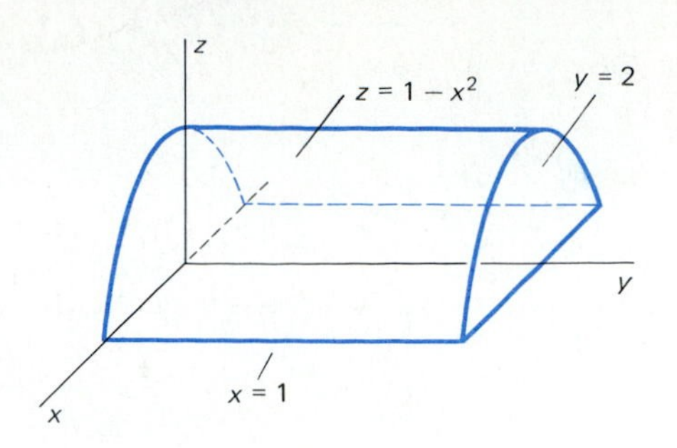

17.28 The region of Example 1

Solution To evaluate the surface integral directly would be quite lengthy. But div $\mathbf{F} = 1 + 1 + 1 = 3$, so the divergence theorem may be applied easily and yields

$$\iint_S \mathbf{F} \cdot \mathbf{n}\, dS = \iiint_T \operatorname{div} \mathbf{F}\, dV = \iiint_T 3\, dV.$$

We examine Fig. 17.28 to find the limits for the volume integral and thus obtain

$$\iint_S \mathbf{F} \cdot \mathbf{n}\, dS = \int_{-1}^{1} \int_0^2 \int_0^{1-x^2} 3\, dz\, dy\, dx = 12 \int_0^1 (1 - x^2)\, dx = 8.$$

EXAMPLE 2 Let S be the surface of the solid cylinder T bounded by the planes $z = 0$ and $z = 3$ and by the cylinder $x^2 + y^2 = 4$. Calculate the outward flux $\iint_S \mathbf{F} \cdot \mathbf{n}\, dS$ given

$$\mathbf{F} = (x^2 + y^2 + z^2)(x\mathbf{i} + y\mathbf{j} + z\mathbf{k}).$$

Solution If we denote by P, Q, and R the component functions of the vector field $\mathbf{F}$, we find that

$$\frac{\partial P}{\partial x} = (2x)(x) + (x^2 + y^2 + z^2)(1) = 3x^2 + y^2 + z^2.$$

Similarly,

$$\frac{\partial Q}{\partial y} = 3y^2 + z^2 + x^2 \quad \text{and} \quad \frac{\partial R}{\partial z} = 3z^2 + x^2 + y^2,$$

so

$$\operatorname{div} \mathbf{F} = 5(x^2 + y^2 + z^2).$$

Therefore, the divergence theorem yields

$$\iint_S \mathbf{F} \cdot \mathbf{n}\, dS = \iiint_T 5(x^2 + y^2 + z^2)\, dV.$$

Using cylindrical coordinates to evaluate the volume integral, we get

$$\begin{aligned}\iint_S \mathbf{F} \cdot \mathbf{n}\, dS &= \int_0^{2\pi} \int_0^2 \int_0^3 5(r^2 + z^2) r\, dz\, dr\, d\theta \\ &= 10\pi \int_0^2 \left[r^3 z + \tfrac{1}{3} r z^3 \right]_{z=0}^{3} dr = 10\pi \int_0^2 (3r^3 + 9r)\, dr \\ &= 10\pi \left[\tfrac{3}{4} r^4 + \tfrac{9}{2} r^2 \right]_0^2 = 300\pi.\end{aligned}$$

EXAMPLE 3 Suppose that the region T is bounded by the closed surface S with a parametrization that gives the outer unit normal. Show that the volume V of T is given by

$$V = \tfrac{1}{3} \iint_S x\, dy\, dz + y\, dz\, dx + z\, dx\, dy. \tag{9}$$

Solution Equation (9) follows immediately from Equation (4) if we take $P(x, y, z) = x$, $Q(x, y, z) = y$, and $R(x, y, z) = z$. For example, if S is the spherical surface $x^2 + y^2 + z^2 = a^2$ with volume V, surface area A, and outer unit normal

$$\mathbf{n} = \langle \cos\alpha, \cos\beta, \cos\gamma \rangle = \left\langle \frac{x}{a}, \frac{y}{a}, \frac{z}{a} \right\rangle,$$

then Equation (9) gives

$$\begin{aligned} V &= \frac{1}{3}\iint_S x\,dy\,dz + y\,dz\,dx + z\,dx\,dy \\ &= \frac{1}{3}\iint_S (x\cos\alpha + y\cos\beta + z\cos\gamma)\,dS \\ &= \frac{1}{3}\iint_S \frac{x^2+y^2+z^2}{a}\,dS = \frac{a}{3}\iint_S 1\,dS = \frac{1}{3}aA. \end{aligned}$$

This result is consistent with the familiar formulas $V = \frac{4}{3}\pi a^3$ and $A = 4\pi a^2$.

EXAMPLE 4 Show that the divergence of the vector field $\mathbf{F}$ at the point P is given by

$$\{\text{div}\,\mathbf{F}\}(P) = \lim_{r \to 0} \frac{1}{V_r} \iint_{S_r} \mathbf{F} \cdot \mathbf{n}\,dS \tag{10}$$

where S_r is the sphere of radius r centered at P and $V_r = \frac{4}{3}\pi r^3$ is the volume of the ball B_r that it bounds.

Solution The divergence theorem gives

$$\iint_{S_r} \mathbf{F} \cdot \mathbf{n}\,dS = \iiint_{B_r} \text{div}\,\mathbf{F}\,dV.$$

Then we apply the average value property of triple integrals, a result analogous to the double integral result of Problem 30 in Section 16.2. This yields

$$\iiint_{B_r} \text{div}\,\mathbf{F}\,dV = V_r \cdot \{\text{div}\,\mathbf{F}\}(P^*)$$

for some point P^* of B_r (here we write $\{\text{div}\,\mathbf{F}\}(P^*)$ for the value of div $\mathbf{F}$ at the point P^*). We assume that the component functions of $\mathbf{F}$ have continuous first-order partial derivatives at P, so it follows that

$$\{\text{div}\,\mathbf{F}\}(P^*) \to \{\text{div}\,\mathbf{F}\}(P) \quad \text{as} \quad P^* \to P.$$

Equation (10) now follows after we divide both sides by V_r and then take the limit as $r \to 0$.

For instance, suppose that $\mathbf{F} = \rho\mathbf{v}$ is a fluid flow vector field. We can interpret Equation (10) as saying that $\{\text{div}\,\mathbf{F}\}(P)$ is the net rate per unit volume that fluid mass is flowing away (or "diverging") from the point P. For this reason the point P is called a **source** if $\{\text{div}\,\mathbf{F}\}(P) > 0$ but a **sink** if $\{\text{div}\,\mathbf{F}\}(P) < 0$.

Heat in a conducting body can be treated mathematically as though it were a fluid flowing through the body. In Miscellaneous Problems 25–27 at the end of this chapter, we ask you to apply the divergence theorem to show this: If $u = u(x, y, z, t)$ is the temperature at the point (x, y, z) at the time t in a body through which heat is flowing, then the function u must satisfy the equation

$$\frac{\partial^2 u}{\partial x^2} + \frac{\partial^2 u}{\partial y^2} + \frac{\partial^2 u}{\partial z^2} = \frac{1}{k} \cdot \frac{\partial u}{\partial t} \tag{11}$$

where k is a constant (the *thermal diffusivity* of the body). This is a *partial differential equation* called the **heat equation.** If both the initial temperature $u(x, y, z, 0)$ and the temperatures on the boundary of the body are given, then its interior temperatures at future times are determined by the heat equation. A large part of advanced applied mathematics consists of techniques for solving such partial differential equations.

Another impressive consequence of the divergence theorem is Archimedes' law of buoyancy; see Problem 21 here and also Problem 22 in Section 17.7.

17.6 PROBLEMS

In each of Problems 1–5, verify the divergence theorem by direct computation of both the surface integral.and the triple integral in Equation (1).

1. $\mathbf{F} = x\mathbf{i} + y\mathbf{j} + z\mathbf{k}$; S is the spherical surface with equation $x^2 + y^2 + z^2 = 1$.

2. $\mathbf{F} = |\mathbf{r}|\mathbf{r}$ where $\mathbf{r} = x\mathbf{i} + y\mathbf{j} + z\mathbf{k}$; S is the spherical surface with equation $x^2 + y^2 + z^2 = 9$.

3. $\mathbf{F} = x\mathbf{i} + y\mathbf{j} + z\mathbf{k}$; S is the surface of the cube bounded by the three coordinate planes and the three planes $x = 2$, $y = 2$, and $z = 2$.

4. $\mathbf{F} = xy\mathbf{i} + yz\mathbf{j} + xz\mathbf{k}$; S is the surface of Problem 3.

5. $\mathbf{F} = (x + y)\mathbf{i} + (y + z)\mathbf{j} + (z + x)\mathbf{k}$; S is the surface of the tetrahedron bounded by the three coordinate planes and the plane $x + y + z = 1$.

In Problems 6–14, use the divergence theorem to evaluate $\iint_S \mathbf{F} \cdot \mathbf{n}\, dS$ where $\mathbf{n}$ is the outer unit normal to the surface S.

6. $\mathbf{F} = x^2\mathbf{i} + y^2\mathbf{j} + z^2\mathbf{k}$; S is the surface of Problem 3.

7. $\mathbf{F} = x^3\mathbf{i} + y^3\mathbf{j} + z^3\mathbf{k}$; S is the surface of the cylinder bounded by $x^2 + y^2 = 9$, $z = -1$, and $z = 4$.

8. $\mathbf{F} = (x^2 + y^2)(x\mathbf{i} + y\mathbf{j})$; S is the surface of the region bounded by the plane $z = 0$ and the paraboloid $z = 25 - x^2 - y^2$.

9. $\mathbf{F} = (x^2 + e^{-yz})\mathbf{i} + (y + \sin xz)\mathbf{j} + (\cos xy)\mathbf{k}$; S is the surface of Problem 5.

10. $\mathbf{F} = (xy^2 + e^{-y}\sin z)\mathbf{i} + (x^2y + e^{-x}\cos z)\mathbf{j} + (\tan^{-1} xy)\mathbf{k}$; S is the surface of the region bounded by the paraboloid $z = x^2 + y^2$ and the plane $z = 9$.

11. $\mathbf{F} = (x^2 + y^2 + z^2)(x\mathbf{i} + y\mathbf{j} + z\mathbf{k})$; S is the surface of Problem 8.

12. $\mathbf{F} = \mathbf{r}/|\mathbf{r}|$ where $\mathbf{r} = x\mathbf{i} + y\mathbf{j} + z\mathbf{k}$; S is the sphere $\rho = 2$ of radius 2 centered at the origin.

13. $\mathbf{F} = x\mathbf{i} + y\mathbf{j} + 3\mathbf{k}$; S is the boundary of the region bounded by the paraboloid $z = x^2 + y^2$ and the plane $z = 4$.

14. $\mathbf{F} = (x^3 + e^z)\mathbf{i} + x^2y\mathbf{j} + (\sin xy)\mathbf{k}$; S is the boundary of the region bounded by the paraboloid $z = 4 - x^2$ and the planes $y = 0$, $z = 0$, and $y + z = 5$.

15. The **Laplacian** of the twice-differentiable scalar function f is defined to be $\nabla^2 f = \operatorname{div}(\operatorname{grad} f) = \nabla \cdot \nabla f$. Show that

$$\nabla^2 f = \frac{\partial^2 f}{\partial x^2} + \frac{\partial^2 f}{\partial y^2} + \frac{\partial^2 f}{\partial z^2}.$$

16. Let $\partial f/\partial \mathbf{n} = \nabla f \cdot \mathbf{n}$ denote the directional derivative of the scalar function f in the direction of the outer unit normal $\mathbf{n}$ to the surface S that bounds the region T. Show that

$$\iint_S \frac{\partial f}{\partial \mathbf{n}}\, dS = \iiint_T \nabla^2 f\, dV.$$

17. Suppose that $\nabla^2 f \equiv 0$ in the region T with boundary S. Prove that

$$\iint_S f \frac{\partial f}{\partial \mathbf{n}} \, dS = \iiint_T |\nabla f|^2 \, dV.$$

(See Problems 15 and 16 for the notation.)

18. Apply the divergence theorem to $\mathbf{F} = f \, \nabla g$ to establish **Green's first identity**

$$\iint_S f \frac{\partial g}{\partial \mathbf{n}} \, dS = \iiint_T (f \, \nabla^2 g + \nabla f \cdot \nabla g) \, dV.$$

19. Interchange f and g in Green's first identity (Problem 18) to establish **Green's second identity**

$$\iint_S \left(f \frac{\partial g}{\partial \mathbf{n}} - g \frac{\partial f}{\partial \mathbf{n}} \right) dS = \iiint_T (f \, \nabla^2 g - g \, \nabla^2 f) \, dV.$$

20. Suppose that f is a differentiable scalar function defined on the region T of space and that S is the boundary of T. Prove that

$$\iint_S f\mathbf{n} \, dS = \iiint_T \nabla f \, dV.$$

(*Suggestion:* Apply the divergence theorem to $\mathbf{F} = f\mathbf{a}$ where $\mathbf{a}$ is an arbitrary constant vector. *Note:* Integrals of vector-valued functions are defined by componentwise integration.)

21. (Archimedes' Law of Buoyancy) Let S be the surface of a body T submerged in a fluid of constant density ρ. Set up coordinates so that $z = 0$ corresponds to the surface of the fluid and so that positive values of z are measured *downward* from the surface. Then the pressure at depth z is $p = \rho g z$. The buoyant force exerted on the body by the fluid is

$$\mathbf{B} = -\iint_S p\mathbf{n} \, dS.$$

(Why?) Apply the result of Problem 20 to show that $\mathbf{B} = -W\mathbf{k}$ where W is the weight of the fluid displaced by the body. Because z is measured downward, the vector $\mathbf{B}$ is directed upward.

22. Let $\mathbf{r} = \langle x, y, z \rangle$, let $\mathbf{r}_0 = \langle x_0, y_0, z_0 \rangle$ be a fixed point, and suppose that

$$\mathbf{F}(x, y, z) = \frac{\mathbf{r} - \mathbf{r}_0}{|\mathbf{r} - \mathbf{r}_0|}.$$

Show that $\operatorname{div} \mathbf{F} = 0$ except at the point $\mathbf{r}_0$.

23. Apply the divergence theorem to compute the outward flux

$$\iint_S \mathbf{F} \cdot \mathbf{n} \, dS,$$

where $\mathbf{F} = |\mathbf{r}|\mathbf{r}$, $\mathbf{r} = x\mathbf{i} + y\mathbf{j} + z\mathbf{k}$, and S is the surface of Problem 8. (*Suggestion:* Integrate in cylindrical coordinates, first with respect to r and then with respect to z. For the latter integration, make a trigonometric substitution and then consult the formula in Equation (8) of Section 9.4 for the antiderivative of $\sec^5 \theta$.)

17.7 Stokes' Theorem

In Equation (9) of Section 17.4, we gave a vector form of Green's theorem,

$$\oint_C P \, dx + Q \, dy = \iint_R \left(\frac{\partial Q}{\partial x} - \frac{\partial P}{\partial y} \right) dA, \tag{1}$$

that amounted to a two-dimensional version of the divergence theorem. There is another vector form of Green's theorem, one that involves the curl of a vector field. Recall that if $\mathbf{F} = P\mathbf{i} + Q\mathbf{j} + R\mathbf{k}$ is a vector field, then curl $\mathbf{F}$ is the vector field given by

$$\operatorname{curl} \mathbf{F} = \nabla \times \mathbf{F} = \begin{vmatrix} \mathbf{i} & \mathbf{j} & \mathbf{k} \\ \dfrac{\partial}{\partial x} & \dfrac{\partial}{\partial y} & \dfrac{\partial}{\partial z} \\ P & Q & R \end{vmatrix}$$

$$= \left(\frac{\partial R}{\partial y} - \frac{\partial Q}{\partial z} \right)\mathbf{i} + \left(\frac{\partial P}{\partial z} - \frac{\partial R}{\partial x} \right)\mathbf{j} + \left(\frac{\partial Q}{\partial x} - \frac{\partial P}{\partial y} \right)\mathbf{k}. \tag{2}$$

Note first that the **k**-component of $\nabla \times \mathbf{F}$ is the integrand of the double integral in Equation (1). We know from Section 17.2 that the line integral in (1) can also be written as

$$\oint_C \mathbf{F} \cdot \mathbf{T}\, ds,$$

where **T** is the positive-directed unit tangent. Consequently, Green's theorem can be rewritten in the form

$$\oint_C \mathbf{F} \cdot \mathbf{T}\, ds = \iint_R (\text{curl } \mathbf{F}) \cdot \mathbf{k}\, dA. \tag{3}$$

Stokes' theorem is the generalization of (3) that we get by replacing the plane region R with a floppy two-dimensional version: an oriented bounded surface S in three-dimensional space with boundary C that consists of one or more simple closed space curves.

Recall that an *oriented* surface is one with a continuous unit normal vector **n**. The positive orientation of the boundary C of an oriented surface S corresponds to the unit tangent vector **T** such that $\mathbf{n} \times \mathbf{T}$ always points *into* S (see Fig. 17.29). You should check that for a plane region with unit normal **k**, the positive orientation of its outer boundary is counterclockwise.

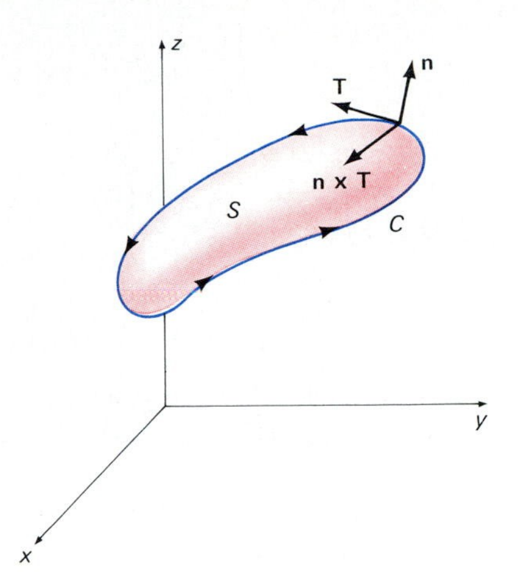

17.29 Vectors, surface, and boundary curve mentioned in the statement of Stokes' theorem

Stokes' Theorem

Let S be an oriented, bounded, and piecewise smooth surface in space with positively oriented boundary C. Suppose that the components of the vector field **F** have continuous first-order partial derivatives in a space region containing S. Then

$$\oint_C \mathbf{F} \cdot \mathbf{T}\, ds = \iint_S (\text{curl } \mathbf{F}) \cdot \mathbf{n}\, dS. \tag{4}$$

Thus Stokes' theorem says that *the line integral around the boundary curve of the tangential component of* **F** *is equal to the surface integral of the normal component of* curl **F**. Compare Equations (3) and (4).

This result first appeared publicly as a problem posed by George Stokes (1819–1903) on a prize examination for Cambridge University students in 1854. It had been stated in an 1850 letter to Stokes from the physicist William Thomson (Lord Kelvin, 1824–1907).

In terms of the components of $\mathbf{F} = P\mathbf{i} + Q\mathbf{j} + R\mathbf{k}$ and those of curl **F**, Stokes' theorem can—with the aid of Equation (14) of Section 17.5—be recast in its scalar form:

$$\oint_C P\, dx + Q\, dy + R\, dz$$
$$= \iint_S \left(\frac{\partial R}{\partial y} - \frac{\partial Q}{\partial z}\right) dy\, dz + \left(\frac{\partial P}{\partial z} - \frac{\partial R}{\partial x}\right) dz\, dx + \left(\frac{\partial Q}{\partial x} - \frac{\partial P}{\partial y}\right) dx\, dy. \tag{5}$$

Here we need the usual stipulation that the parametrization of S corresponds to the given unit normal **n**.

To prove Stokes' theorem, it is therefore enough to establish the equation

$$\oint_C P\,dx = \iint_S \left(\frac{\partial P}{\partial z}\,dz\,dx - \frac{\partial P}{\partial y}\,dx\,dy\right) \tag{6}$$

and the corresponding two equations that are the Q and R "components" of Equation (5). Equation (5) itself then follows by adding the three results.

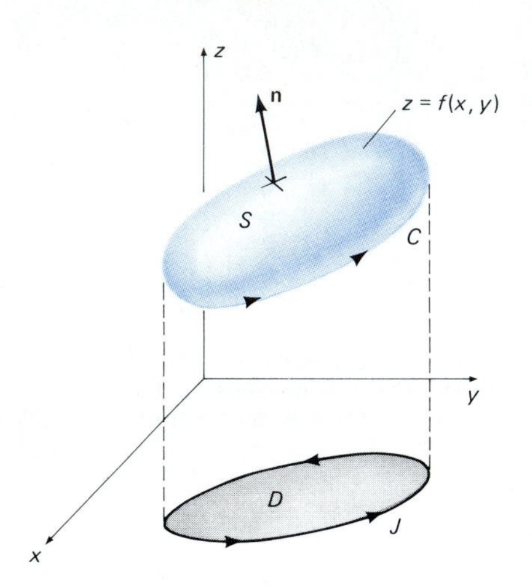

17.30 The surface S

Partial Proof Suppose first that S is the graph of a function $z = f(x, y)$, (x, y) in D, with an upper unit normal and with D a region in the xy-plane bounded by the simple closed curve J (see Fig. 17.30). Then

$$\begin{aligned}\oint_C P\,dx &= \oint_J P(x, y, f(x, y))\,dx\\ &= \oint_J p(x, y)\,dx \qquad (\text{where } p(x, y) \equiv P(x, y, f(x, y)))\\ &= -\iint_D \frac{\partial p}{\partial y}\,dx\,dy \qquad (\text{by Green's theorem}).\end{aligned}$$

We now use the chain rule to compute $\partial p/\partial y$ and find that

$$\oint_C P\,dx = -\iint_D \left(\frac{\partial P}{\partial y} + \frac{\partial P}{\partial z}\frac{\partial z}{\partial y}\right) dx\,dy. \tag{7}$$

Next we use Equation (15) of Section 17.5:

$$\iint_S P\,dy\,dz + Q\,dz\,dx + R\,dx\,dy = \iint_D \left(-P\frac{\partial z}{\partial x} - Q\frac{\partial z}{\partial y} + R\right) dx\,dy.$$

In this equation we replace P by 0, Q by $\partial P/\partial z$, and R by $-\partial P/\partial y$. This gives

$$\iint_S \left(\frac{\partial P}{\partial z}\,dz\,dx - \frac{\partial P}{\partial y}\,dx\,dy\right) = \iint_D \left(-\frac{\partial P}{\partial z}\frac{\partial z}{\partial y} - \frac{\partial P}{\partial y}\right) dx\,dy. \tag{8}$$

Finally, we compare Equations (7) and (8) and see that we have established Equation (6). If the surface S can also be written in the forms $y = g(x, z)$ and $x = h(y, z)$, then the Q and R "components" of Equation (5) can be derived in essentially the same way. This proves Stokes' theorem for the special case of a surface S that can be represented as a graph in all three coordinate directions. Stokes' theorem may then be extended to a more general oriented surface in the now-familiar way: by subdividing it into simpler surfaces, to each of which the above proof is applicable. ■

EXAMPLE 1 Apply Stokes' theorem to evaluate $\oint_C \mathbf{F}\cdot\mathbf{T}\,ds$, where C is the ellipse in which the plane $z = y + 3$ intersects the cylinder $x^2 + y^2 = 1$. Orient the ellipse counterclockwise as viewed from above. Take $\mathbf{F}(x, y, z) = 3z\mathbf{i} + 5x\mathbf{j} - 2y\mathbf{k}$.

Solution The plane, cylinder, and ellipse appear in Fig. 17.31. The given orientation of C corresponds to the upward unit normal $\mathbf{n} = (-\mathbf{j} + \mathbf{k})/\sqrt{2}$ to the elliptical region S in the plane $z = y + 3$ bounded by C. Now

$$\text{curl } \mathbf{F} = \begin{vmatrix} \mathbf{i} & \mathbf{j} & \mathbf{k} \\ \dfrac{\partial}{\partial x} & \dfrac{\partial}{\partial y} & \dfrac{\partial}{\partial z} \\ 3z & 5x & -2y \end{vmatrix} = -2\mathbf{i} + 3\mathbf{j} + 5\mathbf{k},$$

so

$$(\text{curl } \mathbf{F}) \cdot \mathbf{n} = (-2\mathbf{i} + 3\mathbf{j} + 5\mathbf{k}) \cdot \frac{1}{\sqrt{2}}(-\mathbf{j} + \mathbf{k}) = \frac{-3+5}{\sqrt{2}} = \sqrt{2}.$$

Hence by Stokes' theorem,

$$\oint_C \mathbf{F} \cdot \mathbf{T}\, ds = \iint_S (\text{curl } \mathbf{F}) \cdot \mathbf{n}\, dS = \iint_S \sqrt{2}\, dS = \sqrt{2}\, \text{area}(S) = 2\pi,$$

because we can see from Fig. 17.31 that S is an ellipse with semiaxes 1 and $\sqrt{2}$; thus its area is $\pi\sqrt{2}$.

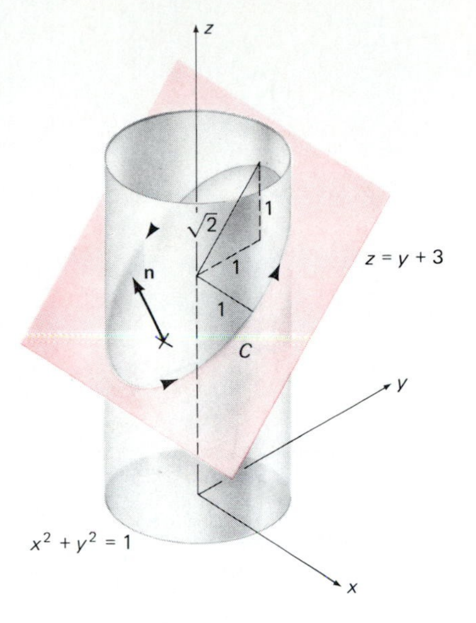

17.31 The ellipse of Example 1

EXAMPLE 2 Apply Stokes' theorem to evaluate $\iint_S (\nabla \times \mathbf{F}) \cdot \mathbf{n}\, dS$ where $\mathbf{F} = 3z\mathbf{i} + 5x\mathbf{j} - 2y\mathbf{k}$ and S is the part of the parabolic surface $z = x^2 + y^2$ that lies below the plane $z = 4$ and has orientation given by the upper unit normal vector (see Fig. 17.32).

Solution We parametrize the boundary circle C of S by $x = 2\cos t$, $y = 2\sin t$, $z = 4$ for $0 \leqq t \leqq 2\pi$. Then $dx = -2\sin t\, dt$, $dy = 2\cos t\, dt$, and $dz = 0$. So Stokes' theorem yields

$$\iint_S (\nabla \times \mathbf{F}) \cdot \mathbf{n}\, dS = \oint_C \mathbf{F} \cdot \mathbf{T}\, ds = \oint_C 3z\, dx + 5x\, dy - 2y\, dz$$

$$= \int_0^{2\pi} (3)(4)(-2\sin t\, dt) + (5)(2\cos t)(2\cos t\, dt) - (2)(2\sin t)(0)$$

$$= \int_0^{2\pi} (-24\sin t + 20\cos^2 t)\, dt$$

$$= \int_0^{2\pi} (-24\sin t + 10 + 10\cos 2t)\, dt$$

$$= \Big[24\cos t + 10t + 5\sin 2t \Big]_0^{2\pi} = 20\pi.$$

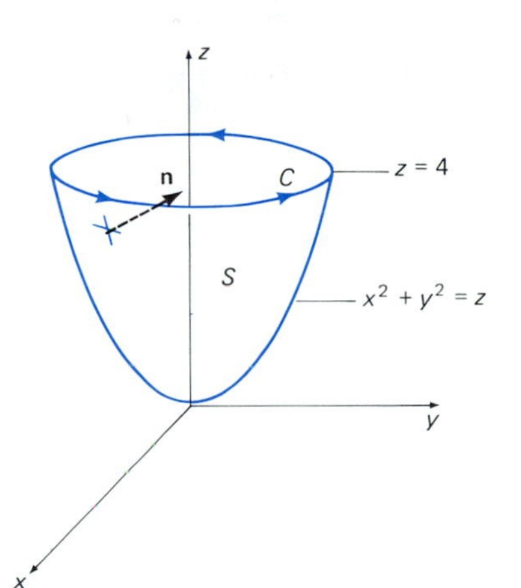

17.32 The parabolic surface of Example 2

Just as the divergence theorem yields a physical interpretation of div $\mathbf{F}$ (the formula in Equation (10) in Section 17.6), Stokes' theorem yields a physical interpretation of curl $\mathbf{F}$. Let S_r be a circular disk of radius r, centered at the point P in space and perpendicular to the unit vector $\mathbf{n}$. Let C_r be the boundary circle of S_r; see Fig. 17.33. Then Stokes' theorem and the average value property of double integrals together give

$$\oint_{C_r} \mathbf{F} \cdot \mathbf{T}\, ds = \iint_{S_r} (\text{curl } \mathbf{F}) \cdot \mathbf{n}\, dS = \pi r^2 \{(\text{curl } \mathbf{F}) \cdot \mathbf{n}\}(P^*)$$

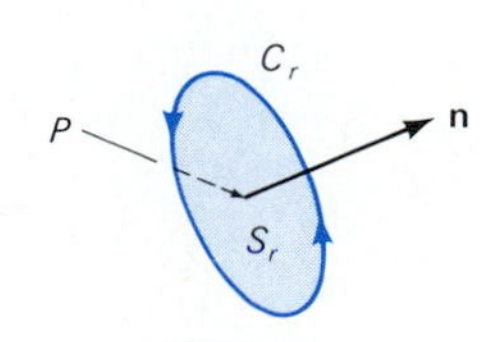

17.33 A physical interpretation of the curl of a vector field

for some point P^* of S_r, where $\{(\text{curl } \mathbf{F}) \cdot \mathbf{n}\}(P^*)$ denotes the value of $(\text{curl } \mathbf{F}) \cdot \mathbf{n}$ at the point P^*. We divide this equality by πr^2 and then take the limit as $r \to 0$. This gives

$$\{(\text{curl } \mathbf{F}) \cdot \mathbf{n}\}(P) = \lim_{r \to 0} \frac{1}{\pi r^2} \oint_{C_r} \mathbf{F} \cdot \mathbf{T}\, ds. \tag{9}$$

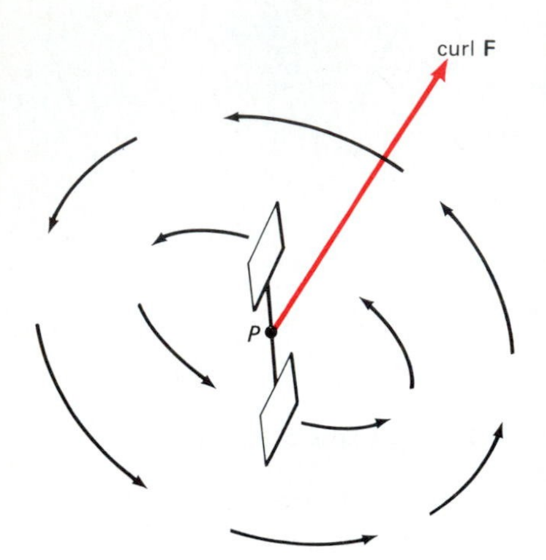

17.34 The paddle-wheel interpretation of curl **F**

This last formula has a natural physical meaning. Suppose that $\mathbf{F} = \rho \mathbf{v}$, where $\mathbf{v}$ is the velocity vector field of a steady-state fluid flow with constant density ρ. Then the value of the integral

$$\Gamma(C) = \oint_C \mathbf{F} \cdot \mathbf{T}\, ds \tag{10}$$

measures the rate of flow of fluid mass *around* the curve C and is therefore called the **circulation** of $\mathbf{F}$ around C. From (9) we see that

$$\{(\text{curl } \mathbf{F}) \cdot \mathbf{n}\}(P) \approx \frac{\Gamma(C_r)}{\pi r^2}$$

if C_r is a circle of very small radius r centered at P and perpendicular to $\mathbf{n}$. If $\{\text{curl } \mathbf{F}\}(P) \neq \mathbf{0}$, it follows that $\Gamma(C_r)$ is greatest (for r fixed and small) when the unit vector $\mathbf{n}$ points in the direction of $\{\text{curl } \mathbf{F}\}(P)$. Hence the line through P determined by $\{\text{curl } \mathbf{F}\}(P)$ is the axis about which the fluid near P is revolving the most rapidly. A tiny paddle wheel placed in the fluid at P (see Fig. 17.34) would rotate the fastest if its axis lay along this line. In Miscellaneous Problem 32 (at the end of this chapter), we ask you to show that $|\text{curl } \mathbf{F}| = 2\rho\omega$ in the case of a fluid revolving steadily about a fixed axis with constant angular speed ω (in radians per second). Thus $\{\text{curl } \mathbf{F}\}(P)$ indicates both the direction *and* rate of rotation of the fluid near P. Because of this interpretation, the notation rot $\mathbf{F}$ for the curl is sometimes seen in older books, an abbreviation that we are happy has disappeared from general use.

If curl $\mathbf{F} = \mathbf{0}$ everywhere, the fluid flow and the vector field $\mathbf{F}$ itself are said to be **irrotational.** An infinitesimal straw placed in an irrotational fluid flow would be translated parallel to itself without rotating. It turns out that a vector field $\mathbf{F}$ defined on a simply connected region D is irrotational if and only if it is conservative, which in turn is true if and only if the line integral $\int_C \mathbf{F} \cdot \mathbf{T}\, ds$ is independent of the path in D. (The region D is said to be **simply connected** if every simple closed curve in D can be continuously shrunk to a point while staying within D. The interior of a torus is an example of a space region that is *not* simply connected. It is true, though not obvious, that any piecewise smooth simple closed curve in a simply connected region D is the boundary of a piecewise smooth oriented surface in D.)

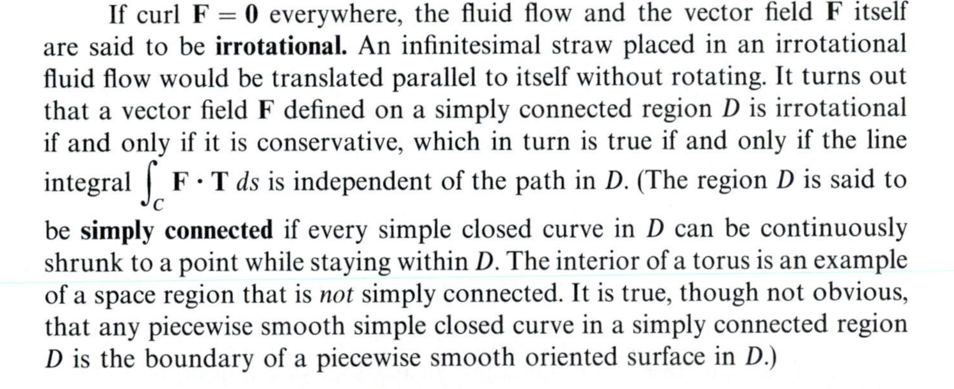

Theorem *Conservative and Irrotational Fields*

Let $\mathbf{F}$ be a vector field with continuous first-order partial derivatives in a simply connected space region D. Then the vector field $\mathbf{F}$ is irrotational if and only if it is conservative; that is, $\nabla \times \mathbf{F} = \mathbf{0}$ if and only if $\mathbf{F} = \nabla \phi$ for some scalar function ϕ defined on D.

Partial Proof A complete proof of the *if* part of the theorem is easy; in Example 7 of Section 17.1 we showed that $\nabla \times (\nabla\phi) = \mathbf{0}$ for any twice-differentiable scalar function ϕ.

Here is a sketch of how one might show the *only if* part of the proof of the theorem. Assume that $\mathbf{F}$ is irrotational. Let $P_0(x_0, y_0, z_0)$ be a fixed point of D. Given an arbitrary point $P(x, y, z)$ of D, we would like to define

$$\phi(x, y, z) = \int_{C_1} \mathbf{F}\cdot\mathbf{T}\,ds \tag{11}$$

where C_1 is a path in D from P_0 to P. But it is essential that we show that any *other* path C_2 from P_0 to P would give the *same* value for $\phi(x, y, z)$.

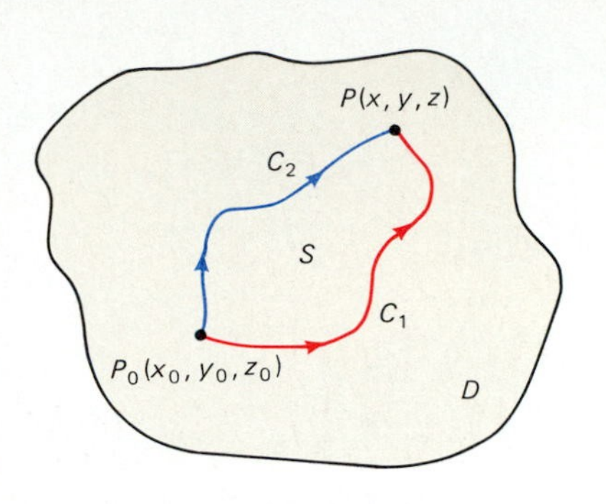

17.35 Two paths from P_0 to P in the simply connected space region D

We may assume, as suggested by Fig. 17.35, that C_1 and C_2 intersect only at their endpoints. Then the simple closed curve $C = C_1 \cup (-C_2)$ bounds an oriented surface S in D, and Stokes' theorem gives

$$\int_{C_1} \mathbf{F}\cdot\mathbf{T}\,ds - \int_{C_2} \mathbf{F}\cdot\mathbf{T}\,ds = \oint_C \mathbf{F}\cdot\mathbf{T}\,ds = \iint_S (\nabla\times\mathbf{F})\cdot\mathbf{n}\,dS = 0$$

because of the hypothesis that $\nabla\times\mathbf{F}\equiv\mathbf{0}$. This shows that the line integral $\int_C \mathbf{F}\cdot\mathbf{T}\,ds$ is *independent of path*, just as desired. In Problem 21 we ask you to complete this proof by showing that the function ϕ of Equation (11) is the one whose existence is claimed in the theorem. That is, $\mathbf{F} = \nabla\phi$. ■

EXAMPLE 3 Show that the vector field $\mathbf{F} = 3x^2\mathbf{i} + 5z^2\mathbf{j} + 10yz\mathbf{k}$ is irrotational. Then find a potential function $\phi(x, y, z)$ such that $\nabla\phi = \mathbf{F}$.

Solution To show that $\mathbf{F}$ is irrotational, we calculate

$$\nabla\times\mathbf{F} = \begin{vmatrix} \mathbf{i} & \mathbf{j} & \mathbf{k} \\ \dfrac{\partial}{\partial x} & \dfrac{\partial}{\partial y} & \dfrac{\partial}{\partial z} \\ 3x^2 & 5z^2 & 10yz \end{vmatrix} = (10z - 10z)\mathbf{k} = \mathbf{0}.$$

Hence the preceding theorem implies that $\mathbf{F}$ has a potential function ϕ. We can apply Equation (11) to actually find ϕ. If C_1 is the straight line segment from $(0, 0, 0)$ to (x_1, y_1, z_1) parametrized by $x = x_1t$, $y = y_1t$, $z = z_1t$ for $0 \leqq t \leqq 1$, then (11) yields

$$\begin{aligned}\phi(x_1, y_1, z_1) &= \int_{C_1} \mathbf{F}\cdot\mathbf{T}\,ds \\ &= \int_{(0,0,0)}^{(x_1,y_1,z_1)} 3x^2\,dx + 5z^2\,dy + 10yz\,dz \\ &= \int_0^1 (3x_1^2t^2)(x_1\,dt) + (5z_1^2t^2)(y_1\,dt) + (10y_1tz_1t)(z_1\,dt) \\ &= \int_0^1 (3x_1^3t^2 + 15y_1z_1^2t^2)\,dt = \Big[x_1^3t^3 + 5y_1z_1^2t^3\Big]_0^1,\end{aligned}$$

and thus

$$\phi(x_1, y_1, z_1) = x_1^3 + 5y_1z_1^2.$$

We may now drop the subscripts because (x_1, y_1, z_1) is actually an arbitrary point of space, and therefore we obtain the scalar potential function

$\phi(x, y, z) = x^3 + 5yz^2$. As a check, we note that $\phi_x = 3x^2$, $\phi_y = 5z^2$, and $\phi_z = 10yz$, so $\nabla\phi = \mathbf{F}$ as desired.

APPLICATION Suppose that $\mathbf{v}$ is the velocity vector field of a steady fluid flow that is both irrotational and incompressible—the density ρ of the fluid is constant. Suppose that S is any closed surface bounding a region T. Then, because of conservation of mass, the flux of $\mathbf{v}$ across S must be zero; the mass of fluid within S remains constant. Hence the divergence theorem gives

$$\iiint_T \operatorname{div} \mathbf{v}\, dV = \iint_S \mathbf{v} \cdot \mathbf{n}\, dS = 0.$$

Because this holds for *any* region T, it follows from the usual average value property argument that $\operatorname{div} \mathbf{v} = 0$ everywhere. The scalar function ϕ provided by the theorem above, for which $\mathbf{v} = \nabla\phi$, is called the **velocity potential** of the fluid flow. We substitute $\mathbf{v} = \nabla\phi$ into the equation $\operatorname{div} \mathbf{v} = 0$ and thereby obtain

$$\operatorname{div}(\nabla\phi) = \frac{\partial^2\phi}{\partial x^2} + \frac{\partial^2\phi}{\partial y^2} + \frac{\partial^2\phi}{\partial z^2} = 0. \tag{12}$$

Thus the velocity potential ϕ of an irrotational and incompressible fluid flow satisfies *Laplace's equation.*

Laplace's equation appears in numerous other applications. For example, consider a heated body whose temperature function $u = u(x, y, z)$ is independent of time t. Then $\partial u/\partial t \equiv 0$ in the heat equation (Equation (15) in Section 17.6) shows that the "steady-state temperature" function $u(x, y, z)$ satisfies Laplace's equation

$$\frac{\partial^2 u}{\partial x^2} + \frac{\partial^2 u}{\partial y^2} + \frac{\partial^2 u}{\partial z^2} = 0. \tag{13}$$

These brief remarks should indicate how the mathematics of this chapter forms the starting point for investigations in a number of areas, including acoustics, aerodynamics, electromagnetism, meteorology, and oceanography, among others. Indeed, the entire subject of vector analysis stems historically from its scientific applications rather than from abstract mathematical considerations. The modern form of the subject is due primarily to J. Willard Gibbs (1839–1903), the first great American physicist, and the English electrical engineer Oliver Heaviside (1850–1925).

17.7 PROBLEMS

In Problems 1–5, use Stokes' theorem for the evaluation of

$$\iint_S (\operatorname{curl} \mathbf{F}) \cdot \mathbf{n}\, dS.$$

1. $\mathbf{F} = 3y\mathbf{i} - 2x\mathbf{j} + xyz\mathbf{k}$; S is the hemispherical surface $z = (4 - x^2 - y^2)^{1/2}$ with upper unit normal.

2. $\mathbf{F} = 2y\mathbf{i} + 3x\mathbf{j} + e^z\mathbf{k}$; S is the part of the paraboloid $z = x^2 + y^2$ below the plane $z = 4$ with upper unit normal.

3. $\mathbf{F} = \langle xy, -2, \arctan x^2 \rangle$; S is the part of the paraboloid $z = 9 - x^2 - y^2$ above the xy-plane with upper unit normal.

4. $\mathbf{F} = yz\mathbf{i} + xz\mathbf{j} + xy\mathbf{k}$; S is the part of the cylinder $x^2 + y^2 = 1$ between the two planes $z = 1$ and $z = 3$ with outer unit normal.

5. $\mathbf{F} = \langle yz, -xz, z^3 \rangle$; S is the part of the cone $z = (x^2 + y^2)^{1/2}$ between the two planes $z = 1$ and $z = 3$ with upper unit normal.

In Problems 6–10, use Stokes' theorem to evaluate

$$\oint_C \mathbf{F} \cdot \mathbf{T}\, ds.$$

6. $\mathbf{F} = 3y\mathbf{i} - 2x\mathbf{j} + 4x\mathbf{k}$; C is the circle $x^2 + y^2 = 9$, $z = 4$ oriented counterclockwise as viewed from above.

7. $\mathbf{F} = 2z\mathbf{i} + x\mathbf{j} + 3y\mathbf{k}$; C is the ellipse in which the plane $z = x$ meets the cylinder $x^2 + y^2 = 4$, oriented counterclockwise as viewed from above.

8. $\mathbf{F} = y\mathbf{i} + z\mathbf{j} + x\mathbf{k}$; C is the boundary of the triangle with vertices $(0, 0, 0)$, $(2, 0, 0)$, and $(0, 2, 2)$, oriented counterclockwise as viewed from above.

9. $\mathbf{F} = \langle y - x, x - z, x - y\rangle$; C is the boundary of the part of the plane $x + 2y + z = 2$ that lies in the first octant, oriented counterclockwise as viewed from above.

10. $\mathbf{F} = y^2\mathbf{i} + z^2\mathbf{j} + x^2\mathbf{k}$; C is the intersection of the plane $z = y$ and the cylinder $x^2 + y^2 = 2y$, oriented counterclockwise as viewed from above.

In each of Problems 11–14, first show that the given vector field $\mathbf{F}$ is irrotational, then apply the method of Example 3 to find a potential function $\phi = \phi(x, y, z)$ for $\mathbf{F}$.

11. $\mathbf{F} = (3y - 2z)\mathbf{i} + (3x + z)\mathbf{j} + (y - 2x)\mathbf{k}$

12. $\mathbf{F} = (3y^3 - 10xz^2)\mathbf{i} + 9xy^2\mathbf{j} - 10x^2z\mathbf{k}$

13. $\mathbf{F} = (3e^z - 5y \sin x)\mathbf{i} + (5 \cos x)\mathbf{j} + (17 + 3xe^z)\mathbf{k}$

14. $\mathbf{F} = r^3\mathbf{r}$, where $\mathbf{r} = x\mathbf{i} + y\mathbf{j} + z\mathbf{k}$ and $r = |\mathbf{r}|$

15. Suppose that $\mathbf{r} = x\mathbf{i} + y\mathbf{j} + z\mathbf{k}$ and that $\mathbf{a}$ is a constant vector. Show that

(a) $\nabla \cdot (\mathbf{a} \times \mathbf{r}) = 0$;

(b) $\nabla \times (\mathbf{a} \times \mathbf{r}) = 2\mathbf{a}$;

(c) $\nabla \cdot [(\mathbf{r} \cdot \mathbf{r})\mathbf{a}] = 2\mathbf{r} \cdot \mathbf{a}$;

(d) $\nabla \times [(\mathbf{r} \cdot \mathbf{r})\mathbf{a}] = 2(\mathbf{r} \times \mathbf{a})$.

16. Prove that

$$\iint_S (\text{curl } \mathbf{F}) \cdot \mathbf{n}\, dS$$

has the same value for all oriented surfaces S that have the same oriented boundary curve C.

17. Suppose that S is a closed surface. Prove that

$$\iint_S (\text{curl } \mathbf{F}) \cdot \mathbf{n}\, dS = 0$$

in two different ways:

(a) by using the divergence theorem, with T the region bounded outside by S;

(b) by using Stokes' theorem, with the aid of a nice simple closed curve C on S.

Line integrals, surface integrals, and triple integrals of vector-valued functions are defined by componentwise integration. Such integrals appear in the following three problems.

18. Suppose that C and S are as in the statement of Stokes' theorem and that ϕ is a scalar function. Prove that

$$\oint_C \phi\mathbf{T}\, ds = \iint_S \mathbf{n} \times \nabla\phi\, dS.$$

(*Suggestion:* Apply Stokes' theorem with $\mathbf{F} = \phi\mathbf{a}$ where $\mathbf{a}$ is an arbitrary constant vector.)

19. Suppose that $\mathbf{a}$ and $\mathbf{r}$ are as in Problem 15. Prove that

$$\oint_C \mathbf{a} \times \mathbf{r} \cdot \mathbf{T}\, ds = 2\mathbf{a} \cdot \iint_S \mathbf{n}\, dS.$$

20. Suppose that S is a closed surface that bounds the region T. Prove that

$$\iint_S \mathbf{n} \times \mathbf{F}\, dS = \iiint_T \nabla \times \mathbf{F}\, dV.$$

(*Suggestion:* Apply the divergence theorem to $\mathbf{F} \times \mathbf{a}$ where $\mathbf{a}$ is an arbitrary constant vector.)

Note that the formulas of Problem 20, the divergence theorem, and Problem 20 of Section 17.6 all fit the pattern

$$\iint_S \mathbf{n} * (\quad)\, dS = \iiint_T \nabla * (\quad)\, dV,$$

where $*$ may denote either ordinary multiplication, the dot product, or the cross product, and either a scalar function or a vector-valued function is placed within the parentheses, as appropriate.

21. Suppose that the line integral $\int_C \mathbf{F} \cdot \mathbf{T}\, ds$ is independent of path. If

$$\phi(x, y, z) = \int_{P_0}^{P} \mathbf{F} \cdot \mathbf{T}\, ds$$

as in Equation (11), show that $\nabla\phi = \mathbf{F}$. (*Suggestion:* Note that if L is the line segment from (x, y, z) to $(x + \Delta x, y, z)$, then

$$\phi(x + \Delta x, y, z) - \phi(x, y, z) = \int_L \mathbf{F} \cdot \mathbf{T}\, ds = \int_x^{x+\Delta x} P\, dx.)$$

22. Let T be the submerged body of Problem 21 in Section 17.6, with centroid

$$\mathbf{r}_0 = \frac{1}{V} \iiint_T \mathbf{r}\, dV.$$

The torque about $\mathbf{r}_0$ of Archimedes' buoyant force $\mathbf{B} = -W\mathbf{k}$ is given by

$$\mathbf{L} = \iint_S (\mathbf{r} - \mathbf{r}_0) \times (-\rho g z \mathbf{n})\, dS.$$

(Why?) Apply the result of Problem 20 of this section to prove that $\mathbf{L} = \mathbf{0}$. It follows that $\mathbf{B}$ acts along the vertical line through the centroid $\mathbf{r}_0$ of the submerged body. (Why?)

CHAPTER 17 REVIEW: Definitions, Concepts, Results

Use the list below as a guide to concepts that you may need to review.

1. Definition and evaluation of the line integral

$$\int_C f(x, y, z)\, ds$$

2. Definition and evaluation of the line integral

$$\int_C P\, dx + Q\, dy + R\, dz$$

3. Relationship between the two types of line integrals, and the line integral of the tangential component of a vector field

4. Line integrals and independence of path

5. Green's theorem

6. Flux and the vector form of Green's theorem

7. The divergence of a vector field

8. Definition and evaluation of the surface integral

$$\iint_S f(x, y, z)\, dS$$

9. Definition and evaluation of the surface integral

$$\iint_S P\, dy\, dz + Q\, dz\, dx + R\, dx\, dy$$

10. Relationship between the two types of surface integrals, and the flux of a vector field across a surface

11. The divergence theorem, in vector and in scalar notation

12. Inverse-square force fields

13. The curl of a vector field

14. Stokes' theorem, in vector and in scalar notation

15. The circulation of a vector field around a simple closed curve

16. Physical interpretations of the divergence and the curl of a vector field

CHAPTER 17 MISCELLANEOUS PROBLEMS

In each of Problems 1–5, evaluate the given line integral.

1. $\int_C (x^2 + y^2)\, ds$, where C is the straight-line segment from $(0, 0)$ to $(3, 4)$.

2. $\int_C y^2\, dx + x^2\, dy$, where C is the graph of $y = x^2$ from $(-1, -1)$ to $(1, 1)$.

3. $\int_C \mathbf{F} \cdot \mathbf{T}\, ds$ where $\mathbf{F} = x\mathbf{i} + y\mathbf{j} + z\mathbf{k}$ and C is the curve $x = e^{2t}$, $y = e^t$, $z = e^{-t}$, $0 \leqq t \leqq \ln 2$.

4. $\int_C xyz\, ds$, where C is the path from $(1, 1, 2)$ to $(2, 3, 6)$ consisting of three straight-line segments, the first parallel to the x-axis, the second parallel to the y-axis, and the third parallel to the z-axis.

5. $\int_C \sqrt{z}\, dx + \sqrt{x}\, dy + y^2\, dz$, where C is the curve $x = t$, $y = t^{3/2}$, $z = t^2$, $0 \leqq t \leqq 4$.

6. Apply Theorem 1 in Section 17.3 to show that the line integral $\int_C y^2\, dx + 2xy\, dy + z\, dz$ is independent of the path C from A to B.

7. Apply Theorem 1 in Section 17.3 to show that the line integral $\int_C x^2 y\, dx + xy^2\, dy$ is not independent of the path C from $(0, 0)$ to $(1, 1)$.

8. A wire shaped like the circle $x^2 + y^2 = a^2$, $z = 0$, has constant density and total mass M. Find its moment of inertia about:

(a) the z-axis; (b) the x-axis.

9. A wire shaped like the parabola $y = \frac{1}{2}x^2$, $0 \leqq x \leqq 2$, has density function $\rho = x$. Find its mass and moment of inertia about the y-axis.

10. Find the work done by the force field $\mathbf{F} = z\mathbf{i} - x\mathbf{j} + y\mathbf{k}$ in moving a particle from $(1, 1, 1)$ to $(2, 4, 8)$ along the curve $y = x^2$, $z = x^3$.

11. Apply Green's theorem to evaluate the line integral

$$\oint_C x^2 y\, dx + xy^2\, dy,$$

where C is the boundary of the region between the two curves $y = x^2$ and $y = 8 - x^2$.

12. Evaluate the line integral

$$\oint_C x^2\, dy,$$

where C is the cardioid $r = 1 + \cos\theta$, by first applying Green's theorem and then changing to polar coordinates.

13. Let C_1 be the circle $x^2 + y^2 = 1$ and C_2 the circle $(x - 1)^2 + y^2 = 9$. Show that if $\mathbf{F} = x^2 y\mathbf{i} - xy^2\mathbf{j}$, then

$$\oint_{C_1} \mathbf{F} \cdot \mathbf{n}\, ds = \oint_{C_2} \mathbf{F} \cdot \mathbf{n}\, ds.$$

14. (a) Let C be the straight line segment from (x_1, y_1) to (x_2, y_2). Show that

$$\tfrac{1}{2}\int_C -y\,dx + x\,dy = \tfrac{1}{2}(x_1y_2 - x_2y_1).$$

(b) Suppose that the vertices of a polygon are (x_1, y_1), $(x_2, y_2), \ldots, (x_n, y_n)$, named in counterclockwise order around the polygon. Apply part (a) to show that the area of the polygon is

$$A = \tfrac{1}{2}\sum_{i=1}^{n} (x_iy_{i+1} - x_{i+1}y_i),$$

where x_{n+1} means x_1 and y_{n+1} means y_1.

15. Suppose that the line integral $\int P\,dx + Q\,dy$ is independent of the path in the plane region D. Prove that

$$\oint_C P\,dx + Q\,dy = 0$$

for every piecewise smooth simple closed curve C in D.

16. Use Green's theorem to prove that $\oint_C P\,dx + Q\,dy = 0$ for every piecewise smooth simple closed curve C in the plane region D if and only if $\partial P/\partial y = \partial Q/\partial x$ at each point of D.

17. Evaluate the surface integral

$$\iint_S (x^2 + y^2 + 2z)\,dS$$

if S is the part of the paraboloid $z = 2 - x^2 - y^2$ above the xy-plane.

18. Suppose that $\mathbf{F} = (x^2 + y^2 + z^2)(x\mathbf{i} + y\mathbf{j} + z\mathbf{k})$ and that S is the spherical surface $x^2 + y^2 + z^2 = a^2$. Evaluate $\iint_S \mathbf{F}\cdot\mathbf{n}\,dS$ without actually performing an antidifferentiation.

19. Let T be the solid bounded by the paraboloids

$$z = x^2 + 2y^2 \quad \text{and} \quad z = 12 - 2x^2 - y^2,$$

and suppose that $\mathbf{F} = x\mathbf{i} + y\mathbf{j} + z\mathbf{k}$. Find (by evaluation of surface integrals) the outward flux of $\mathbf{F}$ across the boundary of T.

20. Give a reasonable definition—as a surface integral—of the average distance of the point P from points of the surface S. Then show that the average distance of a fixed point of a spherical surface of radius a from all points of the surface is $4a/3$.

21. Suppose that the surface S is the graph of the equation $x = g(y, z)$ for (y, z) in the region D of the yz-plane. Prove that

$$\iint_S P\,dy\,dz + Q\,dz\,dx + R\,dx\,dy = \iint_D \left(P - Q\frac{\partial x}{\partial y} - R\frac{\partial x}{\partial z}\right) dy\,dz.$$

22. Suppose that the surface S is the graph of the equation $y = g(x, z)$ for (x, z) in the region D of the xz-plane. Prove that

$$\iint_S f(x, y, z)\,dS = \iint_D f(x, g(x, z), z)\sec\beta\,dx\,dz$$

where $\sec\beta = \sqrt{1 + (\partial y/\partial x)^2 + (\partial y/\partial z)^2}$.

23. Let T be a space region with volume V, boundary surface S, and centroid $(\bar{x}, \bar{y}, \bar{z})$. Use the divergence theorem to show that

$$\bar{z} = \frac{1}{2V}\iint_S z^2\,dx\,dy.$$

24. Apply the result of Problem 23 to find the centroid of the solid hemisphere $x^2 + y^2 + z^2 \leqq a^2$, $z \geqq 0$.

Problems 25–27 outline the derivation of the heat equation for a body having temperature $u = u(x, y, z, t)$ at the point (x, y, z) at time t. Denote by K its heat conductivity and by c its heat capacity, both assumed constant, and let $k = K/c$. Let B be a small solid ball within the body, and let S denote the boundary sphere of B.

25. Deduce from the divergence theorem and Equation (18) of Section 17.5 that the rate of flow of heat across S *into* B is

$$R = \iiint_B k\nabla^2 u\,dV.$$

26. The meaning of heat capacity is that, if Δu is small, then $(c\,\Delta u)\,\Delta V$ calories of heat are required to raise the temperature of the volume ΔV by Δu degrees. It follows that the rate at which the volume ΔV is absorbing heat is $c(\partial u/\partial t)\,\Delta V$ (why?). Conclude that the rate of flow of heat into B is

$$R = \iiint_B c(\partial u/\partial t)\,dV.$$

27. Equate the results of Problems 25 and 26, apply the average value property of triple integrals, and then take the limit as the radius of the ball B approaches zero. You should thereby obtain the heat equation $\partial u/\partial t = k\,\nabla^2 u$.

28. For a *steady-state* temperature function (one that is independent of time t), the heat equation reduces to Laplace's equation

$$\nabla^2 u = \frac{\partial^2 u}{\partial x^2} + \frac{\partial^2 u}{\partial y^2} + \frac{\partial^2 u}{\partial z^2} = 0.$$

(a) Suppose that u_1 and u_2 are two solutions of Laplace's equation in the region T, and that u_1 and u_2 agree on its boundary surface S. Apply Problem 17 in Section 17.6 to the function $f = u_1 - u_2$ to conclude that $\nabla f = \mathbf{0}$ at each point of T.

(b) From the fact that $\nabla f = \mathbf{0}$ in T and $f \equiv 0$ on S, conclude that $f \equiv 0$, so that $u_1 \equiv u_2$. Thus the steady-state temperatures within a region are *determined* by the boundary value temperatures.

29. Suppose that $\mathbf{r} = x\mathbf{i} + y\mathbf{j} + z\mathbf{k}$ and that $\phi(r)$ is a scalar function of $r = |\mathbf{r}|$. Compute:

(a) $\nabla\phi(r)$; (b) $\text{div}[\phi(r)\mathbf{r}]$; (c) $\text{curl}[\phi(r)\mathbf{r}]$.

30. Let S be the upper half of the torus obtained by revolving the circle $(y - a)^2 + z^2 = b^2$ in the yz-plane about the z-axis, with upper unit normal. Describe how to subdivide S to establish Stokes' theorem for it. How are the two boundary circles oriented?

31. Explain why the method of subdivision does not suffice to establish Stokes' theorem for the Möbius strip of Fig. 17.22.

32. (a) Suppose that a fluid or a rigid body is rotating with angular speed ω radius per second about the line through the origin determined by the unit vector $\mathbf{u}$. Show that the velocity of the point with position vector $\mathbf{r}$ is $\mathbf{v} = \boldsymbol{\omega} \times \mathbf{r}$, where $\boldsymbol{\omega} = \omega\mathbf{u}$ is the angular velocity vector. Note first that $|\mathbf{v}| = \omega|\mathbf{r}| \sin\theta$ where θ is the angle between $\mathbf{r}$ and $\boldsymbol{\omega}$.

(b) Use the fact that $\mathbf{v} = \boldsymbol{\omega} \times \mathbf{r}$ (established in part (a)) to show that curl $\mathbf{v} = 2\boldsymbol{\omega}$.

33. Consider an incompressible fluid flowing in space (no sources or sinks) with variable density $\rho(x, y, z, t)$ and velocity field $\mathbf{v}(x, y, z, t)$. Let B be a small ball with radius r and spherical surface S centered at the point (x_0, y_0, z_0). Then the amount of fluid within S at time t is

$$Q(t) = \iiint_B \rho \, dV,$$

and differentiation under the integral sign yields

$$Q'(t) = \iiint_B \frac{\partial \rho}{\partial t} \, dV.$$

(a) Consider fluid flow across S to get

$$Q'(t) = -\iint_S \rho\mathbf{v} \cdot \mathbf{n} \, dS$$

where $\mathbf{n}$ is the outer unit normal to S. Now apply the divergence theorem to convert this into a volume integral.

(b) Equate your two volume integrals for $Q'(t)$, apply the mean value theorem for integrals, and finally take limits as $r \to 0$ to obtain the **continuity equation**

$$\frac{\partial \rho}{\partial t} + \nabla \cdot (\rho\mathbf{v}) = 0.$$

chapter eighteen

Differential Equations

18.1 Introduction

Numerous applications in earlier chapters have illustrated the fact that mathematical models of changing real-world phenomena frequently involve differential equations—that is, equations involving *derivatives* of unknown functions. Several sections have been devoted to the solution of simple types of first-order equations. A *first-order* **differential equation** is one that can be written in the form

$$\frac{dy}{dx} = F(x, y) \tag{1}$$

where x denotes the independent variable and y the unknown function of x. A **solution** of (1) is a function $y = y(x)$ such that $y'(x) = F(x, y(x))$ for all x in an appropriate interval I.

A differential equation of the form in (1) often appears in conjunction with an **initial condition**

$$y(x_0) = y_0. \tag{2}$$

In this case we seek, among the likely multitude of different solutions of (1), that one (if there is only one) which satisfies this initial condition. The differential equation in (1) together with the initial condition in (2) constitute the **initial value problem**

$$\frac{dy}{dx} = F(x, y), \qquad y(x_0) = y_0. \tag{3}$$

The solution of a differential equation inevitably involves the process of integration (or antidifferentiation) to get us from the derivative $y'(x)$ to the solution $y(x)$. For example, consider the very simplest form of differential equation

$$\frac{dy}{dx} = f(x), \tag{4}$$

in which the dependent variable y does not appear on the right. As we saw in Section 4.8, the *general solution* of (4) is obtained by direct integration:

$$y(x) = \int f(x)\,dx = F(x) + C \tag{5}$$

where $F'(x) = f(x)$. That is, every solution of (4) is of the form $y(x) = F(x) + C$ for some specific choice of the *arbitrary constant C*.

Thus we need only find (explicitly) an antiderivative $F(x)$ of the given function $f(x)$. If in addition an initial condition $y(x_0) = y_0$ is given, then we readily determine the appropriate value of the arbitrary constant C by substituting the values $x = x_0$ and $y = y_0$ in the general solution equation

$$y = F(x) + C.$$

The resulting constant $C = y_0 - F(x_0)$ then yields the **particular solution** of the differential equation in (4) that satisfies the initial value problem $dy/dx = f(x, y)$, $y(x_0) = y_0$. We saw in Section 4.9 that this simple technique suffices for the solution of a variety of interesting velocity and acceleration problems stemming from the differential equations

$$\frac{dv}{dt} = a \quad \text{and} \quad \frac{dx}{dt} = v$$

where x denotes position, v velocity, and a acceleration.

EXAMPLE 1 To solve the initial value problem

$$\frac{dx}{dt} = 3x^2, \qquad y(2) = 1,$$

we first integrate to find the general solution

$$y = \int 3x^2\,dx = x^3 + C.$$

Substitution of the values $x = 2$, $y = 1$ yields the equation $1 = 2^3 + C$, and it follows that $C = -7$. Hence the desired particular solution of $dy/dx = 3x^2$ is given by

$$y(x) = x^3 - 7.$$

In Section 6.5 we discussed the solution of a **separable** differential equation

$$\frac{dy}{dx} = g(x)\phi(y), \tag{6}$$

one in which the right-hand side in (1) factors into the *product* of a function of x and a function of y. The method we gave there consisted of multiplying by $f(y) = 1/\phi(y)$ to "separate" the variables in (6), then integrating each side to obtain the general solution in the form

$$\int f(y)\,dy = \int g(x)\,dx + C. \tag{7}$$

If the integrals $F(y) = \int f(y)\,dy$ and $G(x) = \int g(x)\,dx$ can be evaluated, then we can hope (finally) to solve the resulting equation

$$F(y) = G(x) + C \tag{8}$$

algebraically for an *explicit* solution $y = y(x)$ of (6). Otherwise, we settle for x expressed explicitly as a function of y, or at worst for (8) itself as an *implicit* solution of the differential equation. In Sections 7.5 and 7.6 we saw that this method suffices to solve a variety of natural growth and population problems involving the **linear** first-order differential equation

$$\frac{dx}{dt} = ax + b \tag{9}$$

with *constant* coefficients a and b. In Section 2 of this chapter we discuss the solution of a linear equation of the form in (9), though with a and b no longer assumed constant.

EXAMPLE 2 To solve the initial value problem

$$\frac{dy}{dx} = 2x\cos^2 y, \qquad y(0) = \frac{\pi}{4}, \tag{10}$$

we separate the variables and integrate:

$$\int \sec^2 y\,dy = \int 2x\,dx; \qquad \tan y = x^2 + C. \tag{11}$$

Hence a general solution of the differential equation in (10) is given by

$$y = y(x) = \arctan(x^2 + C).$$

Substitution of the values $x = 0$, $y = \pi/4$ in (11) yields $C = 1$, so the desired particular solution of (10) is

$$y(x) = \arctan(x^2 + 1).$$

The solution of separable differential equations (as discussed in Section 6.5) should be considered a prerequisite to this chapter. The problems for this section will provide an opportunity to review the method illustrated in Example 2.

This brief chapter on differential equations will show that a relatively few simple techniques are enough to enable us to solve a diversity of differential equations having an impressive variety of applications. In a single chapter we can only scratch the surface of a vast and venerable area of advanced mathematics whose applications permeate modern science and technology. This subject can be pursued further in C. H. Edwards, Jr. and D. E. Penney, *Elementary Differential Equations with Applications*, second edition (Englewood Cliffs, N.J.: Prentice-Hall, 1989).

18.1 PROBLEMS

Find general solutions (implicit if necessary, explicit if convenient) of the differential equations in Problems 1–16.

1. $\dfrac{dy}{dx} = 2xy$

2. $\dfrac{dy}{dx} = 2xy^{3/2}$

3. $\dfrac{dy}{dx} = y \cos x$

4. $\dfrac{dy}{dx} = 2xe^{-y}$

5. $\dfrac{dy}{dx} = 3x^2 \sec y$

6. $\dfrac{dy}{dx} = \dfrac{1}{2xy}$

7. $\dfrac{dy}{dx} = y + 1$

8. $\dfrac{dy}{dx} = 2y - 1$

9. $\dfrac{dy}{dx} = xy^3$

10. $y\dfrac{dy}{dx} = x(y^2 + 1)$

11. $y^3\dfrac{dy}{dx} = (y^4 + 1) \cos x$

12. $\dfrac{dy}{dx} = \dfrac{1 + \sqrt{x}}{1 + \sqrt{y}}$

13. $\dfrac{dy}{dx} = \dfrac{(x - 1)y^5}{x^2(2y^3 - y)}$

14. $(x^2 + 1)\dfrac{dy}{dx} \tan y = x$

15. $\dfrac{dy}{dx} = 1 + x + y + xy$

16. $x^2\dfrac{dy}{dx} = 1 - x^2 + y^2 - x^2y^2$

Find explicit particular solutions of the initial value problems in Problems 17–30.

17. $\dfrac{dy}{dx} = ye^x; \quad y(0) = 2e$

18. $\dfrac{dy}{dx} = 3x^2(y^2 + 1); \quad y(0) = 1$

19. $2y\dfrac{dy}{dx} = x(x^2 - 16)^{-1/2}; \quad y(5) = 2$

20. $\dfrac{dy}{dx} = 4x^3y - y; \quad y(1) = -3$

21. $\dfrac{dy}{dx} + 1 = 2y; \quad y(1) = 1$

22. $\dfrac{dy}{dx} \tan x = y; \quad y(\pi/2) = \pi/2$

23. $x\dfrac{dy}{dx} - y = 2x^2y; \quad y(1) = 1$

24. $\dfrac{dy}{dx} = 2xy^2 + 3x^2y^2; \quad y(1) = -1$

25. $\dfrac{dy}{dx} = 2x\sqrt{y - 1}; \quad y(0) = 2$

26. $\dfrac{dy}{dx} = 3x^2y; \quad y(0) = 1$

27. $\dfrac{dy}{dx} = 4x\sqrt{y}; \quad y(1) = 100$

28. $\dfrac{dy}{dx} = e^{x-y}; \quad y(0) = 1$

29. $\dfrac{dy}{dx} = \cos x \sec y; \quad y(0) = \pi/2$

30. $\dfrac{dy}{dx} = \tan x \tan y; \quad y(0) = \pi/2$

18.2 Linear First-Order Equations

We have seen how to solve a separable differential equation by integrating *after* multiplying each side by an appropriate factor. For example, to solve the equation

$$\frac{dy}{dx} = 2xy \qquad (y > 0), \tag{1}$$

we multiply each side by the factor $1/y$ to get

$$\frac{1}{y} \cdot \frac{dy}{dx} = 2x; \qquad \text{that is,} \quad D_x(\ln y) = D_x(x^2). \tag{2}$$

Because each side of the equation in (2) is recognizable as a *derivative* (with respect to the independent variable x), all that remains is a simple integration, which yields $\ln y = x^2 + C$. For this reason, the function $\rho = 1/y$ is called an *integrating factor* for the original equation in (1). An **integrating factor** for a differential equation is a function $\rho(x, y)$ such that multiplication of each side of the differential equation by $\rho(x, y)$ yields an equation in which each side is recognizable as a derivative.

With the aid of the appropriate integrating factor, there is a standard technique for solving the **linear first-order equation**

$$\frac{dy}{dx} + P(x)y = Q(x) \tag{3}$$

on an interval where the coefficient functions $P(x)$ and $Q(x)$ are continuous. We multiply each side in Equation (3) by the integrating factor

$$\rho = \rho(x) = e^{\int P(x)\,dx}. \tag{4}$$

The result is

$$e^{\int P(x)\,dx} \frac{dy}{dx} + P(x)e^{\int P(x)\,dx}\, y = Q(x)e^{\int P(x)\,dx}. \tag{5}$$

Because

$$D_x\left[\int P(x)\,dx\right] = P(x),$$

the left-hand side in (5) is the derivative of the *product* $y \cdot e^{\int P(x)\,dx}$, so (5) is equivalent to

$$D_x\left[y(x)e^{\int P(x)\,dx}\right] = Q(x)e^{\int P(x)\,dx}.$$

Integration of both sides of this equation gives

$$y(x)e^{\int P(x)\,dx} = \int \left[Q(x)e^{\int P(x)\,dx}\right] dx + C.$$

Finally solving for y, we obtain the general solution of the linear first-order equation in (3):

$$y = y(x) = e^{-\int P(x)\,dx}\left[\int \left[Q(x)e^{P(x)\,dx}\right] dx + C\right]. \tag{6}$$

The formula in (6) should not be memorized. In a specific problem it is generally simpler to use the *method* by which we developed this formula. Begin by calculating the integrating factor given in Equation (4), $\rho = e^{\int P(x)\,dx}$. Then multiply each side of the differential equation by ρ, recognize the left-hand side as the derivative of a product, integrate both sides, and finally solve for y. Moreover, given an initial condition $y(x_0) = y_0$, we can substitute $x = x_0$ and $y = y_0$ in (6) to solve for the value of C yielding the particular solution of (3) that satisfies this initial condition.

The integrating factor $\rho(x)$ is determined only to within a multiplicative constant. If we replace $\int P(x)\,dx$ by $\int P(x)\,dx + c$ in (4), the result is

$$\rho(x) = e^{\int P(x)\,dx + c} = e^c e^{\int P(x)\,dx}.$$

But the constant factor e^c does not affect the result of multiplying both sides of the differential equation in (3) by $\rho(x)$. Hence we may choose for $\int P(x)\,dx$ any convenient antiderivative of $P(x)$.

EXAMPLE 1 Solve the initial value problem

$$\frac{dy}{dx} - 3y = e^{2x}, \qquad y(0) = 3.$$

Solution Here we have $P(x) = -3$ and $Q(x) = e^{2x}$, so the integrating factor is

$$\rho = e^{\int(-3)\,dx} = e^{-3x}.$$

Multiplication of each side of the given equation by e^{-3x} yields

$$e^{-3x}\frac{dy}{dx} - 3e^{-3x}y = e^{-x},$$

which we recognize as

$$\frac{d}{dx}(e^{-3x}y) = e^{-x}.$$

Hence integration with respect to x gives

$$e^{-3x}y = -e^{-x} + C,$$

so the general solution is

$$y = y(x) = Ce^{3x} - e^{2x}.$$

Substitution of the initial condition $(x_0, y_0) = (0, 3)$ produces the result $C = 4$. Thus the desired particular solution is

$$y(x) = 4e^{3x} - e^{2x}.$$

EXAMPLE 2 Find the general solution of

$$(x^2 + 1)\frac{dy}{dx} + 3xy = 6x.$$

Solution After division of each side of the equation by $x^2 + 1$, we recognize the result

$$\frac{dy}{dx} + \frac{3x}{x^2 + 1}y = \frac{6x}{x^2 + 1}$$

as a first-order linear equation in which we have $P(x) = 3x/(x^2 + 1)$ and $Q(x) = 6x/(x^2 + 1)$. Multiplication by

$$\rho = \exp\left(\int \frac{3x}{x^2 + 1}\,dx\right) = \exp\left(\frac{3}{2}\ln(x^2 + 1)\right) = (x^2 + 1)^{3/2}$$

yields

$$(x^2 + 1)^{3/2} \frac{dy}{dx} + 3x(x^2 + 1)^{1/2} y = 6x(x^2 + 1)^{1/2},$$

and thus

$$D_x[(x^2 + 1)^{3/2} y] = 6x(x^2 + 1)^{1/2}.$$

Integration then yields

$$(x^2 + 1)^{3/2} y = \int 6x(x^2 + 1)^{1/2}\, dx = 2(x^2 + 1)^{3/2} + C.$$

Multiplication of both sides by $(x^2 + 1)^{-3/2}$ now gives the general solution

$$y(x) = 2 + C(x^2 + 1)^{-3/2}.$$

The derivation above of the solution in (6) of the linear first order equation in (3) bears a closer examination. Suppose that the functions $P(x)$ and $Q(x)$ are continuous on the (possibly unbounded) open interval I. Then the antiderivatives

$$\int P(x)\, dx \quad \text{and} \quad \int (Q(x)e^{\int P(x)\, dx})\, dx$$

exist on I. Our derivation of Equation (6) shows that *if* $y = y(x)$ is a solution of Equation (3) on I, *then* $y(x)$ is given by the formula in (6) for some choice of the constant C. Conversely, you may verify by direct substitution (Problem 16) that the function $y(x)$ given in Equation (6) satisfies Equation (3). Finally, given a point x_0 of I and any number y_0, there is (as previously noted) a unique value of C such that $y(x_0) = y_0$. Consequently, we have proved the following existence-uniqueness theorem.

Theorem *The Linear First-Order Equation*

If the functions $P(x)$ and $Q(x)$ are continuous on the open interval I containing the point x_0, then the initial value problem

$$\frac{dy}{dx} + P(x)y = Q(x), \qquad y(x_0) = y_0$$

has a unique solution $y(x)$ on I, given by the formula in Equation (6) with an appropriate choice of the constant C.

MIXTURE PROBLEMS

As a first application of linear first-order equations, we consider a tank containing a solution—a mixture of solute and solvent—such as salt dissolved in water. There is both inflow and outflow, and we want to compute the *amount* $x(t)$ of solute in the tank at time t, given the amount $x(0) = x_0$ at time $t = 0$. Suppose that solution with a concentration of c_i grams of solute per liter of solution flows into the tank at the constant rate of r_i liters per second and that the solution in the tank—kept thoroughly mixed by stirring—flows out at the constant rate of r_o liters per second.

To set up a differential equation for $x(t)$, we estimate the change Δx in x during the brief time interval $[t, t + \Delta t]$. The amount of solute that flows

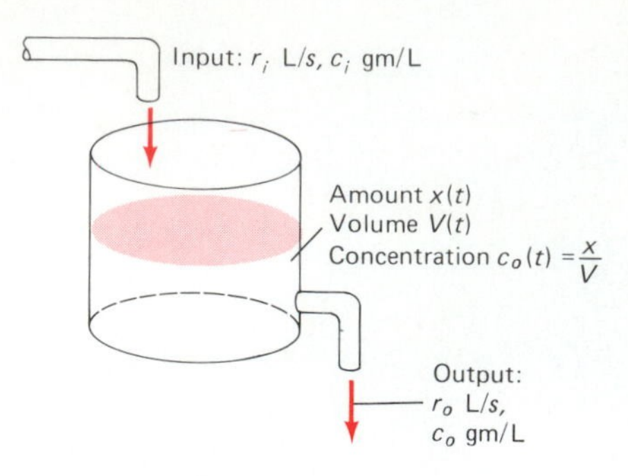

18.1 The single-tank mixture problem

into the tank during these Δt seconds is $r_i c_i \, \Delta t$ grams. To check this, note how the cancellation of dimensions checks our computation:

$$\left(r_i \frac{\text{liters}}{\text{second}}\right)\left(c_i \frac{\text{grams}}{\text{liter}}\right)(\Delta t \text{ seconds})$$

yields a quantity measured in grams.

The amount of solute that flows out of the tank during the same time interval depends upon the concentration $c_o(t)$ in the tank at time t. But, as noted in Fig. 18.1, $c_o(t) = x(t)/V(t)$ where $V(t)$ denotes the volume (not constant unless $r_i = r_o$) of solution in the tank at time t. Then

$$\begin{aligned}\Delta x &= \{\text{grams input}\} - \{\text{grams output}\} \\ &\approx r_i c_i \, \Delta t - r_o c_o \, \Delta t = r_i c_i \, \Delta t - r_o \frac{x}{V} \, \Delta t.\end{aligned}$$

We now divide by Δt:

$$\frac{\Delta x}{\Delta t} \approx r_i c_i - r_o \frac{x}{V}.$$

Finally we take the limit as $\Delta t \to 0$. If all the functions involved are continuous and x is differentiable, then the error in the approximations also approaches zero. Thus we obtain the differential equation

$$\frac{dx}{dt} = r_i c_i - r_o c_o = r_i c_i - \frac{1}{V} r_o x. \tag{7}$$

If $V_0 = V(0)$, then $V(t) = V_0 + (r_i - r_o)t$, so Equation (7) is a linear first-order differential equation for the amount $x(t)$ of solute in the tank.

IMPORTANT Equation (7) is *not* one you should commit to memory. It is the process we used to obtain that equation—examination of the behavior of the system over a short time interval $[t, t + \Delta t]$—that you should strive to understand, for it is an extremely useful tool for obtaining all sorts of differential equations.

EXAMPLE 3 Lake Erie has a volume of 458 km^3, and its rate of inflow and outflow are both 175 km^3/year. Suppose that at time $t = 0$ (years) its pollutant concentration is 0.05%, and that thereafter the concentration of pollutants in the inflowing water is 0.01%. Assuming that the outflow is perfectly mixed lake water, how long will it take to reduce the pollution concentration in the lake to 0.02%?

Solution Here we have $V = 458$ (km^3) and $r_i = r_0 = 175$ (km^3/year). Let $x(t)$ denote the volume of pollutants in the lake at time t. We are given

$$x(0) = (0.0005)(458) = 0.2290 \quad (\text{km}^3),$$

and we want to find when

$$x(t) = (0.0002)(458) = 0.0916 \quad (\text{km}^3).$$

The change Δx in Δt years is

$$\Delta x \approx (0.0001)(175) \, \Delta t - \frac{x}{458}(175) \, \Delta t = [0.0175 - (0.3821)x] \, \Delta t,$$

so the differential equation is

$$\frac{dx}{dt} + (0.3821)x = 0.0175.$$

The integrating factor is $e^{(0.3821)t}$; it yields

$$D_t[e^{(0.3821)t}x] = (0.0175)e^{(0.3821)t};$$

$$[e^{(0.3821)t}]x = (0.0458)e^{(0.3821)t} + C.$$

Substitution of $x(0) = 0.2290$ in the preceding equation gives

$$C = 0.2290 - 0.0458 = 0.1832,$$

so the solution is

$$x(t) = 0.0458 + (0.1832)e^{-(0.3821)t}.$$

Finally, we solve the equation

$$0.0916 = 0.0458 + (0.1832)e^{-(0.3821)t}$$

for

$$t = -\frac{1}{0.3821}\ln\frac{0.0916 - 0.0458}{0.1832} \approx 3.63.$$

The answer to the problem, then, is this: After about 3.63 years.

EXAMPLE 4 A 120-gal tank initially contains 90 lb of salt dissolved in 90 gal of water. Brine containing 2 lb/gal of salt flows into the tank at the rate of 4 gal/min, and the mixture flows out of the tank at the rate of 3 gal/min. How much salt does the tank contain when it is full?

Solution This example has an interesting feature: Because of the differing rates of inflow and outflow, the volume of brine in the tank increases steadily with $V(t) = 90 + t$ (gal) at time t (min). The change Δx in the amount x of salt in the tank from time t to time $t + \Delta t$ is given by

$$\Delta x \approx 4 \cdot 2\,\Delta t - 3 \cdot \frac{x}{90 + t}\,\Delta t,$$

so the differential equation for $x = x(t)$ is

$$\frac{dx}{dt} + \frac{3}{90 + t}x = 8.$$

The integrating factor is

$$\rho = \exp\left(\int \frac{3}{90 + t}\,dt\right) = e^{3\ln(90+t)} = (90 + t)^3,$$

which gives

$$D_t[(90 + t)^3\,x] = 8(90 + t)^3; \qquad (90 + t)^3 x = 2(90 + t)^4 + C.$$

Substitution of $x(0) = 90$ gives $C = -(90^4)$, so the amount of salt in the tank at time t is

$$x(t) = 2(90 + t) - \frac{90^4}{(90 + t)^3}.$$

The tank is full after 30 min, and when $t = 30$ we have

$$x(30) = 2(90 + 30) - \frac{90^4}{120^3} \approx 202$$

lb of salt.

18.2 PROBLEMS

Find general solutions of the differential equations in Problems 1–15. If an initial condition is given, find the corresponding particular solution. Throughout, primes denote derivatives with respect to x.

1. $xy' + y = 3xy;\quad y(1) = 0$

2. $xy' + 3y = 2x^5;\quad y(2) = 1$

3. $y' + y = e^x;\quad y(0) = 1$

4. $xy' - 3y = x^3;\quad y(1) = 10$

5. $y' + 2xy = x;\quad y(0) = -2$

6. $y' = (1 - y)\cos x;\quad y(\pi) = 2$

7. $(1 + x)y' + y = \cos x;\quad y(0) = 1$

8. $xy' = 2y + x^3\cos x$

9. $y' + y\cot x = \cos x$

10. $y' = 1 + x + y + xy;\quad y(0) = 0$

11. $xy' = 3y + x^4\cos x;\quad y(2\pi) = 0$

12. $y' = 2xy + 3x^2\exp(x^2);\quad y(0) = 5$

13. $xy' + (2x - 3)y = 4x^4$

14. $(x^2 + 4)y' + 3xy = x;\quad y(0) = 1$

15. $(x^2 + 1)y' + 3x^3y = 6x\exp(-3x^2/2);\quad y(0) = 1$

16. (a) Show that $y_c(x) = C\exp(-\int P(x)\,dx)$ is a general solution of $y' + P(x)y = 0$.
(b) Show that

$$y_p(x) = e^{-\int P(x)\,dx}\left(\int\left[Q(x)e^{\int P(x)\,dx}\right]dx\right)$$

is a particular solution of $y' + P(x)y = Q(x)$.
(c) Suppose that $y_c(x)$ is any general solution of $y' + P(x)y = 0$ and that $y_p(x)$ is any particular solution of $y' + P(x)y = Q(x)$. Show that $y(x) = y_c(x) + y_p(x)$ is a general solution of $y' + P(x)y = Q(x)$.

17. A tank contains 1000 liters of a solution consisting of 100 kg of salt dissolved in water. Pure water is pumped into the tank at the rate of 5 liters/s, and the mixture—kept uniform by stirring—is pumped out at the same rate. How long will it be until only 10 kg of salt remains in the tank?

18. Consider a reservoir with a volume of 8 billion cubic feet and an initial pollutant concentration of 0.25%. There is a daily inflow of 500 million cubic feet of water with a pollutant concentration of 0.05% and an equal daily outflow of the well-mixed water of the reservoir. How long will it take to reduce the pollutant concentration in the reservoir to 0.10%?

19. Rework Example 3 for the case of Lake Ontario. The only differences are that this lake has a volume of 1636 km^3 and an inflow–outflow rate of 209 km^3/year.

20. A tank initially contains 60 gal of pure water. Brine containing 1 lb of salt per gallon enters the tank at 2 gal/min, and the (perfectly mixed) solution leaves the tank at 3 gal/min. Thus the tank will be empty after exactly 1 h.
(a) Find the amount of salt in the tank after t minutes.
(b) What is the maximum amount of salt ever in the tank?

21. A 400-gal tank initially contains 100 gal of brine containing 50 lb of salt. Brine containing 1 lb of salt per gallon enters the tank at the rate of 5 gal/s, and the mixed brine in the tank flows out at the rate of 3 gal/s. How much salt will the tank contain when it is full of brine?

22. Consider the *cascade* of two tanks shown in Fig. 18.2, with $V_1 = 100$ (gal) and $V_2 = 200$ (gal) the (constant) volumes of brine in the two tanks. Each tank initially contains 50 lb of salt. The three flow rates are each 5 gal/s, with pure water flowing into tank 1.

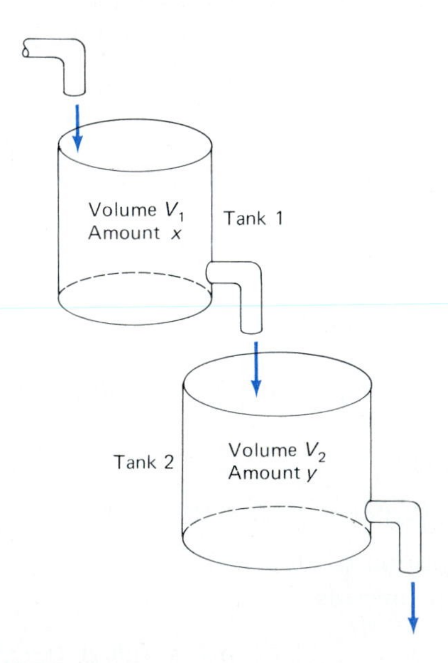

18.2 A cascade of two tanks

(a) Find the amount $x(t)$ of salt in tank 1 at time t.

(b) Suppose that $y(t)$ is the amount of salt in tank 2 at time t. Show first that

$$\frac{dy}{dx} = \frac{5x}{200} - \frac{5y}{200},$$

then solve for $y(t)$, using the value of $x(t)$ found in part (*a*).

(c) Finally, find the maximum amount of salt ever in tank 2.

23. Suppose that in the cascade shown in Fig. 18.2, tank 1 initially contains 100 gal of pure ethyl alcohol and tank 2 initially contains 100 gal of pure water. Pure water flows into tank 1 at 10 gal/min, and the other two flow rates are also 10 gal/min.

(a) Find the amounts $x(t)$ and $y(t)$ of alcohol in the two tanks.

(b) Find the maximum amount of alcohol ever in tank 2.

18.3 Complex Numbers and Functions

We include in this section the elementary algebra and calculus of complex numbers and functions—as much as is needed to solve higher-order differential equations with constant coefficients (Section 18.4). Recall that a **complex number** is one of the form

$$z = a + bi \tag{1}$$

where a and b are real numbers, and i denotes the (so-called imaginary) square root of -1. That is,

$$i = \sqrt{-1}; \qquad i^2 = -1. \tag{2}$$

Complex numbers may appear, for example, when we solve a quadratic equation by use of the quadratic formula.

The **real part** $\text{Re}(z)$ of the complex number $z = a + bi$ is a, and its **imaginary part** $\text{Im}(z)$ is b (not bi):

$$\text{Re}(a + bi) = a, \qquad \text{Im}(a + bi) = b. \tag{3}$$

The **conjugate** of $z = a + bi$ is the complex number $\bar{z} = a - bi$, with the same real part as z but with imaginary part opposite in sign from that of z. Complex numbers are added (and subtracted) by adding their real and imaginary parts:

$$(a + bi) + (c + di) = (a + c) + (b + d)i,$$

$$(a + bi) - (c + di) = (a - c) + (b - d)i. \tag{4}$$

For example, $(2 + 7i) + (3 - 5i) = 5 + 2i$. Complex numbers are multiplied just as binomials are multiplied, using the fact that $i^2 = -1$:

$$(a + bi)(c + di) = ac + (ad + bc)i + bdi^2 = (ac - bd) + (ad + bc)i. \tag{5}$$

To find the quotient of two complex numbers, we can use a technique reminiscent of rationalization:

$$\frac{a + bi}{c + di} = \frac{a + bi}{c + di} \cdot \frac{c - di}{c - di} = \frac{(ac + bd) + (bc - ad)i}{c^2 + d^2}. \tag{6}$$

EXAMPLE 1 Express the complex number $(2 - 3i)/(3 + 4i)$ in the standard form shown in Equation (1).

Solution

$$\frac{2 - 3i}{3 + 4i} = \frac{2 - 3i}{3 + 4i} \cdot \frac{3 - 4i}{3 - 4i} = -\frac{6}{25} - \frac{17}{25}i.$$

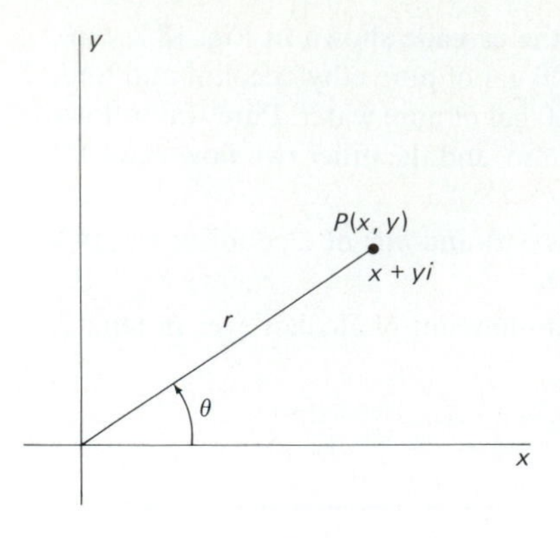

18.3 Geometric interpretation of the complex number $x + yi$

We can represent complex numbers geometrically by regarding the x-axis as the *real axis* and the y-axis as the *imaginary axis*. We then identify the complex number $z = x + iy$ with the point $P(x, y)$ in the plane. In terms of the polar coordinates r and θ of P we have

$$z = x + iy = r(\cos\theta + i\sin\theta). \tag{7}$$

We call the right-hand side in Equation (7) the **polar form** of the complex number z. Note that the polar form is expressed in terms of the **magnitude** $r = |z|$ of the complex number and its **argument** $\theta = \arg z$. From Fig. 18.3 we see that

$$r = |z| = \sqrt{x^2 + y^2} \quad \text{and} \quad \tan\theta = \frac{y}{x} \qquad (\text{if } x \neq 0). \tag{8}$$

In terms of the common and useful abbreviation

$$\operatorname{cis}\theta = \cos\theta + i\sin\theta, \tag{9}$$

the polar form of z is

$$z = r\operatorname{cis}\theta. \tag{10}$$

For example,

$$1 + i = \sqrt{2}\left(\cos\frac{\pi}{4} + i\sin\frac{\pi}{4}\right) = \sqrt{2}\operatorname{cis}\frac{\pi}{4}.$$

The polar form is especially convenient for multiplying and dividing complex numbers. For if

$$z_1 = r_1\operatorname{cis}\theta_1 \quad \text{and} \quad z_2 = r_2\operatorname{cis}\theta_2,$$

then a simple computation (Problem 17) using the addition formula for the sine and cosine shows that

$$z_1 z_2 = (r_1\operatorname{cis}\theta_1)(r_2\operatorname{cis}\theta_2) = r_1 r_2\operatorname{cis}(\theta_1 + \theta_2). \tag{11}$$

Thus the magnitudes are multiplied while the arguments are added. Similarly,

$$\frac{z_1}{z_2} = \frac{r_1\operatorname{cis}\theta_1}{r_2\operatorname{cis}\theta_2} = \frac{r_1}{r_2}\operatorname{cis}(\theta_1 - \theta_2). \tag{12}$$

For example,

$$\frac{1 + i}{1 - i} = \frac{\sqrt{2}\operatorname{cis}\pi/4}{\sqrt{2}\operatorname{cis}(-\pi/4)} = \operatorname{cis}\frac{\pi}{2} = i.$$

From Equation (11) we obtain $(r\operatorname{cis}\theta)^2 = r^2\operatorname{cis}2\theta$ and

$$(r\operatorname{cis}\theta)^3 = (r\operatorname{cis}\theta)(r^2\operatorname{cis}2\theta) = r^3\operatorname{cis}3\theta.$$

We proceed by induction, and find that

$$(r\operatorname{cis}\theta)^n = r^n\operatorname{cis}n\theta \tag{13}$$

for all positive integers n. The special case $r = 1$ is **DeMoivre's formula**

$$(\cos\theta + i\sin\theta)^n = \cos n\theta + i\sin n\theta. \tag{14}$$

EXAMPLE 2 Write $(1 + i\sqrt{3})^{10}$ in standard form.

Solution If $z = 1 + i\sqrt{3}$, then $r = |z| = 2$ and $\theta = \tan^{-1}\sqrt{3} = \pi/3$. Hence (14) gives

$$\begin{aligned}(1 + i\sqrt{3})^{10} &= \left(2 \operatorname{cis}\frac{\pi}{3}\right)^{10} = 2^{10}\operatorname{cis}\frac{10\pi}{3}\\ &= 1024 \operatorname{cis}\left(2\pi + \frac{4}{3}\pi\right) = 1024 \operatorname{cis}\frac{4\pi}{3}\\ &= 1024\left(-\frac{1}{2} - \frac{\sqrt{3}}{2}i\right) \approx -512 - (886.81)i.\end{aligned}$$

DeMoivre's formula can also be used to compute roots of complex numbers. Let $r^{1/n}$ denote the ordinary positive nth real root of the positive real number r. Then Equation (13) gives

$$\left[r^{1/n}\operatorname{cis}\frac{\theta}{n}\right]^n = r \operatorname{cis}\theta.$$

So $r^{1/n}\operatorname{cis}(\theta/n)$ is an nth root of $z = r \operatorname{cis}\theta$. But the number z can also be written in the forms

$$z = r \operatorname{cis}(\theta + 2\pi k), \qquad k = 0, \pm 1, \pm 2, \ldots.$$

The values $k = 0, 1, 2, \ldots, n - 1$ give, as above, the n distinct nth roots

$$z^{1/n} = r^{1/n}\operatorname{cis}\left(\frac{\theta}{n} + \frac{2\pi k}{n}\right), \qquad k = 0, 1, 2, \ldots, n - 1. \tag{15}$$

These complex numbers are, indeed, nth roots of z. But no other nth roots of z can be obtained by using other values of k. For if m is a multiple of n, then

$$r^{1/n}\operatorname{cis}\left(\frac{\theta}{n} + \frac{2\pi(k + m)}{n}\right) = r^{1/n}\operatorname{cis}\left(\frac{\theta}{n} + \frac{2\pi k}{n}\right).$$

The n distinct nth roots of $z = r \operatorname{cis}\theta$ given in (15) are evenly spaced around the circle of radius $r^{1/n}$, the first (corresponding to $k = 0$) having argument θ/n and the arguments of the others differing from θ/n by integral multiples of $2\pi/n$. For instance, the eight eighth roots of 256 are shown in Fig. 18.4.

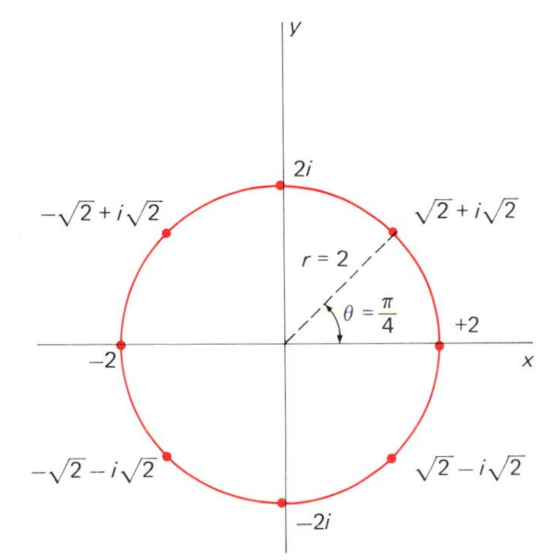

18.4 The eight eighth roots of 256

EXAMPLE 3 Find the three cube roots z_0, z_1, and z_2 of $z = 27i = 27 \operatorname{cis}\pi/2$.

Solution We take $n = 3$, $r = 27$, $\theta = \pi/2$, and $k = 0, 1$, and 2 in (15). Thus we find that

$$\begin{aligned} z_0 &= 3 \operatorname{cis}\frac{\pi}{6} = \frac{3}{2}(\sqrt{3} + i),\\ z_1 &= 3 \operatorname{cis}\left(\frac{\pi}{6} + \frac{2\pi}{3}\right) = 3 \operatorname{cis}\frac{5\pi}{6} = \frac{3}{2}(-\sqrt{3} + i),\\ z_2 &= 3 \operatorname{cis}\left(\frac{\pi}{6} + \frac{4\pi}{3}\right) = 3 \operatorname{cis}\frac{3\pi}{2} = -3i.\end{aligned}$$

These three roots are shown in Fig. 18.5.

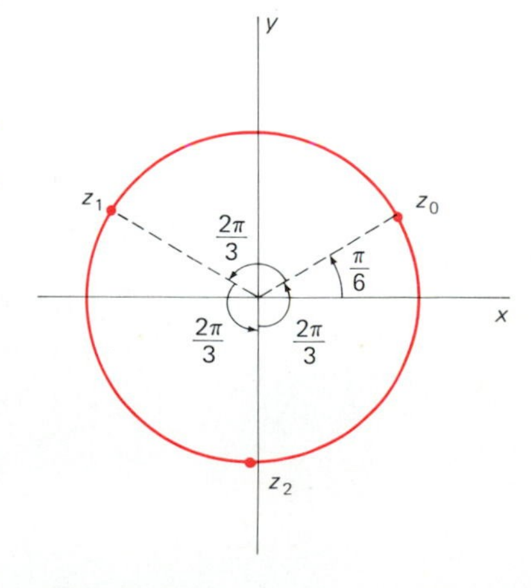

18.5 The three cube roots of $27i$

COMPLEX SEQUENCES AND SERIES

Suppose that $\{a_n + ib_n\}_{n=1}^{\infty}$ is an infinite sequence of complex numbers. We say that

$$\lim_{n\to\infty} (a_n + ib_n) = A + Bi$$

provided that both

$$\lim_{n\to\infty} a_n = A \quad \text{and} \quad \lim_{n\to\infty} b_n = B.$$

When we apply this definition to the sequence of partial sums of an infinite series of complex numbers, we find that we have the definition of the sum of such a series. We thus say that

$$\sum_{n=1}^{\infty} (a_n + ib_n) = A + Bi$$

provided that both

$$\sum_{n=1}^{\infty} a_n = A \quad \text{and} \quad \sum_{n=1}^{\infty} b_n = B.$$

Thus we can sum an infinite series of complex numbers by separately summing the real and imaginary parts of its terms.

For example, if we put $t = i\theta$ in the Taylor series for e^t as a way to *define* $e^{i\theta}$, we obtain

$$\begin{aligned} e^{i\theta} &= \sum_{n=0}^{\infty} \frac{(i\theta)^n}{n!} \\ &= 1 + i\theta - \frac{\theta^2}{2!} - \frac{i\theta^3}{3!} + \frac{\theta^4}{4!} + \frac{i\theta^5}{5!} - \cdots \\ &= \left(1 - \frac{\theta^2}{2!} + \frac{\theta^4}{4!} - \cdots\right) + i\left(\theta - \frac{\theta^3}{3!} + \frac{\theta^5}{5!} - \cdots\right); \end{aligned}$$

$$e^{i\theta} = \cos\theta + i\sin\theta. \tag{16}$$

This result is known as **Euler's formula.** Because of it we *define* the exponential function e^z, for $z = x + iy$, by

$$e^z = e^{x+iy} = e^x e^{iy} = e^x(\cos y + i \sin y). \tag{17}$$

Note that if

$$z_1 = x_1 + iy_1 \quad \text{and} \quad z_2 = x_2 + iy_2,$$

then the formula in Equation (17) implies that

$$\begin{aligned} e^{z_1}e^{z_2} &= e^{x_1}\operatorname{cis} y_1\, e^{x_2}\operatorname{cis} y_2 = e^{x_1}e^{x_2}\operatorname{cis}(y_1 + y_2) \\ &= e^{(x_1+x_2)}e^{i(y_1+y_2)} = e^{z_1+z_2}. \end{aligned} \tag{18}$$

This shows that the complex exponential function satisfies one important law of exponents. By similar methods it can be shown to satisfy the others as well.

In summary, the value of the exponential e^z for $z = x + iy$ is *by definition*

$$e^z = e^x(\cos y + i \sin y). \tag{19}$$

This definition yields the following desirable properties:

1. It agrees with the real case. That is,

$$e^{x+i\cdot 0} = e^x(\cos 0 + i \sin 0) = e^x.$$

2. It agrees with the pure imaginary case in Euler's formula. That is,

$$e^{0+iy} = e^0(\cos y + i \sin y) = \cos y + i \sin y.$$

3. It satisfies familiar laws of exponents like the one in Equation (18).

EXAMPLE 4 By substituting particular values in (19), we get

$$e^{i\pi} = e^0(\cos \pi + i \sin \pi) = -1,$$

$$e^{(1+i\pi)/2} = e^{1/2}\left(\cos\frac{\pi}{2} + i \sin\frac{\pi}{2}\right) = i\sqrt{e},$$

$$e^{2-3i} = e^2(\cos 3 - i \sin 3) \approx -7.3151 - (1.0427)i,$$

$$e^{(2+3i)t} = e^{2t}(\cos 3t + i \sin 3t).$$

Euler's formula leads to important expressions for the sine and cosine functions in terms of the exponential function. The equations

$$e^{ix} = \cos x + i \sin x \quad \text{and} \quad e^{-ix} = \cos x - i \sin x,$$

which follow from (16) (put $\theta = x$, then $\theta = -x$) imply (add them, subtract them) the formulas

$$\cos x = \frac{e^{ix} + e^{-ix}}{2} \quad \text{and} \quad \sin x = \frac{e^{ix} - e^{-ix}}{2i}. \tag{20}$$

COMPLEX-VALUED FUNCTIONS

A **complex-valued function** F of the *real* variable x associates with each real number x in its domain of definition the complex number

$$z = F(x) = f(x) + ig(x). \tag{21}$$

The real-valued functions $f(x)$ and $g(x)$ are called the real and imaginary parts, respectively, of $F(x)$. If they are differentiable, we can define the **derivative** of F to be

$$F'(x) = f'(x) + ig'(x). \tag{22}$$

Thus we merely differentiate the real and imaginary parts of F separately. Similarly, we integrate a complex-valued function by separately integrating its real and imaginary parts, provided the integrals exist. That is,

$$\int_a^b F(x)\,dx = \int_a^b f(x)\,dx + i\int_a^b g(x)\,dx. \tag{23}$$

A particular complex-valued function that plays an important role in the solution of *real* differential equations is the exponential function $F(x) = e^{rx}$ where $r = a + bi$ is a complex number. By (19) we have

$$e^{rx} = e^{(a+bi)x} = e^{ax}(\cos bx + i \sin bx). \tag{24}$$

So the real and imaginary parts of e^{rx} are $e^{ax} \cos bx$ and $e^{ax} \sin bx$, respectively.

EXAMPLE 5 Suppose that $r = a + bi$. Show that

$$D_x e^{rx} = re^{rx}, \tag{25}$$

exactly as if r were a real number.

Solution From (24) we obtain

$$\begin{aligned} D_x e^{rx} &= D_x(e^{ax} \cos bx) + iD_x(e^{ax} \sin bx) \\ &= (ae^{ax} \cos bx - be^{ax} \sin bx) + i(ae^{ax} \sin bx + be^{ax} \cos bx) \\ &= (a + bi)(e^{ax} \cos bx + ie^{ax} \sin bx) = re^{rx}. \end{aligned}$$

THE FUNDAMENTAL THEOREM OF ALGEBRA

The fundamental theorem of algebra, easy to state but difficult to prove, is the underlying reason for the importance—indeed, the very unavoidability—of complex numbers in mathematics. The theorem tells us that every polynomial equation of degree n,

$$p(x) = a_n x^n + a_{n-1} x^{n-1} + \cdots + a_1 x + a_0 = 0, \tag{26}$$

has at least one (complex) root r_1. By the **factor theorem,** we can then write $p(x) = (x - r_1)q(x)$ where $q(x)$ is a polynomial of degree $n - 1$. By induction on n, it then follows that the nth-degree polynomial $p(x)$ has the factorization

$$p(x) = a_n(x - r_1)(x - r_2) \cdots (x - r_n),$$

and the n roots $r_1, r_2, \ldots, r_n$ (not necessarily distinct). Even if the coefficients $a_0, a_1, \ldots, a_n$ of $p(x)$ are all real, the roots will be, in general, complex. This is one reason for the necessity of complex numbers. But if the coefficients of the polynomial $p(x)$ are all real, then it can be shown that the complex roots (if any) occur in **conjugate pairs** of the form $a \pm bi$.

18.3 PROBLEMS

Express the complex numbers given in Problems 1–6 both in the form $x + iy$ and in the form $r \text{ cis } \theta$.

1. $(3 - 4i)(3 + 4i)$

2. $(1 + 2i)(2 - 3i)$

3. $\dfrac{5 + 7i}{3 - 2i}$

4. $\dfrac{(1 + i)^2}{(1 - i\sqrt{3})^2}$

5. $(1 + i)^6$

6. $(-3 + 4i)^3$

7. Find the four fourth roots of -1.

8. Find the three cube roots of 27.

9. Find the six sixth roots of -64.

10. Find the five fifth roots of $-4 - 4i$.

11. Find the four roots of the equation $x^4 - 16 = 0$.

12. Find the four roots of the equation $x^4 - 4x^2 + 6 = 0$.

13. Find the six roots of the equation $x^6 + 4x^3 + 8 = 0$.

14. Expand $(\cos \theta + i \sin \theta)^3$. Then compare the result with that given by DeMoivre's formula, and thereby discover formulas for $\cos 3\theta$ and $\sin 3\theta$ in terms of $\sin \theta$ and $\cos \theta$.

15. Use DeMoivre's formula as in Problem 14 to discover formulas for $\cos 4\theta$ and $\sin 4\theta$ in terms of $\sin \theta$ and $\cos \theta$.

16. Suppose that a is a positive real number. Find the four roots of the equation $x^4 + a^4 = 0$.

17. Prove that $\text{cis}(\alpha + \beta) = \text{cis}\,\alpha\ \text{cis}\,\beta$.

18. If $e^w = z$ then $z^i = e^{wi}$ by definition. Choose w so that $e^w = i$, then conclude that

$$i^i = e^{-\pi/2} \approx 0.20788.$$

19. Obtain formulas for

$$\int e^{ax} \cos bx\, dx \quad \text{and} \quad \int e^{ax} \sin bx\, dx$$

by equating real and imaginary parts in the formula

$$\int e^{(a+bi)x}\, dx = \frac{e^{(a+bi)x}}{a + bi} + C.$$

20. Assume that $(e^z)^r = e^{rz}$ for every complex number $z = x + iy$ and every real number r. Then deduce that DeMoivre's formula in (14) holds for n any *real number.*

21. Let p, q, A, and B be real numbers. Prove that

$$Ae^{(p+iq)t} + Be^{(p-iq)t} = e^{pt}(C \cos qt + D \sin qt),$$

where C and D are complex numbers expressible in terms of A and B.

22. Prove that $e^z \neq 0$ for all complex numbers z.

23. Find the product of the ten tenth roots of 1.

24. Find the sum of the ten tenth roots of 1.

18.4 Linear Second-Order Equations with Constant Coefficients

Aside from second-order equations that can be reduced by elementary substitutions to familiar first-order equations, higher-order differential equations ordinarily require specialized techniques for their solution. Higher-order nonlinear equations can be very challenging; fortunately, many of the important equations arising from real-world applications are linear, and we now turn our attention to these. The general form of a **linear *n*th-order differential equation** is

$$a_n(x)\frac{d^n y}{dx^n} + a_{n-1}(x)\frac{d^{n-1}y}{dx^{n-1}} + \ldots + a_1(x)\frac{dy}{dx} + a_0(x)y = F(x) \tag{1}$$

where the *coefficients* $a_0(x), a_1(x), \ldots, a_n(x)$ and the function $F(x)$ are given. Equation (1) is called *linear* because only the *first* powers of the dependent variable $y(x)$ and of its derivatives are involved. It is linear in y but not necessarily in the independent variable x—the coefficients and the function $F(x)$ need not be linear in x.

The linear differential equation in (1) is called **homogeneous** provided that $F(x) \equiv 0$; otherwise, it is said to be **inhomogeneous.** Thus the general form of a homogeneous linear equation is

$$a_n(x)\frac{d^n y}{dx^n} + a_{n-1}(x)\frac{d^{n-1}y}{dx^{n-1}} + \ldots + a_1(x)\frac{dy}{dx} + a_0(x)y = 0. \tag{2}$$

The homogeneous linear equations that most commonly appear in elementary applications are of order $n = 2$ and have *constant* coefficients. We therefore restrict our attention in this section to the special case of a **homogeneous linear second-order differential equation**

$$ay'' + by' + cy = 0, \tag{3}$$

in which primes denote derivatives with respect to x and the coefficients a, b, and c are (real) constants.

The study of linear differential equations is greatly simplified by the following fact: If $y_1(x)$ and $y_2(x)$ are solutions of the homogeneous equation in (3), then so is the **linear combination**

$$y(x) = C_1 y_1(x) + C_2 y_2(x) \tag{4}$$

for any choice of the *constants* C_1 and C_2. This is so because the linearity of the operation of differentiation implies that

$$y' = C_1 y_1' + C_2 y_2' \quad \text{and} \quad y'' = C_1 y_1'' + C_2 y_2''.$$

Hence it follows that if $y = C_1 y_1 + C_2 y_2$, then

$$\begin{aligned} ay'' + by' + cy &= a(C_1 y_1'' + C_2 y_2'') + b(C_1 y_1' + C_2 y_2') + c(C_1 y_1 + C_2 y_2) \\ &= C_1(ay_1'' + by_1' + cy_1) + C_2(ay_2'' + by_2' + cy_2) \\ &= 0 + 0 = 0 \end{aligned}$$

provided that y_1 and y_2 are both solutions of (3).

EXAMPLE 1 According to the theorem in Section 8.2, the homogeneous second-order linear equation

$$y'' + k^2 y = 0 \tag{5}$$

has general solution

$$y(x) = C_1 \cos kx + C_2 \sin kx. \tag{6}$$

That is, a function $y(x)$ satisfies Equation (5) if and only if it is of the form in (6). Thus the general solution $y(x)$ is a linear combination of the two particular solutions $y_1(x) = \cos kx$ and $y_2(x) = \sin kx$ of the differential equation $y'' + ky = 0$.

Just as the general solution of the equation $y'' + k^2 y = 0$ is a linear combination of the particular solutions $y_1(x) = \cos kx$ and $y_2(x) = \sin kx$, it is known that the general solution of the second-order homogeneous equation $ay'' + by' + cy = 0$ is a linear combination of some two particular solutions. That is, there exist two particular solutions $y_1(x)$ and $y_2(x)$ such that *every* solution $y(x)$ is of the form $y = C_1 y_1 + C_2 y_2$. Our task is to find these two particular solutions. Specifically, in order to construct a general solution of (3), we need to find two particular solutions that are *linearly independent* in the sense that neither is a scalar multiple of the other. [See Section 2.1 of C. H. Edwards, Jr. and David E. Penney, *Elementary Differential Equations with Applications*, second edition (Englewood Cliffs, N.J.: Prentice-Hall, 1989)].

THE CHARACTERISTIC EQUATION

We first look for a *single* solution of Equation (3) and begin with the observation that

$$(e^{rx})' = re^{rx} \quad \text{and} \quad (e^{rx})'' = r^2 e^{rx}, \tag{7}$$

so any derivative of e^{rx} is a constant multiple of e^{rx}. Hence, if we substituted $y = e^{rx}$ in Equation (3), each term would be a constant multiple of e^{rx}, with the constant coefficients dependent on r and the coefficients a, b, and c. This suggests that we try to find a value of r so that these multiples of e^{rx} will have sum zero. If we succeed, then $y = e^{rx}$ will be a solution of Equation (3).

For example, if we substitute $y = e^{rx}$ in the equation

$$y'' - 5y' + 6y = 0,$$

we obtain

$$r^2e^{rx} - 5re^{rx} + 6e^{rx} = 0.$$

Thus

$$(r - 2)(r - 3)e^{rx} = 0.$$

Hence $y = e^{rx}$ will be a solution if either $r = 2$ or $r = 3$. So, in searching for a single solution, we actually have found *two* solutions, $y_1(x) = e^{2x}$ and $y_2(x) = e^{3x}$.

To carry out this procedure in the general case, we substitute $y = e^{rx}$ in Equation (3). With the aid of Equation (7), we find the result to be

$$ar^2e^{rx} + bre^{rx} + ce^{rx} = 0.$$

Because e^{rx} is never zero, we conclude that $y(x) = e^{rx}$ will satisfy the differential equation precisely when r is a root of the algebraic equation

$$ar^2 + br + c = 0. \tag{8}$$

This quadratic equation in r is called the **characteristic equation** of the homogeneous linear differential equation in (3),

$$ay'' + by' + cy = 0.$$

If Equation (8) has two *distinct* (unequal) roots r_1 and r_2, then the corresponding solutions $y_1(x) = e^{r_1x}$ and $y_2(x) = e^{r_2x}$ of (3) are linearly independent (why?). This gives the following result.

Theorem 1 *Distinct Real Roots*

If the two roots r_1 and r_2 of the characteristic equation in (8) are real and distinct, then

$$y(x) = C_1e^{r_1x} + C_2e^{r_2x} \tag{9}$$

is a general solution of Equation (3).

EXAMPLE 2 Find a general solution of

$$2y'' - 7y' + 3y = 0.$$

Solution We can solve the characteristic equation

$$2r^2 - 7r + 3 = 0$$

by factoring:

$$(2r - 1)(r - 3) = 0.$$

The roots $r_1 = 1/2$ and $r_2 = 3$ are real and distinct, so Theorem 1 yields the general solution

$$y(x) = C_1e^{x/2} + C_2e^{3x}.$$

EXAMPLE 3 The differential equation $y'' + 2y' = 0$ has characteristic equation

$$r^2 + 2r = r(r + 2) = 0$$

with distinct real roots $r_1 = 0$ and $r_2 = -2$. Because $e^{0 \cdot x} \equiv 1$, we get the general solution

$$y(x) = C_1 + C_2 e^{-2x}.$$

REMARK Note that Theorem 1 changes a problem in differential equations into a problem involving only the solution of a quadratic equation.

If the characteristic equation in (8) has equal roots $r_1 = r_2$, then we get (at first) only the single solution $y_1(x) = e^{r_1 x}$ of Equation (3). The problem in this case is to produce the "missing" second solution of the differential equation.

A "double" root $r = r_1$ will occur precisely when the characteristic equation is a constant multiple of the equation

$$(r - r_1)^2 = r^2 - 2r_1 r + r_1^2 = 0.$$

Any differential equation with this characteristic equation is equivalent to

$$y'' - 2r_1 y + r_1^2 y = 0. \tag{10}$$

But it is easy to verify by direction substitution (or see Problem 24) that $y = xe^{r_1 x}$ is a second solution (in addition to $y = e^{r_1 x}$) of Equation (10). It is clear that

$$y_1(x) = e^{r_1 x} \quad \text{and} \quad y_2(x) = xe^{r_1 x}$$

are linearly independent functions, so the general solution of the differential equation in (10) is

$$y(x) = c_1 e^{r_1 x} + c_2 x e^{r_1 x}.$$

Theorem 2 *Repeated Roots*

If the characteristic equation in (8) has equal (real) roots $r_1 = r_2$, then

$$y(x) = (C_1 + C_2 x)e^{r_1 x} \tag{11}$$

is a general solution of Equation (3).

EXAMPLE 4 The differential equation

$$y'' - 10y' + 25y = 0$$

has characteristic equation

$$r^2 - 10r + 25 = 0$$

with equal roots $r_1 = r_2 = 5$. Hence the general solution provided by Theorem 2 is

$$y(x) = (C_1 + C_2 x)e^{5x}.$$

The roots of the characteristic equation may be either real or complex. Because Equation (7) holds even when r is complex, it follows that (whether r is real, as before, or complex) e^{rx} will be a solution of the differential equation in (3) if and only if r is a root of its characteristic equation. In the case of a complex conjugate pair of roots $r_1 = \alpha + \beta i$ and $r_2 = \alpha - \beta i$, we get the general solution

$$\begin{aligned} y(x) &= c_1 e^{(\alpha+\beta i)x} + c_2 e^{(\alpha-\beta i)x} \\ &= c_1 e^{\alpha x}(\cos \beta x + i \sin \beta x) + c_2 e^{\alpha x}(\cos \beta x - i \sin \beta x) \\ &= e^{\alpha x}(C_1 \cos \beta x + C_2 \sin \beta x) \end{aligned}$$

where $C_1 = c_1 + c_2$ and $C_2 = (c_1 - c_2)i$. Thus the complex conjugate pair of roots $\alpha \pm \beta i$ leads to the linearly independent *real-valued* solutions

$$y_1(x) = e^{\alpha x} \cos \beta x \quad \text{and} \quad y_2(x) = e^{\alpha x} \sin \beta x.$$

Theorem 3 *Complex Roots*

If the characteristic equation in (8) has complex conjugate roots $\alpha \pm \beta i$ (with $\beta \neq 0$), then

$$y(x) = e^{\alpha x}(C_1 \cos \beta x + C_2 \sin \beta x) \tag{12}$$

is a general solution of Equation (3).

EXAMPLE 5 The characteristic equation of the differential equation

$$y'' + 16y = 0$$

is the equation $r^2 + 16 = 0$ with roots $4i$ and $-4i$. Hence Theorem 3 (with $\alpha = 0$ and $\beta = 4$) yields the general solution

$$y(x) = C_1 \cos 4x + C_2 \sin 4x.$$

In the case of the second-order differential equation in (3), the natural initial value problem has the form

$$ay'' + by' + cy = 0; \qquad y(x_0) = y_0, \qquad y'(x_0) = y_0' \tag{13}$$

where initial values both of the solution $y(x)$ *and* of its derivative $y'(x)$ are prescribed. We can then expect to determine values of the coefficients C_1 and C_2 in a general solution $y(x) = C_1 y_1(x) + C_2 y_2(x)$ so that the two initial conditions in (13) are satisfied.

EXAMPLE 6 Find the particular solution of

$$y'' - 4y' + 5y = 0$$

for which $y(0) = 1$ and $y'(0) = 5$.

Solution The characteristic equation is $r^2 - 4r + 5 = 0$, with roots $2 + i$ and $2 - i$. Hence a general solution is

$$y(x) = e^{2x}(C_1 \cos x + C_2 \sin x).$$

Then

$$y'(x) = 2e^{2x}(C_1 \cos x + C_2 \sin x) + e^{2x}(-C_1 \sin x + C_2 \cos x),$$

and the initial conditions give

$$y(0) = C_1 = 1 \quad \text{and} \quad y'(0) = 2C_1 + C_2 = 5.$$

It follows that $C_2 = 3$, so the desired particular solution is

$$y(x) = e^{2x}(\cos x + 3 \sin x).$$

18.4 PROBLEMS

Find a general solution of each of the second-order differential equations given in Problems 1–14.

1. $y'' - 3y' + 2y = 0$
2. $y'' + 2y' - 15y = 0$
3. $y'' + 5y' = 0$
4. $2y'' + 3y' = 0$
5. $3y'' - 10y' + 3y = 0$
6. $6y'' + 5y' + y = 0$
7. $y'' + 25y = 0$
8. $y'' - 25y = 0$
9. $y'' - 4y' + 8y = 0$
10. $y'' + 10y' + 125y = 0$
11. $9y'' - 6y' + 10y = 0$
12. $25y'' - 10y' + 13y = 0$
13. $y'' = 7y$
14. $y'' = 2y' + 4y$

Solve each of the initial value problems given in Problems 15–20.

15. $y'' - 4y' + 3y = 0;\quad y(0) = 7,\quad y'(0) = 11$
16. $2y'' + y' - y = 0;\quad y(0) = 3,\quad y'(0) = 6$
17. $y'' = 3y';\quad y(0) = 6,\quad y'(0) = 3$
18. $y'' = 4y;\quad y(0) = 0,\quad y'(0) = 8$
19. $y'' - 2y' + 5y = 0;\quad y(0) = 3,\quad y'(0) = 3$
20. $y'' + 8y' + 41y = 0;\quad y(0) = 0,\quad y'(0) = 15$
21. Find a solution $y(x)$ of $y'' + y = 0$ such that $y(0) = 0$ and $y(\pi/2) = 1$.
22. Show that the differential equation $y'' + 4y = 0$ has infinitely many different solutions such that $y(0) = 0$ and $y(\pi/2) = 0$.
23. Show that the differential equation $y'' + 9y = 0$ has *no* solution such that $y(0) = y(\pi/2) = 0$.
24. The characteristic equation of the differential equation
$$y'' - 2ry' + r^2 y = 0$$
has the repeated root r, which yields only one solution $y_1(x) = e^{rx}$. If a second linearly independent solution $y_2(x)$ exists, then $y_2(x) = u(x)y_1(x)$ for some function $u(x)$. Substitute this form for $y_2(x)$ in the differential equation and solve for $u(x)$ to discover a formula for $y_2(x)$.

*18.5 Mechanical Vibrations

18.6 The undamped mass-and-spring system

In Section 8.2 we discussed the motion of a mass m on the end of a spring that exerts a restoring force $F_S = -kx$ on the mass when its displacement from the equilibrium position is x (Fig. 18.6). Newton's law then gives

$$m\frac{d^2x}{dt^2} = F_S = -kx;$$

that is,

$$\frac{d^2x}{dt^2} + \omega^2 x = 0 \qquad \left(\omega^2 = \frac{k}{m}\right). \tag{1}$$

The general solution of (1) is

$$x(t) = A \cos \omega t + B \sin \omega t. \tag{2}$$

If the mass is released from rest at $x = x_0$, then the initial conditions $x(0) = x_0$ and $x'(0) = 0$ give

$$x(t) = x_0 \cos \omega t \tag{3}$$

as the motion that results. This equation describes a periodic oscillation or vibration with **amplitude** x_0, **period**

$$T = \frac{2\pi}{\omega} = 2\pi\sqrt{\frac{m}{k}}, \tag{4}$$

and **frequency** in hertz (cycles per second)

$$f = \frac{1}{T} = \frac{\omega}{2\pi} = \frac{1}{2\pi}\sqrt{\frac{k}{m}}. \tag{5}$$

For a specific particular solution of Equation (1), the values of the coefficients A and B in (2) are readily determined from given initial conditions, typically of the form $x(0) = x_0$, $x'(0) = v_0$. Once this has been done, the identity

$$\cos(\alpha + \beta) = \cos\alpha\cos\beta - \sin\alpha\sin\beta \tag{6}$$

can be used to rewrite the solution in (2) in the form

$$x(t) = C\cos(\omega t - \alpha) \tag{7}$$

where $C = (A^2 + B^2)^{1/2}$ is the **amplitude** of the motion, ω is its **circular frequency** in radians per second, and α is its **phase angle.** This process is illustrated in the following example.

EXAMPLE 1 A body that weighs $W = 16$ lb is attached to the end of a spring which is stretched 2 ft by a force of 100 lb. It is set in motion with initial position $x_0 = 0.5$ (ft) and initial velocity $v_0 = -10$ (ft/s). (Note that these data indicate that the body is displaced to the right and moving to the left at time $t = 0$.) Find the position function of the body, as well as the amplitude, frequency, period of oscillation, and phase angle of its motion.

Solution We take $g = 32$ ft/s^2. The mass of the body is then $m = W/g = 0.5$ (slugs). The spring constant is $k = 100/2 = 50$ (lb/ft), so Equation (1) yields

$$\tfrac{1}{2}x'' + 50x = 0;$$

that is,

$$x'' + 100x = 0.$$

Consequently the circular frequency will be $\omega = 10$ (rad/s). So the body will oscillate with

$$\text{Frequency:} \quad \frac{10}{2\pi} \approx 1.59 \quad \text{(Hz)}$$

and

$$\text{Period:} \quad \frac{2\pi}{10} \approx 0.63 \quad \text{(s)}.$$

We now impose the initial conditions $x(0) = 0.5$ and $x'(0) = -10$ on the general solution

$$x = A\cos 10t + B\sin 10t,$$

and it follows that $A = 0.5$ and $B = -1$. So the position function of the body is

$$x(t) = \tfrac{1}{2}\cos 10t - \sin 10t.$$

Hence its amplitude of motion is

$$C = [(0.5)^2 + 1^2]^{1/2} = \tfrac{1}{2}\sqrt{5} \approx 1.12 \quad \text{(ft)}.$$

To find the phase angle we write

$$x = \frac{\sqrt{5}}{2}\left(\frac{1}{\sqrt{5}} \cos 10t - \frac{2}{\sqrt{5}} \sin 10t\right) = \frac{\sqrt{5}}{2} \cos(10t - \alpha),$$

using the identity in (6). Thus we require $\cos \alpha = 1/\sqrt{5} > 0$ and $\sin \alpha = -2/\sqrt{5} < 0$. Hence α is the fourth-quadrant angle

$$\alpha = 2\pi - \arctan\left(\frac{2/\sqrt{5}}{1/\sqrt{5}}\right) \approx 5.1760 \quad \text{(rad)}.$$

In the form in which the amplitude and phase angle are made explicit, the position function is

$$x(t) \approx \frac{\sqrt{5}}{2} \cos(10t - 5.1760) \approx (1.1180) \cos(10t - 5.1760).$$

DAMPED VIBRATIONS

Now suppose that the mass of the discussion above is attached, as indicated in Fig. 18.7, to a dashpot or shock absorber—as in the suspension system of an automobile—that exerts on it a force F_R of resistance to motion proportional to the velocity v of the mass. Then $F_R = -cv$, where the proportionality constant c will depend on the viscosity of the fluid in the dashpot; the more viscous this fluid is, the larger $c > 0$ is. To take this resistive force into account, we replace Equation (1) by

$$m \frac{d^2x}{dt^2} = F_S + F_R = -kx - c\frac{dx}{dt}.$$

In order to analyze the motion of such a system, we therefore need to solve the homogeneous linear second-order equation

$$mx'' + cx' + kx = 0; \tag{8}$$

alternatively,

$$x'' + 2px' + \omega_0^2 x = 0, \tag{9}$$

where $\omega_0 = \sqrt{k/m}$ is the corresponding *undamped* circular frequency and

$$p = \frac{c}{2m} > 0. \tag{10}$$

The characteristic equation $r^2 + 2pr + \omega_0^2 = 0$ of Equation (9) has roots

$$r_1 = -p + \sqrt{p^2 - \omega_0^2} \quad \text{and} \quad r_2 = -p - \sqrt{p^2 - \omega_0^2}. \tag{11}$$

Whether these roots are real or complex depends upon the sign of

$$p^2 - \omega_0^2 = \frac{c^2}{4m^2} - \frac{k}{m} = \frac{c^2 - 4km}{4m^2}.$$

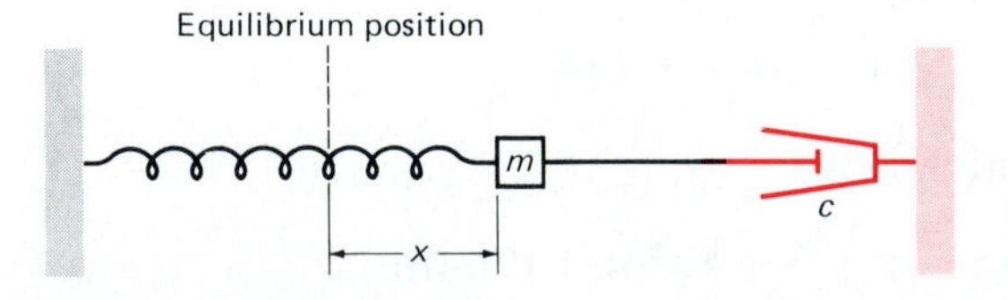

18.7 A mass-and-spring system with damping force provided by a dashpot

The **critical damping** c_{CR} is given by

$$c_{CR} = \sqrt{4km},$$

and we distinguish three cases, according as $c > c_{CR}$, $c = c_{CR}$, or $c < c_{CR}$.

Overdamped Case: $c > c_{CR}$ $(c^2 > 4km)$

Because c is relatively large in this case, we are dealing with a strong resistance in comparison with a weak force or a small mass. Then Equation (11) gives distinct real roots r_1 and r_2, both of which are negative. (Why?) The position function has the form

$$x(t) = c_1 e^{r_1 t} + c_2 e^{r_2 t}. \tag{12}$$

It is apparent that $x(t) \to 0$ as $t \to +\infty$ and that the body settles to its equilibrium position without any oscillations. Figure 18.8 shows some typical graphs of the position function for the overdamped case; we choose x_0 a fixed number and illustrate the effect of varying the initial velocity v_0. In every case the would-be oscillations are damped out.

Critically Damped Case: $c = c_{CR}$ $(c^2 = 4km)$

In this case Equation (11) gives equal real roots $r_1 = r_2 = -p$ of the characteristic equation, so the general solution is

$$x(t) = e^{-pt}(c_1 + c_2 t). \tag{13}$$

Because $e^{-pt} > 0$ for all t and $c_1 + c_2 t$ has at most one positive zero, the mass passes through its equilibrium position at most once, and it is clear that $x(t) \to 0$ as $t \to +\infty$. Some graphs of the motion in the critically damped case appear in Fig. 18.9, and they resemble those of the overdamped case in Fig. 18.8. In the critically damped case the resistance of the dashpot is just large enough to damp out any oscillations, but even a slight decrease in the resistance c will bring us to the remaining case, the one that shows the most dramatic behavior.

Underdamped Case: $c < c_{CR}$ $(c^2 < 4km)$

The characteristic equation now has complex conjugate roots

$$-p \pm i(\omega_0^2 - p^2)^{1/2},$$

and the general solution is

$$x(t) = e^{-pt}(c_1 \cos \omega_1 t + c_2 \sin \omega_1 t), \tag{14}$$

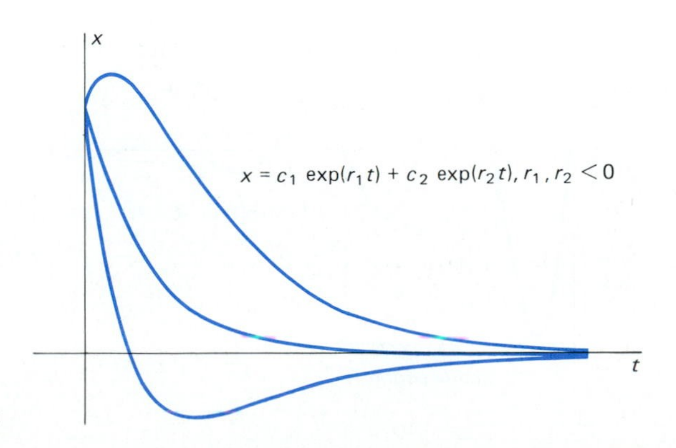

18.8 Overdamped motion

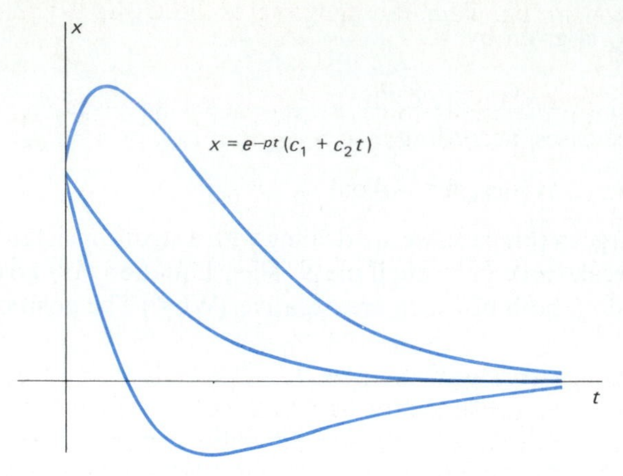

18.9 Critically damped motion

where

$$\omega_1 = \sqrt{\omega_0^2 - p^2} = \frac{1}{2m}\sqrt{4km - c^2}. \tag{15}$$

Using the cosine addition formula as in Example 1, we may rewrite (14) as

$$x(t) = Ce^{-pt}\cos(\omega_1 t - \alpha), \tag{16}$$

where $C = (c_1^2 + c_2^2)^{1/2}$ and $\tan \alpha = c_2/c_1$.

The solution in (16) represents exponentially damped oscillations of the mass about its equilibrium position. The graph of $x(t)$ lies between the curves $x = Ce^{-pt}$ and $x = -Ce^{-pt}$ and touches them when $\omega_1 t - \alpha$ is an integral multiple of π. The motion is not truly periodic, but it nevertheless is useful to call ω_1 its **circular frequency,** $T_1 = 2\pi/\omega_1$ its **pseudoperiod** of oscillation, and Ce^{-pt} its **time-varying amplitude.** Most of these quantities are shown on the typical graph of underdamped motion shown in Fig. 18.10. Note from Equation (15) that in this case ω_1 is less than the undamped circular frequency ω_0, so T_1 is larger than the period T of oscillation of the same mass with-

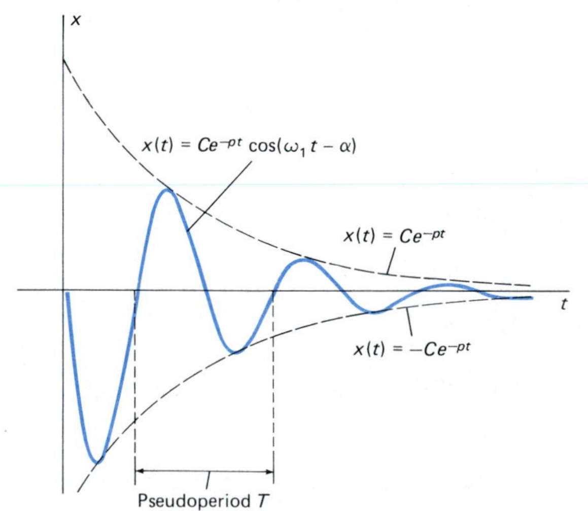

18.10 Underdamped motion

out damping on the same spring. Thus the damping of the dashpot has at least three effects:

1. It exponentially damps the oscillations, in accord with the time-varying amplitude;
2. It slows the motion—that is, the dashpot decreases the frequency of the motion; and
3. It delays the motion—this is the effect of the phase angle in Equation (16).

EXAMPLE 2 The mass-and-spring of Example 1 is now also attached to a dashpot that provides 6 lb of resistive force for each foot per second of velocity. The mass is set in motion with the same initial position $x(0) = 0.5$ (ft) and the same initial velocity $x'(0) = v_0 = -10$ (ft/s). Find the position function of the mass, its new frequency and pseudoperiod, and its phase angle.

Solution Rather than memorizing the various formulas given earlier, it is better practice in any particular case to set up the differential equation and then solve it directly. Recall that $m = 0.5$, $k = 50$, and we are now given $c = 6$ in fps units. Hence Equation (8) is

$$\tfrac{1}{2}x'' + 6x' + 50x = 0;$$

that is,

$$x'' + 12x' + 100x = 0.$$

The roots of the characteristic equation $r^2 + 12r + 100 = 0$ are

$$\frac{-12 \pm \sqrt{144 - 400}}{2} = -6 \pm 8i,$$

so the general solution is

$$x(t) = e^{-6t}(A \cos 8t + B \sin 8t). \tag{17}$$

The new circular frequency is $\omega_1 = 8$ (rad/s) and the pseudoperiod and new frequency are

$$T_1 = \frac{2\pi}{8} \approx 0.79 \text{ (s)} \quad \text{and} \quad \frac{1}{T_1} = \frac{8}{2\pi} \approx 1.27 \quad \text{(Hz)}$$

(in contrast with 0.63 s and 1.59 Hz, respectively, in the undamped case).

From Equation (17) we compute

$$x'(t) = e^{-6t}(-8A \sin 8t + 8B \cos 8t) - 6e^{-6t}(A \cos 8t + B \sin 8t).$$

The given initial conditions then yield the equations

$$x(0) = A = \tfrac{1}{2} \quad \text{and} \quad x'(0) = -6A + 8B = -10,$$

so $A = \frac{1}{2}$ and $B = -\frac{7}{8}$. Thus

$$x(t) = e^{-6t}(\tfrac{1}{2} \cos 8t - \tfrac{7}{8} \sin 8t),$$

so with $C = \sqrt{(\frac{1}{2})^2 + (\frac{7}{8})^2} = \frac{1}{8}\sqrt{65}$, we have

$$x(t) = \frac{\sqrt{65}}{8} e^{-6t}\left(\frac{4}{\sqrt{65}} \cos 8t - \frac{7}{\sqrt{65}} \sin 8t\right).$$

We require $\cos\alpha = 4/\sqrt{65} > 0$ and $\sin\alpha = -7/\sqrt{65} < 0$, so α is the fourth-quadrant angle

$$\alpha = 2\pi - \arctan(\tfrac{7}{4}) \approx 5.2315 \quad \text{(rad)}.$$

Finally,

$$x(t) \approx \frac{\sqrt{65}}{8} e^{-6t}\cos(8t - 5.2315) \approx (1.0078)e^{-6t}\cos(8t - 5.2315).$$

18.5 PROBLEMS

In each of Problems 1–6, a mass m is attached to a spring with given spring constant k, so its position function $x(t)$ satisfies $mx'' + kx = 0$. The mass is set in motion with initial position x_0 and initial velocity v_0. Find $x(t)$, as well as the amplitude, frequency, period of oscillation, and phase angle of the resulting motion.

1. $m = 2, \quad k = 19, \quad x_0 = 5, \quad v_0 = 0$

2. $m = 3, \quad k = 12, \quad x_0 = 0, \quad v_0 = 20$

3. $m = 2, \quad k = 50, \quad x_0 = 5, \quad v_0 = 25$

4. $m = 3, \quad k = 48, \quad x_0 = 3, \quad v_0 = -48$

5. $m = 1/2, \quad k = 18, \quad x_0 = 6, \quad v_0 = 48$

6. $m = 1/2, \quad k = 50, \quad x_0 = 24, \quad v_0 = 100$

7. Determine the period and frequency of the simple harmonic motion of a 4-kg mass on the end of a spring with spring constant 16 N/m (newtons per meter).

8. Determine the period and frequency of the simple harmonic motion of a body weighing 24 lb on the end of a spring with spring constant 48 lb/ft.

9. A mass of 3 kg is attached to the end of a spring that is stretched 20 cm by a force of 15 N. It is set in motion with initial position $x_0 = 0$ and initial velocity $v_0 = -10$ m/s. Find the amplitude, period, and frequency of the resulting motion.

10. A body weighing 8 lb is attached to the end of a spring that is stretched 1 in. by a force of 3 lb. At time $t = 0$ (in seconds), the body is pulled 1 ft to the right (stretching the spring) and set in motion with an initial velocity of 5 ft/s to the left.

(a) Find the displacement $x(t)$ of the body in the form $C\cos(\omega_0 + \alpha)$.

(b) Find the amplitude and period of motion of the body.

11. Consider a floating cylindrical buoy with radius r, height h, and density $\rho \leqq 0.5$ (recall that the density of water is 1 g/cm^3). The buoy is initially suspended at rest with its bottom on the top surface of the water and is released at time $t = 0$. Thereafter, it is acted on by two forces: a downward gravitational force equal to its weight $mg = \rho\pi r^2 hg$, and an upward force of buoyancy equal to the weight $\pi r^2 xg$ of water displaced, where $x = x(t)$ is the depth of the bottom of the buoy beneath the surface at time t. Conclude that the buoy undergoes simple harmonic motion about the equilibrium position $x_e = \rho h$ with period $p = 2\pi\sqrt{\rho h/g}$. Compute p and the amplitude of the motion if $\rho = 0.5$ (g/cm^3), $h = 200$ (cm), and $g = 980$ (cm/s^2).

12. A cylindrical buoy weighing 100 lb floats in water with its axis vertical (as in Problem 11). When depressed slightly and released, it oscillates up and down four times every 10 s. Assume that friction is negligible. Find the radius of the buoy.

The remaining problems in this section deal with free *damped* motion. In Problems 13–17, a mass m is attached both to a spring (with given spring constant k) and a dashpot (with given damping constant c). The mass is set in motion with initial position x_0 and initial velocity v_0. Find the position function $x(t)$, and determine whether the motion is underdamped, critically damped, or overdamped. If it is underdamped, write $x(t)$ in the form $Ce^{-pt}\cos(\omega_1 t - \alpha)$.

13. $m = 1/2, \quad c = 3, \quad k = 4, \quad x_0 = 2, \quad v_0 = 0$

14. $m = 3, \quad c = 30, \quad k = 63, \quad x_0 = 2, \quad v_0 = 2$

15. $m = 1, \quad c = 8, \quad k = 16, \quad x_0 = 5, \quad v_0 = -10$

16. $m = 2, \quad c = 12, \quad k = 50, \quad x_0 = 0, \quad v_0 = -8$

17. $m = 4, \quad c = 20, \quad k = 169, \quad x_0 = -4, \quad v_0 = 16$

18. A 12-lb weight is attached both to a vertically suspended spring that it stretches 6 in. and to a dashpot that provides 3 lb of resistive force for every foot per second of velocity.

(a) Suppose that the weight is pulled down 1 ft below its static equilibrium position and then released from rest at time $t = 0$. Find its position function $x(t)$.

(b) Find the frequency, time-varying amplitude, and phase angle of the motion of the weight.

19. This problem deals with a highly simplified model of an automobile weighing 3200 lb. Assume that its suspension system behaves like a single spring and its shock absorbers like a single dashpot, so that its vertical vibrations satisfy Equation (8) with appropriate values of the coefficients.

(a) Find the stiffness coefficient of the spring if the automobile undergoes free vibrations at 80 cycles per minute when its shock absorbers are disconnected.

(b) With the shock absorbers connected, the automobile is set into vibration by driving it over a bump, and the resulting damped vibrations have a frequency of 78 cycles per minute. After how long will the time-varying amplitude be 1% of its original value?

18.6 Numerical Methods

It is the exception rather than the rule when a differential equation of the general form

$$\frac{dy}{dx} = f(x, y) \tag{1}$$

can be solved exactly by simple methods such as those discussed in Sections 18.1 and 18.2.

Here we discuss **Euler's method** for computing numerical *approximations* to the solution of the initial value problem

$$\frac{dy}{dx} = f(x, y), \qquad y(a) = y_0 \tag{2}$$

on an interval of the form $a \leqq x \leqq b$. We first choose a fixed *step size* $h > 0$ and consider the points

$$a = x_0,\ x_1,\ x_2,\ \ldots,\ x_n,\ \ldots$$

where $x_n = a + nh$, so that $x_n = x_{n-1} + h$ for each $n = 1, 2, 3, \ldots$. Our goal is to find suitable approximations $y_1, y_2, y_3, \ldots$ to the true values of the solution $y(x)$ of Equation (2) at the points $x_1, x_2, x_3, \ldots$. Thus we seek reasonably accurate approximations

$$y_n \approx y(x_n) \tag{3}$$

for each n.

When $x = x_0$, the rate of change of y with respect to x is $y' = f(x_0, y_0)$. If y continued to change at this same rate from $x = x_0$ to $x = x_1 = x_0 + h$, the change in y would be exactly $hf(x_0, y_0)$. We therefore take

$$y_1 = y_0 + hf(x_0, y_0) \tag{4}$$

as our approximation to the true value $y(x_1)$ of the solution at $x = x_1$. Similarly, we take

$$y_2 = y_1 + hf(x_1, y_1) \tag{5}$$

as our approximation to $y(x_2)$. Having reached the nth approximate value $y_n \approx y(x_n)$, we take

$$y_{n+1} = y_n + hf(x_n, y_n) \tag{6}$$

as our approximation to the true value $y(x_{n+1})$.

With h = 0.1:

n	X_n	Y_n	Y actual
0	0.0	1.00000	1.00000
1	0.1	1.10000	1.11034
2	0.2	1.22000	1.24281
3	0.3	1.36200	1.39972
4	0.4	1.52820	1.58365
5	0.5	1.72102	1.79744
6	0.6	1.94312	2.04424
7	0.7	2.19743	2.32751
8	0.8	2.48718	2.65108
9	0.9	2.81590	3.01921
10	1.0	3.18748	3.43656

With h = 0.01:

n	X_n	Y_n	Y actual
0	0.0	1.00000	1.00000
10	0.1	1.10924	1.11034
20	0.2	1.24038	1.24281
30	0.3	1.39570	1.39972
40	0.4	1.57773	1.58365
50	0.5	1.78926	1.79744
60	0.6	2.03339	2.04424
70	0.7	2.31353	2.32751
80	0.8	2.63343	2.65108
90	0.9	2.99727	3.01921
100	1.0	3.40963	3.43656

With h = 0.001:

n	X_n	Y_n	Y actual
0	0.0	1.00000	1.00000
100	0.1	1.11023	1.11034
200	0.2	1.24256	1.24281
300	0.3	1.39931	1.39972
400	0.4	1.58305	1.58365
500	0.5	1.79662	1.79744
600	0.6	2.04315	2.04424
700	0.7	2.32610	2.32751
800	0.8	2.64930	2.65108
900	0.9	3.01699	3.01921
1000	1.0	3.43385	3.43656

18.11 Using Euler's method to approximate the solution of (7)

Algorithm *The Euler Method*

Given the initial value problem

$$y' = f(x, y), \qquad y(a) = y_0, \tag{2}$$

Euler's method with step size h consists in applying the iterative formula

$$y_{n+1} = y_n + hf(x_n, y_n) \qquad (n \geqq 0) \tag{6}$$

to compute successively approximations $y_1, y_2, y_3, \ldots$ to the (true) values $y(x_1), y(x_2), y(x_3), \ldots$ of the (exact) solution $y = y(x)$ at the points $x_1, x_2, x_3, \ldots$, respectively.

Although the most important practical applications of Euler's method are to nonlinear differential equations, we will illustrate the method with the linear first-order initial value problem

$$\frac{dy}{dx} = x + y, \qquad y(0) = 1, \tag{7}$$

in order that we may compare our approximate solution with the exact solution

$$y(x) = 2e^x - x - 1 \tag{8}$$

that is easy to find by the methods of Section 18.2.

EXAMPLE 1 Apply Euler's method to approximate the solution of the initial value problem in (7) on the interval $0 \leqq x \leqq 1$.

Solution Here $f(x, y) = x + y$, so the iterative formula in (6) is

$$y_{n+1} = y_n + h(x_n + y_n). \tag{9}$$

The tables in Fig. 18.11 show the results we obtain, beginning with $y_0 = 1$ and using the step sizes $h = 0.1$, $h = 0.01$, and $h = 0.001$. With $h = 0.1$ the computations can be carried out easily with only a pocket calculator. With $h = 0.01$ and $h = 0.001$ a computer is needed, and we have shown the results only at intervals of $\Delta x = 0.1$. Note in each case that the error $y_{\text{actual}} - y_{\text{approx}}$ increases as x increases—that is, as x gets further and further from x_0. But each error decreases as h decreases. The percentage errors at the final point $x = 1$ are 7.25% with $h = 0.1$, 0.78% with $h = 0.01$, and only 0.08% with $h = 0.001$.

THE IMPROVED EULER METHOD

To increase the accuracy of the basic Euler technique, we can regard the result

$$u_{n+1} = y_n + hf(x_n, y_n) \tag{10}$$

as merely a first attempt at estimating the true value of $y(x_{n+1})$. We then take the average value

$$m_n = \tfrac{1}{2}[f(x_n, y_n) + f(x_{n+1}, u_{n+1})] \tag{11}$$

as an improved estimate of the rate of change of y with respect to x over the interval $[x_n, x_{n+1}]$. This yields

$$y_{n+1} = y_n + hm_n \tag{12}$$

as our $(n + 1)$st approximate value.

The **improved Euler method** consists of applying the formulas in (10) through (12) at each step in the iteration to compute the successive approximations $y_1, y_2, y_3, \ldots$. It is an example of a *predictor-corrector method.* The formula in (10) is the *predictor* used as a first attempt at a next approximation. The formula in (12) is the *corrector* used to improve the value of the first attempt.

EXAMPLE 2 Apply the improved Euler method to approximate the solution of the initial value problem in (7) on the interval $0 \leqq x \leqq 1$.

Solution With $f(x, y) = x + y$, the formulas in (10) through (12) yield

$$u_{n+1} = y_n + h(x_n + y_n),$$

$$m_n = \tfrac{1}{2}(x_n + y_n + x_{n+1} + u_{n+1}),$$

$$y_{n+1} = y_n + hm_n.$$

The tables in Fig. 18.12 show the results we get using the step sizes $h = 0.1$, $h = 0.01$, and $h = 0.001$. The percentage errors at the final point $x = 1$ are 0.244% with $h = 0.1$ and 0.003% with $h = 0.01$, while with $h = 0.001$ the results shown are all accurate to five decimal places.

With h = 0.1:

n	X_n	Y_n	Y actual
0	0.0	1.00000	1.00000
1	0.1	1.11000	1.11034
2	0.2	1.24205	1.24281
3	0.3	1.39847	1.39972
4	0.4	1.58180	1.58365
5	0.5	1.79489	1.79744
6	0.6	2.04086	2.04424
7	0.7	2.32315	2.32751
8	0.8	2.64558	2.65108
9	0.9	3.01236	3.01921
10	1.0	3.42816	3.43656

With h = 0.01:

n	X_n	Y_n	Y actual
0	0.0	1.00000	1.00000
10	0.1	1.11034	1.11034
20	0.2	1.24280	1.24281
30	0.3	1.39970	1.39972
40	0.4	1.58363	1.58365
50	0.5	1.79742	1.79744
60	0.6	2.04420	2.04424
70	0.7	2.32746	2.32751
80	0.8	2.65102	2.65108
90	0.9	3.01913	3.01921
100	1.0	3.43647	3.43656

With h = 0.001:

n	X_n	Y_n	Y actual
0	0.0	1.00000	1.00000
100	0.1	1.11034	1.11034
200	0.2	1.24281	1.24281
300	0.3	1.39972	1.39972
400	0.4	1.58365	1.58365
500	0.5	1.79744	1.79744
600	0.6	2.04424	2.04424
700	0.7	2.32750	2.32751
800	0.8	2.65108	2.65108
900	0.9	3.01921	3.01921
1000	1.0	3.43656	3.43656

18.12 Using the improved Euler method to approximate the solution of (7)

A REAL-WORLD APPLICATION

Suppose that a skydiver steps out of an airplane at an altitude of 10,000 ft, and thereafter is subject both to gravitational acceleration ($g = 32$ ft/s^2) and to air resistance proportional to the square of his velocity v. We denote by $x(t)$ the (downward) distance the skydiver falls in t seconds and assume that $v = dx/dt$ satisfies the differential equation

$$\frac{dv}{dt} = 32 - (0.005)v^2, \tag{13}$$

giving the skydiver's acceleration $a = dv/dt$. In Problem 21 we ask you to separate the variables in (13) to derive the (exact) formulas

$$v(t) = 80 \tanh 0.4t \tag{14}$$

and

$$x(t) = 200 \ln(\cosh 0.4t). \tag{15}$$

Here we want to solve Equation (13) numerically, with initial conditions $x(0) = v(0) = 0$.

Choosing a step size $h > 0$, we want to calculate approximations v_n to $v(t_n)$ and x_n to $x(t_n)$, where $t_n + nh$. We write

$$a_n = 32 - (0.005)v_n^2 \tag{16}$$

for the approximate acceleration at time t_n. Then Euler's method applied to Equation (13) yields

$$v_{n+1} = v_n + ha_n \tag{17}$$

as a first approximation to the skydiver's velocity at time t_{n+1}. We implement the improved Euler method by using the average

$$\bar{v}_n = \tfrac{1}{2}(v_n + v_{n+1})$$

T	APPROX V	ACTUAL V	APPROX X	ACTUAL X
1	30.61	30.40	15.66	15.59
2	53.65	53.12	58.59	58.15
3	67.28	66.69	119.75	118.74
4	74.19	73.73	190.90	189.36
5	77.42	77.12	266.92	265.00
6	78.87	78.69	345.16	343.01
7	79.51	79.41	424.39	422.11
8	79.79	79.73	504.05	501.70
9	79.91	79.88	583.90	581.52
10	79.96	79.95	663.84	661.44
11	79.98	79.98	743.81	741.40
12	79.99	79.99	823.80	821.38
13	80.00	80.00	903.80	901.38
14	80.00	80.00	983.79	981.37
15	80.00	80.00	1063.79	1061.37

18.13 The free-falling skydiver

as a better estimate of the velocity over the interval $[t_n, t_{n+1}]$. This yields

$$x_{n+1} = x_n + \bar{v}_n h = x_n + hv_n + \tfrac{1}{2}h^2 a_n \tag{18}$$

as the approximate position at time t_{n+1}.

The formulas in (16) through (18) were used with $h = 0.1$ to compute the approximate values shown in Fig. 18.13, which also lists for comparison (at intervals of 1 s) the exact values of v and x (obtained from Equations (14) and (15)). Note that after 15 s, the skydiver has fallen 1064 ft (taking the approximate value) and has reached a limiting velocity of 80 ft/s. If the skydiver's parachute fails to open, the skydiver will therefore hit the ground after

$$\frac{10{,}000 - 1064}{80} \approx 112$$

seconds. Thus the total time of descent is about 2 min 7 s.

18.6 PROBLEMS

In each of Problems 1–10, use Euler's method with $h = 0.1$ to approximate the value of the given initial value problem on the interval $[0, 0.5]$. Also find the exact solution, and compare the approximate value of $y(0.5)$ with its exact value.

1. $dy/dx = -y, \quad y(0) = 2$

2. $dy/dx = 2y, \quad y(0) = 0.5$

3. $dy/dx = y + 1, \quad y(0) = 1$

4. $dy/dx = x - y, \quad y(0) = 1$

5. $dy/dx = y - x - 1, \quad y(0) = 1$

6. $dy/dx = -2xy, \quad y(0) = 2$

7. $dy/dx = -3x^2 y, \quad y(0) = 3$

8. $dy/dx = e^{-y}, \quad y(0) = 0$

9. $dy/dx = \frac{1}{4}(1 + y^2), \quad y(0) = 1$

10. $dy/dx = \dfrac{2x}{1 + 3y^2}, \quad y(0) = 1$

11–20. Repeat Problems 1–10, except use the improved Euler method with step size $h = 0.1$.

21. Separate the variables in Equation (13) and then use the integral

$$\int \frac{1}{1 - x^2}\,dx = \tanh^{-1} x + C$$

to derive the formulas in (14) and (15).

22. Suppose that a skydiver steps out of an airplane at an altitude of 8000 ft and falls freely for 30 s with acceleration $a = 32 - (0.005)v^2$. Then he opens his parachute, after which $a = 32 - (0.08)v^2$. Approximate his height and velocity at 1-s intervals after he opens his parachute. When will he reach the ground?

*18.6 Optional Computer Application

The BASIC program shown in Fig. 18.14 was used to compute the approximations that appear in Fig. 18.11.

The function $f(x, y) = x + y$ is defined in line 140 (which can be changed if a different differential equation $y' = f(x, y)$ is to be solved). The input in lines 150–180 specifies the initial data, and how often to calculate and print results. Line 230 is the iterative formula of Euler's method. According to line 250, the values x_n, y_n are printed only if n is an integral multiple of p. For example, the input

```
Initial x,y? 0, 1
Step size h? 0.01
Number of steps? 100
Print step p? 10
```

```
100 REM--Program EULER
110 REM
120 REM--Initialization:
130 REM
140       DEF FNF(X,Y) = X + Y
150       INPUT "Initial x,y"; X,Y
160       INPUT "Step size h"; H
170       INPUT "Number of steps"; K
180       INPUT "Print step p"; P
190 REM
200 REM--Euler iteration:
210 REM
220       FOR N = 1 TO K
230           Y = Y + H*FNF(X,Y)
240           X = X + H
250           IF INT(N/P) = N/P THEN PRINT X,Y
260       NEXT N
270 REM
280       END
```

19.14 Listing of Program EULER

produces the approximate values of y at intervals of $\Delta x = 0.1$ shown in the second table of Fig. 18.11.

In order to compute the improved Euler approximations shown in Fig. 18.12, it is necessary only to replace line 230 of Program EULER with the two lines

```
230  U = Y + H*FNF(X,Y)
235  Y = Y + (H/2)*(FNF(X,Y) + FNF(X + H,U))
```

and then run the modified program three times: with $h = 0.1$, $n = 10$, $p = 1$; with $h = 0.01$, $n = 100$, $p = 10$; and with $h = 0.001$, $n = 1000$, $p = 100$.

CHAPTER 18 REVIEW: Concepts and Methods

Use the list below as a guide to concepts that you may need to review.

1. General and particular solutions of differential equations
2. Solution of separable first-order equations
3. Solution of linear first-order equations
4. Mixture problems
5. Complex numbers and complex-valued functions
6. Euler's formula and DeMoivre's formula
7. Solution of homogeneous linear equations with constant coefficients; the characteristic equation and the solutions corresponding to distinct roots, repeated roots, and complex roots
8. Mechanical vibrations—amplitude, period, frequency, and phase angle
9. Euler's method and the improved Euler method of numerical approximation of solutions

CHAPTER 18 MISCELLANEOUS PROBLEMS

Find general solutions of the differential equations in Problems 1–28. (NOTE Throughout this group of problems, primes denote derivatives with respect to x.)

1. $y' + 2xy = 0$
2. $y' + 2xy^2 = 0$
3. $y' = y \sin x$
4. $(1 + x)y' = 4y$
5. $2\sqrt{x}y' = \sqrt{1 - y^2}$
6. $y' = 3\sqrt{xy}$
7. $y' = (64xy)^{1/3}$
8. $y' = 2x \sec y$
9. $(1 - x^2)y' = 2y$
10. $(1 + x)^2y' = (1 + y)^2$
11. $y' + y = 2$
12. $y' - 2y = 3e^{2x}$
13. $y' + 3y = 2xe^{-3x}$
14. $y' - 2xy = \exp(x^2)$

15. $xy' + 2y = 3x$

16. $xy' + 5y = 7x^2$

17. $2xy' + y = 10\sqrt{x}$

18. $3xy' + y = 12x$

19. $xy' - y = x$

20. $2xy' - 3y = 9x^3$

21. $y'' - 4y = 0$

22. $2y'' - 3y' = 0$

23. $y'' + 3y' - 10y = 0$

24. $2y'' - 7y' + 3y = 0$

25. $y'' + 6y' + 9y = 0$

26. $y'' + 5y' + 5y = 0$

27. $4y'' - 12y' + 9y = 0$

28. $y'' - 6y' + 13y = 0$

Solve the initial value problems in Problems 29–50.

29. $y' = e^x y, \quad y(0) = 2e$

30. $y' = 3x^2(y^2 + 1), \quad y(0) = 1$

31. $2yy' = x(x^2 - 16)^{-1/2}, \quad y(5) = 2$

32. $y' = 4x^3 y - y, \quad y(1) = -3$

33. $1 + y' = 2y, \quad y(1) = 1$

34. $y' \tan x = y, \quad y(\pi/2) = \pi/2$

35. $xy' - y = 2x^2 y, \quad y(1) = 1$

36. $y' = 2xy^2 + 3x^2y^2, \quad y(1) = -1$

37. $xy' + y = 3xy, \quad y(1) = 0$

38. $xy' + 3y = 2x^5, \quad y(2) = 1$

39. $y' + y = e^x, \quad y(0) = 1$

40. $xy' - 3y = x^3, \quad y(1) = 10$

41. $y' + 2xy = x, \quad y(0) = -2$

42. $y' = (1 - y) \cos x, \quad y(\pi) = 2$

43. $xy' = 3y + x^4 \cos x, \quad y(2\pi) = 0$

44. $y' = 2xy + 3x^2 \exp(x^2), \quad y(0) = 5$

45. $y'' = 90x^8, \quad y(1) = 3, \quad y'(1) = 11$

46. $y'' = 10y', \quad y(0) = y'(0) = 1$

47. $y'' = 10y, \quad y(0) = y'(0) = 1$

48. $y'' + 2y' - 8y = 0, \quad y(0) = 8, \quad y'(0) = -2$

49. $9y'' + 6y' + 4y = 0, \quad y(0) = 3, \quad y'(0) = 4$

50. $y'' - 6y' + 25y = 0, \quad y(0) = 3, \quad y'(0) = 1$

References for Further Study

References 2, 3, 7, and 10 may be consulted for historical topics pertinent to calculus. Reference 14 provides a more theoretical treatment of single-variable calculus topics than ours. References 4, 5, 8, and 15 include advanced topics in multivariable calculus. Reference 11 is a standard work on infinite series. References 1, 9, and 12 are differential equations textbooks. References 6 and 13 discuss topics in calculus together with computing and programming in BASIC.

1. BOYCE, W. E., and R. C. DIPRIMA, *Elementary Differential Equations* (4th ed.). New York: John Wiley, 1986.
2. BOYER, C. B., *A History of Mathematics.* New York: John Wiley, 1968.
3. BOYER, C. B., *The History of the Calculus.* New York: Dover Publications, 1959.
4. BUCK, R. Creighton, *Advanced Calculus* (3rd ed.). New York: McGraw-Hill, 1978.
5. COURANT, RICHARD, AND FRITZ JOHN, *Introduction to Calculus and Analysis.* Vol. I. New York: Interscience, 1965. Vol. II. New York: Wiley-Interscience, 1974, with the assistance of Albert A. Blank and Alan Solomon.

6. EDWARDS. C. H., Jr., *Calculus and the Personal Computer*. Englewood Cliffs, N.J.: Prentice-Hall, 1986.

7. EDWARDS, C. H., Jr., *The Historical Development of the Calculus*. New York: Springer-Verlag, 1979.

8. EDWARDS, C. H., Jr., *Advanced Calculus of Several Variables*. New York: Academic Press, 1973.

9. EDWARDS, C. H., Jr., and DAVID E. PENNEY, *Elementary Differential Equations with Applications* (2nd ed.). Englewood Cliffs, N.J.: Prentice-Hall, 1989.

10. KLINE, MORRIS, *Mathematical Thought from Ancient to Modern Times*. New York: Oxford University Press, 1972.

11. KNOPP, KONRAD, *Theory and Application of Infinite Series*. New York: Hafner Press, 1981.

12. SIMMONS, GEORGE F., *Differential Equations with Applications and Historical Notes*. New York: McGraw-Hill, 1972.

13. SMITH, DAVID A., *Interface: Calculus and the Computer* (2nd ed.). Philadelphia: W. B. Saunders, 1984.

14. SPIVAK, MICHAEL E., *Calculus* (2nd ed.). Berkeley: Publish or Perish, 1980.

15. TAYLOR, ANGUS E., and W. ROBERT MANN, *Advanced Calculus* (3rd ed.). New York: John Wiley, 1983.

16. THOMAS, GEORGE B., JR., *Calculus and Analytic Geometry* (3rd ed.). Reading, MA: Addison-Wesley, 1960.

Appendices

Appendix A Review of Trigonometric Functions

In elementary trigonometry, the six basic trigonometric functions of an acute angle θ in a right triangle are defined as ratios between lengths of pairs of sides of the triangle. As in Fig. 1,

$$\cos\theta = \frac{a}{h}, \qquad \sin\theta = \frac{o}{h}, \qquad \tan\theta = \frac{o}{a},$$
$$\sec\theta = \frac{h}{a}, \qquad \csc\theta = \frac{h}{o}, \qquad \cot\theta = \frac{a}{o}. \tag{1}$$

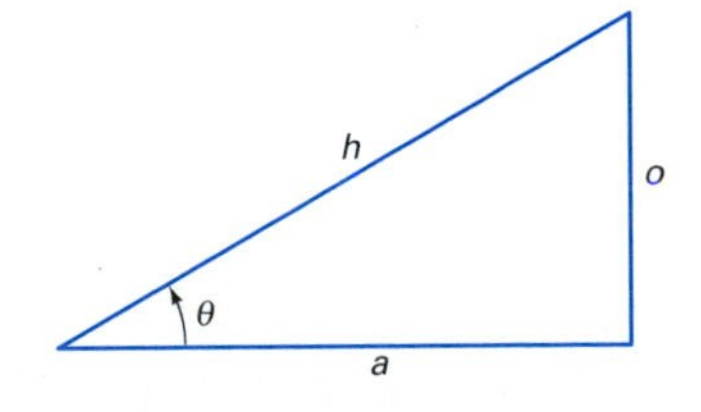

1

We generalize these definitions to *directed* angles of arbitrary size in the following way. Suppose that the initial side of the angle θ is the positive x-axis, so its vertex is at the origin. The angle is **directed** if a direction of rotation from its initial side to its terminal side is specified. We call θ a **positive** angle if this rotation is counterclockwise and a **negative** angle if this rotation is clockwise.

Let $P(x, y)$ be the point at which the terminal side of θ intersects the *unit* circle $x^2 + y^2 = 1$. Then we define

$$\cos\theta = x, \qquad \sin\theta = y, \qquad \tan\theta = \frac{y}{x},$$
$$\sec\theta = \frac{1}{x}, \qquad \csc\theta = \frac{1}{y}, \qquad \cot\theta = \frac{x}{y}. \tag{2}$$

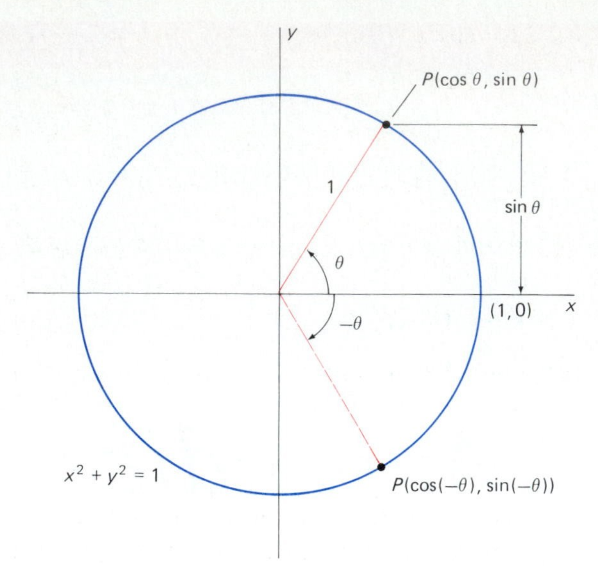

2 Using the unit circle to define the trigonometric functions

Of course we assume that $x \neq 0$ in the case of $\tan\theta$ and $\sec\theta$ and that $y \neq 0$ in the case of $\cot\theta$ and $\csc\theta$. If the angle θ is positive and acute, then it is clear from Fig. 2 that the definitions in (2) agree with the right triangle definitions in (1) in terms of the coordinates of P. A glance at the figure also shows which of the six functions are positive for angles in each of the four quadrants. The diagram in Fig. 3 summarizes this information.

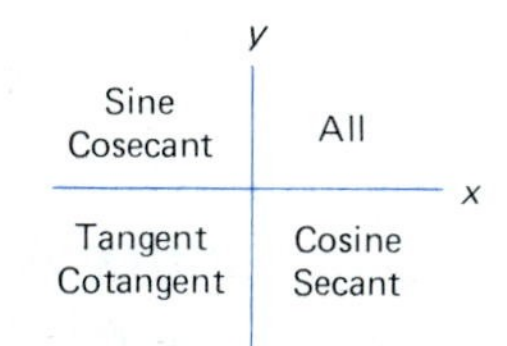

3 The signs of the trigonometric functions

In this appendix we discuss primarily the two most basic trigonometric functions, the sine and the cosine. From (2) we see immediately that

$$\tan\theta = \frac{\sin\theta}{\cos\theta}, \qquad \sec\theta = \frac{1}{\cos\theta},$$
$$\cot\theta = \frac{\cos\theta}{\sin\theta}, \qquad \csc\theta = \frac{1}{\sin\theta}. \tag{3}$$

Next we compare the angles θ and $-\theta$ in Fig. 4. We see that

$$\cos(-\theta) = \cos\theta \quad \text{and} \quad \sin(-\theta) = -\sin\theta. \tag{4}$$

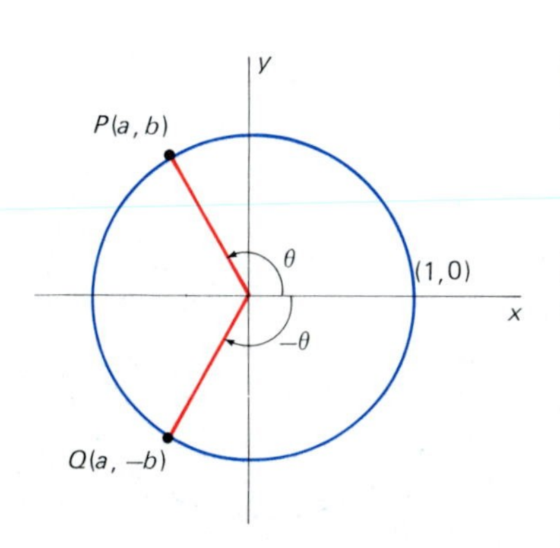

4 Effect of replacing θ by $-\theta$ in sine and cosine

Because $x = \cos\theta$ and $y = \sin\theta$ in (2), the equation $x^2 + y^2 = 1$ of the unit circle translates immediately into the **fundamental identity of trigonometry,**

$$\cos^2\theta + \sin^2\theta = 1. \tag{5}$$

In Problems 21 and 22 of this appendix we outline derivations of the important **addition formulas**

$$\sin(\alpha + \beta) = \sin\alpha\cos\beta + \cos\alpha\sin\beta, \tag{6}$$

$$\cos(\alpha + \beta) = \cos\alpha\cos\beta - \sin\alpha\sin\beta. \tag{7}$$

When we take $\alpha = \theta = \beta$ in (6) and (7), we obtain the **double-angle formulas**

$$\sin 2\theta = 2\sin\theta\cos\theta, \tag{8}$$

$$\cos 2\theta = \cos^2\theta - \sin^2\theta \tag{9}$$

$$= 2\cos^2\theta - 1 \tag{9a}$$

$$= 1 - 2\sin^2\theta. \tag{9b}$$

Equations (9a) and (9b) are obtained from (9) by use of the fundamental identity in (5).

If we solve (9a) for $\cos^2\theta$ and (9b) for $\sin^2\theta$, we get the **half-angle formulas**

$$\cos^2\theta = \tfrac{1}{2}(1 + \cos 2\theta), \tag{10}$$

$$\sin^2\theta = \tfrac{1}{2}(1 - \cos 2\theta). \tag{11}$$

RADIAN MEASURE

In elementary mathematics, angles frequently are measured in *degrees*, with 360° in one complete revolution. In calculus it is more convenient, and often imperative, to measure angles in *radians*. The **radian measure** of an angle is the length of the arc it subtends (cuts off) in the unit circle when the vertex of the angle is at the center of the circle.

Recall that the area A and circumference C of a circle of radius r are given by the formulas

$$A = \pi r^2 \quad \text{and} \quad C = 2\pi r$$

where the irrational number π is approximately 3.14159. Because the circumference of the complete unit circle is 2π and its central angle is 360°, it follows that an angle of radian measure 2π has degree measure 360°, and hence an angle of

$$180° \quad \text{has radian measure} \quad \pi \approx 3.14159. \tag{12}$$

Using (12) we can easily convert back and forth between radians and degrees:

$$1 \text{ radian has measure} \quad \frac{180°}{\pi} \approx 57°17'44.8'', \tag{12a}$$

$$1° \text{ has radian measure} \quad \frac{\pi}{180} \approx 0.01745. \tag{12b}$$

Radians	Degrees
0	0
$\pi/6$	30
$\pi/4$	45
$\pi/3$	60
$\pi/2$	90
$2\pi/3$	120
$3\pi/4$	135
$5\pi/6$	150
π	180
$3\pi/2$	270
2π	360
4π	720

5 Some radian-degree conversions

The table in Fig. 5 shows radian–degree conversions for some frequently occurring angles.

Now consider an angle of θ radians placed at the center of a circle of radius r, as in Fig. 6. Denote by s the length of the arc subtended by θ and by A the area of the sector of the circle bounded by this angle. Then the evident proportions

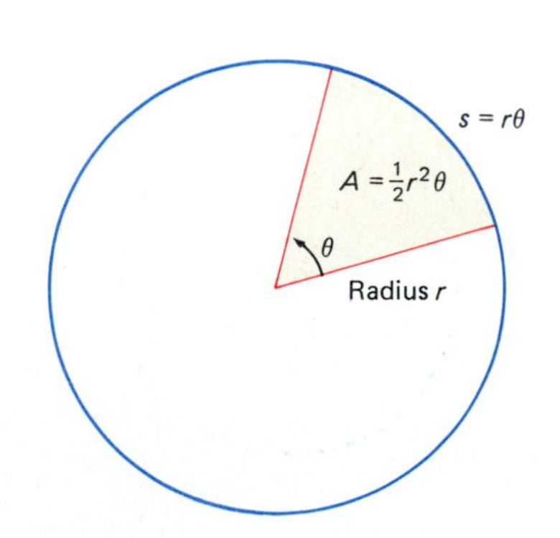

6 Area of a sector and arc length for a circle

$$\frac{s}{2\pi r} = \frac{A}{\pi r^2} = \frac{\theta}{2\pi}$$

give the formulas

$$s = r\theta \qquad (\theta \text{ in radians}) \tag{13}$$

and

$$A = \tfrac{1}{2}r^2\theta \qquad (\theta \text{ in radians}). \tag{14}$$

The definitions in (2) refer to trigonometric functions of *angles* rather than trigonometric functions of *numbers*. Suppose that t is a real number. Then the number $\sin t$ is, *by definition*, the sine of an angle of t radians—remembering that a positive angle is directed counterclockwise from the positive x-axis, while a negative angle is directed clockwise. Briefly, $\sin t$ is the sine of an angle of t *radians*. The other trigonometric functions of the number t have similar definitions. Hence, whenever we write $\sin t$, $\cos t$, and so on, with t a real number, we *always* intend reference to an angle of t *radians*.

When we need to refer to the sine of an angle of t *degrees*, we will henceforth write $\sin t^\circ$. The point is that $\sin t$ and $\sin t^\circ$ are quite different functions of the variable t. For example, you will find that

$$\sin 1^\circ \approx 0.0175 \quad \text{and} \quad \sin 30^\circ = 0.5000$$

with your calculator set in degree mode. But when it is set in radian mode, it will report that

$$\sin 1 \approx 0.8415 \quad \text{and} \quad \sin 30 \approx -0.9880.$$

The relationship between the functions $\sin t$ and $\sin t^\circ$ is this:

$$\sin t^\circ = \sin\left(\frac{\pi t}{180}\right). \tag{15}$$

The distinction extends even to programming languages. In FORTRAN, the function SIN is the radian sine function. In most versions of FORTRAN, one obtains the degree sine function with SIND: $\sin t^\circ =$ SIND(T). In BASIC you must write SIN(PI*T/180) to get the correct value of the sine of an angle of t degrees; SIN(T) yields the value of the radian sine function.

An angle of 2π radians corresponds to one full revolution about the unit circle in Fig. 2. This implies that the sine and cosine functions have **period** 2π, meaning that

$$\sin(t + 2\pi) = \sin t, \qquad \cos(t + 2\pi) = \cos t \tag{16}$$

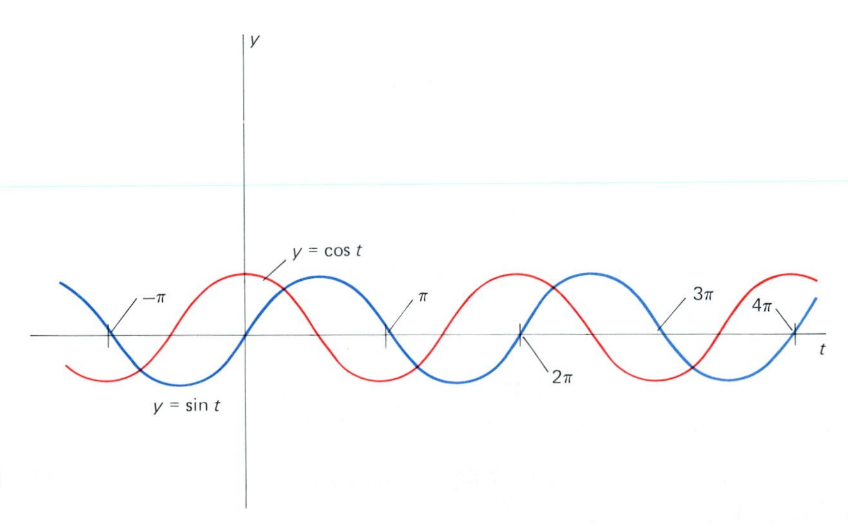

7 Periodicity of the sine and cosine functions

for all real numbers t. It follows from (16) that

$$\sin(t + 2n\pi) = \sin t, \qquad \cos(t + 2n\pi) = \cos t \tag{17}$$

for every integer n (and all real numbers t). This periodicity of the sine and cosine functions is evident in their graphs, which are shown in Fig. 7.

From the definitions of $\sin t$ and $\cos t$ in (2), we see that

$$\sin 0 = 0, \qquad \sin\frac{\pi}{2} = 1, \qquad \sin \pi = 0, \tag{18}$$

$$\cos 0 = 1, \qquad \cos\frac{\pi}{2} = 0, \qquad \cos \pi = -1.$$

The trigonometric functions of $\pi/6$, $\pi/4$, and $\pi/3$ (the radian equivalents of 30°, 45°, and 60°) are easy to read from the well-known triangles of Fig. 8. For example,

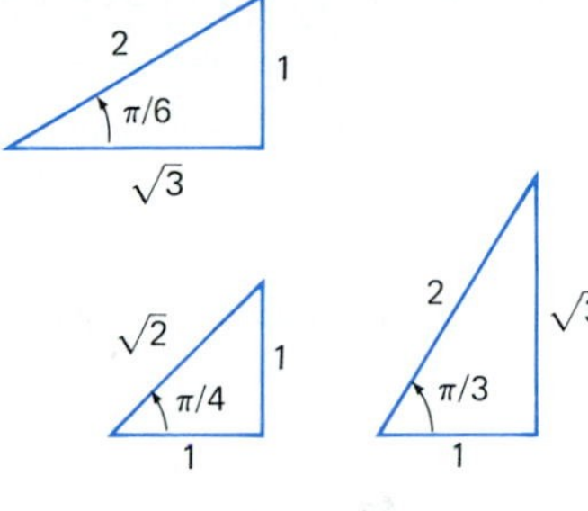

8 Familiar right triangles

$$\sin\frac{\pi}{6} = \cos\frac{\pi}{3} = \frac{1}{2},$$

$$\sin\frac{\pi}{4} = \cos\frac{\pi}{4} = \frac{1}{\sqrt{2}} = \frac{\sqrt{2}}{2}, \quad \text{and} \tag{19}$$

$$\sin\frac{\pi}{3} = \cos\frac{\pi}{6} = \frac{\sqrt{3}}{2}.$$

The following example illustrates the use of periodicity and the addition formulas in (6) and (7) to find sines and cosines of angles larger than $\pi/2$.

EXAMPLE 1

$$\cos\frac{2\pi}{3} = \cos\left(\pi - \frac{\pi}{3}\right) = \cos \pi \cos\frac{\pi}{3} + \sin \pi \sin\frac{\pi}{3}$$

$$= (-1)\left(\frac{1}{2}\right) + (0)\left(\frac{\sqrt{3}}{2}\right) = -\frac{1}{2}.$$

$$\sin\frac{5\pi}{4} = \sin\left(\pi + \frac{\pi}{4}\right) = \sin \pi \cos\frac{\pi}{4} + \cos \pi \sin\frac{\pi}{4}$$

$$= (0)\left(\frac{\sqrt{2}}{2}\right) + (-1)\left(\frac{\sqrt{2}}{2}\right) = -\frac{\sqrt{2}}{2}.$$

$$\sin\frac{17\pi}{6} = \sin\left(2\pi + \frac{5\pi}{6}\right) = \sin\frac{5\pi}{6} = \sin\left(\pi - \frac{\pi}{6}\right)$$

$$= \sin \pi \cos\frac{\pi}{6} - \cos \pi \sin\frac{\pi}{6}$$

$$= (0)\left(\frac{\sqrt{3}}{2}\right) - (-1)\left(\frac{1}{2}\right) = \frac{1}{2}.$$

EXAMPLE 2 Find the solutions (if any) of the equation

$$\sin^2 x - 3\cos^2 x + 2 = 0$$

that lie in the interval $[0, \pi]$.

Solution Using the fundamental identity in (5), we substitute $\cos^2 x = 1 - \sin^2 x$ in the given equation to obtain

$$\sin^2 x - 3(1 - \sin^2 x) + 2 = 0; \qquad 4\sin^2 x - 1 = 0; \qquad \sin x = \pm\tfrac{1}{2}.$$

Because $\sin x \geqq 0$ for x in $[0, \pi]$, we rule out the negative value of $\sin x$. Finally,

$$\sin x = \tfrac{1}{2}$$

for $x = \pi/6$ and for $x = \pi - \pi/6 = 5\pi/6$. These are the solutions of the given equation in $[0, \pi]$.

APPENDIX A PROBLEMS

1. Express the following angles in radian measure.
(a) $40°$; (b) $-270°$; (c) $315°$; (d) $210°$;
(e) $-150°$

2. Express the following angles, given in radian measure, in degrees.
(a) $\dfrac{\pi}{10}$; (b) $\dfrac{2\pi}{5}$; (c) 3π; (d) $\dfrac{15\pi}{4}$;
(e) $\dfrac{23\pi}{60}$

3. Evaluate the six trigonometric functions of x at the following values.
(a) $x = -\dfrac{\pi}{3}$; (b) $x = \dfrac{3\pi}{4}$;
(c) $x = \dfrac{7\pi}{6}$; (d) $x = \dfrac{5\pi}{3}$

4. Find all numbers x such that
(a) $\sin x = 0$; (b) $\sin x = 1$;
(c) $\sin x = -1$.

5. Find all numbers x such that
(a) $\cos x = 0$; (b) $\cos x = 1$;
(c) $\cos x = -1$.

6. Find all numbers x such that
(a) $\tan x = 0$; (b) $\tan x = 1$;
(c) $\tan x = -1$.

7. Suppose that $\tan x = \frac{3}{4}$ and that $\sin x$ is negative. Find the values of the other five trigonometric functions at x.

8. Suppose that $\csc x = -\frac{5}{3}$ and that $\cos x$ is positive. Find the values of the other five trigonometric functions at x.

9. Deduce from the fundamental identity

$$\cos^2\theta + \sin^2\theta = 1$$

and the definitions of the other four trigonometric functions the identities
(a) $1 + \tan^2\theta = \sec^2\theta$;
(b) $1 + \cot^2\theta = \csc^2\theta$.

10. Deduce from the addition formulas for the sine and cosine the tangent addition formula

$$\tan(x + y) = \frac{\tan x + \tan y}{1 - \tan x \tan y}.$$

In Problems 11 and 12, use the method of Example 1 to find the indicated values.

11. (a) $\sin\dfrac{5\pi}{6}$; (b) $\cos\dfrac{7\pi}{6}$;
(c) $\sin\dfrac{11\pi}{6}$; (d) $\cos\dfrac{19\pi}{6}$

12. (a) $\sin\dfrac{2\pi}{3}$; (b) $\cos\dfrac{4\pi}{3}$;
(c) $\sin\dfrac{5\pi}{3}$; (d) $\cos\dfrac{10\pi}{3}$

13. Suppose that $0 < \theta < \pi/2$. Show that
(a) $\sin(\pi \pm \theta) = \mp\sin\theta$;
(b) $\cos(\pi \pm \theta) = -\cos\theta$.

14. Suppose that α is an arbitrary angle. Prove that
(a) $\cos\left(\dfrac{\pi}{2} - \alpha\right) = \sin\alpha$;
(b) $\sin\left(\dfrac{\pi}{2} - \alpha\right) = \cos\alpha$.

In each of Problems 15–20, find all solutions of the given equation that lie in the interval $[0, \pi]$.

15. $3\sin^2 x - \cos^2 x = 2$
16. $\sin^2 x = \cos^2 x$
17. $2\cos^2 x + 3\sin x = 3$
18. $2\sin^2 x + \cos x = 2$
19. $8\sin^2 x\cos^2 x = 1$
20. $\cos 2\theta - 3\cos\theta = 0$

21. The points $A(\cos\theta, -\sin\theta)$, $B(1, 0)$, $C(\cos\phi, \sin\phi)$, and $D(\cos(\theta + \phi), \sin(\theta + \phi))$ are shown in Fig. 9; all are points on the unit circle. Deduce from the fact that the line segments AC and BD have equal length (because they subtend the same angle, $\theta + \phi$) that

$$\cos(\theta + \phi) = \cos\theta\cos\phi - \sin\theta\sin\phi.$$

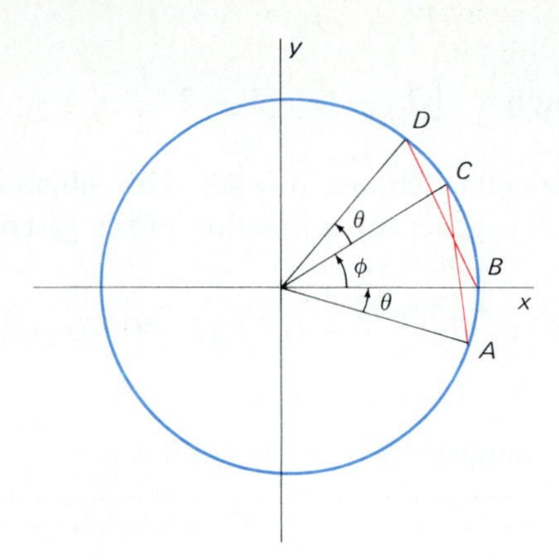

9 Deriving the cosine addition formula (Problem 21)

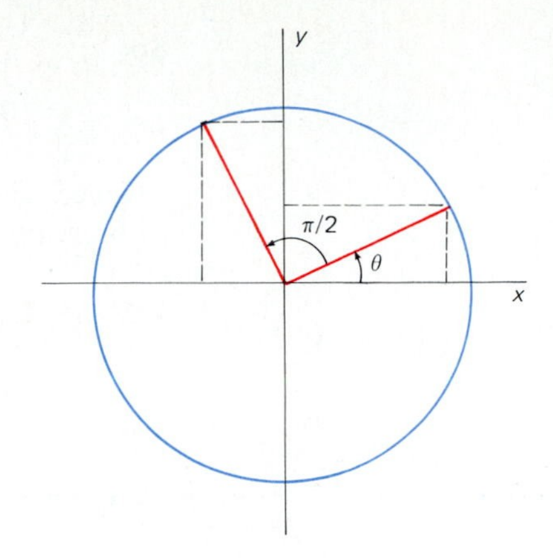

10 Deriving the identities of Problem 22

22. (a) Use the triangles shown in Fig. 10 to deduce that

$$\sin\left(\theta + \frac{\pi}{2}\right) = \cos\theta \quad \text{and} \quad \cos\left(\theta + \frac{\pi}{2}\right) = -\sin\theta.$$

(b) Use the results of Problem 21 and part (a) to derive the addition formula for the sine function.

Appendix B
Proofs of the Limit Laws

Recall the definition of the limit:

$$\lim_{x \to a} F(x) = L$$

provided that, given $\varepsilon > 0$, there exists $\delta > 0$ such that

$$0 < |x - a| < \delta \quad \text{implies} \quad |F(x) - L| < \varepsilon. \tag{1}$$

Note that the number ε comes *first. Then* a value of $\delta > 0$ must be found so that (1) holds. In order to prove that $F(x) \to L$ as $x \to a$, you must, in effect, be able to stop the first person you see on the street and ask him or her to pick a positive number ε at random. Then *you* must *always* be ready to respond with a positive number δ. This number δ must have the property that (1) holds for your δ and the given ε. The *only* restriction on x is the one

$$0 < |x - a| < \delta$$

given in (1).

To do this, you will ordinarily need to give an explicit method—a recipe or formula—for producing the value of δ that works for a given value of ε. As the next few examples show, this method or recipe will depend upon the particular function F involved as well as upon the numbers a and L.

EXAMPLE 1 Prove that $\lim_{x \to 3} (2x - 1) = 5$.

Solution Given $\varepsilon > 0$, we must find $\delta > 0$ such that

$$|(2x - 1) - 5| < \varepsilon \quad \text{if} \quad 0 < |x - 3| < \delta.$$

Now

$$|(2x - 1) - 5| = |2x - 6| = 2|x - 3|.$$

So

$$0 < |x - 3| < \frac{1}{2}\varepsilon \quad \text{implies} \quad |(2x - 1) - 5| < 2 \cdot \frac{1}{2} \cdot \varepsilon = \varepsilon.$$

Hence, given $\varepsilon > 0$, it is sufficient to choose $\delta = \varepsilon/2$. This illustrates the observation that the required δ is generally a function of the given ε.

EXAMPLE 2 Prove that $\lim_{x \to 2} (3x^2 + 5) = 17$.

Solution Given $\varepsilon > 0$, we must find $\delta > 0$ such that

$$0 < |x - 2| < \delta \quad \text{implies} \quad |(3x^2 + 5) - 17| < \varepsilon.$$

Now

$$|(3x^2 + 5) - 17| = |3x^2 - 12| = 3 \cdot |x + 2| \cdot |x - 2|.$$

Our problem, therefore, involves showing that $|x + 2| \cdot |x - 2|$ can be made as small as we please by choosing $|x - 2|$ sufficiently small. The idea is that $|x + 2|$ cannot be too large if $|x - 2|$ is fairly small. For example, if $|x - 2| < 1$, then

$$|x + 2| = |(x - 2) + 4| \leqq |x - 2| + 4 < 5.$$

Therefore,

$$0 < |x - 2| < 1 \quad \text{implies} \quad |(3x^2 + 5) - 17| < 15 \cdot |x - 2|.$$

Consequently, let us choose δ to be the minimum of the two numbers 1 and $\varepsilon/15$. Then

$$0 < |x - 2| < \delta \quad \text{implies} \quad |(3x^2 + 5) - 17| < 15 \cdot \frac{\varepsilon}{15} = \varepsilon,$$

as desired.

EXAMPLE 3 Prove that

$$\lim_{x \to a} \frac{1}{x} = \frac{1}{a} \qquad \text{if} \quad a \neq 0.$$

Solution We take the liberty of considering only the case in which $a > 0$, for that will simplify the notation. The method of proof is similar in the case $a < 0$.

Now suppose that $\varepsilon > 0$ is given. We must find a number $\delta > 0$ such that

$$0 < |x - a| < \delta \quad \text{implies} \quad \left|\frac{1}{x} - \frac{1}{a}\right| < \varepsilon.$$

Now

$$\left|\frac{1}{x} - \frac{1}{a}\right| = \left|\frac{a - x}{ax}\right| = \frac{|x - a|}{a|x|}.$$

The idea is that $1/|x|$ cannot be too large if $|x - a|$ is fairly small. For example, if $|x - a| < a/2$, then $a/2 < x < 3a/2$. Therefore,

$$|x| > \frac{a}{2}, \qquad \text{so that} \quad \frac{1}{|x|} < \frac{2}{a}.$$

In this case it would follow that

$$\left|\frac{1}{x} - \frac{1}{a}\right| < \frac{2}{a^2} \cdot |x - a|$$

if $|x - a| < a/2$. Thus if we choose δ to be the minimum of the two numbers $a/2$ and $a^2\varepsilon/2$, then

$$0 < |x - a| < \delta \quad \text{implies} \quad \left|\frac{1}{x} - \frac{1}{a}\right| < \frac{2}{a^2} \cdot \frac{a^2\varepsilon}{2} = \varepsilon,$$

as desired.

We are now ready to give proofs of the limit laws stated in Section 2.2.

Constant Law

If $f(x) \equiv C$, a constant, then

$$\lim_{x\to a} f(x) = \lim_{x\to a} C = C.$$

Proof Because $|C - C| = 0$, we merely choose $\delta = 1$, regardless of the previously given value of $\varepsilon > 0$. Then, if $0 < |x - a| < \delta$, it is automatic that $|C - C| < \varepsilon$. ■

Addition Law

If $\lim_{x\to a} F(x) = L$ and $\lim_{x\to a} G(x) = M$, then

$$\lim_{x\to a} [F(x) + G(x)] = L + M.$$

Proof Let $\varepsilon > 0$ be given. We must find $\delta > 0$ such that

$$0 < |x - a| < \delta \quad \text{implies} \quad |(F(x) + G(x)) - (L + M)| < \varepsilon.$$

Because L is the limit of $F(x)$ as $x \to a$, there is a number $\delta_1 > 0$ such that

$$0 < |x - a| < \delta_1 \quad \text{implies} \quad |F(x) - L| < \frac{\varepsilon}{2}.$$

Because M is the limit of $G(x)$ as $x \to a$, there is a number $\delta_2 > 0$ such that

$$0 < |x - a| < \delta_2 \quad \text{implies} \quad |G(x) - M| < \frac{\varepsilon}{2}.$$

Let $\delta = \min\{\delta_1, \delta_2\}$. Then

$$0 < |x - a| < \delta \quad \text{implies}$$

$$|(F(x) - G(x)) - (L + M)| \leqq |F(x) - L| + |G(x) - M|$$

$$< \frac{\varepsilon}{2} + \frac{\varepsilon}{2} = \varepsilon,$$

as desired. ■

Product Law

If $\lim_{x \to a} F(x) = L$ and $\lim_{x \to a} G(x) = M$, then

$$\lim_{x \to a} [F(x) \cdot G(x)] = L \cdot M.$$

Proof Given $\varepsilon > 0$, we must find a number $\delta > 0$ such that

$$0 < |x - a| < \delta \quad \text{implies} \quad |F(x) \cdot G(x) - L \cdot M| < \varepsilon.$$

But first, note that the triangle inequality gives the result

$$\begin{aligned} |F(x) \cdot G(x) - L \cdot M| &= |F(x) \cdot G(x) - L \cdot G(x) + L \cdot G(x) - L \cdot M| \\ &\leqq |G(x)| \cdot |F(x) - L| + |L| \cdot |G(x) - M|. \end{aligned} \tag{2}$$

Because $\lim_{x \to a} F(x) = L$, there exists $\delta_1 > 0$ such that

$$0 < |x - a| < \delta_1 \quad \text{implies} \quad |F(x) - L| < \frac{\varepsilon}{2(|M| + 1)}. \tag{3}$$

And because $\lim_{x \to a} G(x) = M$, there exists $\delta_2 > 0$ such that

$$0 < |x - a| < \delta_2 \quad \text{implies} \quad |G(x) - M| < \frac{\varepsilon}{2(|L| + 1)}. \tag{4}$$

Moreover, there is a *third* number $\delta_3 > 0$ such that

$$0 < |x - a| < \delta_3 \quad \text{implies} \quad |G(x) - M| < 1,$$

which in turn implies that

$$|G(x)| < |M| + 1. \tag{5}$$

We now choose $\delta = \min\{\delta_1, \delta_2, \delta_3\}$. Then we substitute (3), (4) and (5) in (2), and finally see that $0 < |x - a| < \delta$ implies

$$\begin{aligned} |F(x)G(x) - LM| &< (|M| + 1) \cdot \frac{\varepsilon}{2(|M| + 1)} + |L| \cdot \frac{\varepsilon}{2(|L| + 1)} \\ &< \frac{\varepsilon}{2} + \frac{\varepsilon}{2} = \varepsilon. \end{aligned}$$

This establishes the product law. The use of $|M| + 1$ and $|L| + 1$ in the denominators takes care of the possibility that L or M might be zero. ■

Substitution Law

If $\lim_{x \to a} g(x) = L$ and $\lim_{x \to L} f(x) = f(L)$, then

$$\lim_{x \to a} f(g(x)) = f(L).$$

Proof Let $\varepsilon > 0$ be given. We must find a number $\delta > 0$ such that

$$0 < |x - a| < \delta \quad \text{implies} \quad |f(g(x)) - f(L)| < \varepsilon.$$

Because $\lim_{y \to L} f(y) = f(L)$, there exists $\delta_1 > 0$ such that

$$0 < |y - L| < \delta_1 \quad \text{implies} \quad |f(y) - f(L)| < \varepsilon. \tag{6}$$

Also, because $\lim_{x \to a} g(x) = L$, we can find $\delta > 0$ such that

$$0 < |x - a| < \delta \quad \text{implies} \quad |g(x) - L| < \delta_1;$$

that is, such that

$$|y - L| < \delta_1$$

where $y = g(x)$. From Equation (6) we see that

$$0 < |x - a| < \delta \quad \text{implies} \quad |f(g(x)) - f(L)| = |f(y) - f(L)| < \varepsilon,$$

exactly as desired. ■

Reciprocal Law

If $\lim_{x \to a} g(x) = L$ and $L \neq 0$, then

$$\lim_{x \to a} \frac{1}{g(x)} = \frac{1}{L}.$$

Proof Let $f(x) = 1/x$. Then, as we saw in Example 3,

$$\lim_{x \to a} f(x) = \lim_{x \to a} \frac{1}{x} = \frac{1}{L} = f(L).$$

Hence the substitution law gives us the result

$$\lim_{x \to a} \frac{1}{g(x)} = \lim_{x \to a} f(g(x)) = f(L) = \frac{1}{L},$$

and this completes the proof. ■

Quotient Law

If $\lim_{x \to a} F(x) = L$ and $\lim_{x \to a} G(x) = M \neq 0$, then

$$\lim_{x \to a} \frac{F(x)}{G(x)} = \frac{L}{M}.$$

Proof It follows immediately from the product and reciprocal laws that

$$\lim_{x \to a} \frac{F(x)}{G(x)} = \lim_{x \to a} F(x) \cdot \frac{1}{G(x)} = \left(\lim_{x \to a} F(x)\right)\left(\lim_{x \to a} \frac{1}{G(x)}\right)$$

$$= L \cdot \frac{1}{M} = \frac{L}{M}. \quad ■$$

Squeeze Law

Suppose that $f(x) \leqq g(x) \leqq h(x)$ in some deleted neighborhood of a and also that

$$\lim_{x \to a} f(x) = L = \lim_{x \to a} h(x).$$

Then

$$\lim_{x \to a} g(x) = L$$

as well.

Proof Given $\varepsilon > 0$, we choose $\delta_1 > 0$ and $\delta_2 > 0$ such that

$$0 < |x - a| < \delta_1 \quad \text{implies} \quad |f(x) - L| < \varepsilon$$

and

$$0 < |x - a| < \delta_2 \quad \text{implies} \quad |h(x) - L| < \varepsilon.$$

Now let $\delta = \min\{\delta_1\ \delta_2\}$. Then $\delta > 0$. Moreover, if $0 < |x - a| < \delta$, then $f(x)$ and $h(x)$ are both points of the open interval $(L - \varepsilon, L + \varepsilon)$. So

$$L - \varepsilon < f(x) \leqq g(x) \leqq h(x) < L + \varepsilon.$$

Thus

$$0 < |x - a| < \delta \quad \text{implies} \quad |g(x) - L| < \varepsilon,$$

as desired. This completes the proof of the squeeze law. ■

APPENDIX B PROBLEMS

In Problems 1–10, apply the definition of the limit to establish the given equality.

1. $\lim_{x \to a} x = a$

2. $\lim_{x \to 2} 3x = 6$

3. $\lim_{x \to 2} (x + 3) = 5$

4. $\lim_{x \to -3} (2x + 1) = -5$

5. $\lim_{x \to 1} x^2 = 1$

6. $\lim_{x \to a} x^2 = a^2$

7. $\lim_{x \to -1} (2x^2 - 1) = 1$

8. $\lim_{x \to a} \dfrac{1}{x^2} = \dfrac{1}{a^2}$ if $a \neq 0$

9. $\lim_{x \to a} \dfrac{1}{x^2 + 1} = \dfrac{1}{a^2 + 1}$

10. $\lim_{x \to a} \dfrac{1}{\sqrt{x}} = \dfrac{1}{\sqrt{a}}$ if $a > 0$

11. Suppose that $\lim_{x \to a} f(x) = L$ and that $\lim_{x \to a} f(x) = M$. Apply the definition of the limit to prove that $L = M$. Thus a limit of a function is unique if it exists.

12. Suppose that C is a constant and that $\lim_{x \to a} f(x) = L$. Apply the definition of the limit to prove that

$$\lim_{x \to a} C \cdot f(x) = C \cdot L.$$

13. Suppose that $L \neq 0$ and that $\lim_{x \to a} f(x) = L$. Use the method of Example 3 and the definition of the limit to show directly that

$$\lim_{x \to a} \frac{1}{f(x)} = \frac{1}{L}.$$

14. In this problem use the algebraic identity

$$x^n - a^n = (x - a)(x^{n-1} + x^{n-2}a + \cdots + xa^{n-2} + a^{n-1}).$$

Show directly from the definition of the limit that $\lim_{x \to a} x^n = a^n$ if n is a positive integer.

15. Apply the identity

$$|\sqrt{x} - \sqrt{a}| = \frac{|x - a|}{\sqrt{x} + \sqrt{a}}$$

to show directly from the definition of the limit that $\lim_{x \to a} \sqrt{x} = \sqrt{a}$ if $a > 0$.

16. Suppose that $\lim_{x \to a} f(x) = f(a) > 0$. Prove that there exists a neighborhood of a on which $f(x) > 0$; that is, prove that there exists $\delta > 0$ such that

$$|x - a| < \delta \quad \text{implies} \quad f(x) > 0.$$

Appendix C
Existence of the Integral

When the basic computational algorithms of the calculus were discovered by Newton and Leibniz in the latter half of the seventeenth century, the logical rigor that had been a feature of the Greek method of exhaustion was largely abandoned. When computing the area A under the curve $y = f(x)$, for example, Newton took it as intuitively obvious that the area function existed, and proceeded to compute it as the antiderivative of the height function $f(x)$. Leibniz regarded A as an infinite sum of infinitesimal area elements, each of the form $dA = f(x)\,dx$, but in practice computed the area

$$A = \int_a^b f(x)\,dx$$

by antidifferentiation just as Newton did—that is, by computing

$$A = \Big[D^{-1} f(x)\Big]_a^b.$$

The question of the *existence* of the area function—one of the conditions that a function f must satisfy in order for its integral to exist—did not at first seem to be of much importance. Eighteenth-century mathematicians were mainly occupied (and satisfied) with the impressive applications of calculus to the solution of real-world problems and did not concentrate on the logical foundations of the subject.

The first attempt at a precise definition of the integral and a proof of its existence for continuous functions was that of the French mathematician Augustin-Louis Cauchy (1789–1857). Curiously enough, Cauchy was trained as an engineer, and much of his research in mathematics was in fields that we today regard as applications-oriented: hydrodynamics, waves in elastic media, vibrations of elastic membranes, polarization of light, and the like. But he was a prolific researcher, and his writings cover the entire spectrum of mathematics, with occasional essays into almost unrelated fields.

Around 1824, Cauchy defined the integral of a continuous function in a way that is familiar to us, as a limit of left-endpoint approximations:

$$\int_a^b f(x)\,dx = \lim_{\Delta x \to 0} \sum_{i=1}^{n} f(x_{i-1})\,\Delta x.$$

Of course, this is a much more complicated sort of limit than the ones we discussed in Chapter 2. Cauchy was not entirely clear about the nature of the limit process involved in the above equation, nor about the precise role that the hypothesis of the continuity of f played in proving that the limit exists.

A complete definition of the integral, as we gave in Section 5.3, was finally produced in the 1850s by the German mathematician Georg Bernhard

Riemann (1826–1866). Riemann was a student of Gauss; he met Gauss upon his arrival at Göttingen for the purpose of studying theology, when he was about 20 years old and Gauss was about 70. Riemann soon decided to study mathematics and became known as one of the truly great mathematicians of the century. Like Cauchy, he was particularly interested in applications of mathematics to the real world; his research particularly emphasized electricity, heat, light, acoustics, fluid dynamics, and—as you might infer from the fact that Wilhelm Weber was a major influence on Riemann's education—magnetism. Riemann also made significant contributions to mathematics itself, particularly in the field of complex analysis. A major conjecture of his, involving the zeta function

$$\zeta(s) = \sum_{n=1}^{\infty} \frac{1}{n^s}, \tag{1}$$

remains unsolved to this day, and has important consequences in the theory of the distribution of prime numbers because

$$\zeta(k) = \prod \left(1 - \frac{1}{p^k}\right)^{-1}$$

where the product is taken over all primes p. (The zeta function is defined in (1) for complex numbers s to the right of the vertical line at $x = 1$ and is extended to other complex numbers by the requirement that it be differentiable.) Riemann died of tuberculosis shortly before his fortieth birthday.

In this section we give a proof of the existence of the integral of a continuous function. We will follow Riemann's approach. Specifically, suppose that the function f is continuous on the closed interval $[a, b]$. We will prove that the definite integral

$$\int_a^b f(x)\,dx$$

exists. That is, we will demonstrate the existence of a number I satisfying the following condition: For every $\varepsilon > 0$ there exists $\delta > 0$ such that, for *every* Riemann sum R associated with *any* partition P with $|P| < \delta$,

$$|I - R| < \varepsilon.$$

(Recall that the mesh $|P|$ of the partition P is the length of the longest subinterval in the partition.) In other words, every Riemann sum associated with every sufficiently "fine" partition is close to the number I. If this happens, then the definite integral

$$\int_a^b f(x)\,dx$$

is said to **exist,** and I is its **value.**

Now we begin the proof. Suppose throughout that f is a function continuous on the closed interval $[a, b]$. Given $\varepsilon > 0$, we need to show the existence of a number $\delta > 0$ such that

$$\left| I - \sum_{i=1}^{n} f(x_i^*)\,\Delta x_i \right| < \varepsilon \tag{2}$$

for every Riemann sum associated with any partition P of $[a, b]$ with $|P| < \delta$.

Given a partition P of $[a, b]$ into n *not necessarily equally long* subintervals, let p_i be a point in the subinterval $[x_{i-1}, x_i]$ at which f attains its minimum value $f(p_i)$; similarly, let $f(q_i)$ be its maximum value there. These numbers exist for $i = 1, 2, 3, \ldots, n$ because of the maximum value property of continuous functions (Theorem 4 in Appendix D).

In what follows we will denote the resulting lower and upper Riemann sums associated with P by

$$L(P) = \sum_{i=1}^{n} f(p_i)\,\Delta x_i \tag{3a}$$

and

$$U(P) = \sum_{i=1}^{n} f(q_i)\,\Delta x_i, \tag{3b}$$

respectively. Then our first lemma is obvious.

Lemma 1

For any partition P of $[a, b]$, $L(P) \leqq U(P)$.

Now a definition. The partition P' is called a *refinement* of the partition P if each subinterval of P' is contained in some subinterval of P. That is, P' is obtained from P by adding more points of subdivision to P.

Lemma 2

Suppose that P' is a refinement of P. Then

$$L(P) \leqq L(P') \leqq U(P') \leqq U(P). \tag{4}$$

Proof The inequality $L(P') \leqq U(P')$ is a consequence of Lemma 1. We will show that $L(P) \leqq L(P')$; the proof that $U(P') \leqq U(P)$ is similar.

The refinement P' is obtained from P by adding one or more points of subdivision to P. So all we really need show is that the Riemann sum cannot be decreased by adding a single point of subdivision. Thus we will suppose that the partition P' is obtained from P by dividing the kth subinterval $[x_{k-1}, x_k]$ of P into two subintervals $[x_{k-1}, z]$ and $[z, x_k]$ by means of the new point z.

The only resulting effect on the corresponding Riemann sum is to replace the term

$$f(p_k) \cdot (x_k - x_{k-1})$$

in $L(P)$ by the two-term sum

$$f(u) \cdot (z - x_{k-1}) + f(v) \cdot (x_k - z),$$

where $f(u)$ is the minimum of f on $[x_{k-1}, z]$ and $f(v)$ is the minimum of f on $[z, x_k]$. But

$$f(p_k) \leqq f(u) \quad \text{and} \quad f(p_k) \leqq f(v).$$

Hence

$$f(u)\cdot(z-x_{k-1})+f(v)\cdot(x_k-z)\geqq f(p_k)\cdot(z-x_{k-1})+f(p_k)\cdot(x_k-z)$$
$$=f(p_k)\cdot(z-x_{k-1}+x_k-z)$$
$$=f(p_k)\cdot(x_k-x_{k-1}).$$

So the replacement of $f(p_k)\cdot(x_k-x_{k-1})$ cannot decrease the sum in question, and therefore $L(P)\leqq L(P')$. Because this is all we needed to show, we have completed the proof of Lemma 2. ■

To prove that all the Riemann sums for sufficiently "fine" partitions are close to some number I, we must give some construction of I. This is accomplished through the next lemma.

Lemma 3

Let P_n denote the regular partition of $[a, b]$ into 2^n subintervals of equal length. Then the (sequential) limit

$$I=\lim_{n\to\infty} L(P_n) \tag{5}$$

exists.

Proof Note first that each partition P_{n+1} is a refinement of P_n, so that (by Lemma 2)

$$L(P_1)\leqq L(P_2)\leqq\cdots\leqq L(P_n)\leqq\cdots.$$

Therefore, $\{L(P_n)\}$ is a nondecreasing sequence of real numbers. Moreover,

$$L(P_n)=\sum_{i=1}^{2^n} f(p_i)\,\Delta x_i\leqq M\sum_{i=1}^{2^n}\Delta x_i=M\cdot(b-a),$$

where M is the maximum value of f on $[a, b]$.

Now Theorem 1 in Appendix D guarantees that a bounded monotone sequence of real numbers must converge. Thus the number

$$I=\lim_{n\to\infty} L(P_n)$$

exists. This establishes Equation (5), and thus the proof of Lemma 3 is complete. ■

It is proved in advanced calculus that if f is continuous on $[a, b]$, then—for every number $\varepsilon>0$—there exists a number $\delta>0$ such that

$$|f(u)-f(v)|<\varepsilon$$

for any two points u and v of $[a, b]$ such that

$$|u-v|<\delta.$$

This property of a function is called **uniform continuity** of f on the interval $[a, b]$. Thus the theorem from advanced calculus that we need to use states that every continuous function on a closed (and bounded) interval is uniformly continuous there.

NOTE That f is continuous on $[a, b]$ means that for each number u in the interval and each $\varepsilon > 0$, there exists $\delta > 0$ such that if v is a number in the interval with $|u - v| < \delta$, then $|f(u) - f(v)| < \varepsilon$. But *uniform* continuity is a more stringent condition; it means that given $\varepsilon > 0$, you can not only find a value δ_1 that "works" for u_1, a value δ_2 that works for u_2, and so on, but more: You can in fact find a universal value δ that works for *all* values of u in the interval. This should not be obvious when you realize it is possible that $\delta_1 = 1$, $\delta_2 = \frac{1}{2}$, $\delta_3 = \frac{1}{3}$, and so on. In any case, it is clear that uniform continuity of f on an interval implies its continuity there.

Remember that, throughout, we have a continuous function f defined on the closed interval $[a, b]$.

Lemma 4

Suppose that $\varepsilon > 0$ is given. Then there exists a number $\delta > 0$ such that if P is a partition of $[a, b]$ with $|P| < \delta$ and P' is a refinement of P, then

$$|R(P) - R(P')| < \tfrac{1}{3}\varepsilon \tag{6}$$

for any two Riemann sums $R(P)$ associated with P and $R(P')$ associated with P'.

Proof Because f must be uniformly continuous on $[a, b]$, there exists a number $\delta > 0$ such that if

$$|u - v| < \delta, \quad \text{then} \quad |f(u) - f(v)| < \frac{\varepsilon}{3(b - a)}.$$

Suppose now that P is a partition of $[a, b]$ with $|P| < \delta$. Then

$$|U(P) - L(P)| = \sum_{i=1}^{n} |f(q_i) - f(p_i)| \, \Delta x_i < \frac{\varepsilon}{3(b - a)} \sum_{i=1}^{n} \Delta x_i = \frac{\varepsilon}{3}.$$

This is valid because $|p_i - q_i| < \delta$, for p_i and q_i both belong to the same subinterval $[x_{i-1}, x_i]$ of P, and $|P| < \delta$.

Now, as shown in the figure below, we know that $L(P)$ and $U(P)$ differ by less than $\varepsilon/3$. We also know that

$$L(P) \leqq R(P) \leqq U(P)$$

for every Riemann sum $R(P)$ associated with P. But

$$L(P) \leqq L(P') \leqq U(P') \leqq U(P)$$

by Lemma 2 because P' is a refinement of P; moreover,

$$L(P') \leqq R(P') \leqq U(P')$$

for any Riemann sum $R(P')$ associated with P'.

The situation is illustrated in the figure: The two numbers $R(P)$ and $R(P')$ both belong to the interval $[L(P), U(P)]$ of length less than $\varepsilon/3$, so Equation (6) follows, as desired. This concludes the proof of Lemma 4. ■

Theorem *Existence of the Integral*
If f is continuous on the closed interval $[a, b]$, then the integral

$$\int_a^b f(x)\,dx$$

exists.

Proof Suppose that $\varepsilon > 0$ is given. We must show the existence of a number $\delta > 0$ such that, for every partition P of $[a, b]$ with $|P| < \delta$, we have

$$|I - R(P)| < \varepsilon,$$

where I is the number given in Lemma 3 and $R(P)$ is an arbitrary Riemann sum for f associated with P.

We choose the number δ provided by Lemma 4 such that

$$|R(P) - R(P')| < \frac{\varepsilon}{3}$$

if $|P| < \delta$ and P' is a refinement of P.

By Lemma 3, we can choose an integer N so large that

$$|P_N| < \delta \quad \text{and} \quad |L(P_N) - I| < \frac{\varepsilon}{3}. \tag{7}$$

Now, given an arbitrary partition P such that $|P| < \delta$, let P' be a common refinement of P and P_N. You can obtain such a partition P', for example, by using all the points of subdivision of both P and P_N to form the subintervals of $[a, b]$.

Because P' is a refinement of both P and P_N and both of the latter have mesh less than δ, Lemma 4 implies that

$$|R(P) - R(P')| < \frac{\varepsilon}{3} \quad \text{and} \quad |L(P_N) - R(P')| < \frac{\varepsilon}{3}. \tag{8}$$

Here, of course, $R(P)$ and $R(P')$ are (arbitrary) Riemann sums associated with P and P', respectively.

Now, given an arbitrary Riemann sum $R(P)$ associated with the partition P having mesh less than δ, we see that

$$\begin{aligned} |I - R(P)| &= |I - L(P_N) + L(P_N) - R(P') + R(P') - R(P)| \\ &\leqq |I - L(P_N)| + |L(P_N) - R(P')| + |R(P') - R(P)|. \end{aligned}$$

In the last sum, each of the last two terms is less than $\varepsilon/3$ by virtue of the inequalities in (8). We also know, by (7), that the first term is also less than $\varepsilon/3$. Consequently,

$$|I - R(P)| < \varepsilon.$$

This establishes the theorem. ■

We close with an example that shows that some hypothesis of continuity is required for integrability.

EXAMPLE Suppose that f is defined for $0 \leqq x \leqq 1$ as follows:

$$f(x) = \begin{cases} 1 & \text{if } x \text{ is irrational;} \\ 0 & \text{if } x \text{ is rational.} \end{cases}$$

Then f is not continuous anywhere. (Why?) Given a partition P of $[0, 1]$, let p_i be a rational point and q_i an irrational point of the ith subinterval of P, for $i = 1, 2, 3, \ldots, n$. As before, f attains its minimum value 0 at each p_i and its maximum value 1 at each q_i. Also

$$L(P) = \sum_{i=1}^{n} f(p_i)\,\Delta x_i = 0, \quad \text{while} \quad U(P) = \sum_{i=1}^{n} f(q_i)\,\Delta x_i = 1.$$

Thus if we choose $\varepsilon = \frac{1}{2}$, there is *no* number I that can lie within ε of both $L(P)$ and $U(P)$, no matter how small the mesh of P. It follows that f is *not* Riemann integrable on $[0, 1]$.

Appendix D The Completeness of the Real Number System

Here we present a self-contained treatment of those consequences of the completeness of the real number system that are relevant to material of this text. Our principal objective is to prove the intermediate value theorem and the maximum value theorem. We begin with the least upper bound property of the real numbers, which we take as an axiom.

Definition *Upper Bound and Lower Bound*

The set S of real numbers is said to be **bounded above** if there is a number b such that $x \leqq b$ for every number x in S. If so, the number b is called an **upper bound** for S. Similarly, if there is a number a such that $x \geqq a$ for every number x in S, then S is said to be **bounded below** and a is called a **lower bound** for S.

Definition

The number λ is said to be a **least upper bound** for the set S of real numbers provided that:

1. λ is an upper bound for S, and
2. If b is an upper bound for S, then $\lambda \leqq b$.

Similarly, the number γ is said to be a **greatest lower bound** for S if γ is a lower bound for S *and* $\gamma \geqq a$ for every lower bound a of S.

Exercise: Prove that if a set S has a least upper bound λ, then it is unique. That is, if λ and μ are both least upper bounds for S, then $\lambda = \mu$.

It is easy to show that the greatest lower bound γ of a set S, if any, is also unique. At this point you should construct examples to illustrate that a set with a least upper bound λ may or may not contain λ, and similarly for its greatest lower bound. We now state the *completeness axiom* of the real number system.

Least Upper Bound Axiom

If the nonempty set S of real numbers has an upper bound, then it has a least upper bound.

By working with the set T consisting of the numbers $-x$ where x is in S, it is not difficult to show the following consequence of the least upper bound axiom: If the nonempty set S of real numbers is bounded below, then S has a greatest lower bound. Because of this symmetry, we need only one axiom, not two; results for least upper bounds also hold for greatest lower bounds provided some attention is paid to getting the inequalities the right way around.

The restriction that S be nonempty is annoying but necessary. For if S is the "empty" set of real numbers, then 73 is an upper bound for S, but S has no least upper bound because $72, 71, 70, \ldots, 0, -1, -2, \ldots$ are all also upper bounds for S.

Definition *Increasing, Decreasing, and Monotone Sequences*

The infinite sequence $x_1, x_2, x_3, \ldots, x_k, \ldots$ is said to be **nondecreasing** if $x_n \leqq x_{n+1}$ for every $n \geqq 1$. This sequence is said to be **nonincreasing** if $x_n \geqq x_{n+1}$ for every $n \geqq 1$. If the sequence $\{x_n\}$ is either nonincreasing or nondecreasing, then it is said to be **monotone**.

The following theorem gives the **bounded monotone sequence property** of the set of real numbers. (Recall that a set S of real numbers is said to be **bounded** if it is contained in an interval of the form $[a, b]$.)

Theorem 1 *Bounded Monotone Sequences*

Every bounded monotone sequence of real numbers converges.

Proof Suppose that the sequence

$$S = \{x_n\} = \{x_1, x_2, x_3, \ldots, x_k, \ldots\}$$

is bounded and nondecreasing. By the least upper bound axiom, S has a least upper bound λ. We claim that λ is the limit of the sequence $\{x_n\}$. For consider an open interval centered at λ; that is, an interval of the form $I = (\lambda - \varepsilon, \lambda + \varepsilon)$ where $\varepsilon > 0$. Some terms of the sequence must lie within I, else $\lambda - \varepsilon$ would be an upper bound for S. But if x_N is within I, then—because we are dealing with a nondecreasing sequence—x_k must also lie in I for all $k \geqq N$. Because ε is an arbitrary positive number, λ is by definition (see Section 12.2) the limit of the sequence $\{x_n\}$. That is, a bounded nondecreasing sequence converges. A similar proof can be constructed for nonincreasing sequences by working with the greatest lower bound. ■

So the least upper bound axiom implies the bounded monotone sequence property of the real numbers. With just a little trouble, you can prove that the two actually are logically equivalent: If you take the bounded monotone sequence property as an axiom, then the least upper bound property follows as a theorem. The **nested interval property** in the next theorem is also equivalent to the least upper bound property, but we shall prove only that it follows from the least upper bound property, because we have chosen the latter as the fundamental completeness axiom for the real number system.

Theorem 2 *Nested Interval Property of the Real Numbers*
Suppose that $I_1, I_2, I_3, \ldots, I_n, \ldots$ is a sequence of closed intervals (so that I_n is of the form $[a_n, b_n]$ for each n) such that

1. I_n contains I_{n+1} for each $n \geqq 1$, and
2. $\lim_{n\to\infty} (b_n - a_n) = 0$

Then there exists exactly one real number c such that c is an element of I_n for each n. Thus

$$\{c\} = I_1 \cap I_2 \cap I_3 \cap \cdots.$$

Proof It is clear from the property in part (2) of the theorem that there is at most one such number c. The sequence $\{a_n\}$ of left-hand endpoints of the intervals is a bounded (by b_1) nondecreasing sequence and thus has a limit a by the bounded monotone sequence property. Similarly, the sequence $\{b_n\}$ has a limit b. Because $a_n \leqq b_n$ for all n, it follows easily that $a \leqq b$. It is clear that $a_n \leqq a \leqq b_n$ for all $n \geqq 1$, so a belongs to every interval I_n; so does b, by the same argument. But then Property (2) above implies that $a = b$, and clearly this common value—call it c—is the number satisfying the conclusion of the theorem. ■

We can now use these results to prove several important theorems used in the text.

Theorem 3 *Intermediate Value Property of Continuous Functions*
If the function f is continuous on $[a, b]$ and $f(a) < K < f(b)$, then $K = f(c)$ for some number c in (a, b).

Proof Let $I_1 = [a, b]$. Suppose that I_n has been defined for $n \geqq 1$. We describe (inductively) how to define I_{n+1}, and this shows in particular how to define I_2, I_3, and so forth. Let a_n be the left-hand endpoint of I_n, b_n its right-hand endpoint, and m_n its midpoint. If $f(m_n) > K$, then $f(a_n) < K < f(m_n)$; in this case, let $a_{n+1} = a_n$, $b_{n+1} = m_n$, and $I_{n+1} = [a_{n+1}, b_{n+1}]$. If $f(m_n) < K$, then let $a_{n+1} = m_n$ and $b_{n+1} = b_n$. Thus at each stage we bisect I_n and let I_{n+1} be the half of I_n on which f takes on values both above and below K. Note that if $f(m_n)$ is ever actually equal to K, we simply let $c = m_n$ and stop.

It is easy to show that the sequence $\{I_n\}$ of intervals satisfies the hypotheses of Theorem 2. Let c be the (unique) real number common to all the intervals I_n. We will show that $f(c) = K$, and this will conclude the proof.

The sequence $\{b_n\}$ has limit c, so (by the continuity of f) the sequence $\{f(b_n)\}$ has limit $f(c)$. But $f(b_n) > K$ for all n, so the limit of $\{f(b_n)\}$ can be no less than K; that is, $f(c) \geqq K$. By considering the sequence $\{a_n\}$, it follows that $f(c) \leqq K$. Therefore, $f(c) = K$. ■

Lemma 1

If f is continuous on the closed interval $[a, b]$, then f is bounded there.

Proof Suppose by way of contradiction that f is not bounded on $I_1 = [a, b]$. Bisect I_1 and let I_2 be either half on which f is unbounded—if f is unbounded on both halves, let I_2 be the left half. In general, let I_{n+1} be a half of I_n on which f is unbounded.

Again it is easy to show that the sequence $\{I_n\}$ of closed intervals satisfies the hypotheses of Theorem 2. Let c be the number common to them all. Because f is continuous, there is a number $\varepsilon > 0$ such that f is bounded on the interval $(c - \varepsilon, c + \varepsilon)$. But for sufficiently large values of n, I_n is a subset of $(c - \varepsilon, c + \varepsilon)$. This contradiction shows that f must be bounded on $[a, b]$. ■

Theorem 4 *Maximum Value Property of Continuous Functions*

If the function f is continuous on the closed interval $[a, b]$, then there exists a number c in $[a, b]$ such that $f(x) \leqq f(c)$ for all x in $[a, b]$.

Proof Consider the set $S = \{f(x) \mid a \leqq x \leqq b\}$. This set is bounded by Lemma 1; let λ be its least upper bound. Our goal is to show that λ is a value $f(c)$ of f.

With $I_1 = [a, b]$, bisect I_1 as before. Note that λ is the least upper bound of the values of f on at least one of the two halves of I_1; let I_2 be that half. Having defined I_n, let I_{n+1} be the half of I_n on which λ is the least upper bound of the values of f. Let c be the number common to all these intervals. It then follows from the continuity of f, much as in the proof of Theorem 3, that $f(c) = \lambda$. And it is clear that $f(x) \leqq \lambda$ for all x in $[a, b]$. ■

The technique we are using in these proofs is called the *method of bisection*. We now use it once again to establish the *Bolzano–Weierstrass property* of the real number system.

Definition *Limit Point*

Let S be a set of real numbers. The number p is said to be a **limit point** of S if every open interval containing p also contains points of S other than p.

Theorem 5 *Bolzano–Weierstrass Theorem*

Every bounded infinite set of real numbers has a limit point.

Proof Let I_0 be a closed interval containing the bounded infinite set S of real numbers. Let I_1 be one of the closed half-intervals of I_0 that contains infinitely many points of S. If I_n has been chosen, let I_{n+1} be one of the closed half-intervals of I_n containing infinitely many points of S. An application of Theorem 2 yields a number p common to all the intervals I_n. If J is an open interval containing p, then J contains I_n for some sufficiently large value of n and thus contains infinitely many points of S. Therefore, p is a limit point of S. ■

Our final goal is in sight: We can now prove that a sequence of real numbers converges if and only if it is a Cauchy sequence.

Definition *Cauchy Sequence*

The sequence $\{a_n\}_1^\infty$ is said to be a **Cauchy sequence** if, for every $\varepsilon > 0$, there exists an integer N such that

$$|a_m - a_n| < \varepsilon$$

for all $m, n \geqq N$.

Lemma 2 *Convergent Subsequences*

Every bounded sequence of real numbers has a convergent subsequence.

Proof If $\{a_n\}$ consists of only finitely many values, then the conclusion of the theorem follows easily. We therefore focus our attention on the case in which $\{a_n\}$ is an infinite set. It is easy to show that this set is also bounded, and thus we may apply the Bolzano–Weierstrass theorem to obtain a limit point p of $\{a_n\}$. For each integer $k \geqq 1$, let $a_{n(k)}$ be a term of the sequence $\{a_n\}$ such that

1. $n(k + 1) > n(k)$ for all $k \geqq 1$, and
2. $|a_{n(k)} - p| < \dfrac{1}{k}$.

It is then easy to show that $\{a_{n(k)}\}$ is a convergent (to p) subsequence of $\{a_n\}$. ■

Theorem 6 *Convergence of Cauchy Sequences*

A sequence of real numbers converges if and only if it is a Cauchy sequence.

Proof It follows immediately from the triangle inequality that every convergent sequence is a Cauchy sequence. Thus suppose that the sequence $\{a_n\}$ is a Cauchy sequence.

Choose N such that

$$|a_m - a_n| < 1$$

if $m, n \geqq N$. It follows that if $n \geqq N$, then a_n lies in the closed interval $[a_N - 1, a_N + 1]$. This implies that the sequence $\{a_n\}$ is bounded, and thus by Lemma 2 it has a convergent subsequence $\{a_{n(k)}\}$. Let p be the limit of this subsequence.

We claim that $\{a_n\}$ itself converges to p. For given $\varepsilon > 0$, choose M such that

$$|a_m - a_n| < \frac{\varepsilon}{2}$$

if $m, n \geqq M$. Next choose K such that $n(K) \geqq M$ and

$$|a_{n(K)} - p| < \frac{\varepsilon}{2}.$$

Then if $n \geqq M$,

$$|a_n - p| \leqq |a_n - a_{n(K)}| + |a_{n(K)} - p| < \varepsilon.$$

Therefore $\{a_n\}$ converges to p by definition. ■

Appendix E Approximations and Riemann Sums

Several times in Chapter 6 our attempt to compute some quantity Q led to the following situation. Beginning with a regular partition of an appropriate interval $[a, b]$ into n equal subintervals each of length Δx, we find an approximation A_n to Q of the form

$$A_n = \sum_{i=1}^{n} g(u_i)h(v_i)\,\Delta x \tag{1}$$

where u_i and v_i are two (generally different) points of the ith subinterval $[x_{i-1}, x_i]$. For example, in our discussion of surface area of revolution that precedes Equation (8) of Section 6.4, we found the approximation

$$\sum_{i=1}^{n} 2\pi f(u_i)\{1 + [f'(v_i)]^2\}^{1/2}\,\Delta x \tag{2}$$

to the area of the surface generated by revolving the curve $y = f(x)$, $a \leqq x \leqq b$, around the x-axis. (In Section 6.4 we wrote x_i^{**} for u_i and x_i^* for v_i.) Note that the expression in (2) is the same as the right-hand side in (1); take $g(x) = 2\pi f(x)$ and $h(x) = \{1 + [f'(x)]^2\}^{1/2}$.

In such a situation we observe that if u_i and v_i were the *same* point x_i^* of $[x_{i-1}, x_i]$ for each i ($i = 1, 2, 3, \ldots, n$), then the approximation in (1) would be a Riemann sum for the function $g(x)h(x)$ on $[a, b]$. This leads us to suspect that

$$\lim_{\Delta x \to 0} \sum_{i=1}^{n} g(u_i)h(v_i)\,\Delta x = \int_a^b g(x)h(x)\,dx. \tag{3}$$

In Section 6.4, we assumed the validity of Equation (3) and concluded from the approximation in (2) that the surface area of revolution ought to be defined to be

$$\begin{aligned} A &= \lim_{\Delta x \to 0} \sum_{i=1}^{n} 2\pi f(u_i)\{1 + [f'(v_i)]^2\}^{1/2}\,\Delta x \\ &= \int_a^b 2\pi f(x)\{1 + [f'(x)]^2\}^{1/2}\,dx. \end{aligned}$$

The following theorem guarantees that Equation (3) holds under mild restrictions on the functions g and h.

Theorem 1 *A Generalization of Riemann Sums*
Suppose that h and the derivative g' are continuous on $[a, b]$. Then

$$\lim_{\Delta x \to 0} \sum_{i=1}^{n} g(u_i)h(v_i)\,\Delta x = \int_a^b g(x)h(x)\,dx, \tag{3}$$

where u_i and v_i are arbitrary points of the ith subinterval of a regular partition of $[a, b]$ into n subintervals each of length Δx.

Proof Let M_1 and M_2 denote the maximum values on $[a, b]$ of $|g'(x)|$ and $|h(x)|$, respectively. Note that

$$\sum_{i=1}^{n} g(u_i)h(v_i)\,\Delta x = R_n + S_n, \quad \text{where} \quad R_n = \sum_{i=1}^{n} g(v_i)h(v_i)\,\Delta x$$

is a Riemann sum approaching $\int_a^b g(x)h(x)\,dx$ as $\Delta x \to 0$, and

$$S_n = \sum_{i=1}^{n} [g(u_i) - g(v_i)]h(v_i)\,\Delta x.$$

To prove (3) it is sufficient to show that $S_n \to 0$ as $\Delta x \to 0$. The mean value theorem gives

$$\begin{aligned} |g(u_i) - g(v_i)| &= |g'(\bar{x}_i)| \cdot |u_i - v_i| \qquad (\bar{x}_i \text{ in } (u_i, v_i)) \\ &\leqq M_1\,\Delta x \end{aligned}$$

because u_i and v_i are both points of the interval $[x_{i-1}, x_i]$ of length Δx. Then

$$\begin{aligned} |S_n| &\leqq \sum_{i=1}^{n} |g(u_i) - g(v_i)| \cdot |h(v_i)|\,\Delta x \leqq \sum_{i=1}^{n} (M_1\,\Delta x) \cdot (M_2\,\Delta x) \\ &= (M_1 M_2\,\Delta x) \sum_{i=1}^{n} \Delta x = M_1 M_2 (b - a)\,\Delta x, \end{aligned}$$

from which it follows that $S_n \to 0$ as $\Delta x \to 0$, as desired. ■

As an application of Theorem 1, let us give a rigorous derivation of the formula in Equation (2) of Section 6.3,

$$V = \int_a^b 2\pi f(x)\,dx, \tag{4}$$

for the volume of the solid generated by revolving the region lying under $y = f(x)$, $a \leqq x \leqq b$, around the y-axis. Beginning with the usual regular partition of $[a, b]$, let $f(x_i^\flat)$ and $f(x_i^\#)$ denote the minimum and maximum values of f on the ith subinterval $[x_{i-1}, x_i]$. Denote by x_i^* the midpoint of this subinterval. From Fig. 1, we see that the part of the solid generated by revolving the region under $y = f(x)$, $x_{i-1} \leqq x \leqq x_i$, contains a cylindrical shell with average radius x_i^*, thickness Δx, and height $f(x_i^\flat)$, and is contained in another cylindrical shell with the same average radius and thickness, but with height $f(x_i^\#)$. Hence the volume ΔV_i of this part of the solid satisfies

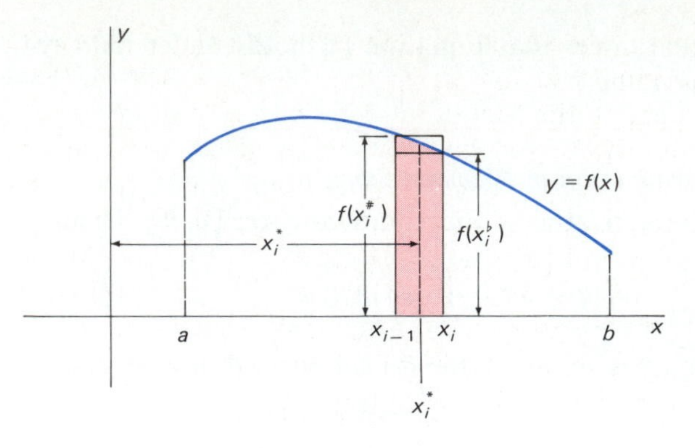

1 A careful estimate of the volume of a solid of revolution around the y-axis

the inequalities

$$2\pi x_i^* f(x_i^\flat)\,\Delta x \leqq \Delta V_i \leqq 2\pi x_i^* f(x_i^\#)\,\Delta x.$$

We add these inequalities for $i = 1, 2, 3, \ldots, n$, and find that

$$\sum_{i=1}^{n} 2\pi x_i^* f(x_i^\flat)\,\Delta x \leqq V \leqq \sum_{i=1}^{n} 2\pi x_i^* f(x_i^\#)\,\Delta x.$$

Because Theorem 1 implies that each of the last two sums approaches $\int_a^b f(x)\,dx$, the squeeze law of limits now implies the formula in (4).

We will occasionally need a generalization of Theorem 1 that involves the notion of a continuous function $F(x, y)$ of two variables. We say that F is *continuous* at the point (x_0, y_0) provided that the value $F(x, y)$ can be made arbitrarily close to $F(x_0, y_0)$ merely by choosing the point (x, y) sufficiently close to (x_0, y_0). We discuss continuity of functions of two variables in Chapter 15. Here it will suffice to accept the following facts: If $g(x)$ and $h(y)$ are continuous functions of the single variables x and y, respectively, then simple combinations such as

$$g(x) \pm h(y), \qquad g(x)h(y), \quad \text{and} \quad \{[g(x)]^2 + [h(y)]^2\}^{1/2}$$

are continuous functions of the two variables x and y.

Now consider a regular partition of $[a, b]$ into n subintervals each of length Δx, and let u_i and v_i denote arbitrary points of the ith subinterval $[x_{i-1}, x_i]$. The following theorem—we omit the proof—tells us how to find the limit as $\Delta x \to 0$ of a sum such as

$$\sum_{i=1}^{n} F(u_i, v_i)\,\Delta x.$$

Theorem 2 *A Further Generalization*

Let $F(x, y)$ be continuous for x and y both in the interval $[a, b]$. Then, in the notation of the preceding paragraph,

$$\lim_{\Delta x \to 0} \sum_{i=1}^{n} F(u_i, v_i)\,\Delta x = \int_a^b F(x, x)\,dx. \tag{5}$$

Note that Theorem 1 is the special case $F(x, y) = g(x)h(y)$ of Theorem 2. Observe also that the integrand $F(x, x)$ on the right in Equation (5) is merely an ordinary function of the (single) variable x. As a formal matter, the integral corresponding to the sum in (5) is obtained by replacing the summation symbol with an integral sign, changing both u_i and v_i to x, replacing Δx by dx, and inserting the correct limits of integration.

For example, if the interval $[a, b]$ is $[0, 4]$, then

$$\begin{aligned}\lim_{\Delta x \to 0} \sum_{i=1}^{n} \sqrt{9u_i^2 + v_i^4}\, \Delta x &= \int_0^4 \sqrt{9x^2 + x^4}\, dx \\ &= \int_0^4 x(9 + x^2)^{1/2}\, dx \\ &= \left[\tfrac{1}{3}(9 + x^2)^{3/2}\right]_0^4 \\ &= \tfrac{1}{3}[(25)^{3/2} - (9)^{3/2}] = \tfrac{98}{3}.\end{aligned}$$

APPENDIX E PROBLEMS

In each of Problems 1–7, u_i and v_i are arbitrary points of the ith subinterval of a regular partition of $[a, b]$ into n subintervals of length Δx each. Express the given limit as an integral from a to b, then compute the value of this integral.

1. $\displaystyle\lim_{\Delta x \to 0} \sum_{i=1}^{n} u_i v_i\, \Delta x;\quad a = 0,\quad b = 1$

2. $\displaystyle\lim_{\Delta x \to 0} \sum_{j=1}^{n} (3u_j + 5v_j)\, \Delta x;\quad a = -1,\quad b = 3$

3. $\displaystyle\lim_{\Delta x \to 0} \sum_{i=1}^{n} u_i\sqrt{4 - v_i^2}\, \Delta x;\quad a = 0,\quad b = 2$

4. $\displaystyle\lim_{\Delta x \to 0} \sum_{i=1}^{n} \frac{u_i\, \Delta x}{\sqrt{16 + v_i^2}};\quad a = 0,\quad b = 3$

5. $\displaystyle\lim_{\Delta x \to 0} \sum_{i=1}^{n} \sin u_i \cos v_i\, \Delta x;\quad a = 0,\quad b = \pi/2$

6. $\displaystyle\lim_{\Delta x \to 0} \sum_{i=1}^{n} \sqrt{\sin^2 u_i + \cos^2 v_i}\, \Delta x;\quad a = 0,\quad b = \pi$

7. $\displaystyle\lim_{\Delta x \to 0} \sum_{k=1}^{n} \sqrt{u_k^4 + v_k^7}\, \Delta x;\quad a = 0,\quad b = 2$

8. Explain how Theorem 1 applies to show that the formula in Equation (8) of Section 6.5 follows from the discussion that precedes it in that section.

9. Use Theorem 1 to derive the formula in Equation (10) of Section 6.4.

Appendix F
Units of Measurement and Conversion Factors

MKS SCIENTIFIC UNITS

Length in meters (m), ***mass*** in kilograms (kg), ***time*** in seconds (s).

Force in newtons (N); a force of 1 N provides an acceleration of 1 m/s^2 to a mass of 1 kg.

Work in joules (J); 1 J is the work done by a force of 1 N acting through a distance of 1 m.

Power in watts (W); a watt is 1 J/s.

BRITISH ENGINEERING UNITS (fps)

Length in feet (ft), ***force*** in pounds (lb), ***time*** in seconds (s).

Mass in slugs; 1 lb of force provides an acceleration of 1 ft/s^2 to a mass of 1 slug. A mass of m slugs at the surface of the earth has a ***weight*** of $w = mg$ pounds, where $g \approx 32.17$ ft/s^2.

Work in ft-lb, ***power*** in ft-lb/s.

CONVERSION FACTORS

1 in. = 2.54 cm = 0.0254 m, 1 m $\approx$ 3.2808 ft

1 mi = 5280 ft; 60 mi/h = 88 ft/s

1 lb $\approx$ 4.4482 N; 1 slug $\approx$ 14.594 kg

1 hp = 550 ft-lb/s $\approx$ 745.7 W

Gravitational acceleration $g \approx 32.17$ ft/s^2 ≈ 9.807 m/s^2

Atmospheric pressure: 1 atm is the pressure exerted by a column of 76 cm of mercury; 1 atm $\approx$ 14.70 lb/in.^{2} $\approx 1.013 \times 10^5$ N/m^2

Heat energy 1 Btu $\approx$ 778 ft-lb $\approx$ 252 cal, 1 cal $\approx$ 4.184 J

Appendix G Formulas from Algebra and Geometry

1. LAWS OF EXPONENTS

$$a^m a^n = a^{m+n}, \qquad (a^m)^n = a^{mn}, \qquad (ab)^n = a^n b^n, \qquad a^{m/n} = \sqrt[n]{a^m};$$

in particular,

$$a^{1/2} = \sqrt{a}.$$

If $a \neq 0$, then

$$a^{m-n} = \frac{a^m}{a^n}, \qquad a^{-n} = \frac{1}{a^n}, \quad \text{and} \quad a^0 = 1.$$

2. QUADRATIC FORMULA

The quadratic equation

$$ax^2 + bx + c = 0 \qquad (a \neq 0)$$

has solutions

$$x = \frac{-b \pm \sqrt{b^2 - 4ac}}{2a}.$$

3. FACTORING

$$\begin{aligned} a^2 - b^2 &= (a - b)(a + b) \\ a^3 - b^3 &= (a - b)(a^2 + ab + b^2) \\ a^4 - b^4 &= (a - b)(a^3 + a^2b + ab^2 + b^3) \\ &= (a - b)(a + b)(a^2 + b^2) \end{aligned}$$

$$a^5 - b^5 = (a - b)(a^4 + a^3b + a^2b^2 + ab^3 + b^4)$$

(The pattern continues.)

$$a^3 + b^3 = (a + b)(a^2 - ab + b^2)$$
$$a^5 + b^5 = (a + b)(a^4 - a^3b + a^2b^2 - ab^3 + b^4)$$

(The pattern continues for odd exponents.)

4. BINOMIAL FORMULA

$$(a + b)^n = a^n + na^{n-1}b + \frac{n(n-1)}{1 \cdot 2} a^{n-2}b^2 + \frac{n(n-1)(n-2)}{1 \cdot 2 \cdot 3} a^{n-3}b^3 + \cdots + nab^{n-1} + b^n$$

if n is a positive integer.

5. AREA AND VOLUME

In the figures below, the symbols have these meanings.

A:	area	b:	length of base	r:	radius
B:	area of base	C:	circumference	V:	volume
h:	height	l:	length	w:	width

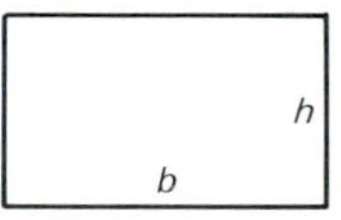

Rectangle

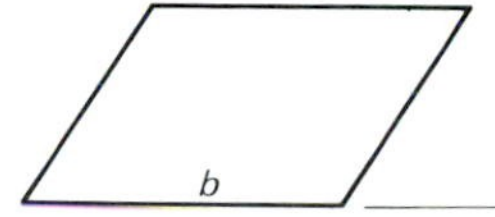

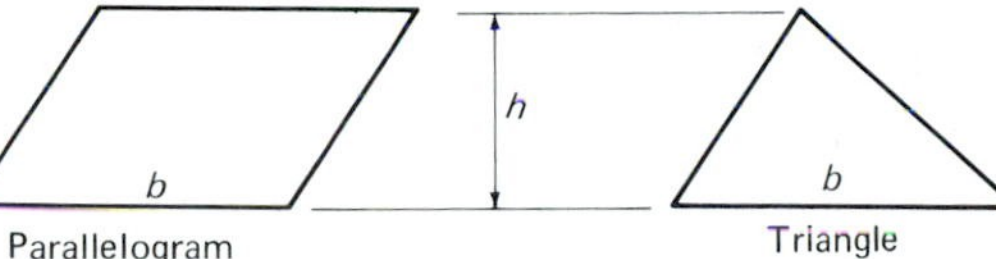

Parallelogram

Triangle

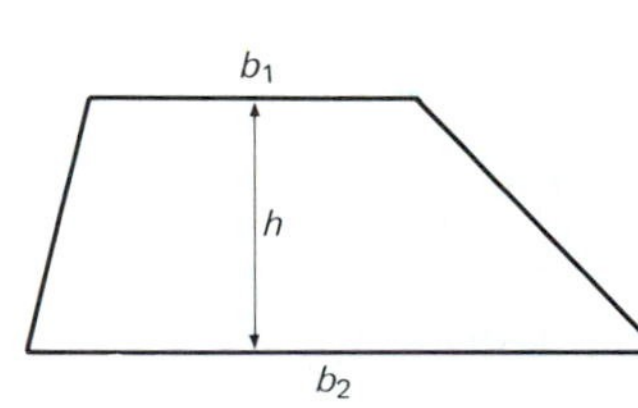

Trapezoid

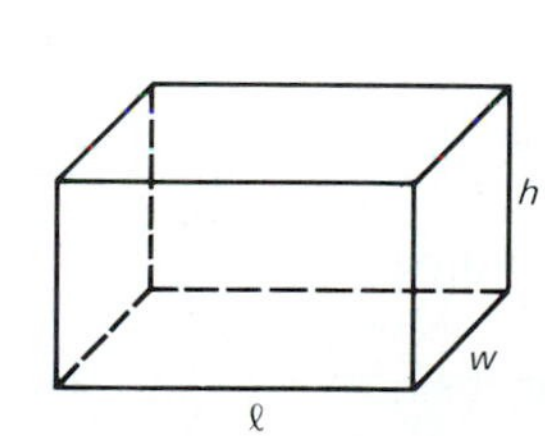

Rectangular parallelepiped

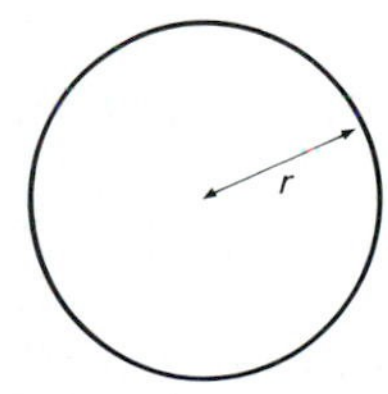

Circle

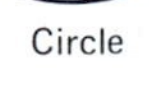

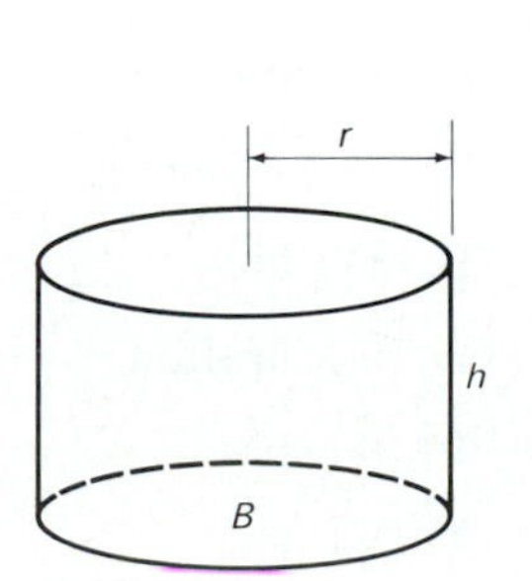

Right circular cylinder

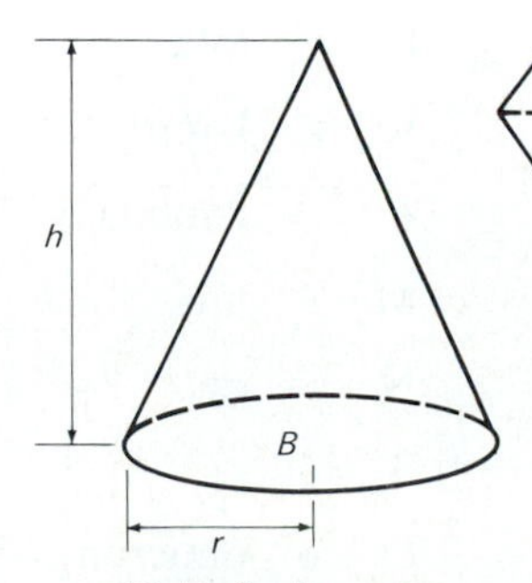

Right circular cone

Pyramid

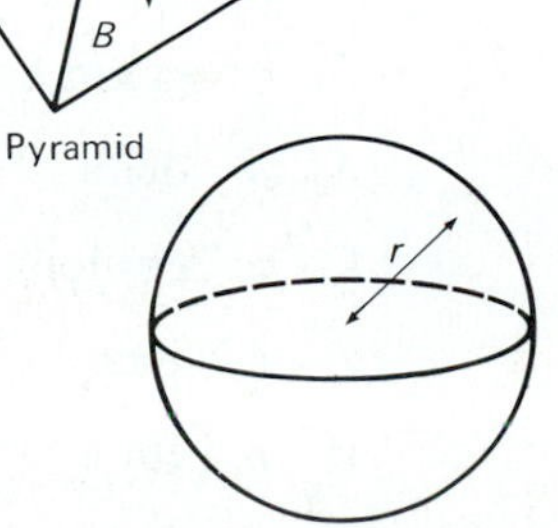

Sphere

Rectangle: $A = bh$
Parallelogram: $A = bh$
Triangle: $A = \frac{1}{2}bh$
Trapezoid: $A = \frac{1}{2}(b_1 + b_2)h$
Circle: $C = 2\pi r$ and $A = \pi r^2$
Rectangular parallelepiped: $V = lwh$

Right circular cylinder: $V = \pi r^2 h = Bh$
Right circular cone: $V = \frac{1}{3}\pi r^2 h = \frac{1}{3}Bh$
Pyramid: $V = \frac{1}{3}Bh$
Sphere: $V = \frac{4}{3}\pi r^3$, $A = 4\pi r^2$

6. PYTHAGOREAN THEOREM

In a right triangle with legs a and b and hypotenuse c,

$$a^2 + b^2 = c^2.$$

Appendix H Formulas from Trigonometry

$$\sin(-\theta) = -\sin\theta, \qquad \cos(-\theta) = \cos\theta$$

$$\sin^2\theta + \cos^2\theta = 1$$

$$\sin 2\theta = 2\sin\theta\cos\theta$$

$$\cos 2\theta = \cos^2\theta - \sin^2\theta$$

$$\sin(\alpha + \beta) = \sin\alpha\cos\beta + \cos\alpha\sin\beta$$

$$\cos(\alpha + \beta) = \cos\alpha\cos\beta - \sin\alpha\sin\beta$$

$$\tan(\alpha + \beta) = \frac{\tan\alpha + \tan\beta}{1 - \tan\alpha\tan\beta}$$

$$\sin^2\frac{\theta}{2} = \frac{1}{2}(1 - \cos\theta)$$

$$\cos^2\frac{\theta}{2} = \frac{1}{2}(1 + \cos\theta)$$

For an arbitrary triangle:

Law of sines $$\frac{\sin A}{a} = \frac{\sin B}{b} = \frac{\sin C}{c}$$

Law of cosines $$c^2 = a^2 + b^2 - 2ab\cos C$$

Appendix I The Greek Alphabet

Α	α	alpha	Ι	ι	iota	Ρ	ρ	rho
Β	β	beta	Κ	κ	kappa	Σ	σ	sigma
Γ	γ	gamma	Λ	λ	lambda	Τ	τ	tau
Δ	δ	delta	Μ	μ	mu	Υ	υ	upsilon
Ε	ε	epsilon	Ν	ν	nu	Φ	ϕ	phi
Ζ	ζ	zeta	Ξ	ξ	xi	X	χ	chi
Η	η	eta	Ο	ο	omicron	Ψ	ψ	psi
Θ	θ	theta	Π	π	pi	Ω	ω	omega

Answers to Odd-Numbered Problems

Section 1.1 (page 10)

1. 14 **3.** $\frac{1}{2}$ **5.** 25 **7.** 27

9. $\frac{22}{7} - \pi \approx 0.00013$ **11.** $(-4, 1)$

13. $(\frac{3}{2}, \frac{11}{2}]$ **15.** $(-1, 4)$

17. $(-\infty, -\frac{1}{3}) \cup (1, +\infty)$ **19.** $[0, 0.4] \cup [1.2, 1.6]$

21. $\dfrac{-1}{a},\quad a,\quad \dfrac{1}{\sqrt{a}},\quad \dfrac{1}{a^2}$

23. $\dfrac{1}{a^2+5},\quad \dfrac{a^2}{5a^2+1},\quad \dfrac{1}{a+5},\quad \dfrac{1}{a^4+5}$

25. $\frac{1}{3}$ **27.** $3, -3$ **29.** 100 **31.** $3h$

33. $2ah + h^2$ **35.** $\dfrac{-h}{a(a+h)}$ **37.** $\{-1, 0, 1\}$

39. $\{-1, 1\}$ **41.** **R** **43.** **R**

45. $x \geqq \frac{5}{3}$ **47.** $x \leqq \frac{1}{2}$ **49.** $x \neq 3$

51. **R** **53.** $0 \leqq x \leqq 16$ **55.** $x \neq 0$

57. $C(A) = 2\sqrt{\pi A},\quad A \geqq 0$

59. $C(F) = \frac{5}{9}(F - 32),\quad F \geqq F_{\min}$

61. $A(x) = x\sqrt{16 - x^2},\quad 0 \leqq x \leqq 4$

63. $C(x) = 3x^2 + \dfrac{1296}{x},\quad x > 0$

65. $A(x) = 2\pi x^2 + \dfrac{2000}{x},\quad x > 0$

67. $V(x) = 4x(25 - x)^2,\quad x \geqq 0$

Section 1.2 (page 18)

1. On a line with slope 1 **3.** Not on a line

9. Slope $\frac{2}{3}$, y-intercept 0

11. Slope 2, y-intercept 3

13. Slope $-\frac{2}{5}$, y-intercept $\frac{3}{5}$

15. $y = -5$ **17.** $y = 2x - 7$

19. $x + y = 6$ **21.** $2x + y = 7$

23. $x + 2y = 13$ **25.** $\frac{4}{13}\sqrt{25}$

33. $K = \dfrac{500F + 229{,}844}{900}$; $F = -459.688$ when $K = 0$.

35. 1136 gal/week

Section 1.3 (page 25)

1. Center (2, 0), radius 2
3. Center (−1, −1), radius 2
5. Center (−0.5, 0.5), radius 2
7. Opens upward, vertex at (3, 0)
9. Opens upward, vertex at (−1, 3)
11. Opens upward, vertex at (−2, 3)
13. Circle, center (3, −4), radius 5
15. There are no points on the graph.
17. The graph is the straight line segment connecting the two points (−1, 7) and (1, −3).
19. Parabola, vertex at (0, 10), opening downward
21.

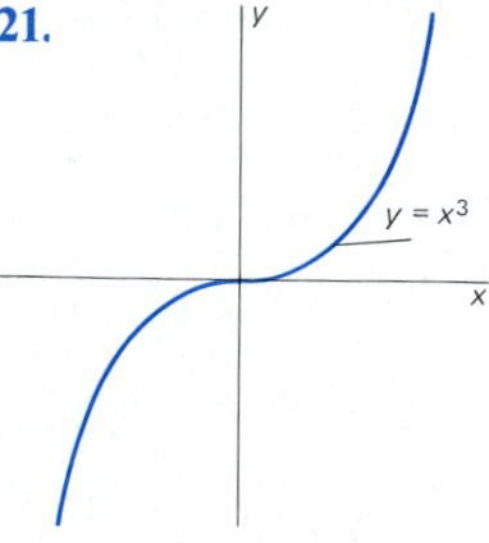

23. Upper half of the circle with center (0, 0) and radius 2
25.

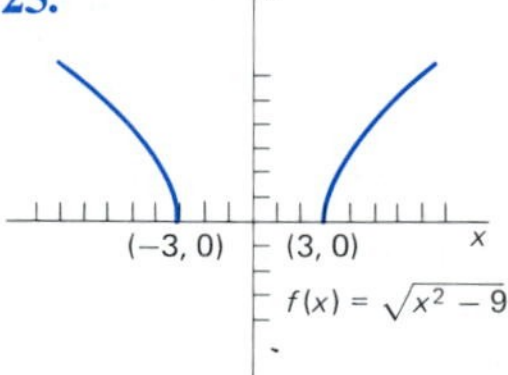

27.

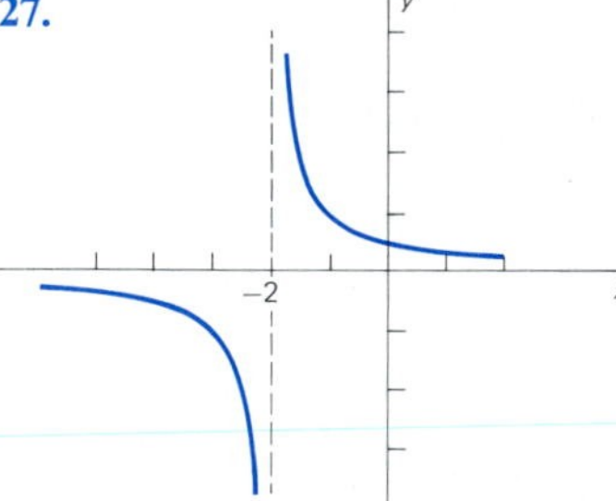

29.

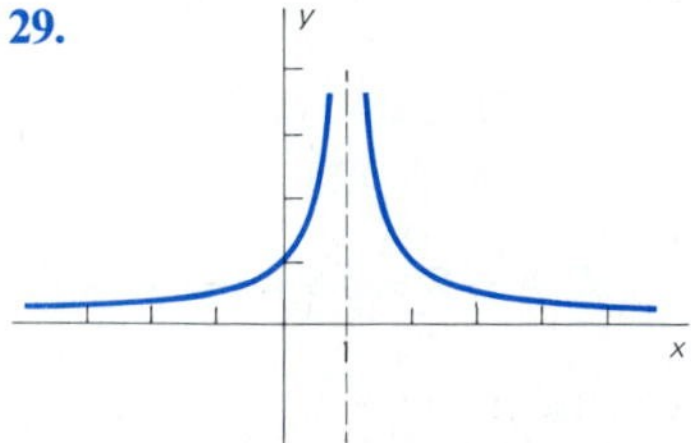

31.

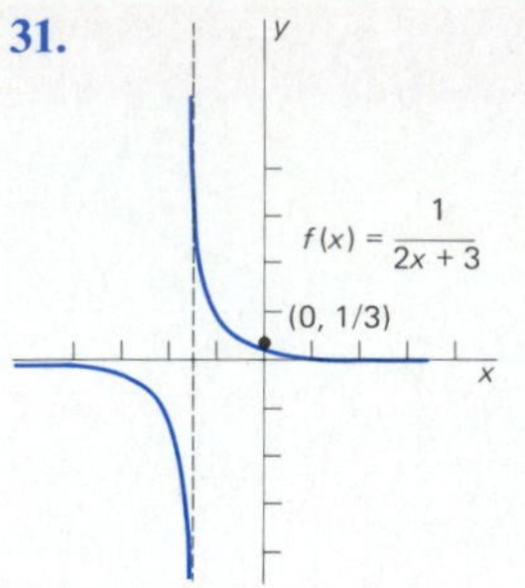

33.

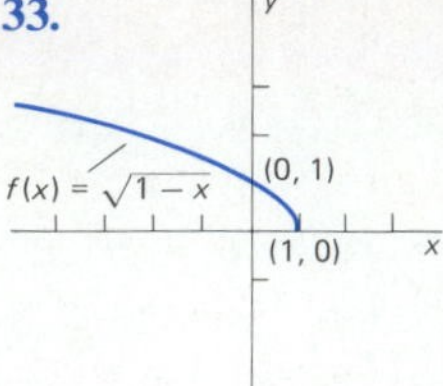

35.

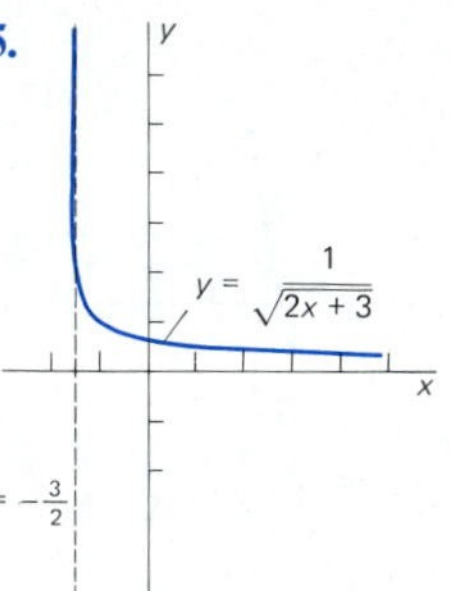

37.

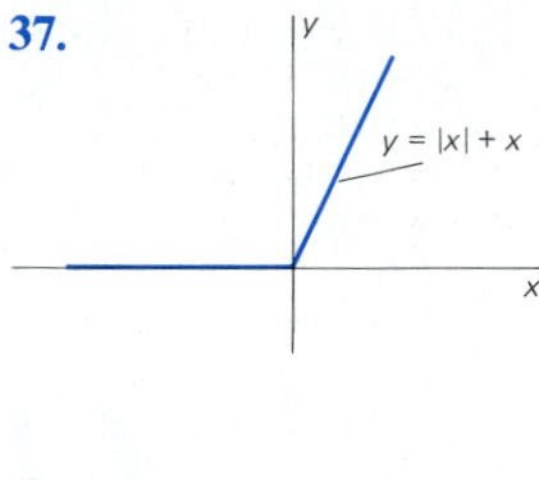

39.

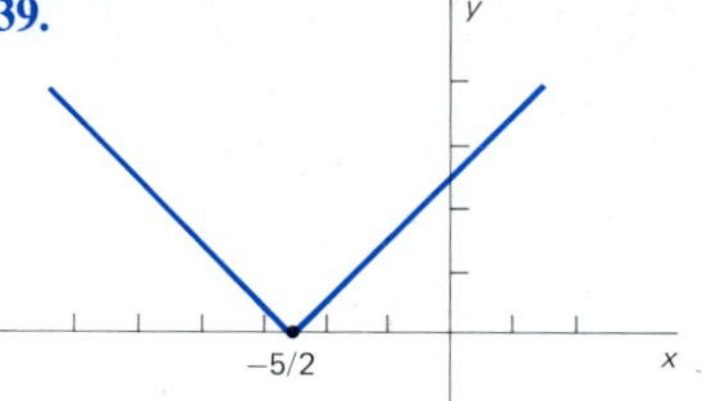

41.

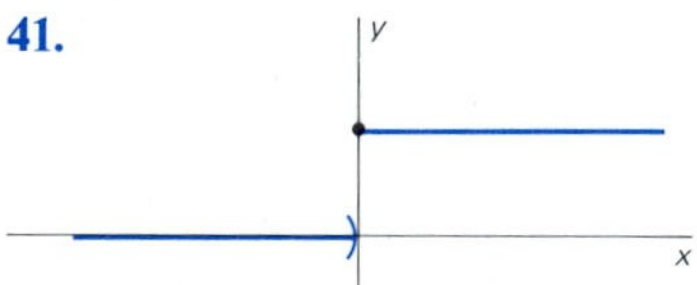

45.

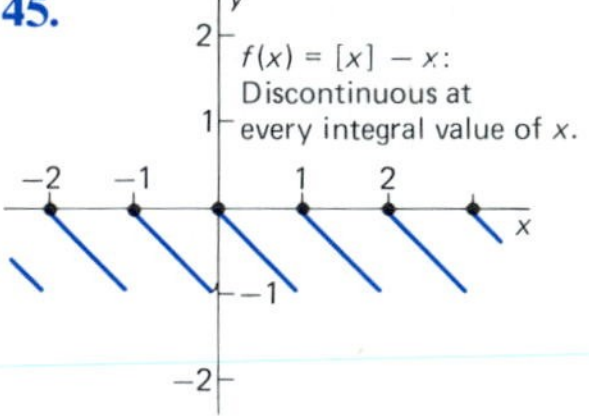

47. 144 ft

49. 625

Chapter 1 Miscellaneous Problems (page 26)

1. $(-\infty, -5)$
3. $(-\frac{15}{7}, \infty)$
5. $(1, 2)$
7. $(-0.5, 2]$
9. $(-\infty, -\frac{1}{8}] \cup [\frac{7}{16}, \infty)$
11. $x \geqq 4$
13. $|x| \neq 3$
15. $x \geqq 0$
17. $x \leqq \frac{2}{3}$
19. **R**
21. $4 \leqq p \leqq 8$
23. $2 < I < 4$
25. $V(S) = \left(\dfrac{S}{6}\right)^{3/2}$
27. $A(P) = \dfrac{P^2\sqrt{3}}{36}$

29. $y = 2x + 11$ **31.** $x - 2y = 10$

33. $x + 2y = 11$

35.

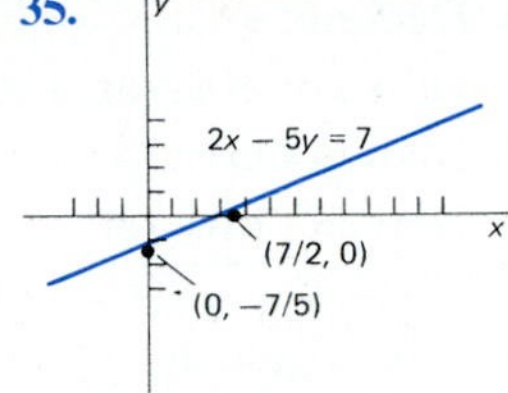

37.

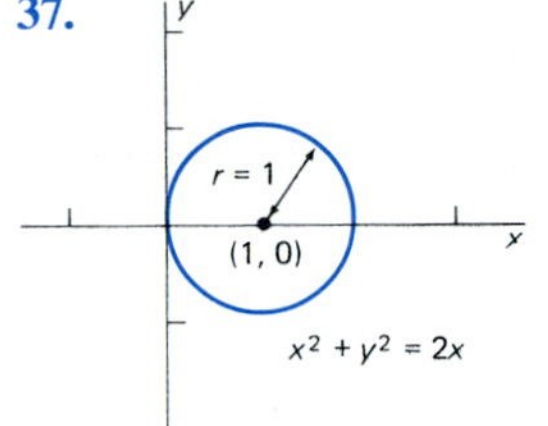

39.

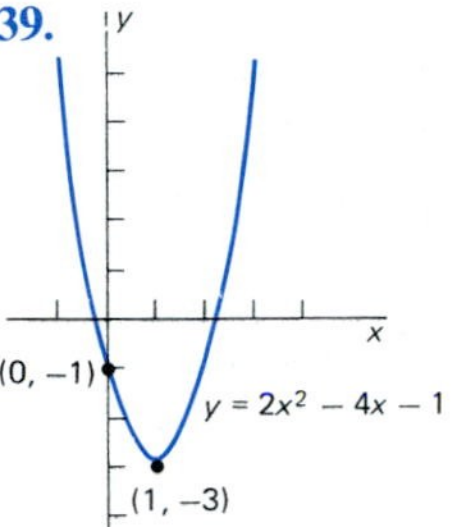

41.

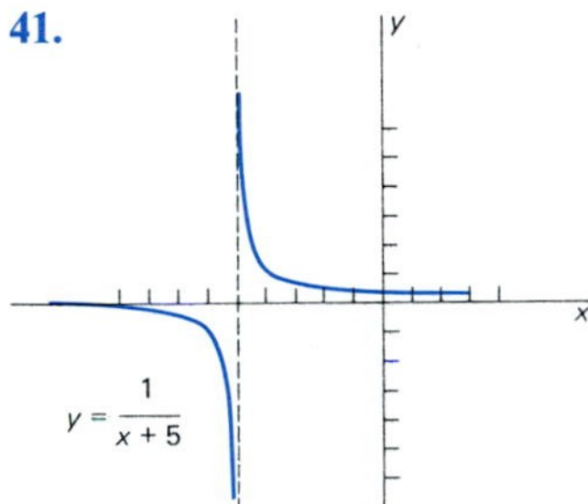

43.

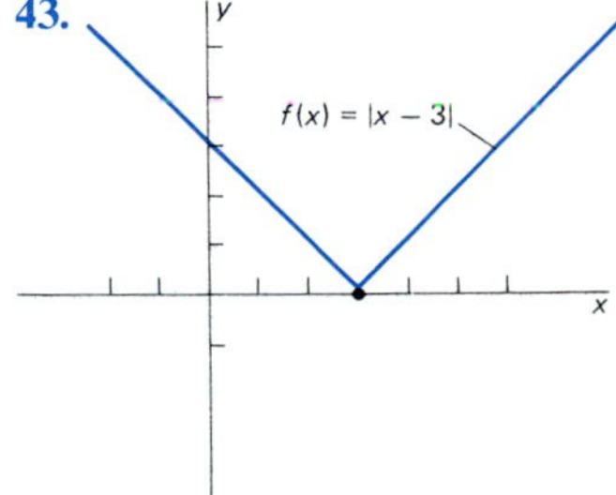

49. $(-\infty, -2) \cup (4, \infty)$

Section 2.1 (page 36)

1. $f'(x) = 0$ **3.** $2x$ **5.** 4

7. $4x - 3$ **9.** $4x + 6$ **11.** $2 - \dfrac{x}{50}$

13. 4 **15.** $(0, 10)$ **17.** $(1, 0)$

19. $(50, 25)$

21. $f'(x) = 3;\quad y - 5 = 3(x - 2)$

23. $f'(x) = 4x - 3;\quad y - 7 = 5(x - 2)$

25. $f'(x) = 2x - 2;\quad y - 1 = 2(x - 2)$

27. $f'(x) = \dfrac{-1}{x^2};\quad x + 4y = 4$

29. $f'(x) = \dfrac{-2}{x^3};\quad x + 4y = 3$

31. $f'(x) = \dfrac{-2}{(x-1)^2};\quad 2x + y = 6$

33. $y - 4 = -4(x + 2);\quad 4(y - 4) = x + 2$

35. $y - 9 = 11(x - 2);\quad -11(y - 9) = x - 2$

37. $y_{max} = 144$ (ft) **39.** 625

41. $y = 12(x - 3)$ **43.** $(1, 1)$

Section 2.2 (page 47)

1. -12 **3.** -1 **5.** 128 **7.** $-\frac{1}{3}$

9. 6 **11.** $16\sqrt{2}$ **13.** 1 **15.** $\frac{1}{4}$

17. $-\frac{1}{54}$ **19.** 2 **21.** 4 **23.** $-\frac{3}{2}$

25. -32 **27.** 1 **29.** 0 **31.** 2

33. $\frac{1}{6}$ **35.** $-\frac{1}{25}$

37. $f'(x) = \dfrac{-1}{2x^{3/2}}$ **39.** $\dfrac{1}{(2x+1)^2}$

41. $\dfrac{x^2 + 2x}{(x+1)^2}$ **43.** $-|x| \leqq g(x) \leqq |x|$

45. $0 \leqq |g(x)| \leqq |\sqrt{x+1} - 1|$ for $-1 < x < 1$

49. If a is not an integer

Section 2.3 (page 54)

1. 3 **3.** Does not exist

5. 0 **7.** 0

9. $+\infty$ (or "does not exist") **11.** -1

13. 1 **15.** -1 **17.** 2 **19.** -1

21. $f(x) \to +\infty$ as $x \to 1^+$, $f(x) \to -\infty$ as $x \to 1^-$.

23. $f(x) \to -\infty$ as $x \to -1^+$, $f(x) \to +\infty$ as $x \to -1^-$.

25. $f(x) \to -\infty$ as $x \to -2^+$, $f(x) \to +\infty$ as $x \to -2^-$.

27. $f(x) \to +\infty$ as $x \to 1$.

29. $f(x) \to -\infty$ as $x \to -2^+$, $f(x) \to +\infty$ as $x \to -2^-$.

31. Both limits are 2.

33. If n is even, then $f(x) \to 1$ as $x \to n^+$ while $f(x) \to -1$ as $x \to n^-$. The situation is reversed for n odd.

35. For each integer n, $f(x) \to 0$ as $x \to n^-$, while $f(x) \to 1$ as $x \to n^+$.

Section 2.4 (page 65)

1. $\{f + g\}(x) = x^2 + 3x - 2$, domain $\mathbf{R}$;
$\{f \cdot g\}(x) = x^3 + 3x^2 - x - 3$, domain $\mathbf{R}$;
$\left\{\dfrac{f}{g}\right\}(x) = \dfrac{x+1}{x^2 + 2x - 3}$, domain $x \neq 1, -3$

3. $\{f+g\}(x)=\sqrt{x}+\sqrt{x-2}$, domain $x \geqq 2$;
$\{f \cdot g\}(x)=(x^2-2x)^{1/2}$, domain $x \geqq 2$;
$\left\{\frac{f}{g}\right\}(x)=\frac{x^{1/2}}{(x-2)^{1/2}}$, domain $x > 2$

5. $\{f+g\}(x)=(x^2-1)^{1/2}+(4-x^2)^{-1/2}$,
$\{f \cdot g\}(x)=\frac{(x^2-1)^{1/2}}{(4-x^2)^{1/2}}$,
$\left\{\frac{f}{g}\right\}(x)=(x^2-1)^{1/2}(4-x^2)^{1/2}$; domain in each case: $(-2,-1] \cup [1,2)$

7. $f(g(x)) \equiv -17$, domain $\mathbf{R}$; $g(f(x)) \equiv 17$, domain $\mathbf{R}$

9. $f(g(x))=1+\frac{1}{(x^2+1)^2}$, domain $\mathbf{R}$;
$g(f(x))=\frac{1}{1+(x^2+1)^2}$, domain $\mathbf{R}$

11. $f(x)=x^2$, $g(x)=2+3x$, etc.

13. $f(x)=x^{1/2}$, $g(x)=2x-x^3$, etc.

15. $f(x)=x^{3/2}$, $g(x)=5-x^2$, etc.

17. $f(x)=x^{-1}$, $g(x)=x+1$, etc.

19. $f(x)=x^{-1/2}$, $g(x)=x+10$, etc.

21. $\mathbf{R}$ 23. $x \neq -3$ 25. $\mathbf{R}$

27. $x \neq 5$ 29. $x \neq 2$ 31. $x \neq 1$

33. $x \neq 0, 1$ 35. $-2 < x < 2$

37. $x=-3$; no

39. $f(x)$ can be made continuous at $x=2$ by defining $f(2)$ to be $\frac{1}{4}$, but f cannot be made continuous at $x=-2$.

41. $x=-1$ and $x=1$; no for both

43. $x=17$; no

45. $f(x)$ can be made continuous at $x=0$ by defining $f(0)$ to be 0.

47. Let $f(x)=x^2-5$. Then f is continuous, $f(2)<0$, and $f(3)>0$.

53. -14, 1, 4, 1, -2, 1, and 16. So there is a root between -3 and -2, between 0 and 1, and between 1 and 2.

55. Begin by letting $f(x)=x^2-a$.

57. At every real number not an integral multiple of 3.

59. Show that f has no limit at the arbitrary real number a.

61. 0.47 (0.4746266176, approximately)

62. 0.75 (0.7548776662, approximately)

Chapter 2 Miscellaneous Problems (page 69)

1. 4 3. 0 5. $-\frac{5}{3}$ 7. -2

9. 0 11. 4 13. 8 15. $\frac{1}{6}$

17. $-\frac{1}{54}$ 19. -1 21. 1

23. Does not exist

25. Does not exist (or $+\infty$)

27. Does not exist (or $+\infty$)

29. Does not exist (or $-\infty$)

31. $f'(x)=4x$; $y-5=4(x-1)$

33. $f'(x)=6x+4$; $y-2=10(x-1)$

35. $f'(x)=4x-1$; $y=3(x-1)$

37. $f'(x)=4x+3$

39. $f'(x)=\frac{1}{(3-x)^2}$

41. $f'(x)=1+\frac{1}{x^2}$

43. $f'(x)=-\frac{2}{(x-1)^2}$

45. $y-14-6\sqrt{4}=(6+2\sqrt{5})(x-3-\sqrt{5})$ and
$y-14+6\sqrt{4}=(6-2\sqrt{5})(x-3+\sqrt{5})$

47. $g(x)=x^2-x$

49. $g(x)=x+1$

51. Either $g(x)=(x^4+1)^{1/2}/2$ or $g(x)=-(x^4+1)^{1/2}/2$

53. f cannot be made continuous at $x=-1$. Define $f(1)$ to be $\frac{1}{2}$ to make f continuous at $x=1$.

55. f cannot be made continuous at $x=-3$. Define $f(1)$ to be $\frac{3}{4}$ to make f continuous at $x=1$.

57. Let $f(x)=x^5+x-1$. Note that $f(0)<0$ and that $f(1)>0$.

59. Let $f(x)=x-\cos x$. Note that $f(0)<0$ and that $f(\pi/2)>0$.

Section 3.1 (page 80)

1. $f'(x)=4$ 3. $25-2z$ 5. $4x+3$

7. $10u+3$ 9. $-10y+17$ 11. $f'(x)=2$

13. $2x$ 15. $-\frac{2}{(2x+1)^2}$

17. $(2x+1)^{-1/2}$ 19. $(1-2x)^{-2}$

21. $s=100$ 23. 99 25. 120

27. $s_{\max}=64$ (ft) 29. 194 ft

31. $\frac{dA}{dC}=\frac{C}{2\pi}$

33. It skids for 10 s, a total distance of 500 ft

35. (a) 2.5 months; (b) 50 rodents per month

37. $v(20) \approx 73$ ft/s, about 50 mi/h; $v(40) \approx 91$ ft/s, about 62 mi/h

41. When $t=30$, $dV/dt=-6.25\pi/3$, so the air is leaking out at approximately 6.545 in.^{3}/s.

43. (a) 144π cm^3/h; (b) 156π cm^3/h

45. $v=0$ (soft touchdown) when $t=2$ (s)

Section 3.2 (page 90)

1. $f'(x) = 6x - 1$
3. $f'(x) = 2(3x - 2) + 3(2x + 3)$
5. $3x^2 + 6x + 3$
7. $(2y - 1)(2y + 1) + 2y(2y - 1)$
9. $(x - 1)^{-2} - (x + 1)^{-2}$
11. $\dfrac{-3(2x + 1)}{(x^2 + x + 1)^2}$
13. $2t(t^3 + t^2 + 1) + (3t^2 + 2t)(t^2 + 1)$
15. $\frac{2}{3}z^{-3} - \frac{1}{2}z^{-2}$
17. $2(3y^2 - 1)(y^2 + 2y + 3) + (6y)(2y)(y^2 + 2y + 3) + (2y + 2)(2y)(3y^2 - 1)$
19. $\dfrac{(t^2 + 2t + 1) - (t - 1)(2t + 2)}{(t^2 + 2t + 1)^2}$
21. $v'(t) = -3(t - 1)^{-4}$
23. $\dfrac{3(x^3 + 7x - 5) - (3x)(3x^2 + 7)}{(x^3 + 7x - 5)^2}$
25. $\dfrac{(2x - 3)(3x^2 - 4x) - 2(x^3 - 2x^2)}{(2x - 3)^2}$
27. $3x^2 - 30x^4 - 6x^{-5}$
29. $2x + 4x^{-2} - 15x^{-4}$
31. $3 + \frac{1}{2}x^{-3}$
33. $-(x - 1)^{-2} - \frac{1}{3}x^{-2}$
35. $\dfrac{x^4 + 31x^2 - 10x - 36}{(x^2 + 9)^2}$
37. $\dfrac{150x^{10} - 240x^5}{(15x^2 - 4)^2}$
39. $\dfrac{x^2 + 2x}{(x + 1)^2}$
41. $12x - y = 16$
43. $x + y = 3$
45. $5x - y = 10$
47. $y - 18x = 25$
49. $3x + y = 0$
51. (a) It contracts; (b) 0.06427 $cm^3/°C$
53. $14{,}400\pi \approx 45.239$ cm^3/cm
55. $y = 3x + 2$
57. Suppose that such a line L is tangent at the points (a, a^2) and (b, b^2). Show that $a = b$ is a consequence.
59. $\left(\dfrac{n - 1}{n}\right) x_0$ for $x_0 \neq 0$
65. $17(x^3 - 17x + 35)^{16}(3x^2 - 17)$

Section 3.3 (page 98)

1. $\dfrac{dy}{dx} = 15(3x + 4)^4$
3. $-3(3x - 2)^{-2}$
5. $3(x^2 + 3x + 4)^2(2x + 3)$
7. $-4(2 - x)^3(3 + x)^7 + 7(2 - x)^4(3 + x)^6$
9. $-\dfrac{6x + 22}{(3x - 4)^4}$
11. $12[1 + (1 + x)^3]^3(1 + x)^2$
13. $\dfrac{dy}{dx} = -x^{-2}(x^2 + 1)^{-1/2}$
15. $24[1 + (4x - 1)^4](4x - 1)$
17. $-4x^{-5}[(1 - x^{-4})^3 - 3x^{-4}(1 - x^{-4})^2]$
19. $-2x^{-3}[2x^{-2}(x^{-2} - x^{-8})^3 + 3x^{-4}(x^{-2} - x^{-8})^2(1 - 4x^{-6})]$
21. $u = 2x - x^2$, $n = 3$; $f'(x) = 3(2x - x^2)^2(2 - 2x)$
23. $u = 1 - x^2$, $n = -4$; $f'(x) = -4(1 - x^2)^{-5}(-2x)$
25. $u = \dfrac{x + 1}{x - 1}$, $n = 7$;
$$f'(x) = -\frac{14(x + 1)^6}{(x - 1)^8}$$
27. $g'(y) = 1 + 10(2y - 3)^4$
29. $F'(s) = 3(s - s^{-2})^2(1 + 2s^{-3})$
31. $f'(u) = 3(1 + u)^2(1 + u^2)^4 + 8u(1 + u)^3(1 + u^2)^3$
33. $h'(v) = -2[v - (1 - v^{-1})]^{-3}[1 + v^{-2}(1 - v^{-1})^{-2}]$
35. $F'(z) = -10(3 - 4z + 5z^5)^{-11}(10z - 4)$
37. $\dfrac{dy}{dx} = 4(x^3)^3(3x^2) = 12x^{11}$
39. $\dfrac{dy}{dx} = 2(x^2 - 1)(2x) = 4x^3 - 4x$
41. $\dfrac{dy}{dx} = 4(x + 1)^3 = 4x^3 + 12x^2 + 12x + 4$
43. $\dfrac{dy}{dx} = -2x(x^2 + 1)^{-2} = -\dfrac{2x}{(x^2 + 1)^2}$
45. $f'(x) = 3x^2 \cos x^3$
47. $g'(z) = 3(\sin 2z)^2(2 \cos 2z)$
49. 40π $in.^2/s$
51. 40 $in.^2/s$
53. 600 $in.^3/h$
55. -18
57. $400\pi \approx 1256.64$ cm^3/s
59. 5 cm
61. Total melting time: $2/(2 - 4^{1/3}) \approx 4.85$ h; completely melted at about 2:50:50 P.M. on the same day.
63. Note that v is also a function of x. Thus $du/dx = (du/dv)(dv/dx)$. But $dv/dx = (dv/dw)(dw/dx)$ by the chain rule.

Section 3.4 (page 105)

1. $f'(x) = \frac{3}{2}x^2(x^3 + 1)^{-1/2}$
3. $f'(x) = 2x(2x^2 + 1)^{-1/2}$
5. $f'(t) = \frac{3}{2}(2t)^{1/2}$
7. $f'(x) = \frac{3}{2}(2x^2 - x + 7)^{1/2}(4x - 1)$
9. $g'(x) = -\frac{4}{3}(1 - 6x^2)(x - 2x^3)^{-7/3}$
11. $f'(x) = (1 - 2x^2)(1 - x^2)^{-1/2}$
13. $f'(t) = -2t(t^2 + 1)^{1/2}(t^2 - 1)^{-3/2}$
15. $f'(x) = 3x^{-4}(x^2 + 1)(x^2 - 1)^2$
17. $f'(v) = -\frac{1}{2}v^{-2}(v + 2)(v + 1)^{-1/2}$

19. $f'(x) = -\frac{2}{3}x(1 - x^2)^{-2/3}$

21. $-2x(3 - 4x)^{-1/2} + (3 - 4x)^{1/2}$

23. $-2x(2x + 4)^{4/3} + \frac{8}{3}(1 - x^2)(2x + 4)^{1/3}$

25. $-2t^{-2}(1 + t^{-1})(3t^2 + 1)^{1/2} + 3t(1 + t^{-1})^2(3t^2 + 1)^{-1/2}$

27. $[2(3x + 4)^5 - 15(3x + 4)^4(2x - 1)](3x + 4)^{-10}$

29. $[(3x + 4)^{1/3}(2x - 1)^{-1/2} - (3x + 4)^{-2/3}(2x + 1)^{1/2}](3x + 4)^{-2/3}$

31. $\dfrac{3y^{2/3}[(1 + y)^{-1/2} - (1 - y)^{-1/2}] - 10[(1 + y)^{1/2} + (1 - y)^{1/2}]}{y^{8/3}}$

33. $\frac{1}{2}[t + (t + t^{1/2})^{1/2}]^{-1/2}[1 + (t + t^{1/2})^{-1/2}(1 + \frac{1}{2}t^{-1/2})]$

35. No horizontal tangent line; vertical tangent line at $(0, 0)$

37. Horizontal tangent line where $x = \frac{1}{3}$, vertical tangent line at $(0, 0)$

39. No horizontal or vertical tangent lines

41. $\pi^2/32 \approx 0.3084$ s/ft

43. $(2/\sqrt{5}, 1/\sqrt{5})$ and $(-2/\sqrt{5}, -1/\sqrt{5})$

45. $a = 2$ is the only real root of the cubic equation. Answer: $x + 4y = 18$.

47. $3x + 2y = 5$ and $3x - 2y = -5$

49. It is permissible to differentiate both sides of the *identity* $f(x) = g(x)$ because the graphs of f and g coincide.

Section 3.5 (page 111)

1. No minimum; maximum: 2
3. No maximum; minimum: 0
5. Maximum: 2; minimum: 0
7. Maximum: 2; minimum: 0
9. Maximum: 4; minimum: $-\frac{1}{6}$
11. -4 and 11
13. -5 and 3
15. 0 and 9
17. -2 and 52
19. 4 and 5
21. Minimum: 1; maximum: 5
23. Minimum: -16; maximum: 9
25. Minimum: -22; maximum: 10
27. Minimum: -56; maximum: 56
29. Minimum: 5; maximum: 13
31. Minimum: 0; maximum: 17
33. Minimum: 0; maximum: $\frac{3}{4}$
35. Minimum: $-\frac{1}{6}$ at $x = 3$; maximum: $\frac{1}{2}$ at $x = -1$
37. Maximum value: $f(1/\sqrt{2}) = \frac{1}{2}$; minimum value: $f(-1/\sqrt{2}) = -\frac{1}{2}$
39. Maximum value: $f(\frac{3}{2}) = 3/(2^{4/3})$; minimum value: $f(3) = -3$

Section 3.6 (page 121)

1. Each is 25
3. 1250
5. 500 in.^{3}
7. 1152
9. 250
11. 11,250 yd^2
13. 128
15. Approximately 3.9665°C
17. 1000 cm^3
19. 0.25 m^3 (all cubes, no open-top boxes)
21. Two equal pieces yield minimum total area 200 in.^{2}; no cut yields a single square of maximum area 400 in.^{2}
23. 30,000 m^2
25. Approximately 9259.26 in.^{3}
27. Five presses
29. The minimizing value of x is $-2 + \frac{10}{3}\sqrt{6}$ in. To the nearest integer, we use $x = 6$ in. of insulation for an annual saving of \$285.
31. Charge either \$1.10 or \$1.15 for the largest revenue.
33. Radius $\frac{2}{3}R$, height $\frac{1}{3}H$
37. $\dfrac{2000\pi\sqrt{3}}{27}$
39. Maximum 4, minimum $16^{1/3}$
41. $\frac{1}{2}\sqrt{3}$
43. Each plank is $\frac{1}{8}(\sqrt{34} - 3\sqrt{2}) \approx 0.19854$ by $\frac{1}{2}(7 - \sqrt{17})^{1/2} \approx 1.69614$
45. Strike the shore $\frac{2}{3}\sqrt{3} \approx 1.1547$ km from the point nearest the island.
47. $\frac{1}{3}\sqrt{3}$
49. The actual value of x is approximately 3.45246.
51. $8/\sqrt{5}$ ft and $10/\sqrt{5}$ ft

Section 3.7 (page 133)

1. $6 \sin x \cos x$
3. $\cos x - x \sin x$
5. $x^{-1}\cos x - x^{-2}\sin x$
7. $\cos^3 x - 2 \cos x \sin^2 x$
9. $4(1 + \sin t)^3 \cos t$
11. $(\sin t + \cos t)^{-2}(\sin t - \cos t)$
13. $2 \sin x - 4x \cos x + 3x^2 \sin x$
15. $3 \cos 2x \cos 3x - 2 \sin 2x \sin 3x$
17. $3t^2 \sin^2 2t + 4t^3 \sin 2t \cos 2t$
19. $-\frac{5}{2}(3 \sin 3t + 5 \sin 5t)(\cos 3t + \cos 5t)^{3/2}$
21. $\dfrac{dy}{dx} = x^{-1/2} \sin x^{1/2} \cos x^{1/2}$

23. $\dfrac{dy}{dx} = 2x\cos(3x^2 - 1) - 6x^3\sin(3x^2 - 1)$

25. $2\cos 2x \cos 3x - 3\sin 2x \sin 3x$

27. $-\dfrac{3\sin 5x \sin 3x + 5\cos 5x \cos 3x}{(\sin 5x)^2}$

29. $4x\sin x^2 \cos x^2$

31. $x^{-1/2}\cos 2x^{1/2}$

33. $\sin x^2 + 2x^2\cos x^2$

35. $\frac{1}{2}(x^{-1/2}\sin x^{1/2} + \cos x^{1/2})$

37. $\frac{1}{2}x^{-1/2}(x - \cos x)^3 + 3x^{1/2}(x - \cos x)^2(1 + \sin x)$

39. $-[\sin(\sin x^2)](2x\cos x^2)$

41. 0

43. 0.5

45. $-\infty$ (or "does not exist")

47. 5

49. Does not exist

51. $\frac{1}{3}$

53. 0

55. 1

57. 0.5

59. 1

63. $\alpha = \pi/4$

65. $\frac{1}{18}\pi \sec^2(5\pi/18) \approx 0.4224$ mi/s

67. $2000\pi/27$ ft/s; that is, about 158.67 mi/h

69. $\theta = \pi/3$

71. $V = \frac{8}{3}\pi R^3$

73. $A = \frac{3}{4}\sqrt{3} \approx 1.299$

75. *Suggestion:* $A(\theta) = \dfrac{s^2(\theta - \sin\theta)}{(2\theta)^2}$

Section 3.8 (page 141)

1. $\dfrac{x}{y}$

3. $\dfrac{-16x}{25y}$

5. $-\sqrt{\dfrac{y}{x}}$

7. $-\left(\dfrac{y}{x}\right)^{1/3}$

9. $\dfrac{dy}{dx} = \dfrac{3x^2 - 2xy - y^2}{3y^2 + 2xy + x^2}$

11. $\dfrac{dy}{dx} = -\dfrac{x}{y}$; $3x - 4y = 25$

13. $\dfrac{dy}{dx} = \dfrac{1 - 2xy}{x^2}$; $3x + 4y = 10$

15. $\dfrac{dy}{dx} = -\dfrac{2xy + y^2}{2xy + x^2}$; $y = -2$

17. $\dfrac{dy}{dx} = \dfrac{25y - 24x}{24y - 25x}$; $3y = 4x$

19. $\dfrac{dy}{dx} = -\left(\dfrac{y}{x}\right)^4$; $x + y = 2$

21. $\dfrac{dy}{dx} = \dfrac{5x^4y - y^2}{3xy - 2x^5}$; slope $\frac{1}{2}$

23. None (dy/dx is undefined at (1, 1); see Problem 63)

25. $(2, 2 + 2\sqrt{2})$ and $(2, 2 - 2\sqrt{2})$

27. $2y = x + 6$, $2y = x - 6$

29. Horizontal tangents at the four points where $x^2 = \frac{3}{8}$ and $y^2 = \frac{1}{8}$, vertical tangents at $(-1, 0)$ and $(1, 0)$.

31. $\dfrac{4}{5\pi} \approx 0.25465$ ft/s

33. $\dfrac{32\pi}{125} \approx 0.80425$ mi/h

35. 20 cm^2/s

37. 0.25 cm/s

39. 6 ft/s

41. 384 mi/h

43. (a) about 0.047 ft/min; (b) about 0.083 ft/min

45. $\frac{400}{9}$ ft/s

47. Increasing at 16π cm^3/s

49. 6000 mi/h

51. (a) $\frac{11}{15}\sqrt{21} \approx 3.36$ ft/s downward; (b) slightly less than 2112 ft/s downward

53. They are closest at $t = 12$ min, and the distance between them is $32\sqrt{13} \approx 115.38$ mi.

55. $-\dfrac{50}{81\pi} \approx -0.1965$ ft/s

57. $\dfrac{10}{81\pi} \approx 0.0393$ in./min

59. $300\sqrt{2} \approx 424.26$ mi/h

61. $\frac{1}{30}$ ft/s

Section 3.9 (page 151)

1. 2.2361

3. 2.5119

5. 0.3028

7. -0.7402

9. 0.7391

11. 1.2361

13. 2.3393

15. 2.0288

17. 2.154435

19. 1.802191

21. (b) 1.25992

23. 0.45018

27. 0.754877666

29. -1.8955, 0, and 1.8955

31. -1.3578, 0.7147, and 1.2570 are the three real solutions.

33. 0.8655

35. 3.4525

37. 0.2261

39. $\alpha_1 \approx 2.0287578 \approx \left(\dfrac{1.29}{2}\right)\pi$,

$\alpha_2 \approx 4.9131804 \approx \left(\dfrac{3.13}{2}\right)\pi$,

$\alpha_3 \approx 7.9786657 \approx \left(\dfrac{5.08}{2}\right)\pi$,

$\alpha_4 \approx 11.085538 \approx \left(\dfrac{7.06}{2}\right)\pi$

Chapter 3 Miscellaneous Problems (page 155)

1. $2x - 6x^{-3}$

3. $\frac{1}{2}x^{-1/2} - \frac{1}{3}x^{-4/3}$

5. $7(x - 1)^6(3x + 2)^9 + 27(x - 1)^7(3x + 2)^8$

7. $4(3x - \frac{1}{2}x^{-2})^3(3 + x^{-3})$

9. $\dfrac{dy}{dx} = -\dfrac{y}{x} = -\dfrac{9}{x^2}$

11. $-\frac{3}{2}(x^3 - x)^{-5/2}(3x^2 - 1)$

13. $\dfrac{4x(1 + x^2)}{(x^4 + 2x^2 + 2)^2}$

15. $\frac{7}{3}[x^{1/2} + (2x)^{1/3}]^{4/3}[\frac{1}{2}x^{-1/2} + \frac{2}{3}(2x)^{-2/3}]$

17. $-(x+1)^{-1/2}[(x+1)^{1/2} - 1]^2$

19. $\dfrac{dy}{dx} = \dfrac{-2xy^2 + 1}{2x^2y - 1}$

21. $\frac{1}{2}\{x + [2x + (3x)^{1/2}]^{1/2}\}^{-1/2} \cdot \{1 + \frac{1}{2}[2x + (3x)^{1/2}]^{-1/2}[2 + \frac{3}{2}(3x)^{-1/2}]\}$

23. $-\left(\dfrac{y}{x}\right)^{2/3}$

25. $-18[1 + 2(1+x)^{-3}](1+x)^{-4}$

27. $\dfrac{\frac{1}{2}[(1+\cos x)/\sin^2 x]^{1/2}[2(1+\cos x)\sin x \cos x + \sin^3 x]}{(1+\cos x)^2}$

29. $-\frac{1}{2}(4 \sin 2x \sin 3x + 3 \cos 2x \cos 3x)(\sin 3x)^{-3/2}$

31. $(6 \sin^2 2x \cos 3x)(\cos 2x \cos 3x - \sin 2x \sin 3x)$

33. $[4 \sin^4(x + x^{-1}) \cos(x + x^{-1})](1 - x^{-2})$

35. $[\cos^2(x^4 + 1)^{1/3}][-\sin(x^4+1)^{1/3}] \cdot (x^4+1)^{-2/3}(4x^3)$

37. $x = 1$ 39. $x = 0$ 41. 0.5 ft/min

43. $\frac{1}{3}$ 45. $\frac{1}{4}$ 47. 0

49. $-x(x^2+25)^{-3/2}$ 51. $\frac{5}{3}(x-1)^{2/3}$

53. $-2x \sin(x^2+1)$ 55. $\dfrac{dV}{dS} = \dfrac{1}{4}\sqrt{\dfrac{S}{\pi}}$

57. $\frac{1}{18}\pi \sec^2 50°$, approximately 0.4224 mi/s

59. The maximum area is R^2.

61. Minimum area: $(36\pi V^2)^{1/3}$, obtained by making *one* sphere of radius $(3V/4\pi)^{1/3}$.
Maximum area: $(8\pi)^{1/3}(3V)^{2/3}$, obtained by making two equal spheres, each of radius $\frac{1}{2}(3V/\pi)^{1/3}$.

63. $\frac{32}{81}\pi R^3$ 65. $\dfrac{M}{2}$

67. 36 ft^3 69. $3\sqrt{3}$

73. 2 mi from the point on the shore nearest the first town

75. (a) The maximum height is $m^2v^2/[64(m^2+1)]$.
(b) The maximum range occurs when $m = 1$, thus when $\alpha = \pi/4$.

77. 2.6458 79. 2.3714 81. −0.3473

83. 0.7402 85. −0.7391 87. −1.2361

89. Approximately 1.54785 ft

91. To five places, the three real solutions are −2.72249, 0.80126, and 2.30998.

97. 4 in.^{2}/s

99. $-\frac{50}{9}\pi \approx -1.7684$ ft/min

101. 1 in./min

Section 4.2 (page 166)

1. $(6x + 8x^{-3})\,dx$

3. $[1 + \frac{3}{2}x^2(4 - x^3)^{-1/2}]\,dx$

5. $[6x(x-3)^{3/2} + \frac{9}{2}x^2(x-3)^{1/2}]\,dx$

7. $[(x^2+25)^{1/4} + \frac{1}{2}x^2(x^2+25)^{-3/4}]\,dx$

9. $-\frac{1}{2}x^{-1/2}\sin(x^{1/2})\,dx$

11. $(2\cos^2 2x - 2\sin^2 2x)\,dx$

13. $(\frac{2}{3}x^{-1}\cos 2x - \frac{1}{3}x^{-2}\sin 2x)\,dx$

15. $(\sin x + x \cos x)(1 - x \sin x)^{-2}\,dx$

17. $f(x) \approx 1 + x$ 19. $f(x) \approx 1 + 2x$

21. $f(x) \approx 1 - 3x$ 23. $f(x) \approx x$

25. $3 - \frac{2}{27} \approx 2.926$ 27. $2 - \frac{1}{32} \approx 1.969$

29. $\frac{95}{1536} \approx 0.06185$ 31. $\dfrac{1 + \pi/90}{\sqrt{2}} \approx 0.7318$

33. $\sin\dfrac{\pi}{2} - \dfrac{\pi}{90}\cos\dfrac{\pi}{2} = 1.000$

35. $\dfrac{dy}{dx} = -\dfrac{x}{y}$ 37. $\dfrac{y - x^2}{y^2 - x}$

41. −4 in.^{2} 43. $-405\pi/2$ cm^3

45. $\frac{1}{2}\sqrt{50}$ (about 3.54 ft) 47. 6 W

49. $25\pi \approx 78.54$ in.^{3} 51. $4\pi \approx 12.57$ m^2

Section 4.3 (page 173)

1. Increasing on the whole real line

3. Increasing for $x < 0$, decreasing for $x > 0$

5. Increasing for $x < 1.5$, decreasing for $x > 1.5$

7. Increasing for $-1 < x < 0$ and for $x > 1$, decreasing for $x < -1$ and for $0 < x < 1$

9. Increasing for $-2 < x < 0$ and for $x > 1$, decreasing for $x < -2$ and for $0 < x < 1$

11. Increasing for $x < 2$, decreasing for $x > 2$

13. Increasing for $x < -\sqrt{3}$, for $-\sqrt{3} < x < 1$, and for $x > 3$; decreasing for $1 < x < \sqrt{3}$ and for $\sqrt{3} < x < 3$

15. $f(0) = 0 = f(2)$, $f'(x) = 2x - 2$, $c = 1$

17. $f(-1) = 0 = f(1)$, $f'(x) = \dfrac{-4x}{(1+x^2)^2}$, $c = 0$

19. $f'(0)$ does not exist. 21. $f(0) \neq f(1)$

23. $c = -\frac{1}{2}$ 25. $c = \frac{35}{27}$

27. The average slope is $\frac{1}{3}$, but $|f'(x)| = 1$ where $f'(x)$ exists.

29. The average slope is 1, but $f'(x) = 1$ wherever it exists.

31. If $g(x) = x^5 + 2x - 3$, then $g'(x) > 0$ for all x in $[0, 1]$ and $g(1) = 0$. So $x = 1$ is the only root of the equation in the given interval.

33. If $g(x) = x^4 - 3x - 20$, then $g(2) = -10$ and $g(3) = 52$. Because

$$g'(x) = 4x^3 - 3 \geqq 4(2^3 - 3) = 29 > 0,$$

g is an increasing function and thus has exactly one zero in the interval $[2, 3]$.

35. Note that $f'(x) = \frac{3}{2}(-1 + \sqrt{x+1})$.

37. Assume that $f'(x)$ has the form

$$a_0 + a_1x + \cdots + a_{n-1}x^{n-1}.$$

Construct a polynomial $p(x)$ such that $p'(x) = f'(x)$. Conclude that $f(x) = p(x) + C$ on the interval $[a, b]$.

Section 4.4 (page 179)

1. Local minimum at $x = 2$

3. Local maximum at $x = 0$, local minimum at $x = 2$

5. No extremum at $x = 1$

7. Local minimum at $x = -2$, local maximum at $x = 5$

9. Local minima where $|x| = 1$, local maximum at $x = 0$

11. Local maximum at $x = -1$, local minimum at $x = 1$

13. Local minimum at $x = 1$

15. Local maximum at $x = 0$

17. -10 and 10 **19.** $(1, 1)$

21. 9 in. wide, 8 in. long, 6 in. high

23. Radius $5\pi^{-1/3}$ in., height $10\pi^{-1/3}$ in.

27. Base 5 in. by 5 in., height 2.5 in.

29. Radius $(25/\pi)^{1/3} \approx 1.9965$ in., height 4 times that radius

31. The nearest points are

$$(\tfrac{1}{2}\sqrt{6}, \tfrac{3}{2}) \quad \text{and} \quad (-\tfrac{1}{2}\sqrt{6}, \tfrac{3}{2});$$

(0, 0) is *not* the nearest point.

33. 8 in.

35. $L = (20 + 12\sqrt[3]{4} + 24\sqrt[3]{2})^{1/2} \approx 8.324$ m

39. Height $(6V)^{1/3}$, base edge $(\tfrac{9}{2}V^2)^{1/6}$.

Section 4.5 (page 186)

1. Parabola, opening upward, global minimum at (1, 2)

3. Increasing for $x < -2$ and for $x > 2$, decreasing for $-2 < x < 2$; local maximum at $(-2, 16)$, local minimum at $(2, -16)$

5. Increasing for $x < 1$ and for $x > 3$, decreasing for $1 < x < 3$, local maximum at (1, 4), local minimum at (3, 0)

7. Increasing for all x, no extrema

9. Decreasing for $x < -2$ and for $-0.5 < x < 1$, increasing for $-2 < x < -0.5$ and for $x > 1$, global minima at $(-2, 0)$ and at (1, 0), local maximum at $(-0.5, 5.0625)$

11. Increasing for $0 < x < 1$, decreasing for $x > 1$, global maximum at (1, 2), no graph for $x < 0$

13. Increasing for $|x| > 1$, decreasing for $|x| < 1$, but with a horizontal tangent at (0, 0); local maximum at $(-1, 2)$, local minimum at $(1, -2)$

15. Decreasing for $x < -2$ and for $0 < x < 2$, increasing for $-2 < x < 0$ and for $x > 2$, global minima at $(-2, -9)$ and at $(2, -9)$, local maximum at (0, 7)

17. Decreasing for $x < 0.75$, increasing for $x > 0.75$, global minimum at $(0.75, -10.125)$

19. Decreasing for $-2 < x < 1$, increasing for $x < -2$ and for $x > 1$, local maximum at $(-2, 20)$, local minimum at $(1, -7)$

21. Local maximum at (0.6, 16.2), local minimum at (0.8, 16.0); increasing for $x < 0.6$ and for $x > 0.8$, decreasing on (0.6, 0.8)

23. Increasing for $-1 < x < 0$ and for $x > 2$, decreasing for $x < -1$ and for $0 < x < 2$; local maximum at (0, 8), local minimum at $(-1, 3)$, global minimum at $(2, -24)$

25. Increasing for $|x| > 2$, decreasing for $|x| < 2$, local maximum at $(-2, 64)$, local minimum at $(2, -64)$

27. Increasing everywhere, no extrema, graph passes through the origin, minimum slope $\tfrac{9}{2}$ occurs where $x = -\tfrac{1}{2}$

29. Global maximum 16 occurs at both points where $x^2 = 2$, local minimum at (0, 0), increasing for $x < -\sqrt{2}$ and for $0 < x < \sqrt{2}$, decreasing for $x > \sqrt{2}$ and for $-\sqrt{2} < x < 0$

31. Increasing for $x < 1$, decreasing for $x > 1$, global maximum at (1, 3), vertical tangent at (0, 0)

33. Decreasing for $0.6 < x < 1$, increasing for $x < 0.6$ and for $x > 1$, local minimum and a cusp at (1, 0), local maximum at (0.6, 0.3257) (ordinate approximate)

Section 4.6 (page 197)

1. $8x^3 - 9x^2 + 6$, $24x^2 - 18x$, $48x - 18$

3. $-8(2x - 1)^{-3}$, $48(2x - 1)^{-4}$, $-384(2x - 1)^{-5}$

5. $4(3t - 2)^{1/3}$, $4(3t - 2)^{-2/3}$, $-8(3t - 2)^{-5/3}$

7. $(y + 1)^{-2}$, $-2(y + 1)^{-3}$, $6(y + 1)^{-4}$

9. $-\tfrac{1}{4}t^{-3/2} - (1 - t)^{-4/3}$, $\tfrac{3}{8}t^{-5/2} - \tfrac{4}{3}(1 - t)^{-7/3}$, $-\tfrac{15}{16}t^{-7/2} - \tfrac{28}{9}(1 - t)^{-10/3}$

11. $3 \cos 3x$, $-9 \sin 3x$, $-27 \cos 3x$

13. $\cos^2 x - \sin^2 x$, $-4 \sin x \cos x$, $4 \sin^2 x - 4 \cos^2 x$

15. $\dfrac{x\cos x - \sin x}{x^2}$,
$\dfrac{2\sin x - x\cos x - x^2\sin x}{x^3}$,
$\dfrac{3x^2\sin x - x^3\cos x + 6x\cos x - 6\sin x}{x^4}$

17. $-\dfrac{2x + y}{x + 2y}$ and $\dfrac{-18}{(x + 2y)^3}$

19. $\dfrac{dy}{dx} = -\dfrac{1 + 2x}{3y^2}$ and
$\dfrac{d^2y}{dx^2} = \dfrac{2(y^{-2} - 21y^{-5})}{9}$

21. $\dfrac{y}{\cos y - x}$ and
$\dfrac{2y\cos y - 2xy + y^2\sin y}{(\cos y - x)^3}$

23. Local minimum value -1 at $x = 2$; no inflection points

25. Local maximum value 3 at $x = -1$, local minimum value -1 at $x = 1$; inflection point at $(0, 1)$

27. No local extrema; inflection point at the origin

29. No local extrema; inflection point at the origin

31. Local maximum value $\frac{1}{16}$ at $x = \frac{1}{2}$, local minimum value 0 at $x = 0$ and at $x = 1$; abscissas of the inflection points are the roots of $6x^2 - 6x + 1 = 0$—the inflection points are approximately $(0.42, 0.06)$ and $(1.58, 0.83)$.

33. Local maximum value 0 at $x = 1$, local minimum value (exactly) -0.03456 at $x = 1.4$; inflection points at $(2, 0)$, $(1.155, -0.015)$, and $(1.645, -0.019)$ (the decimals are approximate values)

45. Increasing for $x < -1$ and for $x > 2$, decreasing for $-1 < x < 2$, local maximum at $(-1, 10)$, local minimum at $(2, -17)$, inflection point at $(0.5, -3.5)$

47. Increasing for $x < -2$ and for $0 < x < 2$, decreasing for $-2 < x < 0$ and for $x > 2$, global maximum value 22 where $|x| = 2$, local minimum at $(0, 6)$, inflection points where $x^2 = \frac{4}{3}$ (and $y = \frac{134}{9}$)

49. Decreasing for $x < -1$ and for $0 < x < 2$, increasing for $-1 < x < 0$ and for $x > 2$, local minimum at $(-1, -6)$, local maximum at $(0, -1)$, global minimum at $(2, -32)$, inflection points with approximate coordinates $(1.22, -19.36)$ and $(-0.55, -3.68)$

51. Local maximum at $(\frac{3}{7}, 0.008)$ (ordinate approximate), local minimum at $(1, 0)$, inflection points at $(0, 0)$ and at $(x, f(x))$ where $(7x - 3)^2 = 2$. The general shape of the graph is shown in the upper left; the lower right shows the behavior of f on and near the interval $[0, 1]$ with the vertical scale greatly magnified.

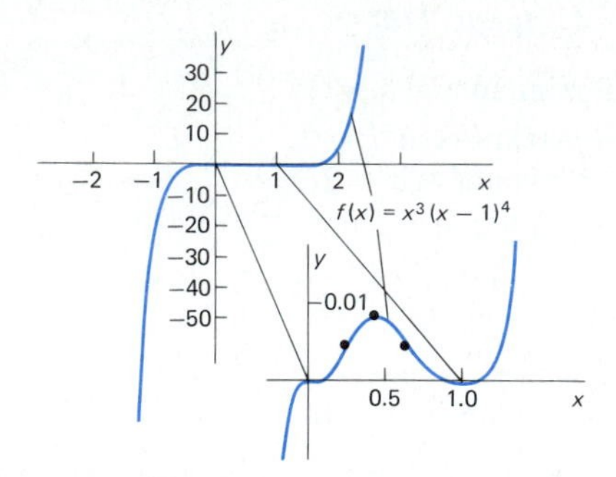

53. The graph is increasing everywhere, with a vertical tangent and inflection point at $(0, 1)$, the other intercept is $(-1, 0)$, and there are no extrema

55. Global minimum at $(0, 0)$, increasing for all $x > 0$, inflection point at $(1, 4)$, concave up for $x > 1$, vertical tangent at the origin

57. Increasing for $x < 1$, decreasing for $x > 1$, vertical tangent and inflection point at $(0, 0)$, another inflection point at $(-2, -7.56)$ (ordinate approximate), global maximum at $(1, 3)$

59. $5(x + 1)^4$, $20(x + 1)^3$, $60(x + 1)^2$, $120(x + 1)$, 120

67. $a = 3PV^2 \approx 3{,}583{,}858.8$, $b = \dfrac{V}{3} \approx 42.7$,
$R = \dfrac{8PV}{3T} \approx 81.80421$

Section 4.7 (page 206)

1. 1 3. 3 5. 2 7. 1
9. 4 11. 0 13. 2

15. $+\infty$ (or "does not exist")

17. No critical points, no inflection points. Vertical asymptote $x = 2$, horizontal asymptote $y = 0$. Sole intercept: $(0, -\frac{2}{3})$.

19. No critical points, no inflection points. Vertical asymptote $x = -2$, horizontal asymptote $y = 0$. Sole intercept: $(0, -\frac{2}{3})$.

21. No critical points, no inflection points.
Vertical asymptote $x = \frac{3}{2}$,
horizontal asymptote $y = 0$.

23. Global minimum at (0, 0)
inflection points where $3x^2 = 1$ (and $y = \frac{1}{4}$),
horizontal asymptote $y = 1$.

25. Local maximum at $(0, -\frac{1}{9})$,
no inflection points.
Vertical asymptotes $x = -3$ and $x = 3$,
horizontal asymptote $y = 0$.

27. Local maximum $(-\frac{1}{2}, -\frac{4}{25})$,
horizontal asymptote $y = 0$,
vertical asymptotes $x = -3$ and $x = 2$,
no inflection points.

29. Local minimum at (1, 2)
local maximum at $(-1, -2)$,
no inflection points,
vertical asymptote $x = 0$.
The line $y = x$ is also an asymptote.

31. Local minimum at (2, 4),
local maximum at (0, 0).
Asymptotes $y = x + 1$, $x = 1$.
No inflection points.

33.

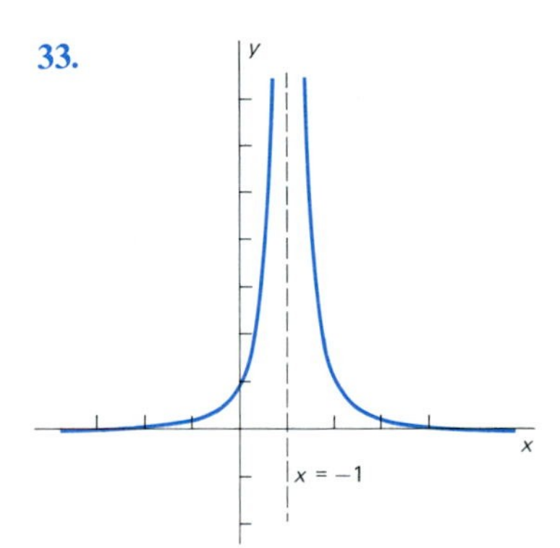

35.

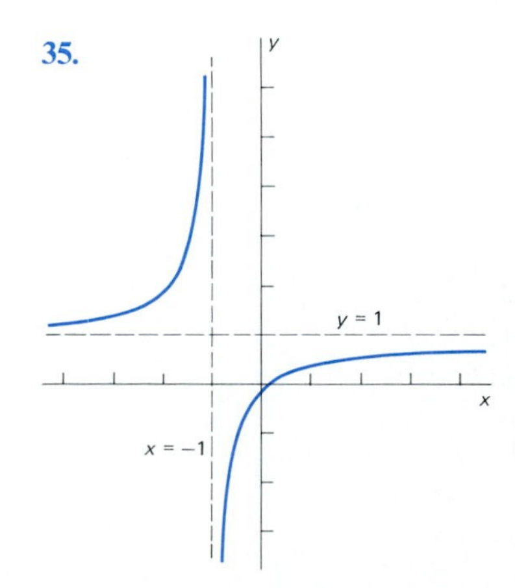

37.

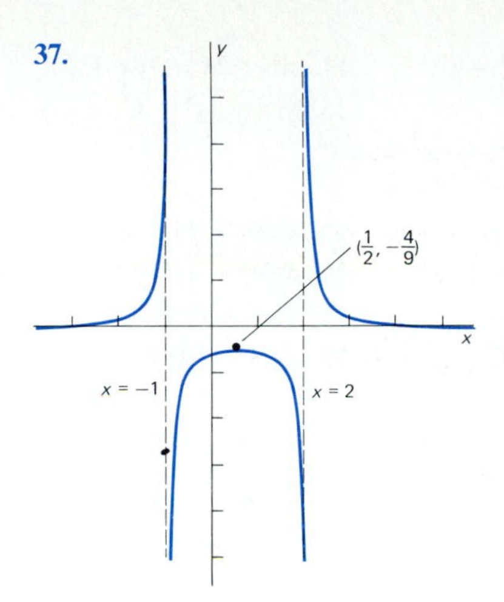

39.

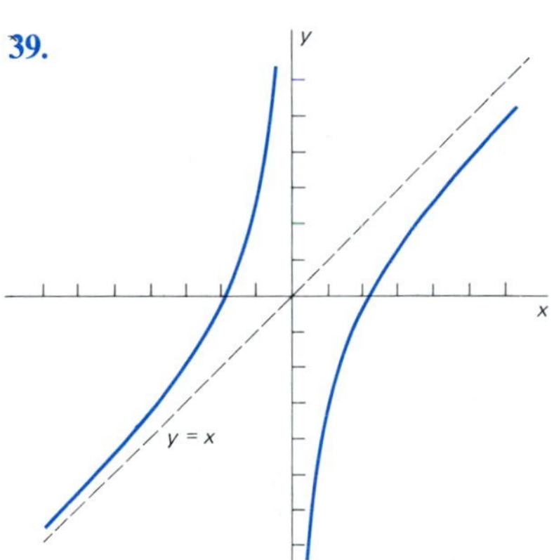

41.

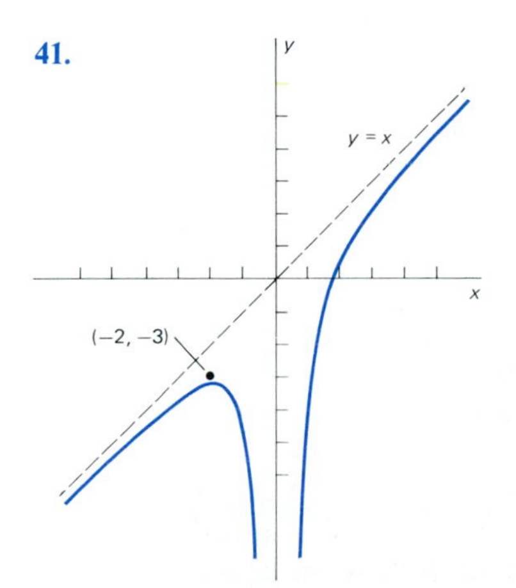

43. Local minimum at (1, 3),
inflection point at $(\sqrt[3]{-2}, 0)$.
Vertical asymptote: $x = 0$.

Section 4.8 (page 213)

1. $x^3 + x^2 + x + C$

3. $x - \frac{2}{3}x^3 + \frac{3}{4}x^4 + C$

5. $-\frac{3}{2}x^{-2} + \frac{4}{5}x^{5/2} - x + C$

7. $t^{3/2} + 7t + C$

9. $\frac{3}{5}x^{5/3} - 16x^{-1/4} + C$

11. $x^4 - 2x^2 + 6x + C$

13. $7x + C$

15. $\frac{1}{5}(x + 1)^5 + C$

17. $-\frac{1}{6}(x - 10)^{-6} + C$

19. $\frac{2}{3}x^{3/2} - \frac{4}{5}x^{5/2} + \frac{2}{7}x^{7/2} + C$

21. $\frac{2}{21}x^3 - \frac{3}{14}x^2 - \frac{5}{7}x^{-1} + C$

23. $\frac{1}{54}(9t + 11)^6 + C$

25. $-7(x + 77)^{-1} + C$

27. $\frac{1}{8}(3t^2 - 1)^8 + C$

29. $\frac{2}{3}(x^2 + 1)^{3/2} + C$

31. $3(x^2 + 4)^{1/2} + C$

33. $\frac{1}{4}(x^3 + 2)^{4/3} + C$

35. $\frac{1}{88}(x^8 + 9)^{11} + C$

37. $\frac{1}{7}x^7 + \frac{3}{5}x^5 + x^3 + x + C$

39. $\frac{1}{2}\sin 10x + 2\cos 5x + C$

41. $\dfrac{3\sin \pi t}{\pi} + \dfrac{\sin 3\pi t}{3\pi} + C$

43. $\frac{1}{6}\sin^6 x + C$

45. $\frac{1}{8}\sin^4 2x + C$

49. $y = f(x) = x^2 + x + 3$

51. $\frac{2}{3}(x^{3/2} - 8)$

53. $2\sqrt{x + 2} - 5$

55. $y = \frac{3}{4}x^4 - 2x^{-1} + \frac{9}{4}$

57. $y = \frac{1}{4}(x - 1)^4 + \frac{7}{4}$

59. $y = 2(x - 13)^{1/2} - 2$

61. $y = -\frac{1}{3}(1 - x^2)^{3/2}$

Section 4.9 (page 220)

1. $x(t) = 25t^2 + 10t + 20$

3. $x(t) = \frac{1}{2}t^3 + 5t$

5. $x(t) = \frac{1}{3}(t + 3)^4 - 37t - 26$

7. $x(t) = \frac{1}{2}(t + 1)^{-1} + \frac{1}{2}(t - 1)$

9. $x(t) = t + 2 - \sin t$

11. 144 ft; 6 s

13. 144 ft

15. 5 s; 112 ft/s

17. $\sqrt{60}$ s (about 7.75 s); $32\sqrt{60}$ ft/s (about 247.87 ft/s)

19. 120 ft/s

21. 5 s; -160 ft/s

23. 400 ft; 10 s

25. $\dfrac{-5 + 2\sqrt{145}}{4} \approx 4.77$ s; $16\sqrt{145} \approx 192.6655$ ft/s

27. $\frac{544}{3}$ ft/s

29. 22 ft/s^2 ($s = -11t^2 + 88t$ is the distance traveled t seconds after the brakes are applied)

31. Approximately 869.434 ft

Chapter 4 Miscellaneous Problems (page 221)

1. $dy = 3(4x - x^2)^{1/2}(2 - x)\,dx$

3. $dy = -2(x - 1)^{-2}\,dx$

5. $dy = (2x\cos x^{1/2} - \frac{1}{2}x^{3/2}\sin x^{1/2})\,dx$

7. $10 + \frac{1}{60} \approx 10.0167$

9. 132.5

11. $2 + \frac{1}{32} \approx 2.0313$

13. 7.5

15. $10\pi \approx 31.416$ (in.^{3})

17. $\frac{1}{96}\pi \approx 0.0327$ s

19. $c = \sqrt{3}$

21. $c = 1$

23. $c = (2.2)^{0.25}$

25. Decreasing for $x < 3$, increasing for $x > 3$, global minimum at $(3, -5)$, concave upward everywhere

27. Increasing everywhere, no extrema

29. Increasing for $x < 0.25$, decreasing for $x > 0.25$, vertical tangent at $(0, 0)$, global maximum at $x = 0.25$

31. $3x^2 - 2$, $6x$, 6

33. $-(t^{-2}) + 2(2t + 1)^{-2}$, $2t^{-3} - 8(2t + 1)^{-3}$, $-6t^{-4} + 48(2t + 1)^{-4}$

35. $3t^{1/2} - 4t^{1/3}$, $\frac{3}{2}t^{-1/2} - \frac{4}{3}t^{-2/3}$, $-\frac{3}{4}t^{-3/2} + \frac{8}{9}t^{-5/3}$

37. $-4(t - 2)^{-2}$, $8(t - 2)^{-3}$, $-24(t - 2)^{-4}$

39. $-\frac{4}{3}(5 - 4x)^{-2/3}$, $-\frac{32}{9}(5 - 4x)^{-5/3}$, $-\frac{640}{27}(5 - 4x)^{-8/3}$

41. $\dfrac{dy}{dx} = -\left(\dfrac{y}{x}\right)^{2/3}$, $\dfrac{d^2y}{dx^2} = \frac{2}{3}y^{1/3}x^{-5/3}$

43. $\dfrac{dy}{dx} = \frac{1}{2}(5y^4 - 4)^{-1}x^{-1/2}$, $\dfrac{d^2y}{dx^2} = -\dfrac{20x^{1/2}y^3 + (5y^4 - 4)^2}{4x^{3/2}(5y^4 - 4)^3}$

45. Global minimum at $(2, -48)$, x-intercepts 0 and 3.1748 (latter approximate), concave upward everywhere, no inflection points, no asymptotes

47. Local maximum at $(0, 0)$, global minima where $x^2 = \frac{4}{3}$ (and $y = -\frac{32}{27}$), inflection points where $x^2 = \frac{4}{5}$ (and $y = -\frac{96}{125}$), no asymptotes

49. Global maximum at $(3, 3)$, inflection points at $(4, 0)$ (also a vertical tangent there) and where $x = 6$; no asymptotes

51. Local maximum at $(0, -0.25)$, horizontal asymptote $y = 1$, vertical asymptotes $|x| = 2$, no inflection points

53. The inflection point has abscissa the only real solution of $x^3 + 6x^2 + 4 = 0$, which is approximately -6.10724

55. 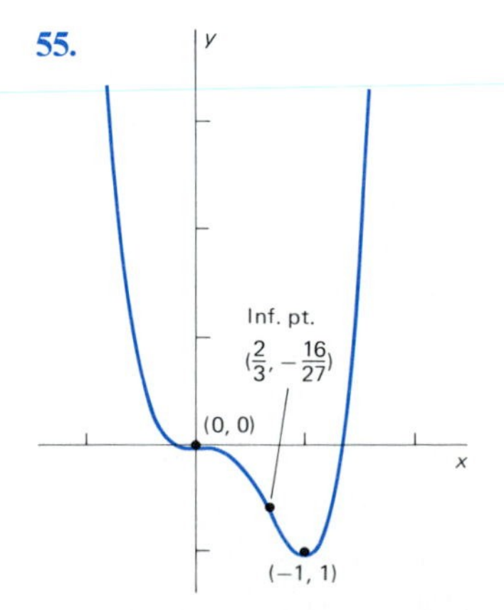

57.

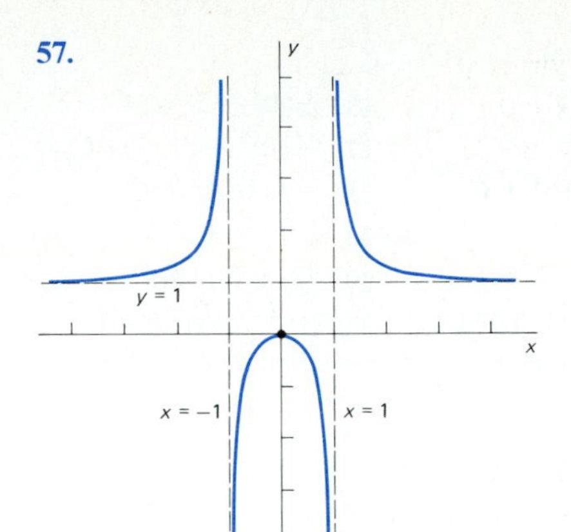

59.

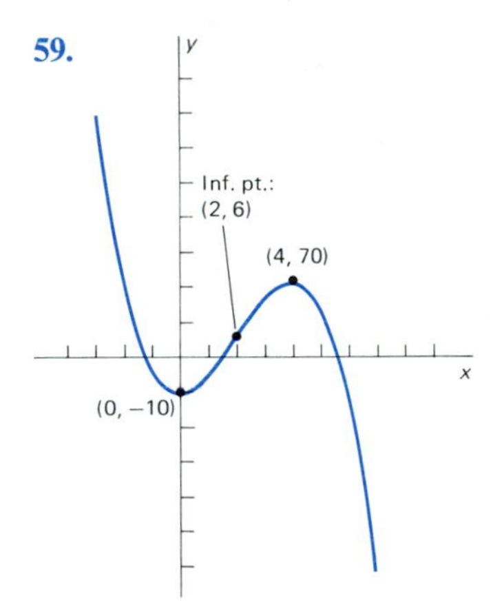

61.

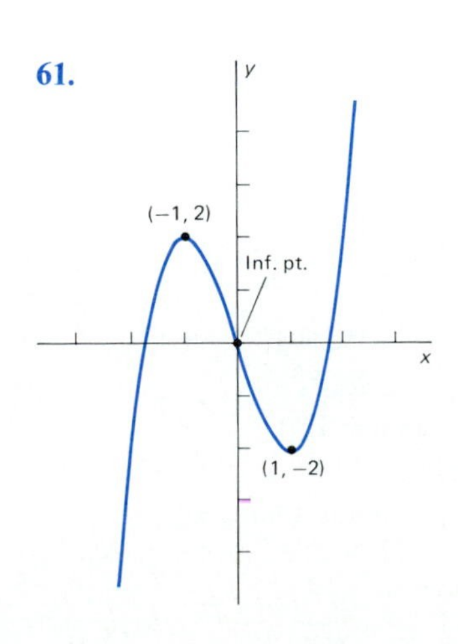

63.

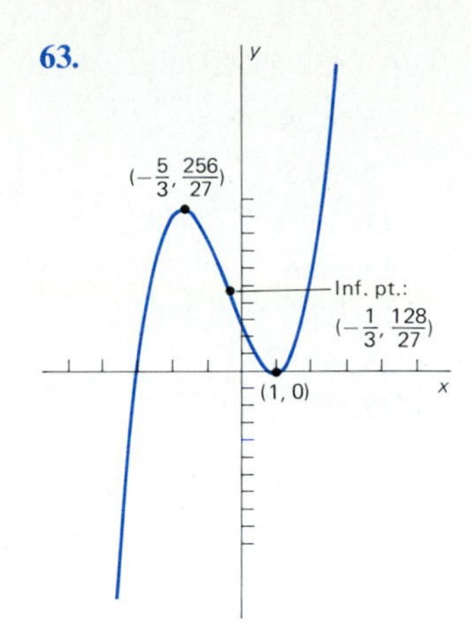

65. $\frac{1}{3}x^3 + 2x^{-1} - \frac{5}{2}x^{-2} + C$

67. $-\frac{1}{30}(1 - 3x)^{10} + C$

69. $\frac{3}{16}(9 + 4x)^{4/3} + C$

71. $\frac{1}{24}(1 + x^4)^6 + C$

73. $-\frac{3}{8}(1 - x^2)^{4/3} + C$

75. $\frac{7}{5}\sin 5x + \frac{5}{7}\cos 7x + C$

77. $y(x) = x^3 + x^2 + 5$

79. $y(x) = \frac{1}{12}(2x + 1)^6 + \frac{23}{12}$

81. $y(x) = \frac{3}{2}x^{2/3} - \frac{1}{2}$

83. Maximum value 1, at $x = -1$

85. The minimizing angle is $\arctan\sqrt[3]{2} \approx 0.9$; that is, about 51.56°. The minimum length is $4(1 + 4^{1/3})^{3/2}$, about 16.65 ft.

87. 6 s; 396 ft

89. 120 ft/s

91. Impact time: $t = 2\sqrt{10} \approx 6.32$ s;
impact speed: $20\sqrt{10} \approx 63.25$ ft/s

93. 176 ft

95. $100(20/9)^{0.4} \approx 54.79$ mi/h

97. Two horizontal tangents, where $x \approx 0.42$, $|y| \approx 0.62$.
Intercepts: (0, 0), (1, 0), and (2, 0).
Vertical tangent at each intercept. Inflection points corresponding to the only positive solution of $3x^2(x - 2)^2 = 4$; that is, $x \approx 2.46789$, $|y| \approx 1.30191$.
No asymptotes.

Section 5.2 (page 233)

1. 190 **3.** 1165 **5.** 224

7. 1834 **9.** 333,833,500

11. 0.5 **13.** 0.2

15. $\frac{1}{3}n(2n - 1)(2n + 1)$ **17.** 24

19. $\frac{1}{3}$ **21.** 132 **23.** 90 **25.** $\frac{1}{3}b^3$

Section 5.3 (page 240)

1. 0.44
3. 1.45
5. 19.5
7. 58.8
9. 0.24
11. $\frac{137}{60} \approx 2.28333$
13. 16.5
15. 26.4
17. 0.33
19. $\frac{6086}{3465} \approx 1.75462$
21. 18
23. 40.575
25. 1.83753
27. 2
29. 12
31. 30

Section 5.4 (page 247)

1. $\frac{55}{12}$
3. $\frac{49}{60}$
5. $-\frac{1}{20}$
7. $\frac{1}{4}$
9. $\frac{16}{3}$
11. 24
13. 0
15. $\frac{32}{3}$
17. 0
19. $\frac{93}{5}$
21. $\frac{17}{6}$
23. $\frac{28}{3}$
25. $\frac{52}{5}$
27. $\frac{122}{9}$
29. $\frac{1}{4}$
31. 0
33. $\frac{1}{2}$
35. $\frac{2}{3}$
37. $\frac{25}{4}\pi \approx 19.635$
41. $1000 + \int_0^{30} V'(t)\,dt = 160$ (gal)
43. Let $Q = \left(\frac{1}{1.2} + \frac{1}{1.4} + \frac{1}{1.6} + \frac{1}{1.8}\right)(0.2) \approx 0.5456$,

    and $I = \displaystyle\int_1^2 \frac{1}{x}\,dx$. Then $Q + 0.1 \leqq I \leqq Q + 0.2$.

    Hence $0.64 < I < 0.75$.

Section 5.5 (page 255)

1. 2
3. $\frac{26}{3}$
5. 0
7. $\frac{125}{4}$
9. $\frac{14}{9}$
11. $\frac{1}{2}$
13. 4
15. $\frac{1}{3}$
17. $-\frac{22}{81} \approx -0.271605$
19. 0
21. $\frac{35}{24} \approx 1.4583$
23. 0
25. 4
27. $\frac{38}{3}$
29. $\frac{61}{3}$
31. $f'(x) = (x^2 + 1)^{17}$
33. $h'(z) = (z - 1)^{1/3}$
35. $f'(x) = -x - x^{-1}$
37. $G'(x) = (x + 4)^{1/2}$
39. $G'(x) = (x^3 + 1)^{1/2}$
41. $f'(x) = 3 \sin 9x^2$
43. $f'(x) = 2x \sin x^2$
45. $f'(x) = \dfrac{2x}{x^2 + 1}$
47. The integral does not exist.
51. Average height: $\frac{800}{3} \approx 266.67$ ft; average velocity: -80 ft/s
53. $\frac{5000}{3} \approx 1666.67$ (gal)

Section 5.6 (page 259)

1. $\frac{1}{6}(1 + x^4)^{3/2} + C$
3. $-2(1 + x^{1/2})^{-1} + C$
5. $\frac{1}{12} \sin 4x^3 + C$
7. $\frac{1}{30}(x^2 + 1)^{15} + C$
9. $\frac{2}{5}(4 - x)^{5/2} - \frac{8}{3}(4 - x)^{3/2} + C$
11. $(1 + x^4)^{1/2} + C$
13. $\frac{1}{24}(4x - 3)^6 + C$
15. $-\frac{1}{9}(2 - 3x^2)^{3/2} + C$
17. $-\frac{1}{9}(x^3 + 5)^{-3} + C$
19. $-\frac{1}{16}(2 - 4x^3)^{4/3} + C$
21. $-3 \cos \frac{1}{3}t + C$
23. $-\frac{3}{8}$
25. 2
27. $\frac{1192}{15} \approx 79.46667$
29. $\frac{15}{128} \approx 0.11719$
31. $\frac{2}{5}(4\sqrt{2} - 1) \approx 1.86274$
33. $\frac{3}{4}(x^4 - 4x)^{1/3} + C$
35. $\frac{4}{15}(6 - t^3)^{5/4} + C$

Section 5.7 (page 267)

1. $\frac{2}{3}$
3. $\frac{16}{3}$
5. $+\frac{1}{4}$
7. $\frac{1}{4}$
9. $+\frac{16}{3}$
11. $\frac{1}{4}$
13. $\frac{1}{20}$
15. $\frac{4}{9}$
17. $\frac{32}{3}$
19. $\frac{128}{3}\sqrt{2}$
21. $\frac{500}{3}$
23. $\frac{64}{3}$
25. $\frac{500}{3}$
27. $\frac{4}{3}$
29. $\frac{1}{12}$
31. $\frac{16}{3}$
33. $\frac{16}{3}$
35. $\frac{27}{2}$
37. $\frac{320}{3}$
39. $\frac{8}{15}$
41. $\frac{37}{12}$
45. 1
49. $f(x) = \sqrt{x}$

Section 5.8 (page 279)

1. $L_4 = 6$, $R_4 = 10$, $T_4 = 8$, true value: 8
3. 0.55, 0.75, 0.65, $\frac{2}{3}$
5. 1.24, 0.72, 0.98, 1
7. $M_4 = 8$
9. 0.6713
11. 1.0115
13. $T_4 = 8.7500$, $S_4 = 8.6667$; true value: $\frac{26}{3}$
15. 0.0973, 0.940, $\frac{3}{32} = 0.09375$
17. 7.3948, 7.3398 (true value: approximately 7.3414)
19. 8.5499, 8.5509 (true value: approximately 8.5507)
21. (a) 3.0200; (b) 3.0717
23. Both answers should be near 2435.
27. 19

Chapter 5 Miscellaneous Problems (page 284)

1. 1700
3. 2845
5. $2(\sqrt{2} - 1) \approx 0.82843$
7. $\frac{2}{3}\pi(2\sqrt{2} - 1) \approx 3.82945$
13. $\frac{2}{3}x\sqrt{2x} + 2(3x)^{-1/2} + C$
15. $-2x^{-1} - \frac{1}{4}x^2 + C$
17. $\frac{2}{3}\sin x^{3/2} + C$
19. $\cos(t^{-1}) + C$
21. $-\frac{3}{8}(1 + u^{4/3})^{-2} + C$
23. $\frac{28}{3}$
25. $\frac{1}{3}x^{-3}(4x^2 - 1)^{3/2} + C$ (use $u = 1/x$)
27. $\frac{1}{30}$
29. $\frac{44}{15}$
31. $\frac{125}{6}$
33. Semicircle, center (1, 0), radius 1: area $\pi/2$
35. $f(x) = (4x^2 - 1)^{1/2}$

37. $n \geqq 9$. $L_{10} \approx 1.12767$, $R_{10} \approx 1.16909$. An estimate: their average is $A = 1.1483 \pm 0.05$.

39. $M_5 \approx 0.28667$, $T_5 \approx 0.28971$. They bound the true value of the integral because the second derivative of the integrand is positive for all $x > 0$.

Section 6.1 (page 291)

1. -320; 320 **3.** -50; 106.25
5. 65; 97 **7.** 1; 1
9. 0; $4/\pi$ **11.** 1
13. $2/\pi$ **15.** 98/3 **17.** 4 **19.** -1

21. $\int_{-1}^{1} [f(x)]^2\,dx$

23. $\int_{-2}^{3} 2\pi x\{1 + [f(x)]^2\}^{1/2}\,dx$

25. 550 gal **27.** 385,000 **29.** 109.5 in.

31. $\frac{3}{4}$

Section 6.2 (page 299)

1. $\dfrac{\pi}{5}$ **3.** 8π **5.** $\frac{1}{2}\pi^2$ **7.** $\dfrac{3\pi}{10}$

9. $\dfrac{512\pi}{3}$ **11.** $\dfrac{16\pi}{15}$ **13.** $\dfrac{\pi}{2}$ **15.** 8π

17. $\dfrac{121\pi}{210}$ **19.** 8π **21.** $\dfrac{49\pi}{30}$ **23.** $\dfrac{17\pi}{10}$

25. 9π **27.** $\frac{4}{3}\pi a^2 b$ **29.** $\frac{16}{3}a^3$

31. $\frac{4}{3}a^3\sqrt{3}$ **37.** $\frac{16}{3}a^3$

Section 6.3 (page 308)

1. 8π **3.** $\dfrac{625\pi}{2}$ **5.** 16π **7.** π

9. $\dfrac{6\pi}{5}$ **11.** $\dfrac{265\pi}{15}$ **13.** $\dfrac{4\pi}{15}$ **15.** $\dfrac{11\pi}{15}$

17. $\dfrac{56\pi}{5}$ **19.** $\dfrac{8\pi}{3}$ **21.** $\dfrac{2\pi}{15}$ **23.** $\dfrac{\pi}{2}$

25. $\dfrac{16\pi}{3}$ **27.** 64π

31. $\frac{4}{3}\pi a^2 b$ **33.** $V = 2\pi^2 a^2 b$

35. $V = 2\pi^2 a^3$ **37.** (a) $V = \frac{1}{6}\pi h^3$

Section 6.4 (page 317)

In 1–19, the integrand is given, followed by the interval of integration.

1. $(1 + 4x^2)^{1/2}$; $0 \leqq x \leqq 1$

3. $[1 + 36(x^4 - 2x^3 + x^2)]^{1/2}$; $0 \leqq x \leqq 2$

5. $(1 + 4x^2)^{1/2}$; $0 \leqq x \leqq 100$

7. $(1 + 16y^6)^{1/2}$; $-1 \leqq y \leqq 2$

9. $x^{-2}(x^4 + 1)^{1/2}$; $1 \leqq x \leqq 2$

11. $2\pi x^2(1 + 4x^2)^{1/2}$; $0 \leqq x \leqq 4$

13. $2\pi(x - x^2)(2 - 4x + 4x^2)^{1/2}$; $0 \leqq x \leqq 1$

15. $2\pi(2 - x)(1 + 4x^2)^{1/2}$; $0 \leqq x \leqq 1$

17. $\pi(4x + 1)^{1/2}$; $1 \leqq x \leqq 4$

19. $\pi(x + 1)(4 + 9x)^{1/2}$; $1 \leqq x \leqq 4$

21. $\frac{22}{3}$ **23.** $\frac{14}{3}$

25. $\frac{123}{32} = 3.84375$

27. $\frac{1}{27}(104\sqrt{13} - 125) \approx 9.25842$

29. $\frac{1}{6}\pi(5\sqrt{5} - 1) \approx 5.3304$

31. $\dfrac{339\pi}{16} \approx 66.5625$

33. $\frac{1}{9}\pi(82\sqrt{82} - 1) \approx 258.8468$

35. 4π

37. 3.8194 (The true value is approximately 3.8202.)

41. Avoid the problem when $x = 0$ as follows:

$$L = 8\int_{1/2\sqrt{2}}^{1} x^{-1/3}\,dx = 6.$$

Section 6.5 (page 323)

1. $y(x) = (\frac{1}{2}x^2 + C)^2$ **3.** $3y^{-2} + 2x^3 = C$

5. $y(x) = 1 + (\frac{1}{2}x^2 + C)^2$

7. $3y + 2y^{3/2} = 3x + 2x^{3/2} + C$

9. $y^3 + y = x - x^{-1} + C$

11. $y(x) = (1 - x)^{-1}$ **13.** $y(x) = (x + 1)^{1/4}$

15. $y(x) = (\frac{1}{2} - \frac{1}{3}x^{3/2})^{-2}$ **17.** $x^2 + y^2 = 169$

19. $y(x) = (3x^3 - 3x + 1)^{1/3}$

21. 20 weeks

23. (a) 169,000; (b) early in 2011

25. $P(t) \to +\infty$ as $t \to 6^-$.

27. 1 h 18 min 40 s after the plug is pulled

29. 16 min 12 s **31.** 14 min 29 s

Section 6.6 (page 331)

1. 30 **3.** 9

5. 0 **7.** 15 ft-lb

9. 2.816×10^9 ft-lb (with $R = 4000$ mi, $g = 32$ ft/s^2)

11. $13{,}000\pi \approx 40{,}841$ ft-lb

13. $\dfrac{125{,}000\pi}{3} \approx 130{,}900$ ft-lb

15. $156{,}000\pi \approx 490{,}088$ ft-lb

17. $4{,}160{,}000\pi \approx 13{,}069{,}025$ ft-lb

19. 8750 ft-lb **21.** 11,250 ft-lb

23. $25{,}000[1-(0.1)^{0.4}]$ in.-lb—approximately 1254 ft-lb

25. 16π ft-lb

27. $1{,}382{,}400\pi$ ft-lb

29. Approximately 690.53 ft-lb

31. 249.6 lb

33. 748.8 lb

35. 19,500 lb

37. $\dfrac{700\rho}{3} \approx 14{,}560$ lb

39. About 32,574 tons

Chapter 6 Miscellaneous Problems (page 334)

1. $-\frac{3}{2}; \frac{31}{6}$

3. 1; 3

5. $\frac{14}{3}$

7. $\dfrac{2\pi}{15}$

9. 12 in.

11. $\dfrac{41\pi}{105}$

13. $10.625\pi \approx 33.379$ g

19. $f(x) = \sqrt{1+3x}$

21. $\dfrac{24-2\pi^2}{3\pi} \approx 0.4521$

23. $\frac{10}{3}$

25. $\frac{63}{8}$

27. $\dfrac{52\pi}{5}$

31. $y(x) = x^2 + \sin x$

33. $y(x) = -1 + (C-x)^{-1}$

35. $y(x) = (1-x^3)^{-1}$

37. $y(x) = (C - 3x^{-1})^{1/3}$

39. $y(x) = (1-\sin x)^{-1}$

41. $x(2y^{1/2}-1) = y(Cx + 2x^{1/2} - 1)$

43. 1 ft

45. $W = 4\pi R^4 \rho$

47. 10,454,400 ft-lb

49. 36,400 tons

51. There is no maximum volume; $c = \frac{1}{3}\sqrt{5}$ minimizes the volume V.

Section 7.1 (page 344)

1. 128

3. 64

5. 1

7. 16

9. 16

11. 4

13. 3

15. 3

17. $3 \ln 2$

19. $\ln 2 + \ln 3$

21. $3\ln 2 + 2\ln 3$

23. $3\ln 2 - 3\ln 3$

25. $3\ln 3 - 3\ln 2 - \ln 5$

27. $2^{81} > 2^{12}$

29. There are three solutions of the equation. The one *not* obvious by inspection is approximately -0.7666647.

31. 6

33. -2

35. 1, 2

37. $81 = 3^4$

39. 0

41. $\dfrac{dy}{dx} = (x+1)e^x$

43. $(x^{1/2} + \frac{1}{2}x^{-1/2})e^x$

45. $x^{-3}(x-2)e^x$

47. $1 + \ln x$

49. $x^{-1/2}(1 + \frac{1}{2}\ln x)$

51. $(1-x)e^{-x}$

53. $\dfrac{3}{x}$

57. $f'(x) = \frac{1}{10}e^{x/10}$

61. $P'(t) = 3^t \ln 3$

63. $P'(t) = -(2^{-t}\ln 2)$

Section 7.2 (page 352)

1. $f'(x) = \dfrac{3}{3x-1}$

3. $\dfrac{1}{1+2x}$

5. $\dfrac{3x^2-1}{3x^3-3x}$

7. $-x^{-1}\sin(\ln x)$

9. $\dfrac{-1}{x(\ln x)^2}$

11. $x^{-1} + x(x^2+1)^{-1}$

13. $-\tan x$

15. $2t\ln(\cos t) - t^2 \tan t$

17. $(\ln t)^2 + 2\ln t$

19. $6(2x+1)^{-1} + 8x(x^2-4)^{-1}$

21. $-x(4-x^2)^{-1} - x(9+x^2)^{-1}$

23. $\dfrac{2}{1-x^2}$

25. $2t^{-1} - 2t(t^2+1)^{-1}$

27. $-x^{-1} + \cot x$

29. $\dfrac{dy}{dx} = \dfrac{y \ln y}{y - x}$

31. $\dfrac{dy}{dx} = \dfrac{y}{-x + \cot y}$

33. $\frac{1}{2}\ln|2x-1| + C$

35. $\frac{1}{6}\ln(1+3x^2) + C$

37. $\frac{1}{4}\ln|2x^2 + 4x + 1| + C$

39. $\frac{1}{3}(\ln x)^3 + C$

41. $\ln|x+1| + C$

43. $\ln(x^2 + x + 1) + C$

45. $\frac{1}{2}(\ln x)^2 + C$

47. $\frac{1}{2}\ln(1 - \cos 2x) + C$

49. $\frac{1}{3}\ln|x^3 - 3x^2 + 1| + C$

51. 0

53. 0

55. 0

59. $m \approx -0.2479, \quad k \approx 291.7616$

65. $y \to 0$ as $x \to 0^+$; also, $dy/dx \to 0$ as $x \to 0^+$. The point (0, 0) is not on the graph. Intercept at (1, 0), global minimum where $x = e^{-1/2}$—the coordinates are approximately (0.61, -0.18). There is an inflection point where $x = e^{-3/2}$—the coordinates are approximately (0.22, -0.07). The figure is *not drawn to scale.*

67. $y \to -\infty$ as $x \to 0^+$; $y \to 0$ as $x \to +\infty$. There is a global maximum at $(e^2, 2e^{-1})$ and an inflection point where $x = e^{8/3}$. The x-axis is a horizontal asymptote and the y-axis is a vertical asymptote. The only intercept is (1, 0). The figure is *not drawn to scale.*

71. Midpoint estimate: approximately 872.47. Trapezoidal estimate: approximately 872.60. The true value of the integral is approximately 872.5174.

Section 7.3 (page 360)

1. $f'(x) = 2e^{2x}$

3. $2xe^{(x^2)} = 2xe^{x^2} = 2x\exp(x^2)$

5. $-2x^{-3}\exp(x^{-2})$

7. $(1 + \frac{1}{2}t^{1/2})\exp(t^{1/2})$

9. $(1 + 2t - t^2)e^{-t}$

11. $-e^{\cos t}\sin t$

13. $-e^{-x}\sin(1 - e^{-x})$

15. $\dfrac{1 - e^{-x}}{x + e^{-x}}$

17. $e^{-2x}(3\cos 3x - 2\sin 3x)$

19. $15(e^t - t^{-1})(e^t - \ln t)^4$

21. $-(5 + 12x)e^{-4x}$

23. $(e^{-t} + te^{-t} - 1)t^{-2}$

25. $(x - 2)e^{-x}$

27. $e^x\exp(e^x) = e^x e^{e^x}$

29. $2e^x\cos 2e^x$

31. $\dfrac{dy}{dx} = \dfrac{e^y}{1 - xe^y} = \dfrac{e^y}{1 - y}$

33. $\dfrac{dy}{dx} = \dfrac{e^x + ye^{xy}}{xe^{xy} - e^y}$

35. $\dfrac{dy}{dx} = \dfrac{e^{x-y} - y}{e^{x-y} + x} = \dfrac{xy - y}{xy + x}$

37. $-\frac{1}{2}e^{1-2x} + C$

39. $\frac{1}{9}\exp(3x^3 + 1) + C$

41. $\frac{1}{2}\ln(1 + e^{2x}) + C$

43. $\frac{1}{2}\exp(1 - \cos 2x) + C$

45. $\frac{1}{2}\ln(x^2 + e^{2x}) + C$

47. $-\exp(-\frac{1}{2}t^2) + C$

49. $2\exp(x^{1/2}) + C$

51. $\ln(1 + e^x) + C$

53. $-\frac{2}{3}\exp(-x^{3/2}) + C$

55. e^2

57. e

59. $+\infty$

61. $+\infty$

63. Global minimum and intercept at (0, 0), local maximum at $(2, 4/e^2)$; inflection points where $x = 2 - \sqrt{2}$ and $x = 2 + \sqrt{2}$. The x-axis is an asymptote. The figure is *not drawn to scale.*

65. Global maximum at (0, 1), the only intercept. The x-axis is the only asymptote. Inflection points at the two points where $x^2 = 2$.

67. $\frac{1}{2}\pi(e^2 - 1) \approx 10.0359$

69. $(e^2 - 1)/(2e) \approx 1.1752$

71. The solution is approximately 1.278464543. Note that if $f(x) = e^{-x} - x + 1$, then $f'(x) < 0$ for all x.

73. $f'(x) = 0$ only for $x = 0$ and $x = n$; f is increasing on $(0, n)$ and decreasing for $x > n$. Thus $x = n$ yields the absolute maximum value of $f(x)$ for $x \geqq 0$. The x-axis is a horizontal asymptote, and there are inflection points with abscissas $x = n + \sqrt{n}$ and $x = n - \sqrt{n}$. The graph is *not drawn to scale.*

77. $y = e^{-2x} + 4x^x$

Section 7.4 (page 367)

1. $10^x\ln 10$

3. $3^x4^{-x}\ln 3 - 3^x4^{-x}\ln 4 = (\frac{3}{4})^x\ln\frac{3}{4}$

5. $-(7^{\cos x})(\ln 7)(\sin x)$

7. $2^{x\sqrt{x}}(\frac{3}{2}\ln 2)\sqrt{x}$

9. $x^{-1}2^{\ln x}\ln 2$

11. $17^x\ln 17$

13. $-x^{-2}10^{1/x}\ln 10$

15. $(2^{2^x}\ln 2)(2^x\ln 2)$

17. $\dfrac{1}{\ln 3}\cdot\dfrac{x}{x^2 + 4}$

19. $\dfrac{\ln 2}{\ln 3} = \log_3 2$

21. $\dfrac{1}{x(\ln 2)(\ln x)}$

23. $\dfrac{\exp(\log_{10}x)}{x\ln 10}$

25. $\dfrac{3^{2x}}{2\ln 3}$

27. $2\dfrac{2^{\sqrt{x}}}{\ln 2} + C$

29. $\dfrac{7^{x^3+1}}{3\ln 7} + C$

31. $\dfrac{(\ln x)^2}{2\ln 2} + C$

33. $\dfrac{dy}{dx}$
$= [x(x^2 - 4)^{-1} + (4x + 2)^{-1}](x^2 - 4)^{1/2}(2x + 1)^{1/4}$

35. $\dfrac{dy}{dx} = 2^x\ln 2$

37. $\dfrac{dy}{dx} = \dfrac{(x^{\ln x})(2\ln x)}{x}$

39. $\frac{1}{3}[(x + 1)^{-1} + (x + 2)^{-1} - 2x(x^2 + 1)^{-1} - 2x(x^2 + 2)^{-1}]\cdot y$

41. $\dfrac{dy}{dx} = (\ln x)^{\sqrt{x}}[\frac{1}{2}x^{-1/2}\ln(\ln x) + x^{1/2}(\ln x)^{-1}]$

43. $\left(\dfrac{3x}{1 + x^2} - \dfrac{4x^2}{1 + x^3}\right)(1 + x^2)^{3/2}(1 + x^3)^{-4/3}$

45. $\left[\dfrac{2x^3}{x^2 + 1} + 2x\ln(x^2 + 1)\right](x^2 + 1)^{(x^2)}$

47. $\frac{1}{4}x^{-1/2}(2 + \ln x)(\sqrt{x})^{\sqrt{x}}$

49. e^x

51. $x^{\exp(x)}e^x(x^{-1} + \ln x)$

57. Note that $\ln\dfrac{x^x}{e^x} = x\ln\dfrac{x}{e}$.

Section 7.5 (page 375)

1. \$119.35; \$396.24
3. Approximately 3.8685 h
5. The sample is about 686 years old.
7. (a) 9.308%; (b) 9.381%; (c) 9.409%; (d) 9.416%; (e) 9.417%
9. \$44.52
11. After an additional 32.26 days
13. Approximately 35 years
15. About 4.2521×10^9 years old
17. 2.40942 min
19. (a) 20.486 in.; 9.604 in.
(b) 3.4524 mi, about 18,230 ft

Section 7.6 (page 380)

1. $y(x) = -1 + 2e^x$

3. $y(x) = \frac{1}{2}(e^{2x} + 3)$

5. $x(t) = 1 - e^{2t}$

7. $x(t) = 27e^{5t} - 2$

9. $v(t) = 10(1 - e^{-10t})$

11. 4,870,328

15. About 46 days after the rumor starts

19. $\dfrac{400}{\ln 2} \approx 577$ ft

23. (b) $1,308,283

57. 20 weeks

59. (a) $925.20; (b) $1262.88

61. About 22.567 h after the power failure; that is, at about 9:34 P.M. on the following evening.

63. (b) $v(10) = 176(1 - e^{-1}) \approx 111.2532$ ft/s, about 75.85 mi/h. The limiting velocity is $a/\rho = 176$ ft/s, exactly 120 mi/h.

Chapter 7 Miscellaneous Problems (page 382)

1. $\dfrac{1}{2x}$

3. $\dfrac{1 - e^x}{x - e^x}$

5. $\ln 2$

7. $(2 + 3x^2)e^{-1/x^2}$

9. $\dfrac{1 + \ln \ln x}{x}$

11. $x^{-1}2^{\ln x} \ln 2$

13. $-\dfrac{2}{(x-1)^2}\exp\left(\dfrac{x+1}{x-1}\right)$

15. $\dfrac{3}{2}\left(\dfrac{1}{x-1} + \dfrac{8x}{3 - 4x^2}\right)$

17. $\dfrac{(\sin x \cos x)\exp(1 + \sin^2 x)^{1/2}}{(1 + \sin^2 x)^{1/2}}$

19. $\cot x + \ln 3$

21. $\dfrac{dy}{dx} = \dfrac{x^{1/x}(1 - \ln x)}{x^2}$

23. $\left(\dfrac{1 + \ln \ln x}{x}\right)(\ln x)^{\ln x}$

25. $-\frac{1}{2}\ln|1 - 2x| + C$

27. $\frac{1}{2}\ln|1 + 6x - x^2| + C$

29. $-\ln(2 + \cos x) + C$

31. $\dfrac{2}{\ln 10}10^{\sqrt{x}} + C$

33. $\frac{2}{3}(1 + e^x)^{3/2} + C$

35. $\dfrac{6^x}{\ln 6} + C$

37. $x(t) = t^2 + 17$

39. $x(t) = 1 + e^t$

41. $x(t) = \frac{1}{3}(2 + 7e^{3t}) + C$

43. $x(t) = \sqrt{2}e^{\sin t}$

45. Horizontal asymptote: the x-axis. Global maximum where $x = \frac{1}{2}$, inflection point where $x = (1 + \sqrt{2})/2$—approximately (1.21, 0.33). Global minimum and intercept at (0, 0), with a vertical tangent there as well. The graph is *not drawn to scale.*

47. Global minimum at $(4, 2 - \ln 4)$, inflection point at (16, 1.23) (ordinate approximate). The y-axis is a vertical asymptote. The graph continues to rise for large increasing x; there is no horizontal asymptote. The graph is *not drawn to scale.*

49. Inflection point at $(0.5, e^{-2})$. The horizontal line $y = 1$ and the y-axis are asymptotes. The point (0, 0) is *not* on the graph. As $x \to 0^+$, $y \to 0$; as $x \to 0^-$, $y \to +\infty$. As $|x| \to \infty$, $y \to 1$. The graph is *not drawn to scale.*

51. Sell immediately!

53. (b) The minimizing value is approximately 10.516. But because the batch size must be an integer, it turns out that 11 (rather than 10) minimizes $f(x)$. Thus the answer to part (c) is $977.85.

Section 8.2 (page 393)

1. $2\cos(2x + 3)$

3. $\frac{2}{3}\sec^2 \frac{2}{3}x$

5. $-\dfrac{1}{x^2}\sec\left(\dfrac{1}{x}\right)\tan\left(\dfrac{1}{x}\right)$

7. $12\sec^2 4x \tan^2 4x$

9. $\cot x$

11. $\dfrac{2}{x}\tan(\ln x)\sec^2(\ln x)$

13. $-\csc x$

15. $-3[\sin^2(\csc x)][\cos(\csc x)](\csc x \cot x)$

17. $\sec x \tan x$

19. $[\exp(\sin x)](1 + \sec x \tan x)$

21. $\frac{1}{3}(\sin x)^{-2/3}\cos x - \frac{1}{3}x^{-2/3}\cos(x^{1/3})$

23. $(\sec^2 x - \sec x \tan x + \sec x)(1 + \sec x)^{-2}$

25. $-12t^3\csc^2(1 + t^4)\cot^2(1 + t^4)$

27. $\dfrac{dy}{dx} = -\dfrac{y}{x}$; $x + y = \sqrt{\pi}$

29. $\dfrac{dy}{dx} = \dfrac{\sin x \cos x}{\sin y \cos y}$; $y = x$

31. $2\tan \frac{1}{2}x + C$

33. $\frac{1}{2}x + \frac{1}{12}\sin 6x + C$

35. $-\cos x + C$

37. $\ln|1 + \sec x| + C$

39. $\frac{1}{2}\sec x^2 + C$

41. $\frac{1}{2}\exp(\sin 2x) + C$

43. $\sec x + C$

45. $\ln|\tan e^x| + C$

47. $\frac{1}{6}\sin^6 x + C$

49. $\frac{1}{8}\sec^4 2x + C$

51. $\frac{1}{11}\sin^{11} x + C$

53. $\frac{1}{4}(1 + \cos x)^{-4} + C$

55. $\frac{1}{2}\sin e^{2x} + C$

57. $\dfrac{1}{\ln 2}2^{\sec x} + C$

59. $\sin(\ln x) + C$

61. $f'(x) = \sec^2 x$, which is always positive. $f''(x) = 2\sec^2 x \tan x$, which is zero at $(0, 0)$, $(\pi, 0)$, $(-\pi, 0)$, and so on; these are all x-intercepts and inflection points. There are vertical asymptotes at the odd integral multiples of $\pi/2$.

63. Minima at $(-3\pi/4, -\sqrt{2})$, $(5\pi/4, -\sqrt{2})$, and so on; maxima at $(\pi/4, \sqrt{2})$, $(9\pi/4, \sqrt{2})$, and so on; x-intercepts at $-\pi/4$, $3\pi/4$, $7\pi/4$, and so on (these are also inflection points); the y-intercept is (0, 1).

65. $\pi/4$

67. $\ln 4$

71. 0.86033359 and 3.42561846 (approximately)

73.

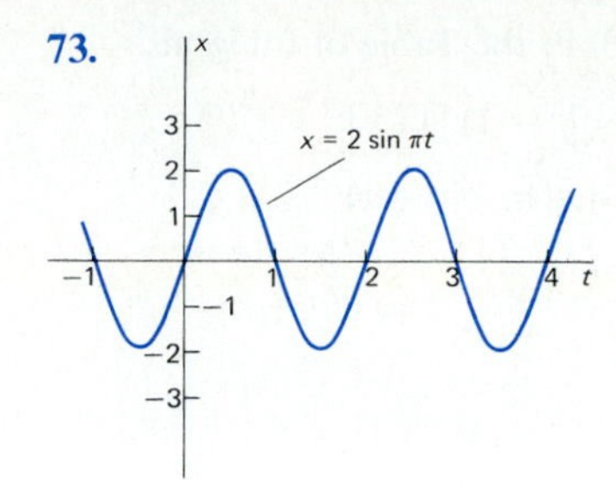

75.

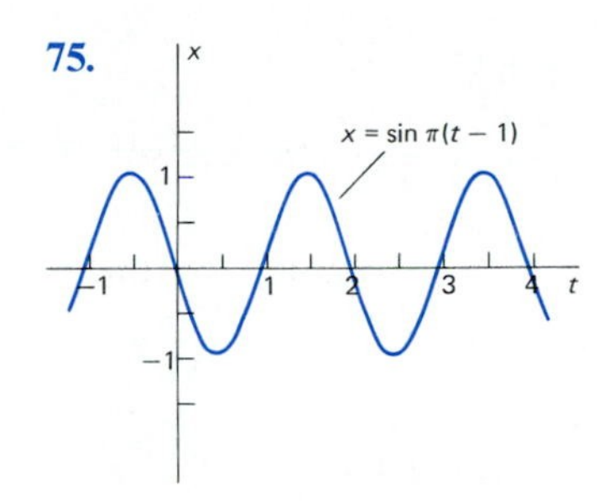

77.

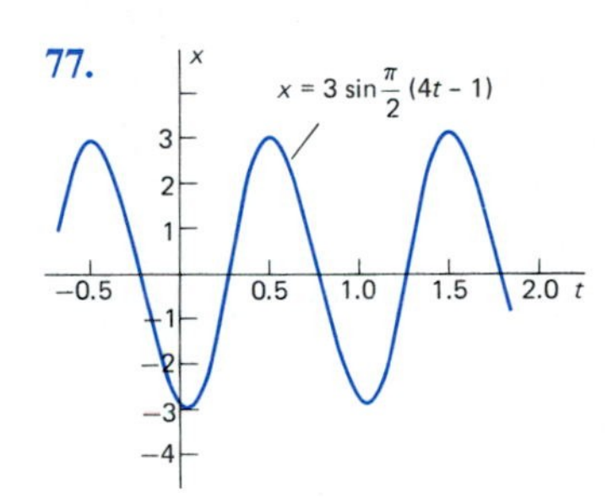

79.

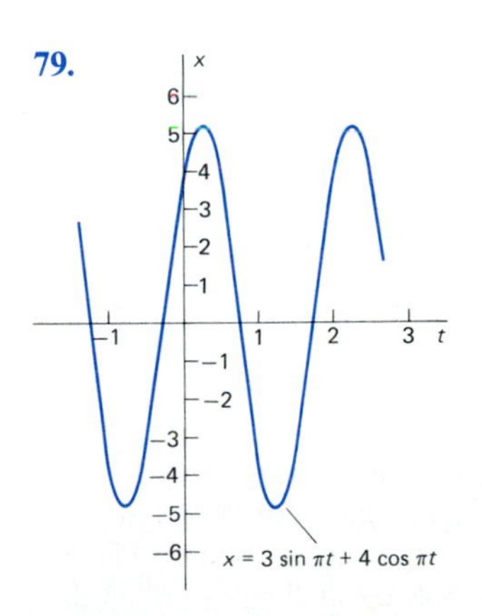

81. 1 **83.** $\frac{1}{2}\sqrt{5}$ **85.** 5π

87. Period 2π, amplitude 5

Section 8.3 (page 401)

1. (a) $\pi/6$, (b) $-\pi/6$, (c) $\pi/4$, (d) $-\pi/3$

3. (a) 0, (b) $\pi/4$, (c) $-\pi/4$, (d) $\pi/3$

5. $100x^{99}(1 - x^{-200})^{-1/2}$

7. $\{x|\ln x|[(\ln x)^2 - 1]^{1/2}\}^{-1}$

9. $(\sec^2 x)(1 - \tan^2 x)^{-1/2}$ **11.** $e^x(1 - e^{2x})^{-1/2}$

13. $-2(1 - x^2)^{-1/2}$ **15.** $-2x^{-1}(x^4 - 1)^{-1/2}$

17. $-[(1 + x^2)(\arctan x)^2]^{-1}$

19. $\{x[1 + (\ln x)^2]\}^{-1}$ **21.** $2e^x(1 + e^{2x})^{-1}$

23. $\dfrac{\cos(\arctan x)}{1 + x^2}$ **25.** $\dfrac{1 - 4x \arctan x}{(1 + x^2)^3}$

27. $\dfrac{dy}{dx} = -\dfrac{1 + y^2}{1 + x^2}$; $x + y = 2$

29. $\dfrac{dy}{dx} = -\dfrac{(1 - y^2)^{1/2} \arcsin y}{(1 - x^2)^{1/2} \arcsin x}$; $x + y = \sqrt{2}$

31. $\dfrac{\pi}{4}$ **33.** $\pi/12$ **35.** $\pi/12$

37. $\frac{1}{2} \arcsin 2x + C$ **39.** $\frac{1}{5} \operatorname{arcsec}|x/5| + C$

41. $\arctan(e^x) + C$ **43.** $\frac{1}{15} \operatorname{arcsec}|x^3/5| + C$

45. Either $\arcsin(2x - 1) + C$ or $2 \arcsin\sqrt{x} + C$

47. $\frac{1}{2} \arctan(x^{50}) + C$ **49.** $\arctan(\ln x) + C$

51. $\pi/4$ **53.** $\pi/2$ **55.** $\pi/12$ **59.** 8 ft

63. $\pi/2$

65. (b) $A = 1 - \frac{1}{3}\pi$, $B = 1 + \frac{2}{3}\pi$

Section 8.4 (page 406)

1. $3 \sinh(3x - 2)$

3. $2x \tanh(1/x) - \operatorname{sech}^2(1/x)$

5. $-12 \coth^2 4x \operatorname{csch}^2 4x$

7. $-e^{\operatorname{csch} x} \operatorname{csch} x \coth x$ **9.** $(\cosh x) \cos(\sinh x)$

11. $4x^3 \cosh x^4$ **13.** $-\dfrac{1 + \operatorname{sech}^2 x}{(x + \tanh x)^2}$

15. $\frac{1}{2} \cosh x^2 + C$ **17.** $x - \frac{1}{3} \tanh 3x + C$

19. $\frac{1}{6} \sinh^3 2x + C$ **21.** $-\frac{1}{2} \operatorname{sech}^2 x + C$

23. $-\frac{1}{2} \operatorname{csch}^2 x + C$ **25.** $\ln(1 + \cosh x) + C$

27. $\frac{1}{4} \tanh x + C$ **33.** $\sinh a$

Section 8.5 (page 410)

1. $\dfrac{2}{(4x^2 + 1)^{1/2}}$ **3.** $\frac{1}{2}x^{-1/2}(1 - x)^{-1}$

5. $(x^2 - 1)^{-1/2}$

7. $\frac{3}{2}(\sinh^{-1} x)^{1/2}(x^2 + 1)^{-1/2}$

9. $(1 - x^2)^{-1}(\tanh^{-1} x)^{-1}$

11. $\operatorname{arcsinh} \frac{1}{3}x + C$ **13.** $\frac{1}{4} \ln \frac{9}{5} \approx 0.14695$

15. $-\frac{1}{2} \operatorname{sech}^{-1}|3x/2| + C$ **17.** $\sinh^{-1}(e^x) + C$

19. $-\operatorname{sech}^{-1}(e^x) + C$

29. Time: approximately 485 s;
impact speed: about 20.656 ft/s

Chapter 8 Miscellaneous Problems (page 411)

1. $\cos \frac{1}{2}\sqrt{x}$
3. $x^{-1}\sec^2(\ln x)$
5. $-10(\csc 2x)(\csc 2x + \cot 2x)^5$
7. $6 \tan 3x$
9. $(1 + x^2)^{-1}\exp(\arctan x)$
11. $\frac{1}{2}x^{-1/2}(1 - x)^{-1/2}$
13. $\dfrac{2x}{x^4 + 2x^2 + 2}$
15. $e^x \sinh e^x + e^{2x}\cosh e^x$
17. 0
19. $\dfrac{x}{|x|(x^2 + 1)^{1/2}}$
21. $\frac{1}{2}\tan x^2 + C$
23. $-\frac{1}{4}\csc^2 2x + C$
25. $x^{-1} + \tan x^{-1} + C$
27. $\frac{1}{2}\ln|\sec(x^2 + 1)| + C$
29. $\sin^{-1}(e^x) + C$
31. $\frac{1}{2}\arcsin(2x/3) + C$
33. $\frac{1}{3}\arctan(x^3) + C$
35. $\sec^{-1}|2x| + C$
37. $\sec^{-1}(e^x) + C$
39. $2\cosh\sqrt{x} + C$
41. $\frac{1}{2}(\tan^{-1} x)^2 + C$
43. $\frac{1}{2}\sinh^{-1}(2x/3) + C$
45. $\pi^2/6$
47. The zeros shown are at (approximately) -0.197, 1.373, and 2.944. The maximum shown is at $(0.588, 13)$ and the minima at $(-0.983, -13)$ and $(2.159, -13)$ (abscissas approximate).
51. $x \approx 4.730041$
53. (d) $\sqrt{Rg}$; about 17,691 mi/h

Section 9.2 (page 416)

1. $-\frac{1}{15}(2 - 3x)^5 + C$
3. $\frac{1}{9}(2x^3 - 4)^{3/2} + C$
5. $\frac{9}{8}(2x^2 + 3)^{2/3} + C$
7. $-2\csc\sqrt{y} + C$
9. $\frac{1}{6}(1 + \sin\theta)^6 + C$
11. $e^{-\cot x} + C$
13. $\frac{1}{11}(\ln t)^{11} + C$
15. $\frac{1}{3}\arcsin 3t + C$
17. $\frac{1}{2}\arctan(e^{2x}) + C$
19. $\frac{3}{2}\arcsin(x^2) + C$
21. $\frac{1}{15}\tan^5 3x + C$
23. $\tan^{-1}(\sin\theta) + C$
25. $\frac{2}{5}(1 + \sqrt{x})^5 + C$
27. $\ln|\arctan t| + C$
29. $\sec^{-1} e^x + C$
31. $\frac{2}{7}(x - 2)^{7/2} + \frac{8}{5}(x - 2)^{5/2} + \frac{8}{3}(x - 2)^{3/2} + C$
33. $\frac{1}{3}(2x + 3)^{1/2}(x - 3) + C$
35. $\frac{3}{10}(x + 1)^{2/3}(2x - 13) + C$
37. $\dfrac{1}{60}\ln\left|\dfrac{3x + 10}{3x - 10}\right| + C$
39. $\frac{1}{2}x(4 + 9x^2)^{1/2} + \frac{2}{3}\ln|3x + (4 + 9x^2)^{1/2}| + C$
41. $\frac{1}{32}x(16x^2 + 9)^{1/2} - \frac{9}{128}\ln|4x + (16x^2 + 9)^{1/2}| + C$
43. $\frac{1}{128}x(32x^2 - 25)(25 - 16x^2)^{1/2} + \frac{625}{512}\arcsin\frac{4}{5}x + C$
45. The substitution $u = e^x$ leads to an integral in the form of (44) in the Table of Integrals (see the endpapers); the answer is
$$\tfrac{1}{2}e^x(9 + e^{2x})^{1/2} + \tfrac{9}{2}\ln(e^x + (9 + e^{2x})^{1/2}) + C.$$
47. With $u = x^2$ and (47) in the Table of Integrals:
$$\tfrac{1}{2}(x^4 - 1)^{1/2} - \operatorname{arcsec} x^2 + C.$$
49. With $u = \ln x$ and (48) in the Table of Integrals:
$$\tfrac{1}{8}([\ln x](2[\ln x]^2 + 1)([\ln x]^2 + 1)^{1/2}$$
$$- \ln|[\ln x] + ([\ln x]^2 + 1)^{1/2}|) + C$$
53. $\sin^{-1}(x - 1) + C$

Section 9.3 (page 420)

1. $\frac{1}{3}\cos^3 x - \cos x + C$
3. $\frac{1}{3}\sin^3\theta - \frac{1}{5}\sin^5\theta + C$
5. $\frac{1}{5}\sin^5 x - \frac{2}{3}\sin^3 x + \sin x + C$
7. $\frac{2}{5}(\cos x)^{5/2} - 2(\cos x)^{1/2} + C$
9. $-\frac{1}{14}\cos^7 2z + \frac{1}{5}\cos^5 2z - \frac{1}{6}\cos^3 2z + C$
11. $\frac{1}{4}(\sec 4x + \cos 4x) + C$
13. $\frac{1}{3}\tan^3 t + \tan t + C$
15. $-\frac{1}{4}\csc^2 2x - \frac{1}{2}\ln|\sin 2x| + C$
17. $\frac{1}{12}\tan^6 2x + C$
19. $-\frac{1}{10}\cot^5 2t - \frac{1}{3}\cot^3 2t - \frac{1}{2}\cot 2t + C$
21. $\frac{1}{4}\cos^4\theta - \frac{1}{2}\cos^2\theta + C$
23. $\frac{2}{3}(\sec t)^{3/2} + 2(\sec t)^{-1/2} + C$
25. $\frac{1}{3}\sin^3\theta + C$
27. $\frac{1}{5}\sin 5t - \frac{1}{15}\sin^3 5t + C$
29. $t + \frac{1}{3}\cot 3t - \frac{1}{9}\cot^3 3t + C$
31. $-\frac{1}{5}\cos^{5/2} 2t + \frac{2}{9}\cos^{9/2} 2t - \frac{1}{13}\cos^{13/2} 2t + C$
33. $\frac{1}{2}\sin^2 x - \cos x + C$
35. $-\cot x - \frac{1}{3}\cot^3 x - \frac{1}{2}\csc^2 x + C$
39. $\frac{1}{4}\cos 2x - \frac{1}{16}\cos 8x + C$
41. $\frac{1}{6}\sin 3x + \frac{1}{10}\sin 5x + C$

Section 9.4 (page 425)

1. $\frac{1}{2}xe^{2x} - \frac{1}{4}e^{2x} + C$
3. $-t\cos t + \sin t + C$
5. $\frac{1}{3}x\sin 3x + \frac{1}{9}\cos 3x + C$
7. $\frac{1}{4}x^4\ln x - \frac{1}{16}x^4 + C$
9. $x\arctan x - \frac{1}{2}\ln(1 + x^2) + C$
11. $\frac{2}{3}y^{3/2}\ln y - \frac{4}{9}y^{3/2} + C$
13. $t(\ln t)^2 - 2t\ln t + 2t + C$
15. $\frac{2}{3}x(x + 3)^{3/2} - \frac{4}{15}(x + 3)^{5/2} + C$
17. $\frac{2}{9}x^3(x^3 + 1)^{3/2} - \frac{4}{45}(x^3 + 1)^{5/2} + C$
19. $-\frac{1}{2}(\csc\theta\cot\theta + \ln|\csc\theta + \cot\theta|) + C$
21. $\frac{1}{3}x^3\arctan x - \frac{1}{6}x^2 + \frac{1}{6}\ln(1 + x^2) + C$
23. $x\operatorname{arcsec} x^{1/2} - (x - 1)^{1/2} + C$
25. $(x + 1)\arctan x^{1/2} - x^{1/2} + C$
27. $-x\cot x + \ln|\sin x| + C$
29. $\frac{1}{2}x^2\sin x^2 + \frac{1}{2}\cos x^2 + C$
31. $-2x^{-1/2}(2 + \ln x) + C$
33. $x\sinh x - \cosh x + C$
35. $x^2\cosh x - 2x\sinh x + 2\cosh x + C$

37. $\pi(e-2) \approx 2.25655$

39. $\frac{1}{2}(x-1)e^x \sin x + \frac{1}{2}xe^x \cos x + C$

47. $6 - 2e \approx 0.563436$

49. $6 - 2e \approx 0.563436$

Section 9.5 (page 434)

1. $\frac{1}{2}x^2 - x + \ln|x+1| + C$

3. $\dfrac{1}{3}\ln\left|\dfrac{x-3}{x}\right| + C$

5. $\dfrac{1}{5}\ln\left|\dfrac{x-2}{x+3}\right| + C$

7. $\frac{1}{4}\ln|x| - \frac{1}{8}\ln(x^2+4) + C$

9. $\frac{1}{3}x^3 - 4x + 8\arctan\frac{1}{2}x + C$

11. $x - 2\ln|x+1| + C$

13. $x + (x+1)^{-1} + C$

15. $\dfrac{1}{4}\ln\left|\dfrac{x-2}{x+2}\right| + C$

17. $\frac{3}{2}\ln|2x-1| - \ln|x+3| + C$

19. $\ln|x| + 2(x+1)^{-1} + C$

21. $\frac{3}{2}\ln|x^2-4| + \frac{1}{2}\ln|x^2-1| + C$

23. $\ln|x+2| + 4(x+2)^{-1} - 2(x+2)^{-2} + C$

25. $\dfrac{1}{2}\ln\left(\dfrac{x^2}{x^2+1}\right) + C$

27. $\dfrac{1}{2}\ln\left(\dfrac{x^2}{x^2+4}\right) + \dfrac{1}{2}\arctan\dfrac{1}{2}x + C$

29. $-\frac{1}{2}\ln|x+1| + \frac{1}{4}\ln(x^2+1) + \frac{1}{2}\arctan x + C$

31. $\arctan\frac{1}{2}x - \frac{3}{2}\sqrt{2}\arctan x\sqrt{2} + C$

33. $2^{-1/2}\arctan 2^{-1/2}x + \frac{1}{2}\ln(x^2+3) + C$

35. $x + \frac{1}{2}\ln|x-1| - 5(2x-2)^{-1}$
$+ \frac{3}{4}\ln(x^2+1) + 2\arctan x + C$

37. $\frac{1}{4}(1-2e^{2t})(e^{2t}-1)^{-2} + C$

39. $\frac{1}{4}\ln|3+2\ln t| + \frac{1}{4}(3+2\ln t)^{-1} + C$

41. $x(t) = \dfrac{2e^t}{2e^t-1}$

43. $x(t) = \dfrac{2e^{2t}+1}{2e^{2t}-1}$

45. $x(t) = \dfrac{21e^t-16}{8-7e^t}$

47. Approximately 153,700,000

49. (a) 0.36 s; (b) 200 g

51. $P(t) = \dfrac{200}{2-e^{t/100}}$;
(a) $t = 100\ln 1.8 \approx 58.8$ (days);
(b) $t = 100\ln 2 \approx 69.3$ (days)

Section 9.6 (page 440)

1. $\arcsin\frac{1}{4}x + C$

3. $-\dfrac{1}{4}\cdot\dfrac{(4-x^2)^{1/2}}{x} + C$

5. $-\frac{1}{8}x(16-x^2)^{1/2} + 8\arcsin\frac{1}{4}x + C$

7. $\frac{1}{9}x(9-16x^2)^{-1/2} + C$

9. $\ln|x+(x^2-1)^{1/2}| - x^{-1}(x^2-1)^{1/2} + C$

11. $\frac{1}{80}[(9+4x^2)^{5/2} - 15(9+4x^2)^{3/2}] + C$

13. $(1-4x^2)^{1/2} - \ln\left|\dfrac{1+(1-4x^2)^{1/2}}{2x}\right| + C$

15. $\frac{1}{2}\ln|2x+(9+4x^2)^{1/2}| + C$

17. $\frac{25}{2}\arcsin\frac{1}{5}x - \frac{1}{2}x(25-x^2)^{1/2} + C$

19. $\frac{1}{2}x(x^2+1)^{1/2} - \frac{1}{2}\ln|x+(1+x^2)^{1/2}| + C$

21. $\frac{1}{18}x(4+9x^2)^{1/2} - \frac{2}{27}\ln|3x+(4+9x^2)^{1/2}| + C$

23. $x(1+x^2)^{-1/2} + C$

25. $\dfrac{1}{256}\left[16x(4-x^2)^{-2} + 6x(4-x^2)^{-1} + 3\ln\left|\dfrac{2+x}{(4-x^2)^{1/2}}\right|\right] + C$

27. $\frac{1}{2}x(9+16x^2)^{1/2} + \frac{9}{8}\ln|4x+(9+16x^2)^{1/2}| + C$

29. $(x^2-25)^{1/2} - 5\,\mathrm{arcsec}\,\frac{1}{5}x + C$

31. $\frac{1}{8}x(2x^2+1)(x^2-1)^{1/2} - \frac{1}{8}\ln|x+(x^2-1)^{1/2}| + C$

33. $-x(4x^2-1)^{-1/2} + C$

35. $-x^{-1}(x^2-5)^{1/2} + \ln|x+(x^2-5)^{1/2}| + C$

37. $\sinh^{-1}\frac{1}{5}x + C$

39. $\cosh^{-1}\frac{1}{2}x - x^{-1}(x^2-4)^{1/2} + C$

41. $\frac{1}{8}x(1+2x^2)(1+x^2)^{1/2} - \frac{1}{8}\sinh^{-1}x + C$

43. $\frac{1}{32}\pi[18\sqrt{5} - \ln(2+\sqrt{5})] \approx 3.8097$

45. $\sqrt{5} - \sqrt{2} + \ln\left(\dfrac{2+2\sqrt{2}}{1+\sqrt{5}}\right) \approx 1.222016$

49. $2\pi[\sqrt{2} + \ln(1+\sqrt{2})] \approx 14.4236$

53. $\$6\frac{2}{3}$ million

Section 9.7 (page 445)

1. $\arctan(x+2) + C$

3. $11\arctan(x+2) - \frac{3}{2}\ln(x^2+4x+5) + C$

5. $\arcsin\frac{1}{2}(x+1) + C$

7. $-2\arcsin\frac{1}{2}(x+1) - \frac{1}{2}(x+1)(3-2x-x^2)^{1/2}$
$-\frac{1}{3}(3-2x-x^2)^{3/2} + C$

9. $\frac{5}{16}\ln|2x+3| + \frac{7}{16}\ln|2x-1| + C$

11. $\frac{1}{3}\arctan\frac{1}{3}(x+2) + C$

13. $\dfrac{1}{4}\ln\left|\dfrac{1+x}{3-x}\right| + C$

15. $\ln(x^2+2x+2) - 7\arctan(x+1) + C$

17. $\frac{2}{9}\arcsin(x-\frac{2}{3}) - \frac{1}{9}(5+12x-9x^2)^{1/2} + C$

19. $\frac{75}{4}\arcsin\frac{2}{5}(x-2) + \frac{3}{2}(x-2)(9+16x-4x^2)^{1/2}$
$+\frac{1}{6}(9+16x-4x^2)^{3/2} + C$

21. $\frac{1}{9}(7x-12)(6x-x^2)^{-1/2} + C$

23. $-(16x^2+48x+52)^{-1} + C$

25. $\frac{3}{2}\ln(x^2+x+1) - \frac{5}{3}\sqrt{3}\arctan(\frac{1}{3}\sqrt{3}[2x+1]) + C$

27. $\dfrac{1}{32}\ln\left|\dfrac{x+2}{x-2}\right| - \dfrac{1}{8}x(x^2-4)^{-1} + C$

29. $\ln|x| - \frac{2}{3}\sqrt{3}\arctan\frac{1}{3}\sqrt{3}(2x+1) + C$

31. $-\frac{5}{4}(x-1)^{-1} - \frac{1}{4}\ln|x-1| - \frac{5}{4}(x+1)^{-1}$
$+ \frac{1}{4}\ln|x+1| + C$

33. $-\frac{1}{4}(x-7)(x^2+2x+5)^{-1} - \frac{1}{8}\arctan\frac{1}{2}(x+1) + C$

37. Approximately 3.69 mi

39. $\ln|x-1| - \frac{1}{2}\ln(x^2+2x+2) + \arctan(x+1) + C$

41. $\frac{1}{2}x^2 + \ln|x-1| + \frac{1}{2}\ln(x^2+x+1)$
$+ \frac{1}{3}\sqrt{3}\arctan\frac{1}{3}\sqrt{3}(2x+1) + C$

43. $\dfrac{1}{4}\ln(x^4+x^2+1) - \dfrac{1}{3}\sqrt{3}\arctan\left(\dfrac{\sqrt{3}}{2x^2+1}\right) + C$

Section 9.8 (page 450)

1. $\frac{2}{729}(3x-2)^{9/2} + \frac{4}{189}(3x-2)^{7/2}$
$+ \frac{8}{135}(3x-2)^{5/2} + \frac{16}{243}(3x-2)^{3/2} + C$

3. $2\sqrt{x} - 2\ln(1+\sqrt{x}) + C$

5. With $u = (x^2-1)^{1/3}$:
$\frac{3}{4}(x^2-3)(x^2-1)^{-1/3} + C$

7. With $x = u^4$: $x - \frac{4}{5}x^{5/4} + C$

9. With $u = 1 + x^3$:
$\frac{2}{9}(1+x^3)^{3/2} - \frac{2}{3}(1+x^3)^{1/2} + C$

11. With $u = x^{1/3}$: $3x^{1/3} - 3\arctan x^{1/3} + C$

13. With $u = (x+4)^{1/2}$: $2(x+4)^{1/2}$
$- 2\ln(1+(x+4)^{1/2}) + C$

15. $\dfrac{\sin\theta - 1}{\cos\theta} + C$

17. $\sqrt{2}\ln\left|\dfrac{\sqrt{2}-1+u}{(1+2u-u^2)^{1/2}}\right| + C$
where $u = \tan\frac{1}{2}\theta$

19. $-\ln|2+\cos\theta| + C$

23. $\frac{8}{15}$

25. $\sqrt{2} + \ln(1+\sqrt{2}) \approx 2.295587$

31. $u + \dfrac{1}{2}\ln\left|\dfrac{u-1}{u+1}\right| + C$, where
$u = (1+e^{2x})^{1/2}$

Chapter 9 Miscellaneous Problems (page 452)

NOTE Different techniques of integration may produce answers that appear to differ from those shown here; if both are correct, they of course differ only by a constant.

1. $2\arctan\sqrt{x} + C$

3. $\ln|\sec x| + C$

5. $\frac{1}{2}\sec^2\theta + C$

7. $x\tan x - \frac{1}{2}x^2 + \ln|\cos x| + C$

9. $\frac{2}{15}(2-x^3)^{5/2} - \frac{4}{9}(2-x^3)^{3/2} + C$

11. $\frac{1}{2}x(25+x^2)^{1/2} - \frac{25}{2}\ln|x+(25+x^2)^{1/2}| + C$

13. $\frac{2}{3}\sqrt{3}\arctan\frac{1}{3}\sqrt{3}(2x-1) + C$

15. $\frac{103}{87}\sqrt{29}\arctan\frac{1}{29}\sqrt{29}(3x-2)$
$+ \frac{5}{6}\ln(9x^2-12x+33) + C$

17. $\frac{2}{3}\arctan(\frac{1}{3}\tan\frac{1}{2}\theta) + C$

19. $\arcsin(\frac{1}{2}\sin x) + C$

21. $-\ln|\ln\cos x| + C$

23. $(1+x)\ln(1+x) + C$

25. $\frac{1}{2}x(x^2+9)^{1/2} + \frac{9}{2}\ln|x+(x^2+9)^{1/2}| + C$

27. $\frac{1}{2}(x-1)(2x-x^2)^{1/2} + \frac{1}{2}\arcsin(x-1) + C$

29. $\dfrac{1}{3}x^3 + 2x - \sqrt{2}\ln\left|\dfrac{x+\sqrt{2}}{x-\sqrt{2}}\right| + C$

31. $\frac{1}{2}(x^2+x)(x^2+2x+2)^{-1} - \frac{1}{2}\arctan(x+1) + C$

33. $\dfrac{1}{2}\tan\theta + C$ or $\dfrac{1}{2}\dfrac{\sin 2\theta}{1+\cos 2\theta} + C$

35. $\frac{1}{5}\sec^5 x - \frac{1}{3}\sec^3 x + C$

37. $\frac{1}{8}x^2[4(\ln x)^3 - 6(\ln x)^2 + 6(\ln x) - 3] + C$

39. $\frac{1}{2}e^x(1+e^{2x})^{1/2} + \frac{1}{2}\ln[e^x + (1+e^{2x})^{1/2}] + C$

41. $\frac{1}{54}\operatorname{arcsec}|\frac{1}{3}x| + \frac{1}{18}x^{-2}(x^2-9)^{1/2} + C$

43. $\ln|x| + \frac{1}{2}\arctan 2x + C$

45. $\frac{1}{2}(\sec x\tan x - \ln|\sec x + \tan x|) + C$

47. $\ln|x+1| - \frac{2}{3}x^{-3} + C$

49. $\ln|x-1| + \ln(x^2+x+1)$
$+ (x-1)^{-1} - 2(x^2+x+1)^{-1} + C$

51. $\dfrac{1}{3}\ln\left|\dfrac{1-\cos\theta+3\sin\theta}{1-\cos\theta-3\sin\theta}\right| + C$

53. $\frac{1}{3}(\arcsin x)^3 + C$

55. $\frac{1}{2}\sec^2 z + \ln|\cos z| + C$

57. $\frac{1}{2}\arctan(e^{x^2}) + C$

59. $-\frac{1}{2}(x^2+1)\exp(-x^2) + C$

61. $-x^{-1}\arcsin x - \ln\left|\dfrac{1+(1-x^2)^{1/2}}{x}\right| + C$

63. $\frac{1}{8}\arcsin x + \frac{1}{8}x(2x^2-1)(1-x^2)^{1/2} + C$

65. $\frac{1}{4}\ln|2x+1| + \frac{5}{4}(2x+1)^{-1} + C$

67. $\frac{1}{2}\ln|e^{2x}-1| + C$

69. $2\ln|x+1| + 3(x+1)^{-1} - \frac{5}{3}(x+1)^{-3} + C$

71. $\frac{1}{2}\ln(x^2+1) + \arctan x - \frac{1}{2}(x^2+1)^{-1} + C$

73. $\frac{1}{45}(x^3+1)^{3/2}(6x^3+4) + C$

75. $\frac{2}{3}(1+\sin x)^{3/2} + C$

77. $\frac{1}{2}\ln|\sec x + \tan x| + C$

79. $-2(1-\sin t)^{1/2} + C$

81. $-2x + \sqrt{3}\arctan\frac{1}{3}\sqrt{3}(2x+1)$
$+ \frac{1}{2}(2x+1)\ln(x^2+x+1) + C$

83. $-x^{-1}\arctan x + \ln|x(1+x^2)^{-1/2}| + C$

85. $\frac{1}{2}\ln(x^2+1) + \frac{1}{2}(x^2+1)^{-1} + C$

87. $\frac{1}{2}(x-6)(x^2+4)^{-1/2} + C$

89. $\frac{1}{3}(1+\sin^2 x)^{3/2}+C$

91. $\frac{1}{2}e^x(x\sin x - x\cos x + \cos x) + C$

93. $-\frac{1}{2}(x-1)^{-2}\arctan x + \frac{1}{2}(x^2+1)(x-1)^{-2} - \frac{1}{4}(x-1)^{-1} + C$

95. $\frac{11}{9}\arcsin\frac{1}{2}(3x-1) - \frac{2}{9}(3+6x-9x^2)^{1/2} + C$

97. $\frac{1}{2}\cos^2\theta + \cos\theta + C$

99. $x\,\text{arcsec}\sqrt{x} - (x-1)^{1/2} + C$

101. $\frac{1}{4}\pi(e^2 - e^{-2} + 4)$

103. (a) $A_b = \pi\left(\sqrt{2} - e^{-b}(1+e^{-2b})^{1/2} + \ln\left[\dfrac{1+\sqrt{2}}{(e^{-b}+(1+e^{-2b})^{1/2}}\right]\right)$;
(b) $\pi[\sqrt{2}+\ln(1+\sqrt{2})] \approx 7.2118$

105. $\frac{1}{2}\pi\sqrt{2}\left[2\sqrt{14} - \sqrt{2} + \ln\left(\dfrac{1+\sqrt{2}}{2\sqrt{2}+\sqrt{7}}\right)\right]$, approximately 11.66353

109. $\frac{5}{4}\pi \approx 3.29699$

111. The value of the integral is $\frac{1}{630}$.

113. $\frac{1}{2}(5\sqrt{6}-3\sqrt{2}) + \frac{1}{2}\ln\left(\dfrac{1+\sqrt{2}}{\sqrt{3}+\sqrt{2}}\right) \approx 3.869983$

115. The substitution is $u = e^x$.
(a) $\frac{2}{3}\sqrt{3}\arctan\frac{1}{3}\sqrt{3}(1+2e^x) + C$

119. $\frac{1}{4}\sqrt{2}\ln\left|\dfrac{1+\tan\theta-(2\tan\theta)^{1/2}}{1+\tan\theta+(2\tan\theta)^{1/2}}\right| - \frac{1}{2}\sqrt{2}\arctan(2\cot\theta)^{1/2} + C$

Section 10.1 (page 459)

1. $x + 2y + 3 = 0$

3. $4y + 25 = 3x$

5. $x + y = 1$

7. Center $(-1, 0)$, radius $\sqrt{5}$

9. Center $(2, -3)$, radius 4

11. Center $(0.5, 0)$, radius 1

13. Center $(0.5, -1.5)$, radius 3

15. Center $(-\frac{1}{3}, \frac{4}{3})$, radius 2

17. The point $(3, 2)$

19. No points

21. $(x+1)^2 + (y+2)^2 = 34$

23. $(x-6)^2 + (y-6)^2 = \frac{4}{5}$

25. Equation: $2x + y = 13$

27. $(x-6)^2 + (y-11)^2 = 18$

29. $\left(\dfrac{x}{5}\right)^2 + \left(\dfrac{y}{3}\right)^2 = 1$

31. $y - 7 + 4\sqrt{3} = (4 - 2\sqrt{3})(x - 2 + \sqrt{3})$,
$y - 7 - 4\sqrt{3} = (4 + 2\sqrt{3})(x - 2 - \sqrt{3})$

33. $y - 1 = 4(x-4)$ and $y + 1 = 4(x+4)$

35. $h = \dfrac{p(e^2+1)}{1-e^2}$, $a = |h^2 - p^2|^{1/2}$,
$b = a(e^2-1)^{1/2}$

Section 10.2 (page 466)

1. (a) $(\frac{1}{2}\sqrt{2}, \frac{1}{2}\sqrt{2})$
(b) $(1, -\sqrt{3})$
(c) $(\frac{1}{2}, -\frac{1}{2}\sqrt{3})$
(d) $(0, -3)$
(e) $(\sqrt{2}, \sqrt{2})$
(f) $(\sqrt{3}, -1)$
(g) $(-\sqrt{3}, 1)$

3. $r\cos\theta = 4$

5. $\theta = \arctan\frac{1}{3}$

7. $r^2\cos\theta\sin\theta = 1$

9. $r = \tan\theta\sec\theta$

11. $x^2 + y^2 = 9$

13. $x^2 + 5x + y^2 = 0$

15. $(x^2+y^2)^3 = 4y^4$

17. $x = 3$

19. $x = 2$; $r = 2\sec\theta$

21. $x + y = 1$; $r = \dfrac{1}{\cos\theta + \sin\theta}$

23. $y = x + 2$; $r = \dfrac{2}{\sin\theta - \cos\theta}$

25. $x^2 + y^2 + 8y = 0$; $r = -8\sin\theta$

27. $x^2 + y^2 = 2x + 2y$; $r = 2(\cos\theta + \sin\theta)$

29. Symmetric about the x-axis

31. Symmetric about the x-axis

33. Symmetric about the x-axis

35. Symmetric about the origin

37. Symmetric about both axes and about the origin

39. Symmetric about the x-axis

41. Symmetric about the y-axis

43. No points of intersection

45. $(0, 0)$, $(1/2, \pi/6)$, $(1/2, 5\pi/5)$, $(1, \pi/2)$

47. Four points: The pole, the point $(r, \theta) = (2, \pi)$, and the two points $r = 2(\sqrt{2}-1)$, $|\theta| = \arccos(3 - 2\sqrt{2})$.

49. (a) $r\cos(\theta - \alpha) = p$

Section 10.3 (page 471)

1. π

3. $3\pi/2$

5. $9\pi/2$

7. 4π

9. $19\pi/2$

11. $\pi/2$ (one of *four* loops)

13. $\pi/4$ (one of *eight* loops)

15. 2 (one of *two* loops)

17. 4 (one of *two* loops)

19. $\frac{1}{6}(2\pi + 3\sqrt{3})$

21. $\frac{1}{24}(5\pi - 6\sqrt{3})$

23. $\frac{1}{6}(39\sqrt{3} - 10\pi)$

25. $\frac{1}{2}(2 - \sqrt{2})$

27. $\frac{1}{6}(20\pi + 21\sqrt{3})$

29. $\frac{1}{2}(2 + \pi)$

31. $\pi/2$

Section 10.4 (page 476)

1. $y^2 = 12x$
3. $(x-2)^2 = -8(y-3)$
5. $(y-3)^2 = -8(x-2)$
7. $x^2 = -6(y+\frac{3}{2})$
9. $x^2 = 4(y+1)$
11. $y^2 = 12x$; vertex $(0, 0)$, axis the x-axis
13. $y^2 = -6x$; vertex $(0, 0)$, axis the x-axis
15. $x^2 - 4x - 4y = 0$; vertex $(2, -1)$, axis the line $x = 2$
17. $4x^2 + xy + 4y + 13 = 0$; vertex $(-0.5, -3)$, axis the line $x = -0.5$
23. About 0.693 days; that is, about 16 h 38 min
27. $\alpha = \pi/12$, $\alpha = 5\pi/12$

Section 10.5 (page 481)

1. $\left(\frac{x}{4}\right)^2 + \left(\frac{y}{5}\right)^2 = 1$
3. $\left(\frac{x}{15}\right)^2 + \left(\frac{y}{17}\right)^2 = 1$
5. $\frac{x^2}{16} + \frac{y^2}{7} = 1$
7. $\frac{x^2}{100} + \frac{y^2}{75} = 1$
9. $\frac{x^2}{16} + \frac{y^2}{12} = 1$
11. $\frac{(x-2)^2}{16} + \frac{(y-3)^2}{4} = 1$
13. $\frac{(x-1)^2}{25} + \frac{(y-1)^2}{16} = 1$
15. $\frac{(x-1)^2}{81} + \frac{(y-2)^2}{72} = 1$
17. Center $(0, 0)$, foci $(\pm 2\sqrt{5}, 0)$, major axis of length 12, minor axis of length 8
19. Center $(0, 4)$, foci $(0, 4 \pm \sqrt{5})$, major axis of length 6, minor axis of length 4
21. About 3466.54 A.U.—that is, about 3.22×10^{11} mi or about 20 light-days
27. $\frac{(x-1)^2}{4} + \frac{y^2}{16/3} = 1$

Section 10.6 (page 489)

1. $\frac{x^2}{1} - \frac{y^2}{15} = 1$
3. $\frac{x^2}{16} - \frac{y^2}{9} = 1$
5. $\frac{y^2}{25} - \frac{x^2}{25} = 1$
7. $\frac{y^2}{9} - \frac{x^2}{27} = 1$
9. $\frac{x^2}{4} + \frac{y^2}{12} = 1$
11. $\frac{(x-2)^2}{9} - \frac{(y-2)^2}{27} = 1$
13. $\frac{(y+2)^2}{9} - \frac{(x-1)^2}{4} = 1$
15. Center $(1, 2)$, foci $(1 \pm \sqrt{2}, 2)$, asymptotes $y - 2 = \pm(x-1)$
17. Center $(0, 3)$, foci $(0, 3 \pm 2\sqrt{3})$, asymptotes $y = 3 \pm x\sqrt{3}$
19. Center $(-1, 1)$, foci $(-1 \pm \sqrt{13}, 1)$, asymptotes $y = \frac{3x+5}{2}$, $y = -\frac{3x+1}{2}$
21. There are no points on the graph if $c > 15$.
25. $16x^2 + 50xy + 16y^2 = 369$
27. The plane is about 16.42 mi north of B and 8.66 mi west of B; that is, it is about 18.56 mi from B at a bearing of 27°48′ west of north.

Section 10.7 (page 494)

1. $2(x')^2 + (y')^2 = 4$: ellipse; origin $(2, 3)$
3. $9(x')^2 - 16(y')^2 = 144$: hyperbola; origin $(1, -1)$
5. The single point $(2, 3)$
7. $4(x')^2 + 2(y')^2 = 1$: ellipse; 45°
9. The two parallel lines (a "degenerate parabola") $(x')^2 = 4$; $\tan^{-1}(\frac{1}{2}) \approx 26.57°$
11. $(x')^2 - (y')^2 = 1$: hyperbola; $\tan^{-1}(\frac{1}{3}) \approx 18.43°$
13. $2(x')^2 + (y')^2 = 2$; ellipse; $\tan^{-1}(\frac{2}{3}) \approx 33.69°$
15. $4(x')^2 + (y')^2 = 4$: ellipse; $\tan^{-1}(\frac{4}{3}) \approx 53.13°$
17. $2(x')^2 + (y')^2 = 4$: ellipse; $\tan^{-1}(\frac{1}{4}) \approx 14.04°$
19. The two perpendicular lines $y' = x'$ and $y' = -x'$ (a "degenerate hyperbola"); $\tan^{-1}(\frac{5}{12}) \approx 22.67°$
21. $25(x'-1)^2 + 50(y')^2 = 50$: ellipse; $\tan^{-1}(\frac{3}{4}) \approx 53.13°$
23. $2(y'-1)^2 - (x'-2)^2 = 1$: hyperbola; $\tan^{-1}(\frac{4}{3}) \approx 53.13°$
25. $(x'-1)^2 - (y')^2 = 1$: hyperbola; $\tan^{-1}(\frac{8}{15}) \approx 28.07°$

Chapter 10 Miscellaneous Problems (page 496)

1. Circle, center $(1, -1)$, radius 2
3. Circle, center $(3, -1)$, radius 1
5. Parabola, vertex $(4, -2)$, opening downward
7. Ellipse, center $(2, 0)$, major axis 6, minor axis 4
9. Hyperbola, center $(-1, 1)$, vertical axis, foci at $(-1, 1 \pm \sqrt{3})$
11. There are no points on the graph.
13. Hyperbola, axis inclined at 22.5° from the horizontal

15. Ellipse, major axis $2\sqrt{2}$, minor axis 1, rotated through the angle $\alpha = \pi/4$, center at the origin

17. Parabola, vertex (0, 0), opening to the "northeast," axis at angle $\alpha = \tan^{-1}(\frac{3}{4})$ from the horizontal

19. Circle, center (1, 0), radius 1

21. Straight line $y = x + 1$

23. Horizontal line $y = 3$

25. Two ovals tangent to each other and to the y-axis at (0, 0)

27. Apple-shaped curve, symmetric about the y-axis

29. Ellipse, one focus at (0, 0), directrix $x = 4$, eccentricity $e = 0.5$

31. $\frac{1}{2}(\pi - 2)$

33. $\frac{1}{6}(39\sqrt{3} - 10\pi) \approx 6.02234$

35. 2

37. $5\pi/4$

39. $r = 2p\cos(\theta - \alpha)$

41. If $a > b$, the maximum is $2a$ and the minimum is $2b$.

43. $b^2 y = 4hx(b - x)$ or, alternatively, $r = b\sec\theta - (b^2/4h)\sec\theta\tan\theta$

45. *Suggestion:* Let θ be the angle that QR makes with the x-axis.

49. The curve is a hyperbola with one focus at the origin, directrix $x = -3/2$, and eccentricity $e = 2$.

45. $\frac{3}{2}$

Section 11.1 (page 505)

1. $\frac{1}{2}$ 3. $\frac{2}{5}$ 5. 0 7. 0
9. $\frac{1}{2}$ 11. 2 13. 0 15. 1
17. 1 19. $\frac{3}{5}$ 21. $\frac{3}{2}$ 23. $\frac{1}{3}$
25. $\dfrac{\ln 2}{\ln 3}$ 27. $\frac{1}{2}$ 29. 1 31. $\frac{1}{3}$
33. $-\frac{1}{2}$ 35. 1 37. $\frac{1}{4}$ 39. $\frac{2}{3}$
41. 6 43. $\frac{4}{3}$ 45. $\frac{2}{3}$ 47. 0

Section 11.2 (page 508)

1. 1 3. $\frac{3}{8}$ 5. $\frac{1}{4}$ 7. 1
9. 0 11. -1 13. $-\infty$ 15. $+\infty$
17. $-\frac{1}{2}$ 19. 0 21. 1 23. 1
25. $e^{-1/6}$ 27. $e^{-1/2}$ 29. 1 31. e
33. $-\infty$

Section 11.3 (page 519)

1. $e^{-x} = 1 - \dfrac{x}{1!} + \dfrac{x^2}{2!} - \dfrac{x^3}{3!} + \dfrac{x^4}{4!} - \dfrac{x^5}{5!} + \left(\dfrac{e^{-z}}{6!}\right)x^6$ for some z between 0 and x.

3. $\cos x = 1 - \dfrac{x^2}{2!} + \dfrac{x^4}{4!} - \left(\dfrac{\sin z}{5!}\right)x^5$ for some z between 0 and x.

5. $(1 + x)^{1/2} = 1 + \dfrac{x}{1!2} - \dfrac{x^2}{2!4} + \dfrac{3x^3}{3!8} - \dfrac{5x^4}{128}(1 + z)^{7/2}$ for some z between 0 and x.

7. $\tan x = \dfrac{x}{1!} + \dfrac{2x^3}{3!} + ((16\sec^4 z\tan z + 8\sec^3 z\tan^3 z)/4!)x^4$ for some z between 0 and x.

9. $\sin^{-1} x = \dfrac{x}{1!} + \dfrac{x^3}{3!}\cdot\dfrac{1 + 2z^2}{(1 - z^2)^{5/2}}$ for some z between 0 and x.

11. $e^x = e + \dfrac{e}{1!}(x - 1) + \dfrac{e}{2!}(x - 1)^2 + \dfrac{e}{3!}(x - 1)^3 + \dfrac{e}{4!}(x - 1)^4 + \dfrac{e^z}{5!}(x - 1)^5$ for some z between 1 and x.

13. $\sin x = \dfrac{1}{2} + \dfrac{\sqrt{3}}{1!2}\left(x - \dfrac{\pi}{6}\right) - \dfrac{1}{2!2}\left(x - \dfrac{\pi}{6}\right)^2 - \dfrac{\sqrt{3}}{3!2}\left(\dfrac{x - \pi}{6}\right)^3 + \dfrac{\sin z}{4!}\left(x - \dfrac{\pi}{6}\right)^4$ for some z between $\pi/6$ and x.

15. $1/(x - 4)^2 = 1 - 2(x - 5) + 3(x - 5)^2 - 4(x - 5)^3 + 5(x - 5)^4 - 6(x - 5)^5 + \dfrac{7}{(z - 4)^8}(x - 5)^6$ for some z between 5 and x.

17. Six-place accuracy

19. Five-place accuracy

21. Problem 17 predicts $e^{1/3} \approx 1.3955286$; the true value is about 1.3956124 (the error is about 0.00003).

25. The third-degree polynomial gives the estimate 0.81915.

27. The third-degree polynomial gives the estimate 0.8829476.

Section 11.4 (page 527)

1. 1 3. $+\infty$ 5. $+\infty$ 7. 1
9. $+\infty$ 11. $-\frac{1}{2}$ 13. $\frac{9}{2}$ 15. $+\infty$
17. Does not exist 19. $2(e - 1)$
21. $+\infty$ 23. $\frac{1}{4}$

Chapter 11 Miscellaneous Problems (page 529)

1. $\frac{1}{4}$ 3. $\frac{1}{2}$ 5. $\frac{7}{45}$ 7. 1
9. $-\infty$ 11. $+\infty$ 13. e^2 15. $-\frac{1}{2}e$
17. $+\infty$ 19. 1 21. 2

23. $e^{x+1} = e + \dfrac{ex}{1!} + \dfrac{ex^2}{2!} + \dfrac{ex^3}{3!} + \dfrac{ex^4}{4!} + \dfrac{ex^5}{5!} + \left(\dfrac{e^{z+1}}{6!}\right)x^6$ for some number z between 0 and x.

25. $e^{-x^2} = 1 - x^2 + \left(\dfrac{(16z^4 - 48z^2 + 12)e^{-z^2}}{4!}\right)x^4$ for some number z between 0 and x.

27. $x^{1/3} = 5 + \dfrac{x - 125}{75} - \dfrac{(x - 125)^2}{28,125} + \dfrac{(x - 125)^2}{6,328,125} - \dfrac{10(x - 125)^4}{243z^{11/3}}$ for some z between 125 and x.

29. $x^{-1/2} = 1 - \dfrac{x - 1}{1!2} + \dfrac{3(x - 1)^2}{2!4} - \dfrac{15(x - 1)^3}{3!8} + \dfrac{105(x - 1)^4}{4!16} - \dfrac{945(x - 1)^5}{5!32z^{11/2}}$ for some z between 1 and x.

31. $x^4 = 1 + 4(x - 1) + 6(x - 1)^2 + 4(x - 1)^3 + (x - 1)^4 + 0$—the remainder ("0") is in fact zero in this case.

33. Five terms of the series for $f(x) = x^{1/10}$, center 1024, give the estimate 1.99526. The true value is approximately 1.995262315.

35. Not even one-place accuracy because the error may be as large as 0.2.

Section 12.2 (page 538)

1. $\frac{2}{5}$ 3. $\frac{1}{2}$ 5. 1 7. Does not converge 9. 0 11. 0 13. 0 15. 1 17. 0 19. 0 21. 0 23. 0 25. e 27. $\dfrac{1}{e^2}$ 29. 2 31. 1 33. Does not converge 35. 0 41. (b) 4

Section 12.3 (page 546)

1. $\frac{3}{2}$ 3. Diverges 5. Diverges 7. 6 9. Diverges 11. Diverges 13. Diverges 15. $\dfrac{\sqrt{2}}{\sqrt{2} - 1}$ 17. Diverges 19. $\frac{1}{12}$ 21. $\dfrac{e}{\pi - e}$ 23. Diverges 25. $\frac{65}{12}$ 27. $\frac{167}{8}$ 29. $\frac{1}{4}$ 31. $\frac{47}{99}$ 33. $\frac{41}{333}$ 35. $\dfrac{314,156}{99,999}$

37. $S_n = \ln(n + 1)$; diverges

39. $S_n = \frac{3}{2} - n^{-1} - (n + 1)^{-1}$; the sum is $\frac{3}{2}$.

45. Computations with S are meaningless because S is not a number.

47. 4.5 s

49. (a) $M_n = (0.95)^n M_0$; (b) 0

51. Peter $\frac{4}{7}$, Paul $\frac{2}{7}$, Mary $\frac{1}{7}$ 53. $\frac{1}{12}$

Section 12.4 (page 554)

1. Diverges 3. Diverges 5. Converges 7. Diverges 9. Converges 11. Converges 13. Converges 15. Converges 17. Diverges 19. Converges 21. Diverges 23. Diverges 25. Converges 27. Converges

29. The terms are not nonnegative.

31. The terms are not monotone decreasing.

33. $n = 100$

37. With $n = 6$, $1.03689 < S < 1.03699$

39. About a million centuries

41. Results with $n = 10$: $S_{10} \approx 1.08203658$ and $3.141566 < \pi < 3.141627$.

Section 12.5 (page 561)

1, 5, 9, 11, 15, 17, 19, 21, 23, 25, 31, and 33: Converge. 3, 7, 13, 27, 29, and 35: Diverge

Section 12.6 (page 568)

1, 3, 5, and 7: Converge

9, 19, 25, and 27: Diverge

11, 15, 17, 21, 29, and 31: Converge absolutely

13 and 23: Converge conditionally

33. $n = 1999$

35. $n = 6$

37. The first six terms give the estimate 0.6065.

39. The first three terms give the estimate 0.0953

41. Results using the first ten terms: $S_{10} \approx 0.8179622$ and $3.1329 < \pi < 3.1488$.

Section 12.7 (page 573)

1. $e^{-x} = 1 - x + \dfrac{x^2}{2!} - \dfrac{x^3}{3!} + \dfrac{x^4}{4!} - \cdots$

3. $e^{-3x} = 1 - 3x + \dfrac{9x^2}{2!} - \dfrac{27x^3}{3!} + \dfrac{81x^4}{4!} - \cdots$

5. $\cos 2x = 1 - \dfrac{4x^2}{2!} + \dfrac{16x^4}{4!} - \dfrac{64x^6}{6!} + \dfrac{256x^8}{8!} - \cdots$

7. $\sin x^2 = x^2 - \dfrac{x^6}{3!} + \dfrac{x^{10}}{5!} - \dfrac{x^{14}}{7!} + \dfrac{x^{18}}{9!} - \cdots$

9. $x - \frac{1}{2}x^2 + \frac{1}{3}x^3 - \frac{1}{4}x^4 + \cdots$

11. $1 - x + \frac{x^2}{2!} - \frac{x^3}{3!} + \frac{x^4}{4!} - \cdots$

13. $(x-1) - \frac{1}{2}(x-1)^2 + \frac{1}{3}(x-1)^3 - \frac{1}{4}(x-1)^4 + \cdots$

15. $\frac{\sqrt{2}}{2} - \frac{\sqrt{2}}{1!2}\left(x - \frac{\pi}{2}\right) - \frac{\sqrt{2}}{6!2}\left(x - \frac{\pi}{2}\right)^2 + \frac{\sqrt{2}}{3!2}\left(x - \frac{\pi}{2}\right)^3 + \frac{\sqrt{2}}{4!2}\left(x - \frac{\pi}{2}\right)^4 - \frac{\sqrt{2}}{5!2}\left(x - \frac{\pi}{2}\right)^5 - \frac{\sqrt{2}}{6!2}\left(x - \frac{\pi}{2}\right)^6 + \frac{\sqrt{2}}{7!2}\left(x - \frac{\pi}{2}\right)^7 + \cdots$

17. $x + \frac{x^3}{3!} + \frac{x^5}{5!} + \frac{x^7}{7!} + \cdots$

19. $1 - (x-1) + (x-1)^2 - (x-1)^3 + (x-1)^4 - \cdots$

21. $\frac{\sqrt{2}}{2}\left[1 + \left(x - \frac{\pi}{4}\right) - \frac{(x-\pi/4)^2}{2!} - \frac{(x-\pi/4)^3}{3!} + \frac{(x-\pi/4)^4}{4!} + \cdots\right]$

29. Seven terms give the estimate 0.2877.

Section 12.8 (page 584)

1. $[-1, 1)$
3. $(-1, 1)$
5. $[-1, 1]$
7. (0.4, 0.8)
9. [2.5, 3.5]
11. Converges only for $x = 0$
13. $(-4, 2)$
15. [2, 4]
17. Converges only for $x = 5$
19. $(-1, 1)$
21. $f(x) = x^2 - 3x^3 + \frac{9x^4}{2!} - \frac{27x^5}{3!} + \frac{81x^6}{4!} - \cdots$; the radius of convergence is $+\infty$.
23. $f(x) = x^2 - \frac{x^6}{3!} + \frac{x^{10}}{5!} - \frac{x^{14}}{7!} + \frac{x^{18}}{9!} - \cdots$; the radius of convergence is $+\infty$.
25. $(1-x)^{1/3} = 1 - \frac{x}{3} - \frac{2x^2}{2!3^2} - \frac{2 \cdot 5x^3}{3!3^3} - \frac{2 \cdot 5 \cdot 8x^4}{4!3^4} - \cdots$; $R = 1$
27. $f(x) = 1 - 3x + 6x^2 - 10x^3 + 15x^4 - \cdots$; $R = 1$
29. $f(x) = 1 - \frac{1}{2}x + \frac{1}{3}x^2 - \frac{1}{4}x^3 + \frac{1}{5}x^4 - \cdots$; $R = 1$
31. $f(x) = \frac{x^4}{4} - \frac{x^{10}}{3!10} + \frac{x^{16}}{5!16} - \frac{x^{22}}{7!22} + \cdots = \sum_{n=0}^{\infty} \frac{(-1)^n}{(2n+1)!(6n+4)} x^{6n+4}$
33. $f(x) = x - \frac{x^4}{4} + \frac{x^7}{2!7} - \frac{x^{10}}{3!10} + \frac{x^{13}}{4!13} - \cdots = \sum_{n=0}^{\infty} \frac{(-1)^n}{n!(3n+1)} x^{3n+1}$
35. $f(x) = x - \frac{x^3}{2!3} + \frac{x^5}{3!5} - \frac{x^7}{4!7} + \frac{x^9}{5!9} - \cdots = \sum_{n=1}^{\infty} \frac{(-1)^{n+1}}{n!(2n-1)} x^{2n-1}$
37. By using six terms, we find that $3.14130878 < \pi < 3.1416744$.

Section 12.9 (page 589)

1. $65^{1/3} = (4 + \frac{1}{64})^{1/3}$. The first four terms of the binomial series give $65^{1/3} \approx 4.020726$; answer: 4.021.
3. Three terms of the usual sine series give 0.479427 with error less than 0.000002; answer: 0.479.
5. Five terms of the usual arctangent series give 0.463684 with error less than 0.000045; answer: 0.464.
7. 0.309
9. 0.174
11. 0.946
13. 0.487
15. 0.0976
17. 0.470
19. 0.747
21. -0.5
23. 0.5
25. 0
31. The first five coefficients are 1, 0, $\frac{1}{2}$, 0, and $\frac{5}{24}$.

Chapter 12 Miscellaneous Problems (page 591)

1. 1
3. 10
5. 0
7. 0
9. No limit
11. 0
13. $+\infty$ (or "no limit")
15. 1

17, 19, 21, 25, and 27: Converge

23 and 29: Diverge

31. $(-\infty, +\infty)$
33. $[-2, 4)$
35. $[-1, 1]$
37. It converges only for $x = 0$.
39. $(-\infty, +\infty)$
41. It converges for *no* x.
43. It converges for all x.
51. Seven terms of the binomial series give 1.084.
53. Six terms of the "obvious" series give 0.747.
55. Fouı terms give 0.444.

Section 13.1 (page 600)

1. $y = 2x - 3$
3. $y^2 = x^3$
5. $y = 2x^2 - 5x + 2$
7. $y = 4x^2$, $x > 0$
9. $\left(\frac{x}{5}\right)^2 + \left(\frac{y}{3}\right)^2 = 1$
11. $x^2 - y^2 = 1$
13. $y - 5 = \frac{9}{4}(x - 3)$; $\frac{d^2y}{dx^2} = \frac{9}{16}t^{-1}$, so the curve is concave upward at $t = 1$.
15. $y = -\frac{1}{2}\pi(x - \frac{1}{2}\pi)$; concave downward.
17. $x + y = 3$; concave downward.

19. $\psi = \dfrac{\pi}{6}$ (constant)

21. $\psi = \dfrac{\pi}{2}$

25. $x = \dfrac{p}{m^2}, \quad y = \dfrac{2p}{m}$

Section 13.2 (page 606)

1. $\frac{22}{5}$
3. $\frac{4}{3}$
5. $\dfrac{1}{2}(1 + e^{\pi}) \approx 12.0703$
7. $\frac{358}{35}\pi \approx 32.13400$
9. $\frac{16}{15}\pi \approx 3.35103$
11. $\dfrac{74}{3}$
13. $\frac{1}{4}\pi\sqrt{2}$
15. $(e^{2\pi} - 1)\sqrt{5} \approx 1195.1597$
17. $\frac{8}{3}\pi(5\sqrt{5} - 2\sqrt{2}) \approx 69.96882$
19. $\frac{2}{27}(13\sqrt{13} - 8) \approx 9.04596$
21. $16\pi^2$
23. $5\pi^2 a^3$
25. (a) $A = \pi ab$; (b) $V = \frac{4}{3}\pi ab^2$
27. $\pi(1 + 4\pi^2)^{1/2} + \frac{1}{2}\ln[2\pi + (1 + 4\pi^2)^{1/2}] \approx 21.25629$
29. $\frac{3}{8}\pi a^2$
31. $\frac{12}{5}\pi a^2$
33. $\frac{216}{5}\sqrt{3}$
35. $\frac{243}{4}\sqrt{3}$

Section 13.3 (page 613)

1. $\sqrt{5}$; $2\sqrt{13}$; $4\sqrt{2}$; $\langle -2, 0\rangle$; $\langle 9, -10\rangle$; no
3. $2\sqrt{2}$; 10; $\sqrt{5}$; $\langle -5, -6\rangle$; $\langle 0, 2\rangle$; no
5. $\sqrt{10}$; $2\sqrt{29}$; $\sqrt{65}$; $3\mathbf{i} - 2\mathbf{j}$; $-\mathbf{i} + 19\mathbf{j}$; no
7. 4; 14; $\sqrt{65}$; $4\mathbf{i} - 7\mathbf{j}$; $12\mathbf{i} + 14\mathbf{j}$; yes
9. $\mathbf{u} = \langle -\frac{3}{5}, -\frac{4}{5}\rangle$, $\mathbf{v} = \langle \frac{3}{5}, \frac{4}{5}\rangle$
11. $\mathbf{u} = \frac{8}{17}\mathbf{i} + \frac{15}{17}\mathbf{j}$, $\mathbf{v} = -\frac{8}{17}\mathbf{i} - \frac{15}{17}\mathbf{j}$
13. $-4\mathbf{j}$
15. $8\mathbf{i} - 14\mathbf{j}$
17. Yes
19. No
21. (a) $15\mathbf{i} - 21\mathbf{j}$; (b) $\frac{5}{3}\mathbf{i} - \frac{7}{3}\mathbf{j}$
23. (a) $\frac{35}{38}\mathbf{i}\sqrt{58} - \frac{15}{58}\mathbf{j}\sqrt{58}$;
    (b) $-\frac{40}{89}\mathbf{i}\sqrt{89} - \frac{25}{89}\mathbf{j}\sqrt{89}$
25. $c = 0$
33. $\mathbf{v}_a = (500 + 25\sqrt{2})\mathbf{i} + 25\mathbf{j}\sqrt{2}$
35. $\mathbf{v}_a = -225\mathbf{i}\sqrt{2} + 275\mathbf{j}\sqrt{2}$

Section 13.4 (page 620)

1. $\mathbf{0}$; $\mathbf{0}$
3. $2\mathbf{i} - \mathbf{j}$; $4\mathbf{i} + \mathbf{j}$
5. $6\pi\mathbf{i}$; $12\pi^2\mathbf{j}$
7. $\mathbf{j}$; $\mathbf{i}$
9. $\frac{1}{2}(2 - \sqrt{2})\mathbf{i} + \mathbf{j}\sqrt{2}$
11. $\frac{484}{15}\mathbf{i}$
13. 11
15. 0
17. $\mathbf{i} + 2t\mathbf{j}$; $t\mathbf{i} + t^2\mathbf{j}$
19. $\frac{1}{2}t^2\mathbf{i} + \frac{1}{3}t^3\mathbf{j}$; $(1 + \frac{1}{6}t^3)\mathbf{i} + \frac{1}{12}t^4\mathbf{j}$
21. $v_0 \approx 411.047$ ft/s
25. (a) $y_m = 100$ ft, $R = 400\sqrt{3}$ ft;
    (b) $y_m = 200$, $R = 800$;
    (c) $y_m = 300$, $R = 400\sqrt{3}$
27. $v_0 = 1056$ ft/s
29. $v_0 = 4\sqrt{4181} \approx 258.64$ ft/s; $\alpha = \tan^{-1}(0.82)$, approximately 39°21′6″
35. Begin with $\dfrac{d}{dt}(\mathbf{v} \cdot \mathbf{v}) = 0$.
37. A repulsive force acting directly away from the origin, with magnitude proportional to distance from the origin.

Section 13.5 (page 627)

1. $\mathbf{v} = a\mathbf{u}_\theta$, $\mathbf{a} = -a\mathbf{u}_r$
3. $\mathbf{v} = \mathbf{u}_r + t\mathbf{u}_\theta$, $\mathbf{a} = -t\mathbf{u}_r + 2\mathbf{u}_\theta$
5. $\mathbf{v} = (12\cos 4t)\mathbf{u}_r + (6\sin 4t)\mathbf{u}_\theta$,
   $\mathbf{a} = (-60\sin 4t)\mathbf{u}_r + (48\cos 4t)\mathbf{u}_\theta$
9. 36.65 mi/s; 24.13 mi/s
11. 0.672 mi/s; 0.602 mi/s
13. About -795 mi—thus it can't be done.
15. About 1.962 h

Chapter 13 Miscellaneous Problems (page 628)

1. The straight line $y = x + 2$
3. The circle $(x - 2)^2 + (y - 1)^2 = 1$
5. Equation: $y^2 = (x - 1)^3$
7. $y - 2\sqrt{2} = -\frac{4}{3}(x - \frac{3}{2}\sqrt{2})$
9. $2\pi y + 4x = \pi^2$
11. 24
13. 3π
15. $\frac{1}{27}(13\sqrt{13} - 8) \approx 1.4397$
17. $\frac{43}{6}$
19. $\frac{1}{8}(4\pi - 3\sqrt{3}) \approx 0.92128$
21. $\dfrac{471{,}295\pi}{1024} \approx 1445.915$
23. $\frac{1}{2}\pi\sqrt{5}(e^{\pi} + 1) \approx 84.7919$
25. $x = a\theta - b\sin\theta$, $y = a - b\cos\theta$
    The graph shows the case $a = 1$, $b = 0.7$.
27. *Suggestion:* Compute $r^2 = x^2 + y^2$.
29. $6\pi^3 a^3$
35. There are two solutions:
    $\alpha \approx 0.033364$ radians (about 1°54′53″) and
    $\alpha \approx 1.29116$ radians (about 73°58′40″).

Section 14.1 (page 638)

1. $\langle 5, 8, -11\rangle$; $\langle 2, 23, 0\rangle$; 4; $\sqrt{51}$; $\frac{1}{15}\sqrt{5}\langle 2, 5, -4\rangle$
3. $2\mathbf{i} + 3\mathbf{j} + \mathbf{k}$; $3\mathbf{i} - \mathbf{j} + 7\mathbf{k}$; 0; $\sqrt{5}$; $\frac{1}{3}\sqrt{3}(\mathbf{i} + \mathbf{j} + \mathbf{k})$

5. $4\mathbf{i} - \mathbf{j} - 3\mathbf{k}$; $6\mathbf{i} - 7\mathbf{j} + 12\mathbf{k}$; -1; $\sqrt{17}$; $\frac{1}{5}\sqrt{5}(2\mathbf{i} - \mathbf{j})$

7. $\theta = \cos^{-1}(-\frac{13}{50}\sqrt{10}) \approx 2.536$

9. $\theta = \cos^{-1}\left(-\dfrac{34}{\sqrt{3154}}\right) \approx 2.221$

11. $\text{comp}_\mathbf{b}\,\mathbf{a} = \frac{2}{7}\sqrt{14}$; $\text{comp}_\mathbf{a}\,\mathbf{b} = \frac{4}{15}\sqrt{5}$

13. $\text{comp}_\mathbf{b}\,\mathbf{a} = 0 = \text{comp}_\mathbf{a}\,\mathbf{b}$

15. $\text{comp}_\mathbf{b}\,\mathbf{a} = -\frac{1}{10}\sqrt{10}$; $\text{comp}_\mathbf{a}\,\mathbf{b} = -\frac{1}{5}\sqrt{5}$

17. $(x + 2)^2 + (y - 1)^2 + (z + 5)^2 = 7$

19. $(x - 4)^2 + (y - 5)^2 + (z + 2)^2 = 38$

21. Center $(-2, 3, 0)$, radius $\sqrt{13}$

23. Center $(0, 0, 3)$, radius 5

25. A plane perpendicular to the z-axis at $z = 10$

27. All points in the three coordinate planes

29. The point $(1, 0, 0)$

31. $\alpha = \cos^{-1}(\frac{1}{9}\sqrt{6}) \approx 74.21^\circ$,
$\beta = \cos^{-1}(\frac{5}{18}\sqrt{6}) \approx 47.12^\circ$,
$\gamma = \beta$

33. $\alpha = \cos^{-1}(\frac{3}{10}\sqrt{2}) \approx 64.90^\circ$,
$\beta = \cos^{-1}(\frac{2}{5}\sqrt{2}) \approx 55.55^\circ$,
$\gamma = \cos^{-1}(\frac{1}{2}\sqrt{2}) = 45^\circ$

35. 48

37. If there's no friction, the work done is mgh.

41. $A = \frac{3}{2}\sqrt{69} \approx 12.46$

43. $\cos^{-1}(\frac{1}{3}\sqrt{3}) \approx 0.9553$, about 54.7356°

49. $2x + 9y - 5z = 23$; the plane through the midpoint of the segment AB perpendicular to AB

51. 60°

Section 14.2 (page 646)

1. $\langle 0, -14, 7\rangle$

3. $\langle -10, -7, 1\rangle$

7. $(\mathbf{a} \times \mathbf{b}) \times \mathbf{c} = \langle -1, 1, 0\rangle$, $\mathbf{a} \times (\mathbf{b} \times \mathbf{c}) = \langle 0, 0, -1\rangle$

11. $A = \frac{1}{2}\sqrt{2546} \approx 25.229$

13. (a) $V = 55$; (b) $V = \frac{55}{6}$

15. 4395.657

17. 31,271.643

21. $\frac{1}{38}\sqrt{9842} \approx 2.6107$

Section 14.3 (page 652)

1. $x = t$, $y = 2t$, $z = 3t$

3. $x = 4 + 2t$, $y = 13$, $z = -3 - 3t$

5. $x = 3 + 3t$, $y = 5 - 13t$, $z = 7 + 3t$

7. $x + 2y + 3z = 0$

9. $x - z + 8 = 0$

11. $x = 2 + t$, $y = 3 - t$, $z = -4 - 2t$;
$x - 2 = -y + 3 = \dfrac{-z - 4}{2}$

13. $x = 1$, $y = 1$, $z = 1 + t$;
$x - 1 = 0 = y - 1$, z arbitrary

15. $x = 2 + 2t$, $y = -3 - t$, $z = 4 + 3t$;
$\dfrac{x - 2}{2} = -y - 3 = \dfrac{z - 4}{3}$

17. $y = 7$

19. $7x + 11y = 114$

21. $3x + 4y - z = 0$

23. $2x - y - z = 0$

25. $\theta = \cos^{-1}\left(\dfrac{1}{\sqrt{3}}\right) \approx 54.736^\circ$

27. The planes are parallel: $\theta = 0$.

29. $\dfrac{x - 3}{2} = y - 3 = \dfrac{-z + 1}{5}$

31. $3x + 2y + z = 6$

33. $7x - 5y - 2z = 9$

35. $x - 2y + 4z = 3$

37. $\frac{10}{3}\sqrt{3}$

41. (b) $D = \dfrac{133}{\sqrt{501}} \approx 5.942$

Section 14.4 (page 658)

1. $\mathbf{v} = 5\mathbf{k}$, $\mathbf{a} = \mathbf{0}$, $v = 5$

3. $\mathbf{v} = \mathbf{i} + 2t\mathbf{j} + 3t^2\mathbf{k}$, $\mathbf{a} = 2\mathbf{j} + 6t\mathbf{k}$,
$v = (1 + 4t^2 + 9t^4)^{1/2}$

5. $\mathbf{v} = \mathbf{i} + 3e^t\mathbf{j} + 4e^t\mathbf{k}$, $\mathbf{a} = 3e^t\mathbf{j} + 4e^t\mathbf{k}$,
$v = (1 + 25e^t)^{1/2}$

7. $\mathbf{v} = (-3\sin t)\mathbf{i} + (3\cos t)\mathbf{j} - 4\mathbf{k}$,
$\mathbf{a} = (-3\cos t)\mathbf{i} - (3\sin t)\mathbf{j}$, $v = 5$

9. $\mathbf{r} = t^2\mathbf{i} + 10t\mathbf{j} - 2t^2\mathbf{k}$

11. $\mathbf{r} = 2\mathbf{i} + t^2\mathbf{j} + (5t - t^3)\mathbf{k}$

13. $\mathbf{r} = (10 + \frac{1}{6}t^3)\mathbf{i} + (10 + \frac{1}{12}t^4)\mathbf{j} + \frac{1}{20}t^5\mathbf{k}$

15. $\mathbf{r} = (1 - t - \cos t)\mathbf{i} + (1 + t - \sin t)\mathbf{j} + 5t\mathbf{k}$

17. $\mathbf{v} = 3\sqrt{2}(\mathbf{i} + \mathbf{j}) + 8\mathbf{k}$, $v = 10$,
$\mathbf{a} = 6\sqrt{2}(-\mathbf{i} + \mathbf{j})$

21. *Suggestion:* Compute $\dfrac{d}{dt}(\mathbf{r} \cdot \mathbf{r})$.

23. Maximum height 200 ft, speed $\sqrt{6425} \approx 80.16$ ft/s

27. $9(2 - \sqrt{3})$ ft

Section 14.5 (page 668)

1. 10π

3. $19(e - 1) \approx 32.647$

5. $2 + \frac{9}{10}\ln 3 \approx 3.5588$

7. 0

9. 1

11. $\dfrac{40\sqrt{2}}{41\sqrt{41}} \approx 0.2155$

13. At $(\frac{1}{2}\ln 2, \frac{1}{2}\sqrt{2})$

15. Maximum curvature $\frac{5}{9}$ at $(5, 0)$ and at $(-5, 0)$, minimum curvature $\frac{3}{25}$ at $(0, 3)$ and at $(0, -3)$

17. $\mathbf{T} = \frac{1}{10}\sqrt{10}(\mathbf{i} + 3\mathbf{j})$, $\mathbf{N} = \frac{1}{10}\sqrt{10}(3\mathbf{i} - \mathbf{j})$

19. $\mathbf{T} = \frac{1}{57}\sqrt{57}(3\mathbf{i} - 4\mathbf{j}\sqrt{3})$, $\mathbf{N} = \frac{1}{57}\sqrt{57}(4\mathbf{i}\sqrt{3} + 3\mathbf{j})$

21. $\mathbf{T} = -\frac{1}{2}\sqrt{2}(\mathbf{i} + \mathbf{j})$, $\mathbf{N} = \frac{1}{2}\sqrt{2}(\mathbf{i} - \mathbf{j})$

23. $a_T = 18t(9t^2 + 1)^{-1/2}$, $a_N = 6(9t^2 + 1)^{-1/2}$

25. $a_T = t(1 + t^2)^{-1/2}, \quad a_N = (2 + t^2)(1 + t^2)^{-1/2}$

27. $\dfrac{1}{a}$

29. $x^2 + (y - \frac{1}{2}) = \frac{1}{4}$

31. $(x - \frac{3}{2}) + (y - \frac{3}{2})^2 = 2$

33. $\frac{1}{2}$

35. $\frac{1}{3}e^{-t}\sqrt{2}$

37. $a_T = 0 = a_N$

39. $a_T = \dfrac{4t + 18t^3}{(1 + 4t^2 + 9t^4)^{1/2}}$, $a_N = \dfrac{2(1 + 9t^2 + 9t^4)^{1/2}}{(1 + 4t^2 + 9t^4)^{1/2}}$

41. $a_T = \dfrac{t}{(t^2 + 2)^{1/2}}$, $a_N = \dfrac{(t^4 + 5t^2 + 8)^{1/2}}{(t^2 + 2)^{1/2}}$

43. $\mathbf{T} = \frac{1}{2}\sqrt{2}\langle 1, \cos t, -\sin t\rangle$, $\mathbf{N} = \langle 0, -\sin t, -\cos t\rangle$; at $(0, 0, 1)$, $\mathbf{T} = \frac{1}{2}\sqrt{2}\langle 1, 1, 0\rangle$, $\mathbf{N} = \langle 0, 0, -1\rangle$

45. $\mathbf{T} = \frac{1}{3}\sqrt{3}(\mathbf{i} + \mathbf{j} + \mathbf{k})$, $\mathbf{N} = \frac{1}{2}\sqrt{2}(-\mathbf{i} + \mathbf{j})$

47. $x = 2 + \frac{4}{13}s, \quad y = 1 - \frac{12}{13}s, \quad z = 3 + \frac{3}{13}s$

49. $x(s) = 3\cos\frac{1}{5}s, \quad y(s) = 3\sin\frac{1}{5}s, \quad z(s) = \frac{4}{5}s$

51. Begin with $\dfrac{d}{dt}(\mathbf{v} \cdot \mathbf{v})$.

53. $|t|^{-1}$

55. $y = 3x^5 - 8x^4 + 6x^3$

Section 14.6 (page 680)

1. A plane with intercepts $(20/3, 0, 0)$, $(0, 10, 0)$, and $(0, 0, 2)$
3. A vertical circular cylinder with radius 3
5. A vertical cylinder intersecting the xy-plane in the rectangular hyperbola $xy = 4$
7. An elliptical paraboloid opening upward from its vertex at the origin
9. A circular paraboloid that opens downward from its vertex at $(0, 0, 4)$
11. Paraboloid, vertex at the origin, axis the z-axis, opening upward
13. Cone, vertex the origin, axis the z-axis (both nappes)
15. Parabolic cylinder perpendicular to the xz-plane, its trace there the parabola opening upward with axis the z-axis and vertex at $(x, z) = (0, -2)$
17. Elliptical cylinder perpendicular to the xy-plane, its trace there the ellipse with center $(0, 0)$ and intercepts $(1, 0)$, $(0, 2)$, $(-1, 0)$, and $(0, -2)$
19. Elliptical cone, vertex $(0, 0, 0)$, axis the x-axis
21. Paraboloid, vertex at the origin, axis the z-axis, opening downward
23. Hyperbolic paraboloid, saddle point at the origin, meeting the xz-plane in a parabola with vertex the origin and opening downward, meeting the xy-plane in a parabola with vertex the origin and opening upward, meeting each plane parallel to the yz-plane in a hyperbola with directrices parallel to the y-axis
25. Hyperboloid of one sheet, axis the z-axis, trace in the xy-plane the circle with center $(0, 0)$ and radius 3, traces in parallel planes larger circles, and traces in planes parallel to the z-axis hyperbolas
27. Elliptic paraboloid, axis the y-axis, vertex at the origin
29. Hyperboloid of two sheets, axis the y-axis
31. Paraboloid, axis the x-axis, vertex at the origin, equation $x = 2(y^2 + z^2)$
33. Hyperboloid of one sheet (see Fig. 14.66) with equation $x^2 + y^2 - z^2 = 1$
35. Paraboloid, vertex at the origin, axis the x-axis, equation $y^2 + z^2 = 4x$
37. The surface resembles a rug covering a turtle. Its highest point is $(0, 0, 1)$; $z \to 0$ from above as $|x|$ or $|y|$ (or both) increase without bound. Its equation is $z = \exp(-x^2 - y^2)$.
39. A circular cone with the z-axis as its axis of symmetry
41. Ellipses with semiaxes 2 and 1
43. Circles
45. Parabolas that open downward
47. Parabolas that open upward if $k > 0$, downward if $k < 0$
51. The projection of the intersection has equation $x^2 + y^2 = 2y$; it is the circle with center $(0, 1)$ and radius 1.
53. Equation: $5x^2 + 8xy + 8y^2 - 8x - 8y = 0$. Because $B^2 - 4AC < 0$, it is an ellipse. In a uv-plane rotated approximately $55°16'41''$ from the xy-plane, the ellipse has center $(u, v) = (0.517, -0.453)$, minor axis 0.352 in the u-direction, major axis 0.774 in the v-direction.

Section 14.7 (page 686)

1. Cylindrical coordinates $(0, 0, 5)$, spherical coordinates $(5, 0, 0)$

3. $\left(\sqrt{2}, \dfrac{\pi}{4}, 0\right)_{\text{cyl}}, \quad \left(\sqrt{2}, \dfrac{\pi}{2}, \dfrac{\pi}{4}\right)_{\text{sph}}$

5. $\left(\sqrt{2}, \dfrac{\pi}{4}, 1\right)_{\text{cyl}}, \quad \left(\sqrt{3}, \tan^{-1}\sqrt{2}, \dfrac{\pi}{4}\right)_{\text{sph}}$

7. $(\sqrt{5}, \tan^{-1}(0.5), -2)_{\text{cyl}}$, $\left(3, \dfrac{\pi}{2} + \tan^{-1}\left(\dfrac{\sqrt{5}}{2}\right), \tan^{-1}(0.5)\right)_{\text{sph}}$

9. $(5, \tan^{-1}(4/3), 12)_{\text{cyl}}$, $(13, \tan^{-1}(5/13), \tan^{-1}(4/3))_{\text{sph}}$

11. Cylinder, radius 5, axis the z-axis

13. The *plane* $y = x$

15. The upper nappe of the cone $x^2 + y^2 = 3z^2$

17. The xy-plane

19. Cylinder, axis the vertical line $x = 0$, $y = 1$; its trace in the xy-plane is the circle

$$x^2 + (y-1)^2 = 1$$

21. The vertical plane having trace the line $y = -x$ in the xy-plane

23. The horizontal plane $z = 1$

25. $r^2 + z^2 = 25$; $\rho = 5$

27. $r(\cos\theta + \sin\theta) = 1$;
$\rho(\sin\phi\cos\theta + \sin\phi\sin\theta + \cos\phi) = 1$

29. $r^2 + z^2 = r(\cos\theta + \sin\theta) + z$;
$\rho = \sin\phi\cos\theta + \sin\phi\sin\theta + \cos\phi$

31. $z = r^2$

33. (a) $1 \leqq r^2 \leqq 4 - z^2$; (b) $\csc\phi \leqq \rho \leqq 2$ (and, as a consequence, $\pi/6 \leqq \phi \leqq 5\pi/6$)

35. About 3821 mi

37. Just under 31 mi

39. $0 \leqq \rho \leqq H\sec\phi$, $0 \leqq \phi \leqq \arctan\left(\dfrac{R}{H}\right)$, θ arbitrary

Chapter 14 Miscellaneous Problems (page 687)

7. $x = 1 + 2t$, $y = -1 + 3t$, $z = 2 - 3t$;
$\dfrac{x-1}{2} = \dfrac{y+1}{3} = \dfrac{2-z}{3}$

9. $-13x + 22y + 6z = -23$

11. $x - y + 2z = 3$

15. 3

17. Curvature $\frac{1}{9}$, $a_T = 42$, $a_N = 1$

21. $3x - 3y + z = 1$

27. $\rho = 2\cos\phi$

29. $\rho^2 = 2\cos 2\phi$; shape: like an hourglass with rounded ends

37. The curvature is zero when x is an integral multiple of π, and reaches the maximum value 1 when x is an odd integral multiple of $\pi/2$.

39. $\mathbf{T} = -\left(\dfrac{\pi}{(\pi^2+4)^{1/2}}\right)\mathbf{i} + \left(\dfrac{2}{(\pi^2+4)^{1/2}}\right)\mathbf{j}$,
$\mathbf{N} = -\left(\dfrac{2}{(\pi^2+4)^{1/2}}\right)\mathbf{i} - \left(\dfrac{\pi}{(\pi^2+4)^{1/2}}\right)\mathbf{j}$

43. $y = \frac{15}{8}x - \frac{5}{4}x^3 + \frac{3}{8}x^5$

Section 15.2 (page 700)

1. All (x, y)

3. Except on the line $x = y$

5. Except on the coordinate axes $x = 0$ and $y = 0$

7. Except on the lines $|y| = |x|$

9. Except at the origin $(0, 0, 0)$

11. The horizontal plane with equation $z = 10$

13. A plane that makes a 45° angle with the xy-plane, intersecting it in the line $x + y = 0$

15. A circular paraboloid opening upward from its vertex at the origin

17. The upper hemispherical surface of radius 2 centered at the origin

19. A circular cone opening downward from its vertex at $(0, 0, 10)$

21. Straight lines of slope 1

23. Ellipses centered at $(0, 0)$, each with major axis twice the minor axis and lying on the x-axis

25. Vertical (y-direction) translates of the curve $y = x^3$

27. Circles centered at the point $(2, 0)$

29. Circles centered at the origin

31. Circular paraboloids opening upward, each with its vertex on the z-axis

33. Spheres centered at $(2, -1, 3)$

35. Elliptical cylinders each of which has axis the vertical line through $(2, 1, 0)$ and with the length of the x-semiaxis twice that of the y-semiaxis

37.

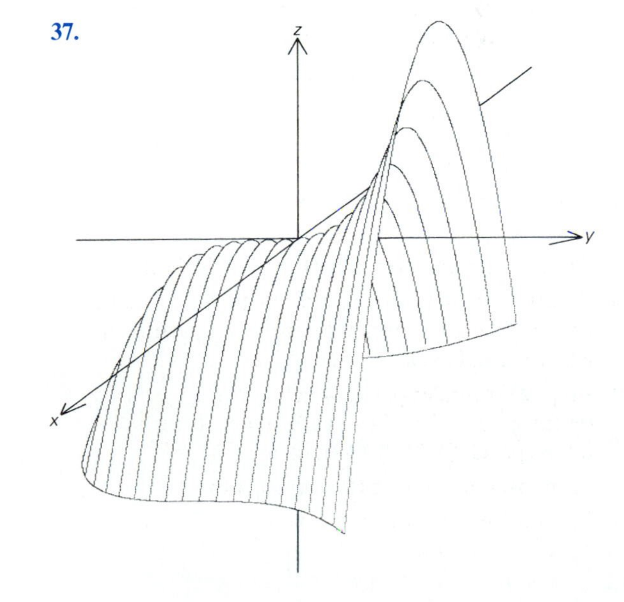

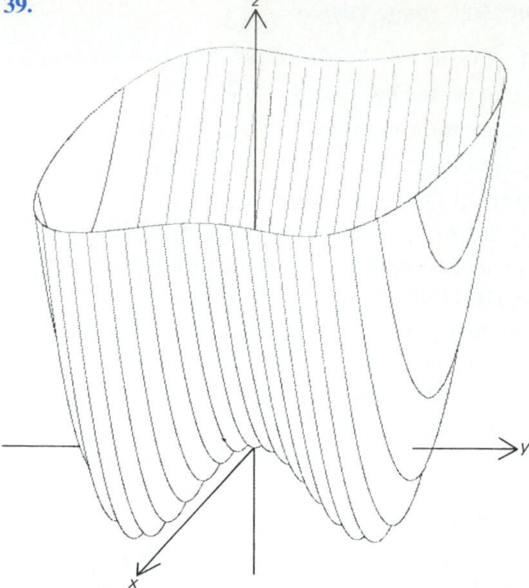

41. Corresponding figures are: 15.22 and 15.30; 15.23 and 15.31; 15.24 and 15.28; 15.25 and 15.29; 15.26 and 15.33; 15.27 and 15.32

Section 15.3 (page 708)

1. 7 **3.** e **5.** $\frac{5}{3}$ **7.** 0
9. e **11.** $-\frac{3}{2}$ **13.** 1 **15.** -4
17. y; x **19.** y^2; $2xy$
29. Note that $f(x, y) = (x - \frac{1}{2})^2 + (y + 1)^2 - \frac{1}{4} \geqq -\frac{1}{4}$ for all x.

Section 15.4 (page 716)

1. $\dfrac{\partial f}{\partial x} = 4x^3 - 3x^2y + 2xy^2 - y^3$,
$\dfrac{\partial f}{\partial y} = -x^3 + 2x^2y - 3xy^2 + 4y^3$

3. $\dfrac{\partial f}{\partial x} = e^x(\cos y - \sin y)$, $\dfrac{\partial f}{\partial y} = -e^x(\cos y + \sin y)$

5. $\dfrac{\partial f}{\partial x} = -\dfrac{2y}{(x - y)^2}$, $\dfrac{\partial f}{\partial y} = \dfrac{2x}{(x - y)^2}$

7. $\dfrac{\partial f}{\partial x} = \dfrac{2x}{x^2 + y^2}$, $\dfrac{\partial f}{\partial y} = \dfrac{2y}{x^2 + y^2}$

9. $\dfrac{\partial f}{\partial x} = yx^{y-1}$, $\dfrac{\partial f}{\partial y} = x^y \ln x$

11. $\dfrac{\partial f}{\partial x} = 2xy^3z^4$, $\dfrac{\partial f}{\partial y} = 3x^2y^2z^4$, $\dfrac{\partial f}{\partial z} = 4x^2y^3z^3$

13. $\dfrac{\partial f}{\partial x} = yze^{xyz}$, $\dfrac{\partial f}{\partial y} = xze^{xyz}$, $\dfrac{\partial f}{\partial z} = xye^{xyz}$

15. $\dfrac{\partial f}{\partial x} = 2xe^y \ln z$, $\dfrac{\partial f}{\partial y} = x^2e^y \ln z$,
$\dfrac{\partial f}{\partial z} = x^2e^y/z$

17. $\dfrac{\partial f}{\partial r} = \dfrac{4rs^2}{(r^2 + s^2)^2}$, $\dfrac{\partial f}{\partial s} = -\dfrac{4r^2s}{(r^2 + s^2)^2}$

19. $\dfrac{\partial f}{\partial u} = e^v + we^u$, $\dfrac{\partial f}{\partial v} = e^w + ue^v$, $\dfrac{\partial f}{\partial w} = e^u + ve^w$

21. $z_{xy} = z_{yx} = -4$
23. $z_{xy} = z_{yx} = -4xy\exp(-y^2)$
25. $z_{xy} = z_{yx} = (x + y)^{-2}$
27. $z_{xy} = z_{yx} = 3e^{-3x}\sin y$
29. $z_{xy} = z_{yx} = -4xy^{-3}\sinh y^{-2}$
31. $6x + 8y - z = 25$ **33.** $z = -1$
35. $27x - 12y - z = 38$ **37.** $x - y + z = 1$
39. $10x - 16y - z = 9$
43. $f_{xyz}(x, y, z) = (x^2y^2z^2 + 3xyz + 1)e^{xyz}$
51. $(10, -7, -58)$
53. (a) A decrease of approximately 2750 cm^3; (b) an increase of approximately 82.5 cm^3

Section 15.5 (page 727)

1. None **3.** $(0, 0, 5)$
5. $(3, -1, -5)$ **7.** $(-2, 0, -4)$
9. $(-2, 0, -7)$ and $(-2, 1, -9)$
11. $(0, 0, 0)$, $(1, 0, 2/e)$, $(-1, 0, 2/e)$, $(0, 1, 3/e)$, and $(0, -1, 3/e)$
13. Minimum value: $f(1, 1) = 1$
15. Maximum value: $f(1, 1) = f(1, -1) = 2$
17. Minimum value: $f(4, -2) = f(-4, 2) = -16$
19. Maximum value: $f(1, -2) = e^5$
21. Maximum value: $f(1, 1) = 3$; minimum value: $f(-1, -1) = -3$
23. Maximum value: $f(0, 2) = 4$; minimum value: $f(1, 0) = -1$
25. Maximum value: $f\left(\dfrac{1}{\sqrt{2}}, \dfrac{1}{\sqrt{2}}\right) = f\left(-\dfrac{1}{\sqrt{2}}, -\dfrac{1}{\sqrt{2}}\right) = 1$;
minimum value: $f\left(\dfrac{1}{\sqrt{2}}, -\dfrac{1}{\sqrt{2}}\right) = f\left(-\dfrac{1}{\sqrt{2}}, \dfrac{1}{\sqrt{2}}\right) = -1$
27. Dimensions: 10 by 10 by 10
29. 10 by 10 by 10 cm
31. Maximum value: $\frac{4}{3}$; $x = 2$, $y = 1$, $z = \frac{2}{3}$
33. Height: 10 ft; front, back 40 ft wide, sides 20 ft deep
35. Base 2 ft by 2 ft, height 3 ft
37. 11,664 in.^3 **39.** $V_{\max} = \frac{1}{2}$
41. Maximum area: 900 (one square); minimum area: 300 (three equal squares)

Section 15.6 (page 735)

1. $dw = (6x + 4y)\,dx + (4x - 6y^2)\,dy$
3. $dw = (1 + x^2 + y^2)^{-1/2}(x\,dx + y\,dy)$
5. $dw = (x^2 + y^2)^{-1}(-y\,dx + x\,dy)$
7. $dw = (x^2 + y^2 + z^2)^{-1}(2x\,dx + 2y\,dy + 2z\,dz)$
9. $dw = (\tan yz)\,dx + (xz\sec^2 yz)\,dy + (xy\sec^2 yz)\,dz$
11. $dw = -e^{-xyz}(yz\,dx + xz\,dy + xy\,dz)$
13. $dw = \exp(-v^2)(2u\,du - 2u^2v\,dv)$
15. $dw = (x^2 + y^2 + z^2)^{-1/2}(x\,dx + y\,dy + z\,dz)$
17. $\Delta f \approx 0.014$ (true value: about 0.01422975)
19. $\Delta f \approx -0.0007$
21. $\Delta f \approx \dfrac{53}{1300} \approx 0.04077$
23. $\Delta f \approx 0.06$
25. 191.1
27. 1.4
29. $x \approx 1.95$
31. 8.18 in.^{3}
33. 0.022 acres
35. The period increases by about 0.0278 s.
37. Approximately 303.8 ft

Section 15.7 (page 742)

1. $-(2t + 1)\exp(-t^2 - t)$
3. $6t^5\cos t^6$
5. $\dfrac{\partial w}{\partial s} = \dfrac{\partial w}{\partial t} = \dfrac{2}{s + t}$
7. $\dfrac{\partial w}{\partial s} = 0, \quad \dfrac{\partial w}{\partial t} = 5e^t$
9. $\dfrac{\partial r}{\partial x} = (y + z)\exp(yz + xy + xz),$
$\dfrac{\partial r}{\partial y} = (x + z)\exp(yz + xy + xz),$
$\dfrac{\partial r}{\partial z} = (x + y)\exp(yz + xy + xz)$
11. $z_x = -\left(\dfrac{z}{x}\right)^{1/3}, \quad z_y = -\left(\dfrac{z}{y}\right)^{1/3}$
13. $z_x = -\dfrac{yz(e^{xy} + e^{xz}) + (xy + 1)e^{xy}}{e^{xy} + xye^{yz}},$
$z_y = -\dfrac{x(x + z)e^{xy} + e^{xz}}{xye^{xz} + e^{xy}}$
15. $z_x = -\dfrac{c^2x}{a^2z}, \quad z_y = -\dfrac{c^2y}{b^2z}$
17. $\dfrac{\partial w}{\partial x} = \frac{1}{4}(x + y)^{1/4}\left(\dfrac{xy}{x - y}\right)^{1/2} + \frac{1}{4}(x - y)^{1/4}\left(\dfrac{xy}{x + y}\right)^{1/2} + \frac{1}{2}(x^2 - y^2)^{1/4}\left(\dfrac{y}{x}\right)^{1/2};$
$\dfrac{\partial w}{\partial y} = -\frac{1}{4}(x + y)^{1/4}\dfrac{xy}{(x - y)^{1/2}} + \frac{1}{4}(x - y)^{1/4}\left(\dfrac{xy}{x + y}\right)^{1/2} + \frac{1}{2}(x^2 - y^2)^{1/4}\left(\dfrac{x}{y}\right)^{1/2}$
19. $\dfrac{\partial w}{\partial x} = (y^3 - 3x^2y)(x^2 + y^2)^{-3} - y,$
$\dfrac{\partial w}{\partial y} = (x^3 - 3xy^2)(x^2 + y^2)^{-3} - x$
21. $x + y - 2z = 7$
23. $5x + 5y + 11z = 31$
27. $\partial w/\partial x = f'(u)(\partial u/\partial x) = f'(u)$, and so on
29. Show that $w_u = w_x + w_y$. Then note that
$$w_{uv} = \frac{\partial}{\partial v} w_u = \frac{\partial w_u}{\partial x}\cdot\frac{\partial x}{\partial v} + \frac{\partial w_u}{\partial y}\cdot\frac{\partial y}{\partial v}.$$

Section 15.8 (page 751)

1. $\langle 3, -7\rangle$
3. $\langle 0, 0\rangle$
5. $\langle 0, 6, -4\rangle$
7. $\langle 1, 1, 1\rangle$
9. $\langle 2, -1.5, -2\rangle$
11. $8\sqrt{2}$
13. $\frac{12}{13}\sqrt{13}$
15. $-\frac{13}{20}$
17. $-\frac{1}{6}$
19. $-6\sqrt{2}$
21. Maximum: $\sqrt{170}$; direction: $\langle 7, 11\rangle$
23. Maximum: $14\sqrt{2}$; direction: $\langle 3, 5, -8\rangle$
25. Maximum: $2\sqrt{14}$; direction: $\langle 1, 2, 3\rangle$
27. $29(x - 2) - 4(y + 3) = 0$
29. $x + y + z = 1$
35. (a) $\frac{34}{3}$°C/ft; (b) 13°C/ft and $\langle 4, 3, 12\rangle$
37. (a) $3x + 2y - 10z = 4$;
(b) 0.44 (true value: 0.448)
39. $x - 2y + z + 10 = 0$
43. $\frac{55}{6}\sqrt{2}$
45. (a) Direction: $\langle -1, -2\rangle$; angle: about 43°44′;
(b) Angle: about 23°10′

Section 15.9 (page 759)

1. Maximum: 4, at $(\pm 2, 0)$, minimum: -4, at $(0, \pm 2)$
3. Maximum: 3, where $y = \pm\sqrt{2}$ and $2x = 3y$;
minimum: -3, where $y = \pm\sqrt{2}$ and $2x = -3y$
5. Minimum: $\frac{126}{49}$, at $(\frac{9}{7}, \frac{6}{7}, \frac{3}{7})$;
no maximum
7. Maximum: 7, at $(\frac{36}{7}, \frac{9}{7}, \frac{4}{7})$;
minimum: -7, at $(-\frac{36}{7}, -\frac{9}{7}, -\frac{4}{7})$
9. Minimum: $\frac{25}{3}$, at $(-\frac{5}{3}, \frac{1}{3}, \frac{7}{3})$
21. Closest points: $(2, -2, 1)$ and $(-2, 2, 1)$
25. $(2, 3)$ and $(-2, -3)$

27. Highest point: $(\frac{2}{5}\sqrt{5}, \frac{1}{5}\sqrt{5}, \sqrt{5} - 4)$;
lowest point: $(-\frac{2}{5}\sqrt{5}, -\frac{1}{5}\sqrt{5}, -\sqrt{5} - 4)$

29. Farthest: $x = -\frac{1}{20}(15 + 9\sqrt{5}), y = 2x,$
$z = \frac{1}{4}(9 + 3\sqrt{5})$;
nearest: $x = -\frac{1}{20}(15 - 9\sqrt{5}), y = 2x,$
$z = \frac{1}{4}(9 - 3\sqrt{5})$

Section 15.10 (page 767)

1. Minimum: $(-1, 2, -1)$, no other extrema

3. Saddle point: $(-\frac{1}{2}, -\frac{1}{2}, \frac{29}{4})$, no extrema

5. Minimum: $(-3, 4, -9)$, no other extrema

7. Saddle point at $(0, 0, 3)$; local maximum at $(-1, -1, 4)$

9. No extrema

11. Saddle point at $(0, 0, 0)$; local minima at $(-1, -1, -2)$ and $(1, 1, -2)$

13. Saddle point at $(-1, 1, 5)$, local minimum at $(3, -3, -27)$

15. Saddle point at $(0, -2, 32)$, local minimum at $(-5, 3, -93)$

17. Saddle point at $(0, 0, 0)$, local minima at the four points where $(|x|, |y|, z) = (1, 2, 2)$

19. Saddle point at $(-1, 0, 17)$, local minimum at $(2, 0, -10)$

21. Local minimum at $(0, 0, 0)$, saddle point at $(2, 0, 4e^{-2})$

23. Local minimum

25. Local maximum

27. Minimum value 3 at $(1, 1)$ and at $(-1, -1)$

29. See Problem 41 of Section 15.5 and its answer.

31. The critical points are of the form (m, n) where m and n are either both even integers or both odd integers. The critical point (m, n) is a saddle point if m and n are both even. It is a local maximum if m and n are both of the form $4k + 1$ or both of the form $4k + 3$. It is a local minimum in the remaining cases.

35. Local minima at $(-1.879385, 0)$ and $(1.532089, 0)$, saddle point at $(0.347296, 0)$

37. Local minima at $(-1.879385, 1.879385)$ and $(1.532089, -1.532089)$;
saddle point at $(0.347296, -0.347296)$

39. Saddle point at $(0, 0)$; local minima at $(3.624679, 3.984224)$ and $(3.624679, -3.984224)$

Chapter 15 Miscellaneous Problems (page 769)

3. On the line $y = x$, $g(x, y) = \frac{1}{2}$, except that $g(0, 0) = 0$.

5. $f(x, y) = x^2y^3 + e^x \sin y + y + C$

7. All points of the form $(a, b, \frac{1}{2})$ (so that $a^2 + b^2 = \frac{1}{2}$) together with $(0, 0, 0)$

9. The normal to the cone at $(a, b, (a^2 + b^2)^{1/2})$ meets the z-axis at $z = 2(a^2 + b^2)^{1/2}$.

15. Base: $24^{1/3}$ by $24^{1/3}$, height $375^{1/3}$

17. 200 ± 2 ohms

19. 3%

21. The six points where $(|x|, |y|, |z|) = (4, 0, 0)$, $(0, 2, 0)$, and $(0, 0, \frac{4}{3})$

25. $3\mathbf{i} + 4\mathbf{j}$; 1

27. The plane tangent to the graph at (a, b, c) has x-intercept $a^{1/3}$.

31. Semiaxes: 1 and 2

33. There is no such triangle of minimum perimeter, unless one is willing to consider as a triangle the figure with all sides of length zero—a single point on the circumference of the circle. The triangle of *maximum* perimeter is equilateral, with perimeter $3\sqrt{3}$.

35. Closest point: $(\frac{1}{3}\sqrt{6}, \frac{1}{6}\sqrt{6})$;
farthest point: $(-\frac{1}{3}\sqrt{6}, -\frac{1}{6}\sqrt{6})$

39. Maximum: 1; minimum: $-\frac{1}{2}$

41. Local minima: -1, at $(1, 1)$ and at $(-1, -1)$; horizontal tangent plane (not extrema) at $(0, 0, 0)$, $(\sqrt{3}, 0, 0)$ and $(-\sqrt{3}, 0, 0)$

43. Local minimum: -8 at $(2, 2)$; horizontal tangent plane at $(0, 0, 0)$

45. Local maximum: $\frac{1}{432}$ at $(\frac{1}{2}, \frac{1}{3})$. The points on the intervals $(-\infty, 0)$ and $(1, +\infty)$ on the x-axis are all local minima (value 0), and those on the interval $(0, 1)$ on the x-axis are all local maxima (value 0). There is a saddle point at $(0, 1, 0)$.

47. Saddle point at $(0, 0, 1)$; each point on the hyperbola $xy = \ln 2$ yields a global minimum.

49. No extrema; saddle points at $(1, 1, 0)$ and at $(-1, -1, 0)$.

Section 16.1 (page 777)

1. 80 **3.** -78 **5.** $\frac{513}{4}$

7. $-\frac{9}{2}$ **9.** 1 **11.** $\frac{1}{2}(e - 1)$

13. $2(e - 1)$ **15.** $2\pi + \frac{1}{4}\pi^4$ **17.** 1

19. $2 \ln 2$ **21.** -32

23. $\frac{4}{15}(9\sqrt{3} - 4\sqrt{2} - 1)$

Section 16.2 (page 782)

1. $\frac{5}{6}$ **3.** $\frac{1}{2}$ **5.** $\frac{1}{12}$ **7.** $\frac{1}{20}$

9. $-\frac{1}{18}$ **11.** $\frac{1}{2}(e - 2)$ **13.** $\frac{61}{3}$

15. $\int_0^4 \int_{-\sqrt{y}}^{\sqrt{y}} x^2 y \, dx \, dy = \frac{512}{21}$

17. $\int_0^1 \int_{-\sqrt{y}}^{\sqrt{y}} x \, dx \, dy + \int_1^9 \int_{(y-3)/2}^{\sqrt{y}} x \, dx \, dy = \frac{32}{3}$

19. $\int_0^4 \int_{2-(4-y)^{1/2}}^{y/2} 1 \, dx \, dy = \frac{4}{3}$

21. $\int_0^{\pi}\int_0^{\pi} \frac{\sin y}{y}\,dx\,dy = 2$

23. $\int_0^1\int_0^x \frac{1}{1+x^4}\,dy\,dx = \frac{1}{8}\pi$

Section 16.3 (page 787)

1. $\frac{1}{6}$ 3. $\frac{32}{3}$ 5. $\frac{5}{6}$ 7. $\frac{32}{3}$
9. $2\ln 2$ 11. 2 13. $2e$ 15. $\frac{1}{3}$
17. $\frac{41}{60}$ 19. $\frac{4}{15}$ 21. $\frac{10}{3}$ 23. 19
25. $\frac{4}{3}$ 27. $\frac{1}{6}abc$ 29. $\frac{2}{3}$
33. $\frac{625}{2}\pi \approx 981.748$
35. $\frac{1}{6}(2\pi + 3\sqrt{3})R^3 \approx (1.913)R^3$
37. $\frac{256}{15}$

Section 16.4 (page 796)

3. $\frac{3}{2}\pi$
5. $\frac{1}{6}(4\pi - 3\sqrt{3}) \approx 1.22837$
7. $\frac{1}{2}(2\pi - 3\sqrt{3}) \approx 0.5435$
9. $\frac{16}{3}\pi$ 11. $\frac{23}{8}\pi$
13. $\frac{1}{4}\pi \ln 2$ 15. $\frac{16}{5}\pi$
17. $\frac{1}{4}\pi(1 - \cos 1) \approx 0.36105$
19. 2π 21. 4π 27. 2π
29. $\frac{1}{3}\pi a^3(2 - \sqrt{2}) \approx (0.6134)a^3$
31. $\frac{1}{4}\pi$ 35. $2\pi^2a^2b$

Section 16.5 (page 805)

1. $(2, 3)$ 3. $(1, 1)$ 5. $(\frac{4}{3}, \frac{2}{3})$
7. $(\frac{3}{2}, \frac{6}{5})$ 9. $(0, \frac{8}{5})$
11. Mass $\frac{1}{24}$, centroid $(\frac{2}{5}, \frac{2}{5})$
13. Mass $\frac{256}{15}$, centroid $(0, \frac{16}{7})$
15. Mass $\frac{1}{12}$, centroid $(\frac{9}{14}, \frac{9}{14})$
17. Mass $\frac{1}{3}$, centroid $(0, \frac{22}{35})$
19. Mass 2, centroid $(\frac{1}{2}\pi, \frac{1}{8}\pi)$
21. Mass a^3, centroid $(\frac{7}{12}a, \frac{7}{12}a)$
23. Mass $\frac{128}{5}$, centroid $(0, \frac{20}{7})$
25. Mass π, $\bar{x} = \frac{\pi^2 - 4}{\pi} \approx 1.87$, $\bar{y} = \frac{\pi}{8} \approx 0.39$
27. Mass $\frac{1}{3}\pi a^3$, $\bar{x} = 0$, $\bar{y} = \frac{3a}{2\pi}$
29. Mass $\frac{2}{3}\pi + \frac{1}{4}\sqrt{3}$, $\bar{x} = 0$,
$\bar{y} = \frac{36\pi + 33\sqrt{3}}{32\pi + 12\sqrt{3}} \approx 1.4034$
31. $I_0 = \frac{2\pi a^{n+4}}{n+4}$
33. $\frac{3}{2}\pi k$ 35. $\frac{1}{9}$
37. $\hat{x} = \frac{2}{21}\sqrt{105}$, $\hat{y} = \frac{4}{3}\sqrt{5}$
39. $\hat{x} = \hat{y} = \frac{1}{10}a\sqrt{30}$
41. $\left(\frac{4r}{3\pi}, \frac{4r}{3\pi}\right)$ 43. $\left(\frac{2r}{\pi}, \frac{2r}{\pi}\right)$
51. (a) $\bar{x} = 0$, $\bar{y} = \frac{4a^2 + 3\pi ab + 6b^2}{12b + 3\pi a}$;
(b) $\frac{1}{3}\pi a(4a^2 + 2\pi ab + 6b^2)$
53. $(1, \frac{1}{4})$ 55. $\frac{484}{3}k$

Section 16.6 (page 812)

1. 18 3. 128 5. $\frac{31}{60}$ 7. 0
9. 12
11. $V = \int_0^3\int_0^{3-(2x/3)}\int_0^{6-2x-2y} 1\,dz\,dy\,dx = 6$
13. $\frac{128}{5}$ 15. $\frac{332}{105}$ 17. $\frac{256}{15}$ 19. $\frac{11}{30}$
21. $(0, \frac{20}{7}, \frac{20}{7})$ 23. $(0, \frac{8}{7}, \frac{12}{7})$
25. $\bar{x} = 0$, $\bar{y} = \frac{44 - 9\pi}{72 - 9\pi}$, $\bar{z} = \frac{9\pi - 16}{72 - 9\pi}$
27. $\frac{8}{7}$ 29. $\frac{1}{30}$ 33. $\frac{2}{3}a^5$ 35. $\frac{38}{45}ka^7$
37. $\frac{1}{3}k$ 39. $(\frac{9}{64}\pi, \frac{9}{64}\pi, \frac{3}{8})$
41. 24π 43. $\frac{1}{6}\pi$

Section 16.7 (page 819)

1. 8π 5. $\frac{4}{3}\pi(8 - 3\sqrt{3})$
7. $\frac{1}{2}\pi a^2h^2$ 9. $\frac{1}{2}\pi a^4h^2$
11. Volume $\frac{81}{2}\pi$, centroid $(0, 0, 3)$
13. 24π 15. $\frac{1}{6}\pi(8\sqrt{2} - 7)$
17. $I_x = \frac{1}{12}\pi\delta a^2h(3a^2 + 4h^2)$
19. $\frac{1}{3}\pi$ 23. $\frac{1}{3}\pi$
25. Volume $\frac{1}{3}\pi a^3(2 - \sqrt{2})$; $\bar{x} = 0 = \bar{y}$,
$\bar{z} = \frac{3}{16}(2 + \sqrt{2})a \approx (0.6402)a$
27. $I_x = \frac{7}{5}ma^2$
29. Description: The surface obtained by rotating the circle in the xz-plane with center $(a, 0)$ and radius a about the z-axis—a doughnut with an infinitesimal hole. Its volume is $2\pi^2a^3$.
31. $I_x = \frac{2}{15}(128 - 51\sqrt{3})\pi a^5$
33. Mass $\frac{37}{48}\pi a^4$, $\bar{z} = \frac{105}{74}a$

Section 16.8 (page 825)

1. $6\pi\sqrt{11}$ 3. $\frac{1}{6}\pi(17\sqrt{17} - 1)$
5. Area $3\sqrt{2} + \frac{1}{2}\ln(3 + 2\sqrt{2}) \approx 5.124$
7. $3\sqrt{14}$ 9. $\frac{2}{3}\pi(2\sqrt{2} - 1) \approx 3.829$
11. $\frac{1}{6}\pi(65\sqrt{65} - 1)$ 15. $8a^2$

Section 16.9 (page 832)

1. $x = \frac{1}{2}(u + v)$, $y = \frac{1}{2}(u - v)$, $J = -\frac{1}{2}$
3. $x = (u/v)^{1/2}$, $y = (uv)^{1/2}$, $J = (2v)^{-1}$

5. $x = \frac{1}{2}(u + v), \quad y = \frac{1}{2}(u - v)^{1/2}, \quad J = -\frac{1}{4}(u - v)^{-1/2}$

7. Area: $\frac{3}{5}$

9. Area: $\ln 2$

11. Area: $\frac{1}{8}(2 - \sqrt{2})$

13. Volume: $\frac{39}{2}\pi$

15. 8

17. S is the region $3u^2 + v^2 \leqq 3$; the value of the integral is $\frac{2}{3}\pi\sqrt{3}(e^3 - 1)e^{-3}$.

Chapter 16 Miscellaneous Problems (page 834)

1. $\frac{1}{3}(2 - \sqrt{2})$

3. $\dfrac{e - 1}{2e}$

5. $\frac{1}{4}(e^4 - 1)$

7. $\frac{4}{3}$

9. Volume 9π, $\bar{z} = \frac{9}{16}$

11. 4π

13. 4π

15. $\frac{1}{16}(\pi - 2)$

17. Mass $\frac{128}{15}$, centroid $(\frac{32}{7}, 0)$

19. Mass $k\pi$, centroid $(1, 0)$

21. $\bar{y} = 4b/3\pi$

23. $(0, \frac{8}{5})$

25. $\frac{10}{3}\pi(\sqrt{5} - 2) \approx 2.4721$

27. $\frac{3}{10}Ma^2$

29. $\frac{1}{5}M(b^2 + c^2)$

31. $\frac{128}{225}\delta(15\pi - 26) \approx (12.017)\delta$ where δ is the (constant) density

33. $\frac{8}{3}\pi$

41. $\frac{18}{7}$

43. $\frac{1}{6}\pi(37\sqrt{37} - 17\sqrt{17}) \approx 81.1418$

47. $4\sqrt{2}$

48. Approximately 3.49608

51. $I_0 = 3\delta$

53. Mass: $\frac{8}{15}\pi abc$

Section 17.1 (page 843)

1.

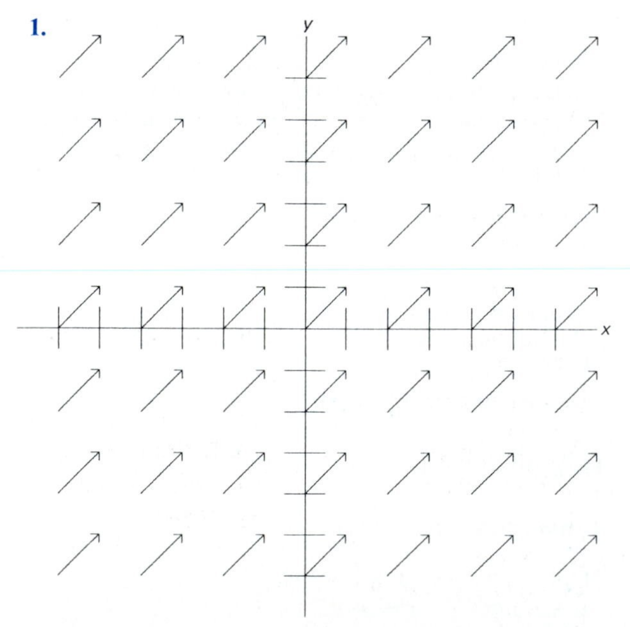

3.

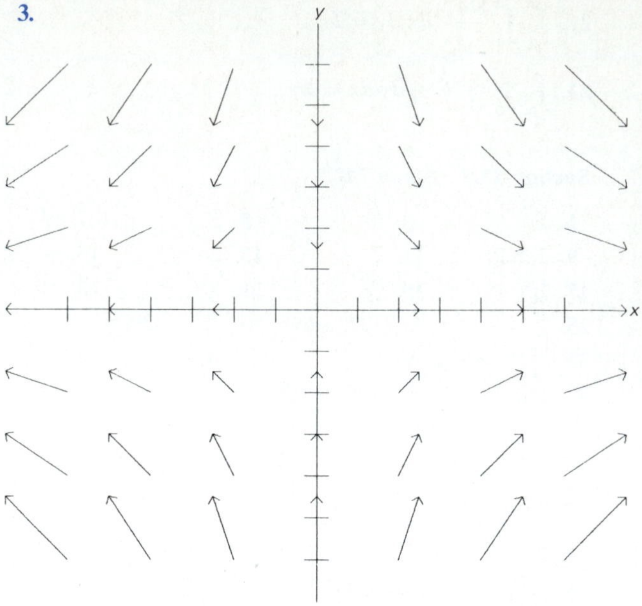

y
x

5.

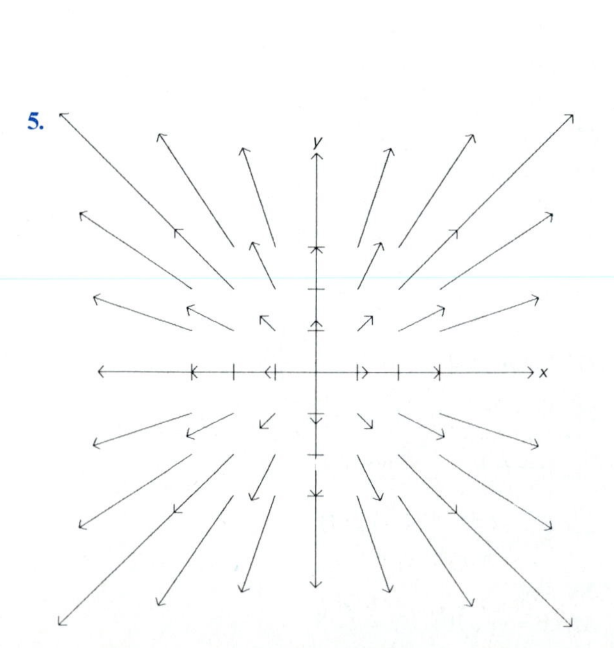

7.

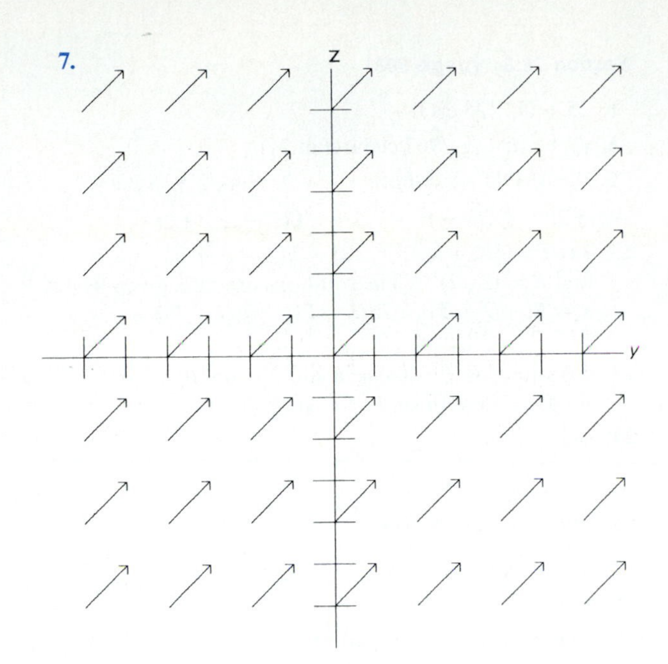

9.

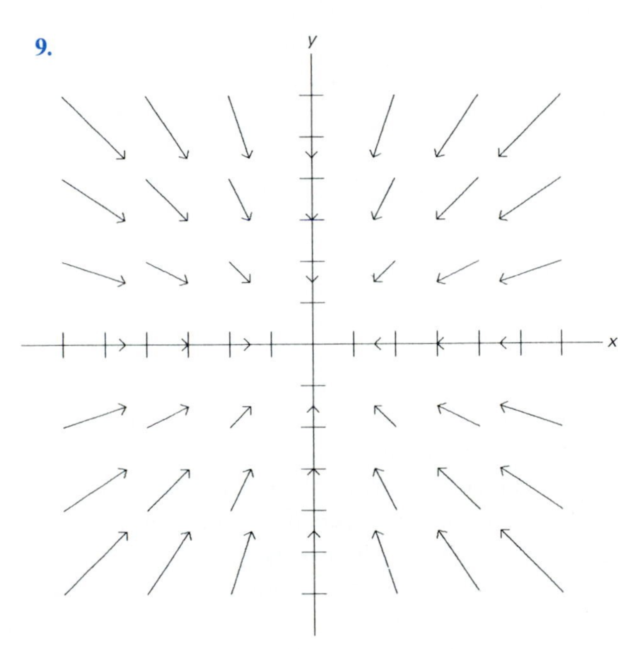

11. 3, $\mathbf{0}$ **13.** 0, $\mathbf{0}$

15. $x^2 + y^2 + z^2$, $\langle -2yz, -2xz, -2xy \rangle$

17. 0, $\langle 2y - 2z, 2z - 2x, 2x - 2y \rangle$

19. 3, $\langle x \cos xy - x \cos xz, y \cos yz - y \cos xy, z \cos xz - z \cos yz \rangle$

Section 17.2 (page 851)

1. $\frac{310}{3}$, $\frac{248}{3}$, 62 **3.** $3\sqrt{2}$, 3, 3

5. $\frac{49}{24}$, $\frac{3}{2}$, $\frac{4}{3}$ **7.** $\frac{6}{5}$ **9.** 315

11. $\frac{19}{60}$ **13.** $\pi + 2\pi^2$ **15.** 28

17. $\frac{1}{6}(14\sqrt{14} - 1) \approx 8.563867$

19. $(0, 2a/\pi)$

21. $10k\pi$; centroid: $(0, 0, 4\pi)$

23. Mass $\frac{1}{2}ka^3$; centroid $(\frac{2}{3}a, \frac{2}{3}a)$; $I_x = I_y = \frac{1}{4}ka^5$

25. (a) $\frac{1}{2}k \ln 2$; (b) $-\frac{1}{2}k \ln 2$

Section 17.3 (page 858)

1. $f(x, y) = x^2 + 3xy + y^2$

3. $f(x, y) = x^3 + 2xy^2 + 2y^3$

5. $f(x, y) = \frac{1}{4}x^4 + \frac{1}{3}y^3 + y \ln x$

7. $f(x, y) = \sin x + x \ln y + e^y$

9. $f(x, y) = x^3y^3 + xy^4 + \frac{1}{5}y^5$

11. $f(x, y) = x^2y^{-1} + y^2x^{-3} + 2y^{1/2}$

17. 6 **19.** $3e^{-1}$

21. $-\pi$ **23.** $f(x, y, z) = xyz$

25. $f(x, y, z) = xy \cos z - yze^x$

27. $\mathbf{F}$ is not conservative on any region containing (0, 0).

29. $Q_z = 0 \neq 2y = R_y$

Section 17.4 (page 866)

1. 0 **3.** 3 **5.** $\frac{3}{10}$ **7.** 2

9. 0 **11.** $\frac{16}{105}$ **13.** πa^2 **15.** $\frac{3}{8}\pi$

25. $\frac{3}{2}$ **31.** 2π

Section 17.5 (page 874)

1. $\frac{1}{120}\sqrt{3}$

3. $\frac{1}{60}\pi(1 + 391\sqrt{17}) \approx 84.4635$

5. $\frac{16}{3}\pi$ **7.** 24π

9. 0 **11.** $(\frac{1}{2}a, \frac{1}{2}a, \frac{1}{2}a)$

13. $\bar{z} = \dfrac{1 + (24a^4 + 2a^2 - 1)(1 + 4a^2)^{1/2}}{10[1 + 4a^2)^{3/2} - 1]}$,

$I_z = \frac{1}{60}\pi\delta[1 + (24a^4 + 2a^2 - 1)(1 + 4a^2)^{1/2}]$

15. $\bar{x} = \dfrac{4}{3\pi - 6} \approx 1.16796$, $\bar{y} = 0$,

$\bar{z} = \dfrac{\pi}{2\pi - 4} \approx 1.13797$

17. Net flux: 0 **19.** Net flux: 1458π

Section 17.6 (page 881)

1. Both values: 4π **3.** Both values: 24

5. Both values: $\frac{1}{2}$ **7.** $\frac{2385}{2}\pi$

9. $\frac{1}{4}$ **11.** $\frac{730,125}{4}\pi$

13. 16π

23. $\frac{1}{48}\pi(482,620 + 29,403 \ln 11)$

Section 17.7 (page 888)

1. -20π 3. 0 5. -52π 7. -8π
9. -2
11. $\phi(x, y, z) = yz - 2xz + 3xy$
13. $\phi(x, y, z) = 3xe^z + 5y\cos x + 17$

Chapter 17 Miscellaneous Problems (page 890)

1. $\frac{125}{3}$
3. $\frac{69}{8}$ (Use the fact that the integral of $\mathbf{F}\cdot\mathbf{T}$ is independent of the path.)
5. $\frac{2148}{5}$
9. Mass: $\frac{1}{3}(5\sqrt{5} - 1) \approx 3.3934$; $I_y = \frac{1}{15}(2 + 50\sqrt{5}) \approx 7.5869$
11. $\frac{2816}{7}$ 17. $\frac{371}{30}\pi$ 19. 72π
29. (a) $\phi'(r)(\mathbf{r}/r)$; (b) $3\phi(r) = r\phi'(r)$; (c) $\mathbf{0}$

Section 18.1 (page 896)

1. $y(x) = C\exp(x^2)$ 3. $y(x) = C\exp(\sin x)$
5. $y(x) = \arcsin(C + x^3)$ 7. $y(x) = Ce^x - 1$
9. $y^2 = (C - x^2)^{-1}$
11. $\ln(y^4 + 1) = 4\sin x + C$
13. $\frac{1}{3}y^{-3} - 2y^{-1} = x^{-1} + \ln|x| + C$
15. $2\ln|1 + y| = 2x + x^2 + C$
17. $y(x) = 2\exp(e^x)$ 19. $y^2 = 1 + (x^2 - 16)^{1/2}$
21. $\ln(2y - 1) = 2(x - 1)$ 23. $\ln|y| = x^2 - 1 + \ln|x|$
25. $y(x) = 1 + (1 + \frac{1}{2}x^2)^2$ 27. $y(x) = (x^2 + 9)^2$
29. $y(x) = \arcsin(1 + \sin x)$

Section 18.2 Miscellaneous Problems (page 902)

1. $xe^{-3x}y = C$; $y \equiv 0$
3. $e^x y = \frac{1}{2}e^{2x} + C$; $y(x) = \frac{1}{2}(e^x + e^{-x})$
5. $y\exp(x^2) = \frac{1}{2}\exp(x^2) + C$; $y(x) = \frac{1}{2}(1 - 5\exp(-x^2))$
7. $(1 + x)y = \sin x + C$; $y(x) = \dfrac{1 + \sin x}{1 + x}$
9. $y\sin x = \frac{1}{2}\sin^2 x + C$
11. $x^{-3}y = \sin x + C$; $y(x) = x^3\sin x$
13. $x^{-3}e^{2x}y = 2e^{2x} + C$
15. $y(x) = [\exp(-\frac{3}{2}x^2)][3(x^2 + 1)^{3/2} - 2]$
17. After approximately 7 min 41 s
19. After approximately 10.85 years
21. 393.75
23. (a) $x(t) = 50e^{-0.05t}$; $y(t) = 150e^{-0.025t} - 100e^{-0.05t}$; (b) 56.25 lb

Section 18.3 (page 908)

1. $25 + 0i$; $25 \text{ cis } 0$
3. $\frac{1}{13} + \frac{31}{13}i$; $\frac{1}{13}\sqrt{962}\text{ cis}(\arctan 31)$
5. $0 + (-8)i$; $8\text{ cis}(\frac{3}{2}\pi)$ 7. $\pm\frac{1}{2}\sqrt{2} \pm \frac{1}{2}i\sqrt{2}$
9. $\pm 2i$, $\pm\sqrt{3} \pm i$ 11. ± 2, $\pm 2i$
13. Let $P = \frac{1}{2}(2 + \sqrt{3})^{1/2}$, $Q = \frac{1}{2}(2 - \sqrt{3})^{1/2}$, and $R = (2\sqrt{2})^{1/3}$. The solutions are then given by $x = R(-Q + Pi)$, $R(Q - Pi)$, $R(Q + Pi)$, $R(-P - Qi)$, and $R(\frac{1}{2}\sqrt{2} \pm \frac{1}{2}i\sqrt{2})$.
15. $\cos 4\theta = \cos^4\theta - 6\cos^2\theta\sin^2\theta + \sin^4\theta$, $\sin 4\theta = 4\cos^3\theta\sin\theta - 4\cos\theta\sin^3\theta$
23. -1

Section 18.4 (page 914)

1. $y(x) = C_1e^x + C_2e^{2x}$ 3. $y(x) = C_1 + C_2e^{5x}$
5. $y(x) = C_1e^{x/3} + C_2e^{3x}$
7. $y(x) = C_1\cos 5x + C_2\sin 5x$
9. $y(x) = e^{2x}(C_1\cos 2x + C_2\sin 2x)$
11. $y(x) = e^{x/3}(C_1\cos x + C_2\sin x)$
13. $y(x) = C_1\exp(x\sqrt{7}) + C_2\exp(-x\sqrt{7})$
15. $y(x) = 5e^x + 2e^{3x}$ 17. $y(x) = 1 + 5e^{3x}$
19. $y(x) = 3e^x\cos 2x$ 21. $y(x) = \sin x$
23. The only solution is the trivial solution $y(x) \equiv 0$.

Section 18.5 (page 920)

1. $x(t) = 5\cos 3t$
3. $x(t) = 5\sqrt{2}\cos(5t - \frac{1}{4}\pi)$
5. $x(t) = 6\cos 6t + 8\sin 6t$
7. Frequency 2 rad/s; that is, $1/\pi$ Hz (cycles/s). Period: π seconds
9. Amplitude 2 m; frequency 5 rad/s; period $2\pi/5$ s
11. Amplitude 100 cm; period approximately 2 s
13. $x(t) = 4e^{-2t} - 2e^{-4t}$; overdamped
15. $x(t) = 5e^{-4t} + 10te^{-4t}$; critically damped
17. $x(t) = \frac{1}{3}\sqrt{313}\,e^{-5t/2}\cos(6t - 0.8254)$
19. (a) $k \approx 7018.39$ lb/ft; (b) After approximately 2.49 s

Section 18.6 (page 924)

1. Approximate value: 1.18098; exact value: 1.21306 (rounded)
3. 2.22102; 2.29744 5. 0.88949; 0.85128
7. 2.73730; 2.64749 9. 1.27847; 1.28743
11. 1.21415; 1.21306 13. 2.29489; 2.29744
15. 0.85255; 0.85128 17. 2.64050; 2.64749
19. 1.28733; 1.28743

Chapter 18 Miscellaneous Problems (page 925)

1. $y(x) = C \exp(-x^2)$
3. $y(x) = C \exp(-\cos x)$
5. $y(x) = \sin(C + x^{1/2})$
7. $y(x) = (2x^{4/3} + C)^{3/2}$
9. $y(x) = \dfrac{C(1 + x)}{1 - x}$
11. $y(x) = 2 + Ce^{-x}$
13. $y(x) = (x^2 + C)e^{-3x}$
15. $y(x) = x + Cx^{-2}$
17. $y(x) = 5x^{1/2} + Cx^{-1/2}$
19. $y(x) = Cx + x \ln x$
21. $y(x) = C_1 e^{2x} + C_2 e^{-2x}$
23. $y(x) = C_1 e^{2x} + C_2 e^{-5x}$
25. $y(x) = C_1 e^{-3x} + C_2 x e^{-3x}$
27. $y(x) = C_1 e^{3x/2} + C_2 x e^{3x/2}$
29. $y(x) = 2 \exp(e^x)$
31. $y^2 = 1 + (x^2 - 16)^{1/2}$
33. $\ln(2y - 1) = 2(x - 1)$
35. $\ln y = x^2 - 1 + \ln x$
37. $y(x) \equiv 0$
39. $y(x) = \frac{1}{2}(e^x + e^{-x})$
41. $y(x) = \frac{1}{2}(1 - 5 \exp(-x^2))$
43. $y(x) = x^3 \sin x$
45. $y(x) = x^{10} + x + 1$
47. $y(x) = \frac{1}{20}(10 + \sqrt{10}) \exp(x\sqrt{10}) - \frac{1}{20}(10 - \sqrt{10}) \exp(-x\sqrt{10})$
49. $y(x) = e^{-x/3}(3 \cos \frac{1}{3}x\sqrt{3} + 2\sqrt{3} \sin \frac{1}{3}x\sqrt{3})$

Index

D

E

J

K

L

M

N

S

T

U

V

X

Y

Z

INVERSE TRIGONOMETRIC FORMS

69 $\int \sin^{-1} u\, du = u \sin^{-1} u + \sqrt{1 - u^2} + C$

70 $\int \tan^{-1} u\, du = u \tan^{-1} u - \frac{1}{2} \ln (1 + u^2) + C$

71 $\int \sec^{-1} u\, du = u \sec^{-1} u - \ln|u + \sqrt{u^2 - 1}| + C$

72 $\int u \sin^{-1} u\, du = \frac{1}{4} (2u^2 - 1) \sin^{-1} u + \frac{u}{4} \sqrt{1 - u^2} + C$

73 $\int u \tan^{-1} u\, du = \frac{1}{2} (u^2 + 1) \tan^{-1} u - \frac{u}{2} + C$

74 $\int u \sec^{-1} u\, du = \frac{u^2}{2} \sec^{-1} u - \frac{1}{2} \sqrt{u^2 - 1} + C$

75 $\int u^n \sin^{-1} u\, du = \frac{u^{n+1}}{n+1} \sin^{-1} u - \frac{1}{n+1} \int \frac{u^{n+1}}{\sqrt{1 - u^2}}\, du + C \quad \text{if } n \neq -1$

76 $\int u^n \tan^{-1} u\, du = \frac{u^{n+1}}{n+1} \tan^{-1} u - \frac{1}{n+1} \int \frac{u^{n+1}}{1 + u^2}\, du + C \quad \text{if } n \neq -1$

77 $\int u^n \sec^{-1} u\, du = \frac{u^{n+1}}{n+1} \sec^{-1} u - \frac{1}{n+1} \int \frac{u^n}{\sqrt{u^2 - 1}}\, du + C \quad \text{if } n \neq -1$

HYPERBOLIC FORMS

78 $\int \sinh u\, du = \cosh u + C$

79 $\int \cosh u\, du = \sinh u + C$

80 $\int \tanh u\, du = \ln (\cosh u) + C$

81 $\int \coth u\, du = \ln|\sinh u| + C$

82 $\int \operatorname{sech} u\, du = \tan^{-1} |\sinh u| + C$

83 $\int \operatorname{csch} u\, du = \ln \left| \tanh \frac{u}{2} \right| + C$

84 $\int \sinh^2 u\, du = \frac{1}{4} \sinh 2u - \frac{u}{2} + C$

85 $\int \cosh^2 u\, du = \frac{1}{4} \sinh 2u + \frac{u}{2} + C$

86 $\int \tanh^2 u\, du = u - \tanh u + C$

87 $\int \coth^2 u\, du = u - \coth u + C$

88 $\int \operatorname{sech}^2 u\, du = \tanh u + C$

89 $\int \operatorname{csch}^2 u\, du = -\coth u + C$

90 $\int \operatorname{sech} u \tanh u\, du = -\operatorname{sech} u + C$

91 $\int \operatorname{csch} u \coth u\, du = -\operatorname{csch} u + C$

MISCELLANEOUS ALGEBRAIC FORMS

92 $\int u(au + b)^{-1}\, du = \frac{u}{a} - \frac{b}{a^2} \ln|au + b| + C$

93 $\int u(au + b)^{-2}\, du = \frac{1}{a^2} \left[\ln|au + b| + \frac{b}{au + b} \right] + C$

94 $\int u(au + b)^n\, du = \frac{(au + b)^{n+1}}{a^2} \left(\frac{au + b}{n + 2} - \frac{b}{n + 1} \right) + C \quad \text{if } n \neq -1, -2$

95 $\int \frac{du}{(a^2 \pm u^2)^n} = \frac{1}{2a^2(n - 1)} \left(\frac{u}{(a^2 \pm u^2)^{n-1}} + (2n - 3) \int \frac{du}{(a^2 \pm u^2)^{n-1}} \right) \quad \text{if } n \neq 1$